Pathology of Childhood and Adolescence

Consolato M. Sergi

Pathology of Childhood and Adolescence

An Illustrated Guide

Volume II

Consolato M. Sergi
University of Alberta
Edmonton, AB
Canada

ISBN 978-3-662-59167-3   ISBN 978-3-662-59169-7 (eBook)
https://doi.org/10.1007/978-3-662-59169-7

© Springer-Verlag GmbH Germany, part of Springer Nature 2020
This work is subject to copyright. All rights are reserved by the Publisher, whether the whole or part of the material is concerned, specifically the rights of translation, reprinting, reuse of illustrations, recitation, broadcasting, reproduction on microfilms or in any other physical way, and transmission or information storage and retrieval, electronic adaptation, computer software, or by similar or dissimilar methodology now known or hereafter developed.
The use of general descriptive names, registered names, trademarks, service marks, etc. in this publication does not imply, even in the absence of a specific statement, that such names are exempt from the relevant protective laws and regulations and therefore free for general use.
The publisher, the authors, and the editors are safe to assume that the advice and information in this book are believed to be true and accurate at the date of publication. Neither the publisher nor the authors or the editors give a warranty, express or implied, with respect to the material contained herein or for any errors or omissions that may have been made. The publisher remains neutral with regard to jurisdictional claims in published maps and institutional affiliations.

This Springer imprint is published by the registered company Springer-Verlag GmbH, DE part of Springer Nature.
The registered company address is: Heidelberger Platz 3, 14197 Berlin, Germany

To my patients and their families, to my family, and to my students and fellows

Foreword

University hospitals, major hospitals, and medical care centers must have a pathological institute, and this is a long-standing requirement. This facility is the prerequisite for the adequate diagnosis of most of the diseases that need to be treated. Since most major hospitals also have pediatric and pediatric and adolescent surgery, the affiliated pathological institute should either have a department for pediatric pathology or at least one or two, better three, designated pathologists of childhood pathology to meet the child-specific pathological diagnosis. It is well known that this is not always the case. There are very few institutes of pathology that have their department or division of pediatric pathology or at least have a pediatric pathologist. The well-known sentence "children are not small adults" is often not taken into account.

At present, at least in the German-speaking world, pediatric pathology is only a subdiscipline of pathology. It is carried out often by general pathologists who are particularly interested with this field, who receive their training on their initiative, by rotating in several renowned institutions and participating to congresses and meetings of the Society for Pediatric Pathology or Pediatric Pathology Society as well as national meetings of pediatric pathology.

What is the challenge for pediatric pathologists? The field of work comprises a vast field of diagnostic tasks, which implies two interlinked fields of knowledge: on the one hand, knowledge of physiological development processes, i.e., the continually changing processes for growth reasons, and typical findings of the various tissues of the human body, starting with the embryo and ending in adolescents and late youth, and on the other hand, knowledge of the possible deviating pathological findings for the respective age group. This means that the ever-changing morphology of healthy development represents an additional dynamic dimension in the evaluation of pathological changes that reduce the diagnostic effort of pediatric pathologists compared to that of general pathologists. An additional, not less critical, area of pediatric pathology is the fetal diagnosis, primarily because it can be used to clarify congenital malformations and intrauterine, perinatal, and postpartum deaths. Finally, placenta diagnosis should not be forgotten, as it can be used to detect relevant evidence not only for intra- and perinatal deaths or possible diseases in the newborn but also for pregnancy complications.

In summary, in addition to diagnostics, the participation of the pediatric pathologists in the context of rare or unusual findings with discussion in the setting of multidisciplinary team meetings with general and subspecialized pediatricians, pediatric surgeons, obstetricians, and human geneticists is very well welcomed. It is surprising that the children's pathology, although its diagnostic-scientific spectrum and its cooperation in the clinical implementation of its findings differ sizably from that of general pathology, is usually covered in very few textbooks. There is virtually no truly comprehensive, clinical, and molecular biology-oriented up-to-date manual that includes detailed information for all stakeholders.

Due to the far-reaching prerequisites mentioned, a great deal of courage is required to commit to the task of writing a handbook for children's pathology because, as mentioned above, children's pathology not only requires a comprehensive spectrum of knowledge per se but also requires solid knowledge of the physiological processes of the growing body and – in connection with diagnostics – also specific ideas regarding pediatric or pediatric surgical treatment options. These requirements make the creation of a manual an almost unmanageable challenge if all these criteria are to be considered efficiently. Consolato Sergi is in the field of pediatric pathology for more than 30 years collecting experiences and acquiring knowledge. He took up this challenge, as he told me about a year ago, after much consideration. His decision for this endeavor was born out of the idea of putting his extensive pediatric pathological knowledge on paper in order to fill the bibliographical gap that exists in this respect.

In this context, I would like to mention a few of my memories that connect me with the author of this book. Between 2004 and 2008, Sergi was employed as a pediatric pathologist at the Pathological Institute of the Medical University of Innsbruck, Austria. As director of the Department of Pediatrics and Adolescent Surgery, I had to deal with him very often, as we had to cope with a quite significant number of surgical patients. On the one hand, we repeatedly discussed not only critical problems of our shared patients (often not sparing any day of the week including Saturday and Sunday) but also scientific projects. On the other hand, we met fortnightly or weekly within the framework of multidisciplinary team meetings. The "case ideas" and discussions were very profitable and sustainable since Dr. Sergi, who is also pediatrician, already had, at that time, an extensive theoretical and clinical knowledge. Therefore, he was able to contribute a lot to the clinical discussions, which was generally well praised.

It was obvious that I remained in contact with him, who has proven himself to this day, after his departure from Innsbruck due to our excellent cooperation. I congratulated him on his decision to write up this textbook, which will complement the library of not only pathologists and pediatric pathologists but also pediatricians, pediatric surgeons, and human geneticists. Because of his knowledge and well-balanced scientific curriculum, I was sure he would be able to complete this project. In reading the table of the contents of his book, I saw my opinion confirmed. He had compiled a textbook covering all relevant pediatric pathological topics. What is important to mention in this context is the fact that not only had he brought his vast wealth of knowledge into text but also didactically presented his theoretical and practical knowledge with a lot of pictorial material so informative that it would benefit not only morphologists but also clinicians. Consequently, it represents for all those who are involved – directly – in the pathological diagnosis with all its facets and in the resulting treatment of sick children or adolescents who are involved in diagnostics and research in the field of pediatric pathology as well as trainees and fellows, an essential and comprehensive source of information.

Medical University of Innsbruck, Innsbruck, Austria Josef Hager

Foreword

Pediatrics, the "Medicine of Children," is an area of medical study that is very broad, diverse, and challenging. Unlike different specialties and subspecialties, pediatrics does not deal with an organ, system, or specialty of function but with a developing organism, the human organism in its total entirety, genetics, embryonal, fetal, neonate to adolescent. The development through this continuum presents with a diversity of pathophysiology. The Department of Pediatrics, at the University of Alberta, has recognized the research and education opportunities to answer clinically important questions that make a difference in the health of children throughout the age spectrum. In collaboration with all our child health colleagues, including Dr. Sergi, the Department of Pediatrics is a leading recipient of national tri-council funding and through research truly making a clinical difference.

Dr. Sergi's book is a compilation of pathological illustrations and molecular biology data put together for the benefit of all who care for the newborn, child, and adolescent, including physicians, trainees, and allied Pediatric health providers. Clinical acumen depends predominantly on logical deduction but also on lateral interaction with disciplines that help to understand the biology of the developing human organism. Pediatric pathology is unique because it helps pediatricians to understand not only the pathological basis of disease but also inquire into molecular biological pathogenesis. Dr. Sergi's marvelous textbook in pediatric pathology is the crown of more than 30 years of the interaction of him with pediatricians. Joining the University of Alberta in 2008 as full professor of pathology and adjunct professor of pediatrics allowed me to interact with him very often as chair of the Department of Pediatrics at the University of Alberta, and importantly as a colleague in hepatology and liver transplantation.

For over a decade I have had the professional opportunity to interact clinically with Dr. Sergi. His clear explanation of concepts of pathophysiology to not only myself but to our multidisciplinary team, and most importantly our trainees, has enriched all of our intellect, and ongoing management of patients. I know that the value of such a compilation relies on extensive and thorough effort, as is witnessed in this textbook. The generosity of Dr. Sergi's dedication, and that of his family, for the advancement of pediatric medicine and pediatric pathology is immeasurable.

I wish all readers, my fellows and colleagues and myself will have the opportunity to interact more with Dr. Sergi and acquire the invaluable steps in progressing the knowledge in pediatrics.

Susan M. Gilmour, MD, MSc, FRCPC
Professor Pediatrics, Pediatric Gastroenterology/Hepatology
Department Chair 2009–2019
University of Alberta
Edmonton
Canada

Preface

The pediatric patient may be mistakenly considered an adult in miniature, but the precise definitions of diseases uniquely appearing in infancy, childhood, and adolescence make this field of medicine astoundingly rich in notions and knowledge ranging from the early intrauterine life to adolescence and youth. Pediatric pathology is one of the fastest growing subspecialties in medicine. Pediatric pathology is also unique because a few diseases that were earlier confined to adults are now occurring in childhood, adolescence, and youth. There may be several reasons, including the optimization of imaging technology, better surgical and clinical pathology criteria, but also some external factors are likely playing some role. The environment has become more and more impregnated with dangerous molecular compounds that are not only carcinogenic but also endocrine and metabolism disruptors. The pediatric pathology clubs, born in the United Kingdom and the United States, have seen growing number of participants in the last few years. The current pediatric pathology society (PPS) and societies of pediatric pathology (SPP) have reached an enormous interest not only in residents in pathology but also in pediatrics, obstetrics, gynecology, as well as other medical and surgical specialties. Long before the founding of the pediatric pathology clubs and their transformation into PPS and SPP, the German embryological schools played an unarguable and incontrovertible role in the expansion of our knowledge of perinatal medicine and congenital defects in the twentieth century.

Three figures have motivated me in collecting and producing this book. They are Guido Fanconi, Klaus Goerttler, and Harald P. Schmitt. Guido Fanconi is regarded as one of the founders of the modern pediatrics for his contributions to biochemistry and how biochemistry helped reshape modern pediatrics. His contributions to the pathology of pediatrics were numerous, and several diseases have been nominated after him. He was head of the *Kinderspital* in Zurich, Switzerland, for about 45 years and recognized Down syndrome years before the chromosomal identification. In 1934, the first patients affected with mucoviscidosis or cystic fibrosis of the pancreas were described in a doctoral thesis written under his direction. His contributions to pediatrics are countless, and some of these are highlighted in the present book. A few years ago, Stephan Lobitz, from the Department of Pediatric Oncology and Hematology, Charité–University Medicine Berlin, Berlin, Germany, and Eunike Velleuer, from the Department of Pediatric Oncology, Hematology and Immunology, University of Düsseldorf, Germany, wrote an outstanding contribution on Professor Fanconi. Most probably, one of the most paramount contributions for understanding the magnitude of Fanconi to pediatrics and the many mentees is a handwritten note kindly made available to Dr. Lobitz and Dr. Velleuer by Fanconi's son. Fanconi wrote "*Forschen: Aufdeckung und Deutung eines neuen Tatbestandes ist an eine Idee gebunden, der eine Unsumme fleissiger Arbeit folgt. Sie setzt einen ideenreichen Kopf und einen fleissigen, systematischen Arbeiter voraus. Ferner ist es wichtig, dass ein gründliches Wissen zugrunde liegt.*" ("Research: the discovery and interpretation of new facts are bound to an idea which is followed by an enormous amount of hard work. It presupposes a creative mind and a diligent, systematic worker. A sound knowledge base is also important.") This sentence has been a path to many pediatric pathologists who contributed to the self-determination, development, and autonomy of pediatric pathology starting about the second half of last century and continuing in this century. The stature of Professor Klaus Goerttler as top-ranking cardiac embryologist and pediatric pathologist has extensively contributed to the German embryology school, pediatric cardiology, and pediatric pathology. The German embryology school laid fundamental notions for the development of numerous concepts of knowledge in pediatric pathology. The clarification of the embryology of heart and its defects by the Heidelberg professor are milestones in the interpretation of the current pediatric cardiology and cardiac surgery as may be identified in his book *Normale und pathologische Entwicklung des menschlichen Herzens* (*Normal and Pathologic Development of the Human Heart*) of 1958. Hypoxia is a dreadful teratogen, causing disruption, particularly of neurulation if it interferes with early stages of embryonic development. Numerous experimental studies performed in amphibian and chick embryos showed that hypobaric-mediated hypoxia determines disruptions of the heart as well as head and brain. Most of the disruptions are induced at the beginning of gastrulation, i.e., before the onset of neurulation, when oxygen consumption is known to be exceptionally high. In these experiments, the underlying developmental mechanisms responsible for the malformations occurring with hypoxia were multiple. They include (1) the disturbance of the migration of blastema, or an alteration of the "*Topogenese*" (topogenesis" of the German embryological school according to Lehmann and of the "integrated cell and tissue movements" according to Gilbert), (2) the decreased induction of the altered blastema, and (3) the disturbance of the organ anlage or primordial organ to further differentiate. These studies continued the Spemann-Mangold experiments successfully on organizers and found that amphibian and chick investigations can also be extrapolated to humans. Professor Schwalbe highly influenced professor Goerttler and early pediatric pathologists worldwide. Ernst Theodor Karl Schwalbe (1871–1920) was a German pathologist, born in Berlin, who specialized in teratology. His medical study crossed three of the most prestigious universities (Strasbourg, Berlin, and Heidelberg). His habilitation (higher Ph.D. degree) was with a thesis on blood coagulation. In Heidelberg, Dr. Schwalbe worked as an assistant under Julius Arnold. As prosector and head of the pathology-bacteriology clinic, he worked at the city hospital in Karlsruhe in 1907/08. He worked for 12 years at the University of Rostock until 1920 when he was killed while serving as a volunteer during the Kapp Lüttwitz Putsch, which aimed to disengage the German Revolution of 1918–1919; overthrow the Weimar Republic, in governmental position after World War I; and establish an autocratic government in its place. In my opinion, Professor Schwalbe is probably one of the founders of pediatric pathology in general, and his studies have helped to shape the birth of several pediatric pathology clubs worldwide. Professor Horst P. Schmitt was my teacher in neuropathology during my residency in pathology. Professor Schmitt allowed me to revise several diseases in neuropathology and stimulated me in broadening my knowledge in developmental biology and pathology not only of humans but also of vertebrates.

The book *Pathology of Childhood and Adolescence* is not only a book of pathology for pathologists or surgeons (adult and pediatric pathologists) or residents in these disciplines but also contains a molecular biology approach to some of the very challenging diseases in pediatrics. This book came after 12 years work and tight cooperation with Springer Publisher. I am very grateful to Mr. Karthik Periyasamy, who looked at the production for several years; all the Springer team; to my colleague, Dr. Atilano Lacson; and my mentors (Francesco Callea, Walter J Hofmann, Herwart F. Otto, Axel von Herbay, Peter Sinn, Philipp Schnabel, Gregor Mikuz, Moorghen Morgan, Josef Hager, Lothar Bernd Zimmerhackl, Ulrich Schweigmann, and Brian Chiu, among others). I am also grateful to numerous adult and pediatric pathology colleagues and clinical colleagues who sent me consultation cases and advised me over the years. In this book, 18 chapters summarize and illustrate the pathology of childhood and youth comprehensively, including diseases of the cardiovascular system and respiratory tract, gastrointestinal tract, liver, pancreas and biliary tract, upper and lower urinary system, gynecological tract, breast, hematolymphoid system, endocrine system, soft tissues, arthro-skeletal system, central and peripheral nervous system, dermatological system, and placenta with pathology of the fetus and newborn. Pediatric pathology is not the search of the curiosities and does not involve the collection of rare specimens or the taking care of a cabinet of teratological cases. The pediatric pathologist tries to address specifically congenital defects in a context that may help to advance medicine and discover new cures. Most of the genes involved in morphogenesis may play a crucial role in carcinogenesis as identified most recently by pharmacology studies.

Apart from being a pathologist, several pediatric pathologists are engaged in various humanitarian missions. Several pediatric pathologists hold a second specialty in pediatrics or gynecology and membership in pediatric or gynecology societies. Our job as pathologist involves an impressive work for advocacy of children and families. Rudolf Virchow, who is probably one of the founders of cellular pathology, was particularly active in social medicine and public health. In 1902, the *British Medical Journal* suggested that Berlin was one of the most hygienic cities in Europe due to Professor Virchow's efforts on social milieu and social reforms in Germany. The increasingly authoritarian nature of his Imperial country moved Virchow to demand democracy and social welfare programs. Following the steps of Virchow's deep sense of humanitarianism, several pediatric pathologists are now engaged in middle- or low-income countries to improve the public health qualities in these geographic areas. Sir William Osler, first Baronet, FRS

FRCP (1849–1919), was a Canadian physician and one of the four founding fathers of Johns Hopkins Hospital in Maryland, United States. As a student, Sir William Osler spent some time in Berlin learning the art of the autopsy from Virchow and, on the pathologist's 70th birthday, emphasized that his consistent humanitarianism amplified Virchow's scientific achievements. Such Canadian esprit has permeated the twenty-first century as well. In 2006, Jean Vanier, a Canadian philanthropist, who recently passed away, gave a talk at Concordia University in the city of Montreal. He argued that the Western culture of the individualism which values beauty, money, and success also creates a gap between the healthy and the disabled.

In this century, healthcare is going to go through financial and technological challenges. It is imperative that the pediatric pathologist is prone to support social initiatives. Pathologists provide strong support to patients and physicians alike as they face the battle with cancer, and the Springer team and I hope that our pages provide the concepts and the foundation for the future of cancer prevention and therapy as we move to the era of personalized molecular oncology. I hope that our book will help the students to learn some pediatric pathology skills and that the physicians are motivated in continuing learning programs. We have employed a coded format that we consider is fruitful and powerful for achieving optimal learning and memorization (DEF, definition; AKA/SYN, also known as/synonyms; EPI, epidemiology; EPG, etio-pathogenesis; CLI, clinics; GRO, gross findings; CLM, conventional light microscopy; TEM, transmission electron microscopy, HSS, histochemical special stains; IHC, immunohistochemistry; FNA, fine-needle aspiration; CMB, cytogenetics and molecular biology; DDX, differential diagnosis; TRT, treatment; PGN, prognosis). We think that the reader will enjoy this book and forgive me if some areas have not been emphasized enough. I trust this book can be a good help for many colleagues in their daily practice. We are eager to share this book with all our health professional colleagues across the world honestly hoping the reader will find the content of pertinence and interest to the care of all children.

Edmonton, AB, Canada   Consolato M. Sergi

Coda
All health is global health, and all medicine is social medicine, and, in this setting, pediatrics plays a significant role for the well-being and health of all children no matter their origin, the skin color they have, the religion they profess, or the orientation their parents have. Pediatric pathology is at the basis of the diseases that can be discovered before birth and after birth. This book is a solid basis that can be useful not only for pediatricians and pathologists involved in the healthcare of developed countries but also for low- and middle-income developing countries. Thus, most of the profit of the author will go to charities involving children. Pediatric pathologists and pediatricians remain essential leaders in the advocacy for pediatric patients and their families. Many of us have been engaged in teaching and providing healthcare services in the setting of the World Health Organization integrated management solutions and nongovernmental organizations, and their efforts have paved the roads for suggesting improvements for better healthcare in low- and middle-income countries. We hope that the application of experience from resource-rich settings to resource-limited settings will be followed by voracious readers able to gather information and notes from this book for the continuing medical education in global health.

Disclaimer
Medicine is an evolving field of knowledge. Thus, please be advised of the following disclaimer. Here, I inform readers of this book that the views and opinions expressed in the text belong solely to the author. They do not necessarily belong to the author's employer, organization, committee, or other group or individual. Moreover, I do not make any warranties about the completeness, reliability, and accuracy of this information. Please verify your current standards of practice in your country of practice. Any action you take upon the information on this book is strictly at your own risk, and by reading and understanding this disclaimer, I will not be liable for any losses and damages in connection with the use of the text of this book.

Contents

Volume I

1 Cardiovascular System 1
  1.1 Developmental and Genetics 2
  1.2 Congenital Heart Disease. 6
      1.2.1 Sequential Segmental Cardio-Analysis 8
      1.2.2 Atrial Septal Defect (ASD), Ventricular Septal Defect (VSD), Atrioventricular Septal Defect (AVSD), Patent Ductus Arteriosus (PDA), and Ebstein's Tricuspid Anomaly (ETA) 11
      1.2.3 Tetralogy of Fallot (TOF) and Double Outlet Right Ventricle (DORV) 18
      1.2.4 Hypoplastic Right Heart Syndrome. 24
      1.2.5 Transposition of the Great Arteries 26
      1.2.6 Common Trunk (Persistent Truncus Arteriosus) 31
      1.2.7 Total Anomalous Pulmonary Venous Return 32
      1.2.8 Scimitar Syndrome. 32
      1.2.9 Hypoplastic Left Heart Syndrome and Coarctation of Aorta 33
      1.2.10 Heterotaxy Syndrome 41
      1.2.11 Common Genetic Syndromes with Congenital Heart Disease 41
  1.3 Myocarditis and Nonischemic Cardiomyopathies 48
      1.3.1 Myocarditis 48
      1.3.2 Dilated Cardiomyopathy 48
      1.3.3 Hypertrophic Cardiomyopathy 66
      1.3.4 Infiltrative/Restrictive Cardiomyopathies 66
      1.3.5 Left Ventricular Non-compaction 78
      1.3.6 Arrhythmogenic Right Ventricular Dysplasia/ Cardiomyopathy (ARVD/C) 79
      1.3.7 Lamin A/C Cardiomyopathy 81
      1.3.8 Mitochondrial Cardiomyopathies 81
      1.3.9 Cardiac Channelopathies and Sudden Cardiac Death. 83
  1.4 Ischemic Heart Disease, Myocardial Dysfunction, and Heart Failure 85
      1.4.1 Ischemic Heart Disease 85
      1.4.2 Myocardial Dysfunction 85
      1.4.3 Heart Failure 85

	1.5	Valvular Heart Disease	87
		1.5.1	Endocarditis and Valvar Vegetation	87
		1.5.2	Mitral Valve Prolapse	91
		1.5.3	Calcific Mitral Annulus and Calcific Aortic Stenosis	91
		1.5.4	Rheumatic Heart Disease (RHD)	91
		1.5.5	Prosthetic Valves	92
	1.6	Transplantation	92
		1.6.1	Allograft Cellular Rejection	92
		1.6.2	Allograft Humoral Rejection	95
		1.6.3	Cardiac Allograft Vasculopathy	97
		1.6.4	Extracardiac Posttransplant Lympho-proliferative Disorders (PTLD) Following HTX	100
	1.7	Tumors	100
		1.7.1	Benign Tumors	102
		1.7.2	Malignant Tumors	109
		1.7.3	Metastatic Tumors	112
	1.8	Pericardial Disease	115
		1.8.1	Non-neoplastic Pericardial Disease	115
		1.8.2	Neoplastic Pericardial Disease	116
	1.9	Non-neoplastic Vascular Pathology	116
		1.9.1	Congenital Anomalies	116
		1.9.2	Fibromuscular Dysplasia	117
		1.9.3	Arteriosclerosis	117
		1.9.4	Large-Vessel Vasculitis	119
		1.9.5	Medium-Vessel Vasculitis	123
		1.9.6	Small-Vessel Vasculitis	125
	Multiple Choice Questions and Answers	125
	References and Recommended Readings	127
2	**Lower Respiratory Tract**	139
	2.1	Development and Genetics	141
	2.2	Dysmorphology and Perinatal Congenital Airway Diseases	145
		2.2.1	Foregut Cysts of Bronchogenic Type	146
		2.2.2	Agenesis, Aplasia, and Hypoplasia Pulmonis (Pulmonary Hypoplasia)	148
		2.2.3	CPAM and Congenital Lobar Emphysema	148
		2.2.4	Tracheal and Bronchial Anomalies	151
		2.2.5	Sequestrations and Vascular Anomalies	152
	2.3	Infantile and Pediatric *NILD*	155
		2.3.1	Neonatal Respiratory Distress Syndrome	157
		2.3.2	"Old" and "New" Bronchopulmonary Dysplasia (OBPD and NBPD)	157
		2.3.3	Atelectasis	158
		2.3.4	Cystic Fibrosis	158
	2.4	Infantile and Pediatric *ILD*	159
		2.4.1	Diffuse Lung Development-Associated ILD	159
		2.4.2	Trisomy 21 (Down Syndrome)-Associated ILD	161

	2.4.3	ILD Related to Other Chromosomal/Genomic Microdeletion Disorders	163
	2.4.4	Neuroendocrine Hyperplasia of Infancy (NEHI)	163
	2.4.5	Pulmonary Interstitial Glycogenosis (PIG)	164
	2.4.6	Surfactant Dysfunction Disorders	164
	2.4.7	Pulmonary Alveolar Proteinosis	166
2.5	"Adult" ILD of the Youth		167
	2.5.1	Diffuse Alveolar Damage (DAD)	167
	2.5.2	Cryptogenic Organizing Pneumonia (COP)	170
	2.5.3	Usual Interstitial Pneumonia/Pneumonitis (UIP)	173
	2.5.4	Nonspecific Interstitial Pneumonia/Pneumonitis (NSIP)	175
	2.5.5	Desquamative Interstitial Pneumonia/Pneumonitis	176
	2.5.6	Respiratory Bronchiolitis ILD	177
	2.5.7	Pulmonary Langerhans Cell Histiocytosis	179
	2.5.8	Lymphoid Interstitial Pneumonia/Pneumonitis (LIP)	180
	2.5.9	Hypersensitivity Pneumonia/Pneumonitis (HSP)	181
	2.5.10	Acute Eosinophilic Pneumonia/Pneumonitis	182
	2.5.11	Progressive Massive Fibrosis	184
2.6	Inflammation: Tracheobronchial		184
	2.6.1	Laryngotracheitis	184
	2.6.2	Acute Bronchitis	184
	2.6.3	Bronchial Asthma	184
	2.6.4	Pulmonary Hyperinflation	188
	2.6.5	Bronchiectasis	189
	2.6.6	Neoplasms	189
2.7	Inflammation: Infectious (Pneumonia/Pneumonitis)		190
	2.7.1	Lobar Pneumonia	190
	2.7.2	Bronchopneumonia	191
	2.7.3	Interstitial Pneumonia	193
	2.7.4	Primary Atypical Inflammation (Pneumonia)	193
2.8	Inflammation: Non-infectious (Pneumonia vs. Pneumonitis)		193
	2.8.1	Aspiration and Chemical Pneumonitis	193
	2.8.2	Lipoid Pneumonia	195
	2.8.3	Diffuse Pulmonary Hemorrhagic Syndromes	196
2.9	Inflammation: Infectious/Non-infectious Granulomatous/Nongranulomatous		197
	2.9.1	Primary Pulmonary Tuberculosis	197
	2.9.2	Sarcoidosis	200
	2.9.3	Aspergillosis	201
	2.9.4	Others	202
2.10	Chronic Obstructive Pulmonary Disease of the Youth		202
	2.10.1	Chronic Bronchitis	203
	2.10.2	Emphysema	203

	2.11	"Adult" Pneumoconiosis of the Youth	204
		2.11.1 Silicosis	205
		2.11.2 Asbestosis	206
		2.11.3 Non-silico Asbestosis-Related Pneumoconiosis	207
	2.12	Pulmonary Vascular Disorders	207
		2.12.1 Pulmonary Congestion and Edema	208
		2.12.2 Pulmonary Embolism/Infarction	208
		2.12.3 Pulmonary Arterial Hypertension (PAH)	210
	2.13	Transplantation-Related Disorders	212
		2.13.1 Acute Rejection	212
		2.13.2 Chronic Rejection	212
	2.14	Pediatric Tumors and Pseudotumors	215
		2.14.1 Pulmonary Teratoma	215
		2.14.2 Inflammatory Myofibroblastic Tumor	215
		2.14.3 Pulmonary Hamartoma	217
		2.14.4 Pulmonary Sclerosing Hemangioma	217
		2.14.5 Pulmonary Carcinoid	217
		2.14.6 Lymphangioma-(Leo)-Myomatosis (LAM)	217
		2.14.7 Kaposiform Lymphangiomatosis	219
		2.14.8 Pleuropulmonary Blastoma	220
		2.14.9 Metastatic Tumors	222
	2.15	"Adult"-Type Neoplasms in Childhood/Youth	222
		2.15.1 "Adult"-Type Preneoplastic Lesions	226
	2.16	Pleural Diseases	230
		2.16.1 Pleural Effusions and Pleuritis	230
		2.16.2 Pneumothorax	234
		2.16.3 Neoplastic Pleural Diseases	235
	2.17	Posttransplant Lymphoproliferative Disorders (PTLD)	237
	2.18	Non-thymic Mediastinal Pathology	240
		2.18.1 Anterior Mediastinal Pathology	241
		2.18.2 Middle Mediastinal Pathology	242
		2.18.3 Posterior Mediastinal Pathology	242
	Multiple Choice Questions and Answers		242
	References and Recommended Readings		244
3	**Gastrointestinal Tract**		255
	3.1	Development and Genetics	256
	3.2	Esophagus	258
		3.2.1 Esophageal Anomalies	258
		3.2.2 Esophageal Vascular Changes	263
		3.2.3 Esophageal Inflammatory Diseases	263
		3.2.4 Esophageal Tumors	274
	3.3	Stomach	277
		3.3.1 Gastric Anomalies	277
		3.3.2 Gastric Vascular Changes	280
		3.3.3 Gastric Inflammatory Diseases	281
		3.3.4 Tissue Continuity Damage-Related Gastric Degenerations	286
		3.3.5 Gastric Tumors	287

	3.4	Small Intestine	304
		3.4.1 Small Intestinal Anomalies	304
		3.4.2 Abdominal Wall Defects (Median-Paramedian)	306
		3.4.3 Continuity Defects of the Intestinal Lumen	312
		3.4.4 Intestinal Muscular Wall Defects	316
		3.4.5 Small Intestinal Dystopias	317
		3.4.6 Composition Abnormalities of the Intestinal Wall	317
		3.4.7 Small Intestinal Vascular Changes	318
		3.4.8 Inflammation and Malabsorption	319
		3.4.9 Short Bowel Syndrome/Intestinal Failure	336
		3.4.10 Small Intestinal Transplantation	336
		3.4.11 Graft-Versus-Host Disease (GVHD) of the Gut	339
		3.4.12 Small Intestinal Neoplasms	340
	3.5	Appendix	343
		3.5.1 Appendiceal Anomalies	343
		3.5.2 Appendiceal Vascular Changes	343
		3.5.3 Appendicitis	349
		3.5.4 Appendiceal Metabolic and Degenerative Changes	351
		3.5.5 Appendiceal Neoplasms	351
	3.6	Large Intestine	353
		3.6.1 Large Intestinal Anomalies	353
		3.6.2 Large Intestinal Vascular Changes	373
		3.6.3 Large Intestinal Inflammatory Disorders	374
		3.6.4 Colon-Rectum Neoplasms	380
	3.7	Anus	387
		3.7.1 Anal Anomalies	388
		3.7.2 Inflammatory Anal Diseases	389
		3.7.3 Benign Anal Tumors and Non-neoplastic Anal Lesions (Pseudotumors)	389
		3.7.4 Anal Pre- and Malignant Lesions	390
	3.8	Peritoneum	393
		3.8.1 Cytology	393
		3.8.2 Non-neoplastic Peritoneal Pathology	394
		3.8.3 Neoplastic Peritoneal Pathology	394
	Multiple Choice Questions and Answers		399
	References and Recommended Readings		402
4	**Parenchymal GI Glands: Liver**		**425**
	4.1	Development and Genetics	426
	4.2	Hepatobiliary Anomalies	430
		4.2.1 Ductal Plate Malformation (DPM)	431
		4.2.2 Congenital Hepatic Fibrosis	435
		4.2.3 Biliary Hamartoma (von Meyenburg Complex)	435
		4.2.4 Caroli Disease/Syndrome	435
		4.2.5 ADPKD-Related Liver Cysts	435
	4.3	Hyperbilirubinemia and Cholestasis	436
		4.3.1 Hyperbilirubinemia	436

	4.3.2	Decreased Bilirubin Conjugation and Unconjugated Hyperbilirubinemia	437
	4.3.3	Conjugated Hyperbilirubinemia	438
4.4	Infantile/Pediatric/Youth Cholangiopathies		439
	4.4.1	Biliary Atresia	440
	4.4.2	Non-BA Infantile Obstructive Cholangiopathies (NBAIOC)	443
	4.4.3	The Paucity of the Intrahepatic Biliary Ducts (PIBD)	444
	4.4.4	Neonatal Hepatitis Group (NAG)	444
	4.4.5	Primary Sclerosing Cholangitis (PSC)	456
	4.4.6	Primary Biliary Cirrhosis (PBC)	458
	4.4.7	Pregnancy-Related Liver Disease (PLD)	459
4.5	Genetic and Metabolic Liver Disease	460
	4.5.1	Endoplasmic Reticulum Storage Diseases (ERSDs)	461
	4.5.2	Congenital Dysregulation of Carbohydrate Metabolism	464
	4.5.3	Lipid/Glycolipid and Lipoprotein Metabolism Disorders	467
	4.5.4	Amino Acid Metabolism Disorders	471
	4.5.5	Mitochondrial Hepatopathies	471
	4.5.6	Peroxisomal Disorders	477
	4.5.7	Iron Metabolism Dysregulation	479
	4.5.8	Copper Metabolism Dysregulation	483
	4.5.9	Porphyria-Related Hepatopathies	485
	4.5.10	Shwachman-Diamond Syndrome (SDS)	485
	4.5.11	Chronic Granulomatous Disease (CGD)	485
	4.5.12	Albinism-Related Liver Diseases	486
4.6	Viral and AI Hepatitis, Chemical Injury, and Allograft Rejection	486
	4.6.1	Acute Viral Hepatitis	486
	4.6.2	Chronic Viral Hepatitis	488
	4.6.3	Autoimmune Hepatitis	489
	4.6.4	Drug-Induced Liver Disease (Chemical Injury) and TPN	490
	4.6.5	Granulomatous Liver Disease	495
	4.6.6	Alcoholic Liver Disease	496
	4.6.7	Non-alcoholic Steatohepatitis	498
	4.6.8	Acute and Chronic Rejection Post-Liver Transplantation	498
4.7	Hepatic Vascular Disorders	503
	4.7.1	Acute and Chronic Passive Liver Congestion	503
	4.7.2	Ischemic Hepatocellular Necrosis	505
	4.7.3	Shock-Related Cholestasis	505
4.8	Liver Failure and Liver Cirrhosis	505
4.9	Portal Hypertension	508
4.10	Bacterial and Parasitic Liver Infections	509

		4.10.1	Pyogenic Abscess	509
		4.10.2	Helminthiasis	510
	4.11	Liver Tumors		510
		4.11.1	Benign Tumors	510
		4.11.2	Malignant Tumors	517
		4.11.3	Metastatic Tumors	534
	Multiple Choice Questions and Answers			538
	References and Recommended Readings			540
5	**Parenchymal GI Glands (Gallbladder, Biliary Tract, and Pancreas)**			**551**
	5.1	Development and Genetics		551
	5.2	Biliary and Pancreatic Structural Anomalies		553
		5.2.1	Choledochal Cysts	553
		5.2.2	Gallbladder Congenital Abnormalities	553
		5.2.3	Pancreas Congenital Anomalies	553
	5.3	Gallbladder and Extrahepatic Biliary Tract		556
		5.3.1	Cholesterolosis and Cholelithiasis	556
		5.3.2	Acute and Chronic Cholecystitis	557
		5.3.3	Gallbladder Proliferative Processes and Neoplasms	558
		5.3.4	Extrahepatic Bile Duct Tumors and Cholangiocellular Carcinoma	561
	5.4	Pancreas Pathology		561
		5.4.1	Congenital Hyperinsulinism and Nesidioblastosis	561
		5.4.2	Acute and Chronic Pancreatitis	563
		5.4.3	Pancreatoblastoma and Acinar Cell Carcinoma in Childhood and Youth	566
		5.4.4	Cysts and Cystic Neoplastic Processes of the Pancreas	567
		5.4.5	PanIN and Solid Pancreas Ductal Carcinoma	571
		5.4.6	Other Tumors	571
		5.4.7	Degenerative Changes and Transplant Pathology	571
	Multiple Choice Questions and Answers			574
	References and Recommended Readings			575
6	**Kidney, Pelvis, and Ureter**			**579**
	6.1	Development and Genetics		581
	6.2	Non-cystic Congenital Anomalies		589
		6.2.1	Disorders of Number	589
		6.2.2	Disorders of Rotation	591
		6.2.3	Disorders of Position	591
		6.2.4	Disorders of Separation	591
	6.3	Cystic Renal Diseases		591
		6.3.1	Classifications	591
		6.3.2	Autosomal Dominant Polycystic Kidney Disease	591
		6.3.3	Autosomal Recessive Polycystic Kidney Disease	596
		6.3.4	Medullary Sponge Kidney	597

	6.3.5	Multicystic Dysplastic Kidney (MCDK)	597
	6.3.6	Hydronephrosis/Hydroureteronephrosis	597
	6.3.7	Simple Renal Cysts	598
	6.3.8	Acquired Cystic Kidney Disease (CRD-Related)	598
	6.3.9	Genetic Syndromes and Cystic Renal Disease	598
6.4	Primary Glomerular Diseases		599
	6.4.1	Hypersensitivity Reactions and Major Clinical Syndromes of Glomerular Disease	599
	6.4.2	Post-infectious Glomerulonephritis	600
	6.4.3	Rapidly Progressive Glomerulonephritis	602
	6.4.4	Minimal Change Disease	604
	6.4.5	(Diffuse) Mesangial Proliferative GN	604
	6.4.6	Focal and Segmental Glomerulosclerosis	604
	6.4.7	Membranous Glomerulonephritis	606
	6.4.8	Membranoproliferative (Membrane-Capillary) Glomerulonephritis	609
6.5	Secondary Glomerular Diseases		610
	6.5.1	SLE/Lupus Nephritis	610
	6.5.2	Henoch-Schönlein Purpura	611
	6.5.3	Amyloidosis	613
	6.5.4	Light Chain Disease	614
	6.5.5	Cryoglobulinemia	614
	6.5.6	Diabetic Nephropathy	614
6.6	Hereditary/Familial Nephropathies		615
	6.6.1	Fabry Nephropathy	615
	6.6.2	Alport Syndrome	615
	6.6.3	Nail-Patella Syndrome	617
	6.6.4	Congenital Nephrotic Syndrome	617
	6.6.5	Thin Glomerular Basement Membrane Nephropathy (TBMN)	617
6.7	Tubulointerstitial Diseases		618
	6.7.1	Acute Tubulointerstitial Nephritis (ATIN)	618
	6.7.2	Chronic Tubulointerstitial Nephritis (CTIN)	622
	6.7.3	Acute Tubular Necrosis	623
	6.7.4	Chronic Renal Failure (CRF)	623
	6.7.5	Nephrolithiasis and Nephrocalcinosis	624
	6.7.6	Osmotic Nephrosis and Hyaline Change	625
	6.7.7	Hypokalemic Nephropathy	625
	6.7.8	Urate Nephropathy	625
	6.7.9	Cholemic Nephropathy/Jaundice-Linked Acute Kidney Injury	625
	6.7.10	Myeloma Kidney	625
	6.7.11	Radiation Nephropathy	626
	6.7.12	Tubulointerstitial Nephritis and Uveitis (TINU)	626
6.8	Vascular Diseases		626
	6.8.1	Benign Nephrosclerosis	626
	6.8.2	Malignant Nephrosclerosis	626
	6.8.3	Renal Artery Stenosis	626

	6.8.4 Infarcts	627
	6.8.5 Vasculitis	627
	6.8.6 Hemolytic Uremic Syndrome	628
	6.8.7 Thrombotic Thrombocytopenic Purpura	630
6.9	Renal Transplantation	630
	6.9.1 Preservation Injury	630
	6.9.2 Hyperacute Rejection	630
	6.9.3 Acute Rejection	631
	6.9.4 Chronic Rejection	632
	6.9.5 Humoral (Acute/Chronic) Rejection	632
	6.9.6 Antirejection Drug Toxicity	633
	6.9.7 Recurrence of Primary Disorder	633
6.10	Hereditary Cancer Syndromes Associated with Renal Tumors	633
	6.10.1 Beckwith-Wiedemann Syndrome	633
	6.10.2 WAGR Syndrome	633
	6.10.3 Denys-Drash Syndrome	634
	6.10.4 Non-WT1/Non-WT2 Pediatric Syndromes	634
	6.10.5 Von Hippel Lindau Syndrome	634
	6.10.6 Tuberous Sclerosis Syndrome	634
	6.10.7 Hereditary Papillary Renal Carcinoma Syndrome	634
	6.10.8 Hereditary Leiomyoma Renal Carcinoma Syndrome	634
	6.10.9 Birt-Hogg-Dube Syndrome	634
6.11	Pediatric Tumors (Embryonal)	634
	6.11.1 Wilms Tumor (Nephroblastoma)	634
	6.11.2 Cystic Partially Differentiated Nephroblastoma (CPDN) and (Pediatric) Cystic Nephroma	645
	6.11.3 Congenital Mesoblastic Nephroma	646
	6.11.4 Clear Cell Sarcoma	646
	6.11.5 Rhabdoid Tumor	647
	6.11.6 Metanephric Tumors	647
	6.11.7 XP11 Translocation Carcinoma	647
	6.11.8 Ossifying Renal Tumor of Infancy	650
6.12	Non-embryonal Tumors of the Young	650
	6.12.1 Clear Cell Renal Cell Carcinoma	650
	6.12.2 Chromophobe Renal Cell Carcinoma	653
	6.12.3 Papillary Adenoma and Renal Cell Carcinoma	653
	6.12.4 Collecting Duct Carcinoma	656
	6.12.5 Renal Medullary Carcinoma	656
	6.12.6 Angiomyolipoma	656
	6.12.7 Oncocytoma	656
	6.12.8 Other Epithelial and Mesenchymal Tumors	659
6.13	Non-neoplastic Pathology of the Pelvis and Ureter	661
	6.13.1 Anatomy and Physiology Notes	661
	6.13.2 Congenital Pelvic-Ureteral Anomalies	661
	6.13.3 Congenital Ureteric Anomalies	661
	6.13.4 Lower Urinary Tract Abnormalities	662

		6.14	Tumors of the Pelvis and Ureter	666
		6.14.1	Neoplasms	666
		6.14.2	Genetic Syndromes	666
	Multiple Choice Questions and Answers			667
	References and Recommended Readings			668
7	**Lower Urinary and Male Genital System**			673
	7.1	Development and Genetics		674
		7.1.1	Urinary Bladder and Ureter	674
		7.1.2	Testis, Prostate, and Penis	675
	7.2	Lower Urinary and Genital System Anomalies		676
		7.2.1	Lower Urinary System Anomalies	676
		7.2.2	Male Genital System Anomalies	678
	7.3	Urinary Tract Inflammatory and Degenerative Conditions		684
		7.3.1	Cystitis, Infectious	684
		7.3.2	Cystitis, Non-infectious	684
		7.3.3	Malacoplakia of the Young	686
		7.3.4	Urinary Tract Infections and Vesicoureteral Reflux	686
		7.3.5	Megacystis, Megaureter, Hydronephrosis, and Neurogenic Bladder	687
		7.3.6	Urinary Tract Endometriosis	688
		7.3.7	Tumorlike Lesions (Including Caruncles)	689
	7.4	Preneoplastic and Neoplastic Conditions of the Urinary Tract		690
		7.4.1	Urothelial Hyperplasia (Flat and Papillary)	690
		7.4.2	Reactive Atypia, Atypia of Unknown Significance, Dysplasia, and Carcinoma In Situ (CIS)	691
		7.4.3	Noninvasive Papillary Urothelial Neoplasms	691
		7.4.4	Invasive Urothelial Neoplasms	693
		7.4.5	Non-urothelial Differentiated Urinary Tract Neoplasms	696
	7.5	Male Infertility-Associated Disorders		696
		7.5.1	Spermiogram and Classification	696
	7.6	Inflammatory Disorders of the Testis and Epididymis		696
		7.6.1	Acute Orchitis	697
		7.6.2	Epidermoid Cysts	697
		7.6.3	Hydrocele	700
		7.6.4	Spermatocele	700
		7.6.5	Varicocele	704
	7.7	Testicular Tumors		704
		7.7.1	Germ Cell Tumors	704
		7.7.2	Tumors of Specialized Gonadal Stroma	720
		7.7.3	Rhabdomyosarcoma and Rhabdoid Tumor of the Testis	727
		7.7.4	Secondary Tumors	727

	7.8	Tumors of the Epididymis	727
		7.8.1 Adenomatoid Tumor	727
		7.8.2 Papillary Cystadenoma	728
		7.8.3 Rhabdomyosarcoma	728
		7.8.4 Mesothelioma	728
	7.9	Inflammatory Disorders of the Prostate Gland	728
		7.9.1 Acute Prostatitis	728
		7.9.2 Chronic Prostatitis	728
		7.9.3 Granulomatous Prostatitis	729
		7.9.4 Prostatic Malakoplakia of the Youth	729
	7.10	Prostate Gland Overgrowths	729
		7.10.1 Benign Nodular Hyperplasia of the Young and Fibromatosis	729
		7.10.2 Rhabdomyosarcoma, Leiomyosarcoma, and Other Sarcomas (e.g., Ewing Sarcoma)	729
		7.10.3 Prostatic Carcinoma Mimickers	730
		7.10.4 Prostatic Intraepithelial Neoplasia (PIN)	732
		7.10.5 Prostate Cancer of the Young	733
		7.10.6 Hematological Malignancies	739
		7.10.7 Secondary Tumors	739
	7.11	Inflammatory and Neoplastic Disorders of the Penis	739
		7.11.1 Infections	739
		7.11.2 Non-infectious Inflammatory Diseases	740
		7.11.3 Penile Cysts and Noninvasive Squamous Cell Lesions	741
		7.11.4 Penile Squamous Cell Carcinoma of the Youth	741
		7.11.5 Non-squamous Cell Carcinoma Neoplasms of the Penis	744
	Multiple Choice Questions and Answers		744
	References and Recommended Readings		746

Volume II

8	**Female Genital System**		757
	8.1	Development and Genetics	758
	8.2	Congenital Anomalies of the Female Genital System	760
	8.3	Ovarian Inflammatory and Degenerative Conditions	761
		8.3.1 Oophoritis	761
		8.3.2 Torsion	761
		8.3.3 Non-neoplastic Cystic Lesions	763
		8.3.4 Non-neoplastic Proliferations	763
	8.4	Infectious Diseases of the Female Genital System	764
		8.4.1 Viral Diseases	764
		8.4.2 Bacterial Diseases	765
		8.4.3 Fungal Diseases	767
		8.4.4 Parasitic Diseases	767
	8.5	Inflammatory, Reactive Changes and Degenerative Conditions of Vulva, Vagina, Cervix, Uterus, and Tuba	767
		8.5.1 Non- and Preneoplastic Vulvar and Vaginal Lesions	767
		8.5.2 Inflammation-Associated Cervical Lesions	769

		8.5.3	Reactive Changes of the Cervix................	770
		8.5.4	Cyst-Associated Cervical Lesions...............	770
		8.5.5	Ectopia-Associated Cervical Lesions............	770
		8.5.6	Pregnancy-Associated Cervical Lesions..........	770
		8.5.7	Cervical Metaplasias........................	771
		8.5.8	Non-neoplastic and Preneoplastic Uterine Lesions.............................	771
		8.5.9	Non-neoplastic Abnormalities of the Tuba........	773
	8.6	Tumors of the Ovary................................		773
		8.6.1	Germ Cell Tumors...........................	774
		8.6.2	Surface Epithelial Tumors.....................	779
		8.6.3	Sex Cord-Stromal Tumors.....................	790
		8.6.4	Other Primary Ovarian and Secondary Tumors.....	798
	8.7	Tumors of the Tuba................................		798
		8.7.1	Benign Neoplasms...........................	798
		8.7.2	Malignant Neoplasms........................	798
	8.8	Tumors of the Uterus................................		798
		8.8.1	Type I/Type II EC...........................	799
		8.8.2	Non-type I/Non-type II EC....................	799
		8.8.3	Uterine Leiomyoma (ULM) and Variants.........	799
		8.8.4	Uterine Leiomyosarcoma (ULMS)..............	799
		8.8.5	Uterine Adenomatoid Tumor...................	804
		8.8.6	Uterine Lymphangiomyomatosis (ULAM)........	804
		8.8.7	Uterine Stroma Tumors.......................	804
		8.8.8	Hematological Malignancies and Secondary Tumors.......................	804
	8.9	Tumors of the Cervix...............................		805
		8.9.1	Benign Neoplasms...........................	805
		8.9.2	Precancerous Conditions and Malignant Neoplasms.................................	805
	8.10	Tumors of the Vagina...............................		811
		8.10.1	Benign Tumors.............................	811
		8.10.2	Precancerous Conditions and Malignant Tumors....................................	813
	8.11	Tumors of the Vulva................................		816
		8.11.1	Benign Neoplasms...........................	816
		8.11.2	Malignant Tumors...........................	818
	Multiple Choice Questions and Answers.....................			823
	References and Recommended Readings...................			825
9	**Breast**...			833
	9.1	Development and Genetics..........................		834
	9.2	Congenital Anomalies, Inflammatory, and Related Disorders................................		836
		9.2.1	Amastia, Atelia, Synmastia, Polymastia, and Politelia................................	836
		9.2.2	Asymmetry, Hypotrophy, and Hypertrophy........	836
		9.2.3	Dysmaturity and Precocious Thelarche...........	838
		9.2.4	Acute Mastitis, Abscess, and Phlegmon..........	838

	9.2.5	Duct Ectasia, Periductal Mastitis, and Granulomatous Mastitis	839
	9.2.6	Necrosis, Calcifications, and Mondor Disease	839
9.3	Pathology of the Female and Young Adult.	840
	9.3.1	Fibrocystic Disease	840
	9.3.2	Soft Tissue Tumors and Hematological Malignancies	840
	9.3.3	Fibroadenoma	840
	9.3.4	Adenoma	841
	9.3.5	Genetic Background of Breast Cancer.	844
	9.3.6	In Situ and Invasive Ductal Breast Carcinoma	847
	9.3.7	In Situ and Invasive Lobular Breast Carcinoma	849
	9.3.8	WHO Variants of the Infiltrating Ductal Carcinoma	849
	9.3.9	Sweat Gland-Type Tumors and Myoepithelial Tumors	850
	9.3.10	Phyllodes Tumor	851
9.4	Cancer Mimickers	852
	9.4.1	Hyperplasia, Ductal and Lobular.	852
	9.4.2	Adenosis.	855
9.5	Male Breast Disease.	855
	9.5.1	Gynecomastia.	855
	9.5.2	Breast Cancer of the Male	855
Multiple Choice Questions and Answers			856
References and Recommended Readings.			857

10 Hematolymphoid System 861

10.1	Development and Genetics		862
10.2	Red Blood Cell Disorders		864
	10.2.1	Anemia.	864
	10.2.2	Polycythemia	867
10.3	Coagulation and Hemostasis Disorders		869
	10.3.1	Coagulation and Hemostasis	869
	10.3.2	Coagulation Disorders	869
	10.3.3	Platelet Disorders.	870
10.4	White Blood Cell Disorders.		871
	10.4.1	Leukocytopenias and Leukocyte Dysfunctionalities	871
	10.4.2	Non-neoplastic Leukocytosis.	871
	10.4.3	Leukemia (Neoplastic Leukocytosis) or *Virchow's "Weisses Blut"*	872
	10.4.4	Myelodysplastic Syndromes	878
	10.4.5	Hodgkin Lymphoma	878
	10.4.6	Non-Hodgkin Lymphomas.	887
	10.4.7	Follicular Lymphoma.	888
	10.4.8	Small Lymphocytic Lymphoma.	890
	10.4.9	Mantle Cell Lymphoma (MCL).	891
	10.4.10	Marginal Cell Lymphoma	891

		10.4.11 Diffuse Large B-Cell Lymphoma (DLBCL)	892
		10.4.12 Lymphoblastic Lymphoma	894
		10.4.13 Burkitt Lymphoma	896
		10.4.14 Peripheral T-Cell Lymphoma	902
		10.4.15 Anaplastic Large Cell Lymphoma	903
		10.4.16 Adult T-Cell Leukemia/Lymphoma	903
		10.4.17 Cutaneous T-Cell Lymphoma (CTCL)	904
		10.4.18 Angiocentric Immunoproliferative Lesions	905
		10.4.19 Extranodal NK-/T-Cell Lymphoma	906
	10.5	Disorders of the Monocyte-Macrophage System and Mast Cells	907
		10.5.1 Hemophagocytic Syndrome	907
		10.5.2 Sinus Histiocytosis with Massive Lymphadenopathy (Rosai-Dorfman Disease)	908
		10.5.3 Langerhans Cell Histiocytosis	908
		10.5.4 Histiocytic Medullary Reticulosis	909
		10.5.5 True Histiocytic Lymphoma	910
		10.5.6 Systemic Mastocytosis	910
	10.6	Plasma Cell Disorders	910
		10.6.1 Multiple Myeloma	911
		10.6.2 Solitary Myeloma	912
		10.6.3 Plasma Cell Leukemia	912
		10.6.4 Waldenstrom's Macroglobulinemia	912
		10.6.5 Heavy Chain Disease	912
		10.6.6 Monoclonal Gammopathy of Undetermined Significance (MGUS)	913
	10.7	Benign Lymphadenopathies	913
		10.7.1 Follicular Hyperplasia	914
		10.7.2 Diffuse (Paracortical) Hyperplasia	918
		10.7.3 Sinus Pattern	919
		10.7.4 Predominant Granulomatous Pattern	920
		10.7.5 Other Myxoid Patterns	921
		10.7.6 Angioimmunoblastic Lymphadenopathy with Dysproteinemia (AILD)	921
	10.8	Disorders of the Spleen	921
		10.8.1 White Pulp Disorders of the Spleen	922
		10.8.2 Red Pulp Disorders of the Spleen	922
	10.9	Disorders of the Thymus	924
		10.9.1 Thymic Cysts and Thymolipoma	924
		10.9.2 True Thymic Hyperplasia and Thymic Follicular Hyperplasia	924
		10.9.3 HIV Changes	925
		10.9.4 Thymoma	925
	Multiple Choice Questions and Answers		927
	References and Recommended Readings		930
11	**Endocrine System**		933
	11.1	Development and Genetics	934
	11.2	Pituitary Gland Pathology	938

		11.2.1	Congenital Anomalies of the Pituitary Gland	939
		11.2.2	Vascular and Degenerative Changes	939
		11.2.3	Pituitary Adenomas and Hyperpituitarism	939
		11.2.4	Genetic Syndromes Associated with Pituitary Adenomas	941
		11.2.5	Hypopituitarism (Simmonds Disease)	942
		11.2.6	Empty Sella Syndrome (ESS)	943
		11.2.7	Neurohypophysopathies (Disorders of the Posterior Pituitary Gland)	943
	11.3	Thyroid Gland Pathology		943
		11.3.1	Congenital Anomalies, Hyperplasia, and Thyroiditis	946
		11.3.2	Congenital Anomalies, Goiter, and Dysfunctional Thyroid Gland	946
		11.3.3	Inflammatory and Immunologic Thyroiditis	948
		11.3.4	Epithelial Neoplasms of the Thyroid Glands	949
	11.4	Parathyroid Gland Pathology		965
		11.4.1	Congenital Anomalies of the Parathyroid Glands	965
		11.4.2	Parathyroid Gland Hyperplasia	965
		11.4.3	Parathyroid Gland Adenoma	966
		11.4.4	Parathyroid Gland Carcinoma	966
	11.5	Adrenal Gland Pathology		967
		11.5.1	Congenital Anomalies of the Adrenal Gland and Paraganglia	967
		11.5.2	Dysfunctional Adrenal Gland	969
		11.5.3	Adrenalitis	969
		11.5.4	Neoplasms of the Adrenal Gland and Paraganglia	974
		11.5.5	Syndromes Associated with Adrenal Cortex Abnormalities	996
	Multiple Choice Questions and Answers			997
	References and Recommended Readings			998
12	**Soft Tissue**			**1003**
	12.1	Development and Genetics		1004
	12.2	Vascular and Inflammatory Changes of Soft Tissue		1005
		12.2.1	Hyperemia	1005
		12.2.2	Necrotizing Fasciitis	1005
		12.2.3	Vasculitis-Associated Soft Tissue Changes	1005
		12.2.4	Miscellaneous	1005
	12.3	Soft Tissue Neoplasms: Scoring		1009
	12.4	Adipocytic Tumors		1010
		12.4.1	Lipoma	1011
		12.4.2	Lipoma Subtypes	1011
		12.4.3	Lipomatosis	1012
		12.4.4	Lipoblastoma	1012
		12.4.5	Hibernoma	1012

		12.4.6 Locally Aggressive and Malignant Adipocytic Tumors.............................. 1013
12.5	Fibroblastic/Myofibroblastic Tumors.................. 1017	
	12.5.1	Fasciitis/Myositis Group 1019
	12.5.2	Fibroma Group............................. 1020
	12.5.3	Fibroblastoma Classic and Subtypes 1021
	12.5.4	Fibrous Hamartoma of Infancy (FHI) 1023
	12.5.5	Fibromatosis of Childhood 1023
	12.5.6	Infantile Myofibroma/Myofibromatosis......... 1027
	12.5.7	Fibroblastic/Myofibroblastic Tumors with Intermediate Malignant Potential.......... 1027
	12.5.8	Malignant Fibroblastic/Myofibroblastic Tumors................................... 1033
12.6	Fibrohistiocytic Tumors........................... 1036	
	12.6.1	Histiocytoma 1036
	12.6.2	Benign Fibrous Histiocytoma 1037
	12.6.3	Borderline Fibrous Histiocytoma.............. 1039
	12.6.4	Malignant Fibrous Histiocytoma 1039
12.7	Smooth Muscle Tumors........................... 1040	
	12.7.1	Leiomyoma 1040
	12.7.2	EBV-Related Smooth Muscle Tumors.......... 1040
	12.7.3	Leiomyosarcoma 1040
12.8	Pericytic Tumors 1041	
	12.8.1	Glomus Tumor............................ 1041
	12.8.2	Glomangiosarcoma 1042
	12.8.3	Myopericytoma 1042
12.9	Skeletal Muscle Tumors........................... 1044	
	12.9.1	Rhabdomyomatous Mesenchymal Hamartoma (RMH) 1044
	12.9.2	Rhabdomyoma............................ 1046
	12.9.3	Rhabdomyosarcoma........................ 1046
	12.9.4	Pleomorphic RMS 1055
12.10	Vascular Tumors................................ 1059	
	12.10.1	Benign Vascular Tumors 1059
	12.10.2	Vascular Tumors with Intermediate Malignant Potential 1062
	12.10.3	Malignant Vascular Tumors.................. 1066
	12.10.4	Genetic Syndromes Associated with Vascular Tumors...................... 1068
12.11	Chondro-Osteoforming Tumors..................... 1069	
	12.11.1	Extraskeletal Chondroma.................... 1069
	12.11.2	Extraskeletal Myxoid Chondrosarcoma......... 1069
	12.11.3	Mesenchymal Chondrosarcoma............... 1070
	12.11.4	Extraskeletal Aneurysmatic Bone Cyst 1070
	12.11.5	Extraskeletal Osteosarcoma (ESOS) 1070
	12.11.6	Extraskeletal Chordoma..................... 1070
12.12	Tumors of Uncertain Differentiation 1071	
	12.12.1	Myxoma................................. 1071
	12.12.2	Myoepithelial Carcinoma.................... 1072

		12.12.3	Parachordoma	1072
		12.12.4	Synovial Sarcoma	1074
		12.12.5	Epithelioid Sarcoma	1076
		12.12.6	Alveolar Soft Part Sarcoma	1076
		12.12.7	Clear Cell Sarcoma	1079
		12.12.8	Extraskeletal Myxoid Chondrosarcoma (ESMC)	1080
		12.12.9	PNET/Extraskeletal Ewing Sarcoma (ESES)	1081
		12.12.10	Desmoplastic Small Round Cell Tumor	1083
		12.12.11	Extrarenal Rhabdoid Tumor	1084
		12.12.12	Malignant Mesenchymoma	1084
		12.12.13	PEComa	1084
		12.12.14	Extrarenal Wilms' Tumor	1085
		12.12.15	Sacrococcygeal Teratoma and Extragonadal Germ Cell Tumor and Yolk Sac Tumor	1085
	Multiple Choice Questions and Answers			1090
	References and Recommended Readings			1091
13	**Arthro-Skeletal System**			**1095**
	13.1	Development and Genetics		1096
	13.2	Osteochondrodysplasias		1097
		13.2.1	Nosology and Nomenclature	1097
		13.2.2	Groups of Genetic Skeletal Disorders	1098
	13.3	Metabolic Skeletal Diseases		1103
		13.3.1	Rickets, and Osteomalacia	1103
		13.3.2	Osteoporosis of the Youth	1104
		13.3.3	Paget Disease of the Bone	1106
		13.3.4	Juvenile Paget Disease	1108
	13.4	Osteitis and Osteomyelitis		1109
		13.4.1	Osteomyelitis	1109
	13.5	Osteonecrosis		1113
		13.5.1	Bony Infarct and Osteochondritis Dissecans	1113
	13.6	Tumorlike Lesions and Bone/Osteoid-Forming Tumors		1115
		13.6.1	Myositis Ossificans	1115
		13.6.2	Fibrous Dysplasia and Osteofibrous Dysplasia	1116
		13.6.3	Non-ossifying Fibroma (NOF)	1118
		13.6.4	Bone Cysts	1120
		13.6.5	Osteoma, Osteoid Osteoma, and Giant Osteoid Osteoma	1124
		13.6.6	Giant Cell Tumor	1127
		13.6.7	Osteosarcoma	1128
	13.7	Chondroid (Cartilage)-Forming Tumors		1136
		13.7.1	Osteochondroma	1137
		13.7.2	Enchondroma	1139
		13.7.3	Chondroblastoma	1140
		13.7.4	Chondromyxoid Fibroma	1143
		13.7.5	Chondrosarcoma	1145
	13.8	Bone Ewing Sarcoma		1147

	13.9	Miscellaneous Bone Tumors 1149

- 13.9 Miscellaneous Bone Tumors 1149
    - 13.9.1 Chordoma.................................. 1149
    - 13.9.2 Adamantinoma............................. 1150
    - 13.9.3 Langerhans Cell Histiocytosis 1151
    - 13.9.4 Vascular, Smooth Muscle, and Lipogenic Tumors................................... 1153
    - 13.9.5 Hematologic Tumors 1153
- 13.10 Metastatic Bone Tumors 1153
- 13.11 Juvenile Rheumatoid Arthritis and Juvenile Arthropathies.................................... 1154
    - 13.11.1 Rheumatoid Arthritis and Juvenile Rheumatoid Arthritis 1154
    - 13.11.2 Infectious Arthritis...................... 1156
    - 13.11.3 Gout, Early-Onset Juvenile Tophaceous Gout and Pseudogout..................... 1157
    - 13.11.4 Bursitis, Baker Cyst, and Ganglion 1159
    - 13.11.5 Pigmented Villonodular Synovitis and Nodular Tenosynovitis 1159
- Multiple Choice Questions and Answers 1161
- References and Recommended Readings................... 1162

14 Head and Neck 1167
- 14.1 Development 1168
- 14.2 Nasal Cavity, Paranasal Sinuses, and Nasopharynx....... 1171
    - 14.2.1 Congenital Anomalies 1171
    - 14.2.2 Inflammatory Lesions 1173
    - 14.2.3 Tumors................................... 1174
- 14.3 Larynx and Trachea 1183
    - 14.3.1 Congenital Anomalies 1183
    - 14.3.2 Cysts and Laryngoceles 1184
    - 14.3.3 Inflammatory Lesions and Non-neoplastic Lesions................................... 1185
    - 14.3.4 Tumors................................... 1186
- 14.4 Oral Cavity and Oropharynx 1189
    - 14.4.1 Congenital Anomalies 1189
    - 14.4.2 Branchial Cleft Cysts..................... 1195
    - 14.4.3 Inflammatory Lesions 1197
    - 14.4.4 Tumors................................... 1197
- 14.5 Salivary Glands 1203
    - 14.5.1 Congenital Anomalies 1203
    - 14.5.2 Inflammatory Lesions and Non-neoplastic Lesions................................... 1203
    - 14.5.3 Tumors................................... 1205
- 14.6 Mandible and Maxilla 1209
    - 14.6.1 Odontogenic Cysts........................ 1210
    - 14.6.2 Odontogenic Tumors 1216
    - 14.6.3 Bone-Related Lesions 1218

	14.7	Ear 1218
		14.7.1 Congenital Anomalies 1218
		14.7.2 Inflammatory Lesions and Non-neoplastic Lesions................................. 1218
		14.7.3 Tumors 1220
	14.8	Eye and Ocular Adnexa 1226
		14.8.1 Congenital Anomalies 1226
		14.8.2 Inflammatory Lesions and Non-neoplastic Lesions................................. 1226
		14.8.3 Tumors 1226
	14.9	Skull... 1230
	Multiple Choice Questions and Answers 1233
	References and Recommended Readings..................... 1235
15	**Central Nervous System**................................. 1243
	15.1	Development: Genetics 1244
		15.1.1 Development and Genetics 1244
		15.1.2 Neuromeric Model of the Organization of the Embryonic Forebrain According to Puelles and Rubenstein 1246
	15.2	Congenital Abnormalities of the Central Nervous System.................................... 1247
		15.2.1 Ectopia.................................... 1247
		15.2.2 Neural Tube Defects (NTDs).................. 1250
		15.2.3 Prosencephalon Defects...................... 1252
		15.2.4 Vesicular Forebrain (Pseudo-aprosencephaly)..... 1254
		15.2.5 Ventriculomegaly/Hydrocephalus 1257
		15.2.6 Agenesis of the ***Corpus Callosum*** (ACC)........ 1257
		15.2.7 Cerebellar Malformations 1258
		15.2.8 Agnathia Otocephaly Complex (AGOTC)........ 1259
		15.2.9 Telencephalosynapsis (Synencephaly) and Rhombencephalon Synapsis 1259
		15.2.10 CNS Defects in Acardia...................... 1261
		15.2.11 CNS Defects in Chromosomal and Genetic Syndromes..................... 1263
		15.2.12 Neuronal Migration Disorders................. 1264
		15.2.13 Phakomatoses............................... 1264
	15.3	Vascular Disorders of the Central Nervous System....... 1267
		15.3.1 Intracranial Hemorrhage 1267
		15.3.2 Vascular Malformations...................... 1272
		15.3.3 Aneurysms................................. 1273
		15.3.4 Thrombosis of Venous Sinuses and Cerebral Veins........................ 1274
		15.3.5 Pediatric and Inherited Neurovascular Diseases ... 1274
	15.4	Infections of the CNS............................... 1275
		15.4.1 Suppurative Infections 1276
		15.4.2 Tuberculous (Lepto-)Meningitis 1279
		15.4.3 Neurosyphilis............................... 1280
		15.4.4 Viral Infections.............................. 1280

		15.4.5 Toxoplasmosis 1282

    15.4.5 Toxoplasmosis 1282
    15.4.6 Fungal Infections 1282
15.5 Metabolic Disorders Affecting the CNS 1282
    15.5.1 Pernicious Anemia 1283
    15.5.2 Wernicke Encephalopathy 1283
15.6 Trauma to the Head and Spine 1284
15.7 Head Injuries 1284
    15.7.1 Epidural Hematoma 1284
    15.7.2 Subdural Hematoma 1284
    15.7.3 Subarachnoidal Hemorrhage 1285
    15.7.4 Spinal Injuries 1285
    15.7.5 Intervertebral Disk Herniation 1286
15.8 Demyelinating Diseases Involving the Central
Nervous System 1286
    15.8.1 Multiple Sclerosis 1286
    15.8.2 Leukodystrophies 1287
    15.8.3 Amyotrophic Lateral Sclerosis 1288
    15.8.4 Werdnig-Hoffmann Disease 1288
    15.8.5 Syringomyelia 1289
    15.8.6 Parkinson Disease and Parkinson
           Disease-Associated, G-Protein-Coupled
           Receptor 37 (GPR37/PaelR)-Related
           Autism Spectrum Disorder 1289
    15.8.7 Creutzfeldt-Jakob Disease (sCJD or
           Sporadic), CJD-Familial and CJD-Variant 1290
    15.8.8 West Syndrome/Infantile Spasms, ACTH
           Therapy, and Sudden Death 1290
15.9 Neoplasms of the Central Nervous System 1291
    15.9.1 Astrocyte-Derived Neoplasms 1291
    15.9.2 Ependymoma 1294
    15.9.3 Medulloblastoma 1295
    15.9.4 Meningioma 1299
    15.9.5 Hemangioblastoma and Filum Terminale
           Hamartoma 1299
    15.9.6 Schwannoma 1300
    15.9.7 Craniopharyngioma 1300
    15.9.8 Chordoma 1300
    15.9.9 Tumors of the Pineal Body 1302
    15.9.10 Hematological Malignancies 1302
    15.9.11 Other Tumors and Metastatic Tumors 1304
Multiple Choice Questions and Answers 1309
References and Recommended Readings 1311

16 Peripheral Nervous System 1321
16.1 Development 1321
16.2 Disorders of the Peripheral Nervous System 1322
    16.2.1 Peripheral Neuropathy 1322
    16.2.2 Traumatic Neuropathy 1323
    16.2.3 Vascular Neuropathy 1323

		16.2.4	Intoxication-Related Neuropathy	1323
		16.2.5	Infiltration (e.g., Amyloid) Related Neuropathy	1323
		16.2.6	Neoplasms of the Peripheral Nervous System	1323
	16.3	Neuromuscular Disorders		1333
		16.3.1	Muscle Biopsy Test	1333
		16.3.2	Neurogenic Disorders	1335
		16.3.3	Myopathic Disorders	1337
		16.3.4	Glycogen Storage Diseases	1339
		16.3.5	Mitochondrial Myopathies	1339
		16.3.6	Inflammatory Myopathies: Non-infectious	1341
		16.3.7	Inflammatory Myopathies: Infectious	1341
	Multiple Choice Questions and Answers			1341
	References and Recommended Readings			1342
17	**Skin**			**1345**
	17.1	Development, General Terminology, and Congenital Skin Defects		1347
		17.1.1	Development	1347
		17.1.2	General Terminology	1348
		17.1.3	Lethal Congenital Contractural Syndromes	1348
	17.2	Spongiotic Dermatitis		1352
		17.2.1	Conventional Spongiosis	1352
		17.2.2	Eosinophilic Spongiosis	1354
		17.2.3	Follicular Spongiosis	1354
		17.2.4	Miliarial Spongiosis	1354
	17.3	Interface Dermatitis		1355
		17.3.1	Vacuolar Interface Dermatitis	1355
		17.3.2	Lichenoid Interface Dermatitis	1360
	17.4	Psoriasis and Psoriasiform Dermatitis		1360
		17.4.1	Psoriasis	1360
		17.4.2	Psoriasiform Dermatitis	1360
	17.5	Perivascular In Toto Dermatitis (PID)		1361
		17.5.1	Urticaria	1361
		17.5.2	Non-urticaria Superficial and Deep Perivascular Dermatitis	1362
	17.6	Nodular and Diffuse Cutaneous Infiltrates		1363
		17.6.1	Granuloma Annulare	1363
		17.6.2	Necrobiosis Lipoidica Diabeticorum (NLD)	1364
		17.6.3	Rheumatoid Nodule	1364
		17.6.4	Sarcoid	1364
	17.7	Intraepidermal Blistering Diseases		1364
		17.7.1	Pemphigus Vulgaris, Pemphigus Foliaceus, and Pemphigus Paraneoplasticus	1364
		17.7.2	IgA Pemphigus and Impetigo	1366
		17.7.3	Intraepidermal Blistering Diseases	1366
	17.8	Subepidermal Blistering Diseases		1367
		17.8.1	Bullous Pemphigoid and Epidermolysis Bullosa	1367
		17.8.2	Erythema Multiforme and Toxic Epidermal Necrolysis	1368

		17.8.3 Hb-Related Porphyria Cutanea Tarda, Herpes Gestationis, and Dermatitis Herpetiformis 1368

17.8.3 Hb-Related Porphyria Cutanea Tarda, Herpes
       Gestationis, and Dermatitis Herpetiformis 1368
17.8.4 Lupus (Systemic Lupus Erythematodes)......... 1369
17.9 Vasculitis 1369
17.10 Cutaneous Appendages Disorders 1369
17.11 Panniculitis... 1369
    17.11.1 Septal Panniculitis 1369
    17.11.2 Lobular Panniculitis......................... 1370
17.12 Cutaneous Adverse Drug Reactions.................... 1370
    17.12.1 Exanthematous CADR....................... 1371
17.13 Dyskeratotic, Non-/Pauci-Inflammatory Disorders 1372
17.14 Non-dyskeratotic, Non-/Pauci-Inflammatory Disorders ... 1372
17.15 Infections and Infestations........................... 1372
17.16 Cutaneous Cysts and Related Lesions 1372
17.17 Tumors of the Epidermis 1373
    17.17.1 Epidermal Nevi and Related Lesions............ 1373
    17.17.2 Pseudoepitheliomatous Hyperplasia (PEH)...... 1376
    17.17.3 Acanthoses/Acanthomas/Keratoses............. 1376
    17.17.4 Keratinocyte Dysplasia 1377
    17.17.5 Intraepidermal Carcinomas 1378
    17.17.6 Keratoacanthoma 1378
    17.17.7 Malignant Tumors 1378
17.18 Melanocytic Lesions 1381
    17.18.1 Lentigines, Solar Lentigo, Lentigo Simplex,
            and Melanotic Macules (Box 17.8) 1381
    17.18.2 Melanocytic Nevi........................... 1381
    17.18.3 Variants of Melanocytic Nevi 1382
    17.18.4 Spitz Nevus and Variants 1384
    17.18.5 Atypical Melanocytic (Dysplastic) Nevi 1385
    17.18.6 Malignant Melanoma and Variants 1385
17.19 Sebaceous and Pilar Tumors 1387
    17.19.1 Sebaceous Hyperplasia 1387
    17.19.2 Nevus Sebaceous (of Jadassohn) (NSJ)......... 1387
    17.19.3 Sebaceous Adenoma, Sebaceoma,
            and Xanthoma 1388
    17.19.4 Sebaceous Carcinoma 1388
    17.19.5 Benign Hair Follicle Tumors 1388
    17.19.6 Malignant Hair Follicle Tumors................ 1390
17.20 Sweat Gland Tumors 1390
    17.20.1 Eccrine Gland Tumors....................... 1391
    17.20.2 Apocrine Gland Tumors...................... 1393
17.21 Fibrous and Fibrohistiocytic Tumors.................. 1394
    17.21.1 Hypertrophic Scar and Keloid 1394
    17.21.2 Dermatofibroma............................ 1394
    17.21.3 Juvenile Xanthogranuloma 1394
    17.21.4 Dermatofibrosarcoma Protuberans............. 1396

	17.22	Vascular Tumors. 1396

- 17.22 Vascular Tumors. 1396
- 17.23 Tumors of Adipose Tissue, Muscle, Cartilage, and Bone 1396
- 17.24 Neural and Neuroendocrine Tumors 1396
  - 17.24.1 Merkel Cell Carcinoma 1396
  - 17.24.2 Paraganglioma 1399
- 17.25 Hematological Skin Infiltrates 1400
  - 17.25.1 Pseudolymphomas......................... 1400
  - 17.25.2 Benign and Malignant Mastocytosis 1400
- 17.26 Solid Tumor Metastases to the Skin. 1400
- Multiple Choice Questions and Answers 1403
- References and Recommended Readings.................... 1405

18 Placenta, Abnormal Conception, and Prematurity 1409
- 18.1 Development and Useful Pilot Concepts and Tables...... 1410
- 18.2 Pathology of the Early Pregnancy 1422
  - 18.2.1 Disorders of the Placenta Formation 1423
  - 18.2.2 Disorders of the Placenta Maturation 1431
  - 18.2.3 Disorders of the Placenta Vascularization 1434
  - 18.2.4 Disorders of the Placenta Implantation Site...... 1434
  - 18.2.5 Twin and Multiple Pregnancies 1442
- 18.3 Pathology of the Late Pregnancy 1449
  - 18.3.1 Acute Diseases............................ 1450
  - 18.3.2 Subacute Diseases 1457
  - 18.3.3 Chronic Diseases 1468
  - 18.3.4 Fetal Growth Restriction 1482
- 18.4 Non-neoplastic Trophoblastic Abnormalities 1485
  - 18.4.1 Placental Site Nodule....................... 1485
  - 18.4.2 Exaggerated Placental Site................... 1485
- 18.5 Gestational Trophoblastic Diseases, Pre- and Malignant .. 1486
  - 18.5.1 Invasive Mole. 1486
  - 18.5.2 Placental Site Trophoblastic Tumor............ 1486
  - 18.5.3 Epithelioid Trophoblastic Tumor.............. 1486
  - 18.5.4 Choriocarcinoma 1486
- 18.6 Birth Defects 1488
  - 18.6.1 Birth Defects: Taxonomy Principles 1490
  - 18.6.2 Birth Defects: Categories.................... 1493
  - 18.6.3 Birth Defects: Pathogenesis (Macro- and Micromechanisms). 1496
  - 18.6.4 Birth Defects: Etiology (Mendelian, Chromosomal, Multifactorial) 1515
- 18.7 Infection in Pregnancy, Prom, and Dysmaturity 1533
  - 18.7.1 Infection in Pregnancy...................... 1533
  - 18.7.2 Premature Rupture of Membranes (PROM) 1541
  - 18.7.3 Fetal Growth Restriction (FGR) and *Dys*maturity.. 1541
- 18.8 IUFD and Placenta................................ 1546

18.8.1 Fetal Death Syndrome (Intrauterine Fetal Demise, IUFD).................................... 1546
  18.8.2 Step-by-Step Approach in the Examination of a Placenta............................. 1548
 Multiple Choice Questions and Answers 1552
 References and Recommended Readings.................... 1554

Index... 1571

Female Genital System

Contents

8.1	**Development and Genetics**	758
8.2	**Congenital Anomalies of the Female Genital System**	760
8.3	**Ovarian Inflammatory and Degenerative Conditions**	761
8.3.1	Oophoritis	761
8.3.2	Torsion	761
8.3.3	Non-neoplastic Cystic Lesions	763
8.3.4	Non-neoplastic Proliferations	763
8.4	**Infectious Diseases of the Female Genital System**	764
8.4.1	Viral Diseases	764
8.4.2	Bacterial Diseases	765
8.4.3	Fungal Diseases	767
8.4.4	Parasitic Diseases	767
8.5	**Inflammatory, Reactive Changes and Degenerative Conditions of Vulva, Vagina, Cervix, Uterus, and Tuba**	767
8.5.1	Non- and Preoplastic Vulvar and Vaginal Lesions	767
8.5.2	Inflammation-Associated Cervical Lesions	769
8.5.3	Reactive Changes of the Cervix	770
8.5.4	Cyst-Associated Cervical Lesions	770
8.5.5	Ectopia-Associated Cervical Lesions	770
8.5.6	Pregnancy-Associated Cervical Lesions	770
8.5.7	Cervical Metaplasias	771
8.5.8	Non-neoplastic and Preoplastic Uterine Lesions	771
8.5.9	Non-neoplastic Abnormalities of the Tuba	773
8.6	**Tumors of the Ovary**	773
8.6.1	Germ Cell Tumors	774
8.6.2	Surface Epithelial Tumors	779
8.6.3	Sex Cord-Stromal Tumors	790
8.6.4	Other Primary Ovarian and Secondary Tumors	798
8.7	**Tumors of the Tuba**	798
8.7.1	Benign Neoplasms	798
8.7.2	Malignant Neoplasms	798
8.8	**Tumors of the Uterus**	798
8.8.1	Type I/Type II EC	799

© Springer-Verlag GmbH Germany, part of Springer Nature 2020
C. M. Sergi, *Pathology of Childhood and Adolescence*,
https://doi.org/10.1007/978-3-662-59169-7_8

8.8.2	Non-type I/Non-type II EC	799
8.8.3	Uterine Leiomyoma (ULM) and Variants	799
8.8.4	Uterine Leiomyosarcoma (ULMS)	799
8.8.5	Uterine Adenomatoid Tumor	804
8.8.6	Uterine Lymphangiomyomatosis (ULAM)	804
8.8.7	Uterine Stroma Tumors	804
8.8.8	Hematological Malignancies and Secondary Tumors	804
8.9	**Tumors of the Cervix**	805
8.9.1	Benign Neoplasms	805
8.9.2	Precancerous Conditions and Malignant Neoplasms	805
8.10	**Tumors of the Vagina**	811
8.10.1	Benign Tumors	811
8.10.2	Precancerous Conditions and Malignant Tumors	813
8.11	**Tumors of the Vulva**	816
8.11.1	Benign Neoplasms	816
8.11.2	Malignant Tumors	818

Multiple Choice Questions and Answers ... 823

References and Recommended Readings ... 825

8.1 Development and Genetics

The ovary is a bilateral organ for reproduction and is made of follicles: primordial, primary, secondary, tertiary, Graafian, and atretic. *Corpora lutea* and *corpora albicantia* are also seen. A *corpus luteum* is a hormone-secreting anatomical structure of the ovary that takes place after an ovum has been discharged. It degenerates after some time (a few days) unless pregnancy has started. The *corpus albicans* represent an atretic *corpus luteum* and is substantially scar tissue composed of connective tissue that forms after the *corpus luteum* degenerates (luteolysis). The *corpus albicans* can be identifiable on the ovary for a few months (Latin, *luteus*: saffron-yellow; *albicans*: white; *corpus*: body). The ovarian stroma is cellular and quite characteristic. Hilus cells (counterpart of Leydig cells of the testis) are closely associated with nervous terminations. Walthard cell nests may be observed as cystic or solid elements, and germ cells arise from endoderm, while the rest of ovary is mesodermal-derived. The vagina is delimited proximally by the uterine cervix and distally by the vulva. Embryologically, it emerges as a paired Müllerian Duct. Histologically, there are mucosa, tunica muscularis, and adventitia. Basal, intermediate, and superficial squamous epitheliums constitute the mucosa, which is responsive to steroid hormones. The lymphatic drainage of the vagina is essential for FIGO and TNM staging of vaginal tumors, which may take place from very young children to elderly. The FIGO or International Federation of Gynecology and Obstetrics, as the acronym of its French name Fédération Internationale de Gynécologie et d'Obstétrique, is a worldwide non-governmental organisation representing obstetricians and gynaecologists worldwide. The TNM Classification of Malignant Tumors (TNM, tumor, node, metastasis) is a globally recognized standard for classifying the extent of spread of malignant tumors. TNM was developed and is maintained by the Union for International Cancer Control (UICC) and is broadly used by the American Joint Committee on Cancer (AJCC) and FIGO. With regard to the lymphatic drainage, *upper anterior wall* lymphatics drain into external iliac nodes, and l*ower anterior wall* lymphatics drain into interiliac, inferior gluteal, and femoral LNs, whereas p*osterior wall* lymphatics drain into deep pelvic, rectal, and aortic nodes. The cervix, which connects the uterus to the vagina via endocervical canal and is also estrogen-responsive, arises as Müllerian Ducts fuse at day 54 postconception and form a uterovaginal canal, lined by Müllerian columnar epithelium.

Box 8.1 Sequential Steps of Sex Organ Development in Humans

1. Germ cell formation in the 6th–10th week in the yolk sac and migration into the gonads.
2. Embryonic kidney-close bulging germ cell formation with sac development with a core.
3a. Germ cells drift to the *cortex* of the gonad and *medulla* degenerates (46, XX).
3b. Germ cells migrate to the *core* of the gonad and *cortex* atrophies (46, XY).
4. Bilateral double ductal system, i.e., Müllerian and Wolffian Ductal Systems, joining at the lower end (cloaca) at 8th–10th week of gestation.
5. *Internal genitalia determination* by ± secretion of Müllerian Duct Inhibitor (MDI).
5a. If XX, (−) MDI ⇒ Müllerian Ductal System ⇒ fallopian tubes, uterus, and upper vagina, while the Wolffian Ductal System disappears under the influence of testosterone.
5b. In XY, (+) MDI ⇒ Müllerian Ductal System disappears, while the Wolffian Ductal System ⇒ epididymis, vas deferens, and sem. Vesicles under the influence of testosterone (the prostate gland develops under the DHT or 17β-OH-5α-androstan-3-one from the conversion of testosterone).
6. *External genitalia determination* arises from the median and paramedian embryonic vestigial.
6a. In XX, (−) DHT ⇒ the tubercle folds, and swellings transform into the *clitoris, labia minora,* and distal vagina, and *labia majora.*
6b. In XY, (+) DHT ⇒ tubercle folds, and swellings transform into glans, penile shaft, and scrotum.
7. *Puberty* (10–12 years in XX, 12–14 years in XY): *Reproductive function*
7a. In XX: FSH/LH ⇒ ovular development, menarche, breast development, pelvis enlargement, pubic, and armpits hair.
7b. In XY: T/DHT ⇒ prostate growth, facial hair, hairline recession, and acne.

Notes: MDI, Müllerian Duct Inhibitor; T, testosterone; DHT, dihydrotestosterone.

The urogenital sinus joins the uterovaginal canal at Müllerian tubercle. This latter structure becomes a vaginal orifice at the hymenal ring. By day 77 on, the epithelium proliferates to become almost purely squamous in the vagina, while between the 91st and 15th day arise specifically endocervical glands and vaginal fornices. *Portio vaginalis* (the portion that protrudes into the vagina) and supravaginal portion constitute the cervix. The squamocolumnar junction is the "meeting point" of squamous and glandular epithelium, usually in the exocervix, while the transformation zone, aka *ectropion*, is located between the original squamocolumnar junction and the border of the metaplastic squamous epithelium. In *ectropion cervicis*, cells from inside the cervix form an inflamed patch on the outside the cervix. Endocrine cells and melanocytes are sporadically seen in the cervix, and multinucleated giant cells may also be observed. Basal cells, aka reserve cells, are cuboidal to low columnar with scant cytoplasm and round-ovoid nuclei and tend to acquire eosinophilic cytoplasm as they mature. Basal cells are positive for LMW-CK and ER but negative for HMW-CK. The vulva is made of *mons pubis, clitoris, labia minora, labia majora*, vulvar vestibule, vestibulovaginal bulbs, urethral meatus, *hymen*, Bartholin's and Skene's glands, and ducts, and *introitus vaginae* (Box 8.1).

Endometrial Dating There is usually a 2-day range, because the secretory phase of the cycle is always 14 days, while the proliferative phase is around 14 days. The proliferative phase of the cycle may vary from individual to individual and cycle to cycle up to 2-3 days. Endometrial dating uses either the 28-day period (Day 1–28) or post-ovulatory day nomenclature (POD 2–14), keeping in mind that the biopsy is performed during the secretory period. Increasing glandular tortuosity characterizes the proliferative phase as progress to the ovulation day with pseudostratified columnar cells containing cigar-shaped nuclei, coarse chromatin, and mitoses. The secretory phase shows secretion-filled tortuous glands with one layer of the cuboidal cell containing round nuclei and fine chromatin.

8.2 Congenital Anomalies of the Female Genital System

The congenital malformations of the ovary are here as follows.

Anomaly of *Number*, e.g., *bilateral gonadal agenesis* in monosomy X0 or Turner syndrome, ataxia-telangiectasia syndrome and *unilateral gonadal agenesis* (abnormalities of the ipsilateral fallopian tube, kidney, and/or ureter), and *supernumerary ovaries* (very rare anomalies that may be present in the pelvis as an attachment to the urinary bladder, omentum, or retroperitoneum). In *streak gonads*, a vestigial remnant of the undifferentiated gonadal ridge, there are dark strands of gonadal stromal cells of irregular shape with dense, round, and irregular nuclei embedded in a stroma constituted of numerous spindle-shaped cells that are often arranged in whorls.

The anomaly of *Composition* is expressed in *pure gonadal dysgenesis* and *mixed gonadal dysgenesis*, e.g., ovotestis or adrenal cortical rests that may be seen in the ovary. In MGD (mixed gonadal dysgenesis or 45,X/46,XY) there is one streak ovary and the other gonad has testicular tissue, a condition that is particularly prone to develop gonadoblastoma. Small duct-like structures constitute mesonephric (Wolffian duct) remnants at ovarian hilus. They may become cystically dilated. Histologic examination shows cuboidal epithelium, mainly non-ciliated, on a well-developed BM unlike paramesonephric (Müllerian) ducts, which have taller epithelium with a mixture of ciliated and non-ciliated cells and an inconspicuous BM.

The anomaly of *Shape* refers to ovarian dysmorphism such as in the case of *splenic-gonadal fusion*, which results from the fusion of the two organs during embryologic development. A complete fusion may be encountered. In most cases, there is a cord-like structure connecting the spleen to the left ovary, or some splenic tissue may be seen on the surface of the ovary. This latter condition should be distinguished from splenosis, which is an acquired condition (e.g., trauma or surgery) and defined as auto-implantation of one or more (often) focal deposits of splenic tissue in different body compartments.

The anomaly of *Location*: *Dystopia/ectopia ovarii* (it may be associated with abnormalities of the ipsilateral fallopian tube, kidney, and ureter).

The peculiarity of *Volume*: Dyscorias are encountered in condition labeled as hypoplastic ovary.

Maldevelopmental cysts of the ovary are paramesonephric, mesonephric, and mesothelial. Although they do not have a clinical significance, it may be opportune to distinguish them. The epithelium is resembling fallopian tube mucosa in the setting of paramesonephric cysts. There is a monolayer of ciliated columnar cells and thin, smooth muscle wall. The mesonephric cysts show a layer of cuboidal to columnar non-ciliated epithelium and a small, smooth muscle wall, while the mesothelial cysts are lined by a monolayer of flat or cuboidal cells with a thin wall with fibrous stroma. Also, the mesothelial cysts lack ciliated cells and are identified exclusively in the paramesonephric cysts.

Cysts

- *Epithelial inclusion cyst*: Cysts lined by squamous epithelium and occurring post-surgery (episiotomy) or trauma.
- *Gartner duct cyst (mesonephric remnants)*: Non-mucin-secreting (mesonephric) cysts lined by cuboidal epithelium and dense eosinophilic PAS+, mucin- intraluminal material and located in the lateral vaginal wall.
- *Müllerian cyst*: Mucin-secreting cysts lined by columnar epithelium ± focal squamous metaplasia.
- *Urothelial cysts*: Transitional epithelium-lined cysts arising from periurethral and Skene's glands.
- *Vaginitis emphysematosa*: Rare (<200 cases), vaginal nodules that produce "popping sound," composed of variably sized cysts with pink, hyaline-like material, foreign body giant cells, and chronic inflammatory infiltrate. This rare

female genital condition is usually self-healing, but septic complications have been reported and a medical and surgical close follow-ups are mandatory.

- *Endosalpingiosis and florid cystic endosalpingiosis*: Disorder of the Müllerian system, which is characterized by benign glands lined by tubal type epithelium (ciliated cuboidal epithelium reminiscent of fallopian tube). The occurrence of cilia is often a useful criterion of a benign lesion and still represents a useful tip to give to trainees and colleagues. Endosalpingiosis involves the peritoneum, subperitoneal tissues, and retroperitoneal LNs and may present as an incidental finding on microscopic examination. In rare occasions, endosalpingiosis may form a cystic mass, which is then termed "florid cystic endosalpingiosis," and its DDX is extraovarian serous cystadenoma.

Neonatal interlabial masses may be of serious concern for pediatricians and need to be carefully identified. Neonatal interlabial masses include Gartner's cyst, genital prolapses, hydrometrocolpos, hymenal cyst, paraurethral (Skene's gland) cyst, prolapsed ectopic ureterocele, sarcoma botryoides, and urethral prolapse.

Congenital Anomalies of the Female Genital Tract
Congenital defects of the female genital tract are deviations from normal anatomy, which result from maldevelopment of the embryonic Müllerian Ducts or paramesonephric ducts. Prevalence of these congenital anomalies is about 5% of the general population. Their knowledge is essential not only from the anatomical point of view but for health (e.g., obstetrical) and sexual (reproductive) aspects. These aspects are more or less relevant depending on the type and the degree of anatomical alteration. Congenital anomalies of the female genital tract should not be managed by a physician only, but need a team including nurses and doctors experienced in this field and the consultation with a psychologist. Three systems are known to describe congenital anomalies of the female genital tract, including the American Fertility Society's presently American Society of Reproductive Medicine system (1988), the embryological clinical classification system of genitourinary malformation (2004, 2011), and a TNM-based system (2005). Recently, a typical working group under the name CONUTA ("CONgenital UTerine Anomalies") has developed a newly updated classification system aiming to combine some positive aspects of the previous classifications and limit the negative elements using a single DELPHI procedure (2013).

8.3 Ovarian Inflammatory and Degenerative Conditions

8.3.1 Oophoritis

Nonspecific, which is a usually acute inflammation associated with salpingitis and characterized by many neutrophils and the formation of a tubo-ovarian abscess that may heal with a scarring process with potentially future fertility problems and possible conversion to an ovarian cyst.

Specific or granulomatous, which is characterized by a chronic inflammation including granulomas with or without caseous necrosis and a chronic inflammatory infiltrate, occurring in TB, actinomycosis, schistosomiasis, *E. vermicularis* (pinworms), sarcoidosis, and Crohn disease.

8.3.2 Torsion

An ovarian torsion belongs to circulatory abnormalities of the ovary and is a dramatic event, which may be secondary to inflammation or tumor of tube or ovary. However, but both anatomical structures are normal in at least 1/5 of cases. Grossly, there is the classic dusky appearance (as described in textbooks of the 20th century) of the ovary, which shows, microscopically, a clear-cut hemorrhagic infarction. According to the time occurring between torsion and reduction, there may be the possibility to salvage the ovary. The ovarian parenchyma may be able to recover in some cases, but it may be unpredictable. Ovarian torsions, which are nonsurgically removed, may calcify.

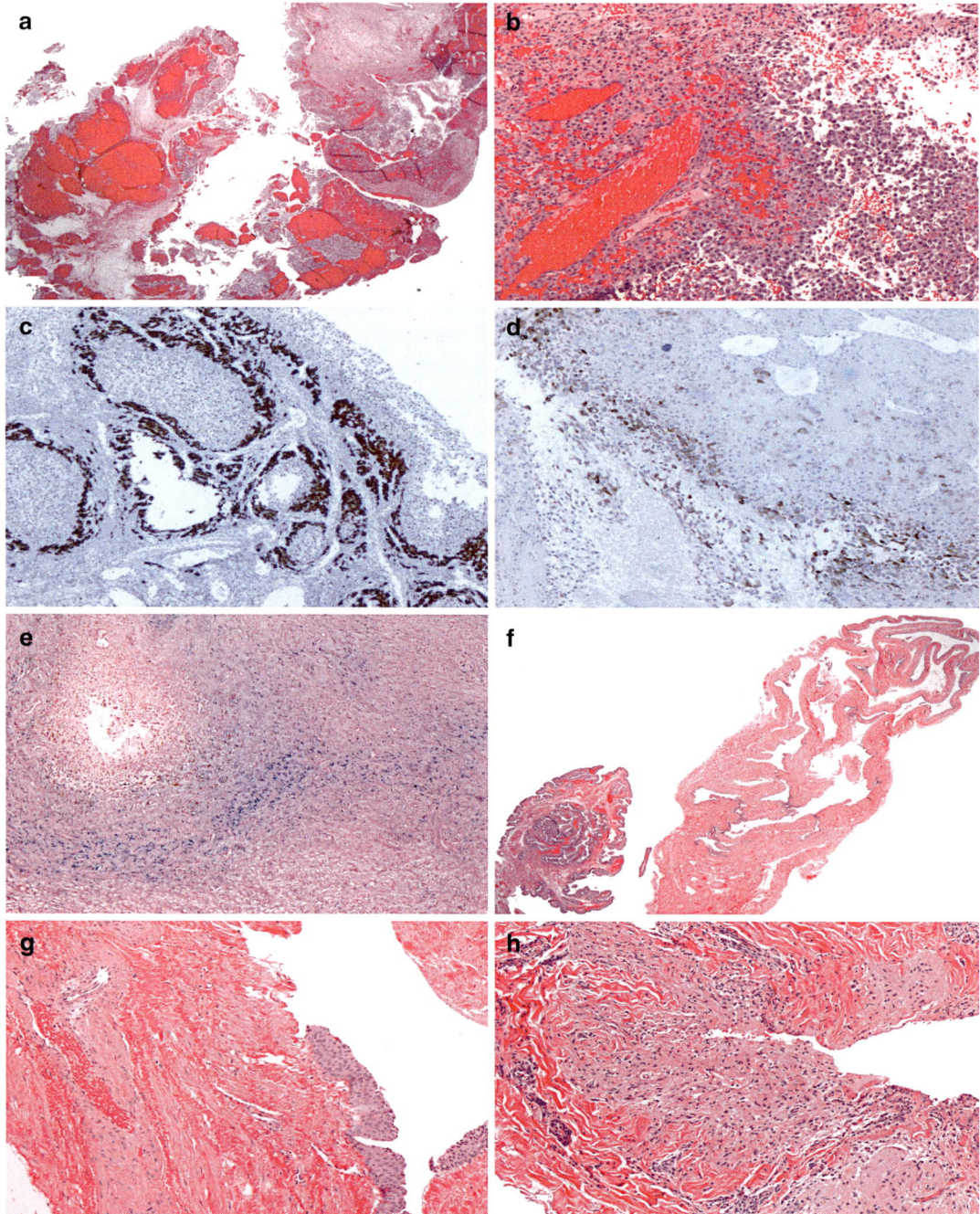

Fig. 8.1 In this panel, the microphotographs (**a**, **b**) show a luteinized cyst (H&E; x12.5 and x100 original magnification), which shows positivity for calretinin (**c**; anti-calretinin immunohistochemistry, avidin-biotin complex, x50) and inhibin A (**d**; anti-inhibin A immunohistochemistry, avidin-biotin complex, x100). In (**e**) is shown an endometriosis cyst with iron deposition (Pearls' Prussian Blue stain, x50); (**f**, **g**) show a paratubal Walthard rest cyst (H&E stain; x100 and x100). In (**h**) is shown the extreme rarity of gliomatosis peritonei in a child with a history of ovarian teratoma (H&E stain, x100). No yolk sac tumor or other malignancies were encountered. Gliomatosis peritonei is a very rare medical condition, which is often associated with immature ovarian teratoma (Norris grade 3) and characterized by mature glial tissue in the peritoneum

8.3.3 Non-neoplastic Cystic Lesions

Non-neoplastic cystic lesions include germinal inclusion cysts, follicular cysts, *corpus luteum* cysts, polycystic change, *hyperreactio luteinalis*, and epidermoid cysts among others (Fig. 8.1).

8.3.3.1 Germinal Inclusion Cysts
Small invaginations of the surface coelomic epithelium, which may become flat and possess tubal metaplasia. They can form cysts, which may be multiple.

8.3.3.2 Follicular Cysts
Distention of developing follicles with a size >2 cm (if smaller "cystic follicles") and lined by theca layer (Ret+ unlike serous cystadenoma), which may be luteinized with or without a granulosa layer.

± McCune-Albright syndrome (*p*olyostotic fibrous dysplasia, *p*atchy *p*igmentation, and *p*recocious *p*uberty) remembering the "p" rule.

8.3.3.3 Corpus Luteum Cysts
Corpus luteum cysts need to be kept differentiated from *corpus luteum* or *corpora lutea* (the Latin word "corpus" is neuter and belongs to the 3rd declension in the Latin grammar). *Corpus luteum*-derived cyst are lined by luteinized granulosa cells surrounded by theca lutein cells (*corpus luteum* cyst has a smooth, yellow lining unlike *corpora lutea*, which has a rimmed coating) with a tendency to rupture into the peritoneal cavity with risk of sepsis.

8.3.3.4 Polycystic Change
Common condition (up to 1/20 women) characterized by multiple subcapsular follicular cysts (usually <1 cm) with nonluteinized granular/hyperplastic luteinized/theca internal and a dense fibrous capsular covering (no *corpora lutea* identifiable). Polycystic change of the ovary determines the medical condition also known as polycystic ovary syndrome or PCOS.

Follicular cysts produce androgens, which are converted to estrogens (androstenedione → estrone with subsequent risk for endometrial hyperplasia and endometrial carcinoma and anovulation due to the suppression of FSH. Two syndromes need to be kept in mind, including the Stein-Leventhal syndrome and the HAIR-AN syndrome.

± Stein-Leventhal syndrome: Polycystic change of the ovary, amenorrhea, and androgen excess with subsequent sterility/infertility.

± HAIR-AN syndrome (*H*yper*A*ndrogenism, *I*nsulin *R*esistance – *A*canthosis *N*igricans). HAIR-AN syndrome is a rare disease which is a subset of PCOS.

8.3.3.5 Hyperreactio Luteinalis
Massive bilateral enlargement of the ovaries due to multiple theca lutein cysts, which are constituted by luteinized theca internal cells ± luteinized granulosa and stromal cells + stroma edema + ↑ HCG (pregnancy or gestational trophoblastic disease to be differentiated on obstetrical and gynecological ground).

8.3.3.6 Epidermoid Cysts
Squamous epithelial-lined cysts, structurally not different from entities found in other systems or organs.

8.3.4 Non-neoplastic Proliferations

Non-neoplastic proliferations of the ovary include endometriosis, endosalpingiosis, stromal hyperplasia/stromal hyperthecosis, hilus cell hyperplasia, (ectopic) decidual reaction, *fibromatosis ovarii*, pregnancy luteoma, and massive edema of the ovary.

8.3.4.1 Endometriosis
It refers to an ectopic proliferation of two out of three components, i.e., glands, stroma, and hemosiderin-laden macrophages (Fe + using special stains, CD68+ using an antibody against a MΦ-myeloid associated antigen (the murine homolog of the human glycoprotein is called macrosialin with 72% identity, useful for vertebrate animal model studies). Endometriosis risk factors are low parity and cervical stenosis. Serum marker is CA-125, and grossly, an ovary with endometriosis is bluish with brown patches (extraovarian locations: fallopian tube, uterus, colon, peritoneum, and lungs). Clinically, there is infertility, and a

"chocolate" cyst due to repeated intralesional hemorrhages is a common surgical term. Complications: decidual change, Arias-Stella reaction, endometrial hyperplasia, clear cell, and endometrioid tumors. Historically considered malignant, the Arias-Stella reaction is a benign change in the endometrium associated with the presence of chorionic tissue due to progesterone primarily.

8.3.4.2 Endosalpingiosis
Proliferation of one-layered tubal epithelium with or without cilia and no stroma (no endometrial stroma as seen in endometrial glands of the cysts) lining an apparent lumen. DDX: noninvasive and invasive ovarian implants (see below) and mesonephric remnants, which have smooth muscle under the epithelial lining.

8.3.4.3 Stromal Hyperplasia and Stromal Hyperthecosis
Stromal hyperplasia and stromal hyperthecosis (± HAIR-AN): Non-neoplastic ovarian proliferations characterized by bilateral and diffuse/nodular clusters or single cells within the cortex and medulla, frequently associated with patchy or extensive luteinization (hyperthecosis or diffuse thecomatosis). In this proliferative background, there is a risk for the development of true thecomas and fibromas, which have more collagen than stromal hyperplasia/hyperthecosis.

8.3.4.4 Hilus Cell Hyperplasia
Hilus cell hyperplasia refers to multiple nodules, which are usually less than 0.2 cm each.

8.3.4.5 Ectopic Decidual Reaction
Ectopic decidual reaction refers to an exaggerated response of the endometrium to progesterone and has been typically observed in the uterus, cervix, and lamina propria of the fallopian tubes and ovaries.

8.3.4.6 Massive Edema of the Ovary
It refers to massive edema of the ovary (MEO) and is an unusual condition characterized clinically by pain and abdominal mass and usually involving one ovary, which presents grossly enlarged, soft, and boggy. Microscopically, there is diffuse edema + cystic follicles. The presence of follicles and their derivatives is indeed essential in the DDX with ovarian fibroma, which lacks follicles and their derivatives. In some cases of MEO, a partial torsion may play an etiologic role as described by some authors.

8.3.4.7 Fibromatosis Ovarii
Fibromatosis of the ovary occurs in a young group of females (13–40 years), a factor that may be relevant in the DDX with fibroma. Grossly, the ovary is firm with a white cut surface and harbors residual follicles (+) and their derivatives (DDX: MEO and ovarian fibroma).

8.3.4.8 Pregnancy Luteoma
Pregnancy luteoma refers to a condition characterized by well-circumscribed nodules occurring in the setting of hCG stimulation and composed of a substantial proliferation of uniform polygonal cells with abundant, eosinophilic granular cytoplasm and reticulin fibers surrounding epithelial clusters differently from luteinized thecoma/fibroma.

8.4 Infectious Diseases of the Female Genital System

8.4.1 Viral Diseases

8.4.1.1 Herpes Genitalis
HSV-2 (Herpes simplex virus, type 2), STD (sexually transmitted disease) is, subclinical in >90% of cases. This STD may manifest clinically as painful 1 mm sized in diameter erythematous vesicles or blisters with focal ulceration on the cervix, vagina, and vulva ± UTI (urinary tract infections) symptoms (fever, dysuria, and malaise), spontaneous resolution (1–3 weeks), but common recurrence. Microscopically, MNGCs with intranuclear inclusions are the hallmark. Complications may include cervical carcinoma. Cesarean delivery for pregnant women is to avoid because of fatal infectious disease of the newborn.

8.4.1.2 Condyloma Acuminatum
HPV, promiscuity, incidental finding on Pap smears, asymptomatic, warty papillary tumors or flat lesions on the cervix, vagina, and vulva.

Microscopically, koilocytosis (cellular ballooning with "crinkled" nuclei).

Complications include coexistent cervical dysplasia (LSIL) with possible evolution to HSIL and cervical carcinoma (LSIL, low-grade squamous intraepithelial lesion; HSIL, high-grade squamous intraepithelial lesion).

8.4.2 Bacterial Diseases

8.4.2.1 Syphilis

Treponema pallidum is a spirochaetal bacterium (subspecies *T.p. pallidum* → syphilis, endemic → bejel or endemic syphilis; *T. carateum* → pinta; *T. pertenue* → yaws), in the group of STD or acting transplacentally (congenital syphilis). Syphilis has three clinicopathologic stages:

1. Primary syphilis with painless, hard-based ulcer or "chancre" at the point of injection (bacterial inoculation) and 3–4 weeks after infection
2. Secondary syphilis: Diffuse skin rash and lymphadenopathy ± *condylomata lata* and general symptoms, fever and malaise, and 1–4 months after primary syphilis
3. Tertiary syphilis: CVS and meningovascular and parenchymatous CNS involvement mostly but also the skin, bone, joint, and liver involvement and very variable and heterogeneous occurrence from 1 year through 30 years but usually 10–25 years after primary syphilis

Microscopically, there are three different lesions:

1. Primary syphilis → "Chancre" (shallow ulcer with an epithelial defect and heavy plasmacytic inflammatory infiltrate and obliterative endarteritis).
2. Secondary syphilis → A lymphocytic and histiocytic bandlike infiltrate is detected in the upper part of the dermis and extended around blood vessels of the deep plexus. Plasma cells are also present in early papular lesions.
3. Tertiary syphilis → Obliterative endarteritis of the *vasa vasorum* of the ascending aorta.

Consequently, there is ischemic medial necrosis and fibrosis of the media and adventitia or syphilitic aortitis with patchy loss of medial elastic fibers and muscle cells ("tree-bark aorta" or wrinkled, nodular change of the aortic intima due to loss of the elastic recoil). Moreover, there is a thoracic aorta aneurysm (if secondary atherosclerosis adds to the injury) as well as aortic insufficiency (proximal extension of the obliterative endarteritis-related scarring process to the aortic valve ring). CNS involvement includes meningovascular syphilis with syphilitic granulomas (*gummae* with granulation tissue including fibroblasts and new capillaries) of the leptomeninges and parenchymatous syphilis with *tabe dorsalis* (progressive granulation tissue-related degeneration of axons and myelin sheaths around the dorsal roots of the lumbar region) and general paresis (atrophy of the frontal gyri due to progressive leptomeningeal infiltration and anoxia-related syphilitic changes with gradual degeneration of neurons).

Laboratory studies include a dark-field examination on chancre material and Dieterle stain on tissue and serology tests, including VDRL, rapid plasma reagin (RPR), treponemal antibody tests (FTA-ABS), *T. pallidum* immobilization reaction (TPI), and syphilis TPHA test. The Venereal Disease Research Laboratory test (VDRL) is a blood test to screen for syphilis with high sensitivity. The fluorescent treponemal antibody absorption (FTA-ABS) test targets the antibodies to this spirochete, while the *Treponema pallidum* hemagglutination assay or TPHA is an indirect hemagglutination assay used for the detection and titration of antibodies against this specific spirochete. In Alberta, Canada, an impressive increase of cases has been recently identified with syphilis cases from 160 cases in 2014 to more than 1,500 in 2018.

8.4.2.2 *Gonorrhea*

N. gonorrheae infection with kidney-shaped, intracellular, a-sporigen, facultatively anaerobic, Gram-negative diplococcus with external pili that allow attachment to the endocervical and tubal epithelium and consequently entry into host cells.

Gonorrhea is an STD characterized by 7–21 days of incubation time. Clinically, gonorrhea can involve pictures ranging from asymptomatic patients to vaginal discharge and dysuria, and PID with acute salpingitis, pelvic peritonitis, and constitutional symptoms (fever, abdominal pain, vomiting) ± gonococcal pharyngitis and proctitis. Microscopically, there is a purulent inflammation of Bartholin's glands, endocervix, and fallopian tubes (pyosalpinx) in females and males.

LAB: Gram staining of direct smears of infected material and bacterial culture (enriched media, such as chocolate agar or Thayer-Martin agar, in a CO_2 atmosphere at 35–37 °C, oxidase and fermenting glucose, but not maltose).

Therapy includes the combination of the antibiotics ceftriaxone (i.v.) and either azithromycin or doxycycline (orally) according to the Centers for Disease Control and Prevention. In Alberta, Canada, an alert was recently distributed, detailing the outbreak. Cases have more than doubled to more than 5,000 cases in 2018 from 1,900 cases in 2014.

Complications include diffuse peritonitis and sepsis as well as decreased fertility or infertility due to adhesions in the tubo-ovarian compartment.

8.4.2.3 *Chlamydia* Infections

Chlamydia genus has three bacterial species, of which one has been found in humans, i.e., *C. trachomatis* (*C. muridarum* in hamsters and mice while *C. suis* is found in swine). *C. trachomatis* is an obligate intracellular, Gram-negative bacterium with an extrachromosomal plasmid that includes three biovars (biovar is a variant prokaryotic strain that shows different features physiologically and/or biochemically from other strains in a particular species):

- Serotypes Ab, B, Ba, and C cause trachoma with potential evolution to blindness.
- Serotypes D–K cause urethritis, cervicitis, and PID (pelvic inflammatory disease) but also pharyngitis and proctitis in females, urethritis, epididymitis, pharyngitis and proctitis in males, and neonatal conjunctivitis and pneumonia in newborns (by vaginal childbirth).
- Serotypes L1-L2-L3 cause lymphogranuloma venereum (LGV), which starts as transient genital and/or anal ulcers (primary stage) with regional lymphangitis and lymphadenitis (secondary stage) following a 2–6 weeks interval showing enlarged lymph nodes (aka buboes) and, histologically, star-shaped abscesses with a central necrotic area surrounded by histiocytes and epithelioid cells (DDX: cat-scratch fever) and fistula formation. Recently, the importance of DNA exchange between different strains of *Chlamydia* has risen significant concerns about potential interspecial diffusion.

Complications include decreased fertility, sterility, and sepsis.

Diagnosis is based on tissue diagnosis and culture, which requires cycloheximide-treated McCoy or HeLa cells, direct fluorescent antibody (DFA) test, and PCR (polymerase chain reaction)-based molecular biology detection of tissue and pus. Recently a fast procedure of real-time PCR (Taqman analysis) has been developed to diagnose LGV.

8.4.2.4 Chancroid

Haemophilus ducreyi is a Gram-negative coccobacillus bacterium, which is responsible for the STD, called universally chancroid. The STD of chancroid type caused multiple painful genital/anal ulcers at the site of injection/inoculation (3–5 days after) (primary stage) and enlarged lymph nodes or buboes showing acute suppurative lymphadenitis (secondary phase) and self-limiting outcome with fibrosis and scarring of the affected lymph nodes. The diagnosis is based on bacterial isolation from an ulcer or bubo.

8.4.2.5 Granuloma Inguinale

Klebsiella (Calymmatobacterium) (Donovania) granulomatis is a Gram-negative, rod-shaped bacterium of the genus *Klebsiella*. It causes the STD of *granuloma inguinale*. This STD is particularly disfiguring and presents with papulae. In the next stage, serpiginous ulcers with satellite lesions (primary stage) take place and regional acute and chronic lymphadenitis. The diagnosis relies on specific detection of bacteria or Donovan bodies by Giemsa stain (secondary phase) and bacterial (microbiological) isolation.

8.4.3 Fungal Diseases

8.4.3.1 Candidiasis

Candida albicans is an opportunistic pathogenic yeast. It is an ordinary member of the human gut flora. It is facultative (e.g., pregnancy, DM, cancer or chemotherapy for cancer, immunosuppression, prolonged antibiotic). Candidiasis is an STD causing red vaginal mucosa with white patches, vaginal secretion, and intense pruritus. The diagnosis is based on cytologic smears with or without use of special stains and clinical/microbiological evaluation.

8.4.4 Parasitic Diseases

8.4.4.1 Trichomoniasis

Trichomonas vaginalis is an anaerobic, flagellated protozoan parasite. It is responsible for the trichomoniasis, which is an STD causing abundant, malodorous, greenish-yellow or grayish vaginal discharge with intense pruritus in females. Trichomoniasis is usually an asymptomatic infection in males.

8.5 Inflammatory, Reactive Changes and Degenerative Conditions of Vulva, Vagina, Cervix, Uterus, and Tuba

8.5.1 Non- and Preneoplastic Vulvar and Vaginal Lesions

Non-neoplastic and preneoplastic vulvar and vaginal lesions include cystic lesions of the vulva, vulvitis, neovagina, urothelial metaplasia, vulvar dystrophies, vaginal adenosis, benign discolored lesions of the vulva, vaginal pseudotumors, and endometriosis of the vulva or vagina or both.

8.5.1.1 Cystic Lesions of the Vulva

Cystic lesions of the vulva may be a significant reason for concern in an adolescent or youth and may be entered in the DDX of malignant tumors as well. The most frequent cystic lesions of the vulva are as follows.

8.5.1.2 Vulvitis, Neovagina, and Urothelial Metaplasia

Vulvitis is an inflammation of the vulva.

Neovagina (aka artificial vagina) is an artificial conduit or vaginoplasty ideated by surgeons as reconstructive plastic surgery procedure for the vaginal canal and its mucous membrane and of vulvovaginal structures owing to agenesis or atresia or because of an acquired cause (e.g., trauma or malignancy). It is created from isolated segments of sigmoid colon or peritoneum and represents a major surgical operation. Large bowel vaginas may show diversion colitis with lymphocytic infiltration and variable acute inflammatory infiltrate with neutrophils in lamina propria, while peritoneum vaginas may display metaplastic squamous epithelium with variable paramesonephric-like glandular elements in both mucosa and submucosa resembling mild vaginal adenosis (*vide infra*).

Urothelial (transitional) metaplasia is considered an incidental microscopic finding, which shows hyperplastic epithelium lacking maturation, composed of spindle nuclei with longitudinal grooves, perinuclear halos, low N/C ratios, and low MI as well as no mitotic figures. Also, no atypia is found (DDX: VAIN, *vide infra*).

8.5.1.3 Vulvar Dystrophies

Most children with this disease continue to show changes consistent with *lichen sclerosus* into adulthood. Although considered part of the disorders of the elderly, they can occur in childhood. The primary symptom of the vulvar dystrophies, which should be defined as non-neoplastic epithelial disorders of the vulvar skin and mucosa, is vulvar pruritus. Treatment is topical steroids, and surgical procedure should not be an option in childhood or youth.

8.5.1.4 Vaginal Adenosis

Vaginal adenosis (VA) was formerly associated with diethylstilbestrol (DES) only (Al Jishi & Sergi 2017). Nowadays, non-induced VA has a reported incidence of about 10% of adult women in some statistics. Although it has also been suggested that VA may be an insignificant coincidental finding, other authorities have pointed out significant implications in having it. Thus it needs to be surveilled because it may be an underrecognized lesion. VA is defined as a Müllerian type glandular epithelium metaplasia in the vagina and may present with excess mucus. Diethylstilbestrol [DES; 4,4-(3E)-hex-3-ene-3,4-diyldiphenol] is a potent synthetic estrogen. For approximately three decades, from the 1940s to the 1970s, pregnant women were administered DES to treat "pregnancy-associated" disturbances, but the link of prenatal DES exposure with adult onset of reproductive abnormalities promoted the ban of DES. DES-associated diseases are VA, cervical ectropion, and clear cell adenocarcinoma as well as thymic atrophy, and apoptosis in T cells. VA occurs in a very broad range from 35 through 90% after in utero exposure to DES. DES adenosis is similar histologically to non-DES adenosis. A mouse model of DES exposure seems to demonstrate VA and structural changes quite similar to actual changes observed in women.

Risk of subsequent clear cell adenocarcinoma is increased in all individuals exposed to DES but usually very small.

- *GRO*: Red granular spots or patches that do not typically stain with Lugol's iodine solution.
- *CLM*: Endocervical-type mucous glands on vaginal surface or in lamina propria, often presenting as cysts or nodules; also tubo-endometrial cells and embryonic columnar cells between lamina propria and squamous epithelium. The finding of intestinal metaplasia remains rare according to numerous authors. There is, often, chronic inflammation and squamous metaplasia, which may be extensive and obliterate glandular lumina. At some point, it may resemble dysplasia/VAIN or VAginal Intra-epithelial Neoplasia. Moreover, microglandular hyperplasia present if patient using oral contraceptives. It may be detected in vaginal smears if present on the vaginal epithelium.
- *HSS*: Mucicarmine (+)
- *DDX*: Clear cell carcinoma (if microglandular hyperplasia present). An important aspect to remember is that transverse ridges (aka cockscomb cervix, rims, collars, hoods, pseudopolyps) are located in the upper vagina or cervix of 25% of women with DES exposure.
- *CLM*: Fibrotic tissue lined by mucinous epithelium or metaplastic squamous epithelium are two extremely important findings. Rarely, the tubal or endometrial epithelium are also observed.

DES-Related VAIN
DES-related VAIN I lesions usually regress after therapy or biopsy, but VAIN II/III lesions tend to persist and recur after biopsy or treatment. These high-grade pathologic conditions are often aneuploid. A molecular biology investigation may be appropriate.

Non-DES-Related Adenosis
Adenosis can also occur in a setting not associated with DES exposure. In this situation, patients have an age range starting in the early adolescence up to the ninth decade of life. It manifests with symptoms in 15% of patients and microscopy is similar to DES-related VA.

8.5.1.5 Benign Discolored Lesions of the Vulva
The following table summarizes all discolored lesions of the vulva that behave benignly. It is of paramount relevance that some lesions need to be taken in the differential diagnosis of melanocytic lesions such as malignant melanoma (Box 8.2). These lesions are benign, and their characterization and identification can be made using appropriate dermatopathology textbooks.

Box 8.2 Benign Discolored Vulvar Lesions
- Lentigo simplex
- Lichen planus
- Lichen sclerosus et atrophicus
- Lichen simplex chronicus
- Melanosis vulvae
- Melanocytic nevi
- Seborrheic keratosis

8.5.1.6 Vaginal Pseudotumors

- *Polyposis vaginalis of pregnancy*: Numerous polyps localized intimately in the vagina and cervical portio that disappear postpartum (DDX: botryoid rhabdomyosarcoma).
- *Postoperative spindle-cell nodule* (POSN): Pseudosarcomatous small friable reddish vaginal lesion, which typically appears within a few weeks of hysterectomy. POSN is characterized by granulation tissue, hypercellular spindle-cell proliferation, typical mitoses, RBC extravasation, and no pleomorphism (DDX: LMS, Kaposi sarcoma).
- *Fibroepithelial polyp*: Pseudosarcomatous fibroepithelial stromal polyp with central fibrovascular core covered by normal appearing intact squamous epithelium and constituted by hypercellularity, bizarre morphology, typical and atypical mitoses (MI > 10/10 HPF). It usually occurs in pregnant women with a wide age range from adolescence through 4th and 5th decades of life. It is probably hormone-induced hyperplasia of connective tissue or the end-stage of granulation tissue. IHC: (+) VIM, (−) DES, and (+) ACT (DDX: botryoid rhabdomyosarcoma, which would show cambium layer, epithelial invasion, rapid growth, cross striations, and (+) Myo-D1).

8.5.1.7 Endometriosis Vulvae Sive Vaginae Vel Vulvovagine

Endometriosis is the presence of endometrial glands and stroma outside of the uterus.

8.5.2 Inflammation-Associated Cervical Lesions

Inflammation-associated cervical lesions include infectious and noninfectious cervicitis, papillary endocervicitis, cervical lymphoma-like lesion, and cervical arteritis.

8.5.2.1 Infective Cervicitis

HPV is the cause of *condyloma acuminatum*, flat condyloma, and papilloma with risk to progress to carcinoma/malignant lesion according to the typing of the virus.

- Low-risk HPV types: 6 and 11
- High-risk HPV types: 16, 18, 31, 33, and 35

HPV is a virus with a high oncogenic risk for some subtypes. High oncogenic risk lasts longer than infection with low oncogenic risk, although 50% of HPV infection are formally cleared within 8 months and 90% within 2 years. Susceptible cells are immature basal cells of the squamous epithelium areas of epithelial breaks and immature metaplastic cells of the squamous epithelium located at the squamocolumnar junction.

Simple koilocytosis refers to a binucleate cell with surrounding halo and denser peripheral cytoplasm and specific nuclear change. There is no disruption or expansion of the basal cell layer, which would point to cervical dysplasia.

Cervix operative procedures are (1) cold knife cone excision, (2) loop electrical excision procedure (LEEP), and (3) large loop excision of the transformation zone.

Herpes simplex virus (HSV) 1 and 2 are two members of the human *Herpesviridae* family, a set of viruses that produce viral infections in the majority of humans. HSV-induced cervicitis can also occur. There is an intense nonspecific chronic and acute inflammation with ulceration. To remember are the characteristic features of multinucleated giant cells (MNGCs) and the molding of nuclei.

Chlamydia trachomatis - induced cervicitis is one of the most common STD, and histological sections of the cervix show chronic nonspecific inflammation, reactive epithelial atypia, and occasional scattered prominent lymphoid follicles, which may raise the suspicion of a lymphomatous lesions at places.

8.5.2.2 Non-infective Cervicitis

Reactive (trauma)-related cervicitis.

8.5.2.3 Papillary Endocervicitis

Papillary endocervicitis refers to papillae of several sizes which are filled with inflammatory cells. It represents a process that may be interpreted either as post-chronic inflammation or hamartomatous in origin, although most of the authors tend to consider it an inflammatory condition. Microscopically, there are edema, chronic inflammation in a fibrotic stroma, and dilated glands underneath of the epithelium. MNGCc may be scattered and present in the stroma.

Other lesions of the cervix include the cervical lymphoma-like lesion and the cervical arteritis.

8.5.3 Reactive Changes of the Cervix

Reactive changes of the cervix include repair atypia, radiation-induced atypia (e.g., post radiotherapy for urinary bladder neoplasms), POSN, inflammatory atypia, and atypia-accompanying atrophic lesions.

8.5.4 Cyst-Associated Cervical Lesions

Cyst-associated cervical lesions include Nabothian cysts, tunnel clusters, microglandular hyperplasia, and mesonephric remnants or hyperplasia. Nabothian cysts refer to dilatation of endocervical glands due to blockage, which is usually secondary to inflammation. Microglandular hyperplasia (MGA) (aka microglandular adenosis) refers to a complex of histologic characteristics including (1) complex proliferation of tightly packed small glands with flat to cuboidal epithelial cells expressing substantially no atypia, (2) squamous metaplasia of some glands and subcolumnar reserve cell hyperplasia, and (3) stroma infiltration with acute and chronic inflammatory cells and intraluminal neutrophils. MGA is seen in young patients with history positive for oral contraceptive pills (OCP) pregnancy, and a history of single/multiple polypoid lesions resembling small cervical polyps. IHC: (−) CEA. DDX: mucinous endocervical adenocarcinoma. The DDX with clear cell carcinoma (CCC) is essential, and in this latter mucin is (+). Clear cell carcinoma, which is an adenocarcinoma, would show tubulocystic and papillary histologic patterns, an irregular infiltration of the stroma, and marked atypia with brisk mitotic figures (↑ MI). Vestigial elements of the mesonephric or Wolffian ducts usually regress during the development of the female genital tract. In the early development of the embryo, two pairs of genital ductular structures form, which are the mesonephric or Wolffian ducts and the paramesonephric or Müllerian Ducts. These ducts parallel one another as they, during the early development of the female genital tract, run cranio-caudally from the mesonephros to the cloaca. Typically, mesonephric rests are located in the lateral wall of the cervix (or in utero) in 1/3 of females, deeply to the endocervical glands. Microscopically, there are a single cell layer, usually cuboidal forming tubules with characteristic - but not pathognomonic - intraluminal eosinophilic homogeneous PAS (+) secretions. Clear cells may be seen, but hobnailing is lacking. Immunohistochemically, the glandular epithelium is Bcl2 (+) and CD10 (+). In particular, both markers may be useful in the DDX with cervical adenocarcinoma, which is Bcl2 (−) and CD10 (−) as well as strongly positive for both p16 and Ki-67 (MIB-1). The latter two markers are usually weakly positive in mesonephric remnants.

8.5.5 Ectopia-Associated Cervical Lesions

Ectopia-associated cervical lesions include *endometriosis cervicis* and *ectopia prostatae* (ectopic prostatic tissue).

8.5.6 Pregnancy-Associated Cervical Lesions

Reactive changes of the cervix include Arias-Stella change or reaction, decidual pseudopolyps, and decidualization as well as a mesodermal stromal polyp or fibroepithelial polyps. Arias-Stella change refers to endometrial and endocervical glandular change associated with the above considered presence of chorionic tissue seen during pregnancy. It is entirely benign, although in the past it has been mis-diagnosed with malignant transformation. Five types are identified, including minimal atypia; early secretory pattern; secretory or hypersecretory pattern; regenerative, proliferative, or nonsecretory pattern; and "monstrous" cell pattern. Microscopically, ASC often displays tortuous glands with clear and hobnailed cells with some cytologic atypia, but – very important – ASC is only focal and always associated with pregnancy, and the decidual reaction is seen in the cervical stroma. Decidual pseudopolyp or decidualization is a decidual reaction to pregnancy, and, grossly, there are multiple, red-yellowish papulae, which may present with a friable aspect, which may represent a pitfall for malignancy.

8.5.7 Cervical Metaplasias

Cervical metaplasia includes squamous metaplasia, tubal metaplasia, tubo-endometrial metaplasia, and transitional cell metaplasia. Two other conditions are part of this paragraph and are cervical ectropion and diffuse laminar endocervical hyperplasia. Squamous metaplasia is a benign change with the retreat of the transitional zone into endocervix, in which squamous cells acquire nucleoli and polygonal appearance on cytology and starts basally on histology. Since it may involve glands, a careful differentiation between pseudo-invasion and invasion is needed. Cervical ectropion refers to the presence of classic glandular epithelium in ectocervix. There is a strong correlation with in utero DES exposure, which also associated with cervical hyperplasia, VA, and CCC. Diffuse laminar endocervical hyperplasia (DLEH) refers to a benign proliferation of endocervical glands with usually an inflammatory infiltrate and a very well-delimitated lower extent of the lesion.

Basal (reserve) cells are essential in dealing with MGA (Box 8.3). Basal cells have some characteristics, including residence in the endocervical glands, squamocolumnar junction, and potential differentiation into either endocervical glandular cells or squamous cells.

8.5.8 Non-neoplastic and Preneoplastic Uterine Lesions

It is important to correctly define the abnormal endometrial bleeding as dysfunctional (anovulatory) uterine bleeding due to an excessive loss of blood without demonstrable cause (e.g., estrogen breakthrough bleeding, estrogen withdrawal bleeding).

8.5.8.1 Dysfunctional Uterine Bleeding

Dysfunctional (anovulatory) uterine bleeding (DUB) refers to excessive bleeding without evident cause. Examples of DUB may be due to estrogen breakthrough bleeding or estrogen withdrawal bleeding.

Bleeding may be labeled as menorrhagia if bleeding is >7 days or > 80 ml flowing at the regular time in the cycle, metrorrhagia when bleeding occurs at irregular intervals, and menometrorrhagia when both menorrhagia and metrorrhagia characteristics do occur, i.e., the female patient is in a setting characterized by prolonged or excessive bleeding outside of regular periodic cycle. Dyssynchrony is also a term indicating a > 4-day variation between glands and stroma maturation.

Typically, DUB causes may be grouped into four categories (mnemonic: COLE):

1. *C*ontraceptive-induced DUB: Discordant appearance or dyssynchrony between glands and stroma showing inactive glands exquisitely intermixed to a mature stroma reminiscent of pregnancy-related decidua.
2. *O*vulation failure DUB: Dyssynchrony showing proliferative glands without secretory activity and stromal breakdown, which can be classified as unscheduled.
3. *L*uteal phase inadequacy: Inappropriate secretory endometrial activity for that determinate POD.
4. *E*ndometrial disorders: This is a very comprehensive term, which may include several heterogeneous processes such as cystic atrophy.

Box 8.3 Emblematic Features Distinguishing Microglandular Adenosis from Clear-Cell Carcinoma of the Cervix

1. Vacuoles at subnuclear location (+)
2. Mucin at intracytoplasmic location (mucin+, CEA−)
3. Lack of glycogen ⇒ PAS (−) glands
4. MI very low or nil
5. No/minimal cell atypia
6. (+) Reserve cell hyperplasia
7. (+) Squamous metaplasia

Notes: MI, mitotic index

8.5.8.2 Endometritis

Acute endometrial inflammation or acute endometritis refers to a critical inflammatory process of the endometrium, which may follow abortion (missed abortion or miscarriage and interruption of pregnancy), delivery, or instrumentation procedures performed in the usually inappropriate way. The main characteristic is tissue neutrophilia. DDX includes as first obviously menstruation, which may be identified from the active secretory glands broken and embedded in a stroma showing obvious decidual changes. To acquire training in this very delicate sector of public health, it may be essential to evaluate the endometrium in adolescent's inquiries following gynecologic visits or during inappropriate post-delivery care.

Chronic endometrial inflammation or chronic endometritis indicates an underlying constant inflammatory process of the endometrium, which may be due to chronic pelvic inflammatory disease (PID), post-delivery or post-abortion, intrauterine device (IUD) manipulation, or TB infection. Microscopic sections show an infiltration of lymphocytes and plasma cells in an edematous stroma. The presence of one plasma cell suffices for the diagnosis of a chronic inflammatory process of the endometrium because they are not present in usual conditions. Clinically, both vaginal bleeding and pelvic pain are present and may be complicated by extensive PID or mucopurulent cervicitis.

Endometrial dating in endometritis is not feasible, because there is gland maturation variability. Therefore, tissue neutrophilia and chronic inflammatory infiltrate should direct the pathologist to liaise with the gynecologist to investigate the possible etiology.

The complications reported by IUD manipulation are endometritis, acute and chronic, stroma hemorrhage, shortening of the secretory phase, squamous metaplasia, atrophy, fibrosis, and actinomycosis.

In mentioning the important gynecologic genes and gene products relevant to diagnostic routine practice, it is essential to recall four members, including *P16, TP53, P57,* and *P63*.

- *P16* gene product is p16, which binds to cyclin-CDK 4/6 complex to intimately regulate the cell cycle at G1-S interphase (cyclin-kinase inhibitor) by preventing the phosphorylation of the retinoblastoma (RB) gene product (pRB or RB1). RB is a tumor suppressor gene, whose function is to prevent excessive cell growth by inhibiting cell cycle progression, other than to recruit several chromatin remodeling enzymes such as methylases and acetylases. In HPV-infected cells, there is overexpression of p16 and pRB, and the target of p16 inhibitory activity is inactivated by the E7 HPV oncoprotein. In practice, p16 is used as a surrogate for HPV infection. Diffuse nuclear and cytoplasmic staining is encountered in HSIL and adenocarcinoma in situ (AIS).
- *TP53* is a tumor-suppressor gene, whose mutational status cause conformational changes and stabilization of the gene product or p53 protein. The protein p53 is involved in regulating cell growth. Gene mutations allow the IHC detection of the protein. A complete absence of signal should point to gene deletion, which is a particularly significant event in some neoplasms.
- *P57* is a parental imprinted gene, which is expressed only when maternal DNA is present, and is a cell cycle inhibitor of cell proliferation. The p57 protein can be detected in decidua and extravillous trophoblast by immunohistochemistry. The marker p57 is particularly crucial in evaluating products of conception to rule out molar pregnancies at the microscopic level.
- *P63* is a transcription factor gene that belongs to the p53 family. The gene product or transcription factor p63 is reactive in both immature basal and reserve squamous cells of the cervix. It is a principal diagnosis for cells of squamoid differentiation that looks like entirely undifferentiated.

8.5.8.3 Endometrial Metaplasias

Metaplasia in the endometrium is not an uncommon finding, although some metaplasia is not usually seen in childhood or youth (Box 8.4).

Box 8.4 Endometrial Metaplasias

Type	Subtypes	Morphology
Squamous	Adenoacanthosis	Acanthosis
	Squamous *morulae*	"Berry"-like clusters
	Ichthyosis uteri	Keratinization
Ciliated cell (tubal)		Ciliated cells
Oxyphilic (eosinophilic)		Oxyphilic epithelium
Mucinous	Endocervical Intestinal	Mucinous epithelium
Mesonephroid		Hobnail and clear cell change
Secretory		Secretory changes
Stromal		Islands of foamy cells, smooth muscle, cartilage, and bone
Syncytial papillary		Papillary growth

Notes: Mnemonic: *SCOMS*, i.e. South Carolina Osteopathic Medical Society

8.5.8.4 Adenomyosis and Endometriosis

Adenomyosis refers to endometrial cells and stroma growing into the uterine wall, while endometriosis is when endometrial cells and endometrial stroma are in a location outside of the uterus. They can occur together, and both can cause pain, but endometriosis does not always cause heavy bleeding. The non-neoplastic endometrial proliferative disorders include atrophy, disordered proliferative endometrium, endometrial hyperplasia, and endometrial polyp. The atrophy is defined as a flat epithelial lining with nuclear pyknosis and cystically dilated glands and spindled stroma.

8.5.9 Non-neoplastic Abnormalities of the Tuba

Non-neoplastic abnormalities of the tuba include Walthard cell nests (benign cluster of epithelial cells, which are most frequently found in the connective tissue of the fallopian tubes but also seen in the *mesovarium, mesosalpinx*, and hilus of the ovary), paratubal cysts, tubal endosalpingiosis, tubal endometriosis, and *salpingitis isthmica nodosa*. Also, an extrauterine pregnancy can occur (tubal pregnancy).

8.6 Tumors of the Ovary

Ovarian neoplasms constitute the fifth cause of cancer-related death in women, although 80% are benign with age range of 15–45 years. Nine out of ten ovarian malignancies are generally of epithelial origin (carcinoma). In 4/5 of cases, there is is spread beyond the ovary at diagnosis.

- *Risk factors*: *IDAHO* mnemonic word including 1) intercourse without parity, nulliparity; 2) drugs (clomiphene and long-term estrogen replacement therapy) – dysgenesis, gonadal; 3) age (older); 4) heritage – genetic background (*BRCA1, BRCA2, HNPCC*), 5) others (other factors). Pregnancy and OCD decrease the risk!
- *BRCA1/BRCA2*: *BRCA1/BRCA2* mutations cause 10% of ovarian carcinomas, of which 16% of women harbor the abnormal genetic background. These patients have ovarian neoplasms, generally serous cystadenocarcinomas, which are occult in 4.5% of prophylactic salpingo-oophorectomies.
- *BRCA1* gene mutation prevalence: 1–2% in US Ashkenazi Jews but ~2.5 per 1000 in other US whites.

In 1–2% of patients, undergoing prophylactic oophorectomy, there is peritoneal adenocarcinoma with clinical and histologic similarities to papillary serous carcinoma of the ovary.

Origin of Ovarian Neoplasms
(a) The coelomic surface epithelium of various organs of Müllerian lineage, including the ovary, endometrium, fallopian tubes, peritoneum, and cervix ⇒ *epithelial neoplasms*
(b) Germ cells migrating to ovary from yolk sac ⇒ *germ cell tumors*
(c) "Specialized" stroma of ovary ⇒ *sex-cord stromal tumors*

The stroma can synthesize hormones, unlike the nonspecific stroma of other epithelial tumors. Clinically, there may be localized pain at the lower abdominal quadrant (right or left or undefined), abdominal enlargement, obstruction signs on neighboring organs (intestinal or ureteral obstruction with hydronephrosis), ascites, or bilateral diffuse uveal melanocytic proliferation, which is considered a paraneoplastic syndrome constituted by an increase of uveal melanocytes. Uveal melanocytic proliferation can cause blindness.

CA-125 (cancer antigen 125 or carbohydrate antigen 125, aka MUC16) is a protein that in humans is encoded by the *MUC16* gene and a member of the mucin family glycoproteins. An elevation of serum levels of CA-125 is somewhat sensitive but not specific for ovarian carcinoma.

Spread of the malignancy can occur toward the contralateral ovary, peritoneal cavity, pelvic and paraaortal LNs (Sister Joseph's nodule: umbilical metastasis, maybe the first manifestation of disease), liver, lung, pleura, omentum, and diaphragm. 5-YSR: ~45%. However, according to the American Cancer Society an early detection (stage 1) typically results in a 5-YSR of 92%.

Features to report in an ovarian tumor are specimen type, tumor site, specimen integrity (intact, ruptured, fragmented, or other), tumor size, typing (histologic type), grading (histologic grade, i.e., benign, borderline, malignant or poor, moderate, or well-differentiated), staging (invasion sites within the ovary, surface involvement, involvement of fallopian tube, opposite ovary, other tissues, implants: noninvasive epithelial, noninvasive desmoplastic, or invasive; *vide infra* for definitions), radicality of surgery (margins), and additional staging features including angiolymphatic space involvement, LNs positive/number of nodes examined, eventual extranodal extension, and the presence of endometriosis or other non-neoplastic findings. The risk of malignancy rises with the increase of solid areas. Accordingly, a generous sampling of the base of papillary formations, which are recognizable in the gross room, and areas adjacent to the ovarian surface is necessary. The designation epithelial-stromal instead of epithelial only has been proposed because most of the epithelial tumors may show a stromal component as well. Ovarian epithelium may undergo metaplasia/neoplasia to bear a resemblance to the tubal, endometrial, or endocervical mucosa (the so-called Müllerian differentiation). Subtyping related to the pattern (cystic, solid, surface), amount of fibrous stroma, and invasiveness/atypia are probably essential factors for ovarian tumors in the youth. Grading of the surface epithelial-stromal tumors is a matter of debate because clear-cut criteria seem to be incomplete or missing for some categories. According to most of the authorities, who are expert in gynecologic pathology, there may be several grading systems different for each tumor type. Tips and pitfalls of the ovarian oncology are presented in Box 8.5.

8.6.1 Germ Cell Tumors

Germ cell tumors constitute about 1/3 of all ovarian tumors, and most entities are observed in children, adolescents, and young adults. Most cases are benign cystic teratomas, but 1/10 of germ cell tumors are of a mixed type, e.g., dysgerminoma with a yolk sac component (non-"seminomatous" component of the testis).

8.6.1.1 Dysgerminoma
- *DEF*: Seminomatous germ cell tumor with uniform cells that resemble primordial germ cells and interpreted as the female counterpart of seminoma of the testis.
- *EPI*: It accounts for less than 1% of all ovarian tumors. Approximately 80% of the patients harboring dysgerminoma are children, adolescents, and young adults (second to third decades of life).
- *CLI*: There are associations with gonadal dysgenesis (Swyer syndrome), androgen insensitivity or pseudohermaphroditism, pregnancy, and hypercalcemia.
- *LAB*: ± ↑ LDH, LDH1, and HCG. Lactate dehydrogenase (LDH) with four distinct enzyme classes catalyzes the interconversion of pyruvate and lactate with concomitant interconversion of NADH (reduced form of nicotinamide adenine dinucleotide) and

8.6 Tumors of the Ovary

Box 8.5 Pearls and Pitfalls
- *IHC-implants/M1*: (+) WT1, ER, PR, CA125; (−) GCDFP-15, CEA (apart from mucinous differentiation) (WT, Wilms tumor; ER, estrogen receptor; PR, progesterone receptor; CA125, cancer antigen 125 aka mucin 16 or MUC16; GCDFP-15, Gross cystic disease fluid protein 15; CEA, carcinoembryonic antigen).
- *Coexistence with uterine carcinoma* may be a diagnostic challenge. It may reflect metastases from uterus or ovary (possibly also contralateral) or two single independent tumors. Some "pearls" pointing to metastasis from uterine origin (uterine carcinoma) are bilateralism, multinodularity, the small size of the ovarian tumor, the involvement of the tubal lumen, deep myometrial invasion, or angioinvasion of the myometrium. Overall, there is, obviously, a better prognosis in independent tumors than in metastatic tumors.
- In *childhood*, most ovarian tumors are benign, and malignant tumors seem to have a favorable outcome with chemotherapy, even after recurrence. Pediatric ovarian tumors are usually unilateral without metastasis at presentation, and pediatric malignant tumors of the ovary are mostly of germ cell type, although sex cord-stromal tumors and epithelial tumors also do occur. Among the benign tumors, mature cystic teratomas are common. In childhood, pediatric ovarian tumors need to be differentiated from other tumors, including neuroblastoma, Ewing sarcoma/PNET, adrenal gland carcinoma, rhabdomyosarcoma, and DSRCT (desmoplastic small round cell tumor).
- *Grading challenge*: Serous carcinomas can be classified using a two-tier system that is related to the tumor biology (low-grade vs. high-grade), although staging (FIGO/WHO) has more weight than typing and grading. Endometrioid carcinoma can be graded using the same three-tier system as used in the uterus, while mucinous carcinomas have a classification according to nuclear atypia in expansile lesions and a three-tier system, comparable to that of the endometrioid type, for infiltrative lesions. Clear cell carcinomas are usually high grade, while transitional cell carcinomas are graded according to a three-tier system based on nuclear morphology.

NAD+ (oxidized form of nicotinamide adenine dinucleotide). Human chorionic gonadotropin (HCG) is a hormone produced by the placenta following uterine implantation.
- *GRO*: Uni-(85%)/bilateral, encapsulated, smooth tumor with a tangled surface and gray-white solid cut surface with minimal/no hemorrhage or necrosis.
- *CLM*: Well-defined nests of clear cells separated by fibrous stroma infiltrated by scattered T lymphocytes. Tumor cells have clear, glycogen-rich cytoplasm (PAS+ cytoplasm), oval, uniform, and mostly central nuclei with 1–2 prominent nucleoli. Scattered HCG+ syncytiotrophoblastic cells (no PGN significance) are seen. If MI ≥ 30/10 HPF is seen, there is an obvious early carcinomatous differentiation of dysgerminoma (anaplastic dysgerminoma), which may potentially aggravate the outcome (Fig. 8.2).
- *IHC*: (+) CD117, OCT 3/4, D2–40, PLAP, (−) CK7 and EMA, but (±) LMW-CK, (−) CK20 and HMW-CK, (−) CD30, and (±) VIM. OCT3/4 or now relabelled OCT-4 is a nuclear POU-domain octamer binding transcription factor specifically expressed in cells with pluripotent capacity. The word POU is derived from the names of three transcription factors in mammals, including the pituitary-specific Pit-1, the octamer-binding proteins Oct-1 and -2, and the neural Unc-86 from *Caenorhabditis elegans*.
- *CGB*: 12p abnormalities in most cases (12p overrepresentation and isochromosome 12p).

a

Germ Cell Tumor FNA Cytology
Mature Teratoma: sheets of epithelial cells forming glands and loosely clustered benign -appearing squamous cells.
Dysgerminoma: dyshesive large polygonal cells with prominent nucleoli.
Embryonal carcinoma: cohesive poorly differentiated pleomorphic epithelial cells in clusters.
Yolk sac tumor: single and clustered middle-sized and large cells with hyaline globules +/- tumor cells forming Schiller-Duval bodies.

b

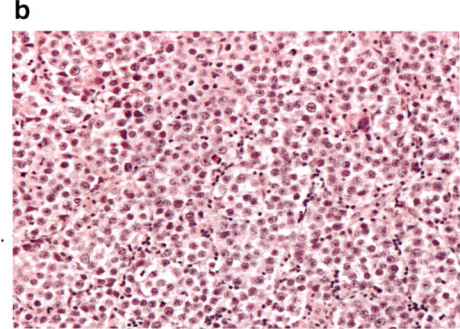

Fig. 8.2 In (**a**) is shown a table with germ cell tumor FNAC features, while (**b**) shows a classic ovarian dysgerminoma (H&E stain, x200 original magnification). The tumor is constituted by cells with clear cytoplasm, centrally located nuclei with coarsely granular chromatin and one or more nucleoli. Lymphocytes are also scattered seen

Box 8.6 Germ Cell Tumors – Differential Diagnosis

Endometrioid carcinoma	Histological growth patterns, INA (−), EMA (+)
ESS (endometrial stromal sarcoma)	"Tongues," estrogenism (−), arterioles (+), INA (−)
Fibroma/thecoma	Reticulin (+)
Gonadoblastoma	Intersex, histological growth patterns
Metastatic breast carcinoma	INA (−), EMA (+), GCDFP-15 (+)
Metastatic melanoma	Intracytoplasmic melanin (+), INA (−), HMB45 (+)
Multiple follicle cysts	Histological features
SCTAT (sex cord stromal tumor with annular tubules)	PJS, multiplicity, histological growth pattern
Small-cell carcinoma	Histological features, hypercalcemia, estrogenism (−)
Undifferentiated carcinoma	INA (−), VIM (−), EMA+

Note: INA, inhibin A; EMA, epithelial membrane antigen; GCDFP-15, gross cystic disease fluid protein 15; HMB45, Human Melanoma Black; VIM, vimentin

- *DDX*: See Box 8.6.
- *TRT*: Surgery + CHT, but no RT, although it may adequate for the male counterpart.
- *PGN*: 5-YSR: 95%, if pure. But mixed germ cell tumor of dysgerminoma (10% of cases) with choriocarcinoma, yolk sac tumor, or embryonal carcinoma have an influence to worse the prognosis. Also, anaplastic dysgerminoma (marked nuclear atypical changes and high MI) may have a worse outcome, but this finding has been disputed in the recent years.

8.6.1.2 Teratoma

Teratoma accounts for 1/5 of all ovarian tumors and mostly in patients younger than 30 years. Most of the cases are benign (98%), unlike the testicular counterpart (Fig. 8.3).

8.6.1.3 Yolk Sac Tumor (YST)/ Endodermal Sinus Tumor (EST)

- *DEF*: Germ cell tumor of children and young adults with an increase of serum alpha-fetoprotein (AFP), which is encoded by the *AFP* gene located on the q arm of chromosome 4 in humans and representing a major plasma protein of the yolk sac and the fetal liver during early embryonic and fetal development.
- *GRO*: There is a smooth and glistening surface.
- *CLM*: A variety of pattern may be observed, including reticular, microcystic, papillary, solid, parietal and endodermal sinus as well as a polyvesicular vitelline pattern, hepatoid pattern, and a glandular pattern. Characteristic are Schiller-Duval bodies, which are glomeruloid structures with bulbous pseudopapillary having a central blood vessel and intracytoplasmic PAS+ hyaline droplets of BM material (+ IV collagen and

8.6 Tumors of the Ovary

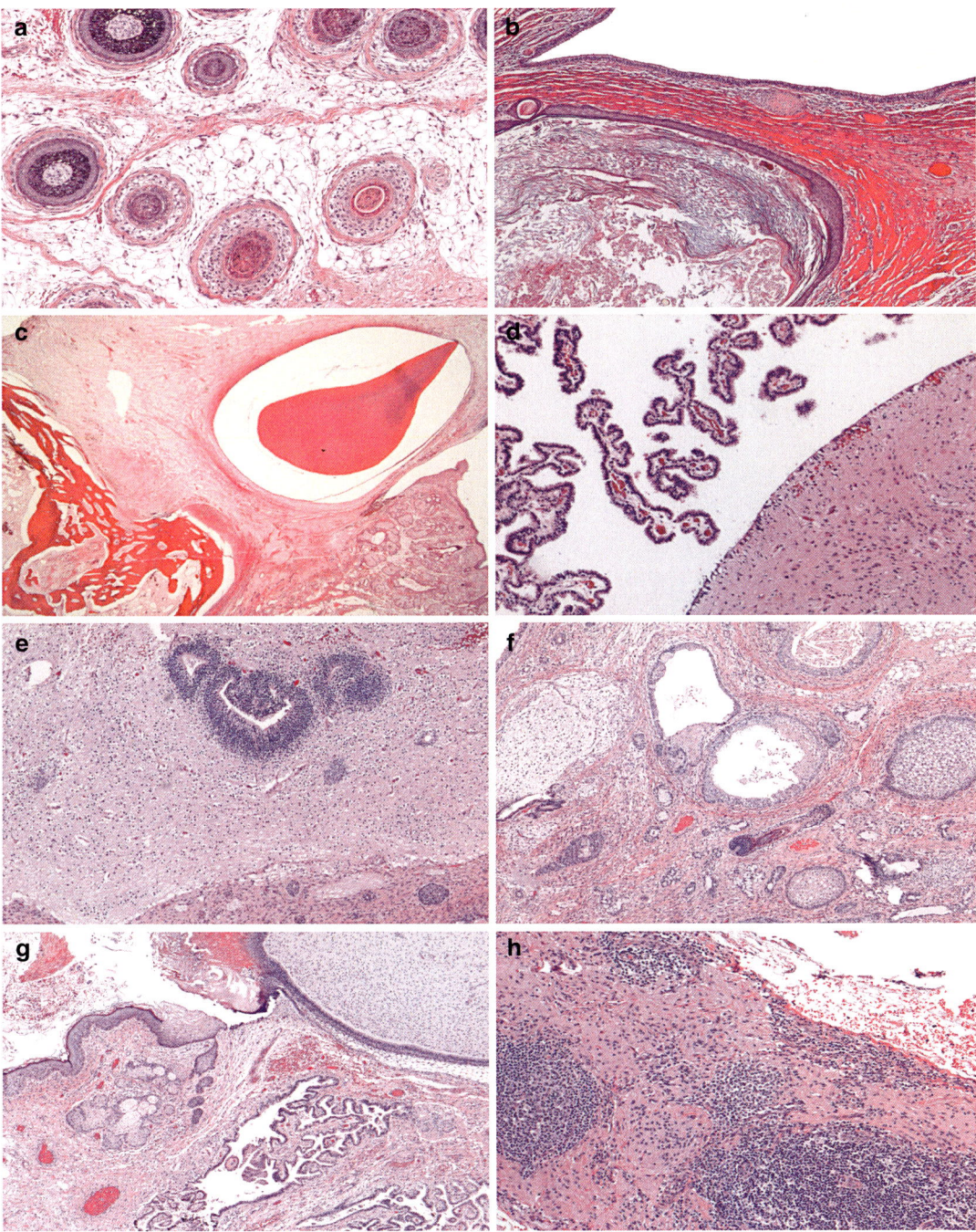

Fig. 8.3 The microphotographs (H&E staining) of this panel show an ovarian teratoma with immature elements (neuroectodermal tube) in (**e**) (x50 original magnification). In (**a**) there is ectodermal differentiation with hair follicles of different age (x100 original magnification). In (**b**) there is also an ectodermal differentiation with dermoid cyst (x50 original magnification), while (**c**) shows bone and a tooth with mesodermal differentiation (x12.5 original magnification). In (**d**) choroidal villi and glial tissue are evident (x100 original magnification). In (**f–g**) all three layers of differentiation seem to be represented (f, x50 original magnification; g, x50 original magnification). In (**h**) (x100 original magnification) the lymph nodal metastasis of teratoma is seen.

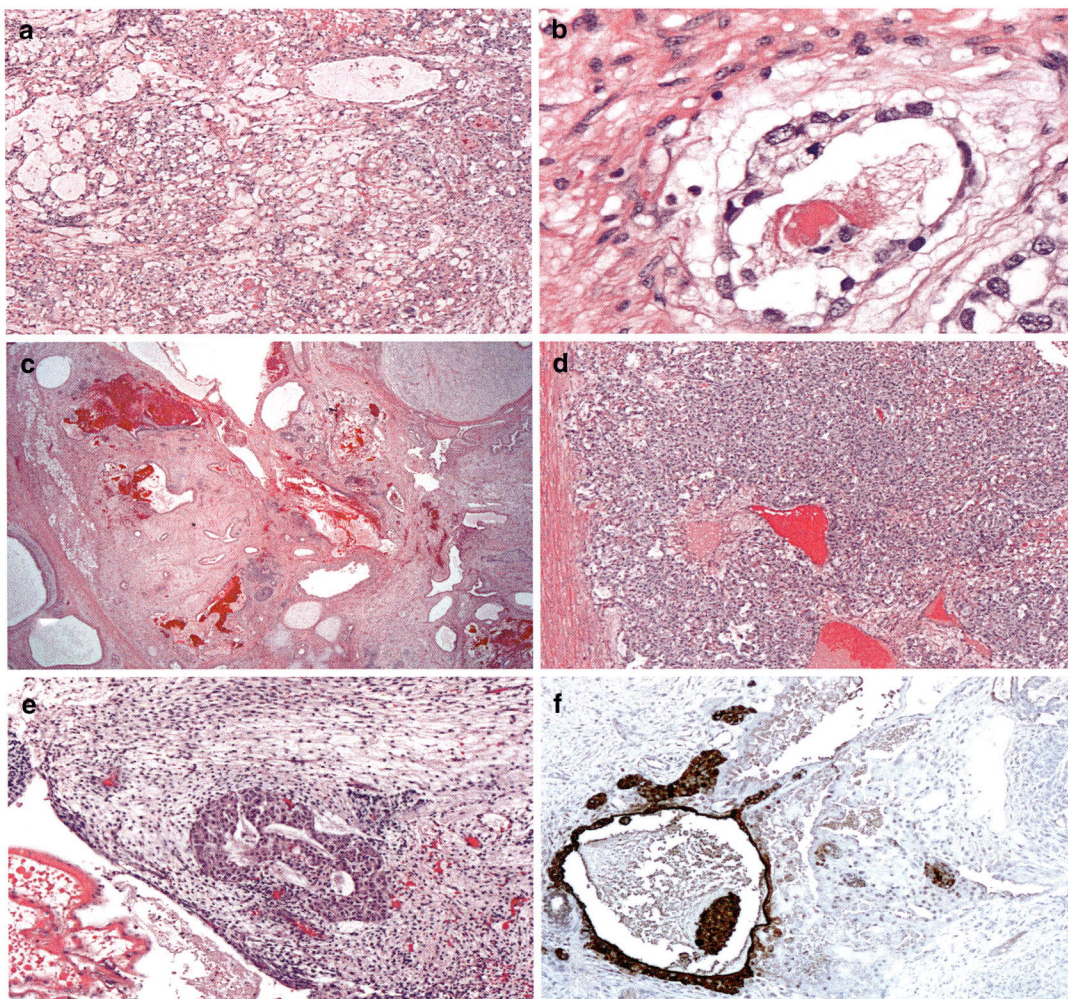

Fig. 8.4 The microphotographs of this panel are characteristic for yolk sac tumor with some different growth patterns of this yolk sac tumor. In (**a**) there is a reticular / microcystic pattern with anastomosizing channels and cysts of variable size (H&E stain, x50). In (**b**) is shown a Schiller-Duval body characterized by tubular structure with papillary accentuation including a vascular core centrally and cuboidal to columnar epithelial cells as lining (H&E stain, x400). In (**c**) a microcystic pattern with aspects of polyvesicular pattern are encountered (H&E stain, x12.5), while a solid and a hepatoid pattern are observed in (**d**) and (**e**), respectively (H&E stain, x50 and x100, respectively). In (**f**) the AFP is positive in this gland with angulated features (anti-AFP immunostaining, ABC method, x100 original magnification)

laminin). In the polyvesicular vitelline pattern, small cystic structures with eccentric constrictions are seen. A dense spindle-cell stroma separates these. The hepatoid pattern looks like hepatoblastoma or HCC (Fig. 8.4).

- *IHC*: (−) CD117, (−) OCT3/4, but (+) AE1–AE3, PLAP, AFP, and Hep par-1 (Hep par-1 recognizes a mitochondrial antigen of hepatocytes), and (−) CK7 and EMA.
- *PGN*: 3-YSR: ~15%.

8.6.1.4 Embryonal Carcinoma

The embryonal carcinoma (aka undifferentiated malignant teratoma) is rare in the ovary, and serum HCG and AFP levels are elevated, and it may present with isosexual precocious puberty or menstrual irregularities (Fig. 8.5).

- *IHC*: (−) CD117, OCT 3/4, D2–40, and PLAP.

8.6 Tumors of the Ovary

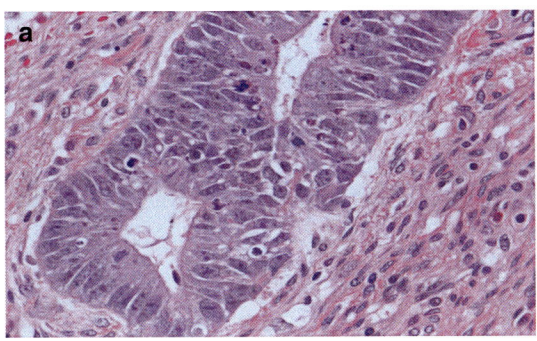 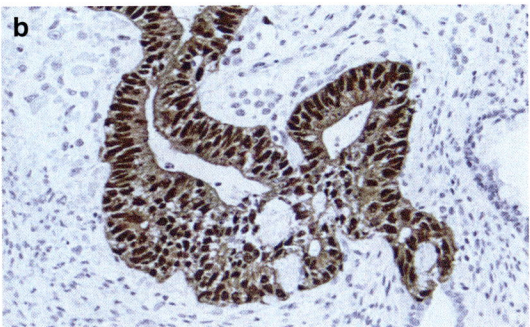

Fig. 8.5 These two microphotographs (**a**, **b**) show an embryonal carcinoma of the ovary with an immunostaining positive for CD30 (**a**, H&E stain, x400 otiginal magnification; **b**, anti-CD30 immunostaining, ABC method, x200 original magnification)

8.6.1.5 Polyembryoma

It is a rarity in the family of the ovarian tumors and the germ cell tumor family. It is constituted by a predominance of embryoid bodies at various stages of differentiation, usually less or equal to 18. There may be malformed embryos or embryonic disc elements with an ectodermal layer of columnar cells and an endodermal layer of cuboidal cells. The embryoid body is considered to have an amniotic cavity-like structure, which is continuous with the intestinal duct, and rarely squamous cell nests, unlike the "yolk sac" which is continuous with the hepatic tissue. Embryoid body seems to be a product of divergent differentiation into the intestine and seems to be derived from the embryonic gut.

8.6.1.6 Choriocarcinoma

Primary ovarian choriocarcinoma is rarer than metastases from uterine tumors. It can present with isosexual precocious puberty, i.e., with the appearance of phenotypically appropriate (girls) secondary sexual characteristics before the age of 8 years. If it is present in a gonad, like an ovary, choriocarcinoma is usually part of an MGCT (mixed germ cell tumor). It has to be differentiated from the scattered HCG+ syncytiotrophoblastic cells of dysgerminoma. In choriocarcinoma, there is a biphasic cell population with cytotrophoblast in addition to syncytiotrophoblasts (+ HPL and HCG).

PGN: Poor, predominantly if the tumor is a pure choriocarcinoma.

> **Box 8.7 Pearls and Pitfalls of Germ Cell Tumors**
> - Accurate, complete, and thorough sampling is of utmost importance in germ cell tumors.
> - Select sampling from various areas mostly from solid areas but without forgetting to sample the cystic changes.
> - Areas with calcification should suggest the pathologist assistant or resident the diagnosis of a gonadoblastoma and additional blocks may be appropriate.

Tips and pitfalls of germ cell tumors are presented in Box 8.7.

8.6.2 Surface Epithelial Tumors

8.6.2.1 Serous Tumors

Serous tumors are ovarian neoplasms, which are often cystic and filled with typically clear serous fluid and characterized by both secretory and ciliated cell differentiation characteristic of the fallopian tube as well as psammoma bodies (concentric calcifications) in carcinomas (Figs. 8.6 and 8.7).

In most of the tumor statistics, surface epithelial tumors represent one-fourth of all ovarian tumors (60% *benign*, 15% *borderline*, 25% *malignant*). Unlike an *in situ* subtype as seen in

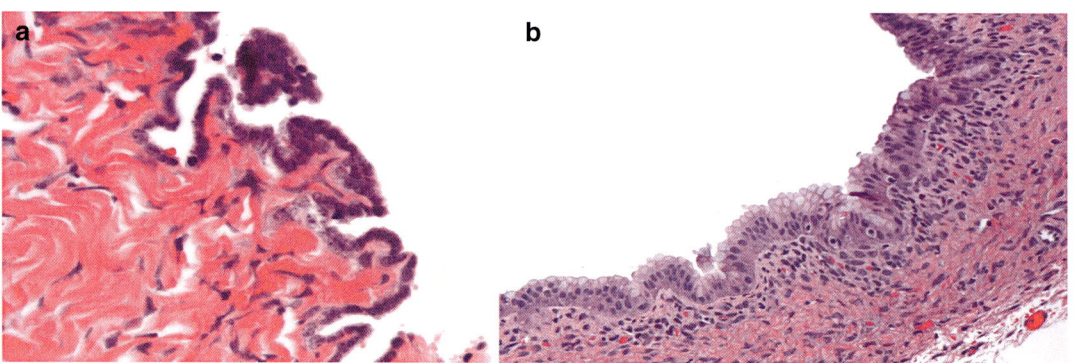

Fig. 8.6 These two microphotographs show a serous cystadenoma (**a**) and a mucinous cystadenoma (**b**) (a, H&E stain, x400; b, H&E stain, x200)

breast and other organs, surface epithelial tumors have been traditionally subdivided in benign, malignant and in a hybrid component, which is called "borderline". This category harbors atypical features of cytology without invasion into the ovarian stroma. The site of origin for serous tumors of pelvic location is the ovarian surface (cortical endosalpingiosis, cortical inclusion cysts), the distal fallopian tube, or the Müllerian epithelial rests in the pelvis. In case neither lesion in the ovary nor the tuba is seen, the tumor is supposed to originate from the peritoneum. If cystic, the prefix "*cysto-*" is added, named as serous "cystadeno-"; if prominent fibrous component ("more than a few papillae having a broad fibrous stromal core"), the suffix "*fibroma*" is added as adenofibroma or adenocarcinofibroma.

- *IHC*: (+) CK7, (−) CK 20; (+) CK8, CK18, CK19, EMA, B72.3, S100; (−) CAL and CEA (+) WT1 and ER and (−) HNF-1β and HER2.

Apart from keratins, B72.3 is also a useful marker. There is a monoclonal antibody that is used in immunohistochemistry. This antibody recognizes a tumor-associated glycoprotein 72 (TAG-72) located on the surface of numerous malignant cells. This antibody is largely used in differentiating lung adenocarcinoma (positive) from malignant mesothelioma (negative). HER2 is a member of the human epidermal growth factor receptor (HER/EGFR/ERBB) family.

HER2 is aka receptor tyrosine-protein kinase erbB-2, CD340, proto-oncogene Neu, HER2/neu, and ERBB2. ERBB is the abbreviation of the words "erythroblastic oncogene B". HNF-1β is hepatocyte nuclear factor 1 homeobox B or transcription factor 2 (TCF2), a protein of the homeobox-containing basic helix-turn-helix family. Borderline (LMP, low malignant potential) tumors are more often in younger women, often pregnant, with unusual malignant behavior and bilateral in 1/3 if include microscopic tumors and their 5-YSR: 100% if confined to the ovary and 90% if the peritoneal surface is involved.

Causes of death may be a bowel obstruction, ureteral obstruction, invasive carcinoma, sepsis, and treatment-related complications.

Grading Serous carcinomas can be graded using a two-tier system that is related to tumor biology (low-grade vs. high-grade), although staging (FIGO/WHO) has more weight than typing and grading according to several authors and cancer agencies.

Serous Cystadenoma/Cystadenofibroma

Grossly, these tumors show a smooth glistening cyst wall with or without papillary projections on the outer surface or protruding into the cystic cavity, and, microscopically, there are usually small, multilocular, simple papillary processes with a single layer (unstratified) of tall, columnar, and ciliated cells resembling normal tubal epithelium. The cells have scant eosinophilic cytoplasm

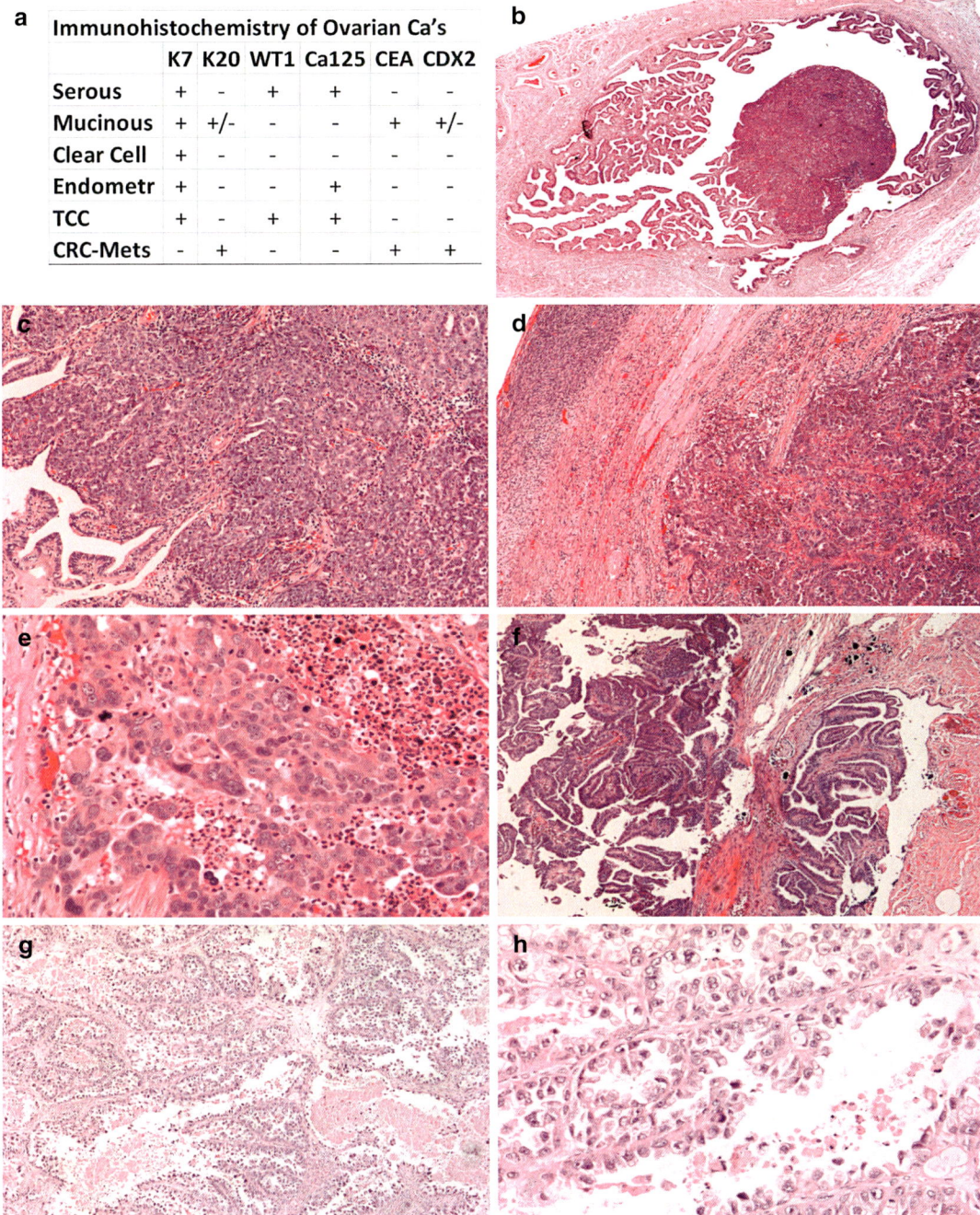

Fig. 8.7 In this panel, (**a**) shows a table with positivity of several markers for different tumors of the ovary. Serous carcinoma from the fallopian tube is shown in (**b**, **c**) (H&E stains, x12.5 and x100, respectively). Serous carcinoma, serous papillary, and clear cell carcinoma are shown in (**d–e**, **f**, and **g–h**, respectively (d, H&E stain, x50; e, H&E stain, x200; f, H&E stain, x50; g, H&E stain, x50; h, H&E stain, x200))

and oval nuclei with bland appearance and small inconspicuous basophilic nucleoli without obvious nuclear/cellular atypia, architectural complexity, or invasion.

Serous Borderline (Proliferative) Tumors (LMP)

The diagnosis of borderline may be subjective, but it is usually based on more than a small focus with the *hierarchical branching* of the papillary structures with clear *stratification* of the epithelial cells and *"budding"* into the extrapapillary space. Grossly, borderline tumors are cystic with "cauliflower-like" papillary projections in the internal cavity wall.

Following the diagnosis of borderline lesions, five features need to be assessed and reported in the final pathology report. These features are of utmost importance and may have clinical and prognostic relevance.

1. Surface involvement
2. Stromal microinvasion
3. Nodal involvement
4. Implants identification
5. Salpingitis

1. Surface involvement refers to *autoimplants*, which are single cells or glands and aggregations of cells with eosinophilic cytoplasm (G1-G2) within dominant fibroblastic stroma and associated with noninvasive peritoneal implants. Autoimplants have no apparent adverse prognostic value in stage I and should not be considered evidence of invasion. In consideration of autoimplants, ovarian clear cell carcinoma with papillary features (unilateral, nonhierarchical branching, monomorphic cells with classic features elsewhere, endometriosis, focal cytologic atypia) and low-grade serous carcinoma (clear-cut destructive stromal invasion) need to be taken into consideration.
2. Stromal microinvasion referred to *microfoci*, which are defined as dishesive cells "percolating" through the papillary stroma and categorized as <10 mm^2 and < 3 mm in the greatest size. Microfoci may be of adverse prognostic significance if associated to other variables, e.g., peritoneal involvement, "invasive" implants, and well-differentiated serous carcinoma. The presence of the micropapillary/micro-cribriform pattern is synonymous with low-grade serous intraepithelial carcinoma and may be associated with invasive implants (*vide infra*) and adverse outcome.
3. Nodal involvement refers to *lymph nodal metastasis*. It has occasionally been associated with a decrease of disease-free survival, but LN positive for metastasis need to be distinguished by Müllerian rests (endosalpingiosis), which are not malignant and may be considered as mimics of nodal involvement.
4. Implants identification refers to either *noninvasive or invasive implants*. Noninvasive and invasive implants are associated with approximately 5% and 30% mortality, respectively.
5. Salpingoliths refer to *intrafallopian concentric calcifications*, which are situated within the lamina propria, and, importantly, a diagnosis of a coexisting borderline tumor of the ovary or peritoneal surface should be considered.

Invasive features are characterized by cytologically atypical cells (increased eosinophilia of the cytoplasm with rounding of nuclei and prominence of nuclei), destructive stromal invasion, stromal reaction, capsular invasion, loss of calretinin-marker for mesothelial cells, loss of stromal CD34 marker+ of fibrocytes around the nests, and presence of α-smooth muscle actin (SMA)+ myofibroblasts as stromal response. The calretinin, CD34, and alpha-SMA have been investigated for the identification of peritoneal invasive implants of serous borderline tumors of the ovary Lee et al. 2006.

Borderline serous tumors with micropapillary/cribriform architecture = Low-grade serous intraepithelial carcinoma. In the micropapillary growth pattern, slender papillae radiate from a large papilla resembling a "Medusa head," whereas in the cribriform, pattern occurs when micropapillae merge. Nuclear cytology shows a striking monomorphism, and cells have scant cytoplasm and hobnailing; columnar or flattened cell morphology may be observed.

The Gorgons were three monsters in Greek mythology, daughters of Echidna and Typhon, including Stheno, Euryale, and Medusa. The most famous Gorgon is indeed Medusa. Medusa's name (Greek Μέδουσα "guardian, protectress") derives from μεδω "to protect", "to rule over", "to exercise absolute authority". Medusa was so gruesome that anyone who gazed upon her was turned to stone. The hero Perseus, the son of Zeus and Danaë, the daughter of Acrisius of Argos, was the slayer of the Gorgon Medusa and the rescuer of Andromeda from a sea monster. Perseus was able to kill Medusa using the reflection in his shield.

Ovarian Implants and Benign Mimics

Implants are deposits of ovarian tumor tissue on peritoneal surfaces, lateral pelvic grooves, diaphragm, or omentum in patients with primary ovarian tumors. It is paramount to emphasize again that benign mimics need to be distinguished from true implants, and implants must be separated between noninvasive and invasive implants that affect the staging and prognosis of these patients. Bell et al.'s (2001) criteria for invasive implants include (1) invasion by proliferative epithelium with DSR (desmoplastic stroma reaction) and infiltrative margin, (2) micropapillary architecture, and (3) clefts-surrounding solid epithelial nests embedded in a fibrous stroma.

If there are multiple peritoneal nodules with features of serous ovarian borderline/proliferative tumor or carcinoma and some minimal ovarian involvement, it is an ovarian carcinoma. In the event that no ovarian involvement is detected, then an extraovarian serous carcinoma or, even, a papillary tumor of peritoneum should be taken into consideration in the differential diagnosis. If a biopsy does not show any underlying tissue, the lesion, whichever morphology shows, has to be classified as "noninvasive" by default and broadly discussed at the multidisciplinary tumor board meetings.

Ovarian Serous (Cyst-) Adenocarcinoma (OSCAC/OSAC)

Clinically, there is a presentation as ovarian enlargement or lymphadenopathy (usually inguinal or supraclavicular), and about 2/3 of cases show bilateral occurrence, but rarely there is AFP production. It may derive from tubal intraepithelial carcinoma (STIC, serous tubal intraepithelial carcinoma).

- *GRO*: The OSCAC is solid, hemorrhagic, and necrotic.
- *CLM*: Branching papillary fronds and glandular complexity ± psammoma bodies (calcium concretions with concentric laminations that may also be intracellular due to autophagocytosis) on low-power magnification and nuclear atypia and stratification, increased MI, and stromal invasion on high-power magnification.
- *IHC*: (+) WT1, ER, CK7, EMA, and CA125 (additional markers may be mesothelin, which may also be positive in lung, pancreas, endometrial adenocarcinomas as well as CK5/6 and D2-40, which are markers that are positive in malignant mesothelioma); (−) CK20, calretinin, VIM, and HNF-1β (positive marker for ovarian clear cell tumors).
- *CGB*: *TP53* and *BRCA1* loci LOH in high-grade tumors.
- *DDX*: Peritoneal mesothelioma (+) h-Cald, calretinin and (−) ER, Ber-EP4, MOC31, and metastatic invasive micropapillary carcinoma of the breast (GCDFP-15+ and WT1−).

Noteworthy, Ber-EP4 antibody is a membrane-enriched fraction derived from the MDF-7 breast cancer cell line reacting with an epitope present on two glycoproteins on epithelial cells except the superficial layers of squamous epithelium, hepatocytes, and parietal cells. Ber-EP4 is a useful antibody to add to a panel for malignant mesothelioma, which shows no Ber-EP4 expression and picking up distal tubules of normal kidney and DSRCT. MOC-31 Mo-Ab is an epithelial glycoprotein 2 on epithelial cells and is expressed by a wide range of malignant epithelial tumors showing some similarities of the epitope with EMA.

DSRCT is an important differential diagnosis in young girls with oncologic peritoneal involvement. DSRCT tumor cells are polyphenotypic, showing features of epithelial/mesothelial, muscle, and neural differentiation, and are positive for epithelial markers, including keratins, Ber-EP4 and MOC-31, and WT1 and CD15, but negative for CK5/6 and CD141 (thrombomodulin).

- *PGN*: 5-YSR, 70%, if confined to the ovary but 25% if involves the peritoneum! The malignant serous ovarian tumor (OSCAC) rarely but occasionally may metastasize to breast or axillary nodes and, in this eventuality, is associated with advanced stage disease.

Low-Grade vs. High-Grade Serous Carcinoma

LG-OSC usually combines both extensive papillary architecture and stromal invasion, and cytology shows cells with eosinophilic cytoplasm and nuclei, which are consistently round or oval. Moreover, chromatin is evenly distributed with or without prominence of nucleoli. MI < 10/10HPF, numerous nests of invasive implants and focal p53 staining, which means no TP53 gene mutations are also distinctive features.

HG-OSC shows a vast spectrum of morphology/histological growth patterns including branching papillary fronds, fenestrations of slit-like type, complex glandular elements, and fibrous/edematous/myxoid/desmoplastic stroma on low-power microscopy. Nuclear atypia (moderate to marked) with marked pleomorphism, prominent nucleoli, nuclear stratification, and high MI are encountered on high-power microscopy. A stromal invasion is also observed.

TP53 Gene Mutational Status and Gene Product Expression (p53 IHC)

There has been some ambiguity in consideration of the *TP53* mutational status gathered from the results of the immunohistochemical staining for the gene product. In particular, some studies have evidenced different results with evident confusion in prognosis and treatment options. This data has generated fruitless discussion at both scientific meetings and clinical meetings (multidisciplinary team meetings). Gene mutational status is particularly efficient, but some factors, including labor-intensive hamper effectiveness, time-consuming, and validation of appropriate molecular biology-trained technologists may play a significant role. Consequently, immunohistochemistry is often used to detect proteins that show gene mutations.

This is the case for MLH1, MSH2, MSH6, and PMS2 for the microsatellite instability patterns of colorectal carcinoma. The wild-type p53 protein is relatively unstable and has a short half-life, which makes it undetectable by immunohistochemistry. Conversely, mutant p53 has a much longer half-life. Consequently, it results in the accumulation in the nucleus. This data makes this mutated protein a stable target for immunohistochemistry-based detection. However, some caveats need to be declared. Two immunohistochemical patterns are essential to gather the *TP53* mutational status from the immunohistochemical findings. Substantially, (1) *overexpression of p53 (>60% tumor cells)* correlates invariably with a *missense* mutation of the *TP53* gene, while (2) *complete lack of immunolabeling* on formalin-fixed and paraffin-embedded tissue corresponds indeed with a *nonsense* mutation leading to protein truncation.

Ovarian Serous Tumors: Variants

Cystadenofibroma: Cystadenofibroma of Borderline Type – Cystadenofibrocarcinoma

In cystadenofibroma, there is a prominent fibrous stromal component surrounding benign glands, while malignant glands are the characteristic feature present in cystadenofibrocarcinoma.

Serous micropapillary carcinoma is aka low-grade serous carcinoma because it behaves like a low-grade adenocarcinoma (10-YSR of 70% vs. 10-YSR of 100% for borderline tumors). This tumor seems that arises from serous borderline tumor or adenofibroma but recurs indeed as classic serous carcinoma. As indicated above, there is a filigree pattern of small, uniform, elongated, stroma-poor, or stroma-free papillae in serous micropapillary carcinoma.

Microinvasive serous carcinoma refers to an ovarian tumor of the serous type with small foci with a size variable, but up to 2 mm constituted by single cells or small clusters of cells, occasionally cribriform or rounded aggregates of papillae and harbors the same prognosis as borderline serous tumors.

Serous psammoma carcinoma is a rare variant of serous carcinoma with multiple and massive psammoma bodies and low-grade cytologic features. The 10-YSR of serous psammoma carcinoma is similar to borderline serous tumors.

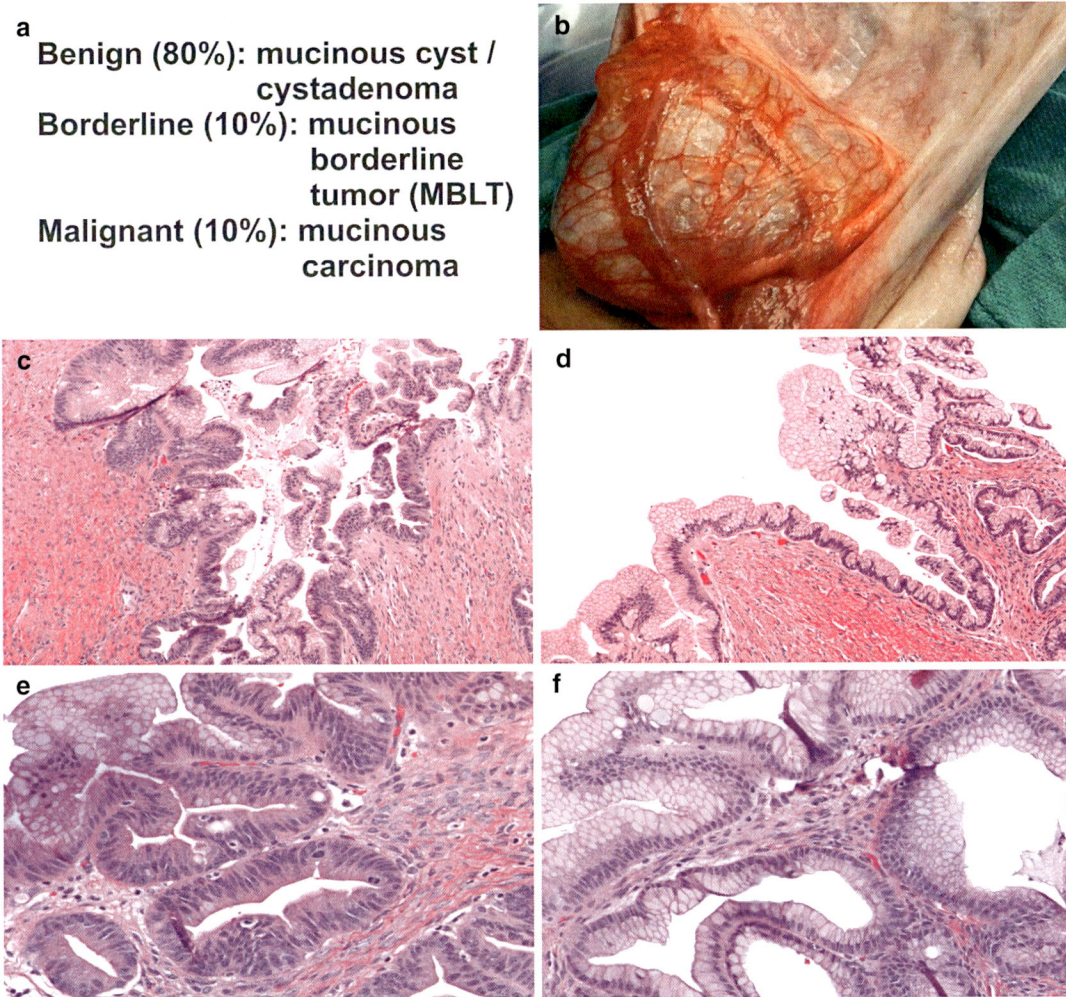

Fig. 8.8 The classification of the mucinous tumor of the ovary can be very challenging. In (**b**) is shown the gross photograph of a very large mucinous borderline tumor of the ovary. The histology of the mucinous borderline tumor is shown in (**c–f**) (c-f, H&E stains, x100, x100, x200, x200, respectively)

8.6.2.2 Mucinous Tumors

Ovarian mucinous tumors may be benign, borderline, or malignant (ovarian mucinous carcinoma or OMC) and represent up to ¼ of ovarian tumors, bilateral up to 1/5 of cases (most commonly when malignant), and most often surface epithelial tumor of children with an association with *teratoma*. Other tumor associations include *Brenner tumor, carcinoid,* and *YST*. The term "mucinous" derives from mucus, which is from the Proto-Indo-European word *mew-k-* ("slimy") and a cognate word include the Greek word μύκης ("mushroom").

- *GRO*: Mucinous tumors tend to grow larger than serous type and have a multiloculated aspect with cysts containing mucinous material, and hints for malignancy include papillae, solid areas ("mural nodules"), necrosis, and hemorrhage.
- *CLM*: Microscopically, the tumor cells are tall, columnar, and non-ciliated. Cells have basal nuclei, abundant intracellular supranuclear mucin, and cellular stroma, which may be luteinized. Mucinous tumors of the ovary, mostly of borderline type (increased layering even up to four cells thick), may be subclassified as an intestinal type or IBMT (+ goblet

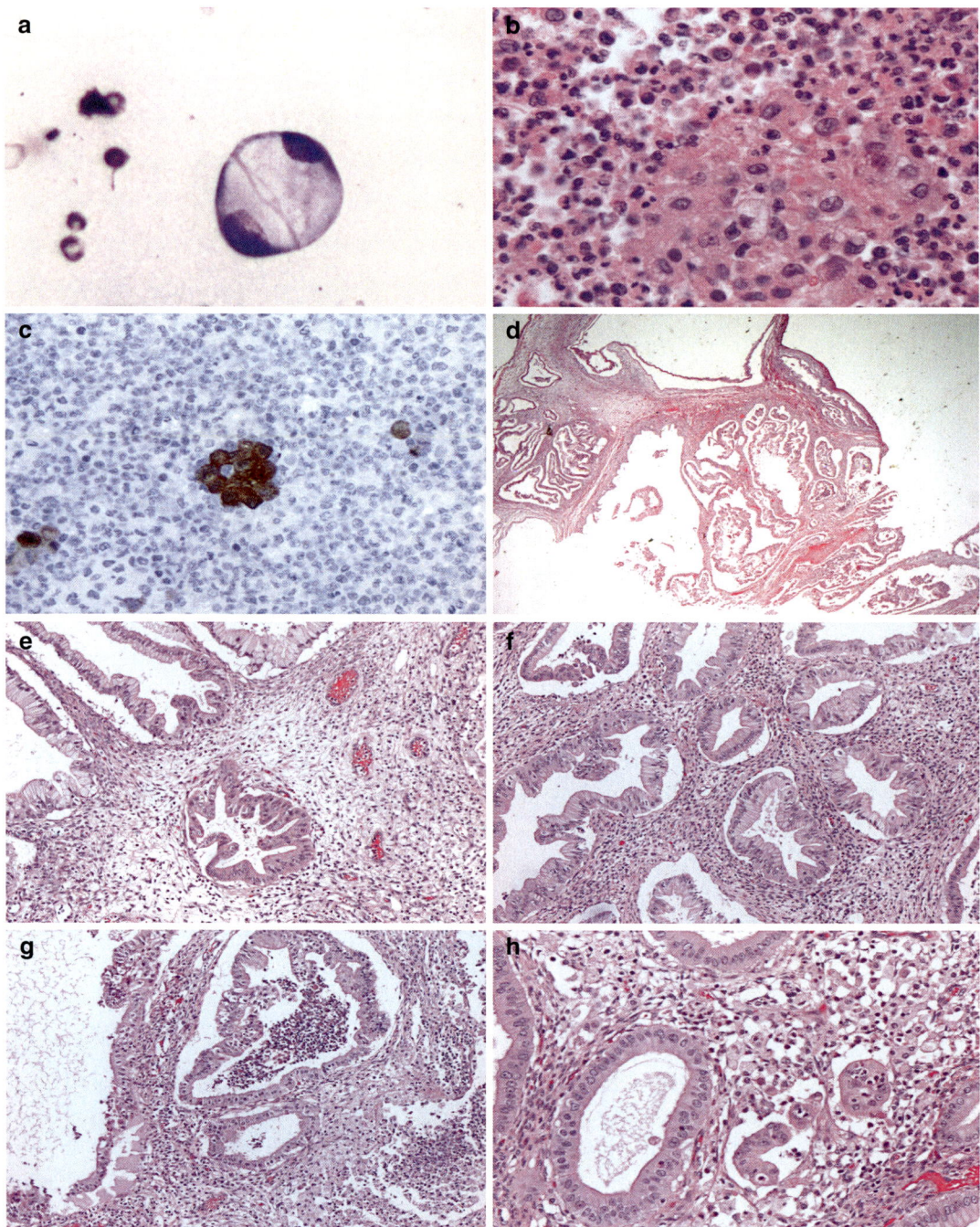

Fig. 8.9 In this panel is shown a mucinous carcinoma with characteristic morphology (**a–h**). In (**a**) is shown the cytology of a peritoneal wash (Diff-Quik stain, x630). In (**b, d–h**), are shown the atypical glandular proliferations of the mucinous carcinoma with necrosis (b, d-h; H&E stains, x400, x12.5, x100, x100, x100, x200, respectively). In (**c**) is shown the peritoneal wash where a cluster results positive for calretinin (Anti-Calretinin immunostaining, ABC method, x400)

cells, Paneth cells) and endocervical kind or EBMT (goblet cells, Paneth cells). IBMT is 85% of cases and tends to be unilateral, while EBMT is 15% of cases and tends to be bilateral. Cell atypia, increased cell layering (>4 cells thick), complex glands, and stroma invasion characterize the deadly form of this kind of tumors. However, stroma invasion is frequently a challenge because of gland complexity (Figs. 8.8 and 8.9). Microinvasive borderline of the thickness of the invasion (θ) < 5 mm and NO pleomorphism/marked atypia are seen, while microinvasive mucinous carcinoma is defined if θ < 5 mm AND marked atypia is detected. Obviously, invasive carcinoma is defined if θ ≥ 5 mm.

Two criteria for OMC:

1. Visible infiltrative glands, cysts, or cell nests with surrounding DSR
2. Solid glandular formations of >10 mm²/area, of which Ø ≥3 mm

Mucinous carcinomas have a grading according to nuclear atypia in expansile lesions and a three-tier system, comparable to that of the endometrioid type, for infiltrative lesions.

In practice, grade 1 shows no substantial areas, in grade 2 up to 50% solid foci, and in grade 3 more than 50% solid foci. However, severe nuclear atypia can raise grade 1 or 2 carcinomas by one degree.

- *IHC*: (+) CK7 and EMA, (±) CK20, (−) WT1, and (±) ER

The primary consideration in DDX is metastatic carcinoma (CRC, appendix), and both histology ("dirty necrosis with basophilic nuclear debris") and IHC (CK7/CK20) may help. The immunohistochemical phenotype of EBMT differs from that of IBMT. Both types of borderline mucinous tumors exhibit (+) CK7, but only IBMT shows positive staining for CK20. (+) CEA is strongly and diffusely in IBMT, while (+) ER and (+) PR are present in EBMT.

- *OMC*: (+) CK7, (±) CK20, (±) CEA, (+) B-CAT$_{(C)}$ and (−) CDX2
- *CRC*: (−) CK7, (+++) CK20, (+++) CEA, (+) B-CAT$_{(N+C)}$ and (+) CDX2

CEA positivity is usually seen in 100% of malignant mucinous tumors but 4/5 of IBMT and 15% of benign mucinous form. β-catenin is usually exclusively cytoplasmic in ovarian mucinous carcinomas, while both cytoplasmic and nuclear in most colon-rectal carcinomas due to the nuclear translocation of the gene product.

- *PGN*: 5-YSR (pure mucinous carcinoma) is about 95%, while it decreases to 87% when the tumor is mixed with other types of cancer.

Criteria of stroma invasion according to WHO are (1) evident invasion, which is obvious as far as desmoplastic adenocarcinoma in other sites; (2) infiltrative pattern due to glands, cords, nests, and/or single cells going into the ovarian or desmoplastic stroma; and (3) solid glandular cluster (>10 mm² area and ≥ 3 mm of two linear dimensions) in >1 of the following "back-to-back" architecture (i.e., little or no stroma), (a) serpiginous shapes, (b) complex glands, (c) complex true papillary, (d) irregular glands with necrotic contents, and (e) cribriform histological growth pattern.

8.6.2.3 Endometrioid Tumors

Endometrioid differentiated tumors account for up to 15% of all ovarian tumors, and about 1/5 of cases are *bilateral*, have concurrent *endometriosis*, and/or *synchronous endometrial carcinoma* of the uterus. The tumoral synchronism may be explained by either a single clonal tumor with concordant genetic alterations using identical LOH, *PTEN* mutation, or sporadic MSI patterns, or a single clonal tumor with genetic progression, or eventually (potentially) ≥ 2 clonal tumors. There is also a possible association with YST.

- *GRO*: Endometrioid tumors frequently show large to huge cysts with some solid, often hemorrhagic areas obviously devoid of papillary structures (grossly).
- *CLM*: The histologic aspect is similar to endometrioid endometrial adenocarcinoma (tubular glands identical to endometrium) with areas of squamous metaplasia ("squamous morules"). Benign (no cytologic atypia, no architectural complexity) and borderline forms (cell clusters, glands, or cysts of endometrioid-type epithelium with atypical features but lacking destructive stromal invasion, glandular confluence, or stromal disappearance) are extremely rare compared with

the carcinoma. Ovarian endometrioid carcinoma (OEC) can be graded using the same three-tier system as used in the uterus. High-grade/grade 3 nuclei are large, pleomorphic with coarse chromatin, large irregular nucleoli, high MI, ± villoglandular architecture (histological growth pattern), and microinvasion, which refers to areas of invasion ≤10 mm^2.

Variants include a granulosa cell tumor-like variant, Sertoliform variant, and spindle-cell variant of endometrioid carcinoma of the ovary.

- *IHC*: (+) CK7, (−) WT1.
- *CGB*: *PTEN* mutations in ~1/5 of cases.
- *PGN*: OEC shows a better survival than serous or mucinous carcinomas.
- *DDX*: OEC vs. OSC.
- *IHC*: WT1 − β -Cat − p16 (p16INK4A):
- OEC is usually WT1−, β-Cat$_n$−, p16−.
- OSC is generally diffusely, and strongly WT1+, shows nuclear translocation of β-Cat (β-Cat$_n$+), and is diffusely and strongly p16+.

8.6.2.4 Clear Cell (Mesonephroid) Tumors
- *DEF*: It is probably a variant of endometrioid carcinoma and indeed may occur with endometrioid carcinoma, and all are (almost) of the *high grade*. Ovarian clear cell (adeno-) carcinoma (OCCC), aka mesonephroid adenocarcinoma, is an epithelial-stromal tumor that may be associated with three clinical conditions, including:
  1. Endometriosis (or endometrioid carcinoma of the ovary)
  2. Hypercalcemia
  3. Thromboembolism

OCCC is typically characteristic of *middle-aged women*, but young females have been reported in the literature.

Pediatric clear cell carcinoma of the ovary (PCCCO) is considered by some authors a separate entity.

- *GRO*: OCCC has a spongy-cystic texture, which corresponds to histology.
- *CLM*: Three histologic patterns are frequently encountered on low-power magnification:
  1. *Tubular-cystic* pattern
  2. *Papillary* pattern with or without prominent hyalinization
  3. *Solid* pattern

On high-power magnification, the tumor is constituted by large cells, some with nuclei that protrude into lumina (hobnail cytologic feature), with clear cytoplasm (mostly glycogen ⇒ PAS+, mucin−, ORO−), and may resemble mucinous tumor cells on intraoperative frozen section.

- *IHC*: (+) CK7 and (−) CK20, (+) EMA, (−) WT1 and ER, (+) HNF1β, but (±) CA-125, CD15, CEA, AFP and (±) PR and p53 (no expression of PR and TP53 gene product in ~90% of cases); CD15 (Leu-M1) is mostly positive in OCCC, but negative cases have been observed in the literature and personal experience.
- *DDX*: It includes epithelial and mesenchymal tumors, although the "specific" immunophenotype with the H&E pattern may help to distinguish OCCC from some other similar tumors.
  1. *O-HGSC*, as these tumors are usually WT1, ER, PR, p53, positive, and negative for HNF1β. Immunohistochemically, OEC is typically WT1 (−), ER/PR +, and p53−. The same immunophenotype as O-HGSC, as described above, is seen in borderline serous tumors except for p53 overexpression.
  2. *YST/EST* (+ AFP, + GPC) and (−) CK7, EMA, CD15.
  3. *Dysgerminoma* (+CD117, OCT 4, D2–40, PLAP).
  4. *JGCT* shows (+) INA, CALR, SF-1, WT-1, and CD 56; (+/-) CKs (dot-like pattern); (-) EMA. Differently from adult-type granulosa cell tumor (AGCT), they do not have *FOXL2* gene mutations.
  Molecular genetic analysis of AGCT has evidenced that over 90% of AGCTs have a missense somatic point mutation in the *FOXL2* gene. JGCT are indistinguishable from AGCT grossly, but histologically, these tumors have solid (diffuse or nodular) and follicular growth. The follicular

growth pattern exhibits follicles of varying size and shape containing eosinophilic or basophilic secretions. Moreover, differently from AGCT, JGCT cells have ample eosinophilic to vacuolated cytoplasm and rounded nuclei without grooves.

5. *Peritoneal malignant mesothelioma* (PMM) and well-differentiated papillary mesothelioma (WDPM) need to be taken into account particularly in women of young age. In WDPM, there is a papillary proliferation with focally hyalinized or foamy macrophages. The fibrovascular cores are lined by a single layer of cuboidal cells and nuclear atypia is minimal with a very low mitotic index (MI) In WDPM, there are no destructive infiltration of the underlying stroma. Conversely, PMM may show focal well-differentiated papillary areas together with other growth patterns (solid, biphasic, sarcomatoid, deciduoid). Moreover, atypia is usually higher than mild or focal. The association of moderate or severe atypia, extensive pseudostratification, or high MI indicate PMM.
6. *Lipid-rich Sertoli cell tumor* shows (+) VIM, *I*NA, *M*ART-1, LMW-CK, CALR, AR, PR, and (−) CK7, WT1, and (±) CD99. SF-1 (adrenal four-binding protein) is a nuclear transcription factor regulating genes that are involved in the development of gonads and adrenal glands, sexual differentiation, steroidogenesis, reproduction, and metabolism.
7. *Clear cell variant of struma ovarii*, which shows no evidence of atypia, and other patterns are also present.
8. *MTX* from numerous sites including the kidney (−CK7, −CK20, ±EMA, +CD10, +RCC-Ag), adrenal, and the gastrointestinal tract.

- PGN: 5-YSR is 40–50%.

8.6.2.5 Brenner (Transitional Cell) Tumors

Brenner tumors or transitional cell-differentiated ovarian tumors account for up to 2% of all ovarian tumors and probably a variant of serous tumors, particularly when high-grade tumors are compared. In ¾ of cases, the presentation is 40+ years, but female adolescents with this kind of tumors have been reported. There is an association with *hyperestrinism*, *mucinous cystadenoma*, *struma ovarii*, and *urothelial carcinoma of urinary bladder*.

Grossly, Brenner tumors are typically unilateral, firm, white, and solid, and, microscopically, there is quite a unique constellation of solid and cystic nests of cells similar to transitional epithelium of the urinary tract with cells (PAS+ histochemically) showing distinct cell borders, oval "coffee-bean" nuclei with small and distinct nucleolus, and longitudinal grooves. The solid and cystic nests of transitional cells are embedded in a dense, fibroma-like stroma.

Benign Brenner tumor does not show atypia or invasion; borderline Brenner tumor has atypia, but no invasion. Malignant Brenner tumor, which needs to harbor a benign Brenner component, shows mucin-containing microscopic spaces, malignant cytology, and obvious stromal invasion. A three-tier system based on nuclear morphology is used for grading. Transitional cell carcinoma of the ovary does not have a benign Brenner component. Differentiating malignant Brenner tumor from transitional cell carcinoma of the ovary is not irrelevant, because the latter has a more aggressive course. Moreover, the transitional cell carcinoma of the ovary may exhibit a better chemosensitivity than malignant Brenner tumor.

- *IHC*: (+) WT1, ER, CK7, EMA, CEA, and CA125 and (±) CK20, URO-III, and CD141

Benign Brenner tumors show a line of some urothelial differentiation (either CK20, or URO-III, or CD141), which is absent in a peculiarly critical differential diagnosis such as transitional cell carcinoma, allowing the differentiation with urothelial carcinoma of the urinary bladder, which is (−) WT1 and (−) ER. Transitional cell carcinoma of the ovary is also (+) WT1 and ER.

Variants:

- *Metaplastic Brenner tumor*: Prominent cystic formations with clear-cut mucinous changes analogous to *cystitis glandularis*

- *Proliferating Brenner tumor*: Papillary tumor with low-grade nuclear atypia similar to low-grade urothelial carcinoma of urinary bladder

DDX includes four entities that need to be ruled out:

(1) *Granulosa cell tumor*, (2) *Carcinoid*, (3) *Mucinous ovarian tumor*, and (4) *Transitional cell carcinoma of the urinary tract*, which shows (+) CK20, URO-III, and CD141. Transitional cell carcinoma of the ovary, which lacks a benign Brenner component, may be impossible to distinguish from metastatic urothelial carcinoma of the urinary tract if an appropriate patient history is not available.

Other tumors include *squamous cell tumors*, *mixed surface epithelial tumors*, and *undifferentiated surface epithelial tumors*. In Box 8.8 there are some tips and pitfalls of the ovarian mucinous tumors.

Box 8.8 Ovarian Mucinous Tumors: Pearls and Pitfalls

If mucin is (+), then think of *primary mucinous carcinoma of the ovary* (+CK7, ±CK20), most *Krukenberg tumors* (CRC-MTX: −CK, +CK20, +CEA, +B-CAT$_{(N+C)}$, +CDX-2), and *mucinous carcinoid tumors* (+CGA and SYN).

If mucin is (−), then think of *primary clear cell ACA of the ovary*, *signet ring stromal tumor*, and *sclerosing stromal cell tumor of the ovary*.

Be aware that ovarian mucinous tumors may be associated with teratomas in children!

OEC needs to be sampled thoroughly, as serous or undifferentiated carcinoma component lowers 5-YSR from 2/3 to ~1/10. Also, a thorough sampling is of utmost importance because a YST component may be present, which makes the tumor's behavior very aggressive.

8.6.3 Sex Cord-Stromal Tumors

Sex cord-stromal tumors (SCSTs) are ovarian neoplastic proliferations derived from the ovarian stroma, which arises from sex cords of embryonic gonad and in turns considered a predecessor of Sertoli, Leydig, granulosa, and theca cells. Sex cord-stromal tumors account for 1/20 of all ovarian neoplasms, and about the same rate has been identified for malignant ovarian neoplasms. It is important to remember that typically estrogens are produced from theca cells, while androgens are produced from Leydig cells. Classification of SCSTs is based on morphology, although clinical and biochemical findings support it.

Characteristic IHC markers useful for SCSTs are in Box 8.9. Negative IHC stains, useful for a combined immunohistochemical panel, are EMA, PLAP, CEA, CA19-9, CA125, as well as CD45 and CGA. PTPRC or protein tyrosine phosphatase, receptor type, C is an enzyme that is encoded by the *PTPRC* gene in humans. Most often, PTPRC is aka CD45 antigen and leukocyte common antigen (LCA).

Typically, WT1 and CD99 are expressed in all SCSTs except steroid cell tumor and fibroma/

Box 8.9 IHC Markers Useful for SCSTs (VINCAP/VIMCAP)

IHC markers	Abbreviation
Vimentin	VIM
Inihin A	INA
Nephroblastoma antigen	WT1
Melanoma antigen recognized by T cells 1	MART-1
Low-molecular weight Cytokeratins	LMW-CK
Cytokeratins 7/8	CAM 5.2
Calretinin	CALR
Androgen receptor	AR
Progesterone receptor	PR
CD99	PNET-Ag

fibrothecoma, respectively, while MART-1 is expressed in steroid cell tumor and Sertoli-Leydig cell tumor only. In the latter entity, MART-1 seems restricted to Leydig cell component only. Moreover, INA and CALR are less frequently expressed in fibroma/fibrothecoma/thecoma group in comparison with the other type of SCST. Recently, SF-1 or adrenal four-binding protein, a nuclear transcription factor that regulates genes involved in steroidogenesis, gonadal and adrenal gland development, sexual differentiation, reproduction, and metabolism, seems to be the most sensitive marker for SCSTs followed by INA, being expressed in 100% of the cases in all types of tumors. Anti-Müllerian hormone staining is also a marker, which is specific for a percentage of granulosa or Sertoli tumor cells. Moreover, a potential pitfall could be some INA positivity that may also occur in luteinized cells from other tumors other than SCSTs.

8.6.3.1 Granulosa Cell Tumor

- *DEF*: Granulosa cell-based SCST (GrCT) with clinical occurrence both pre- (60%) (meno-/metrorrhagia or amenorrhea) and postmenopausal (40%) (postmenopausal bleeding) and 1/20 of cases even prepubertal. Biochemically, there is hyperestrogenism in 3/4 of cases with subsequent estrogenic-related disturbances, including isosexual pubertal precocity, endometrial hyperplasia/carcinoma, meno-/metrorrhagia, and fibrocystic disease of the mammary gland.
- *GRO*: GrCTs (also known as adult granulosa cell tumors or AGCT) are almost constantly unilateral, encapsulated, solid, and lobulated, gray with variegated cut surface with ± cysts with straw-colored fluid.
- *CLM*: Microscopically, there is a variable number of histological patterns, including microfollicular (Call-Exner bodies-rich), macrofollicular, trabecular, insular, solid, and "watered silk" (cell arranged in undulated rows). Call-Exner bodies are small "glands"-like follicles filled with acidophilic material. Tumor cells have "coffee-bean" nuclei with folds and grooves, and bizarre multinucleated cells may be occasionally encountered, which are a sign of degenerative cytologic change rather than malignancy. Theca cell component is often present (up to 90% of cases), and cells may luteinize showing central nuclei, abundant eosinophilic cytoplasm, and well-developed cell borders (Figs. 8.10 and 8.11). Remarkably, reticulin fibers (reticulin stain or silver stain) surround groups of cells, rather than single cells as seen in thecoma or fibroma. This distinguishing feature between GrCT and thecoma/fibroma is particularly important in the evaluation of biopsies. FNAC study shows cells with little cytoplasm and nuclear indentation looking like mesothelial cells.
- *IHC*: (+) VIM, INA, WT1, CAM5.2, CALR, CD56, and (−) EMA but also (+) S100, SMA, DES, desmoplakin, and anti-Müllerian hormone. LMW-CK (CAM 5.2) is present but dot-like. EMA is an important negative marker to rule out possible similarities with ovarian surface tumors and we advise to order systematically for immunohistochemistry.
- *TEM*: Abundant intermediate filaments as common cytoskeletal structural components and desmosomes.
- *CGB*: Monosomy 22 and trisomy 12, although +14, monosomy X, and monosomy 17 have also been observed. Diploidy of the tumor cells is seen in 4/5 of tumors. Diploidy indicates that the tumor cells have the same number of chromosomes as normal, healthy cells (2X sets of 23 chromosomes). This data suggests that these tumors are slower-growing and probably less aggressive than aneuploid malignancies.
- *DDX* includes:
  1. Ovarian, surface epithelial carcinoma, poorly differentiated (→ CK7, EMA)
  2. Carcinoid tumor (CGA and search for neurosecretory granules on TEM)
  3. Endometrial stromal sarcoma
  4. Endometrioid carcinoma (+CK7, −WT1)
  5. Small-cell carcinoma of hypercalcemic type
  6. Pregnancy-related granulosa cell proliferation (multiple, microscopic foci, atretic follicles)
- *PGN* points to the relevance of tumor size and staging with 10-YSR of >90%; most deaths occur after 5 years (up to 20 years later!). In fact, as indicated, most of these tumors are diploid (*vide supra*).

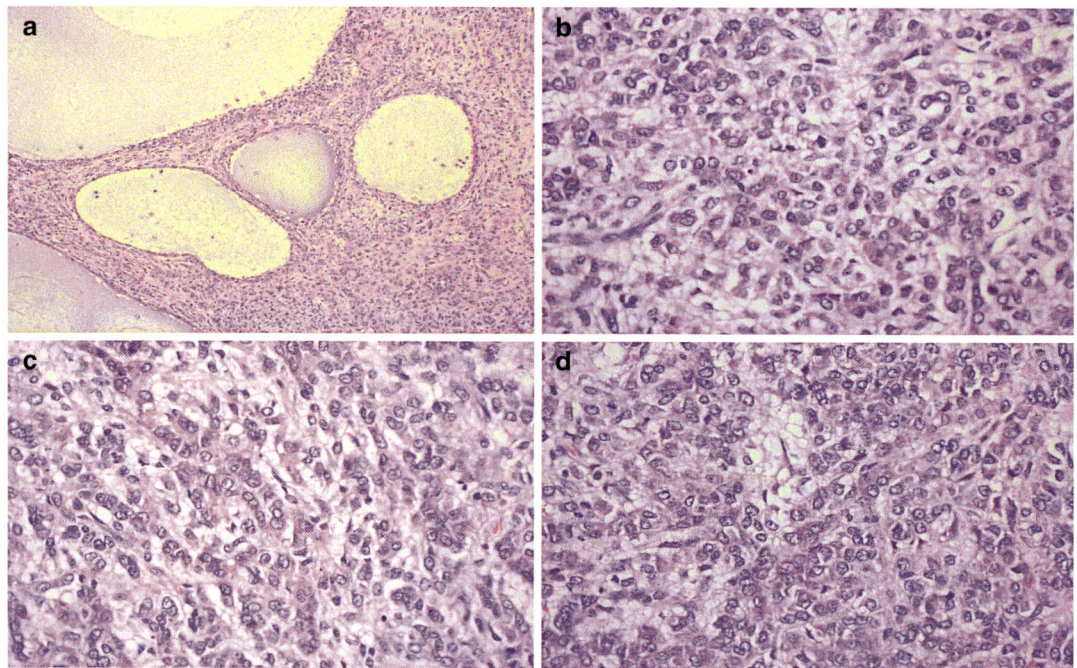

Fig. 8.10 In (**a–d**) are shown the microphotographs of a juvenile granulosa cell tumor with macrocysts (a-d, H&E stain, x40, x400, x400, x400, respectively)

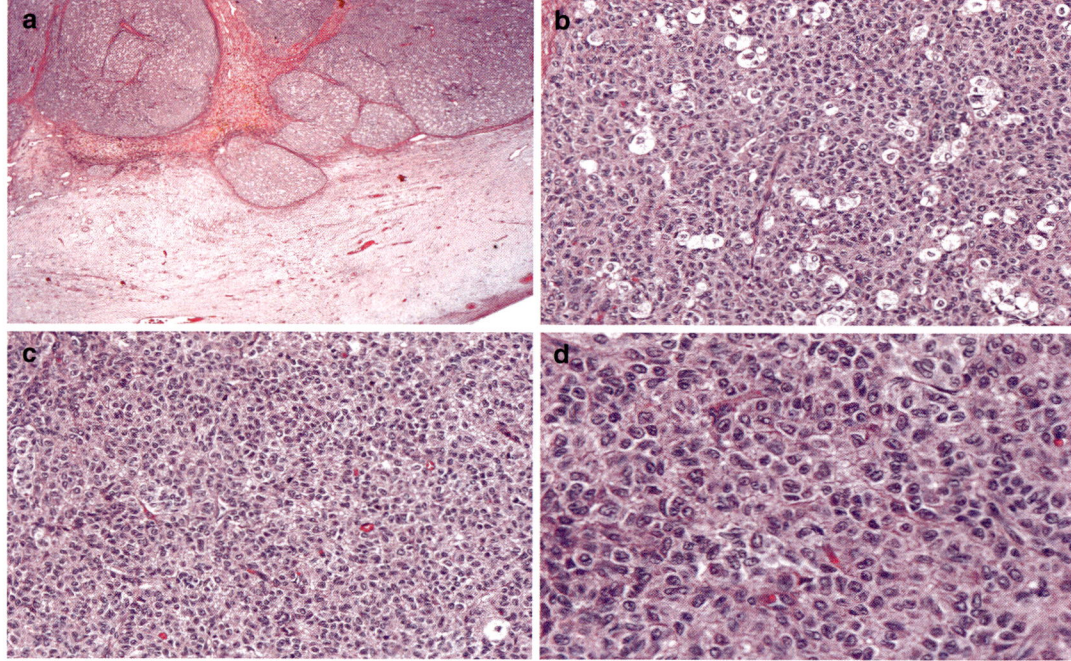

Fig. 8.11 The adult granulosa cell tumor shows small microcysts and Call-Exner bodies (**a–d**, H&E stain, x12.5, x200, x200, and x400 original magnification, respectively)

- Variant: Juvenile granulosa cell tumor (JGCT), which occurs in 4/5 of cases in children and adolescents before the age of 20 years showing a multiloculated architecture with cysts and solid areas with or without hemorrhage and necrosis grossly. JGCT clinical associations are Ollier disease, Maffucci syndrome, Goldenhar syndrome, and ambiguous genitalia/karyotype aberrations. Microscopically, diffuse or macrofollicular patterns with rare or no Call-Exner bodies but mucin (+) in follicular secretions. Tumor cells are large, hyperchromatic round/oval nuclei with few or no nuclear grooves and variable scant (non-luteinized) or large cytoplasm (luteinized). MI is high (~7/10 HPF) (Fig. 8.10).
- *DDX*:
  1. AGCT (microfollicular with intraluminal BM-material, Call-Exner bodies, nuclear grooves, and lack of hyperchromasia)
  2. Thecoma (age > 40 years, no follicles, no atypia, nil or low MI, obvious ret stain pattern)
  3. OCCC (diffuse hobnailing, lack of follicles or JGCT cells, (+) CK7 and EMA, and (−) WT1 and ER)

8.6.3.2 Sertoli-Leydig Cell Tumor

The SCST has a young age group occurrence and mostly androgen-producing features (aka androblastoma, arrhenoblastoma formerly). It accounts for less than 0.2% of ovarian neoplasms. Although it depends from the degree of differentiation, SCST has a relatively good prognosis (Figs. 8.12, 8.13, and 8.14). The Greek word άρρεν (male) is the root for the arrhenoblastoma, which links the ovarian neoplasm to the growth pattern of "testicular-like" tubules and the secretion of male sex hormone, inducing virilization in female individuals. Other names used for this kind of tumors are andreioma, andreoblastoma, androma, and arrhenoma, but they are rarely used. Robert Meyer introduced the term 'arrhenoblastoma' for most widely now as Sertoli–Leydig cell tumor and his subclassification of SLCTs into well-differentiated, intermediate, and poorly differentiated forms remains a milestone in gonadal oncology.

Biochemically, androgens ⇒ erythrocytosis, although testosterone and estradiol are found in both cell types.

Clinical associations: ± FHx (family history is crucial). It has been associated with sarcoma botryoides of the cervix and thyroid disease.

The extended Meyer classification of SLCT includes WD-SLCT, MD-SLCT, PD-SLCT, SLCT with heterologous elements, retiform SLCT, and pure Sertoli cell tumor (WD, well-differentiated; MD, moderately-differentiated; PD, poorly-differentiated).

1. *WD-SLCT* (Meyer type I) (~10%): Hollow/solid tubules lined by Sertoli-like cells separated by cellular (fibrous) stroma with a variable number of Leydig-like cells. PGN: 5-YSR: 100%.
2. *MD-SLCT* (Meyer type II) (~55%): Clusters, cords, or sheets of Sertoli-like cells separated

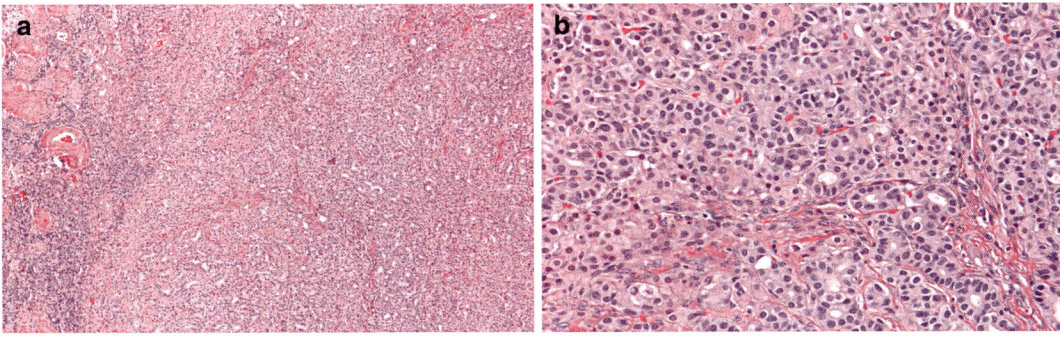

Fig. 8.12 The two microphotographs (**a–b**) show a well-differentiated Sertoli-Leydig cell tumor (a-b; H&E stains, x50 and x200 as original magnifications, respectively)

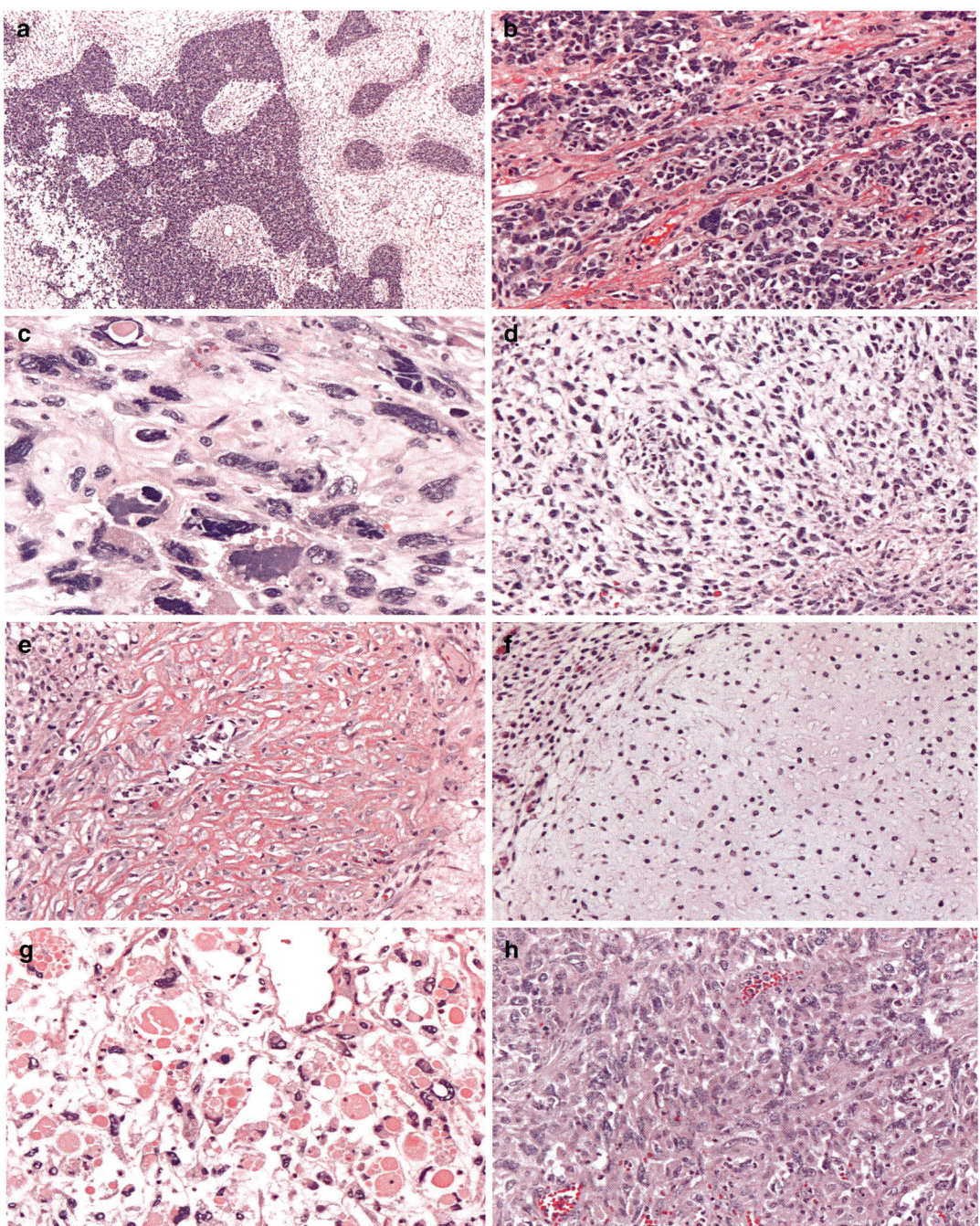

Fig. 8.13 In this panel are shown the microphotographs of a poorly differentiated Sertoli-Leydig cell tumor showing anaplasia and heterologous differentiation (**a–h**; H&E, x100, x200, x400, x200, x200, x200, x200, x200 original magnification, respectively)

by cellular (fibrous) stroma with intermixed Leydig-like cells. PGN: 5-YSR: 90%.

3. *PD-SLCT* (Meyer type III) (~10%): Diffuse growth of undifferentiated cells and/or spindle cells arranged in a sarcomatoid fashion with very few or no Leydig-like cells and mucus-filled epithelial cells. PGN: 5-YSR: 40%.

8.6 Tumors of the Ovary

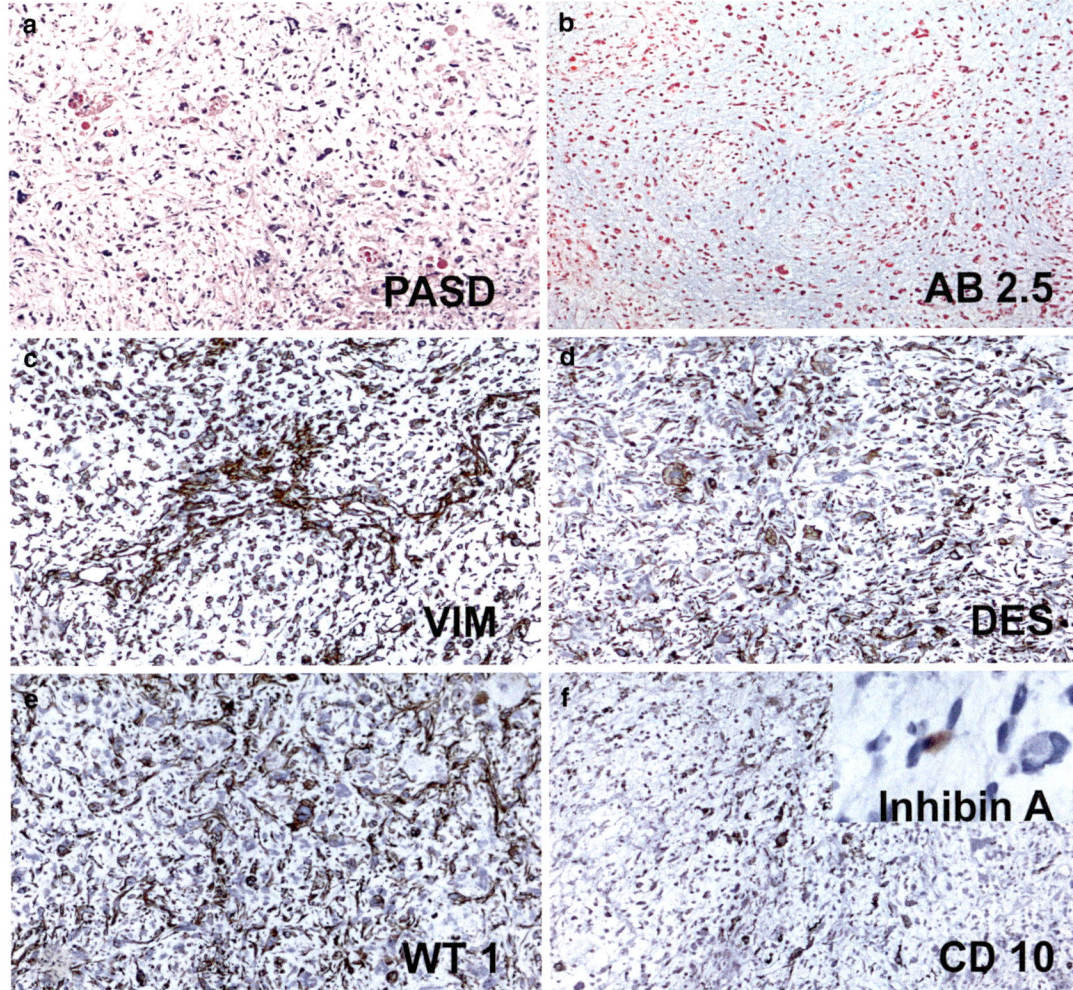

Fig. 8.14 This panel shows some special stains and the positivity of the immunohistochemistry for several markers. We used PASD, AB2.5, VIM, DES, WT1, and CD10 (**a–f**; x100, x100, x100, x100, x100, x100 original magnification, respectively). In (**f**) is also shown the focal positivity for inhibin A (inset, x400 original magnification). Avidin-Biotin Complex was used for immunohistochemical procedures.

4. *SLCT with heterologous elements* (~20%): Mucinous epithelium, hepatic tissue, skeletal muscle, and cartilage (aka "teratoid androblastoma" or "teratoid arrhenoblastoma").
5. *Retiform SLCT* (~10%): Sertoli-Leydig histological elements growing as elongated, branching tubules with low cuboidal epithelium with or without blunted papillae resembling rete testis ± homologous or heterologous differentiation (DDX includes YST, metastatic ACA, carcinosarcoma, carcinoid tumor, serous tumors of borderline or malignant type, endometrioid carcinoma with sex cord-like elements, Wilms' tumor or nephroblastoma).
6. *Pure Sertoli cell tumor* (very rare): ± Peutz–Jeghers syndrome (PJS) (a particularly lipid-rich variant of pure Sertoli cell tumor), premenopausal (precocious pseudopuberty with vaginal bleeding), and postmenopausal (irregular bleeding, solid and hollow tubules lined by Sertoli-like cells with intracytoplasmic lipid (ORO+) identifiable by histochemistry without atypia, excellent prognosis). PJS is an AD medical condition characterized mainly by mucocutaneous pigmentation and intestinal polyposis.

- *IHC*: (+) *V*IM, *I*NA, *N*ephroblastoma-Ag/WT1, *C*K [AE1–AE3, CAM5.2] and *CA*LR, *A*R/T, and *P*R and *P*NET-Ag/CD99 (mnemonic word: VINCAP) as well as (−) EMA, PLAP, CEA, CA19-9, CA125, and, obviously, (−) CD45 and (−) CGA.
- *TEM*: "Sertoli cell-like" ultrastructural features with elongated nuclei, deep indentation *annulate lamellae*, apical microvilli, and frequent desmosomes.
- *PGN*: There are four unfavorable prognostic factors for SLCT, including (1) rupture, (2) extraovarian spread, (3) poor differentiation, and (4) heterologous elements. All four elements need to be reported in the pathological report and discussed at the multidisciplinary tumor board meeting.

8.6.3.3 Steroid/Lipid Cell-Rich Tumor Group

It is a group of SCST of the ovary with steroid/lipid-rich features and includes a few similar entities with different names but that show quite identical clinical and microscopic characteristics, e.g., virilization (amenorrhea or postmenopausal bleeding) and MART-1 (Melanoma-associated Antigen Recognized by T cells - 1) expression (only in steroid cell tumor and restricted to Leydig cell component of Sertoli-Leydig cell tumors).

1. Stromal luteoma.
2. Leydig/hilus cell tumor: Leydig cell tumor, if Reinke crystals ("rod-like" cytoplasmic inclusions) are visible, while in hilus cell tumor, the location of the tumor is at the hilus of the ovary.
3. Steroid cell tumor.

Grossly, unilateral, yellow or yellow-brownish solid mass with vague nodularity is separated by fibrous trabeculae. Microscopically, S/LCRT (Steroid/Lipid Cell-Rich Tumor) shows clusters of large rounded or polyhedral cells with exquisitely identifiable features of steroid hormone secretion (abundant eosinophilic or vacuolated cytoplasm) ± Reinke crystals.

- *IHC*: (+) VIM, CAM 5.2, AE1–3, (±) EMA, and S100.
- *TEM*: SER (smooth endoplasmic reticulum) prominence and tubulovesicular *cristae* in mitochondria.
- *DDX*: neoplasms with the secondary proliferation of steroid hormone-producing cells, including stromal luteoma, luteinized GrCT, fibroma/thecoma, stromal Leydig cell tumor, and focal spread at the periphery of other ovarian tumors.
- *PGN*: Favorable, but unfavorable PGN factors include size, high MI, and nuclear atypia.

8.6.3.4 Fibroma/Thecoma Group

Spindle-cell-based SCSTs with similar gross, histologic, and molecular/chromosomal (12+ by fluorescence in situ hybridization - FISH) features (Fig. 8.15).

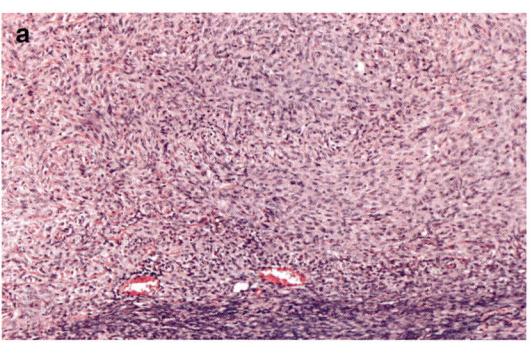 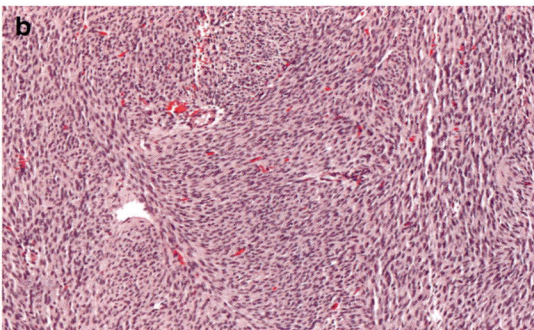

Fig. 8.15 The fibroma (**a**) – fibrosarcoma (**b**) or spectrum – group is shown in this two-microphotograph panel (a-b; H&E stains, x100 original magnification for both microphotographs) (see text for details about definition of both entities)

- Fibroma: age < 40 years, ± hydrothorax (right), Meigs syndrome (+ ascites and pleural effusions), ± Gorlin syndrome or Nevoid Basal Cell Carcinoma Syndrome (NBCCS). NBCCS is a hereditary condition characterized by multiple basal cell skin neoplasms.
- Thecoma: age > 40 years, ± dysfunctional uterine bleeding, pelvic/abdominal pain, or distension and sclerosing peritonitis. Estrogenic-related individualized risk of endometrial hyperplasia/carcinoma
- Fibrothecoma: Generalized term for the combination of both fibroma and thecoma in the same neoplasm
- Fibroma: Unilateral, solid, lobulated, firm, and white, grossly
- Thecoma: Unilateral, solid, lobulated, firm, and yellow, grossly
- Fibroma: Closely packed spindle cells in "feather-stitched" or storiform histological growth pattern, ORO (−), ± hyaline bands/edema, and no atypia
- Thecoma: Spindle cells with pale cytoplasm with reticulin fibers surrounding individual cells, ORO (+), ± hyaline bands/calcification, and no atypia
- Fibroma: (+) WT1, (+) ER-β, (+) PR
- DDX includes:
  1. Cellular fibroma (MI ≤ 3/10 HPF)
  2. Fibrosarcoma (large Ø, hypercellularity, MI > 3/10 HPF, severe nuclear atypia)
  3. Fibromatosis
  4. Sclerosing stromal tumor

8.6.3.5 Sclerosing Stromal Tumor

Fibroma/thecoma group-related SCST of the younger age group (15–30 years) with similar gross, histologic, and molecular (+12 by FISH) features.

Grossly, SST is a unilateral, small, gray-white, and hormonally inactive neoplasm with the variegated cut surface. Microscopically, SST shows a hypercellular pseudo-lobular architectural pattern with dual cell population (collagen-producing spindle cells = fibroblasts + round/ovoidal cells with clear, lipid-rich cytoplasm = luteinized stromal cells) with accompanying prominent, "staghorn"-shaped, ectatic BVs AND intervening, hypocellular edematous zones.

- TEM: Variable cytological configuration of fibroblasts, luteinized theca-like cells, and primitive mesenchymal cells.
- Variant: Signet-ring cell stromal tumor.

8.6.3.6 Sex Cord Tumor with Annular Tubules

- DEF: Sex cord tumor with peculiar histologic annular tubules. It is generally associated with PJS in 1/3 of events (benign fashion: "small and bilateral SCTAT") and hyperestrinism in 1/2 of events (isosexual precocious puberty or menstrual irregularities).
- GRO: Solid mass with distinctive yellow-orange diacritic nuances.
- CLM: There are simple and complex rings ("annular tubules") with basal nuclei-containing cells (antipodal nuclear diacritical arrangement) around a hyalinized luminal "body" of BM material.
- HSS/IHC: (+) PAS, CD10, and INA.
- TEM: Granulosa and Sertoli cell ultrastructural distinctiveness including nuclear indentation, an interdigitating plasmatic membrane with joined desmosomes as well as intracytoplasmic Charcot-Bottcher filaments, which are spindle crystalloids of 10-25 μm in length.
- DDX: Pure Sertoli cell tumor, GrCTs, and gonadoblastoma.
- PGN: Malignant behavior in 1/4 of SCTAT not associated with PJP. Genetic counseling is appropriate and needs to be suggested in the comments section of the pathological report.

8.6.3.7 Gynandroblastoma

SCST showing a mixture of similar amounts of granulosa-thecal elements (GYN component) and Sertoli-Leydig elements (ANDRO component).

The distinction between fibroma, fibrothecoma, and thecoma remains in many centers quite vague and indeed subjective. It has been suggested that thecoma should be designated "fibromatous tumors with large vacuolated or luteinized cytoplasm content and evident band-like hyaline plaques between tumor cell clusters" probably imparting a vague nodularity to the intriguing neoplasm. Incomplete development of thecoma features may suggest the use of fibrothecoma, in which this designation should qualify tumors intermediate between fibroma and thecoma. Gorlin syndrome is an AD-inherited disease characterized by multiple skin carcinomas of basal cell type, odontogenic keratocysts, palmoplantar pits, congenital bony abnormalities, and ectopic calcifications due to mutational inactivation of *PTCH*. The *Drosophila*-homologue specifies polarity. The human *PTCH1* gene, which has been mapped to 9q22.3-31, consists of 23 coding exons (~74 kylobases) encoding a 1447-amino-acid transmembrane glycoprotein. PTCH1 is involved in the Hedgehog (Hh) signaling pathway, which plays a major role in body patterning and organ development during embryogenesis.

8.6.4 Other Primary Ovarian and Secondary Tumors

8.6.4.1 Gonadoblastoma

Gonadoblastoma (aka dysgenetic gonadoma) is an expansion of an SCST with a *germ* cell component and occurs typically in 80% of cases with female phenotype of the patients and 20% with phenotypic males with undescended testes and female internal secondary organs, and in 1/2 of cases, there is a coexistent dysgerminoma (thorough sampling!) and bilaterality in 1/3 of cases. Patient karyotype: gonadal dysgenesis with a Y chromosome, i.e., XV gonadal dysgenesis, XO-XY mosaicism, but not XX gonadal dysgenesis. Gonadal dysgenesis with Y chromosome harbors a neoplastic risk of 25%. Clinical associations: Ataxia-telangiectasia, which is a rare inherited disorder involving early (childhood) difficulty with coordinating movements (ataxia) associated with chorea, myoclonus, neuropathies, and telangiectases of eyes and skin.

- *GRO*: Small solid tumor.
- *CLM*: Tumor cell nests containing Call-Exner-like hyaline "blobs" surrounded by fibrous bands and ± hyalinization and calcification.
- *IHC*: (+) SALL4 and SF1 and anti-Müllerian hormone focally. SALL4, i.e., Sal-like protein 4 is a transcription factor encoded by a member of the Spalt-like (SALL) gene family. SALL4 participates in the Wnt/β-Catenin signaling pathway and regulation of pluripotent stem cells. *SALL4* gene is altered in Duane-Radial Ray Syndrome and Ivic Syndrome.
- *DDX*: It includes SCTAT and MMGCT.

Other entities not described here include the *female adnexal tumor of Wolffian origin (FATWO)*, the *microcystic stromal tumor*, the *ovarian endometrioid stromal sarcoma (OESS)*, the *small-cell carcinoma of hypercalcemic type*, the *small-cell carcinoma of pulmonary type*, the *hepatoid adenocarcinoma*, and *fibromatosis*.

8.7 Tumors of the Tuba

8.7.1 Benign Neoplasms

There are mostly two entities that play a role as benign tumors in the salpinx. These are an adenomatoid tumor and papillary cystadenoma. Leiomyoma and teratomas are rarities.

8.7.2 Malignant Neoplasms

The malignancies that may occur in the salpinx are carcinoma of the fallopian tube, MMMT (Malignant Mixed Müllerian Tumor) or carcinosarcoma, and choriocarcinoma as primary tumors. Moreover, the serous tubal intraepithelial neoplasia (STIC) and secondary tumors need to be taken into account.

8.8 Tumors of the Uterus

Malignant epithelial neoplasm of the endometrial lining with only 1/5 of cases occurring in premenopausal women. Endometrial carcinomas have been identified in the pediatric age group. In particular, children with Turner syndrome receiving unopposed estrogens as well as rarely some genetic syndromes such as Turcot syndrome and Leprechaunism (Donohue syndrome) (Figs. 8.16, 8.17, 8.18, and 8.19) may be affected with this kind of neoplasms.

Risk factors: Mnemonic "IDAHO" acronym word for infertility, diabetes, age > 35 years, hypertension, and obesity.

8.8.1 Type I/Type II EC

1. Type I or *endometrioid* endometrial carcinoma (EC)
2. Type II or *clear cell/serous-papillary* EC
3. Non-type I/non-type II EC

Apart from *type I or endometrioid endometrial carcinoma (EC) and type II or clear cell/serous-papillary EC, there is a non-type I/non-type II EC.* The type I shows closely packed irregular glands lined by atypical columnar epithelium with scant stroma and occurs in women younger than 40 years in less than 5% of cases. Type II is characterized by a clear cell and serous papillary growth pattern, but is inexistent in youth.

8.8.2 Non-type I/Non-type II EC

Non-type I/non-type II endometrial carcinomas are a heterogeneous group including mucinous, squamous cell, transitional cell, mixed, small cell, and undifferentiated. Mucinous endometrial carcinoma refers to carcinoma with >50% of neoplastic cells showing intracellular mucin. Squamous cell carcinoma or epidermoid carcinoma shows squamous differentiated neoplastic cells, and its diagnosis is made only in the absence of squamous cell carcinoma of the cervix. Transitional cell carcinoma requires >90% of neoplastic cells showing urothelial differentiation. Mixed endometrial carcinomas have ≥2 cell types with ≥10% each type. Small-cell carcinoma of the endometrium shows both pan-keratin (AE1–AE3) and neuroendocrine markers (CGA, SYN) expression. Undifferentiated endometrial carcinoma displays no differentiation.

8.8.3 Uterine Leiomyoma (ULM) and Variants

Uterine leiomyoma is a quite frequent finding in the uterus from puberty through middle age. Variants of leiomyoma include cellular, mitotically active, symplastic, and myxoid. The following sections are going to focus on uterine lymphangiomyomatosis, uterine stroma tumors, as well as hematological and secondary malignancies.

8.8.4 Uterine Leiomyosarcoma (ULMS)

It is the malignant counterpart of the uterine leiomyoma. Thus, it is important to assess correctly and carefully mitotic figures as well as cellular atypia. Mitosis criteria include (1) "hairy" extension of chromatin distribution pattern, (2) no nuclear membrane, and (3) exclusion of mitosis mimickers (e.g., lymphocytes, mast cells, stripped nuclei, degenerated cells, and peculiar precipitations of hematoxylin). All three criteria MUST be checked out before the term mitosis is applied! Atypia criteria include an assessment which is based on nuclear pleomorphism, nuclear size, nuclear plasma membrane irregularities, chromatin density, and size and prominence of nucleoli. G1-G2-G3 refers to degrees of differentiation of a tumor (well, moderately, and poorly differentiated, respectively) but may refer – under certain circumstances – to the degree of atypia of the cell type. G1 refers to the uniformity of nuclei ± enlargement, smooth contours of the nuclear membrane, evenly distribution of chromatin pattern, and minimal variation in size and shape. G2 refers to mild dis-uniformity of nuclei (pleomorphism) with enlargement, plumpness, and irregularities, coarse distribution of chromatin, and occasional enlarged abnormal mitoses. G3 refers to marked dis-uniformity of nuclei (nuclear pleomorphism) with enlarged bizarre nuclei, dense chromatin, and atypical prominent nucleoli.

Tumor cell necrosis of coagulative type, i.e., tumor cell necrosis intrinsic to the tumor

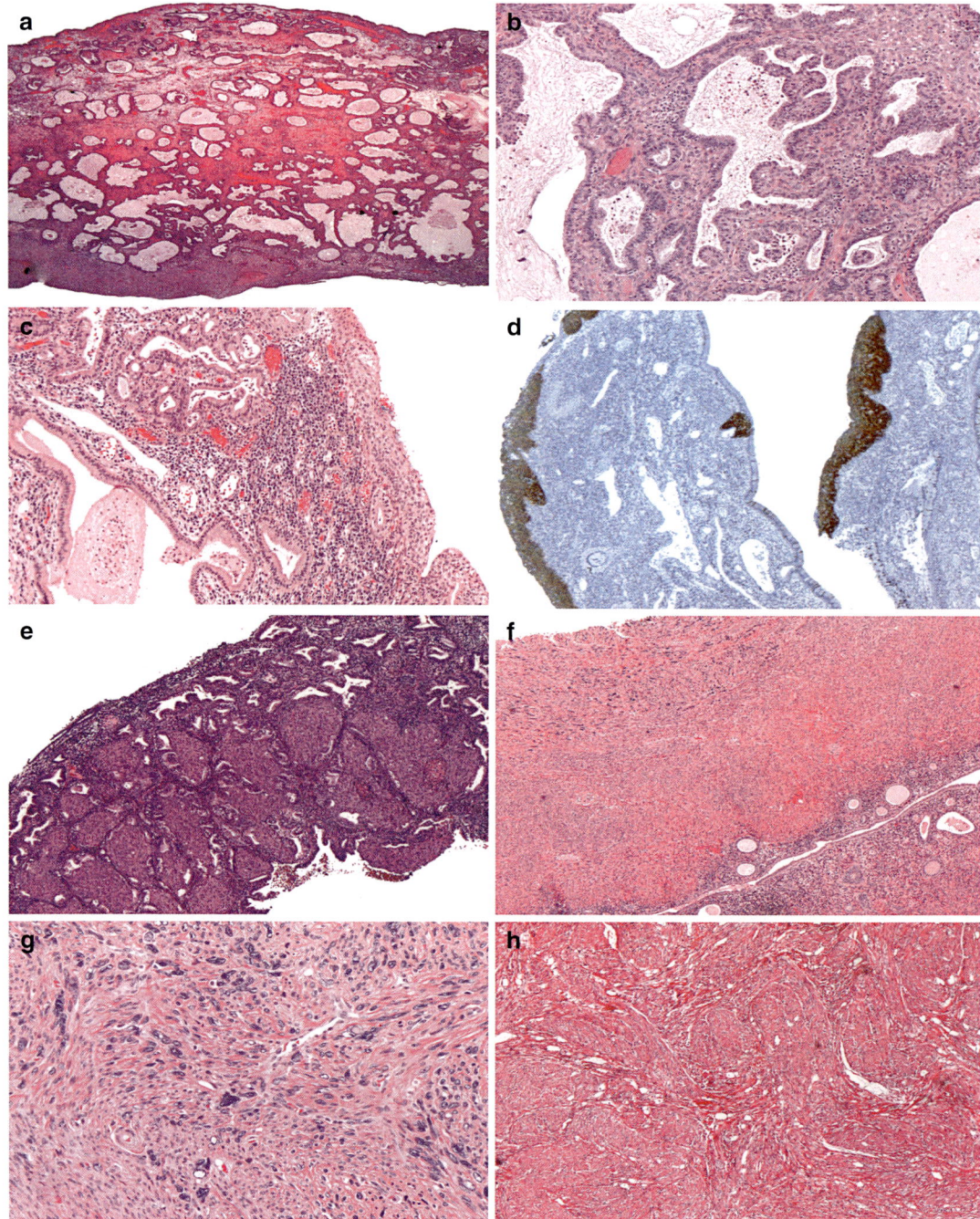

Fig. 8.16 This panel shows benign noninvasive neoplasias of the uterus. In (**a**) and (**b**) the microphotographs show an endometrial polyp with complex atypical hyperplasia progesterone-treated (H&E stains, x12.5 and x100 original magnification, respectively). In (**c**) and (**d**) are shown a high grade superficial intraepithelial lesion with microglandular hyperplasia (H&E stain, x100 and P16 immunostain, ABC method, x100, respectively). In (**e**) is shown an atypical endometrial hyperplasia with morular metaplasia (H&E stain, x50 original magnification). In (**f**–**g**) is shown a symplastic leiommyoma (H&E stain, x50 and x200 original magnification, respectively). In (**h**) is shown an adenomatoid tumor of the uterus (H&E stain, x50 original magnification).

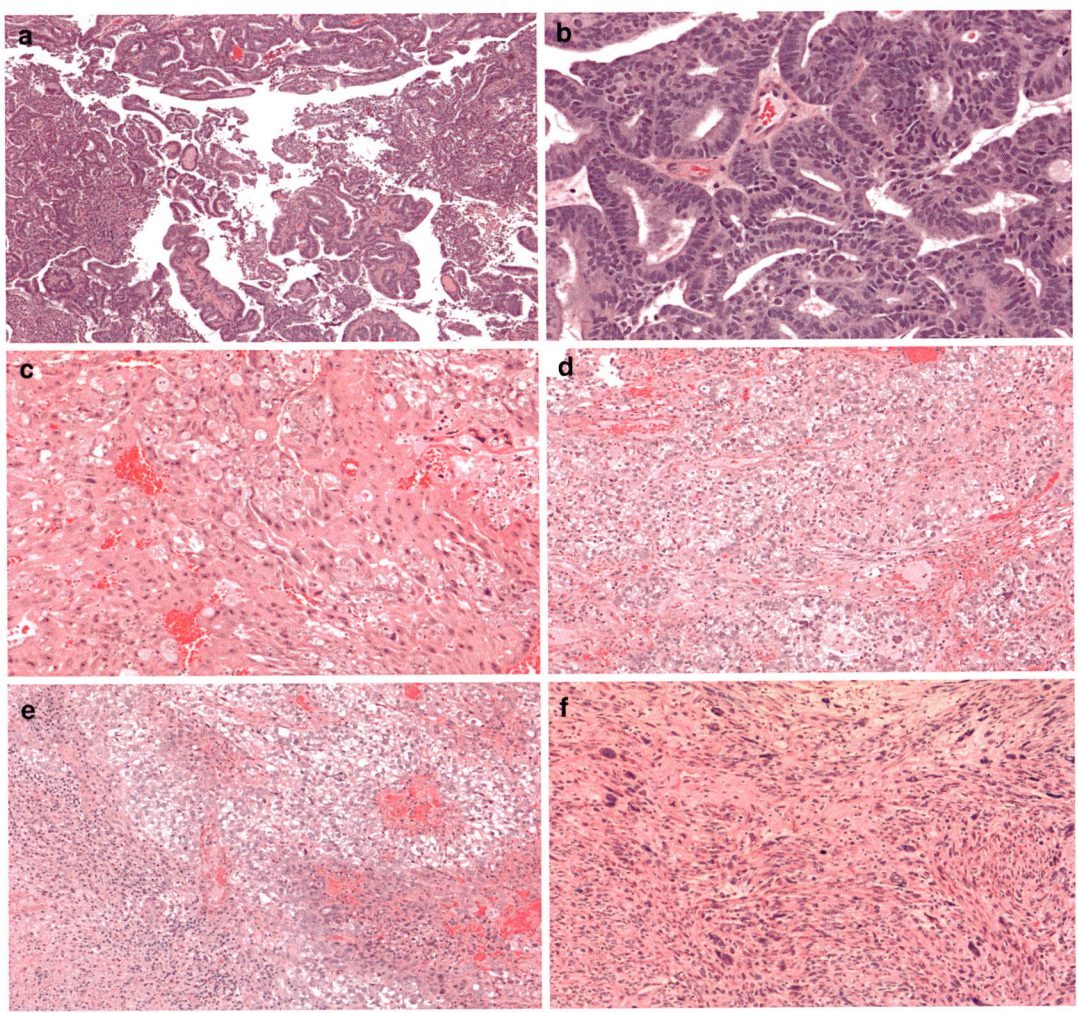

Fig. 8.17 This panel of rare uterine neoplasms include an endometrioid endometrial carcinoma (**a**, **b**), a choriocarcinoma (**c**), a poorly differentiated carcinoma with clear cell areas (**d**, **e**), and a leiomyosarcoma (**f**). All of these tumors occurred in youth. The serous papillary endometrial carcinoma is, conversely, practically inexistent in childhood or youth and reported here for comparison from an older patient (a-f; H&E stains, x100, x200, x100, x100, x100, x200 original magnification, respectively).

and not of ischemic type, needs to be differentiated from hyalinizing patterns of necrosis that may occur for several necrobiotic processes. Once criteria for mitotic figures, atypia, and necrosis type are set and retained with confidence by the gynecologic, pediatric, and general surgical pathologist, the "*If…, then…*" rules to differentiate leiomyoma from atypical leiomyoma and leiomyosarcoma may be applied (Box 8.10).

The concept of STUMP is a controversy. STUMP is an acronym defined as smooth muscle tumor of uncertain malignant potential. Its diagnosis can be applied if all criteria are met in a very rigorous way. The requirements are minimally atypical SM neoplasms with low MI but with uncertain histology AND combination of standard SM differentiation, G3-atypia, low MI, but uncertain necrosis, AND G2-G3 atypia, but with unclear MI.

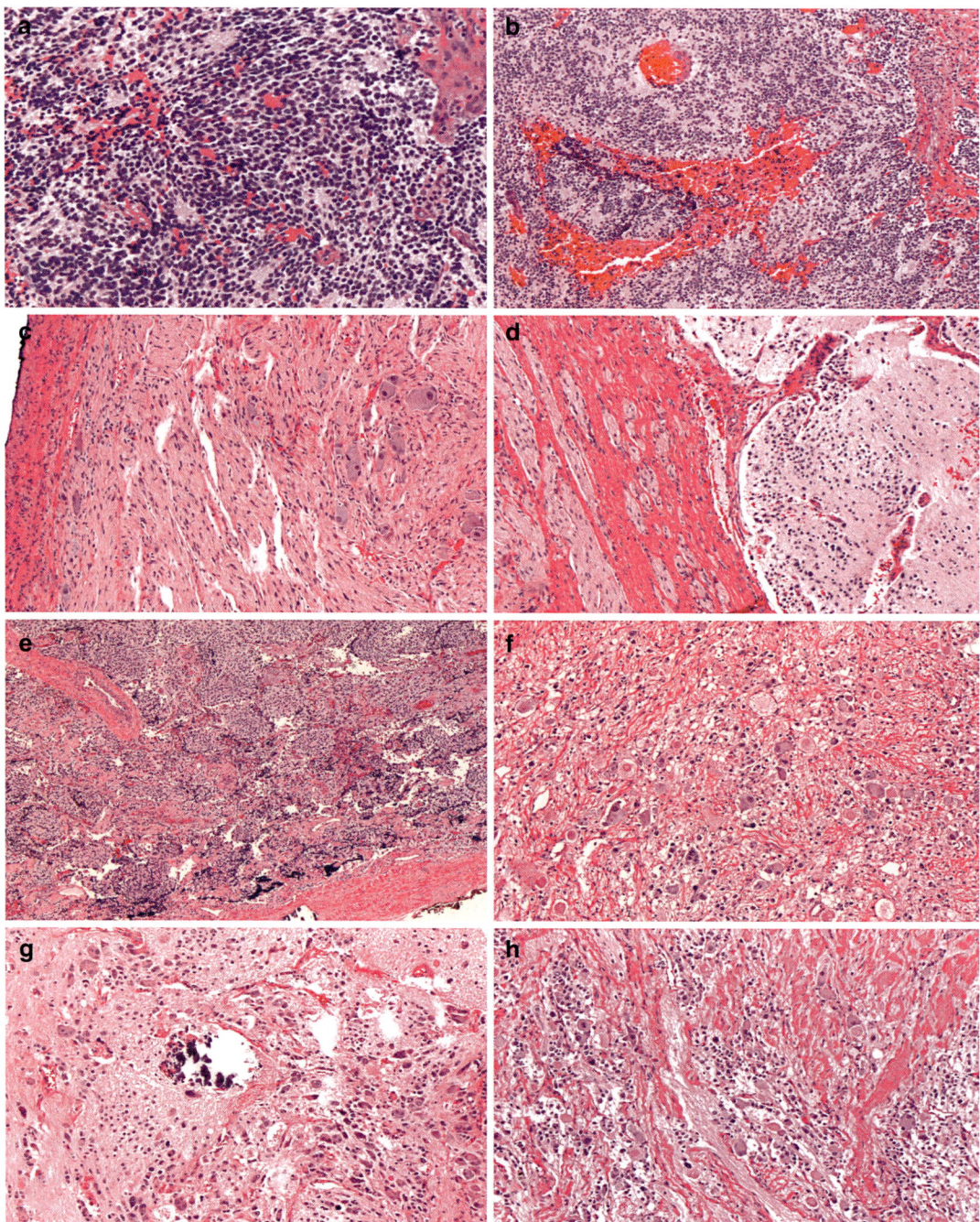

Fig. 8.18 A very rare occurrence of neuroblastoma of the uterine wall with characteristic features infiltrating the wall of the uterus in all the microphotographs presented. In this child, there was no evidence of neuroblastoma in either adrenal gland or paraganglia. It may have arisen from ectopic neuroblasts or choristiae located in the uterus. Extrarenal Wilms' tumor is another rare entity in the uterus with blastemal cells that may share similar pathogenesis. The various locations of extrarenal Wilms' tumor include the retroperitoneum, skin, and thorax, other than uterus (**a–h**; H&E stains, x200, x100, x100, x100, x 50, x 100, x100, x100 original magnification, respectively)

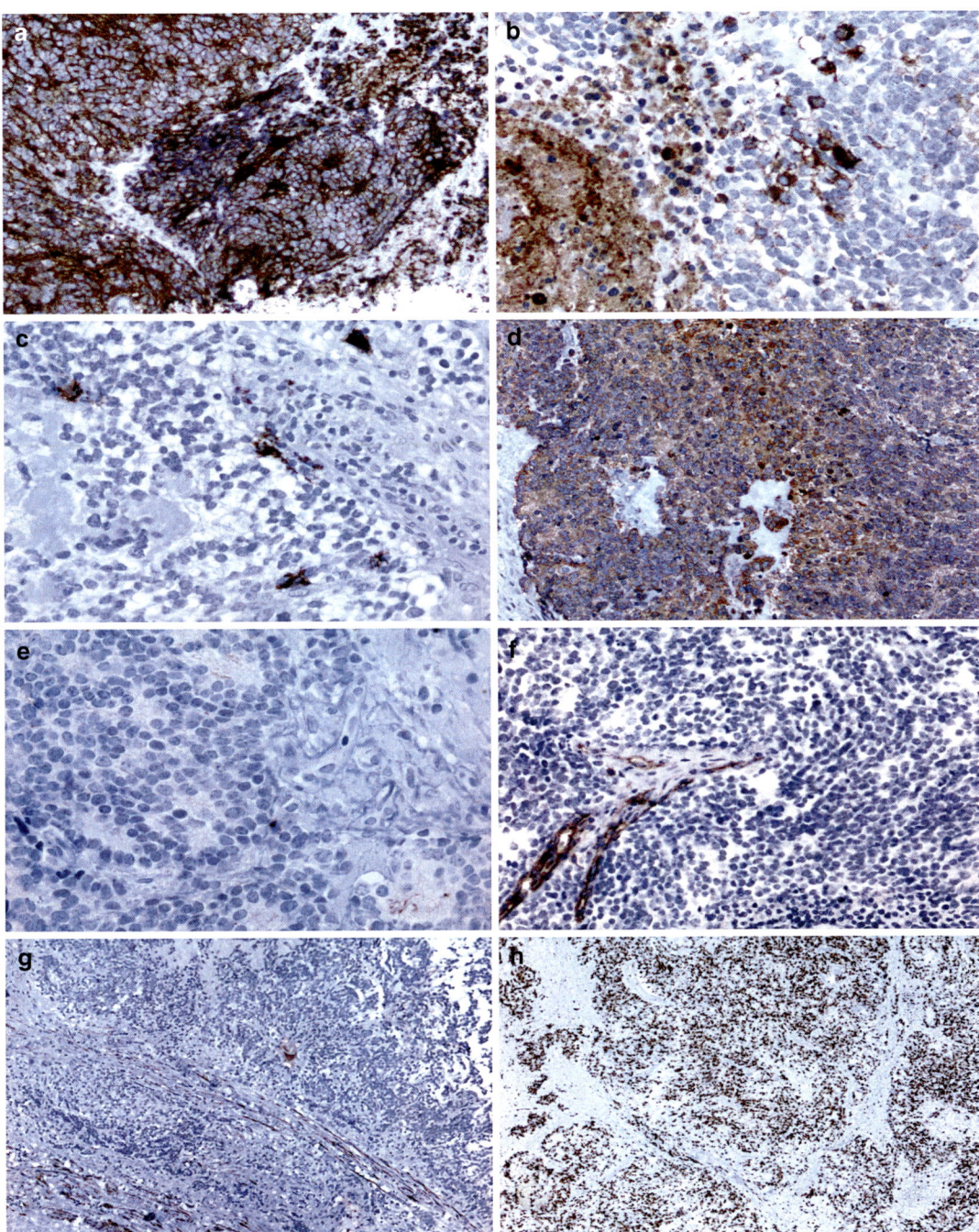

Fig. 8.19 This panel illustrates the immunohistochemistry of the neuroblastoma shown in the previous panel. The immunostaining Avidin-Biotin Complex) of this panel is for CD56, chromogranin A, S100, NSE, neurofilaments, CD99, desmin, and Ki67 (MIB1) in (**a–h**), respectively (a-h; x200, x400, x400, x200, x400, x200, x100, x50 original magnification, respectively)

> **Box 8.10 LM/LMS "If..., then..." Criteria**
> IF "no/mild (G1) atypia" – "tumor cell necrosis," then LM.
> IF "G2-G3 atypia" – "tumor cell necrosis," then assess MI.
> – If MI < 10/10 HPF, then *Atypical* LM.
> – If MI ≥ 10/10 HPF, then LMS.
> IF "G2-G3 atypia" + "tumor cell necrosis," then LMS (whatever MI).
> Notes: LM, leiomyoma; LMS, leiomyosarcoma.

8.8.5 Uterine Adenomatoid Tumor

The adenomatoid tumor is a rare benign tumor of mesothelial origin, which chiefly occurs in the male and female genital tracts. Grossly, there is a single, solid, gray-white mass. Microscopically, there are small cystic spaces of various diameter, lined by flattened to cuboidal and, occasionally, hobnailed cells exquisitely embedded in smooth muscle fascicles.

IHC: (+) Keratins (AE1-3, CAM5.2, K7,8,18,19), CALR, D2-40, WT1, and VIM. Focal (+) ER/PR and Glut-1 may be important pitfalls.

8.8.6 Uterine Lymphangiomyomatosis (ULAM)

Uterine lymphangiomyomatosis (ULAM) refers to a PEComa of reproductive age women. The extra-uterine involved sites include the lungs, axial thoraco-abdominal lymphatic system, and the thoracic duct.

Characteristically, TSC1/TSC2(+) in both tuberous sclerosis (TS)-associated and sporadic ULAM. Moreover, the myomelanocytic differentiation (SMA, SMM, HMB45) ± ER/PR are always present in both types (TS-associated and spradic).

Short spindle cells with slight epithelioid features arranged in short fascicles and large spaces with endothelial lining.

DDX: Leiomyoma/leiomyomatosis peritonei or LMS (proliferation of smooth muscle cells) with variable cellularity, elongated cytological features, abundant eosinophilic cytoplasm, ± necrosis, ± ↑ MI, HMB45 (−) but SMA (+), spindle-cell hemangioma (distal extremities), Kaposi sarcoma (spindle cells with slit-like spaces, hyaline globules, extravasated RBC, HHV8(+), HMB45(−)), and vascular transformation of LN (capillary-sized BV involving the sinuses with preservation of LN architecture, FVIIIRA(+), CD31(+), CD34(+), HMB45(−)).

Features of malignancy include Ø > 5 cm, infiltrative borders, marked pleomorphism, tumor-related necrosis, and high MI (*man*: mitoses, atypia, necrosis).

PGN investigations point to recurrence after lung TX.

8.8.7 Uterine Stroma Tumors

Uterine stroma tumors include stromal nodule, endometrial stromal sarcoma (ESS, formerly endometrial stromal sarcoma of low grade), and undifferentiated stromal sarcoma (USS), formerly endometrial stromal sarcoma of high-grade type (Box 8.11).

8.8.8 Hematological Malignancies and Secondary Tumors

Lynch syndrome (hereditary nonpolyposis colon cancer (HNPCC)) is an essential diagnosis because endometrial carcinomas develop before colon malignancy in more than half of females and HNPCC should always be considered if the patient is younger than 50 years and/or with a family history of colon carcinoma. In these cases, DNA mismatch repair can be performed by IHC using the commercially available kit to MLH1, MLH2, MLH6, and PMS2. The correlation is between loss MLH2 and MLH6. PCR can assess high levels of MSI for defective DNA mismatch repair. Lynch syndrome individuals demonstrate endometrial carcinomas with a wide variety of histologic types, being non-endometrioid endo-

Box 8.11 Stroma Nodule vs. ESS vs. USS

	Stromal nodule	ESS	USS
On LPM			
Invasion	(−)[a]	(+) "tongues"	(+++)
LVI	(−)	(+) "worm-like"	(+)
Extrauterine extension	(−)	(+/−)	(+)
On HPM			
Cellularity	Uniformity	Uniformity	Pleomorphism
MI	<3/10HPF	<10/10HPF	>10/10HPF
CD10	(+)	(+)	(−)
SMA	(+) (focally)	(+) (focally)	(−)
DES	(−)	(−)	(−)
h-CALD	(−)	(−)	(−)
ER/PR	(+)	(+)	(−)
WT1	(+)	(+)	(+/−)
CD34	(−)	(−)	(−)
CD99	(−)	(−)	(−)
EMA/CK/INA	(−)	(+)[b]	(−)

Notes:
[a]In stromal nodules up to three blunt extensions into the myometrium, each up to 3 mm, are permitted, without the need to call them "tongues" as seen for ESS
[b]Sex cord-like areas may show positivity for EMA, CK, and inhibin A. Desmin and h-caldesmon are two better markers than smooth muscle actin to differentiate endometrial stromal tumors from myometrial leiomyoma, atypical leiomyoma, and leiomyosarcoma. ER/PR markers help to distinguish ESS from USS. CD34 and CD99 (together with Bcl2) are usually positive in synovial sarcoma. LPM, low-power magnification by light microscopy; HPM, high-power magnification by light microscopy

metrial carcinomas and higher grades of EC more often described despite the young age of the patients. Report specimen type and what kind of surgical procedure had always been performed, because some supracervical hysterectomy specimens with unsuspected endometrial carcinoma obtained by laparoscopy (morcellation techniques) may harbor venous tumor emboli. These emboli are thought to be iatrogenic, and there is a risk that tumor cells had spread to the pelvis and peritoneal activity and need to be discussed at the multidisciplinary team meetings. In mixed type of endometrial carcinoma, tumors with at least 10% serous component behave like pure serous carcinoma, although other studies point to 1/4 of the tumor as a quote to consider these tumors in a poor prognosis category. It is emphasized that serous carcinomas, although rare in youth, may display a deceiving glandular architecture. If you see gland-forming endometrial carcinoma harboring high-grade nuclear features, then the diagnosis of serous carcinoma should be seriously considered. Carcinosarcoma (aka MMMT) is extremely rare in youth but should be viewed as a high-risk metaplastic carcinoma and staged like endometrial carcinoma schemes according to FIGO and WHO classifications.

8.9 Tumors of the Cervix

8.9.1 Benign Neoplasms

These entities include the endocervical polyp, *condyloma acuminatum*, flat condyloma, and papilloma.

8.9.2 Precancerous Conditions and Malignant Neoplasms

Precancerous conditions represent the primary target of the preventive medicine of the last four decades with outstanding results regarding early diagnosis of cervical cancer.

8.9.2.1 Cervical Dysplasia

Cervical dysplasia denotes an enlargement of cells, an increase of the N/C ratio, irregularity of the nuclear borders, chromatin clumping, and loss of basal cell polarity. It often occurs in the so-called transformation zone and replaces adjacent squamous and glandular epithelium. Graded initially as cervical intraepithelial neoplasia (CIN) I, II, and III according to the degree of dysplasia and thickness of involvement (<1/3, 1/3–2/3, and > 2/3), it is now classified as "low-grade squamous intraepithelial lesion" (LSIL) or "high-grade squamous intraepithelial lesion." (HSIL) The designation of "carcinoma in situ" (CIS) is used when full-thickness changes are associated to no differentiation at any level, although HSIL and CIS may have overlapping features and may be treated similarly. HPV and particularly HPV 16, 18, 31, 33, and 35 are associated with high risk for progression to carcinoma. Cofactors include cigarette smoking, coexisting microbial cervical infections, dietary deficiencies, and hormonal status variations. A controversially discussed cofactor is a p53 Arg72Pro variant, which is particularly interesting because it introduces a specific polymorphism of the tumor suppressor gene *TP53*. *TP53* is a tumor suppressor gene, which is one of the most mutated genes in human cancer. However, other than gene mutations, some polymorphisms of the wild-type *TP53* gene may affect some neoplasms toward cancer progression or have a protective effect. It is important to recall that a mutation is defined as any alteration in the DNA sequence, while polymorphism or better "single nucleotide polymorphism" (SNP) defines a single base pair alteration that is common (i.e., >1%) in the population. The 1% cutoff is arbitrary, although it seems to function in many biological systems. *TP53* gene has 14 polymorphisms, of which p53 Arg72Pro variant (*TP53* rs1042522), i.e., Arg → Pro involving guanine (CGC) to cytosine (CCC) nucleotide exchange on exon four at codon 72, is mostly associated with some cancers, although this data is controversially debated. According to a recent meta-analysis *TP53* rs1042522, it is unlikely to be associated with breast, colorectal, or endometrial cancer. A weak association with lung cancer has been suggested. The use of genetic investigation obtained from tumor tissue for genotyping can still represent a bias in estimating some genetic effect on cancer. Future investigation should avoid the use of tumor tissue as the primary source of genetic material, precisely when the genetic loci of interest are known to demonstrate LOH. About cervical cancer, it seems that p53 Arg72Pro variant is associated with progression of HSIL to cervical carcinoma *only* in the presence of HPV positivity where the abnormal p53 is more susceptible to be degraded by E6 protein. However, it seems that there is no association of p53 Arg72Pro variant with overall risk or initiation of carcinoma in either HPV-(+) or HPV-(−) women. In the malignant transformation, the viral oncogenes E6 and E7 (E stands for early expression) are the key players. Loss of the transcriptional control system leads to high levels of expression of the viral oncogenes throughout the epithelium favoring the malignant progression. E6 has a ubiquitin ligase activity that acts to ubiquitinate p53 leading to its proteasomal degradation. The introduction of viral DNA in the host genome would prevent cell death apoptosis and, finally, cancer progression. E7 acts differently by competing for pRB binding acting on the transcription factor E2F to transactivate its targets. This E7-mediated transactivation of E2F pushes the cell cycle forward leading to p16 protein accumulation. The detection of p16 in HSIL may be interpreted as a surrogate of the HPV test. Koilocytic changes are nuclear changes (enlargement and variation of size and shape, hyperchromasia, coarse chromatin granules) and perinuclear halo.

8.9.2.2 Cervical Epidermoid Carcinoma

Cervical epidermoid carcinoma or squamous cell carcinoma of the cervix (C-SqCC) is the most common malignancy of the female genital tract and epidemiologically the second most common cause of cancer death in North America and most of the Western countries. The distinguishing features between invasion and non-invasion are listed in Box 8.12.

The risk factors to develop a cervical squamous cell carcinoma are listed in Box 8.13.

- *GRO*: C-SqCC may be infiltrative or bulky (exophytic) or both.
- *CLM*: Significant subtypes include large cell nonkeratinizing (~2/3 of cases), keratinizing (~ ¼ of cases), small cell, basaloid, and verrucous histological subtype (microscopic architecture).

8.9 Tumors of the Cervix

Box 8.12 Distinguishing Features Between Invasion and Non-invasion

1. (+) Paradoxic Matur.	(−) Paradoxic Matur.
2. Irregular contour of the invading nest (+)	Irregular contour of the invading nest (−)
3. (+) DSR (+) with metachromatic stain	(−) DSR (+) with metachromatic stain
4. (±) LVI	(−) LVI
5. (±) µ-inv.	(−) µ-inv.

Notes: *Matur.*, maturation; *DSR*, desmoplastic stroma reaction; *LVI*, lympho-vascular invasion; *µ-inv.*, microinvasion (<1 mm).

Box 8.13 Cervical Epidermoid Carcinoma Risk Factors (Mnemonic IDAHO)
1. Intercourse frequency and promiscuity, parity, intraepithelial lesions (e.g., HSIL), and "irrational" living (e.g., smoking)
2. DES-exposure F-Hx. (+) → clear cell type and drugs (OCP, immunosuppressive)
3. Age of first intercourse (*vide supra*)
4. Hypertension, "heritage" (PJS in terms of inheritance → minimal deviation ACA), and HSIL (*vide supra*).
5. Obesity

- *PGN*: Two major significant factors are Ø and stage. LN stations: paracervical, hypogastric, obturator, external iliac, sacral, common iliac, aortic, and inguinal from the first drainage of involvement of LN stations to the latest. M-staging involves M-sites, which are typically lungs and bones. Rare cardiac metastases have been reported.

8.9.2.3 Cervical Adenocarcinoma and Variants
- *DEF*: Cervical adenocarcinoma (C-ACA) is a glandular forming malignant epithelial tumor with either typical mucinous endocervical glandular appearance (endocervical, mucinous type of cervical ACA), or endometrial/intestinal glandular appearance (intestinal, mucinous cervical ACA or endometrioid cervical ACA), or one of the ACA variants. This latter group, which represents about ¼ of all cervical ACA, includes minimal deviation ACA, villoglandular carcinoma, basaloid, clear cell, adenosquamous (mixed), glassy cell, adenoid cystic, and mesonephric carcinoma. The mucinous endocervical ACA accounts for about half of all C-ACA, while intestinal, mucinous cervical ACA or endometrioid cervical ACA makes up the remaining ¼ of all C-ACA.
- *EPG*: RFs include similar factors to cervical epidermoid carcinoma, although HPV 16 and 18, as well as long-term use of OCP, may play a significant role in activating some carcinogenetic pathways.
- N.B. Special stains (AB and mucicarmine), as well as CEA immunostain, are strongly positive intracellularly in almost all cells unlike endometrial adenocarcinoma, which shows focal and superficial staining only. To remember is that in AIS there is minimal intracytoplasmic mucin instead (obviously no DSR!).

The intestinal type of C-ACA may show goblet, argentaffin, and Paneth cells and is (+) for CK7, CK20, CDX2, and p16.

A microinvasion of tumor cells into the underlying stroma can be detected by:

1. Small detached tumor cell nests, which are adjacent to AIS showing DPSR around the nests.
2. Irregular outlines of tightly packed clusters of glandular complexes.
3. Intraglandular cribriform growth pattern.
4. Intraglandular papillary growth pattern.

There are some favorable and some unfavorable variants of cervical adenocarcinoma, and these entities are listed in Box 8.14.

Minimal Deviation Adenocarcinoma
It is not a diagnosis that can be performed with a superficial cervical biopsy, but it requires a deep biopsy or conization specimen. MDACA represents 1% of all C-ACA and is very well-differentiated. MDACA has architectural

> **Box 8.14 Most Favorable and Unfavorable Variants of Cervical Adenocarcinoma**
> Favorable variants of C-ACA:
>
> - Minimal deviation ACA (aka "*adenoma malignum*")
> - Villoglandular carcinoma
> - Basaloid carcinoma
> - Clear cell carcinoma
>
> Unfavorable variants of C-ACA:
>
> - Adenosquamous (mixed) carcinoma
> - Glassy cell carcinoma
> - Adenoid cystic carcinoma
>
> Variable prognosis is found in different carcinomas such as neuroendocrine carcinoma according to the degree of differentiation and probably mesonephric carcinoma.

characteristics of malignancy with a deep location in the cervix but cytologically shows discrepant features of quite a benignancy. It is seen in patients with PJP. Microscopically, there are irregularly shaped glands with DSR constituted by mild atypical cells. The neoplastic glands are more in-depth than other possible mimickers or ectopic endocervical glands. A criterion defines 7 cm from the surface or also outer 1/3 of cervical wall the limit to start to include MDACA in the differential diagnosis.

Villoglandular Papillary Carcinoma

Villoglandular papillary carcinoma (VGPA) is a rare variant of C-ACA mostly seen in young women and characterized by a good long-term prognosis as it infrequently involves LNs. Grossly, VGPA is exophytic mostly. Microscopically, VGPA shows papilla formation with a predominantly exophytic growth pattern for at least 51% of the tumor. An endophytic invasive component with papillary, tubular, microglandular, or mixed patterns may be present but should be 50% or less than 50% of cancer. The cores of the papillae could be thin or thick with or without endocervical-like or desmoplastic stroma reaction. The lining epithelium may be endocervical, endometrial, intestinal, or mixed types. Main DDX is endocervical adenocarcinoma that is typically more infiltrative with severe atypia, a higher MI, and with a higher grade than VGPA.

Basaloid Carcinoma

Basaloid squamous carcinoma of the uterine cervix is a rare tumor subtype of cervical carcinoma, which is characterized by an ulcerated, infiltrating growth pattern, islands or cords of small basaloid cells (hyperchromatic epithelial cells with high nucleus-to-cytoplasm ratio) with prominent peripheral palisading of the tumor cells, and lack of significant stromal reaction. Like its counterpart found in the hypopharynx, the base of the tongue, salivary glands, esophagus, anal canal, prostate, thymus, vulva, and urinary bladder, in the cervix, it is quite rare. It seems that in the cervix, basaloid squamous carcinoma harbors an excellent prognosis. However, a basaloid squamous carcinoma may occur "pure" or in association with squamous dysplasia, in situ SCC, or invasive SCC. IHC shows (±) HMKCK, (+) EMA, and (−) VIM, DES, CGA, and SYN. DDX includes the solid variant of adenoid cystic carcinoma (ACC), small-cell carcinoma, and large cell neuroendocrine carcinoma (LC NEC) of the cervix.

Adenoid Basal Carcinoma

Adenoid basal carcinoma (ABC) shows islands of squamous differentiated epithelial cells with moderate atypia and a prominent peripheral basal cell layer. Some nests of small basaloid cells have cystic change and focal columnar differentiation without DPSR. It is probably virtually related to basaloid carcinoma.

Solid ACC shows BM (+) surrounded by basaloid neoplastic material.

Clear Cell Carcinoma

- *DEF*: Müllerian-derived and glycogen-producing cervical carcinoma with a bimodal distribution, gross appearance of a polypoid lesion or polyp, or polypoid nodule in women, whose 50% have Hx. (+) for DES, presenting with vaginal bleeding or dyspareunia, usually at FIGO I or II.

- *CLM*: Microscopically, three patterns have been described, including tubulocystic, papillary, and solid. Tumor cells are large and have abundant glycogen-filled cytoplasm and hobnailing features.
- *HSS/IHC*: Special stains of interest include (+) PAS but (−) mucin. IHC shows (+) KER (AE1–AE3, CAM 5.2, CK7, CK8, CK18, CK19), EMA, HNF1β, NapsinA, p53 but (−) CEA, CK20, HMW-CK (34βE12), VIM, DES, SMA, CGA, SYN, and CD56. The CEA is particularly useful because it is negative unlike most other types of C-ACA. Napsin A is an aspartic proteinase that is largely expressed in normal lung and kidney tissue. It is often expressed in pulmonary adenocarcinomas.
- *TEM* shows neoplastic cells with continuous lamina densa, abundant mitochondria, copious RER, intracytoplasmic glycogen, blunt microvilli, and vesicular aggregates in the nucleoplasm, perinuclear cytoplasm, or located quite specifically in the nuclear membrane.
- *DDX* includes MGH, ASC, mesonephric hyperplasia/carcinoma, YST, and metastatic RCC.
- *PGN* is good, but late recurrences have been described are common.

Mesonephric Carcinoma

Mesonephric (Wolffian) ducts form epididymis and vas deferens in males and persist in females as rete ovarii, paroophoron, and Gartner's duct. Cervical mesonephric carcinoma (CMC) is a sporadic tumor arising from remnants of mesonephric ducts, variable age range (from 1.5 years up), usually low stage, and in areas adjacent to areas of mesonephric hyperplasia. CMC seems to arise from the "lower zone" of the Wolffian ductal system, contrary to female adnexal tumors of probable Wolffian origin, which occurs from the upper zone. Microscopically, histologic architectural patterns include tubular (ductal), retiform, solid, sex-cord-like and spindled, and endometrioid-like. The presence of eosinophilic secretions has been described as variable. Clear cells are cuboidal or low columnar with atypical nuclei, but no hobnailing and no intracytoplasmic mucin. There is minimal or mild DSR.

- *HSS*: (+) PAS but (−) mucin.
- *IHC*: (+) AE1–AE3, LMWCK, EMA, VIM, CALR, and CD10 but (−) ER, PR, CK20, and mCEA.
- *DDX*: It includes mesonephric hyperplasia, endometrial adenocarcinoma, clear cell carcinoma, serous carcinoma, MMMT, and UTROSCT.

Adenosquamous (Mixed) Carcinoma

Mostly pregnancy-related cervical carcinoma, which may start from subcolumnar reserve cells in the basal layer of endocervix, harbors a worse outcome than pure SqCC.

CLM: Microscopically, there is a biphasic pattern of malignant glandular and squamous components, which are delimitated and identifiable without special stains or immunostains. The gland-forming part is typically endocervical and poorly differentiated with cytoplasmic vacuoles or intraluminal mucin, while the squamous part is quite poorly differentiated.

IHC shows (+) p63 and CK7.

TEM may be particularly useful in quite poorly differentiated tumors showing glandular features including mucous secretory vacuoles, exact luminal formation, scattered glycogen, tonofilaments, and secretory products.

DDX: SqCC with focal mucinous production, adenoid basal carcinoma, endometrial adenocarcinoma with extension into the cervix, but the bulk of tumor in the endometrium and adenocarcinoma with coexisting SIL.

Glassy Cell Carcinoma

Cervical glassy cell carcinoma (C-GCC) is most probably a variant of the adenosquamous carcinoma, being differentiated epidemiologically because the first is most often seen in young females and during pregnancy than the last tumor entity. Microscopically, the tumor shows sheets of large cells with characteristic abundant amphophilic or, sometimes reported, eosinophilic ground glass or finely granular cytoplasm, well-defined cell borders, large nuclei with prominent nucleoli, ↑ MI, and stromal inflammatory infiltration of eosinophils and plasma cells ± rare foci of squamous or glandular differentiation and intracellular mucin. It is regarded as having a poor outcome.

Adenoid Cystic Carcinoma

Cervical adenoid cystic carcinoma (C-ACC) is quite uncommon and most often seen in the older age group of women, characteristically black with multiple pregnancies (multiparity), but can occur in white women younger than 40 years and even nulliparous. C-ACC harbors a poor prognosis due to frequent local recurrences and distant metastases. Grossly, ACC displays an irregular, polypoid, friable mass with an endophytic component on the cervix. Microscopically, islands/clusters of cells in cribriform architectural growth pattern with eosinophilic/hyaline cores, resembling ACC of salivary glands, are classically seen. At least focally there is peripheral cell palisading. There are, however, no myoepithelial cells, unlike the counterpart in the salivary glands. The appearance is quite similar to C-ABC. However, the nuclear atypia is more marked, and the growth pattern is doubtless expansile. DSR ± necrosis are well identifiable, and ↑ MI, angioinvasion (BVs), LVI, and hyalinized stroma are standard features. However, the stromal response may be myxoid, fibroblastic, or hyaline at places. There is a solid variant of C-ACC that may be a diagnostic challenge occasionally.

IHC shows (+) AE1–AE3, MSA (HHF45), and (+) in the extracellular BM for laminin and type IV collagen, and (−) S100 and actin.

TEM may show redundant basal lamina forming pseudocysts, intercellular spaces, and luminal formation with microvilli.

8.9.2.4 Cervical Neuroendocrine Carcinoma (CNEC)

HPV18-related carcinoma of the cervix with neuroendocrine differentiation (+CGA, SYN, NSE, PGP9.5, CD56), tendency to secrete hormones (5-HT-liked carcinoid syndrome, ACTH, insulin, PTH) and to harbor an aggressive behavior with rapid metastases (3-YSR: 10–33%). Neuroendocrine tumors of the cervix are classified as the well-differentiated neuroendocrine tumor (WD-NEC) or carcinoid, atypical carcinoid, and neuroendocrine carcinoma (small cell and large cell type). Survival curves may be similar between neuroendocrine tumors of the cervix, and new strategies to classify these tumors at this location may be needed in the nearest future (Box 8.15).

- *IHC*: (+) CGA, SYN, NSE, PGP9.5, CD56, (±) TTF1 (pulmonary origin), CK20 (DDX: Merkel cell carcinoma), and CD99 (DDX: PNET).
- *TEM*: Dense-core neurosecretory granules.

Box 8.15 CNEC

Criteria	
Tubular/organoid HGP	WD-NEC = carcinoid
Nuclear regularity	
CDP: Fine	
MI: Low	
No necrosis	
Tubular/organoid HGP	MD-NEC = atypical carcinoid
Nuclear regularity	
CDP: Coarse	
MI ↑, but <10/10HPF	
Necrosis ±	
Insular, trabecular, glandular, and solid HGP	PD-NEC, large cell type
Large cells with ↑ N/C	
MI > 10/10 HPF	
Loss of cytoplasmic membrane definition	
Nuclear crowding	
Insular, trabecular, glandular, and solid HGP	PD-NEC, small cell type
Small cells with ↑ N/C	
MI > 10/10 HPF	
Fracture lines with absence of cell polarity and nuclear molding	

Note: Nuclear/cytoplasmic ratio (N/C); HGP, histological growth pattern; CDP, chromatin distribution pattern; HPF, high-power field; NEC, neuroendocrine carcinoma

- *DDX*: Adenocarcinoma, small cell nonkeratinizing SqCC (well-defined nests similar to large cell nonkeratinizing SqCC, p63+, and CGA−), basaloid carcinoma (p63+, CGA−), PNET (CD99 is probably not useful, and FLI and/or t(11;22) detection is necessary), lymphoma (CD45 and B-/ T-cell markers), granulocytic/myeloid sarcoma (+lysozyme, chloroacetate esterase, CD34, CD68, MPO, − CD3, − CD20), melanoma (+S100, HMB45, Mel-A), embryonal rhabdomyosarcoma /RMS (+DES, MYD-1, MYF4), and metastatic carcinoma (lung or other sites). In particular, the expression of myelomonocytic antigens (CD34) and myeloid-associated enzymes (myeloperoxidase) is of enormous help in differentiating GS from malignant lymphoma and acute lymphocytic leukemia (ALL). Both markers can help recognize early myeloid cells. The absence of CD20 and CD3 markers is also necessary to rule out B-cell and T-cell lymphomas, respectively. Some histologic features that are commonly identified in poorly differentiated NEC vs. SqCC include (1) vascular invasion, (2) evident lack of coexisting inflammation, and (3) broad geographic areas of necrosis. Importantly, amphicrine carcinoma are defined tumors with small cell NEC component combined with squamous cell carcinoma or adenocarcinoma and may require caution in the differential diagnosis of this kind of tumors.
- *TRT*: Radical hysterectomy with bilateral lymphadenectomy, radiation therapy, and chemotherapy for poorly differentiated NEC, but PGN remains poor.

Other tumors include *cervical mesenchymal neoplasms* and *hematological malignancies*, as well as *cervical secondary tumors*.

8.10 Tumors of the Vagina

8.10.1 Benign Tumors

The Müllerian and mesonephric structures may be characterized by histochemical stains and are seen in Box 8.16 (Fig. 8.20).

> **Box 8.16 Müllerian vs. Mesonephric Roots**
> *Müllerian Structures*: Mucin (+), PAS (−), DPAS (−).
> *Mesonephric Structures*: Mucin (−), PAS (+), DPAS (−).

8.10.1.1 (Müllerian) Papilloma
Arborizing connective tissue fronds constitute Müllerian papillomas with an overlying superficial epithelium consisting of flat cuboidal or mucinous epithelium.

8.10.1.2 Vaginal Fibroepithelial Polyp
It is a polypoid rubbery lesion covered by classic vaginal mucosa and may protrude, grossly, from the *introitus vaginae* as multiple fingerlike projections, which particularly in infants and children may raise the principal differential diagnosis of embryonal RMS of botryoides subtype (*vide infra*). Microscopically, there is acanthotic squamous epithelium overlying a fibrovascular core consisting of fibrous connective tissue with dilated capillaries and occasional large bizarre cells.

DDX: Botryoid RMS angiomyofibroblastoma, aggressive angiomyxoma, superficial angiomyxoma, *condyloma acuminatum*, neurofibroma (particularly in patients affected with NF1, if histology is spindle and the histological pattern recall neurofibroma and is supported by S100+), and leiomyoma.

8.10.1.3 Vaginal Postoperative Spindle (Cell) Nodule (POSN)
Characteristics of the POSN are Hx. (+) recent surgery, bland cytology, and a prominent delicate network of BVs.

8.10.1.4 Other Vaginal Benign Tumors
Other rare tumors include *tubulovillous adenomas* (arising from cloacal remnants), *leiomyoma*, *rhabdomyoma* (interweaving and haphazardly oriented bundles of spindle- to strap-shaped cells, some with cross striations with focal atypia but rare mitotic figures and no cambium layer),

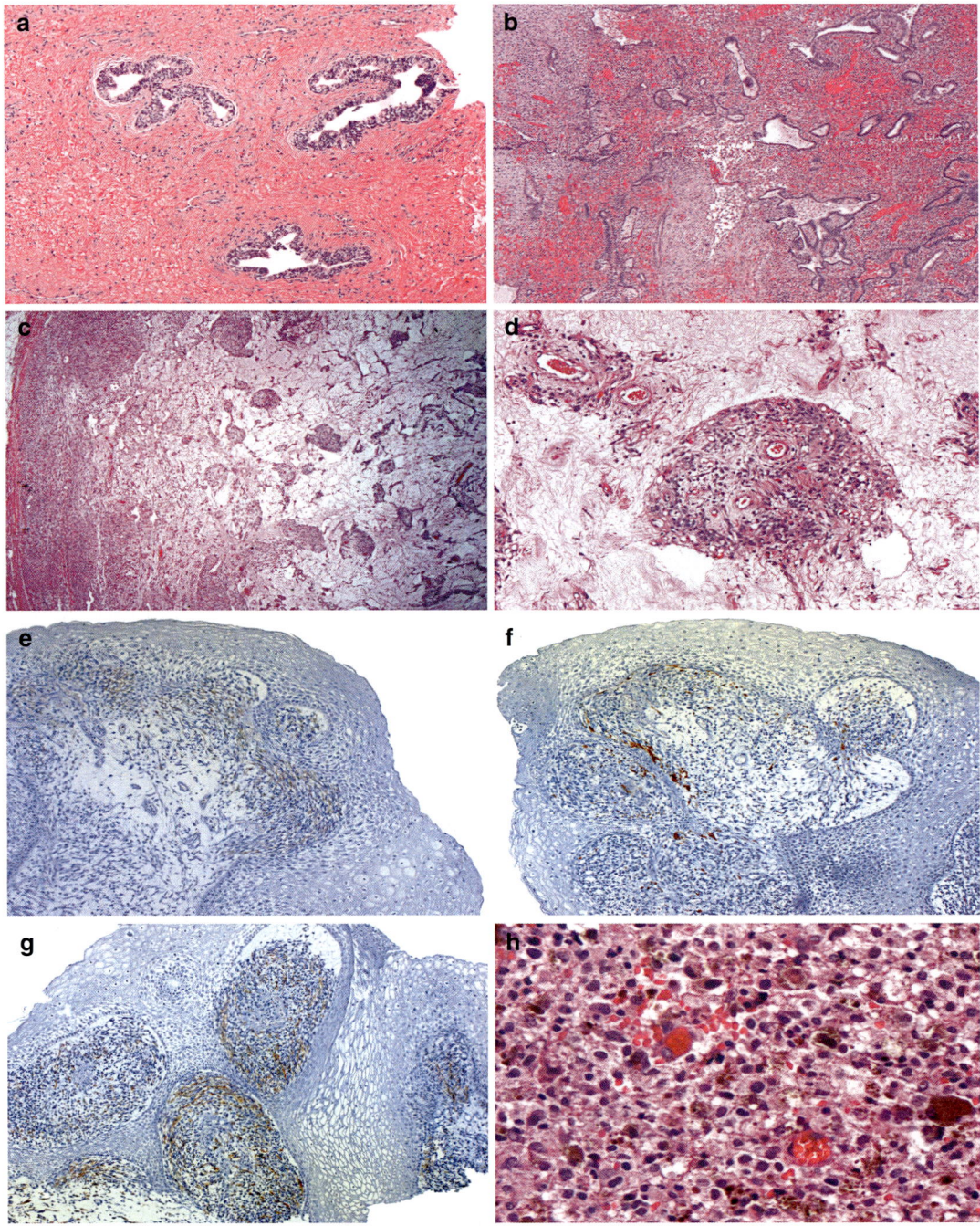

Fig. 8.20 In (**a**) are septum heterotopic endocervical glands (hematoxylin and eosin staining, ×100 original magnification), in (**b**) is a vagina section with progestin-treated endometrial hyperplasia with extensively decidualized stroma in a peri-pubertal child (H&E, 50×), and in (**c, d**) are microphotographs of an angiomyofibroblastoma (H&E stain, ×12.5 and ×100 original magnification, respectively). The microphotographs (**e–g**) show a botryoid rhabdomyosarcoma with cambium layer and positivity for actin, desmin, and myosin, respectively Avidin-Biotin Complex; ×100 original magnification for all three immunostains). Myogenin was also positive (not shown). In (**h**) is the microphotograph of a very rare melanoma of the vagina (hematoxylin and eosin staining, ×400 original magnification)

superficial myofibroblastoma, *Brenner tumor*, and *benign mixed tumor* (aka spindle-cell epithelioma) (DDX: Teratoma). Other rarer tumors are *hymenal polyp* and *polyposis vaginalis*.

8.10.2 Precancerous Conditions and Malignant Tumors

Precancerous conditions of the vagina are collected under the acronym of VAIN (I-II-III), which will be examined in detail in the next paragraph.

8.10.2.1 Vaginal Intraepithelial Neoplasia (VAIN I–III)

VAIN and squamous cell (or epidermoid) carcinoma of the vagina are usually considered rare entities in childhood, but they may indeed occur in women aged less than 25 years. In the Automated Central Tumor Registry (ACTUR), the cancer registry for the US Department of Defense was used by You et al. (2005) to identify children to young adults diagnosed with malignancies of the female genital tract. ACTUR data between 1990 and 2002 were then compared with data obtained from the national Surveillance, Epidemiology, and End Report (SEER) program database, which demonstrated a similar distribution and incidence patterns. Squamous cell carcinoma of the vagina/vulva was about 1/3 of the most common histologic types, and the 21- to 25-year-old age group had the most number of cases for the entire group. Pediatric pathologists should be aware of these rare entities, because of the most efficient and productive application of health maintenance programs for patients in this age group (e.g., pelvic exams and Pap test screening). Unlike in the cervix, the intraepithelial neoplasia of the vagina arise from native squamous epithelium and not from metaplastic epithelium as in cervix. Approximately half of the cases are multifocal and often associated with other neoplasms of the lower genital tract. VAIN is usually located in upper 1/3 of the vagina and may show some tendency to the confluence with similar lesions in the cervix. DDX: urothelial (transitional) metaplasia. VAIN may range from mild dysplasia to carcinoma in situ and is amenable to local treatment, including local excision, partial vaginectomy, CO_2, laser therapy, and 5-FU (Fluorouracil). Risk factors for VAIN or vaginal SqCC are (1) age at first sexual intercourse, (2) # of sexual partners (promiscuity), (3) conditions associated with immunosuppression, (4) tobacco use, (5) HPV status, and (6) malignancy of squamous cell differentiation elsewhere in the female genital tract.

8.10.2.2 Vaginal Epidermoid Carcinoma

Approximately 95% of primary vaginal carcinomas are squamous cell-based but with different degree of squamous differentiation, occurring in the upper posterior vagina of elderly and maybe ± for HPV and/or postradiation locally. A cervical SqCC needs to be ruled out, before making the diagnosis of a primary vaginal SqCC. PGN: 5-YSR, ~40% and related to histology and staging (N1, M1). Therapy includes radiation (external, intracavitary), local excision, or radical surgery for upper 1/3 or posterior wall. If tumor recurs <1 year, worse PGN can be predicted. Moreover, upper lesions recur locally, while lower lesions do distally. *SqCC* usually involves the top third and the posterior wall of the vagina, although most neoplasms may represent an extension of cervical carcinoma. Most of the carcinoma is indeed squamous, but verrucous (*DDX*: Buschke-Loewenstein syndrome) and small cell variants (*DDX*: small blue cell tumors) may occur. FIGO/AJCC staging is related to localization and extension of the tumor with T1 cancer confined to the vagina, T2 invading paravaginal tissues (paracolpium), T3 extending to pelvic wall, and T4 invading urinary bladder, rectum, or extrapelvic. N1 and M1 refer to positive lymph node and distant metastasis, respectively.

- *Verrucous carcinoma*: Well-differentiated variant of SqCC (± HPV), which has a tendency to invade locally (rectum, coccyx) and rarely to metastasize. It needs to be dif-

ferentiated from *condyloma acuminatum*, although this latter rarely involves the vagina.
- *Small-cell carcinoma*: Poor PGN-harboring variant of SqCC, which may be pure or associated with squamous cell or adenocarcinoma.
- *Spindle-cell (sarcomatoid carcinoma)*: DDX is MMMT.

8.10.2.3 Vaginal Adenocarcinoma

There are four subtypes of adenocarcinoma, although the first is most often seen in children or youth (Box 8.17):

- *Clear cell adenocarcinoma* (CCAC): Vaginal clear cell (mesonephroid) carcinoma is an adenocarcinoma that may arise from VA, which is defined as "any Müllerian-type glandular epithelium in the vagina." Typically, the location is at the upper vagina or cervix of children or young adults along Gartner ducts. There is (+) Hx. Prenatal DES exposure (DES was stopped to be distributed in 1971). Malignancy risk for DES exposure is one per 1000. Other DES-related lesions may also be present and include VA, cervical ectropion, and transverse vaginal or cervical ridges. Therapy relies on surgery/radiation according to the size. Grossly, CCAC is generally polypoid and constituted, microscopically, by *tubules/cysts*, *papillary*, and *solid* areas with clear cells, on high magnification, showing abundant PAS+ cytoplasm (glycogen, rarely fat ⇒ ORO−), hobnailing of the epithelium, and low MI.
- *DDX*: Two major non-neoplastic conditions.
  1. MGH, which may present in four growth patterns (glandular, reticular, trabecular, and solid) and shows (−) p53, CEA, and low MI (Ki67/MIB1 < 10%)

> **Box 8.17 Vaginal Adenocarcinoma: Subtypes**
> 1. Clear cell (mesonephroid)
> 2. Endometrioid
> 3. Mucinous
> 4. Urothelial

  2. AS-reaction (pregnancy, progestational agents)
- *PGN*: Relatively good outcome for this kind of tumor and there is no local recurrence if the tumor 1) has been completely excised, 2) was not close to margins, and 3) harbors a depth < 3 mm in extension. MTX: pelvic LN and lungs.

The DDX with YST/EST is discussed below.

- *Endometrioid adenocarcinoma*: AC probably related to a focus of endometriosis
- *Mucinous adenocarcinoma*: AC similar to MAC of the cervix
- *Urothelial carcinoma*: Transitional cell-epithelium differentiating carcinoma, with characteristic (+) CK7 and CK20

8.10.2.4 Vaginal Rhabdomyosarcoma of Botryoid Type (B-ERMS)

The most common malignant mesenchymal tumor in children and adolescents with 1/5 of cases, which occurs in the pelvis of GU tract and typically of embryonal or botryoid subtype, although PGN is relatively good and there is a 10-YSR ranging from ~75% (embryonal) to > 90% (botryoid). Histologically, the embryonal is not different from other embryonal RMS in other locations. The cambium layer characterizes the botryoid subtype. *IHC*: (+) SMA, DES, MYOG (MYF4), and MYO-D1. Myogenin, aka MYOG or MYF-4, is a muscle-specific transcription factor that can induce myogenesis in numerous and different cell types. MYOG belongs to a large family of proteins related by sequence homology (helix-loop-helix or HLH proteins). This transcription factor is involved in the coordination of skeletal muscle development, myogenesis, and repair. MYO-D1 or Myogenic Differentiation 1 is a nuclear protein that belongs also to the HLH family of transcription factors. MYO-D1 regulates muscle cell differentiation and muscle regeneration. In B-RMS, there are, grossly, "grapelike" clusters that fill and project out the vagina, usually anterior vaginal wall, and, microscopically, there are tumor cells localized crowded into the cambium layer beneath the intact vaginal epithelium and perivascular location lying in loose, fibromyxoid stroma with edema and inflammatory cells. The tumor cells have oval nuclei and tennis-racket-like cytoplasm protruding from one end ± rhabdomyoblastic differentiation (cross striations). *DDX*: Polyposis vaginalis of pregnancy (Hx. of pregnancy, benign lesion, and no atypia). Therapy includes surgery, chemotherapy, and radiation therapy.

8.10.2.5 Vaginal Yolk Sac Tumor (YST) or Endodermal Sinus Tumor (EST)

Vaginal YST (aka endodermal sinus tumor) is a tumor of infants (almost exclusively of children < 3 years of age) and is usually located in the posterior wall of vagina or fornices (midline position) and may simulate botryoid rhabdomyosarcoma (ERMS-B) clinically or clear cell adenocarcinoma (CCAC) histologically. The diagnosis is based on histological pattern (reticular or microcystic, macrocystic, endodermal sinus type, papillary, solid, glandular, myxomatous, sarcomatoid, polyvesicular, hepatoid, and/or parietal), characteristic Schiller-Duval bodies, D-PAS (+) hyaline eosinophilic globules, and raised serum AFP, which can also be demonstrated by immunohistochemistry (although relatively nonspecific). The midline position, histology, and (+) AFP point to a germ cell origin and do not show the OCCC IHC triad (−) CK7, EMA, and CD15. Therapy includes multidrug chemotherapy (cisplatinum, etoposide, and bleomycin) and possibly radiation therapy without surgery to preserve sexual and reproductive function. AFP assay helps in preoperative diagnosis, monitoring the effectiveness of the treatment and recurrence. *DDX* includes mostly two tumors:

1. *ERMS of botryoid type*, characterized by subepithelial cambium layer, alternating edematous and cellular areas, Phosphotungstic acid haematoxylin (PTAH)-highlighted rhabdomyoblastic differentiation, (+) DES, MYF4, and MYO-D1. PTAH is a mixture of hematoxylin with phosphotungstic acid, used in histology to visualize better collagen, striated muscle, and glia without using Zenker fixative, which contains mercuric chloride ($HgCl_2$).

2. *CCAC*, characterized by slightly older patients (age > 20 years), (+) CK7, EMA, CD15, while YST is negative for all three markers!

8.10.2.6 Other Vaginal Malignant Neoplasms

Rarely, other tumors may occur in the youth or middle-aged patients or enter in the differential diagnosis of the above listed tumoral entities. These tumors are endometrioid adenocarcinoma, mesonephric adenocarcinoma, malignant fibrous histiocytoma, leiomyosarcoma, malignant melanoma, carcinosarcoma (MMMT), PNET, GIST, DLBCL (diffuse large B-cell lymphoma), granulocytic sarcoma (first clinical manifestation of AML or relapse of AML with (+) Naphthol-AS-D-chloroacetate esterase (NAS-DCE), LYS, MPO, CD68, CD43, CD117, and CD34), and secondary tumors. CAE or chloroacetate esterase, and alternatively naphthol AS-D chloroacetate esterase is useful to demonstrate myeloid differentiation. These latter entities may arise from original tumors of the ovary, uterus, cervix, vulva, colon-rectum, urinary bladder, kidney, and, even, heart. In the setting of AML, the vagina may be the original site of presentation, and BM may disclose the leukemic infiltration. It is not different from granulocytic sarcoma of other locations with a diffuse growth pattern, cords, and/or pseudoacinar spaces and tumor cells showing or not myelocytic differentiation. *HSS/IHC*: (+) NAS-DCE, lysozyme, MPO, CD68, CD43. The *DDX* includes lymphoma, carcinoma, and GrCTs. Other rare tumors include malignant melanoma, which is much more common in the vulva, but it can occur in the vagina (1/20 individuals have melanocytes in the normal vagina). It may present with either vaginal bleeding or vaginal mass and may be associated with melanosis. *PGN*: Poor (local recurrence and/or distant metastases). Therapy may include wide local excision, exenteration with or without LN dissection, RT and/or CHT according to the staging of the lesion. Grossly, *melanomas*: soft blue-black polypoid mass, often ulcerated, and, microscopically, there are epithelioid or spindled cells with clear-cut atypia and junctional activity proving local origin. IHC: (+) S100, HMB45, MART-1, tyrosinase, and MiTF. The *MITF*

> **Box 8.18 Pearls and Pitfalls of the Vagina**
> - Radiation change features include low N/C ratio with vacuolated cytoplasm, multinucleation with smudged chromatin, and lack of mitoses as well as very low or nil MIB1/Ki67.
> - ITC is N1 (TNM, FIGO, and AJCC staging systems).
> - Regional LNs are pelvic or inguinal lymph nodes, and lesions of the upper 2/3 of the vagina may metastasize to the obturator, iliac, and pelvic LNs, while lesions of the lower 1/3 of the vagina may seed to inguinal and femoral LNs.

(Microphthalmia-associated transcription factor) gene is aka class E basic helix-loop-helix protein 32. More vaginal tumors that may be considered in the differential diagnosis are *leiomyosarcoma* (MI ≥ 5/10 HPF; moderate to severe atypia), *lymphoma* (usually DLBCL or SLL/CLL), *MMMT of vagina* (biphasic pattern of malignant glands, epithelial component, and spindle cells, sarcomatous component, and may arise from Gartner ducts or related mesonephric rests), Paget disease of the vagina, and postradiation angiosarcoma. These entities are not usually present in children, but may present theoretically in young adults with underlying genetic predisposition including deletions of tumor suppressor genes. Finally, *metastases to the vagina* include cervix, endometrium, ovary, colon, and kidney carcinomas.

Tips and pitfalls of the vaginal tumor pathology are listed in Box 8.18.

8.11 Tumors of the Vulva

8.11.1 Benign Neoplasms

8.11.1.1 Condyloma Acuminatum

Sexually (HPV6/HPV11) transmitted soft elevated lesion with a tendency to be multiple and coalesce and characterized histologically by papillomatosis, acanthosis with orderly maturation, parakeratosis, hyperkeratosis, prominent granular cell layer, and T-lymphocytic infiltration of the stroma. Koilocytosis is defined as nuclear hyperchromasia, enlargement, pleomorphism, and cytoplasmic cavitation. *IHC*: p16 (−). DDX includes condyloma latum, vulvar/vestibular squamous papillomatosis, warty VIN (+abnormal mitoses, lack of maturation, nuclear pleomorphism, enlargement, and hyperchromasia), and verrucous carcinoma (*vide infra*). Flat condyloma variant is not exophytic and may be more common. Typically, *condyloma acuminatum*/flat shows regression, except in immunocompromised patients, and therapy includes CO_2 laser and podophyllin application. Podophyllin is a cytotoxic drug, which has been used topically in the treatment of genital warts. The active agent is podophyllotoxin. Podophyllin action relies in arresting mitosis in metaphase. This effect is shared with other cytotoxic agents (e.g., vinca alkaloids).

8.11.1.2 Ectopic Mammary Tissue and Hidradenoma Papilliferum

It is a classic boardslide and probably important to remember to avoid mistakes with malignant mimickers. *Hidradenoma papilliferum* is a small, well-delimited subcutaneous nodule covered by normal skin arising from apocrine sweat glands of the vulva or ectopic breast tissue (mammary intraductal papilloma-like). It is often located on *labia majora* or on (uni- or bilateral) interlabial folds. There is no connection with the overlying epidermis. If an epidermis connection is present, the diagnosis of *syringocystadenoma papilliferum* is more appropriate. Grossly, it may ulcerate through skin simulating carcinoma. Microscopically, there are papillary and glandular patterns with stratification on low-power magnification. Columnar secretory bilayered lining including epithelium (+ apocrine decapitation) and myoepithelium are seen on high-power magnification. Some cellular pleomorphism may be present and should not raise any concern if the myoepithelial layer is present. Thus, myoepithelial cell markers are crucial. Ectopic mammary tissue occurs along the primitive milk line and can undergo any changes as seen in normal mammary gland tissue, including lactational changes to malignant neoplasms.

8.11.1.3 Squamous Papilloma, Vulvar/Vestibular Squamous Papillomatosis, and Fibroepithelial Polyp (Acrochordon)

Arborizing connective tissue fronds constitute squamous papillomas with an overlying superficial epithelium consisting of flat epithelium without atypia, ± HPV, but no microscopic features of condyloma (koilocytosis). Vulvar vestibular papillomatosis is a medical condition characterized by multiple papillary vulvar lesions, which are quite similar to *condyloma*, but lack HPV and have slender, "finger-like" cores, contrasting with the typical stalks of condylomas. Fibroepithelial polyp or acrochordon is a polypoid lesion with loose myxoid stroma covered by squamous epithelium with focally bizarre, multinucleated stromal cells, which are (+) DES.

8.11.1.4 Vulvar Melanocytic Nevi

Melanocytic nevi can be overdiagnosed as malignant melanoma, which usually occurs in older patients (age > 50 years), but rarely they may occur in childhood and youth. Melanocytic nevi of the vulva show prominent junctional nests and are common in *labia majora* as intradermal or compound nevi. Dysplastic features may occur (e.g., eosinophilic fibroplasia or lamellar fibroplasia) and should be reported with a comment about the excision and surgical margins.

8.11.1.5 Vulvar Angiomyofibroblastoma (AMF)

Benign vulvar and vaginal tumor with possible/theoretical sarcomatous transformation in the vulva, but usually it does not recur after excision and is composed of alternating hypercellular and hypocellular areas. Spindle cells and plump stroma cells with eosinophilic cytoplasm have a perivascular arrangement with numerous thin-walled and capillary-like vascular structures. Occasionally, there are binucleated or, even, multinucleated cells with mast cells, which can occur. No atypia and low MI.

- *IHC*: (+) VIM, DES, (−) SMA, CK, S100.
- *TEM*: Myofibroblastic differentiation with well-developed RER, Golgi apparatus, intermediate filaments, and pinocytotic vesicles.
- *DDX*: Angiomyxoma (not circumscribed, less cellular, fewer vessels with deep, prominent vascular pattern cuffed by myoid bundles; no bizarre stromal cells, no hyalinization of the BVs, RBC extravasation, and stromal mucin).

8.11.1.6 Vulvar Cellular Angiofibroma (CAF)

Unlike most of the other site-specific vulvovaginal mesenchymal lesions, cellular angiofibroma has a marked preference for the vulva, with only occasional cases described in the vagina. CAF is a benign lesion, which needs to be distinguished from other vulvovaginal mesenchymal neoplasms and particularly smooth muscle neoplasms. However, CAF exhibits fibroblastic rather than myofibroblastic differentiation, based on an absence of positivity for muscle markers and ultrastructural evidence of fibroblastic electron microscopic features. Microscopically, CAF is well-demarcated but unencapsulated lesion with entrapped adipose tissue at the periphery and moderately cellular. At high-power magnification, CAF is composed of a uniform population of bland spindle-shaped cells in a fibrous stroma with bundles of thin collagen. Genetically, 13q loss has been demonstrated similarly to spindle-cell lipoma and extramammary myofibroblastoma, which are also CD34 (+). IHC: (±) SMA, DES, h-Caldesmon, (+) CD34, and (+) ER/PR.

8.11.1.7 Other Benign Lesions of the Vulva

Other benign lesions include benign pilar tumor, syringoma, chondroid syringoma, warty dyskeratoma, angiokeratoma of Fordyce, and leiomyoma. To remember is the occurrence of criteria for malignancy, which are Ø > 5 cm, infiltrative markers, MI > 5/10 HPF, and moderate/severe cytologic atypia. Ultimately, the diagnosis of leiomyoma should be made if only one of these criteria is reached, atypical leiomyoma in case of two definite criteria, and leiomyosarcoma, in

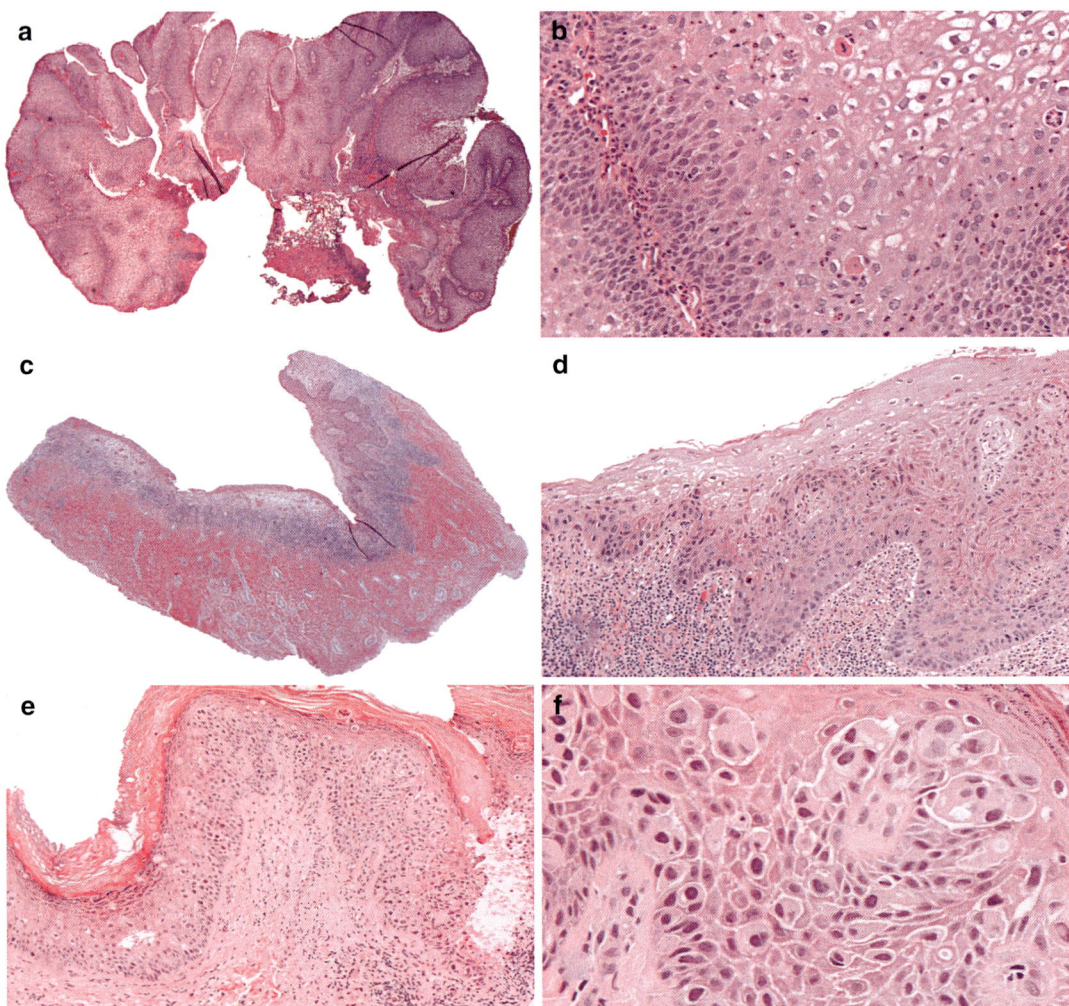

Fig. 8.21 The microphotographs in (**a**, **b**) show a condyloma with cytopathic changes (H&E stains, x12.5 and x200 original magnification, respectively); the microphotographs in (**c**, **d**) show vulvar intraepithelial neoplasia (VIN) III differentiated (H&E stain; x12.5 and x100 original magnifications), while the microphotographs in (**e**, **f**) show an extramammary Paget of the vulva (H&E stain, x100 and x400 original magnification, respectively). Vulvar lichen sclerosus starts as early as the child is 8 years old as recently identified. Vulvar lichen sclerosus can progress to VIN

case of three. Finally, pseudosarcomatous fibroepithelial stromal polyp (pregnancy) or POSNs should be differentiated from malignant mesenchymal tumors (Figs. 8.21 and 8.22).

8.11.2 Malignant Tumors

In childhood, adolescence and youth benign neoplasms of the vulva may occur and need to be taken into account in the *DDX* of vulvar lesions in this age group. In 95% malignancies are of squamous cell carcinoma and in 9 out 10 cases are HPV16 or HPV18 positive. Malignant lesions can be multicentric and associated with vulvar intraepithelial neoplasia (VIN) III (CIS or Bowen disease). Progression of malignancy is commonly seen when the patient is older and/or immunosuppression is present. The metastatic burden is linked to size ($\emptyset <$ or > 2 cm), depth of invasion (θ), and/or ALVI.

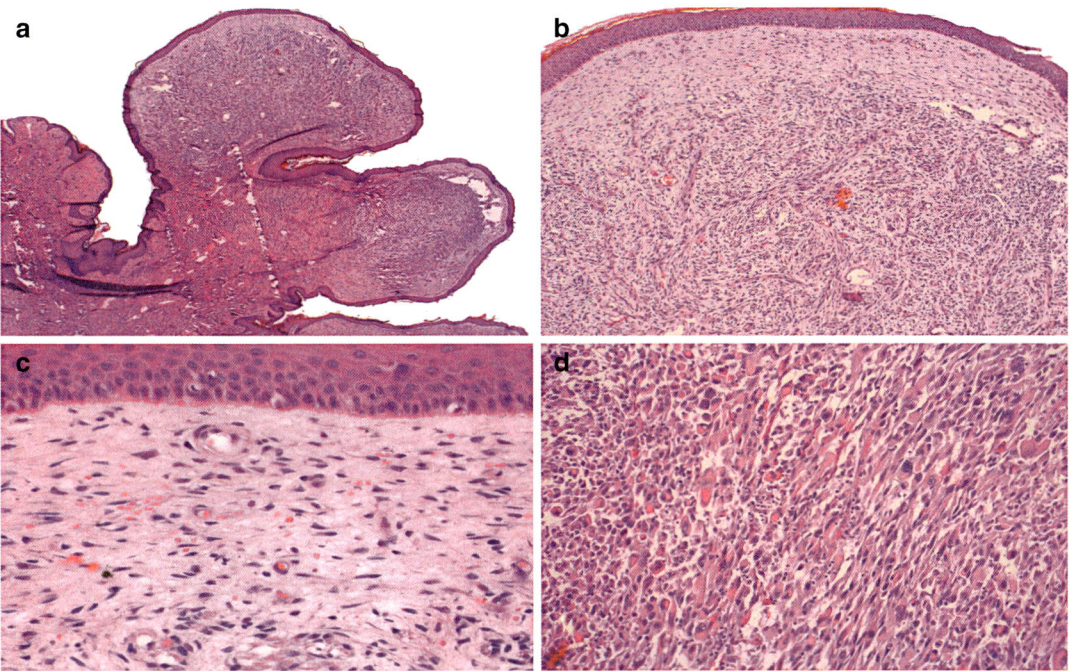

Fig. 8.22 This panel shows another botryoid rhabdomyosarcoma (**a-d**), here located at the vulva with a characteristic cambium layer (H&E stain; x20, x100, x400, x400 original magnification, respectively)

8.11.2.1 Precancerous Malignant Conditions

Precancerous lesions may occur in the youth and particularly important is to remember if the patient is immunocompetent or immunosuppressed or harbors some genetic disease like *xeroderma pigmentosum*.

Vulvar Intraepithelial Neoplasia (VIN)

Premalignant uni-/multifocal lesion, whose incidence has markedly increased recently due to most diffuse screening programs, has become more often diagnosed in young women (20–35 years). Its association with HPV and the early presentation should be a stimulus for pediatric pathologists to investigate these lesions.

VIN-1 is still not correctly defined. VIN-1 includes flat condylomas and acanthotic lesions with mild atypia that are not incorporated into either the *condyloma* or classic VIN categories. It is often used the classification LSIL/HSIL with low-grade squamous intraepithelial lesion (VIN-1) for condyloma and high-grade squamous intraepithelial lesion (HSIL; VIN-2/VIN-3) for classic VIN lesions. VIN3 features include traditional nuclear enlargement, hyperchromasia, and multinucleation in the lower epithelial layers with or without abnormal mitoses and apoptosis with "*corps rounds*". On the surface, there is atypical parakeratosis with or without koilocytotic atypia. Classic VINs has also been subclassified into warty and basaloid types.

Bowenoid papulosis or Bowenoid dysplasia is a type of VIN with multiple small papules, often pigmented, in the vulva of young women, which looks like many warts. Bowenoid papulosis does not typically progress to invasive carcinoma but may recur and needs to be distinguished from Bowen disease or CIS. Microscopically, there are atypical cells in a background of orderly epithelium with sparing of *acrotrichium*, which the uppermost distal part of the infundibular epidermis that traverses the epidermis (Greek ἄκρος, "highest", τρίχωμα, "hair", and -ium).

Bowen disease is CIS (squamous cell carcinoma in situ), i.e., a type of VIN, with full-thickness disorganization of the squamous epithelium with hyperkeratosis, parakeratosis, acanthosis, multinucleated dyskeratotic cells, and abnormal mitoses throughout total thickness with intercellular bridges ± intraepidermal involvement of hair follicle (acrotrichium) but sparing of the sweat gland counterpart (acrosyringium) (HPV+, p16+). John T. Bowen, an American dermatologist, described this lesion in 1912.

Differentiated (simplex) VIN should be considered a form of the preinvasive vulvar disease characterized by maturation, variable hyperplasia, keratinization, and parabasal atypia. Interestingly, nuclear atypia remains *confined to the cells at the epithelial-stromal interface*, and the atypia is not observed in the upper epithelial layers. This situation is different from other HSIL associated with HPV, and it has been suggested that the genetically altered cells retain the capacity to mature. Conservative excision and careful follow-up are part of the management of this lesion. Differentiated (simplex) VIN is not a diagnosis of young and middle-aged women, being more common in older women that share the clinical manifestations with *vulvar lichen sclerosus* and *lichen simplex chronicus*. Patterns of differentiated VIN include lichen sclerosus with basal atypia, prominent basal atypia with a strong lichenoid inflammatory presence, limited basal hyperchromasia with abnormal keratinocyte maturation, and differentiated VIN with variable keratinocyte maturation, hyperkeratosis, and basal atypia, and finally, discrete keratinocytes with dyskeratosis. DDX includes mainly psoriasis, spongiotic dermatitis, Candida vulvitis, which are all conditions that may be present in young females. Atypia in vulvar lesions needs to be carefully ruled out. It occurs as (1) squamous cell hyperplasia, (2) *lichen sclerosus*, (3) intraepithelial neoplasia, VIN I (mild dysplasia), (4) intraepithelial neoplasia, VIN II (moderate dysplasia), (5) intraepithelial neoplasia, VIN III (severe dysplasia) = CIS, (6) vulvar extramammary Paget disease, and (7) *melanoma in situ*.

8.11.2.2 Vulvar Epidermoid Carcinoma

Vulvar epidermoid carcinoma (VEC) accounts for 95% of vulvar carcinomas and is more often seen in the sixth to seventh decades, although rarely may be encountered in female individuals younger than 30 years. Locations are usually *labia majora*, but it may be also found on *labia minora* and *clitoris*. RF include cervical carcinoma, genital granulomatous disease, immunodeficiency, promiscuity, and a tobacco smoking habit.

- *CLM*: VEC is usually well-differentiated, but tumors on the *clitoris* may be more anaplastic, and VIN may be found at margins of the neoplasm. High-grade lesions may show focal glandular differentiation, although it is correct not to overcall adenosquamous subtype when you found yourself in this kind of settings. Depth of invasion characterizes the microinvasive squamous cell carcinoma ($\theta < 5$ mm). The pattern of invasion can be (1) downward stromal extension of irregular nests of neoplastic epithelium, (2) the confluent growth of uneven intersecting clusters of malignant epithelial cells, (3) sharply demarcated bulbous nests with minimal atypia (verrucous carcinoma), and (4) downward stromal extension of irregular nests of keratinized neoplastic epithelium (HPV-negative vulvar carcinoma associated with differentiated (simplex) VIN).
- *CGB*: Loss of 4p13, 3p, and 5q and gains of 3q and 8p have been found.
- *TRT*: It includes wide local excision with 1 cm margins if small size (Ø) of VEC, superficial (3 mm) localized, and WD subtype histologically is diagnosed. Conversely, radical vulvectomy with bilateral iliac LN dissection for advanced cases with or without pelvic exenteration and RT.

About the outcome, SqCC of the vulva has a different 5-YSR according to the FIGO stage, being approximately 70% at stage I, 60% at stage II, 45% at stage III, and less than 10% at stage IV. About 20% of VSqCC metastasize to regional LN and, specifically, labial tumors to superficial inguinal LNs while clitoral tumors to deep inguinal nodes.

In vulvar cancer, independent of the age, the status of the regional represents probably the most crucial factor prognostically and therapeutically. Vulval lymph drainage is to the ipsilateral inguinofemoral nodes for tumors located in the middle of either labium, whereas nodes on both sides are the drainage for perineal tumors. The subsequent station following the inguinofemoral is the deep pelvic iliac and obturator LNs. The risk of node metastasis is related to tumor size (smaller or larger than 2 cm), tumor grade, tumor thickness, fixed or ulcerated LNs, LVI, and older age. Features to include in your pathology report are anatomic site and location, size, depth of invasion (VEC: from epithelial-stromal junction of adjacent most superficial dermal papilla to most profound point of invasion), histologic type and grade, pagetoid spread, type of invasion (infiltrating, pushing, mixed), ALVI, adjacent organs, margins, precancerous lesions if any, vulvar dystrophies if any, descriptive features (ulcers, etc.), LN (total, number definite, location, tumor size), and involvement or not of adjacent tissue.

8.11.2.3 Vulvar Epidermoid Carcinoma Variants

Vulvar verrucous carcinoma is a variant of SqCC cell carcinoma that tends to show no metastasis to regional LNs, and local surgical excision is the treatment of choice, being RT contraindicated. Basal cell carcinoma is another variant, which is also treated by wide local excision.

Melanoma is the most frequent non-squamous vulvar malignant neoplasm comprising about 1/20 of all vulvar malignancies and may occur from the late teens to older women. Melanomas arise from junctional or compound nevi, and pigmented vulvar lesions are usually junctional nevi and require remotion by excision as soon as possible. Clinically, melanomas are brown or blue-black lesions and can be flat or ulcerated with or without darkly pigmented areas (satellite nodules) nearby. Melanoma of superficial spreading type tends to occur in younger patients and shows a better outcome than nodular melanomas. Vulvar melanomas follow the same FIGO classification as indicated for SqCC. Clark levels are also used. *PGN* demonstrates that 5-YSR is ~ 50% overall and seems to be related to tumor invasion (Clark levels) and tumor size (Ø).

8.11.2.4 Extramammary Paget disease (EMPD) of the Vulva or Vulvar Paget Disease (VPD)

Malignant epithelial tumor of the vulva with different origins and tendency to recur (residual disease) and rarely metastasize but can be associated with VIN.

Primary EMPD (usually *labia majora*) is due to a sweat gland carcinoma arising from acrosyringium (intraepidermal spiral duct in the eccrine gland) or basal layer carcinoma with eccrine differentiation, and *secondary EMPD* is due to an epithelial malignancy of the ano-rectum, urothelium, or prostatic gland. MPD usually indicates an underlying carcinoma, and intradermal cells are typically mucin negative, whereas VPD shows a primary in only 15–30% of cases, and the remaining is primary without an underlying carcinoma and is usually mucin positive.

Histogenesis:

1. Epidermotropic spread of *tumor cells from neighborhood organs*
2. Adnexal skin tissue (e.g., *eccrine or apocrine glands*)
3. Ectopic mammary cells (i.e., *Toker cells* of the nipple) or intraepidermal ectopic sweat gland epithelial cells (i.e., *clear cell papulosis*)
4. In situ transformation of *intraepidermal stem cells*

Clinically, there is a crusting, weeping, and oozing lesion with some erythema.

Microscopically, there are large pale-stained cells, usually located in the lower epidermis, in nests, in glandular spaces, or continuously along BM but also in pilosebaceous units and sweat glands. The tumor cells have large nuclei with the prominence of nucleoli and abundant cytoplasm with a separating cleft between tumor cells and neighboring keratinocytes. Potential additional features are intracytoplasmic melanin, lichenoid lymphocytic infiltrate in the papillary dermis, fibroepithelioma-like hyperplasia, and papillomatous hyperplasia.

EMPD secondary to urothelial carcinoma shows anaplasia resembling high-grade urothe-

Box 8.19 IHC Differentiating Panel

	CK7	CK20	MUC1	MUC2	MUC5AC	GCDFP-15	UP-III
MPD	+	−	+	−	−	+	−
VPD	+	+	+	−	+	+	−
AR-EMPD	±	+	±	+	+	−	−
U-EMPD	+	+	+	−	+	−	+

Notes: *EMPD*, extramammary Paget disease; *MPD*, mammary Paget disease; *VPD*, vulvar Paget disease; *AR*, anorectal; *U*, urothelial

lial cells with cellular pleomorphism, coarse chromatin distribution pattern, non-prominent nucleoli, and high MI.

EMPD secondary to anorectal carcinoma is similar to atypical gastrointestinal glands with stratified columnar cells, goblet cells, and intraluminal necrosis.

DDX: MMM, invasive carcinoma, and clear cell papulosis (asymptomatic hypopigmented macules or flat papules distributed in the lower abdomen and along the milk lines of young children with a tendency of regression and no need for treatment) among other entities (Box 8.19).

MPD is CK7+ and CK20−, while EMPD is usually both CK7+ and CK20+ (mainly if there is an underlying anorectal or urothelial neoplasm). MUC2 seems to be more specific than CK20 in identifying the colorectal origin of EMPD. CDX2 may be associated with MUC2 for AR-EMPD, and Uroplakin III may be associated with urothelial carcinoma. Paget cells are also (+) CEA and (+) mucin stains (AB, PAS, and MCM). Mucin stains are variably positive in CCP. In contrast, these stains are consistently (+) in PD and (−) in Toker cells. HMB45 serves to distinguish from MM.

- *TEM*: Evidence of glandular differentiation.
- *PGN*: Good, because it does not metastasize, and if there is no underlying primary carcinoma, therapy consists of excision with lateral and deep margins to include all sweat glands.

8.11.2.5 Bartholin Gland Carcinoma and Other Carcinomas

Arborizing connective tissue fronds constitute Müllerian papillomas with an overlying superficial epithelium consisting of flat cuboidal or mucinous epithelium. Other vulvar carcinomas include adenocarcinoma of mammary gland type, adenocarcinoma of Skene gland origin, and malignant sweat gland tumors. A rarity is Merkel cell carcinoma, which is rare in childhood but may occur in youth, if abnormal immunologic conditions may occur.

8.11.2.6 Vulvar Aggressive Angiomyxoma

Slow-growing tumor of the vulva, vagina, pelvic floor, or perineum with a tendency to recur in ischiorectal or retroperitoneal space (a *tumor with LMP*) and occurring in teens/young adults of both genders. Grossly, AAM is white, soft or gelatinous, and poorly circumscribed and constituted by *spindled and stellate cells in hypocellular myxomatous stroma* with *thick-walled blood vessels with dilated lumina* and occasional hyaline thickening of adventitia, RBC extravasation and vascular pattern cuffed by myoid bundles, and stromal mucin. No atypia and low MI.

- *IHC*: (+) SMA, (−) S100, FVIII-RA, CEA, AE1–AE3.
- *TEM*: primitive mesenchymal cells with fibroblastic features.

8.11.2.7 Vulvar Epithelioid Sarcoma and Other Sarcomas

Epithelioid sarcoma of the vulva is quite more aggressive than the same entity of the extremities (central vs. peripheral epithelioid sarcomas). Malignant rhabdoid features with intracytoplasmic inclusions (*TEM*: whorls of intermediate filaments) may be present and add to the classic elements of sheets of large epithelioid cells with marked cytologic atypia including prominent nucleoli. The classic granuloma-like pattern may be lacking in the vulva.

- *IHC*: (+) VIM, CK, EMA, CD34, and (±) CD99 but (−) CD31 and S100.
- *DDX*: It includes synovial sarcoma (usually biphasic, CD34−), angiosarcoma (malignant angiopoiesis, CD31+, EMA−), malignant melanoma (S100+, HMB45+), MPNST (S100+, Hx. of NF±), extrarenal rhabdoid tumor (INI−), and undifferentiated carcinoma (usually connected to epidermis or adnexae, squamous or glandular differentiation, CD34−).

8.11.2.8 Vulvar Malignant Melanoma

Vulvar malignant melanoma (VMM) is rare or inexistent in children or adolescents, but young people with some immunosuppression or family cancer syndrome may develop this malignant melanocytic lesion. VMM is associated with melanosis. The presentation may be late with lesions at Clark level III or IV, and prognostic factors include the level of infiltration (Clarke, Breslow, and Chung measurements), LN status, and metastases. Grossly, a large and color-variegated nodule may be detectable. Microscopically, VMM with vertical growth phase shows fibroplasia with angiogenesis, while fibroplasia with a plaque-like lymphocytic infiltrate is seen in the radial growth phase of VMM. Therapy consists of radical vulvectomy with bilateral LN dissection or wide local excision according to the size of the neoplasm.

Vulvovaginal mesenchymal lesions include aggressive angiomyxoma, angiomyofibroblastoma, cellular angiofibroma, fibroepithelial stromal polyp, prepuberal vulvar fibroma, superficial myofibroblastoma, and some smooth muscle neoplasms. Box 8.20 lists the most frequent differential diagnoses of vulvo-vaginal lesions.

Vulvar-pigmented lesions include *lentigo simplex*, melanocytic nevi, dysplastic nevi, VIN III, and malignant melanoma. Vulvar depigmented lesions include, conversely, *vitiligo, psoriasis*, chronic dermatitis, *lichen planus, lichen simplex chronicus, lichen sclerosus*, and squamous cell hyperplasia, which are nonmalignant lesions, and three malignant lesions including carcinoma in situ, EMPD of the vulva, and the invasive carcinoma.

> **Box 8.20 Important DDX of Vaginal Lesions**
> - *Aggressive angiomyxoma*: Deep, prominent vascular pattern, myoid bundles cuffing
> - *Angiomyofibroblastoma*: Well-circumscribed subserosal nodule, no atypia, perivascular stromal cells, and delicate wall-harboring BVs
> - *Botryoid embryonal RMS*: Infancy, submucosal hypercellular zone/cambium layer, rhabdomyoblasts, DES+, myoglobin+, MYF4+, MYO-D1+
> - *Cellular angiofibroma*: well circumscribed, less polypoid, diffusely vascular pattern, hyalinized wall-harboring BVs, no atypical stromal cells, DES−
> - *Leiomyosarcoma*: Clear boundary of tumor cells and smooth muscle differentiation
> - *Low-grade ESS*: Spiral arterioles-like BVs, no central blood vascular core, thick bands of collagen ("starburst pattern"), and dot-like staining of DES/AE1–AE3
> - *MPNST*: Perivascular accentuation, S100+, GFAP +
> - *DFSP*: Storiform, CD34+, FXIIIa−, DES−

Finally, hematological malignancies and secondary tumors need to be kept in mind. Secondary tumors may arise from small round blue cell tumors of the abdomen and perineal region in childhood and carcinomas of the cervix, endometrium, gastrointestinal tract, or kidney in adulthood.

Multiple Choice Questions and Answers

- GYN-1 Which of the following conditions is NOT a Müllerian duct anomaly?
  (a) Septate uterus
  (b) Vaginal atresia
  (c) Uterine atresia
  (d) Unicornuate uterus
  (e) Wilms' tumor – aniridia

- GYN-2 Smears of vaginal, cervical, or urethral discharge are useful in diagnosing genital infections. For each microscopic finding indicated below, choose the correct microorganism.
  (a) Flagellate protozoa
  (b) Inclusion bodies (columnar lining cells)
  (c) Gram-negative diplococci
  (d) Budding hyphae
      1. *Neisseria gonorrhoeae*
      2. *Trichomonas vaginalis*
      3. *Candida albicans*
      4. *Chlamydia trachomatis*
         (a)(2) (b)(4) (c)(1) (d)(3)
- GYN-3 Which statement regarding pelvic inflammatory disease (PID) is NOT correct?
  (a) It represents the spread of an infection from the vagina to the cervix, uterus, fallopian tubes, and even peritoneum.
  (b) It occurs most often in women 5–10 years of age.
  (c) Complications include endometritis, salpingitis, parametritis, perihepatitis, and peritonitis.
  (d) It is associated with the use of an intrauterine device (IUD).
  (e) It is associated with subsequent infertility following fallopian tube damage.
- GYN-4 Diethylstilbestrol (DES) was an orally active xenoestrogen prescribed to pregnant women to prevent miscarriages in the past. DES is known as a "biological time bomb." Indeed, long-term effects of DES have been documented in the mothers exposed to DES and their offspring (DES-daughters and DES-sons). Which pathology of the following listed below has NOT been associated with DES exposure in humans?
  (a) Clear cell adenocarcinoma (CCA) of the vagina
  (b) CCA of the cervix
  (c) Carcinoma of the mammary gland
  (d) Cervical intraepithelial neoplasia
  (e) Urinary bladder carcinoma
- GYN-5 A 17-year-old girl presents to her family doctor for evaluation of vaginal discharge. In the presence of a chaperone, the family doctor collects a thick, purulent vaginal discharge, some cervical motion tenderness, and a friability of the cervix. The girl reports that her partner is asymptomatic, and they are using condom protection. She complains painful urination. Which etiology is most likely to be the case in this girl?
  (a) Bacterial vaginosis
  (b) Herpes simplex virus (HSV) infection
  (c) *Treponema pallidum*
  (d) *Chlamydia trachomatis*
  (e) *Neisseria gonorrhoeae*
- GYN-6 What is the definition of an immature teratoma grade III?
  (a) Glial implants
  (b) Prominent neuroepithelium in ≥ 4 low-power fields (LPF) on any slide
  (c) Prominent neuroepithelium in ≥ 3 low-power fields (LPF) on any slide
  (d) Prominent neuroepithelium in ≥ 2 low-power fields (LPF) on any slide
- GYN-7 What is NOT a growth pattern of a yolk sac tumor?
  (a) Reticular/microcystic
  (b) Polyvesicular vitelline
  (c) Solid
  (d) Hepatoid
  (e) Intestinal
  (f) Endometrioid
  (g) Ectodermal body
  (h) Parietal
  (i) Festoon
  (j) Tubular
  (k) Papillary
- GYN-8 A botryoid rhabdomyosarcoma is a histologic variant of embryonal rhabdomyosarcoma that can involve the vulvar region in girls aged 10 years or younger. Which of the following statements is NOT valid for botryoid rhabdomyosarcoma?
  (a) It can mimic a fibroepithelial stromal polyp.
  (b) A rapid growth history characterizes it.
  (c) It contains a "cambium layer" with hyperchromatic cells beneath the epithelium.
  (d) It may show cross striations or strapped cells on histology.
  (e) There is positivity for skeletal muscle markers.

(f) It involves the *PAX3* (paired box gene 3) located at 2q36.1 and the *FOXO1* (forkhead box O1) located at 13q14.11.
- GYN-9 Which criterion is NOT used for invasive ovarian implants?
    (a) Invasion of underlying healthy tissue with infiltrating margin and desmoplastic stroma
    (b) Micropapillary architecture
    (c) Solid epithelial nests surrounded by cleft
    (d) Submesothelial invaginations
- GYN-10 The lymphatic route to regional mesenteric lymph nodes is the most common mode of spread of colon-rectal cancer. Hematogenous spread to the liver and lung and direct extension to adjacent structures are also possible. If the neoplasm has perforated through the serosal surface, metastatic deposits can occur. Which eponym is used for each of the anatomical locations?
    (a) Rectovesical pouch
    (b) Hypoumbilical peritoneum
    (c) Ovary
        1. Sister Joseph's nodule
        2. Blumer shelf
        3. Krukenberg tumor
        (a)(2) (b)(1) (c)(3)

References and Recommended Readings

Aarestrup J, Gamborg M, Ulrich LG, Sørensen TI, Baker JL. Childhood body mass index and height and risk of histologic subtypes of endometrial cancer. Int J Obes. 2016;40(7):1096–102. https://doi.org/10.1038/ijo.2016.56. Epub 2016 Apr 28. PubMed PMID: 27121254; PubMed Central PMCID: PMC4973214

Acién P, Acién MI. The history of female genital tract malformation classifications and proposal of an updated system. Hum Reprod Update. 2011;17(5):693–705. https://doi.org/10.1093/humupd/dmr021. Epub 2011 Jul 4. Review. PubMed PMID: 21727142.

Acién P, Acién M, Sánchez-Ferrer M. Complex malformations of the female genital tract. New types and revision of classification. Hum Reprod. 2004;19(10):2377–84. Epub 2004 Aug 27. PubMed PMID: 15333604.

Agrons GA, Wagner BJ, Lonergan GJ, Dickey GE, Kaufman MS. From the archives of the AFIP. Genitourinary rhabdomyosarcoma in children: radiologic-pathologic correlation. Radiographics. 1997;17(4):919–37. PubMed PMID: 9225391.

Al Jishi T, Sergi C. Current perspective of diethylstilbestrol (DES) exposure in mothers and offspring. Reprod Toxicol. 2017;71:71–7. https://doi.org/10.1016/j.reprotox.2017.04.009. Epub 2017 Apr 28. Review. PubMed PMID: 28461243.

Alonso M, Hamelin R, Kim M, Porwancher K, Sung T, Parhar P, Miller DC, Newcomb EW. Microsatellite instability occurs in distinct subtypes of pediatric but not adult central nervous system tumors. Cancer Res. 2001;61(5):2124–8. PubMed PMID: 11280776

Am J Clin Pathol. 1997;107:12.
Am J Clin Pathol. 2002;117:745.
Am J Surg Pathol. 1994;18:687.
Am J Surg Pathol. 1996;20:1056.
Am J Surg Pathol. 2000;24:1465.
Am J Surg Pathol. 2001a;25:1283.
Am J Surg Pathol. 2001b;25:212.
Am J Surg Pathol. 2006;30:1222.
Am Surg. 2008;74:1062.

Anagnostopoulos A, Ruthven S, Kingston R. Mesonephric adenocarcinoma of the uterine cervix and literature review. BMJ Case Rep. 2012; pii: bcr0120125632. https://doi.org/10.1136/bcr.01.2012.5632. Review. PubMed PMID: 23230242.

Ansari DO, Horowitz IR, Katzenstein HM, Durham MM, Esiashvili N. Successful treatment of an adolescent with locally advanced cervicovaginal clear cell adenocarcinoma using definitive chemotherapy and radiotherapy. J Pediatr Hematol Oncol. 2012;34(5):e174–6. https://doi.org/10.1097/MPH.0b013e318257dc91. PubMed PMID: 22713709.

Arafah M, Zaidi SN. A case of yolk sac tumor of the vagina in an infant. Arch Gynecol Obstet. 2012;285(5):1403–5. https://doi.org/10.1007/s00404-011-2209-y. Epub 2012 Jan 8. PubMed PMID: 22228386.

Arias-Stella J. Atypical endometrial changes associated with the presence of chorionic tissue. Arch Pathol. 1954;58(2):112–28. PMID 13170908.

Arias-Stella J. The Arias-Stella reaction: facts and fancies four decades after. Adv Anat Pathol. 2002;9(1):12–23. PMID 11756756.

Asian Pac J Cancer Prev. 2006;7:234.

Babin EA, Davis JR, Hatch KD, Hallum AV 3rd. Wilms' tumor of the cervix: a case report and review of the literature. Gynecol Oncol. 2000;76(1):107–11. Review. PubMed PMID: 10620450.

Badalyan V, Burgula S, Schwartz RH. Congenital paraurethral cysts in two newborn girls: differential diagnosis, management strategies, and spontaneous resolution. J Pediatr Adolesc Gynecol. 2012;25(1):e1–4. Epub 2011 Nov 16. Review. PubMed PMID: 22088318.

Baker GM, Selim MA, Hoang MP. Vulvar adnexal lesions: a 32-year, single-institution review from Massachusetts General Hospital. Arch Pathol Lab Med. 2013;137(9):1237–46. https://doi.org/10.5858/arpa.2012-0434-OA. PubMed PMID: 23991738.

Bárcena C, Oliva E. WT1 expression in the female genital tract. Adv Anat Pathol. 2011;18(6):454–65. https://doi.org/10.1097/PAP.0b013e318234aaed. PubMed PMID: 21993272.

Baris HN, Barnes-Kedar I, Toledano H, Halpern M, Hershkovitz D, Lossos A, Lerer I, Peretz T, Kariv R, Cohen S, Half EE, Magal N, Drasinover V, Wimmer K, Goldberg Y, Bercovich D, Levi Z. Constitutional mismatch repair deficiency in Israel: high proportion of founder mutations in MMR genes and consanguinity. Pediatr Blood Cancer. 2016;63(3):418–27. https://doi.org/10.1002/pbc.25818. Epub 2015 Nov 6. PubMed PMID: 26544533

Bell KA, Smith Sehdev AE, Kurman RJ. Refined diagnostic criteria for implants associated with ovarian atypical proliferative serous tumors (borderline) and micropapillary serous carcinomas. Am J Surg Pathol. 2001;25(4):419–32.

Biason-Lauber A, Chaboissier MC. Ovarian development and disease: the known and the unexpected. Semin Cell Dev Biol. 2015;45:59–67. https://doi.org/10.1016/j.semcdb.2015.10.021. Epub 2015 Oct 19. Review. PubMed PMID: 26481972

Broome M, Vial Y, Jacquemont S, Sergi C, Kamnasaran D, Giannoni E. Complete maxillo-mandibular syngnathia in a newborn with multiple congenital malformations. Pediatr Neonatol. 2016;57(1):65–8. https://doi.org/10.1016/j.pedneo.2013.04.009. Epub 2013 Jun 15. PubMed PMID: 23778189.

Cancer. 2007;109:1784.

Cancer Epidemiol Biomarkers Prev. 2004;13:2078.

Cao D, Liu A, Wang F, Allan RW, Mei K, Peng Y, Du J, Guo S, Abel TW, Lane Z, Ma J, Rodriguez M, Akhi S, Dehiya N, Li J. RNA-binding protein LIN28 is a marker for primary extragonadal germ cell tumors: an immunohistochemical study of 131 cases. Mod Pathol. 2011;24(2):288–96. https://doi.org/10.1038/modpathol.2010.195. Epub 2010 Nov 5. PubMed PMID: 21057460.

CAP Protocol.

CAP Protocol for endometrium and References from CAP protocol 20–23.

Carcangiu ML, Radice P, Casalini P, Bertario L, Merola M, Sala P. Lynch syndrome–related endometrial carcinomas show a high frequency of nonendometrioid types and of high FIGO grade endometrioid types. Int J Surg Pathol. 2010;18(1):21–6. https://doi.org/10.1177/1066896909332117. Epub 2009 May 14. PubMed PMID: 19443869.

Chauhan S, Nigam JS, Singh P, Misra V, Thakur B. Endodermal sinus tumor of vagina in infants. Rare Tumors. 2013;5(2):83–4. https://doi.org/10.4081/rt.2013.e22. Print 2013 Apr 15. PubMed PMID: 23888222; PubMed Central PMCID: PMC3719117.

Chung-Park M, Zheng Liu C, Giampoli EJ, Emery JD, Shalodi A. Mucinous adenocarcinoma of ectopic breast tissue of the vulva. Arch Pathol Lab Med. 2002;126(10):1216–8. PubMed PMID: 12296762.

Clin Cancer Res. 2008;14:5840.

Creasman WT, Phillips JL, Menck HR. The national cancer data base report on cancer of the vagina. Cancer. 1998;83(5):1033–40. PubMed PMID: 9731908.

Crum CP, Pinto AP, Grantner SR, Peters WA III. Chapter 6 – Squamous neoplasia of the vulva. In: Diagnostic gynecologic and obstetric pathology. Philadelphia: Elsevier; 2018.

Dahabreh IJ, Schmid CH, Lau J, Varvarigou V, Murray S, Trikalinos TA. Genotype misclassification in genetic association studies of the rs1042522 TP53 (Arg72Pro) polymorphism: a systematic review of studies of breast, lung, colorectal, ovarian, and endometrial cancer. Am J Epidemiol. 2013;177(12):1317–25. https://doi.org/10.1093/aje/kws394. Epub 2013 May 31. Review. PubMed PMID: 23729685; PubMed Central PMCID: PMC3676148.

DeLair D, Oliva E, Köbel M, Macias A, Gilks CB, Soslow RA. Morphologic spectrum of immunohistochemically characterized clear cell carcinoma of the ovary: a study of 155 cases. Am J Surg Pathol. 2011;35(1):36–44. https://doi.org/10.1097/PAS.0b013e3181ff400e. PubMed PMID: 21164285.

Demircan M, Ceran C, Karaman A, Uguralp S, Mizrak B. Urethral polyps in children: a review of the literature and report of two cases. Int J Urol. 2006;13(6):841–3. Review. PubMed PMID: 16834678.

Dimashkieh HH, Khazai L, Novak GD, Casey MB. Pathologic quiz case: an 83-year-old woman with a uterine cervical mass. Adenoid cystic carcinoma. Arch Pathol Lab Med. 2004;128(7):817–8. PubMed PMID: 15214839.

Diniz da Costa AT, Coelho AM, Lourenço AV, Bernardino M, Ribeirinho AL, Jorge CC. Primary breast cancer of the vulva: a case report. J Low Genit Tract Dis. 2012;16(2):155–7. https://doi.org/10.1097/LGT.0b013e31823b3bc6. PubMed PMID: 22227843.

Dixit S, Singhal S, Neema J, Soornarayan R, Baboo HA. Adenoid cystic carcinoma of the cervix in a young patient. J Postgrad Med. 1994;40(2):94–5. Review. PubMed PMID: 8737564.

Egan CA, Bradley RR, Logsdon VK, Summers BK, Hunter GR, Vanderhooft SL. Vulvar melanoma in childhood. Arch Dermatol. 1997;133(3):345–8. PubMed PMID:9080895.

Farley-Loftus R, Bossenbroek NM, Rosenman K, Schaffer JV. Clear cell papulosis. Dermatol Online J. 2008;14(10):19. PubMed PMID: 19061618.

Fefferman NR, Pinkney LP, Rivera R, Popiolek D, Hummel-Levine P, Cosme J. Sclerosing stromal tumor of the ovary in a premenarchal female. Pediatr Radiol. 2003;33(1):56–8. Epub 2002 Apr 10. PubMed PMID: 12497242

Fleming NA, Hopkins L, de Nanassy J, Senterman M, Black AY. Mullerian adenosarcoma of the cervix in a 10-year-old girl: case report and review of the literature. J Pediatr Adolesc Gynecol. 2009;22(4):e45–51. https://doi.org/10.1016/j.jpag.2008.06.001. Epub 2009 Jun 2. Review. PubMed PMID: 19493521

Freidrich EG Jr. Vulvar disease. Major Probl Obstet Gynecol. 1976;9:1–217. PubMed PMID: 180357.

Frumovitz M, Bodurka DC. Neoplastic diseases of the vulva. Lichen sclerosus, intraepithelial neoplasia, Paget's disease, and carcinoma. Chapter 30. In: Comprehensive gynecology. Philadelphia: Elsevier Mosby; 2012. p. 685–702.

Gangopadhyay M, Raha K, Sinha SK, De A, Bera P, Pati S. Endodermal sinus tumor of the vagina in children: a report of two cases. Indian J Pathol Microbiol. 2009;52(3):403–4. https://doi.org/10.4103/0377-4929.55008. PubMed PMID: 19679975.

Gerbie MV. Malignant tumors of the vagina. Classification and approach to treatment. Postgrad Med. 1983;73(2):271–82. PubMed PMID: 6681673.

Gevaert O, Daemen A, De Moor B, Libbrecht L. A taxonomy of epithelial human cancer and their metastases. BMC Med Genet. 2009;2:69. https://doi.org/10.1186/1755-8794-2-69. PubMed PMID: 20017941; PubMed Central PMCID: PMC2806369.

Goel V, Verma AK, Batra V, Puri SK. 'Primary extrarenal Wilms' tumour': rare presentation of a common paediatric tumour. BMJ Case Rep. 2014;2014:pii: bcr2013202172. https://doi.org/10.1136/bcr-2013-202172. PubMed PMID: 24907205; PubMed Central PMCID: PMC4054395.

Gray SW, Skandalakis JE. Embryology for surgeons: the embryological basis for the treatment of congenital defects. Philadelphia: W.B. Saunders; 1972. p. 263–82.

Grimbizis GF, Gordts S, Di Spiezio Sardo A, Brucker S, De Angelis C, Gergolet M, Li TC, Tanos V, Brölmann H, Gianaroli L, Campo R. The ESHRE-ESGE consensus on the classification of female genital tract congenital anomalies. Gynecol Surg. 2013a;10(3):199–212. Epub 2013 Jun 13. PubMed PMID: 23894234; PubMed Central PMCID: PMC3718988.

Grimbizis GF, Gordts S, Di Spiezio Sardo A, Brucker S, De Angelis C, Gergolet M, Li TC, Tanos V, Brölmann H, Gianaroli L, Campo R. The ESHRE/ESGE consensus on the classification of female genital tract congenital anomalies. Hum Reprod. 2013b;28(8):2032–44. https://doi.org/10.1093/humrep/det098. Epub 2013 Jun 14. PubMed PMID: 23771171; PubMed Central PMCID: PMC3712660.

Grimsby GM, Ritchey ML. Pediatric urologic oncology. Pediatr Clin N Am. 2012;59(4):947–59. https://doi.org/10.1016/j.pcl.2012.05.018. Epub 2012 Jun 15. Review. PubMed PMID: 22857841.

Groff DB. Pelvic neoplasms in children. J Surg Oncol. 2001;77(1):65–71. Review. PubMed PMID: 11344486.

Guo YY, Zhang JY, Li XF, Luo HY, Chen F, Li TJ. PTCH1 gene mutations in Keratocystic odontogenic tumors: a study of 43 Chinese patients and a systematic review. PLoS One. 2013;8(10):e77305. https://doi.org/10.1371/journal.pone.0077305. eCollection 2013. Review. PubMed PMID: 24204797; PubMed Central PMCID: PMC3804548

Gupta OP, Guirguis MN, Diejomaoh F. Mesonephric adenocarcinoma of cervix: a case report. Int J Gynaecol Obstet. 1988;26(1):137–40. PubMed PMID: 2892730.

Habbous S, Pang V, Eng L, Xu W, Kurtz G, Liu FF, Mackay H, Amir E, Liu G. p53 Arg72Pro polymorphism, HPV status and initiation, progression, and development of cervical cancer: a systematic review and meta-analysis. Clin Cancer Res. 2012;18(23):6407–15. https://doi.org/10.1158/1078-0432.CCR-12-1983. Epub 2012 Oct 12. Review. PubMed PMID: 23065429.

Harel M, Ferrer FA, Shapiro LH, Makari JH. Future directions in risk stratification and therapy for advanced pediatric genitourinary rhabdomyosarcoma. Urol Oncol. 2016;34(2):103–15. https://doi.org/10.1016/j.urolonc.2015.09.013. Epub 2015 Oct 28. Review. PubMed PMID: 26519985.

Harris RM, Waring RH. Diethylstilboestrol – a long-term legacy. Maturitas. 2012;72(2):108–12. https://doi.org/10.1016/j.maturitas.2012.03.002. Epub 2012 Mar 29. Review. PubMed PMID: 22464649.

Hassanein AM, Mrstik ME, Hardt NS, Morgan LA, Wilkinson EJ. Malignant melanoma associated with lichen sclerosus in the vulva of a 10-year-old. Pediatr Dermatol. 2004;21(4):473–6. PubMed PMID: 15283794.

Haupt HM, Stern JB. Pagetoid melanocytosis. Histologic features in benign and malignant lesions. Am J Surg Pathol. 1995;19(7):792–7. PubMed PMID: 7793477.

Hazard FK, Longacre TA. Ovarian surface epithelial neoplasms in the pediatric population: incidence, histologic subtype, and natural history. Am J Surg Pathol. 2013;37(4):548–53. https://doi.org/10.1097/PAS.0b013e318273a9ff. PubMed PMID: 23388124

Heatley MK, Russell P. Florid cystic endosalpingiosis of the uterus. J Clin Pathol. 2001;54(5):399–400. PubMed PMID: 11328842; PubMed Central PMCID: PMC1731417.

Hensle TW, Reiley EA. Vaginal replacement in children and young adults. J Urol. 1998;159(3):1035–8. PubMed PMID: 9474227.

http://www.people.virginia.edu/~rjh9u/sexdev.html

Hubalek M, Smekal-Schindelwig C, Zeimet AG, Sergi C, Brezinka C, Mueller-Holzner E, Marth C. Chemotherapeutic treatment of a pregnant patient with ovarian dysgerminoma. Arch Gynecol Obstet. 2007;276(2):179–83. PubMed PMID: 17342499.

Hum Pathol. 2000;31:1420.

Hum Pathol. 2002.

Hum Pathol. 2005;36:433.

Hum Pathol. 1982;13:190.

Hum Pathol. 1986;17:488.

Ishida H, Nagai T, Sato S, Honda M, Uotani T, Samejima K, Hanaoka T, Akahori T, Takai Y, Seki H. Concomitant sentinel lymph node biopsy leading to abbreviated systematic lymphadenectomy in a patient with primary malignant melanoma of the vagina. Springerplus. 2015;4:102. https://doi.org/10.1186/s40064-014-0773-x. PubMed PMID: 25763309; PubMed Central PMCID: PMC4349903.

J Med Case Reports. 2008;2:7.

Jo W, Sudo S, Nakamura A, Endo D, Konno Y, Ishizu K, Tajima T. Development of endometrial carcinoma in a patient with leprechaunism (Donohue syndrome). Clin Pediatr Endocrinol. 2013;22(2):33–8. https://doi.org/10.1292/cpe.22.33. Epub 2013 Apr 26. PubMed PMID: 23990696; PubMed Central PMCID: PMC3756185

Kairi-Vassilatou E, Dastamani C, Vouza E, Mavrigiannaki P, Hasiakos D, Kondi-Pafiti A. Angiomyofibroblastoma of the vulva: a clinicopathological and immunohistochemical analysis of a rare benign mesenchymal tumor. Eur J Gynaecol Oncol. 2011;32(3):353–5. PubMed PMID: 21797135.

Kazakov DV, Spagnolo DV, Kacerovska D, Michal M. Lesions of anogenital mammary-like glands: an update. Adv Anat Pathol. 2011;18(1):1–28. https://doi.org/10.1097/PAP.0b013e318202eba5. Review. PubMed PMID: 21169735.

Kiratli H, Erkan Balci K, Güler G. Primary orbital endodermal sinus tumor (yolk sac tumor). J AAPOS. 2008;12(6):623–5. https://doi.org/10.1016/j.jaapos.2008.06.011. Epub 2008 Oct 4. PubMed PMID: 18835732.

Koopman P. The curious world of gonadal development in mammals. Curr Top Dev Biol. 2016;116:537–45. https://doi.org/10.1016/bs.ctdb.2015.12.009. Epub 2016 Feb 17. Review. PubMed PMID: 26970639

Kranl C, Zelger B, Kofler H, Heim K, Sepp N, Fritsch P. Vulval and vaginal adenosis. Br J Dermatol. 1998;139(1):128–31. PubMed PMID: 9764164.

Kuo TT, Chan HL, Hsueh S. Clear cell papulosis of the skin. A new entity with histogenetic implications for cutaneous Paget's disease. Am J Surg Pathol. 1987;11(11):827–34. PubMed PMID: 2823621.

Kwon YS, Kim YM, Choi GW, Kim YT, Nam JH. Pure basaloid squamous cell carcinoma of the uterine cervix: a case report. J Korean Med Sci. 2009;24(3):542–5. https://doi.org/10.3346/jkms.2009.24.3.542. Epub 2009 Jun 18. PubMed PMID: 19543425; PubMed Central PMCID: PMC2698210.

Lacy J, Capra M, Allen L. Endodermal sinus tumor of the infant vagina treated exclusively with chemotherapy. J Pediatr Hematol Oncol. 2006;28(11):768–71. PubMed PMID: 17114968.

Lanzillotto MP, Orofino A, Paradies G, Zullino F, Caroppo F, Leggio S. Metrorrhagia in a child with an endodermal sinus tumor of the vagina: a case report. Ann Ital Chir. 2013;84(6):705–9. PubMed PMID: 23103591.

Lataifeh IM, Al-Hussaini M, Uzan C, Jaradat I, Duvillard P, Morice P. Villoglandular papillary adenocarcinoma of the cervix: a series of 28 cases including two with lymph node metastasis. Int J Gynecol Cancer. 2013;23(5):900–5. https://doi.org/10.1097/IGC.0b013e31828efcaa. PubMed PMID: 23552807.

Lebeau S, Braun RP, Masouyé I, Perrinaud A, Harms M, Borradori L. Acquired melanocytic naevus in childhood vulval pemphigoid. Dermatology. 2006;213(2):159–62. PubMed PMID: 16902297.

Lebowitz RL, Olbing H, Parkkulainen KV, Smellie JM, Tamminen-Möbius TE. International system of radiographic grading of vesicoureteric reflux. International Reflux Study in Children. Pediatr Radiol. 1985;15(2):105–9. PubMed PMID: 3975102.

Lee ES, Leong AS, Kim YS, Lee JH, Kim I, Ahn GH, Kim HS, Chun YK. Calretinin, CD34, and alpha-smooth muscle actin in the identification of peritoneal invasive implants of serous borderline tumors of the ovary. Mod Pathol. 2006;19(3):364–72.

Leuschner I, Harms D, Mattke A, Koscielniak E, Treuner J. Rhabdomyosarcoma of the urinary bladder and vagina: a clinicopathologic study with emphasis on recurrent disease: a report from the Kiel pediatric tumor registry and the German CWS study. Am J Surg Pathol. 2001;25(7):856–64. PubMed PMID: 11420456.

Liu Q, Yang J, Tao T, Cao D, Shen K. The clinical features and treatment of endodermal sinus tumor of vagina. Eur J Obstet Gynecol Reprod Biol. 2012;165(1):130–1. https://doi.org/10.1016/j.ejogrb.2012.06.036. Epub 2012 Jul 20. PubMed PMID: 22819272.

Luo LM, Huang HF, Pan LY, Shen K, Wu M, Xu L. [Clinical analysis of 42 cases of primary malignant tumor in vagina]. Zhonghua Fu Chan Ke Za Zhi. 2008;43(12):923–7. Chinese. PubMed PMID: 19134332.

Mahzouni P, Pejhan S, Ashrafi M. Yolk sac tumor of the vagina. Saudi Med J. 2007;28(7):1125–6. PubMed PMID: 17603725.

Mandong BM, Ngbea JA. Childhood rhabdomyosarcoma: a review of 35 cases and literature. Niger J Med. 2011;20(4):466–9. Review. PubMed PMID: 22288325.

Mardi K, Gupta N, Bindra R. Primary yolk sac tumor of cervix and vagina in an adult female: a rare case report. Indian J Cancer. 2011;48(4):515–6. https://doi.org/10.4103/0019-509X.92249. PubMed PMID: 22293272.

Masuguchi S, Jinnin M, Fukushima S, Makino T, Sakai K, Inoue Y, Igata T, Ihn H. The expression of HER-2 in extramammary Paget's disease. Biosci Trends. 2011;5(4):151–5. https://doi.org/10.5582/bst.2011.v5.4.151. PubMed PMID: 21914949.

Matthews-Greer J, Dominguez-Malagon H, Herrera GA, Unger J, Chanona-Vilchis J, Caldito G, Turbat-Herrera EA. Human papillomavirus typing of rare cervical carcinomas. Arch Pathol Lab Med. 2004;128(5):553–6. PubMed PMID: 15086278.

McCluggage WG, White RG. Angiomyofibroblastoma of the vagina. J Clin Pathol. 2000;53(10):803. PubMed PMID: 11064681; PubMed Central PMCID: PMC1731080.

McCluggage WG, O'Rourke D, McElhenney C, Crooks M. Mullerian papilloma-like proliferation arising in cystic pelvic endosalpingiosis. Hum Pathol. 2002;33(9):944–6. PubMed PMID: 12378522.

McCluggage WG, Kennedy K, Busam KJ. An immunohistochemical study of cervical neuroendocrine carcinomas: neoplasms that are commonly TTF1 positive and which may express CK20 and P63. Am J Surg Pathol. 2010;34(4):525–32. https://doi.org/10.1097/PAS.0b013e3181d1d457. PubMed PMID: 20182342.

Meyer R. Pathology of some special ovarian tumors and their relation to sex characteristics. Am J Obstet Gynecol. 1931;22:697–713.

Micheletti L, Preti M, Radici G, Boveri S, Di Pumpo O, Privitera SS, Ghiringhello B, Benedetto C. Vulvar lichen sclerosus and neoplastic transformation: a retrospective study of 976 cases. J Low Genit Tract Dis. 2016;20(2):180–3. https://doi.org/10.1097/LGT.0000000000000186. PubMed PMID: 26882123.

Modlin IM, Shapiro MD, Kidd M. An analysis of rare carcinoid tumors: clarifying these clinical conundrums. World J Surg. 2005;29(1):92–101. Review. PubMed PMID: 15599742.

Moritani S, Ichihara S, Kushima R, Sugiura F, Mushika M, Silverberg SG. Combined signet ring cell and glassy cell carcinoma of the uterine cervix arising in a young Japanese woman: a case report with immunohistochemical and histochemical analyses. Pathol Int. 2004;54(10):787–92. PubMed PMID: 15482569.

Nasioudis D, Alevizakos M, Chapman-Davis E, Witkin SS, Holcomb K. Rhabdomyosarcoma of the lower female genital tract: an analysis of 144 cases. Arch Gynecol Obstet. 2017;296(2):327–34. https://doi.org/10.1007/s00404-017-4438-1. Epub 2017 Jun 20. PubMed PMID: 28634755.

Nielsen GP, Rosenberg AE, Young RH, Dickersin GR, Clement PB, Scully RE. Angiomyofibroblastoma of the vulva and vagina. Mod Pathol. 1996;9(3):284–91. Review. PubMed PMID: 8685229.

Nucci MR, Young RH, Fletcher CD. Cellular pseudosarcomatous fibroepithelial stromal polyps of the lower female genital tract: an underrecognized lesion often misdiagnosed as sarcoma. Am J Surg Pathol. 2000;24(2):231–40. PubMed PMID: 10680891.

Offman SL, Longacre TA. Clear cell carcinoma of the female genital tract (not everything is as clear as it seems). Adv Anat Pathol. 2012;19(5):296–312. https://doi.org/10.1097/PAP.0b013e31826663b1. Review. PubMed PMID: 22885379.

Oppelt P, Renner SP, Brucker S, Strissel PL, Strick R, Oppelt PG, Doerr HG, Schott GE, Hucke J, Wallwiener D, Beckmann MW. The VCUAM (Vagina Cervix Uterus Adnex-associated Malformation) classification: a new classification for genital malformations. Fertil Steril. 2005;84(5):1493–7. PubMed PMID: 16275249.

Park KJ, Kiyokawa T, Soslow RA, Lamb CA, Oliva E, Zivanovic O, Juretzka MM, Pirog EC. Unusual endocervical adenocarcinomas: an immunohistochemical analysis with molecular detection of human papillomavirus. Am J Surg Pathol. 2011;35(5):633–46. https://doi.org/10.1097/PAS.0b013e31821534b9. PubMed PMID: 21490443.

Pavlakis K, Messini I, Yiannou P, Chrissanthakis D, Papaspyrou I, Panoskaltsis T, Stathopoulos EN. A pre-tailored panel of antibodies in the study of cervical mesonephric remnants. Gynecol Oncol. 2010;116(3):468–72. https://doi.org/10.1016/j.ygyno.2009.10.055. Epub 2009 Nov 14. PubMed PMID: 19913894.

Pediatr Blood Cancer. 2005;44:167.

Pediatr Blood Cancer. 2006;46:459.

Pommert L, Bradley W. Pediatric gynecologic cancers. Curr Oncol Rep. 2017;19(7):44. https://doi.org/10.1007/s11912-017-0604-7. Review. PubMed PMID: 28501984.

Rajshekar SK, Guruprasad B, Shakunthala P, Rathod P, Devi U, Bafna U. Malignant mixed Mullerian tumour of the uterus. Ecancermedicalscience. 2013;7:302. https://doi.org/10.3332/ecancer.2013.302. PMID: 23589731; PMCID: PMC3622448

Romao RL, Lorenzo AJ. Vaginectomy and buccal mucosa vaginoplasty as local therapy for pediatric vaginal rhabdomyosarcoma. Urology. 2017;102:222–4. https://doi.org/10.1016/j.urology.2017.02.010. Epub 2017 Feb 21. PubMed PMID: 28232175.

Scheier M, Ramoni A, Alge A, Brezinka C, Reiter G, Sergi C, Hager J, Marth C. Congenital fibrosarcoma as cause for fetal anemia: prenatal diagnosis and in utero treatment. Fetal Diagn Ther. 2008;24(4):434–6. https://doi.org/10.1159/000173370. Epub 2008 Nov 19. PubMed PMID: 19018145.

Schultz KAP, Rednam SP, Kamihara J, Doros L, Achatz MI, Wasserman JD, Diller LR, Brugières L, Druker H, Schneider KA, McGee RB, Foulkes WD. PTEN, DICER1, FH, and their associated tumor susceptibility syndromes: clinical features, genetics, and surveillance recommendations in childhood. Clin Cancer Res. 2017;23(12):e76–82. https://doi.org/10.1158/1078-0432.CCR-17-0629. Review. PubMed PMID: 28620008

Semin Pediatr Surg. 2005;14:100.

Sergi C, Magener A, Ehemann V, De Villiers EM, Sinn HP. Stage IIa cervix carcinoma with metastasis to the heart: report of a case with immunohistochemistry, flow cytometry, and virology findings. Gynecol Oncol. 2000;76(1):133–8. PubMed PMID: 10620458

Seth A, Agarwal A. Adenoid cystic carcinoma of uterine cervix in a young patient. Indian J Pathol Microbiol. 2009;52(4):543–5. https://doi.org/10.4103/0377-4929.56158. PubMed PMID: 19805968.

Shalon L, Markowitz J, Bialer M, Kahn E, Weinblatt M, Giardiello FM, Luce MC, Daum F. Ovarian neoplasm and endometrioid carcinoma in a patient with Turcot syndrome. J Pediatr Gastroenterol Nutr. 1997;25(2):224–7. PubMed PMID: 9252914

Sheibani K, Battifora H, Burke JS, Rappaport H. Leu-M1 antigen in human neoplasms. an immunohistologic study of 400 cases. Am J Surg Pathol. 1986;10(4):227–36. PubMed PMID: 3085523.

Shinkoda Y, Tanaka S, Ijichi O, et al. Successful treatment of an endodermal sinus tumor of the vagina by chemotherapy alone: a rare case of an infant diagnosed by pathological examination of discharged tumor fragment. Pediatr Hematol Oncol. 2006a;23:563–9.

Shinkoda Y, Tanaka S, Ijichi O, Yoshikawa H, Nonaka Y, Tanabe T, Nishikawa T, Ishikawa S, Okamoto Y, Kaji T, Tahara H, Takamatsu H, Nagata K, Kawano Y. Successful treatment of an endodermal sinus tumor of the vagina by chemotherapy alone: a rare case of an infant diagnosed by pathological examination of discharged tumor fragment. Pediatr Hematol Oncol. 2006b;23(7):563–9. PubMed PMID: 16928651.

Singh NP, Singh UP, Nagarkatti PS, Nagarkatti M. Prenatal exposure of mice to diethylstilbestrol disrupts T-cell differentiation by regulating Fas/Fas ligand expression through estrogen receptor element and nuclear factor-κB motifs. J Pharmacol Exp Ther. 2012;343(2):351–61. https://doi.org/10.1124/jpet.112.196121. Epub 2012 Aug 10. PubMed PMID: 22888145; PubMed Central PMCID: PMC3477208.

Soga J, Osaka M, Yakuwa Y. Gut-endocrinomas (carcinoids and related endocrine variants) of the uterine cervix: an analysis of 205 reported cases. J Exp Clin Cancer Res. 2001;20(3):327–34. PubMed PMID: 11718210.

Spence JM, Wright L, Clark VL. Laboratory maintenance of *Neisseria gonorrhoeae*. Curr Protoc Microbiol. 2008; Chapter 4:Unit 4A.1. https://doi.org/10.1002/9780471729259.mc04a01s8. PubMed PMID: 18770539.

Srikantia N, B R, A G R, Kalyan SN. Endometrioid endometrial adenocarcinoma in a premenopausal woman with multiple organ metastases. Indian J Med Paediatr Oncol. 2009;30(2):80–3. doi: 10.4103/0971-5851.60053. PMID: 20596308; PMCID: PMC2885881

Stern JB, Haupt HM. Pagetoid melanocytosis: tease or tocsin? Semin Diagn Pathol. 1998;15(3):225–9. Review. PubMed PMID: 9711673.

Stewart C. Angiomyofibroblastoma of the vagina. Pathology. 2009;41(2):199–200; author reply 200-1. PubMed PMID: 19152197.

Suarez-Quian CA, Dym M. Characterization of Sertoli cell perinuclear filaments. Microsc Res Tech. 1992;20(3):219–31. PubMed PMID: 1543876

Takeuchi K, Oomori S, Oda N, Maeda K, Kaji Y, Maruo T. Coexistence of Mayer-Rokitansky-Küstner-Hauser syndrome and yolk sac tumor of the ovary in a prepubertal girl. Acta Obstet Gynecol Scand. 2006;85(2):245–7. PubMed PMID: 16532924.

Tangour-Bouaicha M, Bel Haj Salah M, Ben Brahim E, Ben Othmène M, Douggaz A, Sassi S, Chatti-Dey S. [Primary peritoneal yolk sac tumour. A case report]. Ann Pathol. 2010;30(5):378–81. https://doi.org/10.1016/j.annpat.2010.07.004. Epub 2010 Oct 16. French. PubMed PMID: 21055525.

Tao T, Yang J, Cao D, Guo L, Chen J, Lang J, Shen K. Conservative treatment and long-term follow up of endodermal sinus tumor of the vagina. Gynecol Oncol. 2012;125(2):358–61. https://doi.org/10.1016/j.ygyno.2011.12.430. Epub 2011 Dec 14. PubMed PMID: 22178761.

Tekgül S, Riedmiller H, Hoebeke P, Kočvara R, Nijman RJ, Radmayr C, Stein R, Dogan HS. European Association of Urology. EAU guidelines on vesicoureteral reflux in children. Eur Urol. 2012;62(3):534–42. https://doi.org/10.1016/j.eururo.2012.05.059. Epub 2012 Jun 5. Review. PubMed PMID: 22698573.

Terenziani M, Spreafico F, Collini P, Meazza C, Massimino M, Piva L. Endodermal sinus tumor of the vagina. Pediatr Blood Cancer. 2007;48(5):577–8. PubMed PMID: 16200632.

The American Fertility Society classifications of adnexal adhesions, distal tubal occlusion, tubal occlusion secondary to tubal ligation, tubal pregnancies, müllerian anomalies and intrauterine adhesions. Fertil Steril. 1988;49(6):944–55. PubMed PMID: 3371491.

Tinari A, Pace S, Fambrini M, Eleuteri Serpieri D, Frega A. Vulvar Paget's disease: review of the literature, considerations about histogenetic hypothesis and surgical approaches. Eur J Gynaecol Oncol. 2002;23(6):551–2. Review. PubMed PMID: 12556103.

Tseng FW, Kuo TT, Lu PH, Chan HL, Chan MJ, Hui RC. Long-term follow-up study of clear cell papulosis. J Am Acad Dermatol. 2010;63(2):266–73. https://doi.org/10.1016/j.jaad.2009.08.056. Epub 2010 Jun 3. PubMed PMID: 20605258.

Udager AM, Frisch NK, Hong LJ, Stasenko M, Johnston CM, Liu JR, Chan MP, Harms PW, Fullen DR, Orsini A, Thomas DG, Lowe L, Patel RM. Gynecologic melanomas: a clinicopathologic and molecular analysis. Gynecol Oncol. 2017;147(2):351–7. https://doi.org/10.1016/j.ygyno.2017.08.023. Epub 2017 Aug 24. PubMed PMID: 28844540.

Uharcek P. Prognostic factors in endometrial carcinoma. J Obstet Gynaecol Res. 2008;34(5):776–83. Review. PubMed PMID: 18958927

Underwood PB Jr, Hester LL Jr. Diagnosis and treatment of premalignant lesions of the vulva: a review. Am J Obstet Gynecol. 1971;110(6):849–57. PubMed PMID: 4327296.

Vang R, Medeiros LJ, Silva EG, Gershenson DM, Deavers M. Non-Hodgkin's lymphoma involving the vagina: a clinicopathologic analysis of 14 patients. Am J Surg Pathol. 2000;24(5):719–25. PubMed PMID: 10800991.

Wani NA, Robbani I, Andrabi AH, Iqbal A, Qayum A. Vaginal yolk sac tumor causing infantile hydrometra: use of multidetector-row computed tomography. J Pediatr Adolesc Gynecol. 2010;23(3):e115–8. https://doi.org/10.1016/j.jpag.2009.10.005. Epub 2009 Dec 5. PubMed PMID: 19963410.

WC MB Jr, Brainard J, Sawady J, Rose PG. Yolk sac tumor of the ovary associated with endometrioid carcinoma with metastasis to the vagina: a case report. Gynecol Oncol. 2007;105(1):244–7. Epub 2007 Feb 21. PubMed PMID: 17316775.

Wee WW, Chia YN, Yam PK. Diagnosis and treatment of vaginal intraepithelial neoplasia. Int J Gynaecol Obstet. 2012;117(1):15–7. https://doi.org/10.1016/j.ijgo.2011.10.033. Epub 2012 Jan 10. PubMed PMID: 22239755.

Wei Q, Zhu Y. Collision tumor composed of mammary-type myofibroblastoma and eccrine adenocarcinoma of the vulva. Pathol Int. 2011;61(3):138–42. https://doi.org/10.1111/j.1440-1827.2010.02642.x. Epub 2011 Jan 26. PubMed PMID: 21355955.

Wepfer JF, Boex RM. Mesonephric duct remnants (Gartner's duct). AJR Am J Roentgenol. 1978;131(3):499–500. PubMed PMID: 98997.

Werner D, Wilkinson EJ, Ripley D, Yachnis A. Primary adenocarcinoma of the vagina with mucinous-enteric differentiation: a report of two cases with associated vaginal adenosis without history of diethylstilbestrol exposure. J Low Genit Tract Dis. 2004;8(1):38–42. PubMed PMID: 15874835.

Wolff JP, Douyon E. Primary cancer of the Vagina. Study of 85 cases examined at the institute Gustave-Roussy from 1949 to 1961. Gynecol Obstet (Paris). 1964;63:565–84. French. PubMed PMID: 14265064.

Yemelyanova A, Vang R, Kshirsagar M, Lu D, Marks MA, Shih IeM, Kurman RJ. Immunohistochemical staining patterns of p53 can serve as a surrogate marker for TP53 mutations in ovarian carcinoma: an immunohistochemical and nucleotide sequencing analysis. Mod Pathol. 2011;24(9):1248–53. https://doi.org/10.1038/modpathol.2011.85. Epub 2011 May 6. PubMed PMID: 21552211.

You W, Dainty LA, Rose GS, Krivak T, McHale MT, Olsen CH, Elkas JC. Gynecologic malignancies in women aged less than 25 years. Obstet Gynecol. 2005;105(6):1405–9. PubMed PMID: 15932836.

Young RH. A brief history of the pathology of the gonads. Mod Pathol. 2005;18(Suppl 2):S3–S17. PubMed PMID: 15529187

Young RH, Clement PB. Endocervical adenocarcinoma and its variants: their morphology and differential diagnosis. Histopathology. 2002;41(3):185–207. Review. PubMed PMID: 12207781.

Zirker TA, Silva EG, Morris M, Ordóñez NG. Immunohistochemical differentiation of clear-cell carcinoma of the female genital tract and endodermal sinus tumor with the use of alpha-fetoprotein and Leu-M1. Am J Clin Pathol. 1989;91(5):511–4. PubMed PMID: 2470246.

Breast

9

Contents

9.1	**Development and Genetics**	834
9.2	**Congenital Anomalies, Inflammatory, and Related Disorders**	836
9.2.1	Amastia, Atelia, Synmastia, Polymastia, and Politelia	836
9.2.2	Asymmetry, Hypotrophy, and Hypertrophy	836
9.2.3	Dysmaturity and Precocious Thelarche	838
9.2.4	Acute Mastitis, Abscess, and Phlegmon	838
9.2.5	Duct Ectasia, Periductal Mastitis, and Granulomatous Mastitis	839
9.2.6	Necrosis, Calcifications, and Mondor Disease	839
9.3	**Pathology of the Female and Young Adult**	840
9.3.1	Fibrocystic Disease	840
9.3.2	Soft Tissue Tumors and Hematological Malignancies	840
9.3.3	Fibroadenoma	840
9.3.4	Adenoma	841
9.3.5	Genetic Background of Breast Cancer	844
9.3.6	In Situ and Invasive Ductal Breast Carcinoma	847
9.3.7	In Situ and Invasive Lobular Breast Carcinoma	849
9.3.8	WHO Variants of the Infiltrating Ductal Carcinoma	849
9.3.9	Sweat Gland-Type Tumors and Myoepithelial Tumors	850
9.3.10	Phyllodes Tumor	851
9.4	**Cancer Mimickers**	852
9.4.1	Hyperplasia, Ductal and Lobular	852
9.4.2	Adenosis	855
9.5	**Male Breast Disease**	855
9.5.1	Gynecomastia	855
9.5.2	Breast Cancer of the Male	855
	Multiple Choice Questions and Answers	856
	References and Recommended Readings	857

© Springer-Verlag GmbH Germany, part of Springer Nature 2020
C. M. Sergi, *Pathology of Childhood and Adolescence*,
https://doi.org/10.1007/978-3-662-59169-7_9

9.1 Development and Genetics

The mammary gland is constituted by a large duct system and terminal duct lobular units (TDLUs) (Fig. 9.1). The TDLU is defined as a mammary gland unit, which is surrounded by specialized, hormone-responsive, and elastic fiber-free myxoid stroma and is the place of origin for fibrocystic disease, most of the malignancies of the breast, and cancer mimickers. The lactiferous ducts are collecting ducts with lactiferous sinuses (fusiform dilatations beneath the nipple) emptying into the nipple which also contain sebaceous glands and smooth muscle bundles. There is one lactiferous duct per lobe for approximately 20 lobes and lactiferous ducts per breast. The TDLU *Intralobular Stroma* (inside of a lobe) is defined as breast-specific hormonally responsive fibroblast-like cells and lymphocytes. The TDLU *Interlobular Stroma* (between lobes) shows dense fibrosis and fat. Glandular structures include:

- One-layered luminal epithelium (large open chromatin, small nucleoli, and abundant cytoplasm)
- One-layered abluminal myoepithelial cell layer (dark compact nuclei and scant cytoplasm)
- Epithelium: (+) keratin, EMA, fat globule membrane Ag, α-lactalbumin
- Myoepithelium: (+) *S*MA and myosin, *C*D10, HMW*C*K (CK5/6), *cal*ponin, *p*63

Pregnancy changes include marked hyperplasia.

Anatomically, the regional LNs are the (1) axillary (interpectoral and LN along the axillary vein and its tributaries: Level I (low axilla), Level II (midaxilla), and Level III (apical axilla)), the (2) infraclavicular, (3) the supraclavicular, and the (4) internal mammary artery LNs.

Bone morphogenetic proteins (BMPs) are members of the transforming growth factor β (TGF β) superfamily. BMPs have been intensively investigated in our cooperation with the University of Minnesota, Minneapolis, USA. BMPs are morphogens that are crucial in the embryonic development. BMPs master the regulation of tissue homeostasis in both physiological and pathological states. Some essential aspects involve cell proliferation, migration, differentiation, apoptosis, and epithelial-stromal interaction. Individually, BMPs play crucial roles in the development of the mammary gland as studied in our animal model. The human mammary gland is a very intriguing organ because of its dynamic structure that may be examined in detail in mice. The embryonic development put the basis for the rudimentary ductal tree system that during puberty, pregnancy, and lactation undergoes morphological changes. Intriguingly, in tissues where BMPs are integral for proper development, BMPs and BMP pathway components are often dysregulated in cancers of that tissue. In the mouse, the mammary gland develops around embryonic day 10.5 (E10.5). At this time, epithelial thickenings form placodes, which further thicken and invaginate as they respond to inductive morphogenetic signals from the underlying mesenchyme at around E13.5. The primary bud epithelium starts proliferating at about E16.5. The bud elongates into the developing fat pad forming a fundamental embryological tree-like structure. At this stage, the development is arrested until puberty. At puberty, terminal end buds (TEBs) form at the leading edge of the duct and are the site of active proliferation and programmed cell death. The TEB consists of bipotent cap cells that produce both keratin 18 (K18) and K14. Both the keratin and cytokeratin terms have been used in this book, because of the different familiarity and/or preference identified in pathologists and molecular biologists. The first is in the luminal, while the second is positive in the myoepithelial cells, which stay in contact with the extracellular matrix (ECM). In the TEB, proliferation creates excess body cells, which generate the mass of the ducts. A mature duct consists of a layer of 1–2 cell thick of epithelial cells expressing K18 surrounded by a layer of myoepithelial cells expressing K14. There is a progressive elongation up to the edge of the fat pad when TEB is reabsorbed and elongation ceases. Together with the elongation, the ductal tree elaborates further developing secondary and tertiary branching. The integration and regulation of morphogens, cytokines, growth factors, and hormones are crucial for the full development of the mature mammary gland.

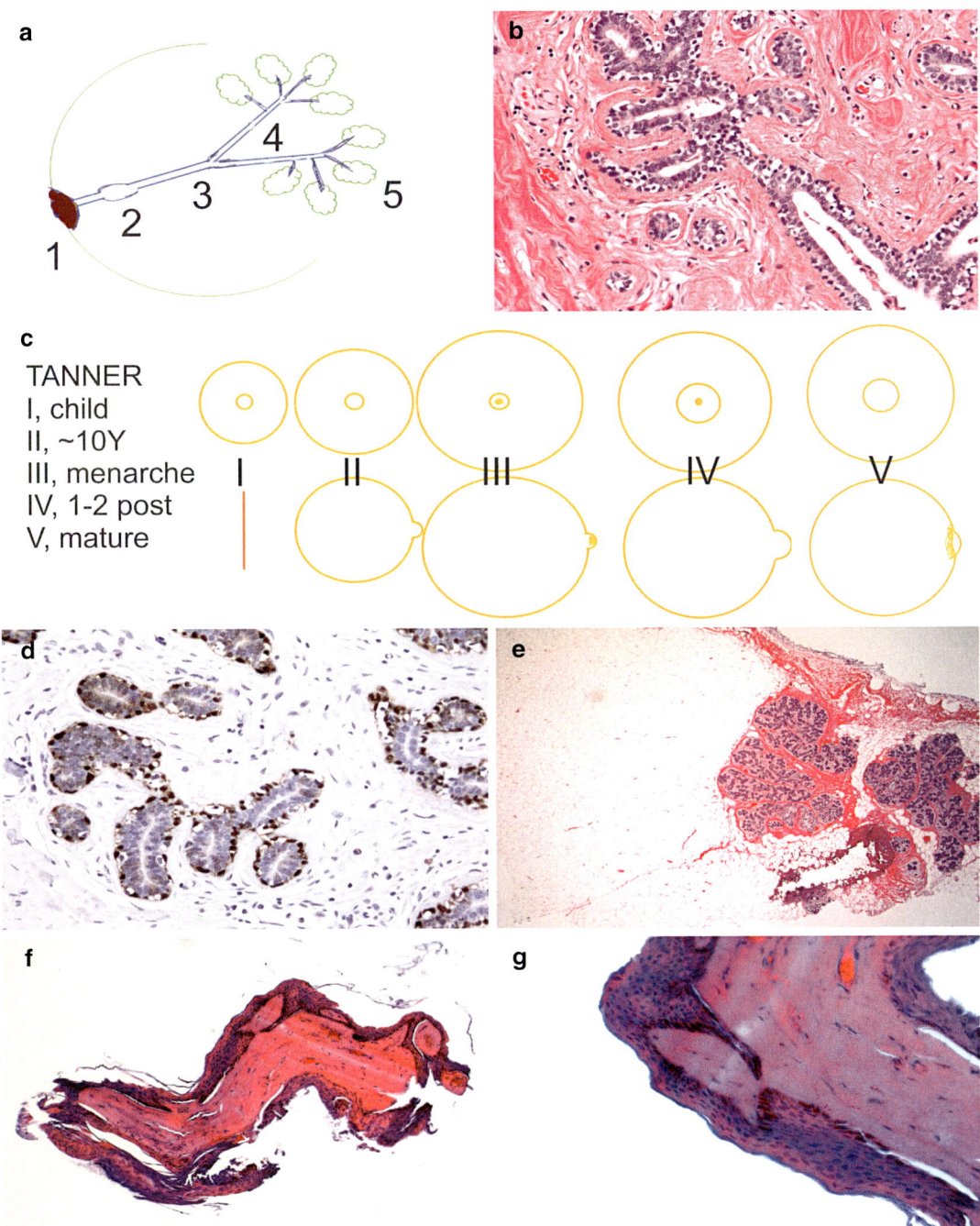

Fig. 9.1 In (**a**) it is shown the anatomy of the mammary gland with (1) the areola, (2) the galactophorous sinus, (3) the excretory duct, (4) a branch, and (5) the loule. In (**b**) the histology of a normal breast is shown with particular prominent myoepithelial cells. In (**c**) is shown the Tanner grading for evaluating the breast development in childhood and adolescence. In (**d**) are highlighted the myoepithelial cells using an antibody against S-100 by immunohistochemistry (anti-S100, 200×), while (**e**) represents a section of a mammary gland reduction for excessive breast fat (H&E stain, 12.5×). The microphotographs (**f**, **g**) show an accessory breast (H&E, 100× and 400×, respectively)

The BMP-specific intracellular signal transducers of BMPs are SMAD1/5/8 proteins, which become phosphorylated upon binding of BMPs to their receptors (BMPR1A, BMPR1B, and BMPR2). Phosphorylated pSMAD1/5/8 binds SMAD4 and translocates to the nucleus where the compound acts as a transcription factor. In the extracellular space, the availability of BMPs to link to their receptors is regulated by the secreted proteins chordin (CHRD), chordin-like1 (CHRDL1), chordin-like2 (CHRDL2), noggin (NOG), and TWSG1 among others. CHRD and NOG are antagonists of BMP action, but TWSG1 has a dual role, acting as either a BMP antagonist as seen in *Danio rerio*, *Drosophila melanogaster*, and *Xenopus laevis*. Mice deficient for *Twsg1* have some developmental defects, including craniofacial malformations and abnormalities of spinal architecture, the kidneys, thymus, and other organs. We found that, at puberty, *Twsg1* is expressed in the myoepithelium and a subset of body cells of the terminal end buds. In the matured duct, the *Twsg1* expression is mainly restricted to the myoepithelial layer. The global deletion of Twsg1 leads to a delay in ductal elongation, reduced secondary branching, enlarged terminal end buds, and occluded lumens. This finding seems to be associated with an increase in the number of luminal epithelial cells and a decrease in programmed cell death. In the mammary gland, pSMAD1/5/8 level and the expression of *BMP* genes are decreased. This data is consistent with a reduction of BMP signaling. Regulation of BMP signaling by TWSG1 is required for normal ductal elongation, branching of the ductal tree, lumen formation, and myoepithelial compartmentalization in the final postnatal mammary gland.

9.2 Congenital Anomalies, Inflammatory, and Related Disorders

9.2.1 Amastia, Atelia, Synmastia, Polymastia, and Politelia

Amastia is probably an infrequent congenital anomaly characterized by the mono- or bilateral lack of the mammary gland. In pediatric age, amastia is a clinical finding of the Poland syndrome, which is a rare birth defect defined by poor development or the absence of the pectoralis on one side of the body. There is also webbing of the fingers (cutaneous syndactyly) of the ipsilateral hand. In the *Poland syndrome*, there is a monolateral mammary gland aplasia associated with hypotrophy of the thoracic cage fascicles of the musculus pectoralis major with or without aplasia of the nervus radialis, syndactyly, and deformity of the anterior portions of the thoracic cage. Both the isolated form and the Poland syndrome require plastic surgery for correction and for the development of substitution of the mammary gland, which may be highly relevant for the psychology of the peripubertal child or female adolescent. In all cases, no plastic surgery is recommended before complete development of the contralateral breast to avoid unesthetical asymmetry. *Atelia* is an extremely rare congenital anomaly characterized by the mono- or bilateral lack of the nipple. *Synmastia* is the coming together of the breast tissue of both breasts across the midline anterior to the sternum. Synmastia is usually surgically corrected by a plastic surgeon. Although rare cases of congenital synmastia exist, most of the circumstances of this condition are iatrogenic, which occurs following breast augmentation. *Polymastia* is a more unusual event and constituted by the supernumerary development of mammary gland tissue, which is still localized along the lactiferous line (Fig. 9.1 f, g). Plastic surgeons commonly see it for esthetic reasons. *Politelia* is an increased number of nipples.

9.2.2 Asymmetry, Hypotrophy, and Hypertrophy

Asymmetry refers to an asymmetry of volume and shape of the breast, which is usually physiological during the pubertal maturation and in the first years after the menarche. If an asymmetry persists in late adolescence or youth, it often required counseling with a plastic surgeon. The acceptance of correct breast shape and volume is part of the complex psychological image, which is found in teenagers. The *pectoralis* muscle is a target in the morphologic anomaly of the breast. *Breast hypotrophy*, either mono- or bilateral, following a com-

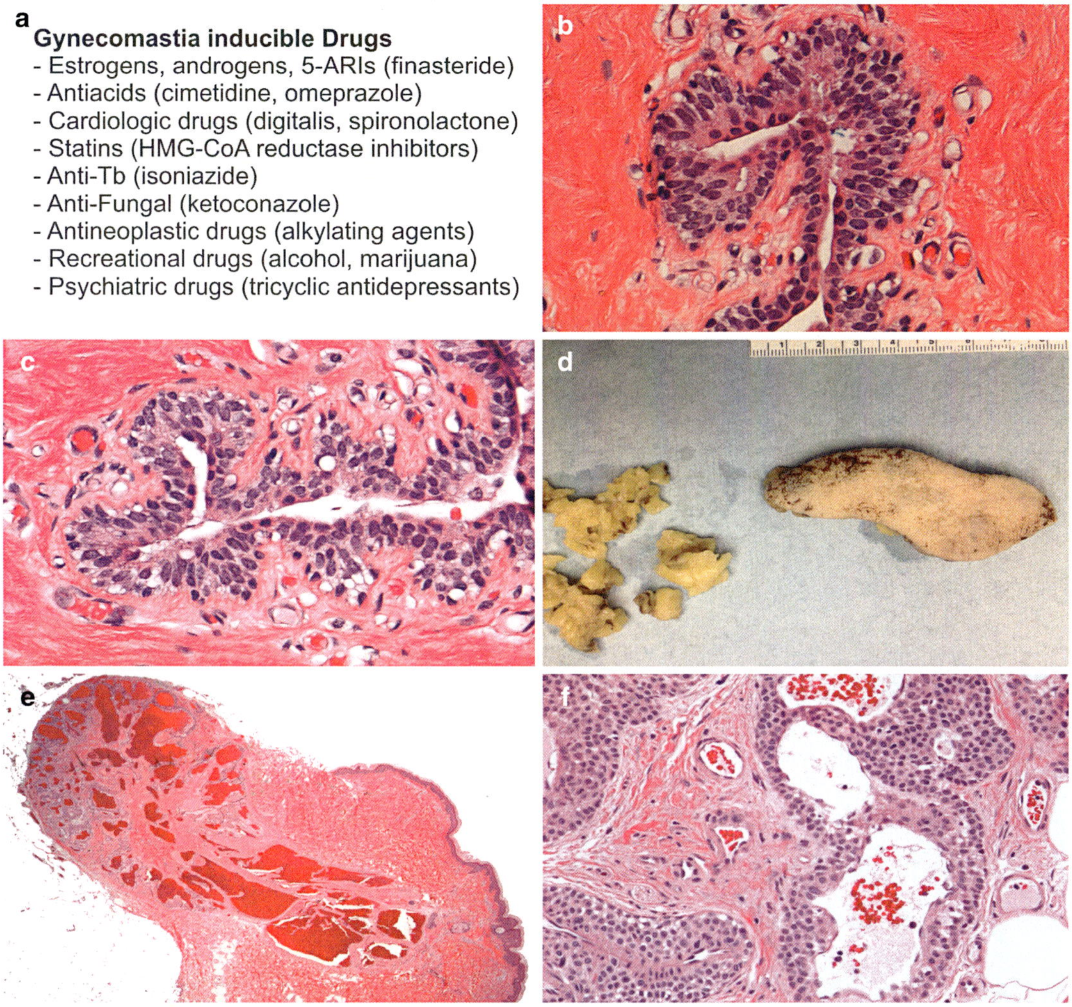

9.2 Congenital Anomalies, Inflammatory, and Related Disorders

a
Gynecomastia inducible Drugs
- Estrogens, androgens, 5-ARIs (finasteride)
- Antiacids (cimetidine, omeprazole)
- Cardiologic drugs (digitalis, spironolactone)
- Statins (HMG-CoA reductase inhibitors)
- Anti-Tb (isoniazide)
- Anti-Fungal (ketoconazole)
- Antineoplastic drugs (alkylating agents)
- Recreational drugs (alcohol, marijuana)
- Psychiatric drugs (tricyclic antidepressants)

Fig. 9.2 In (**a**) is shown a list of drugs inducing gynecomastia. In (**b**, **c**) the histological features of gynecomastia are depicted. They include epithelial hyperplasia and stromal fibrous tissue. Gynecomastia has three phases including florid gynecomastia (ductal epithelial hyperplasia with flat or papillary patterns and increased periductal stromal cellularity, prominent vascularity, and edema), intermediate gynecomastia (florid component and increased fibrosis), and fibrous gynecomastia (epithelial proliferation – less than before – and more collagenous stroma with less edema and vascularity). Florid is up to 1 year of onset, while the fibrous is more than 1 year from the onset. The intermediate is in between and is a transitional phase during few months up to 6 months ((**b**) H&E stain, 400×; (**c**) H&E stain, 400×). In (**d**) is the gross photograph of glomus arterio-vascular malformation of the breast of a pediatric patient with the histology depicted in (**e**, **f**) showing the growth of a solid tumor with thick-walled vessels lined by small tumor cells with eosinophilic cytoplasm ((**e**) H&E stain, 12.5×; (**f**) H&E stain, 200×)

pleted pubertal development, can be seen in a familiar setting and may not require surgery without a documented counseling. In anorexia, there may be a reduction of the fatty tissue. Other causes of hypomastia include the history of radiotherapy for thoracic tumors, transthoracic surgery, or regional burns occurred in childhood. *Breast hypertrophy* is usually associated with the excessive development of the glandular component, and breast reduction surgery is considered. This surgery is not only probably crucial for the psychological evaluation of the child as self-identification but may be indicated because of the abnormalities that may occur in cases the static and dynamic of the body are jeopardized. In children and adolescents as well as youth with overweight, the cause underlying the breast enlargement is the increase of the adipose tissue. Ultrasound may readily distinguish both cases. A breast with reduced thoracic implant site and hypertrophic areola and nipple characterizes *areola hypertrophy*. Plastic surgery may be indicated. Conversely, *nipple hypertrophy* is often seen in normal or almost normal breast and rarely requires the intervention of a plastic surgeon. Axillary tail may be considered a paraphysiological phenomenon and should not need plastic surgery unless it may affect the psychological evolution of the female adolescent or young adult. The "axillary tail of Spence" is a particular accessory axillary breast tissue. Usually, ectopia is present within axillary nodes and along "milk line" (axilla to inguinal region), which is also seen in other mammals. Nipple introflession is rarely an esthetical problem, although it may constitute a problem during breastfeeding. Virginal hypertrophy is very similar to gynecomastia of the male breast (Fig. 9.2), which presents with epithelial hyperplasia and stromal edema or fibrous tissue with atrophic glands according to the phase of development (*vide infra*). Virginal hypertrophy may be uni- or bilateral and is characterized by a benign proliferation of ducts and surrounding stroma with little or no tubular involvement.

9.2.3 Dysmaturity and Precocious Thelarche

Dysmaturity of the mammary gland is constituted by the unilateral mammary button, which may require the attention of the physicians to reassure the parents or a more complex endocrinological evaluation in case the problem persists for a long time after the puberty. Precocious thelarche is part of the precocious puberty. Thelarche is the onset of secondary (postnatal) breast development and typically occurs at the beginning of puberty in girls (etymologically the term comes from Greek θηλή, "nipple" and ἀρχή, "beginning, onset"). Thelarche is part of a group of changes comprising other pubertal signs (adrenarche, gonadarche, menarche, pubarche). In precocious thelarche, the isolated unilateral or bilateral breast development is not accompanied by any other estrogen secretion signs and is usually a benign clinical condition. Precocious thelarche occurs from birth to 3 years of age with spontaneous regression within months or persistence to puberty. Importantly, in isolated precocious thelarche, bone age and growth velocity remain adequate for chronological age, and this aspect needs to be investigated with the pediatrician and pediatric radiologist. The underlying cause of precocious thelarche is not apparent, but baseline serum gonadotropin and estradiol levels are within normal prepubertal range. However, Follicle-stimulating hormone (FSH) levels and inhibin B can be increased, and pelvic ultrasound may be helpful in distinguishing the isolated form of early-stage precocious puberty. FSH is a gonadotropin, which is synthesized and secreted by the gonadotropic cells of the anterior pituitary gland. FSH regulates the development, growth, maturation, and reproductive processes of the human body. Inhibin B is a glycoprotein produced by granulosa cells of the ovary. Inhibin B suppresses synthesis and secretion of FSH. The treatment of the separate form is conservative, but biannual evaluation is mandatory to detect a progression into complete adolescence. Delay of the mammary development is correlated to conditions associated with gonadal dysgenesis, hypogonadotropic hypogonadism, congenital adrenal hyperplasia (3-β-3-OH-steroid dehydrogenase and 17-α-hydroxylase), or constitutional pubertal delay or secondary to other pathologies. Another cause of obstruction in the development of the breast gland is an androgen-secreting tumor.

9.2.4 Acute Mastitis, Abscess, and Phlegmon

Acute mastitis refers to acute inflammation of the mammary gland, which can be localized and circumscribed as in abscess, which is often seen during breastfeeding or poor hygiene conditions. Acute mastitis can diffuse into soft tissue involving several layers of the mammary gland. The involvement of soft tissue is labeled as mammary phlegmon. The signs of acute inflammation are present and evident to the clinical observation and include *rubor*, *calor*, *tumor*, and *dolor* as well as difficulty in secreting or pumping out milk (*functio laesa*). It is almost never brought to the attention of pathologists, except for cases with sepsis and death to septic shock, mainly if the patient is immunodeficient. Cases in maternal death inquiries can be consulted for clinical government purposes (appraisal in the setting of quality assurance and control) and training purposes. A sclerosing lobulitis with dense lymphocytic infiltrate around lobules, and blood vessels with or without atypical stromal cell involvement, is referred to as lymphocytic mastopathy. Clinical settings with predisposing conditions that should keep in mind include DM type I and autoimmune thyroid disease (Hashimoto thyroiditis).

9.2.5 Duct Ectasia, Periductal Mastitis, and Granulomatous Mastitis

Both duct ectasia and periductal mastitis are considered parts of a similar process, which can occur in young female adults. Duct ectasia is not associated with cigarette (tobacco) smoking, while periductal mastitis is often associated with cigarette (tobacco) smoking. Proposed pathogenesis supports the sequence cigarette smoking ⇒ vitamin A deficiency ⇒ keratinizing squamous metaplasia of the nipple ducts with the recruitment of inflammatory cells ⇒ mastitis. Microscopically, there is massive duct dilation with fibrosis of the wall accompanied by squamous metaplasia of the nipple ducts (periductal mastitis only). Significantly, the escape of some secretion material elicits frequently a florid inflammatory response of the surrounding tissue with subsequent calcification, which may become quite common according to diet, climate, and additional underlying pathologic conditions. Granulomatous mastitis is a granulomatous inflammation of the mammary gland with the presence of poorly to well-formed granulomas as seen in tuberculosis infection, foreign body granulomas (e.g., silicone), and sarcoidosis. The review of clinical history is paramount before signing out the pathologic report.

9.2.6 Necrosis, Calcifications, and Mondor Disease

Mammary gland tissue-associated fat necrosis is of liquefactive type and often follows a trauma with primary involvement of the superficial subcutaneous tissue. It may recall the fat necrosis that occurs in any depository of fat tissue in the body, such as retroperitoneum and omentum. In some cases, it may be observed in ruptured ectatic cysts or fibrocystic disease (FCD). Clinically, fat necrosis can be very difficult to diagnose and should be taken into the differential diagnosis of most breast lesions of any female patient of any age. Manifestations include clinically palpable mass, skin retraction, and skin thickening. It may also present with an abnormal mammography finding. Breast infarction may also be the underlying disorder that complicates with necrosis. Calcification refers to the deposition of calcium ions (Ca^{2+}) or more specifically salts of calcium. Calcification in breast tissue may be visible on mammographic imaging. Distinguishing between three types of calcium deposition is important in physiology and pathology. Calcium deposition may include calcium oxalate, calcium phosphate, and Liesegang rings. Calcium oxalate (~10%) is faint or undetectable on H&E, birefringent (under polarized light), and almost always pointing to an underlying benignancy (e.g., ducts with apocrine metaplasia). Calcium phosphate (~90%) is detectable on H&E and potentially may indicate either benignancy or malignancy. Liesegang rings are laminated inclusions containing Ca, Fe, silicone, and sulfur. The College of the American Radiologists proposes the Breast Imaging Reporting and Data System (BI-RADS), which is extremely useful for pathologists. Benignancy-associated imaging patterns include popcorn, milk of calcium, "teacup"-shaped microcalcifications, or large, smooth, and diffuse microcalcifications. Conversely, malignancy-associated imaging patterns include clustering and linear branching-shaped microcalcifications. These following patterns are indeed often observed in comedocarcinoma and cribriform types of ductal carcinoma in situ (DCIS). *Mondor disease* is a rare entity of breast disease referring to sclerosing thrombophlebitis of the subcutaneous veins of the anterior chest wall with breast involvement. Although not usually seen in children or adolescents, it may occur at any age. The abrupt appearance of a subcutaneous cord, which is firstly red and tender and then painless and fibrous with the possible association of tension and skin retraction, should attract our attention into this direction. Mondor disease remains a benign and self-limited pathologic condition of the mammary gland but needs to be taken into consideration in the differential diagnosis of breast cancer.

9.3 Pathology of the Female and Young Adult

9.3.1 Fibrocystic Disease

Fibrous and cystic change (FCD) of the TDLU (FCD) is usually not observed in children but may be detected in young females. It is typically bilateral without a premalignant significance *ab initio*. Of note is that it may be a rare event in women taking long-term contraceptives. Microscopically, FCD is characterized by:

1. Cysts of type 1 with metaplastic apocrine cells (alkaline pH, low Na^+/K^+ ratio) and cysts of type 2 with flattened epithelium (acid pH, high Na^+/K^+ ratio) (type 1 cysts are prone to recur, while type 2 cysts do not show this feature invariably)
2. Fibrosis with mild to severe/dense hyalinization with or without cyst rupture
3. Chronic cellular inflammation with lymphocytes and plasma cells (rupture-related) scattered in the breast tissue
4. Adenomatoid change due to stromal proliferation and slit-like spaces
5. Apocrine metaplasia with eosinophilic, PAS +, and Sudan black + snout cells with prominent nucleoli and abundant granular cytoplasm
6. Epithelial hyperplasia (mild/moderate), including sclerosing adenosis (SA), fibroadenoma (FA), and intraductal papilloma

9.3.2 Soft Tissue Tumors and Hematological Malignancies

These are rare entities in children and youth but need to be kept in mind. These include *pseudoangiomatous stromal hyperplasia* (PASH), *hemangioma*, *angiolipoma*, *angiosarcoma* (freely anastomosing and arborizing vascular channels lined by malignant cells), *stromal sarcoma* (similar to phyllodes tumor but lacking the epithelial component), *granular cell tumor*, *fibromatosis*, *nodular fasciitis*, *malignant lymphoma* (almost always DLBCL, FCL, EN-MZL/MALT lymphoma), *granulocytic sarcoma* (aka myeloid sarcoma or chloroma), and leukemic manifestation with eosinophilic myelocytes or metamyelocytes. Glomus tumor or glomangioma is a benign, subcutaneous neoplasm. It accounts for about 2% of all soft tissue tumors and may be located in virtually any location in the body. Typically, it appears as a painful nodule with pinpoint tenderness and hypersensitivity to cold. The usual site is an area of the skin that is rich in glomus bodies such as the subungual regions of the digits or the deep dermis of the palms, wrists, forearms, and feet. Extra-digital glomus tumors may be located in breast tissue and can occur at any age. Since extra-digital glomus tumors are rare, misdiagnosis of these neoplasms as hemangiomas or arteriovenous malformations or simple cysts can lead to a delay in the correct management of these lesions (Fig. 9.2).

9.3.3 Fibroadenoma

Fibroadenoma (FA) is a TDLU-arising, benign biphasic tumor with epithelial and stromal components. Loose stroma constitutes the stromal element with acid mucopolysaccharides, but no elastic tissue. The FA usually presents as solitary (90%), firm, freely movable breast nodule/mass (usually <3 cm), which is encapsulated and solid, grayish-white cut surface grossly. It is more often in 25–35-year-old women, although female adolescents may also harbor this tumor. There is an increased frequency of FA in Afro-American female adolescents in comparison with Caucasian female adolescents. Characteristically, there is an enlargement of the lesion during pregnancy. It can arise in large or small ducts of the TDLU as intracanalicular (around compressed, cleft-like ducts with stroma > glands and the stroma undoubtedly invaginates glands) or pericanalicular (around tubular ducts with glands > stroma and the glandular configuration seem overall maintained).

Microscopically, FA is circumscribed and composed of both proliferating glandular and stromal elements. The epithelial (glandular) component shows always a myoepithelium (ME) cell layer and no atypia, whereas the stromal component, which should be not very cellular (DDX: phyllodes tumor), may contain cartilage and smooth muscle (hamartoma/choristoma). FCD may also be present, and the simultaneous occurrence is called fibroadenomatosis when FA merges specifically and undoubtedly with surrounding FCD. SA, apocrine metaplasia, and squamous metaplasia may also occur (Fig. 9.3).

Variants:

1. *Juvenile*: Afro-American adolescents, often bilateral and may be very large and have highly cellular stroma and glands (DDX: phyllodes tumor).
2. *Myxoid*: FA with prominent myxoid change ± Carney complex (cardiac and cutaneous myxomas, spotty pigmentation, and endocrine overactivity) ⇒ ↑ risk of recurrence.
3. *Degenerative change-accompanied FA*: FA ± stromal hyalinization, epithelial atrophy, calcification, and ossification.
4. *Fibroadenomatoid change*: Multiple adjacent TDLU with FA-like changes without forming a well-defined nodule or mass.
5. *Complex FA*: It is designated an FA with cysts with Ø > 3 mm, SA, non-SA epithelial hyperplasia, or papillary apocrine metaplasia.
6. *Malignancy-arising FA*: FA with a malignancy arising in a setting of an FA with >50% of lobular type. In most half of the investigated female patients, there is an obvious invasive carcinoma elsewhere in the mammary gland. Malignant transformation: 0.1% → LCIS (good PGN). A sarcomatous transformation still remains a rare event.

Fine needle aspiration cytology (FNAC) plays a major role in breast pathology, and the quality improved enormously in the last two decades. Excellent textbooks of cytopathology can help the non-experienced and experienced pediatric pathologist in dealing with FNAC specimens.

9.3.4 Adenoma

DEF: Any *focal* primary hyperplastic/adenomatous process of the epithelial (glandular) component of the breast, ± associated with pregnancy (lactating adenoma), characterized by a focal increase in the number of acini per lobule. It includes several histologic phenotypes:

9.3.4.1 Tubular Adenoma

Circumscribed (no obviously right or straightforward capsule) firm tan-yellow mass composed of tightly packed regular small tubules with possible luminal secretion, but no cytoplasmic vacuoles. It may be part of a physiologic spectrum including lactating adenoma under different physiology (Fig. 9.3).

9.3.4.2 Lactating Adenoma

Circumscribed gray-tan mass (no obvious and straightforward capsule) of densely packed regular round bilayered tubules with prominent secretory change (cytoplasmic vacuoles and luminal eosinophilic secretion) and scant stroma (no ductal compression), ± ↑ MI, no atypia. DDX: Tubular adenoma (luminal secretion may be present but lack of cytoplasmic vacuoles vs. noticeable lactational change), focal pregnancy-like change (focal changes vs. circumscribed mass or nodule), secretory carcinoma (sheetlike growth with focal infiltration and no myoepithelial cell (ME) layer vs. circumscribed mass of normal round tubules with ME cell layer).

9.3.4.3 Nipple Duct Adenoma (NDA)

Periareolar compact (nodular) proliferation of small bilayered tubules lined by epithelial and myoepithelial cells featuring papillary pattern (intraductal hyperplasia) replacing/destroying the nipple. Fibrosis of adjacent stroma loose and no dense or desmoplastic stroma as seen in syringoadenomatous adenoma or SAN and tubular carcinoma, which shows neither basement membrane (BM) nor ME layer is an important feature of NDA. Other features of NDA include, continuity with the squamous epithelium of epidermis, peripheral clefting, oval nuclei, streaming, epithelial hyperplasia, and sclerosing adenosis foci. Keratin cysts or focal necrosis within proliferating ducts and bland cells in the epidermis that are

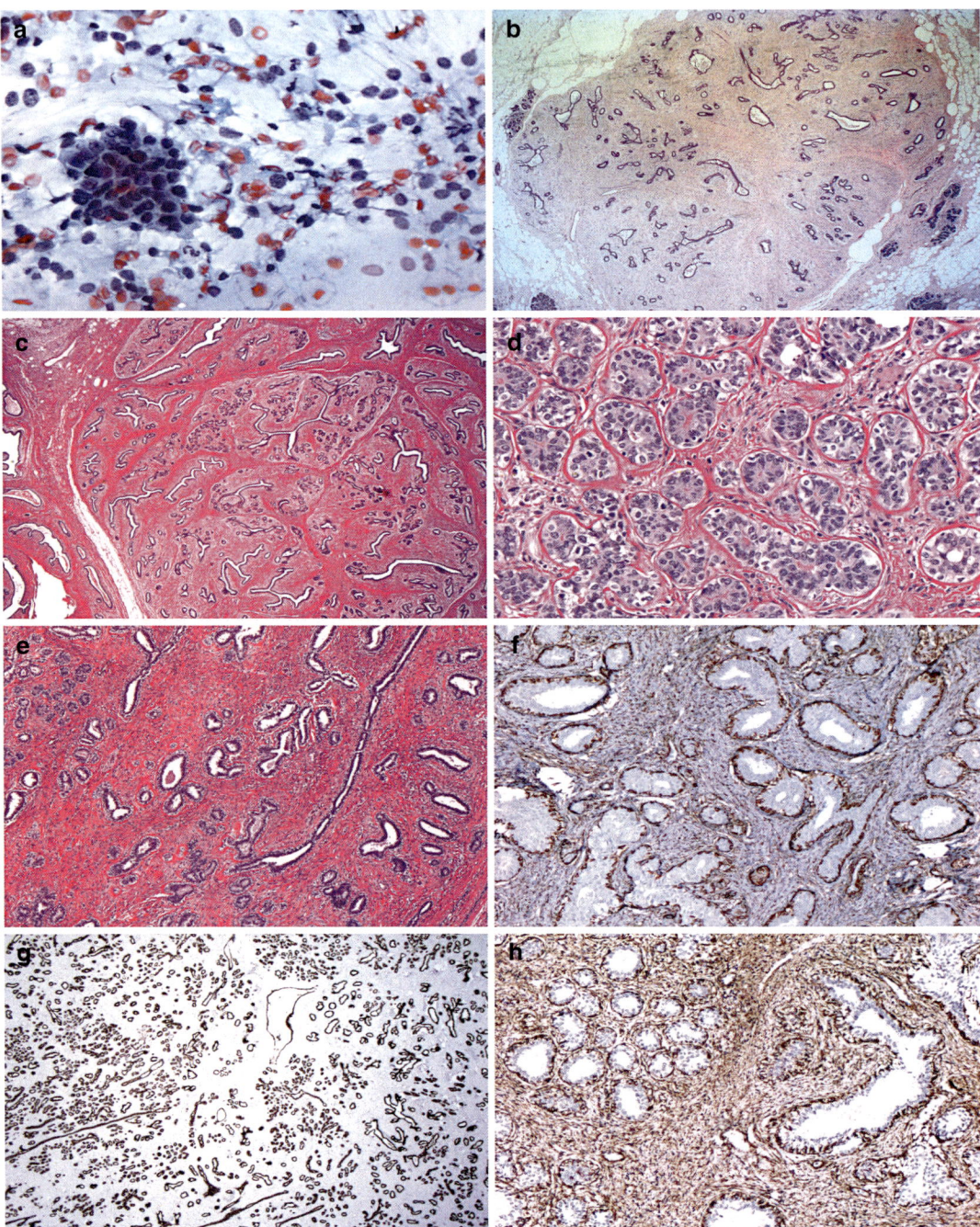

Fig. 9.3 In (**a**) is shown an FNAC of a fibroadenoma (Diff-Quik stain, 400×), in (**b**) an uninodular fibroadenoma (H&E stain, 20×), and in (**c**) a multinodular fibroadenoma (H&E stain, 12.5×) showing intralobular stroma enclosing partly open luminar spaces (pericanalicular) in (**b**) and compressed glandular spaces due to the stroma proliferation in (**c**). In (**d**, **e**) are microphotographs of a mixture of tubular adenoma (**d**) and fibroadenoma (**e**), both H&E stained and taken at 200× and 50× as original magnification. The collision tumor "tubular adenoma-fibroadenoma" is quite rare in childhood and youth. It shows the combined pictures of a tubular adenoma with tightly packed tubules lined by epithelial and myoepithelial cells with little intervening stroma and those of a pericanalicular fibroadenoma showing intralobular stroma (fibrotic) enclosing glandular spaces, mostly with open lumina. The microphotographs (**f**, **g**, **h**) show immunohistochemical findings of the epithelial and stromal components highlighted by actin ((**f**) anti-actin antibody, 100×), CAM 5.2 ((**g**) CAM 5.2, 12.5×), and muscle-specific actin ((**g**) anti-MSA, 100×). CAM 5.2 is a low molecular weight cytokeratin marker that identifies cytokeratins 7 and 8, but not to cytokeratin 18 or 19

scattered or in small aggregates which are CAM5.2+, CK7+, CEA-, and HER2-, no atypia, and no cribriform component are also some additional features of NDA. DDX includes SAN and subareolar sclerosing duct hyperplasia, which is characterized by central sclerosis entrapping and distorting ducts with a pattern similar to radial scar (RS).

9.3.4.4 Syringomatous Adenoma of the Nipple (SAN)

Haphazard proliferation of irregular, angulated, and comma-shaped tubules with compressed and open lumina, which may be widely separated by fibrous stroma, similar to the cutaneous syringoma, with permeative character (not destructive, unlike NDA), ME cell layer, small basophilic cells without intraductal hyperplasia, atypia or ± ↑ MI, and ± squamous metaplasia (unlike tubular carcinoma that lacks squamous metaplasia). DDX: SAN vs. nipple adenoma (SAN permeates the nipple stroma, while NDA replaces the nipple stroma). DDX: Tubular Ca. (single-cell layer, ME markers (−), apical "snouting", no squamous metaplasia).

9.3.4.5 Intraductal Papilloma, Papillomatosis, and "Encysted Intraductal Papillary Carcinoma"

The *intraductal papilloma* can arise in large or small ducts and is not rare in young females. Typically, intraductal papilloma presents with bloody nipple discharge and is solitary in the majority of the cases. However, sometimes more than 2–3 papillomas are found, and the term *papillomatosis* is used. Microscopically, there is an intricate, cellular, "arborescent" pattern which is a feature of all papillomas. Benignancy criteria include small size (<3 cm) well-developed fibrovascular cores, ME layer, normochromatic oval nuclei, nil or low MI, apocrine metaplasia, and lack of cribriform or trabecular pattern. Complications include hemorrhagic infarct, squamous metaplasia, focal necrosis, and pseudoinfiltration at the base of the lesion. The *encysted intraductal papillary carcinoma* needs to be differentiated from the intraductal papilloma. Intracystic malignant variant, which also seems to have a relatively good prognosis, probably better than ductal carcinoma, is not infrequent in the late youth. It is characterized by delicate or absent fibrovascular core recalling the micropapillary variant and atypical epithelium, lack of ME layer, and rare apocrine metaplasia. There is a possible association of histological foci of DCIS in the surrounding tissue.

9.3.4.6 Ductal Adenoma

It is a circumscribed breast mass with a prominent fibrotic capsule and a bilayered tubular component (epithelial and ME cell layers) combining morphologic features of SA and papilloma. Diagnostic criteria include combined morphologic features of SA and papilloma, circumscription, encapsulation, surrounding of an obvious fibrous capsule, prominent hyalinizing fibrosis (often dense in the center), no papillary structure, intermixed bland ducts, and clear-cut lining of epithelial and myoepithelial layers (+ ME-markers). Whereas peripheral ducts are frequently dilated, central ducts may be compressed or obliterated. However, to remember is that there is often apocrine metaplasia with prominent apical snouts on several tissue sections and possible luminal calcification (DDX: DCIS!), myoepithelial overgrowth, or peripheral lymphoid infiltrate.

9.3.4.7 Apocrine Adenoma

It is an adenoma constituted by bland tightly packed glands with flat or papillary lining, apocrine differentiation, and scant stroma. Diagnostic criteria include circumscription of the lesion and apocrine differentiation of the luminal cells tall columnar, granular eosinophilic cytoplasm, GCDFP-15+ (Gross cystic disease fluid protein-15), bland-appearing nuclei located at the base of the cells, apical blebs, no or extremely rare mitoses, and presence of basal cytologically bland myoepithelial cells). ME cells are "SCALP"+, i.e., the cells show positivity for reliable ME markers that are typically *S*MA, *Cal*ponin, and *p*63-nuclear. Although some authors prefer extend the list to smooth muscle myosin heavy chain (SMMHC), VIM, Pan-Keratins, and GFAP with variable reproducibility, "SCALP" markers are indeed the most useful in routine practice.

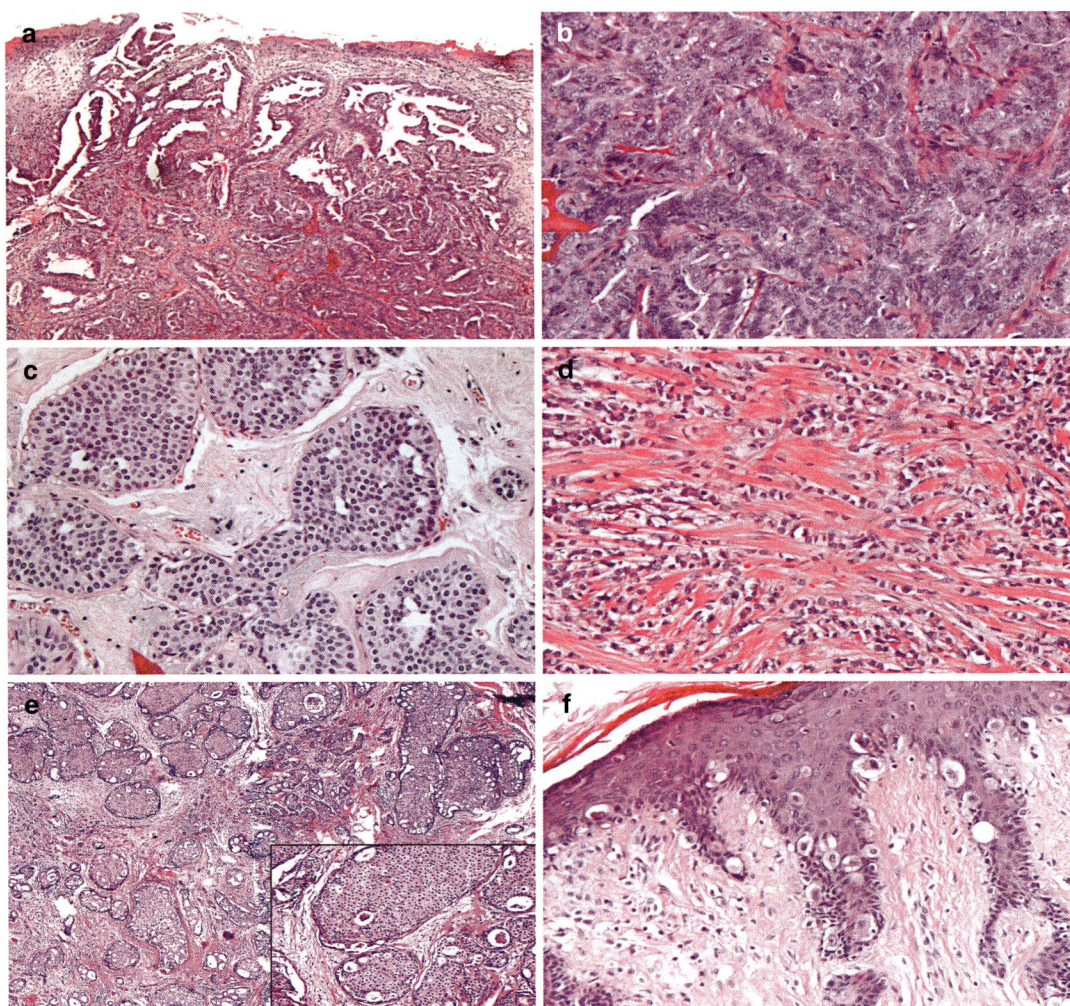

Fig. 9.4 In (**a**, **b**) are microphotographs of a papillary carcinoma of the mammilla in a youth with papillary or solid foci formed by ducts nearly or completely filled by a solid neoplastic proliferation with cells having moderate to abundant cytoplasm, intermediate histologic grade, and no apparently discernible microcalcifications ((**a**) H&E, 50×; (**b**) H&E stain, 200×). The microphotographs (**c**, **d**) show an in situ lobular carcinoma (LCIS) and an invasive lobular carcinoma (ILC), respectively ((**c** and **d**) H&E stain, 200×). In (**c**) the tumor cells have obliterated the lobules without apparent invasion, while in (**d**) the tumor cells grow in single file, linear pattern and also show a loosely dispersed growing pattern throughout the fibrous matrix. In (**e**) is shown a case on an adenoid cystic carcinoma in a youth at 40× (H&E stain) and 200× (H&E stain, inset), while in (**f**) is shown a Paget disease of the breast (H&E stain, 200×). In the adenoid cystic carcinoma, there is a cribriform pattern with true glandular lumina lined by ductal epithelium and eosinophilic "cylinders" with basement membrane material lined by basal/myoepithelial type cells, while Paget disease of the mammilla is characterized by an in situ carcinoma involving the nipple epidermis with cells harboring a copious pale cytoplasm, irregular large nucleus, and prominent nucleoli. An important differential diagnosis is clear cell papulosis. Histological differential diagnoses include the early stages of extramammary Paget disease, pagetoid squamous carcinoma, pagetoid melanoma, and sebaceous carcinoma with malignant cytologic features, whereas the clear cells of clear cell papulosis appear to be benign

9.3.5 Genetic Background of Breast Cancer

The adolescent up to young adult (AYA) is defined, as indicated before, as a patient of 15–39 years of age at original cancer diagnosis. This group of individuals has particular medical needs. There are age-related issues linked to the AYA population, excluding homicide, suicide, or unintentional injury, because the leading cause of

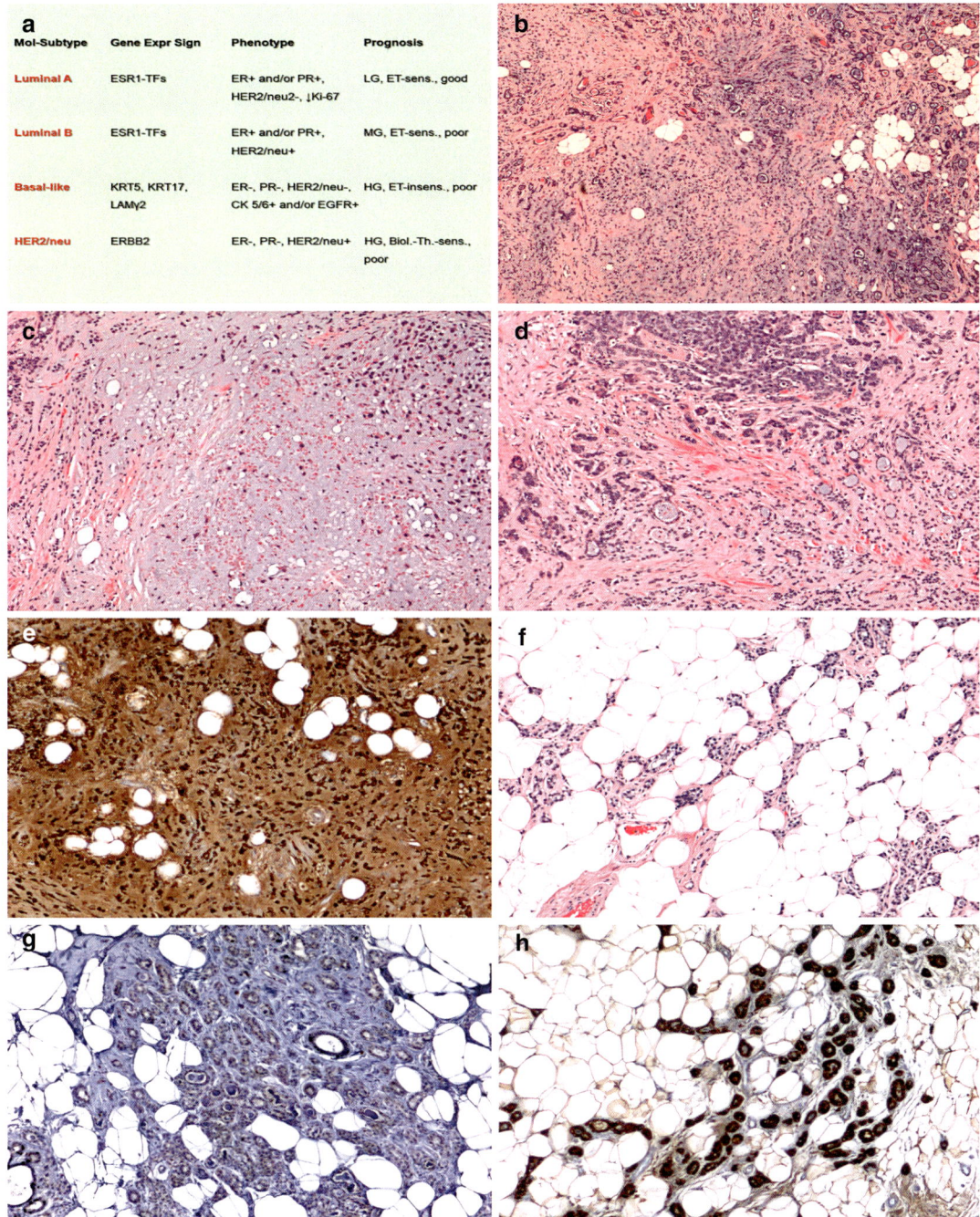

Fig. 9.5 In (**a**) is shown the molecular classification of the breast cancer (see text for details). In (**b–e**) are the microphotographs of a metaplastic breast carcinoma with carcinomatous growth and chondroid portions with S100 positivity ((**b**) H&E stain, 50×; (**c**) H&E stain, 100×; (**d**) H&E stain, 100×; (**e**) anti-S100, 100×). Metaplastic carcinoma can occur in late youth, particularly young females harboring *BRCA1*-mutated gene. In (**f, g, h**) are features of a microglandular adenosis with S100 positivity, an important differential diagnosis for pathologists ((**f**) H&E, 100×; (**g**) anti-p63, 100×; (**h**) anti-S100, 100×). Notes: ESR1, …; KRT5, …; KRT17, …; LAMy2, …; ERBB2, …

death remains cancer. Breast cancer is one of the fatalities accounting with others for 95% of the tumors in this group. Genetic background of breast cancer is now mandatory to know for pathologists, including pediatric pathologists, because of genetic counseling. Interestingly, the most common genomic alteration in both familial and sporadic breast carcinomas resides on chromosome 17: 17p13.1 (*TP53* gene mutations), 17q11–12 (*HER2* gene mutations, human epidermal growth factor receptor 2), and 17q12 (*BRCA1* gene mutations, breast cancer type 1) (Figs. 9.4 and 9.5). In many female adolescents and female youth, the fear of developing breast cancer does not get away with just a simple regulatory palpatory test or regular visits to the family doctor. The reality is that about one in four breast cancers is diagnosed before the age of 50 and in some countries/setting before the age of 40. Numerous early-stage premalignant lesions remain unfortunately undiagnosed and are difficult to catch by conventional imaging. Exposures during childhood and adolescence or previous in the intrauterine environment (e.g., diethylstilbestrol, DES) affect a woman's long-term risk of breast cancer. Al Jishi and I reviewed the DES exposure for public health concerns. Maternal exposure to DES during pregnancy and increased breast cancer risk in daughters are a reality, and family doctors should remember this condition in the differential diagnosis. It is important to remember the high risk and low/intermediate risk of breast cancer in patients with underlying genes that may be mutated in female adolescents (Boxes 9.1 and 9.2). The gene expression profiling of invasive breast carcinoma is shown in Box 9.3.

In *luminal A*, which harbors a better prognosis, the clinical picture is usually characterized by metastases to the bone and pleura, while in luminal B with a worse prognosis, the clinical picture shows metastases typically to the brain and liver. The "basal-like/3-" group of tumors with this characteristic gene expression profiling (~1/5 of all breast carcinomas) is particularly challenging. These breast carcinomas are triple-negative, i.e., they show no expression of *ER* (estrogen receptor), *PR* (progesteron receptor), or *HER2* but an expression of CK5/6 and/or *HER1*. Basal-like/3- breast carcinomas often harbor *BRCA1* mutations

---

Box 9.1 HR Breast Ca Susceptibility Genes

BRCA1 (17q12): Hereditary breast and ovary carcinomas (BC risk 40–90%; basal-like/3- G3)

BRCA2 (13q12.3): Hereditary breast and ovary carcinomas (biallelic germline mut., Fanconi anemia)

TP53 (17p13.1): Li-Fraumeni syndrome (LFS)

PTEN (10q23): E.g., Cowden syndrome

Notes: LFS is BAS; *b*rain, breast, bone, BM, *a*drenal *s*oft tissue; *PTEN*, Phosphatase and tensin homolog (mutated in multiple advanced cancers). Cowden syndrome is PAB: *p*olyposis, *a*ppendiceal skin tumors and *a*kral keratosis, *b*rain and *b*reast ca. BC, breast cancer

---

Box 9.2 LR/MR Breast Ca Susceptibility Genes

STK11 (19p13.3): Peutz-Jeghers syndrome

CDH1 (16q22.1): Hereditary diffuse gastric carcinoma

ATH (11q) Ataxia-telangiectasia

CHEK2 (22q12.1) Li-Fraumeni variant (↑ risk for breast ca. postradiation!)

BRIP1/FANC J (17q1)
PALB2/FANC N (16p)

Notes: CDH1, cadherin 1; CHEK2, checkpoint kinase 2; STK11, serine/threonine kinase 11; BRIP1, BRCA1-interacting protein 1; PALB2, partner, and localizer of BRCA2.

---

and are positive for ME markers (e.g., basal keratins, P-Cad, p63, and laminin ⇒ "SCALP") as well as progenitor cells (stem cells). Most often

> **Box 9.3 Gene Expression Profiling of Invasive Breast Ca, NST**
> *Lum A* (50%): ER+ and/or PR+, HER2−, Ki67 < 15%, WD-C (G1–2), +P53−SM, better PGN
> *Lum B/3+* (15%): (ER, PR, HER2)+, MIB1 > 15%, PD−C (G3), +++P53−SM, worse PGN
> *HER2+* (10%): (ER and PR)−, HER2+
> *Basal-l./3−* (20%): (ER, PR, HER2)−, CK5/6+ and/or HER1+, BRCA1-M
> *"Normal breast"* (5%): ER−, but (+) adipose tissue genes
> Notes: Lum, luminal; WD-C (G1–2), well-differentiated carcinoma (grading 1-2); the other abbreviations are explained in the text

these carcinomas are of medullary, metaplastic (e.g., spindle cell), adenoid cystic type, G3 (poorly differentiated), high MI/MIB1 score, high *TP53* mutations, marked nuclear pleomorphism, pushing borders with prominent lymphocytic infiltrate, syncytial growth pattern, scant stromal content, central necrosis, low lympho-vascular stroma invasion, and low N+ (TNM staging). They show an aggressive course with metastasis to the brain and lung and poor PGN. Other markers that are usually positive in these tumors are CK14, VIM, c-kit, caveolin-1, caveolin-2, nestin, and SMA. BRCA1 carcinomas are often associated with loss of the inactive X chromosome and reduplication of the active X. This aspect results in the absence of the Barr body. Conversely, BRCA2-associated breast carcinomas are luminal A phenotype (ER+/HER2−) (Fig. 9.5).

9.3.6 In Situ and Invasive Ductal Breast Carcinoma

Ductal carcinoma in situ (DCIS) and invasive ductal carcinoma (IDC) of the breast are the most common malignant epithelial tumor of the mammary gland affecting ~1/9 women and being responsible for ~1/5 of the cause of death in women. Ductal carcinoma needs to be differentiated from lobular carcinoma, which harbors a better prognosis (*vide infra*). Ductal and lobular are not only distinguished morphologically but also molecular biologically using transcriptomics or gene expression profiling. Risk factors for breast carcinoma are displayed in Box 9.4.

9.3.6.1 Ductal Carcinoma In Situ

Neoplastic proliferation with malignant cytological and architectural features AND limitation to the ductal compartment of the TDLU (i.e., within spaces bordered by SCALP+ myoepithelial cells and collagen IV+ basement membrane) *without* invasion of the surrounding tissue. It has been demonstrated that 1/3 are multifocal and 1/10 are bilateral. The IHC phenotype is (+) E-Cad, P-Cat$_m$, (−) HMWCK (e.g., CK903), and (+) LMWCK (CK8) except the DCIS of basaloid type (Cad, cadherin; Cat, catenin; m, membranous; HMWCK, high molecular weight cytokeratins; LMWCK, low molecular weight cytokeratins). Microinvasion is defined when the tumor invades with a Ø < 0.1 cm (no invasive foci larger than 1 mm).

Ways to detect DCIS: (1) calcifications, (2) abnormal mammographic density, (3) palpable mass (periductal fibrosis), (4) nipple discharge (DCIS of micropapillary type), and (5) incidentally (upon biopsy for another lesion).

> **Box 9.4 Risk Factors: "IDAHO"**
> - *I*nfertility, encompassing nulliparity, but also early menarche and late menopause
> - *D*iet (high lipids) and EtOH, *D*rugs (Estrogen-only Hormone Replacement Therapy, HRT), ra*D*iation exposure
> - *A*ge (risk increases with advancing age) and *A*typical lesions (UDH, ADH, ULH, ALH)
> - "*H*eritage" (geographic, familiarity, susceptibility genes)
> - *O*besity (postmenopausal, due to the synthesis of estrogens in fat depots)

DCIS Risk for Invasive Carcinoma

1. Low-grade DCIS → invasive Ca. ~1%/year
2. High-grade DCIS → invasive Ca. >1%/year

DCIS Variables for Recurrence (numerical and dichotomic)

1. Grade, size, margins, postoperative radiation therapy and tamoxifen (ER+ DCIS). Tamoxifen ($C_{26}H_{29}NO$) is the oldest and likely most-prescribed selective estrogen receptor modulator (SERM) drug.

DCIS grading systems are based on nuclear grade with or without the parameter of necrosis. The Van Nuys Prognostic Index (VNPI) uses a combination of nuclear grade and necrosis and can be used to localize a subset of patients who are at risk of local recurrence and who may potentially benefit from adjuvant radiotherapy. VPNI uses three categories, high grade with necrosis, non-high grade with necrosis, and non-high grade without necrosis. A combination of nuclear class and cell polarization/differentiation has also been proposed. The European Pathologists' Working Group has made this choice, while the WHO and the "NHS Breast Screening Programme" of the UK suggest using the nuclear grade alone.

The most frequent histology patterns include comedocarcinoma, cribriform, solid, papillary, and micropapillary architecture. *Comedocarcinoma* has high-grade cytology hallmarks with large nuclear size, pleomorphism, and frequent mitoses and necrosis of tumor cells in the central duct space. A stable (uniform) single-cell population with unique punched-out luminal areas, distinct cell borders, and even spaces between them characterizes the *cribriform* pattern. The *solid* variant shows inward growth of large, high-grade pleomorphic tumor cells with a solid filling of the duct with uniform cells, more substantial than LCIS but smaller than comedocarcinoma, with sharp cytoplasmic borders and no necrosis (DDX: neuroendocrine variant). *Papillary* and *micropapillary* show arrangements of the tumor cells in the ducts, including hierarchical with fibrovascular core and nonhierarchical without fibrovascular cores, respectively. Other rarer variants include small cell solid, apocrine, neuroendocrine, signet ring cell, and cystic hypersecretory. Lobular cancerization refers to the presence within a lobule of cytologically ductal carcinoma.

Mammary gland Paget disease (MGPD) is a neoplastic skin lesion of the nipple with eczematous (crusting) gross features and underlying DCIS ± invasion. Microscopically, there are intraepidermal large cells with clear cytoplasm (milk fat globule), usually concentrated along the basal layer with or without small glandular structures. DCIS may be focally continuous with MGPD, and in about 1/3 of cases, metastases are found at the time of diagnosis. *IHC*: (+) CK7, CAM5.2 = LMWCK (CK8, CK18, CK19), EMA, CEA, Her-2, but (−) ER/PR, S100, HMB-45, CD45, CK20. Normal keratinocytes do not express CK8, CK18, and CK19. *DDX*: Toker cells (ER+), melanoma (S100+, HMB-45+), lymphoma (CD45+), Bowen disease (CK20+).

9.3.6.2 Invasive Ductal Carcinoma

- *DEF*: Neoplastic proliferation with malignant architecture and cytological features of the ductal compartment of the TDLU *and* invasion of the surrounding tissue.
- *EPI*: 75% of all invasive ductal carcinomas.
- *GRO*: IDC is a yellowish-gray, gritty tumor with some stellate shape, necrosis, hemorrhage, and cystic degeneration. There may be "chalky streaks" (duct or vessel elastosis) and calcifications in up to 60%.
- *CLM*: Several patterns mirror at least partially the noninvasive form. Some of them have prognostic significance and need to be distinguished carefully (vide infra). Angioinvasion (~1/20), lymphoinvasion (~1/3), and perineural invasion (~1/4) also need to be identified and are used for staging.
- *IHC*: + LMWCK, EMA, milk fat globule; lactalbumin and CEA (+ HMWCK in case of squamous metaplasia).
- GRADING: Scarff-Bloom-Richardson – I–III (SBR, infiltrating ductal carcinoma only) based on the score for each percent of tubule formation, nuclear pleomorphism, and mitotic. The SBR grading system with the Elston–Ellis

modification or Nottingham combined histologic grade, aka Nottingham grading system (NGS), has been validated for its prognostic significance through multiple studies.

- *T*ubule formation (1 point if >75% tubules, 2 points if 10–75% tubules, 3 points if <10% tubules)
- *N*uclear pleomorphism (1 point if minimal, 2 points if moderate, 3 points if marked variation)
- *M*itotic index (1 point, low MI; 2 points, medium MI; 3 points, high MI)

Some tables connecting objective and microscopic type are available. Some conditions need to be respected in assessing the NGS grading. They include the following: (1) ≥10 HPF should be counted, (2) unequivocal mitoses need to be calculated at the periphery of the tumor, and (3) the range of figures that gives each score needs to be correlated with field diameter and field area of the microscope used (e.g., Olympus BX40, HPF (×400 = 0.55 mm Ø)). If 3–5 (low level), then grade I; if 6–7 (intermediate level), then grade II; and if 8–9 (high level), then grade III.

9.3.7 In Situ and Invasive Lobular Breast Carcinoma

Lobular neoplasia has a better prognosis than ductal-arising carcinoma and is typically observed in premenopausal women, often multicentric (3/4 of cases) and bilateral (up to half of the cases), and can occur within FA or SA. The lobular neoplasia is usually not identifiable grossly. Microscopically, LCIS shows filling and distention of lobules by uniformly round, loosely cohesive small cells. The tumor cells have a low nuclear grade (oval normochromatic nuclei, nil or very low MI, nil or minimal atypia) and show some scattered intra-cytoplasmic mucin positivity in most of the cases. The ME layer is usually still present. The risk of invasiveness of lobular neoplasia is up to 30%. The recommended close follow-up for LCIS is generally enough for patients harboring this condition (Fig. 9.4). ILC usually shows 1) a single-cell file pattern with small, uniformly relatively round cells growing singly; 2) mild dense, fibrous stroma with periductal and perivenous elastosis; and 3) scattered *signet ring* cell cytology with intracytoplasmic mucin. Other growth patterns include *alveolar* pattern with sharply outlined groups separated by fibrous septa; *tubulolobular* pattern with typical ILC together with tiny tubules, apparently "closed"; and *pleomorphic* pattern with numerous heterogeneous morphologies.

9.3.8 WHO Variants of the Infiltrating Ductal Carcinoma

- *Tubular*: Stellate growth pattern of size and shape tumoral entity with variable "diploid" glands, which are haphazardly distribute, and show angulated "tear-drop" appearance, open lumina and basophilic secretion, single epithelial layer without either BM or ME, and interglandular desmoplastic stroma reaction. 2/3: (+) DCIS (cribriform or micropapillary). 1/2: multicentric or multifocal or bilateral. PGN: excellent (5-YSR: >95%, but 1/10 prone to metastasize). IHC: (+) ER, EMA, but (−) Her2. DDX: Sclerosing adenosis (SA), radial scar, microglandular adenosis, and micropapillary adenosis.
- *Cribriform*: Tightly linked to a tubular variant, harboring an excellent prognosis and showing infiltrating tubular structures with a prominent cribriform pattern.
- *Mucinous* (aka mucoid, colloid – but only if the non-mucinous component is <10%): Well-circumscribed, "currant jelly"-like tumor with constituted of clusters of carcinoma cells floating in large lakes of mucin (neutral or acid; almost entirely extracellular) IHC: (+) MUC-2 and MUC-6. PGN: good if pure, but late metastases have been recorded for this kind of tumor.
- *Secretory* (or juvenile secretory): Well-circumscribed, small pediatric tumor with central hyalinization, tubule-alveolar and focally papillary growth patterns, and cells showing intra- and extracellular clear cell vacuoles, AB+ and DPAS+. CGB: *ETV6/NTRK3*

fusion gene (+), which represents the translocation of genetic material between chromosomes 12 and 15. A new fusion gene is created by t(12;15), in which the 5′ region of *ETV6* is fused to the 3′ region of *NTRK3*. The chimeric transcript encodes a protein consisting of a helix-loop-helix (HLH) dimerization domain of ETV6 fused to a protein tyrosine kinase (PTK) domain of NTRK3. The *ETV6-NTRK3* gene fusion was identified initially in congenital fibrosarcoma and congenital mesoblastic nephroma of the kidney. PGN: Excellent (5-YSR: 100%).

- *Medullary*: Tumor belonging to basal-like carcinoma group of triple-negative (ER-, PR-, HER2-), *BRCA1* mutation-positive, more often in Asian females, and soft, fleshy, and well-circumscribed grossly. Three distinguishing microscopic features: (1) solid, syncytium-like sheets (>75% of tumor) with no glandular differentiation, indistinct cell borders, cytologically high-grade, large pleomorphic cells, large nucleoli, and high MI, (2) pushing borders, and (3) prominent lymphoplasmacytic cellular infiltrate at periphery (T cell-markers positive; IgA+), often with germinal centers. PGN: Slightly better survival than classic ductal (~80 vs. ~60% in 10-YSR).
- *Papillary*: "Frond"-forming architecture with in situ component, pushing growth pattern, and excellent prognosis.
- *Micropapillary*: Malignant epithelial tufts exhibiting "reverse polarity" portend a poor prognosis.
- *Apocrine*: Large cells with abundant acidophilic, granular PAS+ cytoplasm and decapitation secretion and basally located nuclei.
- *Neuroendocrine*: Invasive ductal carcinoma with quite specific neuroendocrine differentiation (+CGA, SYN, CD56) lacking carcinoid syndrome clinically. Microscopically, there is an organoid pattern with solid nests of cells, ribbons, rosettes, and fibrous stroma. TEM: Dense core neurosecretory granules. PGN: Same as IDC.
- *Metaplastic*: Tumor belonging to the group of triple-negative (ER-, PR-, HER2-) mammary carcinomas. Metaplastic carcinomas are well-circumscribed grossly. Microscopically, there is a mainly sarcomatoid look and there is an incontrovertible VIM (+) mesenchymal immunohistochemical staining. Epithelial markers may be demonstrated using single cytokeratins (CK7, CK19, and CK20) or panels (34BE12 or HMWCK, CAM5.2 or LMWCK, and MNF116, which react with keratins 5, 6, 8, 17 and probably also 19). PGN: 1/4 metastasize thus portend a more aggressive than the classic pattern. Subtypes include monophasic sarcomatoid, which is also known as spindle cell carcinoma and biphasic sarcomatoid carcinoma (aka carcinosarcoma or malignant mixed tumor) (Fig. 9.5).
- *Inflammatory*: Clinically, red and warm breast with edema of skin showing a tumor, which is, microscopically, a carcinoma with extensive dermal lymphatic invasion. PGN: Ominous (pT4d).

9.3.9 Sweat Gland-Type Tumors and Myoepithelial Tumors

Since the mammary gland is a particular sweat gland, some tumors usually occurring in the salivary glands and bronchial system may also be seen in the mammary gland as well. Salivary gland-type tumors of the breast are traditionally pleomorphic adenoma, adenoid cystic carcinoma, acinic cell carcinoma, and mucoepidermoid carcinoma.

Pleomorphic adenoma of the breast (PAB, aka benign mixed tumor) has been reported in the breast but also in other, less common, sites such as the paranasal sinuses, larynx, palate, and nasal septum other than the usual major and minor salivary glands. In the skin, it is known as chondroid syringoma. PAB is a circumscribed lesion grossly that is characterized microscopically by a mixture of epithelial and myoepithelial cells embedded in abundant stroma with ≥1 of the following matrices: (1) myxoid, (2) chondroid, or (3) osseous. The youngest patient that has been reported in the biomedical literature was 23 years old, but a wide age range has been observed according to my records and consultations.

Adenoid cystic carcinoma: Carcinoma with two types of cavity formations, right glands and eosinophilic "cylinders" containing PAS+BM material (collagen IV). It needs to be differentiated from the much more common intraductal cribriform carcinoma. PGN: Poor (LN MTX in 6%; lung MTX in 12%).

Acinic cell carcinoma and *mucoepidermoid carcinoma* are rare entities and similar to the counterparts seen in the salivary glands and bronchial system.

Other tumors include *eccrine spiradenoma* (sharply circumscribed lobules with small lumina and basaloid cells) and *papillary clear cell hidradenoma* (a pedunculated tumor with intracystic and papillary architecture and clear and granular bi-epithelial cell types). A myoepithelial tumor that should be considered in the differential diagnosis is the so-called *adenomyoepithelioma*, which is one of the most common myoepithelial tumors and shares similar features with its counterpart of the salivary glands and bronchial system.

9.3.10 Phyllodes Tumor

- *DEF*: Phyllodes tumor (an early name was "*cystosarcoma phyllodes*") is considered the "malignant" counterpart of fibroadenoma with low- or high-malignancy potential. It is a mesenchymal neoplasm with biphasic components.
- *EPG*: The tumorigenesis of a phyllodes tumor with the involvement of three genes, including *TERT1, MED12,* and *EGFR*. There is a critical role of *TERT* (telomerase reverse transcriptase) promoter mutations, located in 5p15.33, in cooperation with *MED12* (mediator complex subunit 12 genes) mutations, also of epithelial growth factor receptor (*EGFR*), in the development of phyllodes tumors. The most commonly reported chromosomal copy number alterations in phyllodes tumors are 1q gain and 13q, 6q, and 9p losses, and gene amplifications involve *MDM2* (mouse double minute 2 homolog), *MDM4*, *RAF1* (virus-induced rapidly accelerated fibrosarcoma), *PDZD2* (PDZ domain containing 2), *MYC* (from the homologous avian virus, *myelocytomatosis*, or v-myc and its homologous human gene overexpressed in several cancers or *C-MYC*), *EGFR*, *IGF1R* (insulin growth factor 1 receptor), and *TERT* and *PDZ* (for PSD95, discs large, and ZO-1) (gene-related) domains are conserved protein-protein interaction modules. PDZ domains comprised of ~80 to 100 amino acids and recognize specific carboxy-terminal sequences. Typically, they exist in proteins as tandem repeats and act as molecular scaffolds to facilitate the assembly of macromolecular complexes.
- *GRO*: The tumor is somewhat round, quite firm, well-circumscribed, with the solid gray-white cut surface containing cleft-like spaces with or without hemorrhage and necrosis.
- *CLM*: Phyllodes tumors have benign glandular elements with hypercellular, CD34+ stroma ("stroma overgrowth") harboring progesterone receptors (PR) but lacking ER. PR is NR3C3 or nuclear receptor subfamily 3, group C, member 3 as indicated in the official name and is activated by the steroid hormone progesterone. The neoplasm forms cellular leaflike processes protruding into cystic spaces lined by epithelial and myoepithelial cells. The hypercellular stroma may range from a predominantly periductal fibroblastic appearance with foci of mature adipose tissue to marked nuclear atypia and mitoses as well as metaplastic tissue (cartilage, bone, and skeletal muscle). Metaplasia implies the transformation of one type of tissue into another, i.e., to mold into a new form (Greek: μετά- and πλάσσειν, which is the present active infinitive of πλάσσω). Cambium layer is also often present with subepithelial stroma condensation (Box 9.5) (Fig. 9.6). If FA has no infiltration into the surrounding tissue, no cambium layer, no atypia, and low MI, the biphasic neoplasms of low and high grade have a different story.
- *PGN*: It is usually benign with locally aggressive behavior. Although rare in adolescence it is present in young females (female youth). Low grade → tendency for local recurrence, but high grade harbors an increased potential to

Box 9.5 Distinguishing Features of Biphasic Neoplasms

	LG-PT	HG-PT
Size	~ small	~ large
Stroma	Moderate	High
Cambium layer	(+)	(+)
Architecture	Leaflike, cystic	Leaflike, cystic
Atypia	G1-G2	G2-G3
Margins	"Pushing"	Infiltrative
MI	<3/10HPF	>3/10HPF

Notes: FA, fibroadenoma; LG-PT, low-grade phyllodes tumor; HG-PT, high-grade phyllodes tumor; ~ variable; G1-G2, mild to moderate grade of atypia (G1, well differentiated = mild atypia, G2, moderately differentiated = moderate atypia); G3, poorly differentiated = severe atypia. In botany, cambium layer is identified in plants and considered a meristematic tissue (Greek: μεριστός: "divided") between the inner bark or phloem and the wood or xylem. Cambium layer is responsible for the growth of stems and roots by producing new phloem on the outside and new xylem on the inside in stems.

CD10 and CD117 are expressed with increased frequency in HG-PT. HG-PT (benign epithelial component, spindle cell component HMWCK/p63-, and no squamous differentiation) needs to be differentiated from metaplastic carcinoma (malignant epithelial component, spindle cell component HMWCK/p63+, and ± squamous differentiation). *EGFR* gene amplification is seen in HG-PT.

metastasize (3–12% incidence of metastases, usually lung or bones – rarely axillary nodes).

9.4 Cancer Mimickers

9.4.1 Hyperplasia, Ductal and Lobular

Summarizing the degenerative and premalignant disorders, nonproliferative breast changes are, indeed, constituted by three elements that may participate variably to the formation of the pathology. The three elements are cysts, adenosis, and fibrosis. Proliferative breast disease without atypia includes epithelial hyperplasia (mild/moderate), SA, radial scar/complex sclerosing lesion, and papilloma. Proliferative disorders with atypia include atypical ductal hyperplasia and atypical lobular hyperplasia.

Hyperplasia, ductal and lobular, is defined as an epithelial proliferation or "epitheliosis" (according to some other authors) of the TDLU at ductal or lobular level with a slightly increased risk for invasive carcinoma.

Relative Risk of Infiltrating Carcinoma according to Grading	
I Mild hyperplasia	1
II Moderate or florid hyperplasia	1.5
III Atypical ductal or lobular hyperplasia (ADH/ALH) *(aka DIN1b)* (DIN, ductal intraepithelial neoplasia)	4.5
IV In situ ductal/lobular carcinoma (DCIS/LCIS) (aka DIN1c-DIN2-DIN3)	9

Usual ductal hyperplasia (UDH), aka "low-risk ductal intraepithelial neoplasia or DIN" (partly not encompassing with the lesion "flat DIN 1a" or "flat epithelial atypia"), is also classified as:

- *Mild Hyperplasia*, if ≤4 cells thick extending upward from the BM
- *Moderate Hyperplasia*, if >4 cells thick extending upward from the BM and potential some bridging in the lumen
- *Florid (Flourishing) Hyperplasia*, if the lumen is distended and potentially obliterated (a flower may be used to remember this structure with a core and some petals as distending lumens)

However, there is no natural necessity to distinguish moderate from florid hyperplasia, because they are usually grouped to form a single category for risk assessment according to numerous authors. Most probably, it is not only an academic distinction, but it will be important in the future. However, it is imperative to remember that in UDH the luminal spaces observed histologically show preservation of a peripheral par-

9.4 Cancer Mimickers

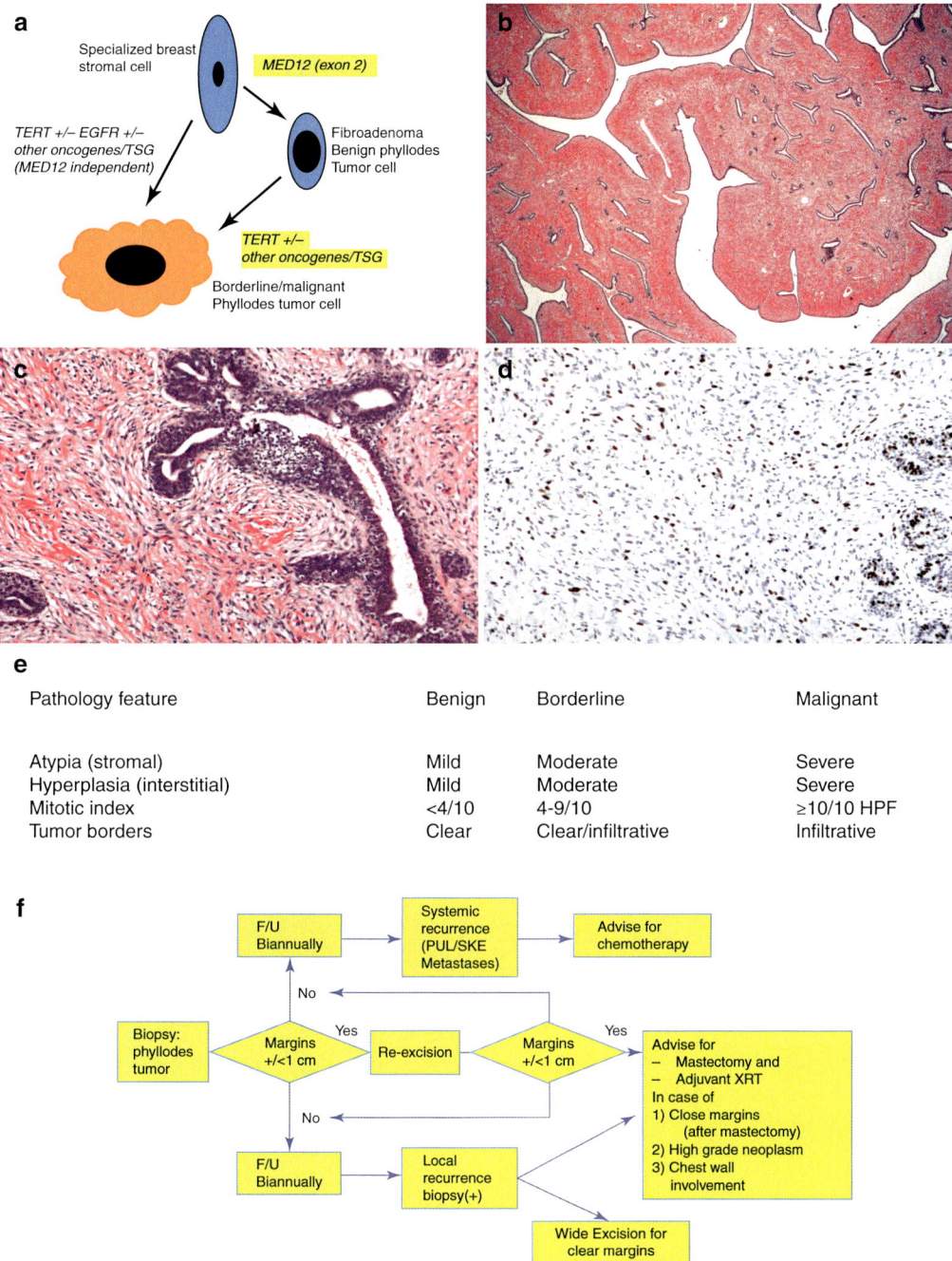

Fig. 9.6 In (**a**) is shown the most current knowledge on tumorigenesis of a phyllodes tumor with involvement of three genes, including *TERT1, MED12,* and *EGFR*. There is a critical role of *TERT* (telomerase reverse transcriptase) promoter mutations, located in 5p15.33, in cooperation with *MED12* (mediator complex subunit 12) gene mutations, in addition to epithelial growth factor receptor (*EGFR*), in the development of phyllodes tumors. The most commonly reported chromosomal copy number alterations in phyllodes tumors are 1q gain and 13q, 6q, and 9p losses, and gene amplifications involve *MDM2, MDM4, RAF1, PDZD2, MYC, EGFR, IGF1R,* and *TERT*. The microphotographs (**b, c, d**) show a borderline phyllodes tumor of the breast ((**b**) H&E stain, 12.5×; (**c**) H&E stain, 100×; and (**d**) anti-Ki67/MIB1, 100×) showing a hypercellular mesenchymal component organized in "leaflike" pattern surrounding benign epithelial- and myoepithelial-lined spaces. In (**e**) the current grading of the phyllodes tumor is presented, while (**f**) gives a potentially useful algorithm for the management of the phyllodes tumors of the breast (see text for detail)

tial ring and irregularity in size and shape and are often found as "slit-like" unlike the "punched-out" lumina of ADH or DCIS.

Moreover, in UDH the cell population is still mixed (LMWCK or CAM5.2 or CK 8, CK18, CK19 vs. HMWCK or 34BE12 or CK1/5/610/14), while in ADH and DCIS, there is almost loss of the HMWCK with uniformity of the luminal epithelial cells for LMWCK, because of the loss of the basal/intermediate epithelial cells. In UDH, basal/intermediate epithelial cells, myoepithelial cells, and lymphocytes are also present. Thus, the population is mixed. An exception is the basaloid DCIS with evidence of HMWCK.

Benignancy Criteria for UDH: "COMBS" (as mnemonic acronym) including

*C*lefting, elongated irregular slit-like fenestrations, observed at the periphery of the duct, and Calcifications, intraluminal or stromal (vs. basal), but no psammoma bodies or necrosis

*O*verlapping, normochromatic oval nuclei (vs. round hyperchromatic)

*M*E cell layer, + basal/intermediate cells and lymphocytes and metaplasia, apocrine-type (complete/incomplete), and MΦ, foamy-type in lumen and associated with epithelium

*B*orders, indistinct (cellular) with acidophilic, finely granular cytoplasm (vs. pale, homogeneous)

*S*treaming of cells and nuclei characteristically creating irregular, *nonrigid* bridges

ADH or ALH are usually observed in a background of FCD and show some but not all of the features of in situ malignancy (DCIS). This definition is far to be perfect but currently is most probably the most applicable description. The word "SAC" may be used to memorize the criteria:

- Size ≤2 mm (or two duct spaces according to the 1992 Paget's definition)
- Architecture: Ducts are filled and exhibit sharp "punched-out" spaces or micropapillae but harboring a lack of uniformity in cytology.
- Cytology: Ducts are made up of cells with uniformity in cytology (DIN1-DIN2) but lack structural characteristics (i.e., only partial filling of ducts or lack of uniformity in revealing sharp "punched-out" spaces, clear-cut micropapillae or microacini).

Conversely, low-grade DCIS, as seen above, shows fulfilling of all three "SAC" criteria!

- Size >2 cm (and involvement of at least *two* ducts).
- Architecture: Complete filling of ducts with sharply "punched-out" space morphology or evidence of micropapillae (Box 9.6).
- Cytology: Uniformity in cellularity and nuclei with lack of a substantial and incontrovertible overlapping, streaming, or columnar cell population.

Distinguishing between ductal and lobular proliferation may also be a challenging task for a non-breast pathologist, and liaise with breast

---

Box 9.6 Distinguishing Features Between UDH and ADH

	UDH (I-II)	Vs.	ADH (DIN 1b, ≤2 mm)
Cellularity	Mixed (two types)		Uniformity
Secondary Lumina	"Slit-like", irregular, peripheral		Rigid "Roman bridges"
Cell border	Indistinct		Distinct
Nuclei	Variability		Uniformity
Nucleoli	−/+		+/−
Necrosis	−/+		+/−
LMWCK (e.g., CAM 5.2)	+		+
HMWCK (e.g., 34BE12)	+		− *

Notes: UDH, usual ductal hyperplasia; ADH, atypical ductal hyperplasia; DIN, ductal intraepithelial neoplasia; *(apart basaloid DCIS)

pathologists is crucial. Features of ductal proliferation include distinct cell borders, secondary lumina, large nuclei (larger than lobular), and even variants, such as stratified spindle cells, and apocrine, while features of lobular proliferations include indistinct cell borders; solid, cohesive growth pattern; intracytoplasmic lumina; and small nuclei (smaller than ductal) apart from the pleomorphic lobular variant. Immunohistochemically, E-Cad appears to be a sensitive marker of ductal differentiation, and its expression is lost in lobularly differentiated proliferations. However, its utility is validated for DCIS vs. LCIS, whereas there is still debate about borderline lesions, and its utility remains currently uncertain.

9.4.2 Adenosis

Any *multifocal* or *diffuse* primary hyperplastic process of the glandular component of the breast, not associated with pregnancy (physiologic adenosis), characterized by an increase in the number of acini per lobule. It includes several histologic phenotypes:

1. *Sclerosing Adenosis* (SA): ↑ of the number of acini and stromal fibrosis (central *hyper*-glandularity) with a peripheral *hypo*-landularity by the preservation of lobular architecture and two-cell layers (+ME markers).
2. *Radial Scar* (RS): A stellate-shaped lesion with lesion with central fibroelastosis and peripheral cystic dilation of glands.
3. *Complex Sclerosing Lesion* (CSL): RS with size diameter >1 cm.
4. *Blunt Duct Adenosis* (BDA): Dilatation of small ducts with blunting of the ends, but no distortion as seen in SA.
5. *Nodular Adenosis*: Combination of BDA and SA.
6. *Microglandular Adenosis* (MGA): Irregular distribution of small, uniform ducts with open lumina containing eosinophilic secretion in fat or fibrous tissue with "infiltrative" character, thick BM, and no ME layer, but not associated with sclerosis or desmoplastic stroma reaction (DDX: SA and tubular carcinoma). IHC: (+) PAN-CK and S100 but (−) EMA, ME markers, ER/PR, Her2, P53, and GCDFP-15.
7. *Adenomyoepithelial Adenosis* (ME-MGA): MGA with variably enlarged glands, incontrovertible apocrine metaplasia, and prominent ME cell layer.
8. *Atypical Apocrine Sclerosing Lesion*: SA with incontrovertible apocrine metaplasia.

DDX: It includes SA with apocrine metaplasia, apocrine adenosis, and apocrine adenoma.

9.5 Male Breast Disease

9.5.1 Gynecomastia

Gynecomastia is a benign disease, characterized by an enlargement of breast tissue in males. Gynecomastia occurs for several reasons. Transiently in newborns, it may also be present during puberty in preadolescent boys. Abnormal conditions associated with gynecomastia are genetic disorders (Klinefelter syndrome or 47, XXY or 46, XY/47, XXY or 47, XXY/48, XXXY by meiotic non-disjunctions), metabolic disorders, side-effect of drugs or medications (marijuana, heroin, anabolics, psychoactive drugs, antiretroviral therapy), or alcohol (Fig. 9.2). Gynecomastia may also be seen because of the natural decrease of testosterone production in older males or result of increased estrogens or decreased androgens. Grossly, there is an oval, disc-shaped mass of elastic consistency, well-circumscribed and characterized, microscopically, by a proliferation of ducts and surrounding fibrotic stroma, which increases with age of lesion (Fig. 9.2). Other benign lesions include duct ectasia, SA, NDA, and myofibroblastoma.

9.5.2 Breast Cancer of the Male

BRCA2-associated breast cancer with an increased incidence in individuals affected with

Klinefelter syndrome (karyotype and genetic counseling may be mandatory in some centers). Papillary carcinoma is the most frequent carcinoma type, and the invasive lobular carcinoma is the most uncommon, although all types can be seen. *PGN*: 10-YSR, 40%. Klinefelter karyotype is in about 80% of the cases XXY, but occasionally XXXY, or mosaics (~20%). It occurs in about 1 out of 850 live births with some increase with maternal age and it is the most common cause of male infertility according to several authors. The phenotype includes an eunuchoid body with hypogonadism and gynecomastia.

Multiple Choice Questions and Answers

- BRS-1 Which Tanner stage of breast maturation corresponds to the following description "breast is slightly elevated, extending beyond the borders of the areola, which continues to expand but remains in outline with surrounding breast"?
  (a) Tanner stage I
  (b) Tanner stage II
  (c) Tanner stage III
  (d) Tanner stage IV
  (e) Tanner stage V
- BRS-2 Which transcription factors play a role in the development of the human mammary gland?
  (a) STAT5A, GATA-3, Elf-5
  (b) TGF-Beta, GATA-3, Elf-5
  (c) BRCA1, GATA-2, Elf-3
  (d) STAT5A, BRCA2, TNF
- BRS-3 What is the role of STAT5A?
  (a) Critical role in the primordial germ cells
  (b) Specification of luminal cell fate in the mammary gland
  (c) Control of the alveolar differentiation
  (d) Critical role in the establishment of luminal alveolar progenitor cells
- BRS-4 Look at the microphotograph arising from a mammary gland resection in a 14-year-old boy with enlargement of both mammary glands. Which of the following statements is NOT correct for the diagnosis of this patient?

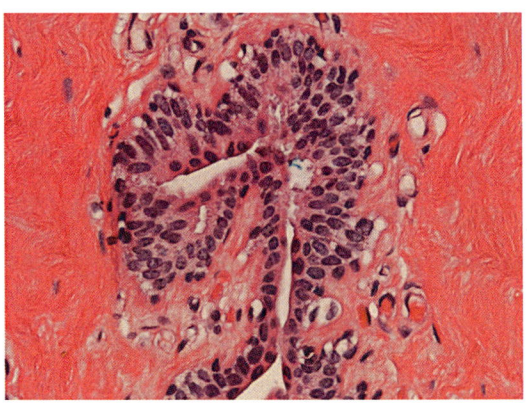

  (a) This condition is defined as enlargement of the male breast, including both epithelial and mesenchymal elements.
  (b) A firm to a rubbery mass of tissue can be palpated beneath the areola.
  (c) It occurs in the setting of hyperprogestinism.
  (d) A variety of drugs have been associated with this condition, including cimetidine, anti-depressants, digitalis, spironolactone, marijuana, levothyroxine, and finasteride.
  (e) Non-pharmacologic causes of gynecomastia include Klinefelter syndrome (47, XXY), alcoholism, long-term use of marijuana, hypogonadism, human chorionic gonadotropin-producing tumors, prolactinomas, hyperthyroidism, chronic renal disease, and chronic pulmonary disease.
- BRS-5 What is the meaning of the Toker cells?
  (a) Inconspicuous cytokeratin 7-positive cells that should be kept separated from intraepidermal involvement by malignant cells seen in Paget disease (PD) of the breast.
  (b) Progenitor cells for PD of the breast that may evolve to PD of the breast.
  (c) Inconspicuous cytokeratin 19-positive cells that should be kept separated from intraepidermal involvement by malignant cells seen in PD of the breast.
  (d) Inconspicuous cytokeratin 20-positive cells that should be kept separated from intraepidermal involvement by malignant cells seen in PD of the breast.

- BRS-6 What is the basis of the traditional Nottingham grading of ductal carcinoma of the breast?
  - (a) Amount of gland formation, nuclear pleomorphism, and mitoses
  - (b) Amount of necrosis, nuclear pleomorphism, and mitoses
  - (c) Amount of calcification, nuclear pleomorphism, and gland formation
  - (d) Amount of necrosis, calcification, and mitoses
- BRS-7 Adenoid cystic carcinoma (AdCC) is a rare type of triple-negative breast cancer (TNBC). Which of the following fusion genes is characteristic of the AdCC?
  - (a) *F3B1-FBXW7* fusion gene
  - (b) *FGFR2, MYB* fusion gene
  - (c) *MYB-NFIB* fusion gene
  - (d) *NOTCH1-MYB* fusion gene
  - (e) *NOTCH2-NFIB* fusion gene
- BRS-8 Which of the following statement regarding HER2/neu is TRUE?
  - (a) *HER2* gene encodes transduction molecule (p185).
  - (b) *HER2* gene encodes an adhesion molecule (p185).
  - (c) *HER2* gene encodes a nuclear receptor (p185).
  - (d) *HER2* gene encodes transmembrane growth factor receptor (p185).
- BRS-9 What is NOT a criterion used for high-grade phyllodes tumor of the breast?
  - (a) Infiltrative margins
  - (b) Marked nuclear atypia
  - (c) Mitotic rate >10/10 HPF
  - (d) Size > 4cm
  - (e) Eosinophilic globules
- BRS-8 Which of the following statements is NOT correct?
  - (a) Mammography represents an investigation of assessing breast masses effectively in a pediatric patient.
  - (b) Over 50% of fibroadenomas resolve spontaneously over 5–10 years.
  - (c) Juvenile fibroadenomas are more common in girls of African American ethnics.
  - (d) Malignant phyllodes tumor is the most common primary pediatric malignancy of the breast.

References and Recommended Readings

1. Abou-Zamzam A, Somers S, Cora C, Pairawan S, Lum S. Percutaneous needle biopsies of the breast in women younger than 35 years: minimally or excessively invasive? Am Surg. 2017;83(10):1019–23. PubMed PMID: 29391087.
2. Al Jishi T, Sergi C. Current perspective of diethylstilbestrol (DES) exposure in mothers and offspring. Reprod Toxicol. 2017;71:71–7. https://doi.org/10.1016/j.reprotox.2017.04.009. Epub 2017 Apr 28. Review. PubMed PMID: 28461243.
3. Bacchi CE, Silva TR, Zambrano E, Plaza J, Suster S, Luzar B, Lamovec J, Pizzolitto S, Falconieri G. Epithelioid angiosarcoma of the skin: a study of 18 cases with emphasis on its clinicopathologic spectrum and unusual morphologic features. Am J Surg Pathol. 2010;34(9):1334–43. https://doi.org/10.1097/PAS.0b013e3181ee4eaf. PubMed PMID: 20697249.
4. Berkey CS, Willett WC, Frazier AL, Rosner B, Tamimi RM, Rockett HR, Colditz GA. Prospective study of adolescent alcohol consumption and risk of benign breast disease in young women. Pediatrics. 2010;125(5):e1081–7. https://doi.org/10.1542/peds.2009-2347. Epub 2010 Apr 12. PubMed PMID: 20385629; PubMed Central PMCID: PMC3075610.
5. Bezić J, Karaman I, Šundov D. Combined fibroadenoma and tubular adenoma of the breast: rare presentation that confirms common histogenesis. Breast J. 2015;21(3):309–11. https://doi.org/10.1111/tbj.12400. Epub 2015 Mar 17. PubMed PMID: 25775939.
6. Boettger MB, Sergi C, Meyer P. BRCA1/2 mutation screening and LOH analysis of lung adenocarcinoma tissue in a multiple-cancer patient with a strong family history of breast cancer. J Carcinog. 2003 Oct 2;2(1):5. PubMed PMID: 14583096; PubMed Central PMCID: PMC239936.
7. Colditz GA, Bohlke K, Berkey CS. Breast cancer risk accumulation starts early: prevention must also. Breast Cancer Res Treat. 2014;145(3):567–79. https://doi.org/10.1007/s10549-014-2993-8. Epub 2014 May 13. Review. PubMed PMID: 24820413; PubMed Central PMCID: PMC4079839.
8. Connor AE, Baumgartner RN, Pinkston C, Baumgartner KB. Obesity and risk of breast cancer mortality in Hispanic and Non-Hispanic white women: the New Mexico Women's Health Study. J Womens Health (Larchmt). 2013;22(4):368–77. https://doi.org/10.1089/jwh.2012.4191. Epub 2013

Mar 26. PubMed PMID: 23531051; PubMed Central PMCID: PMC3627406.

9. D'Angelo P, Carli M, Ferrari A, Manzitti C, Mura R, Miglionico L, Di Cataldo A, Grigoli A, Cecchetto G, Bisogno G, AIEOP Soft Tissue Sarcoma Committee. Breast metastases in children and adolescents with rhabdomyosarcoma: experience of the Italian Soft Tissue Sarcoma Committee. Pediatr Blood Cancer. 2010;55(7):1306–9. https://doi.org/10.1002/pbc.22729. Epub 2010 Aug 20. PubMed PMID: 20730885.

10. Danner-Koptik KE, Majhail NS, Brazauskas R, Wang Z, Buchbinder D, Cahn JY, Dilley KJ, Frangoul HA, Gross TG, Hale GA, Hayashi RJ, Hijiya N, Kamble RT, Lazarus HM, Marks DI, Reddy V, Savani BN, Warwick AB, Wingard JR, Wood WA, Sorror ML, Jacobsohn DA. Second malignancies after autologous hematopoietic cell transplantation in children. Bone Marrow Transplant. 2013;48(3):363–8. https://doi.org/10.1038/bmt.2012.166. Epub 2012 Sep 10. PubMed PMID: 22964594; PubMed Central PMCID: PMC3525761.

11. Djakovic A, Engel JB, Geisinger E, Honig A, Tschammler A, Dietl J. Pleomorphic adenoma of the breast initially misdiagnosed as metaplastic carcinoma in preoperative stereotactic biopsy: a case report and review of the literature. Eur J Gynaecol Oncol. 2011;32(4):427–30. Review. PubMed PMID: 21941969.

12. East EG, Zhao L, Pang JC, Jorns JM. Characteristics of a breast pathology consultation practice. Arch Pathol Lab Med. 2017;141(4):578–84. https://doi.org/10.5858/arpa.2016-0371-OA. PubMed PMID: 28353380.

13. Fallon SC, Hatef DA, McKnight AJ, Izaddoost SA, Brandt ML. Congenital synmastia with concurrent fibroadenomas in a pediatric patient. J Pediatr Surg. 2013;48(1):255–7. https://doi.org/10.1016/j.jpedsurg.2012.10.053. PubMed PMID: 23331826.

14. Farley-Loftus R, Bossenbroek NM, Rosenman K, Schaffer JV. Clear cell papulosis. Dermatol Online J. 2008;14(10):19. PubMed PMID: 19061618.

15. Ferrari A, Casanova M, Massimino M, Sultan I. Peculiar features and tailored management of adult cancers occurring in pediatric age. Expert Rev Anticancer Ther. 2010;10(11):1837–51. https://doi.org/10.1586/era.10.105. Review. PubMed PMID: 21080807.

16. Ferreira CG, de Melo AC, Nogueira-Rodrigues A. The adolescent and young adult with cancer: state of the art--epithelial cancer. Curr Oncol Rep. 2013;15(4):287–95. https://doi.org/10.1007/s11912-013-0322-8. Review. PubMed PMID: 23754487.

17. Forsman CL, Ng BC, Heinze RK, Kuo C, Sergi C, Gopalakrishnan R, Yee D, Graf D, Schwertfeger KL, Petryk A. BMP-binding protein twisted gastrulation is required in mammary gland epithelium for normal ductal elongation and myoepithelial compartmentalization. Dev Biol. 2013;373(1):95–106. https://doi.org/10.1016/j.ydbio.2012.10.007. Epub 2012 Oct 24. PubMed PMID: 23103586; PubMed Central PMCID: PMC3508155.

18. Frank B, Meyer P, Boettger MB, Hemminki K, Stapelmann H, Gast A, Schmitt C, Kumar R, Sergi C, Burwinkel B. ARLTS1 variants and melanoma risk. Int J Cancer. 2006;119(7):1736–7. PubMed PMID: 16646072.

19. Frazier AL, Rosenberg SM. Preadolescent and adolescent risk factors for benign breast disease. J Adolesc Health. 2013;52(5 Suppl):S36–40. https://doi.org/10.1016/j.jadohealth.2013.01.007. PubMed PMID: 23601609.

20. Ghilli M, Mariniello DM, Fanelli G, Cascione F, Fontana A, Cristaudo A, Cilotti A, Caligo AM, Manca G, Colizzi L, Naccarato AG, Roncella M. Carcinosarcoma of the breast: an aggressive subtype of metaplastic cancer. Report of a rare case in a young BRCA-1 mutated woman. Clin Breast Cancer. 2017;17(1):e31–5. https://doi.org/10.1016/j.clbc.2016.08.002. Epub 2016 Aug 30. PubMed PMID: 27697421.

21. Gillcard O, Goodman A, Cooper M, Davies M, Dunn J. The significance of the Van Nuys prognostic index in the management of ductal carcinoma in situ. World J Surg Oncol. 2008;6:61. https://doi.org/10.1186/1477-7819-6-61. PubMed PMID: 18564426; PubMed Central PMCID: PMC2459183.

22. Gupta SS, Singh O. Cystic lymphangioma of the breast in an 8-year-old boy: report of a case with a review of the literature. Surg Today. 2011;41(9):1314–8. https://doi.org/10.1007/s00595-010-4382-1. Epub 2011 Aug 26. Review. PubMed PMID: 21874439.

23. Henderson TO, Amsterdam A, Bhatia S, Hudson MM, Meadows AT, Neglia JP, Diller LR, Constine LS, Smith RA, Mahoney MC, Morris EA, Montgomery LL, Landier W, Smith SM, Robison LL, Oeffinger KC. Systematic review: surveillance for breast cancer in women treated with chest radiation for childhood, adolescent, or young adult cancer. Ann Intern Med. 2010;152(7):444–55; W144-54. https://doi.org/10.7326/0003-4819-152-7-201004060-00009. Review. PubMed PMID: 20368650; PubMed Central PMCID: PMC2857928.

24. Herman JD, Appelbaum H. Hereditary breast and ovarian cancer syndrome and issues in pediatric and adolescent practice. J Pediatr Adolesc Gynecol. 2010;23(4):253–8. PubMed PMID: 20632459.

25. John K, Becker K, Mattejat F. Impact of family-oriented rehabilitation and prevention: an inpatient program for mothers with breast cancer and their children. Psychooncology. 2013; https://doi.org/10.1002/pon.3329. [Epub ahead of print] PubMed PMID: 23760766.

26. Kapila K, Pathan SK, Al-Mosawy FA, George SS, Haji BE, Al-Ayadhy B. Fine needle aspiration cytology of breast masses in children and adolescents: experience with 1404 aspirates. Acta Cytol. 2008;52(6):681–6. PubMed PMID: 19068671.

27. Karadeniz E, Arslan S, Akcay MN, Subaşi ID, Demirci E. Papillary lesions of breast. Chirurgia (Bucur). 2016;111(3):225–9. PubMed PMID: 27452933.

28. Kim SW, Roh J, Park CS. Clear cell papulosis: a case report. J Pathol Transl Med. 2016;50(5):401–3.

https://doi.org/10.4132/jptm.2016.02.16. Epub 2016 May 29. PubMed PMID: 27237133; PubMed Central PMCID: PMC5042892.

29. Knezevich SR, McFadden DE, Tao W, Lim JF, Sorensen PH. A novel ETV6-NTRK3 gene fusion in congenital fibrosarcoma. Nat Genet. 1998;18(2):184–7. https://doi.org/10.1038/ng0298-184. PMID 9462753.

30. Kuijper A, Mommers EC, van der Wall E, van Diest PJ. Histopathology of fibroadenoma of the breast. Am J Clin Pathol. 2001;115(5):736–42. PubMed PMID: 11345838.

31. Kulkarni D, Dixon JM. Congenital abnormalities of the breast. Women's Health (Lond Engl) 2012;8(1):75–86; quiz 87–8. https://doi.org/10.2217/whe.11.84. Review. PubMed PMID: 22171777.

32. La Rocca G, Anzalone R, Corrao S, Magno F, Rappa F, Marasà S, Czarnecka AM, Marasà L, Sergi C, Zummo G, Cappello F. CD1a down-regulation in primary invasive ductal breast carcinoma may predict regional lymph node invasion and patient outcome. Histopathology. 2008;52(2):203–12. https://doi.org/10.1111/j.1365-2559.2007.02919.x. PubMed PMID: 18184269.

33. Laé M, Gardrat S, Rondeau S, Richardot C, Caly M, Chemlali W, Vacher S, Couturier J, Mariani O, Terrier P, Bièche I. MED12 mutations in breast phyllodes tumors: evidence of temporal tumoral heterogeneity and identification of associated critical signaling pathways. Oncotarget. 2016;7(51): 84428–38.https://doi.org/10.18632/oncotarget.12991. PubMed PMID: 27806318; PubMed Central PMCID: PMC5356671.

34. Lee JY, Chao SC. Clear cell papulosis of the skin. Br J Dermatol. 1998;138(4):678–83. PubMed PMID: 9640379.

35. Lemoine C, Mayer SK, Beaunoyer M, Mongeau C, Ouimet A. Incidental finding of synchronous bilateral ductal carcinoma in situ associated with gynecomastia in a 15-year-old obese boy: case report and review of the literature. J Pediatr Surg. 2011;46(9):e17–20. https://doi.org/10.1016/j.jpedsurg.2011.06.010. Review. PubMed PMID: 21929970.

36. Mizutani L, Tanaka Y, Kondo Y, Bando H, Hara H. Glomus tumor of a female breast: a case report and review of the literature. J Med Ultrason (2001). 2014;41(3):385–8. https://doi.org/10.1007/s10396-013-0517-5. Epub 2014 Jan 14. Review. PubMed PMID: 27277916.

37. Mosavel M, Genderson MW. From adolescent daughter to mother: exploring message design strategies for breast and cervical cancer prevention and screening. J Cancer Educ. 2013; 28(3):558–64. https://doi.org/10.1007/s13187-013-0503-z. PubMed PMID: 23813491; PubMed Central PMCID: PMC4046862.

38. Page DL, Rogers LW. Combined histologic and cytologic criteria for the diagnosis of mammary atypical ductal hyperplasia. Hum Pathol. 1992;23(10):1095–7. PubMed PMID: 1328030.

39. Richards MK, Goldin AB, Beierle EA, Doski JJ, Goldfarb M, Langer M, Nuchtern JG, Vasudevan S, Gow KW, Javid SH. Breast malignancies in children: presentation, management, and survival. Ann Surg Oncol. 2017;24(6):1482–91. https://doi.org/10.1245/s10434-016-5747-5. Epub 2017 Jan 5. PubMed PMID: 28058544.

40. Rodríguez Ogando A, Fernández López T, Rodríguez Castaño MJ, Mata Fernández C, Alvarez Bernardi J, Cebollero Presmanes M. Cystosarcoma phyllodes of the breast: a case report in a 12-year-old girl. Clin Transl Oncol. 2010;12(10):704–6. PubMed PMID: 20947486.

41. Sanchez R, Ladino-Torres MF, Bernat JA, Joe A, DiPietro MA. Breast fibroadenomas in the pediatric population: common and uncommon sonographic findings. Pediatr Radiol. 2010;40(10):1681–9. https://doi.org/10.1007/s00247-010-1678-7. Epub 2010 May 7. PubMed PMID: 20449731.

42. Sergi C, Dhiman A, Gray JA. Fine needle aspiration cytology for neck masses in childhood. An illustrative approach. Diagnostics (Basel). 2018;8(2):E28. https://doi.org/10.3390/diagnostics8020028. Review. PubMed PMID: 29690556.

43. Sillesen NH, Hölmich LR, Siersen HE, Bonde C. Congenital symmastia revisited. J Plast Reconstr Aesthet Surg. 2012;65(12):1607–13. https://doi.org/10.1016/j.bjps.2012.08.008. Epub 2012 Sep 29. Review. PubMed PMID: 23026472.

44. Stasik CJ, Davis M, Kimler BF, Fan F, Damjanov I, Thomas P, Tawfik OW. Grading ductal carcinoma in situ of the breast using an automated proliferation index. Ann Clin Lab Sci. 2011;41(2):122–30. PubMed PMID: 21844569.

45. Tang H, Liu F, Li H, Huang X, Zhao T. Pleomorphic carcinoma of breast: a case report and review of literature. Int J Clin Exp Pathol. 2014;7(8):5215–20. eCollection 2014. Review. PubMed PMID: 25197400; PubMed Central PMCID: PMC4152090.

46. Terenziani M, Casalini P, Scaperrotta G, Gandola L, Trecate G, Catania S, Cefalo G, Conti A, Massimino M, Meazza C, Podda M, Spreafico F, Suman L, Gennaro M. Occurrence of breast cancer after chest wall irradiation for pediatric cancer, as detected by a multimodal screening program. Int J Radiat Oncol Biol Phys. 2013;85(1):35–9. https://doi.org/10.1016/j.ijrobp.2012.03.043. Epub 2012 Jun 5. PubMed PMID: 22677366.

47. Testa JR, Malkin D, Schiffman JD. Connecting molecular pathways to hereditary cancer risk syndromes. Am Soc Clin Oncol Educ Book. 2013;2013:81–90. https://doi.org/10.1200/EdBook_AM.2013.33.81. PubMed PMID: 23714463.

48. Tognon C, Knezevich SR, Huntsman D, Roskelley CD, Melnyk N, Mathers JA, Becker L, Carneiro F, MacPherson N, Horsman D, Poremba C, Sorensen PH. Expression of the ETV6-NTRK3 gene fusion as a primary event in human secretory breast carcinoma. Cancer Cell. 2002;2(5):367–76. https://doi.org/10.1016/S1535-6108(02)00180-0. PMID 12450792.

49. Tse GM, Ni YB, Tsang JY, Shao MM, Huang YH, Luo MH, Lacambra MD, Yamaguchi R, Tan

PH. Immunohistochemistry in the diagnosis of papillary lesions of the breast. Histopathology. 2014;65(6):839–53. https://doi.org/10.1111/his.12453. Epub 2014 Oct 30. PubMed PMID: 24804569.
50. Wai DH, Knezevich SR, Lucas T, Jansen B, Kay RJ, Sorensen PH. The ETV6-NTRK3 gene fusion encodes a chimeric protein tyrosine kinase that transforms NIH3T3 cells. Oncogene. 2000;19(7):906–15. https://doi.org/10.1038/sj.onc.1203396. PMID 10702799.
51. Wang Y, Zhu JF, Liu YY, Han GP. An analysis of cyclin D1, cytokeratin 5/6 and cytokeratin 8/18 expression in breast papillomas and papillary carcinomas. Diagn Pathol. 2013;8:8. https://doi.org/10.1186/1746-1596-8-8. PubMed PMID: 23327593; PubMed Central PMCID: PMC3571902.
52. Zils K, Ebner F, Ott M, Müller J, Baumhoer D, Greulich M, Rehnitz D, Rempen A, Schaetzle S, Wilhelm M, Bielack S. Extraskeletal osteosarcoma of the breast in an adolescent girl. J Pediatr Hematol Oncol. 2012;34(6):e261–3. https://doi.org/10.1097/MPH.0b013e31823366b4. PubMed PMID: 22246152.

Hematolymphoid System

Contents

10.1	**Development and Genetics**	862
10.2	**Red Blood Cell Disorders**	864
10.2.1	Anemia	864
10.2.2	Polycythemia	867
10.3	**Coagulation and Hemostasis Disorders**	869
10.3.1	Coagulation and Hemostasis	869
10.3.2	Coagulation Disorders	869
10.3.3	Platelet Disorders	870
10.4	**White Blood Cell Disorders**	871
10.4.1	Leukocytopenias and Leukocyte Dysfunctionalities	871
10.4.2	Non-neoplastic Leukocytosis	871
10.4.3	Leukemia (Neoplastic Leukocytosis) or *Virchow's "Weisses Blut"*	872
10.4.4	Myelodysplastic Syndromes	878
10.4.5	Hodgkin Lymphoma	878
10.4.6	Non-Hodgkin Lymphomas	887
10.4.7	Follicular Lymphoma	888
10.4.8	Small Lymphocytic Lymphoma	890
10.4.9	Mantle Cell Lymphoma (MCL)	891
10.4.10	Marginal Cell Lymphoma	891
10.4.11	Diffuse Large B-Cell Lymphoma (DLBCL)	892
10.4.12	Lymphoblastic Lymphoma	894
10.4.13	Burkitt Lymphoma	896
10.4.14	Peripheral T-Cell Lymphoma	902
10.4.15	Anaplastic Large Cell Lymphoma	903
10.4.16	Adult T-Cell Leukemia/Lymphoma	903
10.4.17	Cutaneous T-Cell Lymphoma (CTCL)	904
10.4.18	Angiocentric Immunoproliferative Lesions	905
10.4.19	Extranodal NK-/T-Cell Lymphoma	906
10.5	**Disorders of the Monocyte-Macrophage System and Mast Cells**	907
10.5.1	Hemophagocytic Syndrome	907
10.5.2	Sinus Histiocytosis with Massive Lymphadenopathy (Rosai-Dorfman Disease)	908
10.5.3	Langerhans Cell Histiocytosis	908
10.5.4	Histiocytic Medullary Reticulosis	909

10.5.5	True Histiocytic Lymphoma	910
10.5.6	Systemic Mastocytosis	910
10.6	**Plasma Cell Disorders**	**910**
10.6.1	Multiple Myeloma	911
10.6.2	Solitary Myeloma	912
10.6.3	Plasma Cell Leukemia	912
10.6.4	Waldenstrom's Macroglobulinemia	912
10.6.5	Heavy Chain Disease	912
10.6.6	Monoclonal Gammopathy of Undetermined Significance (MGUS)	913
10.7	**Benign Lymphadenopathies**	**913**
10.7.1	Follicular Hyperplasia	914
10.7.2	Diffuse (Paracortical) Hyperplasia	918
10.7.3	Sinus Pattern	919
10.7.4	Predominant Granulomatous Pattern	920
10.7.5	Other Myxoid Patterns	921
10.7.6	Angioimmunoblastic Lymphadenopathy with Dysproteinemia (AILD)	921
10.8	**Disorders of the Spleen**	**921**
10.8.1	White Pulp Disorders of the Spleen	922
10.8.2	Red Pulp Disorders of the Spleen	922
10.9	**Disorders of the Thymus**	**924**
10.9.1	Thymic Cysts and Thymolipoma	924
10.9.2	True Thymic Hyperplasia and Thymic Follicular Hyperplasia	924
10.9.3	HIV Changes	925
10.9.4	Thymoma	925
Multiple Choice Questions and Answers		**927**
References and Recommended Readings		**930**

10.1 Development and Genetics

The spleen (adult organ weight, 150 g, 12 cm as adult size in its largest dimension) is a hematopoietic and immunologic paramedian and single organ. It is aimed to filter cells and toxins, act in the immunological system, produce lymphoreticular or hematopoietic cells, and is considered the platelet storage organ for excellence. The cortex consists of red and white pulp. The red pulp has venous sinuses, which seem to be unique in the human body being surrounded by reticulin (Ret) annular rings ("Ret Rings"), and lining cells, called "littoral cells," which show a specific phenotype CD8+, CD31+, FVIIIRA+, but CD34-. These sinuses are separated by pulp cords of Billroth and terminal branches of the arterial system. The white pulp is made of periarteriolar T cells, Malpighian corpuscles including germinal center cells (GCC) (IgD−, BCL2−), mantle small B cells (IgD+, BCL2+), and marginal zone monocytoid B cells (IgD−, BCL2+). Thus, using two antibodies against IgD and BCL2, we may able to differentiate the three zones. In fact, the pulp cord is a meshwork of macrophages, T cells, and B cells.

The thymus is a median lymphoepithelial organ essential for the immunocompetency of the organism.

– Anterior mediastinum, pyramid-shape, thin fibrous capsule, lobular with cortex (dark staining) and medulla (light staining) with interconnecting component of epithelial cells that form the scaffolding for the maturing immune cells (perivascular spaces or PVS are devoid of CK+ epithelial cells but packed with small lymphocytes) and characteristic Hassal's corpuscles with onion-skin layered keratinization.
– It originates from third and fourth pharyngeal pouches, a finding that is essential to correlate for neck remnants.
– Organ weight is steadily increasing after birth until 30–40 g at puberty, and then there is a gradual fatty tissue replacement with weight change.
– Function: T-cell maturation with a selection of cells able to form the repertoire of mature T

cells and deletion of cells that bind with high affinity to self-peptide antigens (Box 10.1).

> **Box 10.1 Immunophenotype of Thymus Cell Population**
> 1. Cortical thymocytes, immature T cells: (+) TdT, CD1a, CD99a AND (+) cCD3 and (−) sCD3
> 2. Medullary thymocytes, mature T cells: (−) TdT, CD1a, CD99a AND (+) s/cCD3
> 3. B cells: (+) CD19, CD20, CD22, CD23; (−) CD5, CD21, CD32, CD35a; (+) cIgD, IgM, IgG
>
> Note: s, surface; c, cytoplasmic; CD, cluster differentiation

TdT may be a quite tricky antigen to be detected, and there are a few positive and negative situations and pitfalls that should be taken into account in the routine of a surgical and clinical pathologist. In fact, TdT usually is positive immunohistochemically in hematogones (B- and T-cell precursors) and cortical thymocytes as well as focally in benign pediatric lymph nodes and pediatric tonsils and needs to be taken into account by the pathologist. This situation may cause possible confusion in patients with acute lymphoblastic leukemia. Pathologic positive staining, i.e., staining that indicates a pathological result and has influence for the proper management of the patient, is seen in acute lymphoblastic leukemia, acute myelogenous leukemia, myeloid sarcoma, plasmacytoid dendritic cell neoplasm (blastic phase), and thymoma (lymphocyte predominant). Another condition, i.e., Merkel cell carcinoma, is also positive, but it is practically inexistent in children and youth. Negative staining should be detected in myeloid cells and developed leukemias and lymphomas (e.g., Burkitt lymphoma, diffuse large B-cell lymphoma, MALT lymphoma, mantle cell lymphoma, and T-cell prolymphocytic leukemia). Moreover, negative staining is seen in sarcomas and lung cell carcinoma, small cell type.

The bone marrow is the central hematopoietic organ of the body and produces a number of approximately 200 billion new erythrocytes every day, along with leukocytes and platelets. Bone marrow also contains mesenchymal and hematopoietic stem cells. Hematopoietic stem cells located in the bone marrow give origin to two main types of cells: myeloid and lymphoid lineages. These include monocytes, macrophages, neutrophils, basophils, eosinophils, erythrocytes, dendritic cells, and megakaryocytes or platelets, as well as T cells, B cells, and natural killer cells. Polymorphonuclear leukocytes (PMN) are usually equivalent to the term of polymorphonuclear neutrophils or granulocytes. There are different types of hematopoietic stem cells, which vary in their regenerative capacity and potency. Some stem cells are multipotent, oligopotent, or unipotent as determined by how many types of cell they can generate. The advent of techniques (e.g., CRISPR-Cas9 of gene editing) and new methods of targeting cancer stem cells have been pillars in new protocols in oncology. The pluripotent hematopoietic stem cells have the properties of both renewal and differentiation. In regeneration, these cells can reproduce another cell identical to themselves, while in differentiation, these cells can generate ≥ 1 subsets of more mature cells. Multipotent hematopoietic stem cells emerge within the delicate environment of human embryonic arteries. This process seems to be, at least partially, independent from the primitive vitelline hematopoiesis and uses a "bed" vascular endothelial cell intermediate. At 24–26 days of early development, the aortic-genital-mesonephric (AGM) primitive zone, i.e., the embryonic region comprising the aorta, genital ridges, and mesonephros, starts to be hemogenic in vitro. This event happens before the appearance of aortic clusters of hematopoietic stem cells. Cell sorting and culture experiments have, indeed, revealed that blood-forming potential in the AGM is strictly confined to the CD34-negative, i.e., non-endothelial, cell fraction. It specifically points to either that ventral aortic endothelial cells switch suddenly to hematopoiesis at day 27 of the embryogenetic development or that more primitive, CD34-angiohematopoietic progenitors migrate through the periaortic mesenchyme and colonize the floor of the aorta at this embryonic age. A body of knoweldge arising from recent studies seems to support that the second hypothesis the second hypothesis as "more" correct. This situation is based on the expression of Fetal Liver Kinase 1 or Flk-1, or its human ortholog vascular endothelial growth factor - receptor 2 (VEGF-R2), in the early human embryo. Some studies of developmental biology have revealed that a population of Flk-1+ CD34 cells migrate from the splanchnopleure into the subaortic mesoderm during the 4th week of embryonic development. Of note, kinase insert domain receptor (KDR), which is a type IV receptor tyrosine kinase and is aka VEGFR-2 is essentially a VEGF receptor. *KDR* is the human gene encoding it. *KDR* has also received a CD label or number (CD309).

10.2 Red Blood Cell Disorders

10.2.1 Anemia

- *DEF*: It is defined as a reduction in the amount of circulating hemoglobin (Hb), red blood cells (RBC), or both. There is an imbalance between one or both the components and there is inability from the body to replace one or both parts through two usually mutually exclusive but possibly combining mechanisms, including:

  1. Excessive blood turnover showing an increased reticulocyte count (↑ Rets).
  2. Failure of blood production showing a decreased reticulocyte count (↓ Rets) (Fig. 10.1).

According to RBC size, anemia is subclassified into microcytic, normocytic, and macrocytic.

According to Hb content, anemia is subclassified in hypochromic and normochromic. Microcytic hypochromic anemia characterizes the well-known condition of sideropenia or iron deficiency.

- *CLI*: It may be remembered clinically as *Pallor cutis et albor membranorum internorum*, a Latin sentence recalling the signs and symptoms of anemia that also accompany tachysphygmia and tachycardia, shortness of breath, palpitations, dizziness, and fatigue. There may be heart failure in case of moderate and severe anemia and fatigue with very few other symptoms, if anemia develops slowly and compensation occurs.

10.2.1.1 Anemia Due to Excessive Blood Turnover

1. Excessive blood turnover is due to bleeding (outside of the vascular system) or hemolysis (inside the vascular system) ⇒ Bone marrow (BM) erythroid hyperplasia with increased Rets in the peripheral blood

   - If haptoglobin is +, then bleeding is the cause of the excessive blood turnover.
   - If haptoglobin is −, then acute hemolysis is the cause of the excessive blood turnover.

This situation is because haptoglobin (Hg) [not to be confused with the symbol of mercury] is a plasma protein able to keep the Fe elaborated from the small amount of Hb, which is usually in the plasma. Thus, Hb released in events of acute hemolysis, binds to Hg within a few hours, and serum Hg is unmeasurable because the complex Hb-Hg is cleared from the circulation by the reticulo-endothelial system (RES). In constant excessive blood turnover, the absence of Fe in the BM aspirate or biopsy indicates blood loss to the outside, whereas Fe deposits are seen in any other type of anemia.

– Hemorrhage:

Hemorrhage-Associated Anemias

- If acute: Normochromic normocytic anemia (⇒ hypovolemic shock with tachycardia, which can progress to death)
- If chronic: Hypochromic anemia (variable etiology: GI, e.g., ulcer, cancer, polyp; GU, e.g., cancer or lithiasis; and GYN, e.g., meno-/metrorrhagia)

– Hemolysis:

Congenital Hemolytic Anemias with Defects in
(a) Hb structure ⇒ hemoglobinopathies, e.g., sickle cell anemia (Glu→Val at position 6 in the β-chain of Hb ⇒ HbS). This anemia is

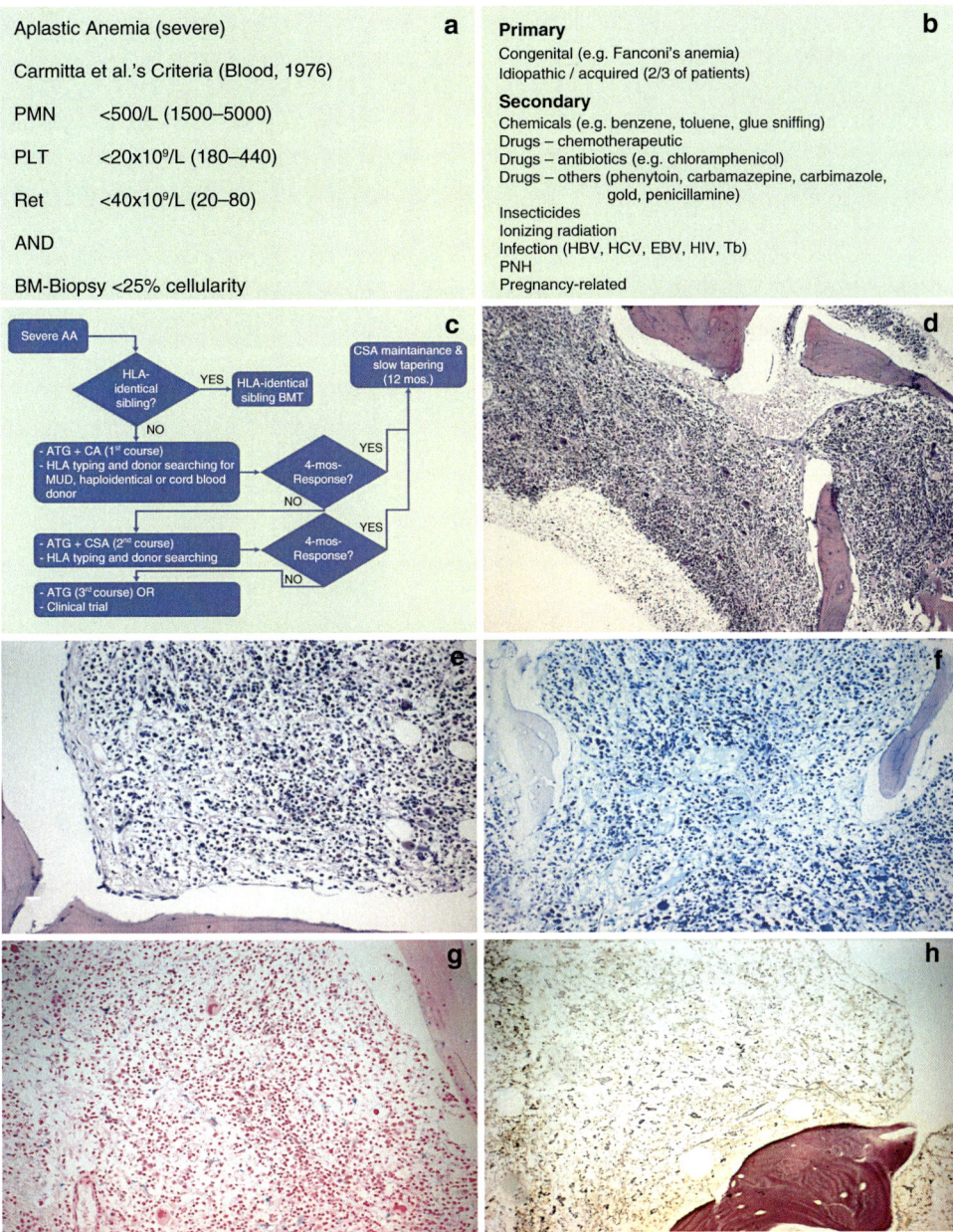

Fig. 10.1 In figure (**a**) are shown the Carmitta et al.'s criteria for aplastic anemia of severe grade. In figure (**b**) the classification of the aplastic anemia is presented. In figure (**c**) an algorithm for aplastic anemia is shown. The bone marrow biopsy shows an aplastic crisis in a young patient with beta-thalassemia intermedia receiving deferasirox. Deferasirox is an oral iron chelator and is largely used to decrease chronic iron overload in patients who are receiving long-term blood transfusions for chronic anemia such as beta-thalassemia. Glucuronidation is the main metabolic pathway to metabolite this drug with subsequent biliary excretion. CYP450-catalyzed (oxidative-type) metabolism of this drug is less than 10% in humans. Aplastic crisis is explained as a rapid fall in Hb levels associated with few or no reticulocytes, indicating a clear-cut failure of the bone marrow to respond to increased cell turnover. The reticulocytopenia occurs without evidence of hemolysis, usually from parvovirus B19 infection in about two thirds of children and adolescents. Figures (**d**) and (**f**) show a quite well preserved level of cellularity at first glance, but the erythroid colonies are very few or nil at paratrabecular level. Some focal iron deposition is seen (**g**) and there is an abnormal elastic fibers appearance (**h**). (**d**, H&E stain, x 100; **e**, H&E stain, x 200; **f**, Giemsa staining x 200; **g**, Pearls' Prussian Blue stain x 200; **h**, Elastic stain x 200) *Note*: *PMN*, polymorphonuclear leukocytes or polymorphonuclear neutrophils; *PLT*, platelets; *Ret*, reticulocytes; *HBV*, hepatitis B virus; *HCV*, hepatitis C virus; *EBV*, Epstein Barr virus; *HIV*, human immunodeficiency virus; *Tb*, tuberculosis; *PNH*, paroxysmal nocturnal hemoglobinuria.

often observed in blacks (0.2% of blacks are homozygous in North America, and 8–10% are carriers of the mutation). It is characterized by an abnormal solubility of Hb with HbS precipitation at low pH and decreased O_2 tension and following deformation of the RBC into sickle cells. The sickle cells are prone to hemolysis and get stuck in the capillaries giving rise to vascular occlusion, ischemia, necrosis and organ dysfunction. Clinically, the patients may present with leg ulcers with bony infarcts and unbearable pain, renal papillary necrosis with hematuria, auto-infarction with atrophy of the spleen, hepatomegaly with jaundice and liver dysfunction, cholelithiasis, as well as pulmonary thrombi (Gr. θρόμβος, clot), and stroke. It is extremely important to emphasize that a sickle cell crisis is a painful event due to acute vascular occlusion and may lead to sudden death.
(b) Hb synthesis ⇒ β-thalassemia is characterized by functionally inadequate Hb due to failure or decrease in the synthesis of the beta chain of Hb and often observed in individuals of the Mediterranean area. RBCs are called hypochromic target cells. Two forms are known such as thalassemia major (Cooley's anemia) or homozygous thalassemia and thalassemia minor or heterozygous form (genetic counseling and mild anemia, otherwise normal life span).
(c) RBC metabolism involving either the membrane (e.g., spectrin in hereditary spherocytosis) or intracellular enzymes with abnormal RBC indices with irregular RBC shape and following RES sequestration.

Acquired Hemolytic Anemias with hemolysis due to

(a) *Autoimmunity*, i.e., autoimmune hemolytic anemia, which is due to IgG antibodies directed against RBC leading to cell lysis directly or following interaction with antigens of the RBC membrane altering the susceptibility of RBC to be entrapped and destroyed in the spleen. Autoimmunity may be either idiopathic or arise secondarily to some underlying diseases, such as Hodgkin lymphoma, chronic lymphocytic leukemia, carcinoma, sarcoidosis, or collagen disorders.
(b) *Blood group (-immune-) incompatibility* occurs in case of fetomaternal incompatibility at Rh locus D, when a D-negative mother following exposure to D-(+) RBC in occasion of her second birth without previous blood transfusion may produce in blood an IgD anti-D antibody. This antibody can easily cross the placenta leading to an inevitable attack of the fetal RBC and subsequent destruction with generalized jaundice of the fetus (erythroblastosis fetalis). ABO incompatibilities may also occur, but jaundice and widespread effects seem to be quite milder than Rh incompatibility.
(c) *C* = καταστρεπτικός (Greek adjective for destructive, disastrous, catastrophic, calamitous) is due to the impact of the RBC destruction following a direct infection of the RBC by malaria parasites (plasmodia).
(d) *D*iaspasis (Greek Διάσπαση = break, disruption, disintegration, breakup, fission, smash) is a mechanical injury-associated anemia (microangiopathy) showing bizarre, fragmented RBC on a peripheral smear and signs of hemolysis. The deformability and twisting of RBCs are quite substantial, but a limiting factor is when these cells pass through blood vessels causing active cell stretching. If shearing forces are applied for any reason, RBC twisting can also occur. Examples of these settings are deformed aortic valves, atrio-ventricular (AV) shunts, ventricular septal defects (VSDs) of the heart, or 1st generation cardiac valvular prostheses.

False causes of MCV (mean corpuscular volume) increase are giant platelets, swollen RBC, cryoglobulins, and in vitro hemolysis.

10.2.1.2 Anemia Due to Failure of Blood Production

2. Failure of blood production is due to either nutritional deficiencies or BM aplasia or replacement (myelophthisis) or systemic disorders.
   – Nutritional deficiency-related anemias are seen in the setting of Fe, folic acid, vitamin

B_{12}, or protein deficiencies. The manifestations of folic acid and vitamin B_{12} deficiencies are megaloblastic anemia, glossitis, lingual atrophy, and subacute combined degeneration of the spinal cord. Subacute combined degeneration of the spinal cord, or Lichtheim disease, refers to degenerative changes of the posterior and lateral columns of the spinal cord as a consequence of vitamin B_{12} deficiency. The adjective "combined" refers to the involvement of both the dorsal columns (*fasciculus gracilis* of Goll and *fasciculus cuneatus* of Burdach) and lateral corticospinal tracts. In tertiary syphilis (tabes dorsalis), the dorsal columns are selectively involved (tabes means wasting disease, or decay from Lat. tabēre to decay). Pernicious anemia is autoimmune in origin and due to antibodies against parietal cells. Following the absence of parietal cells there is a consequent achlorhydria, gastric mucosal atrophy, and no intrinsic factor, which is required for the absorption of vitamin B_{12}. Besides antibodies to parietal cells, these individuals may present with other autoimmune disorders, including hypoadrenalism (Addison disease), and are prone to develop gastric carcinoma even in the youth.

- BM aplasia/replacement-related anemias are due to radiation, drugs (e.g., benzene-derived molecules), antibiotics (e.g., chloramphenicol), viruses (NANB hepatitis viruses), Diamond-Blackfan syndrome, thymoma-associated pure red cell aplasia, and myelophthisis (marrow infiltration, leukoerythroblastic anemia, and splenic extramedullary hematopoiesis (EMH) due to infiltrating neoplasms, lipid storage diseases, osteopetrosi or Albers-Schönberg disease, and several granulomatous diseases). Agnogenic myeloid metaplasia (AMM) or Myelofibrosis (sometimes aka AMM with MF differentiating two processes) is characterized by prominent fibrosis of BM spaces with sparsely isolated residual BM erythropoietic elements. DDX: *Chronic idiopathic myelofibrosis* with (−) Ph-Ch and (−) BCR/Abl hybrid gene and normal or high neutrophilic alkaline phosphatase ALP (LAP) and *CML* with (+) Ph-Ch/BCR/Abl hybrid gene and low neutrophilic ALP (LAP). *BCR-ABL* is a fusion gene made of the *BCR* gene, which is located on chromosome 22, and the *ABL* gene, which is located on chromosome 9. The mutated chromosome 22 is called the Philadelphia chromosome (Ph-Ch) because of the city where investigators first discovered it. The *BCR-ABL* fusion gene is a type of somatic mutation. It is not inherited from parents. DDX: Primary osteo-myelofibrosis with woven bone on spongiosa bone following chemotherapy-related oncological protocols of cytostatic therapy with appositional lamellar bone on spongiosa bone.
- Systemically related anemias include anemias that occur in the setting of endocrine disorders, uremia, chronic low-grade infection, and malignancies (BM infiltrations from endocrinologic tumors (thyroid, adrenal gland) and non-endocrinologic tumors (breast, prostate, kidney, lung, and bone)

In chronic myelofibrosis, megakaryocytes have characteristic "cloud"-like nuclei.

In regular BM, cells have probably the primary source of thymidine for DNA from de novo synthesis directly from deoxyuridine, which requires intact molecules of cobalamin and folate. Thus, if tritium-labeled thymidine (^{3}H-TdR) is carefully administered, <10% is incorporated into DNA. In the setting of megaloblastic BM, there are cobalamin and folate deficiencies, and deoxyuridine cannot be efficiently converted to pyrimidine deoxynucleoside, and more ^{3}H-TdR is incorporated into DNA.

10.2.2 Polycythemia

Erythrocytosis and increase of the RBC mass may be due to either BM-based primary polycythemia or as a reaction to hypoxia (secondary polycythemia).

10.2.2.1 Primary Polycythemia (aka *Polycythemia vera*)

Primary polycythemia is not usually seen in children or young individuals but few cases have been reported in youth. It is a myeloproliferative process and needs to be considered in the case of differential diagnosis or genetic/oncologic counseling/surveillance. Intriguingly, there may be an insidious onset of panmyelopathy (all lines of differentiation - erythro-, myelo-, and thrombo- - involving) with high levels of neutrophilic ALP and TC-III. Both myelofibrosis and leukemia are common in children and youth, who have been treated with alkylating agents or radioactive phosphorus (P^{32}) for other conditions. Thrombocytosis is common in primary polycythemia and may lead to capillary or venule occlusion causing ischemia and necrosis of the organs supplied. An increase in reticulin fibers in pretreatment BM biopsies is, however, not necessarily indicative of the spent phase of the disease. COD (cause of death) in patients with erythrocytosis or polycythemia include heart failure, myocardial infarction, stroke, and SAH (subarachnoid hemorrhage). BM pathology points exquisitely to trilineage hyperplasia with myelofibrosis and increased reticulin and collagen fibers. Commonly, it correlats to the progression of the disease. EMH may be observed in the spleen, liver, and lymph nodes. Iron stain (Perls Prussian Blue Staining) shows no or low hemosiderin deposition, whereas reticulin stain shows an increase of ret fibers, even at the onset of disease in 1/4 of patients. The iron stain is based on the mineral acid hydrolysis, which releases ferric ions (Fe^{3+}) from protein-linked tissue deposits. In the presence of ferrocyanide ions, there is a strong precipitation exhibiting a highly colored and hydro-insoluble complex, which is the potassium ferric ferrocyanide or "Prussian blue" as shown in the following stoichiometric equation: ($FeCl_3 + K_4Fe(CN)_6 = KFeFe(CN)_6^- + 3KCl$). *Polycythemia vera* is one of the common myeloproliferative neoplasms, which are chronic myelogenic leukemia, chronic neutrophilic leukemia, chronic eosinophilic leukemia, polycythemia vera, essential thrombocytopenia, mastocytosis, and primary myelofibrosis. Although myeloproliferative neoplasms are not typically seen in childhood and youth, the occurrence of several genetic syndromes with alteration of both genetic and immunologic inherited platforms by insecticides, pollution, and new microorganisms can accelerate the process and become a differential diagnosis in the nearest future. The genetics of *polycythemia vera* is shown in Box 10.2.

Box 10.2 Genetics of *Polycythemia vera*

V617F mutation	on exon 14 of JAK2	in ~97% of cases
Other *JAK2* gene mutations	on exon 12 of JAK2	in ~3% of cases

10.2.2.2 Secondary Polycythemia

Secondary polycythemia is linked to a variety of hypoxic conditions, including impairment of the ventilation of the lungs and malfunctioning of the cardiac pump. In these settings, there is an increase of the erythropoietin, which stimulates BM cells to produce erythrocytes. Other BM elements are standard in number, function, and morphology. Importantly, there is no BM fibrosis. Secondary polycythemia has been observed in the setting of doping by the utilization of banned performance-enhancing drugs by athletic competitors in young athletes.

In pediatric *polycythemia vera*, features of laboratory testing are similar to those found in adults and include (1) leukocytosis (white blood cells or WBC, $6.7–10.8 \times 10^9$/L), (2) erythrocytosis with a Hb level of 18.0 g/dL and Hct of 53.2–55%, and (3) platelet count of $207–394 \times 10^9$/L. Hyperuricemia and ↑ LDH (lactic acid dehydrogenase) accompany the pediatric process as well. In the pediatric BM of patients with polycythemia vera, there is hypercellularity with erythroid hyperplasia. Moreover, there are atypical megakaryocytes with hyperlobation of nuclei, and some patients show a mild ↑ in reticulin. The revised WHO criteria for polycythemia vera require the *JAK2* mutation and an ↑ red cell mass (RBC). The total blood volume includes RBC, which is commonly referred to as red cell mass (RCM), and the additional part occupied by plasma (i.e., plasma volume). Differently, from the adults, *JAK2* V617F mutations occur only in approximately 1/3 of the pediatric population. Of consequence, the primary diagnostic criteria cannot be fulfilled in most pediatric patients, and two minor rules need to be met to establish a formal diagnosis according to the WHO criteria. All children present with hypercellular bone marrow, but only 1/3 have a low level of serum erythropoietin (EPO) or ↑ endogenous erythroid colony (EEC) growth. Thus, the current diagnostic criteria for polycythemia vera are suboptimal in the pediatric age population as the diagnosis can only be confirmed in about 1/2 of the patients.

10.3 Coagulation and Hemostasis Disorders

10.3.1 Coagulation and Hemostasis

The Virchow's triad is probably one of the first things we learn in medical school. To remember this triad, we can focus on the first three consonants of the surname of Professor Rudolf Virchow, who was one of the fathers of the modern human pathology (VRC). V stands for vessels, R stands for *rouleaux* (the French term for rolls), and C stands for coagulation. Blood vessel rupture or injury (rhexis), the presence of more cells in the lumens of blood vessels in several conditions forming rolls, and clotting deficits may cause the formation of a thrombus in a blood vessel. Rouleaux aggregates form because of the unique discoid shape of the blood cells in vertebrates. This characteristic flat shape of the surface gives RBC a large surface area to make contact and stick to each other. There are underlying conditions, which determine an increase in plasma protein concentration and ESR (erythrocyte sedimentation rate) causing *rouleaux*. These conditions include infections, multiple myeloma, diabetes mellitus, inflammatory and connective tissue disorders, and neoplasms. The perfect balance between pro- and anticoagulative events is at the basis for the correct functioning of all systems and organs. The process of hemostasis relies on the action of platelets (thrombocytes) and of coagulation factors from peripheral blood (plasma, platelets, and other cells) and counterbalancing mechanisms such as the fibrinolysis. Following a break of the endothelial cells of blood vessels, there is an exposition of collagen molecules that cause adhesion of platelets that are also deformed by the collagen itself becoming irregular in shape and emitting pseudopods and simultaneously becoming sticky allowing the aggregation of the same and other platelets. The activation of platelets determines the secretion of ADP (adenosine 5′-diphosphate) and TXA2 (thromboxane A2), which along with thrombin (factor II) increase platelet aggregation, making it irreversible, and PAF (platelet-activating factor) that serves in coagulation and fibrin clot formation.

10.3.2 Coagulation Disorders

Distinguishing between hereditary and acquired clotting disorders is extremely useful and key for the proper management of a patient.

10.3.2.1 Coagulation Disorders
Hereditary disorders: hemophilia A or factor VIII deficiency.

10.3.2.2 Acquired Disorders
1. *Vitamin K deficiency* due to malnutrition, malabsorption, and obstructive cholangiopathy as well as administration of vitamin K antagonists (e.g., drugs such as warfarin) leading to a decrease in serum levels of factors II (prothrombin), VII (proconvertin or stable factor), IX (Christmas factor or plasma thromboplastin), and X (Stuart-Power factor or thrombokinase).
2. Liver disease is leading to a decrease in factors I, II, V, VII, IX, and X.
3. DIC is a life-threatening medical condition in which there is systemic activation of coagulation. DIC results in the formation of intravascular fibrin, which determines intravascular thrombosis as first step. Subsequently. intravascular thrombosis is followed by continuous bleeding due to consumption of coagulation factors. DIC is characterized by a decrease of factors I, II, V, and VIII. In acute DIC, there is

bleeding diathesis, while in chronic DIC, there is thrombotic diathesis. Acute DIC occurs in incompatible blood transfusions, maternal death, and life-threatening peripartum events with characteristic premature separation of the placenta and endotoxic shock. Bleeding diathesis is present because the coagulation cascade is rapidly activated, but the clotting factors are quickly degraded, and fibrinolysis is activated with numerous fragments of fibrinogen and fibrin circulating in the plasma (fibrin split products). Platelets are obviously decreased because they are intimately trapped in areas of coagulation perpetuating bleeding diathesis. Chronic DIC has a similar evolution, although thrombotic diathesis is more common and metastatic malignancies may be observed as underlying disorders.

10.3.3 Platelet Disorders

It is useful to distinguish between thrombocytopenias due to abnormal platelet production or accelerated platelet removal and thrombocytosis of reactive or of idiopathic type as well as mention platelet function defects. Flow cytometry and immunohistochemistry are useful for identifying markers of megakaryoblastic differentiation. These markers are CD41 (glycoprotein IIb/IIIa), CD42 (glycoprotein Ib), and CD61 (glycoprotein IIIa) on flow cytometry. Also, CD41 and CD61 are useful immunohistochemical markers of megakaryoblastic differentiation. We are not discussing the thrombocytopenia due to abnormal platelet production or the thrombocytopenia due to accelerated platelet removal, or either the thrombocytosis of reactive type, but we will focus on essential thrombocythemia.

10.3.3.1 Primary (Essential) Thrombocythemia
- *DEF*: An indolent myeloproliferative disorder with long symptom-free periods interrupted by periods characterized by life-threatening hemorrhage (such as epistaxis [Gr. ἐπίσταξις or "nosebleed"]).

Essential thrombocythemia (ET) is categorized by increased megakaryopoiesis, persistent thrombocytosis, the high risk of vascular complications, and the risk of leukemic transformation. Pediatric essential thrombocythemia occurs approximately in $1:10^7$ children, if considering a cutoff age of 14 years, but higher if deemed the adolescence or even the youth. Like adult ET patients, there is a female predominance. The hereditary or familial forms of thrombocythemia are usually AD-inherited conditions. Moreover, thrombocythemia occurs at a younger age and has been described even in infancy. Almost all patients with ET show a normal karyotype, and *JAK2* V617F mutations are found in 40–50% of adult patients. These mutations are associated with higher Hb levels and PMN counts. A small subset of ET patients has an acquired somatic activating mutation in the thyroid peroxidase (*TPO*) gene receptor, and some *CALR* gene mutations have also been identified as double negative (JAK2−, MPL−) patients of essential thrombocythemia. The *CALR* gene located on chromosome 19 (19p13.13) encodes a protein called calreticulin or calregulin, CRP55, CaBP3, calsequestrin-like protein, and endoplasmic reticulum resident protein 60 (ERp60). Calreticulin is a protein, which inside the endoplasmic reticulum, ensures appropriate folding of newly synthesized glycoproteins and exquisitely modulates calcium homeostasis. Calreticulin is also found in both cellular surface and extracellular compartments. At the extracellular level, calreticulin plays a role in many biologic processes, including cell proliferation, programed cell death, and immunogenic cell death. Calreticulin needs to be confused with calretinin, which is a calcium-binding protein involved in calcium signaling. Two-thirds to three-fourths of children with ET are asymptomatic, and thrombocytosis is discovered as an incidental finding on routine evaluation. If the child is symptomatic, headache is the most common. Splenomegaly is identified in up to about one in five children with essential thrombocythemia. The diagnosis of ET relies essentially on a persistent thrombocytosis (platelet count of $>450 \times 10^9$/L) and the exclusion of (1) reactive thrombocytosis, (2) other diseases that lead to thrombocytosis (e.g., other myeloproliferative neoplasms or MPNs), and (3) specific subclasses of myelodysplastic syndromes or MDS (e.g., 5q-syndrome, refractory anemia with ringed sideroblasts associated with marked thrombocytosis (RARS-T)). The bone marrow histology in pediatric ET is similar to those in adults. It shows a normo- to hypercellular marrow with increased megakaryocytes occurring in clusters, which represents an important clue in the pathologic diagnostic process. Such megakaryocytes are large and have abundant cytoplasm and prominent nuclear lobation (hyperploid nuclei) (the so-called "staghorn-like" megakaryocytes). Minimal or no dysplasia is present, which is important in the differential diagnosis. If it occurs, an MDS with fibrosis should be considered. A standard treatment approach for pediatric ET is challenging. The perusal of the literature exhibits a variegated universal consensus. Typically, approaches target the antiplatelet or cytoreductive therapy using regimens of low-dose aspirin, which is also recommended for low-risk adult patients. Cytoreductive treatment is indicated has been used in about half of the pediatric population. It relies on hydroxyurea and anagrelide followed by single interferon-α (IFN-α) administration or in combination with hydroxyurea. Of note, anagrelide inhibits the maturation of thrombocytes from megakaryocytes. Clinical trials of targeted JAK2 inhibitors (e.g., ruxolitinib) seem encouraging. Giona et al. (2012) addressed the risk for long-term complication in ET of most of the pediatric population with a median follow-up time of over 10 years. In this study, approximately 1/10 children developed thrombosis, all of whom experienced concurrent infections. Splenomegaly with progressive reticulin fibers-rich fibrosis occurred in 9% of children. No patient developed acute leukemia.

10.4 White Blood Cell Disorders

10.4.1 Leukocytopenias and Leukocyte Dysfunctionalities

First, it is quite essential to have in mind the characteristic B- and T-cell markers as well as markers of nonlymphoid white blood cell population. B-cell markers are CD19, cytoplasmic CD79a, and cytoplasmic CD22, while T-cell markers are CD3, CD2, CD5, CD4, and CD8.

Distinguishing between neutropenia and lymphocytopenia is useful. *Neutropenia* is defined as having an absolute neutrophilic count of *<500/μl*, which is often associated with infection. The etiology of neutropenia includes aplastic bone marrow, toxic reactions and side effects of some drugs, autoimmunity, and increased removal by RES (e.g., hypersplenism). Neutropenia constitutes a relevant problem in patients affected with cancer undergoing chemotherapy cycles. *Lymphocytopenia* is defined as having an absolute lymphocyte count of *<1500/μl*, which is often observed post-radiotherapy and in steroid-treated patients and patients affected with malignancies, uremia, and bone marrow aplasia.

Leukocyte dysfunctionalities comprise essential disorders that may play crucial roles in pediatrics and youth. One of the most critical disorders characterized by recurrent pyogenic infections is Chediak-Higashi syndrome (CHS), an AR-inherited disease due to mutations in the *LYST* gene and characterized by white blood cells that contain abnormal granules. Neutrophils show a defective and delayed release of bactericidal substances from the intralysosomal compartment of proteins leading to reduced ability to fight bacterial infections efficiently and effectively. There is also leukocytopenia, which is in addition to defective granulation. Both leukocytopenia and abnormal functionality culminate in a high susceptibility to infections. The underlying defect in CHS is a microtubule polymerization defect, which leads to a decrease in the phagocytic function of white blood cells. Characteristically, neutrophils show abnormalities of nuclear lobation (although not always present) and abnormally large secondary granules on a routine blood smear. Abnormalities of granulation are also observed in cytotoxic T cells leading to cellular immunodeficiency of this lymphocyte population of this lymphocyte population other than increased susceptibility to bacterial infections.

10.4.2 Non-neoplastic Leukocytosis

Also, here are some distinctions of the fractions of the white blood cells useful for the clinician.

Chronic unexplained eosinophilia is often associated with the *FIP1L1-PDGFRA* fusion gene. *FIP1L1-PDGFRA* is an abnormal fusion gene determined by a fusion of platelet-derived growth factor receptor-alpha (*PDGFRA*) to a human gene called FIP1-like-1 (*FIP1L1*). This fusion gene test should also be associated with chromosome analysis, *BCR-ABL1, KIT* mutation, IL-5 for clonal T-cell disease, IgE level, and cardiac tests (e.g., troponin test) to look for evidence of organ damage.

- *Neutrophilia*: Infection (leukemoid reaction, if the neutrophilic count is very high)
- *Eosinophilia*: Allergic conditions (asthma, hay fever, drug allergies), parasitic infestations
- *Lymphocytosis*: Chronic inflammation (e.g., TB, viral infections such as infectious mononucleosis (IM) or measles)
- *Monocytosis*: Chronic inflammation (e.g., TB, fungal infections, listeriosis)

10.4.3 Leukemia (Neoplastic Leukocytosis) or *Virchow's "Weisses Blut"*

Leukemia or *"weisses Blut,"* as a descriptive term suggested by Rudolf Virchow in 1846 in the front page of an old monography, is a condition characterized by an uncontrollable proliferation of leukocytes (granulocytic, lymphocytic, or monocytic) in the peripheral blood and lymphoreticular tissue such as bone marrow, lymph nodes, and spleen and potentiality harboring the ability to infiltrate any organ. The neoplastic cell population has a limited or no capacity for cell differentiation, widespread tendency, and suppressive or impairing action on normal myeloid cell growth (functionally aplastic marrow) with clinical manifestations of anemia, infection, and bleeding. Organ dysfunction is mainly seen in the liver, spleen, lymph nodes, and CNS.

- *DEF*: Diffuse replacement of the bone marrow by neoplastic white blood cells with the presence of numerous and immature white blood cells in peripheral blood ± widespread infiltrates in ontogenetically essential organs, such as the liver and spleen, and lymphatic organs, such as lymph nodes (LNs) and sites throughout the body (e.g., MALT, BALT, etc.) (Fig. 10.2). Leukemia is classified as lymphoid or lymphoblastic and myeloid or myeloblastic according to the white blood cell of origin and acute (acute lymphoblastic leukemia [ALL] or acute myeloblastic leukemia [AML]) and chronic according to the onset and tendency of progression (chronic lymphoblastic leukemia [CLL] or chronic myeloblastic leukemia [CML]).
- *EPI*: Apical cancer death in children <15 years in the USA and Canada (about 1,500-2,000 children of age 0-14 years are diagnosed yearly in Canada).
- *CLI*: Anemia, infections, and thrombocytopenia with aggressive onset (WBC <10 K in 50% of cases and >100 K in 10%) in case of acute leukemias and insidiously with chronic leukemias, which are often detected on the routine physical exam (Box 10.3). Typing of leukemias should follow the division in acute and chronic leukemias and is better done on smears of peripheral blood or BM aspirate.

Box 10.3 Leukemia Changes on Hematopoietic and Non-hematopoietic Organs
1. BM: Hematopoiesis replaced by neoplastic WBC population with the erosion of cancellous and cortical bone ± myelofibrosis (ALL > AML > CLL or CML)
   - Secondary anemia (recovery time: fat cells>erythroids, granulocytes, megakaryocytes)
   - Secondary infections
   - Secondary hemorrhages and hematomas ± infarcts due to vascular sludging
2. Lymphadenopathy, particularly in ALL
3. Hepatomegaly, particularly in CLL
4. Splenomegaly, particularly in CML with the expansion of red pulp and organ enlargement (<1 kg in monocytic, <2.5 kg in CLL)

10.4 White Blood Cell Disorders

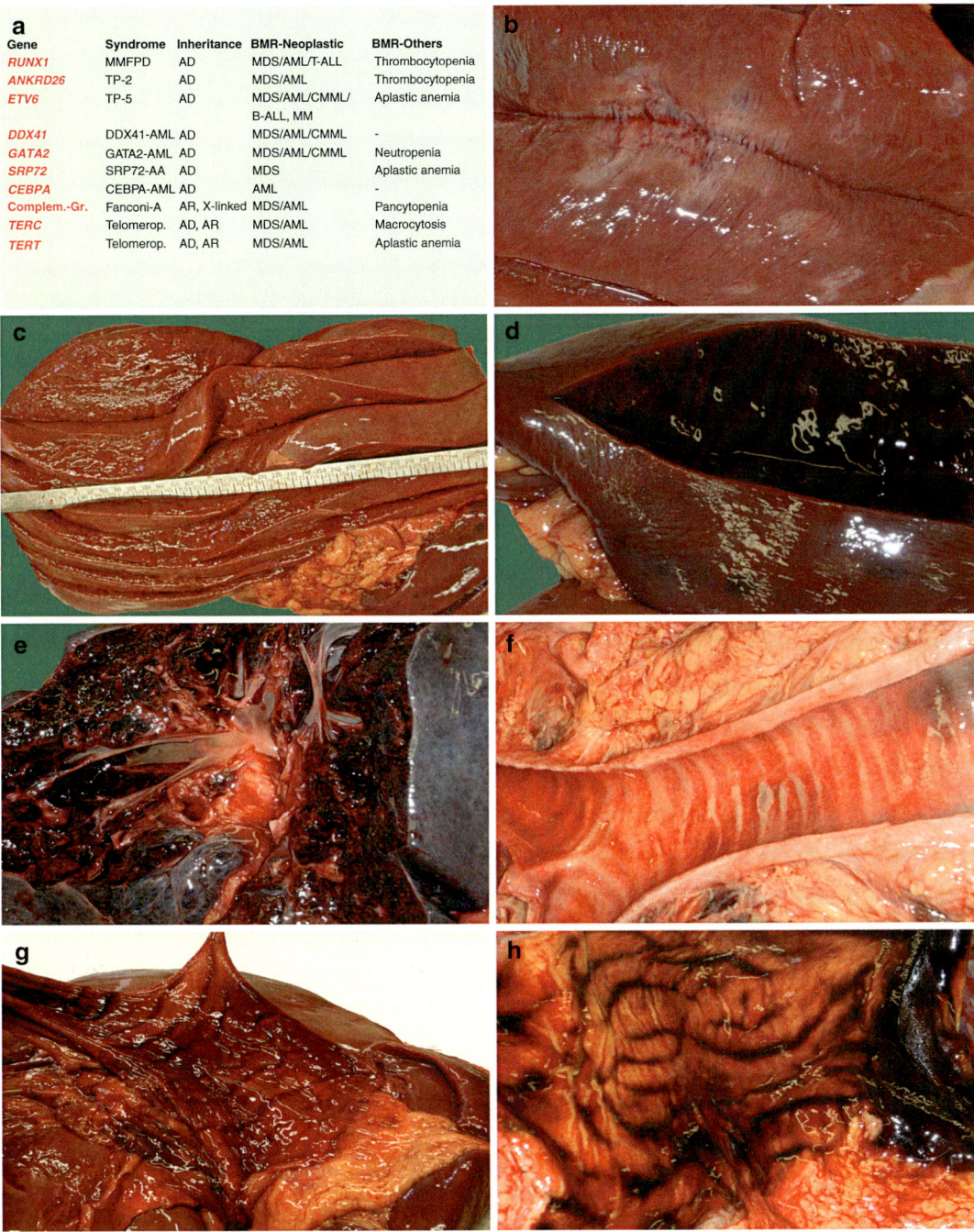

Fig. 10.2 A number of genes associated with some hematologic disorders are listed in figure (**a**). The gross photographs show an infiltration of the kidney with some gray spots surrounded by some hemorrhagic edge (**b**), hepatomegaly (**c**), splenomegaly with colliquated parenchyma (**d**), severe hemorrhage of the lung (**e**), acute tracheitis with histologic evidence of neoplastic infiltration (**f**), and severe gastrointestinal hemorrhage (**g–h**). *Notes*: *RUNX1*, Runt-related transcription factor 1; *ANKRD26*, Ankyrin Repeat Domain-Containing Protein 26; *ETV6*, ETS-transcription factor 6; *DDX41*, DEAD-Box Helicase 41; *GATA2*, Endothelial Transcription Factor of the GATA family of zinc-finger transcription factors, SRP72, Signal Recognition Particle 72; *CEBPA*, CCAAT/enhancer-binding protein alpha; *Complem-Gr*, complementation group (Eight complementation groups have been reported in Fanconi anemia); *TERC*, Telomerase RNA component; *TERT*, Telomerase reverse transcriptase

10.4.3.1 Acute Lymphoblastic (Lymphocytic) Leukemia (ALL)

- *DEF*: B-cell differentiating (80%) leukemia with BM and peripheral blood (PB) as well as L/H/S involvement (L, lymph node; H, liver; S, spleen) with pre-B-cell phenotype (CD19+, CD20−, and Ig−) ± Ph chromosome (t(9;22)(q34;q11)) translocation with *BCR-ABL1* fusion gene) and ± t(1;19).
- *EPI*: Worldwide, children>adults (most frequent cancer <15 years; peak, 4 years), ♂ = ♀.
- *CLI*: Lymphadenopathy, generalized + hepatosplenomegaly, high TdT (terminal deoxynucleotidyl transferase or DNA nucleotidylexotransferase or DNTT or terminal transferase).
- *TNT*: Intense CHT protocols induce remission in >1/2 of cases, and RT reduces cell size becoming close to lymphocytes (CHT, chemotherapy; RT, radiotherapy).
- *PGN*: Good PGN markers are pre-B phenotype and hyperdiploidy (about 1/4 of cases), while poor PGN markers are B/T phenotype, Ph chromosome, and t(1;19).

Crucial is the protein Tax, which is named by the first initials of the words defining this protein, i.e., transactivator from the X-gene region. The HTLV-1 (human T-cell leukemia virus type 1) Tax is known to induce or repress various cellular genes, several of them encoding transcription factors. Tax acts as a transactivator, causing the transcription of viral proteins that are essential for viral replication. One of the most essential steps in medicine was the identification of a consensus to classify leukemia. The French-American-British (FAB) classification is often used.

10.4.3.2 Chronic Lymphocytic Leukemia (CLL)

- *DEF*: Chronic onset of lymphocytic differentiated neoplastic cells with usually B-cell phenotype (CD19+, CD20+), weak sIg, but also CD5+, >15K lymphocytes/mm^3 with ≤10% blasts (in case of 4–15K = early low count CLL, subleukemic CLL).
- *EPI*: Western countries mostly, adults (uncommonly seen with age <35 years), ♂:♀ = 2:1, although rare pediatric cases or cases in young individuals have been reported.
- *CLI*: PB + BM involvement ± lymphadenopathy, hepatosplenomegaly ± accompanying small lymphocytic lymphoma (SLL).
- *CLM*: Homogeneous cell population of small-sized lymphocytes of the mature type with increased reticulin (25%).
- BM: Paratrabecular and interstitial infiltrate.
- LN: Effacement of the architecture with diffuse infiltrate + pseudofollicles or proliferating germinal centers.
- *TNT*: CHT (protocols according to Children's Oncology Group, United Kingdom Children Cancer Study Group, and International Society of Pediatric Oncology).
- *PGN*: ± SLL, ± Richter's transformation (DLBCL, diffuse large B-cell lymphoma).

Pediatric CLL remains extremely rare, and Demir et al., in 2014, could find eight case reports in the English literature. Although half of these reports were written before the 1970s, the illustrations and/or descriptions of CLL seem entirely genuine. However, the analysis of surface membrane immunoglobulins and B-cell monoclonality was carried out in half of these patients. A clonal translocation involving chromosomes 2 and 14 (t(2;14)) was identified in two of this small cohort. A retrospective analysis of SEER (Surveillance, Epidemiology, and End Results) identified only 4 pediatric patients with CLL and 16 patients with SLL with age younger than 25 years.

T-cell CLL is rare with <2%CLL. PB cells show cytoplasmic azurophilic granules (aka large granular lymphocytes) and increase of both beta-glucuronidase and acid phosphatase. Neoplastic cells are CD3+, CD8+ (if CD4+, the T-CLL acquires the label of prolymphocytic sub-variant), CD56−, and harbor *TCR* (T-cell receptor) gene rearrangement, marked lymphocytosis as well as ± skin involvement without epidermotropism. Clinically, T-cell CLL exhibits evidence of neutropenia, pure red cell aplasia, and IF-gamma-induced hypogammaglobulinemia.

NK-cell CLL (rare, <1%CLL, CD2+, CD16+, CD56+, but CD3−, CD4−, CD8−, no TCR gene rearrangement)

Prolymphocytic CLL (larger, mainly immature atypical lymphocytes, B-cell phenotype with strong sIg expression in 80% of cases and T-cell phenotype: CD2+, CD3+, CD4+, CD5+ in 20% of cases). In the PB, there is marked increase of WBC with >55% prolymphocytes. Also, there are splenomegaly without lymphadenopathy, and low MI. Remarkably, PGN exhibits a more aggressive course than non-prolymphocytic CLL

Intermediate CLL (mixed type → CLL/prolymphocytic leukemia (PB: prolymphocytes from one tenth to half of the neoplastic cells)

10.4.3.3 Acute Myeloblastic (Myelocytic) Leukemia (AML)

- *DEF*: Acute onset leukemia with >30% blasts (M3–M5 subtypes include promyelocytes and promonocytes during blastic counting), with the possible oncovirus association.
- *EPI*: Worldwide, 15–60 years, ♂ = ♀.
- *CLM*: Characteristic pentad of the following:
  - (+) Azurophilic granules – modified primary lysosomes (promyelocyte stage)
  - (+) MPO stain → distinguishing myeloblasts from lymphoblasts (MPO, myeloperoxidase)
  - (+) Naphthyl AS-D chloroacetate esterase (NCE) specific for myeloid lineage
  - (+) Alpha-naphthyl acetate esterase (ANE) specific for monocytes
  - (+) Auer bodies (red rod-shaped modified primary lysosomes of both myelocytic and monocytic subtypes)
- *TNT*: Other than the classic CHT regimens, targeted therapy may be used. Midostaurin for patients who have AML with an *FLT3* gene mutation and enasidenib for those who have relapsed or refractory AML with an *IDH2* gene mutation.
- *PGN*: Remission is possible, but relapses are common.

10.4.3.4 Chronic Myelocytic Leukemia (CML)

- *DEF*: Chronically onset leukemia (WBC>10^5) of myelocytic lineage with Philadelphia chromosome (t(9;22), i.e., translocation of 22q11 to 9q34 forming a chimeric protein c-abl/bcr factor *BCR/ABL* tyrosine kinase, breakpoint cluster region), splenomegaly, and BM showing <5% myeloblasts but mostly neutrophils with scattered myelocytes and promyelocytes and increased basophils and decreased alkaline phosphatase activity in leukemic cells.
- *TNT*: CHT induces 2–5 years remission, usually followed by a blast crisis with myeloblasts and 1/3 immature B lymphocytes and subsequent death.
- *PGN*: 2–3 years of survival without treatment, but with therapy, a better outcome is recorded. The development of myelofibrosis may occur and compromise the final outcome.
  - Good PGN factors: Ph (+)
  - Poor PGN factors: Ph (−)

Regarding dynamics, leukemia is a neoplastic process that can go through a transformation: (1) pro-lymphocytoid transformation (dedifferentiation) (5%), which is defined by an increase in prolymphocytes to >20% of BM or PB lymphocytes, often presenting with foci of prolymphocytes surrounded by well-differentiated lymphocytes in BM, characteristic prominent, and single eosinophilic nucleoli; as well as, (2) Richter's transformation (large cell transformation) (3–10%), which is defined by pleomorphic, usually B-immunoblastic lymphoma but also occasionally Hodgkin lymphoma with patients showing reactivation of signs and symptoms, lymphopenia, and invariable downhill disease course. Of note for Richter's transformation, PB may show no lymphoma cells, while BM and LN are probably essential for the diagnosis.

There are several staging systems for leukemia, and the most important is probably the RAI staging system.

10.4.3.5 Myeloid Sarcoma

- *DEF*: Extramedullary tumor mass of neoplastic immature myeloid cells of granulocytic or monocytic lineages, which may present:
  - As an isolated entity or accompanied by an underlying BM disease, such as one of the following: AML M4, AML M5, CML, chronic idiopathic myelofibrosis (agnogenic myeloid metaplasia), hypereosinophilic syndrome, or *polycythemia vera*

- Involving ≥1 bone/soft tissue (skin, soft tissue, viscera, oral mucosa, particularly gingiva, LN, orbit, bone, spinal cord, and mediastinum) with or without BM disease

The translocation t(8;21) is present in pediatric cases.

- *SYN*: Chloroma, granulocytic sarcoma.
- *EPI*: Worldwide, children>adults, ♂ = ♀.
- *GRO*: Papula/nodular lesion, possibly ulcerated on the surface, with green-colored cut surface due to high peroxidase content of the neoplastic cells.
- *CLM*: Three lineages of maturation (singly or relatively mixed) including:
  - *Blastic* type (undifferentiated cells): Myeloblasts with the moderate rim of basophilic cytoplasm, fine nuclear chromatin, and 2–4 nucleoli
  - *Immature* type (intermediate cells): Myeloblasts, promyelocytes, and eosinophilic myelocytes
  - *Differentiated* type (or mature WBCs): Promyelocytes, eosinophilic myelocytes, and more mature WBCs
- *DDX*: Burkitt lymphoma, DLBCL, small round blue cell tumors (SRBCTs).
- *TNT*: CHT regimens like AML or more aggressive regimens and local control by RT and surgery. Otherwise, BM transplantation is an essential and crucial choice.
- *PGN*: The simultaneous presence of isolated tumor and chromosome 7 abnormalities is associated with a worse prognosis.

Box 10.4 illustrates the positivity of myeloid sarcoma for both special stains and immunohistochemistry that can be used as ancillary investigations by the clinical and/or surgical pathologist.

Box 10.5 illustrates some paramount tips on myeloid sarcoma.

10.4.3.6 Hairy Cell Leukemia (HCL)
- *DEF*: Distinct B-cell (+CD19, CD20, CD22, CD79a and −CD5, CD10) leukemia of "hairy cells" with four characteristic diagnostic features, including:

Box 10.4 Myeloid Sarcoma HSS/IHC
- (+) Chloroacetate esterase stain, MPO, Sudan Black on touch preps
- (+) Lysozyme and CD43 (Leu-22)
- (+) MPO and CD117 (myeloid lineage tumor markers)
- (+) CD13 and CD33 (myeloblasts on FC)
- (+) CD68 (monocytic tumors)
- (+) CD11c, CD14, CD56, CD99 (monoblasts markers)
- (+) CD4, CD20, CD79a

Box 10.5 Myeloid Sarcoma Tips
1. Wright stain on touch prep: myeloid granules or Auer roads.
2. Eosinophilic myelocytes are extremely important because their number is proportional to the rise of the differentiation of the myeloid sarcoma lesion.
3. IHC is vital in ruling out potential misdiagnoses (e.g., ALK-1-NHL, DLBCL).
4. Touch preps (imprints) are very useful for molecular biology studies.
5. Subperiosteal bone disease (e.g., skull, sinuses, thoracic cage, pelvis), LN, and skin disease may accompany myeloid sarcoma and, if presenting as first, may be extremely challenging.
6. Subperiosteal disease of myeloid sarcoma of the orbit may prompt unilateral proptosis.

---

1. An oval or cleaved nucleus with absent or inconspicuous nucleoli, abundant pale-blue cytoplasm, and peculiar circumferential cell surface "ruffled" projectors
2. Interstitial or diffuse BM infiltration without discrete nodular clusters and clear cells with "fried egg" (oligodendroglioma-like) or spindled appearance and reticulin fibrosis
3. FC showing clonal B-cell expression of CD11c, CD25, CD103, and CD123
4. IHC showing B-cell expression of *TRAP*, *ANXA1*, and *DNA.44* ("Tad")

Notes: TRAP, tartrate-resistant acid phosphatase, ANXA1, Annexin 1.
- *SYN*: Leukemic reticuloendotheliosis.
- *EPI*: Worldwide, young to middle age, ♂ = ♀.
- *CLI*: Hepatosplenomegaly ± lymphadenopathy (mostly abdominal and retroperitoneal).
- *LAB*: Anemia (Hb ≤12 g/dL), neutropenia (≤1500 × 10^9/L), monocytopenia (<500 × 10^9/L), and "hairy cells" on PB smear examination. Pancytopenia ⇒ anemia, infections, and hemorrhage.
- *CLM*: 10–15 um, ovoid cells with delicate nuclear chromatin and inconspicuous nucleoli with clear to lightly basophilic cytoplasm with numerous cytoplasmic projections and pericellular halos (formalin artifact, similar to oligodendroglioma). In the BM, there is hypercellular marrow, but it may also be hypocellular with loosely packed hairy cells conversely from the tightly packed clusters distinctive of lymphomas, increased reticulin, and erythroid hyperplasia. In the spleen, there is red pulp disease with diffuse infiltration of the red pulp by the monotonous cellular population of the small mononuclear type with common invasion into vascular walls (subendothelial) ⇒ RBC lakes (pseudosinuses or blood lakes).
- *TEM*: Ribosome-lamellar complexes.
- *DDX*: Splenic marginal zone lymphoma (SMZL), HCL variant (HCL-v), chronic lymphocytic leukemia (CLL), and large granular lymphocytic leukemia (LGL).
- *TNT*: Splenectomy, α-IFN, and deoxycoformycin.
- *PGN*: Good, 10-YSR >90%.

HCL may also be subclassified according to the presence or not of leucopenia (Box 10.6).

In childhood, HCL has never been described except a single Congolese child (Gini Ehungu et al. 2013). A 4-year-old child was transferred from Kisangani University Hospital in the Northeast of the Democratic Republic of Congo to the University of Kinshasa for the evaluation of a suspected acute leukemia. The history dates back about 5 weeks, characterized by prolonged fever, fatigue, and weakness. The bone marrow was suggestive of HCL. Bone marrow aspiration showed hypercellular marrow with rare megakaryocytes. In 84% of blast cells, there were the following characteristics: large cells (Ø, 10–25 mm), round or oval, with cytoplasm containing occasional vacuoles, and budding was observed in clumps form of hair that could be quite long and slender. The bulk of chromatin core was oval.

10.4.3.7 Transient Atypical Myelopoiesis (TAM)

AML1 (aka *RUNX1*) gene encodes for core binding factor alpha and is located on chromosome 21. Transient myeloid leukemoid reaction (aka transient atypical myelopoiesis) is a peculiar reaction of the bone marrow occurring in 10% of infants harboring an extra-chromosome 21 (trisomy 21/Down syndrome). It has been estimated that 20–30% of patients with TAM eventually develop AML 1–3 years after the resolution of the diagnosed TAM initially. The prevalence of benign lymphoid aggregates in the BM biopsy may be misleading in some cases because they can be present from 20% to 60% of cases and their presence increases with age. TAM may also be present in other organs than BM.

Box 10.7 illustrates most useful tips and mnemonic helps for the routine practice of pathology.

Box 10.6 Classification
- HCL leukopenic type → WBC<3 K, seen in US and Western countries
- HCL non-leukopenic type → WBC>3 K, seen in Japan

Box 10.7 Useful Tips
- AML with t(9;11) shows a monocytic or myelomonocytic differentiation, and most adult AML with 11q23 abnormalities, including AML cases with t(9;11), express some markers of monocytic differentiation (CD4, CD14, CD11b, CD11c, CD36, and CD64).
- Acute monocytic or acute myelomonocytic leukemia (AMML) shows the fol-

lowing phenotype: bright expression of CD33, CD65, CD4, and HLA-DR and dim expression of CD13, CD34, and CD14 on flow cytometry.
- Adult T-cell leukemia/lymphoma (ATCL) shows in blood the characteristic "flower cell" or "cloverleaf cell," which is quite characteristic for this kind of leukemia/lymphoma.
- HTLV1 is the etiologic agent for ATCL.
- Aggressive diseases with poor prognosis that may be identified on BM biopsy are Erdheim-Chester disease, AML with t(6;9), and AML with inv. (3).
- Evans syndrome has been considered a disorder of immune regulation.
- AML with inv. (3), which is an AML with multilineage dysplasia, shows basophilia and dysplastic megakaryocytes (micro-megakaryocytes) in the BM aspirate, and *FLT3-ITD* gene abnormality plays a vital role for prognosis.
- B-ALL with t(12;21) is typically associated with a core binding factor translocation.
- B-ALL with t(12;21) and B-ALL with hyperdiploid and no structural cytogenetic abnormalities have an excellent prognosis and good cure rate.
- B-ALL with t(9;22) more often expresses the myeloid markers CD13, CD33, as well as CD10.
- B-ALL with t(v;11q23) often demonstrates a translocation involving the MLL gene
- The BM infiltrated by follicular NHL and, occasionally, mantle cell lymphoma show focal, paratrabecular lymphoid infiltrates with discrete clusters of malignant cells molding against bony trabeculae.
- Ipilimumab is an antibody (monoclonal) directed against CTLA-4 (cytotoxic T lymphocytes antigen 4) and is vital for regulating the immune system because CTLA-4 is an "off" switch. The T-cell attack can be turned off by stimulating this receptor.

10.4.4 Myelodysplastic Syndromes

Myelodysplastic syndromes (MDS) represent a problematic topic not only for the student or the pathologist but also for the physician working in hematology and oncology. From the original five categories of MDS in the early 1980s, a morphologic and cytogenetic approach has been used to bring the actual categories of MDS to nine with four additional types combining MDS and myeloproliferative syndromes (MPS). Myelodysplastic syndromes remain difficult to classify. Their heterogeneity probably impacts the taxonomic tools.

10.4.5 Hodgkin Lymphoma

Hodgkin lymphoma (HL) is a neoplastic disease of germinal center origin with chiefly nodal but also extranodal involvement showing Reed-Sternberg cells (RSC) and Reed-Sternberg variants (RSV) as malignant cells. RSC/RCV cell morphology is both aneuploid and clonal and is an intense topic of immunological investigation. HL entails two natures, i.e. (1) *histiocytic* or *IDR cell*, due to lack of the surface markers of B and T cells, and (2) *lymphocytic* origin based on gene rearrangement data (Figs. 10.3, 10.4, 10.5, 10.6, 10.7, 10.8, and 10.9) (IDR cell, interdigitating reticular cell).

- *RSC*: It is a binucleated or multilobated cell with large eosinophilic nucleoli, which can be round or linear in shape and be surrounded by a bright halo and have a clear, abundant cytoplasm.
- IHC: (+) CD15, CD30, CD25 (IL-2R), PAX5, LMP-1 and EBER.
  (−) CD45, (±) CD20, (±) CD79a, (±) BOB1, and OCT2 as well as (−) AE1–AE3, EMA, ALK-1, and S100

CD15 and CD30 have both membranous and Golgi IHC-phenotype pattern, although CD15 may be restricted to the Golgi area only in some cases. PAX5 is a B-cell-specific activator protein. PAX5 is the B-cell marker present in >95% of cases demonstrating the B-cell nature of

Fig. 10.3 B-cell differentiation (normal and abnormal) is presented in figures (**a**) and (**b**)

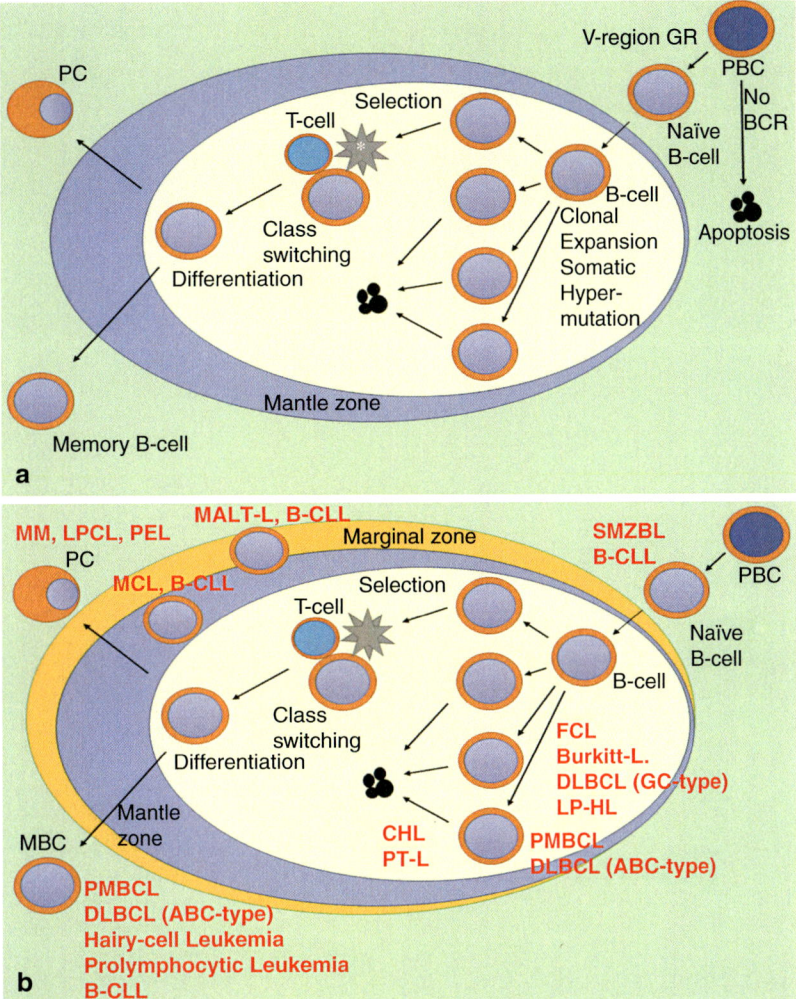

RSC. CD20 and CD79a are usually not expressed or of varied intensity and typically present on a minority of tumor cells. The detection of LMP-1 and EBER is indicative of CHL. The prevalence of EBV in CHL is variable with the highest rate of 75% in MC-CHL and lowest in NS-CHL (see below). The transcription factor OCT2 and its co-activator BOB1 are practically absent in CHL. An important pitfall is that CD30 is normally present in activated immunoblasts BOB1 and OCT2 span all stages of B cell development from precursor B lymphocytes to plasma cells (EBV, Epstein-Barr virus; EBER, EBV-encoded small RNAs; BOB1, B cell specific Octamer Binding protein-1; OCT2, Organic cation transporter 2).

- *RSV*:
  - *Mononuclear* RS (Hodgkin cell): Large mononucleated with large nucleolus and basophilic cytoplasm
  - *Lacunar cell*: Large/giant cell (40–50 um) with mono-/multilobated nucleus, medium-sized nucleoli smaller than classic RSC, and abundant clear cytoplasm (artefactual contraction from formalin fixation)
  - *Popcorn cell* (*LP* or predominant lymphocytic cell, *L&H* or lymphocytic and histiocytic cell) (mnemonics, POP→BOB): Large cells with quite a negligible cytoplasm and large folded or multilobated nucleus to such extent to qualify as popcorn-shaped and multiple, basophilic, and small nucleoli

HL vs. NHL

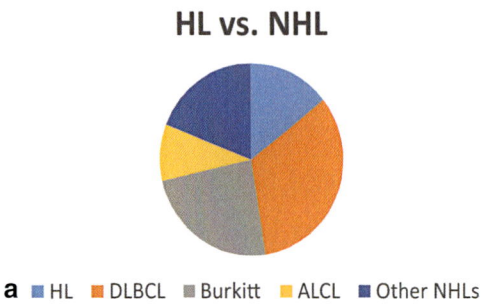

a ■ HL ■ DLBCL ■ Burkitt ■ ALCL ■ Other NHLs

HODGKIN LYMPHOMA vs. NHL
Single axial group vs. multiple peripheral lymph nodes
Contiguous vs. noncontiguous spread
Mesenteric nodes and Waldeyer ring usually not involved vs. involved
Extranodal involvement rare vs. common

b

HODGKIN LYMPHOMA
- cell of origin: germinal centre B-cell
- Hallmark of disease: Reed-Sternberg cells
- Other HL-Cells: RS variants, lacunar and popcorn cells in the affected tissues
- most cells in affected lymph node are polyclonal reactive lymphoid cells, not neoplastic cells

c

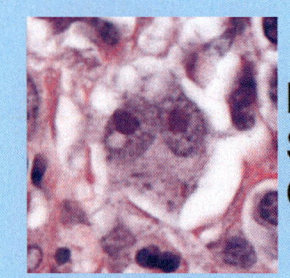 Reed-Sternberg Cell

d

Classical Hodgkin's Lymphoma (CHL):
Most of the HL with a postgerminal center B lymphocyte's genotype.
- Nodular Sclerosis (NSHL): Most of the HL with 1/5 EBV-positive
- Mixed Cellularity (MCHL): HL with 3/4 expressing EBV-positive
- Lymphocyte-Rich HL (LRHL): <5% of all CHL
- Lymphocyte-Depleted HL (LDHL): <5% of all CHL

Nodular Lymphocyte Predominant HL (NLPHL):
5% of all HL with germinal center genotype and usually EBV-negative.

e

X-ray: economic and easy method to assess mediastinal and hilar LN-pathy but are not as accurate as CT.
Lymphangiography: accurate assessment of the lower abdominal LNs and, since of retained contrast material, allows repeated examinations and assessment of the response to therapy.
CT and **US**: the most useful ways to assess abdominal and retroperitoneal LN-pathy.
MRI and **gallium scanning** are not 1st line studies for the assessment of LN-pathy.

f

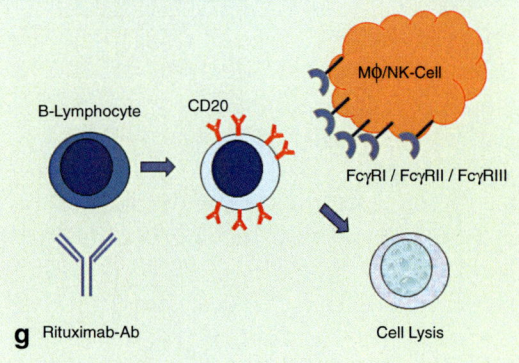

g

Fig. 10.4 Hodgkin lymphoma versus non-Hodgkin lymphomas (NHLs) is a critical differential diagnosis in childhood and youth. In figure (**a**) a pie chart shows the prevalence for Hodgkin lymphoma (HL), diffuse large B-cell lymphoma (DLBCL), Burkitt lymphoma (Burkitt), anaplastic large cell lymphoma (ALCL), and other NHLs. Some differential guidepoints are presented in figure (**b**), and the cells of the HL are shown in the figure (**c**). The hallmark of HL is the Reed-Sternberg cell, which is shown in figure (**d**). The subclassification of the classical HL and nodular lymphocytic predominant HL is shown in figure (**e**), while figure (**f**) summarizes the diagnostic procedures in case of HL. The action of rituximab on B lymphocytes is shown in figure (**g**)

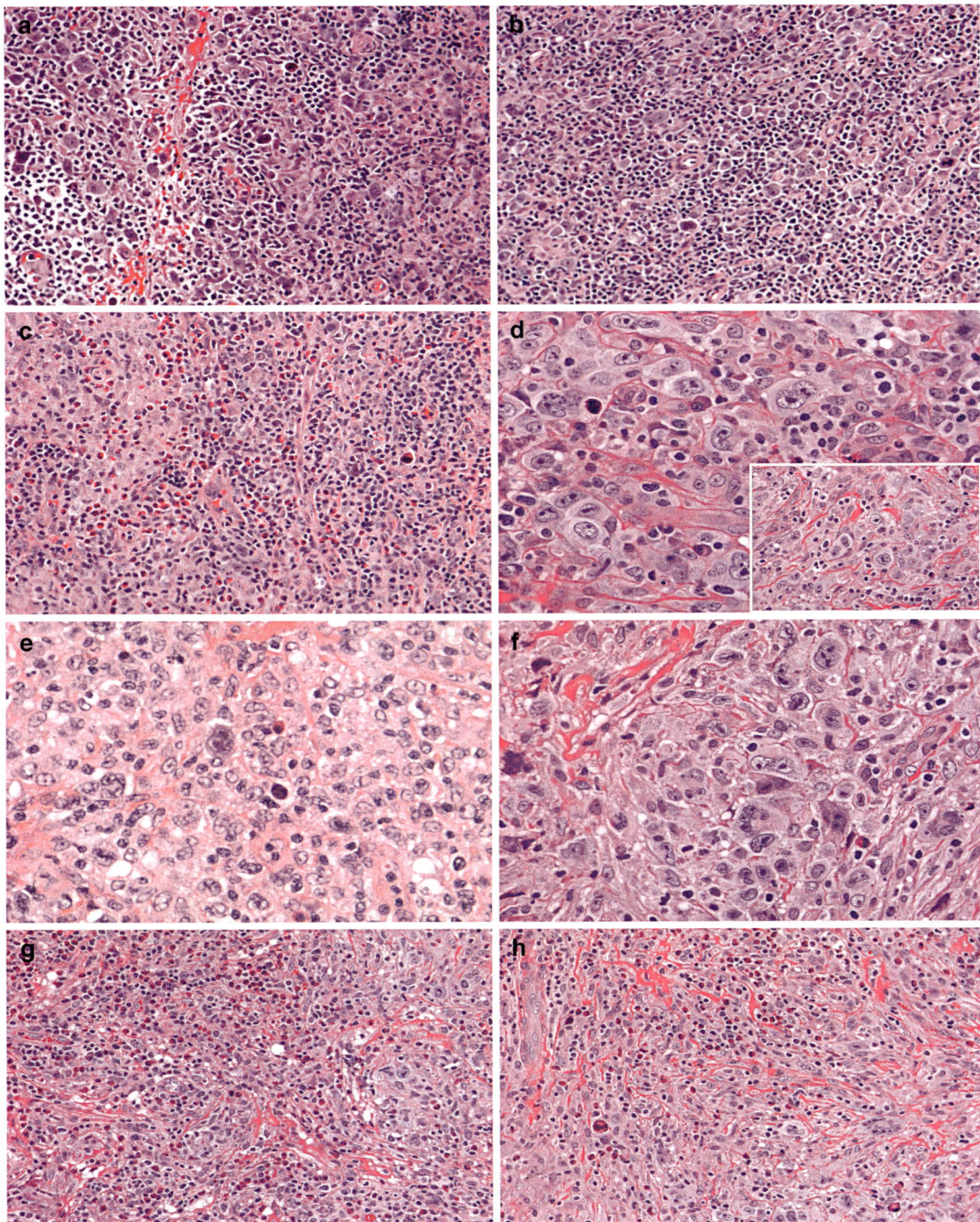

Fig. 10.5 The panel displays the histology of a classical Hodgkin lymphoma (HL) with the different cell variants as described in the text (**a**, x200; **b**, x200; **c**, x200; **d**, x400; **e**, x400; **f**, x400; **g**, x200; **h**, x200). All microphotographs are taken from hematoxylin and eosin stained histological slides from pediatric patients affected lymph nodes with HL

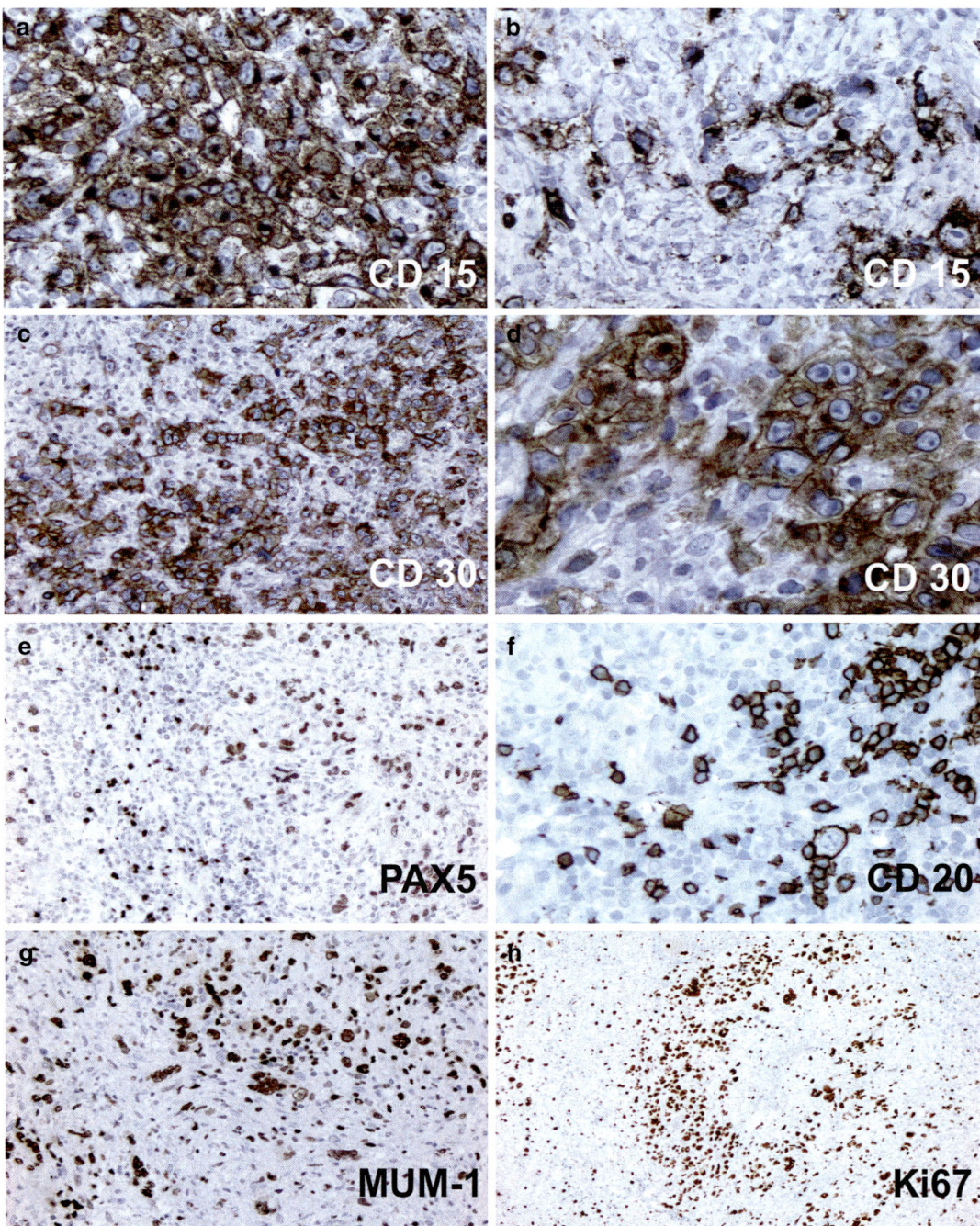

Fig. 10.6 The immunohistochemistry helps confirming the diagnosis of a classical Hodgkin lymphoma with expression of CD15 (**a**, x400; **b**, x400) and CD30 (**c**, x200; **d**, x630). Hodgkin lymphoma shows also positive staining with antibodies against PAX5, a member of the paired box (PAX) family of transcription factors (**e**, x200); CD20, a B-lymphocytic marker (**f**, x400); and MUM-1, a member of the interferon regulatory factor (IRF) family (**g**, x200). The proliferation activity is shown in figure (**h**) using a monoclonal antibody (MIB1) against Ki67 (**h**, x100). All immunohistochemistry procedures have been performed using avidin-biotin complex

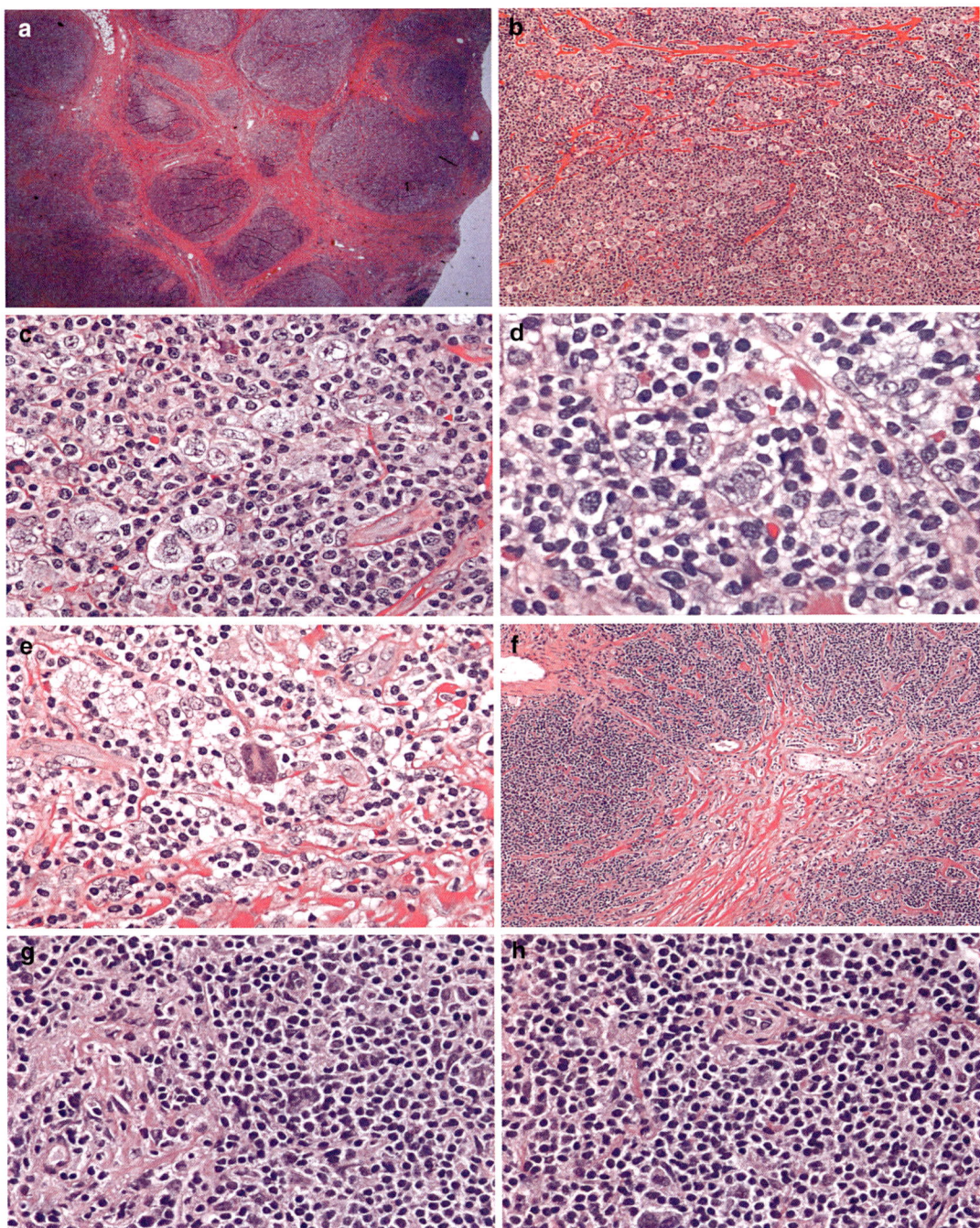

Fig. 10.7 In the classic Hodgkin lymphoma, nodular sclerosis variant, there are bands of sclerosis transetting the lymph node (**a**, x12.5). Hodgkin cells are identified in several fields of a lymph node involved by a neoplastic infiltration of malignant cells (**b**, x100; **c**, x400). Figure (**d**) shows a classic Reed-Sternberg cell (x630), while figure e shows a mummified cell (x400). In figure (**f**), the neoplastic infiltration is disappeared at places leaving the bands of sclerosis on place (**f**, x50), but a closer examination of the lymphoid population reveals hallmark of residual neoplastic infiltration (**g**, x400; **h**, x400). All microphotographs have been taken from histological slides stained with hematoxylin and eosin

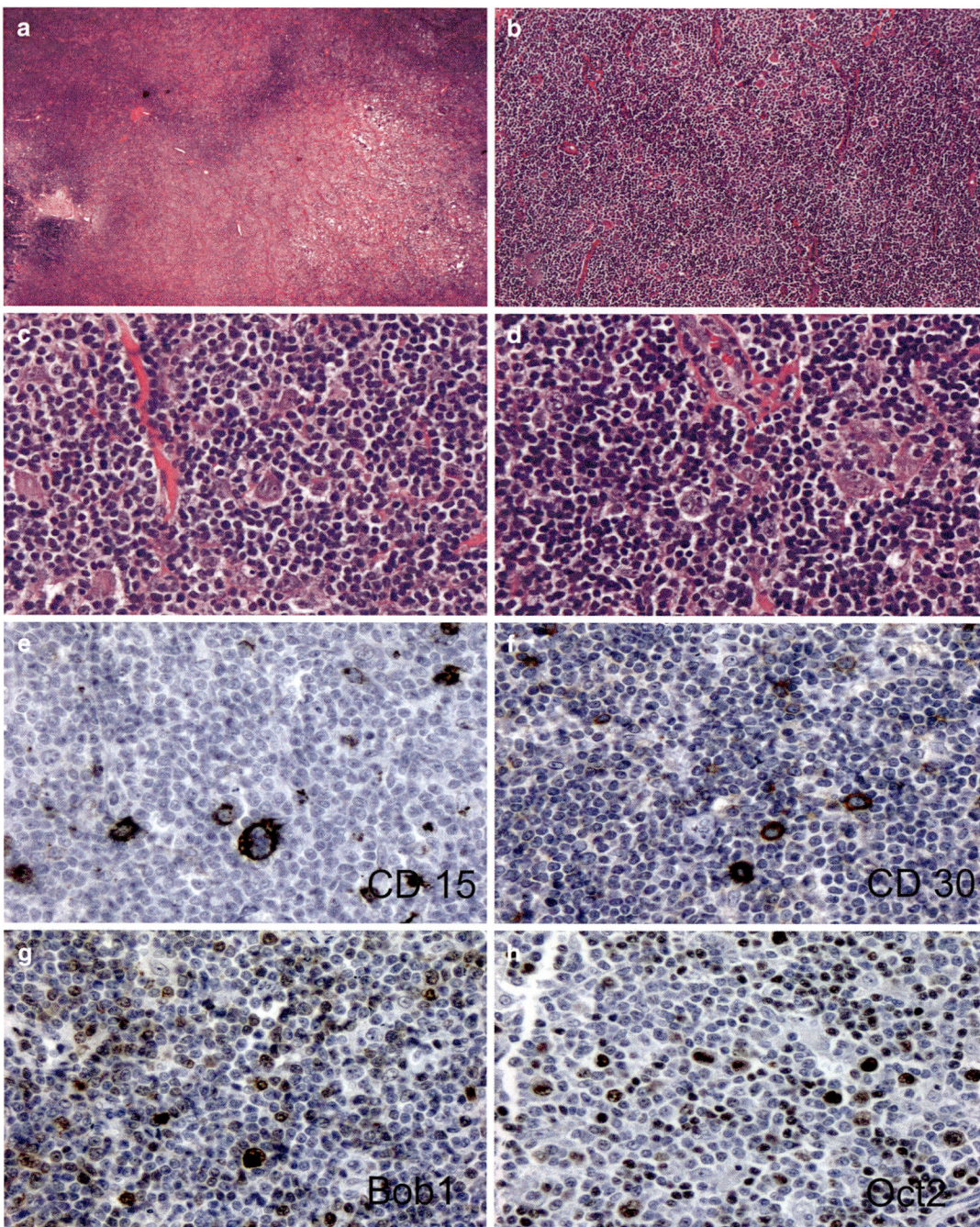

Fig. 10.8 The panel shows microphotographs from a lymph node involved by a nodular lymphocytic predominant Hodgkin lymphoma. There is effacement of the lymph node architecture (**a**, x12.5, hematoxylin and eosin staining). Low- and high-power magnifications reveal the neoplastic infiltration with large, neoplastic cells derived from germinal center B cells, in an inflammatory background (**b**, x100; **c**, x400; **d**, x400). The four microphotographs show positive staining for CD15 (**e**, x400), CD30 (**f**, x400), Bob1 (**g**, x400), and Oct2 (**h**, x400)

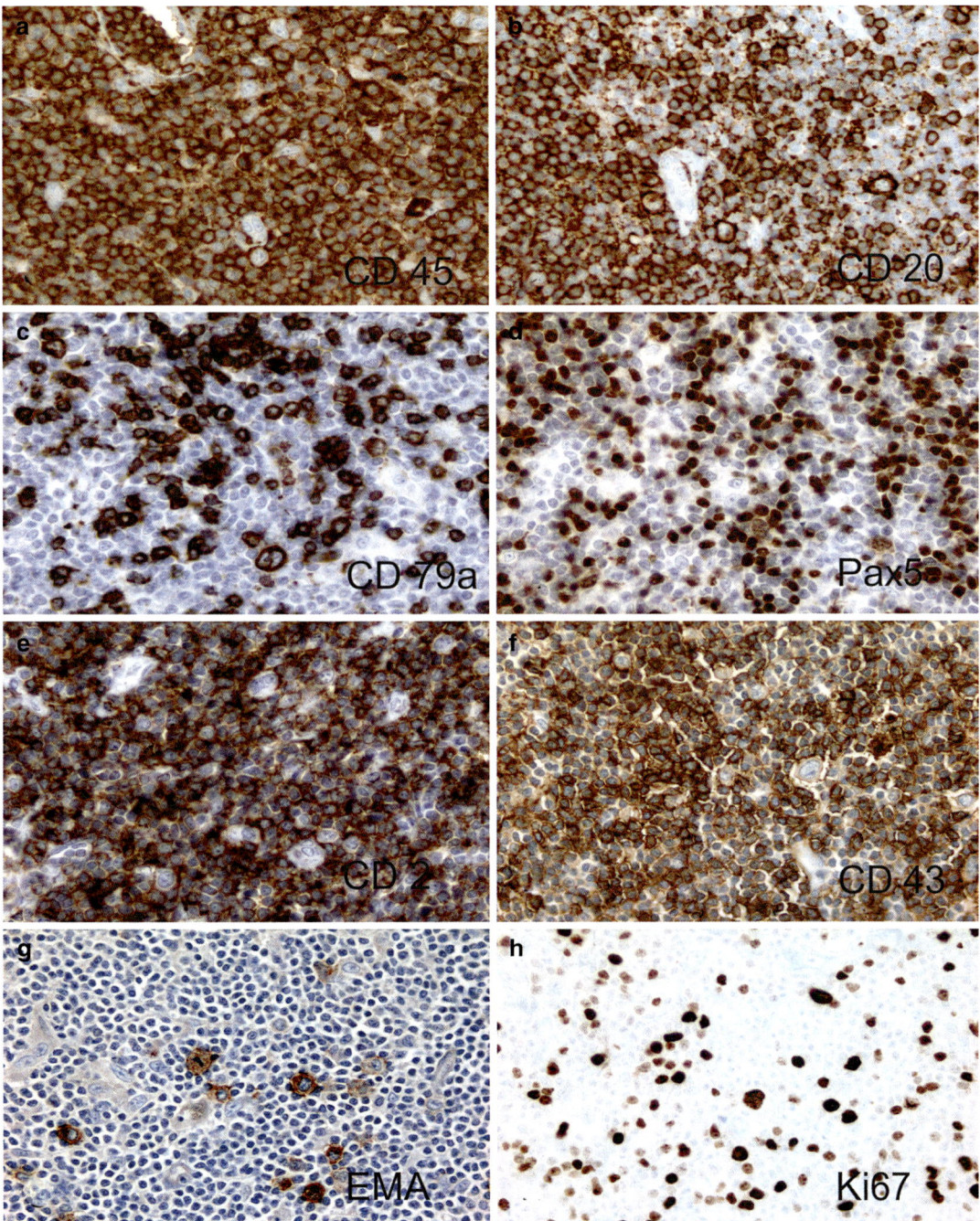

Fig. 10.9 Eight more microphotographs of immunohistochemistry performed on a nodular lymphocytic predominant Hodgkin lymphoma are shown in this panel, which discloses positive staining for CD45 or common leukocyte antigen, which is usually absent in the classical type of Hodgkin lymphoma (**a**, x400); CD20, a B-lymphocytic marker (**b**, x400); CD79a, another B-lymphocytic marker (**c**, x400); PAX5, a member of the paired box (PAX) family of transcription factors (**d**, x400); CD2 (**e**, x400); CD43 (**f**, x400); and epithelial membrane antigen (EMA) (**g**, x400). The microphotograph (h) shows the proliferation rate using a monoclonal antibody (MIB1) against Ki67 (**h**, x400). All immunohistochemical assays have been performed using the avidin-biotin complex

- IHC: (+) BOB1, OCT2, Bcl6, EMA, CD20, and CD45 and (−) CD15, CD30, and EBV

Mononuclear RSV is commonly seen in CHL. The lacunar cells are characteristic of NS-CHL, while popcorn cells are rosetted by NK/T cells, which are CD3+ (more) and CD57+ (less). In contrast to RSC from CHL, the transcription factor OCT2, its co-activator BOB1, and activation-induced deaminase are seen in popcorn cells, which is the substrate of NLPHL (vide infra). It is now known that RSC is pre-apoptotic germinal center B cells, while LP cells are Ag-selected, mutating germinal center B cells.

- Shadow or *mummified cell*: Shrunken nuclei-apoptotic cells with condensed cytoplasm
- Undifferentiated giant multinucleated cell

HL is classified as follows: nodular lymphocyte predominant Hodgkin lymphoma (NLPHL) and classic Hodgkin lymphoma (CHL).

- Nodular sclerosis classical Hodgkin lymphoma (NS-CHL)
- Mixed cellularity classical Hodgkin lymphoma (MC-CHL)
- Lymphocyte-rich classical Hodgkin lymphoma (LR-CHL)
- Lymphocyte-depleted classical Hodgkin lymphoma (LD-CHL)

HL needs to be differentiated from NHL both therapeutically and prognostically. HL is different from NHL owing to:

1. The manner of spread (contiguous vs. simultaneous nodal involvement), i.e., lymph nodal regions are usually involved in HL using a step-by-step fashion from the neck to mediastinum to celiac tripod and extranodal regions, while coincident lymph nodal regions are involved and are usually seen in NHL.
2. Histology (polymorphic vs. monomorphic), i.e., the cell population is in HL polymorphic with many reactive cells, including small lymphocytes, plasma cells, eosinophils, and macrophages, which outnumber RSC/RSV. By contrast, NHL is usually monomorphic, and reactive cells are vastly outnumbered.
3. Age (young vs. older children/adolescents), i.e., younger patients are usually seen in HL, while NHL target most often older children, adolescents, and older individuals.
4. PGN (cure vs. curable/non-curable). Individuals affected with HL harbor usually a better prognosis with 4/5 of cases reaching a virtual cure.

About staging, the Cotswolds revised Ann Arbor staging system is used.

Clinically, NLPHL presents usually with the involvement of localized peripheral LN (stage I or II). CHL manifests with the participation of localized peripheral LN or lymphoid structures (stage I or II), being mediastinal involvement most often seen in the NS-CHL subtype, while abdominal and splenic participation are most frequently seen in the MC-CHL. B symptoms (fever, nights sweats, body weight loss) are observed in up to 40% of patients. HL is accompanied by immunologic dysfunction, being T-cell-mediated responses mostly severely depressed. The consequence of this dysregulation let patients prone to several viral, mycobacterial, and fungal infections.

The four histopathologic subtypes of CHL have different clinicopathologic features, although the number of RSC is roughly correlated with the subtype and prognosis (LD-CHL > MC-CHL > NS-CHL > LR-CHL). The worse prognosis, and high rate of RSC seem to follow LD-CHL. In the vast majority of CHL, the neoplastic cells are derived from mature B cells at the germinal center stage of differentiation and, rarely, from peripheral (post-thymic) T cells.

To remember is that EBV detection may be relevant for the diagnosis of IM, PTLD or other immunodeficiency states, Burkitt lymphoma, NK-cell neoplasm (aggressive NK-cell leukemia and extranodal NK-/T-cell lymphoma, nasal type), and immunoblastic/plasmablastic lymphoma other than HL. Viral-encoded proteins are LMP and EBNA-2 that may be detected by IHC, while EBV-encoded RNA needs to be assessed by in situ hybridizations (ISH). Thus, both IHC and ISH are essential tools in the diagnostic workup of lymphomatous processes.

NLPHL derives from a germinal center B cell at the centroblastic stage of differentiation and is characterized by total or partial replacement of the lymph nodal architecture by a nodular or a nodular and diffuse infiltrate, mainly consisting of small lymphocytes and intermingled LP or popcorn cells. The diffuse areas are usually composed of small lymphocytes with admixed histiocytes that may be single or in clusters. The presence of a single nodule is sufficient to rule out the diagnosis of T-cell/histiocyte-rich large B-cell lymphoma (THRLBCL). At the margins of the nodules of NLPHL, histiocytes and some polyclonal plasma cells can be detected. Reactive follicular hyperplasia accompanied by a progressive transformation of germinal centers (PTGC) adjacent to NLPHL can occasionally be found, and this transformation may either precede or follow a diagnosis of NLPHL. This finding may be significant because PTGC may develop to HL and oncologic surveillance may be advisable. Sclerosis is rarely seen, compared to CHL, and the presence of remnants of small germinal centers may help in the low-power differential diagnosis with LR-CHL.

Genetic Abnormalities and Oncogenes in NLPHL and CHL
NLPHL: Clonal rearrangement of Ig genes (VH region of the *IG*) harboring a high load of somatic mutations as well as *IGH/BCL6* translocation and *BCL6* rearrangements (involving *IG*, *IKAROS*, *ABR*, etc.). Somatic hypermutations of the wrong type have also been found and most often in *PAX5*, *PIM1*, *RhoH/TTF*, and *MYC*.

CHL: Recurrent gains of the chromosomal subregions on 2p, 9p, and 12q and high-level amplifications on 4p16, 4q23-q24, and 9p23-p24 may be detectable by comparative genomic hybridization. In CHL arising in follicular lymphoma, t(14;18) and t(2;5) may be found. Breakpoints in the IGH locus have been observed using interphase cytogenetics.

Accurate diagnosis and staging are critical for the best management of cancer patients. About the treatment effect in HL and NHL, the Deauville 5-point scale or Deauville 5PS plays a significant role. This score is an internationally recommended scale for clinical routine and clinical trials in the initial staging and assessment of treatment response in both HL and some types of NHL. The Deauville score is based on FDG-PET/CT, which is positron emission tomography with 2-deoxy-2-[fluorine-18]F-D-glucose integrated with computed tomography (^{18}F-FDG-PET/CT). FDG-PET/CT has emerged as a powerful imaging tool for the detection of various cancers.

10.4.6 Non-Hodgkin Lymphomas

Non-Hodgkin lymphoma (NHL) is a broad category comprising lymphomas of B- and T-cell type and represents clonal expansions of lymphocytic elements blocked explicitly at particular stages of B-cell and T-cell differentiation. NHLs are common neoplasms in both childhood and adulthood and are second only to acute leukemia as the most common form of tumor in children and youth. In the small bowel, lymphomas constitute 30–50% of all malignant tumors in a range of age comprising from children to adults. Thus, a solid presentation of NHL may not be uncommon and needs to be taken into consideration. Fine needle aspiration (FNA) may be very useful, but some contraindications are bleeding diathesis, suspicion of *Echinococcus granulosus* or liver abscess, carotid body tumor, and pheochromocytoma.

NHL Etiology Directions and Trends
Some factors have been associated or ascribed to be causative in NHL. These factors include alterations in immunoregulatory control mechanisms (congenital and acquired immunodeficiencies), infections, chemicals, and chromosomal changes. A number of molecular compounds have been reviewed by the International Agency for research on Cancer - World Health Organization in the last two decades and associated with NHL in both experimental animals and humans and results of these revisions are published in monographs (https://monographs.iarc.fr/). *Congenital immuno-*

deficiencies such as ataxia-telangiectasia and Wiskott-Aldrich syndrome have an up to 50 times increased incidence of NHL than healthy people. Ataxia-telangiectasia (or Louis-Bar syndrome, OMIM 208900) is a rare inherited multisystem disorder that affects mainly the nervous and immune systems. There is progressive difficulty with coordinating movements (ataxia) beginning in early childhood with associated involuntary jerking movements (chorea), muscle twitches (myoclonus), and disturbances in nerve function (neuropathy). The gene underlying this disorder is called *ATM* and is located on 11q22-q23. ATM mutations prevent repair of broken DNA, increasing the risk of cancer. Wiskott-Aldrich syndrome is characterized by an X-linked congenital immune deficiency and a reduced ability to form blood clots. There are microthrombocytopenia, eczema, and abnormal white blood cells leading to an increased risk of several immune and inflammatory disorders as well as an increased susceptibility to infection and NHL. Mutations in the *WAS* (Wiskott-Aldrich syndrome) gene located on Xp11.4-p11.21, forming a protein called WASp or WASP, cause Wiskott-Aldrich syndrome. WASP is involved in communicating signals from the surface of blood cells to the actin cytoskeleton, and its signaling activates cell adhesion. *Acquired immunodeficiencies* are also conditions predisposing to NHL. An acquired immunodeficiency may develop in SLE, RA, Sjögren syndrome, and Hashimoto thyroiditis, following immunosuppression by transplantation (see posttransplant lymphoproliferative disorders or PTLDs) and aging.

Infections such as HIV, EBV, and HHV8 among viruses are some known examples. EBV is a potent polyclonal stimulator for B-cell proliferation and has been implicated in Burkitt lymphoma as well as some immunoblastic lymphomas. HHV8 may escape HLA-class-I-restricted antigen presentation to cytotoxic T lymphocytes by increasing endocytosis of MHC class I chains from the cell surface. The consequence of this event is the possibility to enable latent infection and immune escape in primary and chronic infection. Both multicentric Castleman disease, a rare lymphoproliferative disorder of the plasma cell type in both HIV-seropositive and HIV-seronegative patients, and pleural effusion lymphoma, or body cavity-based lymphoma, a NHL of B-cell lineage characterized by pleural, pericardial, or peritoneal lymphomatous effusions in the absence of a solid tumor mass, belong to the diseases associated with HHV8 infection. *Chemicals* such as phenytoin altering the surface membrane of some lymphocytes interfere with the immunoregulation and promote cellular proliferation, while *chromosomal translocations* such as t(8;14) involving protooncogenes such as *c-myc* and the heavy chain gene may encourage the development of NHL. Clinically, the immunoregulatory alterations predispose patients harboring NHL to develop infectious diseases. B-NHL usually predisposes patients to bacterial illness because there is increased catabolism of immunoglobulins and suppression of all B-cell functions, and T-NHL predisposes to viral, mycobacterial, or fungal infections.

10.4.7 Follicular Lymphoma

- *DEF*: Germinal center NHL lymphoma with a nodular histologic pattern of growth, which and a "bitonal" taxonomy in the Working Formulation (WF) classification being split with the predominantly small cleaved and mixed cell types of low grade and the primary large cell type of intermediate grade (Fig. 10.10).
- *EPG*: The cell of origin is the follicular center cell (FCC) with B immunophenotype (CD19+, CD20+) and exhibiting the expression of CD10+ and BCL6+ as well as BCL2+ but no expression of CD5 and CD43.
- *EPI*: 1/2 of adult NHL, but pediatric follicular lymphoma has been delineated in the last decade and seem to be better recognized currently. Patients are usually with a median age of 55–65, although this kind of NHLs are not unusual in individuals <40 years of age.
- *CLI*: Stage III or stage IV except for pediatric cases, which present with stage I or II.

In Box 10.8, the distinctive features of follicular lymphoma from reactive follicular hyperplasia are presented.

10.4 White Blood Cell Disorders

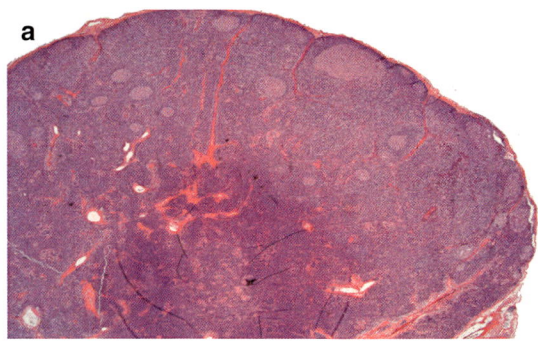
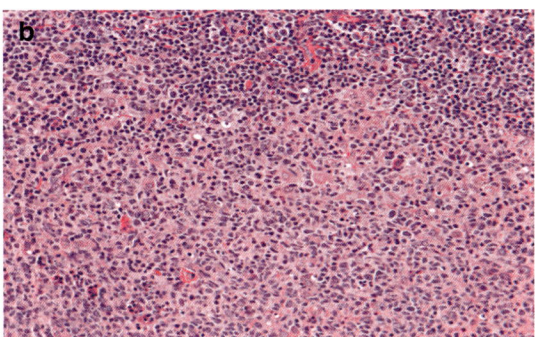
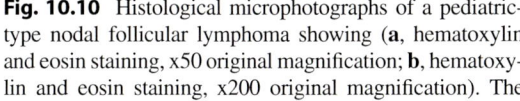

Fig. 10.10 Histological microphotographs of a pediatric-type nodal follicular lymphoma showing (**a**, hematoxylin and eosin staining, x50 original magnification; **b**, hematoxylin and eosin staining, x200 original magnification). The pediatric-type nodal lymphoma is an indolent clonal proliferation in children and young adults with high proliferation index and no BCL2 rearrangement. In this patient, there was positivity for CD20, CD10, and BCL6 (not shown)

> **Box 10.8 Distinctive Features of FCL from Reactive Follicular Hyperplasia**
> 1. Effacement of the LN architecture
> 2. Filling of sinuses and capsular infiltration
> 3. Minimal variation in size and shape of follicles
> 4. Even distribution of follicles throughout the cortex and medulla
> 5. Monomorphism of neoplastic lymphocytes
> 6. Low MI, minimal phagocytosis, no TBM, and BCL2 (+)
> 7. Interfollicular area with monomorphism
>
> Note: *MI*, mitotic index; *TBM*, tingible body macrophages.

Additional tips to the features of Box 10.8 which include the presence of amorphous or better described as eosinophilic (PAS+, PASD+) material extracellularly within follicles, plasmacytosis in the interfollicular regions, and T cells with CD4+ immunophenotype have been described, but controversially weighted in the scientific literature. There are two growth patterns, including the follicular pattern and the diffuse pattern in FCL. The follicular pattern shows firmly packed follicle effacing the LN architecture and disappearance of the centrocytes-centroblasts (CC-CB) polarization of reactive follicles ("mantle zone attenuation") surrounded by a pole rim of marginal zone. The diffuse pattern shows the area of tissue completely lacking follicles and defined by CD21+ and CD23+ FDC (follicular dendritic cells) by immunohistochemistry. To remember is that FDC are cells of the immune system found in lymph follicles of the B cell areas of the lymphoid tissue. One of the challenges is to distinguish FDC from centroblasts. FDC needs to be distinct from CB. FDC have round nucleus, bland, dispersed chromatin, and one small central nucleolus. Conversely, CB has a large nucleus, vesicular chromatin, and ≥1 basophilic nucleus. The grading of FCC is also crucial and is based on WHO: low grade if 0–5 CB/10 HPF, intermediate grade if 6–15 CB/10 HPF, and high grade if >15 CB/10 HPF (HPF, high-power field). However, there is an additional subdivision of high grade with 3A if CCs are present and 3B if solid sheets of CBs are present.

Any area of DLBCL in an FL should be reported, according to the WHO recommendations, as the primary diagnosis and should also be reported an estimate of the proportion of DLBCL and FL present in all histologic tissue sections.

- Variants: Signet ring lymphoma (prominent intracellular Ig, usually IgG of LCs vs. IgM of Russell bodies in lymphoplasmacytic malignancies) (Of note, Russell bodies are homogeneous Ig-containing inclusions representing distended endoplasmic reticulum and showing histochemical positivity for PAS stain and immunohistochemical positivity for both CD38 and CD138 stains)

- BM: Paratrabecular lymphoid aggregates showing cellular monomorphism and increase in reticulin. Interestingly, FCC-NHL tends to be paratrabecular, differently from the non-paratrabecular location of large cell NHL.
- PB: ± involvement ("buttock" cells):
- Splenic involvement: 50%
- Hepatic involvement: 50% (portal triads, involving and crossing the limiting plate)
- CGB: t(14;18)(q32;q21) involving BCL2 on Chr. 18 and IGH on Chr. 14 (~80%).
- PGN: Progression (SC→MC→LC) variable.

Rarely, there may be a blastic transformation with conversion to leukemic phase, in which case the median survival is reduced to 2 months.

- FLIPI (Follicular Lymphoma International Prognostic Index): Remarkably useful index constituted by histologic grade (1–2 vs. 3), ≥6 chromosomal breaks and complex karyotype, TP53 gene mutations, and proliferation of cell fractions (Ki67 immunohistochemistry with internal validation)
- Pediatric FLL: a localized form of FCC-NHL, BCL2-, t(14;18) (−), and grade 3!

10.4.8 Small Lymphocytic Lymphoma

- DEF: Well-differentiated lymphocytic lymphoma of B-cell type with an expression of CD5, CD23, and CD43, although CD23 may also be lacking and 1% of cases are T-cell NHL (immunophenotype-based).
- SYN: WD-lymphocytic lymphoma.
- EPI: Middle-aged to elderly.
- CGB: 12+ (trisomy 12) and 13q−, being trisomy 12 a bad and 13q- a good PGN marker.
- CLM: Effacement of the architecture by diffuse monotonous infiltration of lymphocytes with clumped chromatin pattern, no or inconspicuous nucleoli, few or rare mitoses, PAS+ inclusions either nuclear or cytoplasmic, and capsular invasion. The finding of (+) proliferation centers is crucial (proliferation centers: clusters of blasts and pro-lymphoblasts deter-

> **Box 10.9 Pseudofollicles vs. Follicles**
> - Round nuclei vs. ovoidal nuclei
> - Poor demarcation vs. well demarcation
> - No compression of surrounding Ret fibers vs. compression of surrounding Ret fibers

mining a pseudofollicular appearance with less basophilia) (Box 10.9). In patients with hepatomegaly (~1/10), there are portal triads, typically, which are infiltrated, but sinusoidal involvement is seen in CLL. In patients with splenomegaly (~1/4), the white pulp shows some extension into red pulp with asymmetric, coalescing nodules. In this context and in an appropriate epidemiological background, the DDX with splenic MZL is crucial.
- PGN: "Poor outcome"-indicating markers include the evaluation of ZAP-70 and CD38. Del11q suggests a worse outcome, while del 13q a better result. Tumors with mutated Ig genes behave aggressively.

Richter's transformation: Transformation of a low-grade NHL to a higher-grade NHL (large cell lymphoma) occurring in 1/10 of cases. Another conversion may be prolymphocytic leukemia with BM involvement, usually not paratrabecular. Variant: plasmacytoid (immunocytoma), which is generally associated with monoclonal gammopathy, and up to half of the cases have BM involvement, typically paratrabecular. If most plasma cells are observed, the diagnosis of plasmacytoma (multiple myeloma) should be considered.

The difference between light and heavy chain producers is essential. IgM or heavy chain producers tend to be lymphomas, while IgG or light chain producers tend to be myelomas. If IgM dysproteinemia is present the diagnosis of Waldenstrom's macroglobulinemia should be given. Waldenstrom's macroglobulinemia, is characterized by hyperviscosity, cryoglobulinemia, and Coombs-positive hemolytic anemia. The distinction between Russell and Dutcher bodies is also important. Russell bodies are "cytoplasmic" Ig inclusions, while Dutcher bodies are "nuclear"-located cytoplasmic Ig inclu-

10.4 White Blood Cell Disorders

sions. Frankly speaking, Dutcher inclusions are inclusions of the cytoplasm that are either invaginated into or are covering the nucleus.

10.4.9 Mantle Cell Lymphoma (MCL)

- *DEF*: Intermediate grade B-cell non-Hodgkin lymphoma characterized by overexpression of *cyclin D1* and the t(11;14)(q13;q32) chromosomal translocation arising from naive pre-germinal center B lymphocytes present in primary lymphoid follicles and mantle zones of secondary follicles with non-mutated human immunoglobulin heavy-chain-variable region genes. Clinically heterogeneous, MCL frequently disseminates to extranodal areas with a quite poor overall outcome. MCL shows a monomorphous neoplastic infiltration of small B lymphocytes (formerly MCL was also called *centrocytic lymphoma*) with CD5+, CD23−, CyclinD1+, and t(11;14)(q13;q32) as well as B symptoms (1/3), including fevers, night sweats, and weight loss.
- *EPI*: Middle age, ♂ > ♀, 3–10% of NHL (Western).
- *CLI*: Lymphadenopathy, BM involvement in 2/3 of cases (*paratrabecular* lymphoid aggregates), ~ PB involvement (leukemic phase, usually <2×10^4 absolute lymphocytes), ~ splenomegaly (white pulp involvement), hepatic (periportal hepatic infiltration), gut (lymphomatoid polyposis of small bowel), CNS, and skin involvement (dermal and subcutaneous, perivascular or periadnexal with "grenz zone" sparing of the epidermis and often blastoid).
- *CLM*: Three patterns are recognized, including:
  1. Diffuse replacement of architecture
  2. Mantle zone infiltration and expansion
  3. Vague nodular pattern

Neoplastic lymphocytes are monotonous small lymphocytes (CCs) with scant cytoplasm and cleaved round or slightly irregular nuclei with condensed chromatin. No nucleoli, no large cells (CBs or immunoblasts), no proliferation centers. DDX:. SLL/CLL, prolymphocytic leukemia, FL.

- IHC: (+) CD19, CD20, CD22, and CCND1 (aka cyclin D1, bcl1, PRAD1), but also CD43, CD79a, FMC7, sIgM or IgD, κ/λ, bcl2.
- (−) CD10, bcl6 (unlike FL, which is also CD5-), (−) CD23 (unlike SLL/CLL), CD11c, TdT (unlike lymphoblastic lymphoma), T-cell antigens (unlike many cutaneous T-cell NHL), p27(kip1) (unlike B-CLL, FL, MZBCL, and HCL).
- MZBCL is (−) CD5, CD10, CCND1.
- CCND1 regulates the G1→S cell transition which is not balanced by RB1 and p27Kip1 leading to the development of mantle cells and (+) also in HCL (possible in children and young adults) and myeloma (not typically observed in childhood and youth).
- In situ MCL: CCND1 ISH/IHC restricted to the inner mantle zones only
- *CGB*: t(11;14)(q13;q32) with *IGH/CCND1* gene fusion transcript
- Also, breakpoints at 8q24, 9p22–9p24, and 16q24
- *PGN*: MCL may have variable survival rates but is, in general, more aggressive than SLL/CLL or MZBCL and tends to relapse, but not to transform to DLBCL. Nodular or mantle zone patterns have more extended survival rates (5 years), while blastic variant, high MI, high Ki67/MIB1 index, PB involvement (leukemic phase), and trisomy 12, 3q+, 9q−, p53 mutation/overexpression/loss are associated with shorter survival (3 years). The International Prognostic Index is poorly reproducible, because numerous patients present with high stage disease with multiple foci of EN involvement. Therapy protocols include rituximab (anti-CD20) and bortezomib (proteasome inhibitor) as well as acalabrutinib and lenolidomide as FDA-approved drugs.

10.4.10 Marginal Cell Lymphoma

- *DEF*: Post-germinal center B-cell NHL (~7–8% of all B-cell NHL) strongly associated with AI disorders and composed of morphologically heterogeneous small-sized B lymphocytes with the absence of CD5, CD10, and

CD23 but an expression of BCL2+, CD43 (±), and EMA (±). There is no rearrangement of the *BCL1* or *BCL2* genes, and the course is indolent (slow to disseminate and potential to recur), although there is a potential transformation to DLBCL.
- *SYN*: MALTomas, because of the origin in MALT-harboring organs (e.g., salivary gland, GI tract). MALT or mucosa-associated lymphoid tissue is a diffuse lymphatic system located in the gastrointestinal tract, pharynx and nasopharynx, thyroid, lung, salivary glands, breast, skin, and eye.
- *EPG*: Interestingly, marginal zone cell lymphomas as early lesions are probably antigen-driven (e.g., *H. pylori* in the stomach, *Chlamydia psittaci* in the ocular adnexa, *Campylobacter jejuni* in IPSID, and *Borrelia burgdorferi* as a cutaneous form) and are reversible by removal of the antigen. Subsequent stages including dissemination of the lesion make MZL antigen-independent becoming irreversible. IPSID or immunoproliferative small intestinal disease is an infectious pathogen-associated human NHL.
- *CLM*: Two forms, including the nodular and diffuse forms (Box 10.10). This lymphoma has a centrocyte-like (small cleaved), monocytoid, and/or plasmacytoid cell morphology. In the nodular type, a perifollicular distribution is seen. In this setting, lymphocytes may infiltrate the follicles determining a lymphocytic cuff (mantle zone pattern) or may target germinal centers (follicular colonization).
- *IHC*: (+) CD19, CD20, CD79, and (+) CD21, CD35 (MZ-cell-associated antigens) and (+) BCL2 but ALWAYS (−) CD5 and CYCLIN D1 to rule out MCL and (−) CD10 and BCL6 to rule out FL and BL.

Box 10.10 Marginal Cell Lymphoma
Distinctive Features
1. Lymphoepithelial lesion in the surrounding non-neoplastic area
2. Reactive follicles with germinal centers
3. Cytogenetics (+): abnormal karyotype

- *CMB*: Most often, three translocations are observed.
  - t(1;14) (p22;q32) involving *BCL10/IGH* (transcript dysregulation of BCL10)
  - t(14;18) (q32;q21) involving *IGH/MALT1* (transcript dysregulation of MALT1)
  - t(11;18) (q21;q21) involving *API2/MALT1* (transcript dysregulation of MALT1)
- *PGN*: Since this NHL has an indolent course, it tends to remain localized for long periods before progressing.

Extranodal marginal zone lymphoma (ENMZL) involves the Hashimoto thyroiditis, which may evolve to a thyroid MALToma and Sjögren's syndrome/LESA, which may grow to salivary gland MALToma. LESA is lymphoepithelial sialadenitis, which is aka myoepithelial sialadenitis (MESA).

Splenic MZL (SMZL)
It is a B-cell NHL composed of small lymphocytes, which have a strong tendency to surround and progressively replace the white pulp of GC, efface the mantle of follicles, and, finally, merge with a peripheral zone of large cells (marginal zone) with infiltration of the red pulp. The white pulp nodules show a central dark zone of small lymphocytes, which sometimes surround a residual germinal center and progressively merge into a peripheral area of small- to medium-sized atypical lymphocytes showing more dispersed chromatin and abundant pale cytoplasm and a characteristic targetoid pattern using MIB1 (Ki67).

- *IHC*: (+) CD19, CD20, and CD79, but (−) CD10 and BCL6, (−) CD5 and CD43, and (−) ANXA1.

10.4.11 Diffuse Large B-Cell Lymphoma (DLBCL)

- *DEF*: B-cell neoplasm of either germinal center or post-germinal center (activated) B cells with very heterogeneous presentation and morphology, usually occurring in both children and adults, involving t(14;18) with

IGH/BCL2 (but also t(3;V) and *BCL6* among others) and growing as large bulky fleshy tumor resembling carcinoma and harboring a tendency for EN presentation (~1/2) including GIT, skin, and arthro-skeletal system. The characteristic of this kind of lymphoma is the heterogeneity, which is morphological, biological, and clinical (Figs. 10.11, 10.12, 10.13, and 10.14).

- *SYN*: Diffuse histiocytic lymphoma, reticulum cell sarcoma, and diffuse large cell lymphoma.
- *CLI*: In contrast to FC lymphoma with bilateral sides of diaphragmatic involvement (90%), DLBCL is typically limited to one side of the diaphragm, and BM/Hepatic involvements are rare.
- *CLM*: Large cells with vesicular NN and subtle peripheral nucleoli.
- *IHC*: ~60% B cell (e.g., CD19, CD20), ~10% T cell (e.g., CD3), 5% histiocytic (e.g., CD68), and 25% no markers (mostly B cell by gene rearrangement) (CD10 (+) and BCL6 (+)). In 1/10 of DLBCL cases, CD5 is positive. The CD5+ DLBCL needs to be distinguished from the blastoid variant of MCL by the classic lack of cyclin D1 expression, which is present in MCL by definition!
- *CGB*: Translocations involving *BCL2* and *BCL6* genes.
- *PGN*: Rapid progression and poor outcome, if untreated (high MIB1 rate), but good results if aggressive chemotherapy is started soon after diagnosis (median survival: 1–2 years).

T-cell-rich BCL: Diffuse BCL NHL, sometimes evolving from previous FCC, in which at least 3/4 of the cells are reactive T cells with standard Th/T-suppressor ratio hiding the large malignant B lymphocytes and having an indolent course, probably due to the host response of the T cells.

The most common location for primary CNS DLBCL is supratentorial (~60%).

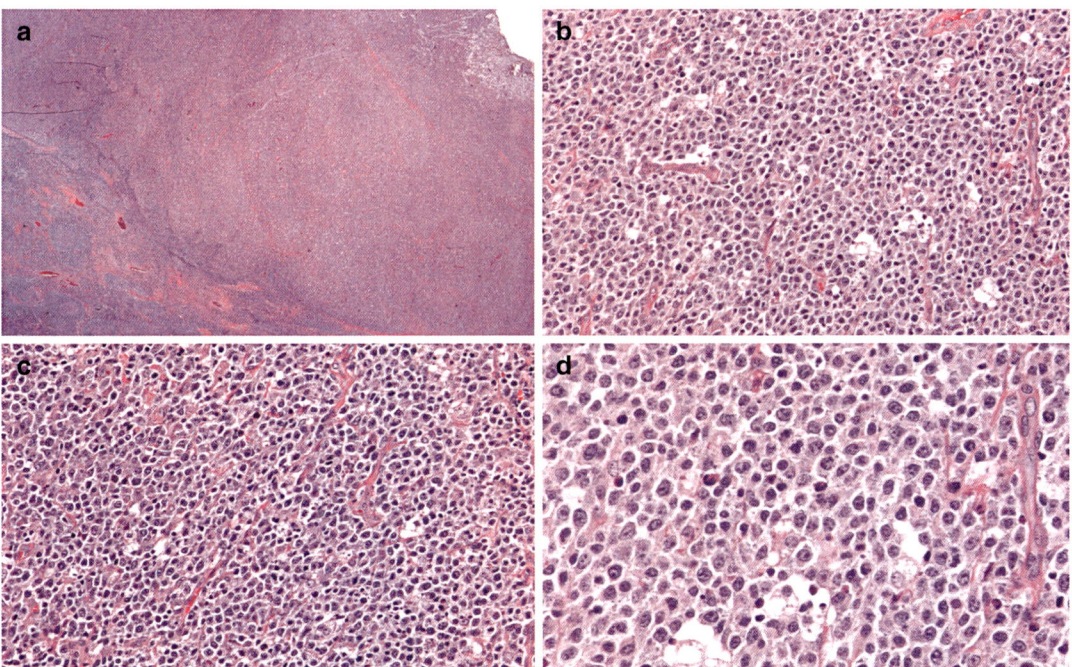

Fig. 10.11 Histological panel of a diffuse large B-cell lymphoma showing a nodal infiltration by an atypical lymphoid population constituted by large lymphoid cells (immunoblasts-like with amphophilic cytoplasm, eccentric nuclei, and one central nucleolus and centroblasts-like with pale or light basophilic cytoplasm, vesicular chromatin and 2–3 nucleoli close to the nuclear membrane) (**a**, hematoxylin and eosin staining, ×12.5 original magnification; **b**, hematoxylin and eosin staining, ×200 original magnification; **c**, hematoxylin and eosin staining, ×200 original magnification; **d**, hematoxylin and eosin staining, ×400 original magnification)

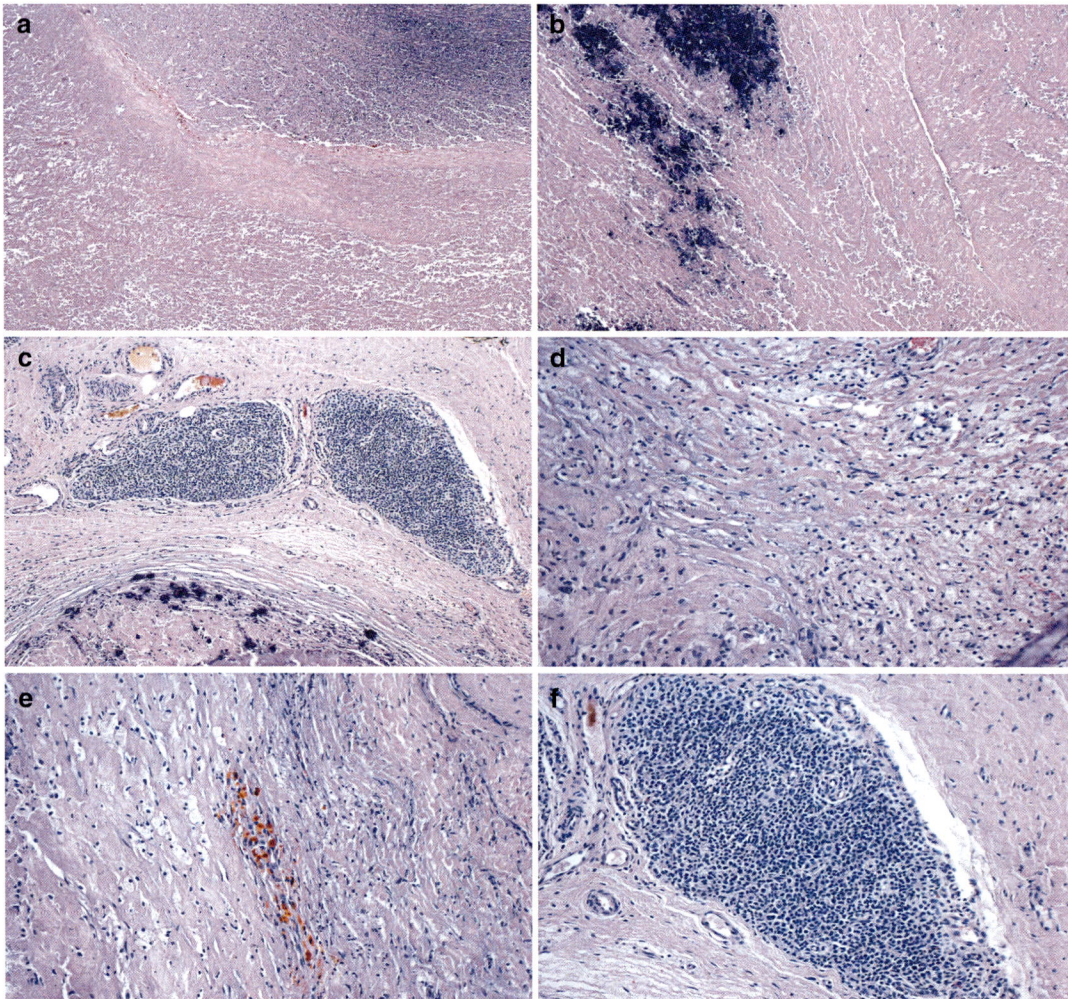

Fig. 10.12 Diffuse large B-cell lymphoma, post-chemotherapy, with numerous regressive changes including pigment deposition, foamy histiocytes, and frank coagulative necrosis (**a**, hematoxylin and eosin staining, ×40 original magnification; **b**, hematoxylin and eosin staining, ×40 original magnification; **c**, hematoxylin and eosin staining, ×100 original magnification; **d**, hematoxylin and eosin staining, ×200 original magnification; **e**, hematoxylin and eosin staining, ×200 original magnification; **f**, hematoxylin and eosin staining, ×200 original magnification;). See text for details

10.4.12 Lymphoblastic Lymphoma

- *DEF*: TdT-expressing NHL of usually T-cell type (4/5) but also B cell (1/5), with T-cell blastic proliferation (CD7+, CD2±) showing a diffuse, monomorphic growth pattern (Figs. 10.15, 10.16, 10.17, and 10.18). In Fig. 10.15 (**a**) is presented the Kobayashi-Iwasaki model. Tax (transcriptional transactivator) (p40) is a crucial factor that intermediated the viral persistence and disease development, oncogenic potential, cell-signaling pathways, the checkpoint control and inhibition of DNA repair, and, finally, the modulation of the miRNAs environment. Three mechanisms can inactivate Tax expression in ATL cells: 1) genetic changes in the *TAX* gene, 2) deletion of the 5' long terminal repeat (LTR) containing the viral promoter, and 3) DNA methylation of the 5' LTR, which leads eventually to promoter inactivation.
- Tax-mediated mutation is considered a crucial event in ATL that triggers the clonal selection of infected T cells as leukemia cells.

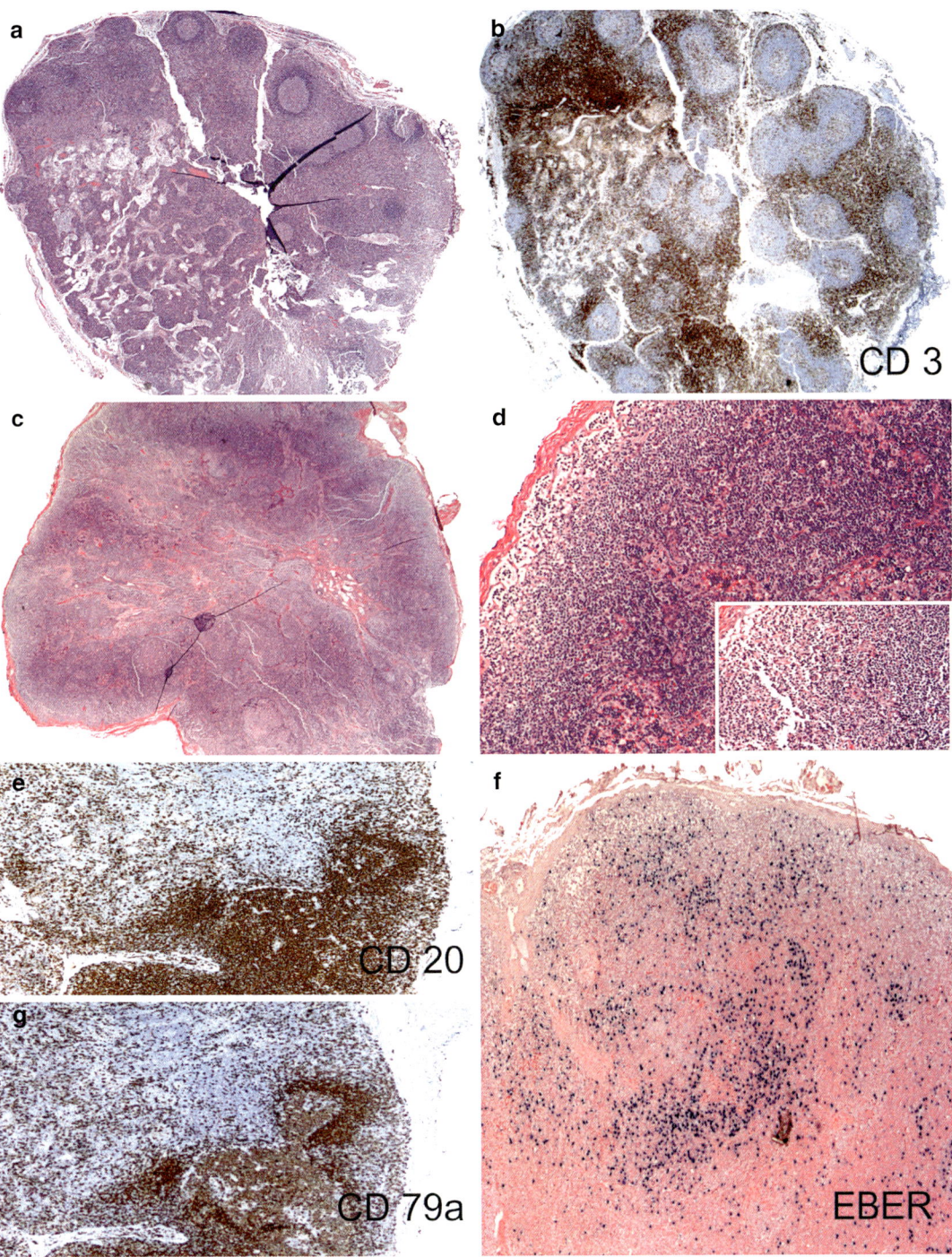

Fig. 10.13 Histological (**a**, **c**, **d**), immunohistochemical (**b**, **e**, **f**), and in situ hybridization (**g**) panel of posttransplant lymphoproliferative disorders (PTLDs). Figure (**a**) (hematoxylin and eosin staining, ×12.5 original magnification) shows an example of early PTLD with CD3 positivity (anti-CD3 immunostaining, avidin-biotin complex, ×12.5 original magnification). Figure (**c**) shows an example of nodal PTLD of polymorphic type (hematoxylin and eosin staining, ×12.5 original magnification). Figure (**d**) and inset of figure (**d**) show a high magnification of a nodal PTLD of polymorphic type. Figures (**e**) and (**f**) show expression of CD20 and CD79a in the polymorphic type of PTLD (anti-CD20 immunostaining, ×50 original magnification; anti-79a immunostaining, ×50 original magnification). The same case of polymorphic type of PTLD shows Epstein-Barr virus RNA by in situ hybridization (EBER, ×50 original magnification)

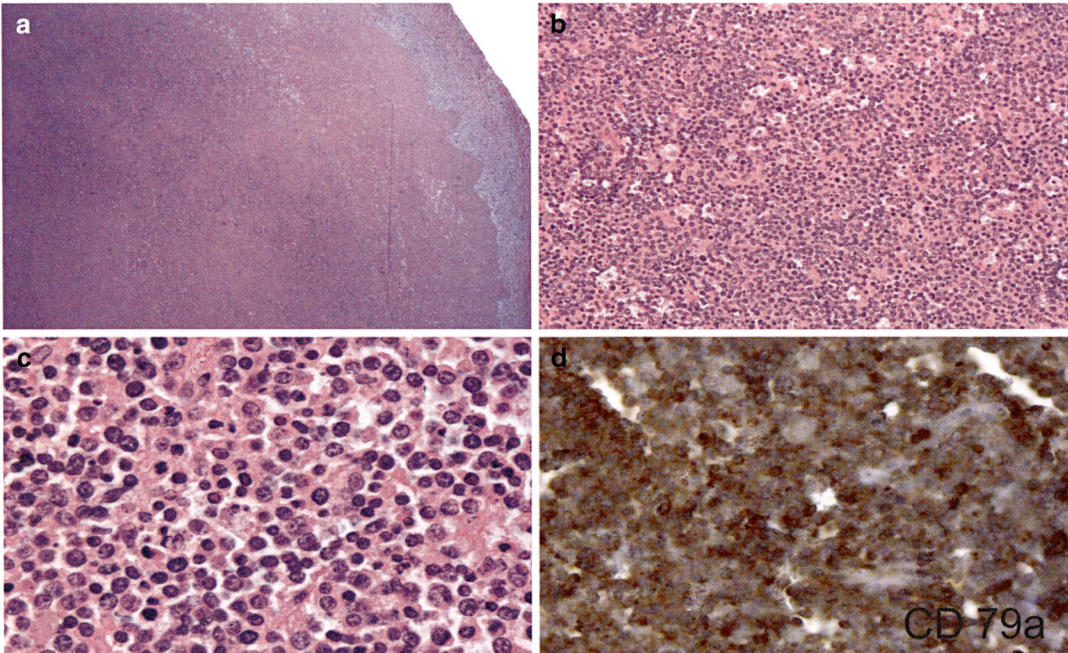

Fig. 10.14 This panel illustrates a diffuse large B-cell lymphoma as posttransplant lymphoproliferative disorder (PTLD) of monomorphic type (**a**, hematoxylin and eosin staining, ×12.5 original magnification; **b**, hematoxylin and eosin staining, ×200 original magnification; **c**, hematoxylin and eosin staining, ×400 original magnification; **d**, anti-CD79a immunostaining, avidin-biotin complex, ×400 original magnification)

- *EPI*: 1/3 of pediatric NHL (3-M's: "male, mediastinum, mitotic active").
- *CLI*: Anterior mediastinal mass with or without CSF and/or skin involvement and mainly paracortical in LNs but also ± BM involvement → acute lymphoblastic leukemia.
- *GRO*: Soft, white mass with foci of hemorrhage and necrosis.
- *CLM*: Diffuse, monomorphic lymphocytic growth with round nuclei, fine convolutions, fine chromatin, small nucleoli, and high MI as well as focal "starry sky" areas (Fig. 10.15).
- *DDX*: Small lymphocytic NHL, Burkitt lymphoma ("small noncleaved cell"), and Hodgkin lymphoma, LP variant.

Pitfall: Acute lymphoblastic lymphoma is known to express CD99 (MIC-2), an antigen usually used in pediatric pathology for the diagnosis of primitive neuroectodermal tumor (PNET) or Ewing sarcoma (mandatory → FLI1 immunohistochemistry!).

10.4.13 Burkitt Lymphoma

- *DEF*: Germinal center B-cell immunophenotypic NHL showing pan-B-cell markers (CD19, CD20, CD22, CD79a, and PAX5) and IgM with light chain restriction, along with CD10, BCL6, CD38, CD77, and CD43 (Fig. 10.19). There are three types of Burkitt lymphoma, including the (1) *Endemic*, Equatorial Africa, jaw, orbit, kidney, adrenals, ovaries, and children (EBV+ (95%)); the (2) *Sporadic*, worldwide, ileocoecum, children (EBV+ (20%)); and the (3) *HIV-associated*, worldwide, multiple sites, and children<adults (EBV+ (25%)).
- *SYN*: "Small noncleaved cell" NHL (WF), diffuse undifferentiated lymphoma.
- *EPG*: The cell of origin is a cell intermediate in size between lymphocytes and histiocytes, usually positive for B-cell markers (sIgM, CD19, CD20) and absence of CD5 and CD23. Neoplastic cells are generally negative for

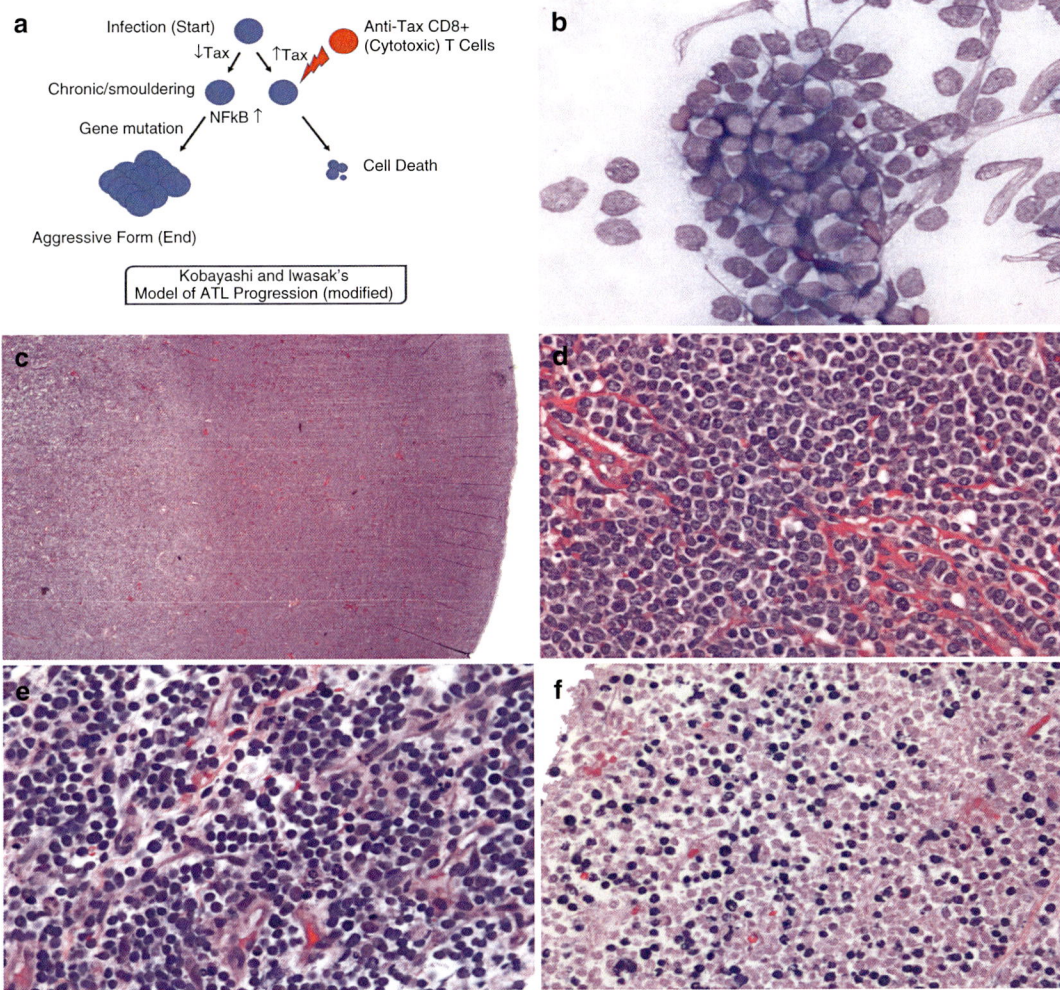

Fig. 10.15 The Kobayashi Iwasaki's model of ATL progression is displayed in figure (**a**), while figure (**b**) shows the characteristic cytologic features of T-lymphoblastic lymphoma (×400). A pediatric T-lymphoblastic lymphoma is presented in figure (**c**) (H&E stain, x12.5), while figures (**d**) (H&E stain, x400) through (**f**) show the intimate relationship of the T-lymphoblastic cells with blood vessels and some areas of necrosis, respectively (×400). Tax (transcriptional transactivator) is a major factor that influences: 1) the viral persistence and disease development. 2) any oncogenic potential, 3) cell-signaling pathways, 4) checkpoint control and inhibition of DNA repair, and 5) the miRNAs environment. Three mechanisms are considered to play a role to inactivate Tax expression in ATL cells: 1) genetic changes (nonsense mutation, deletion, and insertion) in the *TAX* gene, 2) deletion of the 5' long terminal repeat (LTR) containing the viral promoter, and 3) DNA methylation of the 5' LTR, which eventually leads to promoter inactivation. Tax-mediated mutation has been suggested to be a critical event in ATL that triggers the clonal selection of infected T cells as leukemia cells

BCL2 and always negative for TdT. A critical pitfall could be CD10 expression without the characteristic immunophenotype of BL. FL-NHL is another B-cell malignancy showing CD10 (+). In Fig. 10.19 (**a**) there is the schematic representation of the temporal events leading to Burkitt lymphoma. Translocation-inducing factors (EBV, malaria or congenital dysregulated immunity) is the primary event, which through inflammasome dysregulation, genetic changes, and/or cytokine-related events progresses to *IgH MYC* genetic change eventually transforming macrophages into Burkitt lymphoma cells. Double-hit lymphoma (DHL)

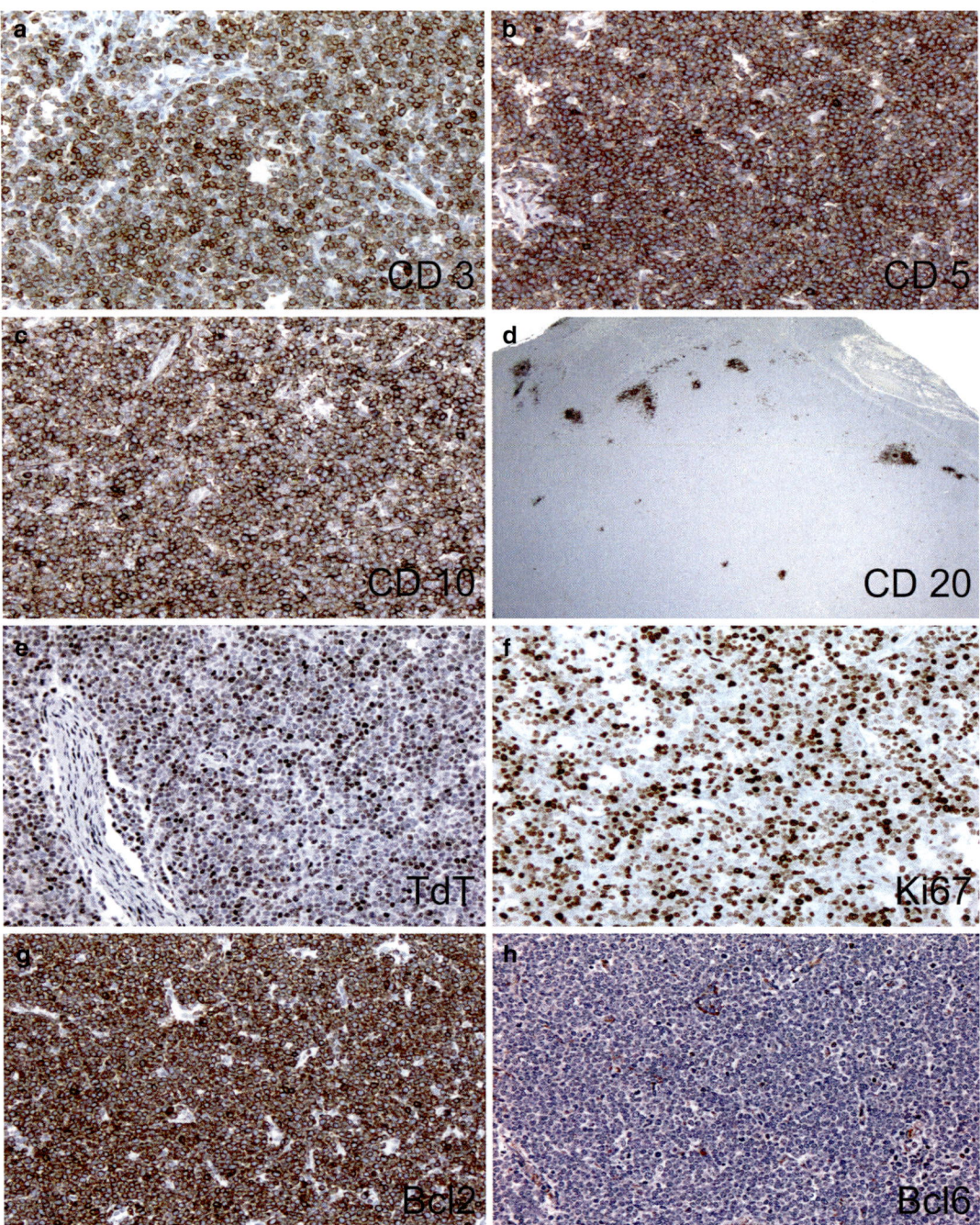

Fig. 10.16 The immunohistochemical panel of T-lymphoblastic lymphoma shows a positivity of the tumor cells for CD3 (**a**, ×200), CD5 (**b**, ×200), CD10 (**c**, ×200), terminal deoxynucleotidyl transferase or TdT (**e**, ×200), and Bcl2 (**g**, ×200), while there is no expression for CD20, a B lymphocytes marker (**d**, ×12.5), and Bcl6 (**h**, ×200). TdT is a nuclear DNA polymerase in thymic and small number of bone marrow cortical lymphocytes with the function to catalyze the addition of deoxynucleotides to 3' OH ends of oligonucleotides or polydeoxynucleotides. CD3, CD5, and CD10 are expressed in T lymphocytes. Bcl2 gene encodes an integral outer mitochondrial membrane protein that has the function to block the apoptotic death of some cells (e.g., lymphocytes). Bcl6 is the protein encoded by a zinc finger transcription factor acting as a sequence-specific repressor of transcription. Bcl6 has been shown to modulate the transcription of STAT-dependent IL-4 responses of B cells. The Bcl6 gene is found to be frequently translocated and hypermutated in DLBCL. All immunostochemistry assays have been performed using the avidin-biotin complex (ABC)

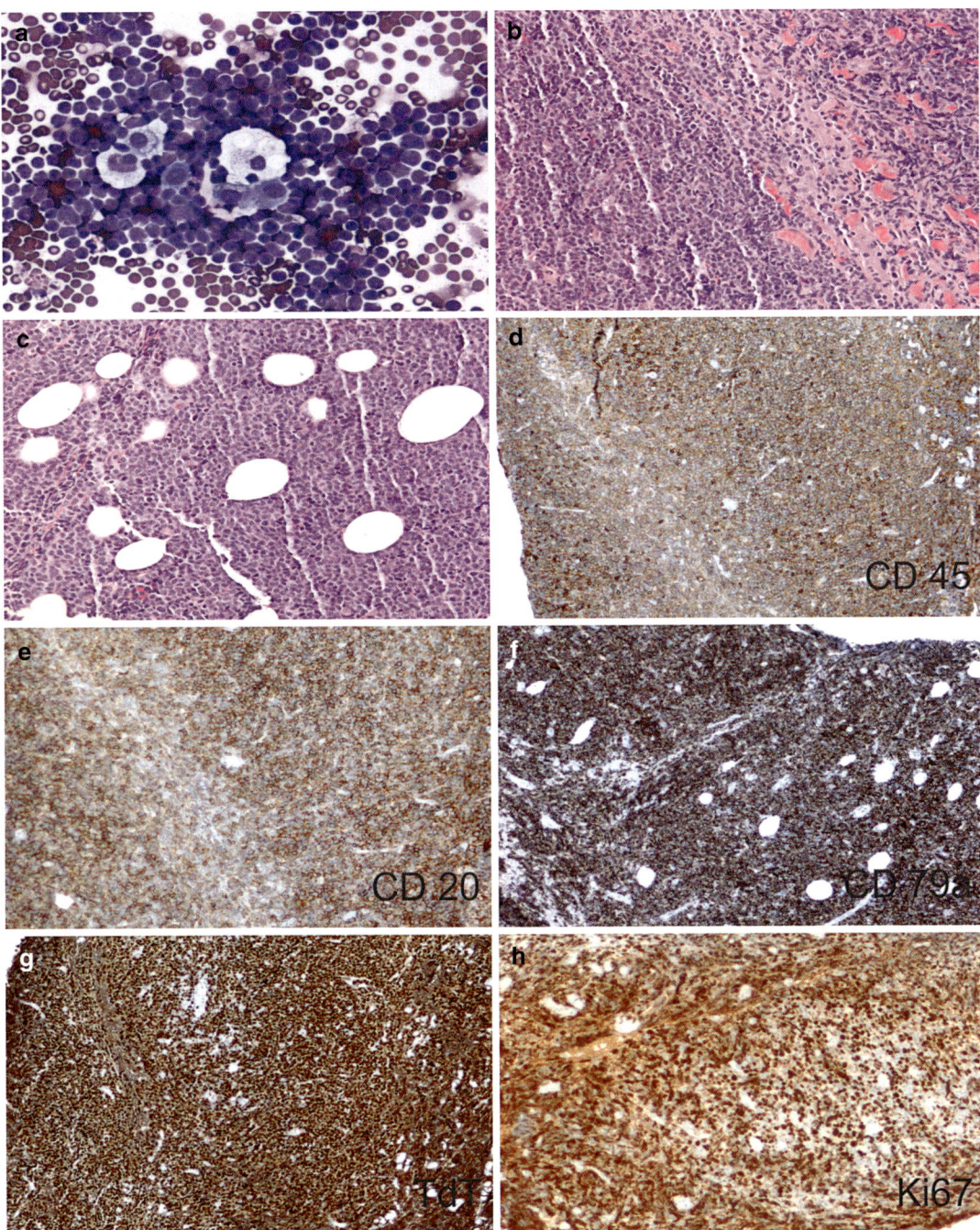

Fig. 10.17 Cytological (**a**), histological (**b–c**), and immunohistochemical (**d–h**) panel of B-lymphoblastic lymphoma of the mediastinum of a child. Cytologically, the atypical lymphoid cells show a high nucleus to cytoplasm ratio (**a**, Diff-Quik staining, ×400 original magnification). Microphotographs of the histology (**b–c**) show the atypical lymphoid cell population with frank fatty tissue and connective tissue infiltration (**b–c**, hematoxylin and eosin staining, ×200 original magnification). Figures (**d**) through (**h**) show the positivity for CD45, CD20, CD79a, terminal deoxynucleotidyl transferase (TdT), and the Ki67 proliferation antigen (immunostaining, avidin-biotin complex, ×100 original magnification). TdT is a nuclear DNA polymerase in thymic and small number of bone marrow cortical lymphocytes with the specific function to catalyze the addition of deoxynucleotides to hydroxyl groups (3'-OH) at the ends of polydeoxynucleotides or oligonucleotides

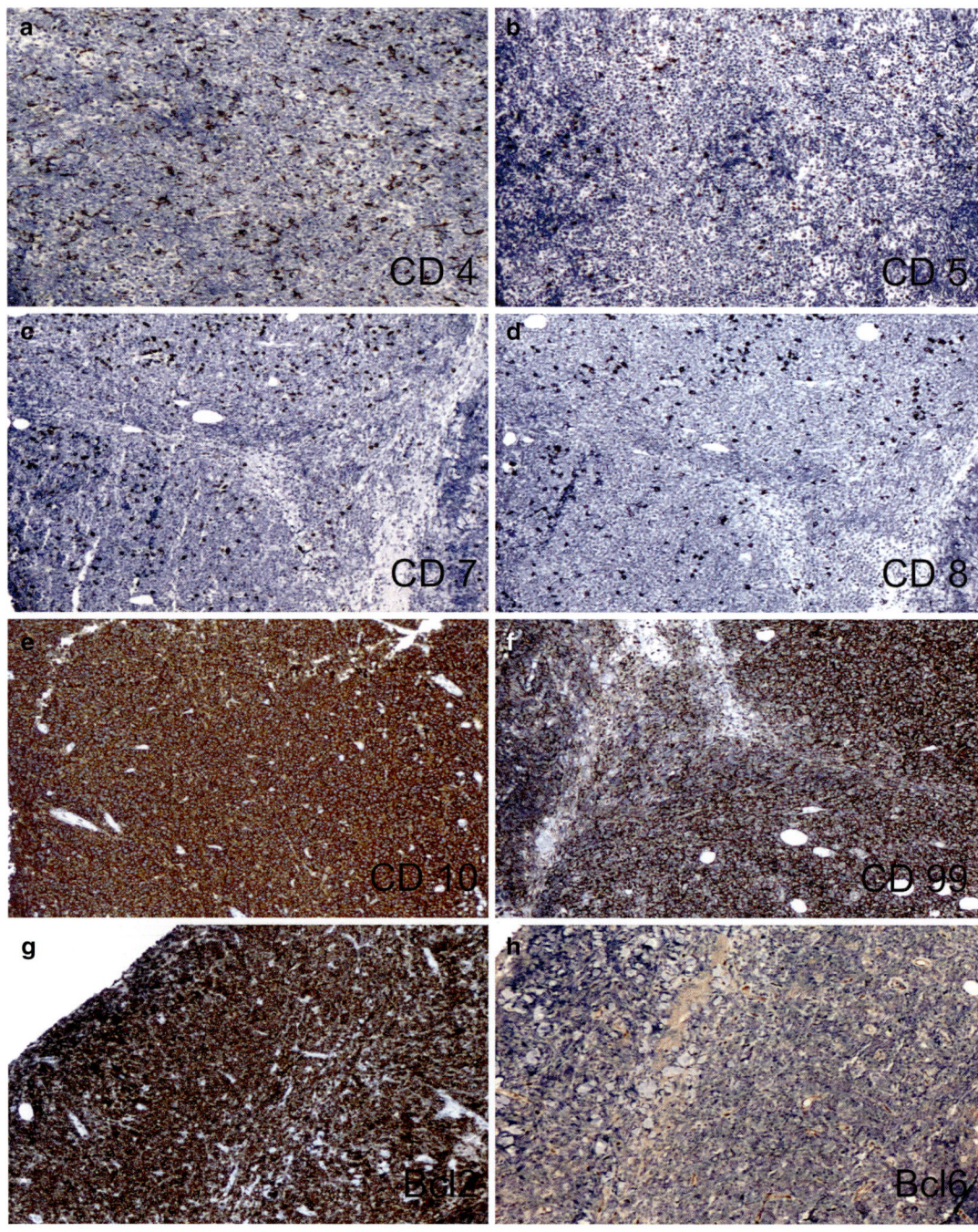

Fig. 10.18 More immunohistochemical images of B-lymphoblastic lymphoma are shown in figures (a) through (h) with immunostaining for CD4, CD5, CD7, CD8, CD10, CD99, Bcl2, and Bcl6 (see text for details). All microphotographs are at ×100 as original magnification. All immunohistochemical procedures have been carried out using the avidin-biotin complex protocol

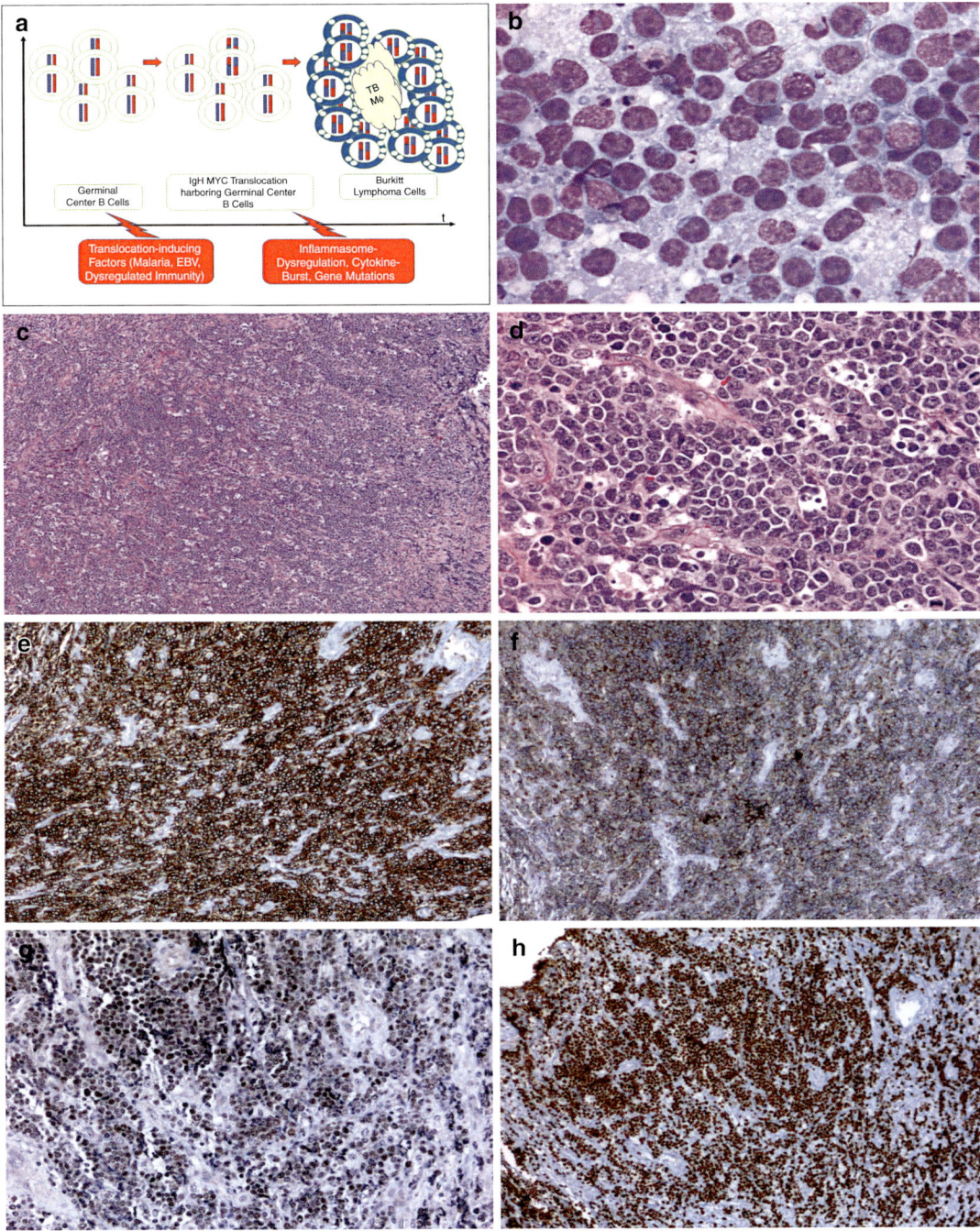

Fig. 10.19 The panel shows a schematic picture of Burkitt lymphoma with a double genetic event (**a**) (see text for details). In the microphotograph (**b**) the cytologic features of the neoplastic cells are shown exhibiting numerous intracytoplasmic vacuoles (Diff-Quick staining, ×630 original magnification). The starry sky pattern of Burkitt lymphoma is shown in the microphotographs (**c**) at low power and (**d**) at moderate power of magnification (hematoxylin and eosin staining, **c**, ×50 original magnification, and **d**, ×400 original magnification). An immunohistochemical panel (**e–h**) discloses the positivity of the tumor cells for CD20, CD10, Bcl6, and Ki67 (**e**, ×100 original magnification; **f**, ×100 original magnification; **g**, ×200 original magnification; **h**, ×100 original magnification). All immunohistochemical studies have been performed using the avidin-biotin complex

is an aggressive type of B-cell NHL characterized by rearrangements in two particular genes, including the *MYC* gene, and the other involves the *BCL2* gene or the *BCL6* gene. DHL has many features with both the diffuse large B-cell lymphoma (DLBCL) and Burkitt lymphoma. In 2016 the World Health Organization categorized DHL as its own group of B-cell NHL.

- *CGB*: t(8;14)(q24;q32) involving the *IGH* gene (80%), t(2;8)(p11;q24) involving the Ig κ light chain locus (15%), and t(8;22)(q24;q11) involving the Ig λ light chain locus (5%).
- *CLI*: In all, but the third type, peripheral lymphadenopathy is quite rare, and BM involvement occurs late.
- *GRO*: Bulky, fleshy mass ± necrotic areas.
- *CLM*: Monotonous pattern of lymphocytes, small size (10–25 μm), round with several prominent basophilic nucleoli and highly mitotic (Ki67, ~100%) with "starry sky" growth pattern with "stars" constituted by TBM. Touch preps, characteristic for intracytoplasmic vacuoles (ORO+).
- *IHC*: (+) CD19, CD20, CD10, BCL6, CD43, and EBER, while (±) BCL2 and (−) CD5 and CD23.
- *TEM*: Atypical lymphocytes showing abundant ribosomes, lipid inclusions, no glycogenic accumulation, and nuclear projections/pockets.
- *CMB*: C-myc from 8q24 to 14q32 of Ig heavy chain genes.
- *TRT*: Chemotherapy-responsive.
- *PGN*: If untreated, rapid dissemination, leukemia, and death within months. If treated, the cure is possible, but relapses common (50%).

Variant: Pleomorphic, older age group, larger tumor cells. More pleomorphism, well-defined cytoplasmic rim, large eosinophilic nucleoli, bi-/multinucleate, BM involvement>GI involvement, CD10-, *c-myc* gene rearrangement rare, *BCL2* rearranged (1/3), and a more aggressive clinical course.

10.4.14 Peripheral T-Cell Lymphoma

- *DEF*: A heterogeneous group of clinically aggressive neoplasms characterized by T-cell immunophenotypic and associated with poor outcome.
- *SYN*: Large cell immunoblastic or diffuse mixed (WF nomenclature).
- *EPI*: Diffuse T-cell NHL, mostly adults (>20 years), 1/3 of diffuse aggressive lymphomas in the USA and Canada, but it is more diffuse in East Asia and Africa.
- *EPG*: The increased incidence of T-/NK-cell lymphomas in East Asia and Africa is probably related to the frequency of endemic human T-leukemia virus 1 (HTLV-1) and EBV infections.
- *CLI*: Generalized lymphadenopathy (~ stage IV) with common B symptoms.
- *CLM*: Small cells (size slightly larger than normal lymphocytes) with highly condensed chromatin, irregular nuclei, small nucleoli ± abundant pale cytoplasm, AND large cells with vesicular nuclei and prominent eosinophilic nucleoli with the presence of nuclear pleomorphism, lobulation, ± multinucleation in both cell size types, AND inflammatory background with PMN, eosinophils, MΦ, and plasma cells. In the cutaneous form, dermis only is typically involved, and PB involvement occurs after skin involvement.
- *IHC*: ≥1 PAN-T-MARKER (CD2,3,4,7), (−) CD1a AND TdT (both markers of immature T cells).

Subtyping: CD4+, CD8+, CD4+ CD8+, CD4− CD8−; in most cases there are α/β T-cell receptors. There is a γ/δ variant, which is an aggressive form with marked hepatosplenomegaly and minimal lymphadenopathy. This variant involves mostly youth and shows common relapses. It is usually fatal.

- *IHC*: CD3+, CD2+, and CD16+, but usually CD4- and CD8−.
- *DDX*: In a partial list of non-cutaneous T-cell NHLs, we should count the T-cell prolymphocytic leukemia/lymphoma. Also, T-cell large granular lymphocytic leukemia/lymphoma, chronic lymphoproliferative disorder of NK cells, aggressive NK-cell leukemia, systemic EBV+ T-cell lymphoproliferative disease of childhood, adult T-cell leukemia/lymphoma,

extranodal NK-/T-cell lymphoma (nasal type), enteropathy-associated T-cell lymphoma, hepatosplenic T-cell lymphoma, angioimmunoblastic T-cell lymphoma, anaplastic large cell lymphoma (ALCL) (ALK+), and anaplastic large cell lymphoma (ALCL) (ALK−) are counted. Clinical features and age need to be taken into account when reading the immunophenotype or the flow cytometry data.

In primary T-cell lymphoma, NOS (not otherwise specified), two most commonly lost pan-T-cell antigens are CD5 and CD7, and this is important to keep in mind during the workup of the immunophenotype of a T-cell NHL. Primary cutaneous CD30+ T-cell lymphoproliferative disorder lymphomatoid papulosis presents spontaneously as regressing papules, which are usually isolated to the limbs. The CD4+CD25+regulatory T cells (Treg) are paramount in immunology. In fact, (1) Treg suppress the activation of self-reactive T cells and prevent autoimmunity, (2) Treg may prevent cytotoxic T cells from operating a killing procedure on tumor cells, and (3) Treg suppress immunologic cells that are reactive to host antigens and to avoid graft-versus-host disease (GVHD).

10.4.15 Anaplastic Large Cell Lymphoma

- *DEF*: Aggressive T-cell (mostly) NHL derived from activated mature CTL that are usually large with abundant pale cytoplasm and pleomorphic nucleus (lobate, "packman nucleus"), showing often a "horseshoe" shape and harboring t(2;5) involving *ALK* and *NPM* genes with ALK1 expression in the nucleus and marked expression of CD30 on cell membrane and Golgi apparatus (Fig. 10.20).
- *SYN*: ALK1 NHL, and Ki-1 Lymphoma.

It is paramount to recall three aspects:

1. Activated CTL with horseshoe-shaped nuclei and large cytoplasm
2. CD30 expression (almost, but not all!)
3. ALK1 expression due to t(2;5)

- *IHC*: (+) CD2, CD5, CD45, CD30, ALK1, and EMA, but (−) CD15 and (±) CD4 and CD3.

(+) Cytotoxic-associated antigens (T-cell intracellular antigen 1/TIA1, granzyme B, and/or perforin)

1. Classic type: younger subjects, often involving the skin, with histology simulating carcinoma, sarcoma, or malignant histiocytosis with the classic immunophenotype, an unusual clinical course.
2. Primary cutaneous type: variant closely related to lymphomatoid papulosis that may be considered the benign counterpart and harboring an abnormal immunophenotype with (−) EMA and ALK1, and no t(2;5).

- *CGB*: Three common translocations:
  1. t(2;5)(p23;q35) involving *ALK* and *NPM* genes.
  2. t(1;2)(q25;p23) involving *TPM3* (encoding a nonmuscular tropomyosin) and *ALK1*.
  3. t(2;3). Variant translocations involving *ALK* and other partner genes on chromosomes 1, 2, 3, 17, 19, 22, and X also may be found. The translocation partners of *ALK* include *TPM3*, *TPM4*, *TFG*, *ATIC*, *MYH9*, *CLTC*, *MSN*, and *ALO17*.

10.4.16 Adult T-Cell Leukemia/Lymphoma

- *DEF*: HTLV-1-associated NHL (HTLV, human T-cell leukemia virus) with distinct T-cell NHL, ♂=♀, youth, acute to subacute course, and often seen in southwestern Japan and the southeastern states of the USA.
- *CLM*: Pleomorphic with multilobated nuclei ("flower cells" or "cloverleaf cells"), medium-sized cell, and mixed large and small cells with large cell and small cell variants.
- *CLI*: Generalized lymphadenopathy, hepatosplenomegaly, leukemia, CSF positive for malignant cells, activated osteoclast-related hypercalcemia and/or bony lesions, and epidermotropism with Pautrier microabscesses.

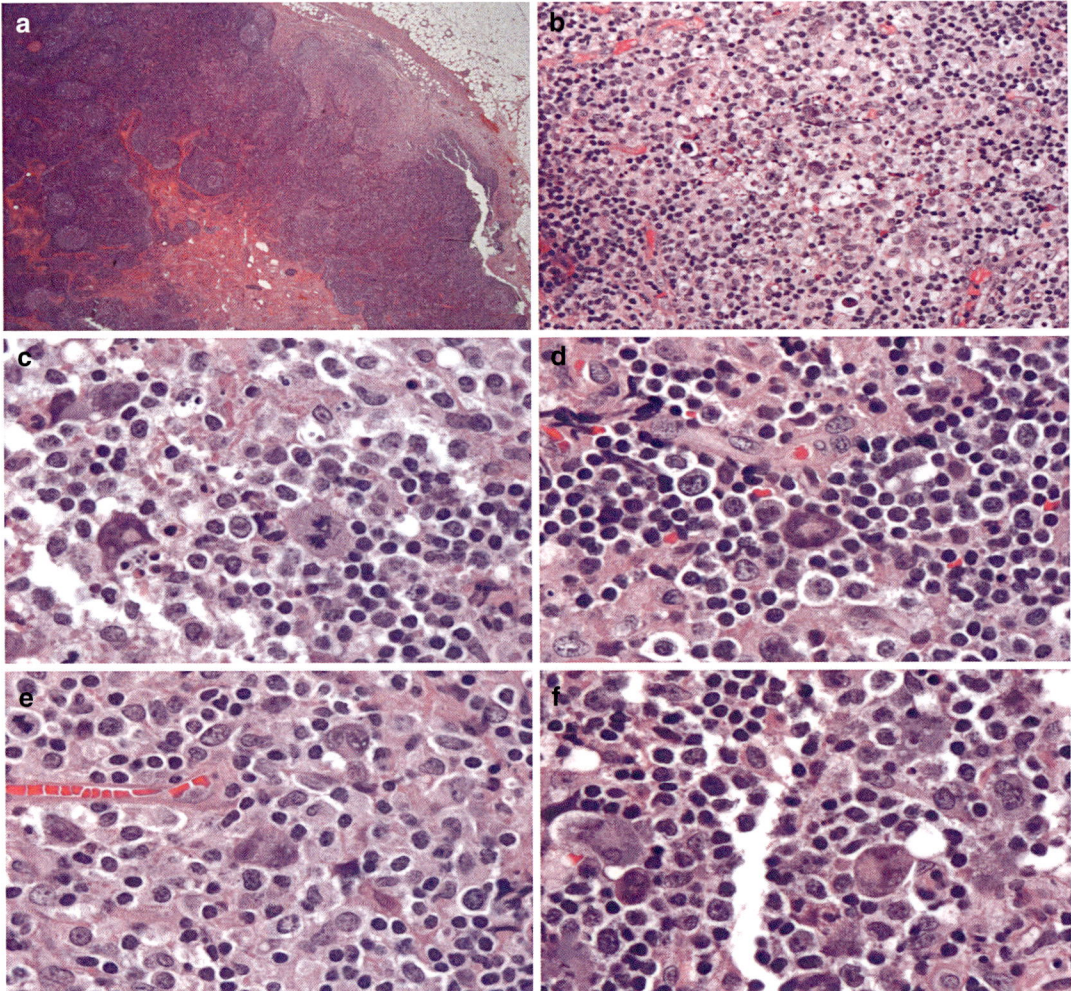

Fig. 10.20 Histological panel with micrographs of a nodal anaplastic large T-cell lymphoma taken at low- and high-power magnification. The lymph node is infiltrated (interfollicular T zones and nodal sinuses) by an atypical lymphoid population constituted by large cells with abundant cytoplasm, wreath-like or multiple nuclei, multiple nucleoli, and nuclear pseudoinclusions (**c–f**). There is a high rate of mitotic figures (**a**, hematoxylin and eosin staining, ×12.5 original magnification; **b**, hematoxylin and eosin staining, ×200 original magnification; **c–f**, hematoxylin and eosin staining, ×630 original magnification)

- *IHC*: CD4+, IL2R+ (CD25+).
- *PGN*: Stage IV at presentation, thus the outcome is poor, independent of the histologic subtype, and the aggressive chemotherapy does not seem to change the prognosis in these patients, being the median survival less than 1 year.

TEM may reveal some cerebriform nuclear irregularity that needs to be distinct from other neoplastic cells. The ultrastructural evidence of extreme cerebriform nuclear irregularity includes mycosis fungoides/Sezary syndrome, DFS, and fibroadenoma of the mammary gland as DDX.

10.4.17 Cutaneous T-Cell Lymphoma (CTCL)

- *DEF*: T-NHL with mainly cutaneous involvement showing a characteristic skin morphology with exocytosis of epidermotropic atypical lymphocytes ("Lutzner cells"), Pautrier microabscesses, and unique immuno-

phenotype (+CD2, CD4, CD5 and −CD30, − cytotoxic molecules).
- *SYN*: Mycosis fungoides, Sezary syndrome.
- *EPG*: The cell of origin of CTCL is CD4+ T cell. CTLA-4 or cytotoxic T lymphocyte antigen 4 is a protein that downregulates the immune system.
- *CLI*: Initially cutaneous involvement only and then progression to another organ involvement.

Skin Disease
0. "Premycotic" stage, erythema, and eczema with focal round-oval circumscribed flat red macules
1. Patch stage, macular coalescence into the formation of patches with scaling and intense pruritus with purple-brown spots with some spontaneous regression, at least partially and focally
2. Plaque stage, tumescence of the macular coalescence into palpable discrete, indurated papules with potential *restitutio ad integrum* centrally and progression/extension peripherally (TNM staging: T1-T2)
3. Nodule stage, nodular tumescence with size increase >1 cm in diameter and inherent tendency to erosion/ulceration

Sezary syndrome: Triad constituted by erythroderma, lymphadenopathy, and atypical lymphocytes in the peripheral blood ("Lutzner cells," cells with cerebriform aspect characterized by marked infolding of nuclear membrane)

- *CLM*: Mnemonic word "API," which is an acronym for:
  – Acanthosis or strophy with focal parakeratosis
  – Pautrier microabscesses or single exocytosis of "Lutzner cells" (epidermotropism)
  – Infiltration of atypical into the superficial dermis with increased Langerhans cells and follicular mucinosis (mucinous degeneration of outer hair shafts)

- LN-3 patterns, including
  – Reactive follicular hyperplasia
  – Malignancy with partial/total effacement of the architecture by replacement of the LN structure by monomorphic infiltrates of atypical lymphocytes ± coagulative necrosis
  – Dermatopathic lymphadenopathy, characterized by paracortical expansion of proliferating histiocytes (TEM: racket-shaped granules → Langerhans cells).

Pitfall: Pautrier microabscesses, although considered almost pathognomonic, may also occur in benign conditions, such as cutaneous lymphoid hyperplasia, lymphomatoid papulosis, and Jessner's lymphocytic infiltration of the skin. Clinical and hematological correlations are needed.

- *IHC*: (+) CD2, CD4, CD5, and CD3, mostly being "helper" CD4 T lymphocytes with IL2R- (CD25-), but (+) TCR β and (−) CD7, CD8, CD30, and cytotoxic molecules (TIA1, granzyme B, and perforin).
- *DGN*: Dense band-like infiltrate of atypical lymphocytes in the papillary dermis with epidermotropism and Pautrier microabscesses (aggregates of atypical lymphocytes with cerebriform nuclei).
- *PGN*: About 1/5 of the patients die of CTCL or related complications. The survival of this neoplasm is associated with skin stage. It seems that the most important clinical predictive factors for survival remain patient age, T classification, and extracutaneous disease.

10.4.18 Angiocentric Immunoproliferative Lesions

EBV is a widespread infection or cohabitation in the human population worldwide. Systemic infection with EBV causes infectious mononucleosis, while local EBV-related conditions with EBV etiology may stay localized, such as hairy leukoplakia or EBV-driven mucocutaneous ulcer of the oral mucosa. It is well known that there are some neoplasms associated with EBV, including African Burkitt lymphoma, nasopharyngeal carcinoma, Hodgkin lymphoma, gastric carcinoma, PTLD, and extranodal NK-/T-cell lymphoma.

An angiocentric immunoproliferative lesion (AIL) is a lesion that should be subdivided into three degrees according to the vascular destruction, luminal compromise, or necrosis present.

AIL grade I or benign lymphocytic vasculitis is an angiocentric mononuclear cellular (lymphocytes, plasma cells, and immunoblasts) that involve blood vessels in an angiocentric fashion, but without evidence of vascular destruction, luminal compromise, or necrosis. AIL-I involves lungs and skin commonly and has the potential to progress to more aggressive forms of AIL.

AIL grade II or lymphomatoid granulomatosis (LYG, aka "polymorphic reticulosis") is an angiocentric mononuclear cellular infiltrate that shows angiodestruction and necrosis. LYG is subdivided in I-II-III degrees according to the angiodestruction and number of EBV positive cells.

- LYG-I shows sparse atypical EBV+ cells, generally <5/HPF, and minimal necrosis.
- LYG-II shows moderate # of EBV+ cells, 5–20/HPF, and more necrosis than LYG-I.
- LYG-III shows numerous EBV+ cells, usually >50/HPF, and prominent necrosis.

There is a natural potential to progression to large cell immunoblastic lymphoma.

LYG is classified under B-cell proliferation of uncertain malignant potential by the WHO, and a spectrum of biologic behavior has been reported in the literature.

AIL grade III corresponds to angiocentric lymphoma with apparent neoplastic nature of the angiocentric mononuclear cellular infiltrate showing atypia in both small and large cells with angioinvasion, destruction, and extensive necrosis and predominantly extranodal (e.g., the spleen with the characteristic periarterial arrangement). The hemophagocytic syndrome may be detected due to lymphokine production by atypical T cells.

EBV may be a challenge in a laboratory, and some tests may be performed (Box 10.11). LMP1 regulates its expression and the expression of human genes, and its expression promotes many of the EBV infection-related changes and activation of primary B cells. A short cytoplasmic N-terminus tail, six transmembrane domains, and a long cytoplasmic C-terminus form the structure of LMP1. The C-terminus form contains three activating domains.

10.4.19 Extranodal NK-/T-Cell Lymphoma

- *DEF*: EBV-driven, mainly extranodal, EN-NHL with NK-/T-cell cytotoxic immunophenotype, which is characterized by angiocentric infiltration AND angiocentric destruction, luminal compromise, and prominent coagulative necrosis.
- *EPI*: Asian and Native American populations from Central and South America, with adult>children, ♂ > ♀, EBV+, and immunosuppression ± (e.g., post-TX).
- EN sites: Upper respiratory and GI tracts, skin, soft tissue, and testis.
- Nodal: ± LN involvement.
- *CLI*: Nasal obstruction, epistaxis, extensive midfacial destruction (aka "Lethal Midline Granuloma").
- *CLM*: Extensive ulceration with diffuse and characteristically permeative neoplastic infiltrate of atypical mononuclear cells (medium-sized cells with irregularly folded nuclei, granular chromatin, inconspicuous nucleoli, and pale to clear cytoplasm showing high MI) + coagulative necrosis + apoptotic bodies.
- *IHC*: Characteristic pattern, including:
  - (+) CD2, CD3+(in T-type)/-(in NK-type), CD56, and cytotoxic granules (TIA1, GRAB, perforin)
  - (+) CD43, CD45RO, HLA-DR, CD25, FAS, and FASL
  - (+) LMP1 and EBER
- *DDX*: Lepra (*M. leprae* infection) and ozena (disease of the nose in which the bony ridges and mucous membranes are progressively destroyed).
- *TNT*: Radiation therapy (external beam radiation therapy) and postradiation chemotherapy (L-asparaginase, etoposide, ifosfamide, cisplatin, and dexamethasone).

Box 10.11 EBV Infection and Lab Tools
LMP1: *L*atent *m*embrane *p*rotein 1
EBER: In situ study for *EB*V-*e*ncoded *R*NA

- *PGN*: Poor outcome with poor PGN factors include stage (III/IV), bone/skin invasion, high levels of circulating EBV-DNA, and BM-EBER+.

10.5 Disorders of the Monocyte-Macrophage System and Mast Cells

This group is quite heterogeneous but is of high relevance to the pediatric pathologist (Box 10.12).

> **Box 10.12 Disorders of the Monocyte-Macrophage System and Mast Cells**
> 10.5.1. Hemophagocytic Syndrome
> 10.5.2. Sinus Histiocytosis with Massive Lymphadenopathy (Rosai-Dorfman Disease)
> 10.5.3. Langerhans Cell Histiocytosis
> 10.5.4. Histiocytic Medullary Reticulosis
> 10.5.5. True Histiocytic Lymphoma
> 10.5.6. Systemic Mastocytosis

10.5.1 Hemophagocytic Syndrome

The hemophagocytic syndrome or hemophagocytic lymphohistiocytosis (HLH) is an unusual immunological disorder first recognized almost 80 years ago. The use of molecular genetics and experimental animals allowed to identify the familial form of HLH which is due to a deficiency of cytotoxic killing. Specific diagnostic criteria for HLH have been revised by the Histiocyte Society and described below.

- *DEF*: An immunological disorder characterized by hyperimmunity, organ damage, and life-threatening cytopenia.
- *SYN*: Hemophagocytic lymphohistiocytosis (HLH).
- *EPG*: Genetic form (rare, infants/young children, rapidly fatal, AR with definite FH or parental consanguinity) or acquired (EBV+ immunocompromised patients), and the pathogenesis may involve defects in the perforin gene, at least for a group of HLH patients.
- *DGN*: The diagnosis of HLH may be established by either a molecular diagnosis consistent with HLH (e.g., pathologic mutations of PRF1, UNC13D, or STX11 are identified) or the fulfillment of five out of the eight criteria listed below. Moreover, in the case of familial HLH, no evidence of malignancy should be present.

The HLH criteria (adapted from Henter et al. 2007) include (1) fever; (2) splenomegaly; (3) cytopenias ($\geq$2/3 cell lineages of the PB) (RBC (Hb <9 g/100 ml (in infants <4 weeks, Hb <10 g/100 ml)), thrombocytopenia (platelets <100 × 10^3/ml), and neutropenia (neutrophils <1 × 10^3/ml)); (4) hypertriglyceridemia (fasting, >265 mg/100 ml) and/or hypofibrinogenemia (<150 mg/100 ml); (5) hemophagocytosis in the BM, spleen, or lymph nodes; (6) low or absent NK-cell activity; (7) hyperferritinemia (ferritin >500 ng/ml); and (8) soluble CD25 (i.e., soluble IL-2 receptor) >2400 U/ml.

- *GRO*: Splenomegaly and bleeding in several organs as well as infections at several sites due to neutropenia.
- *CLM*: Infiltration by non-atypical histiocytes/Kupffer cells containing erythrocytes and occasionally lymphocytes and neutrophils in the setting of a preserved LN/BM/liver architecture.
- *DDX*: Malignant histiocytosis (atypia, effacement of the organ architecture) and peripheral T-cell lymphoma (atypical cells showing a characteristic immunophenotype on both immunohistochemistry and flow cytometry).
- *TRT*: Most probably, hematopoietic cell transplantation (HCT) remains the only long-term curative therapy, but chemo- and/or immunotherapy may be useful in achieving clinical remission of symptoms.
- *PGN*: Since the early 1980s, the prognosis for patients with HLH has improved remarkably, but there is still a lot to do, because HLH will inexorably (and fatally) recur in individuals with intrinsic, severe deficiencies of cytotoxic function. Monitoring of the CBC, soluble CD25, ferritin, spinal fluid, and organ (liver, spleen, bone marrow) function should be routinely assessed.

10.5.2 Sinus Histiocytosis with Massive Lymphadenopathy (Rosai-Dorfman Disease)

Sinus histiocytosis with massive lymphadenopathy or SHML is a benign, self-limiting histiocytic disorder that typically involves the lymph nodes. Rosai and Dorfman described SHML in 1969, and numerous cases have been added to the literature in the last 40–50 years. SHML is characterized by painless, bilateral, cervical lymphadenopathy accompanied by fever, ↑ WBC in the peripheral blood, ↑ ESR, and hypergammaglobulinemia. Extranodal sites include the head and neck, upper respiratory tract, skin, subcutaneous tissue, bone, skeletal muscle, central nervous system, gastrointestinal tract, genitourinary tract, thyroid, breast, liver, kidney, heart, and uterine cervix. Patients presenting with lymphadenopathy are commonly diagnosed as having malignant lymphomas, but patients with extranodal involvement are thought to have various neoplasms, depending on the site of the participation.

10.5.3 Langerhans Cell Histiocytosis

- *DEF*: Oncologic, uni- to a multisystemic disease characterized by an accumulation of Langerhans dendritic cells, which are S100+, CD1a+, and Langerin+ and contain Birbeck granules (racket-shaped electron microscopy detectable inclusions). Clinical picture and organ dysfunction are variable from localized and relatively self-limiting to multiple organ involvement, severe system dysfunction, and life-threatening. Incidence peaks at 1–4 years of age with a prevalence of $9/10^6$, ♂>♀, without evident ethnic association. Treatment may include steroids and chemotherapeutic agents including prednisone, vinblastine, and mercaptopurine. Dendritic cells belong to histiocytes, which also include monocytes and macrophages. All histiocytes are derived from stem cells residing in the BM. Thus, histiocytosis is an abnormal proliferation of any of these types of immune cells. According to the text of the Writing Group of the Histiocyte Society, histiocytosis was classified as class I, which is LCH; class II, which corresponds to histiocytosis of mononuclear phagocytes; and class III, which relates to malignant disorders of the proliferating histiocytes. LCH is caused by an abnormal proliferation and accumulation of Langerhans cells forming granulomatous, yellow-brown lesions. Langerhans cells or epidermal dendritic cells are antigen-presenting cells that the subsequent activation migrates through the epidermis into regional LNs. Langerhans cells can recognize antigens and internalize them by receptor-mediated endocytosis and display a fragment of the antigenic molecule on the MHC class II. This event induces the stimulation of helper T lymphocytes and cytotoxic T lymphocytes to eliminate the antigen. The granulomatous lesions are made up of T cells, macrophages, MNGC, stromal cells, and NK cells with an intricate and orchestrated interaction of multiple cytokines and chemokines (Box 10.13).
- *CLI*: LCH may still be subclassified and is, indeed the case in some centers, in three forms

> **Box 10.13 LCH: Traditional Clinical Forms**
> - Eosinophilic granuloma is the localized or benign form, which is usually limited to skeletal or respiratory system and occurs in children younger than 15 years of age.
> - The LCH-HSC disease is a chronic, multifocal form affecting bones (retro-orbital lesions may cause exophthalmos) and other organs such as the pituitary gland (diabetes insipidus) but also the skin, lymph nodes, liver, spleen, lungs, GI tract, thymus, CNS, and other endocrine glands. The LCH-HSC disease tends to progression and occurs in children younger than 10 years of age.
> - LCH-LS disease is the most severe form of LCH with acute, disseminated, and multisystem presentation and happens in children younger than 2 years of age.

including eosinophilic granuloma, Hand-Schuller-Christian (HSC) disease, and Letterer-Siwe (LS) disease.
- *DGN*: LCH workup includes a full family and medical history, physical examination, complete skeletal bone survey, chest and skull X-rays, comprehensive hematological evaluation (blood count with differentials and ESR and BM biopsy mainly if there is anemia, neutropenia, or thrombocytopenia), liver function tests, electrolytes, and urinalysis as well as a biopsy of the affected area. In case of suspicion of DI-related LCH, serum and urine osmolality after water deprivation, GH and TSH levels are needed.
- *PGN*: Outcome depends on the number of the systems involved and from the degree of organ dysfunction, the age at presentation, and progression rate. Usually, children younger than 2 years of age have the acute and disseminated form of LCH and harbor the lowest rate of long-term survival, although this group may constitute less than 15% of LCH cases.

Skin lesions are erythematous papules with a possible tendency to vesiculation, ulceration, crusting (seborrhea-like eruptions), or bleeding, often localized in the groin, axilla, and scalp (DDX of cradle cap of infants), and tendency to spontaneous regression. Bony lesions are lytic; often affect the skull, lower limbs, ribs, pelvis, and spine; and manifest as painful swelling, dull aching pain, limited range of mobility, and inability to bear weight. LN lesions may be single or multiple, accompany a skin or a bony lesion, and may manifest as painful lymphadenopathy. BM is involved in case of multisystem disease only because Langerhans cells do not usually reside in this region. Manifestations include anemia, neutropenia, and thrombocytopenia as well as fever and splenomegaly. Liver involvement may manifest as ascites, jaundice, hypoproteinemia, and prolonged thrombin time. Liver biopsy shows a range of variable features spanning from mild cholestasis to severe infiltration of portal triads by Langerhans cells with bile duct involvement and lobular inflammatory activity. CNS lesions include a disruption of the hypothalamic-pituitary axis with structural abnormalities, thickening of the pituitary stalk and DI, and dysregulation of GH and TSH production. Neurodegenerative changes may also be observed and include bilateral symmetrical lesions in the cerebellum and basal ganglia with MRI exhibiting white and gray matter changes, extraparenchymal and space-occupying lesions, and atrophy. Cognitive function is impaired in these children, who may show learning deficit, poor performance at school, and emotional disturbances.

- *TRT*: It is variable and depends on the extent and degree of disease and includes regimens differentiated for low- and high-risk patients. Low-risk patients include only system involvement of the skin, bone, lymph nodes, or pituitary gland, while high-risk patients are children with multiple system involvement. Low-risk patients benefit from a combined CHT protocol of vinblastine and prednisone for 6–12 months reaching the goal of 100% survival and 1/5–1/3 risk of recurrence. High-risk patients include several regimens derived from ongoing or completed trials (see COG, SIOP, UKCCSG).

10.5.4 Histiocytic Medullary Reticulosis

- *DEF*: Leukemic reticuloendotheliosis (aka malignant histiocytosis) characterized by progressive systemic proliferation of quite immature, morphologically atypical histiocytes and/or precursors in childhood and youth.
- *CLI*: Patients are acutely ill presenting with fever, lymphadenopathy, hepatomegaly, splenomegaly, and skin involvement. Laboratory parameters exhibit anemia, leukopenia, thrombocytopenia, and increased ferritin. Erythrophagocytosis in BM aspirate is quite common, and increased plasma cells are also a typical finding. Lysozyme (muramidase) and alpha-1-antichymotrypsin are also positive.
- *PGN*: HMR is usually rapidly fatal with about 2/3 of patients succumb within months after diagnosis.

10.5.5 True Histiocytic Lymphoma

This lymphoma is a very rare diagnosis with only very few cases reported in the literature (♂>♀). There are no common histologic features.

10.5.6 Systemic Mastocytosis

- *DEF*: It is mastocytosis with systemic involvement differently from urticarial manifestations or myeloproliferative disorders. Valent et al. classification is that used in pediatrics.
- *EPI*: Systemic mastocytosis is not usually a pediatric disease, although it may occur in children and youth.
- *CLI*: Symptoms are due to release of histamine and heparin as well as other chemokines by mast cell dysregulation. It is important to remember that both mastocytes and basophils have receptors with the ability to bind the Fc portion of immunoglobulin E (IgE) with an extremely high affinity. Thus, a degranulation may be considered explosive with the release of histamine and numerous other potent inflammatory mediators when both mastocytes and basophils are stimulated. This aspect is the occurrence of a specific antigen binding to cell-based IgE-molecules harboring cells. Two dermatologic signs need to be distinguished, the Darier phenomenon and the dermatographism. Darier sign is characterized by swelling of the skin, which becomes itchy and red after cutaneous stroking in consequence of the release of histamine from the neoplastic mast cells invading the skin.
- *CLM*: Mast cells ("Mastzellen" of the German literature) or mastocytes are ovoid cells with uniform centrally located nuclei and characteristic purple "metachromatic" granules, which are positive on Giemsa stain as well as on other special stains and using immunohistochemistry tools. Giemsa stain, Toluidine blue stain, Leder stain, and chloroacetate esterase stain are often listed as special stains (histochemistry). Antibodies against CD117, CD25, and CD2 are considered as appropriate immunohistochemical stains.

In systemic mastocytosis, there are *C-KIT* gene mutations (D816V), which are associated with a gain-of-function effect. Biochemically, there is a serum level of tryptase ≥20 ng/ml. BM aspirate and biopsy show diffuse infiltration of mastocytes. Scattered eosinophils are usually present. The distinction of mastocytosis in cutaneous and systemic is not trivial and needs the activation of the pediatric oncology team. Cutaneous mastocytosis is skin-restricted, while systemic mastocytosis is constituted by cutaneous involvement and ≥ one extracutaneous organ (BM, liver, spleen, and LNs). In cutaneous mastocytosis, the mast cells fill the papillary dermis and extend as sheets and aggregates into the reticular dermis and infiltrate around blood vessels and adnexal tissue. In systemic mastocytosis, LNs are wholly or partially effaced, but the spleen is always involved. Spleen examination shows either ill-defined granuloma-like nodules, scattered throughout parenchyma or angiocentric fibrotic nodules with small assemblages of mast cells embedded within. Skin examination shows multiple and quite widely distributed, round to oval, red-brownish, non-scaling papules and small plaques. The skeletal system may show osteoblastic and osteoclastic lesions. Importantly, BM is the most frequent non-cutaneous site (90%) with a notable increase of reticulin fibers.

- *PGN*: Despite progress in CHT protocols, the outcome is worrisome in systemic mastocytosis, because the outcome may be quite poor.

10.6 Plasma Cell Disorders

In general, plasma cell dyscrasias (Latin, bad mixture, disease, from Greek δυσκρασία or δυσ- (dys-, "bad") + κρασία (krasía, "mixture") refer to a group of monoclonal gammopathies or paraproteinemia. They need two components to be labeled in this way, including an uncontrolled autonomous proliferation of plasma cells or cells related to this origin AND abnormally elevation in blood and/or urine of levels of homogeneous Ig and/or one of its constituent chains. Urinary samples can be positive for Bence Jones proteins, i.e., free light chains that are small enough to be effectively excreted from the urinary system, occurring either alone or in combination with any hyper-Ig syndrome.

10.6.1 Multiple Myeloma

- *DEF*: Post-GC B-cell neoplasm showing plasma cell dyscrasia (aka plasma cell myeloma or plasmacytoma) with numerous masses of immature plasma cells scattered primarily throughout the skeletal system with fulfilled characteristic "CRASH" criteria (hyper<u>c</u>alcemia, <u>r</u>enal insufficiency, <u>a</u>nemia, <u>s</u>keletal lesions, and impaired <u>h</u>ematopoiesis) and relying on signals generated by the hedgehog pathway (self-renewal of abnormal plasma cells) and on IL-6. Although the hematopoiesis does not belong to the famous "CRAB" criteria (hypercalcemia, renal insufficiency, anemia, and bony lesions), we found that an impaired hematopoiesis is almost the rule in patients affected with multiple myeloma.
- *EPI*: Typically, middle age (♂ > ♀), but children/youth have been described with a commonly associated long-standing history of chronic infections (inefficient Ig production).
- *CLI*: Anemia, proteinuria, and recurrent infections (impaired immunity) initially and bony pain, fractures, and renal failure subsequently ± POEMS syndrome (most myelomas associated with POEMS syndrome are osteosclerotic, while the classic non-POEMS myelomas are osteolytic).
- *LAB*: 3 g Ig/100 ml serum OR 6 mg Ig/1000 ml urine.

Ig secretion: IgG:IgA:IgM:IgD:IgE = 60%:15–20%:10–15%:5%:5%

Although IgD are rare, IgD secretors have more aggressive clinical course (see below PGN).

Increased M protein ⇒ rouleaux ⇒ thrombosis

The immunophenotype of normal and atypical plasma cells is slightly different. Normal plasma cells are (+) CD19, (+) CD38, and (+) CD138, while MM "atypical plasma cells" are (−) CD19, (+) CD38, and (+) CD138. Most atypical plasma cells of MM are CD19-, CD20-, but (+) CD38 and (+) CD138, as well as CD56(±), which is considered as aberrant expression (MM, multiple myeloma).

In about 10% of patients, there is evidence of systemic amyloidosis, and X-ray shows clear-cut multiple destructive bony lesions, starting in the medullary cavity and eroding progressively the cortex, forming sharply "punched-out" defects (size, 1–4 cm) with a characteristic of a "soap bubble" aspect, mostly involving the spine (2/3), ribs (~1/2), and skull (~1/2). Other bony lesions involve the pelvis, femur, shoulder girdle, and soft tissue lesions may also appear and affect the spleen, liver, kidneys, lungs, LNs, and stomach. A primary gastric plasmacytoma, as a rare cause of hypertrophic gastritis in an adolescent, has also been reported in the literature. In all soft tissue lesions, the aspect is quite similar with fleshy, red-brown masse, grossly.

- *CLM*: Patchy or diffuse (10–90%) BM/soft tissue infiltration by a variable range of maturity of plasma cells from very immature to mature ("flame cells," Mott cells, cells with inclusions); bi-/trinucleation; PAS+; Russell bodies, which is a form of intracellular accumulation of protein, e.g., alcoholic hyalin or reabsorption droplets; or Dutcher bodies, which are nuclear in location. Other inclusions in the plasma cells include fibrils and crystalline rods. Bone destruction is due to MΦ inflammatory protein 1 α (MIP1α) ⇒ ↑ RANKL, which activates osteoclasts.

BM impaired hematopoiesis ⇒ anemia, infections, thrombosis

Other clinical features are also related to the effects of plasma cellular infiltrates to other non-BM tissues and rely on impaired immunity (pneumonia, abscesses, osteomyelitis) and excess of abnormal Ig productions (myeloma nephrosis and systemic amyloidosis)

Renal pathology in multiple myeloma includes (1) toxic effect of BJP on renal tubular cells, (2) luminal obstruction by Bence Jones proteins (BJP) casts, (3) amyloidosis, (4) light chain nephropathy, and (5) nephrocalcinosis due to hypercalcemia ("C" of CRAB CRASH lesions).

Abnormal interstitial infiltrates of plasma cells + casts into the distal tubules and collecting ducts characterize the myeloma nephrosis ("myeloma kidney"). It is due to mostly BJ protein accumulation, which forms aggregates of eosinophilic material in the lumina.

- *TNT*: Chemotherapy.
- *PGN*: The survival is limited to 1–2 years without treatment. Poor prognostic factors include amyloidosis, anaplastic conversion of plasma cellular infiltrates, extensive/diffuse BM involvement, high immaturity of plasma cellular infiltrate, hyperalbuminemia, hypercalcemia, plasma cell leukemia, serum β-2-microglobulin >4 ng/μl, and renal failure.
- *CGB*: Somatic hypermutation and IGH rearrangements on chromosome 14 include *CCND1*, *CCND3*, *FGFR3*, and *MAF* genes. Good cytogenetics-based prognostic factors are *CCND1*/t(11;14) or *CCND3*/t(6;14) and hyperdiploidy, while poor cytogenetics-based prognostic factors are t(4;14), *MAF*, 17p-/, and *TP53* alterations.

10.6.2 Solitary Myeloma

- *DEF*: Bony or soft tissue-located myeloma with the potential involvement of single soft tissue lesions, such as the lungs, oropharynx, nasal sinuses, and stomach.
- *SYN*: Solitary myeloma (aka solitary plasmacytoma).
- *EPI*: 40 years with occasional cases in childhood (e.g., primary gastric plasmacytoma), ♂>♀.
- *CLI*: >1/2 of patients monoclonal protein in urine and/or serum, but <2 g/100 ml.
- *PGN*: Solitary bony plasmacytoma may progress to MM (25% of solitary plasmacytoma at a skeletal location with dysproteinemia and 10–20% of solitary plasmacytoma at soft tissue location).

Osteosclerotic myeloma → *POEMS syndrome*: Polyneuropathy, organomegaly, endocrinopathy, monoclonal gammopathy, and skin changes. It is a multi-organic disease with plasma cell neoplasm characterized by fibrosis and osteosclerotic modifications in bony trabeculae and LN changes similar Castleman disease. It is thought that it is a plasma cell variant of Castleman disease. Skin changes include hyperpigmentation and hypertrichosis.

10.6.3 Plasma Cell Leukemia

MM variant with plasma cells in peripheral blood >20% or absolute plasma cell count >2000 and very rarely tissue involvement.

10.6.4 Waldenstrom's Macroglobulinemia

- *DEF*: Plasmacytoid lymphoma showing a monoclonal gammopathy (IgM >> IgG or IgA) typically occurring in a setting of a LPD.
- *EPI*: Generally, >60 years, few youth cases.
- *CLI*: General symptoms (weakness, fatigue, weight loss).
- *LAB*: Serum proteins >6.5 g/100 ml with IgM 1–3 g and Ig >15% of all Ig (nl. 5%), BJP (1/3 of cases).
- *CLM*: Diffuse BM infiltration by plasma cells, plasmacytoid lymphocytes, and lymphocytes (all of them monoclonal), although tumor masses or lytic lesions are not seen. Dutcher bodies may be viewed, and similar may be found in the liver, spleen, and LNs.
- *PGN*: Survival 2–5 years.

10.6.5 Heavy Chain Disease

Plasma cell dyscrasias with production only of heavy chains, hepatosplenomegaly, and soft tissue tumors. All ages of individuals may be affected, although young and middle-aged adults are more commonly involved. Some classification according to the location has also been proposed.

- Subtype α ("Mediterranean Lymphoma"): Pediatric HCD with two patterns (pattern 1, massive infiltrate of lamina propria of the intestine by heavy chain-producing plasma cells, villous atrophy, malabsorption, diarrhea, AND abdominal LN involvement, and pattern 2, respiratory tract involvement).

- Subtype γ: Adult HCD with associated TB, RA, and AI disorders and histopathologically more similar to malignant NHL than myeloma (lack of lytic lesions).
- Subtype μ: HCD variant seen in patients affected with CLL.
- *PGN*: The course is variable with months to years of survival. Several studies have been performed and several authors suggest that associated underlying disorders may contribute to the variability of the clinical course.

10.6.6 Monoclonal Gammopathy of Undetermined Significance (MGUS)

It is defined as a: constellation of paraprotein or M protein <30 g/L, abnormal κ/λ ratio (normal: 0.26–1.65), NBM-PC (clonal) <10%, with no "CRAB" lesions, no BCL, no diseases known to produce a secretion of M protein. MGUS occurs middle aged individuals and elderly, but subjects under the age of 40 years have been reported. MGUS prefers African-Caribbean ethnics over Caucasians. Pediatric cases have been associated with EBV and CMV infection. In MGUS, there is an Ig peak in blood, but no apparent associated cellular (plasmacellular) proliferation. BJP is usually not encountered in the workup of the individuals with MGUS. It has been estimated that 1 out of 100 subjects with MGUS develop a related neoplasm.

10.7 Benign Lymphadenopathies

Benign lymphadenopathies may be quite common in childhood and youth. In Box 10.14 are summarized the benign lymphadenopathies that the pathologist should keep in mind. These benign lymphadenopathies are differentiated according to the abnormal histologic pattern seen under the lens (Figs. 10.21 and 10.22).

Box 10.14 Benign Lymphadenopathies
10.7.1. *Follicular Hyperplasia*
    10.7.1.1. Nonspecific Reactive
    10.7.1.2. Toxoplasmosis
    10.7.1.3. Rheumatoid Arthritis and Sjögren's Disease
    10.7.1.4. Systemic Lupus Erythematosus
    10.7.1.5. Necrotizing Lymphadenitis (Kikuchi-Fujimoto Lymphadenitis)
    10.7.1.6. Cat-Scratch Disease
    10.7.1.7. Lymphogranuloma Venereum
    10.7.1.8. Kimura's Disease
    10.7.1.9. Syphilis
    10.7.1.10. Castleman Disease
    10.7.1.11. Progressively Transformed Germinal Centers
    10.7.1.12. AIDS-Related Lymphadenopathy and PAIDS
10.7.2. *Diffuse (Paracortical) Hyperplasia*
    10.7.2.1. Postvaccinal Viral Lymphadenitis
    10.7.2.2. Infectious Mononucleosis
    10.7.2.3. Dermatopathic Lymphadenitis
10.7.3. *Sinus Pattern*
    10.7.3.1. Sinus Histiocytosis
    10.7.3.2. Sinus Histiocytosis with Massive Lymphadenopathy
    10.7.3.3. Lipophagic Reactions
    10.7.3.4. Vascular Transformation of LN Sinuses
10.7.4. *Predominantly Granulomatous Pattern*
    10.7.4.1. Sarcoidosis
    10.7.4.2. Tuberculosis
    10.7.4.3. Atypical Mycobacteriosis
    10.7.4.4. Fungal Infections
    10.7.4.5. Chronic Granulomatous Disease

10.7.5. *Others/Mixed Patterns*
    10.7.5.1. Mucocutaneous LN Syndrome (Kawasaki Lymphadenitis)
    10.7.5.2. Leprosy
    10.7.5.3. Mesenteric Lymphadenitis (Masshoff Lymphadenitis)
10.7.6. *Angioimmunoblastic Lymphadenopathy with Dysproteinemia (AILD)*

The first pattern that catches the eye of the pathologist looking under the microscope for a lymph node with abnormal histology is hyperplastic follicles.

10.7.1 Follicular Hyperplasia

It is essential to distinguish between nonspecific reactive hyperplastic follicles and specific entities, including toxoplasmosis, rheumatoid arthritis, Sjögren disease, systemic lupus erythematosus, necrotizing lymphadenitis (Kikuchi-Fujimoto

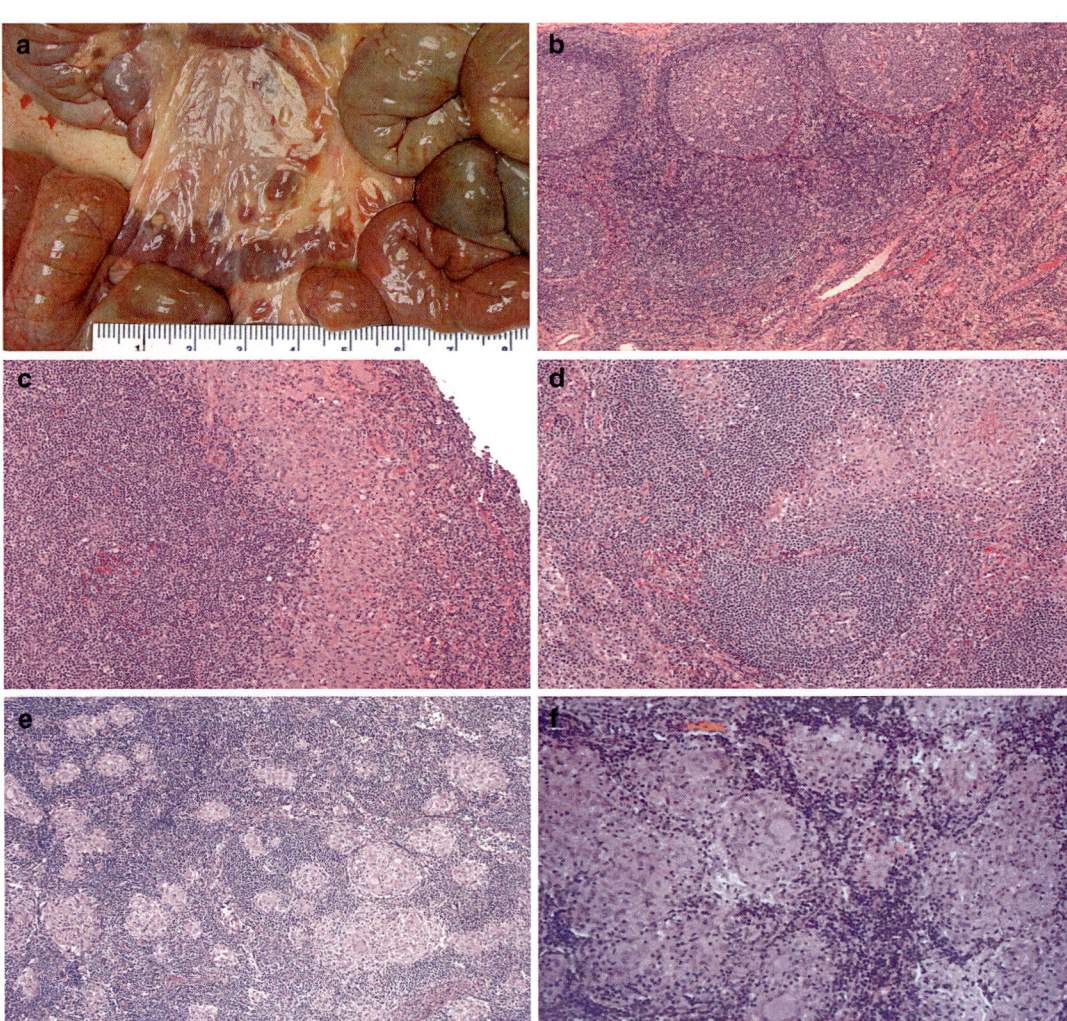

Fig. 10.21 Figure (**a**) shows a post-infectious enlargement of the mesenteric lymph nodes (*Y. enterocolitica* infection). Figures (**b–d**) show a granulomatous lymphadenitis (**b-d**, hematoxylin and eosin staining, ×100 original magnification. Figures (**e–f**) shows a lymph nodal sarcoidosis (**e**, hematoxylin and eosin staining ×100 original magnification; f, hematoxylin and eosin staining, ×200 original magnification)

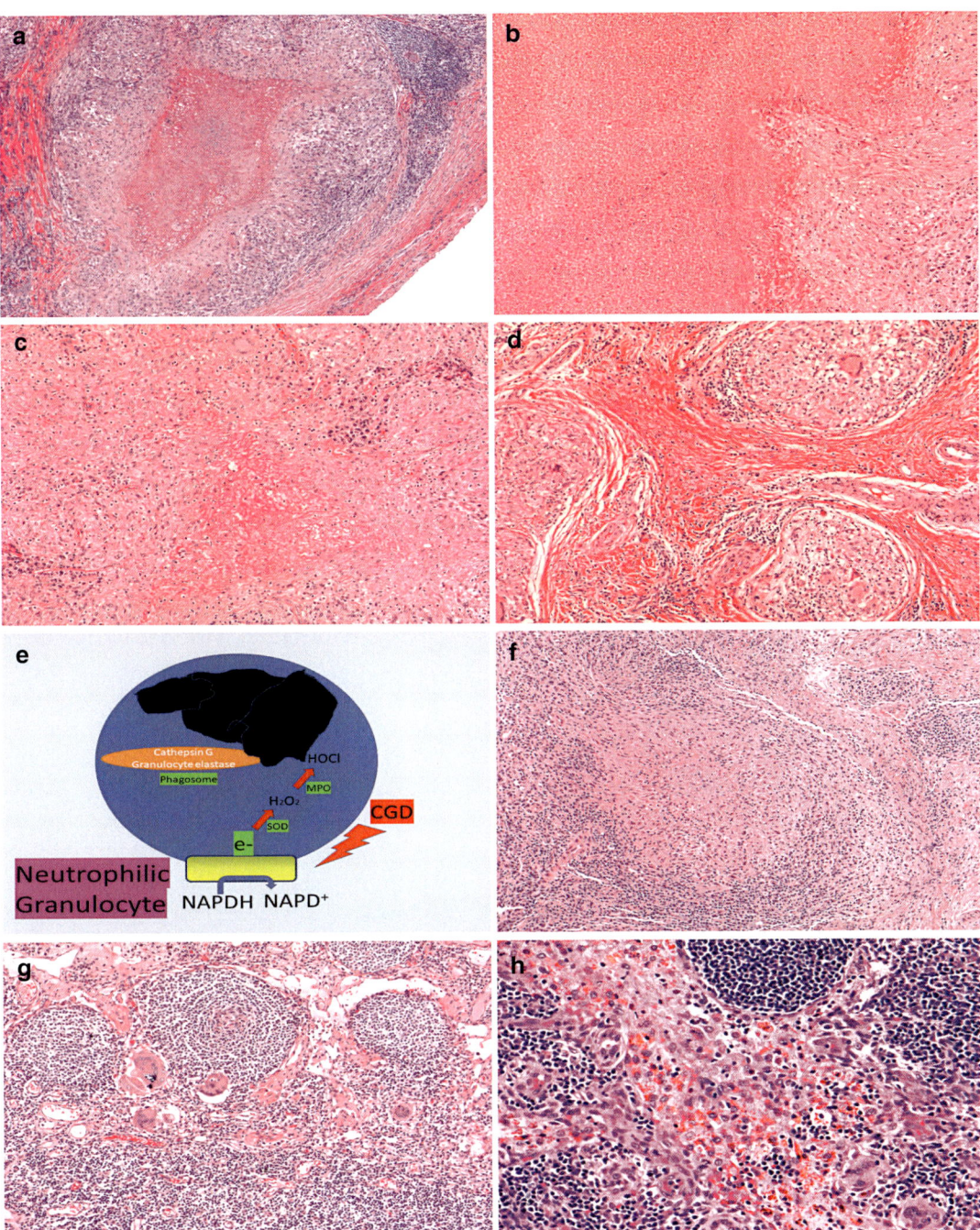

Fig. 10.22 Granulomatous lymphadenitis due to mycobacterial infection with caseous necrosis (**a–d**) (H&E stain; x50, x100, x100, x100, respectively), pathogenetic schema of chronic granulomatous disease (**f**) (H&E stain, x100). Mutations in the *CYBA*, *CYBB*, *NCF1*, *NCF2*, or *NCF4* gene can cause chronic granulomatous disease. The proteins codified by the respective genes are subunits of NADPH oxidase, which plays an essential role in the immune system, particularly in phagocytes. The phagocytes are able to target foreign invaders destroying them. Inside of these cells, NADPH oxidase is involved in the production of superoxide, which is used to generate other toxic substances to kill bacteria and viruses. Neutrophils are also regulated by NADPH. Mutations in any of the five genes result in the production of subfunctional proteins causing chronic granulomatous disease. Crohn disease in a lymph node with granulomas and multinucleated giant cells (**g**) (H&E stain, x100), and pigment (bilirubin)-laden macrophages in a hepatic lymph node of an infant with biliary atresia (**h**) (H&E stain, x200)

lymphadenitis), cat-scratch disease, lymphogranuloma venereum, Kimura disease, syphilis, Castleman disease, progressively transformed germinal centers, and AIDS-related lymphadenopathy in the setting of a possible picture identified in childhood and youth, named "PAIDS."

10.7.1.1 Nonspecific Reactive Lymphadenitis
- Children and adults
- Cortex or throughout
- Size and shape variable follicles with lymphocytes, plasma cells, histiocytes, showing large germinal centers, polarization, well-demarcated mantle zone, expansion through the mantle zone

10.7.1.2 Toxoplasmosis
- *DEF*: Piringer-Kuchinka Toxoplasma gondii-associated Lymphadenitis
- *EPI*: Children and adults (younger women) with often posterior cervical LN involvement
- *CLM*: Characteristic histologic triad, including:
  1. Follicular hyperplasia with high MI and nuclear debris
  2. Small, epithelioid (non-sarcoid-like) giant cell-free granulomas in germinal centers as well as at periphery of germinal centers
  3. Monocytoid B-cell distension of marginal/cortical sinuses and paracortex
- *DDX*: NLPHD, LPHD (in both conditions no granulomas)

10.7.1.3 Rheumatoid Arthritis and Sjögren Disease
RA/SS Lymphadenopathy

- *DEF*: Pediatric and adult lymphadenopathy associated with rheumatoid arthritis or Sjögren disease
  1. Corticomedullary follicular hyperplasia AND (2) marked interfollicular plasmacytosis with cells containing numerous Russell bodies
     - In case of JRA (juvenile RA): (1)+ (2)+ neutrophils
- *PGN*: Long-standing RA is associated with an increased risk for lymphoma!

10.7.1.4 Systemic Lupus Erythematosus (SLE) Lymphadenopathy
- *DEF*: Necrotizing lymphadenitis associated with SLE, inolving mostly the cervical LNs.
- LM: Cortical follicular hyperplasia with interfollicular plasmacytosis ± well-demarcated areas of paracortical necrosis with none or few neutrophils.

10.7.1.5 Necrotizing Lymphadenitis (Kikuchi-Fujimoto Lymphadenitis)
- *DEF*: Self-limited painless lymphadenitis disorder of unknown etiology characterized by systemic symptoms often involving cervical LNs and associated with leukopenia.
- *EPI*: Young women, Japan.
- *CLM*: Well-demarcated necrotizing foci of the paracortex ("zonal necrosis") with scattered fibrin deposits, histiocytes with twisted nuclei surrounded by immunoblasts, and collections of large mononuclear cells and only limited numbers of neutrophils and plasma cells.
- *TEM*: Tubular-reticular inclusions and intracytoplasmic rodlets.

10.7.1.6 Cat-Scratch Disease
- *DEF*: *Bartonella henselae* lymphadenitis (*B. henselae* ⇒ bacillary angiomatosis and peliosis).
- *LAB*: IFA+, ELISA+ (both nonspecific), *B. henselae* in feline RBC
  - Cat to cat (horizontal spread) and cat to human
- *CLM*: Three stages
  1. Early CSD: Follicular hyperplasia and histiocytic proliferation
  2. Intermediate CSD: Granulomatous change with abscesses near germinal centers
  3. Late CSD: Large abscesses with characteristic central stellate necrosis, neutrophils, and palisading histiocytes
- *PGN*: ± Encephalitis AND/OR neuroretinitis

B. henselae infection also causes bacillary angiomatosis, and the LNs shows focal nodal effacement of the architecture by multiple coalescing clusters of small blood vessels, which are lined by endothelial cells with epithelioid aspect showing a pale cytoplasm and focal mild atypia of the identifiable nuclei. The interstitium contains material, which is eosinophilic to amphophilic, sometimes described as amorphous, or granular on H&E showing aggregates of bacteria that may be highlighted by Warthin-Starry staining. PMN can also be seen in the interstitium.

10.7.1.7 Lymphogranuloma Venereum

- *DEF*: *Chlamydia trachomatis* lymphadenitis (LGV).
- *EPI*: Sexually transmitted lymphadenitis with higher prevalence rates in Central and South America than in North America (♂>♀). The main risk factor is HIV-infection.
- *CLM*: Necrotic foci from tiny, often initially, to large stellate abscesses with Langhans giant cells, fibroblasts, and fistulous-harboring histologic images.

10.7.1.8 Kimura Disease

- *DEF*: Florid follicular hyperplasia of the cortical and paracortical type with interfollicular eosinophilia and cortical and paracortical hypervascularity with the proliferation of thin blood vessels and ± soft tissue lesion.

10.7.1.9 Syphilis

- *DEF*: *Treponema pallidum* lymphadenitis.
- *CLI*: Stages 1 and 3 with localized form and stage 2 with generalized lymphadenopathy.

Stage 1
- Capsular and pericapsular inflammation with fibrosis
- Plasmacytic vasculitis with diffuse infiltrate of plasma cells, angio-proliferation with endothelial swelling, and plasma cell-rich inflammation
- Epithelioid histiocytic aggregates, occasionally with accompanying sarcoid-like granulomas

Stage 2
- Florid follicular hyperplasia with large epithelioid histiocyte clusters and sarcoid-like granulomas

- SS: Warthin-Starry → look for spirochetes, particularly in the wall of blood vessels.

10.7.1.10 Castleman Disease

- *DEF*: Plasma cell-rich angiofollicular hyperplasia with or without HHV-8 (+) and ± multisystemic involvement.

Non-HHV-8-associated Castleman disease

- Unicentric, hyaline vascular
- Unicentric, plasma cell-rich
- Multicentric

HHV-8-associated Castleman disease

- Associated with plasmablastic lymphoma (PBL)
- Non-associated with plasmablastic lymphoma (PBL)

PBL is an aggressive NHL often associated with HIV infection, but also encountered in patients with other immunodeficiencies and immunocompetent individuals.

- *SYN*: Giant LN hyperplasia, LN hamartoma, follicular lymphoreticulosis.
- *EPI*: Worldwide, children-adults, ♂ = ♀.

Hyaline vascular (angiofollicular): Large follicles with vascular proliferation, Hassall's bodies-like hyalinization in center of follicles, "lollipop sign" with radially penetrating sclerotic BVs, onion-skin layering of lymphocytes at follicle periphery, and prominent interfollicular stroma with many plasma cells. In the lymphoid subtype, there is an expansion of the mantle zone with small regression of germinal centers.

- Hyaline vascular subtype/variant: Large mass involving ≥ one lymph node with symptoms related to the mass (*tumor quia tumet*) and compression of adjacent structures with consequent shortness of breath due to airway compression or SVC syndrome due to vascular compression or dysphagia or varices due to solid esophageal pressure (SVC syndrome: medical emergency characterized by superior vena cava (SVC) obstruction of blood flow most often in patients with a malignancy located within the thorax).

- *HSS/IHC*: Ig stain→ polyclonal, conversely of monoclonality seen in lymphoma.
- (2) Plasma cell type: Diffuse interfollicular plasma cell proliferation with increased IL-6 and deposition of amorphous acidophilic material in follicles (probably fibrin and Ig) and often systemic symptoms and abnormal lab values (fever, anemia, ↑ ESR, and Ig).

Clinical Presentations
(1) Unicentric form: Round-ovoid mass, well-delimitated, up to 10–15 cm, usually hyaline vascular type (90%), and curable by surgery.
 Mediastinum, lung, axilla, neck, mesentery, retroperitoneum, limbs
 Mediastinum, often hyaline vascular type, while mesentery location, plasma cell type.
(2) Multicentric form
 Lymphadenopathy, generalized ± splenomegaly, ± HIV (+), older patients, systemic symptoms and abnormal lab values (IL-6-associated B symptoms with fever, night sweats), ± POEMS syndrome, ↑ risk of synchronous or metachronous carcinoma/NHL/angiosarcoma.

- *TNT*: Ganciclovir, rituximab.
- *PGN*: Poor.

10.7.1.11 Progressive Transformation of Germinal Centers (PTGC)

- *DEF*: Lymphadenopathy with low malignant potential characterized by large (2–3× standard size) centrally located follicles with germinal centers infiltrated by aggregates of mantle zone lymphocytes showing expression of CD20, CD23, and BCL2 giving the aspect at places of no follicle centers with indistinct margin with the surrounding mantle zone ("Intrafollicular Invasion of Mantle cells").
- *EPI*: Both children (20%, mean age 11 years) and adults (80%, mean age 28 years) are involved (♂:♀ = 3:1).
- *EPG*: PTGC is associated with HL in 1/3 of patients. PTGC may precede, follow, or be concurrent with HL.
- *CLI*: Cervical LNs (50%) with oral cavity involvement.
- *CLM*: There is a nodule, which is 3–5 times the size of a typical reactive follicle, with mantle zone small B cells infiltrating the residual germinal center. Moreover, epithelioid histiocytic clusters, immunoblasts, mononuclear cells, and eosinophils may be found.
- *IHC*: The immunophenotypes of both PTGC and nodular lymphocytic predominant-variant of Hodgkin lymphoma (NLPHL) may be identical (CD45+, CD20+, CD15-, CD30-). Moreover, PTGC is (+) CD23 and (+) BCL2.
- *DDX*: NLPHL. Of note, the nodules in PTGC are relatively well-defined, while they are irregular in NLPHD.
- *TNT*: Follow-up only.
- *PGN*: It has been suggested that it is synchronous or metachronous NLPHL. Recurrent PTGC is more common in children. Although PTGC rarely develops NLPHL, a repeated biopsy is reasonable to exclude this possibility if lymphadenopathy recurs.

10.7.1.12 AIDS-Related Lymphadenopathy

1. Florid reactive hyperplasia, serpentine-like enlarged germinal cells
2. Follicle lysis, showing the collapse of the central portions of the germinal centers with the PTGC-like invasion of mantle lymphocytes
3. Regressive changes with advanced lymphocyte depletion

Ret or CD21 reveal irregular outlines of germinal centers in HIV-associated lymphadenopathy.

Stain for B cells or FDC (CD21 and/or CD35) may highlight germinal centers not identified on routine stainings.

10.7.2 Diffuse (Paracortical) Hyperplasia

Expansion of T-cell regions with effacement of the LN architecture accompanied by hypertrophy of endothelium ("pseudolymphomatous pattern") and usually observed in drug reactions (Dilantin), post-smallpox vaccination, chronic dermatitis, and viral infection. Thus, it is important to distinguish between postvaccinal viral, infectious, and dermatopathic lymphadenitis.

10.7.2.1 ALPS
- *DEF*: The Autoimmune Lymphoproliferative Syndrome (ALPS) is a FAS-ligand/caspase gene-linked AD-inherited genetic disease with florid paracortical hyperplasia with immunoblasts and plasma cells and small hyperplastic lymphoid follicles ± AI anemia, ± AI thrombocytopenia.
- *IHC*: CD45RA (+), CD45RO (−), CD57 (+), CD56 (−), and (−) CD4/CD8 T lymphocytes.

10.7.2.2 Postvaccinal Viral Lymphadenitis
- *DEF*: Diffuse/nodular (paracortical) hyperplasia-based supraclavicular lymphadenitis usually developing 1–10 weeks following vaccination and showing RS-like immunoblast proliferation, vascular proliferation, sinusoidal dilatation, and mixed cellular infiltrate (plasma cells, mast cells, and eosinophils).

10.7.2.3 Infectious Mononucleosis
- *DEF*: Paracortical and follicular hyperplasia-based lymphadenitis with nonspecific effacement of the architecture with infiltration of the trabeculae, capsule, and perinodal fat by immunoblasts, RS-like cells, and numerous plasma cells. The distinguishing features between IM and lymphoma are exhibited in Box 10.15.

Box 10.15 Distinguishing Features Between IM and Lymphoma
- Plasma cell predominance
- Biochemical evidence: Monospot (+), anti-EBV-IgM (+), atypical CD8(+) T cells
- ISH for EBV RNA: EBER (+)
- Lack of sinusoidal distribution
- Lack of numerous large cells with prominent nucleoli

10.7.2.4 Dermatopathic Lymphadenitis
- *DEF*: Itching-/dermatitis-associated lymphadenopathy in drainage areas of disrupted, inflamed, or neoplastic skin (e.g., psoriasis, mycosis fungoides), although on some occasions no identifiable cutaneous lesion is found.
- *SYN*: Lipomelanosis reticularis of Pautrier microabscesses.
- *GRO*: Lipo → centric + melanotic/pigment → periphery of the LN.
- *CLM*: Paracortical hyperplasia with the expansion of the T-dependent zone showing a proliferation of the interdigitating reticulum S100+, CD1a- cells, histiocytes with folded nuclei (Langerhans cells, S100+, CD1a+) compressing the follicles against the capsule, where macrophages with ingested melanin are present (CD68+).
- *DDX*:
  1. Metastatic mycosis fungoides: Lutzner cells with cerebriform nuclei in paracortex, history (+), and FC (+)
  2. LCH: Sinusoidal infiltrate of Langerhans cells, (+) eosinophils, (+) CD1a, particularly strong expression

10.7.3 Sinus Pattern

It is essential to distinguish between sinus histiocytosis, sinus histiocytosis with massive lymphadenopathy (SHML or Rosai-Dorfman disease), lipophagic reactions, and vascular transformation of LN sinuses. Essentially four disorders involve the lymph node with a sinus pattern, including sinus histiocytosis (SH), sinus histiocytosis with massive lymphadenopathy (SHML) or Rosai-Dorfman disease, lipophagic reactions, and vascular transformation of LN sinuses.

10.7.3.1 Sinus Histiocytosis
Benign, nonspecific lymph nodal finding, which is often associated with infections and neoplasia. There is usually enlargement of lymph node, and, microscopically, the sinuses are distended with histiocytes (abundant foamy cytoplasm ± pigment and/or yellow bodies), and there is a ↑ plasma cells. DDX includes SHML (RDD) and dermatopathic lymphadenopathy.

10.7.3.2 Sinus Histiocytosis with Massive Lymphadenopathy

(*Vide supra*)

10.7.3.3 Lipophagic Reactions

- *DEF*: Lymphadenopathy characterized by an accumulation of phagocyted fat in histiocytes including several types. They are mineral oil ingestion (asymptomatic, periportal, and mesenteric LNs), *Tropheryma whipplei* (*T. whipplei*) infection (symptoms related to the organs or systems involved, mesenteric LNs, poorly formed granulomas with ORO+ lipid in the macrophages (CD68+) and PAS+ particles in the cytoplasm by LM, and lymphangiography-related. In the *T. whipplei*-infection, bacilli are seen within the histiocytes by transmission electron microscopy.

10.7.3.4 Vascular Transformation of LN Sinuses

- *DEF*: Lymphadenopathy with vascular proliferation following the sinuses (DDX: Kaposi sarcoma) secondary to extranodal venous outflow obstruction and characterized histologically by subcapsular and interfollicular sinuses with blood-filled endothelial lined spaces, fibrosis, and RBC's extravasation.

10.7.4 Predominant Granulomatous Pattern

It is essential to distinguish between sarcoidosis, tuberculosis, atypical mycobacteriosis, fungal infections, and chronic granulomatous disease (CGD). A more exhaustive list may be found in textbooks of hematology or hematopathology.

Reactive lymphadenopathies of a primary granulomatous PTN:

- Non-infectious (neoplasms, sarcoidosis, berylliosis, Crohn disease)
- Infectious:
    Of nonsuppurative type (Tb, atypical mycobacteriosis, lepra, fungi, pneumocystis)
    Of suppurative type (cat-scratch disease, LGV, Yersinia lymphadenitis)

10.7.4.1 Tuberculosis

M. tuberculosis-related lymphadenitis with or without matting of adjacent LNs similar to the phenomenon seen in LNs adherent to each other following metastatic carcinoma and characterized by small epithelioid granulomas to large caseous masses.

In particular, the term "scrofula" has been used for "matted" cervical lymphadenopathy and "scrofuloderma" for the form draining sinus to the skin.

- SS: AFB on Ziehl-Neelsen or auramine-rhodamine.

10.7.4.2 Atypical Mycobacteriosis

Atypical mycobacteria (Runyon's group-related) lymphadenitis of typically lateral middle neck LNs with foremost involvement of the pediatric age group.

CLM is similar to TB infection, although it has been suggested that the suppurative component is more prominent in the atypical mycobacteriosis than in the TB infection.

10.7.4.3 Sarcoidosis (See Lung) Lymphadenopathy

Noncaseating granulomatous inflammation with back-to-back granulomas and scattered Langhans giant cells, which can be preceded or accompanied by erythema nodosum.

Schaumann bodies (round, concentric laminations with Fe and Ca), asteroid bodies (crisscrossing collagen fibers), and calcium oxalate crystals in cytoplasm of giant cells as well as Kveim test (+) (granulomatous intradermal reaction following inoculation with extract of human spleen involved with sarcoidosis, positive in sarcoidosis in 60–85% of cases)

- *PGN*: 70% recover with minimal sequelae, 20% exhibit permanent loss of lung function, 10% will die for cardiac disease, CNS involvement or progressive pulmonary fibrosis.

10.7.4.4 Fungal Infections

- Suppurative or granulomatous lymphadenitis
- *Histoplasma capsulatum* extensive nodal necrosis accompanied by marked diffuse hyperplasia of sinus histiocytes.

- *Sporotrix*: ± suppurative (*Sporotrix* is an ubiquitous genus of soil-dwelling fungus, whose the best known is *Sporothrix schenckii*, the causative agent of rose handler's disease)

10.7.4.5 Chronic Granulomatous Disease
- Gene-enzyme defect.
- Impossibility to digest microorganisms.
- Granulomas with necrotic, purulent centers.

10.7.5 Other Myxoid Patterns

Other myxoid patterns include mucocutaneous lymph nodal syndrome or Kawasaki lymphadenitis, leprosy, and mesenteric lymphadenitis or Masshoff lymphadenitis.

10.7.5.1 Kawasaki Disease
- *DEF*: Mucocutaneous lymph node syndrome or Kawasaki disease is a pediatric illness including several symptoms, such as fever, cervical lymphadenopathy, pharyngeal and conjunctival inflammation, erythematous skin rash and swelling of the hands and feet.
- *PGN*: Although the acute phase of the condition usually lasts 10–14 days and most children recover fully, Kawasaki disease can lead to death linked to coronary aneurysmatic arterial dilation.
- *LN*: Fibrin thrombi in smaller BVs and patchy infarcts.

10.7.5.2 Leprosy
- Large, pale round histiocytes (CD68+) without granuloma formation.

10.7.5.3 Mesenteric Lymphadenitis
- *DEF*: Yersinia enterocolitis-related lymphadenitis characterized by capsular thickening, immunoblasts, plasma cells in the cortex and paracortex, and germinal centers and one of the significant mimickers of appendicitis.

10.7.6 Angioimmunoblastic Lymphadenopathy with Dysproteinemia (AILD)

Distinguishing between postvaccinal viral, infectious, and dermatopathic lymphadenitis is essential.

AILD
- *DEF*: Polyclonal proliferation of immunoblasts and plasma cells.
- *EPI*: Worldwide, usually adults.
- *CLI*: Fever, anemia, polyclonal hyper-Ig 1/4 of cases following drug administration, especially penicillin.
- Systemic: Involvement of the spleen, liver, LNs, and BM.
- *CLM*: Partial effacement of the LN architecture by obliteration due to eosinophils, plasma cells, immunoblasts, some giant cells, and proliferation of post-capillary venules with accompanying deposition of amorphous eosinophilic interstitial PAS+ material (e.g., cellular detritus) and lack of germinal centers. In BM there is in 1/2–2/3 of cases a focal>diffuse involvement with ↑ reticulin in affected areas.
- *TNT*: Steroids ± combination chemotherapy.
- *PGN*: Death (3/4 of cases). Poor PGN factors: (+) Clones of tightly packed immunoblasts (→ progression to immunoblastic sarcoma, T-cell type).

10.8 Disorders of the Spleen

Spleen disorders are shown in Box 10.16 and congenital anomalies are shown in Box 10.17.

Box 10.16 Disorders of the Spleen
10.8.1. Hypofunction
10.8.2. White Pulp Disorders
10.8.3. Red Pulp Disorders
10.8.4. Splenomegaly
10.8.5. Cysts and Tumors of the Spleen

> **Box 10.17 Congenital Anomalies**
> – Asplenia
> – Hyposplenia
> – Abnormal lobulation
> – Polysplenia
> – Accessory spleens

Hypersplenism, which is defined as an overactive spleen, may result in (1) splenic enlargement; (2) sequestration/loss of circulating RBCs, WBCs, and/or platelets; and (3) resolution of the cellular sequestration/loss following splenectomy.

10.8.1 White Pulp Disorders of the Spleen

Essentially four categories may be delineated, including reactive follicular hyperplasia, reactive non-follicular lymphoid hyperplasia, chronic lymphocytic leukemia, and malignant lymphomas.

10.8.1.1 Reactive Follicular Hyperplasia
- *DEF*: Hyperplasia of the follicles of the white pulp of reactive nature.
- *EPI*: Worldwide, children, ♂=♀.
- *EPG*: Acute infections, ITP, RA, acquired hemolytic anemia, chronic hemodialysis, and Castleman disease.
- *TNT*: None, but it is essential to distinguish between the several and different underlying disorders.
- *PGN*: Usually linked to the underlying disorder.

10.8.1.2 Reactive Non-follicular Lymphoid Hyperplasia
- *DEF*: Hyperplasia of the lymphatic tissue (white pulp) of reactive origin without formation of germinal centers.
- *EPG*: Viral infections, graft rejections, AILD.

10.8.1.3 Chronic Lymphocytic Anemia
It is the sole leukemia, which consistently and selectively targets the white pulp, although the involvement may be quite irregular.

10.8.1.4 Malignant Lymphomas
Involvement of the spleen by either lymphoma confined to the spleen or lymphomas involving other hematopoietic organs.

Four gross patterns:

1. "Miliary small nodules" pattern, which may conceal FL, B-CLL/SLL, LPL, MCL, MZBCL
2. "Solitary or several small clustered nodules" pattern, which usually corresponds to HL
3. "Solitary or multiple large fleshy nodules" pattern, which may represent the gross pattern for DLBCL, FL grade 3, some peripheral TCL, various high-grade lymphomas or HL
4. "Beef-Red" pattern, which may mean leukemia (apart B-CLL/SLL), hepatosplenic TCL, some DLBCL and PTCL, agnogenic myeloid metaplasia, systemic mastocytosis, and LCH.

10.8.2 Red Pulp Disorders of the Spleen

There may be some diseases involving the red pulp, including congestion, infection, leukemia, mastocytosis of systemic type, and histiocytic proliferations exclusively.

10.8.2.1 Congestion
Congestion may occur in hemolytic anemias, such as the case of splenectomy performed in cases of children with spherocytosis. In case of congestion, it is usual to divide into acute and chronic congestions. In the setting of chronic congestion, there is thickening of the splenic capsule and occurrence of Gandy-Gamna bodies, which are organized, fibrosed old hemorrhages with hemosiderin and calcium aggregates.

10.8.2.2 Infection
Two major infections involve the red pulp and are EBV-driven infectious mononucleosis and sepsis. In case of an infection of the red pulp, we consider the process an acute splenitis, which is acute septic splenitis in case of a patient, in almost all cases, at autopsy. No figure is provided with infectious mononucleosis because the patient is not usually undergoing a splenectomy unless the spleen causes major harms to the patient.

10.8.2.3 Leukemia

In case of red pulp involvement by leukemia, the red pulp is strikingly prominent making the white pulp barely discernible. Three conditions need to be kept in mind and are hairy cell leukemia (HCL), chronic myelocytic leukemia (CML), and myeloid metaplasia of idiopathic (agnogenic myeloid metaplasia) or secondary nature. In HCL, blood lakes are frequently found (see below, under leukemia), which makes this pattern quite distinctive. In CML, there is a polymorphous infiltrate with eosinophilic myelocytes, which is quite characteristic, and obviously, clinical history and special stains are essential to review before signing out the case. In myeloid metaplasia, all three cell lines are present in the idiopathic subtype, while the picture may be limited to a single cell line in myeloid metaplasia of secondary type.

10.8.2.4 Histiocytic Proliferations

Some diseases may recruit histiocytes and give the impression of a histiocytosis.

Histiocytosis: LCH, JXG of the spleen, and malignant histiocytosis.

Non-histiocytosis or generic histiocytic proliferations need to be kept separated. In this last group, granulomatous inflammations of the spleen in the setting of several infections, ceroid histiocytosis, Gaucher disease, and Niemann-Pick disease are to be recalled. In ceroid histiocytosis, there is an accumulation of foamy macrophages laden with the wavy material, mainly sphingomyelin, and it may be seen in ITP and CML.

Littoral Cell Angioma

- *DEF*: Endothelial+histiocytic markers-expressing tumor, which seems to derive from the littoral cells lining the venous sinuses of the spleen.
- *CLI*: ± (50%) Anemia, hypersplenism-associated thrombocytopenia, and splenomegaly and ± Crohn disease, colon-rectal carcinoma (CRC), and pancreatic carcinoma.
- *CLM*: Anastomosing vascular channels of a monotonous type resembling splenic sinuses, but lined by tall prominent endothelial cells with LM-identifiable variable hemophagocytosis, channels with irregular lumina, papillary projections, and cystic spaces and detached endothelial cells into the lumina, but neither sclerosis nor atypia is present.
- *IHC*: CD31(+), CD68(+), but CD34(−), CD21(±), and CD163(+). CD21 is the C3d complement receptor and also, very important to remember, the EBV receptor of B lymphocytes and seems to be exclusively positive in SLCA. CD163 is the high-affinity scavenger receptor for the hemoglobin-haptoglobin complex and a marker of cells from the monocyte-macrophage lineage.

Inflammatory Pseudotumor

ALK1(+)-spindle cell proliferation with quite bland cytology mixed with an inflammatory component constituted by plasma cells, lymphocytes, eosinophils, and histiocytes with low malignant potential. ALK1 is, however, positive in about 50% of cases and most of the cases are pediatric.

Hyaline Perisplenitis

Collagenous thickening of the capsule ("*Zuckergussmilz*" of the German literature, spleen with "sugarcoating" or *perisplenitis cartilaginea*), which follows states of portal vein stasis/hypertension. Neither symptoms nor sequelae are documented, apart from the underlying disorder, if noted.

Spontaneous Rupture

It happens exclusively due to underlying pathologies, such as IM, malaria, typhoid fever, CML, and acute splenitis. To list other causes, we should include the littoral hemangioma, angiosarcoma, and hairy cell leukemia. Sequelae are hemorrhagic shock and death. In case of seeding of the peritoneal cavity with splenic tissue, the condition of splenosis may occur.

Peliosis

Widespread, blood-filled cystic spaces due to androgens or chronic wasting diseases (TB or cachexia in the setting of metastatic carcinoma).

- *DDX*: Littoral hemangioma, cavernous hemangioma, hairy cell leukemia, and hemangiosarcoma.

10.9 Disorders of the Thymus

The diseases of the thymus are shown in Box 10.18.

> **Box 10.18 Disorders of the Thymus**
> 10.9.1. Thymic *Cysts and Thymolipoma*
> 10.9.2. *True Thymic Hyperplasia and Thymic Follicular Hyperplasia*
> 10.9.3. *Thymoma*
> 10.9.4. *Thymic Carcinoma*
> 10.9.5. *Thymic Neuroendocrine Carcinoma*
> 10.9.6. *Non-thymic Neoplasms with Primary Intrathymic Onset*
>   10.9.6.1. Hematopoietic
>   10.9.6.2. Germ Cell Tumors
>   10.9.6.3. With Associated Somatic Type Malignancy
>   10.9.6.4. With Associated Hematologic Malignancy
> 10.9.7. *Ectopic Thymic and Related Branchial Tumors*
>   10.9.7.1. Ectopic Thymoma
>   10.9.7.2. Ectopic Hamartomatous Thymoma
>   10.9.7.3. Spindle Epithelial Tumor with Thymus-Like Elements (SETTLE)
>   10.9.7.4. Carcinoma Showing Thymus-Like Elements (CASTLE)

10.9.1 Thymic Cysts and Thymolipoma

The thymic cyst may be unilocular (congenital) or multilocular (acquired). The congenital thymic cysts are a thin-walled cyst filled with serous, limpid fluid and lined by the flat/cuboidal lining. The acquired thymic cysts are a thick-walled cyst filled with torbid, hemorrhagic, dusky fluid and lined by the flat/cuboidal/squamous/columnar/ciliated lining. The thymic cyst is recognizable as a hole cavity in continuity with thymic remnants in the wall containing lymphocytes and cholesterol clefts crystals. The thymic cysts are subdivided into (1) foregut cysts of the mediastinum (branchial or branchiogenic cysts, esophageal cysts, enteric cysts, and duplication cysts) and (2) mesothelial cysts.

- *DDX*: It includes the parathyroid cyst (location, Ca levels, PTH levels), cystic hygroma (age, chromosomes, genetics, dysmorphism), esophageal cyst (radiology, surgical report, histology), bronchial cyst (radiology, surgical report, histology), cystic teratoma (1–3 dermal teratoma), cystic thymoma (age, history, histology, IHC), and cystic degeneration (age, clinical history).

Thymolipoma

It is a rare, benign anterior mediastinal mass of thymic origin displaying both thymic and mature adipose tissue. It is about 5% of all thymic neoplasms and involves children and adults.

10.9.2 True Thymic Hyperplasia and Thymic Follicular Hyperplasia

True thymic hyperplasia (TTH): Thymic enlargement with a specific increase of volume but normal weight. ± AI (e.g., hyperthyroidism, MG, etc.), affecting both children and youth after CHT for malignancy.

- *CLM*: Conservation of the lobular architecture with a normal distribution of lymphocytes and epithelial cells and preservation of corticomedullary differentiation.
- Thymic follicular hyperplasia: Thymus with conserved size and weight, but the presence of numerous follicles with GC in the setting of ± AI (MG, RA, SLE, etc.).

TTH needs to be kept distinct from thymic follicular hyperplasia (TFH) and thymomas. A lobular architecture characterizes the first with a normal distribution of lymphocytes and epithelial

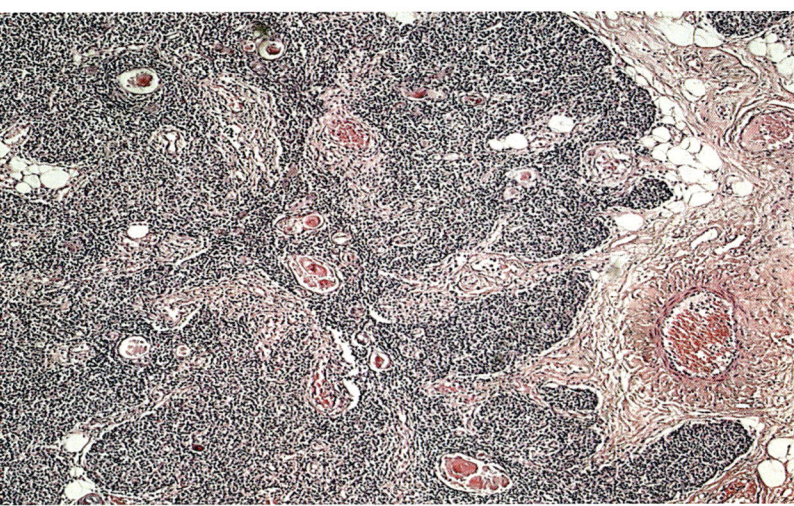

Fig. 10.23 Human immunodeficiency virus changes include regressive changes of the thymus with hypotrophic lymphoid tissue (Modified Periodic acid Schiff with diastase digestion to highlight Hassall's corpuscles, x100 original magnification)

cells and CM differentiation preservation. TFH is characterized by well-formed lymphoid follicles with germinal centers and CD20+ B lymphocytes in the germinal centers, while thymomas show no normal lobulation and absence of corticomedullary differentiation. Moreover, it is useful to distinguish between neoplasms and thymus and non-neoplastic thymic lesions and conditions, which are atrophy, cysts, Castleman disease, dysplasia, and hyperplasia (B cell, GC-typed, rebound, true thymic).

10.9.3 HIV Changes

HIV infects CD4+ T lymphocytes, and the disease is manifested by CD4+ T-cell paucity. Both the thymic epithelial space (TES) and the perivascular space (PVS) are involved in HIV-1-infected children. TES is a lymphoid compartment where thymopoiesis is located, while PVS is a lymphoid compartment where T lymphocytes are pooled. The knowledge of autophagy and apoptosis of the last couple of decades seems to reveal that HIV infection per se is not responsible for dying CD4+ T lymphocytes and anti-retroviral therapy does not instantly reduce the high death rates of CD4+ T lymphocytes. Brunner et al. (2011) found that CD4+ and CD8+ cells were more numerous in PVS than in TES. Moreover, they showed that CD4 + CD8+ cells (immature thymocytes) were 15.4% (age-related control: 80.5%) in thymus cell suspensions. Very few apoptotic CD4+ cells were seen in TES, but very low to absent proliferation activity was demonstrated in both TES and PVS. Brunner et al. (2011) suggest that (1) lymphocyte exhaustion in HIV-1 infection is more pronounced in TES than in PVS, (2) immature thymocytes are not enhanced, and (3) an anti-apoptotic effect appears to be a potential ART mechanism elucidating the CD4+ pool increase (Figs. 10.23 and 10.24).

10.9.4 Thymoma

- *DEF*: Epithelial cell-derived thymic tumor with lymphocytic "innocent bystanders" as immature T cells, conversely from thymic carcinoma where lymphocytes are of B-cell type.
- *CLI*: ± myasthenia gravis, hypogammaglobulinemia, pure red cell aplasia, and connective tissue disorders (RA, SLE, scleroderma, polymyositis).
- *GRO*: Round-oval mass with an appreciable fibrous capsule of variable thickness with tan fleshy lobules delimited by thin/coarse fibrous septa (smooth, bosselated surface) in the form of jigsaw puzzle-like shape ± cysts and H&C (hemorrhage and calcification).

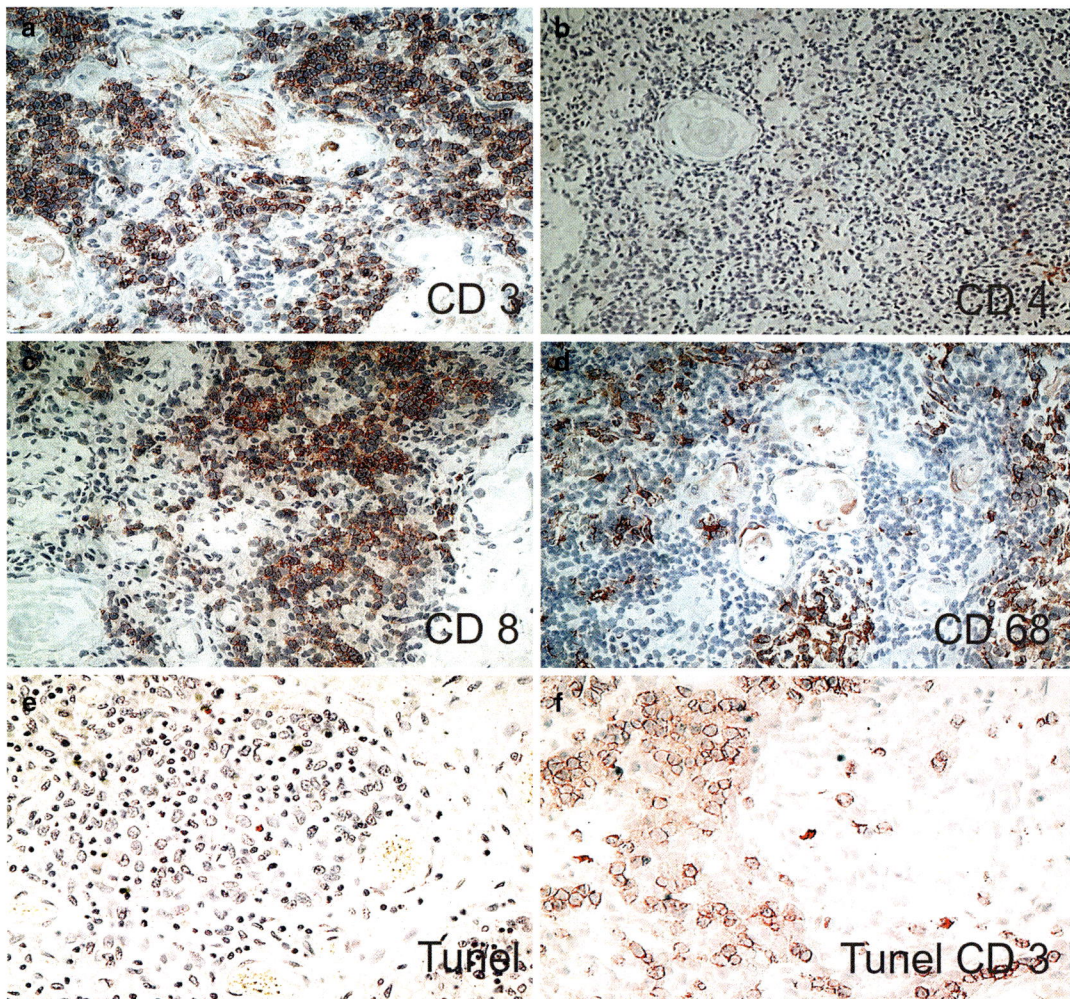

Fig. 10.24 A specific study of the thymus in pediatric acquired immunodeficiency syndrome caused by human immunodeficiency virus shows a depletion of CD4 with an involvement of apoptosis as mechanism of depletion (**a**, anti-CD3 immunostaining, ×100 original magnification; **b**, anti-CD4 immunostaining, ×100 original magnification; **c**, anti-CD8 immunostaining, ×100 original magnification; **d**, anti-CD68 immunostaining, ×100 original magnification; **e**, TUNEL technique of in situ apoptosis detection; **f**, combined TUNEL-CD3 staining, ×100 original magnification)

- *CLM*: A-B1-B3-AB (classification parameters include the type of epithelial cell, organotypic morphology, the relative proportion of associated immature T cells, and degree of epithelial atypia).

A thymoma	Bland, spindle cells (at least focally) with paucity or absence of immature (TdT+) T cells throughout the neoplasm
B1 thymoma	Copiousness of immature T cells, areas of medullary differentiation (the so-called "medullary islands"), scarcity of polygonal or dendritic epithelial cells without clearcut evidence of clustering
B2 thymoma	↑ single or clustered polygonal or dendritic epithelial cells intermixed with copious immature T cells
B3 thymoma	Sheets of polygonal atypical epithelial cells with paucity or absence of intermixed immature (TdT+) T cells
AB thymoma	Bland, spindle cells (at least focally) with copious immature (TdT+) T cells focally or throughout the tumor

Thymic carcinoma is differentiated in low grade (well-differentiated squamous, basaloid, ACA, adenosquamous, mucoepidermoid) and high grade (lymphoepithelial-like, LC undifferentiated, clear cell, sarcomatoid, midline carcinoma with t(15;19) translocation). Thymic NEC is subdivided in well-differentiated or carcinoid and poorly differentiated of small cell/large cell NEC.

- *TEM*: (+) EC features, including tonofilaments, tight intercellular junctions, desmosomes, elongated cytoplasmic processes, and basal lamina (Pitfall! Due to the heterogeneity of the thymic neoplasms and lesions, there is a high potential for sampling error: ergo, TEM interpretation needs to be done only in the context of a contiguous and consonant LM, including H&E and IHC).

Myasthenia Gravis

It is a defect in nicotinic acetylcholine receptor (AChR) present in the subsynaptic membrane of the neuromuscular junction (at motor end plate), which is due to circulating autoantibodies to the receptor. The AChR is also present in normal thymus. Hyperplastic medullary thymic epithelial cells are involved in provoking infiltration, and thymic myoid cells with intact AChR are involved in germinal center formation. Approximately 12% of patients affected with myasthenia gravis have other AI diseases (e.g., Graves' disease, rheumatoid arthritis), 65% thymic hyperplasia, 25% normal thymus, and 10% thymomas. Myasthenia gravis is present in up to 45% of patients with thymomas. Thymectomy is the treatment of choice regardless of the presence of thymoma.

Acute Thymic Involution

Stress (chronic debilitating disease), HIV or other infections, prolonged protein malnutrition and immunosuppressive or cytotoxic drugs, GVHD, infants with chorioamnionitis, and sepsis are subject to involution of the thymus. There is a preservation of lobular architecture and Hassall's corpuscles, but marked lymphocyte depletion (particularly with HIV) and BVs are large compared to the size of lobules, plenty of plasma cells, and fibrohyaline changes of the basement membrane of vessels and thymic epithelium.

The thymic/non-thymic tumors potentially arising in the thymic region are hematopoietic neoplasms (primary mediastinal LBCL, EN-MZL, T-lymphoblastic lymphoma/leukemia), germ cell-type neoplasms (seminoma and non-seminoma germ cell tumor, including embryonal carcinoma, yolk sac tumor, teratoma, choriocarcinoma, mixed germ cell tumor ± associated somatic-type malignancy, e.g., embryonal rhabdomyosarcoma and angiosarcoma, OR ± associated hematologic malignancy (e.g., AML, acute megakaryoblastic leukemia). The thymic neoplasms most commonly associated with cysts or cystic degeneration are a basaloid carcinoma, ENMZL, HL, seminoma, and thymoma. Finally, the ectopic thymic and related branchial tumors are listed as ectopic thymoma (e.g., pleura, pulmonary, pericardium (3Ps)), ectopic hamartomatous thymoma, spindle epithelial tumor with thymus-like elements (SETTLE), and carcinoma showing thymus-like elements (CASTLE).

Multiple Choice Questions and Answers

- HEM-1 A 12-year-old boy presents at the emergency department with pallor, low-grade fever, and enlargement of the spleen. The parents indicate that the child suffers from hereditary spherocytosis. His blood work shows a hemoglobin of 1.86 mmol/L (normal, 7.45–11.17 mmol/L or 12–18 g/dL), reticulocyte count of 2%, white blood cells at 6.0 x 10^9 cells per liter (6,000/mm^3; normal range, 4.3–10.8 x 10^9 cells per liter), and platelet count of 200 x 10^9 cells per liter (200,000/mm^3; normal range, 150–400 x 10^9 cells per liter or 150,000–400,000/mm^3). What is the most likely diagnosis?
  (a) Acute leukemia
  (b) Sepsis
  (c) Aplastic crisis
  (d) Acute splenic sequestration
  (e) Hemolytic crisis
- HEM-2 Fanconi anemia (FA) is a rare inherited recessive disease, which is determined by mutations in one of fifteen genes. These genes

are known to encode FA pathway components. In response to DNA damage, nuclear FA proteins associate into complexes of high molecular weight through a cascade of post-translational modifications and molecular interactions. This results in the repair of damaged DNA. Which of the following features is NOT characteristic of FA?
(a) Chromosome fragility
(b) Infantile hematologic abnormalities
(c) Pancytopenia
(d) Skeletal anomalies
(e) Squamous cell carcinoma of the head and neck

- HEM-3 Which of the following disorders is associated with a factor extrinsic to the red blood cell?
  (a) Sickle cell anemia
  (b) Autoimmune hemolytic anemia
  (c) Glucose-6-phosphate dehydrogenase (G6PD) deficiency
  (d) Hereditary spherocytosis
  (e) Hereditary elliptocytosis

- HEM-4 Which of the following statement on hereditary spherocytosis does NOT match with the disease?
  (a) Mitochondrial DNA pattern of inheritance.
  (b) The defect compromises the scaffolding of the red blood cell leading to a loss of membrane fragments with the formation of microspherocytes.
  (c) The microsporocyte membrane is hugely permeable to sodium.
  (d) Cholelithiasis and cholecystitis may develop in teenagers.
  (e) Splenectomy is the definitive therapy.

- HEM-5 A 12-year-old boy presents with an enlargement of a right-sided lymph node persisting for more than 12 months. The child did not complain about any pain. The lymphadenopathy did not rise in the setting of an infection. There was no history of clinical symptoms, such as fever, malaise, night sweats, or weight loss. On physical examination, the lymph node was mostly soft and mobile. The lymph node was about 5 cm × 3 cm in size without apparent overlying skin changes. Laboratory testing showed a normal white blood cell count with elevated lymphocytes and positivity for Epstein-Barr virus (early immunoglobulin G), while *Bartonella* and *Cytomegalovirus* testing were negative. A biopsy was performed, and the histology of the lymph node is shown here:

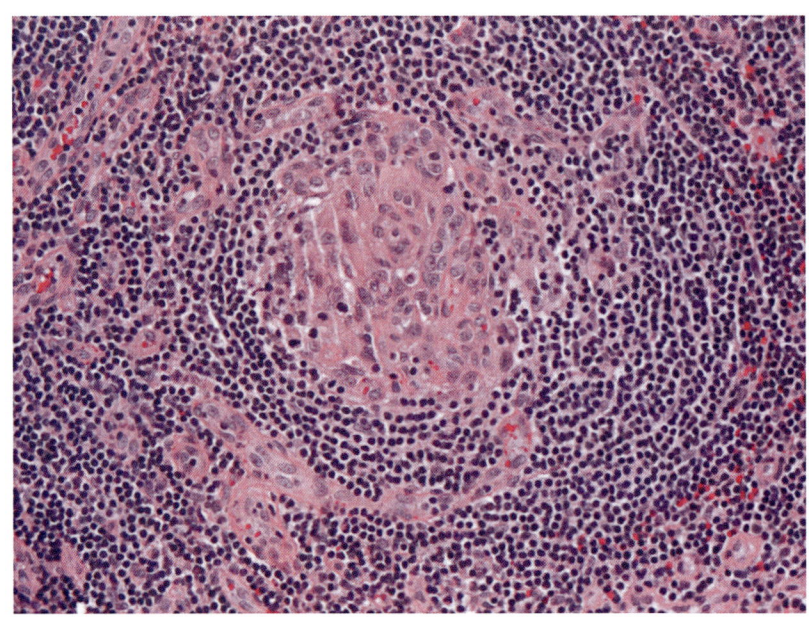

What is the most likely diagnosis?
(a) Castleman disease
(b) Viral lymphadenitis
(c) Langerhans cell histiocytosis
(d) Hodgkin lymphoma
(e) Diffuse large B-cell lymphoma

- HEM-6 A 15-year-old boy develops an enlargement of a cervical lymph node for more than 6 months. Neither weight loss nor other B symptoms are noted. Microbiologically, all tuberculosis and sarcoidosis tests were negative. A biopsy was performed, and the histology of the lymph node is shown here:

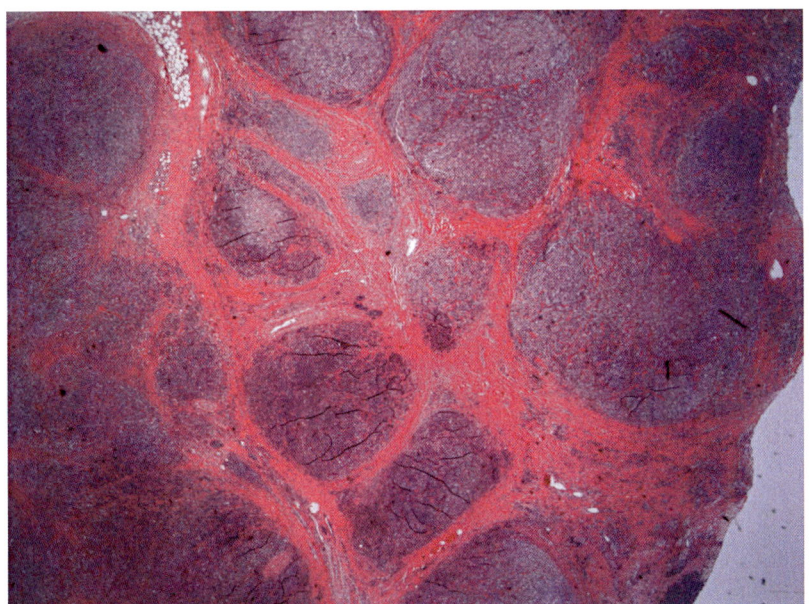

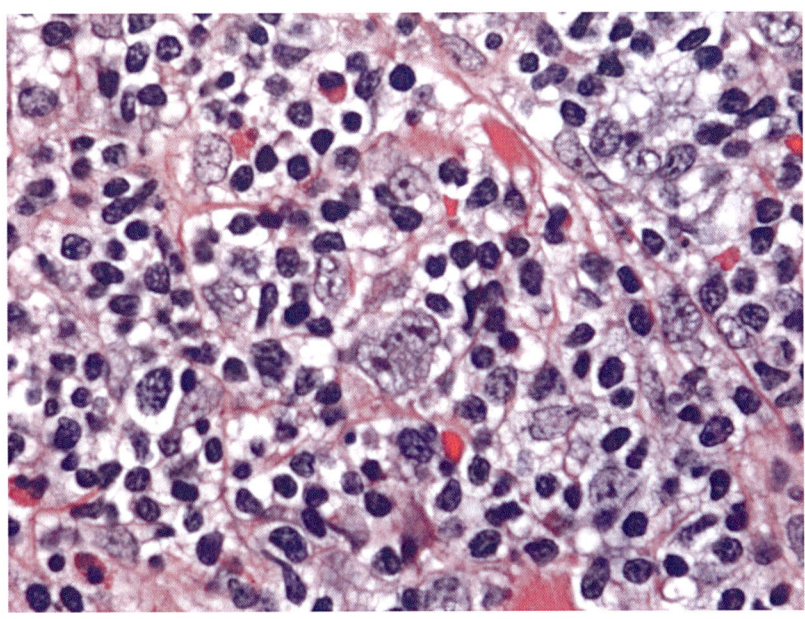

- What is the most likely diagnosis?
  - (a) Hodgkin lymphoma, nodular sclerosis variant
  - (b) Hodgkin lymphoma, depleted lymphocyte variant
  - (c) Diffuse large B-cell lymphoma
  - (d) Mantle cell lymphoma
  - (e) Langerhans cell histiocytosis
- HEM-7 What is the best combination of immunohistochemical markers to diagnose classical Hodgkin cells?
  - (a) AE1–AE3, CD15, CD30, PAX5
  - (b) CD15, CD30, PAX5, MUM-1
  - (c) CD20, CD21, CD15, CD30
  - (d) BOB-1, PAX5, MUM-1
  - (e) CD15, CD30, CD56, CD138
- HEM-8 Which of the following statements about mantle cell lymphoma (MCL) is TRUE?
  - (a) Cyclin D1 is essential for the diagnosis of MCLs.
  - (b) SOX11 is a useful marker in the diagnosis of MCL.
  - (c) The proliferative index recognized by the mouse monoclonal antibody against Ki-67 (MIB-1 immunostaining) has no prognostic relevance.
  - (d) Splenic involvement is rarely encountered as a presentation of patients harboring MCL.
  - (e) B symptoms are common at the presentation of MCL
- HEM-9 A 25-year-old woman who is an intravenous drug user brings to the emergency department her afebrile infant showing severe oral thrush. Upon examination, the infant's blood work shows hypocalcemia on two repeated measurements, but white cell count seems to be within normal limits. What is the most likely diagnosis in this infant of the following ones proposed below?
  - (a) Child abuse by the mother's partner
  - (b) Severe combined immunodeficiency disease with mutations in the tyrosine phosphatase CD45 gene
  - (c) Chronic granulomatous disease
  - (d) DiGeorge syndrome
- HEM-10 Which of the following are the two best antibodies to apply in detecting bone marrow metastases and minimal residual disease of neuroblastoma?
  - (a) PHOX2B and NB84
  - (b) CD56 and CD57
  - (c) NB84 and S100
  - (d) NSE and S100
  - (e) S100 and CD56

References and Recommended Readings

Alinari L, Pant S, McNamara K, Kalmar JR, Marsh W, Allen CM, Baiocchi RA. Lymphomatoid granulomatosis presenting with gingival involvement in an immune competent elderly male. Head Neck Pathol. 2012;6(4):496–501. https://doi.org/10.1007/s12105-012-0378-z. Epub 2012 Jun 19. PubMed PMID: 22711054; PubMed Central PMCID: PMC3500898

Bechan GI, Egeler RM, Arceci RJ. Biology of Langerhans cells and Langerhans cell histiocytosis. Int Rev Cytol. 2006;254:1–43. Review. PubMed PMID: 17147996

Broadbent V, Gadner H, Komp DM, Ladisch S. Histiocytosis syndromes in children: II. Approach to the clinical and laboratory evaluation of children with Langerhans cell histiocytosis. Clinical Writing Group of the Histiocyte Society. Med Pediatr Oncol. 1989;17(6):492–5.

Brunner J, Boehler T, Ehemann V, Kassam S, Otto H, Sergi C. Decreased apoptosis despite severe CD4 depletion in the thymus of a human immunodeficiency virus-1 infected child. Klin Padiatr. 2011;223(4):246–8. https://doi.org/10.1055/s-0030-1270514. Epub 2011 Jan 26. PubMed PMID: 21271506

Chaurasia JK, Singh G, Sahoo B, Maheshwari V. Rosai-Dorfman disease: unusual presentation and diagnosis by fine-needle aspiration cytology. Diagn Cytopathol. 2015;43(9):716–8. https://doi.org/10.1002/dc.23232. Epub 2014 Oct 31.

Demir HA, Bayhan T, Üner A, Kurtulan O, Karakuş E, Emir S, Özyörük D, Ceylaner S. Chronic lymphocytic leukemia in a child: a challenging diagnosis in pediatric oncology practice. Pediatr Blood Cancer. 2014;61(5):933–5. https://doi.org/10.1002/pbc.24865. Epub 2013 Nov 19. PubMed PMID: 24249660.

Drut R, Drut RM. Angiocentric immunoproliferative lesion and angiocentric lymphoma of lymph node in children. A report of two cases. J Clin Pathol. 2005;58(5):550–2. PubMed PMID: 15858132; PubMed Central PMCID: PMC1770650

Gini Ehungu JL, Mufuta JP, Ngiyulu RM, Ekulu PM, Kadima BT, Aloni MN. A rare occurrence of hairy cell leukemia in a congolese child: a presentation and challenge of diagnosis in low resource settings. J Pediatr Hematol Oncol. 2013;35(8):e350–2. https://

References and Recommended Readings

doi.org/10.1097/MPH.0b013e318290b9c7. PubMed PMID: 23652869.

Giona F, Teofili L, Moleti ML, Martini M, Palumbo G, Amendola A, Mazzucconi MG, Testi AM, Pignoloni P, Orlando SM, et al. Thrombocythemia and polycythemia in patients younger than 20 years at diagnosis: clinical and biologic features, treatment, and long-term outcome. Blood. 2012;119:2219–27. https://doi.org/10.1182/blood2011-08-371328.

Grifo AH. Langerhans cell histiocytosis in children. J Pediatr Oncol Nurs. 2009;26(1):41–7. https://doi.org/10.1177/1043454208323915. Epub 2008 Oct 20. PubMed PMID: 18936291

Hait E, Liang M, Degar B, Glickman J, Fox VL. Gastrointestinal tract involvement in Langerhans cell histiocytosis: case report and literature review. Pediatrics. 2006;118(5):e1593–9. Epub 2006 Oct 9. Review. PubMed PMID: 17030599

Henter JI, Horne A, Aricó M, Egeler RM, Filipovich AH, Imashuku S, Ladisch S, McClain K, Webb D, Winiarski J, Janka G. HLH-2004: diagnostic and therapeutic guidelines for hemophagocytic lymphohistiocytosis. Pediatr Blood Cancer. 2007;48(2):124–31. PubMed PMID: 16937360.

Histiocytosis Association of America. The importance of clinical trials in the fight against histiocytosis. 2008. Retrieved 22 Mar 2008, from http://www.histio.org

Hofmann I. Myeloproliferative neoplasms in children. J Hematop. 2015;8(3):143–57. Epub 2015 Aug 2. PubMed PMID: 26609329; PubMed Central PMCID: PMC4655194.

Hoover KB, Rosenthal DI, Mankin H. Langerhans cell histiocytosis. Skelet Radiol. 2007;36(2):95–104. Epub 2006 Oct 7. Review. PubMed PMID: 17028900

Hurford MT, Altman AJ, DiGiuseppe JA, Sherburne BJ, Rezuke WN. Unique pattern of nuclear TdT immunofluorescence distinguishes normal precursor B cells (Hematogones) from lymphoblasts of precursor B-lymphoblastic leukemia. Am J Clin Pathol. 2008;129(5):700–5. https://doi.org/10.1309/ANERT51H38TUEC45. PubMed PMID: 18426728

Jordan MB, Filipovich AH. Hematopoietic cell transplantation for hemophagocytic lymphohistiocytosis: a journey of a thousand miles begins with a single (big) step. Bone Marrow Transplant. 2008;42(7):433–7. https://doi.org/10.1038/bmt.2008.232. Epub 2008 Aug 4. Review. PubMed PMID: 18679369.

Kobayashi S, Iwasaki S. Adult T-cell leukemia/lymphoma (ATL): pathogenesis, treatment and prognosis, clinical epidemiology of acute lymphoblastic leukemia – from the molecules to the clinic, Juan Manuel Mejia-Arangure. IntechOpen. https://doi.org/10.5772/54776. Available from: https://www.intechopen.com/books/clinical-epidemiology-of-acute-lymphoblastic-leukemia-from-the-molecules-to-the-clinic/adult-t-cell-leukemia-lymphoma-atl-pathogenesis-treatment-and-prognosis

Kushwaha R, Ahluwalia C, Sipayya V. Diagnosis of sinus histiocytosis with massive lymphadenopathy (Rosai-Dorfman disease) by fine needle aspiration cytology. J Cytol. 2009;26(2):83–5. https://doi.org/10.4103/0970-9371.55229. PubMed PMID: 21938160; PubMed Central PMCID: PMC3168026.

Makras P, Alexandraki KI, Chrousos GP, Grossman AB, Kaltsas GA. Endocrine manifestations in Langerhans cell histiocytosis. Trends Endocrinol Metab. 2007;18(6):252–7. Epub 2007 Jun 27. Review. PubMed PMID: 17600725

Onciu M, Lorsbach RB, Henry EC, Behm FG. Terminal deoxynucleotidyl transferase-positive lymphoid cells in reactive lymph nodes from children with malignant tumors: incidence, distribution pattern, and immunophenotype in 26 patients. Am J Clin Pathol. 2002;118(2):248–54. PubMed PMID: 12162686

Sidiropoulos M, Hanna W, Raphael SJ, Ghorab Z. Expression of TdT in Merkel cell carcinoma and small cell lung carcinoma. Am J Clin Pathol. 2011;135(6):831–8. https://doi.org/10.1309/AJCPLCB2Q9QXDZAA. PubMed PMID: 21571955

Strauchen JA, Miller LK. Terminal deoxynucleotidyl transferase-positive cells in human tonsils. Am J Clin Pathol. 2001;116(1):12–6. PubMed PMID: 11447741

Sur M, AlArdati H, Ross C, Alowami S. TdT expression in Merkel cell carcinoma: potential diagnostic pitfall with blastic hematological malignancies and expanded immunohistochemical analysis. Mod Pathol. 2007;20(11):1113–20. Epub 2007 Sep 21. PubMed PMID: 17885674

Takeshita M, Akamatsu M, Ohshima K, Suzumiya J, Kikuchi M, Kimura N, Uike N, Okamura T. Angiocentric immunoproliferative lesions of the lymph node. Am J Clin Pathol. 1996;106(1):69–77. PubMed PMID: 8701936

Tavian M, Péault B. Embryonic development of the human hematopoietic system. Int J Dev Biol. 2005a;49(2–3):243–50. Review. PubMed PMID: 15906238.

Tavian M, Péault B. The changing cellular environments of hematopoiesis in human development in utero. Exp Hematol. 2005b;33(9):1062–9. Review. PubMed PMID: 16140155.

Valent P, Horny HP, Escribano L, Longley BJ, Li CY, Schwartz LB, Marone G, Nuñez R, Akin C, Sotlar K, Sperr WR, Wolff K, Brunning RD, Parwaresch RM, Austen KF, Lennert K, Metcalfe DD, Vardiman JW, Bennett JM. Diagnostic criteria and classification of mastocytosis: a consensus proposal. Leuk Res. 2001;25(7):603–25. Review

Writing Group of the Histiocyte Society. Histiocytosis syndromes in children. Lancet. 1987;1(8526):208–9.

Endocrine System

Contents

11.1	**Development and Genetics**	934
11.2	**Pituitary Gland Pathology**	938
11.2.1	Congenital Anomalies of the Pituitary Gland	939
11.2.2	Vascular and Degenerative Changes	939
11.2.3	Pituitary Adenomas and Hyperpituitarism	939
11.2.4	Genetic Syndromes Associated with Pituitary Adenomas	941
11.2.5	Hypopituitarism (Simmonds Disease)	942
11.2.6	Empty Sella Syndrome (ESS)	943
11.2.7	Neurohypophysopathies (Disorders of the Posterior Pituitary Gland)	943
11.3	**Thyroid Gland Pathology**	943
11.3.1	Congenital Anomalies, Hyperplasia, and Thyroiditis	946
11.3.2	Congenital Anomalies, Goiter, and Dysfunctional Thyroid Gland	946
11.3.3	Inflammatory and Immunologic Thyroiditis	948
11.3.4	Epithelial Neoplasms of the Thyroid Glands	949
11.4	**Parathyroid Gland Pathology**	965
11.4.1	Congenital Anomalies of the Parathyroid Glands	965
11.4.2	Parathyroid Gland Hyperplasia	965
11.4.3	Parathyroid Gland Adenoma	966
11.4.4	Parathyroid Gland Carcinoma	966
11.5	**Adrenal Gland Pathology**	967
11.5.1	Congenital Anomalies of the Adrenal Gland and Paraganglia	967
11.5.2	Dysfunctional Adrenal Gland	969
11.5.3	Adrenalitis	969
11.5.4	Neoplasms of the Adrenal Gland and Paraganglia	974
11.5.5	Syndromes Associated with Adrenal Cortex Abnormalities	996
	Multiple Choice Questions and Answers	997
	References and Recommended Readings	998

11.1 Development and Genetics

The *pituitary gland*, aka hypophysis, is a small endocrine bioanatomic structure, weighing 0.35–0.9 g, and located in the sella turcica of the sphenoid bone (neurocranium). The pituitary gland is composed of two portions, which are classified on a sagittal plane, i.e., anterior pituitary gland and posterior pituitary gland, as well as a vestigial intermediate lobe. The anterior portion constitutes about 3/4 of the gland, while the posterior portion represents about 1/4 of the gland. The histology of the pituitary gland shows small nests and cords of round uniform cells finely demarcated by a rich vascular network. The different tinctorial characteristics of the cells of the pituitary gland gave the original name to these cells, using the terms of acidophilic cells, basophilic cells, or chromophobic cells; the latter was used if no staining was present. The use of immunohistochemistry allows the classification of these cells according to the specific hormones they produce. It has been scrutinized that ~1/5 of the cells of the anterior pituitary gland lacks any immunoreactivity. These cells are subclassified as nonsecretory cells. Conversely, the histology of the posterior portion of the pituitary gland reveals an intricate framework of nerve fibers with the support of glial cells. These nerve fibers are the axons of hypothalamic neurons and are unmyelinated. The ultrastructural examination of these unmyelinated nerve fibers discloses membrane-bound secretory granules, which are composed of antidiuretic hormone (ADH) or oxytocin. Neurophysins, which are specific binding proteins, make some complexes with either ADH or oxytocin. The hypothalamus is central in neuroendocrinology and controls both the anterior and the posterior portions of the pituitary gland. The hypothalamic control is, however, different. The anterior portion receives releasing and inhibiting hormones by the hypothalamus using the portal venous system, while the posterior part of the pituitary gland is directly under hypothalamic control because neurons in the hypothalamic cells secrete ADH and oxytocin. These hormones migrate down the axons in the pituitary stalk for storage and then are blood-released by the posterior pituitary. The cytologic diversity of the pituitary gland is impressive and may look like a symphony orchestra, because the several instruments play a role for proper functionality of the organism. The abnormal performance of a single "instrument" may considerably jeopardize the final "sound" of the entire organism (Box 11.1).

Box 11.1 Pituitary Gland Cyto-diversity

Cytologic phenotype	Hormone type	Action
Somato*troph*	GH	Body growth Insulin-antagonist
Cortico*troph*	ACTH	Adrenal gland stimulation
Lacto*troph*	Prolactin	Mammary gland proliferation and function
Thyro*troph*	TSH	Thyroid hormones synthesis and secretion
Gonado*troph*	FSH/LH/ICSH	FSH: pre-ovulatory phase, estrogen secretion, ovulation LH: corpus luteum, progesterone secretion, ovulation
Null	None	
Hypothalamic nuclei	ADH	Distal nephron H_2O resorption and arteriolar constriction
Hypothalamic nuclei	Oxytocin	Smooth muscle contraction in uterus and ducts of the mammary gland

Note: GH or growth hormone is also called somatotropin. During corticotropin synthesis, three additional peptidic hormones are split off, including beta-lipotropic hormone, beta-endorphin, and alpha-melanocyte-stimulating hormone (alpha-MSH). Beta-lipotropic hormone functions. β-endorphin has endogenous opiate action, while alpha-MSH facilitates the dispersion of melanin in the skin. ACTH, adreno-corticotropin hormone; the same gonadotroph cells produce both follicle-stimulating hormone (FSH) and luteinizing hormone (LH). In the male organism, the interstitial cell-stimulating hormone (ICSH) works as a male counterpart for LH. ICSH stimulates the Leydig cells of the testis to secrete androgens. ADH, antidiuretic hormone

11.1 Development and Genetics

The *thyroid gland* is an endocrine organ. It develops from a tubular pouch of the embryonic pharynx (*foramen cecum* of the tongue), the so-called thyroglossal duct that migrates downward into the neck. In this location, the duct is an important bio-anatomic structure in the development of the thyroid gland. Thyroid gland grows downward in front of trachea and thyroid cartilage. The distal end of this primordial organ becomes the adult gland, while the proximal portion regresses progressively by the 5th to 7th week of gestation. This primordium is also joined by tissue derived by the 4th and 5th pharyngeal pouches, which contribute to the parafollicular cells, while the hyoid bones forms from the 2nd branchial arch. Anatomic thyroid gland milestones are, thus, identified the thyroid diverticulum and the thyroglossal duct (stalk), which eventually disappears. In case of persistence of the thyroglossal duct, there will be an ectopic glandular tissue of thyroid. The fate of this tissue is variable. It can stay silent or become cystic, inflamed, as well as, more rarely, neoplastic.

When there is the arrest of the downward migration of the thyroglossal duct, it happens that some ectopic thyroidal tissue is found along this migration (*thyroid gland ectopia*). Some forms are clinically recognized, including *lingual thyroid* and *thyroglossal midline remnant*. The former structure is characterized by the presence of thyroidal tissue at the root of the tongue. The latter structure is a remnant of the thyroglossal duct fund along its pathway of descent. The lingual thyroid is very rare, while thyroglossal duct remnants are more often recognized. The thyroglossal duct remnant may become cystic (*thyroglossal duct cyst*) when epithelial remnants are allowed to secrete with time (late childhood and youth). The lining of the thyroglossal duct cyst is constituted by either squamous or ciliated (respiratory) epithelium, and the wall of the cyst is made up of thyroid tissue. Most commonly, thyroglossal duct cysts are found between the hyoid bone and the isthmus of the orthotopic thyroid gland, and the pediatric surgeon wisely excises the hyoid bone as well to avoid missing ectopic thyroidal tissue. The persistence of a cyst may be prone to be infected and develop into an abscess (*thyroglossal duct abscess*) and the patient can develop local signs of infection but also fever and, potentially, septicemia. Two lateral lobes constitute the thyroid gland, which are joined across the midline by a bridge called isthmus (*isthmus glandulae thyroideae*). Also, there is a small lobe or pyramidal lobe, which extends upward from the isthmus. The pyramidal lobe is the vestigial point of attachment of the thyroglossal duct embryologically digging. At examination, the normal thyroid gland is firm, red-brownish, and smooth with a fibrous capsule that merges with the deep cervical fascia. Histology is variable from the gestational age, but once the thyroid gland has reached the maturity at birth, it is almost indistinguishable from healthy young adult size. The hematoxylin-eosin staining of thyroid gland tissue shows firmly packed follicles separated by a quite rich vascular supply with little intervening stroma. The rich vascular texture is key to the definition of an endocrine gland differently from the poor or normal vascular texture identified in exocrine glands. Endocrine glands secrete into blood, while exocrine gland secrete in a non-blood filled cavity. The follicles are distinctly lined by cuboidal epithelial cells, and their lumens are filled with colloid, which contains thyroglobulin and thyroid hormones in a proteinaceous matrix. Moreover, the parafollicular or C cells are dispersed between the thyroid follicles and secrete calcitonin.

In the thyroid gland, follicles vary in size: ~200 μm, but this size should be considered only cautiously. A single layer of cuboidal-columnar cells is seen, but epithelium may look like quite flat at places. Hürthle or Askanazy cells are acidophilic thyrocytes with abundant mitochondria. Thyroid immunoreactivity displays T3, T4, LMW-CKs, EMA, and VIM. Thyroglobulin (TGB or TG) is a 650 kD protein synthesized and stored extracellularly as "colloid" with or without the presence of calcium oxalate crystals (1/2 of individuals). Tyrosyl residues are iodinated to monoiodotyrosine (MIT) and diiodotyrosine (DIT) and MIT and DIT couple to build T3 and T4, and in response to TSH, T3 and T4 are collected in pinocytotic vesicles and released into the blood. Thyroxine-binding protein (TBP) binds both T3 and T4, although T4 binds tighter, which means, in endocrinological investigations and discussions, is less available in plasma. Parafollicular C cells are neural crest-derived cells, which are located between and within follicles (<10/LPF, LPF = 10X objective, i.e., less than 10 cells per LPF) and found in the upper and middle portions of the lateral lobes, mostly. Immunoreactivity of the parafollicular C cells shows (+) calcitonin, CGA, and CEA (of note immunoreactivity is more prominent in hyperplasia or neoplasia). Parafollicular C cells secrete calcitonin inhibiting osteoclasts in response to hypercalcemia.

The understanding of the thyroid function is vital for the well-being of fetuses, children, and adults. There are "hormone regulation feedback mechanisms" that aim to regulate the concentrations of hormone molecules in the blood and fluids of an organism as an important part of homeostasis. As seen in other endocrine structures, the main body systems involved in hormone regulation by feedback mechanisms are the nervous system, which is directly involved in emitting signals input to and output from the control center or chemical signals, and the endocrine system, which produces and secretes hormones. A feedback mechanism (aka a "feedback system" or a "feedback loop") is central to understand thyroid function as well as other endocrine organs. The feedback loop is a cycle of events in which the state of a specific aspect of the body's condition, which is specifically labeled as a "controlled condition" (e.g., temperature), is continually monitored and accustomed as appropriate to keep the value of that "controlled condition" within a range, which is considered safe from the same organism. Feedback loops are an ideal means of controlling hormone levels because they involve constant monitoring and making adjustments to keep hormone levels stable (e.g., thyroid hormones). This system works perfectly because hormones undergo degradation quite quick. It is important to keep in mind that low concentration of hormones are useful but potentially life-threatening. The pituitary thyrotropin (TSH) and its level control the rate of synthesis of thyroid hormones in the blood. TSH is crucial in regulating all the steps in the synthesis of thyroid hormone. The hypothalamic release of thyrotropin (TSH)-releasing hormone (TRH) stimulates the adenohypophysis directly to release thyroid-stimulating hormone (TSH), which in turn stimulates iodine trapping by follicle cells. In the thyroid gland, the iodide is oxidized to iodine in the follicle cells following trapping. Following oxidation, iodine is incorporated into tyrosine molecules to form monoiodotyrosine (MIT) and di-diiodotyrosine (DIT). Both MIT and DIT are linked to the protein, thyroglobulin, and stored in the colloid. Finally, MIT and DIT are coupled to form triiodothyronine (T_3) and tetra-iodothyronine or thyroxine (T_4) using an enzymatic driven step. Being $T_4:T_3$ ratio about 10:1, T_4 is considered the primary hormone secreted by the thyroid gland. In the plasma, the $T_4:T_3$ rate is 40:1, and T_4 and T_3 are transported in the plasma bound to the proteins, thyroxine-binding globulin (TBG) and thyroxine-binding prealbumin (TBPA). There is an equilibrium between the hormonal forms bounded to proteins and the free molecules. The $T_4:T_3$ ratio shows the higher affinity of the binding proteins to thyroxine and its slower metabolism in comparison with tri-iodothyroxine.

The *parathyroid glands* are at least four in ≥25% of subjects. More than four may be found in some individuals, while less than four is rarer. Each parathyroid gland is 4 × 3 × 1.5 mm with a weight of about 30 mg (together 120 mg). Histologically, two types of cells are seen including chief cells (main cell type) and oxyphil cells (minor cell type). The chief cells have a centrally located nucleus, pale granular cytoplasm, and ill-defined cell borders, while oxyphil cells have more abundant cytoplasm with oncocytic character (mitochondria-rich). Clear cells or "Wasserhelle" cells of the German literature are not present in the normal parathyroid gland. Of note, oxyphil cells are evident soon after puberty and increase in number in adulthood forming islands at the age of 40 and are probably derived from chief cells. The adolescence is also essential because at this age begins fat infiltration of the gland. Embryological studies evidence that the upper pair derives from the 4th branchial cleft, while the lower couple derives from the 3rd branchial cleft. The top pair of parathyroid glands descends into the neck with the thyroid gland and resides in the middle third of posterolateral part of the thyroid gland, while the lower couple descends with thymus to a position near inferior thyroid artery at the lower pole of the thyroid gland. The parathyroid glands produce parathormone (PTH), which regulates calcium metabolism in combination with vitamin D and calcitonin. PTH is secreted in response to hypocalcemia. PTH increases renal excretion of phosphate, increases renal and intestinal reabsorption of calcium, and activates osteoclasts via osteoblasts being the receptors on the surface of the osteoblasts. At places, parathyroid gland architecture may disclose follicles that need to be differentiated from thyroidal follicles. The former are indeed glycogen-rich (PAS (+) of parathyroid follicles) and lack oxalate crystals.

The *adrenal gland* is a bilateral organ essentially for the full intrauterine and extrauterine life of the organism. The adrenal glands are paramedian neuroendocrine capsulated organs (4–6 g, 5 × 3 × 1 cm) constituted by an external structure or *cortex* (three zones) producing aldosterone, cortisol, and testosterone and an internal structure or *medulla* (<10% of gland volume, 1% in newborns) responsible for the production of catecholamines. Vascular supply is from the aorta, inferior phrenic arteries, and renal arteries. Vein drainage is to adrenal veins. Lymphatics are found in the capsule only. The adrenal gland cortex displays three zones ("GFR", which may recall the medical abbreviation for "glomerular filtration rate"):

- *Zona glomerulosa* (ZG, outer, ~10%) constituted by small clusters and short trabeculae of relatively small, well-defined cells with less cytoplasm than other cortical cells, producing mineral-corticoids, lipid-poor (TEM: sparse intracellular lipid, elliptical mitochondria with lamellar or platelike cristae, no/very rare lysosomes, lipofuscin, and smooth endoplasmic reticulum)
- *Zona fasciculata* (ZF, middle, ~80%) with broad bands of large cells with distinct membranes arranged in cords two cells wide, numerous cytoplasmic small lipid vacuoles which may indent the central nucleus and resemble lipoblasts (round/ovoidal/oval mitochondria with short and tubular and/or tubulovesicular cristae, prominent lipid droplets, smooth and rough endoplasmic reticulum, microvillus cytoplasmic projections, and lysosomes)
- *Zona reticularis* (ZR, inner, ~10%) with haphazardly arranged small cells with granular and eosinophilic cytoplasm with lipofuscin but minimal lipid (spherical/ovoid mitochondria with short and long tubular invaginations of the inner membrane, abundant lipofuscin granules, lysosomes and microvilli, sparse lipid droplets)
- *Cortex reactivity*: (+) LMW-CK (CAM 5.2), INA, Melan-A/A103, BCL2, SYN (weak), VIM (~); (−) CGA, epinephrine.
- *Comparative studies*: In the *rat*, ZG and ZF are separated by an undifferentiated zone, while in *mouse* there is no undifferentiated zone separation and capsular mesenchyme cells have properties of adrenocortical stem/progenitor cells.
- *Medulla*: Nests and cords of neural crest-derived *chromaffin cells* or pheochromocytes or medullary polygonal cells (CGA+, S100-) with poorly defined borders, abundant granular and basophilic cytoplasm (TEM: numerous mitochondria, 200–250 nm neurosecretory granules with prominent halo and dense core), and delimited by *sustentacular spindle cells* (CGA-, S100+) (TEM: no neurosecretory granules), and scattered *ganglion cells*, which are the differentiated forms of primitive neuroblasts.
- *Medulla reactivity*: (+) CGA, SYN, NF, TH, epinephrine, S100; (−) LMW-CK (CAM 5.2), VIM (CGA, chromogranin A; LMW-CK, low molecular weight cytokeratins; CAM, cell adhesion molecule; NF, neurofilaments; VIM, vimentin; TH, tyrosine hydroxylase).

Adrenocortical steroidogenesis has the target to produce mineral-corticoid and glucocorticoids and sex steroids. Aldosterone, the principal mineral-corticoid, is controlled by the renin-angiotensin system (RAS), while cortisol, the primary glucocorticoid, is stimulated by the hypothalamus-pituitary system involving CRP and ACTH. Sex steroids, including androgens, estrogens, and progestagens, are a subset of sex hormones with the aim to produce sex differences or support reproduction.

The mesoderm and the neural crest contribute to the development of the adrenal gland which starts during the fourth week (p.c.). The mesodermal ridge that produces the adrenal cortex is in the same embryological field that provides the gonads. Bilateral adrenal primordium develops as cords of large polyhedral cells in coelomic epithelium medial to mesonephros and urogenital ridge by day 25 of gestation. Following the formation of the neural tube, neural crest cells migrate from the CNS and aggregate in a mass that will become the medulla. From the mesodermal ridge and through repeated division cells surround the neural crest cells to form the cortex, which encapsulates the medulla, a process that continues until the end of the 3rd year of age. It is important to remember that replicating paraganglionic cells differentiating into chromaffin cells (first sympathicoblasts) form neuroblastic nodules that peak at weeks 17–20 and after that usually regress. At birth, zona glomerulosa and zona fasciculata are present, but zona reticularis is built during the 3rd year of life. The neonatal adrenal gland shows a rather marked decrease in weight (about 2 g) in the first months of life due to the regression of the fetal (or provisional) cortex. The Wilms' tumor gene (*WT1*) and member 4 of the wingless-type mouse mammary tumor virus integration site family (*WNT4*) play an early role in the ontogenesis of the adrenal gland. Adrenal gland developmental genes include *SF-1, SOX8, SOX10, DAX1,* and *CYP17*. The most significant gene is probably the coding information behind the 53 kDa protein adrenal 4-binding protein (Ad4BP) or *steroidogenic factor 1* (*SF-1*). This site is a classical DNA-binding domain (DBD) characterized by two Cys2-Cys2 zinc fingers properly located in the N-terminal area. SF-1 binds DNA as a monomer and has high homology with the Drosophila Ftz-F1, which has been brilliantly characterized as a transcription factor that controls *fushi tarazu* homeotic gene expression. Other essential genes are constituted by Sry-box (Sox) 8 (*SOX8*), and *SOX10*, which are expressed in the neural crest and the cells migrating to the adrenal gland, *DAX1*, and *CYP17*. *DAX1* which is an acronym for "dosage-sensitive sex reversal, adrenal hypoplasia critical region, on chromosome X, gene 1" is also known as *NROB1* (nuclear receptor subfamily 0, group B, member 1). *DAX1* encodes a nuclear receptor protein, which is expressed in embryonic stem cells, steroidogenic tissues (adrenal glands and gonads), the ventromedial portion of the hypothalamus, and gonadotropic pituitary gland. Individuals harboring *DAX1* mutations develop an X-linked syndrome, which is referred to as adrenal hypoplasia congenita (AHC). CYP17 is an acronym for cytochrome P450 17A1. It is also called steroid 17α-monooxygenase, 17α-hydroxylase, 17,20-lyase, or 17,20-desmolase. CYP17 is an enzyme of the hydroxylase type that in humans is encoded by the *CYP17A1* gene located on chromosome 10. CYP17 is ubiquitously expressed and localizes to the endoplasmic reticulum. The cytochrome P450 superfamily proteins, of which CYP17 is a member, are generally viewed as monooxygenases that catalyze several reactions involved in drug metabolism and synthesis of cholesterol, steroids, and nonsteroid lipids. SHH signaling regulates development of the adrenocortex and identifies progenitors of steroidogenic lineages. The fetal cortex produces a steroid precursor (DEA), which is converted by placenta into estrogen, and fetal adrenal hormones influence remarkably lung maturation. In the 2nd and 3rd trimesters, a steroid precursor dehydroepiandrosterone (DHEA) and sulfated derivative (DHEAS) are produced, which are converted by placenta into estrogen. In summary, the fetal zone is throughout gestation and expresses enzymes required for DHEA-S synthesis, while the transitional zone is initially identical to the fetal region but at 25–30 weeks expresses enzymes for the glucocorticoid synthesis. The definitive zone occurs at 22–24 weeks. After this time there is expression of mineral-corticoid enzymes. Human male neonates produce high levels of DHEA and DHEAS, which decay within some months of birth, due to regression of the fetal zone. In older children and adults, ZR is responsible for the production of DHEA and DHEAS. ZF is accountable for cortisol, corticosterone, and cortisone, while ZG is accountable for aldosterone.

11.2 Pituitary Gland Pathology

Although the pituitary gland seems to play a significant role during the process of ontogenesis, it is now clear that the pituitary gland is essential also for all life functions. It also seems evident that an efficient and effective network of the pituitary

11.2 Pituitary Gland Pathology

> **Box 11.2 Pituitary Gland Pathology**
> 11.2.1. *Congenital Anomalies*
> 11.2.2. *Vascular and Degenerative Changes*
> 11.2.3. *Pituitary Adenomas and Hyperpituitarism*
>   11.2.3.1. Classification of Pituitary Adenomas
>   11.2.3.2. Genetic Alterations of Pituitary Tumors
>   11.2.3.3. Pituitary Tumor Morphology
>   11.2.3.4. Clinical Course of Pituitary Neoplasia
> 11.2.4. *Genetic Syndromes associated with Pituitary Adenomas*
> 11.2.5. *Hypopituitarism (Simmonds Disease)*
> 11.2.6. *Empty Sella Syndrome*
> 11.2.7. *Neurohypophysopathies*
>   11.2.7.1. Diabetes Insipidus
>   11.2.7.2. ADH-Hypersecretion

gland with other endocrine organs as well as with the brain is vital. In almost all cases of anterior pituitary hypersecretion, the primary hyperfunction is due to benign neoplasms (adenoma). The role of malignant tumors (carcinomas) or hyperplasia of pituitary cells is minimal in inducing anterior pituitary hypersecretion. Although rare in children, pituitary adenomas may occur in youth with a peak between 20 and 50 years of age. Curiously, there is a somewhat skewed male prevalence, although the reason for it is not apparent. Pituitary adenomas constitute 1/10 of all primary intracranial neoplasms. In Box 11.2 are listed the topics of pituitary gland pathology that are going to be highlighted in the next paragraphs (Box 11.2).

The abnormalities of the ontogenesis about the pituitary gland are particularly crucial for dysmorphologists as well as endocrine pediatricians or auxologists. Auxology is indeed highly relevant to the pediatric pathologist, targeting the study of all aspects of human physical growth involving physiology, pediatrics, internal medicine, endocrinology, and epidemiology and the involvement of pituitary pathology is the target of several investigations in different centers worldwide.

11.2.1 Congenital Anomalies of the Pituitary Gland

Anomalies of the ontogenesis, mostly involving the absence of the gland.

11.2.2 Vascular and Degenerative Changes

Anomalies of the vascularization and degenerative changes may take place during the peripartum or in states of malnutrition. Interestingly, some correlation with birth delivery has been suggested (Maghnie et al. 1991). In studying 37 patients with idiopathic hypopituitarism by MRI, of whom 12 had multiple pituitary hormone deficiencies (MPHD) and 25 isolated growth hormone deficiency (IGHD), researchers of the Pediatric Department, University of Pavia, IRCCS Policlinico S. Matteo, Italy, defined a group of patients with congenital defects, such as anterior pituitary hypoplasia, stalk agenesis, and ectopic posterior pituitary gland who had breech presentation at birth. This group contained subjects born by breech delivery who presented with MPHD and subjects born by C-section or normal delivery who presented with IGHD only.

11.2.3 Pituitary Adenomas and Hyperpituitarism

The anterior lobe of the pituitary gland represents the most common location of adenomas of the pituitary gland. This benign tumor is the most common etiology of hyperpituitarism, which is manifested by the hypersecretion of the hormones produced from the neoplasia. The hormones produced by the neoplastic cells that can be highlighted by immunohistochemical methods are the basis for the formation of a functional classification of pituitary adenomas, which is routinely used by neurologists and oncologists. The usual clinical constellation of adenomas of the pituitary gland is the secretion of only one hormone, but some pituitary adenomas can be bi-hormonal with GH and prolactin being the most common combination and, in some cases, even pluri-hormonal. In some rare occurrences, pituitary carcinomas and hypothalamic disorders may be at the basis of hyperpituitarism. Rarely, hypopituitarism instead of hyperpituitarism may be encountered. This situation is the case when the tumor invades and destroys the adjacent parenchyma of the anterior pituitary gland. Although the age peak is between the fourth and sixth decades, pituitary adenoma may also occur in children and adolescents. The most recent epidemiological survey for tumors in childhood and adolescence indicates that pituitary tumors are still uncommon. They are <3% of childhood supratentorial tumors and between 3% and 6% of all surgically treated adenomas. The survey studies also contain different rates because the definition of adolescence is variable between studies, with the upper limit for age ranging from 17 to 20 years, without considering the youth that may include young adults up to 40 years of age. Nearly all cases of anterior pituitary hypersecretion (*hyper-secretive adeno-hypophysopathies*) are due to primary hyperfunction caused by a pituitary adenoma. In childhood, the prolactinoma (*vide infra*) is the most common hormonally active pituitary adenoma (Box 11.3).

The following paragraphs will be highlighting the classification of pituitary adenomas the pituitary tumor morphology, and clinical course of pituitary neoplasia.

11.2.3.1 Classification of Pituitary Adenomas

The Box 11.4 is the updated classification of hormone-based pituitary adenomas. In the setting of GH-producing adenoma, two clinical syndromes include either gigantism in children or acromegaly in adults. In the background of hyperproduction of ACTH, either Cushing syndrome or Nelson syndrome may be the consequences. In the case of hyper-gonadotropin production, hypogonadism and hypopituitarism are found and are classified as nonfunctional. Moreover, TSH cell production is the cause of hyperthyroidism, while combined features of GH and prolactin excess are found in mamma-somato-trophic pituitary adenomas. Functional pituitary adenomas include somato-, cortico-, lacto-, mammosomato-, and thyrotrophic. Immunohistochemical tools and electron microscopy play a significant role in the identification of the cell phenotype.

Another classification criterion is the size, which consists in designating *microadenomas*, tumors less than or equal to 1 cm in Ø, and *macroadenomas*, tumors exceeding 1 cm in Ø. Microadenomas or better pituitary "incidentalomas" may be found at

> **Box 11.3 Function-Based Subdivision of Pituitary Adenomas**
> 1. *Functional pituitary adenoma*: Pituitary adenoma associated *with* an excess hormone and related clinical manifestations
> 2. *Nonfunctional pituitary adenoma*: Pituitary adenoma with the immunohistochemical and ultrastructural identification of hormone neo-production at the tissue level, but *without* clinical evidence of hormone excess

> **Box 11.4 Classification of Pituitary Adenomas**
>
> | Hormone | Tumor | Clinical syndrome |
> |---|---|---|
> | GH | GH-producing | Gigantism/acromegaly |
> | ACTH + other POMC-derived peptides | ACTH-producing (basophilic) | Cushing/Nelson |
> | Prolactin | PLT-producing | Galactorrhea, amenorrhea, sexual dysfunction, infertility |
> | Prolactin, GH | Mammasomatotroph | GH + PL symptoms |
> | FSH, LH | Gonadotroph | Hypogonadism and hypopituitarism |
> | TSH | Thyrotroph-producing | Hyperthyroidism |
>
> Notes: *ACTH* adrenocorticotrophic hormone, *FSH* follicle-stimulating hormone, *GH* growth hormone or somatotropin, *LH* luteinizing hormone, *PLT* prolactin, *POMC* proopiomelanocortin, *TSH* thyroid-stimulating hormone

autopsy with a ratio of about 15% for a total number of autopsies as indicated following meta-analysis of very large autopsy studies. Some authors prefer to give the cut off of 10 mm to macroadenoma. Thus, it is very important to correletae with both the radiologist and the endocrinologist. Genetic alterations of pituitary tumors are also significant, and both losses of function and gain of function may be found in pituitary adenomas.

11.2.3.2 Pituitary Tumor Morphology

Pituitary tumors may be quite challenging when investigating the morphology. The characterization of cell diversity requires either immunohistochemical or ultrastructural study. The latter method points to differences of types of granules, while the first method, which is used today, identifies the cell phenotype using monoclonal antibodies raised against a specific hormone.

Differentiating pituitary adenomas from pituitary hyperplasias may be quite challenging. This task is performed using a histochemical special stain, called reticulin. While reticulin displays an intact network with focal enlarged acini in hyperplasia, the acinar architecture is destroyed in pituitary adenomas.

11.2.3.3 Clinical Course of Pituitary Neoplasia

The clinical manifestations seen in the setting of pituitary adenomas may be divided into effects resulting from the local growth of the neoplasm, i.e., the mass effects ("tumor quia tumet"), and those resulting from hormone secretion. The tumor size and the tumor invasive capability are factors influencing the first category, while the latter depends on the type of hormone secreted.

11.2.4 Genetic Syndromes Associated with Pituitary Adenomas

Dysmorphology changed enormously since the human genome has been typed. Numerous syndromes with multiple congenital anomalies (MCAs) have been successfully targeted and genetically characterized. Local effects include standard settings with superior quadrantic bitemporal hemianopia. Of particular importance is the distinction between pituitary adenomas that occur in the context of internal medicine and pituitary adenomas occurring in syndromes with MCAs.

The crucial hypersecretion-related effects of pituitary adenomas are (1) *Hypophysis Apoleia*, i.e., destruction of normal pituitary gland cells with hypopituitarism and diabetes insipidus (the Greek term ἀπώλεια (Gr. ἀπόλλυμι, "cut off") is particularly useful, because it means destruction, causing someone or something to be completely severed – cut off from what could or should have been); (2) *Infra-sellar Expansion*, i.e., sella turcica expansion (X-ray); and (3) *Suprasellar Expansion* (through the sellar diaphragm), promoting:

- Visual field defects by compression of optic chiasm or nerves (e.g., superior quadrantic bitemporal hemianopia or bitemporal hemianopia)
- Diabetes insipidus by compression of the hypothalamus
- ICP (intracranial pressure) increase and hydrocephalus by interference with an outflow of CSF
- A dull headache by compression of the intracranial blood vessels
- Deficits of cranial nerves I–XII, other than optic, by compression of their pathway
- Deficiencies or seizures related to the invasion of brain structures
- Chronic rhinorrhea, sinusitis, and meningitis by invasion into the paranasal sinuses
- Intracranial hemorrhage-related symptoms and thrombosis with orbital edema and congestion due to expansion into the cavernous sinus

Prolactin is the most common hormone produced by a pituitary adenoma. Hyperprolactinemia causes amenorrhea, infertility, and galactorrhea in women and a decrease of libido, impotence, and galactorrhea in men. Galactorrhea, which is specifically defined as milk secretion in the absence of pregnancy in women, may be alarming in youth and quickly denied. Hyperprolactinemia due to hormone excess needs to be distinguished by clinical mimickers, including hypothalamic diseases in which there may be decreased production of prolactin inhibiting factor and adverse drug reactions with toxic response to drugs that block dopaminergic transmission (e.g., methyldopa and reserpine). The differential diagnosis may be quite challenging, mainly because prolactin-secreting tumors are commonly microadenomas. Hypersomatotropinemia causes different phenotypes according to the age. In children, gigantism is the rule and is due to excessive uniform (massive but proportionate) bone growth at the epiphyses, while in adults, the fused growth plates allow generalized enlargement of bones without increase of height. This pattern is known as acromegaly, which is visible mainly in the hands, jaw, and skull. Cartilage of the nose and ears is also involved, and osteoarthritis by the involvement of the vertebral column is also seen. The facial features may be quite distinctive with coarsening of the physiognomy, which is an individual facial feature or expression. Also, enlargement of the internal organs, including the heart, liver, kidneys, adrenals, thyroid, and pancreas, is also observed. Compression-related atrophy of the residual normal (non-neoplastic) pituitary cells is also seen, and patients may show impotence (male), amenorrhea (female), and infertility (male and female). From the metabolic point of view, a frequent problem is secondary diabetes mellitus due to the antagonistic effect of GH on insulin, which is a phenomenon that is correctly identified in about 1/10 of patients. In nearly half of the patients with hypersecretion of GH may be found abnormalities in the oral glucose tolerance test. The lack of suppression of hypersomatotropinemia following glucose administration has a diagnostic value. Hypersecretion of ACTH by a pituitary adenoma has a direct effect on adrenal glands with hyperplasia of the *zona fasciculata* determining increased cortisolemia. The phenotype is Cushing syndrome, which is caused by the lack of feedback inhibition. In these patients, increased skin pigmentation is also seen and linked to increased production of MSH or the melanogenic effect of hyper-corticotropinemia. Diagnostic value relies on serum levels of both cortisol and ACTH. Differential diagnosis should incorporate ectopic ACTH production, e.g., lung carcinoma. Nelson's syndrome is rarely encountered today but remains a critical endocrine imbalance to consider in unique settings. Bilateral adrenalectomy performed in the past for patients believed to have Cushing syndrome of adrenal origin allows a sudden decrease of cortisolemia. It is targeting any residual partial feedback inhibition of the tumor leading to hyperproliferation of pituitary adenoma cells.

TRT: The surgical removal of a pituitary adenoma may be performed through the nasopharynx (transsphenoidal approach) or the skull (craniotomy approach). Following removal of the pituitary adenoma, recurrence is seen in 1/10 of patients. Occasionally, locally aggressive behavior (rarely metastatic) has also been described, and radiation therapy has been advised. The dopamine agonist bromocriptine is a medical therapeutical approach for microadenomas, being this type of pituitary adenomas rarely able to progress into macroadenomas.

11.2.5 Hypopituitarism (Simmonds Disease)

Hypopituitarism remains a rare disease, although in the past when the postpartum hemorrhagic shock was more common, the ischemic necrosis of the gland was seen. This condition, aka Sheehan's syndrome, was due to hyperplasia of the pituitary gland during pregnancy. In developed countries, nonfunctional neoplasms involving the *sella turcica* represent the most prevalent cause of hypopituitarism. Apart from nonfunctional pituitary adenomas, which are rare in pediatrics, craniopharyngioma is the most common neoplasm. If hypopituitarism develops early in life, pituitary dwarfism is seen. The causes of hypopituitarism are sella turcica tumors (nonfunctional adenoma, craniopharyngioma, chordoma, Langerhans cell histiocytosis). Non-neoplastic hypopituitarisms entail (1) ischemic necrosis (postpartum, trauma, angiopathies, diabetes mellitus), (2) chronic inflammation (tuberculosis, syphilitic gumma, neurosarcoidosis), (3) infective causes, (4) infiltrative diseases (genetic hemochromatosis, hemosiderosis, mucopolysaccharidoses, amyloidosis), (5) intrasellar cysts (non-hydatids cysts), (6) empty sella syndrome, and (7) Laron dwarfism (deficiency of end-organ growth receptors). The replacement of the deficient hormones represents the treatment of hypopituitarism, apart from the instances where a neoplastic process is the underlying etiology. In these cases, surgical removal is the therapy of choice. Morris Simmonds was a German physician and pathologist, who was born in St. Thomas (formerly Danish West Indies and now the United States Virgin Islands) and emigrated accompanying his family to Hamburg, Germany.

11.2.6 Empty Sella Syndrome (ESS)

It is shrinkage or flattening of the pituitary gland creating a void in the *sella turcica* with the anatomic structure called *Sella* being occupied by an arachnoid herniation containing cerebrospinal fluid. It is differentiated in primary and secondary ESS.

Primary ESS occurs when an anatomical defect above the pituitary gland increases pressure in the *sella turcica* and causes the gland to flatten and is associated with obesity and blood hypertension in women, possibly representing a sign of idiopathic intracranial hypertension ↑ ICP. *Secondary ESS* follows regression of the pituitary gland within the cavity after an injury, surgery, or radiation therapy. Importantly individuals with secondary ESS due to the destruction of the pituitary gland have symptoms that reflect the loss of pituitary gland functions, such as the ceasing of menstrual periods, infertility, fatigue, intolerance to stress, and infection.

11.2.7 Neurohypophysopathies (Disorders of the Posterior Pituitary Gland)

11.2.7.1 Diabetes Insipidus

The diabetes insipidus is a failure of the hypothalamus and posterior pituitary gland to secrete ADH with consequent deficiency of hydro-resorption in the collecting tubules of the kidney leading to the excretion of the increased amount of urine (polyuria) and hyperosmolality in serum, which points the organism to excessive water intake (polydipsia). Diagnosis is based on the water deprivation test, and therapy relies on the removal of the underlying etiologic disorder. There are some conditions interfering with the hypothalamus-pituitary axis, including hypothalamic neoplasms (primary or secondary), Langerhans cell histiocytosis (Hand-Schüller-Christian disease), neoplastic disruption of the pituitary stalk (metastasis from intra- or extraaxial tumors), pituitary neoplasms (adenomas >> carcinomas), post-infectious disruption of the pituitary stalk (e.g., tuberculous meningitis), and post-traumatic disruption of the pituitary stalk (e.g., sport- or car accident-related). The eponym "Hand-Schüller-Christian disease" is reserved to multifocal LCH, which is usually characterized by a triad of exophthalmos, lytic bone (e.g., skull) lesions, and diabetes insipidus due to the infiltration of the pituitary stalk by Langerhans cells. The American pediatrician Alfred Hand Jr., the Austrian neurologist and radiologist Arthur Schüller, and the American internist Henry Asbury Christian have described the presentation of this disease.

11.2.7.2 ADH Hypersecretion

The hypersecretion of the antidiuretic hormone may have several underlying etiologies and may also play a role in childhood and youth. A high serum level of ADH causes hydro-retention in the collecting tubules of the kidney and excretion of concentrated urine, leading to hypoosmolality in serum (<275 mosm/kg) and hyponatremia. The latter metabolic imbalance causes weakness, confusion, lethargy, convulsions, and coma.

The list of causes of excessive secretion of ADH includes (1) *Schwartz-Batter syndrome* or *syndrome of inappropriate secretion of ADH (SIADH)* (adrenal gland insufficiency, cerebral neoplasms, drugs, head trauma, liver cirrhosis, tuberculosis, pneumonia, and thyroid gland insufficiency) and (2) *paraneoplastic syndromes* or *ectopic ADH neoplastic disorders* (carcinoma of the pancreas and small-cell carcinoma of the lung).

11.3 Thyroid Gland Pathology

Clinical syndromes relative to dysfunctional thyroid gland include hypothyroidism, hyperthyroidism, and goiter. *Hypothyroidism* or "cretinism" as congenital untreated hypothyroidism (physical and mental retardation during development) and *myxedema* (lethargy, cold intolerance, apathy, dry skin, edema in older child/adult) are due to several causes (Figs. 11.1, 11.2 and 11.3). The word "cretinism" arises from

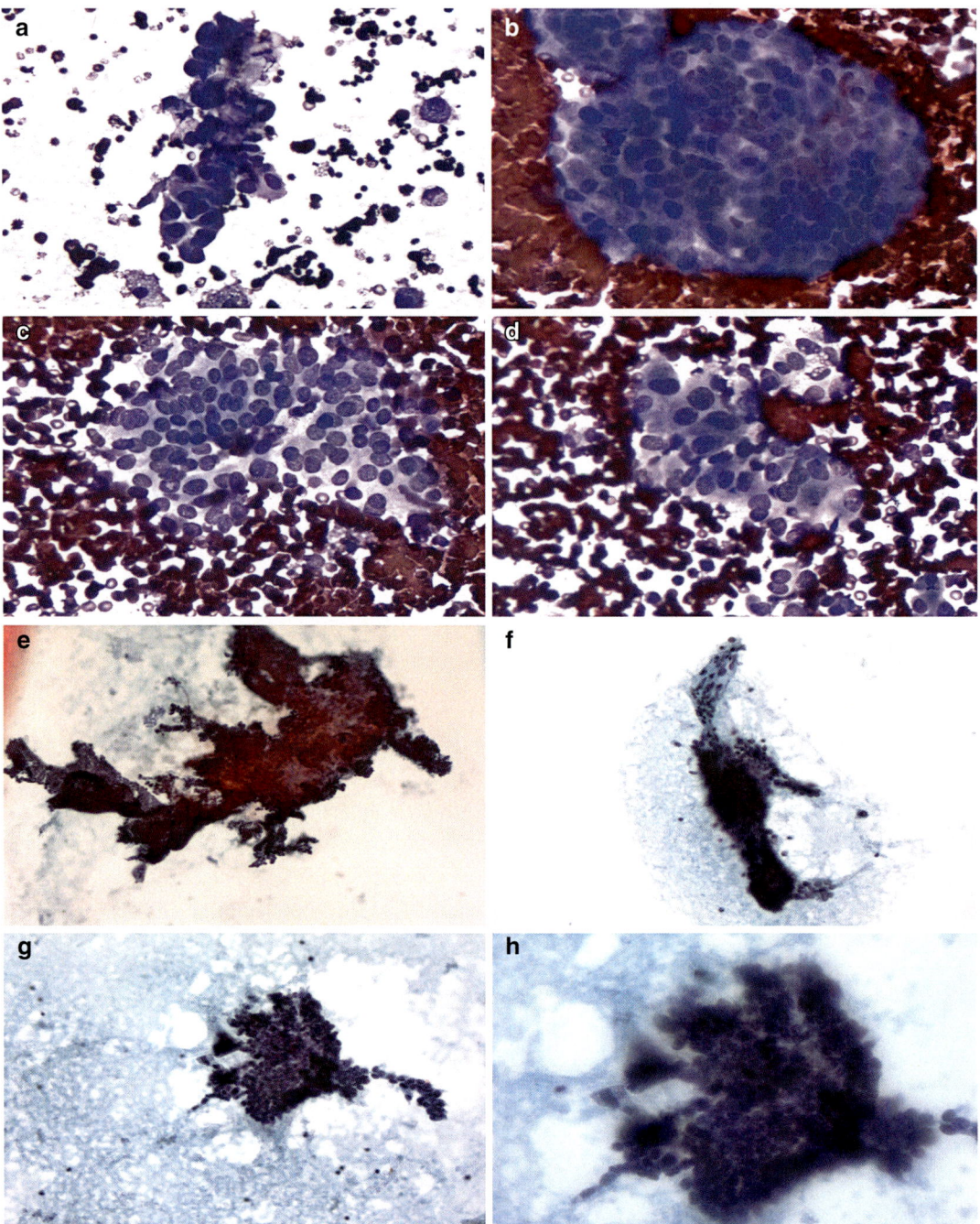

Fig. 11.1 This panel illustrates cytology aspects of a fine needle aspiration cytology of thyroidal conditions in childhood and youth, which may represent a large problem in some countries. Figures (**a**, **b**) display cytology aspirates of a papillary carcinoma, while cytological features of papillary thyroid carcinoma are seen in Figures (**c**) through (**h**) (**a–d**, ×400; **e**, ×100; **f–g**, ×200, **h**, ×400). Figures (**a–d**) are cytological preparations stained with Diff-Quik, a Romanowski-based staining, while figures (**e–h**) are cytological preparations stained with Papanicolau staining

11.3 Thyroid Gland Pathology

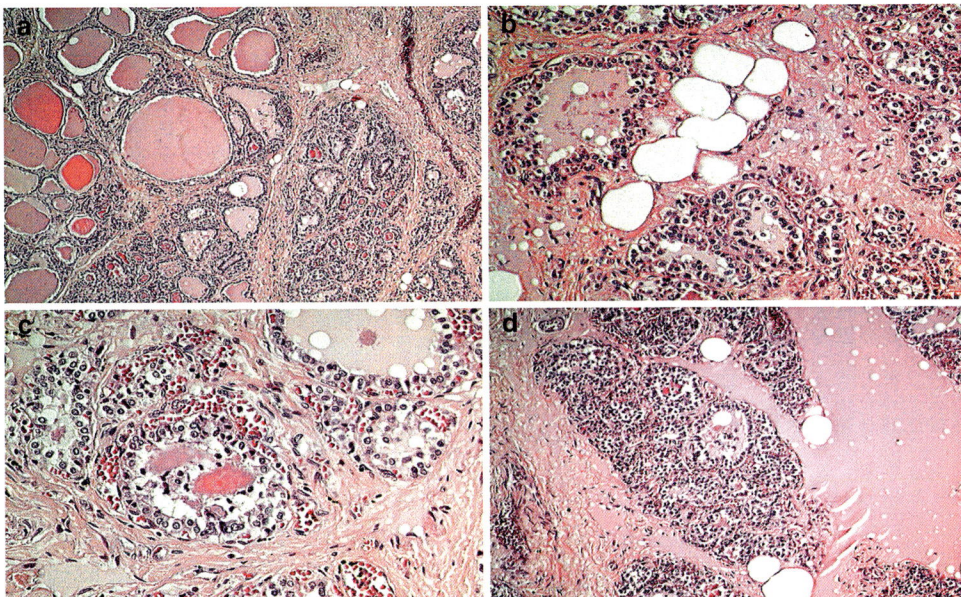

Fig. 11.2 In case of a HIV infection, thyroid changes can be seen, and this aspect affects also children and adolescents as shown in this panel (**a–d**). The thyroid gland shows pronounced fibrosis with inactive and active areas. Some follicles are increased in size, have a flattened epithelium, and have densely eosinophilic intraluminal colloid, whereas some follicles are normal sized, lined by a single layer of columnar follicular cells, and have pale intraluminal colloid with scalloped borders (**a**, ×50, hematoxylin and eosin stain). In some areas fibrotic septa contain collections of lipocytes (**b**, ×160, hematoxylin and eosin stain), intraluminal colloid tends to be broken up in globular formations (**c**, ×200, hematoxylin and eosin stain), or intraluminal colloid is free in the fibrovascular septa or close to the fibrovascular septa (**d**, ×100, hematoxylin and eosin stain). A variable combination of infection, genetic and environmental factors induces an autoimmune response to the thyroid gland, TSH receptor, and several thyroid antigens as suggested by Shukla et al. (2018)

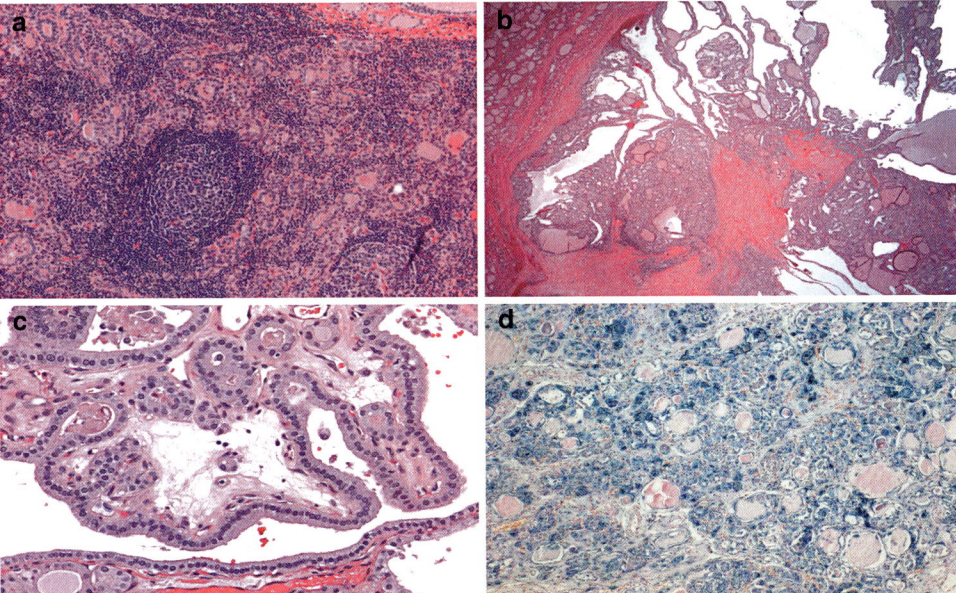

Fig. 11.3 This panel shows the classic features of a Hashimoto autoimmune thyroiditis (**a**, ×100, hematoxylin and eosin staining), adenomatous hyperplasia of the thyroid gland (**b**, ×12.5; **c**, ×200, hematoxylin and eosin staining for both images), and a remarkable case of hemosiderin accumulation of the thyroid gland in a young patient with genetic hemochromatosis (**d**, ×100, Perls Prussian blue staining)

the French word *crétinisme*, dating back to the end of the 18th century.

The etiology of hypothyroidism includes agenesis, autoimmune thyroiditis, diet (e.g., poor nutrition, vitamin B12 deficiency, consumption of thiocyanate-rich cassava plant), hypopituitarism, and post-surgery/postradiation (*ablatio glandulae thyroideae*).

Hyperthyroidism characterized by hyperactivity, tachycardia, exophthalmos, muscle atrophy, and warm skin is also due to several causes, i.e.:

- Non-thyroidal neoplasms (*struma ovarii*, choriocarcinoma)
- Pituitary adenoma
- Thyroid adenoma
- Thyroid carcinoma (functioning)
- Thyroid hyperplasia, diffuse type = Graves disease
- Thyroid hyperplasia, nodular type = multinodular goiter
- Thyroiditis, acute/subacute

Goiter is a visible, usually symmetrical, enlargement of the thyroid gland, which can present as nodular or diffuse and either hypo- or hyperfunctioning (*struma glandulae thyroideae*).

Some thyroid hormone-driven basic energy metabolism of the target cells include (1) ↑ protein synthesis, (2) ↑ oxidative phosphorylation (OXPHOS) in the mitochondria, (3) ↑ cell carbohydrate metabolism, (4) ↑ lipidic metabolism, and (5) Ca^{++} mobilization in the bone. The assessment of the thyroid gland structure is performed at several and well distinct levels as follows:

1. Clinical examination (during swallowing)
2. Radiologic examination (^{125}I) helping to identify "filling defects" or "cold nodules"
3. Ultrasonography assisting in distinguishing cystic from solid thyroid nodules
4. Fine needle aspiration biopsy to evaluate cytologically the thyroid nodules
5. Biopsy (excision/hemithyroidectomy) to assess cytologically suspicious thyroid nodules
6. Thyroidectomy ± lymph nodes to stage thyroid malignancies

The assessment of the thyroid gland function is performed at several levels:

1. Free T4 index and free T3 index
2. Serum TSH (ultrasensitive)
3. TRH stimulation test

11.3.1 Congenital Anomalies, Hyperplasia, and Thyroiditis

Box 11.5 lists the congenital anomalies, hyperplasia, and thyroiditis.

11.3.2 Congenital Anomalies, Goiter, and Dysfunctional Thyroid Gland

Box 11.6 lists the congenital anomalies, goiter, and dysfunctional thyroid gland.

Box 11.5 Congenital Anomalies, Hyperplasia, and Thyroiditis

11.3.1. *Congenital Anomalies, Goiter, and Dysfunctional Thyroid Gland*
    11.3.1.1. Congenital Anomalies of the Thyroid Gland
    11.3.1.2. Dysfunctional Thyroid Gland and Dyshormonogenetic Goiter
    11.3.1.3. Grave Disease (Diffuse Hyperplasia), "Toxic" (Thyrotoxicosis)
    11.3.1.4. Multinodular Goiter (Nodular Hyperplasia), Nontoxic

11.3.2. *Inflammatory and Immunologic Thyroiditis*
    11.3.2.1. Acute Thyroiditis
    11.3.2.2. Granulomatous Thyroiditis
    11.3.2.3. Autoimmune Thyroiditis (Lymphocytic and Hashimoto Thyroiditis)
    11.3.2.4. IgG4-related Thyroiditis and Riedel Thyroiditis

11.3 Thyroid Gland Pathology

> **Box 11.6 Congenital Anomalies, Goiter, and Dysfunctional Thyroid Gland**
> 11.3.1. *Congenital Anomalies, Goiter, and Dysfunctional Thyroid Gland*
> 11.3.1.1. Congenital Anomalies of the Thyroid Gland
> 11.3.1.2. Dysfunctional Thyroid Gland and Dyshormonogenetic Goiter
> 11.3.1.3. Grave Disease (Diffuse Hyperplasia), "Toxic" (Thyrotoxicosis)
> 11.3.1.4. Multinodular Goiter (Nodular Hyperplasia), Nontoxic

11.3.2.1 Congenital Anomalies of the Thyroid Gland

- *DEF*: Anomalies of the ontogenesis, mostly involving the parathyroid gland.
- *Thyroglossal Duct Cyst*: Localized persistence of the embryonic thyroglossal duct, usually in the region of the hyoid bone.

11.3.2.2 Dysfunctional Thyroid Gland and Dyshormonogenetic Goiter

The reader is requested to consult pediatric endocrinology textbooks.

11.3.2.3 Graves Disease (Diffuse Hyperplasia)

- *DEF*: Autoimmune (AI) disease of the thyroid gland characterized by serum autoantibodies of the IgG class directed against the TSH receptor of the thyrocyte with the final event of cytologic hyperstimulation (hyperthyroidism) ± other AI diseases (e.g., pernicious anemia). Graves disease is named for the doctor, Robert J. Graves. This physician first described this disease in 1835. The disease may be written Graves disease or Graves disease because eponymous words are often written nonpossessively. In the European continent, the constellation of symptoms described by Dr. Graves was also reported by Dr. Karl Adolph von Basedow, a German physician and Chief Medical Officer of Merseburg, in 1840 and Basedow disease or Basedow syndrome are more common than Graves disease.
- *EPI*: 15–40 years, ♂:♀ = 1:4; familiarity, (+) HLA-DR3/HLA-B8 in Caucasians.
- *CLI*: Exophthalmos, pretibial myxedema, ↑ thyroid hormones (e.g., T3-Toxicosis), ↓TSH.
- *GRO*: Diffuse symmetrical enlargement of the gland with highly and prominent vascularity.
- *CLM*: Tightly packed thyroid follicles with tall columnar epithelium with papillary infolding, scant pale colloid, and luminal scalloping (rapid thyroglobulin proteolysis) with scattered lymphocytes/lymphoid tissue with germinal centers.
- *DDX*: Other causes of hyperthyroidism.

11.3.2.4 Multinodular Goiter

- *DEF*: Multinodular enlargement of the thyroid gland due to mild deficiency of iodine and hormonal production following an alternating period of hyperplasia and involution according to dietary iodine levels.
- *EPI*: Endemic (e.g., Alps, the Andes, Himalayas), toxic (goitrogen dietary factors, including cabbage or cassava), and sporadic (increased demand such as at puberty or pregnancy and mild, occasionally, a relative deficiency of thyroid hormonal synthesis enzymes) diffuse nontoxic and multinodular goiter.
- *CLI*: From initially painless diffuse enlargement of the thyroid (euthyroid) without dominant nodules through diffuse enlargement of the thyroid with dominant nodule (autonomous hyperplastic nodule → toxic nodular goiter)
- *GRO*: Multinodularity of the thyroid gland with thick gelatinous and glistening cut surface.
- *CLM*: Hyperplasia of the thyroid gland with small follicles lined by tall columnar cells in the early stages with progression to involution with distended follicles with colloid and flat or cuboidal epithelium, once adequate or excess hormone synthesis is reached. Regressive or degenerative changes may be present, and they may range from areas of hemorrhage, cystic degeneration, fibrosis, and calcification.

11.3.3 Inflammatory and Immunologic Thyroiditis

Box 11.7 lists the inflammatory and immunologic thyroiditis.

11.3.3.1 Acute Thyroiditis
It is a severe, inflammation of the thyroid gland with a viral, mostly, or suppurative origin.

11.3.3.2 Granulomatous (Subacute) Thyroiditis
- *DEF*: Inflammatory disease of the thyroid gland often following an upper respiratory infection (adenovirus, mumps virus, echovirus, influenza virus, EBV, Coxsackieviruses).
- *SYN*: Granulomatous (subacute) thyroiditis, DeQuervain thyroiditis.
- *CLI*: Painful enlargement of the thyroid with systemic symptoms (fever, malaise, muscle aches).
- *GRO*: Diffuse enlarged gland, firm and often adherent to the surrounding structures.
- *CLM*: Parenchymal destruction with clusters of MΦ and MNGC around fragments of colloid and marked fibrosis.

11.3.3.3 Autoimmune Thyroiditis
- *DEF*: An autoimmune disease of the thyroid gland characterized by a cytotoxic T-cell-mediated hypersensitivity reaction against the thyroid with several different IgG auto-antibodies, including anti-thyroglobulin (anti-TG), anti-colloid, and anti-microsomal autoantibodies. The most clinically relevant anti-thyroid auto-Abs are anti-thyroid peroxidase (anti-TPO Abs), thyrotropin receptor antibodies, and anti-TG Abs.
- *SYN*: Hashimoto Autoimmune Thyroiditis (HAT). This condition was first described in 1912 by the Japanese physician Hakaru Hashimoto.
- *CLI*: Euthyroid status through hypothyroid status (progressive auto-destruction).
- *GRO*: A variable range from firm and rubbery, diffusely enlarged gland (early) to marked atrophic gland (late).
- *CLM*: Numerous lymphocytes and plasma cells within and around the thyroid follicles (cyto-destructive lymphoepithelial lesion) and scattered small and atrophic follicles with often marked eosinophilia of the cytoplasm (oxyphilic change or Hürthle change).
- *DDX*: Nonspecific lymphocytic thyroiditis (histological changes similar to the HAT, but lack of the clinical background).

Some AI tips and pitfalls are discussed in Box 11.8.

11.3.3.4 IgG4-Related Thyroiditis and Riedel's Thyroiditis
- *DEF*: Immunoglobulin G4-driven inflammation of thyroid gland. IgG4-related disease (IgG4-RD) has been recently delineated, and it is characterized by ↑ serum IgG4 levels and ↑ IgG4-positive plasma cells in involved organs.
- *SYN*: "Woody"/"ligneous" thyroiditis.

Box 11.7 Inflammatory and Immunologic Thyroiditis
11.3.2.1. Acute Thyroiditis
11.3.2.2. Granulomatous Thyroiditis
11.3.2.3. Autoimmune Thyroiditis (Lymphocytic and Hashimoto Thyroiditis)
11.3.2.4. IgG4-related Thyroiditis and Riedel Thyroiditis

Box 11.8 AI Thyroid Pearls and Pitfalls
1. Neonatal hyperthyroidism in mothers suffering from Graves disease is explained by the cross transfer of the maternal IgG autoantibodies stimulating the fetal thyroid gland and disappearance once the baby is delivered.
2. About 1/20 of patients with long-standing Hashimoto autoimmune thyroiditis develop malignancies of the thyroid gland, including malignant B-cell lymphoma and papillary carcinoma of the thyroid gland.

- *EPI*: Practically inexistent in childhood and youth, although HAT in children and adolescents with type 1 DM is associated with elevated IgG4 but not with low vitamin D (Demir et al. 2014).
- *CLI*: Presentation as an aggressive form of Hashimoto thyroiditis.
- *GRO*: Mildly enlarged gland with partial or full replacement by a stony-hard appearance of the glandular tissue, which appears grayish-white ("woody"/"ligneous" appearance).
- *CLM*: Diffuse atrophy of the thyroid follicles with their replacement by dense collagen, which looks like scar tissue, and scattered chronic inflammatory cells (lymphocytes and plasma cells).
- *DDX*: Malignancy!

The incidence of IgG4-related thyroiditis remains poorly known, but Jokish et al. (2016) investigated formalin-fixed and paraffin-embedded thyroid gland samples of 216 patients (191 Hashimoto thyroiditis, 5 Riedel thyroiditis, and 20 goiters, as controls). The study contemplated both clinic evaluation and morphologic/immunohistochemical analysis. IgG4-RD showed a higher IgG4/IgG ratio, a higher median IgG4 count, an association with younger age, and a higher ♂:♀ ratio. In the differential diagnosis, the fibrous variant of Hashimoto thyroiditis was identified in 96% of the IgG4-related cases (96%) and 18% of the non-IgG4-related instances. The incidence of IgG4-related disease (IgG4-RD) of the thyroid gland in Europe is considerably lower than that observed in non-European studies. IgG4-RD of the thyroid gland differs morphologically from the IgG4-RD in other organ systems, exhibiting dense fibrosis without intense tissue eosinophilia or obliterative phlebitis. IgG4-related thyroiditis has been studied by Li et al. (2010) in a surgical series encompassing 70 patients with Hashimoto thyroiditis with age between 40 and 60 years. Nineteen of the 70 patients showed immunostaining positivity for IgG4. Li et al. (2010) found that this subgroup had a higher ♂:♀ ratio, more rapid advance, and a higher level of circulating antibodies, with no clinical evidence of other organ involvement, compared with Hashimoto thyroiditis without immunostaining for IgG4. This Japanese group concluded that IgG4-related thyroiditis constitutes a distinct and more aggressive form of Hashimoto thyroiditis.

11.3.4 Epithelial Neoplasms of the Thyroid Glands

The thyroid carcinoma is the most common malignant neoplasm of the endocrine gland system. It has been estimated that more than 95% of thyroid carcinomas arise from follicular epithelial cells. In most countries, including the USA and Canada, thyroid cancer has increased during the past few decades. Thyroid cancer is now one of the most rapidly growing fatalities, representing a significant cause of morbidity in premenopausal women. The majority of thyroid gland neoplasms can be readily diagnosed using histopathologic criteria. These principles allow the pathologist to discern the malignant from the benign tumors and guarantee a good system for planning the therapy or postsurgical follow-up. In fact, hyperplastic nodules can be diagnosed as nodules of follicles with variable size and association of degenerative changes (cyst, hemorrhage, fibrosis), while follicular adenomas have a monotony of follicular size, being either microfollicular, or macrofollicular, possess a fibrous capsule that separates them from the surrounding thyroidal tissue, and do not evidence any vascular invasion. In some cases, however, the distinction is subtle, and this task for the pathologist is quite challenging. In this case, intradepartmental and interdepartmental consultations are mandatory for the safety of the patients. Indeed, hyperplastic nodules with some monotony, no degenerative changes and disrupted fibrous capsule, as well as atypical follicular adenomas or the controversial entity of hyalinizing trabecular adenoma (HTA) may become challenging diagnostic tasks in the pathology routine. Ancillary techniques and molecular profiling can direct the diagnosis. In most cases, an immunohistochemical panel is very helpful. The diagnosis, which is indeterminate, of follicular neoplasm following fine needle aspiration cytology (FNAC) encompasses some diagnoses with entirely different follow-up. These entities that may all have a FNAC of follicular neoplasm include cellular adenomatoid nodule, follicular adenoma, follicular carcinoma, and follicular variant of papillary thyroid carcinoma (FVPTC). The surgical biopsy is the next procedure for the safety of the patients. It is important to discern among diseases that have a quite different therapeutic approach and follow-up. About the FVPTC, it is important to realize that this diagnosis is one of the most controversial diagnoses in clinical and surgical pathology. It is important not to miss the subtle nuclear features of papillary carcinoma because the misidentification may have significant clinical consequences. Some immunohistochemical markers are available and can be recommended to the routine pathologist. These belong to different categories, including structures involved in cell adhesion (galectin-3, E-cadherin, beta-catenin, fibronectin), receptor signaling (RET), gene transcription control (thyroid transcription factor 1), secretion (thyroglobulin, calcitonin, carcinoembryonic antigen), cell cycle regulation (p27, cyclin D1, p53), subcellular cytoskeleton (K19), and membranous structures of still poorly understood function (Hector Battifora mesothelial cell-1 or HBME-1). *RET* is a proto-oncogene (*c-RET*), which encodes an intracellular signal-transducing tyrosine kinase receptor protein whose ligands belong to the family of the glial cell line-derived neurotrophic factors. A rearranged version of *RET* is *RET/PTC*. In MEN 2, point mutations of *RET* determine in constitutive activation of RET, resulting in medullary thyroid carcinomas. Conversely, *RET/PTC-1, RET/PTC-2,* and *RET/PTC-3* together with other less common genetic rearrangements play a significant role in PTC. Indeed, there is constitutive activation of the tyrosine domain in the carboxyl-terminal end of RET/PTC determining oncogenic effects in thyrocytes as it has been demonstrated in transgenic mice. RET immunostaining is useful in the assessment of thyroid lesions with incomplete and focal or ambiguous features of PTC in which more than half of cases show positivity for this gene product and in line with the morphological features suggestive of PTC. *Cytokeratin or keratin 19* is low molecular weight keratin (LMWK) or intermediate filament of the cytoskeleton. Stratified squamous epithelium shows the expression of high molecular weight keratins (K1, K4, K10, and K13), while simple or glandular epithelium discloses the expression of low molecular weight keratins (K7, K8, K18, K19). PTC shows strong and diffuse immunoreactivity for LMWKs up to 100% of cases. Conversely, K7, K18, and K19 are present in about 50% of poorly differentiated thyroidal carcinomas. These markers are not present in the spindle cell sarcomatoid variant of anaplastic thyroid carcinoma, while some positivity remains in the squamoid and giant cell-solid epithelioid variants of anaplastic thyroid carcinoma. Among the LMWKs, K19 plays a significant role in helping distinguishing follicular adenomas/follicular carcinomas from PTC, because it is less intense and more focal in the former entities. The expertise of the pathologists for endocrine disease is mandatory when the pathologist receives this kind of tissue specimens. Of note, normal follicular epithelium of the thyroid gland is negative for K19, although focal staining has been noted in the compressed thyroid parenchyma surrounding nodules and in lymphocytic thyroiditis. *Galectin-3* is a cell adhesion molecule member of the family of non-integrin β-galactoside-binding lectins, which have related amino acid sequences in the CHO-binding site. Some non-neoplastic cells express galectin-3. PMNs, MΦ, mast cells, and Langerhans cells express galectin-3, which is involved in several biologic and pathologic processes including cell cycle, apoptosis, cell matrix, and cell-to-cell interactions as well as cell adhesion and cell migration. Although downregulated in colorectal and breast cancer, galectin-3 shows strong diffuse cytoplasmic staining in PTC up to 100% of cases, while it is expressed in 45–95% of FTC. MTC, PDTC, and anaplastic carcinomas also show some variable expression, while follicular adenomas are usually positive in about one fifth of cases. There is no expression of galectin-3 in hyperplastic nodules, nodular goiters, and normal follicular epithelium. HBME-1 is a monoclonal antibody with a target in the microvilli of mesothelioma cells; adenocarcinoma of the lung, breast, and pancreas; and normal tracheal epithelium. HBME-1 is positive in follicular-derived thyroid malignancies up to 100% of cases, including well- and poorly differentiated carcinomas, but also PTC in about 80% of cases. Poorly differentiated thyroidal carcinomas and anaplastic carcinomas are positive in about 75% and 25% of cases, respectively. Normal and hyperplastic thyroid gland show no expression of this marker. Other markers that may be considered of some validity include TG; calcitonin; carcinoembryonic antigen (CEA); cyclooxygenase-2 (COX-2); peroxisome proliferator-activated receptor γ; E-cadherin; β-catenin; fibronectin-1; CD44v; thyroid peroxidase (TPO); Cbp/p300-interacting transactivator 1 (CITED1), which is aka melanocyte-specific protein 1; cyclin D1; p27; p53; and thyroid transcription factor 1 (TTF-1), which together with PAX8 controls the expression of TG, TPO, thyrotropin receptor, the Na/iodide symporter, calcitonin, and primary histocompatibility complex class I genes in the thyroid gland. Of note, TTF-1 regulates the expression of surfactant A, B, and C and Clara cell secretory protein genes in the lung. However, reports on these markers may be quite discordant in the literature. It is advisable that a good hematoxylin and eosin staining and pathology expertise in endocrinological pathology are still enough for the diagnosis of most of the neoplastic lesions. In a subset of tumors with follicular architecture without unequivocal evidence of malignancy, the use of ancillary techniques, including immunohistochemistry and molecular biology, may be a good choice. The method of a single immunohistochemical marker is never a good choice in distinguishing thyroid malignancies, while a panel of ≥2 markers is wise. Overall, K19, GAL3, HBME-1, and CITED1 are useful in separating some cases of FVPTC from follicular adenomas.

The *solitary thyroid nodule* (STN) is an intrathyroidal nodule, which standouts during careful palpation of the thyroid gland and may show iodine uptake ("hot" vs. "cold" nodules) using radionuclide scan and usually surgically treatable. A palpation-identified STN is found in ~5% of all thyroid gland clinical examination, and an STN identified at time of autopsy (C5 evidence) is revealed in ~10% of thyroid gland examination grossly. A benign colloid nodule is found in ~ 65% of all STNs, a follicular adenoma STN is found in ~30% of all STNs, and malignancy is found in ~5% of all STNs. In the previous TNM classification (7th edition), the C (1–5) parameter was a modifier of the *Certainty* (quality), which has been removed in the TNM 8th edition. The C parameter or C factor was an optional descriptor that reflected the intensity of the studies and the validity of the classification. The C factor included five categories: C1: evidence from standard diagnostic means (e.g., inspection, palpation, standard radiography, and endoscopy); C2: evidence gathered by special diagnostic means (e.g., radiography with special projections, ultrasonography, lymphography, angiography, scintigraphy, tomography, CT, MRI, PET, PET-CT, biopsy, and cytology); C3: evidence from surgical exploration; C4: evidence of disease extensiveness following surgery and pathological examination of the resected specimen; and C5: evidence from postmortem investigation (autopsy).

Thyroid Scan and Uptake is a particular medical procedure to investigate the thyroid gland in detail. Thyroid Scan and Uptake uses small amounts of radioactive materials ("radiotracers"), a specific camera to collect the signals from the tracers, and a computer to provide information about thyroid gland size, shape, position, and function. Radionuclide scan results can show either a (1) "hot" STN with high iodine uptake portend to a probably benign diagnosis and (2) "cold" STN with low iodine uptake that can foreshadow a probably malignant diagnosis (e.g., carcinoma).

FNA diagnoses include benign colloid nodule, cellular follicular lesion, inconclusive (probably harmless), inconclusive (suspicious), probably malignant, malignant, and inadequate specimen for the cytologic diagnosis (FNAC). Of note, a follicular lesion on smear may be quite challenging, because it may represent a cellular region of nodular goiter, a follicular adenoma, or well-differentiated follicular carcinoma. The sole cytologic examination does not permit to differentiate between these three entities reliably and a surgical procedure is necessary. FNA adequacy is defined when ≥6 groups of ≥10 epithelial cells (EC) in ≥2 slides with two exceptions are met. The first debarment includes the option to have a suggestive cytology of the colloid nodule but nondiagnostic due to lack of EC rather than nonspecific, if colloid ++ but no cells. The second debarment includes the option of a suggestive cytology of thyroiditis, but nondiagnostic due to low # of EC rather than nonspecific, if lymphocytes are present. In case of *colloid nodule/goiter*, there is abundant colloid and scant cellularity, honeycomb sheets of follicular cells (uniform, cohesive with even spacing), degenerative and/or regenerative changes (hemosiderin, foamy MΦ), MNGC, stroma, degenerated RBC, and watery colloid (thin, translucent, pink on Papanicolau stain or PAP and blue on Diff-Quik or DQ stain). The DQ stain is a commercial Romanowsky stain variant widely used in cytopathology and is based on a modification of the Wright-Giemsa stain. The DQ stain has assets over the Wright-Giemsa staining technique in that it reduces the 4-minute process and allows for selective increased eosinophilia or basophilia of the specimens. In case of a *follicular neoplasia*, there is scant colloid and high cellularity, syncytial sheets, and macrofollicles (cells with an iso-nucleosis and cell crowding). In case of *Hashimoto thyroiditis*, there is a mixed population of lymphocytes (also with crushed lymphoid tangle).

Box 11.9 lists the epithelial and non-epithelial neoplasms of the thyroid gland.

11.3.4.1 Follicular Thyroid Adenoma
- *DEF*: Benign epithelial neoplasm of the thyroid gland composed of uniform follicles standing out from the surrounding thyroidal parenchyma (e.g., micro-/macrofollicular adenoma).
- *EPI*: Any age, ♂:♀ = 1:4

> **Box 11.9 Epithelial and Non-Epithelial Neoplasms of the Thyroid Gland**
> 11.3.3.1 *Follicular Thyroid Adenoma*
> 11.3.3.2 *Follicular Thyroid Carcinoma*
> 11.3.3.3 *Papillary Thyroid Carcinoma*
> 11.3.3.4 *Medullary Thyroid Carcinoma*
> 11.3.3.5 *Hürthle Cell Tumors*
> 11.3.3.6 *Poorly Differentiated Thyroid Carcinoma*
> 11.3.3.7 *Anaplastic Thyroid Carcinoma*
> 11.3.3.8 *Other Epithelial Tumors*
> 11.3.3.9 *Non-Epithelial Neoplasms*
> 11.3.3.10 *Secondary (Metastatic) Tumors*

- *CLI*: Euthyroid or rarely hyperthyroidism ("toxic" adenoma).
- *GRO*: Solitary, firm, gray-reddish nodule up to 5 cm in size ± hemorrhage, cystic degeneration, fibrosis, and calcification.
- *FNA*: Cellular smear & +++ microfollicles.
- *CLM*: Nodule of small follicles of different size compact surrounded by a complete fibrous capsule of variable thickness and compressing the non-neoplastic parenchyma (Fig.11.4).
- FTA is differentiated in:
  - *Microfollicular FTA*, when small (fetal) follicles are seen
  - *Macrofollicular FTA*, when large (colloid) follicles are seen
  - *Embryonal FTA*, when solid cords of epithelial cells form rudimentary follicular structures

Variants are listed in Box 11.10.

Carcinoma of the thyroid is a malignant epithelial neoplasm of the thyroid gland arising from either follicular epithelial cell (papillary, follicular, anaplastic) or parafollicular or "C" cell. Thyroid cancer is the most common endocrinologic malignant neoplasm in North America and is one of the most rapidly increasing tumors. In more than 95% of the cases, carcinomas of the thyroid gland originate from follicular epithelial cells, but the pathogenesis is poorly understood. It is important to note that thyroidal carcinoma has increased in the last 50–75 years and exposure to radiation may represent a remarkable risk factor. There have been two types of radiation, which has been put in connection with carcinoma of the thyroid. In particular, two causes of radiation are known to be of some etiologic factor for the onset of carcinoma of the thyroid:

- Iatrogenic radiation, following external neck radiation for treatment of thymic enlargement (erroneously linked to respiratory distress) as used in the 1950s
- Environmental radiation, following exposure to the Hiroshima and Nagasaki atomic bombs, Marshall Islands nuclear tests, and Chernobyl nuclear catastrophe

11.3 Thyroid Gland Pathology

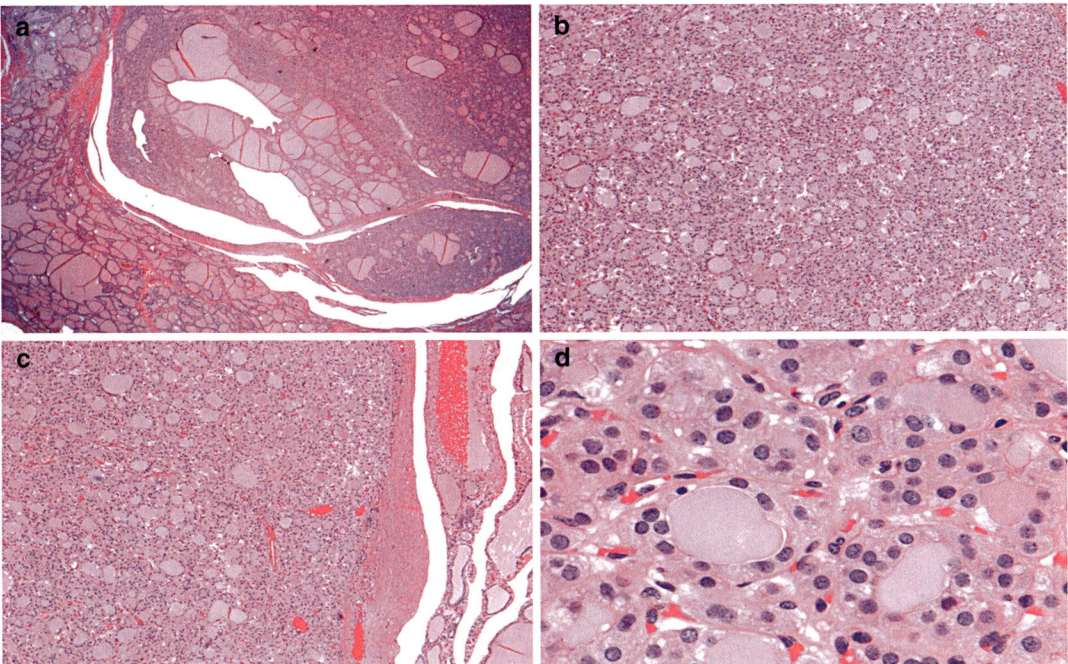

Fig. 11.4 This panel illustrates a thyroid gland with a follicular adenoma in the background of a multinodular goiter. In Figure (**a**) there is the presentation of the multinodular goiter (×12.5), while the Figures (**b–d**) show the microfollicular adenoma without capsular invasion and without vascular invasion (**b**, ×50; **c**, ×50; **d**, ×400). (All images are from hematoxylin and eosin-stained histological slides)

Box 11.10 Follicular Thyroid Adenoma Variants
1. *Hürthle cell Adenoma*: Tumor epithelial cells harboring abundant pink granular cytoplasm
2. *Atypical Adenoma*: Tumor epithelial cells showing cellular pleomorphism and atypia.
3. *Hyalinizing Trabecular Adenoma*: Well-defined organoid tumor with trabeculae simulating paraganglioma (± psammoma bodies), prominent eosinophilic hyaline fibrosis (Congo red negative), and characteristic cytoplasmic MIB1 positivity, and HBME (−)
4. *Clear cell Adenoma*: Tumor epithelial cells with clear cell change (DDX: MTX of RCC)
5. *Adenolipoma*: Adenoma with adipose metaplasia of the stroma

Notes: MTX, metastasis, RCC renal cell carcinoma

In iatrogenic exposure, the risk of thyroid cancer is 5% with infants developing malignancy 15–40 years following radiation, while in nuclear mishaps, the risk is up to 10%, and malignancy may develop 15–40 years following exposure. Epidemiologic details, clinics, gross, and microscopic findings of carcinomas of the thyroid are grouped in the Box 11.11.

In Box 11.12, ten pearls of thyroidal neoplasms are collected.

It is possible to wave the endothelialization of the intravascular tumor seeding if the tumor cluster is attached to the BV wall and associated with the formation of a thrombus. Also, the Sanderson "pollster" needs to be differentiated by papillae. These are collections of small follicles forming a bulge into larger follicles, and the epithelium overlying these collections is columnar and usually negative for Alcian blue (special staining) and EMA (immunohistochemical staining), while true papillae are positive for Alcian blue and EMA. Of course, these pseudo-papillae lack

Box 11.11 Differential Features of the Most Frequently Observed Non-anaplastic Thyroid Carcinomas

	Papillary	Follicular	Medullary
Cell of origin	Follicular	Follicular	Parafollicular
Frequency	~70%	~20%	~5%
Age	15–35 years	>30 years	20–60 years
Gender	♂ < ♀	♂ < ♀	♂ = ♀
Lymphatic	+++	+	+
Hematogenous	+/rare (late)	+++	+
5-YSR	90%	65%	50%
Tumor marker	TG	TG	Calcitonin

Notes: 5-YSR, 5-year survival rate; +/rare (<25% of cases), ++/fair (25–50%), +++/often (50–75%), ++++/very often (>75%); TG, Thyroglobulin (660 kDa, dimeric protein produced by the follicular cells of the thyroid accounting for about 50% of the protein content of the thyroid gland)

Box 11.12 Ten Pearls of Thyroidal Neoplasms
1. Since there is no satisfactory test to allow an absolute differentiation of a follicular adenoma from a follicular carcinoma, rely the DDX on angio-/capsular invasion.
2. Ground-glass change of the nuclei of PTC is often absent in intraoperative frozen sections and cytologic preparations.
3. CK19, GAL3, CITED, HBME-1: useful in extricating some FVPTC from FTA.
4. *Clear nuclei* are not PTC pathognomonic, because benign lesions may display some of them (e.g., nodular hyperplasia, FTA, Graves disease, and Hashimoto thyroiditis); *grooved nuclei* are also not PTC pathognomonic, because they can be seen in solid cell nests, FTA, HTA, poorly differentiated thyroid carcinoma, and adenocarcinomas of non-thyroid origin, while *nuclear pseudoinclusions* are mostly reliable and typical for PTC, although still not pathognomonic.
5. *Capsular invasion*: complete "transgression" of the fibrous capsule (≥10 blocks with the tumor-capsule interface needed for proper handling of the thyroid gland specimens).
6. *Vascular invasion*: BVs tumor cell seeding being located inside or outside of the fibrous capsule AND endothelium-covered intravascular polypoid tumor cell growth.
7. *CD31* and *D2-40* are good markers to highlight the endothelium.
8. MTC is often (+) CEA, differently from stromal carcinoids of the ovary with (−) CEA.
9. Think twice or thrice or more, before diagnosing anaplastic carcinoma of the thyroid gland in a young patient. This entity should be at the bottom of the list following the more frequent diagnoses in this age group (e.g., rhabdomyosarcoma, PNET).
10. It is crucial to differentiate the *FNA track artifact vs. follicular carcinoma true invasion*, and the pathologist can focus to the needle track, cell diffusion, and the end at the capsular site.

the nuclear features of PTC. The term pollster was used by Sanderson-Damberg (1911), while Aschoff (1925) labeled these structures as "Proliferationsknospen." The word "Polster" is of German origin and means a "cushion" or "pillow."

The majority of thyroid tumors can be readily diagnosed using well-established histopathologic criteria. However, in some cases, the separation between benign and malignant lesions can be quite challenging, and the decision for either decision has obvious clinical consequences. There is an immunohistochemical panel that may need to be performed in case of a neoplasm of the thyroid gland, but there is no perfect antibody. A group of pathologists may help to

11.3 Thyroid Gland Pathology

Box 11.13 Immunohistochemical Profile of Thyroid Tumors

	PTC	FTA/FTC	HCT	PDTC	MTC
TG	■	■	■	■	
TTF	■	■	■	■	■
Pan-K	■	■	■	■	*
Calcitonin					■
CGA					■
SYN					■

Notes: The brown color means positive, while light blue negativity. Antibodies include *TG* thyroglobulin, *TTF* thyroid transcription factor, *Pan-K* pan-keratins; calcitonin, *CGA* chromogranin A, *SYN* synaptophysin; tumors included in the box are PTC, papillary thyroid carcinoma, *FTA/FTC* follicular thyroid adenoma and carcinoma, *HCT* Hürthle cell tumors, *PDTC* poorly differentiated thyroid carcinoma, *MTC* medullary thyroid carcinoma. * CAM 5.2 is positive in MTC

shorten the list of differential diagnoses and identify the correct diagnosis. Box 11.13 summarizes the most useful antibodies.

11.3.4.2 Follicular Thyroid Carcinoma
- *DEF*: A malignant epithelial tumor of the thyroid gland with follicular differentiation, no cytologic (nuclear) features of papillary carcinoma (DDX: Follicular variant of papillary thyroid carcinoma, FVPTC), full-thickness capsular ("transgression") and BV invasion, and characterized alternatively by *PAX8/PPAR* gene fusion, *RAS*, or *PTEN* mutations.
- *EPI*: Uncommon in childhood/youth, ~15% of thyroid malignancies, ♂ < ♀.
- *GRO*: Slow-/fast-growing encapsulated fleshy nodule/mass.
- *CLM*: Well-formed follicles ± cribriform areas/trabeculae/solid ± nuclear atypia, ↑ MI (Fig.11.5).
- *IHC*: (+) TG, LMW-CKs, EMA, S100.

Follicular thyroid carcinoma variants include:

1. *Minimally invasive*: grossly encapsulated with microscopic invasion in BVs within or just outside capsule and intravascular tumor nodules eventually covered by endothelium.
2. *Widely invasive*: widespread invasive carcinoma with diffuse BV infiltration and continuous infiltration of the neighboring tissues.

The minimally invasive metastasizes in about 1/20 cases and has a mortality rate of ~3%, while widely invasive has a mortality rate of ~30%. Blood-borne (lungs, bones) metastatic sites are more frequent with shoulder, sternum, skull, and iliac bones more often involved.

11.3.4.3 Papillary Thyroid Carcinoma
- *DEF*: It is a malignant epithelial tumor of the thyroid gland with usually papillary differentiation, obligate cytologic (nuclear) features of PTC (*vide infra*), no malignant relevance of full-thickness capsular and BV invasion, and characterized alternatively by *BRAF* mutations or *RET/PTC1–5* gene fusion (*TRK/TPM3* in radiation-association PTC).
- *EPI*: Childhood (>90%)/youth, ~65% of thyroid malignancies, ♂ < ♀, (+) Hx. CHT, RT.
- *GRO*: Solid, white, firm with gritty cut surface (psammoma bodies, *vide infra*).
- *CLM*: Well-developed branching of complex papillae with edematous/hyalinized fibrovascular cores ± lymphocytes, hemosiderin, follicles, and obligate nuclear features (ground-glass nuclei or orphan Annie eye nuclei, nuclear grooves, and nuclear pseudoinclusions) with low MI but common fibrosis and psammoma bodies (30–50%) (Figs. 11.6, 11.7, 11.8, and 11.9). Harold Gray's figure of Orphan Annie has inspired pathologists, and precisely Dr. Nancy Warner (Hastings Professor of Pathology, Emeritus, University of Southern

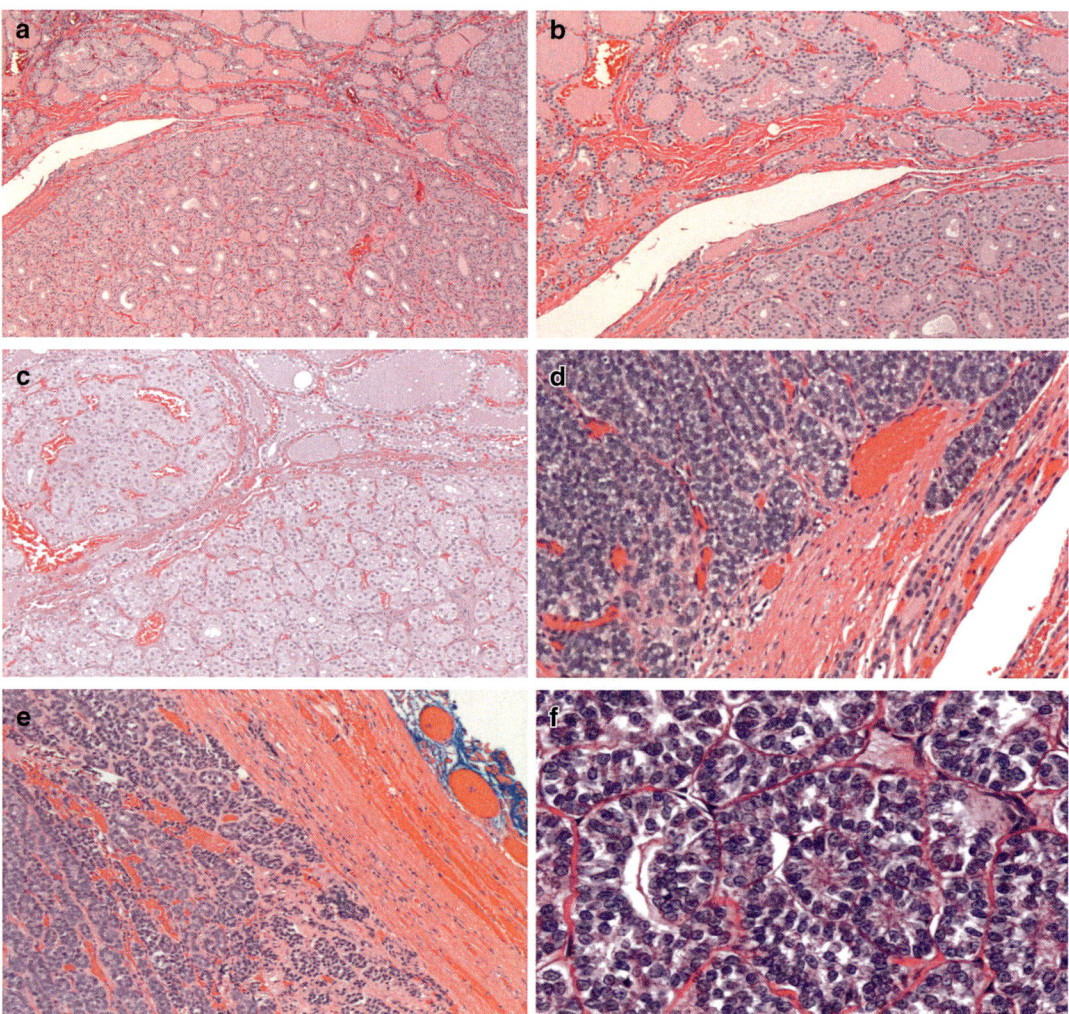

Fig. 11.5 The follicular thyroid carcinoma of this panel discloses clearly capsular and vascular invasion (**a**, ×50; **b**, ×100; **c**, ×100; **d**, ×100; **e**, ×100; **f**, ×400). All images are taken from hematoxylin- and eosin-stained histological slides. The morphology can also be quite variable, but the scrutiny for capsular and vascular invasion is crucial

California) to find a resemblance of the nuclear features of PTC with the eyes of Little Orphan Annie. This was a daily American comic strip and took its name from the 1885 poem "Little Orphant Annie" by James Whitcomb Riley. Its debut came in 1924 using the New York Daily News and ended in 2010. Not inspirational like Orphan Annie's eyes, psammoma bodies are a form of dystrophic calcification and present with round microscopic calcium collections under the lens. Psammoma, which derives from the Greek ψάμμος "sand" is a neoformation containing "grains of sand".

- *IHC*: (+) K7, K19, HMWCKs, TTF-1, TG, EMA, VIM, but also (+) LMWCKs.
- *PGN*: Extrathyroid spread (25%), cervical LNs (+) (~50%), and *AMES* mnemonic (*vide infra*), including the worse features of PTC outcome recorded in the scientific literature (Box 11.14).

Papillary thyroid carcinoma variants include:

(1) *Solid* variant (>50% as solid growth pattern) showing solid growth, islands, and sheets of cells with typical nuclear features of PTC

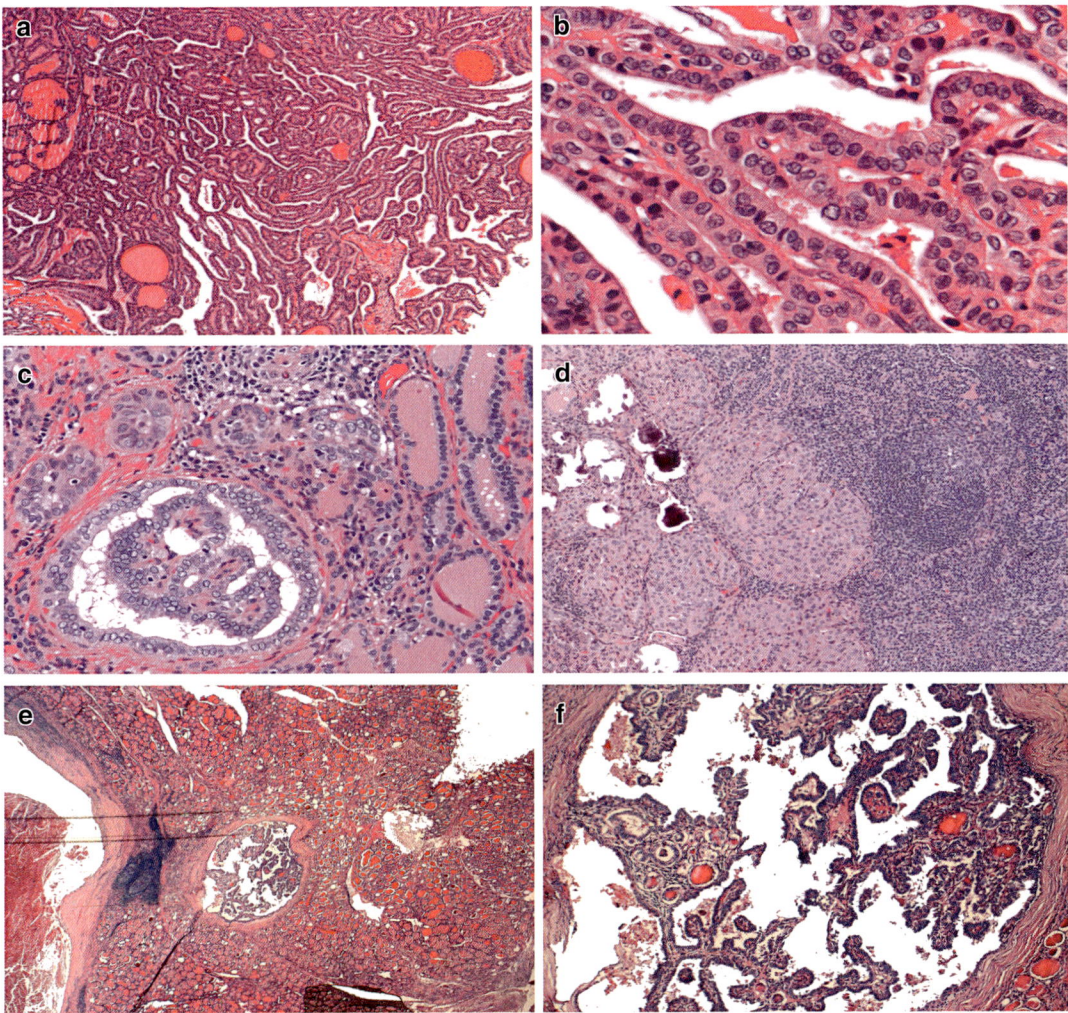

Fig. 11.6 This panel shows several views of papillary thyroid carcinomas (**a**, ×100; **b**, ×400; **c**, ×200; **d**, ×100; **e**, ×20; **f**, ×100). All images are from hematoxylin- and eosin-stained histological slides. The diagnosis of a papillary thyroid carcinoma relies on the nuclear changes (see text for details). Psammoma bodies are clearly identifiable in the (**d**) microphotograph

with distinct and delicate fibrovascular septs ("nuclear incident" PTC)

(2) *Diffuse sclerosing* variant showing dense sclerosis (hypocellular), psammoma bodies, squamous metaplasia, inflammation (heavy), and widespread LN ± lung metastases

(3) *Tall cell* variant showing cells with eosinophilic cytoplasm, height ≥3 times the width, inflammation, and (+) CD15, C-MET, and P53

(4) *Columnar cell* variant displaying tall-cell like morphology, but with chromatin-rich nuclear stratification and histology reminiscent of colon-rectal or endometrioid carcinoma

(5) *Follicular* variant characterized by almost entirely follicles ± focal papillae and nuclear features of PTC ± fibrous trabeculae and psammoma bodies

(6) *Papillary microcarcinoma* identified as Ø ≤ 1 cm (cervical LN MTX in ~1/3 of cases)

(7) *Encapsulated* variant identified as a capsular surrounded PTC

Typically, (1), (2), (3), and (4) variants have worse course than conventional PTC. The variant # (5) behaves as conventional PTC, while (6) and

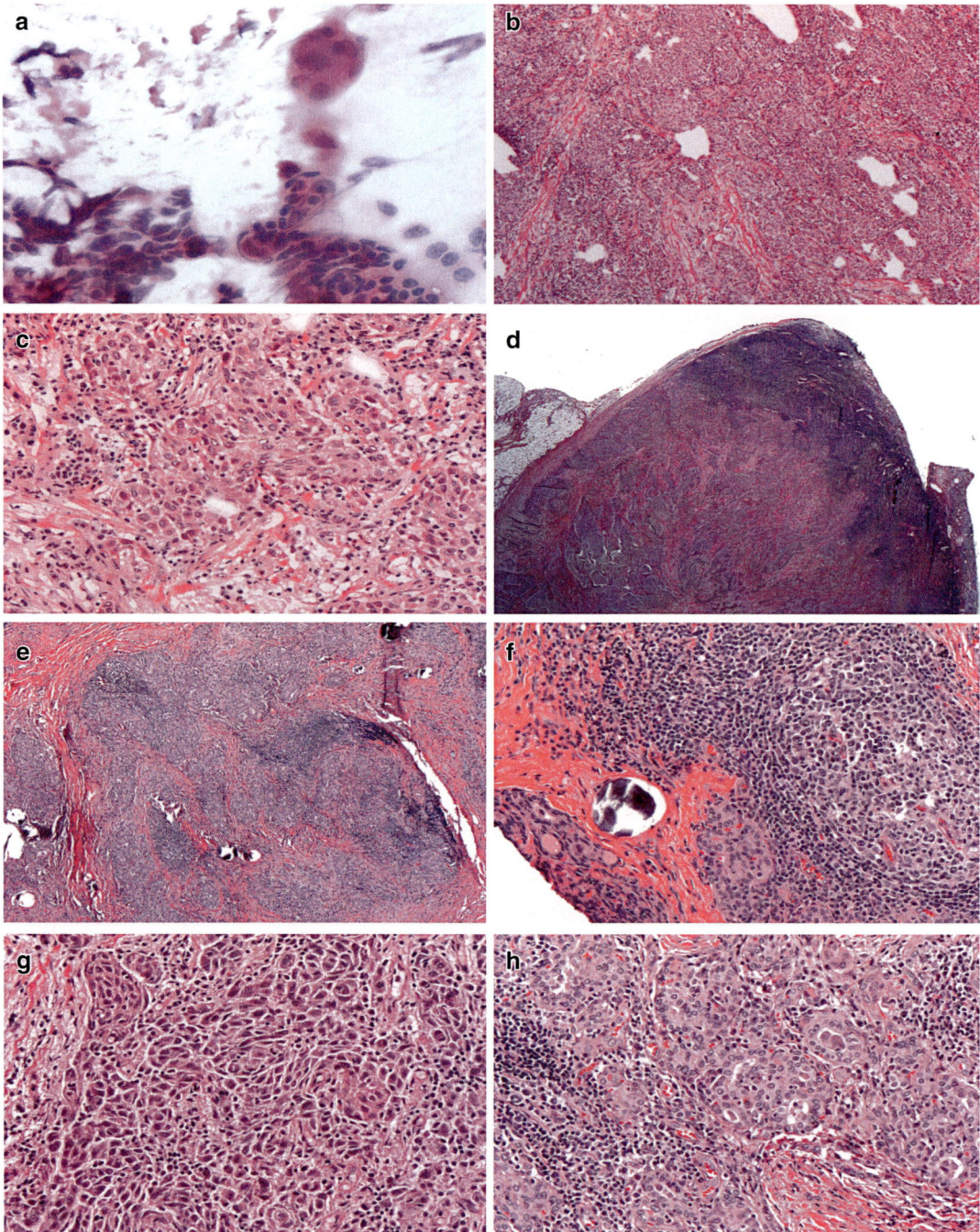

Fig. 11.7 The diffuse sclerosing variant of the papillary thyroid carcinoma is quite often observed in childhood and adolescence. In figure (**a**) are shown some papillae in a Papanicolau stained cytological slide from a fine needle aspiration cytology (×400). The following images (**b–h**) show the characteristic features of a diffuse sclerosing variant of a papillary thyroid carcinoma (**b**, 50; **c**, ×200; **d**, ×12.5; **e**, ×50; **f**, ×200; **g**, ×200; **h**, ×200 as original magnification). Microcalcification is shown in Figure (**f**) (see text for details)

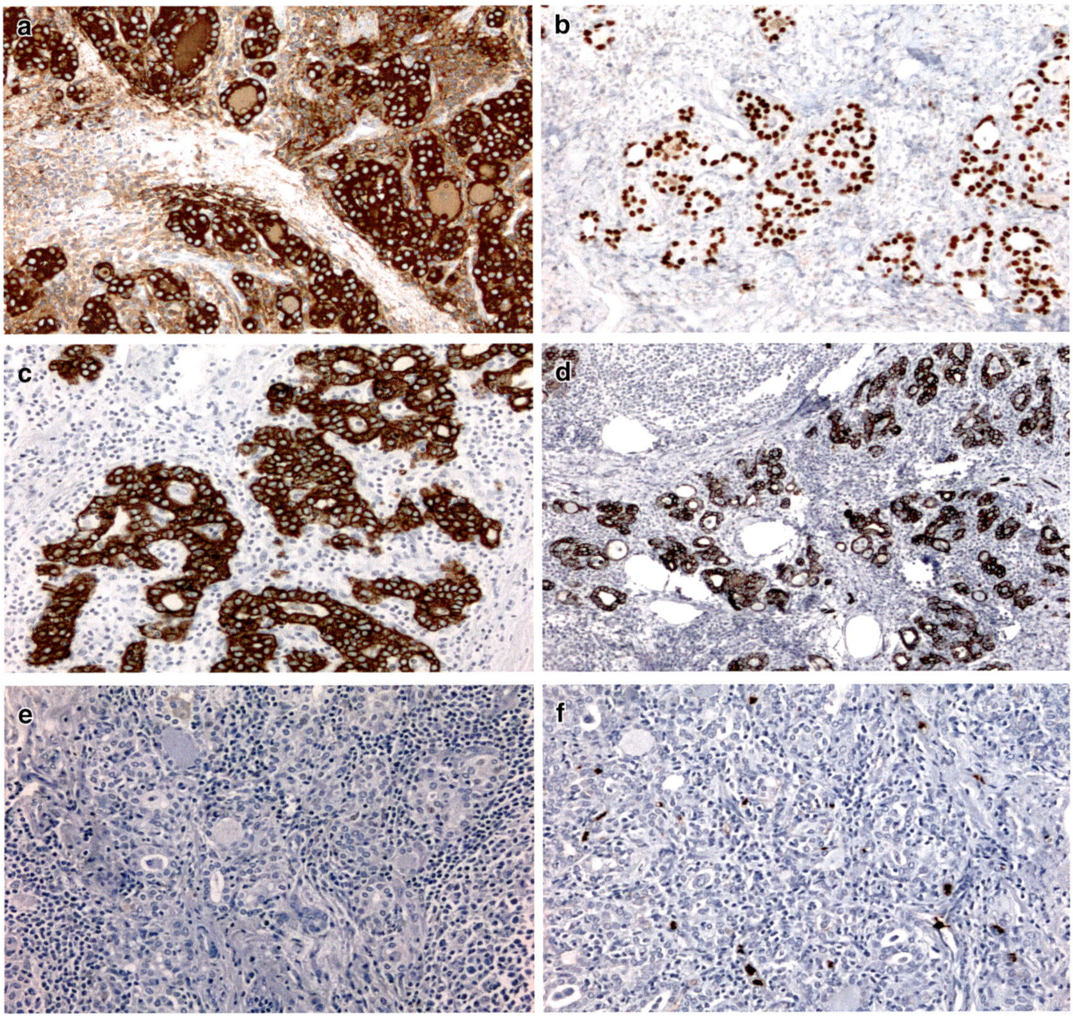

Fig. 11.8 The immunohistochemistry of a diffuse sclerosing variant of a papillary thyroid carcinoma shows the expression of thyroglobulin (**a**, ×200), thyroid transcription factor 1 (**b**, ×200), cytokeratin 7 (**c**, ×200), and cytokeratin 19 (**d**, ×200) but the absence of expression of calcitonin (**e**, ×200), and carcinoembryonic antigen (**f**, ×200), which highlights the C cells of the thyroid gland. Immunohistochemistry following the Avidin Biotin Complex -based method

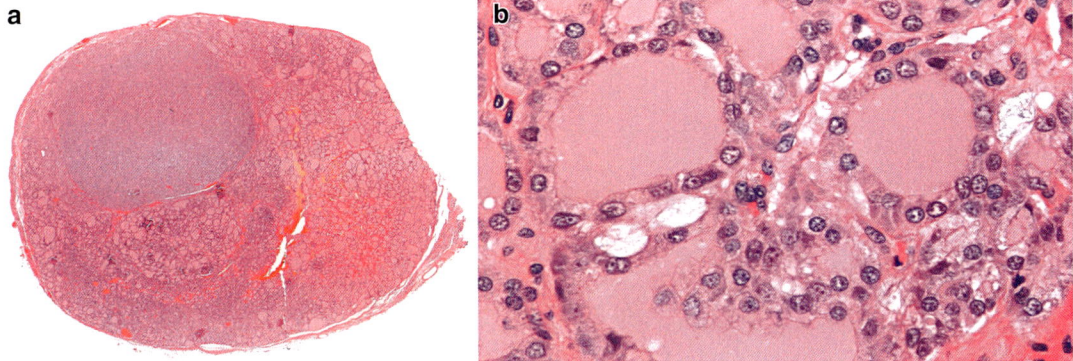

Fig. 11.9 The rare occurrence of a follicular variant of a papillary thyroid carcinoma is shown in Figures **a** (×1) and **b** (×400) (hematoxylin and eosin staining)

> **Box 11.14 AMES: Mnemonic**
> - *Age*: >40 years
> - *Male* gender: ♂ > ♀
> - *Extrathyroidal* extension: ↑ TNM stage
> - *Size-S-phase-Structure*: Ø>4 cm, Aneuploidy, (1-4) variants

(7) variants have an excellent prognosis. Of final note, the oxyphilic variant should be included in the Hürthle cell tumors. The sclerosing variant is mainly restricted to children and young adults.

11.3.4.4 Medullary Thyroid Carcinoma

- *DEF*: Neuroendocrine malignant epithelial tumor with strong *RET* mutations (somatic/germinal), which shows one of the most histologically various growth patterns among thyroid neoplasms. It arises mostly in the upper half of the gland (more C cells) occurring in either *sporadic* form (~80%, solitary cold nodule, ~45 years as mean age) or *familial* form (~20%, multiple/bilateral cold nodules, ~35 years, and C-cell hyperplasia (*vide infra*) in residual gland, MEN II and MEN III, VHL, NF).
- *EPI*: 5–10 of thyroid carcinomas.
- *GRO*: Solid, firm, well delimited, nonencapsulated, and gray-white.
- *CLM*: Variable cytologic appearance showing carcinoid-like nests of round (plasmacytoid) to polygonal cells with granular cytoplasm, trabeculae, glands, or pseudopapillary architecture embedded in highly vascular stroma containing hyalinized collagen and Congo red/crystal violet special stains (+) amyloid (amorphous hyaline material), coarse calcifications, ± psammoma bodies (Fig. 11.10).
- *IHC*: TG (−), but (+) TTF-1 & (+) CGA, SYN, CALC, CEA (± ACTH, VIP).
- *DDX*: PTC, Hürthle cell carcinoma, atypical laryngeal carcinoma, and neuroendocrine lung tumors (*vide infra*) (Box 11.15).
- *PGN*: Hematogenous (lung, liver, bone) and lymphatic spread with neuroendocrine "AMES". AMES mnemonic collects the worse features of MTC outcome:
  - *Age*: >40 years vs. younger individuals, who have a better outcome
  - *Male* gender: ♂ > ♀ vs. female gender, which has better PGN
  - *Extrathyroid* extension, increasing TNM staging vs. confined to the gland (better PGN)
  - *Sporadic-Solid-Stage* (TNM): sporadic occurrence, solid growth, high stage.

C-Cell Hyperplasia
- *DEF*: Precursor lesion of familial MTC (*vide supra*) occurring in MEN2, Hashimoto thyroiditis, hypergastrinemic, and hypercalcemic states and characterized by >50 C cells/follicle with C cells within follicles and between them and mostly located in central part of lateral lobes.
- *IHC*: (+) CALC and CEA. Collagen IV is also useful to highlight the BM, which should be single, complete, continuous without foci of reduplication (DDX: MTC!).
- *DDX*: MTC, cross section of any MTC tumor embolus, intrathyroidal metastasis, "palpation thyroiditis", islands of squamous metaplasia, tangential shave of a healthy follicle, parathyroid gland, and thymic rests. These eight differential diagnoses should be kept in mind at all times and particularly before rendering the diagnosis of neoplasm.

Multiple Endocrine Neoplasias
- *DEF*: AD-inherited genetic syndromes with high degree of penetrance and various neoplasias of the endocrine system harboring from "indigenous" hyperplasias (C cell of the thyroid, adrenal medulla) through frank neoplasias linked to either *MENIN*, a tumor suppressor gene (MEN I), located on chromosome 11q, or *RET*, an oncogene (MEN IIA = MENII and MEN IIB = MEN III), located on chromosome 10q.

MEN Type I
- *DEF*: *MENIN*-related genetic disease
- *SYN*: Wermer syndrome

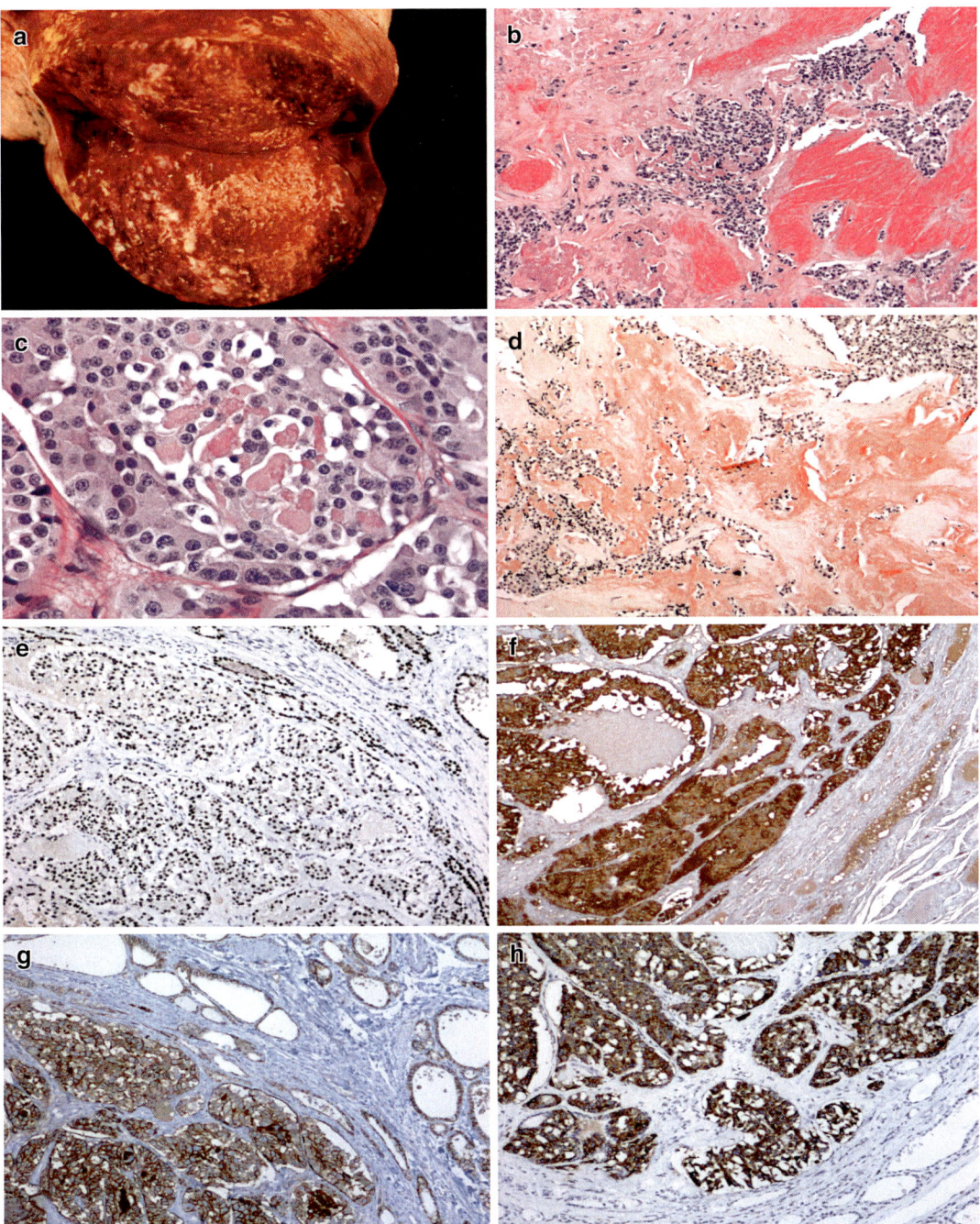

Fig. 11.10 This panel shows the characteristic features of a medullary thyroid carcinoma with gross findings (**a**) including a solid white-yellow nodule with infiltrative margins. The histological examination reveals round, plasmacytoid cells in nests, cords, or follicles (**b**, ×100; **c**, ×400). The tumor cells show round nuclei with finely stippled to, at places, coarsely clumped chromatin and indistinct nucleoli. No nuclear pseudoinclusions are seen, although they may be encountered occasionally. There is a low mitotic index. The stroma has amyloid deposits (**b**), which is Congo red positive (**d**, ×100). The Figures (**b**) and (**c**) are microphotographs taken from histological slides stained with hematoxylin and eosin. In the panel, the four other microphotographs show the immunologic expression of thyroid transcription factor (**e**, ×100), calcitonin (**f**, ×100), CD56 (**g**, ×100), and chromogranin A (**h**, ×100). The immunostaining was performed using the avivin-biotin complex

> **Box 11.15 MTC Differential Diagnosis**
> 1. PTC: both are (+) TTF-1, but MTC is (−) TG and (+) CGA, SYN, CALC.
> 2. Hürthle cell carcinoma: both are (+), but MTC is (−) TG and (+) CGA, SYN, CALC.
> 3. Atypical laryngeal carcinoid (ALC): both are NE tumors and (−) TG and (+) CGA, SYN, CALC, but ALC is (−) TTF-1, while MTC is (+) TTF-1.
> 4. Neuroendocrine tumor of lung (NETL): both MTC and NETL are (+) CGA, SYN, TTF-1, K7, and CAM5.2, but NETL is (−) CALC (usually), and MTC is obviously (+) CALC.
>
> Notes: CALC, calcitonin; CAM 5.2, cell adhesion molecules 5.2; CGA, chromogranin A; K7, (cyto-) keratin7; MTC, medullary thyroid carcinoma; PTC, papillary thyroid carcinoma; SYN, synaptophysin, TTF1 thyroid transcription factor 1, TG, thyreoglobulin.

- *CLM*: *P*ituitary adenoma, *p*arathyroid hyperplasia/adenoma, *p*ancreas endocrine tumor (three Ps as mnemonic help)
  *P*ituitary adenoma
  *P*arathyroid gland chief cell hyperplasia/parathyroid gland adenoma
  *P*ancreas endocrine tumor: G cell (50%), B cell (30%), VIP cell (10%), and others (10%)
  G cells (gastrinoma) ⇒ *Hypergastrinemia* (Zollinger-Ellison syndrome with hypersecretion and peptic ulceration)
  B cells (insulinoma) ⇒ *Hyperinsulinemia*
  VIP cells (VIPoma) ⇒ *WDHA syndrome* or watery diarrhea (pancreatic cholera), hypokalemia, achlorhydria

Also, nodular hyperplasia of the thyroid gland, carcinoid tumors of lung, thymus, GI tract, lipomas, and Menetrier's disease can be observed in patients harbiring MEN type I. The surgical treatment of congenital hyperinsulinism has been recently revised (Adzick 2020).

MEN Type II (IIA)
- *SYN*: Sipple syndrome
- *CLM*: *C*atecholaminergic pheochromocytoma + MT*C* (two Cs) + 1*P*: *P*arathyroid chief cell hyperplasia.
  Also, adrenal cortical tumors can be observed.

MEN Type III (IIB)
- *SYN*: Gorlin syndrome
- *CLM*: *C*atecholaminergic pheochromocytoma + MT*C* (two Cs) + 1*P*: *P*arathyroid chief cell hyperplasia
  + *M*ucosal ganglioneuromas of corneal nerves, lips, and GI tract
- *CLI*: ± *M*arfanoid habitus with skeletal abnormalities
  Also, adrenal cortical tumors and parathyroid tumors can be observed.

11.3.4.5 Hürthle Cell Tumors
Among *Hürthle cell* neoplasms, a significant percentage may show malignant outcome or show destructive tendency locally (5-YSR: 20–40%) along the years as compared to non-Hürthle cell neoplasms. The diagnosis of Hürthle cell cancer is difficult and usually occurs most commonly in subjects above 50 years of age. However, rarely it may be encountered in childhood. Hürthle cell neoplasms are more often seen in females. It appears as a solid, tan, well-vascularized, and usually encapsulated mass grossly. Histologically, they show follicular (more often), trabecular, solid, or papillary growth pattern. Hürthle cell cytology is characterized by cells with granular, acidophilic cytoplasm due to the numerous mitochondria. Malignancy of the Hürthle cell neoplasms is carried out using the same criteria used for follicular lesions (Fig. 11.11).

11.3.4.6 Poorly Differentiated Thyroid Carcinoma
This type of tumor is practically inexistent in children and youth, being reserved to older age group. It exhibits a tremendous high mortality (60%). PDTC harbors *TP53*, *RAS*, and *BRAF* gene mutations. These neoplasms are grossly invasive, and show, histologically, nested (insular) growth pattern (other growth patterns: solid, trabecular) with small uniform cells with convoluted nuclei, necrosis, and high MI (≥3/10 HPF). IHC: (−) TG, CALC (DDX: MTC).

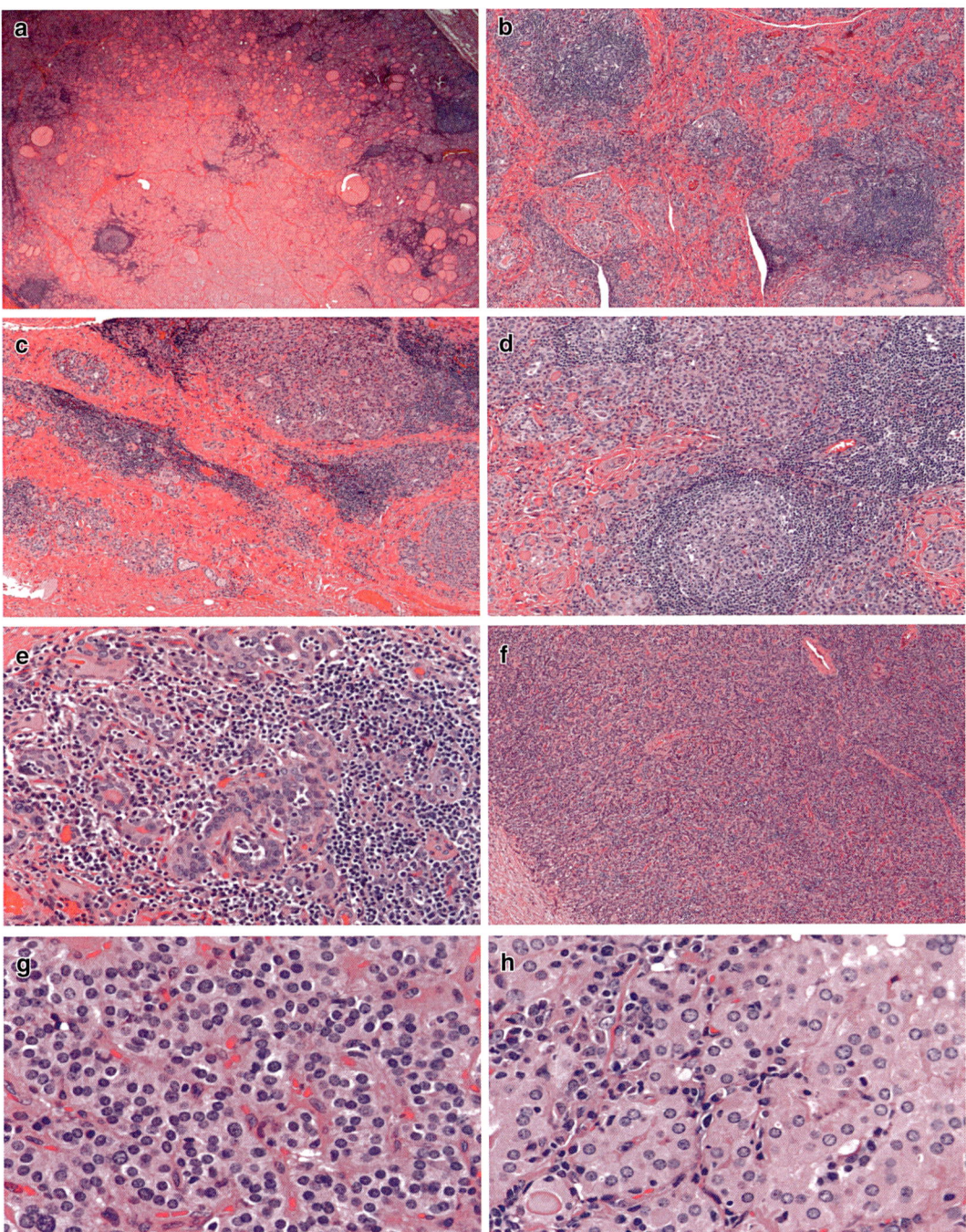

Fig. 11.11 This panel illustrates the occurrence of a Hürthle cell nodule with characteristic oncocytic morphology of the tumor cells (**a**, ×12.5; **b**, ×50; **c**, ×50; **d**, ×100; **e**, ×200; **f**, ×50; **g**, ×400; **h**, ×400; hematoxylin-eosin staining). (See text for details)

11.3.4.7 Anaplastic Thyroid Carcinoma

This type of tumor is also practically inexistent in children and youth, being reserved to older age group. It exhibits the highest mortality (100%) among thyroid gland neoplasms. ATC arises from pre-existing, less aggressive, well-differentiated thyroid carcinomas, specifically papillary thyroid cancer (PTC) and follicular thyroid cancer (FTC), through a process of dedifferentiation, where epithelial and thyroid differentiation markers are lost. This process of dedifferentiations may recall the dedifferentiated liposarcoma. ATC is a rapidly growing neoplasm causing dysphagia/dyspnea and shows grossly hemorrhagic/necrotic areas. Histologically, ATC has three significant growth patterns (squamoid, spindle cell, and giant cell). Immunohistochemically, ATC is (+) VIM, CAM5.2, and EMA but (−) TG and TTF-1. If CALC is (+), the tumor should be considered an anaplastic variant of MTC.

11.3.4.8 Other Epithelial Tumors

Two recently added entities need to be taken into account. They are *spindle epithelial tumor with thymus-like (elements of) differentiation* (SETTLE) and *carcinoma showing thymus-like differentiation* (CASTLE), which need to be considered in the differential diagnosis of *thymoma*, arising from intrathyroid ectopic thymus tissue. Moreover, practically inexistent in childhood and youth are *clear cell tumor* or mucinous carcinoma (DPAS(+), mucicarmine (+)), *squamous cell carcinoma, mucoepidermoid carcinoma, sclerosing mucoepidermoid carcinoma with eosinophilia*, small cell carcinoma, and *paraganglioma*. The paraganglioma may arise from the carotid body (Zellballen and S100 (+) sustentacular cells). The *small cell carcinoma*, has been suggested to be a poorly differentiated medullary carcinoma.

SETTLE occurs in youth and is indolent and may metastasize to the lung and kidney. SETTLE is characterized morphologically by a lobular growth pattern with thick sclerotic fibrous septs as well as spindle (fascicles/storiform) and epithelioid cells with bland cytology and low MI forming papillae, tubules, trabeculae, and solid sheets ± Hürthle cell-like corpuscles) ± cysts from mucinous/respiratory epithelium without immature lymphocytes (DDX: intrathyroidal thymoma, which shows immature lymphocytes). Immunohistochemically, SETTLE exhibits (+) SMA, MSA, CD99 (spindle cells), and (+) keratins/AE1-3 (epithelioid cells) but (−) CD5, CALC, and TG.

CASTLE may occur in youth but is more prevalent in the middle age group and has a sluggish growth with local complications and rare tendency to give metastases. Histologically, lobular and insular patterns with fibrous septs and extrathyroidal spread have been described. Immunohistochemically, CASTLE is (+) CD5 and Bcl2 and (−) CALC and TG. Substantially, CASTLE shows also thymic carcinoma markers, including HMW-CK (high molecular weight keratin), p63, CD117, MCL1 (apoptosis regulator), CEA (carcinoembryonic antigen), GLUT1 (glucose transporter 1), and PAX8 (paired box 8).

11.3.4.9 Non-epithelial Neoplasms

Malignant lymphoma and *teratoma* do play an important role in childhood and youth. *Plasmacytoma* is extremely rare in childhood and adolescence and should be reserved for tumors exclusively composed of atypical plasma cells and may be local or a manifestation of the generalized neoplastic disease. Malignant lymphoma is more often in the middle age but may occurs in childhood. It should be taken into consideration of the differential diagnosis of thyroid gland neoplasms. This is especially important to consider in patients following exposure to radiation therapy. Malignant lymphomas also arise in the setting of lymphocytic and Hashimoto thyroiditis. Two most common types have been described, including the *diffuse large B-cell lymphoma* and the *extranodular marginal zone B-cell lymphoma*. Clinically, the malignant lymphoma presents as one or more cold nodules, and grossly they show a fish-flesh solid tumor mass white cut surface. Morphologically, there are confluent sheets of B cells and immunohistochemically are not different from the respective nodal profile. Prognosis is variable, but it may recur, and if it does, it may recur in the gastrointestinal tract.

Teratoma may occur in infants/children. It is usually cystic and mature with benign, favorable course. Ruling out germ cell tumors and malignant SRBCTs is essential for the oncologist.

Rare other non-epithelial neoplasms include benign and malignant vascular tumors (hemangioma/hemangioendothelioma/angiosarcoma), smooth muscle tumors (leiomyoma, leiomyosarcoma), peripheral nerve sheath tumors (schwannoma, malignant peripheral nerve sheath tumor), solitary fibrous tumor, follicular dendritic cell tumor, Rosai-Dorfman disease, and LCH. Mucin-containing tumors include mucinous carcinoma, mucoepidermoid carcinoma (MEC), sclerosing MEC with eosinophilia, adenosquamous carcinoma, and an amphicrine variant of medullary thyroid carcinoma (mucin-secreting SR-like cells). The sclerosing MEC with eosinophilia has a background of Hashimoto thyroiditis, nests, and cords in both mucoid and epidermoid components, diffuse sclerotic stroma with eosinophils ((−) TG, DDX: Nodular Sclerosis-variant of HL, sclerosing Hashimoto thyroiditis, PTC with squamoid metaplasia, and squamous cell carcinoma).

11.3.4.10 Secondary (Metastatic) Tumors

It is crucial to recall soft tissue tumors, particularly small round blue cell tumors (e.g., rhabdomyosarcoma, embryonal) in childhood, but in the youth, the skin (melanoma), breast, kidney, and lung play a significant role.

11.4 Parathyroid Gland Pathology

Parathyroid gland pathology may manifest with hyper- or hypofunction of the parathyroid gland physiology. Two clinical syndromes may be encountered clinically.

Hyperparathyroidism

Primary hyperparathyroidism is characterized by an increase of PTH secretion due mostly to either adenoma or chief cell hyperplasia. Carcinoma is seen in about 1 out of 20 cases. It is usually encountered in adults but can be familial and seen in MEN-I and MEN-II (or IIa) and has been associated with neck radiation and sarcoidosis. Biochemically, hypercalcemia, hypophosphatemia, and duodenal peptic ulcers are seen. Clinically, Recklinghausen disease (*osteitis fibrosa cystica*) and nephropathic changes are found.

> **Box 11.16 Parathyroid Gland Pathology**
> 11.4.1. Congenital Anomalies of the Parathyroid Glands
> 11.4.2. Parathyroid Gland Hyperplasia
> 11.4.3. Parathyroid Gland Adenoma
> 11.4.4. Parathyroid Gland Carcinoma

1. Recklinghausen disease shows an expansile multilocular bony lesion, most of the jaw with alternating solid and cystic areas, hemosiderin deposition in MΦ (Fe+), and MNGC, which is following the reversible removal of the hyperfunctioning parathyroid glands.
2. Nephropathy includes nephrolithiasis and nephrocalcinosis.

Secondary hyperparathyroidism is delineated by an increase of PTH secretion as a consequence of chronic renal disease (commonly vitamin D-resistant) or intestinal malabsorption. Biochemically, there are hypocalcemia and hyperphosphatasemia. Morphologically, chief cell hyperplasia is noted.

Tertiary hyperparathyroidism is characterized by the autonomy of one or more parathyroid glands in patients with secondary hyperparathyroidism.

Finally, ectopic hyperparathyroidism may originate in renal cell carcinoma and pulmonary small cell carcinoma and need to be considered particularly in young non-smoker females of Chinese origin that may develop lung cancer.

11.4.1 Congenital Anomalies of the Parathyroid Glands

- *DEF*: Anomalies of the ontogenesis, mostly involving the parathyroid gland.
- *Parathyroid Gland Dystopia* (*dystopia glandularum parathyroidarum*) is an unusual location of the parathyroid glands, which may involve one or more parathyroid glands. Anomalous locations include intrathyroidal, carotid sheath, retroesophageal, and anterior mediastinum.

11.4.2 Parathyroid Gland Hyperplasia

- *DEF*: Chief cell hyperplasia or water-clear cell hyperplasia. Chief cell hyperplasia means hyperplasia of chief cells. Water-clear cell hyperplasia means hyperplasia of water-clear cells.
- *Chief Cell Hyperplasia*.

- *SYN*: Primary nodular hyperplasia.
- *GRO*: Enlargement of all glands, usually with tan-reddish discoloration and superior glands larger than inferior glands.
- *CLM*: Lobulated architecture (with enlargement of single lobules) AND intralobular scattered fat cells ± fibrous septs, giant nuclei (DDX secondary hyperparathyroidism – chief cell hyperplasia, characterized by more variability, fewer giant nuclei, and more oxyphil cells).

- *Water-Clear Cell Hyperplasia*.
- *GRO*: Extreme enlargement of all glands (superior glands are more abundant than inferior glands).
- *CLM*: Cells with optically clear cytoplasm with numerous clear vacuoles and size variability without giant nuclei.

11.4.3 Parathyroid Gland Adenoma

- *DEF*: It is a benign epithelial tumor of the parathyroid glands (an inferior gland in ~¾ of cases) with *CYCLIN D1/PRAD1* gene rearrangement, *MEN1* (menin), and, less frequently, *HRPT2* (parafibromin) gene mutations.
- *EPI*: ♂:♀ = 1:3, 2nd-3rd decades of life.
- *GRO*: Encapsulated nodule.
- *CLM*: There is a diffuse growth of predominant chief cells, although other cells may be seen, with marked variability in nuclear size and hyperchromasia favoring benign vs. malignant and rare mitoses or no mitoses = nil mitotic index. Different growth patterns include nesting, follicular, and papillary.

Features favoring parathyroid gland adenoma include one gland enlarged over three normal glands, excess weight (>1 g), lack of adipocytes, thin capsule separating it from the uninvolved gland, "endocrine-type atypia" in oxyphil cells or adjacent to hemorrhage with nil MI. Some authors that just one of these criteria may favor the diagnosis of adenoma over hyperplasia, but other authors require the fulfillment of all criteria and probably the truth is in the middle (*in medias res stat virtus*).

Variants: Oxyphil cell adenoma and lipoadenoma (variable number of adipocytes within a solitary, nonlobulated lesion).
- *TEM*: Glycogen and secretory vacuoles.

11.4.4 Parathyroid Gland Carcinoma

- *DEF*: It is a malignant epithelial tumor of the parathyroid gland with *HRPT2* (parafibromin) gene mutations or, less frequently, *CYCLIN D1/PRAD1* gene rearrangement, *MEN1* gene changes, functioning/nonfunctioning (more aggressive).
- *CLM*: Trabecular arrangement of tumor cells ± spindling AND features favoring carcinoma over adenoma: vascular invasion, capsular invasion (penetration with growth into adjacent tissue), perineural invasion, and/or metastasis. The absence of the above-described features may be suggested, if the following features are seen:
  1. +/++ MI (evident mitotic activity)
  2. Dense fibrous bands (vs. fibrous tissue ± hemosiderin and cysts as seen in adenoma)
- IHC: (+) p105 (pRb) favors carcinoma vs. adenoma, but if (−) p105 → not conclusive!

In some reports, the term atypical adenoma has been used to identify an adenoma with some features of carcinomas of the adrenal gland. They include adherence of the parathyroid gland neoformation to the neighboring structures, increased mitotic activity, fibrosis, trabecular growth pattern, and the presence of tumor cells inside of the capsule but without evidence of invasive growth, including invasion through the capsule, invasion of the blood vessels, and invasion of the perineural space. In carcinomas of the parathyroid gland, there is invasion of the neighboring structures, blood vessel invasion, and perineural infiltrates. A mimicker of the invasion and therefore a caveat may be the condition called parathyromatosis, i.e., the transfer of parts of parathyroid gland due to previous surgery. Thus, clinical history may be determining and the detailed revision of the medical notes is a very helpful clue for the pathologist dealing with an intraoperative frozen section from parathyroid gland.

In parathyroid adenoma, large, pleomorphic, and hyperchromatic nuclei can be seen, but the mitotic index is rare or nil. If numerous mitoses are found, the pathologist should look for the presence of invasion and other parameters that can direct to the diagnosis of parathyroid carcinoma. Parathyroid carcinomas are, however, extremely rare.

Pseudoadenomatous hyperplasia is when one gland looks like larger than others in the setting of parathyroid hyperplasia, while a rarer form is *parathyromatosis*, which is characterized by numerous microscopic foci of hyperplastic parathyroid gland tissue in the neck.

Atypical Adenoma: Parathyroid gland tumor is lacking unequivocal evidence of invasion but showing some other features suspicious for malignancy, including broad fibrous bands with or without hemosiderin depots, conspicuous or accurate mitotic figures, and neoplastic cell groups delimitated by a thickened fibrous capsule ("non-penetrating capsular invasion"). This diagnosis is controversially accepted in the scientific literature.

The features suggesting carcinoma include MI >5 or >10 HPF OR > 5%, metastasis (+), abnormal monotonous cells with prominent nucleoli, ↑ N/C, spindling, coagulative tumor cell necrosis, invasion (angio-/vascular, capsular, perineural, surrounding soft tissue), and a thick hyaline fibrous capsule with internal fibrous bands enclosing nodules. It is important to remember the pseudoinfiltrative pattern of parathyroid tissue implants that can arise from previous surgery or pseudoinfiltrative pattern of clear cells in the wall of a benign parathyroid cyst and about metastasis to rule out renal cell carcinoma and small-cell type of lung carcinoma. Normal/atrophic chief cells need to be distinguished from hyperplastic/neoplastic chief cells. Two essential differential diagnoses are oxyphil parathyroid adenoma vs. thyroid Hürthle cell neoplasm and microfollicular parathyroid adenoma vs. thyroid follicular neoplasm.

11.5 Adrenal Gland Pathology

In Box 11.17 are listed the topics covered in the pathology of adrenal gland and paraganglia.

> **Box 11.17 Pathology of the Adrenal Glands and Paraganglia**
> 11.5.1. *Congenital Anomalies*
> 11.5.2. *Dysfunctional Adrenal Gland*
> 11.5.3. *Adrenalitis*
> 11.5.4. *Neoplasms of the Adrenal Gland and Paraganglia*
>   11.5.4.1. Adrenal Cortical Adenoma
>   11.5.4.2. Adrenal Cortical Carcinoma
>   11.5.4.3. Neuroblastoma
>   11.5.4.4. Pheochromocytoma

11.5.1 Congenital Anomalies of the Adrenal Gland and Paraganglia

- *DEF*: Anomalies of the ontogenesis, mostly involving the adrenal gland.

Developmental Anomalies
A-/Hypoplasia (formation): Complete absence or reduced presence of the adrenal gland as well as small mono- or bilateral adrenal glands. All these anomalies of development are rare events. Additional abnormalities include dystopias (abnormal localization of the adrenal gland or paraganglia) and dysnomia (an unusual number of the adrenal glands and paraganglia). A significant pitfall is the occurrence of paraganglion / paraganglia in the adrenal gland mimicking a carcinoma. Ectopy of the adrenal gland (so-called ectopic adrenal gland) may occur in some anatomic sites. Ectopy is usually observed in retroperitoneum, anywhere along the urogenital ridge from the diaphragm to the pelvis, and is generally composed only of the cortex. However, small rests have also been occasionally found subcapsularly in the kidney, in the hilar region of testis or ovary, or in the sac of a hernia. Two types are distinguished, including the anencephalic type and the cytomegalic one (Fig. 11.12).

Another anomaly is the finding of heterotopic tissue in the adrenal gland. Such a condition may represent an *amartia* or a *coristia*, whereby *amartiae* are heterotopias with tissue homogeneous of the accepting organ, while *coristiae* are

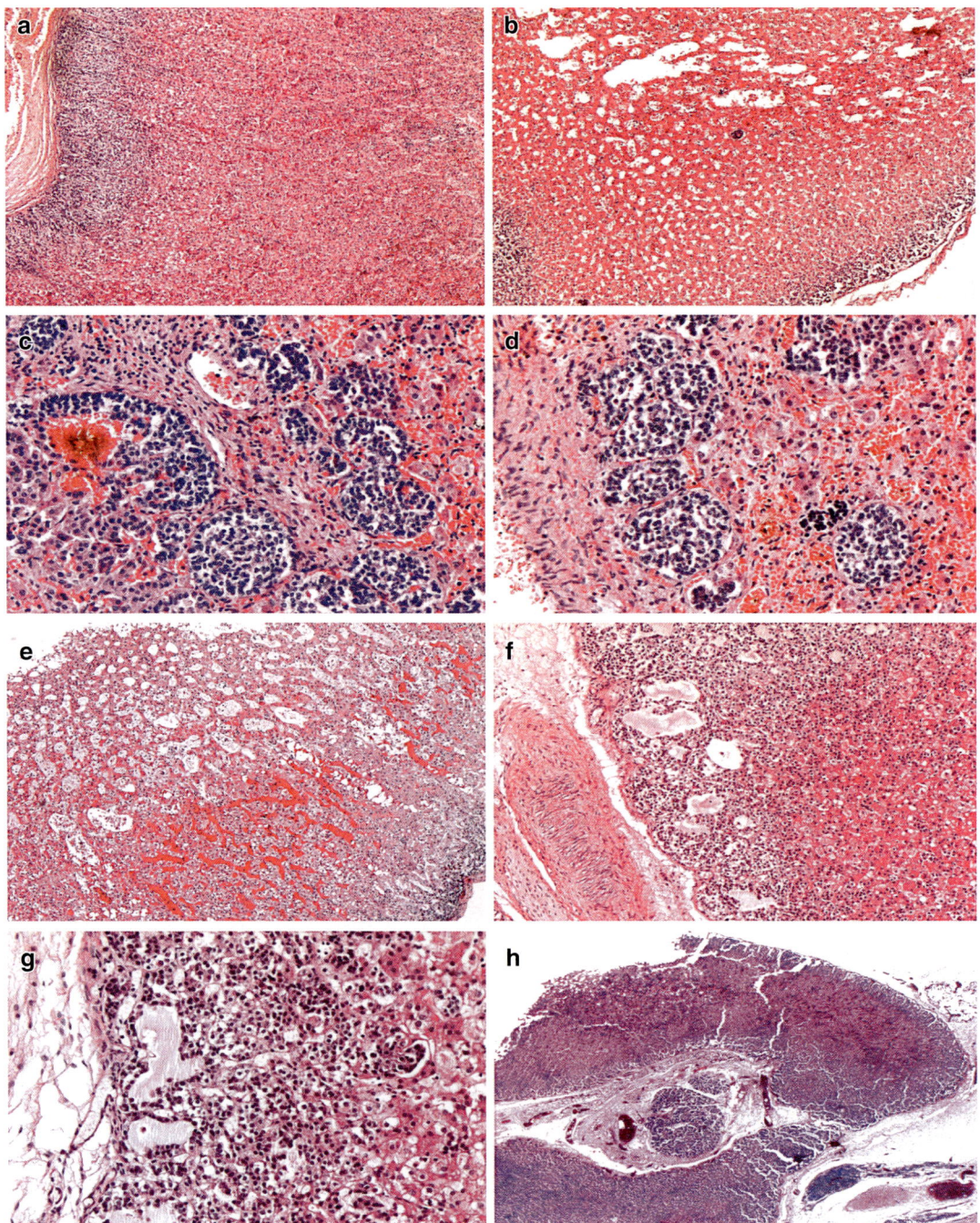

Fig. 11.12 The panel illustrates a normal adrenal gland (**a**, ×50) with loose medulla on postmortem examination (**b**, ×50). Medulla neuroblasts form clusters during the intrauterine life (**c**, ×200; **d**, ×200). Stress- or sepsis-related microcystic degeneration is displayed in Figure **e** (×50), Figure **f** (×50), and Figure **g** (×50). Figure (**h**) shows a hyperplastic nodule (×12.5). (All microphotographs are taken from hematoxylin-/eosin-stained histological slides)

heterotopias with tissue, which is heterogeneous to the accepting organ. In Fig. 11.12f is shown heterotopic pancreas in a fetal adrenal gland.

Embryologically, the heterotopic pancreas is defined as pancreatic tissue that has no contact with the orthotopic pancreas with its ductal draining system and vascular blood supply. About the origin of this congenital anomaly, it is postulated that the pancreatic primordium adhered to the original intestine or penetrates the intestinal wall during the development of the embryo. Heterotopic pancreas is not unusual in the gastric wall, abdominal wall, intestinal wall, and mesenteric wall, because of the original bowel movements linked to the embryogenesis. There is only one additional case of heterotopic pancreas in the adrenal gland described in a 21-year-old woman who presented with chronic lower back pain for a week without urinary disturbance or gastrointestinal pain or discomforts. Ultrasound revealed a cyst in the left kidney, and computed CT scan showed a cyst in the area of the adrenal gland. Removal of the cyst was performed. Histopathologic examination of the removed cyst wall showed heterotopic pancreatic cyst with cystic degeneration. Despite the close anatomic relationship of the adrenal gland and the tail of the pancreas, these two organs have different embryologic origins, and an explanation is still controversially debated for pancreatic *coristiae*.

Agenesis Absence of the adrenal gland(s) for lack of the primordial anlage.

Heterotopy Accessory adrenal tissue can be found in areas of the celiac axis, kidney, broad ligament, adnexa of the testis, spermatic cord, placenta, and, even, liver and lung.

Adrenal Cytomegaly Incidental finding, 3% newborns, 6.5% premature stillbirths, and usually ≤2 months of age, fetal cortex-located large (up to 120 microns in diameter) cells with marked nucleomegaly with pleomorphism and hyperchromasia, containing >25% the normal amount of nuclear DNA, ± occasional nuclear pseudoinclusions due to nuclear indentation or folding with intranuclear protrusion of cell cytoplasm.

Focal "Adrenalitis" Small foci of lymphocytes and plasma cells at the cortico-medullary junction and perivenular may accompany retroperitoneal chronic inflammatory processes.

Ovarian Thecal Metaplasia Small, partially hyalinized, fibroblastic nodules, wedge-shaped, and capsule-attached.

Adreno-Renal Fusion The adrenal glands are united in the midline with some medial deviation of kidneys in most cases. This condition has been associated with Cornelia de Lange syndrome.

11.5.2 Dysfunctional Adrenal Gland

- *DEF*: It is a complex of disorders in which the adrenal glands produce an abnormal amount of sex hormones and cortisol (e.g., Cushing syndrome and Addison disease) (Fig. 11.12).

11.5.3 Adrenalitis

It is an inflammation of the adrenal gland, which is caused by autoimmunopathies or infective events leading to adrenal insufficiency. The most frequent etiology of Addison disease is indeed an autoimmune lymphocytic adrenalitis. In this form, there is lymphoplasmacytic inflammation of the gland leading to the increased destruction of adrenocortical tissue. Another form of adrenalitis is of TB origin with a typical granulomatous change of the gland harboring caseous necrosis. Both types are sporadic in childhood. In both childhood and youth, other bacterial and viral infections play a role. The bacteria *β-Streptococcus*, *N. meningitis*, Epstein-Barr virus (EBV), and cytomegalovirus (CMV) may affect the adrenal glands in patients, specifically with immunodeficiency disorders or at posttransplant time.

Hypoadrenalism
Acute Hypoadrenalism is differentiated in primary (→ prolonged or difficult delivery, sepsis or Waterhouse-Friderichsen syndrome, the stress-related crisis in patients with Addison disease, or too rapid withdrawal of steroids) or secondary (any disorder of hypothalamus or pituitary resulting in decreased levels of ACTH).

Chronic Hypoadrenalism or Addison disease is defined as any condition able to destroy >90% of the adrenal cortex. Etiologies include usually adrenalitis, AI-related atrophy, and tuberculosis, but also amyloidosis, hemochromatosis, and metastases have been described.

Isolated Mineral-Corticoid Deficiency refers to a deficiency of aldosterone production due to impaired release of renin from kidney in patients with DM, AI disease, amyloidosis, sickle cell anemia, heparin administration, and neoplastic proliferations. An idiopathic form is due to an AR-inherited disorder associated with a deficiency in CYP11B2 enzyme able to convert an 18-hydroxyl group to aldehyde at the end of aldosterone biosynthesis, and infants have a failure to thrive, recurrent dehydration, salt wasting.

Primary Chronic Adrenocortical Insufficiency (Addison Disease)
Any conditions which destroy >90% of the adrenal cortex produce the clinical picture of Addison's disease including insidious development of weakness, fatigability, anorexia, nausea, vomiting, weight loss, hypotension, and hyperpigmentation (from elevated proopiomelanocortin peptides).

Two most common causes are idiopathic adrenalitis/atrophy (probably AI-related) and TB. Others include amyloidosis, hemochromatosis, and metastatic carcinoma.

In general, the cortex is atrophic with variable amounts of chronic inflammatory cells.

AI Addison disease is often divided into two types: type I, involving hypoparathyroidism, mucocutaneous candidiasis, and defect in suppresser T-cell function, and type II (Schmidt syndrome) involving AI thyroid disease, and insulin-dependent diabetes mellitus (IDDM), associated with HLA-A1 and B8.

Primary Acute Adrenocortical Insufficiency
It can develop as a crisis in patients with Addison disease (precipitated by stress), following the too rapid withdrawal of steroids (suppressed endogenous production recovers slowly), or result from massive hemorrhage.

In neonates, it can present with an enormous bleeding following prolonged or difficult delivery, presumably secondary to hypoxia or trauma.

Waterhouse-Friderichsen syndrome (WFS): Hemorrhagic destruction of the adrenal glands related to a severe bacterial infection (*N. meningitidis*, *S. pneumoniae*, *S. aureus*, *H. influenzae*) with widespread petechiae, purpura, and hemorrhages throughout body, particularly involving the skin and mucosal surfaces. Adrenals are hemorrhagic and necrotic, and, occasionally, they present as merely sacs full of blood clot.

Secondary Adrenocortical Insufficiency
Any disorder of the hypothalamus or pituitary gland resulting in decreased levels of ACTH is the correct definition for secondary adrenocortical insufficiency. Since tropic hormones are low, we do not usually see any hyperpigmentation. Secondary adrenocortical insufficiency may also be seen in the setting of exogenous corticosteroids.

Hypoadrenalism

- Infants: Adrenal gland hypoplasia (e.g., anencephalic fetus), familial unresponsiveness to ACTH, adrenal hemorrhage, or overwhelming sepsis (WFS).
- Children and youth: Autoimmune adrenal insufficiency (Addison disease) alone or in association with another AI endocrinopathy (e.g., thyroiditis, IDDM) and infection (TB, fungi).
- Adrenoleukodystrophy is a group of X-linked recessive inherited congenital disorders of long-chain fatty acid metabolism with signs of progressive neurologic deterioration in association with adrenal insufficiency.
- Secondary insufficiency due to chronic ACTH deficiency or after withdrawal of pharmacologic steroid therapy as a result of suppression of pituitary ACTH.

Hyperadrenalism
Conn Syndrome → Z.G.-targeting syndrome → ↑ aldosterone levels in the absence of ↑ renin levels. There is no edema, unlike secondary hyperaldosteronism. The underlying cause of Conn syndrome is almost invariably an adrenocortical adenoma with identifiable features,

including Ø<2 cm, bright canary yellow, nonencapsulated neoplasm grossly, and mixed cell types histologically (*vide infra*)

Cushing Syndrome → Z.F.-targeting syndrome → Clinical features include impaired glucose tolerance (overt diabetes in 20%), "moon" facies, "buffalo hump," abdominal striae, loss of libido, acne, vascular fragility with skin hemorrhages, as well as hypertension and muscle weakness.

"Pituitary" Cushing Syndrome (60–70%): Bilateral adrenal hyperplasia due to an elevation of ACTH levels, usually due to a pituitary adenoma (Cushing disease) is the most common etiology in children older than 7 years of age. It is suppressible by high doses of dexamethasone, and the hyperplasia is usually diffuse.

"Adreno-Nodular" Cushing Syndrome (20–25%): Functioning neoplasm (adenoma or carcinoma) of the adrenal cortex with low serum ACTH and symptoms not suppressible by high doses of dexamethasone.

"Ectopic" Cushing Syndrome (10–15%): ACTH (or molecular compound with similar biological activity) is produced by a non-endocrine neoplasm, including bronchogenic carcinoma, teratoma, and thymoma. It is not suppressible by high doses of dexamethasone.

"Iatrogenic" Cushing Syndrome: Patients on chronic steroids, usually transplant recipients or with AI disorders, may present with this kind of syndrome.

Virilism/Feminization of adrenal origin → Z.R.-targeting syndrome → Clinical features are most readily recognized in females due to androgen secretion. The underlying organic cause is an adenoma, which is often pigmented (*vide infra*).

In *Cushing syndrome*: The diagnosis relies on two criteria. The 1st criterion is a 24-h urine free cortisol test (an overnight dexamethasone suppression test administered the night before, which fails to suppress the morning serum cortisol level, is supportive of possible Cushing syndrome). The 2nd criterion is the dexamethasone suppression test in a prolonged fashion (a test including low dose for 2 days and high dose for 2 days, which is needed to differentiate Cushing disease, namely, bilateral adrenal hyperplasia due to pituitary adenoma from Cushing syndrome, namely, adrenal tumor).

Imaging (ultrasound, CT scan, and MRI as well as lung view for staging) is important once there is a suspicion of a tumor.

The differential diagnosis of adrenal cortical enlargement in childhood and youth contains some conditions that may mimic a neoplastic process. These include hemorrhage, infection, WFS, cysts, adrenal cytomegaly, and hyperplasia (CAH, congenital adrenogenital syndrome and ACH or Cushing syndrome).

Inflammation
Adrenalitis is a cause of primary chronic adrenocortical insufficiency or Addison disease (*vide supra*).

Vascular Changes
The adrenal gland is richly vascularized, including arterioles passing through cortex and capillaries from cortex to medulla with a portal-like circulation.

Degeneration
Acute stress reduces lipid content and weight, while prolonged stress induces hypertrophy and hyperplasia and increases the weight of the organ. *Microscopic cystic change*: formation of microcysts in the outer cortex in premature, stillborn, and newborns, who underwent *in utero* stress (Intrauterine Mycrocystic Degeneration of the Adrenal Gland). Amyloidosis, which may be an essential cause for elderly patients, is not seen in children or young patients, but *genetic hemochromatosis* may conversely be a cause of primary chronic adrenocortical insufficiency or Addison disease. However, the deposition of hemosiderin pigment is much more damaging for other organs than the adrenal gland. To keep in mind are also storage disorders, including adrenoleukodystrophy, Pompe disease, and Wolman disease.

Congenital Adrenocortical/Adrenal Hyperplasia (CAH) or Adrenogenital Syndrome
Several inherited disorders of adrenal steroidogenesis enzymes are known and impair cortisol production. The most commonly affected pro-

teins are 21-hydroxylase, 11β-hydroxylase, and 3β-hydroxysteroid dehydrogenase, while the less widely affected are 17-α-hydroxylase/17,20-lyase and cholesterol 20,22-desmolase. In the classical CAH, the androgen excess due to shunting of intermediates to other pathways leads female newborns with external genital ambiguity (virilization: clitoral hypertrophy and pseudohermaphroditism, which may lead to erroneous male sex assignment) and male newborns with the salt-losing crisis within days to weeks after delivery due to decreased synthesis of aldosterone. Infants develop in the first 2–4 weeks of life symptoms of salt-wasting, vomiting, hypotension, and shock and are hypovolemic, hyponatremic, hyperkalemic, hyperreninemic, and acidotic and often hypoglycemic. Postpuberal girls have oligomenorrhea, hirsutism, and cutaneous acne, while boys at peripubertal age show enlargement of external genitalia and precocious puberty.

Classical CAH is rare, affecting only 1 in 14,000 patients, but mild forms of the disease may occur at a remarkable rate of 1:100 live births. The altered enzyme can be entirely or partially impaired, and the degree of enzyme insufficiency determines the severity of the condition. There are several distinct clinical syndromes.

CAH – Diagnosis relies on (1) clinical with measurement of ↑ 17-OH-progesterone in the serum ± ACTH stimulation, (2) immunogenetics with HLA family studies, (3) genetics with molecular biology investigations, and (4) prenatal diagnosis with molecular-biological investigations.

Box 11.18 reveals some tips of pediatric auxologic pathology.

In the classical salt-wasting 21-hydroxylase deficiency, there is an impossibility to gather aldosterone and cortisol, a ↑ precursor of the 21-hydroxylase step, 17-hydroxyprogesterone, and ACTH. The excess of 17-hydroxyprogesterone shunts to androgens, namely, dihydroepiandrostenedione (DHEA) and androstenedione. In 11-hydroxylase deficiency, there is a decreased conversion of 11-deoxycortisol to cortisol with 17-hydroxyprogesterone shunted toward overproduction of androgens, as in 21-hydroxylase deficiency, and no conversion of deoxycorticosterone

> **Box 11.18 Under-/Overtreatment Signs**
> *Undertreatment*: ↑17-OH-progesterone, androstenedione, and renin in the serum, premature sexual hair growth, virilization, accelerated advancement of skeletal maturation, early epiphyseal fusion, and adult short stature.
> *Overtreatment*: Growth suppression and symptoms of hypercortisolism (in fact, the addition of a mineral-corticoid facilitates suppression of adrenal androgens with even very small doses of cortisol).

to corticosterone in the aldosterone pathway. Clinically, there is an overproduction of 11-deoxycorticosterone, which has mineral-corticoid activity, resulting in hypertension and hypokalemia, and overproduction of 11-deoxycortisol with ↑ of androstenedione and testosterone. Renin and aldosterone are suppressed, and tetrahydro-metabolites of 11-deoxycortisol and 11-deoxy-corticosteroids are increased in the urine.

Biochemical Endocrinological View

CAH is a group of biochemical disorders that arise from defective steroidogenesis and the adreno-corticotropic hormone (ACTH) regulates adrenal steroidogenesis via a rate-limiting step that results in the production of *Pregnenolone* promoting StAR protein to transfer free cholesterol to the inner mitochondrial membrane, the site where the first step in steroidogenesis takes place. The central nervous system (CNS) controls the secretion of ACTH, its circadian rhythm, and stress-related variation via the corticotropin-releasing factor (CRF) secreted by the hypothalamus. The adrenal steroidogenesis is part of the hypothalamic-pituitary-adrenal feedback system, which is activated by a variety of circulating levels of plasma cortisol inducing negative feedback of cortisol on CRF and ACTH secretion. An impairment of the enzyme function at each step of the adrenal steroidogenesis determines (1) a unique combination of accumulated precursor molecules in pathways blocked by the enzyme deficiency, (2) overstimulation of the adrenal cortex with hyperplasia of cortical cells, and (3) an excessive synthesis of adrenal products in those pathways unimpaired by the enzyme deficiency.

Adrenal Pseudohermaphroditism
This form of ambiguous genitalia needs to be differentiated by other types because females with classical CAH maintain the potential for fertility. In fact, internal female genitalia, such as the uterus, fallopian tubes, and ovaries, usually develop in contrast to the virilization of the external genitalia. There is neither testicular tissue nor production of an anti-Mullerian hormone (AMH), which is produced by the testicular Sertoli cells. Internal female genitalia are Mullerian derivatives and not responsive to androgens.

21-hydroxylase deficiency is the most common cause of CAH and is caused by genetic changes in the *CYP21A2* gene, which is located on 6p21.33, i.e. the short arm of chromosome 6 at position 21.33.

The most common variations of the *CYP21* gene are the result of either of two types of meiotic recombination events between CYP21 and CYP21P; the latter is a homologue or pseudogene showing 98% of sequence homology in exons: (1) misalignment and unequal crossing over, resulting in large-scale DNA deletions, and (2) transfer to CYP21 of small-scale mutations of the CYP21P pseudogene.

CLI: Corticotropin stimulation test

Prenatal: molecularly (CVS at 9–11 weeks of gestation or amniocentesis in the 2nd trimester of pregnancy, DNA extraction, and CYP21 gene investigation through a combination of molecular techniques to study substantial gene rearrangement and arrays of point mutation) or hormonally (measurement of 17-KS and pregnanetriol in the amniotic fluid).

CAH is AR, i.e., the risk is 1/4 of the fetus being affected with the disease, and 1/8 of the fetus is a female with ambiguous genitalia, and dexamethasone prenatally is used to prevent virilization of the external genitalia.

Preimplantation: There is only one report of preimplantation gene diagnostics utilized in a family where offspring was at risk for CAH. Chronic therapy with glucocorticoids at levels over physiology can result in decreased bone accrual and lead to osteopenia and osteoporosis. Bilateral adrenalectomy is used in rare cases when medical treatment alone showed poor improvement of the symptoms and signs of CAH.

In CAH there is marked hyperplasia of the *zona fasciculata* with the conversion of pale-stained, lipid-rich cells into lipid-depleted cells with dense, eosinophilic cytoplasm, similar to the cells of the *zona reticularis*. Enlarged hyperplastic accessory cortical nodules or their extensions seem common, but true cortical neoplasms are rare. Testicular tumors (Leydig) in CAH may also occur, and even the rare occurrence of adrenal myelolipoma and extra-adrenal osteosarcoma, ES, and astrocytoma have been reported.

BWS or EMG (exomphalos, macroglossia, gigantism) syndrome: Adrenal cytomegaly of the fetal cortex and hyperplastic and inappropriately mature adrenal chromaffin cells.

Carney syndrome

McCune-Albright syndrome

Ectopic ACTH: Bronchial carcinoid tumor, pulmonary and extrapulmonary, small-cell carcinoma, pancreatic islet cell carcinoma, medullary thyroid carcinoma, thymic carcinoid, and pheochromocytoma.

Stress-Related Changes

- Lipid reversion pattern is characterized by the outer aspect of *zona fasciculata* being composed of cells with lipid-depleted, eosinophilic cytoplasm, and the inner zone is composed of pale-stained, lipid-rich cells.
- Tubular degeneration of outer *zona fasciculata* with the conversion of columns and cords of lipid-depleted cells into hollow tubules with rare necrotic cortical cells.
- Intracytoplasmic globules of the *zona fasciculata* in patients dying of streptococcal meningitis, chronic renal failure, pneumonia, and overexposure to cold by overstimulation of the adrenal cortex.

Immunohistochemistry for neurofilament protein (NFP or simply NF) highlights the thin neuritic processes between nests of chromaffin cells as well as the filaments of the nerve bundle. Immunohistochemistry for $P450_{17alpha}$ and $P450_{C21}$ detect cytochromes, which are essential in adrenal steroidogenesis. $P450_{C21}$ detects in all three zones of the adrenal cortex, while $P450_{17alpha}$ is exclusively present in the *zona fasciculata* and *zona reticularis*.

11.5.4 Neoplasms of the Adrenal Gland and Paraganglia

Adrenal Gland Nodules and Diffuse Micronodular Hyperplasia

Adrenal cortical nodules (1+) can be seen in patients without clinical or laboratory evidence of hypercortisolism and can increase with age. They can also be seen in the setting of congenital heart disease, extracorporeal membrane oxygenation (ECMO)-based therapy, and pediatric intensive care unit (PICU) interventional therapy. The cortical nodules are also associated with hypertension and DM. Most small nonfunctioning adrenal cortical adenomas or so-called incidentalomas can produce biologically active steroids. In pituitary-dependent Cushing syndrome (Cushing disease) and the setting of ectopic ACTH syndrome, the adrenal glands may indeed show small areas of nodularity. Occasionally, this nodular aspect might be *diffuse* and *micronodular hyperplasia*, which shows on cut surface numerous nodules in the cortex, some of them projecting just beneath the capsule as an adrenal extension. Some others can be fused with the adrenal capsule. Finally, some may lie free in the adjacent parenchymal connective tissue.

Tumors

Persistent nodules of sympaticoblasts or primitive neuroblasts may be confused with small neuroblastomas. This aspect needs to be kept in mind as an important pitfall in endocrine pathology. *In situ* neuroblastoma is defined as "an adrenal lesion of microscopic size that is cytologically identical to typical neuroblastoma and is detected in infants without demonstrable metastases" by Shimada (2005), who based his definition on the Beckwith and Perrin original description of 1963. Essentially, we need to differentiate between adenoma and carcinoma in the cortex and neuroblastoma and pheochromocytoma in the medulla. Genetic associations include adrenal cortical neoplasm, Wilms tumor, and hepatoblastoma, which have an increased association with both hemihypertrophy and the BWS (11p15 locus). Of note, 11p- may be associated with hepatoblastoma, WT, and RMS. Adrenal cortical neoplasms may also occur in Li-Fraumeni syndrome (SBLA cancer family syndrome – sarcoma, breast, and a brain tumor, leukemia, laryngeal and lung cancer, and ACC) (OMIM#151623) on 17p13.1, which is caused by heterozygous mutations in the *TP53* gene. Congenital anomalies of the gut, hamartomas, and brain tumors, i.e., astrocytoma and medulloblastoma, have been described in patients with virilizing adrenocortical tumors. The most of adrenal cortical adenomas (ACA) is nonfunctional and discovered incidentally often at autopsy of by screening other non-endocrine systems, while adrenal cortical carcinomas (ACC) usually harbor a mass effect ("*tumor quia tumet*") and are found during life. Immunohistochemistry plays a significant role in the identification of the histology of adrenal tumors and their separation from other entities that can present at the same location or contiguous location. As indicated below, a majority of ACC are positive for synaptophysin (SYN) and negative for chromogranin A (CGA). Moreover, the triad "MIC" (*M*elan-A, *I*nhibin, and *C*alretinin) is usually positive and is helpful in distinguishing ACC from other primary or metastatic tumors.

11.5.4.1 Adrenocortical Adenoma
- *DEF*: It is a benign nonfunctioning or functioning epithelial tumor of the adrenal gland cortex.
- *EPI*: Rare in children, often < 50 g, Ø < 5 cm, ♂ < ♀.
- *GRO*: Variable morphology with no confluent necrosis or large hemorrhages.
- *CLM*: Nests and cords of neoplastic cells in a background of eosinophilic and hyalinized matrix.
- *PGN*: Management according to size (Ø <3 cm often monitoring only, 3–6 cm controversial management, >6 cm excision due to ↑ cancer risk).

It is distinguished in three most frequent types according to the cell composition and hormonal production:

Aldosterone-secreting Adenoma (⇒ Conn syndrome) "Bright canary yellow" tumor with ZG-like cells (small darkly stained neoplastic cells with round centrally placed nuclei and characteristic pale amphophilic cytoplasm), ± ZF-like cells, and a hybrid of the two forms. Spironolactone body is a whorled, scroll-like, pink globule surrounded by a clear halo presenting in patients with hyperaldosteronism treated with spironolactone and shows IHC-reactivity with aldosterone. TEM: Few lipid droplets and mitochondria with tubular and lamellar, "plate-like" profile of cristae, and concentric wraps of smooth endoplasmic reticulum (SER). The electron microscopic investigation of the spironolactone body reveals an ultrastructure constituted by a central core with amorphous electron-dense material surrounded by numerous smooth-walled concentric membranes or, better described as laminated "whorls", which are continuous with the SER.

Corticosteroids secreting Adenoma (⇒ Cushing syndrome) "Yellow with brown mottling" tumor with ZF-like cells with pale foamy, lipid-rich, cytoplasm, which may be demonstrated using ORO staining on frozen tissue, and round to slightly vesicular nuclei with small dot-like nucleoli. Additional features include ZR-like cells with compact eosinophilic cytoplasm (lipid depletion), focal nuclear hyperchromasia and cellular pleomorphism, myelolipomatous change (metaplasia), lipochrome pigment deposition, hemorrhage (focal), fibrosis, dystrophic calcification, and bony tissue (metaplasia). TEM: Prominent lipid droplets, mitochondria with a tubular-vesicular profile of cristae, which is typical of steroid-producing cells, small components of rough endoplasmic reticulum (RER) and free polyribosomes, and few primitive intercellular attachments.

Sex Hormones-secreting Adenoma (Virilization or Feminization) "Mahogany brown" cut-surface with ZR-like cells, which exhibit dense, eosinophilic cytoplasm and are often arranged in a diffuse or solid pattern. TEM: ZR-ultrastructure with much more mitochondria in the oncocytic variant.

Distinguishing between adenoma and carcinoma of a cortical tumor is not an easy task and has been highly subjective for years. It is not an easy task because the original parameters of weight and size, taken singly or combined, are not trust-able and no longer used. They can indeed be misleading. The Weiss system and the modified Weiss system do work quite well. The determinant histological features for malignant potential (≥3) include *M*itotic index (MI) >5/50 HPF, *A*typical mitoses, *N*uclear pleomorphism (significant), *N*ecrosis (confluent type), *I*nvasion, *C*apsular/sinusoidal/venous, *C*lear cells ≤25% of total or cystic, *S*oft tissue (peri-adrenal) extension, or "*S*olid" (diffuse) architecture. The mnemonic (acronym) word "MANICS" picking up all the initials of the determinant histological features helps to memorize these features. More or equal three of these features favor a malignant potential, and this system has been validated with a specificity of 96%, a sensitivity of 100%, and a good correlation score ($r = 0.94$). Nevertheless, it is important to emphasize that single features took singly- may not be of any congruent prognostic value. In a few cases, the pathologist should release a diagnosis of indeterminate or borderline tumor. To remember that focal cytologic atypia may be seen in adenomas and atypia is *not* a feature to distinguish between adenomas and carcinomas. Moreover, a careful inspection of the attached adrenal cortex or the contralateral gland may show atrophy. Another "adenoma" feature, which needs to be considered, is the "pushing border" with pseudocapsule. Also, this feature taken singly is not of prognostic value. Finally, there are also some special adenomas, including a pigmented "black" adenoma, which is due to the presence of abundant lipochrome pigment deposition, and a Leydig cell adenoma of the adrenal gland as well as a testosterone-secreting adrenal tumor. Prominent fibrous bands may also be a constant feature of malignancy (in addition to the "MANICS" mnemonics) such as the distinction between parathyroid gland adenoma and carcinoma.

11.5.4.2 Adrenal Cortical Carcinoma

- *DEF*: It is a malignant epithelial tumor of the cortical adrenal gland.
- *EPI*: Children & adults, often >50 g, Ø>5 cm, ♂ < ♀.
- *GRO*: Bulky mass with "coarseness" (coarse nodules), along with areas of confluent necrosis, hemorrhage, and occasionally cystic degeneration, yellow-to-yellow-orange or tan-brownish on the cut surface.
- *CLM*: Three basic microscopic patterns (Fig. 11.13) include (1) trabecular (broad anastomosing epithelial cords or serpentine, or ribbon-like columns of cells with delicate slit-like intervening sinusoids or vascular channels), (2) alveolar (nests of cells), and (3) diffuse (solid arrangement of cells). Cytologically, there are lipid-depleted cells with cytoplasm condensations mimicking hyaline globules. There is, apparently, a mixed microscopic pattern as well. Some more rare features described are a pattern showing free floating nests and a perithelial distribution with intervening zones of necrosis. Intracytoplasmic hyaline globules (round to oval, refractile, intensely eosinophilic, PAS+,

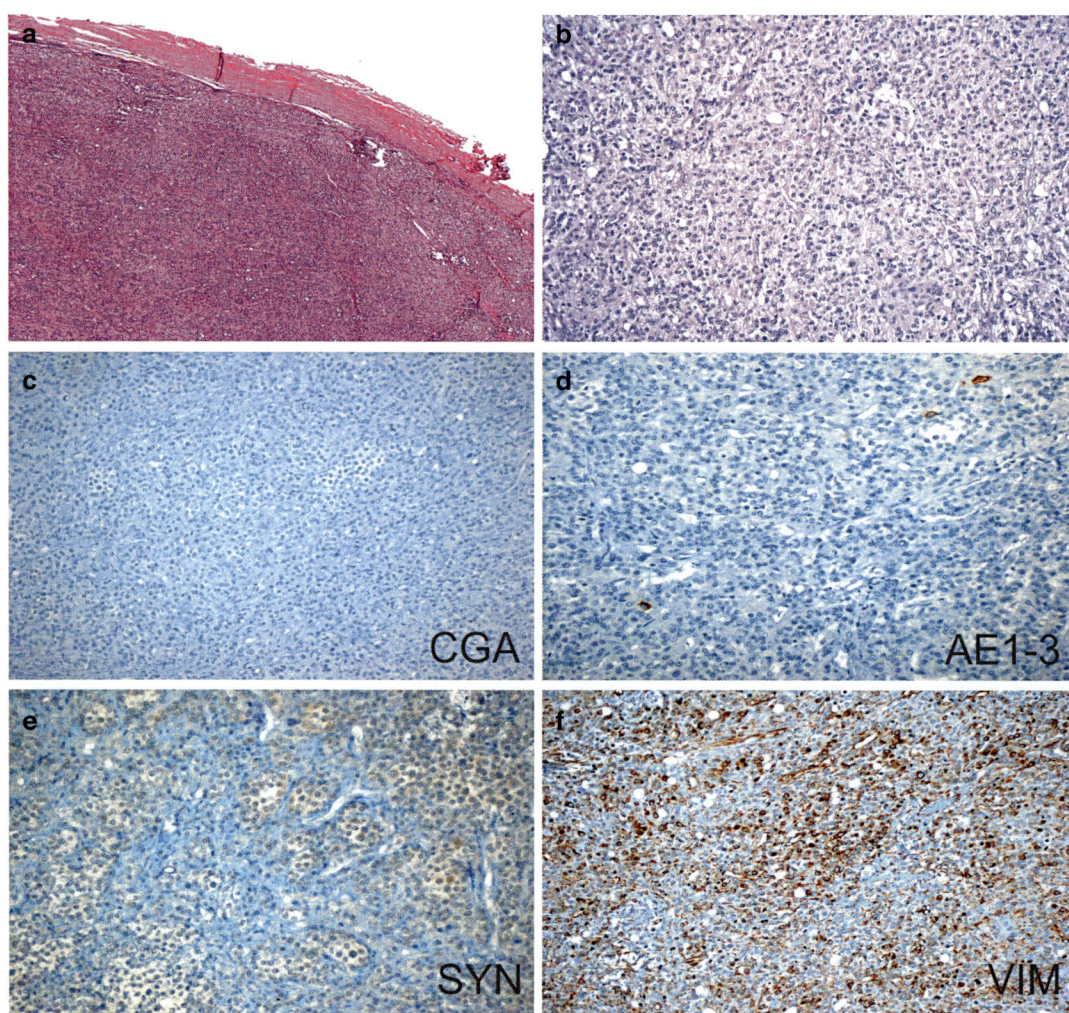

Fig. 11.13 An adrenocortical tumor is shown in this panel at low power (**a**, ×40) and high power magnification (**b**, 200). Both microphotographs are from hematoxylin-/eosin-stained histological slides. The four immunohistochemical microphotographs show no expression of chromogranin A (**c**, ×200), focal expression of pan-cytokeratins (**d**, ×200), no expression of synaptophysin (**e**, ×200), and strong expression of vimentin (**f**, ×200). Immunostaining was performed using the avidin-biotin complex

PASD+) may be a feature of both benign and malignant tumors and not characteristic of malignant potential (Box 11.19).
- *TEM*: RER stacks, abundant SER ("*tangled skein*"), mitochondria with tubulovesicular *cristae* or platelike cristae, variably sized lipid droplets with zonation (strong uniform density with or without an internal round clear zone), lysosomes, and primitive cell junctions, as well as focal dense-core granules constitute identifying ultrastructural elements.
- *PGN*: The prognosis depends on the stage, and about 2/3 of adolescents and young adults present with ACC at stage III or IV disease at the time of the first diagnosis. 5-YSR is about 65% (i.e., ~2/3 of the patients are alive 5 years after the primary diagnosis) if the tumor is strictly contained in the adrenal gland, but 10% in case of ACC with metastatic disease.

From the pearls of the pathology practice, it is essential to mention the features commonly associated with ACC according to Medeiros and Weiss (1992), which include high nuclear grade (according to Fuhrman criteria, Furhman et al. 1982), MI >5/50 HPF, atypical mitoses, cytoplasmic eosinophilia >75% of tumor cells, and diffuse architectural growth pattern in over 33% of the tumor, necrosis, invasion (venous, sinusoidal, and capsular). According to Madeiros and Weiss, the presence of ≤2 features foretells tumor with low-metastatic-potential (LMP), while the presence of >2 elements portends high-metastatic-potential (HMP) and/or recurrence. Although attention should be paid to all nine features, it seems that the three most important criteria are MI (>5/50 HPF), atypical mitotic figures, and venous invasion.

Identification of a Tumor (Usually Carcinoma) as Adrenal-Originating

Adrenal tumors, usually carcinoma, need to be differentiated by mimics originating from the medulla (pheochromocytoma) and other sites, including the liver (hepatocellular carcinoma, HCC) and kidney (clear cell renal cell carcinoma, CCRCC). *M*elan-A/Mart 1, *I*nhibin A/α, and *C*alretinin are positive in ACC and useful for differentiating with pheochromocytoma as well as metastatic carcinoma, which should be negative for the three markers and positive for chromogranin A. Immunoreactivity for the antibodies D11, SF-1, and DAX has been reported as useful in identifying adrenal cortical tumors but is not yet widely used in diagnostic practice and still under intense investigation. Hepatocellular carcinoma is positive for HepPar1 (Hepatocyte Paraffin 1), AFP (α-feto-protein), and AAT ($α_1$-anti-trypsin), while CCRCC is usually positive for CD10. However, as indicated previously, although validated monoclonal or polyclonal antibodies are beneficial for the diagnosis of an adrenal tumor, there is still the necessity to order antibodies in a panel and not singly to avoid abnormal expression, which may lead to a wrong diagnosis and a fatal verdict for the patient. Endocrinologically active ACC may be defined "functional" and subclassified as "pure," if

---

Box 11.19 Features Portend a Diagnosis of ACC
1. MI > 20/50 HPF ⇒ high-grade malignant potential (HMP).
2. ↑Ki67 (MIB1) and/or TP53 expression ⇒ malignant behavior.
3. Confluent Tu.-Necrosis with angioinvasion, and capsular/extra-adrenal invasion are worrisome.
4. CGA (−) (DDX: Pheochromocytoma), but SYN (+), other than (+) Melan-A, INA, Calretinin, VIM and CKs (∗); (−) EMA.

Notes: ∗ CK5 may be just weakly positive after protease digestion. Of note, DNA ploidy is not a reliable predictor of malignancy, and FNAC (fine-needle aspiration cytology) harbors a high rate of false positivity. Thus, it is extremely important to not diagnose ACC on FNAC. Substantially, generous embedding (1 block/cm) remains a key and necessary factor to increase the accuracy for the diagnosis of malignancy. Chessboard-like embedding of tumors with printed or scanned photograph to assess the malignant potential accurately is a wise decision. MI, mitotic index; HPF, high-power field; CGA, chromogranin A; SYN, synaptophysin; INA, inhibin A/α; VIM, vimentin; CK, (cyto-)keratins; EMA, epithelial membrane antigen.

there is Cushing syndrome only, and "mixed" if Cushing syndrome is accompanied by virilization or any biochemical evidence of excessive steroid production.

BWS tumors (incidence): Nephroblastoma > ACC > hepatoblastoma

Adrenocortical blastoma (of infancy): Mixture of immature epithelial and mesenchymal elements with recapitulating morphology of the developing fetal cortex and with focal vascular invasion, but no metastases.

Overall, ACC is rare in infants and young children but may be encountered in the practice of a pediatric pathologist. Occasionally, a newborn can even present with a congenital metastasizing ACC (Sherman et al. 1958).

PGN: Factors able to influence the outcome include age <5 years, complete resection, tumor weight <400 g, low MI, and minimal necrosis. These features are associated with an excellent result. In childhood, there are two clinical groups (infants and adolescents) with better survival in the former group (~50% vs. ~20%, Medeiros and Weiss, 1992). The presence of skin metastases ("blueberry muffin"-like nodules) is a worrisome sign (Lack et al. 1992). Sites of metastases include the lung, liver, peritoneum, regional LNs, kidney, and brain.

11.5.4.3 Neuroblastoma, Ganglioneuroblastoma, and Ganglioneuroma

Neuroblastoma
- *DEF*: The neuroblastoma is the most common solid extracranial and most common neoplasm diagnosed in children before the age of 1 year, derived by sympathetic progenitor cells, and characterized as hallmark by clinical, morphological, and molecular-biological heterogeneity with the likelihood of tumor progression varying widely according to the age at diagnosis, stage of disease, and tumor cell biology (Figs. 11.14, 11.15, 11.16, 11.17 and 11.18).
- *EPI*: 80% < 5 years; 35% in children <2 years; ~ 20 NB cases/10_6 live births/year.

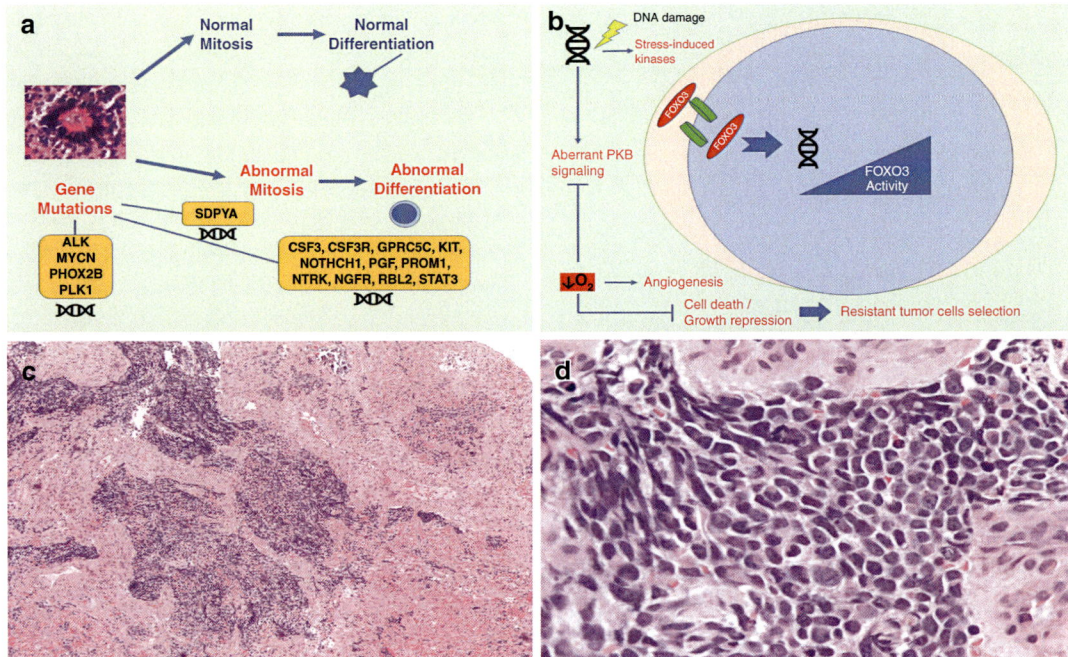

Fig. 11.14 The first two upper microphotographs illustrate some self-explaining pathogenetic mechanism for neuroblastoma (**a**, **b**). The subsequent six photographs illustrate the morphology of neuroblastoma as main component of the category of the small round blue cell tumors (**c–h**: **c**, ×50; **d**, ×400; **e**, ×100; **f**, ×400; **g**, ×50; **h**, ×400; hematoxylin-eosin staining). The tumor cells show a high nucleus to cytoplasma ratio and are aggregated in clusters. Ganglionic differentiation or initial tumor cell maturation is occasionally seen (**h**)

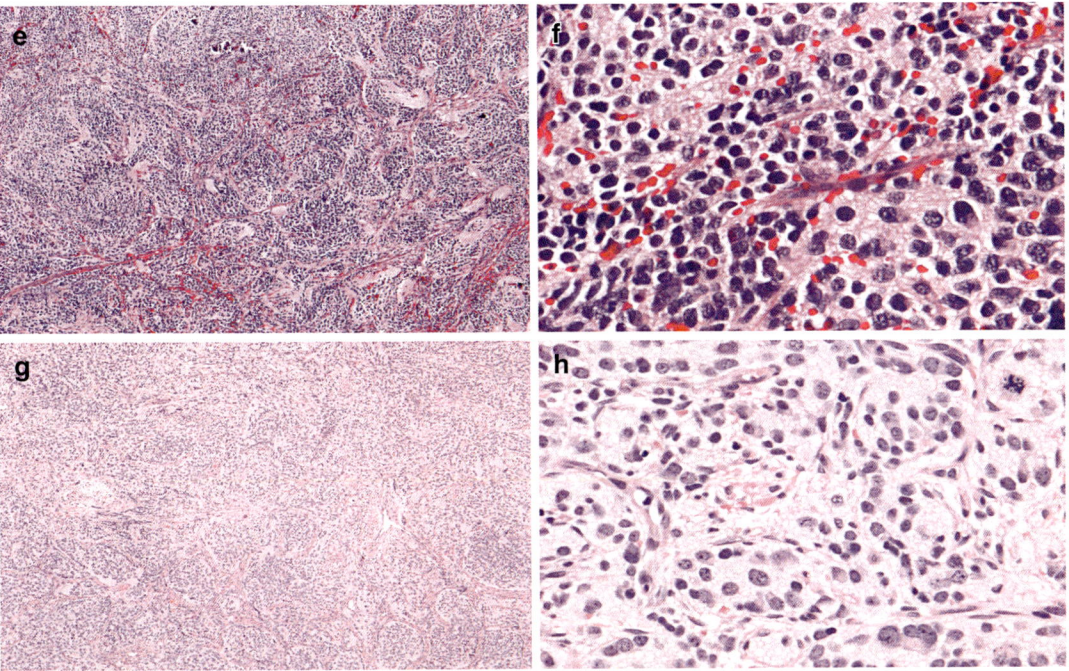

Fig. 11.14 (continued)

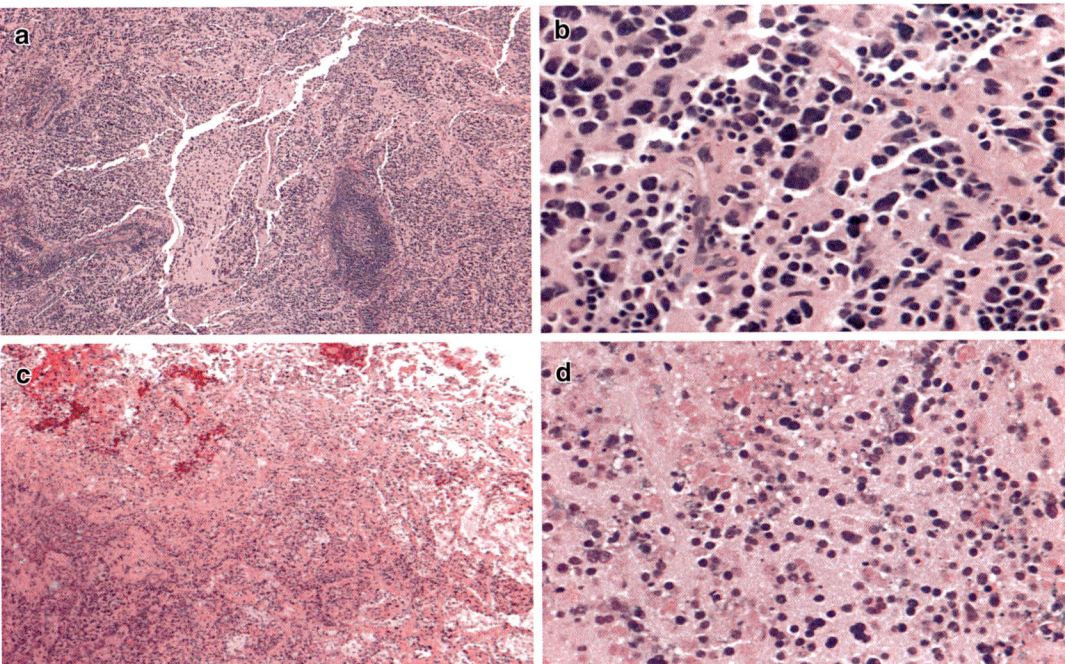

Fig. 11.15 These are additional microphotographs of several neuroblastomas with different degrees of differentiation (**a–h**: x50, x400, x50, x400, x50, x400, x200, and x400 as original magnifications; hematoxylin and eosin staining)

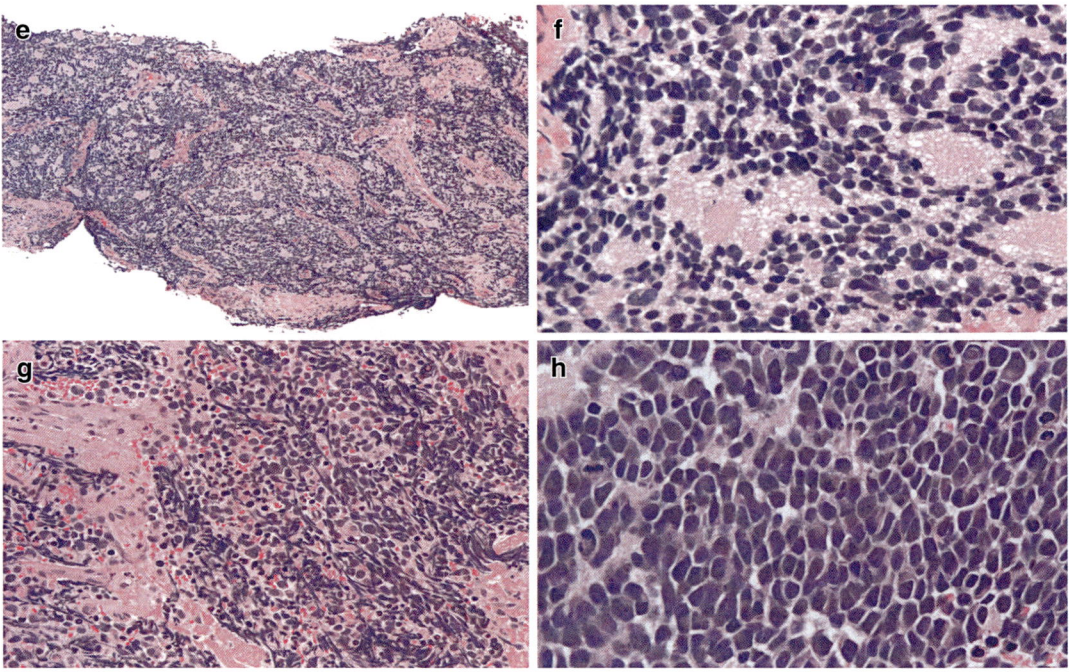

Fig. 11.15 (continued)

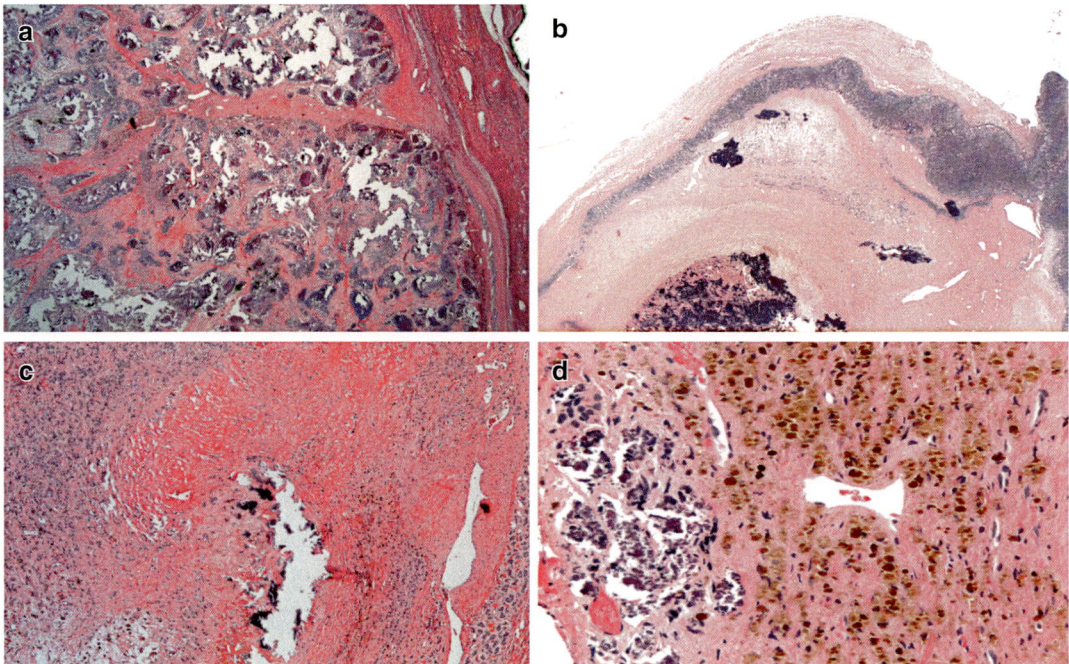

Fig. 11.16 The post-chemotherapy neuroblastoma may also disclose degenerative changes, which may represent a challenge for the diagnosis (**a–h**: 12.5x, 20x, 50x, 200x, 200x, 200x, 400x, 400x, respectively; hematoxylin and eosin staining). Dystrophic calcification, hemorrhage, necrosis, and pigment (hemosiderin)-laden macrophages are identifiable on hematoxylin and eosin stained histological slides. In many cases, a layer of non-neoplastic adrenal gland is recognized. Moreover, clear cell change and ganglionic differentiation can be observed. In some cases, some clusters of neuroblastoma remnants are seen

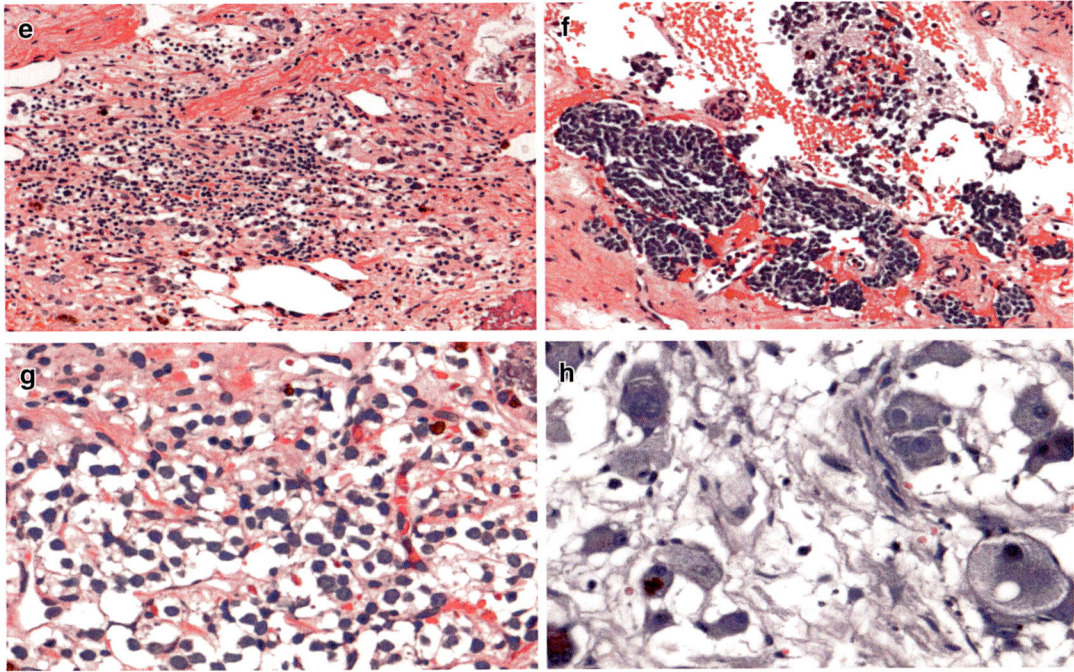

Fig. 11.16 (continued)

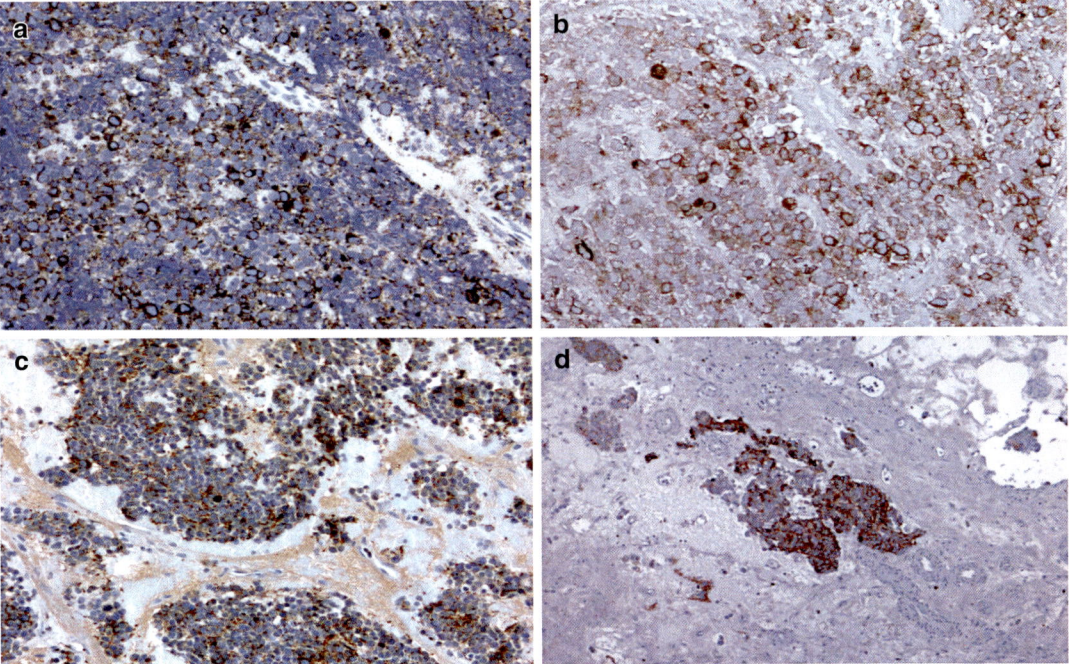

Fig. 11.17 Chromogranin A immunostaining is a useful immunohistochemistry tool for confirming the neuroendocrine nature of the neuroblastoma spanning from undifferentiated through differentiated subtype (**a**–**c**, ×200; **d**, ×100) (Avidin Biotin Complex immunostaining)

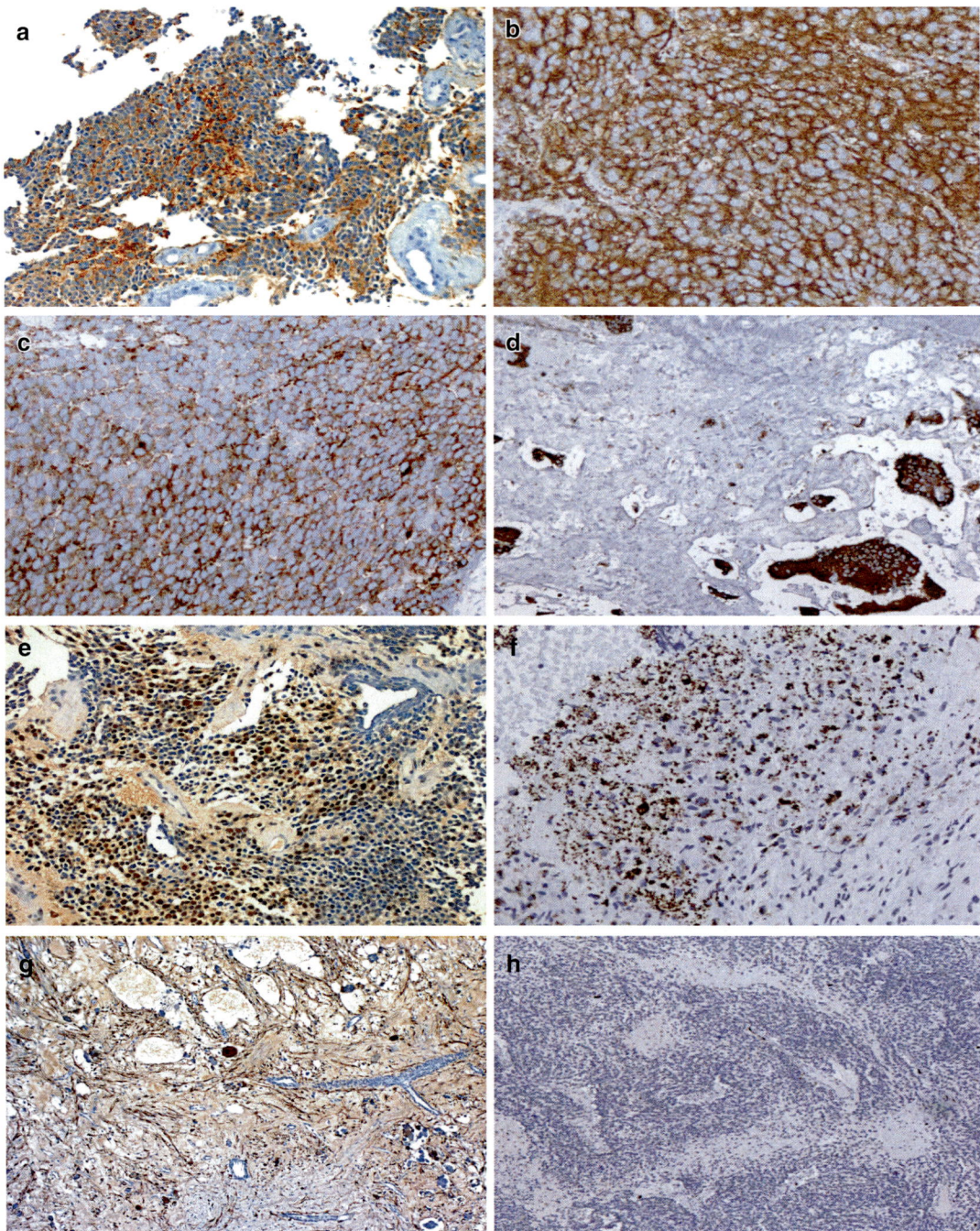

Fig. 11.18 Three immunohistochemical markers help in confirming the diagnosis of neuroblastoma in the group of small round blue cell tumors, including synaptophisin, neurofilaments, and S100. This panel illustrated the expression of synaptophysin (**a–c**, ×200; **d**, ×100), neurofilaments (**e–f**, ×200), and S100 (**g**, ×100; **h**, ×50 as original magnification). In (**h**) there is the absence of Schwannian stroma differentiation (Avidin Biotin complex - immunostaining using monoclonal and polyclonal antibodies against synaptophysin, neurofilament proteins, and S100, which collects a family of low-molecular-weight proteins identified in vertebrates. S100 are characterized by two Ca^{2+}-binding sites that possess helix-loop-helix conformation and considered as Damage-Associated Molecular Pattern molecules or DAMPs)

11.5 Adrenal Gland Pathology

- *CLI*: Presentation sites: abdomen (50–80% in the adrenal gland or adjacent retroperitoneal tissues), posterior mediastinum (respiratory distress, pneumonia, stridor, swallowing difficulties), H&N (Horner syndrome), epidural (posterior growth in dumbbell fashion with back pain and symptoms of cord compression), and skin (bluish cutaneous nodules or aka "blueberry muffin"-like nodules). In 90% ⇒ catecholamines, i.e., urinary ↑ VMA, ↑ serum Ferritin, inhibition of rosette formation by patient T lymphocytes, ↑ NSE (neuron-specific enolase), LDH (lactic dehydrogenase), and, NGF (nerve growth factor). The paraneoplastic syndrome, opsoclonus-myoclonus syndrome (OMS), is found in 2–3% of patients with neuroblastoma and is usually associated with well-differentiated tumors and favorable histology. The degree of maturity of tumor secreted catecholamines, typically vanillylmandelic acid (VMA) and homovanillic acid (HVA), and their ratio mirrors the biologic behavior of the neuroblastoma. The consideration that HVA is an early metabolite of the catecholamine pathway, a VMA / HVA ratio of <1 suggests a more aggressive, biologically primitive type of neoplasm associated with a shorter survival. A detailed list of the clinical presentation of neuroblastoma is shown in Box 11.20.
- *CGB*: Familial NB syndrome (AD, OMIM#), neural crest-related syndromes (NF-1, nesidioblastosis, central hypoventilation syndrome or Ondine course, and Hirschsprung disease) pointing to a complex neurocristopathy or derangement of the healthy development of neural crest tumors, non-neural crest-related syndromes (BWS, DiGeorge syndrome, fetal hydantoin syndrome, fetal alcohol syndrome), and single developmental defects not associated to a specific pattern (Nakissa et al. 1985).
- *GRO*: Large (Ø: 6–8 cm), soft gray-white tumor with lobular cut section and focal to extensive hemorrhage, necrosis, cyst formation, and dystrophic calcification.
- *CLM*: Small cells with homogeneous hyperchromatic nuclei ("neuroblasts") arranged in solid sheets or small nests separated by fibrovascular septa with Homer-Wright (HW) pseudorosettes around eosinophilic centers ("*neuropil*" – tangles of eosinophilic fibrils or neuritic processes arising from the primitive neuroblasts).

Box 11.20 Neuroblastoma Paramount Clinical Details of Presentations

Signs and symptoms	Location
Mass effect, dysphagia, Horner syndrome, breathing difficulty	Neck
Respiratory distress, Horner syndrome	Thorax
Pain, distension, constipation, urinary retention, hypertension	Abdomen/pelvis
Paraplegia/paraparesis, clonus, urinary retention, hypo-/hypertension	Presacral/paraspinal
Myoclonic jerking, random eye movement, cerebellar ataxia	PNS (OMS)
Intractable secretory diarrhea	PNS (VIP)
Hepatomegaly, hyperbilirubinemia, coagulopathy, skin nodules	Stages IV-S or IV
Bone pain, cytopenia, periorbital ecchymoses, fever, weight loss	Metastases (various)

Notes: *PNS* paraneoplastic syndrome, *OMS* opsoclonus-myoclonus syndrome, *VIP* vasoactive intestinal peptide

MKI is the percentage or absolute cell count of mitotic figures and karyorrhectic cells in ten high power fields (HPFs) (10X ocular × 40X, objective). MKI should only be calculated for neuroblastoma and nodules of neuroblasts in ganglioneuroblastoma, but not in ganglioneuroma. The absolute cell count per 5000 neuroblastoma cells is key. If <100 cells in mitosis or karyorrhexis (<2%), 100–200 in mitosis or karyorrhexis (2–4%), or >200 in mitosis or karyorrhexis (>4%), the MKI will be low, intermediate, or high, respectively (*vide infra* with regard to the grading).

- *IHC*: (+) NSE, NF, CGA, SYN, NB84, CD56, CK, and S100 (spindle Schwann cells→stroma) and (−) LCA, CD99, DES, MYG, and VIM
- *DDX*: It includes mostly the pediatric tumoral diagnoses listed in Box 11.21. Of course, an

> **Box 11.21 Pediatric Tumoral DDX of Neuroblastoma**
> - E-RMS/A-RMS (preteens/teens, strap cells with eosinophilic cytoplasm, tapering of cell ends, cytoplasmic cross striations, (+) DES, SMA, MYF4, MYO-D1, MYG)
> - ML (diffuse sheets of atypical quite monotonous "lymphoid" cells without fibrovascular septa, (+) CD45 ± B-cell/T-cell markers, (−) NSE, CGA, and SYN)
> - PNET/ES (sheets of monomorphic round cells with a rim of eosinophilic to clear cytoplasm with PAS+ intracytoplasmic glycogen, (+) CD99 and FLI-1, t(11;22))
> - WT (triphasic pattern, (+) CKs, (+) WT1, but (−) NSE, CGA, and SYN)
>
> Notes: *E-RMS/A-RMS*, embryonal/alveolar rhabdomyosarcoma; *ML*, malignant lymphoma; *PNET/ES*, peripheral neuroectodermal tumor/Ewing sarcoma; *WT*, Wilms' tumor; MYF-4 (myogenin) and MYO-D1 are two important RMS markers in association with desmin.

infectious origin needs to be ruled out primarily.
- *TEM*: Cytoplasmic filaments, dense-core neurosecretory granules, and microtubules are crucial ultrastructural features.

Neuroblastoma Group
This group involves a series of several entities, including undifferentiated neuroblastoma, poorly differentiated neuroblastoma, differentiating neuroblastoma, differentiated neuroblastoma or ganglioneuroblastoma, ganglioneuroma, nodular neuroblastoma, and neuroblastoma in situ. Some of these terms have been widely used in the past, and their designation depends on the variety and degree of differentiation or maturation of the cells composing the tumor. It is important to remember that overlap features exist. It is also paramount to emphasize that this overlapping is kept distinguished because the prognosis is different.

Grading and diagnosis should be made according to the International Neuroblastoma Pathology Classification (INPC) and is based on three crucial parameters to be evaluated microscopically:

(1) *ganglion cell differentiation* (gradual maturation of the neuroblasts toward ganglion cell featuring vesicular nuclear change with peripheral location and prominence of nucleolus), (2) *stroma or Schwann cell differentiation*, and (3) *MKI*.

- Thin H&E section are helpful to identify ganglion cell differentiation!
- S100 immunostaining is useful to distinguish stroma-rich from stroma-poor tumors!
- MKI is based on percentage seen in 10 HPF (=5000 tumor cells) and is classified as low (<2%), intermediate (2–4%), and high (>4%) (~ 1 tumor cell per 100 tumor cells is in either mitosis – non G0 – or shows karyorrhexis) (*vide supra*).

Metastases (1) Nonspecific symptoms, including fever and weight loss; (2) specific symptoms, including BM failure, cortical bone pain, proptosis and periorbital ecchymosis, hepatomegaly, lymphomegaly, and "blueberry muffin" skin nodules; and (3) remote effects, including watery diarrhea, due to differentiated tumor cells secreting VIP and acute myoclonic encephalopathy with opsoclonus (rapid eye movements), myoclonus (OMS, opsoclonus-myoclonus syndrome), and truncal ataxia.

Strong expression of CD57 defines migratory neural crest cells in normal development. CD57 seems to be associated with aggressive neuroblastoma cells *in vitro* and *in vivo*. It is also associated with undifferentiated neuroblastoma cells *in situ*. If compared to CD57 low neuroblastoma cells, CD57 high tumor cells develop neoplasms with decreased latency after orthotopic transplantation into adrenal glands of immunodeficient mice, are more clonogenic, induce more spheres, are less lineage-restricted, and show enhanced invasiveness.

Prenatal Diagnosis The cystic or solid mass, which is usually abutting the upper pole of the kidney is a key prenatal diagnostic feature. Moreover, some other features include, hydrops fetalis, polyhydramnios, and fetal growth restriction (metastases to the umbilical cord can induce fetal demise).

Postnatal Diagnosis Hemolytic Disease of the Newborn (HDN)-like picture, including hydrops, anemia, hepatomegaly and splenomegaly, jaundice, and increase of nucleated RBC on PB smears, hemoperitoneum from spontaneous rupture of tumor during delivery, myoclonic encephalopathy, omphalocele and ileal atresia due to an intestinal involvement of the tumor, intraocular metastases, scrotal mass, and leukoerythroblastic anemia due to BM replacement by neoplasm.

Neuroblastoma In Situ It is neuroblastoma observed in a fetus or a newborn (<4 weeks of age) and possibly associated with several multiple congenital anomalies, congenital heart disease (CHD), tracheoesophageal fistula (TEF), and genitourinary anomalies (GUA) with or without a setting of trisomy 18 syndrome.

Neuroblastoma IV-S (S=Special)
Small or undetectable primary NB with metastases in ≥ 1 site: liver, skin, BM (not bone!).

Spontaneous Regression in Neuroblastoma
It usually occurs at about 3 months of age or after therapy and has the hallmark of spontaneous degeneration characterized by extensive necrosis, hemorrhage, calcification, and natural maturation with ganglion cell differentiation. Spontaneous regression is often encountered in a subset of disseminated *MYCN* single-copy neuroblastomas, i.e., non-amplified *MYCN* gene, which is termed IV-S (stage 4S-NA, 4 special - *MYCN* not amplified). Several investigations have pointed out that other neuroblastomas may experience spontaneous regression occurring in stages I, II, and III (localized-NA, not amplified). Puzzlingly, *MYCN* mRNA and protein levels seem to be higher in localized-NA neuroblastoma with favorable histology, and particularly in stage IV-S NA neuroblastoma than in stage IV-NA neuroblastomas. However, they do not reach the levels seen in clearcut *MYCN* amplified tumors. Of note, neuroblastoma-derived cell lines lacking amplified *MYCN* usually express *C-MYC* rather than *MYCN* and often at higher levels than normal tissues.

Ganglioneuroblastoma (or Differentiating NB)
>5% Maturation of neuroblasts to ganglion cells (round to slightly oval, clear nucleus, and prominent cytoplasm) in a background of delicate fibrillary neuropil, with remaining areas of immature neuroblastoma (Fig. 11.19).

Two types have been identified:

- *Composite*: Compact nodules of neuroblastoma in a ganglioneuroma.
- *Diffuse*: Both elements (differentiated ganglion cells and immature neuroblasts) are distributed throughout the tumor.

The diffuse type of ganglioneuroblastoma harbors a better prognosis than the composite type. In ganglioneuroblastoma, if neuroblasts harbor a favorable histology constituting the nodules of the nodular ganglioneuroblastoma, then the ganglioneuroblastoma has overall favorable histology and good outcome. However, if the buds of the ganglioneuroblastoma have unfavorable histology, then the ganglioneuroblastoma has overall unfavorable histology.

Ganglioneuroma
- *DEF*: Neoplasm that originates from neural crest cells of sympathetic ganglia (posterior mediastinum and retroperitoneum mostly) or adrenal medulla with all mature elements: fibrous and Schwann cell-rich stroma-rich background with scattered ganglion cells and fully differentiated neuroblasts (no immaturity of the neuroblasts). Ganglioneuroma has an excellent prognosis in any case, even it derives from an original ganglioneuroblastoma or neuroblastoma. The proliferation rate of the cells is nil, and MKI scoring does not apply to ganglioneuroma.
- *EPI*: Scholar age (≥5 years old), ♂ = ♀.

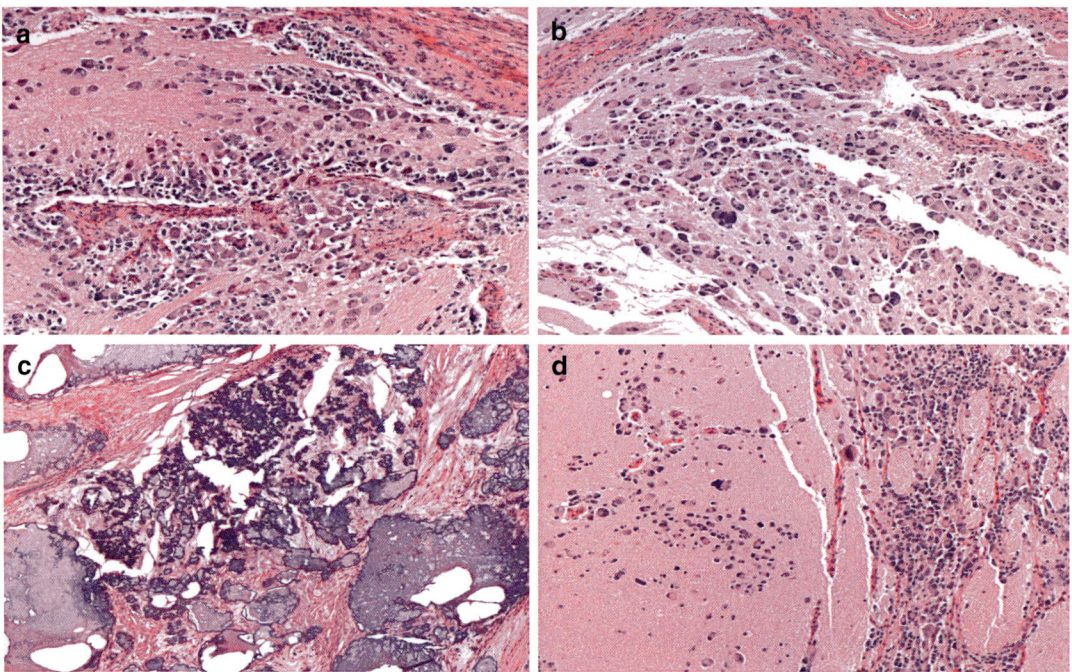

Fig. 11.19 The hematoxylin-eosin stained sections shown in Figures **a**–**d** show a mixture of differentiated neuroblasts and differentiating neuroblasts with or without early ganglionic differentiation (**a**, ×100; **b**, ×100; **c**, ×100; **d**, ×100; all original magnifications)

- *EPG*: Neural crest cells origin with clonal proliferation and genetic predisposition to Turner syndrome and MEN2.
- *CLI*: Tumor compression or asymptomatic (incidental finding), because it is typically nonfunctional.
- *LAB*: Occasionally, catecholamines may induce arterial HTN.
- *IMG*: Well-defined, homogeneous, hypoechogenic mass on the US and well-defined, hypodense, homogeneous, or slightly heterogeneous lesion on CT. The ganglioneuroma may surround the peripheral blood vessels without compression or occlusion and it often harbors punctate calcifications. Contrast medium poorly enhance this tumor. The tumor is homogeneous with signal intensity lower than liver on MRI-T1 and heterogeneous with signal intensity superior to liver on T2 (MRI).
- *GRO*: Well-circumscribed ± true or fibrous capsule with firm texture and gray-white color and slightly whorled or trabecular cut surface.
- *CLM*: Admixture of mature ganglion cells (eosinophilic cytoplasm with distinct cell borders, single eccentric nucleus, prominent nucleolus) and Schwannian cells, which are arranged in small intersecting fascicles. The fascicles are separated by loose myxoid stroma in the Schwannian stroma. Some maturing component (minor component) of scattered collections of differentiating neuroblasts or maturing ganglion cells may be present. However, differently, from the intermixed subtype of ganglioneuroblastoma, these immature foci do not form discrete microscopic nests.
- *IHC*: (+) S100, SYN, and NFP in Schwann cells/stroma and (+) S100, SYN, CGA, NFP, GFAP, PGP 9.5, IV-Coll, and VIP in ganglion cells. (−) KER, EMA, HMB45, WT1, CD99, CD45, DES, and MYF4.
- *TEM*: Admixture of neural bundles and normal appearing ganglion cells with eccentric nuclei and large numbers of cytoplasmic organelles.
- *DDX*: Ganglioneuroblastoma, neuroblastoma and other small round blue cell tumors, neurofibroma, adrenal cortical adenoma, adrenal cortical carcinoma, composite pheochromocytoma, and masculinizing ganglioneuroma.
- *TRT*: Resection.

11.5 Adrenal Gland Pathology

- *PGN*: Excellent outcome is almost the rule, although there is the rare malignant transformation of Schwannian stroma to MPNST (e.g., *de novo* or after abdominal radiation). Variants include the masculinizing ganglioneuroma (ganglioneuroma with admixed Leydig cells containing crystalloids of Reinke) and composite ganglioneuroma (ganglioneuroma and pheochromocytoma).

Neuroblastoma has a prognosis that ranges from spontaneous regression to aggressive metastatic neoplasms. The treatment of this most common extracranial tumor of childhood has served as a strong paradigm for more than two decades incorporating clinical and biological factors to stratify patients and tailor better therapeutic protocols. Currently, patients harboring neuroblastoma can be categorized as low risk (LR), intermediate risk (IR), and high risk (HR) for recurrence. Thus, overall survival (OS) for both LR and IR neuroblastomas is higher than 90%, while the OS of HR neuroblastomas is not more than 50% despite the use of combined treatments and immunotherapy.

Several transformation-linked genetic and epigenetic changes have been identified that have contributed to the first understanding of neuroblastoma tumor biology. Tumor-specific genetic and epigenetic aberrations seem to be highly predictive of treatment response and outcome.

AMES Mnemonics: *A*ge/*a*neuploidy, *M*KI/*M*YC, *E*xtra-adrenal, *S*tage/*S*ecretion/*S*troma

Good Prognostic Factors:

- Age (< 1.5 years), aneuploidy (hyperdiploid or near-triploid)
- MKI < 200/5 × 10^3 cells; No *N-MYC* amplification (the amplification of *N-MYC* does not karyotypically manifest at the resident 2p23–24 site but rather as extra-chromosomal Double Minute Chromatin Bodies or Homogeneously Staining Regions on other chromosomes); No 1p-, 11q-, 17q+; TRK-A expression; Telomerase expression
- Extra-adrenal
- Stage (IVS, I, and II), Secretion (high ratio of urinary VMA: HVA), Stroma/differentiation

Bad Prognostic Factors: High levels of ferritin/NSE; N-myc amplification (>10 copies, ~1/3 of tumors); adrenal; >5 years; TRK-B expression

Assessing differentiation, regression, or response to therapy is vital. Metastases occur early and widely, and particular types include the Hutchinson-type neuroblastoma (extensive bony metastases, particularly to the skull and orbit, producing exophthalmos) and the Pepper-type syndrome with extensive metastases to the liver (DDX: IVS NB). According to the International Neuroblastoma Staging System, stage I corresponds to localized NB with complete gross excision ± R1 (+) on microscopic margins (negative ipsilateral non-adherent LNs, although LNs attached to the tumor and removed with the primary tumor are permitted to be positive). In stage IIA, there is a localized NB with incomplete gross excision (negative ipsilateral non-adherent LNs, although LNs attached to the tumor and removed with the primary tumor are permitted to be positive), while in stage IIB there is a localized NB with complete/incomplete gross excision with (+) ipsilateral, non-adherent LNs or (−) contralateral, non-adherent LNs. A localized NB characterizes stage III, with unilateral OR unresectable unilateral tumor mass infiltrating across the midline OR unresectable midline tumor mass with bilateral extension ± regional LNs OR (+) contralateral regional OR (+) bilateral LNs. Stage IV corresponds to any primary NB with the involvement of distant LNs, bone, BM, liver, skin, and/or other organs (except IVS), while stage IVS (special) corresponds to any primary tumor with involvement of the liver, skin, and/or bone marrow (BM). It is paramount to recall that BM cellularity is <10% in children <1 year of age without bony involvement using meta-iodobenzylguanidine (MIBG) scan.

Some of the new lines of research are focusing on multigene expression profiles, rather than specific candidate genes. These multigene expression profiles may predict outcome and may lead to further refinement of risk categories, and a 14-gene classifier has identified a subset of HR neuroblastoma patients with the worst prognosis. Forkhead box O (FOXO) transcription factors control diverse cellular functions, such as cell death, metabolism, and longevity. In neuroblastoma, caspase-8 is frequently silenced by DNA methylation. Conditional FOXO3 activates caspase-8 gene expression is crucial, but it does not change the DNA-methylation pattern of regulatory sequences in the caspase-8 gene. Conversely, FOXO3 induces phosphorylation of its binding partner ATM and the ATM downstream target cAMP-responsive element-binding protein (CREB). This latter component is critical for FOXO3-mediated caspase-8 expression. Caspase-8 levels above a critical threshold sensitizes neuroblastoma cells to tumor necrosis factor (TNF)-related apoptosis-inducing ligand-induced cell death opening the pathway for new clinical trials. The *ALK* F1174L mutation has a particular relevance in cancer. It is not only because of its role in neuroblastoma but also because it causes resistance to crizotinib in ALK-rearranged tumors in general. Berry et al.'s (2012) murine model of HR neuroblastoma has identified the cooperation between *ALK* F1174L and *MYCN*. They provide an ideal platform for further dissection of oncogenic *ALK* & *MYCN* interactions (Berry et al. 2012).

Germline genetic variants have been identified in neuroblastomas that may predispose to the formation of this embryonal tumor. These variants include single nucleotide polymorphisms in several genes, including *LINC00340*, *BARD1*, *LMO1*, *DUSP12*, *DDX4*, *LIN28B*, *HACE1*, and *TP53*. It seems opportune to consider that the initiation and progression of neuroblastoma may be influenced by the interplay between multiple germline variants and somatic alterations.

Neuroblastoma and Congenital Heart Disease
The association between neuroblastoma and congenital heart disease has been described in the literature. Holzer and Franklin (2002) revised this topic and identified a number of cardiac defects (persistent arterial duct, atrial septal defects, ventricular septal defects, mitral atresia, pulmonary atresia, aortic stenosis, bicuspid aortic valve, coarctation of aorta, anomaly of aortic arch, vascular ring, pulmonary valvar stenosis, *cor triatriatum*, atrioventricular septal defects, transposition of the great arteries, hypoplastic left heart syndrome, tetralogy of Fallot, and more complex cardiac malformations.

11.5.4.4 Pheochromocytoma and Paraganglioma

Pheochromocytoma
- *DEF*: Intra-adrenal or adreno-medullary paraganglioma with usually marked hypertension secondary to catecholamine production (NE > E) and peak in the four to five decades of life. If the pheochromocytoma occurs outside of the adrenal medulla, it is referred to as an *extra-adrenal* paraganglioma.
- *EPI*: <7% of tumors arising from the sympathetic nervous system with incidence rates at $0.3:10^6$ per year. Approximately one-fifth of cases are diagnosed during childhood at an average age of 10 years, with a slight ♂ > ♀ predominance. In children diagnosed with systemic arterial HTN, it has been found that up to 1.7% there is a catecholamine-secreting neoplasm underlying the systemic hypertensive picture. Tumor age peak is located in the 4th decade.
- *EPG*: Von Hippel-Lindau (VHL), multiple endocrine neoplasia type 2 (MEN2), and neurofibromatosis type 1 (NF1) with three gene subunits of the succinate dehydrogenase (SDH) complex, *SDHB*, *SDHC*, and, *SDHD*, are established causes of hereditary pheochromocytoma (paraganglioma). In the recent few years, four new genes (*SDHA*, *SDHAF2*, *MAX*, and *TMEM127*) have been detected to be linked with predisposition to these tumors.
- *CLI*: It relies on 24-h urine ↑ catecholamines (epinephrine and nor-epinephrine) and their metabolites as well as CT scan and MRI using ^{123}I-MIBG depending upon uptake of a radiolabeled analogue of guanethidine collected within the storage granules of the tumor cells.
- *GRO*: Well-demarcated, bulging, pale gray to light brown mass ± hemorrhage, necrosis (slightly mottled to confluent areas), and cysts.
- *CLM*: Three *major* growth patterns ("SAT"):
  – *Solid* (or diffuse)
  – *Alveolar* (or nested, i.e., small clusters or "Zellballen")
  – *Trabecular*

There may be apparent encapsulation with the intermingling of tumor cells with adjacent non-tumorous cortical cells. Other growth patterns include a more marked spindling of the tumor cells, pseudopapillary pattern, and perivascular pattern (DDX: MPNST). Tumor cytology shows mature *chromaffin* polygonal cells with abundant eosinophilic to the amphophilic granular cytoplasm with variable cellular/nuclear pleomorphism, which is separated by thin fibrovascular stroma septs. MI: −/↑ and ± angioinvasion (Figs. 11.20, 11.21, 11.22, and 11.23).

- *HST*: Grimelius stain (+) due to strong argyrophylia of the cytoplasm of the tumor cells and Fontana-Masson silver (FMS) stain (+) due to the argentaffinity of the tumor cells associated to the lability to bleaching procedures using permanganate or picric acid.
- *IHC*: (+) CGA, SYN, TYH, S100 (sustentacular cells) and (−) Melan-A/MART-1, INA, CAM5.2, and LCA, CD99, DES, MYF4, and VIM (~), calretinin (~).
- *TEM*: Numerous dense-core neurosecretory granules, including E-type with narrow, uniform halo and NE-type granules, characterized conversely by a wide eccentric halo between the electron dense-core and limiting membrane.
- *CGB*: 1p-, 3p-, 3q-, 17p, and 22q.
- *DDX*: Adrenal cortical adenoma (ACA), neuroblastoma (NB), rhabdomyosarcoma (RMS), and Wilms tumor (WT)/nephroblastoma.
    (a) *ACA* (golden-yellow grossly, (+) Melan-A/MART-1, SYN, INA, CAM5.2, (−) CGA, S100)
    (b) *NB* (diffuse sheets of small round blue cells with neuropil and Schwannian stroma showing potentially ganglionic cell differentiation, HW pseudorosettes)
    (c) *RMS* (spindle cells ± skeletal muscle differentiation, (+) DES, MYF4, MYOD1)
    (d) *WT* (triphasic pattern consisting of blastema, stroma, and epithelial elements, (+) CKs, but (−) NSE, CGA, and SYN)

In Box 11.22, pearls and pitfalls in diagnosing pheochromocytoma are summarized for the routine pathologist. Morbidity and mortality of pheochromocytoma depend on the severity of the effects of overproduction of catecholamines, including a headache, vomiting, pallor, sweating, and hypertensive encephalopathy. Catecholamines exert a peculiar and essential widespread metabolic effect on glucose metabolism (glycogenolysis and gluconeogenesis) and lipid metabolism (lipolysis) as well as effects on the cardiovascular system. The physiologic effects on cardiac muscle contractility and vasodilation are mediated through α- and β-adrenergic receptors on target cell surfaces. Catecholamine-associated cardiomyopathy (arrhythmia, cardiac failure, sudden death) is characterized by focal degeneration and necrosis of cardiomyocytes along with mild inflammation, mostly involving the inner 2/3 of the left ventricular myocardium. Other aspects that may target the cardiovascular system include fibromuscular dysplasia, renal artery stenosis, and multiple cerebral aneurysms. CODs include hypertensive crisis, due to the invasive procedure (e.g., venography for blood sampling for catecholamines, selective arteriography) without the preventive use of an α-adrenergic blockade, stroke/hypertensive encephalopathy, and shock due to reduced vascular volume after the tumor is removed (lack of aggressive fluid management). In the period of follow-up, postoperative levels of urinary catecholamines and their metabolites should be essentially normal. In case of persistent symptoms, elevated levels of catecholamines and metabolites or new occurrence of increased catecholamines even after many years indicate substantially residual tumor.

Prognostic Factors of Pheochromocytoma
Although features of malignancy (capsular invasion, numerous mitoses, necrosis, solid/diffuse growth pattern without nests or with large nests, spindled cells, aneuploidy) have been proposed, malignant behavior cannot be determined based on morphologic findings, and only presence of distant metastases, which may occur very late in disease, proves definitely malignancy. Malignancy is, indeed, more common (20–40%) in extra-adrenal paragangliomas and in tumors arising in the setting of specific germline mutations. The proliferation antigen Ki67 (MIB1)

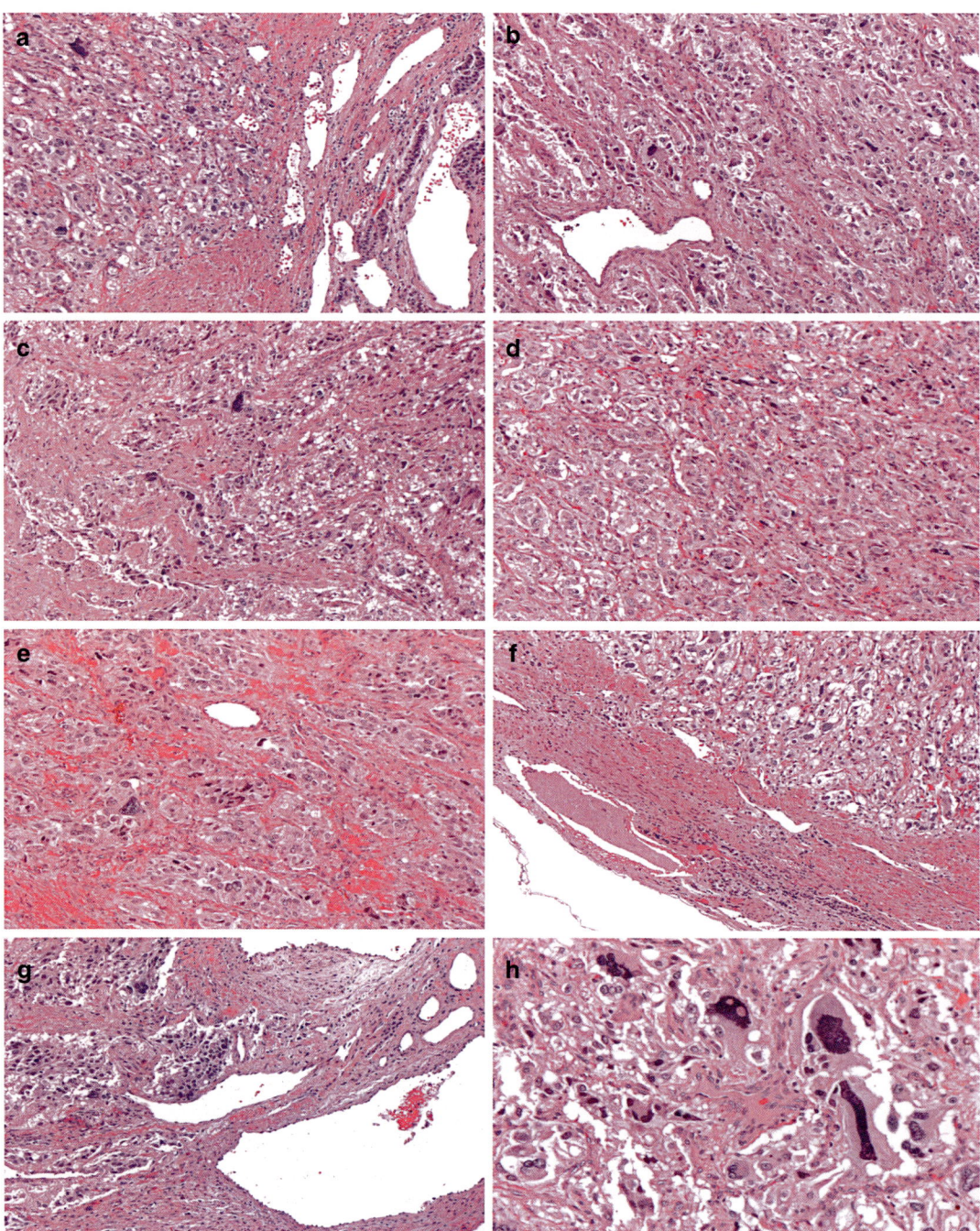

Fig. 11.20 The histological panel shows the characteristic features of a pheochromocytoma with the so-called Zellballen, which is a German name for small nests or alveolar pattern. However, pheochromocytoma can exhibit trabecular or solid patterns of polygonal or spindle cells embedded in a rich vascular network. On high power, it is easy to recognize that the tumor cells have finely granular basophilic or amphophilic cytoplasm, intracytoplasmic hyaline globules, and oval/ovoidal nuclei with some bizarre forms and prominent nucleolus as well as marked pleomorphism (**a**, ×100; **b**, ×100; **c**, ×100; **d**, ×100; **e**, ×100; **f**, ×100; **g**, 100; **h**, ×200). (All micrographs have been taken from hematoxylin-eosin stained histological slides)

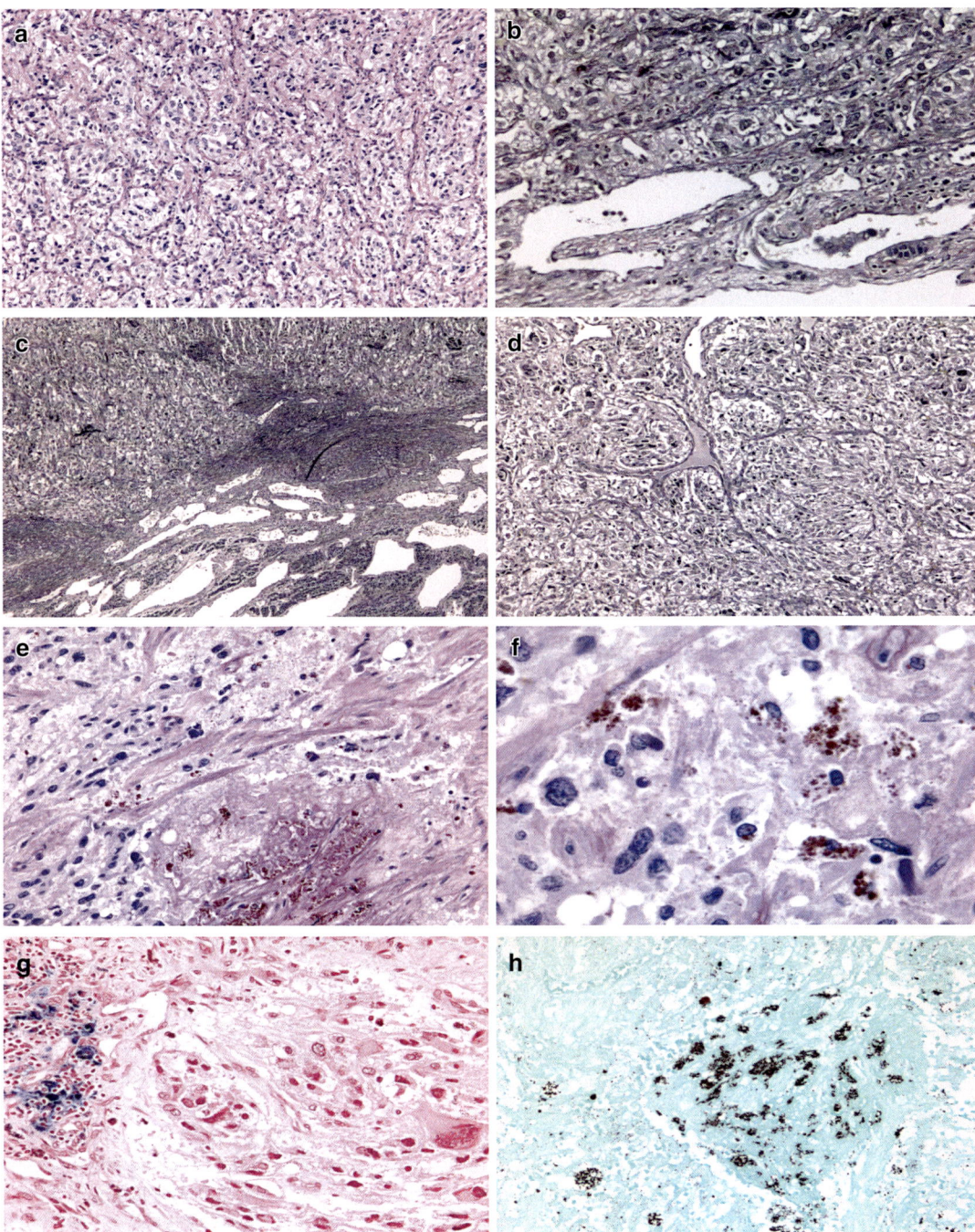

Fig. 11.21 Special stains play a major role in supporting the diagnosis of pheochromocytoma, and the stained slides can arrive to the pathologist desk earlier than the immunostained slides (**a**, ×100, Periodic acid Schiff stain; **b**, ×200, Verhoeff elastica stain; **c**, ×50, Verhoeff elastica stain; **d**, ×100, Verhoeff elastica stain; **e**, ×200, Periodic acid Schiff stain; **f**, ×630, Periodic acid Schiff stain; **g**, ×200, Perls Prussian blue; **h**, ×200, Fontana-Masson stain)

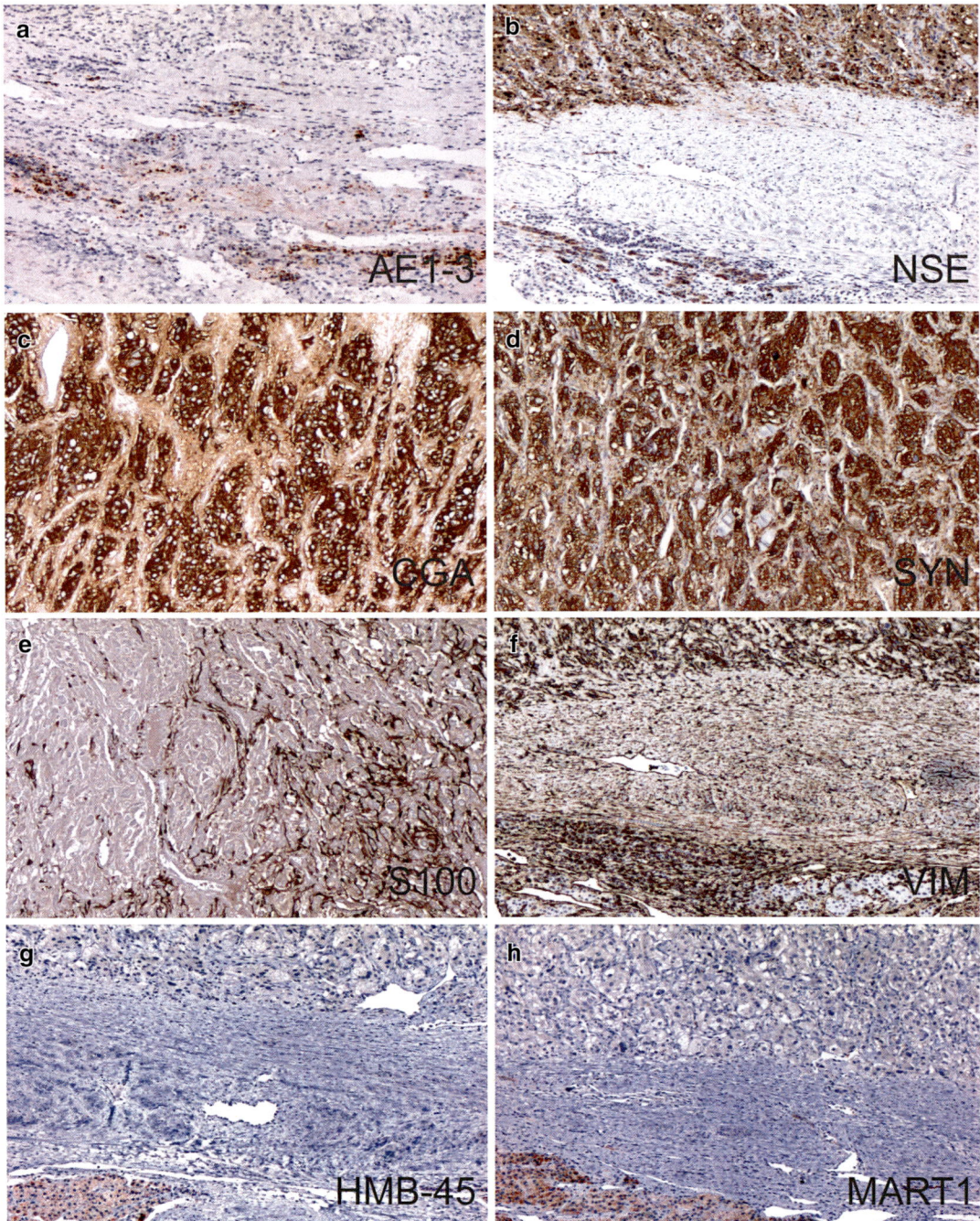

Fig. 11.22 The immunohistochemistry is usually mandatory in confirming the diagnosis of a pheochromocytoma, and this classic panel shows the expression of neuro specific enolase (**b**, x100), chromogranin A (**c**, ×100), synaptophysin (**d**, ×100), and S100 in the sustentacular cells (**e**, ×100). Variable staining is seen with a monoclonal antibody against vimentin (**f**, ×100), while HMB45 (**g**, ×100) and MART-1 (**h**, ×100) are usually negative as well as pankeratins (AE1/AE3, **a**, ×100). Occasionally, a focal expression for HMB45 can be seen. (All immunostainings have been performed using avidin-biotin complex and mono-/polyclonal antibodies against antigens as indicated in the lower right corner)

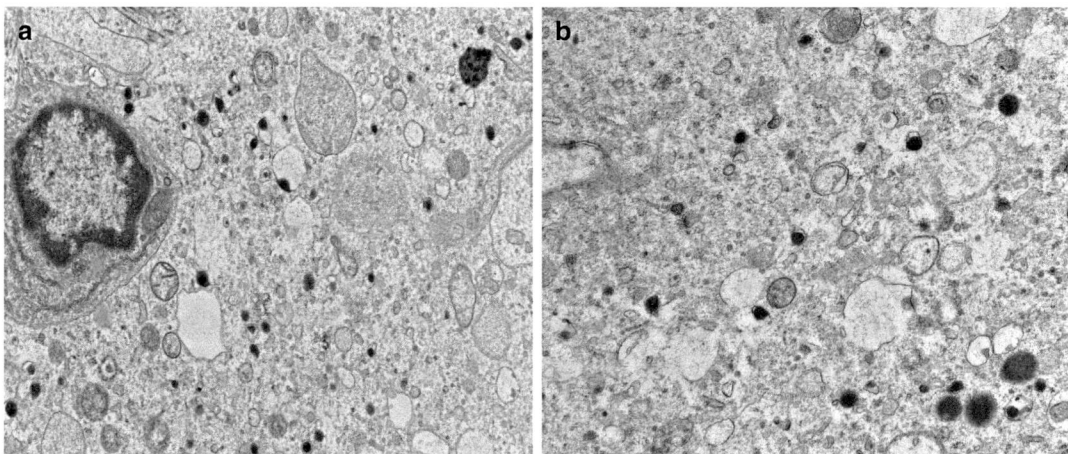

Fig. 11.23 The transmission electron microscopy discloses an abundance of membrane-bound electron-dense granules, tubulolamellar mitochondria, and rough endoplasmic reticulum (**a**, **b**)

Box 11.22 Pheochromocytoma Pearls and Pitfalls
- 10% rule: 10% *extra-adrenal* (e.g., the organs of Zuckerkandl & carotid body), 10% *bilateral*, 10% *biologically malignant* (i.e., metastatic disease), 10% in *childhood and adolescence*, and 10% *not associated with systemic arterial hypertension*[1].
- 50% hereditary: *MEN IIA, MEN IIB, NF-1, VHL, SDHB, SDHC, SDHD*[2].
- MI: −/↑; ± angioinvasion DO NOT imply malignancy!
- Cell Variability: neuronal/ganglion cell morphology
  (a) Bizarre multinucleated tumor cells ± tapering cytoplasm extensions
  (b) Tumor cells with "lavender"-stained cytoplasm with quite often punctate granularity
  (c) Tumor cells with nuclear pseudo-inclusions due to invaginations of the cytoplasm
- Intracytoplasmic hyaline eosinophilic, PAS+ and PASD+ globules, hemorrhage, degenerative changes (e.g., cystic or calcified), and interstitial amyloid deposition can be seen in adrenal cortical neoplasms independently from the outcome (benign/malignant).
- Variants:
  (a) *COMPOSITE* (pheochromocytoma *and* NB/GNB/GN-Group or MPNST)
  (b) *MIXED* (pheochromocytoma + ACA)
  (c) *PIGMENTED* (neuromelanin)
- FMS[3] is also positive in melanin-like pigment due to the common embryogenesis from the neural crest (melanosomes/*pre*melanosomes ultrastructurally).

Notes: [1], ~ 2/3 of patients harboring a pheochromocytoma experience "*paroxysmal*" episodes; [2], at the exception of SDHC, which is associated with paraganglioma, all other genes are usually associated with pheochromocytoma; [3], Fontana-Masson silver special histochemical stain and an immunohistochemical score based on it have been indicated to possess some prognostic relevance by some authors. Although features of malignancy have been suggested (aneuploidy, capsular invasion, conspicuous necrosis, high mitotic index, solid/diffuse growth pattern without nests or with large nests, and spindle cell growth pattern), the only probably satisfied criterion of malignancy seems, in the opinion of most authorities, the presence of metastatic tumor tissue in sites where chromaffin tissue is not typically formed. Complications include catecholamine cardiomyopathy, and patients should have a lifelong cardiological follow-up.

Extra-adrenal Paraganglioma
Same tumor as pheochromocytoma, but occurring outside of the adrenal gland (e.g., perirenal, celiac axis, peri-urinary vesical, and heart). It is generally named according to the site:

1. Chemodectoma: carotid body
2. Jugulotympanic paraganglioma (glomus jugulare; glomus tympanicum)
3. Vagal paraganglioma (glomus vagale)
4. Mediastinal paraganglioma

Composite Pheochromocytoma: Pheochromocytoma (main tumor) + NB/GNB/GN or MPNST/NF/Schwannoma (secondary component)

11.5.4.5 Other Tumors and Secondary Tumors

Other tumors of the adrenal gland include hemangioma, leiomyoma, neurilemmoma (schwannoma), neurofibroma, adenomatoid tumor, ovarian thecal metaplasia (or stroma spindle nodules), primary adrenal granulosa-theca cell tumor, solitary fibrous tumor, inflammatory myofibroblastic tumor, and myelolipoma among benign entities and malignant lymphoma, melanoma, malignant schwannoma (malignant peripheral nerve sheath tumor), angiosarcoma, rhabdomyosarcoma, and extrarenal adrenal Wilms tumor among malignant entities (Fig. 11.24).

Metastatic Tumor The list may be particularly long and is associated with various neoplasms,

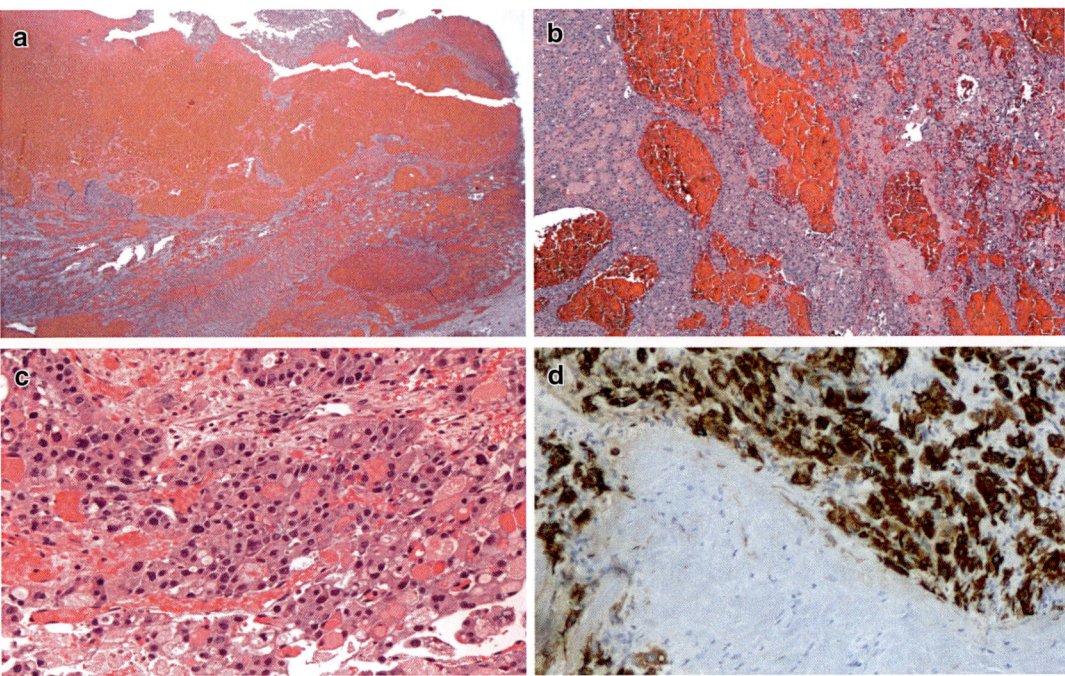

Fig. 11.24 The panel shows the evidence of metastasis in the adrenal gland originating from a fibrolamellar carcinoma of the liver (**a–c**: x12.5, x50, and x200; hematoxylin and eosin staining), which also shows cytokeratin 7 expression immunohistochemically (**d**, anti-CK7, ×100). In the microphotograph (**b**) a peliosis-like architecture is recognized

which may be different according to the age of the subjects. They include lung carcinoma, breast carcinoma, malignant melanoma, gastric carcinoma, colorectal carcinoma, renal cell carcinoma, and Kaposi sarcoma. These neoplasms may be encountered in childhood and youth, mainly if the individual patients harbor a genomic mutation or deletion of tumor suppressor genes.

Adrenal Gland/Periadrenal Tumorlike Lesions

There are a few conditions that may mimic a neoplastic process at the level of the adrenal gland or in the soft tissue surrounding the adrenal glands.

- *Infections*: *H*istoplasmosis (*O*hio and Mississippi rivers areas; PAS stain is very useful to delineate the intracytoplasmic organisms typical for *Histoplasma capsulatum* better), *b*lastomycosis (Southeastern USA including Ala*b*ama), and *c*occidioidomycosis (Southwestern USA including *C*alifornia), and TB with extensive adrenal necrosis (ZN and AR stains).
- *Abscess*: some bacteria are prone to form an abscess, such as *E. coli*, *β-Streptococcus*, and Bacteroides.
- *Hemorrhage-related mass*: WFS (extensive necrosis and hemorrhage of the adrenal gland with hematoma formation and extension into periadrenal adipose tissue). Over time, in survived individuals, the adrenal hematoma may decrease in size or resolve with dystrophic calcification and cystic degeneration and constitute a mass. In neonates, adrenal hemorrhage has been associated with complicated labor or perinatal asphyxia. Apart from the mass effect, other complications of a bleeding are adrenal insufficiency, secondary infection and abscess (*vide supra*), and retroperitoneal bleeding.
- *Cysts*: These pathologies are classified as parasitic (usually *Echinococcus*), epithelial (true glandular or retention cysts, embryonal cysts, and cystic adrenal tumors such as cystic pheochromocytoma), endothelial (lymphangiomatosis and angiomatosis), mesothelial (calretinin +), and "pseudocysts" (the most common).

Myelolipoma

It is a usually asymptomatic and benign and incidental finding not typically present in childhood or youth. It is a sharply circumscribed, nonencapsulated, pale yellow, and cortical located lesion with focal darker areas (hematopoiesis). It is composed by adipocytes admixed with myeloid cells and lymphocytes. No blasts (DDX: Leukemic infiltration → CD34 immunohistochemistry is helpful). Myelolipoma may occur as an incidental finding or in association with endocrine disturbances and congenital adrenal hyperplasia (21-OH and 17-OH). Lipoma is a benign tumor without the hematopoietic areas.

Primary Malignant Melanoma

Pleomorphic cells with aberrant melanin pigment. Melanin or melanin-like pigment has been reported in pheochromocytoma and may be difficult to distinguish from neuromelanin, a pigment related to altered lipofuscin, which has been observed in GNB and GN. Melanocytic melanin pigment shows typical melanosomes or premelanosomes ultrastructurally.

- Primary malignant lymphoma
- Primary vascular tumors: hemangioma, lymphangioma, and angiosarcoma
- Primary muscle or smooth myostructure-forming tumors: leiomyoma, leiomyosarcoma
- Primary nerve-forming tumors: schwannoma, neurofibroma, malignant peripheral nerve sheath tumor

Other Tumors (Including Metastases)

Adenomatoid tumor with (+) CKs and calretinin and ovarian thecal metaplasia. Moreover, Castleman disease (Debatin et al. 1991), retroperitoneal bronchogenic cyst (Foerster et al. 1991), primary PNET (Zhang et al. 2016), primary adrenal extrarenal Wilms tumor (Santonja et al. 1996), and SFT (Prévot et al. 1996) should also be considered. In childhood, metastatic tumors may include Wilms tumor as a first neoplastic entity in this age group. In adults and children or youths with cancer family syndromes, carcinomas from the lung, breast, GI tract, thyroid, and kidney need to be taken into account.

11.5.5 Syndromes Associated with Adrenal Cortex Abnormalities

Several syndromes may manifest clinically with some pathology of the adrenal gland. The pathologist must liaise with the clinician or look into electronic medical/health records for a possible syndrome associated with some adrenal gland pathology. Some of these changes are related to abnormal growth, while others are linked to adrenal insufficiency and adrenal atrophy. We should list X-linked adrenal hypoplasia, congenital, X-linked adrenoleukodystrophy, Wolman disease (WoD), Carney complex (CNC), multiple endocrine neoplasia type 1 (MEN-I), Beckwith-Wiedemann syndrome (BWS), McCune-Albright syndrome (MAS), Carney Complex (CNC), congenital adrenal hyperplasia (CAH), familial adenomatous polyposis (FAP), and Li-Fraumeni syndrome. (LFS), which is a rare type of cancer predisposition syndrome associated with germline *TP53* gene mutations.

In one of the syndromes related to adrenal gland changes, genetic imprinting is highly relevant. Genetic imprinting is the reason for the complexity and variability of a genetic syndrome in some settings. Genetic imprinting is the phenomenon of variable gene expression, which is based on the parent of origin. In other words, if maternal genes are silenced, paternal genes are expressed, while if paternal genes are silenced, maternal genes are expressed. Genetic imprinting characterizes BWS.

BWS may manifest with macroglossia, pre- and postnatal overgrowth, abdominal wall defects (omphalocele, *diastasis recti*, umbilical hernias), genital enlargement, and visceromegaly, which may include hepato-, nephron-, and pancreato-megaly. Prenatal diagnosis include ultrasound and gross features, such as polyhydramnios, placentomegaly, and excessive long umbilical cord. In the placenta, (placental) mesenchymal dysplasia is seen. Mesenchymal dysplasia of the placenta is similar to placenta mole, although there is no trophoblastic hyperplasia. Patients with BWS may show a dome-shaped diaphragm, midface hypoplasia, *nevus flammeus*, and earlobe pits or posterior helical pits. In 10% of patients with BWS, there is a congenital embryonal tumor, including nephroblastoma (Wilms tumor) and hepatoblastoma. However, rhabdomyosarcoma, and adrenal cortical carcinoma can be observed. Tumors seem to be seen more often in patients with hemihypertrophy. The majority of patients with BWS are related to loss of the appropriate imprinting of the genes located in the 11p15 chromosomal region. It seems that the lowest risk of the tumor is linked to loss of imprinting of *LIT1* (risk: 1–5%), while the highest risk of tumors is related to loss of imprinting of *H19* (risk up to 45%).

LFS is a *TP53* germline mutation-based autosomal dominantly inherited the genetic disease. Patients with LFS may have different tumors, including osteosarcoma, soft tissue sarcoma, breast carcinoma, and adrenal cortical carcinoma among the most frequent. *TP53* has been considered the guardian of the genome, because of the tumor suppression gene activities. The activation of *TP53* either arrests the cell cycle in the G1 phase for DNA repair to occur, or it permits the affected cell to go to programmed cell death (apoptosis). *TP53*-inactivating mutations lead to the lack of these pathways in response to DNA damage. The 2nd locus for LFS is the *CHEK2* gene, which is localized on 22q12.1.

CNC is a combination of myxomas (mucosal, cutaneous, cardiac), spotty pigmentation (lips, conjunctiva, inner and outer canthi, and vaginal/penile mucosa), and endocrine overactivity (pituitary adenoma with growth hormone secretion). In about half of the patients, the gene, responsible for CNC, is *PRKAR1A* located on chromosome 17q23-q24, and inactivating mutations are inside of this gene. The most common variation is on exon 5 and is a 2-bp deletion. This gene encodes for the cAMP-dependent protein kinase PKA type I-A regularity subunit (R-Iα). PKA is a 2nd messenger-dependent enzyme that has been associated with a quite wide variety of cellular processes, including cell cycle progression, transcription and cellular apoptosis, as well as cell metabolism. In CNC, hyperplasia and adenoma of the adrenal cortex are seen but also primary pigmented nodular adrenocortical disease (PPNAD), which is considered the characteristic CNC lesion of the adrenal gland. Grossly, adrenal glands have numerous yellow-brownish nodules scattered diffusely throughout both glands. Histologically, nodules are composed of cells that are either with eosinophilic cytoplasm (lipid-poor) or vacuolated cytoplasm with characteristic lipofuscin or lipofuscin-like pigment in the cytoplasm of the cells. The pigment stains with Fontana-Masson special stain (before bleaching only), PAS, and Ziehl-Nielsen. On ultrastructural investigations and histochemical stains, it seems that the pigment is partially melanized (the term "neuro-melanized" pigment has been used by some authors).

Multiple Choice Questions and Answers

- END-1 What is splenogonadal fusion?
  (a) A congenital defect with an abnormal connection between the primitive spleen and gonad that occurs during gestation
  (b) A gene fusion transcript involving the expression of genes of the spleen and the gonads
  (c) A defect involving the genes responsible for the expression of the splenium of the corpus callosum and the developmental genes of the testis
  (d) A combined defect involving the genes responsible for the expression of the splenium of the corpus callosum and the developmental genes of the ovary
- END-2 Which of the following cytologic features does NOT belong to a papillary thyroid carcinoma?
  (a) "Salt-and-pepper" chromatin
  (b) Grooves
  (c) Pseudo-inclusions
  (d) Nucleolus
  (e) Thickening of the nuclear membrane
  (f) Irregularity of the nuclear membrane
- END-3 Some parents are concerned that their 15-year-old child did not start to menstruate. Although in some occasions menstruation may be delayed in healthy children, the child showed short stature, webbing of the neck, and cubitus valgus at the physical examination. Intellectually her school performance was unremarkable. Which of the following investigations will be appropriate at first?
  (a) Serum long-chain fatty acids
  (b) Organic acid analysis of urine
  (c) Muscular biopsy
  (d) Computed tomography of the head with a primary focus on the *sella turcica*
  (e) Karyotype
- END-4 A 17-year-old girl develops a unilateral enlargement of her thyroid gland, and the lump undergoes a fine needle aspiration cytology (FNAC), which is shown here:

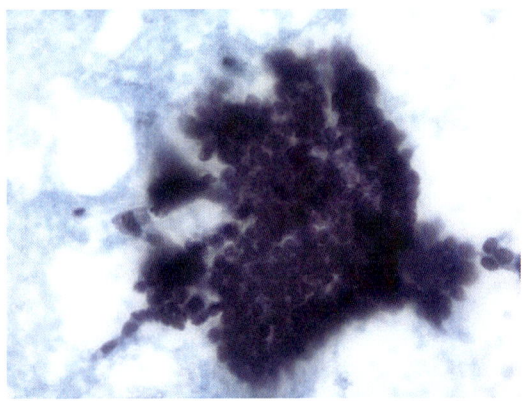

- What is the most likely diagnosis?
  (a) Papillary thyroid carcinoma
  (b) Follicular thyroid carcinoma
  (c) Hashimoto thyroiditis
  (d) Benign nodule
  (e) Graves disease
- END-5 Which of the next features does NOT help in distinguishing a parathyroid carcinoma from parathyroid adenoma?
  (a) Capsular invasion
  (b) Vascular invasion
  (c) Metastases
  (d) Pleomorphism

- END-6 Which of the next statements regarding the Smith-Lemli-Opitz syndrome (SLOS) is NOT correct?
  (a) It is an abnormality in cholesterol metabolism resulting from deficiency of the enzyme 7-dehydrocholesterol (7-DHC) reductase.
  (b) The serum concentration of cholesterol is usually low.
  (c) *DHCR7* gene encodes the 7-DHC reductase.
  (d) The inheritance pattern is autosomal dominant.
  (e) Infants with SLOS show prenatal/postnatal growth retardation, microcephaly, intellectual disability, distinctive facial features, cleft palate, cardiac defects, the underdeveloped external genitalia (males), postaxial polydactyly, and 2–3 syndactyly of the toes.
- END-7 A 1-year-old infant present with neuroblastoma and the anatomo-pathologic diagnosis with mitotic karyorrhectic index (MKI), *MYCN* status, and staging is presented at the pediatric tumor board. Which of the following features that need to be discussed is associated with <u>unfavorable</u> histology in a patient of this age?
  (a) Ganglioneuroblastoma, intermixed
  (b) Differentiating neuroblastoma, high MKI
  (c) Poorly differentiated neuroblastoma, intermediate MKI, MYCN non-amplified
  (d) Hyperdiploid neuroblastoma
  (e) Nodular ganglioneuroblastoma with intramodular low MKI value
- END-8 Which neoplasm reveals a positivity for NSE in tumor cells, negativity for AE1–AE3 (low and high molecular weight keratin group) in tumor cells, and S100 positivity in interstitial ("sustentacular") cells?
  (a) Alveolar soft part sarcoma
  (b) Pheochromocytoma
  (c) Malignant melanoma
  (d) Malignant peripheral nerve sheath tumor
- END-9 Which neoplasm may exhibit mutations in succinate dehydrogenase subunit genes of the mitochondrion?
  (a) Metastatic rhabdomyosarcoma
  (b) Malignant melanoma
  (c) Glioma
  (d) Malignant peripheral nerve sheath tumor
  (e) Pheochromocytoma
- END-10 The immunohistochemistry plays a significant role in the diagnosis of adrenal gland masses. Which immunohistochemical stain of the following list, if positive, would predilect a diagnosis of pheochromocytoma over adrenal cortical carcinoma?
  (a) Calretinin
  (b) Steroidogenic Factor 1
  (c) Chromogranin A
  (d) Inhibin α
  (e) Melan A

References and Recommended Readings

Adzick NS. Surgical treatment of congenital hyperinsulinism. Semin Pediatr Surg. 2020;29(3):150924. https://doi.org/10.1016/j.sempedsurg.2020.150924. Epub 2020 May 17. PMID: 32571515.

Al-Brahim NY, Asa SL. My approach to pathology of the pituitary gland. J Clin Pathol. 2006;59(12):1245–53. https://doi.org/10.1136/jcp.2005.031187. PMID: 17142570; PMCID: PMC1860551.

Arnoux JB, de Lonlay P, Ribeiro MJ, Hussain K, Blankenstein O, Mohnike K, Valayannopoulos V, Robert JJ, Rahier J, Sempoux C, Bellanné C, Verkarre V, Aigrain Y, Jaubert F, Brunelle F, Nihoul-Fékété C. Congenital hyperinsulinism. Early Hum Dev. 2010;86(5):287–94. Epub 2010 Jun 13. Review.

Aschoff L. Lectures on pathology held at the universities and academies of Japan in 1924. Jena: Gustav Fischer; 1925. p. 1–309.

Asgharzadeh S, Pique-Regi R, Sposto R, Wang H, Yang Y, Shimada H, Matthay K, Buckley J, Ortega A, Seeger RC. Prognostic significance of gene expression profiles of metastatic neuroblastomas lacking MYCN gene amplification. J Natl Cancer Inst. 2006;98(17):1193–203. PubMed PMID: 16954472.

Asgharzadeh S, Salo JA, Ji L, Oberthuer A, Fischer M, Berthold F, Hadjidaniel M, Liu CW, Metelitsa LS, Pique-Regi R, Wakamatsu P, Villablanca JG, Kreissman SG, Matthay KK, Shimada H, London WB, Sposto R, Seeger RC. Clinical significance of tumor-associated inflammatory cells in metastatic neuroblastoma. J Clin Oncol. 2012;30(28):3525–32. Epub 2012 Aug 27. PubMed PMID: 22927533; PubMed Central PMCID: PMC3675667.

Aubert S, Wacrenier A, Leroy X, Devos P, Carnaille B, Proye C, Wemeau JL, Lecomte-Houcke M, Leteurtre E. Weiss system revisited: a clinico- pathologic and

immunohistochemical study of 49 adrenocortical tumors. Am J Surg Pathol. 2002;26:1612–9.
Beckwith JB, Perrin EV. In situ neuroblastomas: a contribution to the natural history of neural crest tumors. Am J Pathol. 1963;43:1089–104. PubMed PMID:14099453; PubMed Central PMCID: PMC1949785.
Bellah R, D'Andrea A, Darillis E, Fellows KE. The association of congenital neuroblastoma and congenital heart disease. Is there a common embryologic basis? Pediatr Radiol. 1989;19(2):119–21. PubMed PMID: 2922225.
Berry T, Luther W, Bhatnagar N, Jamin Y, Poon E, Sanda T, Pei D, Sharma B, Vetharoy WR, Hallsworth A, Ahmad Z, Barker K, Moreau L, Webber H, Wang W, Liu Q, Perez-Atayde A, Rodig S, Cheung NK, Raynaud F, Hallberg B, Robinson SP, Gray NS, Pearson AD, Eccles SA, Chesler L, George RE. The ALK(F1174L) mutation potentiates the oncogenic activity of MYCN in neuroblastoma. Cancer Cell. 2012;22(1):117–30. https://doi.org/10.1016/j.ccr.2012.06.001. PubMed PMID: 22789543; PubMed Central PMCID: PMC3417812.
Bolande RP. Neoplasia of early life and its relationships to teratogenesis. Perspect Pediatr Pathol. 1976;3:145–83. Review. PubMed PMID: 184428.
Bolande RP. Neurocristopathy: its growth and development in 20 years. Pediatr Pathol Lab Med. 1997;17(1):1–25. Review. PubMed PMID: 9050057.
Christian H. Defects in membranous bones, exophthalmos, and diabetes insipidus; an unusual syndrome of dyspituitarism. Contrib Med Biol Res. 1919;1:390–401, dedicated to Sir William Osler. New York, P. B. Hoeber. Medical Clinics of North America, Philadelphia, PA., 1920; 3: 849–71.
Clarke MR, Weyant RJ, Watson CG, Carty SE. Prognostic markers in pheochromocytoma. Hum Pathol. 1998;29(5):522–6. PubMed PMID: 9596278.
Deaton MA, Glorioso JE, McLean DB. Congenital adrenal hyperplasia: not really a zebra. Am Fam Physician. 1999;59(5):1190–6, 1172. Review. Erratum in: Am Fam Physician 1999 Sep 15;60(4):1107. PubMed PMID: 10088875.
Debatin JF, Spritzer CE, Dunnick NR. Castleman disease of the adrenal gland: MR imaging features. AJR Am J Roentgenol. 1991;157(4):781–3. PubMed PMID: 1892035.
DeLellis RA. Orphan Annie eye nuclei: a historical note. Am J Surg Pathol. 1993;17(10):1067–8. PubMed PMID: 8372945.
Demir K, Keskin M, Kör Y, Karaoğlan M, Bülbül ÖG. Autoimmune thyroiditis in children and adolescents with type 1 diabetes mellitus is associated with elevated IgG4 but not with low vitamin D. Hormones (Athens). 2014;13(3):361–8. https://doi.org/10.14310/horm.2002.1481. PubMed PMID: 25079460.
Ezzat S, Asa SL. Mechanisms of disease: the pathogenesis of pituitary tumors. Nat Clin Pract Endocrinol Metab. 2006;2(4):220–30. Review. PubMed PMID: 16932287.

Fargion S, Bissoli F, Fracanzani AL, Suigo E, Sergi C, Taioli E, Ceriani R, Dimasi V, Piperno A, Sampietro M, Fiorelli G. No association between genetic hemochromatosis and alpha1-antitrypsin deficiency. Hepatology. 1996;24(5):1161–4. PubMed PMID: 8903392.
Fischer S, Asa SL. Application of immunohistochemistry to thyroid neoplasms. Arch Pathol Lab Med. 2008;132(3):359–72. https://doi.org/10.1043/1543-2165(2008)132[359:AOITTN]2.0.CO;2. Review. PubMed PMID: 18318579.
Flanagan SE, Kapoor RR, Hussain K. Genetics of congenital hyperinsulinemic hypoglycemia. Semin Pediatr Surg. 2011;20(1):13–7. Review. PubMed PMID: 21185998.
Foerster HM, Sengupta EE, Montag AG, Kaplan EL. Retroperitoneal bronchogenic cyst presenting as an adrenal mass. Arch Pathol Lab Med. 1991;115(10):1057–9. Review. PubMed PMID: 1898237.
Fuhrman SA, Lasky LC, Limas C. Prognostic significance of morphologic parameters in renal cell carcinoma. Am J Surg PathoI. 1982;6:655–63.
Geiger K, Hagenbuchner J, Rupp M, Fiegl H, Sergi C, Meister B, Kiechl-Kohlendorfer U, Müller T, Ausserlechner MJ, Obexer P. FOXO3/FKHRL1 is activated by 5-aza-2-deoxycytidine and induces silenced caspase-8 in neuroblastoma. Mol Biol Cell. 2012;23(11):2226–34. Epub 2012 Apr 4. PubMed PMID: 22493319; PubMed Central PMCID: PMC3364184.
Graves RJ. Newly observed affection of the thyroid gland in females archived 2016-03-31 at the Wayback machine. (clinical lectures.). Lond Med Surg J (Renshaw). 1835;7(part 2):516–7. Reprinted in Medical Classics, 1940;5:33–6.
Hagenbuchner J, Rupp M, Salvador C, Meister B, Kiechl-Kohlendorfer U, Müller T, Geiger K, Sergi C, Obexer P, Ausserlechner MJ. Nuclear FOXO3 predicts adverse clinical outcome and promotes tumor angiogenesis in neuroblastoma. Oncotarget. 2016;7(47):77591–606. https://doi.org/10.18632/oncotarget.12728. PubMed PMID: 27769056; PubMed Central PMCID: PMC5363607.
Hand A. Polyuria and tuberculosis. Proc Pathol Soc Phila. 1893;16:282–4. Archives of Pediatrics, New York, 1893: 10: 673–5.
Hashimoto H. Zur Kenntnis der lymphomatösen Veränderung der Schilddrüse (Struma lymphomatosa). Arch Klin Chir. 1912;97:219–48.
Herrmann T, Muckenthaler M, van der Hoeven F, Brennan K, Gehrke SG, Hubert N, Sergi C, Gröne HJ, Kaiser I, Gosch I, Volkmann M, Riedel HD, Hentze MW, Stewart AF, Stremmel W. Iron overload in adult Hfe-deficient mice independent of changes in the steady-state expression of the duodenal iron transporters DMT1 and Ireg1/ferroportin. J Mol Med (Berl). 2004;82(1):39–48. Epub 2003 Nov 15. PubMed PMID: 14618243.
Holzer R, Franklin RC. Congenital heart disease and neuroblastoma: just coincidence? Arch Dis Child.

2002;87(1):61–4. https://doi.org/10.1136/adc.87.1.61. PubMed PMID: 12089127; PubMed Central PMCID: PMC1751155.

Hook S, Spicer R, Williams J, Grier D, Lowis S, Foot A, Sergi C. Severe anemia in a 25-day-old infant due to gastric teratoma with focal neuroblastoma. Am J Perinatol. 2003;20(5):233–7. PubMed PMID: 13680506.

Hundahl SA. Perspective: National Cancer Institute summary report about estimated exposures and thyroid doses received from iodine 131 in fallout after Nevada atmospheric nuclear bomb tests. CA Cancer J Clin. 1998;48(5):285–98. PubMed PMID: 9742895.

Hundahl SA, Fleming ID, Fremgen AM, Menck HRA. National Cancer Data Base report on 53,856 cases of thyroid carcinoma treated in the U.S., 1985-1995 [see comments]. Cancer. 1998;83(12):2638–48. PubMed PMID: 9874472.

James C, Kapoor RR, Ismail D, Hussain K. The genetic basis of congenital hyperinsulinism. J Med Genet. 2009;46(5):289–99. Epub 2009 Mar 1. Review. PubMed PMID: 19254908.

Jokisch F, Kleinlein I, Haller B, Seehaus T, Fuerst H, Kremer M. A small subgroup of Hashimoto's thyroiditis is associated with IgG4-related disease. Virchows Arch. 2016;468(3):321–7. PubMed PMID: 26669779.

Keller M, Scholl-Buergi S, Sergi C, Theurl I, Weiss G, Unsinn KM, Trawöger R. An unusual case of intrauterine symptomatic neonatal liver failure. Klin Padiatr. 2008;220(1):32–6. PubMed PMID: 18172830.

Kirby ML, Waldo KL. Role of neural crest in congenital heart disease. Circulation. 1990;82(2):332–40. Review. PubMed PMID: 2197017.

Lack EE, Mulvihill JJ, Travis WD, Kozakewich HP. Adrenal cortical neoplasms in the pediatric and adolescent age group. Clinicopathologic study of 30 cases with emphasis on epidemiological and prognostic factors. Pathol Annu. 1992;27(Pt 1):1–53. Review. PubMed PMID: 1736241.

Laufer E, Kesper D, Vortkamp A, King P. Sonic hedgehog signaling during adrenal development. Mol Cell Endocrinol. 2012;351(1):19–27. Epub 2011 Oct 13. Review. PubMed PMID: 22020162; PubMed Central PMCID: PMC3288303.

Li Y, Nishihara E, Hirokawa M, Taniguchi E, Miyauchi A, Kakudo K. Distinct clinical, serological, and sonographic characteristics of hashimoto's thyroiditis based with and without IgG4-positive plasma cells. J Clin Endocrinol Metab. 2010;95(3):1309–17. https://doi.org/10.1210/jc.2009-1794. Epub 2010 Jan 22. PubMed PMID: 20097712.

Maghnie M, Larizza D, Triulzi F, Sampaolo P, Scotti G, Severi F. Hypopituitarism and stalk agenesis: a congenital syndrome worsened by breech delivery? Horm Res. 1991;35(3–4):104–8. PubMed PMID: 1806462.

Maris JM, Brodeur GM. Genetics of neuroblastoma. In: Cowell JK, editor. Molecular genetics of cancer. 2nd ed. Oxford: BIOS; 2001. ISBN 10: 185996169X; ISBN 13: 9781859961698; Publisher: Taylor & Francis, 2001.

Maris JM, Matthay KK. Molecular biology of neuroblastoma. J Clin Oncol. 1999;17:2226–79.

Maris JM, Hogarty MD, Bagatell R, Cohn SL. Neuroblastoma. Lancet. 2007;369(9579):2106–20. Review. PubMed PMID: 17586306.

Marquard J, Palladino AA, Stanley CA, Mayatepek E, Meissner T. Rare forms of congenital hyperinsulinism. Semin Pediatr Surg. 2011;20(1):38–44. Review. PubMed PMID: 21186003.

Medeiros LJ, Weiss LM. New developments in the pathologic diagnosis of adrenal cortical neoplasms. A review. Am J Clin Pathol. 1992;97(1):73–83. Review. PubMed PMID: 1728867.

Meguro R, Asano Y, Odagiri S, Li C, Iwatsuki H, Shoumura K. Nonheme-iron histochemistry for light and electron microscopy: a historical, theoretical and technical review. Arch Histol Cytol. 2007;70(1):1–19. Review. PubMed PMID: 17558140.

Meier JJ, Köhler CU, Alkhatib B, Sergi C, Junker T, Klein HH, Schmidt WE, Fritsch H. Beta-cell development and turnover during prenatal life in humans. Eur J Endocrinol. 2010;162(3):559–68. https://doi.org/10.1530/EJE-09-1053. Epub 2009 Dec 18. PubMed PMID: 20022941.

Meissner T, Mayatepek E. Clinical and genetic heterogeneity in congenital hyperinsulinism. Eur J Pediatr. 2002;161(1):6–20. Review. PubMed PMID: 11808881.

Nakissa N, Constine LS, Rubin P, Strohl R. Birth defects in three common pediatric malignancies; Wilms' tumor, neuroblastoma and Ewing's sarcoma. Oncology. 1985;42(6):358–63. PubMed PMID: 2999670.

Nathan BM, Sockalosky J, Nelson L, Lai S, Sergi C, Petryk A. The use of hormonal therapy in pediatric heart disease. Front Biosci (Schol Ed). 2009;1(1):358–75. Review. PubMed PMID: 19482707.

Papaioannou G, McHugh K. Neuroblastoma in childhood: review and radiological findings. Cancer Imaging. 2005;5:116–27.

Prévot S, Penna C, Imbert JC, Wendum D, de Saint-Maur PP. Solitary fibrous tumor of the adrenal gland. Mod Pathol. 1996;9(12):1170–4. PubMed PMID: 8972477.

Rahier J, Guiot Y, Sempoux C. Morphologic analysis of focal and diffuse forms of congenital hyperinsulinism. Semin Pediatr Surg. 2011;20(1):3–12. Review. PubMed PMID: 21185997.

Reiprich S, Claus Stolt C, Schreiner S, Parlato R, Wegner M. SoxE proteins are differentially required in mouse adrenal gland development. Mol Biol Cell. 2008;19(4):1575–86. PMID:18272785.

Sanderson-Damberg E. The thyroid glands from 15–25. Year of life from the North German plain and coastal area as well as from Bern. Frankfurt Z Path. 1911;6:312–34.

Santonja C, Diaz MA, Dehner LP. A unique dysembryonic neoplasm of the adrenal gland composed of nephrogenic rests in a child. Am J Surg Pathol. 1996;20(1):118–24. PubMed PMID: 8540603.

Schlitter AM, Dorneburg C, Barth TFE, Wahl J, Schulte JH, Bruederlein S, Debatin KM, Beltinger C. Strong expression of CD57 is associated with aggressive neuroblastoma cells in vitro and in vivo, and with undifferentiated neuroblastoma cells in situ. [Abstract]. In: Proceedings of the 103rd annual meeting of the American association for cancer research, 2012 Mar 31–Apr 4. Chicago/Philadelphia: AACR; Cancer Res 2012; 72(8 Suppl): Abstract nr. 1438.

Schranz M, Talasz H, Graziadei I, Winder T, Sergi C, Bogner K, Vogel W, Zoller H. Diagnosis of hepatic iron overload: a family study illustrating pitfalls in diagnosing hemochromatosis. Diagn Mol Pathol. 2009;18(1):53–60. PubMed PMID: 19214108.

Schüller A. Über eigenartige Schädeldefekte im Jugendalter («Landkartenschädel»). Fortschr Geb Rontgenstr. 1915–1916;23:12–8.

Sempoux C, Capito C, Bellanné-Chantelot C, Verkarre V, de Lonlay P, Aigrain Y, Fekete C, Guiot Y, Rahier J. Morphological mosaicism of the pancreatic islets: a novel anatomopathological form of persistent hyperinsulinemic hypoglycemia of infancy. J Clin Endocrinol Metab. 2011;96(12):3785–93. Epub 2011 Sep 28. PubMed PMID: 21956412.

Sergi C, Böhler T, Schönrich G, Sieverts H, Roth SU, Debatin KM, Otto HF. Occult thyroid pathology in a child with acquired immunodeficiency syndrome. Case report and review of the drug-related pathology in pediatric acquired immunodeficiency syndrome. Pathol Oncol Res. 2000;6(3):227–32. PubMed PMID: 11033465.

Sergi C, Himbert U, Weinhardt F, Heilmann W, Meyer P, Beedgen B, Zilow E, Hofmann WJ, Linderkamp O, Otto HF. Hepatic failure with neonatal tissue siderosis of hemochromatotic type in an infant presenting with meconium ileus. Case report and differential diagnosis of the perinatal iron storage disorders. Pathol Res Pract. 2001a;197(10):699–709; discussion 711–3. PubMed PMID: 11700892.

Sergi C, Penzel R, Uhl J, Zoubaa S, Dietrich H, Decker N, Rieger P, Kopitz J, Otto HF, Kiessling M, Cantz M. Prenatal diagnosis and fetal pathology in a Turkish family harboring a novel nonsense mutation in the lysosomal alpha-N-acetyl-neuraminidase (sialidase) gene. Hum Genet. 2001b;109(4):421–8. PubMed PMID: 11702224.

Sergi C, Roth SU, Adam S, Otto HF. Mapping a method for systematically reviewing the medical literature: a helpful checklist postmortem protocol of human immunodeficiency virus (HIV)-related pathology in childhood. Pathologica. 2001c;93(3):201–7. Review. PubMed PMID: 11433613.

Sergi C, Dhiman A, Gray JA. Fine needle aspiration cytology for neck masses in childhood. An illustrative approach. Diagnostics (Basel). 2018;8((2). pii):E28. https://doi.org/10.3390/diagnostics8020028. Review. PubMed PMID: 29690556; PubMed Central PMCID: PMC6023333.

Sherman FE, Bass LW, Fetterman GH. Congenital metastasizing adrenal cortical carcinoma associated with cytomegaly of the fetal adrenal cortex. Am J Clin Pathol. 1958;30:439–46.

Shimada H, Ambros IM, Dehner LP, et al. The international neuroblastoma pathology classification (the Shimada system). Cancer. 1999;86:364–72.

Shukla SK, Singh G, Ahmad S, Pant P. Infections, genetic and environmental factors in pathogenesis of autoimmune thyroid diseases. Microb Pathog. 2018;116:279–88. https://doi.org/10.1016/j.micpath.2018.01.004. Epub 2018 Jan 8. Review. PubMed PMID: 29325864.

Simmonds M. Über Hypophysisschwund mit tödlichem Ausgang. Dtsch Med Wochenschr. 1914;40:322–3.

Stefanski H, Wedel L, Haller C, Pierpont ME, Perkins JL. Is there an association between bicuspid aortic valve and neuroblastoma? Pediatr Blood Cancer. 2010;55(2):359–60. https://doi.org/10.1002/pbc.22539. PubMed PMID: 20582975.

Val P, Lefrançois-Martinez A-M, Veyssière G, Martinez A. SF-1 a key player in the development and differentiation of steroidogenic tissues. Nucl Recept. 2003;1(1):8. PMID:14594453.

Verissimo CS, Molenaar JJ, Fitzsimons CP, Vreugdenhil E. Neuroblastoma therapy: what is in the pipeline? Endocr Relat Cancer. 2011;18(6):R213–31. https://doi.org/10.1530/ERC-11-0251. Print 2011 Dec. Review. PubMed PMID: 21971288.

Von Basedow KA. Exophthalmus durch Hypertrophie des Zellgewebes in der Augenhöhle. [Casper's] Wochenschrift für die gesammte Heilkunde, Berlin. 1840;6:197–204; 220–8. Partial English translation in: Ralph Hermon Major (1884–1970): Classic Descriptions of Disease. Springfield, C. C. Thomas, 1932. 2nd edition, 1939; 3rd edition, 1945.

Weiss LM. Comparable histologic study of 43 metastasizing and non metastasizing adrenocortical tumors. Am J Surg Pathol. 1984;8:163–9.

Weiss LM, Medeiros LJ, Vickery AL Jr. Pathologic features of prognostic significance in adrenocortical carcinoma. Am J Surg Pathol. 1989;13:202–6.

Westermann F, Muth D, Benner A, Bauer T, Henrich KO, Oberthuer A, Brors B, Beissbarth T, Vandesompele J, Pattyn F, Hero B, König R, Fischer M, Schwab M. Distinct transcriptional MYCN/c-MYC activities are associated with spontaneous regression or malignant progression in neuroblastomas. Genome Biol. 2008;9(10):R150. https://doi.org/10.1186/gb-2008-9-10-r150. PubMed PMID: 18851746; PubMed Central PMCID: PMC2760877.

Xu X, Sergi C. Pediatric adrenal cortical carcinomas: histopathological criteria and clinical trials. A systematic review. Contemp Clin Trials. 2016;50:37–44. https://doi.org/10.1016/j.cct.2016.07.011. Epub 2016 Jul 14. Review. PubMed PMID: 27424218.

Zhang L, Yao M, Hisaoka M, Sasano H, Gao H. Primary Ewing sarcoma/primitive neuroectodermal tumor in the adrenal gland. APMIS. 2016;124(7):624–9. https://doi.org/10.1111/apm.12544. Epub 2016 May 20. PubMed PMID: 27197626.

Soft Tissue

Contents

12.1	**Development and Genetics**	1004
12.2	**Vascular and Inflammatory Changes of Soft Tissue**	1005
12.2.1	Hyperemia	1005
12.2.2	Necrotizing Fasciitis	1005
12.2.3	Vasculitis-Associated Soft Tissue Changes	1005
12.2.4	Miscellaneous	1005
12.3	**Soft Tissue Neoplasms: Scoring**	1009
12.4	**Adipocytic Tumors**	1010
12.4.1	Lipoma	1011
12.4.2	Lipoma Subtypes	1011
12.4.3	Lipomatosis	1012
12.4.4	Lipoblastoma	1012
12.4.5	Hibernoma	1012
12.4.6	Locally Aggressive and Malignant Adipocytic Tumors	1013
12.5	**Fibroblastic/Myofibroblastic Tumors**	1017
12.5.1	Fasciitis/Myositis Group	1019
12.5.2	Fibroma Group	1020
12.5.3	Fibroblastoma Classic and Subtypes	1021
12.5.4	Fibrous Hamartoma of Infancy (FHI)	1023
12.5.5	Fibromatosis of Childhood	1023
12.5.6	Infantile Myofibroma/Myofibromatosis	1027
12.5.7	Fibroblastic/Myofibroblastic Tumors with Intermediate Malignant Potential	1027
12.5.8	Malignant Fibroblastic/Myofibroblastic Tumors	1033
12.6	**Fibrohistiocytic Tumors**	1036
12.6.1	Histiocytoma	1036
12.6.2	Benign Fibrous Histiocytoma	1037
12.6.3	Borderline Fibrous Histiocytoma	1039
12.6.4	Malignant Fibrous Histiocytoma	1039
12.7	**Smooth Muscle Tumors**	1040
12.7.1	Leiomyoma	1040
12.7.2	EBV-Related Smooth Muscle Tumors	1040
12.7.3	Leiomyosarcoma	1040

© Springer-Verlag GmbH Germany, part of Springer Nature 2020
C. M. Sergi, *Pathology of Childhood and Adolescence*,
https://doi.org/10.1007/978-3-662-59169-7_12

12.8	**Pericytic Tumors**	1041
12.8.1	Glomus Tumor	1041
12.8.2	Glomangiosarcoma	1042
12.8.3	Myopericytoma	1042
12.9	**Skeletal Muscle Tumors**	1044
12.9.1	Rhabdomyomatous Mesenchymal Hamartoma (RMH)	1044
12.9.2	Rhabdomyoma	1046
12.9.3	Rhabdomyosarcoma	1046
12.9.4	Pleomorphic RMS	1055
12.10	**Vascular Tumors**	1059
12.10.1	Benign Vascular Tumors	1059
12.10.2	Vascular Tumors with Intermediate Malignant Potential	1062
12.10.3	Malignant Vascular Tumors	1066
12.10.4	Genetic Syndromes Associated with Vascular Tumors	1068
12.11	**Chondro-Osteoforming Tumors**	1069
12.11.1	Extraskeletal Chondroma	1069
12.11.2	Extraskeletal Myxoid Chondrosarcoma	1069
12.11.3	Mesenchymal Chondrosarcoma	1070
12.11.4	Extraskeletal Aneurysmatic Bone Cyst	1070
12.11.5	Extraskeletal Osteosarcoma (ESOS)	1070
12.11.6	Extraskeletal Chordoma	1070
12.12	**Tumors of Uncertain Differentiation**	1071
12.12.1	Myxoma	1071
12.12.2	Myoepithelial Carcinoma	1072
12.12.3	Parachordoma	1072
12.12.4	Synovial Sarcoma	1074
12.12.5	Epithelioid Sarcoma	1076
12.12.6	Alveolar Soft Part Sarcoma	1076
12.12.7	Clear Cell Sarcoma	1079
12.12.8	Extraskeletal Myxoid Chondrosarcoma (ESMC)	1080
12.12.9	PNET/Extraskeletal Ewing Sarcoma (ESES)	1081
12.12.10	Desmoplastic Small Round Cell Tumor	1083
12.12.11	Extrarenal Rhabdoid Tumor	1084
12.12.12	Malignant Mesenchymoma	1084
12.12.13	PEComa	1084
12.12.14	Extrarenal Wilms' Tumor	1085
12.12.15	Sacrococcygeal Teratoma and Extragonadal Germ Cell Tumor and Yolk Sac Tumor	1085
	Multiple Choice Questions and Answers	1090
	References and Recommended Readings	1091

12.1 Development and Genetics

The embryo grows in the first few weeks exponentially, and the algorithms of the tissue information, so far known, show the complexity of the differentiation of cells to specific organs. Somites characterize the soft tissue. This differentiation is directed by transcription factors inherited via the chromosomes from both parents. Most organs are formed between the 5th and the 8th week of the embryonic life. After the embryogenesis, there is continued growth and development to the time of delivery of the baby at 38–42 weeks of gestation in utero. There are three germ cell layers, which are ectoderm, endoderm, and mesoderm. The ectoderm forms the lining that covers the body of the organism and gives rise to cells in the nervous system. The endoderm is accountable for the formation of the gastrointestinal tract, lung, and the endocrine glands. The mesoderm forms the connective tissues and the soft tissues such as bone, muscle, and fat. Here, the somites are transient, segmentally organized structures that contribute to multiple tissues, including the axial skeleton, skeletal and smooth muscles, dorsal dermis, tendons, ligaments, cartilage, and adipose tissue in the vertebrate embryo. Also, the somites are responsible for the migration pathways of neural crest cells of the trunk and spinal nerve axons. The total number of somites formed is species-specific. Thus, humans have 38–39, while 50 somites are present in chickens and 65 in mice. The dermomyotome contributes specifically to the dorsal dermis and skeletal muscle, while the sclerotome is the source of the axial skeleton, and the myotome contains skeletal muscle precursor cells. The genes that direct cellular proliferation and development in embryologic life are suppressed once their objective is fulfilled in the formation of an adequate organ or structure. In case, some of these genes are inappropriately activated, a neoplastic process can be the result.

> **Box 12.1 Scores Used for the Evaluation of the Outcome of Patients with SNV**
> 1. "5-Factor Score" (French Vasculitis SG, 1996) – Guillevin et al. Medicine 1996
> 2. "Birmingham Vasculitis Activity Score" (BVAS) (1994; 1997; 2009) – Luqmani et al. 1994; Luqmani et al. 1997; Mukhtyar et al. 2009
> 3. "Disease Extent Index" (2001) – De Groot et al. Clin Nephrol 2001
> 4. "Vasculitis Damage Index" (1996) – Exley and Bacon, Curr Opin Rheumatol 1996
> 5. "5-Factor Score" (French Vasculitis SG 2011) – Guillevin et al. Medicine 2011
>
> Notes: SG, the study group; SNV, systemic necrotizing vasculitis

- *PGN*: Rapidly fatal if an appropriate and prompt diagnosis and treatment (IV antibiotics and debridement of necrotic tissue) are not applied.

12.2 Vascular and Inflammatory Changes of Soft Tissue

12.2.1 Hyperemia

Hyperemia can be passive in case of tourniquet compression or active in case of inflammation.

12.2.2 Necrotizing Fasciitis

- *DEF*: Serious infection of the subcutaneous tissue and of the tissue that covers internal organs or fascia caused by several different types of bacteria, and the disease can arise suddenly and spread quickly due to transient or permanent immunosuppression of the individuum.
- *CLI*: Flu-like symptoms and redness and pain around the infection site with the rapid propagation of the infection due to an inefficient removal method by the organism.

12.2.3 Vasculitis-Associated Soft Tissue Changes

Systemic necrotizing vasculitis (SNV) may be associated with a different evolution or damage and harbors an entirely different outcome for patients having the same disease. SNV includes polyarteritis nodosa (PAN), Churg-Strauss syndrome (CSS) or aka eosinophilic granulomatosis with polyangiitis (EGPA), and microscopic polyangiitis (MPA) as well as Wegener granulomatosis (WG). There have been a series of scores (Box 12.1) that may be very useful to determine the outcome of these patients affected by SNV.

12.2.4 Miscellaneous

In this paragraph are simply illustrated very different processes encountered in the soft tissue as iatrogenic or sporadic with or without congenital character (Figs. 12.1, 12.2, 12.3, and 12.4).

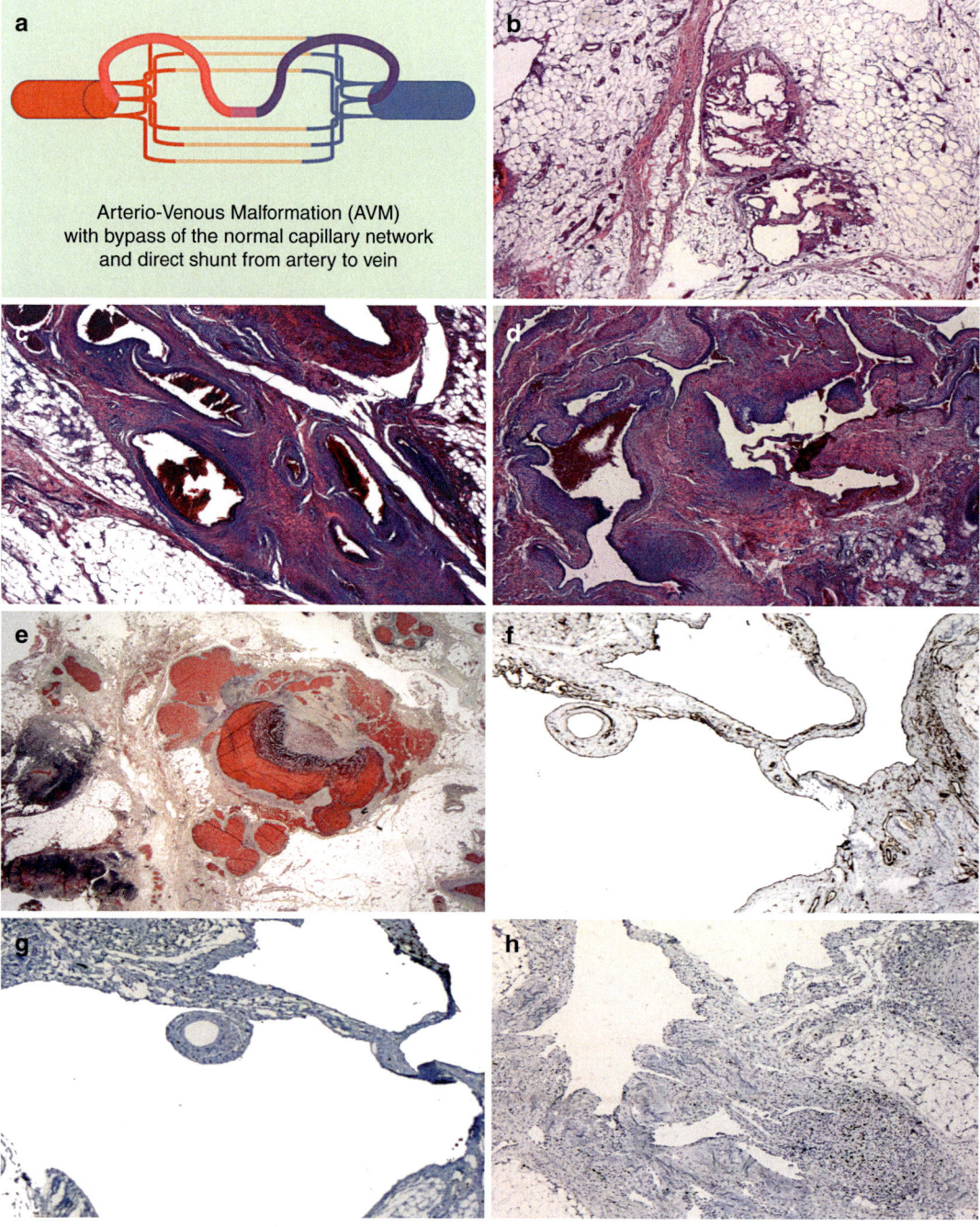

Fig. 12.1 Congenital Anomalies and Inflammation. In (**a**) is a cartoon explaining the underlying pathology of an arteriovenous malformation. The microphotographs (**b**) through (**d**) show an example of an AVM of the soft tissue (**b–d**, H&E, 10× original magnification). In (**e**) is shown an AVM with alteration of the architecture highlighted by the Movat pentachrome stain (50× original magnification). The microphotograph (**f**) shows part of an AVM using an antibody against CD31 (endothelial cell marker) by immunohistochemistry (50× original magnification), while a consecutive section (**g**) shows the absence of staining using an antibody against GLUT1 to differentiate from a capillary hemangioma. GLUT1 is the erythrocyte type glucose transporter protein GLUT-1 derived from *GLUT1* gene. GLUT1 is useful in the differentiation between juvenile hemangioma as it is highly expressed in this lesion, but not in congenital hemangiomas, vascular malformations, pyogenic granulomas or, even, granulation tissue. The microphotograph in (**h**) shows the proliferation activity of the AVM using an antibody against Ki67 (MIB1 immunohistochemistry, avidin-biotin complex) (50× original magnification)

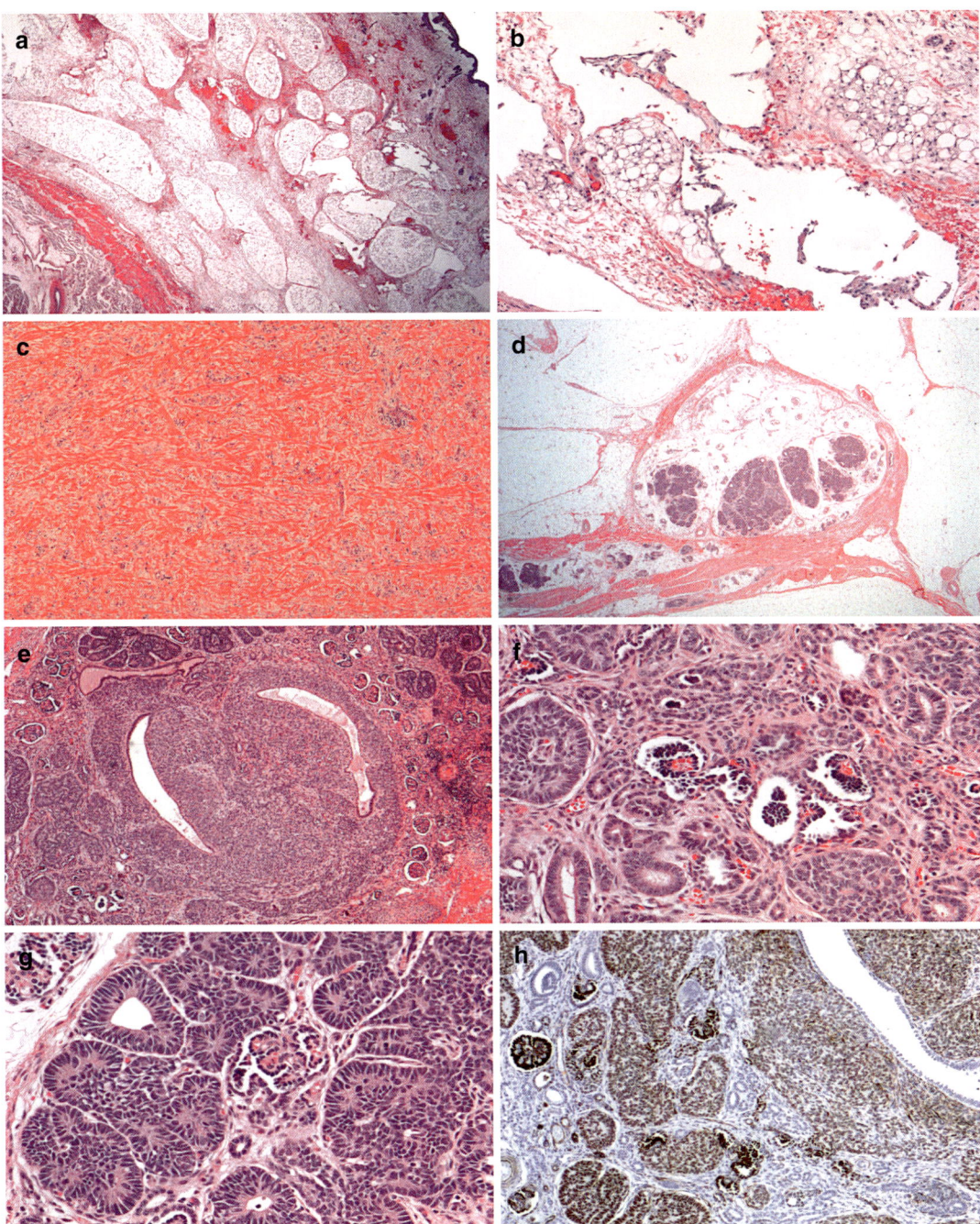

Fig. 12.2 Congenital Anomalies and Inflammation. In (**a** and **b**) are shown two microphotographs of the capillary leak syndrome (H&E stain, 12.5× and 100× original magnification, respectively). The microphotograph (**c**) shows a disorganized soft tissue as histologic correlate of a tuberosity in a patient surgically operated in the suture sagittalis for craniosynostosis (H&E stain, 50× original magnification). In (**d–h**) are shown microphotographs of some ectopic immature renal tissue (d, H&E stain, 12.5× original magnification; **e**, H&E stain, 50×; **f**, H&E stain, 200×; **g**, H&E stain, 200×; **h**, anti-WT1, 100×). The characteristics of a fetal glomerulogenic zone are impressive, and both epithelial and blastema are embedded in an orthotypic stroma. The antibody anti-WT-1 highlights the blastema by immunohistochemistry using the avidin-biotin complex (ABC) method (Bratthauer 2010)

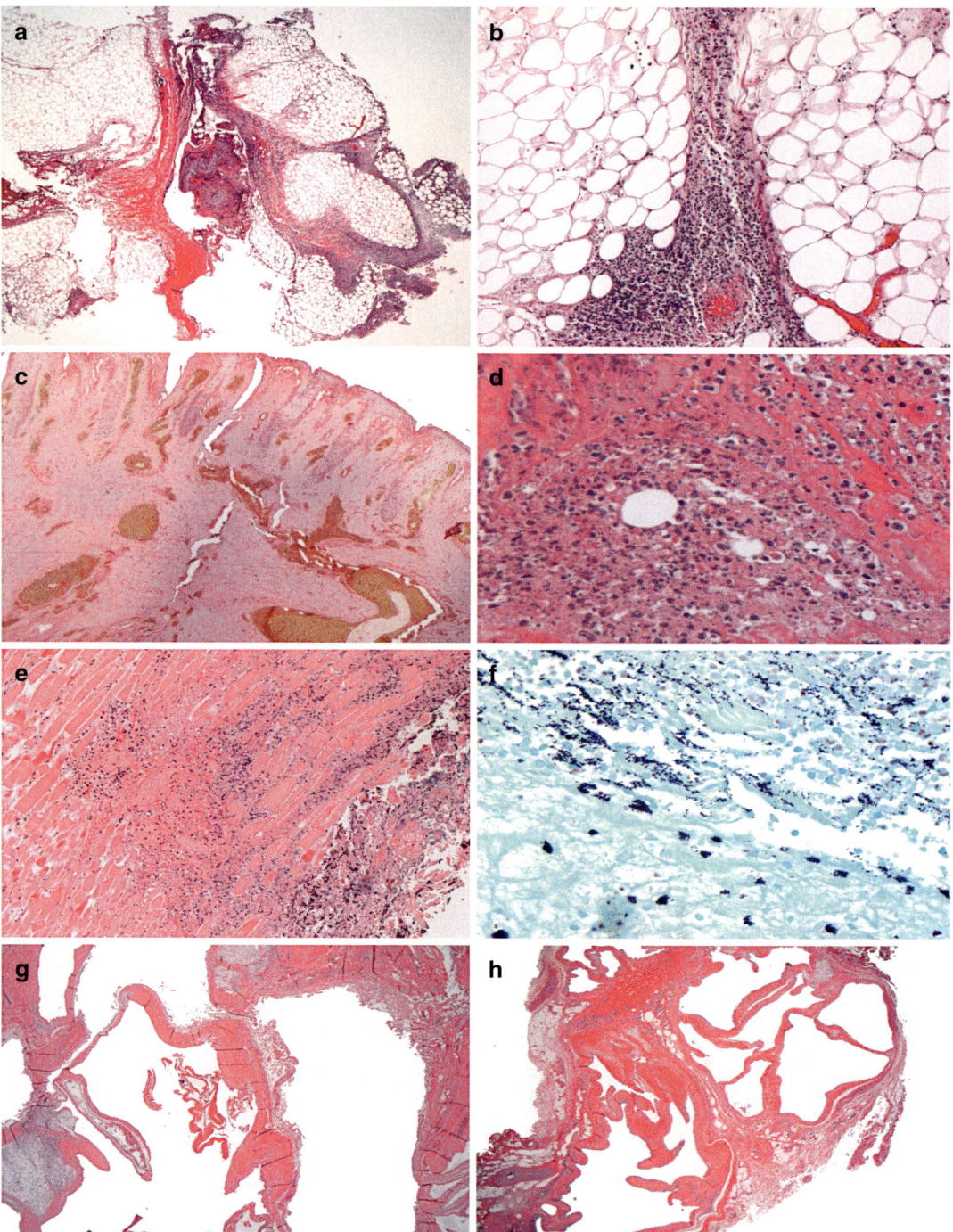

Fig. 12.3 Congenital Anomalies and Inflammation. In (**a**) is shown a leg abscess with panniculitis following superinfection after varicella infection (H&E stain, 12.5× original magnification). The microphotograph (**b**) is taken from the same location by a higher magnification (100×) using the same H&E stain. In (**c** and **d**) is shown *Str. pneumoniae* gangrenous necrosis in an asplenic child (**c**, H&E stain, 100×; **d**, H&E stain, 400×). In (**e**) is shown a necrotizing fasciitis (H&E stain, 100×), and (**f**) is shown the same necrotizing fasciitis with Gram stain showing numerous colonies of Gram-positive bacteria (400×). The microphotographs (**g**) and (**h**) show a ganglion cyst of which (**h**) was located left wrist volar area and excised as a mass (both H&E stains and 50× and 12.5× as original magnifications)

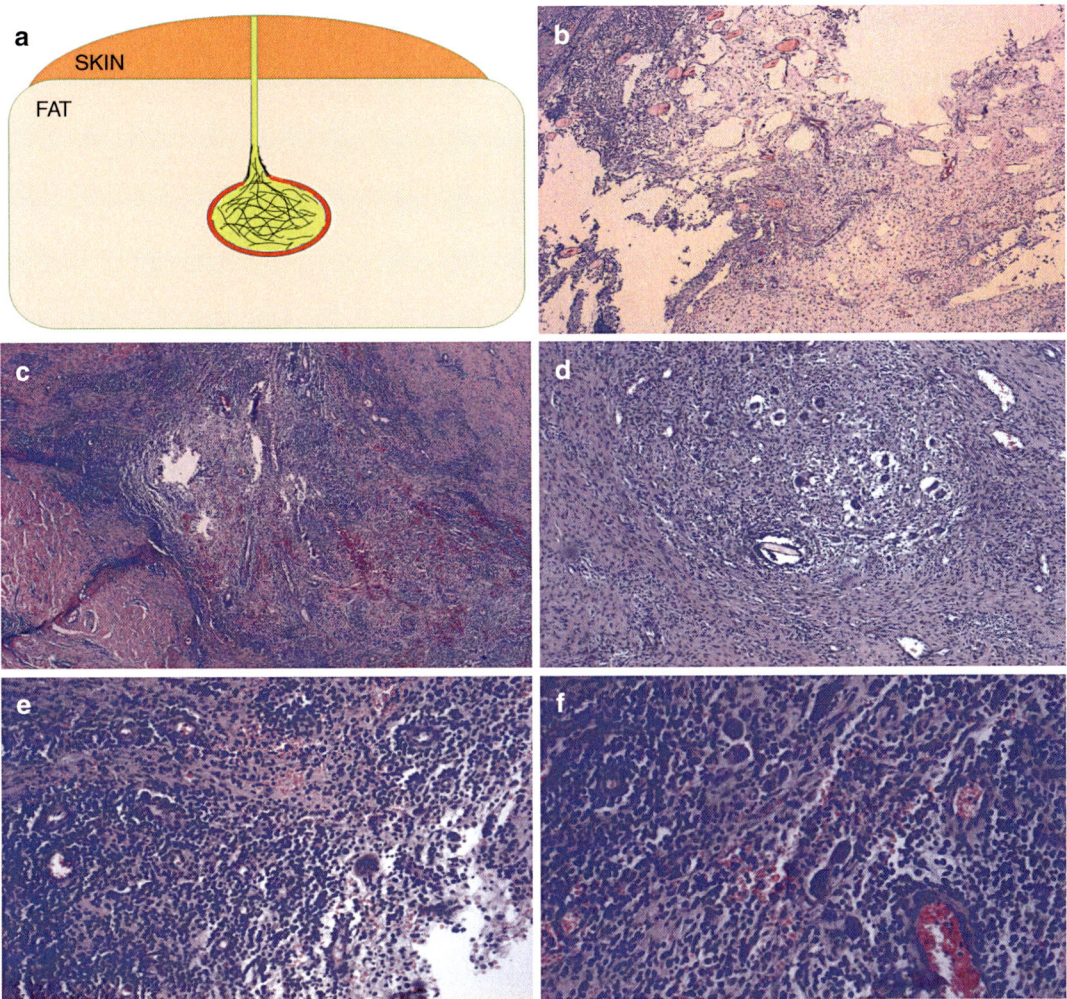

Fig. 12.4 Congenital Anomalies and Inflammation. The figure (**a**) is representing a cartoon of the pilonidal cyst that can become an abscess. The microphotographs from (**b**) through (**f**) evidence a pilonidal cyst at several degrees of magnification in several patients (**b–f**), hematoxylin and eosin staining, ×40, ×40, ×100, ×200, and ×200 as original magnification, respectively)

12.3 Soft Tissue Neoplasms: Scoring

The most recent revisited FFS (5-Factor Score) is based on an original FFS developed in 1996. It includes a series of clinical and pathological parameters based on an extensive database with affected patients between 1957 and 2005. To the best of my knowledge, this probably represents one of the most massive databases of patients with SNV and probably very able to identify the most likely outcome for patients affected by systemic vasculitis.

> **Box 12.2 Non-Age-Related Factors Associated with a Low 5-YSR**
> 1. Cardiac Insufficiency
> 2. GI Involvement
> 3. Renal Insufficiency (creatinine peak ≥150 μmol/L [>1.7 mg/dL])

In Box 12.2, factors associated with a low 5-YSR (apart of the age) are listed. A low 5-YSR is seen in patients affected with a soft tissue sarcoma. All parameters are disease-specific, and

> **Box 12.3 Classification of Soft Tissue Tumors**
> 1. *Benign Tumors*
> 2. *Tumors of Intermediate Malignancy or Tumors with Low-Malignant Potential/Malignant Behavior*
> 3. *Malignant Tumors*

> **Box 12.4 The Cell of Origin of the Soft Tissue Tumors**
>
> | | |
> |---|---|
> | Fatty cell | Adipocytic tumors |
> | Fibroblasts | Fibroblastic/myofibroblastic tumors |
> | Fibrohistiocyte | Fibrohistiocytic tumors |
> | Smooth muscle cell | Smooth muscle tumors |
> | Pericytes | Pericytic tumors |
> | Skeletal muscle cell (myofibers) | Skeletal muscle tumors |
> | Endothelial cell | Vascular tumors |
> | Chondrocytes/osteocytes | Chondro-osseous tumors |
> | ??? | Tumors of uncertain differentiation |

> **Box 12.5 Adipocytic Tumors**
> 12.4.1 Lipoma
> 12.4.2 Lipoma Subtypes
> 12.4.3 Lipomatosis
> 12.4.4 Lipoblastoma
> 12.4.5 Hibernoma
> 12.4.6 Locally Aggressive and Malignant Adipocytic Tumors
>   12.4.6.1 Atypical Lipomatous Tumor/Well-Differentiated Liposarcoma
>   12.4.6.2 Myxoid/Round Cell Liposarcoma
>   12.4.6.3 Pleomorphic Liposarcoma
>   12.4.6.4 Dedifferentiated Liposarcoma

the presence of each was accorded +1 point. FFS of 0, 1, and ≥ 2 are associated with 9%, 21%, and 40% 5-year mortality rates, respectively. Previously, the fifth factor about the ENT (ear, nose, and throat) manifestations was considered, but it is not included anymore in several updated re-evaluations. Interestingly, the better outcome of patients with WG with ENT manifestations, even for them with other visceral involvement, compared with the result for those without ENT manifestations, reflects probably the heterogeneity of the WG phenotype. In the classification of soft tissue tumors, three degrees of severity or malignancy are considered (Box 12.3) and the cell of origin if known (Box 12.4). Coindre and Neuville have revised and updated the grading for soft tissue sarcomas. The French grading system and the National Cancer Institute (NCI) grading systems are the most used scores for soft tissue sarcomas. The French system and the NCI are 3-grade systems. They are predominantly based on histologic type and subtype, tumor necrosis, and mitotic activity. It has been indicated that rules for using grading should be rigorously respected. Moreover, limitations of grading should be well known to obtain the best performance. Of paramount importance is that the most disputable point is the respective values of histologic typing. Histologic typing must be defined before grading is started. Finally, grading should not be applied on tumors belonging to intermediate categories.

12.4 Adipocytic Tumors

Tumors are showing adipocytic differentiation or arising from adipose tissue. Benign tumors include *lipoma, lipomatosis, lipoblastoma, lipoma subtypes (angiolipoma, myolipoma, extrarenal angiomyolipoma, extra-adrenal myelolipoma, chondroid lipoma, spindle cell/pleomorphic lipoma)*, and *hibernoma*. Adipocytic tumors with low-malignant potential (locally aggressive) are a group represented by the *well-differentiated liposarcoma* (aka atypical lipomatous tumor) only. Malignant adipocytic tumors include several variants of *liposarcoma* (dedifferentiated, myxoid, round cell, pleomorphic, mixed-type) and liposarcoma, NOS (not otherwise specified). Most common soft tissue sarcomas of the retroperitoneum, in descending order, are liposarcoma, malignant fibrous histiocytoma, leiomyosarcoma, and rhabdomyosarcoma (Box 12.5).

12.4 Adipocytic Tumors

12.4.1 Lipoma

The lipoma is doubtless the most common soft tissue tumor (~15%) made of mature white adipocytes with uniform nuclei resembling regular white fat with wide age range, usually adults ≥ fifth decade, but children may present with this tumor as well. The lipoma can be associated with genetic syndromes (*NF*, *MEN*, *Bannayan-Riley-Ruvalcaba syndrome*, or macrocephaly, hemangiomas, lipomas, and GI polyposis), one of the four PTEN hamartoma tumor syndromes. The *PTEN* hamartoma tumor syndrome (PHTS) includes Cowden syndrome, Bannayan-Riley-Ruvalcaba syndrome (BRRS), *PTEN*-related Proteus syndrome, and Proteus-like syndrome, all of which may present with intestinal hamartomatous polyposis. Cowden syndrome is a multiple hamartoma syndrome with individuals presenting with a characteristic phenotype, including macrocephaly, trichilemmomas, and papillomatous papules and a high risk for benign and malignant tumors of the thyroid (usually follicular), breast, and endometrium. BRRS is characterized by macrocephaly, lipomas, and pigmented macules of the glans penis. Conversely, subjects with Proteus syndrome may have variable congenital malformations and hamartomatous overgrowth of multiple tissues, as well as connective tissue nevi, epidermal nevi, and hyperostoses. Proteus-like syndrome is currently undefined but should refer to individuals with clinical features similar to Proteus syndrome, although they do not fit the diagnostic criteria for it thoroughly.

- *GRO*: Subcutaneous (unlike liposarcoma, which is deep-seated) bright-yellow homogeneous nodule of the back, shoulder, neck, and proximal extremities, delicately encapsulated in superficial soft tissue with trabeculae, poorly circumscribed when arising deeper with possible infiltrative growth in muscle, greasy cut surface ± areas of infarction, necrosis, calcification; osseous metaplasia can occur.
- *CLM*: Mature white adipose tissue (in presence of a mass!) of large cells (up to 300 μm) without atypia or mitoses with 2–5× variation in cell size (more than standard white adipose tissue), ± intranuclear vacuoles, thickened fibrous septa, intermingled fat necrosis with histiocytes, infarct or calcification, bone or cartilage.
- *IHC*: (+) Vim, S100, CD34 (slender spindle cells), leptin, PAS (capillaries), Ret (surrounding each adipocyte). (−) MDM2, CDK4 and p16.
- *TEM*: Uni-vacuolar adipocytes with peripheral nuclei, pinocytotic vesicles, and pericellular external lamina.
- *CMB*: Up to ¾ rearrangements of *HMGA2*/*HMGIC* at 12q13–15 in solitary lipomas.
- *TRT*: Surgical excision.
- *PGN*: Good, substantially, unless an atypical lipoma or a malignant component is present in the non-submitted tissue specimen for microscopic examination. This event may be the case of deep-seated lipomas that require a satisfactory grossing and microscopic examination.

12.4.2 Lipoma Subtypes

- *Fibrolipoma/myxolipoma*: If fibrous or myxoid tissues are increased, respectively.
- *Angiolipoma*: Vascular component mainly at the periphery with hyaline (fibrin) thrombi common in BVs.
- *Myolipoma*: Mature smooth muscle and mature adipose tissue.
- *Angiomyolipoma (AML), extrarenal*: A combined tumor of vascular, muscular, and adipose tissue.
- *Myelolipoma, extra-adrenal*: Lipoma with extramedullary hematopoiesis.
- *Chondroid lipoma*: Mixture with mature fat, lipoblasts, and chondroid matrix (DDX: myxoid liposarcoma, MLPS, and extraskeletal myxoid chondrosarcoma (ESMC)).
- *Spindle cell/pleomorphic lipoma*: Spindle-cell lipoma and pleomorphic lipoma represent a continuum of benign tumors with most tumors showing some features of both. Microscopically, there is a mixture of mature lipocytes and uniform, primitive, bland, S100 negative spindle cells in a mucinous and fibrous background with many mast cells and hyperchromatic MNGC ("florets with a wreath-like arrangement of nuclei around cell

periphery") as well as ropey collagen bundles. DDX: *Giant cell fibroblastoma/giant cell angiofibroma*, NF1-associated neurofibroma, *WDLS* (deep location, more lipoblasts, variable thick collagen, variable floret MNGC, CD34-), *myxoid liposarcoma* (lipoblasts and pleomorphic spindle cells, prominent plexiform vascular pattern, no thick collagen bundles, CD34-), *SFT* ("patternless" pattern, thin collagen fibers, no prominent adipose component, CD99+, BCL2+, CD34+, STAT6+). Cytogenetically, 13 and 16 chromosomal abnormalities have been observed.

12.4.3 Lipomatosis

Lipomatosis (aka Madelung disease) is a diffuse overgrowth of mature adipose tissue in different clinical settings and anatomic regions probably correlated in some cases to point mutations of mitochondrial genes. Lipomatosis has a wide age range (usually <2 years), but symmetric forms of lipomatosis are known in middle-aged men of Mediterranean region. There may be +Hx. Liver disease, hormonal therapy, and adrenocortical tumors. Moreover, HAART-related HIV lipodystrophy (highly active antiretroviral therapy) should be kept in mind.

12.4.4 Lipoblastoma

Lobulated tumor, either localized (lipoblastoma) or diffusely infiltrating (lipoblastomatosis), resembling fetal adipose tissue of almost exclusively infants and young children (<3 years), preferably male, and located at the proximal regions of upper and lower extremities and consisting of painless superficial lobulated soft tissue mass with mucoid cut surface mostly on the left side of the body. Microscopically, there is adipogenesis in a mixture of stages of differentiation including preadipocytes (spindle or stellate-shaped), uni- or multi-vacuolated lipoblasts and mature adipocytes in a central lobule, prominent separating fibrous septa, plexiform vascular pattern, and abundant myxoid stroma, which may arise the DDX with myxoid LPS. In *lipoblastomatosis, there is* less lobulation and there is infiltration of skeletal muscle.

- *IHC*: Not contributive.
- *TEM*: Primitive cells with high N/C ratio, intracytoplasmic lipid droplets, and few organelles.
- *CMB*: FISH is helpful to confirm diagnosis and rearrangement of the 8q11 approximately q13 region in 82% including the *PLAG1* gene producing *HAS2-PLAG1* and *COL1A2-PLAG1* hybrid genes. Polysomy for chromosome #8 has also been described.
- *DDX*: *Lipofibromatosis/infantile fibromatosis* (fibrous tissue only, but there is an artifact due to the fibrous tissue possibly entrapping mature fat cells, no myxoid stroma, and no plexiform vascular pattern), *myxoid liposarcoma* (adults, no distinct lobulation, more prominent cellularity with giant cells containing pleomorphic nuclei, molecular abnormalities), *ALT/WDLPS* (adults, mature fat with low cellularity and few or no lipoblasts, spindle cells with large, hyperchromatic nuclei and marked nuclear enlargement or pleomorphism (+) MDM2 and CDK4).
- *TRT*: Surgical resection.
- *PGN*: An excellent prognosis accompanies lipoblastoma despite the tumor size at presentation, the rapid growth rate, and its tendency to invade locally. It has been suggested a minimum follow-up of 5 years.

12.4.5 Hibernoma

- *DEF*: Lobulated soft tumor of "brown fat" of the interscapular region, axilla, mediastinum, and retroperitoneum of young male adults (third to fourth decades).
- *GRO*: Well-delineated or encapsulated soft tumor with brown cut surface.
- *CLM*: Organoid arrangement of nests of large cells with central nuclei and cytoplasm filled with many small neutral fat vacuoles, which do not savagely indent the nucleus, unlike liposarcomas.
- *Subtypes*: Classic, lipoma-like, myxoid, spindle cell (CD34+).

- *IHC*: (+) S100 (85%), oil red O/Sudan Black, CD31, uncoupling protein 1/UCP-1, (−) CD34 (usually), P53 (usually).
- *TEM*: Basal lamina, pleomorphic or round mitochondria, undulating plasmalemmal invaginations, micropinocytotic vesicles, and periodic short plasmalemmal densities. Mitochondria are pleomorphic with a very dense matrix or round with transverse lamellar cristae. Mitochondria exhibit an inverse relationship between the lipid droplet size and the number of mitochondria per unit of cytoplasm.
- *CMB*: 11q13-21 rearrangements (also seen in lipomas and liposarcomas).
- *DDX*: WDLS (deep, atypia, different translocations), lipoma (absence of multivacuolation), and residual brown fat (children, not a mass).

12.4.6 Locally Aggressive and Malignant Adipocytic Tumors

Locally aggressive and malignant adipocytic tumors are the most common soft tissue sarcoma of adults (1/5) who are usually in the fifth until seventh decades of life. This kind of tumors are typically located in the proximal extremities (thigh, popliteal fossa) and retroperitoneum but may also be found in internal organs, such as the breast, heart, liver, thorax, testis, and H&N, and are typically *not superficial* (but deep), *not intramuscular*, and *not in the neck* (*vide supra*).

LPS are usually large, well-circumscribed but not encapsulated tumors with three main subtypes including:

(1) Well-differentiated and dedifferentiated liposarcoma (WDLPS/DDLPS)
(2) Myxoid and round cell liposarcoma (M/RC-LPS)
(3) Pleomorphic liposarcoma (P-LPS)

Lipoblasts are relatively specific, resemble fat cells of fetal age, and are small adipocytes with sharply demarcated cytoplasmic small lipid vacuoles, which typically scallop the nucleus (indented nuclei or aka "mulberry cells") with nuclear chromatin spikes projecting between the vacuoles and are (+) S100 and calretinin. Ultrastructurally, lipoblasts have generally lipid droplets, micropinocytotic vesicles, glycogen, external lamina, intermediate filaments, Golgi, RER and SER, and mitochondria.

- *DDX*: It includes a list of benign and malignant conditions, including clear cell melanoma (strong S100+, HMB45+), fat necrosis (circumferential arrangement of histiocytes around lipid droplets, Hx. of trauma, no nuclear indentation), infiltration of fat by non-fatty tumors (standard size and peripheral location of adipocytes), inflammatory lipoma, intramuscular myxoma, lipoblastoma/lipoblastomatosis, lipogranuloma, localized lipoatrophy (insulin injection sites), metastatic signet ring cell carcinoma (infiltration, cytokeratin+, mucin+), pleomorphic/spindle-cell lipoma, and silicon granuloma. PGN factors include size, histology, and location. Retroperitoneal liposarcomas, regardless of type, often recur and eventually dedifferentiate, while pleomorphic and round cell liposarcomas metastasize in 80–90% of cases, usually to the lung.
- *PGN*: 10-YSR, 87% for low-grade sarcomas and 75% for high-grade sarcomas.

Huh et al. (2011) performed a multi-institutional retrospective analysis of medical records for patients younger or equal to 21 years of age at the University of Texas MD Anderson Cancer Center, Houston, TX (USA). They targeted patients presenting with a verified histologic diagnosis of LPS and found 33 patients ($\male:\female$=23:10) with a median age of 17.2 years. The myxoid subtype was present in 24 cases, while pleomorphic subtype was seen in 7 cases. In the children with the myxoid type of LPS, 17 had the tumors located at the extremity. No patients had metastases. In the myxoid subtype, 11 were treated with surgery only, seven with surgery and radiation, and three with surgery, radiation, and chemotherapy. At median follow-up of 4.2 years, two patients exhibited relapse with one death from progressive disease. In the pleomorphic subgroup (7 cases), four patients had the primary mass located at central axial sites. In this subgroup, six patients received multimodal therapy, but they demonstrated relapse of disease and four patients died from progressive disease.

CONCLUSIONS:
Pediatric liposarcoma has a different spectrum of presentation compared to adult cases. Myxoid liposarcoma is the more common subtype, usually occurs in extremities, and has an excellent prognosis. Pleomorphic liposarcoma occurs in axial sites, and despite multimodal therapy, outcome is poor. Further study is needed to identify the optimal therapy for pediatric liposarcoma.

12.4.6.1 Atypical Lipomatous Tumor/Well-Differentiated Liposarcoma

Both terms are equivalent (ALT/WDLPS), although the term well-differentiated liposarcoma (LPS) should be reserved to tumors located in the retroperitoneum or central body sites such as mediastinum or spermatic cord areas. It should occur at sites other than the subcutis of the posterior neck, back, and shoulder and exhibit at least focal cytologic atypia. If the neoplasm is located at the periphery is convenient to use the term "atypical lipomatous tumor" than well-differentiated LPS or atypical lipoma. At the periphery, a complete excision is easy to achieve.

- *GRO*: The tumor size is variable, but these neoplasms are usually medium-sized (Ø <5 cm), yellow-white on cut surfaces without necrosis or hemorrhage. Hemorrhage seems to be present only in large bulging tumors.
- *CLM*: Mature fat, variable-sized adipocytes with atypical cells located in cellular areas (cellular fibrous septa), few lipoblasts (atypical nuclei indented or "scalloped" by perinuclear fat vesicles), and "floret-like" cells are observed. Lipoblasts are sufficient, but not required for diagnosis. Collagen may be subtle or coarse, chronic inflammation may be seen, and the even non-lipogenic component may acquire some features of fibromatosis, low-grade fibrosarcoma or hemangiopericytoma. Heterologous mesenchymal differentiation may include the cartilage, bone, and smooth muscle. If a high-grade sarcoma is found as a component, the process should be designated as dedifferentiated LPS (*vide infra*). Several subtypes: lipoma-like, sclerosing, inflammatory, spindle cell, lipo-leiomyosarcoma, and mixed subtypes, although this subclassification does not seem to have clinical significance. This tumor should be considered grade I by definition.
- *IHC/TEM*: (+) MDM, CDK2+, and p16+ other than S100 (adipocytes) and CD34 (some spindle cells) by immunohistochemistry and TEM shows pseudoinclusions due to invaginations of the nuclear membrane.
- *CMB*: ~4/5 of ALT/WDLPS are characterized by the presence of a supernumerary ring and/or long marker chromosomes. The chromosomal region 12q-13-15 is the target region, which is amplified within these marker chromosomes. Several genes are amplified or rearranged, including the *CDK4* (cyclin-dependent kinases) and the *MDM2* (murine double minutes) families. Both *CDK4* and *MDM2* are proto-oncogenes that permit make ineffective the block operated by the G1-S cell cycle checkpoint on cell proliferation. Moreover, mdm2 protein is known to inhibit transactivation of p53, which in turn results in cell proliferation. Carbossi-Peptidase M (CPM) amplification is also useful in distinguishing WDLPS from lipoma.
- *DDX*: It includes the following entities:
  - *Lipoblastoma*: infancy, (−) MDM2 and CDK4
  - *Lipoma*: no atypical nuclei, MDM2-, CDK4-
  - *Lipomatous AML*: kidney-located, HMB45(+) epithelioid cells, (−) MDM2, (−) CDK4
  - *Malignant fibrous histiocytoma (MFH)*: no adipocytes, (−) MDM2 and (−) CDK4
  - *Myxoid/round cell LPS*: (−) MDM2 and (−) CDK4 and 12q13-15 amplification
  - *Paraffinoma*: Hx. of paraffin injection, no atypia, (−) MDM2 and (−) CDK4
- *TRT*: Retroperitoneal exenterating procedures. PGN is relatively good because it does not differentiate. However, if it is incomplete resected or unresectable or unusually large often raises

concerns to the clinicians or surgeon, because it recurs. No metastases are usually associated with ALT/WDLPS. Retroperitoneal/intra-abdominal location is associated with the worse outcome.

12.4.6.2 Myxoid/Round Cell Liposarcoma

- *DEF*: (50%) Malignant adipocytic tumor with myxoid and round cell morphology characterized by primitive *mesenchymal cells*, signet ring lipoblasts, regular *small round cells* embedded in a *mucopolysaccharide-rich myxoid matrix* (hyaluronidase sensitive) with *prominent arborizing vascularity* (unlike myxoma). Hyaluronidases are a family of enzymes that catalyze the degradation of hyaluronic acid, a constituent of the extracellular matrix (ECM) (hyalurono-glucosidases) with the aim to cleave the (1->4)-linkages between N-acetylglucosamine and glucuronate. They include two classes of *eukaryotic* endo-glycosidase hydrolases and a *prokaryotic* lyase-type of glycosidase. Five functional hyaluronidases (HYAL1-5) are known in humans. By catalyzing the hydrolysis of hyaluronan, hyaluronidase lowers the viscosity of the ECM promoting an increase of the tissue permeability.
- *CLM*: Several histologic patterns are known and include paucicellular (predominantly myxoid, scant cellularity, arborizing vascularity), pulmonary edema-like (numerous microcysts with granular eosinophilic material and virtual absence of lipoblasts), lipoblasts growing in sheets, large fat cells with admixed lipoblasts, and "back-to-back" pattern of uniform round cells without pleomorphism, but focal lipoblasts. Neoplasms with round cell (RC)-histology are more likely to metastasize.
- *IHC*: (+) CD31 and CD34 useful to highlight the vasculature; (+) S100 for lipoblasts.
- *TEM*: Lipid vacuoles and massive proliferation of mitochondria.
- *CMB*: It involves the following translocations:
    - *t(12;16)(q13:p11)* with *FUS-CHOP* as hybrid protein
    - *t(12;22)(q13q12)* with *FUS-EWS* as a hybrid protein with subsequent activation of the PI3K/Akt pathway via activating mutation of PIK3CA, loss of PTEN, or *IGF1R* expression determining the round cell phenotype.
- *DDX*: It includes the following entities:
    - *Ewing sarcoma/PNET*: NE morphology, CD99+, FLI-1+, EWS gene translocations.
    - *Lymphoma*: (+) CD20, CD79a, CD3, CD4.
    - *Melanoma*: Different cytology, prominent nucleoli, (+) S100, HMB45, Melan A, Mart 1.
    - *Met-Ca*: Hx. Ca., (+) CKs, severe pleomorphism.
- *PGN*: Grading (>5% RC areas), necrosis, TP53 overexpression as poor PGN factors. In children and youth, the conventional myxoid LPS is undoubtedly the most common LPS subtype with mostly and worldwide harboring an excellent prognosis. Two growth patterns, apparently novel subtypes of LPS, labeled *pleomorphic* myxoid LPS and *spindle-cell* myxoid LPS, have been described. These two neoplasms show a typical myxoid pattern merging with areas of hypercellularity with marked pleomorphism and increased atypical mitotic figures. Both neoplastic proliferations harbor mortality in 70% of cases within 36 months. The pleomorphic myxoid LPS lacks CHOP translocations of usual myxoid type (useful tip to keep in mind for the differential diagnosis).

12.4.6.3 Pleomorphic Liposarcoma

5–10%, ♂=♀, rare in childhood or youth, ± radiation therapy or NF-I, widespread high-grade, pleomorphic tumor, undifferentiated. Pleomorphic liposarcoma tumor cells are giant cells, often >200 microns in size, admixed with a few lipoblasts but there is no clear-cut evidence of well-differentiated liposarcoma (grade III, high MI). The pleomorphic LPS occurs typically at soft tissue of limbs and limb girdles and is fundamentally aggressive with tendency to early metastasize (5-YSR: 60%). Variant: epithelioid, which may be confused with carcinoma. IHC shows (+) VIM, S100 (useful to highlight lipoblasts in the lipogenic areas), SMA, and CD34. Ultrastructurally, abundant coalescing lipid droplets and numerous cytoplasmic organelles are seen. Molecular cytogenetically, there is a complex karyotype with similar genomic imbalances with myxofibrosarcoma, but the epithelioid variant may have t(12;16)(q13;p11) – TLS-CHOP, more commonly seen in myxoid and round cell LPS.

- *DDX*: It includes the following entities:
  - *Dedifferentiated LPS*: Cellular typology with dimorphism and distinct WDLPS component.
  - *Metastatic carcinoma*: Epithelial differentiation with (+) Pan-CKs, no LPS differentiation.
  - *MFH-pleomorphic type*: No definite lipoblasts (key are both HE/S100 and TEM).
  - *Pleomorphic lipoma*: Back of the neck/shoulder location, giant floret cells, no/↓ MI.
  - *Pleomorphic RMS*: Skeletal muscle differentiation (IHC and/or TEM).

12.4.6.4 Dedifferentiated Liposarcoma

- *DEF*: Well-differentiated LPS with the abrupt transition (dimorphic tumor pattern) to dedifferentiated (high- or low-grade) areas tumor (DD-LPS) with more often retroperitoneal location than external sites, larger size than WD-LPS, firm, and "fish-flesh" cut surface grossly.
- *EPI*: It is extremely rare in the youth.
- *CLM*: DD-LPS characteristics are similarity to MFH, fibrosarcoma, or LMS, MI >5/10, and lack of lipogenesis in the high-grade component (no lipoblasts). Heterologous elements are found in 1/10 of cases. The biphasic nature of DD-LPS may show either adjacent or intermingled components, but WD-LPS should be a ≥ 10x objective field, and the other element *must be* a high-grade sarcoma, being pleomorphic MFH the most common second component. Another option of DD-LPS would be a high-grade sarcoma arising in a location typical of ALT/WD-LPS (e.g., retroperitoneum) that is FISH (+) for *MDM2* gene amplification. Another controversially discussed or accepted option may be an inflammatory myofibroblastic tumor (IMT) as dedifferentiated area, but it does not seem to qualify, because none of the tumors with only IMT in the differentiated component metastasized. Variant: neural-like whirling growth pattern with the metaplastic bone formation. In any retroperitoneal sarcoma, DD-LPS must be ruled out by adequate sampling (e.g., peripheral areas), IHC, and CMB.
- *IHC*: (+) MDM2 and CDK4 in addition to VIM, TP53, RB, and PPAR-γ. PPAR-γ (Peroxisome Proliferator-Activated Receptor gamma), aka the glitazone receptor, or NR1C3 (nuclear receptor subfamily 1, group C, member 3) is a type II nuclear receptor that is encoded by the *PPARG* gene in humans.
- *DDX*: It includes the following entities:
  - *Atypical AML*: Kidney location, abnormal blood vessels, HMB45+, and (−) MDM2/CDK4.
  - *Fibrosarcoma*: No WD-LPS component, (−) MDM2/CDK4, and (−) 12q13–15 amplification.
  - *LMS*: (−) WD-LPS component, (−) MDM2/CDK4, and (−) 12q13–15 amplification.
  - *MPNST*: (−) WD-LPS component, (−) MDM2/CDK4, and (−) 12q13–15 amplification.
  - *P-LPS*: (+) lipoblasts.
  - *RMS*: (−) WD-LPS component, (−) MDM2/CDK4, (−) 12q13–15 amplification, (+) myo markers (e.g., DES, myogenin, MyoD1).
- *CMB*: 12q14 amplification, often due to supernumerary ring or giant chromosome (12).
- *TRT*: Surgery remains the mainstay of treatment for WD/DD LPS.
- *PGN*: Local recurrence rates can be >80%, and DD-LPS is associated with a poor prognosis. Benefit from chemotherapy for has been reported to be limited to DD areas, with response rates of ≤12%. In case a WD-LPS or another tumor undergo dedifferentiation, there is a natural tendency to acquire a metastatic potential. In fact, PGN is poor, although better than P-LPS or P-RMS. DD-LPS metastases contain the high-grade sarcoma component only.

12.5 Fibroblastic/Myofibroblastic Tumors

Benign tumors showing fibroblastic/myofibroblastic differentiation include *nodular fasciitis* group (NOS and subtypes/variants: cranial, intravascular, eosinophilic, proliferative fasciitis, ischemic fasciitis), *fibrous hamartoma of infancy, myofibroma/myofibromatosis, fibromatosis colli, inclusion body fibromatosis, juvenile hyaline fibromatosis, fibroma of tendon sheath, nuchal-type fibroma, Gardner fibroma, elastofibroma, desmoplastic fibroblastoma, mammary-type myofibroblastoma, calcifying aponeurotic fibroma, calcifying fibrous tumor, angiomyofibroblastoma, cellular angiofibroma*, and *giant cell angiofibroma*. Intermediate fibroblastic/myofibroblastic tumors include *fibromatosis* (with subtypes/variants), *solitary fibrous tumor* group (including hemangiopericytoma), *inflammatory myofibroblastic tumor, low-grade myofibroblastic sarcoma, myxo-inflammatory fibroblastic sarcoma*, and *infantile fibrosarcoma*. Malignant fibroblastic/myofibroblastic tumors include *adult fibrosarcoma, myxofibrosarcoma, low-grade fibro-myxoid sarcoma*, and *sclerosing epithelioid fibrosarcoma* (Box 12.6).

However, only a few entities are commonly seen in pediatrics and probably young adults. The group of predominantly pediatric fibroblastic/myofibroblastic lesions includes the following entities:

- Calcifying aponeurotic fibroma
- Fibrous hamartoma of infancy
- Gardner-associated fibroma
- Hyaline fibromatosis
- Inclusion body fibromatosis
- Infantile fibromatosis/lipofibromatosis
- Infantile fibrosarcoma
- Myofibroma

The group of pediatric and adult-type fibromatoses includes the following entities:

- Desmoid fibromatosis
- Desmoid-type fibromatoses
- Fibromatosis colli
- Gingival fibromatosis

Box 12.6 Fibroblastic/Myofibroblastic Tumors
12.5.1 Fasciitis/Myositis Group
    12.5.1.1 Nodular Fasciitis
    12.5.1.2 Cranial Fasciitis
    12.5.1.3 Intravascular Fasciitis
    12.5.1.4 Proliferative Fasciitis/Myositis
    12.5.1.5 Eosinophilic Fasciitis
    12.5.1.6 Ischemic Fasciitis
    12.5.1.7 Myositis Ossificans
12.5.2 Fibroma Group
    12.5.2.1 Nuchal Fibroma
    12.5.2.2 Gardner Fibroma
    12.5.2.3 Nasopharyngeal Fibroma
    12.5.2.4 Ovarian Fibroma
    12.5.2.5 Fibroma of Tendon Sheath
    12.5.2.6 Elastofibroma
    12.5.2.7 Genital Cellular Angiofibroma
    12.5.2.8 Keloid
12.5.3 Fibroblastoma, Classic and Subtypes
12.5.4 Fibrous Hamartoma of Infancy
12.5.5 Fibromatosis of Childhood
    12.5.5.1 Fibromatosis Colli
    12.5.5.2 Infantile Digital Fibromatosis
    12.5.5.3 Infantile Fibromatosis, Desmoid Type
    12.5.5.4 Hyaline Fibromatosis
    12.5.5.5 Gingival Fibromatosis
12.5.6 Infantile Myofibromatosis
12.5.7 Fibroblastic/Myofibroblastic Tumors with Intermediate Malignant Potential
12.5.8 Malignant Fibroblastic/Myofibroblastic Tumors
    12.5.8.1 Adult Fibrosarcoma
    12.5.8.2 Sclerosing Epithelioid Fibrosarcoma
    12.5.8.3 Low-Grade Fibro-Myxoid Sarcoma

> **Box 12.7 β-Catenin Involvements**
> - Cadherin-mediated cellular cohesion
> - Intracellular signaling as a component of the Wnt signaling pathway

- Hyaline fibromatosis
- Inclusion body fibromatosis
- Infantile (lipo-) fibromatosis
- Knuckle pads
- Mesenteric fibromatosis
- Palmar/Penile/Plantar fibromatosis
- Pelvic/retroperitoneal fibromatosis

β-Catenin is a particularly useful marker in soft tissue lesions. β-Catenin is a dual function protein that in humans is encoded by the *CTNNB1* gene. The dual function involves the regulation and coordination of cell-cell adhesion and gene transcription. In Box 12.7, the involvement of β-catenin in cellular processes is summarized.

Wnt/Wingless signaling pathway is essential in both embryonic development and tumorigenesis. β-catenin acts as positive effector of the Wnt signaling pathway, while axin is a negative effector of this pathway. Moreover, β-catenin and GSK-3β form a peculiar complex with axin and the colorectal (tumor suppressor) gene product of the APC gene. Glycogen synthase kinase 3 (GSK-3β) is a serine/threonine protein kinase the function of which is represented by the addition of phosphate molecules onto serine and threonine amino acids. In normal cell metabolism, beta-catenin is regulated by the *APC* (Adenomatous Polyposis Coli) gene and *GSK-3β* gene, which induce phosphorylation of serine/threonine residues with subsequent ubiquitin-mediated degradation of the protein. In case of APC or β-catenin monoallelic gene mutations, there is a loss of β-catenin regulation with subsequent accumulation of β-catenin in the cytosol and its eventual translocation to the nucleus, where it acts as a cofactor with lymphoid-enhancing factor/T-cell factor in upregulating some nuclear oncogenes, including *cyclin D1* and *c-myc*. Thus, nuclear staining only, derived by a single activating mutation, is considered altered and of diagnostic relevance to both research and routine diagnostics (Box 12.8).

> **Box 12.8 Terrific Use of β-Catenin (+ve vs. −ve) in Tumor Cell Nuclei of Fibromatosis and Other Tumors**
> 1. *Superficial fibromatosis* (+/−) vs. scarring and other superficial fibroblastic tumors (−)
> 2. *Mesenteric fibromatosis* vs. GIST or sclerosing mesenteritis (−)
> 3. *Deep fibromatosis* vs. LG-fibroblastic sarcoma, myofibroma, and nodular fasciitis (−)
> 4. *Desmoid-type fibromatosis* vs. desmoplastic fibroma of the bone
> 5. *Tubular adenoma* vs. dysplasia-associated lesion or mass (DALM) of IBD
> 6. *Pancreatic AcCC, pancreatoblastoma, and SPPT* vs. PDCA and pancreatic NETs
> 7. *Hepatoblastoma, hepatic adenoma, and HCC* vs. regenerative nodules (cirrhosis) and FNH
> 8. *Juvenile nasal angiofibroma* vs. hemangioma
> 9. *LG-FLAC/WDFA* vs. HG-FLAC and conventional pulmonary adenocarcinoma
> 10. *Congenital pulmonary blastoma* vs. pulmonary carcinosarcoma
> 11. *HG-Type I-EC* vs. papillary serous EC (type II)
> 12. *Uterine LG-ESS* vs. uterine SM (smooth muscle) tumors
>
> Notes: *AcCC* acinar cell carcinoma, *FLAC/WDFA* low-grade adenocarcinoma of the fetal lung type/well-differentiated fetal adenocarcinoma, *FNH* focal nodular hyperplasia, *PDCA* pancreatic ductal carcinoma, *NET* neuroendocrine tumors, *EC* endometrial carcinoma, *ESS* endometrial stroma sarcoma, *SM* smooth muscle

In tubular adenoma, *CTNNB1* gene mutations are found in about half of cases, being exclusive with *APC* gene, while early *TP53* gene alterations characterize DALM. Thus, the nuclei of tumor cells show accumulation of β-catenin in sporadic adenomas, while p53 in DALM. Superficial fibromatoses do not have *CTNNB1* gene mutations but express a nuclear protein of β-catenin.

12.5.1 Fasciitis/Myositis Group

The fasciitis/myositis group is a slight heterogeneous group of benign fibrous lesions; some of them are paramount to consider benign because they may be mistaken as malignant. In fact, the first entity, nodular fasciitis, is aka "subcutaneous pseudo-sarcomatous fibromatosis" and has been and may be misdiagnosed as fibrosarcoma or liposarcoma.

12.5.1.1 Nodular Fasciitis

- *DEF*: Benign soft tissue, small sized (2–3 cm), rapidly growing, painful lesion of childhood and youth in fascia and subcutis mostly on the flexor forearm, chest, back or H&N, ± Hx trauma-associated.
- *CLM*: Well-circumscribed, unencapsulated lesion with a characteristic zonation, "tissue culture-like" growth, extravasated RBC and inflammatory cells, which may be misdiagnosed as fibrosarcoma or liposarcoma. Zonation consists of zonal growth with a hypocellular center and a hypercellular periphery. There is a "feathery" pattern in the center, which includes cellular tissue culture-like spindle-cell proliferation in a myxoid matrix. The spindle-cell proliferation is made of fibroblasts and myofibroblasts. The MI may be brisk, but nuclear membrane contours are smooth, and chromatin pattern is very delicate. Other characteristics of the lesion include vascular proliferation, extravasated RBC and inflammatory cells (lymphocytes), and wide rolling bands of collagen. This latter component increases with the age of the lesion, and focal metaplastic bone may occasionally be seen. Early lesions tend to have a more myxoid aspect, while old lesions have a more fibrotic character. The outcome of the lesion is also dependent on the size, regressing if less than 3 cm. There is no recurrence, even if incompletely excised.
- *IHC*: (+) VIM, SMA, MSA, calponin (macrophages are CD68+, but CD68 is not specific for histiocytic lesions) and (−) S100, DES, Caldesmon, AE1–3, CD34, ALK, and p53.
- *TEM*: Myofibroblasts (elongated-shaped cells) showing abundant and often dilated RER, numerous cytoplasmic filaments with dense bodies, pinocytotic vesicles, and variable cell junctions.

Nodular Fasciitis Variants

12.5.1.2 *Cranial Fasciitis* is common in children and involves a skull erosion and has been occasionally reported to be associated with birth trauma, although it is controversially discussed in the literature.

12.5.1.3 *Intravascular Fasciitis* is nodular fasciitis, which involves medium-sized BV (walls).

12.5.1.4 *Proliferative Fasciitis/Myositis* refers to a subcutaneous, fascial, or intramuscular proliferation like nodular fasciitis mostly located at the upper/lower extremities of middle-aged and elderly adults. It is characterized by large basophilic *ganglion-like cells (S100) with large nuclei and abundant amphophilic cytoplasm* and *checkerboard pattern* resulting from a peculiar separation of the skeletal muscle cells by the proliferating lesion. There is a degeneration/regeneration of muscle fibers with interstitial inflammation and fibrosis.

- *IHC*: (+) VIM, SMA, MSA, (±) CD68, (−) AE1–3, S100, DES.
- *DDX*: Nodular fasciitis (no ganglion-type cells), ganglioneuroma (different stroma, S100+), RMS (cross-striations, DES+, MYOD1+, MYOGENIN+), and sarcoma (large size, marked pleomorphism with nuclear chromatin and membrane abnormalities, atypical mitoses).

12.5.1.5 *Eosinophilic Fasciitis* is a superficial fasciitis variant of thighs with often symmetrical occurrence and fibrosing inflammation with constant eosinophils and variable lymphocytes and mast cells as well as ± Hx of myalgias (eosinophilia-myalgia syndrome due to a toxin contaminating early preparations of L-tryptophan).

12.5.1.6 *Ischemic Fasciitis* refers to a pseudosarcomatous fibroblastic proliferation that originally occurs usually over osseous prominences and preferentially targeting immobilized (usually nonyoung) patients.

12.5.1.7 *Myositis Ossificans* refers to a solitary, intramuscular nonprogressive reactive/reparative lesion composed of cellular fibrous tissue and metaplastic bone with a radiological pattern of the outer calcific shell and hidden (inner) "air trap". It may be mistaken for osteosarcoma, and this pitfall is particularly treacherous. Myositis ossificans involves flexors of arms and legs and hands, but also mesentery (intra-abdominal myositis ossificans) and subcutis (panniculitis ossificans). There is a Hx. of trauma, and young males are often the affected individuals. If multiple lesions of myositis ossificans are observed, the term "myositis/fibrodysplasia ossificans progressive" (AD, skeletal anomalies, and poor outcome) is used. Microscopically, there is *reversed zonation* with highly cellular stroma with new bone ± cartilage and no inflammation, which inexorably evolves to a cellular core, intermediate zone of osteoid, and outer shell of a calcified bone corresponding to the radiological pattern.

DDX includes fracture callus and paraosteal osteosarcoma.

12.5.2 Fibroma Group

This group contains lesions with quite similar histology that are placed in different locations, including nuchal fibroma, nasopharyngeal fibroma, ovarian fibroma, fibroma of tendon sheath, elastofibroma, and keloid.

12.5.2.1 Nuchal-Type Fibroma (NTF)

Nuchal-type fibroma (NTF): The benign hyalinized fibroblastic proliferation of bundles of thick collagen fibers (aka collagenosis *nuchae*) in the posterior neck entrapping adipose tissue and nerve twigs, of usually middle-aged men with DM (~1/2 of cases) with the possibility of recurrence, but no metastases. Grossly, NTF is Ø <3 cm, hard and white, unencapsulated, poorly demarcated and impressively resembles a traumatic neuroma. IHC: (+) VIM, CD34, CD99, (−) ACT, DES.

- *DDX*: Fibrolipoma (confined, different location), fibromatosis (deep soft tissue, not back of the neck, hypercellular), SFT ("patternless" pattern, hypercellular, "staghorn"-type vessels).

12.5.2.2 Gardner Fibroma

Gardner fibroma is histologically indistinguishable from NFT and occurs in a pediatric age in the paraspinal region, back, chest wall, flank, H&N, and extremities. It is associated with desmoid-type fibromatosis/Gardner syndrome.

12.5.2.3 Nasopharyngeal Fibroma

It is more specifically labeled as nasopharyngeal angiofibroma and is a benign, highly cellular, and heavy vascularized mesenchymal neoplasm that targets the nasopharynx in males at puberty when testosterone spikes. Microscopically, it is characterized by an intact respiratory epithelium overlying a richly vascular neoplastic formation with variably-sized blood vessels surrounded by a highly cellular fibroblastic stroma with collagen production. Endothelial cell markers (FVIIIRA, CD31, CD34) decorate the vascular texture.

12.5.2.4 Ovarian Fibroma

Ovarian fibroma is rare in children. In fact, it usually tends to manifest mostly during perimenopause. The lesion is typically asymptomatic.

12.5.2.5 Fibroma of Tendon Sheath

Fibroma of tendon sheath is a benign fibrous nodule of usually young males (2nd–4th decades), which is well circumscribed, lobulated, attached to tendon. Microscopically, there is a proliferation of haphazardly arranged, spindle-shaped fibroblasts (elongated nuclei) embedded in a collagenous stroma and rare histiocytes, including slit-like vascular channels resembling teno-synovial spaces with no hemosiderin deposition (PPB negative). It may have bizarre tumor cells, but rarely giant cells. FTS may also have extravasated RBC, but neither atypical mitoses, nor necrosis. Also, hyperchromasia is never encountered.

- *IHC*: (+) SMA, (−) DES.
- *DDX*: Some diagnosis of soft tissue lesions, including tenosynovial giant cell tumor (TSGCT) and similar lesions (TSGCT has oval histiocyte-like nuclei, giant cells, foamy histiocytes, hemosiderin deposition (PPB pos-

itive), and the absence of slit-like vascular spaces and is DES ± and SMA−).
- *PGN*: The treatment of choice for FTS is complete surgical excision, but approximately one-fourth of the patients recur after surgery. The recurrence is due to the difficulty in some cases to remove the entire tumor. Thus, the prognosis is usually excellent, but a follow-up of 5 years is suggested to detect recurrences.

12.5.2.6 Elastofibroma

Elastofibroma is a painless, poorly circumscribed; subscapular lesion (aka *elastofibroma dorsi*) of middle-aged to elderly usually Japanese women. Grossly, elastofibroma is made of rubbery, gray-white tissue mixed with yellow streaks occurring. Microscopically, elastofibroma is constituted by *densely collagenized lesion* with collagen I–III (type II typically restricted to articular cartilage and ocular structures) and *coarse enlarged eosinophilic extracellular refractile elastic fibers* ("beads on a string").

- *IHC*: (+) VIM, CD34 (spindle cells); (−) S100, DES, SMA, p53.
- *TEM*: Cylinders composed of immature amorphous elastic tissue, which is removed by elastase digestion.
- *CMB*: Xq12–q22 or #19 gains in ~1/3 of cases.
- *DDX*: Nuchal fibroma (younger age, paravertebral, dense collagen, but no elastic fibers), fibrolipoma (absence of elastic fibers), desmoid fibromatosis (hypercellularity, skeletal muscle infiltration, the lack of elastic fibers).

12.5.2.7 "Genital" Cellular Angiofibroma (GCAF)

- *DEF*: Genital cellular angiofibroma is a benign, densely cellular tumor of the vulva and scrotum/inguinal region of usually middle-aged individuals, but extragenital cases (EGCAF) have been described. Pathogenetically, a link to hormone receptor-positive mesenchymal cells has been suggested.
- *GRO*: Well-confined nodule with soft to rubbery, gray-pink to the brown cut surface.
- *CLM*: Dense fascicles or random pattern of spindle cells with bland cytology (no necrosis and no atypia) having oval to fusiform nucleus, scant, lightly eosinophilic cytoplasm, and ill-defined cell borders and prominent small-to-medium-sized BVs with hyaline fibrosis in the walls. It has been suggested to be related to *spindle-cell lipoma* and, even, to *mammary-type myofibroblastoma*. Additional features include degenerative changes of BVs with fibrin thrombi, intramural inflammation with mast cells, hemosiderin, and fine collagenous fibers in the stroma.
- *IHC*: (+) VIM, CD34, CD31, SMA, ER, PR and (−) S100, DES, and EMA.
- *CMB*: Changes like spindle-cell lipoma with 13q14 involvement.
- *DDX*: It includes the following diagnoses:
  - *Aggressive angiomyxoma*: greater size, deep, hypocellular, infiltrative "aggressive" margin, DES+
  - *Angiomyofibroblastoma*: less uniform cellularity, smaller BVs, DES+.
  - *Leiomyoma*: SMA+, DES+
  - *Perineurioma*: (+) EMA, (±) CD34, and (−) S100
  - *Pleomorphic Hyalinizing Angiectatic Tumor* (PHAT): (±) SMA, but (−) CD31, CD34, DES, S100
  - *Solitary fibrous tumor* (SFT): "patternless," "staghorn" BVs, hyalinized collagen, (+) BCL2, CD34, CD99, STAT6

12.5.2.8 Keloid

Keloid corresponds to hypertrophic scar, which is constituted by thick collagen bands. Please see the skin chapter for more details on this entity.

12.5.3 Fibroblastoma Classic and Subtypes

12.5.3.1 Desmoplastic Fibroblastoma (DFB)

Fibroblastoma of the desmoplastic subtype is a fibroblastic lesion of the subcutaneous tissue with reactive fibroblasts, low cellularity, and abundant collagen deposition (aka collagenous fibroma) of the upper limbs, back, and feet of adult men and "fibronexus" (Fibronexus is a specialized intercellular junction between myofibroblasts and a very subtle communicating structure for providing contact between myofibroblasts and extracellular matrix proteins). Grossly, desmoplastic fibroblastoma is a small (1–4 cm), well-delimitated firm and gray lesion. Microscopically, there is a paucicellular, bland spindled/stellate fibroblasts and myofibroblasts separated by copious collagen with a variable amount of fibro-myxoid stroma. The outcome is wholly benign because it requires conservative excision only and does not recur or metastasize.

- *IHC*: (+) VIM, SMA, and (−) DES, S100, CD34, EMA.
- *TEM*: Myofibroblastic differentiation ("fibronexus" junctions). Fibronexus: Cell surface specialization including extracellular fibronectin filaments and intracellular actin filaments intimately associated with "sub-plasmalemmal plaque material" (Eyden 1993).
- *CMB*: t(2;11)(q31;q12) or 11q12 abnormalities.
- *DDX*: Fibromatosis (e.g., "Lederhosen" fibromatosis or plantar fibromatosis), which is not circumscribed, highly cellular and made of fascicular pattern and prominent BVs.

12.5.3.2 Giant Cell Fibroblastoma (GCFB)

- *DEF*: Pediatric lesion of boys younger than 10 years of age, on back or thigh.
- *SYN*: It has been suggested to be the juvenile version of DFSP.
- *CLM*: The ill-defined proliferation of fibroblasts in collagenized, focally myxoid stroma with hypocellular areas which often alternate with hypercellular regions with thick-walled BVs and characteristic "angiectoid spaces" lined with MNGCs with "floret-like" appearance, which recall some aspects of borderline fibrohistiocytic tumors. GCFB appears to evolve into DFSP by genomic gains of *COL1A1-PDGFB* (same translocation as seen in DFSP).

- *IHC*: (+) VIM, SMA, CD34, CD99 and (−) S100, CD31, DES, AE1–3, HMB45.
- *TEM*: Fibroblastic/myofibroblastic ultrastructural features.
- *CMB*: t(17, 22)(q22;q13) – creating a fusion of collagen type 1 α1 gene and PDGFB chain gene and also supernumerary ring chromosomes derived from t(17;22).
- *DDX*: Angiosarcoma (vascularity, atypia, brisk MI, CD31+), hemangioma (vascular tumor, CD31+), neurofibroma with ancient change (no angiectoid spaces, S100+), and pleomorphic LPS (marked atypia, brisk MI, lipoblasts, no collagenous matrix).

12.5.3.3 Angio-Myofibroblastoma (AMFB)

Angio-myofibroblastoma (AMFB) is a benign, well-circumscribed tumor of myofibroblasts that usually occurs in the vulva but also in the vagina, scrotum, and paratesticular region of young females and males. Grossly, AMFB is a well-demarcated, but not encapsulated, Ø <5 cm, pinkish, and soft nodule. Microscopically, it is characterized by alternating areas of hypercellularity and hypocellularity with thin-walled and ectatic BVs delimitated by a thin fibrous pseudocapsule. Tumor cells are round to spindle-shaped, plasmacytoid/epithelioid/ bi-/multinucleated and have eosinophilic cytoplasm with a characteristic perivascular distribution and admixed mast cells and adipocytes. MI: 0/low, no atypia, no RBC extravasation.

- *IHC*: (+) VIM, SMA, MSA, DES, ER, PR; (±) CD34; (−) S100, AE1–3.
- *TEM*: Myofibroblastic/fibroblastic features.
- *DDX*: Vulvovaginal soft tissue lesions, including aggressive angiomyxoma (not circumscribed, Ø > 5 cm, hypocellularity, thick-walled BVs, (+) RBC extravasation, (+) stromal mucin, (−) SMA), genital cellular angiofibroma (uniform hypercellularity, large, thick-walled BVs and perivascular hyalinization, (−) SMA and DES), epithelioid leiomyoma (hypercellularity, no biphasic pattern, no binucleation, no/rare mast cells), and epitheli-

oid sarcoma (+KER, EMA, VIM, CD34, but ± CD31, ERG, FLI-1; − INI1/hSNF5/SMARCB1, FVIII-RA; abundant intermediate filaments, desmosome-like intercellular junctions, and small intercellular spaces with microvilli).
- *TRT*: Surgery.
- *PGN*: Good.

12.5.3.4 Mammary-Type Myofibroblastoma (MTMFB)

- *DEF*: Benign myofibroblastic subcutaneous lesion of mostly groin, a perineal or abdominal wall with hyalinized collagenous stroma and accompanying fatty tissue, alike to breast lesion of young-middle-aged men that may be discovered incidentally during hernia repair surgery.
- *GRO*: Well-demarcated, but unencapsulated, firm, pink-brown nodule with an astonishingly whirled cut surface.
- *CLM*: Breast tissue-similar myofibroblastic lesion constituted by fascicles of spindle cells with eosinophilic/amphophilic cytoplasm, indistinct cell borders, oval/tapered nuclei with fine chromatin and small nucleoli (myofibroblasts) embedded in a stroma with haphazardly bands of collagen, and ± focal atypia and low MI.
- *IHC*: (+) DES, CD34, SMA.
- *CMB*: 13q and 16q changes.
- *DDX*: Angiofibroma (uniform hypercellularity, large, thick-walled hyalinized BVs and perivascular hyalinization), (−) SMA and DES, lipomatous hemangiopericytoma (staghorn BVs and no myofibroblastic spindle cells), and SFT (*patternless* pattern and staghorn-like BVs).

12.5.4 Fibrous Hamartoma of Infancy (FHI)

- *DEF*: Solitary, poorly circumscribed, dermal to subcutaneous tumor-like (benign) whitish mass of the 1st infancy (<2 years) occurring in boys often at the anterior or posterior axillary fold, shoulder, and upper arm. FHI is characterized, microscopically, by an organoid pattern composed of:
  1. Well-differentiated *dense fibrocollagenous trabeculae*
  2. *Immature mesenchymal areas* with loosely arranged bland primitive spindle cells with scant cytoplasm,
  3. *Mature fat tissue* (Fig. 12.5)
- *IHC*: (+) VIM, (+) SMA in the spindle cells of the WD-dense fibrocollagenous trabeculae.
- *PGN*: Good.

12.5.5 Fibromatosis of Childhood

In the fibromatosis group of childhood, we should list the fibromatosis colli, the infantile digital fibromatosis, the infantile fibromatosis, desmoid type, the juvenile hyaline fibromatosis, and the gingival fibromatosis.

12.5.5.1 Fibromatosis Colli (FCO)

- *DEF*: Bilateral benign soft tissue lesion affecting the lower 1/3 of the sternocleidomastoid muscle, causing thickened muscle ± congenital anomalies (e.g., hip dislocations, breech deliveries), which does not require excision (FNA).
- *GRO*: Tan gritty mass of muscle without hemorrhage or necrosis.
- *CLM*: The intramuscular diffuse proliferation of uniform plump fibroblasts and myofibroblasts and scar-like collagen with entrapped reactive and degenerating skeletal muscle fibers with abnormal sarcolemmas and nuclei (loss of cross-striations, hypercellularity, nuclear enlargement, multinucleation, atrophy).
- *TRT*: It does not require excision (no surgical intervention if FNA is appropriate or considered adequate).
- *PGN*: FCO regresses spontaneously within 3 months from the diagnosis without the need to opting for either a surgical or physical treatment.

12.5.5.2 Infantile Digital Fibromatosis

- *DEF*: Infantile digital (inclusion body) fibromatosis is a benign, possibly recurring but

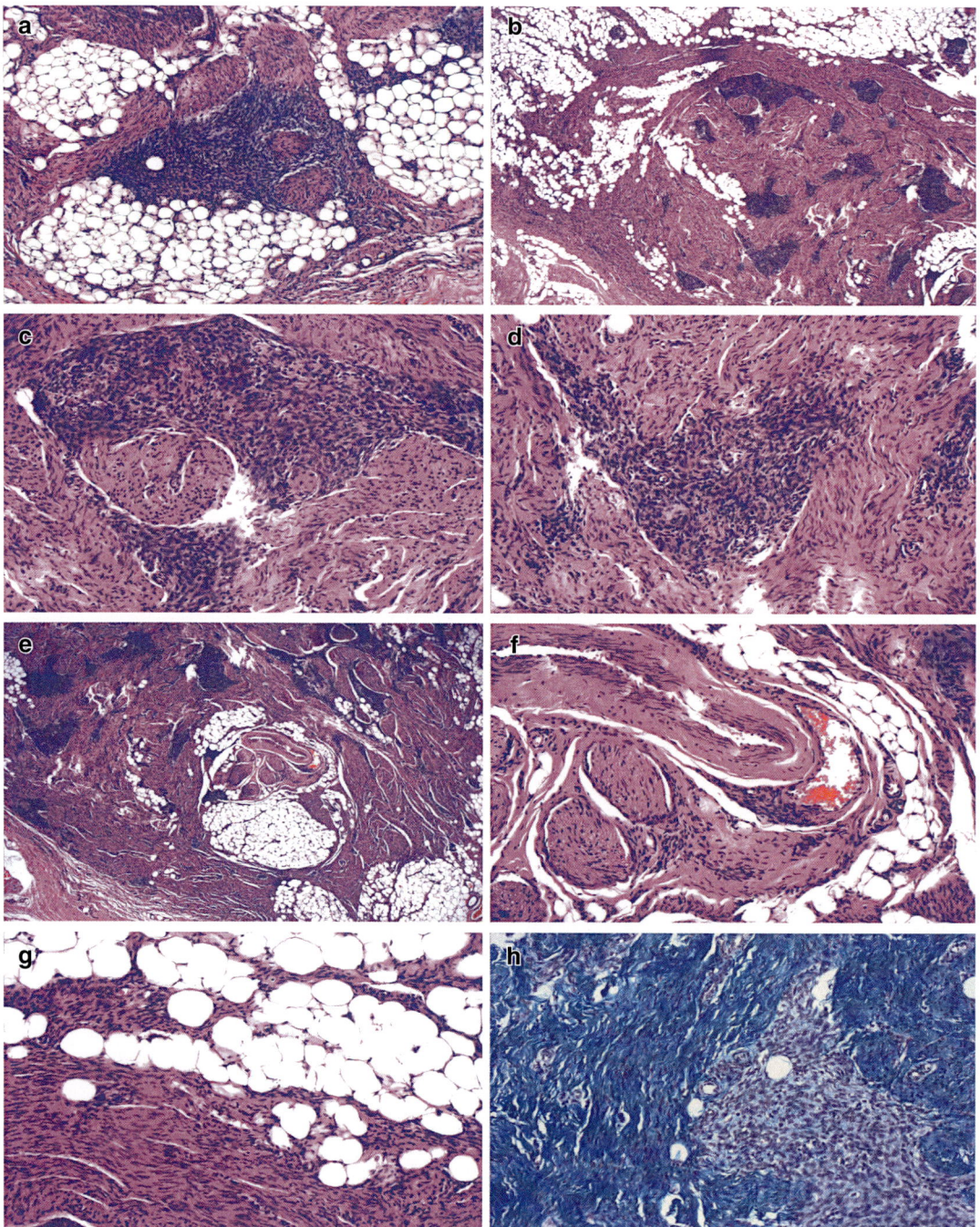

Fig. 12.5 Fibrous Hamartoma of Infancy. The microphotographs show several histological aspects of a fibrous hamartoma of infancy (**a–h**). This soft tissue tumor exhibits characteristic triphasic morphology with an unusual blend of fibroblastic fascicles, nodules of primitive myxoid mesenchyma, and mature adipose tissue (a–g, hematoxylin and eosin staining, h, Celestin-Alcian Blue staining; a–h, ×100, ×40, ×200, ×200, ×40, ×200, ×200, ×200 as original magnification, respectively)

non-metastasizing infantile cellular lesion of the 3rd–4th–5th digits and toes (exterior surface of distal phalanges) showing intersecting fascicles of spindle cells (fibroblasts and myofibroblasts) with scattered RBC-sized eosinophilic PAS-/AB- round eosinophilic inclusions and extracellular collagen. Similar inclusions are also seen in breast FA, breast phyllodes tumor, cervical polyp, and GI leiomyomas (Fig. 12.6).
- *HSS/IHC*: (+) MT, PTAH, ACT, VIM, MSA ("tram track" pattern in spindle cells), calponin, DES, CD99, ± CD117, but (−) PAS, AB, CKs, ER, PR, β-Cat.
- *TEM*: Myofibroblasts with RER and membrane-unbounded inclusions composed of compact masses of actin granules and filaments.
- *DDX*:
  - IFS: no digits, >2 cm, hypercellularity, dense chromatin, high MI, and no inclusions
  - Infantile desmoid fibromatosis: rarely hands, >2 cm, hyper cellularity, and no inclusions
- *TRT*: Excision.
- *PGN*: Good, usually no relapse, if circumscribed and fully resected (*conditio sine qua non*).

12.5.5.3 Infantile Fibromatosis, Desmoid Type

- *DEF*: 1st–2nd infancy-occurring bland fibroproliferative process that invades skeletal muscle subclassified into two types, the *adult-type* desmoid fibromatosis and the *diffuse mesenchymal* type with a remarkable amount of fat (in this last setting some authors have suggested the term "lipo-fibromatosis").
- *SYN*: Aggressive infantile fibromatosis, cellular fibromatosis, congenital fibromatosis, diffuse mesenchymal fibromatosis, fibrosarcoma-like fibromatosis, infantile desmoid-type fibromatosis, lipofibromatosis.
- *EPI*: ≤8 years, ♂ = ♀.
- *CLI*: H&N, thigh.
- *GRO*: Mass infiltrating the skeletal muscle, although the involvement may be superficial.
- *CLM*: Two patterns:
  - *An immature or diffuse* pattern characterized by uniform, bland cells embedded in myxoid stroma and appearing with round/oval/spindled nuclei and scant cytoplasm with the tendency to infiltrate between the muscle fibers, surrounding nerves/nervous structures or bone (rule out malignancy in the DDX!), and fat is commonly present, and having some peripheral inflammation.
  - *Adult-type* pattern with bland spindle cells in the collagenous stroma, variable cellularity, β-Cat-positive, and tendency to destroy the surrounding muscle with invasion.
- *IHC*: β-catenin (adult-type).
- *TEM*: Fibroblastic and myofibroblastic features, including collagen fibers, thin filament bundles, and dense cytoplasmic bodies.
- *PGN*: The tumor may locally recur if inadequately excised.

12.5.5.4 Juvenile Hyaline Fibromatosis

Juvenile hyaline fibromatosis is a hereditary infantile benign "mesenchymal dysplasia" with the extracellular hyaline material in the skin, soft tissue, and bone, due to aberrant fibroblasts harboring mutations in *CMG2* gene on 4q21.21. CMG2 (capillary morphogenesis gene 2) is a transmembrane protein that is activated during capillary morphogenesis and binds laminin and collagen IV using a VWFA (Von Willebrand factor type A) domain. The defect causes the protein to remain in the endoplasmic reticulum. Grossly, there are cutaneous nodules (face and neck), gingival hypertrophy, flexure contractures, and bony lesions. Grossly, nodules are solid, white, and waxy, and, microscopically, they consist of *well-demarcated hypocellular areas* of plump fibroblasts and uniform eosinophilic, non-fibrillar hyaline material without atypia or necrosis.

- *HSS/IHC*: (+) D-PAS, AB, VIM, Coll I and III (−) MSA, Coll. II/IV, S100, β-Cat.
- *TEM*: Fibroblasts with "fibril-filled balls."
- *DDX*: Gingival fibromatosis (limited to gums, collagen-rich fibrous tissue).

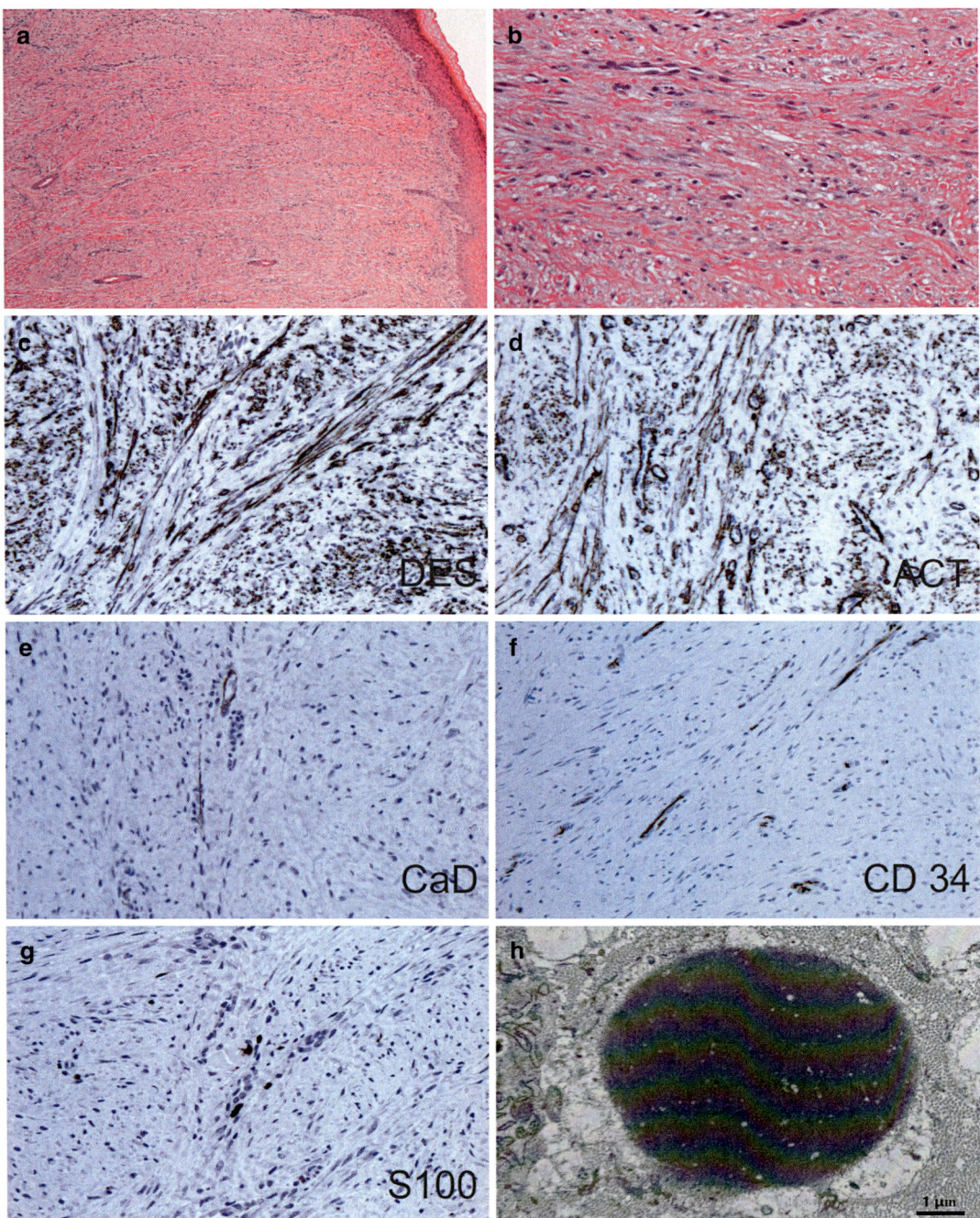

Fig. 12.6 Infantile Digital Fibroma. The microphotographs (**a**) and (**b**) show the features of a nonencapsulated fibroma with a dermal proliferation of hypocellular sheets or fascicles composed of myofibroblasts and fibroblasts with variable collagen and eosinophilic (hyaline) cytoplasmic inclusions with positivity for desmin (DES microphotograph (**c**) and smooth muscle actin (ACT, microphotograph (**d**) and virtual absence of expression of caldesmon (**e**), CD34 (**f**), and S100 (**g**) (a–b, hematoxylin and eosin staining; c-g, immunostainings with antibodies against the corresponding antigens; h, transmission electron microscopy, bar=1 micrometer). This lesion is also called "inclusion body fibroma" or "inclusion body fibromatosis"

- *PGN*: The severity of JHF can vary broadly and overall poor. Children with severe JHF forms frequently do not survive past early childhood.

12.5.5.5 Gingival Fibromatosis
Fibromatosis of the gingival ridge.

- *Calcifying aponeurotic fibroma (CAF)*: Juvenile cellular, ill-defined, slow-growing, painless, usually <3 cm nodule characterized by a diffuse proliferation of fibroblasts with focal calcifications surrounded by chondrocyte-like cells in palisade ± scattered osteoclast-like giant cells and presenting typically in palms, soles, wrists, and ankles.
- *Calcifying fibrous tumor (CFT)*: Juvenile paucicellular, dense collagenous 1–15 cm sized lesion with fibroblasts and scattered chronic inflammation foci (lymphocytes and plasma cells) as well as focal calcifications (dystrophic and psammomatous) presenting anywhere in the body (Gr. ψάμμος "sand"). Associations: Hx. trauma, Castleman disease, IMFT.

Also, calcifying fibrous pseudotumor (CFPT) needs to be distinguished from desmoplastic fibroblastoma, SFT, and GIST. Desmoplastic fibroblastoma has large stellate cells and no calcification, and inflammation is uncommon; SFT has a "patternless pattern" with cellular areas, staghorn-shaped BVs, and no calcification, and inflammation is unusual; and finally GIST is not typically multinodular and possesses sclerotic stroma and no hyalinized stroma, calcification is of dystrophic type and it is not psammomatous, and inflammation is not a typical feature.

12.5.6 Infantile Myofibroma/Myofibromatosis

Infantile myofibroma (IM) as solitary (50%) and infantile myofibromatosis (IM/IM) as multicentric are lesions with similar constitution, but different outcomes. The multicentric type may have extensive visceral involvement and may be fatal (*vide infra*). Myofibroma is a benign fibrous tumor, typically occurring in infancy and early childhood that can arise in the skin, subcutis, muscle, bone, or viscera. Three clinical forms of IM/IM have been described, including *solitary* cutaneous nodules, *multiple* cutaneous nodules, and generalized infantile myofibroma with *visceral* involvement. The most common outcome in IM is a complete spontaneous regression, but the widespread and visceral form tends to carry a poorer prognosis and this aspect is a frequent debate at pediatric tumor boards (Kulkarni et al. 2012). The etiology is unclear, but it seems that a genetic component may be present. On the other hand, conflicting evidence exists about the inheritance pattern. Some familial cases point to an AR inheritance, while other reports indicate an AD pattern with or without variable penetrance or genetic heterogeneity. Grossly, a swirling (spiraling) pattern of fibrous tissue is recognized, and microscopically, there is a proliferation of spindle cells with eosinophilic cytoplasm admixed in a myxoid background. The architecture revealed on low power shows tumor cells, which are arranged in nodules with slit-like vascular spaces or, at places, even a more classic hemangiopericytomatous pattern (Fig. 12.7). Immunohistochemistry-positive evidence using antibodies against several mesenchymal markers (smooth muscle actin [SMA], neuron-specific enolase [NSE]) and no staining for DES, S100, glial fibrillary acidic protein [GFAP], and CD57 confirm the diagnosis of myofibromatosis, which is also supported by TEM showing subcellular myofibroblastic structures.

- *DDX*: Hyaline fibromatosis and neurofibroma.

12.5.7 Fibroblastic/Myofibroblastic Tumors with Intermediate Malignant Potential

Fibroblastic or myofibroblastic tumors with intermediate malignant potential harbor a poorer

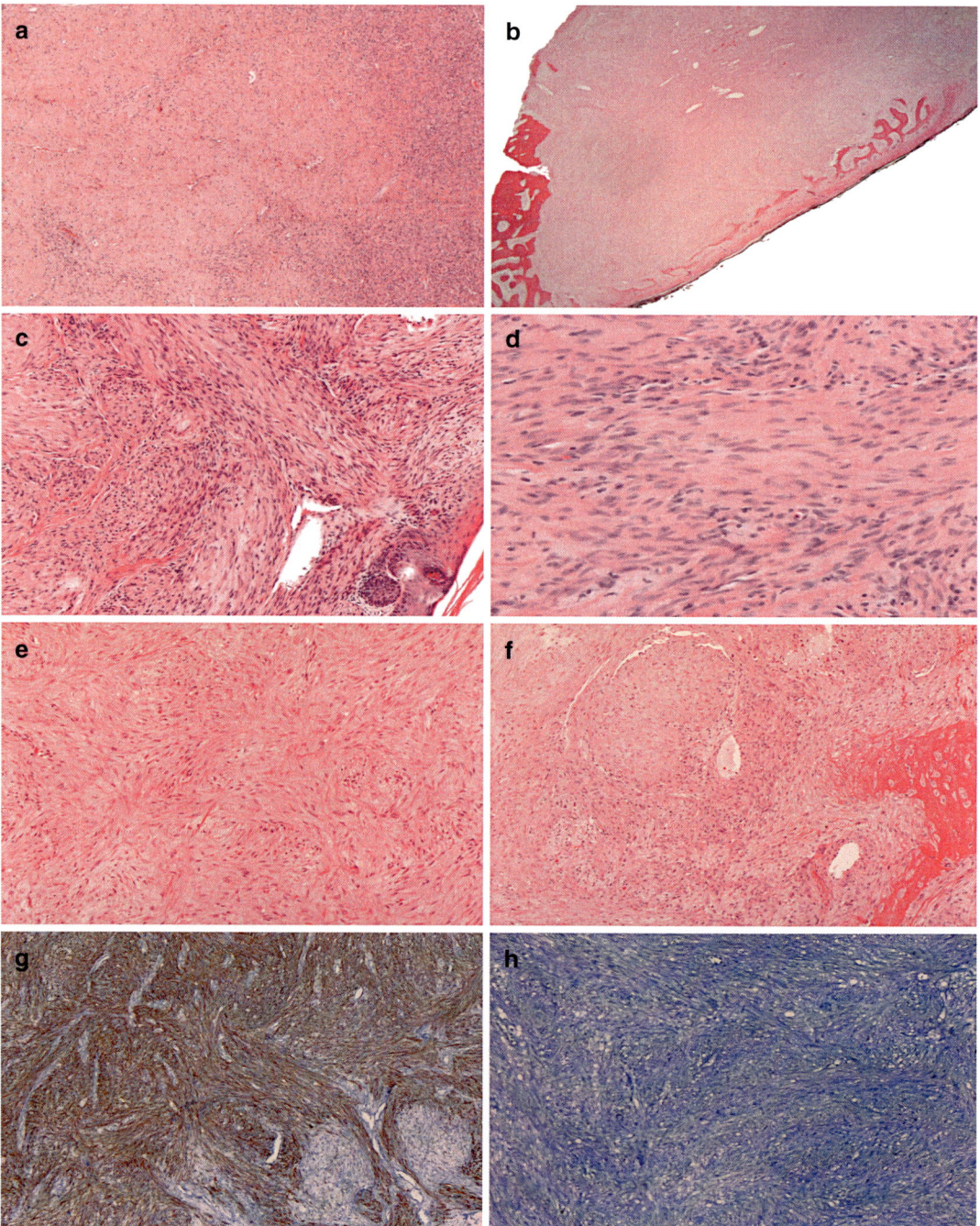

Fig. 12.7 Myofibromatosis. Histologic and immunohistochemical microphotographs of a myofibromatosis showing eosinophilic spindle-cell fascicles and whorls with bland myoid features underneath the skin and adnexal tissue and blending with bone (**a–b**). Two components with markedly varying proportions are seen in these lesions with an additional more cellular component close to the previous more fascicular component. There is an abrupt transition between two components constituted by a more central spindle-cell area and a more primitive cellular area. In the less differentiated area, rounded cells with eosinophilic cytoplasm arranged around small vessels are seen (**c–f**). The presence in some areas of a prominent perivascular arrangement justified in the past the use of the term of "infantile hemangiopericytoma," which may represent part of the spectrum of infantile myofibromatosis. In (**g**) it is easy to see the marked expression of actin, while in (**h**) it is apparent the low proliferation rate (a-f, hematoxylin and eosin staining, ×50, ×12.5, ×100, ×200, ×100, ×100 as original magnification, respectively; g, immunostaining with antibody against smooth muscle actin (Actin), ×50 original magnification; h, immunostaining against Ki67 (MIB1) ×100 original magnification)

12.5.7 Fibroblastic/Myofibroblastic Tumors with Intermediate Malignant Potential

> **Box 12.9 Fibroblastic/Myofibroblastic Tumors with Intermediate Malignant Potential**
> - Superficial Fibromatosis
> - Desmoid-type Fibromatosis
> - Lipofibromatosis
> - Solitary Fibrous Tumor, Hemangiopericytoma, and Giant-Cell-Angiofibroma
> - Inflammatory Myofibroblastic Tumor
> - Low-Grade Myofibroblastic Sarcoma
> - Myxo-Inflammatory Fibroblastic Sarcoma
> - Infantile Fibrosarcoma

prognosis than their benign counterparts, i.e., they are either locally aggressive or rarely metastasizing) (Box 12.9).

12.5.7.1 Solitary Fibrous Tumor, Hemangiopericytoma, and Giant Cell Angiofibroma

- *DEF*: SFT is a mesenchymal tumor with "patternless pattern" of fibroblast-like cells with thin strip-like bands of collagen and gaping BVs. It is similar to the pleural localization of SFT and has a blurred distinction from hemangiopericytoma.
- *EPI-CLI*: Children and adults are presenting with a slow-growing painless mass, potentially presenting with paraneoplastic hypoglycemia due to IGF production by tumor cells.
- *GRO*: Well-circumscribed tumor with multinodular, tan-white, firm, "whorled" cut surface and with or without focal necrosis.
- *CLM*: "Patternless" architecture of hypo- and hypercellular areas with thick, hyalinized collagen with cracking artifact and staghorn-like BVs (hemangiopericytoma-like) on low power of microscopy. This alternating array may be like the *marbled appearance* of SS as well as of other soft tissue sarcomas. In fact, this pattern may suggest erroneous interpretation if overestimated. The *"patternless appearance"* may be found in soft tissue sarcomas and should not be emphasized. High power magnification shows bland and uniform spindle cells with indistinct nucleoli and minimal cytoplasm along thin parallel collagen strands, mast cells, adipose tissue or MNGCs, no atypia, minimal or no pleomorphism, and no/low MI.
- *IHC*: (+) BCL2, CD34, CD99, STAT6 and (−) DES, AE1–3, S100, CD117, CD31, D2-40. The strong nuclear expression of STAT6 differs tremendously from other SFT-mimicking neoplasms, which do not harbor positivitiy for this marker. STAT6 is the signal transducer and activator of transcription 6 (STAT6). The *STAT6* gene is responsible for the production of STAT6 protein. STAT6 is a member of the STAT family of transcription factors and remains a crucial tool for making the diagnosis of SFT.
- *TEM*: Features of myofibroblasts and fibroblasts.
- *CMB*: Molecular aberrations observed in SFTs are *NAB2-STAT6* fusion variants, activation of Akt/mTOR pathway, and ↑ expression of lysine-specific demethylase 1, which is a histone lysine demethylase strongly associated with epigenetic changes.
- *DDX*: Synovial sarcoma, smooth muscle tumors, and benign neural tumors.
  - SS: No or very few thick collagen bands, AE1–3+, CD34− (but CD99 and BCL2+).
  - SM tumors: Fascicular with abundant cytoplasm, blunted nuclei, (+) DES, SMA.
  - Benign neural tumors: S100+.
- *PGN*: prognosticators are the size (cut-off: 10 cm), positive margins (incomplete resection or noncomplete resection), cellular atypia, cellular pleomorphism, and MI ($\geq$5/10-HPF). Substantially, when centrally located and greater than 5.0 cm in diameter, with mitoses greater than 4/10-HPF, increased cytologic atypia, hemorrhage, and/or necrosis, SFT can be considered low-grade malignant neoplasms.

Giant Cell Angiofibroma (GCAF)
- *DEF:* Benign, a slow-growing painful tumor that may be related to SFT and is constituted exquisitely by MNGCs and ectatic BVs. GCAF is seen in young/middle-aged individuals in the eyelid/orbital region.

- *GRO*: Well-circumscribed nodule ± capsule, <5 cm, and hemorrhagic cut surface.
- *CLM*: Bland cellularity of the round to spindle cells with indistinct nucleolus embedded in a myxo-collagenous stroma with MNGCs lining small/medium-sized ectatic BVs.
- *IHC*: (+) VIM and BCL2, CD34, CD99; (−) CD31, CD68, CD117, MSA, DES, S100.
- *TEM*: Fibroblastic (often) and Schwannian (rare) features.
- *CMB*: ± t(12;17)(q15;q23) or 6q13 abnormalities.
- *DDX*: As first, SFT, but if it is extra-orbital, the differential diagnoses must also include giant cell fibroblastoma, synovial sarcoma, fibrosarcoma, malignant fibrohistiocytoma, deep fibrous histiocytoma, hemangiopericytoma, and schwannoma. The giant cells in GCAF look like the mononuclear component and result positive for CD34 and CD99. There is some positive of the giant cells for BCL-2, but they are negative for CD68.
- *TRT*: Therapy consists of excision, and it does not recur.

Other fibromatoses affect the youth and are fibromatoses of the adult. These are characterized by a proliferation of well-defined fibroblasts that may be very cellular, low MI, and harbor some infiltrative growth pattern intermixed with more collagenous areas. There is a variety of names, patterns, and associations, uniformity of cells, and, of course, the rarity of mitoses. *Superficial fibromatoses* include the palmar (Dupuytren) fibromatosis, which is bilateral in 50% and results in fixed painless flexion contractures of digits and the plantar ("Lederhosen", German "leather breeches") fibromatosis, which is bilateral in 10–25% and presents with some degree of pain/paresthesias at the sole of the feet, and the penile fibromatosis (Peyronie), which affects the dorsolateral side of the penis. *Deep fibromatosis (musculoaponeurotic)* include the desmoid, which affects the abdominal wall of women during or following pregnancy and can be associated with Gardner's syndrome and patients harboring Gardner syndrome may also have extra-abdominal or aggressive fibromatoses at the shoulder, chest/back, mesentery, and thigh with increased MI although no atypia. Lipofibromatosis involves the presence of adipose tissue admixed with an extra-abdominal fibromatosis. Idiopathic retroperitoneal fibrosis (aka Ormond disease, sclerosing fibrosis, sclerosing retroperitonitis) is an ill-defined mass composed of mixed inflammatory cells (lymphocytes and germinal centers, plasma cells, eosinophils) in a cellular fibrous stroma with foci of fat necrosis. The Ormond disease surrounds the abdominal aorta and displaces the ureters medially. The Ormond disease is very rare in children.

12.5.7.2 Inflammatory Myofibroblastic Tumor

- *DEF*: Tumor/Pseudotumor (aka inflammatory fibrosarcoma, inflammatory pseudotumor, plasma cell granuloma) constituted by a proliferation of myofibroblasts with plasma cells, lymphocytes and eosinophils that involves retroperitoneum, omentum, and mesentery in children and young adults, but also parenchymal organs such as lung, organs of the GI and GU tracts.
- *CLI*: ±Fever, FTT/weight loss, anemia, thrombocytosis, polyclonal hyper-Ig, and ↑ESR.
- *GRO*: Circumscribed, not encapsulated mass with white-tan whorled fleshy/myxoid cut surface ± H/N/C foci.
- *CLM*: Myofibroblastic and fibroblastic spindle cells with diffusely scattered infiltrates of lymphocytes, plasma cells, eosinophils, and histiocytes in a background of copious BVs. Three growth patterns have recognized, including nodular fasciitis-like, highly cellular, and densely hyalinized stroma. Neither nuclear pleomorphism nor atypical mitoses are seen.
- *IHC*: (+) VIM, SMA, MSA, calponin, ALK1/p80, (±) AE1–3, DES, but (−) S100, CD34, h-Cald, CD117, HHV-8. S100 is important to rule out nerve tumors, which may have some similarity (e.g., ancient schwannoma), CD34 (together with BCL2, CD99 and AE1–3) to rule out SS, h-Cald (together with DES, Myogenin, and MyoD1) to rule out RMS, CD117 (to rule out GIST), and HHV8 (to rule out Kaposi sarcoma).

- *TEM*: Features of myofibroblasts and fibroblasts.
- *CMB*: 2p23 clonal abnormalities, including t(2;5)(p23;q35) involving *ALK* and *NPM* and t(2;17)(p23;q23) involving *ALK* and *CLTC*, and t(2;19)(p23;p13.1) involving *ALK* and *TPM4*.
- *DDX*: It includes calcifying fibrous pseudotumor, nodular fasciitis, low grade myofibroblastic sarcoma, and IgG4 related sclerosing lesion.

- Calcifying Fibrous Pseudotumor (CFPT): calcification, no myofibroblasts, SMA (−)
- Nodular Fasciitis: Small Ø, location, zonation, ALK (−)
- LGMFS: high cellularity, uniformity, hyperchromasia, infiltrating, ALK (−)
- IgG4-Related Sclerosing Lesion: IgG4+ PC and IgG4+/IgG+ PC, obstructive phlebitis, ALK (−)

- *TRT*: Excision, although 1/3 recurs and rarely metastases or multifocality have been recorded.
- *PGN*: It depends on the site and ALK result, being IMT of the abdominal or pelvic sites and/or ALK-negative harboring a poor prognosis. Highly atypical cells with oval nuclei, prominent nucleoli, Reed-Sternberg-like cells, and atypical mitoses have been associated with aggressive behavior.

IMT vs. Pseudosarcomatous Myofibroblastic Proliferation (PMP).

Activation of the anaplastic lymphoma kinase (*ALK*) gene, secondary to chromosomal translocation 2→5, leads to constitutive tyrosine kinase activity, which found in roughly half of the IMT cases. ALK positivity suggests a neoplastic rather than reactive etiology of IMT. The lesion called IMT includes a myofibroblastic proliferation associated with a prominent lymphoplasmacytic infiltrate, which is typically lacking in PMP. Both lesions are quite similar and need to be differentiated from other lesions such as myxomas, angiosarcoma, osteosarcoma, and synovial sarcoma. In PMP, aneuploidy has been found in the heart, although aneuploidy does not seem to indicate destructive behavior, which may be odd to an oncologist. However, this is because some non-neoplastic lesions, such as atherosclerotic plaques or inflammatory disease may harbor numerical chromosomal changes without oncologic significance.

12.5.7.3 Low-Grade Myofibroblastic Sarcoma (LG*MF*S)

- *DEF*: It is a rare tumor of malignant myofibroblasts or distinct atypical myofibroblastic tumor with the mostly deep-seated H&N location of adults more than children and a characteristic tendency to recur locally rather than metastasize.
- *SYN*: Myofibrosarcoma.
- *GRO*: Firm, pale nodule with treacherous unclear margins.
- *CLM*: Hypercellular fascicles of elongated myofibroblasts with ill-defined cell borders and pale eosinophilic cytoplasm and nuclei, which can be either vesicular with inconspicuous nucleoli or spindle-shaped, elongated, and wavy reminding of neural differentiation. There is focal moderate nuclear atypia with hyperchromatic nuclei and irregularities of nuclear membranes, MI = 1–6/10 HPF, and collagenous matrix with prominent hyalinization without histiocytic giant cells or prominent inflammation.
- *IHC*: ≥1 (+) myo-marker (DES, α-SMA, MSA, calponin) and (−) S100, EMA, h-CALD, ALK.
- *TEM*: MF-features (discontinuous basal lamina, thin filaments with focal microdensities, subplasmalemmal attachment plaques, and micropinocytosis).
- *CMB*: Chromosomal imbalances with gains at 1p11 → p36.3 (66%), 12p12.2 → p13.2 (45%), 5p13.2 → p15.3 (31%), and chromosome 22 (28%) and loss at 15q25 → q26.2 (24%) according to Meng et al. who investigated the molecular cytogenetic features of myofibro-

blastic sarcoma. DNA copy number changes was analyzed by comparative genomic hybridization.
- *DDX*: It includes the following diagnoses:
  - Fibromatosis/Myofibromatosis
  - Infantile Fibrosarcoma (IFS): storiform growth pattern, molecular biology characteristics
  - Inflammatory Myofibroblastic Tumor (IMT): heterogeneity, less cellular, hyperchromasia, and infiltrative pattern, and ALK (+)
  - LMS: alternating fascicles
  - Neurofibroma (NF): no infiltrative edges, not deep, no chromosomal anomalies
  - Solitary Fibrous Tumor (SFT): "Patternless," collagen, (+) BCL2, CD34, CD99, STAT6

Notes: *NF* neurofibroma, *SFT* solitary fibrous tumor, *IFS* infantile fibrosarcoma, *IMT* inflammatory myofibroblastic tumor

- *TRT*: Wide surgical excision with removal of the entire lesion is generally suggested. Embolization may be adopted to provide temporary relief and radiation and/or chemotherapy may be added to the therapeutic protocol.
- *PGN*: It is a relatively good, but the outcome depends upon the size of the tumor, staging, tumor location, age, and immunologic status of the individual.

12.5.7.4 Myxo-inflammator Fibroblastic Sarcoma (MIFS)
- *DEF*: Rare alternating dimorphic slow-growing soft tissue fibrosarcoma of the distal extremities (hands and feet) of mostly adults (adults>children) and specific tendency to recur locally more than metastasizing.
- *SYN*: Inflammatory myxohyaline tumor of the distal extremities with virocyte (cell with stick-like protrusions) or Reed-Sternberg-like cells (acral location). In cell culture, the virocyte is a cell with stick-like protrusions, which possesses the ability to infect other cells that remain unprotected by keratinocytes.
- *GRO*: Poorly defined multinodular white and gelatinous tumor mass involving joints and tendons.
- *CLM*: Alternating dimorphic pattern with areas of myxoid tissue and cellular (or solid) areas of neoplastic cells mixed with inflammatory cells. The neoplastic cells may show a different morphology, including spindle cell, lipoblast-like, ganglion-like, and virocyte or Reed-Sternberg-like. MI: 0–1/10 HPF, no atypical forms, no/rare necrosis.
- *IHC*: (±) SMA, (−) DES, (+) VIM, CD34, CD68, CD163, (−) S100, AE1–3, h-Caldesmon.
- *TEM*: Fibroblastic features (abundant RER and mitochondria, IF network ± densely packed perinuclear whorls).
- *CMB*: Complex and heterogeneous karyotypes.
- *DDX*:
  - Hodgkin lymphoma (no ganglion-like or lipoblast-like cells, (+) PAX5, CD15, CD30)
  - Myxofibrosarcoma (rare acral, higher MI, no inflammation)
  - Rosai-Dorfman disease (no intranuclear or cytoplasmic vacuoles, (+) S100)
  - Epithelioid sarcoma (bright eosinophilic cytoplasm, (+) AE1/3)
  - Tenosynovitis (no cellular atypia)

12.5.7.5 Infantile Fibrosarcoma
- *DEF*: Highly cellular aggressive fibroblastic tumor with morphological similarities to the adult counterpart, but with remarkably better outcome, predilected involvement of the extremities or axis in children younger than 5 years of age, and cytogenetically related to congenital mesoblastic nephroma (CMN) of the kidney.
- *GRO*: Large tumor with erythema and erosions/ulcers on the skin overlying the tumor and gray-tan fleshy cut surface with H/N/C and myxoid/cystic degeneration.
- *CLM*: Poorly circumscribed, lobulated mass of small to large spindled cells growing in bundles with a peculiar herringbone pattern with cell crowding, nuclear atypia, and pleomorphism.

Additional features may include hemangiopericytoma-like areas, dystrophic calcification, and extramedullary hematopoiesis. The "herringbone" pattern is constituted by cells in columns of short parallel lines with all the lines in one column inclined one way and lines in adjacent columns leaning the other way.
- *IHC*: (+) VIM, (±) SMA, DES, S100, CD34.
- *TEM*: Fibroblastic and myofibroblastic features.
- *CMB*: t(12;15)(p13;q26) (2/3 of cases) causing *ETV6-NTRK3* gene fusion transcript (ETS variant gene 6 and neurotrophic tyrosine receptor kinase type 3) by FISH/RT-PCR (alike secretory breast carcinoma, secretory salivary gland carcinoma, and CMN). Moreover, additional cytogenetic features include trisomy 8, 11, 17, and 20.
- *DDX*: It includes:
  - Adult-type fibrosarcoma (adolescents and young adults, but lack of t(12;15)).
  - Infantile fibromatosis (no/low MI, no pleomorphism, and lack of t(12;15)).
  - Myofibromatosis (myofibroblastic features and lack of t(12;15)).
- *TRT*: Surgery and chemotherapy.
- *PGN*: Relatively good, although recurrences are common (~1/2 of cases) and only rarely metastases.

12.5.8 Malignant Fibroblastic/Myofibroblastic Tumors

Malignant tumors with fibroblastic and/or myofibroblastic differentiation include essentially four entities as shown below (Box 12.10).

12.5.8.1 Adult Fibrosarcoma (AFS)
- *DEF*: Highly cellular aggressive fibroblastic tumor with morphological similarities to the infantile counterpart, but with worse outcome, occurring in young-/middle-aged adults, the deep soft tissue of lower extremities or trunk and prone both to recur and metastasize (lung and bone). Grossly, AFS is well circumscribed but unencapsulated, fleshy, white-tan with hemorrhages and necrosis. Microscopically, herringbone pattern with cells with tapered

> **Box 12.10 Malignant Fibroblastic/Myofibroblastic Tumors**
> 12.5.8 Malignant Fibroblastic/Myofibroblastic Tumors
>   12.5.8.1 Adult Fibrosarcoma (AFS)
>   12.5.8.2 Myxofibrosarcoma (MFS)
>   12.5.8.3 Low-Grade Fibro-Myxoid-Sarcoma (LGFMS)
>   12.5.8.4 Sclerosing Epithelioid Fibrosarcoma (SEFS)

elongated hyperchromatic nuclei, scant cytoplasm, and variable nucleoli. The MI may be high with atypical mitotic figures, but neither giant cells nor pleomorphism are seen.
- *HSS/IHC*: Pericellular Ret (+), PTAH (+) VIM, type 1-collagen, p53, MIB1/Ki67 +, CD34 ± and (−) for smooth muscle and histiocytic markers.
- *TEM*: Fibroblasts with prominent RER but no myofilaments, no external lamina, no intercellular junction, and no distinct myofibroblasts (DDX: myofibrosarcoma).
- *CMB*: Aneuploidy.
- *DDX*: Other tumors with fibrosarcomatous areas (dedifferentiated liposarcoma, fibromatosis, LGFMS, MFH-pleomorphic, malignant schwannoma or MPNST, and synovial sarcoma).

12.5.8.2 Myxofibrosarcoma
- *DEF*: Spectrum of malignant fibroblastic lesions with multinodular growth pattern, incomplete fibrous septa, prominent myxoid stroma, pleomorphism, and *curvilinear* BVs usually occurring in adult males more than children at the limbs and limb girdles (lower > upper) (aka myxoid MFH with fibroblastic differentiation).
- *GRO*: Multiple mucoid nodules or a single mass with infiltrative margins and tumor necrosis in high-grade lesions.
- *CLM*: Tumor showing (1) *lobularity* with distinctive nodules showing hypocellular (myxo-) and hypercellular (fibro-) areas and incom-

plete septation; (2) characteristic intramyxoidal *curvilinear blood vessels*; and (3) *nuclear pleomorphism*.
- *HSS/IHC*: (+) VIM, CD34, acid mucins, (±) SMA, DES, AE1–3, (−) S100, h-Caldesmon.
- *TEM*: FB-EM features.
- *CMB*: Complex karyotypes.
- *DDX*:
  - Nodular fasciitis (no atypical cells, no atypical mitoses)
  - Myxoma (no atypia, no prominent BV, few mitoses, normal karyotype)
  - Myxoid liposarcoma (lipoblasts with clear ORO+ cytoplasm, plexiform BVs, + S100)
  - Metastatic carcinoma from the kidney or adrenal gland
- *TRT*: Local excision.
- *PGN*: It depends on location (deep-seated ≠ superficial), which adding to high grade, epithelioid differentiation, tumor necrosis, and size (Ø >5 cm) portend to a poor outcome.

12.5.8.3 Low-Grade Fibro-Myxoid Sarcoma (LG*FM*S)

- *DEF*: Low-grade fibro-myxoid sarcoma (LG*FM*S), aka Evans' tumor or hyalinizing spindle-cell tumor with giant rosettes, is a sarcoma with low, bland cellularity, fibro-myxoid areas, and curvilinear BVs. LGFMS is considered a fibrosarcoma variant and at proximal extremities or the trunk in adolescents and young adults with a peculiar tendency to recur more than metastasize (also many years after initial diagnosis).
- *GRO*: LGFMS is well circumscribed and has a "fibro-myxoid" cut surface.
- *CLM*: There is a variable admixture of dense collagenous and myxoid areas of bland spindle cells with whorled arrangement and arcades of characteristic BVs. Two types are differentiated, including the classic one and the LGFMS with *giant collagen rosettes*.
- *HSS/IHC*: (+) VIM, CD99, BCL2 (pitfalls, SS and SFT; however SS is +ve for EMA, AE1–3, and CK7, and SFT is +ve for CD34 other than having a "patternless" pattern and staghorn BV instead of BV arcades). Moreover, LGFMS is ± SMA, but −ve for DES, h-Caldesmon, S100, CKs (DDX→SS), CD34 (DDX→SFT), MDM2 (DDX→LPS), CD117 (DDX→GIST), DOG1 (DDX→GIST), and β-Cat$_n$ (DDX→fibromatosis).
- *TEM*: Fibroblastic differentiation.
- *CMB*: *t(7;16)*(q32–34;p11) ⇒ *FUS-CREB3L2* and *t(11;16)*(p11;p11) ⇒ -*FUS-CRE3L1*.
- *DDX*: It includes the following diagnoses:
  - MFS: Myxoid > fibrous, +++ atypical nuclear changes, and intramyxoidal curvilinear BVs
  - Desmoid fibromatosis: No myxoid areas, slit-like BVs, and (+) β-Cat$_n$
  - Neurofibroma: Cells with wavy nuclei, thick collagen bundles, and (+) S100
  - IFS/AFS: "Herringbone" pattern, no myxoid component, not (7;16) or t(11;16)

12.5.8.4 Sclerosing Epithelioid Fibrosarcoma

- *DEF*: Sclerosing epithelioid fibrosarcoma (SEFS) is a rare, slow-growing sarcoma of deep soft tissue with epithelioid tumor cells in nests and cords and hyalinized fibrous stroma located at the limb/limb girdle, head, and neck, back/chest wall, and the base of the penis of adolescents and young adults.
- *GRO*: SEFS is a poorly circumscribed tumor with firm, gray-white cut surface and may invade the surrounding bone without necrosis.
- *CLM*: Hypo- and hypercellular areas with nests or cords of small-to-medium-sized, round to ovoid, relatively uniform and bland epithelioid cells with clear cytoplasm embedded in a hyalinized fibrous stroma and peculiar infiltrating single rows of cells ("Indian file pattern") resembling carcinoma (grade I vs. grade III showing firmly packed tumor cells with dense chromatin, no/minimal collagenous stroma, and numerous mitotic figures) (Fig. 12.8).
- *HSS/IHC*: (+) VIM, BCL-2, p53, ± EMA, CKs, S100 and (−) CD34 (DDX→SFT), CD45 (DDX→lymphoma), HMB-45, MART-1/Melan-A, Tyrosinase, and MITF (DDX→melanoma), DES, SMA.
- *TEM*: Fibroblastic features with networks of IF that may form perinuclear whorls (epithelioid).

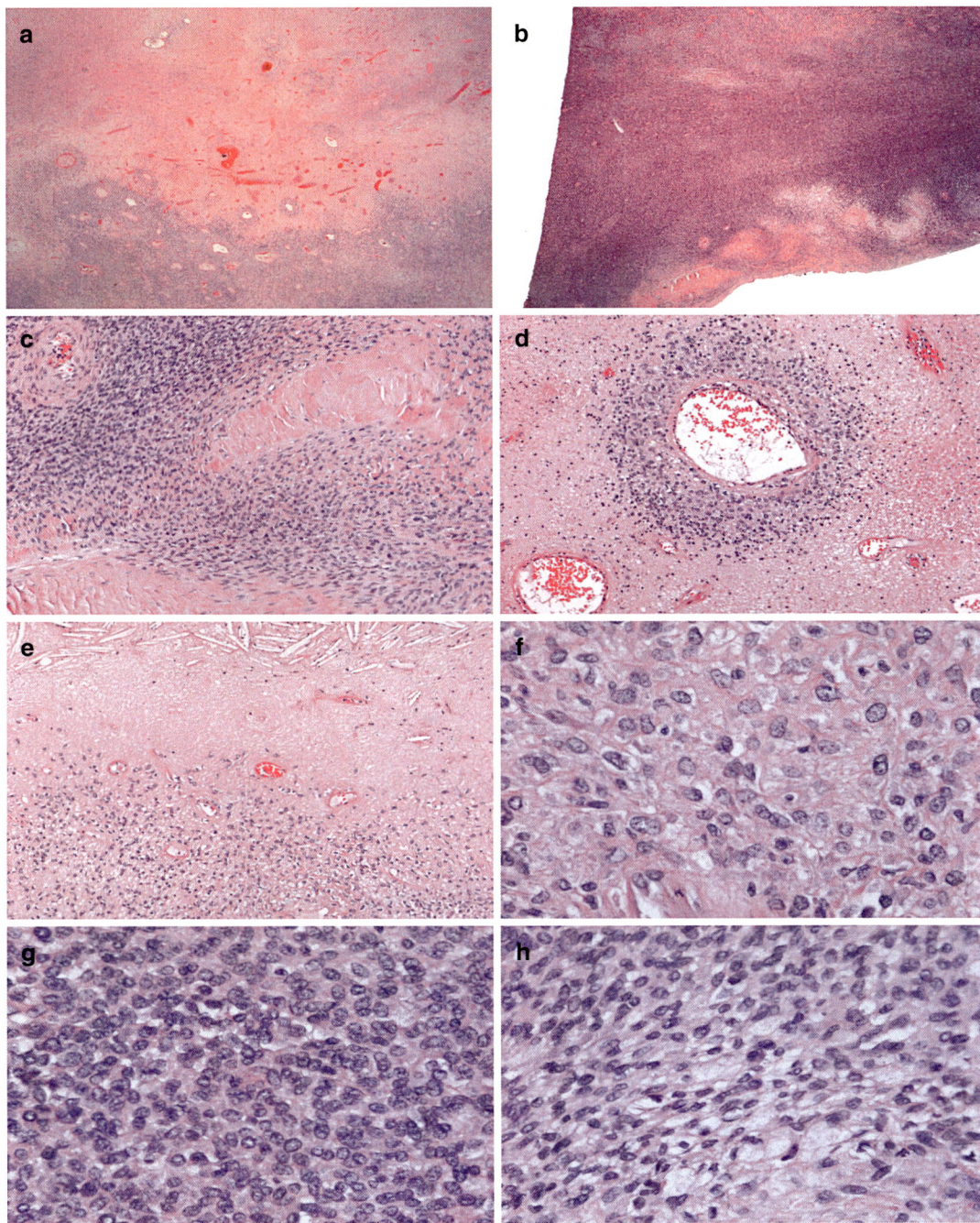

Fig. 12.8 Sclerosing Epithelioid Fibrosarcoma (SEFS). There are hypo- and hypercellular areas with nests or cords of small-to-medium-sized, round to ovoid, relatively uniform and bland epithelioid cells with clear cytoplasm embedded in a hyalinized fibrous stroma and peculiar single-cell file pattern resembling carcinoma

- *CMB*: The majority of LGFMS cases are characterized by a *FUS-CREB3L1* fusion, but both *FUS-CREB3L2* and *EWSR1-CREB3L1* fusions may be encountered in a small number of LGFMS and SEFS/LGFMS hybrid tumors. Thus, generally this tumor is characterized by ± t(7;16)(q32–34;p11) ⇒*FUS-CREB3L2*.
- *DDX*:
    - LGFMS: Monotonous hypocellular with alternating myxoid/collagenous stroma, (+) CD99
    - SS: No extensive sclerosing areas as seen in SEFS, (+) CD99, TLE1, EMA, AE1–3, CK7
    - Sclerosing lymphoma: No extensive sclerosing areas as seen in SEFS and (+) CD45
    - Poorly differentiated carcinoma: (+) EMA, AE1–3, CK7
    - Lobular breast carcinoma: (+) EMA, AE1–3, CK-7 and (+) ER, PR
- *TRT*: It consists of intralesional excision, attempted wide local excision, and amputation with either adjuvant radiation therapy or chemotherapy.
- *PGN*: About half of the patients suffer from a persistent disease or local recurrence, while the other half develops distant metastases. Follow-up of at least 12 months reveals persistent disease or local recurrence in half of the patients and distant metastasis in approximately 90% of the patients.

12.6 Fibrohistiocytic Tumors

Benign entities of this group include the group of the benign fibrous histiocytoma (BFH), including *deep BFH* (counterpart of the superficial BFH or dermatofibroma), *giant cell tumor of tendon sheath* or BFH of the tendon sheath, and the *diffuse-type giant cell tumor* or pigmented villonodular synovitis. Intermediate entities include *plexiform fibrohistiocytic tumor* and *giant cell tumor of soft tissues*. Malignant fibrohistiocytic tumors include *malignant fibrous histiocytoma* with pleomorphic, giant cell, and inflammatory subtypes (Box 12.11).

Box 12.11 Fibrohistiocytic Tumors

12.6.1 Histiocytoma
    12.6.1.1 Juvenile Xanthogranuloma (JXG)
    12.6.1.2 Reticulohistiocytoma
12.6.2 Benign Fibrous Histiocytoma (BFH)
    12.6.2.1 Dermatofibroma
    12.6.2.2 Giant Cell Tumor of Tendon Sheath (GCTTS, TSGCT)
    12.6.2.3 Pigmented Villonodular Synovitis (PVNS)
    12.6.2.4 Cellular Fibrous Histiocytoma
12.6.3 "Borderline" Fibrous Histiocytoma
    12.6.3.1 Dermato-fibrosarcoma Protuberans (DFSP)
    12.6.3.2 Pigmented Dermato-fibrosarcoma (Bednar Juvenile Tumor)
    12.6.3.3 Atypical Fibroxanthoma
12.6.4 Malignant Fibrous Histiocytoma (MFH)
    12.6.4.1 Storiform-Pleomorphic MFH
    12.6.4.2 Myxoid MFH/Myxoid Fibrosarcoma (Myxofibrosarcoma)
    12.6.4.3 Giant Cell MFH
    12.6.4.4 Inflammatory (Malignant Xanthogranuloma)
    12.6.4.5 Angiomatoid MFH

12.6.1 Histiocytoma

Histiocytoma refers polygonal cells tightly packed with minimal stroma, often inflammatory cells and development of fibrosis in older lesions. Two entities will be discussed here, i.e., juvenile xanthogranuloma (JXG) and reticulohistiocytoma.

12.6.1.1 Juvenile Xanthogranuloma

Juvenile xanthogranuloma (JXG) is a benign, self-healing, non-Langerhans cell histiocytosis

(LCH). It is often seen in children of the first infancy (first 2 years of life) and is rarely seen in children older than 3 years of age. Sites are frequently H&N and limbs and in 1/10 of cases are located deeply. Grossly, JXG is papular or nodular in the skin (yellowish asymptomatic papules/nodules) and in extracutaneous location (eyes, lungs, liver, bones, kidneys, pericardium, gastrointestinal tract, testes, ovaries, and central nervous system). Microscopically, there is an ovoid to spindle-cell proliferation of bland histiocytes with some fibrosis and numerous Touton giant cells, foam cells, lymphocytes, and eosinophils. JXG is (+) VIM, CD68, FXIIIa, but (−) CD1a. Prognosis is good without recurrence.

12.6.1.2 Reticulohistiocytoma

Reticulohistiocytoma is a slow-growing non-LCH. Sites are H&N and upper body, and there may be a positive Hx. for arthritis. Grossly, reticulohistiocytoma is characterized by cutaneous or mucosal papulae, which may be associated with severe polyarthritis and arthralgias. Microscopically, there is a uniform cellular population of multinucleated, DPAS+ histiocytes with abundant, "ground glass" cytoplasm and peripherally located nuclei with admixed histiocytes, epithelioid cells, and inflammatory cells (DPAS, diastase resistant periodic acid Schiff histochemical reaction). Reticulohistiocytoma needs to be kept separated from the histiocytic diseases that typically lack the characteristic "ground glass" cytoplasm. Also, CD1a and S100 immunohistochemical studies can be used to exclude LCH and Rosai-Dorfman disease, respectively. JXG shows classic "Touton" giant cells accompanying the lipid-rich histiocytes.

12.6.2 Benign Fibrous Histiocytoma

Benign fibrous histiocytoma (BFH), aka dermatofibroma, is a common finding in the routine practice of pathology. BFH may also include three additional related entities, i.e., giant cell tumor of tendon sheath or BFH of the tendon sheath, and diffuse-type giant cell tumor or pigmented villonodular synovitis, as well as the cellular (or deep) fibrous histiocytoma.

12.6.2.1 Dermatofibroma

Dermatofibroma is a BFH that develops exclusively in the subcutaneous tissue. It is a benign superficial cutaneous, red-brown, painless, and slowly growing nodule. Microscopically, it is constituted by interlacing fascicles of slender spindle cells (storiform pattern) intermingled with foamy histiocytes (xanthoma cells), Touton giant cells, delicate branching vessels, and chronic inflammatory cells with accompanying pseudoepitheliomatous hyperplasia of overlying epidermis, which is useful for the differentiation from DFSP, which bears a deeper localization. No microcysts are seen.

DDX includes:

- Fibromatosis (larger size, infiltration into surrounding soft tissue of the lesion site or neighboring organs, abundant collagen separating spindle cells, and no tissue culture-like appearance)
- Inflammatory MFH (larger size, slower growth, cellular pleomorphism, neutrophils, plasma cells, foam cells and atypical mitoses, no RBC extravasation, no keloid-type collagen)
- IMFT (larger size, no rapid growth, no zonation, no prominent myxoid stroma, and mixed inflammatory infiltrate)
- Myositis ossificans (myocentric reversed zonation with the distinctive zonal pattern showing fasciitis-like features, immature osteoid, or woven bone rimmed by osteoblasts in the center and bone formation at the periphery, initially with woven bone and, subsequently, lamellar bone and calcification, no nuclear atypia)
- Myxofibrosarcoma (hypo-/hypercellular lobularity, curvilinear, regularly arborizing BV, atypia, and cellular pleomorphism)
- Sarcoma NOS (high cellularity, atypia, and cellular pleomorphism)
- Postoperative spindle-cell nodule (positive medico-surgical history, brisk mitotic activity, but no atypia)

12.6.2.2 Giant Cell Tumor of Tendon Sheath (GCTTS)

- *DEF*: BFH that develops in the tendinous tissue of fingers (interphalangeal joints), knees, and hips. GCTTS aka TSGCT, nodular tenosynovitis.

- *GRO/CLM*: Circumscribed (fibrous pseudocapsule) solitary, lobulated, small (0.5–4 cm) slowly growing, painless tumor characterized by a *proliferation of synovial-like mononuclear cells* growing in sheets and accompanied by MNGCs with bland cytology, lipid-laden MΦ with cholesterol clefts, hemosiderin-laden macrophages, and inflammatory cells of the 3th–6th decades of life with rare occurrence in children (♂<♀).
- *IHC*: (+) CD68; (±) MSA (HHF35) and DES in mononuclear cells and (+) CD68, CD45, TRAP in MNGC.
- *TEM*: Cell heterogeneity including features of histiocyte-like cells, fibroblast-like cells, MNGCs, and foamy cells.
- *PGN*: Local recurrence but no malignant potential, although there is the vague concept of neoplasm because of occasional aneuploidy, clonal chromosomal abnormalities, and autonomous growth ability.

12.6.2.3 Pigmented Villonodular Synovitis or Diffuse-Type Giant Cell Tumor

DTGCT (aka pigmented villonodular synovitis, PVNS) is an autonomously growing extra-articular (usually knee), destructive (painful) villonodular hyperplasia with synovial mononuclear cells mixed with MNGCs, foamy cells, siderophages, and inflammatory cells and clonal abnormalities of young adults mostly female and may be classified as the soft tissue counterpart of pigmented villonodular synovitis.

- *GRO*: DTGCT or PVNS is a large (usually Ø > 5 cm), brown-yellow spongy tissue lesion, which is firm and nodular. Microscopically, it consists of diffuse expansive sheets of cells with infiltrative borders and variable cellularity and hyperplastic synovium with papillary projections composed of foamy histiocytes and hemosiderin-laden macrophages, large clefts, pseudoglandular or alveolar spaces lined by synovial cells, osteoclast-like MNGC with numerous nuclei (10–70), and epithelioid cells. Moreover, there are abundant collagen, giant hemosiderotic granules (2–3× diameter of RBC), and brisk mitotic activity (MI > 5/10 HPF). Malignant features include nodular and solid invasive growth, large cells with macronuclei, prominent nucleoli, necrosis, and atypical mitoses.
- *IHC*: (+) CD68 (stromal and giant cells), (±) CD31, CALR, DES, CD14, HAM56, CD45, and LYS (lysozyme); (−) S100, CD45/LCA, EMA, CKs, HMB45, CD34, SMA.
- *TEM*: Ultrastructural features of histiocytes and fibroblasts.
- *CMB*: *CSF1* overexpression or 1p13 (CSF1) rearrangements ± COL6A3 at 2q35.
- *DDX*: Hemosiderotic synovitis (intraarticular bleeding, no mononuclear or giant cell nodular proliferation, hemosiderin primarily in synovial lining cells).
- In the scientific literature, a malignant giant cell tumor of tendon sheath-diffuse type has been described. This lesion seems to be a rare occurrence of tumor showing coexisting prior benign giant cell tumor plus sarcomatous areas, but it is not a WHO-approved diagnosis.

12.6.2.4 Cellular Fibrous Histiocytoma

Plexiform Fibrohistiocytic Tumor

Dermal or subcutaneous, the proliferation of fibrohistiocytic cells growing in small nodules and elongated fascicles with multiple interconnections forming a characteristic plexiform pattern and accompanied by osteoclast-like MNGC and chronic inflammatory infiltrate of the upper extremities (mostly hands and wrists) of children and young adults.

Giant Cell Tumor of Soft Tissues
- *DEF*: Soft tissue (dermis or subcutis) counterpart of giant cell tumor of bone (aka soft tissue giant cell tumor of low-malignant potential) usually found in the extremities but also in the breast, mediastinum, groin, and surgical scars.
- *GRO*: The tumor is a painless growing fleshy, red-brown-gray nodule, which is gritty at periphery due to calcification. Microscopically, it is characterized by fibrous tissue containing siderophages (vascularized

stroma with blood lakes) separating nodules containing round/oval mononuclear cells and osteoclast-like MNGCs, prominent osteoclast-like giant cells and mononuclear cells, but no or mild atypia.
- *IHC*: (+) VIM, SMA, CD68 (MNGC), but also (+) ALP, OPG, RANKL, TRAIL, and TRAP. OPG (Osteoprotegerin, aka osteoclastogenesis inhibitory factor or OCIF or tumour necrosis factor receptor superfamily member 11B - TNFRSF11B), is a cytokine receptor of the TNF receptor superfamily encoded by the *TNFRSF11B* gene.
- *TEM*: Apart from foamy cells, histiocyte-like cells, fibroblast-like cells, and MNGCs are encountered.
- *CMB*: Guo et al. (2005) have described a 12-year-old girl with GCTST of the right leg that metastasized to the lung. G-banded chromosomal analysis from direct and short-term cultures of the primary neoplasm disclosed numerous telomeric associations involving multiple chromosomes.
- *DDX* includes:
  - GCTST-localized: peritendineous, hyalinized stroma, foamy histiocytes, hemosiderin-laden macrophages, but rare bony metaplasia
  - Plexiform fibrohistiocytic tumor: pediatric tumor with plexiform growth pattern and characteristic complex "tentacle-like extensions"
  - MFH-giant cell type: infiltrative, moderate-severe atypia, necrosis, and atypical mitoses
- *TRT*: It consists of complete excision with negative margins, but in 10% recurs (incomplete resection) and, occasionally, metastasizes.

12.6.3 Borderline Fibrous Histiocytoma

Three lesions may enter in this category of fibrous histiocytoma with uncertain malignant potential, including *dermatofibrosarcoma protuberans*, pigmented dermatofibrosarcoma (Bednar tumor), and atypical fibroxanthoma.

12.6.4 Malignant Fibrous Histiocytoma

- *DEF*: High-grade pleomorphic sarcoma with different subtypes that usually occurs in late adulthood (50–70 yrs) representing the second most common sarcoma of retroperitoneum but also occurs in the extremities and the bone (grade III). Apart from myxoid and hyalinized forms, the cells show evidence of histiocytoid differentiation: LYS, FXIIIa, AAT by IHC, and histiocyte-like cells with prominent lysosomes, Golgi, and "surface ruffles" by TEM.

  Storiform-Pleomorphic (70%): Swirling, cartwheel, storiform pattern of polygonal histiocytoid cells, often with bizarre prominent oval nuclei, giant cells with pleomorphism, spindled fibroblasts, myofibroblasts and small undifferentiated mesenchymal cells as well as slit-like vascular spaces. PGN: common metastases at diagnosis in half of the cases.

  Myxoid (Myxofibrosarcoma) (15%): Lose myxoid stroma for more than 50% of the tumor mass with interspersed cellular areas, extensive plexiform vascular network, giant cells focally (DDX: MLPS). PGN: 5-YSR, >50%.

  Giant Cell (10%): Spindled cell morphology with plenty of osteoclast-like giant cells arranged with polygonal cells in vague nodules.

  Inflammatory (5%): Intense inflammatory infiltrate accompanying classic MFH with neutrophils-*phagy* and aggressive course (DDX: sarcomatoid RCC, malakoplakia).

  Angiomatoid (<5%). It is the only variant of MFH with a marked predominance for adolescents and youth, although it is considered clinically and pathologically an utterly different lesion.

- *GRO*: A-MFH is well-circumscribed and located at the extremities.
- *CLM*: Variegated pattern of highly cellular areas with round cells and bland cytology and hemorrhagic cyst-like spaces and chronic inflammatory cells which may show a tendency to form lymphoid follicles. Angiomatoid

("MFH") fibrous histiocytoma (formerly called "angiomatoid" MFH) is a subcutaneous tumor with LMP of the extremities of children and young adults characterized by a multinodular proliferation of eosinophilic, histiocytoid, or myoid-appearing cells, endothelium-lacking intratumoral pseudoangiomatoid spaces, thick pseudocapsule, and marked chronic inflammatory infiltrate at the pericapsular location.

- *IHC*: (+) CALP, (±) CD68, DES, EMA, CD99; (±) SMA; (−) CD21/CD35, S100, HMB-45, CKs, CD34, CD31, FVIIIRA.
- *CMB*:
  – t(12;16)(q13;p11) ⇒ *ATF1-FUS* fusion gene
  – t(12;22)(q13;q12) ⇒ *ATF1-EWSR1* fusion gene (also seen in GI CCS)
  – t(2;22)(q33;q12) ⇒ *CREB1-EWSR1* fusion gene
- *DDX*: It includes:
  – BFH
  – *Aneurysmal MFH* (no thick pseudocapsule, no inflammation, and no significant pleomorphism)
  – *Angiosarcoma* ("endothelial tufting," +CD31, +CD34, −DES)
- *PGN*: 5-YSR, 99% with only 12% of recurrence and 5% metastasis.

12.7 Smooth Muscle Tumors

Smooth muscle tumors include leiomyoma (variants: cutaneous, genital, vascular or angioleiomyoma, and symplastic) and leiomyosarcoma.

12.7.1 Leiomyoma

Leiomyoma is a benign smooth muscle tumor with the highest occurrence in the female genital tract (95%).

12.7.2 EBV-Related Smooth Muscle Tumors

There are some new case reports and case series documenting the occurrence of smooth muscle tumors (SMT) in the pediatric and adult population suffering from immunodeficiency, mainly the acquired immunodeficiency syndrome (Pediatric AIDS/PAIDS) caused by the human immunodeficiency virus (HIV-1). SMTs including leiomyosarcoma are extremely rare in children and adolescents. This population, including the youth, may develop skeletal muscle tumors more often with rhabdomyosarcoma as the culprit of the list. Currently, SMTs appear to be the second most common type of neoplasm arising in children with AIDS. EBV-SMT also occurs in other conditions of immunodeficiency, including posttransplant immunosuppressed patients and children/adolescents with other causes of immunosuppression such as autoimmune disease and common variable immunodeficiency syndrome. Epstein-Barr virus (EBV) infection appears to be a necessary cofactor in the development of these tumors, and the report of EBV-related SMTs in immunocompromised patients was first reported in the early 1990s. Microscopically, tumors had smooth muscle features and are immunoreactive for muscle markers and show the presence of EBV by either immunohistochemistry, in situ hybridization, or real-time PCR. The MI and/or necrosis do not correlate well with clinical outcome (Fig. 12.9). Treatment includes resection primarily. Less often radiotherapy and chemotherapy may be offered. About the outcome, EBV-SMTs appear to have variable aggressiveness exhibiting a more favorable prognosis compared to conventional leiomyosarcoma.

12.7.3 Leiomyosarcoma

- *DEF*: Malignant smooth muscle tumors, which are most often in the extremities, arising from the wall of blood vessels, or retroperitoneal or mesenteric.
- *GRO*: Large tumor, soft with a tendency for hemorrhage, necrosis, and cystic degeneration.
- *CLM*: Fascicular growth pattern with bundles intersecting *at right angles*, unlike the "herringbone" pattern of infantile or adult fibrosarcoma, showing palisading of cigar-shaped, blunt-ended nuclei with possibly brisk MI and

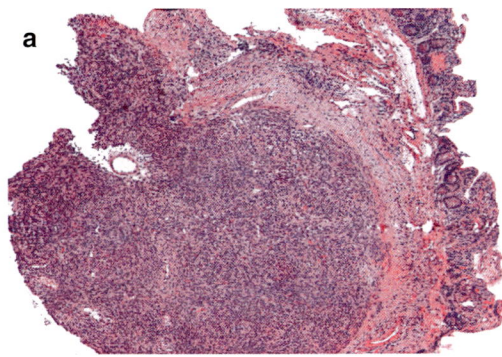
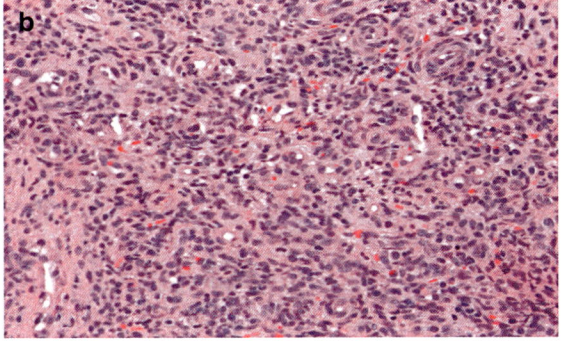

Fig. 12.9 Epstein Barr Virus (EBV)-related smooth muscle tumors. In these microphotographs are depicted a EBV-related smooth muscle tumor with smooth muscle cells with spindle cells with histologic features in keeping with a smooth muscle tumor showing elongated slender blunt-ended nuclei and eosinophilic cytoplasm (**a–b**). Occasionally, intralesional lymphocytes are seen (**b**)

containing antipodal perinuclear cytoplasmic vacuoles with possible nuclear indentation, which is not present in neural lesions.
- *PGN*: Poor factors include location (retroperitoneum and/or mesenteric) and size (Ø > 5 cm).

12.8 Pericytic Tumors

Pericytic tumors include glomus tumor and its variants, its malignant counterpart, and myopericytoma.

12.8.1 Glomus Tumor

Soft tissue neoplasm with a usually benign character that closely looks like the modified smooth muscle cells of the standard glomus body, which is a component of the cutaneous dermis layer the involved in body temperature regulation and made of an arteriovenous shunt surrounded by a capsule of connective tissue and numerously present in the fingers and toes.

- *CLI*: These tumors are mostly in ♀ (apart the multiple lesions → ♂), located in the subungual areas, painful due to abundant nerve fibers, but also to find in unusual locations, such as flexor arm and knee, GI tract, tympanic membrane, and glans penis and occurring in young adults (multiple glomus tumors arise 10–15 years earlier than single neoplasms). In the anterior compartment the superficial muscles include *the flexor carpi ulnaris, palmaris longus, flexor carpi radialis*, and *pronator teres*.
- *Glomangioma*: Glomus tumor with characteristics of a cavernous hemangioma.
- *Glomangiomatosis*: Benign diffuse angiomatosis is resembling angiomatosis with excess glomus cells and needs to be differentiated from angiomatosis with some infiltrative features.
  Familial glomangiomas have been observed in families with individuals harboring a variety of deletions in the *GLMN* gene on chromosome *1p22.1* and responsible for the Glomulin protein, an FKBP associated protein. FKBP (aka FK506 binding protein) is a group of proteins that have prolyl isomerase activity and are related to the cyclophilins in function. These diseases have an AD inheritance pattern with incomplete penetrance. Glomalin a phosphorylated protein that is a member of a Skp1-Cullin-F-box-like complex and its gene has shown interactions with *FKBP52*, *C-Met*, and *FKBP1A*. There is an association of subungual glomus tumor and *NF1* syndrome.
- *GRO*: Ø <1 cm, round, red-blue, firm nodule located under the nail.

- *CLM*: There are *nests of glomus cells surrounded by vasculature*, which can be from delicate to prominent with an obvious tendency to branch (glomangioma). Glomus cells have small, round, central nuclei with the inconspicuous nucleolus and amphophilic to eosinophilic cytoplasm.
- *IHC:* (+) SMA, IV-COLL (pericellular), VIM, H-CALD; ± CD34; (−) CKs, DES, S100. Caldesmon is a calmodulin binding protein. Similar to calponin, CALD tonically inhibits the ATPase activity of myosin in smooth muscle.
- *TEM:* Prominent external lamina, pinocytotic vesicles, and scattered intracytoplasmic microfilaments.
- *PGN:* Usually benign; excision curative.

- *Symplastic glomus tumor:* Glomus tumor with prominent nuclear atypia *as a degenerative feature* but no other features of malignancy, such as large size, deep location, high MI, and necrosis.
- *Glomus tumor of uncertain malignant potential:* Tumor with one atypical feature other than nuclear pleomorphism, but not fulfilling all criteria for malignancy (e.g., MI > 5/50 HPF and superficial location OR Ø > 2 cm only OR deep position only).

12.8.2 Glomangiosarcoma

Glomangiosarcoma is a malignant glomus tumor presenting with *ONE* of the three following features:

1. Tumor Ø > 2 cm AND subfascial/visceral location
2. Atypical mitotic figures
3. Marked nuclear atypia (G2-G3) AND MI > 5/50 HPF

A subdivision of the malignant glomus tumors into three categories based on their histology has been proposed:

1. Locally infiltrative glomus tumors (LIGHT)
2. Glomangiosarcomas arising in benign glomus tumors (GABG)
3. Glomangiosarcomas arising de novo (GADN)

12.8.3 Myopericytoma

- *DEF*: The myopericytoma (MPC) is a benign perivascular myoid tumor occurring in young to middle-aged adults, although pediatric cases have been recorded. It usually affects dermis, subcutis, or soft tissue of the distal extremities and is constituted of *"thin-walled BVs"* with prominent *"gaping"* and *multilayered concentric, perivascular array of the plump spindle to round cells with quite characteristic eosinophilic cytoplasm ("myoid features")*.
- *CLM*: Morphologically, MPC may be similar to infantile myofibroma/myofibromatosis and infantile hemangiopericytoma (Fig. 12.10).
- *IHC*: (+) SMA, h-Cald; but (±) DES, and (−) S100 and CKs.
- *DDX*: An important differential diagnosis is between myopericytoma vs. hemangiopericytoma or other allied lesions. First described by Zimmermann in 1923, pericytes are perivascular cells that are immediately adjacent to capillaries in many tissues. They are continuous with vascular smooth muscle cells of the media. About a half-century later, *myo*pericytes have been identified as transitional cell forms between pericytes and vascular smooth myofibers (Attwell et al. 2016). Histogenetically, perivascular myomas are probably a continuum, including infantile myofibroma/myofibromatosis (MF/MF), glomangiopericytoma (GPC), and myopericytoma (MPC). However, MF/MF is characterized by a biphasic pattern with a mature zone of fascicles of spindle cells resembling smooth myofibers and immature zones with a hemangiopericytic pattern. GPC is distinguished by a biphasic pattern with prominent large dilated blood vessels but lack of myoid nodules. Finally, MPC shows a concentric proliferation of perivascular myoid cells. MPC should also be kept distinct from angioleiomyoma, which is constituted by mature smooth myofiber bundles with abundant vascular channels. MPC is considered a subgroup of

12.8 Pericytic Tumors

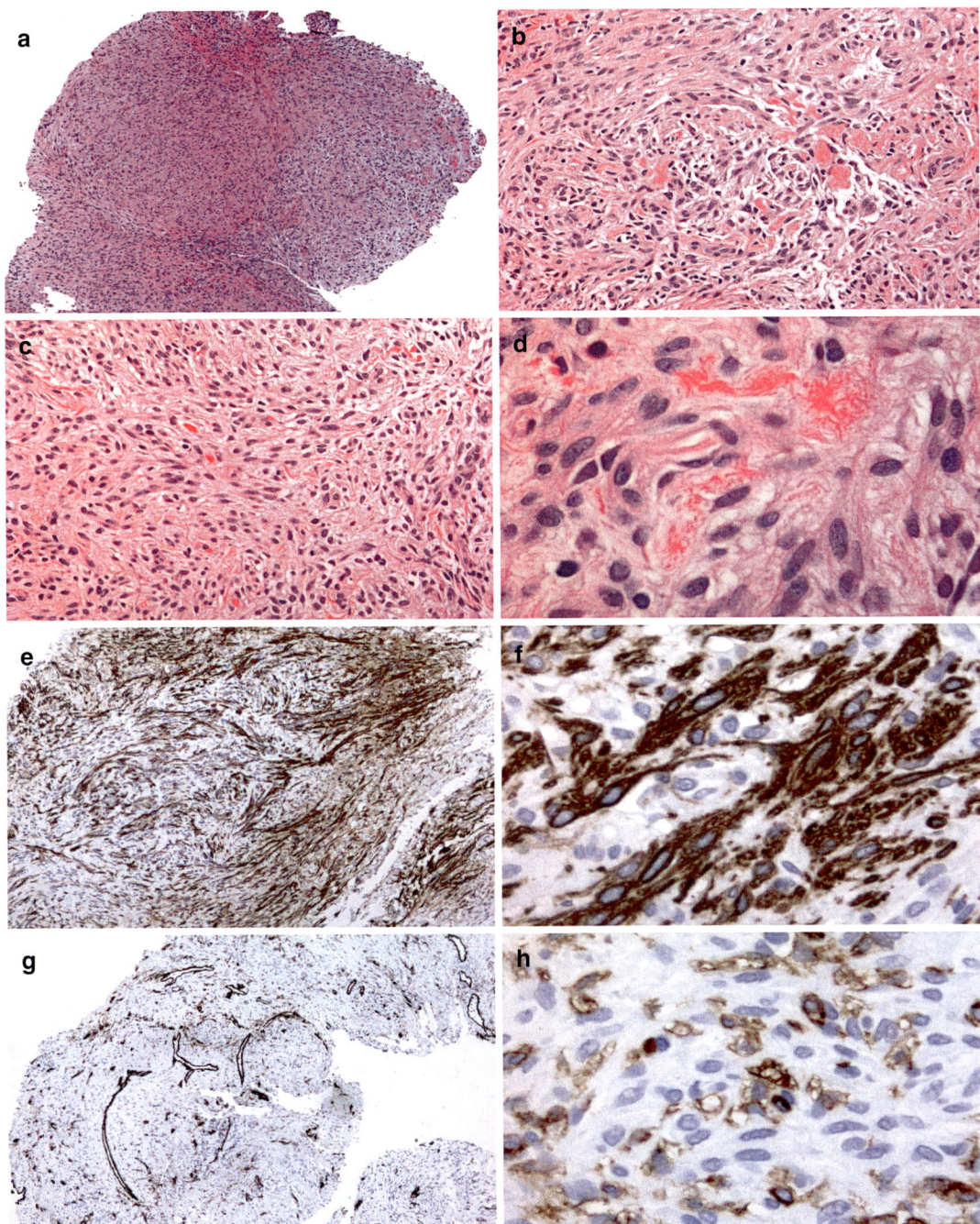

Fig. 12.10 Myopericytoma. Myopericytes are the proliferating cells of this tumor (a, H&E stain 50× original magnification), which are better highlighted at higher magnification (H&E stain and 200× as original magnification for both (b) and (c) microphotographs). In (d) there is a characteristic feature of the myopericytoma, which is the myoid nodules (H&E stain, 630× original magnification). The immunohistochemistry (e) through (h) confirms the histologic diagnosis of myopericytoma using anti-caldesmon (e, 100× original magnification; f, 630× original magnification) and anti-CD31 (g, 50× original magnification; h, 630× original magnification)

perivascular neoplasms with myoid differentiation expressing myogenic markers, i.e., SMA, SMM-HC, and h-Cald, and focally DES. Therefore, MPC shares features of both smooth muscle cells and glomus cells. Progenitor cells of the myopericytes are either myofibroblasts or pericytes. Myofibroblasts are spindle-shaped cells with an elongated nucleus and pale eosinophilic cytoplasm and are SMA+ and DES ± (weakly), while pericytes are contractile arborizing cells with multiple extensions encircling the vasculature and have a particular orientation to BV walls. TEM studies of myopericytes show a cell with a folded nucleus, prominence of RER, thin cytoplasmic filaments with dense focal bodies (myofilaments), subplasmalemmal densities, pinocytotic vesicles, and elongated elongation of cell processes; all features suggestive of differentiation toward myoid subtype. Pinocytosis (Gr. πίνειν 'to drink') is a crucial way of endocytosis. In this process, small particles suspended in extracellular fluid are transferred into the cell through an infolding of the cell membrane. It results in a suspension of the particles within a very small vesicle, which is located within the cell cytoplasm. Subsequently, these vesicles fuse with endosomes to hydrolyze the infolded particles. It is both a key constitutive process and a receptor-mediated process. Four categories of macropinocytosis, caveolae-mediated, clathrin-dependent, and dynamin/clathrin-independent pinocytosis have been delineated (Seto et al. 2002). A major task may differentiate MPC from MF because it depends on the predominant growth pattern. The tumor is considered MPC if the concentric perivascular grouping of plump spindle cells is seen, while the diagnosis of MF relies exquisitely on a zonation/biphasic appearance. MPC/MF spectrum needs to be extended to malignancies composed of small round cells arranged in hemangiopericytoma-like vascular pattern, including Ewing sarcoma (CD99+), poorly differentiated SS (biphasic, EMA+, AE1–3+, bcl-2+, CD99+), mesenchymal chondrosarcoma (cartilaginous islands and S100 (+)), phosphaturic mesenchymal tumor (variable histology, calcification, osteoclast-like giant cells), and angiomatoid MFH.

- *PGN*: Benign, because excision is usually curative.

> **Box 12.12 Skeletal Muscle Tumors**
> 12.9.1 Rhabdomyomatous Mesenchymal Hamartoma
> 12.9.2 Rhabdomyoma
> 12.9.3 Rhabdomyosarcoma (RMS)
>   12.9.3.1 Embryonal RMS
>   12.9.3.2 Botryoid RMS
>   12.9.3.3 Spindle-Cell RMS
>   12.9.3.4 Alveolar RMS
>   12.9.3.5 Pleomorphic RMS

12.9 Skeletal Muscle Tumors

Soft tissue with skeletal muscle differentiation or skeletal muscle tumors includes rhabdomyoma and variants as benign entities as well as rhabdomyosarcoma (and subtypes) as malignant counterparts. The malignant tumor with skeletal muscle differentiation or rhabdomyosarcoma is one of the most frequent and differential diagnoses in pediatric oncology and is one of the small blue round cell tumors (Box 12.12).

12.9.1 Rhabdomyomatous Mesenchymal Hamartoma (RMH)

- *DEF*: Rare benign overgrowth (not correctly tumor) of the skeletal muscle fibers of the skin potentially associated with disorders such as Goldenhar syndrome, amniotic band syndrome, and Delleman syndrome and harboring an excellent prognosis.
- *SYN*: Congenital midline hamartoma of the skin, cutaneous striated muscle hamartoma.
- *EPI*: Infants and children, ♂:♀ = 2:1, all ethnics.
- *CLI*: Painless and solitary and generally present on the midline of the head and neck region including on the nose or chin.
- *GRO*: Cupola-shaped or polyp-shaped lesion.
- *CLM*: It is a disordered and variegated collection of mature adipose tissue, mature skeletal muscle, mature adnexal elements, collagen, and nerve bundles (Fig. 12.11).

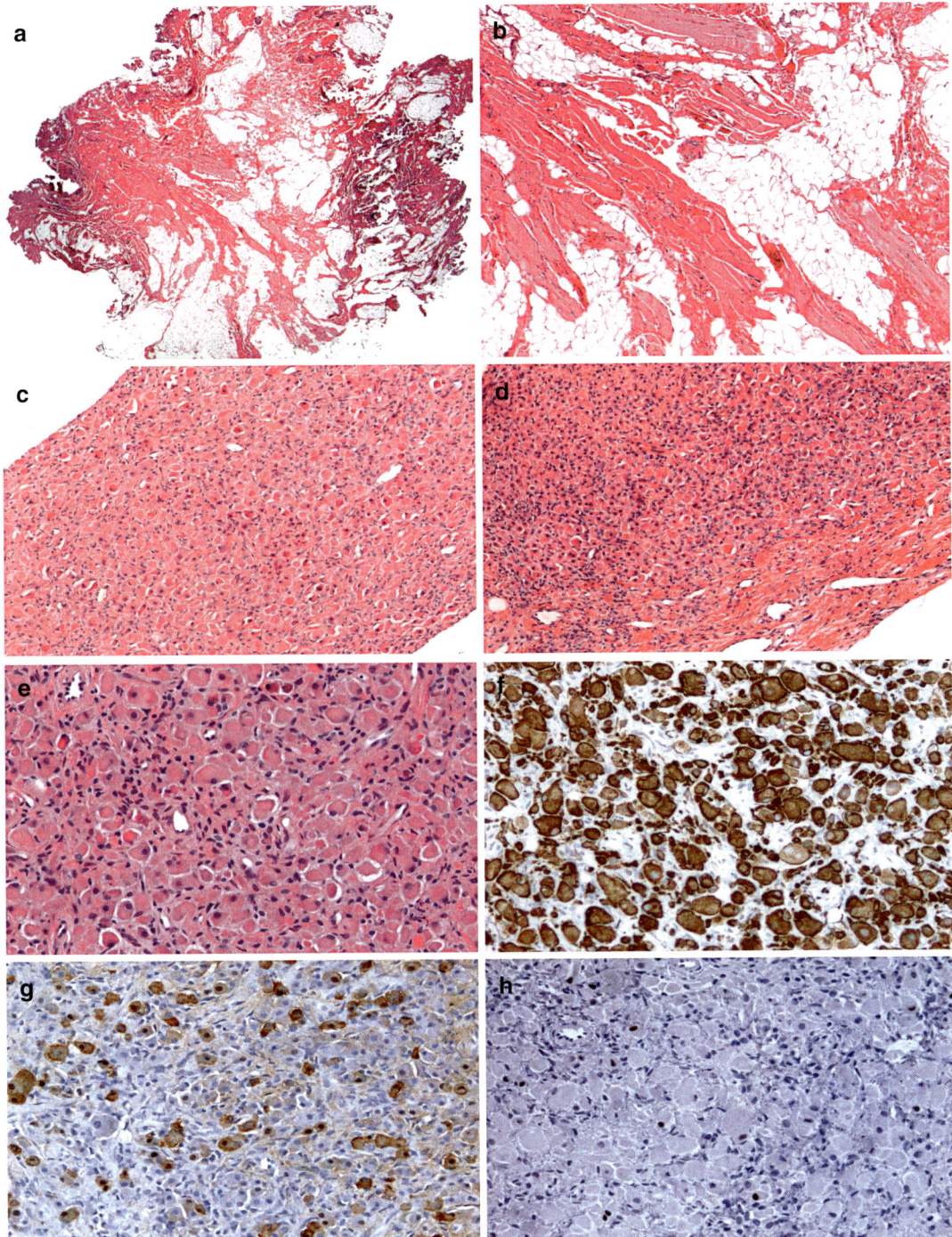

Fig. 12.11 Rhabdomyomatous Mesenchymal Hamartoma (RMH) & Rhabdomyoma. The first two upper microphotographs (**a–b**) show a RMH with its characteristic morphology (disordered and variegated collection of mature adipose tissue, mature skeletal muscle, mature adnexal elements, collagen and nerve bundles) (a, hematoxylin and eosin staining, x12.5 original magnification; b, hematoxylin and eosin staining, ×50 original magnification). The other six microphotographs (**c–h**) show a classic rhabdomyoma (see text for details) (c-e, hematoxylin and eosin staining; f-h, immunohistochemitry with antibodies against desmin (DES), S100, myogenin (MYF4); c-d, ×100 original magnification; e-h, ×200 original magnification)

- *DDX*: Fibroepithelial polyp, *nevus lipomatosus*, embryonal rhabdomyosarcoma, fibrous hamartoma of infancy, benign triton tumor, and fetal rhabdomyoma.
- *PGN*: Excellent, although cosmetic issues have been reported. Moreover, the presence of other genetic disorders should be kept in mind.
- *TRT*: Surgery if necessary from the cosmetic point of view, although other authors suggest a "watchful waiting" approach.

12.9.2 Rhabdomyoma

There is some controversy in considering this lesion a real tumor or a hamartoma. However, it is regarded as a benign tumor in the W.H.O. classification and is presented in this way in this chapter. Several authorities consider cardiac rhabdomyomas and, possibly, also the extracardiac types, hamartomatous lesions, but it may be tissue or organ-based. Cardiac rhabdomyomas have indeed a more hamartomatous character rather than neoplastic because do not show an autonomous growth. Extracardiac types are probably true neoplasms and are divided into two types, *fetal*-type rhabdomyoma and *adult*-type rhabdomyoma (Box 12.13) (Fig. 12.11). The fetal-type rhabdomyoma is also subclassified as classic and intermediate, and the genital location of the fetal-type is almost exclusively seen in middle-aged women.

DDX: It includes benign entities, such as ectopia, fibroepithelial polyp, papilloma (mesonephric or Müllerian), hibernoma, and granular cell tumor (S100+) as well as malignant neoplasms, such as rhabdomyosarcoma, leiomyosarcoma, lymphoma, and melanoma. These neoplasms are most easily identified by the infiltrative growth pattern in addition to the distinguishing and characteristic immunohistochemical phenotypes.

12.9.3 Rhabdomyosarcoma

- *DEF:* Malignant soft tissue tumor with skeletal muscle differentiation. There are several

Box 12.13 Rhabdomyoma

	Fetal-type rhabdomyoma	*Adult*-type rhabdomyoma
Age	Young children and middle-aged adults (♀)	Young- and middle-aged adults
Location	H&N, genital areas	H&N
CLM	Bundles of immature skeletal myofibers intersecting randomly and quite sharply separated by strands of collagen with scattered primitive mesenchymal cells. MI = 0, (−) necrosis, cambium layer, and/or infiltrative growth pattern	Large, well-differentiated, plump cells with cytoplasmic acidophilia (± focal cross-striations) and cytoplasmic clearing (glycogen and/or lipid)

Notes: *H&N*, head and neck; *MI*, mitotic index

classifications, including Horn and Enterline's classification, Intergroup Rhabdomyosarcoma (RMS) Studies I and II, Societe International pour l'Oncologie Pediatrique/International Society for Pediatric Oncology (SIOP), National Cancer Institute, and International Classification of Rhabdomyosarcoma. All of them partly overlap. The *International Classification of Rhabdomyosarcoma* (ICR) seems to be the most recent and the most reproducible way to classify RMS. This classification has been partly modified and introduced in the CAP protocols, which are diffusely used by pathologists not only in North America but worldwide. It includes the following entities:

- *Embryonal*, Botryoid – B-RMS
- Embryonal, not otherwise specified (NOS) – E-RMS (NOS)
- *Alveolar* – A-RMS
- Mixed embryonal and alveolar rhabdomyosarcoma (E-RMS/A-RMS)
- Rhabdoid rhabdomyosarcoma
- Sclerosing rhabdomyosarcoma
- Undifferentiated sarcoma
- Ectomesenchymoma
- Rhabdomyosarcoma, subtype indeterminate

- *EPI:* It is the most common soft tissue sarcoma in children younger than 15 years of age with an incidence of 8% among all cancers. It is one of the most common soft tissue sarcomas in the adolescence and youth. In fact, according to several epidemiological studies, more than half of these neoplasms occur during the first decade of life, while the second peak is located at 15–20 years of age. The early presenting RMS is often of embryonal type, the most common subtype among others. ♂:♀ = 1.3:1.0, although male predominance almost extinguishes toward adolescence and youth. Caucasians ethnic groups are more affected than Afro-American ethnic groups.

- *EPG:* RMS usually occurs at sites of embryonic tissue fusion (H&N, midline), but the postulated cell of origin is the subject of considerable debate for almost a century. It has been argued that there is little to suggest that RMS arises from skeletal muscle cells. RMS is, indeed, found in tissues and organs in which striated muscle tissue is not present (e.g., common bile duct of the extrahepatic biliary system, urinary bladder) or in topographic regions in which striated muscle is scant (e.g., nasal cavity, middle ear, and vagina). However, the developing muscular tissue at 5–8 weeks' gestation of the embryo appears to be similar to the poorly differentiated embryonal type of RMS (E-RMS) showing small, round-ovoid cells with hyperchromatic nuclei and indistinct or scant cytoplasm and absent or very few differentiated rhabdomyoblasts. The developing muscle at 9–15 weeks' gestation would seem the mirroring embryologic image of the well-differentiated E-RMS. It appears that genetic factors and familial cancer syndromes may play a key role, but further investigations are warranted. Congenital retinoblastoma, FAP, multiple lentigines syndrome, NF type I, Costello syndrome, BWS, and a spectrum of other congenital anomalies have been described in conjunction with RMS.

It is probably imperative to keep in mind that alveolar RMS is the most critical subtype to rule out in a tissue specimen received for diagnosis and tumors with *any degree of the alveolar pattern* on histology or cytology portend an *unfavorable prognosis*, regardless of extent. Botryoid and spindle-cell variants of embryonal RMS have, conversely, a better outcome. Pleomorphic areas with focal anaplasia may be found in any subtype of RMS but are more common in E-RMS. Two subtypes need to be considered, although rare in children and youth. These include the pleomorphic RMS, which was not included from the ICR given its extreme rarity in children, and the anaplastic variant of RMS, which has a controversial history and is also rare in children and youth.

- *CLI*: It seems that each subtype often has a site preference. Thus, H&N region (parameningeal, orbit, nasal cavity, nasopharynx, ear, paranasal sinuses, tongue, palate, lips, and soft tissues), which is the most common site harbors mostly ERMS. The genitourinary tract, the second most common site, also shows often ERMS, apart from the paratesticular region with the spindle-cell subtype, being the most frequent RMS subtype in this particular area. Other most common sites are the retroperitoneum and pelvis with mostly ERMS. The botryoid variant of ERMS that has a Greek etymology (Gr. βότρυς 'grapes') has the peculiar gross appearance of multiple nodules of soft, myxoid tissue growing into the lumen of a hollow viscus (*vide infra*). The botryoid RMS is found mainly in the urinary bladder, prostate, vulva, vagina, cervix, uterus, fallopian tube, perineal region, and common bile duct of the hepatobiliary tree. Finally, extremities are involved less commonly than adult soft tissue sarcomas, and in this localization, the alveolar RMS (ARMS) is the most common architectural pattern.

- *GRO*: RMS reflect some intrinsic features, including the degree of cellularity, the amount of collagenous or myxoid stroma, and the presence and/or extent of secondary changes (e.g., bleeding, necrosis, and ulceration) mostly.

- *CLM*: RMS may show both on light and electron microscopy features of skeletal muscle differentiation. Both techniques are usually instrumental in identifying the rhabdomyoblasts. By light microscopy, the elements of striated muscle differentiation show bright cytoplasmic eosinophilia (myoblastic differentiation) with or without cross-striations. Some authorities indicate cross-striations to be a *sine qua non* feature for the diagnosis of RMS, but they are found in about 1/3 of cases. Most importantly, it is the identification of "tadpole cells," i.e., cells with elongated tails of bright eosinophilic cytoplasm, and "strap cells," i.e., cells with cytoplasm in the shape of a ribbon or "strap" (Figs. 12.12, 12.13, 12.14, and 12.15) that play a major role in the diagnosis.

12.9.3.1 Embryonal RMS

Embryonal RMS (E-RMS) shows variable cellularity with pericellular reticulin staining and cells ranging from primitive stellate shape to elongated forms differently showing light or electron microscopy features of striated muscle differentiation. Optional features include myotube-like clusters, "smooth muscle-like spindle forms, myxoid areas and focal anaplasia (singly or clonal clusters). Anaplasia is defined using the Wilms tumor definition of simultaneous occurrence of cellular (nuclear) hyperchromasia, size enlargement (≥ 3 times the size of neighboring nuclei), and atypia (multipolar, polylobated or obvious atypical shape compared with the neighboring cells). Anaplasia may be found in any histologic subtype of RMS and up to 13% of all RMS and is subclassified as focal (group I) or diffuse (group II) anaplasia according to the extension. In patients with tumors in favorable sites, anaplasia seems to be more common, while is less common in younger children and patients with stage or clinical group II or III of disease. The presence of anaplasia, either focal or diffuse, negatively influences the failure-free survival rate and overall survival rates in patients with E-RMS. In contrast to Wilms tumor, anaplasia is not currently used in the stratification of patients. DDX of E-RMS NOS includes the botryoid and spindle cell variants, the solid variant of A-RMS, ectomesenchymoma, undifferentiated embryonal sarcoma of the liver, pleuro-pulmonary blastoma, Wilms tumor with skeletal muscle differentiation, fetal rhabdomyoma, giant cell tumor of tendon sheath (CD68+, Myogenin-), and pseudosarcomatous fibroepithelial polyps of the lower female genital tract. High MI (>15/50 HPF), marked hypercellularity, "cambium layer," and IHC phenotype play a crucial role in addressing the diagnosis correctly.

CMB: Chromosomal gains, including 2, 8, 11, 12, 13, and 20, but the most consistent finding is LOH for multiple closely linked loci at *11p15.5* resulting in activation of tumor suppressor genes. In the 11p15.5, there is imprinting of numerous genes. The allelic loss involves the maternal alleles in the 11p15.5 region. Some important genes involved are *IGF2* (growth factor), *H19* (noncoding RNA), and *CDKN1C* (kinase inhibitor). Other genes implicated are *FGFR1*, *FOXO1*, *PAX3*, and *YBX1*. FOXO1 (Forkhead Box O1) is a transcription factor and is the main target of insulin signaling. FOXO1 regulates metabolic homeostasis in response to oxidative stress. There is substantial evidence that FOXO1 plays important roles in the decision for a preadipocyte to commit to adipogenesis and is linked to fat mass and obesity-associated protein (FTO) (Yang et al. 2019; Peng et al. 2020).

ERMS variants are the botryoid and the spindle-cell variant.

12.9.3.2 Botryoid RMS

Botryoid RMS (B-RMS) is a favorable prognosis subtype characterized by grape-like, polypoid nodules located in sites next an epithelial surface and growing into the lumen of a hollow viscus (e.g., urinary bladder or vagina) and needs to be differentiated from clear cell adenocarcinoma associated or not with DES (Di-Ethyl-Stilbestrol) exposure (Al Jishi & Sergi, 2017). The diagnosis of B-RMS relies on the identification of ≥ 1 microscopic field (any magnification) showing a "cambium layer," which is a *condensed layer* of rhabdomyoblasts underlying a non-neoplastic epithelium. B-RMS seems to account for about 1–20 cases submitted to the Intergroup Rhabdomyosarcoma Study Group (IRSG).

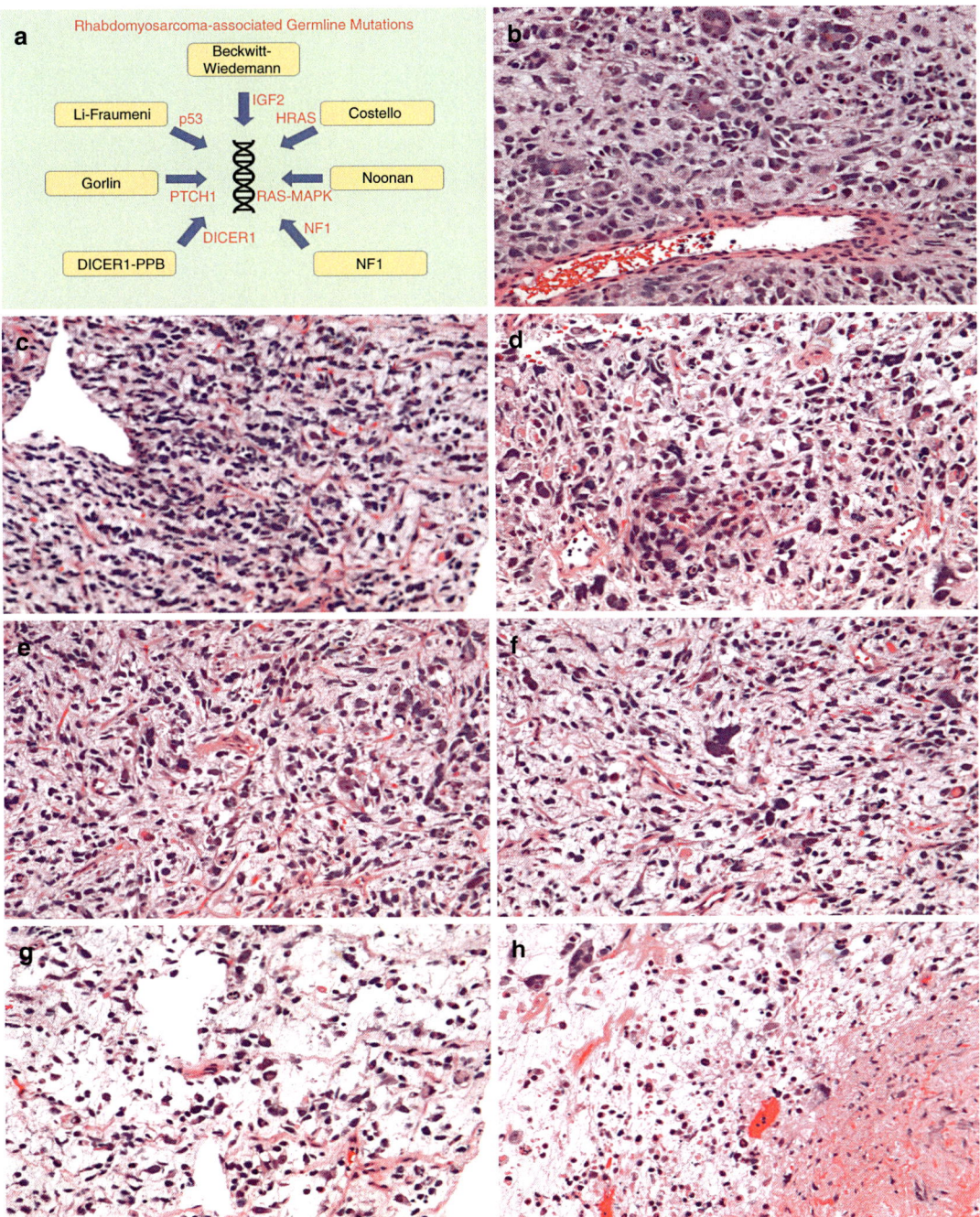

Fig. 12.12 Embryonal Rhabdomyosarcoma. Embryonal rhabdomyosarcoma-associated germline mutations are described in the cartoon in (**a**). The microphotographs (**b**) through (**h**) show an embryonal rhabdomyosarcoma with elements of striated muscle differentiation show bright cytoplasmic eosinophilia (myoblastic differentiation) with or without cross striations (b–h, hematoxylin and eosin staining, ×200, ×200, ×200, ×200, ×200, ×200, ×200 as original magnification, respectively)

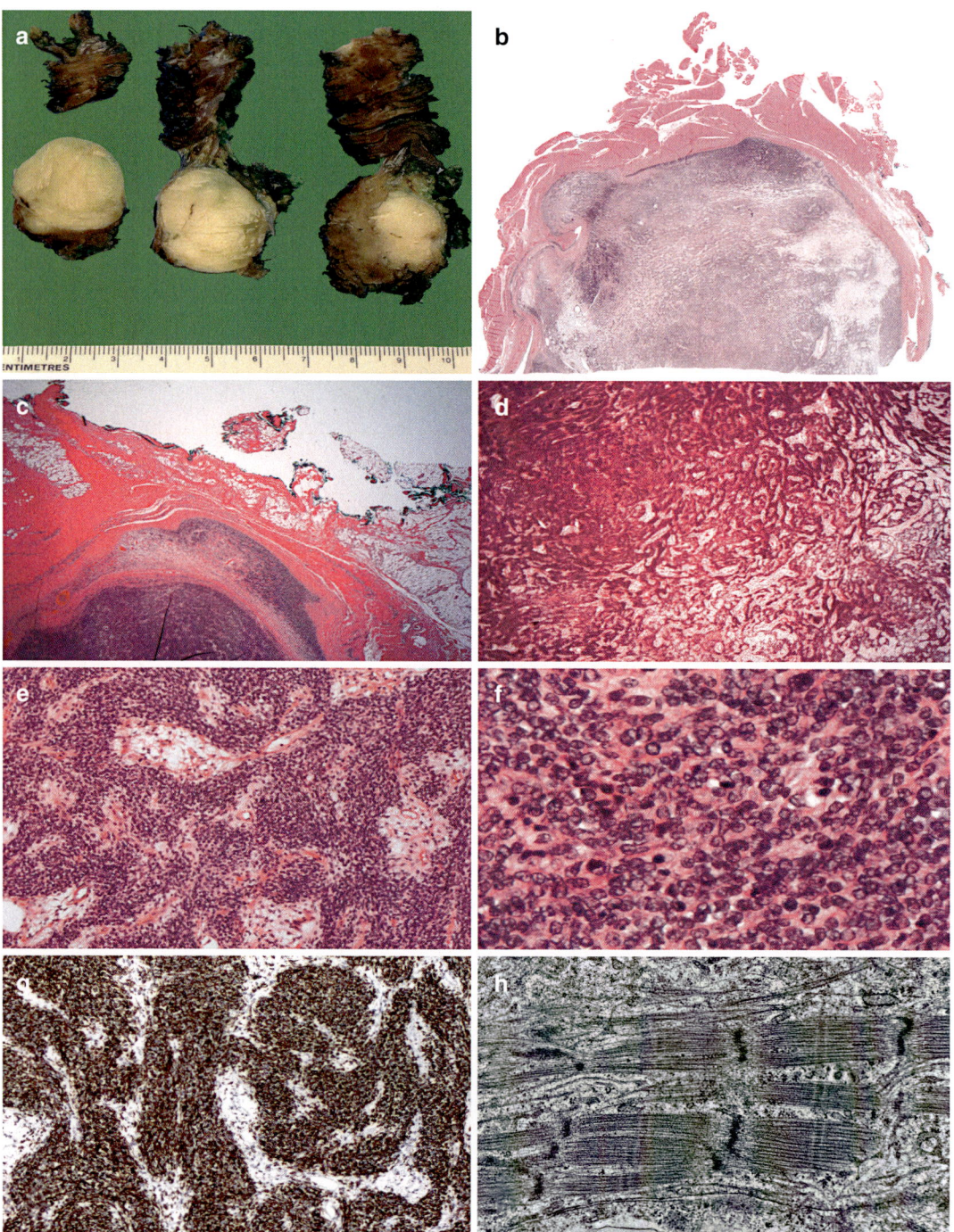

Fig. 12.13 Embryonal Rhabdomyosarcoma. Gross, histology, immunohistochemistry, and ultrastructure of an embryonal rhabdomyosarcoma with features identified in the previous microphotographs. The gross photograph (**a**) show a solid nodule not sharply delimited. The microphotographs (**b–f**) show the characteristic histology of this skeletal muscle differentiating tumor. The microphotograph (**g**) highlights the positivity of this tumor for dermin (anti-desmin immunohistochemistry, avidin-biotin complex). The electron microphotograph (**h**) shows actomyosin filaments, Z-bands, glycogen, and subcellular organelles, including mitochondria (b–f, hematoxylin and eosin staining, g, immunostaining with an antibody against desmin (DES), h, transmission electron microscopy; b, ×0.5 original magnification; c, ×12.5 original magnification; ×12.5 original magnification; ×100 original magnification; ×400 original magnification; g, ×100 original magnification; h, ×18000 original magnification)

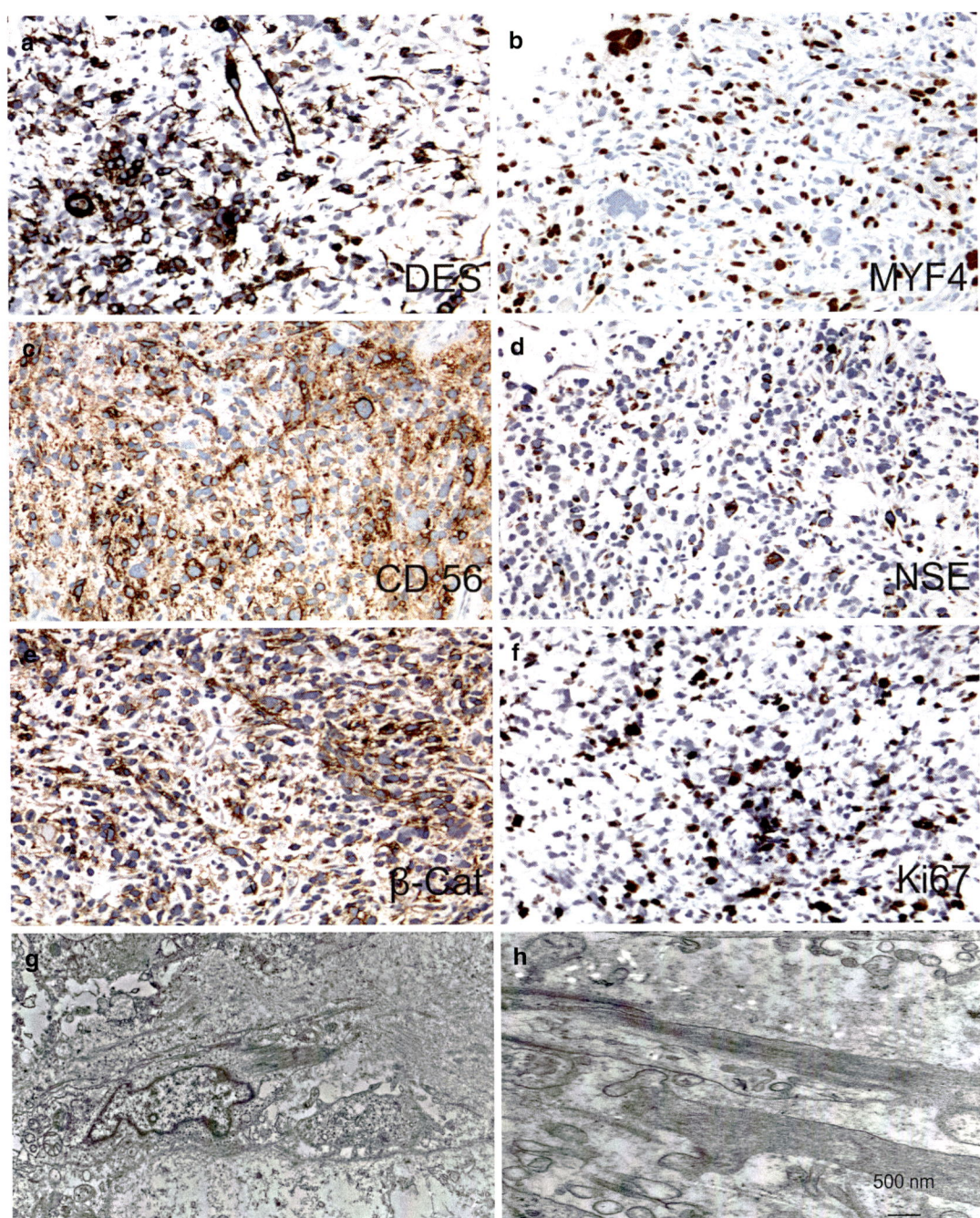

Fig. 12.14 Embryonal Rhabdomyosarcoma. Immunohistochemical panel for the diagnosis of rhabdomyosarcoma with the positivity for myogenic markers (**a**, **b**). The specific antibody used is apparent as abbreviation in the right lower corner of each microphotograph (a, anti-desmin immunostaining; b, anti-myogenin (MYF4) immunostaining). CD56, NSE, and beta-catenin may also be used, although their value is disputable (**c–e**) (c, anti-CD56 immunostaining; d, anti-neuron specific enolase (NSE) immunostaining; e, anti-Beta catenin immunostaining). The proliferation rate of these tumors is quite high (**f**) (f, anti-Ki67 (MIB1) immunostaining). All immuniistochemistry procedures have been performed using the Avidin-Biotin Complex (a-f, x200 as original magnification). The microphotographs (**g**) and (**h**) are transmission electron microscopy photographs and the scale bar in h) corresponds to 500 nanometers

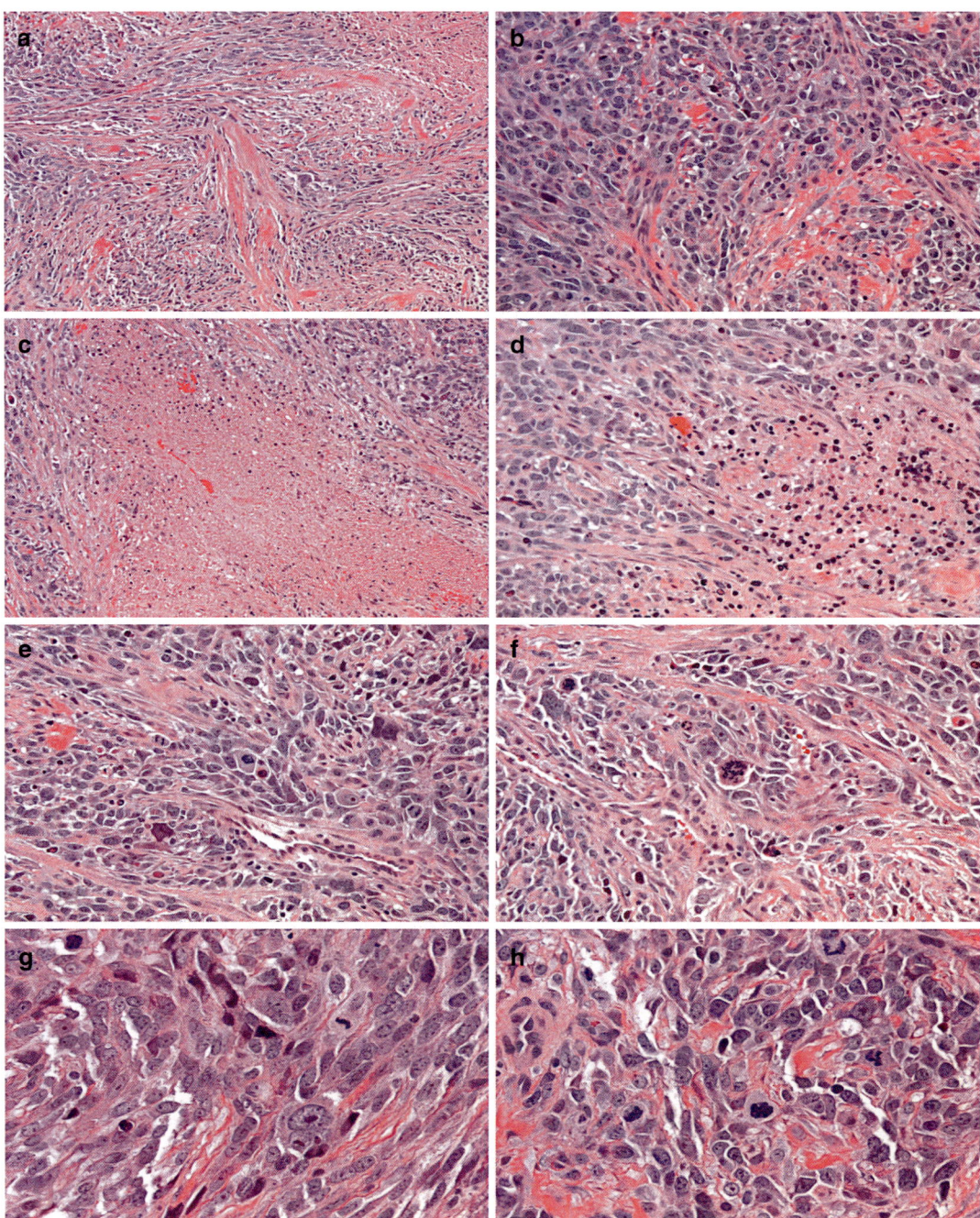

Fig. 12.15 Anaplastic Rhabdomyosarcoma. In this variant of rhabdomyosarcoma necrosis, fibrosis, and anaplasia of the tumor cells are presented (**a–h**). Tumor cells show multipolar mitotic figures, which are particularly evident in the microphotographs e) and f) (a–h, hematoxylin and eosin staining; ×100, ×200, ×100, ×200, ×200, ×200, ×400, ×400 as original magnifications)

12.9.3.3 Spindle-Cell RMS

Spindle-cell RMS (S-RMS) is a favorable prognosis subtype characterized by fascicles of spindled cells with eosinophilic, fibrillar cytoplasm showing a storiform arrangement pattern with a favorite site in the testicular/scrotal region. Spindled cells may be associated with variable collagen deposition, and two recognizable patterns are distinguished. The collagen-rich pattern of S-RMS shows spindle cells with variable myogenic differentiation embedded in a dense collagenous stroma, while the collagen-poor pattern of S-RMS shows spindle cells without significant collagen deposition and having a histologic similarity to a smooth muscle tumor. S-RMS is diagnosed when ≥80% of the neoplasm has elongated spindle cells in at least one of the two architectural patterns. DDX includes E-RMS NOS, fibrosarcoma, MFH, rhabdomyoma, leiomyoma, LMS, and nodular fasciitis. Leiomyoma and LMS are practically reserved to the middle-aged patients as well as MFH. Fibrosarcoma has a quite distinctive herringbone pattern and nodular fasciitis features a classic cell culture-like histology. Both IHC phenotype and ultrastructural features may be pivotal to avoid a misdiagnosis. The favorable outcome is linked to the body site. In fact, extratesticular and extraorbital S-RMS harbor a prognosis similar to E-RMS. S-RMS accounts for about 3% of cases submitted to the IRSG.

12.9.3.4 Alveolar RMS

Alveolar RMS (ARMS) is an unfavorable prognosis subtype of RMS with a 5-YSR of about 50% and characterized typically by reticulin positive, thin fibrous septa, and delimitating cell aggregates with central dyscohesion (the so-called alveolar pattern) (Fig. 12.16). Tumor cells are round with lymphocyte-like nuclei, coarse chromatin and one or more indistinct nucleoli, and a thin rim of eosinophilic cytoplasm. Optional features include focal LM features of striated muscle differentiation, which are often lacking, clear cells, and myxoid areas. ARMS variants are the solid and the mixed subtypes. The solid variant shows no dyscohesion and fibrous septa (retic/MG special stains highlight neoplastic aggregates), while the mixed ARMS/ERMS is a mixed pattern of both essential types with the requirement of more than 50% of the tumor being composed of the alveolar pattern. Extensive sampling is required in case of a solid pattern to identify typical alveolar foci that may help in the classification because the solid variant usually lacks a PAX fusion (*vide infra*). A helpful feature distinguishing the solid variant of A-RMS from a "solid" pattern of E-RMS NOS is the presence of diffuse myogenin expression, which is usually focal in the E-RMS. The presence of anaplasia seems not to affect the outcome in patients with ARMS. DDX includes ASPS that shows, conversely, nests of uniform cells with large eosinophilic cytoplasm and central dyscohesion, ±AB and DPAS, cytoplasmic basophilic crystalloids, +TFE3, ± DES, MSA, NSE, and S100, but – CGA, SYN, HMB45, Melan-A, CK, and EMA, ES/PNET, PD-NB or undifferentiated NB, DSRCT, PD-MSS, DLBCL, Burkitt lymphoma or lymphoblastic lymphoma (Box 12.14).

- *CMB*: The majority of ARMS cases contain one of two recurrent chromosomal translocations: *t(2;13)(q35;q14)* or *t(1;13)(p36;q14)*. The 2;13 and 1;13 translocations rearrange the *PAX3* or *PAX7* gene on chromosome 2 or 1 and the *FOXO1* *(FKHR)* gene on chromosome 13 to generate either *PAX3-FOXO1* or *PAX7-FOXO1* fusion genes. These fusion genes encode *fusion* transcription factors with a PAX3 or PAX7 DNA-binding domain and FOXO1 transactivation domain at nuclear localization. The consequence is an oncogenetic effect with dysregulation of the transcription. ARMS studies identified *PAX3-FOXO1*-positive in ~60% of cases, *PAX7-FOXO1*-positive in ~20%, and

---

Box 12.14 A-RMS Dichotomy and Subtyping
A-RMS (−) t(1;13) OR t(2;13) ⇒ 5% of all A-RMS
A-RMS (+) t(1;13) OR t(2;13) ⇒ 85% of all A-RMS
→ ~30% PAX7-FKHR (+) showing *higher* event-free survival (4-YSR: 75%)
→ ~70% PAX7-FKHR (+) showing *lower* event-free survival (4-YSR: 8%)

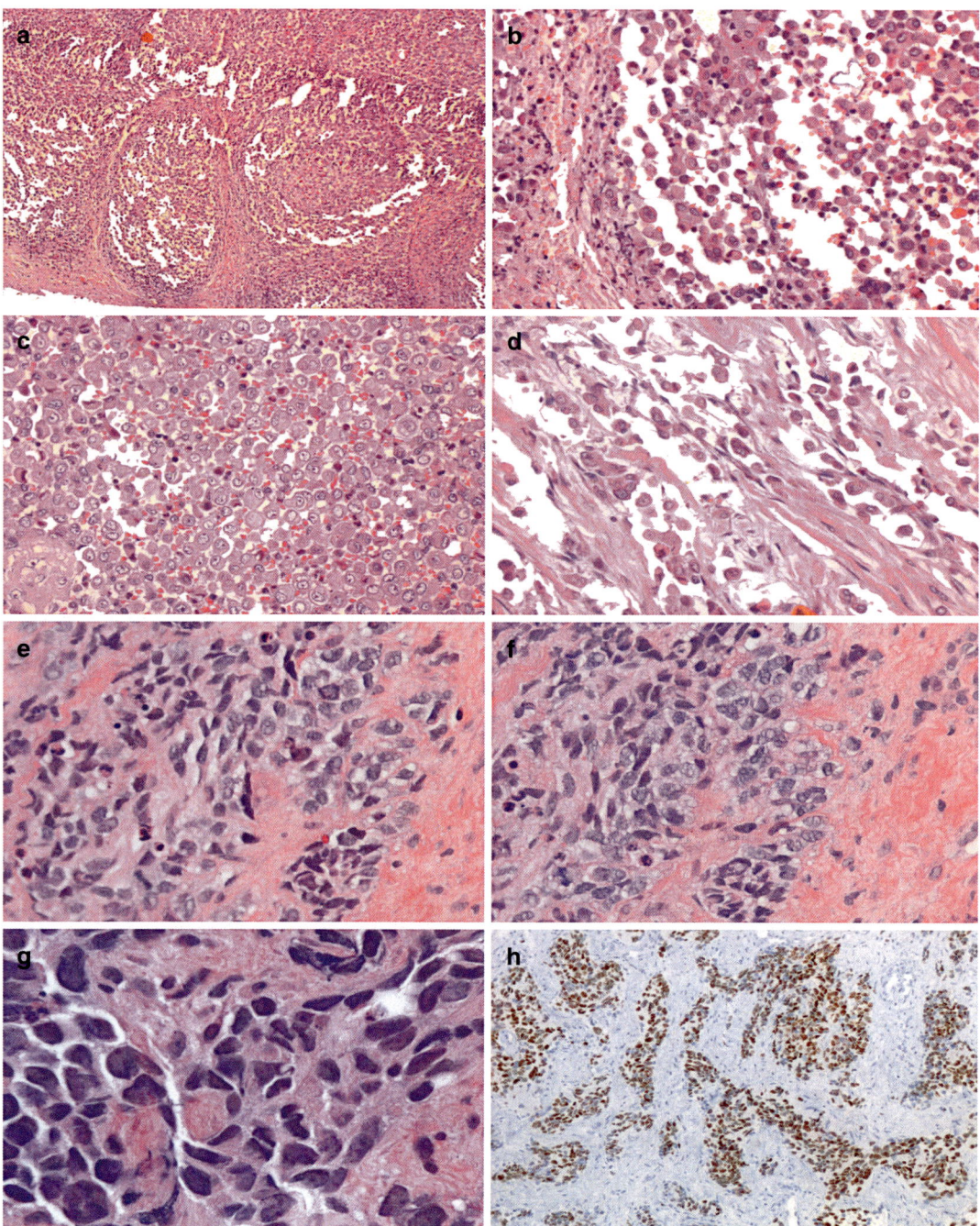

Fig. 12.16 Alveolar Rhabdomyosarcoma. Alveolar Rhabdomyosarcoma with characteristic pseudoalevolar structure filled or partially filled with tumor (rhabdomyosarcoma) cells (**a–h**). MYF4 immunostaining shows a diffuse positivity (**h**) (a–g, hematoxylin and eosin staining; a, ×40, b, ×400, c, ×400, d, ×400, e, ×400, f, ×400, g, ×630 as original magnification respectively; f, anti-myogenin immunostaining, Avidin-Biotin Complex, ×100 original magnification). Myogenin is one of the seven lineage-restricted transcription factors frequently used by soft tissue specialists (myoD1 [MYF3], myogenin [MYF4], FLI1, ERG, Brachyury, SOX10, and SATB2)

no fusion in ~20%. Gene amplification seems to be one of the underlying mechanisms to increase the expression level of the gene fusion in the neoplastic cells. Other genes implicated in ARMS are *DDX1*, *EWSR1*, *FGFR4*, *YBX1* other than *FOXO1* and *PAX3*. CGH studies have highlighted additional cytogenetic events in ARMS, including amplifications involving 12q13–15, 2p24, 1p36, 13q14, 2q34-qter, and 13q31 and numerical chromosomal gains, such as chromosomes 2, 12, and 20.

12.9.4 Pleomorphic RMS

Pleomorphic RMS is rare in childhood, while a tumor with diffuse pleomorphism ± unusual striations, but no alveolar or embryonal areas. PRMS tends to occur in individuals over 45 years at the limbs, especially the thigh. It is characterized by large, atypical tumor cells, often giant cells and difficult to distinguish from liposarcoma or MFH without immunohistochemistry. To make the diagnosis of PRMS, it is necessary to have at least one of two components, including ≥1 SM-IHC markers AND ME-evidence of polygonal rhabdomyoblasts.

The sclerosing variant of RMS is characterized by an unusual hyalinizing, a matrix-rich variant of RMS, which may resemble osteosarcoma, chondrosarcoma, or angiosarcoma. The sclerosing variant has been described in both adults and children, although is a rare entity.

- *IHC*: (+) Myogenin, MyoD1, DES, MSA, and myoglobin. CD99 (±) usually weak granular cytoplasm. Other markers include sarcomeric alpha-actin, myoglobin, fast, slow, and fetal myosin, creatine kinase (isoenzymes MM and BB), beta-enolase, Z-protein, titin, and vimentin.

MyoD1 is a member of the family of myogenic regulatory genes, which includes *myf-4-herculin/myf-6* and *myf-5*. MYO-D1 act as a crucial point for the initiation of skeletal muscle differentiation acting binding to enhancer sequences of muscle-specific genes. Anti-myogenic is better than anti-MYO-D1 because it does not have nonspecific cytoplasmic immunoreactivity, which is sometimes seen with anti-MyoD1. Both myogenin and MYO-D1 are expressed to a higher degree in ARMS than ERMS.

- *TEM*: Hexagonal arrays of thick and thin myofilaments and ribosomal-myosin complexes.
- *DDX*: It includes other SRBCTs, if the tumor is poorly differentiated, muscle infiltration of small round blue cell tumors from a variety of origin, heterologous elements of some other tumors, mixed mesenchymoma, characterized by any combination of ≥2 of osteosarcoma, RMS, or liposarcoma without MPNST, but regardless of any undifferentiated components, including MFH or fibrosarcoma, malignant triton tumor (MTT), or MPNST with admixed RMS elements. The IRSG includes some RMS variants that need to be taken into consideration. These include RMS with rhabdoid features or *rhabdoid RMS* (R-RMS), *sclerosing rhabdomyosarcoma, ectomesenchymoma,* and *undifferentiated sarcoma*. RMS with rhabdoid features shows tumor cells with large copious eosinophilic cytoplasm and intermediate filament globular inclusions similar to the inclusions observed in the malignant rhabdoid tumor (MRT). RMS with rhabdoid features has tumor cells with coarse nuclear chromatin pattern (not vesicular as seen in MRT), inclusions with a characteristic phenotype positive for VIM, DES and peri-inclusion phenotype positive for MSA, P-DES, and nuclear positivity for Myogenin and INI-1, which are negative in MRT. Sclerosing RMS (Scl-RMS) is a subtype of RMS showing a dense hyalinizing collagenous matrix with round tumor cells architecturally forming small nests, single-file rows, and pseudovascular alveolar structures, focal positivity for P-DES and Myogenin in contrast to a diffuse uniform positivity for MyoD1. Scl-RMS has a morphologic overlap with other soft tissue tumors, including sclerosing epithelioid fibrosarcoma, osteosarcoma, and angiosarcoma and may become a challenge for some infiltrating poorly differentiated carcinomas. Ectomesenchymoma is a composite RMS constituted by an RMS component, usu-

ally embryonal, and a neuroblastic component. The RMS component of the ectomesenchymoma is used for the treatment, and age, sex, site distribution, and prognosis are similar to E-RMS. The undifferentiated sarcoma is defined by the absence of any evidence of differentiation and is, historically, included in the RMS classification system and, currently, treated on non-RMS soft tissue tumor regimens in COG protocols.

Current tissue sampling is paramount. In fact, the priority should be given to formalin processing for morphologic evaluation. Individual studies, including RT-PCR and other molecular biology investigations, although critical to the molecular workup of rhabdomyosarcoma, may, at least partially, be performed on FFPE tissue. In case there is enough tissue for morphologic evaluation, and this is not jeopardized, at least 100 mg of viable snap-frozen tissue as the second priority for workup should be set. Translocations may be detected using RT-PCR or FISH on touch preparations from the frozen tissue as well. The procedures that allow tissue to be taken from the patient and sent to the laboratory for processing and evaluation should be discussed at the tumor board with the pathologist and oncologist and include incisional biopsy, through cut, open surgical, excisional biopsy (local, radical, compartmentectomy), amputation, (different types), and other surgical unconventional procedures. Following the type of process, the specimen should include laterality (right, left, midline, indeterminate, not specified), site (urinary bladder, prostate, cranial para-meningeal, extremity, genitourinary-not bladder/prostate, head and neck-excluding para-meningeal, orbit, trunk, retroperitoneum, etc. as well as not specified), greatest dimension (Ø), and additional dimensions in cm, depth (dermal, subcutaneous, subfascial, intramuscular, intra-abdominal, retroperitoneal, intracranial, organ-based, or other/not assessable), the histology type, the presence or lack of anaplasia (focal, diffuse, indeterminate, or not assessable), resection margins with the distance of sarcoma from closest margin in cm or mm if an involvement is identified, the number of lymph nodes sampled, and if lymph nodal metastasis is present. It is generally considered "wide" a resection margin of 2–3 cm of distance from the tumor, but for non-RMS soft tissue sarcomas, 1–2 cm may be adequate for low-grade tumors, while >5 cm may be requested for high-grade tumors. Relevant medical history may be provided to the pathologist and includes any previous therapy, family history of malignancy, and the presence of congenital anomalies. In particular, the assessment of the percentage of necrotic and viable tumor post-therapy is essential at the time of the discussion of the patient at the tumor board. Two additional factors may include extreme cytodifferentiation and the lack or presence of nuclear pleomorphism. It is considered that a genetic predisposition is present up to 1/3 of children with soft tissue sarcomas, and some syndromes may need to be excluded if the tumor occurs in a patient younger than 2 years of age. These syndromes include Li-Fraumeni syndrome, basal cell nevus syndrome, neurofibromatosis, and pleuropulmonary blastoma syndrome. Some congenital anomalies have been described in patients with rhabdomyosarcoma, including CNS anomalies, GU anomalies, GIT anomalies, and CVS anomalies. The tumors with heterologous components of rhabdomyoblastic differentiation are:

- *Epithelial Neoplasms*
  Carcinosarcoma (e.g., breast, urinary bladder), MMMT (uterus, cervix, ovary), thymoma
- *Epithelial-Mesenchymal Neoplasms*
  Wilms tumor
  Hepatoblastoma
  Pleuropulmonary blastoma
- *Germ Cell Tumors or Sex Cord-Stromal Tumors*
  Seminoma/dysgerminoma
  Teratoma
  Sertoli-Leydig cell tumor
- *Mesenchymal Neoplasms (Without Epithelial Component)*
  Dedifferentiated chondrosarcoma
  Dedifferentiated liposarcoma
  Malignant mesenchymoma

- *Neoplasms of Neuroectodermal Derivation*
  Malignant Schwannoma (MPNST)/MTT
  Medulloblastoma
  Medulloepithelioma
  Ectomesenchymoma
  Congenital pigmented nevus or giant nevus

- *PGN*: It is paramount to recall favorable and unfavorable factors for RMS because these are reviewed at the tumor boards during the presentation and discussion of the pathological findings.

Prognostic factors in RMS (favorable +/ unfavorable −) include:

- *Epidemiology-Based/Clinical Variables*: Age at diagnosis (infants/children +) (adults −), anatomic site (orbita, GUT +) (H&N-nonorbital, paraspinal, abdomen, biliary tract, retroperitoneum, perineum, extremities −), distant metastasis at time of diagnosis (PET scan) (if no metastasis +)
- *Surgery/Pathology Team-Linked Variables*: Completeness of resection
- *Pathologist-Linked Variables*: Tumor Ø (≤5 cm +) (>5 cm −), assessment of local invasiveness, histopathology subtype (botryoid, spindle cell +) (alveolar, pleomorphic −), and regional node involvement

Also, unfavorable prognostic factors are a local recurrence, local recurrence during therapy, non-resectability, and diploid content.

According to the histologic subtype, the prognosis would be different. Botryoid and spindle-cell variants harbor a better prognosis than alveolar and undifferentiated sarcoma. The embryonal RMS has an intermediate prognosis.

According to the Intergroup Rhabdomyosarcoma Study (Post-Surgery Clinical Grouping System), four groups are distinct:

- *Group I*
  A: Localized tumor, restricted to the site of origin, wholly resected
  B: Localized tumor, infiltrating beyond the place of birth, wholly resected
- *Group II*
  A: Localized tumor, gross total resection, but microscopic. residuals
  B: Locally large tumor (+) reg. LNs, but completely resected
  C: Locally large tumor (+) reg. LNs, gross total resection, but microscopic residuals
- *Group III*
  A: Localized/locally broad tumor, gross residual after biopsy only
  B: Localized/zonal extensive tumor, gross residual after significant resection (≥50% debulking)
- *Group IV*
  Any size of the primary tumor (±) reg. LNs, but with distant metastases

12.9.4.1 Late Effects of Childhood Cancer Therapy

Surgery, radiation, and multimodal chemotherapy are involved in modern childhood cancer therapy. There is a debate if evidence is sounding or not about an increased risk for offspring of survivors in the absence of genetic syndromes. Some sequelae of the treatment are known and debated. They include poor school performance following CNS irradiation, weak growth and puberty following irradiation of the pituitary gland, and reduced fertility following chemotherapy. It is estimated that approximately 3% of long-term survivors of children with cancer can develop a second tumor within 20 years. Risk factors include familial syndromes, use of alkylating agents, and radiotherapy. Alkylating agents are particularly active in the resting phase of the cell (G_0). Alkylating agents are cell-cycle non-specific and include mustard gas derivatives (mechlorethamine, cyclophosphamide, chlorambucil, melphalan, and ifosfamide), ethylenimines (thiotepa and hexamethylmelamine), alkylsulfonates (busulfan), hydrazines and triazines (altretamine, procarbazine, dacarbazine, and temozolomide), nitrosureas (carmustine, lomustine and streptozocin), and metal salts (carboplatin, cisplatin, and oxaliplatin). Of particular importance is to mention that nitrosureas are unique because they can cross the blood-brain barrier. Box 12.15 illustrates some pearls and pitfalls during the diagnostics of RMS.

Box 12.15 Pearls and Pitfalls

- Degenerated rhabdomyoblasts with a glassy or deeply eosinophilic, hyalinized cytoplasm, and pyknotic nuclei are a frequent feature of ERMS and should not be interpreted as PRMS.
- Cross-striations are rare in the round-ovoid cells of the least differentiated ERMS but are more frequent in the most differentiated areas of ERMS and ERMS with a more prominent spindle-cell component.
- Cross-striations of RMS cells should be differentiated from those seen in residual or entrapped muscle cells by their more accentuated irregularity and incomplete transformation of the cells.
- A prominent degree of cellular pleomorphism or anaplasia in RMS would make the diagnosis difficult with adult PRMS, and the discovery of cross-striations and areas of more typical ERMS would make an asset to make a proper differential diagnosis.
- It is essential to keep separate the solid form of ARMS from the undifferentiated form of ERMS, as the former harbors a less favorable prognosis.
- Clear cell RMS is a rare variant with abundant glycogen-rich cytoplasm, which vaguely looks like clear cell carcinoma or clear cell malignant melanoma, and IHC and MB studies are an asset to make the right diagnosis.
- Polyclonal DES is more sensitive than monoclonal DES in detecting RMS tumor cells and may be the choice to associate to the monoclonal antibodies against the intranuclear myogenic transcription factors MyoD1 and myogenin.

The malignant triton tumor (MTT) or malignant nerve sheath tumor (MPNST) (malignant schwannoma) with rhabdomyoblastic differentiation is a very peculiar tumor with specimens dominated by a sheet-like proliferation of relatively mature rhabdomyoblastic cells with somewhat irregular vesicular nuclei. These tumor cells are smaller than those of an adult rhabdomyoma with admixed smaller rounded cells. Such prominent degree of cytodifferentiation as seen in a triton tumor is never present in a rhabdomyosarcoma (RMS). At places, it is essential to appreciate that there are, usually, small areas in which the tumor cells are basophilic with a high N/C ratio and somewhat elongated nuclei. Although IHC helps in identifying extensive positivity for DES as well as myogenin (MYF-4), i.e., the areas with rhabdomyoblastic differentiation, the hypercellular more spindled areas in the formal excision remain negative for these antigens. Conversely, there is, typically, positivity for both S100 protein and GFAP, which should be present in the hypercellular basophilic areas and also in the subset of the smaller cells scattered between the rhabdomyoblasts.

In RMS the preservation and storage of some tumoral tissue are useful for both diagnosis and prognosis. Tumor tissue may be kept frozen from the tissue used for intraoperative frozen section consultation, which is usually performed to determine sample adequacy and viability or from the tissue resected at the time of surgery and sent fresh to the pathologist. In the first case, at least three tissue slides may be kept at -80 Celsius or lower, while in the second event, ≥ 100 mg of viable tissue may be snap-frozen in iced frozen isopentane and held at -80 Celsius or lower. Isopentane or 2-methylbutane has a high thermal conductivity and does not form a vapor halo. These two qualities are key. Isopentane chilled with liquid nitrogen freezes tissues more effectively and uniformly than putting the tissues samples directly into liquid nitrogen. Molecular biology approach to detect translocations includes RT-PCR or FISH on both frozen or FFPE tumor tissue. Alternatively, air-dried or alcohol-fixed touch preparations may be used, if no tumor tissue is available. To remember is that open incisional biopsy is higher in value than core needle biopsy. Resection tumor tissue specimens may be intralesional, marginal, comprehensive, or radical in extent. Intralesional resections have gross or microscopic residual tumor identifiable at surgical margins; marginal resections show inflammation-involved margin surrounding the tumor, while wide/radical resection involves show surgical margins extending through healthy tissue, which are usually external to the anatomic compartment containing the tumoral tissue.

Diagnosis of botryoid RMS requires ≥1 microscopic field showing a cambium layer, which is a condensed layer of rhabdomyoblasts or spindle cells with focal rhabdomyoblastic differentiation (IHC and/or EM) underlying an intact epithelium (*vide supra*).

Spindle-cell RMS, another good outcome harboring subtype of RMS, requires ≥80% of tumor demonstrating spindle-cell morphology in one of two recognizable patterns (collagen-poor or collagen-rich).

12.10 Vascular Tumors

Vascular tumors include *hemangioma, epithelioid hemangioma, angiomatosis, lymphangioma*, while intermediate vascular tumors include *hemangioendothelioma* (and subtypes), *papillary intralymphatic angioendothelioma*, and *Kaposi sarcoma*. Malignant counterparts are *epithelioid hemangioendothelioma* and *angiosarcoma*.

- *IHC markers of endothelial origin or differentiation*: CD31, CD34, D2-40 but also FVIII and von Willebrand factor. Friend leukemia integration one transcription factor or Fli-1 protein is a member of the ETS family of DNA-binding transcription factors. Fli-1 is involved in cellular proliferation and tumorigenesis. Almost all ES/PNET show t(11;22)(q24;q12) and formation of the *EWS-Fli-1* gene fusion transcript. CD31, CD34, and von Willebrand factor. FLI-1 is a marker for benign or malignant vascular tumors. It has a sensitivity and specificity similar to CD31, CD34, and von Willebrand factor in identifying endothelium and has unambiguous nuclear staining that lacks the cytoplasmic and membranous staining artifacts of other markers as a result of endogenous peroxidases or biotin.

CD141 or thrombomodulin is an integral membrane protein expressed on the surface of endothelial cells and serves as a cofactor for thrombin reducing blood coagulation and is also shown on the human mesothelial cell, monocyte, and a dendritic cell subset. In humans, thrombomodulin is encoded by the *THBD* gene, and gene mutations have also been reported to be associated with the atypical hemolytic-uremic syndrome (aHUS).

D2-40 is a marker of the lymphatic epithelium. It is a monoclonal antibody directed to a 40,000 O-linked sialoglycoprotein that reacts with a fixation-resistant epitope on lymphatic endothelium.

GLUT1 or glucose transporter 1 is a protein that in humans is encoded by the *SLC2A1* gene. GLUT1 does facilitate the transport of glucose across the plasma membranes of mammalian cells. Mutations in the GLUT1 gene are responsible for GLUT1 deficiency, or De Vivo disease, which is an AD inherited disease characterized by hypoglycorrhachia, i.e., a low CSF glucose concentration. GLUT1 is also a receptor used by a virus, the HTLV (Human T-Lymphotropic Virus), to gain entry into target cells. GLUT1 has been observed explicitly in infantile hemangioma allowing the differentiation of capillary hemangioma from pyogenic granuloma.

12.10.1 Benign Vascular Tumors

The W.H.O. classifies in the benign category hemangiomas and their variants as well as lymphangiomas and the condition called angiomatosis. Pyogenic granuloma and capillary hemangioma are the most frequent variants (Figs. 12.17 and 12.18). Pyogenic granuloma is a pedunculated pseudotumor composed of granulation type tissue separated by bands of connective tissue with a collarette appearance at the edges of the lesion and a lobular arrangement of capillaries, similar to capillary hemangioma. The lobules of endothelial cells may have indistinct to prominent lumina with occasional plump endothelial cellular appearance. Pyogenic granuloma is negative for GLUT-1, while capillary hemangioma is positive for this marker (*vide infra*).

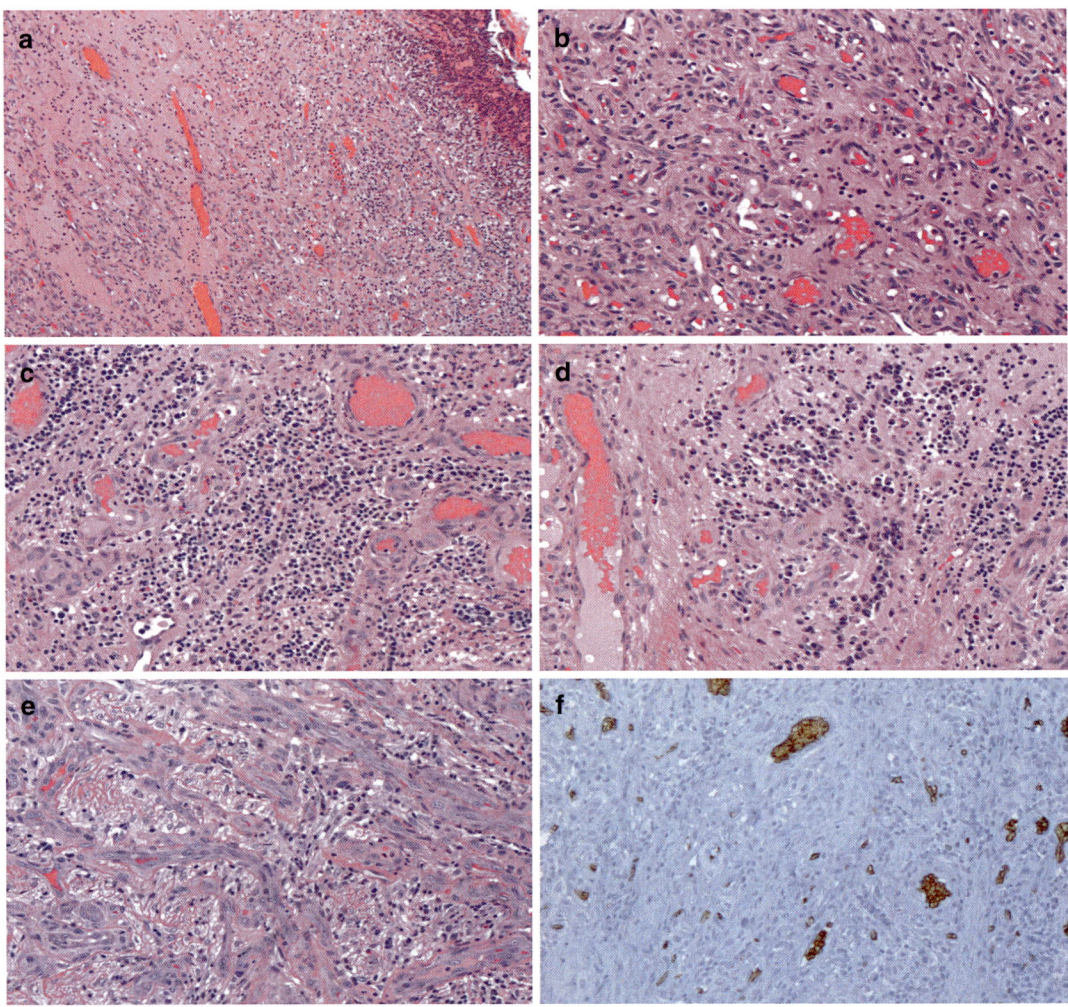

Fig. 12.17 Pyogenic Granuloma. The microphotographs show a pyogenic granuloma with "collarette" and proliferation of capillaries as well as some plump epithelioid cells (**a–d**). In (**e**) the stromal component is quite prominent (a–e, ×100, ×200, ×200, ×200, ×200 as original magnification, respectively). In (**f**) there is positivity for GLUT-1 only in the erythrocytes of preformed blood vessels by immunohistochemistry (Anti-GLUT-1, avidin-biotin complex, 200× original magnification)

12.10.1.1 Hemangioma

- *DEF*: Common benign vascular tumor, particularly often observed in childhood localized either superficially (H&N) or internally (liver) with very rarely malignant transformation. If hemangiomas involve large segments of the body, the term *angiomatosis* is probably appropriated.
- *GRO*: Variably sized vascular channels ± hemorrhage and thrombosis.
- *CLM*: An increased number of BVs, which are lined by a single layer of non-atypical endothelial cells.

12.10.1.2 Hemangioma Variants

- *Capillary*: The skin/subcutaneous tissue (lips, oral cavity)/viscera located hemangioma constituted of firmly packed vascular spaces containing some blood or thrombosed or organized lumina and thin fibrous septa with a tendency to regress by age 7 in ¾ of the cases (Fig. 12.18) (+) GLUT1
- *Cavernous*: The skin/subcutaneous tissue/viscera located hemangioma constituted of mostly firmly packed vascular spaces containing some blood or thrombosed or organized lumina and thick

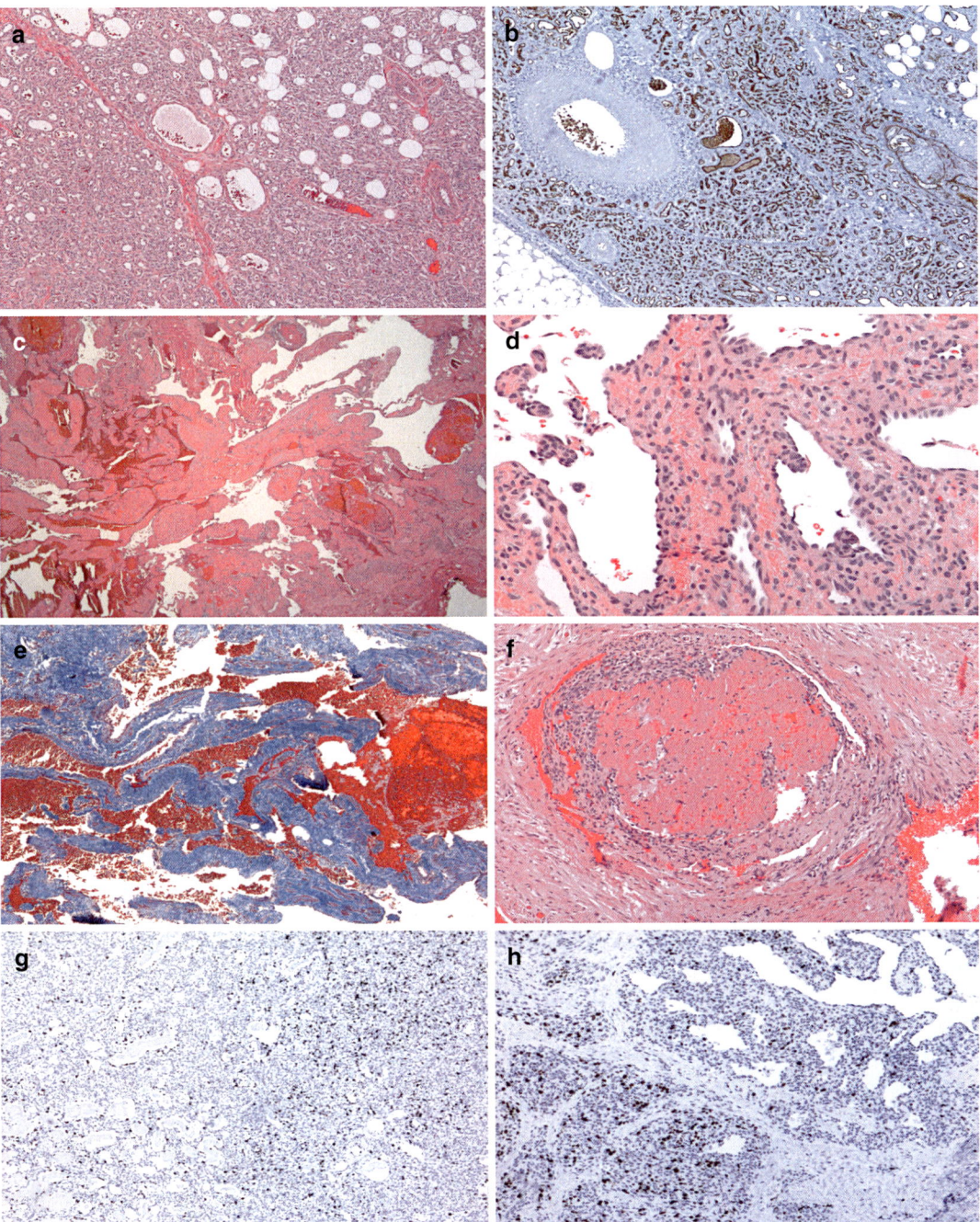

Fig. 12.18 The capillary hemangioma and cavernous hemangioma are shown in the microphotographs (**a**) through (**h**) with positivity for GLUT-1 (**b**) and low proliferation rates (**g** and **h**). In (**f**) an intratumoral mostly thrombosized blood vessel is seen (a, hematoxylin and eosin staining, ×50 original magnification; c, hematoxylin and eosin staining, ×12.5 original magnification; d, hematoxylin and eosin staining, ×200 original magnification; e, Masson trichromic staining, ×50 original magnification; f, hematoxylin and eosin staining, ×100 original magnification)

fibrous septa with no tendency to regress and ± VHLS (*vide infra*)
- *Sinusoidal*: Cavernous hemangiomas with dilated, interconnected, thin-walled channels and sporadic pseudopapillary projections
- *Intramuscular*: Cavernous-like hemangiomas with high cellularity, high MI, intraluminal papillary projections, plump endothelial cells, ± perineurial infiltration, but no atypia

Pyogenic granuloma (aka lobular capillary hemangioma) (Fig. 12.17): Rapidly growing, exophytic red nodule, attached by a stalk to the skin or gingival mucosa with a characteristic collarette with a natural tendency to bleed and/or ulcerate and (−) GLUT1

- *Epithelioid*: ALHE

WT1 gene expression is expressed in different tumors, including vascular neoplasms and can be useful in the distinction of vascular malformations, which are usually negative. The ISSVA or International Society for the Study of Vascular Anomalies issued in 1996 a classification of vascular lesions dividing them into two groups, including vascular tumors and vascular malformations. Vascular proliferation-associated lesions (vascular tumors) comprise hemangioma of infancy, congenital hemangioma, tufted hemangioma, pyogenic granuloma, hemangiopericytoma, kaposiform hemangioendothelioma, and other vascular lesions such as cherry angioma and angiosarcoma. Conversely, vascular malformations are associated with BV with morphologic abnormalities. They are classified according to the distorted BV type and may involve any of the structural components of the vasculature, such as capillaries, veins, arteries, and lymphatics. Both non-involuting congenital hemangioma (NICH) and rapidly involuting congenital hemangioma (RICH), two difficult lesions that may be challenging in the routine practice, are WT1 positive alike other primary tumors. Medical treatment is not useful in vascular malformations, which may benefit from surgical resections, embolization, and sclerotherapy.

12.10.1.3 Lymphangioma
- *DEF*: Benign vascular tumor, which can present in mediastinum or retroperitoneum, associated with thoracic duct anomalies as single neoplasm and, in the diffuse form, as *lymphangio-leiomyomatosis*.
- *CLI*: It may manifest as chylothorax, chylous ascites, and chyluria (Gr. χυλός "animal or plant juice").
- *CLM*: Proliferation of a mixture of BVs and smooth muscle with tumor cells being plumper and paler than usually seen in leiomyoma.
- *IHC*: (+) SMA, DES, HMB45.
- *DDX*: Angioleiomyoma.

12.10.2 Vascular Tumors with Intermediate Malignant Potential

Hemangioendothelioma and Kaposi sarcoma are the essential tumors of this category.

12.10.2.1 Hemangioendothelioma
- *DEF*: Hemangioendothelioma classic pattern and variants except the epithelioid variant belong to the category of vascular tumors with intermediate malignant potential. Hemangioendothelioma is a middle-grade vascular tumor with variable histologic features and clinical behavior ranging from recurrence in about 40% of cases and metastatic rates in about 1/5 of cases, whose 15% of these patients will succumb to this entity. It is not recommended using HE as the abbreviation for hemangioendothelioma, because HE is often used to note hematoxylin and eosin, but HE can be used in combination with other adjectives identifying the variants of this vascular tumor with intermediate malignant potential.

Hemangioendothelioma variants include kaposiform hemangioendothelioma (KHE), retiform (RHE), spindle cell (SHE), endovascular papillary (EVPHE), polymorphous (PHE), and the composite form (CHE). KHE is a locally aggressive vascular tumor of infants and children, with the involvement of the skin, retroperitoneum, and skeleton. It is associated with death due to extensive disease and severe consumption coagulopathy (aka Kasabach-Merritt syndrome), but there is no metastatic potential. Grossly, KHE is usually a cutaneous enlarged and discolored lump located involving an extremity, torso, or the cervicofacial region. Deep KHE lesions may infiltrate skeletal muscle, bone, intrathoracic organs and retroperitoneal cavity. Microscopically, there are infiltrating nodules and sheets of compact spindle cells with peculiar formation of the slit-like lumen as seen in Kaposi sarcoma (*vide infra*).

- *IHC*: (+) CD31, CD34, factor VIII-RA, vWF, VEGFR-3. Vascular endothelial growth factor receptor 3 (VEGFR-3, Flt-4) is the receptor for vascular endothelial growth factors (VEGFs) C and D and is specifically expressed on lymphatic endothelium (Witmer et al. 2001). Folpe et al. (2000) disclosed that anti-VEGFR-3 was positive in 23 of 24 Kaposi sarcomas, 8 of 16 angiosarcomas, 6 of 6 KHE, and 2 of 13 hemangiomas. VEGFR-3 and its ligands VEGF-C and/or VEGF-D lay important roles in cell-to-cell signaling in adult blood vessel angiogenesis.
- *TEM*: Intermediate filaments (nonspecific), pinocytotic vesicles, intracytoplasmic lumina, and pericytic cells are seen.
- *CMB*: The underlying genetic alteration in most EHEs (*vide infra*) is a fusion of the WW domain-containing transcription regulator protein 1 gene (*WWTR1*) with calmodulin-binding transcription activator 1 (*CAMTA1* gene).
- *DDX*: Angiosarcoma and Kaposi sarcoma.

Hemangioendothelioma Variants

CHE is defined by a mixture of benign, low-grade malignant (a vascular tumor with intermediate malignant potential), and destructive vascular components and mainly constituted by epithelioid and retiform growth patterns. CHE remarkably affects the superficial dermis or subcutaneous tissue of the hands and feet of young to middle-aged women. Local recurrence and metastases have been noted in some patients and should be considered at time of the clinical staging.

PHE usually involves lymph nodes and soft tissue with a tendency to recur locally and rarely metastases have been recorded. PHE is characteristically a mixture of solid, primitive vascular, as well as angiomatous patterns with uniform cytology, and spindle cell or angiosarcoma-like features without epithelioid features.

RHE is probably a low-grade variant of AS with limited involvement of the reticular dermis and subcutaneous tissue of distal limbs of adolescents or young adults. RHE tends to recur but has a low rate of metastases. Microscopically, the dominant feature is the retiform architecture, which is net-like, similarly to rete testis. The retiform pattern is characterized by blood vessels that scatter through the reticular dermis and subcutis and is lined by monomorphic hobnail endothelial cells with scant cytoplasm and round, typically naked-type nuclei with or without an accompanying lymphocytic infiltrate. Epithelioid areas or cytoplasmic vacuoles are not found. An essential diagnosis to be ruled out comprises low-grade AS, which demonstrates always areas of severe atypia, cellular pleomorphism, and brisk mitotic activity. The AS typically dissects between individual collagen bundles. Another essential differential diagnosis is constituted by the hobnail hemangioma, which is generally smaller, more superficial and more localized than RHE. Naturally, appropriate tumor tissue sampling is mandatory.

SHE is also localized at dermis or subcutis of distal limbs, but male adults of any age are affected. SHE tends to recur, be multicentric and associated with Maffucci syndrome. Both cavernous hemangioma and Kaposi sarcoma features characterize this kind of tumor.

EPHE (aka Dabska tumor) is a rare variant of hemangioendothelioma occurring in the skin or soft tissue of children with the almost exclusively good outcome. Maria Dabska first described this tumor in 1969. She described 6 cases in her series labeling it as "malignant endovascular papillary angioendothelioma" in children and adults (Dabska 1969). EPHE is characterized by endovascular papillary "tufts" lined by plump endothelial cells (epithelioid- or histiocytic-like) within dilated lumina with or without glomeruloid figures. Folpe et al. (2001) found 4 of 4 Dabska tumors positive with antibodies against the VEGFR-3.

12.10.2.2 Kaposi Sarcoma

Considered very rare before the discovery of the AIDS (acquired immunodeficiency syndrome), Kaposi sarcoma (KS) is a multicentric and reactive hyperplasia or tumor of low-malignant potential rather than a sarcoma (malignant tumor of mesenchymal origin). It is now considered a vascular proliferative disorder mediated by inflammatory cytokines and angiogenic growth factors in patients with HHV-8/Kaposi sarcoma-associated herpesvirus infection and tightly influenced by immunologic status. Thus, although historically valid, the term should be considered a misnomer for this entity. KS may involve mucocutaneous sites and viscera and develops in in one of four different epidemiologic-clinical settings (Box 12.16).

KS seems to arise from cells capable of undergoing lymphatic differentiation based on anti-podoplanin (D2-40) staining, a lymphatic-specific marker. To remember is that HHV-8 is not exclusive of KS, but such viral infection may also be detected in multicentric Castleman disease, primary effusion lymphoma, and some multiple myeloma cases.

- *GRO*: Early, macule/patch; intermediate, plaque; and late, nodule/tumor with multiple red-purple skin plaques or nodules, showing increasing in size and spreading proximally.

Box 12.16 Kaposi Sarcoma (KS) Settings

	Epidemiology	Clinical aspects	Outcome	Notes
Classical	♂ > ♀, 4th–7th, Mediterranean area and Jewish Ashkenazi	Skin (LEX) > mucosa ± viscera	Indolent	Localized
Iatrogenic	♂ = ♀, immune-suppression-linked (AI, drugs, and TX)	Skin (LEX) > mucosa ± viscera	Variable (regression if immunosuppression is stopped)	Localized
African	♂ = ♀, 2nd–5th, Equatorial Africa	Skin (LEX) ± LN	Progressive	Multiple
AIDS-associated	♂ = ♀, 2nd–7th, HIV+ (LGBT, IVD users)	Skin (LEX), mucosa + viscera	Aggressive (ARVT-dependent)	Multiple

N.B. The mnemonic word "CIA" may be used to remember the epidemiologic-clinical settings of Kaposi sarcoma (classical-iatrogenic-African and AIDS-associated). Typically, the higher the immunosuppression (e.g., CD4 T cells <200/mm^3), the more extensive and aggressive KS will be. Visceral involvement includes gastrointestinal tract, lungs, genitals, and lymphatic system mostly. The Metroka-Wems staging system is used for AIDS-associated KS (AIDS Clinical Trials Group issued staging system). T0 refers to KS confined to the skin and LN (lymph nodes) with no or minimal oral involvement, while T1 is used to KS with ulceration or associated edema. Similar staging systems are available for the classical type to assist therapeutic decision-making. AI, autoimmune; ARVT, antiretroviral therapy; IVD, intravenous drug users; LEX, lower extremities; LN, lymph nodes; TX, transplant. Some of the drugs commonly injected include heroin, fentanyl, talwin, buprenorphine, hydrocodone, and oxycodone. Common infections that result from IV drug use are: abscesses, botulism, cellulitis, (bacterial) endocarditis, Hepatitis C Virus, Human Immunodeficiency Virus, necrotizing fasciitis, septic thrombophlebitis, *Staphylococcus* infections, and tetanus. The practice known as "skin popping", i.e., IV drug users who inject subcutaneously, is also associated with a greater risk of developing complications.

All different epidemiologic-clinical forms usually share the same histopathology.

- *CLM*: Proliferation of both spindle cells in fascicles and BV in a sieve-like pattern with slit-like spaces in dermis and subcutis, often admixed with a variable chronic inflammatory infiltrate constituted by lymphocytes, plasma cells, and dendritic cells as well as hemosiderin-laden macrophages and eosinophilic and PAS(+) hyaline globules within lesional cells or extracellularly.
- *IHC*: (+) HHV8, but also (+) CD31 (PECAM-1 or platelet endothelial cell adhesion molecule), CD34, F-VIII-RA(±), D2-40, LYVE-1 (CD44 receptor homologue), VEGFR-3, Bcl2, Prox-1, and FLI-1. Folpe et al. (2001) studied FLI-1 in vascular and non vascular tumors. Fli-1 was expressed by 50 of 53 vascular tumors scored (94%). Vascular neoplasms included 20 of 22 angiosarcomas, 11 of 12 hemangioendotheliomas, 7 of 7 hemangiomas, and 12 of 12 Kaposi sarcomas. Conversely, Fli-1 expression was absent in 68 nonvascular neoplasms, including 45 carcinomas, 16 sarcomas, and 7 melanomas. Folpe et al. concluded that FLI-1 has a high sensitivity (94%) and specificity (100%) allowing the distinction of vascular neoplasms from potential mimics.
- *TEM*: Weibel-Palade bodies and intracellular fragmented RBC.
- *CMB*: The value of HHV8-PCR is inferior to that of LNA-1 (Latent Nuclear Antigen 1) for HHV-8 because contaminating mononuclear cells may harbor this herpesvirus, particularly in HIV-positive patients and may represent a false positive setting.
- *DDX*: Tufted angioma, targetoid hemosiderotic hemangioma, and acroangiodermatitis for KS at plaque-stage, while bacillary angiomatosis, spindle-cell hemangioma, KHE, fibrohistiocytic tumors resolving dermal fasciitis, spindle-cell melanoma, and other spindle-cell mesenchymal neoplasms enter in the differential diagnosis for KS at nodular stage; angiosarcoma (HHV-8 -) needs to be differentiated from KS at advanced stage.
- *TRT*: Series of interventions including symptom palliation with cosmesis improvement, prevention of progression, clearing of associated edema and organ involvement, as well as psychologic support. The treatment for localized lesions includes cryotherapy with liquid nitrogen, radiotherapy, surgery, laser therapy, vinblastine-injection, electrochemotherapy-enhanced bleomycin/cisplatin injection, and topical application of alitretinoin. KS is treated with systemic chemotherapy if >25 lesions are found, extensive oral involvement, marked significant edema or visceral involvement, rapidly progressive disease, and KS flare. The backbone of systemic cytotoxic therapy includes liposomal anthracyclines and taxanes, and novel therapies include IFN-alpha, thalidomide, anti-Herpes, imatinib, and matrix metalloproteinase (MMP)-inhibitors. Tissue inhibitors of metalloproteinases (TIMPs) are the major cellular inhibitors of the MMP sub-family. However, most cells of KS lesions harbor HHV8 in its latent phase.
- *Giant cell angioblastoma* is an essential tumor of pediatric age, because of its clinical implications. GCAB is rare but shows an infiltrative slow-growing pattern and should probably be considered LMP. The GCAB is often considered together with Kaposi sarcoma, because of its high locally destructive fashion and intermediate malignancy without metastatic potential.
- *GRO*: Ulcerated tumor with clear-cut infiltrating character into soft tissue and bone and quite often located in the hand or H&N areas.
- *CLM*: There is a nodular plexiform proliferation of oval-to-spindle undifferentiated mesenchymal cells with prominent, concentric perivascular aggregation with the participation of fibroblasts, myofibroblasts and (+) CD68-MNGC.
- *DDX*: Giant cell fibroblastoma (CD34+, t(17:22)), EHE with osteoclast-like giant cells (Factor VIIIRA+, no concentric perivascular aggregation), plexiform fibrohistiocytic tumor (children/young adults, no concentric perivascular aggregation), myopericytoma (elderly, no giant cells), bacillary angiomatosis (+SS for organisms, +Hx. immunosuppression).

12.10.3 Malignant Vascular Tumors

The WHO incorporates in this section the epithelioid hemangioendothelioma and the angiosarcoma.

12.10.3.1 Epithelioid Hemangioendothelioma

- *DEF*: Angiocentric vascular malignancy that involves soft-tissue and visceral organs such as the liver and lung and composed of epithelioid endothelial cells with an arrangement variable from short cords to nests embedded in a myxohyaline stroma, arising mostly in the extremities, H&N, and mediastinum. Drs. Weiss and Enzinger established this term to describe a soft-tissue vascular tumor with a clinical course in between that of benign hemangioma and that of angiosarcoma. Currently, it is still important to discern EHE from angiosarcoma because patients harboring EHE can survive for a long term.
- *GRO*: Tender gray-white (cutaneous) nodule that may show, at places, some hemorrhage on cut surface.
- *CLM*: The neoplastic endothelial cells show vesicular nuclei and abundant eosinophilic cytoplasm with intracytoplasmic vacuoles ("small intracytoplasmic vascular lumina"), which are mucin negative (Fig. 12.19). Malignant features (~1/3 of the cases) include marked nuclear pleomorphism, MI > 1/10HPF, cell spindling, and necrosis.
- *SS/IHC*: (+) VIM, CD31, vWF, CK (30%, focal), RET (around nests and cords).
- *CMB*: t(1;3)(p36.23;q25.1) is a consistent genetic abnormality in EHEs of different anatomical locations and grades of malignancy with the formation of the gene fusion transcript *WWTR1-CAMTA1*. This recurrent translocation is not present in any of the morphological mimics of EHE, such as epithelioid hemangioma, epithelioid angiosarcoma, or epithelioid sarcoma-like EHE. WWTR1, aka TAZ, is a transcriptional coactivator with a conserved WW domain capable of interacting with the PDZ domain and shares amino acidic homology with YAP (Yes-associated protein). Interestingly, WWTR1 and YAP are downstream effectors of the Hippo pathway, which controls organ size and contact inhibition in mammals by regulating cell proliferation and apoptosis. CAMTA1 belongs to the protein family of calmodulin-binding transcription activators. It is indeed a family of proteins and is a transcription activator potentially involved in cell cycle regulation and possibly interacting with Ca2þ/calmodulin and be engaged in Ca2þ signaling.
- *DDX*: Met.-Ca. (marked atypia, high MI, lack of angiocentricity, keratin+, CD31−), melanoma (S100+, HMB45+, CD31−), epithelioid sarcoma (adolescents and young adults, distal extremities, merging of tumor cells with collagenous stroma and granuloma-like pattern, strong CK+, CD31−), epithelioid angiosarcoma (irregular sinusoidal, interconnected vascular channels, solid cellular sheets harboring marked atypia and high MI, necrosis).
- *PGN:* Unpredictable clinical course, but less aggressive than AS, although 13% will recur, 20–30% metastasize (lung and LNs), and 13% die of the disease (mortality: 65%). It has been suggested that the cutaneous form of EHE has undoubtedly a better prognosis than the deep-localized EHE (Goh & Calonje, 2008).
- High risk EHE (>3 MF/50 HPF and Ø > 3 cm) ⇒ 5-YSR: 59%.
- Low-risk EHE, wide local excision, vs. high-risk EHE, radical local excision ± LN-ectomy.

12.10.3.2 Angiosarcoma

Angiosarcoma (AS) is a malignant endothelial cell-arising vascular tumor of the deep muscles of the lower extremities (~1/2) or other locations, including viscera, with apparent vasoformative growth and complex anastomosing channels that occurs in children and adults with a peak incidence in the seventh decade and harboring a poor prognosis with LN and distant metastases (lungs, liver, bone). Five clinical groups are differentiated:

- Cutaneous AS
- Lymphedema-associated AS
- Radiation-induced AS
- Breast AS
- Deep soft tissue AS

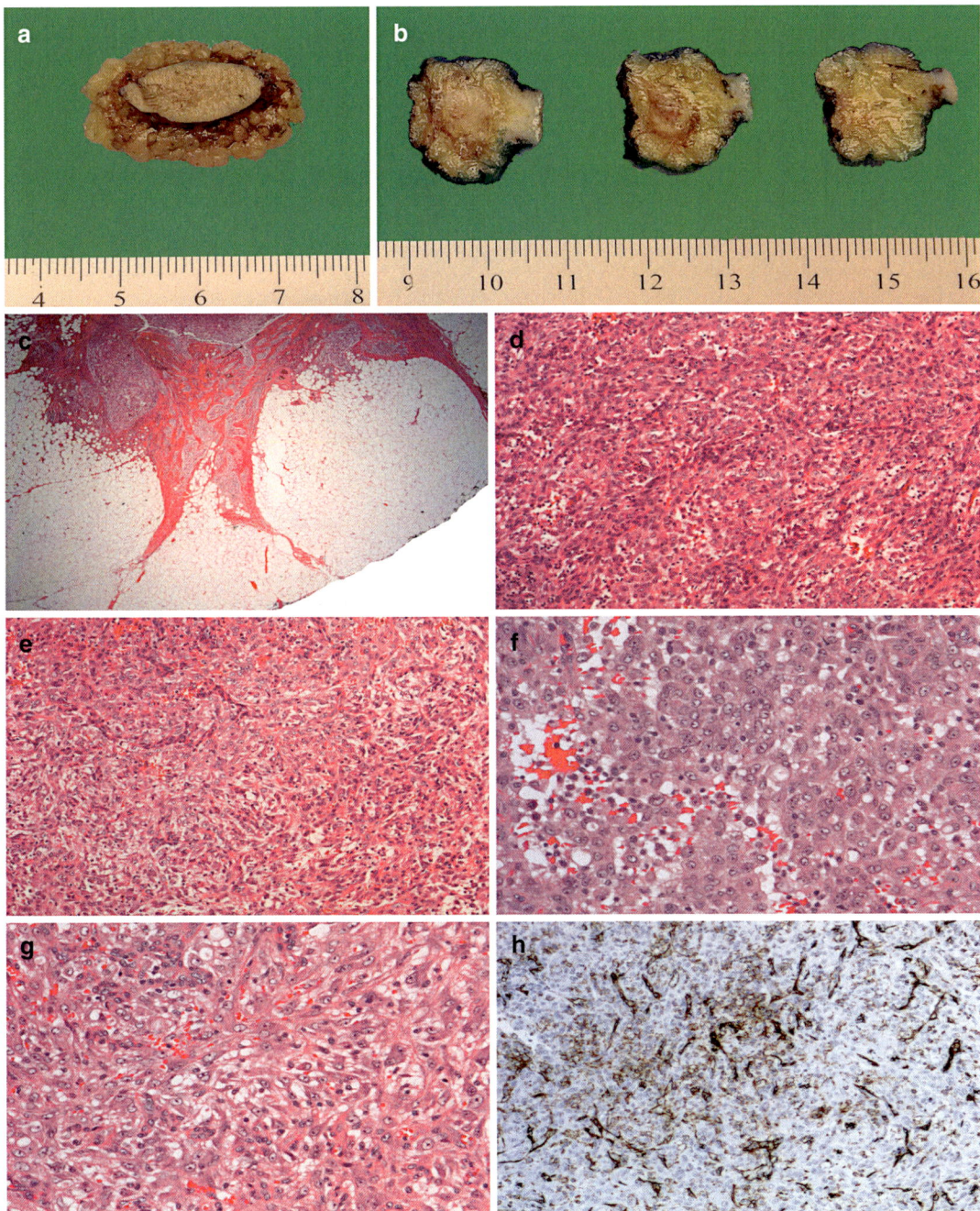

Fig. 12.19 Epithelioid Hemangioendothelioma. An epithelioid hemangioendothelioma is shown grossly (**a** and **b**) and microscopically (**c–h**). In c) is notable the infiltration of the tumor, while the epithelioid nature of the plump atypical endothelial cells is better shown at higher magnification, particularly in f) (c-g, hematoxylin and eosin staining, ×12.5, ×100, ×100, ×200, ×200 as original magnification, respectively). The microphotograph in h) highlights this vascular tumor using an antibody against CD31 by conventional immunohistochemistry. (Anti-CD31 immunostaining, Avidin-Biotin Complex, ×100 original magnification)

However, all forms share similar histology with an infiltrative character, multifocality, and a dissecting pattern of irregular, anastomosing vascular channels between collagen bundles, fatty tissue, and other structures.

- *RF:* Chronic lymphedema (breast), sun exposure, radiation, Thorotrast, polyvinyl chloride (PVC). Epidemiologic studies have identified the vinyl chloride monomer (VCM) as the causative agent (Falk 1987).
- Visceral sites: The breast, liver, lung, and skin (Stewart-Treves syndrome after ~ 10 years post-radical mastectomy for breast cancer-associated lymphedema).
- *GRO:* Early AS ⇒ asymptomatic, small, sharply circumscribed, multiple red nodules vs. late AS ⇒ fleshy, gray-white mass with hemorrhage, necrosis, deeply invasive and painful. AS may also present with coagulability abnormalities, anemia, persistent hematoma, bruisability, and high output cardiac failure.
- *CLM:* Atypical vascular channels lined by neoplastic endothelial cells with evident cytologic atypia, multilayering, intracytoplasmic RBC-containing lumina, high MI, and MN cells with hyaline globules containing AAT and alpha 1-antichymotrypsin (ACT) and necrosis.
- *HSS/IHC:* (+) CD31, CD34, FLI-1, FVIII-RA, CD141, c-kit (50%) with relatively high MIB1 (Ki67).
- *TEM:* (+) Weibel-Palade bodies.
- *DDX:* Hemangiomas (usually Ø <2 cm, well-demarcated, fibrous septa, and thick-walled BV, no infiltrative), angiomatosis, bacillary angiomatosis, A-MFH, epithelioid sarcoma, malignant melanoma, and sarcomatoid carcinoma. Variant: epithelioid AS.
- *TRT:* Early surgery, but 5-YSR is low.

12.10.4 Genetic Syndromes Associated with Vascular Tumors

Several systemic conditions can be linked with vascular neoplasms and the correct diagnosis of these syndromes is essential because they can be related to severe, life-threatening complications. The most relevant syndromes associated with vascular tumors are the following conditions:

1. *Von-Hippel Lindau Syndrome (VHLS):* Clear cell RCC – chromosome 3 (3p25), RAF-1. pVHL inhibits the elongation step of RNA synthesis by interacting with ELONGIN B and C (mnemonics "K-PASH": kidney (CCRCC/renal cysts), pancreas (cysts/serous cystadenoma), adrenal (pheochromocytoma), skin (Cafe au lait spots), hemangioblastoma of retina, and cerebellum (± EPO-induced polycythemia) recalling *K-PAX*, the 2001 American Science Fiction movie with Kevin Spaccy and Jeff Bridges.
2. *Blue Rubber Bleb Nevus Syndrome* (BRBNS): It is a rare disease with an abnormal blood vessel formation and arteriovenous malformation.
3. *Proteus Syndrome* is a syndrome characterized by a mosaic distribution, progressive course, and sporadic occurrence AND one A-lesion (connective tissue nevus) OR two out of three B-lesions (epidermal nevus, disproportionate growth, bilateral ovarian cystadenomas in an individual <20 years, monomorphic adenoma of the parotid gland in an individual <20 years) OR all three C-lesions (lipomas or regional fatty atrophy, vascular malformation, and specific craniofacial dysmorphism, including dolichocephaly, long face, down-slanting palpebrae, ptosis, depressed nasal bridge, anteverted nares, and open mouth position while at rest).
4. *Klippel-Trénaunay-Weber Syndrome* (KTWS) is a constellation composed of varicose veins, single or multiple dysplastic cutaneous hemangioma, and ST/bone hypertrophy).
5. *Kasabach-Merritt Syndrome* (KMS) is a rare, life-threatening coagulopathy of infancy which presents with thrombocytopenia, microangiopathic hemolytic anemia, and consumptive coagulopathy in the background of a fastly enlarging vascular tumor. The vascular neoplasms are almost exclusively kaposiform hemangioendothelioma (KHE) and tufted angioma (TA). An associated coagulopathy may occur in ¾ of patients.

Practically, segmental hemangioma or another vascular tumor may herald underlying and potentially life-threatening structural anomalies of the brain and heart. Other than the above mentioned syndromes, different constellations of signs and symptoms may be necessary to recall. PHACE syndrome is another syndrome, probably relevant to remember. The name is an acronym including posterior fossa defects, hemangiomas, arterial and cardiac defects, and eye anomalies. Neurocutaneous syndromes also include the MCMTC syndrome or macrocephaly-cutis marmorata telangiectatica congenita, which has been renamed to macrocephaly-capillary malformation (M-CM) to reflect the true nature of the syndrome and following the discovery of new genetic mutations (e.g., *RASA-1*). In the therapy that needs to be mentioned, propranolol therapy seems to be highly effective in life-threatening hemangiomas and appears to stop growth and hasten involution. Adverse effects are cardiovascular (bradycardia, hypotension), respiratory (bronchospasm), and metabolic (hypoglycemia). The number of genetic syndromes associated with a vascular tumor may be quite high, and genetic books are probably the best reference. However, the knowledge of these syndromes should be taken into account from a pediatric pathology perspective owing to the involvement of the pediatric pathologist in the multidisciplinary management of these patients.

12.11 Chondro-Osteoforming Tumors

Chondro-osteoforming tumors are mainly *soft tissue chondroma*, *mesenchymal chondrosarcoma*, and *extraskeletal osteosarcoma*, but other entities will also be described briefly.

12.11.1 Extraskeletal Chondroma

- *DEF*: It is a benign mesenchymal tumor constituted of adult-type hyaline cartilage with or without osseous, fibrous, and myxoid stroma in the hands (fingers) and feet (toes).
- *EPI*: Infants to elderly, ♂ = ♀.
- *GRO*: Well-demarcated lobulated, hyaline, and calcified nodule (Ø < 3 cm).
- *CLM*: Plump tumor cells with fine punctate calcification, some hyperchromatic nuclei ± fibrosis, as well as granuloma-like reaction with MNGC and histiocytes.
- *IHC*: (+) S100.
- *DDX*: Chondrosarcoma (rare in hands and feet), calcifying aponeurotic fibroma (different histology in other fields and there is no adult-type hyaline cartilage).
- *TRT*: Complete excision is curative.
- *PGN*: Tendency to recur (20%) if incompletely excised.

12.11.2 Extraskeletal Myxoid Chondrosarcoma

- *DEF*: Malignant chondroid forming mesenchymal tumor that is mostly myxoid and occurs outside of the skeletal system.
- *EPI*: Children to adult, ♂ = ♀.
- *GRO*: Mass with central hardening and "chondroid" or "myxoid" appearance.
- *CLM*: Solid architecture constituted by small-to-medium-sized cells with pale eosinophilic cytoplasm in a myxoid stroma. The tumor cells may harbor occasional vacuoles, but there is no direct evidence of mature chondrocytes.
- *IHC*: (+) S100, CD57, LYS, PAS, acid mucins.
- *TEM*: Intracisternal aggregates or bundles of parallel microtubules, intermediate filaments in the cytoplasm, intracytoplasmic glycogen, lipid droplets, and well-developed Golgi apparatus.
- *CMB*: t(9;22)(q22-31;q11-12): *CHN-EWS* fusion gene.
- *DDX*: Mixed tumor or myoepithelioma of soft tissue, parachordoma, and other myxoid soft tissue tumors (e.g., myxofibrosarcoma, myxoid liposarcoma, and ossifying fibro-myxoid tumor of soft parts). Other tumors that may need to differentially keep distinctly are cellular and poorly differentiated round or epithelioid cell tumors including anaplastic carcinoma, poorly differentiated synovial sar-

coma, epithelioid sarcoma, epithelioid angiosarcoma, and primitive neuroectodermal tumor (PNET).
- *TRT*: Complete resection with wide margins combined to chemotherapy because of potential lung metastases.
- *PGN*: Lung metastases, but less aggressive than bone tumors and usually involving the limbs but the trunk in children as well.

12.11.3 Mesenchymal Chondrosarcoma

- *DEF*: Mesenchymal tumor with biphasic patterns of differentiation (small round blue cells and hyaline cartilaginous islands).
- *EPI*: 3–10% of all chondrosarcoma diagnoses, children and adults.
- *CLI*: Mass presenting in the orbit, dura, trunk, retroperitoneum, extremities, or kidney.
- *CLM*: Clusters of undifferentiated small round blue cells often with hemangiopericytoma appearance mixed with islands of mature-appearing hyaline cartilage (biphasic pattern).
- *IHC*: (+) SOX9 and (−) FLI1.
- *CMB*: Robertsonian translocation der(13;21)(q10;q10) found in both skeletal and extraskeletal cases (*HEY1-NCOA2* fusion gene).
- *DDX*: Small round blue cell tumors, mainly PNET/Ewing sarcoma.
- *PGN*: Poor.

12.11.4 Extraskeletal Aneurysmatic Bone Cyst

- *DEF*: Extraskeletal aneurysmatic bone cyst (EABC) shares features identical to the intraosseous aneurysmal bone cyst and affects children and young adults in the deep soft tissue of upper and lower extremities and characterized clinically by a fast-growing mass without the involvement of adjacent bones.
- *GRO:* Medium-sized mass (<5 cm) with a thin rim of bone and composed of hemorrhagic cystic spaces with discernible fibrous septa.
- *CLM:* Cystic spaces filled with blood; fibrous septa composed of fibroblasts, osteoclast-type giant cells, woven bone present without atypia.
- *CMB*: 46,XY,t(17;17)(p13;q12), similar to intraosseous aneurysmal bone cyst.
- *DDX:* Extraskeletal osteosarcoma.
- *PGN*: High tendency to recur locally if incompletely excised.

12.11.5 Extraskeletal Osteosarcoma (ESOS)

- *DEF*: Extraskeletal osteosarcoma (ESOS) is a malignant mesenchymal tumor of soft tissue (subcutis and dermis) with osteogenic line of differentiation with or without accompanying chondroblastic and fibroblastic cell lines of middle-aged/late adulthood (♂ > ♀), *de novo* (~90%) or postradiation (~10%) in the extremities (thigh), buttock, and shoulder girdle.
- Subtypes: Osteoblastic, chondroblastic, fibroblastic, telangiectatic, well-differentiated (parosteal), and small cell.
- *IHC*: (+) VIM and osteocalcin, (±) SMA, DES, S100, EMA, CK, CD99.
- *TEM*: Osteologic characteristics similar to the bony counterpart.
- *DDX*: Myositis ossificans (no nuclear atypia, a characteristic zonal pattern showing fasciitis-like features, immature osteoid or woven bone rimmed by osteoblasts in the center, and bone formation at the periphery, initially with woven bone and then lamellar bone), other sarcomas producing metaplastic bone (MFH, synovial sarcoma, and fibrosarcoma have specific patterns).
- *PGN*: ESOS harbors a very poor PGN (5-YSR: 25%), and factors are size (Ø < 5 cm/Ø > 5 cm), histology (fibroblastic/chondroblastic have a better outcome than other subtypes), low to high proliferation activity (Ki67 activity).

12.11.6 Extraskeletal Chordoma

- *DEF*: It is a tumor composed of the characteristic physaliphorous cells located outside of the skeleton.

- *IHC*: (+) CK9, S100, brachyury (notochord marker). In humans, brachyury is a protein that is encoded by the *TBXT* gene. Brachyury acts as a transcription factor within the T-box gene family and is key in differentiating chordoma and hemangioblastoma versus neoplastic histological mimickers (Barresi et al. 2014).
- *PGN*: Usually benign.

12.12 Tumors of Uncertain Differentiation

Tumors of uncertain differentiation (TUD) include *myxoma* (and its subtypes and variants), *pleomorphic hyalinizing angiectatic tumor, ectopic hamartomatous thymoma* as benign entities. Intermediate tumoral entities are *angiomatoid fibrous histiocytoma, ossifying fibro-myxoid tumor*, and *mixed tumor* (myoepithelioma and parachordoma). Malignant tumors of uncertain differentiation include *synovial sarcoma* (SS), *epithelioid sarcoma* (EpS), *alveolar soft part sarcoma* (ASPS), *clear cell sarcoma of soft tissue* (CCSS), *extraskeletal myxoid chondrosarcoma* (ESMC), *peripheral neuroectodermal tumor* (PNET), *desmoplastic small round cell tumor* (DSRCT), *extrarenal rhabdoid tumor* (ERRT), *malignant mesenchymoma*, *PEComa*, and *intimal sarcoma* (Box 12.17).

> **Box 12.17 Tumors of Uncertain Differentiation**
> 12.12.1 *Myxoma NOS*
> 12.12.2 *Myoepithelial Carcinoma*
> 12.12.3 *Parachordoma*
> 12.12.4 *Synovial Sarcoma (SSX)*
> 12.12.5 *(Soft Tissue) Epithelioid Sarcoma (STES)*
> 12.12.6 *Alveolar Soft Part Sarcoma (ASPS)*
> 12.12.7 *Clear Cell Sarcoma (CCS)*
> 12.12.8 *Extraskeletal Myxoid Chondrosarcoma (ESMC)*
> 12.12.9 *Extraskeletal Ewing Sarcoma (PNET)*
> 12.12.10 *Desmoplastic Small Round Cell Tumor (DSRCT)*
> 12.12.11 *Extrarenal Rhabdoid Tumor (ERT)*
> 12.12.12 *Malignant Mesenchymoma*
> 12.12.13 *PEComa*
> 12.12.14 *Extrarenal Wilms Tumor*
> 12.12.15 *Sacrococcygeal Teratoma (SCT) Including Extragonadal Germ Cell Tumor and Yolk Sac Tumor*

12.12.1 Myxoma

- *DEF*: Benign soft tissue tumor usually with *intramuscular* location (thigh, buttock, upper arm, and shoulder) and mostly of *female adults*.
- *GRO*: Gelatinous mass with ill-defined margins.
- *CLM*: Uniform and bland spindle and stellate cells embedded in a proteoglycan matrix-rich, slightly basophilic stroma (no endothelial cell proliferation as seen in cardiac myxoma).
- *TEM*: Fibroblastic/myofibroblastic ultrastructural features (dilated RER, Golgi complexes, free ribosomes, pinocytotic vesicles, and occasional filaments) characterize the classic pattern. It may be associated with *GNAS1* (guanine nucleotide-binding alpha-stimulating activity polypeptide 1, i.e., the G-protein that is responsible for the formation of cAMP) mutations with (Mazabraud syndrome) or without polyostotic fibrous dysplasia. *GNAS1* encodes the G-protein alpha stimulatory subunit ($G_s\alpha$), and variations are usually of gain-of-function type interfering with hydrolysis of GTP to GDP with consequent perpetual adenylyl cyclase activation.
- *PGN*: Excellent.
- Myxoma variants and related lesions: *cellular* myxoma (↑ cellularity with ↑ collagen fibers and BVs), *juxta-articular* myxoma ("cellular"-like myxoma associated with ganglion-like cystic change and located close to a large joint), and *deep aggressive angiomyxoma* (uniform and bland spindle and stel-

late ER/PR+ cells embedded in a myxoedematous, loosely collagenous stroma with hyalinized BVs in a locally aggressive form located in the pelvis or perineum of young females).
- *Pleomorphic hyalinizing angiectatic tumor (PHAT)*: Benign soft tissue tumor of uncertain differentiation of usually lower extremities with a mixture of ectasia-prone, thin-walled blood vessels, pleomorphic spindle cells, and inflammation.
- *Ectopic hamartomatous tumor*: Benign soft tissue tumor of probable branchial pouch origin of the lower neck showing a framework of spindle cells, epithelial groups, and adipose cells in highly variable rates. It is a tumor of adulthood with bland cytology, although the "lattice"-like growth of the spindle component and the diverse phenotype of the epithelial groups (squamous islands, syringoma-like tubules, anastomosing networks, glands, and cysts) may raise a potentially long list of differential diagnoses (the lattice is structure consisting of strips of wood or metal crossed and fastened together leaving either diamond- or square-shaped spaces between and often used in gardening for climbing plants).
- *IHC*: (+) CK, HMWCK, (±) SMA, CD34, (−) DES.
- *TEM*: Desmosomes and tonofilaments are often found.

12.12.2 Myoepithelial Carcinoma

- *DEF*: Rare TUD exhibiting myo- and epithelial differentiation.
- *EPI*: Worldwide, all ages (1/5 of cases in childhood), ♂ = ♀.
- *GRO*: Large tumor with hemorrhage and necrosis and marginal irregularity.
- *CLM*: Infiltrative growth pattern of spindle epithelioid cells with fibrillar eosinophilic cytoplasm, distinct vesicular nuclei, and distinct nucleolus. This cell population is also associated with an intimately admixed reactive spindle-cell stroma. Cytologically, the tumor cells have variable atypia, and MI is mild to moderate. Clear cells may be found together with the spindle cells. Areas of necrosis and hemorrhages are not uncommon (Fig. 12.20).
- *IHC*: (+) Calponin, (±) SMA, (+) keratins (CAM5.2, AE1–3), EMA but also (+) S100, GFAP. (+) SOX10 and (−) SMARCB1/INI1.
- *TEM*: Evidence of intercellular junctions and basal lamina.
- *CMB*: *EWSR1* gene rearrangements. Rearrangement of the *EWSR1* gene, encoded on chromosome 22q, is found in many myoepithelial carcinomas of soft tissue, while a subset has alternate *FUS* rearrangement.
- *DDX*: PNET, clear cell RMS, epithelioid sarcoma, extraskeletal myxoid chondrosarcoma (ESMC). Mutations, deletions, and other somatic alterations in INI1 (hSNF5; SMARCB1), which encodes a subunit of the SWI/SNF chromatin remodeling complex, have been first characterized in the malignant rhabdoid tumor of infancy. Currently, numerous INI1-deficient tumors have been delineated. The complete loss group include the malignant rhabdoid tumor (atypical teratoid/rhabdoid tumor), epithelioid sarcoma, renal medullary carcinoma, epithelioid MPNST, myoepithelial tumor, ESMC, pediatric chordoma, pancreatic undifferentiated rhabdoid carcinoma, sinonasal basaloid carcinoma, and rhabdoid carcinoma of the gastrointestinal tract. A mosaic expression group includes schwannomatosis, gastrointestinal stromal tumor, and ossifying fibromyxoid tumor, while the reduced expression group includes synovial sarcoma (Kohashi & Oda 2017; Hollmann & Hornick 2011).
- *PGN*: Local recurrence and metastases each in about half of the patients with about 40% of patients dead after 9 months after diagnosis.

12.12.3 Parachordoma

- *DEF*: Parachordoma is a rare soft tissue tumor resembling ESMC and chordoma, which develops next to tendon, synovium, and bony structures in extremities with wide age range

12.12 Tumors of Uncertain Differentiation

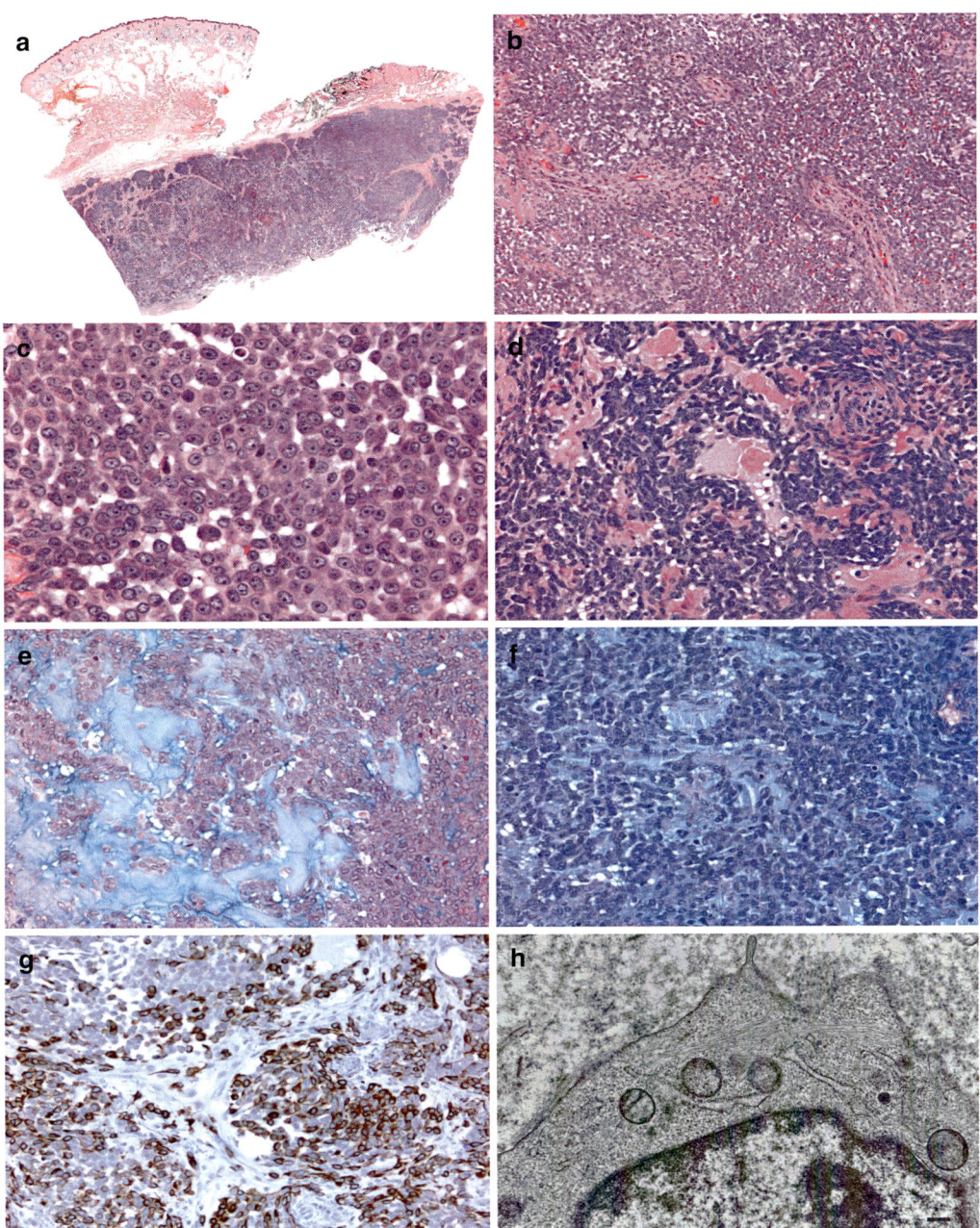

Fig. 12.20 Myoepithelial Carcinoma. There are infiltrating cells of spindle cell type with fibrillar eosinophilic cytoplasm and variable cellular atypia and mitotic activity (**a–c**, hematoxylin and eosin staining, ×0.5, ×100, ×400 as original magnification, respectively). Histochemical stains (**d, e,** and **f**) are positive for the myxoid matrix and some glycogen (d, mucin staining, ×200 original magnification; e, Alcian Blue pH 2.5 staining, ×200 original magnification; f, Alcian Blue Periodic acid Schiff staining, ×200 original magnification). Typically, glycogen is found in clear cells and the myxoid matrix is positive with Alcian blue. Alcian blue denotes any member of a family of polyvalent basic dyes, of which the Alcian blue 8G is the most common and the most reliable member. Alcian blue is used to stain acidic polysaccharides such as glycosaminoglycans (e.g., cartilage), several types of mucopolysaccharides, and sialylated glycocalyx of some cells. In myoepithelial carcinoma, the immunohistochemistry may also play an important role. This tumor is positive for glial fibrillary acidic protein (GFAP) (**g**) (g, anti-GFAP immunostaining, Avidin-Biotin Complex, ×200 original magnification). GFAP is an intermediate filament for astrocytes (both normal, and reactive or neoplastic). The electron microscopy photograph hisghlight the importance of the ultrastructural examination demonstrating tonofilaments, subplasmalemmal bundles of microfilaments with dense bodies, intermediate junctions and poorly developed desmosomes (transmission electron microscopy, ×8000 original magnification)

(3–83 years), mostly young adults. It has been considered to be a tumor originating from ectopic rests of notochord, Schwann cells, myoepithelial cells, or specialized synovial cells, but it does not seem there is a definitive proof for one way or another (Zhang et al. 2013).

- *GRO*: Parachordoma is well-demarcated from the surrounding tissues. The tumor is firm and lobulated with some regions containing translucent, cartilaginous-like tissue with a grayish-white color. Neither necrosis nor hemorrhage is identified.
- *CLM*: Parachordoma shows well-circumscribed lobules of large, round/ovoid, and eosinophilic cells, which may recall, at least focally, physaliferous cells or cartilaginous cells in myxoid to densely hyaline, Alcian blue (AB) 2.5+, negative hyaluronidase matrix. The AB positivity at pH 2.5 (matrix) is eradicated with hyaluronidase predigestion.
- *IHC*: (+) CK 8/18, EMA, S100, VIM, type IV Coll, (−) SMA, GFAP, CK 1/10. Parachordomas only express CK 8/18, but are negative for other keratins, including CK 1/10, CK 7, CK 12-17, CK 19, and CK 20.
- *TEM*: It shows well-developed rough endoplasmic reticulum, abundant intermediate filaments, microvillous cytoplasmic processes, pinocytic vesicles, and desmosome-like junctional structures.
- *CMB*: Trisomy 15 (one case) and monosomy 1, 16, and 17 have been described.
- *DDX*:
  - ESMC: t(9;22)+, type IV Coll (−).
  - Chordoma monosomy 3, 4, 10, or 13 and type IV Coll (−).
- *PGN*: Surgical excision is usually adequate, and the tumor should not recur. However, parachordoma can occur with myoepithelioma and this variant is associated with a completely different outcome.

12.12.4 Synovial Sarcoma

- *DEF*: Morphologically, clinically, and genetically distinct – very peculiar – *carcinosarcoma of soft tissue* with no direct relationship to synovium. SS may involve any site but is mostly a *periarticular* (>80% around the knee and ankle joints) tumor of *young male* adults (15–35 years, ♂ : ♀ = 1.5:1), *deep-seated*, *painful*, and *slow* growing. A single gene, *SS18* (*SYT*), has been incriminated on 18q11.2, while 1 of three related genes, *SSX1*, *SSX2*, or *SSX4*, is usually entangled on Xp11.2.
- *GRO*: Well-circumscribed, firm, gray-pinkish mass.
- *CLM*: *Monophasic or biphasic differentiation* (epithelial and mesenchymal) with reticulin stain highlighting the biphasic nature. The epithelial areas show glands, papillary fronds, or nests of keratinizing squamous epithelium, while mesenchymal areas consist of sarcomatous spindle-cell stroma with short fascicles of plump spindle cells with lobulation, hyalinization, and scattered conventional mast cells. SS ± Calcification → calcifying SS (Fig. 12.21).
- *IHC*: (+) TLE1 (transducin-like enhancer of split belonging to the Groucho/TLE family), VIM, EMA, AE1–3, CK7; (+) CD99, Bcl-2, (−) CD34, SMA, DES, HMB-45; (±) S100. The Drosophila Groucho (Gro) protein is was the key member of the family with transcriptional co-repressor activities in vertebrates (Jennings & Ish-Horowicz 2008). Moreover, ZO-1, claudin1 and occludin are also expressed in SS. The glandular component of most biphasic SS is positive for ZO-1, claudin1 and occludin, while the monophasic SS is positive for ZO-1 and inconstantly with claudin1 and occludin (Billings et al. 2004).
- *TEM*: Short microvilli projecting into the glandular-like lumen, tonofilaments, desmosomes (epithelial component), and prominent RER (mesenchymal component).
- *CMB*: *t(X,18)(p11;q11)* ⇒ *SYT-SSX1* among others including *SYT-SSX2* and *SYT-SSX4* involving genes SS18 and either SSX1, SSX2 or SSX4. The translocation has several effects on oncogenetic pathways, mostly the SWI/SNF chromatin remodeling complex, polycomb repressor complex, and the infamous canonical Wnt pathway (Nielsen et al. 2015).
- *DDX*: MPNST is usually (−) for CK7 and CK19, (−) for SYT-SSX fusion products. Also,

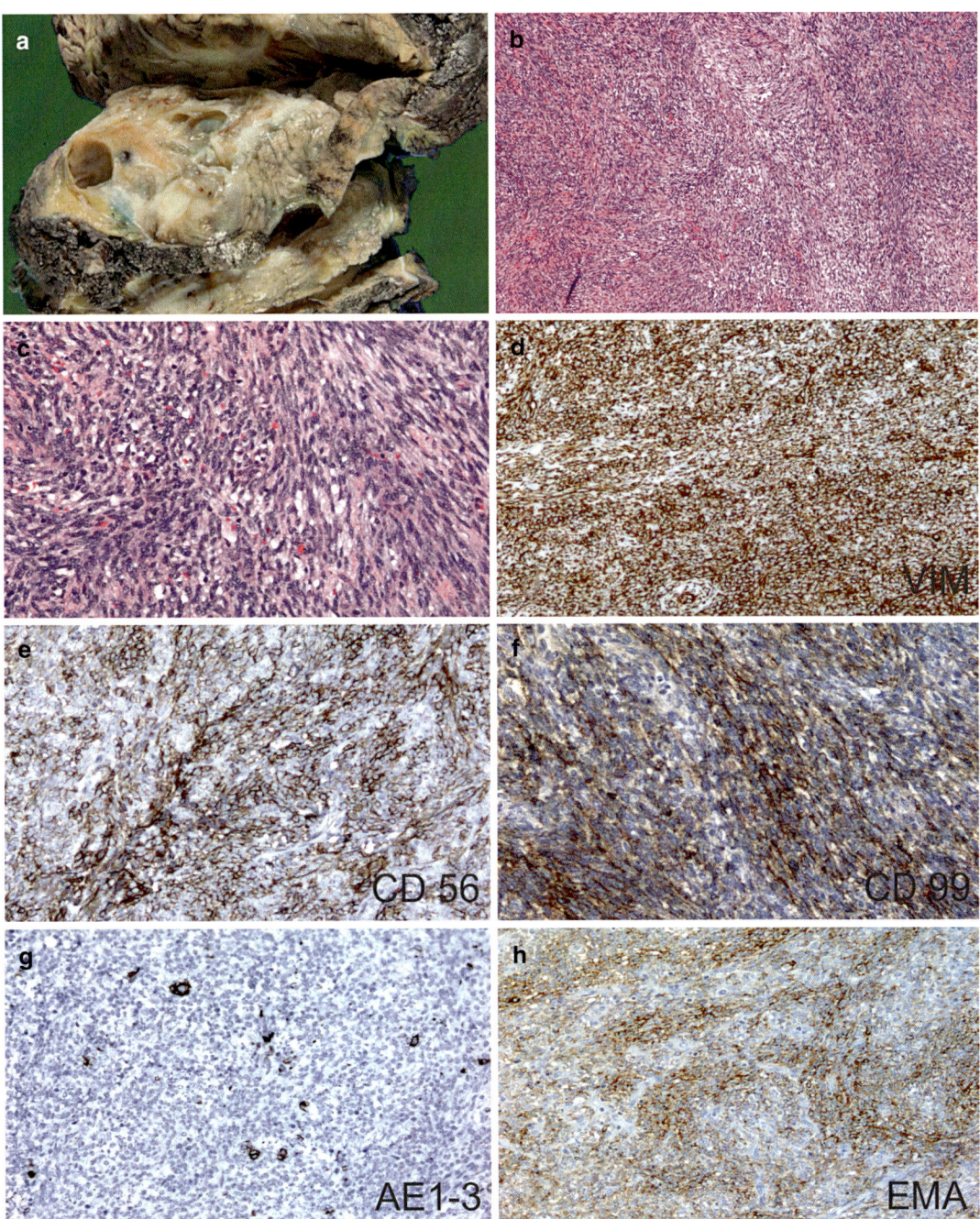

Fig. 12.21 Synovial Sarcoma. Gross photograph (**a**) of a synovial sarcoma with marked heterogeneity and focal cystic formation. The conventional histology (**b**, **c**) shows the infiltration of spindle cells with occasional mitotic figures. There is positivity for immunostaining using vimentin (VIM) (**d**), CD56 (**e**), CD99 (**f**), pan-cytokeratins (AE1-3) (**g**), and epithelial membrane antigen (EMA) (**h**) (b–c, hematoxylin and eosin staining, ×50 and ×200 as original magnifications, respectively; d, anti-VIM immunostaining, ×200 original magnification; e, anti-CD56 immunostaining, ×200 original magnification; f, anti-CD99 immunostaining, ×200 original magnification; g, anti-AE1-3 immunostaining, ×200 original magnification; h, anti-EMA immunostaining, ×200 original magnification; all immunostainigs were performed using the Avidin-Biotin Complex). CD56 is a prototypic marker of Natural Killer cells, but it also regulates homophilic interactions between neurons and, even, between neurons and muscle associated structures with fibroblast growth factor receptor and stimulates tyrosine kinase activity of receptor with the aim to induce neurite outgrowth. Substantially, it contributes to cell-cell or cell-matrix adhesion during development. Conversely, CD99 is a cell surface glycoprotein involved with T cell adhesion, apoptosis of T cells, leukocytic cell migration, and, eventually, differentiation of primitive neuroectodermal cells

fibrous variant resembles other sarcomas and metastatic ACA (if primarily epithelial component).
- *PGN*: (~ from very aggressive to indolent course) 5-YSR, ~50%, and good PGN factors are *child* age (Y/N), *size* (</> 5 cm), *MI* (</> 10/HPF), *necrosis* (N/Y), *calcifying* variant (Y/N), *curative local excision* (Y/N), and *SS18/SSX2 variant* (Y/N).

12.12.5 Epithelioid Sarcoma

- *DEF*: Malignant mesenchymal tissue of the soft tissue with slow, relentless course, multiple focal recurrences, eventually distant metastases (LN and Lung) and death, characterized by INI1 loss, and "poly-phenotypical" differentiation labeling this tumor as "carcinoma of soft tissue," like synovial sarcoma and adamantinoma of soft tissue.
- *INI1* gene (aka *hSNF5* and *SMARCB1*) is a member of SWI/SNF multisubunit chromatin remodeling complex located on 22q11.2. *INI1* action as tumor suppressor gene in exposing DNA to transcription factors (INI1- tumors) is crucial.
- *CLI/EPI*: ES is usually located at the extremities (hands, fingers) of adolescents and young adults.
- *CLM*: Granuloma-like pattern with inconspicuous appearing spindled and polygonal eosinophilic tumor cells with atypical mitoses, typically forming small nests simulating *granuloma annulare* or a *rheumatoid nodule*, and commonly harboring perineural and perivascular infiltration (Fig. 12.22).
- *IHC*: (+) CK (LMWCK and HMWCK), EMA, VIM, (±) CEA, CD34; (−) INI1, S100, CD31. Keratins: CK 8 > CK 19 > CK 14 > CK 7 > CK 20 > CK 5/6 (94%, 72%, 48%, 22%, 15%, 30%, respectively) (Armah & Parwani, 2009).
- *TEM*: Epithelial (desmosomes, tonofilaments, surface microvilli) and mesenchymal differentiation.
- *DDX*: Epithelioid MPNST (S100+, rarely EMA+, CK −), epithelioid angiosarcoma, malignant melanoma (S100+, usually HMB45+). In considering melanocytic lesions, S-100 remains the most sensitive marker. Conversely, markers such as HMB-45, MART-1/Melan-A, Tyrosinase, and MITF (Microphthalmia-associated transcription factor) harbor relatively good specificity but not as good sensitivity as S-100 antibody.
- *PGN*: Very aggressive course with tendency to propagate along fascial planes (5-YSR ~50%) By reviewing the literature, it seems that relatively good PGN factors (*Italic* denotes *good*) are *child* age (Y/N), gender (*female* vs. male), size (Ø </> 5 cm), location (*superficial* vs. deep-seated and *distal* vs. proximal/axial), MI (Ø </> 10/HPF), hemorrhage (N/Y), necrosis (N/Y), *calcifying* variant (Y/N), excision (*curative locally* vs. multiple recurrences, Y/N), rhabdoid features (N/Y), angiolymphatic invasion (N/Y), and LN+ at diagnosis (N/Y).

12.12.6 Alveolar Soft Part Sarcoma

- *DEF*: Alveolar soft part sarcoma (ASPS) is a malignant TUD of deep soft tissues, usually the thigh and leg but also H&N (e.g., orbit, tongue), mostly children and young adults (15–35 yrs.).
- *GRO*: It is a poorly circumscribed tumor with quite often necrosis and hemorrhage.
- *CLM*: Mostly regular nests of 25–50 large (>30 μm), polygonal, loosely cohesive tumor cells with large vesicular nuclei, prominent nucleoli, granular cytoplasm, with irregular (usually "rod-shaped"), +DPAS crystals with thin-walled vascular septs (Fig. 12.23).
- *IHC*: MYO-D1 (cytoplasmic, no nuclear), (+) TFE (TFE also stains Xp11 translocation RCC, focal and less intensely MM, CCS, PEComa, and AML) (RCC, renal cell carcinoma; MM, malignant melanoma; CCS, celar cell sarcoma; PEComa, perivascular epithelioid cell neoplasm; AML, angiomyolipoma), (+) MCT1 and (+) CD147 in the granules (monocarboxylate transporter chaperon-like protein/SLC16A1 located on 1p13.2 and Basigin/EMMPRIN/CD147 located on

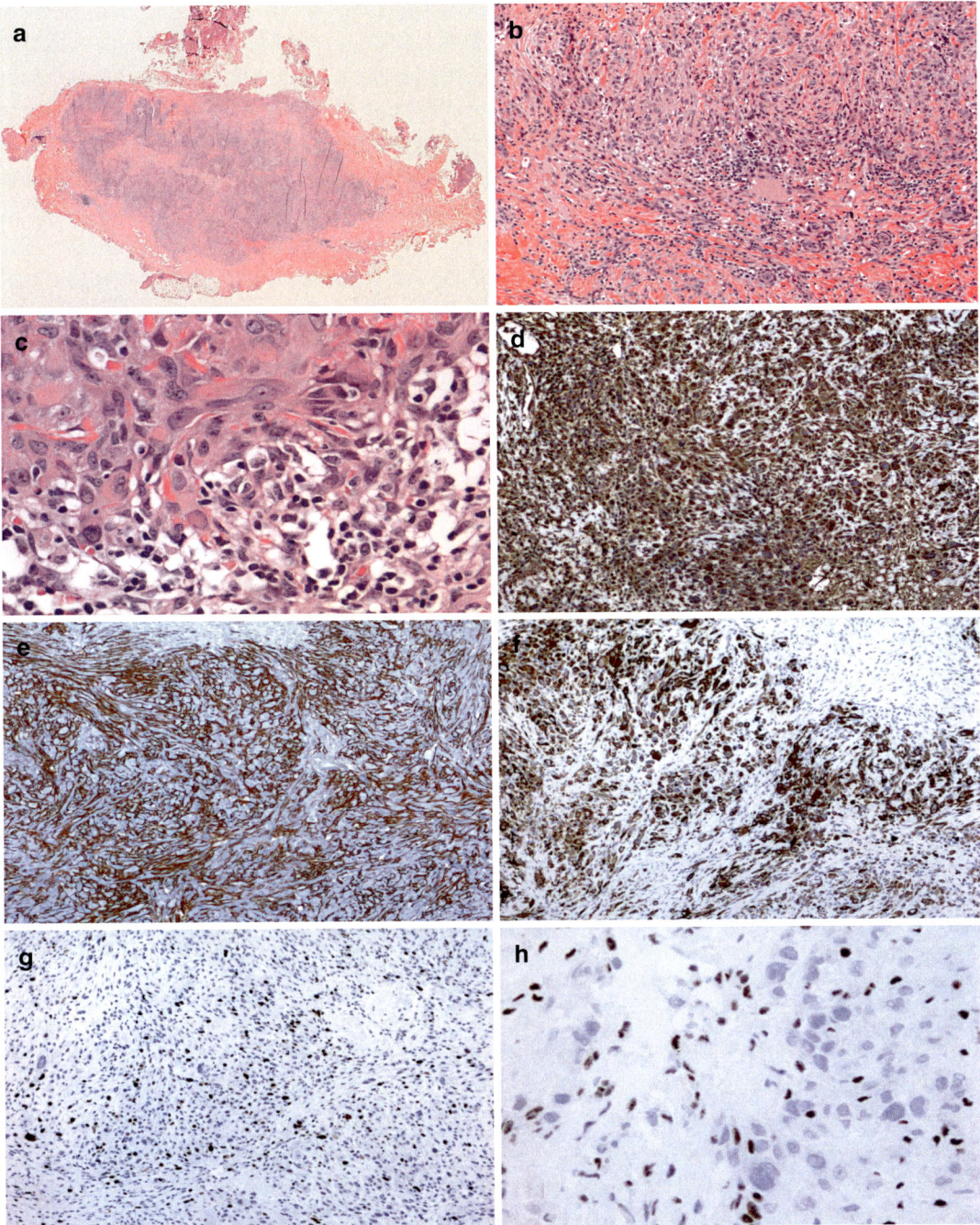

Fig. 12.22 Epithelioid Sarcoma. There is a granuloma-like pattern (**a**) with spindled and polygonal eosinophilic tumor cells with somewhat only occasional prominent nucleolus and atypical mitoses (**b–c**). At places, the cells form small nests simulating *granuloma annulare* or a quite typical rheumatoid nodule. There is immunophenotypical positivity for some markers, which support the diagnosis of epithelioid sarcoma (vimentin, VIM (**d**); actin, ACT (**e**); pan-cytokeratins, AE1-3 (**f**)), increased proliferation rate (**g**) and loss of INI1 (**h**) (a–c, hematoxylin and eosin staining, ×0.5, ×100, ×400 as original magnification, respectively; d, anti-VIM immunostaining, Avidin-Biotin Complex, ×100 original magnification; e, anti-ACT immunostaining, ×100 original magnification; f, anti-AE1-3 immunostaining, Avidin-Biotin Complex, ×100 original magnification; g, anti-Ki67 (MIB1) immunostaining, Avidin-Biotin Complex, ×100 original magnification; h, anti-INI1 immunostaining, Avidin-Biotin Complex, ×400 original magnification). INI1 (hSNF5/SMARCB1) is a member of the SWI/SNF chromatin remodeling complex. The INI1 gene is located on chromosome 22q11.2 and is deleted or/and mutated in malignant rhabdoid tumors (MRT) of infancy, epithelioid sarcoma as well as other malignancies (see text for details)

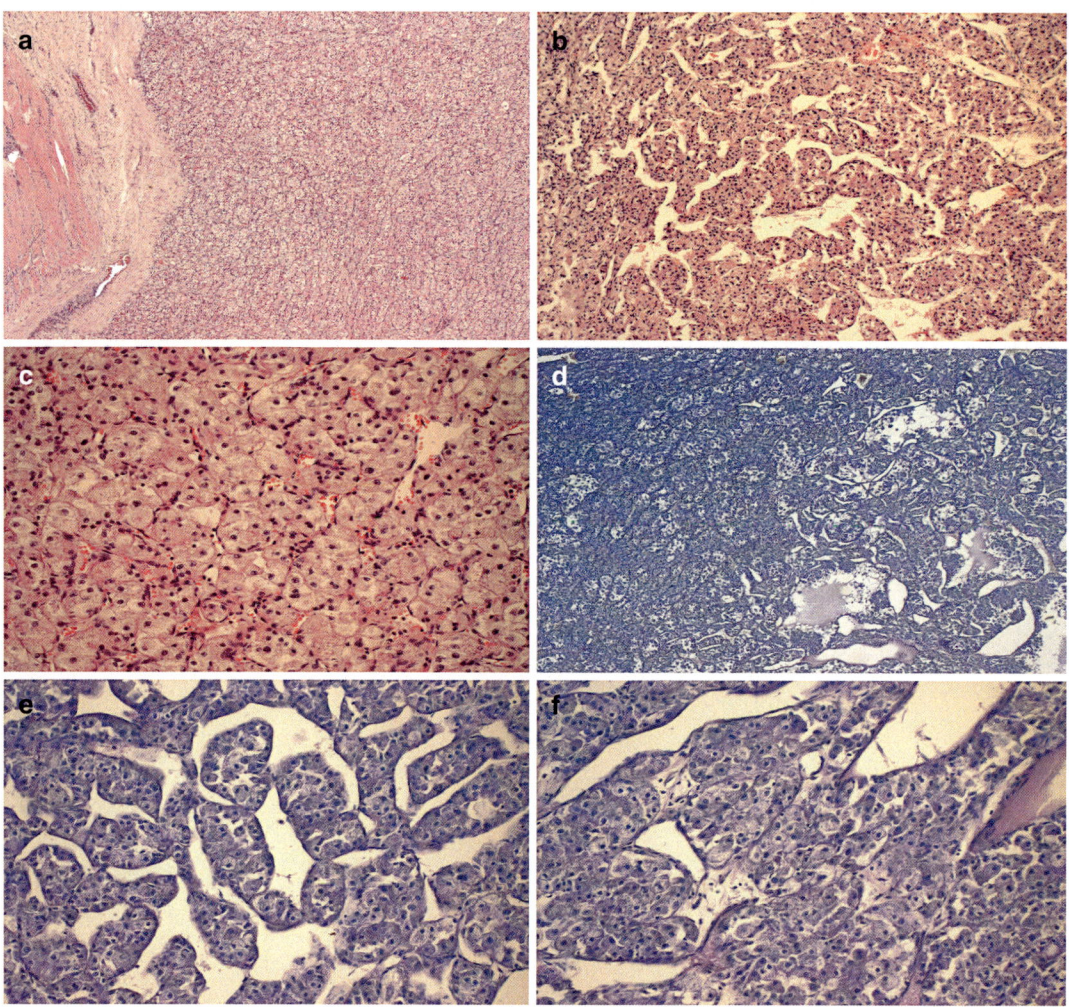

Fig. 12.23 Alveolar Soft Part Sarcoma. There are regular nests of large, polygonal, loosely cohesive tumor cells with large vesicular nuclei, impressively prominent nucleoli, and diffuse granular cytoplasm (**a–f**) (a, hematoxylin and eosin staining, ×40 original magnification; b, hematoxylin and eosin staining, ×100 original magnification; c, hematoxylin and eosin staining, ×200 original magnification; d, periodic acid Schiff, ×40 original magnification; e, periodic acid Schiff, ×200 original magnification; f, periodic acid Schiff, ×200 original magnification)

19p13.3), (−) MYOG (MYF4, myogenin, myogenic factor 4). The DNA-binding factor TFE3 is closely related to MiTF.
- *TEM*: Membrane-bound or membrane-free rhomboid crystals with periodicity of ~10 nm and "cross grid" pattern, numerous electron-dense vesicles near Golgi apparatus, smooth tubular aggregates associated with invaginations of the cytoplasmic membrane, but no features of skeletal muscle differentiation.
- *CMB*: *der(17)t(X;17)*(p11.2;q25) ⇒ ASPL/TFE3.
- *DDX*: ARMS, which shows variably shaped alveolar spaces vs. regular alveolar spaces, fibrous septs vs. thin-walled vascular septs, 10–30 um cellular Ø, cellular pleomorphism, dense nuclei with small nucleoli vs. vesicular nuclei with prominent nucleoli, no crystals, nuclear MYO-D1, (+) MYOG (MYF4, Myogenin, Myogenic Factor 4), t(2;13) or t(1;13) vs. der(17)t(X;17)(p11.2;q25).
- *PGN*: Highly malignant tumor with a tendency to easily metastasize and lethal over a protracted course of 5–10 years with lung, bone,

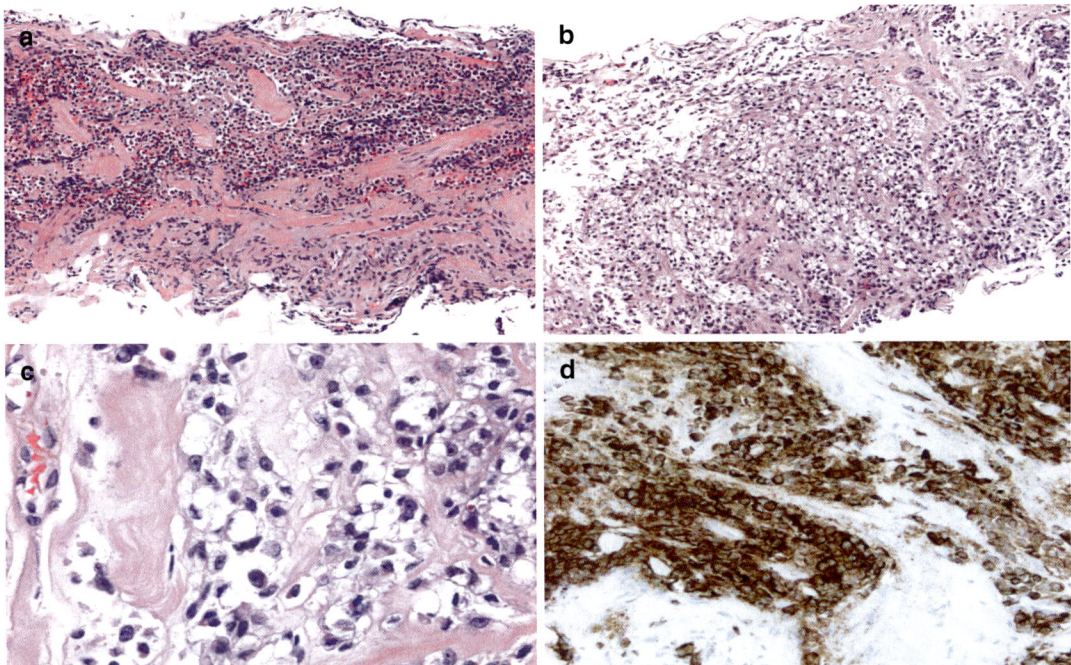

Fig. 12.24 Clear Cell Sarcoma. There are irregular nests of cells separated by fibrous septa as well as short interlacing fascicles. In addition, whorls or solid sheets of tumor cells are also recognized (**a–b**). The tumor cells are round to spindle with hyperchromatic nuclei and clear cytoplasm (**c**). There is immunopositivity for Human Melanoma Black 45 (HMB45) (**d**) (a–c, hematoxylin and eosin staining; a, ×100 original magnification; b, ×100 original magnification; c, ×400 original magnification, d, antii-HMB45 immunostaining, Avidin-Biotin Complex, ×200 original magnification). HMB45 is a monoclonal antibody that recognizes the melanosomal glycoprotein gp100 (Pmel17). HMB45 is required to create and arrange melanosomal fibrils, facilitates maturation of stage I pre-melanosomes to stage II melanosomes, and detects the oncofetal glycoconjugate associated with immature melanosomes

and brain metastases (5-YSR: ~60%). Favorable PGN factors are child age (Y/N), size (Ø</>5 cm), and M (−) at diagnosis (Y/N). Histologic features seem to play no role.

12.12.7 Clear Cell Sarcoma

- *DEF*: Soft tissue sarcoma of young adults with evidence of melanocytic differentiation occurring mostly in large tendons or aponeuroses of the extremities (feet/ankles) of young adults (♂ < ♀) of the third to fourth decades.
- *GRO*: Well-circumscribed, gray-white firm lesion with a gritty texture.
- *CLM*: *Irregular* nests of cells separated by *fibrous* septa as well as short interlacing fascicles, whorls or solid sheets of tumor cells, which are round to spindle with vesicular nuclei and prominent nucleolus, clear to granular cytoplasm associated with wreath-like (or floret-like) MNGC, and occasional melanin pigment and/or Fe-granules (Fig. 12.24).
- *IHC*: (+) S100, HMB45, Melan-A/MART1, MiTF, other than VIM, but also (±) NSE, SYN, CD57, keratins, and (−) SMA, DES, CAM5.2. CAM5.2 is a low-molecular weight marker against keratins 7 and 8 (CK8>CK7), while MITF (MiTF) or Microphthalmia-associated Transcription Factor, which is also often also known as class E basic helix-loop-helix protein 32, is a key marker. Alternatively, MiTF is also known as bHLHe32. MiTF is a transcription factor that in humans is encoded by the *MITF* gene. MITF is a basic leucine

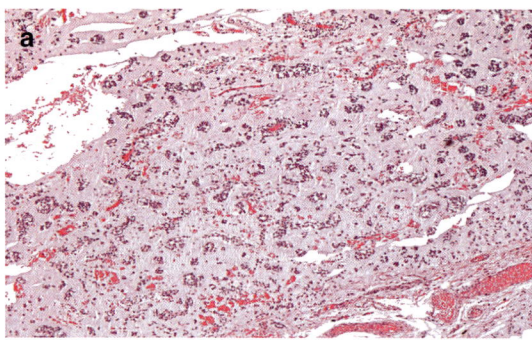 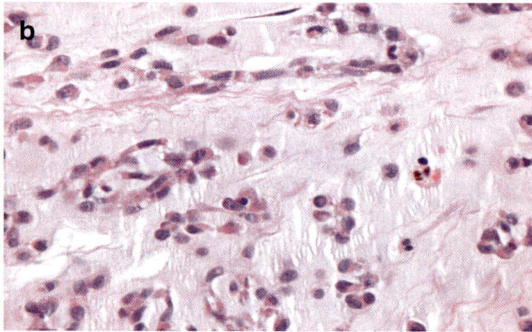

Fig. 12.25 Extraskeletal Myxoid Chondrosarcoma. There are lobules with chondroblast-like cells arranged in cords, clusters, or networks floating in an abundant myxoid matrix (**a–b**). The tumor cells have a straight round to oval nuclei with a small inconspicuous nucleolus. The tumor cells show also a modest amount of deeply eosinophilic cytoplasm (a, hematoxylin and eosin staining, x50 original magnification; b, hematoxylin and eosin staining, x400 original magnification

zipper (helix-loop-helix) transcription factor (Dickson et al. 2011). MITF is engaged in lineage-specific pathway regulation of several types of cells including melanocytes, osteoclasts, and mast cells.
- *TEM*: Stage II/III melanosomes, glycogen, swelling of mitochondria, and basal lamina.
- *CMB*: t(12;22)(q13;q12) – *ATF1* and *EWS* (not seen in melanoma); usually 2n or less aneuploidy than metastatic melanoma to soft tissue. In the GI tract, CCS may show a variant fusion gene *EWSR1-CREB1*.
- *PGN*: Highly malignant tumor with a tendency to readily metastasize: 5-YSR, ~67%, and favorable PGN factors are the *sizes* (Ø</>5 cm), *necrosis* (N/Y), and *local recurrence* (N/Y).

12.12.8 Extraskeletal Myxoid Chondrosarcoma (ESMC)

- *DEF*: Malignant soft tissue tumor of the proximal extremities (thigh) and limb girdles of usually middle-aged adults with very few cases in youth (sixth decade, ♂ > ♀) characterized by extended survival, but high potential for local recurrence and metastases.
- *GRO*: Well-circumscribed tumor with multinodular gelatinous architecture delimitated by a pseudocapsule and fibrous septs, which may be festooned with intratumoral cysts and hemorrhagic areas.
- *CLM*: Multiple lobules with malignant chondroblast-like cells arranged in cords, clusters, or networks floating in the abundant myxoid matrix and increased cellularity at the periphery. The tumor cells have a straight round to oval nuclei with a small inconspicuous nucleolus and a modest amount of deeply eosinophilic cytoplasm, which can exhibit from fine granularity to vacuolation. MI is low (usually <2/10HPF) (Fig. 12.25).
- *IHC*: + VIM, ± S100, EMA, SYN, INI1.
- *CMB*: EWSR1 translocation in ~ ½ of cases that often involves t(9;22)(q31;q12); *EWSR1-NR4A3*; *EWS-CHN* or *TAF2N-CHN* fusion gene transcripts.
- *TRT*: Surgical resection with clear margins with adjuvant chemotherapy.
- *PGN*: Long survival harboring malignant tumor with tendency to local recurrence and metastases (5-YSR, ~90%). Meis-Kindblom et al. (1999) studied 117 cases of ESMC. In their multivariate statistical analysis, these authors identified adverse prognostic factors, which are larger tumor size, proximal tumor location, older patient age, and metastasis. We and others still consider some histologic features also key. In a nutshell, good PGN factors are *age* (young vs. old), *size* (Ø ≤10 cm vs. Ø > 10 cm), *location* (proximal vs. distal; N/Y), and *anaplastic cytology with rhabdoid features* (N/Y).

12.12.9 PNET/Extraskeletal Ewing Sarcoma (ESES)

- *DEF*: Aggressive small round blue cell neoplasm, morphologically similar to Ewing sarcoma of the bone of children and young adults at lower extremities, chest wall (Askin tumor), and paravertebral region with a spectrum of appearance from undifferentiated to forming rosettes (Homer-Wright rosettes) and may represent extension of bone tumor into soft tissue.
- *GRO*: It is a well-circumscribed and whitish to yellowish tumor with scattered hemorrhagic necrosis.
- *CLM*: Small round/oval cells with nuclei exhibiting fine chromatin and small nucleoli and scanty clear cytoplasm containing glycogen (PAS+); peritheliomatous pattern (concentration around blood vessels) (Fig. 12.26). Skeletal Ewing sarcoma cells have been depicted having more evident and substantial cytoplasm than extra-skeletal Ewing sarcoma cells.
- *IHC*: (+) VIM, CD99 ± neuroendocrine lines of differentiation (e.g., S100), (+) DES only in the ectomesenchymoma variant, but it is controversially debated; (−) Pan-Keratins, AE1-3 (usually).
- *TEM*: Abundant cytoplasmic glycogen, poorly developed cell junctions, no neural features.
- *CMB*: The translocations are listed in Box 12.18. Slides are usually cut from paraffin blocks. The slides are deparaffinized in xylene and dehydrated in ethanol. Specimen pretreatment and fluorescence in situ hybridization (FISH) need to be applied. Vysis® (Abbott Molecular Inc.) pretreatment kit and appropriate probes are commonly used. The LSI EWSR1 dual color probe consists of a 500 kb Spectrum-Orange (red) labeled probe that specifically flanks the 5' side of the *EWSR1* gene, and extends inward into intron 4. The second probe is a 1100 kb Spectrum-Green (green) probe, which specifically flanks the 3' side of the *EWSR1* gene. It is known that the known breakpoints within the *EWSR1* gene are restricted to introns 7 through 10. FISH discloses a normal pattern, which includes normal nuclei containing two red/green (yellow) fusion signals. In a normal cell that lacks a t(22q12) in the EWSR1 gene region, a two fusion signal pattern is observed. However, in an abnormal cell with a simple t(22q12), the expected signal pattern will be one fusion, one red and one green signal.

> **Box 12.18 PNET: Chromosomal Translocations**
> - *t(11;22)(q24;q12)* responsible for the fusion transcript of *FLI1-EWS* genes
> - *t(21;22)(q12q12)* creating the gene fusion transcript *ERG-EWS*
> - *t(7;22)(p22;q12)* making the gene fusion transcript *ETV1-EWS* genes
> - *t(17;22)(q12;q12)* producing the gene fusion transcript *E1AF-EWS* genes
> - *t(2;22)(q33;q12)* forming the gene fusion transcript *FEV-EWS* genes

- The lab's normal cut off is established at ≤10% of the cells showing an abnormal signal pattern. In each reaction it is imperative that all appropriate negative and positive controls are used. Any FISH result needs to be interpreted within the context of all available clinical, morphologic, cytogenetic, immunophenotypic, and molecular diagnostic information. The most common translocation found in PNET/ESES is the t(11;22)(q24;q12) resulting in the *EWS/FLI-1* fusion gene found in approximately 90% of cases. The next most common translocation is t(21;22)(q22;q12) resulting in the fusion of *EWSR1* to the *ERG* (21q22). The remaining translocations make up less than 5% of cases.
- *DDX*: The DDX includes a series of tumors that are the following conditions:
  - *Neuroblastoma* (dense chromatin, neurosecretory granules, +CGA, −CD99), *RMS* (nuclear irregularity, dense chromatin, variability in cytoplasmic amount, no rosettes, ±CD99, IHC evidence of myogenic differentiation, t(2;13) or t(1;13) in ARMS)

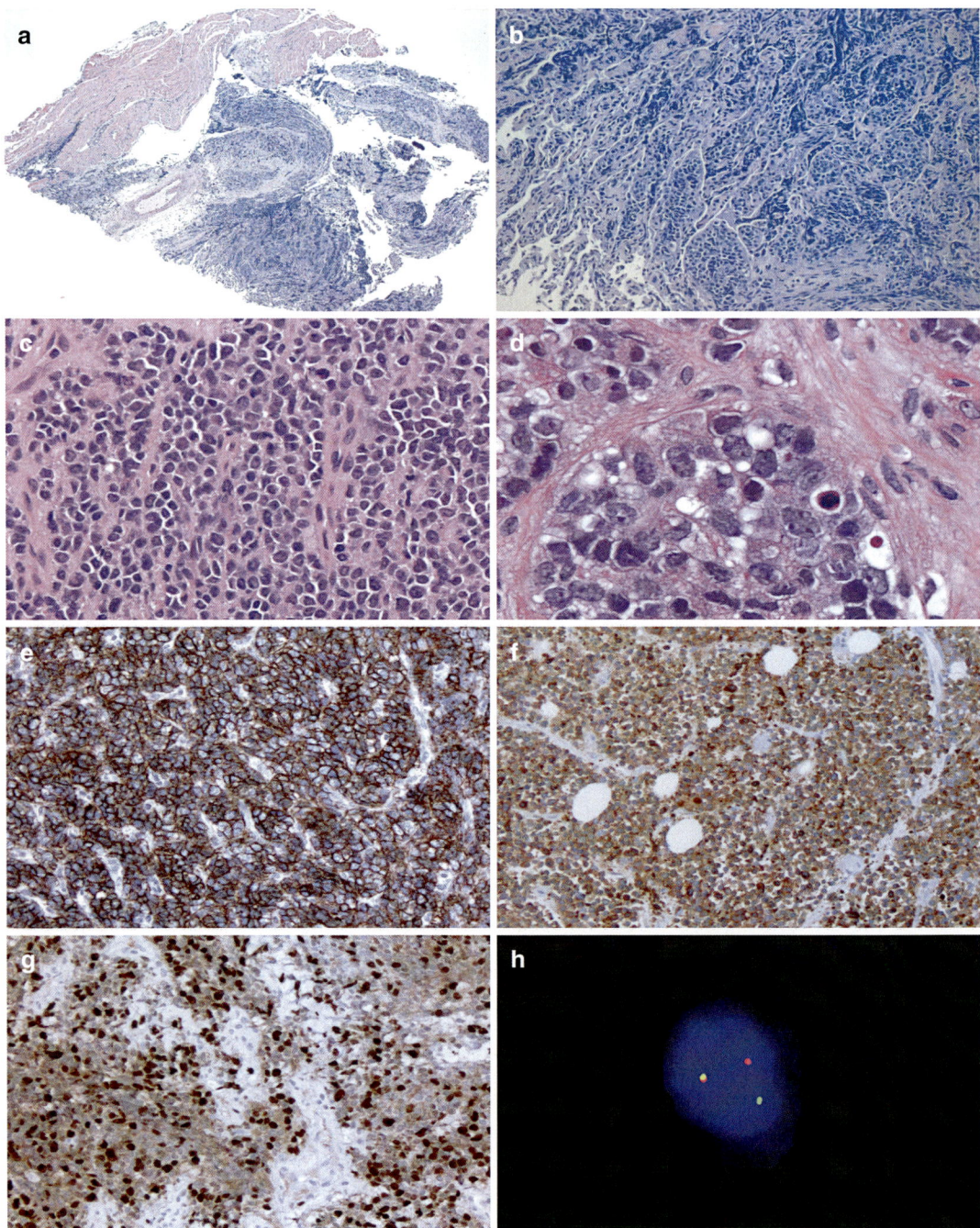

Fig. 12.26 Extraskeletal Primitive Neuroectodermal Tumor – Extraskeletal Ewing Sarcoma (ESES). Primitive neuroectodermal cells growing in lobules where the cells are packed (**a–c**) showing a high nucleus to cytoplasm ratio (**d**). There is immunopositivity for CD99 (**e**) and synaptophysin (SYN) (**f**). The tumor shows also a high proliferation rate (**g**). The immunofluorescent photograph (**h**) demonstrates the expected chromosomal rearrangement (a–d, hematoxylin and eosin staining; ×40, ×200, ×400, ×630 as original magnification, e, anti-CD99 immunostaining, Avidin-Biotin Complex, ×200 original magnification; f, anti-SYN immunostaining, Avidin-Biotin Complex, ×200 original magnification; g, anti-Ki67 immunostaining (MIB1), Avidin-Biotin Complex, ×200 original magnification)

- *Lymphoma* (LCA and B/T lineage positivity) CD99 is not reliable in the differential diagnosis with lymphoproliferative processes, because may be positive up ½ of cases with T-leukemia and T-lymphoblastic lymphoma.
- *DSRCT* (prominent DSR, +WT1, DES, AE1–3, EMA, ± CD99, t(11;22)(p13;q12)
- *Esthesioneuroblastoma* (olfactory neuroblastoma) (S100+ sustentacular cells, − CD99)
- *Small cell osteosarcoma* (malignant osteoid, ± CD99)
- *Mesenchymal CS* (islands of hyaline cartilage, +S100, ± CD99, lack of t(11;22)
- *Wilms tumor* (kidney, ± CD99, lack of t(11;22)
- *Oat cell carcinoma* (older, scant cytoplasm, NN. molding, + AE1–3, ± CD99, lack of t(11;22)

• PGN: Several factors play a role in prognosis (stage, primary tumor site, size, age, and response to therapy). Some authors found that chimeric transcripts in bone marrow of apparently nonmetastatic PNET cases at presentation can be seen in up to 43% of cases. This finding has been associated with an unfavorable outcome. However, other authors have not demonstrated any significant association between chimeric transcript detection at the time of diagnosis and outcome. There is a substantial association between ↑ risk of recurrence and detection of occult tumor cells by RT-PCR during follow-up and patients with more than 90% necrosis, who showed chimeric transcript during follow-up, develop metastasis. Moreover, the risk of local recurrence is associated with the status of the resection margins. The TP53 expression is increased in EWS-FLI1-expressing cells, and cases with p53 > 20% have significantly poorer overall survival among patients with localized disease. In multivariate analysis, p53 > 20% is one of the most influential negative prognostic factors. In ~30% of PNET, there is a homozygous loss of p16, which regulates cell cycle progression. By univariate analysis, a study demonstrated p16/p14ARF deletion alone had only marginal value as a negative factor. Also, by multivariate analysis, p16/p14ARF homozygous deletion is the second most significant factor after p53 mutation (Hense et al. 1999; Eralp et al. 2002; Desai et al. 2010; Verma et al. 2017).

In the context of PNET, it should also be considered the melanotic progonoma or melanotic neuroectodermal tumor of infancy (*Retinal Anlage Tumor*) (Kruse-Losler et al. 2006). Melanotic progonoma occurs in children <6 months old, no gender difference, mostly with anterior maxilla or other head and neck locations. Microscopically, neuroblasts and melanoblasts in nests are observed. The progonoma invades but rarely metastasizes. Since its first clinicopathological description by Krompecher about 100 years ago, more than 360 cases of melanotic progonoma have been reported in the medical literature (Krompecher 1918; Borello & Gorlin 1966; Soles et al. 2018). Apart of the neuroectodermal derivation, it does not seem the two tumors share common cytogenetic or molecular biology pathways. Finally, the pathologist should keep in mind another category labeled as "Ewing-like Sarcomas". This group of tumors are undifferentiated round cell mesenchymal neoplasms that morphologically resemble classical ES, but lack the molecular hallmark for this disease (*EWSR1–ETS* fusion). Four main types of Ewing-like sarcoma have been delineated: *CIC*-rearranged sarcomas, *BCOR*-rearranged sarcomas, sarcomas with a fusion between *EWSR1* and a gene not belonging to the *ETS* family members, and unclassified round cell sarcomas.

12.12.10 Desmoplastic Small Round Cell Tumor

• DEF: Aggressive *small round blue cell tumor* of mostly children, adolescents, and young adults with preferentially intra-abdominal location (multinodular and lobulated) with pain, ascites, visceral obstruction (colon, ureter, and biliary system), keratin and desmin co-expression, and t(11;22)(p13;q12) resulting in a chimeric *EWS-WT1* fusion product.

- *GRO*: Large, multinodular to lobulated mass, often accompanied by several small peritoneal implants with a firm, gray-white cut surface and foci of hemorrhage and necrosis.
- *CLM*: Nests (± confluence) with angulated edges and peripheral palisading feature embedded in a fibroblastic/myofibroblastic desmoplastic stroma. Tumor cells are small, uniformly round, and hyperchromatic with dispersed chromatin, scant cytoplasm, and indistinct cell borders. ± Focal necrosis, cysts, glands, and BV with ectasia.
- *IHC*: (+) VIM, WT1, DES (punctate perinuclear staining), CKs, EMA, NSE, CGA, SYN, ±CD57, CD99; (−) LCA, SMA, HHF-35 (MSA, muscle specific actin), MYOG, MYOD, GFAP, NF, CEA, HMB45, AFP, CK5/6, and CK20. The expression of INI1 is retained.
- *TEM*: Paranuclear bundles of intermediate filaments that can entrap other organelles and few cell junction complexes, including desmosomes. Desmosomes are intercellular junctions that provide strong intercellular adhesion and they also link intracellularly to the intermediate filament cytoskeleton building the strong adhesive bonds in a network that gives mechanical strength to tissues.
- *CMB*: t(11;22)(p13;q12) ⇒ C-terminus DNA-binding domain of WT-1 (11p13) → N-terminus of EWS (22q12).
- *DDX:* It includes mostly RMS and EWS.
- *PGN*: Poor (5-YSR < 10%).

12.12.11 Extrarenal Rhabdoid Tumor

- *DEF*: Extrarenal rhabdoid tumor (ERT) is a highly malignant TUD with similarities to its renal counterpart of infants and children and characterized by rhabdoid features of tumor cells with prominent nucleoli and eosinophilic cytoplasmic inclusions, whorls of intermediate filaments ultrastructurally, and growing in solid sheets exhibiting INI1 loss. It may represent the confluence of an aggressive phenotype of different tumors, such as epithelioid sarcoma, DSRCT, RMS, melanoma, carcinoma.
- *IHC*: (+) VIM, keratins, EMA, but (−) S100 and muscle markers.
- *PGN*: Poor with metastases (lung, liver, nodes) regardless of therapy.

12.12.12 Malignant Mesenchymoma

- *DEF*: Soft tissue malignant mesenchymoma (STMM) is a rare high grade and aggressive soft tissue tumor showing ≥ sarcomatous lines of differentiation (osteosarcoma, chondrosarcoma, leiomyosarcoma, rhabdomyosarcoma, and liposarcoma). There is no fibrosarcoma because some areas are present in most sarcomas and no well-characterized sarcomas with dual differentiation, such as dedifferentiated liposarcoma, dedifferentiated chondrosarcoma, malignant triton tumor, and liposarcoma or chondrosarcoma with myoblastic differentiation. It occurs more often in retroperitoneum or chest wall, and poor PGN factors are young age and rhabdomyosarcomatous component.

12.12.13 PEComa

- *DEF*: Mesenchymal tumor with undefined or better probable malignant potential and characterized by perivascular *clear/epithelioid* cells (*p*erivascular epithelioid *c*ell tumor *-oma*) and characteristic myomelanocytic immunologic differentiation although no known regular counterpart to the perivascular epithelioid cell is known. This class of tumors includes angiomyolipoma (AML), clear cell "sugar" tumor of the lung (CCST), lymphangioleiomyomatosis (LAM), clear cell myomelanocytic tumor of the falciform ligament (CCMMT), and clear cell tumors of other organs, such as the pancreas, rectum, abdominal serosa, uterus, vulva, thigh, and

heart. PEComa involves a wide age range of mostly young to middle-aged females.
- *GRO*: PEComa is sharply demarcated. It shows a lobulated firm gray-tan nodular appearance and focally hemorrhagic areas on the cut surface.
- *CLM*: Perivascular or radial arrangement around BV lumina with epithelioid and spindle features for cells closest to BVs and cells remote from BVs, respectively. Tumor cells have small, central, round/oval NN with small nucleoli and clear to the granular eosinophilic cytoplasm; + MNGC, malignant features (marked atypia, brisk MI, atypical mitoses, and necrosis). A remarkable feature is the small arcing BVs that subdivides the tumor nodules in coarse packets recalling CCSK or RCC (clear cell sarcoma of the kidney or renal cell carcinoma).
- *IHC*: (+) VIM, (+) myo- (SMA, DES, MyoD1), and melanocytic markers (HMB-45, Melan/A, MTF or melano-transferrin, CD63, Tyrosinase); (−) AE1–3, CD117/c-kit, CD34. CD63 (aka NKI-C3, melanoma associated antigen and ME491) is a lysosomal membrane-associated glycoprotein 3 or LAMP-3, which belongs to the tetraspanin superfamily of integral membrane proteins.
- *TEM*: Abundant cytoplasmic glycogen, premelanosomes, thin filaments ± dense bodies, hemidesmosomes, but poorly formed cellular junctions.
- *CMB*: 19-, 16p-, 17p-, 1p-, 18p-, X+, 12q+, 3q+, 5+, 2q+; 16p- (TSC2 gene loss).
- *DDX*: It includes the following diagnoses:
  – Clear cell/epithelioid/ smooth muscle tumors (HMB45-)
  – Clear cell/oxyphilic carcinoma (CKs+)
  – Malignant melanoma (S100+)
  – Pecomatosis (nests of perivascular clear to eosinophilic cells mocking malignant mesothelioma)
  – Pecosis (continuous layer of clear cells with perivascular location remotely from tumor and showing the transition to invasive nests and PEComa)
  – Undifferentiated/high-grade sarcoma.
- *PGN*: Size, infiltrative growth pattern, high nuclear grading, high MI (>1/50HPF), atypical mitoses, and coagulative cell necrosis

12.12.14 Extrarenal Wilms' Tumor

The nephroblastoma or Wilms' tumor is one of the most common childhood solid tumors with three components, including blastema, epithelia (tubules), and stroma. It arises from primitive metanephric cells, but exceptionally it may arise in places other than kidneys. From the oncologic point of view, the extrarenal Wilms' tumor is a rare and challenging neoplasm. Most of the authors support the thesis that the ERWT arises from primitive ectopic nephrogenic rests, but teratoid Wilms' tumor leads to the debate whether this tumor is truly neoplastic or embryonic. A challenging staging complicates the diagnosis. If we consider the National Wilms' Tumor Study (NWTS) recommendations, a nephroblastoma should be considered as stage II or higher as it is beyond the renal capsule. This finding means that the child will need chemotherapy, but most of the case reports illustrate favorable histology and a good outcome with surgery alone (Fig. 12.27).

12.12.15 Sacrococcygeal Teratoma and Extragonadal Germ Cell Tumor and Yolk Sac Tumor

The most common teratoma in children, mainly newborns with female preponderance ♂ > ♀. In 95%, sacrococcygeal teratomas (SCTs) are benign, even when immature elements are present, but the risk of malignancy increases with age of the patient. Teratoma is defined by the presence of tissues from more than one of the three primitive germ cell layers (e.g., teeth, skin, and hair derived from *ectoderm*; cartilage, connective tissue, and bone from *mesoderm*; and intestinal, bronchial, or pancreatic tissue from *endoderm*) (Fig. 12.28). It may be considered a form of parasitic twin

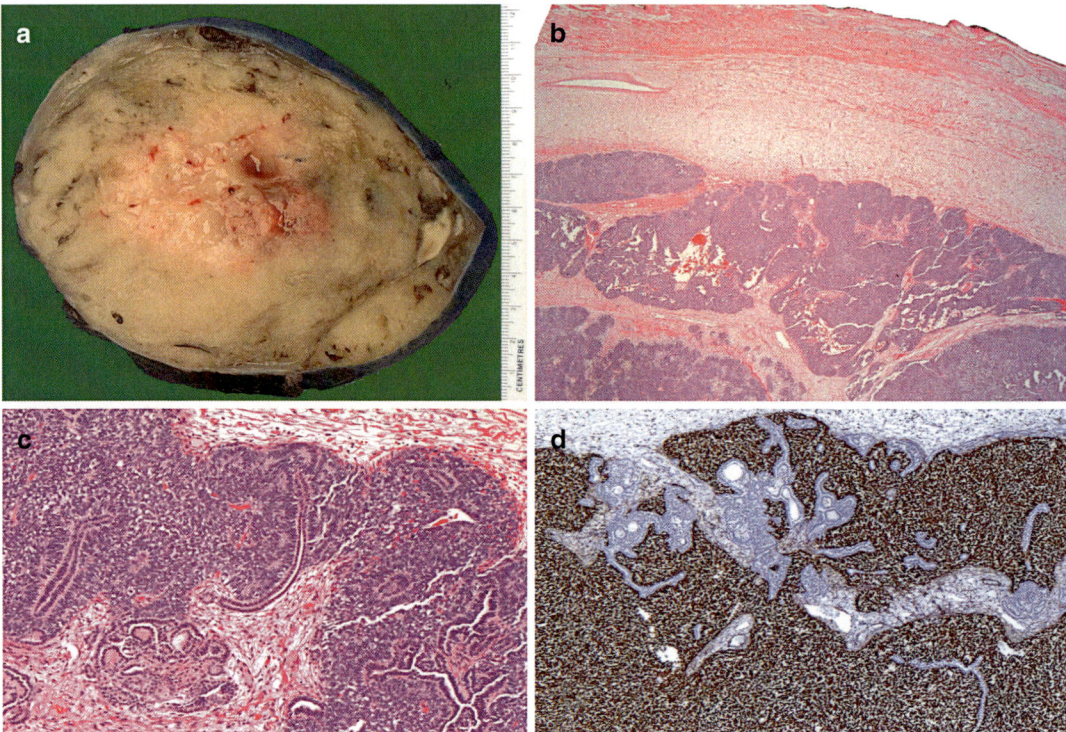

Fig. 12.27 Extrarenal Wilms' Tumor. Gross photograph (**a**) of a solid tumor with soft, brain-tissue like texture showing small round blue cells growing in a solid or serpentine fashion highlighting the blastema character of the tumor cells (**b–c**). In (**d**) there is frank immunopositivity for WT1 (see text for details). (b–c, hematoxylin and eosin staining, ×12.5 and ×100 as original magnification, respectively; d) anti-WT1 immunostaining, Avidin-Biotin Complex, ×50 original magnification)

due to the presence of organs and tissue differentiated toward specific units. In fact, we could find a fully formed eye in a sacrococcygeal teratoma (Sergi et al. 1999). There is a high relevance in considering the possibility to have also malignant components, such yolk sac tumors with polytopic differentiation and choriocarcinoma as well as small round blue cell tumors. One of the challenges of the SCT is that it may be impossible to gross it entirely. The behavior of congenital teratomas remains poorly understood and is frequently considered unpredictable, but studies of large series with follow-up are limited. If it is certain that benign SCTs have a specific potential to malignant manifestations increasing with age, i.e., at the age of 9 months in about 70% of cases of malignant SCTs, it is also stressed that malignant relapses of SCT after primary resection classified as benign initially may develop later on. Moreover, SCTs, dependent on their position, size, and composition can be associated with intrauterine complications, such as urological, anorectal, and osteo-muscular ones, which can be not entirely known at birth. In an extensive study performed at the Children's Hospital, University of Innsbruck, Austria, Hager et al., published in 2012, one of the most extensive reviews of SCTs with follow-up was carried out spanning between January 1968 and December 2011 identifying 29 children who underwent surgical interventions for an

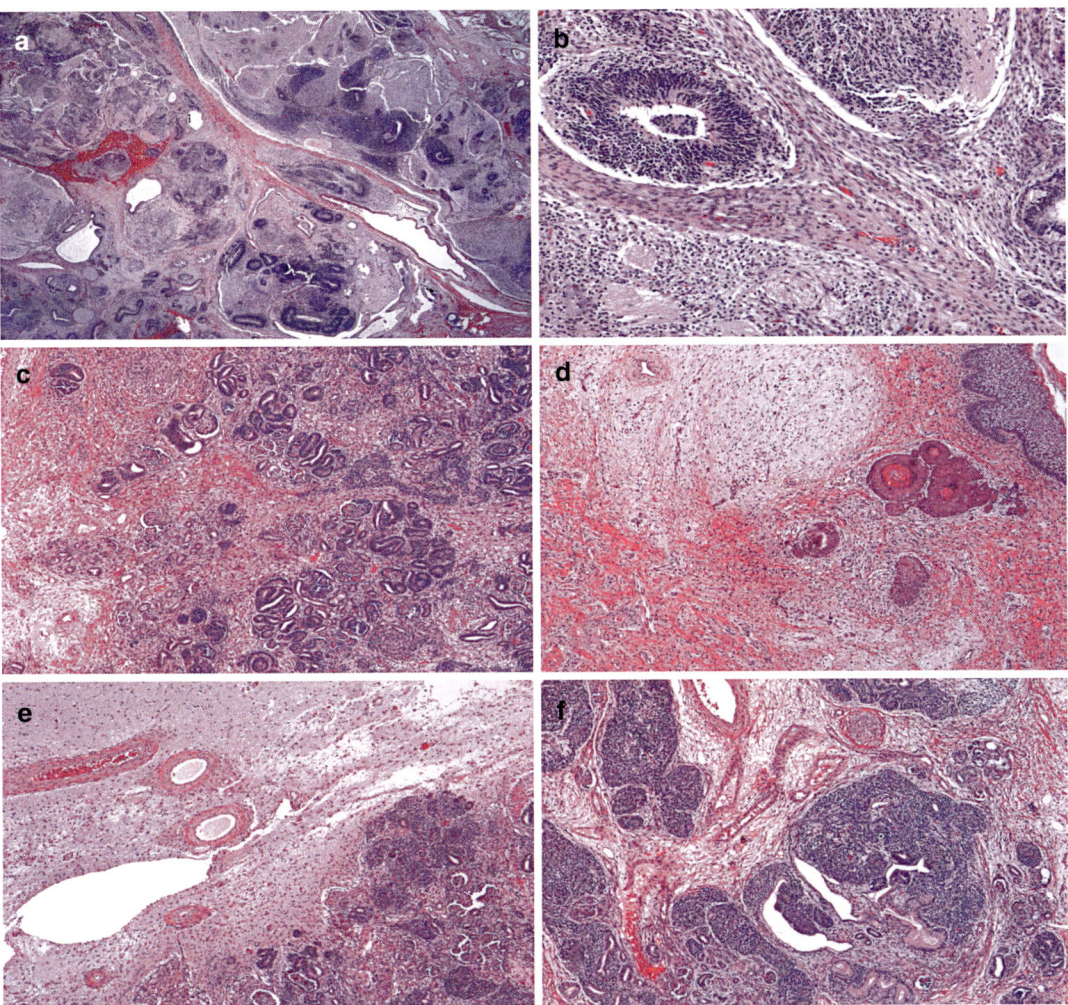

Fig. 12.28 Sacrococcygeal Teratoma. In this teratoma of the sacrococcygeal region, there are lines of differentiation from the endodermal, mesodermal and ectodermal layers of the primitive embryonal differentiation that are similar to other teratomas identified in other body areas (**a–f**). The neuroectodermal tissue, as an example highlighted in (b) is very important for the grading of the immaturity of the teratoma (a–f, hematoxylin and eosin staining, ×12.5, ×100, ×50, ×50, ×50, ×50 as original magnification, respectively)

SCT at the Austrian Children's hospital, which is one of the most famous and high-ranked Children's hospitals worldwide. One more patient was added who was treated for a malignant relapse. Clinical records of 24 children (all survived) and of the girl with the relapse were reviewed, and comparison among three of the most common histological classifications of SCT was made using the histology available of 25 cases. Histological re-evaluation showed no differences between the groupings. Despite the heterogeneity of treatment within the 43 years, 24 patients survived. The histological re-examination of the slides showed no significant differences. However, a yolk sac tumor could be identified in the relapse even if the original tumor seemed to be devoid of a malignant component, and benign components can

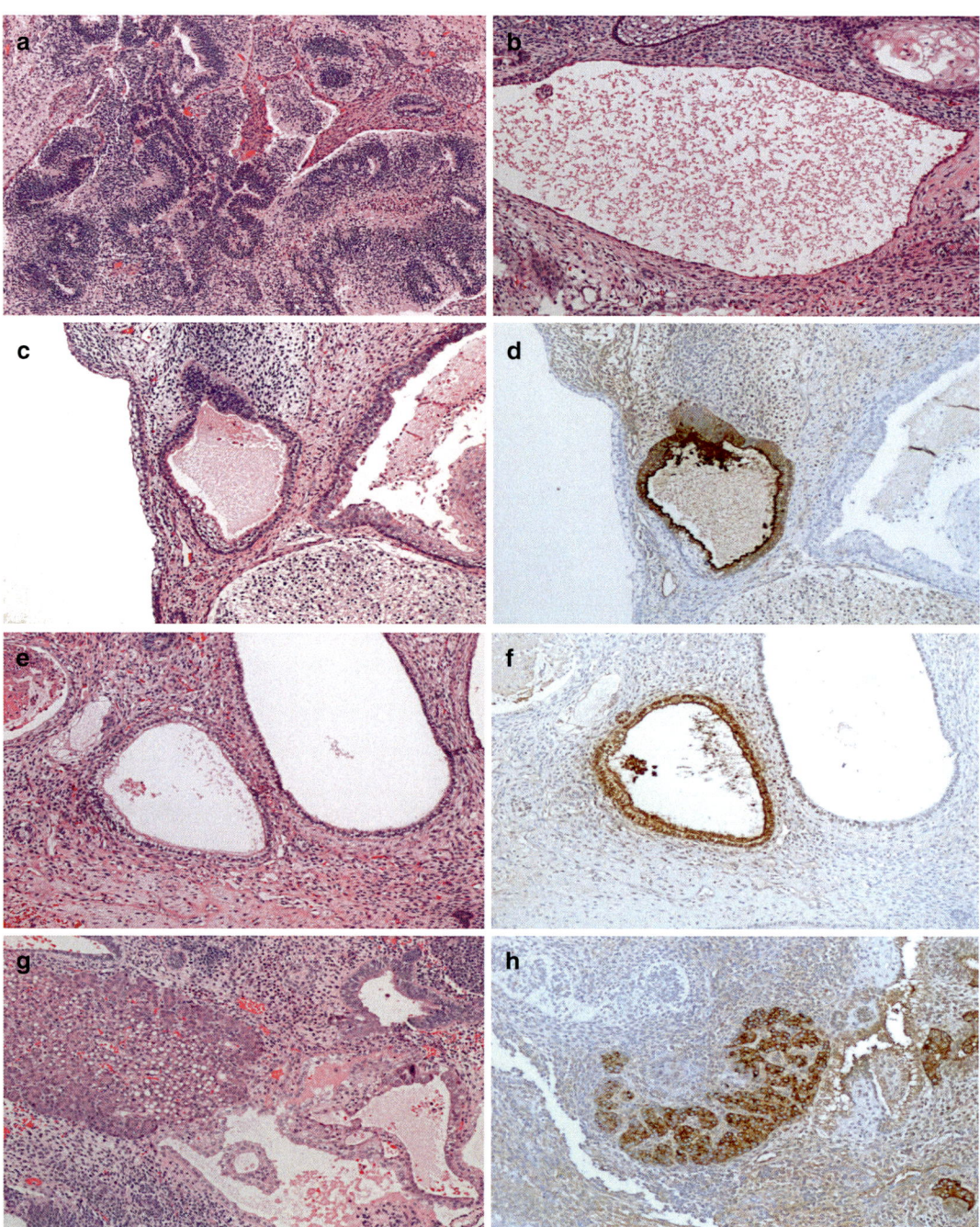

Fig. 12.29 Extragonadal Germ Cell Tumor. In this pediatric tumor, there is a lot of neuroectodermal tissue (**a**) in addition to a component of extragonadal germ cell tumor (**b**). In fact, there are cystic and solid changes characteristic of yolk sac tumor as it is highlighted in (**c**), (**e**) and (**g**), which are accompanied by the corresponding alpha-feto-protein positivity (AFP) by immunohistochemistry (Avidin-Biotin Complex) (**d**), (**f**), and (**h**) (a, b, c, e, and g, hematoxylin and eosin staining, ×50, ×100, ×100, ×100, ×100 as original magnification, respectively; d, f, and h, anti-AFP immunostaining, Avidin-Biotin Complex, ×100, ×100, ×100 as original magnification, respectively)

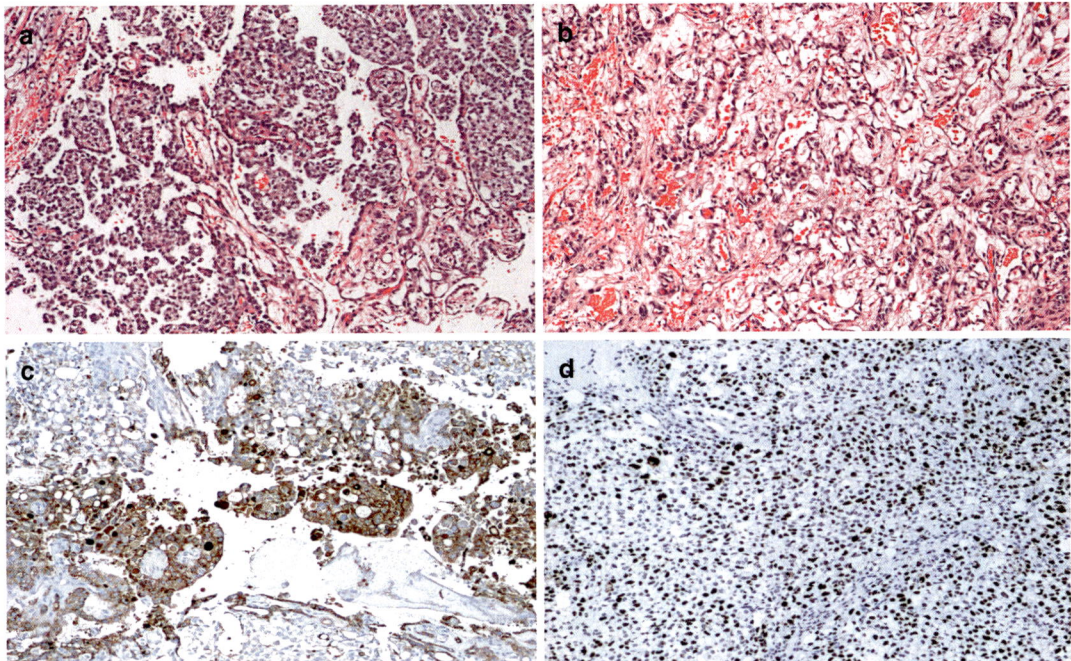

Fig. 12.30 Extragonadal Yolk Sac Tumor. This other extragonadal germ cell tumor of a pediatric patient highlights the histologic and immunohistochemical features of a yolk sac tumor with micropapillary growth pattern (**a**) and microcystic growth pattern (**b**) with immunohistochemical positivity for alpha-fetoprotein (AFP) (**c**) and high proliferation rate (**d**) (a, hematoxylin and eosin staining, ×100 original magnification; b, hematoxylin and eosin staining, ×100 original magnification; c, anti-AFP immunostaining, Avidin-Biotin Complex, ×100 original magnification; d, anti-Ki67 immunostaining (MIB1), Avidin-Biotin Complex, x100 original magnification)

metastasize or relapse as well, even if the coccyx has been resected. Despite accurate resection, annual control investigations should be offered to all patients harboring SCTs, because late recurrence cannot be easily predicted yet. Furthermore, except for incontinence problems in four patients and dissatisfaction with scar formation in five patients, the results were surprisingly good at this institution. Finally, two entities should be considered including the extragonadal germ cell tumor and the extragonadal yolk sac tumor (Figs. 12.29 and 12.30).

Extragonadal germ cell tumor is defined as a germ cell tumor outside of either the testes or ovaries and may be an incidental encounter or manifesting as a tumor. The germ cell tumor is mainly of the teratomatous type with mature and immature components. Yolk sac tumor of extragonadal type is either a component of the germ cell tumor of extragonadal kind or is a pure yolk sac tumor, and occasional cases have been reported in a few organs, such as the heart (Graf et al. 1999).

Multiple Choice Questions and Answers

- SFT-1 A 12-year-old boy presents with a 2.5 cm slow-growing, firm nodule on the scalp. The child started swimming in the pool 3 months ago. There was no clinical history of symptoms, including lethargy, weight loss, or decreased appetite. The school performance was regular and no history of child abuse. Laboratory testing showed a normal blood cell count. A biopsy was performed, and the histology is shown here:

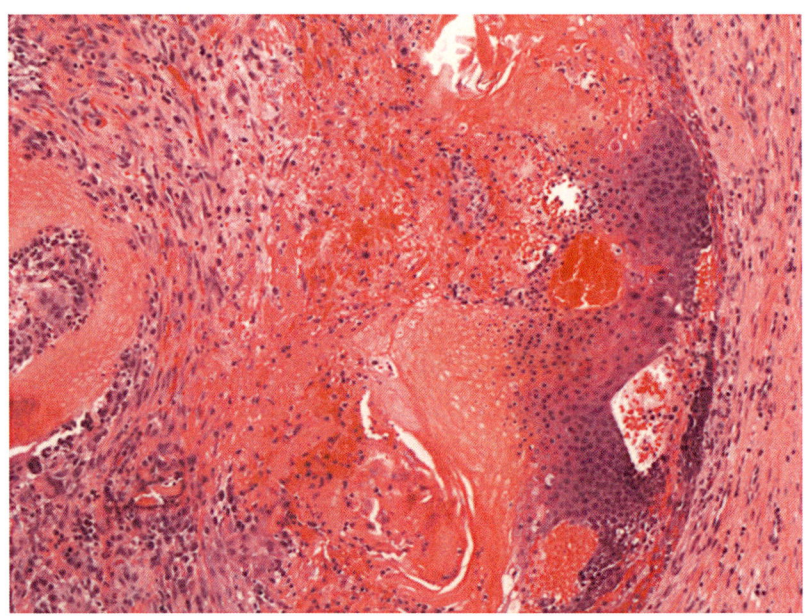

What is the most likely diagnosis?
  (a) Pilomatrixoma
  (b) Polyomavirus infection
  (c) Poxvirus infection
  (d) Molluscum contagiosum
  (e) Trichoblastoma

- SFT-2 Which neoplasm is characterized by the translocation t(X;17)(p11;q25)?
  (a) Neuroblastoma
  (b) Rhabdomyosarcoma
  (c) Synovial sarcoma
  (d) Alveolar soft part sarcoma
  (e) Myoepithelial carcinoma

- SFT-3 Which of the following is CORRECT about the prognosis of a ganglioneuroma?
  (a) It depends on the results of five parameters, including age, stage, MYCN status, histology, and DNA ploidy.
  (b) It depends exclusively on the age of the patient.
  (c) It is good, although a long-term follow-up is required due to the potential malignant transformation of residual tumor leftover following the resection.
  (d) It depends on the mitotic-karyorrhexis index and MYCN status.
  (e) It depends if it occurs sporadically or in a setting of a genetic syndrome.

- SFT-4 A soft tissue tumor biopsy shows sheets and fascicles of relatively uniform spindled cells. The tumor cells harbor a scant cytoplasm and ovoid hyperchromatic nuclei with inconspicuous nucleoli. In the tumor, there are also strands of ropy and wiry collagen as well as bands of hyalinized collagen. The mitotic index reveals 15 mitotic figures per 10 high-power fields. Some focal areas of myxoid change and staghorn-type vascular pattern are also seen. The immunohistochemistry shows positivity for EMA, CD99 (membranous

staining), and transducin-like enhancer of split 1 (TLE1) (nuclear staining), while there is negativity for CD34. Which translocation is expected to be seen in this tumor?
(a) t(2;13)(q35;q14)
(b) t(1;13)(p36;q14)
(c) t(X;2)(q13;q35)
(d) t(21;22)(q22;q12)
(e) t(X;18)(p11.2;q11.2)
(f) t(9;22)(q22;q12)
(g) t(1;22)(q23;q12)

- SFT-5 Which of the following mutations is mostly involved as a somatic mutation in angiosarcoma?
  (a) TP53
  (b) ALK
  (c) BRAF V600E
  (d) SOX10
  (e) PDGFB

- SFT-6 Which of the following statements for a solitary fibrous tumor (SFT) is FALSE?
  (a) SFT is a fibroblastic tumor composed of relatively bland and uniform spindled cells within long, thin, parallel bands of collagen in a "patternless" pattern.
  (b) SFT typically occurs in adults, but young individuals have been reported.
  (c) SFT reveals slow growth and painless masses with low rates of infiltration and metastasis.
  (d) Histologic sections show bland cells with small and elongated nuclei.
  (e) Myxoid change is not a histologic feature.
  (f) Immunohistochemically, SFT is positive for CD34, CD99, bcl-2, and STAT6 but is negative for desmin, keratin, and S100.
  (g) Molecular studies often show NAB2-STAT6 fusion.

- SFT-7 What is TRUE for the MYCN status of neuroblastoma?
  (a) Amplification of the *MYCN* oncogene is extraordinarily crucial for the prognosis.
  (b) When *MYCN* amplification is present, the tumor is elevated to a "middle" risk category.
  (c) *MYCN* amplification is noted in 90% of primary tumors.
  (d) The degree of *MYCN* amplification does not correlate with a worse prognosis.
  (e) The *MYCN* amplification status changes following chemotherapy treatment.

- SFT-8 Which morphologic criterion does NOT help in the diagnosis of lipoblastoma?
  (a) Hypercellular lobules of adipocytes in various stages of differentiation.
  (b) Separation of the lobules by prominent fibrous septa.
  (c) Plexiform vascular pattern and abundant myxoid stroma.
  (d) Prominent extracellular mucinous pools can occur.

- SFT-9 Myxofibrosarcoma is one of the more common subtypes of sarcoma that is very rare in childhood, occurring mostly in middle-aged to elderly adults at the extremities. Myxoid liposarcoma is an intermediate-grade sarcoma that most commonly arises in the deep soft tissue of the extremities. It is also rare in childhood, but it should be kept in the differential diagnosis of myxoid tissue-rich soft tissue sarcomas. Which of the following features is most useful in separating myxofibrosarcoma from myxoid liposarcoma?
  (a) Plenty of Reed-Sternberg-like cells
  (b) *DDIT3* (*CHOP*) translocation
  (c) Lack of *GNAS* mutation
  (d) *MDM2/CDK4* amplification
  (e) Histologic evidence of vascular arcades

- SFT-10 What is the translocation seen in desmoplastic small round cell tumor?
  (a) t(11;22)(p13;q12)
  (b) t(2;13)(q35;q14)
  (c) t(1;13)(p36;q14)
  (d) t(X;2)(q13;q35)
  (e) t(21;22)(q22;q12)

References and Recommended Readings

Akkad T, Sergi C, Gozzi C, Steiner H, Leonhartsberger N, Mitterberger M, Bartsch G, Radmayr C, Oswald J. Metastasizing renal cell carcinoma developing in a congenital ectopic and dysplastic kidney. Urol Int. 2008;81(4):477–9. https://doi.

org/10.1159/000167851. Epub 2008 Dec 10. PubMed PMID: 19077414.

Al Jishi T, Sergi C. Current perspective of diethylstilbestrol (DES) exposure in mothers and offspring. Reprod Toxicol. 2017;71:71–7. https://doi.org/10.1016/j.reprotox.2017.04.009. Epub 2017 Apr 28. PMID: 28461243.

Armah HB, Parwani AV. Epithelioid sarcoma. Arch Pathol Lab Med. 2009;133(5):814–9. https://doi.org/10.1043/1543-2165-133.5.814. PMID: 19415960.

Armstrong SR, Wu H, Wang B, Abuetabh Y, Sergi C, Leng RP. The regulation of tumor suppressor p63 by the ubiquitin-proteasome system. Int J Mol Sci. 2016;17(12):E2041. Review. PubMed PMID: 27929429; PubMed Central PMCID: PMC5187841.

Attwell D, Mishra A, Hall CN, O'Farrell FM, Dalkara T. What is a pericyte?. J Cereb Blood Flow Metab. 2016;36(2):451–455. doi:10.1177/0271678X15610340.

Barresi V, Ieni A, Branca G, Tuccari G. Brachyury: a diagnostic marker for the differential diagnosis of chordoma and hemangioblastoma versus neoplastic histological mimickers. Dis Markers. 2014;2014:514753. https://doi.org/10.1155/2014/514753. Epub 2014. PMID: 24591762; PMCID: PMC3925620.

Billings SD, Walsh SV, Fisher C, Nusrat A, Weiss SW, Folpe AL. Aberrant expression of tight junction-related proteins ZO-1, claudin-1 and occludin in synovial sarcoma: an immunohistochemical study with ultrastructural correlation. Mod Pathol. 2004;17(2):141–9. https://doi.org/10.1038/modpathol.3800042. PMID: 14704716.

Borello ED, Gorlin RJ (1966) Melanotic neuroectodermal tumor of infancy. A neoplasm of neural crest origin. Report of a case associated with high urinary excretion of vanilmandelic acid. Cancer 19: 196–206.

Bratthauer GL. The avidin-biotin complex (ABC) method and other avidin-biotin binding methods. Methods Mol Biol. 2010;588:257–70. https://doi.org/10.1007/978-1-59745-324-0_26. PubMed PMID: 20012837.

Chang MW. Updated classification of hemangiomas and other vascular anomalies. Lymphat Res Biol. 2003;1(4):259–65. Review. PubMed PMID: 15624554.

Coindre JM. Grading of soft tissue sarcomas: review and update. Arch Pathol Lab Med. 2006;130(10) 1448–53. Review. PubMed PMID: 17090186.

Dabska M. Malignant endovascular papillary angioendothelioma of the skin in childhood. Clinicopathologic study of 6 cases. Cancer. 1969;24(3):503–10. https://doi.org/10.1002/1097-0142(196909)24:3<503::aid-cncr2820240311>3.0.co;2-l. PMID: 5343389.

de Groot K, Gross WL, Herlyn K, Reinhold-Keller E. Development and validation of a disease extent index for Wegener's granulomatosis. Clin Nephrol. 2001;55(1):31–8.

Desai SS, Jambhekar NA. Pathology of Ewing's sarcoma/PNET: Current opinion and emerging concepts. Indian J Orthop. 2010 Oct;44(4):363–8. https://doi.org/10.4103/0019-5413.69304. PMID: 20924475; PMCID: PMC2947721.

Dickson BC, Brooks JS, Pasha TL, Zhang PJ. TFE3 expression in tumors of the microphthalmia-associated transcription factor (MiTF) family. Int J Surg Pathol. 2011;19(1):26–30. https://doi.org/10.1177/1066896909352861. Epub 2010 Feb 16. PMID: 20164056.

Eralp Y, Bavbek S, Ba?aran M, Kaytan E, Yaman F, Bilgiç B, Darendeliler E, Onat H. Prognostic factors and survival in late adolescent and adult patients with small round cell tumors. Am J Clin Oncol. 2002;25(4):418–24. https://doi.org/10.1097/00000421-200208000-00020. PMID: 12151977.

Exley AR, Bacon PA. Clinical disease activity in systemic vasculitis. Curr Opin Rheumatol. 1996;8(1):12–8. Review. PubMed PMID: 8867533.

Eyden BP. Brief review of the fibronexus and its significance for myofibroblastic differentiation and tumor diagnosis. Ultrastruct Pathol. 1993;17(6):611–22. Review. PubMed PMID: 8122327.

Falk H. Vinyl chloride induced hepatic angiosarcoma. Princess Takamatsu Symp. 1987;18:39–46. PMID: 3506545.

Folpe AL, Chand EM, Goldblum JR, Weiss SW. Expression of Fli-1, a nuclear transcription factor, distinguishes vascular neoplasms from potential mimics. Am J Surg Pathol. 2001;25(8):1061–6. https://doi.org/10.1097/00000478-200108000-00011. PMID: 11474291.

Frezza AM, Cesari M, Baumhoer D, Biau D, Bielack S, Campanacci DA, Casanova J, Esler C, Ferrari S, Funovics PT, Gerrand C, Grimer R, Gronchi A, Haffner N, Hecker-Nolting S, Höller S, Jeys L, Jutte P, Leithner A, San-Julian M, Thorkildsen J, Vincenzi B, Windhager R, Whelan J. Mesenchymal chondrosarcoma: prognostic factors and outcome in 113 patients. A European Musculoskeletal Oncology Society study. Eur J Cancer. 2015;51(3):374–81. https://doi.org/10.1016/j.ejca.2014.11.007. Epub 2014 Dec 16. PubMed PMID: 25529371.

Graf M, Blaeker H, Schnabel P, Serpi M, Ulmer HE, Otto HF. Intracardiac yolk sac tumor in an infant girl. Pathol Res Pract. 1999;195(3):193–7. https://doi.org/10.1016/S0344-0338(99)80034-X. PMID: 10220801.

Geiger K, Hagenbuchner J, Rupp M, Fiegl H, Sergi C, Meister B, Kiechl-Kohlendorfer U, Müller T, Ausserlechner MJ, Obexer P. FOXO3/FKHRL1 is activated by 5-aza-2-deoxycytidine and induces silenced caspase-8 in neuroblastoma. Mol Biol Cell. 2012;23(11):2226–34. https://doi.org/10.1091/mbc.E11-06-0535. Epub 2012 Apr 4. PubMed PMID: 22493319; PubMed Central PMCID: PMC3364184.

Goh & E Calonje. Cutaneous vascular tumours: an update 2008.

Guillevin L, Lhote F, Gayraud M, Cohen P, Jarrousse B, Lortholary O, Thibult N, Casassus P. Prognostic factors in polyarteritis nodosa and Churg-Strauss syndrome. A prospective study in 342 patients. Medicine (Baltimore). 1996;75:17Y28.

References and Recommended Readings

Guillevin L, Pagnoux C, Seror R, Mahr A, Mouthon L, Le Toumelin P, French Vasculitis Study Group (FVSG). The five-factor score revisited: assessment of prognoses of systemic necrotizing vasculitides based on the French Vasculitis Study Group (FVSG) cohort. Medicine (Baltimore). 2011;90(1):19–27. https://doi.org/10.1097/MD.0b013e318205a4c6. PubMed PMID: 21200183.Al-Dhaybi R et al. J Am Acad Dermatol 2010.

Guo H, Garcia RA, Perle MA, Amodio J, Greco MA. Giant cell tumor of soft tissue with pulmonary metastases: pathologic and cytogenetic study. Pediatr Dev Pathol. 2005;8(6):718–24. Epub 2005 Nov 21. PubMed PMID: 16328671.

Hagenbuchner J, Rupp M, Salvador C, Meister B, Kiechl-Kohlendorfer U, Müller T, Geiger K, Sergi C, Obexer P, Ausserlechner MJ. Nuclear FOXO3 predicts adverse clinical outcome and promotes tumor angiogenesis in neuroblastoma. Oncotarget. 2016;7(47):77591–606. https://doi.org/10.18632/oncotarget.12728. PubMed PMID: 27769056; PubMed Central PMCID: PMC5363607.

Harada O, Ota H, Nakayama J. Malignant myoepithelioma (myoepithelial carcinoma) of soft tissue. Pathol Int. 2005;55(8):510–3. PubMed PMID: 15998380.

Hense HW, Ahrens S, Paulussen M, Lehnert M, Jürgens H. Factors associated with tumor volume and primary metastases in Ewing tumors: results from the (EI) CESS studies. Ann Oncol. 1999;10(9):1073–7. https://doi.org/10.1023/a:1008357018737. PMID: 10572605.

Hollmann TJ, Hornick JL. INI1-deficient tumors: diagnostic features and molecular genetics. Am J Surg Pathol. 2011;35(10):e47-63. https://doi.org/10.1097/PAS.0b013e31822b325b. PMID: 21934399.

Huh WW, Yuen C, Munsell M, Hayes-Jordan A, Lazar AJ, Patel S, Wang WL, Barahmani N, Okcu MF, Hicks J, Debelenko L, Spunt SL. Liposarcoma in children and young adults: a multi-institutional experience. Pediatr Blood Cancer. 2011;57(7):1142–6. https://doi.org/10.1002/pbc.23095. Epub 2011 Mar 10. PubMed PMID: 21394894; PubMed Central PMCID: PMC3134599.

Jennings, B.H., Ish-Horowicz, D. The Groucho/TLE/Grg family of transcriptional co-repressors. Genome Biol 9, 205 (2008). https://doi.org/10.1186/gb-2008-9-1-205.

Kaposiform Hemangioendothelioma: Atypical Features and Risks of Kasabach-Merritt Phenomenon in 107 Referrals … Modern Pathology (2008) 21, 125–130; doi:10.1038/modpathol.3800986.

Kohashi K, Oda Y. Oncogenic roles of SMARCB1/INI1 and its deficient tumors. Cancer Sci. 2017;108(4):547–52. https://doi.org/10.1111/cas.13173. Epub 2017 Apr 12. PMID: 28109176; PMCID: PMC5406539.

Krompecher, E. Zur Histogenese und Morphologie der Adamantnome und Sonstiger Kiefergeschwulste. Bietr. Pathol. 64: 165–97, 1918.

Kulkarni K, Desai S, Grundy P, Sergi C. Infantile myofibromatosis: report on a family with autosomal dominant inheritance and variable penetrance. J Pediatr Surg. 2012;47(12):2312–5. https://doi.org/10.1016/j.jpedsurg.2012.09.046. PubMed PMID: 23217896.

Kulkarni K, Desai S, Grundy P, Sergi C. Infantile myofibromatosis: report on a family with autosomal dominant inheritance and variable penetrance. J Pediatr Surg. 2012;47(12):2312–5. https://doi.org/10.1016/j.jpedsurg.2012.09.046. PubMed PMID: 23217896.

Lau YS, Sabokbar A, Gibbons CL, Giele H, Athanasou N. Phenotypic and molecular studies of giant-cell tumors of bone and soft tissue. Hum Pathol. 2005;36(9): 945–54. PubMed PMID: 16153456.

Luqmani RA, Bacon PA, Moots RJ, Janssen BA, Pall A, Emery P, Savage C, Adu D. Birmingham Vasculitis Activity Score (BVAS) in systemic necrotizing vasculitis. QJM. 1994;87(11):671–8. PubMed PMID: 7820541.

Luqmani RA, Bacon PA, Moots RJ, Janssen BA, Pall A, Emery P, Savage C, Adu D. Birmingham Vasculitis Activity Score (BVAS) in systemic necrotizing vasculitis. QJM. 1994;87(11):671–8. PubMed PMID: 7820541.

Luqmani RA, Exley AR, Kitas GD, Bacon PA. Disease assessment and management of the vasculitides. Baillieres Clin Rheumatol. 1997;11(2):423–46. Review. PubMed PMID: 9220084.

McKillop SJ, Belletrutti MJ, Lee BE, Yap JY, Noga ML, Desai SJ, Sergi C. Adenovirus necrotizing hepatitis complicating atypical teratoid rhabdoid tumor. Pediatr Int. 2015;57(5):974–7. https://doi.org/10.1111/ped.12674. Epub 2015 Aug 19. PubMed PMID: 26508178.

Meng GZ, Zhang HY, Zhang Z, Wei B, Bu H. Myofibroblastic sarcoma vs nodular fasciitis: a comparative study of chromosomal imbalances. Am J Clin Pathol. 2009 May;131(5):701–9. https://doi.org/10.1309/AJCPV6H2WSYXLKFB. PubMed PMID: 19369631.

Meis-Kindblom JM, Bergh P, Gunterberg B, Kindblom LG. Extraskeletal myxoid chondrosarcoma: a reappraisal of its morphologic spectrum and prognostic factors based on 117 cases. Am J Surg Pathol. 1999;23(6):636–50. https://doi.org/10.1097/00000478-199906000-00002. PMID: 10366145.

Mojiri A, Stoletov K, Carrillo MA, Willetts L, Jain S, Godbout R, Jurasz P, Sergi CM, Eisenstat DD, Lewis JD, Jahroudi N. Functional assessment of von Willebrand factor expression by cancer cells of non-endothelial origin. Oncotarget. 2017;8(8):13015–29. https://doi.org/10.18632/oncotarget.14273. PubMed PMID: 28035064; PubMed Central PMCID: PMC5355073.

Mukhtyar C, Lee R, Brown D, Carruthers D, Dasgupta B, Dubey S, Flossmann O, Hall C, Hollywood J, Jayne D, Jones R, Lanyon P, Muir A, Scott D, Young L, Luqmani RA. Modification and validation of the Birmingham Vasculitis Activity Score (version 3). Ann Rheum Dis. 2009;68(12):1827–32. https://doi.org/10.1136/ard.2008.101279. Epub 2008 Dec 3. PubMed PMID: 19054820.

Neuville A, Chibon F, Coindre JM. Grading of soft tissue sarcomas: from histological to molecular assessment. Pathology. 2014;46(2):113–20. https://doi.org/10.1097/PAT.0000000000000048. Review. PubMed PMID: 24378389.

Nielsen TO, Poulin NM, Ladanyi M. Synovial sarcoma: recent discoveries as a roadmap to new avenues for therapy. Cancer Discov. 2015 Feb;5(2):124–34. https://doi.org/10.1158/2159-8290.CD-14-1246. Epub 2015 Jan 22. PMID: 25614489; PMCID: PMC4320664.

Oddone M, Marino C, Sergi C, Occhi M, Negri F, Kotitza Z, De Bernardi B, Jasonni V, Tomà P. Wilms' tumor arising in a multicystic kidney. Pediatr Radiol. 1994;24(4):236–8. PubMed PMID: 7800438.

Osasan S, Zhang M, Shen F, Paul PJ, Persad S, Sergi C. Osteogenic Sarcoma: A 21st century review. Anticancer Res. 2016;36(9):4391–8. Review. PubMed PMID: 27630274.

Peng S, Li W, Hou N, Huang N. A Review of FoxO1-Regulated Metabolic Diseases and Related Drug Discoveries. Cells. 2020;9(1):184. https://doi.org/10.3390/cells9010184. PMID: 31936903; PMCID: PMC7016779.

Purgina B, Rao UN, Miettinen M, Pantanowitz L. AIDS-related EBV-associated smooth muscle tumors: a review of 64 published cases. Pathol Res Int. 2011;2011:561548. https://doi.org/10.4061/2011/561548. PubMed PMID: 21437186; PubMed Central PMCID: PMC3062098.

Renzi S, Anderson ND, Light N, Gupta A. Ewing-like sarcoma: An emerging family of round cell sarcomas. J Cell Physiol. 2019 Jun;234(6):7999–8007. https://doi.org/10.1002/jcp.27558. Epub 2018 Sep 26. Review. PubMed PMID: 30257034.

Santer FR, Bacher N, Moser B, Morandell D, Ressler S, Firth SM, Spoden GA, Sergi C, Baxter RC, Jansen-Dürr P, Zwerschke W. Nuclear insulin-like growth factor binding protein-3 induces apoptosis and is targeted to ubiquitin/proteasome-dependent proteolysis. Cancer Res. 2006;66(6):3024–33. PubMed PMID: 16540651.

Scheier M, Ramoni A, Alge A, Brezinka C, Reiter G, Sergi C, Hager J, Marth C. Congenital fibrosarcoma as cause for fetal anemia: prenatal diagnosis and in utero treatment. Fetal Diagn Ther. 2008;24(4):434–6. https://doi.org/10.1159/000173370. Epub 2008 Nov 19. PubMed PMID: 19018145.

Sergi C, Ehemann V, Beedgen B, Linderkamp O, Otto HF. Huge fetal sacrococcygeal teratoma with a completely formed eye and intratumoral DNA ploidy heterogeneity. Pediatr Dev Pathol. 1999;2(1):50–7. Review. PubMed PMID: 9841706.

Sergi C, Kos M. Bilateral Wilms' tumor in trisomy 18 syndrome: case report and critical review of the literature. Ann Clin Lab Sci. 2018;48(3):369–72. PubMed PMID: 29970442.

Sergi C, Kulkarni K, Stobart K, Lees G, Noga M. Clear cell variant of embryonal rhabdomyosarcoma: report of an unusual retroperitoneal tumor – case report and literature review. Eur J Pediatr Surg. 2012;22(4):324–8. https://doi.org/10.1055/s-0032-1308714. Epub 2012 May 10. Review. PubMed PMID: 22576307.

Sergi C, Zwerschke W. Osteogenic sarcoma (osteosarcoma) in the elderly: tumor delineation and predisposing conditions. Exp Gerontol. 2008;43(12):1039–43. https://doi.org/10.1016/j.exger.2008.09.009. Epub 2008 Sep 25. Review. PubMed PMID: 18845233.

Seto ES, Bellen HJ, Lloyd TE. When cell biology meets development: endocytic regulation of signaling pathways. Genes Dev. 2002;16(11):1314–36. https://doi.org/10.1101/gad.989602. PMID: 12050111.

Shojaeian R, Hiradfar M, Sharifabad PS, et al. Extrarenal Wilms' tumor: challenges in diagnosis, embryology, treatment and prognosis. In: van den Heuvel-Eibrink MM, editor. Wilms tumor [Internet]. Brisbane: Codon Publications; 2016. Chapter 6. Available from: https://www.ncbi.nlm.nih.gov/books/NBK373353/. https://doi.org/10.15586/codon.wt.2016.ch6.

Soles BS, Wilson A, Lucas DR, Heider A. Melanotic Neuroectodermal Tumor of Infancy. Arch Pathol Lab Med. 2018;142(11):1358–1363. https://doi.org/10.5858/arpa.2018-0241-RA. PMID: 30407852.

Verma V, Denniston KA, Lin CJ, Lin C. A Comparison of Pediatric vs. Adult Patients with the Ewing Sarcoma Family of Tumors. Front Oncol. 2017 May 8;7:82. https://doi.org/10.3389/fonc.2017.00082. PMID: 28534008; PMCID: PMC5421143.

Witmer AN, van Blijswijk BC, Dai J, Hofman P, Partanen TA, Vrensen GF, Schlingemann RO. VEGFR-3 in adult angiogenesis. J Pathol. 2001;195(4):490–7. doi: 10.1002/path.969. PMID: 11745682.

Weiss SW, Enzinger FM. Epithelioid hemangioendothelioma: a vascular tumor often mistaken for a carcinoma. Cancer 1982; 50; 970–981.

Yang Y, Shen F, Huang W, Qin S, Huang JT, Sergi C, Yuan BF, Liu SM. Glucose Is Involved in the Dynamic Regulation of m6A in Patients With Type 2 Diabetes. J Clin Endocrinol Metab. 2019;104(3):665–73. doi: 10.1210/jc.2018-00619. PMID: 30137347.

Zimmermann KW. Der feinere Bau der Blutkapillaren. Z Anat Entwicklungsgesch 1923; 68: 29–109.

Zhang J, Wang H, Cheng X, Wang M, Zhu Y. A case of parachordoma on the chest wall and literature review. J Can Res Ther [serial online] 2013 [cited 2020];9:114–7. Available from: http://www.cancerjournal.net/text.asp?2013/9/5/114/119124.

Arthro-Skeletal System

Contents

13.1	**Development and Genetics**	1096
13.2	**Osteochondrodysplasias**	1097
13.2.1	Nosology and Nomenclature	1097
13.2.2	Groups of Genetic Skeletal Disorders	1098
13.3	**Metabolic Skeletal Diseases**	1103
13.3.1	Rickets, and Osteomalacia	1103
13.3.2	Osteoporosis of the Youth	1104
13.3.3	Paget Disease of the Bone	1106
13.3.4	Juvenile Paget Disease	1108
13.4	**Osteitis and Osteomyelitis**	1109
13.4.1	Osteomyelitis	1109
13.5	**Osteonecrosis**	1113
13.5.1	Bony Infarct and Osteochondritis Dissecans	1113
13.6	**Tumorlike Lesions and Bone/Osteoid-Forming Tumors**	1115
13.6.1	Myositis Ossificans	1115
13.6.2	Fibrous Dysplasia and Osteofibrous Dysplasia	1116
13.6.3	Non-ossifying Fibroma (NOF)	1118
13.6.4	Bone Cysts	1120
13.6.5	Osteoma, Osteoid Osteoma, and Giant Osteoid Osteoma	1124
13.6.6	Giant Cell Tumor	1127
13.6.7	Osteosarcoma	1128
13.7	**Chondroid (Cartilage)-Forming Tumors**	1136
13.7.1	Osteochondroma	1137
13.7.2	Enchondroma	1139
13.7.3	Chondroblastoma	1140
13.7.4	Chondromyxoid Fibroma	1143
13.7.5	Chondrosarcoma	1145
13.8	**Bone Ewing Sarcoma**	1147
13.9	**Miscellaneous Bone Tumors**	1149
13.9.1	Chordoma	1149
13.9.2	Adamantinoma	1150
13.9.3	Langerhans Cell Histiocytosis	1151
13.9.4	Vascular, Smooth Muscle, and Lipogenic Tumors	1153
13.9.5	Hematologic Tumors	1153

© Springer-Verlag GmbH Germany, part of Springer Nature 2020
C. M. Sergi, *Pathology of Childhood and Adolescence*,
https://doi.org/10.1007/978-3-662-59169-7_13

13.10	**Metastatic Bone Tumors**...	1153
13.11	**Juvenile Rheumatoid Arthritis and Juvenile Arthropathies**.................	1154
13.11.1	Rheumatoid Arthritis and Juvenile Rheumatoid Arthritis...........................	1154
13.11.2	Infectious Arthritis..	1156
13.11.3	Gout, Early-Onset Juvenile Tophaceous Gout and Pseudogout.................	1157
13.11.4	Bursitis, Baker Cyst, and Ganglion..	1159
13.11.5	Pigmented Villonodular Synovitis and Nodular Tenosynovitis..................	1159

Multiple Choice Questions and Answers... 1161

References and Recommended Readings 1162

13.1 Development and Genetics

At the onset of development of the skeleton, the embryo possesses undifferentiated mesenchymal cells that eventually condense at sites that are predestined to become the future bony structure. These amazing mesenchymal condensations follow precise molecular biology algorithms most of them still mostly unknown to us and acquire the shape of the skeletal elements they foreshadow. The endochondral ossification takes place when a cartilage template is progressively replaced by bone tissue in an exact series of events, in which chondrocytes progress from germinal zone containing stem cells to proliferating chondrocytes, chondrocytes lying in an area of maturation, and then to hypertrophic chondrocytes or hypertrophic states, and the resulting hypertrophic cartilage is then mineralized and replaced by bone. The chondro-osteogenesis steps are as follows.

1. *Segregation of progenitor mesenchymal cells* in the cranial portion of the neural crest (neuroectoderm), mesoderm (*splanchnocranium* or *viscerocranium* and paraxial somites), and lateral plate mesoderm (limbs).
2. *Condensation of the mesenchymal cells.*
3. *Osteochondral differentiation toward chondrocytes* occurs by the action of some transcription factors (SOX5, SOX6, SOX9). SOX9 interacts for COL2A1 expression, which is responsible for the expression of α1 of type 2 collagen and other matrix proteins.
4. *ECM synthesis* occurs involving proteoglycans, aggrecan, decorin, biglycan, fibromodulin, perlecan, and collagens II, IX, and XI.
5. Runx 2 (Cbfa-1) is crucial for the *development of hypertrophic chondrocytes* and the *differentiation of the osteoblast from the primordial osteochondral cell.*
6. *Membranous ossification* (craniofacial bones and portions of the clavicula), which is characterized by the direct differentiation of the common progenitor mesenchymal cells to osteoblasts without an intervening chondroid phase, is induced by Runx 2 and its interaction with members of the TGF-β superfamily (e.g., BMPs, SMADs, WNT, and hedgehog) (Al-Bahrani et al. 2015).
7. *Endochondral ossification* (axial skeleton and limbs) is characterized by the indirect differentiation of the common progenitor mesenchymal cells to osteoblasts with an intervening chondroid phase. It is subdivided in the formation of mid-shaft-located primary centers of ossification, VEGF-induced vascular invasion, periosteal bone collar formation, and induction of secondary centers of ossification, which remain separated from the primary centers by the growth plate where the epiphyseal cartilage proliferates and hypertrophies. They undergo programmed cell death, mostly using Indian hedgehog.
8. *PTH-related peptide* and its COL2A1 promoter and FGFR3-controlled receptor are essential for *bone remodeling*, and activating mutations of this tyrosine kinase receptor at the growth plate are responsible for the development of several forms of skeletal dysplasias.

9. *Osteoclasts* are the BM-derived skeletal cell crucial for bone remodeling, and their malfunction is at the basis of the osteopetrosis ('marble bone disease' or Albers-Schönberg disease), which is is considered to be the prototype of osteosclerosing skeletal dysplasias. In this kind of skeletal dysplasia the bones become denser and harder than normal bones, in striking contrast to osteomalacia, in which the bones progressively soften, or osteoporosis, in which the bones become less dense and more brittle. In osteopetrosis there is indeed a malfunctioning osteoclast, which makes subsequently the bone unable to continue the cycle of deposition and resorption. A range of clinical features and different molecular lesions substantiate the heterogeneity of this skeletal disorder.

In mammals, the long and short bones of mammals are formed mainly by endochondral ossification, but the periosteal bone collar forms via a type of intramembranous ossification without a cartilaginous phase. In birds, the ossification of the long bones takes place via intramembranous ossification of the periosteum. In these animals, the core cartilage undergoes the same steps of endochondral ossification before the mineralization of hypertrophic chondrocytes. However, the matrix of post-hypertrophic chondrocytes is resorbed instead of being mineralized. In joint development, there are three main stages, including interzone formation, cavitation, and final morphogenesis. The study of animals, especially chicks, has been used to understand that the detailed shape of the developing knee joint emerges just following the initiation of muscle contractions.

13.2 Osteochondrodysplasias

13.2.1 Nosology and Nomenclature

The nomenclature of osteochondrodysplasias, including prefixes, suffixes, and categories of limb reduction defects, may be challenging but it is necessary for understanding of numerous pediatric and adult disorders of the skeletal system. The study of the etymology of most of these disorders is particularly useful for pathologists working on skeletal dysplasia interacting closely with pediatricians and pediatric radiologists. It is not only the plethora of genes involved in osteochondrodysplasias but the complex and intricate disturbances of the skeletal development that make the study of osteochondrodysplasias in perinatal and pediatric pathology arduous. One of the most significant challenges is the translation of plain English of the various prefixes and suffixes associated with the dysplasia of the skeletal development. In the 2010 revision of the nosology and classification of genetic skeletal disorders, 40 groups have been assigned including 456 medical conditions challenging our photographic, numerical, and word memory. This nosology should be working in progress because some groups are still waiting for molecular clarification. The criteria used to set up this classification comprise medical conditions including significant bony involvement, which corresponds to osteochondrodysplasias, metabolic bone disorders, dysostoses, skeletal defects *sensu lato*, and/or limb reduction syndromes. Linkage or molecular analysis, genetic counseling data, publications, and OMIM listing supported nosologic autonomy. OMIM or Online Mendelian Inheritance in Man is an authoritative compendium of human genes and clinical genetic phenotypes that comprehensively reviews human and clinical genetics and is freely available and updated daily. Some osteologic nomenclature concepts and definitions are listed below:

- *Acro-* (Greek ἄκρον, "end") refers to the most distal segment or topmost (acromelia is a form of dwarfism in which shortening occurs in the most distal portion of the limbs).
- *Meso-* (Greek μέσος, "middle") refers to the middle segment (mesomelia is a form of dwarfism in which shortening occurs in the forearms and lower legs).
- *Rhizo-* (Greek ρίζα, "middle") refers to the most proximal part or roots (rhizomelia is a form of dwarfism in which shortening occurs in the upper arms and thighs).

- *-melia* relates to limbs (Greek μέλος, "limb").
- *Spondylo-* (σπόνδυλος, "spine") relates to the spine (e.g., spondylolisthesis, which is a forward displacement of a vertebra over a lower vertebra and has been characterized to be either a congenital defect or injury).
- *Epiphysis – Metaphysis – Diaphysis* characterize the several growth portions of the longitudinal bone (Greek φύσις, "growth")
- *Epi-* (ἐπί) refers to above or "on top of" (epiphysis is the rounded end of a long bone near to the joint). *Meta-* (μετά, "after") refers to a middle portion between epiphysis and diaphysis and contains the growth plate, while *Dia-* (δια- and διά) refers to "passing through" (diaphysis is the midsection or shaft of a long bone).
- *-morphic* arises from Greek μορφή, which means form or shape.
- *Amelia* refers to the lack of limbs.
- *Preaxial* refers to the anterior side as seen in aplasia or hypoplasia of thumbs, 1st metacarpal, and radius (radial hypoplasia-aplasia) and/or the absence of hallux, 1st metatarsal, and tibia (tibia hypoplasia-aplasia), but also preaxial polydactyly, when the sixth digit is located near the thumb or hallux.
- *Postaxial* refers to the posterior side, close to the fifth finger or toe (e.g., postaxial polydactyly).
- *Transverse*: loss of distal limb structure with preservation of the proximal structure.

13.2.2 Groups of Genetic Skeletal Disorders

The genetic skeletal disorders according to the most current classification (Warman et al. 2011) include the following 40 groups (adapted classification) (Jung et al. 1998, Sergi et al. 1997, Sergi et al. 1998, Sergi et al. 2001, Sergi et al.2001, Schiffer et al. 2007, Broome et al. 2016, Sergi et al. 2019).

1. *FGFR3 chondrodysplasia group*
2. *Type II collagen group and allied disorders*
3. *Type XI collagen group*
4. *Sulfation disorders group*
5. *Perlecan group*
6. *Aggrecan group*
7. *Filamin group and allied disorders*
8. *TRPV4 group*
9. *Short-rib dysplasia (with or without polydactyly) group*
10. *Multiple epiphyseal dysplasias and pseudoachondroplasia group*
11. *Metaphyseal dysplasias*
12. *Spondylometaphyseal dysplasias*
13. *Spondylo-epi-(meta)-physeal dysplasias*
14. *Spondylodysplastic dysplasias, severe type*
15. *Acromelic dysplasias*
16. *Acromesomelic dysplasias*
17. *Mesomelic and rhizo-mesomelic dysplasias*
18. *Bent bone dysplasias*
19. *Slender bone dysplasia group*
20. *Multiple joint dislocations-associated dysplasias*
21. *Chondrodysplasia punctata (CDP) group*
22. *Osteosclerotic dysplasias, neonatal*
23. *Normoosteomorphic Group with increased bone density*
24. *Dysplasias with increased bone density and meta- and/or diaphyseal involvement*
25. *Dysplasias with decreased bone density, including osteogenesis imperfecta*
26. *Abnormal mineralization group*
27. *Lysosomal storage diseases with skeletal involvement (dysostosis multiplex group)*
28. *Osteolysis group*
29. *Disorganized development of skeletal components group*
30. *Overgrowth syndromes with skeletal involvement*
31. *Genetic inflammatory/rheumatoid-like osteoarthropathies*
32. *Cleidocranial dysplasia and isolated cranial ossification defects group*
33. *Craniosynostosis syndromes*
34. *Dysostoses with predominant craniofacial involvement*
35. *Dysostoses with predominant vertebral with or without the involvement of ribs*
36. *Patellar dysostoses*
37. *Brachydactyly*
38. *Limb (hypoplasia) – reduction defects group*
39. *Polydactyly-syndactyly-triphalangism group*
40. *Defects in joint formation and synostoses*

13.2 Osteochondrodysplasias

Groups 1–8 are built on a common underlying gene or pathway and comprise most common disorders, including thanatophoric dysplasia types 1 and 2 as well as achondroplasia. There are groups involving critical molecular structures in cartilage and bone. In group 1, disorders are associated with mutations in the fibroblast growth factor receptor 3 (*FGFR3*) gene, which has been mapped to 4p16.3. *FGFR3* is part of the tyrosine kinase receptor family and, in normal conditions, is a negative regulator of bone growth. In one of the osteochondrodysplasias (OCDs) of this group, i.e., the thanatophoric dysplasia, activating mutations cause a gain in function, sending negative signals to chondrocytes. In case of ligand binding within the chondrocytes, there is the induction of the receptor homodimerization and heterodimerization, which, in turn, activates tyrosine kinase function potentiating several effects on cell growth and differentiation. Activating mutations in the *FGFR3* lead to the formation of cysteine residues that create disulfide bonds between extracellular domains of mutant monomers. The increased stability by activation of the homodimer receptor complex and promotion of the translocation of the compound into the nucleus affects the interference with terminal chondrocyte differentiation. Thus, there is generalized disorganization of endochondral ossification at the bony growth plate.

13.2.2.1 Thanatophoric Dysplasia

- *DEF*: Most familiar form of OCD divided into two clinical subtypes: thanatophoric dysplasia type I (TDI) and thanatophoric dysplasia type II (TDII). TDI, which is the more common subtype, shows a normal-shaped skull and curved, telephone receiver-shaped long bones, while TDII is associated with a cloverleaf-shaped skull and straight femurs. Clinical overlap is, however, observed between these subtypes.
- *EPI*: 1:20,000–50,000 births, ♂ = ♀.
- *PUS*: Growth deficiency with limb length of less than 5% (by 20 weeks' gestation), macrocephaly, cloverleaf-shaped skull or "Kleeblattschädel", ventriculomegaly, well-ossified skull and spine, platyspondyly (flattening of vertebrae from Greek πλατύς "flat" and σπόνδυλος "spine"), micromelia, bowing of femurs, narrow chest cavity with shortened ribs, and polyhydramnios.
- *EPG*: AD mutations in the *FGFR3* gene with 100% penetrance and currently, all cases are due to de novo mutations in *FGFR3*, although the theoretical possibility of germline mosaicism also needs to be taken into account. Whereas TDI is caused by several different variations that affect either the extracellular or intracellular domains of *FGFR3* with two missense mutations, R248C and Y373C, accounting for as much as 80% of TDI cases, TDII has a single point mutation, SYN K650E, determining an A → G nucleotide transition in the tyrosine kinase domain of *FGFR3*.
- *LAB*: Chromosome analysis and molecular testing for *FGFR3* with targeted and sequence mutation analyses.
- *X-Ray/CT/MRI*: Rhizomelic shortening and irregular metaphysis of the long bones and "telephone receiver-shaped" bowed femurs, platyspondyly with wide intervertebral spacing, macrocephaly with a small foramen magnum, and CNS abnormalities such as hydrocephalus, brainstem hypoplasia, temporal lobe defects, and neuronal migration abnormalities.
- *CLM*: Nonspecific severe disorganization and retardation of the physeal growth plate with disorderly proliferative and hypertrophic chondrocytes, horizontal-oriented band of fibrosis at the periphery of the physis, as well as few, small cartilaginous spicules in the metaphysis.
- *DDX*: Achondrogenesis, achondroplasia, asphyxiating thoracic dystrophy (Jeune syndrome), hypophosphatasia, and osteogenesis imperfecta.
- *PGN*: Lethal, usually at neonatal age (COD is severe respiratory insufficiency from reduced thoracic breathing capacity and hypoplastic lungs or respiratory insufficiency due to brainstem compression).

Groups 1–8 include most common forms, such as achondroplasia and hypochondroplasia that rarely are seen in the morgue by the pediatric pathologist. Achondroplasia is due to a disproportionate dwarfism, and during early fetal development, patients with achondroplasia do not convert all cartilage to bone, and the underlying disorder is due to mutations of the *FGFR3* gene, which instructs the cells to build a protein necessary for bone growth and maintenance. *FGFR3* mutations make the protein overactive and the resulting skeleton unstable. Groups 1–8 also include the sulfation disorders (group 4), which is of particular interest in pediatric pathology. The sulfation disorder group includes achondrogenesis type 1B, atelosteogenesis type 2 (AO2), and diastrophic dysplasia (DD). The growth inhibition of diastrophic dysplasia affects the entire skeleton. There is usually a different expression so that a particularly high degree of growth inhibition determines disproportional dwarfism. We need to distinguish the idiopathic short stature, which is the etiologically unexplained so-called primordial form of dysostotic dwarfism from other forms of short statures and the reader should complete his/her knowledge identifying remarkable textbooks of auxology. For instance, at cranio-mandibular-facial dysmorphic syndromes, called chondrodysplasias osteopetrosis and mucopolysaccharidosis syndromes, a short stature due to hypopituitarism hormonal, hypothyroidism, hyperthyroidism, hypogonadism due and of course also in the context syndromes can occur (see Prader-Willi syndrome). Moreover, the pediatrician interested in OCDs and bone metabolism of extraskeletal disorders can encounter hypoxic dwarfism (cardiac malformations, chronic lung disease hemolytic anemia), hepatic dwarfism (metabolic diseases), intestinal-related dwarfism or dysplasia (e.g. celiac disease), and renal dwarfism (tubular Enzymopathies or chronic glomerulo-tubular global insufficiency). Diastrophic dysplasia occurs as a result of the growth reduction or shortening of limbs. Proximal bones are more involved (shorter) than the distal ones. Also, there is joint luxation that limits the joint mobility. The phalanges of the hands are thick. Usually, the thumb is abducted ("hitchhiker thumbs"). In the development of the child there is kyphoscoliosis (abnormal curvature of the vertebral column in both coronal and sagittal planes from Greek κύφος, "bent" and σκολίωσις, "twisting"). The legs are not only shorter, but also sword-like. The optional cystic deformation in the ears occurs between 1 day after birth and 12 weeks. Often other changes, e.g., cleft palate and hernias, may compromise the life expectancy of these infants harboring this condition. The psychologic development and gender development show a normal course. About the histopathological picture, the diastrophic dysplasia in newborns and children has been reported. Some publications target the fetal age as well. The chondrocytes are pyknotic and eccentric nuclear degenerative changes. The cytoplasm of the chondrocytes is distinguished by a high content of chondroitin sulfate characterizing the extracellular matrix as rarefied (so-called "myxoid degeneration"). The resting cartilage is affected. Moreover, the zone of the columnar cartilage is reduced in length. The differential diagnosis includes diastrophic dysplasia, pseudo-diastrophic dysplasia (PDD), and the atelosteogenesis (AO) type II. In the event of a pseudo-diastrophic dysplasia the metaphysis shows irregular pillars of cartilage cells that are short and far apart standing. Moreover, irregular primary trabeculae of unremarkable resting cartilage can be recognized. During the AO type II (De la Chapelle dysplasia) one can find in the metaphysis of such osteochondrodysplasia, a similar histopathological image as in the diastrophic dysplasia but more pronounced. Also, there is a strong perilacunar staining and numerous cystic areas with radiating arrangement of collagen fibers can be identified. Hästbacka et al. (1996) studied three cultures of fibroblasts from Patients with AO II and found a lack of sulfate transporter enzyme localized the gene on chromosome 5q. Also, a defect in the sulfation of proteoglycans was observed. Mutations of *DTDST* (diastrophic dysplasia sulfate transporter) gene determine how "diastrophic" the patients may present. AO1, the most severe form of AO, has been assigned to the filamin group (group 7) based on molecular genetics. Interestingly, AO2,

13.2 Osteochondrodysplasias

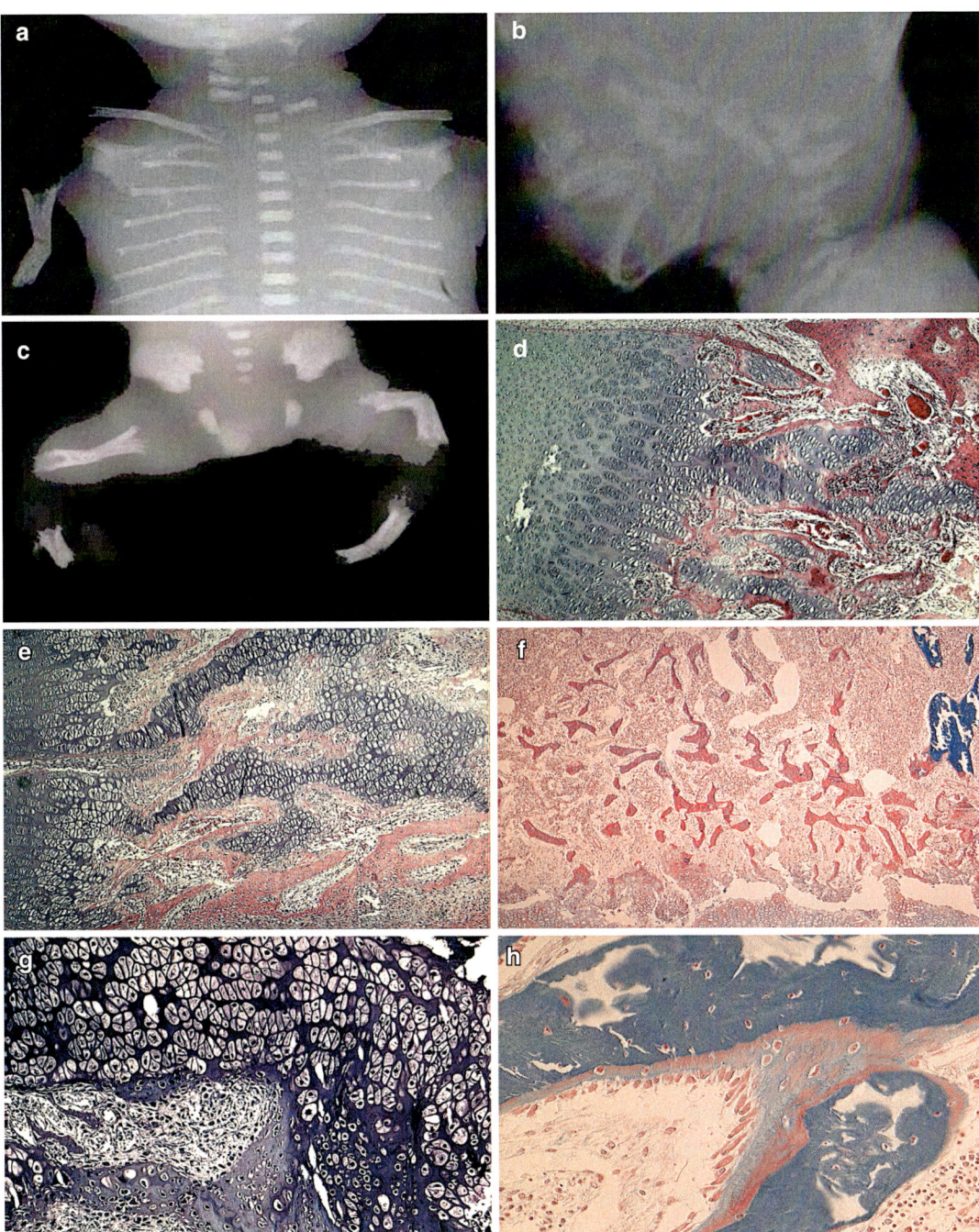

Fig. 13.1 Hypophosphatasia. The radiological images (**a–c**) of hypophosphatasia (HPP) show a generalized decrease in the size of mineralized bones, asymmetric vertebral arch mineralization in the cervical spine and the absence in the remainder of the spine, irregular metaphyseal cupping of the tubular bones of the limbs, bent long bones, short ribs, and a poorly mineralized skull with island-like ossification of frontal and occipital bones. The histological images (**d–h**) of the growth plate in HPP show that the growth plate in the HPP patient is widened and disorganized and large tongues and islands of unmineralized hypertrophic chondrocytes can be seen in metaphyseal bone tissue adjacent to the growth plate embedded in poorly mineralized osteoid ((**d**), hematoxylin and eosin staining, ×50 original magnification; (**e**), hematoxylin and eosin staining, ×50 original magnification; (**f**), Masson Goldner staining, ×25 original magnification; (**g**), Toluidin Blue staining, ×100 original magnification; (**h**), Masson Goldner staining, ×200 original magnification)

DD, AO1B, and other epiphyseal dysplasias share common mutations in the *SLC26A2* gene on chromosome 5q32-q33.1. The *SLC26A2* gene encodes a protein in cartilage that transports inorganic sulfate, and its absence results in hyposulfation of proteoglycans. The filamin group (group 7) is defined by the presence of mutations occurring in the *FLNA* gene on Xq28, which encodes filamin A. OCDs of this group include frontometaphyseal dysplasia, osteodysplasia Melnick-Needles, oto-palato-digital syndrome type 1 and type 2 (OPD1 and OPD2), terminal osseous dysplasia with pigmentary defects, atelosteogenesis type I and type III (AO1 and AO3), Larsen syndrome, spondylo-carpal-tarsal dysplasia, Frank-Ter Haar syndrome, and serpentine fibula-polycystic kidney syndrome. Group 8 includes metatropic dysplasia, spondylo-epimetaphyseal dysplasia, Maroteaux type, spondylometaphyseal dysplasia, Kozlowski type, brachyolmia, and familial digital arthropathy with brachydactyly.

Groups 9–17 involve OCD of significant interest to the pediatric pathologist, because of the differential diagnosis raised at the time of the prenatal ultrasonography and genetic counseling.

Groups 18–20 include bent joint dysplasias, slender bone dysplasias, and OCD associated with multiple joint dislocations. Among these three groups, it is probably essential to emphasize campomelic dysplasia, although bent bones at birth can be observed in a variety of medical conditions, including osteogenesis imperfecta, Antley-Bixler syndrome, cartilage-hair hypoplasia, Cummings syndrome, hypophosphatasia (Fig. 13.1), dyssegmental dysplasia, TD, ATD, and other OCDs. Slender bone dysplasias include 3-M syndrome (3 M1 and 3 M2), Kenny-Caffey dysplasia types 1 and 2, microcephalic osteodysplastic primordial dwarfism, IMAGE syndrome (intrauterine growth retardation, metaphyseal dysplasia, adrenal hypoplasia, and genital anomalies), osteocraniostenosis, and Hallermann-Streiff syndrome. OCD with multiple joint dislocations includes Desbuquois dysplasia and PDD (Sergi et al. 2001, Baumgartner-Sigl et al. 2007).

Groups 21–25 include the *chondrodysplasia punctata* (CDP) group (CDPX1 or brachytelephalangic type, CDPX2 or Conradi-Hunermann type, CHILD syndrome or congenital hemidysplasia, ichthyosis, limb defects, Greenberg dysplasia, rhizomelic CDP types 1–3, CDP tibial-metacarpal type, and Astley-Kendall type), *neonatal osteosclerotic dysplasias* (Blomstrand dysplasia, desmosterolosis, Caffey disease, and Raine dysplasia), *normoosteomorphic dysplasias with increased bone density* but without modification of bone shape (osteopetrosis, osteopoikilosis, melorheostosis, osteopathia striata with cranial sclerosis, dysosteosclerosis, and osteomesopyknosis), *dysplasias with increased bone density and meta- and/or diaphyseal involvement* (craniometaphyseal dysplasia, diaphyseal dysplasia Camurati-Engelmann, hematodiaphyseal dysplasia Ghosal, hypertrophic osteoarthropathy, oculodentooseeous dysplasia, osteoectasia with hyperphosphatasia, sclerosteosis, endosteal hyperostosis, van Buchem type, trichodentooseous dysplasia, craniometaphyseal dysplasia, diaphyseal medullary stenosis with bone malignancy, craniometaphyseal dysplasia, craniometadiaphyseal dysplasia, Wormian bone type, endosteal sclerosis with cerebellar hypoplasia, and Pyle disease), and *dysplasias with decreased bone density* (osteogenesis imperfecta II–VII, Bruck syndrome types 1 and 2, osteoporosis-pseudoglioma syndrome, Calvarial doughnut lesions with bone fragility, idiopathic juvenile osteoporosis, Cole-Carpenter dysplasia, spondylo-ocular dysplasia, osteopenia with radiolucent lesions of the mandible, Ehlers-Danlos syndrome, progeroid form, *geroderma osteodysplasticum*, and *cutis laxa*). *Hypertrophic osteoarthropathy* (HOA) is a clinical triad constituted by periosteal new bone deposition involving the short and long bones of the limbs, arthritis with effusion in adjacent joints, and clubbing, which is a distinctive bulbous enlargement of the tips of the fingers and toes due to edematous fibrovascular hypertrophy of the soft tissue. In pediatrics and youth, clubbing is usually seen in non-treated cyanotic heart disease, cystic fibrosis, and chronic lung infection (e.g., untreated bronchiectasis) differently from the elderly with bronchogenic carcinoma playing an important role.

13.3 Metabolic Skeletal Diseases

Most of the skeletal, metabolic diseases are collected in the Box 13.1. In this chapter, we do not target specific metabolic, skeletal disorders (primary hyperparathyroidism, Gaucher disease, mucopolysaccharidosis group, and osteopetrosis or Albers-Schoenberg marble bone disease) that may be found elsewhere.

> **Box 13.1 Metabolic Skeletal Diseases**
> 13.3.1 Rickets, and Osteomalacia
> 13.3.2 Osteoporosis of the Youth
> 13.3.3 Paget Disease of the Bone
> 13.3.4 Juvenile Paget Disease

> **Box 13.2 Vitamin D Endocrine System Key Elements**
> 1. Dietary intake of vitamin D_3 (prohormone) or cutaneous photo-conversion (ultraviolet) of *7-dehydrocholesterol* to *vitamin D_3* (cholecalciferol).
> 2. Vitamin D_3 metabolism into *$25(OH)D_3$* (calcidiol) by the liver.
> 3. Vitamin D_3 metabolism with conversion of $25(OH)D_3$ into two main dehydroxylated metabolites, namely, *$1a,25(OH)_2D_3$* (calcitriol) and *$24R,25(OH)_2D_3$*.
> 4. Systemic transport of the *$1a,25(OH)_2D_3$* and *$24R,25(OH)_2D_3$* to the target organs.
> 5. Dehydroxylated metabolite binds to a nuclear receptor in the different organs.
> 6. The subsequent generation of various biological responses on:
>   – Bone ⇒ ↑ bone tissue formation by maintaining appropriate Ca/P balance
>   – Kidney ⇒ ↑ Ca^{2+} reabsorption by the tubules
>   – Intestine ⇒ ↑ Ca^{2+} and P absorption and transport
>   – Parathyroid gland ⇒ ↓ PTH secretion
>   – Pancreas ⇒ ↑ insulin secretion
>   – Immune system ⇒ promotion of immunogenic and antitumor activities

Notes: $25(OH)D_3$ represents the dominant form of vitamin D circulating in the blood compartment, and vitamin D-binding protein (DBP) is a plasma protein that carries vitamin D_3 and all its metabolites to their different target organs. $1\alpha,25(OH)_2D_3$ transcriptionally controls the expression of a set of genes mediated through the nuclear vitamin D receptor (VDR) acting as a ligand-inducible factor. The biological actions of $24R,25(OH)_2D_3$ are incompletely investigated, although it seems essential that the combined presence of both dehydroxylated metabolites is required to have the complete spectrum of biological responses attributable to vitamin D.

> **Box 13.3 Etiology of Vitamin D Deficiency**
> 1. ↓ absorption (e.g., malabsorption in IBD, celiac disease, and cystic fibrosis)
> 2. Inadequate exposure (sunscreen with SPF rating ≥8) or intake of sun exposure
> 3. Metabolomics dysregulation
> 4. Resistance to the effects of vitamin D
>
> Notes: IBD, inflammatory bowel disease; SPF, sun protection factor

13.3.1 Rickets, and Osteomalacia

Vitamin D is vital for our organism, and the vitamin D endocrine system has some essential elements, which are summarized in Box 13.2.

13.3.1.1 Vitamin D Deficiency
- *DEF*: Bone mineralization defect (rickets/osteomalacia and, possibly, osteoporosis) due to deficiency of vitamin D by usually either inadequate dietary intake or insufficient exposure to sunlight.

- *EPG*: The causes of vitamin D deficiency are grouped in Box 13.3.

Abnormal metabolism may result from deficiencies in the production of $25(OH)D_3$ or $1,25(OH)_2D_3$, and individuals with chronic renal disorders usually develop rickets or osteomalacia, because of the decrease of $1,25(OH)_2D_3$ production and increase of phosphate levels. In the case of hepatic dysfunction, there is also an interference with the formation of active vitamin D metabolites. Two types of vitamin D-dependent hereditary rickets (VDDR) are known to be caused by mutations in the $1\alpha(OH)$ase and VDR genes, including the *1α(OH)ase* gene, which is responsible for VDDR type I, and *1α,25(OH)D receptor* gene for type II. Both diseases display an autosomal-recessive trait, but clinical features and response to therapy are distinct.

In the healing process of a fracture, first a hematoma forms, and then there is an organization by the entry of capillaries and blood vessels. This step is tightly followed by the absorption of devitalized bone, which begins to be identified about 3 days after the bony fracture. Subsequently, the first callus is formed by the bone growth of intramembranous "pseudosarcomatous" skeletal tissue, which grows across the bony break forming the procallus or primary callus. The second callus replaces, finally, the first callus by lamellar skeletal tissue.

In the differential diagnosis of the four entities described in Box 13.4, it is paramount to rule out parosteal osteosarcoma, which harbors atypical bony structure and nonreactive cellular, spindle-shaped stroma as well as osteoblast rimming. An *exuberant callus* may occur in the setting of infection, inadequate blood supply, delay in the re-adsorbing process (e.g., malnutrition), and insufficient immobilization.

Box 13.4 Bony Fractures: Callus – DDX
- Avascular necrosis
- Myositis ossificans
- Osteomyelitis-linked sub-periosteosis of reactive nature
- Parosteal osteosarcoma

Sequestered bone is dead bone ("*sequestrum*") and may occur in pediatrics in the setting of non-neoplastic and neoplastic disease. Conversely, *involucrum* is new bone formed at a subperiosteal location that encloses an inflammatory focus (reactive periosteal bone). Another aspect to be considered is *foreign material* (e.g., screws) that may become isolated from bone by fibrous tissue which is continuous with periosteum, but no foreign body giant cell reaction occurs. Skeletal and radiographic changes associated with rickets are depicted in Fig. 13.2.

Diagnosis of rickets is based on levels of calcidiol and may be suspected based on any of the following (1) history of inadequate sunlight exposure or dietary intake, (2) symptoms and signs of rickets, osteomalacia, or neonatal tetany, and (3) characteristic bony radiological findings. There is a loss of the sharpness of the diaphyseal ends. They become cup-shaped showing a spotty or fringy rarefaction. Subsequently, the ends of the radius and ulna growing noncalcified and appear radiolucent. The radiology image is striking showing an increased distance between the ends of the radius and ulna and the metacarpal bones (Sergi and Linderkamp 2001, Kato et al. 2002, Nield et al. 2006, Malloy et al. 2014, Zhang et al. 2016).

13.3.2 Osteoporosis of the Youth

- *DEF*: Decrease in the mass of ordinarily mineralized bone differently from osteomalacia, which is an accumulation of non-mineralized bone matrix due to faulty mineralization (Bishop et al. 2014; Saraff et al. 2015; Cascio et al. 2020).
- *EPI*: Usually seen in the elderly as a result of increased absorption, some mild degree of this condition can start early in life.
- *EPG*: Osteoporosis is well known to occur in the postmenopausal period if no estrogen replacement takes place. However, adults experience osteoporosis following endocrine dysfunctions, neoplasms (epithelial and mesenchymal type), and immobilization.

In children, osteoporosis is classified as primary when it occurs in an otherwise healthy child

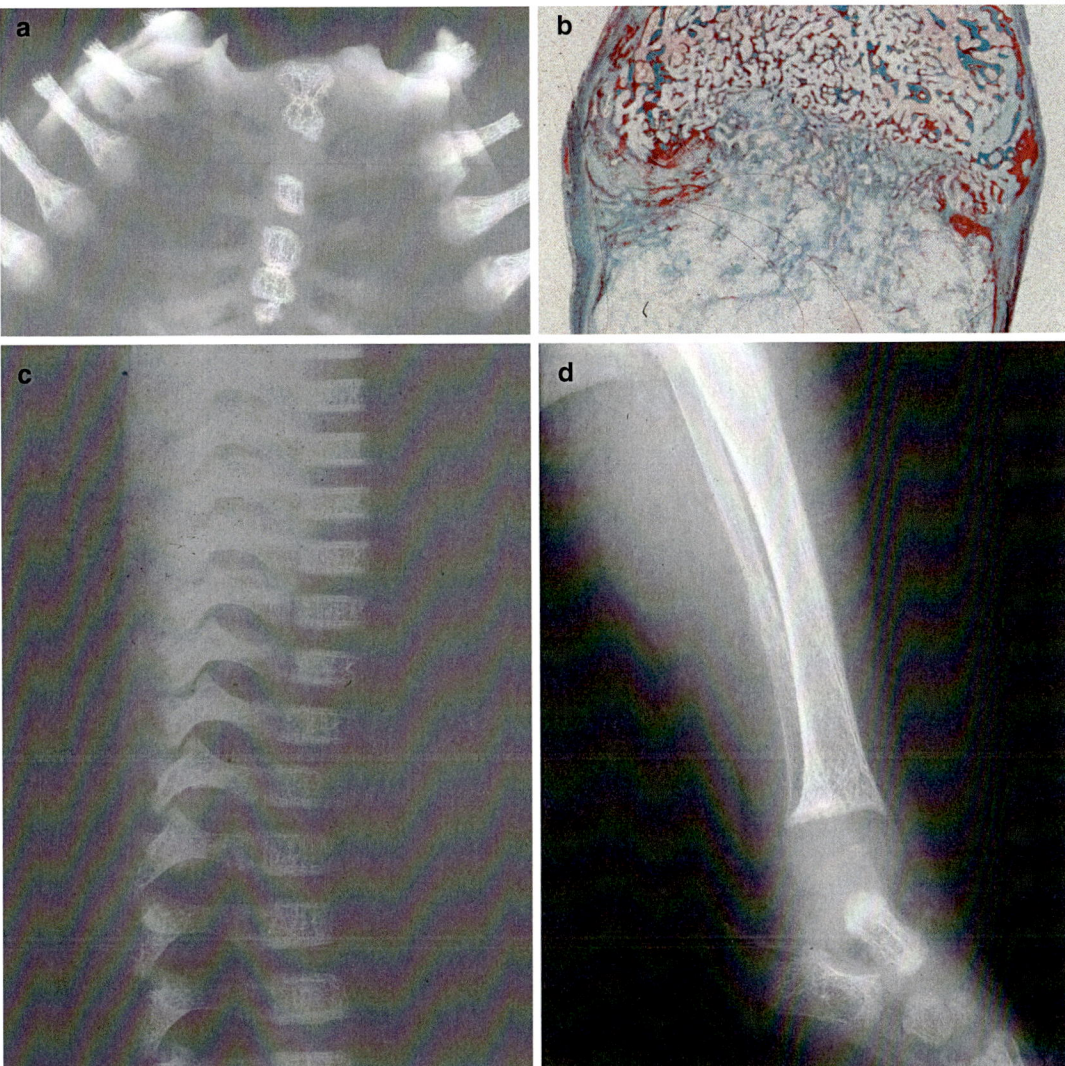

Fig. 13.2 Rickets. In (**a**, **c**, **d**), there are some of the characteristic features of rickets, including enlargements of the costochondral junction in the chest and of the epiphyseal-metaphyseal junctions in long bones, with cupping and fraying of the distal ends accompanied by double contour signs at the vertebral column. The histology (**b**) shows widening and thickening of the physeal growth plate with poor removal of cartilage, persistent hypertrophic chondrocytes in the zone of provisional ossification, disorder of the vascular penetration of cartilage with impairment of the correct chondrocytic proliferation, and deposition of newly formed and poorly defined broad osteoid tissue (glycol methacrylate bone sections without previous decalcification, ×0.5 original magnification)

due to an underlying genetic condition (e.g., osteogenesis imperfecta, cleidocranial dysplasia, Marfan syndrome, Ehlers-Danlos syndrome, and Hajdu-Cheney syndrome). Secondary osteoporosis occurs because of chronic illness or its treatment (e.g., high-dose glucocorticoids, antiepileptic drug therapy, immobility, leukemia, inflammatory conditions – rheumatoid arthritis and Crohn disease, hypogonadotrophic hypogonadism, and poor nutrition – anorexia nervosa). Risk factors for fractures include age, sex, previous fractures, genetic predisposition, poor diet, total body mass, and forceful or lack of physical activity. Mutations in *PLS3* gene, which encodes PLASTIN 3, a bone regulatory protein, have been identified in early-onset X-linked osteoporosis with axial and appendicular fractures developing during childhood. Other forms of early-onset osteoporosis involve the WNT signaling pathway, which is essential for normal bone homeostasis (osteoblast cellular proliferation and differentiation). Defects in this complex signaling pathway chiefly affect bone formation and may involve WNT, low-density lipoprotein receptor-related protein 5 (LRP5), a coreceptor of WNT located on the osteoblast membrane, LGR4, and WNT16. Biallelic mutations in *LRP5* cause osteoporosis-pseudoglioma syndrome (OPPG), and heterozygous *LRP5* mutations cause early-onset osteoporosis.

- *CLI*: History of recurrent low impact fractures or moderate to periods of severe backache. Asymptomatic osteoporosis is being detected through surveillance for vertebral fractures in at-risk children, such as children on high-dose glucocorticoid therapy, or through incidental osteopenia found on X-ray imaging.

A low bone mass and deterioration of bony structure microarchitecture result in increased bone fragility, which is the definition of osteoporosis. The diagnosis of osteoporosis in pediatrics can be confidently made in the presence of:

1. A combination of size-corrected low bone mineral density (BMD) of >2SD below the mean and a substantial history of low-trauma fractures, subjectively defined as the presence of ≥2 long bone fractures by the age of 10 years or ≥3 long bone fractures at any age, up to 19 years.
2. ≥1 vertebral fracture in the absence of high-energy trauma or local disease, independent of BMD.

- *LAB*: Crucial determinations are serum alkaline phosphatase, calcium, phosphate, vitamin D and urinary bone mineral excretion, and dual-energy X-ray absorptiometry (DXA), which remains the technique of choice to measure bony mass. DXA is highly reproducible, commonly available, and inexpensive and exposes the individuals to low radiation. Quantitative computed tomography (QCT) and peripheral QCT (pQCT) possess the benefit of measuring cortical geometry and volumetric densities of both cortical and trabecular bone. These measurements provide information not attainable through DXA, but there is the disadvantage of higher radiation exposure. Other techniques include the digital X-ray *radiogram*-metrics, which estimates BMD by hand radiographs in children, the quantitative ultrasound, magnetic resonance imaging, and the trans-iliac bone biopsy with tetracycline labeling, which is invasive and requires general anesthesia in children. There are several tests to evaluate the functional outcomes, muscle strength, and mobility in children under treatment. The most common functional tests include the 6-minute walk test, Bruininks Oseretsky Test of Motor Proficiency, gross motor function measurement, Childhood Health Assessment Questionnaire score, "faces" pain scale, "chair-rise" test, leg mechanography, and grip force testing by dynamometry. These tests can be further evaluated using textbooks of pediatric orthopedics.
- *TRT*: Biphosphonates, proper nutrition with a 75% implementation of fresh food in the diet, improving muscle strength with regular avoidance of passive hobbies (e.g., video-games, internet, movies, etc.), mobility with avoidance of cars or motorbikes, and rehabilitation are key, but it is paramount to target the primary disorder if it is feasible.

13.3.3 Paget Disease of the Bone

- *DEF*: It is a generalized bone disease with the acme in the adult (middle age) characterized by markedly increased bone turnover (1 mm/month).
- *SYN*: Osteitis deformans.
- *EPI*: 90% of patients are older than 55 years, but some younger patients are present; ♂ = ♀. More common countries are England, Northern European countries, and Australia.
- *EPG*: It is a gene-related disorder with environmental (epigenetic) components. The

genes involved in Paget Disease of the Bone (PDB) are *TNFRSF11A*, *TNFRSF11B*, *SQSTM1*, and valosin-containing protein (*VCP*). *TNFRSF11A* encodes RANK and *Deleted in Colorectal Cancer* (*DCC*), which nonetheless is frequently reduced in expression in osteosarcoma, a neoplasm, which is often seen in patients with PDB. There is nuclear factor kappa B (NF-κB) activation by RANKL signaling (receptor activator for NF-κB ligand, RANKL). Osteoprotegerin, also known as osteoclastogenesis inhibitory factor, or TNF receptor superfamily member 11B, is a protein that is determined by the *TNFRSF11B* gene in humans. In PDB, osteoclasts have more nuclei per single osteoclast, are markedly increased in both number and size, and have an increased rate of cellular proliferation and a significant bone-resorbing capacity per single osteoclast. Nuclear inclusions have been observed in these osteoclasts. These inclusions have been considered *paramyxovirus*-like, which has raised a number of epidemiological and microbiological theories. These specific inclusions seem to be microcylindrical structures, but the true identity has not been established yet. The osteoclast precursors, circulating in the blood, show increased sensitivity to factors that stimulate bone resorption. Osteoblasts may also be the culprit and not bystanders only and several hypotheses have been formulated in the scientific literature. In fact, cultures of osteoblasts from bone lesions are abnormal because they show an increased expression of genes encoding interleukin 1, interleukin 6, and dickkopf 1. The Wnt/Fzd/LRP (frizzled, Fzd; low-density lipoprotein receptor-related protein) complex and the associated β-catenin signaling by treating cells either with Dickkopf-1 (Dkk1), a specific antagonist of the Wnt/β-catenin pathway or LRP5/LRP6 silencing, indeed promote neurite outgrowth by shifting endogenous Wnt activity from the β-catenin pathway to Fzd3-mediated non-canonical pathway that is responsible for neurite formation and this pathway may have been one of the major routes supporting the progression or degeneration of PDB to osteosarcoma.

- *CLI*: Bowing and easily fracture of painful extremities with bony pain, fractures, increased head size, headache with audible bruit over the head ("syndrome of the pagetic vascular steal"), blindness, deafness, leg bowing, kyphosis, and high-output heart failure. High-output congestive heart failure has also other conditions that need to be distinguished clinically. They include anemia, obesity, exacerbation of hyperthyroidism or thyrotoxicosis, liver disease, lung disease, Beriberi heart disease, arterio-venous fistula, sepsis/septic shock, and pregnancy, which may be substantially kept in mind considering two main mechanisms, including the increase in the body's demand for blood from increased metabolism and the bypass of the arteriolar and capillary bed causing an increased blood flow into venous circulation in consequence of a lack of vascular resistance.
- *LAB*: Increased rate of excretion of type I collagen breakdown products hydroxyproline, very high plasma alkaline phosphatase activity, and elevated serum osteocalcin levels. Serum calcium and phosphate are usually normal.
- *IMG*: The radiographs show coarse widening of the diaphyses with cortical thickening and "chalk-stick fracture." Fractures are usually transverse. A circumscribed osteoporosis characterizes the initial phase in the skull and a resorption area with advancing edge in long bones, while blastic bony production along the lines of stress, widening of bones, cortical thickening, and potential fractures are added to the signs of the initial phase in the mid phase. The late phase is characterized by bones showing widening and thickened bone, "cotton wool exudates" of the skull and multiple fractures, postural bowing, kyphosis, and ultimately neoplastic progression to osteosarcoma, chondrosarcoma, or fibrosarcoma among others.
- *GRO*: Many patients have a monostotic and asymptomatic disease, usually affecting the axial skeleton, mainly lumbosacral spine and

pelvis, but other patients have a polyostotic disease.
- *CLM*: Pagetic bone lesions show evidence of ↑ osteoclastic bone resorption, enhanced but disorganized bone formation with ↑ # of osteoblasts and osteocytes (per mm^3) and a ↓ grade of organization of their canalicular network, maintenance of the osteoblastic rimming, fibrosis/fibroplasia of the bone marrow, and ↑ vascularity of bone (numerous dilated capillaries). Three characterized phases are at the basis of the mosaic lamellar bone. The three stages are the osteolytic phase, the mixed (osteoclastic – osteoblastic) phase, and the osteosclerotic phase. Initially osteolytic, the bone is followed by abnormal hyperplasia. There are thick irregular bony trabeculae, disjointed and discontinuous lamellae, and a mosaic of scalloped cement lines with astonishing advancements of the bone of about 1-mm/month.
- *DDX*: Hyperparathyroidism.
- *TRT*: Biphosphonates.
- *PGN*: Mostly, there is an increased risk of osteosarcoma (up to 20% of patients with polyostotic PDB) at the involved sites, including the femur, humerus, tibia, and skull. In the monostotic type, the risk is limited to 10%. Other tumors are also increasingly observed in patients suffering from PDB and include malignant fibro-histiocytoma (MFH), fibrosarcoma, chondrosarcoma, and malignant giant cell tumor.

13.3.4 Juvenile Paget Disease

- *DEF*: AR-inherited generalized bone disease (MIM 239000) with striking similarities to Paget disease of the adult (middle age) (PDB) characterized by markedly increased bone turnover.
- *SYN*: Idiopathic hyperphosphatasia and osteoectasia with hyperphosphatasia.
- *EPI*: About 50 cases have been reported worldwide to date.
- *EPG*: Deletion or mutations of the *TNFRSF11B* gene, a member of the tumor necrosis factor (TNF) receptor superfamily encoding osteoprotegerin (OPG), are considered the culprit in the etiology of Juvenile Paget Disease (JPD).
- *CLI*: Bowing and easily fracture of painful extremities. Patients suffer from short stature, kyphoscoliosis, chest wall deformity, and skull enlargement often leading to cranial nerve deficits similar to the phenotype observed in patients suffering from PBD.
- *LAB*: Increased rate of excretion of type I collagen breakdown products, very high plasma alkaline phosphatase activity, and elevated serum osteocalcin levels are key factors to consider biochemically.
- *IMG*: The radiographs show coarse widening of the diaphysis with cortical thickening.
- *GRO*: The skull shows increased thickness with wide diploic spaces and focal areas of sclerosis and uneven bone mineralization. Vertebral bodies may show osteopenia or sclerosis and sometimes manifest compression fractures. In the pelvis, in addition to thick trabeculae and areas of sclerosis, there can be some areas of rarefaction, and rib changes may incorporate sclerosis, narrowing, or widening.
- *CLM*: Increase of both osteoblasts and osteoclasts in bone tissue. Woven bone is mainly observed in the calvarium, vertebrae, and ribs. The cortex (cortices) of tubular bones are alternatively thinned or thickened along their entire lengths from accelerated endosteal bone remodeling. Conversely, increased bone formation at the periosteum thickens the diaphysis and widens individual bones. It is important to remember that woven bone is hypercellular when compared to lamellar bone on histological ground. This aspect is substantially indifferent if the woven bone is laid down by benign or malignant cells. Osteoid is pink and amorphous and contains collagen by light microscopy. Chondroid is identical to osteoid and not bluish because the content of proteoglycans is quite low differently from the cartilage.
- *DDX*: Two conditions considered in differential diagnosis should be polyostotic fibrous

dysplasia and hereditary hyperphosphatasia. A marked elevation of alkaline phosphatase level is unusual in polyostotic fibrous dysplasia, and radiographs of patients with hereditary hyperphosphatasia do not reveal the typical mosaic pattern observed in patients suffering from JPD.
- *TRT*: Bisphosphonates are often used to reduce the accelerated bone turnover. This drug can impressively meliorate the skeletal phenotype, specifically if started early enough in childhood and continued at least until completion of the bone growth. Studies using recombinant OPG or denosumab also provided favorable results.
- *PGN*: The phenotype tends to be more severe with aging, although mild deformities are recognized after 2 years. Although JPD has some similarities to classical PDB, it is undoubtedly a more severe condition as proven by the early age at onset and the development of marked bone deformity during childhood, adolescence, and youth. Although long-term data may not be extensive, the early-onset of bone disease harbors an increase of the neoplastic potential with malignant transformation into sarcomas.

13.4 Osteitis and Osteomyelitis

One of the most critical challenges for a pathologist is to distinguish between inflammation, inflammatory atypia, and neoplastic atypia. The incisional or excisional biopsy of a bony lesion may represent a challenge at some point. As a standard and good practice for patients' quality assurance, all bone biopsies should be revised by two consultants in pathology, one of which with radiological and histological experience in skeletal and soft tissue pathology. The problems that a pathologist may encounters include the orientation of the biopsy tract that should be longitudinal and carried out in an area that could be easily identifiable radiologically. Thus, computer scanning-guided biopsies are the preferred biopsies for both clinicians and pathologists. In case of an incisional biopsy, following a careful hemostasis, the incision is carried out and that should not go

Box 13.5 Osteitis and Osteomyelitis
13.4.1 Osteomyelitis, pyogenic
13.4.2 Osteomyelitis, tubercular
13.4.3 Osteomyelitis, mycotic
13.4.4 Chronic multifocal recurrent osteomyelitis
13.4.5 Osteitis, sclerosing (Garre Osteomyelitis)

more than one muscle compartment. Both soft tissue and bony tissue need to be sent to the pathology department. Very few fragments should be sent to the microbiology department for bacteriological and mycologic analysis (Box 13.5).

13.4.1 Osteomyelitis

- *DEF*: Bone inflammation with or without evidence of infection with acute hematogenous, subacute focal and chronic types (Box 13.6).
- *EPI*: Most common in infants and children but declining incidence rates have been observed recently in well-developed countries due to better diagnosis and therapy.
- *EPG*: *Staphylococcus aureus* is the most common microbiological agent in all age groups.
- *CLI*: Pain, ↓ ROM (range of mobility), "tumor" (swelling), "rubor" (erythema), fever, malaise, and irritability.
- *IMG*: Delayed changes on imaging (2–4 weeks), lytic lesion with mottled appearance commonly seen in metaphysis, and periosteal reaction.
- *GRO*: Variable gross morphology according to the age of the patient and age of the lesion (acute vs. chronic). Generally, an acute osteomyelitis exhibits pus tracking through bone, periosteal elevation and shell of reactive periosteal bone around a necrotic center, while a chronic disease demonstrates prominent periosteal bone formation without acute inflammatory features.
- *CLM*: Mixed cell population with varying number of acute and chronic inflammatory cells. It is useful to distinguish between

sequestrum or dead cortical bone and *involucrum* or new cortical bone (Figs. 13.3, 13.4 and 13.5). The algorithm approach for diagnostic and prognostic purposes should contain specific items (Box 13.7).
- *TRT*: Following biopsy and culture in case of an acute process IV antibiotics ± surgical debridement are essential, while in the chronic process IV antibiotics are always accompanied by surgical debridement.
- *PGN*: It depends from the state and chronicity of the process because it may be highly for life. The Cierny-Mader staging system is often

> **Box 13.6 Chronologic Classification of Osteomyelitis**
> - Acute (onset <2 weeks) *S. aureus* (mainly)
>   – Hematogenous vs. direct
> - Subacute (onset 2–4 weeks) (*S. aureus*)
> - Chronic (onset >4 weeks) (septic and aseptic)
>   – Chronic sclerosing vs. chronic recurrent multifocal vs. chronic form with uncommon organisms (*M. tuberculosis*, *T. pallidum*)

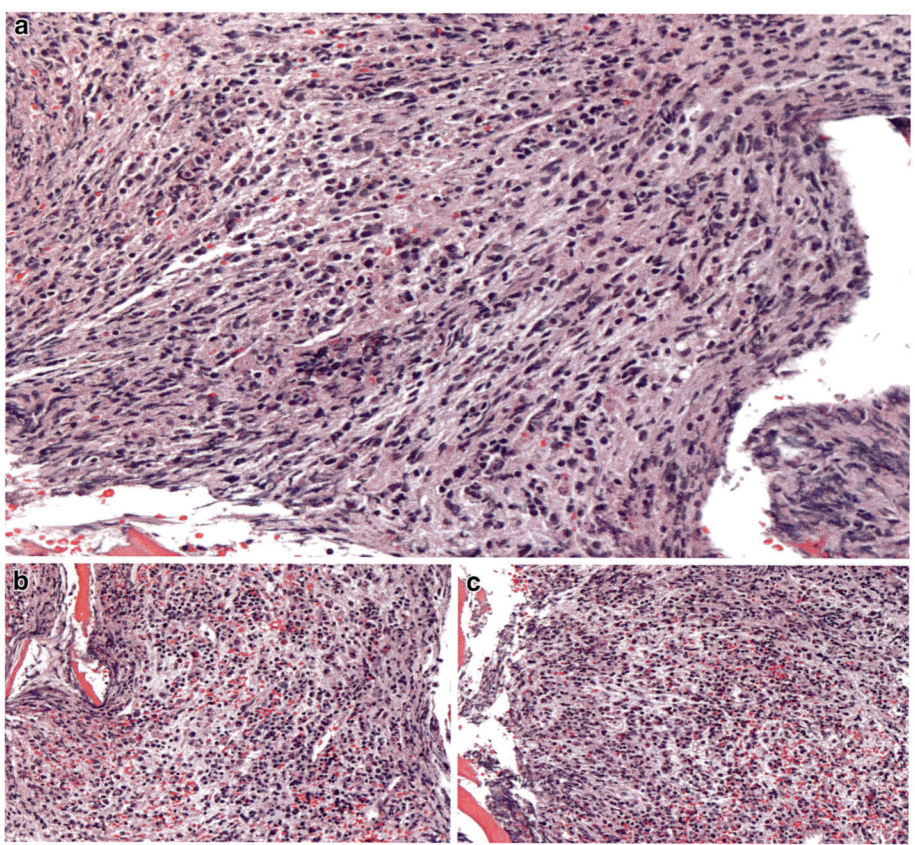

Fig. 13.3 Osteomyelitis, acute and chronic. There is a heavy inflammatory infiltrate of acute and chronic type with neutrophils, lymphocytes, and plasma cells. (**a–c**; H&E staining). Three different fields show the variability of the intensity of inflammatory reaction (**a–c**, hematoxylin and eosin staining, ×200 original magnification for all three microphotographs)

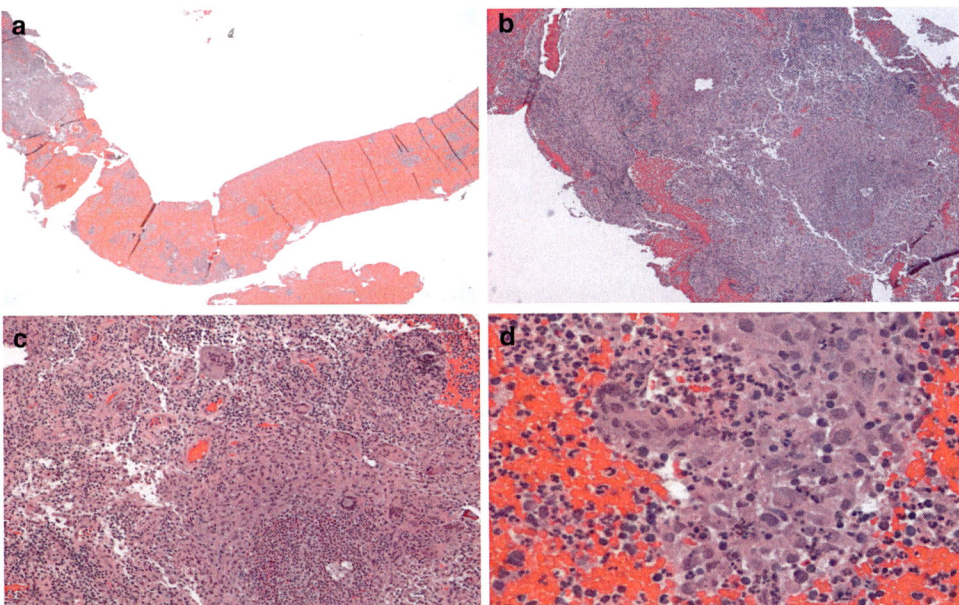

Fig. 13.4 Osteomyelitis, granulomatous. This osteomyelitis has a granulomatous character with multiple granulomas and multinucleated giant cells. (**a–d**; H&E staining. Different magnification of this biopsy show the intense inflammatory reaction with macrophages, epithelioid cells, and multinucleated giant cells (**a–d**, hematoxylin and eosin staining, ×12.5, ×50, ×100, ×400 as original magnification, respectively)

used for staging. It includes medullary osteomyelitis (stage 1), superficial osteomyelitis (stage 2), localized osteomyelitis (stage 3), and diffuse osteomyelitis (stage 4). The factors affecting the immune surveillance, metabolism, and vascularity are shown in Box 13.8.

High temperature on admission also seems to be specific in identifying those patients who may require prolonged treatment. A significantly raised CRP greater than 100 mg/L suggests that more prolonged therapy may be necessary. Also, the CRP responds more rapidly to effective treatment, whereas ESR can remain elevated for several weeks despite a good clinical response. If the clinical picture is one of improvements and the CRP is falling, it is wise to propose that intravenous antibiotics are converted to oral therapy. It has been assessed that approximately 60% of patients can be successfully migrated to oral therapy by day 3 and about 85% can proceed by day 5. In only about 15% of the patients, intravenous therapy is required for 6 days or more Jagodzinski et al. (2009). In the Jagodzinski et al. study, the median duration of inpatient stay was only 5 days. These data show that a significantly shorter period of intravenous therapy is needed than shown in most data published earlier. An initial investigation with blood cultures and hematologic markers (WCC, CRP, and ESR) is, however, necessary to evaluate the progress of the therapy remembering to protect the bone always. Appropriate imaging should also follow the proposed follow-up. Ultrasound should be used and MRI confirmation is advisable in most of the patients. There are studies contemplating series of empirical antibiotics (flucloxacillin and co-amoxiclav) that should be commenced after prompt drainage of joint collections. Sequential alternate-day estimations of hematologic markers should also be taken into consideration. In case that after three full days of I.V. antibiotics, the child is clini-

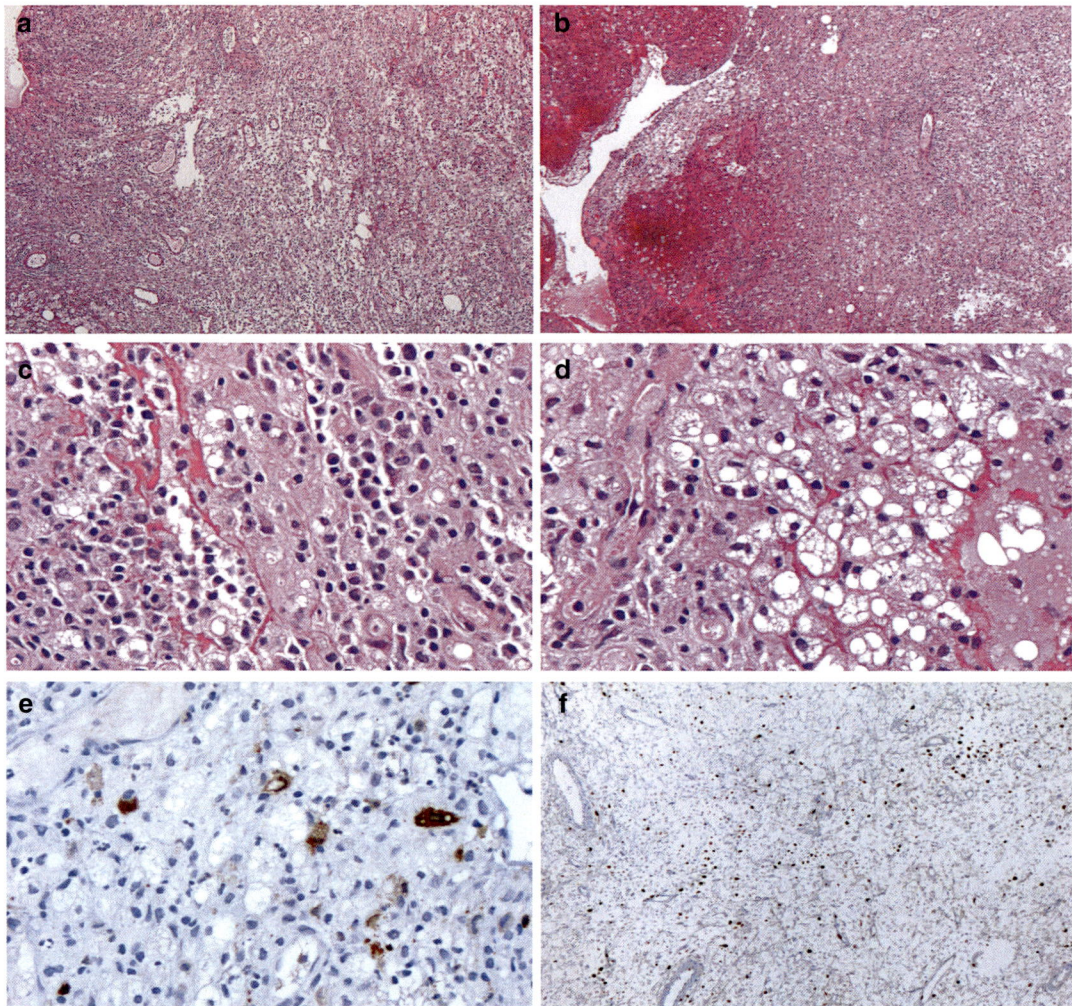

Fig. 13.5 Osteomyelitis, xanthogranulomatous. In this osteomyelitis, there is clear prevalence of foamy osteocytes justifying the term of xanthogranulomatous osteomyelitis (**a–d**, hematoxylin and eosin staining, ×100, ×50, ×400, and ×400 as original magnification, respectively). The osteomyelitis is particularly apparent at high-power magnification (**c**, **d**). The immunohistochemical staining of the osteomyelitis with S100 (**e**) and MIB1 (**f**) shows some positive cells (**e**, ×400, original magnification) and low proliferation activity (**f**) (×100, original magnification). S100, CD1a, and Langerin immunohistochemistry were useful in ruling out Langerhans cell histiocytosis in this child

cally improving, the temperature is settling, and the CRP value is falling, the pediatric patient can be converted to oral antibiotics safely. Meta-analyses investigated numerous studies. It is probably key to use a three-phase bone scintigraphy followed by scintigraphy with mixed WBCs or granulocytes labeled with ^{111}In and, more recently, with ^{99m}Tc-HMPAO (hexamethyl-propyleneamine oxime). HMPAO is often referred to by its pharmaceutical name of exametazime. 18F-FDG (fludeoxyglucose) PET remains the most accurate method (91.9%) for the study of bone infections. The sensitivity is 94%, and the positive predictive value is 94.2%. FDG PET appears to be a valuable tool for the assessment of inflammation during follow-up of secondary osteomyelitis and the study of infections treated with antibiotics.

> **Box 13.7 Algorithmic (step-wise) Approach to Osteomyelitis**
> 1. Identify the organism (blood cultures, open or needle biopsy).
> 2. Drainage of the bone abscess.
> 3. Select the appropriate antibiotic therapy.
> 4. Deliver the appropriate antibiotic therapy making sure that the delivery is complete.
> 5. Stop the tissue destruction – protecting the bone (e.g., nanotechnology).
> 6. Follow up (clinically: local Celsius signs, T°, pain, etc.; laboratory – ESR, CRP, WCC; imaging, X-ray, MRI, bone scan, PET-CT).
>
> Notes: *T°* temperature, *ESR* erythrocyte sedimentation rate, *CRP* C-reactive protein, *WCC* white cells count (neutrophils), *MRI* magnetic resonance imaging, *PET-CT* positron emission tomography and computed tomography scans

> **Box 13.8 Factors Affecting Immune Surveillance, Metabolism, and Vascularity**
> - Systemic
>   Age extremes, chronic hypoxia, diabetes mellitus, hepatic failure, immune deficiency/disease, immunosuppression, malignancy, malnutrition, neuropathy, renal failure
> - Local
>   Blood vessel compromise, chronic lymphedema, neuropathy, radiation-induced fibrosis, scar/keloid of extensive degree, small vessel disease, vasculitis (arteritis), venous stasis

13.5 Osteonecrosis

13.5.1 Bony Infarct and Osteochondritis Dissecans

Necrosis may be the consequence of an infarct or *osteochondritis dissecans*.

13.5.1.1 Bony Infarct
- *DEF*: Triangular/wedge-shaped segment with yellow discoloration and sharply demarcation. In a comprehensive review, the osteonecrosis was reviewed. One hundred and nine studies published since 1985 regarding the results of treatment of osteonecrosis were considered, and 12 major classification systems were cited. The classification system most frequently referred to was that of Arlet and Ficat (1964), Ficat and Arlet (1980), followed by those of the University of Pennsylvania (Steinberg and Steinberg, 2014); Marcus, Enneking, and Massam (1973); the Association Research Circulation Osseous (ARCO), which was originally published in 1991–1993/1994 and carefully revised in 2019 (Yoon et al. 2020); and the 1987/2001 Japanese Investigation Committee. In the University of Pennsylvania Classification of Osteonecrosis the important features are the inclusion of the measurement of lesion size and the accurate extent of joint involvement using Magnetic Resonance Imaging (MRI). In 1987 and 2001, the Japanese Investigation Committee for Avascular Necrosis emphasized the location of lesion as an crucial factor to predict impending collapse. The Japanese Investigation Committee system indicates that lesions progress from medial to lateral as these lesions become progressively larger. In the 2019 ARCO revised classification of osteonecrosis four stages are identified. In stage I, X-ray is normal, but MRI is abnormal with a band lesion of low signal intensity, which is located around the necrotic area. Bone scan exhibits a cold spot. In stage II, X-ray is abnormal and the MRI discloses osteosclerosis, osteoporosis (focal), or cystic changes localized in the femoral head. In stage III, there is subchondral fracture on X-ray or Computer Tomography (CT) scan with fracture in the necrotic portion, and/or flattening of the femoral head. In this stage III, there is a

subclassification with IIIA (early) when the femoral head depression results ≤2 mm, while IIIB (late) when the femoral head depression is >2 mm. In stage IV, there is X-ray osteoarthritis with joint space narrowing, changes of the *acetabulum* bones as large cup-shaped cavity on the anterolateral aspect of the pelvis that articulates with the *caput femoris* to form the hip joint, and definitive bony destruction identified on plain X-rays.

- *SYN*: Avascular (bone) necrosis (AVN), aseptic necrosis, osteonecrosis.
- *EPI*: 1–30/10⁵ children/adolescents, ♂:♀ = 2:1.
- *EPG*: Post-traumatic, alcohol-related (see the Manggold et al. (2002) AVN animal model for aseptic bone necrosis), Gaucher disease (bony infarcts accompanied with Erlenmeyer flask deformity of the femurs, osteonecrosis of the femoral head, and pathologic fractures), decompression sickness or dysbarism (Greek βάρος "weight" and βαρύς "heavy") or "caisson" disease, prolonged (mis-)use of steroidal agents, sickle cell anemia, and chronic pancreatitis.

The "TISH" acronym as mnemonics can be used. In all cases, there is an impairment of the blood supply to a segment of bony tissue and bone marrow.

- *T*: Trauma, alcoholism, dysbarism, radiation, gout, Gaucher, pancreatitis.
- *I*: Infection.
- *S*: Soft tissue damage, including connective tissue disorders, corticosteroids, and tumors.
- *H*: Hemato-vascular, hemoglobinopathies, and coagulopathies.

In any case, there is a disruption of blood supply by intrinsic or extrinsic factors.

- *CLI*: Usually asymptomatic.
- *IMG*: In the first 1–2 weeks, there are no radiological abnormalities. Subsequently, a characteristic radiographic "crescent sign" is seen.
- *GRO*: Lemon-yellow discolored (necrotic) triangular segment with the avital subchondral bony plate as the basis of the triangle and epiphysis as the apex.
- *CLM*: Disorderly mixture of empty or enlarged osteocyte lacunae, karyopyknosis, increased basophilic staining (basophilia) of bony matrix with new bone (woven bone), fibrous and fibro-cartilaginous tissue interposed between the osteonecrotic trabeculae and the surrounding osteosclerotic bone with granulation tissue (inflammatory cells and vascularized reactive stroma) at the periphery of the lesion.
- *TRT*: Observation only. Surgical treatment is not required because it is usually asymptomatic.
- *PGN*: The bony infarct eventually gets osteochondral collapse and secondary arthritis but harbors an increased risk of developing a malignancy.

Kienbock disease is an example of AVN and precisely of the lunate bone (semilunar bone), a carpal bone of the hand (Fig. 13.6). Symptoms include progressive wrist pain and tenderness in the dominant hand of young men (20–45) (bilateral disease, 10%). Imaging (CT, MRI) is diagnostic (sclerosis → cystic changes → fragmentation → collapse). DDX includes dorsal wrist ganglion, synovitis or arthritis, or extensor tendinitis. TRT includes surgery aimed to relieve pressure on the lunate by surgically shortening the radius or lengthening the ulna, implanting a blood vessel or bone graft on a vascular pedicle, free-vascularized bone grafts from the knee, and total wrist arthrodesis (surgical joint fixation, aka artificial ankylosis or syndesis, Greek ἄρθρον "joint" and δέσις "binding" from δέειν "to bind"). PGN: *Sequelae* are lunate bone collapse with the fixed rotation of the scaphoid and subsequent degeneration of the carpal joints.

13.5.1.2 Osteochondritis Dissecans

Post-traumatic (most often) necrosis of the articular cartilage and subchondral bone with separation from the rest of the bone usually located at the medial femoral condyle.

13.6 Tumorlike Lesions and Bone/Osteoid-Forming Tumors

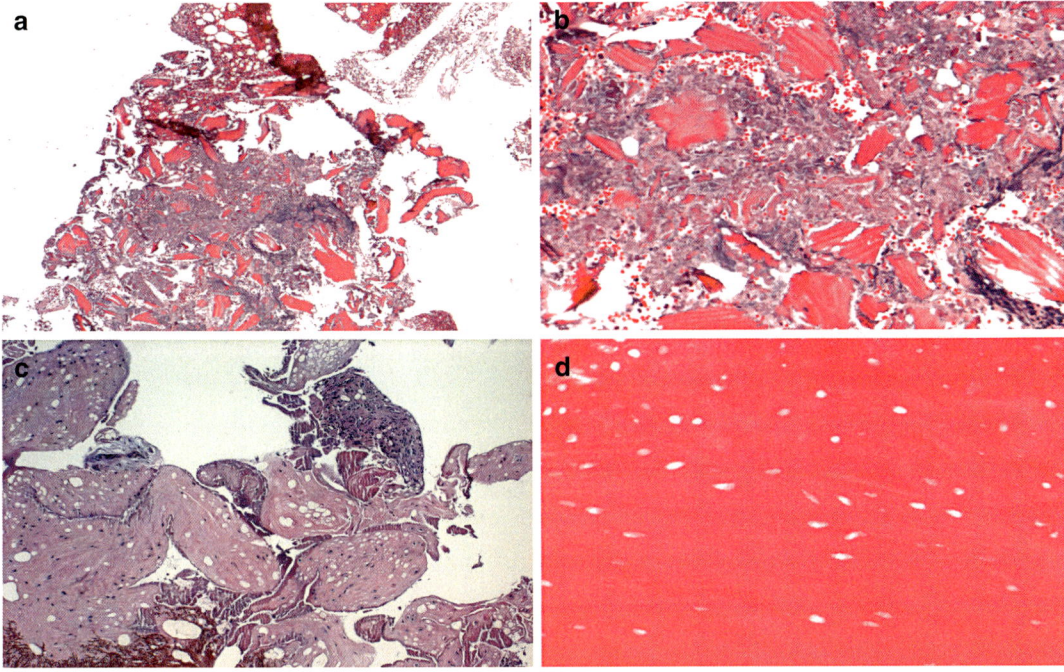

Fig. 13.6 Kienboeck Disease. There is extensive osteonecrosis with broken trabecules (**a–c**) and empty lacunae lacking osteocytes (**d**). All microphotographs are hematoxylin-eosin stained and taken at different magnification (**a–d**, ×50, ×200, ×100, ×100 as original magnification, respectively)

13.6 Tumorlike Lesions and Bone/Osteoid-Forming Tumors

Bone and osteoid-forming tumors may present a true challenge for the pathologist. If the bone can be recognized as well-formed lamellar or cancellous bone, osteoid is non-mineralized bony tissue that needs to be differentiated from reactive processes, including inflammation, fibrosis, and sclerosis (Box 13.9).

Box 13.9 Tumorlike Lesions and Bone/Osteoid-Forming Tumors
13.6.1 Myositis Ossificans
13.6.2 Fibrous Dysplasia and Osteofibrous Dysplasia/Ossifying Fibroma
13.6.3 Non-Ossifying Fibroma/Metaphyseal Fibrous Defect
13.6.4 Unicameral Bone Cyst and Aneurysmal Bone Cyst
13.6.5 Osteoma, Osteoid Osteoma, and Giant Osteoid Osteoma (Osteoblastoma)
13.6.6 Giant Cell Tumor (Osteoclastoma)
13.6.7 Osteosarcoma
13.6.8 Ewing Sarcoma
13.6.9 Miscellaneous Bone Tumors and Metastatic Tumors

13.6.1 Myositis Ossificans

- *DEF*: Solitary, nonprogressive (intramuscular, usually) tumorlike lesion of reactive type.
- *EPI*: Young males with rapid growth.
- *EPG*: Post-traumatic (often).
- *CLI*: Flexors of the arms, quadriceps of the femur, thigh adductors, and soft tissue of hands.
- *IMG*: Periosteal reaction with casing calcification at periphery, mostly 3–6 weeks after trauma.

- *GRO*: It is a well-circumscribed tumor, which shows, on cut surface, a white, soft, and gelatinous aspect in the center. Conversely, a yellow-grayish and most often firm zone with a rough granular surface is identified peripherally.
- *CLM*: Zonation (+)-lesion including three zones in the most mature stage (inner cellular core, an intermediate layer of osteoid, and peripheral shell of highly organized bony tissue). In fact, the bone formation is most bulging at the periphery of the lesion.
- *DDX*: *Fibrodysplasia (myositis) ossificans progressiva* and osteosarcoma. The former lesion shows a progressive fibroblastic proliferation with subsequent calcification and ossification (central) of subcutaneous fat, muscles, as well as tendo-aponeurotic tissue and ligaments. The latter lesion is characterized by atypical cells, no zonation but disorganization with a characteristic "reverse zoning effect," i.e., bone formation in the center of the lesion and immature spindle cells toward the periphery.
- *TRT*: Excision.
- *PGN*: Excellent following removal of the tumor.

13.6.2 Fibrous Dysplasia and Osteofibrous Dysplasia

Fibrous dysplasia and osteofibrous dysplasia are two related lesions. The former is well-recognized by the head and neck surgeons, while the latter is well recognized by pediatric orthopedic surgeons.

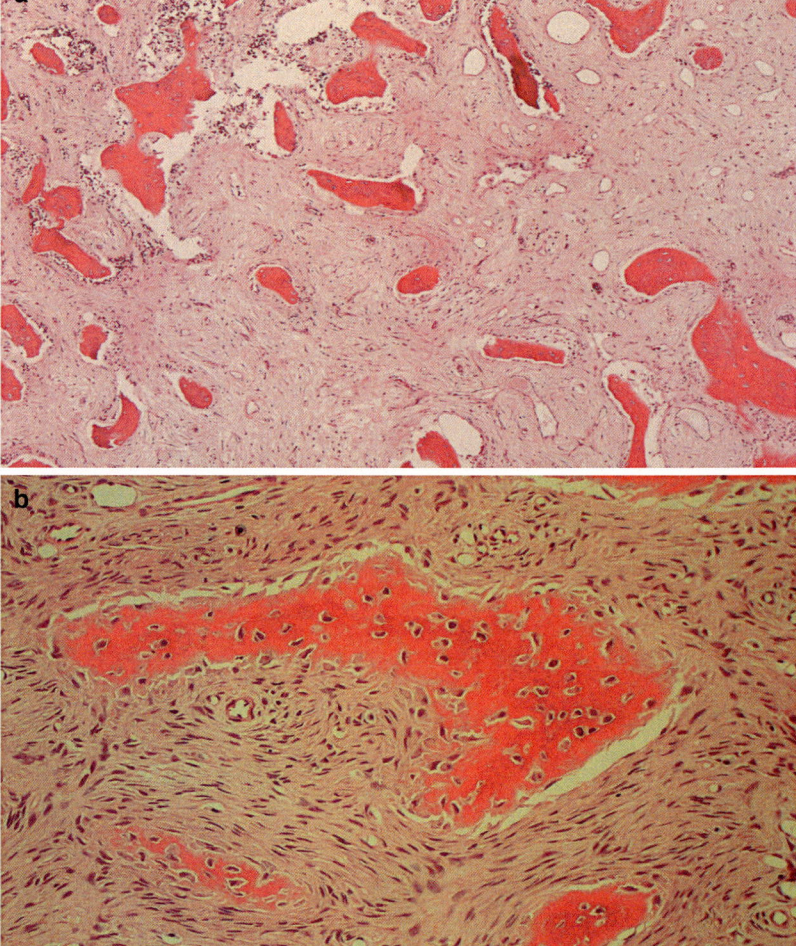

Fig. 13.7 Fibrous Dysplasia. Well-circumscribed lesion with curvilinear trabeculae ("Chinese letters") of metaplastic woven bone in hypocellular (fibroblastic) stroma typically without osteoblastic rimming. No/rare mitotic figures and no atypia (**a–b**, hematoxylin and eosin staining, ×50 and ×200 as original magnification, respectively)

13.6.2.1 Fibrous Dysplasia

- *DEF*: Hamartomatous tumorlike bony lesion derived from osteoprogenitor, fibroblast-like cells and characterized by fibro-osseous metaplasia with monostotic more common than polyostotic involvement (Fig. 13.7).
- *EPI*: 1% of biopsy of bone tumors, 1st-3rd decades of life, ♂:♀ = 1:1.2.
- *EPG*: Activating *GNAS1* gene somatic mutations on chromosome 20q13 with two distinct missense mutations in the α subunit of a G stimulatory protein that accounts for the disease process. These activating mutations activate adenylyl cyclase, and then an obvious rise of cAMP is found. This biochemical setting produces an alteration in the transcription and expression of several downstream target genes, including *c-fos*, which is a proto-oncogene.
- *CLI*: Two patterns of lesions are recognized, including the monostotic type (70%) involving older children and one of the most frequent skeletal sites (splanchnocranium/skull, rib, femur, tibia) and the polyostotic type (30%) involving more than one of the most frequent skeletal sites. The monostotic disease (FD, fibrous dysplasia) can occur incidentally or be associated with pain, swelling, or pathologic fracture, while the polyostotic type is often associated with singular extraskeletal anomalies. Polyostotic FD can be seen without (Jaffe-Lichtenstein syndrome) or with endocrine dysfunction, skin hyperpigmentation ("café au lait" pigmentation with jagged "Coast of Maine"-like borders respecting the midline and Blaschko lines) and *pubertas praecox* (precocious puberty) (=Albright triad: endocrinopathy with pituitary hyperplasia, hyperthyroidism, and adrenal hyperplasia, precocious puberty, and bony lesions). Generally, "cafe au lait" maculae in neurofibromatosis tend to be smaller and have a smooth outline ("Coast of California"); whereas in McCune-Albright syndrome the maculae are characterized by an irregular, jagged outline ("Coast of Maine"). The differentiating radiologic features include permeative ill-defined borders, destroyed cortical outline, speculated periosteal new bone formation, and periodontal ligament space widening. Rarely, FD can occur with soft tissue (intramuscular) myxomas (Mazabraud syndrome). As of 2019, only 106 cases of Mazabraud syndrome have been reported in the scientific literature.
- *IMG*: Bone expansion with ground-glass osteolysis and rim of host bone sclerosis. FD causes facial asymmetry with pathologic features with pain and deformity and leg length discrepancy commonly seen due to the involvement of the upper portion of the femur (the so-called Hockey Stick deformity, Shepherd's crook deformity, and candle flame).
- *GRO*: Fusiform mass of the bone diaphysis arising in cancellous (trabecular, spongy) bone with subsequent thinning of the overlying cortex and variably gritty osseous components.
- *CLM*: "Alphabet soup pattern" or "Chinese characters"-shaped metaplastic bony spicules (narrow, curved, and misshapen bony spicules) mostly devoid of lining osteoblasts immersed in a quite often variably cellular fibrous tissue with or without giant cells. Osteoblasts are interspersed in the woven bone, but usually not lining the trabeculae. Foamy cells can be identified, and fibroblasts have plum, ovoid nuclei without pleomorphism. Multinucleated osteoclast-type giant cells can be seen as well as cartilage with peripheral endochondral ossification. A minimal infiltrative pattern may still be present in FD, but an extensive infiltrative pattern should raise the differential diagnosis with osteosarcoma.
- *IHC*: MIB1/Ki67: low, (−) AE1–3/KER.
- *TEM*: Bone trabeculae of immature woven type and irregular osteoblast lining with cells resembling fibroblasts.
- *DDX*:
  - *OFD* (cortically located, tibia/fibula, age < 5 years, reactive woven bone, and prominent osteoblast rimming)
  - *Osteosarcoma, low-grade parosteal and grade 1 intramedullary types* (anaplasia, lack of osteoblastic rimming, bony trabeculae with the permeation of the tumor into the pre-existing bone, and extensive infiltrative pattern)
  - *Desmoplastic fibroma* (intense collagenization with little or no trabecular bony spiculae)

Box 13.10 Types of Fibrous Dysplasia

	Single bone	Multiple bones	Café au Lait spots	Endocrine disorders	Soft tissue tumors
Monostotic FD	+				
Polyostotic FD		+			
MCAS		+	+	+	
MD		+			+

Note: *FD* fibrous dysplasia, *MCAS* McCune-Albright syndrome, *MD* Mazabraud disease

- *TRT*: Resection is curative although other categories should also be considered (observation, medical therapy, and surgical remodeling) Reconstruction often follows the radical excision of the lesion.
- *PGN*: Polyostotic type may have cosmetic issues, and the endocrinology service needs to be activated to adequately manage the Albright syndrome. Long-standing polyostotic FD may become sarcomatous, which is an eventuality in the absence of radiation therapy. Malignant transformation of FD occurs in <1% of the cases, and osteosarcoma is the most common histologic type. Other tumors include fibrosarcoma, chondrosarcoma, and malignant fibrous histiocytoma.

13.6.2.2 Osteofibrous Dysplasia (OFD)

- *DEF*: Jaw-involving fibrous dysplasia-like fibro-osseous lesion, which appears to be originating from the cortex of either tibia or fibula of children and fully characterized by osteoblasts-lining bony trabeculae (Box 13.10).
- *SYN*: Ossifying fibroma, Campanacci disease, and juvenile adamantinoma.
- *EPI*: 0.2% of primary biopsies of bony tumors, infants, and children <5 years.
- *EPG*: It is a cortically based neoplasm with fibro-osseous histology represented by the ossifying fibroma group of lesions and exhibiting progressive proliferative capabilities with bony expansion and well-defined margins radiologically.
- *CLI*: Tibia and fibula with swelling of the anterior shin and relatively uncommon local pain.
- *IMG*: Blastic pattern with spreading along the long axis of the cortex and potential spreading into the intramedullary bone.
- *CLM*: Fibrous dysplasia-like lesion with prominent osteoblast-lining bony trabeculae (osteoblastic rimming).
- *TRT*: Resection may not be curative, because OFD has a greater tendency to recur.
- *PGN*: Usually considered a self-limited lesion, OFD may show a locally aggressive pattern, and the relationship with the adamantinoma of adulthood has been controversially discussed but unfortunately remains unclear.

13.6.3 Non-ossifying Fibroma (NOF)

- *DEF*: Abnormal ("pseudotumor"-like) connective process constituted by spindle cells of fibroblastic type, collected in storiform bundles located in the flared end of the lower femur and in the flared upper or lower ends of the tibia and harboring a benign character (Fig. 13.8).
- *SYN*: Metaphyseal fibrous defect (MFD), benign fibrous histiocytoma (BFH), fibrous cortical defect (FCD).
- *EPI*: Adolescence, upper tibia or lower femur, (+) bilaterality (~50%) with FCD involving children, 3–5 years of age, and NOF usually children >8 years and adolescents.
- *EPG*: Unknown.
- *CLI*: Usually asymptomatic, "incidentaloma," occasionally painful.
- *IMG*: It is a well-defined radiolucent lesion with self-limiting growth. There are six radiologic principles that need to be known in identifying a lesion as NOF, including incidental finding, small size (1–3 cm in size), eccentric, juxtacortical or intracortical lesion without evidence of intramedullary involvement, metaphyseal or meta-diaphyseal location, lon-

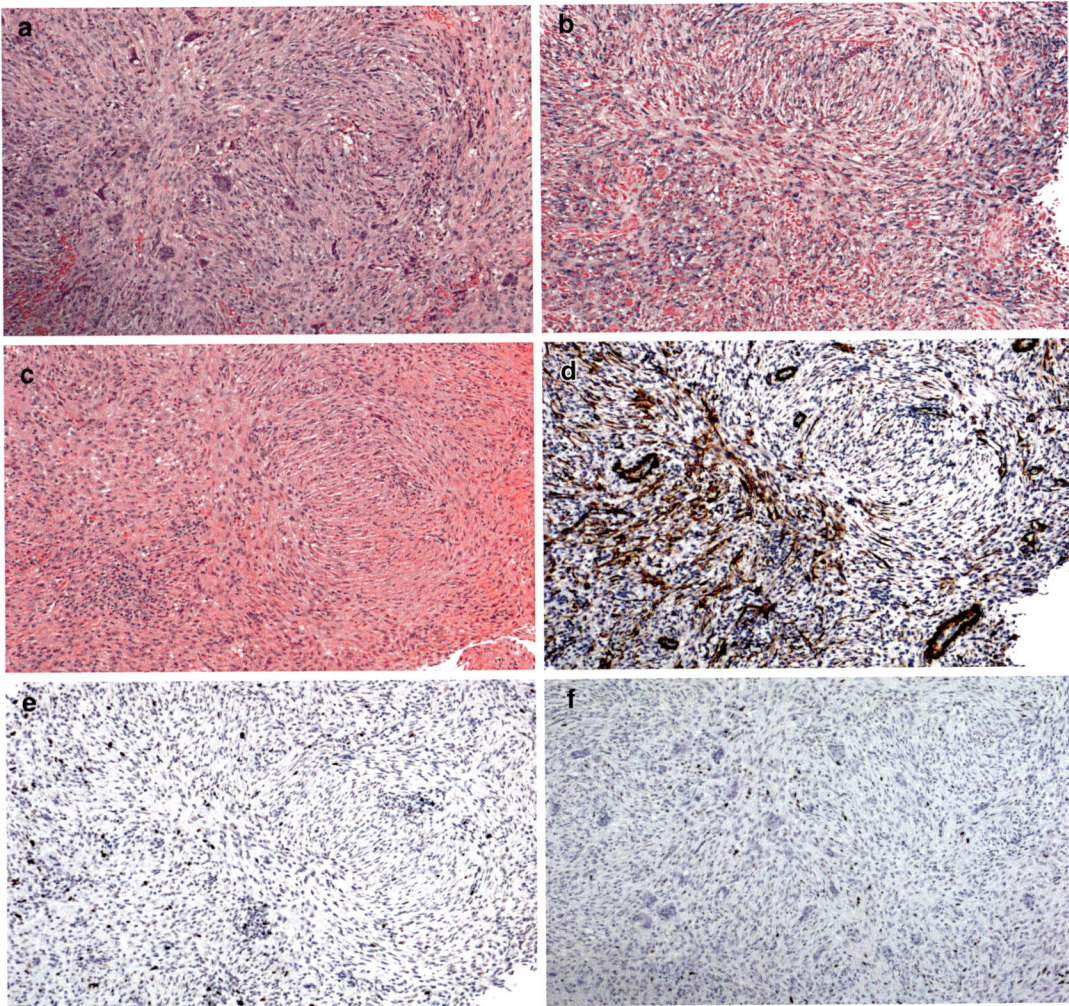

Fig. 13.8 Metaphyseal Fibrous Defect. The histological images exhibit a highly cellular lesion containing spindle-shaped cells on a background of stromal tissue. There is a prominent storiform pattern with foamy histiocytes and multinucleated giant cells and lack of mitotic activity and/or cellular atypia (**a–c**) (×100, ×100, and ×100 as original magnification, respectively). In (**d, e, f**) there is the expression of actin (**d**) and two snapshots of the proliferation activity using the antibody MIB1 against the Ki67 antigen (**e–f**) (avidin-biotin complex immunostaining, ×100 original magnification for all three immunohistochemical microphotographs)

gitudinal Ø > transversal Ø, and centered lucency with a peripheral border of sclerosis.
- *GRO*: Eccentric, sharply delimited peri-epiphyseal tumor.
- *CLM*: Mixture of four components including cellular storiform fibrous tissue, scattered osteoclasts, foamy MΦ, and hemosiderin pigments.
- *IHC*: (+) CD-68, α-1-antichymotrypsin, (±) α-1-antitrypsin, and HLA-DR supporting the histiocytic immunophenotype, but also (+) VIM, MSA indicating some fibroblastic or myofibroblastic characteristics.
- *TRT*: Resection is curative.
- *PGN*: No malignant potential and, most often, NOF disappears spontaneously.

There are several bony cysts, some of them neoplastic, some pseudo-neoplastic, and some harboring a microbiologic character. The simple bone cyst is, probably, the only true cyst of primary intraosseous origin. Although the nature of its layers of lining cells is controversial, it originates within the medullary cavity and is also labeled as a primary bone cyst. The aneurysmal bone cyst (ABC) is a multi-loculated bloody mass whose walls are lined by endothelium. In terms of physiopathology, the ABC has also been considered an AVM secondarily induced by several and different conditions including bony traumas or tumors. Close to the ABC, two cystic lesions need to be considered as follows. The telangiectatic osteosarcoma (TAOS) should be foremost in all of our minds. TAOS is described in the session of the osteosarcoma. Also, the "cystlike" form of fibrous dysplasia should be considered.

The Langerhans cell histiocytosis or eosinophilic granuloma of the bone can also present as a cyst and needs to be distinguished from other entities carefully. Moreover, other cystic lesions include the intrasacral cyst, epidermoid cyst, dermoid cyst, and the echinococcal cyst. The intrasacral cyst is a cyst of meningothelial origin, and the epidermoid and dermoid cysts are lined by epithelium but do not secrete fluid. Finally, the echinococcal cyst is a cyst filled with a serous fluid, but the lining is of extraosseous parasitic origin.

13.6.4 Bone Cysts

13.6.4.1 Simple Bone Cyst

- *DEF*: Benign, solitary fluid-filled (primary) metaphyseal cystic lesion lined by fibrous tissue and a single row of mesothelial-like cells.
- *SYN*: Solitary bone cyst, unicameral bone cyst.
- *EPI*: 3% of primary bone tumors, 1st–2nd decades (85% of patients <20 years), ♂:♀ = 2.5:1.
- *EPG*: It may be considered more a "dysplastic" lesion than a real tumor. SBC shows impairment of venous blood circulation in cancellous bone and bone resorption by ↑ blood pressure and ↑ amount of inflammatory proteins.
- *CLI*: Long bone (upper humerus or femur) and short bone lesion located at the metaphysis migrating away from epiphysis with a propensity to pathologic fracture and recurrence. Typically, SBC presents with pain, but no swelling.
- *IMG*: Central lytic lesion of metaphysis with attenuation and slight expansion of cortex and fluid-filled on CT/MRI. SBC may separate from the growth plate and be found in a diaphyseal location in some cases.
- *GRO*: Unicameral cyst composed of a thin bluish cortex and filled with yellowish fluid.
- *CLM*: Spindle cell-lined cyst (pseudocyst) containing clear or yellow fluid and lined by occasionally brown-laden macrophages (hemosiderin, Fe+) and surrounded by a well-vascularized connective tissue membrane with trapped cholesterol clefts and containing occasionally reactive changes.
- *IHC*: (−) CD31, CD34, FLI-1, ERG, because SBC is not a cyst and has no endothelium.
- *TRT*: Curettage and packing.
- *PGN*: Benign apart for local recurrence if the lining is not destroyed.

13.6.4.2 Aneurysmal Bone Cyst

- *DEF*: Bony lesion representing either a true neoplasm (Ubiquitin-Specific Protease, *USP6* gene rearrangement on chromosome 17) or 1/3 of cases secondary to a pre-existing tumor and showing an exaggerated proliferative response of vascular tissue in bone, lined by fibrous walls and filled with free-flowing blood (Figs. 13.9 and 13.10).
- *SYN*: Multicameral (bloody) bone cyst.
- *EPI*: 1% of primary bone tumors, 1st–2nd decades, ♂:♀ = 1:1.2.
- *ETP*: Exaggerated proliferative response of vascular tissue in bone of primitive (70%) or secondary (30%) origin. Secondary ABC include GCT, chondroblastoma, CMF, NOF, osteoblastoma, fibromyxoma, fibrosarcoma, fibrous histiocytoma, osteosarcoma, and fibrous dysplasia. There is an upregulation of the *USP6* (*Tre2*) gene on 17p13 when combined by translocation with a promoter pairing, which is most commonly described as t(16;17)(q22;p13) leading to juxtaposition of promoter region *CDH11* on 16q22. The first cytogenetic report in ABC occurred in 1999, with a recurrent chromosomal translocation t(16;17)(q22;p13). Subsequently, Oliveira et al. reported the fusion gene partners as *CDH11*

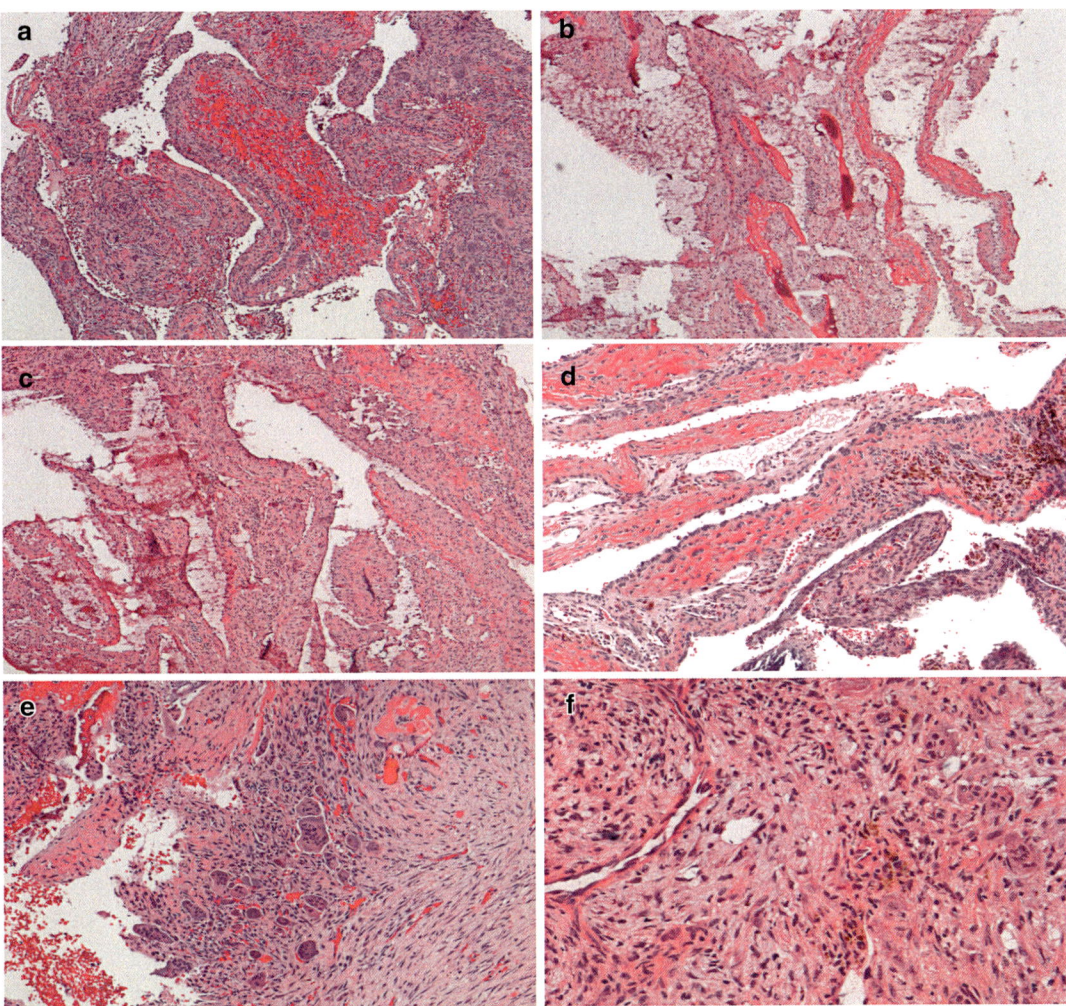

Fig. 13.9 Aneurysmal Bone Cyst. The histology of an aneurysmal bone cyst reveals blood-filled cyst-like spaces which are arranged haphazardly within a moderately cellular stroma (**a–f**). In stroma, both spindle and epithelioid morphology are not anaplastic, but they may show mild nuclear pleomorphism at places (**f**). Osteoclast-like multinucleated cells and scattered mitotic figures are seen (**e**). No necrosis is seen. Moreover, irregularly shaped intralesional fragments of woven bone are seen (**a, b, e**). There is remodeling of trabeculae of lamellar bone at the periphery of the lesion. No malignant osteoid is detected (**a–f**, hematoxylin and eosin staining, ×50, ×50, ×50, ×100, ×100, ×200 as original magnification, respectively)

(osteoblast cadherin 11), on 16q22, and *USP6*, on 17p13. The *CDH11-USP6* fusion transcript was characterized only in primary ABC but not in secondary ABC. Later, additional translocations with up-regulation of other gene promoter swapping were identified, including *ZNF9*, *COL1A1*, *TRAP150*, and *OMD*.

- *CLI*: *LPS* (*l*ong bones, *p*elvis, *s*pine)-located lesion with rapid growth mimicking a malignancy.

Besides pain, the involved area is key. The area affected with ABC is tender on palpation. Intra-operatively, ABC looks like a long-term blood soaked sponge with large pores representing cavernous spaces of the lesion.

- *IMG*: Multiloculated "soap bubble," eccentric radiolucent (osteolytic) expansile (intramedullary, intracortical or intraperiosteal) lesion with thin wall cystic cavities (cortical

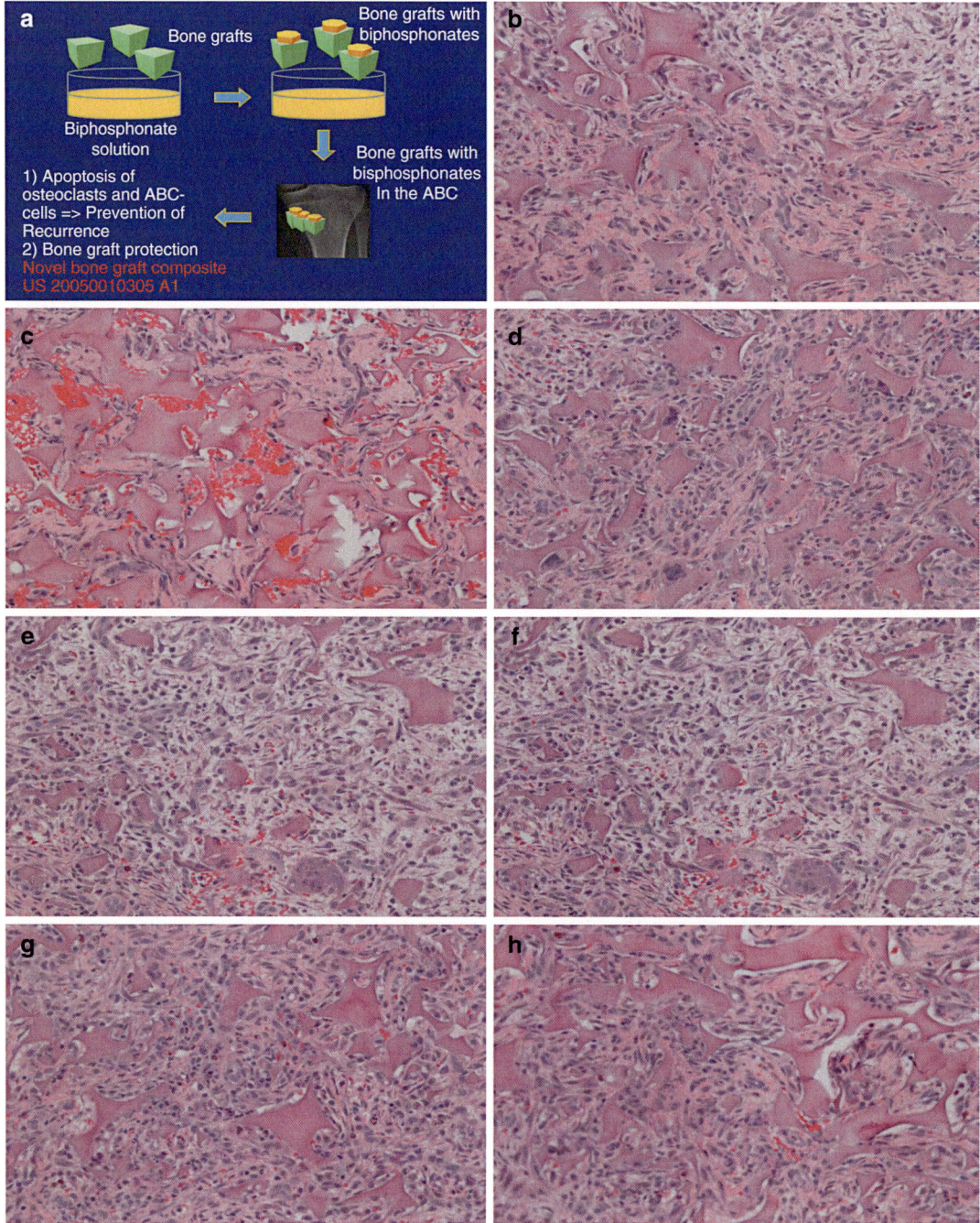

Fig. 13.10 Aneurysmal Bone Cyst. This panel shows the rational to use bone grafting in (**a**). Microphotographs (**b**) through (**h**) show evidence of remodeling with cellular matrix with irregular fragments of promoted new woven bone (**b–h**, hematoxylin and eosin staining, ×200 original magnification for all microphotographs)

attenuation or destruction on imaging), the rim of reactive bone and fluid-fluid levels on MRI.
- *GRO*: Soft, spongelike mass ("*tumor quia tumet*").
- *CLM*: Large, blood-filled spaces without endothelial lining and surrounded by a shell of reactive bone with the rapid tendency to produce an eccentric, often multicystic expansion of the bone, eroding the cortical bone and occasionally extending into the soft tissue. Intralesional components include osteoclasts, reactive bone, and degenerated calcifying fibro-myxoid tissue. In fact, exuberant spindle cells, osteoid and woven bone slivers, and osteoclast-like giant cells characterize the ABC that may look like an aberrant callus. Variant: SABC (Solid ABC): ABC without cystic cavities, mainly located in small bones of hands, jaw, and PS, and constituted by a proliferation of spindle cells with osteoid, neo-osteoid, and giant cells without any evidence of atypia and/or nuclear pleomorphism. SABC has been equated to giant cell reparative granuloma (GCRG).
- *IHC*: (−) CD31, CD34, FLI-1, ERG, because ABC is not a cyst and has no endothelium.
- *DDX*: ABC is easy to be mistaken for a malignant tumor, because of the high rate of growth, incredible bony destruction, and marked cellularity in the early- and mid-phases of development. The appearance of bubbly lytic bone lesions on imaging may portend to another diagnosis other than ABC. Specific lesions need to be kept distinguished including fibrous dysplasia or fibrous cortical defect, osteoblastoma, giant cell tumor, chondroblastoma, chondromyxoid fibroma, hyperparathyroidism (brown tumor), osteomyelitis, non-ossifying fibroma, enchondroma, eosinophilic granuloma, simple bone cyst, and metastases. Histologic ABC-like areas can occur in chondroblastoma, giant cell tumor, fibrous dysplasia, not-ossifying fibroma (NOF), osteoblastoma, hamartoma, and chondrosarcoma.
- *TRT*: ABC is usually treated operatively using bone curettage and bone grafting (a surgical procedure that replaces missing bone to repair complex bone fractures or fractures that fail to heal properly). An alternative treatment is the percutaneous application of fibrosing agents, either as an isolated procedure or as a presurgical option. An aggressive curettage and bone grafting are particularly indicated in symptomatic ABC without acute bony fracture.
- *PGN*: There is a recurrence rate of ~20%, but the tendency to recur is in 1/3 of patients if ABC is treated with curettage alone. Occasionally, spontaneous regression may occur.

Another cystic lesion to list is the *intraosseous ganglion cyst*, which is an intraosseous located pseudocyst with fibrous lining, close to joint space. IGC is much rarer than its similar soft tissue counterpart. It is usually an ankle-located (paraarticular) lesion, typically identified at the distal tibia. Microscopically, there is fibrous tissue in the wall, and the cyst is surrounded by a zone of sclerotic bone with gelatinous content. Cystic lesions need to be carefully differentiated because of the different recurrence rates and medico-surgical approach (Box 13.11).

According to the bone tumor location and grading, several entities should be considered in the differential diagnosis (Box 13.12). Finally, fibro-osseous masses of the maxillofacial region are considered in Box 13.13.

Box 13.11 Cystic Lesions
- Unicameral Bone Cyst
- Aneurysmal Bone Cyst
- Intraosseous Ganglion Cyst

Box 13.13 Fibro-osseous Lesions of the Maxillofacial Region
- Fibrous Dysplasia (monostotic, polyostotic, craniofacial)
- Osseous Dysplasia (periapical, focal, florid, familial "gigantiform")
- Ossifying Fibromas (conventional, juvenile trabecular, juvenile psammomatoid)

Box 13.12 Bone Tumor Location and Grading

Location	Diaphysis	Meta-diaphysis-J	Metaphysis	Epiphysis
	Ewing sarcoma	Fibrous cortical defect	Giant cell tumor	Giant cell tumor
	Adamantinoma	Chondromyxoid fibroma	Enchondroma	Chondroblastoma
		Fibrous dysplasia	Chondrosarcoma	
		Osteoid Osteoma	Osteosarcoma	
		Fibrosarcoma	Osteochondroma	
			Osteoid osteoma	
			Osteoblastoma	
Grading				
	Low grade			
	Grade I	Well-differentiated		
	Grade II	Moderately differentiated		
	High grade			
	Grade III	Poorly differentiated		
	Grade IV	Undifferentiated		

13.6.5 Osteoma, Osteoid Osteoma, and Giant Osteoid Osteoma

13.6.5.1 Osteoma

- *DEF*: Protuberant (bossellated sessile) lesion composed of densely sclerotic well-formed bone projecting out from the cortex and harboring benign features with no tendency to malignant change.
- *EPI*: 4th–5th decades and ♂:♀ = 2:1. Osteoma is the most common tumor of the paranasal sinuses.
- *EPG*: It may have a reactive origin, although it is classified as a benign tumor (three theories have been proposed, including a developmental theory, a traumatic theory, and an infectious theory).
  - Developmental theory: Apposition of membranous and endochondral bony tissues trapping some of the embryonic cells, and ultimately leading to unchecked osseous proliferation.
  - Traumatic theory: Inflammatory process as the inciting force by bony tumor formation.
  - Infectious theory: Osteitis resulting from chronic infection and inciting an initiator other than promotor effect.
- *CLI*: Typically, osteomas are located on the flat bones of skull and face (membranous ossification) with some inherent tendency to protrude into sinuses (sinuses obstruction, impinging on the brain, cosmetical disturb).
- *IMG*: Dense or ivory osteomas appear as radiodense lesions, with a radiodensity similar to the normal cortex, while mature osteomas show a central marrow.
- *GRO*: Protuberant lesion of flat bone.
- *CLM*: Lamellar bone with three histologic patterns, including ivory osteoma with dense bone lacking Haversian system, mature osteoma resembling the normal architecture of bone, and mixed osteoma, which is a mixture of ivory and mature histology.
- *IHC*: MIB1 (Ki67) shows no proliferation activity.
- *TRT*: Excision is curative. No tendency to recur.
- *PGN*: Investigate for Gardner syndrome and liaise with the GI colleagues. In fact, Gardner syndrome is AD-inherited polyposis characterized by the presence of multiple polyps in the colon together with extra-colonic tumors (skull osteomas, thyroid cancer, epidermoid cysts, fibromas, and desmoid tumors).

13.6.5.2 Osteoid Osteoma

- *DEF*: It is a benign skeletal tumor, labeled often as double "O", of less of 2 cm in Ø (typically Ø ≤ 1.5 cm) ("pealike mass") character-

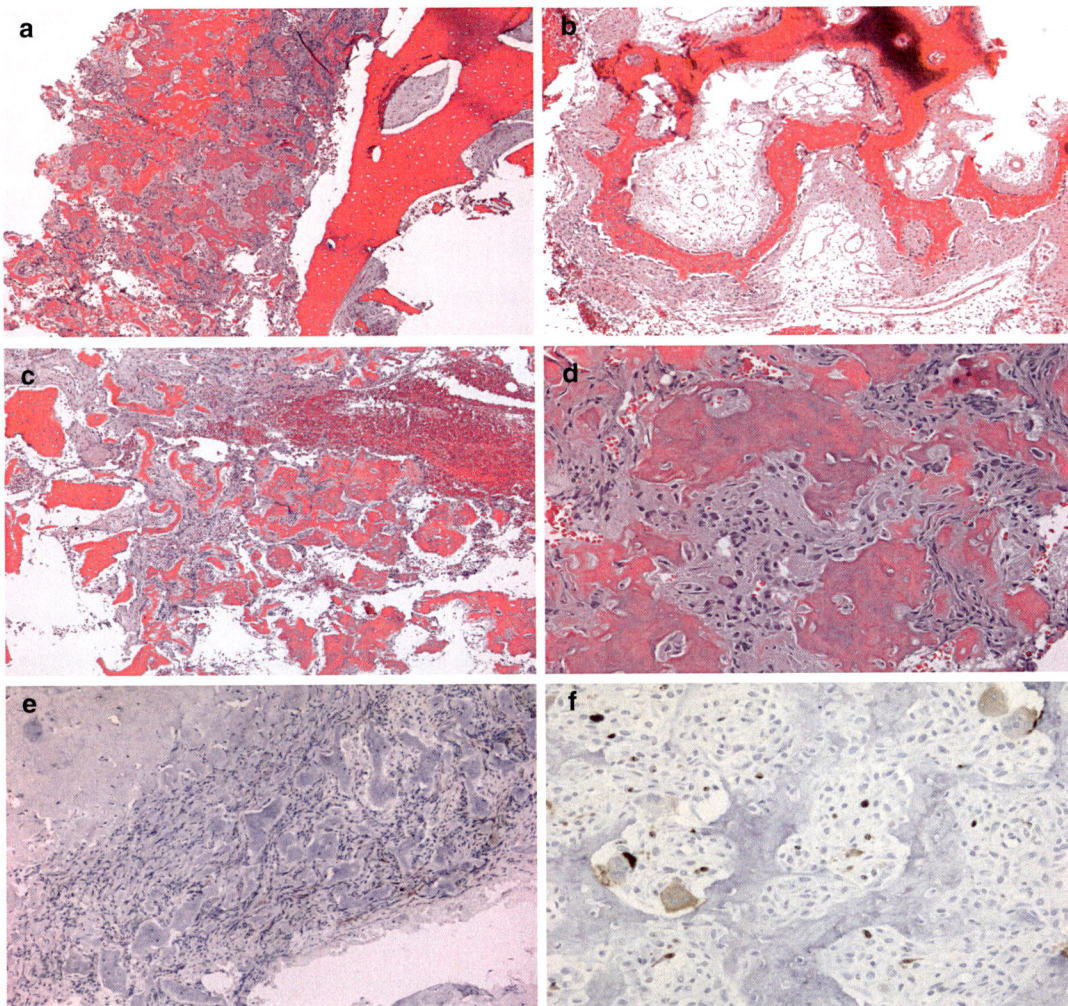

Fig. 13.11 Osteoid Osteoma. This panel shows low and medium power magnification images of the nidus, i.e., the meshwork of dilated vessels, osteoblasts, osteoid and woven bone with some (optional) central region of mineralization, a fibrovascular rim, and surrounding reactive tissue sclerosis (**a–d**, hematoxylin and eosin staining, ×50, ×50, ×50, ×200 as original magnification, respectively). Two immunohistochemical stainings (**e, f**) show the low/no proliferation activity using the antibody MIB1 against Ki67 (**e**, ×100 original magnification; **f**, ×200 original magnification)

ized by abnormal bone (the "nidus"), which is richly innervated by nerve fibers causing considerable pain (Fig. 13.11).

- *EPI*: 3% of primary bone tumors, 1st–3rd decades, ♂:♀-2:1.
- *CLI*: L > PS bones (the femur, tibia, and humerus more frequently involved than pelvis and spine) with intense, well-delimited pain proportionally more than expected in consideration of the size of the lesion. The severe pain classically occurs at night, but it can also be constant at any time of the day, and the chief symptomatology is constituted by dull pain which is non-radiating, persistent with nocturnal exacerbation, and relieved by aspirin in most patients. In some patients, pain may be mistaken for neurotic or psychiatric complaint, though. The pain is due to an ↑ of local concentration of PGE_2 and COX1 and COX2.
- *IMG*: Metaphysis-located cortical tumor (Ø <1.5 or 2 cm) with central radiolucent "nidus". In a CT scan, the nidus is a round

hole, while bone scan reveals the nidus as a hot spot ("double-density sign"). In the femur, OO occurs typically at its upper end, in the neck region, and in the trochanter. In long bones, the lesion is usually in proximity to the cortex (subcortical, intracortical, and intraperiosteal), rather than buried deep within the medullary cavity.

- *GRO*: Multiple bony fragments, some of them tannish or cherry red in the early lesions and yellow-white in the more mature lesions.
- *CLM*: Highly vascular osteoblastic connective tissue with woven bony trabeculae or spiculae and some benign giant cells (nidus) surrounded by a lucent fibrovascular zone and a zone of dense sclerotic bone. The woven bone edges of nidus are sharply circumscribed. The osteoblastic rimming evidences no cytologic atypia. Silver staining reveals numerous non-myelinated axons within the lesion. Thus, the high level of pain that occur in patients harboring OO is probably associated with the rich nerve innervation of this benign tumor.
- *IHC*: MIB1 reveals a low proliferation rate.
- *DDX*: The edges of the nidus remain smooth for all the process, while rough edges denote the involvement of the bone by another process, such as infection, intracortical osteosarcoma, and cortical metastasis. OO, other than osteomyelitis, needs to be distinguished from solitary enostosis, Brodie abscess, eosinophilic granuloma, osteoblastoma, and osteosarcoma.
- *TRT*: NSAIDs such as ibuprofen are used to limit the pain (nonoperative) and percutaneous radiofrequency ablation (operative).
- *PGN*: 10–15% recurrence rate in case of percutaneous radiofrequency ablation.

13.6.5.3 Giant Osteoid Osteoma (GOO)

- *DEF*: Benign bone tissue forming neoplasm producing woven bone spicules delimited by prominent osteoblasts (OO with Ø >2.0 cm ⇒ giant osteoid osteoma).
- *SYN*: Osteoblastoma.
- *EPI*: 1st–3rd decades, ♂:♀ = 2:1; 1% of bone tumors.
- *CLI*: Metaphysis-located medullary tumor, L < PS (pelvis and spine most frequently affected than long bones), pain ± swelling ± pathologic fracture. In the spine, GOO can produce "bent" scoliosis due to a muscle spasm of the muscular groups of the vertebral column.
- *IMG*: Similar to OO ("lytic-blastic"), practically depending upon the ratio of osteoid bone to the woven bone building, commonly expanding bone in a fusiform manner. It is larger than OO (Ø >2 cm) with nidus showing no or minimal rim of sclerotic bone.
- *GRO*: More numerous fragments (Ø > 1.5 cm) with color depending from the osteoid/woven bone ratio and vascularity, being red if richly vascular and osteoid productive, and yellowish/whitish if the bone building is more intense with decreased vascularity.
- *CLM*: Nidus with minimal or no rim of surrounding sclerotic bone. It is mandatory to have fulfilled all conditions before signing out the diagnosis of GOO (osteoblastoma) including (1) lack of unambiguous anaplasia, (2) at least focal osteoblastic rimming, (3) lack of cartilage production, and (4) a sharp circumscription of osteoid and/or woven bone-producing lesion from host lamellar bony tissue (the so-called Bertoni zonation). In the aggressive osteoblastoma atypical cytological features, broad irregular trabeculae lined by osteoblasts and non-trabecular osteoid are seen. In general, the giant OO or osteoblastoma may be difficult to distinguish from osteosarcoma, and the identification of infiltrative margins or cartilage islands suggests malignancy, and a second opinion is warranted. The concept of "pseudoanaplasia" has been suggested for such "pseudomalignant" osteoblastomas. Four clues to the diagnosis of osteosarcoma should not be present, including sheets of elongated spindle cells, a solid stroma appearance, deposits of osteoid between woven bony trabecules, and cartilage production.
- *IHC*: Ki67 (MIB1) score may show a low proliferative rate but may also be misleading in some cases. Thus, scrutiny of all specimens and levels is needed.
- *TRT*: Nonoperative management is rarely a choice because the lesion is going to grow if left untreated. Operative management includes

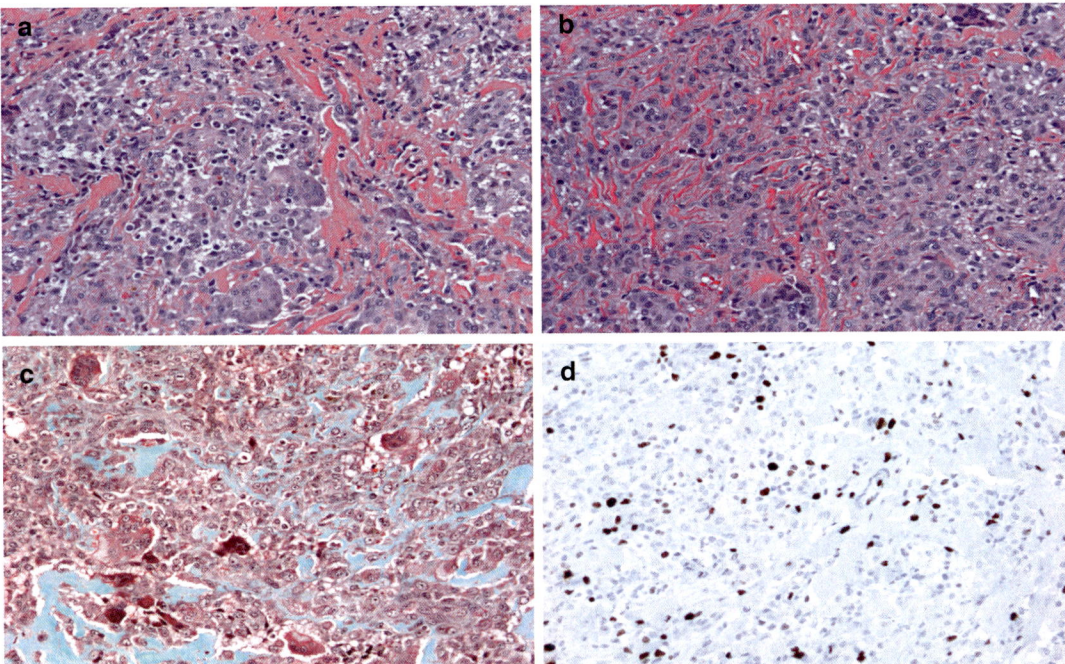

Fig. 13.12 Giant Cell Tumor. The histological panel shows a tumor composed by numerous giant cells with nuclei harboring the same features seen in the stromal cells (**a–b**, hematoxylin and eosin staining, ×200 as original magnification for both microphotographs). There is an open chromatin pattern of the nuclei of the tumor cells (**c**, Masson-Goldner trichromic stain, ×200 original magnification). The proliferation activity is moderate (**d**, avidin-biotin complex with the MIB1 antibody against Ki67, ×200 original magnification)

curettage or marginal excision with bone grafting. There is recurrence rate of 10–20%.
- *PGN*: The osteoblastoma has the same prognosis of OO, but a variant, also known as aggressive osteoblastoma, has more tendency to recur, and its delimitation from the osteosarcoma may be quite challenging in some cases, and second opinion is often necessary.

13.6.6 Giant Cell Tumor

- *DEF*: It is an epiphyseal osteolytic tumor of the long bones characterized by benign appearing osteoclast-like giant cells and stromal cells arising within the epiphysis of adult bones (rarely seen in patients with open growth plates) and harboring an uncertain malignant potential (Fig. 13.12).
- *EPI*: 2nd–3rd decades, ♂ < ♀, Oriental>Western, individuals, 5% of primary bony tumors.
- *EPG*: There are three theories, including an inflammatory theory (Lubarsch-Konjetzny-Oberndorfer) characterized by resorbing giant cell-rich granulation tissue, a dystrophic approach constituted by a circumscribed form of fibrous cystic osteodystrophy, and a neoplastic one (Ewing-Geschickter-Copeland), which regards the GCT as a true neoplasm arising from the fibroblasts of the marrow cavity.
- *CLI*: Tumor of long bones (lower femur, upper tibia, lower radius, humerus, and fibula) with "sign of the pergameneous crepitation" and "soap bubble" aspect.
- *IMG*: Osteolytic, expansile lesion of the epiphyses with or without trabeculated appearance and usually without peripheral sclerosis or periosteal reaction (except a recurrence in soft tissue, when an eggshell of ossification usually surrounds it).
- *GRO*: Variable-sized lesion with solid, tan to light brown, soft with often a hemorrhagic aspect.
- *CLM*: Numerous giant cells (S100-) with many central nuclei regularly distributed in some-

what spindle-shaped mesenchymal (stromal) cells, which form mononuclear cells and then fuse to form giant cells (nuclei are central and similar to the nuclei of mesenchymal/stromal cells) with characteristic spread to the subchondral position. Focal osteoid or bone in 1/3 of the cases. All cells of this neoplasm show nuclear vesicular cytology and do not display any objective anaplasia as seen in TAOS and GCOS. Cells are also differentiated in type I cell (mononuclear stromal cell that look like interstitial fibroblasts or primary neoplastic cells), type II cell (mononuclear cells from peripheral blood and considered precursor of the giant cell), and type III cell (giant cell with the nuclei similar to the stromal cells).

- *CMB*: Type II and III cells have IGF-I and IGF-II activity, and 4/5 of patients have telomeric associations, and the RANK pathway, i.e., receptor activator of nuclear factor-κB or a member of the tumor necrosis factor receptor (TNFR) molecular subfamily, is crucial.
- *DDX*: Osteitis fibrosa cystica (chronic hyperparathyroidism induced osteoclastic destruction of bone with or without cystic degeneration), ABC (when occurs in an uncommon site, more likely is an ABC), telangiectatic OS (malignant cells with osteoid formation), and GC-rich OS. The GCT of hyperparathyroidism is histologically indistinguishable from the non-hyperparathyroidism associated GCT of the epiphysis. Other less common lesions that need to be kept in mind in the differential diagnosis include giant cell-rich fibrosarcoma, malignant fibrous histiocytoma, postirradiation sarcoma, the de novo malignant transformation of a formerly conventional GCT, and the differentiated chondrosarcoma with rich benign osteoclast-like giant cell component.
- *TRT*: It is not a sarcoma, but its uncertain behavioral potential with high local recurrence rate and great bony destruction suggests to the orthopedic surgeon measures similar to those adopted for low-grade malignant tumors such en bloc resection or cryosurgery.
- *PGN*: Uncertain malignant potential with up to 1/2 of the cases showing recurrence and up to 1/10 of the examples showing metastases (lung). Recurrence rate: 1/3 of the cases if curettage is performed, <1/10 of the cases if "*en bloc*" excision is performed. Radiation therapy often induces malignant transformation to osteosarcoma.

13.6.7 Osteosarcoma

Significant relevance for the pathologist, including the pediatric pathologist and the oncologist, is the basic, epidemiological, clinical, radiological, anatomo-pathological, prognostical, and therapeutical knowledge of the osteogenic sarcoma or osteosarcoma. This malignant tumor has a curious bimodal age of incidence, i.e., the adolescence and the elderly sparing the middle age. A similar bimodal age of incidence is also seen in Hodgkin lymphoma, with two peaks in the age groups of 15–34 years and older than 55 years.

- *DEF*: Malignant bone tumor characterized by cells with variable degree of cytologic atypia (minimal [¼ of cases] to frank [¾ of cases] atypia/anaplasia) and osteoid and/or woven (primitive) bone production, which are produced in the course of the evolution either directly from the sarcomatous connective tissue or indirectly through a chondroid/cartilaginous intermediary stage (Figs. 13.13, 13.14, and 13.15). It is essential to keep in mind that osteosarcoma is defined by any osteoid and/or bone that is produced by uncommitted mesenchymal cells without first transit through a cartilaginous phase no matter what the extent of the chondro-sarcomatous elements is. There are several ways to classify osteosarcomas. Indeed, osteosarcomas can be organized. The term juxtacortical, intracortical, and intramedullary refer to the site within bone from which the tumor appears to have arisen.
- Osteosarcoma Taxonomic Criteria: The osteosarcoma (OS) can be classified:
  - According to the site of origin: intramedullary, intracortical (≥90% within the cortex), periosteal (juxtacortical with a variable degree of cortical erosion), and parosteal.

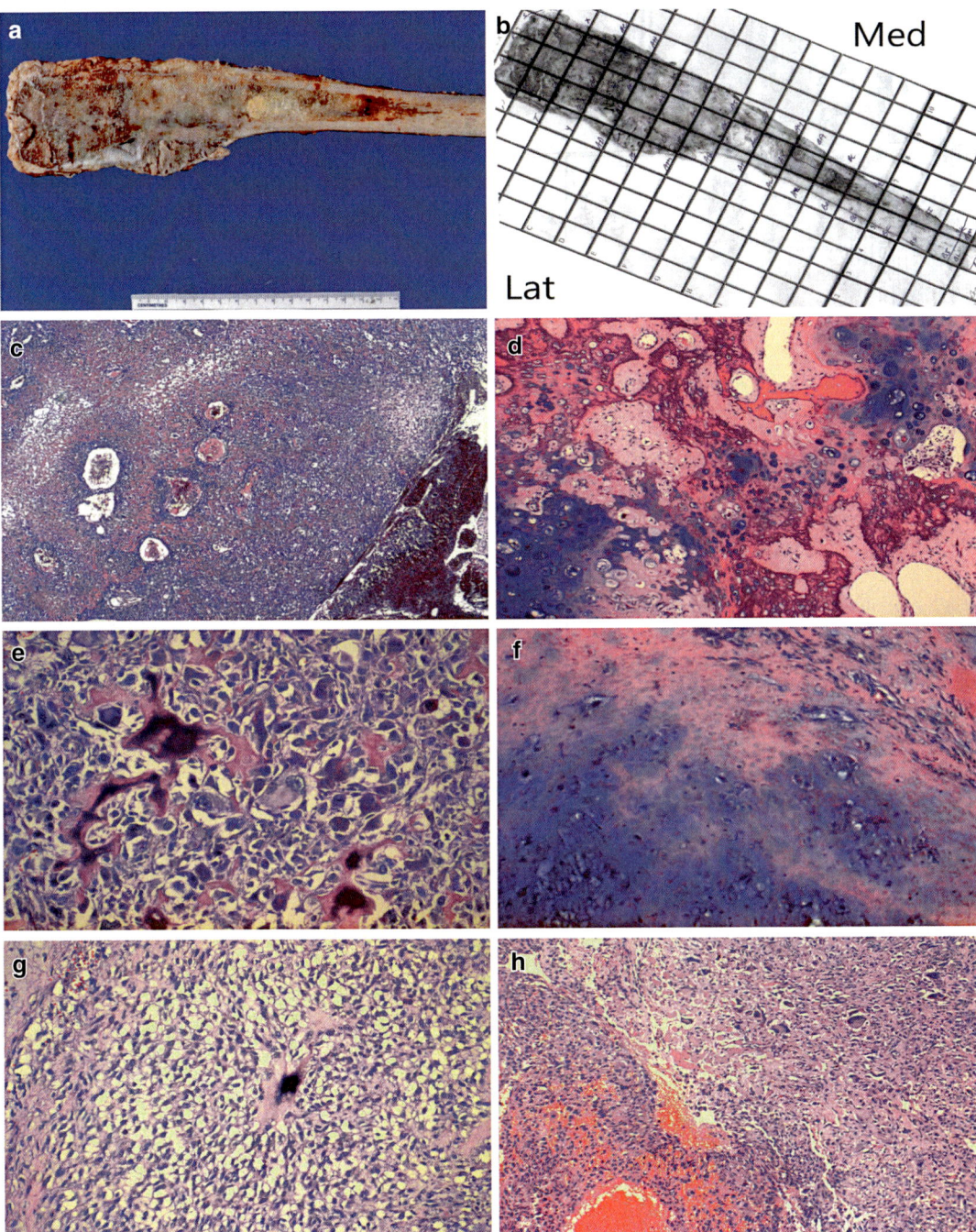

Fig. 13.13 Osteosarcoma. In figure (**a**) it is shown an osteosarcoma of the femur, while the chess board for grossing is used for the correct labeling of the tumor specimens and returning to the sample in case in case this procedure is needed (**b**). The histological images (**c–h**) highlight the presence of malignant osteoid constituted by osteoid and anaplastic cells. There is a difference of chondroid differentiation or bone maturation in the samples (hematoxylin and eosin staining) (**c–h**, ×40, ×100, ×200, ×200, ×200, ×100 as original magnification, respectively)

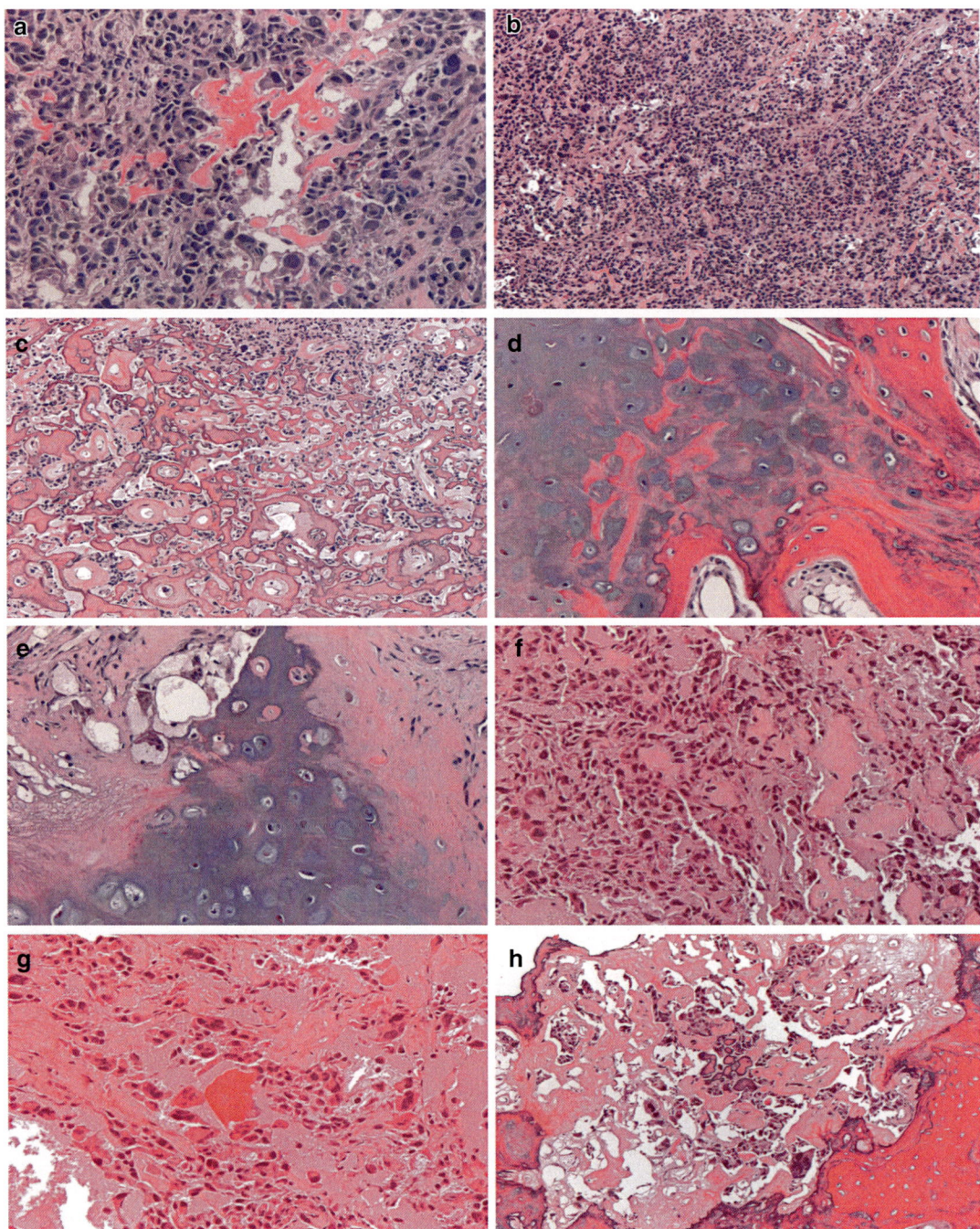

Fig. 13.14 Osteosarcoma. This panel shows different histologic phenotypes of the osteosarcoma with favorable (osteoblastic, chondroblastic, and fibroblastic osteosarcoma). In the panel, there is osteoblastic osteosarcoma showing the conventional and the filigree pattern with osteoid, which comprises thin, randomly arborizing lines of osteoid interweaving between osteosarcoma cells (**a–c**; ×200, 100, and 100, hematoxylin-eosin staining), chondroblastic osteosarcoma with high-grade hyaline cartilage that is intimately associated and randomly mixed with non-chondroid elements (**d–e**, ×200, hematoxylin-eosin staining), and fibroblastic osteosarcoma showing high-grade spindle cell malignancy with only minimal amounts of osseous matrix (**f–h**, ×200, 200, and 100, hematoxylin-eosin staining). The telangectatic osteosarcoma has an unfavorable histology showing blood-filled spaces separated by septa simulating an ABC (not shown)

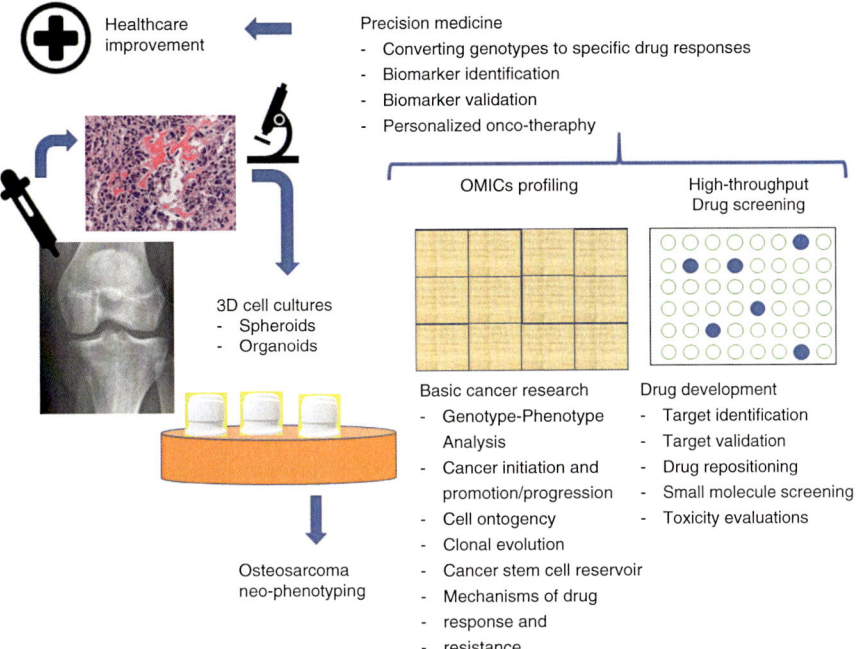

Fig. 13.15 Perspectives in Osteoskeletal Pathology. In this panel is shown an example of an integration of laboratory services, including at first step the radiology with the suspicion of an osteosarcoma and then biopsy with microscopic investigation and confirmation. At time of the biopsy, 3D cell cultures (spheroids and organoids) are created with the intent to have an osteosarcoma neo-phenotyping. Both OMICs profiling and high-throughput drug screening will be useful for a drug repositioning, i.e., the use of already FDA-approved drugs for new purposes. Both OMICs profiling and high-throughput drug screening will help toward precision medicine with converting genotypes to specific drug responses, biomarker identification, biomarker validation, and personalized onco-therapy. FDA, Food and Drug Administration (US Agency)

- According to the number of tumors: unifocal (or solitary) OS and multifocal OS.
- According to the temporal development of multifocal OSs: synchronous (multiple tumors that are diagnosed ≤6 months of each other) and metachronous (various tumors that are diagnosed >6 months of each other).
- According to the histologic growth patterns: bone-rich or "sclerosing," chondroblastic, spindle cell-rich (fibrosarcoma-like), malignant fibro-histiocyte-rich (MFH-like), telangiectatic (ABC-like), small cell, etc.
- According to its inheritance or familiarity: Familial and sporadic osteosarcomas.
- *SYN*: Osteogenic sarcoma.
- *EPI*: 2nd–3rd decade of life and the elderly, ♂ > ♀.

- *CLI:* Local and systemic symptomatology can be encountered in osteosarcoma patients.
  - Local: pain (first mild and intermittent, then severe and persistent), firmness (variable by the intensity of ossification), swelling, sensitiveness to pressure, warm overlying skin, and prominent superficial veins.
  - Systemic: weight loss (specifically with metastatic lung tumor), an increase of ALP (virtual drop after therapy, though not always to the level normal for the age of the patient).
  - History of trauma (?).

Skeletal Distribution: Long bones (metaphysis) of the knee > humerus > pelvis > skull.
- *GEN*: The genetic variations of osteosarcoma are presented in the Box 13.14.
- *IMG*: Radiologic features can be present, but sometimes they may also be absent. In Box 13.15 the radiologic features are summarized.

> **Box 13.14 Genetic Variations of Osteosarcoma**
> 1. Germline mutation and single allelic mutation at 13q14 with "2nd hit" somatic mutation of 13q14 (RB locus) promoting hereditary retinoblastoma and 500 times risk of osteosarcoma.
> 2. *MDM2 amplification* with *P53 inactivation* (Li-Fraumeni syndrome due to an inherited mutant p53 allele and associated with a ↑risk of sarcomas and other malignancies; notably some 500 times ↑ risk of osteosarcoma).
> 3. *LOH on 3, 17, and 18.*

The differential diagnosis of the Codman triangle includes ABC, acute osteomyelitis, hemophilia-related pseudotumor, intraperiosteal hemorrhage (e.g., trauma, parosteal fasciitis), and osteosarcoma.

- *GRO:* Destructive and "permeative" gritty tumor with variable color according to the tissue-producing tumor cells. It includes white-yellowish tumor with gritty to stony in bone-producing regions. It is white and firm in fibrous tissue-producing regions while gray and soft in necrotic areas (particularly crucial after therapy for assessment of the response to treatment → national and international COG protocols), dark red and soft in hemorrhagic regions (particularly telangiectatic OS), and tan in zones richer of osteoclast-like giant cells, as well as bluish and translucent in chondroid or cartilaginous areas. In about 1/5 of cases, a gross pathologic fracture and up to 1/4 of cases extension to the joint are noted. It has been suggested that the color depends on tissue produced from the tumor. Thus, bone regions appear whitish yellow and gritty to stone hard, fibrous areas appear white and firm, necrotic areas appear soft and grey, hemorrhagic portions appear red or dark red, cartilage-bearing parts appear bluish translucent, and former zones of hemorrhage or osteoclast-like giant cells affluent areas appear tan.

> **Box 13.15 Osteosarcoma Radiologic Features**
> - "Fluffy" to "cumulus cloud"-like intramedullary white densities
> - The absence of bone expansion, periosteal bony collar, and sclerotic border of host bone
> - Metaphyseal location
> - Periosteal reactions (e.g., Codman triangle, longitudinal and perpendicular spiculations)
> - Soft tissue component
>
> Notes: The Codman triangle is an isolated cuff of reactive new bone subperiosteally with two components, i.e., an outer homogeneous thick layer of cortical bone with Haversian system and an inner layer characterized by a well-formed network of trabeculae. The OS that manifests before the development of a soft tissue mass is also called "incipient OS" and can be easily mistaken radiologically and pathologically for osteoblastoma or stress fracture. Of note, there is erosion, but not expansion of the cortical bone.

- *CLM*: Several and different histologic patterns ("many faces and disguises")! Growth patterns include fibroblastic (strands and clumps of osteoid with fibrillary pattern), trabecular, filigree, small cell, and telangiectatic. The small cell osteosarcoma can be a very challenging neoplasm because of the broad differential diagnosis related to the small round blue cell tumors. In the small cell growth pattern, sheets of small cells with a size similar to that seen in of Ewing sarcoma, lacy osteoid to primitive filigree bone laid down, and possibly foci of malignant cartilage are seen.
- *IHC*: Immunohistochemical markers of osteoblastic differentiation (e.g., osteocalcin, osteonectin, SATB2) have been suggested to be useful in some difficult cases. However, the use of a good cut H&E slide and Masson trichromic are paramount for the diagnosis.

- *TEM*: There is the extensive presence of dilated, inter-anastomosing, rough endoplasmic reticulum filled with granular material, tight intercellular junctions, and a variety of several cell types.
- *DDX:* Callus, ABC, osteoblastoma, chondroblastoma, chondrosarcoma, mesenchymal chondrosarcoma, fibrosarcoma, giant cell tumor, and Ewing sarcoma (*vide infra*).

Since the OS may have an atypical radiologic presentation, a tremendous variability in histologic patterns and minimal degree of atypia, this tumor is a primary challenge for most of the pathologists. No other tumor is equal to OS in mimicking several and different entities, including GOO, osteoid osteoma, fibrous dysplasia, non-ossifying fibroma, desmoid tumor, fibrosarcoma, malignant fibrous histiocytoma, chondromyxoid fibroma as benign tumors and chondrosarcoma, Ewing sarcoma, giant cell tumor, ABC, and metastatic carcinoma as malignant tumors or tumors of uncertain or low destructive potential.

- **PGN:** High-grade malignancy with a brisk tendency to lung metastasis and a grim prognosis (death $\leq$ 2 years of the diagnosis) if adjuvant chemotherapy using chemotherapy with or without radiotherapy are not implemented. Two groups are identified: (1) patients with good response to chemotherapy ($\geq$90% of diffuse necrosis) and (2) patients with inadequate response to chemotherapy (<90% of focal or diffuse necrosis). The latter group usually requires continued postoperative chemotherapy. Actual 5-YSR is 35–60%, which is enormously improved from the historical range of 12–15% of last century. In most of the patients harboring osteosarcomas (about 85%), micrometastases to the lung are present even before diagnosis and immediate surgery are planned. Along the way for the perfect biomarker, several biomarkers have been correlated to the prognosis of high-grade osteosarcomas, such as *P*-glycoprotein, Her2 and *TP53* gene mutation, and insulin-like growth factor-binding protein 3 (IGFBP-3) (Santer et al. 2006; Ressler et al. 2009; Osasan et al. 2016; Ali et al. 2019). However, to the best of our knowledge, none of these biomarkers has been prospectively validated, and they are not routinely used in clinical practice. To date, the high-grade osteosarcoma is considered a systemic disease, with high probabilities of subclinical metastases upon diagnosis of a localized tumor, and despite its poor prognosis, the cure is actually possible. However, long-term overall survival decreases notably from approximately 60–70% in localized disease to 20–30% if the condition is disseminated. There are many disguises osteosarcoma can appear, and these features need to be kept in mind as indicated above. The diagnosis of osteosarcoma is rarely straightforward, apart obviously of anaplastic osteosarcoma lesions producing large quantities of osseous tissue and/or cartilage. It is controversial to avoid treatment for osteosarcoma with a low degree of metastatic potential, but the intermediate- and high-grade, biologically malignant osteosarcoma need to be treated with systemic chemotherapy or other protocols issued by the Children's Oncology Group (COG), UK Children's Cancer Study Group (UKCCSG), or the International Society of Pediatric Oncology (SIOP) (Redondo et al. 2017).

Osteosarcoma Variants (Other Than Microscopic Patterns)

Parosteal Osteosarcoma (Juxtacortical): 20–60 years, ♂ $\leq$ ♀, metaphysis of long bones (femur, humerus), extracortical calcified mass encircling the bone on imaging and no continuity with bone or BM, slow-growing, and inability to flex the knee or painless swelling over bones. Grossly, it is a firm-to-hard lobulated mass constituted histologically by well-formed bony trabeculae, osteoid ± cartilage and highly cellular spindle cell stroma with mild atypia (0/rare mitoses), numerous myofibroblasts mixed with osteoblasts and fibroblasts with desmosomes identified in undifferentiated cells ultrastructurally. The parosteal osteosarcoma harbors a, very good prognosis with 5-YSR: 80%.

Periosteal Osteosarcoma: 15–25 years, ♂ $\leq$ ♀, diaphysis of long bones (femur, tibia), radiolucent mass on bone surface with bony spicules perpendicular to shaft and penetrating soft tissue with no medullary involvement, bony surface-located tumor constituted by malignant cells of intermediate type producing osteoid and cartilage, intermediate prognosis between parosteal and intramedullary OS. There is 20–35% chance of pulmonary metastasis and the expression of *ABCB1* gene heralds very poor prognosis. Targeting ABCB1 (MDR1) in multi-drug resistant osteosarcoma cells using the new tool based on a bacterial CRISPR-associated protein-9 nuclease (Cas9) from *Streptococcus pyogenes* (CRISPR-Cas9 system) to reverse drug resistance has been recently attempted by Liu et al. (2016). The central causes of MDR is the overexpression of the membrane bound drug transporter protein P-glycoprotein (P-gp), which is the protein product of the MDR gene ABCB1.

Small-cell osteosarcoma (SCOS) is characterized by sheets of small cells like the cells encountered scoping a classic Ewing sarcoma. In fact, Ewing sarcoma is the first differential diagnosis, followed by mesenchymal chondrosarcoma, chondroblastoma, and metastatic neuroblastoma. Age and immunohistochemical phenotype will help characterize these tumors. Also, lymphosarcoma and reticulum cell sarcoma may enter in the differential diagnosis but are extremely rare in children and young adults. SCOS shows small round cells that resemble a primitive neuroectodermal tumor, some variable foci of lacy osteoid to extremely primitive filigree bone deposition between malignant tumor cells, and sporadic foci of chondro-sarcomatous cartilage. SCOS shows a highly lethal course without chemotherapy, although conventional OS chemotherapy protocols are not particularly useful. Specific protocols have been set up for SCOS patients.

Giant cell-rich osteosarcoma contains foci of benign, permeative osteoclast-like giant cells, specifically around hemorrhagic areas. GCOS looks like a giant cell tumor (GCT), and some features may help in the differential diagnosis. One of the most important features include metaphyseal or diaphyseal centering versus epiphyseal centering, but also enlarged and packed stromal cells and primitive deposits of osteoid devoid of osteoblastic rimming help differentiating GCOS from GCT. Tissue submission for histologic evaluation and studies needed for randomized clinical trials may vary and poor (inadequate) sampling may jeopardize the diagnosis. Typically, most tumors require 12–36 tissue blocks, excluding margins (1 block per square cm using a chessboard fashion, Fig. 13.13b). However, tumors showing more significant areas of heterogeneity need a more generous sampling. The usual recommendation of one section per centimeter of maximal dimension is still valid, although extensive tumors showing homogeneity may be assessed correctly with fewer tissue blocks. The concept of taking only one section of the necrotic tumor is not valid anymore, because the assessment of the chemotherapy effect on osteosarcoma and Ewing sarcoma/PNET is crucial for correct reporting to the tumor board, staging, meetings, and multi-disciplinary meetings discussing the follow-up procedures. In fact, gross pathology assessment of necrosis can be misleading, and areas, which appear necrotic, may be myxoid or edematous. Thus, histologic confirmation of the grossly assessed necrosis is critical. In the final surgical pathology report, the highest percentage of necrosis (either grossly or microscopically) should be recorded. It is important to sample for cytogenetics, electron microscopy, and molecular studies. It is recommended that 1 cm^3 of fresh (non-necrotic) tissue should be cut into small, 0.2 cm fragments snap frozen in liquid nitrogen ice-cold isopentane and kept frozen at −70 °C (−94 degrees Fahrenheit). Pragmatically a significant prognostic response in osteosarcoma is when at least 90% of the tumor is necrotic or shows some therapy response, which is associated with a favorable prognosis. The Picci system has been used for assessing Ewing sarcoma/PNET. This system includes grade I with evidence of macroscopic viable tumor, grade II with a microscopic viable tumor, and grade III with no viable tumor. It is recommended to advise the orthopedic surgeon of the challenges of an intraoperative frozen section consultation in case of bone tumors. In the rare cases that it is strongly requested, it is essential to perform the frozen section consultation after the pathologist asks the orthopedic surgeon to give generous tissue samples to avoid artifacts or reactive changes that may mislead the pathologist. The histological diagnosis drives mostly the grading of bone tumors. Typically, Broders advocates a grading system based on cellularity and nuclear features/degree of anaplasia. In the last edition of the AJCC Cancer Staging Manual, a 4-grade system is proposed, including G1 and G2 as low-grade and G3-G4 as high-grade malignant tumors. Low-grade osteosarcomas are parosteal, periosteal and *low-grade* central osteosarcomas. Conventional osteosarcoma, telangiectatic osteosarcoma, small-cell osteosarcoma, secondary osteosarcoma, high-grade surface osteosarcoma, malignant giant cell tumor, Ewing sarcoma/PNET, angiosarcoma, and dedifferentiated chondrosarcoma are high-grade. The conventional chondrosarcoma is graded according to cellularity, cytologic atypia, and mitotic activity. Grade 1 chondrosarcoma is hypocellular with striking similarities to enchondroma, while grade 3 chondrosarcoma is hypercellular, pleomorphic with numerous mitotic figures. Grade 2 chondrosarcoma is between 1 and 2 showing more cellularity than a grade 1 chondrosarcoma with some pleomorphic, hyperchromatic neoplastic chondrocytes. The French Federation of Cancer Centers Sarcoma Group (FNCLCC) grading system is used for mesenchymal chondrosarcoma, fibrosarcoma, leiomyosarcoma, liposarcoma, malignant fibrous histiocytoma, and other soft tissue sarcomas. Chordomas are characterized by a locally aggressive tendency and some inclination for metastasis quite later after surgery. Adamantinomas are low-grade tumors of the bone. The procedures involved in operation are intralesional resection, marginal resection, segmental/wide resection, and radical resection. Each of these procedures has implications for the pathologist and the correct handling of the tissue specimen. A perfect liaison between orthopedic surgeon and pathologist is essential. It is important to remember that for all margins less than 2 cm, the distance of the tumor from the margin is reported in centimeters and all margins need to be specified in case a re-surgery is required. Moreover, margins from bone tumors should be taken as perpendicular margins, and the marrow is scooped out and submitted as a margin, in case the tumor is more than 2 cm from the margin. Osteosarcoma can be a true challenge, and some pearls and pitfalls need to be kept in mind (Box 13.16).

Box. 13.16. Pearls and Pitfalls of OS
1. Second most common primary malignant bone tumor (after multiple myeloma) with challenging pathways.
2. Any uncommitted mesenchymal cells producing osteoid and/or woven bone without a cartilaginous intermediary phase points to the diagnosis of OS no matter how much chondrosarcomatous elements are present.
3. The Codman triangle (*vide supra*) is the most frequent periosteal reaction and is virtually diagnostic of OS, though rare exceptions exist. Before making another diagnosis if Codman triangle is seen, consultation is a wise decision!
4. Osteoid is pink and amorphous and is a specialized form of collagen deposition by osteoblasts (alkaline phosphatase positive). Osteoid and primitive bone have a "crisscross" (or "woven") pattern under polarized light in contrast to no collagen fibers as seen in fibrin and longitudinally oriented in collagen produced by fibroblasts (alkaline phosphatase negative). Moreover, osteoid is less fibrillary, thus more amorphous than collagen, because osteoid is more abundant in proteoglycans Proteoglycans with protein cores having leucine-rich repeat sequences (fibromodulin, and osteoadherin other than decorin and biglycan) are the main form found in

mineralized matrix, although hyaluronan-binding forms (e.g., versican) are present during early stages of osteogenesis. Lamellar bone (compact or cortical and cancellous or spongy or intramedullary) is mature bone and is characterized by a layering of sheets of collagen fibers with a decreased number of osteocytes per unit area as compared to woven or primitive bone, which is hypercellular. Chondroid is also pink, but less fibrillar (less collagenous), and must be distinguished from osteoid because the confusion can have dramatic consequences. Chondroid which is produced by tumor cells of chondroblastoma, chondromyxoid fibroma, and mesenchymal chondrosarcoma other than osteosarcoma may become cartilage (bluish) when the content of proteoglycans is increased (a useful comparison is the cartilage of the mixed tumor of the parotid gland).

5. If chondroid differentiation and no osteoid and/or woven bone production are recognized in a tumor of a patient who is younger than 25 years, do still think of chondroblastic OS and not chondrosarcoma unless the patient has en- or osteochondromatosis.
6. Lamellar bone is never a characteristic histologic feature of OS and prompts the diagnosis in another direction, e.g., callus of ≥6 weeks of age and myositis ossificans.
7. The neoplastic bone of OS lacks a uniform row of osteoblastic rimming, in contrast to reactive periosteal tissues or callus.
8. The "balloon-like expansion," "finger-in-the-balloon" sign, and a collar of periosteal new bone are distinguishing features between ABC and telangiectatic OS (TOS).
9. Three settings are pitfalls in misdiagnosing ABC instead TOS and include (1) a significant amount of hemorrhage and necrosis, (2) smudge nucleus and cytoplasm of tumor cells due to degeneration and ischemia, and (3) intimate insinuation and admixture of tumor cells into reactive cells of the walls.
10. When a lytic lesion or a spongelike tissue presents under the lens with hemorrhage and necrosis, the differential diagnosis between ABC and TOS should be kept in mind. Moreover, hemangioma, lymphangioma, and intraosseous ganglion need to be held in the differential as well.

The fibrosarcoma of bone enters in the differential diagnosis of osteosarcoma rarely in childhood or youth. The radiologic pattern of fibrosarcoma remains nonspecific, although the main feature is osteolysis with other features varying with the degree of malignancy. These features are like other lytic neoplasms of the adult, including malignant fibrous histiocytoma, malignant lymphoma, and metastasis of osteolytic type. Low-grade fibrosarcoma appears often as a sharply edged osteolytic lesion without a sclerotic rim, while high-grade fibrosarcomas arise to produce an extensive area of bony destruction that is lytic partly or entirely.

13.7 Chondroid (Cartilage)-Forming Tumors

This group of tumors is characterized by the production of chondroid and/or cartilage by the tumor cells. Chondroid is a primitive form of cartilage, is rich in collagen, but less fibrillar than osteoid, and is depleted of proteoglycans (determining the pink color similar to osteoid). Conversely, cartilage (mature form), which is rich in proteoglycans, presents under the lens blue in hue. Chondroid is usually well-demarcated and presents in masses under the lens with notably rounded contours. In increasing order of frequency, chondroid is seen in chondromyxoid fibroma (CMF) showing a mixture of chondroid, myxoid and fibroid tissue, osteosarcoma (chondroblastic OS), mesenchymal chondrosarcoma (MCS), and chondroblastoma. It is essential to distinguish mature cartilage into hyaline, which has chondrocytes within lacunae embedded within bluish-pink, homogeneous matrix. Hyaline cartilage is found in the normal articular cartilage, epiphyseal growth plate, maturing callus, enchondroma, chondrosarcoma, and chondroblastic osteosarcoma. The epiphyseal growth plate cartilages have four zones, including resting, germinative, vacuolating, and degenerating zones, and may be found in the osteochondroma caps, *dysplasia epiphysealis hemimelica*, and in some benign bone tumors and parosteal osteosarcoma. Chondrocytes, differently from osteocytes, do not have canaliculi allowing the transfer of nutritive substances from one cell to the other in a densely calcified matrix. Thus, calcification of the matrix means chondrocyte necrosis. A deficiency in the production of collagen fibers with little or disturbance in the production of mucopolysaccharides characterizes the bluish hue and bubbling aspect of the myxoid cartilage, which is usually seen in chondrosarcoma, OS, CMF, parosteal chondroma, and enchondroma. A mixture of chondroid and osteoid (called chondro-osteoid) is found almost exclusively in some OS and callus (Box 13.17).

> **Box 13.17 Chondroid-/Cartilage-Forming Tumors**
> 13.7.1 Osteochondroma and Osteochondromatosis
> 13.7.2 Enchondroma and Enchondromatosis
> 13.7.3 Chondroblastoma
> 13.7.4 Chondromyxoid Fibroma
> 13.7.5 Chondrosarcoma (Chondrogenic Sarcoma)

13.7.1 Osteochondroma

- *DEF*: Most common benign, hamartomatous-like lesion/bony tumor of the metaphysis recapitulating the growth plate of endochondral ossification at metaphysis of mostly long bones (pseudo-neoplastic or single vs. neoplastic or multiple aberrant growths of cartilage) (Fig. 13.16) characterized by the Broca criteria:
  – Site: at the level of the juxtaepiphyseal cartilage
  – Structure: bony at the center, cartilaginous at the periphery
  – Evolution: synchronous with the skeletal growth
  – Multiplicity: more lesions
  – Inheritance: AD pattern with ♂ > ♀

Key is to remember the Bassel-Hagen law, which states that the bone bearing an osteochondroma loses in length what acquires in the exophytic growth.

- *SYN*: Exostosis, ecchondroma.
- *EPI*: 1st–2nd decades, ♂ > ♀.
- *EPG*: Aberrant subperiosteal residuals of growth cartilage (Virchow theory).
- *CLI*: Asymptomatic, slightly esthetic disturbances of "LP" *L*ong bones (lower femur, upper tibia, upper humerus), mostly around the knee + *P*elvis.
- *IMG*: Pedunculated tumor located at metaphysis growing out in a direction opposite that of the adjacent joint.
- *GRO*: Pedunculated tumor with a cap of cartilage, rarely exceeding 1 cm.
- *CLM*: There is a lobulated cap of hyaline cartilage (distal), covered by a fibrous membrane, with an active endochondral ossification (interface), and mature or cancellous bone (proximal to bone metaphysis) beneath, the latter making up the bulk of the lesion. The bone marrow of the diaphysis is *not* connected to the bone marrow of the osteochondroma.
- *DDX*: Parosteal osteosarcoma.
- *PGN*: As the osteochondroma ages, the cap may become thinner and disappear, and the

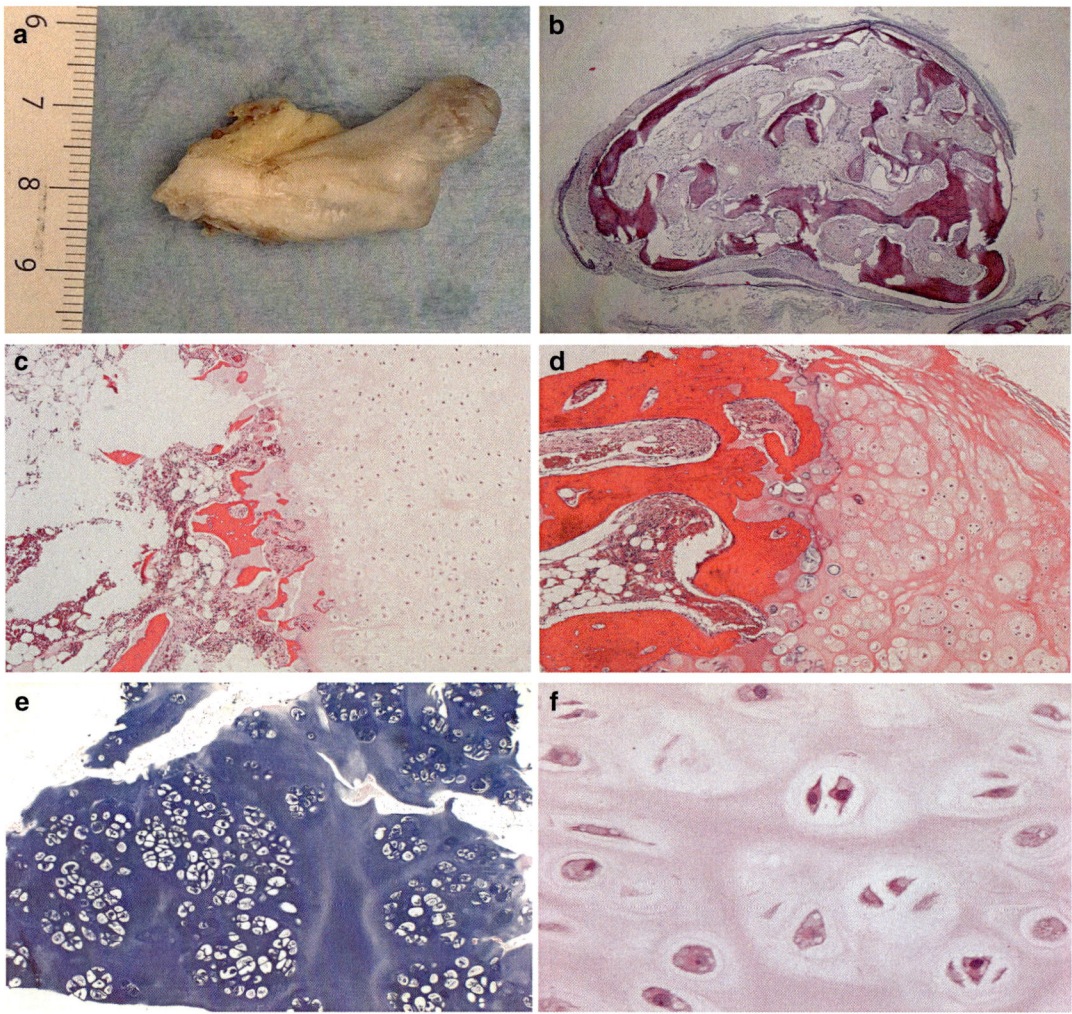

Fig. 13.16 Osteochondroma. This panel illustrates the exostosis resected by the orthopedic surgeon from the metaphysis of a long bone (**a–f**). The histology of the tumor identifies a new cap with a characteristic growth phase without cellular atypia or mitotic figures (**b–f**, hematoxylin and eosin staining, ×40 original magnification; **c**, hematoxylin and eosin staining, ×50 original magnification; d, hematoxylin and eosin staining, ×50 original magnification; **e**, toluidin blue staining, ×100 original magnification; **f**, hematoxylin and eosin staining, ×400 original magnification)

whole lesion may spontaneously regress. However, the osteochondroma harbors a malignant transformation rate of 1–2%, if single. The malignant transformation rate is higher (5–10%) in case of multiple osteochondromas. Worrisome signs portend to a possible/probable malignant transformation: tumor Ø > 8 cm, cap >3 cm, and irregular cap. In case of a developing bursa around head, there is a potentiality to develop chondrosarcoma.

Osteochondroma Variants

Osteochondromatosis (Ehrenfried hereditary deforming chondrodysplasia, Leri disease, Ombredanne disease): *EXT1* or *EXT2* genes-linked, AD-inherited disease with multiple osteochondromas ± Gardner syndrome with a capability to develop into chondrosarcoma (10%). *Dupuytren Subungual Exostoses (DSE)*: Subungual-located osteochondroma with histologic patterns that need to be differentiated from chondrosarcoma. DSE starts as a reactive growth of cellular fibrous tissue and metaplastic cartilage, which undergoes endochondral ossification with postadolescents and young adults mostly affected and usually on the dorsal medial aspect of the great toe. Although a benign lesion, DSE local excision is curative, recurrences are common though (~50%) after incomplete excision or when the lesion has not achieved full maturation.

13.7.2 Enchondroma

- *DEF*: Common asymmetric, unilateral, and painless benign tumor at a medullary or cortical location in the center of the diaphysis of the bones constituted by fundamentally pure cartilage (mature hyaline) (Fig. 13.17).
- *SYN*: Chondroma.
- *EPI*: 2nd–4th decades, 5–10% of benign tumors, ♂ > ♀.
- *EPG*: Hamartomatous origin (the so-called *Hamartiae cartilagineae*). Probably, the enchondromas should be considered dysontogenetic tumors arising from aberrant rests of growing cartilage. Hamartia is a defect of tissue combination during embryo-genetic development with the tissue components usually belonging to the site (Greek: ἁμαρτία, "failure, error", and ἁμαρτάνω, which is the derived verb for "to miss the mark"). Both Greek tragedy and Christian theology have used this term, which has been used since Ancient times. Aristotle was the first using this term in his Poetics. He described "ἁμαρτία" as an error of judgment made by a

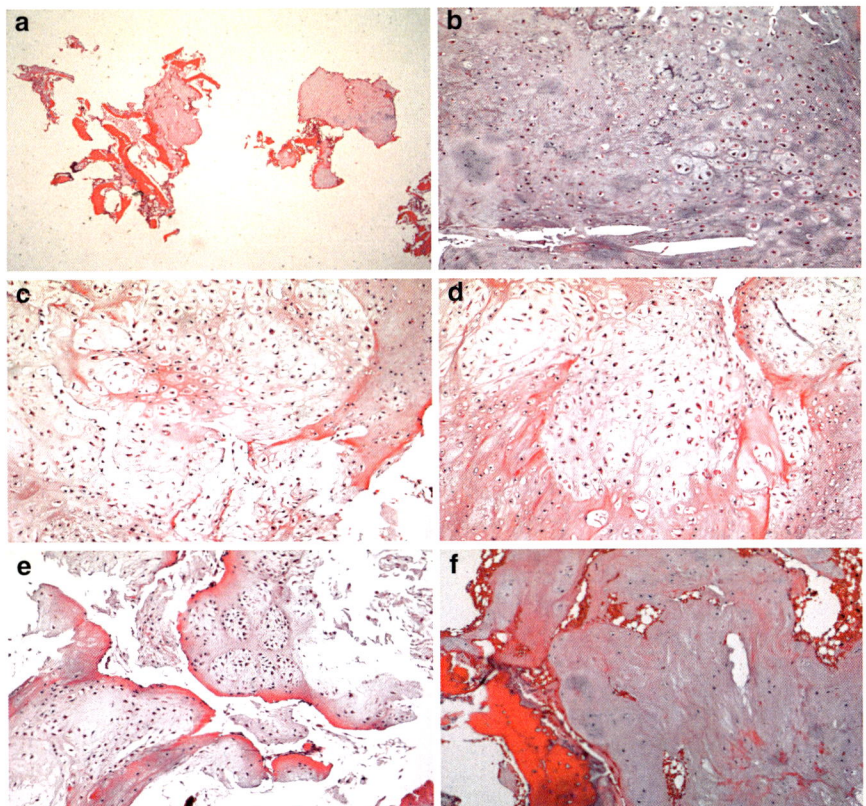

Fig. 13.17 Enchondroma. This panel illustrates an enchondroma with characteristic lobules of hyaline cartilage, which are embedded by bone and covered by fibrous tissue (perichondrium) (**a–f**). The tumor may give the impression of a low-grade chondrosarcoma due to hypercellularity, binucleation of the neoplastic chondrocytes (**b–e**), and occasional prominent myxoid change (**f**) despite a benign appearance radiologically (**a–f**, hematoxylin and eosin staining; ×12.5, ×100, ×100, ×100, ×100, ×100 as original magnification, respectively)

character in a tragedy played in theaters. Subsequently, the Italian (Tuscan) Dante Alighieri relabels *"Hamartia"* a "movement of spirit" played within the sphere of a protagonist. This individual was going to commit actions driving the plot towards its tragic end. When embryonic tissue is found in a non-typical location (heterotopia), the tissue composition derived from it is called choristia (Greek χωριστός "separated, separable").

- *CLI*: Asymptomatic (usually) tumor of the small bones of hands and feet (proximal phalanges), ribs, and long bones. If pain and pathologic fracture occur, then think of low-grade chondrosarcoma.
- *IMG*: Well-circumscribed radiolucency with punctate opacities ("popcorn"-like densities) detected in medullary sites and characterized by sharp margination of the lesion from the surrounding bone without cortical destruction or soft tissue extension, and ring-like or punctate calcifications within the tumor.
- *GRO*: Firm slightly lobulated of gray-blue translucent masses with a Ø often <2 cm. Although enchondromas arise centrally, in their growth, they may erode the adjacent cortex and cause a bone expansion.
- *CLM*: Mature well-circumscribed lobules of hyaline cartilage with or without foci of myxoid degeneration, calcification, and peripheral endochondral ossification. Hyaline cartilage composed of (1) vascularized fibrous stroma and (2) chondrocytes, which appear irregularly dispersed throughout the matrix and entrapped within clearly defined lacunar spaces. Occasionally, atypical chondrocytes with enlarged nuclei are seen.
- *PGN*: Benign but cellular myxoid areas may raise suspicion for low-grade chondrosarcoma. Cellular myxoid regions and double-nucleated lacunae may be seen in the enchondromas of hands and still behave in a benign fashion (intradepartmental consultation is advised).

Risk of malignant transformation: Chondrosarcoma.

If non-hereditary syndromes of multiple enchondromas:

– Ollier disease: multiple enchondromas (unilateral, asymmetric) + ovarian sex-cord stromal tumors.
– Maffucci syndrome: multiple enchondromas + soft tissue hemangiomas.
– Metachondromatosis, which is an AD-inherited osteochrondrodysplasia affecting the growth of bones and harboring multiple enchondromas and osteochondromas.

Variants: Calcifying enchondroma and juxtacortical (periosteal) enchondroma.

Enchondroma: peripheral and subperiosteal chondroma

In general, the lesion is benign regardless of the atypicality of the chondrocytes, mainly if it occurs in the first two decades of life and is located in the small bones of the hands. Ollier and Maffucci diseases/syndromes show enchondromas with more cellularity and greater variability in cell morphology than the sporadic, isolated enchondromas. A chondro-sarcomatous event always needs to be distinguished in these two diseases, and a second opinion is often required.

13.7.3 Chondroblastoma

- *DEF*: Benign, chondroid-forming tumor of the primarily *epiphyseal* growth plate of bones constituted by polygonal chondroblasts, small foci or nodules of the chondroid matrix, osteoclast-like giant cells, and foci of dystrophic calcification with potential extension to metaphysis and even through the physeal plate (Figs. 13.18 and 13.19).
- *EPI*: ~1% of primary bone tumors, ~¾ of patients are in the 2nd decade of life during the active growth of the epiphyseal plate and male (♂:♀ = 3:1).
- *CLI*: Pain and/or swelling at specific sites (*L*ong bones, *P*elvis, *F*eet at epiphyseal location or apophysis or short bone).
- *IMG* – Four features (not simultaneous always present): the *partial occupation* of the epiphysis (usually <50%) by an osteolytic lesion, the *thin border* of sclerosis with *raggedly to wispy*,

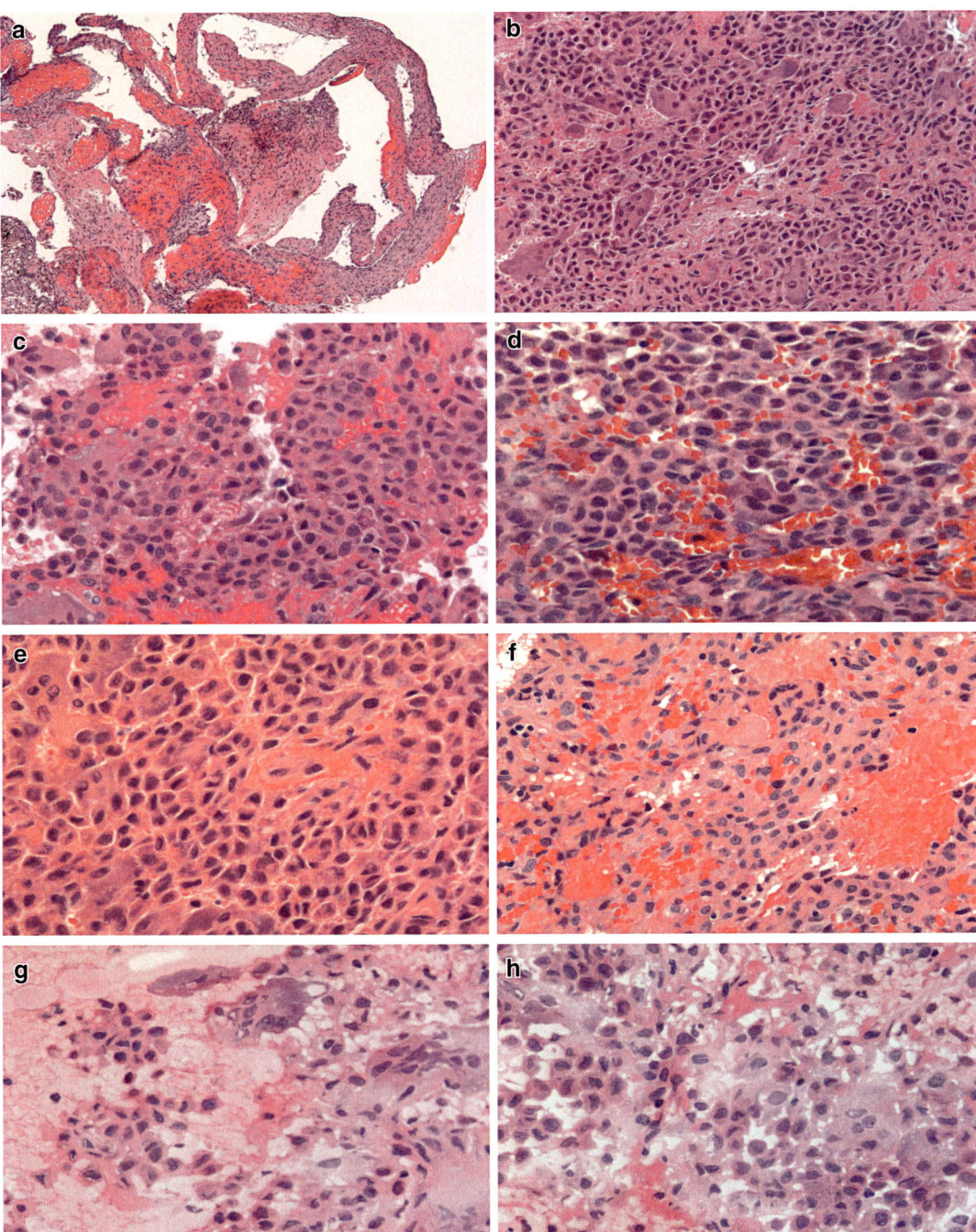

Fig. 13.18 Chondroblastoma. Histological images (**a–h**) of this tumor show plump oval histiocytoid cells with scattered osteoclast-like giant cells embedded in a cellular tissue with occasional islands of chondroid/cartilage. The accompanying eosinophils can be particularly prominent, and the differential diagnosis should include Langerhans cell histiocytosis. All images are microphotographs taken from hematoxylin and eosin-stained slides. Microphotographs (**g**) and (**h**) are intraoperative frozen sections. It is important to compare the different tonalities between intraoperative frozen sections and permanent sections (**a–f**, hematoxylin and eosin staining, ×50, ×200, ×400, ×400, ×400, ×400, ×400, ×400 as original magnification, respectively)

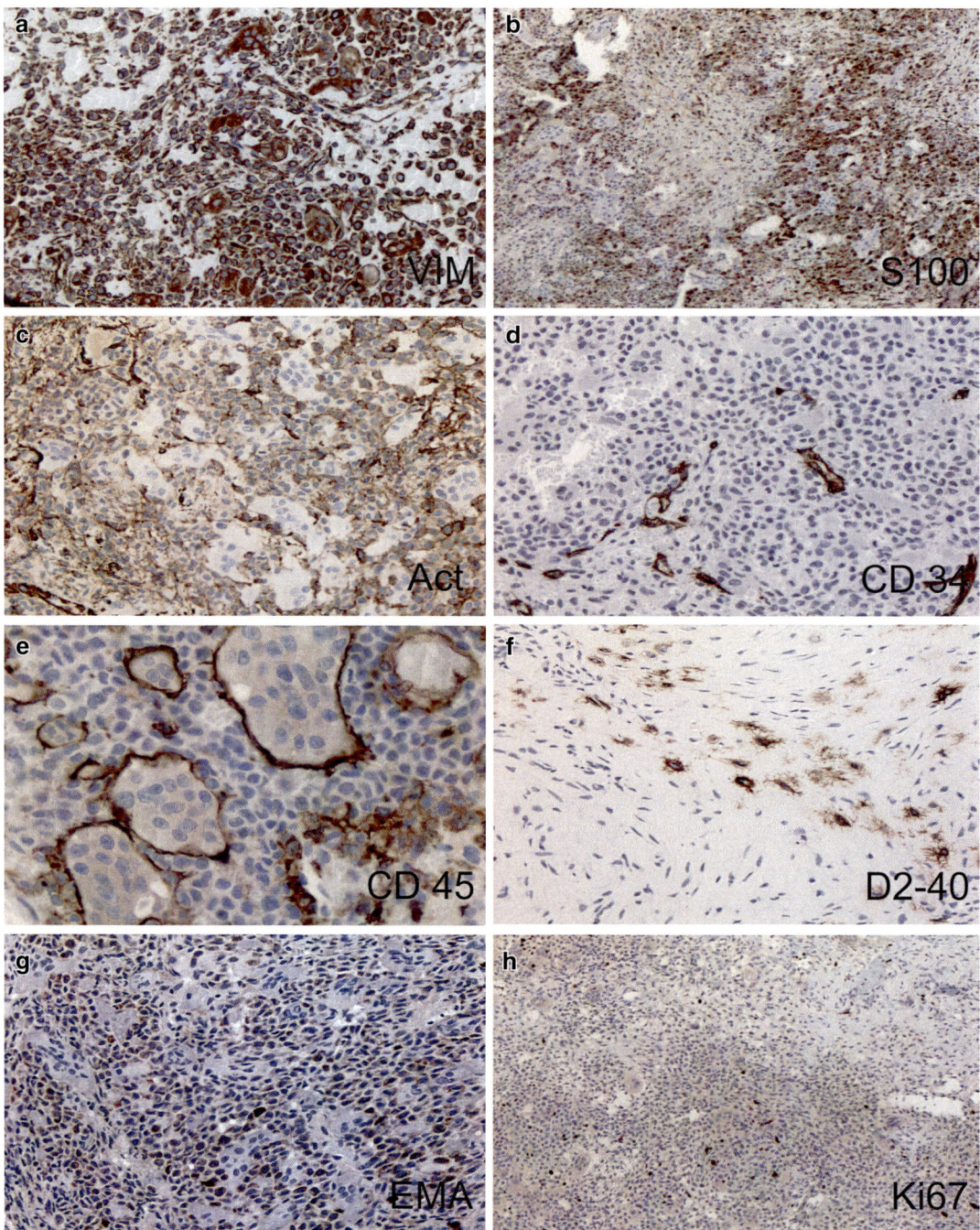

Fig. 13.19 Chondroblastoma. The immunohistochemistry can identify several cell populations with the immunohistochemical markers shown on the lower right corner of each microphotograph in this panel (**a–h**: **a**, anti-vimentin (VIM) immunostaining, Avidin-Biotin Complex, ×200 original magnification; **b**, anti-S100 immunostaining, Avidin-Biotin Complex, ×100 original magnification; **c**, anti-actin (Act) immunostaining, Avidin-Biotin Complex, ×200 original magnification; **d**, anti-CD34 immunostaining, Avidin-Biotin Complex, ×200 original magnification; **e**, anti-CD45 immunostaining, Avidin-Biotin Complex, ×400 original magnification; **f**, anti-D2-40 (Podoplanin) immunostaining, Avidin-Biotin Complex, ×200 original magnification; **g**, anti-epithelial membrane antigen (EMA) immunostaining, Avidin-Biotin Complex, ×200 original magnification; **h**, anti-Ki67 (MIB1) immunostaining, Avidin-Biotin Complex, ×100 original magnification). Note the importance to use CD45 to decorate the osteoclasts in chondroblastoma (**e**), which are not highlighted by actin (**c**)

13.7 Chondroid (Cartilage)-Forming Tumors

fluffy expansion (20%) of the lesion, as well as intralesional 1–2-mm *punctate calcifications* (50%). In CT/MRI the lesion has well-defined borders and fuzzy mottled intratumoral radiodensities.
- *GRO*: Rubbery soft with sharp limits, pink to gray-tan and whitish foci or chalky granules and granular, variegated cut-surface with minimal or no grittiness.
- *CLM*: It includes LM, HSS/IHC, and TEM essential features.

There are four features, including (1) *polygonal stromal cells* (chondroblasts) with distinct cytoplasmic borders, plump nuclei with longitudinally grooves (pseudo-anaplasia may be present showing slightly atypical, pleomorphic nuclei in size and shape, although bland chromatin distribution, small or inconspicuous nucleoli, and no atypical mitoses help in the differential diagnosis with the true anaplasia) and rare mitoses (<1/5 HPF), which are singly encased by reticulin fibers (so-called honeycomb reticular pattern of immature cartilage) and variable "chicken wire" (lacy) pericellular pattern of calcification in which linear calcifications surround individual mononuclear cells, (2) *osteoclast-like giant cells*, (3) *small foci or nodules of pinkish chondroid matrix*, and (4) *foci of dystrophic calcification*.

The mononuclear cells represent embryonic chondroblasts that contain lobulated indented nuclei with some mitotic activity, but without atypical forms.

- *HSS/IHC*: (+) PAS, reticulin, S100, SOX9 (chondrogenesis regulator) and (−) CD1a.
- *TEM*: Six features, including round to polygonal shape, blunted microvillous processes, nuclear irregularity with tendency to multilobulation, heterochromatin condensation with focal discontinuity along the internal surface of the nuclear envelope, quite evident rough endoplasmic reticulum, and granules of intracytoplasmic glycogen (PAS+).
- *DDX*: GCT (different age, radiologic large, >50%, pure lysis of the epiphysis showing a hole to trabeculated image without stippled densities, different cell morphology, osteoid, spindly stroma, no "chicken wire" calcification), OS (site, location, periosteal reactions, soft tissue mass, lack of a ring of host bony sclerosis, true histologic anaplasia, and osteoid), conventional CS (site, location, hyaline cartilage), MCS (dimorphic histologic pattern with undifferentiated stromal cells), CCCS (age, variation from chondroid to hyaline cartilage admixed with bony metaplasia), eosinophilic granuloma or localized LCH (eosinophils, CD1a stain), lymphoma (lack of chondroid, "chicken wire" calcification, scattered osteoclast-like giant cells). Chondroblastoma is not a typical hyaline cartilage-forming tumor. Thus, if I see hyaline cartilage, it is not a chondroblastoma. Conversely, the chondroid matrix is pink.
- *TRT*: Stages 1, 2, and 3 are treated with intralesional excision to segmental resection.
- *PGN*: Almost exclusively benign (99%), although tumor recurrence and perilesional (local) seeding are present in 5% of patients.

13.7.4 Chondromyxoid Fibroma

- *DEF*: Benign, chondroid-forming tumor of *metaphysis* of bones constituted by a mixture of *chondroid*, *myxoid*, and *fibroid* tissues in varying proportions (Fig. 13.20).
- *EPI*: 0.5% of primary bone tumors, ¾ of patients are younger than 30 years, ♂:♀ = 2:1.
- *CLI*: Pain and swelling at specific sites (*L*ong bones and *F*eet at metaphyseal location)
- *IMG:* Metaphyseal location, eccentricity, host bone sclerotic rim, ovoid contour, lobulated architecture to "bubbly" appearance, "bubbling" cortex, periosteal collarette, and the absence of intralesional densities.
- *GRO*: Solid, rubbery, grayish-white to yellow and tan.
- *CLM: Lobular architecture* constituted by a variable pinkish matrix (chondroid) to bluish hypocellular myxoid ("bubbly") matrix with stellate cells with lengthy delicate processes – separated by a highly cellular stroma, particularly prominent at the periphery around chondroid lobules (*peripheral condensation*)

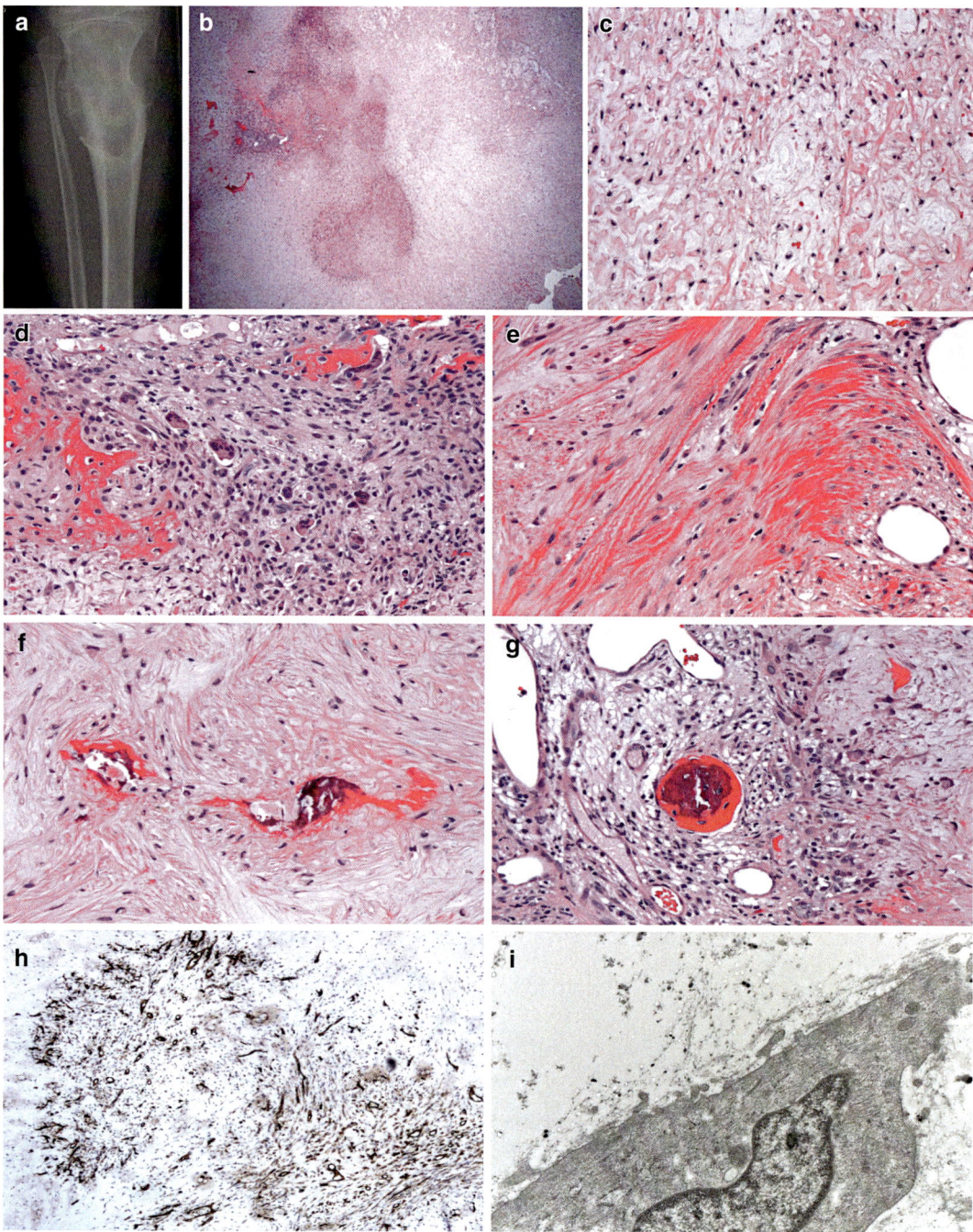

Fig. 13.20 Condromyxoid Fibroma. This panel shows hypocellular lobules of poorly formed hyaline cartilage of a chondromyxoid fibroma. The lobules are composed of chondroblasts with abundant pink cytoplasm and myxoid tissue with fibrous septae, which contain spindle cells and osteoclasts (**a–g**, hematoxylin and eosin staining, ×12.5, ×200, ×200, ×200, ×200, ×200). The microphotograph in (**h**) shows a CD3 immunostaining, while an immature cell is shown on electron microscopy ultrastructural examination (Transmission electron microscope) (**i**)

(fibroid regions) (chondroid + myxoid + fibrous tissues).
- *TEM*: Stellate, ovoid, or elongated cells with abundant intercellular loose matrix with many fine filamentous structures. The nuclei show an indented contour, while the cytoplasm show villous cell processes and intracytoplasmic fine filaments.
- *DDX*: *Enchondroma* (true cartilage, round, and not stellate cells), *chondroblastoma* (epiphyseal, lack of lobular growth with a peripheral increase of cellularity), *chondrosarcoma* (aggressive imaging, hypercellular lobules, and lobular formation are not quite as distinct), and *chondroblastic OS* (atypical cells with osteoid and focal hyaline cartilage) (Box 13.18).
- *PGN*: ¼ patients may show local recurrence, which is probably due to incomplete curetting.

Myxoid cartilage should be differentiated by hyaline cartilage. The former shows a characteristic "bubbly appearance," while the latter is characterized by zonation with the resting zone, germinative zone, vacuolating zone, and degenerating or calcifying zone from the top of physis to its metaphyseal side. Patterns of myxoid cartilage are seen in chondromyxoid fibroma, osteosarcoma, chondrosarcoma, parosteal chondroma, and enchondroma. Patterns of hyaline cartilage are seen in apophyseal plate cartilage, osteochondroma caps, cartilage of benign tumors, benign metaplastically induced cartilage of low-grade osteosarcoma (e.g., parosteal subtype), and the rarest occurrence of an osteochondroma of articular cartilage (the so-called *dysplasia epiphysealis hemimelica*). Chondroid is a primitive form of cartilage, which is rich in collagen and exhausted of proteoglycans. In fact, it is deep pink in sort, as pink as osteoid. The reduced content of proteoglycans makes this kind of cartilage being distinguished from others. Chondroid appears to be deposited in masses with rounded borders, as often seen in chondroblastoma, but also in chondromyxoid fibroma, osteosarcoma, and mesenchymal chondrosarcoma.

13.7.5 Chondrosarcoma

- *DEF*: Malignant chondroid-forming (pure hyaline cartilage) bony tumor with a broad range of age predilection, also occurring in children, although most often observed in middle age and elderly, usually located at flat bones (pelvis and shoulder girdle) and harboring a worrisome prognosis.
- *EPI*: Third most common primary malignancy of bone accounting for about 20% of malignant bone tumors with quite diffuse plateau-like monomodal incidence rate (3rd–6th decades), ♂:♀ = 1.5:1.
- *CLI*: Bony pain and swelling, usually of long duration (months to years) at specific sites (*PS* → *P*elvis (near the acetabulum) and *S*houlder girdle: scapula, sternum, but also ribs and long bones at metaphysis or diaphysis).
- *IMG:* Osteolytic lesion of diaphysis or metaphysis with spotty calcification ("windblown" intramedullary densifications), ill-defined edges, expansive thickening of the shaft, endosteal scalloping, cortical thickening, and soft tissue mass ± areas of perforation or destruction (usually located inside of the periosteum).

> **Box 13.18 CMF: Pearls and Pitfalls**
> 1. When you see myxoid tissue, *rule out first* myxoid chondrosarcoma!
> 2. CMF may show some variability in cellularity and nuclear morphology, and these focal dysplastic features may induce to overcall a lobulated chondroid-forming tumor with benign radiologic signs as chondrosarcoma. However, binucleate or multinucleate cells or multiple cells per lacunae are not seen in CMF!
> 3. *When you see* the radiologic sign of "bubbling out" of the cortex, *then* the separation of CMF from NOF and NFD is crucial.

- *GRO*: The location separates three types:
  - *Central* Chondrosarcoma: Medullary cavity of usually long bones.
  - *Peripheral* Chondrosarcoma: On preexisting osteochondroma or *de novo* with a calcified core.
  - *Juxtacortical* (Periosteal) Chondrosarcoma: Periosteal located at the shaft of long bones.
- *CLM*: Plump, hyperchromatic nuclei, 2+ nuclei per cell, 2+ S100+ cells per lacuna with the permeation of BM and trapping of lamellar bone (No osteoid forming!). Grading is based on cellularity, nuclear size, and hyperchromatic features of the nuclei.
  - Grade 1 (50%): soberly cellular with hyperchromatic slightly enlarged nuclei.
  - Grade 2 (40%): more cellular with an intermediate degree of nuclear atypia.
  - Grade 3 (10%): highly cellular with pleomorphic nuclei and commonly identified mitoses.
- *HSS/IHC*: (+) S100, ER, SOX9 (chondrogenesis regulator), (±) BCL2 (~50%), (−) EZR. Ezrin (EZR), which is aka cytovillin or villin-2 is a member of the ezrin/radixin/moesin (EMR) family of proteins. This family acts as general cross-linking between the actin cytoskeleton and the plasma membrane proteins.
- Conventional Chondrosarcoma: two forms
  - Hyaline – conventional – chondrosarcoma (*vide supra*).
  - Myxoid – conventional – chondrosarcoma: chordoma-like, (−) AE1–3.
- *TEM*: In grade I, there are short branching segments of RER, glycogen, and a prominent Golgi apparatus, while grade II is characterized by a large nucleolus, lipid droplets, dilated RER, and scattered glycogen. Grade III shows cells with pleomorphic nuclei, large nucleolus, and scattered lipid droplets.
- *DDX*: Enchondroma (small size, mild, or no atypia), chondroblastoma (epiphyses, immature skeleton, male predominance, painful, secondary ABC, monomorphic mononuclear cells with indented nuclei, occasional giant cells within a scant immature hyaline type chondroid matrix, "chicken-wire" appearance,

Box 13.19 Chondrosarcoma: Pearls and Pitfalls

1. Chondroid formation in chondrosarcoma = Pure *hyaline* cartilage.
2. *If* imaging points to a chondroid tumor with cortical erosion, cortical thinning/thickening, expansion of the cortical bone, endosteal scalloping (i.e., reabsorption of endosteal cortex), and production of reactive periosteal bone, *then* think of an aggressive tumor.
3. Neoplastic chondrocytes of the chondrosarcoma within lacunae display variable atypia with variation in size and shape and may possess an open-chromatin pattern. Although binucleation is often seen, it must be distinguished from multiple cells within a single lacuna.
4. Chondrosarcomas with myxoid features are classified as grade 2.
5. Differently, from the chondrosarcomas, enchondromas are usually asymptomatic, and commonly occur in the small bones of hands and feet and show an expansile lucent nodular pattern that may have characteristic punctate, stippled or ring mineralization, but no destructive features, as seen in chondrosarcomas.
6. Chondroblastic osteosarcoma is characterized by its predominant chondroid component. The chondroid component can be very extensive and variably represented by high or intermediate grade patterns. By definition, at least a small focus of tumor osteoid must be present to qualify as a chondroblastic osteosarcoma.
7. Specifically, mesenchymal chondrosarcoma harboring small round blue cells it may be a difficult task without ancillary techniques for ruling out "small round blue cell tumors".
8. The prognosis of Dedifferentiated Chondrosarcoma (DCS) is quite miserable.

(+) S100 and SOX9), chondroblastic OS (osteoid formation, age group, (+) EZR), sarcoma (histology, IHC), and small round blue cell tumor (IHC-panel) (Box 13.19).

- *PGN*: Better than OS! It is common to have soft tissue implantation at the biopsy site. Thus *en bloc* excision or removal of biopsy tract is extremely important. 5-YSR: 78%, 53%, and 22% for grade I, II, and III lesions, respectively.

Chondrosarcoma Variants

Clear Cell Chondrosarcoma: Proximal femur/humerus located, lytic, slightly expansile, and sharply located at the edges on imaging chondrosarcoma variant with cells with copious clear cytoplasm, sharply defined cell borders, and interspersed lamellar bony fragments (low grade). *Mesenchymal Chondrosarcoma*: Jaw, pelvis, ribs, femur located chondrosarcoma variant with dimorphic pattern constituted by island of well-differentiated cartilage and highly cellular, undifferentiated stroma. Small round blue cells show rare pleomorphism and mitoses. This tumor has variable grade of malignancy. Typically, it harbors, however, an aggressive clinical course with local recurrence and metastases..

Dedifferentiated Chondrosarcoma: Most malignant variant with <10% long-term survival and characterized by a bimorphic histology located and on imaging showing foci of poorly differentiated sarcomatous elements at the periphery of conventional chondrosarcoma with possible MFH, RMS, fibrosarcoma, and osteosarcoma-like areas (high grade). DCS are bimorphic chondrosarcomas in which chondrosarcoma is adjacent to a malignant spindle cell component, and the two morphologic patterns are sharply demarcated with the cartilaginous component not blending into the spindle cell component as observed in the osteoblastic osteosarcoma.

Most frequent tumors of the ribs include chondrosarcoma (30%), lymphomas (17%), fibrous dysplasia (12%), osteosarcoma (10%), osteochondroma (8%), Ewing sarcoma (6%), and LCH (eosinophilic granuloma) (3%).

13.8 Bone Ewing Sarcoma

- *DEF*: Malignant bone round cell tumor arising in the medullary cavity of diaphysis with at least ten different features at five different tumor sites implicating 15 parameters (relatively independent) comprising a factorial number of 15, which equals 10^9 patterns (Fig. 13.21).
- *EPI*: ~1/20 of malignant bone tumors, sixth most common malignant bone tumor, peak age <30 years (range: 10–25 years); ♂:♀ = 1:1.
- *CLI*: Pain and swelling at specific sites – long bones (femur, humerus, tibia, fibula), pelvis, spine, ribs, jaw, clavicle ± extra-osseous component. Constitutional signs (local heat, prominent subcutaneous BVs, fever, ↓Hb, ↑WBC, and ↑ESR) may suggest osteomyelitis.
- *IMG*: In radiology, BES is considered a nightmare, because it is the most protean of the bony tumors in its radiologic presentation having countless possible radiologic manifestations. BES is intramedullary of diaphysis located tumor with massive extension, usually permeative bone destruction with an invisible boundary between the area of damage and the bone. Other patterns include pinhead-sized holes, moth-eaten, rotten wood, geographic necrosis, cracked ice, honeycomb, and punched out among others. Moreover, there is thickening of the cortex and onion-skin layering as elevates periosteum and simultaneous deposition of new bone. However, Codman triangle is also possible mimicking various malignant and benign bone tumors. BES does not do specific features, including producing an outer rind or shell of host bone sclerosis as seen in NOF, producing smooth bone expansion as observed in ABC, SBC, and other processes, and encircling the soft tissue mass with a shell of bone as seen in ABC.
- *GRO*: White, fleshy tumor, which may present as soft tissue lesion, because permeation of medullary cavity may be nondestructive.
- *CLM*: Peritheliomatous pattern of sheets of small, uniform blue, glycogen-rich (PAS+) cells with indented nuclei and inconspicuous nucleoli (slightly larger than nuclei of lym-

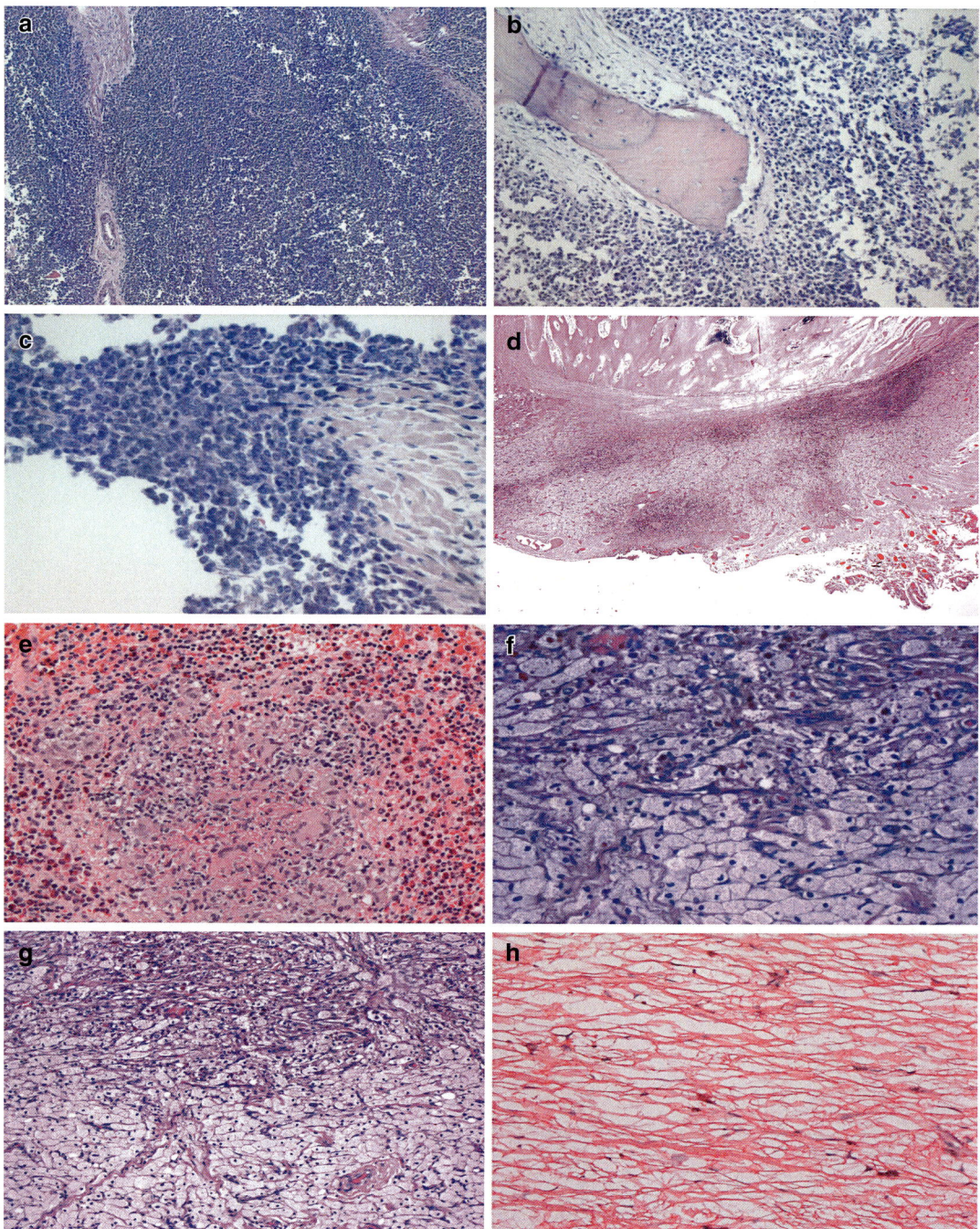

Fig. 13.21 Ewing Sarcoma. The microphotographs (**a–c**) show a small round blue cell tumor with hyperchromatic nuclei and no or minimal discernible cytoplasm. The microphotographs (**d–h**) show post-chemotherapy changes with disappearance of the tumor cells and empty spaces (**a–h**, hematoxylin and eosin staining, ×100, ×200, ×400, ×20, ×200, ×400, ×200, ×200 as original magnification, respectively)

phocytes with stippled chromatin), high N/C, unapparent cell borders, variable number of mitoses, and forming pseudorosettes (Homer-Wright) separated by strands of fibrous stroma.
- *HSS/IHC*: PAS (+), particularly if the bone biopsy is placed in 100% alcohol; (+) CD99, (+) Fli-1, (±) neuroendocrine differentiation (CGA, SYN, etc.).
- *TEM*: Two cell nuclei types, including cells with large nuclei with open chromatin and cells with small contracted nuclei (indented or cleaved) with hyperchromatic chromatin and apoptosis as well as common intracytoplasmic glycogen granules, primitive intercellular junctions, rough endoplasmic reticulum, a few mitochondria in groups, conspicuous Golgi centers, and occasional lipid droplets. Glycogen is both intracytoplasmic and outside of the tumor cells (matrix).
- *CMB*: t(11;22) FLI-1 – EWS (on RT-PCR or FISH).
- *DDX*: *Chronic osteomyelitis* (IHC), *LCH* (+S100, + Langerin), *lymphoma* (CD45, CD20, CD21, CD3, CD5, CD8, CD56), *metastatic neuroblastoma* (CD99-, FLI-1-, but (+) CGA.), *RMS* (+DES, MYF-4, myogenin), *small-cell osteosarcoma* (conventional areas may be present), *mesenchymal chondrosarcoma* (imaging, dimorphic pattern, -CD99), and *small-cell carcinoma of the lung* (age, IHC). When a peritheliomatous pattern of small round blue cells without a dimorphic pattern in the appropriate location and fitting imaging is seen, think to BES. Moreover, more pleomorphism in a supposed BES should raise the possibility of therapy-related changes.
- *PGN:* High-grade lesion ± multiple metastases at presentation (lungs, pleura, other bones, CNS, regional LNs) and death within 2–3 years of initial diagnosis, unless a multistep approach is started including radiotherapy, local resection, and post-adjuvant chemotherapy. 5-YSR with therapy: 75% (without treatment: 5%). Fever has been considered a severe prognostic sign.

13.9 Miscellaneous Bone Tumors

13.9.1 Chordoma

- *DEF*: Fetal notochordal-derived neoplasm with dual epithelial-mesenchymal differentiation. *Ecchordosis physaliphora* is a congenital benign lesion of hamartomatous origin. It is derived from notochord remnants, which are often located in the retroclival prepontine region. These remnants can, however, be detected anywhere from the skull base to the sacrum. *Ecchordosis physaliphora* needs to be kept distinguished from the neoplastic lesion, which is the chordoma. In 1856, Hubert von Luschka (1820–1875), a pathologist of the German School of Pathology, first described the ectopic notochordal tissue at the posterior *clivus*, but Rudolf Virchow coined the term "*chordomata*," and "*ecchondrosis physalifora spheno-occipitalis*," differentiating between neoplasm and hamartomatous lesion. Virchow used the word "physaliphora" (physaliphorous or "having bubbles") in reference to the numerous cytoplasmic vacuoles observed during his microscopic examination of the tissue (Lushka 1857; Virchow 1857; Ho 1985; Rotondo et al. 2007; Beccaria et al. 2015; Sahyouni et al. 2018).
- *EPI*: Any age, but most often around 40s including adolescence and youth, ♂:♀ = 3:1.
- *CLI*: Specific sites 50% sacrococcygeal (elderly), 35% spheno-occipital (children), 15% cervico-thoraco-lumbar spine, usually youth and elderly, but it may occur at earlier age.
- *IMG:* Inter- and intravertebral (discs, bodies) and sacrum-located, solitary, large, central, expansile radiolucency with slow destruction of bone and compression of the nearest structures.
- *GRO*: Soft, gelatinous tumor with areas of hemorrhage and ± bone/cartilage.
- *CLM*: Cords and lobules of physaliferous cells (Gr. φυσαλλίς "bubbles, vacuoles" and φόρος "bearing", φέρω "to bear, to carry") with vesicular nuclei and characteristic bub-

bly cytoplasm separated by a mucoid stroma, "pink cells" characterized by ample, homogeneous deeply pink cytoplasm, devoid of vacuoles or feathery appearance, syncytial arrangement, low MI, no/mild pleomorphism. The chordoma characteristically replaces the space between the bony trabecula as a vacuolated lesion resembling "sweet cheese."
- *HSS/IHC*: Epithelial (AE1–3, EMA) and mesenchymal (VIM, S100, Brachyury) markers of histologic differentiation, CEA. Brachyury is a transcription factor. It is encoded by T, a member of the T-box gene family. Brachyuria (BRA) is required for the formation and differentiation of the posterior mesoderm and for notochord development during early embryogenesis (Barresi et al. 2014).
- *TEM*: Mitochondrial-ER complexes.
- *DDX*: Chondrosarcoma (pelvis & shoulder girdles, D2–40+, EMA+, AE1–AE3-, GFAP-), chordoid meningioma (AE1–AE3-, S100-), myxopapillary ependymoma (AE1–AE3+, S100+, GFAP+, brachyury-), CCRCC (CK7-, AMACR-, CD10+, PAX2+, PAX8+, S100-, brachyury-), and SRCC of rectum (CK7-, CK20+, CDX2+, Mucin+, S100-, brachyury-).

Variant: Chondroid chordoma (spheno-occipital chordoma with abundant cartilaginous component, focal epithelial markers – differentiated and a better prognosis than ordinary chordoma). To remember is that the chordoma never shows an acinar or glandular appearance. This finding helps in the differential diagnosis with carcinomas.

When large cells with bubbly cytoplasm in the appropriate location are recognized, the diagnosis of chordoma should be raised, although the pathologist should also keep in mind that clear cell appearing and papillary tumors (e.g., CCRCC and myxopapillary ependymoma) are also in the differential diagnosis.

- *PGN*: Recurrence common, late metastases (skin, lungs).

13.9.2 Adamantinoma

- *DEF*: Low-grade, diaphysis/metaphysis-based malignant neoplasm of mostly tibia/fibula with characteristic epithelial differentiation and a jaw-located counterpart referred as ameloblastoma that must be carefully differentiated from fibrous dysplasia, chondrosarcoma, and metastatic carcinoma.
- *EPI*: 0.1% of primary bone tumors, 2nd–4th decades, ♂:♀ = 1.7:1.
- *ETP*: In respect to the histogenesis of adamantinoma, the EM, IHC, and cytogenetics proved the epithelial cell origin of the tumor is ruling out angioblastic and synovial cell origin.
- *CLI*: Tibia, but also femur, ulna, and fibula are the main sites. Symptomatology is described as slight to aching pain, swelling, and deformation of the bone for a long period with a positive history of trauma including fracture before the development of the tumor.
- *IMG*: Lytic lesion, poorly defined with a bubbly expansile pattern with or without marked sclerosis demarcating single or multiple lucent areas.
- *GRO*: It is a solid lesion with defects filled with yellowish-brownish firm fibrous tissue.
- *CLM*: Groups of epithelial cells embedded in a fibrous background in a variable microscopic pattern including fibrous dysplasia-like spindle cells surrounding groups of basaloid epithelial cells showing peripheral palisading and occasional squamoid or tubular formation. Based on its histological appearance, Dorfman and Czerniak and others described various histological subtypes/pattern in adamantinomas, like spindle cell, basaloid, tubular (glandular/vascular), squamous, and osteofibrous dysplasia-like forms.
- *IHC*: "KEV (+)": (+) AE1–3/KER (+CK14 and CK19, -CK8 and CK18), EMA, VIM.
- *DDX*: OFD, chondrosarcoma (AE1–3/KER negative), and metastatic carcinoma (IHC).
- *TRT*: Adamantinoma is known to be resistant for both radio- and chemotherapy, and there is no agreement about the optimal therapy. Curettage and spongiosa plastic are not recommended because of the common recur-

rences. The often-suggested amputation, emphasizing the multicentric nature of adamantinoma, is not shared favorably by many centers. Extensive resection in case of a primary tumor, and amputation upon recurrence, have been suggested, but others reported good results after hemicortical resection only. In the Department of Orthopedics, Semmelweis University, wide segment resection is the therapy of choice for both primary tumor and recurrence.

- *PGN*: Low-grade malignant tumor with a tendency to recur locally and occasional late metastases into the lymph nodes and lungs in 15% of the patients (*DDX*, synovial sarcoma). Adverse prognostic factors for recurrences are young age, male sex, history of pain, and aggressive growth, but there is no agreement on prognosis.

13.9.3 Langerhans Cell Histiocytosis

- *DEF*: Inflammatory neoplasm of myeloid precursor cells driven by mutations in the mitogen-activated protein kinase (MAPK) pathway with the diagnosis based on the combination of clinical presentation, histology, and immunohistochemistry (Fig. 13.22).
- *EPI*: 1–3 years, white, ♂:♀ = 2:1.
- *EPG*: Inflammatory myeloid neoplasia driven by activating mutations in the MAPK pathway.
- *CLI*: LCH is characterized by the extent of involvement (single vs. multisystem) and the presence of risk organ involvement (e.g., liver, spleen, or bone marrow).

Histiocytoses are subdivided in several categories:

- *Langerhans cell histiocytosis*
  Erdheim-Chester disease/extracutaneous juvenile Xanthogranuloma.
  Indeterminate cell histiocytosis.
- *Cutaneous and mucocutaneous*
  Juvenile/adult xanthogranuloma, necrobiotic xanthogranuloma, benign cephalic histiocytosis, generalized eruptive histiocytosis, progressive nodular histiocytosis, *xanthoma disseminatum*, solitary reticulohistiocytoma, multicentric reticulohistiocytosis, cutaneous Rosai-Dorfman disease, and cutaneous histiocytoses not otherwise specified.
- *Malignant*
  Histiocytic sarcoma, indeterminate cell sarcoma, Langerhans cell sarcoma, and follicular dendritic cell sarcoma.
- *Rosai-Dorfman disease*
  Rosai-Dorfman disease and non-cutaneous, non-LCH not otherwise specified.
- *Hemophagocytic lymphohistiocytosis and macrophage activation syndrome*
- *IMG*: Lytic lesion which can sharply demarcate or having sclerotic edges. The lesions can extend into soft tissue.
- *GRO*: Soft lesion, dusky with no apparent inhomogeneity.
- *CLM*: There is an inflammatory infiltrate of eosinophils, MΦ, regulatory T lymphocytes (FoxP31 and CD41), and MNGC. LCH cells have a "coffee-bean" cleaved nuclei, rounded shape, and eosinophilic cytoplasm. The MI is low to moderate, and binucleate cells may occasionally be seen. Atypical mitosis and pleomorphism are usually not identified and the detection of such findings should raise suspicion toward an alternative diagnosis (e.g., Langerhans cell sarcoma).
- *IHC*: (+) S100, CD1a, CD207 (Langerin). S100 staining is nuclear and cytoplasmic, while CD1a is expressed mainly on the cell membrane. Conversely, Langerin staining occurs on Golgi apparatus other than the cell membrane of positive cells. Langerin is a Langerhans cell-specific lectin that starts the formation of Birbeck granules and its full expression has been characterized on dendritic cells of the human epidermis, mucosa, lymphoid tissue as well as normal lung.
- *TEM*: Birbeck granules (elongated tennis racket-like electron dense cross striations).
- *TRT*: Treatment of cutaneous single-system LCH relies on topical steroids as first-line therapy for lesions few in quantity, although systemic steroids with vinblastine for

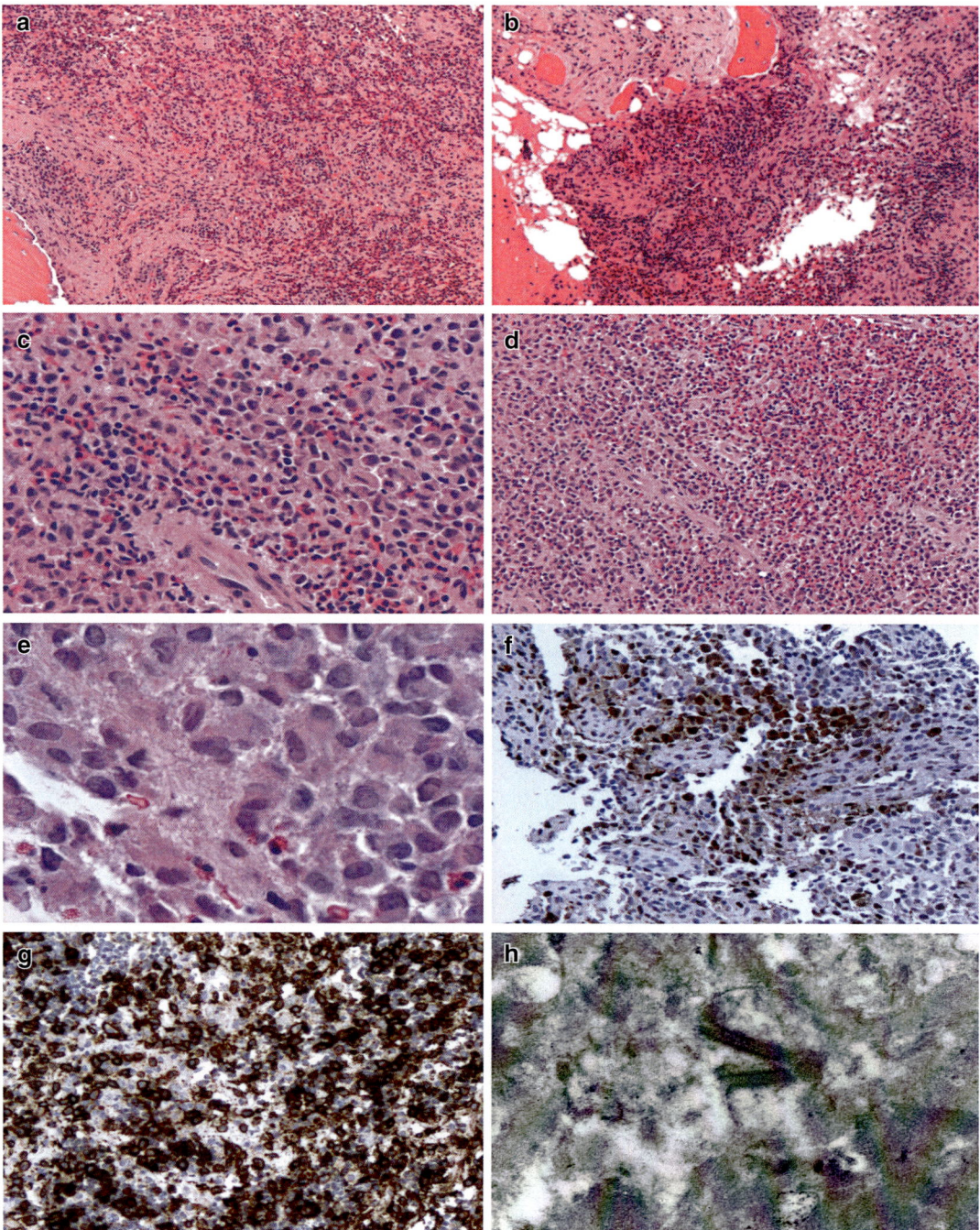

Fig. 13.22 Langerhans Cell Histiocytosis. The histological images (**a–e**) show an infiltration by Langerhans cells. These cells are recognized in microscopic images because they are polygonal with eosinophilic cytoplasm, oval nuclei with longitudinal grooves similar to coffee beans. Also, in the histology, eosinophils, giant cells, foamy cells, and other chronic inflammatory cells may be seen. Mitotic activity can be present, although not brick and focal atypias may be seen (**a–e**, hematoxylin and eosin staining, ×100, ×100, ×400, ×200, ×630 as original magnification, respectively). The immunohistochemistry (**f, g**) shows a positivity of this tumor for S100 and CD1a (**f**, anti-S100 immunostaining, Avidin-Biotin Complex, ×200 original magnification; **g**, anti-CD1a immunostaining, Avidin-Biotin Complex, ×200 original magnification), and the photograph in (**h**) shows the characteristic Birbeck granules using transmission electron microscopy

12 months are first-line therapy for the diffuse disease. Surgical resection of single LCH lesions for diagnostic purposes is usually sufficient for cure. In case of multisystemic disease, 12 months of vinblastine/prednisone as first-line therapy should be started. On the other hand, patients with the low-risk disease have been successfully cured by application of other single drugs or drug combinations (e.g., cladribine, cytarabine, and clofarabine). Options for LCH patients with high-risk disease are a combination of cladribine and cytarabine or a hematopoietic stem cell transplantation.

- *PGN*: The distinction between single-system and multisystem LCH is crucial for the determination of the outcome and treatment. LCH-I has 5-YSR of 62% and 5-YRR of 55%, LCH-II has a 5-YSR of 69% and 5-YRR of 44%, while LCH-III has a 5-YSR of 84% and 5-YRR of 27%. The most frequent permanent consequences include diabetes insipidus, hormonal deficiencies of the anterior pituitary gland, orthopedic disturbances, hearing loss, and progressive neurodegeneration.

13.9.4 Vascular, Smooth Muscle, and Lipogenic Tumors

This group is a collection of entities of quite various and different courses and constituted by hemangioma, hemangiopericytoma, epithelioid hemangioendothelioma, and angiosarcoma. Tumors showing a hematopoietic (plasma cell myeloma, malignant lymphoma), vascular (angiosarcoma), smooth muscle (leiomyosarcoma) or lipogenic (liposarcoma) histogenesis are extremely rare or inexistent at the pediatric age or young adults, while they can be seen in the middle age and elderly apart probably of the benign fibrous histiocytoma.

Benign Fibrous Histiocytoma: → *Metaphyseal Fibrous Defect* (see 13.6.3)

Desmoplastic Fibroma: Long bones, pelvis or lower jaw-located, lytic and honeycomb on imaging tumor with mature fibroblasts and abundant collagen (recurrences, but no metastases).

Malignant Fibrous Histiocytoma: Long bones or jaw-located, ill-defined, etc. on imaging tumor showing atypical spindle cells arranged in fascicles with storiform pattern and large, bizarre, histiocyte-like cells (predisposing conditions, bone infarcts, foreign bodies, history of radiation therapy, Paget disease of bone).

Fibrosarcoma: Long bones (distal femur or proximal tibia), metaphyseal, lytic tumor with "soap bubble" appearance, extension into soft tissues, and harboring cellular areas ± mitoses, cellular pleomorphism and hyperchromatic nuclei that should prompt an MFH diagnosis if these findings are marked (survival correlated with grade of malignancy).

13.9.5 Hematologic Tumors

This is a group with entities of quite various and different course and constituted by lymphomas. Lymphocytic lymphoma of bone may exist in two forms, including the primary lymphocytic lymphoma and the secondary lymphocytic lymphoma. Both NHL and Hodgkin types of lymphomas are rare constituting less than 0.2% of biopsy-analyzed primary bony tumors. *Plasmacytoma* is a sporadic tumor in childhood and youth but occasionally can occur in skeletal and extraskeletal location (Fig. 13.23).

13.10 Metastatic Bone Tumors

Metastatic bony tumors are rare in pediatrics and youth, although neuroblastoma needs to be mentioned and particularly the stage IVS without bone marrow involvement. In contrast to the adult counterpart, in which malignant bone tumors are 25–30 times more common than primary tumors, pediatric metastatic bone tumor remains a rarity. In the adult site, 80% of metastatic tumors originate from the breast, lung, prostate, thyroid gland, and kidney, and 70% involve the axial skeleton, while 30% involve the extremities. The metastatic bony disease is usually osteolytic in the adults but may be osteoblastic (e.g., breast) or mixed. The vertebral disc is rarely destroyed differently from the osteomyelitis.

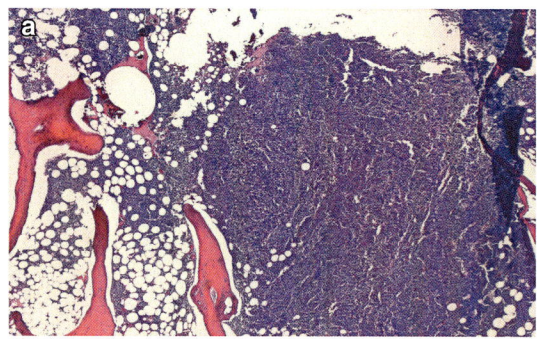
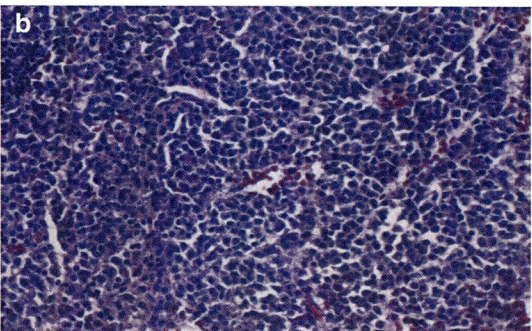
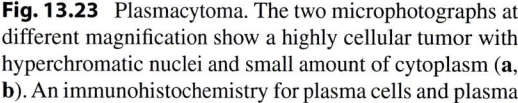

Fig. 13.23 Plasmacytoma. The two microphotographs at different magnification show a highly cellular tumor with hyperchromatic nuclei and small amount of cytoplasm (**a**, **b**). An immunohistochemistry for plasma cells and plasmablasts was positive (not shown) confirming the diagnosis of plasmacytoma, which remains a rarity in childhood and youth (**a–b**, hematoxylin and eosin, ×40 and ×200 as original magnification, respectively).

13.11 Juvenile Rheumatoid Arthritis and Juvenile Arthropathies

The juvenile rheumatoid arthritis and juvenile arthropathies are listed in Box 13.20 (Aletaha et al. 2010).

Small joints include the metacarpophalangeal (MCP), proximal interphalangeal (PIP), 2nd through 5th MCP, and thumb joint IP joints and the wrists. The first carpometacarpal joint, the first MCP, or distal interphalangeal (DIP) are not usually included among the small joints. In fact, these last-mentioned joints are most often affected by osteoarthritis. Large joints are the shoulders, elbows, hips, knees, and ankles.

Chronic inflammatory arthropathies include rheumatoid arthritis, spondyloarthropathies, and other multisystemic rheumatic diseases, including systemic lupus erythematosus (SLE), scleroderma, vasculitis, and others, as well as chronic joint infections.

The diagnosis of arthropathies includes the history (pain, swelling, dysfunction) with details on distribution (monoarthritis, polyarthritis, symmetry) and duration and severity (acute or chronic). The physical examination may include swelling ("*tumor*") variably associated with redness ("*rubor*"), warmth ("*calor*"), pain ("*dolor*"), limitation of motion ("*functio laesa*"), and deformities, as well as the severity of these abnormalities. Another important aspect of the diagnosis of arthropathies includes the radiographic imaging and the joint aspiration. Inflammatory arthropathies are usually associated with increases in joint fluid or effusions. The precise analysis of joint (synovial) fluid may reveal increased numbers of inflammatory cells, bacteria, crystals, and bleeding.

13.11.1 Rheumatoid Arthritis and Juvenile Rheumatoid Arthritis

13.11.1.1 Rheumatoid Arthritis
- *DEF*: Autoimmune disease (HLA-DR1, HLA-DR4, rheumatoid factor or RF+, i.e., IgM autoantibody against the Fc portion of IgG and anti-citrullinated protein antibody or ACPA) of women of 20s–40s characterized by a nonsuppurative proliferation of inflammatory synovium that often progresses to destruction of the articular cartilage, characterized by joint swelling, joint tenderness, and severe disability. Rheumatoid arthritis is associated with dismantling and excruciating morbidity and premature mortality.
- *EPG*: Autoimmunity (Involvement of Th_1, Th_{17}, B cells).

$Th_1 \Rightarrow$ IFN-γ and Th_{17} IL$\Rightarrow$ IL-17 (both IFN-gamma and IL-17 are involved to the upregula-

> **Box 13.20 Juvenile Rheumatoid Arthritis and Juvenile Arthropathies**
> 13.11.1 Rheumatoid Arthritis and Juvenile Rheumatoid Arthritis
> 13.11.2 Infectious Arthritis
> 13.11.3 Gout and Pseudogout
> 13.11.4 Bursitis, Baker Cyst, and Ganglion
> 13.11.5 Pigmented Villonodular Synovitis and Nodular Tenosynovitis

tion of RANKL through IL-1, IL-6, IL-23, TNF, PGE2, NO, GMC-SF, and TGF-β.)
Genetics (HLA-DRB1, *PTPN22*)
Environment (bacteria, viruses, ACPA, or anti-CCP in smokers)
(1) → (6) Pathogenetic Sequence

(1) Lymphocytic infiltration (CD_4Th1/Th17, B cells, PC, DC, MΦ)
(2) Vasodilation and angiogenesis
(3) Fibrin organization and "rice bodies"
(4) Neutrophilic recruitment
(5) Osteoclastic activity (RANKL!)
(6) Pannus (Latin *pannus* "cloth") (skeletal tissue destruction, fibrosis, ankylosis) (Greek ἀγκύλωσις "joint stiffening", ἀγκυλόειν "to bend", ἀγκύλος "bent")

Initial symmetric involvement of small joints of hands and feet is characteristic of this disease. Subsequently, wrists, elbows, knees are involved. The initial classification of RA exposed in 1987 was repeatedly suffering from the lack of sensitivity to detect the early disease of this joint disease. This situation may be catastrophic especially for adolescents and youth. Thus, a joint working group from the American Rheumatism Association (ARA) and the European League Against Rheumatism (ELAR) came together with a new approach to classify RA. In the new criteria set, which define or better classify RA, the diagnosis of RA relies on the confirmation of synovitis in one or more than one joint, the lack of an alternative diagnosis to explain the above confirmed synovitis, and the realization of a total score of ≥6 out of 10 col-

lected in four domains. They include the number and site of involved joints, serologic abnormality, ↑ of an acute-phase response pattern (C-reactive protein, CRP, and erythrocyte sedimentation rate, ESR), and duration of the symptoms.

- *CLM*: Acute inflammation with neutrophils, which proceeds to a chronic phase characterized by marked proliferation of the synovium, lymphocytes, and plasma cells forming the pannus (granulation tissue + inflamed synovium) and accompanied by obvious chondro-lysis and rheumatoid nodules (1/4 of cases) showing palisading histiocytes surrounding an irregular area of fibrinoid necrosis (Fig. 13.24).
- *DDX*: The differential diagnoses of rheumatoid arthritis are in Box 13.21.
- *TRT*: The finest use of disease-modifying antirheumatic drugs (DMARDs), particularly methotrexate, and the availability of new biologic agents have enhanced the management of this chronic inflammatory disease.
- *PGN*: An early therapeutic intervention is crucial in improving the clinical outcome by reducing the accrual of joint damage and disability of the affected individuals.

13.11.1.2 Juvenile Rheumatoid Arthritis

- *DEF*: Autoimmune disease with acute joint involvement, limited to one or few joints, most often involving the knees and usually preceded by a febrile illness with generalized lymphadenopathy and hepatosplenomegaly.
- *SYN*: Still disease.
- *EPI*: ~8/10,000 (prevalence); ♂:♀ = 1:2.
- *EPG*: Early-Onset Pauci-Articular (EOPA) JRA is linked with: HLA-A2, -DR5, -DR8, -DPB1∗0201 and HLA-DR4 is protective, while late pauciarticular JRA is connected with HLA-B27 The polyarticular type of JRA shows an association with HLA-DR1, and HLA-DR4.
- *CLI*: Pauciarticular: early childhood with a peak at 1–2 years of age, while systemic JRA: throughout childhood without peak
- *DDX*: See DDX of rheumatoid arthritis.

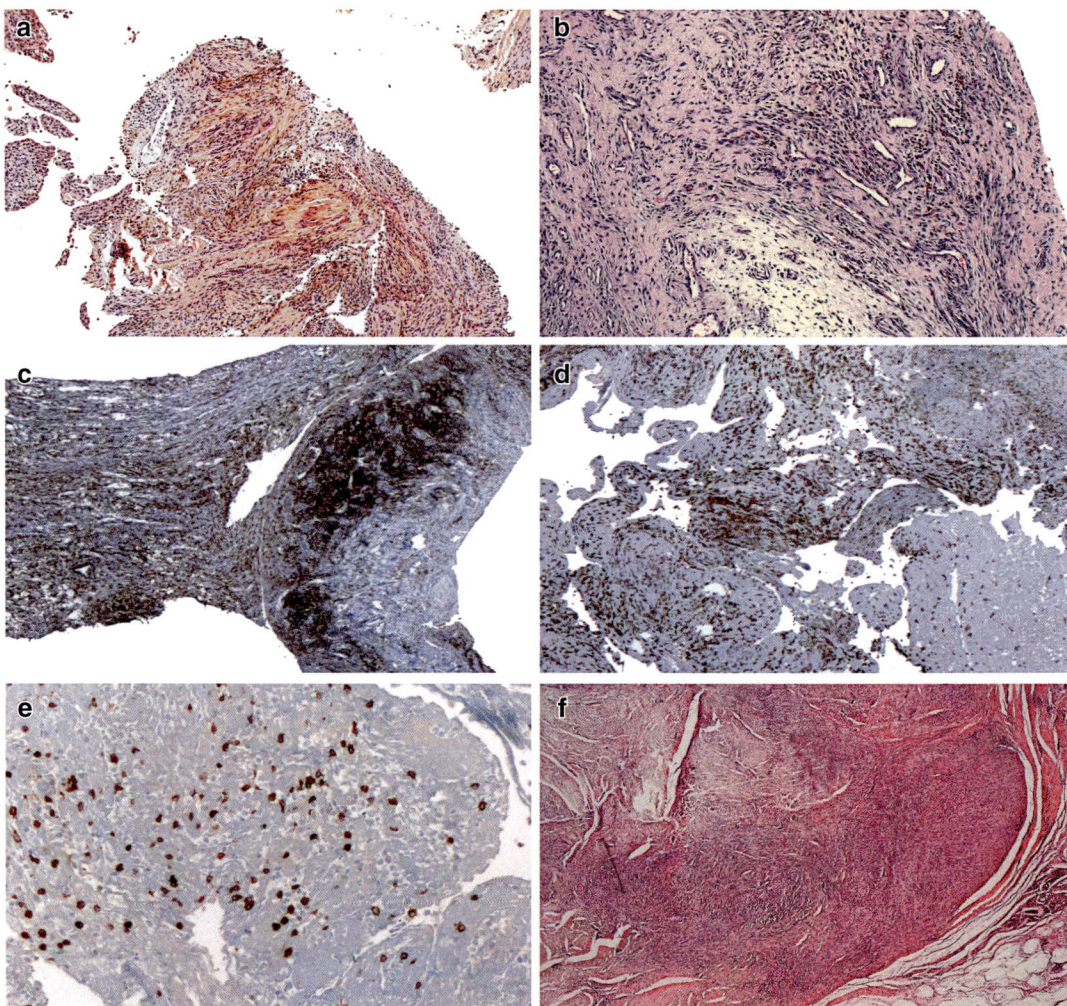

Fig. 13.24 The joint pathology panel displays two histological images of pigmented villonodular synovitis (**a**, ×40 original magnification; **b**, ×100 original magnification) with evidence hemosiderin deposits. The immunohistochemical staining of **c** and **d** (both at ×50 as original magnification) shows the detection of CD4 and CD8 lymphocytes in a 3-year-old child with early synovitis and polyarthropathy with suspicion of juvenile rheumatoid arthritis (positive serology). The CD3 immunostaining of this sample is shown on Figure **e**. Figure **f** shows an early Dupuytren disease with a nodular proliferation of fibroblasts, myofibroblasts, and fibrocytes (hematoxylin and eosin staining, ×100 as original magnification)

13.11.2 Infectious Arthritis

- *DEF*: Infection of joints with pyogenic bacteria.
- *SYN*: Septic arthritis.
- *ETP*: Gram-positive cocci (*Staphylococcus aureus*, *Streptococcus pyogenes*, *Streptococcus pneumoniae*, and *Streptococcus viridans* group), Gram-negative cocci (*Neisseria gonorrheae*, *Neisseria meningitidis*, and *Haemophilus influenzae*), Gram-negative bacilli (*Escherichia coli*, *Salmonella spp.*, and *Pseudomonas spp.*), and mycobacteria and fungi. If Gram-positive cocci are the culprit in about 1/3 of cases in both adults and children, Gram-negative cocci play a role differently in adults and children. *N. gonorrhoeae* and *N. meningitidis* play a role in adults (1/2 of infectious arthritis), and *H. influenzae* is responsible for 40% of the pediatric infectious arthritis.

> **Box 13.21 DDX of Rheumatoid Arthritis**
> - Lyme disease (*Borrelia spp.*)
> - Osteoarthritis
> - Seronegative spondyloarthropathies (ankylosing spondyloarthritis, Reiter arthritis, enteritis-associated arthritis, and psoriatic arthritis)
> - Syphilis (*T. pallidum*)
>
> Notes: In North America, *Borrelia burgdorferi* and *Borrelia mayonii* are the cause of Lyme disease, while in Europe and Asia, the bacteria *Borrelia afzelii* and *Borrelia garinii* are additional etiologic species of Lyme disease.

13.11.3 Gout, Early-Onset Juvenile Tophaceous Gout and Pseudogout

13.11.3.1 Gout

- *DEF*: Crystal-induced arthropathy/arthritis.
- *EPI*: Risk factors for gout are age >30 years, family history positive for gout, alcohol use, obesity, thiazide diuretics administration, and chronic lead intoxication.
- *EPG*: It is caused by a supersaturation of the extracellular fluids concerning monosodium urate. The supersaturation means crystals formation. These crystals eventually induce an acute inflammatory reaction following their ingestion by neutrophils. Apart from the acute inflammation, chronic inflammation also leads to tissue destruction around deposits on monosodium urate crystals, which are called tophi (Plural of *tophus*, the Latin name of stone). At physiologic pH, uric acid is in the monoanionic form. Monosodium urate precipitates when the total urate concentration exceeds 6.5 or 7 mg/100 ml. In case of serum urate >6.5–7 mg/dL ⇒ deposition of monosodium urate (MSU) crystals in joints and viscera and uric acid-based nephrolithiasis with the interstitial renal disease. Hyperuricemia usually occurs because of relatively inefficient excretion of both kidneys. Of note, diuretics and β-blockers in high doses can worsen insulin resistance and atherogenic dyslipidaemia, which may aggravate the physical conditions of subjects with hyperuricemia.
- *CLI*: Acute attacks (acute monoarthritis involvement) are characterized by the subsiding of the inflammatory process after 1–2 weeks if untreated or sooner if treated. There is a recurrence of acute attacks with intervals of weeks to months if no prophylactic treatment is started. Finally, more frequent attacks make the process becoming continuous and persistent with progressive and inexorable bony tissue destruction.
- *IMG*: X-ray changes consistent with longstanding inflammatory arthritis including marked cartilage loss, erosive bony contour changes, and articular subluxation involving multiple joints.
- *CLM*: Uniaxial birefringent materials are classified as either positively or negatively birefringent when, for light (having parallel and perpendicular components) directed to the optic axis, the refractive index of light polarized parallel to the optic axis is either greater or smaller, respectively, than light polarized perpendicularly to the optic axis. Thus, the best way to diagnose gout is by visualization of crystals in a joint aspirate. Gout (monosodium urate) crystals have a needle-shaped appearance under light microscopy. By applying a polarizing light microscope, the needle-shaped crystals of gout are strongly negatively birefringent. This situation means that the crystal appears bright *yellow* when the compensator portion of the microscope is positioned *parallel* to the long axis of the crystal, while the crystals appear bright blue when the compensator is curved perpendicular to the crystal. In case of pseudogout (or calcium pyrophosphate), crystals are rhomboid-shaped and weakly positively birefringent. This situation means that when the compensator is right parallel to the long axis of a pseudogout crystal, the crystals give a faint blue color. When the compensator is right perpendicular to the crystal, the crystal is faintly yellow.

- *TRT*: NSAIDs (cyclooxygenase inhibitors), colchicine aiming to inhibit microtubule function and the phagocytosis of crystals, and glucocorticosteroids with different power that have multiple anti-inflammatory effects are used. The prophylactic treatment of gout seeks to reduce the levels of urate below the solubility of sodium urate. Probenecid enhances the excretion of uric acid by the kidney and allopurinol, which is a xanthine oxidase inhibitor replaces some uric acid with xanthine and hypoxanthine. Both components are more soluble than uric acid.
- *PGN*: Patients with gout exhibit an unusually higher prevalence and more components of metabolic syndrome than age-matched peers. Also, gout or raised serum urate levels may independently increase the risk of nonalcoholic fatty liver disease (NAFLD) and chronic kidney disease. Hyperuricemia and the presence of subcutaneous tophi are reported to increase the risk of mortality in gout patients. This risk is commonly attributed to cardiovascular causes.

13.11.3.2 Early-Onset Juvenile (Tophaceous) Gout (EOJG)

- *DEF*: Although gout onset has shifted to early ages in recent decades, it remains rare before the age of 20 years and is referred to as the early onset of gout in juveniles. Renal involvement in classic gout typically occurs late in the disease process and manifests as urate nephropathy or nephrolithiasis. The combination of hyperuricemia, early-onset renal failure, and EOJG is characteristic of a rare AD-inherited disorder labeled familial juvenile hyperuricemic nephropathy (FJHN), which has been considered synonymous with EOJG by some authors.
- *SYN*: Familial juvenile hyperuricemic nephropathy (*vide supra*).
- *ETP*: Tamm-Horsfall glycoprotein (THP) or uromodulin (UMOD) is the most abundant protein found in normal urine. UMOD is synthesized in the tubular cells of the thick ascending loop of Henle. The *UMOD* gene is found on 16p2.3, which is the short arm of chromosome 16 at position 12.3. There is a substantial genetic heterogeneity as 60 mutations have been reported to contribute to the disease, the majority of which are localized on exon 3 or exon 4. These mutations consist of missense changes or small in-frame deletions. In addition to having genetic variations on the *UMOD*, rare mutations in genes encoding for renin (REN) and hepatocyte nuclear factor 1β (HNF-1β) have been reported in FJHN. Most of the mutations causing UMOD-associated kidney disease may involve addition or deletion of a cysteine residue, changes that alter disulfide bond formation tampering with the process of normal protein folding. The precipitation of the abnormal protein in the endoplasmic reticulum of the tubular cell is the consequence of it. The reduced secretion of UMOD causes early cell death of the tubular cells, which on biopsy is shown as tubular atrophy and abnormal uromodulin deposition in tubular cells morphologically. Lack of uromodulin function induces impairment in tubular function, particularly the urine-concentrating process. The decrease of sodium reabsorption in the ascending loop of Henle is subsequently compensated for by an increase in sodium reabsorption in the proximal tubule, causing a secondary increase in the uric acid reabsorption and the hyperuricemia of FJHN ultimately. In addition to *UMOD*, multiple genetic and clinical factors have been reported for identifying high risks of developing gout. The 869 T/C polymorphism in the *TGF-β* gene was found to be associated with the occurrence of EOJG with tophi.
- *CLI*: Patients with EOJG with tophi exhibit higher serum urate levels and more comorbidities than patients without tophi. In contrast to the middle-age-onset patients, the EOJG patients experience the first gout attack at the ankle, followed by the toe. Regarding the first tophi, the most common site is a finger among the EOJG, different from the middle-age-onset gout patients.
- *IMG*: See gout.
- *CLM*: See gout.
- *DDX*: An enzymatic deficiency of the purine metabolism is often at the basis of the occurrence of hyperuricemia leading to gout in younger populations. Such purine metabolism

deficits are usually hypoxanthine-guanine phosphoribosyltransferase deficiency and phosphoribosyl-1-pyrophosphate synthetase overactivity. Also, mutations in the uromodulin (THP) gene have been found associated with EOJG. Major arthritogenic crystals include monosodium urate (MSU) monohydrate, calcium pyrophosphate dihydrate (CPPD), calcium hydroxyapatite (CHA), corticosteroid esters, calcium oxalate (CaOx), and cholesterol.
- *TRT*: Urate-lowering drugs.

13.11.3.3 Pseudogout

The pseudogout, which is more specifically labeled calcium pyrophosphate deposition (CPPD) disease, is a type of arthritis that causes spontaneous, painful swelling in your joints. CPPD happens when crystals form in the synovial fluid, and the friction of the crystals with the articular structures determines inflammation and pain.

13.11.4 Bursitis, Baker Cyst, and Ganglion

13.11.4.1 Bursitis

It is defined as an inflammation of ≥1 bursae which are anatomic structures located between muscles, tendons, and bony prominences. Bursitis presents with the Galeno's inflammatory characteristics of the *tumor* (swelling), *calor* (warmth), *rubor* (redness), and *dolor* (painful) as well as joint-related inefficiency ("functio laesa"). The bursitis is usually associated with chronic trauma and impossibility of a good joint function ("*functio laesa*"). Bursitis often develops in the shoulders of professional athletes but it is not uncommon in teenagers prone to excessive athletic training.

13.11.4.2 Baker Cyst

It is defined as a fluid-filled synovial cyst that causes a lump at the posterior face of the knee leading to tightness and restricted mobility of the involved lower limb. It can be painful when the extremity is bent or extended. It is due to a herniation of the synovial membrane through the posterior joint capsule, alternatively through escape of synovial fluid from the bursae. Baker cyst is a true cyst and is associated with degenerative joint disease, neuropathic arthropathy, and rheumatoid arthritis.

13.11.4.3 Ganglion

- *DEF*: Fibrous walled pseudocyst (*by definition cyst has an epithelial lining*) containing clear mucinous fluid and, usually, arising on the extensor surfaces in the proximity of the joints of the wrists, hands, and feet.
- *EPG*: It is a mucinous degeneration of the dense fibrous tissue of the joint capsule or tendon sheath, which may be associated with trauma. Although rarely ganglia can communicate with the joint cavity and, occasionally, can become intraosseous following erosion of the adjacent bone.
- *CLM*: The pseudocyst wall is fibrous. In fact, the use of AE1–3 (a pan-keratin immunohistochemical marker), for epithelial lining, CD31 (an immunohistochemical marker for endothelial lining), and D2–40 (an immunohistochemical marker for lymphatic structures) are negative.
- *DDX*: Both the absence of communication of the ganglion with the joint cavity and the lack of a synovial cell lining help distinguishing it from popliteal (Baker) cyst, which is usually a result of arthritis and synovial effusion and represents an enlargement of one or more bursae communicating with the knee joint.
- *TRT*: Excision is usually curative.
- *PGN*: Good.

13.11.5 Pigmented Villonodular Synovitis and Nodular Tenosynovitis

13.11.5.1 Pigmented Villonodular Synovitis (PVNS)

- *DEF*: Villous and/or nodular lesion arising in the synovial lining of the knee or other synovial joints of unknown etiology, asymptomatic to slightly painful, presenting with effusion or a local swelling.
- *GRO*: Multiple sessile or pedunculated tender nodules, which are tan or brown discolored.

- *CLM*: Fibrous nodules, which are associated with villous hypertrophy of the synovial membrane and harboring hemosiderin deposition or hemorrhage (Fig. 13.25).

13.11.5.2 Nodular Tenosynovitis (NTS)

- *DEF*: Solitary, well-circumscribed, tannish, or yellowish discolored (hemorrhage or lipid deposition accordingly) nodule of the flexor tendon sheath of the hand.
- *SYN*: Giant cell tumor of the tendon sheath.
- *EPG*: The inflammatory theory currently prevails over the neoplastic theory.
- *CLI*: Asymptomatic or mildly painful presenting as a local swelling differently from the PVNS of the knee presenting as an effusion.
- *GRO*: Multiple, sessile or pedunculated, fibrous nodules that exhibit either a tan discoloration if an intralesional hemorrhage occurs or a yellow discoloration if an intralesional lipid accumulation occurs.
- *CLM*: Villous or nodular proliferation of fibroblasts or fibroblast-like cells producing collagen, scattered MNGCs, hemosiderin-laden MΦ or lipid-laden MΦ, and chronic inflammatory cells (lymphocytes and plasma cells).

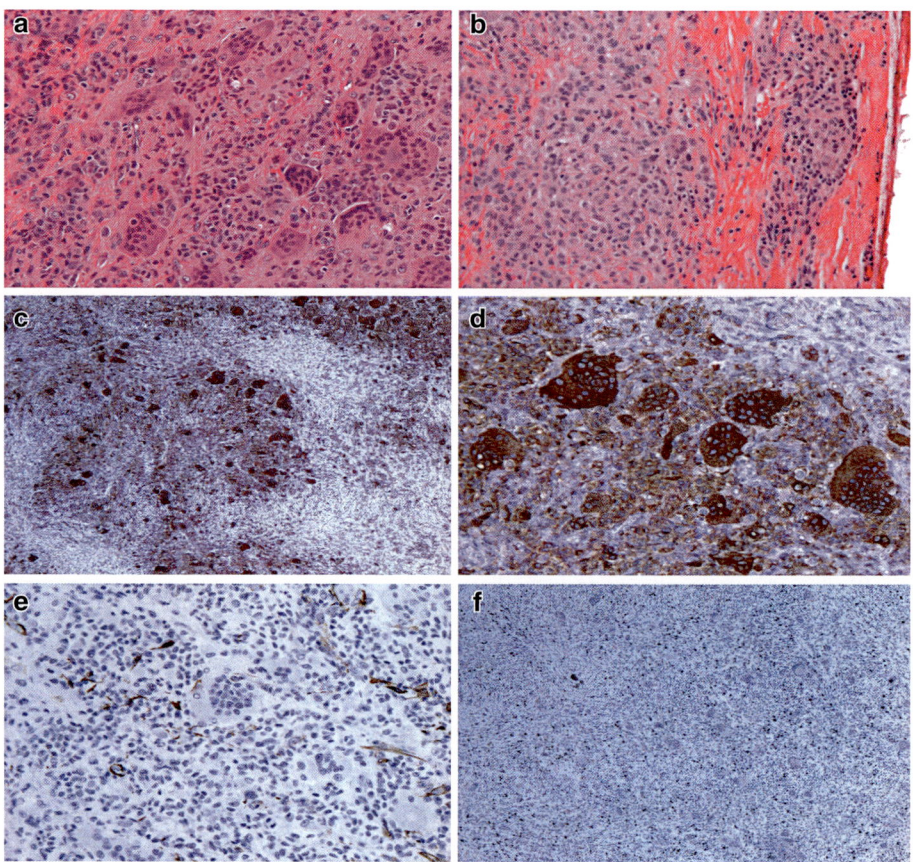

Fig. 13.25 The joint pathology panel displays two histological images (hematoxylin and eosin) of giant cell tumor of the tendon sheath (**a–b**, ×200 as original magnification). The immunostaining (avidin-biotin complex) with an antibody against CD68 highlights the macrophages and the giant cells (**c**, ×50 original magnification; **d**, ×200 as original magnification). The immunostaining (avidin-biotin complex) with an antibody against actin (**e**, ×200 as original magnification), and Ki67 (**f**, ×50 as original magnification) emphasizes the few fibroblasts expressing actin and the low proliferation rate of this tumor

Multiple Choice Questions and Answers

- ASS-1 Which of the following statement is NOT true for thanatophoric dysplasia?
  - (a) Thanatophoric dysplasia is a severe skeletal osteochondrodysplasia characterized by extremely short limbs, folds of redundant skin on the limbs, narrow chest, short ribs, underdeveloped lungs, and an enlarged head with a large forehead and prominent, wide-spaced eyes.
  - (b) Two major forms of thanatophoric dysplasia, type I and type II, have been delineated.
  - (c) It occurs in 1 in 20,000 to 50,000 newborns.
  - (d) Mutations in the *FGFR3* gene cause thanatophoric dysplasia.
  - (e) Thanatophoric dysplasia is inherited with an autosomal recessive pattern.
- ASS-2 What is the gene mutated in fibrous dysplasia?
  - (a) *NF1*
  - (b) *NF2*
  - (c) *PAX7*
  - (d) *GNAS1*
  - (e) *SOX1*
- ASS-3 A 9-year-old boy presents with leg pain near his knee. The child is examined by the pediatrician, who orders some imaging. The roentgenogram demonstrates a 3 cm lytic lesion in the epiphysis of the distal femur. The following biopsy shows pseudo-lobulated sheets of ovoid cells with occasional nuclear grooves, "chicken-wire" calcifications, intermixed giant cells, and fibro-chondroid islands. The tumor cells are positive for vimentin and S100. There is also SOX9 expression. What is the best treatment for such lesion?
  - (a) Curettage and bone graft
  - (b) Curettage with regular radiographs
  - (c) Observation with curettage for symptomatic patients
  - (d) Wide surgical excision with close follow-up
  - (e) Wide surgical resection with prognosis depending on tumor size and margins
- ASS-4 Which of the following genes/syndromes do NOT belong to the osteosarcoma etiopathogenesis?
  - (a) Familial Li-Fraumeni syndrome (germline TP53 inactivation)
  - (b) Hereditary retinoblastoma (germline *RB1* inactivation)
  - (c) Rothmund-Thomson syndrome (germline RECQL4 inactivation)
  - (d) Bloom syndrome (germline *BLM* inactivation)
  - (e) Werner syndrome (germline WRN inactivation)
  - (f) Familial adenomatous polyposis (germline *APC* inactivation)
- ASS-5 Which subtype of osteosarcoma has prognostic relevance?
  - (a) Telangiectatic osteosarcoma
  - (b) Osteoblastic osteosarcoma
  - (c) Chondroblastic osteosarcoma
  - (d) Fibroblastic osteosarcoma
  - (e) Small cell osteosarcoma
- ASS-6 A 17-year-old child undergoes amputation following post-chemotherapy osteosarcoma. According to your assessment of the following photograph, on which ground would you be able to make the diagnosis of residual tumor?
  - (a) Mitoses and pleomorphism
  - (b) Necrosis
  - (c) Apoptosis
  - (d) Macromegaly
  - (e) Osteoid formation
- ASS-7 Which statement on Ewing sarcoma is NOT correct?
  - (a) The tumor cells are arranged in a vasculocentric pattern and form pseudorosettes as well as true Homer-Wright rosettes.
  - (b) Intracytoplasmic glycogen may be identified by periodic acid-Schiff staining.
  - (c) The presence of mesenchymal elements is occasionally reported.
  - (d) The immunohistochemistry with CD99 and FLI1 is contributory for the diagnosis.
  - (e) Two major genetic alterations (fusion genes) have been identified in Ewing sarcoma, including *EWSR1-FLI1* [t(11;22)

(q24;q12)] and *EWSR1-ERG* [t(21;22)(q21q12)].
(f) The overall cure rate of Ewing sarcoma is less than 25%.

- ASS-8 A 14-year-old girl presents to her pediatrician complaining of monoarticular arthritis following a peculiar expanding circular red skin rash with central clearing months ago, which was also preceded by flu-like symptoms. She was wandering in the Rocky Mountains of Canada. Clinical laboratory findings include a positive titer against *Borrelia burgdorferi*. Which diagnosis is the most likely diagnosis in this girl?
    (a) Rocky Mountain spotted fever
    (b) Leptospirosis
    (c) Lyme disease
    (d) Tularemia
    (e) Rheumatoid arthritis
- ASS-9 Match each of the following findings with the proper diagnosis of arthritis subtype.
    (a) Cloudy, hypercellular synovial fluid with crystals that are negatively birefringent in a patient with deficiency of hypoxanthine-guanine phosphoribosyl transferase
    (b) Invasive pannus eroding the cartilage identified at a synovial biopsy and human leukocyte antigen (HLA) DR4
    (c) Polyarthritis following a maculopapular erythematous rash and history of a tick bite
    (d) Seronegative arthritis after an episode of *Chlamydia trachomatis* urethritis and HLA-B27
    (e) Cloudy synovial fluid with 75 x 10^9 /L (98% neutrophils) drained from a warm and tender joint
        1. Rheumatoid arthritis
        2. Reiter syndrome
        3. Gout
        4. Septic arthritis
        5. Lyme disease
- ASS-10 Which morphologic criteria are at the basis for diagnosing a pigmented villonodular synovitis?
    (a) Villous architecture with osteoclast-type giant cells and xanthoma cells
    (b) Lobular architecture with xanthoma cells and iron-laden macrophages
    (c) Villous architecture with xanthoma cells and iron-laden macrophages
    (d) Villous architecture with hemosiderin pigment inside osteoclast-type giant cells

References and Recommended Readings

Al-Bahrani R, Nagamori S, Leng R, Petryk A, Sergi C. Differential Expression of Sonic Hedgehog Protein in Human Hepatocellular Carcinoma and Intrahepatic Cholangiocarcinoma. Pathol Oncol Res. 2015;21(4):901–8. https://doi.org/10.1007/s12253-015-9918-7. Epub 2015 Mar 5. PubMed PMID: 25740074.

Aletaha D, Neogi T, Silman AJ, Funovits J, Felson DT, Bingham CO 3rd, Birnbaum NS, Burmester GR, Bykerk VP, Cohen MD, Combe B, Costenbader KH, Dougados M, Emery P, Ferraccioli G, Hazes JM, Hobbs K, Huizinga TW, Kavanaugh A, Kay J, Kvien TK, Laing T, Mease P, Ménard HA, Moreland LW, Naden RL, Pincus T, Smolen JS, Stanislawska-Biernat E, Symmons D, Tak PP, Upchurch KS, Vencovský J, Wolfe F, Hawker G. 2010 Rheumatoid arthritis classification criteria: an American College of Rheumatology/European League Against Rheumatism collaborative initiative. Arthritis Rheum 2010;62(9):2569–2581. https://doi.org/10.1002/art.27584. PubMed PMID: 20872595.

Ali N, Venkateswaran G, Garcia E, Landry T, McColl H, Sergi C, Persad A, Abuetabh Y, Eisenstat DD, Persad S. Osteosarcoma progression is associated with increased nuclear levels and transcriptional activity of activated β-Catenin. Genes Cancer. 2019;10(3–4):63–79. https://doi.org/:10.18632/genesandcancer.191. PubMed PMID: 31258833; PubMed Central PMCID: PMC6584208.

Arlet J, Ficat RP. Forage-biopsie de la tete femorale dans l'osteonecrose primitive. Observations histo-pathologiques portant sur huit foranes. Rev Rhum. 1964;31:257–64.

Barresi V, Ieni A, Branca G, Tuccari G. Brachyury: a diagnostic marker for the differential diagnosis of chordoma and hemangioblastoma versus neoplastic histological mimickers. Dis Markers. 2014;2014:514753. https://doi.org/10.1155/2014/514753.

Baumgartner-Sigl S, Haberlandt E, Mumm S, Scholl-Bürgi S, Sergi C, Ryan L, Ericson KL, Whyte MP, Högler W. Pyridoxine-responsive seizures as the first symptom of infantile hypophosphatasia caused by two novel missense mutations (c.677T>C, p.M226T; c.1112C>T, p.T371I) of the tissue-nonspecific alkaline phosphatase gene. Bone. 2007;40(6):1655–61. Epub 2007 Feb 14. PubMed PMID: 17395561.

Beccaria K, Sainte-Rose C, Zerah M, Puget S. Paediatric Chordomas. Orphanet J Rare Dis 10, 116 (2015). https://doi.org/10.1186/s13023-015-0340-8.

Bishop N, Arundel P, Clark E, Dimitri P, Farr J, Jones G, Makitie O, Munns CF, Shaw N. Fracture prediction and the definition of osteoporosis in children and adolescents: the ISCD 2013 Pediatric Official Positions. J Clin Densitom. 2014;17:275–80. https://doi.org/10.1016/j.jocd.2014.01.004.

Broome M, Vial Y, Jacquemont S, Sergi C, Kamnasaran D, Giannoni E. Complete Maxillo-Mandibular Syngnathia in a Newborn with Multiple Congenital Malformations. Pediatr Neonatol. 2016;57(1):65–8. https://doi.org/10.1016/j.pedneo.2013.04.009. Epub 2013 Jun 15. PubMed PMID: 23778189.

Cascio A, Colomba C, Di Carlo P, Serra N, Lo Re G, Gambino A, Lo Casto A, Guglielmi G, Veronese N, Lagalla R, Sergi C. Low bone mineral density in HIV-positive young Italians and migrants. PLoS One. 2020;15(9):e0237984. https://doi.org/10.1371/journal.pone.0237984. PMID: 32881882.

Combe B, Landewe R, Daien CI, Hua C, Aletaha D, Álvaro-Gracia JM, Bakkers M, Brodin N, Burmester GR, Codreanu C, Conway R, Dougados M, Emery P, Ferraccioli G, Fonseca J, Raza K, Silva-Fernández L, Smolen JS, Skingle D, Szekanecz Z, Kvien TK, van der Helm-van Mil A, van Vollenhoven R. 2016 update of the EULAR recommendations for the management of early arthritis. Ann Rheum Dis. 2017;76(6):948–59. https://doi.org/10.1136/annrheumdis-2016-210602. Epub 2016 Dec 15.

Cotti E, Sergi C, Bassareo A. Use of a resin-based root canal sealer followed by apicoectomy on two teeth. Dent Today. 2007;26(7):108, 110–1. PubMed PMID: 17708317.

Dal Cin P, Kozakewich HP, Goumnerova L, Mankin HJ, Rosenberg AE, Fletcher JA. Variant translocations involving 16q22 and 17p13 in solid variant and extraosseous forms of aneurysmal bone cyst. Genes Chromosomes Cancer 2000;28:233–4.

Emile JF, Abla O, Fraitag S, Horne A, Haroche J, Donadieu J, Requena-Caballero L, Jordan MB, Abdel- Wahab O, Allen CE, Charlotte F, Diamond EL, Egeler RM, Fischer A, Herrera JG, Henter JI, Janku F, Merad M, Picarsic J, Rodriguez-Galindo C, Rollins BJ, Tazi A, Vassallo R, Weiss LM; Histiocyte Society. Revised classification of histiocytoses and neoplasms of the macrophage-dendritic cell lineages. Blood. 2016;127(22):2672–81. https://doi.org/10.1182/blood-2016-01-690636. Epub 2016 Mar 10. PMID: 26966089; PMCID: PMC5161007..

Ficat RP, Arlet J. Ischemia and necrosis of bone. In: Hungerford DS, editor. Baltimore/London: Williams and Wilkins; 1980.

Forsman CL, Ng BC, Heinze RK, Kuo C, Sergi C, Gopalakrishnan R, Yee D, Graf D, Schwertfeger KL, Petryk A. BMP-binding protein twisted gastrulation is required in mammary gland epithelium for normal ductal elongation and myoepithelial compartmentalization. Dev Biol. 2013;373(1):95–106. https://doi.org/10.1016/j.ydbio.2012.10.007. Epub 2012 Oct 24. PubMed PMID: 23103586; PubMed Central PMCID:PMC3508155.

Gardeniers JWM. ARCO Committee on terminology and staging (report from the Nijmegen meeting). ARCO News Lett 1991;3:153e9.

Gardeniers JWM. ARCO Committee on terminology and staging. Report on the committee meeting at Santiago de Compostella. ARCO News Lett 1993;5:79e82.

Glueckert R, Rask-Andersen H, Sergi C, Schmutzhard J, Mueller B, Beckmann F, Rittinger O, Hoefsloot LH, Schrott-Fischer A, Janecke AR. Histology and synchrotron radiation-based microtomography of the inner ear in a molecularly confirmed case of CHARGE syndrome. Am J Med Genet A. 2010;152A(3):665–73. https://doi.org/10.1002/ajmg.a.33321. PubMed PMID: 20186814.

Hästbacka J, Superti-Furga A, Wilcox WR, Rimoin DL, Cohn DH, Lander ES. Atelosteogenesis type II is caused by mutations in the diastrophic dysplasia sulfate-transporter gene (DTDST): evidence for a phenotypic series involving three chondrodysplasias. Am J Hum Genet. 1996;58(2):255–62. PubMed PMID: 8571951; PubMed Central PMCID: PMC1914552.

Ho KL. Ecchordosis physaliphora and chordoma: a comparative ultrastructural study. Clin Neuropathol. 1985;4(2):77–86.

Jagodzinski NA, Kanwar R, Graham K, Bache CE. Prospective evaluation of a shortened regimen of treatment for acute osteomyelitis and septic arthritis in children. J Pediatr Orthop. 2009;29(5):518–25. https://doi.org/10.1097/BPO.0b013e3181ab472d. PubMed PMID: 19568027.

Johnston J, Al-Bahrani R, Abuetabh Y, Chiu B, Forsman CL, Nagamori S, Leng R, Petryk A, Sergi C. Twisted gastrulation expression in cholangiocellular and hepatocellular carcinoma. J Clin Pathol. 2012;65(10):945–8. Epub 2012 May 25. PubMed PMID: 22639408.

Johnson CA, Gissen P, Sergi C. Molecular pathology and genetics of congenital hepatorenal fibrocystic syndromes. J Med Genet. 2003;40(5):311–9. Review. PubMed PMID: 12746391; PubMed Central PMCID: PMC1735460.

Jung M, Tuischer JS, Sergi C, Gotterbarm T, Pohl J, Richter W, Simank HG. Local application of a collagen type I/hyaluronate matrix and growth and differentiation factor 5 influences the closure of osteochondral defects in a minipig model by enchondral ossification. Growth Factors. 2006;24(4):225–32. PubMed PMID: 17381063.

Jung C, Sohn C, Sergi C. Case report: prenatal diagnosis of diastrophic dysplasia by ultrasound at 21 weeks of gestation in a mother with massive obesity. Prenat Diagn. 1998;18(4):378–83. PubMed PMID: 9602486.

Kato S, Yoshizazawa T, KitanSYN S, Murayama A, Takeyama K. Molecular genetics of vitamin D-dependent hereditary rickets. Horm Res. 2002; 57(3–4):73–8. Review. PubMed PMID: 12006701.

Krooks J, Minkov M, Weatherall AG. Langerhans cell histiocytosis in children: Diagnosis, differential diagnosis, treatment, sequelae, and standardized follow-up. J Am Acad Dermatol. 2018;78(6):1047–56. https://doi.org/10.1016/j.jaad.2017.05.060. Review. PubMed PMID: 29754886.

Krooks J, Minkov M, Weatherall AG. Langerhans cell histiocytosis in children: History, classification, pathobiology, clinical manifestations, and prognosis. J Am Acad Dermatol. 2018;78(6):1035–44. https://doi.org/10.1016/j.jaad.2017.05.059. Review. PubMed PMID: 29754885.

Liu T, Li Z, Zhang Q, De Amorim Bernstein K, Lozano-Calderon S, Choy E, Hornicek FJ, Duan Z. Targeting ABCB1 (MDR1) in multi-drug resistant osteosarcoma cells using the CRISPR-Cas9 system to reverse drug resistance. Oncotarget. 2016;7(50):83502–83513. https://doi.org/10.18632/oncotarget.13148. PubMed PMID: 27835872; PubMed Central PMCID: PMC5347784.

Luschka P. Ueber gallertartige Auswüchse am Clivus Blumenbachii. Arch Pathol Anat Physiol Klin Med. 1857;11:8–12.

Manggold J, Sergi C, Becker K, Lukoschek M, Simank HG. A new animal model of femoral head necrosis induced by intraosseous injection of ethanol. Lab Anim. 2002;36(2):173–80. https://doi.org/10.1258/0023677021912460. PMID: 11943082.

Marcus ND, Enneking WF, Massam RA. The silent hip in idiopathic aseptic necrosis: treatment by bone grafting. J Bone Joint Surg Am. 1973;55:1351–66.

Malloy PJ, Feldman D. Genetic disorders and defects in vitamin d action. Endocrinol Metab Clin N Am. 2010;39(2):333–46, table of contents. https://doi.org/10.1016/j.ecl.2010.02.004. Review. PubMed PMID: 20511055; PubMed Central PMCID: PMC2879401.

Malloy PJ, Tasic V, Taha D, Tütüncüler F, Ying GS, Yin LK, Wang J, Feldman D. Vitamin D receptor mutations in patients with hereditary 1,25-dihydroxyvitamin D-resistant rickets. Mol Genet Metab 2014;111(1):33–40. https://doi.org/10.1016/j.ymgme.2013.10.014. Epub 2013 Nov 4. PubMed PMID: 24246681; PubMed Central PMCID:PMC3933290.

Mazabraud A, Girard J. Un cas particulier de dysplasie fibreuse à localisations osseuses et tendineuses [A peculiar case of fibrous dysplasia with osseous and tendinous localizations]. Rev Rhum Mal Osteoartic. 1957;24(9-10):652–9. French. PMID: 13518962.

Nield LS, Mahajan P, Joshi A, Kamat D. Rickets: not a disease of the past. Am Fam Physician. 2006;74(4):619–26. Review. PubMed PMID: 16939184.

Nowlan NC, Sharpe J, Roddy KA, Prendergast PJ, Murphy P. Mechanobiology of embryonic skeletal development: insights from animal models. Birth Defects Res C Embryo Today. 2010;90(3):203–13. https://doi.org/10.1002/bdrc.20184. Review. PubMed PMID: 20860060; PubMed Central PMCID: PMC4794623.

Oliveira AM, Hsi BL, Weremowicz S, Rosenberg AE, Dal Cin P, Joseph N, Bridge JA, Perez-Atayde AR, Fletcher JA. USP6 (Tre2) fusion oncogenes in aneurysmal bone cyst. Cancer Res. 2004;64(6):1920–3. https://doi.org/10.1158/0008-5472.can-03-2827. PMID: 15026324.

Oliveira AM, Perez-Atayde AR, Inwards CY, Medeiros F, Derr V, Hsi BL, Gebhardt MC, Rosenberg AE, Fletcher JA. USP6 and CDH11 oncogenes identify the neoplastic cell in primary aneurysmal bone cysts and are absent in so-called secondary aneurysmal bone cysts. Am J Pathol. 2004;165(5):1773–80. https://doi.org/10.1016/S0002-9440(10)63432-3. PMID: 15509545; PMCID: PMC3278819.

Oliveira AM, Perez-Atayde AR, Dal Cin P, Gebhardt MC, Chen CJ, Neff JR, Demetri GD, Rosenberg AE, Bridge JA, Fletcher JA. Aneurysmal bone cyst variant translocations upregulate USP6 transcription by promoter swapping with the ZNF9, COL1A1, TRAP150, and OMD genes. Oncogene. 2005;24(21):3419–26. https://doi.org/10.1038/sj.onc.1208506. PMID: 15735689.

Ono K. Diagnostic criteria, staging system and roentgenographic classification of avascular necrosis of the femoral head (steroid induced, alcohol associated or idiopathic nature) (in Japanese) Annual report of Japanese investigation committee for intractable disease, avascular necrosis of the femoral head. Tokyo: Ministry of health and Welfare; 1987.

Osasan S, Zhang M, Shen F, Paul PJ, Persad S, Sergi C. Osteogenic Sarcoma: A 21st Century Review. Anticancer Res. 2016;36(9):4391–8. Review. PubMed PMID: 27630274.

Panoutsakopoulos G, Pandis N, Kyriazoglou I, Gustafson P, Mertens F, Mandahl N. Recurrent t(16;17)(q22;p13) in aneurysmal bone cysts. Genes Chromosomes Cancer 1999;26:265–6.

Prandini N, Lazzeri E, Rossi B, Erba P, Parisella MG, Signore A. Nuclear medicine imaging of bone infections. Nucl Med Commun. 2006;27(8):633–644. https://doi.org/10.1097/00006231-200608000-00006.

Ressler S, Radhi J, Aigner T, Loo C, Zwerschke W, Sergi C. Insulin-like growth factor-binding protein-3 in osteosarcomas and normal bone tissues. Anticancer Res. 2009;29(7):2579–87. PMID: 19596932.

Redondo A, Bagué S, Bernabeu D, Ortiz-Cruz E, Valverde C, Alvarez R, Martinez-Trufero J, Lopez- Martin JA, Correa R, Cruz J, Lopez-Pousa A, Santos A, García Del Muro X, Martin-Broto J. Malignant bone tumors (other than Ewing's): clinical practice guidelines for diagnosis, treatment and follow-up by Spanish Group for Research on Sarcomas (GEIS). Cancer Chemother Pharmacol. 2017;80(6):1113–31. https://doi.org/10.1007/s00280-017-3436-0. Epub 2017 Oct 16. PubMed PMID: 29038849; PubMed Central PMCID: PMC5686259.

Rotondo M, Natale M, Mirone G, Cirillo M, Conforti R, Scuotto A. A rare symptomatic presentation of ecchordosis physaliphora: neuroradiological and surgical management. J Neurol Neurosurg Psychiatry. 2007;78(6):647–649. https://doi.org/10.1136/jnnp.2006.109561.

Sahyouni R, Goshtasbi K, Mahmoodi A, Chen JW. A historical recount of chordoma. J Neurosurg Spine.

2018;28(4):422–428. https://doi.org/10.3171/2017.7.SPINE17668.

Santer FR, Bacher N, Moser B, Morandell D, Ressler S, Firth SM, Spoden GA, Sergi C, Baxter RC, Jansen-Dürr P, Zwerschke W. Nuclear insulin-like growth factor binding protein-3 induces apoptosis and is targeted to ubiquitin/proteasome-dependent proteolysis. Cancer Res. 2006;66(6):3024–33. PubMed PMID: 16540651.

Saraff V, Högler W. Endocrinology and Adolescence: Osteoporosis in children: diagnosis and anagement. Eur J Endocrinol. 2015;173(6):R185-97. https://doi.org/10.1530/EJE-14-0865. Epub 2015 Jun 3. PMID: 26041077.

Schiffer C, Schiesser M, Lehr J, Tariverdian G, Glaeser D, Gabriel H, Mikuz G, Sergi C. Unique occurrence of Brachmann-de Lange syndrome in a fetus whose mother presented with a diffuse large B-cell lymphoma. Pathol Oncol Res. 2007;13(3):255–9. Epub 2007 Oct 7. Review. PubMed PMID: 17922056.

Schiffer C, Tariverdian G, Schiesser M, Thomas MC, Sergi C. Agnathia-otocephaly complex: report of three cases with involvement of two different Carnegie stages. Am J Med Genet. 2002;112(2):203–8. PubMed PMID: 12244557.

Schmutzhard J, Glueckert R, Bitsche M, Abraham I, Falkeis C, Schwentner I, Riechelmann H, Müller B, Beckmann F, Sergi C, Schrott-Fischer A. The cochlea in fetuses with neural tube defects. Int J Dev Neurosci. 2009;(7):669–76. https://doi.org/10.1016/j.ijdevneu.2009.07.008. Epub 2009 Aug 5. PubMed PMID: 19664702.

Schmutzhard J, Glueckert R, Sergi C, Schwentner I, Abraham I, Schrott-Fischer A. Does perinatal asphyxia induce apoptosis in the inner ear? Hear Res. 2009;250(1–2):1–9. https://doi.org/10.1016/j.heares.2008.12.006. Epub 2008 Dec 25. PubMed PMID: 19136052.

Schmutzhard J, Schwentner I, Glueckert R, Sergi C, Beckmann F, Abraham I, Riechelmann H, Schrott-Fischer A, Müller B. Pelizaeus Merzbacher disease: morphological analysis of the vestibulo-cochlear system. Acta Otolaryngol. 2009;129(12):1395–9. https://doi.org/10.3109/00016480802698866. PubMed PMID: 19922087.

Sergi C, Beedgen B, Kopitz J, Zilow E, Zoubaa S, Otto HF, Cantz M, Linderkamp O. Refractory congenital ascites as a manifestation of neonatal sialidosis: clinical, biochemical and morphological studies in a newborn Syrian male infant. Am J Perinatol. 1999;16(3):133–41. PubMed PMID: 10438195.

Sergi C, Graf M, Jung C, Sohn C, Adam S, Krempien B, Otto HF. Ruhender Knorpel und Epiphysenfuge bei diastrophischer Dysplasie. Kasuistik und klinisch-pathologische Charakteristika im Vergleich zur pseudodiastrophischen Dysplasie und zur Atelosteogenese Typ II [Resting cartilage and the growth plate in dystrophic dysplasia: case report and clinicopathologic characteristics as compared to pseudodystrophic dysplasia and type II atelosteogenesis]. Pathologe. 1998;19(5):379–83. German. https://doi.org/10.1007/s002920050301. PMID: 9816594.

Sergi C, Hentze S, Sohn C, Voigtländer T, Jung C, Schmitt HP. Telencephalosynapsis (synencephaly) and rhombencephalosynapsis with posterior fossa ventriculocele ('Dandy-Walker cyst'): an unusual aberrant syngenetic complex. Brain Dev. 1997;19(6):426–32. PubMed PMID: 9339873.

Sergi C, Linderkamp O. Pathological case of the month: classic rickets in a setting of significant psychosocial deprivation. Arch Pediatr Adolesc Med. 2001;155(8):967–8. PubMed PMID: 11483129.

Sergi C, Mornet E, Troeger J, Voigtlaender T. Perinatal hypophosphatasia: radiology, pathology and molecular biology studies in a family harboring a splicing mutation (648+1A) and a novel missense mutation (N400S) in the tissue-nonspecific alkaline phosphatase (TNSALP) gene. Am J Med Genet. 2001;103(3):235–40. PubMed PMID: 11745997.

Sergi C, Penzel R, Uhl J, Zoubaa S, Dietrich H, Decker N, Rieger P, Kopitz J, Otto HF, Kiessling M, Cantz M. Prenatal diagnosis and fetal pathology in a Turkish family harboring a novel nonsense mutation in the lysosomal alpha-N-acetyl-neuraminidase (sialidase) gene. Hum Genet. 2001;109(4):421–8. PubMed PMID: 11702224.

Sergi C, Roth SU, Adam S, Otto HF. Mapping a method for systematically reviewing the medical literature: a helpful checklist postmortem protocol of human immunodeficiency virus (HIV)-related pathology in childhood. Pathologica. 2001;93(3):201–7. PubMed PMID: 11433613.

Sergi C, Shen F, Liu SM. Insulin/IGF-1R, SIRT1, and FOXOs Pathways-An Intriguing Interaction Platform for Bone and Osteosarcoma. Front Endocrinol (Lausanne). 2019;10:93. https://doi.org/10.3389/fendo.2019.00093. eCollection 2019. PubMed PMID: 30881341; PubMed Central PMCID: PMC6405434.

Sergi C, Voigtländer T, Zoubaa S, Hentze S, Meyberg-Solomeyer G, Troeger J, Tariverdian G, Otto HF, Schiesser M. Ellis-van Creveld syndrome: a generalized dysplasia of enchondral ossification. Pediatr Radiol. 2001;31(4):289–93. PubMed PMID: 11321750.

Sergi C, Willig F, Thomsen M, Otto HF, Krempien B. Bronchopneumonia disguising lung metastases of a painless central chondrosarcoma of pubis. Pathol Oncol Res. 1997;3(3):211–4. https://doi.org/10.1007/BF02899923. PubMed PMID: 18470732.

Sergi C, Zwerschke W. Osteogenic sarcoma (osteosarcoma) in the elderly: tumor delineation and predisposing conditions. Exp Gerontol. 2008;43(12):1039–43. https://doi.org/10.1016/j.exger.2008.09.009. Epub 2008 Sep 25. Review. PubMed PMID: 18845233.

Simank HG, Manggold J, Sebald W, Ries R, Richter W, Ewerbeck V, Sergi C. Bone morphogenetic protein-2 and growth and differentiation factor-5 enhance the healing of necrotic bone in a sheep model. Growth Factors. 2001;19(4):247–57. PubMed PMID: 11811780.

Simank HG, Sergi C, Jung M, Adolf S, Eckhardt C, Ehemann V, Ries R, Lill C, Richter W. Effects of local application of growth and differentiation factor-5 (GDF-5) in a full-thickness cartilage defect model. Growth Factors. 2004;22(1):35–43. PubMed PMID: 15176457.

Smolen JS, Aletaha D, McInnes IB. Rheumatoid arthritis. Lancet. 2016;388(10055):2023–2038. https://doi.org/10.1016/S0140-6736(16)30173-8. Epub 2016 May 3. Review. Erratum in: Lancet. 2016 Oct 22;388(10055):1984. PubMed PMID: 27156434.

Spain H, Plumb T, Mikuls TR. Gout as a manifestation of familial juvenile hyperuricemic nephropathy. J Clin Rheumatol. 2014;20(8):442–4. https://doi.org/10.1097/RHU.0000000000000188. PubMed PMID: 25417683.

Steinberg DR, Steinberg ME. The university of pennsylvania classification of osteonecrosis. In: Koo KH, Mont M, Jones L, editors. Osteonecrosis. Springer, Berlin, Heidelberg. First Online 2014. https://doi.org/10.1007/978-3-642-35767-1_25. Print ISBN 978-3-642-35766-4. Online ISBN 978-3-642-35767-1.

Sugano N, Atsumi T, Ohzono K, Kubo T, Hotokebuchi T, Takaoka K. The 2001 revised criteria for diagnosis, classification, and staging of idiopathic osteonecrosis of the femoral head. J Orthop Sci Off J Jpn Orthop Assoc. 2002;7(5):601–5.

Sugano N, Ohzono K. Natural Course and the JIC Classification of Osteonecrosis of the Femoral Head. In: Koo KH, Mont MA, Jones LC, editors. Osteonecrosis. Heidelberg: Springer; 2014. p. 207–10.

Sultan AA, Mohamed N, Samuel LT, Chughtai M, Sodhi N, Krebs VE, Stearns KL, Molloy RM, Mont MA. Classification systems of hip osteonecrosis: an updated review. Int Orthop. 2019;43(5):1089–95. https://doi.org/10.1007/s00264-018-4018-4. Epub 2018 Jun 18. PMID: 29916002.

Sun M, Forsman C, Sergi C, Gopalakrishnan R, O'Connor MB, Petryk A. The expression of twisted gastrulation in postnatal mouse brain and functional implications. Neuroscience. 2010;169(2):920–31. https://doi.org/10.1016/j.neuroscience.2010.05.026. Epub 2010 May 20. PubMed PMID: 20493240; PubMed Central PMCID: PMC2971674.

Szendroi M, Antal I, Arató G. Adamantinoma of long bones: a long-term follow-up study of 11 cases. Pathol Oncol Res. 2009;15(2):209–16. https://doi.org/10.1007/s12253-008-9125-x. Epub 2008 Dec 2. PubMed PMID: 19048403.

Takashima K, Sakai T, Hamada H, Takao M, Sugano N. Which classification system is most useful for classifying osteonecrosis of the femoral head? Clin Orthop Relat Res 2018;476:1240e9.

Toretsky JA, Steinberg SM, Thakar M, Counts D, Pironis B, Parente C, Eskenazi A, Helman L, Wexler LH. Insulin-like growth factor type 1 (IGF-1) and IGF binding protein-3 in patients with Ewing sarcoma family of tumors. Cancer. 2001;92(11):2941–7. PubMed PMID: 11753970.

Virchow R. Untersuchungen über die Entwickelung des Schädelgrundes im gesunden und krankhaften Zustande und über den Einfluss derselben auf Schädelform, Gesichtsbildung und Gehirnbau. 1857, 128 pag. G. Reimer Publisher.

Wagner EF, Karsenty G. Genetic control of skeletal development. Curr Opin Genet Dev. 2001;11(5):527–32. Review. PubMed PMID: 11532394.

Wang L, Zhao P, Ma L, Shan Y, Jiang Z, Wang J, Jiang Y. Increased interleukin 21 and follicular helper T-like cells and reduced interleukin 10+ B cells in patients with new-onset systemic lupus erythematosus. J Rheumatol. 2014;41(9):1781–92. https://doi.org/10.3899/jrheum.131025. Epub 2014 Jul 15. PubMed PMID: 25028374.

Warman ML, Cormier-Daire V, Hall C, Krakow D, Lachman R, LeMerrer M, Mortier G, Mundlos S, Nishimura G, Rimoin DL, Robertson S, Savarirayan R, Sillence D, Spranger J, Unger S, Zabel B, Superti-Furga A. Nosology and classification of genetic skeletal disorders: 2010 revision. Version 2. Am J Med Genet A. 2011;155A(5):943–68. https://doi.org/10.1002/ajmg.a.33909. Epub 2011 Mar 15. PMID: 21438135; PMCID: PMC3166781.

Weir J, Harris N, Sergi C. Pathology quiz case: aneurysmal bone cyst (ABC) of the maxilla. Arch Otolaryngol Head Neck Surg. 2003 Dec;129(12):1345. PubMed PMID: 14676164.

Yoon BH, Mont MA, Koo KH, Chen CH, Cheng EY, Cui Q, Drescher W, Gangji V, Goodman SB, Ha YC, Hernigou P, Hungerford MW, Iorio R, Jo WL, Jones LC, Khanduja V, Kim HKW, Kim SY, Kim TY, Lee HY, Lee MS, Lee YK, Lee YJ, Nakamura J, Parvizi J, Sakai T, Sugano N, Takao M, Yamamoto T, Zhao DW. The 2019 revised version of association research circulation osseous staging system of osteonecrosis of the femoral head. J Arthroplasty. 2020;35(4):933–940. https://doi.org/10.1016/j.arth.2019.11.029. Epub 2019 Nov 27. PMID: 31866252.

Zhang M, Shen F, Petryk A, Tang J, Chen X, Sergi C. "English Disease": Historical Notes on Rickets, the Bone-Lung Link and Child Neglect Issues. Nutrients. 2016;8(11). pii: E722. Review. PubMed PMID: 27854286; PubMed Central PMCID: PMC5133108.

Head and Neck

14

Contents

14.1	**Development**	1168
14.2	**Nasal Cavity, Paranasal Sinuses, and Nasopharynx**	1171
14.2.1	Congenital Anomalies	1171
14.2.2	Inflammatory Lesions	1173
14.2.3	Tumors	1174
14.3	**Larynx and Trachea**	1183
14.3.1	Congenital Anomalies	1183
14.3.2	Cysts and Laryngoceles	1184
14.3.3	Inflammatory Lesions and Non-neoplastic Lesions	1185
14.3.4	Tumors	1186
14.4	**Oral Cavity and Oropharynx**	1189
14.4.1	Congenital Anomalies	1189
14.4.2	Branchial Cleft Cysts	1195
14.4.3	Inflammatory Lesions	1197
14.4.4	Tumors	1197
14.5	**Salivary Glands**	1203
14.5.1	Congenital Anomalies	1203
14.5.2	Inflammatory Lesions and Non-neoplastic Lesions	1203
14.5.3	Tumors	1205
14.6	**Mandible and Maxilla**	1209
14.6.1	Odontogenic Cysts	1210
14.6.2	Odontogenic Tumors	1216
14.6.3	Bone-Related Lesions	1218
14.7	**Ear**	1218
14.7.1	Congenital Anomalies	1218
14.7.2	Inflammatory Lesions and Non-neoplastic Lesions	1218
14.7.3	Tumors	1220
14.8	**Eye and Ocular Adnexa**	1226
14.8.1	Congenital Anomalies	1226
14.8.2	Inflammatory Lesions and Non-neoplastic Lesions	1226
14.8.3	Tumors	1226
14.9	**Skull**	1230
	Multiple Choice Questions and Answers	1233
	References and Recommended Readings	1235

© Springer-Verlag GmbH Germany, part of Springer Nature 2020
C. M. Sergi, *Pathology of Childhood and Adolescence*,
https://doi.org/10.1007/978-3-662-59169-7_14

14.1 Development

The development of the head and neck starts very early in gestation. The head and neck (H&N) early structures include the branchial apparatus, ear primordium, eye primordium, neural tube, oropharyngeal membrane, pharynx primitive, and stomodeum. The stomodeum is the natural oral cavity, lined by the ectoderm, while the neuroectodermal mesenchyme-powered neural tube will form the brain and the spinal cord. The branchial apparatus (*vide infra*) is constituted primarily by four elements/categories, including branchial arches, skeletal elements, muscles, nerves, and arteries; pharyngeal pouches, entoderm; branchial grooves, ectoderm; branchial membranes and, of course, *membrana obturatoria*. The embryological formation of the head and neck is also very complicated. When the head is beginning forming in the protoembryonic structure, the embryo is assumed to be composed of three layers of tissue, the ectoderm, mesoderm, and endoderm. These three germ layers become separate during gastrulation in the 3rd week of development. The neurulation is a process involving the fusion of the neural folds and forming the neural tube. The neurulation is completed in distinct stages that include the forming, shaping, and bending of the neural plate. Then the neural plate will progressively advance in the closing of the neural groove. We can also assume that the early head formation is defined by the migration of neural crest cells that arise from the rhombomeres, which are segments of the forming hindbrain. These structures will give rise to differentiated neurons. The two streams of neural crest cells emerge from the first two rhombomeres. These cell streams collaborate in the building of the face and branchial arch system (*vide infra*). The developing fetal skull is then separated into three sections, including the cranial vault, the cranial base, and the face. During the brain development, a membranous cranium is formed, which becomes the site of cranial (head) osteogenesis. Mesenchymal cells differentiate into osteoblasts without an intermediate cartilage model at week 9 of development. The osteoblasts form an osteoid matrix as well. This osteoid matrix also begins to mineralize. The cranial vault, face, and vomer do not have a cartilage model to follow and are of intramembranous origin. The nasal placodes are paired with prominences of ectodermal origin that arise in the 4th week of human development. A placode (neurogenic or *placodae neurogenicae*) is an epithelial area of condensation of the ectoderm layer of the embryonic head that generate neurons (and other structures) of the sensory nervous system. At the 8th week of embryonic human development, the cheeks and corners of the mouth will progressively form. Once the orbits over the nose form, the face will start to take an identifiable shape. The formation of the sinuses begins then to emerge, but the human development will not be considered completed until into puberty. The splanchnocranium (or visceral skeleton) is the portion of the skull that derives from pharyngeal arches. The splanchnocranium consists of both cartilage and endochondral bone. The pharyngeal or branchial arches begin to form during the 4th week of development. In the human, there are six arches, and they are divided by pharyngeal grooves outwardly and pharyngeal pouches inside. Thus, a groove is constituted of ectoderm unlike its counterpart the pouch on the endodermal side of the developing embryo. Branchial arch musculatures arise from myotomes of somitomeres, which are cell clusters originated from the disengaged masses of paraxial mesoderm that are found at the side of the developing neural tube. The word "groove" derives from the Middle English word "grofe" with the meaning of "cave" and Old English "grōf" with the meaning of "furrow", while the word "pouch" originates from the Old Low Franconian (West Germanic language or Old Dutch) "poka" with the meaning of a pocket or pouch. The derivatives of the branchial apparatus include:

1. *Branchial grooves*
   (a) Branchial groove 1 ⇒ external acoustic meatus
   (b) Branchial grooves 2, 3, and 4 ⇒ no ultimate "grooving" structure after forming:
     1. Cervical sinus
     2. Cervical vesicle
2. *Pharyngeal pouches*
   (a) Pharyngeal pouch 1 ⇒ tympanic cavity of the middle ear and auditory tube

(b) Pharyngeal pouch 2 ⇒ crypts of the palatine tonsils and sinus of tonsils
(c) Pharyngeal pouch 3 ⇒ external parathyroid glands and thymic epithelial cells
(d) Pharyngeal pouch 4 ⇒ internal parathyroid glands
(e) Ultimobranchial body (*Corpus ultimobranchiale*) ⇒ contribution to the thyroid gland (parafollicular or C cells)
3. *Branchial arch musculature*
   (a) Arch 1 gives rise to muscles of mastication (CN V)
   (b) Arch 2 gives rise to muscles of facial expression (CN VII)
   (c) Arch 3 gives rise to *stylopharyngeus* m. (CN IX), which elevates the larynx and the pharynx as well as dilate the pharynx allowing the passage of a food bolus assisting the progress of swallowing
   (d) Arches 4 and 6 give rise to the cricothyroid muscle and intrinsic laryngeal muscles (CN X)

The development of the facial structures (splanchnocranium, contrarily with respect to the neurocranium) is highly sophisticated. The *development of facial features* includes:

- Frontonasal prominence
- Lens placodes
- Mandibular prominence
- Maxillary prominence
- Medial and lateral nasal prominences
- Nasal placodes/pits
- Nasolacrimal grooves

The *development of internal structures of the splanchnocranium* includes:

1. Nasal cavities and palate (primary – secondary)
   (a) Median and lateral nasal processes and (b) nasal septum
2. Oral cavity, teeth, glands, and tongue
   (a) Rostral tongue buds, (b) median tongue bud, (c) caudal tongue bud, (d) hypobranchial eminence, (e) musculature, and (f) innervation

The specialized structures and functions of the oral cavity are as follows.

1. *Taste buds* (specialized nerve endings with afferents in the ninth cranial nerve for the posterior tongue and seventh cranial nerve for the anterior tongue).
2. *Teeth* (structures embedded in the maxilla or upper jaw and mandible or lower jaw with gum, which is the portion of mucosa that is reflected onto the bone).
3. *Salivary glands* constituted by numerous minor salivary glands and three major pairs, including the symmetrical parotid gland entirely serous and draining through Stensen duct (*ductus parotideus*) adjacent to the moral teeth, the seromucous submandibular gland emptying through Warthin's duct into the floor of the mouth, and the seromucous sublingual gland emptying through 10–20 small ducts in the basement of the mouth.
4. Access to nutrients, chewing, lubrification of the bolus, and swallowing of the bolus are the *functions* of the oral cavity (Greek: βῶλος, "lump, cluster, agglomeration").
5. Major *saliva* components: amylase, lysozyme, and secretory IgA.

The milestones of the embryology and anatomy regarding the salivary glands are presented in Box 14.1.

Another notable structure is the thyroid gland. The growth and migration of thyroid tissue are fundamental! Thyroglossal duct (stalk) eventually

---

Box 14.1 Embryology Milestones of the Salivary Glands
- Ectodermal origin
- 4th-6th week of embryonic development
- Serous and mucous cells, arranged in acini, drained by series of small ducts
- Parotid gland: serous acini
- Submandibular gland: serous and mucinous acini
- Minor salivary glands: mucinous acini

disappears. The thyroid gland diverticulum originates from the floor of the pharynx. In case of persistence of the thyroglossal duct, there will be an ectopic glandular tissue of the thyroid, which may become cystic, inflamed, as well as, more rarely, neoplastic. The pituitary gland primordia include (1) the *infundibulum*, which forms neurohypophysis and derives from the floor of the diencephalon, and (2) Rathke's pouch, which forms the adenohypophysis and derives from the roof of the *stomodeum*. The dissection of complex multiple congenital anomaly syndromes such as agnathia-otocephaly syndromes requires a revision of the anatomic structures and the distribution of efferent cranial nerves. The eye and ear are to be covered later, while the brain is included in the section of the neuroanatomy of the respective chapters, and the teeth are referred to in the next paragraph. Odontogenesis is very important for the understanding of the nontumoral and tumoral pathology of the maxillofacial region. Epithelial buds on the alveolar ridge grow down into the primitive stroma, and the lower jaw or mandible grows around the developing teeth. In fact, an invagination forms and deepens, and it is essential to distinguish between epithelium, which is constituted by ameloblasts (forming enamel of tooth), and stroma represented by odontoblasts (forming dentin of tooth). The five milestones of odontogenesis include:

1. Ecto-mesenchymal projections of the *dental lamina*
2. Formation of the *layered cap* (inner/outer enamel epithelium, stratum intermedium, stellate reticulum)
3. *Dentin* secretion by odontoblasts and enamel secretion by ameloblasts
4. *Cementum* secretion by cementoblasts
5. *Periodontal membrane* by fibroblasts

Of note, the dental lamina is a primitive epithelium overlying free margins of jaws and gives the origin of primary and permanent teeth, while epithelial rests are responsible for odontogenic cysts and tumors.

The structures of the head and neck develop primarily from the branchial arches, which are numbered from cranial to caudally. Four branchial arches are well developed and visible on the external surface of the embryo by the 4th week of gestation. The fifth and sixth branchial arches are instead small and located internally. Clefts anatomically separate branchial arches called branchial grooves on the external surface and pharyngeal pouches on the internal surface. Apart from the first branchial groove, all other branchial grooves become obliterated. Histogenetically, each branchial arch consists of mesodermally derived mesenchyme, which is covered externally by ectoderm and internally by endoderm. The several and different mesenchymal structures (bone, muscle, cartilage) of the head and neck arise from the ectodermally derived neural crest cells that migrate into the mesenchyme, proliferate, and induce the formation of the end structures. From the vascular point, each branchial arch is supplied by the correctly numbered aortic arch and has an associated cranial nerve (fifth, seventh, ninth, and tenth, respectively) as follows for the branchial arches. The *mandibular arch* splits into two prominences including the maxillary ("small" prominence) forming the maxilla or upper jaw, zygomatic bone (from Greek zygon "yoke" meaning "to join", also known as cheekbone), and a portion of the temporal bone and the mandibular ("large" prominence) forming the mandible or lower jaw. Of note is that the cartilage of the first arch (Meckel's cartilage) forms malleus and incus. The *hyoid arch* (upsilon-shaped bone) overgrows the third and fourth arches to form the cervical sinus. The hyoid arch also forms a portion of the hyoid bone and the muscles of facial expression, and, mainly, the cartilage of the second arch (Reichert's cartilage) forms the stapes and the styloid. The third arch forms part of the hyoid bone and the stylopharyngeus muscle, while 4) + 6) build laryngeal cartilage and cricothyroid and pharyngeal muscles.

To summarize, the pharyngeal pouch derivatives include:

1. ⇒ External meatus, middle, and inner ear structures, as well as the Eustachian tube.
2. ⇒ Palatine tonsils.
3. ⇒ Inferior parathyroid glands and thymus.
4. ⇒ Superior parathyroid glands.
5. ⇒ It is and remains rudimentary.

The larynx anatomy is particularly complex. The supraglottic portion arises from the third and fourth branchial pouches (buccopharyngeal anlage).

- The glottis and subglottis portions arise from the fifth and sixth branchial pouches (laryngotracheal anlage).
- The cricoid, thyroid, and arytenoid cartilages are of a hyaline type.
- The epiglottic cartilage is of elastic type.
- Thyroid tissue may be found within the fibrous capsule of the larynx and trachea.

Finally, an anatomical note should be considered by the pathologist and surgeons. In the last decades, a few cases of bleeding in the tonsillar bed following tonsillectomy have been reported and the experience of the surgeon has been stressed in the most recent literature. Pediatric tonsillectomy is a common procedure. Tonsillectomy is probably one of the first skills acquired by surgical trainees, but post-tonsillectomy bleeding is one of the most significant complications and can be fatal for the patient. Of importance is the anatomy and, mainly, the blood supply of the tonsils. Blood supply is provided by tonsillar branches of five arteries, including the ascending pharyngeal artery (of the external carotid artery), dorsal lingual artery (of the lingual artery), ascending palatine artery (of the facial artery), tonsillar branch (of the facial artery), and the lesser palatine artery (a branch of the descending palatine artery, which is anatomically a branch of the maxillary artery). These blood vessels arise from the external carotid artery. It has been stressed that during surgery, attention must be paid to the internal carotid artery, lingual artery, and glossopharyngeal nerve, but all five arterial blood vessels need to be tackled properly at time of surgery.

Post-operative hemorrhage remains a frequent complication of tonsillectomy, but a primary hemorrhage, which occurs in the first hours is rapidly dealt with by the surgical team. Conversely, a rash discharge of the patient with a secondary hemorrhage, which typically occurs once the child has returned home, can be fatal if it is not dealt with quickly.

14.2 Nasal Cavity, Paranasal Sinuses, and Nasopharynx

14.2.1 Congenital Anomalies

Congenital anomalies of these three anatomic regions include nasal glioma, nasal encephalocele, nasal dermoid, and epignathus (Fig. 14.1).

Nasal glioma is a glial heterotopia at the nasal location, which presents as a congenital, firm, stable, polypoid mass with an intranasal and an extranasal component and is composed of neuroglial tissue, which is separated from the intracranial contents. Surgery is curative in almost all cases.

Nasal encephalocele is an anterior neural tube defect with herniation of brain tissue through a bony defect in the frontal skull. There is an association with several syndromes. Hydrocephalus and hypertelorism have been observed with nasal encephaloceles. Hypertelorism has also been associated with basal encephaloceles. Corpus callosum agenesis and hydrocephalus are the most often associated anomalies in addition to intracranial cysts, interhemispheric lipomas, and schizencephaly (Tirumandas et al. 2013). Median cleft face syndrome is typically associated with basal encephaloceles.

Nasal dermoid is a midline cyst of hamartomatous or teratomatous nature lined by squamous epithelium. There are several genetic syndromes, sequences, and associations where nasal dermoid can have a clinical feature. Hypertelorism, cleft lip and palate, hemifacial microsomia, aural atresia, branchial sinuses, cardiac, gastrointestinal, genital, and central nervous system (e.g., hydrocephalus) anomalies have been associated with nasal dermoid in up to 41% of cases (Van Wyhe et al. 2016).

Epignathus is an extremely rare congenital teratoma at the pharyngeal location where all three germ layers are mixed arising in the oral cavity, usually in the nasopharynx attaching to the base of the skull, often the hard palate or mandible. Despite the excellent labeling as a benign tumor, it is associated with high mortality and morbidity rates because of severe airway obstruction and other malformations that may coincide with the presentation.

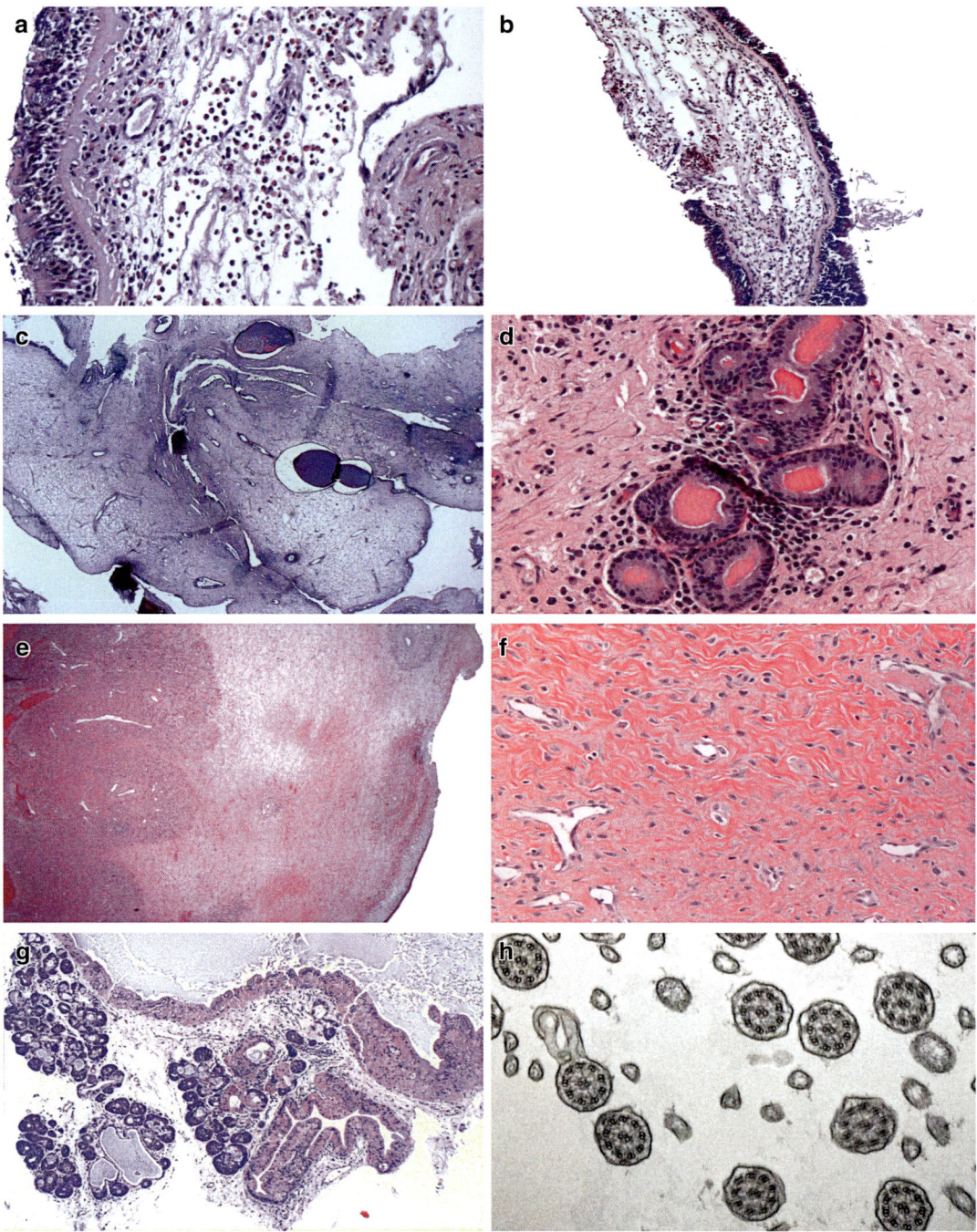

Fig. 14.1 Nasopharynx – congenital and inflammatory lesions with chronic sinusitis (**a**), polyposis nasi et sinuum (**b**), secretion plugs in polyposis of the nose (**c**), inflammatory polyp with dense concretions in the lumina in a patient affected with cystic fibrosis (**d**; hematoxylin and eosin staining, ×200 original magnification), angiofibroma (**e**–**f**; hematoxylin and eosin staining, ×12.5 and ×200 as original magnifications), and oncocytic Schneiderian papilloma (**g**; hematoxylin and eosin staining, ×40 as original magnification). Oncocytic Schneiderian papilloma, fungiform papilloma, and inverted papilloma are three separate tumors arising from the Schneiderian membrane. Classic 9 + 2 combination in normal cilia (**h**; transmission electron microscopy, ×20000 as direct magnification)

- *EPI*: 1 in 35,000–200,000 live births (incidence), ♂ < ♀.
- *ETP*: Teratomas arise from the pluripotent cells of one or more of the three germ layers.
- *CLI*: Unidirectional protruding mass through the oral cavity with rarer intracranial extension, through the sphenoid bone (*os sphenoidale*), and often associated with cleft palate and bifid tongue or nose. The epignathus can present with intrauterine fetal death (IUFD).
- *LAB*: ↑ serum AFP (maternal serum during pregnancy).
- *DGN*: Clinical examination, US, and CT/MRI.
- *DDX*: Hemangioma, lymphangioma, dermoid cyst, encephalocele, and other benign or malignant soft tissue masses of the neonatal period. The differential diagnosis of epignathus includes tumors arising from the neural plate (meningoencephalocele, nasal glioma, olfactory neuroblastoma/esthesioneuroblastoma, primitive neuroectodermal tumor, and retinoblastoma), tumors arising in the neighboring sites (cystic hygroma, ectopic odontogenic tissue, giant epulis, lymphangioma, thyroid gland tissue heterotopia), and rare tumors (e.g., congenital rhabdomyosarcoma).
- *PGN*: A poor outcome has been observed as indicated above and mortality rate is high.

14.2.2 Inflammatory Lesions

Rhinitis: It may be an acute/chronic inflammation of the mucous membrane of the nose (viral, allergic, non-allergic eosinophilic, vasomotor) Vasomotor rhinitis has become quite frequent in industrialized countries in both children and adults. It is characterized by intermittent episodes of sneezing, rhinorrhea, and blood vessel congestion of the nasal mucus membranes. Vasomotor rhinitis seems to be a hypersensitive response to stimuli such as a dry air, pollutants, spicy foods, alcohol, strong emotions, and certain medications.

Mucocele/mucopyocele: Fluid-filled swelling, which is commonly located on the lip or the mouth. The mucocele is a pseudocyst because there is no epithelial lining. It develops when the minor salivary glands of the mouth become plugged with mucus.

Inflammatory (allergic) polyps: Polypoid structures occurring in adolescents ± cystic fibrosis, showing loose edematous stroma with neutrophils, eosinophils, lymphocytes, and plasma cells. Tissue eosinophilia may or may not correlate with plasma eosinophilia (*vide infra*).

14.2.2.1 Nasal Polyposis
- *DEF*: Multiple polyps of the nasal cavity usually occurring in the setting of an underlying local or systemic disease, mostly chronic rhinosinusitis (CRS), but also cystic fibrosis, allergic fungal rhinosinusitis, aspirin-exacerbated respiratory disease, and Churg-Strauss syndrome (eosinophilic granulomatosis with polyangiitis, EGPA).
- *EPI*: 1–4% of the population.
- *GRO*: Polypoid excrescences in the nasal sinuses which may become obstructed.
- *CLM*: Inflammatory polyp or pseudopolyp with an increase of chronic inflammatory cells and associated in some individuals with an abundance of eosinophils. Special stains include PAS, DPAS, and GMS to rule out fungal infections. Nasal polyps can become ulcerated and can also be infected and complicate in an abscess of the nasal sinuses (Fig. 14.1).
- *TRT*: In the setting of CRS, nasal polyps are not likely to be cured by either medical or surgical therapy; however, control is generally achievable. Presently, to the best of our knowledge, we support the use of intranasal corticosteroids for maintenance therapy at least for some periods and, in some settings, short courses of oral corticosteroids for acute exacerbations. There is no clear-cut evidence for short- and long-term antibiotics, but for patients with symptomatic nasal polyposis nonresponsive to medical therapies, functional endoscopic sinus surgery is a valuable option and provides an adjunctive therapeutic route.

Wegener's granulomatosis: T-cell-mediated hypersensitivity reaction, possibly to inhaled infectious or environmental agents in the form of acute necrotizing granulomatous inflammation. In the classic presentation, there is the triad of involvement of:

1. *Upper airways*
2. *Lower airways*
3. *Kidney*

Upper airways may show *chronic sinusitis* and *nasopharyngeal ulcerations*, lower airways may show persistent *pneumonitis*, and in the kidney may be present a *focal and segmental necrotizing glomerulonephritis* (FSNG) at an early stage and crescentic glomerulonephritis at a late stage. Serologically, cytoplasmic antineutrophil cytoplasmic antibodies or c-ANCA (aka PR3-ANCA) may be present. The c-ANCA is a type of autoantibody with a diffusely granular, *cytoplasmic* staining pattern using light microscopy.

Wegener's granulomatosis histologic triad includes:

1. *Parenchymal (basophilic, geographic) liquefactive (suppurative, colliquative) necrosis (In Wegener's granulomatosis, the necrosis is due to vasculitis and similar to bacterial necrosis and EGPA with numerous neutrophils associated with fibrin exudation and abscess formation. This process is quite different from the coagulative necrosis due to hypoxic conditions, such as infarction, although at places both coagulative and colliquative necroses can co-exist in Wegener's parenchymal or geographic necrosis)*
2. *Granulomatous inflammation* (poorly formed granulomas) with eosinophils and multinucleated giant cells (MNGC)
3. *Leukocytoclastic vasculitis* with only scant number of lymphocytes and plasma cells

DDX: Extranodal NK/T-cell lymphoma, which is characterized by an angiocentric infiltrate with mucosal expansion and mucosal gland destruction with prominent necrosis, +CD3, CD56, EBER, and cytotoxic phenotype with (+) perforin, TIA-1, and granzyme B as well as Fas/FasL expression, which is responsible for the prominent necrosis, and P-glycoprotein/MDR1 that may account for the inadequate response to chemotherapy. TIA-1 or T-cell intracellular antigen-1 (aka "granule membrane protein 17") is a 15 kD cytotoxic granule-associated protein. expressed by cytotoxic T-cells. P-glycoprotein 1/ Multidrug resistance protein 1 (MDR1) is an ATP-binding cassette sub-family B member 1 (ABCB1) (aka CD243), which is an important protein of the cell membrane responsible for the excretion of different foreign substances.

Sarcoidosis: This disease is multisystemic and aims to produce noncaseating granulomas with characteristic but not pathognomonic (exclusive) confluent back-to-back organization in several organs, mainly lungs and lymph nodes, but any organ may be involved. T-cell subpopulations with regulatory functions are natural killer T cells (NK T cells) and T regulatory cells (T regs). Both NK T cells and T regs have been described as abnormal in sarcoidosis.

14.2.3 Tumors

The differentiation of the neoplastic processes involves substantially the distinction of benign tumors from malignant tumors (Figs. 14.2, 14.3, and 14.4).

14.2.3.1 Benign Epithelial Tumors

In the nose and paranasal sinuses, there are several essential entities, including *sinonasal papillomas* (inverted or endophytic, cylindrical or oncocytic, and everted or exophytic – mnemonics *ICE*) and *salivary gland-type adenomas* (pleomorphic adenoma or PA, myoepithelioma, and oncocytoma). *Respiratory epithelial adenomatoid hamartoma* (REAH) is a new disease, which is a tumorlike lesion. A classical inflammatory nasal polyp is characterized by edema, goblet cell hyperplasia of the epithelium, BM thickening, and leukocytic inflammatory infiltration, while REAH shows a submucosal adenomatoid proliferation originating from the epithelium and specifically invaginating downward into the submucosa with widely spaced, small to medium-sized pseudoglands, separated by stroma tissue. Everted is aka exophytic or fungi-

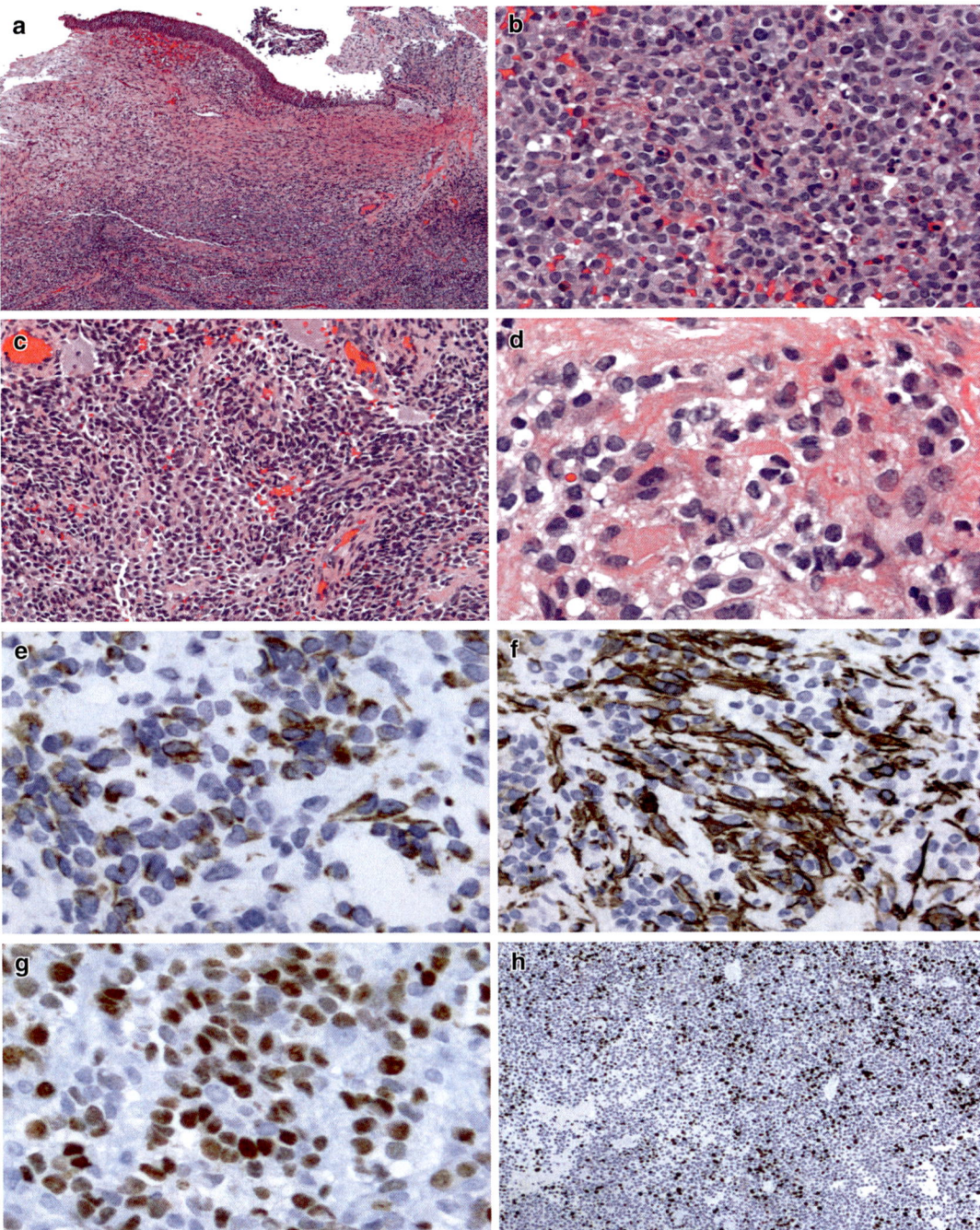

Fig. 14.2 Embryonal rhabdomyosarcoma of the nose showing a dense blue stained infiltration of the tissue deeper than the epithelial layer (**a**), which is constituted by small round blue cell tumors with high nucleus-to-cytoplasm ratio (**b**) with a propensity of a perivascular arrangement (**c**). The examination of the tumor cells at high power reveals some vacuolation of the tumor cells at places (**d**). The immunohistochemical panel (**e–h**) highlights the expression of desmin (**e**), actin (**f**), MYF4 (**g**), and Ki67 (**h**) (**e**, ×600; **f**, ×400; **g**, ×600; **h**, ×100 as original magnifications, avidin-biotin complex immunostaining)

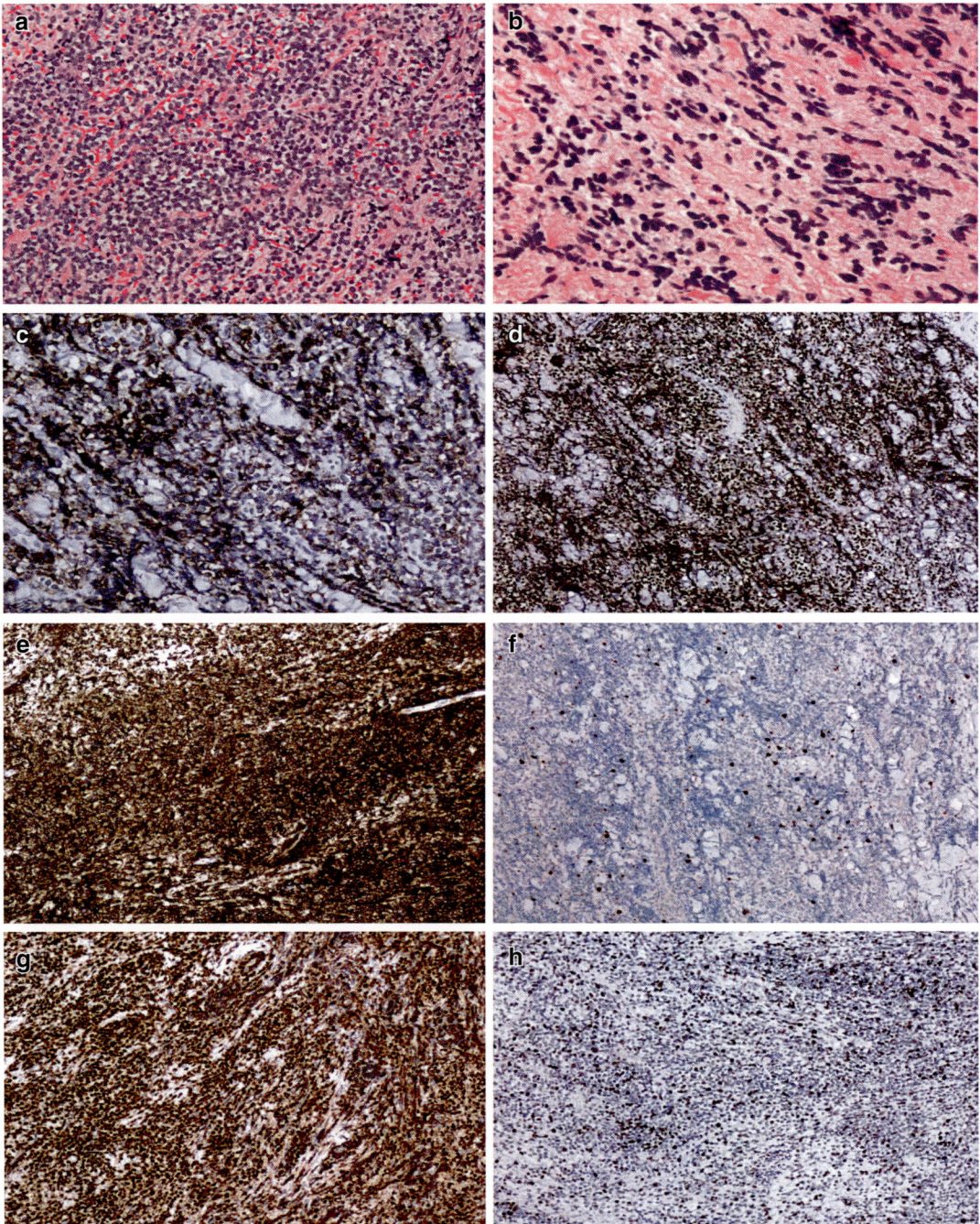

Fig. 14.3 This panel shows a B-cell lymphoblastic lymphoma of the nasopharynx with small round blue cell tumors infiltrating the subepithelial tissue (**a–b**) disguising the hematological nature of the tumor. The immunohistochemical panel highlights the expression of CD79a (**c**, ×200), PAX5 (**d**, ×100), CD10 (**e**, ×100), CD3 (**f**, ×100), TdT (**g**, ×100), and Ki67 (**h**, ×100) (avidin-biotin complex immunostaining)

14.2 Nasal Cavity, Paranasal Sinuses, and Nasopharynx

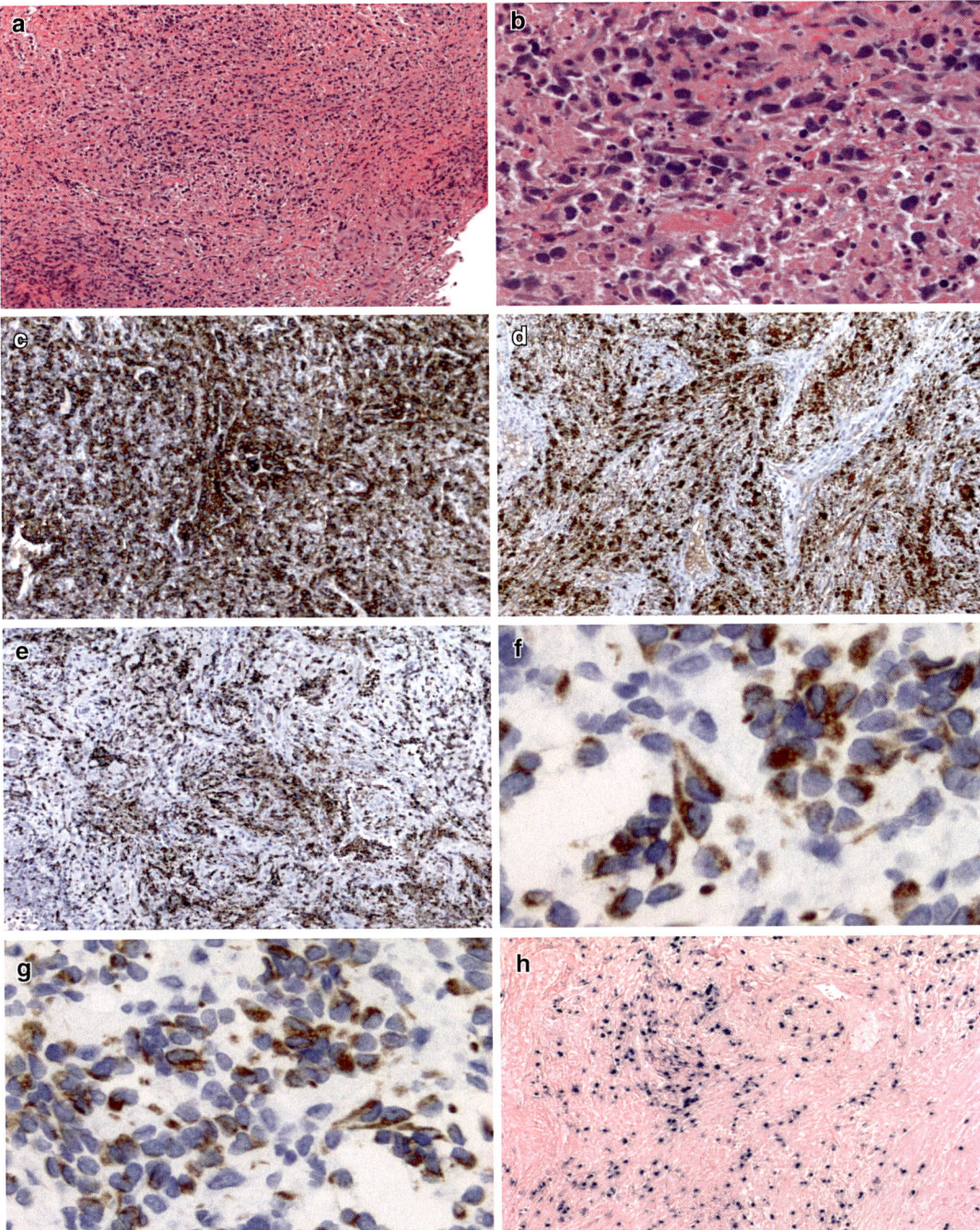

Fig. 14.4 In this panel, the hematoxylin-eosin-stained sections (**a–b**) (**a**, ×100; **b**, ×400 as original magnifications) and the immunostained sections (**c–e**) of a NK-T-cell lymphoma of a young man are shown. The tumor shows medium-sized cells with high nucleus-to-cytoplasm ratio without evidence of rhabdomyoblastic differentiation. The immunohistochemical panel of the tumor cells shows the expression of CD56 (**c**, ×100), granzyme B (**d**, ×100), and TIA-1 (**e**, ×100). In (**h**) there is EBER staining, i.e., the in situ hybridization for the RNA of the Epstein-Barr virus (×100) Extranodal Natural Killer/T-cell lymphoma, nasal type, is a fast-growing and destructive type of NHL, which can start in T cells, but it develops most often in NK cells.

form, most commonly seen. The endophytic growth is characterized by the growth of primarily rounded nests extending into the stroma. The cylindrical papilloma (aka oncocytic Schneiderian papilloma) has sharply defined cell borders, oncocytic cytoplasm, cilia on the surface, mucin droplets, and intraepithelial spaces with inflammatory cells (microabscesses). There are no metastases, but it is very aggressive locally.

In the nasopharynx, the benign epithelial tumors include a few other entities. We list the *hairy or dermoid polyp*, the *Schneiderian-type papilloma* (uncommon, middle age, ♂ > ♀, inverted type), the *squamous papilloma* (larynx papilloma-like), the *ectopic pituitary adenoma* (rare but wide age range), the *salivary gland anlage tumor* or congenital PA (unusual, neonatal age), and the *craniopharyngioma* (exceptionally rare as originating from the nasopharynx or through downward extension from a suprasellar location and harboring an identical histology to the intracranial craniopharyngioma).

Sinonasal Papilloma
- *DEF*: Benign but locally aggressive neoplasm occurring in the nose and paranasal sinuses, usually youth to middle age, but occasionally seen in patients (♂ > ♀) quite young (Box 14.2).
- *CLI*: Epistaxis and mass effect are two classic presentations.
- *CLM*: Proliferating squamous or columnar epithelium intermixed with mucin-containing cells harboring mild to moderate cell atypia.
- *PGN*: Recurrence is frequent if incompletely excised.

Pediatric salivary gland-type adenomas are (1) myoepithelioma, (2) oncocytoma, and (3) pleomorphic adenoma (PA), which are not different from the salivary gland tumors of adulthood.

14.2.3.2 Malignant Epithelial Tumors
Epithelial malignancies include nasopharyngeal carcinoma and sinonasal carcinoma.

Nasopharyngeal Carcinoma
- *DEF*: Carcinoma of the nasopharyngeal mucosa (NPC), mostly arising in the fossa of Rosenmüller and superior posterior wall, with light and/or electron microscopy evidence of squamous differentiation, highly aggressive behavior, extensive locoregional infiltration, early lymphatic spread (jugulodigastric LN), and hematogenous metastases (bone > lung > liver > distant LNs). The fossa of Rosenmüller is a lateral recess of the nasopharynx situated behind the Eustachian tube orifice and also regarded as a partial persistence of the second pharyngeal pouch. The 2005 WHO classification of nasopharyngeal carcinoma includes *keratinizing* squamous cell carcinoma (SqCC), *non-keratinizing* SqCC (differentiated and undifferentiated subtypes), and *basaloid* SqCC.
- *EPI*: <1% (mostly worldwide) to >2% (Southeast Asia, North Africa, and Arctic regions).
- *EPG*: Epstein-Barr Virus (EBV) (↑IgA to EBV with EBER or Epstein–Barr virus-encoded small RNAs present in almost all tumor cells). EBERs are small non-coding RNAs localized in the nucleus of human cells infected with EBV. In carcinogenesis, environmental factors (nitrosamine-rich salted fish/cured meats, smoking, chemical fumes and dust, formaldehyde and radiation exposure) and genetic factors may also play a significant role, which may also be intensified by epigenetic mechanisms. Genetically, LOH on

Box 14.2 Sinonasal Papilloma ("ICE")

Type	HPV6/11	Growth	Site	Epidemiology	PGN
Inverted	+	*Endo*phytic	Middle meatus, sinuses	45% (late youth)	10% cancer
Cylindrical	−	Endo-/exophytic	Lateral nasal wall	5% (late youth)	10% cancer
Everted	+	*Exo*phytic	Nasal septum	50% (early youth)	<1% cancer

3p, 9p, and 14q and inactivation of *P16* (9p21) are often seen.

- *CLI*: A neck mass, nasal features (postnasal drip, discharge, bleeding, and obstruction due to involvement of the Eustachian tube), and aural elements (tinnitus, discharge, earache, deafness) as well as enlarged neck nodes are frequent and associated with IgA against VCA and IgG/IgA against EA of EBV and occasionally dermatomyositis, which belongs to the group of inflammatory myopathies characterized by chronic inflammation and accompanied clinically by muscle weakness.
- *GRO*: NPC may present as a smooth mucosal bulge, a raised nodule, or an infiltrating fungating mass with or without superficial ulceration.
- *CLM*: The three types need to be taken apart, because of a different outcome. Precursor lesion: NPCIS, which shows atypical epithelial change (variable loss of polarity, nuclear enlargement, nuclear crowding, and distinct nucleoli) confined to the surface and/or crypt epithelium.
  - *Keratinizing* SqCC is characterized by irregular islands of tumor cells with clear squamous cell differentiation (intercellular bridges and keratinization on LM and small bundles of tonofilaments and tonofibrils as well as desmosomes on TEM) and desmoplastic stroma with lymphoplasmacytic infiltration. ± EBV (EBER, EBNA-1, LMP-1) according to geographical areas). Latent EBV infection and growth transformation of B lymphocytes is highlighted by EBV nuclear and membrane protein expression (EBV nuclear antigen [EBNA] and latent membrane protein [LMP], respectively). Also, LMP1 is known to be an oncogene in rodent fibroblasts and to induce B-lymphocyte activation and cellular adhesion molecules in the EBV-negative Burkitt lymphoma cell line Louckes. EBV-induced growth transformation requires EBNA-2 (Wang et al. 1990).
- *DDX*: Squamous metaplasia/hyperplasia (stromal invasion, desmoplastic stroma, and nuclear atypia).
  - *Non-keratinizing* SqCC (differentiated and undifferentiated subtypes) is characterized by solid sheets, irregular agglomerated islands, noncohesive sheets, and trabecular arrangement of differentiated (or better "differentiating") and undifferentiated tumor cells. The differentiated subtype shows some cellular stratification and pavements with or without a plexiform growth pattern of plum to slender spindle cells with prominent nucleoli. Occasionally, keratinizing cells, reminding the urothelial carcinoma of the urinary bladder. The undifferentiated subtype shows a syncytial-like arrangement of large cells with indistinct cell borders, round to oval cellular nuclei, and prominent nucleoli. In both subtypes, there is a variable lymphoplasmacytic-rich stroma, which is usually non-desmoplastic, and lymphocytes may border the epithelial nests (*Regaud pattern*) or intermingle with the epithelial cells (*Schmincke pattern*) without prognostic significance.
- *IHC*: (+) p63, CK5/6, CK14, 34βE12 (HMWCKs), but weak and patchy for CAM5.2 (LMWCKs), focal EMA, and (−) CK7 and CK20 for the tumor cells and (+) CD8 > CD4 > CD20 lymphocytes, (+CD138, κ/λ-polyclonal) plasma cells, and S100 dendritic cells!

It has been reported that cases with numerous lymphocytes and dendritic cells as well as few granzyme B-positive cytotoxic T lymphocytes (CTL) have a better PGN. + EBV (EBER, EBNA-1, LMP-1).

- *DDX*: The differential diagnosis of NPC, non-keratinizing, includes the following entities:
  - Benign mimickers: Clusters of germinal center cells, tangentially sectioned crypts, reactive lymphoid hyperplasia, and lymphoid tissue-associated venules
  - Sinonasal undifferentiated carcinoma (SNUC): Sinonasal, EBV (−), (−) keratinization/glandular differentiation/lymphocytes, (−) CK5/6, CK14, 34βE12, but +CK, CK19, EMA, (+) NE markers, including NSE, CGA, and SYN, small

nuclei, low N/C, and (+) CD68. CD99, VIM, muscle markers, hematolymphoid markers, and melanocytic markers need to be associated in a IHC panel and are negative in SNUC.
- Diffuse large B-cell lymphoma (DLBCL): Hematolymphoid blastic morphology, IHC typical for high-grade B-NHL
- Hodgkin lymphoma: Reed-Sternberg and Hodgkin cells, (+) CD15, CD30, and PAX-5
- Sarcoma: Various patterns of IHC, of which paramount is to rule out RMS
- *Basaloid* SqCC is characterized by basaloid differentiation as other tumors of other regions and is accompanied by lower clinical aggressiveness compared to keratinizing and non-keratinizing SqCC. ± EBV (EBER, EBNA-1, LMP-1 according to geographical areas).
- *PGN*: After the therapy with radiation, there is a remarkable 5-YSR of ~75%. Thus, the outcome is relatively favorable. Favorable prognostic factors are younger age, female, low plasma/serum EBV DNA titers, euploidy, and low pre-treatment tumor proliferating fraction. Tumor angiogenesis, c-erbB2, P53, nm23-HI, IL-10, and VEGF have also been indicated to play a role in the outcome. If the tumor is classified as T1, the neoplasm is confined to the nasopharynx or extending to the oropharynx and nasal cavity without parapharyngeal extension, but this last feature with extension beyond the pharyngeal-basilar fascia is characteristic of T2 tumors. In case of bony structures and paranasal sinus involvement, the tumor is classified as T3, while an intracranial extension and involvement of cranial nerves, infratemporal fossa, hypopharynx, orbit, or masticatory space determine the enrollment of the neoplasm as T4. About the LNs, the supraclavicular fossa is the triangle bounded by the superior margin of the sternal head of the clavicle medially, the superior margin of the lateral end of the clavicle laterally, and the point where the neck meets the shoulder posteriorly.

In the nasal cavity and paranasal sinuses, several entities have been described, which are unusual in children, but they may occur in teenagers and young adults. Malignant epithelial tumors of these two regions include *SqCC* (NOS and variants), *lymphoepithelial carcinoma*, *SNUC adenocarcinoma* (intestinal-type and non-intestinal-type), *salivary gland-type carcinomas* (adenoid cystic carcinoma, acinic cell carcinoma, MEC, epithelial-myoepithelial carcinoma, clear cell carcinoma NOS, myoepithelial carcinoma, carcinoma ex-PA, and polymorphous low-grade adenocarcinoma or PLGA), and *neuroendocrine tumors* (typical, atypical and small-cell carcinoma, neuroendocrine type).

14.2.3.3 Benign Mesenchymal Tumors

The following entities we may encounter in the differential diagnosis of benign mesenchymal tumor entities in childhood and youth:

- *Hemangioma, leiomyoma, meningioma, myxoma, nasopharyngeal angiofibroma (NAF), neurofibroma, and schwannoma*

However, NAF represents most likely the soft tissue tumor mostly characteristic for this location, being the other tumors characterized by histology that is not different from other soft tissue locations. In the past, pyogenic granuloma has often been confused both clinically and histologically with infantile lobular hemangioma. The erythrocyte-type glucose transporter protein (GLUT-1) is useful in the differentiation between infantile and juvenile lobular hemangioma as it is highly expressed in these benign vascular tumors, but not in congenital hemangiomas, pyogenic granulomas, granulation tissue, or vascular malformations. Congenital hemangiomas should be kept in a separate group, including two varieties, the rapidly involuting congenital hemangiomas (RICH) and the noninvoluting congenital hemangiomas (NICH) (Leon-Villapalos et al. 2005).

Nasopharyngeal Angiofibroma (NAF)
- *DEF*: Well-demarcated tumor with the lobular arrangement and tannish to a purple-red aspect, which often manifests in adolescent and young males (androgen-dependent) with

nasal obstruction and epistaxis ± associated with FAP.
- *EPI*: Rare (0.5% of all H&N neoplasms), 1:150,000 individuals (incidence), adolescents, and young adults (14–25 years), ♂ > ♀, Indian subcontinent > West countries.
- *CLM*: Antler-shaped (irregular) delicate blood vessels with incomplete muscular coats (thin wall without elastic laminae and minimal smooth muscle), particularly at the periphery, embedded in a collagenized stroma, which can explain the propensity to bleed extensively during surgery. The stroma contains spindle to stellate cells with plump nuclei and, occasionally, MNGC.
- *IHC*: (+) VIM, CD31, CD34, AR, but (±) SMA in the stromal cells, (+) SMA in the perivascular cells only, and (−) ER/PR in the vasculature, perivascular cells, and stromal cells. Movat pentachrome stain is useful to highlight the thin wall without elastic laminae and minimal smooth muscle.
- *TEM*: Tight nuclear RNA-protein complexes implement dense round granules in fibroblasts.
- *DDX*:
  1. *Lobular arranged capillary hemangioma* (LACH) (lobular, (+) CD31, CD34, GLUT1)
  2. *Hemangiopericytoma* (HPC) (Ret-rich, (+) VIM, FXIIIa, CD57, CD99, Bcl2, (−) CD34)
  3. *Solitary fibrous tumor* (SFT) (tangled network of fibroblast-like stromal cells with abundant depots of Ret and collagen fibers, EVG-rich, (+) CD99, Bcl2, CD34, STAT6)
  4. *Vascular angiosarcoma* (VAS) (atypia ± ↑ mitoses)
- *PGN*: After the resection with clear margins, the NAF usually does not recur. However, recurrence is reported in 1/5 of the patients (20%) with incomplete resection. Occasionally, sarcomatoid transformation following RT, and rarely metastases have been observed.

14.2.3.4 Other Tumors

Bone/Cartilaginous Forming Tumors

Benign tumors include giant cell lesion, giant cell tumor, osteoma, chondroma, chondroblastoma, chondromyxoid fibroma, osteochondroma, osteoid osteoma, osteoblastoma, ameloblastoma, and nasal chondromesenchymal hamartoma.

Malignancies include osteosarcoma, chondrosarcoma NOS, mesenchymal chondrosarcoma, and chordoma.

Hematolymphoid Tumors

There are a few entities that may occur in the nasal cavity, sinonasal cavities, and nasopharynx of adolescents and young adults and include extranodal NK/T-cell NHL, DLBCL, Burkitt lymphoma, extramedullary myeloid sarcoma, Langerhans cell histiocytosis (LCH), and juvenile xanthogranuloma (JXG). Other entities, which are pertinent to the youth and middle-aged adults, may include follicular lymphoma, mantle cell lymphoma, extranodal marginal zone B-cell lymphoma, peripheral T-cell lymphoma NOS, extramedullary plasmacytoma, and histiocytic sarcoma. Histiocytic sarcoma that needs to be differentiated from LCH shows large pleomorphic cells with eccentric, indented nucleus and prominent eosinophilic cytoplasm with fine vacuolations showing (+) CD45, CD68, and lysozyme and (−) CD19, CD20, CD22, Cd79a, or pan-B markers, (−) CD3 or pan-T marker, (−) MPO or myeloid marker, (−) CD1a and Langerin, and (−) CD21, CD35, aka FDC markers. Other hematolymphoid tumors that may involve the nasopharynx are *Castleman disease* and *Rosai-Dorfman disease* (sinus histiocytosis with massive lymphadenopathy or SHML).

Extranodal NK/T-Cell Lymphoma (EN-NK/T-NHL)

NHL with extensive effacement (*see hematological chapter*).

Diffuse Large B-Cell Lymphoma (DLBCL)

NHL (*see hematological chapter*).

A DLBCL mimicker is the IM and needs to be suspected in the case of an adolescent or young adult. It is recognized by the presence of a range of large cells showing maturation toward plasma cells, lack of frank cytologic atypia, and polyclonal κ/λ immunophenotypic expression of the large cells.

Extramedullary Myeloid Sarcoma (EMS)

- *DEF*: Myeloid or granulocytic-differentiated sarcoma at extramedullary location (nasal cavity, paranasal cavities, and nasopharynx) constituted by an extramedullary manifestation of acute myeloid leukemia (AML), which can be antecedent, synchronous, or metachronous to EMS. EMS is characterized by a *diffuse chloroacetate esterase reaction* (histochemical special stain) in ~3/4 of cases determining the green color and the old labeling of *chloroma* to this lesion.
- *CLM*: Blast cells with round to ovoid nuclei, fine chromatin, small but distinct nucleoli, and lightly eosinophilic cytoplasm with eosinophilic granules are detected mixed with eosinophilic *myelocytes* and *metamyelocytes*. Giemsa-stained touch preparation is useful to identify cytoplasmic azurophilic granules and Auer rods, when present.
- *IHC*: + MPO, CD13, CD33, CD117, CD68/KP1, neutrophil elastase, and lysozyme. However, EMS with monocytic differentiation is MPO (−), but CD68/PGM1 (+).
- *DDX*: NHL.

Langerhans Cell Histiocytosis (LCH)

- *DEF*: Clonal proliferation (variable) of Langerhans cells in nodal or extranodal H&N tissue.
- *SYN*: Eosinophilic granuloma, histiocytosis X, Langerhans cell granulomatosis.
- *EPI*: Children <15 years (1:200,000–5:10^6) but can also occur more rarely in adults (~1:10^6).
- *ETP*: Neoplastic transformation of Langerhans cells, which are typically present as a few cells in the thymus, lymph nodes, and skin and capture antigens and present them to lymphocytes. Langerhans cells are considered an immune system crucial component. The discovery of recurrent mutations in the mitogen-activated protein kinase (*MAPK*) pathway (i.e., *BRAF* and *MAP2K1* mutations) confirms that it is a neoplastic disease.
- *CLI*: Inconstant clinical picture with single or multiple lesions or disseminated disease.
  - *Hand-Schuller-Christian disease*: Indolent LCH form of children and young adults.
  - *Hashimoto-Pritzker disease*: Congenital, self-limited LCH form.
  - *Letterer-Siwe disease*: Systemic LCH form in infants.
- *GRO*: LN enlargement.
- *CLM*: Partial effacement of the LN architecture node with preservation of follicular centers but distension of nodal sinuses by Langerhans cells, which are 12–15 μm (Ø) large. Neoplastic cells have irregular and elongated nuclei with prominent nuclear grooves and folds, fine chromatin pattern, and indistinct nucleoli. Neoplastic cells are engulfed with abundant, pale eosinophilic cytoplasm. MNGC and eosinophils can be seen, and sinuses may have foci of necrosis. MI is low to moderate.
- *IHC*: (+) S100, CD1a, Langerin, VIM, Fascin, CD68, CD74, HLA-DR, CD45 (IOFS), and (−) CD15, CD21, CD35, CD45RA, CD45RB, EMA, while ± CD2, CD3.
- *TEM*: Birbeck granules (elongated, 200–400 × 33 nm sized, zip-like cytoplasmic inclusion body with double outer sheath).
- *DDX*: Cat-scratch disease, Erdheim-Chester disease, hypersensitivity reaction, Kimura disease, HL, NHL, and SHML (RDD).
- *TRT*: CHT ± surgery.
- *PGN*: Good, if therapy is successful. The recurrence rate is exiguous.

Neuroectodermal Tumors

These tumors constitute a critical category for children, adolescents, and young adults, but they are explained in more detail in other chapters.

PNET/Ewing Sarcoma

See *soft tissue chapter*.

Esthesioneuroblastoma

- *DEF*: Highly malignant neuroblastoma of the olfactory epithelium lining mucosa ± chromosomal translocation of 11–22 (PNET-like).
- *SYN*: Olfactory neuroblastoma.
- *EPI*: Wide age range (3–80 years).
- *GRO*: Reddish gray, highly vascular polypoid mass.
- *CLM*: Small round cells of neural crest origin, with oval nuclei and indistinct cell borders, with

or without Homer-Wright pseudorosettes and, occasionally, large cells growing in solid nests.
- *TRT*: Very sensitive to RT.
- *PGN*: High mortality rate (highly malignant neoplasm with a high level of local recurrence rate and metastatic rate).

Melanotic Neuroectodermal Tumor of Infancy (MNETI)
- *DEF*: Rare pigmented benign neoplasm arising from the neural crest cells.
- *EPI*: <6 months of life, ♂ > ♀ with a typical presentation in craniofacial region (>90%). Locations include the maxilla, mandible, skull, and brain. Extracranial sites are the genitals.
- *CLI*: Painless, nonulcerative, expansile, rapidly growing lesion with intriguing pigmentation and having locally aggressive behavior.
- *DGN*: Clinical assessment, CT/MRI, and histopathology.
- *CLM*: Biphasic tumor, comprising neuroblast-like round cells and melanocytic cells.
- *DDX*: PNET/Ewing sarcoma, desmoplastic small round cell tumor, RMS, peripheral neuroepithelioma, neuroblastoma, NHL, and malignant melanoma.
- *TRT*: Surgical excision, CHT, and RT alone or combined.
- *PGN*: High recurrence rate with metastases and malignant transformation rate.

Mucosal Malignant Melanoma
See *dermatological chapter*.

Germ Cell Tumors
Teratoma, mature and immature and with malignant transformation, yolk sac tumor (aka endodermal sinus tumor, YST/EST), sinonasal teratocarcinosarcoma, mature teratoma, and dermoid cyst represent the germ cell tumors that can occur in the nasal cavity, paranasal sinuses, and nasopharynx. These tumors are similar to the tumors of the male and female genital system.

Secondary Tumors
They are very rare in pediatric groups, although the neuroblastoma may be considered one tumor that is tending to metastasize or localize in the head and neck region.

14.3 Larynx and Trachea

The non-neoplastic and neoplastic lesions of the larynx and trachea are shown in Box 14.3.

> **Box 14.3 Larynx and Trachea Lesions**
> 14.3.1. *Congenital Anomalies*
> 14.3.2. *Cysts and Laryngoceles*
> 14.3.3. *Inflammatory Lesions and Non-neoplastic Lesions*
> 14.3.4. *Tumors*

14.3.1 Congenital Anomalies

14.3.1.1 Thyroglossal Duct Cyst and Laryngeal Disturbances

One of the most common etiologies for a median neck cyst is the thyroglossal duct cyst. Embryologically, the primitive anlage of the thyroid gland descends from the pharyngeal location at the 3rd week of gestation to the front of the pharynx by the 7th week. An enlargement of any persistent portion of this duct and subsequent secretion from the duct epithelium are few of the most common evolutions. Thyroglossal duct remnants are approximately seen in 1:20 individuals of a general population. In about 85% of cases, they are located between the hyoid bone and thyroid gland. In 13–14% of cases, thyroglossal duct remnants are seen at the level of the hyoid bone while in 1–2% are detected at the level of the tongue. A lingual thyroglossal duct cyst is particularly crucial for the weakening of the surrounding tissue of the laryngeal pharynx. This aspect may involve the base of the epiglottis, resulting in fragile laryngomalacia with significant clinical symptoms. An enlarged cyst may press on the epiglottis, irritate it, and become drawn into the vocal cords or ventricle of Morgagni (Morgagni's sinus, *ventriculus laryngis*).

14.3.1.2 Laryngomalacia

In pediatrics, stridor is commonly observed and described as a high-pitched, wheezing sound, which is caused by a disrupted airflow of the trachea. The most common etiology of stridor in babies is laryngomalacia (Greek: μαλακία, "softness"), which affects 1/2–3/4 of all infants with congenital stridor. It is characterized by the inward collapse of laryngeal structures, resulting in a narrow air passage and turbulent airflow during inspiration. Typically, laryngomalacia is characterized by the inner destruction of laryngeal structures. This pathology occurs in a restricted air passage and turbulent airflow during inspiration. Clinically, a high-pitched intermittent inspiratory stridor during the first 2–3 weeks of the neonatal period is observed. The severity of laryngomalacia can be mild, moderate, or severe. In most cases, laryngomalacia is mild, and there is inspiratory stridor with a coordinated suck-swallow-breathe sequence. In these cases, no therapy is required, because the spontaneous resolution is seen in the first infancy in about 2/3 of children. In case of severe laryngomalacia, poor weight gain, feeding difficulties, choking, post-feeding vomiting, cyanosis, dyspnea with intercostal and suprasternal retractions, and failure to thrive may occur. Aryepiglottoplasty is the surgical treatment of choice. It has a high success rate and harbors a low complication rate.

Rilliet and Barthez (1853) were the first to describe congenital stridor, while Sutherland and Lack (1897) published a detailed review including the first description of a congenital laryngeal obstruction. In 1942, Jackson first used the term "laryngomalacia," with the meaning of "soft larynx", to characteristically describe a disorder in which supraglottic tissue collapses onto the laryngeal airway during inspiration. Cartilage floppiness due to developmental abnormalities and immaturity, anatomical variation in the neonate larynx, gastroesophageal reflux, poor neuromuscular control, and hypotonia have been suggested to be etiologic and etiopathogenetic associated with laryngomalacia. However, cartilage is often histologically normal in patients with symptomatic laryngomalacia. Numerous pediatric centers have demonstrated that patients with laryngomalacia have impaired the sensorimotor function of the larynx due to some alteration or adaptation in the laryngeal adductor reflex arch prompting more neurologic research on this topic. Some genetic syndromes have been associated with laryngomalacia, including 22q11.2 microdeletion syndrome, CHARGE syndrome, Cornelia de Lange syndrome, Costello syndrome, Cri du chat syndrome, trisomy 21 (Down) syndrome, oculo-ectodermal syndrome, Sotos syndrome, and the Pierre Robin sequence, which is a set of abnormalities affecting the splanchno- and neurocranium, consisting of micrognathia, glossoptosis, and obstruction of the airways.

Congenital anomalies and genetic disorders occur between 1:10 and 1:5 as estimated incidence. However, the incidence is as high as 40% of infants with severe laryngomalacia who require surgical intervention. Noteworthy, infants with congenital anomalies and genetic disorders may often present with other medical comorbidities such as synchronous airway lesions, cardiac and vascular disease, and neurologic disease. Trisomy 21 syndrome (Down syndrome) appears to be the most commonly reported associated genetic disorder, with 50% who have respiratory symptoms presenting with laryngomalacia.

Benign tumors are granular cell tumor, recurrent respiratory papillomatosis, and squamous papilloma. Dysplastic and invasive squamous lesions are squamous dysplasia, SqCC, early invasive SqCC, and invasive, usual type of SqCC, and SqCC variants (verrucous, papillary, basaloid, spindle cell). Neuroendocrine tumors are carcinoid, atypical carcinoid, small-cell carcinoma (SmCC), and large-cell carcinoma (LCC), while mesenchymal tumors are chondroma and chondrosarcoma.

14.3.2 Cysts and Laryngoceles

According to *DeSanto Classification of Laryngeal Cysts* (1970), two major types are recognized, including saccular cysts and ductal cysts. The frequency is 1/4 for saccular cysts and 3/4 for ductal cysts. Two more rare cysts are the oncocytic cyst and tonsillar cyst. Congenital laryngeal cysts seem to be more common in the British-Pakistani population. The *saccular cyst* arises from a cystic enlargement of the laryngeal saccule and is supraglottic, while the *ductal cyst* originates from dilatation of mucous glands and is located in the true cords or epiglottis. Some neonatal airway obstruction may be associated with saccular cysts. Both squamous and respiratory epithelium can line both cysts. Saccular cysts manifest earlier, often with stridor, while ductal cysts present later with feeding difficulties and failure to thrive. Both cysts are treated with endoscopic marsupialization at the time of diagnostic laryngoscopy in most of the cases without extralaryngeal extension. As indicated above, two additional rarer cysts include the *oncocytic cyst*, as oncocytic metaplasia of a ductal cyst, and the *tonsillar cyst*, which contains thyroid glandular tissue. Unlike saccular cysts, *laryngocele* is an air-containing saccular dilatation, which is located at the appendix of the laryngeal ventricle. Generally, laryngoceles communicate with the lumen of the ventricle through a tiny and narrow pedicle. Laryngocele arises typically from the laryngeal ventricle and extends into peri-laryngeal space.

Although the etiology is unknown, laryngocele seems to be probably related to both congenital and acquired factors. Historically, it seems that Larrey, a very influent Napoleon's surgeon, in 1829, has given the first report on laryngocele. Larrey collected a series of laryngoceles in individuals who apparently would hourly chant the Koran from the minarets. In 1958, laryngocele was analyzed in detail from Burke and Golden. Since then, it seems that a structure to be considered as laryngocele needs to have a Morgagni's ventricle that extends beyond the superior border of the thyroid cartilage. Subsequently, following its distinction from saccules, the definition was changed, being a laryngocele a large Morgagni's ventricle, i.e., an air-containing saccular dilatation, which is located at the appendix of the laryngeal ventricle and explicitly communicating with the lumen. Laryngocele may substantially be due to prolonged periods of increased pressure within the laryngeal lumen as observed in people blowing in the wind musical instrument. It is more frequent in males (5:1) with the highest incidence in the late adolescence/youth, but laryngoceles in children have been described. There are some additional associations, including laryngeal amyloidosis, laryngeal chondroma, history of a tracheotomy, papillomatosis, or ankylosing spondylitis. Patients with an internal laryngocele usually complain of some symptoms, including hoarseness, dyspnea, foreign body sensation, and cough.

- *CLI*: Laryngoceles may manifest with coarse hoarseness, dyspnea, and chronic cough.
- *CLM*: Cystic cavities lined by pseudostratified, columnar, ciliated epithelium with occasional foci of stratified squamous epithelium and a variable mixture of serous and mucous glands characteristically at the submucosal location. In our opinion and that of others, this peculiar composition distinguishes these lesions from laryngeal cysts, which are specially lined entirely by squamous epithelium. In evaluating a laryngocele, the CT scan shows air- or mucus-containing tumor at the level of the false cord. Laryngocele may complicate and have a theoretical risk of malignancy, having been SqCC of the larynx arising in a laryngocele described. This event, although rare, has been reported in a 20-year-old nonsmoker male (Murray et al. 1994).

14.3.3 Inflammatory Lesions and Non-neoplastic Lesions

Non-neoplastic lesions that may be considered are blastomycosis, contact ulcer, laryngeal amyloidosis, laryngeal cysts (ductal, saccular, oncocytic, tonsillar), laryngocele, melanosis, laryngitis, Teflon granuloma, and vocal cord nodule/vocal cord polyp. Laryngocele and laryngeal cysts have been discussed above.

14.3.3.1 Acute Epiglottitis

H. influenzae infection of the epiglottis with cherry red, markedly edematous epiglottis leading often to airway obstruction. It is life-threatening.

14.3.3.2 Chronic Laryngitis

Nonspecific and usually, the result of an infective disease, overuse of the voice, physical/chemical exposure, or irritation by inhalants, alcohol, vaping (e-cigarettes), and tobacco

14.3.3.3 Granulomatous Laryngitis

Granulomas-accompanying chronic laryngitis with granulomas that can be seen in a variable clinical setting, such as tuberculosis, histoplasmosis, blastomycosis, but also Crohn disease, sarcoidosis, and foreign body granuloma (aka "teflonoma"). Also, SHML may also involve the larynx.

14.3.3.4 Laryngeal Nodule

It is a non-inflammatory reaction of the vocal cord, chiefly anterior 1/3, to injury, such as misuse of the voice, and characterized by edema and amorphous eosinophilic material beneath intact epithelium (types/stages: telangiectatic, fibrous, hyaline, and edematous or myxoid).

14.3.3.5 Contact Ulcers of Larynx

Lively granulation tissue or pyogenic granuloma-like tissue lesion occurring on the posterior commissure. Unlike pyogenic granuloma, the lobular pattern is not seen.

14.3.4 Tumors

14.3.4.1 Papilloma (Juvenile Laryngeal Papilloma)

- *DEF*: Juvenile laryngeal human papilloma virus (HPV)-facultative/obligate papilloma with single or multiple benign epithelial histology arising on true cords and optional spread to the false cords, epiglottis, and tracheobronchial tree as well as a tendency to recurrence.
- *ETP*: HPV-6- and HPV-11-related neoplasms.
- *CLM*: Papillary or acanthotic growth of well-differentiated squamous epithelial cells with mild atypia, some mitoses overlying a fibrovascular core and an accompanying mild inflammatory component. Koilocytosis is defined by a cytoplasmic perinuclear cavitation, nuclear hyperchromasia, irregular nuclear border, and nuclear enlargement to at least 3 times the size of a nucleus belonging to an intermediate squamous cell with optional binucleation.
- *DDX*: Adult laryngeal papilloma (♂ > ♀, often solitary, more inflammatory component, usually single, moderate to severe atypia, carcinoma precursor) and verruca vulgaris. The papillomatous structure should be clearly evident to make the diagnosis of a papilloma and this structure needs to be kept differentiated from a polypoid structure.
- *TRT*: Surgery, laser, and cryotherapy.
- *PGN*: Repeated surgical revisions may lead to the progressive and inexorable destruction of the cords. In 2018, Orita et al. studied 77 papilloma patients. In 21 cases with laryngeal papilloma, various types of HPV were identified: 18 (52.9%) were positive of low-risk HPV, 14 cases (41.2%) were positive of high-risk HPV, and 11 (32.4%) were positive of both high-risk HPV and low-risk HPV. Young patients showed a higher rate of HPV infection than old patients. No malignant transformation was observed among the patients with laryngeal papilloma. The immunohistochemical staining with p16 (a tumor suppressor protein, encoded in humans by the CDKN2A gene playing an important role in cell cycle regulation and used as a biomarker to improve the histological diagnostic accuracy of HSIL), i.e., positive expression of p16, was observed in 20 cases. HPV infection and p16-expression were strongly associated with the pathological finding of koilocytosis. It is important to remember that overexpression of p16 is usually caused by the HPV, whereas neoplasms showing p16 downregulation may have other causes.

Juvenile recurrent papillomatosis is seen as an HPV 6-/11-associated, multiple papillary tumors on the true cords, from which they may disseminate to the false cords, epiglottis, and deeper into the trachea and bronchial system. An orderly maturation pattern is characteristic of the benignancy of this lesion, although mitotic activity is present and focal atypia may be present and kept in mind when reporting such lesions. A comment in the histopathological report is often useful to the clinical pediatrician for the follow-up of these patients. There is a low risk to progress to squamous cell carcinoma, differently from the adult laryngeal papilloma showing moderate to severe atypia.

14.3.4.2 Laryngeal Carcinoma

- *DEF*: Malignant epithelial tumor of the larynx. Its delineation as oncological entity is a rarity in children and adolescents (McGuirt et al. 1997). A significant issue remains the diagnosis at late stages because early symptoms are often attributed to inflammatory diseases or developmental anomalies. This issue may be considered a worse-case scenario, but it is similar to the adult counterpart. The flexible laryngoscope allows the first visualization of vocal cords and may be essential to address neoplastic diseases at pediatric age. Juvenile laryngeal papillomatosis is on the top of the list of the neoplastic conditions in children and adolescents. However, rare neoplastic disorders occur, and in most of the cases, there are no established protocols.
- *EPI*: Fifth to seventh decades but also children and adolescents. It is imaginable that the use of vaping (e-cigarettes) may decrease the lower range value of the age incidence for laryngeal cancer in the future. ♂ > ♀.
- *EPG*: Tobacco (probably also e-cigarette and vaping), alcohol, GERD (controversy exists for asbestos and HPV) and Bloom syndrome (predisposition to laryngeal and hypopharyngeal cancer). The knowledge of the genetic syndromes associated with laryngeal carcinoma is essential for the pediatric pathologist. Two other genetic/dysmorphology syndromes, apart from the common cancer genetic syndromes, may be associated with the onset of laryngeal carcinoma. They include the Pierre Robin sequence and the Treacher Collins syndrome. In both conditions, the inflammation may play a role of promoting factor following an initiation process, which can be variable. Bloom syndrome is an autosomal recessive inherited syndrome characterized by short stature, sun-sensitive skin changes, high-pitched voice, long, narrow face with micrognatia (small jaw), large nose, and prominent ears. *BLM* gene mutations have been associated with individuals affected with Bloom syndrome. Treacher Collins syndrome is an autosomal dominant/recessive inherited syndrome characterized by underdeveloped facial bones, specially the cheek, micrognathia, cleft palate, eye abnormalities, and hearing loss. Treacher Collins syndrome may be caused by mutations in the *TCOF1*, *POLR1C*, or *POLR1D* genes. Pierre Robin sequence is a congenital condition characterized by micrognathia, glossoptosis (back-set tongue), and cleft palate associated with *SOX9* gene mutations.
- *CLI*: There is quite a variability, which is correlated to the anatomic site of occurrence. The site can influence the type of symptoms, stage at presentation, treatment, and prognosis. An example is a tumor localized at the base of the epiglottis that is difficult to detect during laryngoscopy (aka "*Winkelkarzinom*" or corner carcinoma). Second primary tumors (SPT) are tumors distinctly separated from the primary or index carcinoma and labeled as synchronous if they are detected within 6 months from the first diagnosis of the primary malignancy. If SPT occurs >6 months, it needs to be classified as a metachronous. neoplasm (Greek: σύγχρονος "contemporaneous" and μετάχρονος "non coeval" from respectively σῠ́ν- "with, together" or μετά "after" with χρόνος "time").
- *CLM*: >95% are squamous cell carcinomas which are either well, moderately, or poorly differentiated.
- *PGN*: The outcome of laryngeal carcinoma is linked to some prognostic factors (Box 14.4). The box contains major useful prognostic factors like those provided from the College of American Pathologists (CAP) simplified to memorize in dealing of laryngeal carcinomas in children and youth. It is important to check clinical history and consult geneticists, because in many countries genetic information is not provided in the electronic medical record or the patient's chart.

> **Box 14.4 Good (Green Highlighted) and Poor (Red Highlighted) Prognostic Factors**
> (1) Small size
> (2) Location (glottic and supraglottic)
> (3) Overexpression of *TP53*
> (1) Extensive cartilage invasion
> (2) TNM stage IV
> (3) Co-morbidities

Remarkably, it seems that in the hypopharynx, the outcome is inversely related to TNM stage and extracapsular spread. Moreover, trachea, mediastinal and distant metastases are very poor prognostic factors.

The histologic variants and subtypes of laryngeal carcinoma are as follows.

- *Verrucous*: Non-metastasizing (good PGN), usually glottic, carcinoma variant characterized by abundant "church-spire" arranged keratosis, broad pushing border of infiltration accompanied by dense lymphocytic and plasmacellular host response, and slow, locally invasive growth.
- *Papillary*: Low-metastasizing (favorable PGN), exophytic papillary growth, usually supraglottic/glottic, carcinoma variant made up of well-defined papillary stalks with thin fibrovascular cores lined by immature basaloid cells ± pleomorphism and minimal keratosis.
- *Spindle cell (sarcomatoid)*: Low-metastasizing (favorable PGN), glottic, polypoid mass with a monoclonal epithelial component with dual/divergent differentiation showing (+) AE1/AE3 and VIM as well as (±) SMA, MSA, and DES.
- *Basaloid*: Aggressive (poor PGN), usually supraglottic, carcinoma characterized by a substantial lobular growth of two components, i.e., basaloid (+ 34βE12 reacting with CK1, 5/6, 10, and 14) and squamous cells, peripheral palisading, comedo-type necrosis, PAS +/AB+ intercellular microcysts, and stromal hyalinization.
- *DDX*: NE tumors (no squamous differentiation, (+) CGA, SYN, NSE, CD56, CD57, PGP9.5), AdCC (myoepithelium), and ASqC (true ductulo-glandular differentiation and intracellular mucin). Although several biomarkers Although several IHC markers have been described to identify a NET, the most widely used and reliable NE markers remain to date CGA and SYN.
- *Acantholytic*: Variant characterized by marked acantholysis of squamous cells forming pseudolumina, anastomosing spaces and channels slightly more aggressive than SqCC. The acantholytic variant may simulate ASqC (+ mucin), AdCC (myoepithelium), MEC (+mucin, lobular arrangement, large clear intermediate cells, limited keratin pearls, and no CIS component), and angiosarcoma (+CD31).

> - *Adenosquamous*: Aggressive (poor PGN) with both squamous and glandular differentiation.

The importance of the surgical report does not need more stress at this time, but often phone calls from the radiotherapists or pediatric oncologists are encountered in the routine practice because the text may be not completely clear and the CAP protocol or minimal dataset has some discordances with the text describing the lesion. Thus, maximal care in reading the reports before signing out is paramount. Critical reports require critical communication (Sergi 2018).

The *surgical pathology report* should include the type of surgical procedure (hemi-, supraglottic, or total laryngectomy); accompanying material, such as neck dissection, thyroid gland, and parathyroid glands; site of origin; size and extent of tumor; histologic type and grade; perineural, lymphovascular, cartilaginous, and/or extralaryngeal invasion; and status of the resection margins. In hypopharyngectomy specimens, it is essential to indicate the laterality of the involvement of piriform sinus, postcricoid area, or posterolateral pharyngeal wall. If we receive a sample from a tracheal resection, the length of the trachea and the eventual intra- or extraluminal involvement also need to be on board of the report.

Neck dissection findings should include the total number of nodes specified for each level, the number of nodes positive for tumor, the size of the largest positive node, and the presence or absence of extracapsular tumor spread.

Other laryngeal tumors include epithelial and mesenchymal neoplasms. Epithelial tumors include lymphoepithelial carcinoma, giant cell carcinoma, malignant salivary gland-like tumors (MEC, AdCC), tumors with neuroendocrine differentiation (typical carcinoid, atypical carcinoid, and small-cell carcinoma), soft tissue tumors (benign and malignant and IMT, which is borderline or LMP), hematolymphoid tumors, tumors of bone and cartilage, mucosal malignant melanoma, and secondary tumors.

The term SILs or squamous intraepithelial lesions has been proposed as an abbreviation including the whole spectrum of epithelial changes, ranging from reactive lesions to carcinoma in situ (CIS). As part of this spectrum, it has been chosen by dysplasia and later also squamous intraepithelial neoplasia (SIN) or, more precisely, laryngeal and oral SIN. All these terms, which have been directly adopted from precursor lesions of the uterine cervix, do not include a spectrum of reactive epithelial changes, as do some systems, for instance, the Ljubljana classification. Comprehensively, all epithelial precursor lesions include squamous cell hyperplasia, mild dysplasia (aka squamous intraepithelial neoplasia/SIN 1, basal/parabasal cell hyperplasia), moderate dysplasia (aka SIN 2, atypical hyperplasia), severe dysplasia (aka SIN 3, atypical hyperplasia), and CIS (aka SIN 3).

14.4 Oral Cavity and Oropharynx

The lesions of the oral cavity and oropharynx are listed in Box 14.5 (Figs. 14.5, 14.6, and 14.7).

> **Box 14.5 Lesions of the Oral Cavity and Oropharynx**
> 14.4.1. Congenital Anomalies
> 14.4.2. Branchial Cleft Cysts
> 14.4.3. Inflammatory Lesions
> 14.4.4. Tumors

14.4.1 Congenital Anomalies

Congenital and developmental anomalies of the oral cavity and oropharynx have been part of the nonmedical literature and inspired several authors. They are itemized here as follows.

1. *Fordyce granules* (ectopic sebaceous tissue on the vermilion rim of lips and oral mucosa)
2. *Lingual thyroid* (thyroid gland tissue at the basis of the tongue)
3. *Nasopalatine duct cyst* (ciliated, mucous, squamous lining cyst)

14.4.1.1 Gingival Cysts

Often observed in newborns with no gender bias with an origin from the dental lamina, these nodules are small, multiple, and whitish located on the crests of the maxillary and mandibular dental ridges.

- *CLM*: Keratin-filled true cysts.
- *TRT*: Not indicated (self-resolution).

14.4.1.2 Epstein Pearls

Asymptomatic nodules of the mid-palatal raphe region along the line of fusion with no gender bias with an origin from entrapment of epithelial remnants. Epstein pearls need to be differentiated from Fox-Fordyce spots, which are characterized by sebaceous glands in the mouth, other than face and genitals. Fox-Fordyce "pearls" are typically small, painless, raised, white spots or bumps of small size (0.1–0.3 cm) and are not associated with any disease or illness.

- *CLM*: Nonodontogenic, keratin-filled cysts.
- *TRT*: Not indicated because of self-resolution.

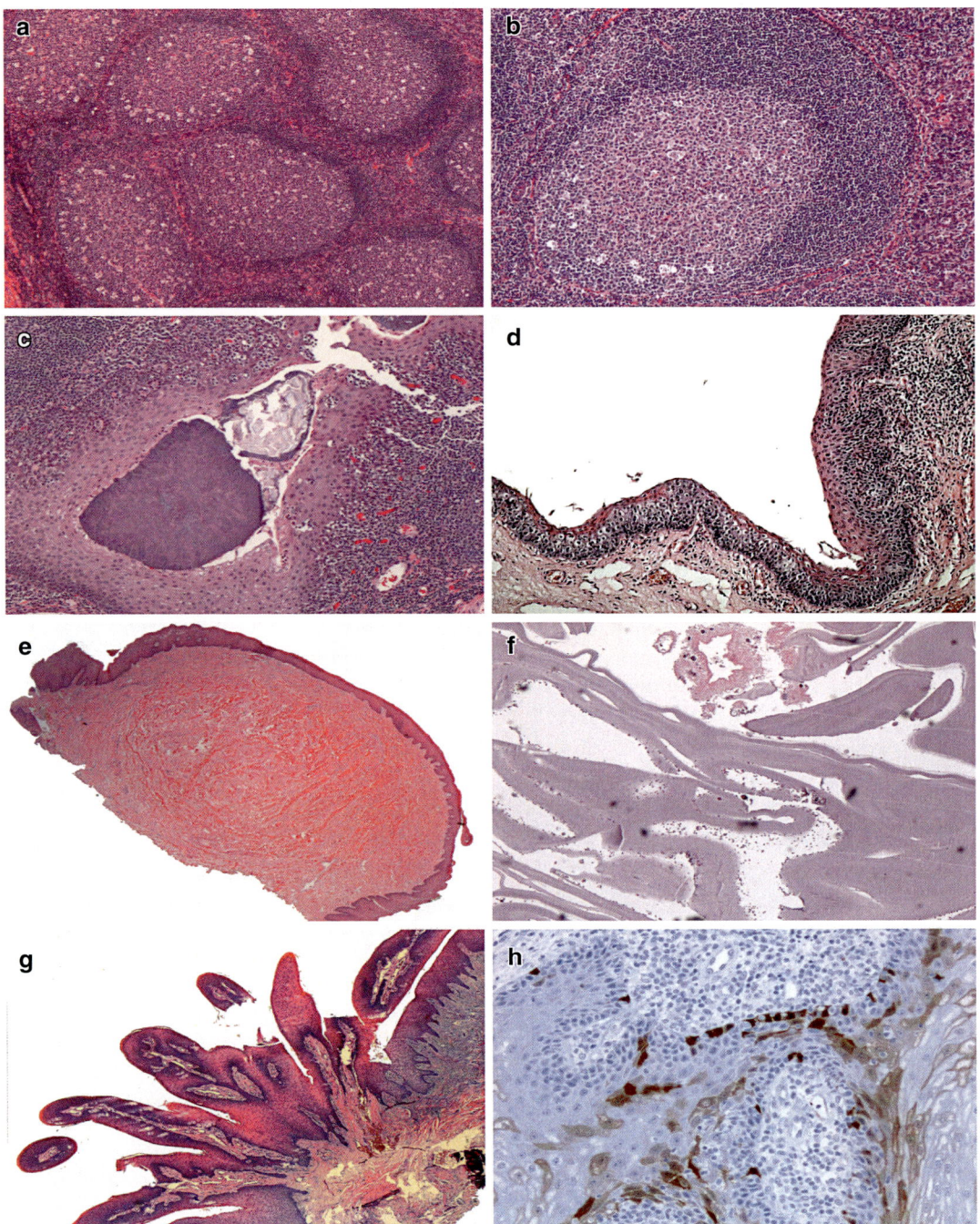

Fig. 14.5 The first three microphotographs show a chronic tonsillitis with normal polarization of the follicles (**a**, ×50; **b**, ×100) and actinomycosis (**c**, ×100). All three microphotographs are from histological slides stained with hematoxylin and eosin. A pauperization of the lymphoid tissue may be present in patients with HIV infection and progression to acquired immunodeficiency syndrome (**d**, ×100, hematoxylin and eosin staining). An irritation fibroma of the mouth is shown in (**e**) (×12.5, hematoxylin and eosin staining). *Echinococcus* cyst can be a sudden discovery in salivary gland mass with a cystic component or a cystic expression as shown in (**f**) (×50, hematoxylin and eosin staining). Figure (**g**) shows a papilloma of the oropharynx in an adolescent (×40, hematoxylin and eosin staining), while figure (**h**) (×200, P16 immunostaining) shows the histology of uvula papillomatosis, human papilloma virus (HPV) driven. P16 is a cyclin-dependent kinase inhibitor that regulates the cell cycle and cell proliferation mediating the inhibition of G1 progression of the cell cycle. P16 is a useful biomarker for transforming HPV infections

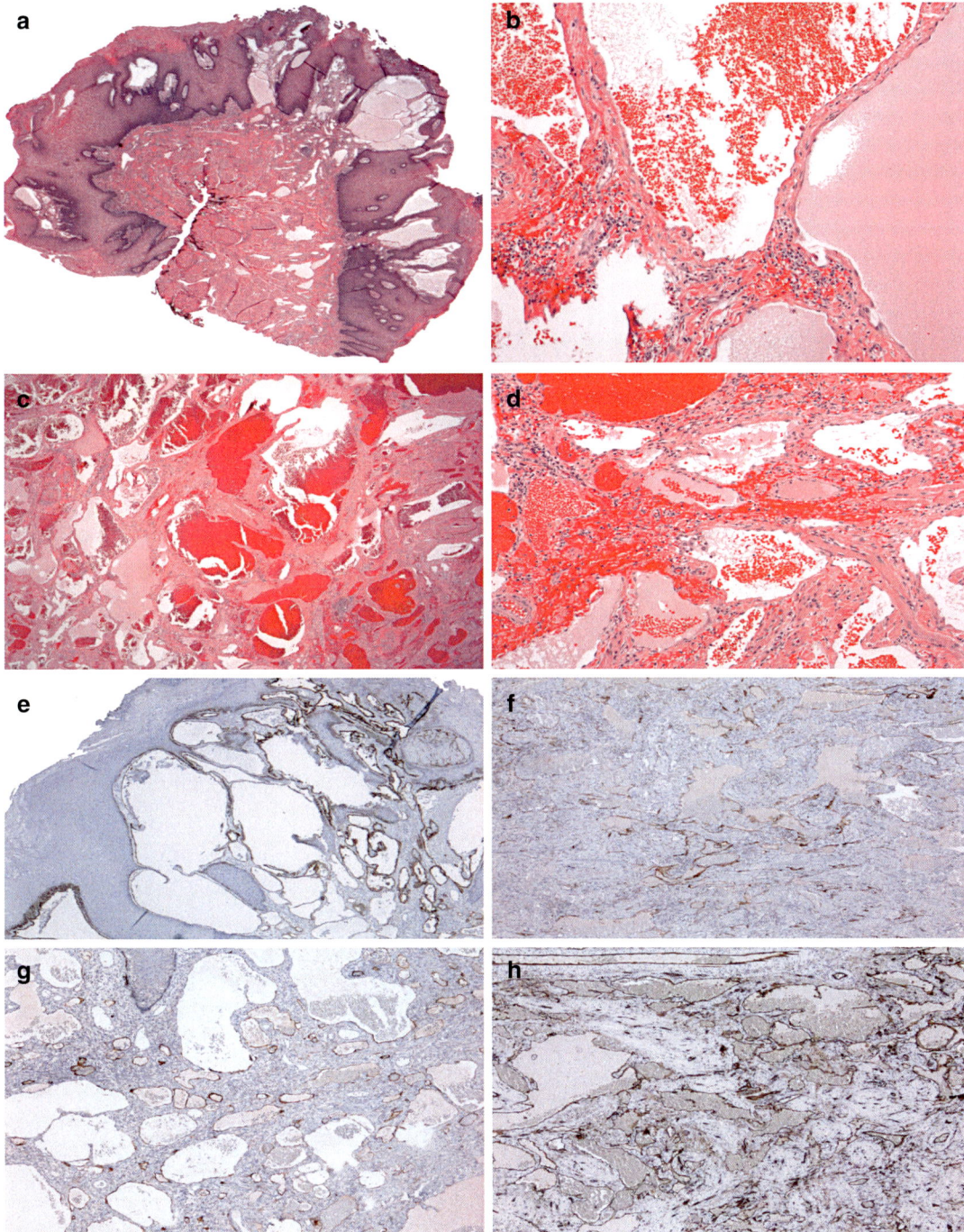

Fig. 14.6 This panel shows a hemo-lymphangioma of the tongue with enlarged cystic spaces, most of them filled with pink eosinophilic secretion (**a**, ×12.5), although some spaces contain also red blood cells (**b**, ×100; **c**, ×12.5; **d**, ×100). The immunohistochemistry panel shows the expression of D2-40 (podoplanin) (**e**, ×50; **f**, ×100; **g**, ×100) and focal expression of CD31 (**h**, ×100). All immunostaining sections have been performed by avidin-biotin complex

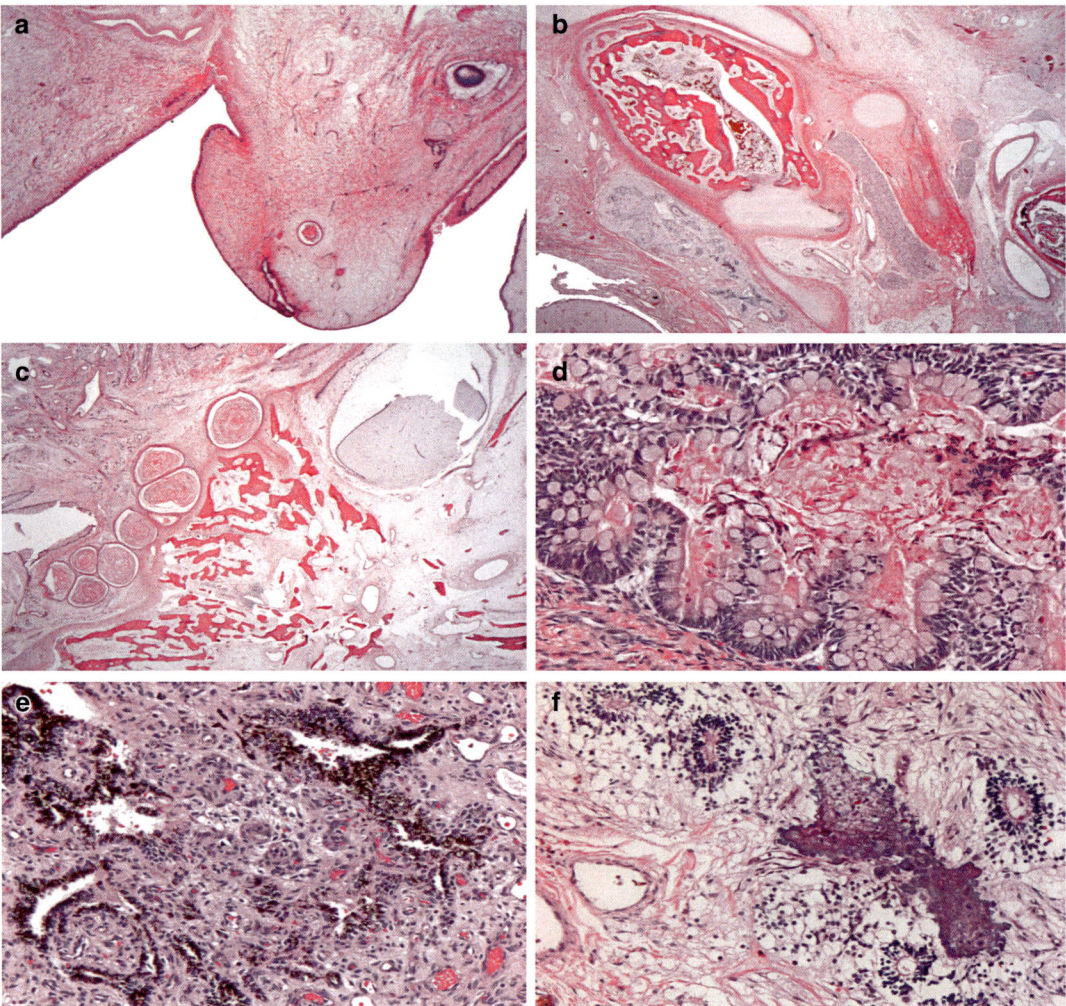

Fig. 14.7 A teratoma with ectodermal, mesodermal, and endodermal differentiation (**a**, ×12.5; **b**, ×12.5; **c**, ×12.5; **d**, ×200; **e**, ×200; **f**, ×200). Pigmented retina epithelium is seen in (**e**). All microphotographs are taken from histological slides stained with hematoxylin and eosin

14.4.1.3 Bohn's Nodules

They are keratin-filled cysts with the prevalence of 47.4% with no apparent gender bias. They are derivatives of palatal salivary gland structures. They clinically appear as numerous nodules along the junction of the hard and soft palate.

- *CLM*: Nonodontogenic, keratin-filled cysts.
- *TRT*: Not indicated (self-resolution).

14.4.1.4 Eruption Cyst

Cysts are originating from degenerative cystic changes in the reduced enamel epithelium or the remnants of the dental lamina due to an impediment of the eruption by overlying dense fibrotic mucosa.

- *CLI*: Bluish, dome-shaped, translucent, compressible mucosal nodules, which overlie the erupting tooth.
- *TRT*: Marsupialization or surgical extraction.

14.4.1.5 Epidermoid and Dermoid Cysts

- *DEF*: True cysts with squamous epithelial lining differentiated by the presence (dermoid) or absence (epidermoid) of adnexal tissue.

- *EPI*: 7% (incidence) in the H&N region.
- *CLI*: Slow-growing asymptomatic unpleasant cysts located commonly in the floor of the mouth and submental region with potential respiratory distress and feeding difficulty in newborns.
- *DGN*: Prenatal/natal US, MRI/CT, FNAC, and histopathology following enucleation.
- *DDX*: Ranula, dermoid cyst, teratoma, heterotopic gastrointestinal cyst, duplication foregut cyst, and lymphatic malformation.
- *CLM*: Epithelial-lined cavities with lumen with or without an adnexal tissue.
- *TRT*: Surgery.
- *PGN*: Recurrence is rare. If it does happen, the pathologist/surgeon needs to review the slides.

14.4.1.6 Maxillary Osteomyelitis of the Newborn

- *DEF*: Life-threatening infection of the maxillary bone observed in the neonatal period.
- *EPI*: 1–7 per 1000 children's hospital admissions with ♂ > ♀ (1.6:1) and preterm predilection.
- *EPG*: Infection due to *S. aureus*, *group B streptococcus*, *E. coli*, and *K. pneumoniae*. RFs are iatrogenic procedures (e.g., catheterization), prolonged hospitalization, parenteral nutrition status, ventilatory support, and nosocomial infections.
- *CLI*: Acute onset of fever followed by edema and redness of cheek, swelling of eyelids with conjunctivitis, and unilateral nasal discharge.
- *DGN*: (+) blood cultures ((+/↑) ESR, CRP, PMNs).
- *TRT*: Antimicrobial drugs ± surgery.
- *PGN*: Poor with high morbidity/mortality (*quoad vitam et valetudinem*).

14.4.1.7 Neonatal Herpes Simplex Virus Infection

- *EPI*: 1 in 3000–20,000 live births.
- *EPG*: HSV1 (orolabial and genital) and HSV2 (genital). The transmission happens during the time of birth delivery and is facilitated by the status of maternal antibody, maternal infection, i.e., primary or recurrent, duration of rupture of membranes, the integrity of mucocutaneous barriers, and mode of delivery, i.e., cesarean or normal. There is an incubation period of 4–21 days after delivery, and symptoms appear between 6 and 21 days.
- *CLI*: Eruptive vesicular rash (one single lesion to clusters) with 1–3 mm lesions involving the face (e.g., mouth), scalp, feet soles, and palms of the hand with a tendency to ulcer within a few days. The rash may be escorted by constitutional symptoms including cyanosis and respiratory distress, seizures, hepatitis, pneumonitis, and DIC.
- *DGN*: Serology, PCR of CSF, and viral cultures.
- *TRT*: Acyclovir (a synthetic nucleoside analogue).
- *PGN*: ~ good (0% mortality rate) in the case of SEM, but 70% for disseminated lesions.

14.4.1.8 Neonatal Candidiasis

- *DEF*: Life-threatening *Candida* infection with disseminated or invasive candidiasis being, probably once, the second most common cause of mortality in the neonatal period.
- *EPI*: 2–20% in preterm newborns.
- *ETP*: *Candida* species with *C. albicans* in 3/4 of cases (1/4: *C. glabrata*, *C. krusei*, *C. tropicalis*, and *C. parapsilosis*). The transmission can be vertical or due to external contaminations. RFs include immaturity of the neonatal immune system, prolonged catheterization and hospital stay, and previous antibiotic treatment of the mother during delivery.
- *CLI*: White plaques on oral mucosa. Complications include meningitis, endophthalmitis, endocarditis, and urinary tract infections.
- *CLM*: Hyphae, epithelial cells, and necrotic tissue (cell debris).
- *DGN*: Blood cultures, urine, and CSF.
- *TRT*: Antifungal treatment.
- *PGN*: If the pediatrician starts a calibrated antimycotic therapy efficiently and without delays, morbidity and mortality rates are remarkably decreased.

14.4.1.9 Mucocele
- *DEF*: Entrapped mucin (pseudocyst) of retention or extravasation type in the oral mucosa.
- *EPI*: 2.7% of patients <1-year-old babies.
- *CLI*: Lower lip, younger infants, trauma-related, bluish, translucent, and fluctuant swelling which may cause mechanical obstruction during feeding.
- *DGN*: FNAC and histopathology. A mucocele (aka sialocele and ptyalocele) is defined as the pooling of mucin in a cystic cavity (cyst or pseudocyst) with two types of mucoceles officially recognized: (1) the retention type, in which the mucin pooling is confined within a dilated excretory duct (cyst), and (2) the extravasation type, in which mucin is spilled into the connective tissues from a ruptured or traumatized duct (pseudocyst).
- *TRT*: Surgical excision.

14.4.1.10 Ranula
- *DEF*: Variant of mucocele found on the floor of the mouth presenting as swelling of connective tissue consisting of collected mucin from a ruptured minor salivary gland caused by local trauma.
- *EPI*: ~7.5 per 1000 live births.
- *CLI*: Swelling in the floor of the mouth and commonly caused by extravasation of mucin.
- *DGN*: FNAC, biopsy with histopathology, and MRI.
- *TRT*: Observation for asymptomatic cases, aspiration, cryosurgery, marsupialization, or surgical excision.

14.4.1.11 Riga-Fede Disease
- *DEF*: Rare benign reactive mucosal disease, trauma-related, with neonatal teeth in newborns occurring in a idiopathic form or associated with other disorders like Riley-Day syndrome, Lesch-Nyhan syndrome, Tourette's syndrome, and cerebral palsy. According to Domingues-Cruz et al. (2007) on the categories of Riga-Fede disease, we can distinguish between "precocious RFD" (younger than 6 months of life with perinatal teeth and no neurological disorders) and "late RFD" (older than 6 months of life with the first dentition association and optional neurological disorders).
- *CLI*: Ulcerated, unifocal/multifocal, and occasionally painful, lesions that often occur on the ventral surface of tongue in the midline region. Other sites include the lip, palate, gingiva, vestibular mucosa, and floor of the mouth.
- *DGN*: Clinical examination and histopathology.
- *DDX*: Ulcers caused by bacterial or fungal infections, immunologic diseases, and neoplasms.
- *TRT*: Dental extraction, corticosteroids, teething rings, oral disinfection, smoothing the incisal edges, and use of protective dental appliances.

14.4.1.12 Neonatal Pemphigus Vulgaris
Rare vesiculobullous disease of newborns caused due to transplacental passage of maternal IgG autoantibodies (mainly class 4) against transmembrane glycoprotein desmoglein 3 (Dsg3) and characterized by multiple cutaneous, mucosal, or mucocutaneous ulcerations after birth.

- *DGN*: Relies on histopathology with the use of immunofluorescence.
- *PGN*: Symptoms resolution within 2–3 weeks.

14.4.1.13 Oral Choristoma
- *DEF*: Histologically normal tissue, presenting in a non-native location and considered to be an aberrant developmental disorder.
- *EPI*: Rare occurrence with ♂ > ♀.
- *CLI*: Most frequent locations include the tongue (lingual choristoma), the floor of mouth, the pharynx, and the hypopharynx. The lesion can present as an asymptomatic mass of variable size causing obstruction in feeding and respiration in newborns.
- *DGN*: Clinical examination, imaging, and histopathology.
- *DDX*: Mucocele of extravasation type, ranula, lymphatic malformation, venous malformation, and dermoid cyst.
- *TRT*: Surgical excision.
- *PGN*: No recurrence is described.

14.4.2 Branchial Cleft Cysts

- *DEF*: Congenital ectoderm-derived (epithelial) cysts. The etiology for the branchial cleft cysts resides in a failure of the obliteration of the second branchial cleft during the development of the embryo. They are located on the lateral part of the neck. From the phylogenetical point of view, the branchial apparatus is intimately related to gill slits. In fact, the name branchial derives from the Greek branchia which means gills. The branchial arches are responsible for the development of the gills in fish and amphibians. Specifically, the event of four branchial clefts at the 4th week of embryonic life marks the formation of five ridges, which form several structures of the head, neck, and upper thorax. The growth of the second arch is caudal and covers the third arch and fourth arch. However, if a portion of the cleft, which is an ectoderm-lined cavity, fails to involute at the 7th week of gestation and persists, there is the generation of a cyst. Like a cyst, i.e., epithelium-lined cavity derives from the ectoderm and harbors commonly squamous epithelium. The entrapped remnant may or may not contain a sinus or fistula tract to the overlying skin.
- *EPI*: Unknown, but it is a common cause of visit and surgery in a department of head, nose, and throat of a children's hospital. In fact, branchial cysts are the most common congenital cause of a neck mass, and bilateralism occurs in 2–3% of cases, and there is a natural predisposition. Moreover, M = F, childhood/youth as presentation, no ethnic preferences.
- *CLI*: The branchial cysts are congenital, but clinical presentation occurs later in life, because of inflammation. In case of a UTI infection, the underlying lymphoid tissue can become hypertrophic, and an abscess can also occur with a tendency to a spontaneous rupture with resulting purulent draining to the skin or the pharynx. The HNT surgeon may identify a tender, enlarged, reddish, and warm neck mass on the lateral neck. Depending on the size, overimposed inflammation, and the extension of the branchial cleft cyst, dysphagia, dysphonia, dyspnea, and stridor may occur.
- *GRO*: Nodules or papulae that can be inflamed acquiring a reddish character and be accompanied by other inflammatory signs and symptoms.
- *CLM*: Squamous epithelium-lined cysts with keratin lamellae in the lumen and sporadic occurrence of respiratory epithelium with or without cartilage. An inflammatory component is often present, although at time of the surgery, it is chronic with lymphocytes and plasma cells. The surgeon tends to treat with antibiotics before a surgical resection is performed. Thus, neutrophils are virtually absent at the time the pathologist receives the specimen.
- *TNT*: Surgery.
- *PGN*: Recurrence is uncommon with an estimated risk of 3%, and the pathologist needs to be contacted about the potential recurrence and revision of the slides coming out from the previous surgery. The risk increases to 20% in case a previous surgery or recurrent infection was present due to the potential persistence of ectoderm-derived stem cells and an inflammatory component. In both cases, numerous schools recommend a postsurgical broad-spectrum antibiotic treatment for both children and adults experiencing a recurrence.

Genetic syndromes involving the head and neck region are extremely important, and a careful examination of the living child or an autopsy can reveal hints to address genetic counseling. Anomalies of the branchial structures include branchial sinuses and branchial cysts. *Branchial sinuses* are due to a failure of obliteration of the second branchial groove and generally manifest with a blind pit and form a "branchial fistula," if there is an extension of the structure into the pharynx. On the other hand, *branchial cysts* are remnants of cervical sinus or second branchial groove and are usually asymptomatic until early adulthood when enlargement occurs due to the accumulation of fluid and cellular debris. The most common genetic syndromes are first arch syndromes or sequences, including *Treacher Collins syndrome*, characterized by an autosomal dominant pattern of inheritance, malar hypoplasia, defects of the lower eyelid, and deformed external ear, and *Pierre Robin sequence* (PRS), characterized by a triad of micrognathia, glossoptosis, and cleft palate. PRS may occur as isolated findings or in association with additional syndromic features, such as ocular and auditory defects. Children harboring PRS may be at risk to variable degrees of upper airway obstruction, which can be life-threatening, and the association of neurologic deficits would probably be an indication for tracheotomy and gastrostomy tube placement to avoid chronic aspiration. Mandibular distraction osteogenesis is the only surgical technique aimed to address the underlying cause of upper airway obstruction in PRS. Congenital thymic aplasia or *DiGeorge syndrome* is due to a failure of differentiation of the third and fourth pouches without evidence of the thymus or parathyroid glands.

The differential diagnosis of branchial cleft cysts is broad and may include neoplastic and non-neoplastic lesions. The non-neoplastic reactive oropharyngeal lesions of the oral cavity that may have relevance in childhood, adolescence, and youth include the following disorders.

1. *Amalgam tattoo* (black-brown granular material embedded in collagen and at perivascular location) DDX: Melanocytic nevi and hemosiderin deposition (Fe-stain)
2. *Lymphoepithelial cyst* (squamous epithelial-lined cyst overlying a dense lymphoid cell population) DDX: Malignant lymphoma
3. *Median glossitis of rhomboid type* (glossitis with elongated rete and loss of papillae and florid *Candida* hyphae)
4. *Necrotizing sialometaplasia* (nests of squamous epithelium with lobular growth with inflammation and focal atypia ⇒ "inflammatory atypia") DDX: squamous cell carcinoma (SqCC) and mucoepidermoid carcinoma (MEC)
5. *Ranula/mucocele* (innocuous collection of mucin, which is either extravasated or trapped in the duct system of the salivary gland system of the oropharynx) DDX: MEC
6. *Teflon FBGC (foreign body giant cell) reaction* (refractive, polarizable, and translucent amorphous material with florid FBGC reaction)
7. *Tonsillar/adenoidal lymphoid hyperplasia* (follicular hyperplasia of the lymphoid tissue of one or both lymphoepithelial organs with benign features, including sharp demarcation to germinal centers, no expansion of the germinal centers, no progressive transformation of the germinal centers, standard mantle zone without increase, and polarization of the germinal cell population) DDX: NHL and carcinoma

Recently, Friedman et al. (2015) proposed a grading system for lingual tonsillar hypertrophy, another situation that may obstruct the airways and may have legal aspects in case of sudden death. The proposed grading system consists of a 0–4 scale. Grade 0 denotes a complete absence of lymphoid tissue in the tongue. Grade 1 comprises lymphoid tissue scattered over the base of the tongue. Grade 2 has lymphoid tissue covering the entirety of the tongue base with limited vertical thickness, while grade 3 consists of significantly raised lymphoid tissue covering the whole of the bottom of the tongue, roughly 5–10 mm in diameter. Finally, grade 4 represents lymphoid tissue ≥1 cm in diameter that rises above the tip of the epiglottis. Montgomery-Downs also modified Brodsky's scale to include midpoint values.

The lingual tonsils are a section of lymphoid tissue in Waldeyer's ring. Waldeyer's tonsillar ring (aka pharyngeal lymphoid ring) is a ringed arrangement of lymphoid tissue in the pharynx consisting of pharyngeal tonsil, two tubal tonsils, two palatine tonsils, and one lingual tonsil. The cause of lingual tonsil hypertrophy has not been identified. Multiple studies have found laryngopharyngeal reflux (LPR) to be correlated with lingual tonsil hypertrophy. Noteworthy, BMI and the metabolism studies behind the growth of a child have also been found to be associated with lingual tonsil hypertrophy. However, a recent survey by Hwang et al. (2015) does not seem to be related to obstructive sleep apnea-hypopnea syndrome (OSAHS) or BMI. High reflux symptom index (RSI), which is collected during patient intake, in younger age, and from the male gender, is associated with increased thickness of the lingual tonsil.

14.4.3 Inflammatory Lesions

Inflammatory lesions of the oral cavity and oropharynx are numerous and may be found in textbooks of head and neck clinics and pathology as well as dermatology. However, it is important to stress here some that may play a role in childhood, adolescence, and youth. The immune-mediated and dermatologic lesions of the oropharynx including oral cavity are grouped together.

1. *Bullous pemphigoid* (bullous inflammatory lesion with infrabasal clefts accompanied by chronic inflammation at the stroma-epithelial interface)
2. *Lichen planus* (band-like lymphoid/lymphocytic subepithelial infiltrate + vacuolar change or liquefaction of the basal layer with a sawtooth pattern of the epithelium)
3. *Pemphigus vulgaris* (bullous inflammatory lesion with acantholysis and suprabasal blisters and richness of "tombstones' row" of basal epithelial cells)
4. *Sjögren's syndrome* (atrophy of the glandular tissue + chronic inflammation with >50 lymphocytes or plasma cells/HPF: (i.e., 40X as objective)

The fibrous and pseudogranulomatous (non-neoplastic) and pseudotumoral lesions of the oropharynx including the oral cavity with some inflammatory character are described here (*vide infra*).

- *Giant cell fibroma*: Nodule of dense collagen scarcely to moderately cellularity and scattered giant cells.
- *Irritation fibroma*: Nodule of paucicellular dense collagen.
- *Lobular capillary hemangioma*: Small, thin-walled BVs with the lobular arrangement and GLUT1 (+).
- *Peripheral giant cell granuloma*: Nodule of spindle cells to ovoid cells with bland cytology and scattered MNGCs, small lymphocytes, and plasma cells.
- *Peripheral ossifying fibroma*: Nodule of high cellularity with fibroblastic stroma and scattered ossification with no atypia or malignant features, e.g., invasion (DDX: Osteosarcoma, fibrous dysplasia, simple bone cyst, and ABC)
- *Pyogenic granuloma*: Small, thin-walled BVs with "collarette", GLUT1 (−).

14.4.4 Tumors

In the following sections are listed tumor lesions that occur more often in late childhood, adolescence, and youth. Most of these lesions do occur in adults but may be present in children and adolescents harboring a predisposition to develop cancer (genetic cancer syndromes) or have been in contact with carcinogens and mutagens in the environment.

14.4.4.1 Epithelial Precursor Lesions

The clinical term "leukoplakia" (Greek: λευκός, "white" and πλάξ, "plate") means "white plaque or plate." It is a premalignant lesion, sharply demarcated that can occur anywhere in oral mucosa, more often seen in late youth in children starting smoking very early. The role of vaping is uncertain at this time. There is an orderly mucosal thickening with epidermal hyperplasia and hyperkeratosis with focal disorderly hyperplasia. Varying degrees of dysplasia up to CIS have been described in late adults only.

14.4.4.2 Benign Epithelial Tumors

Five entities need to be differentiated that are squamous papillomas, lymphangioma, granular cell tumor, congenital granular cell epulis, and keratoacanthoma. However, keratoacanthoma is not seen in youth.

1. *Squamous papilloma* (numerous fibrovascular cores with overlying squamous epithelium with bland cytology)
    DDX: Pseudoepitheliomatous hyperplasia, squamous cell carcinoma, and papillomatosis
2. *Lymphangioma* (thin-walled vascular spaces with a proteinaceous material, with (+) D2–40 and (−) CD31 immunologically on the lining of the vascular areas)
    DDX: Hemangioma (−) D2–40 and (+) CD31
3. *Granular cell tumor* (nodule of cells with granular cytoplasm and small nuclei, charac-

teristically situated underneath of the epithelium and favoring a pseudoepitheliomatous cell reaction and displacing and infiltrating the submucosal tissues, with (+) PAS, S100, PGP 9.5, NSE, CD68, AAT, INA)
DDX: Congenital granular cell epulis
4. *Congenital granular cell epulis* (proliferation of cells with granular cytoplasm admixed with small thin-walled blood vessels and lack of a pseudoepitheliomatous cell reaction, (+) NSE, VIM, but (−) S100
DDX: Granular cell tumor

14.4.4.3 Malignant Epithelial Tumors

Squamous cell carcinoma (SqCC) is sporadic in childhood and youth, but occasional cases may be encountered in immunodeficient or immunosuppressed (e.g., following transplantation) young individuals and should be differentiated from squamous preneoplastic lesions.

Squamous preneoplastic and neoplastic lesions are:

1. Squamous dysplasia vs. reactive atypia (presence of inflammation and downward extension of rete with regular spacing).
2. Proliferative leukoplakia of verrucous type (rare spectrum of changes from simple squamous hyperplasia through verrucous hyperplasia, verrucous carcinoma to SqCC of usual type).
3. Verrucous hyperplasia vs. verrucous carcinoma.
4. Micro- or broadly invasive squamous cell carcinoma.

Squamous dysplasia (SIN, squamous intraepithelial neoplasia) and CIS are:

1. SIN1 = Mild dysplasia ⇒ low-grade dysplastic lesion
2. SIN2 = Moderate dysplasia ⇒ high-grade dysplastic lesion
3. SIN3 = Severe dysplasia ⇒ high-grade dysplastic lesion

As mnemonics technique, the acronym "ASIC" can be used for the architectural disorder, size (nuclear) variability, irregularity of the nuclear contour, and chromatin (nuclear) changes.

The variants of malignant squamous lesions are as indicated below:

- *Squamous cell carcinoma*, early invasive (*vide supra*)
- *Verrucous* (wart-like with characteristic "church-spire" proliferation, no atypia, downward growth with pushing border with DDX including verrucous hyperplasia, proliferative verrucous leukoplakia, and papillary squamous cell carcinoma)
- *Papillary* (long fibrovascular cores of variable complexity lined by atypical squamous epithelium with lack of hyperkeratosis)
- *Basaloid* (basaloid cell proliferation with hyperchromatic nuclei, palisading pattern, and often set in a basement membrane-like hyaline matrix)
- *Adenoid* (squamous cell proliferation with a characteristic pseudoglandular pattern with cleft-like spaces and lack of mucin, goblet cells, and no immunoreactivity for CD31, CD34, and FVIII)
- *Spindle cell/pseudosarcomatous* (exophytic mass of atypical spindle cells with histologic or IHC evidence of distinct keratin production with DDX including fibrosarcoma, and malignant fibrous histiocytoma)
- *Non-keratinizing* (*vide infra*)

The non-keratinizing squamous lesions are classified as follows.

1. *Differentiated* "non-papillary urothelial carcinoma-like tumor" (nests or sheets of cohesive cells with hyperchromatic nuclei and small, dense nucleolar presence)
2. *Undifferentiated* (syncytial growth pattern of cells with pale, vesicular nuclei and prominent nucleoli described as "centroblast-like" cells)
3. *Undifferentiated with lymphoid stroma* (syncytial growth pattern of cells with pale, vesicular nuclei and prominent nucleoli described as "centroblast-like" cells + interspersed EBV+ lymphoid cells ⇒ Schminke's and Regaud patterns)

Non-keratinizing SqCC major DDX is with lymphoma (B-cell lymphoma, (+) CD45, CD20, CD79a) and typically either *DLBCL* (anaplastic, centroblastic, immunoblastic, T-cell histiocyte-rich) or *EN-NK-TCL* (+) CD2, cCD3e, CD56, TIA1, Perforin, EBV, or, even, *malignant melanoma* (+) S100, HMB45, and Melan-A.

14.4.4.4 Soft Tissue Tumors

Benignancy is seen for focal oral mucinosis, congenital granular cell epulis, lymphangioma, and ectomesenchymal chondromyxoid tumor, while malignancy of this region is Kaposi sarcoma (KS).

Congenital Epulis of Newborn

- *DEF*: Rare benign tumors of the oral cavity with remarkable similarity to granular cell tumor of the adults, although distinguished based on its exclusive origin from the neonatal gingiva, the scattered presence of odontogenic epithelium, lack of interstitial cells with angulate bodies, and the more intricate vasculature.
- *EPI*: 6 in 10,000 neonates (incidence rate) with female preference (♂ < ♀) (♂ : ♀ = 1:10).
- *EPG*: There are odontogenic, neurogenic, myogenic, endocrinological, fibroblastic, or histiocytic factors that may play a role.
- *CLI*: The alveolar ridge lesions present as a lobular or ovoid, sessile, or pedunculated swelling of variable size, which is often covered by a smooth reddish mucosa.
- *DGN*: US, CT/MRI, and histopathology.
- *CLM*: Granular cell tumor-similar histopathology.
- *DDX*: Dermoid cysts, teratoma, hemangioma, lymphatic malformations, RMS, and melanotic neuroectodermal tumor of infancy (MNTI).
- *TRT*: Complete surgical excision with no reported cases of recurrence.

Pyogenic Granuloma (Lobular Capillary Hemangioma)

- *DEF*: Inflammatory "tumor" composed of granulation tissue often localized at the maxillary labial gingiva.

Hemangioma

Hemangiomas are most common pediatric vascular benign neoplasms with a specific course including three stages: (1) rapid proliferating phase (0–1 year), (2) involuting phase (1–5 years), and (3) involuted phase (5–10 years).

- *EPI*: 2–3% in neonates and 22–30% preterm infants, ♂ < ♀ (3:1–5:1).
- *CLI*: H&N (60%), commonly the lips, oral mucosa, palate, and uvula, apart from trunk and extremities.
- *RFs*: Gestational hypertension and low birthweight. Kasabach-Merritt syndrome consists of the combination of giant hemangioma, thrombocytopenia, and consumptive coagulopathy. Hemangioma presents as a swiftly growing macule followed by regression into spotted pigments.
- *DGN*: History, FNAC, MRI, color Doppler/US, surgery, histology, and IHC (Glut 1 +, but absent in vascular malformations).
- *DDX*: Hemangioendothelioma (see other chapters).
- *TRT*: Drugs (propranolol, corticosteroids, interferon), lasers (CO_2, diode, flashlamp pulsed dye), and surgery with complete resolution in 3/4 of cases (remnants in 1/4 include telangiectasias, stippled scarring, anetoderma, hypopigmentation, and fibro-fatty residual tissue). The anetoderma, which is rare in childhood, but a frequent complication in adults, occurs when the elastic tissue localized in the dermis is lost, resulting in a progressive depression in the skin. The anetoderma is also known as macular atrophy in some dermatologic literature.

Lymphangioma

- Lymphangiomas are benign neoplasms of the lymphatic channels with 50% discovered at birth and more than 90% <2 years old.
- *EPI*: 1–3:10,000 live births (prevalence), ♂ = ♀, H&N (3/4) (tongue – dorsal face, lips, oral mucosa, soft palate, and floor of the mouth) and 1/4 (trunk, abdomen, and limbs). The etiopathogenesis includes five major theories:
  1. Congenital obstruction of the primitive lymphatic plexus

2. Congenital sequestration of the primitive lymphatic channels
3. Ectopic deposition of lymph tissues
4. Inborn abnormal expression of vascular endothelial growth factor (VEGFR)
5. Incomplete development of the lymphatic channels
- *CLI*: Macroglossia, sialorrhea, dysphagia, abnormal facies, ulcerations, speech and feeding difficulties, respiratory distress, infection, and fever.
- *DGN*: Clinical examination, MRI/CT/color Doppler US, and histopathology.
- *GRO*: Slow, progressive nodule with superficial blue-black or red components.
- *CLM*: Simplex, cavernous, cystic pattern of dilated and proliferated lymphatic channels.
- IHC (+) for D2–40, Prox-1 (Prospero homeobox protein 1), factor VIII-associated antigen, CD-31, LYVE-1, and VEGFR. Lyve-1, aka hyaluronic acid receptor, is lymphatic vessel endothelial hyaluronan receptor 1. Prenatal DGN may reveal abnormalities of the chromosomes 13, 18, 21, X, and Y.
- *TRT*: Surgery, cryotherapy, electrocautery, sclerotherapy, steroids, embolization, and ligation, laser surgery (Nd:YAG or neodymium-doped yttrium aluminum garnet, CO_2), radiofrequency tissue ablation technique, and RT.
- *PGN*: RCR is 39% in case of the tongue, which is followed by the sites of hypopharynx and larynx.

Langerhans Cell Histiocytosis X (LCH)
- *DEF*: Hematologic neoplasm affecting the Langerhans cells, which are a type of dendritic cell which fights infection, and harboring somatic mutations in Langerhans cells involving the *BRAF*, *MAP2K1*, *RAS* and *ARAF* genes.
- *EPI*: 1 in 200,000 children under 15 years (frequency) with ♂ > ♀.
- Classification: *Eosinophilic granuloma*, *Hand-Schuller-Christian disease* (between 3 and 6 years of age), and *Letterer-Siwe disease* (under 2 years of age). This original classification has lost its value, but it is often taught in schools to identify phenotypes. A new revised classification has been proposed (Emile et al. 2016). Often, LCH is simply subdivided in single-system LCH affecting one organ or body system, and multi-system LCH affecting two or more organs or body systems.
- An LCH variant, congenital self-healing reticulohistiocytosis or *Hashimoto-Pritzker disease*, is present at birth and may show complete involution within 2–3 months.
- *CLI*: Erythematous vesicular and pustular lesions with/without crusting and eczematous scaling on the skin. Oral manifestations include petechiae, lytic bone lesions involving posterior mandible, and pain and swelling of gingiva. Bony lytic lesions include the sites of skull, femur, pelvis, and vertebrae. Systemic findings may consist of hepatomegaly, splenomegaly, pulmonary involvement, pancytopenia, CNS involvement, diabetes insipidus, among others, depending upon the severity. Substantially, LCH may affect low-risk organs or high-risk organs/systems. Low-risk organs/systems include the skin, skeleton, LNs, GIT, pituitary gland, and CNS, while high-risk organs/systems include the liver, spleen, and bone marrow.
- *DGN*: Histopathology and IHC ((+) S100, CD1a, Langerin) accompanied by a complete liver and blood chemistry profile and adjuvated by CT/MRI in evaluating the extent of systemic involvement. 18F-FDG (Fluorodeoxyglucose) PET/CT is considered useful for the evaluation of LCH when compared to conventional imaging, probably except in pulmonary manifestations (Albano et al. 2017).
- *DDX*: Seborrheic dermatitis, osteomyelitis of jaw, leukemia, lymphoma, and metastatic tumors.
- *TRT*: Surgical excision, CHT, and RT. High-risk LCH standard of care consists of 1 year of therapy with vinblastine/prednisone/mercaptopurine, based on the LCH-III study, although increasing treatment duration from 6 months in LCH-II to 12 months in LCH-III may be beneficial (Allen et al. 2015).
- *PGN*: It depends upon the age and extent of systemic involvement.

14.4.4.5 Hematolymphoid Tumors, Non-LCH

Non-LCH hematolymphoid tumors of the oral cavity and oropharynx include B-NHL, such as MCL, FCC, Burkitt lymphoma, DLBCL, EN-MALT-L, as well as Hodgkin lymphoma, and T-NHL, such as T-NHL and ALCL. Occasionally, the extramedullary myeloid sarcoma (EMS) occurs in children and adolescents and EMP and FDCS in adults are also observed. These hematolymphoid tumors are also better delineated in the hematological chapter. Of these neoplasms, probably the Burkitt lymphoma has paramount relevance. *Burkitt lymphoma* is characteristically recognized as starry sky pattern with monotonous, medium-sized, "squared-off" tumor cells with coarse chromatin, multiple small nucleoli, scant cytoplasm, and high MKI as well as interspersed histiocytes. IHC: + pan-B markers (CD19, CD20, CD22, CD79a) and + germinal center markers (CD10 and Bcl-6). DDX includes the DLBCL.

Oropharyngeal hematopoietic tumors may be classified according to germinal center origin.

- *Pre-GC neoplasm*:
  - Mantle cell lymphoma
- *GC neoplasms*:
  - Follicular lymphoma
  - Burkitt lymphoma
  - DLBCL
  - Hodgkin lymphoma (NS variant)
- *Post-GC neoplasms*:
  - MZL and ENMZ-BCL (MALT NHL)
  - LPCL
  - CLL/SLL
  - DLBCL
  - PCM
  - EMS

where GC is germinal center; DLBCL, diffuse large B-cell lymphoma; MZL, mantle zone lymphoma; LPCL, lymphoplasmacytic lymphoma; ENMZ-BCL, extranodal marginal zone B-cell lymphoma; MALT, mucosa-associated lymphatic tissue; CLL, chronic lymphocytic lymphoma; SLL, small lymphocytic lymphoma; DLBCL, diffuse large B-cell lymphoma; PCM, plasma cell myeloma; and EMS, extramedullary myeloid sarcoma or granulocytic sarcoma (aka "chloroma").

Extramedullary Myeloid Sarcoma (EMS)

- *DEF*: Granulocytic sarcoma of the oropharynx mostly restricted to the palate and gingiva. EMS is an extramedullary manifestation of acute myeloid leukemia (AML), which can be antecedent, synchronous, or metachronous to EMS. EMS is characterized by a *diffuse chloroacetate esterase reaction (special stain)* in ~3/4 of cases. EMS, also known as "chloroma," was first described by the British physician A. Burns in 1811, although the word "chloroma" appears in 1853. Chloroma is derived from the Greek word χλωρός, which means bright green to yellow. This specific staining pattern is due to the color of the EMS that often exhibits a green tint due to the MPO. However, up to 1/3 of cases these tumors are gray-whitish or brown and more precise term should be granulocytic sarcoma as adopted by many authors.

- *CLM*: *Blast cells* with round to ovoid nuclei; fine chromatin; small, distinct nucleoli; and lightly eosinophilic cytoplasm with more intense eosinophilic granules are seen admixed with eosinophilic *myelocytes* and *metamyelocytes*. Giemsa-stained touch preparation is useful to identify cytoplasmic azurophilic granules and Auer rods, when present (Fig. 14.8).

- IHC: + MPO, CD13, CD33, CD117, CD68/KP1, neutrophil elastase, and lysozyme. However, EMS with monocytic differentiation is -MPO, but +CD68/PGM1. CD68 is the human homologue of mouse macrosialin and is a heavily glycosylated, membrane protein, which is closely related to the family of lysosomal associated, mucin-like membrane proteins (LAMPS). Apart of the lysosomal (cytoplasmic) localization, a small fraction of CD68 is also found on the cell surface. There are two antibodies against CD68, including KP1 and PGM1. In AML, KP1 recognizes M1-M5 types, while PG-M1 recognizes M4 (myelo-monocytic) and M5 (monocytic) only. Moreover, it is known that neutrophils and myeloid precursors are usually detected by KP1, not PGM1, while mast cell and synovial cells react with PGM1 and not KP1.

- *DDX*: NHLs.

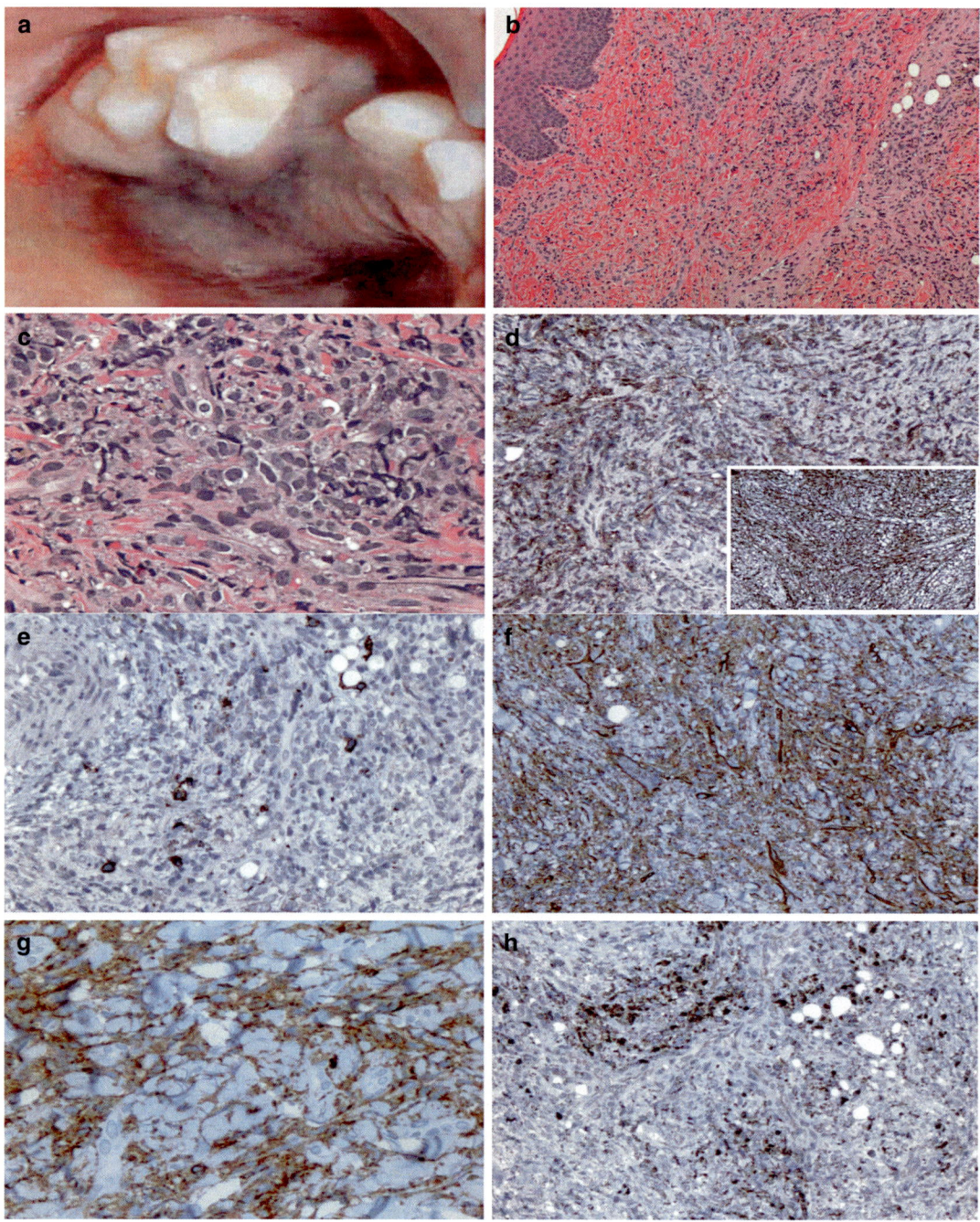

Fig. 14.8 Oral view in a patient presenting with a raised purple lesion of the palate (**a**). The histology of a small incisional biopsy discloses infiltrating tumor cells with various cell size and morphology and hyperchromasia of the nuclei (**b**, ×100; **c**, ×400; both microphotographs are hematoxylin-eosin-stained slides). The immunohistochemical panel highlights the expression of myeloperoxidase (MPO), CD117, CD34, CD45, CD68K (**d**, ×200; **e**, ×200; **f**, ×100; **g**, ×400; **h**, ×200, respectively). All immunostained microphotographs are taken from slides undergoing an avidin-biotin complex (ABC) staining. The inset in (**d**) corresponds to the expression of lysozyme (×100, ABC immunostaining)

Follicular dendritic cell sarcoma (DCS) is a low- to intermediate-grade malignant tumor of FDC, which may evolve from an underlying Castleman disease of hyaline vascular type. Most common sites of the oropharynx include the tonsil and palate. This tumor consists of spindle cells with ill-defined cell borders, elongated nuclei, fine chromatin, small distinct nucleoli, rare nuclear pseudoinclusions, and a large eosinophilic cytoplasm with a "fibrillary quality." Areas with haphazardly clustered nuclei, a sprinkling of lymphocytes, and scattered MNGC may also be seen. Nuclear atypia and pleomorphism may occasionally be present. IHC: +CD21, CD23, CD35, − CKs, and EM: interdigitating lengthy slender cytoplasmic processes and intercellular desmosome junctions.

14.4.4.6 Other Malignant Neoplasms

Neuroblastoma, rhabdomyosarcoma, and malignant melanoma may occasionally arise in the H&N region of a newborn but are often seen later in life (second infancy and childhood) and they are described in other sections of this book.

14.5 Salivary Glands

Salivary gland pathology is not infrequent in the routine of the pediatric pathologist, mainly if fine needle aspiration is added to his or her routine (Box 14.6) (Figs. 14.9, 14.10, 14.11, and 14.12).

> **Box 14.6 Salivary Gland Pathology**
> 14.5.1. *Congenital Anomalies*
> 14.5.2. *Inflammatory Lesions and Non-neoplastic Lesions*
> 14.5.3. *Tumors*

14.5.1 Congenital Anomalies

Congenital anomalies of the salivary glands include agenesis, aplasia, atresia of the excretory ducts, ectopic salivary gland tissue, and Stafne defect, which is an unusual condition due to an ectopic portion of salivary gland tissue which causes the bone of the lower jaw to remodel around the tissue, creating an apparent cyst-like radiolucent area on mandibular radiographs. Since Stafne defect is asymptomatic, it has also been considered to be an anatomic variant rather than a real disease. The difference between agenesis and aplasia is that agenesis means the absence of the salivary gland due to absent primordial tissue, while aplasia implies the lack of the fully developed salivary gland despite the presence of primordial tissue. The lesion with the eponym "Stafne bone defect" is associated with Edward C. Stafne, who first described this entity in 1942.

14.5.2 Inflammatory Lesions and Non-neoplastic Lesions

Inflammatory lesions include sialadenitis of infectious type (e.g., mumps) or non-infectious type (e.g., trauma). Sarcoidosis may also involve the parotid gland alone or with the eyes or uveoparotitis (inflammation of both the parotid and the uvea of the eyes), which occurs in Heerfordt syndrome. Moreover, cheilitis glandularis refers to an inflammation of the minor salivary glands, usually in the lower mouth rim, eversion, and swelling of the lip. Chronic sclerosing sialadenitis is a salivary gland manifestation of the IgG4-related disease. Non-neoplastic lesions, which may have an inflammatory component, but the central element is non-inflammatory, include necrotizing sialometaplasia, mucocele, ranula, Sjögren syndrome, graft-versus-host disease (GVHD), sialolithiasis, sialadenosis, and diverticular disease. Necrotizing sialometaplasia refers to a non-neoplastic lesion that arises from a minor salivary gland on the palate and is due to vascular infarction of the salivary gland lobules. Sialolithiasis is the building of salivary stones due to the altered acidity of saliva, ↓salivary flow rate, abnormal Ca^{2+} metabolism, and abnormalities in the sphincter mechanism of the major or minor salivary duct openings. Sialadenosis (sialosis) is a recurrent swelling of the salivary glands due to irregularities of the neurosecretory control.

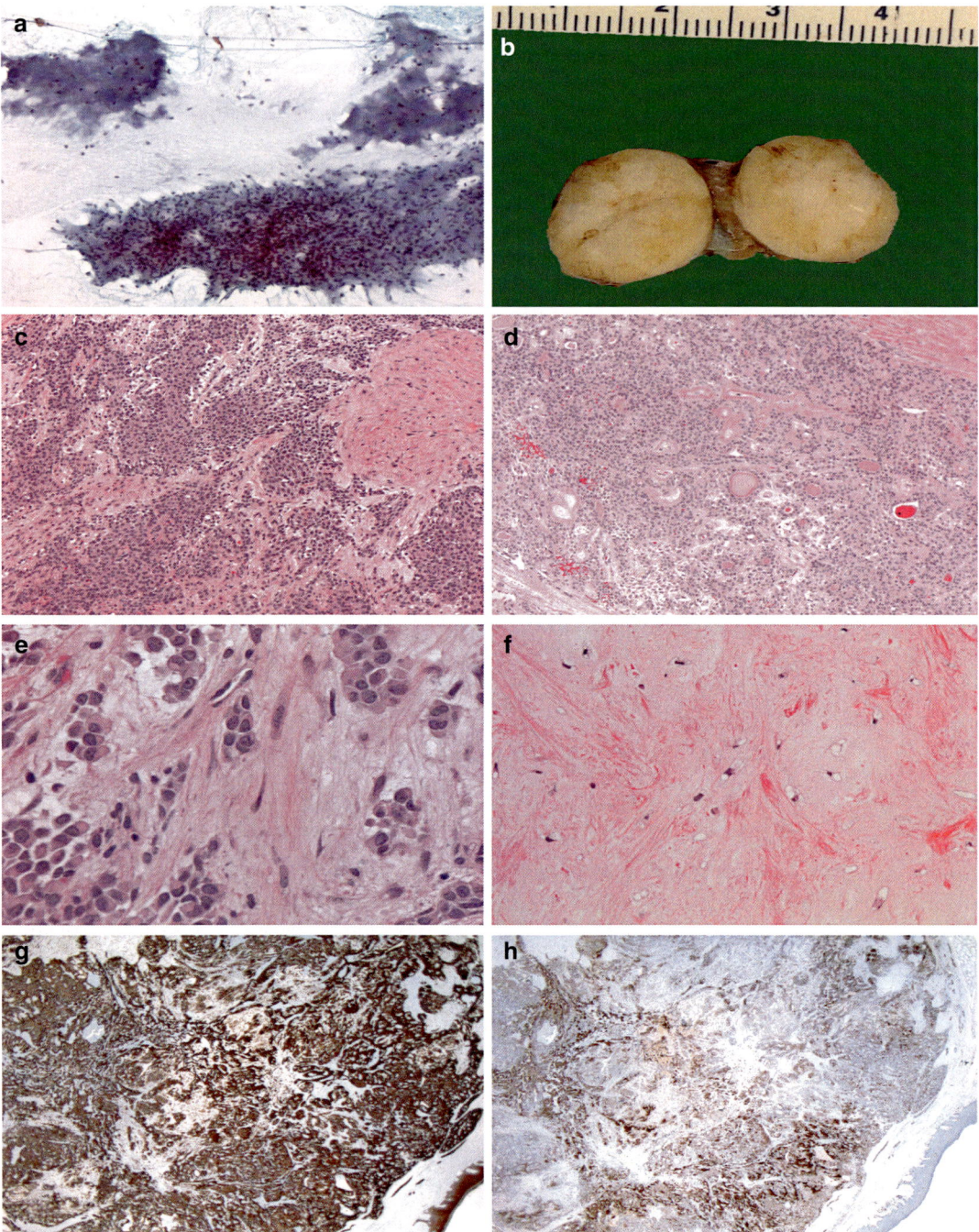

Fig. 14.9 This panel illustrated several features of a pleomorphic adenoma of the oropharynx. In (**a**) is shown the fine needle aspiration cytology (clusters of epithelial cells with blue myxoid matrix with benign features), while in (**b**) is shown the cut surface of a pleomorphic adenoma of a symptomatic child. Grossly, the tumor is well demarcated and partially encapsulated. The cut surface is gray-white with a rubbery aspect. The hematoxylin-eosin-stained histologic sections show in (**c**) through (**f**) the pleomorphic features of a pleomorphic adenoma with a biphasic population of epithelial and mesenchymal cells (**c**, ×100; **d**, ×100; **e**, ×400; **f**, ×200). Figures (**g**) (×12.5) and (**h**) (×12.5) are microphotographs of immunostained slides of a pleomorphic adenoma showing the expression of pancytokeratins (**g**) and S100 (**h**), respectively. The immunohistochemistry was performed using the avidin-biotin complex method

14.5 Salivary Glands

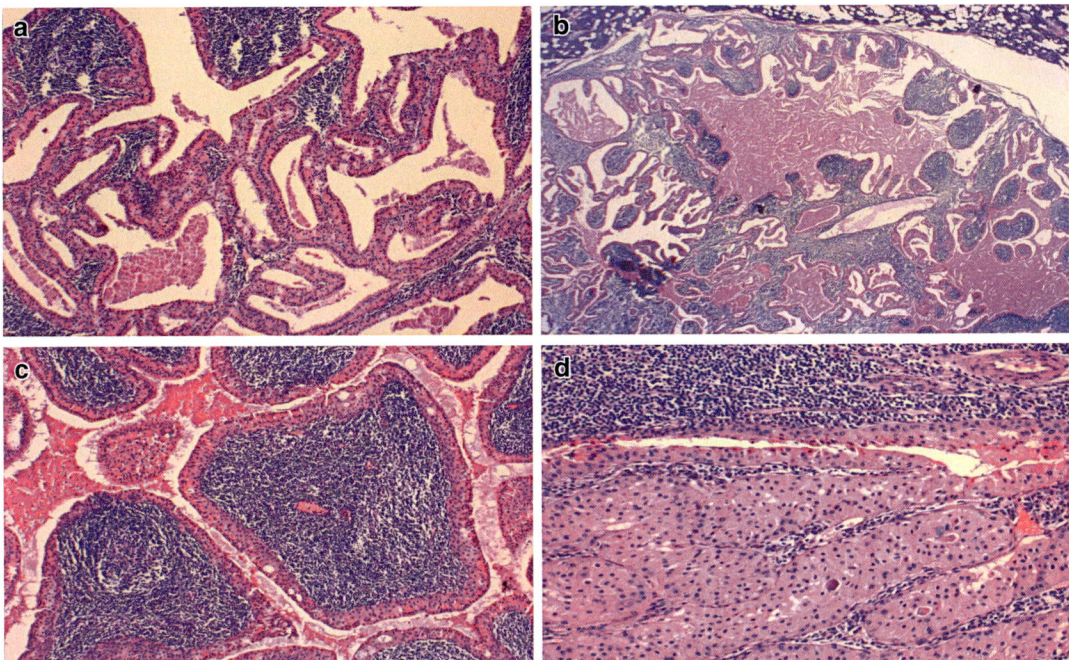

Fig. 14.10 The microphotographs (**a–d**) show the characteristic features that gave the original name to this entity ("*cystadenoma lymphomatosum papilliferum*" or later labeled as adenolymphoma) of an adolescent. There are cystic spaces narrowed by polypoid projections of lymphoepithelial elements, and the epithelial cells form a double layer lying on dense lymphoid stroma with variably activated germinal centers. Warthin tumors are rare in childhood with very few pediatric case reported in English literature (Aoki 2014)

14.5.3 Tumors

14.5.3.1 Benign Tumors

The *benign* salivary gland tumors of the oral cavity and oropharynx include pleomorphic adenoma, myoepithelioma, basal cell adenoma, canalicular adenoma, duct papilloma, and cystadenoma. The most critical salivary gland neoplasms of the oral cavity include:

1. Monomorphic adenoma
2. Canalicular adenoma
3. Pleomorphic adenoma (PA)
4. Polymorphic low-grade adenocarcinoma (PLCA)
5. Acinic cell carcinoma (ACCA)
6. Mucoepidermoid carcinoma (MEC)

14.5.3.2 Malignant Epithelial Tumors

The *malignant* salivary gland tumors of the oral cavity and oropharynx include acinic cell carcinoma, adenoid cystic carcinoma, mucoepidermoid carcinoma, polymorphous low-grade adenocarcinoma, basal cell adenocarcinoma, epithelial-myoepithelial carcinoma, clear cell carcinoma NOS, cystadenocarcinoma, mucinous adenocarcinoma, oncocytic carcinoma, and salivary duct carcinoma.

The category of the malignant epithelial tumors includes carcinoma ex-PA and metastasizing PA, acinic cell carcinoma, adenoid cystic carcinoma, mucoepidermoid carcinoma, PLCA, epithelial-myoepithelial carcinoma, myoepithelial carcinoma, clear cell carcinoma NOS, basal cell adenocarcinoma, sebaceous carcinoma, sebaceous adenocarcinoma, cystadenocarcinoma, low-grade cribriform cystadenocarcinoma, mucinous adenocarcinoma, oncocytic carcinoma, salivary duct carcinoma, adenocarcinoma NOS, carcinosarcoma, SqCC, small-cell carcinoma, large-cell carcinoma, lymphoepithelial carcinoma (LEC), and sialoblastoma. *Carcinoma ex-PA* and *metastasizing PA* are extremely rare in childhood and youth.

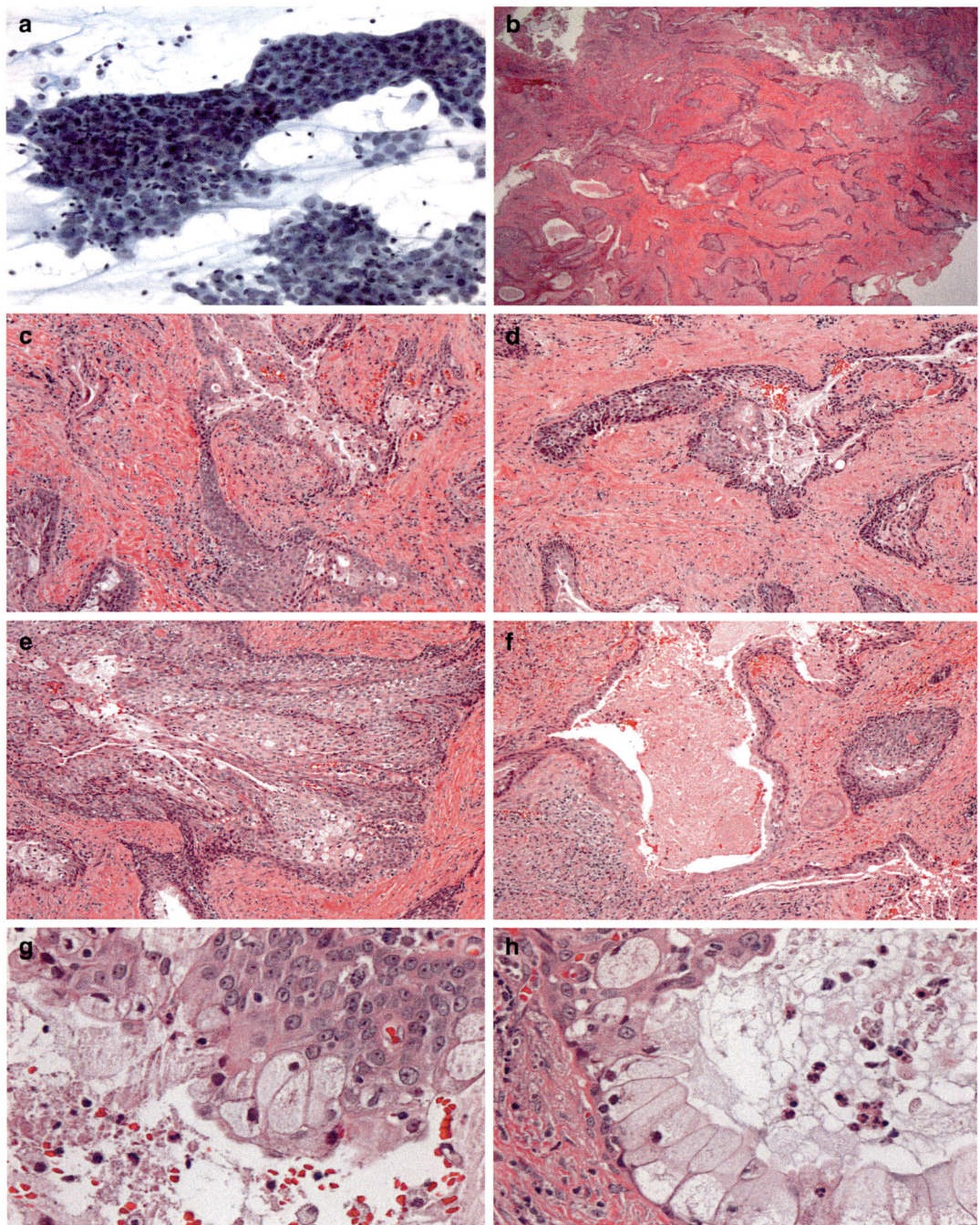

Fig. 14.11 In (**a**) is shown the result of a fine needle aspiration cytology (FNAC) stained with Diff-Quick (×200 as original magnification). The microphotographs (**b**) through (**h**) show the characteristic features of a mucoepidermoid carcinoma with cords, sheets, and groups of mucous and squamous cells with the interposition of intermediate and clear cells. The tumor cells show no marked nuclear atypia (low-grade) (**b**, ×12.5; **c–f**, ×100, **g–h**, ×400 as original magnification). The mucoepidermoid carcinoma is not an infrequent malignant epithelial tumor of childhood and youth and FNAC plays a major role in the 21st century diagnostic approach (Sergi 2018)

14.5 Salivary Glands

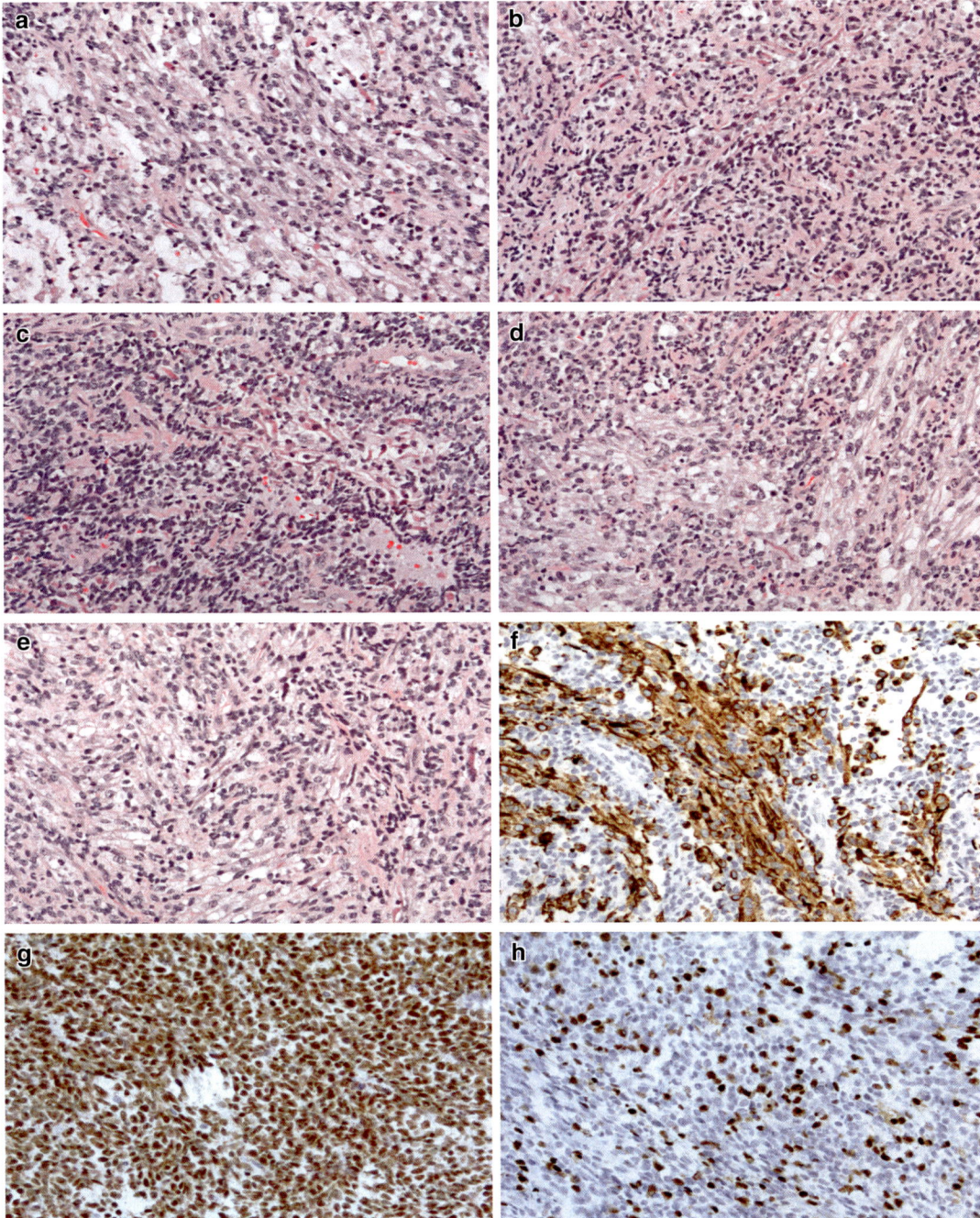

Fig. 14.12 The microphotographs (**a**) through (**e**) show the hematoxylin-eosin-stained sections of a rhabdomyosarcoma of the parotid gland of a child with infiltrating round and spindle tumor cells with mild pleomorphism. Some areas show a particularly marked vacuolation of the tumor cells (**a**), while some areas show particularly basophilic cells with high nucleus-to-cytoplasm ratio (**c**) (**a**, ×200; **b**, ×200; **c**, ×200; **d**, ×200; **e**, ×200, as original magnification). The immunohistochemical panel set shows the expression of desmin (×200) (**f**), Myo-D1 (×200) (**g**), and MYF-4 (×200) (**h**). All stained sections have been carried out using the avidin-biotin complex method

Epithelial-myoepithelial carcinoma and *myoepithelial carcinoma* are mostly seen in middle-aged adults, but they have been described in the pediatric group. The IHC profile should include reactivity to at least one of the myoepithelial markers (SMA, MSA, calponin, P63, and GFAP) other than cytokeratins. *Small-cell carcinoma* is rare and often reserved for middle-aged patients, but can occur in younger patients. Small-cell carcinoma is characterized by sheets, cords, and nests of "anaplastic" cells (2–3× larger than mature lymphocytes) with fine nuclear chromatin, absent or indistinct nucleoli, and scant cytoplasm, ill-defined cell borders, molding of nuclei, numerous mitoses, and extensive necrosis. There is focally squamous differentiation and NE differentiation (+NSE, CGA, SYN, CD56, and CD57 on IHC and membrane-bound neurosecretory granules on TEM) and (+) AE1-3, EMA, and CK20 (paranuclear dot-like pattern) similar to Merkel cell carcinoma, which is a rare type of skin cancer. LEC has a wide age range including children, adolescents, and young adults through the elderly.

Sialoblastoma is a congenital tumor that recalls the primitive salivary anlage and may present as solid nests of basaloid cells.

- *DEF*: Rare salivary gland neoplasm of epithelial origin, which mirrors the developmental phase of the salivary glands.
- *EPI*: Very few cases reported (~30 cases).
- *CLI*: Swelling of variable sizes involving the parotid gland or submandibular gland ± Facial nerve palsy in cases of parotid involvement.
- *TRT*: Surgical excision.
- *PGN*: Local recurrence has been reported, but the long-term course is benign.

Mammary analogue secretory carcinoma (MASC) of salivary glands is a tumor, which has been recently discovered. MASC incidence is unknown and resembles the secretory carcinoma of the breast. Both tumors share the balanced translocation t(12;15)(p13;q25) forming the *ETV6-NTRK6* fusion gene which encodes a chimeric tyrosine kinase. Before its description, MASC was misinterpreted as acinic cell carcinoma, mucoepidermoid carcinoma, and cystadenocarcinoma.

The fine needle aspiration cytology (FNAC) plays a significant role in the diagnosis. FNAC was performed first in 1857, although several techniques and new protocols have transformed this relatively simple technique into a practice that is extraordinarily diffuse currently worldwide. Aside from scarring in some patients, there are important complications due to unnecessary surgery which is well known to many pediatricians. Salivary gland lesions are rare in children, although an increase of such lesions in adolescence and youth has been reported recently. Although salivary gland tumors are considered rare neoplasms in the young population, the incidence of all primary salivary gland carcinomas increased with increasing patient age, particularly in patients younger than 30 years and even younger than 20 years according to most recent statistics and tumor reports. Except for mumps and cytomegaly, lesions of the major cephalic salivary glands remain unusual in children and adolescents. However, these lesions may give rise to some different tentative diagnoses. Since salivary gland tumors are relatively more frequently recognized in younger patients, a practical and safe approach is advisable. Moreover, an intradepartmental consultation is crucial. Cytologic diagnoses of malignant tumors are confirmed histologically in 93% of cases, while benign tumor diagnoses are confirmed on histology in 95% of cases. Inflammatory lesions are confirmed on histology in 73%, while benign salivary gland tissue is ascertained as such histologically in 18% of cases. Malignant salivary gland tumors in childhood, adolescence, and youth are MEC, adenoid cystic carcinoma, acinic cell carcinoma (ACCA), malignant lymphomas, and metastatic tumors from cancer of the H&N region or paraganglia (e.g., neuroblastoma). Benign neoplasms include PA and Warthin tumor. In childhood, 80–90% of all salivary gland malignant neoplasms are represented by three entities, including the mucoepidermoid carcinoma, the adenoid cystic carcinoma, and the ACCA. ACCA is characterized by a hypercellular aspirate with a clean background with tumor cells seen in disorganized clusters. There is a loss of round groupings and lack of an associated ductal epithelium and on higher magnification cells show some uniformity resembling normal serous acinar cells with the cytoplasm being foamy or bubbly and harboring fine dark granules. The examination of the background highlights many naked nuclei. Apart from clear cell tumors, mainly low-grade mucoepidermoid carcinoma and epithelial-myoepithelial carcinoma, oncocytic tumors, sialadenosis, and regular salivary gland tissue need to be kept in the differential diagnosis list. Smears of MEC are usually low in cellularity with a particularly striking dirty background of mucin and debris. In the smears, the presence of scattered cell clusters of intermediate cells with overlapping epithelial groups, some mucin-coated cells (goblet cell-like), and few squamous epithelial cells are particularly evident features for the diagnosis of MEC. Noteworthy, both components may be quite bland and even missing in one slide or all slides brought to the pathologist. According to the grade of differentiation, nuclei may vary from bland to hyperchromatic. There is a high nucleus-to-cytoplasm ratio. Some potential pitfalls in diagnosing MEC are some cystic nature of these tumors, which may harvest only hypocellular (even acellular in some cases) mucoid material. Moreover, extracellular mucin is typically copious mimicking the fibrillary stroma seen in PAs. Interestingly, the mucin observed in PA stains less strongly and had no fibrillary pattern. Lastly, squamous metaplasia is not only characteristic of MEC but may be found in other tumors, such as PA and Warthin tumor. The adenoid cystic carcinoma features are large globules of extracellular matrix with or without surrounding basaloid cells. In FNACs with a predominance of basaloid tumor cells, both benign and malignant salivary gland neoplasms of epithelial differentiation and myoepithelial differentiation should be considered in the differential diagnosis.

14.5.3.3 Soft Tissue Tumors

Soft tissue tumors of the salivary gland are rare and may account in the general population up to 5% of all salivary gland tumors with benign tumors vs. malignant tumors with a ratio variable from 2.5:1 to 18:1, and parotid gland involvement is seen in approximately 85% of cases. There are *vascular* tumors (hemangioma and lymphangioma), *neural* tumors (schwannoma and neurofibroma), *fibroblastic/myofibroblastic* tumors (nodular fasciitis, fibromatosis, myofibromatosis, fibroma, hemangiopericytoma, solitary fibrous tumor, and IMT), *adipocytic* tumors (lipoma and sialolipoma), and miscellaneous, including granular cell tumor, angiomyxoma, glomangioma, myxoma, fibrous histiocytoma, giant cell tumor, and osteochondroma. Malignancies include several sarcomas or tumors with malignant potential, and potentially almost any type of sarcoma can present in the salivary gland. The most frequent malignant soft tissue tumors are "hemangiopericytoma," which should probably be considered of uncertain malignant potential, malignant peripheral nerve sheath tumor (MPNST), which may occur in patients with neurofibromatosis type 1 (von Recklinghausen Disease), fibrosarcoma, and malignant fibrous histiocytoma (MFH). MFH (aka undifferentiated pleomorphic sarcoma) of the head and neck is an uncommon malignancy. Although MFH was once considered the most common soft tissue sarcoma in adults, its existence as a distinct pathologic entity has been questioned. It seems to be exceedingly rare in the pediatric population. There are two well documented MFHs of the parotid gland in children, which are described in the English scientific literature (Kariya et al. 2003; Chang et al. 2007).

14.5.3.4 Hematolymphoid Tumors

Rarely hematolymphoid neoplasms may involve the salivary glands and include HL and NHL, including DLBCL and EN-MZBCL.

14.5.3.5 Secondary Tumors

Rarely, metastases from PNET/Ewing sarcoma or neuroblastoma as well as rhabdomyosarcoma can be encountered in childhood and youth.

14.6 Mandible and Maxilla

The approach to odontogenic lesions is not unusual to the pediatric pathologist, being odontogenic lesions common in childhood and youth. However, in many university centers, there is significant support from odontopathologists or maxillofacial pathologists that may help in the consultation. The mandible and maxillary lesions are grouped in Box 14.7. The approach to the odontogenic lesions is depicted in Box 14.8 (Figs. 14.13, 14.14, 14.15, 14.16, 14.17, and 14.18).

> **Box 14.7 Mandible and Maxilla Lesions**
> 14.6.1. *Odontogenic Cysts*
> 14.6.2. *Odontogenic Tumors*

> **Box 14.8 Approach to Odontogenic Lesions**
> - Complete history
>   – ± Pain, loose teeth, (mal-)occlusion, tissue swellings, dysthesias, and delay in tooth eruption
> - Data from physical examination
>   – Clinical suspicion
> - Plain radiographs
>   – Panorex, historical dental radiographs, and historical/current comparison
> - CT imaging for large and aggressive lesions
> - Review of the supposed differential diagnosis
> - Determination of the tissue obtained and sent to the laboratory
>   – FNAC – to rule out vascular lesions, inflammatory, and neoplastic
>   – Excisional biopsy – smaller cysts and unilocular tumors
>   – Incisional biopsy – larger lesions performed before definitive therapy usually to identify or confirm of benignancy or rule out malignancy

14.6.1 Odontogenic Cysts

A simple classification of the odontogenic cysts includes inflammatory (radicular or paradental) and developmental (dentigerous, developmental lateral periodontal, odontogenic keratocyst, glandular odontogenic). The nonodontogenic cysts are incisive canal cyst, Stafne bone cyst, traumatic bone cyst, and surgical ciliated cyst (of the maxilla).

A cyst is a benign pathologic cavity with lumen filled with fluid, lined by epithelium, and surrounded by a connective tissue wall. Odontogenic cysts occur around teeth, mostly jaw-located, and are derived from remnants of the dental lamina epithelium. They are lined by hyperplastic squamous to thin squamous or cuboidal epithelium, which is derived from odontogenic epithelium hence the term odontogenic cyst. Odontogenic cysts are often histologically quite similar to epidermal inclusion cysts (see skin chapter) and subclassified as developmental or inflammatory. Location except for odontogenic keratocyst typically names odontogenic cysts. Pulp treatment of primary teeth requires close follow-up radiographic examination, even though the treated tooth may become an asymptomatic and radiographic assessment of pulp-treated deciduous teeth within 1 year after the treatment is recommended. The treatment modalities of cysts include enucleation for small cysts, marsupialization for larger cysts with the objective to alleviate pressure through an accessory cavity, marsupialization followed by enucleation, and enucleation with curettage. Enucleation and curettage may result in neurosensory dysfunction or predispose the patient to an increased rate of pathologic fracture and need care. It is also important to remember the children's bone growth potential and the paramount preservation of the underlying permanent tooth. An obturator is used to prevent the formation of fibrous healing tissue and to promote the decompression of the cystic lesion as well as to avoid the entry of food debris into the cystic cavity. Pseudocysts (lack of lining epithelium) or nonodontogenic cysts are simple bone cyst (traumatic bone cyst), aneurysmal bone cyst, mucous retention cyst, and Stafne bone cyst (aka Stafne bone defect). Soames and Southam (1998) collected the incidence of cystic lesions of the jaw and found the following percentages.

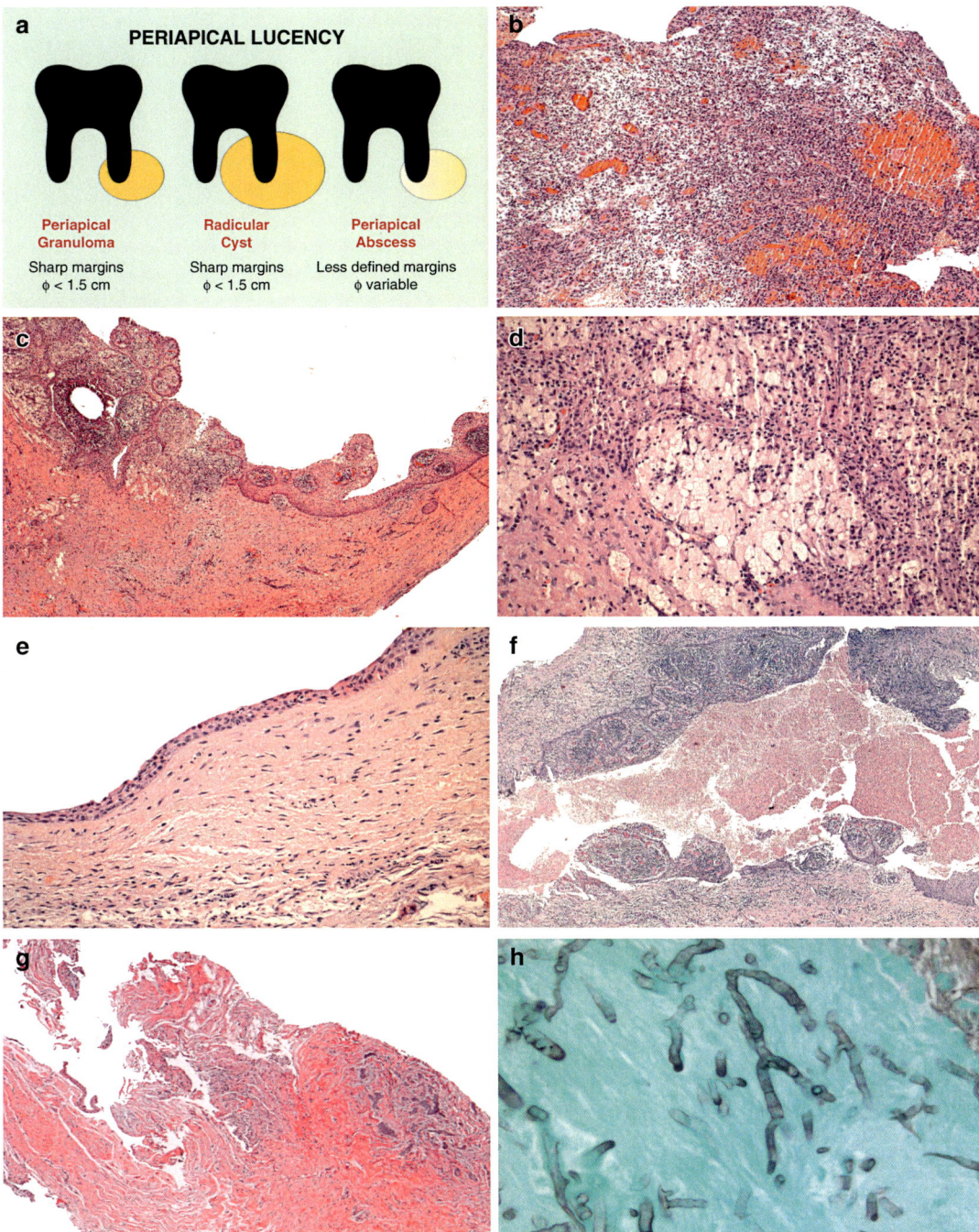

Fig. 14.13 In (**a**) is shown a schema differentiating periapical granuloma from radicular cyst and a periapical abscess. The microphotographs (**b**) through (**h**) show an apical granuloma (**b**, ×10), a radicular cyst (**c**, ×10; **d**, ×40), a follicular cyst (**e**, ×100), a radicular cyst with a prominent inflammatory component (**f**, ×40), a dentigerous cyst (**g**, ×50), and aspergillus infection in an immunocompromised patient (**h**, ×630). All microphotographs but (**h**) are from histological slides stained with hematoxylin and eosin. Grocott methenamine silver was used for (**h**)

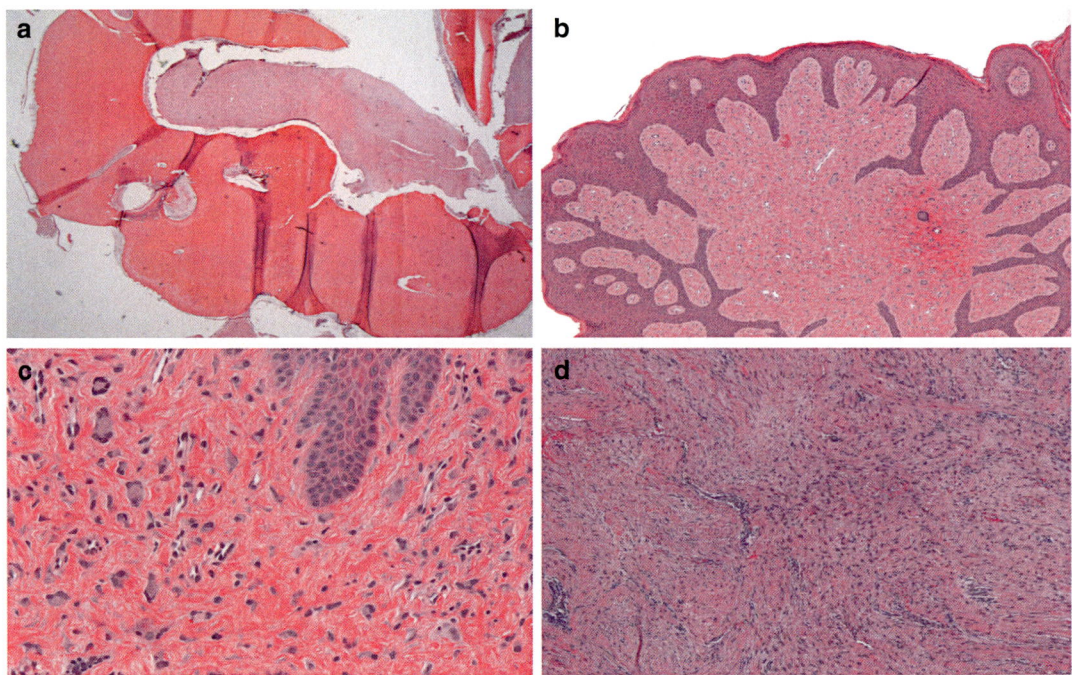

Fig. 14.14 Microphotographs showing a complex odontoma (**a**, ×12.5), a giant cell fibroma (**b**, ×50; **c**, ×200), and a desmoplastic fibroma (**d**, ×50). All microphotographs are stained with hematoxylin and eosin. See text for details

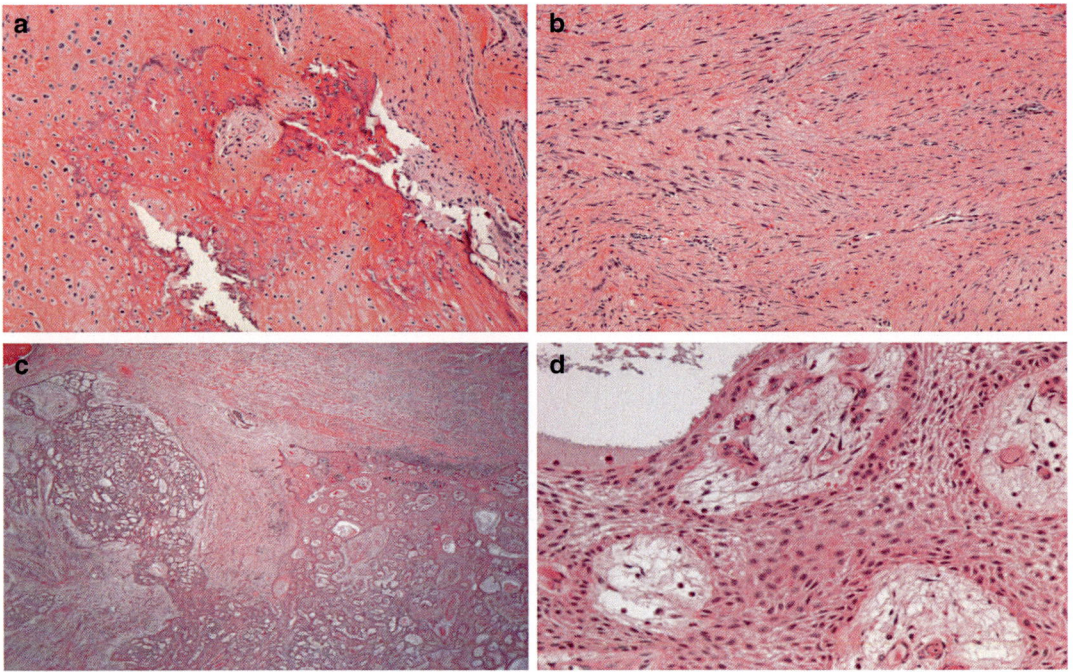

Fig. 14.15 The microphotographs (**a**) and (**b**) show the histological features of an ameloblastic fibroma (**a**, ×100; **b**, ×100), while the microphotographs (**c**) through (**f**) show the histological features of an ameloblastoma (**c**, ×12.5; **d**, ×200; **e**, ×200; **f**, ×400). See text for details

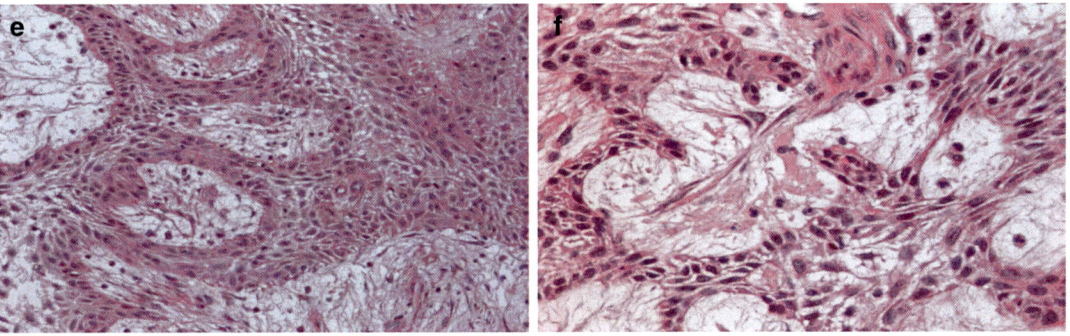

Fig. 14.15 (continued)

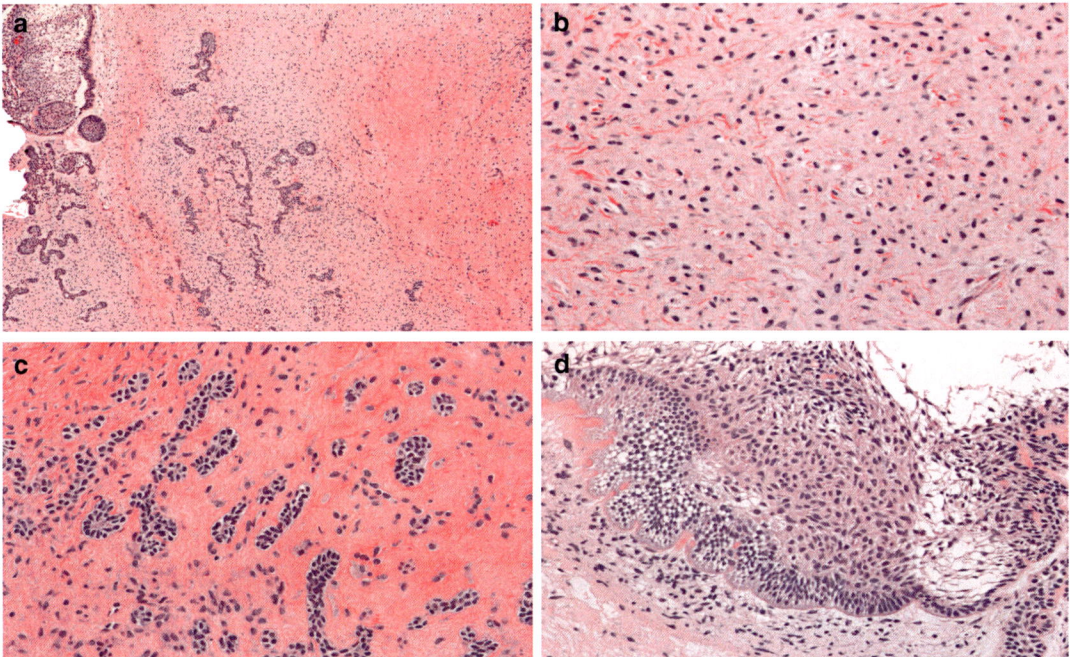

Fig. 14.16 This panel shows microphotographs of an ameloblastic fibro-dentinoma (**a**, ×50; **b–d**, ×200). All microphotographs are from hematoxylin and eosin-stained histological slides. See text for details

Radicular cysts	60–70%
Dentigerous cysts	10–15%
Nasopalatine cysts	5–12%
Keratocysts	5–10%
Paradental cysts	3–5%
Other (odontogenic cysts)	<1%

More than 20 years ago, the histogenesis of the cyst formation has been brilliantly elucidated by Benn and Altini (1996). The histogenesis may include:

- Developmental dentigerous cysts originate from the dental follicle.
- Periapical inflammation originates from a non-vital deciduous tooth or other source and, subsequently, spreads to involve follicle of a permanent successor (dentigerous cyst of inflammatory origin).
- Radicular cysts arise at the apex of the non-vital deciduous tooth.

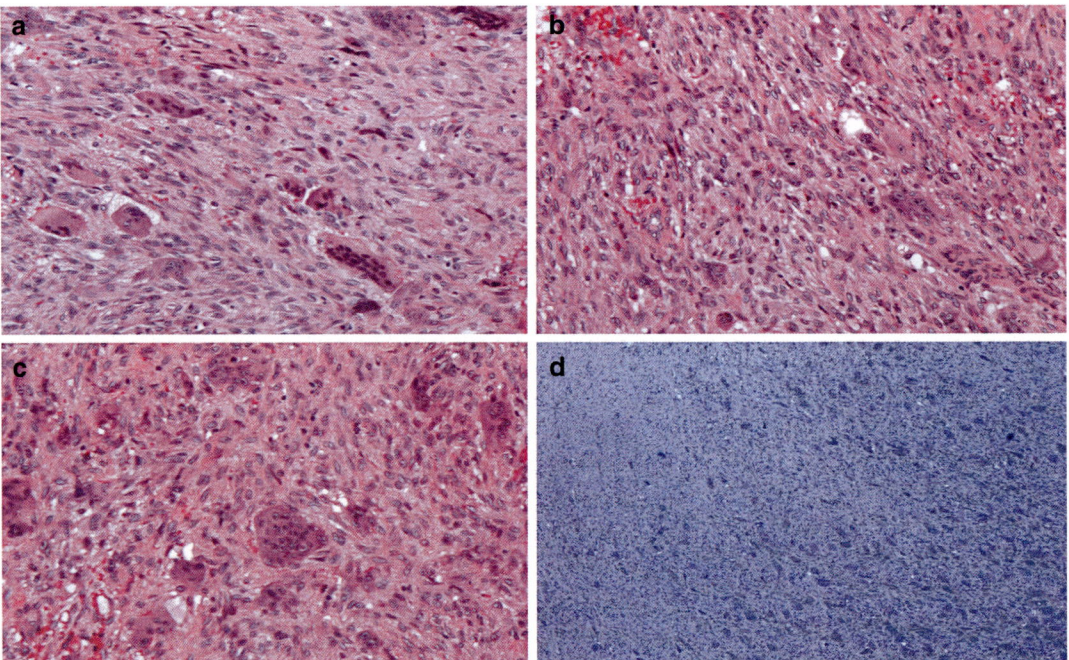

Fig. 14.17 This panel illustrates the histology of this apparently difficult to classify lesion of the jaw (similar lesions have been observed in craniofacial bones and small bones of hands and feet). The former name of "reparative granuloma" may be misleading, because some patients having this lesion show a lesion, which behaves aggressively. Histologically, there is a stroma of fibrillar connective tissue type with embedded small oval and spindle mononuclear cells, which are mixed with clusters of multinuclear giant cells, hemorrhage, hemosiderin, reactive bony spicule with osteoblastic rimming, and neo-angiogenesis (**a–c**; ×200 original magnification, hematoxylin and eosin staining). There are no mitotic figures with a low or nil proliferation rate (**d**), using anti-Ki67 immunohistochemistry and avidin-biotin complex. There is no evidence of cellular pleomorphism, which is an important feature

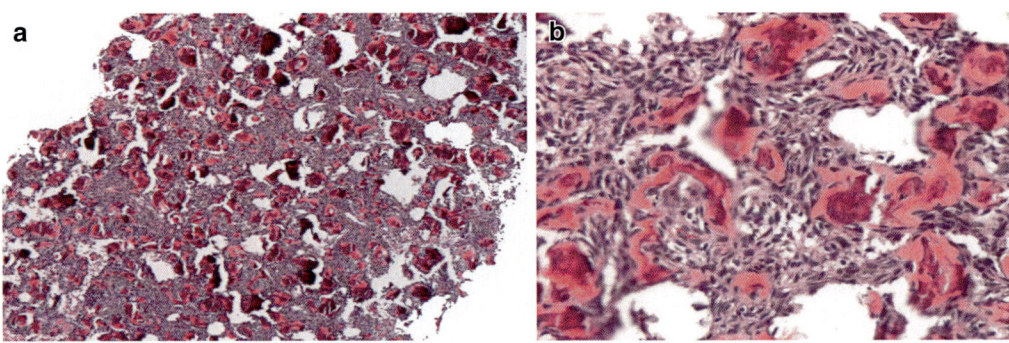

Fig. 14.18 Ossifying fibroma microphotographs showing (**a**) ×50, hematoxylin and eosin staining; (**b**) ×200, hematoxylin and eosin staining; (**c**) ×100, anti-S100 immunostaining; (**d**), ×100 anti-muscle specific actin immunostaining; (**e**), ×100, anti-actin immunostaining; and (**f**), ×100, anti-Ki67 immunostaining. There are haphazardly distributed lamellated bony spicules embedded in a background of fibrous stroma

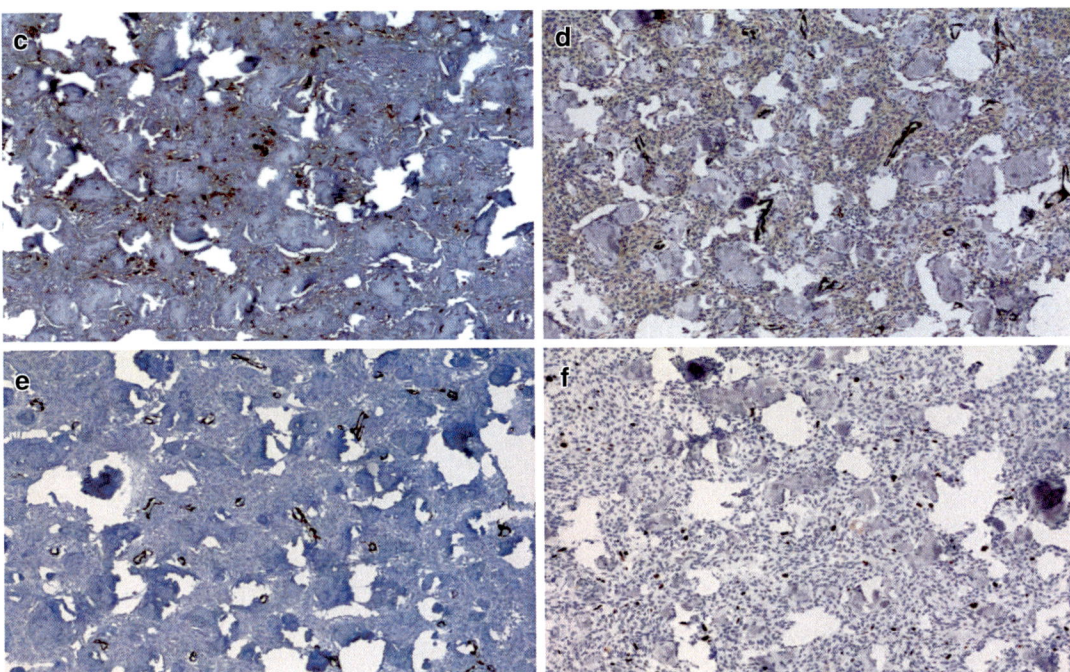

Fig. 14.18 (conrinued)

In dentigerous (follicular) cysts, a permanent tooth may erupt into a radicular cyst associated with the deciduous predecessor.

14.6.1.1 Radicular Cyst
- *DEF*: Inflammatory-based odontogenic cyst at the periapical location, usually maxillary molars at the apex of tooth root due to severe pulp inflammation/death from infection or trauma, which stimulates the proliferation of the rest of Malassez.
- *EPI*: Most common odontogenic cyst (65% of cysts).
- *EPG*: Proliferation and cystic degeneration of epithelial cell rest of Malassez as a response to inflammation.
- *CLI*: Pulpless, non-vital tooth.
- *IMG*: Small, well-defined periapical radiolucency.
- *CLM*: True cyst lined by the stratified squamous epithelium of variable thickness, often with scattered ciliated cells with an acute inflammatory cell infiltrate, Rushton hyaline bodies (eosinophilic, rectilinear to the crescent-shaped bodies), and a fibrous capsule of variable thickness with lymphocytes and plasma cells and cholesterol clefts.
- *PGN*: Good, following extraction and identification of the root canal.

14.6.1.2 Paradental Cyst
- *DEF*: Inflammatory-based cyst associated with partially impacted third molars.
- *EPI*: 0.5–5% of cysts.
- *EPG*: Result of inflammation of the gingiva over an erupting molar.
- *IMG*: Radiolucency in the apical portion of the root.
- *PGN*: Good, following enucleation.

14.6.1.3 Dentigerous (Follicular) Cyst
- *DEF*: Developmental odontogenic cyst that arises by separation of the dental follicle from around the crown of an unerupted tooth.
- *EPI*: Second most common odontogenic cyst, but most common developmental cyst (25%).

- *EPG*: It surrounds the crown of an unerupted permanent tooth and has either a destructive or a deforming character and is accompanied by fluid between tooth crown and reduced enamel epithelium.
- *CLI*: Asymptomatic, if it is identified on routine radiographs, but symptomatic if inflammation is present.
- *IMG*: Radiologic unilocular radiolucency with well-defined sclerotic margins.
- *CLM*: Inflamed dentigerous cyst includes fibrous connective tissue, hyperplastic non-keratinized epithelium with elongated interconnecting rete ridges, chronic inflammatory cells, cholesterol clefts, Rushton bodies, scattered mucous or ciliated or sebaceous cells, and odontogenic epithelial rests, small, inactive-appearing. Non-inflamed dentigerous cyst includes fibrous to fibromyxoid connective tissue, lack of rete ridges, flat interface and lining epithelium of 2–4 layers of cuboidal epithelium, devoid of superficial keratinization, and occasional mucous cells and odontogenic epithelial rests. Some lesions submitted as dentigerous cysts are partially lined with a thin, fragmented layer of eosinophilic columnar cells/low cuboidal epithelium representing the post-functional ameloblastic layer of the reduced enamel epithelium. Many of these lesions probably do not represent true cysts but probably just hyperplastic connective tissue dental follicles.
- *TRT*: Enucleation and decompression.

14.6.1.4 Odontogenic Keratocyst
- *DEF*: Squamous-lined cyst with parakeratosis and palisading basal cells with an aggressive or destructive character and high RCR.
- *EPI*: 10% of jaw cysts.
- *CLI*: Mimicker of any other maxillary/mandibular cysts and most often in mandibular ramus and angle.
- *IMG*: Well-delimited, radiolucent cyst which may be pericoronal, interradicular, or multilocular.
- *CLM*: Thin epithelial lining with underlying connective tissue (collagen and epithelial nests) with secondary inflammation.
- *TRT*: Enucleation with complete removal of cyst wall and peripheral ostectomy in case of a large cyst. It is recommended to have 1 cm of bony margins.
- *PGN*: High RCR (up to 2/3) with mandatory long-term F/U (5–10 years).

Other cysts include the developmental lateral periodontal cyst, the glandular odontogenic cyst, the incisive canal cyst, the Stafne bone cyst, the traumatic bone cyst, and the surgically ciliated cyst, which play minimal or no role at pediatric age. Odontogenic cysts are also classified as an eruption cyst, primordial cyst, paradental cyst, lateral periodontal cyst, calcifying odontogenic cyst (Gorlin cyst), and gingival cyst. Fissural cysts are also subdivided into the nasopalatine cyst, which may occur in the 30s as earliest age, the median palatine cyst, the nasoalveolar cyst (nasolabial), and the globule-maxillary cyst.

14.6.2 Odontogenic Tumors

Box 14.9 illustrates a simplified, pathogenesis-based classification of odontogenic tumors with an additional benignancy/malignancy sub-criterion. This classification is based on WHO classification of the odontogenic tumors but implemented with pathogenesis and mnemonic tools.

Box 14.9 Odontogenic Tumors
Odontogenic tumors with OE + fibrous stroma – OEM: OE, odontogenic epithelium; OEM, odontogenic ectomesenchyme

1. Ameloblastoma
2. Odontogenic tumors
   - Squamous
   - Calcifying epithelial (Pindborg)
   - Adenomatoid
   - Keratocystic

Odontogenic tumors with OE + OEM (± hard tissue)

1. Ameloblastic fibroma/FA/FO
2. Odontoma (complex/compound)

3. Odonto-ameloblastoma
4. Calc.-cystic odontogenic tumor
5. Ghost cell dentinogenic tumor

Odontogenic tumors with M and/or OEM ± OE

1. Odontogenic fibroma
2. Odontogenic myxoma/myxofibroma
3. Cementoblastoma

Bone-related lesions

1. Ossifying fibroma
2. Fibrous dysplasia
3. Osseous dysplasia
4. Central GC lesion
5. Cherubism
6. ABC/SBC

Odontogenic carcinomas

1. Metastasizing ameloblastoma
2. Ameloblastic carcinoma
3. Primary intraosseous SqCC
4. CC odontogenic carcinoma
5. Ghost cell odontogenic carcinoma

Other tumors

- MNETI

Notes: ABC, aneurysmal bone cyst; SBC, single bone cyst; CC, clear cell; GC, giant cell; MNETI, melanotic neuroendocrine tumor of infancy.

14.6.2.1 Ameloblastoma (Adamantinoma)

- *DEF*: Most common odontogenic tumor (uni-/multicystic, solid/cystic) with aggressive potential and characteristically located in the posterior mandible (80%) in adolescence and youth (30s) and associated with a dentigerous cyst or impacted tooth.
- *EPI*: Adolescence and youth, ♂ = ♀.
- *CLM*: Biphasic tumor with peripheral tall columnar cells with reversed polarity and central stellate cells lying in a fibrous stroma. These cells have been recognized to arise from cell rests of enamel organ epithelium or basal layer of the oral mucosa or dentigerous cysts.
- Variants: Follicular, plexiform, acanthomatous or keratotic, granular, and basal cell.
- *TRT*: Complete surgical excision with peripheral ostectomy if extension through cyst wall is noted. RT is contraindicated since it too frequently induces malignant transformation.
- *PGN*: Benign, but locally invasive. Poor prognostic factors include ↑MI, atypia, cell crowding ± necrosis, clear cells, and root resorption on X-ray.

14.6.2.2 Calcifying Epithelial Odontogenic Tumor

- *DEF*: An aggressive tumor of epithelial derivation from impacted tooth, mandible body/ramus with cortical expansion.
- *SYN*: Pindborg tumor.
- *EPI*: 30–40 years old, but adolescents may also be affected.
- *CLI*: Pain ±.
- *IMG*: Expanded cortices with radiolucent appearance.
- *CLM*: Islands of eosinophilic epithelial cells and cells infiltrating the bony trabeculae dominate the scene. Nuclear hyperchromatism and pleomorphism, as well as psammoma-like calcifications (Liesegang rings), can be identified.
- *TRT*: En bloc resection, hemimandibulectomy, and partial maxillectomy.
- *PGN*: Low RCR.

14.6.2.3 Odontogenic Fibroma

- *DEF*: Benign mesenchymal odontogenic tumor composed of mature highly cellular fibrous tissue with variable amounts of inactive-appearing odontogenic epithelium with two variants "intraosseous or central odontogenic fibroma" and "extraosseous or peripheral odontogenic fibroma."
- *EPI*: Children and young adults.
- *CLI*: Most commonly seen in the lower jaw.
- *CLM*: Very cellular fibrous connective tissue with plump fibroblasts and nests/strands of odontogenic epithelium.
- *DDX*: Odontogenic myxoma (honeycomb or multilocular, with delicate myxoid stroma comprising scattered odontogenic epithelial rests which resemble immature dental pulp).

Other odontogenic tumors include odontogenic myxoma, cementoma, adenomatoid odontogenic tumor, and squamous odontogenic tumor, among others.

Mixed Odontogenic Tumors

Ameloblastic fibroma, ameloblastic fibrodentinoma, ameloblastic fibro-odontoma, and odontoma should be included in the category as mixed odontogenic tumors. Both epithelial and mesenchymal cells dominate the scene, and they can mimic the differentiation of developing tooth. The treatment includes enucleation with thorough curettage with the extraction of the impacted tooth. Ameloblastic fibrosarcoma is malignant, and treatment needs to be aggressive with "en bloc" resection. Related jaw giant cell and non-giant cell lesions are listed in Box 14.10.

> **Box 14.10 Related Jaw Giant Cell and Non-giant Cell Lesions**
> - Giant cell lesions
>   - *Aneurysmal bone cyst*
>   - *Brown tumor*
>   - *Central giant cell granuloma*
> - Condensing osteitis
> - Fibro-osseous lesions
> - Fibrous dysplasia
> - Ossifying fibroma

These lesions are described in the arthroskeletal chapter of this book. Lesions of the osseous jaw are osteomyelitis and central giant cell granuloma, while the fibro-osseous lesions are cemento-osseous dysplasia, fibrous dysplasia, and osteosarcoma.

14.6.3 Bone-Related Lesions

In this paragraph, seven entities need to be taken into consideration, including ossifying fibroma, fibrous dysplasia (Box 14.11), osseous dysplasia, central giant cell lesion (granuloma), cherubism, aneurysmal bone cyst, and simple bone cyst.

> **Box 14.11 Fibrous Dysplasia Classification**
> - Monostotic (e.g., jaws, skull)
> - Polyostotic
>   - *McCune-Albright syndrome* (polyostotic fibrous dysplasia + cafe au lait spots with jagged "coast of Maine" borders and hyperfunctioning endocrine glands, leading to precocious puberty)
>   - *GNAS1 mosaicism* (*g*uanine *n*ucleotide-binding protein, *a*lpha-*s*timulating activity polypeptide 1) mutation leading to downregulation of cAMP of the G-signaling.

Central giant cell granuloma is a tumorlike reactive proliferation, which affects children and young adults, ♂ < ♀, mainly at mandible location, and presenting as an asymptomatic lesion, if the CGCG is an expansile tumor with slow growth, while is symptomatic or painful if the CGCG has a rapid extension. Surgical excision with thorough curettage is mandatory, and intralesional corticosteroids may be applicable for extensive disease. RT is contraindicated because of the risk of sarcoma. Radiologically, CGCG is uni-/multilocular radiolucent lesion with well-defined or irregular borders. Histologically, MNGC dispersed throughout a fibrovascular stroma are seen.

14.7 Ear

14.7.1 Congenital Anomalies

External congenital anomalies include dysplastic ears, low-set and malrotated ears, and syndromes such as leprechaunism. In the following pages, the inflammatory lesions and neoplastic lesions of the ear are described.

14.7.2 Inflammatory Lesions and Non-neoplastic Lesions

In the external ear, it is crucial to list *keratinous cysts* (cyst that may be related to branchial cleft, lined by keratinizing squamous epithelium, and of pilar type for periauricular cysts), *cauliflower ear* (auricular deformity following trauma-induced cartilage degeneration, such as in boxers and wrestlers), and *chondrodermatitis nodularis chronica helicis* or CNCH. The CNCH is a small painful nodular lesion of the helix with a raised center containing a crust, or scale occurs in patients over 40 years old and needs to be differentiated by BCC and actinic keratosis, particularly in individuals with genetic cancer syndromes or subjects with primary or secondary immunodeficiency (e.g., posttransplant). Although not a pediatric lesion, CNCH needs to be considered in the differential diagnosis, particularly in genetic syndromes with accelerated decline (e.g., progeria). The histologic features and clinical characteristics suggest that CNCH is an actinically induced perforating necrobiotic granuloma. CNCH is characterized by keratin epithelium-filled crater with a cleft in the subcutis toward cartilage with chronic inflammation and pseudoepithelial hyperplasia. *Relapsing polychondritis* (acute autoimmune inflammatory destruction of the cartilage of the helix ± polymyalgia rheumatica), *necrotizing external otitis* (*Pseudomonas* spp.-associated inflammation of diabetic and immunocompromised patients with possible progression to brain abscess, osteomyelitis, sepsis, and death), and *gout* (tophaceous granulomatous inflammation ± ulceration) are extremely rare in childhood, but occasionally reported in youth.

Lesch-Nyhan syndrome or juvenile gout is due to a deficiency of enzyme hypoxanthine-guanine phosphoribosyltransferase (HGPRT) produced by mutations of the HPRT gene located on X chromosome. The drug allopurinol may be particularly useful. The allopurinol is a drug used to decrease high blood uric acid levels (hyperuricemia) and is correctly applied to prevent gout, to prevent specific types of kidney stones (uric acid rich), and for the high uric acid levels that occur with chemotherapy protocols as well as in Lesch-Nyhan syndrome. The drug blocks the conversion of oxypurines into uric acid avoiding the progression of joint and soft tissue tophi as well as uric acid nephropathy and consequent nephrolithiasis. In the middle ear, acute inflammation or *acute otitis media* may be associated with infection or *S. pneumoniae* or *H. influenzae* and showing a bulging hyperemic tympanic membrane. The chronic form or chronic otitis media is characterized by persistent drainage, tympanic membrane perforation, and polypoid granulation tissue (or *otic polyp*), which may evolve to cholesteatoma (*vide infra*).

14.7.2.1 Acute Otitis Media (AOM)
- *DEF*: Acute inflammation of the tympanic membrane (eardrum).
- *SYN*: Infectious myringitis, acute myringitis, bullous myringitis.
- *EPG*: The eardrum is a three-layered structure, and the layer facing the ear canal is lined by a thin layer of keratinizing squamous epithelium, while the layer on the other side of the eardrum is a mucous membrane identical to the lining of the middle ear, mastoid, and Eustachian tube. A segment that provides the original integrity of the eardrum is in between these two layers. Any infectious process of the middle ear or external ear can indeed lead to inflammation of the tympanic membrane. The early finding in an AOM is an intense inflammation of the tympanic membrane, whereas bullous myringitis is a variant of infectious myringitis, whereby blisters form on the drum. Previously, this pathology was thought uniquely secondary to *Mycoplasma pneumoniae*, but it may be seen with other bacteria as well. Most often, the etiologic microorganisms include *S. pneumoniae*, *H. influenzae*, and *M. catarrhalis* other than *Mycoplasma* spp.; *P. aeruginosa* or *S. aureus* bacteria are more often seen starting from the external ear canal. Viruses that attack mucous membrane include adenovirus, rhinovirus, coxsackievirus, influenza virus, and parainfluenza viruses. Rarely, tuberculosis (*Mycobacterium tuberculosis*) can cause an eardrum infection, which may be associated with multiple small perforations in the drum. *M. tuberculosis* is a bacterium that causes tuberculosis in humans. Although tuberculosis is a disease that primarily affects the lungs, it can attack other parts of the body and the eardrum is not an exception.
- *CLI*: Ear pain ± fever and hearing loss with bloody and pus-fluid draining from the ear canal and tenderness around the opening to

the ear canal, and, on examination, the tympanic membrane is intensely red with increased prominence of the small blood vessels irrigating the tympanic membrane, and purulent debris may be visible in the external canal. Chronic myringitis usually has a granular, red aspect on the drum surface covered by infected material. The principal risks of myringitis are perforation of the tympanic membrane or significant scarring of the eardrum that could result in hearing loss.
- *GRO* and *CLM*: Do not play any role. They are never biopsied fresh in a living patient. An exception may be the investigation of SIDS or legal cases, where the examination of the eardrum and ear canal are often routinely performed.
- *TRT*: Antibiotics or antifungal therapies are indicated in the case of a specific infection, but the purely viral form of infectious myringitis will resolve spontaneously in an immunocompetent patient. Ear canal anesthetics or oral analgesics are used to enhance patient comfort. Chronic myringitis or chronic otitis media (COM) treatment, as a complication of AOM, involves antibiotic ear drops but may respond to acetic acid ear canal irrigations followed by application of steroid cream.

Cholesteatoma is a tumorlike lesion of the middle ear or mastoid cells of the mastoid bone (mastoiditis) with a pearl-like external appearance of the adolescent and young adult and characterized histologically by a keratinizing squamous epithelium-lined cyst filled with desquamated keratin debris and cholesterol crystals. There are a congenital form, which arises from squamous rests, and an acquired form originating from ingrowth of squamous epithelium into middle ear following chronic otitis media with drum perforation. Complications of cholesteatoma include meningitis and brain abscess. Cholesteatoma needs to be differentiated from *cholesterol granuloma*, which presents as cholesterol clefts with surrounding foreign body giant cell response of the host. The *inner* ear is dominated by the otosclerosis, which is an AD-inherited disease with variable penetrance with a tendency to cause a conductive hearing loss in young adults. There is hard tissue deposition around stapes causing fixation to the oval window (Figs. 14.19 and 14.20). Historically, cholesteatoma was first described as a "steatoma" in 1683 by De Verney in France (De Verney, 1863). Subsequently, Cruveilhier illustrated in 1829 as a "pearl tumor" of the temporal bone (Cruveilhier, 1829). In 1838, Müller introduced the term "cholesteatoma" (Müller, 1838). A few years later, Virchow categorized cholesteatoma under epidermoid carcinoma and atheroma (Virchow, 1855), but, ultimately, Von Troeltsch in 1861 considered its epidermal origin and in the studies by Gruber, Wendt and Rokitansky (1855 through 1888), the pathology of cholesteatoma was categorized as a malpighian metaplasia ofthe middle-ear mucosa in response to chronic inflammation. Bezold and Habermann suggested in 1889 and 1891 that cholesteatoma was determined by migration of external auditory canal skin to the middle ear, induced by chronic inflammation (Habermann, 1889; Bezold, 1891; Soldati and Mudry, 2001; Nevoux et al. 2010).

14.7.3 Tumors

The most recent WHO classification of tumors of the ear subdivides according to the regional location (external, middle, and inner ear).

14.7.3.1 Tumors of the External Ear

External ear neoplasms include *benign tumors of the ceruminous glands* (adenoma or ceruminoma, chondroid syringoma, and syringocystadenoma papilliferum) and *corresponding malignancies* (adenocarcinoma, adenoid cystic carcinoma, mucoepidermoid carcinoma), which may occur with a wide age range (adolescents and young adults to elderly), *cylindroma*, *squamous cell carcinoma*, *angiolymphoid hyperplasia with eosinophilia* (ALHE), soft tissue tumors (*embryonal RMS*), and bony tumors (*fibrous dysplasia*, *osteoma*, and *exostosis*).

Ceruminoma is unencapsulated tumor composed of bilayered oxyphilic gland epithelium with focal decapitation secretion (+CK7) and outer myoepithelial layer (+CK5/6, p63). *Chondroid syringoma* is similar to PA of salivary glands.

14.7 Ear

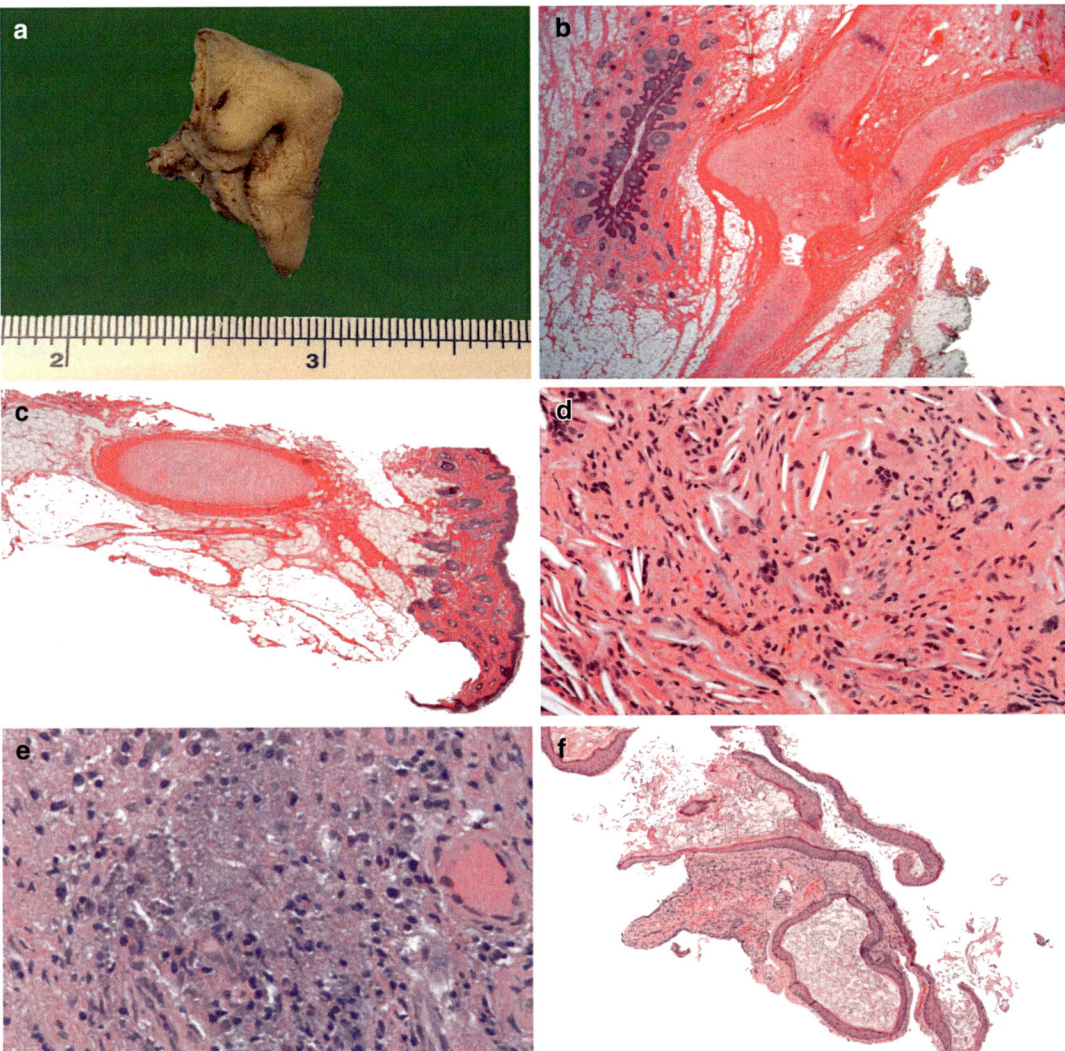

Fig. 14.19 This panel illustrates some pathology of the ear with the gross photograph of an accessory tragus (**a**) and its microscopic examination findings (**b**) ×12.5 original magnification, hematoxylin and eosin staining). In figure (**c**) is shown a cartilaginous remnant of the ear in a periauricular region that may be interpreted as a primordial accessory tragus (×12.5 original magnification, hematoxylin and eosin staining). Figure (**d**) shows the histology of a chronic middle ear infection with chronic inflammatory cells, cholesterol crystals, and foreign body giant cells (cholesterol granulomas) that may be associated with haphazard glandular metaplasia, cilia, fibrosis, hemorrhage, calcification, and reactive bone formation (×200 original magnification, hematoxylin and eosin staining). In figure (**e**) there is an area of basophilic necrosis and acute and chronic inflammatory cells (×400 original magnification, hematoxylin and eosin staining). Figure (**f**) shows a cholesteatoma at low-power magnification (×50 original magnification) with inflamed middle ear mucosa, sac-like accumulation of keratin, and keratinous stratified squamous epithelium with no atypia at high-power magnification (hematoxylin and eosin staining). Figure (**g**) shows another microscopic view of a cholesteatoma (×100 original magnification, hematoxylin and eosin staining), while figure (**h**) shows a rare case of myringosclerosis (×100 original magnification, hematoxylin and eosin staining). Myringosclerosis is a healed recurrent otitis media showing scar tissue and presenting clinically as whitish, sclerotic plaques in some areas of the tympanic membrane

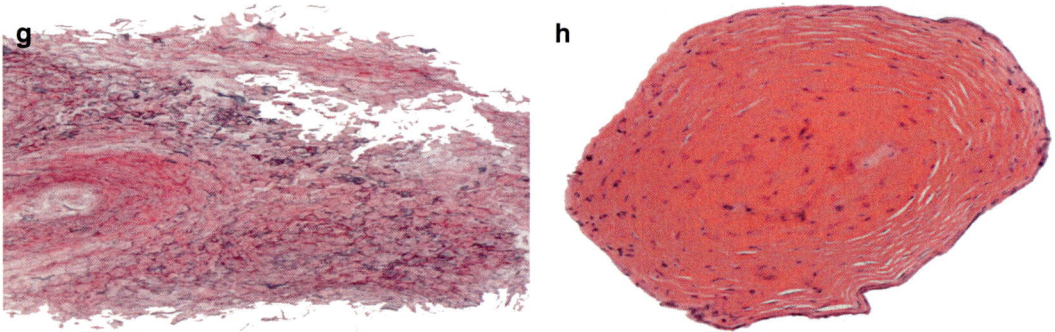

Fig. 14.19 (continued)

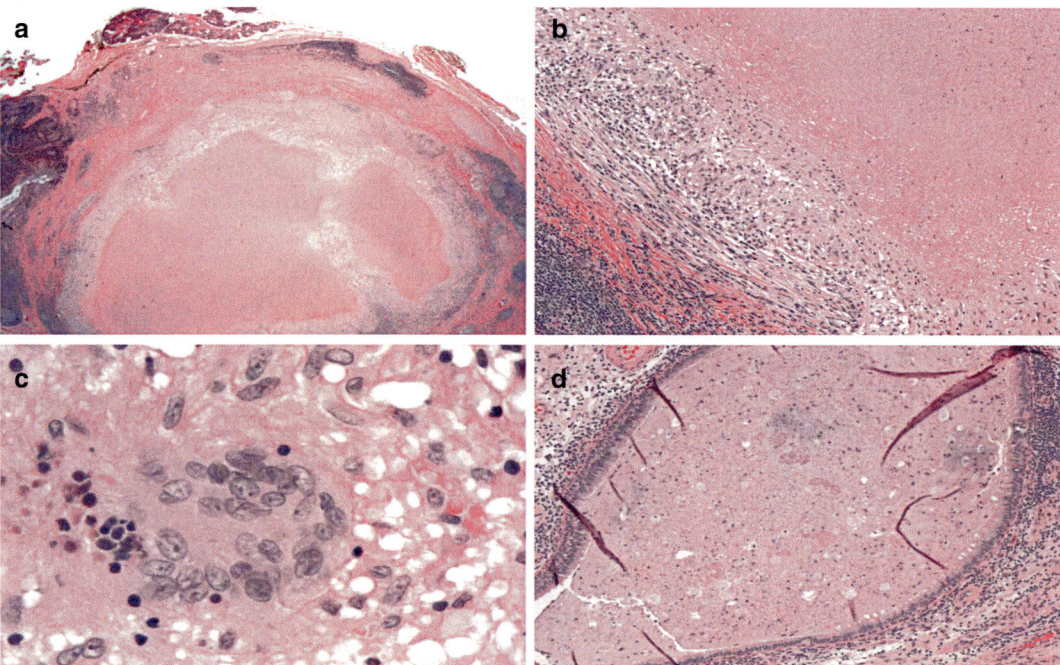

Fig. 14.20 This panel shows some other pathologies of the ear or periauricular region with a *M. tuberculosis* infection of a retroauricular lymph node with characteristic central caseous necrosis (**a**, ×12.5 original magnification, hematoxylin and eosin staining; **b**, ×100 original magnification, hematoxylin and eosin staining) and Langhans giant cells (**c**, ×630 original magnification, hematoxylin and eosin staining). Figure (**d**) shows a fibroepithelial polyp of the meatus in a child affected with cystic fibrosis (**d**, ×100 original magnification, hematoxylin and eosin staining). A hyperemic inflammatory polyp of the middle ear is shown in figure (**e**) (×100 original magnification, hematoxylin and eosin staining), while a capillary lobular hemangioma of the middle ear is shown in (**f**) through (**h**) (**f**, ×50 original magnification, hematoxylin and eosin staining; **g**, ×100 original magnification, anti-CD31 immunostaining; **h**, ×100 original magnification, anti-Glut1 immunostaining). GLUT1 or glucose transporter 1, which is also known as solute carrier family 2 (SLC2A1), is a uniporter protein that is encoded by the SLC2A1 gene in humans. As specified in its name, GLUT1 facilitates the transport of glucose across the plasmatic membranes of mammalian cells and is expressed in almost of all capillary lobular hemangiomas

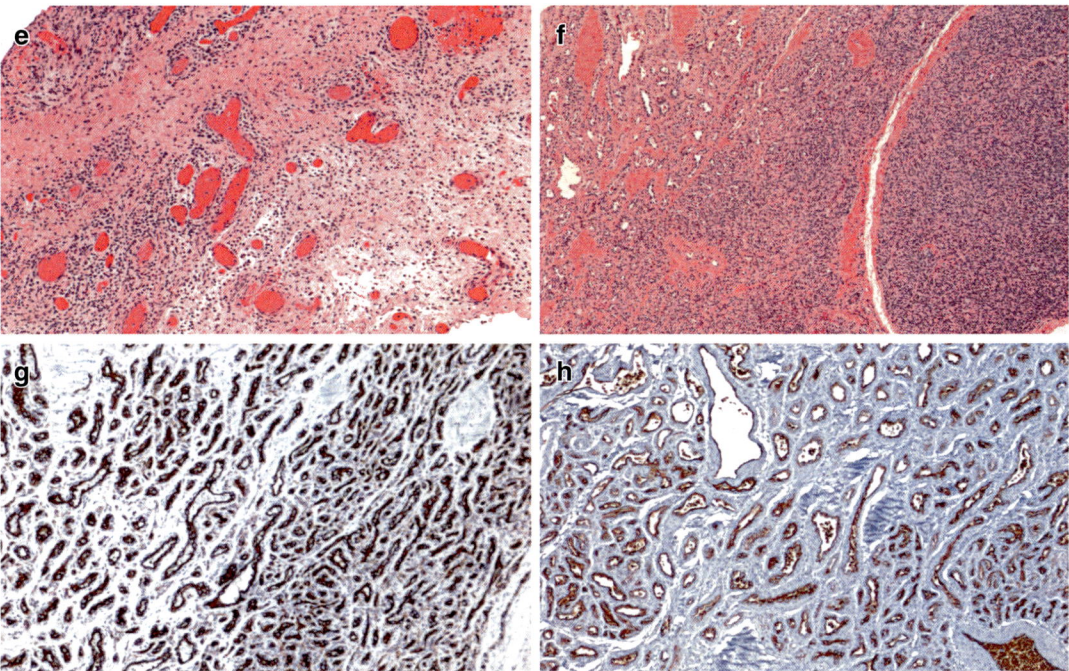

Fig. 14.20 (continued)

Syringocystadenoma papilliferum is characterized by papillae lined by bilayered glandular epithelium with decapitation secretion projecting into a cystic lumen.

Adenocarcinoma of the ceruminous glands (low- and high-grade) is characterized by an infiltrating biphasic neoplastic proliferation separated by desmoplastic stroma.

Adenoid cystic carcinoma is characterized by the cribriform ("Swiss cheese") pattern. *Mucoepidermoid carcinoma* shows epithelial cells and mucous cells and the morphology recapitulates that of the corresponding tumor of the salivary glands.

Cylindroma is characterized by a "jigsaw"-like pattern of small, darkly stained cell groups with extracellular hyaline globules surrounded by a pink-stained hyaline basement membrane. Unlike elderly, *squamous cell carcinoma* is not seen in children or young adults. Actinic keratosis may precede the squamous cell carcinoma of the external ear.

Angiolymphoid hyperplasia with eosinophilia (ALHE) or epithelioid hemangioma is a vasoformative reactive/neoplastic process that needs to be differentiated from Kimura disease (KD); although in both diseases the prognosis is good, keeping in mind that relapses may be encountered.

Here are summarized the clinical and microscopical differentials between ALHE and KD.

- *ALHE*: ♂ < ♀, worldwide, second to fifth decades, ± blood eosinophilia, negative hyper-IgE, red, subcutaneous nodule, forehead, scalp, ear-LN involvement.
- *KD*: ♂ > ♀, Asia, children to fifth decades, ++ blood eosinophilia, positive hyper-IgE, large deep mass, submandibular and parotid glands, pre-auricular region, + LN involvement.
- ALHE: Small blood vessel proliferation with prominent endothelial cells ("histiocytoid" appearance) with mast cells in a lobular arrangement, ± tissue eosinophilia, and ± lymphoid follicles with activated GC without IgE on FDCs.
- KD: High endothelial venule-like small blood vessel proliferation in a mast cell-rich edematous stroma, + tissue eosinophilia, and ++ lymphoid follicles with polykaryocytes and IgE on FDCs.

RMS and Bony Lesions

The rhabdomyosarcoma (RMS) may occur in the middle ear because H&N is a significant area for this kind of tumor (palate, middle ear, orbit) other than extremities. RMS include in theory the embryonal and the alveolar subtype that recapitulate the histology encountered in other organs and systems. Most RMS may arise in the middle ear and protrude into the external canal as an "aural polyp." The embryonal type remains currently the only subtype occurring in this location. Mutations have been identified on 11p15 among others. Bony lesions include fibrous dysplasia, osteoma, and osteocartilaginous exostosis that may occur in children and adolescents.

Fibrous dysplasia is a benign tumorlike lesion *intramedullary proliferation of osteoblastic rimming-free trabecular woven bony trabeculae ("Chinese ideograms") embedded in cellular fibrous tissue ± MNGCs ± cementum*-like spherules localized in the face/skull, rib, femur, and tibia (X-ray: diaphyseal fusiform expansion of a mass arising in cancellous bone with thinning of overlying cortex) and occurring either in a mono- (70%) or polyostotic (30%) form. The polyostotic form (POFD) may occur in the setting of McCune-Albright syndrome, caused by (somatic) activating mutations of the GNAS gene on chromosome 20 and characterized by *ca*fe au lait hyperpigmentation with jagged "coast of Maine" borders respecting the midline and Blaschko's lines:

- *F*ibrous dysplasia of polyostotic type (femur>skull>tibia>jaw>humerus)
- *E*ndocrine dysfunction (pituitary hyperplasia or GH-secreting pituitary adenoma, hyperthyroidism, adrenal hyperplasia, as well as the early start of puberty) (mnemonic: *CAFE*)

GNAS activation results in an abnormal version of the G protein that causes the adenylate cyclase enzyme to be always (constitutively) activated leading to excess cAMP. Fibrous dysplasia and soft tissue myxomas characterize Mazabraud syndrome.

Osteoma is a bosselated lesion composed of dense, mature, chiefly lamellar bone, generally localized on the flat bone of the face/skull and ± protrusion into sinuses, and seen in patients with Gardner syndrome, characterized by:

- *D*esmoids and AD inheritance (APC gene on 5q21 with variable expressivity)
- *O*steomas
- *S*ebaceous cysts
- *T*hyroid carcinoma/teeth abnormalities
- *O*steochondromas
- *P*olyposis of colonic localization (mnemonic: *DO STOP*)

Osteochondroma or exostosis is characterized by a lobulated cap of cartilage, covered by a fibrous membrane with cancellous bone, active endochondral ossification at the interface, and hyaline cartilage that may occur with multiple lesions in Gardner syndrome as well.

14.7.3.2 Tumors of the Middle Ear

Middle ear neoplasms include middle ear adenoma (or carcinoid of the middle ear), papillary tumors (aggressive papillary tumor, Schneiderian papilloma, inverted papilloma), squamous cell carcinoma, and meningioma. Closely apposed small glands characterize *middle ear adenoma* with "back-to-back" pattern without desmoplastic stroma. Tumor cells are regular, cuboidal to columnar with bland cytology, rare nucleoli and no mitoses encircling some +PAS/AB luminal secretion and + NE markers (CGA, SYN) other than Pan-CK. Grossly, a gray-white, firm lesion producing a conductive hearing loss is seen. This kind of tumors are usually benign, but may be locally aggressive with bone destruction. Moreover, papillary tumors, squamous cell carcinoma, and meningioma have been reported, but these neoplasms are rare in childhood and youth.

14.7.3.3 Tumors of the Inner Ear

Inner ear neoplasms include *vestibular schwannoma* (or "acoustic neuroma"), *lipoma, hemangi-*

oma, and *endolymphatic sac tumor*. *Vestibular schwannoma* is not seen in children or adolescents without the genetic setting of neurofibromatosis type II. VS is a benign tumor of Schwann cells of the eighth cranial nerve and usually located in the region of the internal auditory canal and in the area of the cerebellopontine angle. Vestibular schwannoma is often bilateral and shows Antoni A and B areas and Verocay bodies.

Neurofibromatosis I (NF I): AD-inherited disease with mutations on *NF1* tumor suppressor gene located on chromosome 17 (17q11.2) is linked to neurofibromin protein and downregulates p21 ras (penetrance 100%, but variable expressivity). The diagnosis is based on *c*afe au lait spots ("California coast" borders); *a*xillary, inguinal freckling; *f*ibroma (neurofibromas constituted by loose proliferation of neurites, Schwann cells, and fibroblasts in a myxoid stroma); *e*ye lesions (Lisch nodule or pigmented hamartoma of iris); *s*keletal deformities (bowing leg, etc.); *p*edigree/positive family history; and *o*ptic *t*umor (glioma). There is 2–4× risk to develop pheochromocytoma, WT, AML/CML, RMS, optic glioma, and meningiomas.

Neurofibromatosis II (NF II): AD inherited with the gene located on chromosome 22 (22q12), which codes for Merlin (Moezin-Ezrin radixin like protein)/schwannomin. In NF II there are typically bilateral vestibular schwannomas and meningiomas but also other intracerebral and intracranial tumors. Neurofibromas are dermal. Lipoma and hemangioma are similar to lipomas and hemangiomas elsewhere.

Endolymphatic sac tumor (ELST) is a tumor typically associated with VHL disease (hemangioblastomas of the cerebellum, retina or brainstem, pancreatic cysts, and hemangiomas of several organs (liver, kidneys, epididymis), pheochromocytoma, and ↑risk of RCC and testicular cystadenoma). ELST can also arise sporadically.

- *DEF*: Petrous temporal bone tumor classified as mastoid papillary tumors of unknown origin, identified as a distinct clinic-pathological entity by Heffner in 1989, and associated with von Hippel-Lindau (VHL) disease (VHL-ELST) or not (sporadic ELST). Although initially considered as a low-grade papillary adenocarcinoma, the bland histologic appearance and the probably apparent lack of metastatic potential have since convinced most pathologists and ENT surgeons to reclassify ELSTs as papillary adenomas.
- *EPI*: Rare (~175 cases reported), ♂ < ♀.
- *EPG*: VHL is an AD-inherited disease with the VHL gene located on chromosome 3 (3p25–26) encoding pVHL. VHL is a tumor suppressor gene. The VHL gene product, pVHL, forms a multiprotein complex that contains elongin B, elongin C, Cul-2, and Rbx1.
- *CLI*: Pulsatile tinnitus, aural fullness, imbalance, otalgia, otorrhea, vertigo, and facial paresis.
- *IMG*: MR T1-weighted axial images of the brain at the level of the endolymphatic sac and internal auditory canal show moderate expansion of the endolymphatic sac and duct.
- *CLM*: A highly vascular neoplasm with papillary cystic structures that are lined with a simple cuboidal or columnar epithelium. Siderophages and cholesterol clefts are seen, as are clear, vacuolated cells with mild or minimal nuclear pleomorphism and rare mitoses.
- *IHC*: (+) CKs, VIM, EMA, ± GFAP, and VEGF: PAS (+), DPAS (−).
- *DDX*: Intrinsic temporal bone neoplasms (most often paraganglioma), metastatic papillary thyroid carcinoma (+TGB), metastatic renal cell carcinoma, and choroid plexus papilloma (+transthyretin).
- *TRT*: Surgical resection.
- *PGN*: No metastases, but recurrence may take place at high rate.

In addition to these regional specific tumors, there are rare hematolymphoid tumors (mostly *B-CLL/SLL* and histiocytic/dendritic cell neoplasms chiefly *LCH*), *secondary tumors* metastatic to temporal bone (mostly breast, H&N, lower airways, prostate gland, thyroid gland, and malignant melanoma), and *choristoma* of the middle ear (salivary gland or glial tissue), which is a tumorlike lesion and not a neoplasm.

14.8 Eye and Ocular Adnexa

14.8.1 Congenital Anomalies

Congenital eye anomalies include slanting, hypertelorism, hypotelorism, telecanthus, epicanthus, epiblepharon, congenital ptosis, blepharophimosis, lid coloboma, ankyloblepharon, anterior segment dysgenesis, Peters anomaly, persistent pupillary membranes, congenital cysts of the pupil margin, anisocoria, congenital cataract, persistent hyperplastic primary vitreous, microphthalmia and coloboma, optic nerve hypoplasia, peripapillary pigmentary abnormalities, tilted discs, cupping, "morning glory" disc, and pseudopapilledema (Levin 2003).

14.8.2 Inflammatory Lesions and Non-neoplastic Lesions

The essential inflammatory lesions are conjunctivitis, *Herpes simplex* keratitis, infectious keratitis, retinitis, and uveitis. Among the non-neoplastic lesions, it is necessary to mention the dermoid cyst, which has well-described features in other chapters and paragraphs (Fig. 14.21).

14.8.3 Tumors

Two critical tumors need to be taken in mind, including retinoblastoma and malignant melanoma, of which only the retinoblastoma is described here.

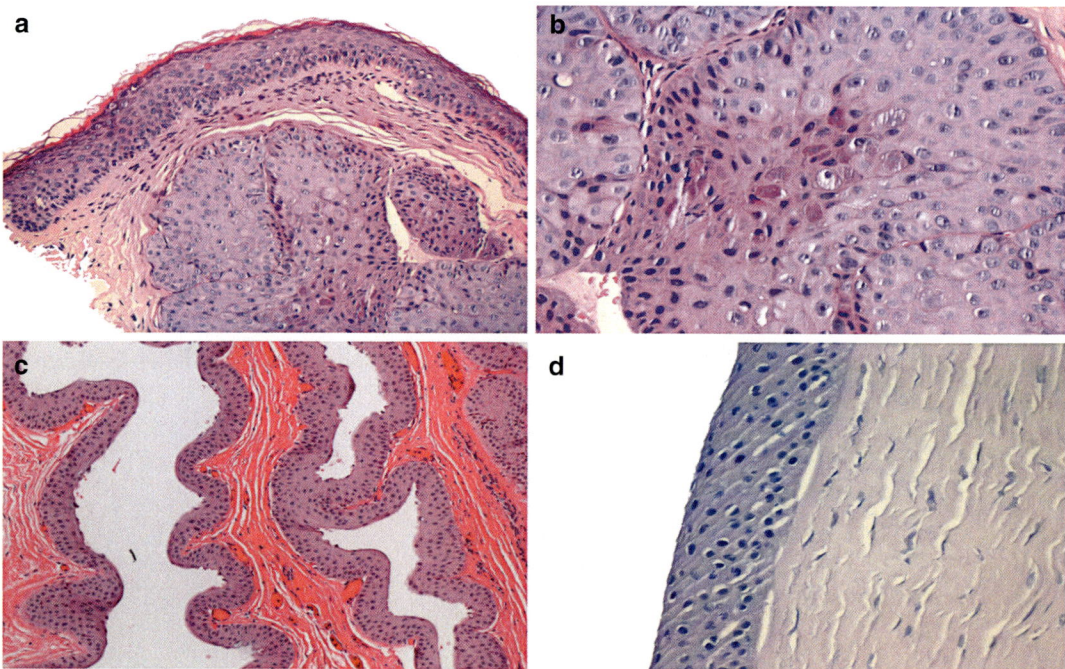

Fig. 14.21 In (**a**) and (**b**) are shown a moderate- (×200) and high-power (×400) magnification of an ocular molluscum (contagiosum) (**a**, hematoxylin and eosin staining; **b**, hematoxylin and eosin staining), while figure (**c**) shows a conjunctival cyst (×100 original magnification, hematoxylin and eosin staining). Molluscum contagiosum is a viral infection often encountered in children, and the diagnosis is usually clinical with pearl-like and dome-shaped bumps with a central crater (the so-called umbilication). The distinction of true conjunctival cysts with large pseudoglands of Henle (invaginations of surface epithelium forming tubular and microcystic structures) may be challenging. Figure (**d**) is a keratoconus, which shows thinning and fibrosis of cornea with numerous breaks in Bowman's layer and devoid of inflammation or neovascularization (×400 original magnification, hematoxylin and eosin staining)

14.8.3.1 Retinoblastoma and Related Lesions

- *DEF*: An embryonal tumor of the retina with variable laterality, uni- or multifocality, and genetics and characterized by an improved outcome and the diagnosis of which, differently of other pediatric malignancies, is made without tissue acquisition (Figs. 14.22 and 14.23).
- *EPI*: 1:20,000 live births, yearly incidence: 10–14/10^6 for children with age <5 years, and ~300 new cases each in the USA. It accounts for ~10% of malignancies in the first year of life, and the average age at diagnosis is 18 months.
- *EPG*: Genetically, RB may be uni- or bilateral, uni- or multifocal, and hereditary or non-hereditary. Hereditary tumors are usually bilateral and multifocal and occur in the first year of life and account approximately for 40% of cases. Non-hereditary tumors are typically unilateral and unifocal and occur in second and third years of life. Tumor development can be either syn- or metachronous. According to Knudson's "2-hit hypothesis," the first mutation can be either germline or somatic and the second somatic in the retinoblasts. *RB1* gene (27 exons, 110 kD nuclear phosphoprotein with 928 amino acids or pRB) is located on chromosome 13q14. RB protein acts as a tumor suppressor gene. RB1 is inherited in an AD pattern and has a high penetrance (~95%). A genetic syndrome labeled 13q deletion syndrome is associated with an increased risk of RB and includes several features, such as microcephaly, broad frontonasal bones, hypertelorism, microphthalmia, epicanthic folds, ptosis, micrognathia, and hypoplasia or absence of thumbs. Environmental and demographic include increased paternal age, peculiar parental employment (military, metal manufacturing, welder machinist), maternal use of steroid hormones, and high birth weight.
- *CLI*: Leukocoria, strabismus, decreased visual acuity, inflammation, hyphema, and vitreous hemorrhage (i.e., "black pupil") are the most common features.
- *CLM*: The tumor is constituted mainly by undifferentiated embryonal cells of the nuclear layers of the retina. The cells have nuclei with hyperchromasia and scanty cytoplasm and show little evidence of photoreceptor differentiation. Three forms of rosettes can usually be identified, including Flexner-Wintersteiner rosettes (FWR), Homer-Wright rosettes (HWR), and fleurettes. FWRs are true rosettes with radially arranged cuboidal or columnar cells with basal and cylindrical nuclei and cytoplasmic extensions into the lumen. Homer-Wright rosettes are pseudorosettes with cells lining up around a tangle of fibrils as seen in neuroblastoma. Fleurettes are composed of cells with small hyperchromatic nuclei and abundant pale eosinophilic perikaryon. In some cases, the cells seem much more differentiated toward neuronal cells and are known as bipolar-like cells, because of the similarity to the bipolar cells of the normal retina. These bipolar-like cells usually proliferate in clusters or vague nests and are surrounded by retinoblastoma cells. High mitotic index, high level of Ki67/MIB1 labeling, and apoptosis are typically found. Patterns of spread are intra- and extraocular. Direct infiltration via the optic nerve to the central nervous system or via the choroid to the orbit constitutes patterns of intraocular spread. Conversely, dispersion of tumor cells through the subarachnoid space may reach the contralateral optic nerve or through the cerebrospinal fluid dissemination to the CNS. Hematogenous dissemination is seen in the lung, bone, and brain. In case of lymphatic spreading tumor cells can disseminate anteriorly into the conjunctivae, eyelids, or extraocular tissue. The diagnosis relies on a combination of an ophthalmologic examination (under anesthesia or sedation) together with retinal camera imaging, ultrasound, CT, and MRI. A few systems have been proposed, but three have reached most consensus in the scientific literature (Box 14.12).

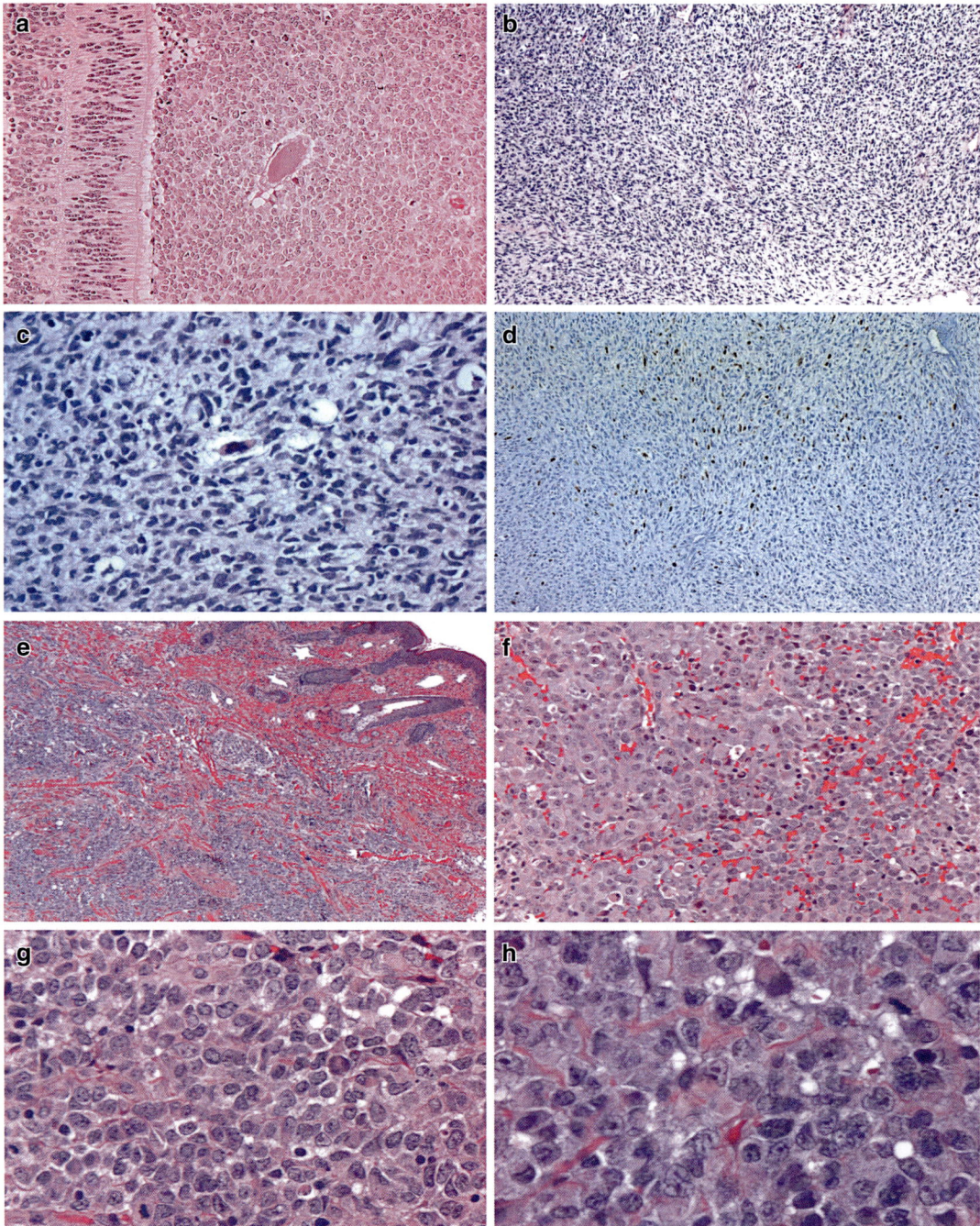

Fig. 14.22 Figure (**a**) shows the histology of a retinoblastoma with (×100 original magnification, hematoxylin and eosin staining), while figures b through d show the microscopic view of rhabdomyosarcoma of the orbita with small to moderately large spindle basophilic tumor cells, which are densely packed (**b**, ×100, hematoxylin and eosin staining) and show focal rhabdomyoblastic differentiation (**c**, ×400, periodic acid Schiff special stain). Figure (**d**) shows the expression of myogenin of this orbital rhabdomyosarcoma (anti-myogenin immunostaining, avidin-biotin complex). An extrarenal rhabdoid tumor of the orbita is shown in the hematoxylin-stained microphotographs (**e**) through (**h**) with moderately large cells with prominent pleomorphism, quite densely packed ((**e**) ×100 original magnification; (**f**) ×200 original magnification; (**g**) ×400 original magnification; (**h**) ×630 original magnification)

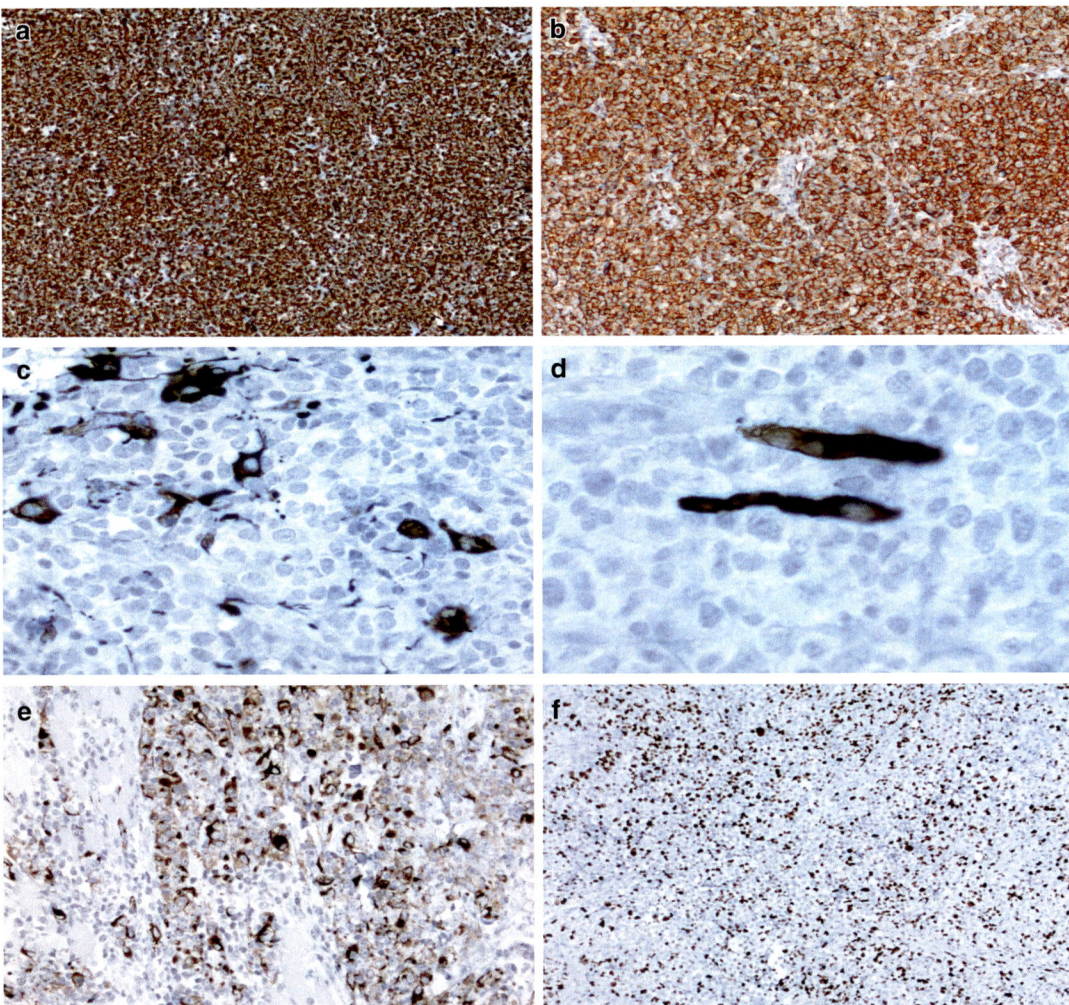

Fig. 14.23 The immunohistochemical panel confirms the diagnosis of the extrarenal rhabdoid tumor ((**a**) anti-vimentin immunostaining, ×100 original magnification; (**b**) anti-actin immunostaining, ×100 original magnification). Some focal desmin immunostaining is seen in this extrarenal rhabdoid tumor (anti-desmin immunostaining, ×400 original magnification). INI1 was lost in this extrarenal rhabdoid tumor (not shown). The orbital rhabdomyosarcoma shows desmin positivity in rhabdomyoblasts (anti-desmin immunostaining, ×630 original magnification). Figure (**e**) shows the cytokeratin positivity in the extrarenal rhabdoid tumor (anti-AE1-3 immunostaining, ×200 original magnification). Figure (**f**) shows the high proliferation rate of the extrarenal rhabdoid tumor (anti-Ki67 immunostaining using MIB1 monoclonal antibody, ×100 original magnification). All immunostaining procedures have been carried out using the avidin-biotin complex method

---

Box 14.12 Classification Systems
1. *Reese-Ellsworth Classification*, with the focus aiming to predict the PGN after radiation
2. *ABC Classification* proposed for the possibility of preserving the eye
3. *International Classification System for Intraocular Retinoblastoma (ICSIR)*

Notes: In ABC all most modern techniques are applied (Linn Murphree 2005). ICSIR covers the whole range of the disease, from intra- to extraocular extension (Aerts et al. 2006).

Enucleating the eye harboring a retinoblastoma is usually performed when there is no chance for useful vision even if the whole tumor is destroyed and when high-risk features for the development of extraocular or metastatic disease are present, including anterior chamber seeding, choroidal involvement, and tumor beyond the *lamina cribrosa*, as well as intraocular hemorrhage or scleral and extra-scleral extension. Retinocytoma is a benign variant of retinoblastoma without evidence of poor outcome. There are bland-looking cells with a high degree of photoreceptor differentiation.

- *PGN*: Poor prognostic factors include >2/3 of the globe filled with tumor, tumor in the anterior segment or anterior to the vitreous, tumor in or on the ciliary body, iris neovascularization, neovascular glaucoma, opaque media from hemorrhage, tumor necrosis with aseptic orbital cellulitis, and *phthisis bulbi*. In situations where a fetus is known to harbor the *RB1* gene defect, it has been advised to screen the conceptus carefully. Since retinoblastoma can have calcification, ultrasonography of the fetus or MRI focusing on ocular structures may be used. In Canada, some centers suggest early induction at 36 weeks to allow for early diagnosis and management of small tumors, but other countries, including the USA, do not favor this approach. We consider that genetic counseling is an integral part of the therapy and prognostic evaluation and risks for the offspring may be quite variable and genetic counseling plays a significant role in the twenty-first century. In patients with heritable RB, there is also a risk for a second malignant neoplasm, which is usually *osteosarcoma*, *soft tissue sarcoma*, and *malignant melanoma*. The highest risk is when treatment comprised full-dose and external beam radiotherapy without the use of conformational fields in infants of the 1st year of life. The 50-year risk is roughly 50% for those patients treated with radiation. In patients with unilateral RB, the risk is approximately 5% only. Finally, trilateral retinoblastoma is a rare variant form of retinoblastoma with bilateral retinoblastoma and embryonic (neuroblastic) tumor of the pineal gland with poor prognosis.

Other tumors include melanocytic tumors that are like the adult counterpart and usually do not occur before 30 years of age and adnexal tumors that are like other adnexal tumors and are also described in the dermatological chapter.

14.9 Skull

The skull may be involved of congenital processes that are described in the CNS chapter. Inflammatory processes may be found in the scalp and subscalp soft tissue. Neoplastic proliferations may be part of both dermatological competence and soft tissue competence (Figs. 14.24, 14.25, and 14.26).

14.9 Skull

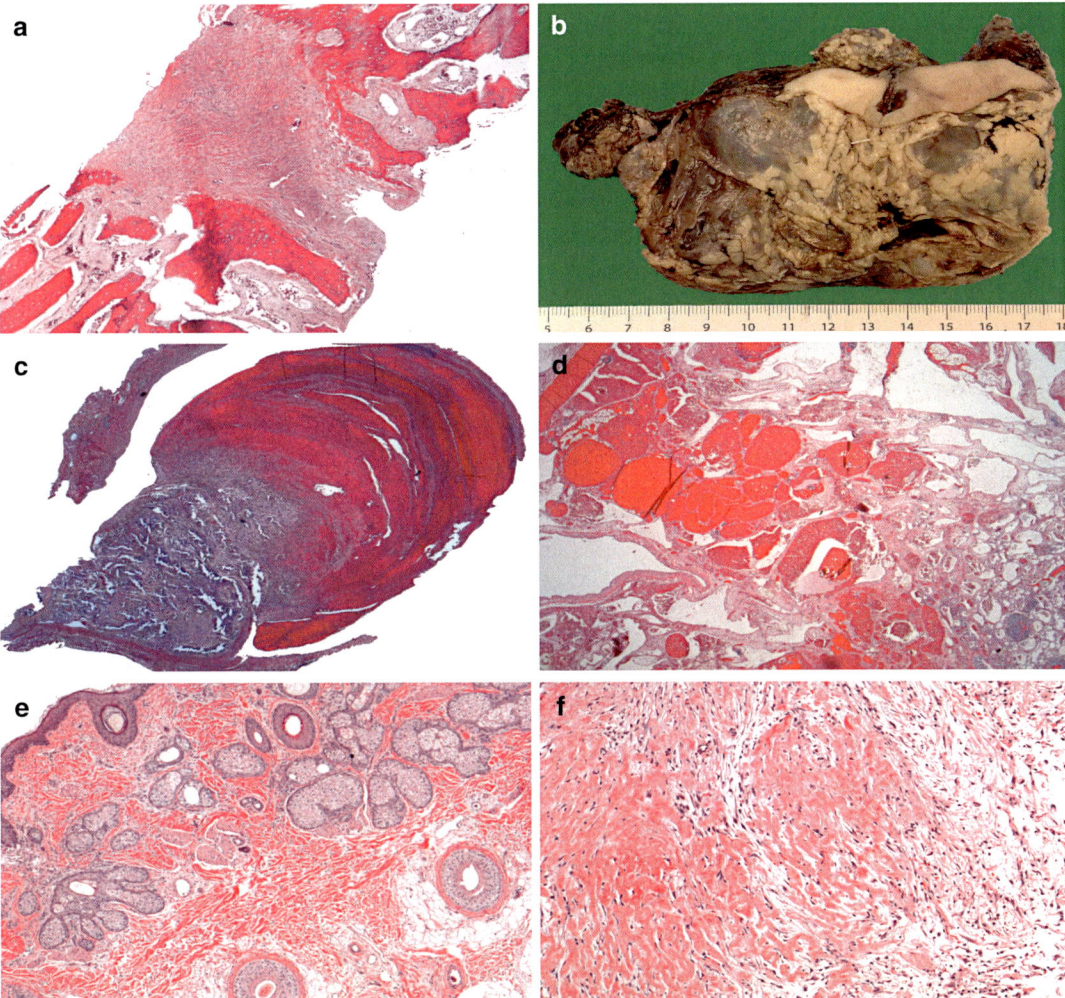

Fig. 14.24 The microphotograph in (**a**) is a fibrous sagittal synostosis, which was interpreted originally as pseudotumor of the sagittal suture (×50, hematoxylin and eosin staining). The gross photograph presented in (**b**) shows a bosselated mass corresponding to arteriovenous malformation of the skull in a child. The arteriovenous malformation is shown in the microphotographs in (**c**) and (**d**) (×12.5, hematoxylin and eosin staining). The microphotographs in (**e**) are a nevus sebaceous of Jadassohn of the skull (×50) and a nodular fasciitis of the skull (×200) with a tissue-like pattern. The nevus sebaceous of Jadassohn is not uncommon in a pediatric pathology practice. It is a hamartoma and is composed of large sebaceous and heterotopic apocrine glands, defective hair follicle formation, acanthosis, and epithelial papillomatosis. All microphotographs have been taken from histological slides stained with hematoxylin and eosin. Nodular fasciitis is discussed in detail in the soft tissue chapter of this book.

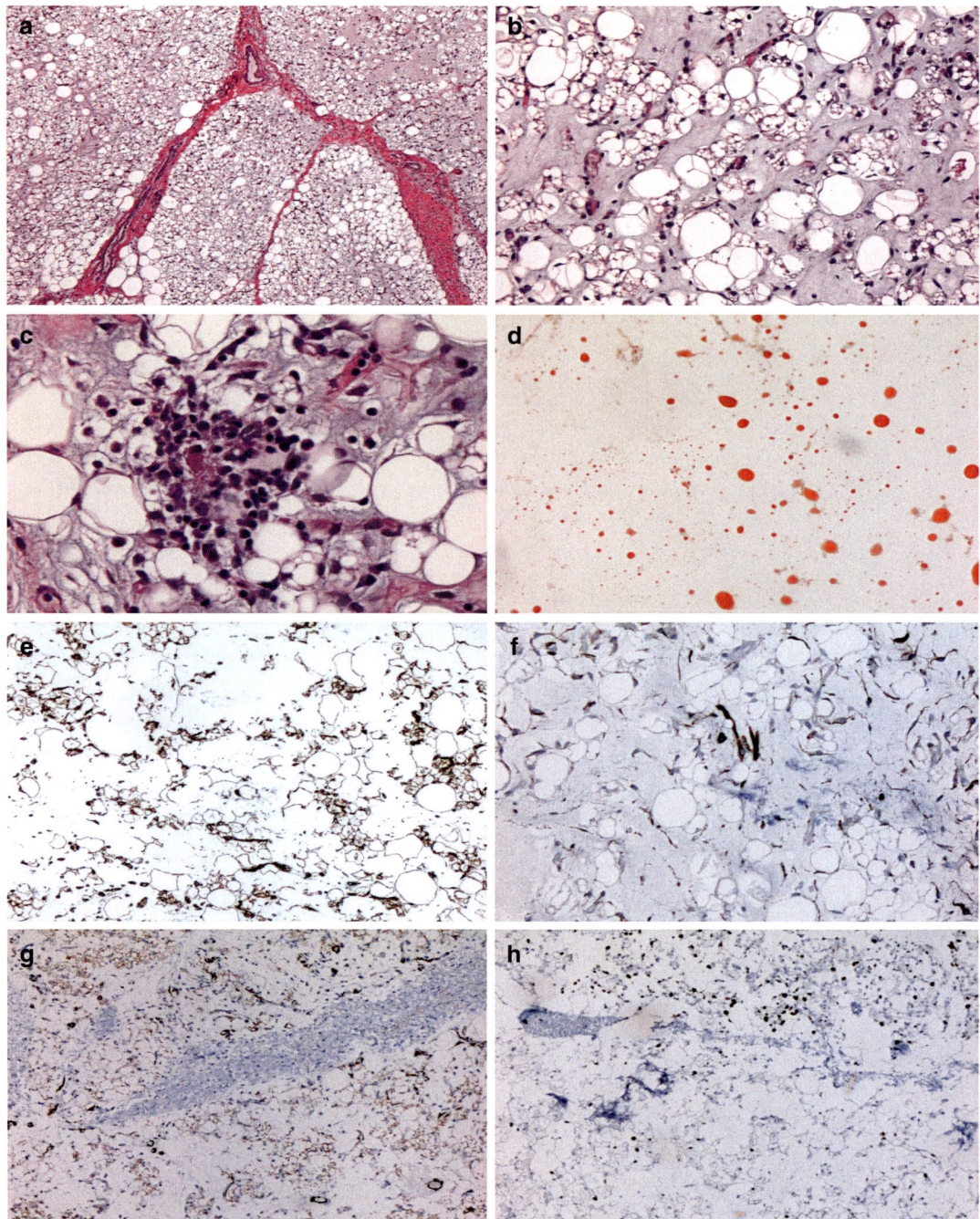

Fig. 14.25 The microphotographs display the characteristic features of a lipoblastoma with hypocellular lobules of adipocytes in various stages of differentiation (**a–c**). There are preadipocytes (spindle- or stellate-shaped), uni- or multi-vacuolated lipoblasts, and classic mature adipocytes. The lobules are separated by prominent fibrous septa. These may be quite cellular. A plexiform vascular pattern and abundant myxoid stroma with prominent extracellular mucinous pools can be seen as well (**a**, ×50, hematoxylin and eosin staining; **b**, ×200, hematoxylin and eosin staining; **c**, ×400, hematoxylin and eosin staining; **d**, ×100, oil red O staining; **e**, ×100, anti-vimentin immunostaining; **f**, ×200, anti-desmin immunostaining; **g**, ×100, anti-actin immunostaining; **h**, ×100, anti-Ki67 immunostaining). All immunostained slides have been carried out using the avidin-biotin complex method

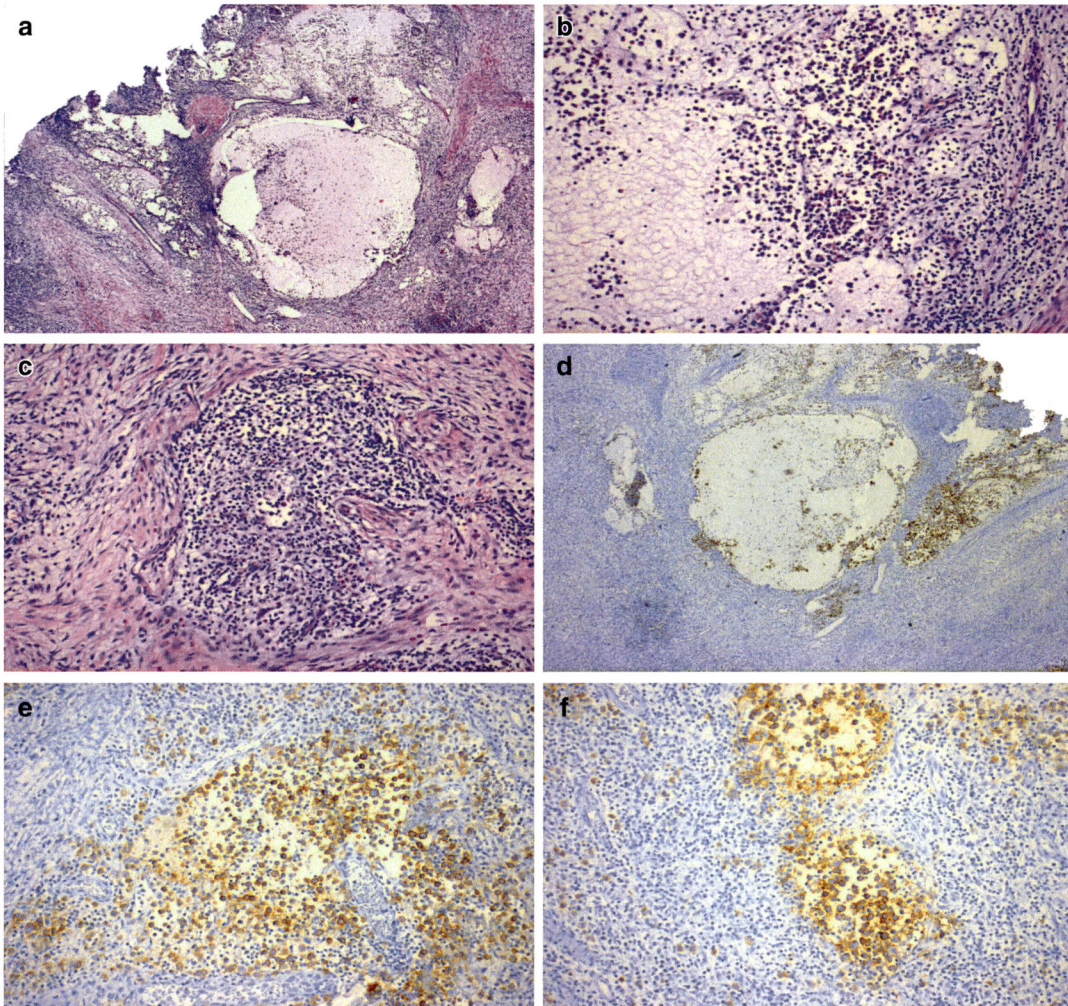

Fig. 14.26 The microphotographs show hematoxylin and eosin-stained infiltration of the skull by Langerhans cells with a microcystic appearance (**a**, ×40; **b**, ×200; **c**, ×200). The microphotographs (**d**) through (**f**) confirm the diagnosis of Langerhans cell histiocytosis using a monoclonal antibody against CD1a (**d**, ×40; **e**, ×200; **f**, ×200). This tumor showed also Birbeck granules on transmission electron microscopy (not shown in this panel)

Multiple Choice Questions and Answers

- HNK-1 During an otoscopic examination of a 36-month-old infant, a bulging tympanic membrane is noted, while the child is crying. What are the two most common bacterial infections that can cause otitis media?
    (a) *Streptococcus pneumoniae* and *Haemophilus influenzae*
    (b) *Streptococcus pyogenes* and *Moraxella catarrhalis*
    (c) *Staphylococcus aureus* and *Streptococcus pneumoniae*
    (d) *Streptococcus pyogenes* and *Staphylococcus aureus*
    (e) *Methicillin-resistant* and *Methicillin-sensitive Staphylococcus aureus*
- HNK-2 Match the sinonasal papilloma with the corresponding features.
    (a) Inverted sinonasal papilloma
    (b) Cylindrical sinonasal papilloma
    (c) Everted sinonasal papilloma

1. Lateral nasal wall, late youth, 10% cancer predisposition
2. Middle meatus (and sinuses), late youth, 10% cancer predisposition
3. Nasal septum, early youth, <1% cancer predisposition
(a)(2) (b)(1) (c)(3)

- HNK-3 Which one of the following neoplasms is associated with Epstein-Barr virus (EBV)?
  (a) Kaposi sarcoma
  (b) Pindborg tumor
  (c) Nasopharyngeal carcinoma
  (d) Osteosarcoma

- HNK-4 Which differential diagnosis does NOT belong in the list of a pre-auricular neck mass in childhood?
  (a) Hemangioma
  (b) Vascular malformation
  (c) Type I branchial cleft cyst
  (d) Reactive lymphadenopathy/lymphadenitis
  (e) Parotitis
  (f) Thyroglossal duct cyst
  (g) Pilomatrixoma
  (h) Salivary gland tumor

- HNK-5 A 12-year-old child complains to be unable to breathe properly. The pediatric otolaryngologist excised a round, smooth, soft, translucent, pale, glistening mass from the prolapsed lining of the ethmoid sinus. The histology reveals a polypoid structure without cellular atypia and abnormal epithelial growth. What is the most likely diagnosis?
  (a) Nasopharyngeal carcinoma
  (b) Anthro-choanal polyp
  (c) Ordinary nasal polyp
  (d) Inverted papilloma
  (e) Sinusitis

- HNK-6 A child develops a cyst that sheds layers of old skin that builds up inside the ear following recurrent infections. Histological examination reveals squamous epithelium and keratin lamellae without signs of cellular atypia or tissue dysplasia. Which of the following statement is TRUE for the diagnosis considered in this case?
  (a) It is mostly caused by a repeated infection that determines an ingrowth of the skin of the eardrum.
  (b) It may metastasize to distant sites.
  (c) It is treated medically.
  (d) Hearing loss, dizziness, and facial muscle paralysis are common findings.

- HNK-7 What is the most common cause of bilateral conductive deafness in a child?
  (a) Congenital cholesteatoma
  (b) Otitis media with effusion
  (c) Otosclerosis
  (d) Acute otitis media
  (e) Chronic suppurative otitis media

- HNK-8 Which of the following causes is NOT an etiology of congenital stridor?
  (a) Laryngomalacia
  (b) Acute epiglottitis
  (c) Laryngeal web
  (d) Subglottic stenosis
  (e) Vascular malformation

- HNK-9 A 12-year-old boy develops a rapidly growing painless mass in his posterior neck. A biopsy of the mass shows a small round basophilic cell population with focal skeletal muscle-like cross striations. The tumor shows an immunohistochemical positivity for vimentin, myosin, desmin, myoglobin, MyoD1, and myogenin. The MyoD1 family of myogenic nuclear regulatory proteins includes MyoD1/myf3 and myogenin/myf4. Which of the following statements is NOT correct for the diagnosis considered in this case?
  (a) CT of the chest is required for the staging of the disease.
  (b) Most lesions occur in the head/neck, genitourinary system, or retroperitoneum.
  (c) MRI reveals the tumor bright on T1, dark on T2.
  (d) A subtype of this tumor expresses *PAX3-FKHR* or *PAX7-FKHR* gene fusions resulting from t(2;13) or t(1;13) translocations.
  (e) Sentinel lymph node biopsy may be considered as part of treatment because of this tumor propensity to get nodal metastasis.

- HNK-10 Which of the following statements is NOT correct regarding retinoblastoma?
  (a) *RB1* is a large gene with 27 exons, encoding a 4.7 kb mRNA that translates into a 928 amino acid protein, pRB.

(b) pRB is a cell cycle regulator that binds to E2F transcription factors to repress cell proliferation-related genes.

(c) The International Retinoblastoma Staging System (IRSS) focused on overall staging (different again from the standard TNM).

(d) There are four different mostly used classifications, including the International Intraocular Retinoblastoma Classification according to Murphree (2005), the International Intraocular Retinoblastoma Classification according to Shields (2006), the Children Oncology Group classification, and the Tumor, Node, Metastasis (2010).

(e) Chondroid differentiation is a useful recently discovered feature to distinguish among several subtypes of retinoblastoma.

(f) Staining of specific markers for photoreceptor cells (the cone-rod homeobox transcription factor; CRX) and for tumor cells (N-glycosylated ganglioside; NeuGc-GM3) helps in detecting the tumor cells in settings where artifacts may jeopardize the diagnosis.

References and Recommended Readings

Abrahams JM, McClure SA. Pediatric odontogenic tumors. Oral Maxillofac Surg Clin North Am. 2016;28(1):45–58. https://doi.org/10.1016/j.coms.2015.08.003. Review. PubMed PMID: 26614700.

Adisa AO, Lawal AO, Effiom OA, Soyele OO, Omitola OG, Olawuyi A, Fomete B. A retrospective review of 61 cases of adenomatoid odontogenic tumour seen in five tertiary health facilities in Nigeria. Pan Afr Med J. 2016;24:102. https://doi.org/10.11604/pamj.2016.24.102.9400. eCollection 2016. PubMed PMID: 27642441; PubMedCentral PMCID: PMC5012829.

Aerts I, Lumbroso-Le Rouic L, Gauthier-Villars M, Brisse H, Doz F, Desjardins L. Retinoblastoma. Orphanet J Rare Dis. 2006;1:31. https://doi.org/10.1186/1750-1172-1-31. Review. PubMed PMID: 16934146; PubMed Central PMCID: PMC1586012.

Akdogan O, Ibrahim O, Selcuk A, Dere H. The association of laryngoceles with squamous cell carcinoma of the larynx presenting as a deep neck infection. B-ENT. 2007;3(4):209–11. PubMed PMID: 18265728

Albano D, Bosio G, Giubbini R, Bertagna F. Role of (18)F-FDG PET/CT in patients affected by Langerhans cell histiocytosis. Jpn J Radiol. 2017;35(10):574–83. https://doi.org/10.1007/s11604-017-0668-1. Epub 2017 Jul 26. PubMed PMID: 28748503

Allen CE, Ladisch S, McClain KL. How I treat Langerhans cell histiocytosis. Blood. 2015;126(1):26–35. https://doi.org/10.1182/blood-2014-12-569301. Epub 2015 Mar 31. PubMed PMID: 25827831; PubMed Central PMCID: PMC4492195

Allison JR, Garlington G. The value of cone beam computed tomography in the Management of Dentigerous Cysts – a review and case report. Dent Update. 2017;44(3):182–4, 186–8. Review. PubMed PMID: 29172322.

Aoki M, Nishihori T, Obara N, Kato H, Mizuta K. Pediatric Warthin tumor of parotid gland. Int J Clin Pediatr. North America, 3, 2014. Available at: https://theijcp.org/index.php/ijcp/article/view/149/120. Accessed 23 Sep 2019.

Arce K, Streff CS, Ettinger KS. Pediatric odontogenic cysts of the jaws. Oral Maxillofac Surg Clin North Am. 2016;28(1):21–30. https://doi.org/10.1016/j.coms.2015.07.003. Review. PubMed PMID: 26614698.

Báez A. Genetic and environmental factors in head and neck cancer genesis. J Environ Sci Health C Environ Carcinog Ecotoxicol Rev. 2008;26(2):174–200. Review. PubMed PMID: 18569329.

Bambakidis NC, Megerian CA, Ratcheson RA. Differential grading of endolymphatic sac tumor extension by virtue of von Hippel-Lindau disease status. Otol Neurotol. 2004;25(5):773–81. PMID 15354010.

Barnes C, Sexton M, Sizeland A, Tiedemann K, Berkowitz RG, Waters K. Laryngo-pharyngeal carcinoma in childhood. Int J Pediatr Otorhinolaryngol. 2001;61(1):83–6. Review. PubMed PMID: 11576635.

Barrett AW, Sneddon KJ, Tighe JV, Gulati A, Newman L, Collyer J, Norris PM, Coombes DM, Shelley MJ, Bisase BS, Liebmann RD. Dentigerous cyst and ameloblastoma of the jaws. Int J Surg Pathol. 2017;25(2):141–7. https://doi.org/10.1177/1066896916666319. Epub 2016 Sep 24. PubMed PMID: 27621276.

Baylis A. Head and neck embryology: an overview of development, growth and defect in the human fetus. Honors Scholar Theses, 105. 2009. http://digitalcommons.uconn.edu/srhonors_theses/105

Becker M, Stefanelli S, Rougemont AL, Poletti PA, Merlini L. Non-odontogenic tumors of the facial bones in children and adolescents: role of multiparametric imaging. Neuroradiology. 2017;59(4):327–42. https://doi.org/10.1007/s00234-017-1798-y. Epub 2017 Mar 13. Review. PubMed PMID: 28289810; PubMed Central PMCID:PMC5394153.

Benn A, Altini M. Dentigerous cysts of inflammatory origin. A clinicopathologic study. Oral Surg Oral Med Oral Pathol Oral Radiol Endod. 1996;81(2):203–9. PubMed PMID: 8665316

Bezold F. Cholesteatom, Perforation der Membrana Flaccida Schrapnelli und Tubenverschluss: eine ätiologische Studie. Z Ohrenheilkd. 1889;20:5–28.

Boedeker CC, Neumann HP, Offergeld C, Maier W, Falcioni M, Berlis A, Schipper J. Clinical features of paraganglioma syndromes. Skull Base. 2009;19(1):17–25. PubMed PMID: 19568339; PubMed Central PMCID: PMC2637571.

Bragulla H. DVM, Ph.D. Thesis. 2005.

Brodsky L. Modern assessment of tonsils and adenoids. Pediatr Clin N Am. 1989;36(6):1551–69. Review. PubMed PMID: 2685730.

Burke EN, Golden JL. External ventricular laryngocele. Am J Roentgenol Radium Therapy, Nucl Med. 1958;80(1):49–53. PubMed PMID: 13545459

Canalis RF. Laryngeal ventricle. Historical features. Ann Otol Rhinol Laryngol. 1980;89(2 Pt 1):184–7. PubMed PMID: 6989307.

Celin SE, Johnson J, Curtin H, Barnes L. The association of laryngoceles with squamous cell carcinoma of the larynx. Laryngoscope. 1991;101(5):529–36. Review. PubMed PMID: 2030634

Cesmebasi A, Gabriel A, Niku D, Bukala K, Donnelly J, Fields PJ, Tubbs RS, Loukas M. Pediatric head and neck tumors: an intra-demographic analysis using the SEER∗ database. Med Sci Monit. 2014;20:2536–42. https://doi.org/10.12659/MSM.891052. PubMed PMID: 25473782; PubMed Central PMCID: PMC4266203.

Chan CC, Chan YY, Tanweer F. Systematic review and meta-analysis of the use of tranexamic acid in tonsillectomy. Eur Arch Otorhinolaryngol. 2013;270(2):735–48. https://doi.org/10.1007/s00405-012-2184-3. Epub 2012 Sep 21. Review. PubMed PMID: 22996082.

Chang RC, Dave SP, Robinson PG. Undifferentiated pleomorphic sarcoma of the parotid gland: a rare pediatric case. Head Neck. 2008;30(7):970–3. PubMed PMID: 18098306

Chantada G, Doz F, Antoneli CB, Grundy R, Clare Stannard FF, Dunkel IJ, Grabowski E, Leal-Leal C, Rodríguez-Galindo C, Schvartzman E, Popovic MB, Kremens B, Meadows AT, Zucker JM. A proposal for an international retinoblastoma staging system. Pediatr Blood Cancer. 2006;47(6):801–5. PubMed PMID: 16358310.

Chaudhary S, Sah JP. Hypercalcemia due to nasopharyngeal carcinoma. JNMA J Nepal Med Assoc. 2017;56(205):182–5. PubMed PMID: 28598460.

Chetty R. Familial paraganglioma syndromes. J Clin Pathol. 2010;63(6):488–91. Review. PubMed PMID: 20498024.

Cohen D, Dor M. Morbidity and mortality of post-tonsillectomy bleeding: analysis of cases. J Laryngol Otol. 2008;122(1):88–92. Epub 2007 Mar 12. PubMed PMID: 17349099.

Cohen EG, Yoder M, Thomas RM, Salerno D, Isaacson G. Congenital salivary gland anlage tumor of the nasopharynx. Pediatrics. 2003;112(1 Pt 1):e66–9. Review. PubMed PMID: 12837908.

Costacurta M, Maturo P, Docimo R. Riga-Fede disease and neonatal teeth. Oral Implantol (Rome). 2012;5(1):26–30. Published online 2012 Jul 17. PMCID: PMC3533976

Cruveilhier J. Anatomie pathologique du corps humain. Paris: Baillière; 1829.

Dalpiaz G, Cancellieri A, editors. Atlas of diffuse lung diseases: a multidisciplinary approach. Cham, Switzerland: Springer; 2017. ISBN: 978-3-319-42752-2

De Verney JG. Traité de l'organe de l'ouïe. Paris. Paris: E. Michallet; 1683.

Dedhia P, Dedhia S, Dhokar A, Desai A. Nasopalatine duct cyst. Case Rep Dent. 2013;2013:869516. https://doi.org/10.1155/2013/869516. Epub 2013 Nov 4. PubMed PMID: 24307954; PubMed Central PMCID: PMC3834977

DeSanto LW, Devine KD, Weiland LH. Cysts of the larynx – classification. Laryngoscope. 1970;80(1):145–76. PubMed PMID: 5411821.

Dhupar A, Yadav S, Dhupar V, Mittal HC, Malik S, Rana P. Bi-maxillary dentigerous cyst in a non-syndromic child – review of literature with a case presentation. J Stomatol Oral Maxillofac Surg. 2017;118(1):45–8. https://doi.org/10.1016/j.jormas.2016.12.001. Epub 2017 Feb 3. PubMed PMID: 28330574.

Diaz RC. Head and neck: ear: endolymphatic sac tumor (ELST). Atlas Genet Cytogenet Oncol Haematol. 2009. http://AtlasGeneticsOncology.org/Tumors/EndolymphaticSacTumID5096.html

Din NU, Fatima S, Kayani N. Mammary analogue secretory carcinoma of salivary glands: a clinicopathologic study of 11 cases. Ann Diagn Pathol. 2016;22:49–53. https://doi.org/10.1016/j.anndiagpath.2016.04.003. Epub 2016 Apr 11. PubMed PMID: 27180060.

Domingues-Cruz J, Herrera A, Fernandez-Crehuet P, Garcia-Bravo B, Camacho F. Riga-Fede disease associated with postanoxic encephalopathy and trisomy 21: a proposed classification. Pediatr Dermatol. 2007;24(6):663–5. PubMed PMID: 18035997

Donegan JO, Strife JL, Seid AB, Cotton RT, Dunbar JS. Internal laryngocele and saccular cysts in children. Ann Otol Rhinol Laryngol. 1980;89(5 Pt 1):409–13. PubMed PMID: 7002004.

Doyle LA, Vivero M, Fletcher CD, Mertens F, Hornick JL. Nuclear expression of STAT6 distinguishes solitary fibrous tumor from histologic mimics. Mod Pathol. 2014;27(3):390–5. https://doi.org/10.1038/modpathol.2013.164. Epub 2013 Sep 13. PubMed PMID: 24030747

Emile JF, Abla O, Fraitag S, Horne A, Haroche J, Donadieu J, Requena-Caballero L, Jordan MB, Abdel-Wahab O, Allen CE, Charlotte F, Diamond EL, Egeler RM, Fischer A, Herrera JG, Henter JI, Janku F, Merad M, Picarsic J, Rodriguez-Galindo C, Rollins BJ, Tazi A, Vassallo R, Weiss LM, Society H. Revised classification of histiocytoses and neoplasms of the macrophage-dendritic cell lineages. Blood. 2016;127(22):2672–81. https://doi.org/10.1182/blood-2016-01-690636. Epub 2016 Mar 10. PMID: 26966089; PMCID: PMC5161007

Eppsteiner RW, Smith RJ. Genetic disorders of the vestibular system. Curr Opin Otolaryngol Head Neck Surg. 2011;19(5):397–402. Review. PubMed PMID: 21825995.

References and Recommended Readings

Evrard SM, Meilleroux J, Daniel G, Basset C, Lacoste-Collin L, Vergez S, Uro-Coste E, Courtade-Saidi M. Use of fluorescent in-situ hybridisation in salivary gland cytology: a powerful diagnostic tool. Cytopathology. 2017;28(4):312–20. https://doi.org/10.1111/cyt.12427. Epub 2017 May 15. PubMed PMID: 28503786.

Ferreira MA, Feiz-Erfan I, Zabramski JM, Spetzler RF, Coons SW, Preul MC. Endolymphatic sac tumor: unique features of two cases and review of the literature. Acta Neurochir. 2002;144(10):1047–53. PMID 12382133.

Friedman M, Yalamanchali S, Gorelick G, Joseph NJ, Hwang MS. A standardized lingual tonsil grading system: interexaminer agreement. Otolaryngol Head Neck Surg. 2015;152(4):667–72. https://doi.org/10.1177/0194599815568970. Epub 2015 Jan 27. PubMed PMID: 25628371.

Gale N, Zidar N, Poljak M, Cardesa A. Current views and perspectives on classification of squamous intraepithelial lesions of the head and neck. Head Neck Pathol. 2014;8(1):16–23. https://doi.org/10.1007/s12105-014-0530-z. Epub 2014 Mar 5. PubMed PMID: 24595419; PubMed Central PMCID: PMC3950392.

Ghosh A, Saha S, Pal S. Myoepithelial neoplasm of nasal cavity: an uncommon tumor presenting with an unusual clinical presentation. Kulak Burun Bogaz Ihtis Derg. 2014;24(1):42–5. https://doi.org/10.5606/kbbihtisas.2014.00243. PubMed PMID: 24798439.

Goette DK. Chondrodermatitis nodularis chronica helicis: a perforating necrobiotic granuloma. J Am Acad Dermatol. 1980;2(2):148–54. PubMed PMID: 7364972.

Goldman JL, Baugh RF, Davies L, Skinner ML, Stachler RJ, Brereton J, Eisenberg LD, Roberson DW, Brenner MJ. Mortality and major morbidity after tonsillectomy: etiologic factors and strategies for prevention. Laryngoscope. 2013;123(10):2544–53. https://doi.org/10.1002/lary.23926. Epub 2013 Apr 17. PubMed PMID: 23595509.

Gombos DS. Retinoblastoma in the perinatal and neonatal child. Semin Fetal Neonatal Med. 2012;17(4):239–42. https://doi.org/10.1016/j.siny.2012.04.003. Epub 2012 May 22. Review. PubMed PMID: 22622484.

Grecchi E, Borgonovo AE, Re D, Creminelli L, Grecchi F. Aneurismal bone cyst: a conservative surgical technique. A case report treated with a small access osteotomy. Eur J Paediatr Dent. 2016;17(2):100–3. PubMed PMID: 27377106.

Habermann J. Zur Entstehung des Cholesteatoms des Mittelohres. Arch Ohrenheilkd 1888;27:43–51

Habermann J. Cholesteatom des Mittelohres, seine Entstehung. Z Ohrenheilkd. 1889a;19:348.

Habermann J. Zur Entstehung des Cholesteatoms des Mittelohrs. Arch Ohrenheilkd. 1889b;27:42–50.

Harney M, Patil N, Walsh R, Brennan P, Walsh M. Laryngocele and squamous cell carcinoma of the larynx. J Laryngol Otol. 2001;115(7):590–2. PubMed PMID: 11485599

Harvey RT, Ibrahim H, Yousem DM, Weinstein GS. Radiologic findings in a carcinoma-associated laryngocele. Ann Otol Rhinol Laryngol. 1996;105(5):405–8. PubMed PMID: 8651636

He W, Hashimoto H, Tsuneyoshi M, Enjoji M, Inomata H. A reassessment of histologic classification and an immunohistochemical study of 88 retinoblastomas. A special reference to the advent of bipolar-like cells. Cancer. 1992;70(12):2901–8. Erratum in: Cancer 1993;71(8):2697. PubMed PMID: 1451072.

Holinger LD, Konior RJ. Surgical management of severe laryngomalacia. Laryngoscope. 1989;99:136–42.

Holinger LD, Barnes DR, Smid LJ, Holinger PH. Laryngocele and saccular cysts. Ann Otol Rhinol Laryngol. 1978;87(5 Pt 1):675–85. PubMed PMID: 718065.

Hwang MS, Salapatas AM, Yalamanchali S, Joseph NJ, Friedman M. Factors associated with hypertrophy of the lingual tonsils. Otolaryngol Head Neck Surg. 2015;152(5):851–5. https://doi.org/10.1177/0194599815573224. Epub 2015 Mar 9. PubMed PMID: 25754182.

Inaba T, Fukumura Y, Saito T, Yokoyama J, Ohba S, Arakawa A, Yao T. Cytological features of mammary analogue secretory carcinoma of the parotid gland in a 15-year-old girl: a case report with review of the literature. Case Rep Pathol. 2015;2015:656107. https://doi.org/10.1155/2015/656107. Epub 2015 Mar 1. PubMed PMID: 25815230; PubMed Central PMCID: PMC4359859.

Isaacson G. An approach to congenital malformations of the head and neck. Otolaryngol Clin N Am. 2007;40(1):1–8, v. Review. PubMed PMID: 17346558.

Ito Y, Ishibashi K, Masaki A, Fujii K, Fujiyoshi Y, Hattori H, Kawakita D, Matsumoto M, Miyabe S, Shimozato K, Nagao T, Inagaki H. Mammary analogue secretory carcinoma of salivary glands: a clinicopathologic and molecular study including 2 cases harboring ETV6-X fusion. Am J Surg Pathol. 2015;39(5):602–10. https://doi.org/10.1097/PAS.0000000000000392. PubMed PMID: 25651470.

Izumo T. Oral premalignant lesions: from the pathological viewpoint. Int J Clin Oncol. 2011;16(1):15–26. https://doi.org/10.1007/s10147-010-0169-z. Epub 2011 Jan 14. Review. PubMed PMID: 21234636.

Jackson C, Jackson CL. Diseases and injuries of the larynx. New York: Macmillan; 1942. p. 63–9.

Johnson Chacko L, Wertjanz D, Sergi C, Dudas J, Fischer N, Eberharter T, Hoermann R, Glueckert R, Fritsch H, Rask-Andersen H, Schrott-Fischer A, Handschuh S. Growth and cellular patterning during fetal human inner ear development studied by a correlative imaging approach. BMC Dev Biol. 2019;19(1):11. https://doi.org/10.1186/s12861-019-0191-y. PubMed PMID: 31109306; PubMed Central PMCID: PMC6528216

Kamil AH, Tarakji B. Odontogenic Keratocyst in children: a review. Open Dent J. 2016;10:117–23. https://doi.org/10.2174/1874210601610010117. eCollection 2016. PubMed PMID: 27335612; PubMed Central PMCID: PMC4891985.

Kansu L, Aydin E. Atypical presentation of antrochoanal polyp in a child. Turk J Pediatr. 2011;53(3):320–4. PubMed PMID: 21980816.

Kariya S, Aoji K, Kuyama K, Akagi H, Fukazawa M, Nishizaki K. Malignant fibrous histiocytoma of the parotid gland. Auris Nasus Larynx. 2003;30(3):315–8. Review. PubMed PMID: 12927301

Khurram SA, Sultan-Khan J, Atkey N, Speight PM. Cytogenetic and immunohistochemical characterization of mammary analogue secretory carcinoma of salivary glands. Oral Surg Oral Med Oral Pathol Oral Radiol. 2016;122(6):731–42. https://doi.org/10.1016/j.oooo.2016.07.008. Epub 2016 Jul 20. PubMed PMID: 27720350.

Kim WY, Kaelin WG. Role of VHL gene mutation in human cancer. J Clin Oncol. 2004;22(24):4991–5004. (Review) PMID 15611513.

Kim HJ, Butman JA, Brewer C, Zalewski C, Vortmeyer AO, Glenn G, Oldfield EH, Lonser RR. Tumors of the endolymphatic sac in patients with von Hippel-Lindau disease: implications for their natural history, diagnosis, and treatment. J Neurosurg. 2005;102(3):503–12. PMID 15796386.

Kölle G. [Hyperuricemia and gout in childhood (Lesch-Nyhan syndrome)]. Med Klin. 1971;66(17):626–30. German. PubMed PMID: 4251840.

Krause KA, Butler SL. Koilocytosis. [Updated 2018 Dec 1]. In: StatPearls [Internet]. Treasure Island (FL): StatPearls Publishing; 2019 Jan. Available from: https://www.ncbi.nlm.nih.gov/books/NBK532958/

Kunisch E, Fuhrmann R, Roth A, Winter R, Lungershausen W, Kinne RW. Macrophage specificity of three anti-CD68 monoclonal antibodies (KP1, EBM11, and PGM1) widely used for immunohistochemistry and flow cytometry. Ann Rheum Dis. 2004;63(7):774–84. PubMed PMID: 15194571; PubMed Central PMCID: PMC1755048

Lanzkowsky P. Retinoblastoma. In: Manual of pediatric hematology and oncology: Academic Press/Elsevier. https://doi.org/10.1016/B978-0-12-375154-6.00026-4.

Larrey DJ. Du goitre aerien ou vesiculaire: Clinic chirugicale, Exercée particulierement dans les hospitaux militaires, depuis 1792 jusqu'en 1836. 1829;2:81(Nov.).

Lee JI, Kang SJ, Jeon SP, Sun H. Stafne bone cavity of the mandible. Arch Craniofac Surg. 2016;17(3):162–4. https://doi.org/10.7181/acfs.2016.17.3.162. Epub 2016 Sep 23. PMID: 28913275; PMCID: PMC5556806

Leon-Villapalos J, Wolfe K, Kangesu L. GLUT-1: an extra diagnostic tool to differentiate between haemangiomas and vascular malformations. Br J Plast Surg. 2005;58(3):348–52. PubMed PMID: 15780229

Levin AV. Congenital eye anomalies. Pediatr Clin N Am. 2003;50(1):55–76. Review. PubMed PMID: 12713104.

Linn MA. Intraocular retinoblastoma: the case for a new group classification. Ophthalmol Clin N Am. 2005;18(1):41–53, viii. Review. PubMed PMID: 15763190

Lonser RR, Kim HJ, Butman JA, Vortmeyer AO, Choo DI, Oldfield EH. Tumors of the endolymphatic sac in von Hippel-Lindau disease. N Engl J Med. 2004;350(24):2481–6. PMID 15190140.

López F, Suárez V, Costales M, Rodrigo JP, Suárez C, Llorente JL. Endoscopic endonasal approach for the treatment of anterior skull base tumours. Acta Otorrinolaringol Esp. 2012;63(5):339–47. https://doi.org/10.1016/j.otorri.2012.02.002. Epub 2012 Apr 10. English, Spanish. PubMed PMID:22498372.

Lowe LH, Booth TN, Joglar JM, Rollins NK. Midface anomalies in children. Radiographics. 2000; 20(4):907–22; quiz 1106–7, 1112. Review. Erratum in: Radiographics 2000 Sep–Oct;20(5):1494. PubMed PMID: 10903683

Lu S. [A case report of childhood primary gout]. Zhongguo Dang Dai Er Ke Za Zhi. 2015;17(8):884–5. Chinese. PubMed PMID: 26287359.

Luz J, Zweifel D, Hüllner M, Bühler M, Rücker M, Stadlinger B. Oral manifestation of Langerhans cell histiocytosis: a case report. BMC Oral Health. 2018;18(1):106. https://doi.org/10.1186/s12903-018-0568-5. PubMed PMID: 29884166

Makhasana JA, Kulkarni MA, Vaze S, Shroff AS. Juvenile nasopharyngeal angiofibroma. J Oral Maxillofac Pathol. 2016;20(2):330. https://doi.org/10.4103/0973-029X.185908. PubMed PMID: 27601836; PubMed Central PMCID: PMC4989574.

Maksimović Z, Rukovanjski M. Intracranial complications of cholesteatoma. Acta Otorhinolaryngol Belg. 1993;47(1):33–6. PubMed PMID: 8470548.

McGinnis LM, Nybakken G, Ma L, Arber DA. Frequency of MAP2K1, TP53, and U2AF1 mutations in BRAF-mutated Langerhans cell histiocytosis: further characterizing the genomic landscape of LCH. Am J Surg Pathol. 2018;42(7):885–90. https://doi.org/10.1097/PAS.0000000000001057. PubMed PMID: 29649018.

McGuirt WF Jr, Little JP. Laryngeal cancer in children and adolescents. Otolaryngol Clin N Am. 1997;30(2):207–14. Review. PubMed PMID: 9052665

McKnight CD, Parmar HA, Watcharotone K, Mukherji SK. Reassessing the anatomic origin of the juvenile nasopharyngeal angiofibroma. J Comput Assist Tomogr. 2017;41(4):559–64. https://doi.org/10.1097/RCT.0000000000000566. PubMed PMID: 28632604.

Megerian CA, McKenna MJ, Nuss RC, Maniglia AJ, Ojemann RG, Pilch BZ, Nadol JB Jr. Endolymphatic sac tumors: histopathologic confirmation, clinical characterization, and implication in von Hippel-Lindau disease. Laryngoscope. 1995;105(8 Pt 1):801–8. PMID 7630290.

Megerian CA, Haynes DS, Poe DS, Choo DI, Keriakas TJ, Glasscock ME 3rd. Hearing preservation surgery for small endolymphatic sac tumors in patients with von Hippel-Lindau syndrome. Otol Neurotol. 2002;23(3):378–87. (Review) PMID 11981399.

Mitroi M, Căpitănescu A, Popescu FC, Popescu C, Mogoantă CA, Mitroi G, Surlin C. Laryngocele associated with laryngeal carcinoma. Romanian J Morphol

Embryol. 2011;52(1):183–5. PubMed PMID: 21424053.

Modh A, Gayar OH, Elshaikh MA, Paulino AC, Siddiqui F. Pediatric head and neck squamous cell carcinoma: patient demographics, treatment trends and outcomes. Int J Pediatr Otorhinolaryngol. 2018;106:21–5. https://doi.org/10.1016/j.ijporl.2017.12.032. Epub 2018 Jan 3. PubMed PMID: 29447885.

Montgomery-Downs HE, Ramadan HH, Clawges HC, McBean AL, Insana SP, Santy EE. Digital oral photography for pediatric tonsillar hypertrophy grading. Int J Pediatr Otorhinolaryngol. 2011;75(6):841–3. https://doi.org/10.1016/j.ijporl.2011.03.022. Epub 2011 Apr 22. PubMed PMID: 21514678.

More CB, Bhavsar K, Varma S, Tailor M. Oral mucocele: a clinical and histopathological study. J Oral Maxillofac Pathol. 2014;18(Suppl 1):S72–7. https://doi.org/10.4103/0973-029X.141370. PMID: 25364184; PMCID: PMC4211243

Morrison, A. Odontogenic cysts. PathologyOutlines.com website. http://www.pathologyoutlines.com/topic/mandiblemaxilladentigerous.html. Accessed 15 June 2018.

Müller J. Über den feinern Bau und die Formen der krankhaften Geschwülste. Berlin: G. Reimer; 1838.

Murphree AL. Intraocular retinoblastoma: the case for a new group classification. Ophthalmol Clin N Am. 2005;18(1):41–53, viii. Review. PubMed PMID: 15763190.

Murray SP, Burgess LP, Burton DM, Gonzalez C, Wood GS, Zajtchuk JT. Laryngocele associated with squamous carcinoma in a 20-year-old nonsmoker. Ear Nose Throat J. 1994;73(4):258–61. PubMed PMID: 8020424

Nagori SA, Jose A, Bhutia O, Roychoudhury A. Large pediatric maxillary dentigerous cysts presenting with sinonasal and orbital symptoms: a case series. Ear Nose Throat J. 2017;96(4–5):E29–34. Review. PubMed PMID: 28489242.

Nevoux J, Lenoir M, Roger G, Denoyelle F. Ducou Le Pointe H, Garabédian EN. Childhood cholesteatoma. Eur Ann Otorhinolaryngol Head Neck Dis. 2010;127(4):143–50. https://doi.org/10.1016/j.anorl.2010.07.001. Epub 2010 Aug 11. Review. PubMed PMID: 20860924

Ngouajio AL, Drejet SM, Phillips DR, Summerlin DJ, Dahl JP. A systematic review including an additional pediatric case report: pediatric cases of mammary analogue secretory carcinoma. Int J Pediatr Otorhinolaryngol. 2017;100:187–93. https://doi.org/10.1016/j.ijporl.2017.07.004. Epub 2017 Jul 6. Review. PubMed PMID: 28802370.

Orita Y, Gion Y, Tachibana T, Ikegami K, Marunaka H, Makihara S, Yamashita Y, Miki K, Makino T, Akisada N, Akagi Y, Kimura M, Yoshino T, Nishizaki K, Sato Y. Laryngeal squamous cell papilloma is highly associated with human papillomavirus. Jpn J Clin Oncol. 2018;48(4):350–5. https://doi.org/10.1093/jjco/hyy009. PubMed PMID: 29447361

Oza N, Sanghvi K, Shet T, Patil A, Menon S, Ramadwar M, Kane S. Mammary analogue secretory carcinoma of parotid: is preoperative cytological diagnosis possible? Diagn Cytopathol. 2016;44(6):519–25. https://doi.org/10.1002/dc.23459. Epub 2016 Mar 4. PubMed PMID: 26945684.

Patel NP, Wiggins RH 3rd, Shelton C. The radiologic diagnosis of endolymphatic sac tumors. Laryngoscope. 2006;116(1):40–6. PMID 16481807.

Patil S, Rao RS, Majumdar B, Jafer M, Maralingannavar M, Sukumaran A. Oral lesions in neonates. Int J Clin Pediatr Dent. 2016;9(2):131–8. https://doi.org/10.5005/jp-journals-10005-1349. Published online 2016 Jun 15. Review. PubMed PMID: 27365934; PubMed Central PMCID: PMC4921882

Prowse S, Knight L. Congenital cysts of the infant larynx. Int J Pediatr Otorhinolaryngol. 2012;76(5):708–11. Epub 2012 Feb 28. PubMed PMID: 22376997.

Raney RB, Meza J, Anderson JR, Fryer CJ, Donaldson SS, Breneman JC, Fitzgerald TJ, Gehan EA, Michalski JM, Ortega JA, Qualman SJ, Sandler E, Wharam MD, Wiener ES, Maurer HM, Crist WM. Treatment of children and adolescents with localized parameningeal sarcoma: experience of the Intergroup Rhabdomyosarcoma Study Group protocols IRS-II through -IV, 1978–1997. Med Pediatr Oncol. 2002;38(1):22–32. PubMed PMID: 11835233.

Raubenheimer EJ, Noffke CE, Boy SC. Osseous dysplasia with gross jaw expansion: a review of 18 lesions. Head Neck Pathol. 2016;10(4):437–43. Epub 2016 May 9. PubMed PMID: 27161103; PubMed Central PMCID: PMC5082044.

Reese AB, Ellsworth RM. The evaluation and current concept of retinoblastoma therapy. Trans Am Acad Ophthalmol Otolaryngol. 1963;67:164–72. PubMed PMID: 13973597.

Rilliet F, Barthez E. Traite Clinique et Pratique des Maladies des Enfants. Paris: Germer Bailliere; 1853. p. 484–8.

Rodriguez DP, Orscheln ES, Koch BL. Masses of the nose, nasal cavity, and nasopharynx in children. Radiographics. 2017;37(6):1704–30. https://doi.org/10.1148/rg.2017170064. Review. PubMed PMID: 29019747.

Rodriguez-Galindo C, Wilson MW, Chantada G, Fu L, Qaddoumi I, Antoneli C, Leal-Leal C, Sharma T, Barnoya M, Epelman S, Pizzarello L, Kane JR, Barfield R, Merchant TE, Robison LL, Murphree AL, Chevez-Barrios P, Dyer MA, O'Brien J, Ribeiro RC, Hungerford J, Helveston EM, Haik BG, Wilimas J. Retinoblastoma: one world, one vision. Pediatrics. 2008;122(3):e763–70. Review. PubMed PMID: 18762512; PubMed Central PMCID: PMC2844325

Rokitansky K. Handbuch der allgemeinen pathologischen Anatomie. Vienna: Braumüller & Seidel; 1846.

Rokitansky K. Neubildung von äusserer Haut, Schleim und seröser Haut. In: Lehrbuch der pathologischen Anatomie. Vienna: W. Braumüller; 1855.

Sabageh D, Solaja TO, Olasode BJ. Malignant tumors of the upper aerodigestive tract as seen in a Nigerian tertiary health institution. Niger J Clin Pract. 2015;18(2):231–5. https://doi.org/10.4103/1119-3077.151050. PubMed PMID: 25665998.

Saha D, Sinha R, Pai RR, Kumar A, Chakraborti S. Laryngeal cysts in infants and children – a pathologist's perspective (with review of literature). Int J Pediatr Otorhinolaryngol. 2013;77(7):1112–7. https://doi.org/10.1016/j.ijporl.2013.04.012. Epub 2013 May 15. Review. PubMed PMID: 23684174.

Satarkar RN, Srikanth S. Tumors and tumor-like conditions of the nasal cavity, paranasal sinuses, and nasopharynx: a study of 206 cases. Indian J Cancer. 2016;53(4):478–82. https://doi.org/10.4103/ijc.IJC_551_16. PubMed PMID: 28485333.

Scott AR, Tibesar RJ, Sidman JD. Pierre Robin sequence: evaluation, management, indications for surgery, and pitfalls. Otolaryngol Clin N Am. 2012;45(3):695–710, ix. https://doi.org/10.1016/j.otc.2012.03.007. Review. PubMed PMID: 22588044.

Sergi C. Promptly reporting of critical laboratory values in pediatrics: a work in progress. World J Clin Pediatr. 2018;7(5):105–10. https://doi.org/10.5409/wjcp.v7.i5.105. eCollection 2018 Nov 12. PubMed PMID: 30479975; PubMed Central PMCID: PMC6242778

Sergi C, Dhiman A, Gray JA. Fine needle aspiration cytology for neck masses in childhood. An illustrative approach. Diagnostics (Basel). 2018;8(2):pii:E28. https://doi.org/10.3390/diagnostics8020028. Review. PubMed PMID: 29690556; ; PubMed Central PMCID: PMC6023333.

Sergi C, Dhiman A, Gray JA. Fine needle aspiration cytology for neck masses in childhood. An illustrative approach. Diagnostics (Basel). 2018a;8(2):28. https://doi.org/10.3390/diagnostics8020028. PMID: 29690556; PMCID: PMC6023333

Settipane RA, Peters AT, Chiu AG. Chapter 6: Nasal polyps. Am J Rhinol Allergy. 2013;27(Suppl 1):S20–5. https://doi.org/10.2500/ajra.2013.27.3926.

Skalova A, Sima R, Bohus P, Curik R, Lukas J, Michal M. Endolymphatic sac tumor (aggressive papillary tumor of middle ear and temporal bone): report of two cases with analysis of the VHL gene. Pathol Res Pract. 2008;204(8):599–606. PMID 18423895.

Soames JV, Southam JC. Oral pathology. Oxford University Press; 1998. Medical – 340 pages.

Soldati D, Mudry A. Knowledge about cholesteatoma, from the first description to the modern histopathology. Otol Neurotol. 2001;22(6):723–30. PubMed PMID: 11698787

Soluk Tekkesin M, Tuna EB, Olgac V, Aksakallı N, Alatlı C. Odontogenic lesions in a pediatric population: review of the literature and presentation of 745 cases. Int J Pediatr Otorhinolaryngol. 2016;86:196–9. https://doi.org/10.1016/j.ijporl.2016.05.010. Epub 2016 May 11. Review. PubMed PMID: 27260607.

Stafne EC. Bone cavities situated near the angle of the mandible. J Am Dent Assoc. 1942;29:1969–72.

Sturgis EM, Potter BO. Sarcomas of the head and neck region. Curr Opin Oncol. 2003;15(3):239–52. Review. PubMed PMID: 12778019.

Subramanyam R, Varughese A, Willging JP, Sadhasivam S. Future of pediatric tonsillectomy and perioperative outcomes. Int J Pediatr Otorhinolaryngol. 2013;77(2):194–9. https://doi.org/10.1016/j.ijporl.2012.10.016. Epub 2012 Nov 16. Review. PubMed PMID: 23159321.

Sutherland GA, Lack HL. Congenital laryngeal obstruction. Lancet. 1897;2:653–5.

Swibel Rosenthal LH, Caballero N, Drake AF. Otolaryngologic manifestations of craniofacial syndromes. Otolaryngol Clin N Am. 2012;45(3):557–77, vii. https://doi.org/10.1016/j.otc.2012.03.009. Review. PubMed PMID: 22588037.

Szeremeta W, Parikh TD, Widelitz JS. Congenital nasal malformations. Otolaryngol Clin N Am. 2007;40(1):97–112, vi–vii. Review. PubMed PMID: 17346563

Taweevisit M, Tantidolthanes W, Keelawat S, Thorner PS. Paediatric oral pathology in Thailand: a 15-year retrospective review from a medical teaching hospital. Int Dent J. 2018; https://doi.org/10.1111/idj.12380. [Epub ahead of print] PubMed PMID: 29377105.

Thompson LD. Mucocele: retention and extravasation types. Ear Nose Throat J. 2013;92(3):106–8. PubMed PMID: 23532645

Tirumandas M, Sharma A, Gbenimacho I, Shoja MM, Tubbs RS, Oakes WJ, Loukas M. Nasal encephaloceles: a review of etiology, pathophysiology, clinical presentations, diagnosis, treatment, and complications. Childs Nerv Syst. 2013;29(5):739–44. https://doi.org/10.1007/s00381-012-1998-z. Epub 2012 Dec 18. Review. PubMed PMID: 23247827

Tiwari PK, Teron P, Saikia N, Saikia HP, Bhuyan UT, Das D. Juvenile nasopharyngeal angiofibroma: a rise in incidence. Indian J Otolaryngol Head Neck Surg. 2016;68(2):141–8. https://doi.org/10.1007/s12070-015-0898-4. Epub 2015 Sep 16. PubMed PMID: 27340627; PubMed Central PMCID: PMC4899360.

Tkaczuk AT, Bhatti M, Caccamese JF Jr, Ord RA, Pereira KD. Cystic lesions of the jaw in children: a 15-year experience. JAMA Otolaryngol Head Neck Surg. 2015;141(9):834–9. https://doi.org/10.1001/jamaoto.2015.1423. PubMed PMID: 26248292.

Tuchtan L, Torrents J, Lebreton-Chakour C, Niort F, Christia-Lotter MA, Delmarre E, Nicollas R, Piercecchi-Marti MD. Liability under post-tonsillectomy lethal bleeding of the tonsillar artery: a report of two cases. Int J Pediatr Otorhinolaryngol. 2015;79(1):83–7. https://doi.org/10.1016/j.ijporl.2014.11.006. Epub 2014 Nov 16. PubMed PMID: 25464852.

Van Wyhe RD, Chamata ES, Hollier LH. Midline craniofacial masses in children. Semin Plast Surg. 2016;30(4):176–80. https://doi.org/10.1055/s-0036-1593482. PMID: 27895540; PMCID: PMC5115923

Virchow R. Ueber Perlgeschwulste. Arch Anat Physiol Klin Med. 1855;8:371–418.

Walner DL, Karas A. Standardization of reporting post-tonsillectomy bleeding. Ann Otol Rhinol Laryngol.

2013;122(4):277–82. Review. PubMed PMID: 23697327.

Wang F, Gregory C, Sample C, Rowe M, Liebowitz D, Murray R, Rickinson A, Kieff E. Epstein-Barr virus latent membrane protein (LMP1) and nuclear proteins 2 and 3C are effectors of phenotypic changes in B lymphocytes: EBNA-2 and LMP1 cooperatively induce CD23. J Virol. 1990;64(5):2309–18. PubMed PMID: 2157887; PubMed Central PMCID: PMC249392

Weir J, Harris N, Sergi C. Pathology quiz case: aneurysmal bone cyst (ABC) of the maxilla. Arch Otolaryngol Head Neck Surg. 2003;129(12):1345. PubMed PMID: 14676164

Whitaker JA, Shaheedy M, Baum J, James J, Flume JB. Gout in childhood leukemia. Report of a case and concepts of etiology. J Pediatr. 1963;63:961–6. PubMed PMID: 14071051

Wilkins BS, Jones DB. Heterogeneity of expression of CD68 and other macrophage-associated antigens in human long-term bone marrow culture. Biologicals. 1996;24(4):333–7. PubMed PMID: 9088549

Willner JP. Genetic evaluation and counseling in head and neck syndromes. Otolaryngol Clin N Am. 2000;33(6):1159–69. Review. PubMed PMID: 11449781.

Windfuhr JP. Malpractice claims and unintentional outcome of tonsil surgery and other standard procedures in otorhinolaryngology. GMS Curr Top Otorhinolaryngol Head Neck Surg. 2013;12:Doc08. https://doi.org/10.3205/cto000100. Review. PubMed PMID: 24403976; PubMed Central PMCID: PMC3884543.

Woo J, Seethala RR, Sirintrapun SJ. Mammary analogue secretory carcinoma of the parotid gland as a secondary malignancy in a childhood survivor of atypical teratoid rhabdoid tumor. Head Neck Pathol. 2014;8(2):194–7. https://doi.org/10.1007/s12105-013-0481-9. Epub 2013 Aug 7. PubMed PMID: 23921806; PubMed Central PMCID: PMC4022939.

Xie CM, Liu XW, Mo YX, Li H, Geng ZJ, Zheng L, Lv YC, Ban XH, Zhang R. Primary nasopharyngeal non-Hodgkin's lymphoma: imaging patterns on MR imaging. Clin Imaging. 2013;37(3):458–64. https://doi.org/10.1016/j.clinimag.2012.07.009. Epub 2012 Oct 3. PubMed PMID: 23041160.

Xu B, Aneja A, Ghossein R, Katabi N. Salivary gland epithelial neoplasms in pediatric population: a single-institute experience with a focus on the histologic spectrum and clinical outcome. Hum Pathol. 2017;67:37–44. https://doi.org/10.1016/j.humpath.2017.07.007. Epub 2017 Jul 21. PubMed PMID: 28739497.

Xu N, Huang XM, Fang WG, Zhang Y, Qiu ZQ, Zeng XJ. [Glycogen storage disease type Ia: a rare cause of gout in adolescent and young adult patients]. Zhonghua Nei Ke Za Zhi. 2018;57(4):264–269. https://doi.org/10.3760/cma.j.issn.0578-1426.2018.04.007. Chinese. PubMed PMID: 29614584.

Yagasaki H, Makino K, Goto Y, Suzuki T, Oyachi N, Obana K, Ko J, Komai T. Thyroglossal duct cyst accompanied by laryngomalacia and pulmonary sequestration. Pediatr Int. 2014;56(3):e7–10. https://doi.org/10.1111/ped.12309. PubMed PMID: 24894942.

Yi Z, Fang Z, Lin G, Lin C, Xiao W, Li Z, Cheng J, Zhou A. Nasopharyngeal angiofibroma: a concise classification system and appropriate treatment options. Am J Otolaryngol. 2013;34(2):133–41. https://doi.org/10.1016/j.amjoto.2012.10.004. Epub 2013 Jan 16. PubMed PMID: 23332298.

Zanetta A, Cuestas G, Méndez Venditto N, Rodríguez H, Tiscornia C, Magaró M, Magaró S. Cáncer de laringe en niños. Caso clínico [Laryngeal cancer in children: case report]. Arch Argent Pediatr. 2012;110(3):e39–42. Spanish. PubMed PMID: 22760757.

Zeng K, Ohshima K, Liu Y, Zhang W, Wang L, Fan L, Li M, Li X, Wang Z, Guo S, Yan Q, Guo Y. BRAFV600E and MAP2K1 mutations in Langerhans cell histiocytosis occur predominantly in children. Hematol Oncol. 2017;35(4):845–51. https://doi.org/10.1002/hon.2344. Epub 2016 Sep 6. PubMed PMID: 27597420.

Central Nervous System

Contents

15.1	**Development: Genetics**	1244
15.1.1	Development and Genetics	1244
15.1.2	Neuromeric Model of the Organization of the Embryonic Forebrain According to Puelles and Rubenstein	1246
15.2	**Congenital Abnormalities of the Central Nervous System**	1247
15.2.1	Ectopia	1247
15.2.2	Neural Tube Defects (NTDs)	1250
15.2.3	Prosencephalon Defects	1252
15.2.4	Vesicular Forebrain (Pseudo-aprosencephaly)	1254
15.2.5	Ventriculomegaly/Hydrocephalus	1257
15.2.6	Agenesis of the *Corpus Callosum* (ACC)	1257
15.2.7	Cerebellar Malformations	1258
15.2.8	Agnathia Otocephaly Complex (AGOTC)	1259
15.2.9	Telencephalosynapsis (Synencephaly) and Rhombencephalon Synapsis	1259
15.2.10	CNS Defects in Acardia	1261
15.2.11	CNS Defects in Chromosomal and Genetic Syndromes	1263
15.2.12	Neuronal Migration Disorders	1264
15.2.13	Phakomatoses	1264
15.3	**Vascular Disorders of the Central Nervous System**	1267
15.3.1	Intracranial Hemorrhage	1267
15.3.2	Vascular Malformations	1272
15.3.3	Aneurysms	1273
15.3.4	Thrombosis of Venous Sinuses and Cerebral Veins	1274
15.3.5	Pediatric and Inherited Neurovascular Diseases	1274
15.4	**Infections of the CNS**	1275
15.4.1	Suppurative Infections	1276
15.4.2	Tuberculous (Lepto-)Meningitis	1279
15.4.3	Neurosyphilis	1280
15.4.4	Viral Infections	1280
15.4.5	Toxoplasmosis	1282
15.4.6	Fungal Infections	1282
15.5	**Metabolic Disorders Affecting the CNS**	1282
15.5.1	Pernicious Anemia	1283
15.5.2	Wernicke Encephalopathy	1283
15.6	**Trauma to the Head and Spine**	1284

© Springer-Verlag GmbH Germany, part of Springer Nature 2020
C. M. Sergi, *Pathology of Childhood and Adolescence*,
https://doi.org/10.1007/978-3-662-59169-7_15

15.7	**Head Injuries**	1284
15.7.1	Epidural Hematoma	1284
15.7.2	Subdural Hematoma	1284
15.7.3	Subarachnoidal Hemorrhage	1285
15.7.4	Spinal Injuries	1285
15.7.5	Intervertebral Disk Herniation	1286
15.8	**Demyelinating Diseases Involving the Central Nervous System**	1286
15.8.1	Multiple Sclerosis	1286
15.8.2	Leukodystrophies	1287
15.8.3	Amyotrophic Lateral Sclerosis	1288
15.8.4	Werdnig-Hoffmann Disease	1288
15.8.5	Syringomyelia	1289
15.8.6	Parkinson Disease and Parkinson Disease-Associated, G-Protein-Coupled Receptor 37 (GPR37/PaelR)-Related Autism Spectrum Disorder	1289
15.8.7	Creutzfeldt-Jakob Disease (sCJD or Sporadic), CJD-Familial and CJD-Variant	1290
15.8.8	West Syndrome/Infantile Spasms, ACTH Therapy, and Sudden Death	1290
15.9	**Neoplasms of the Central Nervous System**	1291
15.9.1	Astrocyte-Derived Neoplasms	1291
15.9.2	Ependymoma	1294
15.9.3	Medulloblastoma	1295
15.9.4	Meningioma	1299
15.9.5	Hemangioblastoma and Filum Terminale Hamartoma	1299
15.9.6	Schwannoma	1300
15.9.7	Craniopharyngioma	1300
15.9.8	Chordoma	1300
15.9.9	Tumors of the Pineal Body	1302
15.9.10	Hematological Malignancies	1302
15.9.11	Other Tumors and Metastatic Tumors	1304
	Multiple Choice Questions and Answers	1309
	References and Recommended Readings	1311

15.1 Development: Genetics

15.1.1 Development and Genetics

There are four significant steps of the neurodevelopment, including (1) the closure of the neural tube at the cephalic end, (2) the prosencephalic cleavage, (3) the formation of the neoencephalic commissures, and (4) the development of the cerebral mass with the molding of the telencephalic structures and the differentiation of the cortical plate. The completion of the first process with the closure of the anterior neuropore between days 22 and 24 of gestation determines a *primordium cerebri* of the neuraxis consisting of three enlargements: the prosencephalic, the mesencephalic, and the metencephalic vesicles. At approximately day 28, the olfactory placodes appear, and on the 34th day of gestation (CRL: 7–9 mm), the hemispheric or telencephalic vesicles, including the olfactory placodes, evaginate *dorsolaterally* and symmetrically from the originate prosencephalic vesicle. Both hemispheric vesicles grow very rapidly. The diencephalon expands backward and downward to form the temporal lobes (Greek διά "through, across" and ἐγκέφαλος "brain"). Between days 38 and 42, the primordial areas of the olfactory bulb rostrally and the tuberculum olfactorium more caudally appear, whereas on day 42 the olfactory nerve enters the brain. The *anteromedial* part of the primary prosencephalic vesicle lies in the depth of the developing interhemispheric fissure, an area extending from the chiasmal primordium up to the *velum transver-*

sum to the roof of the 3rd ventricle. At ca. 50 days of gestation, the intermediate part of this area becomes enfolded in the midline, forming a sagittal groove, the *sulcus medianus telencephali medii* of Hochstetter (1919). This sulcus deepens gradually, the walls of which (*Laminae reunientes* of His [1904]) are separated by specific connective tissue, which is the extension of the primitive meninx. Subsequently, these laminae become thickened and finally fuse. The area of fusion, the commissural mass, is a sort of crossing-over of the telencephalic fibers. The periventricular zone of the prosencephalic vesicle undergoes a marked germinal activity gradually. The medial and lateral striatal ridges bulge into the floor of the telencephalic evaginations. These structures are continuous across the rostral prosencephalic midline through a transverse germinal bar. From these striatal ridges starts the development of the striatum, amygdala, *nucleus accumbens septi*, pyriform cortex, and septum. It is still controversially discussed whether the septum pellucidum derives from the primary prosencephalic vesicles or represents thinned parts of the medial hemispheric walls. By approximately day 45, most of the major components of the forebrain have appeared and are undergoing rapid development. However, growth is especially striking in the phylogenetically older parts of the brain, i.e., the diencephalon, whose role is the maintenance of life by control of reflex, vegetative autonomic, and endocrine functions. The molding of the cerebral hemispheres begins covering the thalami by the 3rd month, an extension to the quadrigeminal plate by the 4th, and growth of the occipital lobes by the 6th month. The thickness of the cerebral mantle begins from the 4th month (six-layered pattern) and begins to show regional differences from the 5th month onward (Boxes 15.1 and 15.2). As historical note, the name rolandic was first applied by Leuret in 1839 in the book "Anatomie Comparée du Systéme Nerveux" because the Italian anatomist Luigi Rolando (1773–1831) drew attention to the sulcus. On the other hand, this fissure was observed and recorded by Félix Vicq d'Azyr (1746–1794) in his Traité d'Anatomie et de Physiologie in 1786.

Box 15.1 CNS: Normal Anatomy
- *Cerebrum* division (2X): Frontal, temporal, parietal, and occipital lobes
- *Sulcus Rolandi* (Rolando fissure) separates the frontal from parietal lobes and is bounded by the precentral motor cortex anteriorly and the postcentral sensory cortex posteriorly
- *Fissura Sylvii* (Sylvian fissure) splits the upper border of the temporal from the parietal lobes.
- *Convolutions*: High level of cerebral architecture (*gyri* for cerebrum and *folia* for cerebellum)
- *Stria Gennarii*: A prominent horizontal layer of myelinated fibers in the visual cortex
- *Cranial nerves*: 12 pairs with significant coating division:
  - Oligodendrocytes myelinate 1st–2nd cranial nerves, while Schwann cells do the 3rd–12th nerves.
- *Cytoarchitecture*:
  - *Perikarya*: Cell bodies of neurons
  - *Bergmann Glia*: Elongated astrocytes located in the cerebellum
  - *Rosenthal Fibers*: Intracytoplasmic corkscrew shaped hyaline fibers of pilocytic astrocytes.
  - *Gemistocytic Astrocytes*: Bright pink cytoplasm
- *Meninges* (Greek: μῆνιγξ "membrane"), rostrocaudally (external → internal), subdivided in:
  - Dura mater (Pachymeninx): Collagenous for the large part
  - Arachnoid mater: Extremely delicate web-like membrane
  - Pia mater: Innermost thin layer of the meninges covering the surface of the brain and permeable to water and small solutes.

> **Box 15.2 CNS: Normal Anatomy – Spinal Cord**
> - *Portio ventralis medullae spinalis*: predominantly motor with α-motoneurons
> - *Portio dorsalis medullae spinalis*: predominantly sensory
> - *Canalis centralis medullae spinalis*, ependymal lined, which becomes discontinuous by pubertal age
> - *Filum terminale*: Tapering of the caudal end of the cord to form a fibrous extension containing nests of glia and ependymal cells and fatty tissue (1/10 of healthy individuals)

15.1.2 Neuromeric Model of the Organization of the Embryonic Forebrain According to Puelles and Rubenstein

The neuromeric model of the organization of the embryonic forebrain remains a milestone in the developing embryo. Prosencephalic brains have fascinated numerous developental biologists and developmental pathologists. Sergi and Schmitt (2000) disclosed the results of an investigation according to the neuromeric model of the organization of the embryonic forebrain of Puelles and Rubenstein assuming that this model applies to the man. This study was based on expression of genes containing multiple homeodomains in mouse and chicken embryos at various stages of embryogenesis and may provide an anatomic framework for studies of forebrain development in several species. This model divides the forebrain of mammalians into six transverse domains, called prosomeres. Prosomeres p1–3 destiny is to become the diencephalon, while prosomeres p4–6 are projected to become the secondary prosencephalon, where the ventral area consists of the hypothalamus, and the dorsal region comprises of the telencephalic vesicles.

The *corpus callosum* is derived from the midline prosencephalic commissural plate and is formed between the 7th and 20th week of gestation. In adults, the *corpus callosum* has a crescent-shaped structure and reaches the size of about 10 cm in length. It is subdivided into four parts representing the *rostrum, genu, truncus,* and *splenium* from anterior to posterior side. The *corpus callosum* allows the transfer across the hemispheres of the motor, sensorial, and cognitive information and is essential in inhibiting or regulating some concurrent activity of one hemispheric territory to the other.

The *cerebellum* is the most prominent posterior fossa midline structure constituted by three-layered cortex of cellular constitution and well-differentiated fiber connections to the brain and spinal cord. Cerebellar functions are not only limited to coordinate motor activity but also to work with the cerebrum in motor learning, cognition, as well as linguistic, perceptual, and affective functions. The development starts in the early embryonic period until the first few years after birth.

Moreover, *falx cerebri* and *tentorium cerebelli* are collagenous separations made of reduplication of *dura mater* separating the brain in the midline and rostral to the cerebellum separating the cerebrum from the cerebellum. *Dura mater* contains essentially branches of the middle meningeal artery. *Corpora amylacea* are polyglucosan protein globules constituted by α-β-crystallin + HSP27 + ubiquitin, located in the terminal processes of astrocytes at the subpial and perivascular level and associated with degenerative change. They are argyrophilic and basophilic other than PAS-positive and (+) ubiquitin. Ubiquitin is a small regulatory protein found in most tissues of eukaryotic organisms with four genes in the human genome coding for this protein: *UBB, UBC, UBA52,* and *RPS27A*. HSP27 is a chaperone of the small heat shock protein group including ubiquitin, α-crystallin, and others. HSP27 functions include chaperone activity, thermotolerance, inhibition of programmed cell death or apoptosis, regulation of cell development, and cell differentiation. Commonly found after the age of 40 years, *corpora amylacea* can occur earlier than this time in the event of degenerative processes. It is crucial that the reader consults texts of neuropathology and clinical neuroradiology for completeness of all concepts illustrated in this chapter.

The cerebral spinal fluid (CSF) is basilar in clinical pathology. Through the lumbar access, it is possible to obtain this fluid and receive a lot of information useful for pediatric neurologists. CSF is produced in the choroidal plexus, which is a specialized tissue projecting into the lateral, 3rd and 4th ventricle. It is crucial to keep in mind the following few notes of neuropediatrics.

– Liquor System: CSF reabsorbed at the level of the meninges at the level of the Pacchioni structures
– Liquor System: The ventricular system is constituted by two lateral ventricles connecting to the 3rd ventricle via the *foramina of Monro* and the 3rd ventricle connected to the 4th ventricle via the *aqueductus Sylvii*.
– Lack of a lymphatic system, thus, there is no possibility of the brain to drain any excess of fluid, such as in case of edema, making the brain a sensitive organ to edema.

Box 15.3 summarizes some cytology characteristics, while Box 15.4 lists the most useful neuropathology special stains.

> **Box 15.3 CNS Cytology Characteristics**
> - Neurons: CNS cells with large nuclei, prominent nucleolus, and cytoplasmic neurofilaments
> - Astrocytes: Glial ("cerebral glue") cells with large, vesicular nuclei and indistinct cytoplasm ($\neq$ reactive astrocytes, which show a distinct cytoplasm in the setting of gliosis)
> - Oligodendrocytes: Glia cells with small round nuclei with surrounding halo (artifact, but not present on intraoperative frozen section) giving a "fried egg" appearance under the lens
> - Microglia: Glia cells with elongated cigar-shaped nuclei
> - Ependyma: Epithelium lining the ventricular system

> **Box 15.4 Neuropathology Special Stains**
> - LFB: Luxol fast blue
> - MB: Modified Bielschowsky
> - BSS: Bodian silver stain
> - Reticulin
> - Congo red (amyloid)
> - Cajal Gold Stain
> - PTAH: Phosphotungstic acid hematoxylin
>
> Notes: PTAH stain aims the astrocytes and is useful for astrocytosis/gliosis, although an antibody against the glial fibrillary acidic protein (GFAP) by immunohistochemistry has replaced it in numerous laboratories worldwide.

15.2 Congenital Abnormalities of the Central Nervous System

The diagnosis of CNS anomalies is important for the assessment of long-term outcomes of congenital disabilities affecting this complex system. The pathologist should be able to provide the neonatology and pediatric team with appropriate information. The CNS is highly susceptible to damage. This damage can be *genetic*, *chemical*, or *physical* in origin (Figs. 15.1 and 15.2).

15.2.1 Ectopia

- *DEF*: An ectopic nodular slow-growing tissue of mature type with predominantly astrocytic elements usually localized between the eye browses on the frontal site and is improperly called "nasal glioma" (other sites include the occipital bone among others).
- *SYN*: "Nasal glioma."
- *EPI*: It is a sporadic congenital lesion with no familial tendency or gender predilection.
- *DDX*: Dermoid cyst, pilomatrixoma, melanoma among others. Other ectopic tissues are also possible in the CNS (e.g., thyroidal tissue in

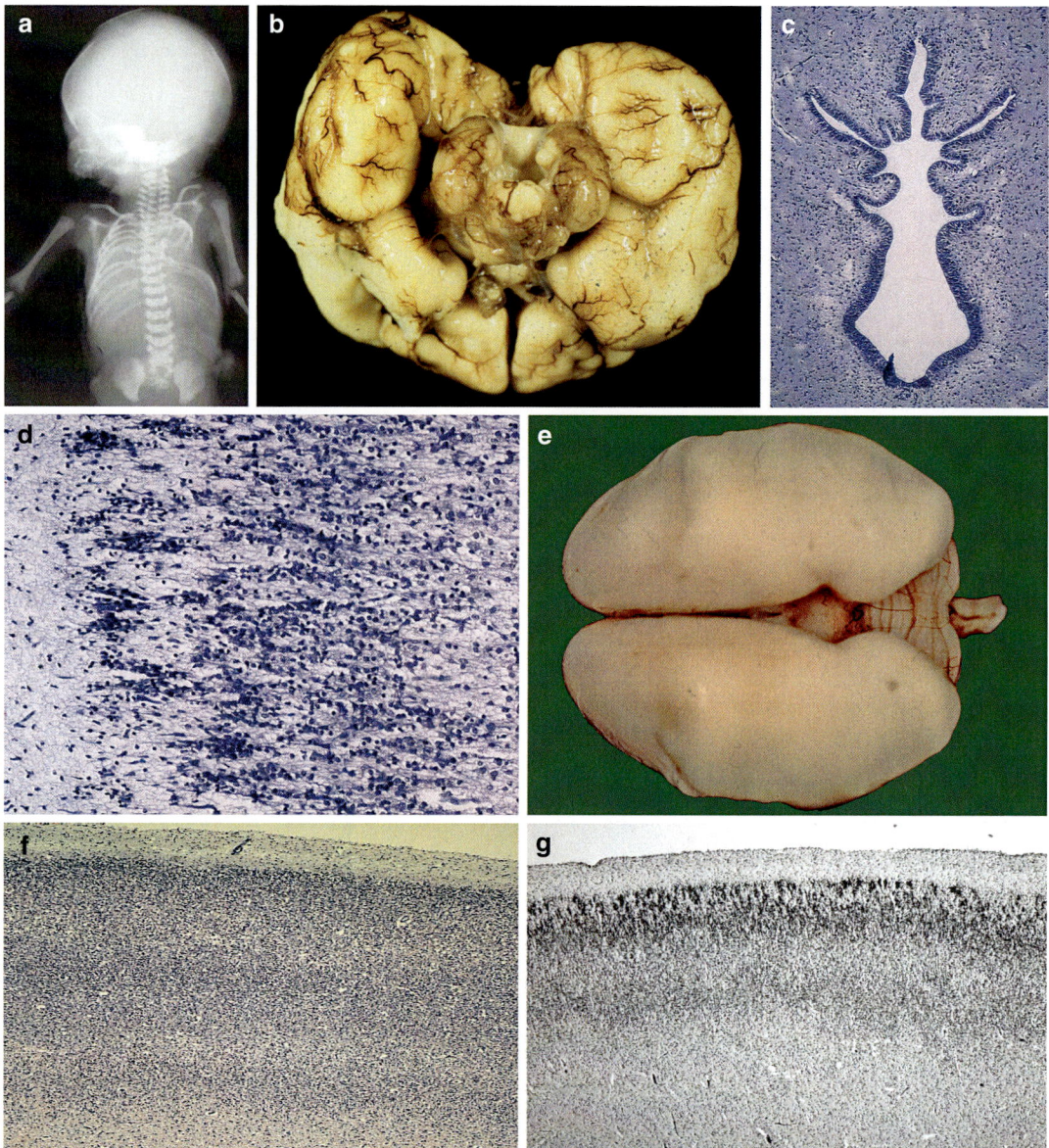

Fig. 15.1 This panel shows some congenital abnormalities of the CNS. Figure **a** is a close-up on the postmortem radiography of a nonviable baby with agnathia-otocephaly syndrome, which shows the lack of the lower jaw and some rib fusion. Note the fusion of the 6th and 7th, 8th and 9th, and 10th and 11th ribs on the right and of the 1st through 6th, 7th through 10th, and 11th and 12th ribs on the left. Patients with this condition may harbor some defects of the development of the prosencephalon. Figure **b** is the gross photograph of a brain of an infant with Dandy-Walker malformation (see text for details). Figure **c** is shown a cross section of the forking area (×60, Kluver-Barrera stain). Figure **d** shows the histology of a patient with lissencephaly (×160, Kluver-Barrera stain). Figure **e** shows the gross photograph of a baby with WDR62 deficiency with marked smoothness and no gyri (there is an almost completely smooth surface of the hemispheres). The microphotographs (**f** and **g**) show the histologic features of the brain (cortical plate, area 4) of a patient affected with mutations of the WDR62 gene compared with an age-matched control. The age-matched control is shown on (**f**) showing and a six-layered cortex, while the *WDR62* gene mutated brain is shown on (**g**) showing only a four-layered cortex (Luxol-Fast-Blue staining, x50 as original magnification for both **f** and **g**). Luxol fast blue stain or LFB, is a commonly used stain to detect myelin under light microscopy. It has been created by Heinrich Klüver and Elizabeth Barrera in 1953 and is also called Klüver-Barrera staining

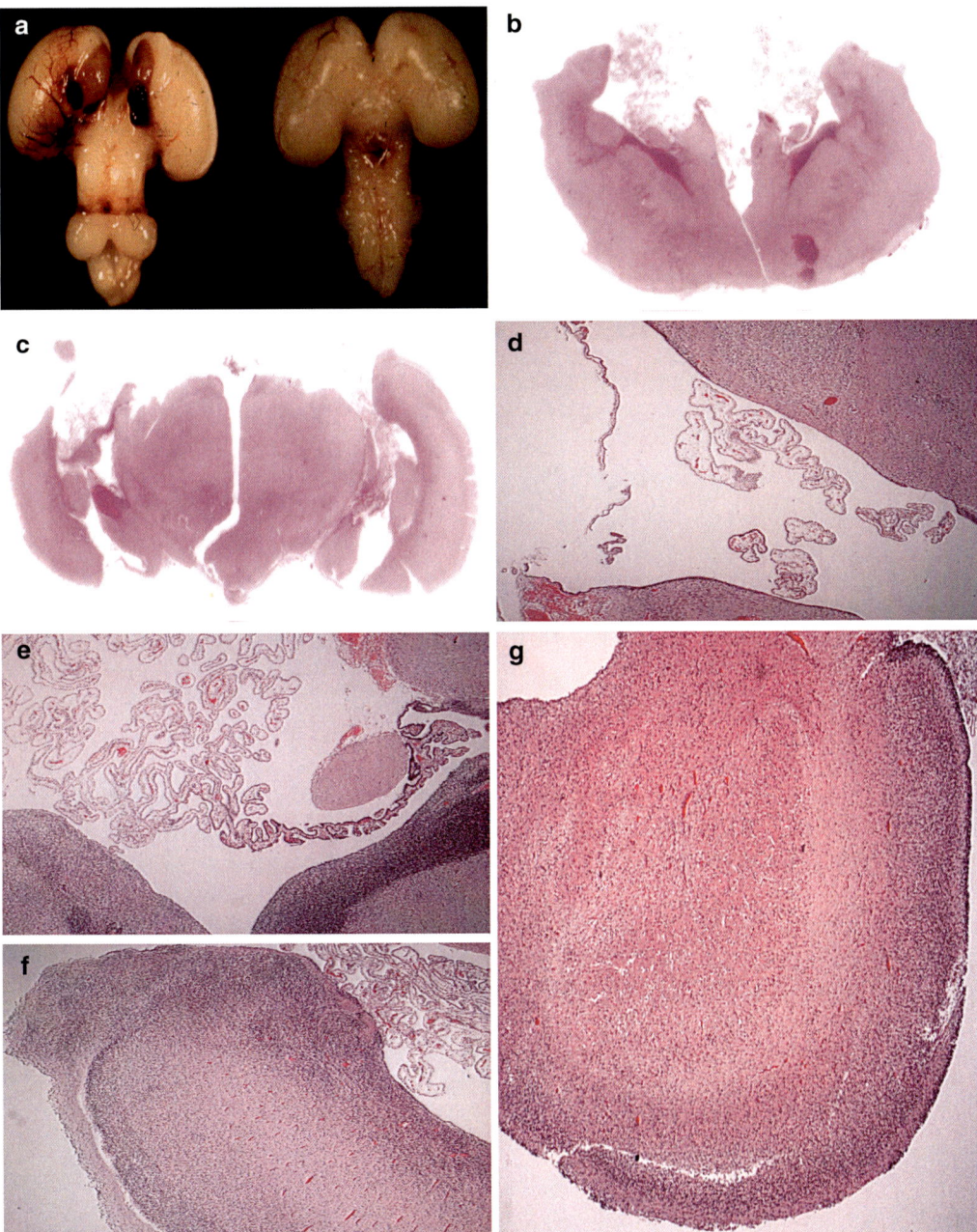

Fig. 15.2 This panel illustrates the abnormalities identified in a multiple congenital anomaly syndrome labeled Seckel syndrome. In figure (**a**) (gross photograph of the brain with superior and inferior views), the brain shows severely hypoplastic telencephalic hemispheres and comparatively normal-shaped lower brainstem and cerebellar anlage. (**b**, **c**) Double-hemispheric microscopic coronal sections at the level of the striatum (**b**) and the thalamus (**c**). Note the severe midline defect with missing medial hemispheric walls, open lateral ventricles and protruding choroid plexus in (**d**, **e**), and the thalamic fusion and missing hippocampal differentiation in (**f**); matrix cell heterotopias at the base of the right striatum in (**f**). Microphotograph (**g**) shows a region-matched microscopic coronal sections from a normally developed brain at 19 weeks of gestation for comparison. Hippocampal differentiations is clearly visible (H&E, **d–g**, x50 original magnification)

sella region, "*ectopia glandulae thyroideae in sella turcica cerebri*"; salivary gland tissue in cerebellopontine angle, "*ectopia glandulae salivaris in angulo cerebello-pontino cerebri*").

- *CLI*: Firm, red to bluish skin-covered tumor with no pulsation or size increase using the Valsalva scheme or compression of the ipsilateral jugular vein (Furstenberg sign), if outside of the skull, but large, firm, submucosal ones if located inside of the nose. The evolution may involve obstruction of the nasal passage, respiratory distress, obstruction of the nasolacrimal duct with epiphora (Greek: ἐπῐ̓φορά, in medicine with the meaning of excessive tear production), cerebrospinal rhinorrhea, meningitis, and epistaxis (Greek: ἐπίσταξις, "nosebleed" from ἐπί or above and στάζω or to drip).
- *CLM*: Dysplastic glial tissue, which is best categorized as heterotopia or heterotopic tissue.
- *TNT*: Surgical excision.
- *PGN*: Resection is curative.

A J-shaped sella is a morphologic pattern variant of the *sella turcica*. In the J-shaped sella, the *tuberculum sellae* is flattened corresponding to the straight edge of the "J" letter, while the *dorsum sellae* stays rounded forming the loop of the "J" letter. Apart from being a standard variant, some conditions may present with a J-shaped conformation of sella, which occur in the hydrocephalus of chronic type, mucopolysaccharidosis, optic glioma, osteochondrodysplasia of achondroplasia type, osteochondrodysplasia of osteogenesis imperfecta type, and neurofibromatosis 1, aka von Recklinghausen disease, which is a genetic disorder characterized by the progressive development of multiple tumors of nerves and skin (neurofibromas) and areas of abnormal skin pigmentation.

The most common causes of the destruction of the *sella turcica* include:

- *Developmental destruction*: Maldevelopment of the enlarged 3rd ventricle
- *Benign Tumors or pseudotumors-related destruction*: Mucocele of the sphenoid sinus, pituitary adenoma, chordoma, giant cell tumor, aneurysmal bone cyst (ABC)
- *Malignancies-linked destruction*: Suprasellar carcinoma of sphenoid and posterior ethmoid sinus, nasopharyngeal carcinoma, osteosarcoma, plasmacytoma, metastasis (PNET, neuroblastoma, RMS, Wilms' tumor, and thyroid cancer)

In the rostral area of the 3rd ventricle, cysts may be located and may constitute the etiology for acute hydrocephalus by obstruction of the *foramina of Monro*, brain herniation, and rapid death. They occur in approximately $3:10^6$ per year, and clinical symptoms may be intermittent, self-resolving, and probably most often nonspecific with most cases arising between the third and fifth decade of life. CT imaging may show a rounded mass in the anterior aspect of the 3rd ventricle, hyperdense, while MR imaging is variable. The origin of the colloid cysts is still a matter of debate. Its origin is attributed to either a diencephalic vesicle or the persistence of embryonic paraphysis. It is important to note that colloid cysts of the third ventricle are a rare event in children. Grossly, the cysts are smooth-walled and spherical (Ø: 0.3–4 cm). Microscopically, there is a simple to pseudostratified epithelial lining with intermingled mucous goblet cells and ciliated cells. Underneath of the epithelium, there is a thin layer of connective tissue, and the cyst content is PAS-(+) and consists of an amorphous material with the inclusion of very few cholesterol clefts and necrotic white blood cells.

15.2.2 Neural Tube Defects (NTDs)

- *DEF*: Posterior midline lesions (*SYN*: Dysraphias, Greek δυσ- distorted, and ραφή suture, seam) resulting from the defective closure of dorsal midline structures of developing embryo during early gestation. Several sites may be involved such as the skull, brain, spinal cord, skin, soft tissue, spine, and meningeal layer, and severity and symptoms are proportional to the localization, size, and depth of the defect.

15.2.2.1 Spina Bifida

- *DEF*: It is the most common defect among NTDs and consists of an absence of arches and dorsal spines of ≥1 vertebral units with 70% located at the level of the lumbosacral vertebral column and the remaining mostly in the cervical and occipital area (Box 15.5).
- *EPI*: Worldwide, 1:1000–2000 live newborns, ♂ = ♀. *Spina bifida occulta* (occult spina bifida) is seen in 1–25% of the average population.
- *CLI*: Malformations in the face (oral clefts), in the musculoskeletal system, in the renal system, and in the cardiovascular system are the most common other anomalies associated with this disorder. Motor disability and urinary or anal sphincter incontinence are the most frequent manifestations of this neurologic disorder. There is also an association with hydrocephalus and Arnold-Chiari malformation (*vide infra*).
- *DDX*: Dermoid cyst, *sinus pilonidalis*, teratoma (e.g., sacrococcygeal teratoma).
- *TNT*: Ongoing surgery, medications, and physical and behavioral therapy, depending on the type and severity of the defect.
- *PGN*: It is quite variable according to the extension of the defect and associated malformations with a combined survival rate decreased of 1/3 of the non-affected individuals.

15.2.2.2 Meningocele/Meningomyelocele

Both meningocele and meningomyelocele are herniation of spinal arachnoid and dura through a defect of the spine, whereby in meningomyelocele a spinal root or part of the spinal cord are included in the herniation. Grossly, a saclike bulge is seen beneath the skin in the dorsal midline position. Complications include CSF leakage and suppurative leptomeningitis. Syringomyelocele is a form of spina bifida in which the fluid of the syrinx of the spine is increased. There is an expansion of the cord tissue into a thin-walled sac that in turn expands through the vertebral defect.

15.2.2.3 Arnold-Chiari Malformation (ACM)/Chiari II Malformation

- *DEF*: It is a caudal displacement of the medulla of brainstem and cerebellum into the cervical region of the vertebral canal and is frequently seen with an additional lumbar NTD (spina bifida, meningocele, myelomeningocele) and hydrocephalus.
- *CLI*: Tonsillar herniation with compression of the aqueduct with deformity of the dorsal midbrain.
- *DDX*: Chiari I malformation does not have a myelomeningocele, Chiari III has an occipital and/or high cervical encephalocele, while Chiari IV has severe cerebellar hypoplasia without displacement of the cerebellum through the foramen magnum. Chiari II malformation is considered a more severe form of the more common Chiari I malformation.
- *TNT*: Myelomeningocele repair and management of neurogenic bladder, ventricular shunting to overcome the hydrocephalus, and craniovertebral decompression if hydrocephalus is absent or if symptoms and signs do not improve with shunting.
- *PGN*: Life-threatening, if surgical intervention is not carried out.

15.2.2.4 Anencephaly

- *DEF*: Most severe dysraphic defect with an absence of the *neurocranium* in the presence of the *splanchnocranium* and lack of the cerebral hemispheres, diencephalon, and midbrain. Instead of these encephalic structures, there is a mass of undifferentiated neurovascular tissue, called *substantia cerebrovasculosa*. The absence of the hypophysis is at the basis

---

Box 15.5 Spina Bifida Classification
- Meningocele
- Meningomyelocele
- Syringomyelocele
- Lipomeningomyelocele
- Myelocele
- Encephalocele

of the bilateral adrenal gland hypoplasia that can be found in fetuses affected with anencephaly.
- *EPI*: 1:8000 live births.
- *CLI*: No brain with an opening skull.
- *DDX*: Encephalocele.
- *TNT*: No intervention is possible.
- *PGN*: Incompatible with life.

Midline craniospinal dimples or pits are relatively conventional dimples or pits of the midline, which can be subclassified as coccygeal dimple and dermal sinus tract dimple. Coccygeal dimples are dimples located at or in the proximity of the tip of the coccyx within the gluteal cleft and visible on inspection only when the buttocks is parted and occur in 4% of the general population and do not have usually associated abnormalities (e.g., hairs, skin markings). Conversely, spinal dermal sinus tract (SDST) occurs in 1:2500 and is a lesion located in the lumbosacral spine overlying the flat portion of the *os sacrum* cranially to the gluteal cleft. There is an association of SDST with hairs inside of the dimple, skin tags, *nevus flammeus*, or a subcutaneous dermoid or epidermoid cyst. SDST may have some associated complications. The revision of the scientific literature revealed that symptoms and signs include signs of suppurative infection, localized neurological disturbances, orthopedic foot deformities, back pain, and scoliosis. MRI is handy to identify the lesion, restrict the level of the *conus medullaris*, and find other malformations. DST also has a cranial position. Nasofrontal DST may be observed between the glabella and the tip of the nose with potential intracranial extension through the *foramen cecum* at the skull basis, and the dura (high risk of suppurative intracranial infections) or the occipital DST is located in the occiput potentially extending intradurally about the occipital bones or cerebellum. Anatomically, the small indentation which is transformed within a *foramen* through articulation with the ethmoid bone, where the frontal crest of the frontal bone ends inferiorly is the structure called *foramen cecum*. Some SDST complications include:

- Meningitis of aseptic type due to epithelial debris within the cerebrospinal fluid
- Meningitis or suppurative infections due to the entrance of bacteria
- Spinal cord/nerve roots compression due to the dermoid/epidermoid cyst
- Tethering of the spinal cord

Notes: In infants younger than 1 year harboring an SDST, half of these patients present with focal deficits, while infants and children older than 1 year harboring an SDST have similar symptoms in nine out of ten cases.

> Ultrasound from 7 weeks on may be diagnostic and may show a "No divided leaves and half leaf type." This appearance includes microcephaly with a single, sickle, or horseshoe-like ventricle, ventricular drainage into the backpack together, a hemispheric fissure, absence of *septum pellucidum*, absence of midline echo, a thin cortex margin, single and fused thalami (Greek θάλαμος, "chamber") - *fusio thalamorum*, expansion of ventricular system, hypotelorism to cyclopia, as well as central facial cleft, cleft lip, and cleft palate. The "leaf type" shows frontal ventricle fusion and posterior horn of lateral ventricle visible separation.

15.2.3 Prosencephalon Defects

There is a well known classification, which is associated with the similarly famous De Myer and Zeman's sentence "the face predicts the brain" where facial dysmorphism was correlated to abnormal divisions of the prosencephalon, which occurs at 4–6 weeks' gestation. In fact, the term holoprosencephaly or HPE (Greek ὅλος "whole", πρόσω "forwards", ἐγκέφαλος "brain" + –y, which is the suffix denoting a condition) is used to define a dorsoventral disruptive disorder of the CNS, in which the prosencephalon fails to develop into two hemispheres and the "*median*

Yakovlev's holosphere" with a single ventricular cavity as an alternative of two hemispheres with symmetrical lateral ventricles takes place. The worldwide famous neuropathologist Paul Yakovlev, an immigrant from Russia, noted a specimen of semilobar holoprosencephaly in which "the alveus and the fimbria of the hippocampal formations formed a massive fornical system… arching over the thalamus." The resulting anomaly shows a dorsal cyst, particularly in association with non-cleaved thalami (thought to be an expansion of partially blocked dorso-posterior third ventricle), a variable degree of fusion of diencephalon and basal ganglia and thalamus with incorporation into upper brainstem and hypoplasia of the neocortex. The clinical severity relates to the degree of hemispheric and deep gray nuclei fusion. Facial anomalies may include midline cleft, premaxillary agenesis, absent superior frenulum, central incisor, proboscis, single nare, single nasal bone, and absent internasal sutures and caudal metopic suture. Here cannot be less stressed and remembered that the main product of the embryonic diencephalon is indeed the *thalami*, as first identified in 1893 by Wilhelm His Senior, who was a Swiss anatomist and professor world famous for his invention of the microtome.

HPE is not purely a failure of hemispheric cleavage, but severe hypoplasia of the neopallium is a well-known feature in most severe cases such as the alobar type, which shows abnormal cortex and abnormalities of the circle of Willis (*Circulus arteriosus cerebri Willisi*), constituted of the anterior communicating artery (median), anterior cerebral arteries (left and right), internal carotid arteries (left and right), posterior communicating arteries (left and right), and posterior cerebral arteries (left and right). The incidence of HPE is quite variable with ~50–90/10^6 in live-born babies but about 4000/10^6 in miscarriages or terminations of pregnancies in different tertiary healthcare centers. HPE is found in other species and is etiologically heterogeneous. Nongenetic causes include environmental factors such as maternal diabetes, alcohol consumption, toxoplasmosis, syphilis, and rubella. An association with alkaloids has been observed in lambs, probably associated with an abnormal sonic hedgehog pathway. Chromosomal aberrations are described in HPE cases and include mostly trisomy 13 and trisomy 18 syndromes as well as triploidy syndrome. Multiple loci for HPE have been mapped. In some instances of HPE, specific chromosomal sites and molecular defects have been identified. Holoprosencephaly-1 (*HPE1*) maps to chromosome 21q22.3 (OMIM 236100) and HPE2 (OMIM 157170) is caused by a mutation in the *SIX3* gene (OMIM 603714), which maps to 2p21. Mutations in the sonic hedgehog gene (*SHH*; OMIM 600725), which is situated at 7q36 cause HPE3 (OMIM 142945). HPE4 (OMIM 142946) is associated with mutations in the *TGIF* gene (OMIM 602630), which maps to 18p11.3. HPE5 (OMIM 609637) is caused by a mutation in the *ZIC2* gene (OMIM 603073), which is located at 13q32. *HPE6* (OMIM 605934) maps to 2q37.1, while *HPE7* (OMIM 610828) maps to 9q22.3 and mutation in the *PTCH1* gene (OMIM 601309) have been linked. *HPE8* (OMIM 609408) does map to 14q13. HPE9 (OMIM 610829) is caused by a mutation in the *GLI2* gene (OMIM 165230) and maps to 2q14, while *HPE10* (OMIM 612530) maps to 1q41-q42, and, finally, HPE11 (OMIM 614226) is caused by a mutation in the *CDON* gene (OMIM 608707) on 11q23–11q24. Systemically, in approximately 3/4 of HPE patients, mainly if a chromosomal aberration is an underlying disorder, there are abnormalities of cardiac, genitourinary, and gastrointestinal development. The single median maxillary central incisor is one of several microforms of HPE. There are a few classifications for holoprosencephaly. The most common forms of holoprosencephalies are those classified by DeMyer and Zeman in alobar, semilobar, and lobar types depending on the degree of fusion of the prosencephalic structures. Probst provided an extended classification with four categories including six subgroups. They were the (I) *dorsal sac category*, types A–C; (II) *intermediate category*, types A–C; (III) *pseudo-hemispheric category*; and (IV) *partial prosencephalic category*. More recently, four major subtypes of HPE seem to be delineated. The spectrum includes the classic HPE subtypes in the De Myer and Zeman's classification and the most recent fourth subtype aka a middle interhemispheric variant of HPE (MIH) or syntelencephaly. The Sylvian angle of Barkovich are lines drawn tangentially through Sylvian fissures. Anteriorly displaced Sylvian fissures give rise to an increased Sylvian angle. It is important to remember that the larger the Sylvian angle is, the more severe the frontal lobe hypoplasia is.

Cortex histology is variable, but the degree of disturbance is typically directly proportional to the severity of the gross malformation. From the historical point of view, some classifications may still be used and are derived from the Isidore Geoffroy Saint-Hilaire's original investigations of 1832. This author classified the cyclocephalians as follows:

- Two closely set orbital fossae
  - *Ethmocephalus*: Two closely-set but distinct eyes, atrophic nose, proboscis above the eyes
  - *Cebocephalus*: Two closely set, distinct eyes, atrophic nasal apparatus, no proboscis
- Single orbital fossae
  - *Rhinocephalus*: Two fused eyes or one double eye in the midline, atrophic nose, proboscis
  - *Cyclocephalus*: Two fused eyes or one double eye, atrophic nose, no proboscis
  - *Stomocephalus*: Two fused eyes or one double eye, rudimentary mandible, very abnormal or absent mouth

The same author classified the otocephalians as follows:

- *Sphenocephalus* (ears nearly fused in the midline, basic lower jaw, mouth, two discrete eyes)
- *Otocephalus* (ears nearly fused in the midline, basic lower jaw, mouth, cyclopia)
- *Edocephalus* (ears nearly fused in the midline, basic lower jaw, cyclopia, nasal proboscis)
- *Opocephalus* (ears nearly fused in the midline, basic lower jaw, cyclopia)
- *Triocephalus* (ears nearly fused in the midline, basic lower jaw)

In all cases, the lower jaw is rudimentary and basic showing an embryonic or similar embryonic structure.

15.2.4 Vesicular Forebrain (Pseudo-aprosencephaly)

This defect is a very rarity representing probably the missing link in the teratogenetic spectrum of the defective brain anlage and its discrimination from aprosencephaly. Grossly, fetuses show microcephaly and missing forebrain, appearing to be cases of aprosencephaly. However, in pseudo-aprosencephaly, glio-mesenchymal membranes with an ependymal outline, consistent with the microscopic appearance of the dorsal sac membrane in holoprosencephaly, are found. This finding highlights the concept that these structures represent remnants of a collapsed primitive prosencephalic vesicle. The microscopic specification of a primitive prosencephalic vesicle does not justify the diagnosis of atelencephaly/aprosencephaly because the prosencephalon is not missing, i.e., this condition should better be labeled "pseudo-aprosencephaly." In fact, the prosencephalic anlage is formed but remains vesicular without further differentiation of a holospheric brain mantle as observed in common holoprosencephaly ("vesicular forebrain"). Pseudo-aprosencephaly represents probably the most primitive form of holoprosencephaly, in which the forebrain remains as a complete sac, linking classical holoprosencephaly with "true" aprosencephaly, which is a defective prosencephalic anlage due to developmental arrest. The pseudo-aprosencephaly or "vesicular forebrain" allows the extension of the Probst classification of prosencephalic defects by an additional category. This new category might also be termed "complete sac category," intercalated between the dorsal sac category and "true" atelencephaly/aprosencephaly. Prosencephalic malformations are thought to result from incomplete cleavage of the embryonic forebrain arising after interference with the induction and/or topogenesis (field segregation) of the prechordal mesoderm during gastrulation.

In the Probst classification, the essential criterion for the most primitive category is a dorsal sac representing the cystic roof of the 3rd ventricle with an ependymal outline and continuous with the meninges. The sac is attached posteriorly to transverse hippocampal ridges and communicates with a single bowl-shaped holosphere. The most severe forms of prosencephalic developmental failure are a complete absence of the forebrain (aprosencephaly) or the telencephalon (atelencephaly, atelencephalic microcephaly). Other authors suggested summarization of aprosencephaly and holoprosencephaly under the term of "midline cerebral dysgenesis." Grossly, there is an entire but microcephalic cranial vault, which differentiates atelencephaly/aprosencephaly from anencephaly (cranioschisis). A continuous cohort of 41 patients with prosencephalic malformations was examined from 1975 through 1996 at the Department of Neuropathology of the University of Heidelberg (Sergi and Schmitt 2000). Thirty-three cases could be classified regarding DeMyer and Zeman as alobar (10 cases), semilobar (18), and lobar (5) holoprosencephaly. According to Probst, 6 cases were labeled "dorsal sac, type A"; 12 "dorsal sac, type B"; 6 "dorsal sac, type C"; 3 "intermediate, type A"; 3 "intermediate type B"; and 1 "pseudohemispheric." In two cases, the left half of the prosencephalon had remained utterly cystic, while the right half showed a highly abnormally differentiated brain mantle. Six cases were atypical and difficult to place in any of the DeMyer and Zeman or Probst categories. Two microcephalic instances presented with complete absence of the forebrain on gross inspection resembling aprosencephaly. However, remnants of a vesicle wall, which showed microscopically features of a developmental forebrain bleb were found in of them. It suggested that the prosencephalon anlage had been present but not further developed. The other case showed, on the other side, neither such remnants nor signs of secondary encephaloclastic damage. It appeared justified to label this case as a true atelencephaly/aprosencephaly, probably due to a developmental arrest. In this unique study, the detailed description of two original patients with prosencephalon maldevelopment have been illustrated demonstrating the fundamental difference between pseudo-aprosencephaly and aprosencephaly.

In one patient, the cerebellum was also missing, and a diaphragmatic defect, an omphalocele, and incomplete bowel malrotation were found. Both cases represent malformative (genetic) or disruptive (environmental) inhibition of the prechordal progenitor field with the failure of the forebrain to form (atelencephaly/aprosencephaly) because of the absence of signs of bleeding, vascular damage, or other regressive changes. To the best of knowledge of the authors, the first case appears to be *monotopic*, while the second case presents *polytopic* developmental field defects, involving the anterior notochord-induced hindbrain with the failure of the cerebellar anlage and the lateroventral mesoderm, resulting in a defective ventral body wall. In the first case, microscopic examination revealed convolutions of a thin membrane composed of meningovascular and glial tissue with a cuboidal cellular outline at the inner surface. The fibrovascular layer was continuous with the leptomeninges of the lower brainstem, while the ependymal layer outlined the rostrodorsal surface of the poorly organized midbrain. There were no residues of degenerated glio-neuronal tissue, calcifications, or remnants of old hemorrhage as indicators of damage to a previously existing brain mantle. Lack of signs of hypoxic damage divided the present condition from hydranencephaly, which is a condition in which the cerebral hemispheres are lacking to a great degree and the cranial cavity is filled with CSF, and hydrocephalus, which is an accumulation of CSF in excess in the brain ventricles. There was a monovesicular state without further development of a holospheric brain mantle, and the forebrain vesicle had collapsed, more likely, at the time on the opening of the cranial cavity. It grossly suggested a complete absence of the prosencephalon. The demonstration of remnants of the vesicle wall does not allow to classify this case as atelencephaly/aprosencephaly in the strict sense of the term (*sensu stricto*). This consideration is due to the formation of the brain anlage, which had advanced to the three-vesicle stage (rhombencephalic, mesencephalic, and prosencephalic bleb) before the development was interrupted. This case should be termed *"vesicular forebrain"* or *"pseudo-aprosencephaly"* since remnants of a collapsed prosencephalic vesicle could be demonstrated. The vesicular forebrain or pseudo-aprosencephaly appears to be a missing link between dorsal sac alobar holoprosencephaly and "true" atelencephaly/aprosencephaly in the pathogenetic spectrum of the inhibited brain anlage. Ultimately, case 1 of the paper represents an extension of the Probst classification, or the spectrum of midline cerebral anomalies of Leech and Shuman, by a further "early" category even more primitive than the dorsal sac category, that is, a *"complete sac category."* In the second case, the authors described neither damaged brain tissue nor remnants of a forebrain vesicle. Such structures could not be demonstrated making it likely that a prosencephalic bleb had never been formed in the early development of the brain anlage. Therefore, this case was indeed, labeled atelencephaly/aprosencephaly. Cohen summarized 12 cases of atelencephaly/aprosencephaly from literature reports. Siebert et al. (1986), Goldsmith et al. (1993), and Harris et al. (1994) added further observations making a total of about 22 cases shortly before 2000 at the time of the publication in *Acta Neuropathologica*. It has been suggested that holoprosencephaly, atelencephaly/ aprosencephaly, and anencephaly might be related to teratogenic interference with different developmental fields of the neural plate. In embryonic development, the brain anlage with its neuromeric organization forms out of the anterior part of the neural plate which develops into the neural tube between days 18 and 28 of gestation. While the posterior part of the neural plate including the hindbrain rhombomeres (r1–r8) and the midbrain mesomeres (m1, m2) become induced by the notochord, the anterior part of the neural plate receives impact to form the forebrain (prosomeres p1 p6) from the prechordal mesoderm. In one patient, impairment of the brain anlage had probably two components. First, development became arrested at the three-vesicle stage without further cleavage of the prosencephalon. Second, induction of the histogenesis of a brain mantle would have resulted in one of the different forms of common holoprosencephaly, the forebrain remaining at a monovesicular stage. In the second case, anterior neural tube development had probably ceased after formation of the rhomboid- and mesencephalon. This arrest resulted in a defect of the prosomeric part of the brain anlage which is induced by the prechordal mesoderm. Ultimately, the interference of secondary regressive mechanisms with the early development of the subarachnoid space and meninges cannot be excluded given the sclerotic meningeal and fibrous plate covering the hindbrain structures that were identified at the time of the autopsy.

Neurulation is controlled by a complex interaction of genetic and epigenetic (environmental) factors. Several regulatory genes, such as homeobox (Hox), "zinc finger" (Krox-20), paired box (Pax), wingless (Wnt), Engrailed (En), fibroblast growth factor (Fgf), open brain (Ob), sonic hedgehog (Shh), limbless (Lim), Otx, *Drosophila* distal-less (Dlx), and others have been demonstrated to be involved in the control of neurulation by forming individual patterns of expression for each neuromere. The expression of sonic hedgehog (Shh), Lim-1, and Otx-2 are crucial for the development of the forebrain and mesencephalon. Mice knocked out for Lim-1 and Otx-2 show deletion of the rostral brain anterior to rhombomere R3. Holoprosencephaly syndromes are frequently associated not only with trisomy 13–15, 18, and 21 but also with a variety of chromosomal aberrations such as del (2) (p21), dup (3pter), del (7) (q36), del (13q), del (18p), and del (21) (q22.3). The two cases presented in 2000 in detail suggest that the classical holoprosencephalies and the "true" atelencephaly/aprosencephaly (i.e., the developmentally arrested formation of the prosencephalon) are linked by a *"complete sac"* category of prosencephaly in which the forebrain remains vesicular (*vesicular forebrain*) appearing to be missing only on gross inspection (pseudo-aprosencephaly). As in "true" atelencephaly/aprosencephaly, the closed microcephalic neurocranium appears empty at autopsy, except for hindbrain structures at the base.

15.2.5 Ventriculomegaly/Hydrocephalus

It is a quite common condition and particularly heterogeneous being associated with other CNS defects as well. The CSF is under pressure in the ventricular system of the brain determining a progressive dilatation of the system and reduction of the peri-luminal brain substance (aka "internal hydrocephalus"). Clinically, the non-closure of the fontanels and bony sutures of early infancy is the condition predisposing to the abnormal enlargement of the head (macrocephaly) and the infant can show a downward gaze ("setting-sun" sign). This condition is life-threatening and needs a shunt to drain the excessive CSF within the cavities of the ventricular system such as ventriculoperitoneal shunt. Morphologically, there is a massive brain with dilatation of the ventricular system, flattening of the gyri, narrowing of sulci, thinning of the cerebral hemispheres, atrophy of the central white matter, and compression of the basal ganglia and thalamus. Hydrocephalus is subdivided into noncommunicating and communicating according to the presence or absence of obstruction of the flow of CSF.

The obstruction may be located either in the intraventricular or subarachnoid space in proximity of the exit from the 4th ventricle. In communicating hydrocephalus, the etiology may include malformation of subarachnoid spaces, overproduction of fluid by the choroid plexus, or abnormal filtration through the Pacchioni granulations of the arachnoid or arachnoid villi. These structures (aka pacchionian bodies) are small protrusions of the 2nd layer covering the brain through the dura mater protruding into the venous sinuses of the brain allowing CSF to exit the subarachnoid space and entering the blood. Arachnoid granulations are named after Antonio Pacchioni (1665–1726), an Italian scientist and anatomist, who mainly focused on the meningeal layers of the brain. Pacchioni studied medicine at the University of Reggio Emilia and graduated in 1688, but left for Rome in 1689. In 1703, Pacchioni published his famous studies on meninges in *Acta Eruditorum*, which is considered the first scientific journal of the German-speaking lands of Europe, published entirely in Latin from 1682 through 1782.

15.2.6 Agenesis of the *Corpus Callosum* (ACC)

The *corpus callosum* is the most significant of the midline commissures of forebrain, and ACC is an etiologically heterogeneous condition with no symptoms or minor to more severe psychiatric alterations characterized by complete (agenesis) or partial absence (hypogenesis) of the *corpus callosum* with resulting abnormal connection of the hemispheres. The only relationship that may be left between the two hemispheres is at the brainstem level. There are also some genetic syndromes associated with ACC, and this aspect should be considered during the clinical or pediatric pathological evaluation. Described initially by Reil in 1812, the prevalence of ACC is now reported in the general population as 1.8–7 cases per 10^5 births. It is up to 2–3% in neuropediatric patients with mental retardation. Two major theories have proposed to explain the abnormal formation of the callosal commissure, including either the *failure of the anterior neuropore to properly close* with the subsequent absence of the commissural plate and lamina terminalis or as *result of the diencephalic roof plate expansion with hydrocephalus*, with an abnormal dilated third ventricle interfering with the growth of the callosal commissure. The complete absence of this midline commissure is probably an effect of abnormal embryogenesis, while the partial formation of this structure may be considered as a result from a disruptive event at any time during pregnancy with inflammatory or vascular pathogenesis under intense debate. Apart from some chromosomal disorders involving the triplication of the chromosomes 8, 13, or 18, there are some genetic syndromes, inborn defects of the metabolisms, and environmental disruptions that may present with ACC (Box 15.6). Ultrasound identifies a dark band of uniform thickness throughout its length. Sonographic features consistent with ACC include dilatation and upward displacement of the third ventricle, dilatation of the lateral ven-

> **Box 15.6 Nonchromosomal Etiology of ACC**
> - *Genetic Syndromes*
>   - Aicardi syndrome (ACC, chorioretinal lacunae, infantile spasms)
>   - Andermann syndrome (ACC, peripheral neuropathy)
>   - Acrocallosal syndrome (ACC, hallux duplication, polydactyly)
>   - Calloso-genital dysplasia (ACC, coloboma, amenorrhea)
>   - Shapiro syndrome (ACC, ataxia, drowsiness)
> - *Inborn Defects of Metabolism*
>   - Glutaric aciduria type II
>   - Neonatal adrenoleukodystrophy
>   - Nonketotic hyperglycemia
>   - Pyruvate dehydrogenase deficiency
> - *Environmental Disruptions*
>   - Fetal alcohol syndrome

tricles and occipital horns (aka "tear-drop ventricles" or colpocephaly), separation of the frontal horns and widening of the interhemispheric fissure, the concave medial border of the frontal horns, and absence of the septum pellucidum Colpocephaly (Greek κόλπος "hollow space") is a congenital brain abnormality with disproportionately enlarged occipital horns of the lateral ventricles due to an absence or maldevelopment of the white matter in this location. Fetal MRI may detect by ultrasound occult abnormalities, including sulci formation and gyration aberrations, periventricular nodular heterotopia, interhemispheric cysts, Dandy-Walker malformation (DWM), and cerebellar and brainstem abnormalities.

Isolated ACC is seen in about half of the patients, while the other half presents with various CNS and extra-CNS abnormalities. CNS abnormalities include periventricular nodular heterotopia, patchy- or polymicrogyria, hypothalamic hypoplasia, DWM, cerebellar vermis agenesis, and pontine hypoplasia. Extra-CNS abnormalities include abnormalities of the visceral- and neurocranium (macro- or microcephaly, hypertelorism, and cleft palate), cardiac defects (ASD, VSD, PDA, and Ebstein anomaly), skeletal deformities (clubfoot, brachy- and syndactyly, scoliosis), and urogenital abnormalities. Evaluating prognosis for ACC may be challenging without an efficient and effective multidisciplinary teamwork because of the associated CNS and extra-CNS anomalies. Experienced clinicians should help parents to address correctly and promptly health-related issues in individuals with ACC (feeding, delayed toilet training, sleep disorders, self-injurious behavior and temper tantrums, ranging from whining and crying to screaming, kicking, hitting, and breath holding).

15.2.7 Cerebellar Malformations

Cerebellar abnormalities may occur as an isolated event or be associated with CNS and extra-CNS defects and have a significant impact on the long-term neurodevelopmental prognosis. Chromosomal aberrations (such as trisomy 13 and trisomy 18 syndromes), genetic syndromes (such as Joubert syndrome), as well as metabolic disorders (mitochondrial cytopathies) may be associated with cerebellar abnormalities. Some definitions of dysontogenetic cerebellar entities are important.

Cerebellar agenesis	Complete absence of the cerebellum
Cerebellar hypoplasia	Incomplete development of the cerebellum
Vermis hypoplasia	Incomplete development of the cerebellar vermis
Dandy-Walker malformation	Vermis agenesis with cystic dilatation of the 4th ventricle with enlarged posterior fossa and superior displacement of the cerebellar hemispheres

Cerebellar agenesis is extremely rare and is associated with developmental delay, cognitive impairment, ataxia, and dysmetria. Ataxia is lack of coordination and can include gait and ocular abnormalities and speech changes (Greek α- and -τάξις "order", i.e., "lack of order" considering "α-" a common negative prefix in medical terminology), while dysmetria (Greek: δυσ- "hard, difficult" and μετρέω "measure") is considered an finding of ataxia, in which the individual has an inability to control the distance, intensity, and/or speed of an act. Cerebellar hypoplasia is associated with developmental delay, cognitive impairment, ataxia, hypertonia, and impaired coordination. Other CNS (cerebral atrophy, periventricular leukomalacia, neuronal migration disorders, and abnormalities of the callosal commissure) and extra-CNS anomalies (renal and skeletal malformations) have also been described in association with cerebellar hypoplasia. Moderate-to-severe developmental delay, severe disruption in language skills, delayed motor development, and abnormal behavior modulations have been observed in the long-term neurodevelopmental outcome of children with cerebellar vermis hypoplasia. DWM is associated with poor neurodevelopmental outcome, because of the vermis absence and associated CNS (such as ventriculomegaly and ACC) and extra-CNS abnormalities (such as congenital heart defects, renal defects, facial dysmorphism, and single umbilical artery).

15.2.8 Agnathia Otocephaly Complex (AGOTC)

The complex defined by combined agnathia and otocephaly is a rare, almost always lethal, condition characterized by mandibular hypoplasia or agenesis (agnathia, from Gr. a- gnathos), ventromedial malposition of the ears or melotia and/or auricular fusion or synotia, as well as microstomia (Gr. μικρός "small" and στόμα "mouth") with hypo-/aglossia. Holoprosencephaly is the most commonly identified CNS association, but cardiovascular, genitourinary, and skeletal anomalies have been observed. Missense mutations in *PRRX1* (paired-related homeobox gene 1) were detected in two sporadic cases of otocephaly, and *OTX2* gene mutations have been found in patients with isolated severe ocular and pituitary malformations. In a patient with AGOTC reported by Schiffer et al. (2002), Sergi and Kamnasaran (2011) detected a heterozygous loss-of-function mutation in the *PRRX1* gene (F113S). The role of *PRRX1* in the mandibular-facial development was key in selecting this gene for molecular genetics studies. Chassaing et al. (2012) reported the identification of a deleterious mutation in *OTX2* in a large French family in which variable expressivity extends from micro-/anophthalmia to otocephaly. The inheritance manner was dominant among the four generations. Chassaing et al. (2012) performed a molecular screening of *OTX2* in nine additional, nonrelated otocephalic patients and identified a second sibship with an *OTX2* mutation. Moreover, these authors suppressed otx2 during zebrafish development and determined that (1) otx2 is also necessary for the correct formation of the mandible and (2) otx2 can interact genetically with other loci to modulate the severity of mandibular deformities.

15.2.9 Telencephalosynapsis (Synencephaly) and Rhombencephalon Synapsis

The agenesis of the cerebellar vermis with a fusion of the cerebellar hemispheres is an uncommon malformation of the CNS. This abnormality was first described by Heinrich Obersteiner from Vienna (Austria), in 1916, as an incidental autopsy finding in a patient who had not exhibited clinical signs of cerebellar pathology. Subsequently, on reviewing paleocerebellar agenesis including Obersteiner's observation, De Morsier (1955) coined the term "*rhombencephalosynapsis*" for the latter, as opposed to "*rhombencephaloschizis*" which represents the more common situation in which cerebellar vermian agenesis is associated with midline dysraphia. Later, additional cases of rhombencephalosynapsis were described. In one patient, cystic enlargement of the fourth ventricle (ventriculocele) was reported (Sergi et al. 1997). Further supratentorial abnormalities may occur with rhombencephalosynapsis. They include agenesis or deficiency in the development of the septum pellucidum, hypoplasia of the commissural system, the optic nerves, the chiasm and the optic tracts, fusion of the thalami and the colliculi of the quadrigeminal plate, and agenesis of the posterior pituitary lobe. In *Brain Development's* reported patient, (2012) a complex CNS malformation was illustrated in detail at 23 weeks' gestation. (Sergi et al. 1997). The CNS abnormality included telencephalosynapsis (synencephaly) with telencephalic ventricular aplasia and paleocerebellar agenesis with rhombencephalosynapsis and cystic enlargement of the 4th ventricle. In both telencephalosynapsis and rhombencephalosynapsis, the neurocranium was microcephalic and displayed a partial premature closure of the cranial sutures with slight steplike protrusion of the occiput at the lambda suture or outward convex protuberance of midportion of the occipital bone just above the torcular Herophili (bathrocephaly). On gross examination, the brain showed a smooth surface with an absence of the interhemispheric cleft and Sylvian fissures. Coronal sections confirmed the complete fusion of the brain hemispheres including the most rudimentary anlage of the basal ganglia and thalami. While the 3rd ventricle appeared to be missing, the lumina of the small lateral ventricles were not filled with blood but were absent. There were only ependymal rosettes embedded in a medium compact central germ cell layer. Similar situation was seen for the 3rd ventricle and the aqueduct. Between the central germ cell layer in the depth of the brain and the disorganized cortex, the *centrum semiovale* displayed, in places, garland-like heterotopias. The cerebellum was severely hypoplastic. Substantially, the cerebellar hemispheres were flattened and fused in the midline, and there was no vermis of the cerebellum. A further unusual feature was the association of the agenesis of the vermis with a cystic dilatation of the 4th ventricle, appearing as a median posterior fossa cyst, outlined by an ependymal-glio-mesenchymal membrane. The malformation complex may be enlisted into two major teratogenetic spectra: (1) the (holo-) prosencephaly-arhinencephaly complex and (2) the spectrum of paleocerebellar agenesis. Kepes et al. (1969) reported slowly progressive hydrocephalus in a 6-month-old infant with congenital thalamic fusion and atresia of the third ventricle, the aqueduct, and the upper portion of the fourth ventricle. There was also a fusion of the colliculi of the quadrigeminal plate along with *rhombencephalosynapsis* and incomplete *synencephaly*. Considering that the cerebellar anlagen are first distinguished at the 5.7–7.3 mm embryonic stage (corresponding to 38 days of gestation) and fuse in the midline at the 24–27 mm crown-rump length (CRL) (~60 days) and the dentate nuclei differentiate at about 16th gestational week, the vermis is almost fully developed by the 20th gestational week, the fusion of the cerebellar hemispheres is complete by the 20th–24th gestational week, a teratogenetic stimulus was able to operate and induce vermal agenesis around 8–12 weeks of gestation, and *rhombencephalosynapsis* becomes completed around 16–20 weeks of gestation. The combination of *rhombencephalosynapsis* and *synencephaly* is a further unusual feature of the malformation complex described in 1997, the teratogenetic termination period of which could be assessed concerning the synencephaly around the fourth gestational week, after the closure of the neural tube, at the earliest. The defective cerebellar vermis links the malformation complex with the spectrum of paleocerebellar agenesis which is roughly summarized in Box 15.7. The *rhombencephalosynapsis* precludes the simple classification of the patient published in *Brain & Development* in 1997 with the Dandy-Walker syndrome associated with additional forebrain malformation. The authors considered the malformation complex to be associated with maternal diabetes. In an infant, a possible teratogenetic influence to determine *rhombencephalosynapsis* was found in her drug prescription. In fact, the mother had taken phencyclidine in the first 6 weeks of pregnancy (Michaud et al. 1982). In teratogenesis, of importance are the etiologic factor, the time of exposure to the offending agent, the route of exposure, and the genotype of the embryo and parents. Different patterns of malformation may be attributed to phenylketonuria, diabetes mellitus, and some chemical substances such as ethanol, cannabis, and therapeutic drugs. Although confounding factors cannot always be ruled out, there is a strong body of evidence that cannabis has also teratogenic effects on the developing brain following perinatal exposure.

> **Box 15.7 Paleocerebellar Agenesis (Adapted)**
> - *Partial/total agenesis of the cerebellar vermis with midline defect*
>   (Rhombencephaloschisis)
>   – With rhombencephalic ventriculocele (cystic enlargement of the 4th ventricle), posterior fossa enlargement with elevated torcular, and hydrocephalus (Dandy-Walker syndrome)
>   – Without rhombencephalic ventriculocele (common partial/total agenesis of the vermis)
>   Sporadic with or without associated malformations
>   Familial (Joubert syndrome)
>
> - *Agenesis of the vermis and midline fusion of cerebellar hemispheres and dentate nuclei*
>   (Rhombencephalosynapsis)
>   – With cystic 4th ventricle (posterior fossa cyst of Dandy-Walker type), telencephalosynapsis and forebrain ventricular aplasia (Sergi et al. 1997)
>   – Without cystic 4th ventricle (ordinary form)
> - *Agenesis of the vermis with encephalocele*
>   – "Inverse cerebellum" or "tectocerebellar dysraphia" with occipital encephalocele
>   – Agenesis of the vermis with encephalocele, type 2 lissencephaly hydrocephalus, and ocular dysgenesis in Walker-Warburg syndrome

15.2.10 CNS Defects in Acardia

Twin-twin transfusion syndrome (TTTS) is still one of the most dreaded complications of monochorionic twin pregnancy although therapy has improved the outcome. TTTS harbors in several countries an extremely high pre- and perinatal morbidity and mortality. Net transfusion from the umbilical artery of the donor twin to the umbilical vein of the recipient twin categorizes TTTS. This blood transfusion occurs in a villous zone of overlapping perfusion of the placenta. Transfusion from the donor to the recipient via placental arterio-arterial anastomoses in monochorionic gestations is at the basis of the twin reversed arterial perfusion (TRAP) sequence, which was delineated by Van Allen in 1983, but brilliantly described by Schwalbe in 1907. The arterial supply into the placenta by the donor or pump twin can overcome the blood pressure of the recipient twin to perfuse retrogradely the recipient twin through the umbilical arteries and, via the iliac arteries, the aorta. The recipient twin presents with different degrees of developmental arrest of internal organs. The label *"acardius"* was coined with accompanying adjectives such as *anceps*, *acephalous*, *amorphous*, *acormus*, etc. demonstrating the different degrees of reduction of anatomic structures. The occurrence of well-defined malformations and various chromosomal abnormalities in some cases of TRAP sequence has led to an alternative hypothesis that the recipient twins are constitutionally abnormal. Although many cases with TRAP sequence have been reported, details about the nervous system remain elusive. In two cases of *acardius anceps*, gross descriptions of brains affected with holoprosencephaly were provided by Chi (1989), but no histopathology was carried out. In 2000, a neuropathological investigation was reported on *acardius anceps* with a thorough gross and light microscopic examination of the rudimentary brain, which displayed a combination of deformity and encephaloclastic destruction. Although some features were compatible with a holoprosencephaly complex of malformation, the non-cystic parts of the brain mantle and the basal ganglia exhibited marked changes of (secondary) encephaloclastic nature, such as depletion of the neuronal matrix with gliosis, calcifications, and scar formation. Also, the irregular distribution of the cystic parts of the brain mantle is not compatible with a dorsal sac of cystic holoprosencephaly. Conversely, they correspond to the areas of the internal carotid artery blood supply. The authors suggested that cyst formation in the brain was more likely the result of vascular (hypoxic) damage to the brain mantle than of the initial configuration of a dorsal sac in a prosencephalic brain. Overall, the changes of the skull and brain in the *acephalous anceps* suggest, indeed, that circulatory failure with hypoxic encephaloclasia and cyst formation had been superimposed on a primarily malformed brain of prosencephalic type. Briefly, "hydranencephaly met holoprosencephaly". In a sample of ten TRAP sequences, Chi (1989) found holoprosencephaly in two cases of acardius anceps. A gross illustration convincingly described one of the patients as prosencephalic brain. On the other side, since the authors did not perform microscopic examinations, it remains open, whether hypoxic damage complicated the holoprosencephaly. In the case of Isoda and Hamamoto (1980), the rudimentary brain also stayed without detailed gross or microscopic descriptions. As early as in 1859, Claudius raised the hypothesis that reversed feto-fetal perfusion via placental anastomoses in twin pregnancies may be the cause of acardia in one of the twins. Claudius conceived that primarily the anlage of both twins was normal, but, subsequently, the previously formed heart in the recipient twin would undergo regression because of the reversed perfusion with increased mechanical stress due to continuous pumping against the stronger heart of the donor twin, whose heart was the motor of both their circulations. Later studies support that concept by demonstration of the placental anastomoses. Consequently, the currently used designations TTTS for artery-to-vein and TRAP sequence for the artery to the artery and vein-to-vein shunts in closely apposed umbilical cords were coined and have as base the impressive work of Claudius. Other authors like Marchand opposed the concept of Claudius proposing that the acardiac twin pregnancy is the result of a primary anlage defect started at the zygote level, e.g., asymmetric cleavage in monozygotic twins. General signs of initial developmental arrest, an about 10% incidence rate of malformations occurring in the co-twin (in the setting of a proposed multiple congenital anomaly syndromes), and the proof of a variety of different chromosomal aberrations in acardia, appeared to favor of Marchand's hypothesis. Interestingly, almost 100 years after Marchand's suggestion, Fisk et al. (1996) could exclude polar body fertilization as a cause of TRAP sequence using molecular biologic investigations. On the other hand, both hypotheses might be correct to a convinced degree, probably. An early occurring common cause, such as hypoxia due to reversed placental perfusion, may be responsible for inducing both developmental arrest in early embryogenesis and later secondary vascular damage ("angiolesive factors") to either defective or developed organs and tissues. In fact, the brain of the acardius anceps supports this concept by presenting signs of both types of changes, developmental anomalies (prosencephaly, etc.), and hypoxic damage (encephaloclastic hydranencephaly). Anoxia has been shown to be a very useful teratogen. Anoxia/hypoxia causes disruption, particularly of neurulation, if it interferes with early stages of embryonic development. Experimentally, hypoxia induced by exposure to hypobaric pressure conditions revealed in amphibian and chick embryos mainly disruptions of the skull and brain, as well as of the heart. It has been indicated that the most severe brain and head changes result if the hypoxia is induced at the start of gastrulation, i.e., before the onset of neurulation, when O_2 consumption is known to be especially high. The developmental abnormalities included different degrees of reduction of the brain anlage and the eyes before the onset of neurulation. In this setting, synophthalmia, cyclopia, anophthalmia, anencephaly, platyneuria, or even the entire head anlage may be missing. Congenital defects of the heart, aorta, and pulmonary artery were often seen, as were defects of the kidneys (e.g., horseshoe kidney) and of the limbs. If the hypoxia occurred after the onset of neurulation, developmental changes of the CNS were less severe with histogenetic disturbance such as brain cortical disarrangement. The basic developmental mechanisms responsible for the defects occurring in the setting of anoxia/hypoxia in these experiments are considered to be (1) disturbance of the migration of blastemas (Lehmann's "topogenesis" and Gilbert "integrated cell and tissue movements"), (2) decrease of the inductive capacity of the altered blastemas, and (3) disturbance of further differentiation of the organ anlage. These radical experiments visibly showed that hypoxia must occur before the onset of the formation of the organ anlage starts to induce severe developmental deviation. The results obtained by hypoxia in amphibians and chicks are also basically valid for mammals and can be extrapolated to man as well. They match the findings observed in dissection studies of acardia to a high degree in both the human species and other vertebrate animals. The severe malformation or absence of the heart cannot be a significant prerequisite of the origin of a TRAP sequence, but it is probably the result of the action of a teratogen, such as oxygen deficiency, in early embryogenesis. Disturbances of placenta formation and/or insertion, including developmental anomalies of the umbilical vessels, such as anastomoses (artery-to-artery and vein-to-vein) with a reversal of placental circulation, would be able to cause oxygen deficiency at the required early stage of development in TRAP sequences. The earlier the disturbance occurred, the more severe the resulting body defects would be concerning the acardiac phenotype (acardius anceps, acephalous, amorphous). This concept has been reiterated here but has its roots in the impressive work of Claudius as indicated above. Friedrich Matthias Claudius was born in Lübeck, Germany, in 1822. His publications signed using the middle name, Matthias, given to him in honor of his grandfather and of his uncle, the German poet Matthias Claudius, are pillar in anatomy and pathology. He completed studies in Jena, Göttingen, and Kiel. After receiving his doctorate in Göttingen in 1844, he volunteered as physician in the army of Schleswig-Holstein from 1848 to 1850 and was appointed to the zoological museum at Kiel in 1849 and as docent for anatomy in 1854. Five years later, he became professor and director of the Institute for Anatomy at the University of Marburg. Claudius established a collection of morphological specimens. In 1862, he fell severely ill and died in Kiel seven years later.

15.2.11 CNS Defects in Chromosomal and Genetic Syndromes

It is a quite a long list of chromosomal and genetic syndromes showing CNS defects. For instance, isolated agenesis of the corpus callosum/callosal commissural agenesis (ACC) may be inherited, but no loci have been mapped. Nonsyndromic genetic transmission is rare. CNS defects may be associated with trisomy 13 and trisomy 18 as well as autosomal and X-linked syndromes with multiple congenital anomalies. Moreover, ACC is part of the fetal alcohol syndrome and is observed in lactic acidosis and non-ketotic hyperglycinemia. Cornelia De Lange syndrome may be an example of CNS defects in genetic syndromes. A detailed study of genetic syndromes with CNS defects will not be examined in this book. It is crucial that the reader liaise with pediatric neuropathologists and consult pediatric neuropathology books for an extensive presentation of the multiple congenital anomaly syndromes with CNS defects.

15.2.11.1 Cornelia De Lange Syndrome (CDLS)

- *DEF*: Multisystemic developmental disorder affecting several and different parts of the body
- *SYN*: Brachmann-De Lange Syndrome (BDLS), De Lange Syndrome
- *EPI*: Worldwide, 1:10,000–1:30,000 newborns, ♂ = ♀.
- *GEN*: The underlying defect in about 3/4 of patients is found in five genes [*NIPBL* (Nipped-B-like protein), *SMC1A* (structural maintenance of chromosomes 1A), *SMC3* (structural maintenance of chromosomes 3), *RAD21* (human homolog of Schizosaccharomyces pombe radiation sensitive mutant 21), and *HDAC8* (histone deacetylase 8)]. These genes regulate the structure and organization of chromosomes and DNA repair mechanisms and play a significant role in the development of several somatic structures.
- *CLI*: Pre- and postnatal slow growth, microcephalic intellectual disability with often autism

with or without seizures, bony deformities involving both the arms and the hands, and distinctive facial dysmorphism (arched eyebrows ± synophrys or a single eyebrow when the two eyebrows meet in the middle above the bridge of the nose, long eyelashes, low-set ears, small, upturned nose, and abnormal primary and permanent dentition with widely spaced teeth) and other somatic features, including cleft palate, congenital heart defects, and anomalies of the gastrointestinal tract. Congenital heart defects occur in one in five patients with CDLS/BDLS (OMIM 122470) including usually a structural heart defect and an isolated non-obstructive hypertrophic cardiomyopathy. In patients with heart defects an isolated obstructive anomaly may include pulmonary stenosis, isolated left to right shunt due to atrial and/or ventricular septal defects, and a combination of structural anomalies. Isolated late-onset mitral or tricuspid valve dysplasia have been described.

- *PGN*: Intellectual disability and supervised living are common in patients with CDLS/BDLS. The life expectancy is assessed to be 10–20 years shorter compared with the general population. This shorter survival is associated with the complications present in several organs and tissues. Apart of the seizures, COD include respiratory diseases including aspiration pneumonia (or reflux pneumonia), gastrointestinal infarction due to obstruction and volvulus, diaphragmatic defects, and complex heart defects (more or less inoperable).

15.2.12 Neuronal Migration Disorders

The failure of migrating neurons to reach their destination in the cerebral cortex is collected in the group of neuronal migration disorders. The defects may be focal or diffuse. Thus, if neurons fail to leave the ventricular zone of the brain, periventricular heterotopias may occur. In addition, if they fail to fulfill their migration paths in the cortex, the neurodevelopmental disorder that ensues is called lissencephaly (Greek λισσός "smooth") or "smooth brain." On the other hand, nodular or band heterotopias are the result if only a subpopulation of neurons are affected, while the remaining neurons complete the migration to the cortex. Agyria (or lissencephaly) means the complete absence of gyri, while pachygyria (Greek παχύς "thick") describes a reduced number of broadened and flat gyri with less collapsible of the cortex than usual. In type I lissencephaly, the cortex is thick with the white matter forming a thin rim along the ventricles. Infants with type I lissencephaly are typically separated into two groups, including a non-dysmorphic group (majority) with isolated lissencephaly sequence and a dysmorphic group (majority) with the characteristic features of Miller-Dieker syndrome. Miller-Dieker syndrome is associated with deletions of 17p13.3, a region which includes the *LIS1* gene, while the non-dysmorphic group is a heterogeneous group with about half of the patients harboring a deletion of, or mutations within, the *LIS1* gene. Mutations also cause lissencephaly if a second gene located on the X chromosome, doublecortin (*DCX*) is involved.

15.2.13 Phakomatoses

The term phakomatoses (Greek φακός "spot, lens", suffix-ωμα and suffix –ωσις "process, disease" from –όω stem verbs + –σις) was coined by van der Hoeve in the 20s of last century. In his paper, he was concerned with the similarities between Bourneville's tuberous sclerosis and von Recklinghausen neurofibromatosis. Jan van der Hoeve (1878–1952), a Dutch ophthalmologist, graduated from the University of Leiden (Netherlands) and received his doctorate at the University of Bern (Switzerland). Van der Hoeve became a professor of ophthalmology at the University of Groningen and, later, transferred at the University of Leiden, became member of the Royal Netherlands Academy of Arts and Sciences, and, finally, president of the Physical Section of the Institute in 1932. By definition, phakomatoses are a group of neuro-oculocutaneous syndromes involving structures arising from the embryonic ectoderm (CNS, eyes, and skin) with variable severity. Later, it has been observed that mesodermal and endodermal tissues are also involved. Phakomatoses include *ataxia telangiectasia*, *incontinentia pigmenti*, neurofibromatosis type 1 (von Recklinghausen disease) and type 2 (central neurofibromatosis), nevoid basal cell carcinoma syndrome (Gorlin-Goltz syndrome), Sturge-Weber syndrome, Bourneville-Pringle disease or tuberous sclerosis, Wyburn-Mason syndrome (Bonnet-Dechaume-Blanc syndrome), and von Hippel-Lindau disease.

15.2.13.1 Tuberous Sclerosis Complex (TSC)

Tuberous sclerosis (aka Bourneville's disease or phakomatosis, epiloia, or cerebral sclerosis) is a relatively frequent, autosomal dominant inherited genetic syndrome with an incidence of 1 in 6000 individuals. Sherlock coined the term "epiloia". It includes the clinical triad of *epi*lepsy, *lo*w intelligence, and *a*denoma sebaceum. Mutations in the *TSC1* gene located on *9q34* or *TSC2* gene situated on *16p13.3* can cause tuberous sclerosis complex. *TSC1* and *TSC2* gene products make the proteins *hamartin* and *tuberin*, respectively. Both proteins work together within the cells helping regulate cell growth and size and act as tumor suppressors. Patients affected with TSC show an inherited mutation. A second mutation located in either gene typically occurs over an affected person's lifetime. The loss of either hamartin or tuberin in different types of cells leads to the growth of non-malignant tumors in several organs such as the brain, skin, kidneys, heart, lungs, and pancreas. Tuberous sclerosis complex also causes neurodevelopmental symptoms, including epileptic seizures and mental retardation. Behavioral problems such as hyperactivity and aggression as well as intellectual disability or learning problems have been described in patients affected with TSC. Moreover, some children may also show the characteristic features of autism. In the brain, the distinguishing morphologic feature is a *"tuber"* or triangular-shaped smooth nodule with the apex pointed towards the ventricles consisting of glial fibers and malformed astrocytes and neurons in the wall of the ventricular system and cerebral gyri. The tubers represent foci of abnormal neuronal migration. Also, *subependymal nodules* in the walls of ventricles and *subependymal giant cell astrocytoma* (SEGA) are other two conditions found in the brain of individuals affected by TSC. Subependymal nodules are abnormal, swollen glial cells and bizarre multinucleated cells with a tendency to calcify as the patient grows up. A SEGA usually develops in the region of the foramen of Monro and patients with SEGA harbor a risk to develop obstructive hydrocephalus. On MRI, there may be abnormal neuron migration showing radial white matter tracts hyperintense on T2WI (T2 weighted image) and heterotopic gray matter. Extracranial manifestations of TSC include renal angiomyolipoma (rarely RCC and oncocytoma), cardiac rhabdomyoma (2nd half of pregnancy and postpartum regression), pulmonary lymphangioleiomyomatosis (LAM), cutaneous facial angiofibroma ("*adenoma sebaceum*"), peri-ungueal fibroma, hypopigmented macules, forehead plaques, shagreen patches, skin tags, café au lait spots, and poliosis, as well as astrocytic hamartomas (or "phakomas") of the retina (non-retinal lesions associated with TSC also include coloboma, angiofibroma of the eyelids, and hydrocephalus-related papilledema).

15.2.13.2 Neurofibromatosis (NF)

Neurofibromatosis (NF) is another phakomatosis like TSC and aka *von Recklinghausen disease*. NF is an AD-inherited genetic disease with variable expressivity and approximately half of patients harbor *de novo* mutations.

Neurofibromatosis type 1 (aka "von Recklinghausen disease") is the most common form of NF, accounting for up to 90% of all cases (one in 4000), and is due to mutations of *Neurofibromin 1* gene on chromosome *17q11.2*. Neurofibromin is a regulator of the GTPase activating enzyme (GAP) and acts as TSG whose target is p21 ras oncoprotein. Inability to suppress this protein leads to tumor proliferation. The problem is the inability to inactivate GTP due to a defective GTPase. In the absence of this tumor suppressor's inhibitory control on the Ras oncoprotein, cellular proliferation is promoted. The diagnosis of NF1 is based on when any two of the following nine criteria are met: (1) café au lait spots (pigmented, light brown cutaneous macules with smooth-edged, "coast of California"-shaped birthmarks) being ≥6 measuring 5 mm in greatest Ø in prepubertal children and >15 mm in greatest Ø in postpubertal individuals; (2) *a*xillary or inguinal freckling; (3) *f*ibromas of neural type / ≥2 neurofibromas (cutaneous or subcutaneous) or ≥1 plexiform neurofibroma; (4) *e*ye hamartomas (Lisch nodules); (5) *s*keletal abnormalities (e.g., sphenoid bone dysplasia or cortical thinning of long bones or tibial pseudo-arthrosis); (6) *p*ositive family history (a first-degree relative with NF-1); and (7) *o*ptic *t*umor (optic nerve glioma). Other features may include macrocephaly, kyphoscoliosis, epilepsy, juvenile posterior lenticular opacity, learning disabilities, hydrocele, and early puberty. *Neurofibromatosis type 2* (aka "central neurofibromatosis") is the result of mutation of the NF2 gene located on *22q12*, which encodes the protein *Merlin* or neurofibromin 2. It accounts for only 10% of all patients affected with NF (1:45,000). NF2 criteria are the Baser criteria (2011) which are a revision of the Manchester Criteria. NF2 features include bilateral acoustic "neuromas" (schwannomas of the vestibule-cochlear nerve) and brain tumors (gliomas, meningiomas, ependymomas) and need 1/3 to diagnose NF-2. *Schwannomatosis is* showing multiple schwannomas with segmental schwannomatosis in 1/3 of the cases. In these patients, schwannomas are limited to a single part of the body, such as an arm, a leg, or the spine. *Neurofibromatosis type 3A*, *type 4,* and *type 5* are segmental neurofibromatoses.

15.2.13.3 Von Hippel-Lindau Syndrome (VHLS)

Von Hippel-Lindau syndrome is an AD-inherited multiorgan genetic condition without gender preference characterized by CNS hemangioblastomas, retinal capillary hemangiomas/hemangiomatosis, endolymphatic sac tumor of the inner ear, as well as various solid and cystic visceral hamartomas, hamartia, and malignant neoplasms such as renal cell carcinoma (RCC) and pheochromocytoma. The median age at detection of the first clinical features of VHLS, which are the capillary hemangiomas of the retina, is 20–25 years. Molecular biological studies have localized the *VHL* gene to chromosome 3p25.3. This gene has the official name of "von Hippel-Lindau tumor suppressor" or "E3 ubiquitin protein ligase" but also different names, such as elongin binding protein, pVHL, and VHL1. The *VHL* gene is part of a complex called VCB-CUL2 that targets some proteins to be degraded by the cells when they are no longer useful. One of these protein targets of VCB-CUL2 is indeed hypoxia-inducible factor 2-alpha (HIF-2α), which plays an essential role in the body's ability to adapt to changing O_2 levels. In fact, HIF controls several genes involved in cell division, neo-angiogenesis, and erythropoiesis through erythropoietin. Extraocular features are hemangioblastomas (capillary hemangiomas) of the brain and spinal cord; RCC, pheochromocytoma, and islet cell carcinoma of the pancreas; and cystadenomas of the pancreas and epididymis. Genetic counseling and periodic physical examinations (Triennial MRI of CNS, annual ultrasound of abdomen, and urogenital system, a yearly 24-h urine collection for vanillyl-mandelic acid) are decisive. The COD of these patients is linked to the bleeding or mass effect of the intracranial hemangiomatous lesion and the progression of the RCC.

15.2.13.4 Sturge-Weber Syndrome (SWS)

Sturge-Weber syndrome (SWS) is a neuro-oculo-dermatologic syndrome characterized by a *nevus flammeus* of the skin along the distribution of the fifth cranial nerve (trigeminal nerve), usually the first branch, ipsilateral diffuse cavernous hemangioma of the choroid, and ipsilateral meningeal hemangiomatosis, which are congenital in origin. Although no gender preference or racial predilection has been identified, the exact frequency of SWS is unknown owing to the numerous "*formes frustes*" or mild forms. Dermatopathological investigation displays a flat to moderately thick zone of dilatated telangiectatic capillaries of the dermis covered by a single endothelial cell layer. The CNS lesion may cause atrophy of the cortical parenchyma of the brain, seizures, and mental retardation, while glaucoma and exudative retinal detachment may complicate the ocular injury.

COD (prognosis *quoad vitam*) of patients affected with SWS is due to intractable seizures or status epilepticus. Mental retardation influences prognosis *quoad valetudinem*.

15.2.13.5 Wyburn-Mason Syndrome (WMS)

Wyburn-Mason syndrome (WMS) is an uncommon phakomatosis characterized by arteriovenous malformations (AVMs) of the retina and ipsilateral brain. It is not universally accepted as phakomatosis, because the abnormal lesions are not distinct tumors, but somewhat anomalous AV communications. The retinal and intracranial AVMs of WMS are congenital, and MRI and/or MR angiography of the ipsilateral orbit and brain is used. COD is linked to spontaneous bleeding from the intracranial lesions.

15.3 Vascular Disorders of the Central Nervous System

15.3.1 Intracranial Hemorrhage

There are four anatomic compartments in the intracranial space, and each chamber has distinctive features. The four compartments are the brain parenchyma, the subarachnoid space, the subdural space, and the epidural space (Figs. 15.3 and 15.4). The neonatal conditions affecting the

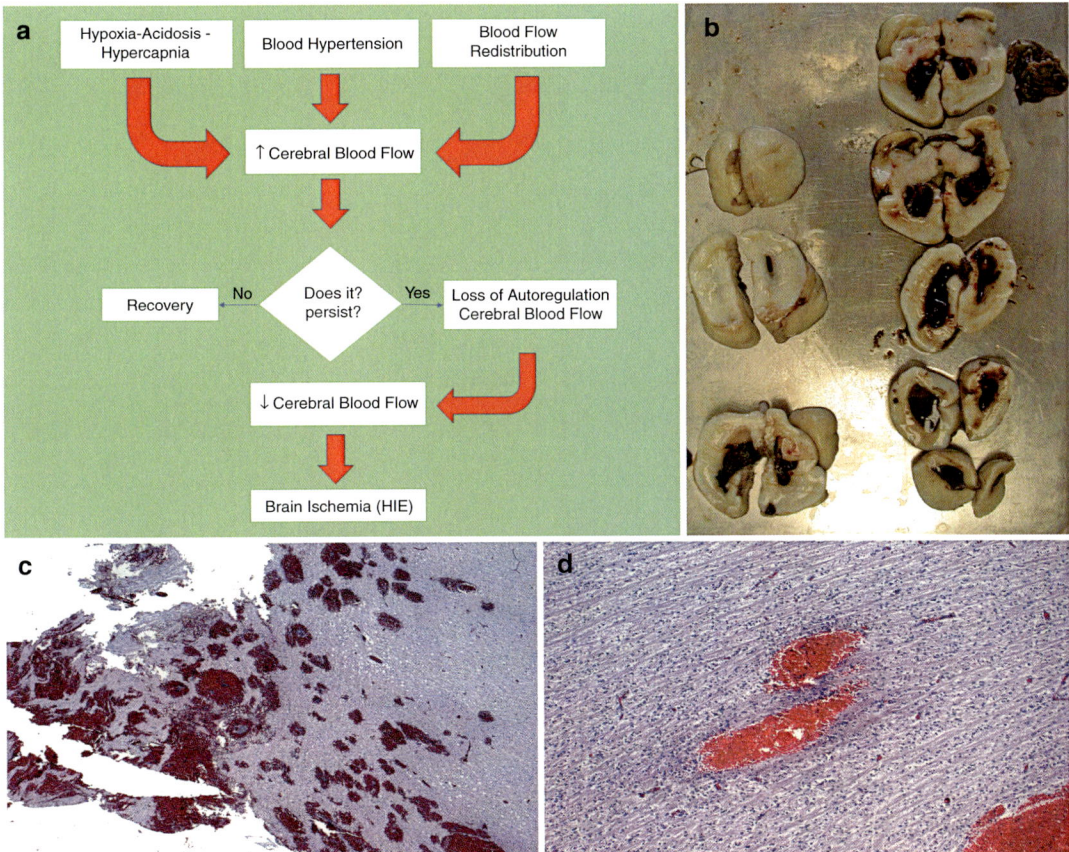

Fig. 15.3 One of the most accepted mechanism for hypoxic ischemic encephalopathy is presented in figure (**a**). Cross sections of a premature baby with grossly identifiable extensive intraventricular hemorrhage (hematocephalus internus (figure **b**). Figures **c** and **d** show microphotographs of a premature baby suffering from hypoxic ischemic encephalopathy with periventricular hemorrhage and brain necrosis (**c**, 40×, hematoxylin and eosin staining; **d**, 100×, hematoxylin and eosin staining)

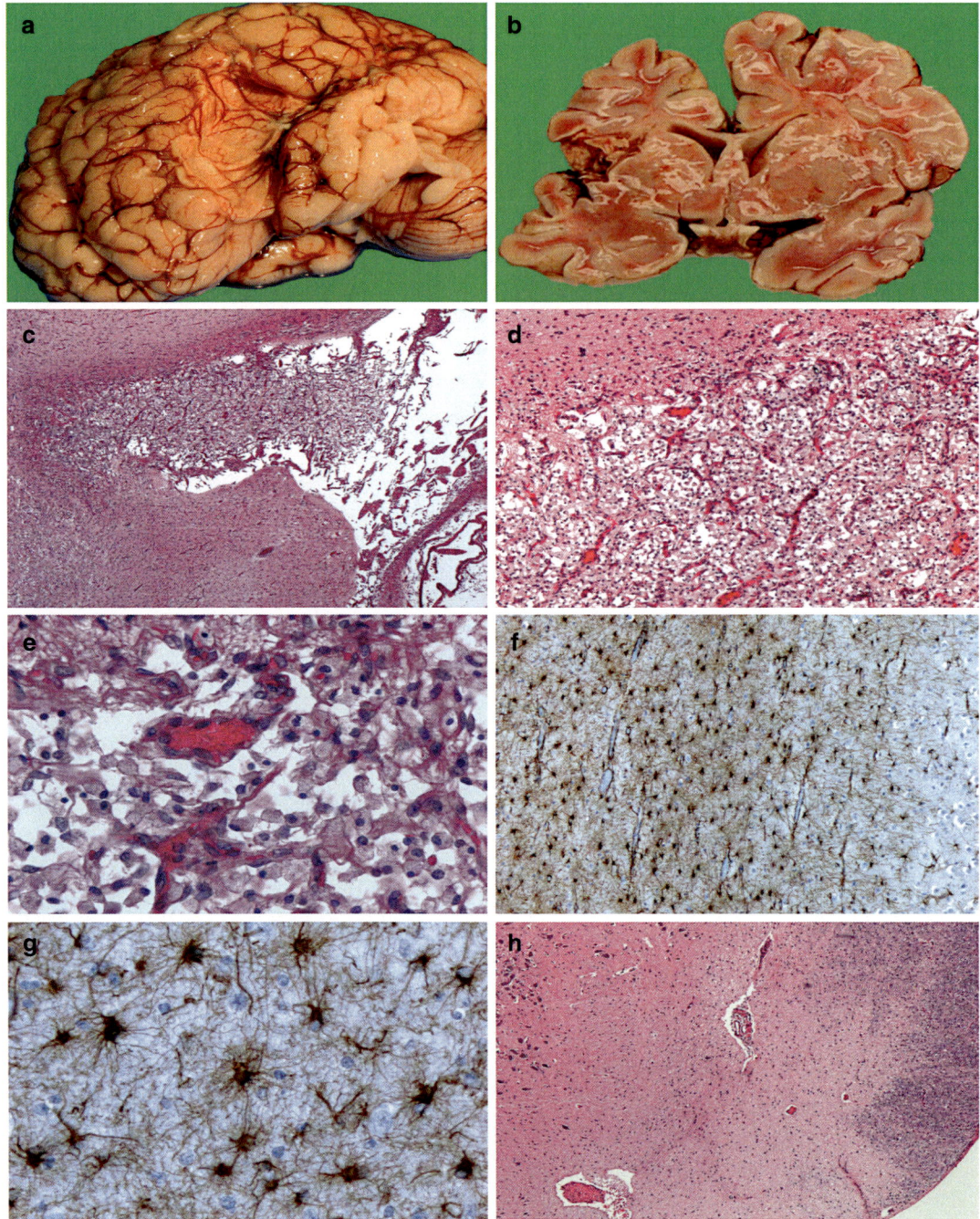

Fig. 15.4 Hemorrhage in the brain can change its shape as noted in this old infarct grossly (**a** and **b**). The microphotographs of this infarct show coagulative necrosis with cavitation eosinophilic neurons, neovascularization, and astrogliosis (**c**, ×20; **d**, ×100; **e**, ×400, hematoxylin and eosin staining). Some fresh vascular thrombosis is seen in (**e**). The astrogliosis is particularly highlighted if an immunostaining is performed with an antibody against glial fibrillary acid protein (GFAP) ("white matter gliosis") (**f**, ×100; **g**, ×400, avidin-biotin complex). The microphotograph (**h**) shows a degeneration of the corticospinal tract at the midbrain (×50, modified Luxol Fast Blue staining)

CNS may be delineated as below, and vascular disorders play a major role because of the frailty of the germinal matrix and of the cerebrum at this time of fetal development.

Prenatal
- Genetic disorders
- Intrauterine infections
- Intrauterine vascular
- Malformations/multiple congenital anomalies
- Maternal disease
- Pregnancy-related disorders

Perinatal
- Birth complication
- Prematurity

15.3.1.1 Intraparenchymal Hemorrhage (IPH)

Etiologically heterogeneous, IPH may be due to hypertension, leukemia-associated hematomas obstructing the small blood vessels, rupture of berry aneurysms and AV malformations, intratumoral bleeding of primary or metastatic neoplasms, trauma, coagulation disorders, and vasculitis. Clinically, the patient having a hemorrhagic stroke suffers from a sudden loss of consciousness and trauma-related fall, thrown back head, congested face, difficulty in breathing, and hemiplegia as well as urinary and fecal incontinence. Convulsions may develop if the bleeding occurs on or in the proximity of the brain surface. Coma and death may occur in a range between a few hours or some days. If the bleeding is small and able to be reabsorbed, *restitutio ad integrum* (restoration to original condition) or, alternatively, a neurologic defect may be the fate of IPH. Grossly, there is acute massive hemorrhage in the brain (basal ganglia > pons > cerebellum), swollen (edematous) cerebral hemispheres, and flattening of the gyri. The bleeding may invade the ventricular system filling its cavities. If the patient survives, the clot shows shrinkage, change in chocolate-color, and disappearance of edema. If the heme, i.e., the complex consisting of Fe as ion coordinated to a porphyrin acting as a tetradentate ligand, and to one/ two axial ligands, in the hemoglobin becomes oxidized, it may become methemoglobin, which is chocolate-like or brownish. Methemoglobin, which has the iron ion in Fe^{3+} state, not the Fe^{2+} of normal hemoglobin, is unable to bind O_2. Consequently, there is no possibility to carry O_2 to tissues. Microscopically, erythrocytic extravasates are the rule replacing the destroyed tissue with the recruitment of microglial phagocytes within 3 days, which ingest degenerating or degenerated RBC, and hemosiderin deposition within 6–10 days. The late fate is a cystic space, which corresponds to the central area of destruction when the phagocytic activity clears out the area of damage. Said that, we also need to emphasize that pathological assessment of timing of traumatic events remains associated with error, which is linked to inter-individual variability, the lack of precision of gross and microscopic changes, and the primitive nature of the techniques in general use (gross observation, histopathology, Prussian special stains). Ultimately, all these factors suggest caution when attempting to assign a time interval between hemorrhage and presentation or death (Castellani et al. 2016).

Risk pregnancies include low and high maternal age, drug addiction, alcohol use/abuse, consanguinity, and hereditary disorders, while maternal complications include diabetes mellitus (DM), hypertonia, preeclampsia, and prenatal infections (e.g., TORCH). The well-known acronym "TORCH" stands for the etiologic pathogens of congenital infections, including *Toxoplasma gondii*, "*o*thers" (including *Treponema pallidum*, *Listeria monocytogenes*, *Varicella Zoster Virus*, and *Parvovirus B19*), *Rubella Virus*, *Cytomegalovirus*, and *Herpes Simplex Virus*. Pregnancy-related problems include prematurity, fetal growth restriction (formerly known as intrauterine growth retardation), placental dysfunction, prolonged labor, breech presentation, twins, and premature rupture of membranes (PROM). In Box 15.8 the CNS disorders observed in prematurity are listed. The mnemonic word "PIRI" may be used. Piri is the Swahili word for pepper, and Piri Piri is a one of the sources of chili pepper that grows in Africa, mostly, but also worldwide.

> **Box 15.8 Prematurity: Observable CNS Disorders**
> - *P*eriventricular leukomalacia
> - *I*mmaturity-based white matter damage
> - *R*etinopathy of prematurity (ROP)
> - *I*ntraventricular hemorrhage

15.3.1.2 Subarachnoidal Hemorrhage (SAH)

SAH is etiologically more homogeneous, being caused in the majority of cases by trauma or by rupture of a berry aneurysm. Clinically, there are headache, nuchal rigidity, a very broad range of alterations of the mental status, and hydrocephalus.

15.3.1.3 Subdural Hemorrhage (SDH)

SDH usually results from rupture of the superior cortical veins leading to the superior sagittal sinus (venous blood).

15.3.1.4 Epidural Hemorrhage (EDH)

EDH usually results from tearing of the middle meningeal artery (arterial blood).

The most important cerebrovascular accidents include atherosclerosis, embolism, intracerebral hemorrhage, and SAH. CNS, and particularly the brain, has an absolute requirement for oxygen. In fact, an arrest of blood flow to the brain may have devastating consequences, although some factors (e.g., body and external temperature) may influence brain damage. Thoresen pioneer works have helped to identify how to preserve the preterm brain using cooling devices. Blood flow arrest and brain function are paralleled as below.

→	11 s	Unconsciousness
→	40 s	Polygraphy with flattening of electroencephalogram (EEG)
→	180 s (3 min)	Irreversible damage, because of the pauperization of glucose
→	5–7 min	Total consumption of tissue adenosine triphosphate (ATP)
→	>7 min	Death

Sensitivity to ischemia is different in several and different cell types, and different regions of the brain show also different sensitivities to global ischemia. Ischemia-related brain sensitivity areas include the cerebellum (Purkinje cells), the hippocampus (CA1 + CA4 > CA2 + CA3), the inferior olivary neurons, the neocortex, layer 3 (external pyramidal layer), and the subthalamic nucleus. Layer 3 and layer 5 represent the two layers of the neocortex that show pyramidal neurons, while stellate cells are seen in layer 4 of the neocortex. The anoxic-ischemic injury may be evidenced on tissue sections by histologic examination. Neuronal anoxic-ischemic injury changes include:

1. Acidophilic (eosinophilic) degeneration
2. Glassy change of the cytoplasm
3. Hyperchromasia of nuclei
4. Loss of intracellular Nissl substance
5. Neuronal (cytologic) shrinkage (↓)
6. Pericellular (perineuronal) space ↑

An acute neuronal injury displays shrinkage of the cell body, pyknosis, nucleolus disappearance, vanishing of the Nissl substance, and cytoplasmic eosinophilia, while an axonal reaction consists of rounding of the cell body with central chromatolysis (loss of central Nissl bodies) and retraction of presynaptic terminals. Factors that impair regeneration of the axons include lack of matrix proteins (e.g., laminin and fibronectin), lack of growth factors (e.g., GAP 43, Growth Associated Protein 43), the formation of inhibitory proteins (e.g., oligodendroglial glycoproteins), and the perinatal and postnatal creation of glial scars. The regions of elective parenchymal necrosis include Purkinje cells (cerebellar cortex), pyramidal cells of the cortex, hippocampus, striatal neurons, and thalamic neurons, and significant causes of elective parenchymal necrosis include anoxia/hypoxia (e.g., cardiac arrest), anemia, CO intoxication, pulmonary disease, and hypoglycemia. Consequences of infarction in the brain are different from any other organs, because of the almost unique cellular and tissue organization. Infarct may be anemic or hemorrhagic, depending upon whether no blood flow is present or circulation is reestablished through either internal carotid arteries or the vertebrobasilar system at the level of the *circulus cerebri* (*circulus Willisii*). The cavitation process is progressive with a speed of 1 cm³/3 months.

Types of Brain Herniation
- *R*etrograde cerebellar herniation (infratentorial → supratentorial)
- *U*ncus or orthograde herniation (supratentorial – infratentorial)
- *S*ubfalcial (cingulate) herniation involving the R&L supratentorial cavities
- *T*onsillar herniation (infratentorial → spinal cord)

The mnemonic acronym "RUST" may also be used (rust is an iron oxide).

A "watershed infarction" of the brain is defined as an infarct of the brain in very specific areas, e.g., between anterior and middle cerebral arterial circulation. It means that there is a susceptibility of endoarterial vascular supply (i.e., tissue will be damaged at the boundaries of limited blood flow as seen in the large bowel in the ileocolic region and splenic flexure). Thrombosis of venous sinuses and their branches causes congestion, bleeding, and necrosis of brain tissue with obvious disastrous consequences. Thrombosis of the superior sagittal sinus causes parasagittal venous infarcts. The causes of venous thrombosis are varied and include inherited thrombophilia, use of oral contraceptives, cancer, and dehydration (mostly in infants). The etiopathogenesis and characteristics of venous infarcts include:

1. *Thrombosis of sinuses of dura mater or cerebral blood vessels* (e.g., in case of infection or dysregulation of the Virchow's triad)
2. *Hemorrhage*, bilateral with white matter over gray matter contribution
3. *Superior sagittal sinus thrombosis*, which induces a parasagittal hemorrhagic infarction

The hypertensive cerebrovascular disease can predominantly manifest under two critical forms (multi-infarction-related disease and Binswanger disease) with three anatomic-pathologic aspects, including:

1. *Lacunar Infarcts* (small, usually ≤1 cm, a subcortical gray matter of basal ganglia, brainstem, and deep white matter, HTN-associated, multiple and bilateral)
2. *Slit Hemorrhages* (bleeding of the subcortical area)
3. *Hypertensive Encephalopathy* (cerebral atherosclerosis in the setting of generalized atherosclerosis, cerebral arteriolosclerosis in the context of chronic HTN or DM, and thrombotic states or post-embolization from carotid arteries or directly from the heart)

In more than 60% of cases, hypertensive hemorrhage occurs in the basal ganglia, although other familiar locations are thalamus, pons, and midbrain.

Multi-infarct disease is characterized by mainly microinfarcts and lacunar infarcts (*vide infra*) of cortex and white matter, while Binswanger disease is a subcortical arteriosclerotic encephalopathy, which is a form of hypertension-based encephalopathy involving broad areas of subcortical basal ganglia and white matter with both myelin and axon loss. Although rare, such cerebrovascular disease may have an encephaloclastic component or be the consequence of a perinatal insult. There are different types of intracranial hemorrhage:

1. Intracerebral (intraparenchymal) hemorrhage (hypertension, Charcot-Bouchard microaneurysms, cerebral amyloid angiopathy)
2. Subarachnoidal hemorrhage (Berry aneurysm rupture, AV malformations, traumas, and stroke)

Charcot-Bouchard aneurysms are microaneurysms of the cerebral vascularity, particularly of the small blood vessels (Ø < 300 μm) and often located in the basal ganglia (lenticular-striate) and associated with chronic hypertensive states of the patient, and represent a common etiology of stroke. A specific pitfall in examining a brain is "fat embolism syndrome," which shows white matter petechial hemorrhages. Another aspect to ponder is the location of typical "contrecoup lesions," which are typically observed in the inferior frontal lobes and anterior temporal lobes. A common mnemonic word for causes of SAH is "BATS" (*B*erry aneurysm, *A*V malformation/ADPKD, *T*rauma, and *S*troke). Aneurysms may be of mycotic (septic) nature or arteriosclerotic origin, other than "berry" aneurysms. CADASIL (a cerebral AD-inherited arteriopathic disease with subcortical infarcts and leukoencephalopathy) is a progressive neurodegenerative condition, associated with mutations in the *NOTCH3* gene. Patients experience migraine with aura, mood disorders, gradual cognitive decline, subcortical ischemic strokes, dementia, and premature death. CADASIL is rare in childhood, but not absent. MRI reveals subcortical foci of increased T2 hyperintensity and needs to be kept in mind in case of deficits in several pathological aspects with regard to executive functioning, and verbal learning difficulties (Box 15.9).

Duret hemorrhage refers to a secondary brainstem hemorrhage generally due to an acute uncal herniation for a supratentorial mass resulting in punctate hemorrhages throughout the tectum of the pons in consequence of stretching of perforating arterioles and subsequent ischemia. Grossly, there are small-sized areas of bleeding in the ventral and paramedian part of the upper brainstem, involving the midbrain and pons.

Box 15.9 CADASIL
- Cerebral *AD a*rteriopathy characterized by *s*ubcortical *i*nfarcts and *l*eukoencephalopathy
- *NOTCH3* gene mutation on 19q ⇒ abnormal folding of the ectodomain
- Childhood- and young-onset dementia (± pseudobulbar palsy, ± hemorrhages)
- CARASIL: *R*ecessive inheritance of CADASIL
- Concentric thickening of media and adventitia with deposition of granular material (basophilic, PAS+, and osmiophilic)
- Diagnosis: biopsy of the skin/skeletal muscle AND/OR blood test (molecular genetic testing)

15.3.2 Vascular Malformations

Vascular malformations include chiefly arteriovenous malformations (AVM), telangiectasias, venous angioma, and cavernous angioma/malformations. The cavernous sinus is an important structure that receives blood through the ophthalmic artery using the superior orbital fissure and from superficial cortical veins and drains into the basilar plexus of veins posteriorly. Each cavernous sinus contains vertically, from superior to inferior the oculomotor nerve (3rd cranial nerve, CN III), the trochlear nerve (4th cranial nerve, CN IV), the ophthalmic branch (V_1) of the trigeminal nerve (5th cranial nerve, CN V), and the maxillary branch (V_2) of the ipsilateral trigeminal nerve (CN V). Horizontally, from medial to lateral, three structures are present, and they include the internal carotid artery (and sympathetic plexus). The topography mnemonic acronym is OTOM CAT, i.e., O, oculomotor nerve; T, trochlear nerve; O, ophthalmic branch; M, maxillary branch; C, internal carotid artery; A, abducent nerve; and T, trochlear nerve.

15.3.2.1 Arteriovenous Malformations (AVM)
- *DEF*: Tangle of large, blood-filled, and thin-walled vessels representing the most significant aberration of angiogenesis and located mostly along the distribution of the middle cerebral artery. It has been suggested that an *intravitam* prerequisite for the diagnosis is the presence of congenital, low-resistance AV shunts that siphon blood from the adjacent parenchyma (Fig. 15.5).
- *SYN*: Vascular pseudotumor
- *EPI*: Worldwide, 1.5% of intracranial "tumors."
- *CLI*: Seizures, neurologic deficits, and chronic mass effects.
- *GRO*: Tangles of abnormally developed arteries and veins without intervening capillaries.
- *CLM*: Irregular thickening of the blood vessel wall with the elastin stain (Elastica van Gieson staining) highlighting the duplication (or even triplication) of the internal elastic lamina. Such pattern is virtually always accompanied by areas where the elastic lamina is wholly missing.

15.3 Vascular Disorders of the Central Nervous System

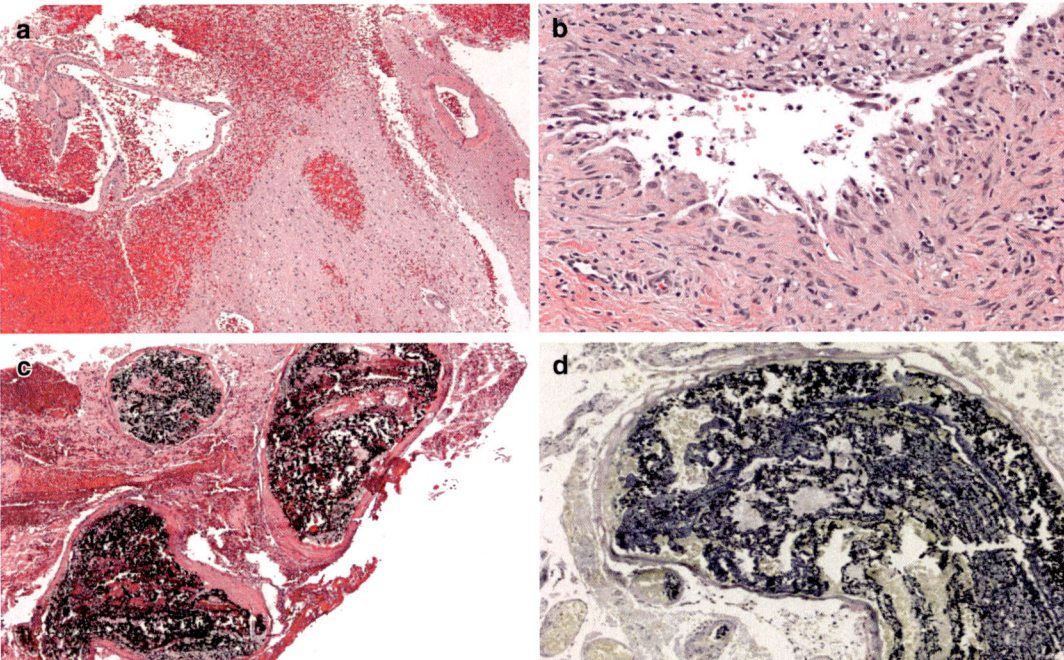

Fig. 15.5 An arteriovenous malformation (AVM) is shown in this panel (**a**–**d**) with tangles of abnormally developed arteries and veins with intimal hyperplasia, without intervening capillaries and surrounding hemorrhage in the brain parenchyma (**a**), granulation tissue in areas with repair changes (**b**), and some of the attempted therapy modalities (**c**, **d**). Different treatments may be proposed that are based on size, location and patient age, including "watch and see" with periodic observation, embolization, sclerotherapy, radiofrequency ablation, and even wide surgical excision (**a**, x50; **b**, x200; **c**, x50; **d**, x100 as original magnifications).

- *DDX*: Aneurysm, hemangioma, angiosarcoma, Kaposi sarcoma.
- *TRT*: Surgical resection if it is feasible.
- *PGN*: It depends on the mass effect and of the vascular connection and is exceptionally anatomically dependent. Apparent tendency to enlarge and eventually rupture, which may cause combined IPH and SAH in about 2/3 of cases, SAH in about 1/4, and IPH in the remaining patients.

Differential telangiectasias, venous hemangioma, and cavernous hemangioma need to be considered as vascular pseudotumors and, of course, benign tumors.

15.3.3 Aneurysms

Aneurysms may be subdivided into berry aneurysms, mycotic aneurysms, and arteriosclerotic aneurysms.

15.3.3.1 Berry Aneurysms

Berry aneurysms are small saccular aneurysms arising from a structural weakness of the branch point (blood vessel wall defect ≠ AVM pathology) of a massive cerebral artery usually participating to the constitution of the circle of Willis. The most frequent localizations are the following:

1. Anterior communicating artery (30%)
2. Posterior communicating artery (25%)
3. Middle cerebral artery at the bifurcation site (20%)

- *EPI*: 5% of the healthy population but 1/4–1/5 of these individuals have multiple berry aneurysms.

If multiple berry aneurysms are seen, then think of:

1. ADPKD
2. Ehler-Danlos syndrome IV

3. Neurofibromatosis type I
4. Marfan syndrome
5. Fibromuscular dysplasia
6. Coarctatio aortae

Multiple berry aneurysms are seen in 1/5 of cases. The underlying defect is a media defect in the wall of the blood vessel at the bifurcation of an artery with degeneration of the internal elastic lamina. The critical radius of an aneurysm is determined by the differential expression of the volume distensibility evaluated at 90° and considering the elastic modulus of an aneurysm (elastin and collagen), the wall thickness, and the systolic pressure. Estimated size of a berry aneurysm between 4 and more than 10 mm (1 cm: 0.393701 inches) seems the most accepted in the literature to predict a rupture. Five methods have been described to determine this risk. The first method involves a direct approach addressing the follow-up of an aneurysm by angiography. An indirect method evaluates a previous ruptured aneurysm comparing with the unruptured one, while the third method is based on epidemiological data. The fourth method uses a decision analysis root using stochastic processes, and the final one applies a biomathematical model based on Laplace's law.

- *PGN*: Only $12/10^5$ aneurysms rupture per year and mortality is ranked 30% within the first 24 h and doubles within the early 30 days.

15.3.3.2 Mycotic Aneurysms
Mycotic aneurysms are common in subacute bacterial endocarditis. There is a weakening of the walls of large or small arteries by infected emboli and may cause single massive or multiple petechial hemorrhages anywhere in the brain. Other fates of septic aneurysms include subarachnoidal space invasion and intraparenchymal extension with the formation of an abscess.

15.3.3.3 Arteriosclerotic Aneurysms
Fusiform type of aneurysms that are usually located on the internal carotid arteries or vertebrobasilar arterial system.

15.3.4 Thrombosis of Venous Sinuses and Cerebral Veins

Cerebral venous disorders, in the form of thrombosis, target poorly nourished, children affected with severe anemia, young individuals with severe acute or chronic infections, and postpartum women. These thromboses reflect either venous wall defects or abnormalities of the coagulation, and the most affected sites are the superior sagittal sinus and superior cerebral veins. Grossly, there is distension of veins, which are quite firm. Accompanying features are hemorrhages and marked congestion of the leptomeninges and cerebrum. Microscopically, two common elements are multiple pericapillary hemorrhages and encephalomalacia, characterized by coagulation necrosis, microglial infiltration, astrocytic proliferation, and ultimate cyst formation.

15.3.5 Pediatric and Inherited Neurovascular Diseases

Congenital vascular lesions of the blood vessels of the CNS account for 20–25% of CNS vascular malformations in children. Some diseases have been found in youth as well and show familiarity. These diseases include cavernous cerebral malformations (CCM) and hereditary hemorrhagic telangiectasia mainly.

15.3.5.1 Cerebral Cavernous Malformations (CCM)
- *DEF*: Cerebrovascular lesions predisposing to chronic headaches, epilepsy, and hemorrhagic stroke and inherited forms have been associated with four genes. Apart from the sporadically acquired, the inherited CCM are mostly caused by mutations in *KRIT1* (*CCM1*) or in *Malcavernin* (*CCM2*) which is the murine ortholog. KRIT1 is a protein that is involved in regulating cell adhesion and migration via its interaction with β-1 integrin. The familiar form is inherited as AD pattern with incomplete penetrance.

- *EPI*: 0.5% or 5:1000 or 1:20 individual of the general population and up to 20% of cerebrovascular malformations in all age groups. Up to 3:4 of multiple cavernous malformations are familiar.
- *IMG*: Gradient-echo MRI, which is the most sensitive technique to detect these lesions, although T2WI MRI is also a proper technique.
- *CLI*: Seizures (~50%) and cerebral hemorrhage (~40%) are the most common symptoms followed by focal sensory and motor neurologic deficits, visual field deficits, and nonmembranous headache.
- *GRO*: "Raspberry"-like collections of sinusoidal-shaped vascular structures without intervening brain parenchyma.
- *CLM*: Sinusoidal vascular channels with moderate irregularity of the wall and lined by a single layer of endothelium (CD31+, D2-40−).
- *TRT*: Microsurgery or surgery.
- *PGN*: In the familiar form, aggressive behavior is found in patients of Hispanic origin. Cavernous malformations in childhood occur four times less often than in adulthood but have a more aggressive growth pattern and clinical behavior.

15.3.5.2 Hereditary Hemorrhagic Telangiectasia (HHT)

AVM of several solid organ and telangiectasis of the mucosal membranes and dermis, with an AD inheritance with high penetrance and variable expressivity and two genes including endoglin gene (*ENG*) causing HHT1 and activin receptor-like kinase 1 (*ALK1* on 12q13.13) causing HHT2. Both ENG and ALK1 are associated with the regulation of the TGF-β family. ALK-1 regulates the angiogenesis positively, and ENG increases the presence of ALK1 in endothelial cells determining that mutations affecting either ENG or ALK1 will possess a similar phenotype. Diagnosis is based on 3/4 fulfilled criteria, including spontaneous, recurrent epistaxis, mucocutaneous telangiectasia, AVMs of internal organs, usually solid organs (CNS, lung, liver, and GI tract), and 1st degree relative with HHT.

- *EPI*: 1–2:10,000, no ethnic or geographic differences.
- *CLI*: Epistaxis, telangiectasias, stroke, TIAs (Transient Ischemic Attacks), and extraneurologic bleeding.
- *IMG*: CNS with brain/spine MRI with or without gadolinium.
- *TRT*: Surgery or stereotactic radiosurgery.
- *PGN*: It depends on the extension of the disease.

PHACE(S) syndrome is a neurocutaneous disease with strong female preponderance (phacomatosis) including malformations of the posterior fossa (from cerebellar hypoplasia to the DWS), facial hemangiomas (not restricted to cutaneous distribution of the trigeminal nerve), arterial and cerebrovascular abnormalities, cardiovascular defects, and eye abnormalities and with or without sternal cleft or supraumbilical raphe. Up to 4/5 of PHACE(S) patients present with posterior fossa abnormalities, which tend to occur ipsilateral to the facial hemangioma. Congenital vascular defects may occur in PHACES which include an aberrant origin or course of primary cerebrovascular structures, agenesis of the arteries, saccular aneurysms, arterial dysplasia, and persistence of fetal anastomoses. Progressive vasculopathy may progress to moyamoya-like progressive vasculopathy with resultant ischemic strokes. In PHACES have also been described some lesions, such as pachygyria, polymicrogyria, cortical thickening, heterotopic gray matter, and cerebral volume loss.

15.4 Infections of the CNS

Infection can get the brain through essentially four ways:

1. Direct extension from neighboring structures (e.g., acute otitis media)
2. Retrograde extension from other neural structures (e.g., otic neuritis)
3. Hematogenous dissemination (e.g., sepsis)
4. Inoculation following trauma or iatrogenic (e.g., skull basis fracture)

15.4.1 Suppurative Infections

Suppurative infections of the CNS include acute suppurative leptomeningitis, brain abscess, and septic thrombophlebitis (Figs. 15.6 and 15.7).

15.4.1.1 Acute Suppurative (Lepto-) Meningitis

- *DEF*: It is a principal diagnosis that needs to be addressed quite quickly to avoid life-threatening conditions and death. In children and youth, a suppurative CNS infection mainly occurs secondarily to diseases of the middle ear such as otitis media involving the related cavities of the middle ear, diseases of the nasal sinuses and mastoid cell cavities, and diseases of the throat and thoracic organs. A liquor fistula or a trauma involving the cribroid lamina of the ethmoid bone may also be an underlying condition evolving to a suppurative CNS infection. These last conditions may generally be associated with cranial traumas or traumas to the spinal cord. Rarely, CNS infection of suppurative type may be due to the hematogenous spread of disease elsewhere.

Leptomeningitis is defined as acute suppurative meningitis and is the most frequent pyogenic infection of the CNS and, although it may occur at any age, is most commonly observed in childhood. The etiology is different according to the age of children. Group B streptococci, *E. coli*, and *Listeria monocytogenes* are mostly seen in newborns and infants, while meningococci, pneumococci, streptococci, and *Hemophilus* spp. are more often seen in older children and young adults. Staphylococci and Gram-negative bacteria are also seen in individuals with immunodeficiency, either primary or secondary in type. The mechanism of invasion of the meninges is direct from an infected sinus or ear or indirect in a setting of sepsis. This last condition is observed in children hospitalized in PICU (pediatric intensive care unit) of Children's Hospitals.

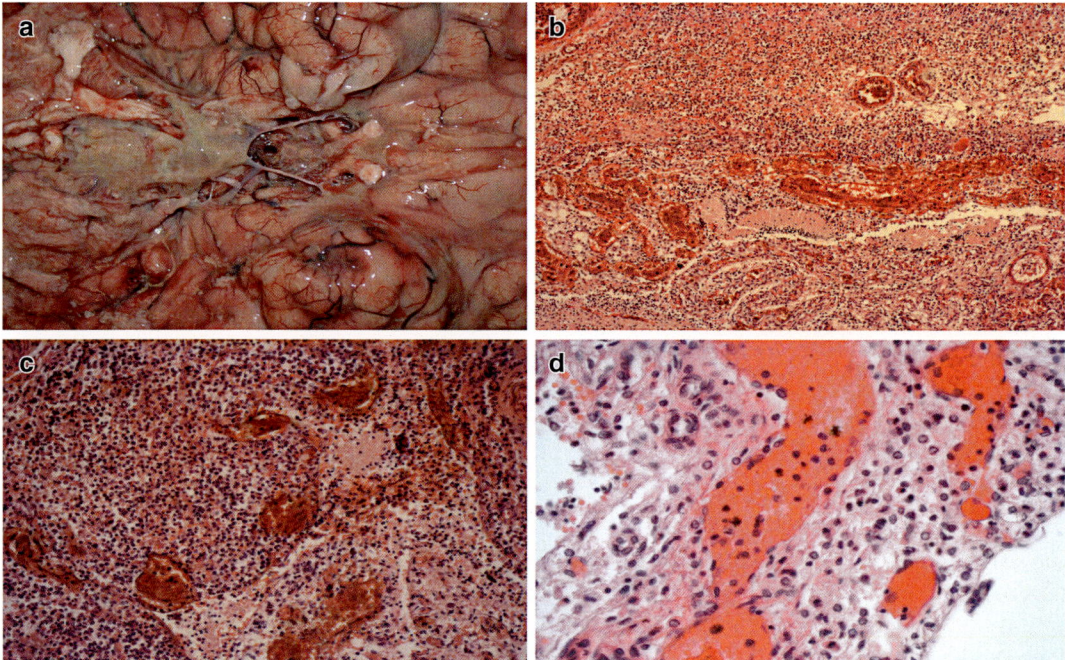

Fig. 15.6 Meningoencephalitis with purulent exudate at the basis of the brain (**a**) and microphotographs (**b–d**) showing an infiltration of acute inflammatory cells in the leptomeninx and in the blood vessels (**b–d**, hematoxylin and eosin staining; **b**, ×50; **c**, ×100; **d**, ×200)

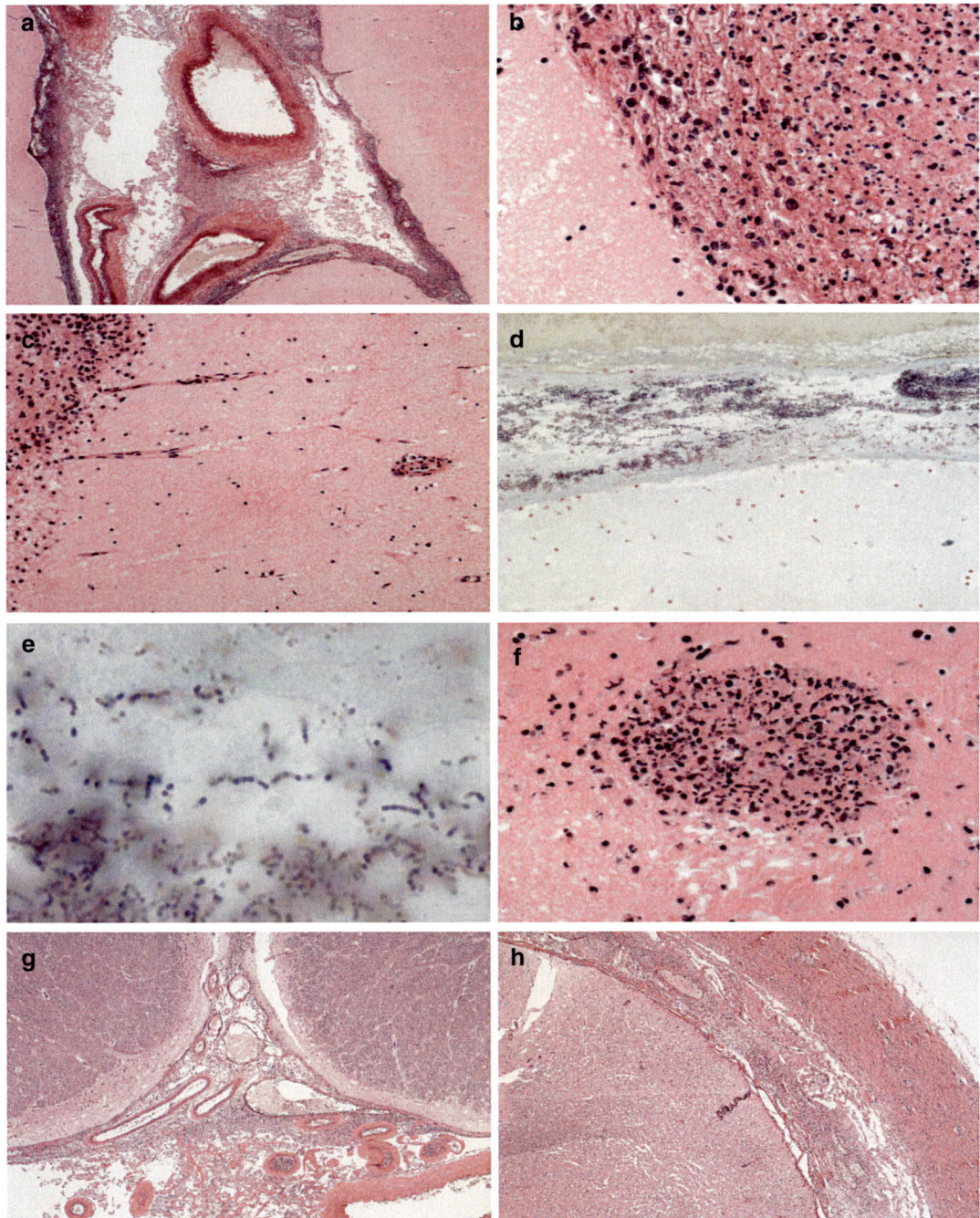

Fig. 15.7 This panel illustrates the presence of bacteria (purulent) (**d**, **e**) leptomeningitis in several areas of the central nervous system (**b**, **c**, **h**) with inflammatory cells filling up the perivascular regions (**a**, **g**) and creating microabscesses (**f**) (**a–c** and **f–h**, H&E staining, **d–e**, Gram staining; **a**, ×20; **b**, ×200; **c**, ×100, **d**, ×200; **e**, ×600; **f**, ×200, **g**, ×20, **h**, ×20 as original magnifications)

- *CLI*: There is a unique picture of a headache, stiff neck, fever, increased intracranial pressure, and vomiting. Seizures and motor disabilities may complicate this clinical picture and are indicative of an irritation slightly spreading to the cortex. Lumbar puncture shows a clouded CSF with neutrophilia, hyperprotidorrachia, and hypoglycorrhachia. Microbiological investigations may show pathogenic organisms. Symptoms and signs of a rapidly expanding intracranial lesion may add an altered mental status and more significant sensorimotor deficits to the above-described clinical picture.
- *GRO*: The exudate is seen in the subarachnoid space, overlying the vault of the brain, more than the basis, and spinal cord.
- *CLM*: There are neutrophils, fibrin, and bacteria in the subarachnoid space, and the underlying brain and cord show edema and congestion. A neutrophilic infiltration of the brain and spinal cord characterizes the direct involvement and the designation as meningoencephalitis would be appropriate in this context. Lymphocytes and monocytes/macrophages will quickly populate the neutrophilic granulocytic rich exudate in 24 hours. The activity of macrophages is crucial if the patient recovers. In fact, a complete resolution of the inflammatory process (*restitutio ad integrum*) is standard, but temporary or permanent disabilities may appear and include localized hemi- or paraplegy, speech defect, or mental retardation in childhood. Complications include septicemia and hydrocephalus, which is due to an obstruction of the CSF outflow from the ventricular system when low-grade infection persists followed by progressive fibrosis and sclerosis with narrowing or localized obliteration of the subarachnoid space.
- *PGN*: Complications include communicating hydrocephalus, sepsis and septic shock, multiorgan failure (MOF) (>3 organ/systems), and death. The occurrence of communicating hydrocephalus following meningitis is due to the process of fibrosis, which may be a natural consequence of an inflammatory process, mainly if the inflammation is severe. Fibrosis blocks CSF resorption through the Pacchioni arachnoidal granulations.

15.4.1.2 Brain Abscess

- *DEF*: Pseudocystic cavity arising in necrosis results from encephalitis, often 1–2 weeks following infarction by direct implantation, local extension, or hematogenous spread.
- *EPI*: Brain abscesses represent 1/5 of the incidence of the suppurative leptomeningitis, and 4/5 of the cases are associated with extramural pyogenic infections, anatomic anomaly, penetrating cranial trauma, and previous neurosurgery.
- *ETP*: Common bacteria include staphylococci, pneumococci, and streptococci. Gram-negative bacilli, including aerobic microorganisms such as *Proteus* spp., *E. coli*, *Klebsiella* spp., *Haemophilus* spp., and *Bacteroides* are more rarely seen. *Actinomyces israeli* and *Nocardia* spp. are usually observed in debilitated and immunosuppressed patients. Sources for the development of a brain abscess include otitis media, mastoiditis, sinusitis of frontal location, lung abscess, pleural empyema, and bacterial endocarditis. A direct extension to cephalic structures or a spreading as infected emboli in the bloodstream is the pathogenesis. If most multiple abscesses are seen, embolic pathogenesis is most likely. Pediatric acute mastoiditis differs significantly between age groups. Typically, acute mastoiditis is most common in children younger than two years of age. The infants show more rapid progress of symptoms and more distinct signs of acute inflammation of the mastoid bone. In combination with the virulence of the bacteria the anatomy of the air cell system of the mastoid bone including its tympano-mastoid suture and vascular channels in the cribriform area play a major role for determining a higher rate of acute mastoiditis in infants than older children (Groth et al. 2012).

Hematogenous-driven brain abscess includes acute bacterial endocarditis, congenital heart disease with R → L shunt, chronic pneumonia, sepsis, and immunodeficiency/immunosuppression (e.g., primary immunodeficiency, transplantation).

- *CLI*: There is often a clinical picture of mass effect ("*tumor quia tumet*") due to a rapidly expanding intracranial lesion possibly associated with infection-related symptoms and signs with stereotyped topographic presentations. History is essential, and the source of infection (e.g., otitis media) remains key in the era of precision-medicine of the 21st century. Lumbar puncture shows an increased pressure of CSF, which also indicates inflammatory cells and hyperprotidorrachia, but glycorrhachia is within the normal range. Generally, no pathogenic microorganisms are easily identified, unless the brain abscess has ruptured and leptomeningitis ensues.
- *IMG*: X-ray characteristics are
  1. Ring enhancement due to a developing pseudo-capsule
  2. "Budding" of "daughter" lesions
  3. Hypodensity of adjacent white matter, which underlies the surrounding edema
- *GRO*: Typically located in the temporal lobe and cerebellum when observed as a single lesion or in the brain located at the intersection of the gray and white matter when multiple abscesses occur, the brain abscess is a pseudocyst without an epithelial lining. It is characterized as a cavity with filled with putrid tissue containing a thick exudate surrounded by a narrow marginal band of whitish (granulation tissue) or intensely congested reddish brain tissue. The surrounding brain tissue displays marked swollen white matter with edema.
- *CLM*: There is necrotic tissue centrally, which is separated from neighboring white matter by granulation tissue constituted by an angioblastic and fibroblastic zone of activity with accompanying gliosis. The absence of an epithelial lining is conformed by the negativity of the immunohistochemical investigation with antibodies against keratins (e.g., AE1/3).
- *PGN*: Complications depend on the virulence of the microorganism, patient's immunocompetency, and resistance or compactness of the abscess wall. In case of high virulent bacteria, immunodeficiency or immunosuppression, or inadequate development of the abscess wall or its early fragility, a spreading suppurative leptomeningitis and sepsis may ensue.

15.4.1.3 Septic Thrombophlebitis

Septic thrombosis/thrombophlebitis is a condition occurring in patients affected with middle ear disease or mastoiditis, localized osteomyelitis, and epidural abscess and develops in transverse sinuses. Septic thrombosis/thrombophlebitis may also occur in the cavernous sinus near to the sella turcica as drainage location following suppurative infections of the face, particularly of the upper lip, with the retrograde involvement of the angular and ophthalmic veins.

15.4.2 Tuberculous (Lepto-) Meningitis

Tuberculous leptomeningitis is a rare complication in immunocompetent individuals but is the most common form of TB infection in the nervous system in immunodeficient or immunosuppressed patients. Two pathogenetic ways are observed, including either miliary dissemination (particularly immunodeficient or immunosuppressed patients) or secondary to infection in the mediastinal or mesenteric glands, bones, joints, lungs, or GU tract. Clinically, there is an insidious onset with 2–3 weeks of anorexia, weight loss, and change of disposition. A specific lucid interval precedes a period of drowsiness with or without delirium. Lumbar puncture shows chronic inflammatory cells (lymphocytes), hyperprotidorrachia, hypoglycorrhachia, and acid-fast bacilli, which may be detected using either staining methods (Ziehl-Nielsen stain, auramine-rhodamine stain) or microbiological investigations. The auramine-rhodamine staining or Truant stain is a histological technique used to detect acid-fast bacilli (genus: *Mycobacterium*) using fluorescence microscopy. It is a mixture of auramine O and rhodamine B. Caution should be taken when using this mixture because of its carcinogenicity. Grossly, a white or gray-whitish, lacy exudate is mainly seen at the base of the brain, conversely from the acute suppurative leptomeningitis (*vide supra*) and, particularly, in all basilar cisterns as well as in the Sylvian fissure. The exudate is marginalized by sharply outlined, round white nodules ("tubercles"). Microscopically, there is a granulomatous exudate with or without caseous necrosis and numerous lymphocytes, mononuclear cells, and single multinucleated giant cells of Langhans type. Special stains, including Ziehl-Nielsen and auramine-rhodamine (*vide supra*), allow highlighting the acid-fast bacilli. It is advisable to take, at time of neurosurgery or autopsy, some tissue for microbiological investigations. Both classic special stains and microbiology may show no bacilli, owing to previous therapy or the presence of atypical mycobacteria or artifact procedures. In case of the second event, Wade-Fite stain may be useful to detect atypical mycobacteria. The special stain "AFB" uses carbol-fuchsin to stain the lipid walls of acid-fast organisms that are present in *M. tuberculosis*. As indicated above, the most commonly used method is the Ziehl-Neelsen method, but there is also the Kinyoun method. The special stain, known as Wade-Fite or, simply, Fite stain, has a weaker acid for supposedly more fragile atypical mycobacteria, such as *M. leprae* among others. Atypical mycobacteria are much less acid- and alcohol-fast as compared to *Mycobacterium tuberculosis*. The mycolic acid coat of atypical mycobacteria is less strong than *Mycobacterium tuberculosis*. Thus, the mycolic acid coat is quickly decolorized by the standard Ziehl-Neelsen technique. In the Fite stain, peanut oil is used with the deparaffinizing solvent (xylene) to minimize the exposure of the bacterial cell wall to organic solvents allowing the preservation of the acid-fastness property. ZN and Fite stains are usually frustrating for the pathologist, because the search for acid-fast bacilli may take hours. This aspect is because much of the lipid in mycobacteria is removed by tissue processing, which causes extensive searches. Overall, the most sensitive stain for mycobacteria is doubtless the auramine-rhodamine stain, which requires a fluorescence microscope for viewing. Essential caveats for the pathologist are false-positive results that may occur in infection or infestations with cryptosporidium, isospora, and cysticerci (hooklets). The role of next-generation sequencing is being investigated to deliver a fast diagnosis and also to rule out laboratory-derived contamination.

15.4.3 Neurosyphilis

Neurosyphilis represents the tertiary stage of an infection due to *Treponema pallidum*. It develops in approximately 2% of syphilis-infected patients, and two forms are specifically delineated that are not mutually exclusive. Neurosyphilis is subdivided in meningovascular and parenchymatous neurosyphilis. Cerebral cortex involvement (e.g., impaired intellectual efficiency, memory, and judgment as well as delusions) with bedriddenness and incontinence.

- *CSF*: Hypercellularity (neurosyphilis-induced pleocytosis) (≥20 cells/µL) and hyperprotidorrachia with positive serology (reactive CSF VDRL and/or a positive CSF intrathecal *T. pallidum* antibody index).
- *GRO/CLM*: Mostly frontally located, atrophy of cerebral gyri showing mild to marked lymphocytic and plasmacellular leptomeningeal infiltration and gradual capillary hypoxia-derived degeneration of neurons.

15.4.4 Viral Infections

Meningoencephalitis is a combining clinical syndrome with pathologic correlate, which is often characteristic for virus infection. It targets both meningeal and neural areas, but the route of infection is heterogenous varying from gastrointestinal (poliovirus), an arthropods-born insect bite (equine encephalitis), mucocutaneous pathway (HSV), ascending neural propagation (rabies) to hematogenous spread. Grossly, edema may be present, but an exudate is usually lacking if no suppurative component is concomitantly associated. Microscopically, there is a typically perivascular lymphocytic/mononuclear cuffing associated with neuronal degeneration and associated gliosis with accompanying intracellular (intranuclear/intracytoplasmic) inclusion bodies in both neural and glial cells.

15.4.4.1 Poliomyelitis

The etiologic agent is poliovirus (neurotropic RNA virus, sewage-contaminated water, ± GI symptoms) that induces 2nd motor neuron lesional signs with caudocranial damage to the anterior horn cells of the spinal cord. The pathologic process is characterized by initial hemorrhage and congestion and subsequent neutrophilic infiltration, degeneration, and gliosis. The clinical picture is sequential: Infection ⇒ lower limbs to brainstem ⇒ paraplegia to respiratory paralysis. Clinically, an upper motor neuron disease manifests with more muscle reflexes (hyperreflexia), more muscle contraction (spasticity), more muscle tone (hypertonicity), and disuse (atrophy), while a lower motor neuron disease manifests with less muscle reflexes (hyporeflexia), less muscle contraction (flaccidity), less muscle tone (hypotonicity), and loss of innervation (devervation atrophy). The Babinski sign shows toes pointing up in upper motor neuron diseases, while it shows toes pointing down in lower motor neuron diseases.

15.4.4.2 Rabies

Rabies: is caused by *Lyssaviruses*, including the rabies virus and Australian bat *Lyssavirus*. The Rhabdovirus is an RNA virus. Rabies starts with bites from wild animals and dogs, and affected individuals experience restlessness and hydrophobia. An early death is the consequence of 1st and 2nd motor neuronal damage throughout the nervous system with or without subsequent gliosis depending from the course of disease. The Negri bodies are pathognomonic inclusions of Rabies infection. They are eosinophilic, sharp intracytoplasmic bodies (Ø: 2–10 μm) found in neurons, particularly pyramidal cells within Ammon's horn (hippocampus). Also, they can be found in Purkinje cells of the cerebellar cortex. Negri bodies are ribonuclear proteins produced by the Rhabdovirus.

15.4.4.3 Herpes Simplex Encephalitis

Herpes simplex encephalitis is an acute necrotizing hemorrhagic inflammation of the brain due to HSV. Fever, lethargy, and coma are due to necrosis and softening of the brain parenchyma (typically frontal and temporal lobes) grossly and neuronal destruction and perivasculitis (Cowdry A inclusions and perivascular lymphocytic infiltrates) initially, and later gliosis microscopically. Complications include sepsis and MOF. Herpes zoster is, conversely, associated with a unilateral vasculopathy (granulomatous arteritis) causing contralateral hemiparesis. There is some confusion about Cowdry A and B inclusions. Cowdry A inclusions are eosinophilic amorphous intranuclear inclusions surrounded by a clear halo with marginalization of chromatin on the nuclear membrane and composed of nucleic acid and protein observed in cells infected with HSV, VZV, and CMV. Cowdry B inclusions may represent the initial form of type A and characterized by inclusions without chromatin marginalization and found in neurons infected with Poliovirus and cells infected with Adenovirus.

15.4.4.4 Subacute Sclerosing Panencephalitis (SSPE)

SSPE is defined as post-sequela of measles infection with the onset of involuntary muscular movements and dementia leading to death within 1–2 years and showing grossly multifocal areas of neuronal damage and microscopically viral inclusions, in oligodendrocytes and neurons, inflammation, myelin degeneration, and gliosis.

The viral inclusions in nerve cells may be located in the nucleus and/or in the cytoplasm:

- Nucleus (Cowdry type A)
    (i) Cytomegalovirus
    (ii) Herpes simplex virus (HSV) /herpes zoster virus
    (iii) Papova/JC (progressive multifocal leukoencephalopathy or PML)
    (iv) Paramyxoviridae (measles)
- Cytoplasm
    (v) Negri/lyssa bodies (rabies)

15.4.4.5 Progressive Multifocal Leukoencephalopathy (PML)

- *DEF*: Papovavirus (usually JC virus)-linked multifocal demyelination with oligodendroglia as target containing intranuclear inclusions and presenting with multiple sclerosis-like white

matter irregular areas of granularity and clinically as visual/motor deficit with or without dementia, particularly in the setting of an underlying disease or condition (*T*ransplant, TB, *I*mmunosuppression such as following natalizumab for multiple sclerosis or Crohn disease, *S*LE, and *H*ematological malignancy such as CLL → mnemonic word: "TISH", which the pronunciation of the word "table" in German ("Tisch"), but also "table" in Yiddish ("טיש") and reminds the gathering of Hasidim around their Rebbe. In such events, speeches on Torah topics and singing of melodies and hymns are accompanied by refreshments). Natalizumab is a humanized monoclonal antibody against the cell adhesion molecule α4-integrin.
- *CLM*: Oligodendrocytic demyelinization with ground-glass oligodendroglial nuclei (virion filling), large pleomorphic astrocytes, minimal or mild perivascular chronic inflammation, and minimal damage to neurons. JC virus is SV40+, because of immunohistochemical cross-reaction (BK virus, JC virus, and Simian Virus 40 are polyomaviruses).
- *TEM*: "spaghetti with meatballs"-virions.

15.4.4.6 Human Immunodeficiency Virus (HIV): Infection

Human immunodeficiency virus may damage the CNS both directly and indirectly (Box 15.10).

15.4.5 Toxoplasmosis

T. gondii infestation as a parasite (protozoa) is not infrequent and may occur either transplacentally or through dissemination from an asymptomatic infection. Congenital toxoplasmosis: hydrocephalus, bilateral chorioretinitis (focal), intracerebral calcifications, convulsions, and ocular palsies. Grossly, depressed, soft, sharply circumscribed yellowish areas on the surfaces of the cerebral hemispheres with typical stenosis of the aqueduct of Sylvius (mesencephalon) are found. Microscopically, sharply outlined necroinflammatory lesions of the brain parenchyma with associated secondary leptomeningitis and Toxoplasma cysts destroying both neural and glial structures are identified. PGN is variable from early death to severe deficits due to neural damage (e.g., chronic hydrocephalus, seizures, blindness, and mental retardation). An acquired toxoplasmosis may occur in immunodeficient hosts, particularly with AIDS.

15.4.6 Fungal Infections

Cryptococcal infection and coccidiomycosis play a major role. Actinomycosis is not a fungal infection, because the etiologic agent is *A. israeli* or other actinomycetes, which are bacteria and harbor the misnomer as mycetes.

15.5 Metabolic Disorders Affecting the CNS

Neurometabolic disorders have four characteristics, including neonatal presentation or metabolic crisis in older children (e.g., hypoglycemia), regression, and dysmorphic features. Neurometabolic disorders include disorders of carbohydrate, lipid, amino acid, and organic acid metabolism as well as mitochondrial disorders, lysosomal disorders, and peroxisomal disorders. Neonatal symptoms in case of intoxication include feeding problems, vomiting, lethargy, acidosis, hypotonia, seizures, and, potentially, coma. In the case of organic aciduria, urea cycle disorders need to be considered. Energy depletion neonatal symptoms include hypoglycemia, lethargy, hypothermia, jitteriness, and seizures. Lysosomal disorders of neurometabolic category include most commonly gangliosidosis, Niemann-Pick type A and B, metachromatic leu-

> **Box 15.10 HIV and CNS/PNS**
> 1. HIV-direct (early-onset dementia and dementia-like cerebropathies, vacuolar myelopathy, peripheral neuropathy)
> 2. Opportunistic infections (toxoplasmosis, tuberculosis, *Cytomegalovirus*, cryptococcosis, JC virus)
> 3. Neoplasms (e.g., B-non-Hodgkin malignant lymphoma)

kodystrophy, Gaucher's disease, and Krabbe's disease. Peroxisomal disorders most widely include X-linked adrenoleukodystrophy and Zellweger syndrome.

15.5.1 Pernicious Anemia

Pernicious anemia is *vitamin B12 deficiency-related megaloblastic anemia* with *subacute combined degeneration of the spinal cord* constituted by the combined involvement of posterior (Goll fascicles and Burdach fascicles) and lateral (spinocerebellar fascicles ± cortico-spinal fascicles) columns of the spinal cord. Subacute combined degeneration (SCD) is the most common metabolic myelopathy, caused by vitamin B12 (cobalamin) or folic acid deficiency. The disease has been associated with autoantibody production that is directed against the parietal cells of the stomach, lack of intrinsic factors, malabsorption (gastrointestinal disorders), malnutrition, extreme diet, an overdose of antacids, multiple genetic defects, and the use of N_2O anesthetic. The deficit of the posterior (Goll-Burdach tract) and lateral columns of the spinal cord is at the basis for the proprioceptive sensory loss and corticospinal symptoms. The classical symptoms include proprioceptive sensory loss and spinal ataxia, Lhermitte's sign, and spastic paraparesis or quadriparesis. Furthermore, autonomic dysfunction, paresthesia on the trunk, polyneuropathy, optic nerve lesion, psychosis, or early dementia/mental retardation may also develop. The peripheral signs of polyneuropathy may mask corticospinal tract involvement, which may make the diagnosis more difficult. Other accompanying signs are signs of megaloblastic anemia seen in the blood and the bone marrow. Cheilitis, glossitis, atrophic gastritis, achlorhydria, and later secondary stomach cancer may develop. If subacute combined degeneration is suspected, the following tests should be performed: blood tests (blood count, vitamin B12, folic acid, homocysteine and methylmalonic acid levels), Schilling test, and gastroscopy. McNeil et al. (2014) described an infant with vitamin B12 deficiency. The 7-months-old infant presented with feeding intolerance, poor growth, and developmental delay. The boy had macrocytic anemia, a markedly low serum vitamin B12 level, and elevated homocysteine and methylmalonic acid levels. Moreover, there were antibodies to intrinsic factor. Although rare, children presenting with failure to thrive and neurological symptoms should prompt the pediatrician to investigate for vitamin B12 deficiency.

15.5.2 Wernicke Encephalopathy

Thiamine (vitamin B1) deficiency causes Wernicke encephalopathy. Although it is often considered to be a disease of adult alcoholics, it occurs in childhood and nonalcoholic conditions. The global prevalence rates of hunger, poverty, and resultant nutrient deprivation have decreased in the twenty-first century, but Wernicke encephalopathy is still present in underdeveloped areas as well as in children and adolescents affected with malignancies, prolonged intensive care unit stays, and surgical procedures for the treatment of obesity (Cefalo et al. 2014; Lallas et al. 2014; Bhat et al. 2017). Other predisposing conditions include magnesium deficiency and defects in the *SLC19A3* gene causing thiamine transporter-2 deficiency. The classic triad consists of encephalopathy, oculomotor dysfunction, and gait ataxia, although it is rarely seen in all patients. MRI shows symmetric T2 hyperintensities in the dorsal medial thalamus, mammillary bodies, periaqueductal gray matter, and tectal plate. Grossly, marked congestion and multiple petechial hemorrhages of the periventricular gray matter, around the 3rd ventricle, the mesencephalic aqueduct of Sylvius (*aqueductus mesencephali cerebri, aqueductus Sylvii*), and the floor of the 4th ventricle are found. Microscopically, there are endothelial hypertrophy and hyperplasia of capillaries with acute necro-ischemic damage of neurons supplied by these affected blood vessels. Treatment should include vitamin replacement and symptomatologic or physiological promptness to salvage the organs. Wernicke encephalopathy is a medical emergency. Delay in its recognition and treatment leads to significant morbidity, irreversible neurological damage, or even death.

Brain edema may have several causes, including acute liver failure, diabetic ketoacidosis, dialysis disequilibrium syndrome, high-altitude exposure, and salicylate poisoning. Tissue ischemia in these conditions induce compensatory mechanisms with the aim to enhance oxygen delivery to cells and carbon dioxide removal such as hyperventilation, increased 2,3-P2-glycerate concentration (reducing the affinity of Hb for oxygen), and capillary vasodilation. Perivascular edema is seen at autopsy.

15.6 Trauma to the Head and Spine

Box 15.11 summarizes the anatomo-pathologic aspects of injuries to the head and spine. Trauma has a multi-etiologic origin, including motor-vehicular accidents (MVA), falls, blows on the chin, and childbirth. Clinical correlation is necessary. Head injuries result in intracranial hemorrhage (epidural, subdural, subarachnoid), concussion, contusion, and brain laceration.

Box 15.11 Trauma to the Head and Spine
15.7.1 *Head Injuries*
    15.7.1.1 Epidural Hematoma
    15.7.1.2 Subdural Hematoma
    15.7.1.3 Subarachnoidal Hemorrhage
    15.7.1.4 Brain Concussion
    15.7.1.5 Brain Contusion
    15.7.1.6 Brain Laceration
15.7.2 *Spinal Injuries*
    15.7.2.1 Meningeal Hemorrhages
    15.7.2.2 Hematomyelia
    15.7.2.3 Spinal Cord Concussion
    15.7.2.4 "Spinal Cord Crush" (Compression associated with Contusion)
    15.7.2.5 Spinal Cord Laceration
15.7.3 *Intervertebral Disk Herniation*

15.7 Head Injuries

Head injuries may manifest as epidural or subdural as well as subarachnoid bleeding or as a brain concussion, contusion, or laceration.

15.7.1 Epidural Hematoma

- *DEF*: Hemorrhage located between the skull and the dura mater of the meninges.
- *EPG*: Laceration of the middle meningeal artery (e.g., skull fracture).
- *EPI*: The least common of traumatic bleeding with ~2% of pediatric closed head injury admissions and male predominance ($\male:\female = 2:1$).
- *CLI*: Loss of consciousness after a "lucid interval" follows the skull injury in consequence of an apparent post-traumatic recovery and relapse with signs of ↑ICP, followed by respiratory depression and death unless the bleeding is promptly and correctly drained.
- *IMG*: "Epidural ellipse" with large angle edges.
- *CLM*: Fresh clot, without usually granulation tissue in contrast to subdural bleeding (*vide infra*).
- *TNT*: High pressure based bleeding with no alternatives to drainage. The epidural hematoma/hemorrhage is a life-threatening neurosurgical emergency, which requires urgent surgical evacuation. The goal is the prevention of irreversible neurological injury and, subsequently, death secondary to hematoma expansion and herniation.
- *PGN*: It depends from the localized area and focal cerebral compression, degree of ↑ICP, and subtentorial herniation, which is COD in untreated cases.

15.7.2 Subdural Hematoma

- *DEF*: Hemorrhage located between the dura mater and the arachnoid spaces of the meninges.
- *EPG*: Laceration of the bridging veins (e.g., *contrecoup* or birth injury at the frontal or occipital region), which lie unprotected in the subdural space.

- *EPI*: Quite common traumatic cause of bleeding.
- *CLI*: Gradual decline in the level of consciousness over days to weeks but two phases may be seen, although without a well-defined "lucid interval" as in the epidural bleeding and the clot may progressively be reabsorbed.
- *IMG*: "Sharp angle bridge" lesion.
- *CLM*: Proliferation of capillaries, fibroblasts, and hemosiderin.
- *TNT*: Low-pressure bleeding with alternatives to drainage.
- *PGN*: It depends from the localized area and focal cerebral compression, but it may be self-contained, and clot eventually reabsorbed leaving a yellow-stained "membrane" constituted by a thin inner membrane and a thick outer membrane strongly attached to the dura. Usually, ↑ICP and subtentorial herniation may occur but are rare occurrences. They need to be evaluated progressively.

15.7.3 Subarachnoidal Hemorrhage

SAH: *Vide supra*.

15.7.3.1 Brain Concussion

There is widespread paralysis of brain function without visibly identifiable changes and a strong tendency to spontaneous recovery. The pathogenesis seems to be related to acceleration effects and shearing (rotational) strains on the brain. Clinically, dazing and transitory unconsciousness are seen first, and subsequently, there is temporary impairment of higher functions of brain activity. In severe cases, prolonged unconsciousness, arterial hypotension, bradycardia, bradypnea, and muscular flaccidity are seen. Also, headache, vomiting, and delirium may occur upon return of consciousness. Complete recovery is common.

15.7.3.2 Brain Contusion

Brain bruising is consequent to a blow on the calvarium and is probably due to capillary tearing following to stress on the vascular network. The lesion may be located directly beneath the area of the skull trauma or opposite it, the so-called *contrecoup* lesions. In most cases, the frontal and temporal lobes are involved. Clinically, unconsciousness is at first observed, and progression to coma and death may follow in severe cases. Grossly, swollen gyri with multiple petechial hemorrhages are seen in early lesions, while cysts represent the characteristic feature of old contusions. Microscopically, there is edema of the cortex and subcortical white matter with numerous fresh pericapillary hemorrhages, such as Duret hemorrhages of the brainstem in the early lesions, while gliosis is the frequent finding of older contusions. Duret hemorrhage needs to be differentiated from Kernohan notch, which is an indentation of the cerebral peduncle associated with some forms of transtentorial herniation (uncal herniation). The syndrome of the shaken-baby syndrome is a severe brain injury caused by forcefully and violently shaking a baby (abusive head trauma) with severe brain damage.

15.7.3.3 Brain Laceration

Injury causes laceration of the brain with wounds penetrating the brain parenchyma, and symptoms are related to the site of the lesion, and, microscopically, there is the complete destruction of all neural, glial, and vascular elements in the line of the tear.

15.7.4 Spinal Injuries

Spinal injuries include meningeal hemorrhages, hematomyelia, spinal cord concussion, crush, and laceration. We need to distinguish several forms, including meningeal hemorrhages, hematomyelia, spinal cord concussion, spinal cord crush (compression and contusion), and spinal cord laceration, which may be revised in specific forensic and neuropathology books. Briefly, spinal injuries include hemorrhages in the meninges. Epidural hemorrhages follow vertebral fractures and spinal cord concussion with hematomyelia (e.g., post-spinal flexion or following a severe blow to the back, cord crush).

Compression and contusion of the spinal cord follow a vertebral dislocation (e.g., a lower cervical region in MVA or lower dorso-lumbar region following a massive blow across the lower back or a fall from height), and spinal cord laceration is seen following stab wounds or bullet wounds with hemisecting or fully sectioning the spinal cord.

15.7.5 Intervertebral Disk Herniation

Intervertebral disk herniation occurs following extrusion of the nucleus pulposus usually in men of the 3rd–5th decades of life and usually located in the lumbar or lumbosacral spinal region (L4-L5 or L5-S1) and presenting as low-back pain (sciatica) following mild to moderate trauma to the back possibly associated with a prolonged period of dehydration. The nucleus pulposus extrudes in the annulus fibrosus, and a tear may be demonstrated pathologically. Symptomatic lumbar intervertebral disc herniation is rare in children and adolescents. Karademir et al. (2017) studied 70 pediatric patients and proposed surgical treatment for patients only with persistent low back pain or radicular pain with a duration of more than 6 weeks, despite rest and medication.

15.8 Demyelinating Diseases Involving the Central Nervous System

A characteristic feature of demyelinating diseases is the involvement of the white matter in which the myelin sheath of neurons is damaged. Demyelinating diseases exclude the diseases in which axonal degeneration occurs first and degradation of myelin is secondary (Love 2006). Apart of the PML caused by a papovavirus described above, demyelinating CNS diseases have probably a critical autoimmune etiologic component.

Primary myelin diseases include the following disorders:

- Allergic diseases
- Infectious diseases
- Nutritional diseases
- Toxic diseases
- Traumatic diseases
- Vascular diseases

Acquired demyelinating disorders include the following:

- Acute disseminated encephalomyelitis (ADEM)
- Demyelinating viral infections
- Guillaume-Barre syndrome
- Multiple sclerosis

15.8.1 Multiple Sclerosis

- *DEF*: It is a demyelinating disease of the white matter with AI etiology with CD4, Th1, and Th17 cells against self-myelin antigens on a genetic susceptibility background (25% concordance between monozygotic twins). In some animal models, mice lacking the Th1 cytokine IFN-γ are not protected and have enhanced susceptibility to disease. IL-17-producing CD4+ effector cell lineage (Th17) are induced in parallel to Th1, and may have the potentiality to induce inflammation and autoimmune diseases (Damsker et al. 2010).
- *EPI*: Worldwide with prevalence in youth (onset 20–30s), 1:500, ♂ < ♀.
- *CLI*: Several subtypes with relapsing episodes of paresthesias, visual disturbances, and incoordination, including neuromyelitis optica (Devic), classical (Charcot), acute (Marburg), diffuse sclerosis (Schilder), and concentric (Balo).
- *LAB*: Albumino-cytologic dissociation (CSF protein is moderately elevated with mild mononuclear pleocytosis. However, if total protein >110 mg/dl and cell counts >50/mm^3 the diagnosis of MS unlikely).
- *IMG*: Old plaques are hyperintense on T2-weighted and FLAIR (fluid-attenuated inversion recovery) MRI studies, while new (active) plaques show gadolinium enhancement.

- *GRO*: Sharply demarcated gray plaques in the periventricular region, subcortical areas, optic pathways, brainstem, and spinal cord with the characteristic feature that white matter without myelin looks like gray matter ("white turns gray" phenomenon).
- *CLM*: Using LFB-PAS and Bielschowski stains as well as IHC stains, there is a triad of:
  1. Perivascular cuffing of T-/B-lymphocytes
  2. Colonization of macrophages (Mφ)
  3. Gliosis

  Loss of myelin with relative sparing of axons ≠ infarcts (PAS+ myelin debris in Mφ). Luxol fast blue stain or LFB, is a commonly used stain to detect myelin under light microscopy. It has been created by Heinrich Klüver and Elizabeth Barrera in 1953 and is also called Klüver-Barrera staining.
- *DDX*: Benign focal inflammatory demyelination (BFID) (optic neuritis, partial cord syndromes, etc.); non-benign, non-disseminated syndromes (transverse myelitis, Devic disease); disseminated diseases (acute disseminated encephalomyelitis and acute haemorrhagic encephalomyelitis); and, less likely, toxins (e.g., tobacco amblyopia and methanol), vitamin B12 deficiency, other inflammatory disorders (e.g., sarcoidosis, vasculitis, and SLE), infections, ischemia, and local tumors.
- *TNT*: It involves treating relapses (acute episodes of inflammation) with high doses of steroids, but also disease-modifying therapies to prevent relapses and disease progression, physical therapy and other types of rehabilitative therapeutic protocols in addition to a healthy lifestyle.
- *PGN*: Urinary tract infections, decubitus ulcers, and pneumonia are three essential complications that may aggravate the outcome (*quoad valetudinem et quoad vitam*).

15.8.2 Leukodystrophies

Leukodystrophies are a group of diseases involving the white matter (Greek λευκός "white") with abnormalities of myelin metabolism leading to ineffective myelination and subsequent demyelination, and most of these disorders have an AR inheritance.

The typing of demyelination can be:

- Diffuse (e.g., metabolic, toxic)
- Multifocal (e.g., inflammatory)
- Unifocal (e.g., traumatic, neoplastic)

15.8.2.1 Metachromatic Leukodystrophy

Cerebroside sulfatase (gene, chromosome) deficiency-based leukodystrophy leading to accumulation of sulfatides within the neurons and in the perineuronal areas with patients (epidemiology) experiencing progressive motor impairment.

15.8.2.2 Globoid Cell Leukodystrophy (Krabbe Disease)

Galactocerebroside B galactoside (gene, chromosome) deficiency-based leukodystrophy leads to accumulation of unmetabolized lipids within the neurons and in the perineuronal areas with patients experiencing rigidity and decreased alertness by 6 months of age. This disease is progressive and fatal by age 2. It is due to mutations in the *GALC* gene on 14q31 causing a deficiency of galactosylceramidase. Rarely, Krabbe disease may be caused by a lack of active saposin A. There are demyelination and multinucleated histiocytes at perivascular sites (aka "globoid cells").

15.8.2.3 Adrenoleukodystrophy

ABCD1 (X-ALD) deficiency-based leukodystrophy with lack of a peroxisome membrane protein necessary for β-oxidation of VLCFA (very long chain fatty acid) leading to accumulation of VLCFAs and progressive loss of myelin with patients experiencing Addison disease (concurrent adrenal gland insufficiency) in addition to a demyelinating syndrome. This disease is fatal by age 3. There are large plaques of demyelination with perivascular lymphocytic infiltrates progressively seen. VLCFA is a fatty acid with 22 or more carbons and its biosynthesis occurs in the endoplasmic reticulum of the cell.

Rosenthal fibers are thick, elongated, wormlike eosinophilic bundles that are found on H&E staining of the brain in specific conditions (Box 15.12).

> **Box 15.12 Rosenthal Fibers**
> - Genetic 1: Alexander disease
> - Genetic 2: Giant axonal neuropathy
> - Reactive 1: Perineoplastic (e.g., around craniopharyngioma)
> - Reactive 2: Perilesional (e.g., around plaques of multiple sclerosis)
> - Neoplastic: Pilocytic astrocytoma

- *Alexander Disease:* Diffuse Rosenthal fibers, which may be secondary to GFAP gene mutations.
- *Canavan Disease:* Megalencephaly with spongiform degeneration of the white matter and AZII astrocytosis of the gray matter.
- *Cockayne Disease:* Patchy myelin loss, basal ganglia calcification, and fibrous thickening of the leptomeninges.

15.8.3 Amyotrophic Lateral Sclerosis

- *DEF*: Primarily motor pathway-affecting CNS disease with neuronal loss of the lower and upper motoneurons (both 1st and 2nd alpha-motor neurons) with little if any sensory deficits leading to death in 1.5 years after onset with bulbar involvement or a median survival of 3 years without.
- *EPG*: AD inheritance (10% of cases, familial ALS or FALS) with involvement of chromosome 21 (In 1991 positional cloning identified linkage of FALS to the *SOD1* locus on 21q22 indicating genetic locus heterogeneity) (Siddique et al. 1991, Siddique and Siddique 2008).
 ALS-linked Genes include mostly
 – *SOD1* (Cu-Zn superoxide dismutase)
 – *DYNACTIN*
 – *ALSIN*
 – *VAMP-B*
- *CLI*: Signs of 1st and 2nd alpha-motor neurons with the weakness of the extremities (pyramidal tracts), muscular atrophy (hands as first), labio-glosso-laryngeal paralysis (bulb involvement) with hyperreflexia (exaggerated deep tendon reflexes), and fasciculations.
- *GRO*: Atrophy of the anterior spinal roots.
- *CLM*: Progressive swelling, pyknosis, and loss of alpha-motor neurons, Bunina bodies (small eosinophilic inclusions in the surviving lower motor neurons), demyelination in the white matter, and neurogenic pattern of muscle fiber atrophy. TAR (trans-activation response)-DNA-binding protein 43 (TDP-43) is the primary component of intraneuronal inclusions in both genetic and sporadic ALS. TDP-43 is a 43 kDa protein that in humans is encoded by the *TARDBP* gene. TDP-43 represses HIV-1 transcription, regulates alternate splicing of the *CFTR* gene as a splicing factor binding to the intron8/exon9 junction of the *CFTR* gene, is a splicing factor to the intron2/exon3 region of the *APOA-2* gene. It has been demonstrated that TDP-43 is a low molecular weight neurofilament mRNA-binding protein in human spinal motor neurons (Strong et al. 2007).

15.8.4 Werdnig-Hoffmann Disease

- *DEF*: AR-inherited rapidly progressive muscular weakness secondary to loss of the 2nd (lower) motor neurons of the anterior horns of the spinal cords leading to the "floppy infant syndrome," which may also be aspecific.
- *EPI*: 1 in 10,000 children in Caucasian populations with 1/40 as a carrier.
- *EPG*: SMA is due to the lack of the survival motor neuron (SMN) protein due to gene alteration on SMN1. If both copies of the *SMN1* gene are defective and SMN protein is absent, spinal neurons are going to die, leading to muscle weakness in the affected patients. The *SMN* genes are about 20 kb pairs long and both located on chromosome 5. *SMN1* is situated on the telomeric portion, while *SMN2* is centromeric. It is now known that only mutations in *SMN1* can cause SMA, while mutations in the centromeric portion result to be "modifiers" of SMA without the potentiality to cause the neuro-

logic disease. Both *SMN1* and *SMN2* are part of a 500 kb inverted duplication on chromosome 5q13. Since this region is duplicated, it has numerous repetitive elements, harboring the consequence that an increased probability of rearrangements and deletions results in copy number variation (CNV). It is assumed that this is the reason why SMA occurs randomly at such a high frequency in several populations.

SMA has a broad clinical spectrum, and five types (SMA 0–4) have been delineated based on clinical findings (ability to stand or walk, arthrogryposis, lack of breathing):

- SMA 0: Prenatal SMA (arthrogryposis)
- SMA 1: Acute infantile SMA (Werdnig-Hoffmann disease)
- SMA 2: Chronic infantile SMA (Werdnig-Hoffmann disease)
- SMA 3: Chronic juvenile SMA (Kugelberg-Welander disease)
- SMA 4: Adult SMA

15.8.5 Syringomyelia

A rare condition in which a cyst or channel (the "syrinx") forms within the spinal cord (Greek σῦριγξ "channel, tube"). As this fluid-filled cyst expands and lengthens over time, it provokes compression and damages to part of the spinal cord in a centro-peripheral fashion.

15.8.6 Parkinson Disease and Parkinson Disease-Associated, G-Protein-Coupled Receptor 37 (GPR37/PaelR)-Related Autism Spectrum Disorder

Parkinsonism is a clinical syndrome constituted by tremor, rigidity, brady- or akinesia, and postural instability. "*TRAP*"-Tetrad including (pill-rolling) tremor, rigidity, a-/hypokinesis (slowed voluntary movements), and postural (expressionless facies) caused by numerous factors, including adverse drug reactions and post-encephalitis event or be a genetic disease associated to a few genes, mainly α-synuclein aggregation-triggered misfolded protein/stress response.

The most frequent genes and proteins in Parkinson disease are:

α-SYNUCLEIN
LARRK2 (leucine-rich repeat kinase 2)
PARKIN
DJ-1
PINK1

Protein deglycase DJ-1 is also known as Parkinson disease protein 7. PINK1- or PTEN-induced putative kinase 1 is a mitochondrial serine/threonine-protein kinase. PINK1 is encoded by the *PINK1* gene and protect cells from stress-induced mitochondrial dysfunction.

- *EPI*: Worldwide, 4th–6th decade, 1–2/1000 population, ♂ > ♀. Juvenile-onset disease can occur rarely and can be caused by a single gene defect that is inherited in an AD, AR, or X-linked manner. Suleiman et al. (2018) reported on a 10-year-old child who had juvenile-onset Parkinson disease. Whole exome sequencing showed compound heterozygosity for two previously unreported novel mutations in *ATP13A2* (*PARK9*).
- *GRO*: There are developmental abnormalities in the *archicortex*, subcortical structures, as well as the cerebellum and brainstem. The occipital circumference and the brain size are increased compared to age-matched control children. This increase may be due to an increased number of neurons or an increased amount of neuropil.
- *CLM*: Cortical and non-cortical areas evidence region-specific defects in neuronal morphology and cytoarchitecture. There are findings located at the prefrontal cortex, fusiform gyrus, fronto-insular cortex, cingulate cortex, hippocampus, and amygdala. The cerebellum and brainstem are also involved. Loss of pigmented neurons from the *substantia nigra* in the midbrain and locus caeruleus in the upper pons, being the lateral *substantia*

nigra most severely affected (lateral → least neuronal pigment). Lewy body (LB): round, concentrically laminated, pale eosinophilic cytoplasmic inclusion, silver stain (−), but ubiquitin (+), and α-synuclein (+), conversely to the Pick bodies, which are silver stain (+).
- *TEM*: A core with densely packed fine filaments and loose at the periphery is observed.
- *DDX*: Parkinsonism with numerous etiologic factors.
- *PGN*: If LB is found in the cortex → diffuse Lewy body disease with patients affected with dementia, while if Alzheimer disease change are seen → Lewy body variant of Alzheimer disease (β-APP gene mutations).

15.8.7 Creutzfeldt-Jakob Disease (sCJD or Sporadic), CJD-Familial and CJD-Variant

- *DEF*: Subacute spongiform encephalopathy (85% sporadic, 10% familial, and 5% iatrogenic) related to "kuru" and the sheep model "scrapie" due to a conformational change of proteinase K-resistant prion ("proteinaceous infectious agent") protein (PrP) acting as a β-pleated sheet (amyloidogenic) localized on 20p constituting the basis for abnormal protein folding (rod polymerization with amyloid features), abnormal ubiquitin, and intracellular trafficking. In fact, PrP 33–35 sc is degraded only to a 27–30 kD protein (partial digestion) by proteinase K digestion in affected individuals, conversely to the complete proteinase K digestion.
- *CLI*: Generalized bilateral short interval periodic sharp wave complexes (~2/3 cases).
- *IMG*: Diffusion-weighted MRI (DWI) and FLAIR MRI sequences show high-signal abnormalities situated in the caudate nucleus and putamen ≥2 cortical regions (temporal, parietal-occipital). In the setting of vCJD, patients have high-intensity signal in the posterior thalamus, which is also known as pulvinar.
- *CLM*: Vacuolar spongiform degeneration of the neuropil with progressive and inexorable neuronal loss ("punched-out" vacuoles in gray matter–cortex and/or basal nuclei) and gliosis (glial proliferation) as well as amyloid plaques ± kuru plaques (10%) in the cerebellar granular cellular layer.
- *PGN*: Fatal in all cases, usually within a half year.

In 1936, Cuille` and Chelle demonstrated the transmissible nature of prion diseases experimentally through the intraocular administration of scrapie-infected spinal cord to a goat, while 30 years later, kuru, a condition of the South Fore people of New Guinea was transmitted to chimpanzees by Gajdusek, the co-recipient of the Nobel Prize in Physiology or Medicine in 1976 for work on kuru, the first known human prion disease. This population was practicing funerary cannibalism.

Prion diseases can be distinguished in CJD forms and non-CJD prion diseases. Other clinical phenotypes of prion disease include kuru, fatal familial insomnia (FFI), and Gerstmann-Straussler-Sheinker syndrome (GSS). Particular precaution needs to be taken into account in performing the autopsies of cases with potential CJD. Following brain removal with CJD or CJD suspicion, 10% NaClO solution (bleach) or NaOH (caustic soda) is used to ensure decontamination of surfaces and equipment. These guidelines are like Rabies Autopsy Procedure that can be accessed online at the CDC or Centers for Disease Control and Prevention, which is the leading national public health institute of the United States of America. The CDC is a US federal agency under the Department of Health and Human Services. The CDC has innumerable guidelines providing paths for many agencies worldwide. It is headquartered in Atlanta, Georgia, USA. Ultimately, it should be mentioned the association with early Alzheimer disease and trisomy 21 syndrome (Down syndrome), which may present with a myoclonus that is different from the CJD-myoclonus.

15.8.8 West Syndrome/Infantile Spasms, ACTH Therapy, and Sudden Death

West syndrome/infantile spasms (WS/IS) is a severe, tragic epilepsy syndrome consisting of the triad of infantile spasms, an interictal EEG pattern labeled "hypsarrhythmia," and mental retardation, but two out of three criteria are enough to make this diagnosis. The onset is between 4 and 7 months of age, and the prevalence is 1:2000 to 1:6000 live births. The etiology is multifactorial, including hypoxic-ischemic injury, congenital infections, cortex development defects, tuberous sclerosis complex (Bourneville-Pringle disease), inborn errors of metabolism, genetic multiple congenital anomaly (MCA) syndromes, and chromosomal abnormalities as well as idiopathic. ACTH is the drug of choice for treatment of WS/IS and is quite safe, but some side effects need to be taken into consideration. These include Cushingoid features, an increase of appetite and weight gain, irritability, immunosuppression, an increase of risk for joint infections (pneumonia, GU and GI infections), hypertension, glucosuria/diabetes mellitus, hypokalemia, gastritis, and cerebral ventriculomegaly. Thus, a death occurring in a child under treatment of ACTH (or steroids) needs to be examined under medico-legal aspects and take into account septicemia, shock, and brain pathology.

15.9 Neoplasms of the Central Nervous System

In approximately one in ten cancer deaths, a CNS neoplasm is the cause underlying the fatal course. Age is important in CNS neoplasms because it affects the distribution and favors specific histologic types. In fact, pediatric brain tumors are in about ¾ of cases localized below the tentorium, while brain tumors are with the same rate above the tentorium in adults. Furthermore, pediatric brain tumors are about 20% of all pediatric tumors just second to leukemias, and adult brain tumors are quite rare if compared with tumors of other organs. The most common pediatric brain tumors are gliomas of the astrocytic type followed by medulloblastomas, ependymomas, and craniopharyngiomas, while the most common adult brain tumors following the astrocytic nature are metastases and meningiomas. Males are more affected than females. Neoplasms may derive from all cell types, although criteria of benignancy and malignancy are not entirely straightforward as seen in other organs. Separating the tumors between glial and nonglial neoplasms is useful. Glial neoplasms show sponge-like quality and a tendency to infiltrate the surrounding tissue, two features that allow categorizing this kind of tumor as biologically malignant (exceptions apply, *vide infra*). Nonglial neoplasms have a more robust structure and grow by expansion enabling a better resection. CNS tumors rarely metastasize outside the CNS. It is also imperative to distinguish between direct and secondary pathophysiologic effects of CNS neoplasms. Direct effects involve several invariable functions of the area where the tumor is located and also affect the inhibitory influence on afferent impulses of local neurons rendering the neurons biophysically electrically unstable with the consequence of a liability to seizures. Indirect effects include the edema of the surrounding tissue, circulatory effects due to compression of the brain and spinal cord, compression and "*functio laesa*" of adjacent vital structures, herniation of the brain and cerebellar tonsils with Duret hemorrhage, as well as interference with CSF circulation and progressive development of hydrocephalus and papilledema. Clinically, headache, nausea, and vomiting may be accompanied by seizures, palsy, aphasia, blindness, deafness, lethargy, or abnormal behavior. Herniation of brain and cerebellar tonsils induce loss of consciousness, central respiratory failure, and death. CSF cytology is crucial in the initial diagnostic approach and following treatment (Fig. 15.8).

15.9.1 Astrocyte-Derived Neoplasms

It is essential to differentiate between *astrocytoma, pilocytic astrocytoma, glioblastoma multiforme,* and *oligodendroglioma.*

15.9.1.1 Astrocytoma

Astrocytomas represent 1/3 of all gliomas and occur more frequently in the cerebellar hemispheres in children and the central white matter of the brain in adults. There are two types

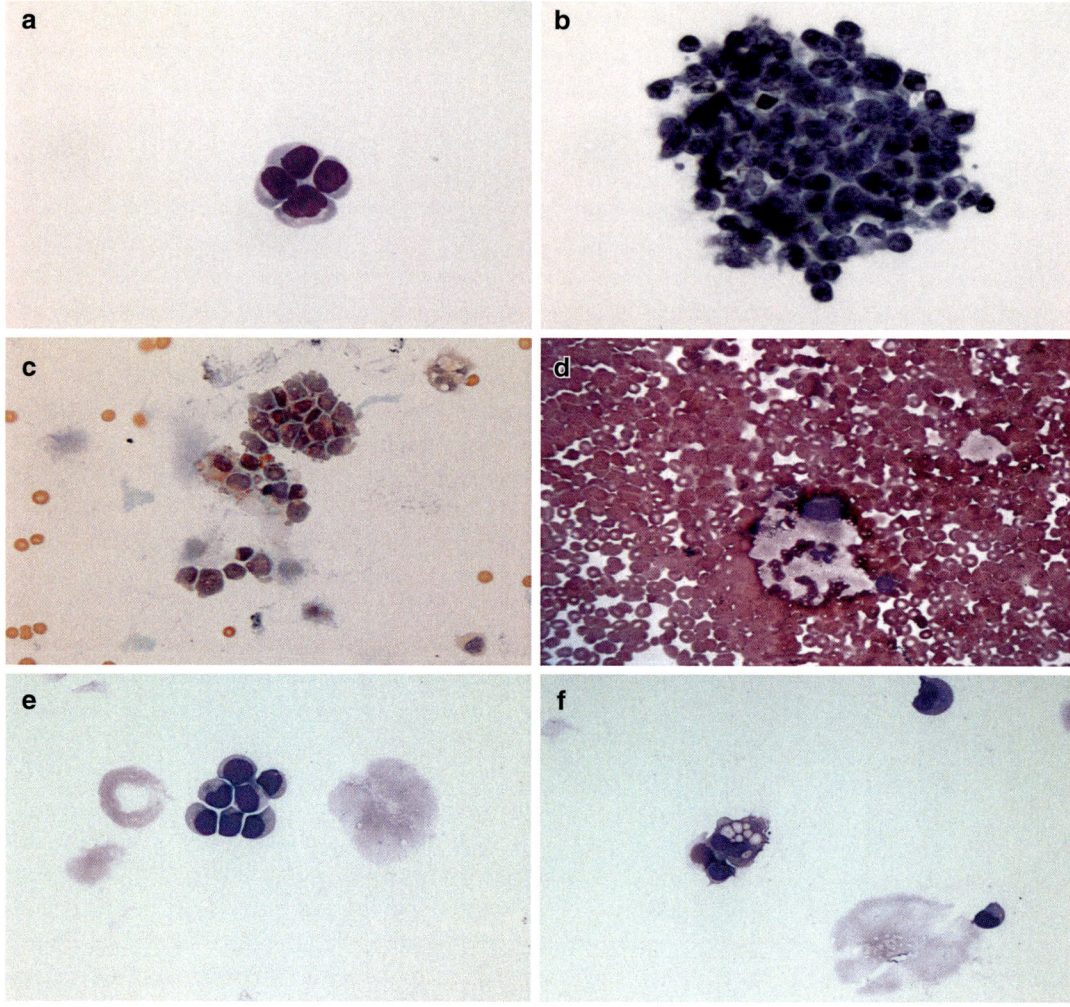

Fig. 15.8 This panel illustrated a cluster of four malignant cells in the CSF in Figure **a** (Diff-Quik staining, ×630 original magnification). The chromatin is clumped and the nuclear contour is irregular. A cluster of numerous cells is shown in Figure **b** (Diff-Quik staining, ×630 original magnification). This cluster shows a high nucleus to cytoplasm ratio with very dense chromatin and hyperbasophilic nucleus. Malignant cells are also shown in Figure **c** (Papanicolaou staining, ×400 original magnification). Leptomeningitis proof is shown in Figure **d** (Papanicolaou staining, ×400 original magnification). A cluster of acute myeloblastic leukemic cells is shown in Figures **e** and **f** (Giemsa staining, ×400 original magnification) Diff-Quik is a commercial Romanowsky stain variant, which is often used to rapidly stain cytological preparations. It differentiates a variety of blood films and cytopathological smears

distinguished in protoplasmic and fibrillary according to the kind of astrocytes that predominate:

- *GRO*: Gray-whitish or white, firm, poorly demarcated with difficulty to appreciate the resection margins and may harbor tiny cysts filled with clear yellow fluid.

- *CLM*: The architecture is typical with tumors varying from highly to sparsely cellular and highly fibrillary matter. If only the edge of the lesion is biopsied, it is essential to distinguish a glial neoplasm from reactive gliosis. Reactive gliosis shows astrocytes that are evenly dispersed within the tissue, at approximately equal distances from each other.

Interestingly, in infarctions and demyelinating disease, there will be a mosaic of reactive astrocytes and macrophages, which is no present in glial neoplasms. In subacute reactive peri-abscessual gliosis, there will be few macrophages, but the astrocytes are scattered in the background neuropil. An *irregular distribution* of the astrocytes and disorganization with a "*tendency to touch each other or clash*" is characteristic of glial neoplasms. Moreover, reactive astrocytes have long, tapering, starlike cytoplasmic processes, while neoplastic astrocytes display short and irregular blunted processes and gemistocytic features. A useful IHC stain is glial fibrillary acidic protein (GFAP), which may strikingly reveal the differences between reactive and neoplastic astrocytes. Moreover, reactive astrocytes do not show nuclear atypia and high cellular density. An important caveat can be radiation therapy (external beam or stereotactic), which can cause significant cytologic atypia in astrocyte. However, true hypercellularity is lacking. Importantly, macrophages, mainly bland necrosis devoid of cellular response, fibrinoid vascular necrosis, and hyalinized and altered blood vessels, are not features of a neoplastic process, and therapy may be held. Conversely, gemistocytic (Greek: γέμιζω 'to fill up') astrocytes, but not reactive astrocytes, scattered with astrocytes that show angular hyperchromatic nuclei and nuclear anaplasia and pleomorphism with or without straight necrosis point to a neoplastic relapse needed to get more effective therapy.
- *PGN*: It depends from the grade (see most recent WHO classification). Survival has a variable range from 1 to 10 years.

In 2016 WHO update of the classification of tumors of the central nervous system, there is a combination of histology with tumor genetic information to create a specific diagnostic algorithm. In the first place, the histologic information is used to address a diffuse glioma/tumor grade. Second, the isocitrate dehydrogenase - *IDH* gene mutation status is tested using immunohistochemistry (IHC) with an antibody against IDH1 R132H (~90% of all *IDH* mutations). Mutations in *IDH1* and *IDH2* occur in most of low grade gliomas and secondary high grade gliomas. These mutations have been found to occur early in gliomagenesis and are able to change the function of the enzymes, inducing them to produce 2-hydroxyglutarate, which is a potential oncometabolite. In case of the negative result by IHC, genetic sequencing is performed to identify minor *IDH1/IDH2* mutations (~10% of all *IDH* mutations). In the case of *IDH*-mutant II–III grade gliomas, a set of genetic parameters can differentiate astrocytomas from oligodendrogliomas. In case of oligodendrogliomas, the proof of loss of heterozygosity due to chromosomal arms 1p and 19q deletion is required, which is a molecular hallmark of oligodendroglioma. Fluorescence in situ hybridization (FISH) is used to identify 1p19q codel. *IDH*-mutant astrocytomas (~80% of all astrocytomas) are distinct by mutations of *ATRX* and *TP53* (and lack of 1p19q codel) because *ATRX* and *TP53* mutations are not present in oligodendrogliomas. In case of GBM in individuals with age >55 years, sequencing is not recommended because of the very low likelihood for detecting other IDH mutations in this subgroup.

15.9.1.2 Pilocytic Astrocytoma
- *DEF*: "Hairlike"-astrocytoma with lengthy, bipolar processes and excellent prognosis (WHO grade I)
- *EPI*: Most common CNS tumor of childhood, 1:100,000 (incidence), 1st–2nd decades, ♂ = ♀
- *CLI*: Midline structures in posterior fossa including cerebellum, 3rd ventricle, thalamus, hypothalamus, neurohypophysis, optic nerve, optic chiasm, hypothalamus, cerebral hemispheres, brainstem, and spinal cord with the potential to be multicentric.
- *IMG*: Well circumscribed and contrast enhancing, often cystic tumor, in which the invasion of subarachnoid space and endothelial proliferation are not considered poor prognostic factors.
- *GRO*: Discrete, cystic (microcystic/macrocystic) mass.

- *CLM*: Biphasic pattern with varying proportions of piloid areas alternating with spongy areas. Piloid regions are constituted by bipolar neoplastic cells with elongated "hairlike" cytoplasmic processes that are arranged in parallel bundles and look like rugs of hair, Rosenthal fibers (tapered twisted shaped, brightly eosinophilic, hyaline masses), eosinophilic protein droplets ± mural nodule, and calcifications. Spongy areas are constituted by multipolar cells (protoplasmic astrocytes) associated with microcysts and eosinophilic granular bodies embedded in a loose texture. Malignant transformation if hypercellularity, mitotic figures, and necrosis are present (leptomeningeal extension does not portend a sign of malignancy). However, it needs to be distinguished from degenerative changes (hyalinized blood vessels, infarct-like necrosis, degenerative nuclear atypia, calcification, hemosiderin deposits, and lymphocytic infiltrates) and vascular modifications (glomeruloid proliferation and vascular hyalinization). Rarely, a diffuse growth pattern is seen.
- *HSS/IHC*: (+) GFAP, PTAH, PAS (protein droplets), A1ACT (protein droplets).
- *DDX*: Diffuse astrocytoma (WHO grades II–IV), gliosis, hemangioblastoma, pleomorphic xanthoastrocytoma, ganglioglioma, and piloid gliosis (hypocellular with numerous Rosenthal fibers, but lack of spongy areas).
- *TRT*: Resection and adjuvant radiochemotherapy for tumors of the optic pathway and hypothalamic region.
- *PGN*: 5-YSR: >90% (10-YSR: 100% if supratentorial and gross total resection, but 74% if subtotal resection).

15.9.1.3 Glioblastoma Multiforme (GBM)

- *DEF*: High-grade glioma with rapid growth, variable symptomatology with increased ICP.
- *EPI*: Rare in pediatrics and youth.
- *GRO*: Large, well-demarcated, multicolored, soft tumor in the central white matter of the brain with surrounding edema inciting marked compression, distortion, and displacement of adjacent structures potentially crossing the midline through a direct invasion of the corpus callosum.
- *CLM*: Highly cellular and hypervascular architecture with extensive necrosis, prominent atypia and pleomorphism, and intense endothelial hyperplasia (all three features are necessary histologic features for a diagnosis of glioblastoma multiforme).
- *TEM*: Tumor cells are swollen variably and contain a large rounded nucleus with marginalized chromatin. The cytoplasm contains mitochondria and well-developed ER. In the extracellular space, myelin sheaths and cell processes are usually observed.
- *CMB*: Abnormalities of chromosomes 7, 9, and 10 with overexpression of the oncogene *c-sis* in more than half of the cases and *c-erb B* in approximately 1/3 of the cases.
- *PGN*: Poor, although adjuvant radiochemotherapy seems to have increased some survival.

15.9.1.4 Oligodendroglioma

- *DEF*: Oligodendroglioma accounts for approximately 1/20 of all intracranial tumors and is the most common glioma of the cerebral hemispheres in adulthood.
- *CLI*: There is an indolent clinical course, and grossly, these tumors are typically well-circumscribed in the white matter, although they may break through into the cortex cerebri.
- *CLM*: Uniform architecture of a honeycomb or "fried egg" appearance of the tumor cells (small rounded nuclei surrounded by clear spaces, although this feature is not seen on frozen section) is observed accompanied by prominent calcification foci. To remember is that the distinction between oligodendroglioma and astrocytoma on intraoperative frozen section is not typically necessary.
- *PGN*: 5-YSR of ~50%.

15.9.2 Ependymoma

- *DEF*: Tumor-derived from the ependymal lining of the ventricular system and the central canal of the spinal cord. About half of the

tumors are below the tentorium, and nearly half of the tumors are above. A few ependymomas occur in the spinal cord or *filum terminale*.
- *EPI*: All age groups with peak at age of 3–4 years.
- *CLI*: Papilledema and headache.
- *GRO*: It is a soft tan mass with a distinct edge from surrounding brain or spinal cord with characteristic exophytic growth into 4th ventricle and ability to fill this ventricular cavity. If the ependymoma fills the 4th ventricle and exits out through foramina of Luschka or Magendie, the ependymoma is called of plastic type.
- *CLM*: There is a classic picture of a combination of perivascular pseudorosettes (more common) and ependymal rosettes (less common) with a monomorphic nuclear morphology. Perivascular pseudorosettes are formed by tumor cells arranged in a radial fashion around blood vessels, while ependymal rosettes are formed by cylindrical cells arranged radially around central lumina resembling ependymal canals. Regressive changes may be encountered and include hemorrhage, calcification, myxoid degeneration, and, rarely, cartilage and bone islands.
- *IHC*: (+) GFAP, S100, VIM, EMA, ± AE1–3. EMA is positive along the lumen of ependymal rosettes and as dot-like cytoplasmic vacuoles illustrative of early microlumens.
- *CMB*: In 1/3–2/3 of ependymomas there are deletions of the chromosome 22, inactivation of *NF2* or loss of expression of the protein in about 1/3 of cases. It is common to observe the loss of protein 4.1B and 4.1R in childhood ependymomas. Protein 4.1G deletions have been associated with a more aggressive outcome.
- *PGN*: The outcome is directly linked to the location of the tumor, carrying the infratentorial ependymomas a more benign outcome than the supratentorial ependymomas.

Histopathologic variants of ependymomas include cellular, papillary, clear cell, tanycytic, ependymoma with lipomatous differentiation, giant cell ependymoma of the *filum terminale*, ependymoma with extensive tumor cell vacuolation, melanotic ependymoma, and ovarian ependymoma. Two essential entities are the myxopapillary ependymoma and the anaplastic ependymoma. Myxopapillary ependymoma is a slow-growing ependymoma composed of cells arranged in a papillary pattern around vascularized myxoid cores and located almost solely near *conus medullaris*, *cauda equina*, and *filum terminale* of spinal cord (Fig. 15.9). The anaplastic ependymoma is a mainly pediatric ependymoma with localization at the brain or cerebellum and tendency to infiltrate leptomeninges and spreading like medulloblastoma (*vide infra*).

15.9.3 Medulloblastoma

- *DEF*: Medulloblastoma is one small-cell tumor of the CNS and PNS and both clinically and pathologically malignant tumors with an occurrence usually below pubertal age (Figs. 15.10 and 15.11).
- *EPG*: Four genetic groups: (1) WNT activated, (2) SHH activated (either *TP53* wild type or *TP53* mutated), non-WNT/non-SHH, (3) medulloblastoma group 3, and (4) medulloblastoma group 4.
- *EPI*: It is the most common embryonal brain tumor of childhood after the pilocytic astrocytoma. The classic type has a midline location, while the desmoplastic/nodular medulloblastoma predilects the cerebellar hemispheres and midline with a bimodal age distribution. Gorlin syndrome may be present in some patients harboring medulloblastoma.
- *GRO*: Homogenous, reddish-gray, well-defined tumors arising in the midline posteriorly (e.g., the roof of the fourth ventricle) with an affinity to spread along the CSF pathway.
- *CLM*: Highly cellular, perivascular with cellular monomorphism ("carrot-shaped blue cells") and high N/C ratio.
- *IHC*: (+) SYN, NeuN, SMARCB1/INI1 and SMARCA4, SOX11, PAX5, TTF1, and ISL1. In medulloblastoma with melanocytic differentiation, HMB45 and melanA are positive, while

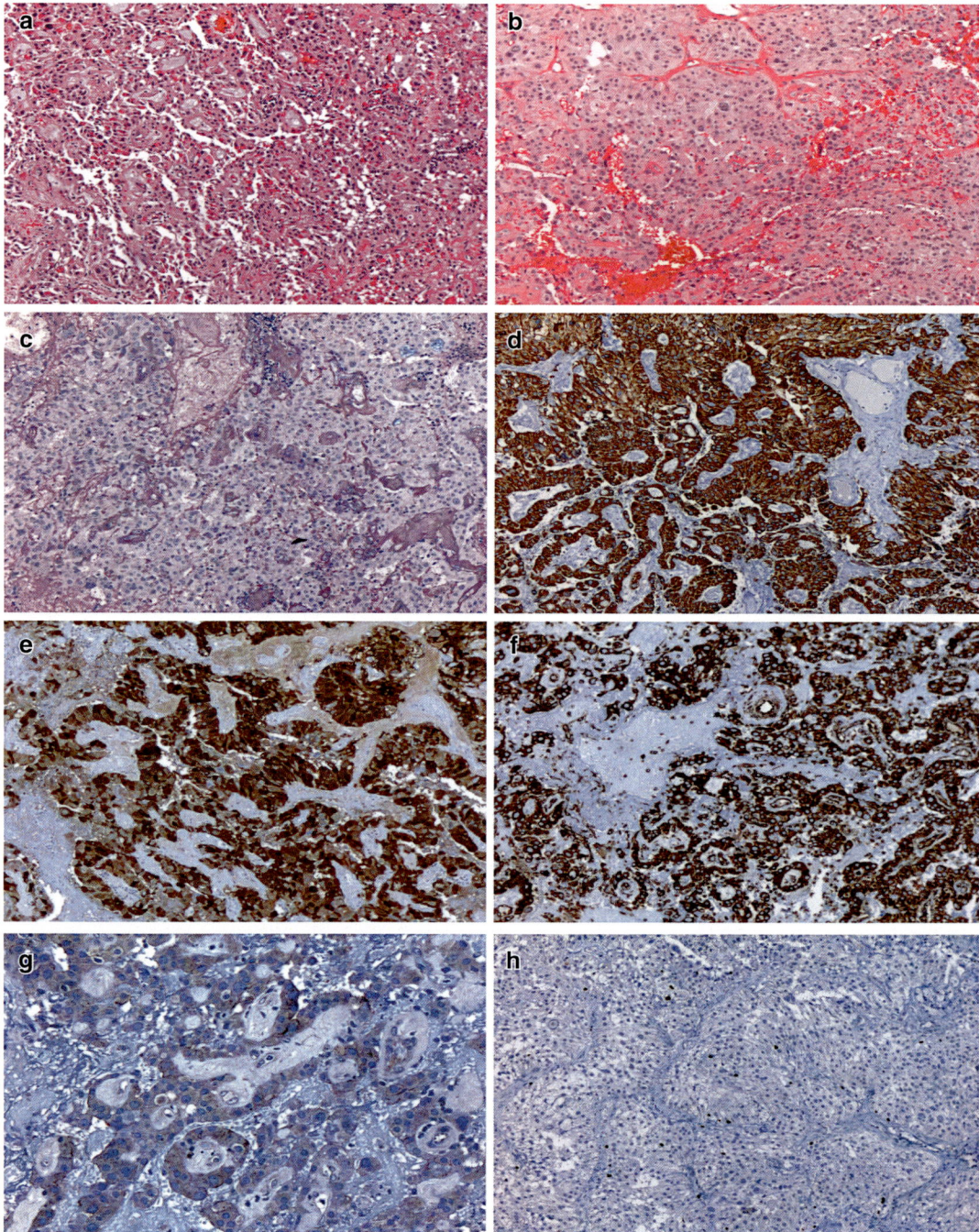

Fig. 15.9 Myxopapillary ependymoma is a tumor which is located in the sacrococcygeal area of children and young adults at intradural level. The microphotographs stained with hematoxylin and eosin are shown in Figures **a** and **b** (**a**, ×100; **b**, ×100, both original magnifications) disclosing cuboidal to mildly elongated cells with a radial orientation around vascularized myxoid cores. In figure **c** (Alcian-Blue-periodic acid Schiff staining, ×100 original magnification), the myxoid cores are highlighted using a special stain. The following five immunohistochemical microphotographs show the expression of glial fibrillary acidic protein (GFAP) (**d**, ×100 original magnification), S100 (**e**, ×100 original magnification), vimentin (**f**, ×100 original magnification), epithelial membrane antigen (**g**, ×100 original magnification), and Ki67 (**h**, ×100 original magnification). All immunostainigs have been performed using the avidin-biotin complex method. GFAP is an intermediate filament protein identified in numerous cell types of the central nervous system including ependymal cells and astrocytes during development. The proliferation antigen Ki67 is a cellular marker for proliferation (MIB1 monoclonal antibody)

15.9 Neoplasms of the Central Nervous System

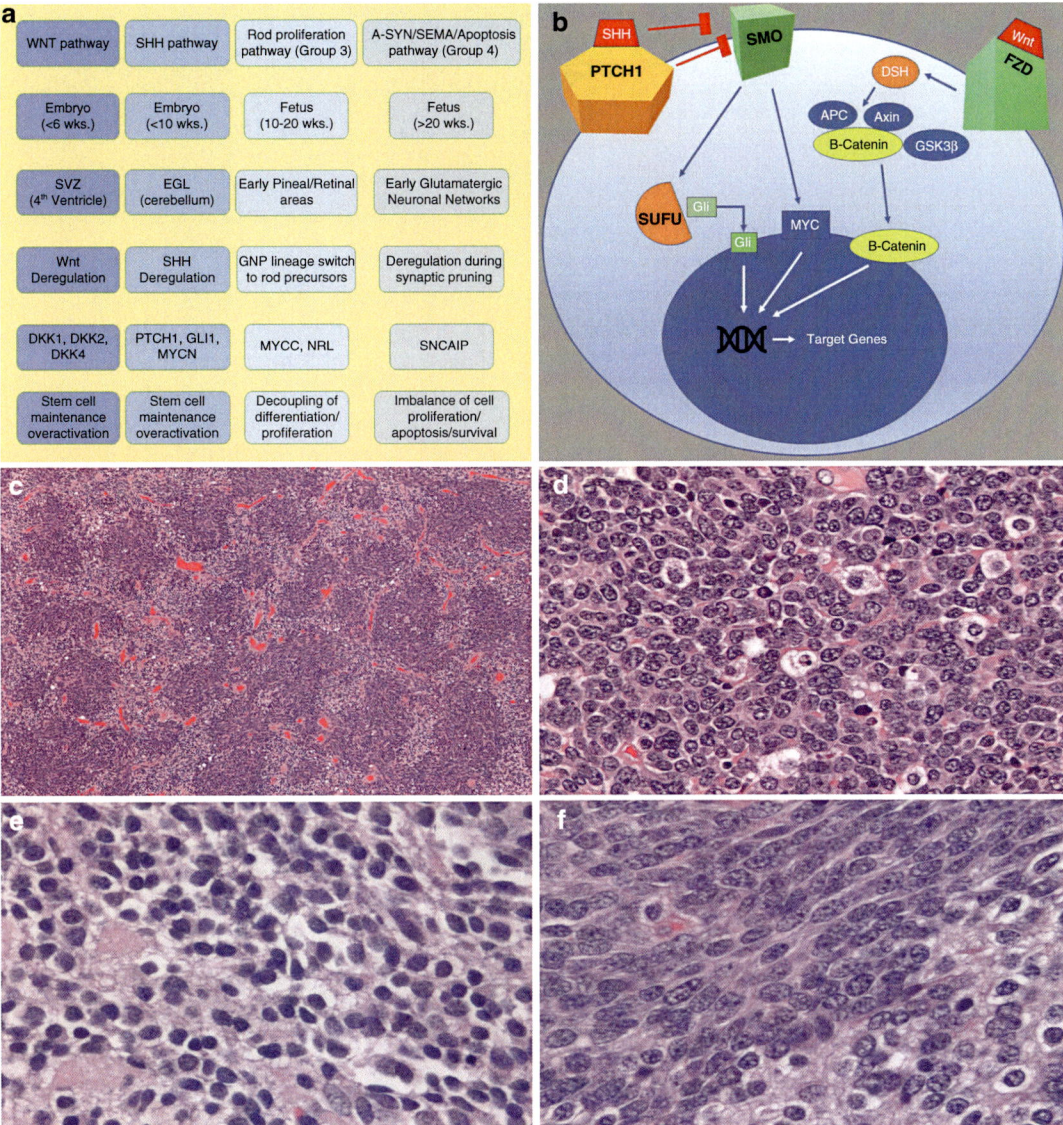

Fig. 15.10 In figure **a** are shown the subtypes of medulloblastomas, while Figure **b** illustrates the pathogenic mechanisms seen in medulloblastoma. The following four microphotographs illustrate the microscopic structure of this tumor of the central nervous system that in its classic representation discloses small round blue cell tumor, "carrot" shaped at places, with a syncytial arrangement exhibiting mitotic figures, apoptotic bodies, and Homer-Wright rosettes (**c**, ×50; **d**, ×400; **e**, ×630; **f**, ×630 as original magnifications). All four microphotographs are from hematoxylin-eosin stained histological slides

in medulloblatoma with myogenic differentiation, DES and MYOGENIN are positive.
- *CMB*: The immunologic positivity for GAB1 and YAP1 suggests the SHH-activated group of medulloblastoma, while their immunologic negativity indicates a medulloblastoma belonging to non-WNT/non-SHH group.
- *TRT*: Surgery and radio-chemotherapy.
- *PGN*: Although considered grade IV tumors harboring a poor prognosis, medulloblastoma with extensive nodularity and the desmoplastic type in young age have a relatively good outcome. On the other hand, *TP53* (p53) mutants bear a poor prognosis.

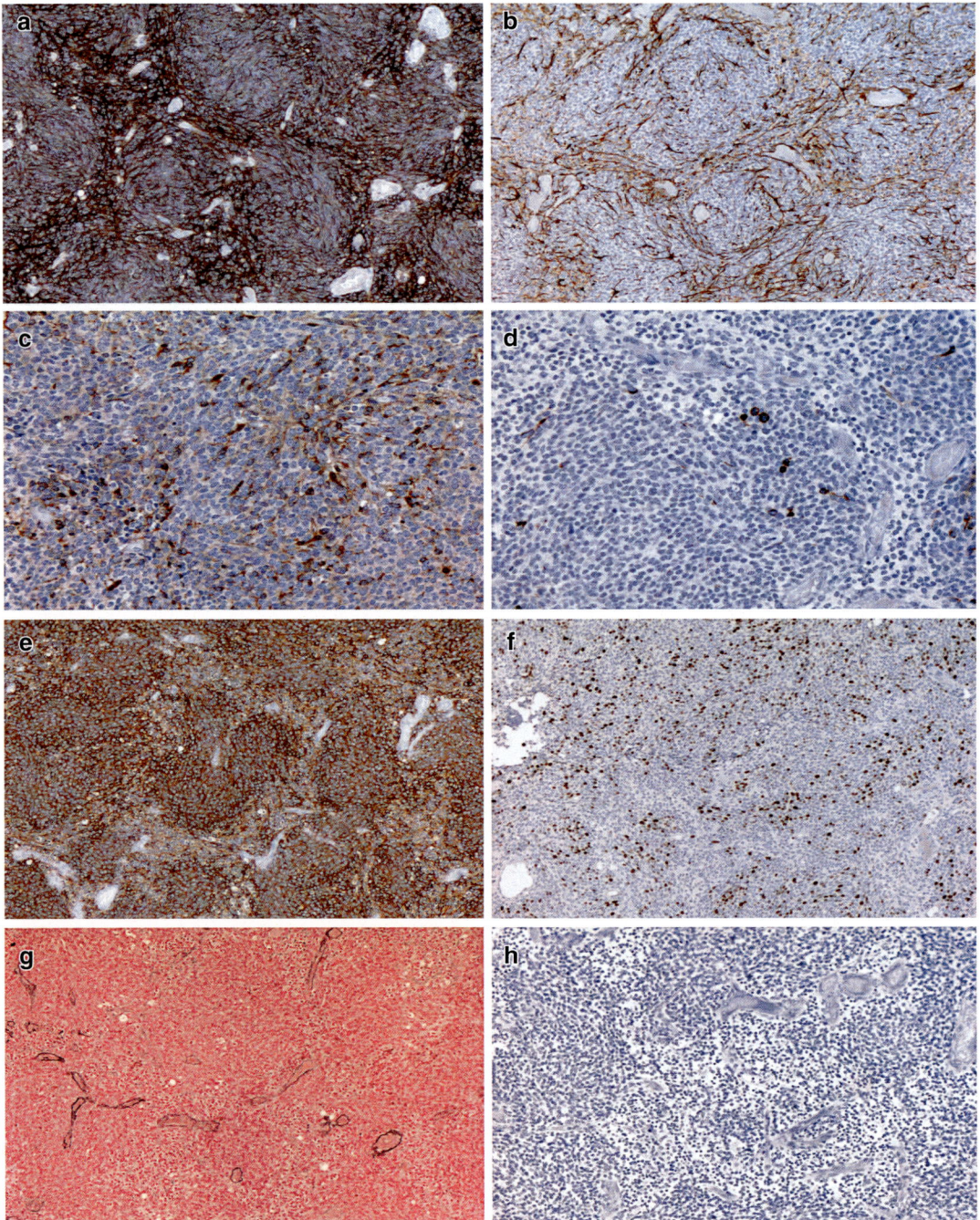

Fig. 15.11 The immunohistochemistry of medulloblastoma has distinctive features including the expression of CD56 (**a**, ×100 original magnification), glial fibrillary acidic protein (GFAP) (**b**, ×100 original magnification), neuron-specific enolase (NSE) (**c**, ×100 original magnification), neurofilaments (**d**, ×100 original magnification), synaptophysin (**e**, ×100 original magnification), and Ki67 (**f**, ×100 original magnification). The special stain reticulin is shown in Figure **g** (×100 original magnification). No expression of cytokeratins uses a pan-cytokeratin monoclonal antibody AE1-3 (×100 original magnification) (**h**). All immunohistochemistry microphotographs have been carried out using antibodies and the avidin-biotin complex method

15.9.4 Meningioma

Typically, meningiomas are slow-growing benign tumors attached to the dura mater, which are made up of neoplastic meningothelial cells and account for 1/5 of all intracranial tumors in adults with female preference.

- *EPI*: Meningiomas are uncommon in childhood or youth.
- *CLI*: Silent or mild neurologic deficits by compression of adjacent structures.
- *GRO*: Well-outlined, firm to rubbery, gray-white tumors with a gritty red appearance that are located in approximately half of the cases near the vertex and the midline with a propensity to push inward rather than invade the surrounding tissue.
- *CLM*: There is a cellular uniformity with various histologic patterns. Tumor cells have ovoid nuclei and whorls, and psammoma bodies are common.
  The WHO classification of meningioma entails low and high risk of aggressive behavior:
  – *Low-risk* meningiomas (WHO grade I):
  Meningothelial, fibroblastic, transitional, psammomatous, angiomatous, microcystic, secretory, lymphoplasmacyte-rich, metaplastic meningioma
  – *High-risk* meningioma (WHO grade II) or atypical meningioma:
  Clear cell meningioma
  Chordoid meningioma
  – *High-risk* meningioma (WHO grade III) or anaplastic meningioma:
  Papillary meningioma
  Rhabdoid meningioma
- *IHC*: (+) VIM, S100, PR, (±) EMA, (−) GFAP, OCT4.
- *CMB*: Cytogenetically and on a molecular level, meningiomas may show mutations or deletions of *NF2* gene and monosomy of the chromosome 22.
- *PGN*: It depends on the location and WHO grade.

15.9.5 Hemangioblastoma and Filum Terminale Hamartoma

- *DEF*: Neoplasm arising from blood vessels and almost exclusively located exquisitely in the cerebellum, constituting up to one in ten of primary posterior fossa tumors, and being associated with *VHLS* gene mutations.
- *CLI*: Ataxia and abnormal locomotion are observed. Polycythemia may be encountered, and CSF analysis may show erythropoietin.
- *GRO*: Well-demarcated, purplish-red or brown, and spongy-like large cysts.
- *CLM*: There is a vascular pattern with capillaries and large anastomosing cavernous channels confluent to more cellular areas with large polygonal cells containing lipids.
- *IHC*: (+) NSE, ORO (at intraoperative frozen section), CD34, VEGF, INA, other than reticulin, (±) EPO, GFAP, S100.
- *CMB*: Loss of VHL induces an increased production of VEGF and EPO.
- *DDX*: Posterior fossa tumors in adults vs. child include:

Adulthood

- *Extraaxial*: Acoustic neuroma (=schwannoma, 80–90% of cases), meningioma, chordoma, choroid plexus papilloma, and epidermoid cyst
- *Intraaxial*: Metastasis (lung, breast), hemangioblastoma, lymphoma, lipoma, and glioma

Childhood

- *Extraaxial*: Schwannoma
- *Intraaxial*: Posterior fossa astrocytoma, pilocytic astrocytoma, brainstem glioma, medulloblastoma, ependymoma, atypical teratoid/rhabdoid tumour (AT/RT), haemangioblastoma, and teratoma

Posterior fossa cystic malformations vs. cystic mass in the cerebellum include:

1. *Posterior Fossa Cystic Malformation:* Dandy-Walker Malformation, Dandy-Walker Variant, Mega cisterna magna, and arachnoid poach
2. *Cerebellar Cystic Mass:* Hemangioblastoma, aneurysm, medulloblastoma of lateral location, choroid plexus papilloma, metastasis
- PGN: Excellent, because solitary hemangioblastomas can be straightforwardly shelled out

Filum terminale hamartoma is a hamartoma (pseudotumor) compressing the filum terminale (Fig. 15.12).

15.9.6 Schwannoma

Neurilemmomas or schwannomas are tumors of nerve sheath origin accounting for 1 in 10 of all intracranial tumors. Typically, they occur in middle-aged adults, but younger patients have been described, and the most common location is the 8th cranial nerve (acoustic "neuroma"). CLI: Hearing loss, disequilibrium, and headache have been noted. They can be associated with NF of Recklinghausen.

- GRO: The tumor is spherical or oblongated and grows externally to a conjoined nerve, differently from neurofibroma, in which the nerve penetrates cancer. On cut section, the tumor can be either homogeneous or variegated, and consistency is usually cystic or granular.
- CLM: The schwannoma is biphasic with uniformly spindled Schwann cells with Antoni A (cellular or fascicular) and Antoni B (myxoid or vacuolated) areas.
- PGN: The outcome is excellent, if the tumor may be fully resected. Otherwise, recurrence is the rule.

15.9.7 Craniopharyngioma

Craniopharyngioma derives from epithelial nests arisen from Rathke's pouch. It is a pediatric tumor, which shows a calcified suprasellar mass giving rise to visual problems and hypothalamic disturbances clinically.

- GRO: The craniopharyngioma is multiloculated encapsulated tumors with pushing tendency into the surrounding brain tissue. On cut section, dense material oozes (aka "machine-oil" of the craniopharyngioma) with numerous suspended cholesterol crystals, whereas the solid portions of the tumor are granular or crumbly.
- CLM: The craniopharyngioma is multicystic with numerous foreign-body giant cell reaction fields in solid areas, fibrous tissue, mineralization, and bony islands. Solid regions show an epithelial differentiation looking like tooth bud or squamous epithelium, occasionally quite like ameloblastomas.
- PGN: The outcome is excellent, if the tumor may be fully resected. Otherwise, recurrence is the rule.

15.9.8 Chordoma

Chordoma is a slow-growing tumor of the sacrococcygeal region (60%) or sphenoid clivus (30%) with a male preference and peaks between 3rd and 5th decades of life and an aggressive tendency to invade the surrounding bony tissue (spine and basal structures of the skull). X-ray shows axial location, geographic lytic bony destruction, soft tissue mass, and shards of bony detritus.

- GRO: Gray-white, gelatinous, highly friable, and quite lobulated neoplasm.
- CLM: Physaliphorous cells (Greek φυσαλλίς "bladder, bubble" and φόρος "carrying", which derives from φέρειν "to carry") bearing neoplasm with large, water-clear cells (PAS+) embedded in an amphiphilic matrix. These cells are sometimes confused with cartilaginous tumor cells, cells from clear cell renal cell carcinoma (CCRCC), or liposarcomas and need to be carefully differentiated.
- IHC: (+) Brachyury, S100, KER (CK8/18, CK19, AE1/AE3), EMA, 5′ nucleotidase, N-Cad, (±) VIM, CK903, CEA, and lysozyme.

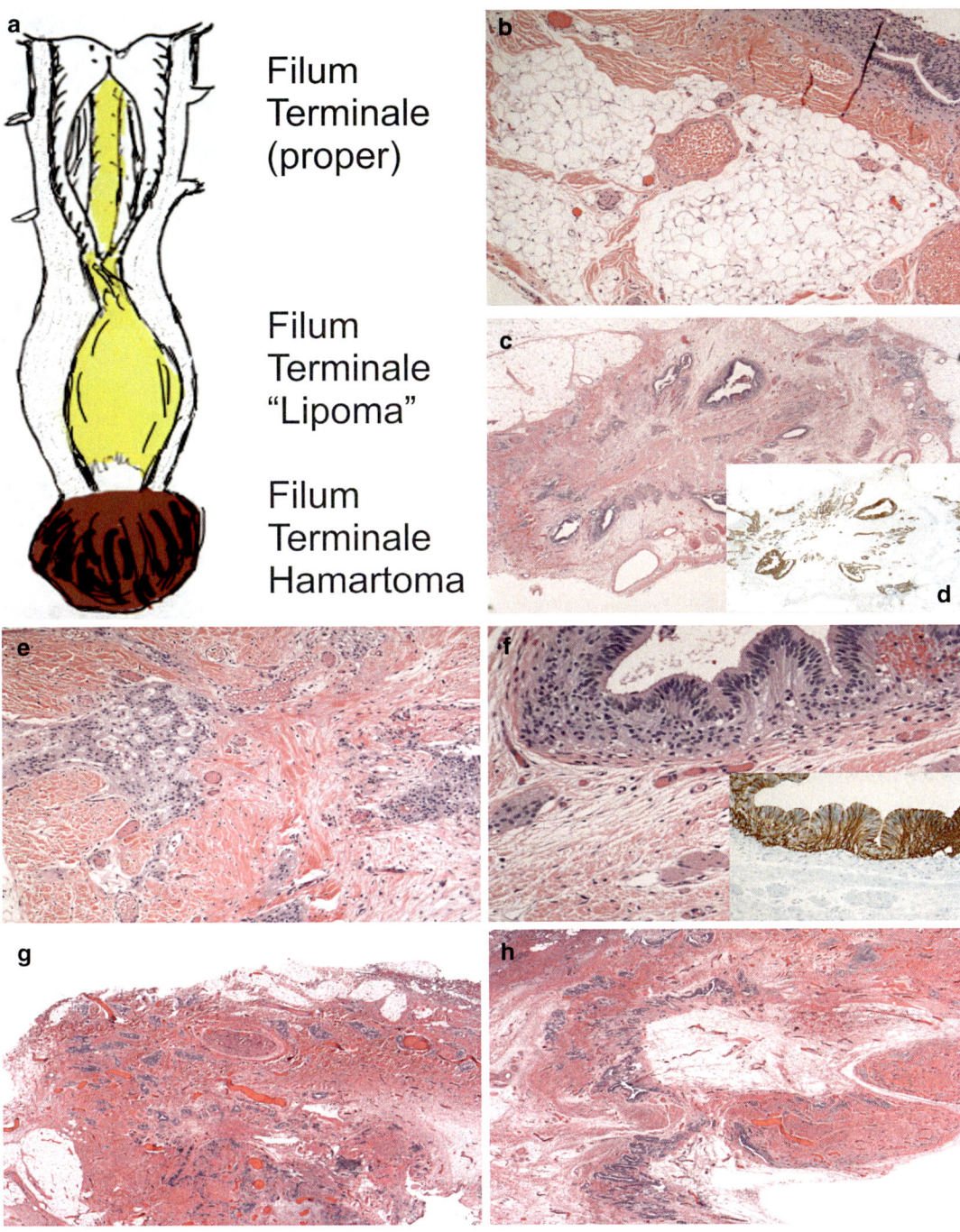

Fig. 15.12 The filum terminale representation with its "tumors" is shown in Figure **a**. The microphotographs **b**, **c**, **e**, **f**, **g**, and **h** show the histology of the filum terminale hamartoma, which is defined by an abnormal mixture of tissues and cells (normal) from the area in which it grows (autotochtonous growth) with dense connective tissue, fatty tissue, meningoepithelial proliferation, and occasional hemorrhage and calcification (hematoxylin-eosin staining). The inset **d** of Figure **c** and the inset of Figure **f** show the immunohistochemical expression of glial fibrillary acidic protein (GFAP) using avidin-biotin complex at different degrees of magnification (**b**, ×50; **c**, 20; **d**, ×20; **e**, ×50; **f**, ×100 with inset at ×400; **g**, ×50; **h**, ×50)

Brachyury may distinguish chordoma from histologic mimickers, including CCRCC, chondrosarcoma, and chordoid meningioma.
- *PGN*: The outcome is poor due to the local aggressivity of this kind of tumor.

15.9.9 Tumors of the Pineal Body

- *DEF*: These pediatric tumors are mostly teratoid and identical to germ cell tumors of the genital reproductive organs.
- *CLI*: Signs of increased ICP due to CSF flow obstruction at the level of the mesencephalic aqueduct or posterior 3rd ventricle, where the pineal gland is located.
- *GRO*: Variegated appearance with possible cystic change.
- *CLM*: Germ-cell histology is noted.
- *PGN*: The outcome is good for seminoma-based pineal body tumor, because of the radiosensitivity. Otherwise, poor survival is the rule.

15.9.10 Hematological Malignancies

Primary hematological solid malignancies of the brain and spinal cord (primary central nervous system lymphoma or PCNSL) are extranodal NHL of the brain, leptomeninges, eyes, or spinal cord without evidence of systemic disease. PCNSLs are relatively uncommon because they do not usually involve children, adolescents, or young adults. Conversely, leukemic infiltration of the brain in the setting of acute lymphoblastic leukemia may be observed in both childhood and youth (*vide infra*). Primary CNS NHL may, however, be found in younger patients, if they are immunocompromised (e.g., renal transplant recipients, AIDS). In fact, CNS lymphomas seem to be the most common primary brain neoplasm in immunocompromised patients from the epidemiological point of view. The lifetime risk of developing CNS lymphoma is approximately 1/5 of cases in patients with AIDS, which also show positive markers for EBV. Congenital immunodeficiency syndromes that predispose to NHL are ATS (Ataxia–Telangiectasia syndrome or Louis–Bar syndrome), WAS (Wiskott–Aldrich syndrome), SCID (Severe Combined Immuno-Deficiency), and CVID (Common Variable Immuno-Deficiency). Autoimmune conditions that may influence to NHL are RA, SLE, Sjögren syndrome, myasthenia gravis, sarcoidosis, and vasculitis.

DLBCL represents <1% of all NHL and 2–3% of all brain tumors. DLBCL is heterogeneous both clinically and morphologically and is classified into two groups, either germinal center B-cell-like (GCB) group or activated B-cell group according to the immunophenotype (Box 15.13). This distinction is not only academic because the ABC subgroup of DLBCL is characterized by constitutive activation of the NFkB pathway, which may represent a target for treatment strategy. In fact, DLBCL of the CNS has a poor prognosis relative to other extranodal DLBCL.

The ABC phenotype is predominant for PCNSL in both WHO classification of lymphoid malignancies and the literature covering this topic. The role of *BCL-2* as an outcome predictor in systemic DLBCL is controversial, but the BCL-2 expression has been indicated to be associated with the significant adverse outcome on overall survival within the ABC subgroup. Activation of the NFkB pathway or 18q21 amplification seems to be responsible for the upregulation of Bcl-2 expression in the ABC subgroup. The protein, called "signal transducer and activator of transcription 3" (STAT3) has been studied in PCNSL-DLBCL. Although the results are not concordant among several studies, it seems that PCNSL is a distinct subset of ABC DLBCL with low STAT3 expression and non-NFkB/STAT3 mechanisms may play a role in the upregulation of bcl-2.

> **Box 15.13 GCB/ABC Grouping of DLBCL of the Brain**
> - Subgroup *GCB*,
>   if CD10 (+) with or without BCL-6 (+) OR CD10 (−), BCL-6 (+), and MUM-1 (−)

- Subgroup *ABC*,
  if both CD10 (−) and BCL-6 (−) OR CD10 (−), BCL-6 (+), and MUM-1 (+).

  Notes: CD10 and BCL-6 are markers of GCB cells, while MUM1 is expressed in the later stages of B-cell development and plasma cells.

- *CLI*: A subtle onset (few months) with focal neurologic deficits associated with alterations of higher neural functions (neuropsychiatric symptoms and personality changes), headache, nausea, vomiting, seizures, and ocular symptoms is seen.
- *GRO*: There is a *multi-* more than unifocal and *cerebral* (supratentorial, noncerebellar) localized tumor, which is usually located in *periventricular areas* involving the thalamus, basal ganglia, and corpus callosum. Occasionally, an encephalitis-like picture with associated leptomeningeal and spinal cord involvement may dominate the presentation. The tumor is firm, homogeneously gray-tan or brownish and centrally necrotic with focal areas of hemorrhage. In case of diffuse infiltration, the term of *lymphomatosis cerebri* is used.
- *CLM*: Angiocentric pattern (encephalitis-like) with neoplastic cuffs of centroblasts or immunoblasts with high N/C ratio and numerous mitotic figures around cerebral blood vessels and also intraluminal. Reticulin stain may highlight the concentric perivascular arrangement of the reticulin fibers. Brain tissue is infiltrated as both small groups of tumor cells and individual cells. The diffuse growth pattern is the only pattern observed so far. Reactive astrocytic and microglial responses and accompanying reactive lymphocytes (CD4+ T cells) may be observed. PCNSL are mostly diffuse large B-cell lymphomas (DLBCL), and IHC shows (+) CD45 and (+) B-cell markers (CD20, CD79a). CD3 (T-cell marker) detects small benign admixed T-lymphocytes, and GFAP highlights the reactive gliosis. BCL2 is also positive. CD10, BCL6 (a Zn-finger transcriptional repressor required for the arrangement of the germinal center), and MUM-1 (a marker of germinal center/early post-germinal center B lymphocytes) helped to subdifferential DLBCL into germinal center B-cell and activated B-cell type. T-cell, low-grade *ALCL* (+ CD45, CD30, T-cell markers, EMA, ALK-1 and t(2;5)(p23;q35) ⇒ *NPM/ALK*) and *Hodgkin lymphoma* have been observed.
- *CMB*: Cytogenetically, gains of chromosomes 1, 7, 12, and 18 by CGH have been found in DLBCL, which are also characterized by homozygous deletion and promoter hypermethylation of CDKN2a, which produces p14ARF differently from systemic NHL. *ALK* on 2p23 produces a glycoprotein, which is a specific membrane-associated tyrosine kinase receptor, while *NPM1* on 5q35 is an RNA-binding nucleolar phosphoprotein involved in pre-ribosomal assembly. The hybrid gene (5' *NPM*-3' *ALK* on der(5)) is an 80 kDa protein of 680 amino acids that is localized both in the cytoplasm and in the nucleus. Also, the oncogenesis is linked to the kinase function activated by oligomerization of NPM-ALK mediated by the NPM part.
- *DDX*: GBM and metastases (e.g., small blue cell tumors) as malignant diseases and subacute infarctions, demyelinating diseases (axons are preserved in demyelinating disorders, while they are destroyed in infarctions), or infectious/parasitic-associated space-occupying lesions as benignancy. Useful histology patterns include the degree of cellular and nuclear pleomorphism, border infiltration, angio-proliferation, necrosis with pseudopalisading of tumor cells, tumor cell cohesiveness, and macrophagic-richness. Useful stains include reticulin stains, GFAP, CD68, pan-keratins (e.g., AE1/3), CD45, and B-cell markers as well as microbiological studies for *T. gondii* in immunocompromised patients.
- *PGN*: Some progress has been reached almost a century following PCNSL first description by Bailey in 1929. About outcome, 5-YSR is moderately good using combined radiation and chemotherapy. Good PGN factors for

DLBCL include unifocality, well circumscription, no meningeal or periventricular component, no immunodeficiency, and younger age.

15.9.11 Other Tumors and Metastatic Tumors

Other tumors may be overwhelming because of the heterogeneity. We may include here just three entities, e.g., occipital neuroblastoma (Fig. 15.13), the rhabdoid tumor (Fig. 15.14), and the leptomengeal melanomatosis (Fig. 15.15), among other entities (Fig. 15.16), which need a neuropathology consultation. The reader is advised to consult books of neuropathology for further information on neuro-oncological diseases.

One fourth or one-fifth of all brain tumors are metastatic neoplasms. Metastasis occurs through the bloodstream or direct invasion and diffuse spread by a paracranial tumor. In children, particularly relevant to consider are rhabdomyosarcoma, PNET, and Burkitt lymphoma or nasopharyngeal carcinoma. In adults, lung carcinoma, leukemias, intestinal carcinoma, clear cell renal cell carcinoma, and breast carcinomas are most often seen. Less frequently, malignant melanoma, follicular thyroid carcinoma, and sarcoma are observed. Clinically, there are signs and symptoms of increased ICP, as well as other signs of mass effect are found. Grossly, most metastatic neoplasms are located at the junction of the gray and white matter, which is a particular region quite poor in blood vessels, and the circulation is slower than in other parts of the CNS. Microscopically, most metastatic tumors show perivascular proximity, seem to be discretely separated from the surrounding brain, and show features of the original tumor.

Leukemia is the most common form of neoplasm in childhood and the advances in imaging techniques, and chemotherapeutic protocols have prolonged survival indeed. CNS is the "sanctuary" for blasts of ALL. These premature cells are protected by the blood-brain barrier and cannot be eradicated with systemic chemotherapy alone, and cranial irradiation is often necessary. The side effect is the increase of CNS complications evidenced in numerous recent studies. CNS complications can be divided into (1) *Direct* effects or (2) *Indirect* effects, i.e., from underlying leukemia or the anti-leukemic therapy. Leukemic direct effect-related CNS sites are leptomeninges, brain parenchyma, and/or blood vessels, while treatment-related CNS sites involve white matter, small and large blood vessels, brain parenchyma, and ventricular system. Moreover, therapy-related effects may also include the onset of secondary tumors and infections.

Leukemia-related CNS complications include:

- Leptomeningeal leukemia, which may show also an infiltration of adjacent bone and scalp
- Myeloid sarcoma (*chloroma*) of the orbit
- Posterior reversible leukoencephalopathy syndrome (the so-called PRES)
- Bilateral retinal hemorrhages
- Intraparenchymal hemorrhage

Of note, PRES is a syndrome characterized by headache and confusion, which may be accompanied by seizures and visual loss. PRES may occur due to several factors, including malignant hypertension and eclampsia, and may also be drug-related.

It is important to recall that cytologic investigation is necessary for the diagnosis of leukemic meningitis, but repeated analysis of CSF is often requested owing to the high incidence of false-negative cytologic findings. Myeloid sarcoma occurs in 3–8% of pediatric patients with AML and may be detected in the orbits, temporal bone, cerebellopontine angle, and spinal canal, but it can occur anywhere throughout the body. This diagnosis needs to be considered a priority (a common medical abbreviation for urgent or rush is "STAT" from the Latin word *statum*, meaning 'immediately.'). The oncologist needs to act fast because myeloid sarcomas are radiosensitive. In this setting, radiotherapy and chemotherapy with or without surgery are warranted.

DIC-related therapy or leukemia-based with hemorrhages in small or large areas of the brain parenchyma may also occur, and retinal hemorrhages, usually bilateral and posteriorly located,

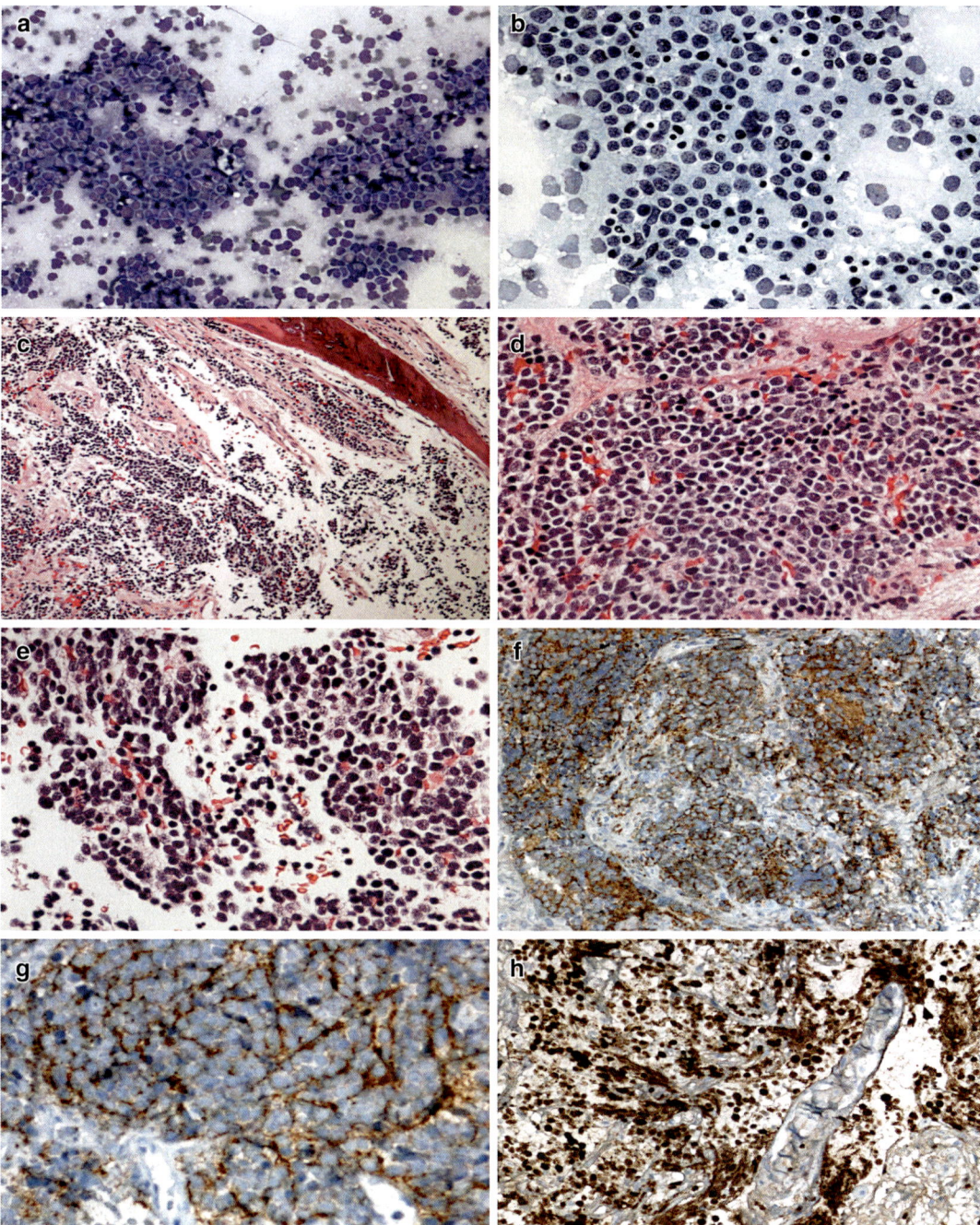

Fig. 15.13 This panel shows the rare occurrence of an occipital neuroblastoma that needs to be taken into the differential diagnosis with medulloblastoma. The cytologic features of this neuroblastoma are shown in Figures **a** and **b** (Diff-Quik staining and Giemsa staining, respectively), while the microphotographs (**c**–**e**) show the characteristic histology of the neuroblastoma (small round blue cell tumor) with Homer-Wright rosettes with pink neuropil surrounded by tumor cells (**d**) (**a**, ×200; **b**, ×400; **c**, ×100; **d**, ×200; **e**, ×200). The immunohistochemical expression of neuron-specific enolase (NSE), synaptophysin (SYN), and Ki67 (proliferation antigen) is shown in Figures **f**, **g**, and **h**, respectively (**f**, ×200; **g**, ×200; **h**, ×100)

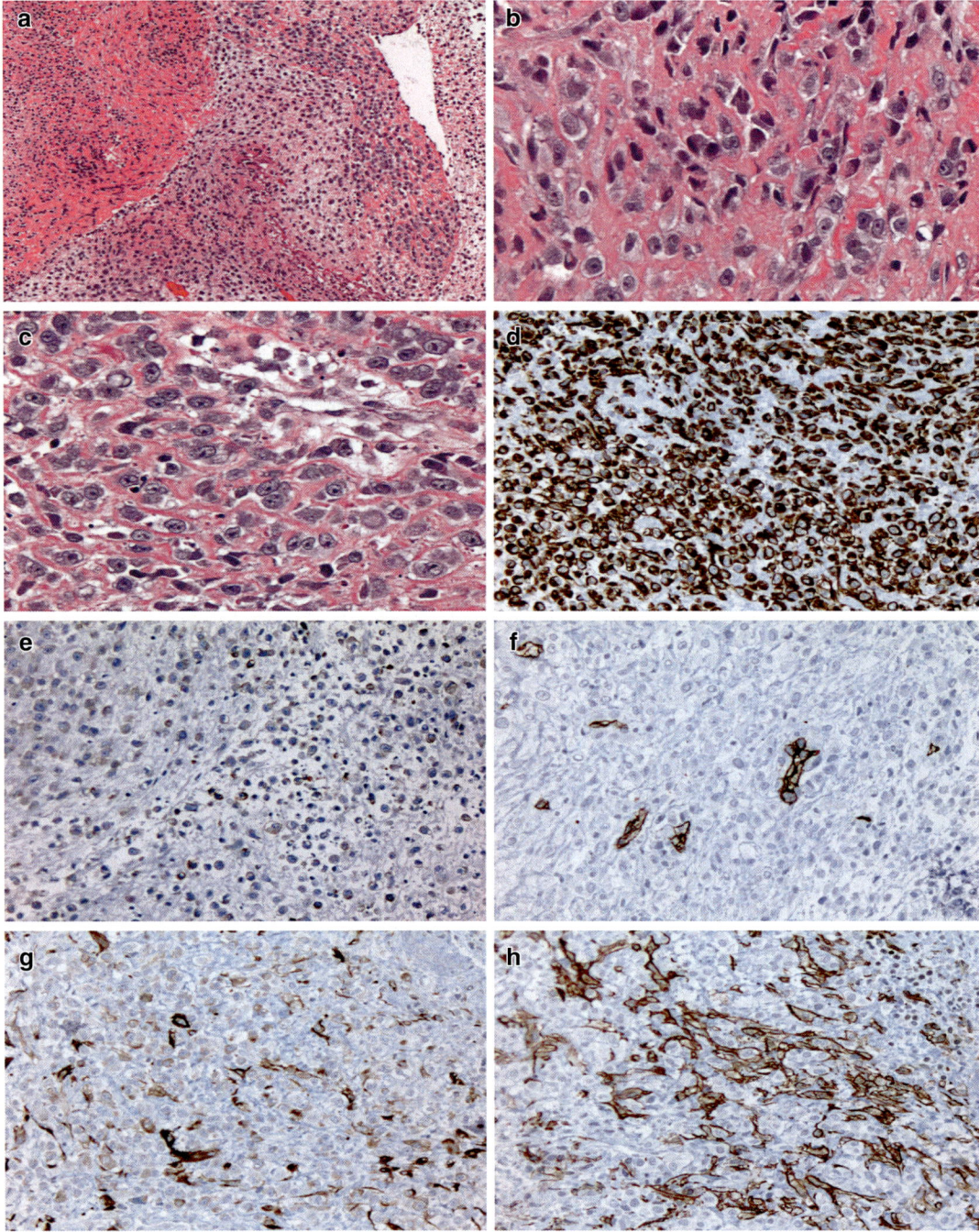

Fig. 15.14 The first three microphotographs illustrate the characteristic of this malignant embryonal tumor of the central nervous system that is composed of poorly differentiated cells (rhabdoid cells) and loss of INI1. Either SMARCB1 or SMARCA4 genes need to be lost for the diagnosis of atypical teratoid/rhabdoid tumor (ATRT) (**a**, ×100; **b**, ×400; **c**, ×400; original magnifications). The immunohistochemical microphotographs (**d**–**h**) show the expression of vimentin (**d**, ×200), epithelial membrane antigen (**e**, ×200), CD34 (**f**, ×200), desmin (**g**, ×200), and actin (**h**, ×200). All immunohistochemical assays have been performed using the avidin-biotin complex

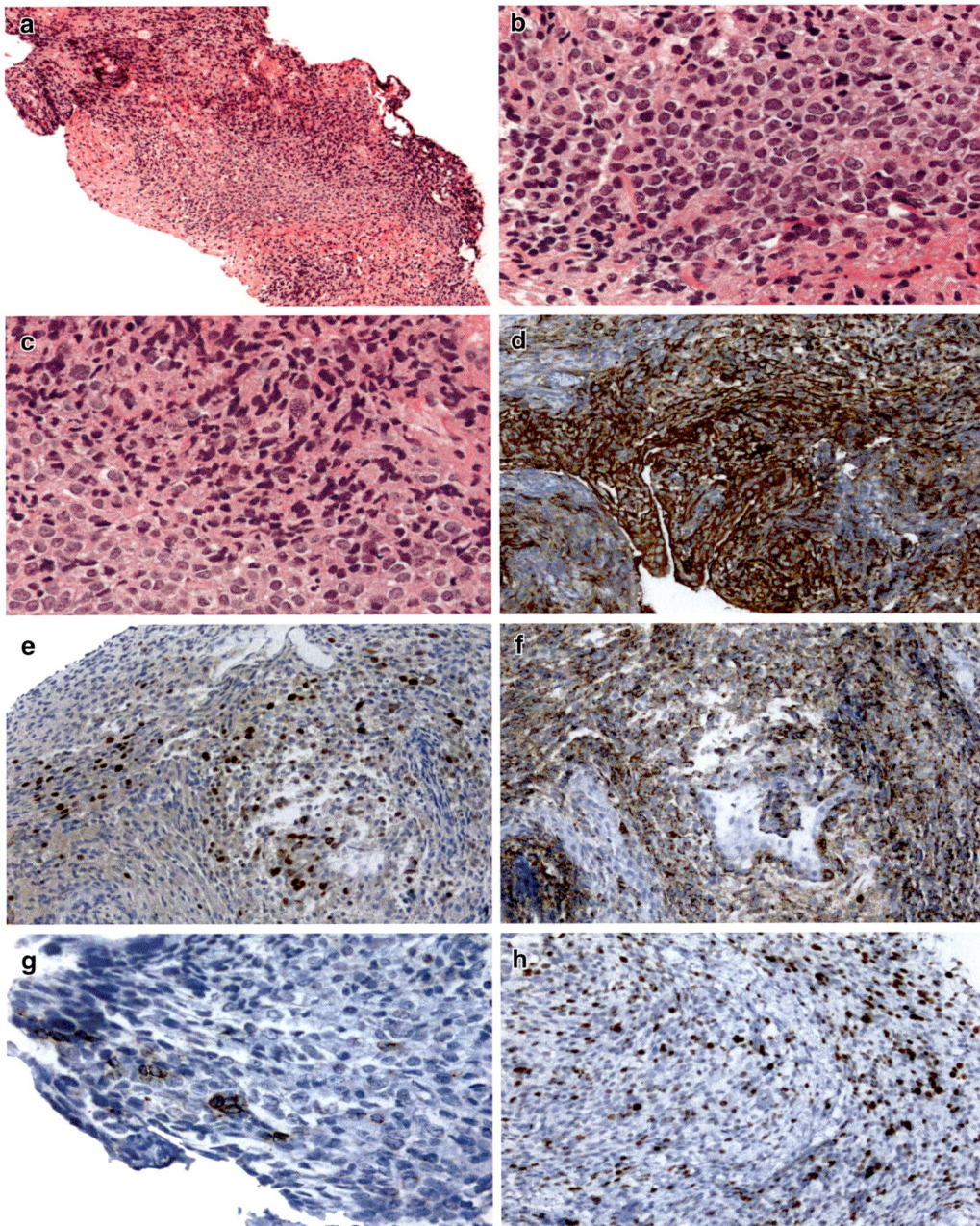

Fig. 15.15 Leptomeningeal melanomatosis is an aggressive form of meningeal melanocytosis and primary melanocytic tumors of the central nervous system with proliferation of melanocytic cells within the subarachnoid space and a strong association with cutaneous melanocytic lesions. Figures **a–c** show the dense infiltration of the leptomenges by densely packed with high nucleus to cytoplasm ratio exhibiting at places a prominent nucleolus. Cellular atypia, high mitotic rate, and necrosis are not uncommon (hematoxylin and eosin staining, **a**, ×100; **b**, ×400; **c**, ×400, original magnifications). The subsequent five microphotographs (**d–h**) show the expression of vimentin (**d**, ×200, original magnification), S100 (**e**, ×200, original magnification), HMB45 (**f**, ×200, original magnification), MART-1 (**g**, ×400, original magnification), and Ki67 proliferation antigen (**h**, ×200, original magnification). All immunohistochemistry slides have been performed using the avidin-biotin complex method. The HMB45 or Human Melanoma Black 45 is a monoclonal antibody directed against the melanosomal glycoprotein gp100 (Pmel17). MART-1 (or MelanA, a synonymous) is an antigen recognized by tumor infiltrating cytotoxic T cells originally identified from a melanoma patient (melanoma antigen recognized by T cells 1, melanoma antigen)

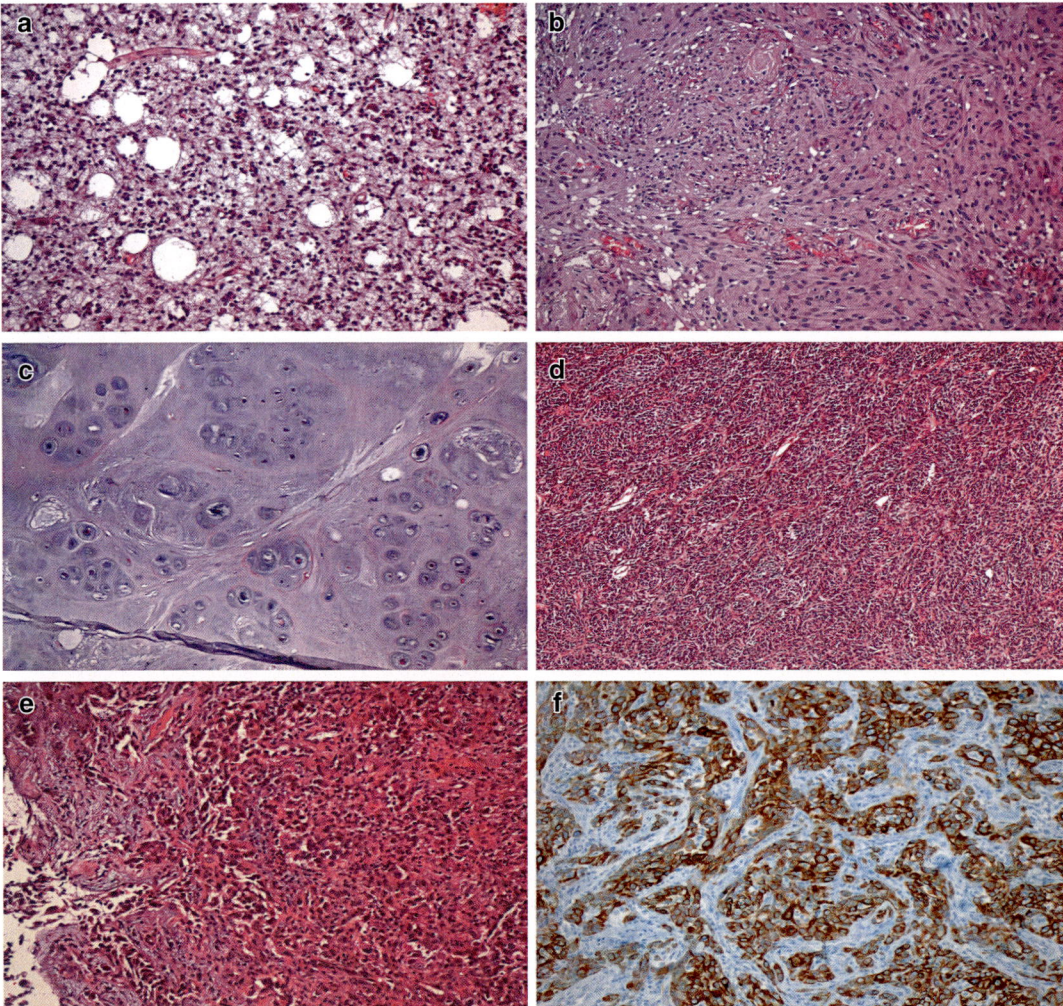

Fig. 15.16 This six-photograph panel illustrated the histology of an astrocytoma (**a**, ×200, original magnification, hematoxylin and eosin staining), meningioma (**b**, ×200, original magnification, hematoxylin, and eosin staining), chondroma (**c**, ×200, original magnification, hematoxylin, and eosin staining), mesenchymal tumor of the central nervous system not otherwise specified (**d**, ×100, original magnification, hematoxylin and eosin staining), the metastasis of an adenocarcinoma (**e**, ×200, original magnification, hematoxylin and eosin staining), and the corresponding identification of the tumor cells using a monoclonal antibody against cytokeratin 7 (**f**, ×200, immunohistochemistry using avidin-biotin complex)

are the most frequent finding, being the retina the most frequently involved structure. Thrombosis is also a complication of remission induction with the combination of prednisone-vincristine-asparaginase. Drug-induced PRES with high-intensity lesions on T2-weighted images of the cortex and subcortical white matter can also occur. Drugs associated with PRES include cyclosporine, tacrolimus, anti-retroviral therapy, erythropoietin, IFN-alpha, corticosteroids, and chemotherapeutic agents such as asparaginase, cisplatin, cytarabine, methotrexate, and vincristine. The most common secondary tumors post-radiotherapy include sarcoma (especially osteosarcoma of the skull) and meningioma, rarely glioblastoma multiforme. Radiotherapy side effects also include necrotizing diffuse leukoencephalopathy, mineralizing microangiopathy, a volume loss of brain parenchyma ("secondary atrophy"), and cryptic vascular malformations. *Candida* and *Aspergillus* species are the organisms most often observed in large series. Endocrinopathies caused by the late effects of the treatment can also occur. Finally, neurodevelopmental sequelae of childhood need to be taken into consideration, and ALL is associated with long-term neurodevelopmental sequelae causing a decrease in global IQ of about 10 IQ points following cranial irradiation. Risk factors include young age, female, time since, and cranial irradiation dosage. Conversely, using chemotherapy alone does not typically compromise global IQ, although some subtle deficits may be identified such as arithmetic performance, attention, processing of the information, visual-motor integration, excellent motor skills, and verbal memory. Also, in this case, young age, female, and treatment intensity are risk factors. Relatively rare are treatment-related neurodevelopmental sequelae in AML, although AML itself and related cerebrovascular complications can represent findings seen as long-term neurodevelopmental deficits in these patients.

Multiple Choice Questions and Answers

- CNS-1 Which of the following risk factors are NOT modifiable risk factors of neural tube defects?
  (a) Genetic defects
  (b) Maternal diabetes
  (c) Maternal obesity
  (d) Maternal hyperthermia
  (e) Valproate
  (f) Inadequate maternal nutritional status
- CNS-2 What is the most appropriate definition of *Rhombencephalosynapsis*?
  (a) Bilateral brain malformation characterized by a fusion of the cerebellar hemispheres.
  (b) Bilateral brain malformation characterized by missing cerebellar vermis and fusion of the cerebellar hemispheres.
  (c) Midline brain malformation characterized by missing cerebellar vermis with a fusion of the cerebellar hemispheres.
  (d) Midline brain malformation characterized by a fusion of the cerebellar hemispheres.
- CNS-3 An 18-year-old boy suffers from a head injury playing at home with an older sibling with mental disabilities. Parents are consanguineous (first-cousins) and are recently immigrants from Middle East. The boy develops the symptoms of the syndrome of inappropriate anti-diuresis hormone (ADH) or vasopressin following an abnormally high level of its secretion. Which signs are going to be be experienced by the attending physician during the physical examination and lab work-up of this boy?
  (a) Hyponatremia due to direct inhibition of ADH on distal tubular sodium resorption
  (b) Hyponatremia due to the dilutional effect of ADH-induced hydric retention in the collecting tubules
  (c) Hypernatremia due to direct stimulation of ADH on distal tubular sodium resorption
  (d) Hypernatremia due to the concentrating effect of ADH-induced hydric retention in the collecting tubules
  (e) Normal plasmatic levels of natremia, because the dilutional effect of the ADH action is counterbalanced by a direct stimulatory effect of ADH on distal tubular resorption of the sodium
- CNS-4 Which of the following statements does NOT belong to Tay-Sachs disease?
  (a) Mutations in the *HEXA* gene cause Tay-Sachs disease.
  (b) It is more common among Ashkenazi Jews than other ethnic groups.
  (c) It is a lysosomal storage disease.
  (d) Accumulation of GM2 gangliosides characterizes Tay-Sachs disease.
  (e) It has an autosomal dominant pattern of inheritance.
- CNS-5 Match the correct definition of each of the three types of Arnold-Chiari malformation?
  (a) Chiari type 1
  (b) Chiari type 2
  (c) Chiari type 3

1. Isolated protrusion of the cerebellar tissue and hydrocephalus
2. Isolated protrusion of the cerebellar tissue, hydrocephalus, and *cranium bifidum* with optional encephalocele (protrusion of cerebral tissue)
3. Isolated protrusion of the cerebellar tissue

   (a)(3), (b)(1), (c)(2)

- CNS-6 A young woman cannot move either the right side of her face or her left extremities. She had right-sided ptosis of the eyelid. The right eye deviated laterally. After light examination, the right eye did not respond and did not show accommodation. Where is the lesion most likely located?
  (a) Telencephalon
  (b) Posterior lobes
  (c) Midbrain
  (d) Pons
  (e) Medulla

- CNS-7 A 2-year-old boy develops acute hemiplegia as a result of an internal carotid artery thrombosis. Which of the following differential diagnosis is NOT appropriate?
  (a) Periarteritis nodosa
  (b) Sickle cell anemia
  (c) Systemic lupus erythematosus
  (d) Todd paralysis
  (e) Hemiplegic migraine
  (f) Bell paralysis

- CNS-8 A 6-year-old girl received no varicella vaccination and got infected developing full disease. During the recovering time, she develops gait ataxia. The child is examined by the pediatrician and reveals having difficulty reaching for objects due to an intention tremor. The girl is alert and afebrile, does not have meningismus, and has no headache. At the examination, she had deep tendon reflexes, and her plantar responses are flexor. A wide-based gait is noted when she walks and sways while standing still. What is the most likely diagnosis?
  (a) Reye syndrome
  (b) Postinfectious encephalomyelitis
  (c) Varicella encephalitis
  (d) Ataxia-telangiectasia
  (e) Guillain-Barré syndrome

- CNS-9 Medulloblastoma (MB) is one of the most common malignant brain tumors in childhood. Surgical resection, craniospinal irradiation, and adjuvant chemotherapy represent the current standard treatment regimen. There are four molecular subgroups. Which of the following corresponds to the four subgroups?
  (a) TP53, WNT, sonic hedgehog (SHH), and RET
  (b) WNT, sonic hedgehog (SHH), Group 3, and Group 4
  (c) RET, TP53, Frizzled (FRZ), and SHH
  (d) TP53, WNT, Group 3, and Group 4

- CNS-10 Spinal muscular atrophies are mostly autosomal recessive disorders with rare occurrences of autosomal dominant and X-linked patterns of inheritance. Match the right presentation with three different types of spinal muscular atrophy.
  (a) Spinal muscular atrophy, type I (Werdnig-Hoffmann disease)
  (b) Spinal muscular atrophy, type II
  (c) Spinal muscular atrophy, type III (Kugelberg Welander disease)
     1. The onset of weakness occurs after walking has been established (even in adolescence).
     2. The onset of weakness occurs after 4–8 months of age when the child can sit.
  3. The onset of weakness occurs at birth or shortly after that.

     (a)(3) (b)(2) (c)(1)

References and Recommended Readings

Abe T, Nishi S, Furukawa T, Ajioka Y, Masuko M, Fuse I. Acute renal failure induced by adenovirus after stem cell transplantation. 2011; https://doi.org/10.5772/21995.

Acampora D, Mazan S, Lallemand Y, Avantaggiato V, Maury M, Simeone A, Brulet P. Forebrain and midbrain regions are deleted in $Otx2^{-/-}$-mutants due to a defective anterior neuroectoderm specification during gastrulation. Development. 1995;121:3270–90.

Ackerman LL, Menezes AH. Spinal congenital dermal sinuses: a 30-year experience. Pediatrics. 2003;112(3 Pt 1):641–7. PubMed PMID: 12949296.

Aicardi J. Malformations of the CNS. In: Aicardi J, editor. Diseases of the nervous system in childhood. London: Mc Keith Press; 1992. p. 108–203.

Allen LH, Miller JW, de Groot L, Rosenberg IH, Smith AD, Refsum H, Raiten DJ. Biomarkers of nutrition for development (BOND): vitamin B-12 review. J Nutr. 2018;148(suppl_4):1995S–2027S. https://doi.org/10.1093/jn/nxy201. Review. PubMed PMID: 30500928; PubMed Central PMCID: PMC6297555.

Ang SL, Jin O, Rinn M, Daigle N, Stevenson L, Rossant J. A targeted mouse *otx2* mutation leads to severe defects in gastrulation and formation of axial mesoderm and to deletion of rostral brain. Development. 1996;122:243–52.

Ardern-Holmes S, Fisher G, North K. Neurofibromatosis type 2: presentation, major complications, and management, with a focus on the pediatric age group. J Child Neurol. 2017;32(1):9–22.

Armao D, Castillo M, Chen H, Kwock L. Colloid cyst of the third ventricle: imaging-pathologic correlation. AJNR Am J Neuroradiol. 2000;21(8):1470–7. PubMed PMID: 11003281.

Bailey P. Intracranial sarcomatous tumours of leptomeningeal origin. Arch Surg. 1929;18:1359–402.

Balkan W, Phillips LS, Goldstein S, Sadler TW. Potential role of somatomedin inhibitors in the production of diabetic embryopathies. Teratology. 1988;37:271–82.

Barbanti-Brodano G, Sabbioni S, Martini F, et al. BK virus, JC virus and Simian virus 40 infection in humans, and association with human tumors. In: Madame Curie bioscience database [Internet]. Austin: Landes Bioscience; 2000–2013. Available from: https://www.ncbi.nlm.nih.gov/books/NBK6100/

Barnes JD, Crosby JL, Jones CM, Wright CV, Hogan BL. Embryonic expression of *Lim-1*, the mouse homolog of Xenopus *Xlim-1*, suggests a role in lateral mesoderm differentiation and neurogenesis. Dev Biol. 1994;161:168–78.

Bartelmez GW, Dekaban AS. The early development of the human brain. Contrib Embryol Carnegie Instn. 1962;37:13–32.

Baser ME, Friedman JM, Joe H, Shenton A, Wallace AJ, Ramsden RT, Evans DG. Empirical development of improved diagnostic criteria for neurofibromatosis 2. Genet Med. 2011;13(6):576–81. https://doi.org/10.1097/GIM.0b013e318211faa9. PubMed PMID: 21451418.

Baumgartner-Sigl S, Haberlandt E, Mumm S, Scholl-Bürgi S, Sergi C, Ryan L, Ericson KL, Whyte MP, Högler W. Pyridoxine-responsive seizures as the first symptom of infantile hypophosphatasia caused by two novel missense mutations (c.677T>C, p.M226T; c.1112C>T, p.T371I) of the tissue-nonspecific alkaline phosphatase gene. Bone. 2007;40(6):1655–61. Epub 2007 Feb 14. PubMed PMID: 17395561.

Bennetto L, Scolding N. Inflammatory/post-infectious encephalomyelitis. J Neurol Neurosurg Psychiatry. 2004;75(Suppl 1):i22–8. Review. PubMed PMID: 14978147; PubMed Central PMCID: PMC1765651.

Bhagavathi S, Wilson JD. Primary central nervous system lymphoma. Arch Pathol Lab Med. 2008;132(11):1830–4. https://doi.org/10.1043/1543-2165-132.11.1830. Review. PubMed PMID: 18976024.

Bhat JI, Ahmed QI, Ahangar AA, Charoo BA, Sheikh MA, Syed WA. Wernicke's encephalopathy in exclusive breastfed infants. World J Pediatr. 2017;13(5):485–8. https://doi.org/10.1007/s12519-017-0039-0. Epub 2017 May 24. PubMed PMID: 28540694.

Bitsche M, Schrott-Fischer A, Hinterhoelzl J, Fischer-Colbrie R, Sergi C, Glueckert R, Humpel C, Marksteiner J. First localization and biochemical identification of chromogranin B- and secretoneurin-like immunoreactivity in the fetal human vagal/nucleus solitary complex. Regul Pept. 2006;134(2–3):97–104. Epub 2006 Mar 10. PubMed PMID: 16530281.

Boettger MB, Kirchhof K, Sergi C, Sakmann C, Meyer P. Colobomas of the iris and choroid and high signal intensity cerebral foci on T2-weighted magnetic resonance images in Klinefelter's syndrome. J Pediatr Ophthalmol Strabismus. 2004;41(4):247–8. PubMed PMID: 15305539.

Boltenstern M, Konrad A, Jost W, Uder M, Kujat C. Rhombenzephalosynapsis. Rofo Fortschr Geb Rontgenstr Neuen Bildgeb Verfahr. 1995;163:91–3.

Boragina M, Cohen E. An infant with the "setting-sun" eye phenomenon. CMAJ. 2006;175(8):878. https://doi.org/10.1503/cmaj.060507. PMID: 17030938; PMCID: PMC1586074.

Bordarier C, Aicardi J, GoutiŠres F. Congenital hydrocephalus and eye abnormalities with severe developmental brain defects: Warburg's syndrome. Ann Neurol. 1984;16:660–5.

Boyle MI, Jespersgaard C, Brøndum-Nielsen K, Bisgaard AM, Tümer Z. Cornelia de Lange syndrome. Clin Genet. 2015;88(1):1–12. https://doi.org/10.1111/cge.12499. Epub 2014 Oct 28. Review. PubMed PMID: 25209348.

Buntinx JM, Bourgeoise N, Buyaert PM, Dumon JE. Acardiac amorphus twin with prune belly sequence in the co-twin. Am J Med Genet. 1991;39:453–7.

Butler AB, Hodos W. Segmental organization of the head, brain, and cranial nerves. In: Butler AB, Hodos W, editors. Comparative vertebrate neuroanatomy. Evolution and adaptation. New York: Wiley-Liss; 1996. p. 120–32.

Cairncross JG, Macdonald DR. Lumbar puncture. In: Wittes RE, editor. Manual of oncologic therapeutics. Philadelphia: Lippincott; 1991. p. 305–7.

Castellani RJ, Mojica G, Perry G. The role of the Iron stain in assessing intracranial hemorrhage. Open Neurol J. 2016;10:136–42. https://doi.org/10.2174/1874205X01610010136. PMID: 27857815; PMCID: PMC5090778.

Cavanagh JB. Corpora-amylacea and the family of polyglucosan diseases. Brain Res Brain Res Rev. 1999;29(2–3):265–95. Review. PubMed PMID: 10209236.

Cefalo MG, De Ioris MA, Cacchione A, Longo D, Staccioli S, Arcioni F, Bernardi B, Mastronuzzi A. Wernicke encephalopathy in pediatric neuro-oncology: presentation of 2 cases and review of literature. J Child Neurol. 2014;29(12):NP181–5. https://doi.org/10.1177/0883073813510355. Epub 2013 Nov 28. Review. PubMed PMID: 24293308.

Chassaing N, Sorrentino S, Davis EE, Martin-Coignard D, Iacovelli A, Paznekas W, Webb BD, Faye-Petersen O, Encha-Razavi F, Lequeux L, Vigouroux A, Yesilyurt A, Boyadjiev SA, Kayserili H, Loget P, Carles D, Sergi C, Puvabanditsin S, Chen CP, Etchevers HC, Katsanis N, Mercer CL, Calvas P, Jabs EW. OTX2 mutations contribute to the otocephaly-dysgnathia complex. J Med Genet. 2012;49(6):373–9. https://doi.org/10.1136/jmedgenet-2012-100892. Epub 2012 May 10. PubMed PMID: 22577225.

Chatkupt S, Hol FA, Shugart YY, Geurds MP, Stenroos ES, Koenigsberger MR, Hamel BC, Johnson WG, Mariman EC. Absence of linkage between familial neural tube defects and PAX3 gene. J Med Genet. 1995;32:200–4.

Chavrier P, Janssen-Timmen U, Mattei MG, Zerial M, Bravo R, Charnay P. Structure, chromosome location, and expression of the mouse zinc finger gene *Krox 20*: multiple gene products and coregulation with the proto-oncogene *c-fos*. Mol Cell Biol. 1989;9:787–97.

Chi JG. Acardiac twins – an analysis of 10 cases. J Korean Med Sci. 1989;4:203–16.

Chisaka O, Musci TS, Capecchi MR. Developmental defects of the ear, cranial nerves and hindbrain resulting from targeted disruption of the mouse homeobox gene *Hox-1.6*. Nature. 1992;355:516–20.

Cincotta RB, Fisk NM. Current thoughts on twin-twin transfusion syndrome. Clin Obstet Gynecol. 1997;40:290–302.

Claudius FM. Die Entwicklung der herzlosen Missgeburten. Kiel: Schwers'sche Buchhandlung; 1859. p. 7, 23.

Cohen MM. An update on the holoprosencephalic disorders. J Pediatr. 1982;101:865–9.

Cohen MM. Perspectives on holoprosencephaly: part III. Spectra, distinctions, continuities, and discontinuities. Am J Med Genet. 1989;34:271–84.

Cohen MM Jr, Jirásek JE, Guzman RT, Gorlin RJ, Peterson MQ. Holoprosence-phaly and facial dysmorphia: nosology, etiology and pathogenesis. Birth Defects Orig Artic Ser. 1971;7:125–35.

Cohen AL, Holmen SL, Colman H. IDH1 and IDH2 mutations in gliomas. Curr Neurol Neurosci Rep. 2013;13(5):345. https://doi.org/10.1007/s11910-013-0345-4. PMID: 23532369; PMCID: PMC4109985.

Cowdry EV. The problem of intranuclear inclusions in virus diseases. Arch Pathol. 1934;18:527–42.

Crossley PH, Martin GR. The mouse *Fgf8* gene encodes a family of polypeptides and is expressed in regions that direct outgrowth and patterning in the developing embryo. Development. 1995;121:439–51.

Crossley PH, Martinez S, Martin GR. Midbrain development induced by *Fgf8* in the duck embryo. Nature. 1996;380:66–8.

Damsker JM, Hansen AM, Caspi RR. Th1 and Th17 cells: adversaries and collaborators. Ann N Y Acad Sci. 2010;1183:211–21. https://doi.org/10.1111/j.1749-6632.2009.05133.x. PMID: 20146717; PMCID: PMC2914500.

Danielian PS, McMahon AP. Engrailed-1 as a target of the Wnt-1 signalling pathway in vertebrate midbrain development. Nature. 1996;383(6598):332–4. PubMed PMID: 8848044.

De Morsier G. Etudes sur les dysraphies cranio-encéphaliques. II: Agénésie du vermis cérébelleux Dysraphies rhomboencéphalique médiane (rhomboschizis). Monatsschr Psychiatr Neurol. 1955;129:321–34.

De Myer W, Zeman W. Alobar prosencephaly (arhinencephaly) with median cleft lip and palate: clinical, electroencephalographic and nosologic considerations. Confin Neurol. 1963;23:1–36.

Deacon JS, Machin GA, Martin JME, Nicholson S, Nwankwo DC, Wintemute R. Investigation of acephalus. Am J Med Genet. 1980;5:85–99.

Dekaban A. Brain dysfunction in congenital malformations of the nervous system. In: Gaull GE, editor. Biology of brain dysfunction, vol. III. New York/London: Plenum Press; 1975. p. 381–423.

Demaerel P, Kendall BE, Wilms G, Halpin SF, Casaer P, Baert AL. Uncommon posterior cranial fossa anomalies: MRI with clinical correlation. Neuroradiology. 1995;37:72–6.

DeMyer W. The median cleft face syndrome. Differential diagnosis of cranium bifidum occultum, hypertelorism, and median cleft nose, lip, and palate. Neurology. 1967;17(10):961–71. PubMed PMID: 6069608.

DeMyer W. Median facial malformations and their implications for brain malformations. Birth Defects Orig Artic Ser. 1975;11(7):155–81. Review. PubMed PMID: 764897.

DeMyer W, Zeman W. Alobar holoprosencephaly (arhinencephaly) with median cleft lip and palate: clinical, electroencephalographic and nosologic considerations. Confin Neurol. 1963;23:1–36.

DeMyer W, Zeman W, Palmer C-G. The face predicts the brain: diagnostic significance of median facial anomalies for holoprosencephaly (arhinencephaly). Pediatrics. 1964;34:256–63.

Dobrozsi S, Flood VH, Panepinto J, Scott JP, Brandow A. Vitamin B12 deficiency: the great masquerader.

Pediatr Blood Cancer. 2014;61(4):753–5. https://doi.org/10.1002/pbc.24784. Epub 2013 Sep 21. PubMed PMID: 24115632.

Dong WF, Heng HH, Larsky R, Xu Y, DeCocteau JF, Shi XM, Tsui LC, Minden MD. Cloning, expression, and chromosomal localization to 11p12-13 of a human LIM/HOMEOBOX gene, hLim-1. DNA Cell Biol. 1997;16:671–8.

Dubourg C, Bendavid C, Pasquier L, Henry C, Odent S, David V. Holoprosencephaly. Orphanet J Rare Dis. 2007, 2:8. Review. PubMed PMID: 17274816; PubMed Central PMCID: PMC1802747.

Edwards BO, Fisher AQ, Flannery DB. Joubert syndrome: early diagnosis by recognition of the behavioral phenotype and confirmation by cranial sonography. J Child Neurol. 1988;3:247–9.

Egger J, Bellman MH, Ross EM, Baraitser M. Joubert-Boltshauser syndrome with polydactyly in siblings. J Neurol Neurosurg Psychiatry. 1982;45:737–9.

El-Dib M, Inder TE, Chalak LF, Massaro AN, Thoresen M, Gunn AJ. Should therapeutic hypothermia be offered to babies with mild neonatal encephalopathy in the first 6 h after birth? Pediatr Res. 2019;85(4):442–8. https://doi.org/10.1038/s41390-019-0291-1. Epub 2019 Jan 16. PubMed PMID: 30733613.

Erickson JD. Risk factors for birth defects: data from the Atlanta birth defects case-control study. Teratology. 1991;43:41–51.

Fedrizzi T, Meehan CJ, Grottola A, Giacobazzi E, Fregni Serpini G, Tagliazucchi S, Fabio A, Bettua C, Bertorelli R, De Sanctis V, Rumpianesi F, Pecorari M, Jousson O, Tortoli E, Segata N. Genomic characterization of nontuberculous mycobacteria. Sci Rep. 2017;7:45258. https://doi.org/10.1038/srep45258. PubMed PMID: 28345639; PubMed Central PMCID: PMC5366915.

Fisk NM, Ware M, Stanier P, Moore G, Bennett P. Molecular genetic etiology of twin reversed arterial perfusion sequence. Am J Obstet Gynecol. 1996;174:891–4.

Florell SR, Townsend JJ, Klatt EC, Pysher TJ, Coffin CM, Wittwer CT, Viskochil DH. Aprosencephaly and cerebellar dysgenesis in sibs. Am J Med Genet. 1996;63:542–8.

Friede RL. Uncommon syndromes of cerebellar vermis aplasia. II: tectocerebellar dysraphia with occipital encephalocele. Dev Med Child Neurol. 1978;20:746–72.

Friede RL, Boltshauser E. Uncommon syndromes of cerebellar vermis aplasia. I. Joubert syndrome. Dev Med Child Neurol. 1978;20:758–63.

Gallagher ER, Evans KN, Hing AV, Cunningham ML. Bathrocephaly: a head shape associated with a persistent mendosal suture. Cleft Palate Craniofac J. 2013;50:104–8.

Garcia CA, Duncan C. Atelencephalic microcephaly. Dev Med Child Neurol. 1977;19:227–32.

Gilbert SF. Developmental biology. 5th ed. Sunderland: Sinauer; 1997.

Gilland E, Baker R. Longitudinal and tangential migration of cranial nerve efferent neurons in the developing hindbrain of *Squalus acanthias*. Biol Bull. 1992;183:356–8.

Goldsmith CL, Tawagi GF, Carpenter BF, Speevak MD, Hunter AG. Mosaic r(13) in an infant with aprosencephaly. Am J Med Genet. 1993;47(4):531–3. PubMed PMID: 8256818.

Gross H. Die Rhombencephalosynapsis, eine systemisierte Kleinhirnfehlbildung. Arch Psychiatr Nervenkr. 1959;199:537–52.

Gross H, Jellinger K, Kaltenback E. Rhombencephalosynapsis eine seltene Dysraphieform. Zentralbl Allg Pathol Pathol Anat. 1978;122:577.

Groth A, Enoksson F, Hultcrantz M, Stalfors J, Stenfeldt K, Hermansson A. Acute mastoiditis in children aged 0-16 years – a national study of 678 cases in Sweden comparing different age groups. Int J Pediatr Otorhinolaryngol. 2012;76(10):1494–500. https://doi.org/10.1016/j.ijporl.2012.07.002. Epub 2012 Jul 23. PubMed PMID: 22832239.

Gunn AJ, Thoresen M. Neonatal encephalopathy and hypoxic-ischemic encephalopathy. Handb Clin Neurol. 2019;162:217–37. https://doi.org/10.1016/B978-0-444-64029-1.00010-2. PubMed PMID: 31324312.

Gurrieri F, Trask BJ, van den Engh G, Krauss CM, Schinzel A, Pettenati MJ, Schindler D, Dietz-Band J, Vergnaud G, Scherer SW, Tsui LC, Muenke M. Physical mapping of the holoprosencephaly critical region on chromosome 7q36. Nat Genet. 1993;3(3):247–51. PubMed PMID: 8485580.

Halsey JH, Allen N, Chamberlin HR. The morphogenesis of hydranencephaly. J Neurol Sci. 1971;12:187–217.

Hameister H, Schulz WA, Meyer J, Thoma S, Adolph S, Gaa A, von Deimling O. Gene order and genetic distance of 13 loci spanning murine chromosome 15. Genomics. 1992;14:417–22.

Hamilton M, Mrazik M, Johnson DW. Incidence of delayed intracranial hemorrhage in children after uncomplicated minor head injuries. Pediatrics. 2010;126(1):e33–9. https://doi.org/10.1542/peds.2009-0692. Epub 2010 Jun 21. PubMed PMID: 20566618.

Hargitai B, Szabó V, Cziniel M, Hajdú J, Papp Z, Szende B, Sergi C. Human brain of preterm infants after hypoxic-ischaemic injuries: no evidence of a substantial role for apoptosis by using a fine-tuned ultrasound-guided neuropathological analysis. Brain Dev. 2004;26(1):30–6. PubMed PMID: 14729412.

Harris CP, Townsend JJ, Norman MG, White VA, Viskochil DH, Pysher TJ, Klatt EC. Atelencephalic aprosencephaly. J Child Neurol. 1994;9:412–6.

Hart MN, Malamud N, Ellis WG. The Dandy-Walker syndrome. A clinicopathological study based on 28 cases. Neurology. 1972;22:771–80.

Hartig PC, Hunter ES III. Gene delivery to the neurulating embryo during culture. Teratology. 1998;58:103–12.

Hartley J, Westmacott R, Decker J, Shroff M, Yoon G. Childhood-onset CADASIL: clinical, imaging, and neurocognitive features. J

Child Neurol. 2010;25(5):623–7. https://doi.org/10.1177/0883073810361382. Epub 2010 Mar 1. PubMed PMID: 20197270.

Hdeib A, Cohen AR. Hydrocephalus in children and adults. Chapter 6. In: Principles of neurological surgery. 3rd ed. Philadelphia: Elsevier; 2012. p. 105–27.

Hein PR, van Groeninghen JC, Puts JJ. A case of acardiac anomaly in the cynomolgus monkey (Macaca fascicularis): a complication of monozygotic monochorial twinning. J Med Primatol. 1985;14:133–42.

Hermanto Y, Takagi Y, Yoshida K, Ishii A, Kikuchi T, Funaki T, Mineharu Y, Miyamoto S. Histopathological features of brain arteriovenous malformations in Japanese patients. Neurol Med Chir (Tokyo). 2016;56(6):340–4. https://doi.org/10.2176/nmc.oa.2016-0032. Epub 2016 Apr 6. PubMed PMID: 27053330; PubMed Central PMCID: PMC4908077.

His W. Vorschläge zur Einteilung des Gehirns. Arch. Anat. Entwickelungsgesch. 1893;3:172–9.

His W. Die Entwicklung des menschlichen Gehirns. Leipzig: Hirzel; 1904.

Hoa M, Slattery WH 3rd. Neurofibromatosis 2. Otolaryngol Clin N Am. 2012;45(2):315. Source: http://books.google.com/books?hl=en&lr=&id=kwdw7RQHeKcC&oi=fnd&pg=PA315&dq=nf2,+NIH,+Natural+History+Study&ots=1JzDOLvv__&sig=9mPRcLKT7aqMbU14304wCtlmaf8|; https://doi.org/10.1016/j.otc.2011.12.005

Hochstetter F. Beitrage zur Entwicklungsgeschicbte desmenschlichen Gehirns. 1. Part. Vienna: Deuticke; 1919.

Hochstetter F. EntwickJungsgeschichte der Ohrmuschel und des ausseren Gehrganges des Menschen. Denkscbr Akad Wissensch Wieri Matb-Naturwiss Klasse. 1948;108:1–50.

Holmes LB. Fetal environmental toxins. Pediatr Rev. 1992;13:364–9.

Holmqvist AS, Olsen JH, Mellemkjaer L, Garwicz S, Hjorth L, Moëll C, Månsson B, Tryggvadottir L, Hasle H, Winther JF; ALiCCS study group. Autoimmune diseases in adult life after childhood Cancer in Scandinavia (ALiCCS). Ann Rheum Dis 2016;75(9):1622–1629. doi: https://doi.org/10.1136/annrheumdis-2015-207659. Epub 2015 Nov 10. PubMed PMID: 26555403.

Hsu SM, Raine L, Fanger H. Use of avidin-biotin-peroxidase complex (ABC) in immunoperoxidase techniques: a comparison between ABC and unlabeled antibody (PAP) procedures. J Histochem Cytochem. 1981;29:577–80.

Hunter ES 3rd, Phillips LS, Goldstein S, Sadler TW. Altered visceral yolk sac function produced by a low-molecular-weight somatomedin inhibitor. Teratology. 1991;43:331–40.

Iivanainen M, Haltia M, Lydecken K. Atelencephaly. Dev Med Child Neurol. 1977;19:663–8.

Isaac M, Best P. Two cases of agenesis of the vermis of cerebellum, with fusion of the dentate nuclei and cerebellar hemispheres. Acta Neuropathol (Berl). 1987;74:278–80.

Isoda K, Hamamoto Y. An autopsy case of incomplete acardius acephalus. Bull Osaka Med Sch. 1980;26:27–32.

Jea A, Vachhrajani S, Widjaja E, Nilsson D, Raybaud C, Shroff M, Rutka JT. Corpus callosotomy in children and the disconnection syndromes: a review. Childs Nerv Syst. 2008;24(6):685–92. https://doi.org/10.1007/s00381-008-0626-4. Epub 2008 Mar 29. Review. PubMed PMID: 18373102.

Jeanes C, O'Grady J. Diagnosing tuberculosis in the 21st century - Dawn of a genomics revolution? Int J Mycobacteriol. 2016;5(4):384–91. https://doi.org/10.1016/j.ijmyco.2016.11.028. Review. PubMed PMID: 27931678.

Johnston JB. The morphology of the forebrain vesicle in vertebrates. J Comp Neurol Psychol. 1909;19:457–539.

Jones KL. Holoprosencephaly sequence. In: Jones KL, editor. Smith's recognizable patterns of human malformation. 5th ed. Philadelphia: Saunders; 1997. p. 605–6.

Joseph FG, Scolding NJ. Neurosarcoidosis: a study of 30 new cases. J Neurol Neurosurg Psychiatry. 2009;80(3):297–304. https://doi.org/10.1136/jnnp.2008.151977. Epub 2008 Oct 31. PubMed PMID: 18977817.

Josephson J, Turner JM, Field CJ, Wizzard PR, Nation PN, Sergi C, Ball RO, Pencharz PB, Wales PW. Parenteral soy oil and fish oil emulsions: impact of dose restriction on bile flow and brain size of parenteral nutrition-fed neonatal piglets. JPEN J Parenter Enteral Nutr. 2015;39(6):677–87. https://doi.org/10.1177/0148607114556494. Epub 2014 Oct 17. PubMed PMID: 25326097.

Joubert M, Eisenring JJ, Robb JP, Andermann F. Familial agenesis of the cerebellar vermis. Neurology. 1969;19:813–25.

Karademir M, Eser O, Karavelioglu E. Adolescent lumbar disc herniation: impact, diagnosis, and treatment. J Back Musculoskelet Rehabil. 2017;30(2):347–52. https://doi.org/10.3233/BMR-160572. PubMed PMID: 27858699.

Karch SB, Urich H. Occipital encephalocele: a morphological study. J Neurol Sci. 1972;15:89–112.

Kastury K, Druck T, Huebner K, Barleta C, Accampora D, Simeone A, Faiella A, Bonicelli E. Chromosome locations of human EMX and OTX genes. Genomics. 1994;22:41–5.

Kepes JJ, Clough C, Villanueva A. Congenital fusion of the thalami (atresia of the third ventricle) and associated abnormalities in a 6-month-old infant. Acta Neuropathol (Berl). 1969;13:97–104.

Khairat A, Waseem M. Epidural Hematoma. [Updated 2018 Nov 15]. In: StatPearls [Internet]. Treasure Island: StatPearls Publishing; 2019 Jan-. Available from: https://www.ncbi.nlm.nih.gov/books/NBK518982/

Khera KS. Maternal toxicity in humans and animals: effects on fetal development and criteria for detection. Teratog Carcinog Mutagen. 1987;7:287–95.

Kim TS, Cho S, Dickson DW. Aprosencephaly: review of the literature and report of a case with cerebellar hypoplasia, pigmented epithelial cyst and Rathke's cleft cyst. Acta Neuropathol (Berl). 1990;79:424–31.

Klüver H, Barrera E. A method for the combined staining of cells and fibers in the nervous system. J Neuropathol Exp Neurol. 1953;12(4):400–3. PubMed PMID: 13097193.

Kojima Y, Kawata K. Morphological observation on two cases of Acardius Amorphus in Holstein-Friesian cattle. Jpn J Vet Res. 1960;8(1–4):261–70.

Kommareddi S, Abramowsky CR, Swinehart GL, Hrabak L. Nontuberculous mycobacterial infections: comparison of the fluorescent auramine-O and Ziehl-Neelsen techniques in tissue diagnosis. Hum Pathol. 1984;15(11):1085–9. PubMed PMID: 6208117.

Lallas M, Desai J. Wernicke encephalopathy in children and adolescents. World J Pediatr. 2014;10(4):293–8. https://doi.org/10.1007/s12519-014-0506-9. Epub 2014 Dec 17. Review. PubMed PMID: 25515801.

Larroche J-C, Encha-Razari F, DeVries L. Central nervous system. In: Gilbert-Barness E, editor. Potter's pathology of the fetus and infant, vol. 2. St. Louis: Mosby; 1997. p. 1028–99.

Lazjuk GI, Lurie IW, Cherstvoy ED. Genetic syndromes of multiple congenital malformations. Arkh Patol. 1977;39:3–11.

Lee SC. Diffuse gliomas for nonneuropathologists: the new integrated molecular diagnostics. Arch Pathol Lab Med. 2018;142(7):804–14. https://doi.org/10.5858/arpa.2017-0449-RA. Epub 2018 May 18. PubMed PMID: 29775073.

Lee AC. Macropolycyte in pediatrics. J Pediatr Hematol Oncol. 2019;41(2):137. https://doi.org/10.1097/MPH.0000000000001344. PubMed PMID: 30475299.

Leech RW, Shuman RM. Holoprosencephaly and related midline cerebral anomalies. A review. J Child Neurol. 1986;1:3–18.

Lehmann FE. Die embryonale Entwicklung. Entwicklungsphysiologie und experimentelle Teratologie. In: Büchner F, Letterer E, Roulet F, editors. Handbuch der allgemeinen Pathologie, vol. VI/1. Berlin/Göttingen/Heidelberg: Springer; 1955. p. 1–53.

Lehmann FE. Zellbiologische und biochemische Probleme der Morphogenese. In: Induktion und Morphogenese. 13. Colloquium der Gesellschaft für Physiologische Chemie. Berlin/Göttingen/Heidelberg: Springer; 1963. p. 1–20.

Lehmann FE. Die embryonale Entwicklung. Entwicklungsphysiologie und experimentelle Teratologie. In: Büchner F, Letterer E, Roulet F, editors. Handbuch der allgemeinen Pathologie, vol. VI/1. Berlin/Göttingen/Heidelberg: Springer; 1983. p. 1–53.

Lemire RJ, Loeser JD, Leech RW, Alvord EC. Normal and abnormal development of the human nervous system. Hagerstown: Harper & Row; 1975a. p. 2–13.

Lemire RJ, Loeser JD, Leech RW, Alvord RC Jr. Normal and abnormal development of the human nervous system. Hagerstown: Harper and Row; 1975b. p. 319–36.

Lichstein J, Solis-Cohen L. Familial tuberous sclerosis (EPILOIA) without adenoma Sebaceum: report of two cases. JAMA. 1943;122(7):429–32. https://doi.org/10.1001/jama.1943.02840240019007.

Litherland J, Ludlam A, Thomas N. Antenatal ultrasound diagnosis of cerebellar vermian agenesis in a case of rhombencephalosynapsis. J Clin Ultrasound. 1993;21:636–8.

Love S. Demyelinating diseases. J Clin Pathol. 2006;59(11):1151–9. https://doi.org/10.1136/jcp.2005.031195. PMID: 17071802; PMCID: PMC1860500.

Lurie IW, Nedzved MK, Lazjuk GI, Kirillova IA, Cherstvoy ED. Aprosencephaly-atelencephaly and the aprosencephaly (XK) syndrome. Am J Med Genet. 1979;3:301–9.

Machin GA. Multiple pregnancies and conjoined twins. In: Gilbert-Barness E, editor. Poter's pathology of the fetus and infant, vol. 1. St Louis: Mosby; 1997. p. 281–321.

Mallam E, Scolding N. The diagnosis of MS. Int MS J. 2009;16(1):19–25. PubMed PMID: 19413922.

Martinez-Frias ML, Frias JL, Opitz JM. Errors of morphogenesis and develop-mental field theory. Am J Med Genet. 1998;76:291–6.

McKechnie L, Vasudevan C, Levene M. Neonatal outcome of congenital ventriculomegaly. Semin Fetal Neonatal Med. 2012;17(5):301–7. https://doi.org/10.1016/j.siny.2012.06.001. Epub 2012 Jul 20. Review. PubMed PMID: 22819382.

McKillop SJ, Belletrutti MJ, Lee BE, Yap JY, Noga ML, Desai SJ, Sergi C. Adenovirus necrotizing hepatitis complicating atypical teratoid rhabdoid tumor. Pediatr Int. 2015;57(5):974–7. https://doi.org/10.1111/ped.12674. Epub 2015 Aug 19. PubMed PMID: 26508178.

McMahon AP, Bradley A. The Wnt-1 (int-1) protooncogene is required for the development of a large region of the mouse brain. Cell. 1990;62:1073–85.

McNeil K, Chowdhury D, Penney L, Rashid M. Vitamin B12 deficiency with intrinsic factor antibodies in an infant with poor growth and developmental delay. Paediatr Child Health. 2014;19(2):84–6. PubMed PMID: 24596481; PubMed Central PMCID: PMC3941681.

Michaud J, Mizrahi EM, Urich H. Agenesis of the vermis with fusion of the cerebellar hemispheres, septo-optic dysplasia and associated anomalies. Report of a case. Acta Neuropathol. 1982;56(3):161–6. PubMed PMID: 7072487.

Miller NJ. NF2 storyboard. CDMRP. National Institute of Health. 2010. Source: http://cdmrp.army

Misser SK. Paediatric neuro-imaging: diagnosis. S Afr J Rad. 2013;17(2):72–3. https://doi.org/10.7196/SAJR.889.

Moore CA, Buehler BA, McManus BM, Harmon JP, Mirkin LD, Goldstein DJ. Acephalus-acardia in twins with aneuploidy. Am J Med Genet Suppl. 1987;3:139–43. Review. PubMed PMID: 3130847.

Murakami U, Kameyama Y. Vertebral malformation in the mouse foetus caused by maternal hypoxia during early stages of pregnancy. J Embryol Exp Morphol. 1963;11:107–18.

Murakami U, Kameyama Y, Nogami H. Malformation of the extremity in the mouse foetus caused by X-radiation of the mother during pregnancy. J Embryol Exp Morphol. 1963;11:549–69. PubMed PMID: 14061959.

Murphy P, Davidson DR, Hill RE. Segment-specific expression of homoeobox-containing gene in the mouse hindbrain. Nature. 1989;341:156–9.

Naidich TP, Chakera TM. Multicystic encephalomalacia: CT appearance and pathological correlation. J Comput Assist Tomogr. 1984;8(4):631–6. PubMed PMID: 6736360.

Nerlich A, Wisser J, Draeger A, Nathrath W, Remberger K. Human acardiac anomaly: a report of three cases. Eur J Obstet Gynecol Reprod Biol. 1991;38(1):79–85. PubMed PMID: 1988330.

Neuromuscular Disease Center, Washington University, St. Louis. http://neuromuscular.wustl.edu/index.html.

Noden DM. Vertebrate craniofacial development: the relation between ontogenetic process and morphologic outcome. Brain Behav Evol. 1991;38:190–225.

Noetzel H. Rhombencephalosynapsis (Kleinhirn ohne Wurm). In: Ule G. Pathologie des Nervensystems, Doerr W, Seifert G, Uehlinger E, editors. Spezielle pathologische Anatomie, vol. 13/II. Berlin/Heidelberg/New York: Springer; 1983. p. 189.

Nonchev S, Maconochie M, Vesque C, Aparicio S, Ariza-McNaughton L, Manzanares M, Maruthainar K, Kuroiwa A, Brenner S, Charnay P, Krumlauf R. The conserved role of Krox-20 in directing Hox gene expression during vertebrate hindbrain segmentation. Proc Natl Acad Sci U S A. 1996;93(18):9339–45. PubMed PMID: 8790331; PubMed Central PMCID: PMC38429.

Norman MG. Bilateral encephaloclastic lesions in a 26 week gestation fetus: effect on neuroblast migration. Can J Neurol Sci. 1980;7(3):191–4. PubMed PMID: 7192593.

O'Reilly SA, Toffol GJ. Adult Arnold-Chiari malformation: a postpartum case presentation. J Am Osteopath Assoc. 1995;95(10):607–9. PubMed PMID: 8557551.

Obersteiner H. Ein Kleinhirn ohne Wurm. Arb Neurol Inst Univ Wien. 1916;21:124–36.

Obladen M. From monster to twin reversed arterial perfusion: a history of acardiac twins. J Perinat Med. 2010;38(3):247–53. https://doi.org/10.1515/JPM.2010.043. PubMed PMID: 20121538.

Okamoto K, Mizuno Y, Fujita Y. Bunina bodies in amyotrophic lateral sclerosis. Neuropathology. 2008;28(2):109–15. Epub 2007 Dec 5. Review. PubMed PMID: 18069968.

Online Mendelian Inheritance in Man, OMIM®. McKusick-Nathans Institute of Genetic Medicine, Johns Hopkins University (Baltimore, MD), {date}. World Wide Web URL: https://omim.org/.

Padget DS, Lindenberg R. Inverse cerebellum morphogenetically related to Dandy-Walker and Arnold-Chiari syndromes: Bizzarre malformed brain with occipital encephalocele. Johns Hopkins Med J. 1972;131:228–46.

Parsons DW, Jones S, Zhang X, Lin JC, Leary RJ, Angenendt P, Mankoo P, Carter H, Siu IM, Gallia GL, Olivi A, McLendon R, Rasheed BA, Keir S, Nikolskaya T, Nikolsky Y, Busam DA, Tekleab H, Diaz LA Jr, Hartigan J, Smith DR, Strausberg RL, Marie SK, Shinjo SM, Yan H, Riggins GJ, Bigner DD, Karchin R, Papadopoulos N, Parmigiani G, Vogelstein B, Velculescu VE, Kinzler KW. An integrated genomic analysis of human glioblastoma multiforme. Science. 2008;321(5897):1807–12. https://doi.org/10.1126/science.1164382. Epub 2008 Sep 4. PubMed PMID:18772396; PubMed Central PMCID: PMC2820389.

Payson RA, Wu J, Liu Y, Chiu IM. The human FGF-8 gene localizes on chromosome 10q24 and is subjected to induction by androgen in breast cancer cells. Oncogene. 1996;13(1):47–53.

Pestronk A. Acquired immune and inflammatory myopathies: pathologic classification. Curr Opin Rheumatol. 2011;23(6):595–604. https://doi.org/10.1097/BOR.0b013e32834bab42. Review. PubMed PMID: 21934500.

Pié J, Gil-Rodríguez MC, Ciero M, López-Viñas E, Ribate MP, Arnedo M, Deardorff MA, Puisac B, Legarreta J, de Karam JC, Rubio E, Bueno I, Baldellou A, Calvo MT, Casals N, Olivares JL, Losada A, Hegardt FG, Krantz ID, Gómez-Puertas P, Ramos FJ. Mutations and variants in the cohesion factor genes NIPBL, SMC1A, and SMC3 in a cohort of 30 unrelated patients with Cornelia de Lange syndrome. Am J Med Genet A. 2010;152A(4):924–9. https://doi.org/10.1002/ajmg.a.33348. PubMed PMID: 20358602; PubMed Central PMCID: PMC2923429.

Pierquin G, Deroover J, Levi S, Masson T, Hayez-Delatte F, Van Regemorter N. Dandy-Walker malformation with postaxial polydactyly: a new syndrome? Am J Med Genet. 1989;33:483–4.

Porto L, Kieslich M, Schwabe D, Zanella FE, Lanfermann H. Central nervous system imaging in childhood leukaemia. Eur J Cancer. 2004;40(14):2082–90. PubMed PMID: 15341983.

Price M, Lazzaro D, Pohl T, Mattei MG, Ruther U, Olivo JC, Duboule D, DiLauro R. Regional expression of the homeobox gene *Nkx-2.2* in the developing mammalian forebrain. Neuron. 1992;8:241–55.

Probst EP. The prosencephalies. Morphology, neuroradiological appearance and differential diagnosis. Berlin/Heidelberg/New York: Springer; 1979.

Puelles L, Rubenstein JL. Expression patterns of homeobox and other putative regulatory genes in the embryonic mouse forebrain suggest a neuromeric organization. Trends Neurosci. 1993;16(11):472–9. Review. PubMed PMID:7507621.

Raimondi AJ, Mullan S, Evans JP. Human brain tumors: an electron-microscopic study. J Neurosurg. 1962;19:731–53.. PubMed PMID: 14489938.

Rakic P, Yakovlev PI. Development of the corpus callosum and cavum septi in man. J Comp Neurol. 1968;132(1):45–72. PubMed PMID: 5293999.

Raybaud C. Destructive lesions of the brain. Neuroradiology. 1983;25:265–91.

Raybaud C, Levrier O, Brunel H, Girard N, Farnarier P. MR imaging of fetal brain malformations. Childs Nerv Syst. 2003;19(7–8):455–70. Epub 2003 Jul 17.PubMed PMID: 12879341.

Reece AS. Chronic toxicology of cannabis. Clin Toxicol (Phila). 2009;47(6):517–24. https://doi.org/10.1080/15563650903074507. Review. PubMed PMID: 19586351.

Reil JC. Mangel des mittleren und freyen Theils des Balkens im Menschengehirn. Arch Physiol. 1812;11:314–44.

Rice CM, Cottrell D, Wilkins A, Scolding NJ. Primary progressive multiple sclerosis: progress and challenges. J Neurol Neurosurg Psychiatry. 2013;84(10):1100–6. https://doi.org/10.1136/jnnp-2012-304140. Epub 2013 Feb 16. Review. PubMed PMID: 23418213.

Robbe C, Paraskeva C, Mollenhauer J, Michalski JC, Sergi C, Corfield A. DMBT1 expression and glycosylation during the adenoma-carcinoma sequence in colorectal cancer. Biochem Soc Trans. 2005;33(Pt 4):730–2. Review. PubMed PMID: 16042587.

Rodeck C, Deans A, Jauniaux E. Thermocoagulation for the early treatment of pregnancy with an acardiac twin. N Engl J Med. 1998;339(18):1293–5. PubMed PMID: 9791145.

Rollins N. Semilobar Holoprosencephaly seen with diffusion tensor imaging and Fiber tracking. Am J Neuroradiol. 2005;26(8):2148–52.

Rosman NP, Donnelly JH, Braun MA. The jittery newborn and infant: a review. J Dev Behav Pediatr. 1984;5(5):263–73. Review. PubMed PMID: 6149233.

Rubenstein JLR, Puelles L. Homeobox gene expression during development of the vertebrate brain. Curr Top Dev Biol. 1994;29:1–63.

Rubenstein JL, Martinez S, Shimamura K, Puelles L. The embryonic vertebrate forebrain: the prosomeric model. Science. 1994;266(5185):578–80. Review. PubMed PMID: 7939711.

Rübsaamen H. Uber die teratogenetische wirkung des sauerstoffmangels in der frühentwicklung: ein Beitrag zur Kausalgenese der Missbildungen bei Mensch und Tier. Naturwissenschaften. 1955;42:319–25.

Ryan M, Heverin M, McLaughlin RL, Hardiman O. Lifetime risk and heritability of amyotrophic lateral sclerosis. JAMA Neurol. 2019; https://doi.org/10.1001/jamaneurol.2019.2044. [Epub ahead of print] PubMed PMID: 31329211; PubMed Central PMCID: PMC6646974.

Saint-Hilaire IG. Histoire générale et particulière des anomalies de l'organisation chez l'homme et les animaux [General and detailed history of anomalies of the organization of the human being and animals] (French). Paris J.B. Baillière et fils; 1832.

Sasai Y. Identifying the missing links: genes that connect neural induction and primary neurogenesis in vertebrate embryos. Neuron. 1998;21:455–8.

Satta G, Lipman M, Smith GP, Arnold C, Kon OM, McHugh TD. Mycobacterium tuberculosis and whole-genome sequencing: how close are we to unleashing its full potential? Clin Microbiol Infect. 2018;24(6):604–9. https://doi.org/10.1016/j.cmi.2017.10.030. Epub 2017 Nov 3. Review. PubMed PMID: 29108952.

Savolaine ER, Fadell RJ, Patel YP. Isolated rhombencephalosynapsis diagnosed by magnetic resonance imaging. Clin Imaging. 1991;15:125–9.

Schachenmayr W, Friede R. Rhombencephalosynapsis: a Viennese malformation. Dev Med Child Neurol. 1982;24:178–82.

Schacht J, Hawkins JE. Sketches of otohistory part 4: a cell by any other name: cochlear eponyms. Audiol Neurootol. 2004;9(6):317–27. Epub 2004 Oct 1. PubMed PMID: 15467285.

Scherer HJ. A critical review: the pathology of cerebral gliomas. J Neurol Psychiatry. 1940;3(2):147–77. PubMed PMID: 21610973; PubMed Central PMCID: PMC1088179.

Schiffer C, Tariverdian G, Schiesser M, Thomas MC, Sergi C. Agnathia-otocephaly complex: report of three cases with involvement of two different Carnegie stages. Am J Med Genet. 2002;112(2):203–8. PubMed PMID:12244557.

Schild RL, Plath H, Födisch HJ, Bartmann P, Hansmann M. Triplet pregnancy with acardius acranius after preimplantation. Fertil Steril. 1998;70:1167–8.

Schmitt HP. "Hindbrain upside-down" in occipital encephalocele – an alternative to the syndrome of Padget and Lindenberg. In: Voth ID, Glees P, editors. Spina bifida – neural tube defects. Basic research, interdisciplinary diagnostics and treatment, results and prognosis. Berlin/New York: de Gruyter; 1986. p. 61–6.

Schmitt HP, Born IA. Pseudoaprosencephaly: aplasia of the forebrain in a median facial cleft syndrome with arhinia and anophthalmia. Cong Anom. 1988;28:169–77.

Schmitt HP, Fehlbildungen d. Zentralnervensystems im Bereich der hinteren Schädelgrube. In: Neuhäuser G, editor. Entwicklungsstörungen des Zentralnervensystems. Stuttgart/Berlin/Köln/Mainz/ Kohlhammer: Ursachen und Folgen; 1986. p. 74–84.

Schmitt HP, Sergi C. The central nervous system in microcephalic primordial dwarfism: is there a characteristic developmental brain pathology in Seckel or Seckel-like syndrome? Congenit Anom. 2000;40(1):32–9.

Schmutzhard J, Glueckert R, Sergi C, Schwentner I, Abraham I, Schrott-Fischer A. Does perinatal asphyxia induce apoptosis in the inner ear? Hear Res. 2009;250(1–2):1–9. https://doi.org/10.1016/j.heares.2008.12.006. Epub 2008 Dec 25. PubMed PMID: 19136052.

Schwalbe E. Acardii und Verwandte. In: Schwalbe E, editor. Die Morphologie der Mißbildungen des Menschen und der Tiere, II. Teil:Die Doppelbildungen. Jena: Gustav Fischer; 1907. p. 133–74.

Schwechheimer K, Schmitt HP. Ventrikelerweiterung und Hirnmanteldefekte bei Prosencephalie: Abgrenzung gegen Porencephalie und Hydranencephalie. In: Voth D, Glees P, editors. Hydrocephalus im frühen Kindesalter. Stuttgart: Ferdinand Enke; 1983. p. 94–7.

Scolding N. The differential diagnosis of multiple sclerosis. J Neurol Neurosurg Psychiatry. 2001;71(Suppl 2):ii9–15. Review. PubMed PMID: 11701778; PubMed central PMCID: PMC1765571.

Scolding N. Devic's disease and autoantibodies. Lancet Neurol. 2005;4(3):136–7. PubMed PMID: 15721820.

Scolding N. Acute disseminated encephalomyelitis and other inflammatory demyelinating variants. Handb Clin Neurol. 2014;122:601–11. https://doi.org/10.1016/B978-0-444-52001-2.00026-1. Review. PubMed PMID: 24507537.

Selicorni A, Colli AM, Passarini A, Milani D, Cereda A, Cerutti M, Maitz S, Alloni V, Salvini L, Galli MA, Ghiglia S, Salice P, Danzi GB. Analysis of congenital heart defects in 87 consecutive patients with Brachmann-de Lange syndrome. Am J Med Genet A. 2009;149A(6):1268–72. https://doi.org/10.1002/ajmg.a.32838. PubMed PMID: 19449412.

Sergi C. EPAS 1, congenital heart disease, and high altitude: disclosures by genetics, bioinformatics, and experimental embryology. Biosci Rep. 2019;39(5) https://doi.org/10.1042/BSR20182197. pii: BSR20182197. Print 2019 May 31. PubMed PMID: 31015364; PubMed Central PMCID: PMC6509053.

Sergi C, Kamnasaran D. PRRX1 is mutated in a fetus with agnathia-otocephaly. Clin Genet. 2011;79(3):293–5. https://doi.org/10.1111/j.1399-0004.2010.01531.x. PubMed PMID: 21294718.

Sergi C, Schmitt HP. The vesicular forebrain (pseudo-aprosencephaly): a missing link in the teratogenetic spectrum of the defective brain anlage and its discrimination from aprosencephaly. Acta Neuropathol. 2000a;99(3):277–84. PubMed PMID: 10663970.

Sergi C, Schmitt HP. Central nervous system in twin reversed arterial perfusion sequence with special reference to examination of the brain in acardius anceps. Teratology. 2000b;61(4):284–90. PubMed PMID: 10716747.

Sergi C, Weitz J, Hofmann WJ, Sinn P, Eckart A, Otto G, Schnabel PA, Otto HF. Aspergillus endocarditis, myocarditis and pericarditis complicating necrotizing fasciitis. Case report and subject review. Virchows Arch. 1996;429(2–3):177–80. Review. PubMed PMID: 8917720.

Sergi C, Hentze S, Sohn C, Voigtländer T, Jung C, Schmitt HP. Telencephalosynapsis (synencephaly) and rhombencephalosynapsis with posterior fossa ventriculocele ('Dandy-Walker cyst'): an unusual aberrant syngenetic complex. Brain Dev. 1997;19(6):426–32. PubMed PMID: 9339873.

Sergi C, Beedgen B, Kopitz J, Zilow E, Zoubaa S, Otto HF, Cantz M, Linderkamp O. Refractory congenital ascites as a manifestation of neonatal sialidosis: clinical, biochemical and morphological studies in a newborn Syrian male infant. Am J Perinatol. 1999;16(3):133–41. PubMed PMID: 10438195.

Sergi C, Grischke EM, Schnabel PA, Sippel F, Adam S, Krempien B, Otto HF. Akardius oder "twin-reversed arterial perfusion" Sequenz. Bericht uber 4 Geminigraviditaten und Ubersicht uber den aktuellen stand der therapeutischen Moglichkeiten [Acardius or "twin-reversed arterial prefusion" sequence. Report of four cases and review of current therapeutic possibilities]. Pathologe. 2000a;21(4):308–14. Review. German.

Sergi C, Zoubaa S, Schiesser M. Norman-Roberts syndrome: prenatal diagnosis and autopsy findings. Prenat Diagn. 2000b;20(6):505–9. PubMed PMID: 10861718.

Sergi C, Penzel R, Uhl J, Zoubaa S, Dietrich H, Decker N, Rieger P, Kopitz J, Otto HF, Kiessling M, Cantz M. Prenatal diagnosis and fetal pathology in a Turkish family harboring a novel nonsense mutation in the lysosomal alpha-N-acetyl-neuraminidase (sialidase) gene. Hum Genet. 2001;109(4):421–8. PubMed PMID: 11702224.

Shawlot W, Behringer RR. Requirement for *Lim1* in head-organizer function. Nature. 1995;374:425–30.

Sherlock EB. The feeble minded. London: Macmillan Company, Ltd.; 1911. p. 239.

Shirahata M, Ono T, Stichel D, Schrimpf D, Reuss DE, Sahm F, Koelsche C, Wefers A, Reinhardt A, Huang K, Sievers P, Shimizu H, Nanjo H, Kobayashi Y, Miyake Y, Suzuki T, Adachi JI, Mishima K, Sasaki A, Nishikawa R, Bewerunge-Hudler M, Ryzhova M, Absalyamova O, Golanov A, Sinn P, Platten M, Jungk C, Winkler F, Wick A, Hänggi D, Unterberg A, Pfister SM, Jones DTW, van den Bent M, Hegi M, French P, Baumert BG, Stupp R, Gorlia T, Weller M, Capper D, Korshunov A, Herold-Mende C, Wick W, Louis DN, von Deimling A. Novel, improved grading system(s) for IDH-mutant astrocytic gliomas. Acta Neuropathol. 2018;136(1):153–66. https://doi.org/10.1007/s00401-018-1849-4. Epub 2018 Apr 23. PubMed PMID: 29687258.

Siddique N, Siddique T. Genetics of amyotrophic lateral sclerosis. Phys Med Rehabil Clin N Am. 2008;19(3):429–39, vii. https://doi.org/10.1016/j.pmr.2008.05.001. PMID: 18625408; PMCID: PMC2553626.

Siddique T, Figlewicz DA, Pericak-Vance MA, et al. Linkage of a gene causing familial amyotrophic lateral sclerosis to chromosome 21 and evidence of genetic-locus heterogeneity. N Engl J Med. 1991;324(20):1381–4.

Siebert JR, Warkany J, Lemire RJ. Atelencephalic microcephaly in a 21-week human fetus. Teratology. 1986;34:9–19.

Siebert JR, Cohen MM Jr, Shaw C-M, Sulic KK, Lemire RJ. Holoprosencephaly. New York: Alan R Liss; 1990a.

Siebert JR, Cohen MM Jr, Sulik KK, Shaw C-M, Lemire RJ. Holoprosencephaly: an overview and atlas of cases. New York: Wiley-Liss; 1990b.

Simmons G, Damiano TR, Truwit CL. MRI and clinical findings in rhombencephalosynapsis. J Comput Assist Tomogr. 1993;17:211–4.

Sipe JC, Herman MM, Rubinstein LJ. Electron microscopic observations on human glioblastomas and astrocytomas maintained in organ culture systems. Am J Pathol. 1973;73(3):589–606. PubMed PMID: 4358392; PubMed Central PMCID: PMC1904091.

Smith MT, Huntington HW. Inverse cerebellum and occipital encephalocele. A dorsal fusion defect uniting the Arnold-Chiari and Dandy-Walker spectrum. Neurology. 1977;27:246–51.

Soejima H, Fujimoto M, Tsukamoto K, Matsumoto KI, Yoshiura KI, Fukushima Y, Jinno Y, Niikawa N. The novel PAX3 mutations observed in patients with the Waardenburg syndrome type 1. Hum Mutat. 1997;9:177–80.

Starck D. Embryologie. Ein Lehrbuch auf allgemein biologischer Grundlage. 2nd ed. Stuttgart: G Thieme; 1965.

Stephens TD. Muscle abnormalities associated with the twin reversed-arterial-perfusion (TRAP) sequence (acardia). Teratology. 1984;30(3):311–8. PubMed PMID: 6515559.

Stoll C, Alembik Y, Dott B. Associated malformations in cases with neural tube defects. Genet Couns. 2007;18(2):209–15. PubMed PMID: 17710873.

Strong MJ, Volkening K, Hammond R, Yang W, Strong W, Leystra-Lantz C, Shoesmith C. TDP43 is a human low molecular weight neurofilament (hNFL) mRNA-binding protein. Mol Cell Neurosci. 2007;35(2):320–7. Epub 2007 Mar 20. PubMed PMID: 17481916.

Suleiman J, Hamwi N, El-Hattab AW. ATP13A2 novel mutations causing a rare form of juvenile-onset Parkinson disease. Brain Dev. 2018;40(9):824–6. https://doi.org/10.1016/j.braindev.2018.05.017. Epub 2018 Jun 11. PubMed PMID: 29903538.

Sulik KK, Cook CS, Webster WS. Teratogens and craniofacial malformations: relationship to cell death. Development. 1988;103(Suppl):219–31.

Sun M, Forsman C, Sergi C, Gopalakrishnan R, O'Connor MB, Petryk A. The expression of twisted gastrulation in postnatal mouse brain and functional implications. Neuroscience. 2010;169(2):920–31. https://doi.org/10.1016/j.neuroscience.2010.05.026. Epub 2010 May 20. PubMed PMID: 20493240; PubMed Central PMCID: PMC2971674.

Swendeman SL, Spielholz C, Jenkins NA, Gilbert DJ, Copeland NG, Sheffery M. Characterization of the genomic structure, chromosomal location, promoter, and development expression of the alpha-globin transcription factor CP2. J Biol Chem. 1994;269(15):11663–71. PubMed PMID: 8157699.

Takashima S, Becker LE, Chan F, Takada K. A Golgi study of the cerebral cortex in Fukuyama-type congenital muscular dystrophy, Walker-type 'lissencephaly', and classical lissencephaly. Brain Dev. 1987;9:621–6.

Temming P, Jenney ME. The neurodevelopmental sequelae of childhood leukaemia and its treatment. Arch Dis Child. 2010;95(11):936–40. https://doi.org/10.1136/adc.2008.153809. Review. Erratum in: Arch Dis Child. 2011;96(8):787. PubMed PMID: 20980277.

The physics of cerebrovascular diseases: biophysical mechanisms of development, diagnosis and therapy hardcover – Nov 20 1997 by George J. Hademenos (Author), Tarik F. Massoud (Author), F. Vinuela (Foreword) Hardcover: 319 pages Publisher: American Institute of Physics; 1998 edition (Nov. 20 1997) Language: English ISBN-10: 1563965585 ISBN-13: 978-1563965586.

Thomas KR, Cappecchi MR. Targeted disruption of the murine *int-1* proto-oncogene resulting in severe abnormalities in midbrain and cerebellar development. Nature. 1990;346:847–50.

Töndury G. Entwicklungsstörungen durch chemische Faktoren und Viren. Naturwiss. 1955;42:312–9.

Truwit CL, Barkovich AJ, Shanahan R, Maroldo TV. MR imaging of rhombencephalosynapsis: report of three cases and review of the literature. AJNR Am J Neuroradiol. 1991;12(5):957–65. Review. PubMed PMID: 1950929

Tubbs RS, Loukas M, Shoja MM, Mortazavi MM, Cohen-Gadol AA. Félix Vicq d'Azyr (1746-1794): early founder of neuroanatomy and royal French physician. Childs Nerv Syst. 2011;27(7):1031–4. https://doi.org/10.1007/s00381-011-1424-y. PubMed PMID:21445631.

Ulu EM, Töre HG, Bayrak A, Güngör D. Coşkun M. MRI of central nervous system abnormalities in childhood leukemia. Diagn Interv Radiol. 2009;15(2):86–92. PubMed PMID: 19517377.

Van Allen MI, Smith DW, Shepard TH. Twin reversed arterial perfusion (TRAP) sequence: a study of 14 twin pregnancies with acardius. Semin Perinatol. 1983;7:285–93.

Van Allen MI, Kalousek DK, Chernoff GF, Juriloff D, Harris M, McGillivray BC, Yong SL, Langlois S, MacLeod PM, Chitayat D, Friedman JM, Wilson RG, McFadden D, Pantzar J, Ritchie S, Hall JG. Evidence for multi-site closure of the neural tube in humans. Am J Med Genet. 1993;47(5):723–43. Review. PubMed PMID: 8267004.

Van der Hoeve J. Eye symptoms in tuberous sclerosis of the brain. Trans Ophthalmol Soc UK. 1920;40:329–34.

Van der Hoeve J. Eye diseases in tuberous sclerosis of the brain and in Recklinghausen's disease. Trans Ophthalmol Soc UK. 1923;43:534–41.

Vanaman MJ, Hervey-Jumper SL, Maher CO. Pediatric and inherited neurovascular diseases. Neurosurg Clin N Am. 2010;21(3):427–41. https://doi.org/10.1016/j.nec.2010.03.001. Review. PubMed PMID: 20561493.

Vasudevan C, McKechnie L, Levene M. Long-term outcome of antenatally diagnosed agenesis of corpus callosum and cerebellar malformations. Semin Fetal Neonatal Med. 2012;17(5):295–300. https://doi.org/10.1016/j.siny.2012.07.001. Epub 2012 Jul 25. Review.

Verity C, Firth H, Ffrench-Constant C. Congenital abnormalities of the central nervous system. J Neurol Neurosurg Psychiatry. 2003;74(Suppl 1):i3–8. Review. PubMed PMID: 12611928; PubMed Central PMCID: PMC1765611.

Vicq d'Azyr F. Traité d'anatomie et de physiologie—avec des planches colorës représentant au naturel les divers organes de 'Homme et des Animaux. Paris: F.A. Didot; 1786.

Wassink G, Davidson JO, Dhillon SK, Zhou K, Bennet L, Thoresen M, Gunn AJ. Therapeutic hypothermia in neonatal hypoxic-ischemic encephalopathy. Curr Neurol Neurosci Rep. 2019;19(2):2. https://doi.org/10.1007/s11910-019-0916-0. Review. PubMed PMID: 30637551.

Weir PE, Ratten GJ, Beischer NA. Polyhydramnios a acomplication of monozygous twins. Br J Obstet Gynaecol. 1979;86:849–53.

Weprin BE, Oakes WJ. Coccygeal pits. Pediatrics. 2000;105(5):E69. PubMed PMID: 10799633.

Werthemann A. Allgemeine Teratologie mit besonderer Berücksichtigung der Verhältnisse beim Menschen. 1. Zwillinge, Mehrlinge und Doppelbildungen in ihrer Beziehung zum Organisationsfeld. In: Büchner F, Letterer E, Roulet F, editors. Handbuch der allgemeinen Pathologie, vol. VI/1. Berlin/Göttingen/Heidelberg: Springer; 1955. p. 100–6.

White RA, Dowler LL, Angeloni SV, Pasztor LM, MacArthur CA. Assignment of FGF8 to human chromosome 10q25-q26: mutations in FGF8 may be responsible for some types of acrocephalosyndactyly linked to this region. Genomics. 1995;30:109–11.

Wildhardt G, Winterpacht A, Hilbert K, Menger H, Zabel B. Two different PAX3 gene mutations causing Waardenburg syndrome type 1. Mol Cell Probes. 1996;10:229–31.

Wilkins DE, Hallett M, Berardelli A, Walshe T, Alvarez N. Physiologic analysis of the myoclonus of Alzheimer's disease. Neurology. 1984;34(7):898–903. PubMed PMID: 6234478.

Williams RS, Swisher CN, Jennings M, Ambler M, Caviness VS. Cerebro-ocular dysgenesis (Walker-Warburg syndrome): neuropathologic and etiologic analysis. Neurology. 1984;34:1531–41.

Wright CVE. Vertebrate homeobox genes. Curr Opin Cell Biol. 1991;3:976–82.

Wyllie DH, Robinson E, Peto T, Crook DW, Ajileye A, Rathod P, Allen R, Jarrett L, Smith EG, Walker AS. Identifying mixed Mycobacterium tuberculosis infection and laboratory cross-contamination during mycobacterial sequencing programs. J Clin Microbiol. 2018;56(11) https://doi.org/10.1128/JCM.00923-18. pii: e00923-18. Print 2018 Nov. PubMed PMID: 30209183; PubMed Central PMCID: PMC6204665.

Yakovlev PI. Pathoarchitectonic studies of cerebral malformations. J Neuropathol Exp Neurol. 1959;18:22–55.

Yakovlev PI, Wadsworth RC. Double symmetrical Porencephalies (Schizencephalies). Tr Am Neurol A. 1941;67:24–9.

Yakovlev PI, Wadsworth RC. Schizencephalics. A study of the congenital clefts in the cerebral mantle. Clefts with fused lips. J Neuropath Exp Neurol. 1946;5:116–30.

Yan H, Parsons DW, Jin G, McLendon R, Rasheed BA, Yuan W, Kos I, Batinic-Haberle I, Jones S, Riggins GJ, Friedman H, Friedman A, Reardon D, Herndon J, Kinzler KW, Velculescu VE, Vogelstein B, Bigner DD. IDH1 and IDH2 mutations in gliomas. N Engl J Med. 2009;360(8):765–73. https://doi.org/10.1056/NEJMoa0808710. PubMed PMID: 19228619; PubMed Central PMCID: PMC2820383.

Yazigi F, Kahwash BM, Al Sufiani F, Conces M, Prasad V, Kahwash SB. Histopathologic identification and pattern recognition of common viral infections in the general pathology practice: an illustrated review. Ibnosina J Med Biomed Sci. 2016;8:28. https://doi.org/10.4103/1947-489X.210214.

Yoganathan S, Thomas MM, Mathai S, Ghosh U. Neuroregression as an initial manifestation in a toddler with acquired pernicious anaemia. BMJ Case Rep. 2015;2015 https://doi.org/10.1136/bcr-2015-213540. pii: bcr2015213540. PubMed PMID: 26678841; PubMed Central PMCID: PMC4691932.

Yoshiura K, Leysens NJ, Chang J, Ward D, Murray JC, Muenke M. Genomic structure, sequence, and mapping of human FGF8 with no evidence for ist role in craniosynostosis/limb defect syndromes. Am J Med Genet. 1997;72:354–62.

Yu TW, Mochida GH, Tischfield DJ, Sgaier SK, Flores-Sarnat L, Sergi CM, Topçu M, McDonald MT, Barry BJ, Felie JM, Sunu C, Dobyns WB, Folkerth RD, Barkovich AJ, Walsh CA. Mutations in WDR62, encoding a centrosome-associated protein, cause microcephaly with simplified gyri and abnormal cortical architecture. Nat Genet. 2010;42(11):1015–20. https://doi.org/10.1038/ng.683. Epub 2010 Oct 3. PubMed PMID:20890278; PubMed Central PMCID: PMC2969850.

Zamorano L, Chuaqui B. Teratogenetic periods for the principal malformations of the central nervous system. Virchows Arch A Pathol Anat Histol. 1979;384(1):1–18. Review. PubMed PMID: 159540.

Zupanc ML. Infantile spasms. Expert Opin Pharmacother. 2003;4(11):2039–48. Review. PubMed PMID: 14596657.

Peripheral Nervous System

Contents

16.1	**Development**	1321
16.2	**Disorders of the Peripheral Nervous System**	1322
16.2.1	Peripheral Neuropathy	1322
16.2.2	Traumatic Neuropathy	1323
16.2.3	Vascular Neuropathy	1323
16.2.4	Intoxication-Related Neuropathy	1323
16.2.5	Infiltration (e.g., Amyloid) Related Neuropathy	1323
16.2.6	Neoplasms of the Peripheral Nervous System	1323
16.3	**Neuromuscular Disorders**	1333
16.3.1	Muscle Biopsy Test	1333
16.3.2	Neurogenic Disorders	1335
16.3.3	Myopathic Disorders	1337
16.3.4	Glycogen Storage Diseases	1339
16.3.5	Mitochondrial Myopathies	1339
16.3.6	Inflammatory Myopathies: Non-infectious	1341
16.3.7	Inflammatory Myopathies: Infectious	1341
	Multiple Choice Questions and Answers	1341
	References and Recommended Readings	1342

16.1 Development

The formation of the muscular system begins about the 4th week of embryonic development, and the primitive cells are called myoblasts and are derived directly from the mesodermal germ layer. The muscles arise from the fusion of myoblasts to form a multinucleated syncytium or syncytial unit called a myotube which stimulates the formation of myofibrils. Exceptions to the mesodermal germ layer origin are the smooth muscles from the ocular pupil, cutaneous sweat glands, and a mammary gland that derive from ectoderm. Skeletal muscles originate from paraxial mesoderm, which forms somites from the occipital to the sacral regions and somitomeres in the head. Cardiac muscles originate from splanchnic mesoderm surrounding the heart tube, while the smooth muscles differentiate from splanchnic mesoderm surrounding the gut. In a cranio-caudal fashion, somites form and differentiate into sclerotome, dermatome, and two-muscle-forming

© Springer-Verlag GmbH Germany, part of Springer Nature 2020
C. M. Sergi, *Pathology of Childhood and Adolescence*,
https://doi.org/10.1007/978-3-662-59169-7_16

regions, of which one is located in the dorsolateral area of the somite providing progenitor cells for limb and body wall musculature (the so-called hypomeric musculature). The other muscle-forming region is located in the dorsomedial region and builds the myotome (the so-called epimeric musculature). It is by the end of the 3rd month that myofibrils appear in the sarcoplasm of the myoblasts and cross-striation appears in the skeletal muscle. When the muscles become innervated, diffusely distributed acetylcholine receptors become concentrated at motor end plate. With the conclusion of the 5th week of embryonic development, the primitive musculature is divided into epimere (small dorsal portion), which is innervated by the primary dorsal ramus, and hypomere, which is innervated by the primary ventral ramus. The epimere forms the extensor muscle of the vertebral column, while from the hypomere originate the muscles of the limbs and body wall. With the elongation of the limb buds, the muscle tissue splits into flexor and extensor parts. The myofibril is composed of alternating bands. The I-bands (isotropic in polarized light) appear light in color, and the A-bands (anisotropic in polarized light) are dark in color. The alternating pattern of these bands results in the striated appearance of skeletal muscle. The Z-lines (*Zwischenschieben*) bisect the I-bands, while a light band called the H-band (*Heller*) is located within each A-band and the M-line (*Mittelschiebe*) bisects each A-band and, of course, also each H-band. Each myofibril is a series of contractile units (sarcomeres) that contains two types of filaments: thick filaments, composed of myosin, and thin filaments, composed of actin. The thick filaments are a bipolar array of polymerized myosin motors, while the thin filaments are attached to a disc-like zone that appears histologically as the Z-line. Skeletal muscles are separated into two muscle fiber types: slow-twitch (type I) and fast-twitch (type II). Type I muscle fibers contract more slowly, rely on aerobic metabolism, and have many mitochondria and myoglobin, which gives the reddish color of the fibers ("red fibers"). Type I muscles can maintain continuous contraction and are useful in activities such as the maintenance of posture. Type II muscle fibers contract more rapidly due to the presence of a faster myosin and are subdivided into fibers that have large amounts of mitochondria and myoglobin and those that have few mitochondria and little myoglobin. In the former subtype, aerobic metabolism plays a major role, while the anaerobic glycolysis plays a major role in the latter. The lack of myoglobin results in a paler color than the slow-twitch muscles ("white fibers"). These muscles are essential for intense but sporadic contractions. Some special stains are useful in pathology. NADH is a stain for lipid and is not as useful as the ATPase reaction since denervation causes both fiber types to stain dark. Nonspecific esterase reaction stains type I fibers slightly darker than type II but stains denervated fibers of either type very dark, owing to the diffuse distribution of acetylcholine esterase activity. Alkaline phosphatase reaction stains regenerating fibers.

16.2 Disorders of the Peripheral Nervous System

16.2.1 Peripheral Neuropathy

Causes of peripheral neuropathy include diabetes mellitus, alcohol, nutrition-related vitamin B deficiency, Guillain-Barré syndrome, trauma, hereditary (e.g., Friedreich ataxia, Charcot-Marie-Tooth syndrome), environmental toxins/drugs (e.g., metronidazole, Fe, Pb), paraneoplastic, amyloidosis, porphyria, inflammatory, syphilis, and tumor. Mononeuropathy indicates a focal process, while polyneuropathy and polyneuritis point to a widespread degeneration of the peripheral nerves. Any combination of peripheral nerves may be affected, although the distal portions of the upper and lower limbs seem most often the target of polyneuropathies. No gross abnormalities are seen, but, microscopically, there is swelling of the myelin sheaths with fragmentation and globalization (aka Wallerian degeneration). These myelin sheaths stain progressively poorly and then disappear, and, later, axons show also swelling, fragmentation, and disappearance. *Wallerian degeneration* is a process of progressive demyelinization and disintegration of the distal axonal segment following the damage to the neuron. There are a few causes associated with Wallerian degeneration which include trauma, cerebral infarction, hemorrhage, necrosis, and focal demyelinization. Four stages have been delineated:

- Stage I: The physical disintegration of the axons and myelin sheaths.
- Stage II: The brisk destruction of the myelin fragments with central involvement of breakdown of the myelin fats in single lipids and neutral fat.
- Stage III: Glia proliferation (gliosis) develops in the areas of degenerated axons and myelin sheaths.
- Stage IV: Volume loss, which is the final consequence of the atrophy of the white matter.

16.2.2 Traumatic Neuropathy

Traumatic neuropathy may occur by compression (e.g., neoplasm, callus, aneurysm, prolonged pressure, necrosis, and bandages), tension or stretching (e.g., overextension of a limb, bony fracture), and disconnection (e.g., knife bullet, jagged edge of skeletal fracture). Regeneration of the peripheral nerves is possible after injury, although the degree of disruption, the degree of blood supply, and simultaneous infections of the injury site may compromise the regeneration process.

16.2.3 Vascular Neuropathy

Vascular neuropathy may occur in either vasculitis or DM, and ischemic pathogenesis is the underlying condition for the neuropathy.

16.2.4 Intoxication-Related Neuropathy

Intoxication-related neuropathies may occur in the setting of arsenic (As), lead (Pb), and ethanol (CH_3CH_2OH) intoxication or poisoning.

16.2.5 Infiltration (e.g., Amyloid) Related Neuropathy

Infiltrative conditions include most often amyloidosis.

16.2.6 Neoplasms of the Peripheral Nervous System

Neoplasms include *schwannomas* or neurilemmomas (Fig. 16.1), *neurofibromas* (Figs. 16.2 and 16.3), as well as *malignant peripheral nerve sheath tumor* (MPNST or malignant schwannomas) (Figs. 16.4, 16.5, and 16.6). To mention also is the neuroblastoma that may develop from the paraspinal ganglia (Fig. 16.7) and the ganglioneuroma (Fig. 16.8) that are described elsewhere in this book.

16.2.6.1 Schwannoma
- *DEF*: Biphasic tumor with highly ordered cellular component (Antoni A areas) that palisades (Verocay bodies) plus myxoid hypocellular component (Antoni B areas) associated with a perceptible nerve and strong S100 immunoreactivity. The schwannoma of cellular type is usually a centrally or axially located benign neoplasm with no tendency to malignant transformation (Fig. 16.1).
- *SYN*: Neurilemmoma, neurinoma.
- *EPG*: In familial tumor syndromes (e.g., NF2, schwannomatosis, or Carney complex) or sporadically. There is a loss of function of the tumor suppressor gene merlin (Schwannomin) or direct genetic change involving the *NF2* gene or secondarily to merlin inactivation.
- *EPI*: Vast age range (second to fourth decade), ♂ = ♀, 90% as a sporadic tumor, while the remaining 10% as a tumor in NF-2, schwannomatosis, and meningiomatosis ± NF-2 type 2.
- *CLI*: Upper limbs but also head and neck area, posterior mediastinum, retroperitoneum, gastrointestinal tract, and endocrine glands. More rare locations may also occur. According to the site, there may be pain and neurological symptoms being the tumor waxing and wan-

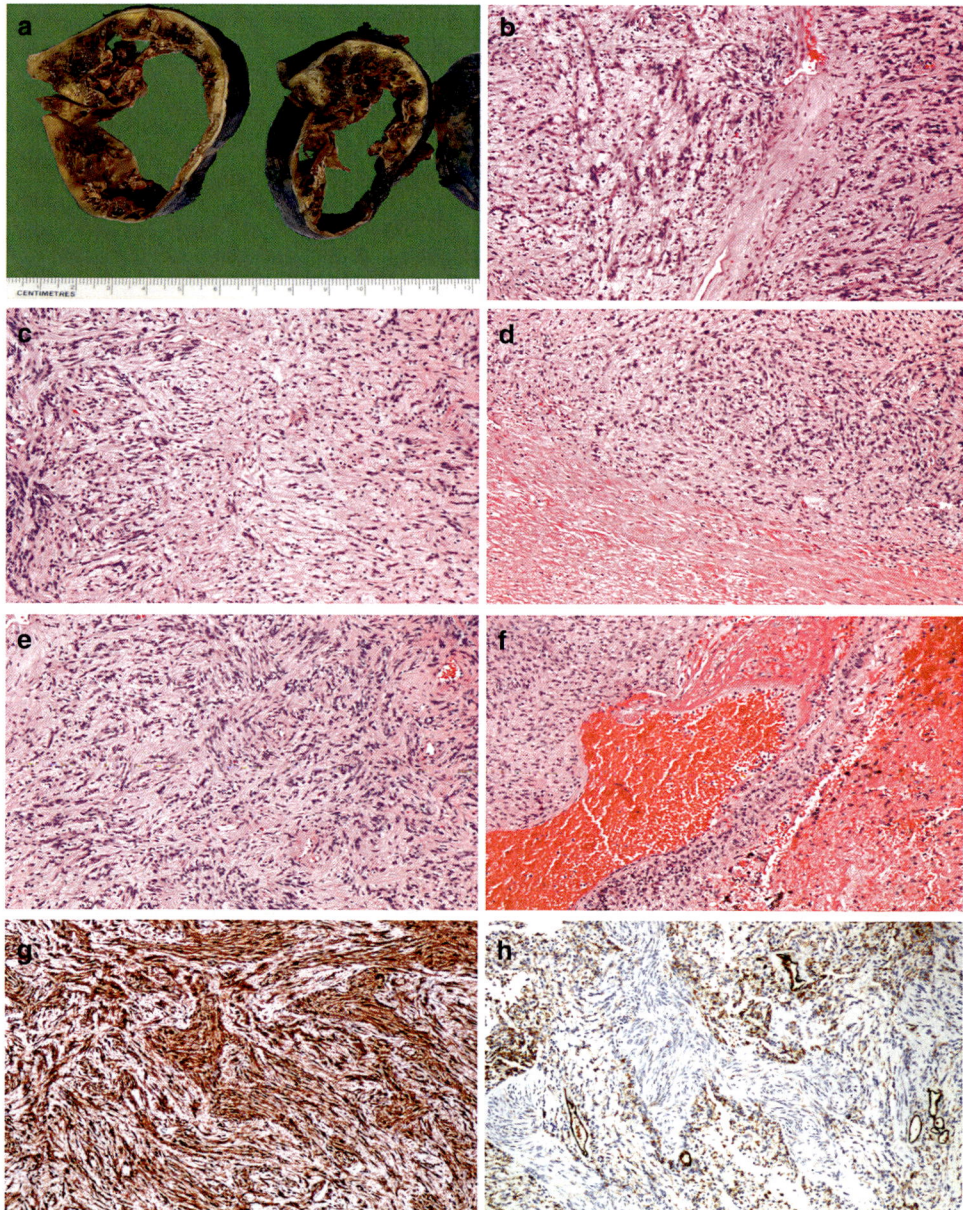

Fig. 16.1 Schwannoma. The gross photograph of an unusual cystically degenerated and hemorrhagic schwannoma is presented (**a**). The microphotographs (**b–f**) show the microscopic appearance of this schwannoma with well-differentiated Schwann cells (hematoxylin-eosin staining). The tumor shows an alternating representation of areas with compact spindle cells (Antoni A) and hypocellular less orderly areas (Antoni B) with myxoid and microcystic appearance. Antoni A areas show the characteristic nuclear palisading. The alternating rows of palisading nuclei and intervening nuclei-free stroma constitute the Verocay bodies (**b**, ×100 original magnification; **c**, ×100 original magnification; **d**, ×100 original magnification; **e**, ×100 original magnification; **f**, ×100 original magnification). Two immunohistochemical microphotographs (**g**, ×100 original magnification; **h**, ×100 original magnification) disclose the expression of S100 and CD31, an endothelial cell marker. Typically, S100 proteins are mostly expressed in cells derived from the neural crest (Schwann cells, and melanocytes), chondrocytes, adipocytes, myoepithelial cells, macrophages, Langerhans cells, and dendritic cells. All immunohistochemistry stains have been performed using the avidin-biotin complex method

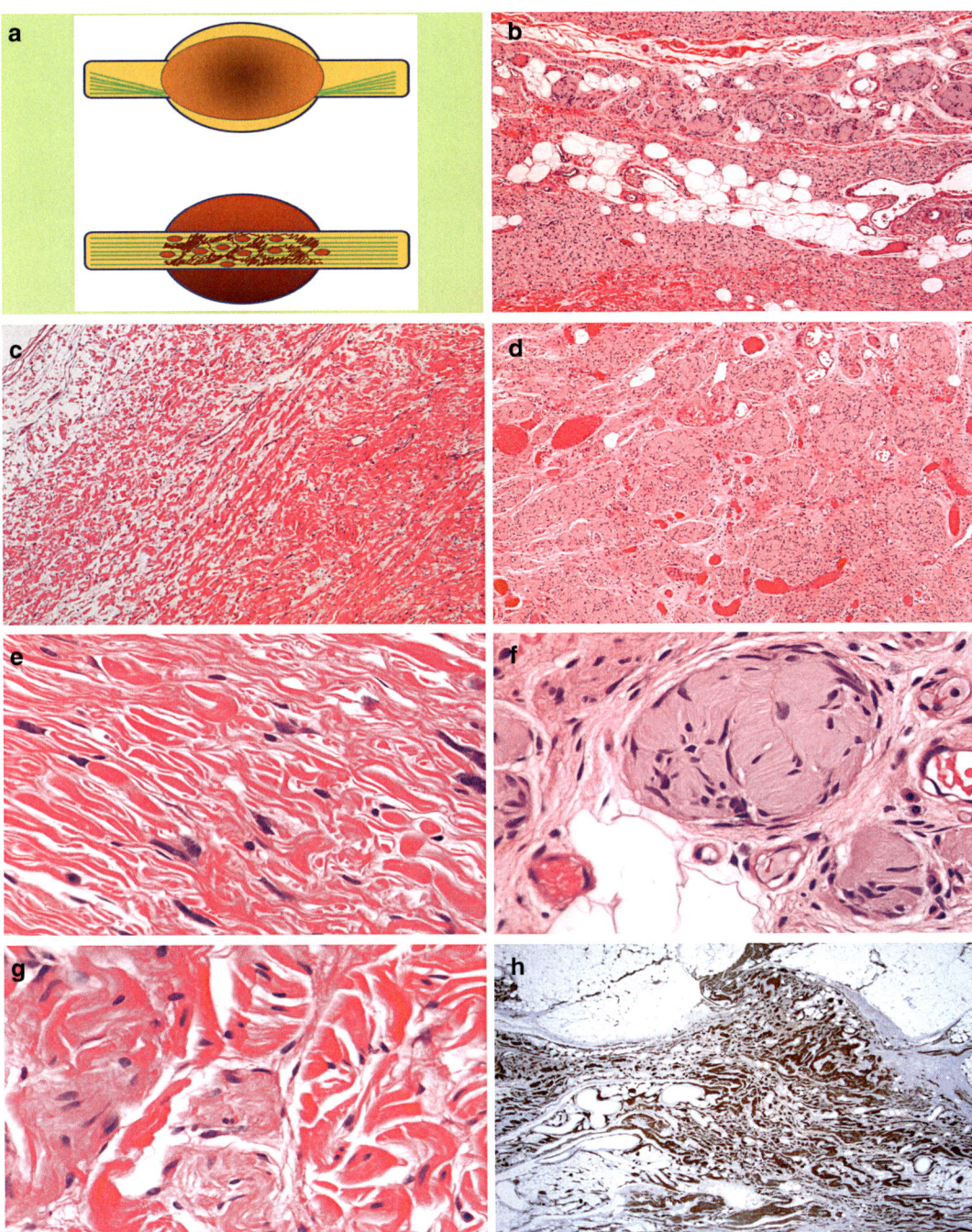

Fig. 16.2 Neurofibroma. In (**a**) the schematic differentiation between schwannoma (upper tumor colored in brown not infiltrating the nerve) and neurofibroma (lower tumor colored in brown infiltrating the nerve). The hematoxylin-eosin microphotographs (**b**) through (**g**) show the characteristic features of a neurofibroma with all elements of peripheral nerves that proliferate in this tumor. There are Schwann cells with wire-like collagen fibrils showing wavy serpentine nuclei and harboring pointed ends. Also, Wagner-Meissner corpuscles (**g**) and Pacinian corpuscles can be seen. Stromal mucosubstances, mast cells, axons, fibroblasts, and collagen may complement the histologic appearance (**b**, ×50 original magnification; **c**, ×50 original magnification; **d**, ×50 original magnification; **e**, ×400 original magnification; **f**, ×400 original magnification; **g**, ×400 original magnification). (**h**) The immunohistochemical expression of S100 (×12.5 original magnification) using the avidin-biotin complex method

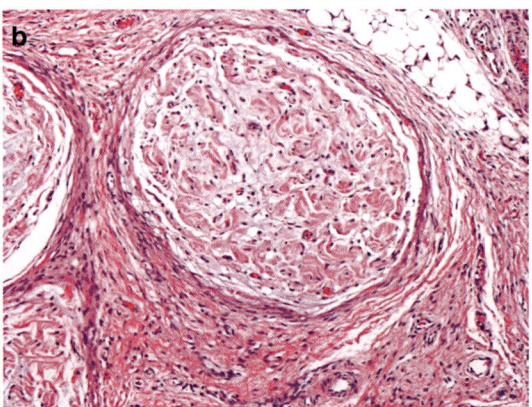
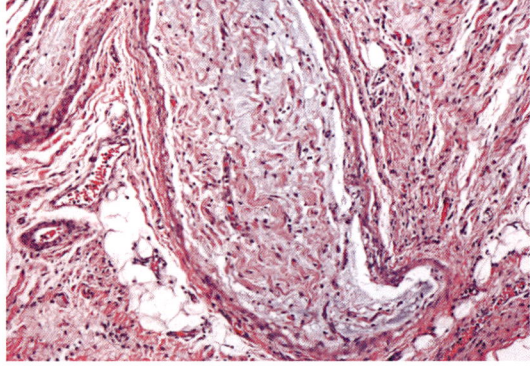

Fig. 16.3 Plexiform Neurofibroma. Neurofibromatosis I and II belong to the group of the RASopathies (see text). In (**a**) are shown the criteria for the diagnosis of this disease. The microphotographs (**b**) and (**c**) are from hematoxylin-eosin-stained slides of a plexiform neurofibroma showing a distorted and tortuous appearance of a neurofibroma (**b**, **c**, ×50 original magnification)

ing in size, which may be related to the amount of cystic degeneration it encompasses.
- *IMG*: Well-circumscribed masses displacing adjacent structures without direct invasion ± cystic and fatty degeneration, hemorrhage, and calcification.
- *GRO*: The tumor is often eccentric to the nerve. In fact, the nerve of origin may be present at the periphery but does not penetrate the substance of tumor.
- *CLM*: Characteristic fusiform growth pattern composed of Schwann cells. There are Antoni A areas with cellular fascicular spindled areas, wavy and slightly plump nuclei, pink cytoplasm, and eosinophilic hyalinization of stroma and around vessels, with or without the well-formed Verocay bodies of classic schwannoma. Moreover, there are Antoni B areas with the hypocellular myxoid (probably neurofibroma-like) growth pattern. Histologic features of malignant change (marked pleomorphism, atypical mitoses, or geographic necrosis) are not observed and need to be carefully identified and need to be distinguished from some good degenerative changes, including cystic degeneration, necrosis, cytologic atypia, and mild increase of the mitotic figures. Reactive xanthoma cell inflammation is typically seen in the setting of central degeneration.

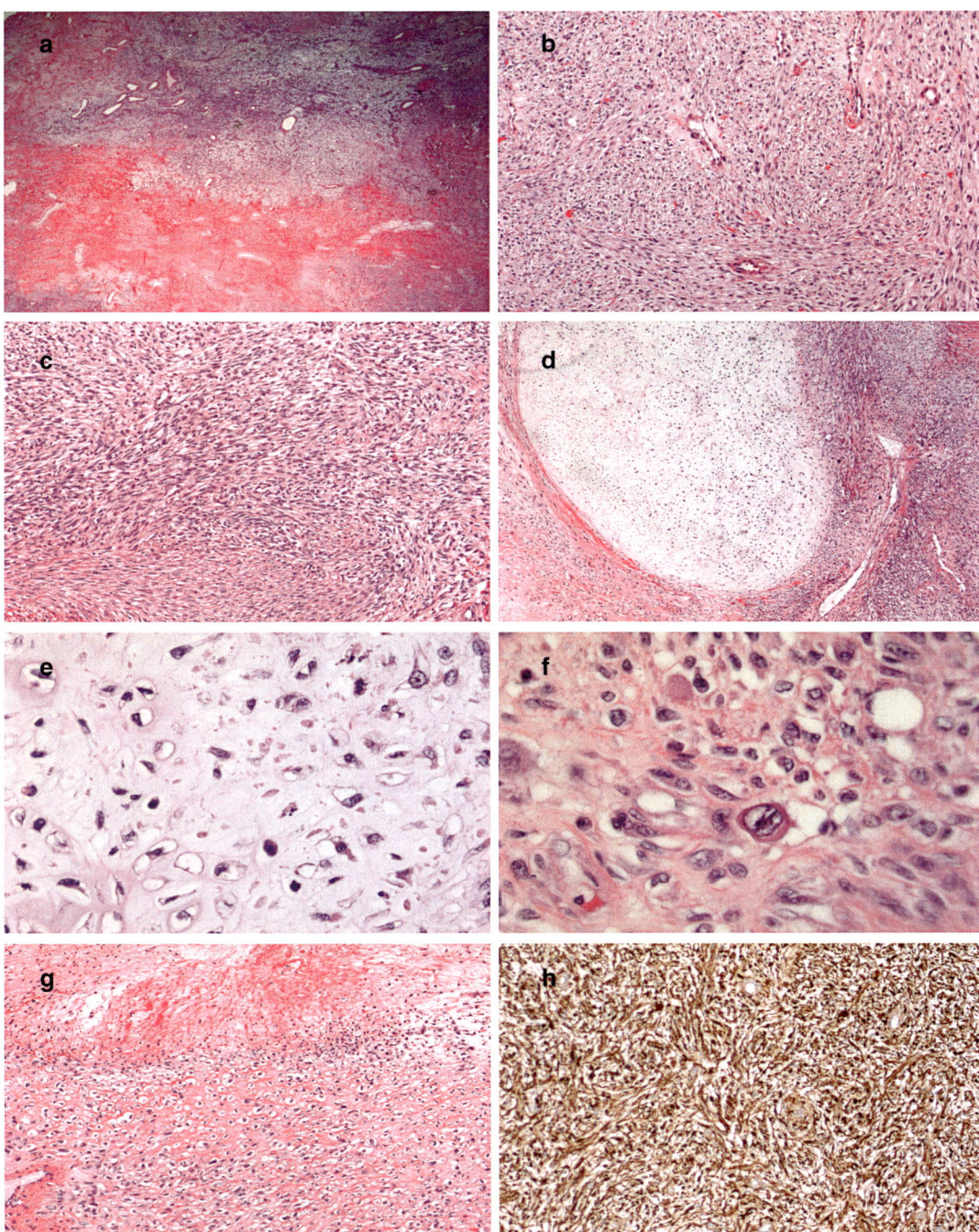

Fig. 16.4 Malignant Peripheral Nerve Sheath Tumor. The malignant peripheral nerve sheath tumor (MPNST) can exhibit a disguising histology and a number of differential diagnoses need to be kept in mind. The microphotographs (**a–g**) show the hematoxylin-eosin-stained histology of an MPNST (**a**, ×12.5 original magnification; **b**, ×100 original magnification; **c**, ×100 original magnification; **d**, ×50 original magnification; **e**, ×400 original magnification; **f**, ×630 original magnification, **g**, ×100 original magnification). The tumor shows a heterogeneity with variation in cellularity and growth pattern. It can be recognized from tightly packed spindle cells, which can be arranged in either a herringbone or interwoven fasciculated growth pattern (**b** and **c**). In (**d**) and (**g**), a chondroid differentiation may be recognized. The microphotograph (**h**) shows the immunohistochemical expression of S100 (×100, original magnification). The avidin-biotin complex has been used for the immunostaining of this tumor

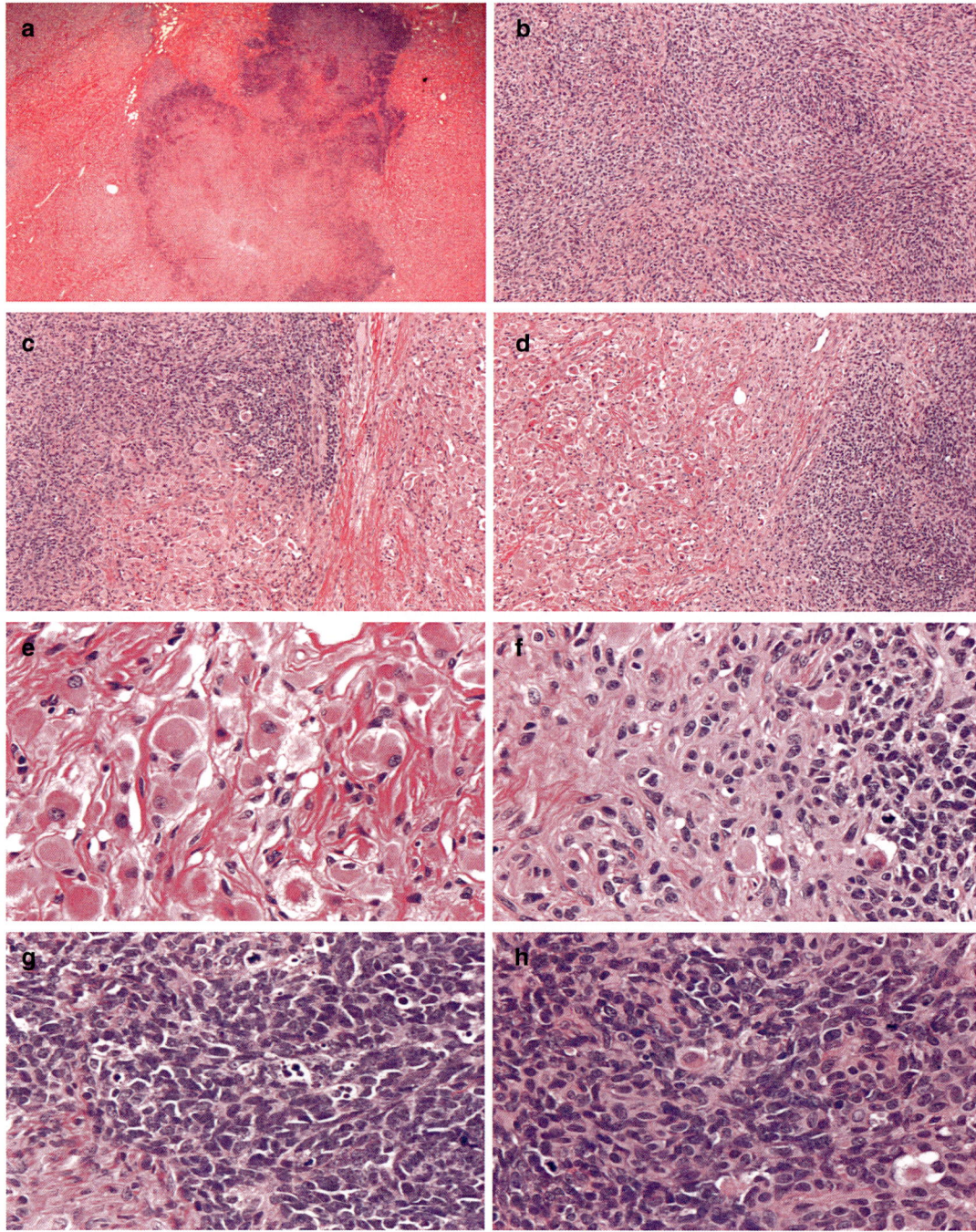

Fig. 16.5 The malignant triton tumor is a specific subtype of malignant peripheral nerve sheath tumor (MPNST) with rhabdomyoblastic differentiation (**e–h**) in addition to a a classic morphology of MPNST (**a–d**) (**a**, hematoxylin-eosin staining, ×12.5 original magnification, **b**, hematoxylin-eosin staining, ×100 original magnification; **c**, hematoxylin-eosin staining, ×100 original magnification; **d**, hematoxylin-eosin staining, ×100 original magnification; **e–h**, hematoxylin-eosin staining, ×400 original magnification)

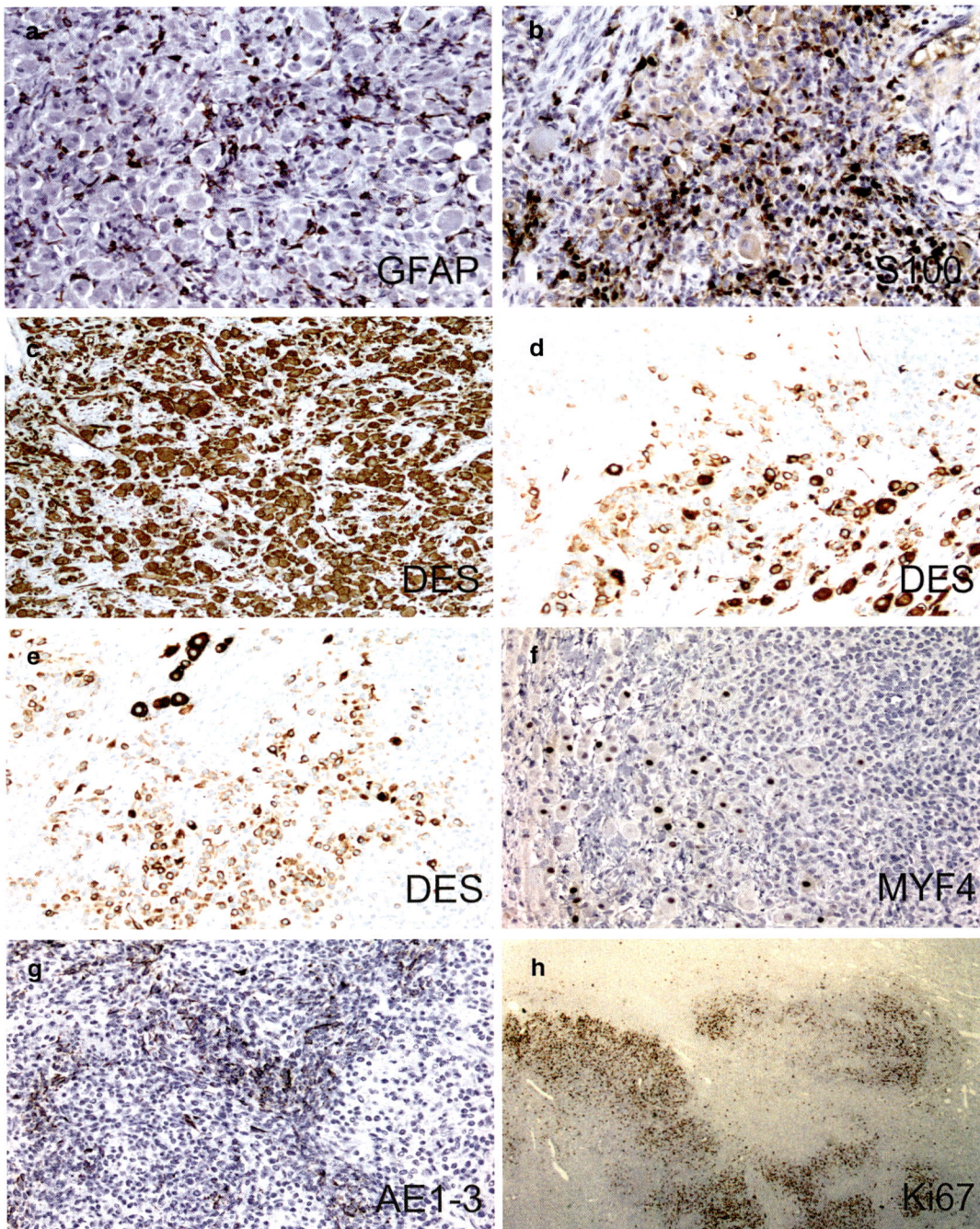

Fig. 16.6 Malignant Peripheral Nerve Sheath Tumor. This immunohistochemical panel of a malignant peripheral nerve sheath tumor (MPNST) shows the expression of GFAP (**a**, ×200 original magnification), S100 (**b**, ×200 original magnification), DES (**c–e**, ×200 original magnification), MYF4 (**f**, ×200 original magnification), AE1-3 (**g**, ×200 original magnification), and of the proliferation antigen Ki67 (**h**, ×12.5 original magnification). All immunohistochemical stainings have been performed using the avidin-biotin complex method

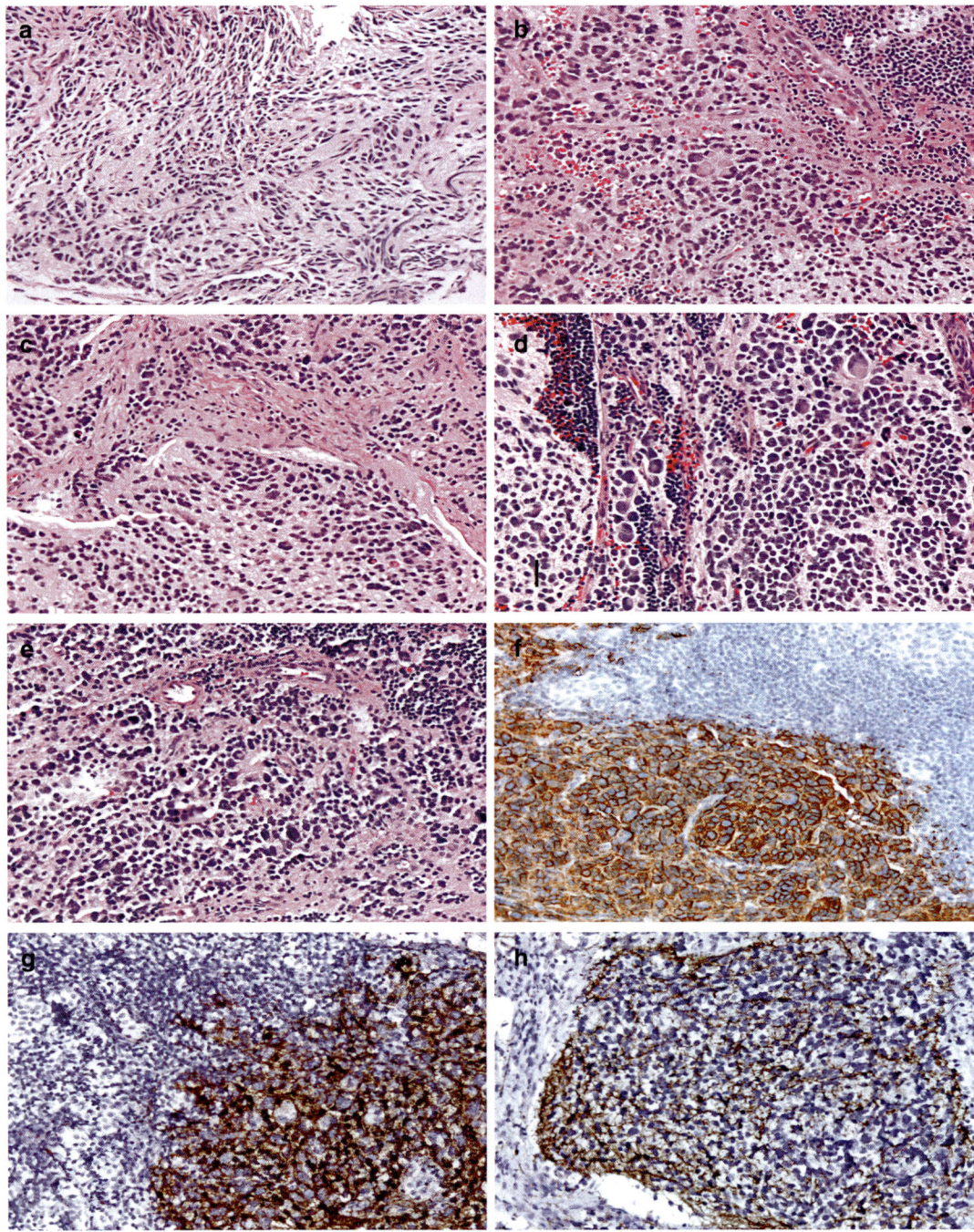

Fig. 16.7 Paraspinal Neuroblastoma. The panel illustrates paraspinal neuroblastoma with five hematoxylin and eosin-stained histologic slides showing a classic poorly differentiated neuroblastoma with small round blue cell tumors with high nucleus to cytoplasm ratio and less discernible cytoplasm (**a**, ×200 original magnification; **b**, ×200 original magnification; **c**, ×200 original magnification; **d**, ×200 original magnification; **e**, ×200 original magnification). Three microphotographs from immunohistochemical-stained slides show the expression of CD56, CD57, and neurofilaments (**f**, ×200 original magnification; **g**, ×200 original magnification; **h**, ×200 original magnification). The tumor was also positive for synaptophysin (SYN) and chromogranin A (CGA) (not shown)

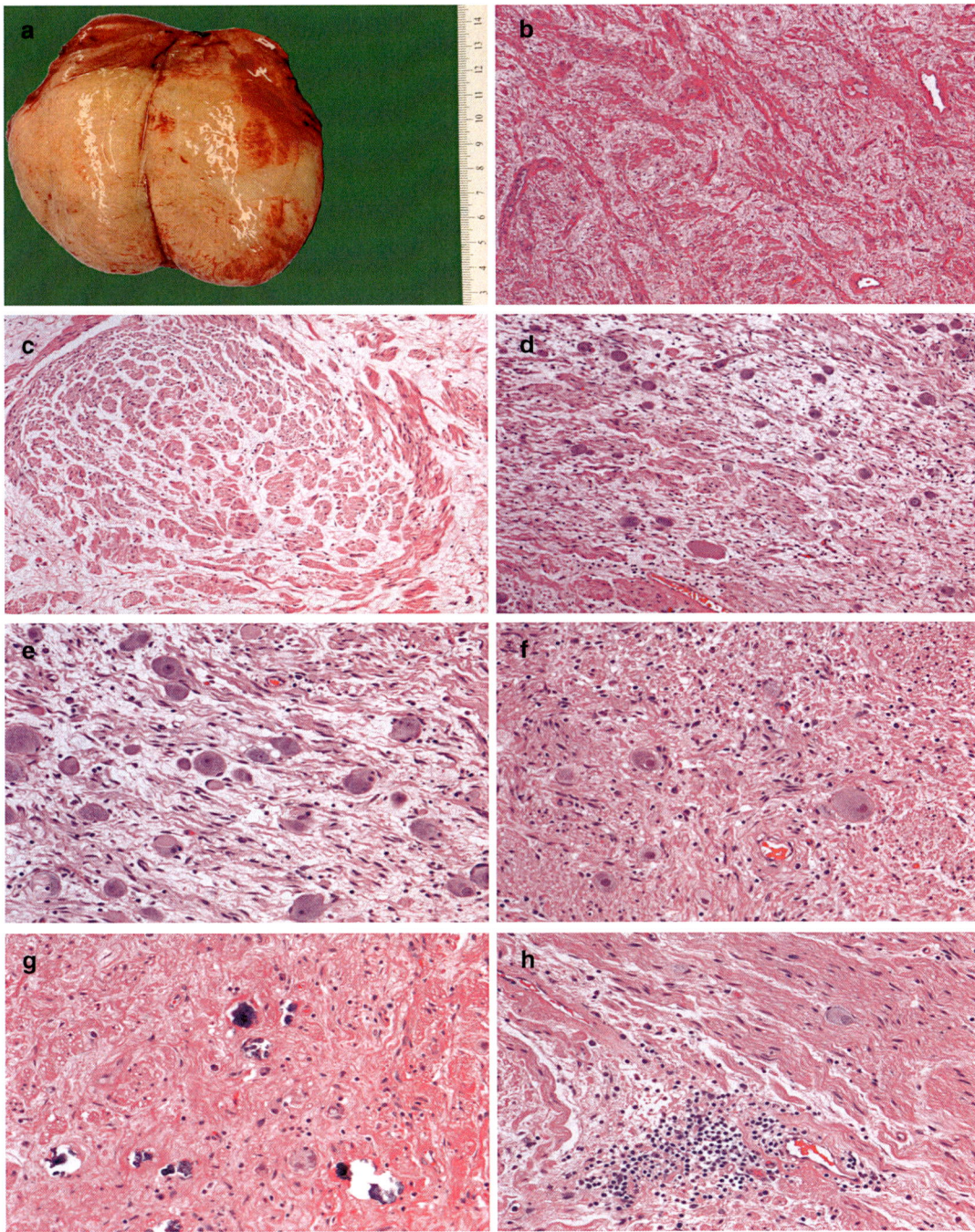

Fig. 16.8 Ganglioneuroma. The gross photograph of this ganglioneuroma discloses the gross appearance of this tumor, which is well circumscribed with a firm texture, gray-white on cut surface and harboring a trabecular or whorled appearance (**a**). The hematoxylin-eosin microphotographs (**b**–**h**) show a mixture of Schwann cells and ganglion cells (**d**–**f**) and the whorled appearance can be recognized microscopically as well (**c**). Occasionally, calcification (**g**) and scattered chronic inflammatory cells (**h**) can be seen (**b**, ×50 original magnification; **c**, ×100 original magnification; **d**, ×100 original magnification; **e**, ×200 original magnification; **f**, ×200 original magnification; **g**, ×200 original magnification; **h**, ×200 original magnification)

- *IHC*: (+) S100, GFAP (focal), AE1–AE3 (focal, cross-reactivity), EMA (capsular area with staining of the perineurial or epineurial cells) but (−) CD34, NFP, HMB45, SMA, and DES. Immunohistochemically, S100 and calretinin can help in the diagnosis. S100 is a protein belonging to a family of low molecular weight found in vertebrates and characterized by two Ca-binding sites of the helix-loop-helix ("EF-hand type") configuration and consists of two identical polypeptides linked by non-covalent bonds. S100 is expressed in cells arising from the neural crest (Schwann cells, melanocytes, and glial cells), chondrocytes, fat cells, myoepithelial cells, macrophages, Langerhans cells, dendritic cells, and keratinocytes. S100 is involved in protein phosphorylation, transcription factor regulation, Ca^{2+} homeostasis, cytoskeleton dynamics, cell growth and differentiation, and inflammation. Calretinin is a Ca-binding protein belonging to the family of EF-hand proteins and is expressed in some neurons of the CNS and PNS, as well as several other cells, including mesothelial cells, and is useful in differentiating mesothelioma from adenocarcinoma, ovarian sex cord-stromal tumors, Leydig cell tumors of the testis, adrenal cortical tumors, and adenomatoid tumors. S100 is expressed in both tumors, although it is higher in schwannomas than neurofibromas, while calretinin is usually detected in schwannomas and very few neurofibromas.
- *DDX*: Six diagnoses need to be kept in mind – *neurofibroma* (nerve-arising tumor with internally located entering and exiting nerve with a central more neural-like and peripheral more edematous aspect, (+) S100, CD34, NFP, EMA, (−) GFAP), *SFT* (eosinophilic appearance with haphazard plump cells, scant mitotic activity, (+) CD34, (−) S100), *synovial sarcoma* (hemangiopericytoma-like vasculature, ovoid cells, and denser cellularity than cellular schwannoma, geographic necrosis, and S100 limited to residual intratumoral wavy neurons), *MPNST* (high-grade sarcoma with invasion of surrounding structures or pseudocapsule, perivascular tumor sparing and geographic necrosis, high N/C imparting a blue tumor appearance, (+) S100, GFAP, but (−) AE1–AE3, DES, HMB45, CD34), *sarcomatoid carcinoma* (malignant morphologic features with clues of origin, e.g., primary pulmonary adenocarcinoma (CK18 and TTF1 +) or squamous cell carcinoma (CK5/6 +)), and *spindle cell melanoma* (prominent cherry red nucleoli and/or intranuclear cytoplasmic inclusions and often atypical mitotic activity, marked cytologic atypia and geographic necrosis, negativity of melanoma markers, including HMB45, Melan-A, tyrosinase, and MITF, but (+) SOX10, NGFR, KBA.62, and clusterin).
- *TEM*: The tumor cells simulate the appearance of differentiated Schwann cells differently from the neurofibromas, which have a mixture of different tumor cells, including Schwann cells, perineurial cells, and endoneurial fibroblasts. In TEM, elongated cells with continuous basal lamina, thin cytoplasmic processes, aggregates of microfibrils in the cytoplasm, intracytoplasmic lamellar bodies, and long-spacing (exceeding 100 nm) extracellular collagen "bodies" (Luse bodies) are observed.
- *PGN*: There is a low tendency for malignant transformation, but signs that portend into this direction include (1) sudden progressive increase in size, (2) epithelioid features in cellular schwannoma, (3) melanotic (pigmented) schwannoma, and (4) angiosarcoma-like areas.

16.2.6.2 Neurofibroma
(Figs. 16.2 and 16.3)

- *DEF*: Benign, common, spindle cell tumor ± NF-1, in the case of multiple tumors, with a malignant transformation more common in NF-1 neurofibromas than in sporadic neurofibromas. Subtypes include plexiform neurofibroma with irregularly expanded nerve bundles with nodular appearance and prominent myxoid matrix; associated with NF-1, diffuse cutaneous, focal cutaneous, and intraneural (Figs. 16.2 and 16.3).

- *CLI*: Small, pedunculated nodules over the skin if superficial or tumor-mass effect if deeply located and large.
- *GRO*: Not encapsulated with a softer consistency than schwannoma.
- *CLM*: It is characterized by the proliferation of all fundamental components of peripheral nerves (Schwann cells with wavy serpentine nuclei, pointed ends, stromal mucosubstances, mast cells, Wagner-Meissner corpuscles, Pacinian corpuscles, axons) with perineurial cells in plexiform types, low MI, and a focally infiltrative character with myxoid areas and potential epithelioid morphology. There are neither Verocay bodies nor nuclear palisading or hyalinized thickening of vessel walls.
- *HSS/IHC*: (+) Silver or AcHE stain, NSE, NF, S100, CD34 (focal), FXIIIa (focal), EMA (plexiform neurofibromas only).
- *TEM*: Schwann cells enclose axons in invaginations of plasmalemmal type.
- *DDX*: Myxoid lipoma and myxoid liposarcoma.
- *TRT*: Resection.
- *PGN*: Meager chance of malignant transformation for sporadic neurofibromas, but a tendency to transform in MPNST is present in multiple neurofibromas.

16.2.6.3 Malignant Peripheral Nerve Sheath Tumor (Figs. 16.4, 16.5, 16.6, 16.7, and 16.8)

- *DEF*: It is a large deep-seated tumor arising from significant nerves in the neck, buttock, and limbs and associated with NF-1, previous radiation, or malignant transformation of a neurofibroma or plexiform neurofibroma.
- *SYN*: Malignant schwannoma.
- *CLI*: Apart from the tumor-mass effect, signs, and symptoms related to NF-1.
- *GRO*: It is a large mass-producing fusiform expansion of a major nerve.
- *CLM*: Monomorphic serpentine cells, palisading, large gaping vascular spaces, perivascular plump tumor cells, necrosis, imparting a "geographic" appearance with tumor palisading at edges with atypical mitoses and 1/10 cases harboring metaplastic cartilage, bone, and muscle with not uncommon glandular differentiation and intracellular melanin pigmentation (Figs. 16.4, 16.5, and 16.6).
- *IHC*: (+) S100, CD99, CD57, PGP 9.5, p53, but also (+) KER, EMA, CEA, and CGA (glandular).
- *TEM*: Infoldings of the cellular membrane with the lamellar arrangement, intermittent basal lamina, obvious intercellular junctions, and dense-core granules (occasionally).
- *CMB*: t(X;18) negative.
- *DDX*: MFH, pleomorphic liposarcoma, and synovial sarcoma as well as neuroblastoma and ganglioneuroma that are described elsewhere in this book (Figs. 16.7 and 16.8).
- *TRT*: Resection with safe margins.
- *PGN*: MPNST can recur locally, even if wholly excised, and distant metastases are frequent. The plexiform variant in children seems to have a better prognosis.

16.3 Neuromuscular Disorders

16.3.1 Muscle Biopsy Test

The interpretation of diagnostic muscle biopsies has as a condition that we understand and remember the skeletal muscle histology. This information is basilar in differentiating several patterns that may be associated with congenital hypotonia of the infant. Skeletal muscle fibers or aka myofibers are delimited by sarcolemma (cell membrane) composed of sarcoplasmic reticulum (particularly important for transport and storage of calcium) and myofibrils, which are arrays of actin (thin) and myosin (thick) filaments surrounded by sarcoplasmic reticulum (Box 16.1). Multiple such displays may be present per myofiber. The functional unit of the skeletal muscle cell is the sarcomere, which is the collection of actin and myosin filaments between two Z-bands. Sarcomere banding includes five bands labeled with letters. Two of these letters are A and I, which stand for anisotropic and isotropic, i.e., they can rotate plane polarized light, although both bands are anisotropic, being the A-band more anisotropic than I-band. Muscle contraction is based on sarcomere shortening, which consists of Z-bands approaching each other. During the sarcomeric shortening, the A-band remains the same, and I- and H-bands become shorter than before the contraction. When evaluating muscle biopsies, it is important to discern between neuropathic or myopathic, although in chronicity, there can be a mixture of both features. Myopathic processes usually demonstrate coexistent degeneration and regeneration. Of note, a characteristic feature of the regenerating fibers is their tendency to show rounded edges and central nuclei with a prominent nucleolus and be strongly stained with the alkaline phosphatase reaction. In the case of a myopathic process, it is essential to distinguish between inflammatory and non-inflammatory processes. The identification of neutrophils and mononuclear cells within the muscle with or without an accompanying vasculitis is crucial. Non-inflammatory processes show coagulative necrosis and "ragged-red" fibers. Moreover, variation in fiber size and internal nuclei are standard at the site of tendinous insertion. Rhabdomyolysis is characterized by diffuse destruction or lysis of skeletal muscle, which may be acute, subacute, or chronic. The subsequent massive myoglobinuria can induce renal failure.

Box 16.1 Muscle Fiber Types

	Type I	Type II
Color	Red	White
Speed	Slow	Fast
Duration/fatigue/use	Protracted/slow/posture	Swift/rapid/strength
Muscle prevalence	~ 30%	~70%
Metabolism	O_2-based (aerobic)	Non-O_2-based (anaerobic)
Glycogen content	+	+++
Lipid content	+++	+
Mitochondria/myoglobin	+++/High	+/Low

The two types of muscle fibers can be differentiated using a series of histochemical stains (Box 16.2).

Special invaginations of the sarcolemma, aka transverse tubules, extend to the sarcoplasmic reticulum of each myofibril of a myofiber and allow myofiber depolarization, i.e., the rapid transmission of an electrical stimulus. Connective tissue has three different sheath functional roles, including epimysium surrounding entire muscle, perimysium separating muscle fascicles (groups of myofibers), and endomysium surrounding each myofiber. The motor unit is a single lower motor neuron of the anterior horn of the spinal cord connected to myofibers that innervate. The

Box 16.2 Histochemical Stains for Skeletal Muscle Fibers

ATPase, acid (pH 4.6)		
ATPase, basic (pH9.4)		
NADH-TR		
Non-specific esterase		

Notes: Nonspecific esterase reaction is peculiarly useful in detecting denervated fibers, while alkaline phosphatase reaction is particularly useful in highlighting regenerating myofibers, light (IIA)/intermediate (IIB). Nonspecific esterase shows stronger stain in denervated myofibers. Dark is brown color, while light is green color with intermediate in yellow. No stain is represented by no color. NADR-TR, nicotinamide adenine dinucleotide tetrazolium reductase.

number of myofibers per motor neuron is variable between less than 10 for fine motor control and more than 1000 for large leg muscles. Except for the extraocular muscles, any myofiber is innervated by a single motor neuron. Skeletal muscle histogenesis is based on a *syncytial* process, which refers to the fusion of embryonic myoblasts to form a multinucleated syncytium, aka myotube, able to stimulate the formation of myofibrils, and on *innervation* process intimately linked to the distribution of acetylcholine receptors at motor end plate. The evaluation of muscle biopsies progresses with diagnostic degrees of certainty, including first if changes are neuropathic or myopathic and then in the case of a myopathic process if an inflammatory component is at the basis of the phenotypically based clinical abnormalities. However, in chronic processes, which are usually rare in youth, a mixture of both neuropathic and myopathic may co-exist. Inflammatory myopathies often show neutrophils, and mononuclear cells within the myofibers and between the bundles and vasculitis may also be an accompanying feature. Non-inflammatory myopathies show, instead, coagulative necrosis of the myofibers and "ragged-red" fiber morphological patterns. Three patterns need to be distinguished in the evaluation of skeletal muscle biopsies, in particular, regeneration, which shows rounded edges-associated myofibers with central nuclei and strongly positive for alkaline phosphatase; rhabdomyolysis, which refers to complete and diffuse destruction of skeletal myofibers with acute, subacute, or chronic progression and complicated by myoglobinuria-associated renal failure; and *ex non-usu* atrophy with selective loss of type II fibers. Patients using high doses of steroids may show disuse atrophy as well.

16.3.2 Neurogenic Disorders

Neurogenic disorders that will be treated are denervation and some diseases associated with denervation process and *myasthenia gravis*, which is rare in pediatric age, but is related to autoimmune dysregulations and may be necessary from a pathophysiology perspective.

16.3.2.1 Denervation
(Figs. 16.9 and 16.10)

A loss of myofibrils or atrophy is the consequence of denervation, and all the myofibers in the motor unit associated with the lower motor neuron are lost (type II fibers first, then type I). According to the extension of the damage or loss of individual neurons, there will be small group atrophy for individual neurons and considerable group atrophy for larger nerves. Histologically, there is angulation of the myofibers on cross-section and nuclear aggregation into hyperchromatic clumps. Two additional histologic patterns may be seen. A fiber with central pallor with condensed eosinophilic or hyper-eosinophilic rim surrounded by normal-appearing muscle is labeled target fibers. A renervation pattern is recognized when the previously randomly distributed type I and type II fibers are converting into type "grouping" fibers and usually coincides with denervation. Two conditions may need to be recalled, including Werdnig-Hoffmann disease (WHD) and amyotrophic lateral sclerosis (ALS). WHD or infantile spinal muscular atrophy is an AR congenital hypotonia showing progressive group atrophy with decline and death within 1 year of age. The histology of the spinal cord shows loss of lower motor neurons of the anterior horn and extensive group atrophy. ALS is a group of rare neurological diseases that involve the neurons responsible for controlling voluntary muscle movement. The disease is progressive with early symptoms in the 30s with a median survival of 3 years. Homozygous loss-of-function mutations in the *ALS2* gene determine the diagnosis of juvenile-onset ALS, one of the several neurological conditions having overlapping symptoms with different neurological phenotypes. In case of involvement of both upper and lower motor neurons, there are fasciculations and group atrophy. In the setting of tumors, the traumatic neuroma and the granular cell tumor have been described in association with denervation, at least partly and are described elsewhere in this book (Figs. 16.9 and 16.10).

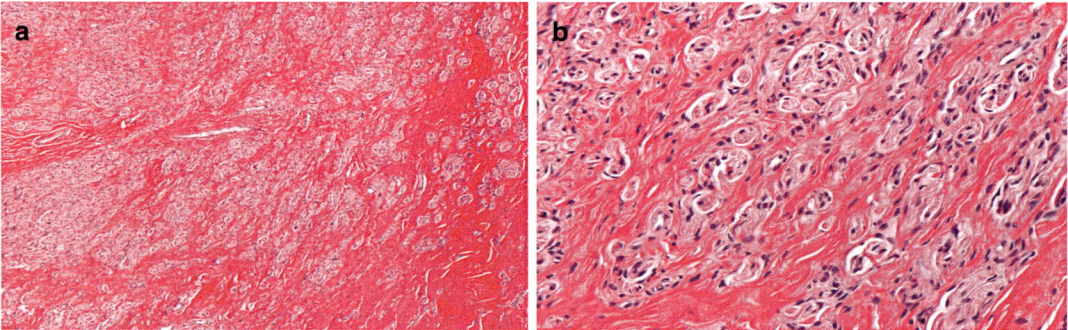

Fig. 16.9 Traumatic Neuroma. The two microphotographs of hematoxylin and eosin-stained histological slides show the typical presentation of a traumatic neuroma, which is a non-neoplastic lesion (**a**, ×50 original magnification; **b**, ×200 original magnification). The histologic examination reveals an irregular arrangement of nerve fascicles embedded in fibrous scar tissue. Sometimes, there is a concentric condensations of fibrous tissue around individual fascicles with the suggested appearance of multiple separate nerves

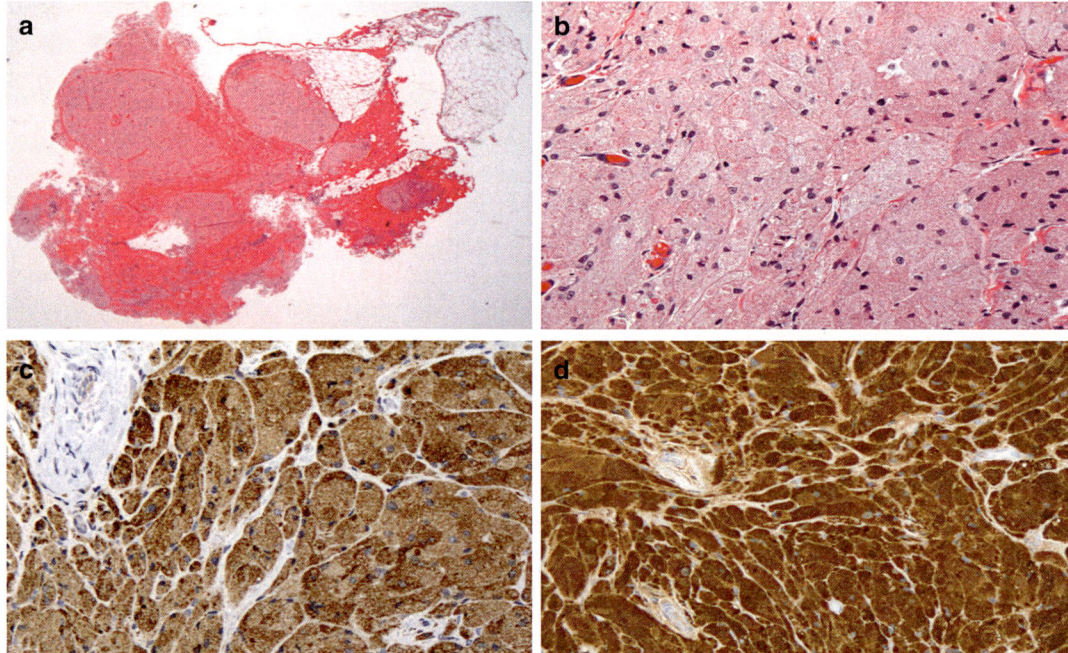

Fig. 16.10 Granular Cell Tumor. This panel illustrates the histology of a granular cell tumor with the first two microphotographs (**a**, **b**) showing a poorly circumscribed tumor with large cells harboring a highly granular cytoplasm (**a**, hematoxylin and eosin, ×12.5 original magnification; **b**, hematoxylin and eosin staining, ×200 original magnification). Two additional microphotographs show the positivity of this tumor for neuron specific enolase (NSE) (**c**, ×200 original magnification) and S100 (**d**, ×200 original magnification). The immunohistochemistry was performed using the avidin-biotin complex

16.3.2.2 Myasthenia Gravis

MG is an autoimmune disorder affecting usually young women in the third to fourth decades of life showing auto-Ab against the neuromuscular receptors of acetylcholine, but pediatric cases have been reported and the term of Juvenile Myasthenia Gravis (JMG) is often in use in the literature. JMG is a rare condition, which is defined as MG in children younger than 18 years of age (O'Connell et al. 2020). Although clinical phenotypes are similar to the adult counterpart, there are a number of situations (pitfalls) that may influence management. These aspects include a broader differential diagnosis, a higher rate of spontaneous remission, and the requirement to start appropriate treatment early with the aim to avoid the both long-term physical and psychosocial morbidities. There are antibodies against the acetylcholine (ACh) receptors (AChR) at the post-synaptic membrane. Both antibodies block stimulation by acetylcholine and induce a down-regulation of the AChR. It may be associated with AI thyroid disorders, rheumatoid arthritis, SLE, and pernicious anemia. IgG and complement may interfere with neural transmission. Clinically, there is weakness relieved by rest and worsened by exercise with mostly involvement of cranial and adjacent extraocular muscles. There is positivity for the edrophonium/neostigmine test. There is a fluctuant course with partial remissions, followed by relapses and progression. EMG shows the decrement of the responses to repeated nerve stimulation. In up to 4/5 of the patients, there is thymic hyperplasia, and 1/10–1/3 patients exhibit a thymoma. Microscopically, there is lymphocytic infiltration with hemorrhages around degenerating myofibers on light microscopy, and abnormal myoneural junctions with widened clefts are seen on TEM. Ultrastructurally, there are irregular motor end plates with the simplified post-synaptic region. The outcome is poor with death due to recurrent infections of the lower respiratory tract (pneumonia with respiratory, muscular weakness), although thymectomy has been correlated with remission of the myasthenia. Although up to 1/3 of patients may develop thymoma, the rest may have thymic hyperplasia. Myasthenia gravis may be life-threatening in case the respiratory function is compromised. Eaton-Lambert syndrome is a paraneoplastic, myasthenia-like syndrome, which occurs in about 2/3 of patients suffering from small cell carcinoma of the lung.

16.3.3 Myopathic Disorders

Myopathic disorders may be subclassified in muscular dystrophies and congenital myopathies. Muscular dystrophies are common in the young adult, while the congenital myopathies may start early in life. These include muscular dystrophies and congenital myopathies (Fig. 16.11).

16.3.3.1 Muscular Dystrophies
Muscular dystrophies are hereditary diseases of the neuromuscular and motor unit with the progressive tendency. The prototype of muscular dystrophies is the X-linked recessively inherited Duchenne (pseudo-hypertrophic) muscular dystrophy, affecting mainly boys. Clinically, weakness (proximal > distal, symmetric, legs and arms, mostly *adductor magnus* in legs and relatively spared the muscles *gracilis* and *Sartorius*) and falling off the child beginning to walk due to the involvement of the pelvic girdle are encountered by clinical history and physical examination. Subsequently, there may be the participation of other muscular groups and a very high level of CK is detected in serum. Other clinical features include DCM of the heart (>15 years), mental retardation, night blindness, and pseudo-obstruction as well as gastric dilatation. The diagnosis is based on the lack of the gene product of dystrophin. The *DYSTROPHIN* gene is on Xp21.2-p21.1 with 96% with a frameshift mutation. Most often, 30% demonstrate new variations. In particular, 10–20% of new mutations are of a gonadal mosaic type. Grossly, there are atrophy and yellow discoloration of the muscles. Microscopically, there is a pattern characterized by variable fiber size with small rounded fibers and hypercontracted (opaque) muscle fibers and the tendency to cluster necrotic muscles (myopathic grouping) with muscle fiber degeneration and regeneration (particularly at early stages). The muscle fiber internal architecture is standard or shows some immature features. Dystrophin immunostaining is absent, whereas other membrane proteins, such as sarcoglycans and aquaporin-4, are reduced in staining. Fibrosis of the endomysia is a characteristic of the late stages. In case of absent dystrophin, DMD phenotype is severe; there is a reduction in sarcoglycans and other proteins in the dystrophin-glycoprotein complex. In case of functional consequences of loss of dystrophin on muscle fibers, there is an increased movement of membrane-impermeant molecules into and out of muscle cells, decreased force production, and decreased hypersensitive to lengthening or eccentric contraction. Force decrement with eccentric contraction typically correlates with acutely increased sarcolemmal permeability. Death is most commonly observed between 15 and 25 years and is due to cardiac or respiratory failure, although life may be prolonged with respiratory support.

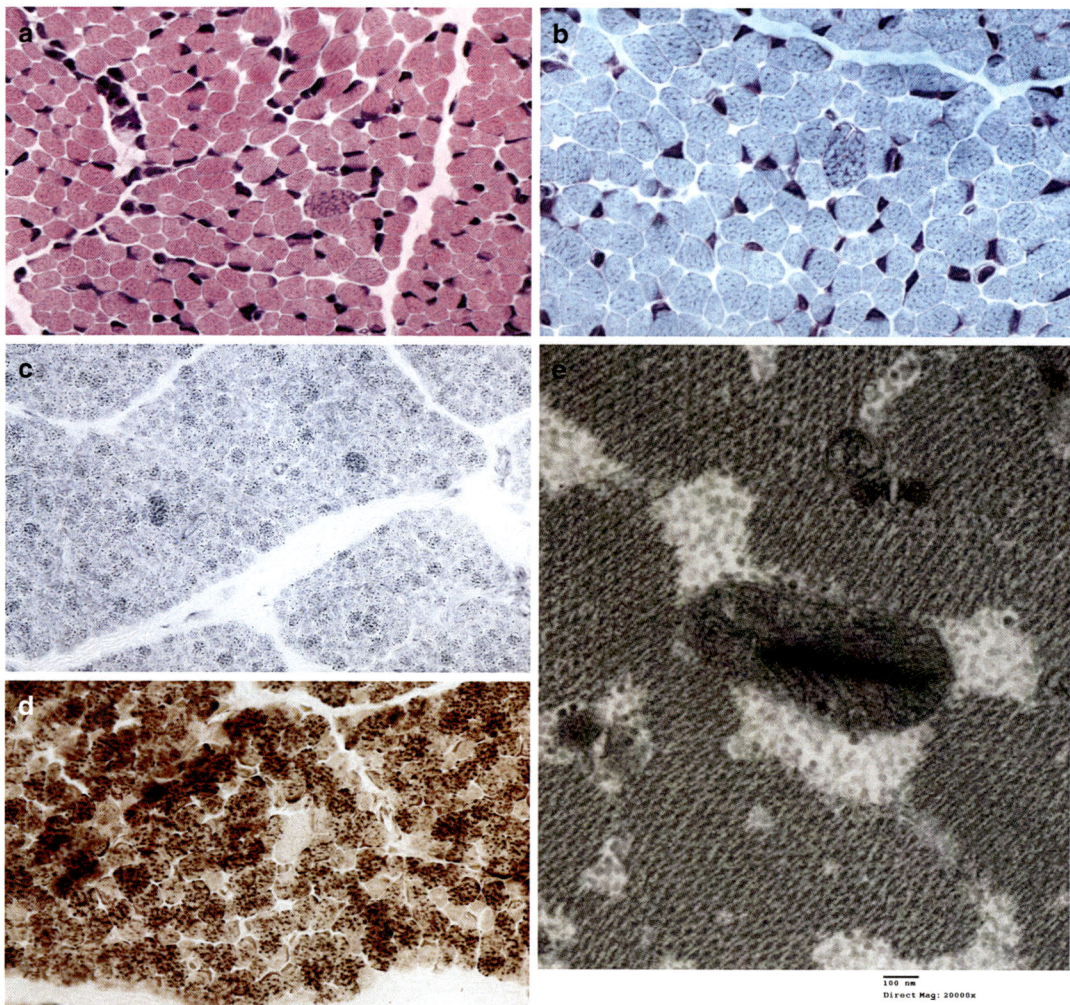

Fig. 16.11 Mitochondrial cytopathy panel with excessive blue staining due to increased mitochondria (**a**), excessive red staining due to increased mitochondria (**b**), scattered darker fibers on succinate dehydrogenase histochemical staining as ragged red fiber equivalents (**c**), scattered negative fibers on cytochrome oxidase (**d**), inclusion body in mitochondria (**e**) (**a**, hematoxylin and eosin staining, ×200 original magnification, **b**, modified muscle trichrom, ×200 original magnification, **c**, succinate dehydrogenase (SDH) staining, ×100 original magnification; **d**, cytochrome oxidase (COX), ×200 original magnification; **e**, transmission electron microscopy, bar = 100 nm)

16.3.3.2 Congenital Myopathies

Congenital myopathies include a group of disorders which are not progressive but restrict the activity of the patient and are complicated by secondary skeletal problems such as kyphoscoliosis. Clinically, they start at birth with hypotonia, hyporeflexia (↓ deep tendon reflexes), and low muscle mass. The progression of the disease shows that motor milestones are delayed. Skeletal muscle biopsy reveals that morphologic abnormalities are restricted to the type I fibers, which predominate, while the type II fibers are decreased in number but structurally normal. Central core disease is a sporadic, AR/AD-inherited congenital myopathy with type I fibers showing a central core of lucency running the length of the cell due to a loss of membranous organelles and often with disorganization of the surrounding myofibrils. The NADH-TR (nicotinamide adenine dinucleotide dehydrogenase tetrazolium reductase) reaction is advantageous to demonstrate the cores. Nemaline (rod) myopathy exhibits inclusions within the type I myofibers of rod-shaped structures which arise from the Z-bands. They are stainable with trichrome stain or with PTAH.

16.3.4 Glycogen Storage Diseases

Glycogen storage disorders may accompany a presentation of hypotonia and need to be kept in the differential diagnosis. For the discussion of this group, please refer to the liver chapter.

16.3.5 Mitochondrial Myopathies (Fig. 16.11)

Mitochondriopathies (MCP) are usually multisystemic disorders with involvement of visceral organs, such as endocrine organs, heart, gastrointestinal tract, liver, kidney, or hematopoietic system, other than central and peripheral nervous system, eyes, or ears (Finsterer 2006, Mohammed et al. 2012, Cave et al. 2013, Hsu et al. 2016). The diagnosis of MCP may be a real challenge, because of their extensive, variable, and diversified clinical and genetic heterogeneity. The specific diagnostic determination depends on clinical, blood chemical, electrophysiological, imaging, histological, immunological, biochemical, polarographic, magnetic spectroscopic, and genetic investigations. Human mtDNA is a 16.5 kb circular mini-chromosome consisting of two complementary strands (H and L strand). MtDNA contains 13 mitochondrial genes encoding for subunits of the respiratory chain complexes I (NADH dehydrogenase (ND)1–4, ND4L, ND5–6), III (cytochrome b), IV (cytochrome-c-oxidase (COX)I-III), and V (ATPase 6, ATPase 8) and 24 syn genes encoding for 22 tRNAs (ribonucleic acid) and 2 rRNAs. Only the D-loop is a noncoding stretch. This portion contains the promoters for the transcription of L and H strands. Since the mtDNA genetic code differs from the universal code, the expression of mtDNA genes relies on the specific mitochondrial protein synthesis, depending on the interplay between nuclear-encoded transcriptional and translational factors with mitochondrial tRNAs and rRNAs. Neuropathologists use the term mitochondrial cytopathies. The mitochondrial cytopathies have distinct histology, although they represent a heterogeneous group of disorders which primarily affect the neuromuscular systems. The mitochondrial cytopathies are caused either by mutations in the mitochondrial genome or by mutations in the nuclear genome (Fig. 16.5).

Genetics of mitochondria differs from the genetics of nuclei. In particular, some features can be better delineated as follows:

1. Mitochondrial DNA is maternally inherited.
2. Polyploidy is a characteristic feature of mitochondria, containing 2–10 mtDNA copies per organelle (each cell includes hundreds of mitochondria).
3. All mtDNA copies are identical (homoplasmy), but there is the tendency of mtDNA to mutate randomly, and the coexistence of wild-type mtDNA and mutant mtDNA in a single cell and organ is called "heteroplasmy."
4. There is a stochastic distribution of mitochondria and mutant mtDNA to daughter cells, which gives rise to a changing mutation load in a different generation and consequent phenotypic variation of MCP.
5. Phenotypic expression is based on a threshold effect because of mitotic segregation and polyploidy. Phenotype expression of mtDNA mutations often requires the influence of nuclear modifier genes, environmental factors, or the presence of mtDNA haplotypes (polymorphisms).

Chronic progressive external ophthalmoplegia (CPEO) is the most prevalent manifestation of mtDNA rearrangements and often associated with ptosis. Subsequently, cataract, retinitis pigmentosa, deafness, fatigue, ataxia, limb weakness, neuropathy, arrhythmias, or renal insufficiency may develop. The clinical course is usually benign because organ failure is mild. CPEO is due to mtDNA deletions or mutations in the *tRNALeu, tRNAIle, tRNALys,* or *tRNAAsn* genes.

Kearns-Sayre syndrome (KSS) is characterized by retinitis pigmentosa, arrhythmias, cerebellar ataxia, raised CSF protein, and onset before the age of 20. Proximal myopathy develops with the progression of the disease, and other features include mental retardation, deafness, syncope, bulbar symptoms such as dysphagia, stroke-like episodes, and endocrine dysfunction, such as delayed puberty, primary amenorrhea, or diabetes mellitus, sideroblastic anemia, and lactic acidosis. The prognosis of KSS is poor because organ failure is severe and the patients do not survive the third decade. KSS is due to sporadic, single or multiple large-scale deletions or from mtDNA duplications.

Mitochondrial, encephalomyopathy, lactic acidosis, and stroke-like episodes (MELAS) is the commonest of the encephalomyopathies and is characterized by stroke-like episodes with hemisyndrome, migraine, nausea, or vomiting. Other features include deafness, DM, seizures, dementia, ataxia, cortical blindness, optic atrophy, retinitis pigmentosa, DCM, exercise intolerance, and short stature. The course is slowly progressive, and the prognosis relies mostly on the cardiac involvement. MELAS is due to mtDNA point mutations in *tRNA, COX3,* or *ND5* genes or due to small-scale mtDNA deletions. A frequently detected variation in MELAS patients is the transition A3242G.

Myoclonic epilepsy with ragged red fiber (MERRF) syndrome presents between childhood and early adulthood with photosensitive general tonic-clonic seizures, myopathy, including ptosis and ophthalmic paralysis, cerebellar ataxia, dementia, and deafness. Other features include stroke-like episodes, optic atrophy, dorsal column loss, cardiomyopathy, heart block, heart failure, respiratory failure, paralytic ileus, pancytopenia, lipomatosis, *pes cavus*, and polyneuropathy. MERRF is caused by mtDNA tRNALys point mutations or multiple mtDNA deletions resulting from nDNA mutations. A variation frequently found in MERRF patients is the mtDNA transition A8344G.

Other mitochondriopathies include maternally inherited diabetes and deafness (MIDD), Leber's hereditary optic neuropathy, neuropathy, ataxia, and retinitis pigmentosa (NARP), and maternally inherited Leigh syndrome (MILS).

Nonsyndromic, mitochondrial MCPs with visceral manifestations include myopathy associated with cardiomyopathy due to *tRNALeu* mutations and hypertrophic cardiomyopathy due to *12rRNA, tRNASIle, tRNALys, tRNAGly* or cytb mutations among others.

Syndromic, nuclear mitochondriopathies with visceral manifestations include nuclear Leigh syndrome, myo-neuro-gastrointestinal encephalopathy (MNGIE), also termed POLIP (polyneuropathy, ophthalmoplegia, leukoencephalopathy, intestinal pseudo-obstruction), diabetes insipidus, DM, optic atrophy, deafness (Wolfram syndrome) or DIDMOAD syndrome, mitochondrial DNA depletion syndrome, and Barth syndrome.

The *mtDNA depletion syndrome* has mtDNA depletion of various degrees, which leads to a fatal multisystem infantile disorder, characterized by weakness, muscle hypotonia, CPEO, and severe lactic acidosis. Other features include hepatopathy, Fanconi syndrome, encephalopathy, seizures, cardiomyopathy, and cataract. Total mtDNA levels in these patients are below 35% of those in controls. Although no mtDNA mutations are found, the underlying defect is an impaired replication or maintenance of mtDNA due to nuclear DNA (nDNA) mutations in the thymidine kinase or deoxyguanosine kinase gene, causing gastrointestinal abnormalities; deoxyguanosine kinase (*DGUOK*) gene, resulting in encephalo-myopathy and hepatopathy; and polymerase gamma (*POLG*) gene causing hepatopathy with lactic acidosis.

Barth syndrome is a rare X-linked disease, characterized by the triad dilated cardiomyopathy, skeletal myopathy, and neutropenia. Other features include growth retardation and 3-methylglutaconic aciduria. Barth syndrome is due to mutations in the *Tafazzin* gene, which is located on chromosome Xq28. The *Tafazzin* gene is suspected to encode for one or more acyltransferases, resulting in reduced cardiolipin synthesis and thus cardiolipin deficiency in the skeletal muscle, myocardium, and platelets. Other syndromes include Friedreich ataxia, Mohr-Tranebjaerg syndrome, nuclear chronic progressive external ophthalmoplegia, and nuclear myoclonic epilepsy with ragged red fibers.

Nonsyndromic, nuclear mitochondriopathies with visceral manifestations include several disorders, i.e., fatal, multisystem complex I deficiency due to mutations in the *NDUFS4* gene, familial idiopathic cardiomyopathy due to multiple, secondary mtDNA, and thiamine-responsive megaloblastic anemia among others.

16.3.6 Inflammatory Myopathies: Non-infectious

These include polymyositis/dermatomyositis, viral myositis, and inclusion body myositis. They are presented in other chapters.

16.3.7 Inflammatory Myopathies: Infectious

Several microorganisms may be associated with infective myositis. Infectious myopathies are infections-related inflammatory myopathies, which have, typically, a good prognosis. However, the outcome is strictly dependent on the immune response of the child or young adult. Numerous viral infections can cause temporary inflammatory myopathies and be part of a fleeting visit to clinics. HIV myopathy may occur early in the HIV infection. However, it is currently more often observed as a complication of AIDS. The influenza virus myositis is more severe in adults than in children. In patients affected by epidemic myalgias, group B coxsackievirus has been isolated from the striated muscle of the patients. Parasitic infections of the tissue include trichinosis, toxoplasmosis, and cysticercosis. The most common parasitic myositis is due to the roundworm *Trichinella spiralis* or *Trichinella* spp. Trichinosis symptoms include diarrhea, abdominal cramps, muscle pain, and fever. Individuals acquire trichinosis by eating raw or undercooked contaminated meat. Pyomyositis is a located zone of suppuration of muscle due to *Staphylococcus* in about 90% of the cases.

Multiple Choice Questions and Answers

- PNS-1 Which statement regarding acetylcholinesterase (AChE) is TRUE?
  (a) It is a neurotransmitter.
  (b) It catalyzes the breakdown of acetylcholine and of some other choline esters that function as neurotransmitters.
  (c) It has enhanced activity from organophosphorus compounds.
  (d) It does not belong to carboxylesterase family of enzymes.
- PNS-2 Which statement regarding the enteric nervous system is FALSE?
  (a) The myenteric plexus is located between the longitudinal and circular layers of muscle in the tunica muscularis.
  (b) The submucous plexus is embedded in the submucosa and plays a role in sensing the environment within the lumen, regulating blood flow and controlling cell function.
  (c) In the enteric plexuses, there are three types of neurons (motor neurons, sensory neurons, and interneurons), most of which are bipolar.
  (d) Sympathetic stimulation determines the inhibition of gastrointestinal secretion and motor activity. It also acts to contract gastrointestinal sphincters and blood vessels.
  (e) Parasympathetic stimulation typically stimulates the gastrointestinal secretion and motor activity, and contraction of gastrointestinal sphincters and blood vessels.
- PNS-3 Which statement regarding myasthenia gravis is TRUE?
  (a) Myasthenia gravis is a neurodegenerative disease.
  (b) Myasthenia gravis is associated to an inadequacy of the organism to release enough amounts of acetylcholine.

(c) An anti-AChE is not efficacious in myasthenia gravis.
(d) Most patients affected with myasthenia gravis have weakness in the muscles of the eyes, eyelids, and face.
(e) Thymectomy is not recommended for myasthenia gravis.

- PNS-4 Verocay bodies are histologic areas lacking nuclei between areas of nuclear palisading. In which tumor are they found characteristically?
  (a) Glioblastoma multiforme
  (b) Schwannoma
  (c) Neurofibroma
  (d) Plexiform neurofibroma
  (e) Ependymoma

- PNS-5 Which of the following statements regarding the traumatic neuroma is FALSE?
  (a) It is a reparative lesion at the site of traumatic injury of peripheral nerves.
  (b) Wallerian degeneration is present initially after nerve interruption, which is followed by an exuberant regeneration of Schwann cells, axons, and fibrous cells.
  (c) Histologic sections show haphazard presentation of nerves within a collagenous scar with entrapped smooth muscle elements.
  (d) The lesion is immunohistochemically S100 negative, but neurofilaments are positive.

- PNS-7 Which of the following statements regarding classic neurofibromatosis (von Recklinghausen's disease) is FALSE?
  (a) Malignant peripheral nerve sheath tumor (MPNST) can occur.
  (b) Iris hamartomas may be present.
  (c) Hemangioblastomas of the cerebellum can occur later in life.
  (d) Acoustic neuroma is an important finding.
  (e) Both meningiomas and pheochromocytoma can be found.

- PNS-8 Which of the following statements of MPNST is FALSE?
  (a) MPNST occurs in the setting of NF1 probably due to a biallelic inactivation of the NF1 gene.
  (b) MPNST often displays disruption of the RB1 pathway.
  (c) CDKN2B and RB1 tumor suppressor genes may be inactivated in a subset of MPNSTs.
  (d) Dominantly acting genes in the RB1 pathway (precisely CDK4, MDM2, and CCND2) may be overexpressed or amplified in a subset of patients harboring MPNSTs.
  (e) The TP53 gene is biallelically affected in most of the MPNSTs.

- PNS-9 In the case of a young woman affected with a mitochondrial disorder, who will most likely inherit the disorder?
  (a) All her daughters, but none of her sons
  (b) All her sons, but none of her daughters
  (c) All her children
  (d) All her grandsons
  (e) All her granddaughters

- PNS-10 Environmental pollution has an enormous burden currently. Thus, some diseases may occur at an early age than before. A 25-year-old man presents to the family doctor with weakness and cramping of both hands. Upon examination, the physician identifies atrophy of the muscles of both hands, hyperresponsive "deep tendon" reflexes, muscular fasciculation involving both arms and legs, and an abnormal positive Babinski response. The neurological examination of the sensation reveals normal feeling in both arms and legs. What is the most likely diagnosis in this patient?
  (1) Wilson disease
  (2) Guillain-Barré syndrome
  (3) Huntington disease
  (4) Amyotrophic lateral sclerosis
  (5) Metachromatic leukodystrophy

References and Recommended Readings

Attarian S, Azulay JP. Myopathies infectieuses [Infectious myopathies]. Rev Prat. 2001;51(3):284–88. Review. French. PubMed PMID: 11265425.

Bahitham W, Liao X, Peng F, Bamforth F, Chan A, Mason A, Stone B, Stothard P, Sergi C. Mitochondriome and cholangiocellular carcinoma. PLoS One. 2014;9(8):e104694. https://doi.org/10.1371/journal.pone.0104694. eCollection 2014. PubMed PMID: 25137133; PubMed Central PMCID: PMC4138114.

Cave D, Ross DB, Bahitham W, Chan A, Sergi C, Adatia I. Mitochondrial DNA depletion syndrome-an unusual reason for interstage attrition after the modified stage 1 Norwood operation. Congenit Heart Dis. 2013;8(1):E20–E23. https://doi.org/10.1111/j.1747-0803.2011.00569.x. Epub 2011 Oct 20. PubMed PMID: 22011012.

Chiu B, Jantuan E, Shen F, Chiu B, Sergi C. Autophagy-inflammasome interplay in heart failure: a systematic review on basics, pathways, and therapeutic perspectives. Ann Clin Lab Sci. 2017;47(3):243–252. Review. PubMed PMID: 28667023.

Finsterer J. Overview on visceral manifestations of mitochondrial disorders. Neth J Med. 2006;64(3):61–71. Review. PubMed PMID: 16547358.

Gray SW, Skandalakis JE. Embryology for surgeons: the embryological basis for the treatment of congenital defects. Philadelphia: WB Saunders Co; 1972. p. 263–82.

Hsu YH, Yogasundaram H, Parajuli N, Valtuille L, Sergi C, Oudit GY. MELAS syndrome and cardiomyopathy: linking mitochondrial function to heart failure pathogenesis. Heart Fail Rev. 2016;21(1):103–116. https://doi.org/10.1007/s10741-015-9524-5. Review. PubMed PMID: 26712328.

Katona I, Weis J. Diseases of the peripheral nerves. Handb Clin Neurol. 2017;145:453–474. https://doi.org/10.1016/B978-0-12-802395-2.00031-6. Review. PubMed PMID: 28987189.

Mohammed S, Bahitham W, Chan A, Chiu B, Bamforth F, Sergi C. Mitochondrial DNA related cardiomyopathies. Front Biosci (Elite Ed). 2012;1(4):1706–1716. Review. PubMed PMID: 22201986.

O'Connell K, Ramdas S, Palace J. Management of Juvenile Myasthenia Gravis. Front Neurol. 2020;11:743. https://doi.org/10.3389/fneur.2020.00743. PMID: 32793107; PMCID: PMC7393473.

Parasca I, Damian L, Albu A. Infectious muscle disease. Rom J Intern Med. 2006;44(2):131–141. Review. PubMed PMID: 17236294.

Pestronk A. Acquired immune and inflammatory myopathies: pathologic classification. Curr Opin Rheumatol. 2011;23(6):595–604. https://doi.org/10.1097/BOR.0b013e32834bab42. Review. PubMed PMID: 21934500.

Röhrich M, Koelsche C, Schrimpf D, Capper D, Sahm F, Kratz A, Reuss J, Hovestadt V, Jones DT, Bewerunge-Hudler M, Becker A, Weis J, Mawrin C, Mittelbronn M, Perry A, Mautner VF, Mechtersheimer G, Hartmann C, Okuducu AF, Arp M, Seiz-Rosenhagen M, Hänggi D, Heim S, Paulus W, Schittenhelm J, Ahmadi R, Herold-Mende C, Unterberg A, Pfister SM, von Deimling A, Reuss DE. Methylation-based classification of benign and malignant peripheral nerve sheath tumors. Acta Neuropathol. 2016;131(6):877–87. https://doi.org/10.1007/s00401-016-1540-6. Epub 2016 Feb 8. PubMed PMID: 26857854.

Schulz A, Grafe P, Hagel C, Bäumer P, Morrison H, Mautner VF, Farschtschi S. Neuropathies in the setting of neurofibromatosis tumor syndromes: complexities and opportunities. Exp Neurol. 2018;299(Pt B):334–344. https://doi.org/10.1016/j.expneurol.2017.06.006. Epub 2017 Jun 3. Review. PubMed PMID: 28587874.

van der Knaap MS, Bugiani M. Leukodystrophies: a proposed classification system based on pathological changes and pathogenetic mechanisms. Acta Neuropathol 2017;134(3):351–382. https://doi.org/10.1007/s00401-017-1739-1. Epub 2017 Jun 21. Review. PubMed PMID: 28638987; PubMed Central PMCID: PMC5563342.

Skin 17

Contents

17.1	**Development, General Terminology, and Congenital Skin Defects**	1347
17.1.1	Development	1347
17.1.2	General Terminology	1348
17.1.3	Lethal Congenital Contractural Syndromes	1348
17.2	**Spongiotic Dermatitis**	1352
17.2.1	Conventional Spongiosis	1352
17.2.2	Eosinophilic Spongiosis	1354
17.2.3	Follicular Spongiosis	1354
17.2.4	Miliarial Spongiosis	1354
17.3	**Interface Dermatitis**	1355
17.3.1	Vacuolar Interface Dermatitis	1355
17.3.2	Lichenoid Interface Dermatitis	1360
17.4	**Psoriasis and Psoriasiform Dermatitis**	1360
17.4.1	Psoriasis	1360
17.4.2	Psoriasiform Dermatitis	1360
17.5	**Perivascular In Toto Dermatitis (PID)**	1361
17.5.1	Urticaria	1361
17.5.2	Non-urticaria Superficial and Deep Perivascular Dermatitis	1362
17.6	**Nodular and Diffuse Cutaneous Infiltrates**	1363
17.6.1	Granuloma Annulare	1363
17.6.2	Necrobiosis Lipoidica Diabeticorum (NLD)	1364
17.6.3	Rheumatoid Nodule	1364
17.6.4	Sarcoid	1364
17.7	**Intraepidermal Blistering Diseases**	1364
17.7.1	Pemphigus Vulgaris, Pemphigus Foliaceus, and Pemphigus Paraneoplasticus	1364
17.7.2	IgA Pemphigus and Impetigo	1366
17.7.3	Intraepidermal Blistering Diseases	1366
17.8	**Subepidermal Blistering Diseases**	1367
17.8.1	Bullous Pemphigoid and Epidermolysis Bullosa	1367

© Springer-Verlag GmbH Germany, part of Springer Nature 2020
C. M. Sergi, *Pathology of Childhood and Adolescence*,
https://doi.org/10.1007/978-3-662-59169-7_17

17.8.2	Erythema Multiforme and Toxic Epidermal Necrolysis.........................	1368
17.8.3	Hb-Related Porphyria Cutanea Tarda, Herpes Gestationis, and Dermatitis Herpetiformis..	1368
17.8.4	Lupus (Systemic Lupus Erythematodes)..	1369
17.9	**Vasculitis**...	1369
17.10	**Cutaneous Appendages Disorders**...	1369
17.11	**Panniculitis**...	1369
17.11.1	Septal Panniculitis..	1369
17.11.2	Lobular Panniculitis..	1370
17.12	**Cutaneous Adverse Drug Reactions**..	1370
17.12.1	Exanthematous CADR..	1371
17.13	**Dyskeratotic, Non-/Pauci-Inflammatory Disorders**.......................	1372
17.14	**Non-dyskeratotic, Non-/Pauci-Inflammatory Disorders**................	1372
17.15	**Infections and Infestations**..	1372
17.16	**Cutaneous Cysts and Related Lesions**..	1372
17.17	**Tumors of the Epidermis**...	1373
17.17.1	Epidermal Nevi and Related Lesions...	1373
17.17.2	Pseudoepitheliomatous Hyperplasia (PEH)...	1376
17.17.3	Acanthoses/Acanthomas/Keratoses..	1376
17.17.4	Keratinocyte Dysplasia...	1377
17.17.5	Intraepidermal Carcinomas...	1378
17.17.6	Keratoacanthoma...	1378
17.17.7	Malignant Tumors...	1378
17.18	**Melanocytic Lesions**...	1381
17.18.1	Lentigines, Solar Lentigo, Lentigo Simplex, and Melanotic Macules.........	1381
17.18.2	Melanocytic Nevi..	1381
17.18.3	Variants of Melanocytic Nevi...	1382
17.18.4	Spitz Nevus and Variants..	1384
17.18.5	Atypical Melanocytic (Dysplastic) Nevi...	1385
17.18.6	Malignant Melanoma and Variants...	1385
17.19	**Sebaceous and Pilar Tumors**...	1387
17.19.1	Sebaceous Hyperplasia...	1387
17.19.2	Nevus Sebaceous (of Jadassohn) (NSJ)..	1387
17.19.3	Sebaceous Adenoma, Sebaceoma, and Xanthoma..................................	1388
17.19.4	Sebaceous Carcinoma...	1388
17.19.5	Benign Hair Follicle Tumors..	1388
17.19.6	Malignant Hair Follicle Tumors...	1390
17.20	**Sweat Gland Tumors**...	1390
17.20.1	Eccrine Gland Tumors..	1391
17.20.2	Apocrine Gland Tumors...	1393
17.21	**Fibrous and Fibrohistiocytic Tumors**...	1394
17.21.1	Hypertrophic Scar and Keloid..	1394
17.21.2	Dermatofibroma..	1394
17.21.3	Juvenile Xanthogranuloma...	1394
17.21.4	Dermatofibrosarcoma Protuberans...	1396
17.22	**Vascular Tumors**..	1396
17.23	**Tumors of Adipose Tissue, Muscle, Cartilage, and Bone**.............	1396
17.24	**Neural and Neuroendocrine Tumors**..	1396
17.24.1	Merkel Cell Carcinoma..	1396
17.24.2	Paraganglioma..	1399

17.25	**Hematological Skin Infiltrates**...	1400
17.25.1	Pseudolymphomas...	1400
17.25.2	Benign and Malignant Mastocytosis..	1400
17.26	**Solid Tumor Metastases to the Skin**..	1400
	Multiple Choice Questions and Answers..	1403
	References and Recommended Readings...	1405

17.1 Development, General Terminology, and Congenital Skin Defects

17.1.1 Development

The development of the tegument in humans originates at the early gastrula from the juxtaposition of two embryonic elements, including the ectoderm and mesoderm. The mesoderm is essential for the differentiation of adnexal epidermal structures such as hair follicles and participates in the maintenance of the adult epidermis. At the 3rd week of embryonic development, there is a single layer of glycogen-filled cells. These entirely undifferentiated cells are called periderm. At 4–6 weeks old, two layers are present, i.e., the periderm or epithelium and the germinative layer. The periderm is a purely embryonic structure that will not be present later. At 8–11 weeks, a middle layer is formed, and the glycogen is still quite copious in all segments. The embryo is at this specific time 26–50 mm as crown-rump length (CRL). At 12–16 weeks (CRL 69–102 mm), one or more intermediate layers of cells appear with a more complex ultrastructural compartmentalization, including mitochondria, Golgi apparatus, few tonofilaments, and numerous microvilli. It is at this determined stage that a dome-shaped bleb starts to project from the center of the periderm. At 16–26 weeks, there is an increase of intermediate layers. At 21 weeks, keratohyalin granules appear in the uppermost layer, and at 24 weeks, the periderm is continuously recycled in the amniotic fluid and caseous vernix of the intrauterine fetus. The proteins of desmosomes are detected by the 10th week gestation and the basal keratins demonstrable by the 14th week. Hair follicles and apocrine glands appear first at the 9th week. The pregerminative or pregerm stage is characterized by the crowding of nuclei at the basal layer of epidermis hair germ. The germ stage is characterized by elongation of the basal cells and their nuclei and the growth toward the dermis. At this time, mesenchymal cells and fibroblasts increase in number to form the hair papilla beneath the germ with the outer cells of hair peg becoming more cylindrical and arranged radially according to the long axis. There is also an oblique growth downward with the advancing axis growing bulbous gradually enveloping the mesodermal papilla. In many hair follicles, a third pulp appears above the sebaceous rudiment to form the apocrine gland. The peg growth is accompanied by the development of the cells of the inner root sheath above the matrix, the downward extension of the matrix, and the upward growth of the internal cells to form the hair canal. Sebaceous glands form as solid hemispherical protuberances on the posterior surface of the hair peg, while the eccrine glands start to develop at about 12 weeks of gestation. Nails grow at about 12 weeks as well. Melanocytes develop from the neural crest and travel to the mesenchyme before 4–6 months of gestation when they travel to the basal layer. Langerhans cells are derived from monocyte-macrophagic lineage and enter the epidermis at 12 weeks gestation as well, while Merkel cells penetrate the epidermis at 16 weeks of pregnancy. The dermis originates from the ventrolateral part of the somite dermatome and is very cellular up to the 2nd month. Regular bundles of collagen appear at the end of the 3rd month, and by the 5th month, papillary and reticular dermis become distinguishable. Elastic fibers start at 22nd week gestation. In the dermis, three types of cells are seen at 14 weeks of embryonic development, including stellate, macrophages, and granular secretory cells (melanoblasts or mast cells). At 14–21 weeks, there is the clear-cut appearance of fibroblasts, perineural cells, pericytes, mast cells, histiocytes, Langerhans cells, and Merkel cells. The dermo-epidermal junction *lamina densa* becomes distinct at the 2nd month, while hemidesmosomes are definitely seen at the 3rd month.

17.1.2 General Terminology

The following definitions according to major dermatology books may promote a proper interdisciplinary communication.

Type of skin lesions	
Macula	Flat discoloration ≤1 cm
Patch	Flat discoloration >1 cm
Papula	Solid elevated lesion ≤1 cm
Plaque	Solid elevated flat lesion >1 cm
Nodule	Solid elevated round lesion >1 cm
Vesicle	Small fluid-filled blister ≤1 cm
Bulla	Large fluid-filled blister >1 cm
Telangiectasia	Dilated capillaries
Scale	Cluster of cornified cells
Crust	Scale subtype with serum

Type of skin patterns	
Acanthosis	Thickening of the epidermal layer
Lichenification	Thickening of the skin due to chronic rubbing
Hyperkeratosis	Increase of the thickness of the cornified layer
Parakeratosis	Residual pyknotic nuclei in an increased or not cornified layer
Hypogranulosis	Thin granular cell layer due to fast cell turnover (e.g., psoriasis)
Hypergranulosis	Thick granular cell layer due to slow cell turnover
Acantholysis	Loss of cell-cell adhesion
Ballooning	Intracellular edema
Spongiosis	Intercellular edema
Vacuolar degeneration	Cell vacuoles separating the epithelium from the upper dermis
Dyskeratosis	Premature (abnormal) keratinization in the squamous epithelium
Papillomatosis	Elongation of dermal papillae

17.1.3 Lethal Congenital Contractural Syndromes

Free fetal movements are necessary for a healthy intrauterine development of the fetus. In fetuses with reduced movements, a complex of severe consequences occurs. This complex of consequences has been termed *fetal akinesia deformation sequence*, a condition which has to be distinguished from the scleredema and sclerema of premature infants. *Sclerema neonatorum* and scleredema are special forms of edema characterized by an extensive induration of the skin with (*scleredema*) or without edema (*sclerema*). In scleredema, the skin is hard and taut, while in *sclerema* the skin is hard and dried out from a clinical point of view. Both the proportions between the palmitic and stearic acids (saturated fats) and the oleic acid (unsaturated fat) and the hypothermia and the circulatory weakness of the premature babies are considered as etiopathogenetic factors. An increase of the lipid peroxidation and a decrease of the superoxide dismutase activity have been suggested. In subcutaneous fat necrosis, the areas of subcutaneous induration are sharply defined and nonelevated and appear in large, well-developed, otherwise healthy infants. To the phenotype of fetal akinesia deformation sequence belong restrictive dermopathy (RD), Pena-Shokeir syndromes, and the Neu-Laxova syndrome primarily. Differently from RD, which shows no visceral defect, the reduction of the body movements in Pena-Shokeir syndromes and Neu-Laxova syndrome, which do not present tautness of the skin, is due to a primary defect of the central nervous system (e.g., ischemic-anoxic brain damage, brain anlage disorders, lack of anterior horn cells that should be reside in the spinal cord). Regarding the Pena-Shokeir syndrome, three main types are known. Pena-Shokeir syndrome type I (MIM 208150) shows multiple ankyloses, camptodactyly, facial anomalies, and pulmonary hypoplasia. In 1983, it was suggested that the Pena-Shokeir type I phenotype is not specific. It is rather the result of a deformation sequence caused by fetal akinesia. In 1986, Hall could identify at least five specific subgroups among the 16 multiplex families by distinguishing features. Pena-Shokeir syndrome type II (MIM 214150) or cerebro-oculo-facial-skeletal syndrome (COFS) is an autosomal recessive (AR) inherited degenerative disease of the CNS leading to brain atrophy with calcifications, bilateral congenital cataract, microcornea, optic atrophy, progressive joint contractures, and growth failure (Le Van Quyen et al. 2020; Faridounnia et al. 2015; Suzumura and Arisaka 2010). A further main type has been described, and it is a phenotype intermediate between the COFS syndrome and the Neu-Laxova syndrome suggesting that the two last conditions may represent different degrees of severity of the same AR mutation in the homozygote state. Concerning the Neu-Laxova syndrome, it has been described as a unique entity characterized by severe intrauterine growth retardation, a peculiar cerebro-arthro-digital phenotype, ectodermal dysplasia, and edema. In 1982, Curry suggested dividing the Neu-Laxova syndrome into two types: the first one showing no edema and no increased fat layer, the second one showing an increased layer of subcutaneous adipose tissues with hypertrophy of fat cells. Other syndromes presenting with congenital contractural syndromes include the *lethal multiple pterygium syndrome* and the *aplasia cutis congenita*. In addition to the classic multiple pterygium syndrome, Hall (1984) commented on two other possible distinct forms: the first one with spinal fusion and the second one with congenital bone fusions. However, Aslan et al. (2000) could identify at least 15 different entities characterized by multiple pterygia or web of the skin and multiple congenital anomalies and proposed one of their own. The relationship between RD and *aplasia cutis congenita* (MIM 107600, congenital skull and scalp defect; MIM 207700, congenital scalp defect only) was described in a rather confusing way in the past. In an invited editorial comment, Toriello indicated that five infants with extensive skin erosions and contractures had been initially diagnosed as *aplasia cutis congenita*. It is, however, known that *aplasia cutis congenita* may occur as an isolated defect or associated with different phenotypes. These phenotypes include gastrointestinal atresia (Carmi syndrome), cone-rod dysfunction with high myopia and congenital nystagmus, epibulbar dermoids and cutaneous hyperpigmentation (oculo-ectodermal syndrome), intestinal lymphangiectasis, type I hereditary motor and sensory neuropathy (X-linked Charcot-Marie-Tooth peroneal muscular atrophy), distal limb reduction anomalies (Adams-Oliver syndrome), coarctation of the aorta, hereditary symmetrical bitemporal skin aplasia ("forceps marks" or Setleis syndrome), and ear abnormalities with unilateral facial palsy and ear abnormalities with amastia and/or athelia (Finley-Marks syndrome). In the 1st year of life, pathological hair loss is a rare event per se but is a potential leading symptom of congenital diseases. Congenital should be differentiated from acquired hair loss. Age of onset, history, and associated symptoms are important components to be considered to reach the diagnosis. The most frequent etiologies of hair loss among 1–3-year-old children are hair shaft anomalies (genotrichoses) with or without associated defects, alopecia areata, and mycotic scalp infestation (e.g., *Tinea capitis*). Between 4 and 11 years, the most significant causes are alopecia areata, loose anagen hair, artificial hair loss (trichotillomania, traction alopecia), and hair loss of infectious origin (*Microsporidia*, *Tinea capitis*, folliculitis). Adolescents (12–18 years) may suffer from alopecia areata and androgen hair loss or androgenetic alopecia. The age group of 1–3 years may be the most challenging because the hair shaft anomalies may be associated with other defects. There may be congenital localized or diffuse hypotrichosis. Alopecia areata and mycotic infection of the hair shaft are more often seen in children older than 3 years. At all times, both the dermatologist and pediatric pathologist should search for characteristic features of genetic syndromes in this age group. Figure 17.1a–c shows a case of congenital alopecia in a toddler. One of the most important differential diagnoses is congenital triangular alopecia (CTA), which is an uncommon disorder and no gender preference. Most cases present in 2–9-year-old patients. In 15% of cases, CTA is associated with other diseases, including pigmented-vascular phakomatosis, trisomy 21 syndrome (Down syndrome), leukonychia, hip dislocation, mental retardation, epilepsy, and tracheoesophageal fistula. Diagnosis is mainly clinical, and microscopic examination shows a normal number of follicular units, the majority being vellus-type hairs. The main differential diagnosis remains alopecia areata, although other forms of non-scarring alopecia should also be investigated. In Fig. 17.1d–e, the characteristic features of RD with an age-matched related newborn are presented (Fig. 17.1f). In Box 17.1 are shown the most common lethal congenital contractural syndromes by analysis of the phenotype. However, a complete clinical-pathological workup needs also to exclude the *ichthyosis congenita of the harlequin fetus type*, by which the entire body of the infant is covered with thick, horny plaques separated by fissures and ectropion of the eyes, nostrils, and mouth is seen. In contrast to the RD, in the *ichthyosis congenita of the harlequin fetus type*, there is marked hyperkeratosis of the skin with papillomatosis and nonobstructive plugging of hair follicles and sweat ducts. A *collodion baby* phenotype clinically manifests the *lamellar ichthyosis* also occurring at birth. The *lamellar ichthyosis* is characterized by erythema, generalized scales of plate-like type, and ectropion. An entity which should be cited here is the *collodion baby* phenotype associated to a rare subset of infants with *type 2 Gaucher disease*, an acute neuronopathic form culminating in early death as a result of the severe neurological disease resulting from the inherited deficiency of the lysosomal enzyme glucocerebrosidase. Finally, two other pathologies should be included in the differential diagnosis of RD: the "*stiff skin syndrome*" and the "*Parana hard skin syndrome*." The former shows that the skin of the entire body is thickened and indurated, and there is a limitation of joint mobility with flexion contractures. The latter shows that the skin of the entire body becomes progressively thicker, and "freezing" of all joints occurs. A few hypotheses about the pathogenesis of this rare genodermatotic disease have been proposed. The clinical and histological features of RD were variably interpreted either as the expression of an ectodermal dysplasia or as a more specific morphogenetic alteration of the epidermis and dermis. Holbrook et al. (1987) found an abnormal differentiation of keratins with arrested development of hair follicles and eccrine sweat glands and suggested an underlying disorder of skin differentiation, involving dermal-epidermal interactions. Impaired skin maturation was also demonstrated by the weak or no labeling of cytokeratin with antibodies. Witt et al. (1986) found a decreased amount of high molecular weight keratins in the skin of a brother and a sister with RD. Dean et al. (1993) pointed out that RD is a disorder of skin differentiation due to impaired cell adhesiveness but assumed to be due to abnormal integrin expression. However, Sillevis Smitt et al. (1998) did not show any abnormal $\alpha 1 \beta 1$ and $\alpha 2 \beta 1$ integrin expression in fibroblast cultures taken from five patients. No preference for any race is known because RD occurs in almost all ethnics. Consanguinity was demonstrated in only a few cases. The survival time was not correlated to the birth weight or to the time of delivery (Sergi et al. 2001). The values of the time of delivery (weeks of gestation) were distributed along a regression line with a significant slope (r^2, 0.27; p-value, 0.0084; significant p-value <0.05), i.e., close weeks of gestation showed different birth weights. By using the Kaplan-Meier estimator, a median survival of 132 h (5.5 days) was found. The longest survival time was 2880 h (about 4 months). Although the pulmonary insufficiency is the common cause of death in these infants, the lungs are compressed, but not dysplastic. About the course of these infants, it is difficult to say whether the outcome would have been any different if any therapeutic option had been proposed. First, until the genetic defect is identified, the gene cloned and the ability to express it demonstrated, a role for gene therapy seems far away. However, since skin biopsy specimens from affected fetuses up to 20 weeks of gestation have been reported as normal, a prenatal diagnosis in time for consideration of therapeutic abortion is not feasible. The possible role of skin transplantation in genodermatosis has not been clearly and fully defined yet. Autologous skin transplantation remains the favorite form of treatment in patients with significant skin loss, but there are commercially available allogeneic products that can be used to replace the skin temporarily or permanently. Early surgical intervention is revolutionizing our therapeutic concepts. A 130 cm^2 lesion has been covered in a newborn affected with aplasia cutis congenita by using allogeneic dermis and cultured epithelial autografts. After 2.5 weeks 90% of the wound had healed, and at 27 months, the grafted area was smooth and pliable with normal skin texture. Acellular allogenic grafts are real options in burn surgery and acne scars. Moreover, cultured epithelial autografts contain keratinocyte stem cells. These cells can be stably transduced with retroviral vectors. These cells are attractive targets for the gene therapy of genodermatoses.

17.1 Development, General Terminology, and Congenital Skin Defects

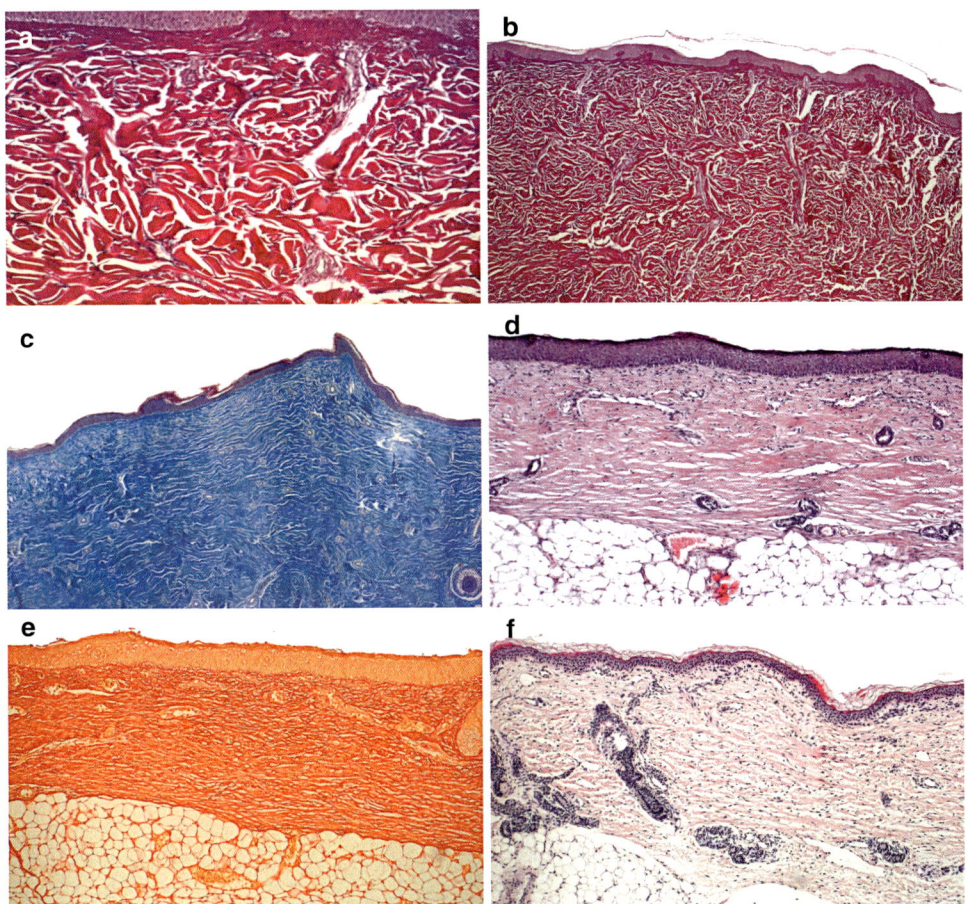

Fig. 17.1 In **a–c** are shown microscopic photographs of congenital alopecia in a toddler with absence of hair shaft (**a**, EvG stain, 200×; **b**, EvG stain, 40×; **c**, CAB stain, 40×). In **d–e** are shown the microphotographs of a newborn with restrictive dermopathy (**d**, H&E stain, 100×; **e**, EvG stain, 100×), while **f** is an age-matched newborn with the same skin location showing a normal skin histology (H&E stain, 100×)

Box 17.1 Most Common Lethal Congenital Contractural Syndromes

Disease	Inheritance	External features	Skin
Pena-Shokeir syndrome type I	AR (50%)	Hypertelorism, micrognathia, low-set ears, depressed nose tip, hip and knee ankyloses, camptodactyly, club feet	Hypo-/aplasia of dermal ridges
Pena-Shokeir syndrome type II	AR	Microcephaly/microophthalmia, large ear pinna, large and prominent root of nose, flexion contractures, coxa valga, camptodactyly, rocker-bottom feet with grooves on soles	Hirsutism
Neu-Laxova syndrome	AR	Microcephaly, exophthalmos, sloping forehead, large ears, flat nose, flexion contractures with pterygia, puffiness of hands and feet, syndactyly	Thin, transparent, scaling skin and edema, ichthyosis (+/-)
Lethal multiple pterygium syndrome	AR/X--linked	Hypertelorism, epicanthal folds, low-set ears, flat nose, microstomia, flexion contractures with pterygia	Hypoplasia of dermal ridges

The use of allogeneic dermis and retrovirus-infected cultured epithelial autografts may be supposed to treat infants affected with RD.

17.2 Spongiotic Dermatitis

Spongiosis refers to intercellular edema, and under light microscopy, the keratinocytes appear stretched apart accentuating the intercellular bridges or spinous processes highlighting the desmosomal character of the squamous epithelial cells. In 1864, a young Italian pathologist, Giulio Bizzozero (1846–1901), described these structures in detail. In his original examination of the *stratum spinosum* of the epidermis, Bizzozzero noted small dense nodules at the contact points between adjacent cells, aka "nodes of Bizzozero" later. Thus, the diagnosis of spongiotic dermatitis relies on the identification of the intercellular edema as seen more than a century ago. Four types of spongiosis should be carefully differentiated, including conventional spongiosis, eosinophilic spongiosis, follicular spongiosis, and miliarial spongiosis.

17.2.1 Conventional Spongiosis

Conventional spongiosis may be encountered in numerous conditions and textbooks of dermatopathology may be the correct source (Barnhill and Crowson 2004; Elston et al. 2013; Massi 2016). However, some histologic features accompanying the intercellular edema may address the etiology correctly. *Angiodermatitis* and *chronic stasis dermatitis* show proliferation of the superficial vascular plexus, thickening of the blood vessel walls, some lymphocytic infiltrate, hemorrhage, and hemosiderin-laden macrophages. *Pigmentary purpura* (eczematoid-like purpura of Doucas and Kapetanakis) is characterized by superficial perivascular lymphocytic infiltrate with hemorrhage and hemosiderin-laden macrophages in addition to spongiosis. A shallow and deep tightly cuffed perivascular lymphoid infiltrate is seen in *figurate erythemas*, dyskeratotic cells in *polymorphous light reaction* and *photo drug reaction*, and many dyskeratotic cells with atypia in *phototoxic reaction*. A wedge-shaped superficial/deep perivascular lymphocytic infiltrate with eosinophils may point to a *hypersensitivity reaction to arthropod attacks*. A superficial perivascular lymphoid infiltrate with intraepidermal/intracorneal neutrophils indicates *seborrheic dermatitis, early psoriasis, treated psoriasis, Sneddon-Wilkinson disease*, and *contact dermatitis*. A superficial perivascular lymphoid infiltrate with intraepidermal vesicles and with or without eosinophils is characteristic of *pompholyx, allergic contact dermatitis*, and *dyshidrotic eczema*. A superficial perivascular lymphoid infiltrate raises the possibility of *chronic superficial dermatitis, pityriasis lichenoides chronica, pruritic urticarial papules and plaques of pregnancy, Gianotti-Crosti syndrome*, and early *mycosis fungoides*. A superficial perivascular lymphoid infiltrate with or without eosinophils may indicate *drug hypersensitivity reaction, atopic dermatitis, nummular dermatitis, pityriasis rosea, Grover disease, Id reaction* (autosensitization), and *hypersensitivity reaction to arthropod attacks*. Intracorneal hyphal forms with spongiosis may raise the diagnosis of *dermatophytosis*, while intracorneal Gram-positive cocci *impetigo* and plasma cells and endothelial swelling may point to the determination of secondary *syphilis*. In childhood and youth, eosinophilic spongiosis is seen mainly as atopic dermatitis and nummular (discoid) dermatitis among other disorders (*vide infra*). Finally, two lesions that show some commonalities in histology need to be carefully differentiated particularly for the relevance concerning therapy and outcome. These two lesions are *erythema neonatorum*, which is self-limited, and *incontinentia pigmenti*, which is an X-linked genetic disorder with multisystemic involvement.

17.2.1.1 Atopic Dermatitis
- *DEF*: Spongiotic dermatitis common in all ages usually starts at around 6 weeks of age on H&N and progresses to flexural aspects of limbs with symmetrical character and associated with FHx positive for atopy, asthma, and hay fever.
- *CLI*: There is marked pruritus associated with a variety of lesions according to the stage, including erythematous scaling areas, edematous, oozing, weepy regions, and lichenified areas with dry fissured scaly areas.

17.2 Spongiotic Dermatitis

- Three stages:
  - Acute: Spongiosis (± vesiculation) + superficial perivascular lymphocytic and eosinophilic infiltrate
  - Subacute: Thickening of the epidermis with psoriasiform character + spongiosis + superficial perivascular lymphocytic and eosinophilic infiltrate +/- fibrosis of papillary dermis
  - Chronic: Psoriasiform hyperplasia, + superficial perivascular lymphocytic and eosinophilic infiltrate +/- fibrosis of papillary dermis
  - DDX: Other types of spongiotic dermatitis

The diagnosis is mainly based on the clinicopathologic correlation.

17.2.1.2 Allergic Contact Dermatitis
- *DEF*: Type IV hypersensitivity-related spongiotic dermatitis with a variable age range (>5 years) and showing an eczematous rash at the site of exposure usually in the previously sensitized host (allergic, e.g., poison ivy, nickel, or irritants, e.g., soaps) with three stages:
  - *Acute*: Spongiosis + vesiculation + superficial perivascular lymphocytic/eosinophilic infiltrate + intraepidermal inflammatory cells
  - *Subacute*: Thickening of the epidermis + spongiosis + superficial perivascular lymphocytic and eosinophilic infiltrate +/- variable exocytosis
  - *Chronic*: Psoriasiform hyperplasia, + superficial perivascular lymphocytic and eosinophilic infiltrate
- *DDX*: Other types of spongiotic dermatitis, particularly irritant and photoallergic reactions, scabies, parasites, drug reactions, and bullous pemphigoid

17.2.1.3 Seborrheic Dermatitis
- *DEF*: Spongiotic dermatitis, psoriasis-like, with prominent spongiosis targeting infants (aka Leiner disease) and early adolescents with manifestation onward on the scalp (dandruff), forehead, cheeks, upper chest, and back and evolving with greasy, scaly, red-brown, often pruritic patches.
- *CLM*: Psoriasiform hyperplasia with hyper- and parakeratosis + spongiosis + neutrophilic exocytosis + superficial perivascular lymphocytic and eosinophilic infiltrate.
- *DDX*: Other types of spongiotic dermatitis (e.g., contact dermatitis), psoriasis, impetigo, drug reaction, dermatophytosis, secondary syphilis, *pityriasis rosea*.
- *PGN*: In childhood and youth, there is a strong association of seborrheic dermatitis with epilepsy, obesity, and malabsorption and HIV infection.

17.2.1.4 Nummular Dermatitis
- *DEF*: Spongiotic dermatitis with pruritic coin-shaped plaques on lower legs, backs of hands, and forearms and often seen in 15–30 years old individuals with ♂ < ♀ tendency.
- *SYN*: Discoid dermatitis.
- *CLM*: Three stages as described above as seen in another spongiotic dermatitis (*vide supra* for histological details).

17.2.1.5 Stasis Dermatitis
- *DEF*: Spongiotic dermatitis of lower legs of usually middle-aged women with increased venous hypertension (HTN) and showing red, edematous, scaling lesions with variable oozing on legs (especially medially/internally) and evidence of stasis, such as brown skin pigmentation, ulcers, varicosity, and atrophic scars.
- *CLM*: Acute, subacute, and chronic spongiotic dermatitis pattern with the lobular proliferation of capillaries + hemorrhage + hemosiderin ± fibrosis. Other forms of spongiotic dermatitis of childhood and youth include *pityriasis alba*, *pityriasis rosea*, papular acrodermatitis of Gianotti-Crosti, *erythema neonatorum*, and *incontinentia pigmenti*.

17.2.1.6 Pityriasis Alba
- *DEF*: Spongiotic dermatitis of children (3–16 years of age) with >1 hypopigmented maculae (average # of melanocytes) on face > trunk/limbs with spontaneous remission and re-pigmentation <1–2 years.
- *CLM*: Hyper-/parakeratosis, spongiosis (including follicular spongiosis), superficial perivascular lymphocytic infiltrate + capillary ectasia.
- *DDX*: *Pityriasis versicolor* (+ yeast and hyphae) (Greek πιτυρίασις or Latin pityriāsis from πίτυρον "bran" and versicolōrus from versō "turn" + color "color") and *Vitiligo* (loss of melanocytes).

17.2.1.7 Pityriasis Rosea
- *DEF*: Self-limited spongiotic dermatitis of children through middle-aged adults with original herald patch followed by a generalized "fir tree"-like ("Christmas tree"-like) eruption on the trunk with multiple salmon-colored maculae with fine scale.
- *CLM*: Spongiosis + superficial perivascular lymphocytic infiltrate with variable exocytosis.
- *DDX*: Other spongiotic dermatitis, drug reactions, pityriasis lichenoides, guttate psoriasis, and dermatophytosis.

17.2.1.8 Papular Acrodermatitis of Gianotti-Crosti
- *DEF*: Virus-related (e.g., HBV, EBV, *Coxsackieviruses*, CMV, echovirus, poliovirus, RSV, parainfluenza, and HAV) spongiotic dermatitis of children (6 months–12 years) characterized by an erythematous, nonpruritic, papular eruption on the face, buttocks, and limbs with an acute phase of about 50–60 days length.
- *CLM*: Spongiotic dermatitis with superficial perivascular lymphocytic infiltrate, which is not specific (keys are FHx and lab values).
- *DDX*: Viral exanthemas, drug reactions, dermatophytosis, and hypersensitivity reactions.

17.2.1.9 Erythema Neonatorum
- *DEF*: Self-limited spongiotic dermatitis of newborns due to the high viscosity of the ECM of neonatal age and mild, self-limited acute GvHD of the skin from maternal to fetal lymphocytic transfer which resolved within 2–7 days and characterized by erythematous maculae and/or papulae mostly on the trunk.
- *CLM*: There are dermal edema and perivascular eosinophilic infiltrate without clear-cut spongiosis.
- *DDX*: Incontinentia pigmenti.

17.2.1.10 Incontinentia Pigmenti
- *DEF*: An X-linked multisystemic disorder with three types of cutaneous lesions, including erythematous lesions with vesicles mostly on limbs and according to Blaschko lines arranged (first stage), linear verrucous lesions mostly on arms and legs (second stage), and, then, pigmentation following Blaschko lines mainly on the trunk.
- *CLM*: There is a very large intraepidermal vesicle containing eosinophils in the ventricular stage. The extracutaneous lesions involve the teeth, eyes, and CNS signs and symptoms (seizures, spina bifida, and mental retardation) are common.
- *DDX*: *Erythema neonatorum* (eosinophilic vesicles, but no spongiosis).

17.2.2 Eosinophilic Spongiosis
- *DEF/DDX*: Spongiosis (intercellular edema) with eosinophils into the edematous intercellular spaces of the epidermis. The differential diagnosis of eosinophilic spongiosis includes allergic contact dermatitis, drug eruptions, reactions to arthropod attacks, some blistering diseases (e.g., bullous pemphigoid, herpes gestationis, pemphigus vegetans), early stages of incontinentia pigmenti, and an idiopathic eosinophilic spongiosis, NOS. In childhood and youth, eosinophilic spongiosis is seen mainly as *allergic contact dermatitis, drug eruptions, reactions to arthropod attacks*, and early stages of *incontinentia pigmenti*.

17.2.3 Follicular Spongiosis
- *DEF/DDX*: Spongiosis (intercellular edema) localized in the follicular epithelium and includes some diseases in the differential diagnosis (follicular eczema, follicular mucinosis, eosinophilic folliculitis, infectious folliculitis, infundibulo-folliculitis, *pityriasis alba*, apocrine miliaria, and impetigo of Bockhart), which are well treated in dermatology and dermatopathology books.

17.2.4 Miliarial Spongiosis
- *DEF*: Spongiosis (intercellular edema) of the acrosyringium (i.e., the intraepidermal part of the duct of the eccrine sweat gland nearest to the surface of the skin). Miliarial spongiosis is carefully sub-differentiated in:
  – *Inflammatory dermatitis*, which involves the epidermis and the acrosyringium

- *Miliaria crystallina* showing a subcorneal blister filled with fluid and less than 50% peri- and intraductal neutrophils
- *Miliaria profunda* with spongiosis of the lower half of the acrosyringium and peri- and intraductal lymphocytic infiltrate
- *Miliaria rubra* showing spongiosis of the upper half of the acrosyringium and peri- and intraductal lymphocytic infiltrate

17.3 Interface Dermatitis

The diagnosis of interface dermatitis relies on an obligate, but the variable combination of leukocytic infiltration of the dermis ("dermatitis"), vacuolar change of the basal layer of the epidermis, melanophages accumulation in the upper dermis, and necrosis of keratinocytes. Interface dermatitis may be subclassified as vacuolar or lichenoid interface dermatitis

17.3.1 Vacuolar Interface Dermatitis

- *DEF*: Interface dermatitis with vacuolar degeneration of the basal layer refers to basal keratinocyte cytoplasmic vacuolation ± paucicellular lymphoid infiltrate at DEJ (dermo-epidermal junction) and with or without pigmentary incontinence. The vacuolar-type of interface dermatitis may be observed in lupus erythematodes (LE), dermatomyositis, *Pityriasis Lichenoides Et Varioliformis Acuta*, *vide infra* (PLEVA), Toxic Epidermal Necrolysis, *vide infra* (TEN), GvHD, erythema multiforme, fixed drug eruption, and secondary syphilis.

17.3.1.1 Lupus: SLE/Discoid Lupus
- *DEF*: Discoid lupus is the cutaneous form of SLE without accompanying systemic manifestations, with erythematous macules and patches seen on the face, earlobes, neck, and scalp with a vacuolar and interface dermatitis.
- SLE skin features include annular to plaques, photosensitive distribution, scales, atrophy/scarring (late), follicular plugging (late), dermal edema and/or mucin deposits, telangiectases, and lichenoid to superficial and deep perivascular lymphoid infiltrate.
- The acute form shows a lichenoid perivascular inflammation, mostly superficial, hydropic degeneration of the basal layer, and edema of the papillary dermis, while the chronic form is characterized by eosinophilic homogenization of the papillary dermis, thick BM, epidermal atrophy, and hyperkeratosis with follicular plugging.
- *DIF*: Granular band of Ig and C3 (IgM + IgG) along the DEJ, i.e., (+) lupus band test, unlike dermatomyositis, which exhibits a negative lupus band test. SLE is a multisystemic autoimmune disease of the late 2nd to 4th decades of life with female preference, butterfly rash, multi-organ involvement (serositis, oral ulcers, arthritis, photosensitivity, blood disease, renal disease, antinuclear antibodies (ANA), immunologic features, neurologic disease, malar rash, and discoid rash), and histology similar to discoid lupus. However, patients affected with systemic LE show a granular band of Ig and C3 also in the clinically normal skin (the so-called "expanded positivity of lupus band test").
- *DDX* of lichenoid lesions include lichenoid photodermatitis (superficial and deep dermatitis with spongiosis), lichenoid solar keratosis (atypical budding and alternating ortho-hyperkeratosis and parakeratosis), lichenoid LE (vacuolar change and dermal mucin), *lichen aureus* (pigmented purpuric dermatosis and Fe-laden macrophages), *mycosis fungoides* (epidermotropism and lamellar fibrosis), and *lichen planus*-like keratosis (usually solitary, peripheral).

17.3.1.2 Dermatomyositis
- *DEF*: Syndrome characterized by variably distributed erythematous macules and plaques with photo-accentuation, mostly indistinguishable purely histologically from LE. Increased mucin deposition in the dermis is seen in both conditions. However, there is no band, i.e., the lupus band test (DIF) is negative.
- *DDX*: Lupus (SLE or discoid lupus).

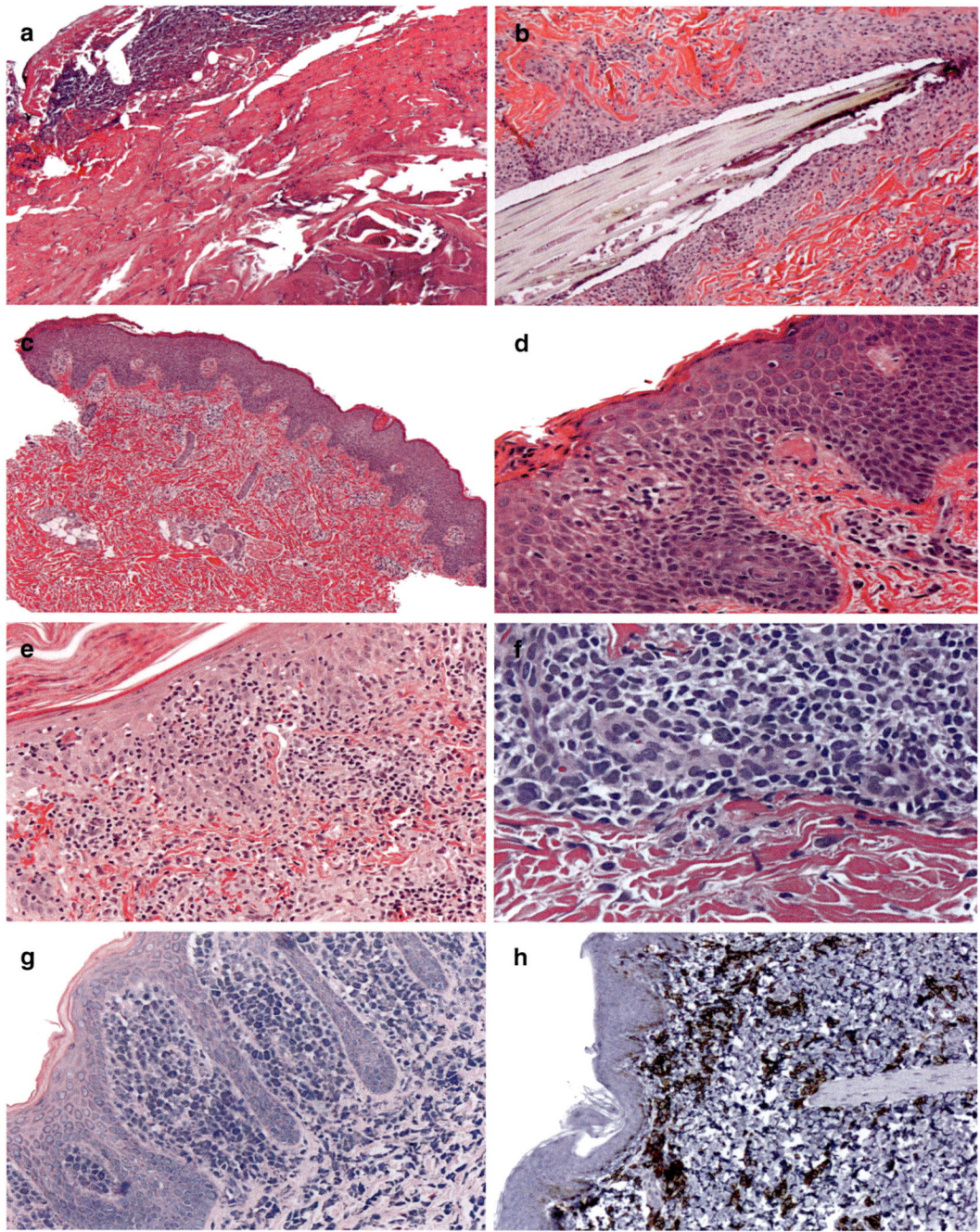

Fig. 17.2 This panel shows (**a**) third-degree burn (H&E stain, 100×), (**b**) inflammation and fibrosis accompanying a dermatitis produced by a wood splitter (H&E stain, 100×), (**c**) and (**d**) spongiotic dermatitis in small bowel transplant recipient (**c**, H&E stain, 50×; **d**, H&E stain, 200×), (**e**) pityriasis lichenoides et varioliformis acuta (H&E stain, 200×), (**f**–**h**) show an example of urticaria pigmentosa with numerous mast cells between the rete pegs proliferating in papillary dermis (**f**, H&E, 400; **g**, Giemsa, 200×; **c**-kit immunohistochemistry, 100×)

17.3.1.3 Pityriasis Lichenoides Et Varioliformis Acuta (PLEVA)

- *DEF*: Lympho-exocytosis-dominated interface dermatitis with abrupt onset of erythematous/pink papulae on the trunk and limbs, central vesiculation, ulceration without scarring of late childhood to early adulthood with male preference (Fig. 17.2).
- *SYN*: Mucha-Habermann disease.
- *EPG*: It has been proposed the concept that PLEVA and pityriasis lichenoides chronica (PLC) are parts of the same disease with a variable but overlapping clinical and histologic features.
- *CLM*: Ulcero-necrotic lesions with wedge-shaped central confluent keratinocyte necrosis, parakeratotic scale with neutrophils, spongiosis, and exocytosis of lymphocytes (CD3+, CD4+) and RBCs with perivascular lymphoid infiltrate in the dermis.
- *DDX*: PLC, syphilis, drug eruption, viral exanthema, papulonecrotic tuberculid, arthropod-bite reaction, polymorphous light eruption, connective tissue disorders (CTD), and *erythema multiforme*.
- PLC = *Pityriasis lichenoides chronica* is a pityriasis form of dermatitis of children and adolescents with a gradual onset of erythematous papulae with the dusky center. PLC shows an interface dermatitis, which is less pronounced than in PLEVA but includes hyperkeratosis with parakeratotic scale, mild vacuolar degeneration of the basal layer, and superficial perivascular lymphocytic infiltrate.

17.3.1.4 Erythema Multiforme

- *DEF*: Self-limited, relapsing, cytotoxic cell-mediated hypersensitivity-associated vacuolar interface dermatitis with CD8+ (but also CD4+ and NK+ cells) recruitment and auto-Abs to desmoplakin I and II (dominant form).
- *EPI*: Youth.
- *EPG*: Etiology does not seem to be fully known but EM can be caused/triggered by infectious agents (e.g., HSV, *M. pneumoniae*), drugs (e.g., β-lactam, NSAIDs, sulphonamides), collagen vascular diseases, and malignancies. Tissue damage seems to be due to both humoral and cell-mediated immunity.
- *CLI*: Symmetric acral groups of maculopapulae, often with annular configuration ("target lesions") with dusky centers mostly on palms and soles.
- *CLM*: Vacuolar interface dermatitis with CD8+ (but also CD4+ and NK+ cells) recruitment and auto-Abs to desmoplakins I and II. There is the "constellation" of *NESVI*, including *n*ecrosis (single and clusters of necrotic keratinocytes), *e*xocytosis, *s*pongiosis, *v*acuolation (basal), and *in*flammation (upper dermal interstitial perivascular infiltrate of lymphocytes). TEM of EM may show intraepidermal, but mostly *subepidermal blistering* with the roof of the bulla only partially necrotic. The ultrastructural studies evidence collagen fibers and individual fibrils, which are swollen. This aspect of fibers and fibrils is accompanied by deposition of proteins arising from plasma, which can be found on and between the collagen fibers. There is also thinning of the basement membrane, which is followed occasionally by actual disruption. In the healing phase there is evidence of reduplication and layering of the membrane. At places, focal degeneration of epidermal cells is found in regions of severe dermal lesions (Caulfield and Wilgram, 1962).
- *DIF*: IgM and C3 in the walls of superficial dermal blood vessels.
- *DDX*: TEN/SJS, which has been considered, probably inappropriately, a severe form of EM with mucosal involvement, conjunctivitis, high fever, and death in 5% of patients.
- *PGN*: Good (self-limited).

17.3.1.5 Toxic Epidermal Necrolysis (TEN)/Stevens-Johnson Syndrome (SJS)

- *DEF*: Severe mucocutaneous syndromes with tenderness, hemorrhagic erosions, erythema, and epidermal detachment presenting as blisters and areas of denuded skin with essentially a spectrum of lesions variable according to the cutaneous involvement. There is a variable widespread tender erythematous rash with full-thickness epidermal denudation, mucosal and conjunctiva participation, and positive Nikolsky sign (epidermal dislodgement by a shearing force demonstrating a plane of cleavage at the dermal-epidermal junction of the skin). In 1922, Stevens and Johnson described first an acute mucocutaneous syndrome in two young male children affected with severe conjunctivitis, stomatitis with extensive mucosal necrosis, and macules with purpuric character. Alan Lyell described in 1956 four patients with an eruption, which looked like scalding of the skin and was called by him *toxic epidermal necrolysis* (TEN). TEN and SJS are now considered the ends of a spectrum of adverse cutaneous drug eruptions divergent only by the extent of cutaneous detachment. Pyotr Vasilyevich Nikolsky (1858–1940) was a Russian professor of dermatology at the Imperial University of Warsaw and later created the Department of Dermatology and Venerology in Rostov.

- *EPG*: Drugs in most cases with drug-specific cytotoxicity by T-lymphocytes against autologous cells. "High"-risk drugs triggering TEN/SJS include allopurinol, trimethoprim-sulfamethoxazole, and other sulfonamide-based antibiotics and drug formulations, aminopenicillins, cephalosporins, quinolones, carbamazepine, phenytoin, phenobarbital, and NSAIDs of the *oxicam* type. Other triggers are *Mycoplasma pneumoniae* infection, HSV, lupus erythematosus, and post-allogeneic hematopoietic stem cell transplantation for severe aplastic anemia.
- *CLM*: Full-thickness epidermal necrosis with *subepidermal blistering*, vacuolar interface dermatitis, and mild dermal perivascular inflammatory infiltrate.
- *DDX*: SSSS, acute GVHD, and drug eruption. Most importantly, TEN must be differentiated by SSSS (staphylococcal scalded skin syndrome). Both conditions may present in children and adults in a very similar way with extensive or diffuse blistering, and denudation and both states harbor a terrible prognosis because they are both life-threatening diseases. However, both illnesses require different lines of treatment. Frozen section diagnosis is often required from PICU or ICU clinical colleagues for rapid therapeutic intervention. To remember is that SSSS shows an intraepidermal cleavage placed exclusively at the uppermost epidermis (at the granular layer). In the blister roof, there is a stratum corneum/corneal layer (in contrast to TEN with full-thickness epidermal necrosis), subcorneal blisters, and acantholysis. Characteristically, an intraoperative frozen section of the peeled skin confirms the site of cleavage as superficial in SSSS, while TEN frozen section shows deeper cleavage (subepidermal cleavage). Other differential diagnoses include blistering autoimmune diseases such as IgA dermatosis and paraneoplastic *pemphigus* but also *pemphigus vulgaris*, bullous pemphigoid, acute generalized exanthematous pustulosis (AGEP), and disseminated fixed bullous drug eruption.
- *PGN*: SCORTEN is a validated disease severity scoring system, which was developed by Bastuji-Garin and co-workers to help predict mortality. It distinctly evaluates the following parameters: age, malignancy, tachycardia, epidermal detachment degree (%BSA), serum urea, serum glucose, and bicarbonate. It is to recall the SCORe of TEN (SCORTEN) by memorizing the acronym "CABST" targeting some key parameters including cancer/malignancy, involved area of the body greater than 10% TBSA (total body surface area), age older than 40 years, bicarbonate < 20 mEq/L, blood urea nitrogen (BUN) > 20 mg/dL, glycemia > 250 mg/dL, and tachycardia (beats per minute greater than 120). Precisely, the identification of a score of 0-1 indicates dramatically a mortality risk of 3.2%; a score of 2, 12.1%; score of 3, 35.3%; while a score of 4, 58.3%; and, finally, a score of 5 or more, 90%. Mortality is variable from 1% of SJS up to 35% of TEN and can be even higher in immunocompromised patients or older patients. Management involves a multidisciplinary approach with rapid diagnosis, SCORTEN evaluation, identification and interruption of the triggering drug, and specialized supportive care ideally PICU/ICU, with or without the introduction of immunomodulating agents (e.g., high-dose IV immunoglobulin therapy).

17.3.1.6 Graft Versus Host Disease (GVHD)

- *DEF*: Medical complication following the receipt of transplanted tissue from a genetically different person. In graft versus host disease (GVHD), the skin is the earliest and most commonly involved organ leading to address promptly patient management following an accurate interpretation of skin biopsy specimens. Erythematous macule/patches develop typically within the 1st month after bone marrow transplantation (BMT) as a manifestation of the acute disease (Box 17.2).

Chronic GVHD, which may arise after acute GVHD or as *de novo*, is divided into lichenoid form (>60 days post-BMT, LP-like histology) and sclero-dermoid form of GVHD (>100 days post-BMT, progressive superficial toward deep thickening of collagen bundles from papillary dermis unlike PSS and morphea, which show upward dermal sclerosis). In the sclero-dermoid form of chronic GVHD, epidermal subset atrophy, rete ridges abolition, loss of adnexal structures, and scarring of superficial and deep dermis are often recognized.

17.3.1.7 Fixed Drug Eruption

A fixed drug eruption is seen following trimethoprim-sulfamethoxazole, ASA, tetracycline, and barbiturates and is characterized clinically by a variable number of well-defined erythematous to purple edematous plaques with central duskiness and blistering tendency, usually recurring at the same site with each drug administration. Microscopically, all elements of interface dermatitis are observed in a fixed drug eruption. Compared to interface dermatitis of EM or TEN, fixed drug eruption contains usually more neutrophils, involves the deep vascular plexus, and harbors melanophages accumulation or pigment incontinence.

17.3.1.8 Lichen Sclerosus et Atrophicus (LSA)

- *DEF*: A rare event occurring in females >435 years of age on genitals and characterized by white or ivory angulated maculae or papulae, often with an edematous center, and over time, dermal sclerosis may ensue, resulting in the clinical term of scleroderma, but rarely in adolescence. On the penis, this lesion is called balanitis xerotica obliterans (BXO), which is most commonly observed in childhood.
- *SYN*: Balanitis xerotica obliterans (penis).
- *CLM*: There are hyperkeratosis, epidermal atrophy, prominent homogeneous eosinophilic hyalinization of the papillary dermis, and a subjacent band-like lymphoid infiltrate.
- *PGN*: Telangiectasia can be seen, and there is a SqCC risk up to 5%, which may mean one case out of 20. Thus, all cases need to be microscopically evaluated.

17.3.1.9 Infections

- *DEF*: Infections associated with interface dermatitis that needs to be mentioned here are secondary syphilis, *acrodermatitis chronica atrophicans*, and HIV (human immunodeficiency virus, types I and II). Secondary syphilis may present, clinically, with a wide range of clinical findings, but the mostly papulosquamous eruption of variable distribution, particularly palms and soles.
- *CLM*: Interface dermatitis with a superficial and deep perivascular and interstitial lymphoid-plasma cellular infiltrate. Psoriasiform acanthosis, spongiosis, exocytosis, PMN ± dyskeratosis, basal vacuolation, edema, lymphoplasmacytic infiltrate, and perivascular lymphocytes and plasma cells.
- *CLI*: Genital *chancre*, 20–30 days post-*T. pallidum* exposure, painless ulcer, which resolves by itself and secondary cutaneous lesions.
- *Acrodermatitis chronica atrophicans* occurs following infection with *B. afzelii* subspecies of *B. burgdorferi*, which is most commonly seen in Europe, and begins, clinically, as erythematous patches on arms and legs with pro-

Box 17.2 Acute GVHD: Grading

Grade 1	Vacuolization of the basal layer
Grade 2	Vacuolization + dyskeratosis, exocytosis, necrotic keratinocytes
Grade 3	"Subepidermal clefting" of the skin
Grade 4	Bulla and loss of the epidermis

gressive atrophy of epidermis, dermis, and adnexal structures with an upper dermis plasma cellular infiltrate, which helps in the DDX with LE and LSA.
- HIV-interface dermatitis is characterized by erythematous patches and plaques, which become purple over time and result in hyperpigmentation, and has been correlated to HIV drug therapy, which targets the skin and is prone to photosensitization.

17.3.2 Lichenoid Interface Dermatitis

Interface dermatitis with lichenoid leukocytic infiltrate refers to basal keratinocyte necrosis with band-like lymphocytic infiltrate and colloid Civatte bodies, the irregular border of the basal membrane, and lymphocytic exocytosis with or without pigmentary incontinence. The lichenoid type of interface dermatitis may be encountered in *lichen planus*, lichenoid drug eruptions, *lichen nitidus*, and other conditions with vacuolar changes (*vide supra*). Here, the *lichen planus* (LP) is described.

- *Lichen planus* is the idiopathic, self-limited prototype of lichenoid interface dermatitis and is characterized by *p*ruritic, *p*urple, *p*olygonal, planar *p*apulae, and *p*laques with white dots (fine scale) mostly on limbs (wrist, elbow, glans penis).
- *EPI*: Adolescents and young adults but also children.
- *CLM*: Histology characterized by the tetrad, including:
  (1) Hyperplasia of *stratus granulosum* (slow turnover)
  (2) Hydropic change/liquefaction of the basal layer with eosinophilic "colloid or Civatte bodies"
  (3) "Saw-toothing" phenomenon of rete ridges (irregular epidermal hyperplasia)
  (4) Tight superficial dermal band-like chronic lymphocytic (CD3+, CD4+) inflammation, which fills the papillary dermis entirely and obscures incontrovertibly the DEJ
- Variant: *Lichen plano-pilaris*, which preferentially affects the epithelium of hair follicles.

- *DDX*: Other oral vesiculo-ulcerative conditions such as *pemphigus vulgaris* and benign mucous membrane pemphigoid, SLE, chronic ulcerative stomatitis, frictional keratosis, and *morsicatio buccarum* (persistent cheek biting), oral leukoplakia, and oral candidiasis.

17.4 Psoriasis and Psoriasiform Dermatitis

17.4.1 Psoriasis

- *DEF*: Dermatitis with regular elongation of the epidermal rete ridges and affects up to 2% of the general population with a chronic relapsing clinical feature. There is evidence of inheritance or abnormal genetics in psoriasis and linkage disequilibrium with HLA-B13, and HLA-Bw17 have been described as well as a genetic susceptibility locus (*PSORS1*). Typically, psoriasis is a disease of young adults (3rd decade) with the involvement of the elbows, knees, and scalp. Three stages may be differentiated:
  (1) Plaque: sharply demarcated, scaly surface with positive Auspitz sign
  (2) Gutta: eruptive lesion on the trunk and proximal extremities
  (3) Pustule: multiple sterile pustular lesions over trunk and limbs following fever and surrounding generalized erythema
- *CLM*: Three main features characterize the histology:
  (1) Hyper-/parakeratosis with hypogranulosis (*fast* cell turnover)
  (2) Acanthosis with long rete ridges and thinning of epidermis over dermal papillae
  (3) Munro microabscesses with neutrophils within the squamous layer
- Variants: Pustular psoriasis and spongiform pustule of Kogoj

17.4.2 Psoriasiform Dermatitis

Psoriasiform dermatitis is seen in several conditions with different etiologies and clinical manifestations. A familiar psoriasiform pattern of inflammation of the dermis is seen. Psoriasiform dermatitis is seen in *Reiter syndrome* (urethritis, arthritis, and conjunctivitis), *pityriasis rubra pilaris*, *lichen simplex chronicus*, *parapsoriasis en plaques*, subacute and chronic spongiotic dermatitis (or *eczematous dermatitis*), *erythroderma* and *mycosis fungoides*, *pityriasis rosea*, *dermatophytosis*, *scabies*, inflammatory linear verrucous epidermal nevus (*ILVEN*), *Bazex syndrome* or *acrokeratosis neoplastica*, *pellagra*, *acrodermatitis enteropathica*, *glucagonoma syndrome*, *syphilis* at secondary stage, *Bowen disease*, *clear cell acanthoma*, and *lamellar ichthyosis*. Reiter disease lesions are crusted erythematous to pustular papules and plaques on the feet, genitalia, buttocks, and scalp.

- *CLM*: There is psoriasiform hyperplasia with parakeratosis and numerous neutrophils, often forming microabscesses, practically indistinguishable from pustular psoriasis without clinical knowledge.

Inflammatory linear verrucous epidermal nevus, or shortened as *ILVEN*, is a form of hamartoma of the epidermis, which occurs as a sporadic or familial lesion in children usually up to 5 years of age in 75% of cases. Clinically, there is a linear arrangement of grouped or coalescent lichenoid, psoriasiform, or verrucous papulae in the lower limbs, characteristically along the Blaschko lines. Skeletal malformations may be observed in some patients.

- *CLM*: There is an alternation of portions of hyperkeratosis and underlying hypergranulosis and portions of hypokeratosis with underlying hypogranulosis, thickening of the rete ridges, slight spongiosis, and focal neutrophilic exocytosis. The upper dermis shows a perivascular lymphocytic infiltrate.

Acrodermatitis enteropathica is an AR-inherited disease with mutations of the *SLC39A4* gene on chromosome 8q24.3. The *SLC39A4* gene encodes a transmembrane Zn-transport protein that serves as a zinc uptake protein. The disease usually starts as an infant is weaned from breast milk. Since Zn supplementation may reverse the disease, the defect is not complete.

17.5 Perivascular In Toto Dermatitis (PID)

The diagnosis of perivascular *in toto* (superficial and deep) dermatitis (PID) includes urticaria and non-urticarial dermatitis with superficial and deep lymphocytic infiltrate.

17.5.1 Urticaria

- *DEF*: It is a disease of the 3rd and 4th decades of life with individual lesions (diffuse or scattered edematous papulae or wheals, often annular with itching) that typically develop and fade within 24 h, when new lesions appear elsewhere and characterized by acute, chronic, symptomatic (dermatographism), and physical forms (cold, pressure, solar, and cholinergic).
- *CLM*: The histology (Fig. 17.2) shows:
  (1) Superficial and deep perivascular chronic inflammatory cellular infiltrates of mixed type (lymphocytes, neutrophils, and eosinophils)
  (2) Dermal edema constituted by separation of collagen bundles
  (3) Dilation of lymphatic channels in the absence of epidermal changes
  (4) Mast cells with granules (Giemsa stain for mast cells is a crucial stain!) and degranulated within the interstitium.
- *DDX*: Insect bite reaction, persistent dermal hypersensitivity to drugs, cutaneous contact hypersensitivity, viral exanthemas, gyrate erythema, and vasculitides involving small blood vessels.

Urticaria, which is also subdivided into acute and chronic, needs to be differentiated by urticarial-like conditions. Acute urticaria of ordinary type, from a few hours to 6 weeks after onset (time 0), while chronic urticaria of usual kind, more than 6 weeks after start (time 0). Infection, food components, or drugs induce acute urticaria of typical kind and are self-limiting, while idiopathic is chronic urticaria of usual kind, often considered of autoimmune type. Angioedema may accompany both types. Physical urticaria is a response to an external factor and is subclassified in dermographism, cholinergic urticaria, cold urticaria, contact urticaria, actinic urticaria, and delayed pressure urticaria according to the responsible agent.

17.5.2 Non-urticaria Superficial and Deep Perivascular Dermatitis

This heterogeneous group includes *toxic erythema of pregnancy*, *gyrate erythema*, *pityriasis lichenoides*, *perniosis*, *rickettsia infections* and *viral infections*, *polymorphous light eruption* (papular, papulovesicular, plaque eruption, or diffuse type occurring on the sun-exposed skin within 24 h post-actinic exposure with the postpubertal onset and nonspecific histology), and *photosensitive eruptions*. Superficial and deep perivascular dermatitis may also be seen in *discoid LE*, *leprosy*, *syphilis*, *borreliosis*, *leukemia*, and, of course, *urticaria pigmentosa*.

Urticarial skin lesions include several entities as follows:

- *Insect bites* (± sting, exposed sites, summer/fall, central blister, asymmetrical distribution of groups of lesions, days to weeks, brown discoloration later, spongiosis of epidermis + inflammatory infiltrate including eosinophils)
- *Urticarial dermatitis* (± CADR, persistent red itchy plaques with either a smooth or dry scratched surface, symmetrical distribution of the lesions on trunk, upper arms, and thighs, dermal inflammatory infiltrate + minimal spongiosis of the epidermis)
- *Contact dermatitis* (± identification of the provoking factor, ± allergic contact dermatitis, clearing up of lesions over days to weeks, ± (+) patch test)
- *Erythema multiforme* (urticaria-like plaques with concentric rings with lesions persisting for 1–3 weeks, ± HSV infection, characteristic histology)
- *Urticaria-like drug eruption* (<2 weeks of a new medication with exception of drug hypersensitivity syndrome with <8 weeks of a new medication, + rechallenge, symmetrical rash of red maculae/patches more severe on the trunk, ± itchy, fever, dark marks, peeling after fading of the lesions, apoptotic keratinocytes of epidermis + inflammatory infiltrate with lymphocytes and eosinophils)
- *Urticaria pigmentosa* (infants, few to many papulae/wheals on trunk and limbs ± systemic involvement with flushing and faintness, massive mast cell infiltrate)
- *AI blistering diseases* (symmetrical distribution of urticarial lesions on trunk and skin folds with subclassification on bullous pemphigoid, herpes gestationis, linear IgA bullous dermatosis, and *epidermolysis bullosa acquisita*, +DIF)
- *PUPPP* (pruritic urticarial papules and plaques of pregnancy, last few weeks of first pregnancy, spots on *striae gravidarum* that clear up in a few weeks following delivery)
- *Annular erythema* (small red spot forming a slowly enlarging ring shape with central clearing, asymptomatic)
- *AI progesterone dermatitis* (childbearing women with a recurrent itchy rash of the 2nd half of the menstrual cycle, autoimmune)
- *Interstitial granulomatous dermatitis* (persistent skin eruption of urticarial type ± AI disorders, including seronegative polyarthritis or malignancy/drugs/infections)
- *Wells syndrome* (red/purple itchy papulae/plaques for days/weeks on one site or generalized, ± persistent brown marks, eosinophilia (tissue + blood) + flame figures)
- *Neutrophilic eccrine hidradenitis* (+ synchronous/metachronous AML, ± fever, perieccrine adnexal neutrophilic colonization)

The urticaria in systemic diseases should discern between the following conditions:

- Scombroid fish poisoning (Scombroid toxicity, also known as histamine poisoning, arises when the individuals ingest fish incorrectly held at warm temperatures after capture or during subsequent handling and storage)

- Urticarial vasculitis
- Sweet syndrome (acute neutrophilic dermatosis) (with blisters/mucosal involvement)
- Neutrophilic urticarial dermatosis (no blisters/mucosal participation)
- Autoinflammatory syndromes (AIS)

17.5.2.1 Autoinflammatory Syndromes

IL-1β pathway hyperactivation is a driver of a heterogeneous group of monogenic diseases with recurrent episodes of multisystemic inflammatory pathology and accompanied clinically by skin lesions, fever, and arthralgia/arthritis. Urticarial-like lesions may be notably observed in cryo-pyrinopathies and familial Mediterranean fever (FMF). AIS histology shows perivascular neutrophilic infiltrate ± lymphocytes. Other than FMF it is important to mention TNF Receptor–Associated Periodic Syndrome or TRAPS, hyper-IgD-emia with periodic fever syndrome, and cryo-pyrinopathies, which belong to AIS. Cryo-pyrinopathies include also familiar cold autoinflammatory syndrome (FCAS), Muckle-Wells syndrome (MWS), and neonatal-onset multisystem inflammatory diseases (NOMID) or aka chronic infantile neurologic cutaneous and articular syndrome, or CINCA syndrome. Other conditions that have been described as AIS are Blue syndrome, pyogenic arthritis-*pyoderma gangrenosum*-acne (PAPA) syndrome, chronic recurrent multifocal osteomyelitis (CRMO), hereditary angioedema, Behçet disease, Gaucher disease, and gout. In gout, monosodium urate crystals are linked to NALP3 inflammasome, which stimulates IL-1β secretion. The NALP3 inflammasome (NACHT, LRR and PYD domains containing protein) is a key regulator of IL-1β secretion.

Major systemic urticarial syndromes include (1) vasculitis-associated SUS, including urticarial vasculitis, SLE, Sjogren syndrome, dermatomyositis, mixed connective tissue disease (MCTD), juvenile rheumatoid arthritis, eosinophilic granulomatosis with polyangiitis (EGPA, Churg-Strauss disease), WG, PAN, and neutrophilic urticarial dermatosis. Moreover, (2) hematologic disorders include NHL (B-cells), Waldenström macroglobulinemia, Schnitzler syndrome, monoclonal gammopathy of undetermined significance (MGUS), cryoglobulinemia, hypereosinophilic syndromes, episodic angioedema with eosinophilia or Gleich syndrome, and *polycythemia vera*, and (3) Autoinflammatory Syndromes (AIS, *vide supra*). In neutrophilic urticarial dermatosis, there is transient urticarial eruption with underlying perivascular and interstitial neutrophilic dermal infiltrate. Schnitzler syndrome is chronic urticaria associated with IgM gammopathy, fever, and arthralgia.

17.6 Nodular and Diffuse Cutaneous Infiltrates

Nodular and diffuse infiltrates of the skin may be organized according to the predominant type of cutaneous inflammation. Eight categories are seen, including granulomatous infiltrates, palisaded granulomatous infiltrates, suppurative granulomatous infiltrates, diffuse histiocytic infiltrates, neutrophilic infiltrates, lymphocytic-plasma cellular infiltrates, mast cell infiltrates, and infiltrates with eosinophilic predominance. Four diseases will be discussed in detail.

17.6.1 Granuloma Annulare

- *DEF*: Infiltrative dermatitis belonging to the category of palisaded granulomatous infiltrates, mostly superficially located, and identified on back of upper limbs of diabetic female youth. Four types are known, including:
  (1) Localized
  (2) Generalized (DM-associated)
  (3) Perforating (keen to pierce the skin)
  (4) Subcutaneous (pediatric form)
- *CLI*: There are >1 skin-colored to red papules in an annular, arcuate, or polycyclic pattern with a tendency to enlarge centrifugally and coalesce. GA resolves spontaneously within 2 years but may recur.

- *CLM*: Discrete *circular* palisading granulomas, generally superficial with a *clear* zone of marked disintegration of collagen and mucin with positive special stains (*colloidal Fe and mucin*).
- *DDX*: Tinea corporis, *pityriasis rosea*, nummular eczema, psoriasis, or *erythema migrans* of Lyme disease.
- *TNT*: Because localized GA is self-limited and asymptomatic, no therapy is necessary. Many patients remain troubled by the appearance and want some treatment. Thus, intralesional corticosteroid has been indicated. Systemic therapy is required for the disseminated form of GA.
- *PGN*: Good in localized but guarded for the disseminated form.

17.6.2 Necrobiosis Lipoidica Diabeticorum (NLD)

Necrobiosis lipoidica diabeticorum (NLD) also belongs to the category of palisaded granulomatous diseases, mostly symmetrically and superficially located, and identified on lower limbs of diabetic female youth. Clinically, there is a sclerotic, round, circumscribed, slightly elevated plaque, which is initially red-brown and then yellow with a purple border. Microscopically, horizontally and parallel arranged palisading granulomas, generally superficial, with a poorly defined zone of the disintegration of collagen, but no central mucin, with accompanying infiltration of plasma cells, NLD has been considered a primary vascular disease or a primary disease of collagen with secondary vascular changes.

17.6.3 Rheumatoid Nodule

Rheumatoid nodule belongs to granulomatous disease with deep localization and develops at sites of trauma or pressure points (elbows, knuckles, Achilles tendons) in about 1/3 of patients with RA. Clinically, there are erythematous plaques or nodules, which are fixed to fascia or bone. Microscopically, subcutaneous palisades of histiocytes ± MNGC with central eosinophilic fibrinoid necrosis (no central mucin!) (Fig. 17.3).

17.6.4 Sarcoid

Sarcoid represents the cutaneous involvement of sarcoidosis in about 1/3 of patients affected with sarcoidosis. Clinically, there is a diffuse, asymptomatic maculopapular cutaneous eruption with an associated increase in lung fibrosis. Lupus pernio defines a chronic purple plaque, typically on the nose. Microscopically, subcutaneous non-necrotizing granulomata (lobular panniculitis) are seen. Small lymphocytes with Langhans (Pirogov-Langhans cells) multinucleated giant cells (MNGC) are also observed. The overlying skin is stretched but intact.

17.7 Intraepidermal Blistering Diseases

Intraepidermal blistering diseases include mostly three common forms of *pemphigus* (*vulgaris*, *foliaceus*, and *paraneoplasticus*), *IgA pemphigus*, *impetigo*, and three "diseases" (*Darier disease*, *Grover disease*, and *Hailey-Hailey disease*).

17.7.1 Pemphigus Vulgaris, Pemphigus Foliaceus, and Pemphigus Paraneoplasticus

Pemphigus refers to an Ancient *Greek* term "πέμφιξ," which means bubble or blister and is a mucocutaneous autoimmune disease that causes vesicles (blisters).

17.7.1.1 Pemphigus Vulgaris
- *DEF*: It is an AI disease (autoantibodies against DSG1 and DSG3) and the most common type of intraepidermal bullous disorders.
- *CLI*: Mucocutaneous fragile blisters with early rupture with following painful crusted

17.7 Intraepidermal Blistering Diseases

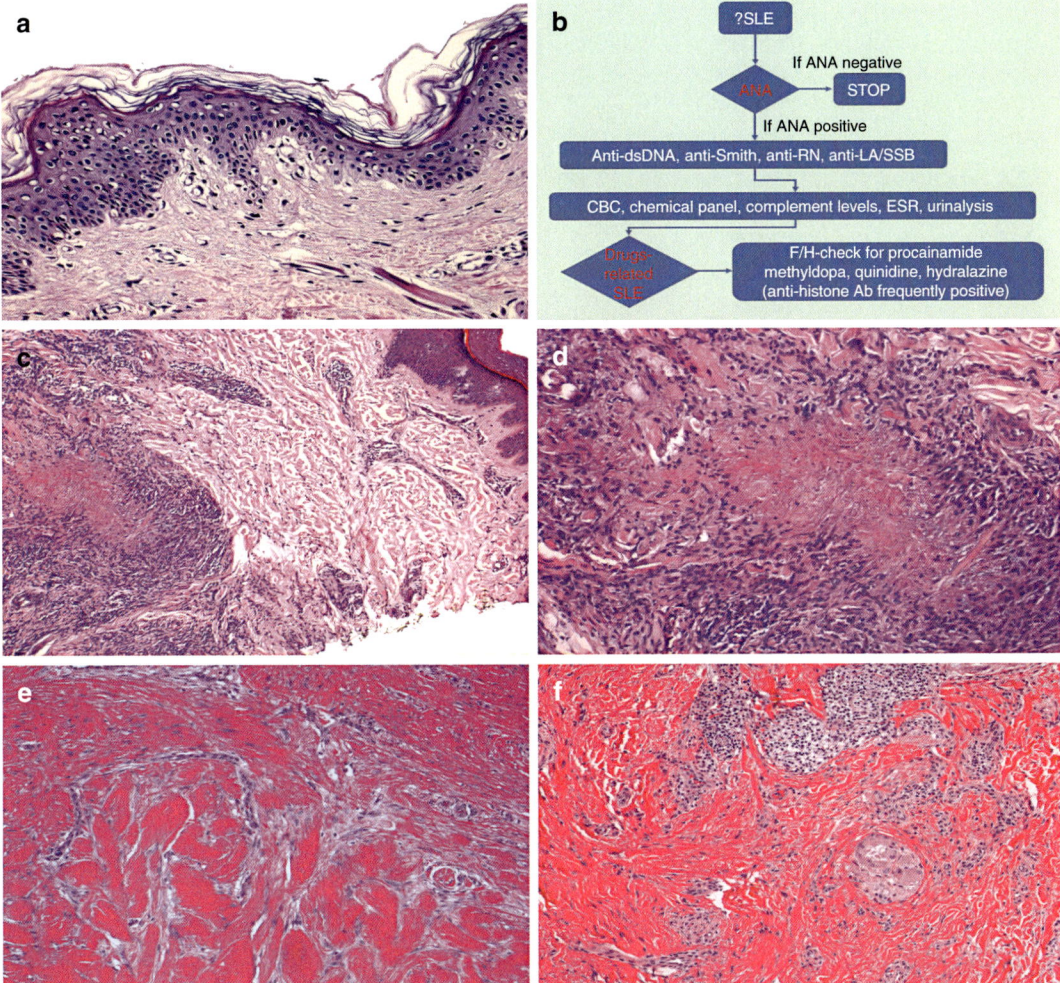

Fig. 17.3 This panel illustrates in (**a**) the microphotograph of SLE with some vacuolar degeneration of the basal epithelium as well as scattered lymphocytes in the papillary dermis with increase of collagen deposition in the papillary dermis (H&E stain, 200×). In (**b**) is the flow chart for a suspicion of SLE with antinuclear antibodies, anti-ds DNA, anti-Smith, anti-RN, anti-LA/SSB as first two initial diagnostic steps. In (**c** and **d**) are two microphotographs showing a rheumatic nodule with central necrotic collagen bundles with fibrin surrounded by inflammatory cells ("palisading granuloma"). The differential diagnosis is mostly with granuloma annulare and necrobiosis lipoidica. In granuloma annulare, there is mucin in the core of the granuloma (negative in our case, not shown) (**c**, H&E stain, 50×; **d**, H&E stain, 200×). The microphotographs (**e** and **f**) show keloid scar not inflamed (**e**, H&E stain, 100×) and an inflamed scar (**f**, H&E stain, 100×)

blood erosions are seen on the scalp, face, axilla, groin, and trunk.
- *CLM*: Suprabasal cleft with acantholytic cells situated in the blister cavity and a "row of tombstones" as remaining basal cells tombstone above an elongated epidermal rete ridge.
- *DIF*: IgG deposition is visualized as a reticular or fishnet-like pattern (aka "chicken-wire" pattern of IF in intraepidermal blistering diseases).
- *TNT*: High-dose corticosteroids have reduced the mortality. The *vegetans* variant of pemphigus has large moist warty plaque studded with a pustule and microscopic findings like pemphigus vulgaris, although overlying epidermal hyperplasia is also present. Rituximab is used

in the treatment of childhood and juvenile pemphigus. Rituximab is an anti-CD20 monoclonal antibody, which attaches to the CD20 receptor and cause the cells to lyse (disintegrate).
- *PGN*: Death can occur because of infections (e.g., pneumonia and septicemia) and cardiovascular diseases.

17.7.1.2 Pemphigus Foliaceus

Pemphigus foliaceus is also an AI disease (auto-Abs against DSG1 only – DSG1 is a desmosomal cadherin with higher concentration in the upper layers of the epidermis explaining the limited subcorneal-granular cleavage plane of the blister) and represents probably a more benign form and is endemic in South America. Clinically, flaccid, fragile blisters rupture to leave shallow crusted erosions and crusted patches and plaques on the scalp, face, H&N, proximal limbs, and trunk of adult subjects. The variant of pemphigus erythematosus is a variant of *pemphigus foliaceus* with features combining *pemphigus foliaceus* and LE and is restricted to the malar region of the face.

17.7.1.3 Pemphigus Paraneoplasticus

This pemphigus type is a generalized polymorphous bullous eruption with lichenoid papules with painful oral ulcers (DDX: SJS) in association with HL, thymoma, CLL, and other hematological malignancies as well as lung cancer and retroperitoneal sarcoma. Microscopically, suprabasal acantholysis and interface dermatitis of vacuolar or lichenoid type. DIF: Intercellular IgG ± C3 and BMZ IgG/IgM/C3 (in both linear and granular patterns). The prognosis shows an aggressive blistering disease with refractoriness to treatment.

17.7.2 IgA Pemphigus and Impetigo

17.7.2.1 IgA Pemphigus
- *DEF*: Pruritic, subcorneal/intraepidermal vesiculopustular eruption occurring worldwide in both children and young adults, without gender preference, and involving the axilla, trunk, and limbs.
- *CLM*: A subcorneal or an intraepidermal pustule with intercellular IgA in the epidermis (DIF) with two variants:
  - SPD, subcorneal pustular dermatosis type IgA pemphigus
  - IEN, intraepidermal neutrophilic type Ig A pemphigus
- *DDX*: Impetigo, other forms of pemphigus, SSSS, and SPD.

17.7.2.2 Impetigo
- *DEF*: Common superficial bacterial infection, typically of children, with localization of "honey-colored" crusted lesions, particularly peri-oral and around the nose and at the groin and due to *S. aureus* or *S. pyogenes* following mild skin abrasions or insect bites and favored by poor hygiene, warm, humid weather, and skin occlusion. The name "ecthyma" (Greel εκθύω "break out" composed of εκ- "out" and θύω "seethe") is used if ulceration on impetigo is present.
- *CLM*: Subcorneal vesicles splitting beneath stratum granulosum and filled with neutrophils and acantholytic cells, ± Gram (+) cocci ± variable sparse dermal infiltrate.
- *DDX*: Numerous other diseases with subcorneal pustules (although clinical information is pivotal!) as well as *pemphigus foliaceus* and *erythematosus*.
- *TNT*: Topical mupirocin or oral antibiotics (e.g., cephalexin, erythromycin, dicloxacillin).
- *PGN*: ± Progression to cellulitis, glomerulonephritis, *erythema nodosum*, or *erythema multiforme*.

17.7.3 Intraepidermal Blistering Diseases

Three diseases need to be kept in mind, particularly, Darier disease, Grover disease, and Hailey-Hailey disease.

17.7.3.1 Darier Disease
- *DEF*: AD-inherited genodermatosis (aka keratosis follicularis) with *ATP2A2* gene mutations (12q23–q24.1) coding for a sarco-/endoplas-

mic reticulum Ca^{2+}-ATPase isoform 2 with a characteristic "corp ronds" and "corp grains."
- *CLI*: Plaques of greasy, crusted yellow-brown papules on the scalp, forehead, back, and upper chest possibly accompanied by palmar pits and longitudinal nail splitting.
- *CLM*: Acantholytic dyskeratosis with a suprabasal split, *corps ronds*, *corps grains* and parakeratosis, as well as nonspecific granular C3 staining at the DEJ (basement membrane) by DIF.
- *DDX*: Hailey-Hailey disease, Grover disease, and pemphigus vulgaris.

17.7.3.2 Grover Disease
- *DEF*: Transient acantholytic dermatosis with marked self-limited acantholysis of unknown etiology showing small foci of acantholysis and four patterns (*vide infra*).
- *EPI*: Worldwide, youth, ♂: ♀ = 2.4: 1.
- *CLI*: Intensely pruritic linear arrangement of papulae, vesicles, plaques, and excoriations on the chest, back, and thighs.
- *CLM*: Suprabasal splitting with four histologic patterns (≥1 histologic pattern in the same biopsy!), including:
  - Darier disease-like focal acantholytic dyskeratosis pattern
  - Hailey-Hailey disease-like pattern
  - *Pemphigus vulgaris*/foliaceus-like pattern
  - Spongiotic dermatitis-like pattern
- *DIF*: Nonspecific pattern.
- *DDX*: Darier disease, Hailey-Hailey disease, pemphigus vulgaris/foliaceus, and spongiotic dermatitis.
- *PGN*: Self-limited (benign, but may last for a while from several weeks to several years).

17.7.3.3 Hailey-Hailey Disease
- *DEF*: Benign familial chronic pemphigus with AD inheritance (*ATP2C1* gene on 3q coding a P-type Ca^{2+} transport ATPase) and onset in adolescence with characteristic "Dilapidated Brick Wall" pattern and common fungal or bacterial superinfection.
- *EPI*: Worldwide, youth, ♂ = ♀.
- *CLI*: Blisters at friction sites (neck, axillae, groin).
- *CLM*: Supra-basal acantholysis of severe degree giving rise to the characteristic "Dilapidated Brick Wall" histologic pattern and nonspecific DIF pattern.
- *DDX*: Grover disease, Darier disease, and pemphigus vulgaris.
- *PGN*: Worsening of symptoms by trauma or sun and common superinfection by staphylococci or *Candida* spp.

17.8 Subepidermal Blistering Diseases

Subepidermal blistering diseases include several diseases, including bullous pemphigoid and epidermolysis bullosa, erythema multiforme and toxic epidermal necrolysis, Hb-related porphyria cutanea tarda (PCT), herpes gestationis, dermatitis herpetiformis, and lupus (systemic lupus erythematodes, SLE).

17.8.1 Bullous Pemphigoid and Epidermolysis Bullosa

17.8.1.1 Bullous Pemphigoid
- *DEF*: Subepidermal non-acantholytic blistering disease due to anti-BM Ab in serum (mostly IgG) resulting in a linear DIF pattern for Ig and C3 at the DEJ (anti-BPAg2 antibodies) with "ribbon candy" aspect (roof of the blister) and usually harboring an increased risk of other autoimmune disorders.
- *EPI*: Worldwide, children < adults (middle age to elderly), ♂ = ♀.
- *CLI*: Large, tense blisters on an erythematous base, some with blood on "normal" skin of mostly extremities (inner thighs, flexor surface of arms, groin, axilla) sparing face and scalp.
- *CLM*: Subepidermal non-acantholytic blister with dense inflammatory infiltrate rich in eosinophils (inflammatory cell-rich and cell-poor variants) with some edema in the dermis and no necrosis of overlying epidermis.
- *DIF*: Linear DIF pattern for Ig and C3 at the DEJ (anti-BPAg2 antibodies) with "ribbon candy" aspect (roof of the blister with salt-split skin).
- *DDX*: Epidermolysis bullosa, herpes gestationis, bullous arthropod bite, and bullous drug eruption.

17.8.1.2 Epidermolysis Bullosa

- *DEF*: Hereditary subepidermal blistering disease with variable onset and considered a heterogeneous family of disorders, which have in common an etiology due to structural disorders of the BMZ.
- *EPI*: The incidence and prevalence of AD dystrophic epidermolysis bullosa are 2.12 and 1.49 cases per 1 million live births, respectively. The incidence and prevalence of AR dystrophic epidermolysis bullosa are 3.05 and 1.35 cases per 1 million live births.
- *CLM*: Non-inflammatory subepidermal blister with blood or fibrin within the vesicles, dermal papillae retained, and lack of dermal edema due to several defects from CK5 and CK14, to reduced hemidesmosomes and anchoring filaments, to auto-Ab to collagen VII and IgG deposition at the floor of the blister (Fig. 17.4).
- *DDX*: *Porphyria cutanea tarda*, acquired *epidermolysis bullosa*, traumas, penicillamine dermopathy, cell-poor bullous pemphigoid, and TEN.

17.8.2 Erythema Multiforme and Toxic Epidermal Necrolysis

See Sects. 17.3.1.4 and 17.3.1.5.

17.8.3 Hb-Related Porphyria Cutanea Tarda, Herpes Gestationis, and Dermatitis Herpetiformis

These disorders belong to the "H"-group involving three heterogeneous diseases, including *porphyria cutanea tarda*, *herpes gestationis*, and

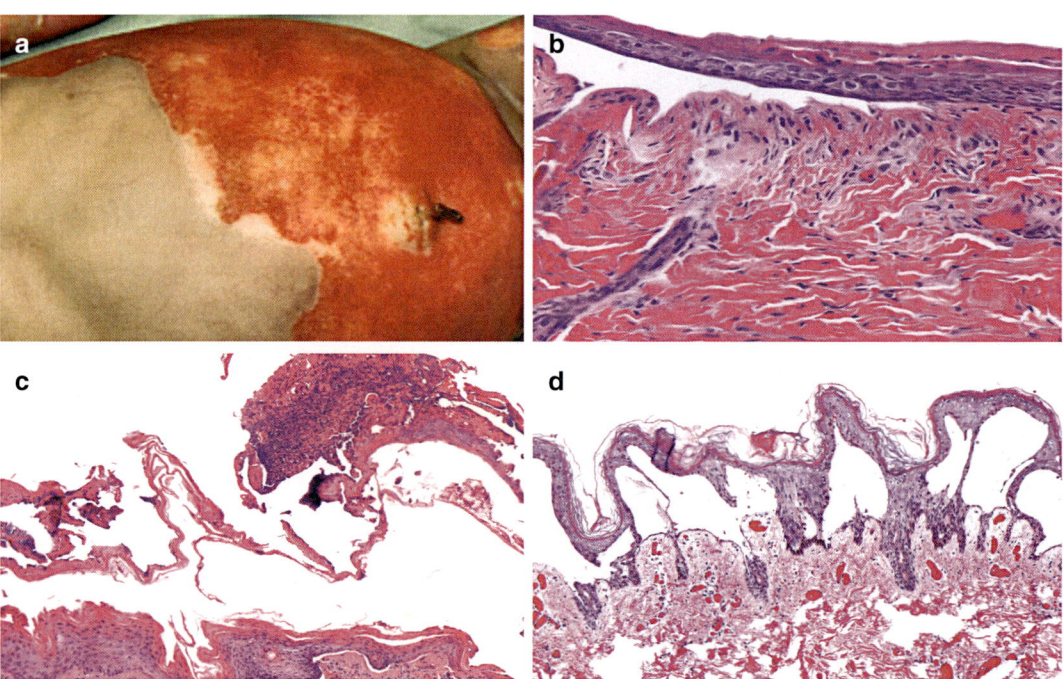

Fig. 17.4 The (**a**) gross photograph shows the extensive skin damage found in an infant affected with molecular genetics confirmed epidermolysis bullosa. The microphotographs (**b**–**d**) (H&E staining with **b** at 200×, **c** at 100×, and **d** at 100×) show the cleavage planes at subepidermal level (epidermolysis bullosa) (**b**), at subepidermal level (epidermolysis bullosa) with some fibrosis of the upper dermis (**c**) highlighted by collagen IV expression (not shown), and at suprabasal level (suprabasal blister) as differential diagnosis (e.g., *pemphigus vulgaris*, which shows suprabasal acantholysis from IgG against desmosomes and production of plasminogen activator)

dermatitis herpetiformis, although all of them share the subepidermal blistering formation. *Dermatitis herpetiformis* is an itchy, blistering, skin rash, occurring on the elbows, knees, scalp, back, and buttocks. *Dermatitis herpetiformis* indicates gluten intolerance, which may be related to celiac disease.

17.8.4 Lupus (Systemic Lupus Erythematodes)

See paragraph of the upper urogenital tract.

17.9 Vasculitis

Vasculitis is an inflammation of a blood vessel and has been discussed in other chapters of the book (Box 17.3).

17.10 Cutaneous Appendages Disorders

Disorders of the dermal appendages involve several diseases including alopecia, chromhidrosis, folliculitis, Fox-Fordyce disease, neutrophilic eccrine hidradenitis, palmoplantar eccrine hidradenitis, and sweat gland necrosis. The reader should refer to specific texts of dermatopathology for the single diseases.

17.11 Panniculitis

Panniculitis = inflammation of the subcutaneous adipose tissue, which has several and different causes with mostly the same clinical appearance. Skin biopsy is essential, and each disease has characteristic microscopic features that need to be correlated with the clinical setting. The following boxes separate panniculitis according to the localization of the inflammatory infiltrate either in a septal way or the fat lobules. In either condition, this is not absolute, and some inflammatory cells of mostly septal panniculitis may be seen in the lobules and vice versa. Another way to classify *panniculitis* is summarized in Box 17.4.

- *CLI*: Thickening of the skin, woody ± discoloration of the overlying skin (red/brown) and tenderness of the area; alternatively raised nodules or lumps under the skin can occur, or even a plaque or large flat area of thickened skin. Following settled inflammation, a depression of the surface may be observed.
- *TNT*: It is based on the treatment of the underlying cause (e.g., drug, withdrawal, change of dosage, resolution of bacterial infection) rest and elevation of the affected body area, compression hosiery (legwear) (18–25 mm Hg pressure), and in the case it is less tolerated, pain relief medications are used like NSAIDs, systemic steroids, anti-inflammatory antibiotics (e.g., tetracycline or hydroxychloroquine), and potassium iodide. In dramatic cases, a surgical removal in the presence of persistence or ulcerated lesions may be considered mandatory.

17.11.1 Septal Panniculitis

17.11.1.1 Erythema Nodosum
- *DEF*: Septal panniculitis *without* vasculitis often associated with infections, drug administration, IBD, some malignancies, sarcoidosis, and pregnancy.
- *EPI*: Worldwide, 20–50 years, ♂:♀ = 1:9.
- *CLI*: Dome-shaped slightly polypoid reddish nodule of the subcutis, >1 cm, tender/painful,

---

Box 17.3 Vasculitides
17.9.1 Leukocytoclastic Vasculitis (LCV)
17.9.2 Polyarteritis Nodosa (PAN)
17.9.3 Wegener Granulomatosis (WG)

Box 17.4 Panniculitis
17.11.1 Septal Panniculitis
17.11.2 Lobular Panniculitis
17.11.3 Mixed Septal and Lobular Panniculitis

self-limited on anterior and lateral lower legs. If pyrexia, malaise, and joint pain are present, it is also often the case of a *granulomatous fibrosing septal panniculitis*.
- *CLM*: Septal inflammation of neutrophils ± eosinophils, then lymphocytes, edematous and thickened septa with the peripheral involvement of fat lobules ± granulomatous inflammation with giant cells (aka Miescher's granulomas). Miescher's radial granulomas (MRGs) are "radial" nodules consisting of small histiocytes, "radially" placed around a central cleft and mainly located in the interlobular septa and in the deeper layers of the skin (Sánchez Yus et al. 1989).
- *DDX*: Other septal panniculitides (sclerosing panniculitis, connective tissue diseases).

17.11.2 Lobular Panniculitis

17.11.2.1 Vasculitis-Associated Lobular Panniculitis (Erythema induratum)
- *DEF*: Lobular panniculitis *with* vasculitis, characteristic of "tender, dusky, bluish nodules" of the posterior calves of adult obese women with a controversial relationship to tuberculosis ("tuberculide").
- *SYN*: "Deep inflammatory necrotizing nodular disease of the calves."
- *CLI*: Erythematous, tender, dusky nodule.
- *CLM*: Extensive intralobular inflammatory infiltration of lymphocytes, MNGCs, and epithelioid histiocytes with necrotizing vasculitis of medium-sized blood vessels, which may become occluded with a subsequent ulcer, necrosis, and lipophagic granulomas.
- *DDX*: Other lobular panniculitides with vasculitis.

17.12 Cutaneous Adverse Drug Reactions

One of the most challenging tasks for a pathologist is to liaise with therapy-related diseases, e.g., cutaneous adverse drug reactions (CADRs). CADRs or cutaneous adverse drug reactions are a common diagnosis in clinical pediatrics and youth and may represent a challenge for some pathologists who are not aware of the numerous new drugs that every year are added in the repertoire of the pharmacopeia of both pediatricians and adult medical internists. It seems that a good starting figure to tackle this important, but difficult, chapter is to say that a rate of 5.5% of ADRs occurs per drug exposure (about one in 20 drugs used clinically), of which CADRs are 2.2% (about one in 40 drugs used clinically). Although CADRs are substantially half of the ADRs, they are rarely considered severe. The parents or relatives have, nevertheless, significant concerns, and CADRs account for a good number of clinic visits. Thus, CADRs may also represent a topic for healthcare econometrics. In assessing a CADR, a relationship of causality needs to be started. Causality assessment utilizes investigation tools aiming to weigh the biological plausibility or "culpability" of a drug causing a reaction. Naranjo et al.'s (1981) criteria for the "definitive culpability" in ADRs setting include (1) temporal sequence between drug administration and assessed drug level in body fluids, (2) recognition of a well-determined response to the suspected drug from literature data or clinical records, (3) improvement following drug withdrawal, and (4) ADRs' reappearance following drug re-exposure. We consider it appropriate to state that fullfilled criteria 1 and 3 determine a reaction as "probable," while criterion 1 only determines a reaction as "possible" only. ADRs may be classified as type A and type B ADRs, accounting type A reactions for the majority of ADRs. *Type A ADR*: Reactions related to the well-known pharmacologic effects of a drug, dose-dependent, predictable, and mild-moderate in the degree of clinical seriousness. *Type B ADR*: Reactions NOT related to the well-known pharmacologic effects of a drug, dose-independent, unpredictable, and moderate-severe in the degree of clinical seriousness. Although 29 drug-related CADRs have been identified, most of them are extremely rare, and only 5 are common in children and youth.

17.12 Cutaneous Adverse Drug Reactions

17.12.1 Exanthematous CADR

- *CLI*: Morbilliform OR scarlatiniform maculopapular exanthema.
- *DDX*: Viral exanthems or scarlet fever.
- *DAT-CADR Time*: <2 weeks of drug therapy with recurrence on rechallenge.
- *Drugs*: Carbamazepine, aminopenicillin, NSAIDs, sulfonamides, anti-TB drugs, and phenobarbital.
- *PGN*: Following discontinuation, oral antihistamines and systemic corticosteroids, as well as the use of emollients and avoidance of sunlight exposure, fading of erythema with common superficial desquamation and transient hyperpigmentation.

Notes: DAT-CADRT – Drug administration time – cutaneous adverse drug reaction time.

17.12.1.1 Urticaria-Like CADR

UCADR is caused by early type I hypersensitivity or drug stimulation of mast cells.

- *CLI*: Edematous, pruritic papulae and plaques (wheals) with duration of the individual lesion not more than 24 h in one single cutaneous location.
- *DDX*: Acute urticaria, allergies to foods, additives, or excipients, and viral infections.
- *DAT-CADRT*: < hours to days of drug administration.
- *Drugs*: NSAIDs, aminopenicillins, penicillin, cephalosporins, sulfonamides, anti-TB drugs, phenytoin, carbamazepine, histamine-releasing drugs (morphine, quinine, IV radiocontrast dye, etc.).
- *PGN*: Following drug discontinuation, oral histamines and systemic corticosteroids administration, as well as the use of emollients, the disappearance of the lesions in 24 h. Rechallenge is advised according to the severity of the reaction.

17.12.1.2 Fixed Drug Eruptions

- *CLI*: Erythematous to purple, edematous, and distinct round, 2–10 cm-sized lesions often fading to gray or light-brown patches over time.
- *DDX*: Bruises, insect bites, erythema multiforme, viral infections, child abuse.
- *DAT-CADRT*: 30 min–8 h.
- *Drugs*: NSAIDs, sulfonamides, ciprofloxacin, metronidazole, penicillins, anti-TB drugs, phenytoin, phenolphthalein-containing laxatives, pseudoephedrine, and acetaminophen.
- *PGN*: Following drug discontinuation, erythema fades becoming gray-brown ("gray-brown pigmented patches") recurring at the same location for months or years.

Notes: Pathogenesis is unknown.

17.12.1.3 Photosensitivity-Driven CADR (P-CADR)

- *CLI*: Phototoxic (sunburn-like) OR photoallergic (eczema-like) reactions in sun-exposed areas of the skin ± pruritus.
- *DDX*: Acute urticaria, allergies to food, additives, excipients, and viral infections.
- *DAT-CADR*: Immediate with drug administration and sun exposure in the setting of phototoxic P-CADR and requires sensitization in the photoallergic subtype.
- *Drugs*: Tetracyclines, fluoroquinolones, amiodarone, psoralens, griseofulvin, diuretics (furosemide and thiazides), NSAIDs, antipsychotic agents (chlorpromazine, prochlorperazine), St. John's wort, topical (furocoumarins from lime, lemon, celery, parsley, and figs) for phototoxic P-CADRs and sunscreens, fragrances, antibacterial agents, latex, NSAIDs, thiazide diuretics, griseofulvin, quinidine, sulfonamides, sulfonate-ureas, and pyridoxine (vitamin B6) for photoallergic P-CADRs.
- *PGN*: Following drug discontinuation, oral administration of antihistamines and systemic corticosteroids as well as emollients, hyperpigmentation takes place and resolves over months to years.

UV radiation interacts with chemical compounds in the skin-forming free radicals, which start local cytotoxic effects for phototoxic type, while a type IV variant reaction underlies a photoallergic kind, in which light enables the pharmacologic substance to conjugate with a carrier protein. Atopy (allergy, asthma) is a risk factor.

17.12.1.4 Serum Sickness- Like CADR (SSLR)

- *CLI*: Systemic and cutaneous symptomatology constituted by fever, urticarial eruptions, pruritis, and arthralgias recurring on rechallenge.
- *DDX*: Serum sickness, erythema multiforme, and viral infections.
- *DAT-RO*: 1–3 weeks.
- *Drugs*: Cefaclor, ciprofloxacin, macrolides, penicillin, sulfonamides, tetracyclines, rifampin, bupropion, and fluoxetine.
- *PGN*: Following drug discontinuation, use of oral antihistamines and systemic corticosteroids, and IV Ig, resolution of the symptomatology. An SSLR lacks the impoverishment of complement (hypocomplementemia), vasculitis, and renal lesions and requires the typical target-like lesions with multiple centers.
- *DAT-RO*: Drug administration time-reaction onset

Dangerous drug eruptions are, substantially, drug hypersensitivity syndrome (DHS), Stevens-Johnson syndrome (SJS), and toxic epidermal necrolysis (TEN). In DHS mucosal lesions are infrequent when compared with SJS and TEN.

DHS – Features include severe exanthemata-like rash, exfoliative dermatitis, ± fever, lymphadenopathy, hepatitis, nephritis, carditis, eosinophilia, "atypical lymphocytes" (1/2 of cases); TTO (time to onset), 1–6 weeks; drugs, phenytoin, carbamazepine, phenobarbital, sulfonamides, allopurinol, minocycline, nitrofurantoin, and terbinafine. DDX: cutaneous lymphoma.

SJS – Features include atypical targets ≤10% BSA with mucous membrane involvement (conjunctivae, mouth, and genitals in ≥2 sites) with high fever, sore throat, rhinorrhea, and cough, 4–28 days; drugs, antiseizures, sulfonamides, allopurinol, and NSAIDs.

TEN – Features include confluent and extensive epidermal detachment ≥30% BSA with mucous membrane involvement (conjunctivae, mouth, and genitals in ≥2 sites) with fever, sore throat, headache, leukopenia, and lesions of the respiratory and/or GI tracts, 4–28 days; drugs, antiseizures, sulfonamides, allopurinol, and NSAIDs.

DDX: SSSS, Kawasaki disease (exanthemata's stage).

17.13 Dyskeratotic, Non-/Pauci-Inflammatory Disorders

Dyskeratotic, non-/pauci-inflammatory disorders are a heterogeneous group of cutaneous diseases, which are characterized by a modest inflammatory infiltrate. This group includes ichthyosis, warts, and the very rare actinic keratosis, cutaneous horn, and seborrheic keratosis.

17.14 Non-dyskeratotic, Non-/Pauci-Inflammatory Disorders

Non-dyskeratotic, non-/pauci-inflammatory disorders are a heterogeneous group of cutaneous diseases, which are characterized by a modest inflammatory infiltrate. This group comprises collagen/elastin disorders, deposition disorders, ectopic tissues, and pigmentation disorders.

17.15 Infections and Infestations

Infections are a common cause of presentation of a sick child to a pediatrician or a GP. Naturally, most of the diseases are quite well manageable with medical therapy and do not require a biopsy to be sent to a pathologist. However, sometimes this is the case, especially if the process is chronic and do not heal after changing one or more courses of therapy. Some cutaneous infections include common bacterial infections, Treponema and Rickettsia infections, viral infections, fungal infections, protozoal and algal infections, and helminthic infestations (Fig. 17.5). The reader should refer to dermatopathology and clinical microbiology books for these topics.

17.16 Cutaneous Cysts and Related Lesions

Cutaneous cysts may be a significant reason for concern, particularly in infancy. Epidermal inclusion cyst is a cyst with squamous epithelial lining and keratin lamellae in the lumen. Milia are multiple tiny whitish cysts located typically on the nose and cheeks due to keratin entrapment

17.17 Tumors of the Epidermis

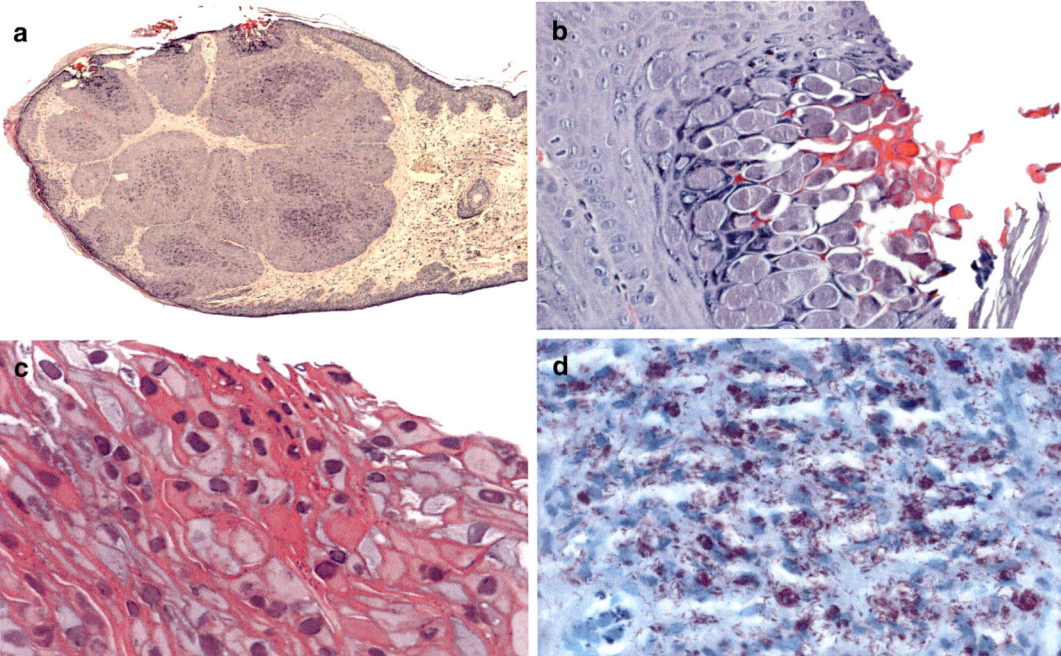

Fig. 17.5 The tetrad of these microphotographs shows molluscum contagiosum (**a**, H&E, 40×; **b**, H&E, 400×), virus cytopathic changes in verruca vulgaris (**c**, H&E stain, 400×) and atypical mycobacterial infection (**d**, Fite staining, 630×). In (**b**) a lobule of a molluscum contagiosum is highlighted. The lobules contain hyalinized molluscum bodies, also known as Henderson-Paterson bodies, and the molluscum contagiosum body is an intracytoplasmic inclusion body. In (**d**) the atypical mycobacteria are seen as red bacilli

beneath the surface of the skin. Trichilemmal cyst or pilar cyst is a smooth, mobile cyst that forms from a hair follicle and usually located on the scalp. Microscopically, the pilar cyst is lined by stratified squamous epithelium. It is filled with keratin and does not show a granular layer. Pseudocysts do have an epithelial lining (e.g., mucocele) (Fig. 17.6).

17.17 Tumors of the Epidermis

Contrarily to the significant consideration that only elderly may be affected by tumors of the epidermis, epidermal tumors do indeed occur in childhood and adolescence as well as young adults and need to be differentiated by the most common cutaneous adnexal tumors. Although the number of these lesions may be quite overwhelming in recalling the several and quite diverse entities, only a few are characteristically seen in the routine of dermatology and pathology, and that occur in childhood and youth. Moreover, a plethora of overlapping names and synonyms are often used.

17.17.1 Epidermal Nevi and Related Lesions

Nevi and angiomas are often grouped, because they may have some clinical appearance.

Nevi of *Epithelial* Derivation

(a) *Epidermal nevi* include (1) *nevus verrucosus*, (2) *nevus unicus lateralis*, (3) inflammatory linear verrucous epidermal nevus (ILVEN), and (4) epidermal nevus syndrome.
(b) *Nevi of the cutaneous appendages* include (1) *nevus comedonicus*, (2) *nevus eccrinus*, (3) nevus sebaceous of Jadasshon, (4) nevus syringe-cysticus adenomatous, (5) and trichofolliculoma (*Nevus folliculi piliferi*).

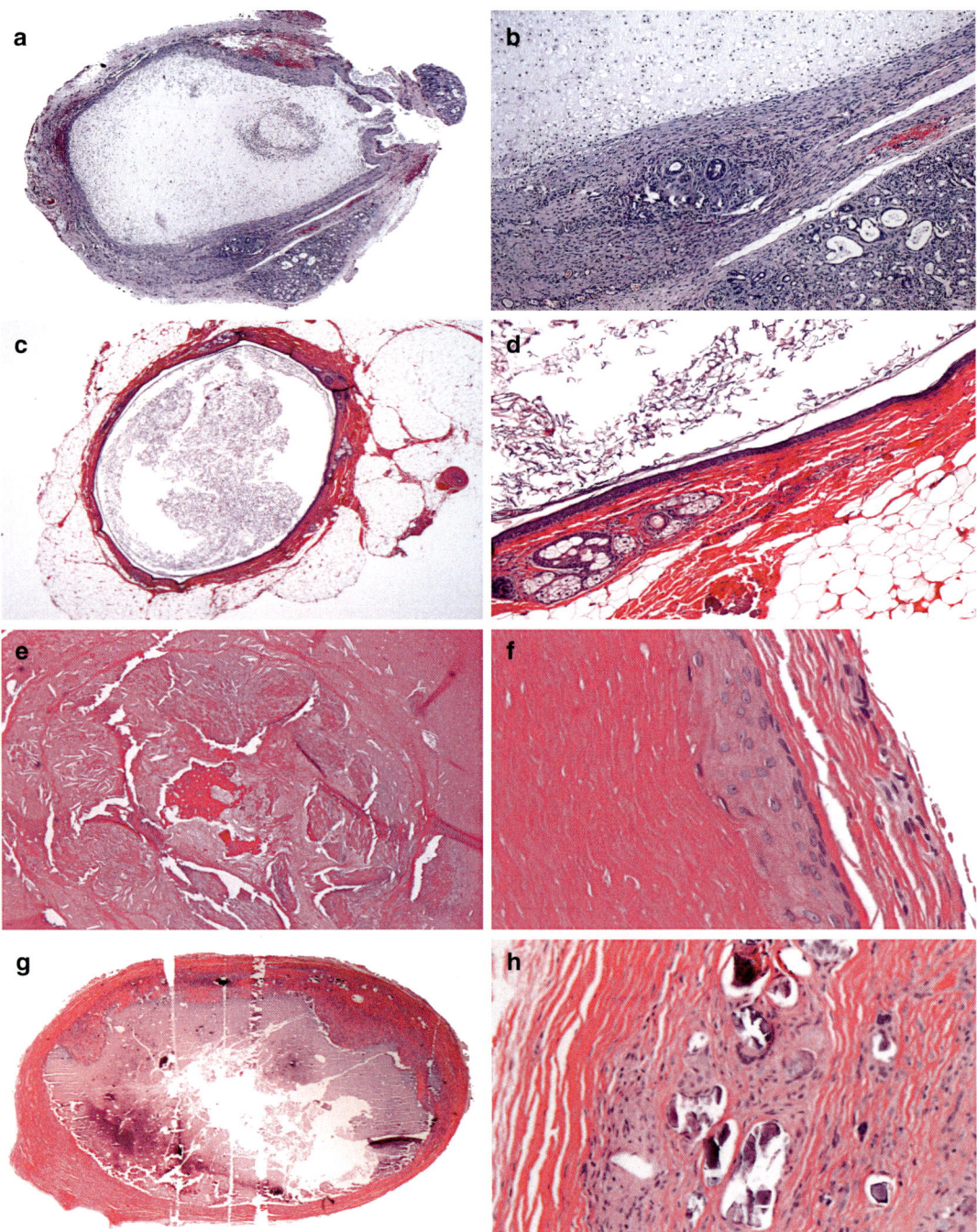

Fig. 17.6 This panel shows cysts and pseudocysts encountered in pediatric dermatopathology. Pseudocysts are retention cysts (**a** and **b**) and digital myxoid ganglion cyst (**g** and **h**). In all four microphotographs, no epithelial lining is seen. Mucin is present in a and b and myxoid tissue is present in g and h. In this latter pseudocyst, dystrophic calcification is also seen (**a**, H&E stain, 20×; **b**, H&E stain, 100×; **g**, H&E stain, 12.5×; **h**, H&E stain, 200×). The microphotographs (**c** and **d**) are taken from a dermoid cyst showing adnexal (sebaceous) cyst associated with the epithelium (**c**, H&E stain, 12.5×; **d**, H&E stain, 50×), while the microphotographs (**e** and **f**) illustrate a trichilemmal or pilar cyst with keratin lamellae, but lack of granular layer (**e**, H&E stain, 50×; **f**, H&E stain, 400×)

Nevi of *Dermal* Derivation

(a) *Connective nevi* include (1) nevus of Lewandowski, (2) nodular connectival nevus, and (3) cutis zigrinata lipomatosis.
(b) *Vascular nevi* include (1) nevus telangiectasia (sive macularis sive planus), (2) nevus angiomatosis (sive tuberosus sive cavernosus), and (3) *nevus anemicus*.
(c) *Melanocytic epidermal lesions* comprise *lentigo simplex*, ephelides, and nevus of Becker
(d) *Nevocellular nevi* comprise nevus of junctional type, nevus of compound type, nevus of intradermal type, and halo nevus of Sutton.
(e) Congenital melanocytic nevus.
(f) *Nevus spilus* (café au lait maculae nevus or speckled lentiginous nevus, which is a hyperpigmented lesion due a focal proliferation of melanocytes along the basal layer of the epidermis within a café-au-lait spot).

Melanocytic Dermal Lesions

(a) Spitz Nevus
(b) Mongol Macula
(c) Nevus of Ota
(d) Nevus of Ito

Nevus Achromicus

Epidermal nevi are a cellular proliferation of hamartomatous character. An epidermal nevus (non-melanocytic nevus) is an overgrowth of the epidermis that can be present at birth or develop during infancy. The epidermal nevus arises from a defect in the ectoderm. Epidermal nevi may occur alone or be associated with other malformations or genetic syndromes. In the last event, epidermal nevi constitute the sign of several "epidermal nevus syndromes." The linear epidermal nevus is not uncommon and represents one lesion that a pathologist can see 2–3 times a year in a busy pediatric pathology practice. It is essential to recall the Blaschko lines because the linear epidermal nevi tend to fall along these skin tension lines. The histopathology examination of the linear epidermal nevi shows benign growths comprised of keratinocytes and a variable mixture of apocrine, eccrine, and sebaceous glands, as well as components of pilosebaceous units.

The four elements of an epidermal nevus are:

(1) Epidermal orthohyperkeratosis, irregular
(2) Hypergranulosis
(3) Papillomatosis
(4) Basal layer hyperpigmentation

No gender differences have been noted. Epidermal nevi tend to manifest between birth and adolescence, although they are usually asymptomatic. An exception is represented by the specific subtype called *inflammatory linear verrucous epidermal nevus* (ILVEN), which can appear and be accompanied by pruritus, erythema, and scaling. Six different epidermal nevus syndromes have been described:

- *Nevus sebaceous*: 50% of epidermal nevi, at birth with thickening and papillomatosis changes at puberty and after puberty potential association with basal cell carcinoma with CNS involvement.
- *CHILD syndrome*: This is an acronym composed by the letters meaning congenital hemidysplasia with ichthyosiform nevus and limb defect. It is X-linked dominant and manifests as unilateral inflammatory erythematous plaques with overlying yellow scale and ipsilateral limb hypoplasia.
- *Nevus comedonicus*: It represents a localized collection of dilated follicles containing keratin, often found on the face, trunk, and upper extremities. There is a potential occurrence of follicular tumors, EEG changes, skeletal abnormalities, and cataracts.
- *Becker nevus*: It is a focal area of hyperpigmentation with hypertrichosis of late childhood or adolescence with or without a pseudo-Darier sign when stroked. Becker nevus can be associated with skeletal abnormalities, hypoplasia of breast/musculature, and other congenital defects.
- *Proteus syndrome*: Syndrome characterized by a complex syndrome including overgrowth of multiple tissues with epidermal nevi.
- *Phacomatosis pigmentokeratotica*: Syndrome including speckled lentiginous nevi, hemiatrophy with muscle weakness, and epidermal

nevi, in association with other criteria. There is some practical more than the academic difference between hyperplastic scar and keloid, and this is quite important for the decision that needs to be taken from the family.

A nevus is often not resected completely if it is too large and scar formation may be found at the time of the second resection (Box 17.5).

However, much more important is not to miss the diagnosis of fibromatosis, which is not a tumor able to metastasize, but there is a quite apparent tendency to recur if the excision is not complete.

Fibromatosis
Round plump spindle cells in collagenous stroma with a tendency to hyalinize. See chapter of the soft tissue.

Dermatofibroma
See other soft tissue chapter.

17.17.1.1 Morphea
- *DEF*: Localized scleroderma without associated systemic involvement (no systemic sclerosis).
- *CLI*: There are white, indurated areas with a smooth surface with a natural ivory ± violaceous border and presented in four forms, including plaque, linear, segmental, and generalized, (+) ANA, (+) Scl70, ± CREST (anticentromere). Scl70 (topoisomerase 1) is a 100-kD nuclear and nucleolar enzyme and antibodies against Scl70 are considered to be specific for scleroderma and are detectable in up to 60% of patients with this connective tissue disorder. The acronym "CREST" refers to the predominant features of this disorder: *C*alcinosis, *R*aynaud phenomenon, *E*sophageal dysmotility, *S*clerodactyly, and *T*elangiectasia.
- *CLM*: Pronounced sclerosis of the dermis with hyalinization and epidermal sparing but extending into fibrous septa of deeper adipose tissue and accompanied by a superficial and deep perivascular lymphoid-plasma cellular infiltrate.

Particularly noteworthy is the identification of cutaneous pre-cancerous conditions, which may play an essential role in the routine of the pediatrician or medical internist other than a dermatologist.

17.17.2 Pseudoepitheliomatous Hyperplasia (PEH)

- *DEF*: Hyperplasia of the epithelium with no signs of malignancy, but associated with chronic insult/inflammation (e.g., at sites of trauma, chronic irritation, ulcers) and can be seen during fungal infections.
- *CLM*: Squamous epithelium with thin elongated anastomosing ridges showing maturation from basaloid cells to squamous cells, inflammatory cells, and dermal vascular proliferation.
- *DDX*: SqCC (no basaloid-squamous cell maturation) and fibroepithelioma of Pinkus.

17.17.3 Acanthoses/Acanthomas/Keratoses

Keratoses or acanthosis or acanthomas are all synonyms indicating proliferation of some part of the epidermis.

17.17.3.1 Acrochordons
- DEF: Fibroepithelial papilloma/polyp constituted by normal exuberant skin ("skin tag").

Box 17.5 Scar Versus Hyperplastic Scar Versus Keloid
- *Scar* = Usual collagenization of the connective tissue following a lesion with loss of continuity
- *Hyperplastic Scar* = Increase of thin collagen fibers following a lesion with loss of continuity
- *Keloid* (aka exuberant scar) = Predominant large intensely eosinophilic collagen fibers with very few scattered fibroblasts following a lesion with loss of continuity

17.17.3.2 Fibrokeratoma, Acquired, Digital

- DEF: Interphalangeal collagenous protrusion with the hyperkeratotic epidermis.

17.17.3.3 Lichenoid Keratosis

- DEF: Lichen planus-like solitary (differently from lichen planus), asymptomatic, slightly verrucous or scaly, bright red to brown lesion located on the sun-exposed skin with or without eosinophils or plasma cells and focal areas of parakeratosis and epithelial atypia.

17.17.3.4 Cutaneous Horn

- *DEF*: A protuberant column of keratin, usually in older adults, but also late childhood and youth with immunosuppression (Pointdujour-Lim et al. 2017).

17.17.3.5 Acanthosis Nigricans

- *DEF*: Dark, velvety discoloration of the skin in body folds and creases that occurs in both children (4/5 of cases) and adults (1/5) and constituted by hyperkeratosis and papillomatosis. This lesion may point to underlying malignancy, commonly gastric carcinoma, in adults.

17.17.3.6 Leukoplakia

- *DEF*: Nonspecific skin lesion constituted by hyperkeratosis, irregular acanthosis, hyalinization of the papillary dermis, and mild chronic inflammation with or without cell atypia.

17.17.3.7 Seborrheic Keratosis

- *DEF*: Sharply demarcated, round, tan to black plaque, which occurs on the face, arms, and upper trunk of middle-aged to elderly (aka basal cell papilloma) and characterized by full-thickness proliferation of keratinocytes (acanthosis) with hyperkeratosis, horn cysts of orthokeratin type (different from the parakeratin type of horn cysts or squamous pearls seen in SqCC), thick basal cell layer, and "flat bottom." Variants include keratotic (papillomatous), adenoid, acanthotic, and inverted follicular. A case of seborrheic keratosis has been described in a 1-years-old infant presenting with seborrheic keratosis in the external auditory canal. The patient was treated successfully with surgical excision (Ozbay et al. 2012).
- Leser-Trelat sign: Metrical and numerical increase of seborrheic keratosis at sudden onset and associated with internal malignancy.

17.17.3.8 Actinic (Solar) Keratosis

- *DEF*: Yellow-brown keratosis (scaly lesion) on sun-exposed skin showing keratinocyte dysplasia (especially lower 1/3 of the epidermis) with a small proportion progression to SqCC and constituted by:
  (1) Parakeratotic scale with alternating parakeratotic/orthokeratosis horn and agranulosis
  (2) Basal cell proliferation, focally budding into the dermis
  (3) Basophilic degeneration of collagen ("actinic elastosis") with chronic inflammation of the papillary dermis
- Variants: hyperplastic, atrophic, Bowenoid, and bowenoid actinic keratosis (BAK), which is similar to, but smaller than, Bowen disease (see below, 17.5.1). BAK occurs on the sun-exposed skin and presents without involvement of the full thickness of the skin and adnexal structures, unlike Bowen disease that affects indeed the adnexal structures. Xeroderma pigmentosum (XP) is a rare AR genodermatosis associated with hypersensitivity to UV radiation (UVR). It is due to defects involving the nucleotide excision repair pathway and XP patients are prone to develop multiple cutaneous neoplasms including non-melanoma skin cancers and melanoma. Collision tumors have also been described in XP patients and include actinic keratosis, BCC, SqCC, and in situ melanoma (Bostanci et al. 2020).
- *PGN*: A small proportion of actinic keratosis progresses to invasive SqCC.

17.17.4 Keratinocyte Dysplasia

- *DEF*: It may be defined as intraepidermal nests or cords of enlarged atypical keratinocytes with an unaffected underlying basal layer.

17.17.5 Intraepidermal Carcinomas

Three most essential lesions need to be discussed and differentiated. These include Bowen disease, erythroplasia of Queyrat, and Bowenoid papulosis. The latter condition is different from the Bowenoid actinic keratosis that has been described as a variant of the actinic keratosis (*vide supra* under 17.3.8). In the setting of *in situ carcinoma*, it should also be mentioned the cutaneous Paget disease. Moreover, neoplastic transformation of nevus sebaceous of Jadassohn has been described in the literature and has been considered an *in situ carcinoma* form from some authors.

17.17.5.1 Bowen Disease
- *DEF*: Squamous cell carcinoma in situ with full-thickness dysplasia ± As-Hx. exposure.
- *EPI*: Middle-aged to elderly, but younger in specialized settings (e.g., immunosuppression), ♂ > ♀.
- *CLI*: Scaly, erythematous sharply demarcated papula showing enlargement slowly and occurring anywhere on mucocutaneous sites, but devoid of adnexae.
- *CLM*: Full-thickness dysplasia of the squamous epithelium showing nuclear hyperchromasia, moderate to marked cytoplasmic vacuolization, multinucleated keratinocytes, dyskeratosis, parakeratosis and increased MI (Ki67 score).
- *PGN*: 1/20 ⇒ invasive SqCC.

17.17.5.2 Erythroplasia of Queyrat
- *DEF*: Squamous cell carcinoma in situ with full-thickness dysplasia of the glans penis of uncircumcised men and presenting as asymptomatic, erythematous, sharply defined papula.

17.17.5.3 Bowenoid Papulosis
- *DVEF*: Bowen disease-like multiple erythematous lesions of uncircumcised glans penis/vulva, rapidly developing of youth (20–40 years) showing similarities to Bowen disease, but more orderly cellular background with the presence of some noticeable maturation of the surface cells.

17.17.6 Keratoacanthoma

- *DEF*: "Benign" skin tumor of pilosebaceous origin on sun-exposed skin of the face, back of hands, wrists, and forearms with male predominance (♂ : ♀ = 3:1), showing rapid growth and usually spontaneous regression (resolution).
- *CLI*: Round, red nodule with characteristic central cup-shaped crater filled with keratin.
- *CLM*: Symmetrical and signs of maturation with well-differentiated squamous cells ± perineural invasion, which may be a potential pitfall.
- *IHC*: (+) Anti-DSG1, (+) Anti-DSG2, (+) E-Cad.
- *TNT*: Full excision with clear margins.
- *PGN*: Typically, good outcome following full excision.

Keratoacanthoma in children and youth can arise on nevus sebaceous and in patients with *xeroderma pigmentosum* (XP), which is a rare AR inherited disorder caused by defective DNA repair in various cutaneous and ocular cells following exposure to sunlight. XP occurs in about 1/65,000–1/100,000, with 3/4 of patients diagnosed before age 4.

17.17.7 Malignant Tumors

Probably it is quite challenging to think about malignant tumors of the epidermis in childhood and youth, but they indeed occur. The reasons for an early appearance of this unusual tumor in childhood and adolescence may be linked to an abnormal genetic status or abnormal immunologic surveillance as well as the increase of environmental factors, sunbed culture, and inappropriate solar protection as well as nutrition, which may be not healthy. It is known that "*loci of minoris resistentiae*" may be skin exposed inadequately to sun, to thermal burns, to chemicals, and to PUVA (renal transplantation). PUVA is a combination treatment consisting of Psoralens (P) and then exposing the affected skin to UVA (long wave ultraviolet radiation). Chemicals may induce skin cancer in a wide variety of experimental animals and humans as well. The most common benign experimental tumors are papillomas and keratoacanthomas (*vide supra*), while common malignancies are carcinomas and melanomas. The carcinogen-induced deoxyribonucleic acid damages lead to a genetic change. The most common carcinogens are polycyclic aromatic hydrocarbons, alkylating agents, and nitrosamines. Historically, the *Harvey ras* gene is the prototype of genes involved in the initiation stage, which is a gene involved in epidermal proliferation. In the second stage or tumor promotion, there is a repeated exposure of the initiated skin. Non-carcinogenic promoting agents include phorbol esters, benzoyl peroxide, anthralin, or some halogenated aromatic hydrocarbons, which create a tissue environment that is permissive to the selective clonal outgrowth of the initiated cell population. In the third stage of carcinogenesis, premalignant cells go further to progress to frank malignancy. This third stage may occur either spontaneously or following exposure to mutagens that also include several initiating agents. One of the problems of our civilization is indeed the exposure to some poisons that are invisible for the most of cases. Arsenic (As) exposure may be acute or chronic, and this latter may lead to benign skin changes, skin malignancies, and internal cancers. High levels of As in drinking water have been identified in various studies of epidemiology from all over the world constituting a significant problem in regions of the world with poor environmental sensitivity. In New Zealand, an area with a high risk for overexposure to UVA, it is estimated that about 50,000 new cases of nonmelanoma skin cancer occur each year. Chronic skin disease, such as discoid lupus erythematosus, may also be a predisposing factor for the development of squamous cell carcinoma of the epidermis. *Carcinoma duplex* is a term used to define the combination of a basal cell carcinoma with a squamous cell carcinoma.

17.17.7.1 Invasive Squamous Cell Carcinoma

- *DEF*: Squamous cell-differentiated malignant tumor of the epidermis.
- *SYN*: Epidermoid carcinoma.
- *EPI*: 2nd most common cutaneous malignancy following basal cell carcinoma (BCC).
- *GRO*: Protuberant skin lesion with or without ulceration.
- *CLM*: There is an invasive solid growth of cells, which show compactness (Box 17.6).
- Variants: Well-differentiated, spindle cell, acantholytic (aka adenoid or pseudoglandular), verrucous, and clear cell.
- Broder classification: differentiation and keratinization entail the following categories:
  (1) >75% of keratinization present ⇒ well-differentiated SqCC
  (2) 25–75% of keratinization present ⇒ moderately differentiated SqCC
  (3) <25% of keratinization present ⇒ poorly differentiated SqCC
  (4) No keratinization ⇒ undifferentiated carcinoma ("SqCC")
- *TEM*: Prominent desmosomes in grade 1 but decreasing in grades 2–3.
- *DDX*: The following diagnoses need to be taken into consideration:
  (1) Pseudoepitheliomatous hyperplasia
  (2) Inverted follicular keratosis
  (3) Metatypical basal cell carcinoma (BCC)
  (4) Melanoma, spindle cell type
  (5) Pilar tumor, proliferating type
- *TNT*: Excision with wide margins ± adjuvant radiochemotherapy.
- *PGN*: The outcome of the SqCC is related to the depth of invasion and tumor thickness, being tumors with diameter < 2 cm having less 1/20 probability of having metastases to regional lymph nodes. High-risk features include depth/invasion (>4 mm of thickness, Clark level IV, LVI (+) and perineural tissue), anatomy (ear and non-glabrous lip), and, of course, the degree of differentiation (G3-G4 > G1-G2).

Box 17.6 Low-Grade (LG) Versus High-Grade (HG) SqCC

- LG is a SqCC, which does not invade beyond the level of the eccrine glands.
- HG is a SqCC, which invades beyond the level of the eccrine glands.

Neoplastic transformation of nevus sebaceous of Jadassohn (NSJ) is rare but has been reported in an 11-year-old girl (Belhadjali et al. 2009) and in a 15-year-old girl (Hidvegi et al. 2003), and altogether only five well-documented cases of SqCC arising in an NSJ have been described, thus pointing to a rare, but potential occurrence of SqCC transformation of NSJ.

17.17.7.2 Basal Cell Carcinoma

- *DEF*: Most common skin cancer on the mainly sun-exposed skin and proportionally to the number of pilosebaceous units, characterized by the slow and indolent growth of synchronous or metachronous appearance.
- *SYN*: "Basalioma" (this term should not be used because it may indicate a benign nature, instead of the true malignancy behind basal cell carcinoma. However, this term is still in use in some countries).
- *EPI*: Worldwide, ♂ > ♀, in childhood and youth, mainly seen in transplant patients.
- *CLI*: Papule or nodule with telangiectasias, which may be eroded or ulcerated (*ulcus rodens*), and pigmented BCC may simulate melanocytic lesions.
- *CLM*: The neoplastic proliferation of basal cells with peripheral palisading, characteristic separation artifact with the surrounding tissue, stromal mucin, and incomplete differentiation toward adnexal structures, particularly pilar adnexae.
- Variants include baso-squamous, clear cell, diffuse sclerosing (morphea-type), localized, cystic/ulcerative, and superficial.
- *IHC*: Bcl-2 (+), CD10 (+), Ki67 > 20% (unrelated to PGN).
- (−) CK20, (−) CD15, (−) CD34, (−) EMA, (−) CEA, but (+) stromelysin-3 (stroma).
- *DDX*: SqCC, which is usually CD10 (−).
- *PGN*: Rare metastases (2/10,000), but if (+) ⇒ fatal within 1 year. Incompletely excised BCC does recur in 1/3 of cases.

In addition to the above-described variants, two more conditions need some words, particularly fibroepithelial tumor of Pinkus and Gorlin syndrome or basal cell nevus syndrome (BCNS). Fibroepithelial tumor of Pinkus is a "BCC" of the back, abdomen, and thigh with a "skin tag"-like shape and constituted by 2–3 cell thick basaloid epithelial strands, which peculiarly anastomize to divide the fibrous stroma in compartments. BCCS or Gorlin syndrome is an AD inherited genetic disease characterized by multiple BCCs starting when the patients are quite young and accompanied by jaw cysts, bony abnormalities, skull anomalies (e.g., *Falx cerebri* calcification), and palmar pits. An important differential diagnosis for the pathologist is distinguishing BCC from *trichoepithelioma*, which shows lobules of basaloid cells disconnected from the epidermis (Box 17.7).

Box 17.7 BCC Versus Trichoepithelioma

- (−) "Follicle-like structures"/(+) "follicle-like structures"
- (+) "Separating cleft"/(−) "separating cleft"
- (+) Programmed cell death/(−) programmed cell death
- (+) Melanin pigmentation/(−) melanin pigmentation

Notes: Follicle-like structures are abortive hair-papillae. Moreover, trichoepithelioma is (−) Bcl-2 and (+) CD34 in the stroma. In addition, there are more "tricho"-features which are absent or not well represented in BCC, including well-delimited organoid architecture (no infiltrative growth pattern), unfolding of the stroma into the nests or cell clusters (e.g., "papillary mesenchymal bodies,") eosinophilic collagenous cuticle surrounding the cell clusters, distinct nucleoli, and IHC. The IHC panel required to differentiate between the two tumors is CD10, CK15, CK20, and D2–40 (all four leading to trichoepithelioma) and low MI (Ki67 20%), (±) Bcl-2, which may show some expression limited to peripheral basal cells, unlike the usual diffuse staining pattern as seen in BCC.

17.17.7.3 Paget Disease

- *DEF*: Mammary Paget is associated with underlying carcinoma, and the most common extramammary sites are perineum (vulva, penis, scrotum, anus), axillae, umbilicus, and eyelids.
- *CLM*: Paget cells are dispersed singly or in clusters at an epidermal location with PAS (+), eosinophilic cytoplasm, and vesicular nuclei and show a characteristic IHC phenotype: CK7 is very useful because it is not expressed by the surrounding epidermal cells. However, normal Toker cells and Merkel cells are positive for CK7 and must be distinguished morphologically (atypia). GCDFP15 (BRST2) and CK20 are useful for separating primary from secondary Paget disease, being GCDFP15 (+) in primary Paget disease (GCDFP15 is low expressed in secondary Paget) and CK20 (+) in secondary Paget disease (CK20 is low expressed in primary Paget). The nature of the underlying carcinoma will also dictate the phenotype in secondary Paget disease. Most cases with underlying rectal adenocarcinoma are CK7+ and CK20+, albeit 90% of rectal adenocarcinomas are CK7– and CK20+. Since S100 can be expressed by Paget cells, MelanA should be used in the differential diagnosis with melanoma.

17.18 Melanocytic Lesions

The diagnosis of CNS anomalies goes further the general implications for the parental decision to carry on pregnancy or terminate it. Parental expectations are important, but long-term outcomes of birth defects affecting the CNS are vital to be able to provide the neonatology and pediatric team with appropriate information. The CNS is highly susceptible to damage.

Three cell types need to be kept separate:

- *Melanocytes*
- *Melanoblasts*
- *Melanophages*

Melanocytes are neuroectodermally derived cells with melanin granules that can be detected using Fontana-Masson stain, but also other silver stains, and immunohistochemically using antibodies against S-100 and VIM and are positive using DOPA reaction. Melanocytes are negative for antibodies against keratins, neurofilaments, and GFAP. These findings are useful for the DDX of carcinoma, neuroblastoma, NET, and glial tumors. The immature forms are called *melanoblasts*, while *melanophages* are macrophages (anti-CD68+), which have phagocytosed melanin pigment. According to the location of the melanin-containing cells, there is a different color by inspection of the skin. In the epidermis, the color is brown, and in the superficial dermis, the nevus will assume a black color, while nevus cells located in the deep dermis will give a blue nevus.

17.18.1 Lentigines, Solar Lentigo, Lentigo Simplex, and Melanotic Macules (Box 17.8)

17.18.2 Melanocytic Nevi

DEF: Any localized benign abnormality of melanocytes with nesting (theques) with initial occur-

---

Box 17.8 Five Benign Disorders of Pigmentation

1. *Albinism*: ↓ Melanin (hypopigmentation), melanocytes present, but not melanin-producing
2. *Freckles* (aka *Ephelis*): ↑ Melanin deposition, melanocyte hypertrophy with normal number of melanocytes, sunlight-reactive, 1–10 mm, irregular borders
3. *Melasma gravidarum*: ↑ Melanin deposition, melanocyte hypertrophy with normal number, pregnancy-associated (often)
4. *Vitiligo*: ↓ Melanin (hypopigmentation), absence of melanocytes
5. *Lentigo simplex*: Melanocyte hyperplasia and hyperpigmentation, but no melanocyte nesting, sunlight-unresponsive, 5–10 mm, well-demarcated borders, common in children

rence at 2–6 years of age until age 20 (Box 17.9) (Figs. 17.7 and 17.8).

17.18.3 Variants of Melanocytic Nevi

Melanocytic nevi variants include the following:

Blue Nevus
Deep located melanocytic nevus with a band of uninvolved dermis between the epidermis and

Box 17.9 Melanocytic Nevi

	Junctional	Intradermal	Compound
Gross	Flat/slightly elevated	Papillary, pedunculated, flat	J + ID features
		Brown-colored	~ Pigmentation
Nesting	DEJ	Intradermal	DEJ + Intradermal
Risk	MM	Nil	MM

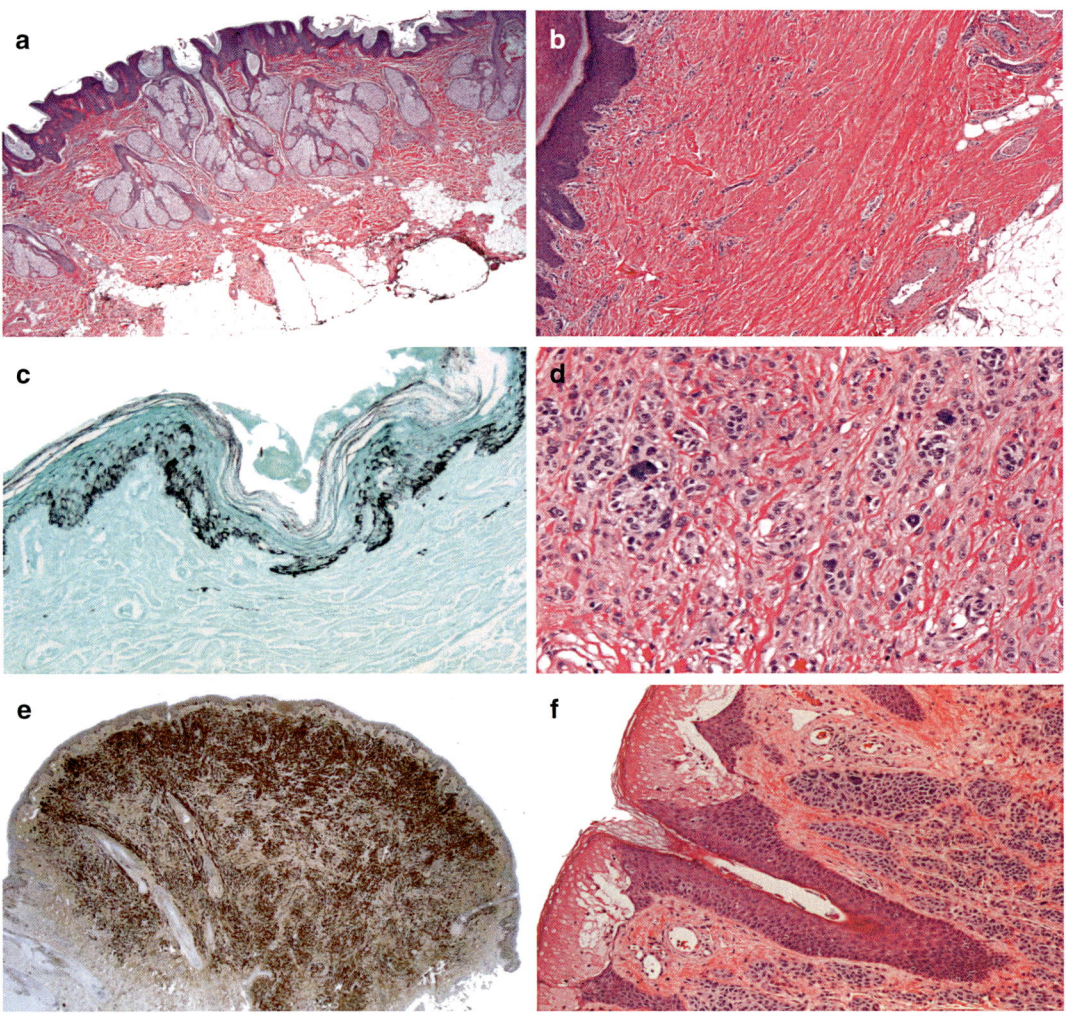

Fig. 17.7 This dermatopathology panel illustrates the first panel of non-melanocytic and melanocytic nevi and melanoma with a nevus sebaceous of Jadassohn in (**a**) (H&E stain, 12.5×), a connective tissue nevus in (**b**) (H&E stain, 50×), hyperplasia of nevus cells in (**c**) (Fontana Masson stain, 50×), an intradermal nevus with melanocytes single and in cluster in the papillary dermis in (**d**) (H&E stain, 200×), an intradermal nevus in (**e**) (antibody anti-S100 immunostain, 50×), and excoriated intradermal nevus in (**f**) (H&E stain, 100×)

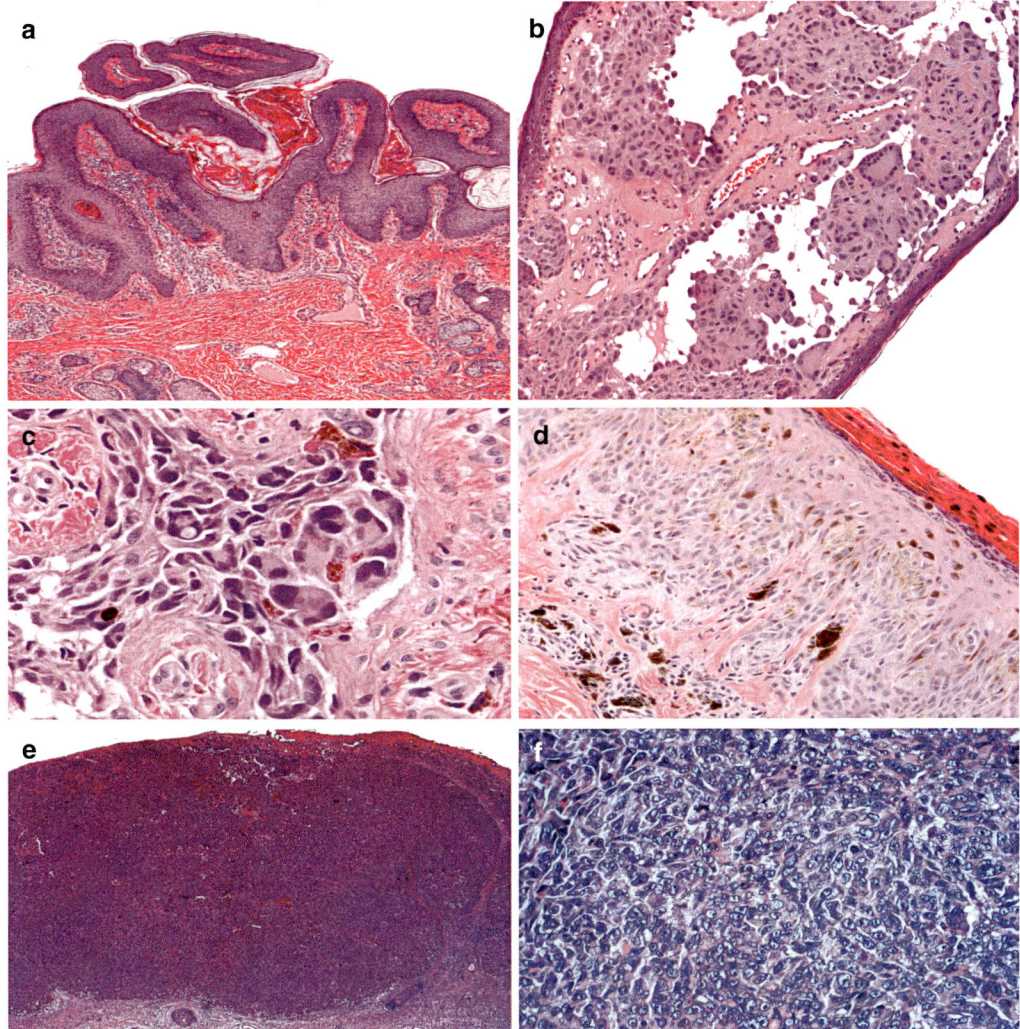

Fig. 17.8 The second panel of nevi and melanocytic lesions shows an ILVEN in (**a**) (H&R stain, 50×), a cavitated intradermal epithelioid Spitz nevus in (**b**) (H&R stain, 100×), a Spitz nevus compound with atypical features in (**c**) (H&R stain, 400×), a Reed nevus or pigmented spindle cell nevus of Reed in (**d**) (H&R stain, 50×), a malignant melanoma with atypical epithelioid nevus cells with nodular architecture and cytologically a vesicular nucleus with prominence of nucleoli in (**e** and **f**) (H&E stain for both (**e**) and (**f**) with (**e**) taken at 50× and (**f**) taken at 400×)

lesion usually situated at H&N or upper limbs with natural abundant melanin pigment.

Unusual forms of blue nevus are the following lesions:

- Mongolian spot: Sacral area
- Nevus of Ota: Melanocytic blue nevus (*fusoceruleus*) at ophthalmic-maxillary site
- Nevus of Ito: Melanocytic blue nevus (*fusoceruleus*) at acromion-deltoideus site

Cellular blue nevus: Deep melanocytic nevus with a band of uninvolved dermis between the dermis and the lesion, which is usually located at the buttock, lower back, and back of hands or feet (3 "b": buttock, back, and back of hands and feet). It shows high cellularity, pushing margins, dermal sclerosis, and elongated dendritic process with fine melanin granules without junctional activity or inflammation (benign, 1/4 of cases: congenital).

Deep penetrating nevus: Melanocytic nevus, usually located on the face and upper trunk, showing nevus cell nesting with scattered melanocytes filling the dermis and tendency to infiltrate the *erector pili* muscles as well as through the dermis into the subcutis (favorable outcome, 10–30 years, not prone to recurrence or metastases).

Congenital nevus (non-giant): Melanocytic nevus with the involvement of reticular dermis, subcutaneous tissue, and skin adnexa (benign, not prone to recurrence or metastases).

Giant congenital (pigmented) nevus (GCPN): Melanocytic nevus with the involvement of the reticular dermis, subcutis, and skin adnexa and specific bathing-trunk distribution and large, coarse skin with folds and satellite lesions (benign, but requiring multiple surgeries → scar tissue in 1+ resections).

Halo nevus: Melanocytic nevus showing non-pigmented skin secondary to regression.

Balloon nevus: Melanocytic nevus showing large foamy macrophages.

Bulky perineal neurocytoma: Uncommon congenital variant of GCPN showing innate features (*vide supra*) with or without cystic structures lined by nevus cells and large numbers of small bodies (Messner bodies-like), S100+, HMB-45+ (cave), usually large, polypoid or pedunculated and hyperpigmented located in the genital area, often obscuring the genitals (aggressive histology, but rare malignant transformation <6 months of age).

17.18.4 Spitz Nevus and Variants

- *DEF*: Particularly worrying melanocytic lesion by inspection and by the first glance on histology (aka "spindle cell and/or epithelioid cell nevus" or "juvenile melanoma"). Half of the cases occur before puberty and clinically as pink-red papule on the face and almost always with good fashion (Internal 2nd opinion or external consultation are usually required from internal policies or pediatric dermatologists). The term of "Reed nevus" is used for a pigmented spindle cell nevus.
- *EPG*: MM genes *CDKN2A*, inactivating mutations (tumor suppressor protein p16), and *BRAF* gene, activating mutations (mostly V600E) → good! (utility of BRAF inhibitors, personalized medicine!), *NRAS*, *KIT*, and *GNAQ*.
- *CLM*: Hypopigmented symmetrical melanocytic nevus (compound, 70% of cases; intradermal, 20% of cases; and junctional, 10% of cases) with spindle and epithelioid cytologic features, epidermal scatter of nests and single melanocytes, intraepidermal hyaline bodies (Kamino bodies), sharp lateral demarcation, flat bottom, and maturing tendency of nevus cells, telangiectasia, lymphocytic infiltrates, and overlying epidermal hyperplasia. The deep maturation of nevus cells is an essential diagnostic histologic feature. As the nevus cells extend more deeply into the dermis, or even into the subcutis, acquire a more neural looking. The deep penetrating nevus cells should show characteristics of maturation by both getting smaller and looking much more like "ordinary" nevus cells of classic melanocytic nevi and simultaneously displaying a tendency to arrange in small aggregates.
- MM in dark-skinned individuals is mostly located in acral and mucosal sites!
- MM distinguishing features (different from nevi with small and uniform melanocytic nests):
  - Bridging of nests (confluent growth) + pagetoid spread (single melanocytes in the epidermis) + epidermal thinning and consumption (loss of rete ridges with attenuation of basal and suprabasal epidermal layers) + diffuse, sheet-like growth pattern with lack of maturation of the invasive dermal component. The Breslow index and Clark levels are used to determining the depth of invasion. The Breslow index is the determination of the degree of MM infiltration measuring depth in mm from overlying stratum granulosum or ulcer bed to a most in-depth extension in the primary tumor. The central thickness (Breslow thickness index) is compared with the histologic most in-depth infiltration (Clark histologic levels).
- *DDX*: MM, nodular type, intradermal melanocytic nevus, Spitz nevus, clear cell sarcoma, DFSP. Spitzoid (malignant) melanoma. Characteristics of MM are present and allow

the distinction from Spitz nevi in most cases. They include asymmetry, irregularity of lateral borders, uneven base, pagetoid scatter of melanocytes ("malignant scatter") with the prominent epidermal spread of melanocytes, uneven nesting, lack of maturation, the absence of perpendicular direction of spindle cell to the surface, and ulcerated surface ± severe cytologic atypia.

- *TRT*: Resection with adequate clear margins ± chemotherapy.
- *PGN*: Spitz nevus is a benign lesion. However, very few atypical Spitz tumors may progress to melanoma. If the atypical Spitz tumors are not removed, it swiftly can lead to dramatic health risks. If an atypical Spitz tumor becomes malignant, it is often called "spitzoid melanoma."

17.18.5 Atypical Melanocytic (Dysplastic) Nevi

DEF: Melanocytic nevi showing some characteristic "ABCDE" features, including size >0.5 cm, border irregularity, variegated appearance, and "pebbly" surface and considered precursor for MM. The "ABCDE" rule includes Asymmetry, Border irregularity, Color not uniform, Diameter >6 mm, and Evolving size, shape or color. Microscopically, atypical melanocytes (nuclear hyperchromasia, prominent nucleoli, dusty melanin pigment), lymphocytic infiltrate in the papillary dermis, bridging of rete ridges by fusion of melanocytic nests, intradermal extending junctional component, dermal fibroplasia with eosinophilic concentric, or lamellar fiber deposition. Activating mutations in *CDKN2A, CDK4, NRAS, and BRAF* have been described associated with atypical melanocytic (dysplastic) nevi!

17.18.6 Malignant Melanoma and Variants

Malignant melanoma does occur in childhood and youth and the misconception that a juvenile melanoma harbors a better outcome than melanoma in adulthood has persevered for many years. The same name "juvenile melanoma" was sometimes used for Spitz tumor.

MM in situ → Lentigo maligna (aka "Hutchinson freckle"): Sun-exposed skin areas (e.g., cheek), tan to black, slow-growing melanocytic lesion of elderly with individual ± nesting with confluence of atypical junctional activity of melanocytes with epidermal atrophy ("consumption of epidermis"), solar elastosis, and no or minimal transepidermal migration or adnexal involvement. PGN: Invasion in 30–40% of cases.

Malignant melanoma (invasive) → MM with invasion (transepidermal migration): Melanocytic lesions with significant propensity to occur in sunlight exposed skin areas, red-haired whites, 1/5 of MM originating from pre-existing melanocytic nevus with a common junctional component (if intradermal, MM is probably metastatic).

Predisposing factors: Inherited genes (*vide infra*) + sunlight exposure.

Melanin (special) stains: Fontana-Masson, DOPA-oxidase, and Schmorl stains. The DOPA-oxidase reaction involves the conversion of L-dopa (L-3,4-dihydroxyphenylalanine) to dopachrome (red).

IHC markers include:

- S100 (nuclear and cytoplasmic): sensitive, but not specific!
- HMB-45 (cytoplasm): sensitive and specific, but (−) in the desmoplastic MM (D-MM).
- Melan A/MART1: sensitive, but not specific being (+) in the ovary, testis, and adrenal gland.
- Tyrosinase: sensitive, but also stains PNST and NET.
- MTF (microphthalmia transcription factor): sensitive, but also stains DF and SMT. MTF is negative in D-MM.

MM altered pathways include (1) RAS-RAF-MEK-ERK (MAPK: mitogen-activated protein kinase) with activating mutations of *NRAS*, (2) *p16-CDK4-RB*, (3) *ARF-TP53*, (4) *KIT* (activating mutations), (5) *PTEN* (epigenetically silenced), (6) *CDKN2A* gene mutations (three TSGs including p14, p15, and p16), *CMM1* gene mutations, MSI, and multiple chromosomal genes and losses.

MM-subtypes other than conventional type:
With "radial growth phase" (MM in situ with ≥3 rete ridges beyond the infiltrative component):

1. *Lentigo maligna melanoma* (LMM) (15% – all ages): Better PGN-harboring MM-subtype associated with sun exposure and elderly arising on *lentigo maligna* in hair follicles and sweat ducts with atypical cells on superficial dermis (macular→ papular→ nodular).
2. *Superficial spreading melanoma* (60% – all ages): Most common form on the trunk and legs and sun exposure-related with both macular and nodular appearance clinically, blue hue with admixed tan and brown (early stage) but white, pink, and blue when areas of regression are present.
3. *Acral-lentiginous melanoma* (10% – all ages): Intraepidermal MM mostly on palms, soles, nails, and genitals, particularly in dark-skinned individuals and characterized by the hyperplastic epidermis.

With "vertical growth phase" (≥1 dermal cluster more substantial than the most massive epidermal group and/or MI > 1 in the dermis):

1. *Nodular melanoma* (15% – all ages): Young age, raised lesion, short duration, symmetrical ± amelanotic, and no intraepidermal component.

Desmoplastic MM: MM-variant with fascicles of spindle cells, (+) S100, (+) TIL.

Breslow index and Clark levels are useful for the grading of MM.

Breslow index entails the tumor invasion/thickness (θ) and grades I if $\theta \leq 0.75$ mm; II, if θ, 0.75–1.5 mm; III if θ, 1.5–2.25 mm; IV if θ, 2.25–3 mm; and V if $\theta > 3$ mm.

Clark levels use the anatomic level of invasion:

- Level I: Intraepidermal localization (only!) = MM in situ
- Level II: Papillary dermis infiltration
- Level III: Papillary dermis (fully) filling and expansion (only!)
- Level IV: Reticular dermis infiltration
- Level V: Subcutis infiltration

DDX: Dysplastic nevi!

Dysplastic nevi – Features include some characteristics grouped with the acronym "BAJIF," including (B) bridging of nests, (A) atypia of melanocytes, (J) junctional shouldering, (I) inflammation, and (F) fibrosis of lamellar type.

The following features are useful to distinguish DYSPLASTIC NEVI from MM using the acronym "TAMM": (T) transepidermal migration of melanocytes, (A) angio-invasion of melanocytes, (M) mitotic activity, and (M) maturation of descending melanocytes.

PGN: See Box 17.10 for favorable prognostic factors.

Regression: It is essential to ponder that a primary lesion can spontaneously regress, while metastases do not. By clinical inspection, regression is white, often with a bluish hue. Microscopically, there are irregularities along the DEJ, focal necrotic cells, band-like lymphocytic infiltrate, melanophages, subepidermal fibrosis, and atrophy of rete ridges. The presence of regression is NOT a good prognostic sign, because it indicates that the lesion was once more abundant. The tumor regression distinguishing features include (1) dermal fibrosis of nonlaminated type with scattered inflammatory cells, (2) epidermal attenuation, (3) melano-phagocytosis, (4) replacement of tumor cells by lymphocytic infiltrate, and (5) telangiectasia.

Box 17.10 MM: Prognosis

Factors → *Favorable PGN*	
(1) Gender, age	♀, youth
(2) Location (central body vs. limbs)	Limbs
(3) Tumor depth (θ = thickness)	< 1.5 mm
(4) Tumor regression	(−)
(5) TILs (tumor infiltrating lymphocytes)	(+)
(6) MI	0/low

Malignant features include the "trailing off" phenomenon (i.e., a lateral extension of individual cells), ↓ nesting cohesiveness, adnexal epithelium extension, fine dusty melanin granules, ↑ MI, intratumoral lymphocytes, lack of maturation, large Ø, melanocyte necrosis, poorly circumscription, prominent nucleoli, and size/shape variation.

In Box 17.11, some pearls and pitfalls of MM are listed.

> **Box 17.11 Pearls and Pitfalls**
> - To distinguish between pT1a and pT1b is of primary importance because sentinel node examination should be performed on pT1b and above according to AJCC recommendations, 6th edition, TNM classification system.
> - The pN status is evaluated on the lymph nodes: <1 LN, between 2 and 3, ≥4.
> - Ulceration and MI are two critical staging factors and need to be evaluated carefully. MM-ulceration is defined as (1) full-thickness defect of the epidermis with the absence of corneal layer and BM, (2) reactive changes (i.e., fibrin exudation, neutrophils), and (3) changes of the surrounding epidermis, including thinning, effacement, or reactive hyperplasia in the absence of trauma or recent surgery.
> - Microsatellitosis is defined as tumor nests >50 μm in Ø located in the Breslow index tissue slide section either in the reticular dermis, panniculus subcutaneous, or blood vessels beneath the primary infiltrating tumor but separated from it by ≥300 μm.
> - In the case of sentinel node examination, the following advice should be followed:
>   – If LN <0.5 cm → embed the LN whole.
>   – If LN > 0.5 cm → bisect the LN and/or slice it at 2 mm intervals.

17.19 Sebaceous and Pilar Tumors

The diagnosis of sebaceous tumors is quite impressive because it is mainly linked to possible clinical manifestations.

17.19.1 Sebaceous Hyperplasia

Sebaceous hyperplasia is unusual in children and youth but mostly present on the face of older individuals as yellowish dome-shaped papule and constituted, microscopically, by hyperplastic glands in the upper dermis that drain into the central duct.

17.19.2 Nevus Sebaceous (of Jadassohn) (NSJ)

- *DEF*: Benign, congenital hamartoma of the follicular-sebaceous apocrine unit and epidermis presenting clinically as a single, round, yellow melanocytic, pebble-like, papule or plaque located on H&N with a warty tendency (aka organoid nevus).
- *CLM*: Hamartomatous lesion composed of large sebaceous glands, heterotopic apocrine glands, defective hair follicles, acanthosis, and epithelial papillomatosis and hyperkeratosis. The nevus sebaceous of Jadassohn often presents at birth, appears to regress in childhood, and grows during puberty, pointing to possible hormonal control. NSJ may develop syringocystadenoma papilliferum, benign hair follicle tumors, and BCC. The differential diagnosis includes sebaceous hyperplasia and seborrheic keratosis, both of them typical lesions of older patients.

As indicated above nevus sebaceous is a hamartoma with sebaceous, follicular, and apocrine gland in addition to epidermal abnormalities. Nevus sebaceous is well known as a background for the development of cutaneous neoplasms such as trichoblastoma, syringocystadenoma papilliferum, basal cell carcinoma and keratoacanthoma, and rarely squamous cell carcinoma.

Although most of these cutaneous neoplasms develop in patients older than 16 years of age, some are present in childhood and need to be identified at the time of diagnosis for proper follow-up from the pediatric or adult dermatologist. Since follicular, sebaceous, and apocrine elements have different distribution depending on the area, nevi sebacei of the face or scalp have different biological behaviors. In fact, nevus sebaceous of the scalp is more often complicated by tumors than nevus sebaceous of other localizations, although only the face is the site where seems to occur keratoacanthoma on nevus sebaceous. The neoplastic transformation of NSJ is extremely rare, but it has been reported as mentioned above.

17.19.3 Sebaceous Adenoma, Sebaceoma, and Xanthoma

Sebaceous adenoma is a rare benign epithelial tumor looking like a BCC clinically as yellow papule and showing, microscopically, individual lobules, collagen-formed pseudocapsule and peripheral small germinative cells with round-to-ovoid vesicular nuclei and eosinophilic cytoplasm. The term *"adenoma sebaceum,"* which occurs in tuberous sclerosis, is something else, i.e., is an angiofibroma.

Sebaceoma or sebaceous epithelioma is a tumor with focal sebaceous differentiation showing more or less mature sebaceous cells, which are mixed randomly with basophilic, immature epithelial cells.

Xanthoma is an accumulation of fat-laden macrophages in dermis or subcutis located on tendons, the synovium, or bone.

17.19.4 Sebaceous Carcinoma

Sebaceous carcinoma is not usually seen in children or young individuals, although familial cancer syndromes and individuals with congenital or acquired immune deficiencies may theoretically develop this kind of tumors.

- *DEF*: An aggressive periocular variant with a relatively less aggressive extraocular form showing an *irregular lobular pattern* in upper dermis with *basophilic sebaceous cells* containing small granular nuclei with eosinophilic nucleoli. Mebazaa et al. (2007) reported a recent sebaceous carcinoma in a 12-year-old male child on the eyebrow and reviewed the literature identifying seven more cases described in the English literature. Age varied from 3.5 to 16 years, and sites included upper eyelid or eyelid (4 cases), eyebrow (1), Meibomian gland (1), buccal mucosa (1), and thorax (1). Following surgical excision, recurrence was observed in one case only, although short follow-up times and no specific follow-up were found in the other cases.
- *DDX*: The differential diagnosis of poorly differentiated sebaceous carcinoma includes cutaneous clear cell tumors, chiefly squamous cell carcinoma with sebaceous differentiation, clear cell acanthoma, clear cell hidradenoma, clear cell hidradenocarcinoma, cutaneous clear cell carcinoma, tricholemmoma, pilar tumor, balloon cell melanoma, and metastasis from an RCC. The clear cell acanthoma is a well-demarcated tumor that may arise from intraepidermal eccrine ducts. It presents with a solitary (raised moist red) lesion on lower limbs of middle-aged to elderly, and histologically there is plenty of glycogen.

17.19.5 Benign Hair Follicle Tumors

In the hair follicle can develop tumors that may harbor the tendency to recurrence, if not completely excised (Fig. 17.9).

17.19.5.1 Pilar Sheath Acanthoma
Pilar sheath acanthoma (PSA) and dilated pore of Winer (DPW) are two related lesions, which belong to benign hair follicle tumors. PSA is a solitary tumor with *central pore with the keratinaceous material* in the upper lip and constituted by a *multiloculated lobulated neoplastic masses* radiating into surrounding dense stroma. DPW shows a more dilated pore than present in PSA showing an enlarged open follicular cystic cavity filled with keratin and irregularly proliferating rete ridges.

17.19 Sebaceous and Pilar Tumors

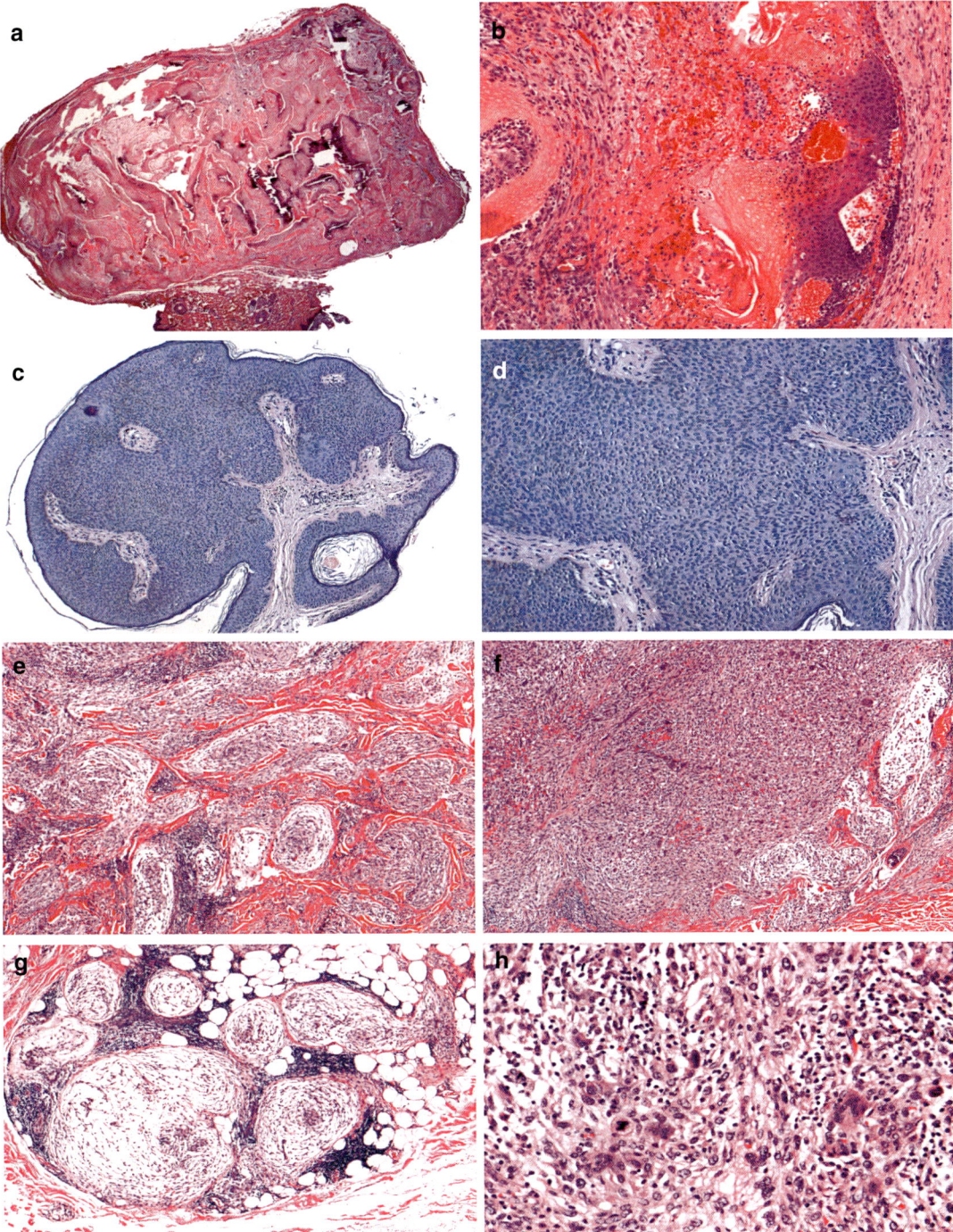

Fig. 17.9 This panel illustrates three non-melanocytic tumors in children and youth. In (**a**) and (**b**), a pilomatrixoma is shown. In (**c** and **d**), an eccrine poroma is shown, and in (**e**) through (**h**), a rare plexiform fibrohistiocytic tumor. The pilomatrixoma (**a**, H&E, 12.5×; **b**, H&E, 100×) shows a basaloid proliferation with similarity to hair matrix cells, which matures into structureless eosinophilic cells without nuclei (the so-called ghost cells). The eccrine poroma (**c**, H&E, 100×; **d**, H&E, 200×) shows a nodular proliferation of cords and nests of small keratinocytes attached to the epidermis, while the plexiform fibrohistiocytic tumor (**e**, H&E, 50×; **f**, H&E, 50×; **g**, H&E, 50×; **h**, H&E, 100×) shows a plexiform proliferation of fibrohistiocytic cells with minimal atypia, osteoclast-like giant cells, and a lymphocytic infiltrate

17.19.5.2 Trichilemmoma

Trichilemmoma may be associated to *Cowden syndrome* and is a small warty or smooth papule on the face of older individuals showing a proliferation of outer root sheath with uniform small cells with variably clear cytoplasm forming one or more sharp-delimited lobules arising from the epidermis and keeping *extensive* connections with the epidermis, peripheral palisading, and thickened BM.

17.19.5.3 Trichofolliculoma

Trichofolliculoma is a dome-shaped lesion with a central pore with silky white threat-like hairs and showing a dilated hair follicle with numerous secondary follicles of the mature and immature type with "*caput medusae*" appearance arising from its stratified squamous wall. Two characteristic features are the infundibular squamous epithelium with prominent granular layer and *tight* connections with the epidermis.

17.19.5.4 Trichoepithelioma

Trichoepithelioma is a small, flesh-colored cutaneous nodule showing lobules of basaloid cells *disconnected* from the epidermis with follicle-like structures, horn cysts, peripheral palisading without clefting, and cellular fibrotic stroma surrounding the lobules. Most importantly, the primary differential diagnosis of trichoepithelioma is BCC, which shows no follicle-like structures, but clefting around the peripheral palisading of the basaloid cells.

17.19.5.5 Pilomatrixoma

Pilomatrixoma (aka calcifying epithelioma of Malherbe) is a slowly growing, firm nodule with chalky consistency usually located on the face and showing a specific *biphasic cell population of basaloid cells and ghost cells* (hair matrix differentiation) and prominent tendency to *calcify* in at least ¾ of cases.

17.19.5.6 Piloleiomyoma

Piloleiomyoma is a small, firm, intradermal tumor constituted by several nodules which are located as groups or linear arrays on the trunk or limbs.

17.19.6 Malignant Hair Follicle Tumors

Malignant hair follicle tumors are very rare malignancies of the adnexa of the skin that have a preference for the head and neck region. These tumors can be listed into several different types. The reader should refer to texts of dermatopathology for this topic.

17.20 Sweat Gland Tumors

Adnexal cutaneous tumors can be differentiated toward the ductal and or/glandular portion of the eccrine or apocrine glands, although they are often seen as a mixture of patterns on the histological ground. Sweat gland with ductal differentiation show some glandular/duct structures and/or intracytoplasmic glandular lumens. Special stains, including PAS, and immunohistochemical stains, including EMA and CEA, may highlight true ducts and intracytoplasmic lumens, while apocrine differentiation may show a decapitation secretion pattern with ducts showing "apocrine snouts" (Box 17.12). These "snouts" represent

---

Box 17.12 Sweat Gland (Eccrine and Apocrine) Tumors

17.20.1 *Eccrine Gland Tumors*
    17.20.1.1 Cylindroma
    17.20.1.2 Syringoma
    17.20.1.3 Chondroid Syringoma
    17.20.1.4 Eccrine Poroma
    17.20.1.5 Eccrine Acrospiroma
    17.20.1.6 Eccrine Spiradenoma
    17.20.1.7 Eccrine Carcinoma
17.20.2 *Apocrine Gland Tumors*
    17.20.2.1 Syringocystadenoma Papilliferum (Apocrine Nevus)
    17.20.2.2 Apocrine Tubular Adenoma
    17.20.2.3 Hidradenoma

accumulations of secreted granules in the apical cytoplasm spilling into the lumen of the ducts.

Apocrine glands should be considered "mammary glands" or "mammary"-like glands of the skin, especially in the anogenital and axillary regions. Both eccrine and apocrine glands express EMA, CEA, and LMWCK (e.g., CAM5.2). Thus, these three markers are not used to differentiate between eccrine and apocrine differentiation. The expression of CK14 of the eccrine glands is characteristically seen in the and also due to acrosyringium, which is the spiraled intraepidermal ductal component.

17.20.1 Eccrine Gland Tumors

Eccrine gland tumors include cylindroma, syringoma, chondroid syringoma, eccrine poroma, eccrine acrospiroma, eccrine spiradenoma, and eccrine carcinoma.

17.20.1.1 Cylindroma
Cylindromas are benign adnexal tumors showing an eccrine and an apocrine differentiation and often as a single lesion. If multiple lesions ("turban tumors") are observed, cylindromas are frequently linked to the *Brooke-Spiegler syndrome*, an irregular AD syndrome with variable penetrance caused by mutations in the *CYLD* gene on 16q12-q13. Other adnexal tumors, such as trichoepitheliomas and spiradenomas have also been observed. Brooke-Spiegler syndrome (BSS), familial cylindromatosis (FC), and multiple familial trichoepitheliomas (MFT), initially described as distinct syndromes, share overlapping clinical features. *CYLD* inhibition increases resistance to apoptosis, leading to oncogenesis. Brooke-Spiegler syndrome manifests in adolescence. Cylindromas can reach considerable size, and malignant transformation has been described. In almost 100% of the cases, cylindromas are located on H&N (scalp) with a female preference, are slow-growing tumors and show, microscopically, *epidermal disconnected multiple lobules in a jigsaw pattern in the upper dermis*. In each lobule, a thick PAS + -BM, hyperchromatic basal cells, and central cells with vesicular nuclei are seen. No regional nodal involvement has been described for cylindromas, but the differential diagnosis includes AdCC. However, signs of malignant transformation include ulceration, bleeding, sudden growth, and color change.

17.20.1.2 Syringoma
Syringomas are multiple symmetrical small papules on lower eyelids with pubertal appearance and female preponderance. Microscopically, the acrosyringium is involved, and there are small cysts/bilayered ducts with comma tails (some of the tadpole shape) with lumina lined by cuticle and filled with eosinophilic granular material. Clear cell variant of syringoma may occur in the setting of DM.

17.20.1.3 Chondroid Syringoma
Chondroid syringoma (aka benign mixed tumor of the skin) is a firm, solitary, lobulated nodule in dermis or fat located on H&N with no symptoms. Two characteristic features are (1) well-delimitated, multilobulated mass with nests or cords of polygonal cells with basophilic nuclei and a significant amount of eosinophilic cytoplasm and (2) pseudo-cartilagineous/chondroid stroma.

17.20.1.4 Eccrine Poroma
- *DEF*: Eccrine glandular-arising benign adnexal cutaneous tumor (outer acrosyringium: upper dermal eccrine duct) showing "anastomosing ribbons."
- *EPI*: Worldwide, children and adults, ♂ = ♀.
- *CLI*: Solitary, scaly red nodule of soles at the sides of feet.
- *CLM*: Epidermis-replacing tumor with anastomosing ribbons, growing down into the dermis and sharp demarcated to the not affected skin and constituted by monomorphic, cuboidal cells and occasional duct differentiation.
- *DDX*: Both benign and malignant diagnoses should enter in the differential:
  – Benign DD include eccrine acrospiroma and clonal seborrheic keratosis.
  – Malignant DD include BCC, SqCC in situ, Paget disease, melanoma, clonal

seborrheic keratosis, acrospiroma, and porocarcinoma.
- *PGN*: Malignant eccrine poroma is a rare variant and shows marked nuclear pleomorphism.

17.20.1.5 Eccrine Acrospiroma
- *DEF*: Eccrine gland-arising adnexal cutaneous tumor with large intradermal lobules with ductal structures that are accompanied by cystic spaces and hyalinized stroma.
- *SYN*: Eccrine hidradenoma, clear cell hidradenoma.
- *EPI*: Worldwide, any age, ♂ < ♀.
- *CLI*: Papula/nodule of almost any location.
- *CLM*: Large intradermal (superficial dermis) lobules of monomorphic cuboidal cells with a basophilic cytoplasm mixed with clear cells showing ductal structures, cystic spaces, and hyalinized stroma (Fig. 17.10).

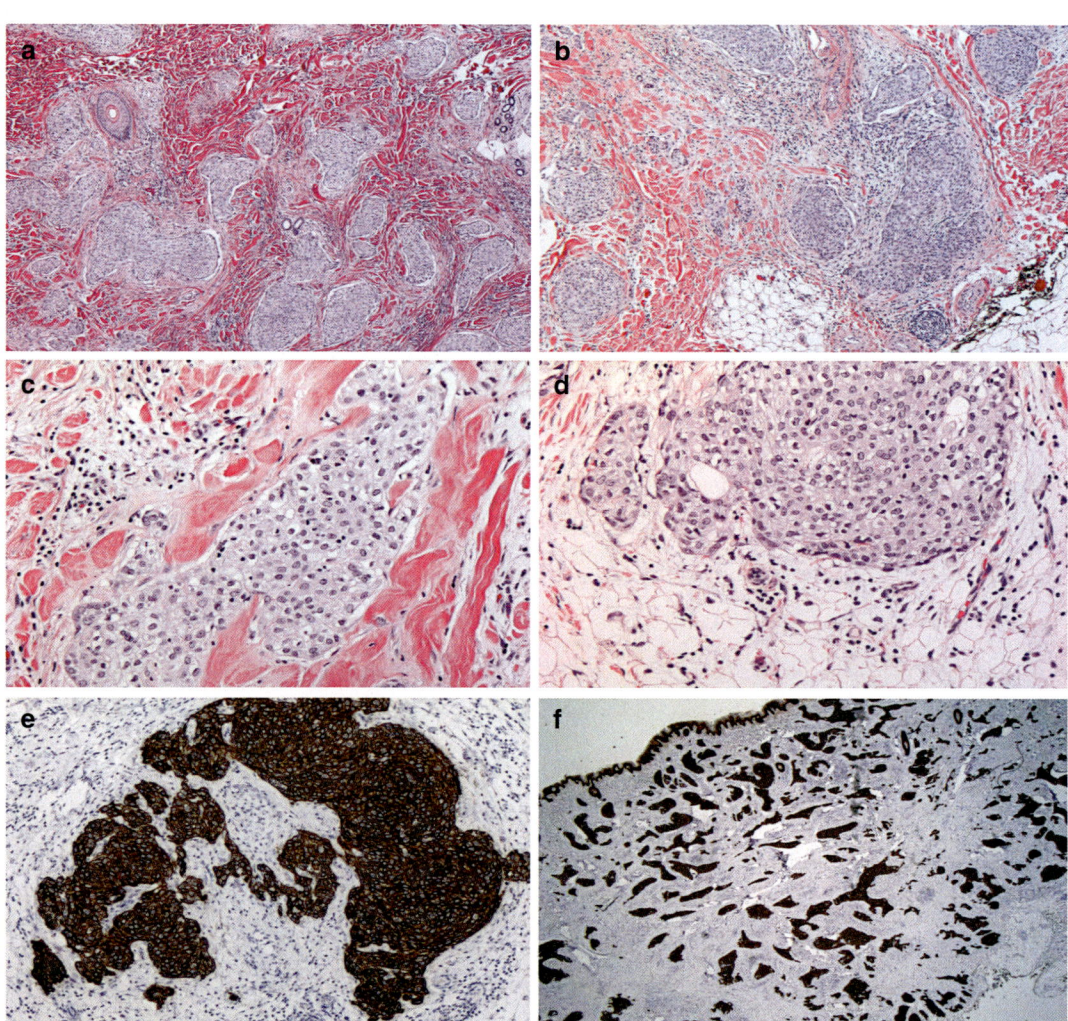

Fig. 17.10 This panel shows a malignant eccrine acrospiroma (MEAS) (**a**, H&E, 50×; **b**, H&E, 50×; **c**, H&E, 200×; **d**, H&E, 200×; **e**, Anti-CK5/6, 100×; **f**, Anti-CK5/6, 12.5×). The MEAS shows solid lobules of tumor cells located in the dermis and extending into the subcutaneous tissue without connection between the epidermis and the tumor. Some of the lobules contain small cystic spaces filled with homogenous, clear or eosinophilic material. Interstitial thick hyaline collagenization is seen around lobules. The cell population is made up of two cells types, clear cells and spindle cells

- *DDX*: Glomus tumor, apocrine-mixed carcinoma, and metastasis of carcinoma, mostly renal cell carcinoma.

17.20.1.6 Eccrine Spiradenoma

- *DEF*: Benign eccrine gland-arising cutaneous adnexal tumor with a characteristic organoid pattern of intradermal cannon blue balls enriched with hyaline droplets, blood vessels, and lymphocytes.
- *EPI*: Worldwide, youth, ♂ = ♀.
- *CLI*: Solitary, bluish or skin-colored, usually painful nodule on ventral surfaces of the upper body.
- *CLM*: Rounded intradermal lobules "cannon blue balls" without epidermal connection or jigsaw puzzle pattern, delimited by BM-material and hyaline droplets, rich vascular supply, and intralesional scattered lymphocytes.
- *DDX*: BCC, eccrine cylindroma, glomus tumor, other vascular tumors. Skin lesions presenting as painful nodules need to be taken into consideration (Box 17.13).
- *PGN*: Good.

Variants include the Ancell-Spiegler syndrome and the multifocal spiradenomatosis.

17.20.1.7 Eccrine Carcinoma

There are several variants, including classical, syringoid, microcystic adnexal, mucinous, and adenoid cystic variants, all of them including nuclear pleomorphism and more or less obvious features of malignancy.

17.20.2 Apocrine Gland Tumors

Apocrine gland tumors include *syringocystadenoma papilliferum*, apocrine tubular adenoma, and *hidradenoma papilliferum*.

17.20.2.1 Syringocystadenoma Papilliferum

- *DEF*: Benign apocrine glandular-arising cutaneous adnexal tumor with "EPIDERMAL cystic invagination, papillae with plasma cell rich fibrovascular core, and bilaminar apocrine epithelium with decapitation snouts."
- *EPI*: Worldwide, youth, ♂ = ♀.
- *CLI*: Solitary, gray to brown, moist nodule on the scalp.
- *CLM*: Epidermal cystic invagination with papillae showing a plasma cell-rich fibrovascular core and a bilayered lining of apocrine epithelium with decapitation secretion.
- *DDX*: Nevus sebaceus, hidradenoma papilliferum.
- *PGN*: Good.

17.20.2.2 Apocrine Tubular Adenoma

- *DEF*: Benign apocrine glandular-arising cutaneous adnexal tumor characterized by lobular masses of tubular structures and apocrine secretion pattern.
- *EPI*: Worldwide, youth, ♂ = ♀.
- *CLI*: Dome-shaped nodule on the scalp.
- *CLM*: Poorly circumscribed intradermal tumor with lobular masses of tubular structures showing columnar epithelium, apocrine secretion, and myoepithelial layer within foci of ductular communication with the epidermis.
- *DDX*: Papillary eccrine adenoma, hidradenoma papilliferum.
- *PGN*: Good.

17.20.2.3 Hidradenoma Papilliferum

- *DEF*: Benign apocrine glandular-arising cutaneous adnexal tumor with intradermal cyst

---

Box 17.13 Skin Lesions Presenting as Painful Nodules

- *Blue Rubber Bleb Nevus* (= asyndromic vascular nevus)
- *Cutaneous Tumors*: Dermatofibroma, eccrine spiradenoma, pilomatrixoma
- *Vascular Tumors*: Angioleiomyoma, angiolipoma, glomus tumor/glomangioma
- *Nerve Tumors*: Schwannoma, neurofibroma, neuroma (traumatic), granular cell tumor
- Endometriosis

intersected by papillary folding of cells with plasma cell-rich fibrovascular core and bilaminar epithelium of apocrine type with decapitation snouts.
- *EPI*: Worldwide, youth, ♂ < ♀.
- *CLI*: Solitary, gray to brown, dome-shaped nodule on the perineum and genital region.
- *CLM*: Intradermal cyst with papillae showing a plasma cell-rich fibrovascular core and a bilayer lining of apocrine epithelium with decapitation secretion.
- *DDX*: Syringocystadenoma papilliferum, apocrine tubular adenoma.
- *PGN*: Good.

17.21 Fibrous and Fibrohistiocytic Tumors

Fibrous and fibrohistiocytic tumors are listed in their complexities in textbooks of dermatopathology. Here, we will discuss the scar, dermatofibroma or benign fibrous histiocytoma (BFH), dermatofibrosarcoma protuberans (DFSP), and juvenile xanthogranuloma (JXG). Other entities are described in other chapters or dermatopathology books.

17.21.1 Hypertrophic Scar and Keloid

Both hypertrophic scar and keloid are the result of a cutaneous injury. Both hypertrophic scars and keloids are manifestations of the same fibroproliferative skin disorder, and they may differ in the intensity and duration of inflammation (*vide supra*).

17.21.2 Dermatofibroma

- *DEF:* It is a benign fibrous tumor of the *upper trunk or neck* of adolescents and young adults presenting as pale pink or hypopigmented plaque composed of bundles of myoid-appearing spindle cells in the reticular dermis (Fig. 17.11).
- *SYN*: Benign fibrous histiocytoma (BFH).
- *CLM*: Plaque-like growth pattern of fascicles of myoid-appearing spindle cells located in the reticular dermis with a Grenz zone sparing the superficial subepithelial dermis (DDX: Dermatofibroma showing a storiform pattern and "encapsulation" of single collagen bundles) and lacking a significant extension into fat and subcutaneous tissue or CD34 staining (DDX: DFSP).

17.21.3 Juvenile Xanthogranuloma

- *DEF*: Benign tumor of dendritic cell origin with potential spontaneous regression.
- *SYN*: Nevoxanthoendothelioma.
- *EPI*: Uncommon, infants, ♂ > ♀.
- *CLI*: The skin of the face or trunk but all sites can be affected. Rarely, non-cutaneous sites are involved (the subcutis, skeletal muscle, eye, peripheral nerve, testis, liver). Multiple lesions in 1/5 patients. There is a clinical association with glaucoma and amblyopia (iris and ciliary body involvement), NF-I, NPD, urticaria pigmentosa, and CMV infection.
- *GRO*: Yellow-red, papula/nodule.
- *CLM*: Dense poorly circumscribed, the lymphohistiocytic proliferation of dermis with foamy and Touton giant cells, short fascicles of spindle cells, fibrohistiocytic cells and fibrosis with the thin epidermis, elongated rete ridges, preservation of adnexa, lymphocytes, and eosinophils as well as no or scattered mitotic figures (Fig. 17.11).
- *IHC*: (+) CD68, A1ACT, lysozyme, VIM, FXIIIa, (−) S100, CD1a.
- *TEM*: Cytoplasmic lipid (+), but Birbeck granules (−).
- *DDX*: Epidermoid cyst (cyst of keratinized epithelium with distinct granular layer containing lamellated keratin), BFH (dense collagenous stroma, storiform growth pattern, pseudoepitheliomatous hyperplasia), xanthoma (tumor of uniform foamy histiocytes), LCH (nuclear grooves, S100+, CD1a+, Langerin+. Birbeck granules by TEM), lipoma

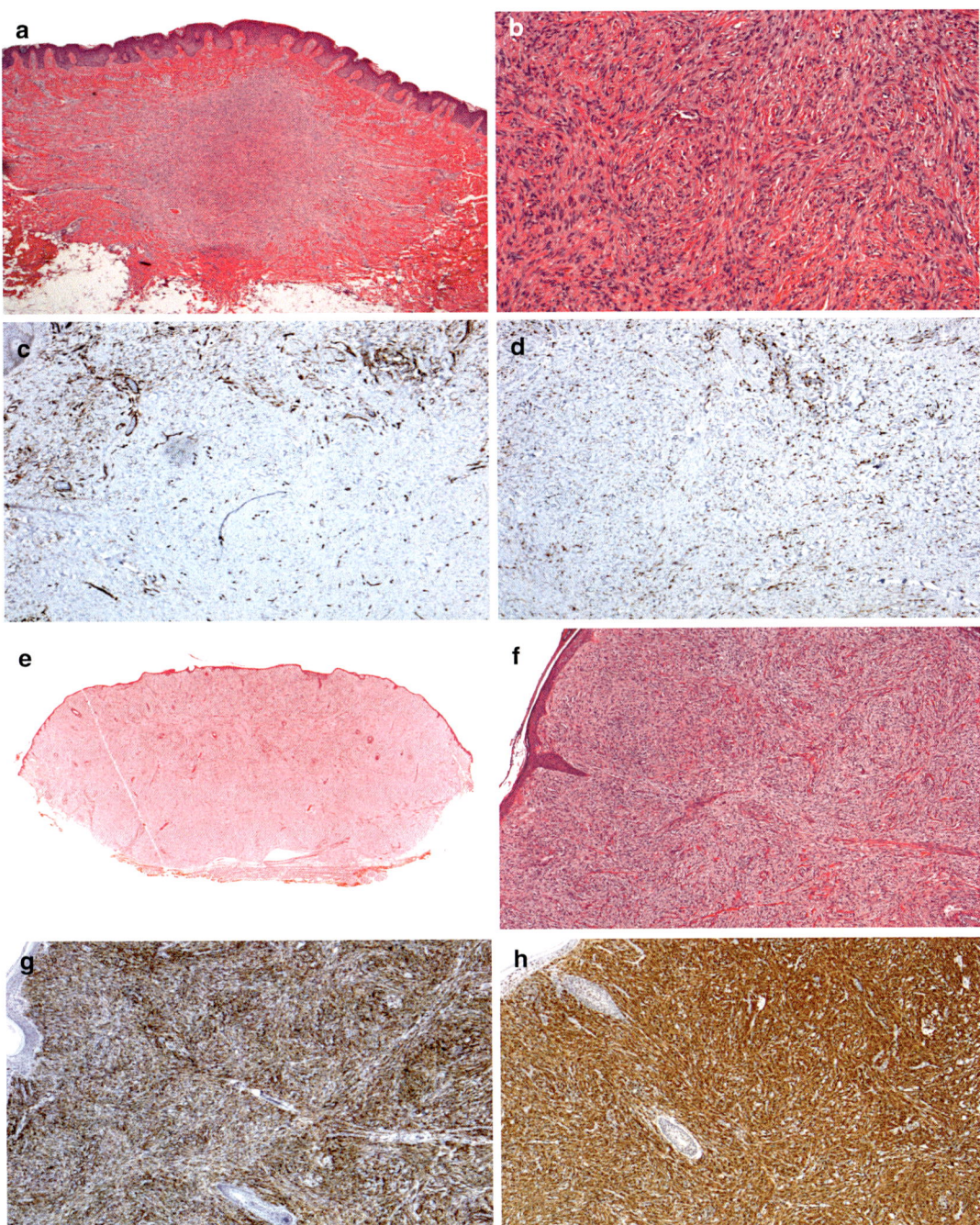

Fig. 17.11 This panel shows a dermatofibroma (**a–d**) and a juvenile xanthogranuloma (**e–h**). In the dermatofibroma, there is a nodular proliferation of fibroblasts and myofibroblasts with some whorling but no atypia (**a**, H&E, 12.5×; **b**, H&E, 100×; **c**, Anti-CD34, 50; **d**, Anti-F13, 50×). In the juvenile xanthogranuloma, there is dense lymphohistiocytic proliferation of dermis with Touton giant cells (**e**, H&E, 2×; **f**, H&E; 50×; **g**, Anti-CD68 immunostain, 50×; **h**, Anti-F13 immunostain, 50×)

(tumor of mature adipocytes), and reticulohistiocytoma (random distribution of MNGC with eosinophilic or ground glass cytoplasm).
- *TRT*: Excision.
- *PGN*: Although spontaneous regression is potential, newborns may develop the systemic disease and liver failure. Recurrence rate: 5–10%.

17.21.4 Dermatofibrosarcoma Protuberans

- *DEF*: Fibrohistiocytic tumor with high rate of recurrence rate and prone to metastasis if incompletely excised. DFSP carries a t(17;22) (q22;q13), which results in the fusion gene *COL1A1-PDGFB*.
- *EPI*: Rare in children but seen in youth.
- *GRO*: Red-brown raised tumor located in the trunk (e.g., shoulder) and limbs.
- *CLM*: Poorly/uncircumscribed, highly cellular tumor with tight storiform pattern prone to infiltrate deeply into subcutaneous tissue entrapping adipocytes and forming characteristic honeycomb pattern (Fig. 17.12).
- *IHC*: (+) VIM, CD34, (±) SMA, Bcl2, CD99, NKI-C3, ApoD, (−) FXIIIa, AE1–3, EMA, DES, S-100, HMB-45, CD117.
- *TRT*: Excision with negative margins.
- *PGN*: Metastasis in 5% of the cases.

17.22 Vascular Tumors (Fig. 17.13)

Vascular tumors are mainly described in Chap. 1 (*vide supra*) but also in some other sections.

17.23 Tumors of Adipose Tissue, Muscle, Cartilage, and Bone

These tumors are described in the chapter of soft tissue. The autochthonous tumors need to be differentiated from skin infiltration from contiguous lipogenic neoplasms, adjacent myogenic tumors, contiguous cartilaginous neoplasms, and skin infiltration from contiguous osteogenic neoplasms.

17.24 Neural and Neuroendocrine Tumors

Neural and neuroendocrine tumors of the skin are extremely rare entities but have occasionally been described, mainly, in immune-deficient patients and need to be taken into consideration during the diagnostic procedure of blue cell tumors with cutaneous localization or neuro-differentiated neoplasms.

17.24.1 Merkel Cell Carcinoma

- *DEF*: Highly aggressive malignant neuroendocrine tumor of the skin at probable viral etiology, i.e., Merkel cell polyomavirus (MCPyV), with a characteristic perinuclear dot-like CK20 stain.
- *SYN*: Trabecular carcinoma, neuroendocrine carcinoma.
- *EPG*: Merkel cell polyomavirus (MCPyV) is a DNA polyomavirus that has been identified as the probable etiological agent of Merkel cell carcinoma (MCC), which is a particularly aggressive cutaneous carcinoma of neuroendocrine origin. Polyomaviruses encode for large and small T-antigens, which bind to host proteins facilitating both replications of the viruses and inactivation of p53 and pRb. Interestingly, the integration of MCPyV preceded the metastatic spreading of the carcinoma. MCPyV is also responsible for high specific seroprevalence, starting in childhood and reaching up to 90% in adults, a finding that has been observed in both blood donors and individuals from Western countries. Recently, Martel-Jantin et al. (2013) emphasize that most primary MCPyV infections occur before the age of 6 years! Mother-fetus transmission is not a significant route of MCPyV transmission, but MCPyV seems to be transmitted from mother to child through close interpersonal contact involving saliva and/or the skin.
- *CLI*: Pink, firm, raised, painless nodule located on H&N and limbs.
- *CLM*: Neural crest-derived Merkel cell NET of the skin, which shows a predominantly dermal

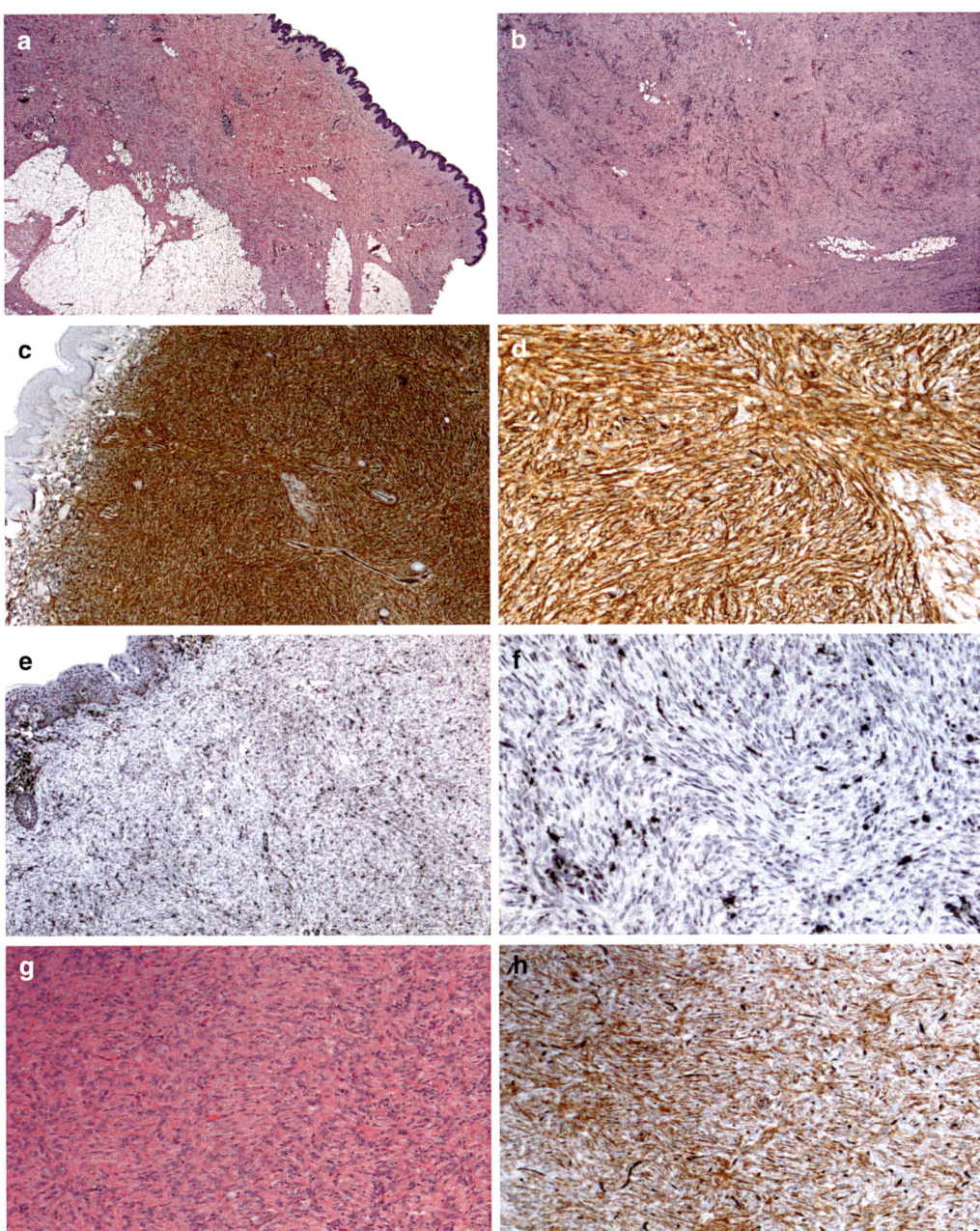

Fig. 17.12 This panel illustrates a dermatofibrosarcoma protuberans with highly cellular, tight storiform pattern with cells radiating in spokes and infiltrates deeply into subcutaneous tissue entrapping fat cells (**a–f**) (**a**, H&E, 20×; **b**, H&E, 20×; **c**, Anti-CD34 immunostain, 100×; **d**, Anti-CD34 immunostain, 200×; **e**, Anti-F13a, 50×; **f**, Anti-F13a, 200×). The two microphotographs (**g**) and (**h**) illustrate a hybrid form of giant cell fibroblastoma and dermatofibrosarcoma protuberans (**g**, H&E, 100×; **h**, Anti-CD34 immunostain, 100×)

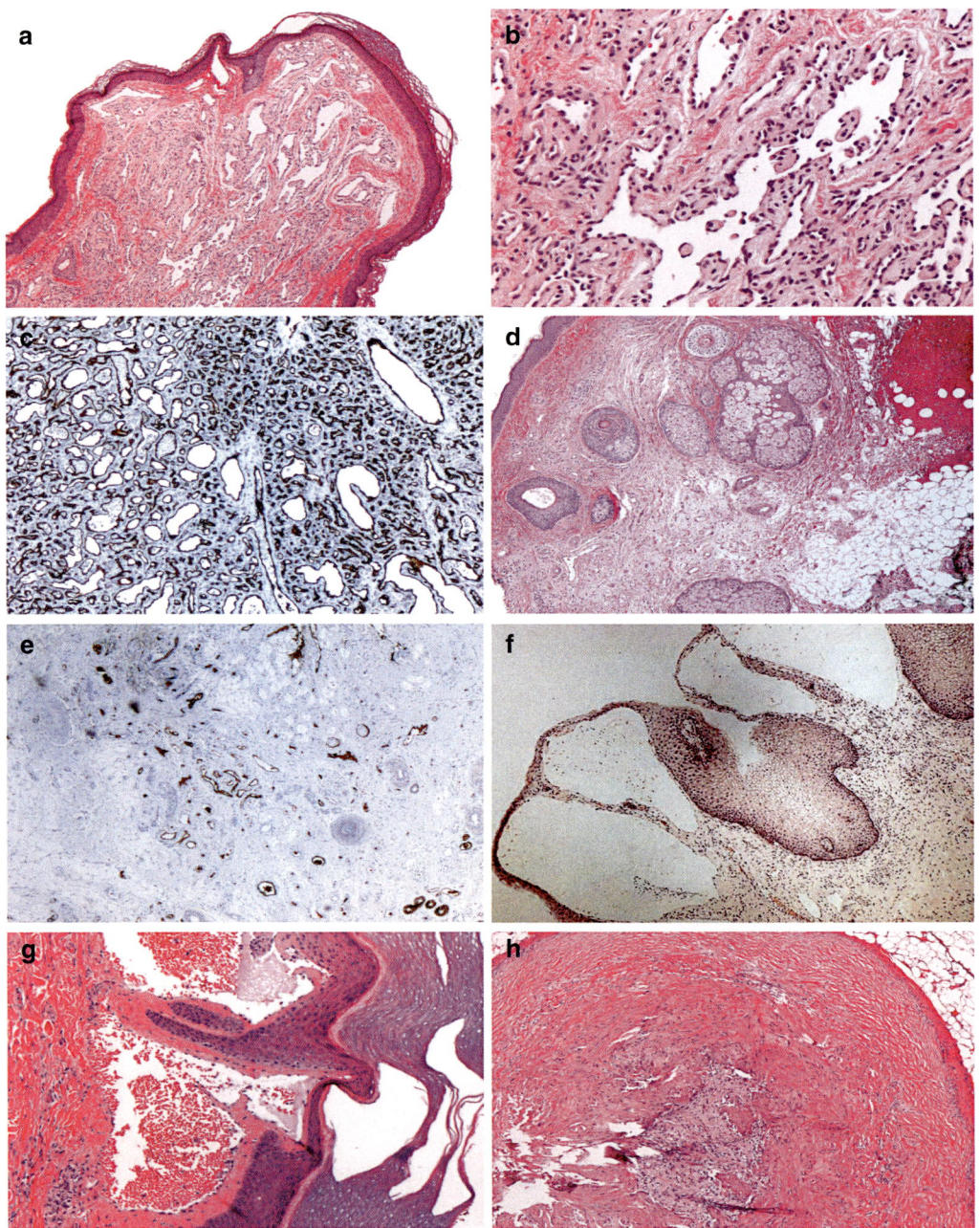

Fig. 17.13 In (**a**–**c**), there is capillary hemangioma (**a**, H&E stain, 50×; **b**, H&E stain, 200×; **c**, Anti-CD31 immunostain, 100×). In (**b**) it is possible to identify some plump endothelial cells, but no atypia. The microphotographs (**d**) and (**e**) show an involuted hemangioma (**d**, H&E stain, 50×; **e**, Anti-Glut-1 immunostain, 50×). The microphotograph (**f**) shows a lymphangioma circumscriptum (H&E stain, 50×). The microphotograph (**g**) shows an angiokeratoma (H&E stain, 100×). The microphotograph (**h**) shows an angioleiomyoma (H&E stain, 50×)

localization with sparing of the epidermis ("Grenz" zone) with trabecular growth pattern and collagen-dissecting features by small basophilic cells with high N/C ratio, inconspicuous eccentric nucleolus, and nuclear molding.
- *IHC*: (+) CD57, NSE, CGA, SYN, and characteristic perinuclear dot-like CK20 positivity, unlike another small cell carcinoma except for the small cell carcinomas of the salivary gland tissue as well as (±) CD99, FLI1, and CD117.
- *TEM*: 100–150 nm neurosecretory granules and paranuclear filament whorls.
- *TNT*: Surgery and chemotherapy, but therapy options and recommendations may vary accordingly to site of origin.
- *PGN*: Recurrences and metastases are up to 1/3 of cases.

17.24.2 Paraganglioma

- *DEF*: Neuroendocrine neoplasm arising parallel the sympathetic chain ganglion in the thoracic-lumbar region of the vertebral column and parasympathetic nervous system in the cranial and sacral areas and classified as a neoplasm of uncertain malignant potential. A rich neural network is present in the skin, but ganglia are not present.
- *EPI*: Rare tumor in childhood.
- *CLI*: Asymptomatic or as a painless mass, but up to 3% of the paragangliomas show symptoms related to the secretion of hormones from neurosecretory granules.
- *CLM*: Characteristic "Zellballen" constituted by aggregates of NSE+, SYN+ neuroendocrine cells separated by fibrovascular stroma (S100+ sustentacular cells).
- *TEM*: Electron-dense cytoplasmic granules with characteristic surrounding halo may be present in some tumor cells.
- *DDX*: Granular cell tumor, hibernomas, epithelioid Spitz nevi, cutaneous meningiomas, juvenile xanthogranulomas, and ectopic neural hamartomas (Box 17.14).
- *TNT*: Surgery and radiation therapy, but therapy options and recommendations vary accordingly to site of origin.
- *PGN*: Tumor with uncertain malignant potential.

Box 17.14 Paraganglioma of Head and Neck, Differential Diagnosis

Granular cell tumor (polygonal cells with eosinophilic granular cytoplasm, central oval nuclei, S100+, EM of granules→ degenerated lysosomes)

Hibernoma (small round adipocytes with granular eosinophilic cytoplasm interspersed with large cells with both eosinophilic and vacuolated cytoplasm and mature adipocytes)

Spitz nevus, epithelioid type (nests and lobules of large polygonal cells with enlarged nuclei, prominent nucleoli, and copious eosinophilic cytoplasm, S100+, HMB-45+, MART-1+)

Meningioma, cutaneous (whorl arrangement of spindle cells and sheets and nests of polygonal or spindle cells, psammoma bodies = laminated calcified bodies, intranuclear pseudoinclusions, EMA+, VIM+, S100)

LCH (Langerhans cells, eosinophils, S100+, CD1a+, Langerin+, Birbeck granules on TEM)

JXG (mixed inflammatory infiltrate of lymphocytes, plasma cells, neutrophils, eosinophils, and many large histiocytes with foamy or granular eosinophilic cytoplasm, Touton-type giant cells, CD68+, FXIII)

Neuroblastoma (small round blue cells, (+) CGA, SYN, CD56, (−) CD99, neurosecretory granules)

PNET (small to the medium-sized spindle to more polygonal cells with the perivascular arrangement, hyperchromasia, CD99+)

Neurothekeoma (multinodular mass, myxoid matrix, peripheral fibrosis, (+) VIM, imentin, NKI-C3, CD10, MITF, (−) S100, GFAP, MelanA)

Notes: Gr. θήκη, sheath; LCH, Langerhans cell histiocytosis; JXG, juvenile xanthogranuloma; PNET, primitive neuroectodermal tumor.

Two cases of paraganglioma of the scalp have been described in the English literature. DDX is important, and neuroblastoma, Langerhans cell histiocytosis, and lymphoma should be considered first. The rare occurrence of paragangliomas in the skin of the head is intriguing and may be argued that this is due to aberrant migration of neural crest cells.

17.25 Hematological Skin Infiltrates

Some conditions may be seen in the skin and be associated with or originated from a neoplastic involvement of the hematopoietic system. Tumors and related lesions of the hematopoietic and lymphoid systems include:

- Cutaneous infiltration of leukemia
- Cutaneous pseudolymphomas
- Langerhans cell histiocytosis
- Malignant lymphomas
- Mastocytosis maligna
- Mycosis fungoides
- Sezary syndrome
- Urticaria pigmentosa

The following is the list of the entities that have been discussed in the chapter dedicated to the hematopathology (Box 17.15).

17.25.1 Pseudolymphomas

This is a very heterogeneous group including conditions that have not been fully understood or diseases with poor description. The group includes perivascular lymphocytic infiltrates of Jessner-Kanoff of benign nature, sarcoid of Spiegler-Fendt, lymphomatoid papulosis, lymphomatoid granulomatosis, and localized forms of cutaneous reticulosis that may be in situ forms of Langerhans cell histiocytosis (*vide infra*).

17.25.2 Benign and Malignant Mastocytosis

Urticaria pigmentosa (infants, few to many papulae/wheals on the trunk and limbs ± systemic involvement with flushing and faintness, +++ mast cell infiltrate).

17.26 Solid Tumor Metastases to the Skin

Cutaneous metastases can occur in the setting of a pediatric (children) or adult (youth) tumor. Pediatric neoplasms include neuroblastoma, PNET/Ewing sarcoma, rhabdomyosarcoma, lymphoblastic lymphoma, other non-common SRBCTs, malignant rhabdoid tumor, congenital fibrosarcoma, and epithelioid sarcoma. Youth neoplasms include PNET/Ewing sarcoma, breast carcinoma, lung carcinoma, gastrointestinal carcinoma, carcinoma of the oral cavity and sinus-nasal carcinomas, urinary tract tumors, and tumors of the genital tract. Special stains are useful in distinguishing the nature of "clearing" of tumor cells, being intracytoplasmic lipid and glycogen (PAS+ and DPAS-) in CCRCC. Urothelial carcinomas are usually strongly positive for both CK7 and CK20, which are also both positive in most of the pancreatic carcinomas and some ovarian mucinous carcinomas.

Box 17.15 Hematological Infiltrates of the Skin

17.25.1 Cutaneous T-Cell Lymphomas
17.25.2 Cutaneous Non-T-Cell Lymphomas (Fig. 17.14)
17.25.3 Cutaneous Pseudolymphomas
17.25.4 Leukemia with Skin Infiltration and Myeloid Sarcoma
17.25.5 Langerhans Cell Histiocytosis (Fig. 17.15)
17.25.6 Benign and Malignant Mastocytosis

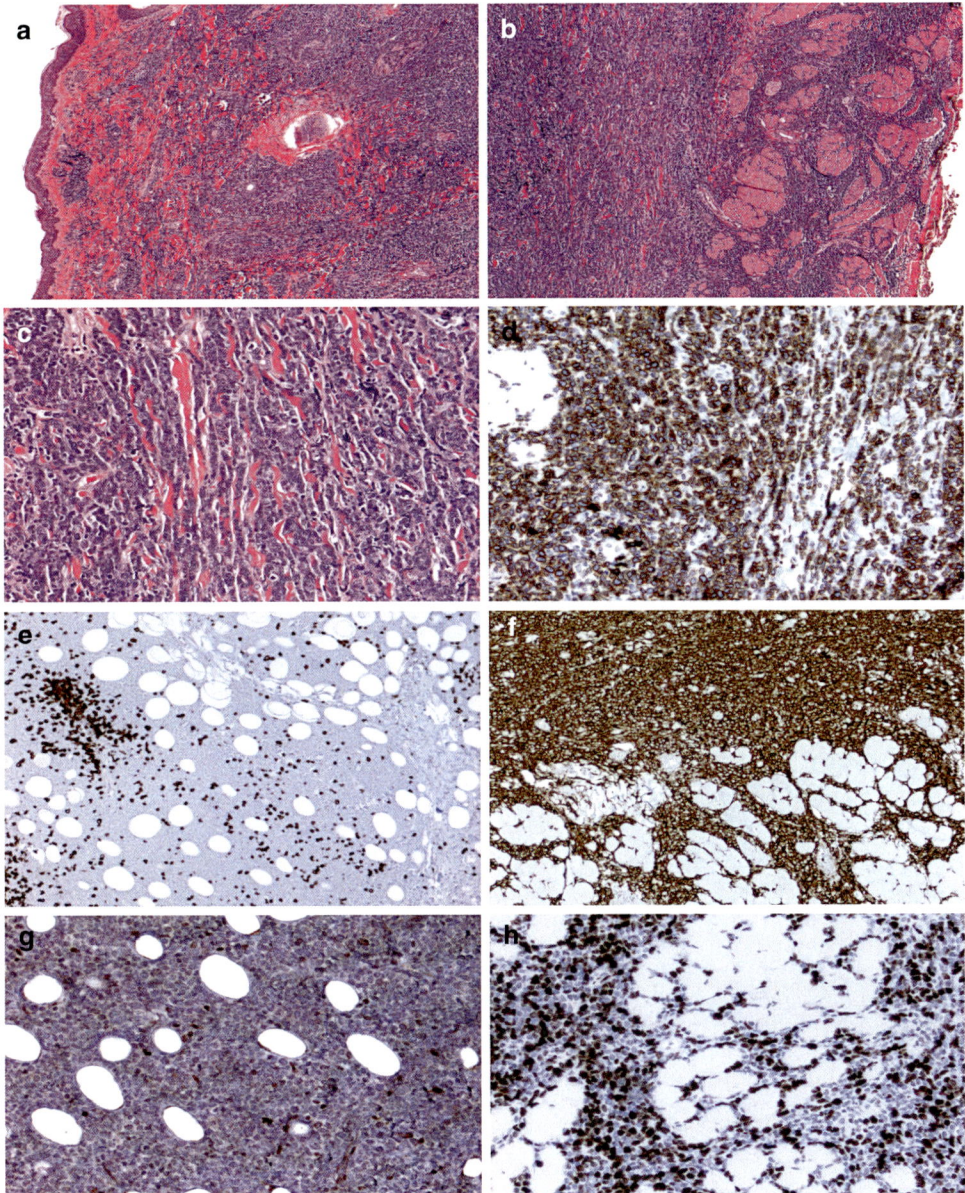

Fig. 17.14 DLBCL. This panel shows a rare diffuse large B-cell lymphoma (see hematopathology chapter for detail) (**a**, H&E, 50×; **b**, H&E, 50×; **c**, H&E, 200×; **d**, Anti-CD79a immunostain, 200×; **e**, Anti-CD3 immunostain, 100×; **f**, Anti-CD10 immunostain, 100×; **g**, Anti-Bcl6 immunostain, 50×; **h**, Anti-Ki67 immunostain, 200×)

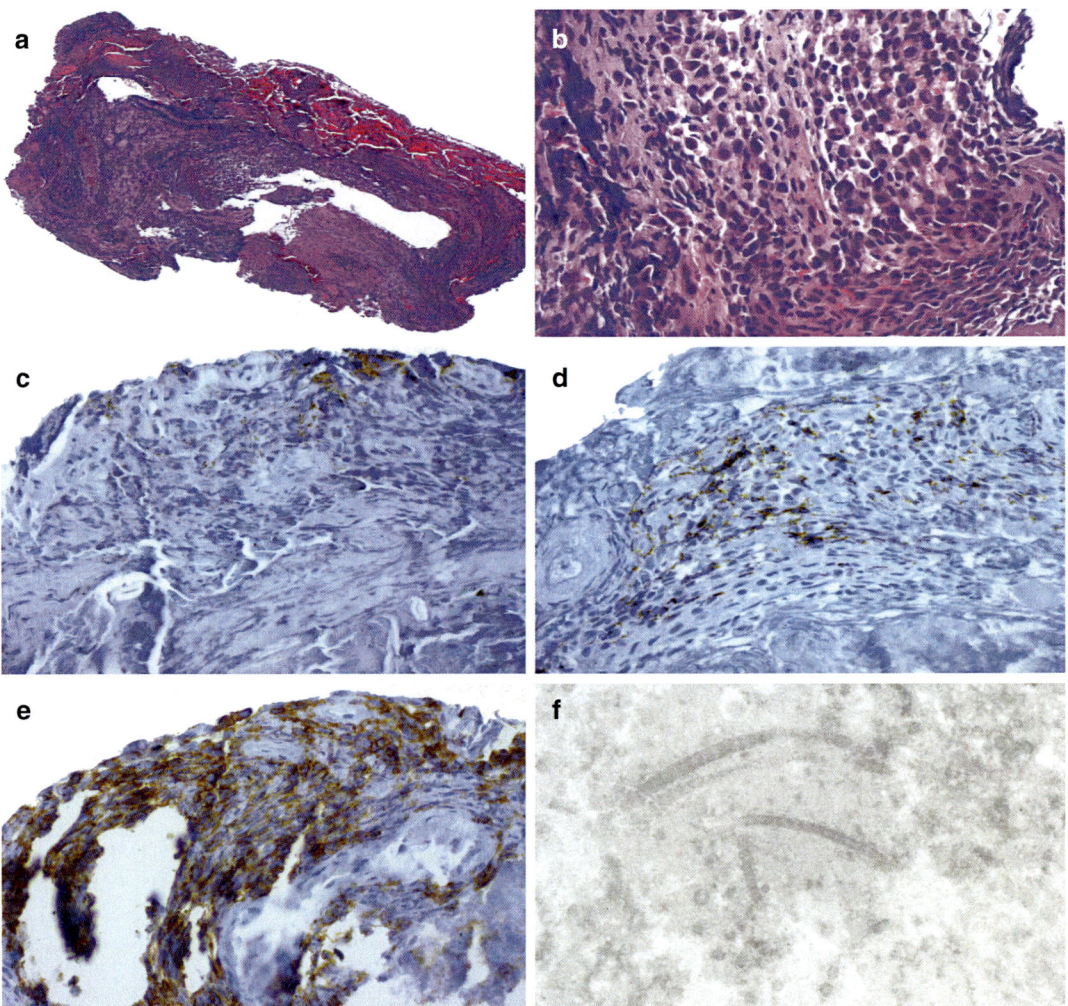

Fig. 17.15 Subungual Langerhans cell histiocytosis. Nail involvement is distinctly uncommon in Langerhans cell histiocytosis (LCH), which may present with longitudinal grooving, purpuric striae, hyperkeratosis, subungual pustules, deformity, loss of nail plate, paronychia, onycholysis, and pitting. This panel shows the microphotographs of pediatric case of subungueal LCH. In (**a**, **b**) there is an effacement of the subungual plate and an infiltration with Langerhans cells in combination with other leukocytes, mainly lymphocytes and eosinophils (**a**, H&E stain, 40×; **b**, H&E stain, 400×). Langerhans cells have abundant, pale eosinophilic cytoplasm, irregular and elongated nuclei with nuclear grooves and folds, fine chromatin and indistinct nucleoli. The microphotographs (**c**), (**d**), and (**e**) show the immunostains with antibodies against CD45 (400×), CD3 (400×), and CD1a (400×), respectively. An electron microscopy investigation of the formalin-fixed and paraffin-embedded block identified Birbeck granules, which present as structures with a size variable between 200 and 400 nm × 33 nm and double outer contour

Multiple Choice Questions and Answers

- DER-1 An 18-year-old girl surviving osteosarcoma of the right leg presents with a slowly growing firm nodule near the site of a recent ear-piercing. The nodule is excised, and the histology is shown here:

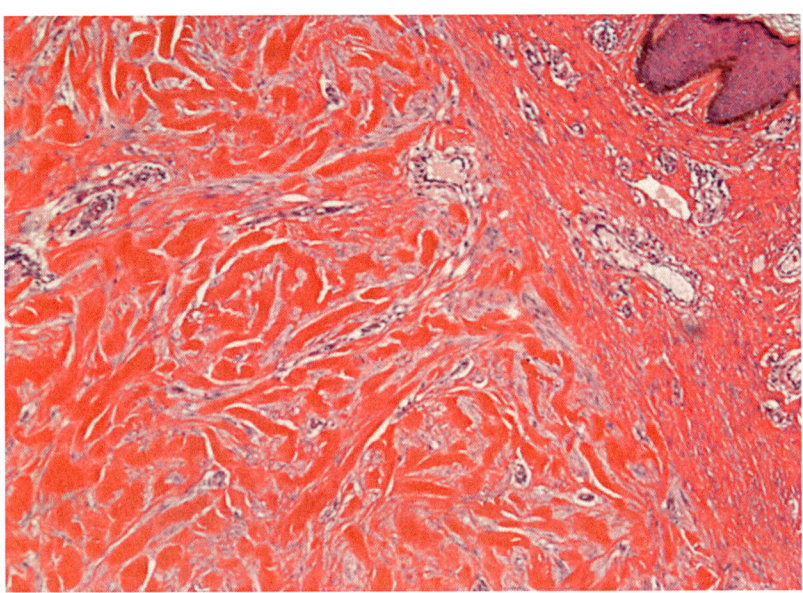

What is the most likely diagnosis?
  (a) Metastasis of an osteosarcoma
  (b) Metastasis of a synovial sarcoma
  (c) Hamartoma
  (d) Teratoma
  (e) Keloid

- DER-2 A 2-year-old boy presented with a nodule on the right leg. The nodule is pink-colored with a dimple in the center, firm, and 3 millimeters in diameter. In the last week, the nodule became itchy, sore, and red. He started swimming in the pool a few months ago. The nodule is excised, and the histology is shown here:

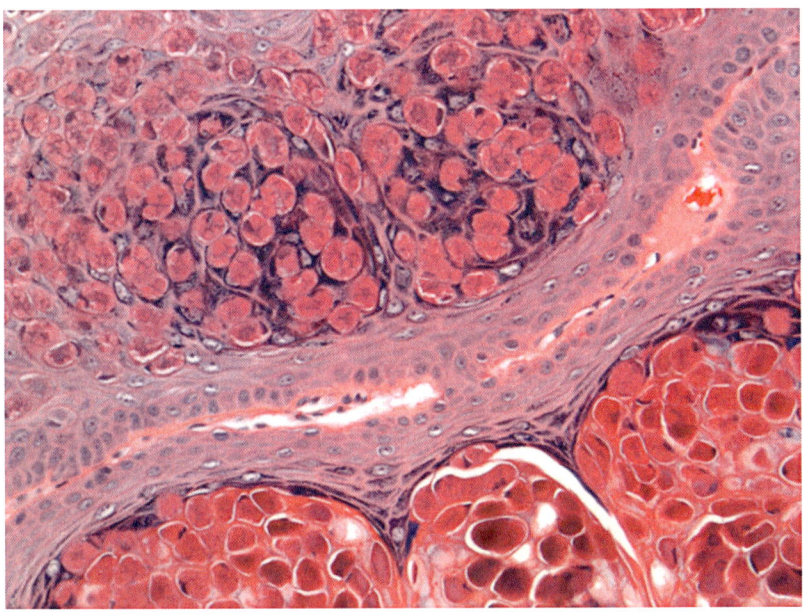

What is the most likely diagnosis?
- (a) Molluscum contagiosum
- (b) Juvenile xanthogranuloma
- (c) Pilomatrixoma
- (d) Scar
- (e) Ecthyma

- DER-3 What is the exact proof of tuberculosis infection in an active stage?
  - (a) Acid-fast staining on tissue
  - (b) Ziehl-Neelsen stain
  - (c) Tuberculin test
  - (d) Microbiological isolation of *M. tuberculosis*

- DER-4 A 6-month-old female infant developed otitis media for which she was given a course of antibiotics. A few days later, the infant developed widespread peeling of the skin. Upon examination, the pediatrician does not note any mucosal lesions, and the baby was not "toxic". What is the most likely diagnosis in this baby?
  - (a) Infantile pemphigus
  - (b) Toxic epidermal necrolysis
  - (c) Steven Johnson syndrome
  - (d) Staphylococcal scalded skin syndrome

- DER-5 An 18-month-old developmentally normal boy was brought to the pediatrician by the parents. He showed generalized itching and symmetrical skin lesions predominantly on the extensor aspect of all four extremities for the 1-week duration. The symptomatology was evident 3 days following DPT (diphtheria, pertussis, tetanus) immunization and vaccination against poliomyelitis with some increases of the body temperature that subsided 3 days later. There was no history of drug intake or atopy. The lesions started to progress from both upper limbs to the buttocks, lower limbs, and face, which was accompanied by generalized lymphadenopathy. On examination, the rash consisted of multiple, small, skin-colored nontender, pruritic papules. What is the most likely diagnosis in this infant?
  - (a) Gianotti-Crosti syndrome
  - (b) Acrodermatitis enteropathica
  - (c) Erythema infectiosum
  - (d) Erythema multiforme
  - (e) Henoch-Schönlein purpura
  - (f) Kawasaki disease

- DER-6 A 19-year-old girl suffering from inflammatory bowel disease for more than 10 years presents to the family doctor with dome-shaped slightly polypoid reddish, painful nodules located on anterior and lateral lower legs. She is not pregnant and denies any drug administration. Septal panniculitis is diagnosed following a skin biopsy. What is the most likely diagnosis?
  - (a) Alpha-1-antitrypsin deficiency
  - (b) Non-Hodgkin lymphoma
  - (c) Lupus panniculitis
  - (d) Erythema nodosum
  - (e) Nodular fat necrosis
  - (f) Necrobiosis lipoidica
  - (g) Necrobiotic xanthogranuloma
  - (h) Scleroderma
  - (i) Subcutaneous granuloma

- DER-7 Epidermal nevi are a cellular proliferation of hamartomatous character. An epidermal nevus (non-melanocytic nevus) is an overgrowth of the epidermis that can be present at birth or develop during infancy. Which of the following features does NOT belong to epidermal nevi?
  - (a) Epidermal ortho-hyperkeratosis
  - (b) Hypergranulosis
  - (c) Hyperpigmentation
  - (d) Papillomatosis
  - (e) Basal layer hyperpigmentation

- DER-8 Which of the following conditions predispose to squamous cell carcinoma in childhood?
  - (a) Nevus sebaceous
  - (b) Down syndrome
  - (c) Patau syndrome
  - (d) Pilomatrixoma
  - (e) Spitz nevus

- DER-9 What is the most common type of malignant melanoma?
  - (a) Acral lentiginous melanoma
  - (b) Mucosal melanoma
  - (c) Nodular melanoma
  - (d) Juvenile melanoma
  - (e) Polypoid melanoma
  - (f) Superficial spreading melanoma

- DER-10 Which of the following pathways is NOT altered in malignant melanoma?
  (a) RAS-RAF-MEK-ERK (MAPK (mitogen-activated protein kinase)) with activating mutations of NRAS
  (b) p16-CDK4-RB
  (c) ARF-TP53
  (d) FAP
  (e) KIT (activating mutations)
  (f) PTEN (epigenetically silenced)
  (g) CDKN2A gene mutations (three TSGs including p14, p15, and p16)
  (h) CMM1 gene mutations, MSI, and multiple chromosomal genes and losses

References and Recommended Readings

Abedi Kiasari B, Vallely PJ, Klapper PE. Merkel cell polyomavirus DNA in immunocompetent and immunocompromised patients with respiratory disease. J Med Virol. 2011;83(12):2220–4. https://doi.org/10.1002/jmv.22222. PubMed PMID: 22012732.

Agelli M, Clegg LX. Epidemiology of primary Merkel cell carcinoma in the United States. J Am Acad Dermatol. 2003;49(5):832–41. Erratum in: J Am Acad Dermatol. 2004;50(5):733. PubMed PMID: 14576661.

Altaykan A, Ersoy-Evans S, Erkin G, Ozkaya O. Basal cell carcinoma arising in nevus sebaceous during childhood. Pediatr Dermatol. 2008;25(6):616–9. https://doi.org/10.1111/j.1525-1470.2008.00726.x. PubMed PMID: 19067866.

Ashena Z, Alavi S, Arzanian MT, Eshghi P. Nail involvement in langerhans cell histiocytosis. Pediatr Hematol Oncol. 2007;24(1):45–51. PubMed PMID: 17130113.

Aslan Y, Erduran E, Kutlu N. Autosomal recessive multiple pterygium syndrome: a new variant? Am J Med Genet. 2000;93(3):194–7. PubMed PMID: 10925380.

Banerjee SS, Agbamu DA, Eyden BP, Harris M. Clinicopathological characteristics of peripheral primitive neuroectodermal tumour of skin and subcutaneous tissue. Histopathology. 1997;31(4):355–66. PubMed PMID: 9363452.

Barnhill RL. Childhood melanoma. Semin Diagn Pathol. 1998;15(3):189–94. Review. PubMed PMID: 9711668.

Barnhill RL, Crowson AN. Textbook of Dermatopathology, Volume 355. McGraw-Hill, Medical Pub. Division, 2004 - Medical - 1093 pages

Bastuji-Garin S, Rzany B, et al. Clinical classification of cases of toxic epidermal necrolysis, Stevens-Johnson syndrome, and erythema multiforme. Arch Dermatol. 1993;129(1):92–6. PubMed PMID: 8420497.

Belhadjali H, Moussa A, Yahia S, Njim L, Zakhama A, Zili J. Simultaneous occurrence of two squamous cell carcinomas within a nevus sebaceous of Jadassohn in an 11-year-old girl. Pediatr Dermatol. 2009;26(2):236–7. https://doi.org/10.1111/j.1525-1470.2009.00895.x. PubMed PMID: 19419489.

Bialasiewicz S, Lambert SB, Whiley DM, Nissen MD, Sloots TP. Merkel cell polyomavirus DNA in respiratory specimens from children and adults. Emerg Infect Dis. 2009;15(3):492–4. https://doi.org/10.3201/eid1503.081067. PubMed PMID: 19239774; PubMed Central PMCID: PMC2681122.

Bizzozero G. Delle cellule cigliate, del reticolo Malpighiano dell'epidermide. Annal Univ Med. 1864;190:110–8.

Bostanci S, Akay BN, Kirmizi A, Okcu Heper A, Farabi B. Basosquamous carcinoma and melanoma collision tumor in a child with xeroderma pigmentosum. Pediatr Dermatol. 2020;37(2):390–392. https://doi.org/10.1111/pde.14097. Epub 2020 Jan 19. PubMed PMID: 31957124.

Broders AC. Microscopic grading of cancer. In: Treatment of cancer and allied diseases. p. 9.

Broders AC. The grading of carcinoma. Minn Med. 1925;8:726.

Broders AC. Grading of Cancer. In: Treatment of Cancer and Allied Diseases. Pack GT, Livingston EM. ed. 1, New York. Paul B. Hoeber, Inc., 1940, pp. 19–41.

Brunner J, Freund M, Prelog M, Binder E, Sailer-Hoeck M, Jungraithmayr T, Huemer C, Sergi C, Zimmerhackl LB. Successful treatment of severe juvenile microscopic polyangiitis with rituximab. Clin Rheumatol. 2009;28(8):997–9. https://doi.org/10.1007/s10067-009-1177-0. Epub 2009 Apr 24. PubMed PMID: 19390907.

Busam KJ, Mentzel T, Colpaert C, Barnhill RL, Fletcher CD. Atypical or worrisome features in cellular neurothekeoma: a study of 10 cases. Am J Surg Pathol. 1998;22(9):1067–72. PubMed PMID: 9737238.

Calkins CC, Setzer SV. Spotting desmosomes: the first 100 years. J Invest Dermatol. 2007;127:E2–3.

Caulfield JB, Wilgram GF. An electron microscopic study of blister formation in erythema multiforme. J Invest Dermatol. 1962;39:307–316. DOI: 10.1038/jid.1962.118.

Chen T, Hedman L, Mattila PS, Jartti T, Ruuskanen O, Söderlund-Venermo M, Hedman K. Serological evidence of Merkel cell polyomavirus primary infections in childhood. J Clin Virol. 2011;50(2):125–9. https://doi.org/10.1016/j.jcv.2010.10.015. Epub 2010 Nov 19. PubMed PMID: 21094082.

Cohen MC, Kaschula RO, Sinclair-Smith C, Emms M, Drut R. Pluripotential melanoblastoma, a unifying concept on malignancies arising in congenital melanocytic nevi: report of two cases. Pediatr Pathol Lab Med. 1996;16(5):801–12. PubMed PMID: 9025878.

Comar M, Cuneo A, Maestri I, Melloni E, Pozzato G, Soffritti O, Secchiero P, Zauli G. Merkel-cell polyomavirus (MCPyV) is rarely associated to B-chronic lymphocytic leukemia (1 out of 50) samples and occurs late in the natural history of the disease. J Clin

Virol. 2012;55(4):367–9. https://doi.org/10.1016/j.jcv.2012.08.011. Epub 2012 Sep 7. PubMed PMID: 22959215.

Cordon M. Extrait d'une lettre au sujet de trois enfants de la meme mere avec partie des extremites denuee de peau. J Med Chir Pharm. 1767;26:556–7.

Cribier B, Scrivener Y, Grosshans E. Tumors arising in nevus sebaceus: a study of 596 cases. J Am Acad Dermatol. 2000;42(2 Pt 1):263–8. PubMed PMID: 10642683.

Curry CJ. Further comments on the Neu-Laxova syndrome. Am J Med Genet. 1982;13(4):441–4. PubMed PMID: 6891563.

Dasgupta T, Wilson LD, Yu JB. A retrospective review of 1349 cases of sebaceous carcinoma. Cancer. 2009;115(1):158–65. https://doi.org/10.1002/cncr.23952. PubMed PMID: 18988294.

Dean JC, Gray ES, Stewart KN, Brown T, Lloyd DJ, Smith NC, Pope FM. Restrictive dermopathy: a disorder of skin differentiation with abnormal integrin expression. Clin Genet. 1993;44(6):287–91. PubMed PMID: 8131298.

Demmel U. Clinical aspects of congenital skin defects. I. Congenital skin defects on the head of the newborn. Eur J Pediatr. 1975;121(1):21–50. Review. PubMed PMID: 765133.

El Hachem M, Diociaiuti A, Latella E, Zama M, Lambiase C, Giraldi L, Surrenti T, Callea F. Congenital myxoid and pigmented dermatofibrosarcoma protuberans: a case report. Pediatr Dermatol. 2013;30(5):e74–7. https://doi.org/10.1111/pde.12131. Epub 2013 Mar 28. PubMed PMID: 23534369.

Elston D, Ferringer T, Ko C, Peckham S, High W, DiCaudo D. Dermatopathology, 2nd Edition. Saunders Ltd. 2013:1–464. eBook ISBN: 9780702055287. eBook ISBN: 9780702055294

Erovic I, Erovic BM. Merkel cell carcinoma: the past, the present, and the future. J Skin Cancer. 2013;2013:929364. https://doi.org/10.1155/2013/929364. Epub 2013 Apr 16. PubMed PMID: 23691324; PubMed Central PMCID: PMC3652192.

Faridounnia M, Wienk H, Kovačič L, Folkers GE, Jaspers NG, Kaptein R, Hoeijmakers JH, Boelens R. The Cerebro-oculo-facio-skeletal Syndrome Point Mutation F231L in the ERCC1 DNA Repair Protein Causes Dissociation of the ERCC1-XPF Complex. J Biol Chem. 2015 Aug 14;290(33):20541-55. https://doi.org/10.1074/jbc.M114.635169. Epub 2015 Jun 17. PubMed PMID: 26085086; PubMed Central PMCID: PMC4536458.

Faust H, Andersson K, Ekström J, Hortlund M, Robsahm TE, Dillner J. Prospective study of merkel cell polyomavirus and risk of merkel cell carcinoma. Int J Cancer. 2014;134(4):844–8. https://doi.org/10.1002/ijc.28419. Epub 2013 Aug 29. PubMed PMID: 23922031.

Frieden IJ. Aplasia cutis congenita: a clinical review and proposal for classification. J Am Acad Dermatol. 1986;14(4):646–60. Review. PubMed PMID: 3514708.

Gallager RL, Helwig EB. Neurothekeoma – a benign cutaneous tumor of neural origin. Am J Clin Pathol. 1980;74(6):759–64. PubMed PMID: 7446487.

Gorlin RJ, Cohen MM, Levin LS. Syndromes of head and neck. 3rd ed. New York: Oxford University Press; 1990. p. 417–9.

Hall JG. The lethal multiple pterygium syndromes. Am J Med Genet. 1984;17(4):803–7. PubMed PMID: 6539071.

Hall JG. Analysis of Pena Shokeir phenotype. Am J Med Genet. 1986;25(1):99–117. Review. PubMed PMID: 3541610.

Hamm H, Höger PH. Skin tumors in childhood. Dtsch Arztebl Int. 2011;108(20):347–53. https://doi.org/10.3238/arztebl.2011.0347. Epub 2011 May 20. Review. PubMed PMID: 21655460; PubMed Central PMCID: PMC3109276.

Harr T, French LE. Toxic epidermal necrolysis and Stevens-Johnson syndrome. Orphanet J Rare Dis. 2010;5:39.

Heelan K, Shear NH. Cutaneous drug reactions in children: an update. Paediatr Drugs. 2013;15(6):493–503. https://doi.org/10.1007/s40272-013-0039-z. Review. PubMed PMID: 23842849.)

Hidvegi NC, Kangesu L, Wolfe KQ. Squamous cell carcinoma complicating naevus sebaceous of Jadassohn in a child. Br J Plast Surg. 2003;56(1):50–2. PubMed PMID: 12706152.

Holbrook KA, Dale BA, Witt DR, Hayden MR, Toriello HV. Arrested epidermal morphogenesis in three newborn infants with a fatal genetic disorder (restrictive dermopathy). J Invest Dermatol. 1987;88(3):330–9. PubMed PMID: 2434579.

Horn TD. Interface dermatitis. In: Barnhill R, Crowson AN, eds. Textbook of Dermatopathology. 2nd ed. New York, NY: McGraw-Hill Co; 2004:35–60.

Jacobson-Dunlop E, White CR Jr, Mansoor A. Features of plexiform fibrohistiocytic tumor in skin punch biopsies: a retrospective study of 6 cases. Am J Dermatopathol. 2011;33(6):551–6. https://doi.org/10.1097/DAD.0b013e318206a648. PubMed PMID: 21697703.

Jafarian F, McCuaig C, Kokta V, Hatami A, Savard P. Plexiform fibrohistiocytic tumor in three children. Pediatr Dermatol. 2006;23(1):7–12. Review. PubMed PMID: 16445402.

Joshi RR, Nepal A, Ghimire A, Karki S. Eccrine poroma in neck of a child – a rare presentation. Nepal Med Coll J. 2009;11(1):73–4. PubMed PMID: 19769246.

Kazakov DV, Sima R, Vanecek T, Kutzner H, Palmedo G, Kacerovska D, Grossmann P, Michal M. Mutations in exon 3 of the CTNNB1 gene (beta-catenin gene) in cutaneous adnexal tumors. Am J Dermatopathol. 2009;31(3):248–55. https://doi.org/10.1097/DAD.0b013e318198922a. PubMed PMID: 19384065.

Kempf W, Kazakov DV, Belousova IE, Mitteldorf C, Kerl K. Paediatric cutaneous lymphomas: a review and comparison with adult counterparts. J Eur Acad Dermatol Venereol. 2015;29(9):1696–709. https://doi.org/10.1111/jdv.13044. Epub 2015 Feb 25. Review. PubMed PMID: 25715748.

Kim LIJ, Park MC, Kim JH, Lim H. Cutaneous paraganglioma of the vertex in a child. J Craniofac Surg. 2012;23(4):e338–40. https://doi.org/10.1097/SCS.0b013e3182564b3a. Review. PubMed PMID: 22801173.

Koens L, Qin Y, Leung WY, Corver WE, Jansen PM, Willemze R, Vermeer MH, Tensen CP. MicroRNA profiling of primary cutaneous large B-cell lymphomas. PLoS One. 2013;8(12):e82471. https://doi.org/10.1371/journal.pone.0082471. eCollection 2013. PubMed PMID: 24358187; PubMed Central PMCID: PMC3865085.

Köksal Y, Toy H, Talim B, Unal E, Akçören Z, Cengiz M. Merkel cell carcinoma in a child. J Pediatr Hematol Oncol. 2009;31(5):359–61. https://doi.org/10.1097/MPH.0b013e3181984f6b. PubMed PMID: 19415020.

Kose D, Ciftci I, Harmankaya I, Ugras S, Caliskan U, Koksal Y. Pilomatrixoma in childhood. J Cancer Res Ther. 2014;10(3):549–51. https://doi.org/10.4103/0973-1482.137918. PubMed PMID: 25313737.

Kulkarni K, Desai S, Grundy P, Sergi C. Infantile myofibromatosis: report on a family with autosomal dominant inheritance and variable penetrance. J Pediatr Surg. 2012;47(12):2312–5. https://doi.org/10.1016/j.jpedsurg.2012.09.046. PubMed PMID: 23217896.

Küster W, Traupe H. Klinik und Genetik angeborener Hautdefekte [Clinical aspects and genetics of congenital skin defects]. Hautarzt. 1988;39(9):553–63. Review. German. PubMed PMID: 3053531.

Kwan TH. Spongiotic dermatitis. In: Barnhill R, Crowson AN, Busam K, Granter S ed's. Textbook of Dermatopathology. New York: McGraw-Hill Co. 1998:17–32.

Le Van Quyen P, Calmels N, Bonnière M, Chartier S, Razavi F, Chelly J, El Chehadeh S, Baer S, Boutaud L, Bacrot S, Obringer C, Favre R, Attié-Bitach T, Laugel V, Antal MC. Prenatal diagnosis of cerebro-oculo-facio-skeletal syndrome: Report of three fetuses and review of the literature. Am J Med Genet A. 2020 Feb 13. https://doi.org/10.1002/ajmg.a.61520. [Epub ahead of print] PubMed PMID: 32052936.

Leclerc-Mercier S, Pedeutour F, Fabas T, Glorion C, Brousse N, Fraitag S. Plexiform fibrohistiocytic tumor with molecular and cytogenetic analysis. Pediatr Dermatol. 2011;28(1):26–9. https://doi.org/10.1111/j.1525-1470.2010.01370.x. Epub 2011 Jan 25. PubMed PMID: 21261704.

Levi N, Bastuji-Garin S, Mockenhaupt M, Roujeau JC, Flahault A, Kelly JP, et al. Medications as risk factors of Stevens–Johnson syndrome and toxic epidermal necrolysis in children: a pooled analysis. Pediatrics. 2009;123(2):e297–304.

Litt J. Drug eruption reference manual. New York: Parthenon; 2000.

Loh J, El-Hakim H, Sergi CM, Fiorillo L. Branchiooculofacial syndrome and bilateral ectopic thymus: report of a family. Pediatr Dermatol. 2012;29(6):759–61. https://doi.org/10.1111/j.15251470.2012.01877.x. PubMed PMID: 23106675.

Lyell A. Toxic epidermal necrolysis: an eruption resembling scalding of the skin. Br J Dermatol. 1956;68:355–61.

Mandt N, Vogt A, Blume-Peytavi U. Differential diagnosis of hair loss in children. J Dtsch Dermatol Ges. 2004;2(6):399–411. Review. PubMed PMID: 16281597.

Martel-Jantin C, Pedergnana V, Nicol JT, Leblond V, Trégouët DA, Tortevoye P, Plancoulaine S, Coursaget P, Touzé A, Abel L, Gessain A. Merkel cell polyomavirus infection occurs during early childhood and is transmitted between siblings. J Clin Virol. 2013;58(1):288–91. https://doi.org/10.1016/j.jcv.2013.06.004. Epub 2013 Jul 2. PubMed PMID: 23829968.

Massi D (Ed.). Dermatopathology. Springer. 2016.

Mataix J, Bañuls J, Botella R, Laredo C, Lucas A. Sindrome de Brooke-Spiegler: una entidad heterogenea [Brooke-Spiegler syndrome: an heterogeneous entity]. Actas Dermosifiliogr. 2006;97(10):669-72. Review. Spanish. PubMed PMID: 17173833.

Maximova N, Granzotto M, Kiren V, Zanon D, Comar M. First description of Merkel cell polyomavirus DNA detection in a patient with Stevens-Johnson syndrome. J Med Virol. 2013;85(5):918–23. https://doi.org/10.1002/jmv.23550. PubMed PMID: 23508917.

Mebazaa A, Boussofara L, Trabelsi A, Denguezli M, Sriha B, Belajouza C, Nouira R. Undifferentiated sebaceous carcinoma: an unusual childhood cancer. Pediatr Dermatol. 2007;24(5):501–4. PubMed PMID: 17958796.

Mori O, Hashimoto T. Plexiform fibrohistiocytic tumor. Eur J Dermatol. 2004;14(2):118–20. PubMed PMID: 15197003.

Nair PS. A clinicopathologic study of skin appendageal tumors. Indian J Dermatol Venereol Leprol. 2008;74(5):550. PubMed PMID: 19086136.

Naranjo CA, Busto U, Sellers EM, Sandor P, Ruiz I, Roberts EA, Janecek E, Domecq C, Greenblatt DJ. A method for estimating the probability of adverse drug reactions. Clin Pharmacol Ther. 1981;30(2):239–45. PubMed PMID: 7249508. Saadat P, Cesnorek S, Ram R, Kelly L, Vadmal M. Primary cutaneous paraganglioma of the scalp. J Am Acad Dermatol. 2006;54(5 Suppl):S220–3. PubMed PMID: 16631945.

Orlandi C, Arcangeli F, Patrizi A, Neri I. Eccrine poroma in a child. Pediatr Dermatol. 2005;22(3):279–80. PubMed PMID: 15916587.

Ozbay M, Kiniş V, Firat U, Bakir S, Yorgancilar E. Seborrheic keratosis of the external auditory canal in a 1-year-old boy. Turk J Pediatr. 2012;54(5):543–4. PubMed PMID: 23427523.

Pandey P, Dixit A, Chandra S, Tanwar A. Cytological features of malignant eccrine acrospiroma presenting as a soft tissue mass axilla: a rare sweat gland tumor with histologic correlation. Int J Appl Basic Med Res. 2015;5(2):145–8. https://doi.org/10.4103/2229-516X.157173. PubMed PMID: 26097826; PubMed Central PMCID: PMC4456892.

Pointdujour-Lim R, Marous MR, Satija CE, Douglass AM, Eagle RC Jr, Shields JA, Shields CL. Cutaneous Horn of the Eyelid in 13 Cases. Ophthalmic Plast Reconstr Surg. 2017;33(4):233–236. https://doi.org/10.1097/IOP.0000000000000816. PubMed PMID: 27811637.

PubMed PMID: 21697703.

Salamanca J, Rodríguez-Peralto JL, de la Torre JP G, López-Ríos F. Plexiform fibrohistiocytic tumor without multinucleated giant cells: a case report. Am J Dermatopathol. 2002;24(5):399–401. PubMed PMID: 12357200.

Sánchez Yus E, Sanz Vico MD, de Diego V. Miescher's radial granuloma. A characteristic marker of erythema nodosum. Am J Dermatopathol. 1989;11(5):434–42. PubMed PMID: 2679196.

Schmid C, Beham A, Feichtinger J, Auböck L, Dietze O. Recurrent and subsequently metastasizing Merkel cell carcinoma in a 7-year-old girl. Histopathology. 1992;20(5):437–9. PubMed PMID: 1587495.

Segal AR, Doherty KM, Leggott J, Zlotoff B. Cutaneous reactions to drugs in children. Pediatrics. 2007;120(4):e1082–96. Review. PubMed PMID: 17908729.

Sergi C, Beedgen B, Kopitz J, Zilow E, Zoubaa S, Otto HF, Cantz M, Linderkamp O. Refractory congenital ascites as a manifestation of neonatal sialidosis: clinical, biochemical and morphological studies in a newborn Syrian male infant. Am J Perinatol. 1999;16(3):133–41. PubMed PMID: 10438195.

Sergi C, Kahl P, Otto HF. Immunohistochemical localization of transforming growth factor-alpha and epithelial growth factor receptor in human fetal developing skin, psoriasis and restrictive dermopathy. Pathol Oncol Res. 2000;6(4):250–5. PubMed PMID: 11173656.

Sergi C, Penzel R, Uhl J, Zoubaa S, Dietrich H, Decker N, Rieger P, Kopitz J, Otto HF, Kiessling M, Cantz M. Prenatal diagnosis and fetal pathology in a Turkish family harboring a novel nonsense mutation in the lysosomal alpha-N-acetyl-neuraminidase (sialidase) gene. Hum Genet. 2001;109(4):421–8. PubMed PMID: 11702224.

Sergi C, Poeschl J, Graf M, Linderkamp O. Restrictive dermopathy: case report, subject review with Kaplan-Meier analysis, and differential diagnosis of the lethal congenital contractural syndromes. Am J Perinatol. 2001;18(1):39–47. PubMed PMID: 11321244.

Sharma VK, Dhar S. Clinical pattern of cutaneous drug eruption among children and adolescents in North India. Pediatr Dermatol. 1995;12(2):178–83. PubMed PMID: 7659648.

Shin HT, Chang MW. Drug eruptions in children. Curr Probl Pediatr. 2001;31(7):207–34. Review. PubMed PMID: 11500668.

Sia PI, Figueira E, Allende A, Selva D. Malignant hair follicle tumors of the periorbital region: a review of literature and suggestion of a management guideline. Orbit. 2016;35(3):144–56. https://doi.org/10.1080/01676830.2016.1176048. Epub 2016 May 12. Review. PubMed PMID: 27171562.

Sillevis Smitt JH, van Asperen CJ, Niessen CM, Beemer FA, van Essen AJ, Hulsmans RF, Oranje AP, Steijlen PM, Wesby-van Swaay E, Tamminga P, Breslau-Siderius EJ. Restrictive dermopathy. Report of 12 cases. Dutch Task Force on Genodermatology. Arch Dermatol. 1998;134(5):577–9. Review. PubMed PMID: 9606327.

Singh DD, Naujoks C, Depprich R, Schulte KW, Jankowiak F, Kübler NR, Handschel J. Cylindroma of head and neck: review of the literature and report of two rare cases. J Craniomaxillofac Surg. 2013;41(6):516–21. https://doi.org/10.1016/j.jcms.2012.11.016. Epub 2012 Dec 21. PubMed PMID: 23260808.

Sourvinos G, Mammas IN, Spandidos DA. Merkel cell polyomavirus infection in childhood: current advances and perspectives. Arch Virol. 2015;160(4):887–92. https://doi.org/10.1007/s00705-015-2343-0. Epub 2015 Feb 10. Review. PubMed PMID: 25666196.

Stevens AM, Johnson FC. A new eruptive fever associated with stomatitis and ophtalmia: report of two cases in children. Am J Dis Child. 1922;24:526–33.

Suzumura H, Arisaka O. Cerebro-oculo-facio-skeletal syndrome. Adv Exp Med Biol. 2010;685:210-4. Review. PubMed PMID: 20687508.

Sybert VP. Aplasia cutis congenita: a report of 12 new families and review of the literature. Pediatr Dermatol. 1985;3(1):1–14. Review. PubMed PMID: 3906608.

Tebcherani AJ, de Andrade HF Jr, Sotto MN. Diagnostic utility of immunohistochemistry in distinguishing trichoepithelioma and basal cell carcinoma: evaluation using tissue microarray samples. Mod Pathol. 2012;25(10):1345–53. https://doi.org/10.1038/modpathol.2012.96. Epub 2012 Jun 8. PubMed PMID: 22684216.

Terrier-Lacombe MJ, Guillou L, Maire G, Terrier P, Vince DR, de Saint Aubain Somerhausen N, Collin F, Pedeutour F, Coindre JM. Dermatofibrosarcoma protuberans, giant cell fibroblastoma, and hybrid lesions in children: clinicopathologic comparative analysis of 28 cases with molecular data – a study from the French Federation of Cancer Centers Sarcoma Group. Am J Surg Pathol. 2003;27(1):27–39. PubMed PMID: 12502925.

Ujiie H, Kato N, Natsuga K, Tomita Y. Keratoacanthoma developing on nevus sebaceous in a child. J Am Acad Dermatol. 2007;56(2 Suppl):S57–8. PubMed PMID: 17224391.

Vanchinathan V, Marinelli EC, Kartha RV, Uzieblo A, Ranchod M, Sundram UN. A malignant cutaneous neuroendocrine tumor with features of Merkel cell carcinoma and differentiating neuroblastoma. Am J Dermatopathol. 2009;31(2):193–6. https://doi.org/10.1097/DAD.0b013e31819114c4. PubMed PMID: 19318809.

Wartchow EP, Goin L, Schreiber J, Mierau GW, Terella A, Allen GC. Plexiform fibrohistiocytic tumor: ultrastructural studies may aid in discrimination from cellular neurothekeoma. Ultrastruct Pathol. 2009;33(6):286–92. https://doi.org/10.3109/01913120903348860. PubMed PMID: 19929176.

Witt DR, Hayden MR, Holbrook KA, Dale BA, Baldwin VJ, Taylor GP. Restrictive dermopathy: a newly recognized autosomal recessive skin dysplasia. Am J Med Genet. 1986;24(4):631–48. PubMed PMID: 2426945.

Zheng JF, Mo HY, Wang ZZ. Clinicopathological characteristics of xeroderma pigmentosum associated with keratoacanthoma: a case report and literature review. Int J Clin Exp Med. 2014;7(10):3410–4. eCollection 2014. PubMed PMID: 25419376; PubMed Central PMCID: PMC4238483.

Placenta, Abnormal Conception, and Prematurity

18

Contents

18.1	**Development and Useful Pilot Concepts and Tables**	1410
18.2	**Pathology of the Early Pregnancy**	1422
18.2.1	Disorders of the Placenta Formation	1423
18.2.2	Disorders of the Placenta Maturation	1431
18.2.3	Disorders of the Placenta Vascularization	1434
18.2.4	Disorders of the Placenta Implantation Site	1434
18.2.5	Twin and Multiple Pregnancies	1442
18.3	**Pathology of the Late Pregnancy**	1449
18.3.1	Acute Diseases	1450
18.3.2	Subacute Diseases	1457
18.3.3	Chronic Diseases	1468
18.3.4	Fetal Growth Restriction	1482
18.4	**Non-neoplastic Trophoblastic Abnormalities**	1485
18.4.1	Placental Site Nodule	1485
18.4.2	Exaggerated Placental Site	1485
18.5	**Gestational Trophoblastic Diseases, Pre- and Malignant**	1486
18.5.1	Invasive Mole	1486
18.5.2	Placental Site Trophoblastic Tumor	1486
18.5.3	Epithelioid Trophoblastic Tumor	1486
18.5.4	Choriocarcinoma	1486
18.6	**Birth Defects**	1488
18.6.1	Birth Defects: Taxonomy Principles	1490
18.6.2	Birth Defects: Categories	1493
18.6.3	Birth Defects: Pathogenesis (Macro- and Micromechanisms)	1496
18.6.4	Birth Defects: Etiology (Mendelian, Chromosomal, Multifactorial)	1515
18.7	**Infection in Pregnancy, Prom, and Dysmaturity**	1533
18.7.1	Infection in Pregnancy	1533
18.7.2	Premature Rupture of Membranes (PROM)	1541
18.7.3	Fetal Growth Restriction (FGR) and *Dys*maturity	1541

© Springer-Verlag GmbH Germany, part of Springer Nature 2020
C. M. Sergi, *Pathology of Childhood and Adolescence*,
https://doi.org/10.1007/978-3-662-59169-7_18

18.8	IUFD and Placenta	1546
18.8.1	Fetal Death Syndrome (Intrauterine Fetal Demise, IUFD)	1546
18.8.2	Step-by-Step Approach in the Examination of a Placenta	1548

Multiple Choice Questions and Answers 1552

References and Recommended Readings 1554

18.1 Development and Useful Pilot Concepts and Tables

A variety of non-neoplastic lesions can present clinically and radiologically as uneventful or life-threatening conditions and may result in a hysterectomy. In such situations, the pediatric pathologist, general pathologist and/or gynecologic pathologist has an important role to play in determining the nature of these lesions. Awareness of the development of the embryo at an early stage and entities that can mimic malignancies will assist the pathologist in the diagnosis of the most common lesions with confidence. The placenta was probably one of the very few underutilized or poorly evaluated assessment of pathological routine. The roots of the problems were lying in (1) poor adaptability and familiarity of the clinical or surgical pathologist with placental lesions, (2) variable acceptance of the placental diagnoses in clinics, and (3) relative nonuniform terminology and standard diagnostic criteria used in different centers. Although this was the rule in the 1980s and 1990s, the situation changed in the twentieth century, when the pathologist correlated the gross and histopathologic data with the functionality of the placenta during pregnancy and the cardiotocography and the evolution of the birth with the neurological status of the newborn. Reasons for undervaluation of the placenta have been ascribed to some historical "incompetence" or low confidence of some pathologists in dealing with placenta specimens but also to the extreme variable heterogeneity of samples that may lead to different data and inconclusive or too speculative directions. There has been a huge advancement in the last couple of decades, which has delineated important factors that may be clinically relevant for both mother and fetus. Today, a detailed examination of the placenta provides, in most cases, a wealth of information. This information may help gynecologists in treating maternal conditions in the post-delivery period and perinatologists for the cure or treatment of the sick infant. In case of the unfortunate event of fetal or neonatal death, pathologic examination of the placenta is an essential component of the postmortem investigation, often giving useful insights for the diagnosis of the sick baby. In the past two decades, all three problems have been targeted by perinatal pathologists. Currently, there is more familiarity of the general pathologist with placental lesions, more understanding of the placental diagnoses by obstetricians and gynecologists, and there is a better terminology and fined criteria for most of the placental lesions. Moreover, the placenta may become a "gold" organ sometimes in the right hands; it was once said. In fact, some people would suggest that we might make this field quite lucrative, although it may be argued that it would be considered unethical. Dr. Drucilla Roberts' categories of situations where a pathologic examination of the placenta may be clinically, genetically, or forensically relevant include (1) legal issues regarding the presence and relevance of acute vs. chronic perinatal stress, (2) legal questions regarding the significance of potential acute vs. chronic perinatal insults, (3) diagnosis of fetal/neonatal/maternal diseases in the setting of abnormal pregnancies, (4) determination of zygosity in multiple pregnancies, (5) OB/GYN advise for the management of subsequent pregnancies, and (6) genetic counseling of the family and risk assessment for newborn and mother.

In Fig. 18.1 are pictured some usual early appearances of the placenta. The pregnancy outcome for the conceptus is as follows: about 15% of the oocytes undergo no fertilization, about 15% of the fertilized oocytes undergo no implantation, about 15% of the implanted blastocysts do not survive the end of the 2nd week, about 15% of the survived embryos harbor fatal malformations, and about 40% of the survived embryos are

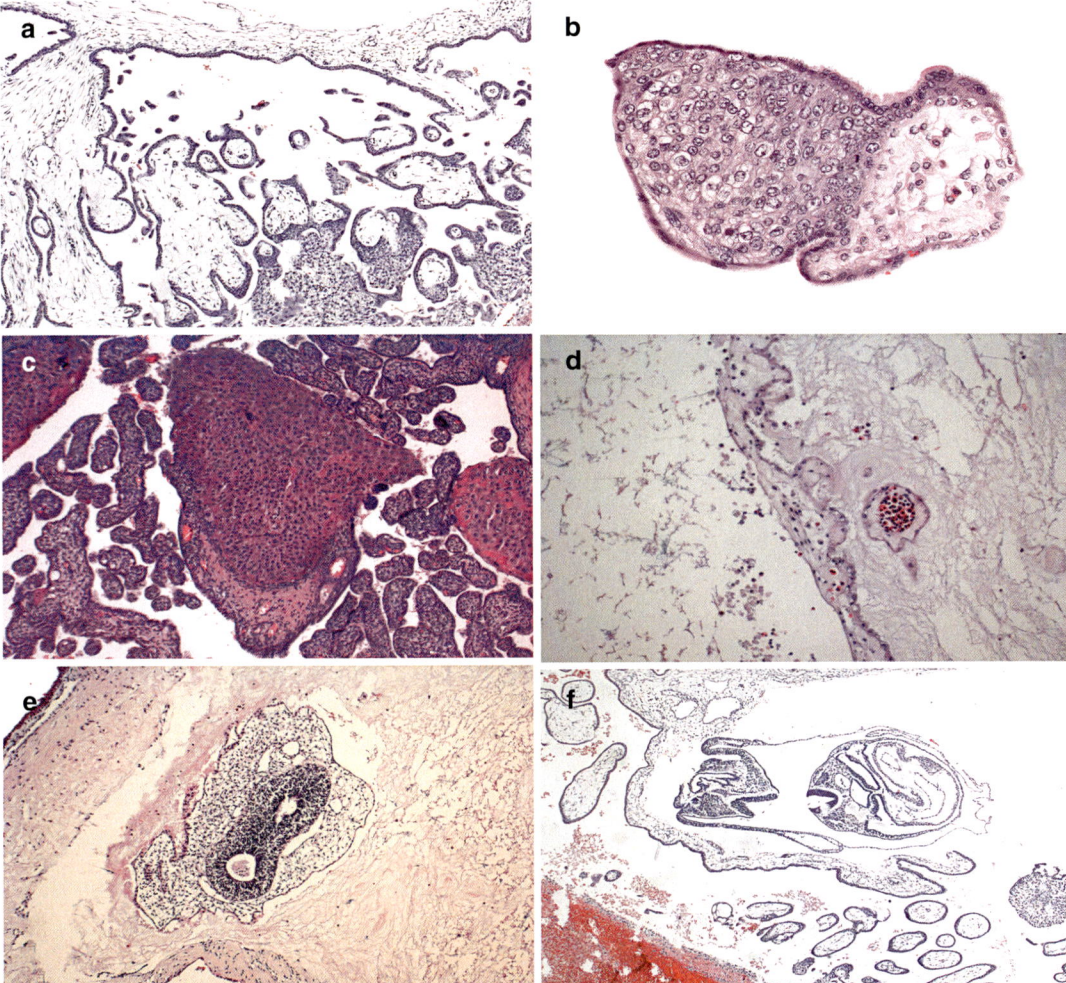

Fig. 18.1 Development of the placenta at early gestation. (**a**) From the chorion plate or primitive allantois, there is a centrifugal growth and proliferation of intermediate trophoblast (dark cells located in the lower right corner). (**b**) A close-up of the early placenta proliferations illustrates the polarized growth of the intermediate trophoblast. (**c**) In other sections of the placenta, the intermediate trophoblast may create an calyx shape on the top of the polarized villus (the renal calyx is a peculiar chamber in the kidney parenchyma that surrounds the apex of the renal pyramids). (**d**) A characteristic feature of the early gestation is the presence of nucleated red blood cells (NRBC) or progenitors of the erythrocytes that populate the cross sections of the blood vessels in the 2nd trimester. (**e**) This microphotograph sh ows the early embryo with the formation of the primitive neural tube. (**f**) This microphotograph shows the neural tube in a more advanced stage, although it has been transected during the preparation of the slide. All photographs are from H&E stained histological slides (a, original magnification x100; b, original magnification x200; **c**, original magnification x100; d, original magnification x200; e, original magnification x100; f, original magnification x100)

alive up to the second/third trimester or birth healthy or show some congenital defects.

"We ought not to set them aside with idle thoughts or idle words about 'curiosities' or 'chances'. Not one of them is without meaning; not one that might not become the beginning of excellent knowledge, if only we could answer the question – why is it rare, or being rare,why did it in this instance happen?" This sentence was written by James Paget at the end of the nineteenth century emphasizing how important is to identify the anomalous pathways behind the congenital or perinatal anomaly detected in life (Fig. 18.2). The main milestones of the embryo

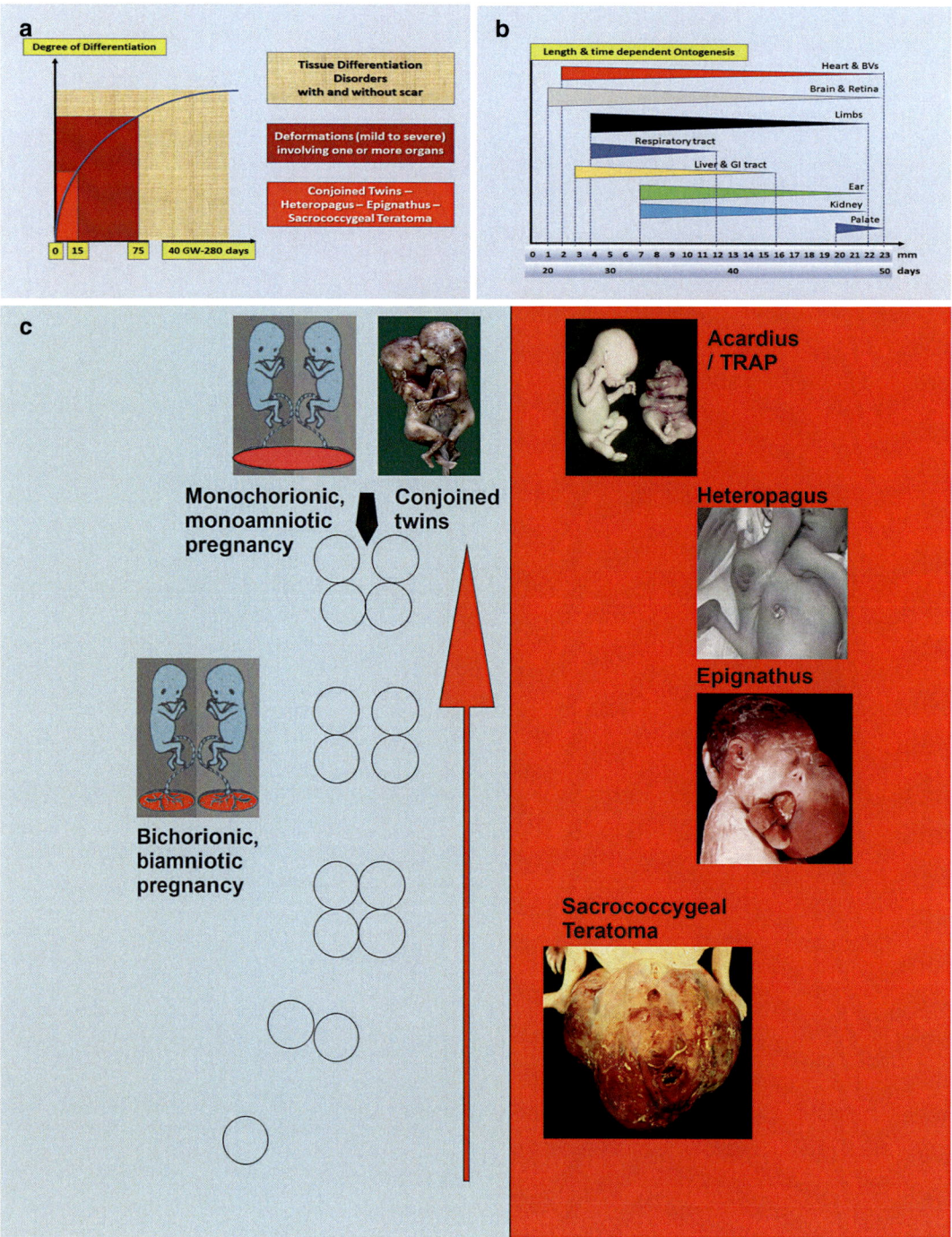

Fig. 18.2 Ontogenesis and early defects. In (**a**) is shown a graphic with the time on the X-axis and the degree of differentiation on the Y-axis. The graphic illustrates the severity of the birth defects in the early gestation with conjoined twins, heteropagus, epignathus, sacrococcygeal teratoma (red columns), and less severe disorders (yellow columns) later in the pregnancy. The figure (**b**) illustrates the same concept specifying the sensitive periods for the single systems of the body. In (**c**) is illustrated the formation of the morula and the respective major malformations that can originate in the early pregnancy

18.1 Development and Useful Pilot Concepts and Tables

Box 18.1 Main Milestones of the Embryo Development (First Embryonic Development)

- *1st week of the tubal migration*
  - Stage 1: Fertilization
  - Stage 2 (2nd–3rd day): Furrowing (2 up to 2^5 cells) and differentiation
  - Stage 3 (4th–5th day): Free blastocyst
- *2nd week: embryo implantation and two-layered germinal disc*
  - Stage 4 (5th–6th day): Attachment and implantation collapse
  - Stage 5 (7th–12th day): Two-layered germinal disc, amniotic cavity, and primary yolk sac
- *3rd week: three-layered germinal disc*
  - Stage 6 (13th–15th day): Extra-embryonal mesoderm, chorionic cavity, villi, primitive streaks
  - Stage 7 (15th–17th day): Chorda extension, stalk, allantois pouch, hemo- and angiogenesis
  - Stage 8 (17th–19th day): Primitive pit, chorda channel, axial channel
  - Stage 9 (19th–21st day): Neural folds, heart anlage, yolk sac folding
- *4th week: embryo folding*
  - Stage 10: Fusion of the neural folds, two pharyngeal arches, and eye furrows
  - Stage 11: Closure of the neuroporus anterior and optical vesicles
  - Stage 12: Closure of the neuroporus posterior, three pharyngeal arches, and arm buds
  - Stage 13: Leg buds, lens placode, otic vesicles

Box 18.2 Placenta Development: Villous Tree Branching

Villous branching	Gradual ↓ ∅ (day 15 p.c. → term)
Connective tissue	
2-years Villus	(++/+++ mesenchymal substance)
3-years Villus	(+ mesenchymal substance)
Hofbauer cells	(↓ # as gestation proceeds)
Vasculature (weeks – gestational age):	
0–6 weeks:	No capillary lumina
6–8 weeks:	Lumina + 100% nucleated hematologic precursors
10–12 weeks:	10% nucleated hematologic precursors
>12 weeks:	Lack of nucleated hematologic precursors

Notes: ∅ diameter, # number

Box 18.3 Placental Barrier (From External-Intervillous Space to Internal-Villus)

1. The outer layer of syncytiotrophoblast (STB) (continuous)
2. The inner layer of cytotrophoblast (CTB) (dis-/continuous)
3. Trophoblastic basal lamina
4. Connective tissue
5. Fetal endothelium

development in the first 4 weeks are summarized in Box 18.1. The embryogenesis entails 8 weeks after the fertilization. In comparison to the embryo of the 8 weeks, the fetus during the rest of the 38-week gestation has more identifiable external features and a quite more complete set of developing organs.

In Box 18.2 the placental development concerning the villous tree branching, which is a fundamental unit of the placenta, is summarized.

The placenta barrier concerning the exchange of oxygen between maternal and fetal blood is presented in Box 18.3. This concept is particularly important for the understanding of the function of the placenta and to direct new research topics in this very precious and fragile barrier.

The trophoblast changes during gestation. By the start of the 2nd month, the trophoblast is categorized by an increased number of second- and third-order villi that give rise to his unique radial appearance. With the progression of the placenta maturation, several settings are characterized. These include the chorionic plate, the outer cytotrophoblast shell, the lacunar or intervillous spaces, and the syncytial knots. Regarding stages of development, in the placenta, there is both a prelacunar and a lacunar stage. The prelacunar stage entails the apposition (days 7–8 postconception), syncytiotrophoblast (cellular fusion), and cytotrophoblast (stem cell-like) formation. The lacunar stage involves the formation of the lacunae (days 8–13 postconception), primary chorionic plate, lacunar system, trophoblastic shell, and cytotrophoblast and syncytiotrophoblast (day 12 postconception). The *chorion frondosum* and *decidua basalis* are also a critical component of the placenta and need to be adequately investigated. By the 4th month, the placenta has two components: fetal portion and maternal portion, i.e., chorionic and decidual plates, junctional zone, and decidual septa and cotyledons.

Mature Placenta

At term, the placenta is a discoid organ with a mean diameter of about 15–25 cm, is about 3 cm thick, and weighs about 500–600 g (Box 18.4).

The margin of coincidence between fetal and maternal surfaces is essential for the assessment of the presence or not there is a circumvallate or circummarginate placentation (vide infra).

Box 18.4 Placenta: Gross Main Features
- Discoid shape (15–25 cm in Ø, 3 cm in θ, ~500 g of weight)
- Two surfaces (fetal surface or chorionic plate, smooth, and maternal surface, rough)
- One insertion of the umbilical cord to the fetal surface
- Two arteries and one vein inside of the umbilical cord
- One margin with the coincidence of the diameters between fetal and maternal surfaces
- Placental membranes: amnion, chorion, and decidua

Moreover, the most critical aspect of the placenta and its pathologies is to understand the fetoplacental circulation.

Fetoplacental Circulation

1. Two circulatory systems, including (A) maternal blood that runs from the spiral arteries of decidua around villous bodies, intervillous space (through the syncytial-capillary membranes), and back via sinuses and (B) fetal blood that runs from the umbilical cord (vein), blood vessels of the chorionic plate on surface (in front of the fetus and antipodal to the maternal side), down stem villi, and then secondary and tertiary villi and reverse to the umbilical cord (arteries).
2. The intervillous space of the placenta at term (mature placenta) contains ~150 ml of blood, which is satisfactorily replenished ~3–4 times per minute.

Amniotic cells produce the clear, watery fluid, which fills the amniotic cavity. The amniotic fluid is derived primarily from maternal blood. The functions of the placenta are essentially an exchange of metabolic and gaseous products between maternal and fetal blood streams, exchange of gases, exchange of nutrients and electrolytes, transmission of maternal antibodies, and production of hormones.

In the placenta, the normal trophoblast is made up of cytotrophoblast (CTB), syncytiotrophoblast (STB), and intermediate trophoblast (IT). STB invades the endometrium allowing the implantation of the blastocyst and produces hCG, while CTB acts as a supplier to the STB with cells progressively becoming the chorionic villi covering the chorionic sac. Thus, the functional placenta is formed by the villous chorion adjacent to the endometrium and one basal layer of the endometrium together. The IT is localized in the villi, implantation site, and, apparently, in the chorionic sac.

Placenta Weight

Placental weight is a critical parameter to evaluate the functionality of the placenta during pregnancy. It was considered to having such an importance in the ancient time, and we are still

Box 18.5 Placental Weight Normality and Abnormalities

16 weeks of gestation – placenta:	60 g	fetus: ~60 g (×1)
25 weeks of gestation – placenta:	200 g	fetus: ~600 g (×3)
40 weeks of gestation – placenta:	500 g	fetus: ~3500 g(×7)
If PW at term >650 g ⇒ placenta >90th centile		
If PW at term <350 g ⇒ placenta <5th centile		

Notes: *PW* placental weight

Box 18.6 "Too Short" Cord: Seven Major Risk Factors

1. Alcohol consumption during pregnancy
2. Chromosomal abnormalities of the fetus
3. Chronic hypertension of the mother
4. Fetal growth restriction (FGR)
5. Gestational diabetes of the mother
6. Polyhydramnios
7. Tobacco consumption during pregnancy (probably also cannabis and vaping)

relying on this important parameter. At term, the placenta is in average 22 cm in diameter (Ø) and has a thickness (θ) of about 3 cm, and the weight is about 500 g (Box 18.5).

In the McNamara et al.'s study (2014), chronic hypertension was associated with low placental weight (PW) with a relative risk (RR) of 2.1 [95% confidence interval (CI) 1.8, 2.4] and 1.8 [95% CI 1.5, 2.1] before and after accounting for birth weight (BW). Conversely, preeclampsia was associated with low PW before, but not after adjustment for BW. Anemia and gestational diabetes have been associated with high PW (RRs 1.2–1.4, respectively) before and after adjustment for BW, while smoking was linked with high placental weight only after adjustment for BW (RR 1.4 [95% CI 1.3, 1.5]). In that Canadian study, placental and cord determinants of high PW included chorioamnionitis, chorangioma/chorangiosis, the placenta of circumvallate type, marginal cord insertion, and other cord abnormalities. The significant factors responsible for a small placenta at less than 10th percentile of the normal PW include chronic hypertension, chronic inflammation, and placental tissue malperfusion. The significant factors for a weight over 90th percentile rely on diabetes, anemia (maternal), abnormal storage disorder (e.g., lysosomal), genetic/chromosomal syndrome, and tumor (e.g., chorangioma).

Umbilical Cord

The umbilical cord (UC) is vital for the flow of blood between fetus and mother. Its length ranges between 45 and 55 cm (35–70 cm as range). UC spiraling is left-to-right spiral twist 4–7:1 with ~0.2 coils/cm (0.2=1/5). If its length is less than 35 cm, the UC is too short, while if the length is more than 70 cm, the UC is too long. The cord of the placental disc contains two arteries and one vein. It provides a connection between maternal and fetal blood circulation. As indicated above, the normal function of the placenta is to supply nutrients and O_2 to the fetus from the mother's blood and remove wastes from the fetal body. Shortening of the cord has been associated with fetal distress, cord rupture, and bleeding. It can also cause low Apgar scores. In Box 18.6 are the seven major risk factors.

The consequences of cord shortening are decreased fetal motion, FGR, perinatal morbidity (e.g., neurologic abnormalities, seizures, low intelligence quotient), and perinatal mortality. Conversely, the consequences of cord elongation are congestion and thrombosis (Virchow's triad), perinatal morbidity (e.g., hyperactivity syndromes), and perinatal mortality. Other than too short or too long cords, the cord may show several anomalies of its insertion to the chorionic plate (Box 18.7). Para-orthologue insertion of the umbilical cord (UC) should not give any consequences for the fetus or the mother, while potential implications for fetus and mother accompany pathologic inserts of the UC.

In the case of membranous or velamentous insertion, there is a high risk of compression injury and association with multiple gestations and congenital syndromes. Moreover, mothers may also have diabetes mellitus, be older than 35 years of age, and have a habit of smokers

> **Box 18.7 UC-Insertion Anomalies**
>
> 1. *Orthologue*: Central insertion of the umbilical cord
> 2. *Paralogue (Para-Orthologue)*: Slight eccentric insert (<1/4 placental disc)
> 3. *Heterologous (Pathologic) Cord Insertions*:
>   – Moderate-strong eccentric (<1/2 placental disc)
>   – Extremely eccentric (<3/4 placental disc)
>   – Marginal (external edge of the placental disc)
>   – Velamentous (unprotected blood vessels in the membranes)
>   – Furcate (a division of the UC a few cm before the attachment to the chorionic plate joining the placental disc singly and often gel unprotected)

(tobacco, cannabis, vaping). Both perinatal morbidity (neurologic abnormalities as well as hyperactive syndromes) and mortality are recorded. Catastrophic blood loss may occur in the fetus with rupture, often by rupturing vasa praevia (i.e., velamentous cord blood vessels overlying cervical os), and trauma causes fetal exsanguination and death. A true knot may occur in 1% of regular deliveries without having specific consequences. However, it needs to be differentiated from false knots. True cord features are the following ones. The true knot, when untied, shows compression with a groove at the knot site, partial or total loss of Wharton's jelly, and persistence of structural change after being untied. There may be optional edema, venous drainage, and vascular congestion distal to the knot location. An occlusive knot may induce intrauterine fetal demise (IUFD) due to thrombosis or hemorrhage. The dentation when untied plays a major role particularly in medicolegal cases, when untightening would mask a wrong diagnosis of no knots. Moreover, there is often evidence of edematous cord portion on the placental side due to cord vein obstruction. Moreover, the cord may show a single umbilical artery, the persistence of vitelline blood vessels, cord angioma, and tumors. In case of a single umbilical artery, which is found in 1% of deliveries, there is an association with genito-urinary tract anomalies of the fetus and significant congenital (general non-urological and non-genital) anomalies, chromosomal abnormalities, fetal growth restriction (FGR), and preterm delivery. Also, a single umbilical artery has been associated with stillbirth. Moreover, it is essential to identify the extrinsic compression. In the event of external compression, it is critical to look for occlusion venous thrombi, squeezing between the fetus and the maternal bony pelvis (cord prolapse) entanglement around fetal body parts. Intrauterine abnormalities include hypercoiling and torsion in association with anomalous cord insertion (marginal, membranous, furcate). Hypercoiling is defined as an excessive twisting of the cord. There are numerous great articles on cord coiling or "twisting." It has been argued that some patterns of coiling are associated with a worse prognosis for the fetus. The coiling index has been defined as the number of twists (coils) per 10 cm of cord in any placenta. A standard coiling index is between 1 and 3 coils/10 cm of UC, while hypocoiling and hypercoiling go below or above this range. Thus, hypocoiling is defined as <1 coil/10 cm, while hypercoiling is defined as >3 coils/10 cm. RFs for hypercoiling include a decrease of Wharton's jelly, an increase of cord length, a decrease of the amniotic fluid volume, frequent changes in fetal position, and fetal thrombophilia. However, 10–20% of apparently normal placentas show some areas of hypercoiling. Coiling of the cord is the result of fetal movement. In fact, focal areas of hypercoiling and localized strictures can be identified, even in normal placentas. The significance of cord strictures in cases of prolonged intrauterine fetal demise / death (IUFD) remains controversial, and several schools have interpreted this finding differently. However, it may be important to say that the interpretation of hypercoiling and intrauterine fetal demise should be reviewed in the setting of full clinical and obstetrical histories and following the perinatal pathology conference with physicians of the fetal medicine division. The outcome may invoke a dramatic reduction of the blood (O$_2$-rich) flow from mother to the fetus. Additional pathologic evidence of mechanical obstruction to fetal blood flow should be ruled out before making the hypercoiling as the cause of IUFD. It is important to stress that both hypocoiling and hypercoiling are also associated with an increased risk of FGR, fetal distress, cerebral palsy, premature or breech delivery, intrapartum death, and perinatal death in some older placental investigations. On the other hand, some studies diverge from these conclusions suggesting that a poor clinical outcome is rarely seen or be coincidental. In fact, some apparently "normal" placenta may also show hypocoiling or hypercoiling without evidence of abnormal pathology for the fetus. We and others suggest always to report abnormal coiling in the placenta pathology report. The gross and histopathological reports need to be clinically correlated. Some tips and clues of the UC torsion are presented in Box 18.8. Just before cessation of blood flow and IUFD, high resistance to the flow of both arterial and venous blood vessels is followed by absent diastolic flow and subsequent reverse flow.

Cord Infections

Infections of the UC may show acute funisitis, although various blood vessels may be associated with chorioamnionitis. Infective diseases of the umbilical cord may also show long-standing processes that go through calcification and then necrosis. *Candida* funisitis is the presence of *Candida spp.* in the cord and may be observed in women with diabetes mellitus (type I/type II) or immunosuppressive statuses. An etiology of syphilis (Treponema pallidum)or herpes simplex virus (HSV) may be encountered in case of a lymphoplasmacytic funisitis.

In the setting of an extrachorial placentation, the circummarginate placenta with a frequency of 3–25% needs to keep separated from the circumvallate placenta with a frequency of 2–20% (Box 18.9). Multiparity is a recognized risk factor for the circumvallate placenta. In case of a circummarginate placenta, fetal malformations have been described, while fetal growth restriction and preeclampsia are two medical conditions not rarely associated with the circumvallate placenta. Dr. Arizawa from Tokyo Metropolitan Ohtsuka Hospital, Japan, examined 1549 placentas in his practice and established the incidence of 14.9% (231/1549) of partial circummarginate (CM) and circumvallate (CV) and the incidence of 5.9% (92/1549) of total CM and CV.

The amniotic band is an intrauterine and intraamniotic fibrotic band able to entrap fetal parts leading to mutilations or death of the fetus. It is part of amniotic band syndrome, which is also known under other names in several countries worldwide, including amniotic band complex, ADAM (Amniotic Deformity & Adhesions Mutilations) complex, amniotic band sequence, and congenital constriction bands or constriction ring syndrome (CRS). CRS is the congenital defect due to a constriction ring, which is located around mostly a limb. Sequelae of the ongoing loss of much fluid include amniotic bands, fetal

---

Box 18.8 Pearls on Umbilical Cord Torsion

1. Check Doppler state for variable decelerations and "late" component!
2. Document all pathologic abnormalities (RF) that may predispose to UC integrity damage.
3. In case of cord torsion, first signs of chronic fetal compromise may be evident a few weeks before fetal death.
4. In case of suspicion of FGR, oligohydramnios, and chronic placental insufficiency, a Doppler flow velocimetry of the umbilical artery may be paramount.

Box 18.9 Circummarginate Versus Circumvallate Extrachorial Placentation

	Circummarginate	Circumvallate
Early fluid loss	No	Yes
Chronic siderosis	No	Yes[a]
Decidual necrosis	Yes	Yes

Notes: [a]Chronic hemosiderosis is due to bleeding during pregnancy

parts entrapped in webs of amniochorion, fetal amputation, fetal malformations, and fetal deformations. In Box 18.10, some features of the amniotic band complex are summarized.

Focal lesions of the amnion surface over the placenta and umbilical cord include *amnion nodosum*, squamous metaplasia, and *Candida* amnionitis. The first two entities are seen as whitish dots grossly, while *Candida* amnionitis are seen as yellowish dots. Amnion nodosum infers severe oligohydramnios of the pregnancy as observed in cases with renal agenesis, renal dysgenesis, and cystic renal disease and is associated with poorly developed fetal lung. *Amnion nodosum* is a consequence of severe oligo-/anhydramnios, which is due to prolonged, premature rupture of membranes and marked decreased urine production or output. If oligohydramnios is due to the fetus (no or marked ↓ of urine production), there are most often the congenital absence of kidneys, ureters, urethra, and urethral valves (posterior type). Other causes of oligo-/anhydramnios may also be causes of *amnion nodosum* with placental dysfunction among others (Box 18.11).

The word oligo- means few and a- is alpha privative, and hydramnios is the amniotic fluid (Gr. ὀλίγος 'few', ὕδρο- 'water', and ἀμνίον 'fetal membrane'). The word ending in -s is sometimes substituted with the ending in -n. Both words are similar, but the first is nominative, while the second is accusative. The consequences of amnion nodosum are impairment of fetal lung development (poor fetal lung development), ulceration of amnion, deposits of squamous lamellae and vernix on denuded amnion, intrauterine fetal death (IUFD), and perinatal morbidity (respiratory insufficiency ± renal failure).

Life-threatening placental lesions may be recognized grossly and potentially interpretable easily by naked eye exam following appropriate pediatric pathology fellowship training. The etiology of IUFD is summarized in Box 18.12. The tight nuchal cord is often associated with long cords (>90th percentile of the cord range for the gestational age). A giant chorangioma or multiple

> **Box 18.11 Etiology of Oligo-/Anhydramnios Other than Placenta Insufficiency**
>
> - Obstructive uropathy (e.g., posterior urethral valves)
> - Renal agenesis/dysgenesis ⇒ Nonfunctioning fetal kidneys (e.g., MCKD, ARPKD)
> - Genetic/chromosomal abnormalities
> - Preterm rupture of membranes (PROM)
> - Viral infections
>
> Notes: Viral infections can also cause polyhydramnios. Placenta insufficiency is a pregnancy complication when the placenta is unable to deliver an adequate supply of nutrients and O_2 to the fetus

> **Box 18.10 Amniotic Band Complex**
>
> - If fluid loss is mild and discontinuous ⇒ No sequelae, mostly!
> - If fluid loss is continuous ⇒ Sequelae, mostly!

> **Box 18.12 Nontumoral and Tumoral Etiology of Placenta-Related IUFD**
>
> 1. Placental abruption/infarction, extensive
> 2. UC, harboring ≥1 right knot(s), coarctation, thrombi
> 3. UC of tight nuchal type, furcation-building, or with velamentous insertion
> 4. Vasa praevia rupture
> 5. Chorionic vascular thrombosis, extensive
> 6. Terminal villous insufficiency, severe
> 7. Chorangiomata (multiple chorangiomas) or giant chorangioma
> 8. Choriocarcinoma
> 9. Non-choriocarcinoma placental tumors with mass effect
>
> Notes: *IUFD* intrauterine fetal death, *UC* umbilical cord

chorangiomatas may result in IUFD if they usually comprise more than one-third of the total placental volume. Similarly, non-choriocarcinoma placental tumors may be a cause of IUFD, if unusually large and also one-third of the total placental volume is considered sufficient to elicit the baby's death.

Meconium Release Discoloration Changes of Membranes and Cord

Meconium release discoloration changes of membranes and cord are due to the release of meconium in consequence of fetal distress, and the correct determination which membranes and cord are discolored is not only crucial for the neonatologist, but it may have critical medicolegal consequences and needs to be correctly recorded in the gross placenta examination. It is paramount to recall that in case of an extended period of exposure to meconium, eventually, meconium may not be easily discernible (Box 18.13).

The apoptosis (programmed cell death) observed in the myocytes of chorionic blood vessels exposed to meconium shows classic features of apoptosis with pyknosis of the nuclei. This phenomenon was described by Altshuler in 1989 and later observed and functionally correlated by King and Redline in 2004. All these reports are significant for the obstetricians and gynecologists (OBGYN). According to the color of the liquor, a decision is taken by the midwife, obstetrician, or general practitioner. The etiology of the meconium passage entails (1) intrauterine stress causing fetal hypoxia, asphyxia, and acidosis considering that hypoxia causes increased gastrointestinal peristalsis and relaxed anal sphincter tone and (2) mature (postdate) babies which involve myelination of nerve fibers, an increase in parasympathetic tone, an increase in the concentration of motilin, and vagal stimulation produced by a cord or head compression leading to intrauterine fetal stress. After the meconium passage, there is a medical decision according to the grade of staining. In case of liquor stained light green/yellow (Grade 1), it has been suggested to let labor progress. In case of liquor stained dark green (Grade 2), there is fetal distress (labor must be allowed only exceptionally). In case of meconium dominance over liquor with semisolid black paste (Grade 3), immediate delivery is the only option.

Hematoidin and Hemosiderin

Particular importance needs to be addressed in differentiating hematoidin correctly from hemosiderin, having two different meanings for the neonatologist. The "ABCDE" rule of an examination of the placenta may be applied (Box 18.14). What are the most critical aspects in assessing and recording placentae?

There are also some tips as well that need to be kept in mind from both OBGYN personnel and pathologist (Box 18.15).

Additional tips before discussing everything in detail are summarized in the numerous lawsuits or medicolegal case reviews. Placenta under the lens includes the examination of chronic fibrinoid deposition, chronic inflammation, infections (maternal/fetal/maternal and fetal), placenta delivery failure, reactive changes (pigment due to meconium or hemosiderin or calcium as well as upstream occlusions and downstream effects), and structural changes (abruptions, infarcts, thrombi, hemorrhages). Subacute/chronic hypoxia is responsible for about 15% of cases, while chronic placental dysfunction is seen in about 20% of well investigated. Siderosis may be seen in macrophages in 1–3 days, while 3–8 days (about 1 week) are necessary for Fe impregnation in other tissues. In nearly half of the cases, there is 1+ lesion. Babies delivered at term but affected

Box 18.13 Meconium Discoloration/Constitutional Changes of Membranes and UC

1.	Green amnion	following 1–3 h of exposure to meconium
2.	Green chorion	following 3–6 h of exposure to meconium
3.	Green decidua	following 3–6 h of exposure to meconium
4.	Green cord and CP	following >6 h of exposure to meconium
5.	SMC apoptosis (UC)	following 16 h of exposure to meconium

Notes: *h* hours, *SMC* smooth muscle cells, *UC* umbilical cord

> **Box 18.14 The "ABCDEFG" Rule**
>
> - *A*ssess the clinical information correctly and thoroughly with peer review!
> - *B*e free from preconceptions or providing unavoidably a diagnosis, if you don't have one!
> - *C*luster your findings into clinicopathologic syndromes following reading the OBGYN report, EFM traces, and Doppler velocimetry data (if available)
> - *D*ifferentiate peculiar/characteristic patterns one from each other (e.g., villous agglutination/maturity/syncytial knots, perivillous fibrin/fibrinoid deposition, processes affecting large blood vessels of fetal CP and decidua – *vide infra*)
> - *E*nhance your service with *e*fficiency (output of path reports) and *e*ffectiveness (the outcome of path reports) *e*mphasizing the relationship of your findings to *clinically relevant queries* interacting with OBGYN colleagues promptly!
> - *F*oster a healthy work relationship with the obstetrician and/or midwife explaining your pathology data in a simple and unambiguous vocabulary!
> - *G*ather a multidisciplinary team meeting with obstetricians and gynecologists, midwives, geneticists, neonatologists, and academic personnel of fetal medicine unit regularly (fortnightly or monthly)!

> **Box 18.15 Pearls and Pitfalls**
>
> - PW is highly relevant and may be the first indicator as an underlying pathology.
> - PW of fixed placenta = PW of fresh placenta + 10% (i.e., the addition of formalin increases the PW of about 10%, but leaving the placenta unfixed determines a PW loss, which is 4% of PW <12 hours, 6% <24 hours, and 10% <48 hours) (Popek & Collins, 2017).
> - Cord abnormalities are highly relevant for causes underlying fetal distress and IUFD
> - Placental malperfusion remains currently the second most common encountered pathology following ACA.
> - There is either fibrin or fibrinoid in the placenta, but the term "fibrosis" should be avoided.
>
> Notes: *UC* umbilical cord, *ACA* acute chorioamnionitis, *PW* placental weight

with cerebral palsy may show clinical/sentinel events during pregnancy (~20% of cases); severe, large fetoplacental vascular lesions (~1/3 of cases); and subacute/chronic hypoxia and chronic placental dysfunction (~1/3 of cases). In about 10% of cases, the etiology remains unknown (idiopathic cases). If a neurologic impairment is seen in infants ≥37 weeks with placental pathology, the etiology may include infection and inflammation, in the form of severe fetal chorioamnionitis (severe fetal chorionic vasculitis with subintimal expansion and dissolution of individual smooth muscle cells, meconium-associated fetal vascular necrosis, or smooth muscle cell necrosis); hemosiderosis of diffuse, chorioamniotic type; chronic peripheral separation; focal or diffuse Fe-laden macrophages; and avascular villi of quite extensive type.

The Ultrasound Biometric Report

Talking with OBGYN personnel or reading correctly their reports is here strongly emphasized and some explanation and notes to be taken in mind are necessary. The gestational sac is measured identifying the dark area from outer to outer – most significant measurement without including the white ring – and the diameter is $L \times B \div 2$, while the volume is $L \times B \times H \times 0.523$ corresponding to the volume of a sphere. If the gestational sac is visible with the yolk sac, the gestation is around 5 weeks, while the presence of an embryo dates to 6 weeks. The viability is determined by the simultaneous presence of gestational sac, embryo, and heart flicker. Fetal heart rate should be assessed using M-mode as far as possible mainly in the first trimester of the pregnancy. The crown-rump length (CRL) measures the embryo or embryonic pole in a sagittal section. CRL is measured from the tip of the cephalic pole (crown) to the tip of the caudal pole (rump). CRL is the most accurate dimension for determining gestation if adequately measured. However, several pitfalls need to be taken in mind, including fetal flexion or hyperextension, improper oblique section, and the inclusion of yolk sac in CRL measurement in early pregnancy. Biparietal diameter (BPD) demonstrates the fetal skull in a transverse section, aka occipito-transverse section, and should show all the following structures: midline, thalami, basal cisterns, and the cavum of the septum pellucidum; BPD pitfalls include fetal position, incorrect measurement level, and abnormal fetal head shape (e.g., dolichocephaly, brachycephaly). Dolichocephaly (long flat skull) in embryos and fetuses may be found in oligohydramnios, PPROM, and microcephaly, while brachycephaly (short, round skull) may be observed in breech anomalies (Greek δολιχός 'long' and βραχύς 'short'). In these cases, it is also important to measure the occipitofrontal diameter (OFD) from occiput to front (Box 18.16).

An embryonic cranial index (ECI) is crucial in defining dolichocephalic fetuses from brachycephalic fetuses. ECI = BPD ÷ OFD × 100 = 78.3%, but if ECI ≤75%, the fetus is dolichocephalic, while if ECI ≥85%, the fetus is brachycephalic. The head circumference (HC) is an indirect estimate of the size of the fetal brain, in consideration that skull sutures have not yet closed (no craniosynostosis or σύν- + ὀστέον 'bone' + -osis, which is the fusion of adjacent bones by the pathologic growth of a bony tissue material) and may be measured from 15 weeks of gestation on. The abdominal circumference (AC) is the circumference of the abdomen representing an indirect estimate of the fetal liver size and glycogen storage/nutritional status of the fetus. AC is usually measured between 15 and 23 weeks of gestation and shows a cross section through the fetal abdomen and is AC = 3.14 {(APAD ÷ 2 + TAD ÷ 2)} (APAD antero-posterior diameter, TAD, transverse abdominal diameter). Another critical measurement is the femur length (FL), which means the length of the femur shaft. This measurement is also performed between 15 and 23 weeks of gestation. It measures from the greater tubercle of the femur to its distal end. If it is too close to the BPD, we may erroneously measure the humerus and careful inspection is necessary. In assessing the amniotic fluid, four quadrants need to be summated, and if single pocket >8 cm, it means polyhydramnios, while a single pocket <2 cm indicates oligohydramnios.

Correlations with abnormal development occur in case of premature aging (higher than average grade), which may be associated with FGR or PE, and persistence of Grade 0 into the third trimester may be singularly associated with cardiovascular or renal abnormalities of the fetus.

Pregnancy Complications

Pregnancy can be complicated, and OB/GYN physicians like to divide the complications according to the delivery time.

- *Antepartum* (early <24 weeks, late >24 weeks)
- EARLY: Miscarriage, ectopic, and mole
- LATE: Antepartum hemorrhage (APH)
- *Intrapartum* (during labor)

> **Box 18.16 Gestational Age-Related Biometry Data and Ultrasound Embryology**
> - LMP
> - <6 weeks (early gestation): gestation sac and content
> - 6–14 weeks: CRL embryo fetus
> - 14–23 weeks: BPD, HC, TCD, AC, FL
> - Chorionic sac visible in the uterus
> - Chorionic sac visible in the endometrium lateral of the midline
>
> Notes: *AC* abdominal circumference, *BPD* biparietal diameter, *CRL* crown-rump length, *FL* femur length, *HC* head circumference, *LMP* last menstrual period, *TCD* transverse cerebellar diameter

- Fetal distress, abnormal fetal presentation
- Maternal exhaustion, uterine bleeding, uterus rupture
- *Postpartum* (from delivery until 6 weeks later)
- Sepsis, shock, cardiac failure, renal failure

Antepartum hemorrhage (APH) is a medical emergency. APH is genital bleeding that occurs during pregnancy between the 20th and 24th weeks of pregnancy and up to delivery. It may be life-threatening and can be linked with reduced fetal BW. The prevention of APH has been targeted in several studies, and the use of aspirin before 16 weeks of pregnancy to prevent preeclampsia appears effective at preventing APH.

18.2 Pathology of the Early Pregnancy

It is important to emphasize that appropriate placental investigation leads to better patient care and each gross examination finding needs to be correlated with microscopic results and clinicopathological correlated. The recognition, staging, and grading of placental lesions are essential and are at the basis of specific clinicopathologic syndromes. The utility of a detailed placental examination has been emphasized in previous workshops and reviews. Particular pathogens (e.g., *Listeria, Treponema, Toxoplasma, Candida*), fetal diseases (e.g., inborn errors of the metabolism, coagulopathies), and maternal illnesses (e.g., placental retention, proliferative and neoplastic gestational trophoblastic disorders) are specifically targeted during the placental examination. A recurrence risk needs to be considered and discussed with the obstetrician or family doctor, who will liaise with the mother. Recurrence risks include some clinicopathologic events, such as maternal floor infarction, chronic intervillositis of histiocytic type, diffuse chronic villitis, thrombophilia-associated vasculopathy of maternal origin, and chorioamnionitis with decidual plasma cells. The indications for a placental investigation adapted by the College of American Pathologists recommendations may be:

Placenta-based

- When there are abnormalities of shape, color, and odor
- When there are masses or hemorrhages
- When there is an abnormal cord insertion or there are cord anomalies or lesions
- When the cord is too short (i.e., <32 cm) or too long (>100 cm)
- When the placenta is SGA (i.e., <350 g) or LGA (>650 g)

Baby-based

- When there is birth distress (i.e., pH < 7.0, Apgar < 7, Hct < 35, assisted ventilation >10 min)
- When there is IUFD or early neonatal death
- When the baby has been admitted to NICU
- When birth is part of multiple pregnancies
- When the baby is SGA (i.e., <10th percentile) or LGA (i.e., >90th percentile)
- When neonatal seizures have been recorded
- When suspicion for infection has been raised
- When there is *hydrops fetalis*
- When MCA (multiple congenital anomalies) or single anomalies have been noted

Mother-based

- When preterm delivery occurred (i.e., ≤36 weeks)
- When there is maternal fever or suspicion for an infection
- When vaginal bleeding occurred
- When abnormal antenatal testing occurred leading to intervention
- When systemic diseases are recorded (e.g., hypertension, diabetes (T1DM, T2DM), autoimmune disorders)
- When meconium is thick or viscid
- When oligo-/anhydramnion or polyhydramnios have been recorded in the maternal chart
- When history is positive for substance abuse

The pathology of the early pregnancy can be subdivided as indicated in Box 18.17.

18.2 Pathology of the Early Pregnancy

> **Box 18.17 Pathology of the Early Pregnancy**
> 18.2.1. *Disorders of the Placenta Formation*
> 18.2.1.1. GD1 or Anembryonic Pregnancy
> 18.2.1.2. GD2-4 or Embryonic Miscarriage
> 18.2.1.3. Partial Mole (Partial Molar Degeneration)
> 18.2.1.4. Complete Mole (Complete Molar Degeneration)
> 18.2.2. *Disorders of the Placenta Maturation*
> 18.2.3. *Disorders of the Placenta Vascularization*
> 18.2.4. *Disorders of the Placenta Implantation Site*
> 18.2.5. *Inflammation*
> 18.2.6. *Twin and Multiple Pregnancies*

18.2.1 Disorders of the Placenta Formation

The risks may also be associated with preterm birth, intrauterine growth restriction (IUGR), and neurodisability. Placental lesions associated with preterm delivery may include acute chorioamnionitis, maternal vascular underperfusion, marginal abruption, chronic abruption, and chronic deciduitis. Placental lesions associated with IUGR of the fetus may include maternal vascular underperfusion, chronic villitis, perivillous fibrin deposition, and fetal vasculopathy of thrombotic origin of the fetus. Neurodisability is also significantly associated with fetal chorioamnionitis with or without associated thrombi in the chorionic plate, severe villous edema, and multiple placental lesions for preterm babies and with fetal thrombotic vasculopathy, chronic villitis, meconium-associated placental lesions of necrotic type, chorioamnionitis, and numerous placental lesions (Figs. 18.3 and 18.4).

Embryo and First-Trimester Specimens
The embryo is defined as a conceptus until the end of the 8th week, and death of the embryo characterizes an early spontaneous abortion, while the fetus is described as a conceptus starting with the 9th week of gestation until birth. In this section, it is important to remind that embryonal weeks or weeks of embryogenesis are different from the gestational age, which is calculated from the last menstrual period (LMP). Thus, embryonic age is a gestational age less 2 weeks. This calculation is particularly important when assessing the degree of differentiation of tissues and organs of the conceptus. The embryonic age or developmental age is dated from fertilization, while gestational age is dated from LMP (Box 18.18). The dilatation and evacuation (D&E) specimens are usually not sent to laboratories, because they are voluntary interruptions of pregnancy.

Causes of miscarriage or spontaneous abortion in the first trimester may be entirely different and be grouped into categories if a well-developed embryo is identified by the perinatal pathologist. Pregnancy loss in the first trimester may be due to *corpus luteum* (ovary)-dependent low progesterone production, non-hypoprogestinic estrogen/progesterone balance, non-estrogenic/progestin hormonal dysfunction, endometrial insufficiency or abnormal endometrial stroma or milieu (multiple causes), chromosomal abnormalities or lethal genetic mutations, dysmorphic and other disruptive events, teratogens, autoimmune and alloimmune abnormalities, specific parental human leukocyte antigens (HLA) sharing or excessive genomic sharing, infections, abnormal uterine anatomy, and ectopic pregnancy.

Multiple causes may be the underlying factors linked to an abnormal endometrial lining and a previous intrauterine device for contraception, among others. Less commonly, the abnormal anatomy of the human gynecologic system or an ectopic pregnancy has been considered by the first-trimester miscarriages. Excessive genomic sharing or parental HLA (human leukocyte antigen) sharing may be seen in conceptuses between parents, who are first cousins or even in legal causes of incest. The HLA is the human version of the major histo-compatibility complex (MHC), which specifies a gene group that occurs in many species.

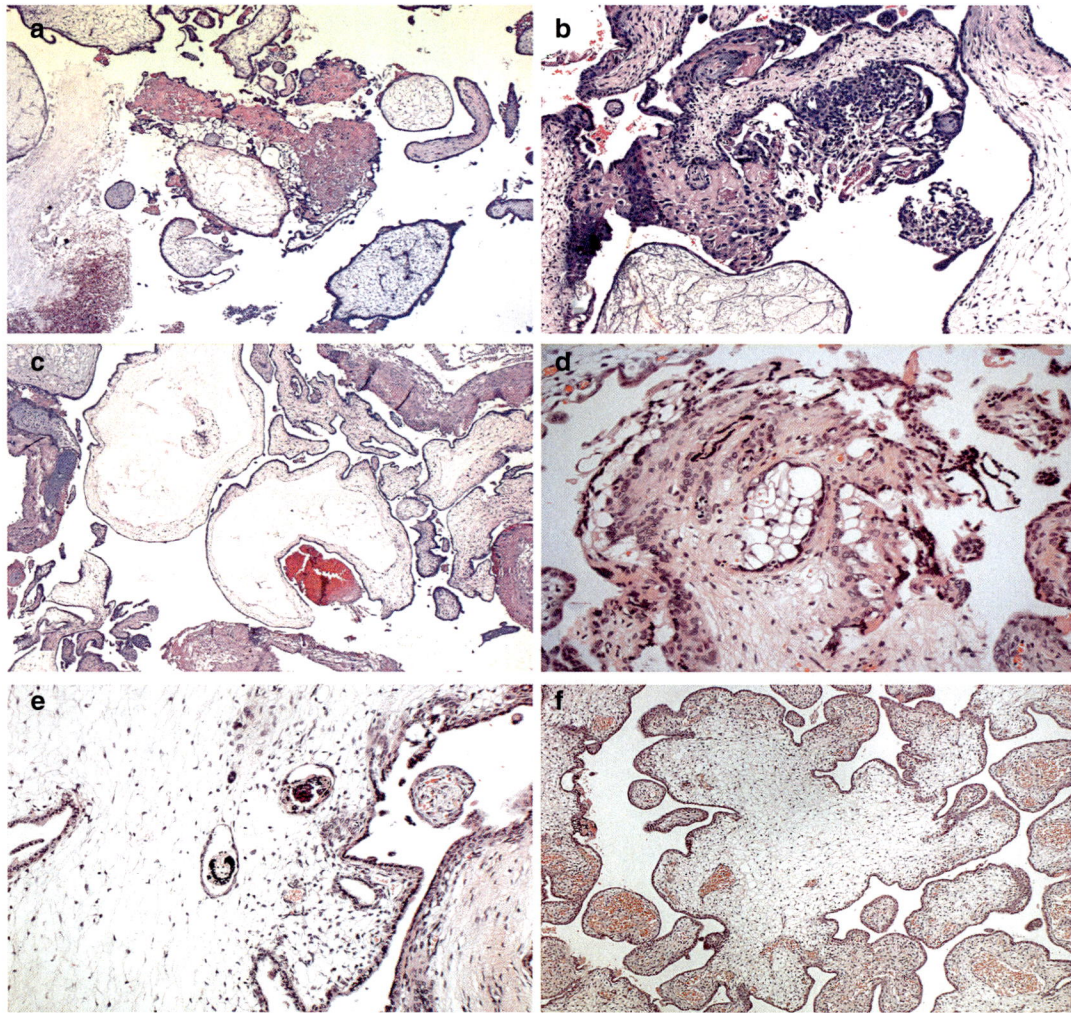

Fig. 18.3 Partial mole degeneration. (**a**) through (**f**) evidentiate the degeneration identified in partial mole with cystic trophoblast hyperplasia (**a**), non-cystic trophoblast hyperplasia at places (**b**), cistern formation (**c**), with trophoblast inclusions in the stroma of the villi (**d**, **e**), and abnormal (geographical) shape of the stem villi (**f**). All microphotographs are from H&E stained histological slides (a, original magnification x40; b, original magnification x100; c, original magnification x40; d, original magnification x200; e, original magnification x200; f, original magnification x40)

The major landmarks of the development (DPC: days postconception) of the early fetus are as below:

35–37	DPC	Retinal pigment identification
42	DPC	Separation of the common aortic-pulmonary trunk
51	DPC	Distinct elbow and developing eyelids
56	DPC	Lack of tail, but the presence of fingernails and well-defined neck
56–70	DPC	Scalp vascular plexus and intestines into the umbilical cord
70–84	DPC	A fetus, no longer an embryo

The aortic-pulmonary trunk separation at 42 days postconception means individualization of the aorta and pulmonary artery as separate blood vessels and is a typical landmark of both ultrasound and anatomy.

There are a few causes that need to be kept in the mind of the pediatric pathologist in performing the fetal autopsy of a stillborn. The concealed etiology of IUFD on second-trimester gestation include:

- Atresia of the *ductus arteriosus* (Botalli)
- Atresia of the *foramen ovale*

18.2 Pathology of the Early Pregnancy

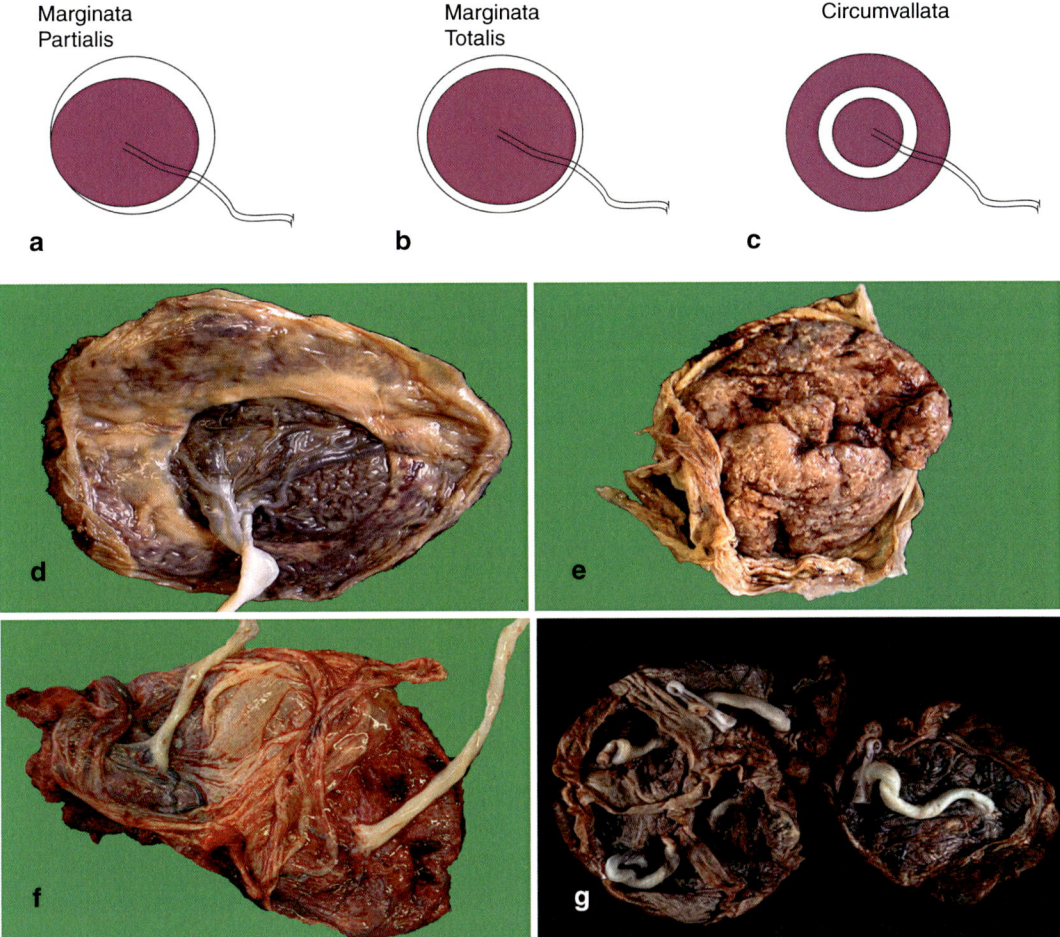

Fig. 18.4 Placenta formation disorders. The schema with the cartoon between (**a**) and (**c**) illustrates some abnormal shape of the placenta indicated as marginata partialis (**a**), marginata totalis (**b**), and circumvallata (**c**). In all three placentas, the edge is abnormally placed (see text for details). In (**d**) is the gross photograph of a circumvallate placenta. Gross photograph (**e**) shows the basal plate of a circumvallate placenta that may disclose some disruption or loss of continuity. The gross figures (**f**) and (**g**) show twinning with a monochorionic diamniotic placenta (**f**) and a pentaplet pregnancy (g) with two separated chorion plates, the one with four amniotic cavities identified by four umbilical cords and the other with only amniotic cavity identified by a single umbilical cord

---

Box 18.18 First-Trimester Specimens
1. Embryos without apparent anomalies
2. Embryos with growth disorganization (GD)
3. Fetuses from the early fetal period after embryogenesis (0–8 weeks), i.e., 9 – 12 + 0 weeks
4. Dilatation and evacuation (D&E) specimens

Ductal atresia is seen on the surgical bench between 14 and 16 weeks' gestation and may be commonly missed as a pathologic diagnosis in these small fetuses.

There are also a few terms and definitions that are commonly used in OBGYN – perinatal pathologist talks or multidisciplinary team meetings that are important to improve interdisciplinary communication.

- *Complete Specimen*: An intact chorionic sac that may be empty or contain various embryonic or extraembryonic tissues
- *Incomplete Specimen*: An opened or ruptured chorionic sac without an identifiable embryo
- *First Trimester*: Period encompassing the first 12 weeks of pregnancy
- *Second Trimester*: Period covering 13th to the 24th week of pregnancy
- *Third Trimester*: Period covering the period between 25th and the 40th weeks of pregnancy
- *Chromosomal Defects*: 5–10% of stillbirths
- *Confined Placental Mosaicism*: Chromosomal abnormalities in the extraembryonic tissue which are absent from the fetal tissue (~1–2% of CVS at 9–12 weeks) and subtyped in type 1 CPM (CPM1: cytotrophoblast only), type 2 CPM (CPM2: mesenchymal core of the chorionic villi), and type 3 CPM (CPM3: cytotrophoblast and mesenchymal core) (Totain et al. 2018)
- *Early Neonatal Period*: First 7 days of life
- *Early Spontaneous Abortion*: It occurs in the embryonic period
- *Embryonic Period*: Conception to the end of the 8th-week postfertilization (FA)
- *Fetomaternal Hemorrhage* (FMH): Entry of fetal blood into the maternal circulation $\leq$ delivery with antenatal FMH harboring a wide spectrum of clinical variation
- *FGR* (fetal growth restriction) = *IUGR* (intrauterine growth retardation)
- *Gestational Age* (GA): Age at birth measured from the 1st day of the LMP
- *Late Spontaneous Abortion*: Between 9th and 20th weeks
- *Low Birth Weight*: Birth weight less than the 5th percentile expected for gestational age
- *Major Malformations*: They are identified in about 5-25% of stillbirths
- *Multiple Pregnancies*
- *Neonatal Period*: First 28 days of life
- *Perinatal Death*: Stillbirths and early neonatal deaths <7 completed days from birth
- *Perinatal Period*: 6-week period including prenatal, natal, and postnatal time
- *Postconceptional Age*: Total age of the conceptus measured from the estimated day of conception and including the postnatal period (vide infra for fertilization age and fetal age)
- *Postmaturity*: Delivery $\geq$42 weeks of gestation
- *Prematurity*: Delivery at less than 37 weeks of gestation
- *Prenatal*: Before birth
- *Stillbirth*: Delivery of an infant with no signs of life between 20 weeks of gestation

Gestational age (GA) is a measure of the age of a pregnancy. GA is taken from the woman's last menstrual period (LMP). It includes adding 14 days to a known duration since fertilization (as is possible in in vitro fertilization (IVF)) or by obstetric ultrasonography. We often use such a definition of GA, because menstrual periods are typically always noticed, while there is a lack of a comfortable way, although some women can identify very well, to discern when ovulation or fertilization occurs. The *fertilization age* or FA (also called embryonic age and, later, probably improperly fetal age) is the time from the fertilization. The fertilization typically occurs within 1 day of ovulation, which, in turn, occurs on average 14.6 days after the beginning of the preceding LMP.

According to the American College of Obstetricians and Gynecologists, the primary methods to calculate GA are:

1. Directly counting the days since the start of the LMP
2. Early obstetric ultrasound, comparing the size of the conceptus to that of a reference group of pregnancies of known GA
3. In the case of IVF, calculating the number of days since oocyte retrieval or co-incubation with spermatozoa and adding 14 days

Thus, by convention, GA = FA + 2 weeks or, otherwise, FA = GA – 2 weeks.

Fetal age is not calculated from the date of the LMP but the time of conception. Thus, fetal age is practically equal to fertilization age.

Usually, menstruation starts 14 days after ovulation. When menstruation is missing, the woman uses a pregnancy test to verify the pregnancy. Then, it should start from the LMP.

Ovulation and conception usually occur in the middle of the menstrual cycle, considering that the ovulation cycle lasts for 28 days. In conclusion, fetal age and FA are less than the GA. The WHO defines the perinatal period as the period starting at 22 completed gestational weeks (154 days) of pregnancy and ending 7 completed days (1 week) after birth. Perinatal mortality is the death of fetuses/neonates during the perinatal period. If only a small proportion of births occur before 24 completed weeks of gestation (probably about 1 per 1000), survival is quite rare. In this setting, most of these babies are either fetal deaths or live births followed by a neonatal death, subsequently.

Grannum Score
One of the fascinating scores in placenta maturation and probably one of the most debated scores in obstetrics and gynecology is the Grannum score (1979, 1982). According to the Grannum classification, this score refers to an ultrasound grading system of the placenta based on its maturity. This score primarily involves the extent of calcifications. Although in some countries the use of the placental grading has obviously fallen out of obstetric practice due to a weak correlation with adverse perinatal outcome, it may be found in other countries. Although limitations need to be taken into account the Grannum score may be considered as an approximative orientation on the maturation of the placenta (Ragozzino et al. 1983; Hills et al. 1984; Dudley et al. 1993; Merz and Bahlmann, 2005; Fadl et al. 2017).

An early progression to a grade III placenta is concerning and may be associated with placental insufficiency, a setting which occurs in mothers who smoke or are affected with chronic hypertension, SLE, or diabetes mellitus (T1DM, T2DM). There are 11 times greater risk of IUFD in the case of multiple pregnancies. IUFD is usually during the second trimester. The mortality rate of the third trimester in case of multiple pregnancies is also increased. Multiple pregnancies account for 7–15 of stillbirths. It is 2–3 times higher for monozygous than dizygous twins. The uteroplacental insufficiency is a common cause of asymmetric fetal growth restriction (FGR). There is a relatively large head compared to the body size. Chromosomal abnormalities are typically associated with symmetric growth restriction. Multiple pregnancies and fetal growth restriction (FGR) may have a bunch of causes, including:

- ↑ brain/liver ratio (3:1 ⇒ 6:1)
- Ascending infection
- Early placental separation
- Fetal anomalies
- Incompetent cervix
- Premature dilation
- Uteroplacental insufficiency

Disorders involving an abnormal formation of the placenta include an anembryonic pregnancy, a so-called embryonal mole or early miscarriage and the molar degenerations of partial and complete type, which are subclassified more in detail in the section of the gestational trophoblastic diseases. A very useful classification used by both gynecologists and perinatal pathologists is the "growth disorganization" (GD) classification (Philipp et al. 2003). The identification of the implantation site changes is critical in ruling out ectopic pregnancy, and the gold standard is the presence of chorionic villi. Arias-Stella reaction is a phenomenon seen in case of pregnancy and is characterized microscopically by cellular stratification, enlargement of cells with bizarre forms, and hyperchromatic nuclei. Arias-Stella reaction may be seen with both intrauterine and ectopic pregnancies. Thus, it is crucial to identify chorionic villi for an orthotopic pregnancy in the uterus. In the absence of chorionic villi, the immunohistochemistry staining of intermediate trophoblast cell populations surrounding hyalinized blood vessels may be a possible sign of intrauterine gestation.

Genomic imprinting is a genetic situation by which precise genes result to be selectively expressed in a parent-of-origin-specific manner. The imprinted or differentially expressed genes or alleles are practically muted or better silenced according to the genetic nomenclature, such that the genes are either expressed only from the non-imprinted allele inherited from the other parent (e.g., *CDKN1C* gene) or the non-imprinted from the allele of the same parent (e.g., IGF-2 gene). It is also known that DNA methylation and histone modifications are essential in regulating genomic imprinting in the healthy development of organs and tissues. However, genetic diseases associated with imprinting defects are also known, including Beckwith-Wiedemann syndrome, Silver-Russell syndrome, Angelman syndrome, and Prader-Willi syndrome. The *CDKN1C* gene has a gene product p57kip2. *CDKN1C* is a paternally (differentially) imprinted, maternally expressed gene (11p15) and showed unique and well-delineated patterns of expression in the trophoblastic population. The staining is nuclear only, as it is to expect from a nuclear gene product (Box 18.19). Growth disorganization (GD) without chromosomal abnormalities may disclose submicroscopic lethal genetic defects are preventing normal embryogenesis or teratogenic effects interfering with normal embryogenesis, and clinical and environmental history of the parents is extremely important.

Growth disorganization (GD) classification includes:

- GD 1 – Intact chorionic or amniotic sac with no evidence of an embryo or yolk sac
- GD 2 – Fragment of embryonic tissue (1–4 mm) with neither recognizable external structures nor cephalic/caudal pole
- GD 3 – Disorganized embryo (5–10 mm) with both identifiable cephalic and caudal poles but morphology not distinguishable (±retinal pigment)
- GD 4 – Major distortion of body shape always involving the head (3–17 mm), but not consistent with any known stage of development

Gestational trophoblastic diseases (GTD) are a group of conditions showing a variable range of biologic behavior and potential for metastasis and life-threatening illness. GTDs include hydatidiform mole (partial and complete), invasive mole, choriocarcinoma, and placental site trophoblastic tumor, being the last three properly tumors, because of the metastatic potential. The incidence of the hydatidiform mole is 0.6–1.1 per 1000 pregnancies in the Western countries, but the rate is notably higher in Asia. The incidence of choriocarcinoma is 1 per 20,000–40,000 pregnancies, although ethnics also play a role. Correct identification of the products of conception is crucial for both pathologists and gynecologists. Hydatidiform mole or hydatidiform molar degeneration is classified as complete (CHM) or partial mole (PHM) using clinical, pathological (surgical and clinical pathology tools), and genetic investigations.

18.2.1.1 Partial Mole (Partial Molar Degeneration)

Partial hydatidiform mole (PHM) is identified histologically as patchy villous *hydrops gestationis* with scattered abnormally shaped irregular villi and patchy trophoblast hyperplasia with scalloped villous outlines giving rise to intravillous invaginations of trophoblast with pseudoinclusions associated with a background of normal-appearing, spongy, villous component (*dimorphic villous population*) (Fig. 18.3).

18.2.1.2 Complete Mole (Complete Molar Degeneration)

Complete hydatidiform mole (CHM) is identified histologically as abnormal budding villous structure showing frequently circumferential trophoblast hyperplasia, intravillous cistern formation, karyorrhectic stromal debris, and the collapse of the villous blood vessels (*uniform villous population*)

Box 18.19 Cyto-/Syncytiotrophoblast and Intermediate Trophoblast: IHC Panel

	AE1-3	HCG	HPL	p57^{kip2}
Cytotrophoblast	++	−	−	+
IT	++	+	++	+
Syncytiotrophoblast	++	++	+	+

Notes: *IHC* immunohistochemistry, *IT* intermediate trophoblast, *AE1-3* pan-keratins of low and high molecular weight, *HPL* human placental lactogen, *HCG* chorionic gonadotropin

Cisterns are usually defined as acellular, fluid-filled intravillous spaces, which are exactly demarcated by villus stroma..

18.2.1.3 Hydropic Abortus/Hydropic Molar Degeneration (HMD)

The hydropic abortus (HAM) or hydropic molar degeneration (HMD) or non-molar miscarriage (NMM) presents as a missed or spontaneous abortion and less tissue on curettage and typically occurs at 6–14 weeks of gestation. The evaluation for β-HCG is paramount and shows no increase, and the uterine size is either average or low for gestational age. Histologically, there is edema with vascularized villi that have a smooth rounded contour, different from the classic histology of PHM and CHM. Occasionally, trophoblastic hyperplasia can be detected but shows a polar arrangement. Circumferential trophoblastic hyperplasia is never seen. However, trophoblastic inclusions are sometimes observed, although they are typically absent. No embryonic elements are noted, although degenerative or autolytic foci may be seen. Hydropic molar degeneration (HMD) or better labeled NMM may be challenging in differentiating it from PHM. Immunostaining for $p57^{kip2}$, the product of *CDKN1C* (expressed by the maternal allele), is extremely useful, being detectable as nuclear staining of cytotrophoblast and villous mesenchyme but lacking in the androgenetic CHM. Flow cytometry or ISH are helpful in distinguishing diploid from triploid products of conception. However, both techniques are not useful in the distinction between CHM from diploid NM miscarriage or molar vs. non-molar triploidy (Box 18.20).

Some authors differentiate between early complete mole and CHM. Early complete mole (ECM) is characterized by an early hydatidiform degeneration of the villous parenchyma that, if left *in situ*, would develop into a full-size CHM. ECM is evacuated before all characteristics of CHM are reached. It is that ECM lacks florid trophoblastic hyperplasia and degeneration of the fetal blood vessels, but the specimen is grossly more substantial than one would expect for the given gestational age. Some swollen villi are, however, seen. Main features of ECM include abnormally shaped villi, which may be branching or polypoid, stromal mucin, recognition of stromal vessels, and stromal nuclear debris. The nature of the villous and extravillous trophoblastic cell population makes ECM distinguishable from NMM or hydropic abortion. In fact, the excessive proliferation, even focally seen, is to begin in ECM. Characteristically, the cytotrophoblast is more stratified than expected, and the overlying STB forms lacy surface projections.

Cytogenetics of a typical GTD of complete molar type (CM) has an androgenetic, diploid 46, XX karyotype (90%), and there is no maternal DNA. In the remaining 10%, a 46, XY karyotype is observed. The androgenetic karyotype is deriving from dispermy, i.e., two separate spermatozoa fertilizing an "empty" ovum lacking maternal chromosomes. Conversely, GTD of partial molar type (PM) is classically triploid, which shows two sets of paternal chromosomes (diandric) and one haploid maternal set. The most common composition is 69, XXY, but 69, XXX and 69, XYY can rarely be observed. Both PM and CM disclose additional genetic aberrations, including CM appearing triploid and PM appearing diploid, haploid, tetraploid, and aneuploid variants. Immunohistochemistry is crucial in the evaluation for $p57^{kip2}$, a cell cycle inhibitor, which is expressed in the maternal allele in most tissues. Since CM contain only paternally derived genes, $p57^{kip2}$ is absent or uttered faintly. The immunohistochemistry for the antigen shows absent or weak expression just in cytotrophoblast and villous stroma of CM, while the CTB and stroma of PM or hydropic abortion harboring a maternal genetic contribution will show a remarkably very strong $p57^{kip2}$ expression. This

---

Box 18.20 $p57^{kip2}$ Immunostaining in Normal Conceptus, PHM, and CHM

	Normal	PHM	CHM
Villous cytotrophoblast	+	+	−
Villous stromal cells	+	+	−
IT islands (intervillous)	+	+	+
Decidualized stromal cells	+	+	+

immunohistochemical result can be considered very useful in confirming the diagnosis of CHM (absent or weak expression) versus PHM (stable expression). On the other hand, the antibody is not perfect and this aspect should be kept in mind in forensic medico-legal cases. Maternally derived, decidualized stromal cells will react strongly with the antibody, and intervillous trophoblast islands in GTD-CM have been shown to express $p57^{kip2}$ often causing equivocal interpretation. In the end, $p57^{kip2}$ cannot be used without using proper, wise hematoxylin and eosin investigation and cytogenetic results. The diagnostic procedure should not rely exclusively on $p57^{kip2}$. Thus, it has been suggested that in case of a possible mole, $p57^{kip2}$ staining is crucial. If it is positive, it is advised to perform a molecular genotyping to distinguish between androgenic diploidy, diandric triploidy, and biparental diploidy. The identification of a diandric triploidy indicates a PHM pregnancy, while androgenic diploidy would mean a CHM. Biparental diploidy suggests a non-molar pregnancy. In case of a negative $p57^{kip2}$ and definite pathology, a CHM is still higher on the table, while an equivocal pathology should prompt the pathologist to perform a molecular genotyping and carry on with the distinction between androgenic diploidy, diandric triploidy, and biparental diploidy.

- *CLI*: Vaginal bleeding, uterus enlargement, and high levels of β-hCG are seen in CHM, while PHM is often present in women with vaginal bleeding, small uterus, and relatively low levels of β-hCG with symptoms of an incomplete or missed abortion. Cytogenetically, CHM has usually paternally derived chromosomes with 46, XY androgenetic karyotype with a sperm cell fertilizing an empty oocyte and then duplicates its DNA, but about one-tenth of CHM are 46, XY (androgenetic diploid-dispermia). Genetic studies identified a remarkable clustering of mutations in the leucine-rich region of NLRP7 at chromosome 19q13.3–13.4 in both CHM and PHM. PHM harbors a fertilization error instead, in which two spermatozoa fertilize a normal ovum with final triploid karyotype (69, XXY). Flow cytometry may help to differentiate the two molar degenerations of the product of conception. By ultrasound, CHM shows a vascular pattern ("snowstorm") in the first trimester before vaginal spotting or the cervical passage of vesicles ("a bunch of grapes") without a recognizable embryo or fetus. PHM may show focal cystic spaces of the stroma, an increase in the transverse diameter of the gestational sac as well as a visible embryo or fetus.

- *TRT*: The evacuation of a mole includes a ≥10% increase of β-hCG levels, β-hCG of three stable values over 2 weeks, an β-hCG rebound, histologic diagnosis of choriocarcinoma or PSTT (placental site trophoblastic tumor), detection of metastasis, β-hCG >20,000 mIU/mL more than 4 weeks post-evacuation, and persistence of elevated β-hCG 6 months post-evacuation. GTD is separated in low (good PGN)-, moderate-, and high-risk (poor PGN) group using several scoring systems, such as WHO Classification, Charing Cross Hospital Trophoblast Disease Centre, and FIGO (Eysbouts et al. 2017). Using the WHO score, a value of ≤4 identified low-risk patients, and 5–7 are patients in the moderate-risk group, while ≥8 are high-risk patients.

In Box 18.21, the distinguishing features of NMM, PHM, ECM, and CHM are summarized allowing a better approach in analyzing these specimens in the usual pathological routine. NMM (non-molar miscarriage) is a term, which may be exchangeable with hydropic abortus or non-molar hydropic abortion or hydropic molar degeneration by some authors (Sharami et al. 2020). To remember, mostly villi are large and more magnificent than two times a low-power microscopic field using a 10X objective. Cistern formation is variable in PHM and more consistent in CHM. The ECM has a mild increased volume, but nothing compared to the CHM, probably similar to PHM. Features are pretty like PHM, but the villous pattern has been described harboring a "cauliflower villi" pattern. Cytologic atypia can be absent or rarely

18.2 Pathology of the Early Pregnancy

Box 18.21 Distinguishing Features of NMM, PHM, and CHM

	NMM	PHM	CHM
Feature – gross			
Tissue – volume	Normal	↑	↑↑↑
Tissue – vesicles	No/+	+/++	+++
Feature – micro			
Uniformity of the hydrops	−	−	+
Villous Ø (max.)	~3*10^3 μm	~5*10^3 μm	~7*10^3 μm
Villous Dimorphism	No	+	−
Hypercellularity	−	−	+
Cytologic atypia	−	−	+
Fetal body elements	+	+	−
THC	−	+	++
VSK	−	−	+
Villous BVs	±	+	±
ISTA	−	±	+
Flow cytometry			
DNA content	Diploid (often)	Triploid	Di-/tetraploid
Genetics			
Karyotype	46, ± ♂ and ♀	69, ♂ and ♀	46, ♂
Outcome			
Neoplastic GTD risk	0%	2.5%	15%

Notes: GTD, gestational trophoblastic disease ISTA implantation site trophoblast atypia, THC trophoblast hyperplastic change, VSK villous stromal karyorrhexis, ♂ paternal, ♀ maternal

present and focal. DNA content and karyotype are variable. There is a risk of neoplastic GTD of 5–10%, which is intermediate between PHM and CHM.

There are, apparently, chemotherapy protocols that need to be assessed according to decisions performed at individual institutions.

Chemotherapy Protocols

LR: Single-agent chemotherapy, either methotrexate or dactinomycin.

MR: Multi-agent chemotherapy, either MAC or EMA, or single-agent chemotherapy.

HR: Multi-agent chemotherapy (e.g., EMA/C) – etoposide, dactinomycin, and methotrexate alternated at weekly intervals with vincristine and cyclophosphamide with selective use of surgery and radiotherapy. Salvage chemotherapy with EP/EMA (etoposide, cisplatin, etoposide, methotrexate, dactinomycin) plus surgery should be the line of choice in resistant disease.

18.2.2 Disorders of the Placenta Maturation

Placenta maturation is probably one of the most critical histologic features to assess in the diagnosis of hypoxic vascular events but also for chromosomal aberrations as well. Although it represents a task to determine a catastrophic event, it is currently quite a topic of intense investigation, particularly for the novelty of findings related to genomics and proteomics, but probably also after the most recent discoveries of epigenomics. Placenta maturation may be quite difficult to assess and also harbors a subjective component, which is not indifferent. Histopathology plays a crucial role in the determination of the maturation and clarification of intrauterine fetal growth restriction and stillbirth (Fig. 18.5). As pregnancy advances to term, there are an increasing number of terminal villi and syncytial knots, a decrease of Hofbauer cells, the disappearance of the villous cytotrophoblast and vascular profiles, and extra-

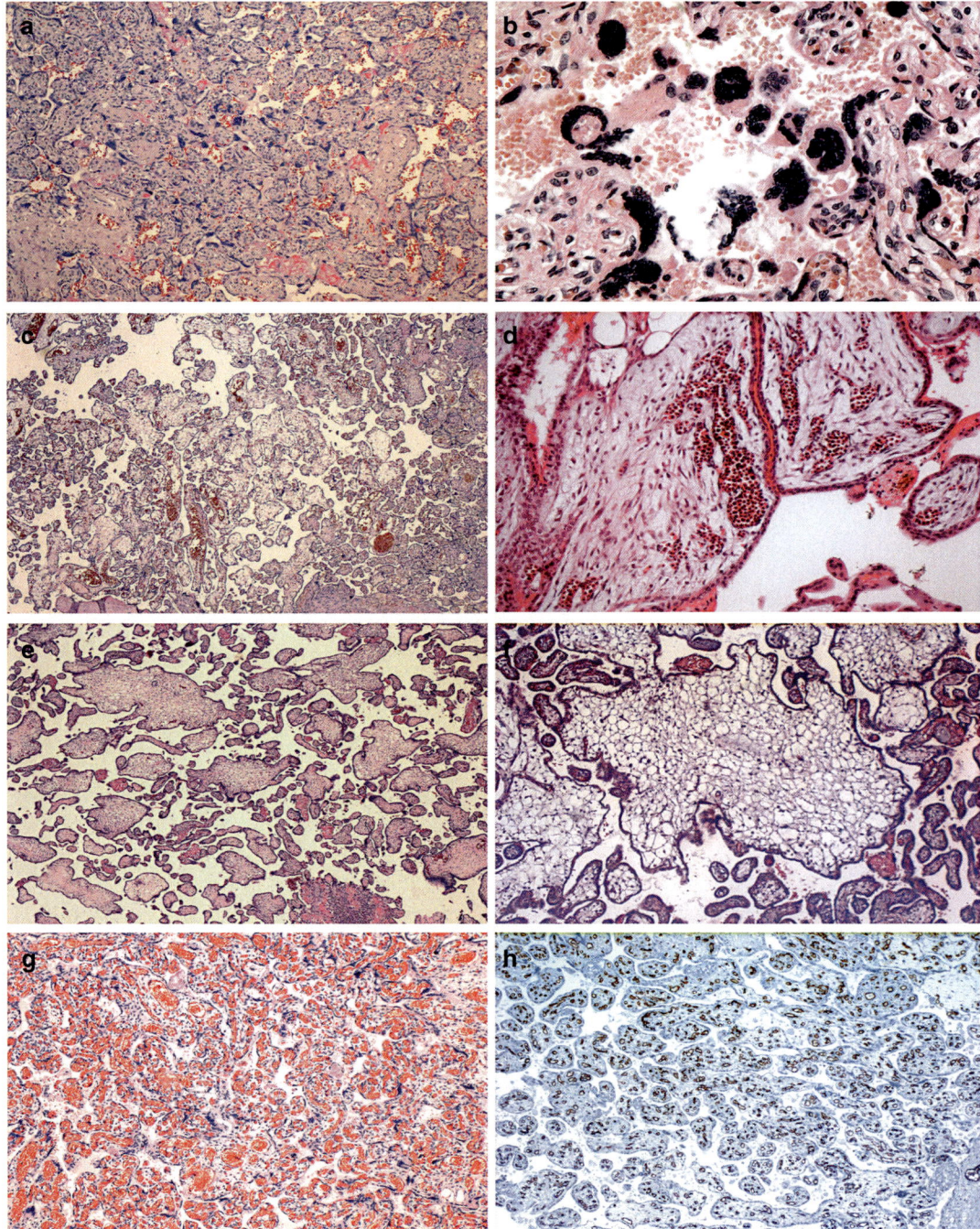

Fig. 18.5 Placenta maturation disorders. The eight microphotographs show several examples of disorders of the maturation of the placenta, namely, syncytial knots zonal aggregation (**a**), syncytial knots close-up (**b**), villi with lower density of the stromal cellularity identifying a partial delay in the maturation (**c**), nucleated red blood cells in the cross sections of the villous capillaries in the late pregnancy, which is an abnormal finding at this time of pregnancy, (**d**), fibrotic villi without vascularity (**e**), stem villi with abnormal shape without discernible capillaries (**f**), and chorangiosis with >10 capillaries in 10 villi taken at 10× magnification as objective (**g–h**) (see text for definition and details). All microphotographs are H&E stained apart from **h**, which is avidin-biotic complex immunostained slide using an antibody against the endothelial cells (anti-CD34 antibody) (a, original magnification x40; b, original magnification x200; c, original magnification x40; d, original magnification x200; e, original magnification x40; f, original magnification x100; g, original magnification x100; h, original magnification x100)

cellular matrix composition changes of the villous stroma. All of these characteristics may harbor an unusually high subjective component in the interpretation of findings, and several tools may be applied, including immunohistochemistry and special stains (histochemistry). Site heterogeneity further complicates subjective evaluation, i.e., better oxygenated central areas of cotyledons, which are also known as placentones (maternofetal circulation units). They are, indeed, more primordial (less mature) than less oxygenated peripheral regions. In fact, this site is where blood returns toward the uterus. Thus, a unquestionable picture of the homogeneity of placental maturation is entirely unrealistic in a typical setting scoring two sections only of a placenta. The finding that may be under the lens may be hypomature placenta or hypermature placenta. Indeed, placental hypermaturation is better appreciated in preterm babies, while the hypomature placenta is more natural to see in full-term babies. Hypervascularity of villi as defined according to the Altshuler's rule ($10 \times 10 \times 10$) is more difficult to appreciate in some settings, mainly if the placenta storage has not been adequate. Thus, CD31 immunostaining targeting the endothelial cells may be extremely useful. Villous hypo-/avascularity can be diffuse, such as in prolonged stillbirth, or focal, such as in FTV. A decrease of the ECM of chorionic villi can be seen in hypoxia, while an increased ECM may be seen as intense eosinophilia of the villous cores and may be noted in cases of intervillous hyperoxemia associated with intravillous hypoxemia. The CTB of the villi is easily identified up to the end of the second trimester. The CTB from the third trimester on is present only in cases of chronic hypoxia. In fact, a state of chronic hyperoxemia, such as in preeclampsia, inhibits the fusion and differentiation of the trophoblast and stimulates proliferative activities instead. Villous CTB is readily detectable using a double stain combining E-cadherin, an adhesion molecule, and Ki67, a proliferation marker, which highlights the cytoplasmic membrane and the nuclei, respectively. Pregnancy hypertension may also promote syncytial knots, i.e., transcriptionally stable, condensed nuclear structures of STB, without nucleoli, which is different from transcriptionally active syncytial sprouting with the prominence of nucleoli of the first trimester. The number of syncytial knots is directly and positively correlated with the length of time and the severity of the hypertensive state. Villous Hofbauer cells, which can be highlighted in a magnificent way using an antibody against the antigen CD68, are present up to the end of the second trimester and are essential for the synthesis of proteins able to allow a villous proliferation in response to hypoxia. The so-called obliterative endarteritis (OEA) is a misnomer due to a herniation of smooth muscle cytoplasm into the vascular luminal spaces showing a vascular occlusion due to a pseudo-swelling of the endothelial layer. OEA is associated with preeclampsia, FGR, and umbilical cord and hypercoiling. In Box 18.22 are listed the disorders involving an abnormal maturation of the placenta.

Placentomegaly is usually classified as mild-moderate, if it is in the weight range of 650–800 g, while a severe placentomegaly occurs when placental weight is over 800 g. Otherwise, a weight of 700 g or higher at term may be considered "large placenta," while the word of placentomegaly may be used in a case of a placenta with two standard deviations more substantial than the 90th percentile. It is crucial to recognize placental villous edema because it had implications for the fetal and newborn's well-being. If placental villous edema is severe and diffuse,

---

Box 18.22 Etiology of Placentomegaly and/or Distal Villous Immaturity

1. *Maternal Factors*: Diabetes mellitus, obesity ab initio, excessive weight gain, maternal anemia (severe)
2. *Placental Factors*: Hydrops fetalis universalis, TORCHES microbiological complex, placental mesenchymal dysplasia, metabolic storage diseases (e.g., mucolipidosis, sialidosis), multiple placental chorangiomata
3. *Fetal Factors*: Overgrowth syndromes (e.g., Beckwith-Wiedemann syndrome, Sotos syndrome)

> **Box 18.23 Distal Villous Hypoplasia (aka Terminal Villous Deficiency)**
> - Small placenta
> - Absent or reversed end-diastolic flow
> - ↓ # & Ø of distal villi
> - Filiform unbranched villous pattern
> - Intervillous spaces devoid of cellular components
> - Villous vascular dysgenesis

babies will experience hypoxia with low Apgar scores at 5 and 10 min, difficulty in resuscitation at birth, neonatal respiratory distress, an increase of postnatal morbidity and mortality rates, and an increased frequency of long-term neurologic impairments. Takemoto et al. (2018) found also a low BW/PW ratio ("relatively large placenta") in preterm babies with cerebral palsy. They argue that the suboptimal condition of the fetus may induce compensatory placental enlargement and a predisposition to preterm birth.

Distal villous immaturity (Box 18.23) may be the disorder underlying the condition of placentomegaly (mild/moderate to severe), but it is not a synonymous and is characterized by ↑ villous stroma, centrally located intravillous capillaries, and ↓ SVM (syncytial-vascular membranes). Distal villous maturity may be seen in DM or abnormal glucose tolerance and stillbirth, other than placentomegaly. Stillbirth may be explained by an increase of placental demand and decrease of placental function with an increased risk of adverse outcomes.

18.2.3 Disorders of the Placenta Vascularization

Disorders involving an abnormal vascularization of the placenta include chorangiosis, chorangioma, chorangiomatosis, and angiosarcoma. Chorangioma represents a benign neoformation of proliferating capillaries behaving benignly, i.e., compressing the surrounding parenchyma. It is different from the angiosarcoma, which would destroy or infiltrate the surrounding parenchyma and induce neoplastic necrosis. In the chorangioma, the neoplasm presents with a bulging red to white mass according to the homogeneity and ratio of neo-formation of blood vessels and supporting connective tissue (Fig. 18.6). The chorangioma is different from chorangiosis and chorangiomatosis which are usually not identifiable grossly. Histologically, chorangiomas are composed of proliferating fetal blood vessels and a variable amount of fibrous and cellular stroma. Chorioangiomas are rare vascular lesions but are seen more commonly in multiple gestations and newborns with congenital anomalies. Clinically, the presence of chorangioma and its clinical consequence is often associated with the size of the lesion. Thus, it may entail fetal hydrops, stillbirth, FGR, fetal anemia, fetal thrombocytopenia, fetal congestive heart failure, disseminated intravascular coagulation (DIC), premature delivery, abruption, and preeclampsia.

Substantially and recapitulating the above-exposed data, the *autonomous disorders of the placenta vascularization* are constituted by the following entities:

- *Chorangiosis* (capillaries surrounded by a distinct basement membrane without loose reticulin bundles or circumferential lining of pericytes following Altshuler criteria)
- *Chorangioma* (well-circumscribed nodular lesion showing capillary hyperplasia surrounded by trophoblastic cells)
- *Chorangiomatosis* (non-expansile heterogeneous lesion between chorangiosis and chorangioma showing capillary proliferation in normal stem villi, related to hypoxia, and divided into focal, multifocal, and diffuse types)
- *Angiosarcoma* (a malignant vascular tumor with neoangiogenesis, high-grade cellular atypia, and metastatic potential)

18.2.4 Disorders of the Placenta Implantation Site

Disorders involving an abnormal maturation of the placenta include deviations of intrinsic or extrinsic type. Placenta implantation anomalies

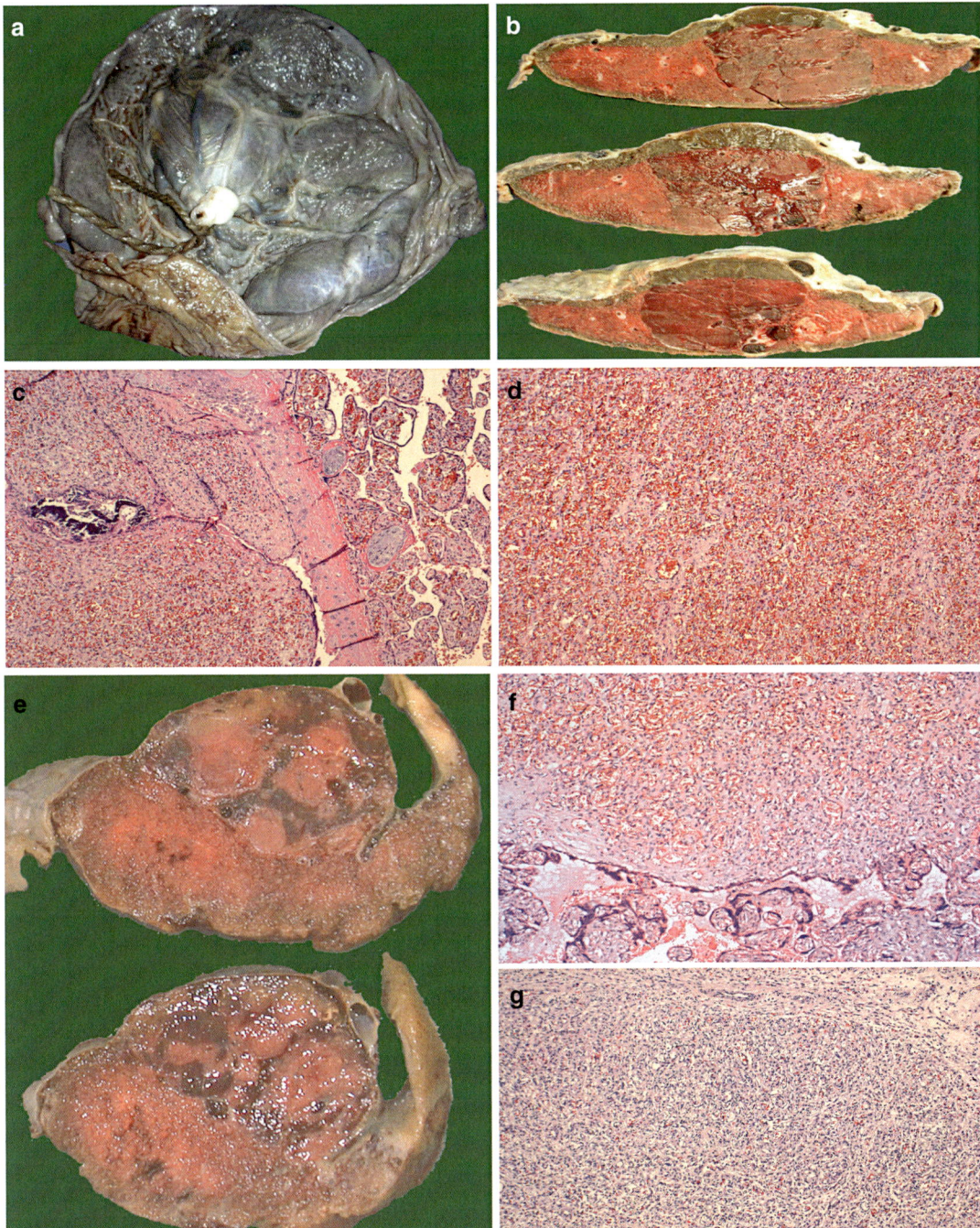

Fig. 18.6 Chorangioma. The gross and microscopic figures taken from (**a**) through (**g**) show a placenta vascularization disorder with chorangioma formation. In particular, the gross photographs (**a**, **b**, and **e**) show the extension of the chorangioma with compression of the placenta parenchyma. This aspect is highly relevant because there is a restriction of the placental reservoir to react for placenta insults in the late pregnancy. Microphotographs (**c**, **d**, **f**, and **g**) show predominantly capillary type proliferative blood vessels (**c**, **d**, **f**, and **g**, ×100).

may play a significant role in perinatal pathology, ranging from some minor difficulties in labor to perinatal death during labor. Perinatal implantation anomalies may also play a role before birth inducing IUFD or preterm delivery with prematurity-related complications in the newborn baby. The classification of the placenta implantation anomalies may be quite confusing. The introduction of intrinsic and extrinsic placenta implantation deviations may help to clarify some background and helps in memorizing names and conditions. Placental implantation anomalies of primary type involve anomalies that are strictly related to the placental disc itself; placental implantation anomalies include abnormalities of the placental disc about the uterine cavity (Box 18.24).

In the placenta bi- or multilobate, the incisions or divisions of the placenta are present at the edge of the placenta, while membranes, which define specific areas, support the placenta bi or tripartite the division. Conversely, *placenta succenturiata* is the term used for accessory placental discs. *Placenta membranacea* should be used when the placenta has a thickness of less of 0.5 cm. *Placenta fenestrata* occurs when in the placental disc, some areas are lacking villous parenchyma identifying windows of membranes (*chorion laeve*) only (fenestra = window). *Placenta zonaria* or *anularis* is a normal placenta seen in mammals when the placenta goes around the chorioallantois. Three more variants (other than *placenta diffusa* or homogeneous placenta) are the *placenta cotyledonaria*, when only cotyledons not fused together can be appreciated, the *placenta discoidalis* or disc-shaped, and the bi-discoidal (placenta bi-discoidalis) when two discs can be seen. In the event of a placenta extrachorialis, the chorionic plate or part of it has a diameter, which is inferior to the basal plate. The chorionic villous plate is in the extrachorial placenta larger than the chorionic roof, and it may be recognized grossly that the membranes reflect off of the chorionic roof before the periphery of the placental discoid edge. The extrachorial part of all or some part of the chorionic plate can be covered or not entirely with membranes. There is a discrepancy between external placental diameter and surface of the chorionic plate and attachment of the membranous layers on the placental disc. In the event of a *placenta marginata*, the extrachorial placenta is found to be beyond the point of membrane insertion. In the fact of a *placenta vallata*, there is an extra-membranous placental plate tissue around all or part of the peripheral discoid with subchorionic fibrin deposited beneath the intramembranous roof determining a folding over with plication of the membranes from the Latin term *vallate* which means having a ridge (Figs. 18.7 and 18.8). The disorders of the placenta implantation site may be seen in Boxes 18.25 and 18.26.

Placenta Previa

Placenta previa is a medical and surgical emergency putting the mother and the conceptus at high risk of death and needs to be correctly managed to avoid both maternal/perinatal morbidity/mortality. In the 1990s, the introduction of confidential inquiries into maternal deaths has notably reduced maternal mortality worldwide. It is vital in case of confidential inquiries into maternal deaths or in case of bleeding beyond the first trimester to read the medical notes carefully and reconstruct the events. A speculum examination with ultrasound need to have been done. Also, no digital examination should have been performed for the risk of provoking life-threatening hemorrhage. Moreover, uterine activity monitoring gives information

---

Box 18.24 Placenta Implantation Deviations of Intrinsic Type
- Minor anomalies of placental shape (e.g., reniform, heart shape-like, seeds-like)
- P. bi-/multilobata vs. bi-/tripartita vs. succenturiata
- P. extrachorialis circummarginata totalis sive partialis (*placenta marginata*)
- P. extrachorialis circumvallata totalis sive partialis (*placenta vallata*)
- P. membranacea/P. fenestrata/P. zonaria

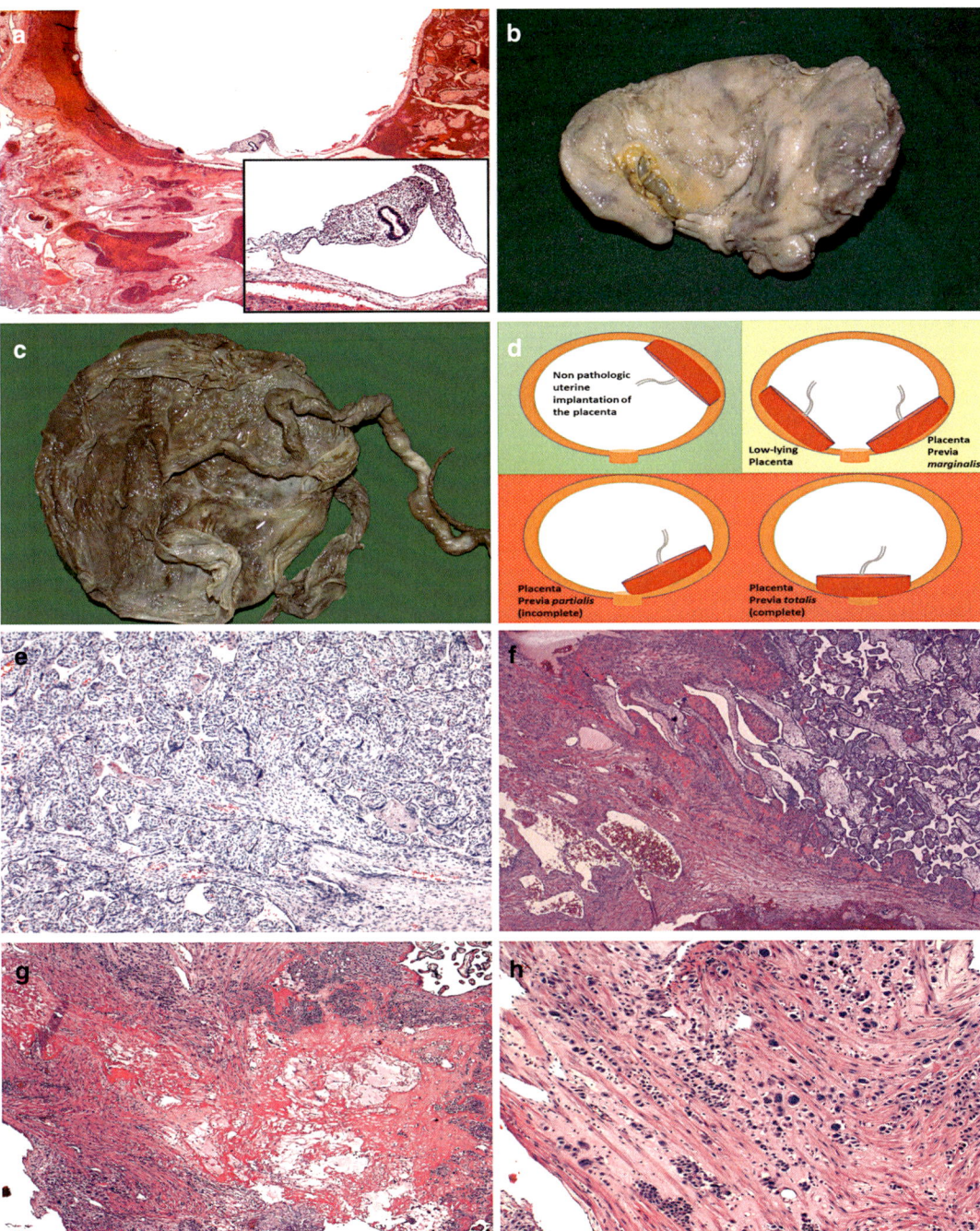

Fig. 18.7 Placenta implantation disorders. Placenta implantation disorders can have several morphologies and consequences for the placenta itself but also for the pregnancy (fetus and mother). In (**a**) is a cross section of an extrauterine (tubal) pregnancy with the early neurulation of the embryo (inset). The figures (**b**, **c**) identify a lithopedion or "petrified" fetus and its placenta. The cartoon in (**d**) separates the different degrees of placenta previa with the most troubling situation in the right lower corner of (**d**) with a complete occlusion of the os (see text for details). The microphotographs from (**e**) through (**h**) show the different morphology of the placental accretion (a, inset, e–h, H&E; a, original magnification x40; inset, original magnification x100; e, original magnification x100; f, original magnification x40; g, original magnification x40; h, original magnification x100)

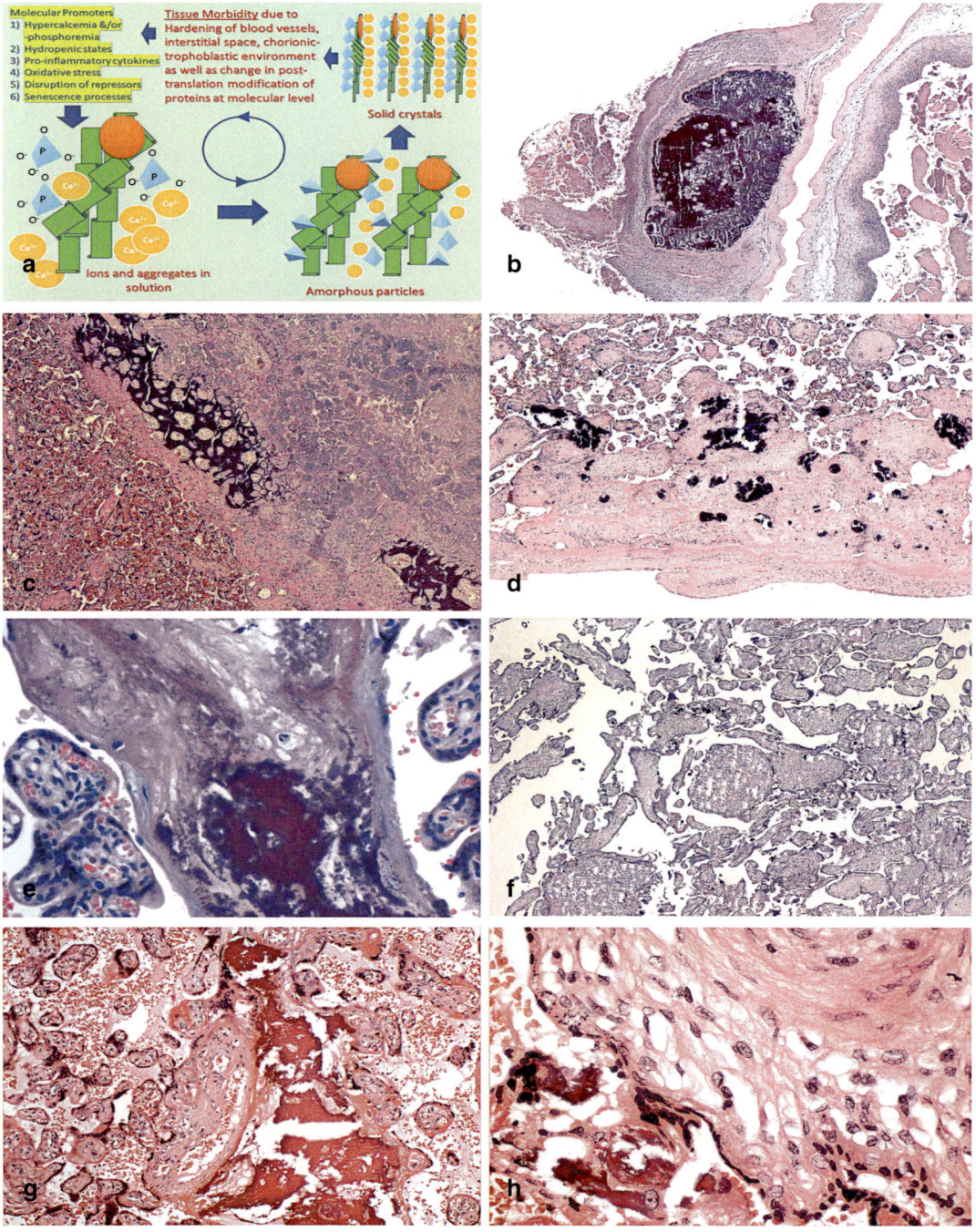

Fig. 18.8 Calcifications may be a typical event at the end of pregnancy, but if it appears early in gestation or is massive in the late pregnancy, it may have consequences for the adequate function of the placenta and growth of the fetus. In (**a**) is shown the formation of the crystallization of calcium aggregates. Calcification may present in the membranes (**b**), at the basal plates (**c**, **d**), and in the stem villi at different weeks of gestation (**g**, **h**). All microphotographs are taken from histological slides stained with hematoxylin and eosin (b, original magnification x40; c, original magnification x40; d, original magnification x40; e, original magnification x400; f, original magnification x40; g, original magnification x100; h, original magnification x400)

Box 18.25 Disorders of the Placenta Implantation Site

- Placenta implantation site disorders *without* non-neoplastic trophoblastic abnormalities
  - Ectopic pregnancy
  - *Placenta previa*
  - Placenta with *vasa praevia*
  - *Placenta accreta* (villous adherence to myometrium due to lack of decidua)
  - *Placenta increta* (villous adherence into myometrium)
  - *Placenta percreta* (villus adherence to through the myometrium)
- Placenta implantation site disorders *with* non-neoplastic trophoblastic abnormalities
  - *Placental site nodule (PSN)*
  - *Exaggerated placental site (EPS)*

Notes: Placental site nodule and the exaggerated placental site will be described in the non-neoplastic trophoblastic abnormalities. A placenta may be considered accreta if villous adherence to the internal myometrial surface occurs with attenuated decidua

Box 18.26 Anomalies of Extrinsic Placental Implantation (EPIA)

1. *Placental implantation to a quite, nearly, or clear malicious site*
   - Deep-seated placenta
   - *Placenta previa marginalis* (placenta approaching the border of the os)
   - *Placenta previa partialis* (placenta partially covering the os)
   - *Placenta previa totalis* (placenta covering the os completely)
2. *Placenta implantation with abnormal uterine penetration*
   - Placenta *accreta*
   - Placenta *increta*
   - Placenta *percreta*

Notes: EPIA, extrinsic placental implantation anomalies. In the event of the *accreta* type, villi are immediately adjacent to the myometrium without interposing decidua, while *increta* type is marked by villi penetrating the myometrial layer of the uterus, and the third type or *percreta* type is characterized by villi penetrating into the myometrial layer up to the serosal surface.

about the 1/5 of the pregnant women who have concurrent contractions with their bleeding. Sexual intercourse should also be investigated as a potential cause of hemorrhage in the setting of placenta *previa*. From the medicolegal point of view, the primary issue is a failure to diagnose this condition. In fact, any second-trimester ultrasound should be able to discover a previa placentation; then there is inadequate preparation and counseling. Thus, adequate documentation of both training of the woman and professional advice is mandatory.

- *DEF*: Placenta situated partially or wholly within the lower segment AND ≥ 28 weeks of gestation.
- *EPI*: Incidence, 0.3–0.5% deliveries (approximately 1 in 200 pregnancies as the highest incidence rate), but *placenta previa*
- accounts for 13% of deaths secondary to hemorrhage. More than 150 years ago, the maternal mortality rate was up

to 1/3 of deliveries, but in the industrialized countries, the rate is practically nil, probably following the introduction of systematic Confidential Enquiries into Maternal Death.
- *EPG*: Although the etiology is unknown, it has been suggested that an abnormal vascularization of the endometrium caused by scarring or atrophy from previous trauma, surgery, or infection is at the origin of the *placenta previa*
- . There are risk factors, which may reduce the differential growth of the lower segment, resulting in a less upward shift in placental position as pregnancy proceeds to the end. Risk factors: Multiparity, >35 years of age, twin placentation, endometrial lining damage (D&C, abortion, RPOC, ± TOP), preterm delivery, previous CS, erythroblastosis, smoking, and cocaine use.
- *DGN*: Antepartum hemorrhage (APH) and breech presentation or transverse lie and/or imaging (ultrasound/MRI). Conventionally, four grades of *placenta previa* have been used, but it is more common currently to merely differentiate between "major" and "minor" cases. The major *placenta previa* is in the lower uterine segment, and the lower edge does cover the internal os, while the minor placenta is in the lower uterine portion, but the lower side does not include the internal os. Moreover, *placenta previa* may be classified as complete, when the placenta covers the cervix completely; partial, when the placenta incompletely covers the cervix; and marginal, when the placenta ends close to the edge of the cervix uteri, about 2 cm from the internal cervical os (vide supra).
- *GRO/CLM*: Marginal/velamentous cord insertions, succenturiate, and bipartite.
- *PGN*: Maternal mortality 0% (in contrast to 30% maternal mortality more than one and a half century ago) AND perinatal mortality 8–80/1000 live births (direct correlation with gestational age at delivery and amount of antepartum blood loss – APH). Recurrence rate: 4–8% in the case of previous *previa* placentation. Placental extension ≥40 mm beyond the internal *os* in the setting of a *placenta previa* is a valid predictor of pregnant women who will require a peripartum hysterectomy. As a result, these women had longer surgical times but fortunately not increased operative or postoperative complications.

Ectopic Pregnancy

The ectopic pregnancy is a life-threatening condition, and ER, as well as OBGYN personnel, may be primarily involved requiring the maximal care to avoid the death of the mother. There is no option to save the baby's life because it occurs very early in gestation. The blastocyst or fertilized ovum is implanted outside the endometrial lining of the uterine cavity. In particular, it is crucial not to forget this condition, if the patient is in childbearing age and comes to the emergency department with an acute abdomen symptomatology. Moreover, some teenagers have sexual intercourses quite early and the increase of sexual intercourses prior of 18 years of age should not be overlooked during the interpretation of the clinical history, ultrasound, and lab work. There is also a mandatory and ethical obligation to avoid missing child abuses and all physicians need to be involved in this setting. Previous genital infection should be considered, but they do always not necessarily evolve with chronic inflammation. The risk factors underlying an ectopic pregnancy are as follows:

High Risk: Tubal pathology (mostly inflammatory: salpingitis), previous tubal surgery (e.g., abnormal sex development or tubal sterilization), S/P (status/post) ectopic pregnancies, in utero exposure to diethylstilbestrol (DES), and S/P use of intrauterine contraceptive device (IUD)

Moderate Risk: Infertility, S/P genital infection, and promiscuity

Mild Risk: S/P pelvic/abdominal surgery, tobacco smoking, vaginal douching, and history of sexual intercourse before the age of 18 years with some cases in patients with age less than 15 years

Ectopic pregnancies represent circa 1/50 of all pregnancies in the USA and most industrialized countries. There have been increasing rates of incidence in most countries because of the increasing prevalence of STIs (sexually transmitted infections or diseases), early diagnosis, contraception that predisposes failure to be ectopic, use of tubal sterilization techniques, use of assisted reproductive technologies, and tubal surgery, such as salpingotomy and tuboplasty (mainly due to sexual differentiation disorders). To keep in mind is that ectopic pregnancy is the most frequent cause of *maternal mortality* within the first trimester. The incidence in non-Caucasian women is about 1.5 times higher than in Caucasian women.

The implantation can be at different level:

Interstitial: Implantation in the interstitial portion of the fallopian tube

Abdominal: Implantation within the peritoneal cavity

Angular: Implantation with extension well beyond the interstitium into the uterus

Vaginal: Implantation in the vagina

Broad ligament: Implantation in the *ligamentum latum uteri, i.e., the peritoneal fold, which exactly passing from the lateral margin of the uterus to the wall of the pelvis bilaterally encloses the uterine tubes and the ovaries.*

To note is that an abdominal ectopic pregnancy can also occur secondary to tubal pregnancy. However, the tubal site remains the most frequent site of ectopic pregnancy (99%), and the ampulla is the most frequent location of implantation (2/3 of tubal cases). The symptoms start about 7 weeks after LMP with abdominal pain and vaginal bleeding, accompanied by abdominal tenderness, adnexal tenderness, amenorrhea (fundamental question in the Emergency Department), early pregnancy symptoms, Cullen sign (periumbilical bruising), as well as nausea, vomiting, diarrhea, and dizziness. Other signs may be quite nonspecific and include tachycardia, low-grade fever, Chadwick sign (cervix and vaginal cyanosis), Hegar sign (softened uterine isthmus), hypoactive bowel sounds, cervical motion tenderness, enlarged uterus, tender pelvic or adnexal mass, 'cul-de-sac' fullness, and decidual cast (passage of decidua in one piece only). Finally, three signs may represent red flags in the medical and nursing personnel and are at 6–12 weeks' gestation severe abdominal tenderness with rebound, guarding (perforated appendicitis-like) in the history of amenorrhea, and orthostatic hypotension. All three are suggestive of ruptured ectopic pregnancy. In a pediatric emergency room, it is important to remember the potential pitfalls and the differential diagnosis of ectopic pregnancy should include extrauterine etiologies such as appendicitis, an ovarian pathology (ruptured cyst, ovarian torsion), inflammation (PID/Pelvic Inflammatory Disease, salpingitis, endometritis), urinary tract infection (UTI), nephrolithiasis, diverticulitis, and mesenteric lymphadenitis (e.g., infection with *Yersinia* spp.).

Placenta accreta: Abnormal placentation with a peculiar invasion of the chorionic villi through the uterine wall. It is subclassified according to the degree of invasion, being defined as properly *placenta accreta* when chorionic villi are adjacent or adherent to the myometrial layer of smooth muscle fibers, placenta increta when invasion of the myometrial thickness by chorionic villi occurs, and *placenta percreta* in case of chorionic villi penetrating through the myometrium with following rupture of the uterus. The incidence of peripartum hysterectomy secondary to *placenta accreta* and increta is between 0.2 and 1.5 per 1000 deliveries, while the mortality in this condition has been assessed ranging from 2% to 7%. In the past half-century, the incidence of *placenta accreta* has been seen increasing more and more literally. Probably, the improper use (or 'abuse') of cesarean delivery may have contributed to this increase. Women who have two or more cesarean sections or central placenta previa may have an unusual risk of 40% to develop a kind of abnormal placentation. Apart of the multiparity and placenta previa, other causes of *placenta accreta* list S/P myomectomy for different reasons, Asczation of leiomyomas, and a maternal age more than 35 years. Moreover, it is also important to recall the settings when accretion aspects of placentation take place (Box 18.27) Asherman syndrome is constituted by the scar tissue formation inside of the uterine cavity that usually develops after uterine surgery.

18.2.5 Twin and Multiple Pregnancies

Disorders involving abnormal placentation include twin and multiple pregnancies (Boxes 18.28 and 18.29) (Figs. 18.9, 18.10, 18.11, 18.12, and 18.13). We define dizygotic twins (fraternal twins), twins that originate from two zygotes, and monozygotic twins (identical twins), twins that arise from one zygote.

> **Box 18.27 Pearls in Dealing with Accretion Aspects of Placentation**
> 1. Ultrasound features of *placenta accreta* include deficiency of retroplacental sonolucent zone, multiple vascular lacunae, thinning of the myometrium, and abnormal urinary bladder.
> 2. Tissue sampling from the pathology assistant or the pathologist must be generous.
> 3. To distinguish smooth muscle fibers from fibrin fragment or eosinophilic intermediate trophoblast cells ("X cells"), the use of immunohistochemistry with antibodies against smooth muscle actin is strongly indicated. The immunohistochemistry seems to be more helpful than Masson trichrome special stain.
> 4. Uterine rupture secondary to placenta *percreta* is a severe, life-threatening condition leading to intra-abdominal bleeding and death, if not dealt with hysterectomy performed in an emergency. Occasionally pediatric pathologists may be called in court addressing this particular issue.

> **Box 18.28 Twin and Multiple Pregnancies**
> - Dizygotic twins
> - Monozygotic twins
> - Monozygotic twins with TTTS
> - Multiple pregnancies
> - Twin reversal arterial perfusion (TRAP) sequence or *acardius*
> - *Pagus* type of multiple pregnancies

> **Box 18.29 Ink-Injection Patterns of Anastomoses in Monochorionic Placentation/Placentae (MP)**
>
> | | |
> |---|---|
> | A-A anastomoses (on the chorionic surface): | ~3/4 of injected MP |
> | V-V anastomoses (on the chorionic surface): | ~1/2 of injected MP |
> | Villous capillary bed anastomoses (third circulation): | ~1/2 of injected MP |
>
> Notes: Vascular anastomoses are present in almost all monochorionic placentae (MP), although they show variable number, size, and shunt direction

Acute and chronic disturbances of the arterial or venous blood flow in the maternal or fetal circulation of the placenta are at the basis of the two major categories, including materno-placental vascular perfusion disturbances and feto-placental vascular perfusion disturbances.

It is important to differentiate an infarct from a thrombus (or a pseudo-infarct) because the clinical relevance of the outcome for the newborn and the recurrence risk are entirely different. A placenta infarction is a territorial delimited villous parenchymal necrosis due to an interruption of the arterial maternal blood flow. The so-called '*Gitterinfarkt*' of the German literature, which corresponds to "maternal floor infarction" of the English literature, corresponds to dysfunctional foci of the villous parenchyma of different size without apparent relation to fetomaternal placental units. This entity overlaps with pseudo-infarct,

18.2 Pathology of the Early Pregnancy

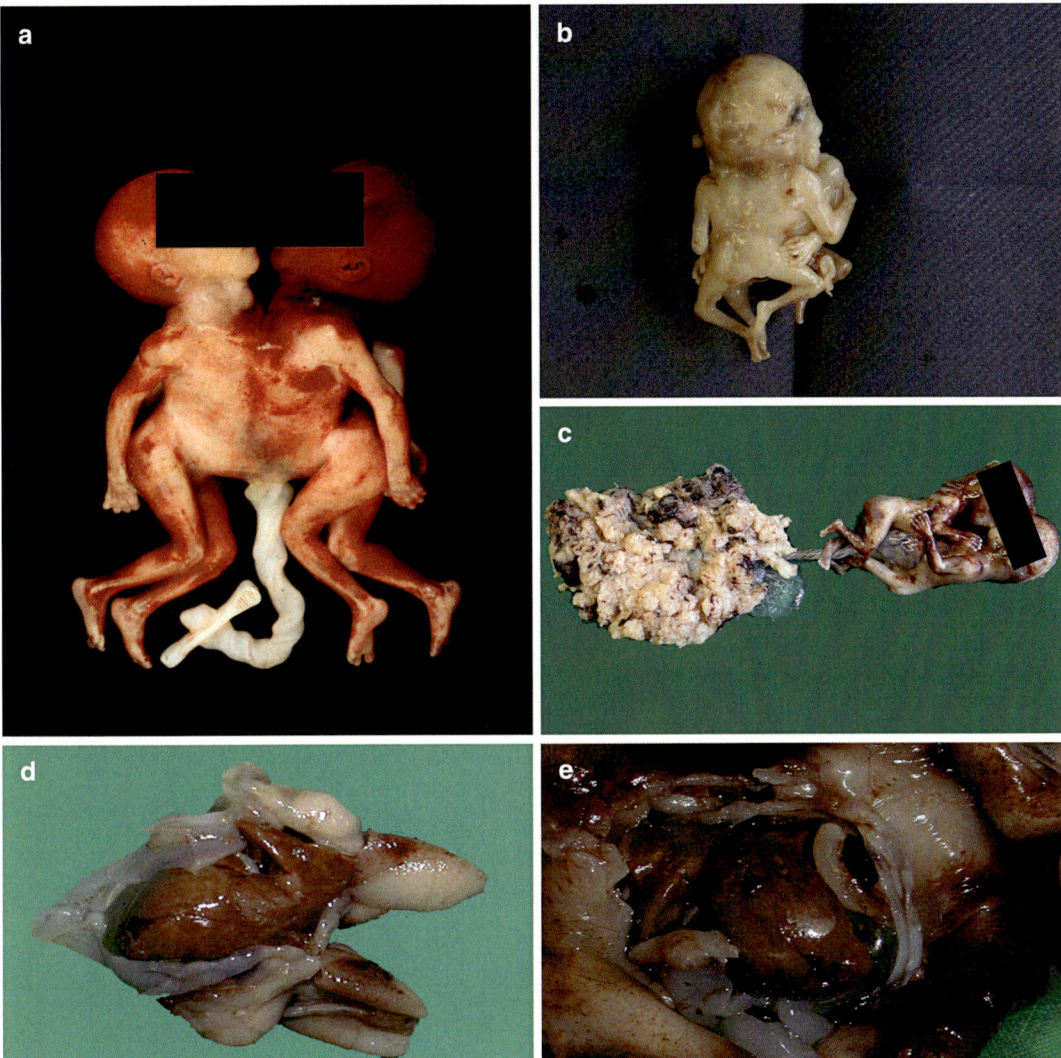

Fig. 18.9 Congenital twinning. Conjoined twins with different morphology and week of gestation. In (**a**) there is thoraco-omphalopagus symmetricus of the late 2nd trimester of pregnancy. In (**b**) there is cephalothoracopagus of the early pregnancy. In (**c**), a thoraco-omphalopagus of the early gestation is shown with the monochorionic, monoamniotic placenta. A close-up of the fused organs of this thoracopagus identifies a fusion of the liver (**d**) and bowel (not shown). In (**e**), is shown a close-up of the fusion

which is characterized by increased fibrin deposition in the intervillous space. Two other non-necrotic vascular pathology-related phenomena are the intervillous thrombus and the placental hematoma. The intervillous thrombus is a thrombus in the intervillous space due to an alteration of one or more factors of the Virchow's triad. The placental hematoma is defined as a collection of fresh blood outside of the intervillous space. The degree of the organization of the blood collection may be necessary to determine for both hospital and legal cases. In addition to the above categories, some bleeding phenomena may occur inside of the villous stroma of some villi or some groups of villi. Fibrosis of the villous stroma is associated with poor capillarization of the villi, while a fetal thrombus is an expression of the macro- and/or microcirculation. In the case of multiple pregnancies, the number of amnions and chorions and their fusion need to be noted.

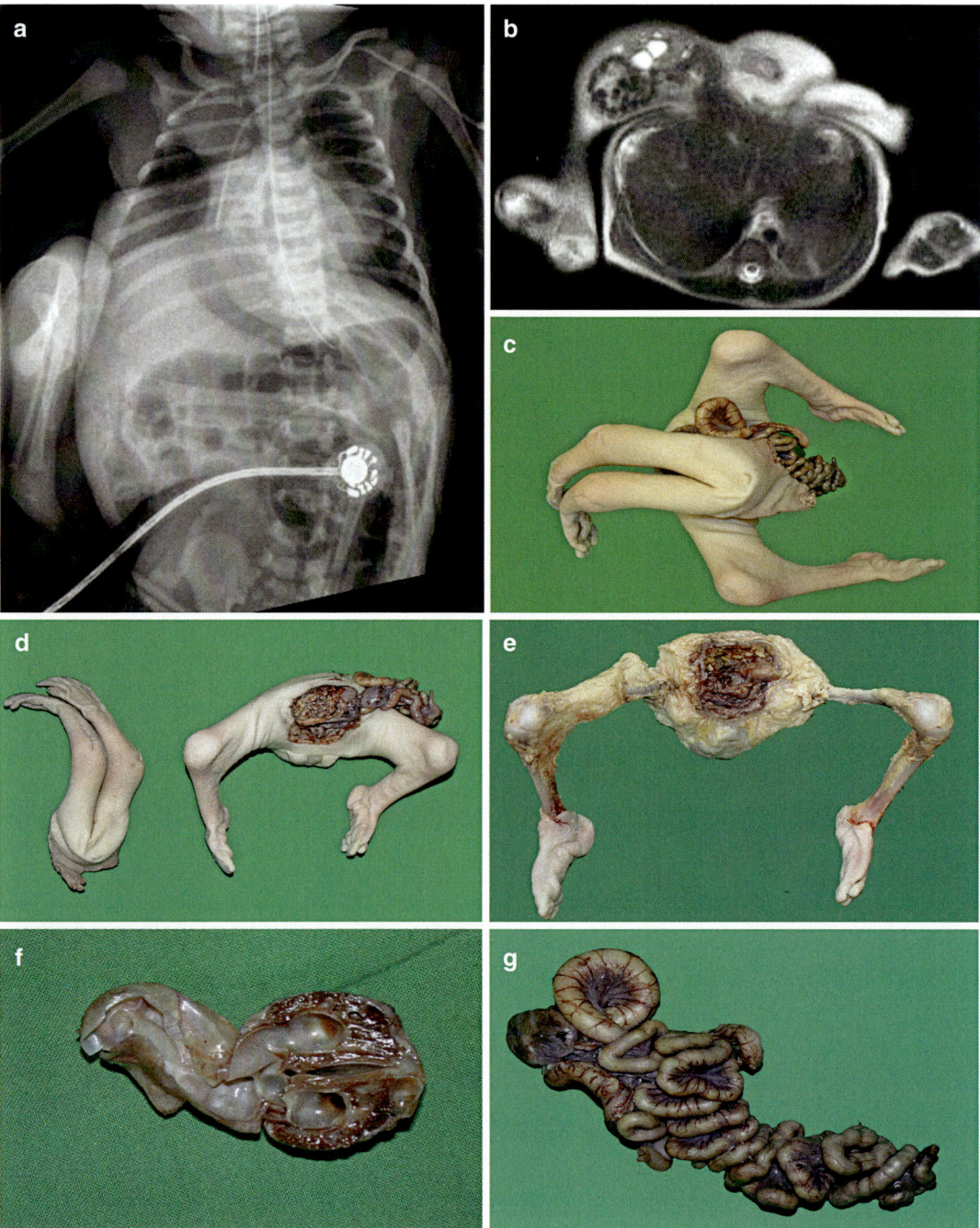

Fig. 18.10 Heteropagus twinning. Heteropagus with X-ray of the autosite twin showing the bony structures of the pelvis and lower limbs of the parasite (twin) attached to the autosite (**a**). Magnetic resonance imaging of the abdomen (**b**) showing the incomplete conjoined twin malformation. It shows the parasite connected with the abdominal cavity of the autosite. No evidence of common blood vessel communication was found (**b**). The figures (**c**, **d**) show the surgical specimen of the parasite (incomplete twin malformation) with pterygia between the thighs and legs. Figure (**e**) shows the delicate dissection of the surgical specimen identifying femurs, tibias, and fibulas as well as normally conformed feet. Figure (**f**) shows the cystic renal disease of the kidneys due to a cloacal malformation with obstruction/atresia of the lower urinary tract. Figure (**g**) shows the shorted bowel of the parasite with fewer loops than expected in a term newborn

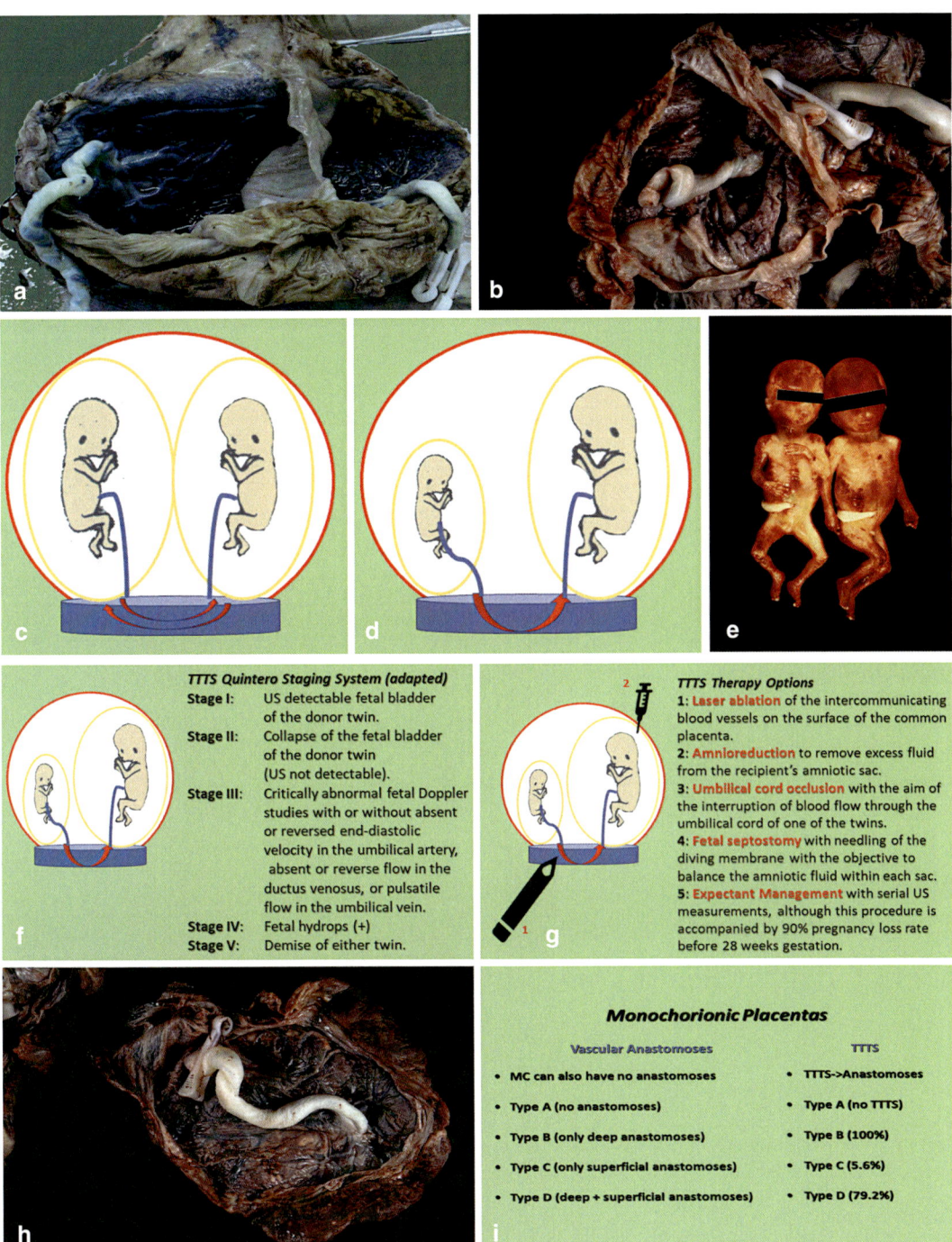

Fig. 18.11 Twin-to-Twin transfusion syndrome (TTTS). Twin-twin transfusion syndrome may be alarming for both mother and fetus. Figures (**a**) and (**b**) show the chorion plates of two TTTS placentas with identified anastomoses (not shown) but different distance between the insertions of the umbilical cords. Figure (**c**) is a TTTS with no apparent imbalance of the blood flow between the two fetuses, while figure (**d**) discloses the imbalance of the TTTS with a fetus twin (donor) giving blood to another fetus twin (acceptor) with growth difference that may be seen in (**e**) in a real TTTS. Figure (**f**) presents the Quintero staging system (adapted) and (**g**) the therapy options for the parents and physicians. Figure (**h**) shows the chorion plate of a monochorionic placenta, while (**i**) shows the vascular anastomoses identified in different degrees of severity of TTTS

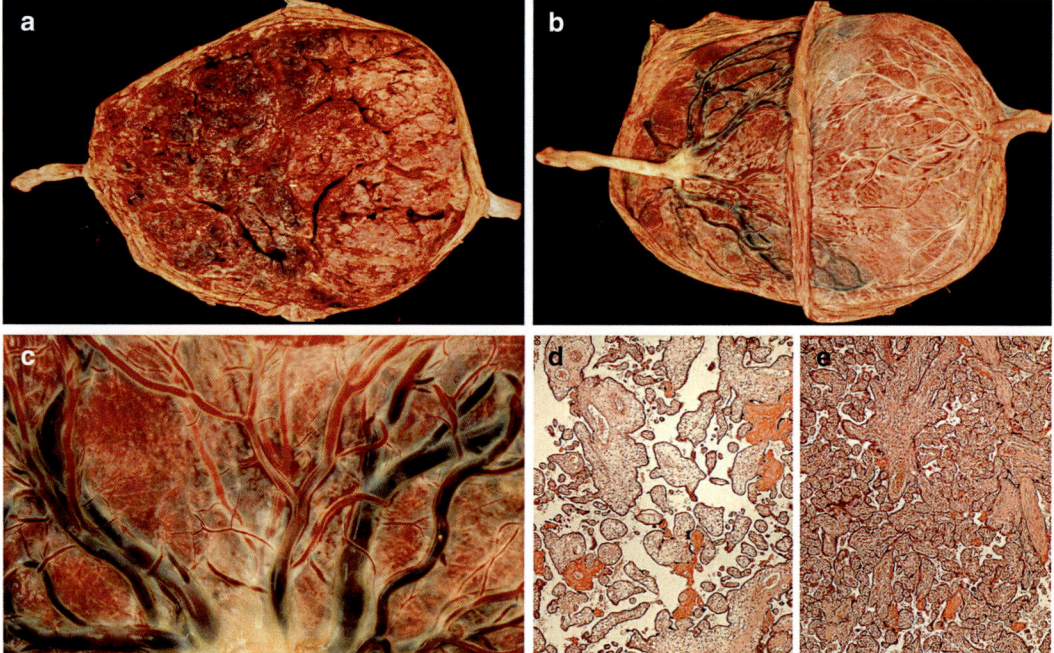

Fig. 18.12 Twin-to-Twin transfusion syndrome (TTTS). TTTS with gross (**a–c**) and microscopic photographs (**d–e**) showing the hyperemia of one side (left) and pallor of the other side (right). The superficial anastomosis is identified using contrast medium or milk instillating it through an umbilical cord and demonstrating the retrograde flow into the other half of this monochorionic placenta (**c**). The microscopic photographs taken from the two placentas show a quite different histology in term of paleness and hyperemia (**d–e**) (H&E stain, x100)

We distinguish as follows:

(A) Amnions (2) + central fused chorions (2): dichorionic diamniotic mono-/dizygotic
(B) Amnions (2) and single disc: monochorionic diamniotic monozygotic ("2 in 1")
(C) Amnion (1) + "artifactual peeling": monochorionic monoamniotic monozygotic

In case of monochorionic twin placenta, sequential injection of arteries needs to be performed. The arteries joining the thinner umbilical cord are examined first. Then, the arteries of the thicker umbilical cord are studied, and finally any vascular anastomosis is recorded by photographic documentation, which is attached to the requisition form or sent electronically to the signing out pathologist.

Twin-to-Twin Transfusion Syndrome (TTTS)
Twin-to-twin transfusion type of twin pregnancies is a challenge for the obstetrician and gynecologist as well as for the NICU team.

- *DEF*: A peculiar condition characterized by an unbalanced blood supply through placental anastomoses in diamniotic monochorionic twin gestations (~5-25%, i.e., ranging between 1:20 and 1:4).
- *EPG*: Vascular shunts between the circulation of one twin (the donor) and that of its co-twin (the recipient) lead to circulatory disequilibrium.
- *CLI*: Oliguria in the donor twin and polyuria in the recipient twin are the main clinical features. The fate of the TTTS depends substantially on the volume of blood transferred in the anastomoses, although other factors may play a significant role as well. Major polyhydramnios may cause extreme premature labor, and in the most severe cases, the recipient twin may die from cardiac failure whereas the donor twin from anemia and hypoxia. Alternatively, both twins may die as a result of premature labor and preterm-associated pathologies.

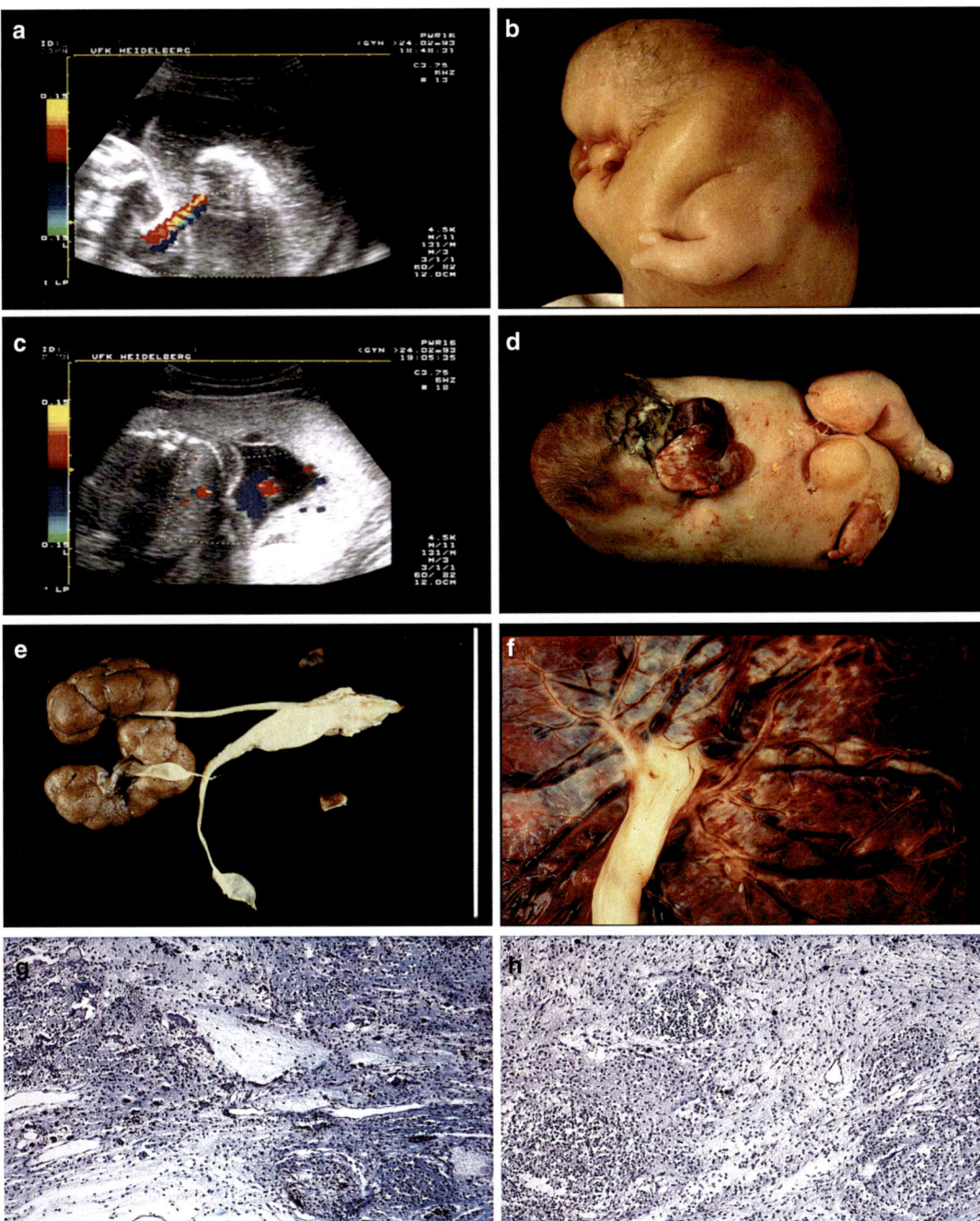

Fig. 18.13 Twin reversed arterial perfusion (TRAP) sequence. TRAP sequence should be considered the end of a TTTS with an amorphic twin labeled differently across centuries. Figures (**a**) and (**c**) show the abnormal flow of the blood in a typical TRAP sequence. Figures (**b**) and (**d**) show an acardius anceps in a front and lateral view. Acardius is another synonym for TRAP. The number of defects can be quite variable, but the urogenital tract is often preserved. Here an exception with low-joined kidneys (**e**). Figure (**f**) shows the extreme closeness of the cord insertions. Figures (**g**) and (**h**) show encephaloclastic features in the brain of an acardius (see text for details). Both microphotographs (**g**, **h**) are stained with Kluver-Barrera staining (x40)

- *GRO/CLM*: The unbalanced blood supply through placental anastomoses induces growth restriction, renal tubular dysgenesis (RTD), and oliguria in the donor twin and visceromegaly and polyuria in the recipient twin.
- *PGN*: Over the last two decades, the survival rate of newborns with TTTS has increased notably owing to improved neonatal resuscitation and to prenatal therapy based on amniotic fluid drainage, coagulation of the intertwine placental vascular shunts, selective feticide, and amniotomy. The perinatal mortality is still high with rates variable worldwide, but it may be as high as 50% with a short-term significant neurological morbidity up to 1/5 cases, unavoidably underscoring the need for further research. Cardiac disease of the recipient twin has also been emphasized mainly affecting the right ventricle and pulmonary artery, which can result in neonatal morbidity and mortality. Fetuses may show tricuspid valve regurgitation *in utero* already. Increased afterload may be the most likely mechanism. Increased pressure in the right ventricular outflow tract may lead to obvious cardiomegaly and tricuspid valve incompetence. The chronic decrease of blood flow through the pulmonary valve has the consequence of narrowing of this valve and finally its anatomic stenosis.

RTD may be studied by IHC using anti-CD15, which is a marker of proximal tubules, and anti-EMA, which is a marker for distal tubules and collecting ducts. The kidney of donor twins may show no or only very occasional tubules of the deep cortex, while strong and uniform tubular apical labeling recognizes distal tubules and collecting ducts. The kidney of recipient twins shows a normal tubular differentiation with the positive CD15 labeling of proximal segments and EMA labeling of distal and collecting tubules.

The putative etiologic factors of cardiac dysfunction of the recipient twin include:

(A) The increase of the preload
(B) Primary cardiac pathology (congenital cardiomyopathy)
(C) The rise of the afterload

TRAPS or Acardius/Acardia

Twin reversed arterial perfusion sequence (TRAPS) type of twin pregnancies may represent a challenge for the perinatal pathologist regarding the proper classification (Boxes 18.30, 18.31, 18.32, and 18.33). TRAPS is a form of conjoined twins in which the conjunction is via large A-A anastomoses that determine the premature circulatory failure of a twin and have been labeled *acardia*, because of the lack of the heart, for many centuries (Fig. 18.13). Four TRAPS seem to be group 99.99% of all TRAP sequences, although it has been clearly stated that any *acardius* should be properly reported in a scientific journal as a case report because there is a lot of knowledge that can be shared and is crucial for the progressive understanding of the early damage. Among several etiopathogenetic theories, the ischemic argument is the most relevant and explicative theory.

However, the history and teratogenesis play a significant role in describing and interpreting the origin of *acardia*. Although to date very few communities may treat unwell a woman giving birth to an *acardius*, there are still misconceptions about acardia and the pondering, if they should be considered embryos, malformed embryos, or amorphous structures. In Box 18.34, some historical theories of the origin of acardia are summarized.

By ultrasound, some signs can be detected to identify TRAPS, including twin gestation with discordance, twin gestation with "bizarre" (not-

---

Box 18.30 Acardius Acormus (Literally "Without Lower Rump")

- The rarest form of acardia (ca. 5% of all acardia).
- It is an embryo constituted by a well-distinct head with a caudal part.
- The caudal portion has an attached umbilical cord.
- Skeleton: Visceral-/neurocranium (± present), short spine (caudally), rudiments of ribs.
- Autopsy: Differentially (craniocaudally) development with head structures, but there are no rudiments or any anlage of organs of chest, abdomen, or pelvis.

> **Box 18.31 Acardius Amorphous (Literally "Amorphous" Means Amorphous "Without Form")**
> - A most underdeveloped form of acardia (ca. 20% of all *acardia*).
> - It is an embryo constituted by edematous tissue without a visible organ formation.
> - Skeleton: Disorganized bony elements.
> - Autopsy: No organ formation, their rudiments or organ anlage
> - but there is a mixture of connective, muscular, fatty, and bone tissue.

> **Box 18.32 Acardius Anceps (Literally Anceps Means "Dubious")**
> - A most well-developed form of acardia (ca. 10% of all acardia).
> - It is an embryo constituted by well-identifiable head, chest, and abdomen.
> - Skeleton: Splanchnic-/neurocranium, low number of ribs, abnormal, but present limbs.
> - Autopsy: Lack of most organs or presence of their rudiments, although gastrointestinal and urogenital tract seems to be the most differentiated structures.

> **Box 18.33 Acardius Acephalous (Literally Acephalus Means "Without a Head")**
> - A most frequent form of acardia (ca. 60–75%).
> - It is an embryo characterized by lack of head and the upper limb girdle.
> - Skeleton: Vertebral column blind ending cranially with some rudiments of the skull basis, ribs (low number), and some well-structured development of pelvis and lower limbs.
> - Autopsy: Lack of the head (DDX, anencephalic defect).
> - Autopsy: Urogenital tract quite well-developed.

> **Box 18.34 Historical Theories of Origin of the Acardius**
> - 1533: Benedetti reported the first case of acardius
> - 16th, 17th, and 18th centuries: Disturbed imaginings or vivid imagination of pregnant mothers
> - 1775: Defect of "germ plasm"
> - 1859: Abnormal circulation of a twin pregnancy with AA anastomoses and atrophy of the heart of the recipient from disuse ("*atrophia ex non usu*")
> - 1964: Aberrant germ cells involving the consideration of a "teratoma of the placenta"
> - 1972: "Rejection reaction" by the mother or regular twin
>
> Notes: *AA* arterio-arterial

classifiable) malformations, and twin gestation with retrograde blood flow in one fetus (acardiac twin) demonstrated by pulsed/color flow Doppler. Anencephaly, cystic hygroma, conjoined twins, twin demise, and intraamniotic placental tumors are differential diagnoses to be ruled out at all times.

18.3 Pathology of the Late Pregnancy

Disorders of the late pregnancy may involve events that occur within hours from the birth or peripartum, days or recent, and weeks from the delivery or isolated incidents. The most important events are summarized in Box 18.35.

In case of a circumvallate placenta, the membranes double back for a short distance over the fetal surface when the chorionic plate is too small. This abnormality of placentation is associated with an increase of fetal demise. Cardiotocography (CTG) is a technical instrument of recording the fetal heartbeat and the uterine contractions during pregnancy and is particularly useful in late pregnancy. Hon, Bradfield, and Hess (1961) invented first fetal monitoring, but Konrad Hammacher (1962) developed later a refined version (antepartum, noninvasive, beat-to-beat) version (cardiotocograph) (Fig. 18.14).

> **Box 18.35 Clinicopathologic Fetoplacental Events**
> *Acute* events (hours) → *Peripartum* ("ABC")
>
> - *A*bruptio placentae (placental abruption)
> - *B*leeding of the fetus ("fetal/baby hemorrhage")
> - *C*ord occlusion
>
> *Subacute* events (days) → *Recent* (4 "Is")
>
> - *I*nfection-associated events
> - "*I*nk" of the fetus-related or meconium-associated events
> - *I*schemia-related events
> - *I*ncreased nucleated RBC
>
> *Chronic* events (weeks) → *Remote* (5 "Vs")
>
> - *V*ascular underperfusion of maternal origin
> - *V*asculopathy of thrombotic type (of fetal origin)
> - *V*illitis of chronic type
> - Peri*v*illous fibrin deposition
> - Circum*v*allate placenta and related lesions (chronic abruption)

18.3.1 Acute Diseases

Acute diseases consist of abruption ("A"), umbilical cord integrity damage ("C"), and fetal hemorrhage ("B").

18.3.1.1 Abruptio Placentae

> - *DEF*: Sudden catastrophic uteroplacental separation, which is usually centrally located and involving the rupture of maternal spiral arteries (Box 18.36) (Figs. 18.15, 18.16, and 18.17).
> - *SYN*: "Acute" placental abruption, acute placental insufficiency.
> - *EPI*: Risk factors include HTN, drug abuse, advanced maternal age, low pregnancy weight gain, multiparity, and exhausting labor.
> - *EPG*: (1) *Abnormal vascular wall*, such as acute atherosis in preeclampsia; (2) *mechanical injury* due to severe HTN, trauma, placenta previa-related cervical dilation, or attempted vaginal delivery following a "recent" previous C-section; and (3) *ischemia-reperfusion*

Acute peripheral separation (aka "marginal abruption") is usually associated with low implantation, PROM, chorioamnionitis and abnormal marginal anatomy. An alteration of the uterine geometry or an increase of maternal venous pressure may be two crucial underlying factors. The usual presentation includes vaginal bleeding and precipitous delivery. Apart from the prematurity association, the recurrence risk is rarely similar to the central abruption. Preeclampsia, a "primigravida notable risk", is considered a specific maternal disorder of late pregnancy defined explicitly by onset (*de novo!*) of hypertension and proteinuria secondary to endothelial injury caused by factors released by the trophoblast of the placenta. Preeclampsia mothers have underlying genetic disorders (increased HLA-sharing paternal constellation and positive family history) with or without clinical risk factors, including obesity, essential hypertension, hyperlipidemia, insulin resistance, clotting disorders, renal diseases, connective tissue disorders, placental hypertrophy, and maternal vascular underperfusion (MVUP).

In a study of 1,570,635 women of which 3496 (0.22%) experienced a placental abruption, Ruiter et al. (2015) found that the risk of placental abruption in a subsequent pregnancy was significantly higher in women with a previous placental abruption compared with women without (5.8% vs. 0.06%). Women with a placental abruption that occurred at term in their first preg-

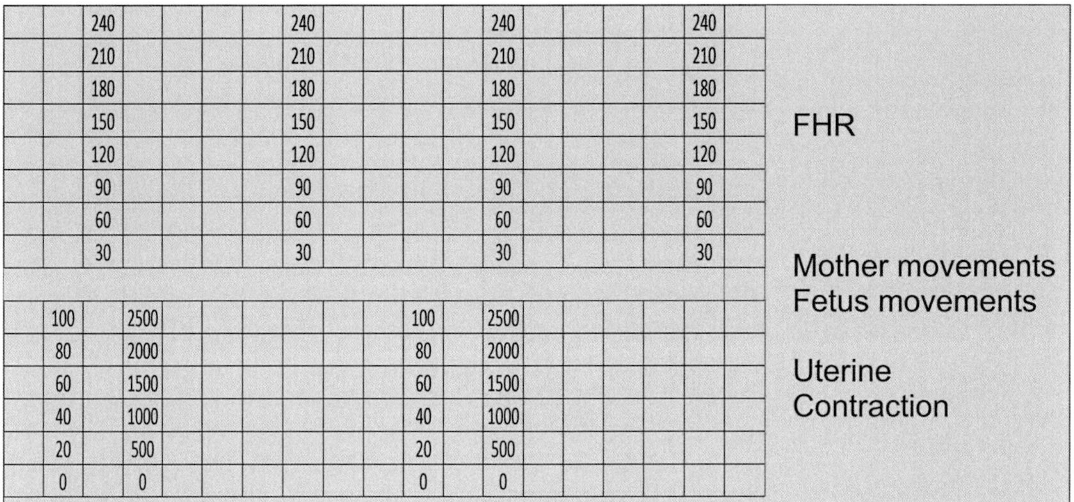

18.3 Pathology of the Late Pregnancy

	240			240			240			240		
	210			210			210			210		
	180			180			180			180		
	150			150			150			150		FHR
	120			120			120			120		
	90			90			90			90		
	60			60			60			60		
	30			30			30			30		

Mother movements
Fetus movements

100	2500					100	2500					Uterine
80	2000					80	2000					Contraction
60	1500					60	1500					
40	1000					40	1000					
20	500					20	500					
0	0					0	0					

Objectives:
1. To monitor fetal progression and viability with regard to the uterine contraction
2. To determine life-threatening risks and act tempestively
3. To provide a standardized terminology with the appropriate intervention
4. To enlist correctly pregnancy in risk categories and obstetrical documentation

Uterine Contraction
Normal: ≤ 5 contractions / 10 min (averaged over a 30-min window)
Tachysystole: ≥ 5 contractions / 10 minutes (averaged over a 30-min window)

Fetal heart rate (FHR): 100-160 bpm (over a 5-10 min period excluding accelerations and decelerations).

Accelerations
Hallmark of fetal health with transient increase in FHR of ≥ 15-bpm and lasting ≥ 15 s and ≥ 2 accelerations in a 20 min period (reactive trace), but the lack of them in an otherwise normal CTG does not indicate acidosis.
Baseline variability (10-25 bpm) excluding accelerations and decelerations is an index of the integrity of the autonomic nervous system, but reduced variability period can occur and associated with fetal sleeping.
Reduced baseline variability: hypoxia, prematurity, tachycardia, fetal infection, drugs (e.g. sedatives), fetal anemia, congenital malformations, and cardiac arrhythmia.

Deceleration: Transient episode of FHR slowing < the baseline level.
Early deceleration occurs in the 2nd stage of labour,
while late deceleration occurs after the uterine contraction (e.g. placental abruption).
Variable deceleration vary in shape, size, and in timing with respect to each other and is a manifestation of compression of the umbilical cord.

Fig. 18.14 Cardiotocography (CTG) essentials. CTG-essential knowledge requested for the pediatric pathologist with objectives and significance of accelerations and decelerations and to interpret correctly the pathologic findings of the placenta at the multidisciplinary team meetings (see text for details)

Box 18.36 Abruptio Placentae (Abruption or Central Retroplacental Hemorrhage)
- *DEF*: Premature utero-separation of a typically implanted placenta (late pregnancy).
- *EPI*: History of preeclampsia, physical trauma, and/or use of vasoactive drugs.
- *EPG*: I/R-I with rupture of a spiral artery.
- *CLI*: Vaginal bleeding, uterine rigidity/pain/tenderness, abdominal pain, hypotension, hemorrhagic shock, anemia, DIC.
- *GRO*: RPH or basal plate excavation in RPH-missing cases adherent to and compressing the maternal surface.
- *CLM*: Fresh hematoma or hematoma with different degree of organization.
- *PGN*: FGR, preterm birth, prematurity complications, neurodisability, FMH. RR: 5.8%.

Notes: *DIC* disseminated intravascular coagulation; *FGR* fetal growth restriction, *FMH* feto-maternal hemorrhage, *I/R-I* ischemia-reperfusion injury, *RPH* retroplacental hematoma *RR* recurrence risk

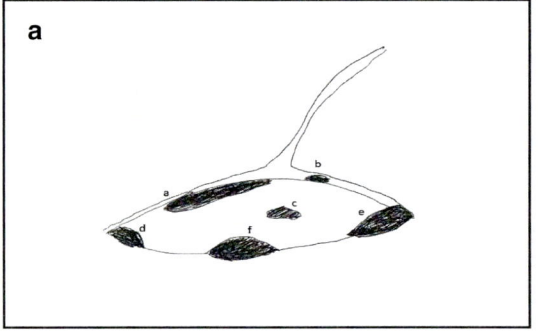

a: Massive subchorionic thrombosis
b: Demarcated subchorionic hematoma
c: Intervillous thrombus
d: Chorionic plate-sparing marginal hematoma
e: Chorionic plate-involving marginal hematoma
f: Retroplacental hematoma

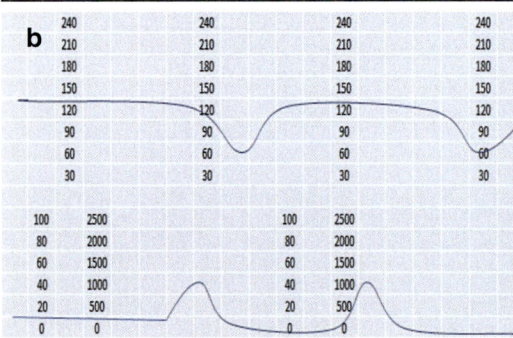
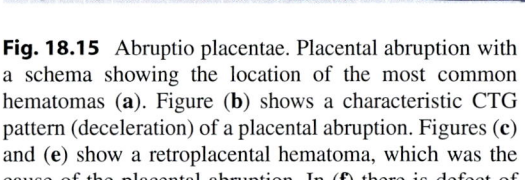

Fig. 18.15 Abruptio placentae. Placental abruption with a schema showing the location of the most common hematomas (**a**). Figure (**b**) shows a characteristic CTG pattern (deceleration) of a placental abruption. Figures (**c**) and (**e**) show a retroplacental hematoma, which was the cause of the placental abruption. In (**f**) there is defect of the basal plate evidencing the site of a hematoma. The hematoma was not delivered to the pathologist at the time of the placenta examination. In (**g**) there is a chorionic plate-sparing marginal hematoma, and (**h**) shows a bulging mass under the chorionic plate, which corresponds to a massive subchorionic hematoma

18.3 Pathology of the Late Pregnancy

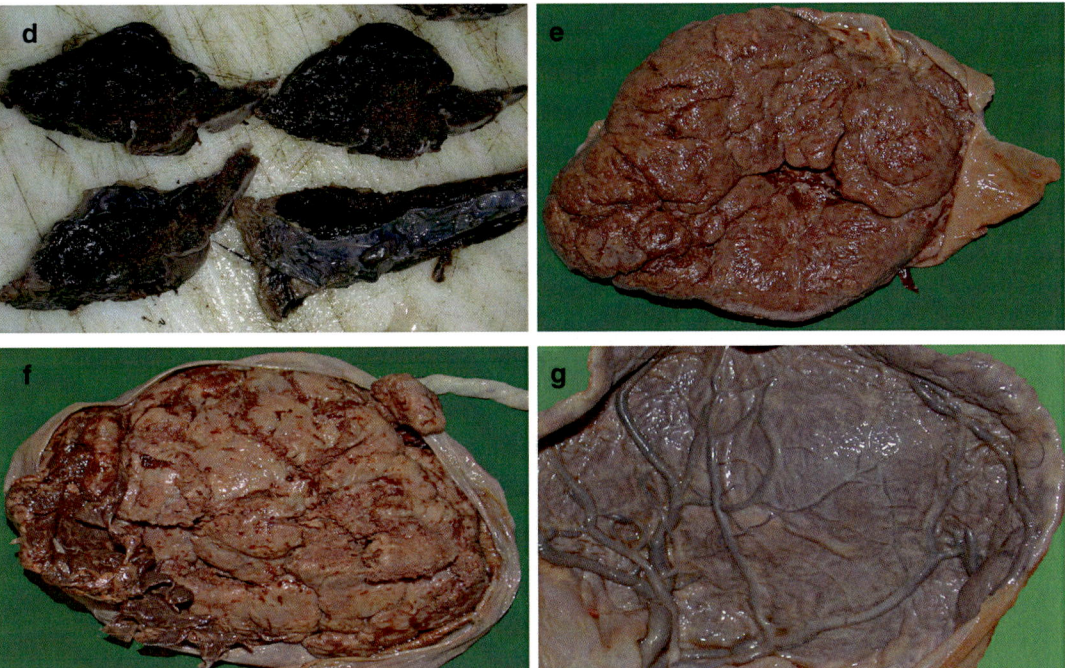

Fig. 18.15 (continued)

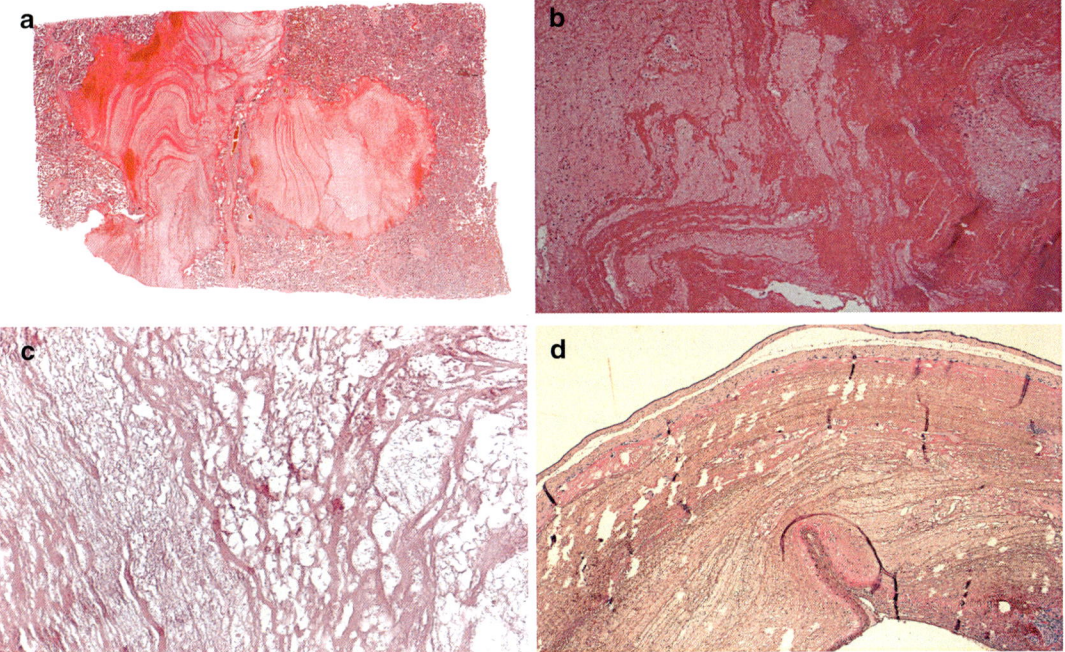

Fig. 18.16 Abruptio placentae. Microphotographs showing areas of thrombosis at different location (**a** and **b** in the parenchyma, **c** in the retroplacental area (necrosis), and **d** is subchorionic). The microphotographs **a** and **b** evidence the important differential diagnosis of fetomaternal hemorrhage with intervillous thrombi. All microphotographs are taken from histological slides stained with hematoxylin and eosin (a, x2.5; b, x40; c, x100; d, x40)

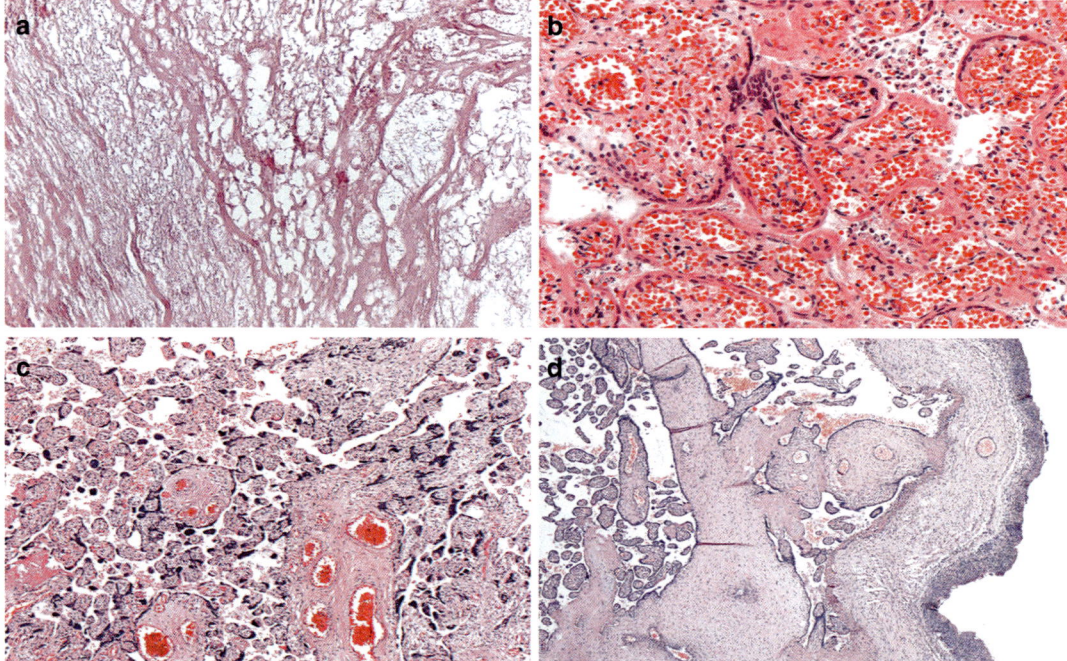

Fig. 18.17 Abruptio placentae. Microphotographs of placental abruption and perilesional changes with necrosis (**a**), acute hemorrhagic infarction (**b**), prominence of syncytial knots at the periphery (**c**), and accompanying inflammation (**d**) (x100). Microphotograph (**a**) of Fig. 18.17 is the same of the microphotograph (**c**) of Fig. 18.16 reproduced for comparison

injury, such as vasospasm-associated drug (e.g., cocaine) or heavy tobacco use (moderate smoker: 11-19 cigarettes per day, while heavy smoker: ≥ 20 cigarettes per day).
- *GRO/MIC*: *Retro*placental hematoma with indentation of the central part of the placental disc ("crater") or *intra*placental extension by direct evidence OR interstitial hemorrhage of the basal plate, diffuse retro-membranous hemorrhage, and ischemic changes of the overlying placenta with a recent villous infarction or villous stromal hemorrhage by indirect evidence.
- *PGN*: Vaginal bleeding, abdominal pain, and *rigor uteri* (maternal); preterm delivery, FGR, stillbirth, and HIE/Hypoxic-Ischemic Encephalopathy (neonatal); and neurodisability with or without cerebral palsy.

nancy were more at risk for recurrence (adjusted odds ratio [AOR], 188; 95% CI, 116–306) than women with a preterm (AOR, 52; 95% CI, 25–111) or early preterm (<32 weeks of gestation) placental abruption in their first pregnancy (AOR, 39; 95% CI, 13–116). Moreover, placental abruption was more frequent among women with hypertensive disorder compared with normotensive women.

A variant of the placental abruption located centrally or near central is the acute peripheral separation of marginal placental abruption (Box 18.37).

18.3.1.2 Bleeding of the Fetus (Fetal Maternal Hemorrhage, FMH)

The Kleihauer-Betke (KB) test, which was originally first described in 1957 by Enno Kleihauer and Klaus Betke, relies on the less solubility of fetal hemoglobin compared to maternal hemoglobin in an acid milieu. Maternal blood films are air-dried, fixed, eluted in the buffer, and carefully stained. At the microscope, fetal RBCs

Box 18.37 Acute Peripheral Separation (Marginal Placental Abruption)
- *EPG*: Abnormality involving the marginal anatomy of the placental disc with alteration of the geometry AND ↑ maternal venous pressure.
- *CLI*: Low implantation, PROM, chorioamnionitis, vaginal bleeding, and precipitous labor.
- *GRO*: Marginal hematoma.
- *CLM*: Fresh hematoma or hematoma with different degree of organization.
- *PGN*: Preterm birth (RR: rare).

Notes: *PROM* premature rupture of membranes

- *DEF*: Bleeding with fetal blood entering the maternal circulation with positive Kleihauer-Betke test or flow cytometric test identifying fetal RBC in the maternal circulation and accompanied with potential life-threatening evolution for the fetus (IUFD, stillbirth) or severe anemia of the newborn (Fig. 18.18).
- *SYN*: Fetal maternal hemorrhage (FMH).
- *EPI*: FMH is a cause of fetal death in 1:2000 deliveries with a significant fetal hemorrhage, defined as 20 ml/kg, occurring in about 5% of births.
- *EPG*: Mechanical blood loss of fetal blood into the maternal circulation (intervillous space), which occurs when the membrane of cytotrophoblast and syncytiotrophoblast ceases to function as a barrier and fetal cells, may reach the maternal area and enter the maternal vessels in the decidua/endometrium. The etiology may be recognized (rarely) or may remain unknown (most). Known causes are maternal trauma, placental abruption, external cephalic version, trauma-related amniocentesis, chorangioma, and choriocarcinoma.
    1. Maternal factors, e.g., physical trauma, amniocentesis
    2. Placental factors, e.g., *abruptio placentae*, chorangioma, choriocarcinoma
    3. Fetal factors, e.g., external cephalic version
- *CLI*: Decreased fetal movement and sinusoidal fetal heart rate pattern with positive Kleihauer-Betke test or flow cytometry test (false-negative tests may occur when ABO incompatibility resulting in rapid lysis of fetal cells does happen). Low neonatal hematocrit in case of neonatal survival and a sustained increase in the count of nucleated RBCs in the first few days following birth.
- *GRO*: Pale placenta ± intervillous thrombi (placenta-fetal → maternal intervillous space)
- *CLM*: Pale, hydropic placenta (nongestational age-related) with intervillous thrombi and increase of nucleated RBCs into the fetal circulation of the placenta.
- *PGN*: IUFD, stillbirth, neonatal anemia. Cerebral infarcts (palsy) and microcephaly have also resulted in FMH, which may rely on an acute hypotension-related mechanism.

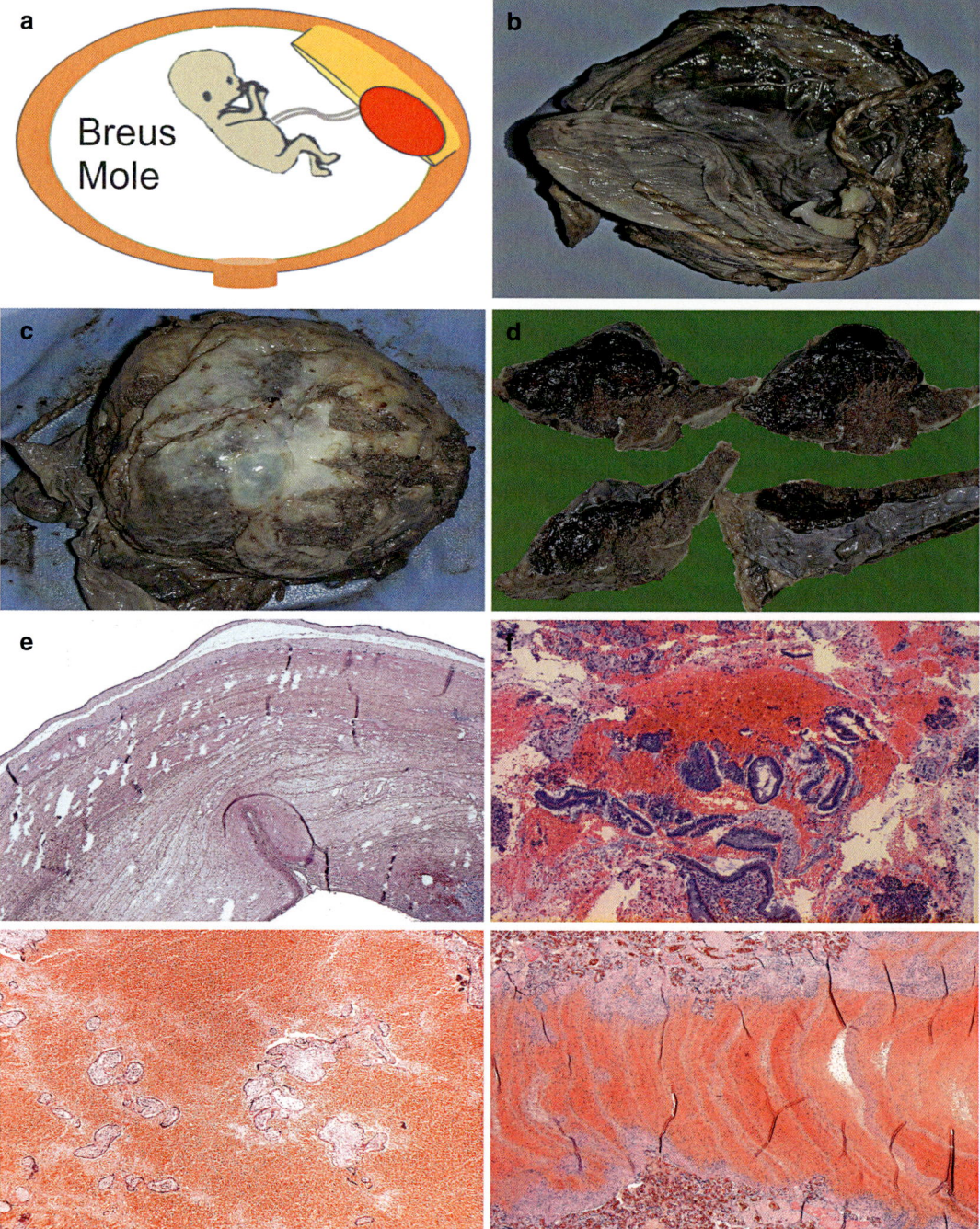

Fig. 18.18 Breus mole is shown in (**a**), which shows a cartoon of a massive, subchorionic hematoma from maternal blood in in pregnancy and first described by Karl Breus in 1892. The placenta harboring a Breus mole is shown in (**b–d**) with chorionic plate showing a bulging mass, which is more evident in cut sections. In (**c**) the hematoma is displayed following trimming of the placental disc. The microphotograph in (**e**) shows fetal hemorrhage, while the microphotograph (**f**) shows an intervillous thrombus. All microphotographs are taken from histological slides stained with hematoxylin and eosin (**e–f**, original magnification ×40)

preserve their color. Conversely, the maternal hemoglobin is mainly responsible for the mere shadow appearance of the maternal RBCs. The test is performed by counting fetal and maternal cells in 10 fields at 25× (objective) × 10× (ocular) magnification (250× magnification) or a total of 2000 cells. The results are stated as a rate of fetal RBC in the maternal circulation with an estimated blood volume of 5 L (5000 mL) with or without adding the maternal and fetal hematocrit. The final equation is the following Fetal Cells % = Number of RBCs × 100/Total Number RBCs and % fetal cells × 5000 mL = Amount (mL) of fetal blood in maternal circulation or, in case of the hematocrit (HCT) correction, % fetal cells × 5000 mL × maternal HCT (typically 35)/fetal HCT (typically 45). If the observer counts 20 fetal cells in 2000 as total, fetal cells are 1%. Thus, 1% × 5000 mL = 50 mL of whole fetal blood and 50/30 = 1.7 doses of Rh-Ig, because 300 μg of Rh-Ig are required for 30 ml of fetal blood implying two vials (1.7 + 0.3) as a safety measure. The KB test may result in false positive in the case of maternal persistence of fetal hemoglobin or other maternal hemoglobinopathies that occur in the setting of an elevated HbF. In this case, flow cytometry must be used to correctly quantitate the amount of fetal hemorrhage in maternal circulation. Of note, the blood volume of an infant is about 80 ml per kg body weight. Then, the volume of hemorrhage relative to the total fetal blood volume can easily be determined.

18.3.1.3 Cord Integrity Damage

- *DEF*: Any damage to the integrity of the umbilical cord (UC) with consequent fetal thrombotic vasculopathy (FTV) and catastrophic consequences to the fetus. Although CID may be associated with FTV, it is not synonymous with FTV (Figs. 18.19 and 18.20).

- *EPG*: UC compression or obstruction (e.g., excessively long, short, or twisted cords, cord entanglements, velamentous vessels, knots, prolapse, hematoma, tumor) altering the blood flow through the cord. Any hypercoagulable state influences on thrombosis and any condition leading to mechanical obstruction or ↑venous pressure are prone to block the venous return from the placenta, endothelial injury, and thrombosis. In fact, venous thrombosis is more commonly associated with cord problems, while thrombosis in the arterial circulation is more often associated with abnormalities in coagulation. Moreover, any condition causing direct endothelial injury prompts for thrombosis, such as chorionic vasculitis in amniotic sac infection.
- *CLI*: FTV-clinical features, IUFD, stillbirth.
- *GRO*: Placenta with surface thrombi due to the FTV.
- *CLM*: FTV lesions, including thrombosis, mural or occlusive, intimal fibrin cushions, avascular villi, fibromuscular sclerosis, and villous stromal-vascular karyorrhexis (hemorrhagic endovasculitis/endovasculopathy/endovasculosis) (see Sect. 18.3.3.4).
- *PGN*: IUFD, stillbirth.

18.3.2 Subacute Diseases

Subacute events prolonging for days are infection-associated events (chorioamnionitis), the discharge of the special fetal *ink* or meconium-associated disorders, and ischemic-hemorrhagic-related events (aka fetomaternal hemorrhage, FMH - *vide supra*) with or without an increase of nucleated RBCs (erythrocytic progenitors).

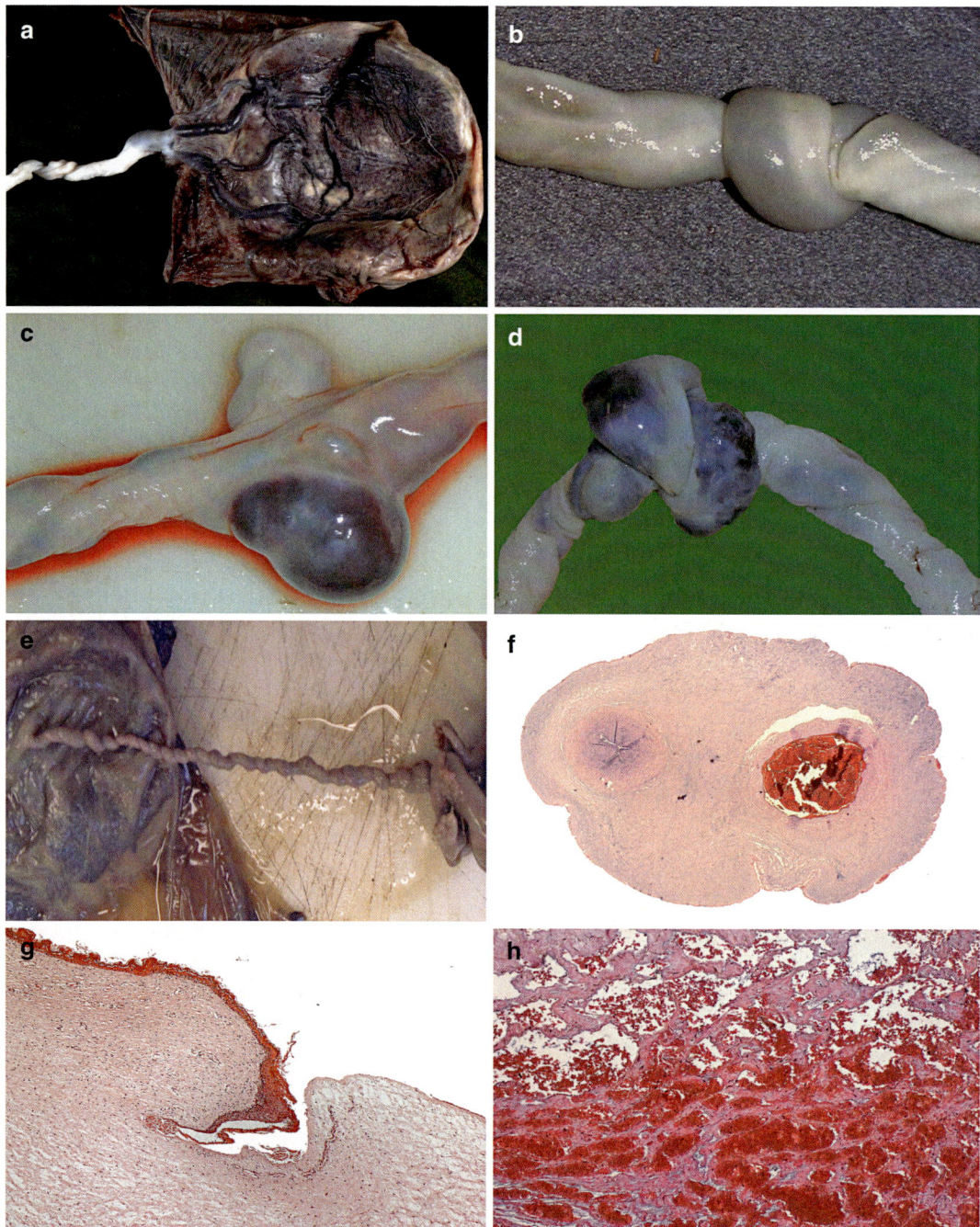

Fig. 18.19 Umbilical cord abnormalities. Gross photographs of cord integrity damage patterns with marginal insertion with vasa praevia (**a**), true knot (**b**), false knots (**c**, **d**), torsion (**e**), single umbilical artery (**f**), prominent squamous metaplasia (**g**), hemorrhage (**h**). All microphotographs are taken from histological slides stained with hematoxylin and eosin (f–h, x20, x100, x100)

18.3 Pathology of the Late Pregnancy

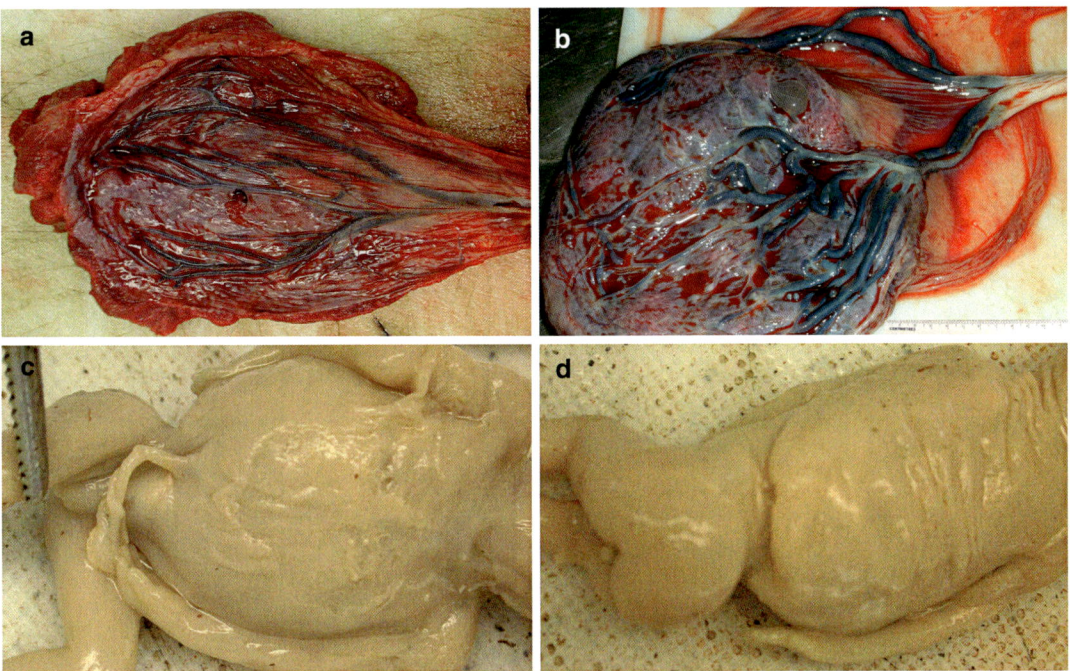

Fig. 18.20 Umbilical cord abnormalities. Gross photographs with velamentous insertion (**a**) and furcate (**b**) insertion of the umbilical cords. Gross photographs (**c**) and (**d**) show some external features of an ADAM sequence (see text for details)

18.3.2.1 Amniotic Fluid Infection (AFI)/Acute Chorioamnionitis (ACA)

- *DEF*: Inflammation of the chorionic plate and amnion (Fig. 18.21).
- *SYN*: Amniotic fluid infection syndrome.
- *EPI*: Maternal age (e.g., <21 years old), low socioeconomic status, primipara, abnormal cervix, prolonged labor, cervical dilatation secondary to PROM and ROM at an early gestational age, multiple vaginal examinations during labor, pre-existing infections of the lower genital tract (e.g., bacterial vaginosis, group B streptococci – GBS infection), foreign bodies (e.g., cerclage, IUD), as well as internal fetal and uterine monitoring.
- *EPG*: Infection through different pathways (ascending, hematogenous, or contiguous) and etiologic microorganisms include the healthy flora of the cervicovaginal tract (e.g., *Mycoplasma* spp., GBS, *E. coli*, *L. monocytogenes*, *Campylobacter fetus*, other aerobic and anaerobic bacteria, and *C. albicans*). ACA occurring before 18 weeks of gestation is often related to a maternal cause, while an ACA occurring at or after 18 weeks of pregnancy may originate from the mother and/or infant. The more advanced is the gestational age, more likely is that etiology is maternal.
- *CLI*: Preterm labor (PTL), PROM, maternal fever (T > 37.8 °C × 2 weeks/1 h interval or T > 38.3 °C ×1) or leukocytosis (PMN > 11×10^3/mm), abdominal pain, maternal tachycardia (>100/min), fetal tachycardia (>160/min), and foul odor of the amniotic fluid.

- *CLM*: 2 Major categories of response, including;
    1. *Maternal inflammatory response*: >14 weeks, diffuse subchorionic fibrin and neutrophilic (PMN) band at the choriodecidual junction accentuated at the site of rupture ± acute deciduitis and acute marginal hemorrhage in PTL.
    2. *Fetal inflammatory response*: >20 weeks, term > preterm, non-cord chorionic vasculitis, and umbilical vasculitis (V > A), which is more often seen at the fetal end of the UC.
- *PGN*: Prematurity, neurodisability with cerebral palsy, and early-onset sepsis about the fetal inflammatory response and with the cascade of cytokines in the fetal circulation.
- Prognostic salient features include amnion necrosis, subchorionic microabscesses, concentric umbilical perivasculitis (aka sub-necrotizing funisitis), umbilical arteritis or perivasculitis, severe chorionic vasculitis, and chorionic vasculitis with recent nonocclusive mural thrombosis. The relative risk (RR) is high (following VLBW delivery) because predisposing factors are likely to remain (VLBW, very low birth weight).

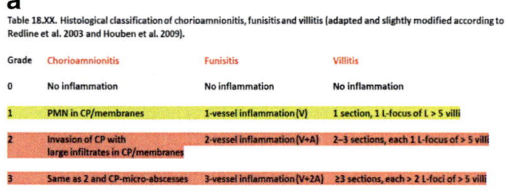

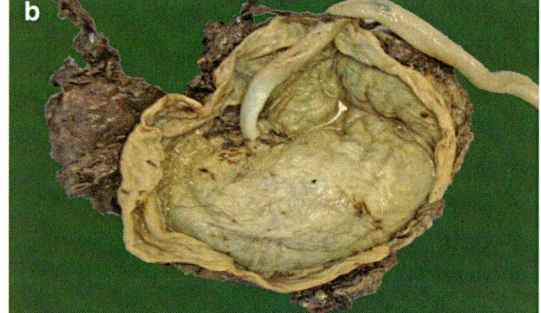

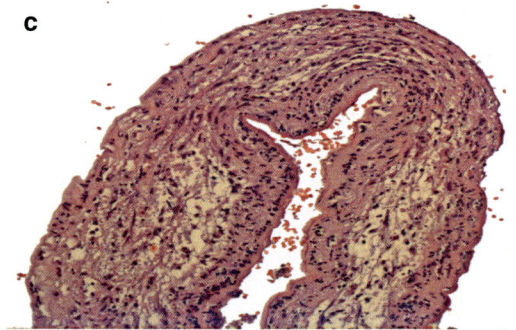

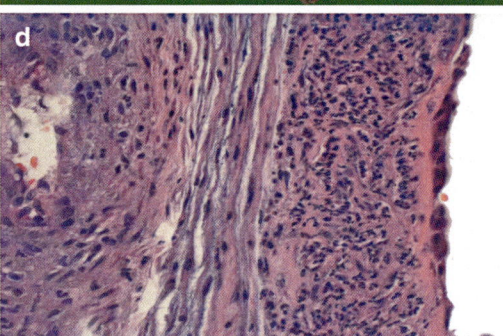

Fig. 18.21 Amnion fluid infection. In (**a**) is the current classification of chorioamnionitis, funisitis, and villitis with degrees from 0 to 3, while (**b**) shows a gray-yellowish change of the color of the membranes with opacification and thickening, a gross feature of chorioamnionitis. In (**c**) and (**d**) are microphotographs of a chorioamnionitis with severe inflammation constituted by a marked infiltration of neutrophils. Figures (**e**) and (**f**) show the histopathology of a funisitis with infiltration of the wall of the blood vessels as well as Wharton's jelly by neutrophils. Figures (**g**) and (**h**) show a necrotic chorioamnionitis. All photographs are taken from histologic slides stained by hematoxylin and eosin (**c–h**, original magnification x200, x400, x100, x200, x20, and x40)

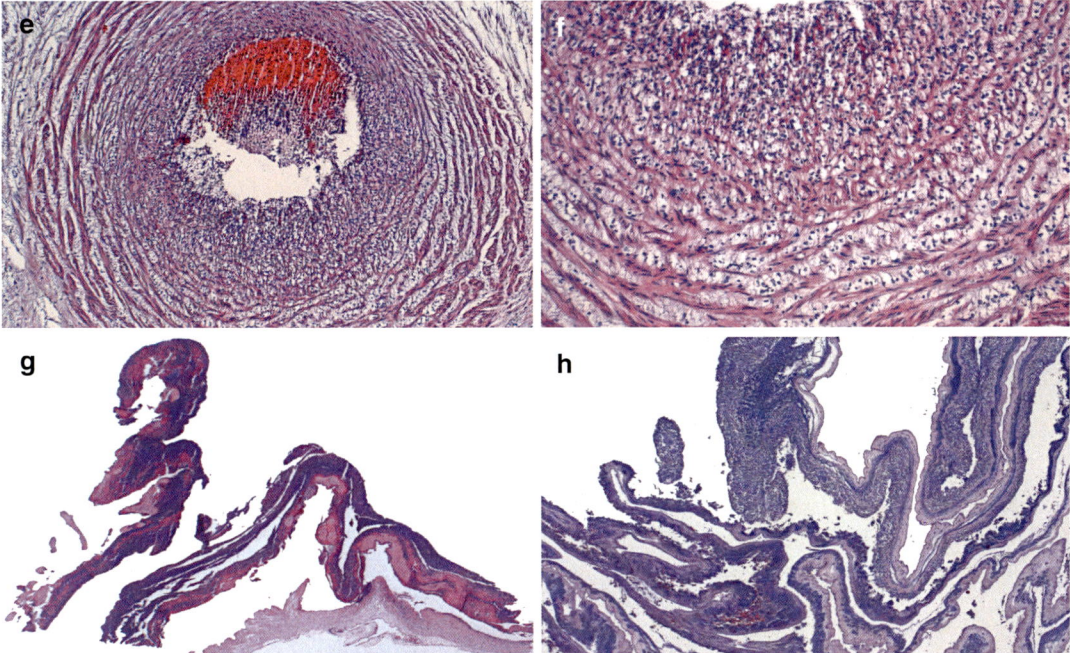

Fig. 18.21 (continued)

Acute chorioamnionitis is suspected in a pregnant woman with a fever (e.g., in case of the premature rupture of membranes or PROM). In studies involving chorioamnionitis, acute inflammation is defined by evidence of amniotic fluid infection/acute inflammatory pathology. Maternal acute inflammation is identified by neutrophilic infiltration of chorion (stage 1), amnion (stage 2), and necrotizing chorioamnionitis (stage 3). Neutrophils' diapedesis recognizes fetal involvement through the wal (subchorionic abscesses)l ls or umbilical vein (stage 1), umbilical artery (stage 2), and necrotizing funisitis (stage 3) defined by neutrophils' karyorrhexis in a band-like configuration within Wharton's jelly. Chronic inflammation is defined as the presence of significant persistent (lymphocytic or histiocytic) infiltrates in the membranes (chorion and/or amnion), chorionic villi, intervillous space, or basal plate. Chronic villitis was defined as lymphocytes or histiocytes infiltrating the chorionic villi and is graded as low (few, small foci) or high (multiple, large foci). Chronic intervillositis is identified when a lymphohistiocytic infiltrate is present in the intervillous space without an identifiable villous infiltrate. Basal chronic intervillositis is considered a diagnostic term for chronic deciduitis when plasma cells are identified within the chronic inflammatory infiltrate. The usua staging and grading of chorioamnionitis and the fetal inflammatory response syndrome (FIRS) are very crucial aspects. If the inflammatory process involves the umbilical cord (umbilical vein, umbilical artery, and Wharton's jelly), the term acute funisitis is used and is the histologic counterpart of the FIRS).

Staging and Grading of Chorioamnionitis (CAI) are in blue and Fetal Inflammatory Response Syndrome (FIRS) are in red.

Stage = Duration ±Fetal Maturity on
1) Subchorionitis/Chorionitis,
2) Acute chorioamnionitis as a combination of chorion and amnion involvement
3) Necrotizing Chorioamnionitis

Grade = intensity: mild degree (subchorionic PMN) and severe degree (subchorionic abscesses)
FIRS stages:

1) Cord phlebitis (stage 1/3)/ arteritis (stage 2/3)/ perivasculitis (concentric) (stage 3/3), AND
2) Necrotizing funisitis or concentric umbilical perivasculitis (>25% amniocytes)
Grade = Intensity:mild degree (subchorionic PMN) & severe degree (subchorionic abscesses)
Note: PMN, polymorphonucleates; UC, umbilical cord.

The histopathology is key for the diagnosis and differential diagnosis (Fig. 18.22). Special stains and culture are used in the routine for placental pathology and include Gram stain mostly for *L. monocytogenes* and Group B Strep (streptococcus) (GBS); silver stains (Dieterle, Warthin-Starry, Steiner) for anaerobic bacteria, *C. fetus*, *T. pallidum*, and other bacteria; and PAS and GMS for *Candida* spp. and other fungi. The differential diagnosis includes laminar (ischemic) necrosis and T-cell-driven eosinophilic vasculitis. The former is characterized by necrotic tissue localized in the decidua capsularis, which is often escorted by varying numbers of neutrophils with a variable degree of cellular degradation. Laminar necrosis has been interpreted as maternal small-vessel disease or meconium toxicity-related. The latter is substantially an infiltration of chorionic and major fetal stem villous blood vessels by T lymphocytes (CD3-positive by immunohistochemistry) and eosinophils (Luna special stain by histochemistry). The T-cell-driven eosinophilic vasculitis is not polarized toward the cavity of the amnion differently from the fetal inflammatory response seen in acute chorioamnionitis. *Listeria monocytogenes* is a Gram-positive bacillus found in soil, sewage, and animal silage which is a type of fodder made from green foliage crops. The crops are preserved by acidification, which is achieved through fermentation.. Periodically, there are outbreaks of listeriosis, and the food industry is particularly aware of this infection of meat and meat-related products as well as dairy products. In fact, *L. monocytogenes* is well known and contaminated food source for deli foods, which are particularly contraindicated in pregnancy. Lunch meats, cold cuts, or other deli meats, may contain this bacterium and should not be eaten by pregnant women unless they are heated to an internal temperature of 165°F or 74°C before serving.Listeriosis tends to occur in clusters, and notifications by public health agencies (e.g., FDA and the Canadian Food Inspection Agency or Health Canada) are routinely issued yearly. An intervillositis is commonly seen in cases of listeriosis of the placenta. However, the same histology may be encountered with infections by *Campylobacter fetus*, *Chlamydia psittaci*, *Francisella tularensis*, *Klebsiella pneumoniae*, and *Coccidioides immitis*. Occasionally, other bacteria can be behind this histologic pattern. *M. tuberculosis* has been indeed detected in the absence of any granulomatous reaction. The listeriosis histopathology slides show a variable mixture of acute intervillositis with intervillous abscesses and chorioamnionitis. The histopathology may also include the so-called septic infarcts that may be grossly appreciable as irregular, firm, yellow opacities on the placenta cut surface. However, more often, there are gross cut surface opacities and green discoloration of the membranes as well as acute deciduitis of the basal plate showing a pallor change of the basal plate. In the intervillous space other than the neutrophilic exudate in the form of small clusters or diffusely there may be villous

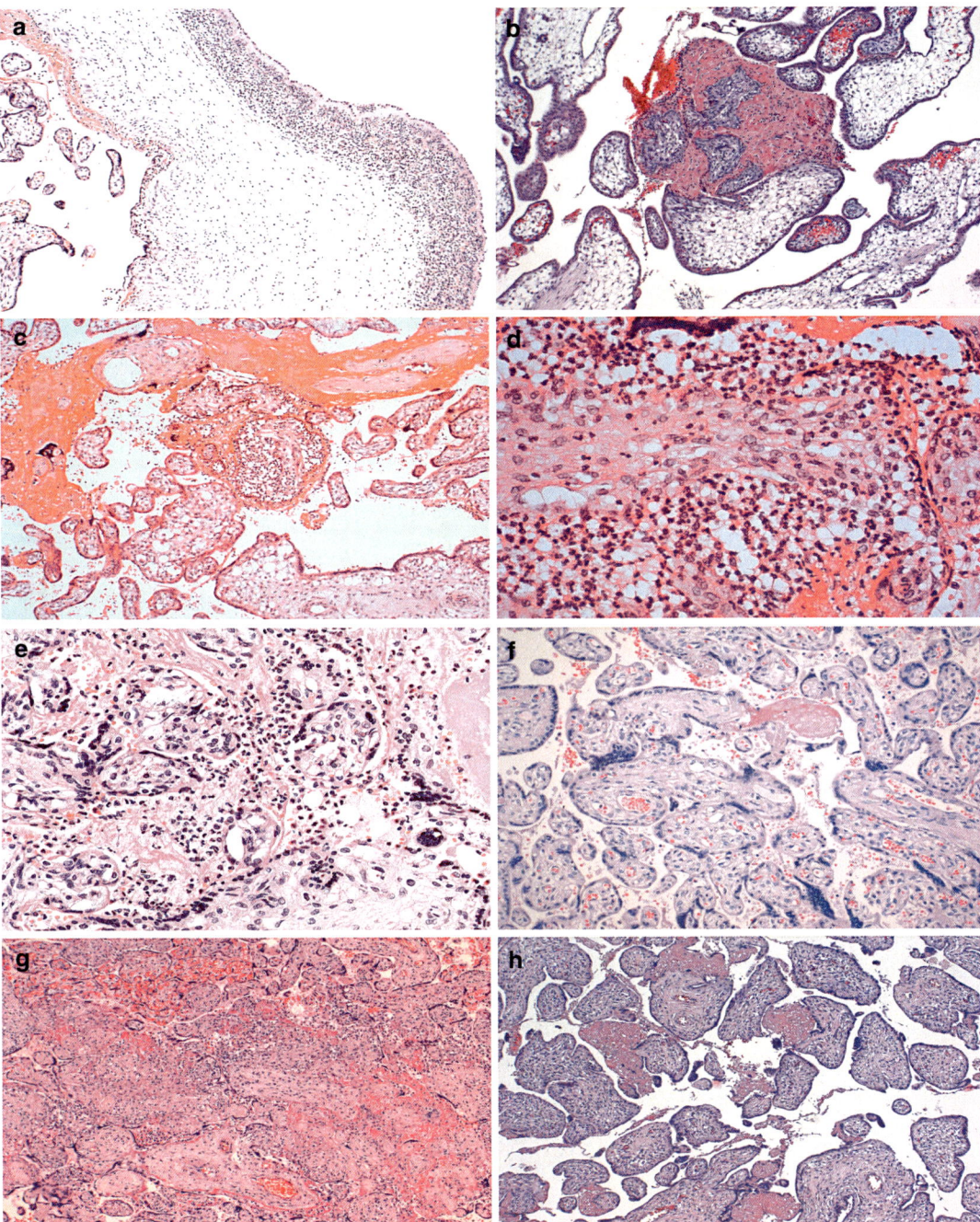

Fig. 18.22 Villitis and perivillitis. Figures (**a**) through (**h**) show villitis and perivillitis of inflammatory character with neutrophils infiltrating the syncytiotrophoblast and the perivillous space. In (**c**) and (**d**), the etiology of the inflammation is an infection with *Listeria monocytogenes* (**a–h**, H&E stain; a, original magnification x100; **b**, original magnification x100; **c**, original magnification x100; **d**, original magnification x200; **e**, original magnification x200; **f**, original magnification x200; **g**, original magnification x100; **h**, original magnification x100)

necrosis and perivillous fibrin deposition. Importantly, neutrophils are not accompanied by other chronic inflammatory cells, including lymphocytes, eosinophils, plasma cells, and giant cells. Acute villitis, i.e., neutrophils inside the villi, may infrequently be seen in conjunction with acute intervillositis. The finding of the acute villitis in the setting of a listeriosis infection is interpreted as secondary spread to villi by a primary acute intervillositis process. Similar to the acute villitis, another infrequent finding is the vasculitis of the stem villi. Gram staining of the placenta at the time of delivery may disclose the nature of the bacteria. *L. monocytogenes* has been named after Joseph Lister and is a facultative anaerobic bacterium, which can rapidly grow and reproduce inside the host's cells. Currently, *L. monocytogenes* is considered one of the most virulent foodborne pathogens. It has been assessed that up to 30% of foodborne listeriosis infections in high-risk individuals may result in the death of the host. Listeriosis grades third in a total number of deaths among foodborne bacterial pathogens. Its fatality rates exceeding even *Salmonella* and *Clostridium botulinum* infections in both the USA/Canada and European Union. One of the most frightening landscapes of this bacterium is its ability to grow at 0 °C (32 °F). Thus, the standard refrigerator with a temperature of 4–10 °C (39.2–50 °F) is optimal for its growth and multiplication inside of the food. Soft cheeses (e.g., Brie, Camembert, feta, and queso blanco fresco) are frequently contaminated with this bacterium. As indicated above, listeriosis outbreaks can also occur by consuming deli meats, frankfurters, tacos, and a variety of other foods with meat. Moreover, celery, sprouts, and cantaloupe are also often contaminated. *L. monocytogenes* can move within eukaryotic cells by polymerization of the actin filaments. *L. monocytogenes* is the third most common cause of meningitis in newborns after group B streptococci (GBS; *Streptococcus agalactiae*) and *E. coli*. *L. monocytogenes* can infect the brain, the spinal cord membranes, and/or the bloodstream of the host through the ingestion of contaminated food (e.g., unpasteurized dairy or raw foods) found to the table or in the crib. *Fusobacterium nucleatum* is an anaerobic bacterial organism, which is now recognized as a relatively common cause of acute chorioamnionitis and preterm labor. *F. nucleatum* chorioamnionitis may originate from fecal flora contamination but also from oral cavity-contaminated food. Secondary hematogenous dissemination following maternal periodontal disease should also be considered. *Fusobacterium* histopathology includes acute chorioamnionitis, with or without funisitis, and copious filamentous organisms, which coat the surface of the amnion. Other than H&E with numerous filamentous bacterial organisms coating the surface of the partially denuded amnion, Giemsa stain and Warthin-Starry silver stain can be used to detect the microorganism. In the differential diagnosis, *L. monocytogenes* and *E. coli* should be considered, but acute villitis and perivillositis are useful features because they are not typical features of *F. nucleatum* infection but more pointing to listeriosis or *E. coli* infection. *Candida* infection results in acute chorioamnionitis with peripheral microabscesses on the umbilical cord. In the differential diagnosis, *Corynebacterium kutscheri*, *H. influenzae*, and *L. monocytogenes* should be considered. In case of ACI, it is important to assess the extent of cord involvement, i.e., if it is segmental, regional, or total and which blood vessels involve (funicular phlebitis, funicular arteritis, or funisitis). The term funisitis should be left to the umbilical cord involvement of blood vessels (≥ 1) and Wharton's jelly stroma. Thus, it is recommended to take three blocks with cross sections of the umbilical cords (three centimeters from the cord insertion to the placental disc, three centimeters from the distal end, and three blocks with cross sections of the central region of the umbilical cord).

18.3.2.2 Meconium-Stained Amniotic Fluid-Related Disorders (MAFD)

Meconium aspiration syndrome (MAS) is a chemical pneumonitis which occurs in 1:20–1:10 live births with evidence of meconium staining. Babies show respiratory distress syndrome (RDS) requiring oxygen supply and display an abnormal chest X-ray. Babies are at risk of multiple thrombi, extrapulmonary air leaks, pulmonary hypertension-associated PFC, and neonatal encephalopathy.

Subacute hypoxia changes can occur in MAFD and include presence or increase of eryth-

- *DEF*: Transient fetal hypoxia-related greenish staining of the amniotic fluid (AF) with universal variable CTG deceleration (Fig. 18.23).
- *SYN*: "Green water" syndrome.
- *CLI*: Postdates (>42 weeks), macrosomia (LGA), oligohydramnios.
- *EPG*: Umbilical cord occlusion, vagal stimulation, and reflex peristalsis.
- *CLM*: According to the contact time of meconium with the amniotic epithelium, several degrees may be identified, including:
    1. *Early stage* (<6 h): Pseudo-columnar change, vacuolated pigment-laden macrophages, and pigment-laden macrophages in amnion (>1 h) and decidua (>3 h).
    2. *Intermediate/late stage* (>6–12 h): Green-staining of the umbilical cord, chorionic plate meconium, chorionic plate meconium with necrosis of fetal BVs, and apoptotic vascular smooth muscle cells (cytoplasmic eosinophilia, nuclear pyknosis, cellular dehiscence) due to the caustic effects of bile acids and phospholipases of the AF on myocytes.
- *PGN*: Neurodisability, IUFD, neonatal death, meconium aspiration syndrome (MAS), or persistent fetal circulation (PFC). RFs: LGA, hypercoiled, entangled, or tethered cord, peripheral cord insertion, and ↓ AF.

rocytic progenitors or nucleated red blood cells (NRBC) defined by >10 NRBC/10 HPF (40×) and are indicative of significant hypoxia at intrauterine onset, which leads to the release of erythropoietin and choragiosis or hyper-capillarization of the villous tree. The choragiosis, which is a nonspecific finding, is defined by Altshuler's rule of >10 capillaries/villus in >10 villi in >10 fields of the placenta at 10×. Prussian blue special stain (Perl's stain) may be useful to separate the deposition of hemosiderin granules from meconium pigment.

18.3.2.3 Fetomaternal Hemorrhage (FMH) ± Intervillous Thrombus(i)

- *DEF*: Fetal bleeding/villous rupture (± villous immaturity) into maternal circulation (↓fetal blood supply) associated with intervillous thrombus development, ↑ intravillous NRBC, ± fetal hypovolemia.
- *EPI*: 1–3:1000 deliveries (RFs: genetic factors/MCA, maternal trauma, OB/GYN procedures).
- *EPG*: Idiopathic (>4/5) or linked to placental abruption, choriangioma, choriocarcinoma, preeclampsia, maternal trauma, and OB/GYN procedures (e.g., external cephalic version or amniocentesis).
- *CLI*: ↓ fetal movement (US/CTG), sinusoidal fetal heart rate, NIHF, (+) Kleihauer-Betke test (HbF resists the cell elution in an acid medium) (+) HbF detection by flow cytometry.
- *GRO/CLM*: Damage of the trophoblastic coat of the placental villi with the escape of fetal blood into the maternal intervillous space with resulting FMH. At some point, a breach of the placental villi coat may either heal with perivillous fibrin deposition or result in massive acute blood loss and fetal death. Perivillous thrombosis may be a protective mechanism to limit the extent of FMH. An ↑ of NRBC occurs in case of profound/sustained hypoxia and is a bone marrow response. It needs >6–12 h to develop (Fig. 18.18).
- *PGN*: In case of small FMH (0.5–40 ml), there is minimal sequela, but large GMH (>40 ml) are associated with neurodisability and IUFD. There is no apparent recurrence risk.

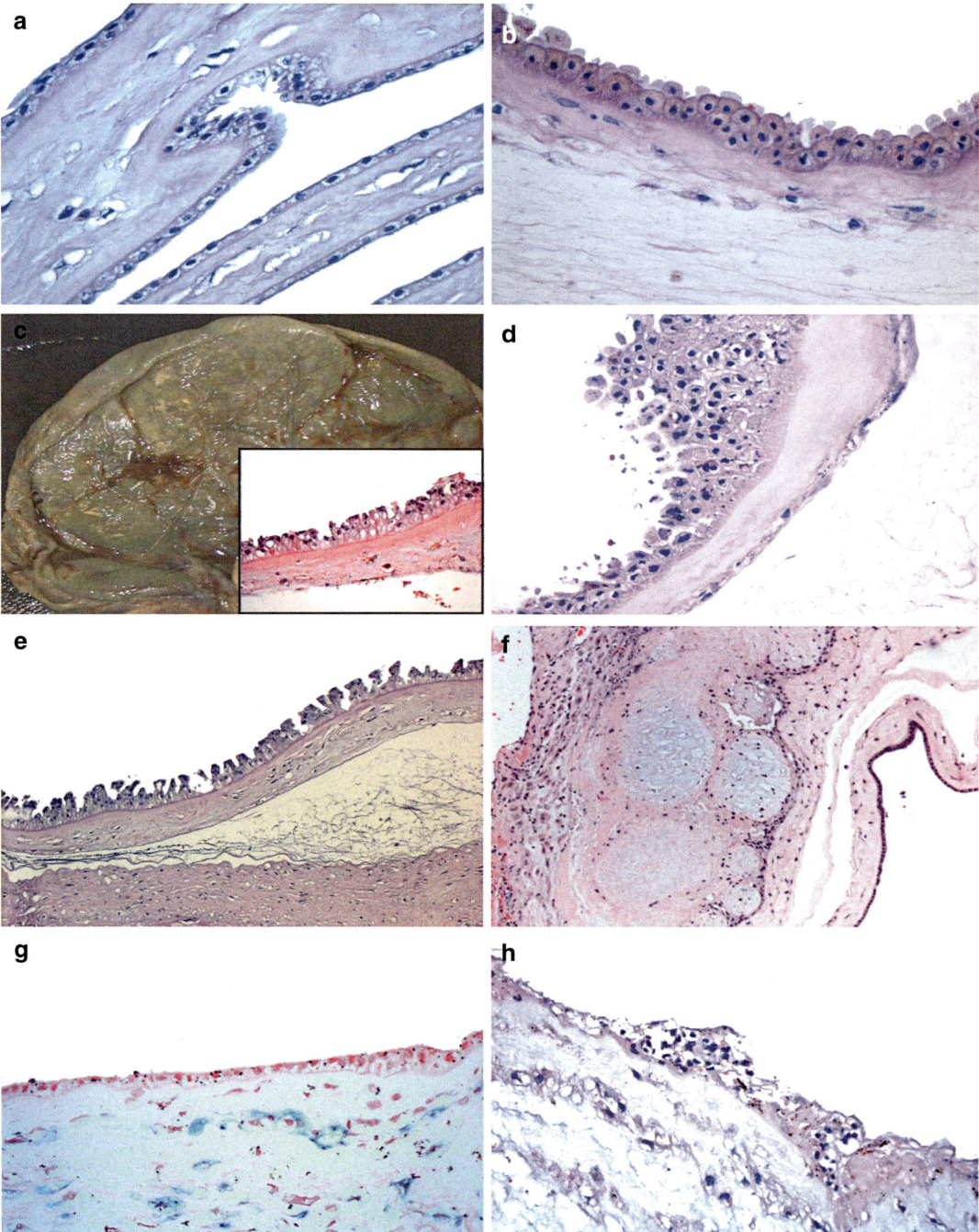

Fig. 18.23 Membranes changes. The figures (**a**) through (**h**) show the changes observed in presence of meconium exposure with cylindrical appearance of the amniocytes, papillary foldings, and bulbous change as well as incorporation of athetoid changes (**f**), hemosiderin (**g**), and frank necrotic of the epithelium (**h**). Figure (**c**) gross photograph corresponds to the inset with cylindrical change of the epithelium. All microphotographs have been taken from histological slides stained with hematoxylin and eosin apart from (**g**), in which a Pearls blue stain for iron was carried out (**a**, original magnification x400; **b**, original magnification x400; **c**, original magnification x400 [inset]; **d**, original magnification x400; **e**, original magnification x100; **f**, original magnification x100; **g**, original magnification x200; **h**, original magnification x400)

In details, clinical features include reduced fetal movements with CTG abnormalities, including decreased variability and sinusoidal rhythm with variable fetal outcome dependent from the rate of bleeding and increase in maternal hematocrit as a subtle sign of FMH. The acute form is characterized by rapid blood loss, which is followed by perinatal hypoxia and intrauterine death or severe anemia and hypoxia at birth. The chronic form shows compensatory mechanisms of increased hemopoietic activity revealed by increased erythroblasts and reticulocytes in the peripheral smear and active erythropoiesis in the fetal liver at time of the autopsy. The sequela for the inadequate compensatory reaction is hydrops. Fetal growth and ultrasound are usually within the normal range. As explained earlier, the Kleihauer-Betke test is useful in women presenting with a decrease of fetal movements or decreased variability and sinusoidal pattern. This test is a convenient tool to assess FMH. This investigation takes into account that adult hemoglobin is more readily eluted through the plasmatic membrane in the presence of acid (e.g., citrate-phosphate buffer) than is fetal hemoglobin (HbF). It occurs after fixation on a slide with ethanol 80% and finally stained with erythrosine B. The number of fetal cells and maternal cells is counted, and this rate is directly proportional to the estimate of fetal blood volume (ml) to maternal blood volume (ml). Massive FMH are often defined by reaching of the thresholds of 80 ml and 150 ml of fetomaternal bleeds, respectively. Moreover, a 250-ml fetal blood loss or a 5% fetal Hb level in the maternal circulation may be associated with fetal death. Apt test or flow cytometry which means quantifying the number of fetal cells present by measuring the intensity (fluorescence) of specific monoclonal antibodies binding to HbF is also used in cases of suspicious FMH. The crucial mimickers of the NRBCs are hypoxic-ischemic disorders (hours-days, normoblasts), fetal blood loss (days-weeks, normoblasts in most terminal villi), and fetal anemia/hydrops (weeks/months, pre-normal blast precursors and EMH). The pitfalls of the Kleihauer-Betke test(KBT) are distinct in false positive and false negative. False-positive KBT (may) occur in maternal HbF cells, maternal thalassemia minor, maternal sickle cell anemia, and hereditary persistence of HbF. False-negative KBT (may) occur in case of ABO incompatibility and Rh incompatibility.

Substantially, to remember is that:

- CTG abnormalities are late signs and often associated with the reduced fetal outcome if delivery is performed outside of the hospital.
- Consider FMH in cases of unexplained IUFD, fetal distress, NIHF, neonatal shock, and non-hemolytic neonatal anemia.
- Timing is essential to save the fetus and prompt delivery followed by neonatal transfusion or in the case of a fetus cord sampling and intrauterine transfusion.
- Settings where an FMH test is appropriate to include, in order of severity, are unexplained stillbirth, hydrops, the persistent maternal perception of decreased fetal movements, neonatal anemia, as well as the unexplained elevation of the middle cerebral artery Doppler consistent with anemia. In fact, ultrasound evaluation of the peak systolic blood flow velocity through the fetal middle cerebral artery is critical to identify fetal anemia. In case of fetal anemia, there is an increase of cerebral blood flow to preserve oxygen to the brain, and this can be demonstrated by ↑ peak systolic blood flow (>1.5 multiples of the median → red flag!). FMH may recur in subsequent pregnancies, but the risk and management are unclear as our knowledge of FMH remains far from complete as well as the long-term neurodevelopmental consequences of the surviving neonates.

18.3.2.4 S/P Prolonged/Repetitive Hypoxia

- *DEF*: Neuro-placental concept combining specific patterns of brain injury of the newborn with prolonged partial asphyxia emphasized by sustained or repetitive periods of significant hypoxia or other systemic stress affecting the fetus over time of many hours to days.
- *SYN*: Prolonged hypoxia-cerebral palsy.

- *CLI*: Cerebral palsy and patterns of neurodisability.
- *EPG*: Severe fetal acute vasculitis, prolonged meconium exposure with fetal vascular necrosis, chronic intermittent umbilical cord occlusion, subacute abruption (*abruptio placentae*), and fetomaternal hemorrhage (FMH).
- *GRO/CLM*: Histology characteristic of the specific etiopathogenic factors.
- *PGN*: Several meta-analyses support the association of the above-delineated risk factors with subsequent cerebral palsy and neurodisability of the newborn.

18.3.3 Chronic Diseases

Chronic diseases of the placenta include maternal vascular underperfusion (MVUP), vasculopathy of thrombotic type of fetal origin, villitis of chronic type, perivillous fibrin deposition (PVFD), and circumvallate placenta and related lesions. In case of the identification of retroplacental hemorrhage, placental abruption as a whole, marginal abruption, and chronic abruption need to be distinguished.

18.3.3.1 Maternal Vascular Underperfusion (MVUP)
(Figs. 18.24, 18.25, 18.26, and 18.27)

- *DEF*: Multifactorial placental syndrome due to inadequate tissue perfusion and cellular oxygenation affecting the placenta as a whole.
- *SYN*: "Chronic placental insufficiency."
- *EPI*: Risk factors include primipara, FHx (+), and Hx (+) of underlying disease (e.g., metabolic syndrome, renovascular or autoimmunity, thrombophilia).
- *CLI*: Preeclampsia, IUGR, PTL, PROM, HELLP (hemolysis, elevated liver enzymes, low platelets) (Figs. 18.24, 18.25, 18.26, and 18.27).
- *EPG*: Perfusion may be decreased either systemically (e.g., hypotension, intravascular volume expansion failure) or limited to regional maldistribution (failure of appropriate trophoblastic invasion and consequent remodeling of the uterine spiral arteries, anatomical abnormalities of the pelvic and uterine blood supply, infection of the female genital tract). Regardless of etiology or severity, all forms of MVU have the common property of perfusion, which is inadequate to meet metabolic demands at the cellular level. Decreased organ perfusion leads naturally to tissue hypoxia, anaerobic metabolism, activation of the inflammatory cascade, and organ dysfunction.
- *GRO/CLM*: Gross and microscopic features are identifiable and include:
  1. *Placental hypotrophy* (decrease of placental weight and increase of fetoplacental weight ratio)
  2. *Decidual arteriopathy* (i.e., mural hypertrophy of decidual arteries, acute atherosis, or fibrinoid necrosis of the vascular smooth muscle wall and scattered clusters of lipid-laden macrophages)
  3. *Villous/intervillous changes* (placental villous tree immaturity with distal villous hypoplasia, an increase of the intervillous fibrin deposition, villous agglutination, and increase in syncytial nuclear clumping and basophilia by accelerated syncytiotrophoblast turnover, which is the so-called "Tenney-Parker" or knotting changes of the terminal villi)
  4. *Placental infarction*
  5. Nonspecific findings, such as thin umbilical cord and laminar membrane necrosis
- *PGN*: Prematurity, IUGR, adult cardiovascular disease, recurrence in subsets (severe preterm birth, underlying disease).

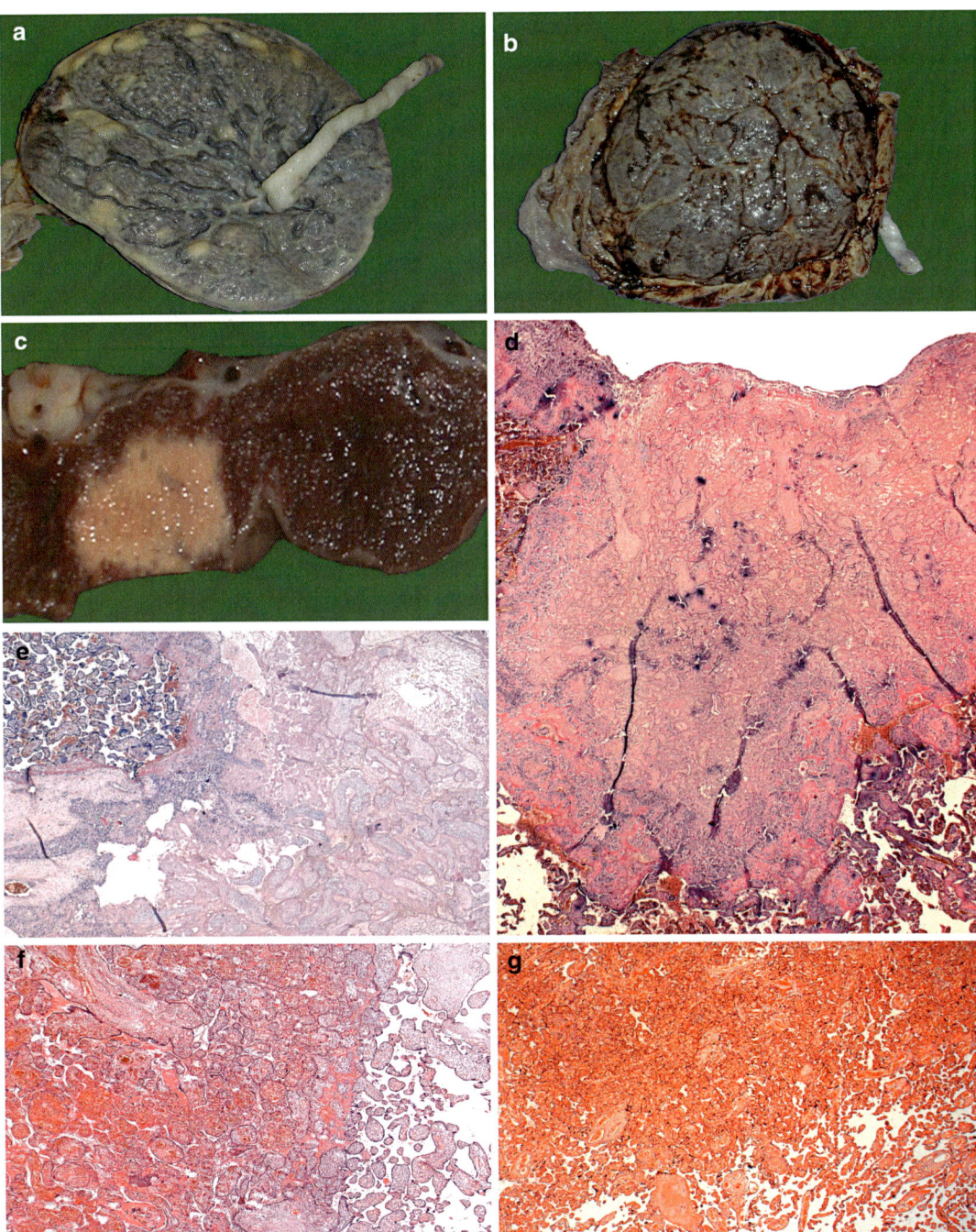

Fig. 18.24 Maternal vascular underperfusion. Maternal vascular underperfusion shows fibrotic changes on the chorionic plate (**a**) as well as at the basal plate (**b**). Figure (**c**) is an old infarct with necrosis of the placental parenchyma, and its histopathology is shown in Figure (**d**). Another old infarct is shown in Figure (**e**). Figures (**f–g**) show fresh acute placenta infarctions. All microphotographs have been taken from histological slides stained with hematoxylin and eosin (d, original magnification x20; e-g, original magnification x40)

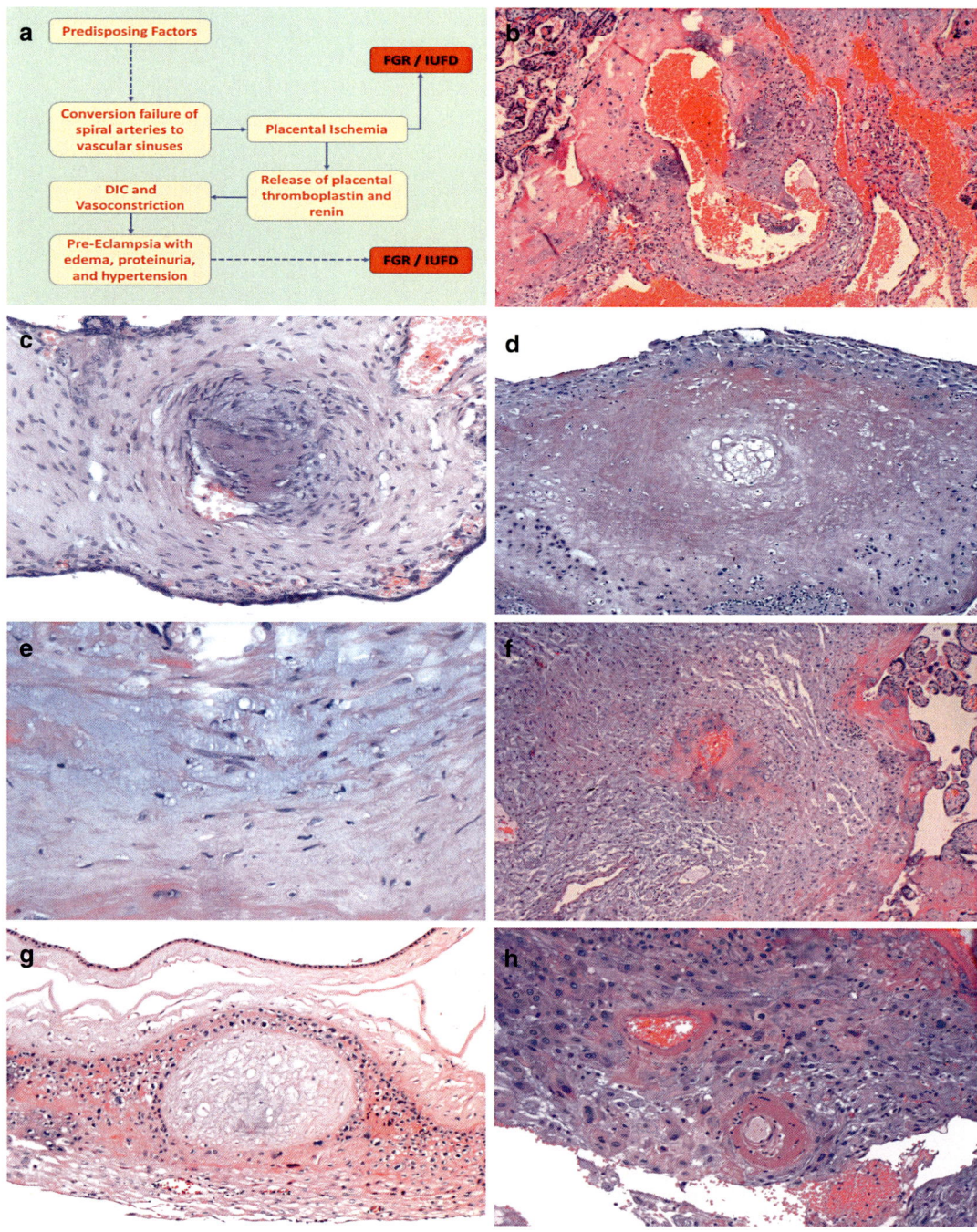

Fig. 18.25 Pre-Eclampsia. In (**a**) is shown the physiopathology concepts evolving to fetal growth restriction or intrauterine fetal demise in consequence of a conversion failure of spiral arteries to vascular sinuses, which the histopathologic correlates for preeclampsia. Figures (**b–c**) show different degrees of preeclampsia histopathology changes with atherosis and fibrinoid necrosis. All microphotographs have been taken from histological slides stained with hematoxylin and eosin (b, original magnification x100; c, original magnification x200; d, original magnification x100; e, original magnification x400; f, original magnification x100; g, original magnification x100; h, original magnification x200)

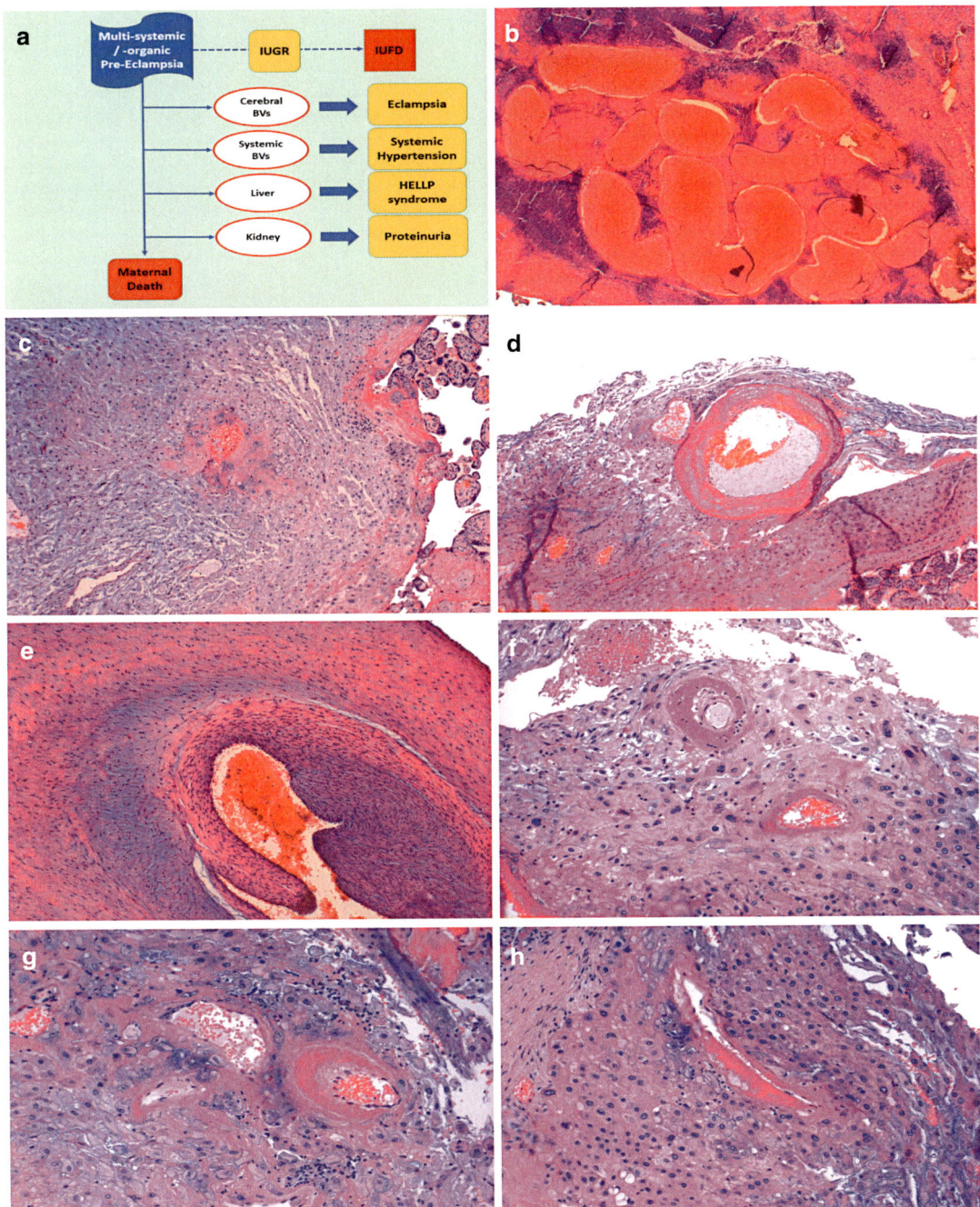

Fig. 18.26 HELLP (hemolysis, elevated liver enzymes, and a low platelet count) syndrome. HELLP syndrome may be dreadful complication of a multisystemic or multi-organic preeclampsia with liver involvement. The histopathology of preeclampsia is shown in these figures (see text for details), and fibrinoid necrosis is more diffuse than a classic preeclampsia. All microphotographs have been taken from histological slides stained with hematoxylin and eosin (**b**, original magnification x20; **c**, original magnification x100; **d**, original magnification x100; **e**, original magnification x40; **f**, original magnification x200; **g**, original magnification x200; **h**, original magnification x200; the microphotograph **f** is the same of the microphotograph **h** of Fig. 18.25 reproduced due to a patient with initial pre-eclampsia developing later HELLP syndrome)

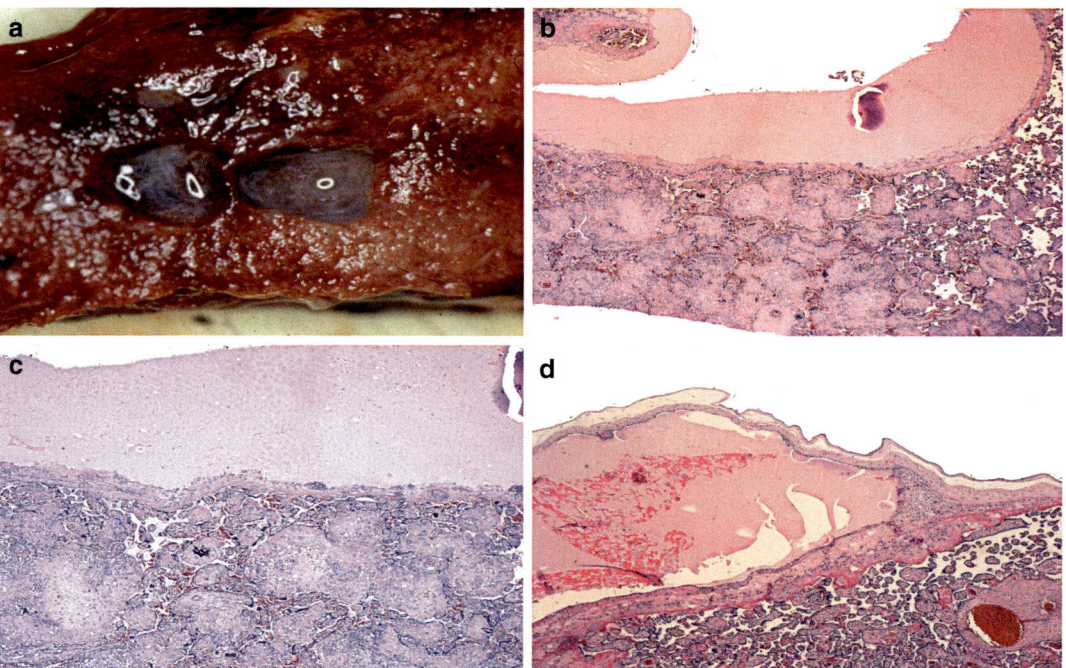

Fig. 18.27 Cysts. Cysts may be an intriguing encounter by the grossing of a placenta (**a**), and the histopathology frequently involves the presence of intermediate trophoblast lined cavities (**b, c**). In (**d**), a cyst is seen at level of the chorionic plate. All microphotographs have been taken from histological slides stained with hematoxylin and eosin apart c, which is periodic acid Schiff stained (b, original magnification x20; c, original magnification x40; d, original magnification x40)

In studies involving MVUP, the target is recording pathologic findings in the maternal vasculature of the parietal and basal decidua (vessel changes), which included mural fibrinoid necrosis/acute atherosis, muscularized basal plate arteries, and mural hypertrophy of membrane arteries (MHMA). Moreover, villous hypoxic lesions (villous changes) including infarcts, increased syncytial knots, villous agglutination, increased perivillous fibrin, distal villous hypoplasia/small terminal villi need to be documented. The degree of MVUP may be graded as severe if one or more vascular lesions are present, one or more villous lesions are seen, and the placental weight is <10th percentile for GA. If findings of MVUP are current but did not meet all these criteria, well-defined by Redline, a grade of mild MVUP should be assigned. Preeclampsia is associated with the maternal HELLP syndrome (hemolysis, elevated liver enzymes, low platelets) (Fig. 18.26). Epidemiologically speaking, MVUP occurs in 1–3:1000 deliveries (Manitoba experience with ≥80 ml hemorrhage), and to emphasize hypoperfusion again and cellular ischemia are considered necessary, but not sufficient to cause MVUP and, apparently, clinical relevance. These factors are probably only the triggers for a complex physiologic cascade. Cellular hypoxia predisposes tissues to "reperfusion injury" leading to local vasoconstriction, thrombosis, regional malperfusion, the release of superoxide radicals, and direct cellular damage. Subsequently, there is activation of neutrophils and the release of proinflammatory cytokines. Some risk factors are associated and need to be taken into account. They include the initial exposure to fetoplacental antigens in first pregnancies or anti-endothelial cell antibodies (immunological factors), genetic polymorphisms of the renin-angiotensin system (genetic factors), or factors damaging the integrity of the blood vessels such as diabetes mellitus type I, connective tissue disorders, vasculitides, chronic renal insufficiency, essential hypertension, antiphospholipid syndrome, and maternal thrombophilia as well as coagulopathies. The final consequences of such malperfusion vary from patient to patient depending upon the degree and duration of hypoperfusion, the extension of the process in the placenta, and the presence of chronic diseases of the mother. Mostly, the neonatal outcome will include FGR with a decrease of the fetoplacental growth and increase of the fetoplacental weight ratio, preterm birth, abruption, and preeclampsia. In general, vasculitides are inflammatory diseases affecting variably sized blood vessels (large, medium, or small), caused by various immunological processes and possibly triggered by infectious agents.

MVUP grading deploys as follows considering some villous/intervillous changes (distal villous hypoplasia, numerous syncytial knots or ↑ # for estimated gestational age, villous agglutination, and ↑ intervillous fibrin deposition), the extension at cut surface (CS), and placental weight (PW):

Grade	Features
(1)	Villous compartment with ≥2 villous/intervillous signs of MVUP in <1/3 CS
	AND
	≤1 non-marginal villous infarct
(2)	PW <10th percentile AND ≥
	either villous parenchyma with ≥2 villous/intervillous signs of MVUP in >1/3 CS
	or >1 non-marginal villous infarct(s)

The presence of infarcts infers an acute vascular occlusion by some process, although central and floor infarctions have more relevance than peripheral perimeter infarctions. Moreover, the age of the infarctions is essential, i.e., if they are of same (one episode of inadequate vascular perfusion or underperfusion) or variable (chronic vascular underperfusion), the geography, i.e., sharply delineation, the color (red means fresh and pale brown to yellowish means older infarctions), and subchorial or intervillous and fetomaternal hemorrhage, should be taken into consideration. Breus' mole is a massive subchorionic hematoma, which has been associated with fetal growth restriction (FGR). The differential diagnosis includes intervillous or perivillous fibrin thrombosis that may also rim and outline a cotyledonous lobule ("*Gitterinfarkt*"), X-cell island nodules, which are fibrin islands containing intermediate trophoblast where fibrin accumulates in the centers and centers may become cystic and intraplacental vascular thromboses. MVUP showing arterial remodeling deficiency (ARD) changes include (1) ↓ placental weight/↑ fetoplacental weight ratio, (2) villous/intervillous changes, (3) villous infarction(s), (4) decidual arteriopathy, (5) thin umbilical cord, and (6) laminar membrane necrosis. As above indicated, villous/intervillous changes comprise distal villous hypoplasia, ↑ syncytial knots for estimated gestational age ("Tenney-Parker changes"), villous agglutination, and ↑ intervillous fibrin deposition.

Placental cysts should be differentiated carefully (Fig. 18.27). Amniotic placental cysts are cysts that protrude from amnion surface into the amniotic cavity. Superficial cysts occur when present at the maternal surface and are frequently associated with subchorionic fibrin and rimed by subchorionic trophoblasts. Intraplacental or chorionic cysts represent the cystic cavitation of X-cell islands showing a gelatinous fluid and be located beneath the chorionic roof (subchorionic) or intra-chorionic (septal) and may interrupt the blood flow of a roof blood vessel or the blood flow of a blood vessel at cord insertion site. The superficial or maternal cysts of the placenta, when extensive, may be associated with FGR.

Villous cells are quite characteristics and possess the following histologic, histochemical and immunohistochemical qualities.

- Cytotrophoblast: single-nucleated trophoblast, cuboid, polyhedral-shaped, PAS+
- Decidual cells: large epithelioid cells located externally to the trophoblast
- Endothelial cells: cells lining villous blood vessels, CD31+, and CD34+
- Hofbauer cells: ovoidal or reniform macrophages, CD68+
- Lymphocytes: small round cells with scant or no discernible cytoplasm, CD3+
- Mast cells: cells with granular cytoplasm, Giemsa+ and derived from CD34+, CD117/c-kit+, CD13+ hematopoietic cell precursor
- Plasma cells: cells with coarse nuclear chromatin and RNA-rich cytoplasm, CD138+

- Stromal cells: stellate, fibroblasts or fibroblast-like cells, actin-positive
- Syncytiotrophoblast: multinucleated trophoblast

Some controversies about placental insufficiency have not been clarified notwithstanding decades of research. MVUP needs to be kept distinguished by perivillous fibrin plaques and intervillous thrombi. Perivillous fibrin plaques are a classical mimicker of placental infarction but are not accompanied by villous agglutination, trophoblast karyorrhexis with nuclear fragmentation, and the collapse of the intervillous space. Intervillous thrombi need to be also distinguished by placental infarcts and may be recognized by the sphericity of non-basal lesions, smoothness of texture, and focal compression of the villous tree in adjacent fields. Substantially, placental insufficiency is a disorder affecting the placenta that may interfere with its function. In turn, it may harm the fetus. The fetus may experience a reduction in both O_2 and nutrient delivery, which eventually will affect growth with following FGR.

Features that are accompanying or underlying the placental insufficiency (concomitant or subsequent) are as follows: (1) abnormal or non-reassuring fetal monitoring tests, (2) abnormal Doppler flow velocimetry waveform analysis, (3) oligohydramnios, and (4) birth weight <10th percentile for the GA.

Factors affecting the impact on outcome are (1) *Severity* AND/OR (2) *Duration*.

Factors to assess carefully with the OBGYN and report adequately in the pathology report are:

(5) Meconium staining, which may occur if fetal hypoxemia is severe enough
(6) Doppler flow velocimetry, which includes analysis of the end-diastolic flow in the cord artery

If ≥1 of (1)–(6) criteria are seen, then the placenta should be sent for pathologic analysis to a pathologist with experience in pediatric pathology and/or perinatal pathologist. Not with standing these findings are suggestive, it has been suggested that they need clinical correlation and are not diagnostic by themselves.

18.3.3.2 Fetal Thrombotic Vasculopathy (FTV)

- *DEF*: Vasculopathy of thrombotic origin involving fetal compartments and sub-classified in:
  1. *Primary large-vessel* FTV (adequately named FTV)
  2. *Secondary large-vessel* FTV
  3. *Secondary villous lesions of thrombo-occlusive character*
- *SYN*: Feto-vascular thrombo-occlusive disease (FVTOD), thrombo-occlusive vasculopathy, hemorrhagic endovasculitis (HEV), primary fetal thrombo-occlusive vasculopathy.
- *CLI*: ↓ fetal movement (hypokinesis) by US/CTG, neonatal HIE, and ↓ neonatal platelet count.
- *EPG*: Cord obstruction, DM, fetal thrombophilia, maternal antiplatelet Abs, and hyperviscosity. *Virchow's* triad (*V*ascular – endothelial damage OR "*R*ouleaux" of blood cells due to blood flow stasis OR *C*oagulation abnormalities (hypercoagulability): cord obstruction, DM, fetal thrombophilia, and maternal antiplatelet antibodies.
- *CLM*: The correspondent pathological findings identified under the above-designated subclassification include:
  1. Large vessel thrombi AND/OR intimal fibrin cushions (recent or remote), due to ↑ fetal venous pressure, myofibroblast-derived matrix secretion, and intramural incorporation of old thrombi.
  2. Fibromuscular sclerosis, recent and distant, due to ↓ fetal blood flow.
  3. Hyalinized avascular villi AND/OR villous stromal-vascular karyorrhexis.
- *PGN*: Neurodisability (cerebral palsy), IUGR, neonatal thromboembolic disease.

FTV is a chronic lesion characterized in the placenta by thrombosis of large vessels and downstream distal villous vascular degenerative changes. FTV is strongly associated with neonatal brain injury, including cerebral palsy. The most common pathogenic factor for FTV remains the vascular stasis related to chronic partial or intermittent cord obstruction. FTV is paramount in studies involving the placenta and was well-defined by Redline and his group. FTV lesions include the presence of thrombi within chorionic, stem villous, or umbilical vessels. Avascular villi are identified as two or more terminal villi showing a total loss of villous capillaries and uniform fibrosis of the villous stroma. A diagnosis of FTV is made when multifocal avascular villi are present (>15 villi involved/slide).

FTV is graded 1 if >1 focus of some affected villi/some slides <15 and grade 2, if >1 focus of some affected villi/some slides ≥15 with type (a) involving large foci while type (b) only small foci (<10 villi).

Non-FTV vasculopathies may include:

- Vasculitis as part of an ACA.
- Cases of villitis of unknown etiology (VUE) with obliterative fetal vasculopathy, i.e., vasculitis and perivasculitis with downstream presence of a constellation of avascular terminal villi, meconium-induced vasospasm, single/multiple thromboses in chorionic blood vessels of unknown etiology, and terminal villous deficiency, although some authors do not classify this entity as vasculopathy. Meconium-induced vasospasm may be recognized by vascular myonecrosis or "myo-apoptosis."

18.3.3.3 Chronic Villitis of Infectious Type (CVI) and Villitis of Unknown Etiology (VUE)

- *DEF*: ↑ of T lymphocytes and Mɸ in placental villi at preterm in CVI or near term in VUE.
- *CLI*: Maternal illness and fetal infection in CVI (hepatosplenomegaly, pneumonitis, coagulopathy, cytopenias, hydrops fetalis, organism-specific IgM), but no in VUE.
- *EPG*: TORCH in CVI (fetal infection with mixed maternal and fetal response), but unknown in VUE.
- *GRO/CLM*: The extent of involvement is UC, chorionic plate, and membranes in CVI, while terminal and stem villi are the only involved compartment in VUE. The pattern of the participation is diffuse with a variable degree of severity in CVI and focal/patchy in VUE. The duration of the involvement is long-standing with fibrosis and calcification, chronic chorion-deciduitis, and increased nucleated erythro-progenitors in CVI and recent with fibrin and necrosis in VUE. VUE may also be associated with decidual plasma cells (usually "basal villitis"), extensive perivillous fibrin (see pitfalls below), obliterative fetal vasculopathy (stem villous perivasculitis and vascular obliteration and chronically inflamed avascular distal villi), and active intervillositis.
- *PGN*: Prematurity, IUGR, neurodisability, IUFD, and congenital anomalies in CVI, while IUGR, IUFD, and neurodisability (specifically if obliterative fetal vasculopathy is present) in VUE (prematurity if the basal pattern occurs). RR: occasional in CVI, while up to 1/4 in VUE, particularly if diffuse villitis with extensive perivillous fibrin is noted.

Non-TORCH or "O"-organisms include *Treponema pallidum*, *Trypanosoma cruzi*, EBV, VZV, smallpox/vaccinia (cowpox), *Borrelia burgdorferi*, parvovirus B19, enterovirus, and mumps virus. Previous classification of chronic villitis may have contributed to some confusion in terms of placental taxonomy, a T-lymphocytic villous inflammation due to maternofetal cell trafficking with a localized host vs. graft reaction in the villous tree and lymphoplasmacytic infiltration of the basal plate (chronic deciduitis).

> **Box 18.38 TORCH-CVI (Typical + Specific Features)**
> Typical features
>
> - Diffuse *histiocytic villitis* accompanied by:
>   *Fibrosis, calcification,* chronic *chorio-deciduitis*, and ↑ *NRBC*
>
> Specific features
>
> - *Toxoplasmosis* (*T. gondii*): Focal necrotizing/granulomatous villitis with cord pseudocysts
> - *Syphilis* (*T. pallidum*): Necrotizing cord periphlebitis/periarteritis at stem villi site
> - *CMV*: Endothelialitis, viral inclusions, and villous plasma cells
> - *HSV/VZV*: Necrosis, calcification, and viral inclusions (Cowdry A inclusions)
>
> Notes: *CMV* cytomegalovirus, *HSV* herpes simplex virus, *NRBC* nucleated red blood cells, *VZV* varicella zoster virus

VUE: localized <10 villi randomly or basally and diffuse when ≥10 villi are involved, stem villous vasculitis, and perivasculitis. Typical and atypical features of chronic villitis of infectious type are depicted in Box 18.38.

Chronic villitis of idiopathic type may present differently.

- *CLI*: Non-hypertensive FGR at term
- *CLM*: Villi (moderately/heavily) engulfed by maternal T lymphocytes
- *PGN*: FGR, prematurity, and neurodisability (RR: 10–25%)

Box 18.39 presents some "pearls" associated to VUE diagnostics.

VUE grading is graded as low, if ≤10 villi/focus in case of focal (≤3 foci) and multifocal (>3 foci), and as high, if >10 villi/focus in case of patchy (≤3 foci) and diffuse (>5% of all villi). Some common pitfalls of the VUE diagnostics due to VUE mimickers are shown in Box 18.40.

Chronic histiocytic intervillositis (HIV) is a rare placental lesion which is often associated with autoimmune disorders of the mother. There is diffuse infiltration of the intervillous space by mononuclear inflammatory cells, i.e., macrophages (CD68+) ± fibrin exudation occurring in spontaneous recurrent miscarriage, FGR, and IUFD with a high perinatal mortality rate up to 3/4 of cases. Most often, there is a positive family history of autoimmunity and steroid response.

> **Box 18.39 VUE Pearls**
>
> - *When you see* endothelialitis, viral inclusions, and villous plasma cells, *then think of* CMV
> - *When you see* necrotizing cord periphlebitis or stem villous periarteritis, *then think of* syphilis
> - *When you see* necrosis, calcification, and inclusions, *then think of* HSV/VZV
> - *When you see* umbilical cord pseudocysts, *then think of* toxoplasmosis

> **Box 18.40 Pitfalls VUE Histologic Mimickers**
> 1. *Maturitas tarda* (delayed villous maturation with increased intravillous Hofbauer cells)
> 2. *Maternal floor infarction* (fibrinoid, intermediate trophoblast, no inflammation)
> 3. *Chronic histiocytic intervillositis* (CHIV)
> 4. *Fetal thrombotic vasculopathy* (no lymphocytes)
> 5. *Villous agglutination-associated Terrey-Parker phenomenon* (↑ syncytial knots)
> 6. *Villous stromal-vascular karyorrhexis* (a secondary thrombo-occlusive villous lesion)

18.3.3.4 Massive Perivillous Fibrin Deposition (MPFD) and "Maternal Floor Infarction"

- *DEF*: Massive deposition of fibrin or fibrinoid at the basal plate or in the perivillous space (Figs. 18.28 and 18.29).
- *EPI*: 0.1–0.5% pregnancies. RFs: autoimmunity, thrombophilia, and HTN.
- *EPG*: ↓ perfusion, hypercoagulability, and excessive matrix secretion are leading to a massive (i.e., strikingly significant) deposition of perivillous fibrin or fibrinoid ⇒ ↓ perfusion, hypercoagulability, and excessive matrix secretion.
- *CLI*: Oligohydramnios, SGA, and fetal distress.
- *GRO/CLM*: Two radically alternative histologic patterns constitute the morphologic basis for MPFD, which are fibrinoid with intermediate trophoblast ("X cells") and villous degeneration with bright red fibrin at the perivillous location.
- *PGN*: PTL, FGR, neurodisability, IUFD, recurrent miscarriage (RR: 1/10–8/10).

MPFD mimickers, so-called the non-massive depositions, may be found in a series of conditions including (1) MVU with ↑ intervillous fibrin (subchorionic, basal, stem villi located, or marginal atrophy) and (2) nonspecific perivillous (intervillous) fibrin plaques (term placentas without clinical correlates or adverse outcomes). Of note, localized intervillous fibrin (marginal atrophy) induces diffuse intervillous fibrin deposition with X-cells (uteroplacental insufficiency), which in turn leads to an increased basal perivillous fibrin with X-cells (FGR, IUFD). A perivillous fibrin plaque induces a "localized" maternal floor infarction, which in turn leads to "diffuse" maternal floor infarction (perinatal mortality and recurrence).

18.3.3.5 "Chronic" Placental Abruption (CPA)

Abortion and miscarriage may be used interchangeably, but the origin of the former is Latin (abortus), while the source of the latter is Russian. In both cases, the event is a termination of pregnancy up to 28 weeks of pregnancy, i.e., in the period when the fetus is unable to survive without mechanical support outside the mother's body. The interruption of pregnancy can be spontaneous or artificial. In case of community-acquired miscarriage, it is important to differentiate among different types of miscarriages according to the dynamics of the miscarriage (Box 18.41).

Each form of abortion corresponds to a specific clinical picture. *Threatening miscarriage* is characterized by the mother feeling a small, occasionally emerging, pain in the lower abdomen (contractions), slight bleeding as blood stains on the strip, and the external orifice of the uterus closed. In *starting miscarriage*, more frequent and severe pain in the lower abdomen occurs and there is an obvious beginning separation of chorionic villi, which manifests with more fresh blood. Vaginal examination shows an external orifice of the uterus with a finger passing the cervical canal. In both conditions, pregnancy can be saved if there are strict bed rest, analgesia, progesterone, and oxygen inhalation. If this therapeutic regimen fails, *miscarriage* is *in progress*. Pain and bleeding are more accentuated, and the fertilized egg can be palpable at the vaginal examination. In the event of an *incomplete miscarriage*, which is the most common clinical form, there is profuse uterine bleeding. In the blood clots, a fertilized egg can be visualized. *Complete abortion* is characterized by the complete detachment of the chorionic villi from his uterine bed without ensuring that they were cut out and did not stay in the uterus at the site of implantation. Uterine curettage is always justified, although a lack of pain and bleeding characterizes this form. *Missed miscarriage* (delayed miscarriage, silent miscarriage, or missed abortion) is characterized by pregnancy signs in the mother's body accompanied by the death of the embryo or fetus, although a *miscarriage* did not occur yet. The chorionic villi are the specific functional unit of the placental disc.

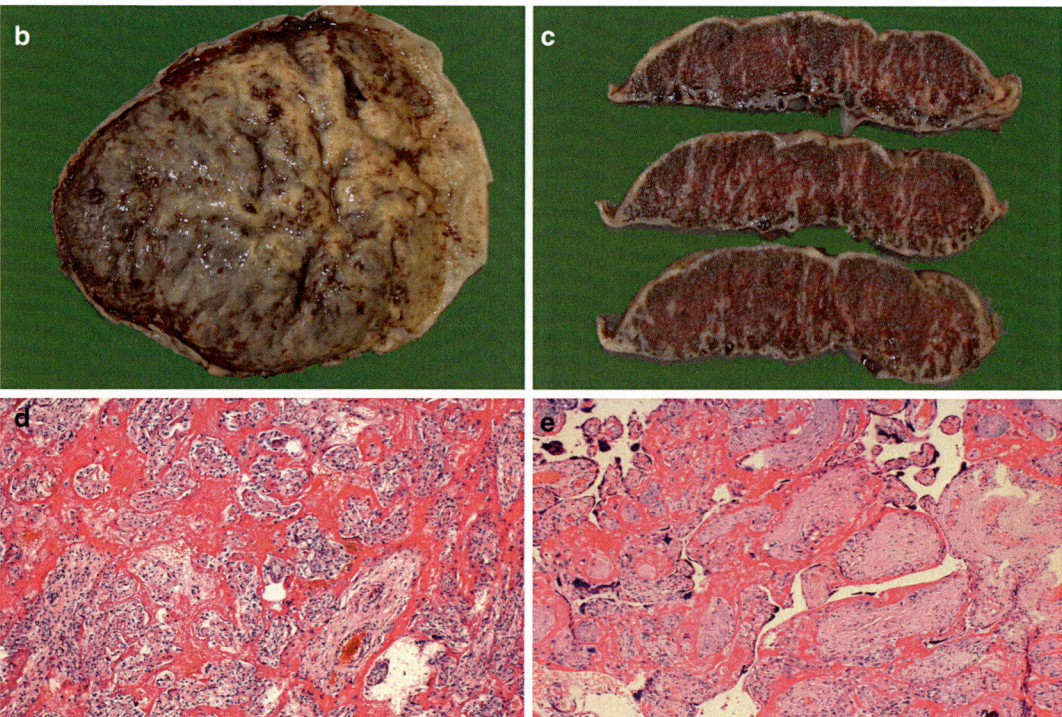

Fig. 18.28 Maternal perivillous fibrin deposition and maternal floor infarction. Maternal perivillous fibrin deposition (MPVFD) and maternal floor infarction (MFI) are part of the same process involving an impressive amount of fibrin deposition in the intervillous space obturating the space that should allow for exchange of oxygen and nutrients. Figure (**a**) lists the current criteria, while (**b**) and (**c**) show gross photographs of MPVFD with features reaching the MFI with more than 50% of basal villi associated with hypovascular or fibrotic villi (**d** and **e**). All microphotographs have been taken from histological slides stained with hematoxylin and eosin (d-e, original magnification x100)

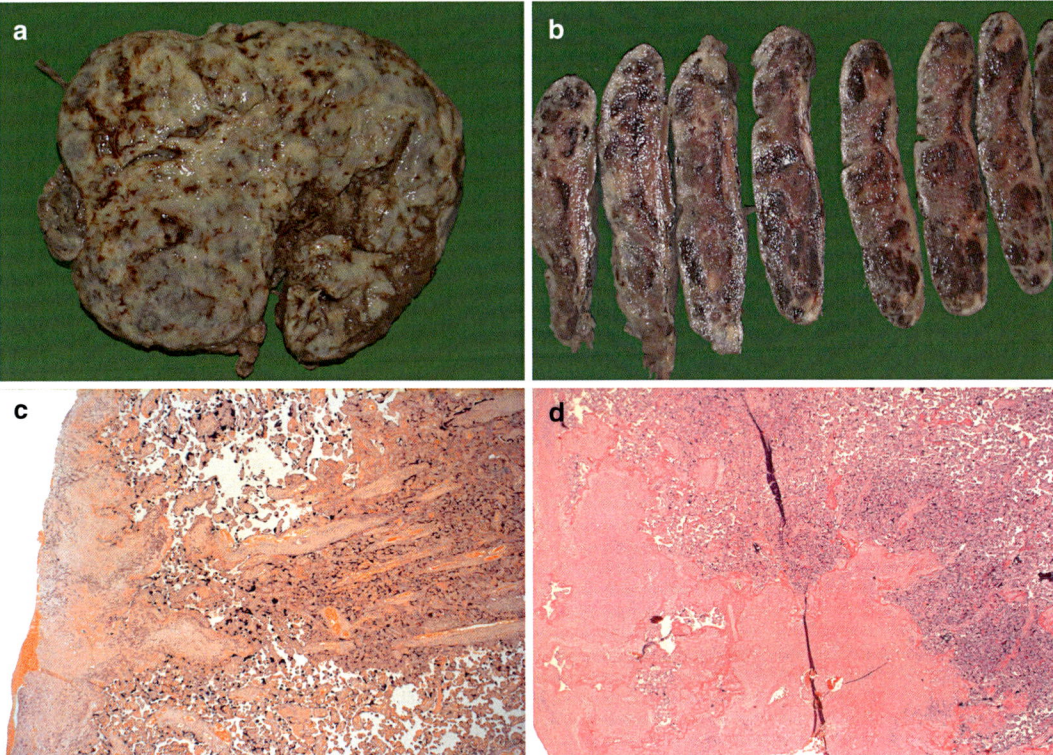

Fig. 18.29 Maternal floor infarction. A maternal floor infarction (MFI) is depicted in (**a**) and (**b**), and the histology of (**c**) and (**d**) is the characteristic of MFI with clearly more than 50% of the basal villi being hypovascular or fibrotic. All microphotographs have been taken from histological slides stained with hematoxylin and eosin (c-d, original magnification x20)

- *DEF*: Chronic peripheral (marginal) separation of the placenta from the uterus.
- *EPG*: Marginal uteroplacental separation due to inadequate support by the surrounding endometrium or when subjected to elevated intramural pressure. There is an abnormality involving the marginal anatomy with alteration of the geometry of the placenta located in utero AND ↑ maternal venous pressure. Both the geometric aberration and the increased maternal venous pressure are not relevant enough to determine an acute event of peripheral abruption, yet chronic.
- *CLI*: Chronic vaginal bleeding (continued/intermittent dark brown vaginal spotting) and oligohydramnios.
- *GRO*: Peripheral (marginal) hematoma and/or circumvallate placenta.
- *MIC*: Old blood clot with chorionic/amniotic hemosiderin and/or Hb-related pigment (biliverdin) deposition and/or histology of circumvallate placenta.
- Pearls: When you see pigment in a premature placenta, then think of chronic abruption!
- When you see pigment ≥37 weeks placenta, then think of meconium release!
- *PGN*: Threatened miscarriage (*abortus imminens*) (see below) in early pregnancy or vaginal bleeding in late pregnancy, preterm delivery, atypical form of neonatal lung disease, and neurological impairment of the newborn (neurodisability), such as cerebral palsy. RF: multiparity, PROM, smoking, oligohydramnios, excessive low uterine implantation. RR: rare/minimal.

> **Box 18.41 Abortus/Miscarriage Classification**
> 1. Threatening miscarriage (*abortus imminens*)
> 2. Starting miscarriage (*abortus incipiens*)
> 3. Miscarriage in progress (*abortus protragens*)
> 4. Incomplete miscarriage (*abortus incompletus*)
> 5. Complete miscarriage (*abortus completus*)
> 6. Missed miscarriage or missed abortion (*abortus amissus*)

These villi provide oxygen and sustenance to the fetus and serve as an excretory unit. The histological appearance of chorionic villi varies with the gestational age and with the stage of development and maturation of the villous tree. Normal villi contain numerous thin-walled patent vascular channels, and the examination of several cross sections of chorionic villi can bring the average count to 8–10 blood vessels. Avascular or hypovascular villi indicate poor blood supply to the fetus and result into FGR. In the examination of spontaneous miscarriages, the pathologist can identify a reduced number of blood vessels (up to 97%), stromal fibrosis (~85%), fibrinoid degeneration (~75%), and prominent Hofbauer macrophages (~67%). Avascular villi point to an obliterative fetal vasculopathy characterized by diffuse perivillous fibrin deposition. Redline and Pappin have demonstrated that when avascular villi account for more than 2.5% of the placental parenchyma, a few events can occur. They include an increased rate of FGR, abnormal fetal Doppler, oligohydramnios, and maternal coagulopathy. Placentas with avascular villi indicate fetal thrombotic vasculopathy (FTV) and may coexist with chronic villitis, membrane hemosiderin, meconium in all three membrane layers, and villous chorangiosis. A Virchow's triad associated FTV may be due to severe fetal inflammation and any damage involving the integrity of the blood vessels as vascular damage. A prolonged maternal cord obstruction, an increased central venous pressure, and an altered hematocrit have been ascribed to intraluminal abnormalities. Clotting anomalies include hemoglobinopathies, such as platelet disorders, maternal DM, and thrombophilia, including protein C, protein S, Factor V Leiden (FVL), prothrombin gene (PT G20210A) and methylenetetrahydrofolate reductase (MTHFR) C677T and A1298C polymorphisms. Pediatric pathologists and embryologists often work in the same environment and frequently join quality assurance or clinicopathological conferences (aka multidisciplinary meetings) in the same institution. Pediatric pathologists are also asked about their interpretation in different conditions. The assessment of the embryo vitality is probably one of the most important and challenging tasks in IVF. Microscopy was traditionally the assessment method of choice until a few years ago. It provided a relatively simple and somewhat reliable method for this aim. However, numerous factors contributing to the oocyte and/or embryo quality are often not picked up by morphology only, and new methods have been proposed in the recent years. In the range of the secretome, which is the total amount of all compounds that are secreted by the oocyte and/or embryo into its environment, the metabolome is probably the fascinating procedure that gained interest in the last few years. It refers to the study of several factors present in the culture medium that are secreted by the oocyte and/or embryo into the environment because of metabolic processes that occur at the cellular level or due to leakage from cell damage. Metabolomics can identify several processes of the preimplantation development, a period when a major metabolic pathway is a switch in ATP synthesis from carboxylic acid to glucose metabolism around the time of compaction. Visual grading of embryos is likely to be highly subjective, and intra- and interobserver inconsistency has been reported and controversially debated. Metabolomic profiling of embryos has stated to be the object of extensive investigation, although it is mostly impractical for a clinical setting. In IVF, metabolomic profiling is usually less challenging due to the longer culture periods during which several metabolites can accumulate and therefore detectable in the medium. There are additional methods of analysis, and specific reviews have been a study object.

18.3.3.6 Villous Capillary Proliferation Disorders (VCPD)

- *DEF*: Primary and secondary lesions characterized by an intravillous proliferation of capillaries.
- *EPI*: Risk factors include DM, BWS, high altitude, smoking, twin pregnancy, and preeclampsia (Beckwith-Wiedemann syndrome, BWS).
- *EPG*: Unknown in the autonomous forms but associated with MVU lesions as reactive (subacute hypoxia and associated with growth factors and other cytokines). There is often Hx. (+) for DM, BWS, high altitude, and smoking in chorangiosis and an association of twins and preeclampsia in chorangioma.
- *CLI*: Variable according to the underlying lesion with a spectrum ranging from asymptomatic to hydrops fetalis and IUFD.
- *CLM*:
  1. Chorangiosis (common or reactive), i.e., >10 capillaries/villus in >10 villi in several areas.
  2. Chorangioma (single or multifocal), i.e., autonomous proliferation (placental hemangioma).
  3. Chorangiomatosis (localized OR diffuse), i.e., large areas of the placenta or diffuse autonomous proliferation (immature placenta).
- *PGN*: Adaptive when accompanying chronic lesions while consumptive vasculopathy, A-V shunt with hydrops, and IUGR in case of chorangioma or chorangiomatosis.

Vascular-syncytial or syncytial-vascular membranes (SVMs) is a characteristic feature of the placenta at term with syncytial trophoblasts and absent cytotrophoblast. Syncytial trophoblasts evidence nuclear knots and attenuated cytoplasm over the villous surface. If a cross section is performed on the last (terminal) villous tree, the histological slides will show most of the cross section occupied by capillary profiles with and one or more peripheral capillaries bulging under attenuated syncytiotrophoblastic cytoplasm. There is typically no stroma between maternal (intervillous) and fetal (villi) blood cells. In case there is an obstruction between maternal and fetal blood, chronic hypoxia is the consequence and fetal death at the final event. Thus, the determination of the satisfactory presence of SVMs is essential to the obstetrician and perinatologist.

18.3.3.7 Distal Villous Immaturity (DVI)

- *DEF*: Immaturity of the distal villous completion for an appropriate placental function.
- *EPI*: Prepregnancy (+) glucose tolerance test, metabolic disorders leading to hyperglycemia (*vide supra*).
- *CLI*: Gestational DM, T1DM, (+) glucose tolerance test, maternal obesity, excessive weight gain, TORCH infections, severe maternal anemia, hydrops fetalis, fetal overgrowth syndromes (e.g., BWS), metabolic storage diseases.
- *GRO/CLM*: Placentomegaly (>90th centile) (<800 g mild-moderate/≥800 g severe) due to an increase of villous stroma, central capillaries, and the consequent increase of the gestational age-related ratio of the vascular-syncytial membrane (number of all intravillous capillaries/number of capillaries juxtaposed to the trophoblastic epithelium).
- *PGN*: The increase of placental demand and the decrease of the placental function are directly proportional to an increased risk of adverse outcome, even stillbirth.

Notes: *DM* diabetes mellitus, *T1DM* type 1 diabetes mellitus

18.3.4 Fetal Growth Restriction

- *DEF*: Functional status of the fetus, who is unable for one or several reasons to achieve the genetically determined potential size. FGR is different from SGA, which is a fetus with growth ≤10th percentile for weight at that gestational age, because not all SGA fetuses are growth restricted, because some of them may be constitutionally small. Likewise, not all fetuses lacking to meet genetic growth potential are ≤10th percentile for estimated fetal growth (EFG).
- *SYN*: Fetal growth restriction (FGR) or intrauterine growth restriction (IUGR) are synonymous and have replaced the term intrauterine growth retardation.
- *PGN*: Fetuses situated at or below the 10th percentile can be in one of three categories:
  - Category A (~40%), fetuses at high risk of potentially preventable perinatal death (not only morbidity).
  - Category B (~40%), fetuses who are constitutionally small and many of them will go through high-risk protocols for preterm delivery with potential iatrogenic prematurity.
  - Category C (~20%), SGA fetuses with pathology secondary to chromosomal (e.g., trisomy syndromes) or multifactorial etiology, including environmental factors (e.g., intrauterine infections or fetal alcohol syndrome).

Maternal-based etiology of FGR comprises:

1. Group of *vascular-related etiologies*, including pregnancy-associated hypertension, chronic hypertension, thrombophilia, autoimmune disorders affecting BV leading to the placenta, and uterine malformation involving abnormal blood supply
2. Group of *hypoxia-related etiology*, including cyanotic heart disease, hemoglobinopathies, tobacco smoking (but also other causes such as cannabis), and prolonged exposure to wildfires
3. Group of *metabolic-related etiologies*, including malnutrition (either low in proteins or low in calories) and diabetes mellitus
4. Group of *iatrogenic causes/multifactorial etiology*, including alcohol, drugs (e.g., cocaine), smoking (e.g., tobacco, cannabis), extreme age groups (young age and advanced age), short stature/thinness in maternal prepregnancy, nulliparity, failure to receive adequate medical care during pregnancy, lower socioeconomic status, Afro-American ethnicity (USA), multiple gestations, polyhydramnios, and preeclampsia

Placental-based etiology of FGR comprises:

- Umbilical cord-related factors, including cord anomalies and abnormal cord insertion
- Disc-related factors, including abnormal placentation (e.g., twin pregnancy, TTTS) and placenta implantation disorders (e.g., placenta previa)

The pathologist approaching the autopsy of a fetus with FGR and the examination of its placenta needs to acquire fetal karyotype, fetal blood and swabs as well as maternal serology for infectious diseases, and detailed environmental exposure history. Fetal mortality and nonlethal perinatal complications can accompany FGR. Nonlethal perinatal complications of FGR are fetal morbidity events (e.g., NEC, thrombocytopenia, renal failure, and temperature control instability), prematurity and related complications, fetal compromise in labor with or without the need for induction of labor, and cesarean section and its related comorbidities. A partially controversially discussed concept of growth is its symmetry, which may direct to some possible etiologies. Campbell and Thomas' (1977) concept of symmetric vs. asymmetric growth reveals a symmetrical low growth, mostly due to early global insult (e.g., aneuploidy, virus, FAS), and an asymmetrical low growth, primarily due to nutrient and/or gas exchange-related growth restriction. Although this concept is not a law, it is quite well accepted that symmetrically grown babies who were SGA have outcomes very similar to the babies who are AGE and very different result than the asymmetrically SGA fetuses. Long-term outcome of infants with IUGR is determined by some factors including neurovascular stability, pancreatic β-cell mass, etc. Diagnosis and surveillance of FGR require some activities to be put on the place, including biometry and measurement of amniotic fluid volumes, uterine artery Doppler measurement, umbilical artery Doppler measurement, middle cerebral artery Doppler, venous Doppler waveforms, and 3D ultrasonography. Flow patterns regarding maternal uterine arteries are probably to reflect most of the placentation on maternal circulation, and umbilical artery resistance is a useful parameter to evaluate for uteroplacental insufficiency. In measuring the GA-specific UA resistance, the systolic-to-diastolic ratio of flow is primarily used. Changes are seen from a baseline value to an estimated value with disease showing worsening features. The current management of FGR at 32 weeks of gestation or older is complex. The current management of FGR at ≥32 weeks according to most O&G Soc's is as follows:

- If UA Doppler and BPP (Bio-Physical Profile) weekly are regular, then delivery at term
- If UA Doppler and BPP weekly are regular but oligohydramnios (+), then daily BPP and delivery if abnormal
- IF UA Doppler and BPP weekly show reverse end-diastolic flow (EDF), then prompt delivery
- IF UA Doppler and BPP weekly show absent EDF and oligohydramnios (+), then a quick delivery is needed
- IF UA Doppler and BPP weekly show absent EDF and ≥33 weeks, then quick delivery
- IF UA Doppler and BPP weekly show absent EDF and <33 weeks, then daily full BPP and delivery if BPP is abnormal

The biophysical profile (BPP) includes fetal heart rate, fetal breathing movements, fetal body movements, fetal muscle tone, and amount of amniotic fluid.

In considering delivery, the obstetrician and midwife do not consider delivery if there are normal umbilical artery flow, fetal movements, and normal amniotic fluid, but deliver promptly if CTG shows late deceleration, abnormal venous flow 2–3 standard deviation (SD), reversed EDF by 31 weeks, absent in diastolic flow by 33 weeks, and ↓EDF by 36 weeks. In any case, antenatal steroids before 34–(36) weeks' gestation are highly recommended.

Disorders of deep placentation are now recognized as a significant mechanism at the basis of disease for several pregnancy-related complications. Preterm labor occurs in pregnant women with uteroplacental ischemia, but the precise mechanism is under investigation. Briefly, the renin-angiotensin system is probably based on this cascade. Fetal membranes are associated with a functional renin-angiotensin system. Uterine ischemia leads to increased renin production. Angiotensin II can provoke myometrial contractility either directly or through the release of prostaglandins. If the uteroplacental ischemia is severe enough to induce decidual necrosis and hemorrhage, thrombin activates the labor cascade. The initial event leading to acute abruption is postulated to be an ischemic lesion of the decidual tissue. This event has as consequence decidual necrosis, vascular disruption, and then bleeding, which occurs along a decidual plane and placental separation with retroplacental blood accumulation. A higher rate of spontaneous abortion, FGR, preeclampsia, preterm delivery with prematurity, and death are also associated with the primary antiphospholipid syndrome. MVU shows a constellation of pathologic findings that are accentuated in several and different cases. They include placental infarction, thrombosis, fibrinoid necrosis, and atherosis. These lesions may also be seen in primary antiphospholipid syndrome, and placental bed biopsies may be quite useful in patients with antiphospholipid antibodies, who show a high density of inflammatory cells, mainly macrophages. All three factors, including inadequate angiogenesis, thrombosis, and/or inadequate or insufficient physiologic transformation of the spiral arteries, can induce ischemia of the placenta. This situation is critical with regard to the fetal viability in the setting of placental vascular pathology. It has been demonstrated in experimental models that decreased blood supply in the placenta can unavoidably lead to the demise of the fetus and/or its growth restriction, maternal hypertension, and preterm delivery. It is essential to reiterate that the diagnosis of deep placentation disorders needs placental bed biopsies, which are characterized by the presence of interstitial trophoblast (keratins+).

Maternal floor infarction (MFI) and maternal perivillous fibrin deposition (MPVFD) may share the same etiologic factors, including acute chorioamnionitis (infection), autoimmunity (particularly antiphospholipid and anti-urokinase auto-Ab), coagulation disorder, host-placenta interaction dysregulation (idiopathic), and toxicity (e.g., pregnancy-associated major basic protein). Finally, the clinicopathologic context is essential in operating a differential diagnosis between pathologic fibrinoid material (MFI or MPVFD) from nonspecific or secondary fibrinoid material deposition. This latter may be a consequence of prolonged placental retention, after fetal death, infarctions, or inflammatory processes. Moreover, marginal and subchorionic fibrinoid material may be in most cases "physiologic" and be the result of eddying of maternal blood in the perivillous intervillous space or from fetus-related minor traumas. Katzman-Genest categories are presented in Box 18.42.

FGR has a definition, which is intimately linked to utero-placental interface perfusion imbalance (*UPIPI*). It comprises the rate of fetal growth that is less than usual for the growth potential of a specific infarct, and the most common cause is, apparently, fetal undernutrition, which is linked to an insufficiency of the placental functions.

> **Box 18.42 Katzman-Genest Categories**
> 1. *Classic MFI*: Maternal floor basal villi encased by perivillous fibrinoid at variable extension but limited to $\theta \geq 3$ mm on ≥ 1 histologic tissue slide.
> 2. *Transmural MPVFD*: Perivillous fibrinoid with a fetomaternal surface transmural extension (process encasing $\geq 50\%$ of the total villous body on ≥ 1 histologic tissue slide).
> 3. *Borderline MPVFD*: Perivillous fibrinoid with a distribution like fetomaternal surface transmurality but encasing 25–50% of the entire villous body on ≥ 1 histologic tissue slide.
>
> Notes: θ thickness, *MFI* maternal floor infarction, *MPVFD* massive perivillous fibrin deposition

The factors underlying placenta malperfusion due to a compromised uteroplacental interface perfusion include extrinsic factors (e.g., decidual vasculopathy, placental abruption) and intrinsic ones (e.g., abnormalities of the villous tree development, infarcts). Decidual vasculopathy is recognized by a series of non-necessarily concomitant factors including mural hypertrophy of the blood vessels of the decidua capsularis, muscularization of basal plate arteries (absence of remodeling), acute atherosis of the blood vessels located in decidua and chorion laeve or derivatives, mural fibrinoid change of the blood vessels, and thrombosis of the blood vessels found in decidua and chorion leave or derivatives.

Another argument of controversy is the hemodynamical relevance of placental infarction. It is considered appropriate to label as relevant when the infarctions are $\geq 15\%$ of the total placental volume (or cut surface) OR if $5\% < \emptyset < 15\%$ of the volume (or cut surface), if randomly distributed or involving central areas (particularly near the cord insertion).

18.4 Non-neoplastic Trophoblastic Abnormalities

Non-neoplastic trophoblastic abnormalities are disorders involving an abnormal formation of the products of conception that include the so-called molar degeneration, which includes two major categories, the partial and the complete molar degeneration, that we discussed in the section of the pathology of the early pregnancy and two other conditions, including placental site nodule and exaggerated placental site.

18.4.1 Placental Site Nodule

Placental site nodule (PSN) is a non-neoplastic gestational trophoblastic disease of the chorionic type showing small, well-circumscribed nodular cellular aggregates of intermediate trophoblast (chorionic type) embedded in a hyalinized stroma forming a "plaque." PSN is a benign entity, but it is important to distinguish it from other placental lesions like decidua, placental polyp, exaggerated placental site (EPS), and placental site trophoblastic tumor (PSTT). Although rarely, PSN may also be histologically confused with squamous cell carcinoma (SCC) of the cervix, and patients with human papillomavirus (HPV) infection are particularly at risk as a mimicker or confounding factor. PSNs are immunoreactive for inhibin-α and cytokeratin-18, while squamous cell carcinoma is negative. There is no recurrence or malignant potential following the diagnosis of PSN. However, there have been described patients harboring "atypical" PSNs and PSNs coexistent with epithelioid trophoblastic tumor (ETT) and PSTT. Although rare and controversially discussed, it has been argued that some trophoblasts of PSNs retain differentiation plasticity and can differentiate to form other trophoblasts proliferations, even with recurrence or malignant potential. Thus, it is wise to closely follow up cases of PSN or "atypical" PSN.

18.4.2 Exaggerated Placental Site

Exaggerated placental site (EPS) is a non-neoplastic gestational trophoblastic disease of implantation site showing cells of intermediate trophoblast infiltrating the endometrium as well as underlying myometrium (Fig. 18.30). EPS can occur after not only first-trimester abortion or molar pregnancy but also even following normal pregnancy. As reviewed by Takebayashi et al. (2014), EPS is an entity consisting of an excessive number of intermediate trophoblast of the implantation site in the decidua and uterine myometrium. EPS may be considered as the end of a physiological process rather than a right lesion, and the difference between a regular placental site and EPS is arbitrary. There are no reliable data quantifying the amount and extent of trophoblastic infiltration at different stages of normal gestation. Takebayashi et al. (2014) could find only ten cases reported in English (1990–2013). Molar pregnancy, cervical pregnancy, abortion or induced abortion of early pregnancy, IUFD at 24 weeks of gestation, and term pregnancy have been settings where EPS was detected. No bleeding was noted in a patient harboring an EPS as a small lesion in the uterine cavity and in three patients with a relatively large injury on the uterine wall. Five women with a polypoid lesion disclosed continuous blood spotting or short-term active bleeding. Two patients presented with severe uterine bleeding with atonic uterus without macroscopic lesions. The treatment was hysterectomy in nine women to prevent progressive gestational trophoblastic disease or to control severe uterine bleeding. The remaining lesions were resected in two patients during caesarian section. In consideration of the literature, it seems that most cases of EPS appear to develop a mass, although it should currently be considered the end of a physiological process rather than a true neoplastic or preneoplastic lesion.

18.5 Gestational Trophoblastic Diseases, Pre- and Malignant

Gestational trophoblastic diseases, both premalignant and malignant, include four conditions, i.e., invasive mole, placental site trophoblastic tumor (PSTT), epithelial trophoblastic tumor (ETT), and choriocarcinoma. Risk needs to be assessed following any pregnancy. β-human chorion gonadotropin (β-hCG) is key in the diagnosis and for the monitoring of therapeutic effects. Since a definitive diagnosis cannot often be achieved by histology, persistent or recurrent disease is critically diagnosed by elevated or persistent levels of β-hCG in the serum (Stevens et al. 2015).. Risks for the development of a complete hydatidiform mole are the extremes of maternal age and prior molar pregnancies, while the risk for choriocarcinoma is linked to a history of complete hydatidiform mole, the ethnicity (Asian, American Indian, Afro-American), long-term OCP, and the blood group A.

18.5.1 Invasive Mole

Disorders involving an abnormal formation of the products of conception include the so-called invasive mole. The *invasive mole* (IM, aka *chorioadenoma destruens*) is an HM showing growth into the muscular layer of the uterus developing from either CHM or PHM with risk being much higher in CHM. The risk of developing an IM is also increased if there is a marked enlargement of the uterus, age >40 years, (+) GTD-Hx. In the complete mole, the villi usually penetrate the myometrium or the vessels of the myometrium. IM occurs in 16% of all moles, and serosa is generally intact. It is paramount to realize that vascular invasion may lead to embolization of villi. Radio-chemotherapy has been proposed as a line of treatment.

18.5.2 Placental Site Trophoblastic Tumor

Placental site trophoblastic tumor (PSTT) is a neoplastic gestational trophoblastic disease of the implantation site, showing a well-localized but ill-defined myometrial mass and histologically representative monomorphism of intermediate trophoblastic cells with polygonal or round cellular shape with or without marked atypia, but with a dissecting growth pattern of invasion into the underlying myometrium. PSTT follows a spontaneous miscarriage (1/4 of cases) or a healthy pregnancy (3/4 of cases). PSTT is locally invasive but typically self-limited with a 10% risk of metastases and lethality.

The extrauterine deciduosis is a real diagnostic challenge because it may be interpreted grossly as peritoneal carcinomatosis. Microscopically, EUD shows decidualized stromal cells, usually without endometrial glands, although endometriosis needs to be taken into account as well (Fig. 18.30). EUD is found on the ovary and cervix but also in the bowel, peritoneum, vagina, lungs, pleura, retroperitoneal LNs, and skin.

No surgical intervention is usually necessary, because the lesions spontaneously involute within the first 1–2 months after birth. The theory recalls the endometriosis theories considering that subcoelomic mesenchymal cells may undergo temporary progesterone-induced metaplasia and scattered distribution of decidual cells in the peritoneum. The source of progesterone in nonpregnant women is either exogenous (drugs) or endogenous (corpus luteum and adrenal cortex)

18.5.3 Epithelioid Trophoblastic Tumor

Epithelioid trophoblastic tumor (ETT) is a neoplastic gestational trophoblastic disease of the chorionic type showing relatively representative monomorphism of intermediate trophoblastic cells of the chorionic kind with eosinophilic cytoplasm arranged in nests or cords closely associated with an eosinophilic (hyaline-like), fibrillar material as well as necrotic debris.

18.5.4 Choriocarcinoma

Choriocarcinoma (CCA) is a neoplastic gestational trophoblastic disease showing bilaminar, dimorphic, or biphasic malignant epithelial tumor cells with marked atypia and unique alter-

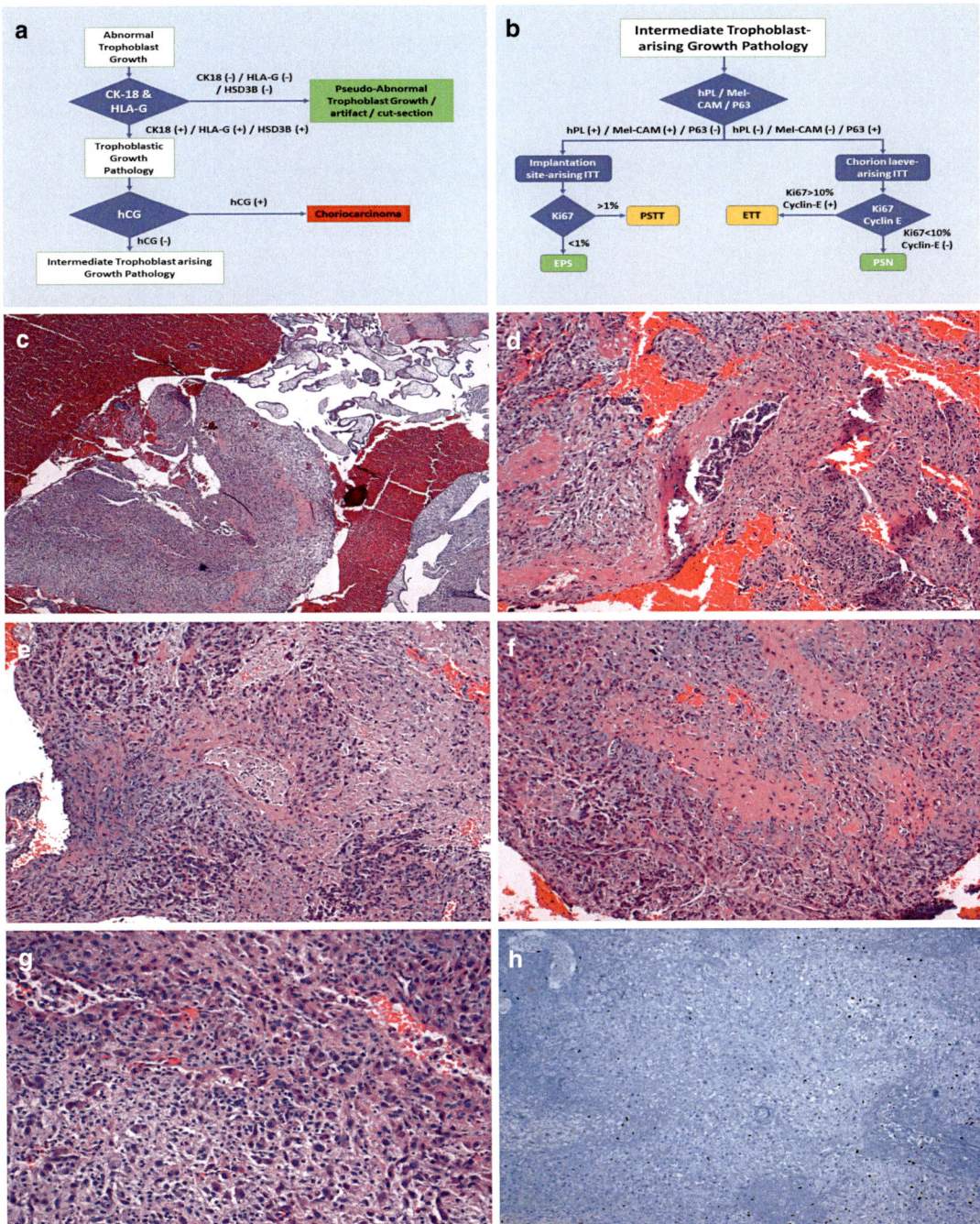

Fig. 18.30 Exaggerated Placental Site - Placental Site Trophoblastic Tumor. Figure (**a**) and (**b**) show the algorithms for abnormal trophoblast growth and the intermediate trophoblast growth pathology, respectively. Figures (**c**) through (**h**) show the classic histopathology of exaggerated placental site and placenta site trophoblastic tumor (see text for detail). All microphotographs have been taken from histological slides stained with hematoxylin and eosin, apart from (**h**), which is from an immunohistochemical slide stained with antibody against Ki67 (MIB1, avidin-biotin complex) (c, original magnification x40; d, original magnification x20; e, original magnification x100; f, original magnification x100; g, original magnification x200; h, original magnification x100)

nating arrangement of mononucleate cytotrophoblast (CT) cells and syncytiotrophoblast (ST) cells. Intermediate trophoblast (IT) and villi should not be present (Fig. 18.31).

Both CC and PSTT may develop from a CHM, PHM, normal pregnancy, or pregnancy that ends early (e.g., miscarriage, elective abortion). Unlike CC, which can metastasize to lungs or other organs, PSTT usually does not spread to other sites in the body but shows a brisk tendency to invade the muscular layer of the uterus at the site of implantation of the placenta to the womb. Most likely, the improvement of the socioeconomic conditions and dietary lifestyles has decreased the incidence of molar pregnancy in South Korea from 4.4 cases of the 1960s to 1.6 cases per 1000 pregnancies in the 1990s. CC usually develops in about 1 in 50,000 pregnancies, while PSTT is probably detected in 0.2% of cases of GTD.

- *RF*: menarche age >12 years, abnormal menstrual flow, previous contraceptive history, reduced consumption of dietary carotene and animal fat as well as advanced maternal age. PHM, CHM, and CC arise from villous trophoblast, while PSTT is a tumor derived from the interstitial (intermediate) trophoblast. PSTT arises from extravillous, interstitial implantation site-like trophoblast (intermediate trophoblast).

Most commonly, CC occurs following a complete mole (50%); a quarter occurs following an abortion and the other quarter following a healthy pregnancy. Unfortunately, CC metastasizes quite early to lung, vagina, bone, and other organs. Since it is chemosensitive, CC has improved survival from <20% to nearly 100%. It is crucial to address the type of GTD correctly, and a few immunohistochemical markers may be extremely helpful.

Other tumors or non-neoplastic lesions of proliferative type are rare but need to be considered. They include choriangioma and choriangiomatosis as well as placenta mesenchymal dysplasia.

Placenta Mesenchymal Dysplasia
- *DEF*: Disorder of abnormal placental development characterized by an enlargement of the placenta due to overgrowth and edema of the chorionic villi, which present with an abnormal vascular growth.
- *CLI*: There is fetal overgrowth syndromes, omphalocele, or mosaicism.
- *GRO*: A "*bed of yarn*" aspect of the maternal surface and abundant grape-like cysts and dilated blood vessels on the chorionic plate, which may be thrombosed or show aneurysms.
- *CLM*: There are thick-walled blood vessels with fibromuscular hyperplasia on the chorionic plate, enlarged and hypervascular stem villi with hydropic stroma, and mole-like central cisterns with the intraluminal gelatinous material. There is no trophoblastic hyperplasia (unlike molar degeneration), intermediate villi with choriangiomatosis-like changes with abnormal vascular branching, and/or dilatation and immature or normal terminal villi.

18.6 Birth Defects

Disruptions and chromosomal aberrations represent probably a regular diagnostic routine in the practice of a pediatric pathologist in many centers. The most frequent cytogenetic syndromes that should be known to a pathologist and particularly to a pediatric pathologist are trisomy 21 syndrome (Down syndrome), trisomy 18 syndrome (Edwards syndrome), trisomy 13 syndrome (Patau syndrome), 5p- syndrome (aka "cri du chat syndrome"), monosomy X0 (Turner syndrome), XXY syndrome (aka Klinefelter syndrome), XXX (aka multi-X syndrome), XYY syndrome, as well as the conditions known as true and pseudo-hermaphroditism. The study of congenital disabilities is one of the most complexes in medicine because the underlying etiology is often very heterogeneous (Box 18.43).

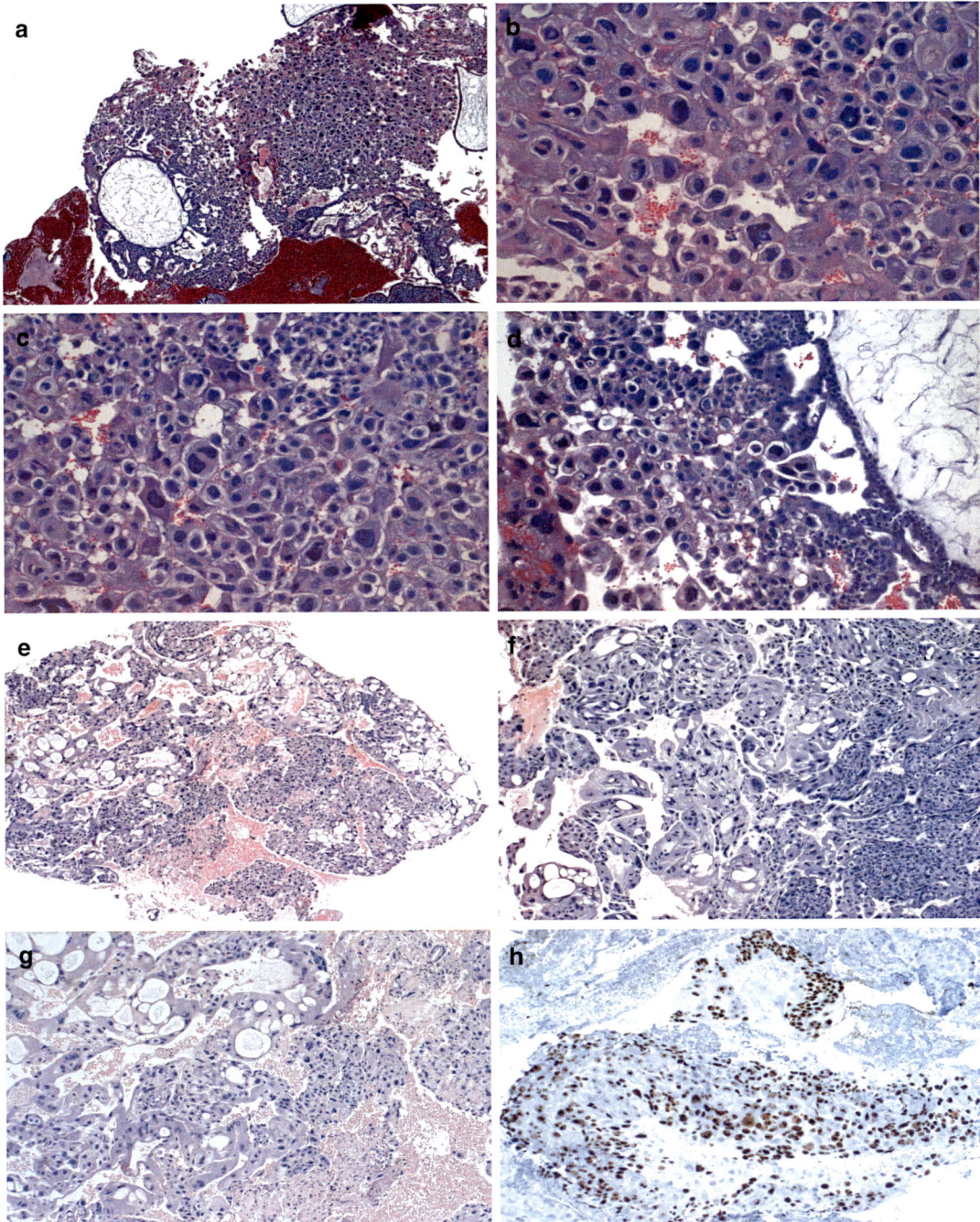

Fig. 18.31 Complete Mole and Choriocarcinoma. Figure (**a**) shows the circumferential abnormal growth of trophoblast of a complete mole. The histopathology of the trophoblast shown in (**b–d**) shows cells with high nucleus-to-cytoplasm ratio, hyperchromatic nuclei, and increased rate of normal mitoses. Atypias may be quite impressive, and, in some cases, the label of invasive mole has been used for preneoplastic stages. Figures (**e–h**) show a characteristic pattern of choriocarcinoma with clear-cut evidence of malignancy, high-grade atypias, necrosis, and high proliferation rate (Anti-Ki67 antibody immunostain, avidin-biotin-complex). All microphotographs have been taken from histological slides stained with hematoxylin and eosin, apart from (**h**) (**a**, original magnification x40; **b**, original magnification x200; **c**, original magnification x200; **d**, original magnification x200; **e**, original magnification x40; **f**, original magnification x100; **g**, original magnification x100; **h**, original magnification x100)

> **Box 18.43 Birth Defects**
> 18.6.1. *Birth Defects: Taxonomy Principles*
> 18.6.2. *Birth Defects: Categories*
>   18.6.2.1. Isolated Birth Defects (Minor and Major Anomaly, Sequence)
>   18.6.2.2. Multiple Birth Defects (Syndromes, PDFD, Associations)
> 18.6.3. *Birth Defects: Pathogenesis* (Macro- and Micromechanisms)
> 18.6.4. *Birth Defects: Etiology* (Mendelian, Chromosomal, Multifactorial)
>
> Notes: *PDFD* polytopic developmental field defects

> **Box 18.44 Developmental Stages and Abnormality Categories (Dys-embryology)**
>
> | Developmental stages | Category of abnormalities |
> |---|---|
> | *Gametogenesis* | ⇒ *Gametopathies* (*dys*-spermatogenesis/*dys*-oogenesis) |
> | *Blastogenesis* | ⇒ *Blastopathies* (fertilization → end of 2nd developmental week) |
> | *Embryogenesis* | ⇒ *Embryopathies* (1st and 2nd embryonic periods until stage 23) |
> | *Fetogenesis* | ⇒ *Fetopathies* (growth, differentiation, and maturation, 9th week-birth) |

We usually think that a pregnancy is a straightforward process, but it is not. The etiology of the human congenital anomalies is in 50% unknown, while four etiologic factors may play a causal role in the other half of cases. These include gene mutations, chromosomal, environment-related, and multifactorial.

18.6.1 Birth Defects: Taxonomy Principles

A variety of disruptions may occur during the process going through from the fertilized ovum to birth. Dys-embryological categories include gametopathies, blastopathies, embryopathies, and fetopathies.

In Box 18.44 of this chapter are listed the processes involved in defective embryology including the fetal development, which evolves from the 9th-week post-ovulation until birth. During the gametogenesis, both *dys*-spermatogenesis and *dys*-oogenesis can occur. These first two abnormal processes mean an abnormal production of germinal cells, i.e., abnormal spermatozoa or oocytes from spermatogonia and oogonia, respectively. The anomalies involved during the process, which runs from fertilization through the end of the 2nd developmental week, are blastopathies. Embryopathies include the abnormalities that occur in the 1st and 2nd early period until Carnegie stage 23 (approximately between the 3rd week and the full 8th-week post-ovulation). Stage 23 corresponds to 56th-day post-ovulation. Fetopathies occur during the fetogenesis, which regards growth, differentiation, and maturation of the conceptus. This process is the most substantial period and runs from the 9th week until birth. Fertilization is *"Befruchtung"* in the German literature and reminds the process of formation of fruit, particularly of the product of conception. This part of the book will also present some German names. This note is linked to the enormous contribution of the German scientists to the embryology. In particular, Spemann was probably one of the largest contributors to German embryology. In embryology, the Spemann-Mangold organizer or Spemann organizer is a cluster of cells in the developing embryo of an amphibian that induces development of the central nervous system. The first two to three decades of the twentieth century have been full of numerous embryological experiments that promoted our current understanding of the embryology and the anomalies that occur during the embryogenesis and fetogenesis. Hans Spemann experimented himself and led his graduate students in conducting experiments with South African clawed frog embryos (*Xenopus laevis*) and embryos of newt (*Triturus taeniatus* and *Triturus cristatus*), which is a salamander in the subfamily Pleurodelinae. During these experiments, Spemann realized the need to create microtools able to perform correctly and with high reproducibility his investigations. The German scientist developed glass needles and micropipettes. Mangold, a graduate student, joined Spemann's lab in the spring of 1921. Mangold's task was to conduct transplant across species using blastopore lips between different new species. Since the tissue was differently colored between species, this quality allowed Mangold to detect whether the features that developed originated from transplanted or host tissue. Spemann's microtools were key, and Mangold excised the blastopore lip of the unpigmented *Triturus cristatus* egg and transplanted it under the ectoderm of a pigmented *Triturus taeniatus* newt egg. The rank of the Spemann-Mangold organizer experiment is the discovery that a part of the mesoderm influences the ectoderm as the ectoderm differentiates into central nervous system tissue. In 1924, Spemann and Mangold published their finding of the amphibious organizer in the research article, *Über Induktion von Embryonalanlagen durch Implantation artfremder Organisatoren (Induction of Embryonic Primordia by Implantation of Organizers from a Different Species)*.

Laminopathies are a consequence of a numerical aberration of the chromosomes during the meiosis of the germinal cells, while blastopathies are a consequence of damage of the germinal disc (aka *disco-blastula* or "Keimscheibe" of the German Embryology School). Conversely, embryopathies are structural deviations due to damage during the reorganization of the *disco-blastula* (or "Umgestaltung" of the original German embryologists) from germinal disc to embryo, a phase where the organs are formed (organogenesis). Thus, blastopathies are a consequence of damage to the *disco-blastula* during the early development (blastogenesis). Implications for the embryo may be entirely different according to the degree of damage, quantifiable regarding a substance, amount, and time of action (Box 18.45).

The most famous conjoined twins were the brothers Bunker, Chang and Eng Bunker, who lived 63 years (1811–1874). These Thai brothers were born in Siam, now Thailand, and were labeled as the Siamese twins as they engaged themselves traveling with P.T. Barnum's circus for many years.

Pagus Type of Twin Pregnancies

Conjoined twins, also known as Siamese twins, have captivated people since the time man began to ponder the etiology of fetal malformations. In fact, Leon Diakonos, mentions in 974/975 AD, having seen a *thoracopagus* conjoined twin pair on several times. Most probably, this pair is the same twin couple described by two major Byzantine scientists and authors, Leon Grammatikos and Theodore Daphnopates, in

Box 18.45 Mild Through Severe Damage During Blastopathies

Mild insult	⇒ *Reconstitution* ("Wiederherstellung" of the Early German School of Embryology) due to the absolute potential of the surviving cells of the blastocyst
Severe insult	⇒ *Death* of the zygote or blastocyst (early miscarriage)
	Disturbances of the total embryonic cell complex
	("*Störungen des Ganzen*"):
	1. Monochorionic monozygotic twins with complete separation ("*Eineiige Zwillinge bei vollständiger Trennung*" of the German Embryology School)
	2. Doubled defects of twin pregnancy with incomplete separation ("*unvollständige Trennung*" of the German Embryology School)
	Symmetric type (Siamese twins) ("*Doppelfehlbildung*")
	Asymmetric type (autosite and parasite twins)

Constantinople during the second half of the tenth century. Constantinople, which changed the name to Istambul in 1930, was the capital city of the Roman Empire (330–395 AD), the Eastern Roman (Byzantine) Empire (395–1204 AD and 1261–1453 AD), the Latin Empire (1204–1261 AD), and, finally, the Ottoman Empire (1453–1923 AD).This historiography describes the surgical separation of a *thoracopagus symmetricus* twin pair. It has been described that one infant survived, while the co-twin died 3 days later following a many hours long surgical operation. Although we may inquire about several surgical procedures, the Byzantine culture and surgery was very well-developed illustrating the high standard of Byzantine medicine. This twin pair and the surgical procedure were described again 150 years later by Johannes Skylitzes (Geroulanus et al. 1993). Since then, several cases have been illustrated, and, currently, reports of conjoined twins are presented in media. Conjoined twins are a rare and often catastrophic obstetrical event that occurs once in every 50,000–100,000 births worldwide. They are usually classified according to Harper et al. (1980) and Gore et al. (1982) into complete or symmetrical and into incomplete or asymmetrical sets. We distinguish the following categories:

- *Craniopagus*: at the level of the head, specifically on the skull (neurocranium fusion)
- *Syncephalus*: at the level of the head, specifically on the face (splanchnocranium fusion)
- *Thoracopagus*: at the level of the ventrolateral chest (thoracic cage)
- *Thoraco-omphalopagus*: at the level of the ventrolateral chest (thoracic cage) and omphalocele
- *Xiphopagus*: at the level of the xiphoid process
- *Rachipagus*: at the level of the suprasacral vertebral column
- *Pygopagus*: at the level of the rump (coccyx) or sacrum (sacral bone)
- *Ischiopagus*: at the level of the ischiatic bone

The cardiovascular and gastrointestinal systems are traditionally adjoined. The severity of the visceral malformations depends inversely on the extent of duplication (complete or partial). The most significant abnormalities involve the cardiovascular and central nervous systems because of their obvious medical and surgical relevance.

Prenatal echocardiography of thoracic-abdominally conjoined twins is extremely helpful for evaluating the feasibility of separation. Surgical options for such twins are primarily dependent on the anatomy of the cardiovascular system, and the extent of cardiac duplication and the severity of the associated cardiac defects primarily determine the suitability of the successful separation of the infants. The anatomical connections are assessed prenatally using magnetic resonance imaging (MRI) in association with fetal ultrasonography. In 1989, MRI was used for the first time in the preoperative planning to separate omphalopagus conjoined twins (Donaldson et al. 1990). Today, the information supplied by MRI is an essential part of the preoperative workup of all types of conjoined twins. If a postmortem examination of conjoined twins is undertaken, a detailed anatomic investigation of the various developmental malformations is typically performed. MRI findings may be used by the pathologist for autopsy planning, especially with a demonstration of critical clinic-topographic correlations (Sergi et al. 1998). Asymmetrical conjoined twinning is a sporadic condition and constitutes only 1–2% of all conjoined twins, i.e., one twinning per one or two million live births. This anomaly, called heteropagus twinning, is characterized by an incompletely formed twin, the parasite, which is attached to a region of the well-developed autosite (Hager et al. 2007). One of the common areas in which such a parasite can be connected is the epigastrium (Fig. 18.10). Although imaging diagnostics using X-ray, ultrasonography, CT scan, MRI of the abdomen, and echocardiogram are usually performed, the anatomic variability of both twins may jeopardize the successful surgical separation of conjoined twins. In fact, pediatric surgeons may be required to change their surgical approach and the planned dissection techniques used for separation in situ to achieve a good result. Thus, while MRI is essential when planning separation strategies for symmetrical conjoined twins, it becomes of paramount importance for asymmetrical twins. Embryopathies may encompass a quite broad spectrum of anomalies. In summary, congenital structural deviation due to damage during the reorganization of the germinal disc ("*Keimscheibeumgestaltung*" of the German Embryology School) to the embryo (organogenesis), depending on the time of the processes of differentiation ("*Differenzierungsvorgänge*" of the German Embryology School), arises a disturbance of the differentiation ("*Fehlbildung*" of the German Embryology School), and, finally, the sooner the differentiation processes are disturbed, the more massive and more extensive is/are the malformation(s) of various organs (see, e.g., the neural tube defects in Box 18.46).

Rachischisis
Rachischisis (Greek: "ῥάχις," spine, and "σχίσις," split) is a congenital developmental disability involving the neural tube. This anomaly occurs in utero when the posterior neuropore of the neural tube fails to close by the 27th intrauterine day, and in Box 18.46, the neural tube defects of lower type (spina bifida or rachischisis) are summarized.

18.6.2 Birth Defects: Categories

A variety of non-neoplastic lesions can present clinically and radiologically as uneventful or life-threatening conditions and may result in a hysterectomy. In such situations, the pediatric pathologist or gynecologic pathologist has an essential role to play in determining the nature of these lesions. Awareness of the development of the embryo at an early stage and entities that can mimic malignancies will assist the physician in the correct diagnosis of these lesions. Genetics is part essential of the study and practice of the pediatric pathologist. A pediatric pathologist should not limit his or her activity to some parts only of the pediatric pathology knowledge. All components may be useful in an integrative concept of pediatrics. Neoplastic proliferation may be intimately linked to embryogenesis, and a pediatric pathologist should practice or maintain skills in both perinatal pathology and histopathology of the child developing into an adult. Thus, it is important to remind that more than 1/10th of newborns have one minor defect or malformation, mostly in the head, neck, or hands, and 0.75% newborns present with two minor abnormalities, while only 1:2000 of newborns have three errors. The word malformation is synonymous with deficiency, although defect is preferable when addressing and talking to parents. In embryo-pathology, malformation is a synonym of the defect. The organ changes and birth anomalies need to be strictly defined.

Definitions of organ changes include the following categories:

Agenesis/agenesia ⇒ Lack of an organ due to the absence of the anlage of the specific organ

Aplasia ⇒ Absence of the development of the anlage of the particular organ

---

Box 18.46 Neural Tube Defects
Neural tube defects (NTD) = *Dysraphias*

1. Spinal cord: 5th–8th week
   Spina bifida occulta (5th–8th developmental week): no bulging rachischisis covered by skin
   (e.g., hypertrichosis, areas with hyper-/hypopigmentation, *naevus vasculosus*, lipoma)
   Spina bifida cystica sive aperta: bulging rachischisis ± skin covering
2. *Canalization defect – Spina bifida* (5th–8th developmental week), if covered with skin
3. *Neurulation defect – Spina bifida* (<5th developmental week), if covered with connective tissue

Atresia	⇒ Incomplete development of a cavitary organ by lack of luminal formation
Hypoplasia	⇒ Arrest of the development of a specific organ
Hypertrophy	⇒ Enlargement of an organ/tissue by enlargement of the single cells
Hyperplasia	⇒ Enlargement of an organ/tissue by an increase in the number of cells
Atrophy	⇒ Acquired size reduction of an organ/tissue by ↓ of either cell size or cell number

Definitions of birth anomalies include the following categories:

Malformation	⇒ Congenital intrinsic structural abnormality of the human development occurring in (usually)≤4% of the general population in consequence of a defective process of development of a specific organ (entirely/partially) or a part of the developing embryo (< 1/20 people)
Variation	⇒ Congenital or postnatal intrinsic structural abnormality of the human development occurring in (usually) >4% of the general population associated or not with gene polymorphisms (standard variant) (> 1/20 people)
Disruption	⇒ Congenital extrinsic structural abnormality of the human development occurring usually in consequence of secondary destruction (interference) of any standard process of differentiation (e.g., amniotic bands-derived disruptions)
Deformation	⇒ Congenital extrinsic structural abnormality of the human development occurring in the general population in consequence of the modification of the intrauterine environment (e.g., oligohydramnios-derived deformations)
Dysplasia	⇒ Congenital intrinsic abnormal organization of cells in tissue and their consequences

Anomalies may be considered singly or grouped in constellations that may have critical prognostic relevance. As above emphasized, the subdivision and definitions of birth anomalies are crucial in the pediatric pathology routine. The term dysplasia regarding dysmorphogenesis needs to be kept separated from the dysplasia concept as preneoplastic condition (e.g., Barrett dysplasia as a predisposing factor for the development of esophageal carcinoma). Dysplasia as a preneoplastic condition should be considered mostly as an extrinsic abnormality, in contrast to the dysplasia as the abnormal development of a tissue, which possesses exquisitely an intrinsic quality. One of the most important examples of disruption is the ADAM sequence. Although genetic factors are also involved, a disruption definition relies on these multiple congenital anomalies.

ADAM Sequence

- *DEF*: Amniotic deformity, adhesion, and mutilation (ADAM) sequence is the acronym illustrating a different series of defects with a broad spectrum of anomalies.
- *EPI*: ~1:10^5 births.
- *EPG*: Three theories are explaining the pathogenesis, the germplasm theory of Streeter (intrinsic) (Streeter, 1930), the early amnion rupture theory of Torpin (extrinsic) (Torpin, 1965), and the vascular disruption theory of Van Allen (Van Allen et al. 1987). They include deficiency of germplasm (developing germinal disc), and disturbance of threshold boundaries of morphogenetic substances may alternate with causes of extrinsic type. The extrinsic theory is the most accepted model for ADAM, although vascular disorders may be responsible for some craniofacial and abdominal wall defects, difficult to quickly explain using the extrinsic model only. LBWC includes the presence of body wall defects with the evisceration of thoracic and/or abdominal organs, limbs defects, and neural tube defect of a lower type. A subgroup of

18.6 Birth Defects

ADAM sequence, i.e., LBWC, may or may not be added to the ADAM cases making the frequency of this sequence different in several studies.
- *PGN*: Mostly fatal due to the complexity of MCA. RFs: High altitude (hypoxia), tobacco, drugs, and DM.

Anomalies may be considered as single elements or seen in a constellation of several and different units of a more complex group. The most current definition of *multiple congenital anomalies* is also key in pediatric pathology.

Syndrome	⇒ Group of congenital anomalies occurring together (not necessarily all simultaneously), which present in different organs or systems not explainable by a single event but thought to be pathogenetically related (e.g., Wolf-Hirschhorn syndrome)
Sequence	⇒ Group of congenital anomalies occurring together (usually simultaneously), which present in different organs or systems and produced by a variety of causes producing a pattern of abnormalities in consequence of a single known structural defect or a mechanic factor (e.g., oligohydramnios sequence)
Polytopic Field Defect	⇒ Group of congenital anomalies occurring together (usually simultaneously) resulting from disturbance (interference) of development (*mal*development) of a single region or developmental field (e.g., Midline Field Defect)
Association	⇒ Group of congenital anomalies occurring in a nonrandom fashion and more than two individuals without corresponding or known syndromes, sequences, or polytopic field defects (e.g., VATER association)
Phenocopy	⇒ Group of congenital anomalies occurring together (not necessarily, but usually simultaneous), which present in different organs or systems resulting from the action of environmental factors on developing embryo or fetus

Note: VATER is an acronym for vertebral, anal, tracheoesophageal, and renal anomalies

Congenital syndromes may be associated with differential expression of genes. More definitions include the following ones:

- *Pleiotropism*: A single gene with noticeable multiple effects on embryo and future neonate
- *Genetic Heterogeneity*: Setting where underlying the same phenotype there are multiple genotypes
- *Variable Expressivity*: Setting in which a variable phenotype occurs in different individuals with the same genotype
- *Penetrance*: The same phenotype is underlying the same genotype but expressed to a varying degree in different individuals

A classic and vital example to understand the concept of *pleiotropism* is found in the *fibrillin* gene, which causes Marfan syndrome. Skeletal, ocular, cardiovascular changes may occur by mutations determining an abnormal gene with a structural protein common to tissues present in these organs. In fact, pleios- means from the Greek πλείων, which means "more," and τρόπο, which means "turn," i.e., "more turns," is a central term in developmental genetics. In pleiotropism, a single gene affects several phenotypic traits in the same organism. These effects often seem to be unrelated to each other. *Genetic heterogeneity* is seen in conditions where different mutations determine the production of a given trait. An example is Charcot-Marie-Tooth disease, which is a "heterogenous" hereditary motor and sensory neuropathy affecting more than 2.5 million people worldwide with mutations in different genes producing similar clinical symptoms.. Different genetic mutations can cause the same hereditary neuropathy. *Variable expressivity* shows a variable range of different clinical or pathological situations different from patient to patient with the same autosomal dominant (AD)-inherited disease (e.g., patients with myotonic dystrophy may present with variable expressivity of the disease). The word expression derives from the Latin *exprimere* "a pressing out, a projection." *Penetrance* represents the percentage of individuals carrying an AD inherited gene and expressing the trait (i.e., individuals with 100% penetrance mean that all these individuals that carry the specific gene will have the disease). Chromosomal variation may be necessary to categorize. To keep in mind is that abnormalities of the sex chromosomes are better tolerated than autosomal defects and are also more common in the routine of a perinatal pathologist. The most current cytogenetic definitions are specifically designed to help the pathologist to interact at the multidisciplinary team meetings.

- *Metacentric Chromosome*: Chromosome with centromere near the center.
- *Submetacentric Chromosome*: Chromosome with centromere eccentrically placed.
- *Acrocentric Chromosome*: Chromosome with centromere near the end.
- *Reciprocal Translocation*: Translocation between chromosomes with single breaks in two chromosomes with the balanced exchange of genetic material and without loss.
- *Robertsonian Translocation*: Translocation between chromosomes with a fusion of extended portions of two acrocentric chromosomes ("acrocentric fusion") with a small chromosome formed by two small parts, which is often lost.
- *Isochromosome*: Centromere separates in transverse planes with subsequent one chromosome with two p-arms and one chromosome with two q-arms, i.e., there is a chromosome in which one arm is practically missing, and the other arm is duplicated.
- *Inversion*: Chromosomes that show both break sites on the same side determining a para-centric inversion or on opposite sides of the centromere determining a peri-centric inversion of the chromosomes.
- *Ring Chromosome*: Fusion of some "sticky ends" of a chromosome, which has lost telomeres of p and q arms (circularization of the remaining chromosome).
- *Lyonization*: Inactivation of all but one X chromosome.

18.6.3 Birth Defects: Pathogenesis (Macro- and Micromechanisms)

A variety of disruptions may occur during the process going through from the fertilized ovum to birth. Agnathia-Otocephaly Complex (AOC) or Syndrome (AOS), which occurs at 4th–5th weeks of the embryonic development, results from a defect of the embryo characterized by the absence of the low jaw (mandible), microstomia, aglossia, and abnormal positioning of the ears in the midline (Schiffer et al., 2002; Sergi and Kamnasaran, 2011; Dubucs et al. 2020).

Prosencephalon Defects

DeMyer suggested that "the face mirrors the brain," but a few years later, it has been evident that "the face does not mirror the brain," being the possibility to have some several variations. To date, it is possible to state, probably with some certainty, that "the face usually mirrors the brain, but NOT always." (DeMyer 1967; DeMyer 1975; Sadove et al. 1989) Holoprosencephaly (HPE) also derives from Greek ὅλος ("whole"), πρόσω ("forwards"), and ἐγκέφαλος ("brain"), i.e., pros or onwards and encephalon or brain, "proso-enképhalos"). HPE is about $0.5/10^4$ live births with about 2/3 of cases showing an extra-chromosome 13 and rarely other chromosomal abnormalities (Urioste et al. 1988; Gekas et al. 2012; Chassaing et al. 2012; Sergi et al. 2012; Sun et al. 2010; Sergi and Schmitt, 2000 (a); Sergi and Schmitt 2000 (b))).

Prosencephalon Defects (4th–5th Embryonic Week) "Brain Defects"

- Incomplete division of the prosencephalon derived from interference with the induction and/or topogenesis of the prechordal mesoderm.
- According to the defective developmental stages of the cerebral lobes, prosencephalon developmental defects or holoprosencephaly (HPE) may be differentiated in:
  - *Alobar* HPE, in which there is no separation of the brain
  - *Semilobar* HPE, in which there is division in the occipital region only
  - *Lobar* HPE, showing two hemispheres with some of the rostral structures in the middle

According to the face defects, there may be quite a surprise in the ultrasound/MRI or at the

18.6 Birth Defects

time of the autopsy. However, most of the cases may have some common associations:

- *Cyclopy*: Alobar holoprosencephaly (HPE)
- *Ethmocephaly*: Alobar HPE
- *Cebocephaly*: Alobar HPE
- *"Median cleft lip"-type*: Alobar HPE
- *Mild facial dysmorphism*: Semilobar/lobar HPE

Another defect of separation is represented by the *caudal regression syndrome* (CRS) (Fig. 18.32). CRS is a congenital malformation syndrome with variable degrees of failure in the

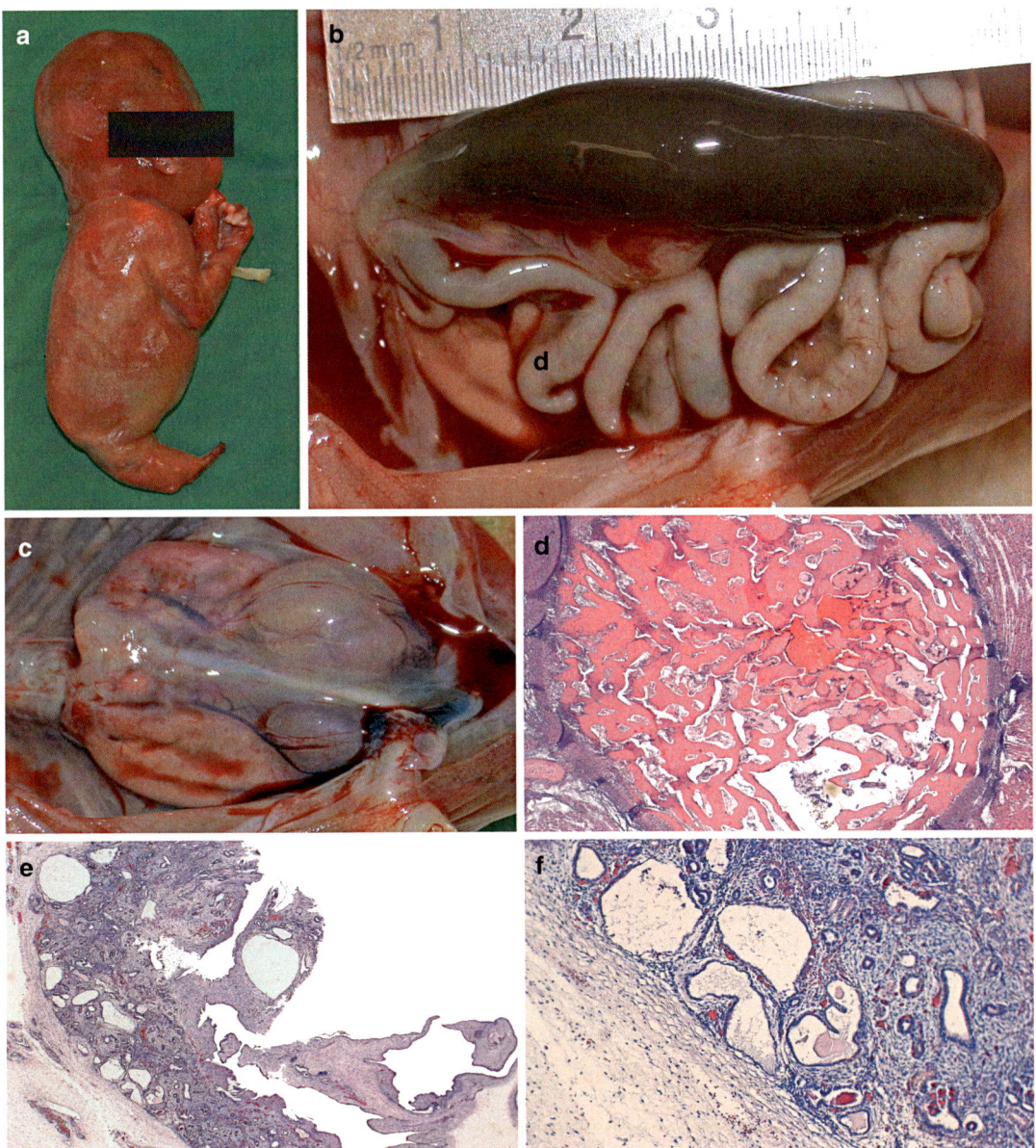

Fig. 18.32 Caudal Regression Syndrome. Caudal regression syndrome (CRS) is depicted in the figures a through f with sirenomelia (**a**), atresia of the distal gastrointestinal tract (**b**), dysplasia of the kidneys, single femur (**d**), and abnormal renal histopathology with cystic dysplasia associated with cloaca malformation (**e–f**). All microphotographs have been taken from histological slides stained with hematoxylin and eosin (d, original magnification x40; e, original magnification x20; f, original magnification x100)

early gestational development. CRS is also known as sacral agenesis or caudal dysplasia. CRS may involve thoracic, lumbar, coccygeal spine, and lower extremities. The cause of this syndrome is a defect of the neurula (error in the induction of the caudal elements) at the 28th day of the embryonic development. In addition to motor and sensory deficits of the lower extremities, there are cardiac diseases, gastrointestinal disorders, neural tube defects, and genitourinary anomalies. Uncontrolled maternal diabetes mellitus (DM), genetic predisposition, and vascular hypoperfusion are probable risk factors, but the actual pathogenesis is unclear. CRS incidence is 0.1–0.25 per 10,000 pregnancies but is 1% in the case of DM. The embryologic insult occurs at the mid-posterior axis of the mesoderm which causes the lack of the progression of the mesoblastic caudal bud. It seems that one or more processes of first streak migration, primary or secondary neuralization, or differentiation are compromised by investigating the axial mesoderm patterning of the early gestation. Sirenomelia, best defined as "a limb anomaly in which the normally paired lower limbs are replaced by a single midline limb," (Stevenson 2006) is an extremely rare congenital malformation disorder probably the severe end of a spectrum of CRS ranging from ectopic anus to sirenomelia. A single midline lower limb, and an aberrant abdominal umbilical artery have been considered the main anatomic findings that distinguish sirenomelia from CRS, but the pathogenesis remains a matter of dispute (or a "vascular steal" dispute) with CRS, as indicated above, hypothesized to arise from primary deficiency of caudal mesoderm. Renshaw (1978) classified the spectrum of CRS into 5 types based on type of defect and articulation between the skeletal segments. In type 1, there is a total or partial unilateral sacral agenesis, in type II there is a variable lumbar and total sacral agenesis and the iliac bones articulates with the sides of the lowest vertebral bodies, while type III is characterized by a variable lumbar and total sacral agenesis and the caudal end plate of the lowest vertebral body rests above fused iliac bones or an iliac amphi-arthrosis, and type IV shows a fusion of soft tissues in both lower limbs, while, finally, type V, also known as sirenomelia, harbors a fusion of the bones of lower limbs (Seidahmed et al. 2014). Stocker and Heifetz (1987) used a 7-tier classification for sirenomelia, including type I with all thigh and leg bones present, type II with a single fibula, type III with absent fibulae, type IV with partially fused femurs and fused fibulae, type V with partially fused femurs, absent fibulae, type VI with single femur and single tibia, and type VII with a single femur, but absent tibiae.Padmanabhan (1998) performed an experimental study using the Hsd/Ola:TO strain of mice and retinoic acid. He showed the role of retinoic acid in producing CRS in the mouse fetus. Agenesis of the tail, caudal vertebral defects, spina bifida, imperforate anus, rectourethral fistula or recto-vesicle, renal malformations, cryptorchidism, gastroschisis, and limb defects, including sirenomelia, were characteristic features of this animal model. Several craniofacial abnormalities were also seen in the GD 8 treatment group. Another study by Tse et al. (2005) studying a mouse model of retinoic acid-induced CRS found that the nephrogenic defect was intrinsic to the kidney and that it resided in the metanephric mesenchyme. The transcription factor Wilms' tumor 1 (WT1), which is indispensable for kidney development, failed to express in the metanephros of retinoic acid-exposed embryos. The vascular steal mechanism seems currently to have been well consolidated as a key factor exquisitely linked to sirenomelia which results in severe ischemia of the caudal spinal segment.

Congenital Heart Disease
Congenital heart disease is primarily discussed in Chap. 1.

Gastrointestinal Anomalies
In the 6th week (post-ovulation or embryonic), several insults may act on developing embryo, and some insults can determine *defects of fusion of coelomic type*. Coelom is the body cavity in metazoans, located between the intestinal canal and the body wall, and coelomate animals have a body cavity with a complete lining called peritoneum derived from mesoderm.

Defects of Fusions of Coelomic Type (6th Week)

Diaphragmatic defects (aka – although not correctly – diaphragmatic hernias) are defects of the fusion of the following four structures that may act either singly or in a combined way:

– Septum transversum
– Pleuro-peritoneal membranes
– Dorsal mesentery of the esophagus
– Muscles of the lateral chest (thoracic cage)

EPI: 1:3000 live births.

PGN: The best prenatal method to assess the severity of a congenital diaphragmatic defect or hernia is probably the observed-to-expected lung-head ratios using sonography and total fetal lung volumes using MRI. The fetal liver position is also a suitable method.

Congenital diaphragmatic defects (CDD) are a challenge for the perinatology team (Fig. 18.33). Newborns with CDD require intensive cardiopulmonary support after birth, which includes immediate endotracheal intubation, ventilation, fluid, and inotropic support. One of the major complications of these infants is the occurrence of pulmonary hypertension, which may require the use of pulmonary vasodilators in addition to prostaglandin E1 and milrinone, which is a phosphodiesterase 3 inhibitor with the aim to increase the heart's muscular contractility and, simultaneously, decrease pulmonary vascular resistance. In severe CDDs, extracorporeal life support may be mandatory. The physiologic stability of the infant is a conditio sine qua non for the pediatric surgical team to attempt an open surgical repair. The aim is to stabilize the newborn in the first 2 weeks of life to avoid cardiovascular and pulmonary complications. Neurodevelopmental testing should be offered to the family always. In the 7th week (postfertilization), recanalization defects can also occur.

Canalization and Recanalization Steps and Related Defects (7th Week)

1. Physiologic initial closure of the future stomach-bowel (Stage 17) through the proliferation of the endodermal epithelium up to the closing of the lumen
2. Recanalization of the stomach and bowel (Stage 21) through the formation of microlumina that are going to join ("spongelike"-primitive digestive tract) forming, in the end, a solitary lumen
3. Digestive tract with a single lumen (Stage 23)

Recanalization defects:

- Atresia, when some or all of the digestive tract shows no lumen formation
- Septs, when some portions of the "spongelike"-septation are still in place
- Cysts, when a part or a microlumen becomes separated from the rest of the digestive tract showing a lumen not communicating with the central lumen of the formed digestive tract
- Stenosis, when some portions of the initially physiologically closed stomach-bowel complex remain closed determining a small central lumen (there is no extrinsic, outside of the digestive tract, which is a situation determining compression of the track itself)

Between the 5th- and the 6th-week postovulation, it is also possible to have rotation anomalies. A defect of canalization is due to the obstructive form of fetal megacystis (Fig. 18.34). One of the organs that can be visualized by ultrasound early during the pregnancy is the fetal urinary bladder. The fetal bladder appears in 50% of fetuses without congenital defects at 10 weeks of gestation and in 100% of fetuses without congenital defects at 13 weeks of gestation. The ultrasound is also a pillar in identifying some urinary defects. Fetal megacystis is the presence of a distended bladder in a fetus and is defined specifically according to the gestational age. In the first trimester, the longitudinal bladder diameter should be ≥ 7 mm. In the second trimester, the longitudinal bladder diameter should be ≥ 30 mm, and in the third trimester, the longitudinal bladder diameter should be ≥ 60 mm. There is an association of the fetal megacystis with ↑ nuchal translucency and ↑ incidence of chromosomal defects (e.g., trisomy 13 or 18). The *cloaca* is the poste-

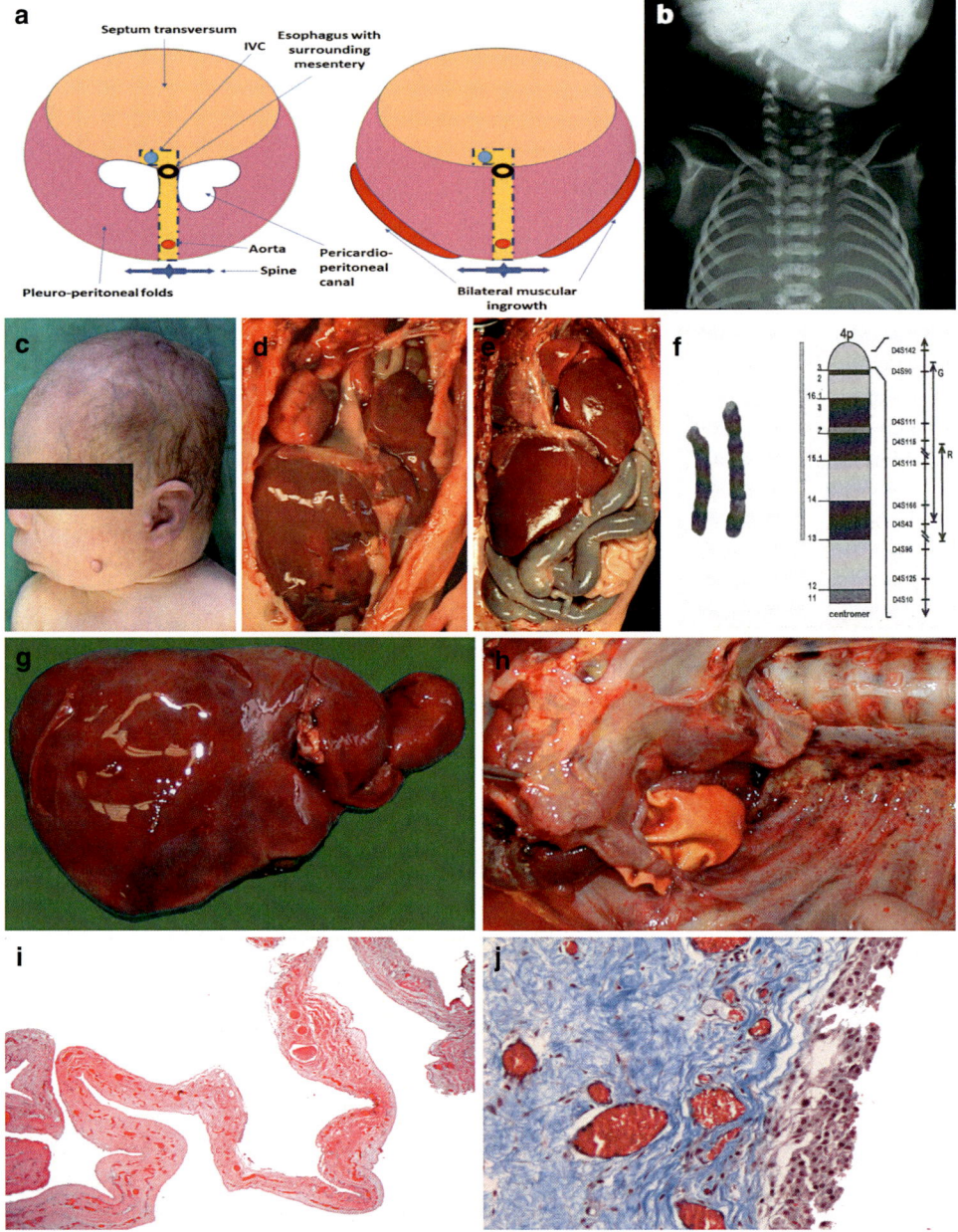

Fig. 18.33 Wolf Hirschhorn Syndrome. Figure (**a**) shows the components that participate to the formation of the diaphragm (see text for details). Figure (**b**) shows the X-ray of a patient with Wolf-Hirschhorn syndrome with midline vertebral defect; Figure (**c**) shows fibroepithelial polyps and accessory tragus identified in Goldenhar syndrome. Goldenhar syndrome and other genetic syndromes or multiple congenital anomalies syndromes may present with diaphragmatic defect and genetic counseling should be considered appropriate. A hypoplasia of the left diaphragm leaves the intestinal organs located in the left hemithorax. In most of these cases, there is a hypoplasia of both lungs (Figures **d** and **e**). Figure (**f**) shows the characteristic genetic variation of chromosomes with WHS caused by a missing piece of genetic material near the end of the short arm of chromosome 4 (4p-). The liver shape is also abnormal in case of a diaphragmatic defect, because the liver is compressed in the left hemithorax. Figure (**h**) shows the repair of congenital diaphragmatic defect performed at the Innsbruck Children's Hospital, Austria. The study of the rest of the diaphragm may disclose a marked fibrosis with hyperemia of blood vessels and mesothelial cell hyperplasia (**i** and **j**). Figure (**i**) is taken from histological slides stained with hematoxylin and eosin, while figure (**j**) is taken from histological slides stained with Masson's trichrome for collagenous tissue (blue staining) (i, original magnification x12.5; j, original magnification x200)

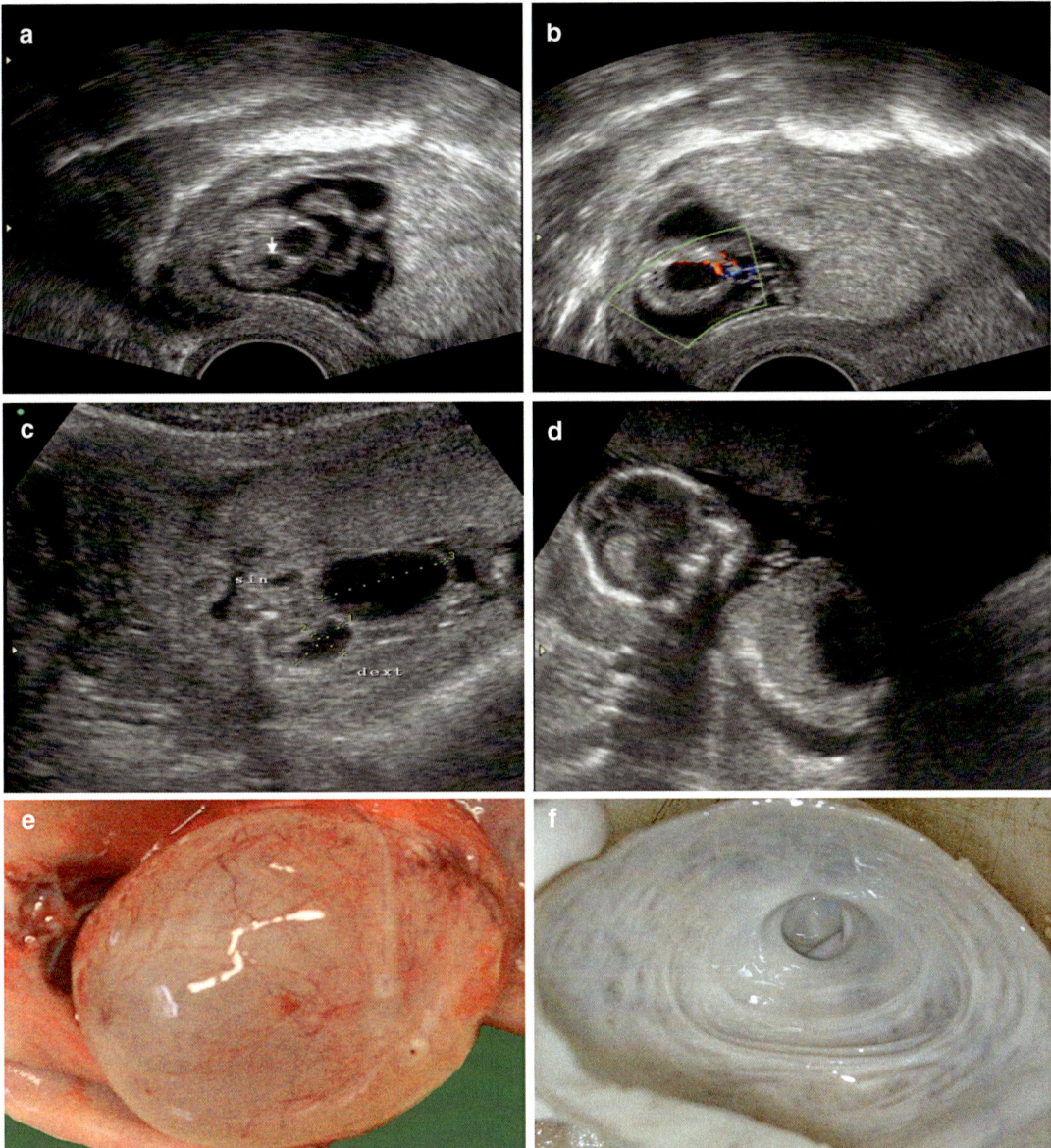

Fig. 18.34 Fetale Megacystis. The examination of the urinary bladder is essential in the prenatal ultrasonography and to rule out urogenital defects. Figures (**a–d**) show the ultrasound of a case with fetal megacystis (see text for detail). The gross preparation at the postmortem of a urinary bladder discloses a large urinary bladder filled with prenatal urinary output. The view of the urinary bladder after evacuation of the prenatal urine discloses changes of the internal surface with marked hypertrophy of the underlying musculature (**e, f**)

rior orifice and cavity at the end of the intestinal tract for the release of both excretory and genital products in certain vertebrates and invertebrates. In zoology, the cloaca is present in birds, reptiles, amphibians, most fish, and monotremes, which are the only mammals that lay eggs. At 4–7 weeks of gestation, the cloaca is divided by the cloacal septum into the *ventral urogenital sinus* and in the *dorsal rectum*. The sequelae of the urogenital sinus are specific. The upper part will form the urinary bladder, the middle or pelvic will originate the prostatic and the membranous urethra in the male and the entire urethra in the female, while the caudal phallic part will form the penile urethra in the male and the vaginal vestibule in the female. A fetal megacystis has two leading causes. The obstruction-underlying megacystis is due to maldevelopment of the urethra, which determines an obstruction to the flow of the urinary function. The non-obstruction-underlying megacystis is part of a heterogeneous group including chromosomal defects (e.g., trisomy 13 and trisomy 18) and genetic syndromes or multiple congenital anomalies, such as "prune belly syndrome" and megacystis-microcolon. In the setting of the obstructive causes, there is a spectrum of abnormalities from complete atresia through the formation of valves or diaphragms in male fetuses. Female fetuses have a more complex defect of the development of the urogenital system, which is labeled as "cloacal plate abnormalities." Firstly, the urinary bladder is in continuity with the allantois, but ultimately the lumen of allantois becomes restrained, and a thick fibrous cord develops that is known as urachus. The caudal portion of the mesonephric ducts become incorporated into the dorsal wall of the urinary bladder contributing to the mucosa of the trigone (trigonum vesicae urinariae). The urethral valves can originate following an anomalous insertion of secondary valves into the mesonephric duct in the cloaca, probably in the first few weeks of intrauterine development. There are three types of posterior urethral valves: type I, which originates from the *verumontanum* and develops down and laterally, coinciding on the anterior wall of the urethra; type II, which arises from the *verumontanum* and develops progressively up inserting into the bladder neck; and, finally, type III, which is constituted by a full diaphragm pierced in the center. The diagnosis relies on dilated ureters and hydronephrosis of variable degrees. There is a key-hole sign, which is an enlarged prostatic urethra, frequently with distension of the prostatic utricle and thickening of the bladder neck. Renal agenesis and dysgenesis (Figs. 18.35 and 18.36) are defects of migration and defects of differentiation of the renal blastema. Unilateral renal agenesis (URA) is the absence of one single kidney, while bilateral renal agenesis (BRA) is the bilateral absence of organs. Both URA and BRA occur in ≤1% of births annually. They happen when the uretic bud fails to successfully develop at early gestation.

Rotation Anomalies of the Intestine (5th–6th Week)

Rotation of the umbilical loop: counterclockwise by 270° of the superior mesenteric artery formed axis

In non-rotation, the colon is on the left and the small intestine on the right. Malrotation is the most frequent incomplete rotation with caecum located in front of the duodenum.

In the setting of gastrointestinal anomalies, the ultrasound examination of the fetus is essential, and a sonolucent cystic structure is seen in the upper left quadrant of the abdomen of the conceptus from 9 weeks of gestation. An ordinarily uniform echogenicity of the bowel is recognized until the third trimester. At this time, prominent meconium-filled loops of large bowel are usually seen. The vast supply of oxygenated blood is responsible for the size of the fetal liver, which fills most of the upper abdomen, and for the largest dimension of the left liver, which will be not the case after birth. By ultrasound, the gallbladder is represented as an ovoidal cystic structure located in the right upper abdominal quadrant and located below the intrahepatic portion of the umbilical vein. To the left of the fetal stomach, it is possible to visualize the spleen. The abdominal circumference of the fetus is measured using a cross section of the abdomen visualizing the stomach and the portal sinus of the liver. In the case of *situs viscerum solitus*, the relative positions of the stomach, hepatic vessels,

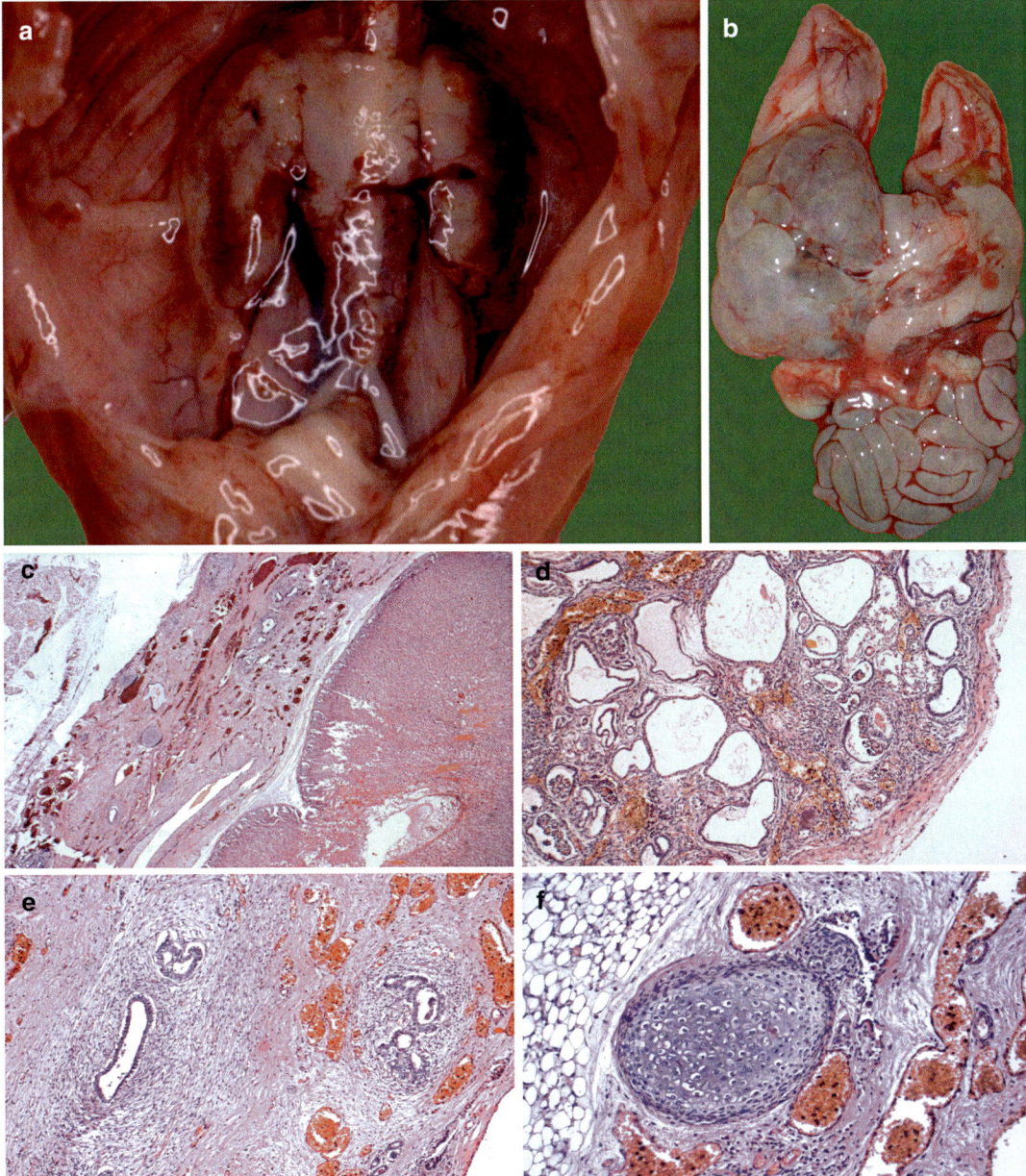

Fig. 18.35 Renal Agenesis & Cysts. The examination of the kidneys and urogenital tract is crucial to identify cases of oligohydramnios. In Figure (**a**) is a patient with renal agenesis with two relatively normal adrenal glands lying on both psoas muscles. Figure (**b**) depicts unilateral cystic dysplasia on the right side, which is accompanied by an agenesis of the left kidney. Figures (**c–f**) show the histopathology of renal aplasia with minimal renal blastema (**c**), cystic renal dysplasia (**d**), and multicystic dysplastic kidney (**e, f**) with primitive ducts surrounded by early mesenchyma in Figure (**e**) and primitive cartilage (island) in Figure (**f**). All microscopic figures have been taken from histological slides stained with hematoxylin and eosin (c-f, original magnification x40)

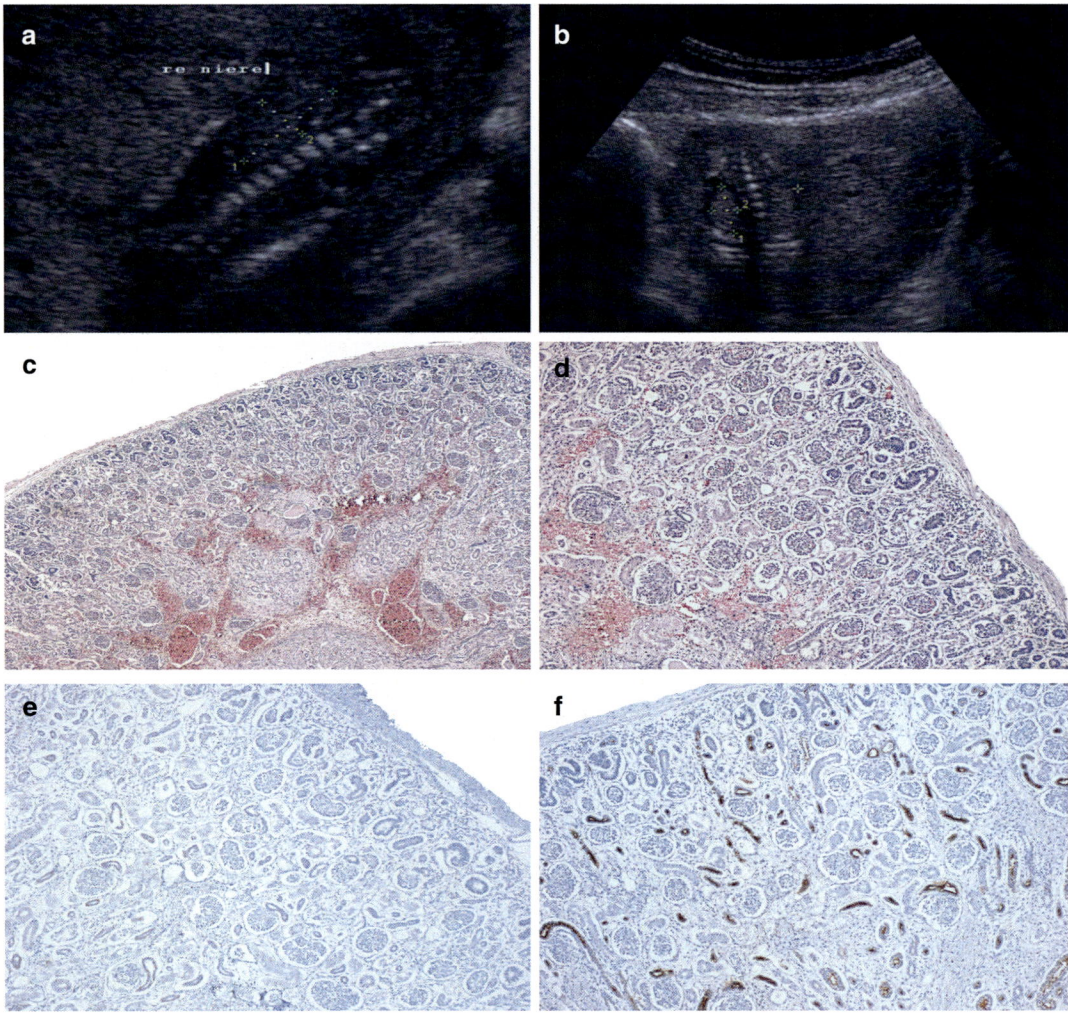

Fig. 18.36 Renal Tubular Dysgenesis. This Figure discloses the characteristic ultrasound and histopathology of renal tubular dysplasia (see text for details). All microscopic figures have been taken from histological slides stained with hematoxylin and eosin, apart from e (anti-CD15, avidin-biotin complex) and f stained by immunohistochemistry

abdominal aorta, and inferior vena cava (IVC) are recorded in the OB/GYN report.

Congenital Hand Anomalies

Congenital hand anomalies have been classified variably and mostly according to the observed form of the hand. However, a classification has no sense, if there is no connection to either mechanistic or surgical aim. Several names combining Greek words for the etymology have been used and are often appropriately used. Some examples are the following, e.g., brachydactyly for short digits, polydactyly for multiple fingers, macrodactyly for big digits, ectrodactyly or oligodactyly for defective digits, and syndactyly for fused fingers. The Greek words are βραχύς for short, πολύς for multiple, μακρός for large, ἐκτρώμα for miscarriage, ὀλίγος for few, σύν for fusion, and δάκτυλος for fingers. There are various types of ectrodactyly or split hand, cleft hand, including those with constriction rings; those amputation-like; those with bony, hypo-, or aplasia; and those with defects on the radial or ulnar side of the upper limb. Ectrodactyly involves the absence or deficiency of one or more central digits or toes of the hand or foot, respectively. Moreover, ectrodactyly is known as split hand/split foot malformation (SHFM).The teratogenic insults that cause the effects (abnormal hand) are not often evident at the time of the clinical observation. A quite useful classification has been proposed in Japan. There are at least four diverse types of teratogenic mechanisms of congenital defects of the digits.

1. Longitudinal deficiencies due to mesenchymal cell death in an early developmental stage
2. Cleft or split hand due to abnormal induction of digital rays in the hand plate
3. Constriction band syndrome due to amputation of digital radiations (ADAM or Amniotic Deformity, Adhesions, Mutilations Complex)
4. Transverse deficiency type due to a still unknown teratogenic event

ADAM complex is also reported in the literature with various names, including amniotic band syndrome, amniotic band sequence, amniotic bands, amnion rupture sequence, congenital constriction rings, constriction band syndrome, Streeter anomaly, Streeter bands, Streeter dysplasia, and TEARS (The Early Amnion Rupture Spectrum). It has also labeled LBWC or Limb body wall complex, although it seems to be a different condition. Pseudoainhum is also a term used for constrictions bands. In the African Yorub language, ainhum means "to saw" and, in the Brazilian patois, means "fissure" (Brodell and Helms, 2012).The gradients of signaling molecules in three spatial dimensions, proximo-distal (shoulder-finger direction), anteroposterior (thumb-little finger direction), and dorsoventral (back-palm direction), are at the basis of developmental patterning of the limbs. Three specialized cell clusters of fundamental importance are the apical ectodermal ridge (AER), the progress zone (PZ), and the territory or zone of polarizing activity (ZPA). The production of signaling molecules is crucial in determining the fate of neighboring cells to proliferate or to differentiate into a cell type. Truncations of all skeletal elements of the limb (stylopod, zeugopod, autopod) originate from a failure to initiate the AER. Genetic anomalies, as well as environmental factors, may cause ectrodactyly following interference with AER. Environmental factors that are known to induce ectrodactyly in rodents include retinoic acid, cadmium, hydroxyurea, cytarabine, methotrexate, ethanol, caffeine, cocaine, valproic acid, acetazolamide, and methoxyacetic acid. There are several factors, including fibroblast growth factors (FGFs), bone morphogenetic proteins (BMPs), WNT signaling molecules, and homeobox-containing proteins, such as MSX1 and MSX2. Mesodermal signaling induces the formation of AER to the overlying ectoderm. Several FGFs are restricted to the AER: FGF4, FGF8, FGF9, and FGF17. These AER-FGFs are crucial for limb development. In mice, simultaneous conditional ablation of Fgf4 and Fgf8 is compatible with normal AER initiation. However, there is defective gene expression in the underlying mesenchyme. Up to E11.25, the AER is maintained; since then the AER begins progressively to degenerate. The Fgf4 and Fgf8 double-knockout animals present with aplasia of both proximal and distal limb elements. This finding may be clarified by a reduction of mesenchymal cells in the limb bud. Molecular compounds widely diffuse worldwide may be associated with teratogenic defects, and those defects that are visualized on the limbs may be more recognizable than others that need imaging.

Migration Defects

Migration defects of ganglion cells usually occur in the 11th-week post-ovulation. The most critical migration defect of ganglion cells occurs distally in the rectosigmoidal region in about 70% of cases of Hirschsprung disease. In 2016, a particular and noteworthy anniversary recurred that had an enormous impact on clinical pediatrics, pediatric surgery, and pediatric pathology. Professor Harald Hirschsprung (1830–1916), a Danish pediatrician, who first described the congenital megacolon, died about one century ago (Sergi 2015). Hirschsprung is especially acknowledged as the writer of the first description of two children who died of intestinal obstruction called "congenital megacolon," which is now known as Hirschsprung disease (HSCR). Dr. Hirschsprung was working as a pediatrician at Queen Louise Children's Hospital in Copenhagen, Denmark, when he came across two patients with this disease. The examination of the scientific literature seems to indicate that about 20 similar cases were recorded in the nineteenth century. However, Hindu surgeons of prehistoric India had considerable knowledge about HSCR. Currently, HSCR is the most common cause of neonatal lower intestinal obstruction occurring in 1:5000 live birth newborns. HSCR involves in about three-quarters of cases male children, and its incidence is variable according to ethnics. The caudal migration of the primordial neural crest cells starts at the upper end of the gut following the vagal fibers progressively toward the distal end. A delay or arrest in this cellular migration induces failure of the neural crest cells to reach the desired distal bowel. The fate of this non-migration is an abnormal nerve innervation of the bowel. In HSCR, there is a caudo-cranial severity, which means from the internal anal sphincter extending proximally for a variable length of the gut. In HSCR, the proximal bowel is dilated and progresses to an abrupt or gradual transition to a normal calibrated distal bowel. Conversely, the distal bowel segment shows a funnel-like or cone-shaped zone in between (the so-called "transition zone"). Moreover, there is proximal muscle hypertrophy. The bowel becomes distended with thickening of its wall, and the degree of dilatation and hypertrophy depends intricately upon both the time and degree of obstruction and, indirectly, to the age of the patient. Clinical presentation settings include failure to pass meconium within the 24 h of life considering that 98% of newborns pass meconium in less than 24 h of age, neonatal intestinal obstruction syndrome (abdominal distension, refusal to be fed, and vomiting of bilious type), and recurrent enterocolitis (mainly <3 months of life), toxic megacolon, spontaneous perforation, and chronic constipation. Toxic megacolon includes fever, abdominal distension, bile-stained vomiting, explosive diarrhea, dehydration, and shock. History comprises failure to pass meconium, painless abdominal distension, and, naturally, constipation. The myenteric plexus (or Auerbach's plexus) provides motor innervation to both muscular layers of the tunica muscularis of the gut, harboring both parasympathetic and sympathetic input, while the submucous plexus has only parasympathetic fibers. It provides secretomotor innervation to the mucosa of the bowel. Box 18.47 lists the migration defects of ganglion cells.

A crucial pediatric pathology aspect to keep in mind is the maturity of the ganglion cells. The ultrashort Hirschsprung disease (US-HSCR) is a very controversial issue. It is a very short segment of aganglionosis extending 2–4 cm proximal to the internal anal sphincter. In US-HSCR the degree of constipation may be less severe, and the complications of growth retardation and enterocolitis are rare. The diagnosis of US-HSCR is established by taking two biopsies, of which one should be just proximal to the dentate line showing aganglionosis. The lack of ganglion cells at this location helps to differentiate US-HSCR from internal anal sphincter achalasia. This last condition has similar findings on anorectal manometry, but in achalasia ganglion cells are present. The second biopsy should be gathered approximately 4 cm above the internal sphincter. In US-HSCR, this second biopsy shows normal ganglion cells. The presence of ganglion cells at this location helps to differentiate the US-HSCR from classic HSCR, in which ganglion cells would be absent.

Another migration defect significant in perinatal and pediatric pathology is migration anomalies of the pancreas, which has been an object of study by several scientists of the German School of Embryology (Box 18.48).

Resorption defects may occur at the 5th-week post-ovulation (Box 18.49).

The allantois is a part of the developing conceptus consisting of a small sac-like structure between the first urinary bladder and the chorionic plate of the placenta. It is an endodermal outpouching of the developing fetus, which becomes surrounded by the mesodermal connecting stalk. The urachus allows the communication from the urinary bladder to the allantois, which

Box 18.47 Migration Defects of Ganglion Cells (11th Week)
Hirschsprung disease (HSCR): Incomplete migration and colonization of the parasympathetic ganglion cells in the plexus of the submucosa and myenteric plexus of the digestive tract

- Zeulzer-Wilson syndrome (total colonic aganglionosis)
- Long-segment HSCR (aganglionosis distally continuing up to splenic flexure [mostly])
- Short-segment HSCR (aganglionosis distally involving up to sigmoid colon)
- Ultrashort-segment HSCR (aganglionosis below the rectosigmoid junction)

Box 18.48 Pancreas Ontogenesis and Pancreas Anomalies (5th–7th Weeks)
1. Formation of two endodermal buds ("*Knospen*") from the duodenum
2. Migration of the ventral bud ("*ventral Knospenwanderung*") toward the dorsal side and fusion ("*Verschmelzung der wandernden ventralen Knospen mit der dorsalen Pankreasanlage*") of the first German Embryological School
3. Joining of the pancreatic ducts ("*Vereinigung der Pankreasgänge*") with the formation of two significant ducts, Wirsung duct or *ductus pancreaticus major*, which represents the joining of the dorsal and ventral ducts, and Santorini's duct or *ductus pancreaticus accessorius*, which is the proximal dorsal duct
4. *Pancreas anulare*, when there is an abnormal migration of the ventral bud

Box 18.49 Resorption Defects (5th Week)
- *Ductus omphalo-entericus* (5th week)
- *Meckel diverticulum*: Incomplete resorption of the omphalos-enteric duct

collects nitrogenous waste. Allantois regresses and stays as a vestigial organ in the umbilical cord containing the umbilical arteries and the umbilical vein. Humans and other mammals are identified as amniotes due to the presence of the allantois, amnion, and chorion. However, *amphibians and fishes* lack this structure among vertebrates (Box 18.50).

Fetopathies
A large chapter of embryology is the development of the fetuses following the embryonal stages. Abnormalities occurring at this stage of the human development are summarized and collected under the heading of fetopathies. Box 18.51 summarizes some of the most important features regarding fetopathies. In this definition, two conditions need to be present to be called fetopathies, i.e., (1) completed embryogenesis and (2) an insult or damage, which occurs in the fetogenesis. Fetopathies may also be classified according to the etiology. The most frequent etiological factors are infections of the fetus with protozoa, bacteria, or viruses, metabolic diseases (e.g., diabetes mellitus), hemolytic disease of the

Box 18.50 Resorption Defects of the Allantois (8th Week) and Urachus Anomalies
Allantois-diverticulum: Extrinsic evagination (outpouching) of the yolk sac ("*Dottersackausstülpung*") that may be considered a rudiment in humans in the form of the urachus or umbilical ligament (*ligamentum umbilicale medianum*)

- Urachus-cyst, when the urachus closes at both the umbilicus and the urinary bladder but remains patent between these two endpoints
- Urachus-diverticulum, when the urachus closes at one end (umbilicus/urinary bladder) and remains patent
- Urachus-fistula, when the urachus remains open at both the umbilicus and the urinary bladder and remains patent between these two endpoints

> **Box 18.51 Fetopathies**
> - Abnormal development of the fetus in consequence of an insult/damage during the fetogenesis following a completed and theoretically unremarkable embryogenesis
> - Fetogenesis: Detailed differentiation of the single organs/tissues of the completed embryo ("*Organsystemausdifferenzierung*" of the German embryological school)
> - Forms of reactions of the fetus to insults/damaging factors
> - The primitive mesenchymal response of defense (e.g., granulation tissue, migration, and colonization of macrophages, fibroblasts)

newborn (aka "*Morbus hemolyticus neonatorum*"), and toxic factors (e.g., alcohol, cocaine).

Cerebral Cortex Maturation Defects

Maturation defects of the cerebral cortex are discussed in detail in the nervous system chapter.

Maturation defects of the cerebral cortex include:

- *Agyria*: Absence of the gyri ("*Großhirnwindungen*" of the German embryology school), such that the brain surface is smooth (aka "lissencephaly") (Greek ἀ- "not, without" and γῦρος "circle").
- *Pachygyria*: Low developed brain gyri are showing a ↓ # of sulci and gyri (Greek παχύς "thick" and γῦρος "circle").
- *Micropolygyria*: Atypical and cytoarchitectonic (four-layered) abnormal brain structure (Greek μικρός "small", πολύς "many", and γῦρος "circle").
- *Ulegyria* (aka encephaloclastic microgyria): secondary, scar-like, lamellar ↓ # of the gyri in consequence of a peri- or postnatal insult/damage (Greek οὐλή and οὐλαί "scar, wound scarred over" combined with γῦρος "circle").

Lung Maturation Defects

In the human embryo, development of the lung initiates as early as four weeks of embryonic life. This process does not stop at birth but continues into postnatal life 2 years up to 8 years. The structural and vascular development of the lung is closely related to this development of the architecture. The antenatal growth and development of the human lung are crucial for understanding numerous respiratory pathologies of the newborn and infant. There are five distinct but overlapping stages of lung development. Major features related to the air-blood barrier (ABB) may be compromised in more than one stage.

Stage of lung development	Major features
Embryonic stage (4–7 weeks)	Organogenesis start – major airways
Pseudoglandular stage (5–17 weeks)	Formation of the bronchial tree – acinus
Canalicular (16–26 weeks)	Completion of conducting airways
	Epithelial differentiation, first ABB, surfactant
Saccular stage (24 weeks-term)	Expansion of airspaces
Alveolar stage (36 weeks–2 years after birth)	Alveolarization (secondary septa)

The lung starts with the formation of a groove in the ventral lower pharynx, which is known as the sulcus laryngotrachealis (stage 10, ~ 28 days). A bud forms after a couple of days - from the lower part, which is universally considered the true lung primordium (primordium pulmonale). The lung progressively subdivides into the two main bronchi (stage 14, ~ 33 days) with some right-left asymmetry due to the topographically relevant cardiac development, being the left budding more laterally than the right bud running caudally more parallel to the esophagus. The endodermal branches with their subsequent divisions also occur asymmetrically with the right bud forming three areas of development, while the left bud forming two areas of development. The developmental disorders of the lung can be divided into (1) tracheobronchial anomalies, (2) anomalies of lung parenchyma, and (3) anomalies of the pulmonary vasculature.

Liver Maturation Defects
Through an orderly process of selection and deletion, the fundamental fetal biliary structure or ductal plate is remodeled during intrauterine life into the excretory system of bile ducts. Morphometric data indicate that a "slowdown" period of the progressive ramification of the intrahepatic biliary tree occurs between the 20th week and the 32nd week of gestation (Sergi et al. 2000 a-d). The growth of the portal tracts, described regarding cross-section circumference and encircled area, slows for a time interval between the end of the second trimester and the beginning of the third trimester. Accordingly, the development of bile ducts showed also that an acceleration of the process occurs before 20 weeks and after 32 weeks of gestation. The interim hematopoietic function of the liver and, in a special way, the intraportal granulopoiesis may play a role for this "slow down" of the liver development. Liver erythropoiesis dominates between 12 weeks of gestation and the beginning of the third trimester (25 weeks of gestation) when 50% of the blood cells are formed in the liver and 50% in the bone marrow. However, until the 32nd week of gestation, the liver plays a major role in both the hematopoiesis and the maternofetal exchange. It is now understood that only at this time do the hematopoietic cells in the liver begin to form islands out of a previously diffuse distribution. After birth, the bone marrow becomes the major site of production of both red and white cells series. In this study, an investigative approach including both an immunohistochemical procedure using antibodies directed against keratins and a computer-based image analysis system was set up to more carefully evaluate the development and the maturation of the primitive intrahepatic biliary system. A mixture of monoclonal antibodies AE1 and AE 3 should theoretically stain both biliary epithelial cells and hepatocytes because AE 1 reacts with CK 19 and AE 3 reacts with both CK 7 and CK 8 (Boxes 18.52 and 18.53).

> **Box 18.52 Type I or Acidic Cytokeratins or Keratins**
> - CK10, CK12-19
> - 40–56.5 kDa (LMW)
> - Chromosome gene location at 17q21.2

> **Box 18.53 Type II or Neutral-Basic Cytokeratins or Keratins**
> - CK1–CK8
> - 53–67 kDa (HMW)
> - Chromosome gene location at 12q13.13

In formalin-fixed tissue, the AE 1–3 cocktail reacts only with biliary cells. Probably, this is due to formaldehyde fixation of the tissue. In fact, the epitope recognized by AE 3 is changed by formalin fixation in a way that is less available for binding. Further, it differs from other keratin-related epitopes because it can only partially be "restored" by antigen-unmasking techniques. The development of bile ducts also showed that an acceleration of the process occurs before the 21st week of gestation and after the 32nd week of gestation. If a third-order polynomial function was used, both turning points (20 weeks and 32 weeks) of the curve were confirmed. Polynomial regression ($y = -d + cx - bx^2 + ax^3$) is $y = -1.2 + 0.19x - 0.009x^2 + 0.0001x^3$ and r^2: 0.83 and $p < 0.05$. Each cell type in the vertebrate body possesses cilia, which are highly conserved organelles projecting from the surface ubiquitously across species from nematodes to ancient protozoa. Cilia are complex and dynamic structures and classified as motile and non-motile subtypes. Both subtypes share a 25-μm micrometer cytoskeletal scaffold, the axoneme, which contains nine peripheral microtubule doublets, consisting of A and B tubules, either surrounding a central pair of microtubules (9 + 2 pattern) or lacking the central pair (9 + 0 pattern). The 9 + 0 motile and non-motile cilia exist as single monocilia on the cell surface, while motile 9 + 2 cilia exist as multiple cilia (multi-cilia). Practically, all cells of the human body have a single non-motile (primary or sensory) cilium, whereas multi-cilia are generated by specialized cells only. The term "ciliopathies" has been used to label a group of diseases or complex multisystem human disorders with cilia as etiopathogenetic momentum (driving power). Momentum refers to the quantity of motion that a specific object possesses. If an object is on the move (in motion), then it has momentum. There exist cilia with involvement of all the major organs, and they contribute subtypes to many common diseases such as retinal dystrophy and kidney disease, but ciliopathies are an expanding group including many syndromes such as primary ciliary dyskinesia (PCD), autosomal dominant and recessive polycystic kidney diseases (ADPKD, ARPKD), nephronophthisis (NPHP), Leber congenital amaurosis (LCA), and Bardet-Biedl syndrome (BBS), Senior-Løken syndrome (SLS), Joubert syndrome (JBTS), Jeune syndrome (or asphyxiating thoracic dystrophy, JATD), short rib-polydactyly syndrome (SRPS), Meckel-Gruber syndrome (MGS), and lethal and nonlethal oro-facial-digital (OFD) syndromes.

Dysgenesis, Dysgenetic Proliferations, and Embryonic Neoplasms

At the beginning of the chapter, we tried to address the importance of knowledge of embryogenesis for the histopathologist. In Box 18.54, there is some representation of the continuum between dys-embryogenesis or altered embryogenesis and neoplasms or new proliferations considering the concepts and teaching of the German School of Embryology and Pediatric Pathology ("*dysgenetische Geschwulst*") (Fig. 18.37).

The etymology of the single words may be helpful in the process of recalling the different concepts regarding genetic counseling or treatment. It is important to stress that a teratoma is a type of germ cell tumor composed of tissue derived from at least two germ cell layers.

Box 18.54 Dysgenesis, Dysgenetic Proliferations, and Embryonic Neoplasms

- *Hamartia*: Autochthonous tissue defect of formation ("*Gewebsfehlentwicklung*")
- *Hamartoma*: Tumorlike further autonomous development ("*Gewebsfehlentwicklung*" with "*Weiterentwicklung*") of the autochthonous tissue in the new surrounding area
- *Hamartoblastoma*: The neoplastic proliferation of hamartia or hamartoma in the sense of a dysregulated, disinhibited ("*enthemmten*"), autonomous growth of autochthonous tissue
- *Choristia*: Scatter of embryonal primordium ("*Versprengung embryonal Anlagen*")
- *Choristoma*: Tumorlike further autonomous (independent) development of scattered primordial tissue in the new surrounding area
- *Choristoblastoma*: The neoplastic proliferation of a choristia or choristoma in the sense of a dysregulated, disinhibited ("*enthemmten*"), autonomous growth of scattered primordial tissue
- *Terata*: Simple or complex malformations ("*einfache bzw. komplexe Fehlbildungen*")
- *Teratoma*: Tumorlike autonomous development of abnormal primordial tissue in the new surrounding area composed of different germ layers (mono-/bi-/triphasic teratomas)
- *Teratoblastoma*: Teratoma with a malignant component

Notes: A "monophasic" teratoma, which by definition does not exist may be considered a choristoma

18.6 Birth Defects

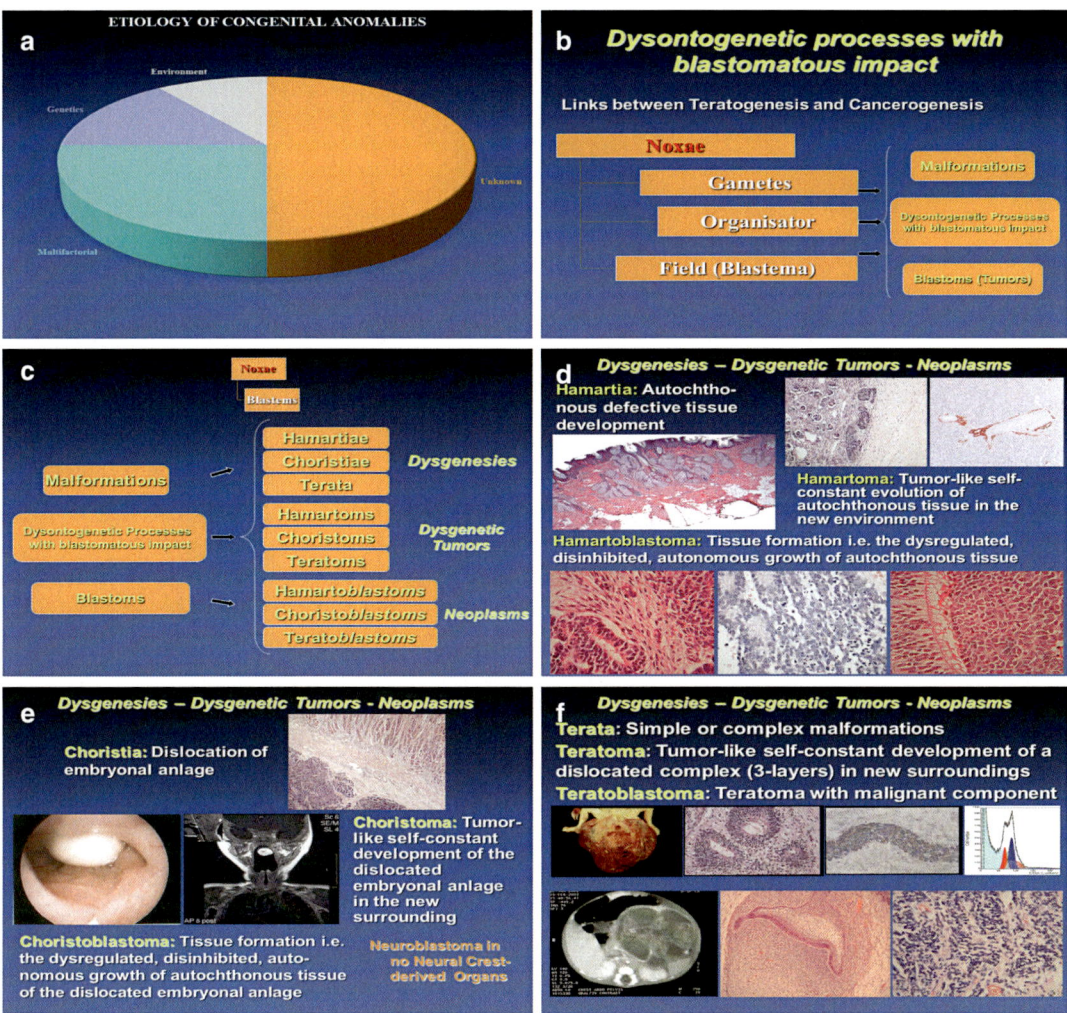

Fig. 18.37 Teratogenesis and Oncogenesis. Teratogenesis addresses the differences identified between defects and proliferative processes with or without a malignant component (see text for detail)

Each combination (ectoderm + endoderm OR mesoderm + ectoderm OR mesoderm + endoderm) is valid. More often, all three germ cell layers are involved (ectoderm, mesoderm, and endoderm). Teratomas can occur everywhere. They are usually found in the ovary or testis and may be either benign or malignant. It is called malignant teratoma, a solid tumor harboring a worrisome prognosis and composed of immature embryonal or extra-embryonal elements derived from all three germ layers characterized by malignant histologic features. Teratoma is a true neoplasm in which none of its tissue needs to be native to the area in which it occurs (Figs. 18.38, 18.39, and 18.40). Also, teratoma differs in prognosis according to the organ involved and, specifically, to the degree of maturation of the tissues. Remarkably, teratomas in the ovary are usually benign dermoid cysts, while those in the testis may portend a

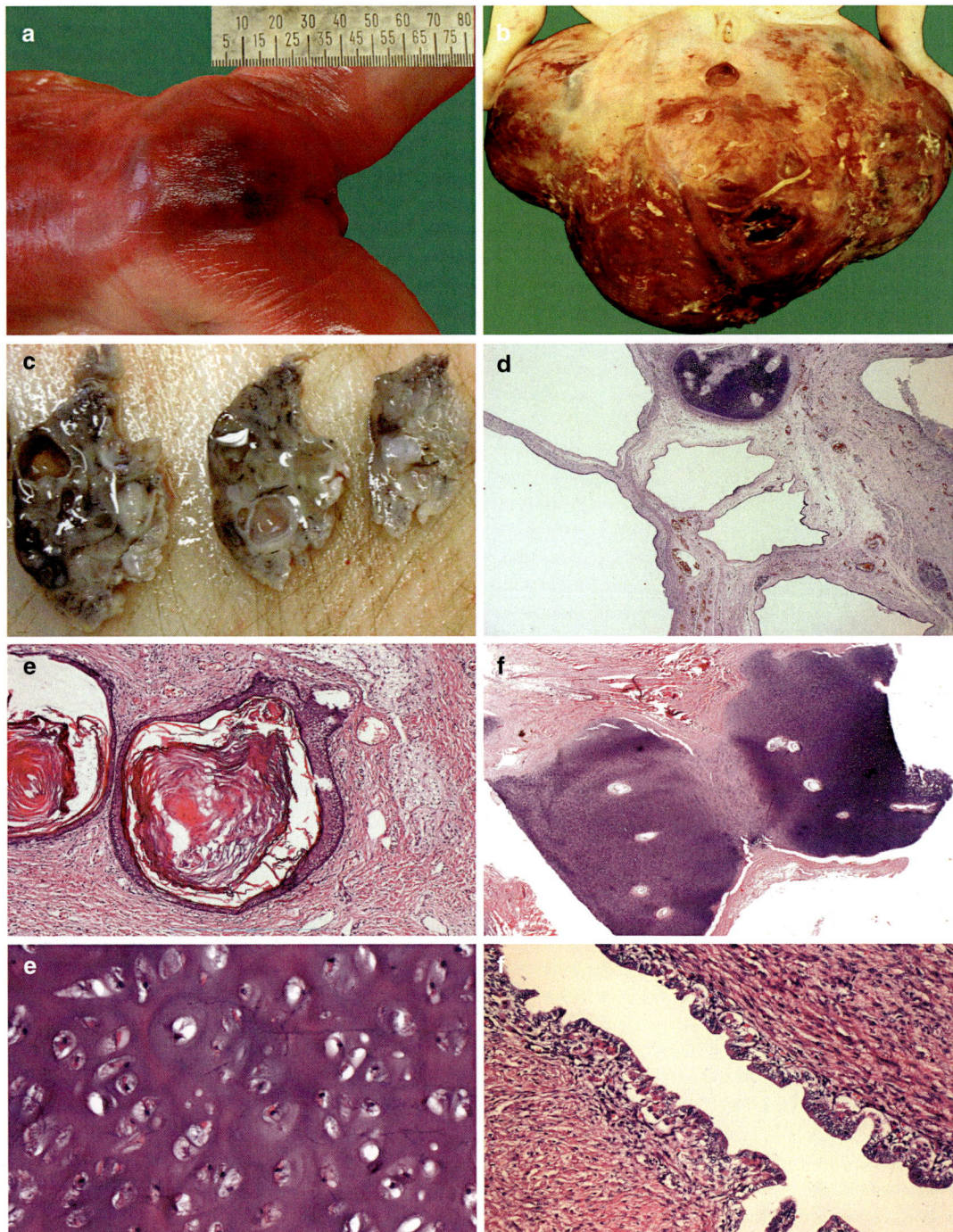

Fig. 18.38 Sacrococcygeal Teratoma. Minimal bulging of sacrococcygeal teratoma (**a**) and very large sacrococcygeal teratoma (**b**). The tumor of (**a**) shows a cystic degeneration grossly in (**c**) and ectodermal, mesodermal, and endodermal derivatives in (**d**) through (**h**). All microscopic figures have been taken from histological slides stained with hematoxylin and eosin (d, original magnification x20; e, original magnification x100; f, original magnification x20; g, original magnification x400; h, original magnification x200)

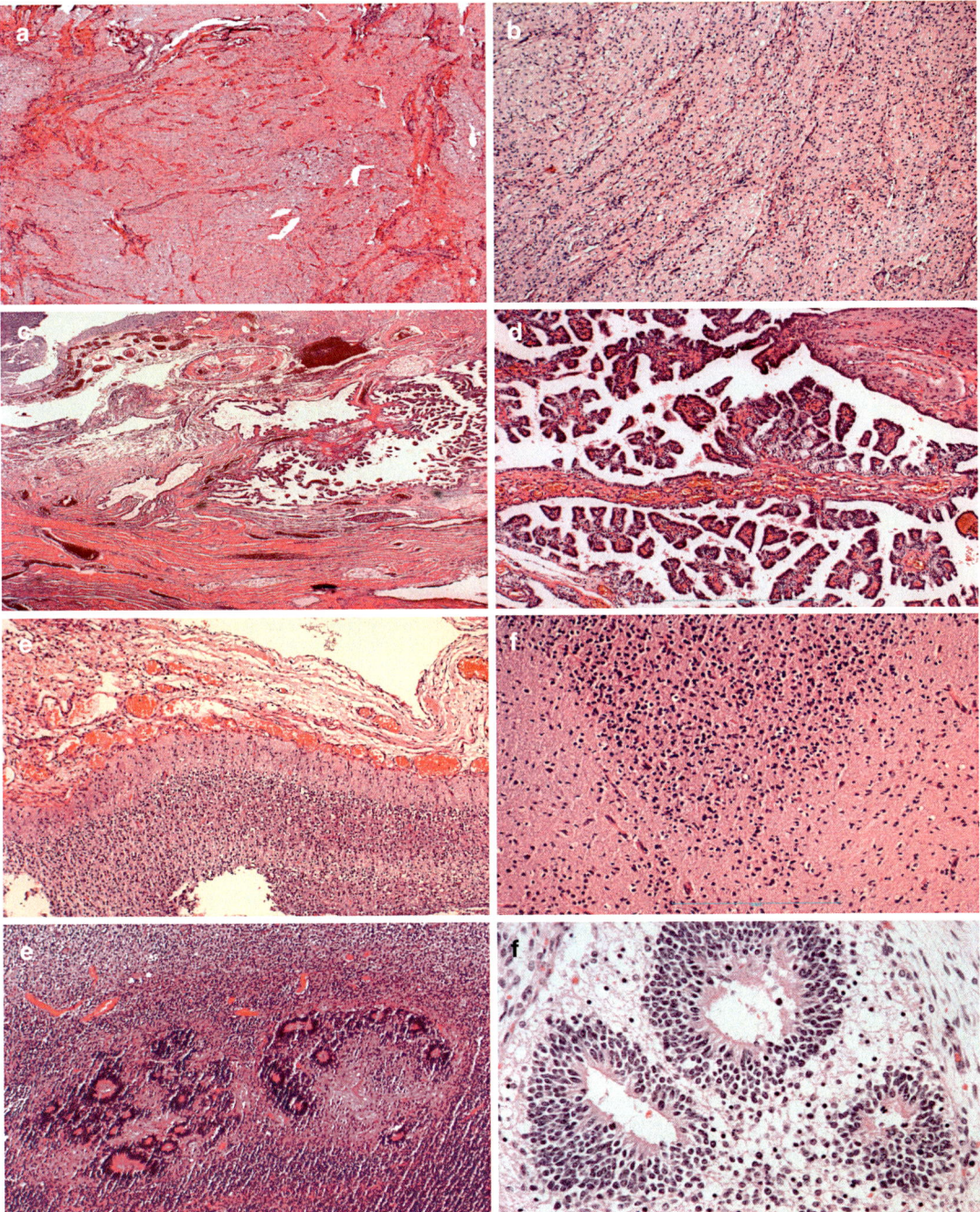

Fig. 18.39 Sacrococcygeal Teratoma. Teratoma with glia and nervous tissue (**a** and **b**), choroidal plexus (**c** and **d**), brain (**e** and **f**), and primitive neural tube (**g** and **h**) (a, original magnification x40; b, original magnification x100; c, original magnification x20; d, original magnification x100; e, original magnification x100; f, original magnification x200; g, original magnification x100; h, original magnification x400)

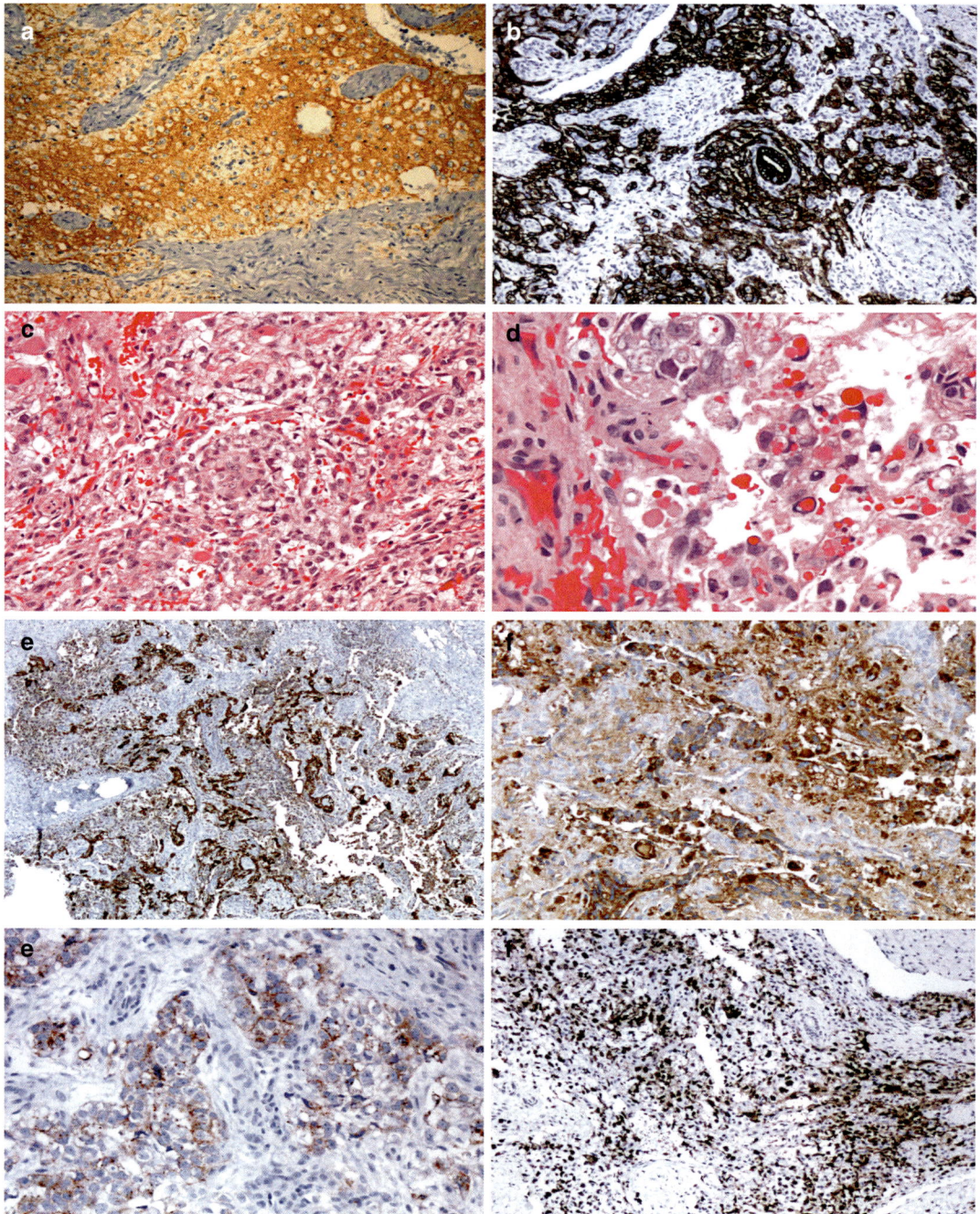

Fig. 18.40 Sacrococcygeal Teratoma. Teratoma with several components including yolk sac tumor. In (**a**) there is neuroglial tissue stained with GFAP by immunohistochemistry. In (**c**) and (**d**), there is yolk sac tumor (H&E stain), which is positive for cytokeratin (AE1-3) in (**b**), and for alpha-feto-protein in (**e**) and (**f**), as well as for CD117 (c-kit) in (**g**). In (**h**) is shown the proliferation rate of the yolk sac component (anti-Ki67; all immunohistochemical microphotographs were carried out using the avidin-biotin complex) (a, original magnification x100; b, original magnification x100; c, original magnification x200; d, original magnification x400; e, original magnification x50; f, original magnification x200; g, original magnification x200; h, original magnification x100)

less favorable outcome. By light microscopy, the cysts are lined by either stratified squamous epithelium or simple ciliated epithelium, and skin appendages. Other tissues, such as cartilage, bone, and mature brain, can also be seen. Teratogenesis and oncogenesis are two closely related biologic events. Some agents or first hits may induce both malformations and tumors. One of the most important aspects of neonatal neuroblastoma is the association in up to 1/3 of cases with congenital malformations, such as Beckwith-Wiedemann syndrome, esophageal atresia, tracheoesophageal fistula, hypoplastic left heart, Ondine curse, Hirschsprung disease, and von Recklinghausen neurofibromatosis. Berry et al. (1970) have given a challenge to the concept of a strict relationship between neoplasia and malformation. They found no significant association of malformation with neuroblastoma, Wilms' tumor, hepatoblastoma, and teratoma. Probably, the relationship between neoplasia and malformation is more complicated than simply thought. Although the molecular biology of the events inducing malformations/neoplasia remains to be fully investigated, some authors have postulated on the intriguing role of some oncogenes, anti-oncogenes, transcription factors, and growth factors. Another interesting and debated aspect of congenital tumors is in some cases the relatively benign outcome of infants harboring a histologically immature or even malignant appearing neoplasm. The spontaneous regression of neuroblastoma IVS is well known. To explain the enigmatic features of some congenital tumors, specifically regressive influences existing in the fetus and young baby, may modify or suppress the progression of unfavorable tendency. Lesser aggressive congenital teratomas, neuroblastomas, and sarcomas were evidenced in neonates than in older children. Interestingly, this representation of the proliferation, originally taught from Professor Goerttler on the teaching of early embryologists and pediatric pathologists or anatomists in Baden-Württemberg (Germany), is still truly actual in the genomic era. Neoplasm means neo-, i.e., new proliferation, and should be considered as such if an autonomous growth is intrinsically present in the neoplastic cells.

18.6.4 Birth Defects: Etiology (Mendelian, Chromosomal, Multifactorial)

According to the etiology, it is useful to classify the congenital disabilities as Mendelian, chromosomal, and multifactorial. Causes of malformations are mostly single gene disorders in 8% of the cases, chromosomal in 6%, multifactorial in 20%, and environmental in 5% of the cases.

Mendelian-Based Birth Defects

Mendelian-based birth defects are linked to either autosomal dominant or recessive inherited diseases due to alteration of genes located in nonsexual chromosomes or X-linked disorders due to change of genes located on heterochromosomes. Numerous genetic syndromes are included in atlases of dysmorphology. Some genetic syndromes are depicted in this book and include Cornelia de Lange syndrome (Fig. 18.41), Smith-Lemli-Opitz syndrome (Fig. 18.42), Fryns syndrome (Fig. 18.43), limb-body wall complex (Fig. 18.44), and Fraser syndrome (Fig. 18.45).

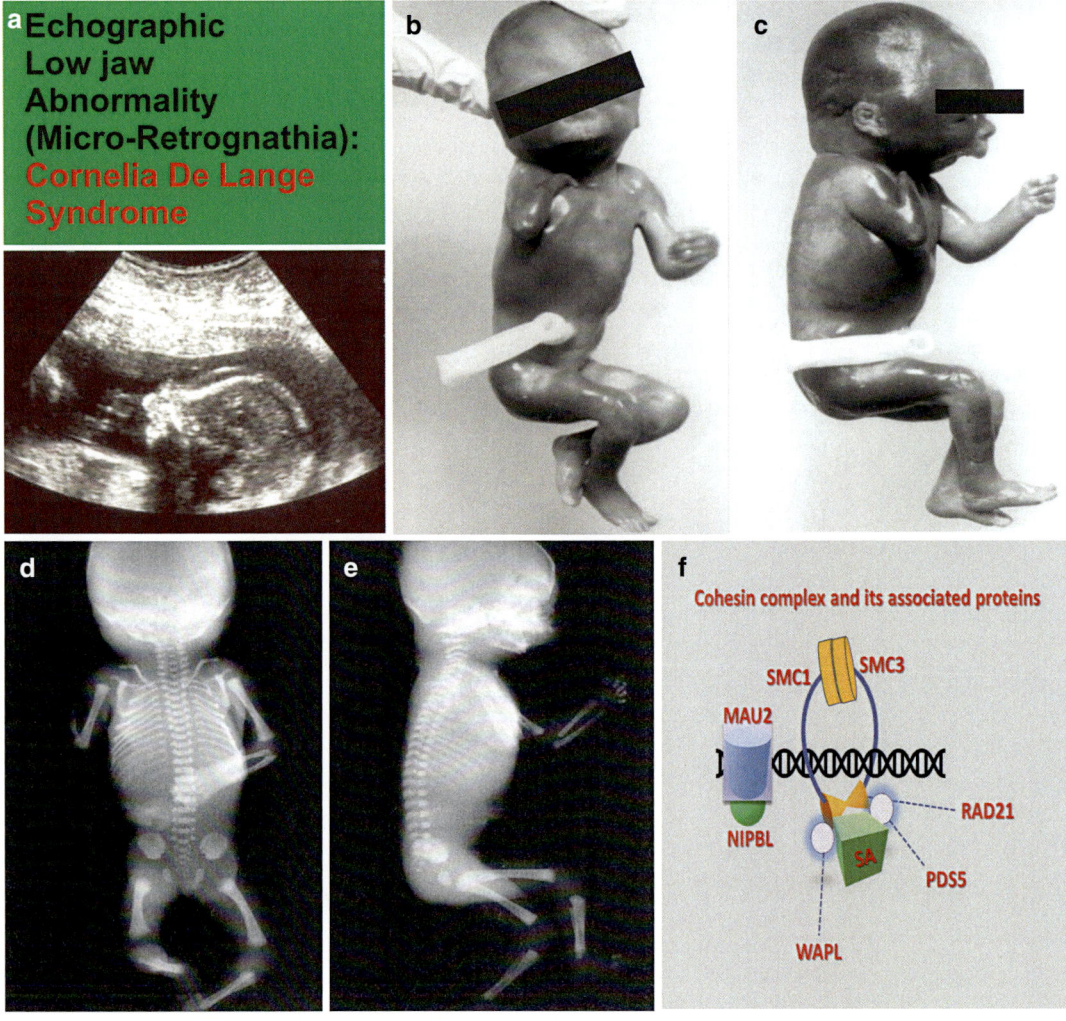

Fig. 18.41 Cornelia De Lange Syndrome. This Figure showing ultrasound, gross, and radiological pictures of a patient with Cornelia de Lange syndrome with low jaw microretrognathia (**a**), upper limb shortening and malformation (**b** and **c**), radiological changes of the limb abnormalities and micro- and retrognathia, and the cohesin complex and its associated proteins (**f**)

18.6 Birth Defects

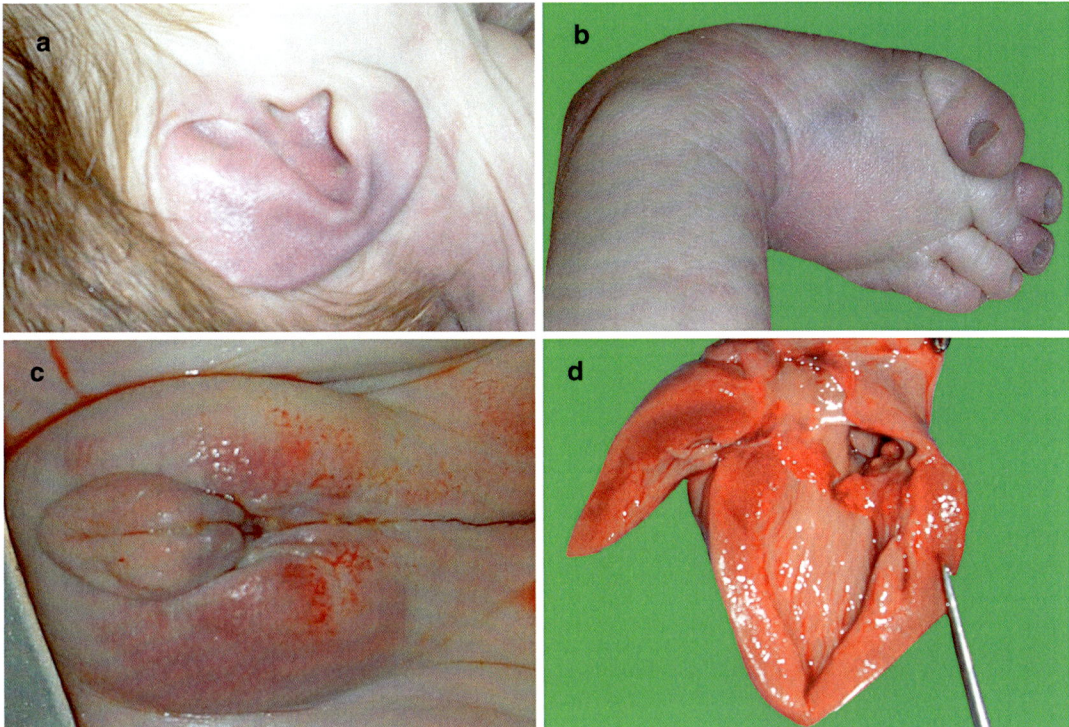

Fig. 18.42 Smith-Lemli-Opitz syndrome. Patient with characteristic features of Smith-Lemli-Opitz syndrome, which is an inborn error of cholesterol synthesis, with autosomal recessive inheritance, caused by a mutation in the enzyme 7-dehydrocholesterol reductase, or DHCR7, with dysplastic ears (**a**), syndactyly of the toes (**b**), dysplastic external genitals (**c**), and atrioventricular septal defect of the heart (**d**)

Autosomal Dominant (AD) Diseases

- AD gene mutations affect either structural or regulatory active gene products.
- 1/2 of offspring inherits the gene carrying the mutation.
- Both penetrance and expressivity may be variable.
- Clinical features may start later than clinical characteristics of individuals carrying gene mutations for autosomal recessive (AR) diseases.
- Categories of affected proteins: Regulation of complex metabolic pathways, critical structural proteins, and proteins harboring a gene mutation-related gain-of-function effect.
- Example: Marfan syndrome.

Marfan Syndrome

- AD-inherited disease characterized by genetic pleiotropism involving a disorder of connective tissues of the body, mainly skeleton, eyes, and cardiovascular system.
- Prevalence: 1:5000 (men/women of all ethnic groups).
- Familiarity rate: up to 85% of cases with great variability in the gene expression.
- Gene 1: Fibrillin-1 (*FBN1*) – >500 mutations (mostly missense, a type of non-synonymous substitution).
- Gene 2: Fibrillin-2 (*FBN2*) – less common.
- Protein (structure): Extracellular, glycoprotein, a major component of microfibrils of extracellular matrix (ECM).

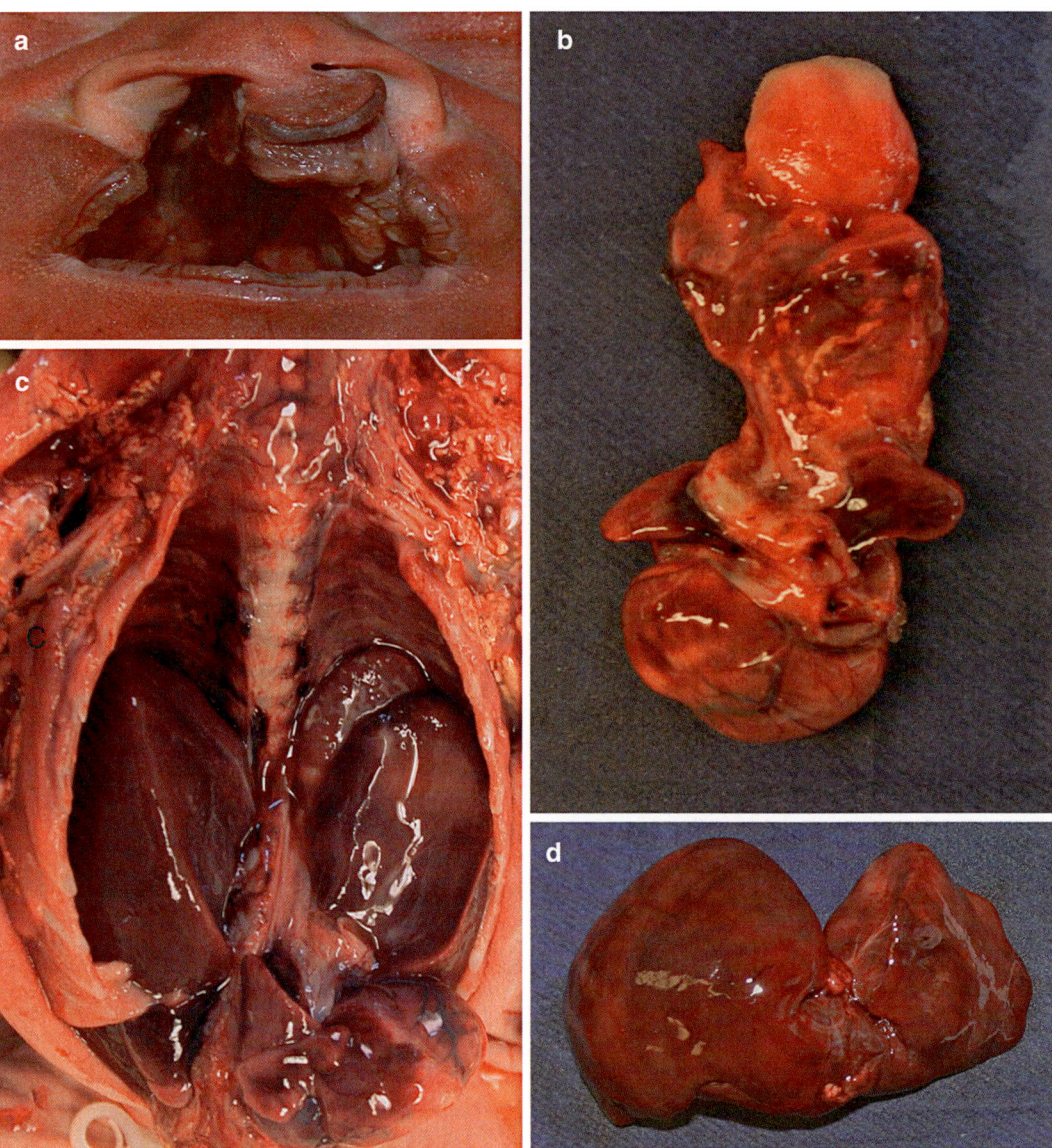

Fig. 18.43 Fryns Syndrome. Patient with Fryns syndrome, multiple congenital anomaly syndrome characterized mostly by dysmorphic facial features, congenital diaphragmatic hernia, pulmonary hypoplasia, and distal limb hypoplasia showing cleft lip and palate (**a**), pulmonary hypoplasia bilaterally (**b**), congenital diaphragmatic defect (**c**) with location of the abdominal organs in the thorax, and consequent abnormal shape of the liver (**d**) with potential obstruction of the excretory function of the biliary system. In Fryns syndrome, additional anomalies include gastrointestinal malrotation, anal atresia, omphalocele, urogenital (renal cysts, ureteral dilation, cryptorchidism), and musculoskeletal (talipes, broad clavicles) and brain malformations (ventricular dilation, hydrocephalus). COD is mainly due to complications of congenital diaphragmatic defect and pulmonary hypoplasia. Several chromosomal aberrations including microdeletions involving chromosome bands 15q26.2 and 8p23.1 have been reported in probands with FS

18.6 Birth Defects

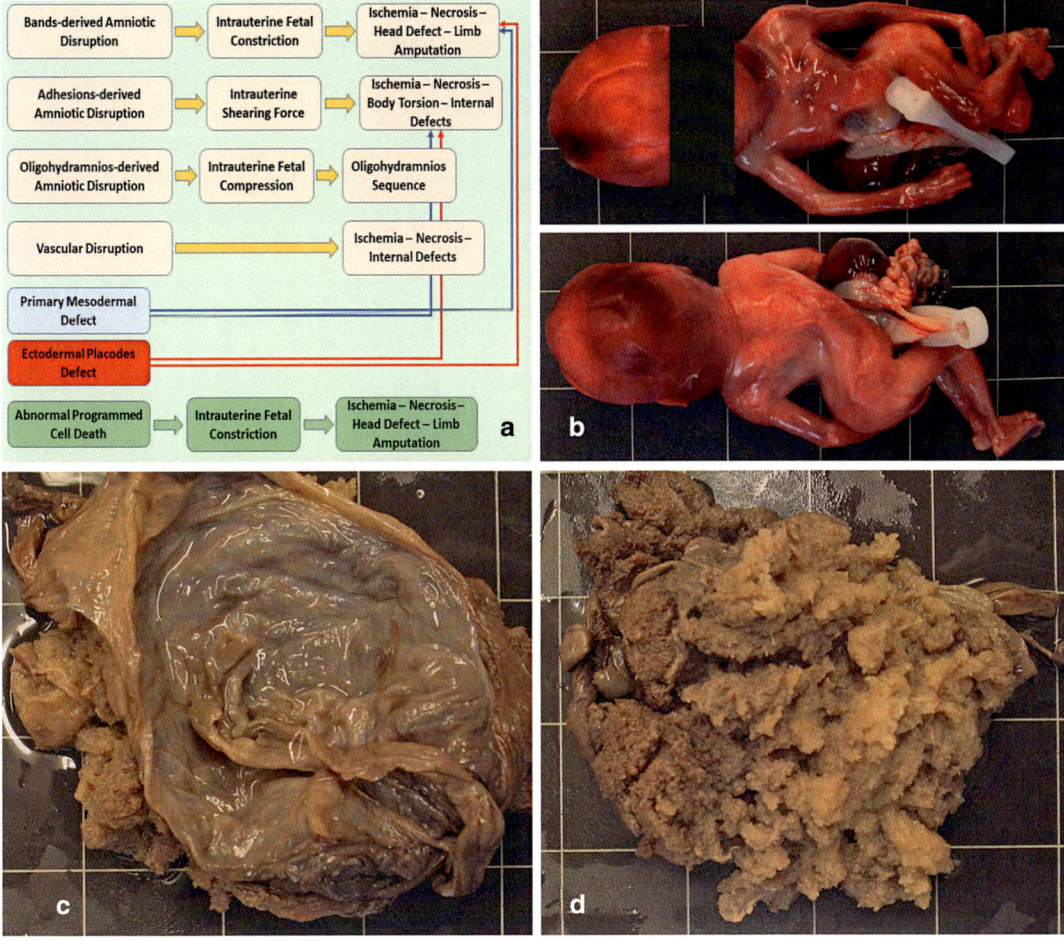

Fig. 18.44 Limb Body Wall Complex. Limb Body Wall Complex (LBWC) is a condition characterized by multiple, severe congenital abnormalities in a fetus. They result in openings in the anterior body wall and defects of the limbs with additional features including facial clefts, short or missing umbilical cord, scoliosis, neural tube defects, and abnormalities of the urogenital organs. In Figure (**a**) there is a pathophysiologic flowchart. In (**b**) and (**c**), a fetus with LBWC is shown, while (**d**) and (**e**) show the placenta with short umbilical cord and disrupted basal plate

- Protein (action): Scaffolding unit on which topoelastin is deposited forming elastic fibers.
- *FBN2* gene mutations ⇒ congenital pattern with fingers (and hands) abnormally long and slender with fingers (and hands) abnormally long and slender (arachnodactyly, Greek ἀράχνη "spider").
- Arthro-skeletal: tall individuals with long extremities (CRL/CHL lower than the norm), lax ligaments (double-jointed), kyphosis, scoliosis, chest deformities (pectus excavatum or pigeon breast).
- Ocular: Lens subluxation or dislocation, bilaterally.
- Cardiovascular: mitral valve prolapse, ascending aorta aneurysm (AAA).
- COD: Rupture of the AAA with internal hemorrhagic shock, cardiac failure due to aortic valve incompetence, and left ventricular insufficiency, sudden cardiac death (SCD), traumas, and car accident.

Of note, the aneurysm of the ascending aorta may become dissecting, and dilation may involve

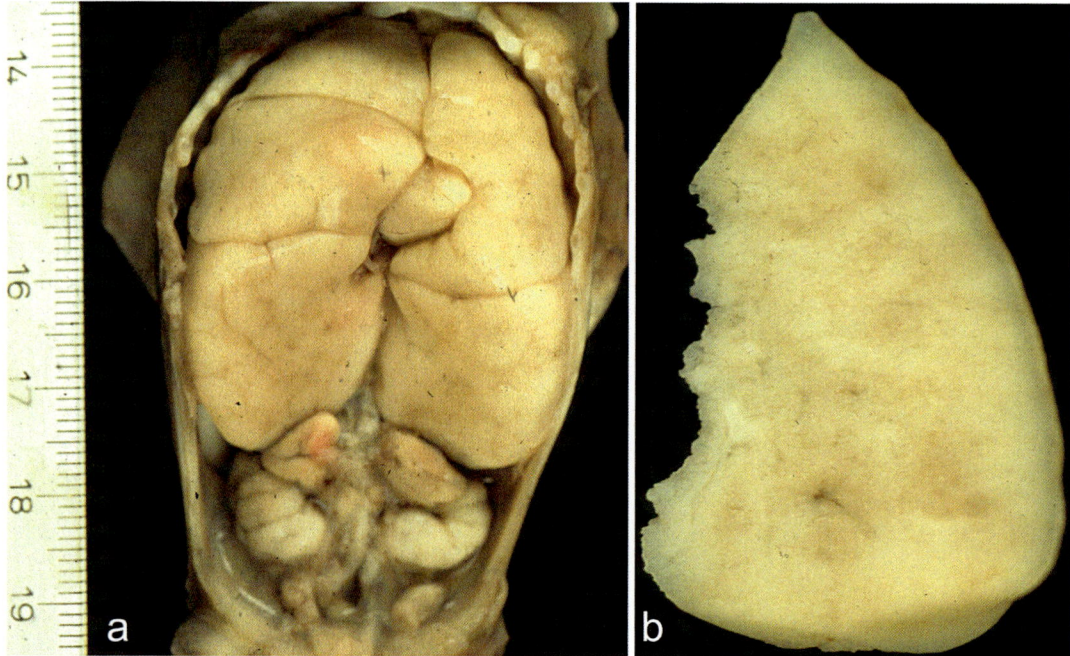

Fig. 18.45 Fraser syndrome. Fraser syndrome is rare genetic disorder characterized by partial webbing of the fingers and/or toes (partial syndactyly), kidney (renal) abnormalities, genital malformations, and/or cryptophthalmos as well as pulmonary hyperplasia (a-b)

the aortic ring determining incompetence of the aortic valve.

Autosomal Recessive (AR) Diseases
- The largest category of Mendelian disorders.
- AR gene mutations generally affect gene products that act as enzymes (biochemistry).
- 1/4 of offspring inherits the gene from both parents on both alleles carrying the mutation.
- AR pattern: both alleles at a given locus show gene mutations.
- Onset: Early in life (typically).
- Penetrance: Complete (commonly).
- Expressivity: Uniform (usually).
- Familiarity: high with more often seen in consanguineous marriages.
- Examples: phenylketonuria, lysosomal storage diseases.

Metabolic diseases play a significant role in the routine of neonatologists and pediatricians and are crucial for genetic counseling as well. Here, some metabolic disorders are briefly described.

Phenylketonuria
- Deficiency of phenylalanine hydroxylase
- Hyperphenylalaninemia
- At birth: Standard, but then with the development there are rising levels of phenylalanine
- ↑ phenylalanine ⇒ brain development impairment and mental retardation (MR)
- Diagnosis: Routine screening in the postnatal period
- Therapy: Strict diet from birth

Lysosomal Storage Diseases
Lysosomal storage diseases as Mendelian-inherited defects diagnosed at birth, infancy, childhood, or youth represent probably a unique biochemical repertoire of conditions with a very characteristic abnormality of the subcellular organelle, the lysosome, which shows essential features on electron microscopy allowing the pathologist to reach the diagnosis in many situations (Box 18.55).

18.6 Birth Defects

> **Box 18.55 Lysosomal Storage Diseases**
> - Inheritance of deficiency of lysosomal enzymes essential for biochemical reactions
> - Accumulation of undigested substrate in lysosomes identifiable on TEM
> - Dysregulation of cell functions
> - Cellular degeneration identifiable in several tissues by CLM (vacuolar, foamy, etc.)
> - Organ enlargement (e.g., hepatomegaly, splenomegaly, cardiomegaly)
> - Principles of categorization of the storage disorders according to the accumulated macromolecule
>   - Sphingolipidosis
>   - Mucopolysaccharidosis
>   - Mucolipidosis
>
> Notes: *CLM* conventional light microscopy, *TEM* transmission electron microscopy

Sialidosis (Fig. 18.46)
Sialidosis (MIM 256550) is a rare, AR-inherited disease, caused by α-N-acetyl neuraminidase deficiency resulting from a mutation in the alpha-N-acetyl-neuraminidase gene (*NEU1*), located on 6p21.33. This genetic alteration leads to abnormal intracellular accumulation as well as urinary excretion of sialyloligosaccharides. As recently reviewed, there are 40 mutations of *NEU1* reported in the literature. An association exists between the individual gene mutations and the severity of the presentation of sialidosis (Khan and Sergi 2018). According to the symptoms, sialidosis has been separated into two subtypes with different ages of onset and severity, including sialidosis type I (normomorphic or mild form) and sialidosis type II (dysmorphic or severe form). Sialidosis II is subdivided into (i) congenital, (ii) infantile, and (iii) juvenile. A complete understanding of the underlying pathology is still challenging. This challenge limits the development of effective therapeutic strategies. In 2001, a report presented the accumulation of sialidosis in detail (Sergi et al. 2001) (Fig. 18.46). Conventional light microscopic (CLM) examination revealed a large micro- and macro-vacuolation in the cytoplasm of the cells of numerous organs that were investigated. In the liver, the hepatocytes contained mostly large empty vacuoles. The Kupffer cells showed a foamy pattern caused by the presence of vacuoles of variable size, and this pattern was notably confirmed using an antibody against CD68 for macrophages by immunohistochemistry. The intra-acinar fields showed diffuse extramedullary hematopoiesis and no ductal plate malformation of the intrahepatic biliary system in the liver. Fibrosis was absent in the portal tracts. CLM of the other organs showed a moderate foam cell alteration, including bone marrow, spleen, thymus, the proximal convoluted tubules of the kidney, the interstitial Leydig cells of the testes, and in the cortex of the adrenal gland and adenohypophysis. In the brain, microvacuolar alterations were found in the neurons of the brain stem nuclei and the *plexus choroideus* of the ventricular system. Tissues were partly stained with PAS, whereas Sudan black was practically absent from frozen sections. Acid phosphatase staining showed a very marked lysosomal reaction. Transmission electron microscopy (TEM) of fibroblasts, liver, and kidney samples identified membrane-bound vacuoles containing fine reticule-granular material. In the perinuclear cytoplasm of the fibroblasts, there were membrane-bound inclusions. These inclusions contained polymorphic material of granular-fibrillar and multilamellar content, whirls of membrane-like structures, and electron-dense tubular formations. The placenta showed a generalized delay of maturation of the villous tree and numerous calcium aggregates located in the terminal villous ends. Foamy cell change of the cytoplasm of the trophoblastic cells with marked acid phosphatase staining was seen. TEM of the placenta showed distension of the syncytiotrophoblast with storage material consisting of two different types, including both fine mono- or bilamellar structures and finely granular structures. In the Hofbauer cells, predominant storage of the granular type was recognized. DNA sequencing analysis showed that both the proband and the third sibling had a novel homozygous nonsense point mutation at position 87 in exon 1

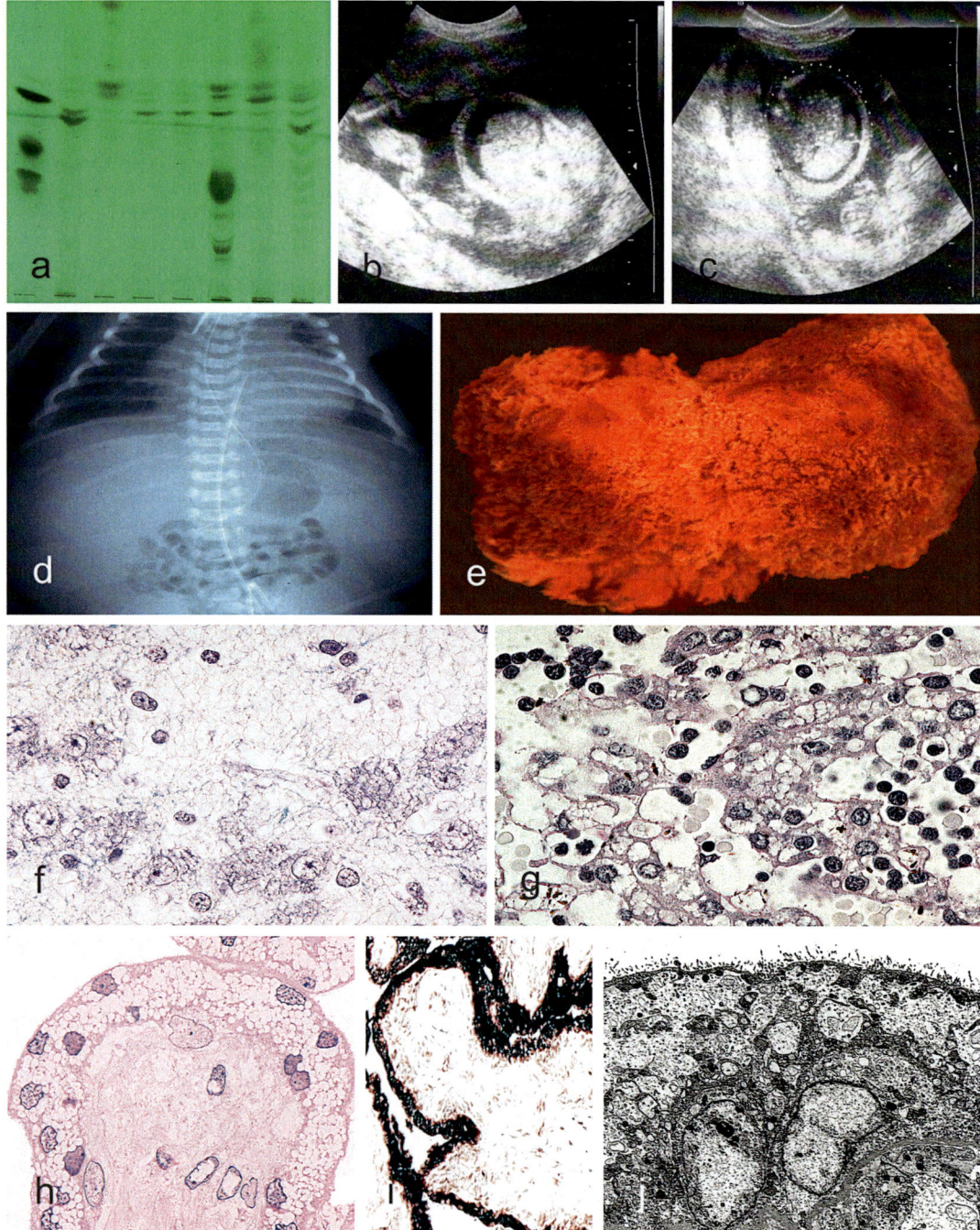

Fig. 18.46 Sialidosis. Sialidosis is a rare metabolic disorder with urine chromatographic pattern (**a**), hydrops fetalis (**b** and **c**), hepatomegaly on X-ray (**d**), and grossly (**e**) as well as vacuolation of the brain (**f**), liver(**g**), and placenta (**h-i**). Microphotograph (**i**) is the acid phosphatase staining of the chorionic villus of a placenta affected with sialidosis. ultrastructural study of the placenta (**j**) shows cytosolic expansion of the trophoblastic cells by fine and reticular material (f, original magnification x630; g, original magnification x630; h, original magnification x630; i, original magnification x200; j, original magnification x1100)

of the *NEU* gene. This mutation caused tryptophan at codon 29 (TGA) to be substituted by a termination codon (W29X).

Complex Lipids

Lecithin or phosphatidylcholine results from the esterification of two out of three alcoholic ossidriles of the glycerol with a molecule of fatty acids and with phosphorylcholine.

Sphingomyelin: N-acyl-sphingosine + phosphorylcholine

Sphingolipidosis

Sphingolipidoses are a class of lipid storage disorders relating to sphingolipid metabolism involving mainly the central nervous system but also other organs. The main sphingolipidoses are Niemann-Pick disease, Fabry disease, Krabbe disease, Gaucher disease, Tay-Sachs disease, and metachromatic leukodystrophy.

Tay-Sachs Disease

- *SYN*: GM2 gangliosidosis.
- *EPG*: Genetic alteration of *HEXA* ⇒ enzyme deficiency, β-hexosaminidase A ⇒ ↑ GM2 gangliosides ⇒ cell toxicity ⇒ organ failure (GM2 ganglioside accumulation with progressive cell destruction and tissue architecture destruction).
- *CLI*: Motor deterioration ~3–6 months, a cherry red spot of the retina with seizures, vision and hearing loss, intellectual disability, and paralysis.
- *CLM*: ballooning of neurons with cytoplasmic vacuoles (ORO+).
- *TEM*: Lysosomes with the whorled configured accumulated material.

Of note, ORO or oil red O stain is a fat lipid special stains, which may be substituted by Sudan black or other fat lipid stains

The *HEXA* gene instructs how to build part of β-hexosaminidase A, an enzyme, which plays a paramount role in the brain and spinal cord.

Mucopolysaccharidoses

Other inborn errors of metabolism that need to be taken into consideration are the mucopolysaccharidoses (*MPS*), which are a group of closely related syndromes resulting from genetically determined deficiencies of specific lysosomal enzymes. All these enzymes are responsible for the degradation of mucopolysaccharides, and all syndromes are inherited in an autosomal recessive manner except Hunter syndrome (MPS-II), which is X-linked recessive. Mucopolysaccharidoses are usually progressive diseases and characterized clinically by coarse facial features, corneal clouding, joint stiffness, and mental retardation. The accumulated macromolecules are typically found in mononuclear phagocytic cells, endothelial cells, smooth muscle cells, and fibroblasts. The cell morphology is affected in these cells. There is ballooning on light microscopy and minute vacuoles on electron microscopy. Clinically, several organs involved by the accumulation of mucopolysaccharides determine a phenotype characterized by hepatosplenomegaly, bony deformities, valvular lesions of the heart, and subendothelial arterial deposits.

Glycogen Storage Diseases

Glycogen storage diseases (GSD) are diseases characterized by the accumulation in the cell of glycogen in a way, which is morphologically evident. In Canada, the incidence is 1:43,000 births, while in the USA it is 1:20,000–1:25,000 births. The accumulated glycogen can be part of the conventional or abnormal structure. The cells showing glycogenic accumulation belong to the liver, kidney, spleen, heart, and skeletal muscle. The accumulated glycogen, examined in tissue sections fixed with formalin or other aqueous fixatives, shows an increase of cellular volume, which has been considered comparable to vegetal or plant cells. The nucleus is not displaced from its position but has a central location, while the cytoplasm appears empty, and the structured parts of the cells are indeed reduced to thin segments forming a tenuous network around the empty spaces. This aspect is due to the accumulation of glycogen, which is soluble in water and, thus, also in hydro-based fixatives. If the tissue is fixed in alcohol, in which the glycogen is practically insoluble, the accumulated glycogen may be highlighted using special stains. Special stains include iodide iodate solution (red-brown), best carmine (red carmine), and Hotchkin (periodic acid Schiff) stain (red). The most important features of the 14 forms of GSD are shown in Box 18.56. There are additional forms of GSD, which will not be presented in this box, being extremely rare. The inheritance is mostly autosomal recessive. Biochemically, in the hepatic forms, the deficiency of glucose-6-phosphatase determines an inadequate conversion in the liver of glucose-6-phosphate to glucose, and patient suffers hypoglycemia and has an enlarged liver. In the myopathic forms (e.g., type V), the enzymes that fuel the glycolytic pathway are deficient, and glycogen stores in the muscle increase. The patient experiences muscular cramping after exercise and a failure of exercise-induced lactic acidosis. In the GSD type I, which is the prototype of GSD, hypoglycemia is associated biochemically with ketosis, hyperuricemia, and hyperlipidemia.

GSD is not a rare event in the practice of a perinatal and pediatric pathologist. In the cases of sudden death needs to be added Pompe disease (GSD II), and in cases related to traumas with medicolegal consequences where hypoglycemia has been considered, GSD should also be considered together to cases of congenital or acquired hyperinsulinism. Moreover, some GSD has different forms, e.g., infantile, juvenile, and adult (late-onset), and some GSD has different subtypes, e.g., 1a, 1b, etc. In GSD type 0, there is no accumulation of extra glycogen, but it is classified among the GSD because it is another defect of glycogen storage causing similar glycogenosis disturbances. GSD type VIII was considered a distinct condition, but it is now classified with GSD type VI or GSD IXa1. GSD type XI is also known as Fanconi-Bickel syndrome or hepatorenal glycogenosis with renal Fanconi syndrome and is no longer classified as a glycogen storage disease. GSD type XIV has also been removed from the GSD group and is now recognized as the congenital

Box 18.56 Glycogen Storage Diseases (GSDs)

Type, eponym	Enzyme	Organs	TEM	Clinics
0	Glycogen synthase	Muscle	–	Hypoglycemia
I – von Gierke	Glu-6-phosphatase	Liver, kidney	↑ glycogen, nl.	Hepatomegaly
II – Pompe	α-1-4-Glucosidase	HEP, KNY, COR, CNS, muscle, LN, histiocytes	↑↑↑ glycogen, nl.	Cardiac failure
III – Forbes	α-1-6-Glucosidase (debranching)	Liver, heart, muscle	↑ glycogen, short branches	Hepatomegaly
IV – Andersen	Branching (α-1-4 → α-1-6)	Liver (Kupffer), spleen, LN, CNS	Glycogen long branches	Liver cirrhosis
V – McArdle	Myophosphorylase	Muscle	↑ glycogen, nl.	↓ physical activity
VI – Hers	Hepatic phosphorylase	Liver	↑ glycogen, nl.	Hepatomegaly
VII – Tarui	Myo-PFK	Muscle	↑ glycogen, nl.	↓ physical activity
IX	Phosphorylase kinase	Liver	↑ glycogen, nl.	Hepatomegaly
X	Enolase 3	–	–	–
XI	Muscle LDH	–	–	–
XII	Aldolase A	Muscle	–	Exercise intolerance, hemolytic anemia
XIII	B-Enolase	Muscle	–	Exercise intolerance
XV	Glycogenin-1	Muscle	–	↓ physical activity

disorder of glycosylation type 1, which affects the phosphoglucomutase enzyme derived from the *PGM1* gene. Of note, Lafora disease is considered a complex neurodegenerative disease and also a glycogen metabolism disorder but is not individualized among the GSD group.

X-linked diseases are mostly X-linked or recessive type. They are fully expressed in males and females, which are carriers, and may partially express the condition. X-linked disorders are based on the Lyon hypothesis, which is the random inactivation of either paternally or maternally inherited X chromosome occurring in each cell of females. Thus, healthy females are mosaics. It is known that The *XIST* gene located on Xq13 regulates X inactivation. However, the inactivated X is reactivated in germ cells. X-linked dominant disorders essential to recall are Rett syndrome and vitamin D resistant rickets. In Rett syndrome, mental retardation is seen in infertile females and prenatally lethal males. Vitamin D resistant rickets shows a typical production and metabolism of vitamin D, but the renal tubular loss of phosphate seen in this disease interferes with skeletal ossification. X-linked recessive disorders include Lesch-Nyhan syndrome and Duchenne muscular dystrophy among others.

Lesch-Nyhan Syndrome
- X-linked inheritance
- *HPRT1* gene defect ⇒ ↓ hypoxanthine phosphoribosyltransferase 1 to recycle purines
- 1: 380,000 births
- Hyperuricemia ⇒ lack of dopaminergic terminals and cells
- Abnormal involuntary muscle movements (dystonia), jerking movements (chorea), and flailing of the limbs (ballismus), self-injury, and self-mutilation

Of note, HPRT has other names (aliases) including HGPRT or hypoxanthine-guanine phosphoribosyltransferase, HGPRTase, HOX5.4, and IMP pyrophosphorylase.

HPRT is involved in the salvage pathway for purine synthesis, and it salvages purine bases from the breakdown of nucleic acids. In case of HPRT deficiency, there is an increased balance of purine nucleotides *de novo* with subsequent hyperuricemia. Quite often, a differential diagnosis or genetic disease is Duchenne muscular dystrophy, which may also be encountered from the perinatal pathologist. In HPRT deficiency, uric acid, a waste product of purine breakdown, progressively accumulates in the body. This accumulation can determine gouty arthritis and nephro-/urolithiasis. It seems undefined how HPRT deficiency causes the neurological and behavioral symptomatology characteristic of Lesch-Nyhan syndrome.

Mitochondrial disorders are a group of chronic genetic disorder that occurs when the mitochondria of the cells fail to produce enough aerobic energy for cell, tissue, or organ function. The incidence is about 1:4000 individuals in the USA and 1:5000 in Canada with similar rates in most of the other Western countries (Box 18.57).

Chromosomal-Based Birth Defects
The essential chromosomal-based birth defects include trisomy 13 (Patau syndrome) (Fig. 18.47), trisomy 18 (Edwards syndrome) (Fig. 18.48), trisomy 21 (Down syndrome) (Fig. 18.49), XXY or

---

Box 18.57 Mitochondrial Disorders
- *MELAS*: Mitochondrial encephalomyopathy – lactic acidosis – stroke-like episodes
- *MERRF*: Myoclonus, epilepsy, ragged red fibers
- *LHON*: Leber hereditary optic neuropathy
- *Leigh syndrome*: Subacute, symmetric, and necrotizing encephalomyelopathy
- *MMCM*: Mitochondrial myopathy and cardiomyopathy
- *Wolfram syndrome*: Diabetes insipidus, diabetes mellitus, optic atrophy, and deafness
- *KSS* (Kearns-Sayre syndrome): Progressive external ophthalmoplegia, retinal pigmentary degeneration, cerebellar ataxia, deafness, and cardiac conduction defects

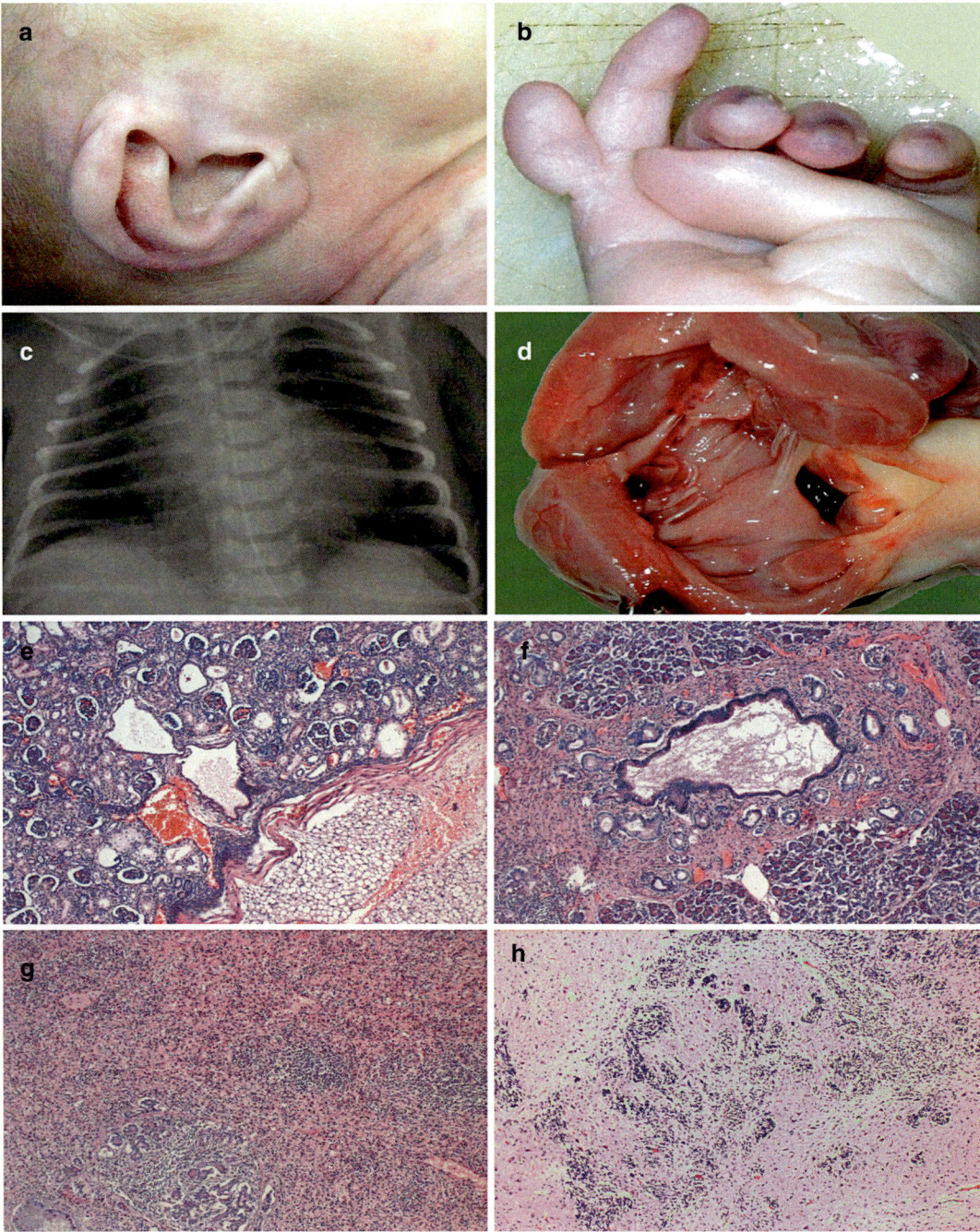

Fig. 18.47 Trisomy 13 Syndrome. Trisomy 13 syndrome in a patient showing dysplastic low-set ears (**a**), polydactyly (**b**), tetralogy of Fallot with characteristic X-ray (**c**) and gross heart preparation with ventricular septal defect (**d**), dysplastic kidney with cortical cysts or Potter IV (**e**), cystic pancreas dysplasia (**f**), splenopancreatic fusion (**g**), and cerebellar heterotopia, which may be considered similar to periventricular neuronal heterotopia (PNH). PNH is characterized by unorganized islands of neurons under the ependyma of the lateral ventricles. During embryogenesis the neurons fail to migrate and differentiate in their original positions. Unilateral PNH occurs frequently with other brain malformations. Bilateral PNH (BPNH) is caused, commonly, by mutations of the FLNA gene, located on Xq28. The product of this gene, filamin A, is an actin-binding protein that specifically cross-links actin filaments. Filamin A is important for the integrity of the cytoskeleton and cellular movement (e-h, H&E stain, original magnification x100)

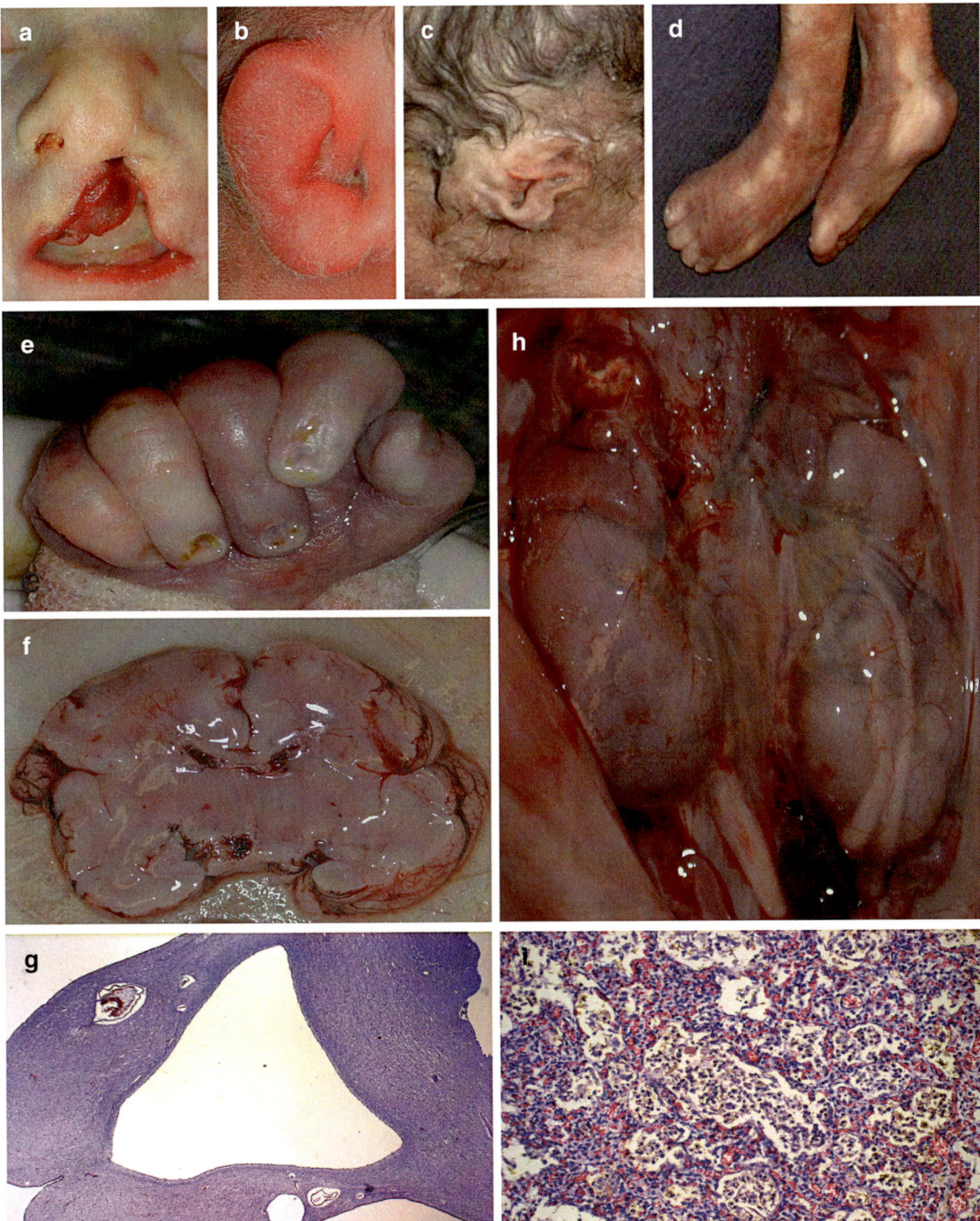

Fig. 18.48 Trisomy 18 Syndrome. Trisomy 18 syndrome with unilateral cleft lip and palate (**a**), dysplastic ears (**b, c**), talipes equinovarus (**d**), abnormal position of the fingers (**e**), basal ganglia fusion (**f**), septum pellucidum cyst (**g**), renal ptosis with double ureter (**h**), and lung immaturity with pneumonia (**i**) (g, H&E stain, original magnification x20; i, H&E stain, original magnification x100)

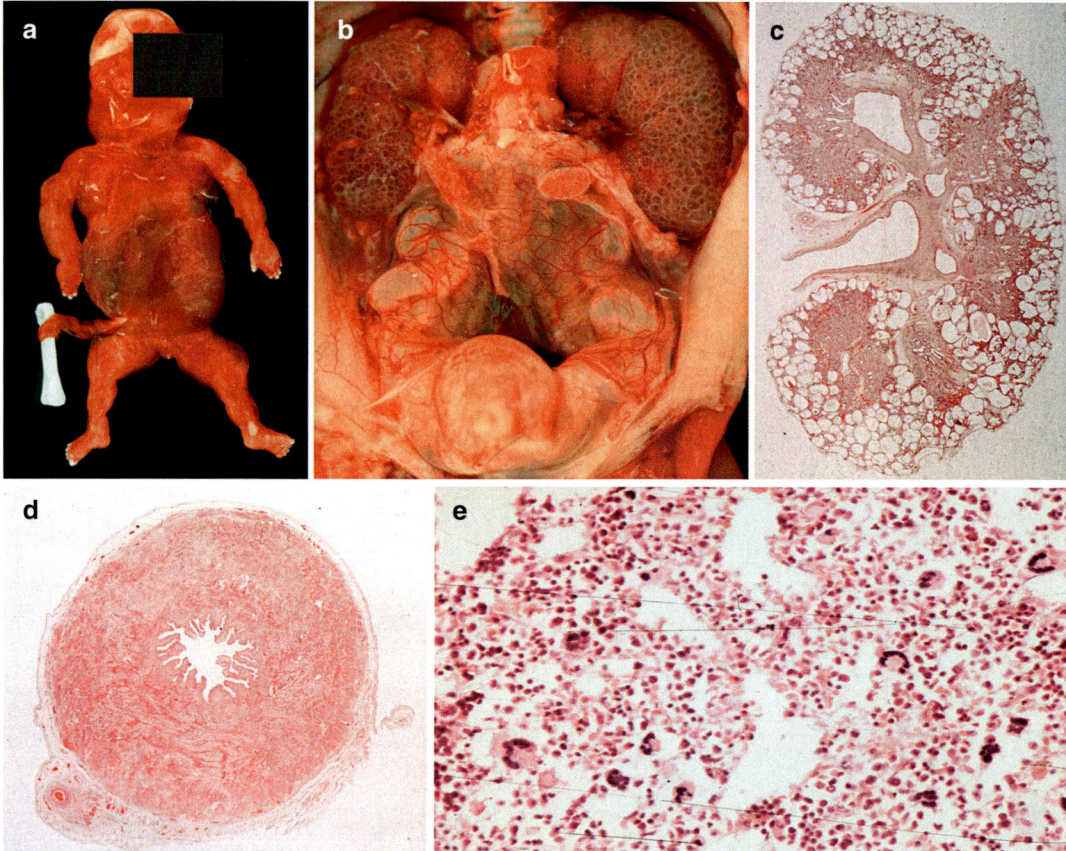

Fig. 18.49 Trisomy 21 Syndrome Trisomy 21 syndrome with non-immunologic hydrops fetalis (**a**), Potter IV renal cortical cysts (**b**, **c**), obstruction of the lower urinary system (**d**), and transient megakaryoblastic leukemia reaction of the liver (**e**) (**c**, original magnification x2.5; **d**, original magnification x20; **e**, original magnification x200)

Klinefelter syndrome, Turner syndrome or monosomy X0 (Fig. 18.50), and microdeletion syndromes as well as triploidy syndrome (Fig. 18.51).

Trisomy 21 (Down Syndrome)
- *DEF*: Syndrome with multivisceral involvement and cognitive impairment in individuals harboring an additional chromosome 21 in all cells or some (mosaic) of the cells of the individuum (Fig. 18.49). Of note, mosaicism is when an affected individual harbors two or more genetically different sets of cells in his/her body.
- *EPI*: 1:1000 newborns.
- *EPG*: Maternal age-linked risk (~1:1500 in women <20 years and then 1:25 in >45 years).
- True trisomy (meiotic nondisjunction)–Robertsonian translocations (95%) (extra-chromosome 21 fused to another acrocentric chromosome, e.g., 14, 22, etc.) and mosaicism (4%) (two different cell lines). Cytogenetic abnormality includes 21q22 band harboring *GART* gene, which is involved in purine metabolism. The protein encoded by the GART gene is a trifunctional polypeptide. This enzyme has phosphoribosyl-glycinamide

18.6 Birth Defects

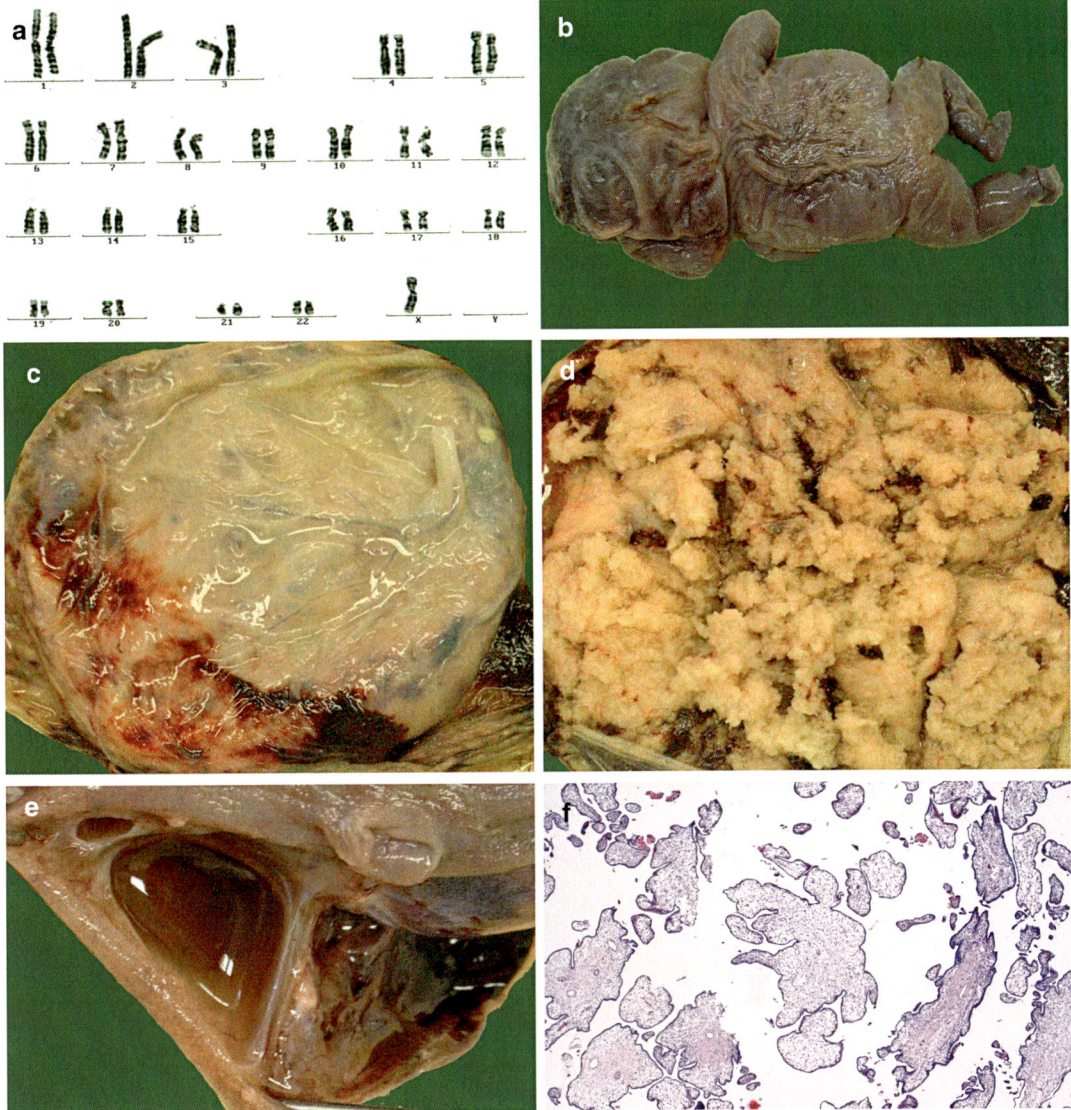

Fig. 18.50 Monosomy X0 (Turner) Syndrome. Monosomy X0 showing the karyotype of the missing chromosome X (**a**), non-immunological hydrops fetalis (**b**), hydropic placenta (**c, d**), cystic hygroma of the neck (**e**), and plump villi with abnormal configuration of the placenta (**f**) (**f**, H&E stain, original magnification x40)

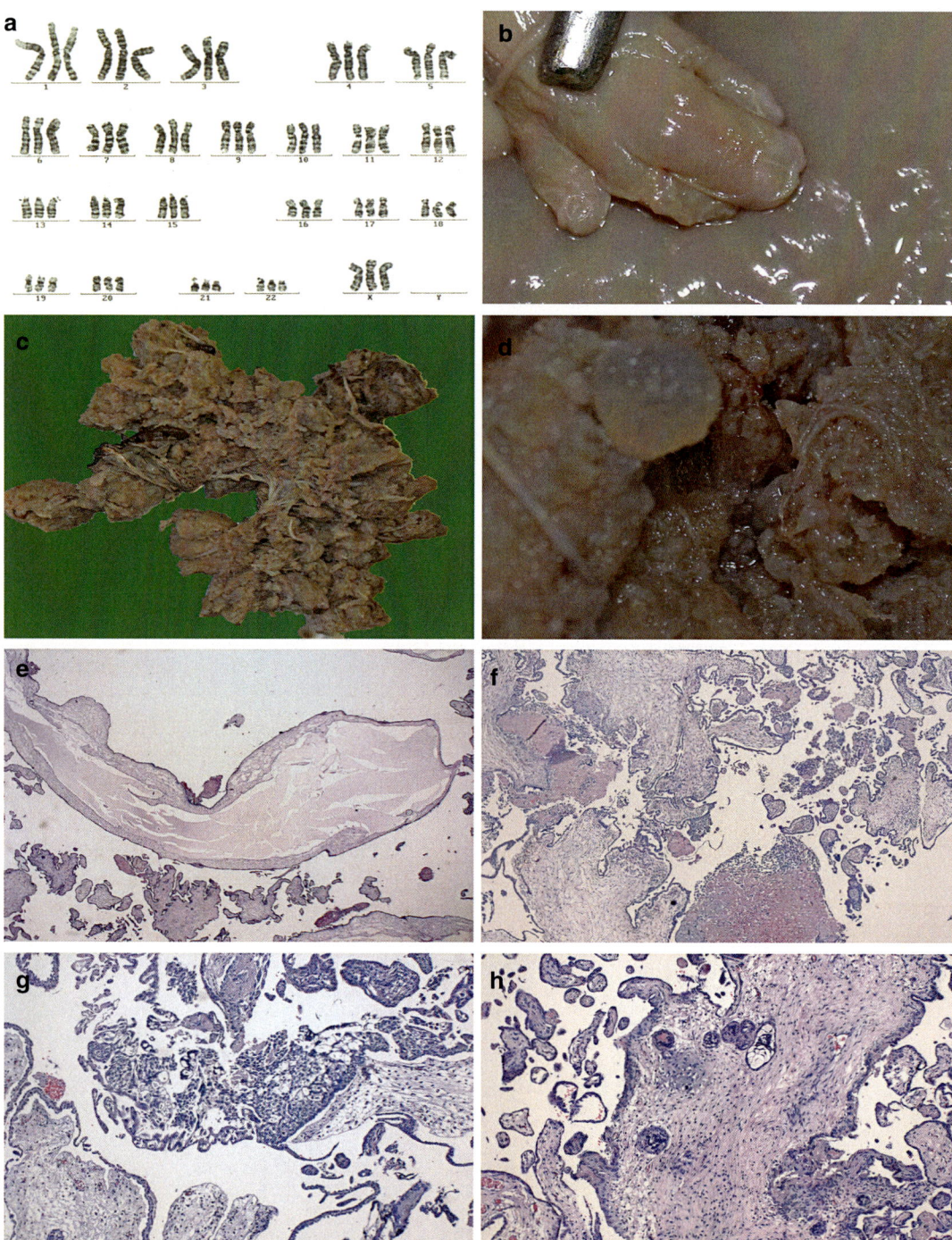

Fig. 18.51 Triploidy Syndrome. Triploidy with characteristic triploid set of chromosomes (69), syndactyly of the 3rd and 4th fingers (**b**), dystrophic placenta with multiple cysts (**c**, **d**), partial molar degeneration (**e–h**) with cisterns €, trophoblast proliferations (**f**, **g**), and trophoblast inclusions (**h**) (e-h, H&E stain; e, original magnification x20; f, original magnification x40; g, original magnification x40; h, original magnification x100)

formyltransferase, phosphoribosyl-glycinamide synthetase, phosphoribosyl-aminoimidazole synthetase activity which is required for de novo purine biosynthesis. The GART enzyme is highly conserved in vertebrates. There is a common alternative splicing of this gene resulting in two transcript variants encoding different isoforms. Diseases associated with GART abnormality include nondisjunction and ureter leiomyoma. Gene ontology annotations related to GART gene include phosphoribosyl-amine-glycine ligase activity and methyltransferase activity.

- *CLI*: Flat face, thick nuchal fold, upward slanting palpebral fissures, depressed nasal bridge.
- *GRO/CLM*: 10/20-fold amplified risk of acute leukemias (transient myeloproliferative disorder, TMD; acute megakaryoblastic leukemia, AMKL; and acute lymphoblastic leukemia, ALL) and a markedly decreased incidence of solid tumors. A recent survey of the Danish Cytogenetic Register revealed that the risk of all major groups of solid tumors is decreased in individuals affected with trisomy 21 syndrome, except testicular cancer (Hasle et al. 2016).
- *PGN*: (1) Mental retardation, which is highly variable, (2) risk of neoplasia, and (3) early Alzheimer disease, i.e., the onset of Alzheimer disease at an earlier age than general population without the extra-chromosome 21. It has been suggested that a dysfunction of endosomes is one of the earliest shared phenotypes of trisomy 21 syndrome (Down syndrome) and Alzheimer disease, when Aβ levels are relatively low (Weick et al. 2016). The retrograde signaling of neurotrophins (e.g., nerve growth factor) through the endosomal trafficking has been exquisitely implicated in the neuronal cell death in both conditions.

Pregnancy panel: First 3 months period (↑ hCG, ↓ PAPPA); second 3 months period (↓ AFP, ↓ unconjugated-Estriol, ↑ hCG, ↓ inhibin A/INA). Pregnancy-associated placental protein-A (PAPP-A) is a metalloprotease. PAPP-A circulates as an hetero-tetramer in maternal blood. PAPP-A levels increase throughout normal pregnancy. PAPP-A rates do not exhibit such increase in trisomy 21. In fact, PAPP-A decreases during the first trimester. PAPP-A levels are also decreased in trisomy 13 and in trisomy 18, no matter the gestational age (Leguy et al. 2014).

There is a protective effect of DS genetic background against the development of intestinal tumors as demonstrated in Ts65Dn- *Apc*^{Min} mice, which are trisomic for mouse orthologues of about 1/2 the human chromosome 21 genes and crossed to *Apc*^{Min} mice. Two other genes studied are *ETS2* and copper-zinc superoxide dismutase as well as *COL18A1* coding for endostatin, a cleavage product of collagen XVIII with anti-angiogenesis effects. Less consumption of tobacco and alcohol may also contribute to the relatively low incidence of solid cancer in individuals harboring an extra-chromosome 21. Gross photographs of a fetus with trisomy 21 and their defects are presented in Fig. 18.49. Lung pathology in trisomy 21 is quite distinctive and has been described in Chap. 2. Substantially, infants with trisomy 21 syndrome and structural CHD may develop earlier and progress more rapidly to irreversible pulmonary vascular changes than infants without trisomy 21 syndrome. The more rapid development of pulmonary hypertension in infants with trisomy 21 syndrome is due to some reasons, including chronic upper airway obstruction (e.g., tracheomalacia and tracheal bronchus), recurrent pulmonary infection, and alveolar hypoventilation. The development of progressive pulmonary hypertension in trisomy 21 syndrome is associated with the failure of neonatal pulmonary vascular remodeling in such infants. Criteria of PPHN (Persistent Pulmonary Hypertension of the Newborn) include hypoxemia refractory to O_2 therapy or lung recruitment strategies, which are achieved by conventional ventilation and high positive end-expiratory pressure (PEEP) formerly and now replaced by high-frequency oscillatory ventilation later, as well as the presence of extrapulmonary shunt at an atrial or ductal level in the absence of severe lung disease. Early development of the pulmonary vascular disease is seen in 1/10 to almost all children with trisomy 21 syndrome and affected with structural CHD. One of the major problems of neonates, infants, and then children with DS is their dysregulated immunological mechanisms.

Trisomy 21 syndrome-associated immunological deficiencies include:

1. ↓ T-cell numbers, mild to moderate
2. ↓ B-cell numbers, mild to moderate
3. Lack of normal lymphocyte expansion in infancy
4. ↓ size of the thymus as compared to age-matched control children
5. ↓ naive T-cell percentages, mild to moderate
6. ↓ T-cell "excision"
7. ↓ immunization response to antibody
8. ↓ total and specific IgA production in saliva
9. ↓ chemotaxis by neutrophils

Marker Chromosome and Microdeletion Syndromes

Marker chromosome and microdeletion syndromes are additional events associated with congenital disabilities. In Box 18.58 the most critical microdeletion syndromes are listed.

Velocardiofacial syndrome, also known as VCFS or as DiGeorge syndrome (DGS) or as a 22q11.2 syndrome, is the most common syndrome associated with cleft palate. It has been estimated that about 1 in 2000–5000 infants are born with VCFS, which has some characteristic features, including a long face with a prominent upper jaw among others.

Box 18.58 Some Microdeletion Syndromes

Chromosome	Syndrome
1p36	Monosomy 1p36
4p16	Wolf-Hirschhorn
5p15.2	Cri du chat
7q11.2	Williams
15q11-q13	Angelman
15q11-q13	Prader-Willi
17p11.2	Smith-Magenis
17p13.3	Miller-Dieker
22q11.2	DGS, VCFS

Notes: *DGS* DiGeorge syndrome, *VCFS* velocardiofacial syndrome

Triploidy Karyotype Syndrome (Fig. 18.51)

The concept of triploid karyotypes is tightly associated with chromosome imprinting. Chromosome imprinting determines the chromosome's function or lack of function at well-specified times in development allowing a functional difference of the maternal and/or paternal haploid contributions during embryogenesis. Mouse experiments have evidenced that paternal contribution to the zygote is crucial for the healthy development of trophoblastic tissue, i.e., placental tissue, while maternal contribution is essential for the embryonic development. Triploidy is 3X the haploid set of chromosomes $\Rightarrow$ 69n, which is usually 69, XXX or 69, XXY or 69, XX. It may occur as a result of fertilization of a standard haploid set by two normal spermatozoa (dispermy) or fertilization of a diploid oocyte in consequence of a failure of meiosis I or II by a haploid spermatozoon or fertilization of a haploid oocyte by a diploid spermatozoon (meiosis I or II failure). The McFadden-Kalousek-Rehder fetal phenotypes are key in perinatal pathology. (Fig. 18.51). In triploidy Type I or DI-ANDRIC (extra-haploid set: paternal), there is a large cystic placenta with microscopic evidence of PHM (focal hydrops of the villi, focal intravillous cistern change, focal hyperplasia of the trophoblast epithelium) showing a trophoblast with a scalloped or lacey appearance. In Type I, there is a well-grown fetus showing proportionate head or slight microcephaly. Triploidy Type II or DI-GYNIC (extra-haploid set: maternal) shows small non-cystic placenta with non-molar degeneration and growth-restricted fetus showing relative macrocephaly consistent with the phenomenon called "head-sparing growth retardation." Some gross features common to both phenotypes of triploidy include syndactyly, adrenal hypoplasia, and congenital heart defects, among others. Several studies have identified dispermy as the most common mechanism, and two fetal phenotypes may be encountered in the pediatric pathology routine, other than partial hydatidiform molar degeneration. Interestingly, the placental phenotype is determining for the fetal phenotype in humans in a way similar to that identified by blastocyst manipulation studies of the mouse. Moreover, most of the diandric triploid conceptuses are aborted in the first trimester due to the cystic degeneration of the placenta. Confined placental mosaicism plays a role in determining triploid conceptuses to reach later stages of pregnancy, being the cytotrophoblast mosaic diploid/triploid in more than some occasions. PHM is seen in a few triploid conceptuses, and the low incidence of PHM supports the results of investigations that have demonstrated, using analysis of molecular polymorphisms, that digyny is responsible for most conceptuses with triploidy.

Multifactorial-Based Birth Defects

Multifactorial-based congenital disabilities may be quite valuable to study because they may be prevented. A classic example is the Erin Brockovich movie. In this film, directed by Steven Soderbergh and written by Susannah Grant, the dramatization of the true story of Erin Brockovich is depicted. The protagonist actress (Julia Roberts) is an unemployed single mother of three. She is desperate to find a job, she takes finally a relatively minor role as a legal assistant in a California law firm for low wages. The movie becomes peculiarly interesting when she begins to investigate a suspicious real estate case involving a private company, which was trying quietly to buy land that was contaminated by hexavalent chromium, a deadly toxic waste that the company had improperly and illegally dumped. The consequence was a continuous poisoning of the residents in that area. Multiple abortions and congenital disabilities were occurring in concomitance with the onset of cancer of the adult. The groundwater contamination incident was spectacularly narrated. Apart from the theatricality there are some notes that need to be taken into account., Multifactorial inheritance results from combined actions of the environment, and ≥2 mutant genes and features include the risks presented in Box 18.59.

Multifactorial-based birth defects include diabetes mellitus, congenital heart disease, cleft lip and/or palate, hypertension, and gout as

> **Box 18.59 Multifactorial Risks**
> – Risk of expressing the disorder influenced by the # of mutant genes inherited
> – Risk of recurrence remains the same for all first-degree relatives
> – Risk of both identical twins getting disorder is <100%, but apparently and tragically greater than the risk of both fraternal twins getting the disorder

most often considered examples among geneticists and genetic counselors Numerous environmental toxins may be associated with birth defects. However, it is not fully clear how many birth defects are exclusively related to environmental exposures, such as drugs, chemicals, and ionizing radiation. It is known that some endocrine-disrupting chemicals, including dioxins, polychlorinated biphenyls (PCBs), and several pesticides, have been linked to defects of the nervous system and psycho-motoric developmental problems. The surrounding of the future mother is crucial and living near a hazardous waste site has been recognized as a possible risk factor for birth defects including cleft lip or palate, body wall defects (gastroschisis), hypospadias, cardio-vascular defects, neural tube defects, and chromosomal anomalies. Pregnant women, who are exposed to disinfection by-products in drinking water such as trihalomethanes may exhibit an increased risk of birth defects affecting the brain and spinal cord, the heart, and the urinary tract (Kaufman et al. 2020).

18.7 Infection in Pregnancy, Prom, and Dysmaturity

18.7.1 Infection in Pregnancy

A variety of non-neoplastic lesions can present clinically and radiologically as uneventful or life-threatening conditions and may result in a hysterectomy. In such situations, the pediatric pathologist or gynecologic pathologist has an essential role to play in determining the nature of these lesions. Awareness of the development of the embryo at an early stage and entities that can mimic malignancies assists the physician in the correct diagnosis of these lesions (Boxes 18.60, 18.61, and 18.62) (Figs. 18.52, 18.53, and 18.54).

> **Box 18.60 Infection in Pregnancy**
> 18.7.1. *Infection in Pregnancy*
> 18.7.2. *Perinatal Infection*
> 18.7.3. *Premature Rupture of Membranes (PROM)*
> 18.7.4. *Dysmaturity*

> **Box 18.61 Prenatal Transmission**
> - *Cytomegalovirus* (CMV) (Fig. 18.52)
> - *Herpes simplex virus* (HSV)
> - *Parvovirus B19* (Fig. 18.54)
> - *Rubella virus*
> - *Varicella zoster virus* (VZV)
> - *Zika virus*

> **Box 18.62 Perinatal Transmission**
> - *Cytomegalovirus* (CMV)
> - *Hepatitis B virus* (HBV)
> - *Hepatitis C virus* (HCV)
> - *Herpes simplex virus* (HSV)
> - *Human immunodeficiency virus* (HIV)
> - *Papilloma virus* (HPV)
> - *Varicella zoster virus* (VZV)
> - *Zika virus*

TORCH is an acronym for *Toxoplasma*, "other microorganisms," rubella, cytomegalovirus, and herpes virus. A TORCH screen is a panel of serologic investigations for detecting infections in pregnant women. Diseases may be passed on to a fetus during pregnancy with consequences ranging from malformations to intrauterine fetal death (IUFD). Early detection and therapy of a disease can prevent several complications in fetuses and newborns.

Toxoplasmosis

Toxoplasmosis is an infective disease, which recurs as host microorganisms in birds, terrestrial animals, and the general population. Although *Toxoplasma gondii*, the etiologic agent for toxoplasmosis, does not cause serious health problems, it is a high risk for the pregnant woman's growing baby, because this parasite can cause brain damage and vision loss. There is an incidence of congenital toxoplasmosis of ~1:10,000 live births in the USA, and similar rates are seen in other well-developed countries due to the use of domestic animals. Although multifactorial in etiology, maternal infection is primarily attributed to the consumption of contaminated meat or water. Routine universal screening should not be carried out for pregnant women at low risk, but serologic testing should be offered to pregnant women harboring a risk of *T. gondii* infection. If the acute infection is suspected, tests should be repeated within 2–3 weeks, but starting therapy immediately with spiramycin should be a high priority without waiting to repeat the test results. Amniocentesis should be offered to identify the microorganism in the amniotic fluid by PCR if the primary maternal infection is diagnosed, if serology is questionable for an acute infection, or in the presence of abnormal ultrasound findings (intracranial calcification, microcephaly, hydrocephalus, ascites, hepatosplenomegaly, or severe FGR/IUGR). False-negative results need to be considered for amniocentesis <18 weeks' gesta-

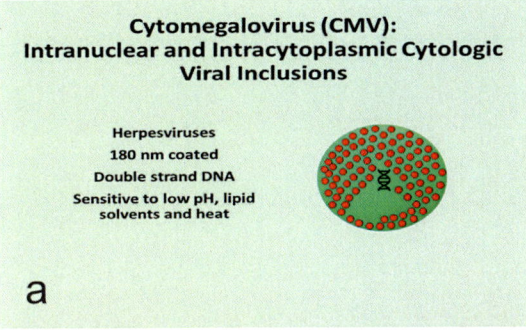

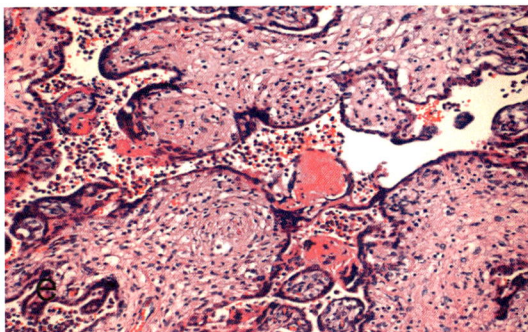

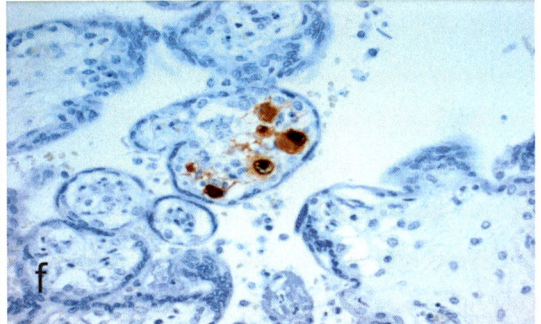

Fig. 18.52 Cytomegalovirus Infection. Cytomegalovirus (CMV) data (**a**) with its relevance in pregnancy (**b**), CMV encephalitis with characteristic inclusions (**c**), the importance of different tests according to the age of the infection (**d**), plasma cell-rich intervillositis (**e**) with immunohistochemical confirmation of CMV infection inside of some villi (**f**) (**c**, H&E stain, original magnification x630; **e**, H&E stain, original magnification x200; **f**, anti-CMV immunostaining, original magnification x400)

tion and <4 weeks after suspected acute maternal infection. If the maternal infection is confirmed, but the fetus may not be infected, spiramycin should be started for fetal prophylaxis as soon as possible. A combination of pyrimethamine, sulfadiazine, and folinic acid should be offered as a treatment for women in whom fetal infection has been confirmed or suspected. Pregnant women with HIV infection or immunosuppressed should be routinely screened for the risk of reactivation and toxoplasmosis encephalitis. There is a 6-month interval for a nonpregnant woman to become pregnant after confirmation of a *T. gondii* infection.

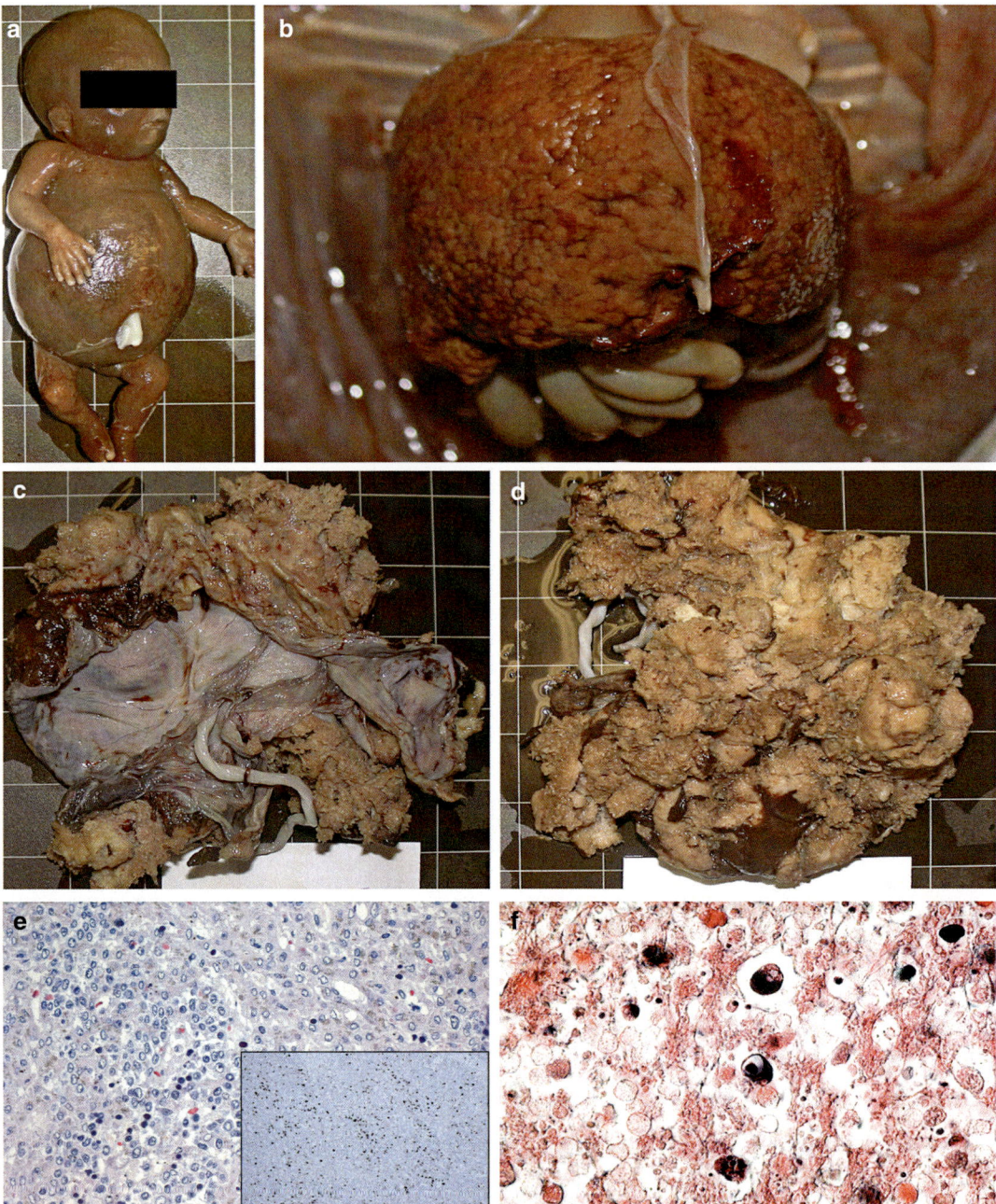

Fig. 18.53 Parvovirus B19 Infection. Parvovirus B19-induced non-immunological hydrops fetalis (**a**) with early cirrhosis of the liver (**b**), hydropic placenta (**c**, **d**), liver infection with immunohistochemical confirmation (**e** with inset), and inclusions in a placenta of a macerated twin fetus (**f**) (e, H&E stain, original magnification x200; inset, anti-Parvovirus B19 immunostaining, avidin-biotin complex, original magnification x100; h, H&E stain, original magnification x400)

Mosquito
- **Exogenous sexual phase**
  (Sporogony) (*7-20 days*)

Human
- **Endogenous asexual phase**
  (Schizogony)
  Liver, spleen, RBC (*48 hrs.*)

a

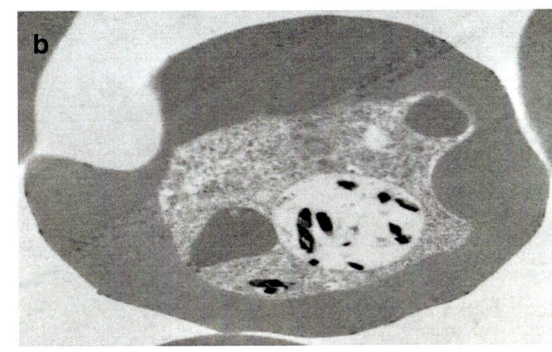

Sequestration
RBC - *PfEMP molecule*
Endothelium *CSA / HA receptors*
Interruption of blood supply
IUGR - 1st / 2nd trimester
Prematurity - 3rd trimester
Infant Mortality
Low Birth weight (<2500g)

c

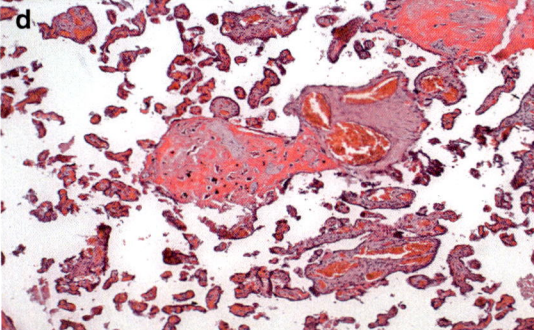

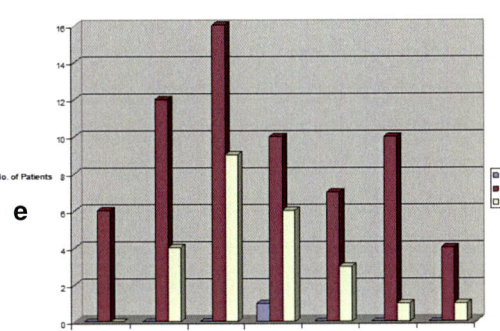

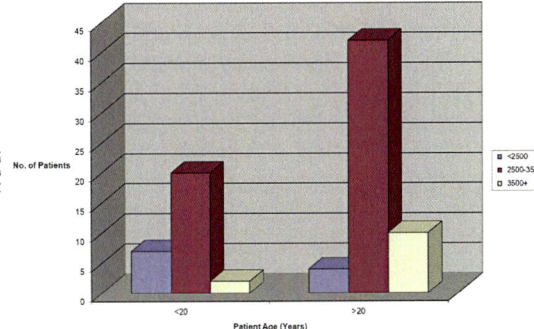

e

Fig. 18.54 Malaria Infection in Pregnancy. *Plasmodium* infection in the erythrocytes (**a**) as well as plasmodial pigments in a placenta (**b**) (d, H&E stain, original magnification x100). Data on mosquito, sequels, and malaria infection in pregnancy are also provided (a, **c**, and e) (see text and references for details)

Cytomegalovirus (Fig. 18.52)

- *DEF*: Enveloped, icosahedral, double-stranded DNA virus of the *Herpesviridae* family.
  - Family: *Herpesviridae*
  - 180 nm Ø enveloped virus
  - Double-stranded DNA virus
  - Sensitive to low pH, lipid solving liquid, and heat
  - Nucleus: Basophilic/amphophilic Cowdry type A single inclusion ("owl's eye cells")
  - Cytoplasm: Infected cells with up to four times ↑ of the cell size (cytomegaly)
  - Cytoplasm: Dense eosinophilic/basophilic multiple inclusions (± GMS and PAS)
- *EPI*: West countries, ~50% (prevalence), while low socioeconomic status countries, ~98% (prevalence). Double-peak infection in infancy and adolescents (early sexual contacts).
- *SYN*: Human herpesvirus 5 (HHV-5).
- *EPG*: Although the virus is ubiquitous worldwide, cytomegalovirus (CMV) can establish a latent infection in the host and undergo subsequent reactivation. CMV targets both epithelial and endothelial cells and pregnant women, who are infected with CMV and are at risk to transmit the virus across the placenta to the fetus. Another transmission from mother to baby can occur at the time of the delivery or by breastfeeding.
- CMV – infection routes are:
  - Primary infection
  - Latent infection (fibroblasts, epithelial cells, endothelial cells, smooth muscle cells, monocytes, and macrophages)
  - Reactivation (immunosuppression, pregnancy, neuropsychiatric disorders, drug abuse)
  - Chronic infection (prenatal and perinatal infection)
  - Exogenous reinfection (immunodeficiency, graft acceptors in organ transplants)
- CMV – transmission are:
  - Permanent intermittent virus elimination (saliva, genital secretions, urine)
  - Blood transfusion, transplantation
  - Trans-placenta or intrauterine (congenital)
  - Vaginal secretion, breast milk (perinatal, postnatal)
- *CLI*: CMV can be a challenge in some setting as follows:
  - Immunocompetent host: asymptomatic up to mononucleosis-like infection
  - Immunosuppressed host: severe disease (pneumonia, retinitis, enterocolitis, encephalitis)
  - Intrauterine infection: an acute illness with permanent damage to the product of conception, including microcephaly, encephalitis, chorioretinitis, pneumonia, hepatitis, bone marrow hypo-/aplasia, damage of the ear function, IUFD
  - Perinatal infection: chorioretinitis, pneumonia, hepatitis, bone marrow hypo-/aplasia, damage of the ear function, sepsis, shock, neonatal death
- Retinitis is the most common disease in coinfected HIV patients. Retinitis shows retinal necrosis, vasculitis, and choroiditis.
- CMV – intrauterine infection
  - ~50% transmission to the embryo (40% in the first and second trimesters and 70% in the third trimester)
  - ~1% of all newborns (West countries): CMV infection, but 5–10% show disease
  - ~10% of all newborns with CMV infection ⇒ will die for disease complications
- CMV – perinatal infection
  - Icterus, petechiae, retinitis, pneumonia (early)
  - Hepatomegaly, splenomegaly, thrombocytopenia (late)
- CMV avidity IgG/avidity index (AI) is as follows:

<40% AI	Low avidity	Acute infection ≤12 weeks
40–60% AI	Medium avidity	Recent infection
>60% AI	High avidity	Exclusion of a primary infection ≤20 weeks

Material for detecting CMV infection includes amniotic fluid, blood cells (e.g., viremia), breast milk, cervical secretion, sperm, throat swabs, tracheal secretion (e.g., pneumonia), umbilical cord, and urine.

CMV remains one of the most challenging opportunistic infections in immunocompromised hosts. In fact, it is known that several CMV gene products are devoted to escaping from the host immune response through which this herpesvirus establishes a balance in the host that sustains viral persistence with consequent sporadic shedding throughout the life of the host. Some characterized immunomodulatory gene products help CMV to escape the effectors of the innate and adaptive immune response, but few described functions target initiating events in the immune response are known. The first immunological steps of the host include cell-intrinsic alarm signals such as apoptosis or programmed cell death and activation of interferon (IFN) α/β. CMV suppresses apoptosis, which is conserved in CMV, but modulation of the IFN response by this virus is being discovered quite recently (Cui et al. 2014). A specific CMV virion protein can modulate the rapid induction of an IFN-like response in cells that follows virus binding and penetration. The use of functional genomics allowed to identify the role of pp65 (ppUL83) in counteracting this response. The mechanism entails a differential impact of pp65 on the regulation of interferon response factor 3 (IRF-3). It has been demonstrated that pp65 are sufficient to prevent the IRF-3 activation when introduced alone into cells. The crucial role of pp65 is played in the inhibition of the nuclear accumulation of IRF-3 triggered by pp65. By subverting IRF-3, the virus escapes the mechanism of the host to regulate the immune response properly. Diagnostics relies on serology (high prevalence), IgM antibodies, DNA-PCR and real-time PCR, antigen test (pp65), and virus isolation. CMV inclusions need to be differentiated from other inclusion bodies associated with other viruses, including HSV, VZV, adenovirus, BKV, and JCV. Although some of them do not play any or little role in pregnancy and infancy, they need to be kept in mind. The use of immunohistochemistry or in situ hybridization and correlation with culture and qualitative/quantitative PCR results are beneficial in confirming the diagnosis of CMV infection. Pasteurization is the destruction of all disease-producing bacteria and about 90–99% of all other bacteria, which may be responsible in affecting the product quality. Two common types of pasteurization techniques are used for pasteurizing breast milk, i.e., the low temperature long time (LTLT) and the high temperature short time (HTST). The LTLT involves heating the milk to 63 °C/145 °F and holding this for 30 min. A Holder variant of pasteurization is the Vat pasteurization, which is also known as batch pasteurization and heats every single particle of milk or cream in optimally designed and operated equipment at 63 °C/145 °F for 30 min. Milk banks mostly use LTLT. The HTST requires heating the milk faster and to a higher temperature than used with the Holder method. There is a classic HTST, which is the flash-heating pasteurization. HTST is a high-tech commercially used dairy pasteurization method, of which the heating of the milk reaches 72 °C/161 °F for 15 s only.

Parvovirus (Fig. 18.53)
Parvovirus B19 is a small single-stranded DNA virus with the associated *erythema infectiosum*, which is one of the exanthemata in childhood. In adults, anemia with transient aplastic crisis and persistent sequela, as well as arthropathy, can be seen. The overall seroprevalence of parvovirus B19 in serum from adult individuals in Western countries is ~70%. In pregnancy, acute infection with parvovirus B19 can cause several severe complications in the fetus, such as anemia, hydrops, neurological complications, and IUFD. Parvovirus B19 infection is classified as non-immunologic hydrops fetalis (NIHF) and probably is one of the significant differentials for non-immunologic causes of hydrops in fetuses. Early diagnosis and therapy of intrauterine parvovirus B19 infection are crucial for preventing fetal complications. Testing maternal serum for IgM antibodies against parvovirus B19 and DNA detection by PCR can confirm maternal infection, and ultrasound investigation of the fetus is routinely applied for diagnosis of fetal hydrops. The highest rate of fetal hydrops is observed if maternal infection occurs in the first 20 weeks of gestation and intrauterine transfusion might be useful for the treatment of fetal anemia. In fact, the proportion of fetuses with severe hydrops that survives following fetal transfusion is ~85%. During the second trimester of pregnancy, the fetus is particularly susceptible to parvovirus B19 infec-

tion, because at this stage, erythroid progenitor cells derived from the fetal liver are preferred targets of the virus. Microscopically, the characteristic feature of this infection is evident in the identification of eosinophilic intranuclear inclusion bodies in circulating normoblasts and their precursors in fetal organs. The histologic diagnosis is also feasible in stillbirth fetuses with maceration. Thus, a thorough microscopic investigation is crucial. Serologically, 1 week after maternal infection, peak viral load occurs, and peak IgM levels are seen after another week. Vertical transmission occurs 1–3 weeks after maternal infection, suggesting that fetal infection occurs during the maternal peak viral load. Parvovirus B19-derived fetal hydrops occurs in 95% of cases within 12 weeks of maternal infection and, currently, viral DNA detection in maternal blood is the best diagnostic sensitivity for identifying maternal parvovirus B19 infection.

Malaria (Fig. 18.54)

Malaria is the fifth most important cause of illness in sub-Saharan Africa. Four plasmodia are sensible to human, i.e., *Plasmodium falciparum*, *Plasmodium vivax*, *Plasmodium ovale*, and *Plasmodium malariae*. *P. falciparum* stands out as the deadliest of the four malaria parasites that infect humans. Maternal anemia, fetal growth restriction, and preterm labor are conditions associated with low birth weight and can complicate malaria infection contributing to fetal morbidity and mortality. Perinatal mortality associated with malaria infection ranges between 25 and 80 per 10^3 infants per year where the virus is endemic. However, malaria can be asymptomatic, and the link between the disease and perinatal morbidity is not always apparent in areas of stable endemic malaria where pregnant women have acquired immunity. Although malaria parasites do not cross the placental barrier under normal conditions, infection of the fetomaternal interface may play a crucial role in determining perinatal morbidity. Since malaria parasites accumulate within the placental tissues, the placenta is the ideal site to study delivery outcomes in women with malaria infection. The placental infection has been linked to up to 35% of preventable cases of low birth weight in areas where the disease is endemic. The prediction of fetal morbidity is safely based on placental studies more than maternal peripheral blood. Numerous studies have revealed that a microscopic examination of the placenta is more sensitive and accurate than maternal peripheral blood. Placenta malaria infection is based on the presence or absence of parasites and of malaria pigment in monocytes or fibrin deposits and the scrupulous examination of the intervillous space (IVS) is key.

The placenta may be infected from malaria parasites. Pathological characteristics of malaria infection on placenta can be detected on:

- Conventional light microscopy
    - Intervillositis with intra-syncytial pigment and scattered necrosis
    - RBC and macrophages containing parasites in the IVS and blood vessels
    - Cytotrophoblast proliferation and ↑ thickness of trophoblast basement membrane (TBM)
- Electron microscopy
    - Pleomorphism of macrophages containing parasites
    - Attenuation/loss of surface microvilli of the syncytiotrophoblast
    - Trophoblast necrosis and abnormal TBM

Malaria in pregnancy is categorized by the accumulation of infected erythrocytes in the placenta, and the critical ligand is a *P. falciparum* erythrocyte membrane protein 1 (PfEMP1) family member, termed VAR2CSA (Variant surface antigen 2-chondroitin sulfate A). This molecule seems to be the principal ligand responsible for adhesion to chondroitin sulfate A (CSA). Yosaatmadja et al. (2008) characterized VAR2CSA-deficient *P. falciparum*-infected erythrocytes selected for adherence to the BeWo placental cell line. These authors found that infected erythrocytes with characteristics similar to the CS2KO (a parasite line CS2 carrying a disrupted *var2csa* gene) have a limited role in the pathogenesis of placental malaria. VAR2CSA appears to be the significant ligand for placental adhesion and could be the basis for a vaccine against pregnancy malaria. Obstacles to vaccine development include the large size, high cysteine content, and sequence variation of VAR2CSA. Two products seem to be ready to enter human clinical studies shortly based on N-terminal VAR2CSA fragments that have a high binding affinity for CSA. Glycosaminoglycan (GAG) consists of repeating disaccharide units with variable type and level of sulfation. Thrombomodulin is a transmembrane thrombin-binding protein with one CSA chain and is considered to play a role in the adhesion of infected RBCs to CSA both *in vivo* and *in vitro*. Although the pattern of expression of thrombomodulin and the cytologic adhesion pattern of infected RBCs on the syncytiotrophoblast are different, 80–90% of the total thrombomodulin is in the chondroitin sulfate form, and the remaining 10–20% lacks chondroitin-4-sulfate chains. High levels of low sulfated CSA proteoglycans have been detected in the placental IVS and support the adhesion of infected RBCs, thus suggesting that *in vivo* adhesion of IE to CSA occurs in the IVS and not to the syncytiotrophoblast layer.

A variety of non-neoplastic lesions can present clinically and radiologically as uneventful or life-threatening conditions and may result in a hysterectomy. In such situations, the pediatric pathologist or gynecologic pathologist has an essential role to play in determining the nature of these lesions. Awareness of the development of the embryo at an early stage and entities that can mimic malignancies assists the physician in the correct diagnosis of these lesions.

18.7.2 Premature Rupture of Membranes (PROM)

- *DEF*: Pregnant women who complain of watery vaginal discharge need to be investigated immediately for premature rupture of membranes (PROM). The diagnosis relies on the vaginal pool or noticeable fluid leakage from the cervix. The fluid can be tested for its neutral or alkaline pH (different from the acidic cervix environment) with phenaphthazine (nitrazine paper) or by the identification of the characteristic "arborization" or "ferning." The fern-like pattern occurs when high levels of estrogen (e.g., just before ovulation, amniotic fluid) induce sodium chloride within the cervical mucus to crystallize on mucus fibers. In case of absent fluid, several procedures may take place, ranging from less invasive, such as a vaginal tampon, or more invasive techniques, using a transabdominal injection of a dye (e.g., indigo carmine, Evans blue, fluorescein). The use of methylene blue is not endorsed because it may cause fetal methemoglobinemia. Spontaneous PROM often leads to the onset of labor, and management issues are paramount.
- Prolonged PROM is defined by a latency period longer than 24 h. Epidemiologically speaking, PROM is around 3% before 37 weeks of gestation, but the reality is that between 30% and 40% of preterm babies are born to women who have PROM. Thus, PROM is the identifiable leading cause of preterm delivery worldwide.
- *PGN*: Apart from preterm delivery, there is an infection in approximately 30% of women who have PROM, and these patients are highly likely to present with chorioamnionitis, endometritis, and neonatal septic and septic shock than women who have a negative amniotic fluid culture on admission. Etiology of the infection includes mycoplasmas (*Ureaplasma urealyticum* and *Mycoplasma hominis*), *Streptococcus agalactiae* (group B streptococcus), *Fusobacterium*, and *Gardnerella vaginalis* as well as *H. influenzae*, although this last microorganism is quite uncommon in the female genital tract.

18.7.3 Fetal Growth Restriction (FGR) and *Dys*maturity

As we defined in Sect. 18.3.4, FGR is a functional status of the fetus, who is unable for one or several reasons to achieve the genetically determined potential size. FGR is different from SGA, which is a fetus with growth ≤10th percentile for weight at that gestational age, because not all SGA fetuses are growth restricted, because some of them may be constitutionally small. Likewise, not all fetuses lacking to meet genetic growth potential are ≤10th percentile for estimated fetal growth (EFG).

- *SYN*: Intrauterine growth restriction (IUGR).
- *EPI*: Fetuses situated ≤10th percentile can be in one of three categories:
  - Category A (~40%): High risk of potentially preventable perinatal death (not only morbidity)
  - Category B (~40%): Constitutionally small fetuses with high risk to preterm labor (prematurity)
  - Category C (~20%): SGA with pathology linked to chromosomal or multifactorial etiology

The pediatric pathologist approaching the autopsy of a fetus with FGR and the examination of its placenta needs to acquire fetal karyotype, fetal blood and swabs as well as maternal serology for infectious diseases, and detailed environmental exposure history.

A partially controversially discussed concept of growth is its symmetry, which may direct to some possible causes. *Prematurity-related complications* (Figs. 18.55 and 18.56) include:

- Respiratory distress syndrome and hyaline membrane disease due to lung immaturity (most babies have mature lungs by 36 weeks of gestation) and respiratory distress syndrome (RDS) (lack of surfactant in the lungs that helps prevent the alveoli from collapsing)
- Transient tachypnea (recovery within 3 days and no needed treatment)
- Bronchopulmonary dysplasia (BPD) (type I and type II)
- Pneumonia (inflammation with inadequate amounts of O_2)
- Apnea/bradycardia
- Infection/sepsis
- Jaundice (liver immaturity, unconjugated hyperbilirubinemia)
- Intraventricular hemorrhage (IVH) with a ↑ risk of bleeding in the brain and liquor system if the baby is <34 weeks of gestation with neurological complications such as cerebral palsy, mental retardation, and learning difficulties
- °C/°F control instability (low fat and thin skin of premature babies)
- Immature gastrointestinal system (immaturity to absorb nutrients effectively) with potential need of total parenteral nutrition (TPN) and potential therapy-refractory anemia
- Patent ductus arteriosus (PDA) with breathing difficulties after delivery because of the patency of the communication between the pulmonary artery and aorta
- Retinopathy of prematurity (ROP), which affects most premature (24–26 weeks of gestation) babies
- Necrotizing enterocolitis (NEC) due to a complex interaction between poor blood flow in the bowel, tissue immaturity, and abnormal intestinal flora

Dysmaturity

Dysmaturity is a variety of non-neoplastic lesions that can present with babies born before 37 weeks or after 41 weeks of gestation.

Asphyxia is one of the most dreaded complications of prematurity. Asphyxia prevalence ranges between 1 and 7 per thousand live-born babies. The Greek etymology means no pulse, and there are various definitions (ἀσφυξία: ἀ- "absent" and σφύξις "heartbeat"). It is often seen as an abnormal cardiotocogram (CTG) and an abnormal scalp pH/lactate. The babies with asphyxia do not breathe within 30 s and do not scream within 60 s. In the medical charts or electronic medical records, it is often recorded as an Apgar score of 0–3 at 5 min. The words "Appearance (skin color), Pulse (heartbeat), Grimace, Activity, and Respiration" compose the acronym "APGAR" that involves scoring each parameter to check a baby's health. Each of the five parameters is scored twice or three times on a scale of 0 to 2, with 2 being the best score. Moreover, the pH is usually less than 7 (acidosis). In a term child, neuronal damage of the gray matter is found with localization of the cortical and subcortical areas

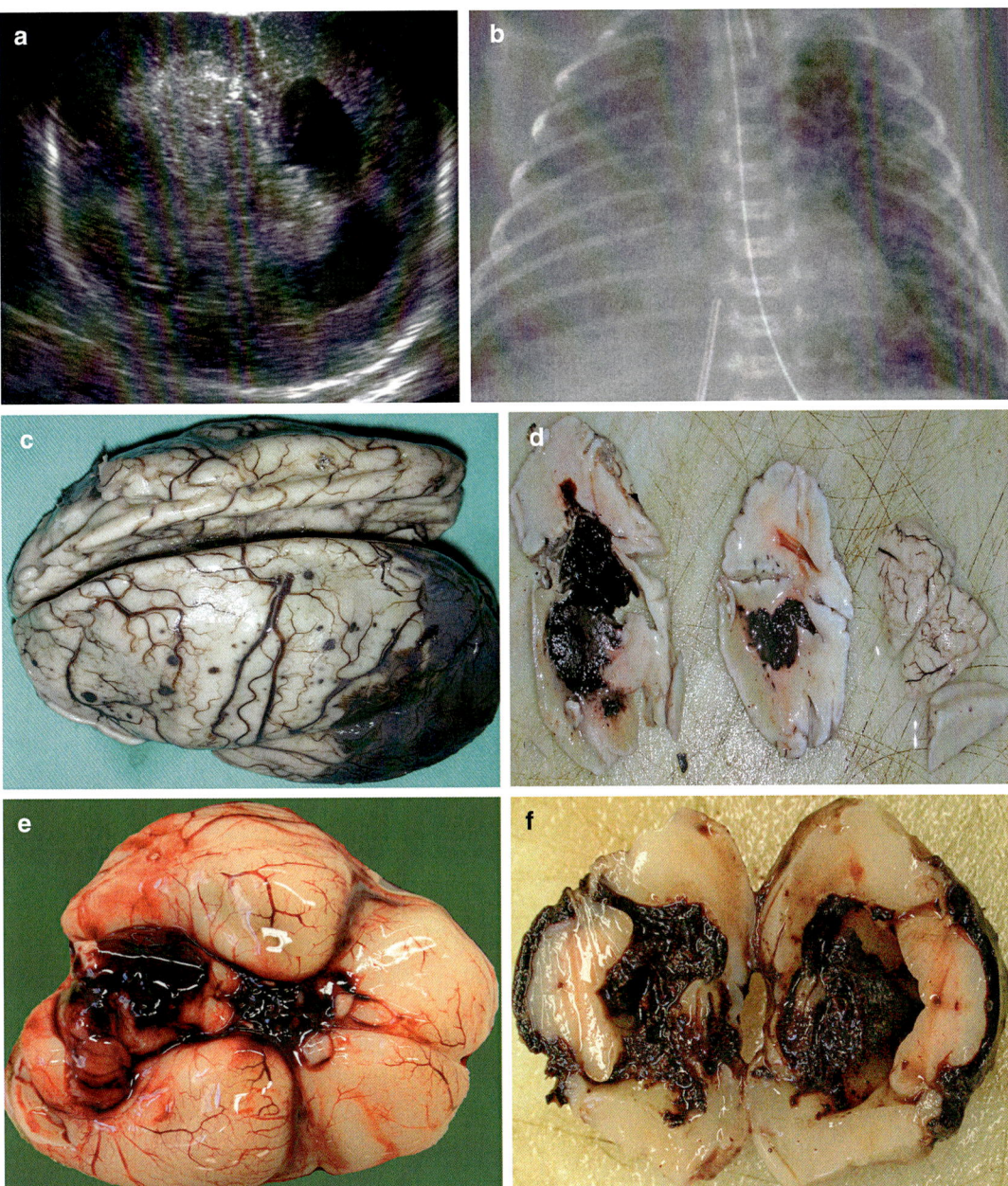

Fig. 18.55 Prematurity Complications. Prematurity complications with brain hemorrhage (**a**), hyaline membrane disease with respiratory distress syndrome (**b**), occipital hemorrhage with involvement of the ventricular system (**c**, **d**), and basal brain hemorrhage with involvement of the ventricular system (**e**, **f**)

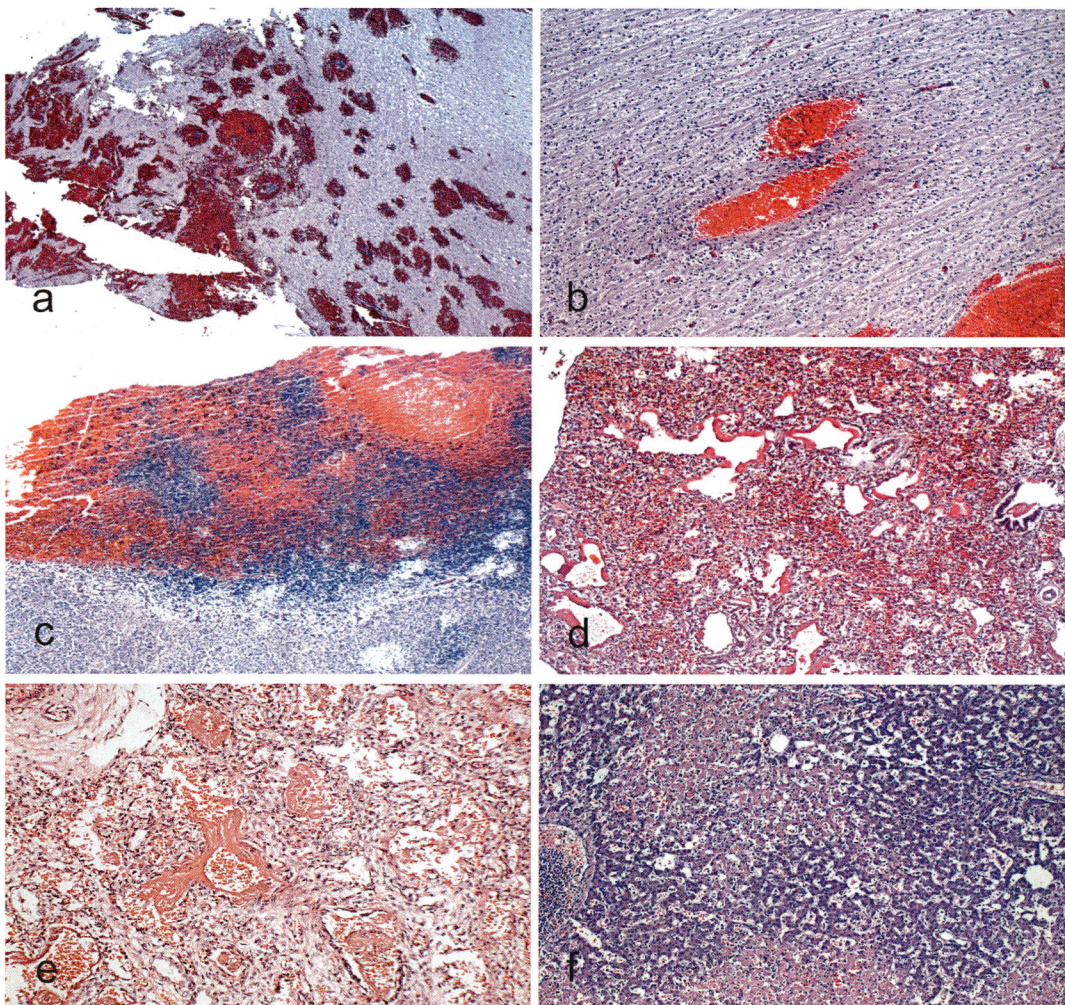

Fig. 18.56 Prematurity Complications. Prematurity complications with brain hemorrhage in the periventricular (**a**), cortical (**b**), and germinal matrix (**c**), hyaline membrane disease of the lung (**d**, **e**), and confluent pericentral necroses (**f**)(a-f, H&E stain, a, original magnification x40; b, original magnification x100; c, original magnification x40; d, original magnification x100; e, original magnification x100; f, original magnification x100)

and basal ganglia. The term for this neuropathology is hypoxic-ischemic encephalopathy (HIE). Lesions that may be found in a baby who died for asphyxia are parasagittal infarctions and intracerebral bleeding. The clinical characteristics of HIE can be described as mild, moderate, or severe, and the most common classification or staging system has been suggested by the Sarnat scale developed by Sarnat and Sarnat in 1976. The Sarnat score combines clinical and EEG findings (I, II, III = mild, moderate, severe). The grading system presented here is called the modified Sarnat score, and the different stages are detailed as below.

Histopathology of hypoxic-ischemic encephalopathy includes:

- *Gray matter lesions*
- Neuronal necrosis of high energy requiring areas, cerebral cortical necrosis, thalamus and basal ganglia necrosis, mesencephalon and brain stem necrosis, pontosubicular necrosis

(PSN), global cerebral necrosis, spinal cord and cerebellum necrosis
- *White matter lesions*
- Periventricular leukomalacia and cerebral white matter gliosis
- *Combined gray and white matter lesions*
- Parasagittal cerebral injury, bilateral necrosis of the cerebral cortex and subjacent white matter (watershed areas between the three major cerebral arteries), and focal/multifocal infarcts
- *HIE late stages*
- Ulegyria, sclerotic gyri, multifocal cystic encephalopathy, marmorated status, unifocal pseudocyst (cystic periventricular leukomalacia, white matter hypoplasia, porencephaly, and hydranencephaly/"basket brain")

In infants with a delivered age ≥29 gestational weeks (GW), neuronal karyorrhexis is restricted to the pons and subiculum (PSN). Conversely, in infants <29 GW, neurons in other brain regions, such as the basal ganglia, thalamus, inferior olivary nucleus, cerebellum, and cerebral cortex, are also involved (Sohma et al. 1995).

Hydranencephaly occurs often as result of a severe hypoxic-ischemic damage occurring toward the end of the second trimester, and preterm infants develop periventricular leukomalacia. A classic pattern of full-term infants is the bilateral necrosis of the cerebral cortex and subjacent white matter, which is maximally recognized in the parasagittal area. Cerebral palsy (CP) is a term collecting group of neurological disorders with potential neuropathologic correlates. It involves a disturbance of movement as well as of motor function. It should be considered permanent, but it would be not correct if considered stable because it is changing. CP is due to a nonprogressive interference of the neural circuits, lesion, or abnormality. CP-associated anomalies include cognitive, visual, hearing, and nutritive and may be associated with epilepsy. The prevalence is 2:1000 live-born babies, and it may be classified as prenatal, perinatal, and postnatal. The most common etiologic factors are at prenatal onset. Prenatal causes include malformations or multiple congenital anomalies syndromes, intrauterine infection, vascular lesions (e.g., bleeding, infarction), and periventricular leukomalacia. Perinatal factors include HIE, intracranial hemorrhage, intraventricular hemorrhage (*hematocephalus internus*), infections of the CNS, and kernicterus. Postnatal factors include asphyxia due to drowning/near-drowning, cardiovascular surgery, infections of the CNS, intracranial hemorrhage, and infarctions. Clinically, CP can be spastic, dyskinetic, or choreoathetosis. Postmaturity-associated pathology comprises a series of diseases that occur when the baby is born after 42 weeks of gestation. However, the postmaturity syndrome is not confined to postdate infants but may arise from 40 weeks or even earlier. Fetal implications of postmaturity include fetal ossification of the skull (difficult baby's molding), macrosomia, cephalic-pelvic disproportion, shoulder dystocia, cephalohematoma, subdural hematoma, fractures, palsies, hypoglycemia, polycythemia, and hypothermia due to low-fat stores.

Persistent Pulmonary Hypertension of the Newborn (PPHN)

The condition called as "persistent pulmonary hypertension of the newborn (PPHN)" is persistent fetal circulation, i.e., a clinical syndrome resulting from the failure of the transition from the normal fetal to neonatal circulation, accompanied by a substantial burden of infant mortality, and perinatal and child morbidity. The risk factors for PPHN are emphasized in Italic in Box 18.63. The pulmonary flow of the human fetus is characterized by a relatively high vascular resistance, which causes the oxygenated blood from the placenta to go through the *ductus arteriosus* and the *foramen ovale* in the fetal systemic circulation. At birth and following the birth, lungs expand; there is a release of vasodilatory factors and fall of the pulmonary vascular resistance. In consequence, the right-to-left fetal hemodynamic shunt ends, and pulmonary blood flow increases.

PPHN features include:

> **Box 18.63 Risk Factors Associated with PPHN**
> - ♂ vs. ♀ of the infant
> - *Black/Asian* vs. Caucasian Ethnicity
> - *High BMI* (>27) (prepregnancy) vs. BMI <20
> - *Maternal asthma* vs. non-asthma
> - *Maternal DM* vs. non-DM
> - *BW >90th percentile* vs. BW 10th–90th percentile
> - *Cesarean section* vs. vaginal delivery
> - *Prematurity of late preterm type* (newborns 34–37 weeks of gestation)
> - *Postmaturity* (newborns >41 weeks of gestation)
>
> Notes: *BMI* body mass index, *DM* diabetes mellitus, *BW* birth weight

EPI: 1-2: 10^3 live births

CLI: Severe respiratory failure a few hours post-birth requiring intubation, NO administration, mechanical ventilation, and/or ECMO

PGN: Death (10–20%), chronic lung disease, seizures, and neurodevelopmental disabilities

Notes: *NO* nitric oxide, *ECMO* extracorporeal membrane oxygenation

The outcome may be quite dramatic with long-term sequelae because of both PPHN hypoxemia and the aggressive therapy used to contrast the condition. Some risk factors may weigh more than others. It seems that the "cesarean section" might affect babies with a rate of about 1 per 100 births.

18.8 IUFD and Placenta

Perinatal death is a dramatic event for families and practitioners. In the era of genomics and proteomics, the placenta examination is still crucial in determining the cause of death in case of miscarriage or stillbirth.

18.8.1 Fetal Death Syndrome (Intrauterine Fetal Demise, IUFD)

Fetal death syndrome or better known as intrauterine fetal demise (IUFD) is the death of a product of conception or fetus occurring during a pregnancy, which is usually or apparently "normal and uncomplicated." IUFD has a specific period, which is after the 20th week of gestation (or after the 22nd week according to the American College of Obstetricians and Gynecologists). Moreover, IUFD has a weight limit, which is the body weight of the fetus. This weight should be ≥500 g (or ≥350 g in some states or countries). Miscarriage is the death of a product of conception or fetus before 20 weeks of gestation. *Intrauterine fetal demise or death* (IUFD) is a death that usually occurs after 20 weeks' gestation. Before 20 weeks' gestation, the demise is typically classified as a spontaneous abortion. Fetal death is reported differently not only in several countries but also in different states of the USA. However, the definitions vary among countries and even among states or provinces. IUFD occurs in about 1% of pregnancies. The fetal death is about 6–7:1000 total births and accounts for approximately half the perinatal mortality in most industrialized countries. This rate varies obviously and is probably highly dependent on the quality of medical care available in the land of the pregnant woman. Fetal death is defined by the World Health Organization (WHO) as the intrauterine death of a fetus, which occurs at any time during pregnancy, while the term "stillbirth" is applied when two conditions occur, including the has no signs of life and the event occurs at or after 28 weeks of gestation (2018).

Categories of signs and symptoms that may accompany miscarriage or IUFD are crucial, and an electron fetal monitoring (EFM) may be very helpful.

Definitions of Electrical Fetal Monitoring
Decelerations: Decelerations-transient slowing of fetal heart rate (FHR) being below the baseline level of >15 bpm and lasting for at least 15 s and classified as:

- *EARLY*, if the deceleration begins with the onset of contraction, but it returns to baseline as the contraction ends. Etiology is often centered to head compression. Early decelerations should not be disregarded if they appear "early" in labor.
- *LATE*, if there is a uniform periodic slowing of FHR with the onset of the contractions after the peak (and after the length of the uterine contraction), often with a slow return to the baseline, and being repetitive late decelerations increase the risk of umbilical artery acidosis and Apgar score <7 at 5 min with an increased risk of cerebral palsy. Etiology: Acute and chronic fetoplacental vascular insufficiency. Late decelerations are precipitated by hypoxemia and are associated with respiratory and metabolic acidosis. Late decelerations are frequent in women with pregnancy-induced hypertension (PIH), diabetes mellitus (DM), fetal growth restriction (FGR), or another form of placental insufficiency. PIH considerably affects the placenta by reducing its weight and dimensions. These morphological changes may induce placental insufficiency as a result of compromised utero-placental blood flow (Salmani et al. 2015).
- *VARIABLE*, if the decelerations show a variable intermittent periodic slowing of FHR with rapid onset recovery and isolation. It may look like that it comprises different types of deceleration in timing and shape. Atypical variable decelerations are associated with an increased risk of umbilical artery acidosis and Apgar score <7 at 5 min and are worrying when the rule of 60 is exceeded (i.e., a decrease of 60 bpm or rate of 60 bpm and more prolonged than 60 s). Meaning: Cord compression of the umbilical cord, oligohydramnios with or without ROM.
- *PROLONGED*, when FHR drop is 30 bpm or decelerations lasting for >2 min and are pathological when crosses two contractions, i.e., 3 min. There is a reduction in O_2 transfer to the placenta, and prolonged decelerations are associated with poor neonatal outcome.

  Etiology: Cord prolapse, maternal hypertension, and uterine hypertonia.

The factors linked to IUFD and stillbirth are important to keep in mind when doing an autopsy. The most common "clear" causes of IUFD include cord accidents, decompensation of the functional capacity of the placenta, *hydrops fetus et placentae*, infection (chorioamnionitis, omphalitis, villitis), multiple congenital anomalies (MCA) incompatible with "intrauterine" life, multiple pregnancies-associated complications (TTTS, TRAP, pagus), and obstetric complications (abnormal labor, trauma, drugs).

The most common maternal factors associated with stillbirth include diabetes mellitus, hypertensive disease, maternal age <20 years or >35 years, autoimmunity (e.g., systemic lupus erythematosus), and maternal drug abuse.

A thorough examination of the placenta may more easily let clarify the placental causes of IUFD. There are some nonspecific risk factors, which are known to ↑ the risk of IUFD. These include specific ethnics (e.g., Afro-American), age (early or advanced maternal or paternal age), abnormal body mass index or rheologic status (e.g., obesity and maternal hemoconcentration), history (IUFD, maternal infertility, genetic or familial background, SGA in previous pregnancies), abnormal placentation (e.g., multiple pregnancy), and infection (e.g., maternal colonization with specific pathogens such as GBS or *Ureaplasma urealyticum*).

Complications may include disseminated intravascular coagulation (DIC), shock, or death, mainly if uterine rupture is associated. In all cases of baby loss, both parents, particularly the mother, will go through a period of grief, which is the specific process of adapting to such a loss. There are feelings of guilt and helplessness, and, in case of a postmortem investigation, a meeting with the pathologist may be essential to facilitate the end of the period of grief. It is important to emphasize the relationship between obstetrician and gynecologist, and pediatric pathologist is necessary, and the prenatal healthcare support team also shares the emotions of baby loss. It is also crucial to perform a full investigation of the mother, fetus, and placenta for subsequent pregnancies, and healthcare will be individualized depending on the clarifications obtained from these investigations and on the needs of the woman. In most cases, more frequent prenatal visits need to be scheduled for these couples, and specialized laboratory examinations may be recommended to provide some reassurance to achieve a healthy outcome of next pregnancy.

18.8.2 Step-by-Step Approach in the Examination of a Placenta

Placental gross examination starts with reading the requisition and checking the data provided in the specimen container and the requisition match. Then, it is essential to evaluate the adequacy of the sending (i.e., if enough formalin has been sent or in case fresh specimens have been sent to fill with formalin). It is wholly subject to centers and labs if the placenta would need a fixation or can be grossed fresh. In many centers, the placenta is refrigerated for 1 week at 4 °C. If an obstetrician or a perinatologist requests a placental examination, the specimen is obtained from the fridge and examined; otherwise, it should be incinerated. The placenta should not be frozen, because freezing distorts the villous architecture, obscures the correct interpretation of meconium staining, and may compromise some placental findings. Fixation in buffered formalin may cause an increase in placental weight corresponding to 10-12%. The fixation of placenta prevents any microbiological tests to be successful, determines the distension of blood vessels, and causes congested thrombotic lesions to appear hemorrhagic. The following steps target the exam of the placenta following optional photographic documentation.

1. Maternal surface
2. Umbilical cord
3. Fetal surface
4. Membranes
5. Cut surface

Maternal surface needs to be evaluated first and especially if it is complete or fragmented, weighed (after removal of membranes and cord), noted lobation (accessory lobes, indentation >50%), atrophy, retroplacental hemorrhage or hematoma (extent in cm and marginal vs. central), as well as a cavum or texture deficiency (so-called compression crater).

The second step refers to the *umbilical cord* that needs to be examined about the length (short <35 cm, long >85 cm), Ø (<0.8 cm or thin cord), and distance to nearest edge and type of insertion (marginal, membranous, furcated, amnion web). In addition, it should be noted some qualities of the cord, including the color (yellow, gray, green, red), twisting (max number of diagonal coils per 10 cm of length), number of vessels and presence or absence of thrombi, knots, abnormal surface, hemorrhage, and abrupt changes.

Fetal surface and *membranes* need to be examined concerning color (blue-gray, gray-whitish, green, brown), membranous insertion (e.g., marginal, circumvallate), completeness with or without a note on the point of rupture, amniotic hemorrhage, amnion nodosum, and amnion bands. Amnion nodosum needs not to be confused with yolk sac remnant on the fetal surface of the placenta at term, which has no clinical significance. Amnion nodosum is constituted by many gray-tan nodular swellings, which are a localized accumulation of *vernix caseosa* with embedded desquamated fetal cutaneous cells. Amnion nodosum is indicative of oligohydramnios or anhydramnios.

The *cut surface* needs to be investigated following thin slices of 0.5–1 cm of thickness with or without a magnification lens. Solid lesions include *infarcts* (number, size, location, percentage to the full placental surface, and color with red for recent or fresh infarcts and gray-white for old infarcts as well as *non-infarcts solid lesions*. Photographic documentation showing the percentage of the affected area is beneficial at the time of sign out and clinic-pathological correlation. Non-infarcts solid lesions include fetal thrombi or regions of avascular villi, perivillous fibrin in the form of plaques or extensive ("maternal floor infarction"), and chorioangiomas. Nonsolid lesions include hemorrhagic lesions or thrombi, which are subtyped according to the location (intervillous, subchorionic, and basal), and cystic lesions, including septal cysts and mesenchymal dysplasia.

Guidelines on Grossing and Histology of Placenta

The *membranes* are handled in rolls, and two rolls in one cassette for one paraffin block. Two cassettes are prepared, one for membranes close to the edge of the disc and one at the site of rupture. The histology of the membranes may disclose abnormalities of the maternal arterioles (e.g., atherosis, mural hypertrophy), acute or chronic inflammation (e.g., neutrophils, lymphocytes, plasma cells), exogenous pigments (e.g., meconium, hemosiderin), laminar necrosis due to maternal small-vessel disease, hemorrhage due to marginal abruption, and plaques that may indicate vanishing twin or accessory lobe. A vanishing twin is defined in occasion of one of a set of twin or multiple fetuses that disappear in the uterus during pregnancy. A vanishing twin occurs as result of a miscarriage of one twin. The other twin, placenta or, even, the mother, may absorb the fetal tissue.

The *umbilical cord* is examined with cross sections, and two cross sections are put in one block. Two cassettes are prepared, being one for cross sections proximal to the cord insertion and one for cross sections distal to the cord insertion and close to the fetus. The histology of the membranes may confirm the number of the blood vessels, disclose any inflammation (veins, arteries, Wharton jelly, and periphery), and determine the Ø (< or ≥0.8 cm), microorganisms (e.g., *T. gondii*, *T. pallidum*, *Candida* spp.), necrosis (e.g., meconium related or syphilis associated), and mineralization (thrombi, hemorrhage and its age, and inflammation). It is wise to sample the cord at ≥3 cm from its insertion to avoid an incorrect diagnosis of the single umbilical artery because umbilical cord vascular anastomoses may occur near the placental disc.

The *placenta* is sectioned in slices of 0.5–1 cm of thickness, and two full-thickness blocks 2 cm distant from the cord insertion are taken. Both blocks should not show any evidence of pathology, including infarcts, thrombi, etc. Additional blocks are taken from any other abnormal change in the parenchyma. The chorion plate needs to be studied for *leukocytes*, *thrombi*, *pigment* (meconium, iron, calcium), and *necrosis*. The stem villi may show *thrombi, vasculitis, cysts*, or *vascular anomalies*. The distal villi are pointing to the maturation of the villous tree. A decrease of the number of distal villi to stem villi may indicate *accelerated maturation* (aka distal villous hypoplasia, "*maturitas praecox*"), while an increase of the diameter of the distal villi with central capillaries would point to immaturity (aka *delayed maturation*, "*maturitas tarda*"). The trophoblast is investigated mainly to identify neoplastic and non-neoplastic lesions, such as *syncytial knots* and *villous agglutination*. The stroma of the villi may show *edema, fibrosis, hemorrhage*, and/or *lymphocytes* or other inflammatory cells. Capillaries may show nucleated RBCs and a variation in their number. A decrease of capillaries is characteristic of *avascular villi*, while an increasing amount is indicative of *chorangiosis*. Before 6 weeks of gestation, very few or no capillary lumina are identified. By about 8 weeks of pregnancy, only nucleated precursors of erythrocytes are found in the villous capillary spaces. These nucleated precursors will decrease progressively, and only 1/10 of the red blood cells are nucleated between 10 and 12 weeks of gestation and virtually none after the first trimester. The IVS is locked for *hemorrhage* (age), *leukocytes*, and *fibrin* and/or *fibrinoid*. The basal plate is studied for *bleeding* (age), *chronic inflammatory cells* (specifically plasma cells), and *accretion*, such as placenta increta, accreta, and percreta (Box 18.64).

At term, average length of the cord is 50–60 cm, and the diameter is 1–1.5 cm. Thus, any cord with length <40 cm short and <35 cm is abnormally short! Any cord with length >70–80 cm is abnormally long! At 18 weeks, the cord should be 22–23 cm, while at 34 weeks, the cord should reach 50–55 cm. At least 32 cm of cord are required for a term delivery. Moreover, abnormal umbilical cord coiling is associated with adverse perinatal outcomes as indicated by Machin et al. (2000). Practically, 0.2 coils/1 cm length (1 full coil/5 cm) need to be present. Thus, hypocoiled umbilical cord is defined by a coiling index (CI) of

> **Box 18.64 Light Microscopy Target Features**
> 1. Chorioamnionitis, funisitis (omphalitis), meconium staining of the membranes
> 2. Cord vasculitis, amnion nodosum, and hydrops (hydrops fetus et placentae)
> 3. Discolorations-related lesions (e.g., infarction, thrombi, cysts)
> 4. Maternal floor infarction, intervillous fibrin deposition, and perivillositis
> 5. Villous maturity (e.g., avascularity, chorangiosis, fetal NRBC, intravillous hemorrhage, and hemosiderin deposition)
> 6. Villitis of specific type (e.g., proliferative, granulomatous, necrotizing)
> 7. Villitis of unknown etiology
> 8. Microorganisms or degenerated products of them (e.g., *P. malariae*)
> 9. Tumors (chorangioma, choriocarcinoma, etc.)

> **Box 18.65 Categories of Placental Diagnoses**
> 1. *Amniotic fluid sterility*
>   – Chorioamnionitis/acute villitis/intervillositis/intervillous abscesses
> 2. *Leukocytic status (inflammation)*
>   – Infectious (specific villitis)/idiopathic (nonspecific villitis)
> 3. *Maternal vasculature*
>   – Abnormal vascular development/occlusion/integrity loss
> 4. *Fetal vasculature*
>   – Abnormal vascular development/occlusion/integrity loss
> 5. *Intervillous fibrin deposition*
>   – Massive perivillous fibrin deposition/maternal floor infarction/plaques
> 6. *Fetal-based lesions*
>   – Nucleated red blood cell precursors/meconium

<1 coil/5 cm length. Hypercoiled umbilical cord has a CI >0.3/5 cm length. Cord consequences can be dramatic. The pathology report should indicate to the clinicians about the following minimal six categories (Box 18.65) that may be relevant for the health of both mother and baby. Each pathological lesion should be graded and staged for extent (diffuse, patchy, multifocal, focal), intensity (grade), duration (stage), character (cellular composition), and activity (active or chronic). Of note, placental gross examination starts with reading the requisition and checking the data provided by the obstetrician and gynecologist. Synoptic data sheet for placenta examination (gross and histology) are displayed in Box 18.66.

In a typical pathology report, the following diagnosis may be used:

PLACENTA, UMBILICAL CORD, AND FETAL MEMBRANES, BIRTH:

- PLACENTAL DISC WITH A CENTRAL THROMBUS (4.5 CM MAX. DIMENSION)
- THREE-VESSEL UMBILICAL CORD WITHIN NORMAL LIMITS
- FETAL MEMBRANES WITHIN NORMAL LIMITS

PLACENTA, UMBILICAL CORD, AND FETAL MEMBRANES, BIRTH:

- THREE-VESSEL UMBILICAL CORD WITHIN NORMAL LIMITS
- FETAL MEMBRANES WITHIN NORMAL LIMITS
- PLACENTAL DISC WITH THIRD TRIMESTER VILLI, TWO SMALL PLACENTAL INFARCTS (1.2 CM AND 0.4 CM IN MAX. DIMENSION) AND PERILESIONAL PROMINENCE OF SYNCYTIAL KNOTS

Box 18.66 Synoptic Data Sheet for Placenta Examination (Gross and Histology)
Unit No.: ....
    Family Name – First Name: ....
    DOB: .... Gender: ....
    Location: ....
    Pathology No.: ....
    Physician: .... Delivery Unit No.: ....
    Previous Specimen/s: YES/NO (..........................)
Administrative QA
    Date of Delivery: ....
    Date Received: ....
    Date Grossed: ....
    Date and Time of Dictation: ....
    Date and Time of Typing: ....
    Print Date: ....
Clinical Information
    IUGR (BW below 2.5 kg or third centile): .... Maternal pyrexia: ....
    Two-vessel cord: .... PROM: ....
    Abruption: .... Fetal hydrops: ....
    Abnormal placenta shape: .... Gestational diabetes: ....
    Morbidly adherent placenta: .... Isoimmunization – Rh/ABO/Kell/Other ...*
    Preeclampsia: .... Maternal Group B Streptococcus
    Prematurity (<37/40): Stillbirth – antepartum/intrapartum
    Fetal abnormality Severe fetal distress (requiring NICU admission)
    Apgar scores: 1′ 5′ 10′ Other (must specify): ....
    Multiple pregnancy (specify): Monochorionic/dichorionic/multiple
    Complications/interventions (specify): ....
    Relevant previous medical/OBGYN Hx. (specify): ....
Diagnoses: Singleton placenta ...
Gross Description Data
    Cord ... × ... cm, cord pieces: .... Insertion: ....
    Blood Vessels (#): ....
    Membranes: complete/incomplete, narrowest width (... cm), clear/opaque, meconium (old/recent/none) Vascular thrombi (Yes/No and # ....) Calcification (Yes/No)
    Disc Dimensions: ... × ... × ... cm
    Parenchyma (color and consistency): red/pale/friable Abruption (Yes/No): hemorrhage/"crater"
    Infarcts (Yes/No, #, age, percentage to the cut surface parenchyma): ....
    Other: ....
Microscopic Findings
    Description by region: chorionic plate, stem villi, distal villi maturity, trophoblast changes, villous stroma, capillaries, intervillous space, and basal plate with decidua evaluation.
    Routine and Special Stains (underline where appropriate): H&E, PAS, Masson's trichrome, EVG, Gram, GMS, ZN, Auramine-Rhodamine, WS, other (............)
    Immunohistochemistry (..........................)

COMMENT:

There is no decidual or intravillous hemorrhage. The prominent syncytial knots are a nonspecific finding suggestive of (focal) ischemia.

PLACENTA, UMBILICAL CORD, AND FETAL MEMBRANES, BIRTH:

- THREE-VESSEL UMBILICAL CORD WITHIN NORMAL LIMITS
- FETAL MEMBRANES WITHIN NORMAL LIMITS
- PLACENTAL DISC WITH THIRD TRIMESTER VILLI, THREE INTERVILLOUS THROMBI (3.1 CM, 2.8, AND 0.5 CM IN MAX. DIMENSION) TOTALIZING MORE THAN 15% OF THE CUT SURFACE AND FOCAL PROMINENCE OF SYNCYTIAL KNOTS

COMMENT:

There is focal intervillous hemorrhage, and the presence of more than 15% of the cut surface showing intervillous thrombi is indicative of fetal villous bleeding. The prominent syncytial knots are a nonspecific finding suggestive of (focal) ischemia. Clinical correlation is needed.

Multiple Choice Questions and Answers

- PLA-1 Some viruses have cytoplasmic replication, while others have nuclear replication. In which of the following viruses is the replication not in the cytoplasm?
  (a) Herpes simplex virus
  (b) Paramyxovirus
  (c) Picornavirus
  (d) Poxvirus
  (e) Reovirus
- PLA-2 Rubella is a usually mild viral disease that typically occurs in childhood in children who did not get vaccines against this virus. Rubella virus crosses the placenta of infected women during pregnancy. In the first trimester, the rubella virus causes miscarriage, or congenital rubella syndrome can develop. Typically, congenital rubella syndrome includes cardiac, auditory, sensorineural, and ocular abnormalities. What is the best method to diagnose congenital rubella?
  (a) Detection of rubella virus IgM in fetal blood or viral genome in amniotic fluid, fetal blood, or chorionic villus biopsies
  (b) Detection of rubella virus IgG in fetal blood or viral genome in amniotic fluid, fetal blood, or chorionic villus biopsies
  (c) Detection of rubella virus IgG in fetal blood of the newborn
  (d) Detection of rubella virus in the urine of the mother
- PLA-3 Congenital syphilis became a rare disease in Western countries, but it became more often discussed in the differential diagnosis following the recent increased migratory flows. Infected infants may harbor severe sequelae, including hydrocephalus, cerebral palsy, musculoskeletal deformity, and sensorineural hearing loss. What is the best method to diagnose congenital syphilis?
  (a) Dark-field microscopy examination of desquamative or ulcerative skin lesions, nasal discharge ("snuffles"), and placenta
  (b) FTA-ABS IgM test (fluorescent treponemal antibody absorption immunoglobulin M test)
  (c) AgNO3 staining of spirochetal bacteria
  (d) Long bone radiography
  (e) Updated Wassermann reaction of antiphospholipid antibodies including the introduction of a sample of blood to the antigen of cardiolipin extracted from the bovine muscle or heart
- PLA-4 What is the best presumptive procedure in case of a suspect of congenital syphilis?
  (a) An infant's nontreponemal titer (VDRL or RPR) is twofold higher than that of the mother when both blood samples are drawn at the time of delivery.
  (b) An infant's nontreponemal titer (VDRL or RPR) persists or increases after birth.
  (c) An infant's treponemal antibody titer (FTA-ABS or MHA-TP) remains positive at 6 to 9 months of age.

(d) An infant's treponemal antibody titer (FTA-ABS and MHA-TP) remains positive at 6 to 9 months of age.
- PLA-5 Schwalbe's and Goerttler's deterministic periods indicate which period is more susceptible to give rise to congenital defects by environmental factors. Which period is most vulnerable for the human embryo (postfertilization day (PFD))?
  (a) PFD 1–14
  (b) PFD 15–60
  (c) The early second trimester
  (d) The late second trimester
  (e) Between the 22nd week and the 32nd week of pregnancy
- PLA-6 Which of the following viruses/bacteria are associated with intranuclear inclusions in precursors of the red blood cells?
  (a) *Fusobacterium nucleatum*
  (b) Cytomegalovirus
  (c) Parvovirus B19
  (d) Herpes simplex virus
  (e) Varicella-zoster virus
- PLA-7 Placenta thrombi may be particularly damaging for the correct hemodynamics of the placenta and determine the death of the fetus if they are compromising the nutritional function of the placenta. An intravascular thrombus can occur in a series of conditions. Which one of the following states is unlikely to start an intravascular thrombogenesis and form a thrombus?
  (a) Hyperviscosity syndromes
  (b) Hypercoagulable states
  (c) Endothelial damage
  (d) Blood stasis
  (e) Warfarin therapy
- PLA-8 Which of the following statements is NOT correct regarding the massive perivillous fibrin deposition (MPVFD)?
  (a) MPVFD is defined as excessive fibrin deposition ensnaring villi and entirely obliterating the intervillous space.
  (b) When the fibrin deposition mainly involves the basal plate, it is often referred to as maternal floor infarct.
  (c) MPVFD has a chance of recurrence in future pregnancies.
  (d) MPVFD is associated with decreased alpha-fetoprotein in the maternal serum.
  (e) The differential diagnosis includes chronic villitis of unknown etiology and chronic histiocytic villitis.
- PLA-9 Twin-reversal arterial perfusion sequence (TRAP) or acardia is defined by which of the following definition?
  (a) Arterial-arterial (A-A) anastomosis, with retrograde perfusion of poorly oxygenated blood from the healthy twin or pump twin to the acardiac twin
  (b) Arterial-venous (A-V) anastomosis, with retrograde perfusion of poorly oxygenated blood from the healthy twin or pump twin to the acardiac twin, and venous-arterial (V-A) anastomosis is carrying blood back from the acardiac to the donor twin
  (c) Arterial-arterial (A-A) anastomosis, with retrograde perfusion of poorly oxygenated blood from the healthy twin or pump twin to the acardiac twin, venous-venous (V-V) anastomosis carrying blood back from the acardiac to the donor twin, and circulatory failure of the acardiac twin
  (d) Venous-venous (V-V) anastomosis carrying blood back from the acardiac to the donor twin, and circulatory failure of the acardiac twin
- PLA-10 Which of the following statements on choriocarcinoma of the placenta is TRUE?
  (a) Gestational choriocarcinoma occurs in 1 in 40,000 pregnancies.
  (b) It is a highly aggressive malignant tumor of the amnion epithelium in association with any form of gestation.
  (c) Of all forms of choriocarcinoma, placental choriocarcinoma is the most frequent.
  (d) Gestational choriocarcinoma is rarely diagnosed in symptomatic patients with metastases.
  (e) Metastases to the lung and brain usually are seen in both fetus and mother.

References and Recommended Readings

Abate DA, Watanabe S, Mocarski ES. Major human cytomegalovirus structural protein pp65 (ppUL83) prevents interferon response factor 3 activation in the interferon response. J Virol. 2004;78(20):10995–11006. PubMed PMID: 15452220; PubMed Central PMCID: PMC521853.

Al Jishi T, Sergi C. Current perspective of diethylstilbestrol (DES) exposure in mothers and offspring. Reprod Toxicol. 2017;71:71–7. https://doi.org/10.1016/j.reprotox.2017.04.009

Altemani AM. Thrombosis of fetal placental vessels. A quantitative study in placentas of stillbirths. Pathol Res Pract. 1987;182:685–9.

Altshuler G. Placenta within the medicolegal imperative. Arch Pathol Lab Med. 1991a Jul;115(7):688–95.

Altshuler G. Placenta within the medicolegal imperative. Arch Pathol Lab Med. 1991b Jul;115(7):688–95.

Altshuler G. A conceptual approach to placental pathology and pregnancy outcome. Semin Diagn Pathol. 1993a Aug;10(3):204–21. Review

Altshuler G. Some placental considerations related to neurodevelopmental and other disorders. J Child Neurol. 1993b Jan;8(1):78–94. Review

Altshuler G. A conceptual approach to placental pathology and pregnancy outcome. Semin Diagn Pathol. 1993c Aug;10(3):204–21. Review

Altshuler G. Some placental considerations related to neurodevelopmental and other disorders. J Child Neurol. 1993d Jan;8(1):78–94. Review

Altshuler G. Role of the placenta in perinatal pathology (revisited). Pediatr Pathol Lab Med. 1996a Mar-Apr;16(2):207–33. Review

Altshuler G. Role of the placenta in perinatal pathology (revisited). Pediatr Pathol Lab Med. 1996b Mar-Apr;16(2):207–33. Review

Altshuler G, Hyde S. Meconium-induced vasocontraction: a potential cause of cerebral and other fetal hypoperfusion and of poor pregnancy outcome. J Child Neurol. 1989a Apr;4(2):137–42.

Altshuler G, Hyde S. Meconium-induced vasocontraction: a potential cause of cerebral and other fetal hypoperfusion and of poor pregnancy outcome. J Child Neurol. 1989b Apr;4(2):137–42.

Altshuler G, Hyde SR. Clinicopathologic implications of placental pathology. Clin Obstet Gynecol. 1996a Sep;39(3):549–70. Review

Altshuler G, Hyde SR. Clinicopathologic implications of placental pathology. Clin Obstet Gynecol. 1996b Sep;39(3):549–70. Review

Ameis D, Khoshgoo N, Keijzer R. Abnormal lung development in congenital diaphragmatic hernia. Semin Pediatr Surg. 2017;26(3):123–8. https://doi.org/10.1053/j.sempedsurg.2017.04.011

Amundson R. The changing role of the embryo in evolutionary thought. Cambridge: Cambridge University Press; 2005.

Ananthan A, Nanavati R, Sathe P, Balasubramanian H. Placental findings in singleton stillbirths: a case-control study. J Trop Pediatr. 2018; https://doi.org/10.1093/tropej/fmy006

Ariel I, Anteby E, Hamani Y, Redline RW. Placental pathology in fetal thrombophilia. Hum Pathol. 2004;35(6):729–733. PubMed PMID: 15188139.

Arizawa M. Frequency of circumvallate and circummarginate. Abstracts/Placenta 36 (2015) A1eA14 (JPA2015–20).

Avagliano L, Bulfamante GP, Massa V. Cornelia de Lange syndrome: To diagnose or not to diagnose in utero? Birth Defects Res. 2017;109(10):771–7. https://doi.org/10.1002/bdr2.1045

Balduf K, Kumar TKS, Boston U, Sathanandam S, Lee MV, Jancelewicz T, Knott-Craig CJ. Improved Outcomes in Management of Hypoplastic Left Heart Syndrome Associated With Congenital Diaphragmatic Hernia: an Algorithmic Approach. Semin Thorac Cardiovasc Surg. 2018 Summer;30(2):191-196. https://doi.org/10.1053/j.semtcvs.2018.02.010. Epub 2018 Feb 12. PubMed PMID: 29448010. .

Balinsky BI. An introduction to embryology. Philadelphia: W. B. Saunders; 1970.

Baxi L, Warren W, Collins MH, Timor-Tritsch IE. Early detection of caudal regression syndrome with transvaginal scanning. Obstet Gynecol. 1990;75(3 Pt 2):486–489. PubMed PMID: 2406664.

Becattini S, Littmann ER, Carter RA, Kim SG, Morjaria SM, Ling L, Gyaltshen Y, Fontana E, Taur Y, Leiner IM, Pamer EG. Commensal microbes provide first line defense against Listeria monocytogenes infection. J Exp Med. 2017 Jul 3;214(7):1973–89. https://doi.org/10.1084/jem.20170495. Epub 2017 Jun 6. 28588016. PMC5502438

Beebe LA, Cowan LD, Hyde SR, Altshuler G. Methods to improve the reliability of histopathological diagnoses in the placenta. Paediatr Perinat Epidemiol. 2000a Apr;14(2):172–8.

Beebe LA, Cowan LD, Hyde SR, Altshuler G. Methods to improve the reliability of histopathological diagnoses in the placenta. Paediatr Perinat Epidemiol. 2000b Apr;14(2):172–8.

Beeksma FA, Erwich JJ, Khong TY. Placental fetal vascular thrombosis lesions and maternal thrombophilia. Pathology. 2012;44:24–8.

Behrendt N, Galan HL. Twin-twin transfusion and laser therapy. Curr Opin Obstet Gynecol. 2016;28(2):79–85. https://doi.org/10.1097/GCO.0000000000000247

Beiler HA, Sergi C, Wagner G, Zachariou Z. Accessory liver in an infant with congenital diaphragmatic hernia. J Pediatr Surg. 2001;36(6):E7. https://doi.org/10.1053/jpsu.2001.24020

Berry CL, Keeling J, Hilton C. Coincidence of congenital malformation and embryonic tumours of childhood. Arch Dis Child. 1970a Apr;45(240):229–31. https://doi.org/10.1136/adc.45.240.229. 4315971, PMC2020237.

Berry CL, Keeling J, Hilton C. Coincidence of congenital malformation and embryonic tumours of child-

References and Recommended Readings

hood. Arch Dis Child. 1970b Apr;45(240):229–31. https://doi.org/10.1136/adc.45.240.229. 4315971, PMC2020237.

Berry CL, Keeling J, Hilton C. Coincidence of congenital malformation and embryonic tumours of childhood. Arch Dis Child. 1970c Apr;45(240):229–31. https://doi.org/10.1136/adc.45.240.229. 4315971, PMC2020237.

Blauth W, Gekeler J. Symbrachydaktylien; Beitrag zur Morphologie, Klassifikation und Therapie. Handchirurgie. 1973;5(3):121–74. German. PubMed PMID: 4598758.

Boer LL, Boek PLJ, van Dam AJ, Oostra RJ. History and highlights of the teratological collection in the museum Anatomicum of Leiden University, the Netherlands. Am J Med Genet A. 2018;176(3):618–637. https://doi.org/10.1002/ajmg.a.38617.

Bohrer JC, Kamemoto LE, Almeida PG, Ogasawara KK. Acute chorioamnionitis at term caused by the oral pathogen fusobacterium nucleatum. Hawaii J Med Public Health. 2012;71:280–1.

Bouchahda H, El Mhabrech H, Hamouda HB, Ghanmi S, Bouchahda R, Soua H. Prenatal diagnosis of caudal regression syndrome and omphalocele in a fetus of a diabetic mother. Pan Afr Med J. 2017;27:128. https://doi.org/10.11604/pamj.2017.27.128.12041. eCollection 2017. PubMed PMID: 28904658.; PubMed Central PMCID: PMC5568004.

Bras A, Rodrigues AS, Gomes B, Rueff J. Down syndrome and microRNAs. Biomed Rep. 2018;8(1):11–6. https://doi.org/10.3892/br.2017.1019

Robert T. Brodell; Stephen E. Helms in Fitzpatrick's Dermatology in General Medicine, 8e. Ainhum and Pseudoainhum. Lowell A, Goldsmith, Stephen I Katz, Barbara A Gilchrest, Amy S Paller, David J Leffell, Klaus Wolff (2012) McGraw Hill, New York, United States.

Broome M, Vial Y, Jacquemont S, Sergi C, Kamnasaran D, Giannoni E. Complete Maxillo-mandibular Syngnathia in a newborn with multiple congenital malformations. Pediatr Neonatol. 2016;57(1):65–8. https://doi.org/10.1016/j.pedneo.2013.04.009

Campbell S, Thoms A. Ultrasound measurement of the fetal head to abdomen circumference ratio in the assessment of growth retardation. Br J Obstet Gynaecol. 1977 Mar;84(3):165–74. https://doi.org/10.1111/j.1471-0528.1977.tb12550.x

Canadian Congenital Diaphragmatic Hernia Collaborative. Diagnosis and management of congenital diaphragmatic hernia: a clinical practice guideline. CMAJ. 2018;190(4):E103–12. https://doi.org/10.1503/cmaj.170206. Review.. 29378870. PMC5790558

Carlson TL, Daugherty R, Miller A, Gbulie UB, Wallace R. Successful separation of conjoined twins: the contemporary experience and historic review in Memphis. Ann Plast Surg. 2018; https://doi.org/10.1097/SAP.0000000000001342.

Casella E, Giunta G, Ventura B, Cariola M, Lo Presti L, Gallo C, Tomaselli CF, Sapia F, Comito C, Castellano LM, Di Simone G, Caruso S. Fetal megacystis: aetiology and management. Giorn It Ost Gin CIC Edizioni Internazionali. 2013;35(5):669–673. ISSN: 0391-9013.

Cave D, Ross DB, Bahitham W, Chan A, Sergi C, Adatia I. Mitochondrial DNA depletion syndrome-an unusual reason for interstage attrition after the modified stage 1 Norwood operation. Congenit Heart Dis. 2013;8(1):E20–E23. https://doi.org/10.1111/j.1747-0803.2011.00569.x. Epub 2011 Oct 20. PubMed PMID: 22011012.

Chan JS, Baergen RN. Gross umbilical cord complications are associated with placental lesions of circulatory stasis and fetal hypoxia. Pediatr Dev Pathol. 2012;15:487–94.

Chandrasekharan PK, Rawat M, Madappa R, Rothstein DH, Lakshminrusimha S. Congenital diaphragmatic hernia – a review. Matern Health Neonatol Perinatol. 2017;3:6. https://doi.org/10.1186/s40748-017-0045-1

Chassaing N, Sorrentino S, Davis EE, Martin-Coignard D, Iacovelli A, Paznekas W, Webb BD, Faye-Petersen O, Encha-Razavi F, Lequeux L, Vigouroux A, Yesilyurt A, Boyadjiev SA, Kayserili H, Loget P, Carles D, Sergi C, Puvabanditsin S, Chen CP, Etchevers HC, Katsanis N, Mercer CL, Calvas P, Jabs EW. OTX2 mutations contribute to the otocephaly-dysgnathia complex. J Med Genet. 2012 Jun;49(6):373-379. https://doi.org/10.1136/jmedgenet-2012-100892. Epub 2012 May 10.PMID: 22577225.

Chisholm KM, Heerema-McKenney A. Fetal thrombotic vasculopathy: significance in liveborn children using proposed society for pediatric pathology diagnostic criteria. Am J Surg Pathol. 2015;39:274–80.

Chisholm KM, Norton ME, Penn AA, Heerema-McKenney A. Classification of preterm birth with placental correlates. Pediatr Dev Pathol. 2018.; https://doi.org/10.1177/1093526618775958. PMID: 29759046.

Colvin KL, Yeager ME. What people with down syndrome can teach us about cardiopulmonary disease. Eur Respir Rev. 2017;26(143) https://doi.org/10.1183/16000617.0098-2016

Commander SJ, Jacques SJ, Lloyd MS, Rushing A, Karlberg H, Buchanan EP. Bioinformatics associated with conjoined twin separation. J Craniofac Surg. 2018;29(1):109–11. https://doi.org/10.1097/SCS.0000000000004067

Committee Opinion No 611. Obstet Gynecol. 2014;124(4):863. https://doi.org/10.1097/01.AOG.0000454932.15177.be

Cooney TP, Wentworth PJ, Thurlbeck WM. Diminished radial count is found only postnatally in Down's syndrome. Pediatr Pulmonol. 1988;5(4):204–9.

Costello RA, Nehring SM. Disseminated Intravascular Coagulation (DIC). StatPearls: Treasure Island; 2018.

Craig M. Pocket guide to ultrasound measurements: JB Lippincott Company. 1988:152–4.

Crowther CA, Aghajafari F, Askie LM, Asztalos EV, Brocklehurst P, Bubner TK, Doyle LW, Dutta S, Garite TJ, Guinn DA, Hallman M, Hannah ME, Hardy P, Maurel K, Mazumder P, McEvoy C, Middleton PF,

Murphy KE, Peltoniemi OM, Peters D, Sullivan L, Thom EA, Voysey M, Wapner RJ, Yelland L, Zhang S, PRECISE Study Group. Repeat prenatal corticosteroid prior to preterm birth: a systematic review and individual participant data meta-analysis for the PRECISE study group (prenatal repeat corticosteroid international IPD study group: assessing the effects using the best level of evidence) – study protocol. Syst Rev. 2012;1:12. https://doi.org/10.1186/2046-4053-1-12. Review. PubMed PMID: 22588009; PubMed Central PMCID: PMC3351733.

Cui J, Chen Y, Wang HY, Wang RF. Mechanisms and pathways of innate immune activation and regulation in health and cancer. Hum Vaccin Immunother. 2014;10(11):3270-3285. https://doi.org/10.4161/21645515.2014.979640. PMID: 25625930; PMCID: PMC4514086.

Cyr PR. Diagnosis and management of granuloma annulare. Am Fam Physician. 2006;74(10):1729–1734. Review. PubMed PMID: 17137003.

de Almeida V, Bowman JM. Massive fetomaternal hemorrhage: Manitoba experience. Obstet Gynecol. 1994;83(3):323–328. PubMed PMID: 8127519.

De Robertis EM. Spemann's organizer and self-regulation in amphibian embryos. Nat Rev Mol Cell Biol. 2006;7:296–302.

Delgado AR. Listeriosis in pregnancy. J Midwifery Womens Health. 2008;53:255–9.

DeMyer W. The median cleft face syndrome. Differential diagnosis of cranium bifidum occultum, hypertelorism, and median cleft nose, lip, and palate. Neurology. 1967 Oct;17(10):961–71. https://doi.org/10.1212/wnl.17.10.961

DeMyer W. Median facial malformations and their implications for brain malformations. Birth Defects Orig Artic Ser. 1975;11(7):155–81.

Diana A, Epiney M, Ecoffey M, Pfister RE. "White dots on the placenta and red dots on the baby": Congenital cutaneous candidiasis-a rare disease of the neonate. Acta Paediatr. 2004;93:996–9.

DiGiulio DB. Diversity of microbes in amniotic fluid. Semin Fetal Neonatal Med. 2012;17:2–11.

Donaldson JS, Luck SR, Vogelzang R. Preoperative CT and MR imaging of ischiopagus twins. J Comput Assist Tomogr. 1990 Jul-Aug;14(4):643–6. https://doi.org/10.1097/00004728-199007000-00024

Dorn L, Menezes LF, Mikuz G, Otto HF, Onuchic LF, Sergi C. Immunohistochemical detection of polyductin and co-localization with liver progenitor cell markers during normal and abnormal development of the intrahepatic biliary system and in adult hepatobiliary carcinomas. J Cell Mol Med. 2009;13(7):1279–90. https://doi.org/10.1111/j.1582-4934.2008.00519.x

Drucilla J. Roberts (2008) Placental Pathology, a Survival Guide. Archives of Pathology & Laboratory Medicine: April. 2008;132(4):641–51.

Dubucs C, Chassaing N, Sergi C, Aubert-Mucca M, Attié-Bitach T, Lacombe D, Thauvin-Robinet C, Arpin S, Perez MJ, Cabrol C. Aziza J, Colin E, Martinovic J, Calvas P, Plaisancié J. Re-focusing on Agnathia-Otocephaly complex. Clin Oral Investig: Chen CP; 2020 Jul 9. https://doi.org/10.1007/s00784-020-03443-w. Epub ahead of print.

Dudley NJ, Fagan DG, Lamb MP. Short communication: ultrasonographic placental grade and thickness: associations with early delivery and low birthweight. Br J Radiol. 1993;66(782):175–7. https://doi.org/10.1259/0007-1285-66-782-175

Duijf PHG, van Bokhoven H, Brunner HG. Pathogenesis of split-hand/split-foot malformation. Hum Mol Genet. 2003;12(suppl_1):R51–60. https://doi.org/10.1093/hmg/ddg090.

Edwards L, Hui L. First and second trimester screening for fetal structural anomalies. Semin Fetal Neonatal Med. 2018;23(2):102–11. https://doi.org/10.1016/j.siny.2017.11.005

Elsasser DA, Ananth CV. Prasad V, Vintzileos AM; New Jersey-placental abruption study investigators. Diagnosis of placental abruption: relationship between clinical and histopathological findings. Eur J Obstet Gynecol Reprod Biol. 2010;148(2):125–130. https://doi.org/10.1016/j.ejogrb.2009.10.005. Epub 2009 Nov 7. PubMed PMID: 19897298; PubMed Central PMCID: PMC2814948.

Enders M, Weidner A, Enders G. Current epidemiological aspects of human parvovirus B19 infection during pregnancy and childhood in the western part of Germany. Epidemiol Infect. 2007;135(4):563–9. Epub 2006 Oct 26. PubMed PMID.

Enders M, Weidner A, Zoellner I, Searle K, Enders G. Fetal morbidity and mortality after acute human parvovirus B19 infection in pregnancy: prospective evaluation of 1018 cases. Prenat Diagn. 2004;24(7):513–518. PubMed PMID: 15300741.

Ernst LM, Chou D, Parry S. Fetal thrombotic vasculopathy in twin placentas with complete hydatidiform mole. Pediatr Dev Pathol. 2009;12:63–7.

Ernst LM, Grossman AB, Ruchelli ED. Familial perinatal liver disease and fetal thrombotic vasculopathy. Pediatr Dev Pathol. 2008;11:160–3.

Ernst LM, Minturn L, Huang MH, Curry E, Su EJ. Gross patterns of umbilical cord coiling: correlations with placental histology and stillbirth. Placenta. 2013;34:583–8.

Expert Panel on Women's Imaging:, Glanc P, Nyberg DA, Khati NJ, Deshmukh SP, Dudiak KM, Henrichsen TL, Poder L, Shipp TD, Simpson L, Weber TM, Zelop CM. ACR Appropriateness Criteria(®) Multiple Gestations. J Am Coll Radiol. 2017 Nov;14(11S):S476-S489. https://doi.org/10.1016/j.jacr.2017.08.051. PubMed PMID: 29101986.

Eysbouts YK, Ottevanger PB, Massuger LFAG, IntHout J, Short D, Harvey R, Kaur B, Sebire NJ, Sarwar N, Sweep FCGJ, Seckl MJ. Can the FIGO 2000 scoring system for gestational trophoblastic neoplasia be simplified? A new retrospective analysis from a nationwide dataset. Ann Oncol. 2017 Aug 1;28(8):1856-1861. https://doi.org/10.1093/annonc/mdx211. PMID: 28459944; PMCID: PMC5834141.

Fadl S, Moshiri M, Fligner CL, Katz DS, Dighe M. Placental Imaging: Normal Appearance with Review of Pathologic Findings. Radiographics : a review publication of the Radiological Society of North America, Inc. 37;(3);979-998. https://doi.org/10.1148/rg.2017160155.

Fariba G, Ayatollahi A, Hejazi S. Pemphigus foliaceus. Indian Pediatr. 2012;49(3):240–241. PubMed PMID: 22484744.

Fotaki A, Novaes J, Jicinska H, Carvalho JS. Fetal aortopulmonary window: case series and review of the literature. Ultrasound Obstet Gynecol. 2017;49(4):533–9. https://doi.org/10.1002/uog.15936

Fried M, Duffy PE. Designing a VAR2CSA-based vaccine to prevent placental malaria. Vaccine. 2015;33(52):7483–7488. https://doi.org/10.1016/j.vaccine.2015.10.011. Epub 2015 Nov 26. Review. PubMed PMID: 26469717; PubMed Central PMCID: PMC5077158.

Froeling FE, Seckl MJ. Gestational trophoblastic tumours: an update for 2014. Curr Oncol Rep. 2014;16(11):408. https://doi.org/10.1007/s11912-014-0408-y. Review. PubMed PMID: 25318458.

Fulton C, Klein AO. Explorations in developmental biology. Cambridge: Harvard University Press; 1976.

Garcia-Delgado R, Garcia-Rodriguez R, Romero Requejo A, Armas Roca M, Obreros Zegarra L, Medina Castellano M, Garcia Hernandez JA. Echographic features and perinatal outcomes in fetuses with congenital absence of ductus venosus. Acta Obstet Gynecol Scand. 2017 Oct;96(10):1205-1213. https://doi.org/10.1111/aogs.13176. Epub 2017 Jul 18. PubMed PMID: 28574580.

Garcia-Flores J, Cruceyra M, Canamares M, Garicano A, Nieto O, Tamarit I. Candida chorioamnionitis: report of two cases and review of literature. J Obstet Gynaecol. 2016;36:843–4.

Gauthier S, Tetu A, Himaya E, Morand M, Chandad F, Rallu F, Bujold E. The origin of fusobacterium nucleatum involved in intra-amniotic infection and preterm birth. J Matern Fetal Neonatal Med. 2011;24:1329–32.

Geirsson RT. Ultrasound instead of last menstrual period as the basis of gestational age assignment. Ultrasound Obstet Gynecol. 1991;1(3):212–219. https://doi.org/10.1046/j.1469-0705.1991.01030212.x. PMID 12797075.

Gekas J, Sergi C, Kamnasaran D. Molecular prenatal diagnosis of a sporadic alobar holoprosencephalic fetus: genotype-phenotype correlations. J Prenat Med. 2012 Jul;6(3):36–9. PMID: 23181171 Free PMC article

Geroulanos S, Jaggi F, Wydler J, Lachat M, Cakmakci M. Thoracopagus Symmetricus. Zur Trennung von siamesischen Zwillingen im 10. Jahrhundert n. Chr. durch byzantinische Ärzte [Thoracopagus symmetricus. On the separation of Siamese twins in the 10th century A. D. by Byzantine physicians]. Gesnerus. 1993;50 (Pt 3-4):179-200. German.

Gerulath AH, Ehlen TG, Bessette P, Jolicoeur L, Savoie R; Society of obstetricians and gynaecologists of canada; gynaecologic oncologists of canada; society of canadian colposcopists. Gestational trophoblastic disease. J Obstet Gynaecol Can. 2002 May;24(5):434-446. English, French. PubMed PMID: 12196865.

Giacoia GP. Severe fetomaternal hemorrhage: a review. Obstet Gynecol Surv. 1997;52(6):372–380. Review. PubMed PMID: 9178311.

Giannubilo SR, Pasculli A, Cecchi A, Biagini A, Ciavattini A. Fetal Intra-abdominal Umbilical Vein Aneurysm. Obstet Gynecol Surv. 2017;72(9):547–52. https://doi.org/10.1097/OGX.0000000000000476

Given JE, Loane M, Garne E, Nelen V, Barisic I, Randrianaivo H, Khoshnood B, Wiesel A, Rissmann A, Lynch C, Neville AJ, Pierini A, Bakker M, Klungsoyr K, Latos Bielenska A, Cavero-Carbonell C, Addor MC, Zymak-Zakutnya N, Tucker D, Dolk H. Gastroschisis in Europe - A Case-malformed-Control Study of Medication and Maternal Illness during Pregnancy as Risk Factors. Paediatr Perinat Epidemiol. 2017 Nov;31(6):549-559. https://doi.org/10.1111/ppe.12401. Epub 2017 Aug 25. PubMed PMID: 28841756.

Glueckert R, Rask-Andersen H, Sergi C, Schmutzhard J, Mueller B, Beckmann F, Rittinger O, Hoefsloot LH, Schrott-Fischer A, Janecke AR. Histology and synchrotron radiation-based microtomography of the inner ear in a molecularly confirmed case of CHARGE syndrome. Am J Med Genet A. 2010 Mar;152A(3):665-673. https://doi.org/10.1002/ajmg.a.33321. PubMed PMID: 20186814.

Goldenberg PC, Adler BJ, Parrott A, Anixt J, Mason K, Phillips J, Cooper DS, Ware SM, Marino BS. High burden of genetic conditions diagnosed in a cardiac neurodevelopmental clinic. Cardiol Young. 2017 Apr;27(3):459-466. https://doi.org/10.1017/S104795111600072X. Epub 2016 Sep 19. PubMed PMID: 27641144.

Goncalves LF, Chaiworapongsa T, Romero R. Intrauterine infection and prematurity. Ment Retard Dev Disabil Res Rev. 2002;8:3–13.

Gore RM, Filly RA, Parer JT. Sonographic antepartum diagnosis of conjoined twins. Its impact on obstetric management. JAMA. 1982 Jun 25;247(24):3351–3.

Grannum PAT, Berkowitz RL, Hobbins JC. The ultrasonic changes in the maturing placenta and their relation to fetal pulmonic maturity. Am J Obstet Gynecol. 1979a;133:915–22.

Grannum PA, Berkowitz RL, Hobbins JC. The ultrasonic changes in the maturing placenta and their relation to fetal pulmonary maturity. American Journal of Obstetrics and Gynecology. 1979b;133:915–22.

Grannum PA, Hobbins JC. The placenta. Radiologic Clinics of North America. 1982;20:353–65.

Groom KM, David AL. The role of aspirin, heparin, and other interventions in the prevention and treatment of fetal growth restriction. Am J Obstet Gynecol. 2018;218(2S):S829–S40. https://doi.org/10.1016/j.ajog.2017.11.565

Haberer K, Buffo-Sequeira I, Chudley AE, Spriggs E, Sergi C. A case of an infant with compound heterozygous mutations for hypertrophic cardiomyopathy producing a phenotype of left ventricular noncompaction. Can J Cardiol. 2014;30(10):1249 e1–3. https://doi.org/10.1016/j.cjca.2014.05.021.

Hager J, Sanal M, Trawöger R, Gassner I, Oswald E, Rudisch A, Schaefer G, Mikuz G, Sergi C. Conjoined epigastric heteropagus twins: excision of a parasitic twin from the anterior abdominal wall of her sibling. Eur J Pediatr Surg. 2007a Feb;17(1):66-71. Review. PubMed PMID: 17407026.

Hager J, Sanal M, Trawöger R, Gassner I, Oswald E, Rudisch A, Schaefer G, Mikuz G, Sergi C. Conjoined epigastric heteropagus twins: excision of a parasitic twin from the anterior abdominal wall of her sibling. Eur J Pediatr Surg. 2007b Feb;17(1):66–71. https://doi.org/10.1055/s-2007-964951

Hamadeh S, Addas B, Hamadeh N, Rahman J. Spontaneous intraperitoneal hemorrhage in the third trimester of pregnancy: clinical suspicion made the difference. J Obstet Gynaecol Res. 2018;44(1):161–4. https://doi.org/10.1111/jog.13479

Hamburger V. The heritage of experimental embryology, Hans Spemann and the organizer. New York: Oxford University Press; 1988.

Hammacher K. New method for the selective registration of the fetal heart beat. Geburtshilfe Frauenheilkd. 1962;22:1542–3.

Hampton MM. Congenital toxoplasmosis: a review. Neonatal Netw. 2015;34(5):274–278. https://doi.org/10.1891/0730-0832.34.5.274. Review. PubMed PMID: 26802827.

Han YW. Fusobacterium nucleatum: a commensal-turned pathogen. Curr Opin Microbiol. 2015;23:141–7.

Han YW, Fardini Y, Chen C, Iacampo KG, Peraino VA, Shamonki JM, Redline RW. Term stillbirth caused by oral fusobacterium nucleatum. Obstet Gynecol. 2010;115:442–5.

Hancks DC. A role for retrotransposons in Chromothripsis. Methods Mol Biol. 2018;1769:169–81. https://doi.org/10.1007/978-1-4939-7780-2_11

Hanzlik E, Gigante J. Microcephaly. Children (Basel). 2017;4(6) https://doi.org/10.3390/children4060047

Harper RG, Kenigsberg K, Sia CG, Horn D, Stern D, Bongiovi V. Xiphopagus conjoined twins: a 300-year review of the obstetric, morphopathologic, neonatal, and surgical parameters. Am J Obstet Gynecol. 1980 Jul 1;137(5):617-629. https://doi.org/10.1016/0002-9378(80)90707-3.

Hasbun J, Sepulveda-Martinez A, Haye MT, Astudillo J, Parra-Cordero M. Chorioamnionitis caused by listeria monocytogenes: a case report of ultrasound features of fetal infection. Fetal Diagn Ther. 2013;33:268–71.

Hasle H, Friedman JM, Olsen JH, Rasmussen SA. Low risk of solid tumors in persons with Down syndrome. Genet Med. 2016 Nov;18(11):1151-1157. https://doi.org/10.1038/gim.2016.23. Epub 2016 Mar 31.

Hay WW Jr, Brown LD, Rozance PJ, Wesolowski SR, Limesand SW. Challenges in nourishing the intrauterine growth-restricted foetus – lessons learned from studies in the intrauterine growth-restricted foetal sheep. Acta Paediatr. 2016;105(8):881–889. https://doi.org/10.1111/apa.13413. Epub 2016 May 10. Review. PubMed PMID: 27028695; PubMed Central PMCID: PMC5961494.

Heider A. Fetal Vascular Malperfusion. Arch Pathol Lab Med. 2017;141(11):1484–9. https://doi.org/10.5858/arpa.2017-0212-RA

Heller DS, Moorehouse-Moore C, Skurnick J, Baergen RN. Second-trimester pregnancy loss at an urban hospital. Infect Dis Obstet Gynecol. 2003;11:117–22.

Hernández-Díaz S, Van Marter LJ, Werler MM, Louik C, Mitchell AA. Risk factors for persistent pulmonary hypertension of the newborn. Pediatrics. 2007;120(2):e272–e282. PubMed PMID: 17671038.

Hertig AT, Rock J, Adams EC. A description of 34 human ova within the first 17 days of development. Amer J Anat. 1956;98:435–93.

Hertig AT, Rock J, Adams EC, Mulligan WJ. On the preimplantation stages of the human ovum: a description of four normal and four abnormal specimens ranging from the second to the fifth day of development. Carnegie Instn Wash Publ 603, Contrib Embryol. 1954;35:199–20.

Hills D, Irwin GA, Tuck S, Baim R. Distribution of placental grade in high-risk gravidas. AJR Am J Roentgenol. 1984;143(5):1011–3.

Holtfreter J. Neural induction in explants which have passed through a sublethal cytolysis. In: Fulton C, Klein AO, editors. Explorations in developmental biology. Cambridge: Harvard University Press; 1976. Abridged and originally published in. J Exp Zool. 1947;106:197–22.

Hon EH, Bradfield A, Hess OW. The electronic evaluation of the fetal heart rate. V. The vagal factor in fetal bradycardia. American journal of obstetrics and gynecology. 1961;82:291–300.

Hoorsan H, Mirmiran P, Chaichian S, Moradi Y, Hoorsan R, Jesmi F. Congenital malformations in infants of mothers undergoing assisted reproductive technologies: a systematic review and meta-analysis study. J Prev Med Public Health. 2017;50(6):347–60. https://doi.org/10.3961/jpmph.16.122

Huls CK, Detlefs C. Trauma in pregnancy. Semin Perinatol. 2018;42(1):13–20. https://doi.org/10.1053/j.semperi.2017.11.004. PMID: 29463389.

Hunter LA. Issues in pregnancy dating: revisiting the evidence. A simple solution to dating discrepancies: the rule of eights. J Midwifery Womens Health. 2009;54(3):184–90. https://doi.org/10.1016/j.jmwh.2008.11.003

Hutcheon JA, McNamara H, Platt RW, Benjamin A, Kramer MS. Placental weight for gestational age and adverse perinatal outcomes. Obstet Gynecol. 2012;119(6):1251–1258. https://doi.org/10.1097/AOG.0b013e318253d3df. PubMed PMID: 22617591.

IJsselstijn H, Gischler SJ, Wijnen RMH, Tibboel D. Assessment and significance of long-term outcomes in pediatric surgery. Semin Pediatr Surg. 2017;26(5):281–5. https://doi.org/10.1053/j.sempedsurg.2017.09.004

In SH, Porter R. Hutchinson dictionary of scientific biography. Oxford: Helicon Publishing. 2005:566–9.

Ito F, Okubo T, Yasuo T, Mori T, Iwasa K, Iwasaku K, Kitawaki J. Premature delivery due to intrauterine candida infection that caused neonatal congenital cutaneous candidiasis: a case report. J Obstet Gynaecol Res. 2013;39:341–3.

Jain V, Yadav DK, Kandasamy D, Gupta DK. Hepatopulmonary fusion: a rare and potentially lethal association with right congenital diaphragmatic hernia. BMJ Case Rep. 2017;2017. https://doi.org/10.1136/bcr-2016-218227

James D, Steer P, Weiner C, Gonik B, Robson S. (Eds.). High-Risk Pregnancy: Management Options. Cambridge: Cambridge University Press. https://doi.org/10.1017/CBO9781108349185

Johnson CA, Gissen P, Sergi C. Molecular pathology and genetics of congenital hepatorenal fibrocystic syndromes. J Med Genet. 2003;40(5):311–9.

Kalyani R, Bindra MS. Twin reversed arterial perfusion syndrome (TRAP or Acardiac twin)-a case report. J Clin Diagn Res. 2014;8(1):166–167. https://doi.org/10.7860/JCDR/2014/7012.3965. Epub 2014 Jan 12. PubMed PMID: 24596758; PubMed Central PMCID: PMC3939538.

Kardon G, Ackerman KG, McCulley DJ, Shen Y, Wynn J, Shang L, Bogenschutz E, Sun X, Chung WK. Congenital diaphragmatic hernias: from genes to mechanisms to therapies. Dis Model Mech. 2017 Aug 1;10(8):955-970. https://doi.org/10.1242/dmm.028365. Review. PubMed PMID: 28768736; PubMed Central PMCID: PMC5560060.

Kassam SN, Nesbitt S, Hunt LP, Oster N, Soothill P, Sergi C. Pregnancy outcomes in women with or without placental malaria infection. Int J Gynaecol Obstet. 2006;93(3):225–232. Epub 2006 Apr 13. PubMed PMID: 16626713.

Katzman PJ, Genest DR. Maternal floor infarction and massive perivillous fibrin deposition: histological definitions, association with intrauterine fetal growth restriction, and risk of recurrence. Pediatr Dev Pathol. 2002;5(2):159–164. Erratum in: Pediatr Dev Pathol. 2003 Jan-Feb;6(1):102. PubMed PMID: 11910510.

Kaufman JA, Wright JM, Evans A, Rivera-Núñez Z, Meyer A, Narotsky MG. Disinfection By-Product Exposures and the Risk of Musculoskeletal Birth Defects. Environ Epidemiol. 2020 Feb 13;4(1):https://doi.org/10.1097/EE9.0000000000000081. https://doi.org/10.1097/EE9.0000000000000081. PMID: 32154492; PMCID: PMC7061532.

Khan A, Sergi C. Sialidosis: a review of morphology and molecular biology of a rare pediatric disorder. Diagnostics (Basel). 2018;8(2). https://doi.org/10.3390/diagnostics8020029. pii: E29. Review. PubMed PMID: 29693572.

Khan A, Sergi C. Sialidosis: A Review of Morphology and Molecular Biology of a Rare Pediatric Disorder. Diagnostics (Basel). 2018 Apr 25;8(2). pii: E29. https://doi.org/10.3390/diagnostics8020029. Review. PubMed PMID: 29693572; PubMed Central PMCID: PMC6023449.

Khushdil A, Niaz H, Ahmed Z. Epigastric Heteropagus conjoined twins. J Coll Physicians Surg Pak. 2018;28(3):S42–3. https://doi.org/10.29271/jcpsp.2018.03.S42

Kim A, Economidis MA, Stohl HE. Placental abruption after amnioreduction for polyhydramnios caused by chorioangioma. BMJ Case Rep. 2018;2018. https://doi.org/10.1136/bcr-2017-222399

Kimball JW. Organizing the Embryo: the central nervous system. Biology Pages, last modified May. 2011;17. http://users.rcn.com/jkimball.ma.ultranet/BiologyPages/S/Spemann.html. Accessed 28 Nov 2011

King EL, Redline RW, Smith SD, Kraus FT, Sadovsky Y, Nelson DM. Myocytes of chorionic vessels from placentas with meconium-associated vascular necrosis exhibit apoptotic markers. Hum Pathol. 2004a;35(4):412–417. PubMed PMID: 15116320.

King EL, Redline RW, Smith SD, Kraus FT, Sadovsky Y, Nelson DM. Myocytes of chorionic vessels from placentas with meconium-associated vascular necrosis exhibit apoptotic markers. Hum Pathol. 2004b Apr;35(4):412–7.

King EL, Redline RW, Smith SD, Kraus FT, Sadovsky Y, Nelson DM. Myocytes of chorionic vessels from placentas with meconium-associated vascular necrosis exhibit apoptotic markers. Hum Pathol. 2004c Apr;35(4):412–7.

Kingdom JC, Audette MC, Hobson SR, Windrim RC, Morgen E. A placenta clinic approach to the diagnosis and management of fetal growth restriction. Am J Obstet Gynecol. 2018;218(2S):S803–S17. https://doi.org/10.1016/j.ajog.2017.11.575

Kleihauer E, Braun H, Betke K. Demonstration von fetalem Hämoglobin in den Erythrocyten eines Blutausstrichs [Demonstration of fetal hemoglobin in erythrocytes of a blood smear]. Klin Wochenschr. 1957 Jun 15;35(12):637-8. German. https://doi.org/10.1007/BF01481043.

Klosowska A, Cwiklinska A, Kuchta A, Berlinska A, Jankowski M, Wierzba J. Down syndrome, increased risk of dementia and lipid disturbances. Dev Period Med. 2017;21(1):69–73.

Ko HS, Cheon JY, Choi SK, Lee HW, Lee A, Park IY, Shin JC. Placental histologic patterns and neonatal seizure, in preterm premature rupture of membrane. J Matern Fetal Neonatal Med. 2017 Apr;30(7):793-800. https://doi.org/10.1080/14767058.2016.1186634. Epub 2016 Jul 20. PubMed PMID: 27145920.

Kontopoulos E, Chmait RH, Quintero RA. Twin-to-twin transfusion syndrome: definition, staging, and ultrasound assessment. Twin Res Hum Genet. 2016;19(3):175–83. https://doi.org/10.1017/thg.2016.34

Kosasa TS, Ebesugawa I, Nakayama RT, Hale RW. Massive fetomaternal hemorrhage preceded by decreased fetal movement and a nonreactive fetal heart rate pattern. Obstet Gynecol. 1993;82(4 Pt 2 Suppl):711–714. Review. PubMed PMID: 8378023.

Kovo M, Schreiber L, Ben-Haroush A, Asalee L, Seadia S, Golan A, Bar J. The placental factor in spontaneous preterm labor with and without premature rupture of membranes. J Perinat Med. 2011;39(4):423–429. https://doi.org/10.1515/JPM.2011.038. PubMed PMID: 21526977.

Kridin K, Sagi SZ, Bergman R. Mortality and cause of death in patients with pemphigus. Acta Derm Venereol. 2017;97(5):607–611. https://doi.org/10.2340/00015555-2611. PubMed PMID: 28093595.

Krywko DM, Shunkwiler SM. Kleihauer Betke Test. [Updated 2017 May 1]. In: StatPearls [Internet]. Treasure Island: StatPearls Publishing; 2018. Available from: https://www.ncbi.nlm.nih.gov/books/NBK430876/

Lalani SR. Current genetic testing tools in neonatal medicine. Pediatr Neonatol. 2017;58(2):111–21. https://doi.org/10.1016/j.pedneo.2016.07.002

Lattanzi W, Barba M, Di Pietro L, Boyadjiev SA. Genetic advances in craniosynostosis. Am J Med Genet A. 2017;173(5):1406–29. https://doi.org/10.1002/ajmg.a.38159

Law MA, Mohan J. Right Aortic Arches. [Updated 2019 Jun 4]. In: StatPearls [Internet]. Treasure Island (FL): StatPearls Publishing; 2019 Jan-. Available from: https://www.ncbi.nlm.nih.gov/books/NBK431104/.

Lazo-Langner A, Al-Ani F, Weisz S, Rozanski C, Louzada M, Kovacs J, Kovacs MJ. Prevention of venous thromboembolism in pregnant patients with a history of venous thromboembolic disease: a retrospective cohort study. Thromb Res. 2018;167:20–25. https://doi.org/10.1016/j.thromres.2018.05.005. PMID: 29772489.

Leguy MC, Brun S, Pidoux G, Salhi H, Choiset A, Menet MC, Gil S, Tsatsaris V, Guibourdenche J. Pattern of secretion of pregnancy-associated plasma protein-A (PAPP-A) during pregnancies complicated by fetal aneuploidy, in vivo and in vitro. Reprod Biol Endocrinol. 2014 Dec 28;12:129. https://doi.org/10.1186/1477-7827-12-129.

Lepais L, Gaillot-Durand L, Boutitie F, Lebreton F, Buffin R, Huissoud C, Massardier J, Guibaud L, Devouassoux-Shisheboran M, Allias F. Fetal thrombotic vasculopathy is associated with thromboembolic events and adverse perinatal outcome but not with neurologic complications: a retrospective cohort study of 54 cases with a 3-year follow-up of children. Placenta. 2014;35:611–7.

Li X, Liu Y, Yue S, Wang L, Zhang T, Guo C, Hu W, Kagan KO, Wu Q. Uniparental disomy and prenatal phenotype: Two case reports and review. Medicine (Baltimore). 2017 Nov;96(45):e8474. https://doi.org/10.1097/MD.0000000000008474. PMID: 29137034; PMCID: PMC5690727.

Li Z, Zeki R, Hilder L, Sullivan EA. Australia's Mothers and Babies 2010. Perinatal statistics series no. 27. Cat. no. PER 57. Australian Institute of Health and Welfare National Perinatal Statistics Unit, Australian Government; 2012. Retrieved 4 July 2013.

Liu X, Dong K, Zheng S, Xiao X, Shen C, Dong C, Zhu H, Li H, Bi Y, Ma R. Separation of pygopagus, omphalopagus, and ischiopagus with the aid of three-dimensional models. J Pediatr Surg. 2018 Apr;53(4):682-687. https://doi.org/10.1016/j.jpedsurg.2017.06.016. Epub 2017 Jun 27. PubMed PMID: 28688793.

Loh TJ, Lian DW, Iyer P, Lam JC, Kuick CH, Aung AC, Chang KT. Congenital gata1-mutated myeloproliferative disorder in trisomy 21 complicated by placental fetal thrombotic vasculopathy. Hum Pathol. 2014;45:2364–7.

Machin GA, Ackerman J, Gilbert-Barness E. Abnormal umbilical cord coiling is associated with adverse perinatal outcomes. Pediatr Dev Pathol. 2000 Sep-Oct;3(5):462–71.

Maienschein J. Whose view of life? Cambridge, MA: Harvard University Press; 2003.

Maki Y, Fujisaki M, Sato Y, Sameshima H. Candida chorioamnionitis leads to preterm birth and adverse fetal-neonatal outcome. Infect Dis Obstet Gynecol. 2017;2017:9060138.

Marks WA, Leech RW, Altshuler GP. Placental examination in perinatal asphyxia. J Child Neurol. 1989a Apr;4(2):124.

Marks WA, Leech RW, Altshuler GP. Placental examination in perinatal asphyxia. J Child Neurol. 1989b Apr;4(2):124.

Masek J, Andersson ER. The developmental biology of genetic notch disorders. Development. 2017;144(10):1743–63. https://doi.org/10.1242/dev.148007

Mayer-Pickel K, Eberhard K, Lang U. Cervar-Zivkovic M. Pregnancy outcome in women with obstetric and thrombotic Antiphospholipid syndrome-a retrospective analysis and a review of additional treatment in pregnancy. Clin Rev Allergy Immunol. 2017;53(1):54–67. https://doi.org/10.1007/s12016-016-8569-0

McFadden DE, Kalousek DK. Two different phenotypes of fetuses with chromosomal triploidy: correlation with parental origin of the extra haploid set. Am J Med Genet. 1991 Mar 15;38(4):535-538. PubMed PMID: 2063893.

McFadden DE, Kwong LC, Yam IY, Langlois S. Parental origin of triploidy in human fetuses: evidence for genomic imprinting. Hum Genet. 1993 Nov;92(5):465-469. PubMed PMID: 7902318.

McFadden DE, Pantzar JT. Placental pathology of triploidy. Hum Pathol. 1996 Oct;27(10):1018-1020. PubMed PMID: 8892584.

McHugo J, Whittle M. Enlarged fetal bladders: aetiology, management and outcome. Prenat Diagn. 2001;21(11):958–963. Review. PubMed PMID: 11746149.

McLean K, Cushman M. Venous thromboembolism and stroke in pregnancy. Hematology Am Soc Hematol Educ Program. 2016;2016(1):243–50. https://doi.org/10.1182/asheducation-2016.1.243

McNamara H, Hutcheon JA, Platt RW, Benjamin A, Kramer MS. Risk factors for high and low placental weight. Paediatr Perinat Epidemiol. 2014;28(2):97–105. https://doi.org/10.1111/ppe.12104. Epub 2013 Dec 20. PubMed PMID: 24354883.

Mei Y, Lin Y. Clinical significance of primary symptoms in women with placental abruption. J Matern Fetal Neonatal Med. 2018;31(18):2446–9. https://doi.org/10.1080/14767058.2017.1344830

Mekinian A, Costedoat-Chalumeau N, Masseau A, Botta A, Chudzinski A, Theulin A, Emmanuelli V, Hachulla E, De Carolis S, Revaux A, Nicaise P, Cornelis F, Subtil D, Montestruc F, Bucourt M, Chollet-Martin S, Carbillon L, Fain O; SNFMI and the European Forum of APS. Chronic histiocytic intervillositis: outcome, associated diseases and treatment in a multicenter prospective study. Autoimmunity. 2015 Feb;48(1):40-45. https://doi.org/10.3109/08916934.2014.939267. Epub 2014 Jul 16. PubMed PMID: 25028066.

Melo A, Dinis R, Portugal A, Sousa AI, Cerveira I. Early prenatal diagnosis of parapagus conjoined twins. Clin Pract. 2018;8(2):1039. https://doi.org/10.4081/cp.2018.1039

Merz E, Bahlmann F. Ultrasound in obstetrics and gynecology. Thieme Medical Publishers. (2005) ISBN:1588901475.

Miny P, Koppers B, Dworniczak B, Bogdanova N, Holzgreve W, Tercanli S, Basaran S, Rehder H, Exeler R, Horst J. Parental origin of the extra haploid chromosome set in triploidies diagnosed prenatally. Am J Med Genet. 1995 May 22;57(1):102-106. PubMed PMID: 7645587.

Miralles-Gutierrez A, Narbona-Arias I, Gonzalez Mesa E. Neurological complications after therapy for fetal-fetal transfusion syndrome: a systematic review of the outcomes at 24 months. J Perinat Med. 2017; https://doi.org/10.1515/jpm-2017-0217

Mohammed S, Bahitham W, Chan A, Chiu B, Bamforth F, Sergi C. Mitochondrial DNA related cardiomyopathies. Front Biosci (Elite Ed). 2012;4:1706–1716. Review. PubMed PMID: 22201986.

Mohangoo AD, Blondel B, Gissler M, Velebil P, MacFarlane A, Zeitlin J. International comparisons of fetal and neonatal mortality rates in high-income countries: should exclusion thresholds be based on birth weight or gestational age? PLoS One. 2013;8(5):e64869. https://doi.org/10.1371/journal.pone.0064869. Bibcode:2013PLoSO..864869M. PMC 3658983 Freely accessible. PMID 23700489.

Molnar-Nadasdy G. Altshuler G. Perinatal pathology casebook. A case of twin transfusion syndrome with dichorionic placentas. J Perinatol. Nov-Dec. 1996a;16(6):507–9.

Molnar-Nadasdy G. Altshuler G. Perinatal pathology casebook. A case of twin transfusion syndrome with dichorionic placentas. J Perinatol. Nov-Dec. 1996b;16(6):507–9.

Montelongo EM, Blue NR, Lee RH. Placenta accreta in a woman with escherichia coli chorioamnionitis with intact membranes. Case Rep Obstet Gynecol. 2015;2015:121864.

Morin L, Lim K. No. 260-ultrasound in twin pregnancies. J Obstet Gynaecol Can. 2017;39(10):e398–411. https://doi.org/10.1016/j.jogc.2017.08.014

Mous DS, Kool HM, Wijnen R, Tibboel D, Rottier RJ. Pulmonary vascular development in congenital diaphragmatic hernia. Eur Respir Rev. 2018;27(147) https://doi.org/10.1183/16000617.0104-2017

Mousavi AS, Hashemi N, Kashanian M, Sheikhansari N, Bordbar A, Parashi S. Comparison between maternal and neonatal outcome of PPROM in the cases of amniotic fluid index (AFI) of more and less than 5 cm. J Obstet Gynaecol. 2018:1–5. https://doi.org/10.1080/01443615.2017.1394280

Muhelo AR, Montemezzo G, Da Dalt L, Wingi OM, Trevisanuto D, Gamba P, Pizzol D, Cavaliere E. Successful management of a parasitic ischiopagus conjoined twins in a low-income setting. Clin Case Rep. 2018 Jan 10;6(2):385-390. https://doi.org/10.1002/ccr3.1374. PMID: 29445482; PMCID: PMC5799649.

Nardozza LM, Araujo Júnior E, Caetano AC, Moron AF. Prenatal diagnosis of amniotic band syndrome in the third trimester of pregnancy using 3D ultrasound. J Clin Imaging Sci. 2012;2:22. https://doi.org/10.4103/2156-7514.95436. Epub 2012 Apr 28. PubMed PMID: 22616039. PMC3352605.

Nimrodi M, Kleitman V, Wainstock T, Gemer O, Meirovitz M, Maymon E, Benshalom-Tirosh N, Erez O. The association between cervical inflammation and histologic evidence of HPV in PAP smears and adverse pregnancy outcome in low risk population. Eur J Obstet Gynecol Reprod Biol. 2018;225:160–165. https://doi.org/10.1016/j.ejogrb.2018.04.023. PMID: 29727786.

O'Driscoll M. The pathological consequences of impaired genome integrity in humans; disorders of the DNA replication machinery. J Pathol. 2017;241(2):192–207. https://doi.org/10.1002/path.4828

Ogino T. Current classification of congenital hand deformities based on experimental research. In: Saffar P, Amadio CP, Foucher G, editors. Current practice in hand surgery. London: Martin Dunitz; 1997. p. 337–41.

Ogino T. Clinical features and teratogenic mechanisms of congenital absence of digits. Develop Growth Differ. 2007;49:529–31.

Ogino T. Classification of congenital differences of the upper extremity. Locomotor Syst. 2012;19(1+2)

Ohyama M, Itani Y, Yamanaka M, Goto A, Kato K, Ijiri R, Tanaka Y. Re-evaluation of chorioamnionitis and funisitis with a special reference to subacute chorioamnionitis. Hum Pathol. 2002;33(2):183–190. PubMed PMID: 11957143.

Oneda B, Rauch A. Microarrays in prenatal diagnosis. Best Pract Res Clin Obstet Gynaecol. 2017;42:53–63. https://doi.org/10.1016/j.bpobgyn.2017.01.003

Padmanabhan R. Retinoic acid-induced caudal regression syndrome in the mouse fetus. Reprod Toxicol. 1998;12(2):139–151. PubMed PMID: 9535508.

Paget J. The Bradshawe lecture on some rare and new diseases. Lancet. 1882;2:1017–21.

Paquet C, Yudin MH, Society of Obstetricians and Gynaecologists of Canada. Toxoplasmosis in pregnancy: prevention, screening, and treatment. J Obstet Gynaecol Can. 2013;35(1):78–81. English, French. PubMed PMID: 23343802.

Parast MM, Crum CP, Boyd TK. Placental histologic criteria for umbilical blood flow restriction in unexplained stillbirth. Hum Pathol. 2008;39:948–53.

Perrot LJ, Williamson S, Jimenez JF. The caudal regression syndrome in infants of diabetic mothers. Ann Clin Lab Sci. 1987;17(4):211–220. PubMed PMID: 3304122.

Philipp T, Philipp K, Reiner A, Beer F, Kalousek DK. Embryoscopic and cytogenetic analysis of 233 missed abortions: factors involved in the pathogenesis of developmental defects of early failed pregnancies. Hum Reprod. 2003 Aug;18(8):1724–32. https://doi.org/10.1093/humrep/deg309

Pinar H. Pathological case of the month. Arch Pediatr Adolesc Med. 1998;152(2):199–200.

Pinar H, Goldenberg RL, Koch MA, Heim-Hall J, Hawkins HK, Shehata B, Abramowsky C, Parker CB, Dudley DJ, Silver RM, Stoll B, Carpenter M, Saade G, Moore J, Conway D, Varner MW, Hogue CJ, Coustan DR, Sbrana E, Thorsten V, Willinger M, Reddy UM. Placental findings in singleton stillbirths. Obstet Gynecol. 2014 Feb;123(2 Pt 1):325–36. https://doi.org/10.1097/AOG.0000000000000100. 24402599, PMC3948332.

Popek EJ, Collins KA. Examination of the Placenta (Chapter 18). In: Autopsy Performance & Reporting, College of American Pathologists, Collins KA (editor). Northfield, IL: CAP Press. p. 2017.

Pramanick A, Hwang WS, Mathur M. Placental site nodule (PSN): an uncommon diagnosis with a common presentation. BMJ Case Rep. 2014. https://doi.org/10.1136/bcr-2013-203086. pii:bcr2013203086. PubMed PMID: 24695661; PubMed Central PMCID: PMC3987568.

Puvabanditsin S, Savla J, Garrow E, Kierson ME, Rosenbaum RD, Brandsma E. Symmetrical upper limb peromelia and lower limb amelia associated with persistent omphalomesenteric duct: a case report. Clin Dysmorphol. 2011;20(2):102–106. https://doi.org/10.1097/MCD.0b013e3283439657. PubMed PMID: 21278573.

Rabin KR, Whitlock JA. Malignancy in children with trisomy 21. Oncologist. 2009;14(2):164–173. https://doi.org/10.1634/theoncologist.2008-0217. Epub 2009 Jan 28. Review. PubMed PMID: 19176633; PubMed Central PMCID: PMC2761094.

Raghu S. Disorders of lung development. J NTR Univ Health Sci [serial online] 2015 [cited 2019 Sep 1];4:65-74. Available from: http://www.jdrntruhs.org/text.asp?2015/4/2/65/158571.

Ragozzino MW, Hill LM, Breckle R, Ellefson RD, Smith RC. The relationship of placental grade by ultrasound to markers of fetal lung maturity. Radiology. 1983;148(3):805–7.

Ram G, Chinen J. Infections and immunodeficiency in Down syndrome. Clin Exp Immunol. 2011 Apr;164(1):9-16. https://doi.org/10.1111/j.1365-2249.2011.04335.x. Epub 2011 Feb 24. Review. PubMed PMID: 21352207; PubMed Central PMCID: PMC3074212.

Rasmussen S, Ebbing C, Linde LE, Baghestan E. Placental abruption in parents who were born small: registry-based cohort study. BJOG. 2018;125(6):667–74. https://doi.org/10.1111/1471-0528.14837

Redline RW. Clinical and pathological umbilical cord abnormalities in fetal thrombotic vasculopathy. Hum Pathol. 2004a;35(12):1494–1498. PubMed PMID: 15619208.

Redline RW. Placental inflammation. Semin Neonatol. 2004b;9(4):265–274. Review. PubMed PMID: 15251143.

Redline RW. AP115 Placental pathology: keeping it simple and focusing on what matters. 2005a College of American Pathologists.

Redline RW. Severe fetal placental vascular lesions in term infants with neurologic impairment. Am J Obstet Gynecol. 2005b;192:452–7.

Redline RW. Inflammatory response in acute chorioamnionitis. Semin Fetal Neonatal Med. 2012;17(1):20–25. https://doi.org/10.1016/j.siny.2011.08.003. Epub 2011 Aug 23. Review. PubMed PMID: 21865101.

Redline RW, Ariel I, Baergen RN, Desa DJ, Kraus FT, Roberts DJ, Sander CM. Fetal vascular obstructive lesions: nosology and reproducibility of placental reaction patterns. Pediatr Dev Pathol. 2004;7(5):443–452. Epub 2004 Jul 30. PubMed PMID: 15547768.

Redline RW, Boyd T, Campbell V, Hyde S, Kaplan C, Khong TY, Prashner HR. Waters BL; Society for Pediatric Pathology, perinatal section, maternal vascular perfusion nosology committee. Maternal vascular underperfusion: nosology and reproducibility of placental reaction patterns. Pediatr Dev Pathol. 2004;7(3):237–249. Epub 2004 Mar 17. PubMed PMID: 15022063.

Redline RW, Minich N, Taylor HG, Hack M. Placental lesions as predictors of cerebral palsy and abnormal neurocognitive function at school age in extremely low birth weight infants (<1 kg). Pediatr Dev Pathol. 2007;10(4):282–292. PubMed PMID: 17638433.

Redline RW, O'Riordan MA. Placental lesions associated with cerebral palsy and neurologic impairment following term birth. Arch Pathol Lab Med. 2000;124:1785–91.

Redline RW, Pappin A. Fetal thrombotic vasculopathy: the clinical significance of extensive avascular villi. Hum Pathol. 1995;26:80–5.

Redline RW, Wilson-Costello D, Borawski E, Fanaroff AA, Hack M. Placental lesions associated with neurologic impairment and cerebral palsy in very low-birth-weight infants. Arch Pathol Lab Med. 1998;122(12):1091–1098. PubMed PMID: 9870858.

Renshaw TS. Sacral agenesis. J Bone Joint Surg Am. 1978 Apr;60(3):373–83.

Richardson RJ, Applebaum H, Taber P, Woolley MM, Chwals WJ, Warden MJ, Dietrich R. Use of magnetic resonance imaging in planning the separation of omphalopagus conjoined twins. J Pediatr Surg. 1989 Jul;24(7):683-684; discussion 684-5. PubMed PMID: 2754585.

Roberge S, Bujold E, Nicolaides KH. Meta-analysis on the effect of aspirin use for prevention of preeclampsia on placental abruption and antepartum hemorrhage. Am J Obstet Gynecol. 2018;218(5):483–9. https://doi.org/10.1016/j.ajog.2017.12.238

Robinson HP, Fleming JEE. A critical evaluation of sonar "crown-rump length" measurements. BJOG. 1975;82(9):702. https://doi.org/10.1111/j.1471-0528.1975.tb00710.x

Rode ME, Morgan MA, Ruchelli E, Forouzan I. Candida chorioamnionitis after serial therapeutic amniocenteses: a possible association. J Perinatol. 2000;20:335–7.

Rogers BB, Alexander JM, Head J, McIntire D, Leveno KJ. Umbilical vein interleukin-6 levels correlate with the severity of placental inflammation and gestational age. Hum Pathol. 2002;33(3):335–340. PubMed PMID: 11979375.

Rogers BB, Momirova V, Dizon-Townson D, Wenstrom K, Samuels P, Sibai B, Spong C, Caritis SN, Sorokin Y, Miodovnik M, O'Sullivan MJ, Conway D, Wapner RJ. Avascular villi, increased syncytial knots, and hypervascular villi are associated with pregnancies complicated by factor V Leiden mutation. Pediatr Dev Pathol. 2010 Sep-Oct;13(5):341–7. https://doi.org/10.2350/09-05-0657-OA.1. Epub 2010 Feb 1. PubMed PMID: 20121426. PMC3161512.

Romero R, Kusanovic JP, Chaiworapongsa T, Hassan SS. Placental bed disorders in preterm labor, preterm PROM, spontaneous abortion and abruptio placentae. Best Pract Res Clin Obstet Gynaecol. 2011;25(3):313–327. https://doi.org/10.1016/j.bpobgyn.2011.02.006. Epub 2011 Mar 8. Review. PubMed PMID: 21388889; PubMed Central PMCID: PMC3092823.

Ross MG, Mansano RZ. Fetal growth restriction. Medscape Reference: Drugs. Diseases and Procedures. Emedicine.medscape.com/article/261226_overview. Accessed February 20, 2013

Rouse DJ, Keimig TW, Riley LE, Letourneau AR, Platt MY. Case records of the Massachusetts general hospital. Case 16-2016. A 31-year-old pregnant woman with fever. N Engl J Med. 2016;374:2076–83.

Royal College of Obstetricians and Gynaecologists placenta praevia, placenta praevia accreta and vasa praevia: diagnosis and management (Green-top Guideline No. 27). https://www.rcog.org.uk/en/guidelines-research-services/guidelines/gtg27/

Royal College of Obstetricians; Gynaecologists UK (April 2001). Further Issues Relating to Late Abortion, Fetal Viability and Registration of Births and Deaths. Royal College of Obstetricians and Gynaecologists UK. Archived from the original on 5 November 2013. Retrieved 4 July 2013.

Rubod C, Deruelle P, Le Goueff F, Tunez V, Fournier M, Subtil D. Long-term prognosis for infants after massive fetomaternal hemorrhage. Obstet Gynecol. 2007;110(2 Pt 1):256–260. PubMed PMID: 17666598.

Ruiter L, Ravelli AC, de Graaf IM, Mol BW, Pajkrt E. Incidence and recurrence rate of placental abruption: a longitudinal linked national cohort study in the Netherlands. Am J Obstet Gynecol. 2015;213(4):573.e1–8. https://doi.org/10.1016/j.ajog.2015.06.019. Epub 2015 Jun 10. PubMed PMID: 26071916.

Ruiz-Cordero R, Birusingh RJ, Pelaez L, Azouz M, Rodriguez MM. Twin reversed arterial perfusion sequence (TRAPS): an illustrative series of 13 cases. Fetal Pediatr Pathol. 2016;35(2):63–80. https://doi.org/10.3109/15513815.2015.1131785

Sadove AM, Eppley BL, DeMyer W. Single stage repair of the median cleft lip deformity in holoprosencephaly. J Craniomaxillofac Surg. 1989 Nov;17(8):363-366. https://doi.org/10.1016/s1010-5182(89)80107-6.

Salafia CM, Minior VK, Lopez-Zeno JA, Whittington SS, Pezzullo JC, Vintzileos AM. Relationship between placental histologic features and umbilical cord blood gases in preterm gestations. Am J Obstet Gynecol. 1995;173:1058–64.

Saleemuddin A, Tantbirojn P, Sirois K, Crum CP, Boyd TK, Tworoger S, Parast MM. Obstetric and perinatal complications in placentas with fetal thrombotic vasculopathy. Pediatr Dev Pathol. 2010;13:459–64.

Salmani D, Purushothaman S, Somashekara SC, Gnanagurudasan E, Sumangaladevi K, Harikishan R, Venkateshwarareddy M. Study of structural changes in placenta in pregnancy-induced hypertension. J Nat Sci Biol Med. 2014 Jul;5(2):352-355. https://doi.org/10.4103/0976-9668.136182. PMID: 25097413; PMCID: PMC4121913.

Sarantopoulos GP, Natarajan S. Placenta accreta. Arch Pathol Lab Med. 2002;126(12):1557–1558. PubMed PMID: 12503589.

Sarnat HB, Sarnat MS. Neonatal encephalopathy following fetal distress. A clinical and electroencephalographic study. Arch Neurol. 1976 Oct;33(10):696-705. PubMed PMID: 987769.

Sasai Y, Lu B, Steinbeisser H, Geissert D, Gont LK, De Robertis EM. Xenopus Chordin: a novel Dorsalizing factor activated by organizer-specific Homeobox genes. Cell. 1994;79:779–90.

Saxena AK. Surgical perspectives regarding application of biomaterials for the management of large congenital diaphragmatic hernia defects. Pediatr Surg Int. 2018;34(5):475–89. https://doi.org/10.1007/s00383-018-4253-1

Saxena D, Srivastava P, Tuteja M, Mandal K, Phadke SR. Phenotypic characterization of derivative

22 syndrome: case series and review. J Genet. 2018;97(1):205–11.

Schalasta G, Schmid M, Lachmund T, Enders G. LightCycler consensus PCR for rapid and differential detection of human erythrovirus B19 and V9 isolates. J Med Virol. 2004;73(1):54–59. PubMed PMID: 15042648.

Schiesser M, Sergi C, Enders M, Maul H, Schnitzler P. Discordant outcomes in a case of parvovirus b19 transmission into both dichorionic twins. Twin Res Hum Genet. 2009;12(2):175–9. https://doi.org/10.1375/twin.12.2.175

Schiffer C, Tariverdian G, Schiesser M, Thomas MC, Sergi C. Agnathia-otocephaly complex: report of three cases with involvement of two different Carnegie stages. Am J Med Genet. 2002 Oct 1;112(2):203–8. https://doi.org/10.1002/ajmg.10672

Schloo BL, Vawter GF, Reid LM. Down syndrome: patterns of disturbed lung growth. Hum Pathol. 1991 Sep;22(9):919-923. PubMed PMID: 1833304.

Schmidt P, Skelly CL, Raines DA. Placental Abruption (Abruptio Placentae). 2019 May 23. StatPearls [Internet]. Treasure Island (FL): StatPearls Publishing; 2019 Jan-. Available from http://www.ncbi.nlm.nih.gov/books/NBK482335/ PubMed PMID: 29493960.

Schorling S, Schalasta G, Enders G, Zauke M. Quantification of parvovirus B19 DNA using COBAS AmpliPrep automated sample preparation and LightCycler real-time PCR. J Mol Diagn. 2004;6(1):37–41. 14736825, PMC1867462.

Seckl MJ, Sebire NJ, Berkowitz RS. Gestational trophoblastic disease. Lancet. 2010;376(9742):717–729. https://doi.org/10.1016/S0140-6736(10)60280-2. Epub 2010 Jul 29. Review. PubMed PMID: 20673583.

Seidahmed MZ, Abdelbasit OB, Alhussein KA, Miqdad AM, Khalil MI, Salih MA. Sirenomelia and severe caudal regression syndrome. Saudi Med J. 2014 Dec;35 Suppl 1(Suppl 1):S36-43. PMID: 25551110; PMCID: PMC4362094.

Sergi C. Hirschsprung's disease: Historical notes and pathological diagnosis on the occasion of the 100(th) anniversary of Dr. Harald Hirschsprung's death. World J Clin Pediatr. 2015 Nov 8;4(4):120-125. https://doi.org/10.5409/wjcp.v4.i4.120. PMID: 26566484; PMCID: PMC4637802.

Sergi C, Adam S, Kahl P, Otto HF. Study of the malformation of ductal plate of the liver in Meckel syndrome and review of other syndromes presenting with this anomaly. Pediatr Dev Pathol. 2000a;3(6):568–583. PubMed PMID: 11000335.

Sergi C, Adam S, Kahl P, Otto HF. The remodeling of the primitive human biliary system. Early Hum Dev. 2000b;58(3):167–178. PubMed PMID: 10936437.

Sergi C, Beedgen B, Kopitz J, Zilow E, Zoubaa S, Otto HF, Cantz M, Linderkamp O. Refractory congenital ascites as a manifestation of neonatal sialidosis: clinical, biochemical and morphological studies in a newborn Syrian male infant. Am J Perinatol. 1999a;16(3):133–141. PubMed PMID: 10438195.

Sergi C, Beedgen B, Kopitz J, Zilow E, Zoubaa S, Otto HF, Cantz M, Linderkamp O. Refractory congenital ascites as a manifestation of neonatal sialidosis: clinical, biochemical and morphological studies in a newborn Syrian male infant. Am J Perinatol. 1999b;16(3):133–41.

Sergi CM, Caluseriu O, McColl H, Eisenstat DD. Hirschsprung's disease: clinical dysmorphology, genes, micro-RNAs, and future perspectives. Pediatr Res. 2017 Jan;81(1-2):177-191. https://doi.org/10.1038/pr.2016.202. Epub 2016 Sep 28.

Sergi C, Daum E, Pedal I, Hauröder B, Schnitzler P. Fatal circumstances of human herpesvirus 6 infection: transcriptosome data analysis suggests caution in implicating HHV-6 in the cause of death. J Clin Pathol. 2007;60(10):1173–1177. Epub 2007 Jun 1. PubMed PMID: 17545558; PubMed Central PMCID: PMC2014822.

Sergi C, Dörfler A, Albrecht F, Klapp J, Jansen O, Sartor K, Otto HF. Utilization of magnetic resonance imaging in autopsy planning with specimen preservation for thoraco-omphalopagus symmetricus conjoined twins. Teratology. 1998a Sep-Oct;58(3-4):71-75. PubMed PMID: 9802185.

Sergi C, Dörfler A, Albrecht F, Klapp J, Jansen O, Sartor K, Otto HF. Utilization of magnetic resonance imaging in autopsy planning with specimen preservation for thoraco-omphalopagus symmetricus conjoined twins. Teratology. 1998b Sep-Oct;58(3-4):71–5. https://doi.org/10.1002/(SICI)1096-9926(199809/10)58:3/4<71::AID-TERA1>3.0.CO;2-C

Sergi C, Gekas J, Kamnasaran D. Holoprosencephaly-polydactyly (pseudotrisomy 13) syndrome: case report and diagnostic criteria. Fetal Pediatr Pathol. 2012a;31(5):315–8. https://doi.org/10.3109/15513815.2012.659390

Sergi C, Gekas J, Kamnasaran D. Holoprosencephaly-polydactyly (pseudotrisomy 13) syndrome: case report and diagnostic criteria. Fetal Pediatr Pathol. 2012b Oct;31(5):315-318. https://doi.org/10.3109/15513815.2012.659390. Epub 2012 Mar 20.

Sergi C, Kahl P, Otto HF. Contribution of apoptosis and apoptosis-related proteins to the malformation of the primitive intrahepatic biliary system in Meckel syndrome. Am J Pathol. 2000;156(5):1589–98. 10793071, PMC1876920.

Sergi C, Kamnasaran D. PRRX1 is mutated in a fetus with agnathia-otocephaly. Clin Genet. 2011 Mar;79(3):293–5. https://doi.org/10.1111/j.1399-0004.2010.01531.x

Sergi C, Penzel R, Uhl J, Zoubaa S, Dietrich H, Decker N, Rieger P, Kopitz J, Otto HF, Kiessling M, Cantz M. Prenatal diagnosis and fetal pathology in a Turkish family harboring a novel nonsense mutation in the lysosomal alpha-N-acetyl-neuraminidase (sialidase) gene. Hum Genet. 2001a;109(4):421–428. PubMed PMID: 11702224.

Sergi, C., Penzel, R., Uhl, J., Zoubaa, S., Dietrich, H., Decker, N., Rieger, P., Kopitz, J., Otto, H. F., Kiessling,

M., Cantz, M. Prenatal diagnosis and fetal pathology in a Turkish family harboring a novel nonsense mutation in the lysosomal alpha-N-acetyl-neuraminidase (sialidase) gene. Hum. Genet. 109: 421-428, 2001b. [PubMed: 11702224].

Sergi C, Penzel R, Uhl J, Zoubaa S, Dietrich H, Decker N, Rieger P, Kopitz J, Otto HF, Kiessling M, Cantz M. Prenatal diagnosis and fetal pathology in a Turkish family harboring a novel nonsense mutation in the lysosomal alpha-N-acetyl-neuraminidase (sialidase) gene. Hum Genet. 2001c Oct;109(4):421–8.

Sergi C, Penzel R, Uhl J, Zoubaa S, Dietrich H, Decker N, et al. Prenatal diagnosis and fetal pathology in a Turkish family harboring a novel nonsense mutation in the lysosomal alpha-N-acetyl-neuraminidase (sialidase) gene. Hum Genet. 2001;109(4):421–8. https://doi.org/10.1007/s004390100592

Sergi C, Poeschl J, Graf M, Linderkamp O. Restrictive dermopathy: case report, subject review with Kaplan-Meier analysis, and differential diagnosis of the lethal congenital contractural syndromes. Am J Perinatol. 2001;18(1):39–47.

Sergi C, Schiesser M, Adam S, Otto HF. Analysis of the spectrum of malformations in human fetuses of the second and third trimester of pregnancy with human triploidy. Pathologica. 2000;92(4):257–63.

Sergi C, Schmitt HP. Central nervous system in twin reversed arterial perfusion sequence with special reference to examination of the brain in acardius anceps. Teratology. 2000a Apr;61(4):284-290. https://doi.org/10.1002/(SICI)1096-9926(200004)61:4<284::AID-TERA7>3.0.CO;2-T. PMID: 10716747.

Sergi C, Schmitt HP. The vesicular forebrain (pseudo-aprosencephaly): a missing link in the teratogenetic spectrum of the defective brain anlage and its discrimination from aprosencephaly. Acta Neuropathol. 2000b Mar;99(3):277–84. https://doi.org/10.1007/pl00007438

Sergi C, Serpi M, Müller-Navia J, Schnabel PA, Hagl S, Otto HF, Ulmer HE. CATCH 22 syndrome: report of 7 infants with follow-up data and review of the recent advancements in the genetic knowledge of the locus 22q11. Pathologica.1999 Jun;91(3):166-172. Review. PubMed PMID: 10536461.

Shah PS, Hellmann J, Adatia I. Clinical characteristics and follow up of Down syndrome infants without congenital heart disease who presented with persistent pulmonary hypertension of newborn. J Perinat Med. 2004;32(2):168-170. PubMed PMID: 15085894.

Sharami SRY, Saffarieh E. A review on management of gestational trophoblastic neoplasia. J Family Med Prim Care. 2020 Mar 26;9(3):1287-1295. https://doi.org/10.4103/jfmpc.jfmpc_876_19. PMID: 32509606; PMCID: PMC7266251.

Shih IM. The role of CD146 (Mel-CAM) in biology and pathology. J Pathol. 1999;189:4–11.

Shih IM. Trophogram, an immunohistochemistry-based algorithmic approach, in the differential diagnosis of trophoblastic tumors and tumorlike lesions. Ann Diagn Pathol. 2007;11(3):228–234. Review. PubMed PMID: 17498600.

Shih IM, Kurman RJ. Ki-67 labeling index in the differential diagnosis of exaggerated placental site, placental site trophoblastic tumor, and choriocarcinoma: a double immunohistochemical staining technique using Ki-67 and Mel-CAM antibodies. Hum Pathol. 1998;29:27–33.

Sills A, Steigman C, Ounpraseuth ST, Odibo I, Sandlin AT, Magann EF. Pathologic examination of the placenta: recommended versus observed practice in a university hospital. Int J Women's Health. 2013;5:309–312. https://doi.org/10.2147/IJWH.S45095. Print 2013. PubMed PMID: 23788842; PubMed Central PMCID: PMC3684225.

Sinkin JA, Craig WY, Jones M, Pinette MG, Wax JR. Perinatal outcomes associated with isolated Velamentous cord insertion in singleton and twin pregnancies. J Ultrasound Med. 2018;37(2):471–8. https://doi.org/10.1002/jum.14357

Sivanathan J, Thilaganathan B. Book: genetics for obstetricians and gynaecologists: chapter: genetic markers on ultrasound scan. Best Pract Res Clin Obstet Gynaecol. 2017;42:64–85. https://doi.org/10.1016/j.bpobgyn.2017.03.005

Smith WC, Harland RM. Expression cloning of noggin, a new Dorsalizing factor localized to the Spemann organizer in Xenopus embryos. Cell. 1992;70:829–40.

Smith J, Treadwell MC, Berman DR. Role of ultrasonography in the management of twin gestation. Int J Gynaecol Obstet. 2018;141(3):304–14. https://doi.org/10.1002/ijgo.12483

Snijders RJ, De Courcy-Wheeler RH, Nicolaides KH. Intrauterine growth retardation and fetal transverse cerebellar diameter. Prenat Diagn. 1994;14(12):1101–1105. PubMed PMID: 7899277.

Sohma O, Mito T, Mizuguchi M, Takashima S. The prenatal age critical for the development of the pontosubicular necrosis. Acta Neuropathol. 1995;90(1):7–10. https://doi.org/10.1007/BF00294453

Spemann H, Mangold H. Induction of embryonic primordia by implantation of organizers from a different species. Int J Dev Biol. 2001;45:13–38. Originally published in Archiv für Mikroskopische Anatomie und Entwicklungsmechanik, 100 (1924): 599–63.

Srinivasan A, Graves L. Four true umbilical cord knots. J Obstet Gynaecol Can. 2006;28(1):32–35. https://doi.org/10.1016/S1701-2163(16)32053-9. PubMed PMID: 16533453.

Stanek J. Association of coexisting morphological umbilical cord abnormality and clinical cord compromise with hypoxic and thrombotic placental histology. Virchows Arch. 2016;468(6):723–32. https://doi.org/10.1007/s00428-016-1921-1

Stanek J. Placental examination in nonmacerated stillbirth versus neonatal mortality. J Perinat Med. 2018a;46(3):323–31. https://doi.org/10.1515/jpm-2017-0198

Stanek J. Placental pathology varies in hypertensive conditions of pregnancy. Virchows Arch. 2018b;472(3):415–23. https://doi.org/10.1007/s00428-017-2239-3

Stevens FT, Katzorke N, Tempfer C, Kreimer U, Bizjak GI, Fleisch MC, Fehm TN. Gestational Trophoblastic Disorders: An Update in 2015. Geburtshilfe Frauenheilkd. 2015 Oct;75(10):1043-1050. https://doi.org/10.1055/s-0035-1558054. PMID: 26556906; PMCID: PMC4629994.

Stevenson RE. Limbs. In: Stevenson RE, Hall JG, editors. Human malformations and related anomalies. 2nd ed. New York (NY): Oxford University Press; 2006. p. 835–925.

Stiller AG, Skafish PR. Placental chorioangioma: a rare cause of fetomaternal transfusion with maternal hemolysis and fetal distress. Obstet Gynecol. 1986;67(2):296–298. PubMed PMID: 3945441.

Stirnemann JJ, Chalouhi G, Ville Y. Twin-to-twin transfusion syndrome: from observational evidence to randomized controlled trials. Twin Res Hum Genet. 2016;19(3):268–75. https://doi.org/10.1017/thg.2016.22

Stocker JT, Heifetz SA. Sirenomelia. A morphological study of 33 cases and review of the literature. Perspect Pediatr Pathol. 1987;10:7–50.

Streeter G. Focal deficiencies in fetal tissues and their relation to intrauterine amputations. Contrib Embryol Carnegie Inst. 1930;22:1–46.

Studd J. Placenta praevia. Progress in Obstetrics & Gynaecology, vol. 11.

Subtil D, Cosson M, Houfflin V, Vaast P, Valat A, Puech F. Early detection of caudal regression syndrome: specific interest and findings in three cases. Eur J Obstet Gynecol Reprod Biol. 1998;80(1):109–12.

Sugibayashi R, Ozawa K, Sumie M, Wada S, Ito Y, Sago H. Forty cases of twin reversed arterial perfusion sequence treated with radio frequency ablation using the multistep coagulation method: a single-center experience. Prenat Diagn. 2016;36(5):437–43. https://doi.org/10.1002/pd.4800

Sun M, Forsman C, Sergi C, Gopalakrishnan R, O'Connor MB, Petryk A. The expression of twisted gastrulation in postnatal mouse brain and functional implications. Neuroscience. 2010 Aug 25;169(2):920-931. https://doi.org/10.1016/j.neuroscience.2010.05.026. Epub 2010 May 20.

Swanson AB. A classification for congenital limb malformations. J Hand Surg. 1976;1A:8–22.

Tachibana M, Nakayama M, Miyoshi Y. Placental examination: prognosis after delivery of the growth-restricted fetus. Curr Opin Obstet Gynecol. 2016;28(2):95–100. https://doi.org/10.1097/GCO.0000000000000249

Takahashi Y, Iwagaki S, Chiaki R, Asai K, Matsui M, Kawabata I. Ultrasonic identification of pump twin by dual-gate Doppler in a monochorionic-triamniotic triplet twin reversed arterial perfusion sequence before preventative radiofrequency ablation: a case report. J Med Ultrason (2001). 2018;45(1):185–7. https://doi.org/10.1007/s10396-017-0792-7

Takano M, Nakata M, Murata S, Sumie M, Morita M. Chorioamniotic membrane separation after Fetoscopic laser photocoagulation. Fetal Diagn Ther. 2018;43(1):40–4. https://doi.org/10.1159/000472713

Takawira C, D'Agostini S, Shenouda S, Persad R, Sergi C. Laboratory procedures update on Hirschsprung disease. J Pediatr Gastroenterol Nutr. 2015 May;60(5):598–605. https://doi.org/10.1097/MPG.0000000000000679

Takebayashi A, Kimura F, Yamanaka A, Takahashi A, Tsuji S, Ono T, Kaku S, Kita N, Takahashi K, Okabe H, Murakami T. Exaggerated placental site, consisting of implantation site intermediate trophoblasts, causes massive postpartum uterine hemorrhage: case report and literature review. Tohoku J Exp Med. 2014;234(1):77–82. Review. PubMed PMID: 25186195.

Takemoto R, Anami A, Koga H. Relationship between birth weight to placental weight ratio and major congenital anomalies in Japan. PLOS ONE. 2018a;13(10):e0206002. https://doi.org/10.1371/journal.pone.0206002

Takemoto R, Anami A, Koga H. Relationship between birth weight to placental weight ratio and major congenital anomalies in Japan. PLOS ONE. 2018b;13(10):e0206002. https://doi.org/10.1371/journal.pone.0206002

Takemoto R, Anami A, Koga H. Relationship between birth weight to placental weight ratio and major congenital anomalies in Japan. PLOS ONE. 2018c;13(10):e0206002. https://doi.org/10.1371/journal.pone.0206002

Taney J, Anastasio H, Paternostro A, Berghella V, Roman A. Placental abruption with delayed fetal compromise in maternal acetaminophen toxicity. Obstet Gynecol. 2017;130(1):159–62. https://doi.org/10.1097/AOG.0000000000002089

Tantbirojn P, Saleemuddin A, Sirois K, Crum CP, Boyd TK, Tworoger S, Parast MM. Gross abnormalities of the umbilical cord: related placental histology and clinical significance. Placenta. 2009;30:1083–8.

Teixeira AB, Lana AM, Lamounier JA, Pereira da Silva O, Eloi-Santos SM. Neonatal listeriosis: the importance of placenta histological examination-a case report. AJP Rep. 2011;1:3–6.

Thia E, Thain S, Yeo GS. Fetoscopic laser photocoagulation in twin-to-twin transfusion syndrome: experience from a single institution. Singap Med J. 2017,58(6):321–6. https://doi.org/10.11622/smedj.2016067

Thurn L, Wikman A, Lindqvist PG. Postpartum blood transfusion and hemorrhage as independent risk factors for venous thromboembolism. Thromb Res. 2018;165:54–60. https://doi.org/10.1016/j.thromres.2018.03.002

Tonkin MA. Classification of congenital anomalies of the hand and upper limb. J Hand Surg Eur Vol. 2017;42(5):448–56. https://doi.org/10.1177/1753193417690965

Tonni G, Granese R, Martins Santana EF, Parise Filho JP, Bottura I, Borges Peixoto A, Giacobbe A, Azzerboni A, Araujo Júnior E. Prenatally diagnosed fetal tumors of the head and neck: a systematic review with antenatal and postnatal outcomes over the past 20 years. J Perinat Med. 2017 Feb 1;45(2):149-165. https://doi.org/10.1515/jpm-2016-0074. Review.

Torpin R. Amniochorionic Mesoblastic fibrous strings and amniotic bands. associate constricting fetal malformations or fetal death. Am J Obstet Gynecol. 1965;91:65–75.

Toutain J, Goutte-Gattat D, Horovitz J, Saura R. Confined placental mosaicism revisited: Impact on pregnancy characteristics and outcome. PLoS One. 2018a Apr 12;13(4):e0195905. https://doi.org/10.1371/journal.pone.0195905. 29649318, PMC5897023.

Toutain J, Goutte-Gattat D, Horovitz J, Saura R. Confined placental mosaicism revisited: Impact on pregnancy characteristics and outcome. PLoS One. 2018b Apr 12;13(4):e0195905. https://doi.org/10.1371/journal.pone.0195905. 29649318, PMC5897023.

Toutain J, Goutte-Gattat D, Horovitz J, Saura R. Confined placental mosaicism revisited: Impact on pregnancy characteristics and outcome. PLoS One. 2018c Apr 12;13(4):e0195905. https://doi.org/10.1371/journal.pone.0195905. 29649318, PMC5897023.

Triebwasser JE, Treadwell MC. Prenatal prediction of pulmonary hypoplasia. Semin Fetal Neonatal Med. 2017;22(4):245–9. https://doi.org/10.1016/j.siny.2017.03.001

Tse HK, Leung MB, Woolf AS, Menke AL, Hastie ND, Gosling JA, Pang CP, Shum AS. Implication of Wt1 in the pathogenesis of nephrogenic failure in a mouse model of retinoic acid-induced caudal regression syndrome. Am J Pathol. 2005;166(5):1295–307. 15855632, PMC1606386.

Tunon K, Eik-Nes SH, Grøttum P, Von Düring V, Kahn JA. Gestational age in pregnancies conceived after in vitro fertilization: a comparison between age assessed from oocyte retrieval, crown-rump length and biparietal diameter. Ultrasound Obstet Gynecol. 2000;15(1):41–6. https://doi.org/10.1046/j.1469-0705.2000.00004.x

Turner JM, Wales PW, Nation PN, Wizzard P, Pendlebury C, Sergi C, Ball RO, Pencharz PB. Novel neonatal piglet models of surgical short bowel syndrome with intestinal failure. J Pediatr Gastroenterol Nutr. 2011 Jan;52(1):9–16. https://doi.org/10.1097/MPG.0b013e3181f18ca0

Uhl J, Penzel R, Sergi C, Kopitz J, Otto HF, Cantz M. Identification of a CTL4/Neu1 fusion transcript in a sialidosis patient. FEBS Lett. 2002a;521(1–3):19–23.

Uhl J, Penzel R, Sergi C, Kopitz J, Otto HF, Cantz M. Identification of a CTL4/Neu1 fusion transcript in a sialidosis patient. FEBS Lett. 2002b Jun 19;521(1-3):19–23.

Ul Haque A, Siddique S, Jafari MM, Hussain I, Siddiqui S. Pathology of chorionic villi in spontaneous abortions. Int J Pathol. 2004;2(1):5–9.

Urioste M, Valcarcel E, Gomez MA, Pinel I, Garcia de Len R, Diaz de Bustamante A, Tebar R, Martinez-Frias ML. Holoprosencephaly and trisomy 21 in a child born to a nondiabetic mother. Am J Med Genet. 1988a;30:925–8.

Urioste M, Valcarcel E, Gomez MA, Pinel I, Garcia de Len R, Diaz de Bustamante A, Tebar R, Martinez-Frias ML. Holoprosencephaly and trisomy 21 in a child born to a nondiabetic mother. Am J Med Genet. 1988b;30:925–8.

Urioste M, Valcarcel E, Gomez MA, Pinel I, Garcia de León R, Diaz de Bustamante A, Tebar R, Martinez-Frias ML. Holoprosencephaly and trisomy 21 in a child born to a nondiabetic mother. Am J Med Genet. 1988 Aug;30(4):925–8. https://doi.org/10.1002/ajmg.1320300408

Van Allen MI, Curry C, Walden CE, Gallagher L, Patten RM. Limb body wall complex.11. Limb and spine defects. Am J Med Genet. 1987;28:549–65.

van Lier MG, Lopriore E, Vandenbussche FP, Streekstra GJ, Siebes M, Nikkels PG, Oepkes D, van Gemert MJ, van den Wijngaard JP. Acardiac twinning: High resolution three-dimensional reconstruction of a low resistance case. Birth Defects Res A Clin Mol Teratol. 2016 Mar;106(3):213-7. https://doi.org/10.1002/bdra.23477.Epub 2015 Dec 21.

Vander Haar EL, So J, Gyamfi-Bannerman C, Han YW. Fusobacterium nucleatum and adverse pregnancy outcomes: epidemiological and mechanistic evidence. Anaerobe. 2018;50:55–9.

Velez-Perez A, Younes P, Tatevian N. Placental fetal thrombotic vasculopathy occurring in association with Megacystis-microcolon-intestinal Hypoperistalsis syndrome: a case report. Ann Clin Lab Sci. 2017;47(3):357–61.

Vinay K, Kanwar AJ, Sawatkar GU, Dogra S, Ishii N, Hashimoto T. Successful use of rituximab in the treatment of childhood and juvenile pemphigus. J Am Acad Dermatol. 2014;71(4):669–75. https://doi.org/10.1016/j.jaad.2014.05.071. Epub 2014 Jul 9

Vinkesteijn AS, Mulder PG, Wladimiroff JW. Fetal transverse cerebellar diameter measurements in normal and reduced fetal growth. Ultrasound Obstet Gynecol. 2000;15(1):47–51.

Waddington CH, Needham J, Needham DM. Physico-chemical experiments on the amphibian organizer. In: Fulton C, Klein AO, editors. Explorations in developmental biology. Cambridge: Harvard University Press; 1976. Originally published in Nature 132; 1933. p. 239.

Wang Y, Cao L, Liang D, Meng L, Wu Y, Qiao F, Ji X, Luo C, Zhang J, Xu T, Yu B, Wang L, Wang T, Pan Q, Ma D, Hu P, Xu Z. Prenatal chromosomal microarray analysis in fetuses with congenital heart disease: a prospective cohort study. Am J Obstet Gynecol. 2018 Feb;218(2):244.e1-244.e17. https://doi.org/10.1016/j.ajog.2017.10.225. Epub 2017 Nov 8. Erratum in: Am J Obstet Gynecol. 2018 Apr 12;:.

Weick JP, Kang H, Bonadurer GF 3rd, Bhattacharyya A. Gene Expression Studies on Human Trisomy 21 iPSCs and Neurons: Towards Mechanisms Underlying Down's Syndrome and Early Alzheimer's Disease-Like Pathologies. Methods Mol Biol. 2016;1303:247–65. https://doi.org/10.1007/978-1-4939-2627-5_15

Weiner E, Barber E, Feldstein O, Dekalo A, Schreiber L, Bar J, Kovo M. Placental Histopathology Differences and Neonatal Outcome in Dichorionic-Diamniotic as Compared to Monochorionic-Diamniotic Twin Pregnancies. Reprod Sci. 2018 Jul;25(7):1067–72. https://doi.org/10.1177/1933719117732163. Epub 2017 Oct 2

Weiner E, Mizrachi Y, Grinstein E, Feldstein O, Rymer-Haskel N, Juravel E, Schreiber L, Bar J, Kovo M. The role of placental histopathological lesions in predicting recurrence of preeclampsia. Prenat Diagn. 2016 Oct;36(10):953–60. https://doi.org/10.1002/pd.4918. Epub 2016 Sep 28

Wenger TL, Chow P, Randle SC, Rosen A, Birgfeld C, Wrede J, Javid P, King D, Manh V, Hing AV, Albers E. Novel findings of left ventricular non-compaction cardiomyopathy, microform cleft lip and poor vision in patient with SMC1A-associated Cornelia de Lange syndrome. Am J Med Genet A. 2017 Feb;173(2):414–20. https://doi.org/10.1002/ajmg.a.38030. Epub 2016 Nov 7. Review.

Witchel SF. Disorders of sex development. Best Pract Res Clin Obstet Gynaecol. 2018;48:90–102. https://doi.org/10.1016/j.bpobgyn.2017.11.005

Wolfe B, Wiepz GJ, Schotzko M, Bondarenko GI, Durning M, Simmons HA, Mejia A, Faith NG, Sampene E, Suresh M, Kathariou S, Czuprynski CJ, Golos TG. Acute Fetal Demise with First Trimester Maternal Infection Resulting from Listeria monocytogenes in a Nonhuman Primate Model. MBio. 2017 Feb 21;8(1). pii: e01938-16. https://doi.org/10.1128/mBio.01938-16. PubMed PMID: 28223455; PubMed Central PMCID: PMC5358912.

World Health Organization. ICD-11 for Mortality and Morbidity Statistics (ICD-11 MMS) 2018 version. https://icd.who.int/browse11/l-m/enAccessed on July. 24:2020.

Wortman AC, Schaefer SL, McIntire DD, Sheffield JS, Twickler DM. Complete placenta Previa: ultrasound biometry and surgical outcomes. AJP Rep. 2018;8(2):e74–8. https://doi.org/10.1055/s-0038-1641163. Epub 2018 Apr 20. PubMed PMID: 29686936; PubMed Central PMCID: PMC5910059

Wylie BJ, D'Alton ME. Fetomaternal hemorrhage. Obstet Gynecol. 2010;115(5):1039–51. https://doi.org/10.1097/AOG.0b013e3181da7929. Review.

Yaguchi C, Itoh H, Tsuchiya KJ, Furuta-Isomura N, Horikoshi Y, Matsumoto M, Jeenat FU, Keiko MK, Kohmura-Kobatashi Y, Tamura N, Sugihara K, Kanayama N.Placental pathology predicts infantile physical development during first 18 months in Japanese population: Hamamatsu birth cohort for mothers and children (HBC Study). PLoS One. 2018 Apr 10;13(4):e0194988. https://doi.org/10.1371/journal.pone.0194988. eCollection 2018. PubMed PMID: 29634735; PubMed Central PMCID: PMC5892873.

Yamamoto R, Ishii K, Muto H, Ota S, Kawaguchi H, Hayashi S, Mitsuda N. Incidence of and risk factors for severe maternal complications associated with hypertensive disorders after 36 weeks' gestation in uncomplicated twin pregnancies: A prospective cohort study. J Obstet Gynaecol Res. 2018a Jul;44(7):1221-1227. https://doi.org/10.1111/jog.13650. Epub 2018 Apr 19.

Yamamoto R, Ishii K, Muto H, Ota S, Kawaguchi H, Hayashi S, Mitsuda N. Incidence of and risk factors for severe maternal complications associated with hypertensive disorders after 36 weeks' gestation in uncomplicated twin pregnancies: a prospective cohort study. J Obstet Gynaecol Res. 2018b; https://doi.org/10.1111/jog.13650

Yamamoto R, Ishii K, Nakajima E, Sasahara J, Mitsuda N. Ultrasonographic prediction of antepartum deterioration of growth-restricted fetuses after late preterm. J Obstet Gynaecol Res. 2018; https://doi.org/10.1111/jog.13626

Yeniel AÖ, Ergenoğlu AM, Sağol S. Prenatal diagnosis of caudal regression syndrome without maternal diabetes mellitus. J Turk Ger Gynecol Assoc. 2011;12(3):186–188. https://doi.org/10.5152/jtgga.2011.43. eCollection 2011. PubMed PMID: 24591990; PubMed Central PMCID: PMC3939279.

Yosaatmadja F, Andrews KT, Duffy MF, Brown GV, Beeson JG, Rogerson SJ. Characterization of VAR2CSA-deficient Plasmodium falciparum--infected erythrocytes selected for adhesion to the BeWo placental cell line. Malar J. 2008;7:51. https://doi.org/10.1186/1475-2875-7-51. PubMed PMID: 18364051; PubMed Central PMCID: PMC2329659.

Yu TW, Mochida GH, Tischfield DJ, Sgaier SK, Flores-Sarnat L, Sergi CM, Topçu M, McDonald MT, Barry BJ, Felie JM, Sunu C, Dobyns WB, Folkerth RD, Barkovich AJ, Walsh CA. Mutations in WDR62, encoding a centrosome-associated protein, cause microcephaly with simplified gyri and abnormal cortical architecture. Nat Genet. 2010 Nov;42(11):1015-1020. https://doi.org/10.1038/ng.683. Epub 2010 Oct 3. PubMed PMID: 20890278; PubMed Central PMCID: PMC2969850.

Zanardo V, Trevisanuto D, Cosmi E, Chiarelli S. Chorioamnionitis and cerebral palsy: a meta-analysis. Obstet Gynecol. 2010;116(6):1454.; author reply 1454. https://doi.org/10.1097/AOG.0b013c3181fd343a

Zarrei M, Hicks GG, Reynolds JN, Thiruvahindrapuram B, Engchuan W, Pind M, Lamoureux S, Wei J, Wang Z, Marshall CR, Wintle RF, Chudley AE, Scherer SW. Copy number variation in fetal alcohol spectrum disorder. Biochem Cell Biol. 2018 Apr;96(2):161-166. https://doi.org/10.1139/bcb-2017-0241. Epub 2018 Mar 13.

Zaw W, Stone DG. Caudal regression syndrome in twin pregnancy with type II diabetes. J Perinatol. 2002;22(2):171–4.

Zepeda-Mendoza CJ, Bardon A, Kammin T, Harris DJ, Cox H, Redin C, Ordulu Z, Talkowski ME, Morton CC. Phenotypic interpretation of complex chromosomal rearrangements informed by nucleotide-level resolution and structural organization of chromatin. Eur J Hum Genet. 2018 Mar;26(3):374-381. https://doi.org/10.1038/s41431-017-0068-0. Epub 2018 Jan 10. PubMed PMID: 29321672; PubMed Central PMCID: PMC5838977.

Zhong QY, Gelaye B, Fricchione GL, Avillach P, Karlson EW, Williams MA. Adverse obstetric and neonatal outcomes complicated by psychosis among pregnant women in the United States. BMC Pregnancy Childbirth. 2018;18(1):120. https://doi.org/10.1186/s12884-018-1750-0. 29720114, PMC5930732.

Zhong QY, Gelaye B, Smoller JW, Avillach P, Cai T, Williams MA. Adverse obstetric outcomes during delivery hospitalizations complicated by suicidal behavior among US pregnant women. PLoS One. 2018;13(2):e0192943. https://doi.org/10.1371/journal.pone.0192943

Zimmerman LB, De Jesus-Escobar JM, Harland RM. The Spemann organizer signal noggin binds and inactivates bone morphogenetic Protein-4. Cell. 1996;86:599–606. 17064457; PubMed Central PMCID: PMC2870617.

Index

A
ABC classification, 1229
ABCD1 (X-ALD) deficiency-based leukodystrophy, 1287
ABCDE Rule of examination of placenta, 1419
Abdominal circumference (AC), 1421
Abdominal wall defects
  gastroschisis, 312
  omphalocele, 306
Abetalipoproteinemia, 326
Abnormal vascularization of placenta, 1434
Abortion, 1477
Abortus/Miscarriage Classification, 1480
Abruptio placentae, 1450–1455
Abruption/central retroplacental hemorrhage, 1452
Abscess, breast, 838
Abscess-pneumonia, 192
Acalculous cholecystitis, 557
Acantholytic, 1188
Acanthosis nigricans, 1377
Acardia, 1261–1263
Acardius, 1447
Acardius acephalous, 1449
Acardius acormus, 1448
Acardius amorphous, 1449
Acardius anceps, 1449
Achalasia, 262
Achondroplasia, 1100
Acidic mucins, 266
Acinar cell carcinomas (ACCs), 567, 851
Acquired anomalies of low/eumotility, 261, 262
Acquired anomalies with dysmotility, 262, 263
Acquired atelectasis, 158
Acquired atresia and stenosis of the colon, 354, 356
Acquired cystic kidney disease, 598
Acquired hemolytic anemias, 866
Acquired immunodeficiencies, 888
Acquired toxoplasmosis, 1282
Acral-lentiginous melanoma, 1386
Acrocentric chromosome, 1496
Acrochordon, 817, 1376
Acrodermatitis enteropathica, 327, 1361

Acromegaly, 942
ACTH therapy and sudden death, 1291
Actinic (solar) keratosis, 1377
Actinomycosis, 1282
Acute (tubulo-) interstitial nephritis (ATIN)
  acute hypersensitivity nephritis, 618
  acute pyelonephritis, 618, 620
  pyonephrosis, 618
  renal papillary necrosis, 618–622
Acute bilirubin elevation (ABE), 436
Acute bronchitis, 184
Acute cellular rejection (ACR), 337
Acute chorioamnionitis (ACA), 1461, 1462, 1464, 1484
Acute eosinophilic pneumonia, 182–184
Acute epiglottitis, 1185
Acute fibrinous and organizing
  pneumonia (AFOP), 172
Acute funisitis, 1461
Acute gastritis, 281
Acute hypersensitivity nephritis (AHN), 618
Acute intestinal ischemia, 318
Acute lymphoblastic (lymphocytic) leukemia (ALL), 739, 874
Acute lymphoblastic lymphoma, 896
Acute mastitis, 838
Acute myeloblastic (myelocytic) leukemia (AML), 875
Acute nephrotic syndrome, 603
Acute otitis media (AOM), 1219–1220
Acute pancreatitis, 563–564
Acute peripheral separation, 1450, 1455
Acute pyelonephritis, 620
Acute rejection, 212, 632
Acute suppurative (lepto-) meningitis, 1276, 1278
Acute thymic involution, 927
Acute viral hepatitis, 486, 488
Adamantinoma, 1150
ADAMTS13, 374
Adenocarcinoma, 277, 352
  gastric tumors, 293
  small intestinal neoplasms, 341
Adenoid basal carcinoma (ABC), 808
Adenoid cystic carcinoma (ACC), 810, 851

Adenoma, 275
  apocrine adenoma, 843
  ductal adenoma, 843
  intraductal papilloma, 843
  lactating adenoma, 841
  nipple duct adenoma, 841, 843
  papillomatosis, 843
  small intestinal neoplasms, 341
  syringomatous adenoma of nipple, 843
  tubular adenoma, 841
Adenomas, 559
Adenomatoid tumor, 727, 995
Adenomatosis, 562
Adenomatous, 290
Adenomatous, hyperplasia, 732
Adenomyoepithelial adenosis (ME-MGA), 855
Adenomyoepithelioma, 851
Adenomyosis, 773
Adenosine monophosphate deaminase-1 (AMPD1) gene, 59
Adenosquamous (mixed) carcinoma, 809
ADH-hypersecretion, 943
Adipocytic tumors
  atypical lipomatous tumor/well-differentiated liposarcoma, 1014
  dedifferentiated liposarcoma, 1016
  hibernoma, 1012–1013
  lipoblastoma, 1012
  lipoma subtypes, 1011–1012
  lipomatosis, 1012
  locally aggressive and malignant, 1013
  myxoid/round cell liposarcoma, 1015
  pleomorphic liposarcoma, 1015–1016
Adipose tissue tumors, 1396
Adnexal cutaneous tumors, 1390
Adolescent up to young adult (AYA), 844, 846
ADPKD-associated cysts, 432, 433, 435
Adrenal 4 binding protein (Ad4BP), 938
Adrenal cortical adenomas (ACA), 974
Adrenal cortical carcinoma (ACC), 974, 976–978
Adrenal gland
  abscess, 995
  adrenal cortex abnormalities, 996
  adrenal cortical carcinoma, 976–978
  adrenal gland/ periadrenal tumor-like lesions, 995
  adrenocortical adenoma, 974, 975
  congenital heart disease, 988
  cysts, 995
  diffuse micronodular hyperplasia, 974
  extraadrenal paraganglioma, 994
  ganglioneuroblastoma, 985
  ganglioneuroma, 985–988
  hemorrhage-related mass, 995
  infections, 995
  metastatic tumor, 995
  myelolipoma, 995
  neuroblastoma, 978–982, 988
    clinical presentation, 983
    grading, 984
    in situ, 985
    IV-S, 985
    metastases, 984
    pediatric tumoral DDX, 984
    postnatal diagnosis, 985
    prenatal diagnosis, 985
    spontaneous regression, 985
  nodules, 974
  pheochromocytoma, 988, 989
    pearls and pitfalls, 993
    prognostic factors, 989
  primary malignant lymphoma, 995
  primary malignant melanoma, 995
  tumors, 974
  zona fasciculata, 937
  zona glomerulosa, 937
  zona reticularis, 937
Adrenal hypoplasia congenita (AHC), 938
Adrenocortical adenoma, 974, 975
Adrenocortical steroidogenesis, 937
Adrenoleukodystrophy, 1287, 1288
Adria cell, 66
Adult fibrosarcoma (AFS), 1033
Adult granulosa cell tumor, 792
Adult ILD of youth
  acute eosinophilic pneumonia, 182–184
  COP, 170, 172
  DAD, 167, 168, 170
  desquamative interstitial pneumonia/-ITIS, 176, 177
  HSP, 181, 182
  LIP, 180, 181
  NSIP, 175, 176
  PMF, 184
  pulmonary langerhans cell histiocytosis, 179, 180
  respiratory bronchiolitis-ILD, 177, 179
  UIP, 173, 175
Adult pneumoconiosis, 204, 205
  asbestosis, 206, 207
  non-silico-asbestosis-related pneumoconiosis, 207
  silicosis, 206
Adult T-cell leukemia/lymphoma (ATCL), 878, 903, 904
Adult-type neoplasms, childhood/youth, 222, 226–230
Adult-type preneoplastic lesions, 227–230
Adult-type rhabdomyoma, 1046
a-dystrobrevin gene, 64, 65
Aganglionosis, 366, 367
Agenesis of the appendix, 343
Agenesis of the corpus callosum (ACC), 1257, 1258
Agnathia-otocephaly complex/syndrome (AOC/AOS), 1259, 1496
AI blistering diseases, 1362
AI pancreatitis–granulocytic epithelial lesion type, 566
AI pancreatitis–lympho-plasmacytic sclerosing type, 564–566
AI progesterone dermatitis, 1362
AIDS-related lymphadenopathy, 918
AKT1, 229
Alagille syndrome, 444
Albinism-related liver diseases, 486

Albright's syndrome, 1118
Alcoholic embryopathy, 47
Alcoholic liver disease, 496–498
Alcoholic steatohepatitis, 499
Aldosterone-secreting adenoma, 975
Allergic contact dermatitis, 1353
Allograft cellular rejection, 92, 95
Allograft humoral rejection, 95, 97
Alpers-Huttenlocher syndrome, 473
Alpha 2C-adrenoceptor gene, 56
Alpha-1-antitrypsin deficiency (AATD), 461–464
Alpha-melanocyte stimulating hormone (alpha-MSH), 934
Alport syndrome, 615
Altered embryogenesis and neoplasms, 1510
Alveolar capillary dysplasia (ACD), 161
Alveolar RMS (ARMS), 1053–1055
Alveolar soft part sarcoma (ASPS), 1076, 1078, 1079
Amanita genus mushroom, 507
Amastia, 836
Ambiguous situs, 8
Ameloblastoma (adamantinoma), 1217
Amino acid metabolism disorders, 471
AML1 gene, 877
Amnion nodosum, 1418
Amniotic band, 1417
Amniotic cells, 1414
Amniotic deformity, adhesion, and mutilation (ADAM) sequence, 1494, 1495
Amniotic fluid infection (AFI) syndrome, 1459–1462, 1464
Ampullary carcinomas, 560
Amyloidosis, 327, 613, 614, 690
Amyotrophic lateral sclerosis (ALS), 1288, 1335
Anal transitional zone (ATZ), 257
Anaplastic large cell lymphoma (ALCL), 903, 904
Anaplastic lymphoma kinase (ALK) 1, 218
Anaplastic rhabdomyosarcoma, 1052
Anaplastic seminoma, 707
Androgenic diploidy, 1430
Anembryonic pregnancy, 1427
Anemia
    acquired hemolytic anemias, 866
    CLI, 864
    congenital hemolytic anemias, 866
    definition, 864
    failure of blood production, 867
        BM aplasia/replacement, 867
        nutritional deficiencies, 867
        systemic disorders, 867
    hemorrhage-associated anemias, 864
Anencephaly, 1251, 1252, 1449
Aneurysmal bone cyst (ABC), 1120–1124
Angiocentric immunoproliferative lesions (AIL), 905, 906
Angiocentric vascular malignancy, 1066
Angiodysplasia, 374
Angioimmunoblastic lymphadenopathy with dysproteinemia (AILD), 921

Angioleiomyoma, 1398
Angiolipoma, 840
Angiomatosis, 1059
Angio-myo-fibroblastoma (AMFB), 1022
Angiomyolipoma, 517, 656
Angiosarcoma (AS), 109, 840, 1066, 1068, 1434
Angiotensin-converting enzyme (ACE) gene, 54, 55
Angiotensin-II type 1 receptor (AGTR1) gene, 55
Angiotensinogen (AGT) gene, 55
Anisakiasis, 330
Annular erythema, 1362
Annular pancreas, 555–556
Anorchia/anorchism, 679
Antepartum hemorrhage (APH), 1422, 1440
Antepartum, pregnancy, 1421
Anterior heart field (AHF), 2
Anterior mediastinal pathology, 241
Anti-diuretic hormone (ADH), 934
Anti-endomysial antibody (IgA EMA), 323
Anti-mesenteric (Meckel's) diverticulum, 316
Anti-tissue transglutaminase (IgA TTG), 323
Aorta coarctation, 39, 41
Aorto-pulmonary collateral arteries (APCAs), 21
Apical Na+-dependent bile transporter (ASBT), 450
Aplastic anemia, 865
Apocrine adenoma, 843
Apocrine gland tumors, 1393, 1394
Apocrine tubular adenoma, 1393
Apolipoprotein E (ApoE) gene, 58
Apoptosis, 1419
Appendiceal neoplasms, 351–353
Appendicitis, 345, 349, 351
Appendix, 343, 349
    anomalies, 343
    appendiceal neoplasms, 351–353
    appendicitis, 349–351
    appendix anomalies, 343
    duplication, 343
    intussusception, 350
    metabolic and degenerative changes, 351
    neuroma, 350
    vascular changes, 343
Archipelagineous architecture, 659
Areola hypertrophy, 838
Arias-Stella change (ASC), 770
Arias-Stella reaction, 1427
Arnold-chiari malformation (ACM), 1251
Arrhythmogenic right ventricular cardiomyopathy (ARVC), 52, 79, 81, 82
Artemin, 581
Arterial and venous blood vessels, 1417
Arterial fibromuscular dysplasia, 120
Arterial maternal blood flow, 1442
Arterial remodeling deficiency (ARD) changes, 1473
Arterial stimulation with venous sampling (ASVS), 562
Arterial switch operation (ASO), 31
Arterio-portal hypertension, 509
Arteriosclerosis, 117
Arteriosclerotic aneurysms, 1274

Arterio-venous malformation (AVM), 116, 1273
Arthropathies, 1154
Arthro-skeletal system
    adamantinoma, 1150
    baker cyst, 1159
    bone Ewing sarcoma, 1147–1149
    bursitis, 1159
    chondroid (cartilage)-forming tumors
        chondroblastoma, 1140–1143
        chondromyxoid fibroma, 1143–1145
        chondrosarcoma, 1145–1147
        enchondroma, 1139, 1140
        osteochondroma, 1137, 1138
    chordoma, 1149–1150
    development and genetics, 1096
    early-onset Juvenile (Tophaceous) gout, 1158–1159
    ganglion, 1159
    gout, 1157–1158
    hematologic tumors, 1153, 1154
    infectious arthritis, 1156
    Juvenile rheumatoid arthritis, 1155
    Langerhans cell histiocytosis, 1151–1153
    metabolic diseases
        juvenile paget disease, 1108, 1109
        osteomalacia, 1104
        osteoporosis, youth, 1104, 1106
        paget disease of bone, 1106, 1107
        vitamin D, 1103
        vitamin D-dependent hereditary rickets, 1104
    metastatic bony tumors, 1153
    nodular tenosynovitis, 1160
    osteitis, 1109
    osteochondrodysplasias
        genetic skeletal disorders, 1098–1100, 1102
        nosology and nomenclature, 1097
    osteomyelitis, 1109–1113
    osteonecrosis
        bony infarct, 1113–1114
        osteochondritis dissecans, 1114
    pigmented villonodular synovitis, 1159–1160
    pseudogout, 1159
    Rheumatoid arthritis, 1154–1157
    tumor-like lesions and bone/osteoid-forming tumors
        aneurysmal bone cyst, 1120–1124
        fibrous dysplasia, 1116–1118
        giant cell tumor, 1127–1128
        giant osteoid osteoma, 1126–1127
        myositis ossificans, 1115–1116
        non-ossifying fibroma, 1118–1120
        osteo-fibrous dysplasia, 1118
        osteoid osteoma, 1124–1126
        osteoma, 1124
        osteosarcoma, 1128–1136
        simple bone cyst, 1120
    vascular, smooth muscle, and lipogenic tumors, 1153
Asbestosis, 184, 206, 207
Ask-Upmark kidney, 623
Aspartylglycosaminuria, 467
Aspergillosis, 201, 202
Asphyxia, 1542
Aspiration, 193–195
Aspiration pneumonia, 194
Asthma, 186
Astrocytomas, 1291, 1293
Asymmetric unit membrane, 674
Ataxia telangiectasia, 888
Atelectasis, 158, 188
Atelia, 836
Atelosteogenesis (AO) type II, 1100
Atherosclerosis, 626
Atopic dermatitis, 1352, 1353
Atresia, 151, 312, 316
Atrial septal defect (ASD), 11, 13
Atrioventricular node (AVN), 108
Atrio-ventricular septal defect (AVSD), 13, 14, 16
Atrophic Fundal type, 274
Atrophy, 731
Atypia, 731, 1489
Atypia of unknown significance, 691
Atypical adenoma, 967
Atypical apocrine sclerosing lesion, 855
Atypical hemolytic-uremic syndrome (aHUS), 1059
Atypical lipomatous tumor, 1014–1015
Atypical melanocytic (dysplastic) nevi, 1385
Atypical mycobacteriosis, 920
Autochthonous tumors, 1396
Autoimmune enteropathy (AIE), 331, 332
Autoimmune hepatitis, 489, 490
Autoimmunity, 1484
Autoinflammatory syndromes, 1363
Autonomous disorders of placenta vascularization, 1434
Autosomal dominant (AD) diseases, 1517
Autosomal dominant polycystic kidney disease (ADSPKD), 591, 595, 596
Autosomal recessive (AR) diseases, 1520
Autosomal recessive polycystic kidney disease, 596, 597
AV connections, 9
A-V-Concordance, 26
Avidin-biotin-complex method, 1307
Axillary tail of Spence, 838

B

Babygram, 598
Bacillus Calmette–Guérin (BCG) therapy, 695
Baker cyst, 1159
Balanitis circumscripta plasmacellularis (Zoon's Balanitis, BCP), 740
Balanitis xerotica obliterans (BXO), 740
Balanoposthitis, 739–740
Balloon atrial septostomy (BAS), 31
Barrett dysplasia, 274
Barrett esophagus, 272
Barrett metaplasia, 272
Barth's syndrome, 1340
Bartholin gland carcinoma, 822
Bartonella henselae lymphadenitis, 916
Basal cell carcinoma (BCC), 1380
Basal cell hyperplasia, 263, 732
Basaloid squamous carcinoma, 277, 808

Bassel-Hagen law, 1137
Bat's wings pattern, 166
BCL-1 gene, 892
BCL-2 genes, 892
BCL6 rearrangements, 887
Becker nevus, 1375
Beckwith-Wiedemann syndrome (BWS), 633, 635, 644, 645, 996
"Beef-Red" pattern, 922
Benign and malignant mastocytosis, 1400
Benign discolored vulvar lesions, 768
Benign epithelial neoplasms, 287, 292
Benign epithelial tumors, 1174–1178, 1197–1198
Benign esophageal tumors, 275
Benign fibrous histiocytoma (BFH), 1153
    cellular fibrous histiocytoma, 1038–1039
    dermatofibroma, 1037
    DTGCT, 1038
    giant cell tumor of tendon sheath, 1038
Benign lymphadenopathies, 914
    angioimmunoblastic lymphadenopathy with dysproteinemia, 921
    diffuse hyperplasia
        dermatopathic lymphadenitis, 919
        infectious mononucleosis, 919
        postvaccinial Viral lymphadenitis, 919
    follicular hyperplasia
        AIDS-related lymphadenopathy, 918
        Castleman disease, 917, 918
        cat-scratch disease, 916, 917
        Kimura disease, 917
        lymphogranuloma venereum, 917
        necrotizing lymphadenitis, 916
        nonspecific reactive lymphadenitis, 916
        progressive transformation of germinal centers, 918
        rheumatoid arthritis, 916
        Sjogren disease, 916
        SLE-lymphadenopathy, 916
        syphilis, 917
        toxoplasmosis, 916
    leprosy, 921
    mesenteric lymphadenitis, 921
    mucocutaneous lymph nodal syndrome, 921
    Predominant granulomatous pattern, 920
    sinus pattern
        lipophagic reactions, 920
        sinus histiocytosis, 919, 920
        vascular transformation of LN sinuses, 920
Benign lymphocytic vasculitis, 906
Benign mesenchymal tumoral entities in childhood and youth, 1180
Benign mesothelioma, 235
Benign nephrosclerosis, 626
Benign nodular hyperplasia, 723, 729
Benign salivary gland tumors, 1205
Berry aneurysms, 116, 596, 1273, 1274
Beta-catenin, 1018
Beta-endorphin, 934
Beta-hemolysin, 85
Beta-lipotropic hormone, 934
Bi-directional Glenn operation, 39
Bifid scrotum, 683
Bilateral adrenal primordium, 938
Bilateral fibromuscular uterine remnants, 689
Bilateral gonadal agenesis, 760
Bilateral PNH (BPNH), 1526
Bile acid synthesis disorders (BASD)
    ASBT, 450
    3β-δ5-C27-steroid oxidoreductase, 451
    bile acid conjugation defects, 452
    cerebrotendinous xanthomatosis/sterol 27-hydroxylase, 449, 451
    cholesterol 7alpha-hydroxylase, 451
    clinical and biochemical features, 445
    CYP8B1, 449
    deconjugation products, 450
    Δ4-3-oxosteroid-5β-reductase (AKR1D1/SRD5B1), 452
    farnesoid X receptors, 450
    inadequate bile flow, 445
    2-methyl acyl-CoA racemase, 452
    neutral pathway, 445
    primary physiologic functions, 449
    urinary cholanoids, 451
Bile duct adenoma (BDA), 512, 515
Bile duct hamartoma (BDH), 515
Biliary and pancreatic structural anomalies
    agenesis/aplasia/hypoplasia, 553
    annular pancreas, 555–556
    choledochal cysts, 553
    congenital cysts of pancreas, 556
    cystic dysplasia, 556
    ductal abnormalities, 555
    gallbladder congenital abnormalities, 553, 554
    heterotopia pancreatis, 555
    restricted exocrine hypoplasia, 553–555
Biliary atresia (BA), 440, 442, 443
Biliary cystadenoma (BCA), 515
Biliary intraductal papillary mucinous neoplasm, 559
Biliary papillomatosis, 515–516
Biophysical profile (BPP), 1483
Biparental diploidy, 1430
Biphasic neoplasms, 851, 852
Birbeck granules, 908
Birmingham Vasculitis Activity Score (BVAS), 1005
Birth defects
    categories, 1493–1496
    etiology, 1515, 1517, 1520, 1521, 1523–1525, 1531–1533
    pathogenesis
        canalization and re-canalization, 1499, 1502
        gastrointestinal anomalies, 1498, 1499
        prosencephalon defects, 1496, 1498
        rotation anomalies of intestine, 1502, 1504
    taxonomy principles, 1490–1493
Birt-Hogg-Dube syndrome, 634
Bladder augmentation, 678
Bladder exstrophy, 662
Blastocyst implantation, 1414

Blastopathies, 1491
Bleeding diathesis, 870
Blue rubber bleb nevus syndrome (BRBNS), 1068
Blunt duct adenosis (BDA), 855
BNH/PNH-small gland type, 732
Bohn's nodules, 1192
Bone Ewing sarcoma, 1147–1149
Bone morphogenetic protein-10 (BMP10), 55
Bone morphogenetic proteins (BMPs), 834
Bone remodeling, 1096
Bone tumors, 1396
Bone-/cartilagineous forming tumors, 1181
Bone-related lesions, 1218
Bony fractures–callus– DDX, 1104
Bony infarct, 1113–1114
Borderline MPVFD, 1484
Borderline fibrous histiocytoma, 1039
Borrelia burgdorferi, 892
Botrioyd rhabdomyosarcoma, 819
Botryoid RMS (B-RMS), 1048, 1059
Bowel duplication, 305
Bowenoid papulosis, 741, 1378
Bowen's disease, 1378
Brachyury, 365
BRAF gene, 228
Brain abscess, 1278, 1279
Brain concussion, 1285
Brain contusion, 1285
Brain herniation, 1271
Brain laceration, 1285
Branchial arch musculatures, 1168
Branchial arches, 1170, 1171
Branchial grooves, 1170
Branchial sinuses, 1195
BRCA1 gene mutations, 845, 846
Breast
    abscess, 838
    acute mastitis, 838
    adenoma
        apocrine adenoma, 843
        ductal adenoma, 843
        encysted intraductal papillary carcinoma, 843
        intraductal papilloma, 843
        lactating adenoma, 841
        nipple duct adenoma, 841, 843
        papillomatosis, 843
        syringomatous adenoma of nipple, 843
        tubular adenoma, 841
    amastia, 836
    areola hypertrophy, 838
    asymmetry, 836
    atelia, 836
    "basal-l./3-" group of tumors, 846
    calcification, 839
    cancer mimickers
        adenosis, 855
        hyperplasia, 852
    DCIS grading systems, 848
    duct ectasia, 839
    ductal carcinoma in situ, 847, 848
    dysmaturity, 838
    fibroadenoma, 840–842
    fibrocystic disease, 840
    gene expression profiling of invasive Ca-susceptibility genes, 847
    granulomatous mastitis, 839
    gynecomastia, 837, 855
    hematological malignancies, 840
    HR-breast Ca-susceptibility genes, 846
    hypertrophy, 837
    infiltrating ductal carcinoma, 849, 850
    invasive ductal carcinoma, 848, 849
    invasive lobular breast carcinoma, 844, 849
    LR/MR-breast Ca-susceptibility genes, 846
    male, cancer in, 855
    mammary gland Paget disease, 848
    Mondor disease, 839
    myoepithelial tumor, 851
    necrosis, 839
    nipple hypertrophy, 838
    periductal mastitis, 839
    phlegmon, 838
    phyllodes tumor, 851, 852
    politelia, 836
    polymastia, 836
    precocious thelarche, 838
    soft tissue tumors, 840
    sweet-gland type tumors, 851
    synmastia, 836
Breast Imaging Reporting and Data System (BI-RADS), 839
Brenner tumors, 789, 790
Breslow level, 1384
Broad-based triangular atrial appendage, 8
Broca's criteria, 1137
Brodsky's scale, 1196
Bronchial asthma, 184, 187, 188
Bronchiectasis, 189
Bronchogenic cyst, 147
Bronchopneumonia, 191
Brugada syndrome, 84
Bruininks Oseretsky test, 1106
b-sarcoglycan (SGCB), 59, 60
Budd Chiari syndrome, 504
Bullous pemphigoid, 1367
Burkitt lymphoma, 300, 313, 727, 888, 896, 901, 902
Bursitis, 1159
Byzantine culture and surgery, 1492

C

CADASIL, 1272
Calcific aortic stenosis, 91
Calcific mitral annulus, 91
Calcification, breast, 839
Calcified amorphous tumor, 108
Calcifying aponeurotic fibroma (CAF), 1027
Calcifying epithelial odontogenic tumor, 1217
Calcifying fibrous pseudotumor (CFPT), 1027
Calcifying fibrous tumor (CFT), 1027

Calcium deposition, 839
Calcium pyrophosphate deposition (CPPD) disease, 1159
Calculus cholecystitis, 557
Calyceal diverticula, 661
Cambium layer, 851
Ca++ mobilization in the bone, 946
Campylobacter jejuni, 892
Canals of Herring, 427
Candida albicans, 767
Candida amnionitis, 1418
Candida-pseudomonas-infection, 358
Candidiasis, 767
Cannabis (Marijuana)-associated CHD, 48
CAP tumor checklist, 296
Capillariasis, 327
Capillary hemangioma, 1059, 1398
Capillary leak syndrome, 85, 1007
Carcinoid tumor, 296, 352–353
Carcinoma in situ (CIS), 691
Carcinoma of the ampulla of vater, 341, 342
Carcinoma of the nasopharyngeal mucosa (NPC), 1178–1180
Carcinoma showing thymus-like elements of differentiation (CASTLE), 927, 964
Cardiac allograft vasculopathy (CAV), 97, 99, 100
Cardiac angiosarcoma, 113
Cardiac catheterization, 31
Cardiac channelopathies, 83–85
Cardiac crescent, 2, 5
Cardiac epithelioid hemangioendothelioma, 112
Cardiac hamartoma, 106
Cardiac myxoma, 101
Cardiac troponin I (TNNI3) gene, 57, 85
Cardiac troponin T (TNNT2) gene, 57
Cardiac tumors, 100
    benign, 102
        AVN, congenital polycystic tumor, 108
        calcified amorphous tumor, 108
        cardiac hamartoma, 106
        hemangioma, 105
        HICM/Purkinje cell hamartoma, 108
        inflammatory myofibroblastic tumor, 105
        lipoma, valve fibrolipoma and LHIAS, 104
        MICE, 108
        mural thrombi, 108
        myxoma, 102, 104
        paraganglioma, 105
        PFE and fibroma, 104, 105
        rare non-cardiac benign tumors, 109
        rhabdomyoma, 105
    malignancies
        angiosarcoma, 109
        epithelioid hemangioendothelioma, 109
        rhabdomyosarcoma, 109
        synovial sarcoma, 109–110
Cardiac/junctional type, 274
Cardiomyocytes, 77
    degeneration, 54
    necroses, 86
Cardiomyopathic phenotypes, 83

Cardiotocography, 1410
Cardiovascular system
    ARVD/C, 79, 81
    cardiac channelopathies & sudden cardiac death, 83–85
    congenital heart disease, 6
        aorta coarctation, 39, 41
        ASD, 11, 13
        AVSD, 13, 16
        congenital syndromes with, 41, 44, 47, 48
        DORV, 22, 24
        ETA, 16, 18
        great arteries, transposition of, 26, 31
        heterotaxy, 41
        HLHS, 33, 39
        hypoplastic right heart syndrome (*see* Hypoplastic right heart syndrome)
        PDA, 16
        persistent truncus arteriosus, 31, 32
        POF, 22
        Scimitar syndrome, 32, 33
        sequential segmental cardio-analysis, 8–11
        TOF, 18, 21
        total anomalous pulmonary venous return, 32
        VSD, 13
    developmental and genetics, 2, 3, 5, 6
    dilated cardiomyopathy, 48, 54
        ACE gene, 54
        ACTA1, 64
        a-dystrobrevin gene, 64
        AGT gene, 55
        AGTR1 gene, 55
        alpha2C-adrenoceptor gene, 56
        alpha-cardiac actin gene, 57
        AMPD1 gene, 59
        autosomal dominant DCM, locus on chromosome 6q12-q16 for, 64
        b1-AR, 56
        β2-adrenoceptor gene, 56
        beta-myosin heavy chain gene, 56
        BMP10, 55
        chemotherapy-related cardiotoxicity, 65, 66
        CTF1, 63
        Cypher/ZASP gene, 59
        desmin gene, 63
        dystrophin, 61
        ETA and ETB, 57, 58
        G4.5 gene, 62
        heart calcineurin pathway, PPP3CA, PPP3CB, GATA4, NFATC4, 61
        HLA gene, 56
        HS426 gene, 62
        lamin A/C gene, 58, 59
        mtDNA abnormalities, 65
        MyBP-C gene, 57
        myotrophin gene, 58
        NOS3 gene, 63, 64
        PABP2 gene, 60
        PAF-AH, 61
        PLN, 55

Cardiovascular system (cont.)
    SCN5A gene, 58
    SGCB and SGCD, 59, 60
    SOD2 gene, 65
    TGF-β1, 62
    titin/connectin gene, 55
    TNF-α gene, 55
    TNNI3 gene, 57
    vinculin and metavinculin genes, 58
  heart failure, 85
  hypertrophic cardiomyopathy, 66
  infiltrative/restrictive cardiomyopathies, 66, 71, 77
  ischemic heart disease, 85
  lamin A/C cardiomyopathy, 81
  left ventricular non-compaction, 78, 79
  metastatic tumors, 112, 115
  mitochondrial cardiomyopathies, 81, 83
  myocardial dysfunction, 85
  myocarditis, 48
  non-neoplastic vascular pathology
    arteriosclerosis, 117
    congenital anomalies, 116
    fibromuscular dysplasia, 117
    large-vessel vasculitis, 119, 123
    medium-vessel vasculitis, 123, 125
    small-vessel vasculitis, 125
  pericardial disease
    neoplastic pericardial disease, 116
    non-neoplastic pericardial disease, 115, 116
  transplantation (see Transplantation)
  tumors (see Cardiac tumors)
  valvular heart disease
    calcific mitral annulus and calcific aortic stenosis, 91
    endocarditis and valvar vegetation, 87
    mitral valve prolapse, 91
    prosthetic valves, 92
    RHD, 91
Cardiovascular/renal abnormalities of fetus, 1421
Carney complex (CNC), 996, 997
Carnitine deficiency, 475
Caroli disease, 434, 435, 553
Cartilage tumors, 1396
Caseous pericarditis, 116
Castleman disease, 912, 917, 918
Catastrophic blood loss, 1416
Catch 22 syndrome, 46
Catecholaminergic Polymorphic Ventricular Tachycardia (CPVT), 84, 85
Cat-Scratch disease, 916, 917
Caudal dysgenesis syndrome (CDS), 365, 366
Caudal regression syndrome (CRS), 1497
Cauliflower ear, 1219
Cavernous Hemangioma, 516
CCAM, 148, 151
C-cell hyperplasia, 960
CD141 or thrombomodulin, 1059
CD15, 878
CD30, 879
CD31 immunostaining, 1433
CD57, 984
CD8+ IELs, 285
CD99 (MIC-2) antigen, 896
CDKN1C gene, 1428
Cell carbohydrate metabolism, 946
Cellular fibrous histiocytoma, 1038–1039
Cellular hypoxia, 1472
Cementum-secretion, 1170
Central chondrosarcoma, 1146
Central giant cell granuloma, 1218
Central motility disorders, 262
Central nervous system (CNS)
  cerebellar abnormalities, 1258
  congenital abnormalities, 1259
    ACC, 1257, 1258
    AGOTC, 1259
    CNS defects in Acardia, 1261–1263
    CNS defects in chromosomal and genetic syndromes, 1263, 1264
    ectopia, 1247, 1250
    neuronal migration disorders, 1264
    NTD, 1250–1252
    phakomatoses, 1264–1267
    prosencephalon defects, 1252–1254
    telencephalosynapsis (synencephaly) and rhombencephalosynapsis, 1259, 1260
    ventriculomegaly/hydrocephalus, 1257
    vesicular forebrain (pseudo-aprosencephaly), 1254–1256
  demyelinating and degenerative diseases, childhood and youth
    amyotrophic lateral sclerosis, 1288
    leukodystrophies, 1287, 1288
    MS, 1286, 1287, 1290
    Parkinson's disease and Parkinson disease-associated, G-protein-coupled receptor 37 (GPR37/PaelR) related Autism Spectrum Disorder, 1289–1290
    syringomyelia, 1289
    variant Creutzfeldt-Jacob disease, 1290
    Werdnig-Hoffmann disease, 1288, 1289
    West syndrome/infantile spasms, ACTH therapy and sudden death, 1291
  development and genetics, 1244
    cerebellum, 1246
    CNS cytology characteristics, 1247
    corpus callosum, 1246
    neuropathology special stains, 1247
    normal anatomy, 1245
  head and spine trauma
    head injuries, 1284, 1285
    intervertebral disk herniation, 1286
    spinal injuries, 1285, 1286
  infections
    fungal, 1282
    neurosyphilis, 1280
    suppurative, 1276, 1278, 1279
    toxoplasmosis, 1282
    tuberculous (lepto-) meningitis, 1279, 1280
    viral, 1281, 1282

    metabolic disorders
        pernicious anemia, 1283
        Wernicke's encephalopathy, 1283
    neoplasms
        astrocytic-derived neoplasms, 1291, 1293, 1294
        chordoma, 1300, 1302
        craniopharyngioma, 1300
        ependymomas, 1294, 1295
        hemangioblastoma and filum terminale hamartoma, 1299–1300
        hematological malignancies, 1302–1304
        medulloblastoma, 1295, 1297
        meningioma, 1299
        neurilemmoma (Schwannoma), 1300
        pineal body tumors, 1302
    vascular disorders
        aneurysms, 1273, 1274
        in children, 1274, 1275
        inherited neurovascular diseases, 1274, 1275
        intracranial hemorrhage, 1267, 1269–1272
        thrombosis of venous sinuses and cerebral veins, 1274
        vascular malformations, 1272, 1273
Centrocytic lymphoma, see Mantle cell lymphoma
Cerebellar agenesis, 1258
Cerebellar hypoplasia, 1259
Cerebral cavernous malformations (CCM), 1274, 1275
Cerebral cortex maturation defects, 1508
Cerebral palsy, 1417
Cerebral spinal fluid (CSF), 1247
Cerebro-oculo-facial-skeletal syndrome (COFS), 1348
Ceroid histiocytosis, 923
Cervical adenocarcinoma (C-ACA), 807
Cervical dysplasia, 805
Cervical epidermoid carcinoma, 806, 807
Cervical metaplasia, 771
Cervical neuroendocrine carcinoma (CNEC), 810, 811
Cervicitis, 769
Cervix tumors, 806
Chamber formation, 2
Chancroid, 766
Charcot-Bouchard aneurysms, 1271
CHARGE syndrome, 47
Chediak-Higashi syndrome (CHS), 486, 871
Chemical pneumonitis, 193–195
Chemotherapy protocols, 1431
Chemotherapy-related cardiotoxicity, 65, 66
CHILD syndrome, 1375
Childhood cancer therapy, 1057–1059
Childhood CLL, 874
Chlamydia infections, 766
Chlamydia lymphadenitis, 917
Chlamydia psittaci, 892
Chlamydia trachomatis, 769
Chloroquine-toxicity, 77
Cholangiocellular carcinoma (CCA), 529–531, 561
Cholangiopathies
    biliary atresia, 440, 442, 443
    etiology and clinical-radiologic-biochemical investigations, 441
    metabolic investigations, 439
    NBAIOC, 443–444
    Neonatal Hepatitis Group (NAG) (see Neonatal Hepatitis Group)
    PIBD, 444, 445
    pregnancy-related liver disease, 459, 460
    primary biliary cirrhosis, 458, 459
    primary sclerosing cholangitis, 456–458
    symptoms, 439
Choledochal cysts, 553, 555
Choledochal system, 553
Cholelithiasis, 556–557
Cholelithiasis-related chronic inflammation, 558
Cholemic nephropathy, 625
Cholestasis, 437
Cholestatic hepatitis, 494
Cholesteatoma, 1220
Cholesterol granuloma, 1220
Cholesterol polyps, 558
Cholesterolosis, 557, 558
Chondroblastoma, 1140–1143
Chondrodermatitis nodularis chronica helicis (CNCH), 1219
Chondrodysplasia punctata (CDP), 1102
Chondrodysplasias osteopetrosis, 1100
Chondroid (cartilage)-forming tumors
    chondroblastoma, 1140–1143
    chondromyxoid fibroma, 1143–1145
    chondrosarcoma, 1145–1147
    enchondroma, 1139, 1140
    osteochondroma, 1137, 1138
Chondroid syringoma, 1391
Chondromyxoid fibroma (CMF), 1137, 1143–1145
Chondro-osteoforming tumors
    extraskeletal aneurysmatic bone cyst, 1070
    extraskeletal chondroma, 1069, 1070
    extraskeletal osteosarcoma, 1070
    mesenchymal chondrosarcoma, 1070
Chondrosarcoma, 1145–1147
Chorangioma, 1434
Chorangiomatosis, 1434
Chorangiosis, 1434
Chordoma, 942, 1149–1150, 1300, 1302
Chorioamnionitis, 1460
Choriocarcinoma (CCA), 712, 717–720, 1486, 1488
Chorion frondosum and *decidua basalis*, 1414
Chorionic villi, 1477
Chromaffin cells, 938
Chromosomal and genetic syndromes, 1263, 1264
Chromosomal translocations, 888
Chromosomal variation, 1495
Chromosomal/genomic microdeletion disorders, 163
Chromosomal-based birth defects, 1525, 1531–1533
Chromosome imprinting, 1532
Chronic (tubulo-) interstitial nephritis (CTIN)
    Ask-Upmark kidney, 623
    malakoplakia, 622
    pelvic lipomatosis, 622
    tuberculosis, 622
    xanthogranulomatous pyelonephritis, 622

Chronic appendicitis, 350
Chronic bilirubin elevation (CBE), 437
Chronic bronchitis, 203
Chronic calcifying pancreatitis, 564
Chronic cholecystitis, 558
Chronic gastritis, 281
Chronic granulomatous disease (CGD), 327, 485–486, 729, 921
Chronic GVHD, 339
Chronic histiocytic intervillositis (HIV), 1476
Chronic idiopathic myelofibrosis, 867, 875
Chronic inflammatory arthropathies, 1154
Chronic intestinal ischemia, 318, 319
Chronic intestinal pseudo-obstruction, 370–372
Chronic laryngitis, 1185
Chronic lymphocytic leukemia (CLL), 874, 875, 922
Chronic myelocytic leukemia (CML), 875
Chronic obstructive pancreatitis, 564
Chronic obstructive pulmonary disease, 202
   chronic bronchitis, 203
   emphysema, 203, 204
Chronic pancreatitis, 564–566
Chronic peripheral (marginal) separation, 1479
Chronic placental abruption (CPA), 1479, 1480
Chronic progressive external ophthalmoplegia (CPEO), 81, 1339
Chronic prostatitis, 728
Chronic rejection, 212, 213, 338
Chronic renal failure (CRF), 623
Chronic ulcerative jejunoileitis, 326
Chronic villitis of infectious type (CVI), 1475, 1476
Chronic viral hepatitis, 488, 489
Churg-Strauss syndrome (CSS), 1005
Circumvallate placenta, 1417
Circumvallate/circum-marginate placentation, 1414
Citrin deficiency, 476
C-KIT gene mutations, 910
Clark index, 1384
Classic Hodgkin-type PTLD, 338
Classic MFI, 1484
Classical Hodgkin's lymphoma, 881–883
Clear cell (mesonephroid) tumors, 788, 789
Clear cell carcinoid, 352
Clear cell carcinoma, 808–809
Clear cell cribriform hyperplasia, 732
Clear cell sarcoma, 646, 1079–1080
Cloacal malformities, 687
Clotting anomalies, 1480
CMV encephalitis, 1535
Coagulation disorders, 1484
   acquired disorders, 869, 870
   hereditary disorders, 869
Coal workers pneumoconiosis (CWP), 184, 207
Coarctation of the aorta, 39, 41
Codman triangle, 1132
Collodion baby phenotype, 1350
Colloid nodule/goiter, 952
Colon, 257
Colon diverticulosis, 372
Comedo-carcinoma, 848

Common atrioventricular canal, 13
Common bile duct, 555
Communicating cavernous ectasias, 553
Community-acquired miscarriage, 1477
Complete abortion, 1477
Complete hydatiform mole (CHM), 1428
Complete molar degeneration, 1428
Complex sclerosing lesion (CSL), 855
Composite form hemangioendothelioma (CHE), 1062, 1063
Composite pheochromocytoma, 994
Conception, 1427
Concordant bi-ventricular AV connection, 9
Condensation of the mesenchymal cells, 1096
Condyloma acuminatum, 764–765, 816
Confined placental mosaicism, 1533
Congenital abnormalities, CNS, 1248
Congenital acinar dysgenesis/dysplasia, 160, 161
Congenital adrenal hyperplasia (CAH), 973, 996
Congenital airways diseases, 145
   agenesis, aplasia and hypoplasia pulmonis, 148
   CPAM and congenital lobar emphysema, 148, 150, 151
   foregut cysts, 146, 148
   sequestrations and vascular anomalies, 152–155
   tracheal and bronchial anomalies, 151
Congenital alopecia, 1351
Congenital alveolar dysgenesis/dysplasia, 161
Congenital anomalies, 116, 344
Congenital atresia–stenosis of the colon, 353, 354
Congenital cysts of pancreas, 556
Congenital diaphragmatic defects (CDD), 1499
Congenital disorders of glycosylation (CDG), 466
Congenital diverticula, 343
Congenital dysregulation carbohydrate metabolism diseases (CCMDD), 465
Congenital ectoderm-derived (epithelial) cysts, 1195
Congenital Epulis of newborn, 1199
Congenital hand anomalies, 1504, 1505
Congenital heart disease, 6, 23, 25, 27, 38
   aorta coarctation, 39, 41
   ASD, 11, 13
   AVSD, 13, 16
   congenital syndromes with, 41, 44, 47, 48
   DORV, 22, 24
   ETA, 16, 18
   great arteries, transposition of, 26, 31
   heterotaxy, 41
   HLHS, 33, 39
   hypoplastic right heart syndrome (*see* Hypoplastic right heart syndrome)
   PDA, 16
   persistent truncus arteriosus, 31, 32
   POF, 22
   Scimitar syndrome, 32, 33
   sequential segmental cardio-analysis, 8–11
   TOF, 18, 21
   total anomalous pulmonary venous return, 32
   VSD, 13
Congenital hemolytic anemias, 866

Congenital hepatic fibrosis (CHF), 435, 597
Congenital hyperinsulinism (CHI), 561–563
Congenital hypertrophic pyloric stenosis (CHPS), 280
Congenital immunodeficiencies, 887–888
Congenital infantile enteropathies, 330
Congenital lobar emphysema, 151
Congenital megacolon, 366
Congenital mesoblastic nephroma, 645, 646
Congenital myopathies, 1338
Congenital nephrotic syndrome, 617
Congenital syndromes, 41, 44, 47, 48
Congenital uterine anomalies (CONUTA), 761
Congenitally corrected TGA, 26
Conjoined twin malformation, 1444
Conjoined twins, 1443, 1449, 1491, 1492
Conn syndrome, 975
Constipation, 370–372
Constrictive pericarditis, 116
Contact dermatitis, 1362
Contact ulcers of larynx, 1186
Conventional spongiosis, 1352
    allergic contact dermatitis, 1353
    atopic dermatitis, 1352, 1353
    erythema neonatorum, 1354
    incontinentia pigmenti, 1354
    nummular dermatitis, 1353
    papular acrodermatitis of Gianotti-Crosti, 1354
    pityriasis alba, 1353
    pityriasis rosea, 1354
    spongiotic dermatitis, 1353
    stasis dermatitis, 1353
Copper metabolism dysregulation, 483, 484
Cord infections, 1417–1419
Cord integrity damage patterns, 1457, 1458
Cord shortening, 1415
Cornelia De Lange syndrome, (CDLS), 1263, 1264, 1515, 1516
Corpus luteum cysts, 763
Cortex histology, 1254
Cortical tumor, 975
Corticosteroids-secreting adenoma, 975
Costello syndrome, 667, 711
Council for International Organization of Medical Sciences (CIOMS), 494
Cowper glands, 730
Cranial fasciitis, 1019
Craniopharyngioma, 942, 1300
Crazy paving pattern, 166
Cretinism, 943
Creutzfeldt-Jacob disease (sCJD or sporadic), 1290
Cribriform pattern, 848
Crigler-Najjar syndrome, 437
Crohn colitis, 359
Crohn disease, 327, 333–335
Cryoglobulinemia, 614
Cryopreserved homografts, 92
Cryptococcal infection and coccidiomycosis, 1282
Cryptogenic organizing pneumonia (COP), 170–172
Cryptorchidism, 680, 682
CT-1 (CTF1), 63

CTG abnormalities, 1467
Cuboidal/columnar epithelium, 559
Cushing's syndrome, 942, 975
Cutaneous adverse drug reactions (CADR's)
    exanthematous, 1371–1372
    fixed drug eruptions, 1371
    P-CADR, 1371
    SSLR, 1372
    UCADR, 1371
Cutaneous appendages disorders, 1369
Cutaneous bronchogenic cysts, 146
Cutaneous cysts, 1372
Cutaneous horn, 1377
Cutaneous mastocytosis, 910
Cutaneous T-cell lymphoma (CTCL), 904, 905
CYCLIND1/PRAD1 gene rearrangement, 966
Cylindromas, 1391
CYP17 enzyme, 938
CYP17A1 gene, 938
Cypher/ZASP, 59
Cystadenofibroma, 784
Cystadenoma/cystadenofibroma, 780–782
Cyst-associated cervical lesions, 770
Cystic dysplasia, 556
Cystic dysplasia-associated with cloaca malformation, 1497
Cystic fibrosis, 158, 159, 453–455
Cystic hygroma, 1449
Cystic nephroma, 645
Cystic pancreatic ductal adenocarcinoma (CPDA), 567
Cystic partially differentiated nephroblastoma (CPDN), 645
Cystitis
    infectious, 684, 685
    non-infectious, 684–686
Cystitis cystica, 689
Cystitis glandularis, 689
Cystocele, 678
Cytochrome c oxidase, 69
Cytogenetic syndromes, 1488
Cytogenetics, GTD of complete molar type, 1429
Cytokeratin, 950
Cytolysin, 85
Cytomegalovirus (CMV)
    colitis, 359
    infection, 1538, 1539
Cytotrophoblast (CTB), 1414

D

Dairy pasteurization method, 1539
Dandy-Walker malformation (DWM), 1259
Danio rerio, 836
Darier disease, 1366–1367
Darier phenomenon, 910
Darier sign, 910
De la Chapelle dysplasia, 1100
Deamidated gliadin peptide antibodies, 323
Deauville five-point scale, 887
Decidual vasculopathy, 1484

Decidualized stromal cells, 1430
Dedifferentiated chondrosarcoma, 1147
Dedifferentiated liposarcoma, 1016
Deep fibromatosis (musculoaponeurotic), 1030
Defective embryology, 1490
Dehydroepiandrosterone sulfate (DHEAS), 938
Delayed miscarriage, 1477
Dental lamina, 1170
Dentigerous (follicular) cyst, 1215, 1216
Dentin-secretion, 1170
Denys-Drash syndrome, 634, 635
Dermatitis herpetiformis, 1368–1369
Dermatofibroma, 1037, 1376, 1394, 1395
Dermatofibrosarcoma protuberans (DFSP), 1396, 1397
Dermatomyositis, 1355
Dermatopathic lymphadenitis, 919
Dermatopathology, 1382
DeSanto Classification of Laryngeal Cysts, 1184
Desmin gene, 63
Desmin-related myopathy, 63
Desmoid-type fibromatosis, 1029
Desmoplastic fibroblastoma (DFB), 1022
Desmoplastic fibroma, 1153
Desmoplastic MM, 1386
Desmoplastic small round cell tumor, 1083–1084
Desquamative interstitial pneumonia (DIP), 160, 176, 177
Dexamethasone suppression test (DST), 971, 1252
Diabetes mellitus, 571–573
Diabetic nephropathy, 615
Diandric triploidy, 1430
Diaphragma, 278
Diastrophic dysplasia sulfate transporter (DTDST), 1100
Di-diiodotyrosine (DIT), 936
Diffuse (paracortical) hyperplasia
    dermatopathic lymphadenitis, 919
    infectious mononucleosis, 919
    postvaccinial viral lymphadenitis, 919
Diffuse alveolar damage (DAD), 168–170
Diffuse hyperplasia, 947
Diffuse laminar endocervical hyperplasia (DLEH), 771
Diffuse large B cell lymphoma (DLBCL), 300, 301, 397, 892–896, 964, 1181
Diffuse lung development-associated ILD, 159, 160
Diffuse micronodular hyperplasia, 974
Diffuse pulmonary hemorrhagic syndromes, 196, 197
Diffuse type carcinoma, 293
Diffuse-type giant cell tumor (DTGCT), 1038
DiGeorge syndrome, 1196
Dilatation and evacuation (D&E) specimens, 1423
Dilated cardiomyopathy, 48, 54
    ACE gene, 54
    ACTA1, 64
    a-dystrobrevin gene, 64
    AGT gene, 55
    AGTR1 gene, 55
    alpha2C-adrenoceptor gene, 56
    alpha-cardiac actin gene, 57
    AMPD1 gene, 59
    autosomal dominant DCM, locus on chromosome 6q12-q16 for, 64
    b1-AR, 56
    β2-adrenoceptor gene, 56
    beta-myosin heavy chain gene, 56
    BMP10, 55
    chemotherapy-related cardiotoxicity, 65, 66
    CTF1, 63
    Cypher/ZASP gene, 59
    desmin gene, 63
    dystrophin, 61
    ETA and ETB, 57, 58
    G4.5 gene, 62
    heart calcineurin pathway, PPP3CA, PPP3CB, GATA4, NFATC, 61
    HLA gene, 56
    HS426 gene, 62
    lamin A/C gene, 58, 59
    mtDNA abnormalities, 65
    MyBP-C gene, 57
    myotrophin gene, 58
    NOS3 gene, 63, 64
    PABP2 gene, 60
    PAF-AH, 61
    PLN, 55
    SCN5A gene, 58
    SGCB and SGCD, 59, 60
    SOD2 gene, 65
    TGF-β1, 62
    titin/connectin gene, 55
    TNF-α gene, 55
    TNNI3 gene, 57
    vinculin and metavinculin genes, 58
Dilated cardiomyopathy (DCM), 53
Dilated Pore of Winer (DPW), 1388
Disco-blastula, 1491
Discoid Lupus, 1355
Discoidin domain receptor 2 (DDR2), 229
Discordant bi-ventricular AV connection, 9
Disease extent index, 1005
Disease-modifying anti-rheumatic drugs (DMARDs), 1155
Disorders of deep placentation, 1483
Disorders of the placenta implantation site (PISD), 1434, 1436, 1439, 1441
Disorders of the placenta vascularization (PVD), 1434
Disruptions and chromosomal aberrations, 1488
Dissecting aortic aneurysma, 121
Distal villous hypoplasia, 1434
Distal villous immaturity (DVI), 1434, 1481
Distopias, 277
Diverticula, 316, 317
Dizygotic twins (fraternal twins) twins, 1442
DNA fragmentation index (DFI), 696
DNA-binding domain (DBD), 938
Double outlet right ventricle (DORV), 22, 24
Double ureters or duplication, 676
Double-inlet ventricle (DIV), 9
Down syndrome, 161–163
Drosophila, 836
Drug-induced liver disease
    allograft transplantation, 494

chemical-induced damage, 491
cholestatic hepatitis, 494
direct hepatotoxins, 491
histopathology patterns, 492
indirect hepatotoxicity, 492
inflammation-associated hepatocellular necrosis, 494
inflammation-free cholestasis, 494
inflammation-free hepatocellular necrosis, 494
inflammation-free microvesicular steatosis, 494
intrinsic hepatotoxins, 491
postoperative cholestasis, 495
total parenteral nutrition, 493, 495
d-sarcoglycan (SGCD) genes, 59, 60
Dubin-Johnson syndrome (DJS), 438
Duct ectasia, 839
Ductal abnormalities, 555
Ductal adenoma, 843
Ductal atresia, 1425
Ductal carcinoma in situ, 847, 848
Ductal communication variations, 555
Ductal cysts, 1185
Ductal plate (DP), 428
Ductal plate malformation (DPM), 431, 435
    cystic abnormalities, extrahepaic biliary tract, 434
    cystic biliary structures, 431
    glomeruloid, 432
    immunohistochemical procedure, 432
    polypoid lesions, 432
    type 1 and 2, 431
Duodenal atresia, 316
Dupuytren (subungual) exostoses (DSE), 1138
Duret hemorrhage, 1272
Dutcher bodies, 890
Dys-embryogenesis, 1510
Dysfunctional (anovulatory) uterine bleeding (DUB), 771
Dysfunctional ABCA3-related ILD, 165
Dysfunctional adrenal gland, 968
Dysfunctional SP-B-related ILD, 165
Dysfunctional SP-C-related ILD, 165
Dysgenesis, 1510, 1515
Dysgenetic proliferations, 1510, 1511, 1515
Dysgerminoma, 774–776
Dyshormonogenetic goiter, 947
Dyskeratotic, non-/pauci-inflammatory disorders, 1372
Dysmaturity, 1542, 1544
Dysmaturity, mammary gland, 838
Dysphagia lusoria, 261
Dysplasia, 559
Dysplasia epiphysealis hemimelica, 1145
Dysplasias with decreased bone density, 1102
Dysplasias with increased bone density, 1102
Dysplastic Nevi, 1386
Dystopias, 317, 591
Dystrophin gene, 61

E

Eagle-Barrett syndrome, 687
Ear
    congenital anomalies, 1218

    inflammatory lesions & non-neoplastic lesions, 1218–1220
    tumors
        external neoplasms, 1220, 1223, 1224
        inner neoplasms, 1224, 1225
        middle neoplasms, 1224
Early complete mole (ECM), 1429
Early pregnancy
    PISD, 1434, 1436, 1439–1441
    placenta formation
        complete mole (complete molar degeneration), 1428
        embryo and 1st trimester specimens, 1423, 1426
        GA, 1426, 1427
        GD1/anembryonic pregnancy, 1427
        Grannum score, 1427, 1428
        HMD, 1429, 1430
        partial mole (partial molar degeneration), 1428
    PMD, 1431
    PTMP, 1442, 1446, 1448, 1449
    PVD, 1434
Early pregnancy, pathology, 1422, 1423
Early-onset Juvenile (Tophaceous) gout (EOJG), 1158–1159
Early-type non-destructive PTLD, 338
Eaton-Lambert syndrome, 1337
Ebstein's tricuspidal anomaly, 17
Eccondroma, 1140
Eccrine acrospiroma, 1392, 1393
Eccrine carcinoma, 1393
Eccrine gland tumors, 1391, 1393
Eccrine glandular-arising benign adnexal cutaneous tumor, 1391
Eccrine poroma, 1391, 1392
Eccrine spiradenoma, 851, 1393
Echinococcal cyst, 154
ECM synthesis, 1096
Ectopia, 1247, 1250
Ectopia/heterotopia, 261, 278
Ectopia-associated cervical lesions, 770
Ectopic ADH neoplastic disorders, 943
Ectopic decidual reaction, 764
Ectopic hamartomatous tumor, 1072
Ectopic pregnancy, 1440, 1441
Elastofibroma, 1021
Electron fetal monitoring (EFM), 1546–1548
Electron transport (respiratory chain) defects
    Alpers-Huttenlocher syndrome, 473
    laboratory tools, 472
    mitochondrial DNA depletion syndrome (mtDNA-DS), 473
    Navajo neuro-hepatopathy, 474
    neonatal liver failure, 473
    oral glucose tolerance test, 472
    Pearson syndrome, 473
    villous atrophy syndrome, 474
Electronic medical devices, 84
Ellis-van Creveld syndrome (ECS), 47
Embolization, 101
Embryo development, 1413

Embryology, 1490
Embryonal carcinoma, 703, 711–713, 778, 779
Embryonal mole/early miscarriage, 1427
Embryonal rhabdomyosarcoma, 561, 1048–1051
Embryonal RMS (E-RMS), 1048, 1224
Embryonal tumor, malignant of CNS, 1306
Embryonic cranial index (ECI), 1421
Embryonic tissue fusion, 1047
Embryonic/developmental age, 1423
Embryopathies, 1490, 1491
Emphysema, 203, 204
Emphysematous cystitis, 684
Empty Sella syndrome (ESS), 943
Empyema-based bacterial producing gas, 235
Enchondral ossification, 1096
Enchondroma, 1139, 1140
Encrusted cystitis, 684
Encysted intraductal papillary carcinoma, 843
Endocarditis, 87–90
Endocrine cell dysplasia, 296
Endocrine cell neoplasia, 296
Endocrine dysregulation-based neonatal hepatitis (EDNH), 455–456
Endocrine system
    Adrenal gland (*see* Adrenal gland)
    parathyroid gland (*see* Parathyroid gland)
    pituitary (*see* Pituitary gland)
    thyroid gland (*see* Thyroid gland)
Endodermal sinus tumor (EST), 715–717, 776–778, 815
Endogenous erythroid colony (EEC) growth, 869
Endo-lymphatic sac tumor (ELST), 1225
Endometrial dating, 759
Endometrial heterotopia, 317–318
Endometrial metaplasias, 772–773
Endometriosis, 317–318, 763–764, 773
Endometritis, 772
Endoplasmic reticulum storage diseases (ERSD's), 461, 462, 464
Endosalpingiosis, 764
Endoscopically suspected esophageal metaplasia, 265
Endothelial nitric oxide synthase (NOS3) gene, 63, 64
Endothelin-1, 57
Endothelin-A (ETA), 57
Endothelin-B (ETB) receptor genes, 57
Endovascular papillary hemangioendothelioma (EVPHE), 1062, 1063
Enterobius vermicularis, 346
Enteropathy associated T-cell lymphoma (EATL), 301, 302
Eosinophilia, 872
Eosinophilic appendicitis, 350
Eosinophilic cholecystitis, 558
Eosinophilic cystitis, 686
Eosinophilic esophagitis, 267–269
Eosinophilic fasciitis, 1019
Eosinophilic gastritis, 285
Eosinophilic gastroenteritis, 326
Eosinophilic granuloma, 908, 909, 942
Eosinophilic granulomatosis with polyangiitis (EGPA), 125
Eosinophilic granulopoiesis, 430
Eosinophilic pneumonia, 182, 183
Eosinophilic spongiosis, 1354
Eparterial broncho-arterial crossing, 144, 145
Ependymoma, 1294, 1295
Epidermal nevi, 1375, 1376
Epidermoid & dermoid cysts, 1192–1193
Epidermoid cysts, 697, 699, 763
Epidermolysis bullosa, 1368
Epididymis
    adenomatoid tumor, 727
    mesothelial cells, 728
    papillary cystadenoma, 728
    rhabdomyosarcoma, 720, 721, 728
Epidural hematoma, 1284
Epidural hemorrhage (EDH), 1270–1272
Epiglottitis, 1185
Epignathus, 1171
Epithelial branching regulation, 587, 588
Epithelial inclusion cyst, 760
Epithelial neoplasms, 1056
    anaplastic thyroid carcinoma, 963, 964
    CASTLE, 964
    follicular thyroid adenoma, 952, 953, 955
    follicular thyroid carcinoma, 955, 956
    Hürthle cell tumors, 962, 963
    keratin 19, 950
    malignant lymphoma, 964
    medullary thyroid carcinoma, 960, 961
        C-cell hyperplasia, 960
        differential diagnosis, 962
        multiple endocrine neoplasias, 960, 962
    papillary thyroid carcinoma, 955–960
    plasmacytoma, 964
    poorly differentiated thyroid carcinoma, 962
    secondary (metastatic) tumors, 965
    SETTLE, 964
    solitary thyroid nodule, 951
    teratoma, 964
Epithelial precursor lesions, 1197
Epithelial tumors, 1189
Epithelial tumors, malignant, 1198, 1199
Epithelial-mesenchymal neoplasms, 1056
Epithelial-mesenchymal transition (EMT), 175
Epithelioid AML (EAML), 656
Epithelioid hemangioendothelioma, 109, 1066
Epithelioid sarcoma, 1076
Epithelioid trophoblastic tumor (ETT), 1486
Epithelioid sarcoma, 1076
Epstein pearls, 1189
Epstein-Barr virus (EBV), 100, 312, 888, 902, 905
Eruption cyst, 1192
Erythema multiforme, 1357, 1362, 1368
Erythema neonatorum, 1354
Erythema nodosum, 1369–1370
Erythrophagocytosis, 909
Erythroplasia of Queyrat, 1378
Erythropoietic protoporphyria (EPP), 485
Esophageal anomalies, 258
    acquired anomalies of low- or eumotility, 261, 262

acquired anomalies with dysmotility, 262, 263
EA-TEF, 258, 260
ectopia/heterotopia, 261
esophageal duplication, 260, 261
obstructive disorders, 261
Esophageal Atresia ± Tracheo-Esophageal Fistula (EA-TEF), 258, 260
Esophageal carcinoma, 276
Esophageal duplication, 260, 261
Esophageal dysmotility, 262
Esophageal inflammatory diseases
eosinophilic esophagitis, 267–269
esophagitis, 270
GVHD, 272
infectious esophagitis, 269
inflammation-associated lining changes, 272, 274
injury-associated esophagitis, 269
reflux esophagitis, 263, 265–267
Esophageal tumors
benign esophageal tumors, 275
squamous cell carcinoma, 275, 277
Esophagitis, 270
Esophagus, 256, 258
acquired anomalies of low- or eumotility, 261, 262
acquired anomalies with dysmotility, 262, 263
EA-TEF, 258, 260
ectopia/heterotopia, 261
esophageal duplication, 260
esophageal inflammatory diseases, 263, 265–269, 272, 274
esophageal tumors, 274, 275, 277
esophageal vascular changes, 263
obstructive disorders, 261
Essential thrombocythemia, 870
Esthesioneuroblastoma, 1182
Etiopathogenetic theories, 1448
ETV6/NTRK3 fusion gene, 850
European Pathologists' Working Group, 848
Evolution of birth, 1410
Ewing's sarcoma, 896
EWS-Fli-1 gene fusion transcript, 1059
Exaggerated placental site (EPS), 1485
Exanthematous CADR, 1371–1372
Excessive genomic sharing, 1423
Experimental discordant xenografts, 97
Exstrophy/congenital eversion, 678
External ear neoplasms, 1220, 1223, 1224
Extraadrenal paraganglioma, 994
Extracellular matrix (ECM), 175
Extrachorial placentation, 1417
Extradigital glomus tumors, 840
Extragonadal germ cell tumor, 1085–1087, 1089
Extragonadal yolk sac tumor, 1085–1087, 1089
Extrahepatic bile ducts, 561
Extrahepatic iron storage disorder, 480–182
Extramammary Paget disease (EMPD), 821, 822
Extramedullary myeloid sarcoma (EMS), 1182, 1201
Extranodal marginal zone B-cell lymphoma of MALT (ENMZL/MALT), 301, 892, 964

Extranodal NK/T cell Lymphoma (EN-NK/T-NHL), 906, 907, 1181
Extrarenal rhabdoid tumor (ERT), 696, 1084
Extraskeletal aneurysmatic bone cyst (EABC), 1070
Extraskeletal chondroma, 1069, 1070
Extraskeletal myxoid chondrosarcoma (ESMC), 1069–1070, 1080–1081
Extraskeletal osteosarcoma (ESOS), 1070
Extrinsic placental implantation (EPIA), 1439
Eye and ocular adnexa
congenital anomalies, 1226
inflammatory lesions & non-neoplastic lesions, 1226
tumors RB, 1227, 1230

F

Fabry disease, 73
Fabry nephropathy, 615, 616
5-factor-score (FFS), 1005, 1009
Falx cerebri, 1246
Familial adenomatous polyposis (FAP), 996
Familial glomangiomas, 1041
Familial Juvenile hyperuricemic nephropathy, 1158
Familial mediterranean fever (FMF), 613
Farnesoid X receptors (FXRs), 450
Fasciitis/myositis group
cranial fasciitis, 1019
eosinophilic fasciitis, 1019
intravascular fasciitis, 1019
ischemic fasciitis, 1019
myositis ossificans, 1020
nodular fasciitis, 1019
proliferative fasciitis/myositis, 1019
Feedback loops, 936
Feedback mechanism, 936
Female genital system
bacterial diseases
chancroid, 766
chlamydia infections, 766
gonorrheae, 765
granuloma inguinale, 766
syphilis, 765
benign discolored vulvar lesions, 768
cervical metaplasia, 771
cervix tumors
benign neoplasms, 805
cervical adenocarcinoma, 807
cervical dysplasia, 805, 806
cervical epidermoid carcinoma, 806, 807
cervical neuroendocrine carcinoma, 810, 811
congenital malformations
of female genital tract, 761
of ovary, 760, 761
cyst-associated cervical lesions, 770
cystic lesions of vulva, 767
development and genetics, 758, 759
ectopia-associated cervical lesions, 770
endometrial dating, 759
fungal diseases, 767
inflammation-associated cervical lesions

Female genital system (*cont.*)
  infective cervicitis, 769
  non-infective cervicitis, 769
  papillary endocervicitis, 769
 microglandular hyperplasia, 771
 neovagina, 767
 non-neoplastic cystic lesions
  corpus luteum cysts, 763
  ednometriosis cyst, 762
  epidermoid cysts, 763
  follicular cysts, 763
  germinal inclusion cysts, 763
  gliomatosis peritonei, 762
  hyperreactio luteinalis, 763
  luteininc cyst, 762
  paratubal Waldhard rest cyst, 762
  polycystic change, 763
 non-neoplastic ovary proliferations
  ectopic decidual reaction, 764
  endometriosis, 763–764
  endosalpingiosis, 764
  fibromatosis ovarii, 764
  Hilus cell hyperplasia, 764
  massive edema of the ovary, 764
  pregnancy luteoma, 764
  stromal hyperplasia, 764
  stromal hyperthecosis, 764
 oophoritis, 761
 ovarian torsion, 761
 ovary tumors
  BRCA1/2 mutations, 773
  germ cell tumors (*see* Germ cell tumors)
  gonadoblastoma, 798
  mucinous cystoadenoma, 780
  ovarian neoplasms, 773–775
  risk factors, 773
  serous carcinoma, 781
  serous cystoadenoma, 780
  sex cord-stromal tumors (SCSTs) (*see* Sex cord-stromal tumors)
  surface epithelial tumors, 779, 780, 782–790
 parasitic diseases, 767
 pregnancy-associated cervical lesions, 770–771
 reactive changes of cervix, 770
 sex organ development, 759
 tuba tumors, 798
 urothelial (transitional) metaplasia, 767
 uterine lesions
  adenomyosis, 773
  dysfunctional (anovulatory) uterine bleeding (DUB), 771
  endometrial metaplasias, 772–773
  endometriosis, 773
  endometritis, 772
  non-neoplastic abnormalities of tuba, 773
 uterus tumors
  hereditary nonpolyposis colon cancer (HNPCC), 804
  Lynch syndrome, 804
  non-type I/non-type II endometrial carcinomas, 799
  type I/type II EC, 799–803
  uterine adenomatoid tumor, 804
  uterine leiomyoma, 799
  uterine lymphangiomyomatosis, 804
  uterine stroma tumors, 804, 805
 vagina tumors
  adenocarcinoma, 814
  endodermal sinus tumor, 815
  malignant neoplasms, 815, 816
  Mullerian papilloma, 811
  pathology, 816
  post-operative spindle (cell) nodule, 811
  vaginal epidermoid carcinoma, 813–814
  vaginal fibroepithelial polyp, 811
  vaginal intraepithelial neoplasia, 813
  vaginal rhabdomyosarcoma of Botryoid type, 814, 815
  vaginal YST, 815
 vaginal adenosis, 767, 768
 vaginal pseudotumors, 769
 viral diseases
  condyloma acuminatum, 764–765
  herpes genitalis, 764
 vulva tumors
  Bartholin gland carcinoma, 822
  EMPD, 821, 822
  epithelioid sarcoma, 822
  malignant melanoma (*see* Vulva tumors)
  vulvar aggressive angiomyxoma, 822
  vulvar paget disease, 821, 822
  vulvar verrucous carcinoma, 821
 vulvar dystrophies, 767
 vulvitis, 767
Femur length (FL), 1421
Fertilization, 1490
Fertilization age (FA), 1426
Fetal age, 1426
Fetal blood bleeding, 1455, 1457
Fetal death syndrome, 1546–1548
Fetal distress, 1417
Fetal growth restriction (FGR), 1427, 1482–1484, 1541, 1542, 1545
Fetal growth restriction and pre-eclampsia, 1417
Fetal growth restriction/intrauterine fetal demise, 1470
Fetal laterality, 8
Fetal lung development, 1418
Fetal maternal hemorrhage (FMH), 1454, 1455, 1457
Fetal megacystis, 681
Fetal or neonatal death, 1410
Fetal thrombotic vasculopathy (FTV), 1474, 1475
Fetal-type rhabdomyoma, 1046
Feto-maternal hemorrhage (FMH) ± intervillous thrombus, 1465, 1467
Fetopathies, 1490, 1507, 1508
Fetoplacental circulation, 1414
 intervillous space of placenta, 1414
 maternal blood, 1414
FGFR1, 229
Fibrinoid necrosis, 626
Fibrinous pericarditis, 116
Fibroadenoma (FA), 840, 842, 904

Fibroadenomatosis, 841
Fibroblast growth factor receptor 3 (*FGFR3*) gene, 1099
Fibroblastic/myofibroblastic tumors
    β-catenin, 1018
    β-catenin involvements, 1018
    fasciitis/myositis group
        cranial fasciitis, 1019
        eosinophilic fasciitis, 1019
        intravascular fasciitis, 1019
        ischemic fasciitis, 1019
        myositis ossificans, 1020
        nodular fasciitis, 1019–1020
        proliferative fasciitis/myositis, 1019
    fibroblastoma classic and subtypes
        angio-myo-fibroblastoma, 1022
        desmoplastic fibroblastoma, 1022
        giant cell fibroblastoma, 1022
        mammary-type myo-fibroblastoma, 1023
    fibroma group
        elastofibroma, 1021
        Gardner fibroma, 1020
        genital cellular angiofibroma, 1021
        keloid, 1021
        nasopharyngeal fibroma, 1020
        nuchal-type fibroma, 1020
        ovarian fibroma, 1020
        tendon sheath fibroma, 1020
    fibromatosis of childhood
        fibromatosis colli, 1023
        gingival ridge, 1027
        infantile digital (inclusion body) fibromatosis, 1023, 1025
        infantile fibromatosis, desmoid type, 1025
        Juvenile hyaline fibromatosis, 1025
    fibrous hamartoma of infancy, 1023, 1024
    infantile fibrosarcoma, 1032–1033
    infantile myofibroma, 1027, 1028
    inflammatory myofibroblastic tumor, 1030–1031
    with intermediate malignant potential
        desmoid-type fibromatosis, 1029
        giant cell angio-fibroma, 1029, 1030
        hemangiopericytoma-like, 1029
        lipo-fibromatosis, 1029
        solitary fibrous tumor, 1029
        superficial fibromatosis, 1029, 1030
    low-grade myo-fibroblastic sarcoma, 1031–1032
    malignant tumors
        adult fibrosarcoma, 1033
        low-grade fibro-myxoid sarcoma, 1034
        myxo-fibrosarcoma, 1033–1034
        sclerosing epithelioid fibrosarcoma, 1034–1036
    myxo-inflammatory fibroblastic sarcoma, 1032
Fibroblastoma classic and subtypes
    angio-myo-fibroblastoma, 1022
    desmoplastic fibroblastoma, 1022
    giant cell fibroblastoma, 1022
    mammary-type myo-fibroblastoma, 1023
Fibrocystic disease, 840
Fibrocystic disorders, 428
Fibroelastic hyperplasia, 626
Fibroepithelial polyp, 689, 769
Fibrohistiocytic tumors
    borderline fibrous histiocytoma, 1039
    cellular fibrous histiocytoma, 1038–1039
    dermatofibroma, 1037
    DTGCT, 1038
    giant cell tumor of tendon sheath, 1038
    histiocytoma, 1036
    malignant fibrous histiocytoma, 1039
Fibrokeratoma, acquired, digital, 1377
Fibroma, 105
Fibroma group
    elastofibroma, 1021
    Gardner fibroma, 1020
    genital cellular angiofibroma, 1021
    keloid, 1021
    nasopharyngeal fibroma, 1020
    nuchal-type fibroma, 1020
    ovarian fibroma, 1020
    tendon sheath fibroma, 1020
Fibromatosis, 840, 1376
    of childhood
        fibromatosis colli, 1023
        gingival ridge, 1027
        infantile digital (inclusion body) fibromatosis, 1023, 1025
        infantile fibromatosis, desmoid type, 1025
        Juvenile hyaline fibromatosis, 1025
    colli, 1023
    ovarii, 764
Fibromuscular dysplasia, 117, 626
Fibrosarcoma, 1136, 1153
Fibrotpiethelial polyps/accessory tragus, 1500
Fibrous and fibrohistiocytic tumors, 1394, 1396
Fibrous and pseudogranulomatous (non-neoplastic) lesions of oropharynx, 1197
Fibrous collagenous silicosis, 205
Fibrous dysplasia, 1116–1118, 1224
Fibrous hamartoma, 102
Fibrous hamartoma of infancy (FHI), 1023, 1024
Fibrous obliteration of the appendix tip, 350
Filum terminale Hamartoma, 1300
Fine-needle aspiration cytology (FNAC), 949
FISH sperm test, 696
Fissural cysts, 1216
Fixed drug eruption, 1359, 1371
Flat dysplasia, 559
Florid cystic endosalpingiosis, 761
Florid hyperplasia, 852
Fluid-filled synovial cyst, 1159
Focal & segmental glomerulosclerosis, 604, 606, 607
Focal nodular hyperplasia (FNH), 510–514
Focally enhanced gastritis (FEG), 283
Follicle lysis, 918
Follicle-stimulating hormone (FSH), 934
Follicular cholecystitis, 558
Follicular cystitis, 684
Follicular cysts, 763
Follicular Dendritic Cell Sarcoma (DCS), 1203
Follicular hyperplasia

AIDS-related lymphadenopathy, 918
Castleman disease, 917, 918
cat-scratch disease, 916, 917
Kimura disease, 917
lymphogranuloma venereum, 917
necrotizing lymphadenitis, 916
nonspecific reactive lymphadenitis, 916
progressive transformation of germinal centers, 918
rheumatoid arthritis, 916
Sjogren disease, 916
SLE-lymphadenopathy, 916
syphilis, 917
toxoplasmosis, 916
Follicular lymphoma, 299, 300, 888–890
Follicular Lymphoma International Prognostic Index (FLIPI), 890
Follicular neoplasia, 952
Follicular spongiosis, 1354
Follicular thyroid adenoma, 952–955
Follicular thyroid carcinoma, 115, 956
Follicular variant of papillary thyroid carcinoma (FVPTC), 950
Fontan operation, 39
Fontana-Masson special stain, 997
Food protein-induced enterocolitis syndrome (FPIES), 332, 333
Fordyce granules, 1189
Foregut cysts, 146
Fraser syndrome, 1515, 1520
Free fetal movements, 1348
French-American-British (FAB) classification, 874
Friend leukemia integration one transcription factor or Fli-1 protein, 1059
Fryns syndrome, 1515, 1518
FSGS-collapsing glomerulopathy, 605
FSH-stimulated spermatogenesis, 675
Fucosidosis, 467
Functional pituitary adenoma, 940
Fundic gland polyps, 292
Fungal infection, 920
Fusiform-type of aneurysms, 1274
Fusion defects of coelomic type, 1498, 1499

G
Galactocerebroside B galactoside (gene, chromosome) deficiency-based leukodystrophy, 1287
Galactosemia, 466
Galectin-3, 950
Gallbladder and extrahepatic biliary tract pathology
   acute cholecystitis, 557–558
   cholelithiasis, 556–557
   cholesterolosis, 557
   chronic cholecystitis, 558
   embryonal rhabdomyosarcoma, 561
   eosinophilic cholecystitis, 558
   extrahepatic bile ducts, 561
   follicular cholecystitis, 558
   proliferative processes and neoplasms, 558–561
   xanthogranulomatous cholecystitis, 558
Gallbladder carcinoma, 559–561
Gallbladder congenital abnormalities, 553, 554
Gandy-Gamna bodies, 922
Gangliocytic paraganglioma, 342
Ganglion cell differentiation, 984
Ganglion cells, 355
Ganglioneuroblastoma, 985, 986
Ganglioneuroma, 985, 986
Gangrenous cystitis, 684
Gardner fibroma, 1020
Gartner duct cyst, 760
Gastric agenesis, 277
Gastric anomalies
   acquired anomalies with dysmotility, 280
   ectopia/heterotopia, 278
   gastric agenesis and distopias, 277
   gastric duplications, 278
   obstructive disorders, 280
Gastric hyperplasia, 286, 287
Gastric inflammation
   acute gastritis, 281
   chronic gastritis, 281
   eosinophilic gastritis, 285
   granulomatous gastritis, 285
   Helicobacter heilmannii gastritis, 285
   lymphocytic gastritis, 284, 285
Gastric metaplasia, 559
Gastric mucosa, 261
Gastric teratoma, 289
Gastric tumors
   benign epithelial neoplasms, 287, 292
   Burkitt lymphoma, 300
   diffuse large B-cell lymphoma, 300, 301
   EATL, 301, 302
   ENMZL/MALT, 301
   follicular lymphoma, 299, 300
   gastrointestinal B-NHL, 298, 299
   gastrointestinal T-NHL, 301
   lymphoproliferative lesions, 297
   malignant epithelial neoplasms, 292–294
   mantle cell lymphoma, 299
   neuroendocrine tumors, 294–297
   non-nasal NK/T-cell NHL, 302, 303
   non-stroma/smooth muscle differentiating tumors, 304
   secondary tumors, 304
   stroma tumors and smooth muscle differentiating tumors, 303, 304
Gastric ulcers, 286
Gastrinoma, 327
Gastroesophageal reflux (GER), 263
Gastrointestinal B-NHL, 298, 299
Gastro-intestinal stromal tumor (GIST), 303
Gastrointestinal T-NHL, 301
Gastrointestinal tract
   appendix, 343
      appendiceal neoplasms, 351–353
      appendicitis, 349–351
      appendix anomalies, 343
      metabolic and degenerative changes, 351
      vascular changes, 343

development and genetics, 256, 257
esophagus, 258
    acquired anomalies of low- or eumotility, 261, 262
    acquired anomalies with dysmotility, 262, 263
    EA-TEF, 258, 260
    ectopia/heterotopia, 261
    esophageal duplication, 260
    esophageal inflammatory diseases, 263, 265–269, 272, 274
    esophageal tumors, 274, 275, 277
    esophageal vascular changes, 263
    obstructive disorders, 261
large intestine
    inflammatory disorders, 374–380
    large intestine anomalies, 353, 354, 365–372
    vascular changes, 373, 374
small intestine, 304
    dystopias, 317
    GVHD, 339, 340
    inflammation and malabsorption, 319, 320, 322–336
    intestinal lumen, continuity defects of, 312, 316
    intestinal muscular wall defects, 316, 317
    intestinal wall, composition abnormalities of, 317, 318
    short bowel syndrome/intestinal failure, 336
    small intestinal neoplasms, 340–343
    small intestine anomalies, 304, 306
    small-intestinal transplantation, 336–339
    vascular changes, 318
stomach, 277
    gastric anomalies, 277, 278, 280
    gastric inflammation, 281, 284–286
    gastric tumors, 287, 292–304
    tissue continuity damage-related degenerations, 286, 287
    vascular changes, 280, 281
Gastroschisis, 311
GATA4, 61
Gaucher disease, 469
GDNF/c-Ret/WntI pathway, 585
GDNF/Ret signaling, 587
G4.5 gene, 62
Gene therapy of genodermatoses, 1350
Genetic and metabolic liver disease
    albinism-related liver diseases, 486
    amino acid metabolism disorders, 471
    chronic granulomatous disease, 485–486
    congenital dysregulation of carbohydrate metabolism
        CDG, 466
        galactosemia, 466
        glycogen storage diseases, 464
        hereditary fructose intolerance, 466
        Lafora progressive myoclonic epilepsy, 465, 466
    copper metabolism dysregulation, 483, 484
    ERSD, 461, 462, 464
    hepatomegaly, 460
    iron metabolism dysregulation, 479, 482, 483
    lipid/glycolipid and lipoprotein metabolism disorders (see Lipid/glycolipid and lipoprotein metabolism disorders)
    microvesicular steatosis, 460
    mitochondrial hepatopathies (see Mitochondrial hepatopathies)
    pericellular fibrosis, 460
    peroxisomal disorders, 477–479
    porphyria-related hepatopathies, 485
    primary tumor infiltration, 460
    Shwachman-Diamond syndrome, 485
Genetic cancer syndromes with renal tumors
    Beckwith-Wiedemann syndrome, 633
    Birt-Hogg-Dube' syndrome, 634
    Denys-Drash syndrome, 634
    hereditary leiomyoma renal carcinoma syndrome, 634
    hereditary papillary renal carcinoma syndrome, 634
    non-WT1/WT2 pediatric syndromes, 634
    tuberous sclerosis syndrome, 634
    Von Hippel Lindau syndrome, 634
    WAGR syndrome, 633
Genetic heterogeneity, 1495
Genetic imprinting, 996
Genetic/hereditary hemochromatosis, 483
Genital cellular angiofibroma (GCAF), 1021
Genomic imprinting, 1427
Germ cell tumors
    anaplastic seminoma, 707
    choriocarcinoma, 712, 717–720, 779
    differential diagnosis, 776
    dysgerminoma, 776
    embryonal carcinoma, 703, 711–713, 778, 779
    endodermal sinus tumor, 715–717, 776–778
    ITGCN, 705–706
    polyembryoma, 717, 779
    seminoma, 702, 706
    seminoma with syncytiotrophoblastic cells, 706
    spermatocytic seminoma, 707
    teratoma, 776, 777
    testicular teratomas, 713–715
    yolk sac tumor, 715–717, 776–778
Germ cell tumors, 1056, 1183
German embryology, 1490
Germinal inclusion cysts, 763
Gestational age (GA), 1426
Gestational alloimmune liver disease (GALD), 481
Gestational trophoblastic diseases, 1427
    CCA, 1486, 1488
    choriocarcinoma, 1486
    ETT, 1486
    invasive mole, 1486
    pre-malignant and malignant, 1486
    PSTT, 1486
Gestational trophoblastic diseases (GTD), 1428
Ghadially's myelinosomes, 466
Giant cell angioblastoma, 1065
Giant cell angio-fibroma (GCAF), 1029, 1030
Giant cell cystitis, 686
Giant cell fibroblastoma (GCFB), 1022

Giant cell rich osteosarcoma (GCOS), 1134, 1135
Giant cell tumor (GCT), 1127–1128
Giant cell tumor of tendon sheath (GCTTS), 1038
Giant osteoid osteoma (GOO), 1126–1127
Giardia lamblia, 308
Giardiasis/lambliasis, 326, 328
Gilbert syndrome, 438
Gingival cysts, 1189
Gingival ridge, 1027
Glassy cell carcinoma (GCC), 809
Glial cell line-derived neurotrophic factor (GDNF), 581
Glioblastoma multiforme (GBM), 1294
Globoid cell leukodystrophy (Krabbe disease), 1287
Glomangioma, 840
Glomangiosarcoma, 1042
Glomerulonephritis
    definition, 599
    membranoproliferative (membrane-capillary) glomerulonephritis, 609, 610
    membranous glomerulonephritis, 606, 608, 609
    post-infectious glomerulonephritis, 600–602
    rapidly progressive glomerulonephritis, 602–604
Glomerulosclerosis, 604, 606, 607
Glomerulus, molecular patterning and vascularization, 588, 589
Glomus tumor, 840, 1041
Glucose transporter 1 (Glut1), 1059
Gluten-sensitive enteropathy, 311, 312, 322–325, 327
Glycogen storage diseases (GSD), 464, 1523–1525
Glycogen storage disorders, 1339
Glycogenosis cardiomyopathy, 72
Glycosylation, 332
Goiter, 946
Gonadal dysgenesis, 679
Gonadoblastoma, 798
Gonorrheae, 765
Goodpasture syndrome, 196, 602
Gorlin syndrome, 962
Gout, 1157–1158
Graft versus host disease (GVHD), 272, 339, 340, 903, 1359
Grannum score, 1427
Granular cell tumor, 559, 840
Granulation tissue/pyogenic granuloma-like tissue lesion, 1186
Granulocytic sarcoma, 840
Granuloma annulare, 1363–1365
Granuloma inguinale, 766
Granulomas-accompanying chronic laryngitis with granulomas, 1186
Granulomatosis with polyangiitis, 197
Granulomatous appendicitis, 349
Granulomatous cystitis, 684
Granulomatous gastritis, 285
Granulomatous liver disease, 495, 496
Granulomatous lymphadenitis, 915
Granulomatous mastitis, 839
Granulosa cell tumor, 715, 716, 791, 793
Graves' disease, 947
Great arteries transposition, 26, 31

Group B streptococcal (GBS) infection, 85
Grover disease, 1367
Growth disorganization (GD), 1427, 1428
Gynandroblastoma, 797
Gynecomastia, 837, 855

H
Haemophilus ducrey, 766
Hailey-Hailey disease, 1367
Hair follicle tumors
    benign, 1388, 1390
    malignant, 1390
Hairy cell leukemia (HCL), 876, 877
Hamartomatous polyps, 288, 292
Hamman-Rich disease, 170
Hand-Schiller-Christian (HSC) disease, 909, 942, 943
Hashimoto autoimmune thyroiditis, 945
Hashimoto's thyroiditis, 888, 892, 949, 952
Hassal's corpuscles, 862
Hb-related Porphyria cutanea tarda, 1368–1369
HBV hepatitis, 487
Head and Neck (H&N)
    branchial apparatus, 1168
    development
        branchial arch musculature, 1169
        branchial grooves, 1168
        facial structures, 1169
        in gestation, 1168
        neurulation, 1168
        pharyngeal pouches, 1168
        splanchnocranium, 1169
    ear
        congenital anomalies, 1218
        inflammatory lesions & non-neoplastic lesions, 1218–1220
        ear primordium, 1168
        tumors, 1220, 1223–1225
    eye and ocular adnexa
        congenital anomalies, 1226
        inflammatory lesions & non-neoplastic lesions, 1226
        primordium, 1168
        tumors, 1226, 1227, 1230
    larynx & trachea
        congenital anomalies, 1183, 1184
        cysts & laryngoceles, 1184, 1185
        inflammatory lesions & non-neoplastic lesions, 1185, 1186
        tumors, 1186–1189
    mandible & maxilla
        odontogenic cysts, 1210, 1215, 1216
        odontogenic tumors, 1216–1218
    nasal cavity, paranasal sinuses & nasopharynx
        congenital anomalies, 1171, 1173
        inflammatory lesions, 1173, 1174
        tumors, 1174, 1178–1183
    neural tube, 1168
    oral cavity & oropharynx
        branchial cleft cysts, 1195, 1196

congenital anomalies, 1189, 1193, 1194
  tumors, 1197–1201, 1203
 oropharyngeal membrane, 1168
 pharynx primitive, 1168
 salivary glands
  congenital anomalies, 1203
  inflammatory lesions & non-neoplastic lesions, 1203
  tumors, 1205, 1208, 1209
 skull, 1230
 stomodeum, 1168
Head circumference (HC), 1421
Head injuries, 1284, 1285
Heart calcineurin pathway, 61
Heart failure, 85
Heart transcriptome, 5
Heavy chain disease, 912
Hector Battifora Mesothelial Cell-1 (HBME-1), 950
Hedgehog pathway, 911
Helicobacter heilmannii gastritis, 285
HELLP syndrome, 1471
Helminthiasis, 510
Hemangioblastoma, 1299, 1300
Hemangioendothelioma, 1062–1064
Hemangioma, 105, 840, 1060, 1061, 1199
Hematogenous-driven brain abscess, 1278
Hematoidin, 1419, 1420
Hematologic tumors, 1153, 1154
Hematological malignancies, 739, 840
Hematological skin infiltrates, 1400
Hematological solid malignancies of the brain and spinal cord, 1302, 1303
Hematolymphoid neoplasms, 1209
Hemato-lymphoid system
 benign lymphadenopathies, 914 (*see* Benign lymphadenopathies)
 bone marrow, 863
 coagulation (*see* Coagulation disorders)
 granulomatous lymphadenitis, 915
 hemostasis, 869
 monocyte-macrophage system (*see* Monocyte-macrophage system)
 plasma cell dyscrasias (*see* Plasma cell disorders)
 platelet disorders, 870, 871
 RBC (*see* Red blood cell disorders)
 spleen, 862 (*see* Spleen disorders)
 TdT, 863
 thymus, 862 (*see* Thymus disorders)
 WBC (*see* White blood cell disorders)
Hemato-lymphoid tumors, 1181, 1182
Hematolymphoid tumors of oral cavity and oropharynx, 1201
Hematopoietic tumors, 112
Hemolytic uremic syndrome, 628, 631
Hemophagocytic lymphohistiocytosis (HLH), 907, 1151
Hemophagocytic syndrome, *see* Hemophagocytic lymphohistiocytosis
Hemorrhage, brain, 1268
Hemorrhagic cystitis, 684
Hemorrhagic pericarditis, 116
Hemostasis, 869
Henoch-Schönlein purpura (HSP), 611–613
Hepatic angiosarcoma (HAS), 531–533
Hepatic leiomyosarcoma (HLMS), 533
Hepatic rhabdomyosarcoma (HRMS), 533
Hepatic vascular disorders
 acute and chronic passive liver congestion, 503, 504
 ischemic hepatocellular necrosis, 505
 liver cirrhosis, 508
 peliosis hepatis, 504
 shock-related cholestasis, 505
Hepatoblastoma, 517–526
Hepatocellular carcinoma (HCC), 521, 522, 526–529
Hepatocellular dysplasia (HCD), 515
Hepato-cerebral phenotypes, 83
Hepatomegaly, 460
Hepcidin, 479
Her-2 gene mutations, 846
Hereditary C4 deficiency (C4def), 608
Hereditary fructose intolerance, 466
Hereditary hemochromatosis (HFE), 480
Hereditary hemorrhagic telangiectasia, 1275
Hereditary leiomyoma renal carcinoma syndrome, 634
Hereditary nonpolyposis colon cancer (HNPCC), 804
Hereditary papillary renal carcinoma syndrome, 634
Hereditary/familial nephropathies
 Alport syndrome, 615
 congenital nephrotic syndrome, 617
 Fabry nephropathy, 615, 616
 Nail-Patella syndrome, 617
 thin glomerular basement membrane nephropathy (TBMN), 617
Hermansky-Pudlak syndrome (HPS), 486
Herpes genitalis, 764
Herpes gestationis, 1368–1369
Herpes simplex encephalitis, 1281
Herpes simplex virus infection, 271
Heterotaxy, 41, 42
Heterotopia, 683
 mimickers, 683
 pancreatis, 555
Heyde syndrome, 373
Hiatal hernia, 261
Hibernoma, 1012–1013
Hidradenoma papilliferum, 816, 1393–1394
High-grade fibrosarcomas, 1136
Hilus cell hyperplasia, 764
Hirschsprung disease, 582
Histiocytic medullary reticulosis, 909
Histiocytic proliferations, 923
Histiocytoid CM (HICM), 108
Histoplasma capsulatum, 920
Histoplasmosis, 327
HIV-causing AIDS, 888
Hodgkin lymphoma, 880–885
Holt-Oram syndrome (HOS), 47
Hormone regulation feedback mechanisms, 936
Horton arteritis, 123
Host-placenta interaction dysregulation, 1484
HS426 (nebulette) gene, 63

Human congenital anomalies, 1490
Human immunodeficiency virus (HIV), 925, 926, 1282
Human leukocyte antigen (HLA) gene, 56
Human T-cell leukemia virus (HTLV), 903
Human T-cell leukemia virus type 1 (HTLV-1), 874, 902
Hürthle cell tumors, 962, 963
Hürthle/Askanazy cells, 935
Hyaline arteriolosclerosis, 626
Hyaline cartilage, 1137
Hyaline perisplenitis, 923
Hyalinized fibrous tissue, 234
Hyalinizing trabecular adenoma (HTA), 949
Hydranencephaly, 1545
Hydrocalyx/hydrocalycosis, 661
Hydrocele, 700
Hydrocephalus, 1257
Hydronephrosis, 597–598, 687
Hydropic abortion, 1429
Hydropic abortus (HAM), 1429–1431
Hydropic molar degeneration (HMD), 1429
Hydro-ureteronephrosis, 597–598
Hyparterial broncho-arterial crossing, 144, 145
Hyperacute rejection, 97
Hyperammonemia, 475
Hyperbilirubinemia
  abnormal intracellular binding, 437
  acute bilirubin elevation, 436
  apical membrane molecules, 436
  basolateral membrane molecules, 436
  bilirubin production, 436
  cholestasis, 437
  chronic bilirubin elevatio, 437
  clinicopathological condition, 436
  conjugated hyperbilirubinemia, 438, 439
  Crigler-Najjar syndrome, 437
  Dubin-Johnson syndrome, 438
  Gilbert syndrome, 438
  intracellular transfer protocol, 436
  Rotor syndrome, 438
  subtle bilirubin encephalopathy, 437
  unconjugated hyperbilirubinemia, 436
Hyper-calcemia, renal insufficiency, anemia, bony
    lesions (CRAB) criteria, 911
Hypercoiling, 1416, 1417
Hyperemia, 1005
Hypereosinophilic syndrome, 875
Hyperlipidemic diet, 498
Hyperparathyroidism
  primary hyperparathyroidism, 965
  secondary hyperparathyroidism, 965
  tertiary hyperparathyroidism, 965
Hyperpituitarism, 940
Hyperplasia
  adenomatous, hyperplasia, 732
  adenosis, sclerosing, 731
  basal cell hyperplasia, 732
  BNH/PNH-small gland type, 732
  clear cell cribriform hyperplasia, 732
Hyperplastic, 290
Hyperplastic arteriolitis, 626

Hyperprolactinemia, 941
Hyperreactio luteinalis, 763
Hyper-secretive adeno-hypophysopathies, 940
Hypersensitivity pneumonia (HSP), 181, 182
Hyper-somatotropinemia, 942
Hypersplenism, 922
Hyperthyroidism, 946
Hypertrophic cardiomyopathy, 66, 68
Hypertrophic osteoarthropathy (HOA), 1102
Hypertrophic scar and keloid, 1394
Hypochondroplasia, 1100
Hypocoiling, 1417
Hypokalemic nephropathy, 625
Hypophosphatasia (HPP), 1101
Hypophysis-Apoleia, 941
Hypopituitarism, 942
Hypoplastic right heart syndrome
  HRHTA, 26
  PAVSD, 24
  PSIVS/PAIVS, 24, 26
Hypoplastic right heart with tricuspid atresia (HRHTA), 26
Hypothalamic neoplasms, 943
Hypothalamus, 934, 937, 938, 941, 943, 970, 972
Hypothyroidism, 943
Hypoxia-related etiology, 1482
Hypoxic ischemic encephalopathy, 1267, 1544

I

Iatrogenic causes/multifactorial eti, 1482
IBAFIS, 213
Idiopathic muscular hypertrophy, 262
Idiopathic pulmonary fibrosis (IPF), 173
Idiopathic pulmonary hemosiderosis, 196
IgA Pemphigus, 1366
IgG4-related thyroiditis, 948, 949
IGH/BCL6 translocation, 887
IgM dysproteinemia, 890
IL-1 beta pathway-hyperactivation, 1363
Ileal atresia, 316
Image J-based Automated Fibrosis Index Score, 213
Immune dysregulation, 331
Immune-mediated and dermatologic lesions of the
    oropharynx, 1197
Immunohistochemistry for antigen, 1429
Immunohistochemistry staining of intermediate
    trophoblast cell populations, 1427
Immuno-inflammatory inflammatory enteropathies, 331
Immunoproliferative small intestine disease (IPSID), 301
Impetigo, 1366
In situ ductal/lobular carcinoma (DCIS/LCIS), 844, 852
Incomplete miscarriage, 1477
Incontinentia pigmenti, 1354
Infantile colic, 304
Infantile digital (inclusion body) fibromatosis, 1023, 1025
Infantile fibromatosis, desmoid type, 1025
Infantile fibrosarcoma, 1032–1033
Infantile Hemangioendothelioma (IHE), 516
Infantile ILD, 159

chromosomal/genomic microdeletion disorders, 163
diffuse lung development-associated ILD, 159–161
NEHI, 163, 164
PIG, 164
pulmonary alveolar proteinosis, 166, 167
surfactant dysfunction disorders, 164–166
trisomy 21-associated ILD, 161
Infantile myofibroma (IM), 1027, 1028
Infantile myofibromatosis (IM/IM), 1027, 1028
Infantile NILD, 155, 157
    atelectasis, 158
    cystic fibrosis, 158, 159
    neonatal respiratory distress syndrome, 157
    OBPD and NBPD, 157
Infarcts, 627, 628
Infection in pregnancy, 1534, 1538–1541
Infection-related neonatal hepatitis, 454
Infections, 338
Infections of umbilical cord, 1417–1419
Infections-associated with interface dermatitis, 1359, 1360
Infectious arthritis, 1156
Infectious esophagitis, 269
Infectious mononucleosis, 919
Infectious pneumonia, 190
    bronchopneumonia, 191, 193
    interstitial pneumonia, 193
    lobar pneumonia, 190, 191
    primary atypical inflammation, 193
Infectious/noninfectious granulomatous/nongranulomatous, 197
    aspergillosis, 201, 202
    primary pulmonarytuberculosis, 197, 199
    sarcoidosis, 200, 201
Infective cervicitis, 769
Infiltrating ductal carcinoma, 845
Infiltration related neuropathy, 1323
Infiltrative diseases, 942
Infiltrative/restrictive cardiomyopathies, 66, 71, 77
Inflammation, 189, 1463
    gastrointestinal tract, 319, 320
    infectious pneumonia (see Infectious pneumonia)
    infectious/noninfectious granulomatous/nongranulomatous (see Infectious/noninfectious granulomatous/nongranulomatous)
    non-infectious (see Non-infectious pneumonia)
    tracheo-bronchial, 184, 187, 188
Inflammation and fibrosis, 1356
Inflammatory (allergic) polyps, 1173
Inflammatory arthropathies, 1154
Inflammatory bowel disease (IBD), 324
Inflammatory disorders
    large intestine, 374
    penis
        infections, 739, 740
        non-infectious inflammatory diseases, 740, 741
    prostate gland, 728
    testis
        epidermoid cysts, 697, 699
        hydrocele, 700
        orchitis, 697, 710
        spermatocele, 700
        varicocele, 704
Inflammatory fibroid polyps, 275, 292
Inflammatory linear verrucous epidermal nevus (ILVEN), 1361, 1375
Inflammatory myofibroblastic tumor, 105, 215, 218, 1030–1031
Inflammatory polyps, 559
Inflammatory pseudotumor, 923
Inflammatory-based cyst, 1215
Infra-sellar expansion, 941
Injury-associated esophagitis, 269
Inner ear neoplasms, 1224, 1225
Insulin-like growth factor 1 (IGF-1) receptor, 229
Interface dermatitis with lichenoid leukocytic infiltrate, 1360
Interface dermatitis with vacuolar degeneration, 1355
Intermediate trophoblast (IT), 1414
International Classification System for Intraocular Retinoblastoma (ICSIR), 1229
Interstitial (Hunner's) cystitis, 686
Interstitial cell-stimulating hormone (ICSH), 934
Interstitial granulomatous dermatitis, 1362
Interstitial pneumonia, 174, 193
Interval appendicitis, 350
Intervertebral disk herniation, 1286
Intervillous thrombi, 1474
Intestinal AMR, 338
Intestinal atresia, 358
Intestinal epithelial dysplasia, 330
Intestinal epithelial dysplasia/tufting enteropathy (IED-TE), 331
Intestinal lumen
    atresia, 312, 316
    meconium plug syndrome, 316
    stenosis, 316
Intestinal lymphoma, 326
Intestinal metaplasia, 559
Intestinal muscular wall defects, 316, 317
Intestinal neuronal dysplasia (IND), 370
Intestinal type carcinoma, 293
Intestinal wall composition abnormalities, 317, 318
Intimal fibroplasia, 117
Intoxication-related neuropathies, 1323
Intra cytoplasmic sperm injection (ICSI), 696
Intraamniotic placental tumors, 1449
Intracavitary growth, 100
Intractable diarrhea of infancy, 330–333
Intraductal papillary mucinous neoplasm (IPMN), 569
Intraductal papilloma, 843
Intraepidermal blistering diseases, 1364, 1366, 1367
Intraepidermal carcinomas, 1378
Intraepithelial eosinophils, 263
Intraepithelial lymphocytes, 263
Intramyocardial growth, 101
Intraosseous ganglion cyst, 1123
Intraparenchymal hemorrhage (IPH), 1269
Intrapartum death and perinatal death, 1417

Intrapartum, pregnancy, 1421
Intrasellar cysts, 942
Intratubular germ cell neoplasia (ITGCN), 705–706
Intrauterine and ectopic pregnancies, 1427
Intrauterine fetal death, 1418
Intrauterine fetal demise (IUFD), 1416, 1546–1548
Intrauterine fetal stress, 1419
Intrauterine growth restriction (IUGR), 1423
Intrauterine stress, 1419
Intravascular fasciitis, 1019
Intussusception, 317
Invasive ductal carcinoma, 848, 849
Invasive lobular breast carcinoma, 849
Invasive lobular carcinoma (ILC), 844
Invasive squamous cell carcinoma, 1379, 1380
Invasive urothelial carcinoma variants, 695
Invasive urothelial neoplasms, 693–695, 742
Inversion, 1496
Inverted papilloma, 693
IPEX syndrome, 331, 332
Ipilimumab, 878
Iron metabolism dysregulation
    extrahepatic iron storage disorder, 480–482
    genetic/hereditary hemochromatosis, 483
    gestational alloimmune liver disease, 481
    hepcidin, 479
    HFE, 480
    neonatal hemochromatosis, 479, 482, 483
Ischemia, 504
Ischemic acute tubular necrosis (ATN), 623
Ischemic colitis, 373
Ischemic fasciitis, 1019
Ischemic heart disease, 85
Ischemic hepatocellular necrosis, 505
Ischemic necrosis, 942
Isochromosome, 1496
Isomerism, 144
Ivemark syndrome, 42, 43
Ivy renal dystopia, 591

J
Jacobsen syndrome, 47
JAK2 V617F mutation, 868, 870
Jatene's procedure, 31
Johanson-Blizzard syndrome, 555
Juvenile granulosa cell tumor (JGCT), 792, 793
Juvenile hyaline fibromatosis (JHF), 1025
Juvenile laryngeal papilloma, 1186
Juvenile paget disease, 1108, 1109
Juvenile recurrent papillomatosis, 1186
Juvenile Rheumatoid arthritis, 1155
Juvenile canthogranuloma (JXG), 1036, 1394–1396
Juxtacortical (periosteal) chondrosarcoma, 1146
Juxtaglomerular cell tumor, 660

K
K19, 950
Kaposi sarcoma (KS), 110, 1064
Kaposiform hemangioendothelioma (KHE), 1062
Kaposiform lymphangiomatosis, 219, 220
Kasabach-Merritt syndrome (KMS), 1068
Katzman-Genest categories, 1484
Kawasaki disease, 123, 124
Kearns-Sayre syndrome (KSS), 81, 1340
Keloid, 1021
Keratinocyte dysplasia, 1377
Keratinous cysts, 1218
Keratoacanthoma, 1378
Kidney development stages, 582–584
Kidneys examination, 1503
Kienbock disease, 1114, 1115
Kikuchi-Fujimoto lymphadenitis, 916
Kimura disease, 917
Kleihauer-Betke (KB) test, 1454
Klinefelter syndrome, 856
Klippel-Trénaunay-Weber syndrome (KTWS), 1068
Kobayashi-Shigeki Iwasak's model, 897
Kwashiorkor, 470

L
Lacerations, 262
Lactating adenoma, 841
Lactiferous ducts, 834
Lafora progressive myoclonic epilepsy, 465, 466
Lambliasis, 328
Lamin A/C cardiomyopathy, 81, 82
Lamin A/C gene, 58, 59
Laminopathies, 1491
Langerhans cell histiocytosis (LCH), 179, 908, 909, 942, 943, 1151–1153, 1182, 1200, 1402
Large bowel carcinoma, 363
Large cell carcinoma, 297
Large intestine
    inflammatory disorders, 374–380
    large intestine anomalies, 353, 354, 365–372
    vascular changes, 373, 374
Large intestine anomalies
    acquired atresia and stenosis of the colon, 354, 356
    CDS, 365, 366
    colon diverticulosis, 372
    congenital atresia–stenosis of the colon, 353, 354
    constipation and chronic intestinal pseudo-obstruction, 370–372
    intestinal atresia, 358
    non-Hirschsprung-recto-colonic dysmotility syndromes, 369, 370
    recto-colonic dysmotility syndromes, 366–369
Large-vessel vasculitis, 119, 123
Laron Dwarfism, 942
Laryngeal carcinoma, 1187, 1188
Laryngeal nodule, 1186
Laryngeal web, 151
Laryngitis, 1185
Laryngocele, 1185
Laryngomalacia, 1184
Laryngotracheitis, 184
Larynx & trachea

congenital anomalies, 1183, 1184
cysts & laryngoceles, 1184, 1185
inflammatory lesions & non-neoplastic lesions, 1185, 1186
tumors, 1186–1188
Last menstrual period (LMP), 1423
Late pregnancy
acute diseases
abruptio placentae, 1450, 1452, 1455
cord integrity damage, 1457
fetus bleeding, 1454, 1455, 1457
chronic diseases
CPA, 1477, 1479, 1480
CVI and VUE, 1475–1477
DVI, 1481
FTV, 1474, 1475
MVUP, 1468, 1472–1474
villous capillary proliferation disorders, 1481
fetal growth restriction, 1482–1484
subacute diseases
AFI-C, 1459, 1461, 1462, 1464
insular RBC-progenitors' accumulation, 1467, 1468
IVT/IVTs, 1465, 1467
MAFD, 1465
Leber's hereditary optic neuropathy (LHON), 82
Lecithin or phosphatidylcholine, 1523
Left isomerism, 8
Left-ventricle non-compaction (LVNC), 78–80
Legitimate genetic syndromes, 511
Leigh syndromes, 82
Leiomyoma, 1040
Leiomyomatosis of the esophagus, 262
Leiomyomatous-like hyperplasia, 688
Leiomyosarcoma, 730, 1040–1041
Lentigines, 1381
Lentigo maligna melanoma (LMM), 1386
Lentigo Simplex, 1381
Leprosy, 921
Leptomeningeal melanomatosis, 1307
Leptomeningitis, 1277
Lesch-Nyhan syndrome/juvenile gout, 1219, 1525
Lethal congenital contractural syndromes, 1348–1351
Letterer-Siwe (LS) disease, 909
Leukemia, 1304, 1309
acute lymphoblastic leukemia, 874
acute myeloblastic (myelocytic) leukemia, 875
chronic lymphoblastic leukemia, 874, 875
chronic myelocytic leukemia, 875
hairy cell leukemia, 876, 877
hematopoietic and non-hematopoietic organs, 872
myeloid sarcoma, 875, 876
transient atypical myelopoiesis, 877
Leukemic reticuloendotheliosis, 877
Leukocyte-dysfunctionalities, 871
Leukodystrophies, 1287
Leukoplakia, 1377
Leydig cell adenoma, 975
Leydig cell tumor (LeCT), 713, 720
LH-stimulated Testosterone production, 675

Lichen planus, 1360
Lichen sclerosus, 767
Lichen Sclerosus et Atrophicus (LSA), 1359
Lichenoid interface dermatitis, 1360
Lichenoid keratosis, 1377
Liesegang rings, 839
Life-threatening defect, 33
Li-Fraumeni syndrome, 974, 996
Light chain disease, 614
Limb-body-wall complex, 1515
Linear heart, 2
Linear hyperplasia, 296
Lingual thyroid, 935, 1189
Lipid storage disorders, 327
Lipid/glycolipid and lipoprotein metabolism disorders
aspartylglycosaminuria, 467
fucosidosis, 467
Gaucher disease, 469
mannosidosis, 467
metabolic placenta, 470
mucolipidoses, 469, 470
Niemann-Pick disease, 469
sphingolipidoses, 469
sphingomyelins, 467, 469
sphingosine, 467
Lipidic metabolism, 946
Lipid-laden macrophage index (LLMI), 195
Lipoblastoma, 1012
Lipoblastomatosis, 1012
Lipo-fibromatosis, 1029
Lipofibromatosis /infantile fibromatosis, 1012
Lipoid pneumonia, 195
Lipoma, 104, 1011
Lipomatosis, 1012
Lipomatous hypertrophy of the interatrial septum (LHIAS), 104
Lipomatous pseudohypertrophy of the pancreas, *see* Restricted exocrine hypoplasia
Lipophagic reactions, 920
Littoral cell angioma, 923
Littoral cells, 862
Liver cell adenoma (LCA), 514–515
Liver cirrhosis, 508
Liver maturation defects, 1509, 1510
Liver tumors
benign tumors
angiomyolipoma, 517
bile duct adenoma, 512, 515
bile duct hamartoma, 515
biliary cystadenoma, 515
biliary papillomatosis, 515–516
cavernous hemangioma, 516
focal nodular hyperplasia, 510–514
hemangioma, 511
hepatocellular dysplasia, 515
infantile hemangioendothelioma, 511, 516
legitimate genetic syndromes, 511
liver cell adenoma, 514–515
macro-regenerative nodule, 514
mesenchymal hamartoma, 513, 516–517

Liver tumors (cont.)
    nodular regenerative hyperplasia, 510
    partial nodular transformation, 510
    peliosis hepatis, 516
    teratoma, 517
  malignant tumors
    cholangiocellular carcinoma, 529–531
    hepatic angiosarcoma, 531–533
    hepatoblastoma, 517–526
    hepatocellular carcinoma, 521, 522, 526–529
    HLMS, 533
    HRMS, 533
    rhabdoid tumor, 533, 534
    UELS, 524, 533
  metastatic tumors, 534–538
LKB1, 229
Lobar pneumonia, 190, 191
Lobular cancerization, 848
Lobular neoplasia, 849
Lobular panniculitis, 1370
Locally aggressive and malignant adipocytic tumor, 1013
Long QT syndrome (LQTS), 83, 84
Louis-Bar syndrome, 888
Low molecular weight keratin (LMWK), 950
Low-density lipoprotein receptor-related protein 5 (LRP5), 1106
Lower respiratory tract
  adult ILD of youth
    acute eosinophilic pneumonia, 182–184
    COP, 170, 172
    DAD, 167, 168, 170
    desquamative interstitial pneumonia/-ITIS, 176, 177
    HSP, 181, 182
    LIP, 180, 181
    NSIP, 175, 176
    PMF, 184
    pulmonary langerhans cell histiocytosis, 179, 180
    respiratory bronchiolitis-ILD, 177, 179
    UIP, 173, 175
  adult pneumoconiosis, 204, 205
    asbestosis, 206, 207
    non-silico-asbestosis-related pneumoconiosis, 207
    silicosis, 206
  adult-type neoplasms, childhood/youth, 222, 226–230
  chronic obstructive pulmonary disease, 202
    chronic bronchitis, 203
    emphysema, 203, 204
  congenital airways diseases, 145
    agenesis, aplasia, and hypoplasia pulmonis, 148
    CPAM and congenital lobar emphysema, 148–151
    foregut cysts, 146, 148
    sequestrations and vascular anomalies, 152–155
    tracheal and bronchial anomalies, 151
  development and genetics, 141, 144
    eparterial vs. hyparterial broncho-arterial crossing, 144, 145
    inhaled particles, 145
  infantile and pediatric ILD, 159
    chromosomal/genomic microdeletion disorders, 163
    diffuse lung development-associated ILD, 159–161
    NEHI, 163, 164
    PIG, 164
    pulmonary alveolar proteinosis, 166, 167
    surfactant dysfunction disorders, 164–166
    trisomy 21-associated ILD, 161, 163
  infantile and pediatric NILD, 155, 157
    atelectasis, 158
    cystic fibrosis, 158, 159
    neonatal respiratory distress syndrome, 157
    OBPD and NBPD, 157
  infectious pneumonia, 190
    bronchopneumonia, 191, 193
    interstitial pneumonia, 193
    lobar pneumonia, 190, 191
    primary atypical inflammation, 193
  infectious/noninfectious granulomatous/nongranulomatous, 197
    aspergillosis, 201, 202
    primary pulmonarytuberculosis, 197, 199
    sarcoidosis, 200, 201
  inflammation, 184, 188, 189
  non-infectious pneumonia, 193
    aspiration and chemical pneumonitis, 193–195
    diffuse pulmonary hemorrhagic syndromes, 196, 197
    lipoid pneumonia, 195
  nonthymic mediastinal pathology, 240
    anterior mediastinal pathology, 241
    middle mediastinal pathology, 242
    posterior mediastinal pathology, 242
  pediatric tumors and pseudotumors (see Pediatric tumors and pseudotumors)
  pleural diseases, 230, 234–237
  PTLD, 237, 238, 240
  pulmonary vascular disorders, 207
    PAH, 210, 212
    pulmonary congestion and edema, 208
    pulmonary embolism / infarction, 208
  transplantation related disorders, 212
    acute rejection, 212
    chronic rejection, 212, 215
Lower urinary system
  anomalies, 677
    diverticula, 678
    double ureters or duplication, 676
    exstrophy or congenital eversion, 678
    urachus, 678
  pre-neoplastic and neoplastic conditions
    atypia of unknown significance, 691
    carcinoma in situ, 691
    dysplasia, 691
    invasive urothelial neoplasms, 693–695, 742
    non-invasive papillary urothelial neoplasms, 691, 697, 698
    non-urothelial differentiated benign lesions, 696, 699, 743
    reactive atypia, 691

urothelial hyperplasia (flat and papillary), 690–691
Low-grade fibro-myxoid sarcoma (LGFMS), 1031–1032, 1034
Low-grade fibrosarcoma, 1136
Low-grade myo-fibroblastic sarcoma, see Low-grade fibro-myxoid sarcoma
Low-grade *vs.* high-grade serous carcinoma, 784
Low-risk appendicitis rule, 351
Lumbar renal dystopia, 591
Lung carcinogenesis, 216
Lung cytology, 239
Lung development, 142, 143
Lung maturation defects, 1508
Lupus (systemic lupus erythematodes), 1369
Lupus nephritis, 610, 611
Luteinizing hormone (LH), 934
Lutzner cells, 904, 905, 919
Lymphangiectasia, 220
Lymphangiectasis, 326
Lymphangioma, 315, 1062, 1199, 1200
Lymphangioma-(LEO)-myomatosis (LAM), 217
Lymphatic malformations, 220
Lymphoblastic lymphoma, 894, 896–900
Lymphocyte-depleted classical hodgkin lymphoma (LD-CHL), 886
Lymphocyte-rich classical hodgkin lymphoma (LR-CHL), 886
Lymphocytic gastritis, 283–285, 311
Lymphocytic lymphoma, 1153
Lymphocytopenia, 871
Lymphocytosis, 872
Lymphogranuloma venereum, 740, 917
Lymphoid follicles, 285
Lymphoid interstitial pneumonia (LIP), 180, 181
Lymphomas, 694, 695, 723, 727, 745
Lymphomatoid granulomatosis (LYG), 181, 906
Lymphoproliferative lesions, 297
Lynch syndrome, 804
Lysinuric protein intolerance (LPI), 165–166
Lysosomal storage diseases, 1520, 1521

M

Macrocephaly-capillary malformation (M-CM), 1069
Macrocephaly-cutis marmorata telangiectatica congenita (MCMTC), 1069
Macroglobulinemia, 327
Macrophage activation syndrome, 1151
Macrophage inflammatory protein 1 alpha (MIP1alpha), 911
Macro-regenerative nodule (MRN), 514
Madelung disease, 1012
Major systemic urticarial syndromes, 1363
Malabsorption
    gastrointestinal tract, 319, 320
    non-GSE causes of, 328, 330
Malakoplakia, 622, 686, 729
Malaria, 1540, 1541
Male breast, 856
Male genital system
    anomalies
        anorchia/anorchism, 679
        bifid scrotum, 683
        cryptorchidism, 680, 682
        heterotopias and heterotopia-mimickers, 683
        numerical abnormalities, 684
        paraphimosis, 683
        penis size abnormalities, 684
        phimosis, 683
        polyorchism, 683
        prostate gland agenesis/dysgenesis, 683
        TRS/VRS, 680
        urethral meatus, 683
    spermiogram and classification, 696
Malignant eccrine acrospiroma (MEAS), 1392
Malignant epithelial neoplasms, 292–294
Malignant epithelial tumors, 1178–1180, 1198
Malignant fibrous histiocytoma, 1153
Malignant hair follicle tumors, 1390
Malignant histiocytosis, 923
Malignant lymphoma, 840, 922, 964
Malignant melanoma and variants, 1203, 1385
    prognosis, 1386
    subtypes, 1386, 1387
Malignant mesothelioma, 236, 237
Malignant nephrosclerosis, 626
Malignant salivary gland tumors of oral cavity and oropharynx, 1205, 1208, 1209
Malignant tumors of epidermis, 1378–1381
Malignant fibrous histiocytoma, 1039
Mallory hyaline (MDBs), 500
Mallory-Denk body, 500
Mallory-hyaline, 459
Malrotation, 317
Mammary analogue secretory carcinoma (MASC) of salivary glands, 1208
Mammary gland
    anatomy, 834, 835
    BMPs, 834, 836
    See also Breast
Mammary gland Paget disease (MGPD), 848
Mammary-type myo-Fibroblastoma (MTMFB), 1023
Mandible & maxilla
    odontogenic cysts, 1210, 1215, 1216
    odontogenic tumors, 1216–1218
Mandibular arch, 1170
Manganese superoxide dismutase (MnSOD), 65
Mannosidosis, 467
Mantle cell lymphoma (MCL), 299, 891
Mantle small B-cells, 862
Marfan syndrome, 47, 119, 1517, 1520
Marginal abruption, 1450
Marginal cell lymphoma, 891
Marginal placental abruption, 1455
Marginal uteroplacental separation, 1479
Marker chromosome and microdeletion syndromes, 1532
Massive edema of the ovary (MEO), 764
Massive perivillous fibrin deposition (MPFD), 1477
Mast cells, 910, 919

Maternal floor infarction (MFI), 1442, 1477–1479, 1484
Maternal mortality, 1441
Maternal perivillous fibrin deposition (MPVFD), 1478, 1484
Maternal vascular under-perfusion (MVUP), 1468, 1469, 1472–1474
Maternal-based etiology of FGR, 1482
Maturation defects of cerebral cortex, 1508
Mature (post-date) babies, 1419
Mature teratoma, 395
Maxillary osteomyelitis of the newborn, 1193
Mayer-Rokitansky-Kuster-Hauser syndrome [MRKH], 688
McCune-Albright syndrome (MAS), 996
Mechanical obstruction to fetal blood flow, 1417
Meconium aspiration syndrome (MAS), 156, 1464
Meconium plug syndrome, 316
Meconium release discoloration changes of membranes and cord, 1419
Meconium-stained amniotic fluid related disorders (MAFD), 1465
Media dysplasia, 117
Medial dissection, 117
Medial fibroplasia, 117
Medial hyperplasia, 117
Medium-vessel vasculitis, 123, 125
Medullary sponge kidney (MSK), 597
Medullary thyroid carcinoma, 961, 965
Medulloblastoma, 1295, 1297, 1298
Megacystis, 687
Megacystis-microcolon-intestinal-hypoperistalsis syndrome (MMIHS), 316, 687
Megaureter, 687
Melan-A, inhibin, and calretinin (MIC), 974
Melanocytic lesions
    malignant melanoma and variants, 1385
    melanoblasts, 1381
    melanocytes, 1381
    melanocytic nevi variants, 1382
    melanophages, 1381
    Spitz Nevus and Variants, 1384–1385
Melanocytic lesions atypical melanocytic (dysplastic) nevi, 1385
Melanocytic lesions malignant melanoma and variants, 1385–1387
Melanocytic nevi, 817, 1381–1382, 1384
Melanoma, malignant, 1203
Melanotic macules, 1381
Melanotic neuroectodermal tumor of infancy (MNETI), 1183
Membranoproliferative (membrane-capillary) glomerulonephritis, 609, 610
Membranous glomerulonephritis, 606, 608, 609
Membranous ossification, 1096
Mendelian-based birth defects, 1515, 1517, 1520
Menetrier's disease, 286, 287
Meningiomas, 1299
Meningo encephalitis with purulen exsudate, 1276
Meningocele, 1251
Meningococcal sepsis, 85

Meningoencephalitis, 1280
Meningomyelocele, 1251
Menkes syndrome, 484
Menstruation, 1427
Merkel cell carcinoma, 1396, 1399
Merkel cell polyoma-virus (MCPyV), 1396
Mesenchymal chondrosarcoma, 1147
Mesenchymal dysplasia, 996
Mesenchymal hamartoma (MH), 513, 516–517
Mesenchymal neoplasms (without epithelial component), 1056
Mesenchymal tumors, benign, 1180
Mesenchymal chondrosarcoma, 1070
Mesenchymal-to-epithelial transdifferentiation (MET), 588
Mesenteric diverticulum, 316
Mesenteric lymphadenitis, 921
Mesonephric carcinoma, 809
Mesonephric gland remnants, 730
Mesonephric remnants, 760
Mesothelial/monocytic incidental excrescences (MICE), 108
Mesothelioma, 205, 728
Meta- and/or diaphyseal involvement, 1102
Metabolic and Degenerative changes, 351
Metabolic diseases
    juvenile paget disease, 1108, 1109
    osteomalacia, 1104
    osteoporosis, youth, 1104, 1106
    paget disease of bone, 1106, 1107
    vitamin D, 1103
    vitamin D-dependent hereditary rickets, 1104
Metabolic placenta, 470
Metabolic-related etiologies, 1482
Metabolomics, 1480
Metacentric chromosome, 1496
Metachromatic leukodystrophy, 1287
Metanephric kidney, 585
Metanephric tumors, 647, 649
Metanephrogenic blastema, 584, 585
Metanephrogenic cap mesenchyme, 586
Metaplasia, 287, 559
Metastasis, 727
Metastatic bony tumors, 1153
Metastatic lung tumor, 224, 225
Metastatic neoplasms, 1304
Metastatic tumors, 112, 115, 222
Metavinculin genes, 58
Micro epididymal sperm aspiration (MESA), 696
Microadenomas, 940
Microglandular adenosis (MGA), 855
Microglandular hyperplasia (MGA), 771
Microinvasive carcinoma, 362
Microinvasive serous carcinoma, 784
Micropapillary/cribriform architecture, 782
Microscopic polyangiitis (MPA), 125, 1005
Microvillous inclusion disease (MVID), 330, 331
Middle mediastinal pathology, 242
Middle-ear neoplasms, 1224
Midline craniospinal dimples or pits, 1252

Midostaurin, 875
Migration defects of ganglion cells, 1505–1507
Mild hyperplasia, 852
Miliarial spongiosis, 1354, 1355
Miliary small nodules pattern, 922
Mimickers, 272
Minimal deviation adenocarcinoma (MDACA), 807–808
Minimum pathology dataset, 296
Misalignment of pulmonary veins (MPV), 161
Miscarriage, 1423
Missed abortion, 1477
Missed miscarriage, 1477
Mitochondrial cardiomyopathies, 81, 83
Mitochondrial disorders, 1525
Mitochondrial DNA (mtDNA) defects, 83
Mitochondrial DNA (mtDNA) depletion syndromes (MDS), 472
Mitochondrial DNA cardiomyopathies, 70
Mitochondrial DNA depletion syndrome (mtDNA-DS), 473
Mitochondrial encephalopathy, lactic acidosis, and stroke-like episodes (MELAS), 81
Mitochondrial hepatopathies
    classification, 471
    electron transport (respiratory chain) defects, 471, 473, 474
    fatty acid oxidation and transport defects, 474–476
    secondary mitochondrial hepatopathies, 477
Mitochondriopathies (MCPs), 81, 1339–1341
Mitotic index, 967, 975
Mitral valve prolapse, 91
Mixed Cellularity Classical Hodgkin Lymphoma (MC-CHL), 886
Mixed function oxidase (MFO) system, 492
Mixed gonadal dysgenesis, 760
Mixed odontogenic tumors, 1218
Moderate hyperplasia, 852
Molecular genotyping, 1430
Molluscum contagiosum, 365
Molluscum contagioum, 1373
Mönckeberg arteriosclerosis, 117
Mondor disease, 840
Monochorionic twin placenta, 1446
Monoclonal gammopathy of undetermined significance (MGUS), 913
Monocyte-macrophage system
    hemophagocytic syndrome, 907
    histiocytic medullary reticulosis, 909
    Langerhans cell histiocytosis, 908–909
    sinus histiocytosis with massive lymphadenopathy, 908
    systemic mastocytosis, 910
    true histiocytic lymphoma, 910
Monocytopenia, 877
Monocytosis, 872
Monoiodotyrosine (MIT), 936
Monolayer teratomas, 699
Monomorphic PTLD, 338
Monosomy X0 syndrome, 42, 1488, 1529
Monozygotic twins (identical twins), 1442
Morphea, 1376
mtDNA depletion syndrome, 83, 1340
Mucinous carcinoid, 352
Mucinous carcinoma, 964
Mucinous cystadenocarcinoma, 352
Mucinous cystadenoma, 351–352, 780
Mucinous cystic neoplasms (MCN), 567, 569
Mucinous gland metaplasia, 731
Mucinous tubular and spindle cell carcinoma (MTS-RCC), 659
Mucocele/hydrops, 559, 1194
Mucocele/mucopyocele, 1173
Mucocutaneous lymph nodal syndrome/Kawasaki lymphadenitis, 921
Mucoepidermoid carcinoma (MEC), 851, 964
Mucolipidoses, 469, 470
Mucopolysaccharidosis syndromes, 1100, 1523
Mucosa-associated lymphoid tissue (MALT) lymphoma, 727
Mucosal homeostasis critical factors (MHCF), 321
Mucosal malignant melanoma, 1183
Müllerian anomalies, 688
Müllerian aplasia, 688
Mullerian cyst, 760
Mullerian differentiation, 774
Müllerian lesions, 690
Multicystic dysplastic kidney (MCDK), 594, 597
Multifactorial placental syndrome, 1468
Multifactorial-based congenital disabilities, 1533
Multi-infarct disease, 1271
Multi-nodular goiter, 947
Multiple berry aneurysms, 1274
Multiple congenital anomalies (MCAs), 941
Multiple endocrine neoplasia type 1 (MEN-I), 996
Multiple endocrine neoplasias, 960, 962
Multiple myeloma, 911, 912
Multiple sclerosis (MS), 1286, 1287
Multipotent hematopoietic stem cells, 863
Multi-visceral graft (MVTx), 336
Mural thrombi, 108
Muscle tumors, 1396
Muscular dystrophies, 1337
Muscularis propria, 675
Muscularis propria invasion, 695
Myasthenia gravis, 927, 1336–1337
MYCN gene, 985
Mycosis fungoides, 904
Mycotic aneurysms, 1274
Myelodysplastic syndromes (MDS), 878
Myeloid sarcoma, 875, 876
Myelolipoma, 995
Myeloma kidney, 625
MYO5B gene, 331
Myocardial dysfunction, 85
Myocarditis, 48
Myoclonic epilepsy-ragged red fiber (MERRF) syndrome, 81, 1340
Myocytes of chorionic blood vessels, 1419
Myoepithelial carcinoma, 1072
Myoepithelial tumor, 851

Myofibroblasts, 175
Myo-neuro-gastrointestinal encephalopathy (MNGIE), 1340
Myopathic phenotypes, 83
Myopathies of distal type (MPDs), 60
Myopathy, Encephalopathy, Lactic Acidosis, Stroke-like episodes (MELAS), 1340
Myopericytoma (MPC), 1042–1044
Myosin binding protein-C (MyBP-C), 57
Myositis ossificans, 1020, 1115–1116
Myotrophin gene, 58
Myxedema, 943
Myxo-fibrosarcoma, 1033–1034
Myxoid degeneration, 1100
Myxoid liposarcoma, 1012
Myxoid/round cell liposarcoma, 1015
Myxo-inflammatory fibroblastic sarcoma (MIFS), 1032
Myxoma, 102, 104, 1071
Myxo-papillary ependymoma, 1296

N
Nail-Patella syndrome, 617
Nasal cavity, paranasal sinuses & nasopharynx
  congenital anomalies, 1171, 1173
  inflammatory lesions, 1173, 1174
  tumors, 1174, 1178–1181
Nasal dermoid, 1171
Nasal encephalocele, 1171
Nasal glioma, 1171
Nasal polyposis, 1173–1174
Nasopharyngeal angiofibroma (NAF), 1180, 1181
Nasopharyngeal carcinoma, 1178–1180
Nasopharyngeal fibroma, 1020
National Wilms' Tumor Study (NWTS), 1085
Navajo neuro-hepatopathy, 474
Necrobiosis lipoidica, 1365
Necrobiosis lipoidica diabeticorum (NLD), 1364
Necrosis, 839
Necrotizing capillaritis, 197
Necrotizing enterocolitis, 319–322
Necrotizing external otitis, 1219
Necrotizing fasciitis, 1005
Necrotizing lymphadenitis, 916
Needle stick injury, 489
NEN III lesions, 297
Neonatal atelectasis, 158
Neonatal candidiasis, 1193
Neonatal cholestasis, 440
Neonatal hemochromatosis, 479, 482, 483
Neonatal hepatitis group (NAG)
  abnormalities of biliary tract, 446
  BASD
    ASBT, 450
    3β-δ5-C27-steroid oxidoreductase, 451
    bile acid conjugation defects, 452
    cerebrotendinous xanthomatosis/sterol 27-hydroxylase, 451
    cholesterol 7alpha-hydroxylase, 451
    clinical and biochemical features, 445

CYP7A1, 449
CYP8B1, 449
deconjugation products, 450
Δ4-3-oxosteroid-5β-reductase (AKR1D1/SRD5B1), 452
farnesoid X receptors, 450
inadequate bile flow, 445
2-methyl acyl-CoA racemase, 452
neutral pathway, 445
oxysterol 7α-hydroxylase, 451
primary physiologic functions, 449
urinary cholanoids, 451
bile canaliculus, 446
cholestatic jaundice, 456
CK7 expression, 447, 448
conjugated bilirubin, intracellular transport, 452–454
endocrine dysregulation-based neonatal hepatitis, 455–456
infection-related neonatal hepatitis, 454
non-BASD metabolic dysregulations-related neonatal hepatitis, 454–455
Stauffer syndrome, 456
Trisomy 13/18/21 syndromes, 456
Neonatal herpes simplex virus infection, 1193
Neonatal iron storage disorder, 482
Neonatal liver failure (NLF), 473
Neonatal osteosclerotic dysplasias, 1102
Neonatal pemphigus vulgaris, 1194
Neonatal respiratory distress syndrome, 157
Neontal HSV-liver failure, 506
Neoplasms, 189, 559
Neoplastic disorders, penis
  non-squamous cell carcinomas, 744
  penile cysts, 736, 737, 741
  penile non-invasive squamous lesions, 738, 741
  PeSCCY, 739, 741, 743, 744
Neoplastic leukocytosis, *see* Leukemia
Neoplastic pericardial disease, 116
Neoplastic pleural diseases, 235–237
Neoplastic transformation of nevus sebaceous of Jadassohn (NSJ), 1380
Neovagina, 767
Nephroblastoma/Wilms' tumor, 1085
Nephrocalcinosis, 624, 625
Nephrogenic adenoma/metaplasia, 689, 731
Nephrolithiasis, 624
Nephron differentiation, 586
Nephropathy, 965
Nephrotoxic acute tubular necrosis (ATN), 623
Nesidioblastosis, 562
Neu-Laxova syndrome, 1349
Neural and neuroendocrine tumors of skin, 1396, 1399, 1400
Neural crest migration, 367–369
Neural tube defects (NTD's), 1251, 1252, 1493
Neurilemmomas/schwannomas, 1300
Neuroblastoma, 978–982, 1203, 1209
  clinical presentation, 983
  grading, 984
  in situ, 985

Index

IV-S, 985
  metastases, 984
  pediatric tumoral DDX, 984
  postnatal diagnosis, 985
  prenatal diagnosis, 985
  spontaneous regression, 985
Neuroblastoma cells, 394
Neuroblastoma IVS, 534
Neuro-developmental delay (NDD), 321
Neuroectodermal derivation, 1057
Neuroectodermal tumors, 1182, 1183
Neuroendocrine carcinoma, 364
Neuro-endocrine hyperplasia of infancy (NEHI), 163, 164
Neuroendocrine tumors, 294–297
Neurofibroma, 559, 1332
Neurofibromatosis (NF), 1265, 1266
Neurofibromatosis I (NF I), 1225
Neurogenic bladder, 688
Neurogenic muscle weakness, ataxia, retinitis pigmentosa (NARP), 82
Neuro-hypophysopathies, 943
Neuromeric model, embryonic forebrain organization according to Puellas and Rubenstein, 1246, 1247
Neuronal migration disorders, 1264
Neurophysins, 934
Neurosyphili, 1280
Neurturin (NTN), 581
Neurulation process, 1168, 1256
Neutral mucin, 266
Neutropenia, 871, 877, 909
Neutrophilia, 872
Neutrophilic eccrine hidradenitis, 1362
Nevi and angiomas, 1373
Nevi and melanocyici lesions, 1383
Nevus comedonicus, 1375
Nevus sebaceous, 1375
Nevus sebaceous (of Jadassohn) (NSJ), 1387, 1388
New bronchopulmonary dysplasia (OBPD), 158
NFATC4, 61
NHS Breast Screening Programme, 848
Niemann-Pick disease (NPD), 469
Nipple duct adenoma (NDA), 841, 843
Nipple hypertrophy, 838
Nipple introflession, 838
Nodular adenosis, 855
Nodular and diffuse infiltrates, 1363, 1364
Nodular fasciitis, 840, 1019–1020
Nodular hyperplasia, 296
Nodular lymphocyte predominant Hodgkin lymphoma (NLPHL), 886, 887, 918
Nodular melanoma, 1386
Nodular regenerative hyperplasia (NRH), 510
Nodular sclerosis classical hodgkin lymphoma (NS-CHL), 886
Nodular tenosynovitis (NTS), 1160
Non melanocytic tumors in children and youth, 1389
Non-AATD endoplasmic reticulum storage diseases, 464
Non-achalasia acquired anomalies, 262

Non-alcoholic steato-hepatitis (NASH), 498, 499
Non-BA infantile obstructive cholangiopathies (NBAIOC), 443–444
Non-BASD metabolic dysregulations-related neonatal hepatitis, 454–455
Non-cystic congenital anomalies
  disorders of number, 589, 590
  disorders of rotation, 591
  disorders of separation, 591
  dystopia, 591
Non-dyskeratotic, non-/pauci-inflammatory disorders, 1372
Nonepithelial neoplasms, 570
Non-FTV vasculopathies, 1475
Nonfunctional adenoma, 942
Non-functional pituitary adenoma, 940
Non-Hirschsprung-recto-colonic dysmotility syndromes, 370
Non-histiocytosis/generic histiocytic proliferations, 923
Non-Hodgkin Lymphoma (NHL), 887, 888
Non-infectious pneumonia, 193
  aspiration and chemical pneumonitis, 193–195
  diffuse pulmonary hemorrhagic syndromes, 196, 197
  lipoid pneumonia, 195
Non-invasive papillary urothelial neoplasms, 691, 697, 698
Non-molar miscarriage (NMM), 1429
Non-molar pregnancy, 1430
Non-nasal NK/T-cell NHL, 302, 303
Non-nec vascular necrosis, 318
Non-necrotic vascular pathology-related phenomena, 1443
Non-neoplastic diseases, 231
Nonneoplastic lesions, 1533
Non-neoplastic lesions, 1410
Non-neoplastic leukocytosis, 871
Non-neoplastic pericardial disease, 115, 116
Non-neoplastic trophoblastic abnormalities
  EPS, 1485
  molar degeneration, 1485
  PSN, 1485
Non-neoplastic vascular pathology
  arteriosclerosis, 117
  congenital anomalies, 116
  fibromuscular dysplasia, 117
  large-vessel vasculitis, 119, 123
  medium-vessel vasculitis, 123, 125
  small-vessel vasculitis, 125
Non-occlusive, 318
Non-ossifying fibroma (NOF), 1118–1120
Non-Palatine duct cyst, 1189
Non-pediatric lung tumors, 226
Non-rotation, 317
Nonselective proteinuria, 604
Non-specific interstitial pneumonia (NSIP), 160, 175, 176
Nonspecific reactive lymphadenitis, 916
Non-squamous cell carcinomas, 744
Non-steroidal anti-inflammatory drug-induced gut lesions, 336

Non-stroma/smooth muscle differentiating tumors, 304
Non-syndromic HSCR, 367
Nonthymic mediastinal pathology, 240
    anterior mediastinal pathology, 241
    middle mediastinal pathology, 242
    posterior mediastinal pathology, 242
Nontumoral and tumoral etiology of placenta-related IUFD, 1418
Noonan syndrome, 47
Normoosteomorphic dysplasias with increased bone density, 1102
Nuchal-Type Fibroma (NTF), 1020
Nuclear receptor subfamily 0, group B, member 1 (NROB1), 938
Nummular dermatitis, 1353

O

Obstructed megaureter, 687
Obstructive disorders, 261, 280
Occipital neuroblastoma, 1305
Occlusive, 318
Odontogenesis, 1170
Odontogenic cysts, 1210, 1215
Odontogenic fibroma, 1217
Odontogenic keratocyst, 1216
Odontogenic tumors, 1216–1218
Oil Red O (ORO), 195
OLD bronchopulmonary dysplasia (OBPD), 158
Oligodendroglioma, 1294
Omphalocele, 306, 311
Oncocytoma, 656
Ontogenesis, 1412
Oophoritis, 761
Oral cavity and oropharynx
    branchial cleft cysts, 1195, 1196
    congenital anomalies, 1189, 1193, 1194
    saliva components, 1169
    salivary glands, 1169
    taste buds, 1169
    teeth, 1169
    tumors, 1197–1201, 1203
Oral choristoma, 1194
Oral glucose tolerance test (OGTT), 472
Orchitis, 697, 710
Ossifying renal tumor, 650
Osteitis, 1109
Osteochondral differentiation, 1096
Osteochondritis dissecans, 1114
Osteochondrodysplasias
    genetic skeletal disorders, 1098–1100, 1102
    nosology and nomenclature, 1097
Osteochondroma, 1137, 1138
Osteochondroma/exostosis, 1224
Osteochondromatosis, 1138
Osteoclasts, 1097
Osteo-fibrous dysplasia (OFD), 1118
Osteogenesis, 1168
Osteoid matrix, 1168
Osteoid osteoma, 1124–1126
Osteoma, 1124
Osteomyelitis (OM), 1109–1113
Osteonecrosis
    bony infarct, 1113–1114
    osteochondritis dissecans, 1114
Osteoporosis-pseudoglioma syndrome (OPPG), 1106
Osteosarcoma (OS)
    chondro-sarcomatous elements, 1128
    classification, 1128
    Codman triangle, 1132
    differential diagnosis, 1132, 1134
    genetic variations, 1132
    giant cell rich osteosarcoma, 1134, 1135
    juxtacortical, intracortical and intramedullary, 1128
    parosteal osteosarcoma (juxtacortical), 1133
    pearls and pitfalls, 1135, 1136
    periosteal osteosarcoma, 1133
    radiologic features, 1131
    small cell osteosarcoma, 1134
Ovarian dysmorphism, 760
Ovarian fibroma, 1020
Ovarian serous (cyst-) adenocarcinoma (OSCAC/OSAC), 783–784
Ovarian serous tumors, 784
Ovarian torsion, 761
Ovulation, 1427
Oxidative phosphorylation (OXPHOS), 946
Oxyphil cells, 936

P

Paget's disease, 1381
Pagus type of twin pregnancies, 1491–1493
Paleocerebellar agenesis, 1261
Pancarditis, 89
Pancreas divisum, 555
Pancreas endocrine neoplasms, 571, 573
Pancreas heterotopia, 279
Pancreas pathology
    acinar cell carcinomas, 567
    acute pancreatitis, 563–564
    chronic pancreatitis, 564–566
    congenital hyperinsulinism, 561–563
    cysts and cystic neoplastic processes, 567
    diabetes mellitus, 571–573
    nesidioblastosis, 562
    pancreas endocrine neoplasms, 571, 573
    pancreas transplant, 573–574
    pancreatoblastoma, 567, 568
    PanIN, 571
    solid pancreas ductal carcinoma, 571
Pancreas transplant, 573–574
Pancreatic intraepithelial neoplasia (PanIN), 567, 571
Pancreatic mucosa, 261
Pancreatic neoplasms, 573
Pancreatic pseudocysts (PPC), 567–571, 689
Pancreatoblastoma, 567, 568
Panniculitis, 1369, 1370
Papillary clear cell hidradenoma, 851
Papillary cystadenoma, 728

Papillary endocervicitis, 769
Papillary fibroelastoma (PFE), 103, 104
Papillary hyperplasia, 691
Papillary neoplasms, 692
Papillary thyroid carcinoma, 957–959
Papillary-polypoid cystitis, 689
Papilloma, 273, 559, 1186
Papillomatosis, 559, 843
Papovavirus (usually JC virus)-linked multifocal demyelination with oligodendroglia, 1281
Papular acrodermatitis of gianotti-crosti, 1354
Parachordoma, 1072
Paradental cyst, 1215
Paraduodenal pancreatitis, 566
Parafollicular C cells, 936
Paraganglioma, 105, 559, 964, 994, 1399, 1400
Paraganglioma of Head and Neck, Differential Diagnosis, 1399
Paraganglionic cells, 938
Paraneoplastic syndromes, 943
Paraphimosis, 683
Parathormone (PTH), 937
Parathyroid gland
    adenoma, 966
    carcinoma, 966, 967
    congenital anomalies, 965
    hyperparathyroidism, 965
    hyperplasia, 966
    PTH, 937
Parenchymal GI glands
    acute and chronic rejection post-liver transplantation, 498, 500–502
    acute viral hepatitis, 486–488
    alcoholic liver disease, 496–498
    arterio-portal hypertension, 509
    autoimmune hepatitis, 489, 490
    bacterial and parasitic liver infections, 509, 510
    bile duct development, 428–430
    biliary and pancreatic structural anomalies
        agenesis/aplasia/hypoplasia, 553
        annular pancreas, 555–556
        choledochal cysts, 553
        congenital cysts of pancreas, 556
        cystic dysplasia, 556
        ductal abnormalities, 555
        gallbladder congenital abnormalities, 553, 554
        heterotopia pancreatis, 555
        restricted exocrine hypoplasia, 553–555
    cholangiopathies (see Cholangiopathies)
    chronic viral hepatitis, 488, 489
    classical lobule, 426
    development and genetics, 551–553
    drug-induced liver disease (see Drug-induced liver disease)
    gallbladder and extrahepatic biliary tract pathology
        acute cholecystitis, 557–558
        cholelithiasis, 556–557
        cholesterolosis, 557
        chronic cholecystitis, 558
        embryonal rhabdomyosarcoma, 561
        eosinophilic cholecystitis, 558
        extrahepatic bile ducts, 561
        follicular cholecystitis, 558
        proliferative processes and neoplasms, 558–561
        xanthogranulomatous cholecystitis, 558
    granulomatous liver disease, 495, 496
    hepatic vascular disorders (see Hepatic vascular disorders)
    hepato-biliary anamolies
        ADPKD-associated cysts, 432, 433, 435
        biliary hamartoma, 435
        Caroli disease, 434, 435
        congenital hepatic fibrosis, 435
        ductal plate malformation, 431, 435
    hepatocytic nuclear size, 426
    hyperbilirubinemia (see Hyperbilirubinemia)
    immunophenotype, 427
    intrahepatic biliary system, 427–430, 432, 434
    Kupffer cells, 427
    liver biopsy, 428
    liver failure, 505–508
    liver tumors (see Liver tumors)
    non-alcoholic steato-hepatitis, 498, 499
    pancreas pathology
        acinar cell carcinomas, 567
        acute pancreatitis, 563–564
        chronic pancreatitis, 564–566
        congenital hyperinsulinism, 561–563
        cysts and cystic neoplastic processes, 567
        diabetes mellitus, 571–573
        nesidioblastosis, 562
        pancreas endocrine neoplasms, 571, 573
        pancreas transplant, 573–574
        pancreatoblastoma, 567, 568
        PanIN, 571
        solid pancreas ductal carcinoma, 571
    portal hypertension, 508
    portal lobule, 426
    Rappaport zones, 426
Parental human leukocyte antigens (HLA) sharing, 1423
Parkinson's disease and Parkinson disease-associated, G-protein-coupled receptor 37 (GPR37/PaelR) related autism spectrum disorder, 1289, 1290
Parkinsonism, 1289
Parosteal osteosarcoma (juxtacortical), 1133
Partial hydatiform mole (PHM), 1428, 1430
Partial molar degeneration, 1424, 1428
Partial nodular transformation (PNH), 510
Parvovirus B19, 1539, 1540
Parvovirus B19-induced non-immunological hydrops fetalis, 1536
Patent ductus arteriosus (PDA), 16
Paucity of intrahepatic biliary ducts (PIBD), 444, 445
Pautrier microabscesses, 903
PAX5, 878
1p36 deletion syndrome, 44
Pearson syndrome, 473
PEComa, 1084, 1085
Pectoralis muscle, 836
Pediatric and adult-type fibromatoses, 1017

Pediatric and inherited neurovascular diseases, 1274, 1275
Pediatric embryonal tumors
  clear cell sarcoma, 646
  congenital mesoblastic nephroma, 645, 646
  cystic nephroma, 645
  cystic partially differentiated nephroblastoma, 645
  metanephric tumors, 647–649
  ossifying renal tumor, 650
  rhabdoid tumor, 647
  Wilms' tumor, 634–645
  XP11 translocation carcinoma, 647, 650
Pediatric essential thrombocythemia, 870
Pediatric ILD, 159
  chromosomal/genomic microdeletion disorders, 163
  diffuse lung development-associated ILD, 159–161
  NEHI, 163, 164
  PIG, 164
  pulmonary alveolar proteinosis, 166, 167
  surfactant dysfunction disorders, 164–166
  Trisomy 21-associated ILD, 161
Pediatric liver cirrhosis, 459
Pediatric NILD, 155, 157
  atelectasis, 158
  cystic fibrosis, 158, 159
  neonatal respiratory distress syndrome, 157
  OBPD and NBPD, 157
Pediatric tumors and pseudotumors, 215
  inflammatory myofibroblastic tumor, 215
  kaposiform lymphangiomatosis, 219, 220
  LAM, 217
  metastatic tumors, 222
  pleuro-pulmonary blastoma, 220, 222
  pulmonary carcinoid, 217
  pulmonary hamartoma, 217
  pulmonary sclerosing hemangioma, 217
  pulmonary teratoma, 215
Peliosis, 923
Peliosis hepatis (PH), 504, 516
Pelvic lipomatosis, 622
Pelvic renal dystopia, 591
Pelvic-ureteral junction (PUJ) stenosis, 661
Pelvis and ureter
  genetic syndromes, 666, 667
  neoplasms, 666
  non-neoplastic pathology
    anatomy and physiology, 661
    calyceal diverticula, 661
    congenital ureteric anomalies, 661–665
    hydrocalyx/hydrocalycosis, 661
    lower urinary tract abnormalities, 662, 666
    partial duplication of pelvis and ureter, 661
    PUJ stenosis, 661
Pemphigus foliaceus, 1366
Pemphigus paraneoplasticus, 1366
Pemphigus vulgaris, 1364–1366
Pena-Shokeir syndrome, 1348
Penile cysts, 736–739, 741
Penile intraepithelial neoplasia (PeIN), 741
Penile malignancies, 744
Penile non-invasive squamous lesions, 738, 741
Penile squamous cell carcinoma of the youth (PeSCCY), 739, 741, 743, 744
Penis, 676
  infections, 739, 740
  neoplastic disorders
    non-squamous cell carcinomas, 744
    penile cysts, 736, 737, 741
    penile non-invasive squamous lesions, 738, 741
    PeSCCY, 739, 741, 743, 744
  non-infectious inflammatory diseases, 740, 741
Penis size abnormalities, 684
Pentalogy of Fallot (POF), 22
Peptic ulceration-associated diseases, 335, 336
Percutaneous epididymal sperm aspiration (PESA), 696
Periampullary myoepithelial hamartoma/adenomyoma, 342
Periarterial fibroplasia, 117
Pericardial disease
  neoplastic pericardial disease, 116
  non-neoplastic pericardial disease, 115, 116
Pericardial effusion, 115
Pericarditis, 116
Pericytic tumors
  glomangiosarcoma, 1042
  glomus tumor, 1041
  myopericytoma, 1042–1044
Periductal mastitis, 839
Perimedial fibroplasia, 117
Perinatal implantation anomalies, 1436
Perinatal mortality, 1427
Periodontal membrane by fibroblasts, 1170
Periosteal osteosarcoma, 1133
Peripheral chondrosarcoma, 1146
Peripheral nervous system (PNS)
  development, 1321, 1322
  infiltration related neuropathy, 1323
  intoxication-related neuropathies, 1323
  malignant peripheral nerve sheath tumor, 1333
  neurofibroma, 1332, 1333
  neuromuscular disorders
    congenital myopathies, 1338
    glycogen storage disorders, 1339
    histochemical stains, 1334
    infectious inflammatory myopathies, 1341
    inflammatory myopathies, 1335
    mitochondriopathies, 1339–1341
    muscle biopsy test, 1333
    muscle fiber types, 1334
    muscular dystrophies, 1337
    neurogenic disorders, 1335, 1337
    non-infectious inflammatory myopathies, 1341
    non-inflammatory myopathies, 1335
    skeletal muscle histogenesis, 1335
  peripheral neuropathy, 1322
  Schwannoma, 1323–1332
  traumatic neuropathy, 1323
  vascular neuropathy, 1323
Peripheral neuropathy, 1322
Peripheral T-cell lymphoma, 727, 902, 903

Peripheral tubular structures (PTS), 430
Perisplenitis cartilaginea, 923
Perivascular in toto (superficial and deep) dermatitis
    non-urticaria superficial and deep perivascular dermatitis, 1362, 1363
    urticaria, 1361
Perivascular space (PVS), 925
Periventricular neuronal heterotopia (PNH), 1526
Perivillous fibrin plaques, 1474
Pernicious anemia, 867, 1283
Peroxisomal disorders, 477–479
Persephin, 581
Persistent pulmonary hypertension of the newborn (PPHN), 1545, 1546
Persistent truncus arteriosus, 31, 32
Peutz-Jeghers polyp, 361
Peyronie disease, 741
PHACE(S) syndrome, 1069, 1275
Phacomatosis pigmentokeratotica, 1375
Phakomatoses, 1264–1267
Phenylketonuria, 1520
Pheochromocytoma, 988–993
Phimosis, 683
Phlegmon, 838
Phospholamban (PLN), 55
Photosensitivity-driven CADR (P-CADR), 1371
Phyllodes tumor, 851–853
Picci system, 1134
Pierre-Robin sequence (PRS), 1196
Pigmented villonodular synovitis (PVNS), 1038, 1159–1160
PIK3CA, 228
Pilar sheath acanthoma (PSA), 1388
Pilocytic astrocytoma, 1293, 1294
Piloleiomyoma, 1390
Pilomatrixoma, 1390
Pineal body tumors, 1302
Piringer-Kuchinka's lymphadenitis, 916
Pituitary gland
    congenital anomalies, 939
    cyto-diversity, 934
    Empty Sella syndrome, 943
    hyperpituitarism, 940
    hypopituitarism, 942
    hypothalamic control, 934
    neuro-hypophysopathies (*see* Neuro-hypophysopathies)
    pathology, 939
    pituitary adenomas, 939
        ACTH, 942
        classification, 940
        hyperprolactinemia, 941
        hypersecretion, 941
        MCAs, 941
        pituitary neoplasm, 941
        prolactin, 941
        tumor morphology, 941
    vascular and degenerative changes, 939
Pituitary neoplasia, 941
Pituitary thyrotropin (TSH), 936
Pityriasis alba, 1353
Pityriasis Lichenoides Et Varioliformis Acuta (PLEVA), 1357
Pityriasis rosea, 1354
Placenta accreta, 1441, 1442
Placenta development at early gestation, 1411
Placenta formation disorders, 1425
    embryo, 1423
    1st trimester specimens, 1423, 1424, 1426
    GA, 1426, 1427
    Grannum score, 1427, 1428
    neurodisability, 1423
Placenta implantation deviations of intrinsic type, 1436
Placenta implantation disorders, 1437
Placenta malaria infection, 1540
Placenta maturation disorder (PMD), 1431–1434
Placenta mesenchymal dysplasia, 1488
Placenta pathology report, abnormal coiling, 1417
Placenta praevia, 1436, 1439
Placenta vascularization disorder with chorangioma formation, 1435
Placental abruption, 1452, 1454
Placental abruption and perilesional changes with necrosis, 1454
Placental cysts, 1473
Placental development, villous tree branching, 1413
Placental diagnoses, 1410
Placental gross examination, 1548
    cut surface, 1548
    fetal surface and membranes, 1548, 1549
    light microscopy target features, 1550
    maternal surface, 1548
    synoptic data sheet, 1551, 1552
    umbilical cord, 1548, 1549
Placental infarction, 1484
Placental insufficiency, 1474
Placental lesions, 1410
    adaptability and familiarity, 1410
    associated with IUGR of fetus, 1423
    associated with preterm delivery, 1423
Placental site nodule (PSN), 1485
Placental site trophoblastic tumor (PSTT), 1486
Placental weight (PW), 1414, 1415
Placental-based etiology of FGR, 1482
Placentomegaly, 1433
Plasma cell disorders
    heavy chain disease, 912, 913
    monoclonal gammopathy of undetermined significance, 913
    multiple myeloma, 911, 912
    plasma cell luekemia, 912
    solitary myeloma, 912
    Waldenstrom's macroglobulinemia, 912
Plasma cell leukemia, 912
Plasmablastic lymphoma (PBL), 917
Plasmacytoma, 727, 964, 1153
Plasmodium infection in the erythrocytes, 1537
Platelet disorders, 870, 871
Pleiotropism, 1495
Pleomorphic adenoma of the breast (PAB), 850

Pleomorphic cells, 995
Pleomorphic hyalinizing angiectatic tumor (PHAT), 1072
Pleomorphic liposarcoma, 1015–1016
Pleomorphic RMS, 1055, 1057
Pleural diseases
 neoplastic pleural diseases, 235–237
 pleural effusions and pleuritis, 230
 pneumothorax, 234, 235
Pleural effusions, 230
Pleural neoplastic diseases, 233
Pleural pleuritis, 230
Pleuro-pulmonary blastoma, 220, 222
Plexiform fibrohistiocytic tumor, 1038
Plummer-Vinson syndrome, 262
1p36 monosomy, 44
PNET/Ewing sarcoma, 1182, 1209
PNET/extra-skeletal ewing sarcoma (ESES), 1081–1083
Pneumoconiosis, 205
Pneumocystis jerovici, 239
Pneumothorax, 234, 235
POEMS syndrome, 911, 912
Poland syndrome, 836
Poliomyelitis, 1281
Politelia, 836
Pollard willow pattern, 428
Polyarteritis nodosa (PAN), 123, 125, 1005
Polycystic change, 763
Polycythemia
 primary, 867, 868
 secondary, 868, 869
Polycythemia vera, *see* Primary polycythemia
Polyembryoma, 717, 779
Polyendocrinopathy, 331
Polymastia, 836
Polymorphic PTLD, 338
Polymorphic reticulosis, *see* Lymphomatoid granulomatosis
Polymorphous hemangioendothelioma (PHE), 1062, 1063
Polyorchism, 683
Polypoid mucosal prolapse, 292
Polyposis vaginalis of pregnancy, 769
Pompe's disease, 72
Porphyria cutanea tarda (PCT), 485
Porphyria-related hepatopathies, 485
Portal hypertension (PHT), 508
Posterior mediastinal pathology, 242
Posterior urethral valves (PUV), 666, 687, 688
Post-infectious disruption of the pituitary stalk, 943
Post-infectious glomerulonephritis, 600–602
Post-mortem urothelial sloughing, 680
Postnatal-onset male hypogonadism, 679
Post-operative spindle (cell) nodule (POSN), 690, 769, 811
Postpartum, pregnancy, 1422
Post-transplant lymphoproliferative disoders (PTLDs), 100, 112, 237, 238, 240, 291, 500, 895, 896
Post-traumatic disruption of the pituitary stalk, 943
Postvaccinial viral lymphadenitis, 919
Potter´s microdissection studies, 599

PPP3CA, 61
PPP3CB, 61
Pre- and post-fundoplication motility disorder, 262
Precocious thelarche, 838
Predominant granulomatous pattern
 atypical mycobacteriosis, 920
 chronic granulomatous disease, 921
 fungal infections, 920
 sarcoidosis lymphadenopathy, 920
 tuberculosis, 920
Predominantly pediatric fibroblastic/myofibroblastic lesions, 1017
Preeclampsia, 1450
Pregnancy
 complications, 1421, 1422
 infections, 1533–1535, 1538–1541
 luteoma, 764
 outcome for conceptus, 1410
Pregnancy-associated cervical lesions, 770–771
Pregnancy-related liver disease (PLD), 459, 460
Premature rupture of membranes (PROM), 1541
Premature/breech delivery, 1417
Prematurity–observable CNS disorders, 1270
Prematurity complications with brain hemorrhage, 1543
Prematurity-related complications, 1542
Prenatal echocardiography of thoracic-abdominally conjoined twins, 1492
Preservation injuries, 337
Preterm birth, 1423
Primary (essential) thrombocythemia, 870, 871
Primary adrenal extrarenal Wilms tumor, 995
Primary atypical inflammation, 193
Primary biliary cirrhosis (PBC), 458, 459
Primary empty Sella syndrome, 943
Primary glomerular diseases
 allergy-related (immediate) hypersensitivity, 599
 clinical syndromes, 600
 cytotoxic hypersensitivity, 599
 delayed hypersensitivity, 600
 (diffuse) mesangial proliferative GN, 604–606
 focal & segmental glomerulosclerosis, 604, 606, 607
 immune complex hypersensitivity, 599
 membranoproliferative (membrane-capillary) glomerulonephritis, 609, 610
 membranous glomerulonephritis, 606, 608, 609
 minimal change disease, 604
 pathogenic mechanisms, 599
 post-infectious glomerulonephritis, 600–602
 rapidly progressive glomerulonephritis, 602–604
Primary hyperoxalurias, 624
Primary hyperparathyroidism, 965
Primary hypogonadism, 678
Primary malignant melanoma, 995
Primary ovarian choriocarcinoma, 779
Primary pigmented nodular adrenocortical disease (PPNAD), 997
Primary polycythemia, 868
Primary pulmonary tuberculosis, 197, 199
Primary sclerosing cholangitis (PSC), 456–458
Primitive neuroectodermal tumor (PNET), 185, 896

Primordial form dysostotic dwarfism, 1100
PRKAR1A, 996
Probst classification, 1256
Progenitor mesenchymal cells, 1096
Progressive familial intrahepatic cholestasis (PFIC-1-3), 452, 453
Progressive massive fibrosis (PMF), 184
Progressive multifocal leukoencephalopathy (PML), 1281
Progressive systemic sclerosis, 262
Progressive transformation of germinal centers (PTGC), 887, 918
Prolactin, 941
Prolamins, 323
Proliferationsknospen, 954
Proliferative fasciitis/myositis, 1019
Prolonged stillbirth, 1433
Prolymphocytic CLL, 875
Prosencephalon defects, 1252–1254, 1496–1498
Prostate, 675–676
Prostate gland
    acute prostatitis, 728
    atrophy, 731
    atypia, 731
    benign nodular hyperplasia, 723, 729
    chronic granulomatous disease, 729
    chronic prostatitis, 728
    Cowper glands, 730
    hematological malignancies, 739
    hyperplasia
        adenomatous, hyperplasia, 732
        adenosis, sclerosing, 731
        basal cell hyperplasia, 732
        BNH/PNH-small gland type, 732
        clear cell cribriform hyperplasia, 732
    inflammation, 731
    leiomyosarcoma, 730
    malakoplakia, 729
    mesonephric gland remnants, 730
    mimickers, 729, 730
    mucinous gland metaplasia, 731
    nephrogenic adenoma / metaplasia, 731
    paraganglion, 730
    PNET or Ewing sarcoma, 730
    prostatic intraepithelial neoplasia, 730, 732, 733
    rhabdomyosarcoma, 725, 726, 729
    secondary tumors, 739
    seminal vesicles/ejaculatory ducts, 730
    Veru montanum mucosal gland hyperplasia, 731
    young, 733–739
Prostate gland agenesis/dysgenesis, 683
Prostatic carcinoma mimickers, 727
Prostatic intraepithelial neoplasia (PIN), 730, 732, 733
Prostatic-type polyp, 690
Prosthetic valves, 92
Proteus syndrome, 1068, 1375
Protracted conjugated hyperbilirubinemia, 440
PRSS1-related hereditary pancreatitis, 564
Prune Belly syndrome (PBS), 662, 682, 687
Pruritic urticarial papulae and plaques of pregnancy (PUPPP), 1362

Pseudoangiomatous hyperplasia (PASH), 840, 967
Pseudo-aprosencephaly, 1256
Pseudocarcinomatous hyperplasia, 690
Pseudocysts, 569, 1374
Pseudo-diastrophic dysplasia (PDD), 1100
Pseudo-epitheliomatous hyperplasia (PEH), 1376
Pseudofollicles *vs.* follicles, 890
Pseudogout, 1159
Pseudolymphomas, 303, 1400
"Pseudosarcomatous" skeletal tissue, 1104
PSIVS/PAIVS, 24, 26
Psoriasiform dermatitis, 1360, 1361
Psoriasis, 1360
4p-syndrome, 44
5p- syndrome, 46, 1488
PTH-related peptide, 1096
Pubertal maturation, 678
Pulmonary alveolar proteinosis, 166, 167
Pulmonary arterial hypertension (PAH), 210, 212
Pulmonary atresia with VSD (PAVSD), 24
Pulmonary carcinoid, 217
Pulmonary congestion, 208
Pulmonary edema, 208
Pulmonary embolism/infarction, 208
Pulmonary hamartoma, 217
Pulmonary homografts, 92
Pulmonary hyperinflation, 188, 189
Pulmonary hypertension, 21, 211
Pulmonary interstitial glycogenosis (PIG), 164
Pulmonary Langerhans cell histiocytosis, 179, 180
Pulmonary mycotoxicosis, 207
Pulmonary sclerosing hemangioma, 217
Pulmonary teratoma, 215
Pulmonary vascular disorders, 207
    PAH, 210, 212
    pulmonary congestion and edema, 208
    pulmonary embolism / infarction, 208
Pure gonadal dysgenesis, 760
Purkinje cell hamartoma, 108
Pyogenic abscess, 509–510
Pyogenic granuloma (lobular capillary hemangioma), 1059, 1060, 1199
Pyonephrosis, 618

Q

22q11.2 microdeletion syndrome, 44
Queyrat's erythroplasia, 741
Quilty lesions, 92

R

Rabies, 1281
Rachischisis, 1493
Radial alveolar count (RAC), 141, 144
Radial scar (RS), 855
Radiation cystitis, 686
Radiation enteritis, 336
Radiation nephropathy, 626
Radicular cyst, 1215

Ranula, 1194
Rapidly progressive glomerulonephritis, 602–604
Rare exocrine neoplasms, 570
Rare non-cardiac benign tumors, 109
Rare vesiculobullous disease of newborns, 1194
Rastelli repair, 31
Reactive atypia, 691
Reactive follicular hyperplasia, 922
Reactive inflammatory cells, 238
Reactive non-follicular lymphoid hyperplasia, 922
Reciprocal translocation, 1496
Recklinghausen's disease, 965
Rectal biopsy, 370
Recto-colonic dysmotility syndromes, 366–369
Recurrent respiratory papillomatosis, 275
Red blood cell disorders
    anemia, 864
        acquired hemolytic anemias, 866
        CLI, 864
        congenital hemolytic anemias, 866
        definition, 864
        failure of blood production, 866, 867
        hemorrhage-associated anemias, 864
    primary, 867, 868
    secondary, 868, 869
Red pulp disorders, spleen
    congestion, 922
    histiocytosis, 923
    hyaline perisplenitis, 923
    infection, 922
    inflammatory pseudotumor, 923
    leukemia, 923
    littoral cell angioma, 923
    non-histiocytosis, 923
    peliosis, 923
    spontaneous rupture, 923
Reed-Sternberg cells (RSC), 878, 883
Reed-Sternberg variants (RSV), 878
Reese-Ellsworth classification, 1229
Reflux esophagitis, 263, 265–267
Relapsing polychondritis, 1219
Remodeled bile ducts (RBD), 428, 429
Remodeling ductal plate (RDP), 428
Renal and ureteral congenital defects, 589
Renal artery stenosis, 626
Renal cystic diseases
    acquired cystic kidney disease, 598
    ADSPKD, 591, 596
    autosomal recessive polycystic kidney disease, 596, 597
    classifications, 591, 596
    cystic renal changes, 593
    encephalocele, 593
    fibrocystin, 592
    genetic syndromes, 598, 599
    hydronephrosis, 597
    hydro-ureteronephrosis, 597
    medullary sponge kidney, 597
    multicystic dysplastic kidney, 597
    multicystic kidney dysplasia, 594
    nephrocystin-1, 593
    nephromegaly, 593
    polycystin-1, 592
    polycystin-2, 592
    segmental renal dysplasia, 595
    simple renal cysts, 598
Renal dysplasia, 591, 598
Renal morphogenesis, 581
Renal papillary necrosis, 618–622
Renal pathology in multiple myeloma, 911
Renal transplantation
    acute, 631, 632
    anti-rejection drug toxicity, 633
    chronic, 632
    humoral, 632, 633
    hyperacute, 630
    preservation injury, 630
Renal tubular dysplasia, 1504
Rendu-Osler-Weber disease (ROWD), 155
Renomedullary interstitial cell tumor, 660
Resorption defects, 1506, 1507
    of allantois, 1507
    of urachus-anomalies, 1507
Respiratory bronchiolitis, 177–179
Restricted exocrine hypoplasia, 553–555
Restrictive cardiomyopathy, 78
Reticulin (Ret) annular rings ("Ret Rings"), 862
Reticulohistiocytoma, 1037
Retiform hemangioendothelioma (RHE), 1062, 1063
Retinoblastoma (RB) and related lesions, 1227, 1230
Retro-placental hematoma, 1454
Rhabdoid tumor, 533–534, 647, 698
Rhabdomyoma, 105, 1046
Rhabdomyomatous mesenchymal hamartoma (RMH), 1044–1046
Rhabdomyosarcoma (RMS), 109, 717, 718, 720, 721, 725–729, 1203, 1209
    alveolar pattern, 1047
    alveolar RMS, 1053–1055
    botryoid RMS, 1048
    diagnosis, 1047, 1048
    embryonal RMS, 1048
    Horn and Enterline's classification, 1046
    incidence, 1047
    pearls and pitfalls, 1057, 1058
    pleomorphic RMS, 1055, 1057
    spindle cell RMS, 1053
    unfavorable prognosis, 1047
Rhabdovirus, 1281
Rheumatic fever, 51
Rheumatic heart disease (RHD), 91
Rheumatoid arthritis, 916, 1154–1157
Rheumatoid nodule, 1364
Rhinitis, 1173
Rhombencephalon-synapsis, 1260
Richter's transformation, 890
Rickets, 1105
Ridson-Hill lesions, 652
Riedel's thyroiditis, 949
Riga-Fede disease, 1194
Right isomerism, 8

Right middle lobe syndrome (RMLS), 187, 188
Ring chromosome, 1496
Risk pregnancies, 1269
Robertsonian translocation, 1496
Rokitansky-Aschoff sinuses, 428
Rolling (paraesophageal) a hernia, 261
Rosai-Dorfman disease, *see* Sinus histiocytosis with massive lymphadenopathy (SHML)
Rosenthal fibers, 1287, 1288
Rotor syndrome, 438
Rubeola embryopathy, 48
RUNX1 gene, *see* AML1 gene
Russell bodies, 889, 890, 911, 916

S

Saccular cysts, 1184
Sacrococcygeal teratoma, 1085–1087, 1089, 1512
Salivary glands
    embryology and anatomy, 1169
    pathology
        benign, 1205
        congenital anomalies, 1203
        inflammatory lesions & non-neoplastic lesions, 1203
        malignant, 1205, 1208, 1209
    tissue, 261
Sanfilippo cardiomyopathy, 74
Santorini/accessory duct, 555
Sarcoid, 1364
Sarcoidosis, 71, 200, 201, 310, 1174
Sarcoidosis lymphadenopathy, 920
Sarcomatoid carcinoma, 1332
Sarnat score, 1544
Scar *vs.* hyperplastic scar *vs.* Keloid, 1376
Schatzki's rings, 261
Schaumann bodies, 920
Schistosomiasis, 327
Schwannoma, 1300, 1323–1332
Schwartz-Batter syndrome/syndrome of inappropriate secretion of ADH (SIADH), 943
Scimitar syndrome, 32, 33
Sclerosing adenosis (SA), 855
Sclerosing epithelioid fibrosarcoma (SEFS), 1034–1036
Sclerosing mucoepidermoid carcinoma with eosinophilia, 964
Sclerosing RMS (Scl-RMS), 1055
Sclerosing stromal tumor, 797
Sclerosisng cholangitis, 502
Sclerotherapy/banding therapy, 454
SCN5A gene, 58
Sebaceoma/sebaceous epithelioma, 1388
Sebaceous adenoma, 1388
Sebaceous carcinoma, 1388
Sebaceous hyperplasia, 1387
Seborrheic keratosis, 1377
Secondary (metastatic) tumors, 965
Secondary empty Sella syndrome, 943
Secondary glomerular diseases
    amyloidosis, 613, 614
    cryoglobulinemia, 614
    diabetic nephropathy, 614, 615
    Henoch-Schönlein purpura, 611–613
    light chain disease, 614
    SLE/lupus nephritis, 610, 611
Secondary hypogonadism, 679
Secondary mitochondrial hepatopathies, 477
Secondary motility disorders, 262
Secondary tumors, 304, 1183
Segmental arterial mediolysis, 626
Segmental dilatation, 317
Segmental renal dysplasia, 595
Sella turcica, 1250
Sella turcica-tumors, 942
Seminal vesicles/ejaculatory ducts, 730
Seminoma, 702, 706
Seminoma with syncytiotrophoblastic cells, 706
Sepsis-carditis, 90
Septal panniculitis, 1369–1370
Septation, 2
Septic infarcts, 1462
Septic shock erosive gastritis, 282
Septic thrombosis/thrombophlebitis, 1279
Septum interventriculare cordis, 13
Sequential segmental analysis (SSA), 8
Sequential segmental cardio-analysis, 8–11
Sequestrations, 152, 153
Serous carcinoma, 781
Serous cystoadenoma, 780
Serous microcystic adenoma (SMAP), 567
Serous micropapillary carcinoma, 784
Serous pericarditis, 116
Serous psammoma carcinoma, 784
Sertoli cell tumors (SeCT), 714, 722
Sertoli cells, 675
Sertoli-Leydig cell tumor, 793–796
Serum sickness-like CADR (SSLR), 1372
Severe mucocutaneous syndromes, 1357
Sex cord-stromal tumors (SCSTs)
    with annular tubules, 797
    classification, 790
    fibroma/thecoma group, 796, 797
    granulosa cell tumor, 791–793
    gynandroblastoma, 797
    IHC markers, 790
    sclerosing stromal tumor, 797
    sertoli-Leydig cell tumor, 793–796
    steroid/lipid cell rich tumor-group, 796
Sex hormones-secreting adenoma, 975
Sex organ development, 674
Sexual development disorder, 679
Sexual intercourse, 1439
Sezary syndrome, 905
Sheehan's syndrome, 942
Shock-related cholestasis, 505
Short QT syndrome (SQTS), 84
Shwachman-Diamond syndrome (SDS), 485
Sialidosis, 468, 1521–1523
Sialoblastoma, 1208
Siamese twins, 1491

Signet ring cell cytology, 849
Signet ring lymphoma, 889
Silent miscarriage, 1477
Silicosis, 184, 204
Simmond's disease, *see* Hypopituitarism
Simple bone cyst (SBC), 1120
Simplified Scheuer system (SSS), 489
Single nucleotide polymorphisms (SNP), 54
Sinonasal papilloma, 1178
Sinus histiocytosis, 919
Sinus histiocytosis with massive lymphadenopathy (SHML), 908
Sinus pattern, 920
Sipple syndrome, 962
Situs, 8
Situs ambiguous, 41
Situs inversus, 8, 41
Situs solitus, 8, 41
Sjogren disease, 916
Sjogren syndrome, 888
Skeletal muscle alpha-actin gene (ACTA1), 64
Skeletal muscle tumors
    rhabdomyoma, 1046
    rhabdomyomatous mesenchymal hamartoma, 1044
    rhabdomyosarcoma
        alveolar pattern, 1047
        alveolar RMS, 1053–1055
        botryoid RMS, 1048
        diagnosis, 1047, 1048
        embryonal RMS, 1048
        Horn and Enterline's classification, 1046
        incidence, 1047
        pearls and pitfalls, 1057, 1058
        pleomorphic RMS, 1055, 1057
        spindle cell RMS, 1053
        unfavorable prognosis, 1047
    RMH, 1044–1046
Skin
    CADR's, 1370–1372
    cutaneous appendages disorders, 1369
    cutaneous cysts and lesions, 1372
    development, 1347, 1348
    dyskeratotic, non-/pauci-inflammatory disorders, 1372
    fibrous and fibrohistiocytic tumors, 1394, 1396
    hematological skin infiltrates, 1400
    infections and infestations, 1372
    interface dermatitis
        lichenoid interface dermatitis, 1360
        vacuolar interface dermatitis, 1355, 1357–1360
    intraepidermal blistering diseases, 1364, 1366, 1367
    lethal congenital contractural syndromes, 1348–1352
    malignant tumors, 1378–1381
    melanocytic lesions, 1381–1387
    neural and neuroendocrine tumors, 1396, 1399, 1400
    nodular and diffuse infiltrates, 1363, 1364
    non-dyskeratotic, non-/pauci-inflammatory disorders, 1372
    panniculitis, 1369, 1370
    PID, 1361–1363
    psoriasiform dermatitis, 1361
    psoriasis, 1360
    sebaceous tumors, 1387, 1388, 1390
    solid tumor metastases, 1400
    spongiotic dermatitis
        conventional spongiosis, 1352–1354
        eosinophilic spongiosis, 1354
        follicular spongiosis, 1354
        milarial spongiosis, 1354, 1355
    sub-epidermal blistering diseases, 1367–1369
    Sweat gland (eccrine and apocrine) tumors, 1390, 1391, 1393, 1394
    tumors of adipose tissue, muscle, cartilage and bone, 1396
    tumors of epidermis, 1373, 1375–1381
    vascular tumors, 1396
    vasculitis, 1369
Skin lesions, types, 1348
Skin patterns, types, 1348
Skin transplantation in genodermatosis, 1350
Skull, 1230
Sliding hernia, 261
Small bowel atresia, 305
Small cell carcinoma, 277
Small cell carcinoma/poorly differentiated NET, 297
Small cell osteosarcoma (SCOS), 1134
Small intestinal neoplasms, 340–342
Small intestine, 304
    anomalies, 304, 306
        abdominal wall defects, 306, 312
        infantile colic, 304
    dystopias, 317
    GVHD, 339, 340
    inflammation and malabsorption, 319, 320, 322–336
    intestinal lumen, continuity defects of, 312, 316
    intestinal muscular wall defects, 316, 317
    intestinal wall, composition abnormalities of, 317, 318
    short bowel syndrome/intestinal failure, 336
    small intestinal neoplasms, 340–343
    small-intestinal transplantation, 336–339
    vascular changes, 318
Small intracytoplasmic vascular lumina, 1066
Small lymphocytic lymphoma, 890, 891
Small-intestinal transplantation (SITx), 336–339
Small-vessel vasculitis, 125
Smith-Lemli-Opitz syndrome, 1515, 1517
Smooth muscle actin (SMA), 303
Smooth muscle differentiating tumors, 303, 304
Smooth muscle tumors, 304
Soft tissue
    adipocytic tumors, 1010
        atypical lipomatous tumor/well-differentiated liposarcoma, 1014
        dedifferentiated liposarcoma, 1016
        hibernoma, 1012–1013
        lipoblastoma, 1012
        lipoma, 1011
        lipoma subtypes, 1011–1012
        lipomatosis, 1012

locally aggressive and malignant, 1013
  myxoid/round cell liposarcoma, 1015
  pleomorphic liposarcoma, 1015–1016
 cell of origin, 1010
 chondro-osteoforming tumors
  extraskeletal aneurysmatic bone cyst, 1070
  extraskeletal chondroma, 1069, 1070
  extraskeletal osteosarcoma, 1070
  mesenchymal chondrosarcoma, 1070
 classification, 1010
 development and genetics, 1004
 fibroblastic/myofibroblastic tumors (see fibroblastic/myofibroblastic tumors)
 fibrohistiocytic tumors
  borderline fibrous histiocytoma, 1039
  cellular fibrous histiocytoma, 1038–1039
  dermatofibroma, 1037
  DTGCT, 1038
  giant cell tumor of tendon sheath, 1038
  histiocytoma, 1037
  malignant fibrous histiocytoma, 1039
 pericytic tumors
  glomangiosarcoma, 1042
  glomus tumor, 1041
  myopericytoma, 1042, 1043
 skeletal muscle tumor (see Skeletal muscle tumors)
 smooth muscle tumors
  EBV-SMT, 1040, 1041
  leiomyoma, 1040
  leiomyosarcoma, 1040–1041
 uncertain differentiation
  alveolar soft part sarcoma, 1076, 1078, 1079
  clear cell sarcoma, 1079–1080
  desmoplastic small round cell tumor, 1083–1084
  epithelioid sarcoma, 1076, 1077
  extragonadal germ cell tumor, 1085–1089
  extragonadal yolk sac tumor, 1085–1087, 1089
  extrarenal rhabdoid tumor, 1084
  extraskeletal myxoid chondrosarcoma, 1080–1081
  myoepithelial carcinoma, 1072
  myxoma, 1071
  nephroblastoma/Wilms' tumor, 1085, 1086
  parachordoma, 1072
  PEComa, 1084, 1085
  PNET/extra-skeletal ewing sarcoma, 1081–1083
  sacrococcygeal teratoma, 1085–1087, 1089
  soft tissue malignant mesenchymoma, 1084
  synovial sarcoma, 1074–1076
 vascular and inflammatory changes
  hyperemia, 1005
  iatrogenic/sporadic with or without congenital character, 1005–1008
  necrotizing fasciitis, 1005
  vasculitis-associated soft tissue changes, 1005
 vascular tumors (see Vascular tumors)
Soft tissue malignant mesenchymoma (STMM), 1084
Soft tissue tumors, 840, 1199
Soft tissue tumors of salivary gland, 1209
Solar lentigo, 1381

Solid (and cystic) pseudopapillary tumor of the pancreas, 570–572
Solid pancreas ductal carcinoma, 571
Solid tumor metastases to skin, 1400
Solitary fibrous tumor (SFT), 235, 1029
Solitary myeloma, 912
Solitary/multiple large fleshy nodules pattern, 922
Solitary/several small clustered nodules pattern, 922
Solitary thyroid nodule, 951
Somatotropin, 934
SOX2, 229
Soy protein reaction, 327
Specialized gonadal stroma
 gonadoblastoma, 723
 granulosa cell tumor, 715, 716
 Leydig cell tumor, 713, 720
 rhabdoid tumor of testis, 719, 727
 sertoli cell tumors, 714, 722
Spemann's microtools, 1491
Sperm chromatin structure assay (SCSA), 696
Spermatocele, 700
Spermatocytic seminoma, 707
Spermatogenesis, 675
Sperm-cervical mucus contact (SCMC) test, 696
Sphincter of Oddi, 555
Sphingolipidoses, 469, 1523
Sphingomyelins, 467, 469, 1523
Sphingosine, 467
Spina bifida, 1251
Spinal cord, 1246
Spinal injuries, 1285, 1286
Spindle and epithelioid cell tumor with thymus-like elements of differentiation (SETTLE), 964
Spindle cell (sarcomatoid), 1188
Spindle cell carcinoma, 276
Spindle cell hemangioendothelioma (SHE), 1062, 1063
Spindle cell melanoma, 1332
Spindle cell RMS (S-RMS), 1053
Spindle epithelial tumor with thymus-like elements (SETTLE), 927
Spitz nevus and variants, 1384, 1385
Spleen disorders
 red pulp disorders (see Red pulp disorders, spleen)
 white pulp disorders (see White pulp disorders, spleen)
Splenic marginal zone lymphoma (SMZL), 877, 892
Spongiosis (intercellular edema), 1354
Spongiotic dermatitis, 1353
Spontaneous abortion, 1423
Spontaneous intestinal perforation (SIP), 321
Spontaneous miscarriages, 1480
Spontaneous rupture, 923
S/P prolonged/repetitive hypoxia, 1467
Squamous cell carcinoma, 275, 277, 964
Squamous intraepithelial lesions (SILs), 1189
Squamous metaplasia, 559, 690
Squamous papilloma, 275
Squamous preneoplastic and neoplastic lesions, 1198
"Staghorn-like" megakaryocytes, 871
Starting miscarriage, 1477

Stasis dermatitis, 1353
Stauffer syndrome, 456
Stenosis, 316
Steroidogenic factor 1 (SF-1), 938
Stevens-Johnson syndrome (SJS), 1357, 1358
Stomach, 256, 277
    gastric anomalies, 277, 278, 280
    gastric inflammation, 281, 284–286
    gastric tumors, 287, 292–304
    tissue continuity damage-related degenerations, 286, 287
    vascular changes, 280, 281
Straddling, 10
Stress ulcers, 286
Stroma/Schwann cell differentiation, 984
Stroma tumors, 303, 304
Stromal hyperplasia, 764
Stromal hyperthecosis, 764
Stromal microinvasion, 782
Stromal sarcoma, 840
Strongyloidiasis, 326
Sturge-Weber syndrome (SWS), 1266
Subacute hypoxia, 1464
Subacute sclerosing panencephalitis (SSPE), 1281
Subarachnoidal hemorrhage (SAH), 1270, 1285
Subdiaphragmatic renal dystopia, 591
Subdural hematoma, 1284, 1285
Subdural hemorrhage (SDH), 1270
Sub-epidermal blistering diseases
    bullous pemphigoid, 1367
    epidermolysis bullosa, 1368
    erythema multiforme and toxic epidermal necrolysis, 1368
    Hb-related porphyria cutanea tarda, 1368–1369
    herpes gestationis, and dermatitis herpetiformis, 1368–1369
    lupus (systemic lupus erythematodes), 1369
Submetacentric chromosome, 1496
Subtle bilirubin encephalopathy (SBE), 437
Succinate dehydrogenase, 69
Sudden cardiac death, 83–85
Sudden catastrophic uteroplacental separation, 1450
Sulfated musin, 266
Sulfation disorder, 1100
Superficial fibromatosis, 1029, 1030
Superficial spreading melanoma, 1386
Supernumerary ovaries, 760
Suppurative infections of CNS
    acute suppurative leptomeningitis, 1276, 1278
    brain abscess, 1276, 1278, 1279
    septic thrombophlebitis, 1276, 1279
Suppurative/granulomatous Lymphadenitis, 920
Supraaortic metastasis, 115
Supra-sellar expansion, 941
Surface epithelial tumors
    Brenner tumors, 789, 790
    clear cell (mesonephroid) tumors, 788, 789
    endometrioid tumors, 787, 788
    mucinous tumors, 785–787
    serous tumors
    benign mimics, 783
    borderline (proliferative) tumors (LMP), 782–783
    cystadenofibroma, 784
    cystadenoma/cystadenofibroma, 780–782
    grading, 780
    IHC, 780
    low-grade vs. high-grade serous carcinoma, 784
    microinvasive serous carcinoma, 784
    micropapillary/cribriform architecture, 782
    ovarian implants, 783
    ovarian serous (cyst-) adenocarcinoma (OSCAC/OSAC), 783, 784
    ovarian serous tumors, 784
    serous micropapillary carcinoma, 784
    serous psammoma carcinoma, 784
    site of origin, 780
    TP53 gene mutational status and gene product expression (p53 IHC), 784
Surfactant dysfunction disorders, 164–166
Surfactant protein C deficiency (*SFTPB*), 165
Sweat gland (eccrine and apocrine) tumors, 1390, 1391, 1393, 1394
Symmetrical conjoined twins, 1493
Symplastic glomus tumor, 1042
Syncytiotrophoblast (STB), 1414
Syndromic diarrhea of infancy (SDI), 330–332
Syndromic HSCR, 366
Synmastia, 836
Synovial sarcoma, 109–110, 1074–1076, 1332
Syphilis, 740, 917
Syringocystadenoma papilliferum, 1393
Syringomas, 1391
Syringomatous adenoma of nipple (SAN), 843
Syringomyelia, 1289
Syringomyelocele, 1251
Systemic lupus erythematosus (SLE), 916
Systemic mastocytosis, 910
Systemic necrotizing vasculitis (SNV), 1005
Systemic/constitutional symptoms, 101

T
Takayasu aortitis, 122
Takayasu arteritis, 119, 123
Tamponade, 116
Tartrate-resistant acid phosphatase (TRAP), 877
Taussig-bing anomaly, 11
Tax, 874
Tay-Sachs disease, 1523
T-cell CLL, 874
T-cell/histiocyte-rich large B-cell lymphoma (THRLBCL), 887
TDLU intralobular stroma, 834
Telencephalosynapsis (synencephaly), 1259, 1260
Tendon sheath fibroma, 1020
Tentorium cerebelli, 1246
Teratogenesis, 1448, 1511
Teratoma, 111, 517, 776, 777, 964, 1514
Teratoma with glia and nervous tissue, 1513
Terminal duct-lobule units (TDLUs), 834

Terminal villous deficiency, 1434
Testicular regression syndrome (TRS), 679
Testicular septa, 675
Testicular sperm extraction (TESE), 696
Testicular teratomas, 713–716
Testicular tissue, 676
Testicular tumors
   germ cell tumors
      anaplastic seminoma, 707
      embryonal carcinoma, 703, 711–713
      endodermal sinus tumor, 715–717
      ITGCN, 705–706
      polyembryoma, 717
      seminoma, 702, 706
      seminoma with syncytiotrophoblastic cells, 706
      spermatocytic seminoma, 707
      testicular teratomas, 713–716
      yolk sac tumor, 715–717
   metastasis, 727
   rhabdomyosarcoma, 717, 718, 727
   specialized gonadal stroma
      gonadoblastoma, 723
      granulosa cell tumor, 715, 716
      leydig cell tumor, 713, 720
      sertoli cell tumors, 714, 722
Testis, 675
   epidermoid cysts, 697, 699
   hydrocele, 700
   orchitis, 697, 710
   rhabdoid tumor, 719, 727
   rhabdomyosarcoma, 717, 718, 727
   spermatocele, 700
   varicocele, 704
Tetralogy of fallot (TOF), 18, 20, 21
Thalassemia major, 71, 77
Thanatophoric dysplasia, 711, 1099–1102
Thelarche, 838
Thiamine (vitamin B1) deficiency, 1283, 1284
Thin glomerular basement membrane nephropathy (TBMN), 617
Threatening miscarriage, 1477
Thrombocytosis, 868, 870, 871
Thrombosis of venous sinuses and cerebral veins, 1274
Thrombotic thrombocytopenic purpura, 630
Thymectomy, 927
Thymic carcinoma, 926
Thymic cyst, 924
Thymic epithelial space (TES), 925
Thymic follicular hyperplasia (TFH), 924
Thymic/non-thymic tumors, 927
Thymolipoma, 924
Thymoma, 964
   acute thymic involution, 927
   myasthenia gravis, 927
Thymus disorders
   HIV, 925, 926
   thymic cyst, 924
   thymic follicular hyperplasia, 924
   thymolipoma, 924
   thymoma, 925–927

   acute thymic involution, 927
   myasthenia gravis, 927
   true thymic hyperplasia, 924
Thyroglobulin (TGB), 935
Thyroglossal duct abscess, 935
Thyroglossal duct cyst, 935, 947
Thyroglossal midline remnant, 935
Thyroid gland
   anatomy, 935
   congenital anomalies of thyroid gland, 947
   diverticulum, 1170
   dysfunctional thyroid gland, 947
   dyshormonogenetic goiter, 947
   ectopia, 935
   epithelial neoplasm (*see* Epithelial neoplasms)
   fine needle aspiration cytology, 944
   Graves' disease, 947
   HIV infection, 945
   hormone regulation feedback mechanisms, 936
   hyperplasia, 946
   hyperthyroidism, 946
   hypothyroidism, 946
   inflammatory and immunologic thyroiditis
      acute thyroiditis, 948
      autoimmune thyroiditis, 948
      granulomatous thyroiditis, 948
      IgG4-related thyroiditis, 948, 949
      Riedel's thyroiditis, 949
   MIT and DIT, 936
   multi-nodular goiter, 947
   thyroglossal duct abscess, 935
   thyroglossal duct cyst, 935
   thyroid gland ectopia, 935
   thyroiditis, 946
Thyroid tissue, 1169
Thyroid transcription factor (TTF1), 230
Thyroidal neoplasms, 953
Thyroid-stimulating hormone (TSH), 936
Thyronin binding protein (TBP), 935
Thyrotropin/TSH-releasing hormone (TRH), 936
Thyroxine-binding globulin (TBG), 936
Thyroxine-binding pre-albumin (TBPA), 936
Tissue continuity damage-related degenerations
   gastric hyperplasia, 286, 287
   gastric ulcers, 286
Titin/connectin gene, 55
Tonsillar cyst, 1185
TORCH-CVI, 1476
Total anomalous pulmonary venous return, 32
Total parenteral nutrition hepatopathy, 493
Toxic epidermal necrolysis (TEN), 1357, 1358, 1368
Toxoplasma, "other microorganisms", rubella, cytomegalovirus, and herpes (TORCH) virus, 1534
Toxoplasmosis, 916, 1282, 1534, 1535
TP53 gene, 5, 846, 996
Trachea-bronchial system, 184, 187–189
Tracheal stenosis, 151
Tracheo-esophageal fistula (TEF), 151, 259
Transforming growth factor–β1 (TGF- β1) gene, 62

Transient atypical myelopoiesis (TAM), 877
Transmural MPVFD, 1484
Transplantation
    allograft cellular rejection, 92, 95
    allograft humoral rejection, 95, 97
    cardiac allograft vasculopathy, 97, 99, 100
    PTLD, 100
Transplantation related disorders, 212
    acute rejection, 212
    chronic rejection, 212, 215
Transplant-vasculopathy, 98
Traumatic neuropathy, 1323
Treacher Collins syndrome, 1195–1196
Treponema pallidum, 765
Trichilemmoma, 1390
Trichoepithelioma, 1380, 1390
Trichofolliculoma, 1390
Trichomonas vaginalis, 767
Trichomoniasis, 767
Tricuspid regurgitation in utero, 1448
1st trimester specimens, 1425
Triple-A syndrome, 267
Triploid karyotype syndrome, 1532, 1533
Triploidy, 1530
Trisomy 13 syndrome, 42, 456, 1488, 1526
Trisomy 18 syndrome, 42, 456, 1488, 1527
Trisomy 21 (Down Syndrome), 42, 456, 1488, 1528, 1531
Trisomy 21-associated ILD, 161, 163
Tritium-labeled thymidine (3H-TDR), 867
Trophoblast changes during gestation, 1413
Tropical sprue, 330
True and pseudo-hermaphroditism, 1488
True histiocytic lymphoma, 910
True thymic hyperplasia, 924
Trypanosoma cruzii, 267
Tuberculosis, 622, 920
Tuberculous leptomeningitis, 1279, 1280
Tuberous sclerosis complex (TSC), 1265
Tuberous sclerosis syndrome, 634
Tubular adenocarcinoid, 352
Tubular adenoma, 841
Tubulocystic carcinoma, 659
Tubulo-interstitial disease
    acute (tubulo-) interstitial nephritis, 618, 622
    acute hypersensitivity nephritis, 618
    acute pyelonephritis, 618
    acute tubular necrosis, 623
    cholemic nephropathy, 625
    chronic (tubulo-) interstitial nephritis, 622, 623
    CRF, 623
    hyaline change, 625
    hypokalemic nephropathy, 625
    jaundice-linked acute kidney injury, 625
    myeloma kidney, 625
    nephrocalcinosis, 624
    nephrolithiasis, 624
    osmotic nephrosis, 625
    primary hyperoxalurias, 624
    radiation nephropathy, 626
    tubulointerstitial nephritis and uveitis, 626
    urate nephropathy, 625
Tubulointerstitial nephritis and uveitis (TINU), 626
Tumor necrosis factor-α (TNF-α) gene, 55
Tumor-like lesions and bone/osteoid-forming tumors
    aneurysmal bone cyst, 1120–1124
    fibrous dysplasia, 1116–1118
    giant cell tumor, 1127–1128
    giant osteoid osteoma, 1126–1127
    myositis ossificans, 1115–1116
    non-ossifying fibroma, 1118–1120
    osteo-fibrous dysplasia, 1118
    osteoid osteoma, 1124–1126
    osteoma, 1124
    osteosarcoma, 1128–1136
    simple bone cyst, 1120
Tunica vaginalis, 728
Turner syndrome, 666
Twin and multiple pregnancies (PTMP), 1442, 1446, 1448, 1449
Twin demise, 1449
Twin gestation with "bizarre" (not-classifiable) malformations, 1448–1449
Twin gestation with discordance, 1448
Twin reversed arterial perfusion (TRAP) sequence, 1261, 1447
Twin-reversal arterial perfusion sequence (TRAPS)-type of twin pregnancies, 1448
Twin-to-twin transfusion syndrome (TTTS), 1261, 1445, 1446, 1448
TX hepatopathy, 501
TX SV40 (Polyoma)-nephropathy, 632
Type I/acidic cytokeratins/keratins, 1509
Type II/neutral-basic cytokeratins/keratins, 1509
Tyrosinemia, 471

U

Ulcerative colitis, 359
Ullrich-Turner syndrome, 42
Ultrashort Hirschsprung disease (US-HSCR), 1506
Ultrasound Biometric Report, 1420, 1421
Umbilical cord (UC), 1415
    compression/obstruction, 1457
    torsion, 1417
Umbrella cells, 674
Unconjugated hyperbilirubinemia, 436
Undifferentiated embryonal liver sarcoma (UELS), 524, 533
Unicameral bone cyst, 1123
Unilateral gonadal agenesis, 760
Unilateral PNH, 1526
Univentricular AV connection, 10
Urachal remnants, 678
Urachus remnant, 678
Urate nephropathy, 625
Urea cycle enzyme defects, 475, 476
Ureteral bud, 581
Ureteral dilatation, 687
Ureteric bud, 584

Urethral atresia/stenosis, 687
Urethral meatus abnormalities, 683
Urinary and genital ontogeny, 584
Urinary bladder diverticula, 678
Urinary bladder examination, 1501
Urinary tract endometriosis, 688–689
Urinary tract inflammatory and degenerative conditions
    cystitis, infectious, 684, 685
    cystitis, non-infectious, 684–686
    hydronephrosis, 687
    infections, 686
    malakoplakia, 686
    megacystis, 687
    megaureter, 687
    neurogenic bladder, 688, 694
    tumor-like lesions
        amyloidosis, 690
        cystitis cystica, 689, 690
        cystitis glandularis, 689
        fibroepithelial polyp, 689
        Müllerian lesions, 690
        nephrogenic adenoma, 689
        papillary-polypoid cystitis, 689
        postoperative spindle cell nodule, 690
        prostatic-type polyp, 690
        pseudocarcinomatous hyperplasia, 690
        squamous metaplasia, 690
        Von Brunn's nests, 689
    urinary tract endometriosis, 688–689
    vesicoureteral reflux, 686, 687
Urogenital tract examination, 1503
Urothelial (transitional) metaplasia, 767
Urothelial carcinoma, 694, 695
Urothelial cysts, 760
Urothelial hyperplasia (flat and papillary), 690–691
Urothelium, 674
Urticaria pigmentosa, 1362, 1400
Urticarial dermatitis, 1362
Urticarial skin lesions, 1362
Urticaria-like CADR (UCADR), 1371
Urticaria-like drug eruption, 1362
Usual ductal hyperplasia (UDH), 852
Usual interstitial pneumonia/pneumonitis (UIP), 160, 173, 175
Uterine adenomatoid tumor, 804
Uterine leiomyoma (ULM), 799
Uterine leiomyosarcoma (ULMS), 799
Uterine lymphangiomyomatosis (ULAM), 804
Uterine stroma tumors, 804, 805
Utero-placental interface perfusion imbalance (UPIPI), 1484
Uterus tumors, 804

V

VACTERL association, 45, 47
Vacuolar interface dermatitis, 1355, 1357–1360
Vaginal adenosis (VA), 767, 768
Vaginal epidermoid carcinoma, 813–814
Vaginal intraepithelial neoplasia (VAIN I-III), 813

Vaginal pseudotumors, 769
Vaginal rhabdomyosarcoma of Botryoid type (B-ERMS), 814, 815
Vaginitis emphysematosa, 760
Valvar vegetation, 87
Valve dysplasia, 90
Valve fibrolipoma, 104
Valvular heart disease
    calcific mitral annulus and calcific aortic stenosis, 91
    endocarditis and valvar vegetation, 87
    mitral valve prolapse, 91
    prosthetic valves, 92
    RHD, 91
Van nuys prognostic index (VNPI), 848
Vanishing regression syndrome (VRS), 679
Variable expressivity, 1495
Varices, 262
Varicocele, 704
Vascular anomalies, 152–155
Vascular disease
    benign nephrosclerosis, 626
    hemolytic uremic syndrome, 628, 631
    infarcts, 627, 628
    malignant nephrosclerosis, 626
    renal artery stenosis, 626
    thrombotic thrombocytopenic purpura, 630
    vasculitis, 627, 629, 630
Vascular neuropathy, 1323
Vascular transformation of LN sinuses, 920
Vascular tumors, 1396
    benign
        hemangioma variants, 1060, 1061
        lymphangioma, 1062
    blue rubber bleb nevus syndrome, 1068
    endothelial origin/differentiation, IHC markers, 1059
    with intermediate malignant potential
        hemangioendothelioma, 1062
        Kaposi sarcoma, 1064
    Kasabach-Merritt syndrome, 1068
    Klippel-Trénaunay-Weber syndrome, 1068
    malignant, 1066, 1068
    proteus syndrome, 1068
    Von-Hippel Lindau syndrome, 1068
Vascular, smooth muscle, and lipogenic tumors, 1153
Vascular-related etiologies, 1482
Vascular-syncytial or syncytial-vascular membranes (SVMs), 1481
Vasculitides, 119
Vasculitis, 627, 629, 630, 1369
Vasculitis damage index, 1005
Vasculitis-associated hemorrhage, 197
Vasculitis-associated lobular panniculitis (*Erythema induratum*), 1370
Vasculitis-associated soft tissue changes, 1005
Velocardiofacial syndrome (VCFS), 1532
Veno-occlusive disease, 213
Ventricular septal defect (VSD), 13
Ventricular septum, 68
Ventriculomegaly, 1257
Vermis hypoplasia, 1259

Verrucous carcinoma, 276
Vertebral, anal, tracheo-esophageal and renal (VATER) anomalies, 1495
Veru montanum mucosal gland hyperplasia (VMGH), 731
Vesicoureteral reflux (VUR), 686, 687
Vesicular forebrain (pseudo-aprosencephaly), 1254–1256
Vesiculobullous disease of newborns, 1194
Vestibular schwannoma, 1224, 1225
Villitis of unknown etiology (VUE), 1476
Villoglandular papillary carcinoma (VGPA), 808
Villous atrophy syndrome, 474
Villous capillary proliferation disorders (VCPD), 1481
Villous cells, 1473
Villous cytotrophoblast, 1433
Villous tree branching, placental development, 1413
Vinculin genes, 58
Virchow's triad, 869, 1443
Virginal hypertrophy, 838
Vitamin B12 deficiency-related megaloblastic anemia with subacute combined degeneration of spinal cord, 1283
Vitamin D deficiency, 1104
Vitamin D-dependent hereditary rickets (VDDR), 1104
VOLVULUS, 317
Von Brunn's nests, 689
Von Hippel Lindau syndrome (VHLS), 634, 728, 1068, 1266
von Meyemburg complex, 435
von Recklinghausen disease, 1265
Vulva tumors
  benign neoplasms
    botrioyd rhabdomyosarcoma, 819
    condyloma, 818
    condyloma acuminatum, 816
    ectopic mammary tissue, 816
    fibroepithelial polyp/acrochordon, 817
    Hidradenoma papilliferum, 816
    melanocytic nevi, 817
    squamous papilloma, 817
    vulvar angiomyofibroblastoma, 817
    vulvar cellular angiofibroma, 817
    vulvar vestibular papillomatosis, 817
  malignant tumors
    differentiated (simplex) VIN, 820
    vulvar epidermoid carcinoma, 820, 821
    vulvar intraepithelial neoplasia, 819–820
Vulvar aggressive angiomyxoma, 822
Vulvar angiomyofibroblastoma (AMF), 817
Vulvar cellular angiofibroma (CAF), 817
Vulvar dystrophies, 767
Vulvar epidermoid carcinoma (VEC), 820, 821
Vulvar intraepithelial neoplasia (VIN), 819–820
Vulvar Paget disease (VPD), 822
Vulvar verrucous carcinoma, 821
Vulvar vestibular papillomatosis, 817
Vulvitis, 767

W
WAGR syndrome, 633, 635
Waldenstrom's macroglobulinemia, 890, 912
Waldeyer's spaces, 687
Wallerian degeneration, 1322
"Wasserhelle" cells, 936
Water-clear cell hyperplasia, 966
Watershed infarction of brain, 1271
WDR62 deficiency, 1248
Webs, 261
Wegener's granulomatosis(WG), 125, 197, 1005, 1173
Weisses Blut, see Leukemia
Well-differentiated liposarcoma, 1014
Well-Differentiated NET, 297
Wells syndrome, 1362
Werdning-Hoffman disease (WHD), 1288, 1289, 1335
Wermer syndrome, 960
Wernicke encephalopathy, 1283, 1284
West syndrome/infantile spasms (WS/IS), 1290
Whipple disease, 326, 330
White blood cell disorders
  adult T-cell leukemia/lymphoma, 903, 904
  anaplastic large cell lymphoma, 903, 904
  angiocentric immunoproliferative lesions, 905, 906
  Burkitt lymphoma, 896, 901, 902
  cutaneous T-cell lymphoma, 904, 905
  diffuse large B-cell lymphoma, 892–896
  extranodal NK-T-cell lymphoma, 906, 907
  follicular lymphoma, 888–890
  Hodgkin lymphoma, 878–885
    classification, 886
    lacunar RS, 879
    mononuclear RS, 879
    nodular lymphocyte predominant Hodgkin lymphoma, 887
    vs. non-Hodgkin-lymphoma, 886
    popcorn cell, 879
    RSC/RCV, 878
  leukemia (*see* Leukemia)
  leukocyte-dysfunctionalities, 871
  leukocytopenias, 871
  lymphoblastic lymphoma, 896–900
  lymphocytopenia, 871
  mantle cell lymphoma, 891
  marginal cell lymphoma, 891, 892
  myelodysplastic syndromes, 878
  neutropenia, 871
  non-Hodgkin lymphoma, 887, 888
  non-neoplastic leukocytosis, 871
  peripheral T-cell lymphoma, 902, 903
  small lymphocytic lymphoma, 890, 891
White pulp disorders, of spleen
  chronic lymphocytic anemia, 922
  malignant lymphomas, 922
  reactive follicular hyperplasia, 922
  reactive non-follicular lymphoid hyperplasia, 922
Williams Beuren syndrome, 47
Wilms' tumor, 634–637, 639–645, 995
Wilms tumor gene (*WT1*), 938
Wilson disease, 484
Wirsung's/main duct, 555
Wiskott-Aldrich syndrome, 888
Wnt/Wingless signaling pathway, 1018, 1106

Wolffian duct (WD), 581
Wolf-Hirschhorn syndrome (WHS), 598, 1500
Wyburn-Mason syndrome (WMS), 1267

X
Xanthogranulomatous cholecystitis, 558
Xanthogranulomatous pyelonephritis, 622
Xanthoma, 1388
Xenopus, 836
X-linked diseases, 1525
XP11 translocation carcinoma, 647, 650
XXX (multi-X syndrome), 1488
XXY syndrome, 1488
XYY syndrome, 1488

Y
Yersinia enterocolitis-related lymphadenitis, 921
Yolk sac tumor (YST), 715–717, 776–778
Young non-embryonal tumors
    angiomyolipoma, 656–658
    chromophobe renal cell carcinoma, 653
    clear cell renal cell carcinoma, 650–653
    collecting duct carcinoma, 656
    cystic nephroma, 659
    juxtaglomerular cell tumor, 660
    mixed epithelial and stromal tumor, 659
    mucinous tubular and spindle cell carcinoma
        (MTS-RCC), 659
    oncocytoma, 656
    papillary adenoma and renal cell carcinoma, 653–655
    renal medullary carcinoma, 656
    renomedullary interstitial cell tumor, 660
    tubulocystic carcinoma, 659

Z
Zellballen, 990
Zellweger syndrome, 478, 479
Zollinger-Ellison syndrome, 287
Zona fasciculata, 937
Zona glomerulosa, 937
Zona reticularis, 937